Freshwater and Terrestrial Radioecology

Freshwater and Terrestrial Radioecology

A Selected Bibliography

Alfred W. Klement, Jr.
and
Vincent Schultz

Dowden, Hutchinson & Ross, Inc.
Stroudsburg Pennsylvania

Library of Congress Catalog Card Number: 80-22169
ISBN: 0-87933-389-8

82 81 80 1 2 3 4 5
Manufactured in the United States of America.

Library of Congress Cataloging in Publication Data
Klement, Alfred W.
Freshwater and terrestrial radioecology.
1. Radioecology—Bibliography. I. Schultz, Vincent, joint author.
II. Title.
Z5322.R2K57 [QH543.5] 016.5745′222 80-22169
ISBN 0-87933-389-8

Distributed world wide by Academic Press,
a subsidiary of Harcourt Brace Jovanovich,
Publishers.

Preface

An increasing interest in the relatively new field of radioecology began about 1958. The use of radionuclides and ionizing radiation in investigative biology had become well known. They were seen as useful tools for more detailed studies in ecology as well as other environmental sciences. The biological aspects of radioactive fallout from nuclear weapons tests and peaceful uses of nuclear explosives became of interest, both from the standpoint of possible effects on humans and their environment and the use of the widespread radiotracers as sources of radiation produced by these activities. Further interest in natural radioactivity was stimulated also. Disposal of radioactive wastes from the various nuclear activities was and is a major environmental concern. More recently, the nuclear power industry has become of considerable interest from an environmental point of view. These aspects of nuclear energy continue to be useful tools for research and for solving environmental problems. The authors felt, as did others, that a comprehensive list of publications in radioecology would be of value. Here an attempt was made to list those publications pertaining to or related to freshwater and terrestrial radioecology.

The authors began compiling such a list of publications in 1959 and issued the first bibliography in 1962. This bibliography was reprinted in Schultz and Klement, *Radioecology* (The First National Symposium on Radioecology, held at Fort Collins, Colorado, September 10-15, 1961), published in 1963 by the American Institute of Biological Sciences and Reinhold Publishing Corp. Since the original bibliography was issued, we have compiled and issued twelve supplements, the last in 1978. The supplies of nearly all of these issues have been exhausted for some years, although they are available in some form in a limited number of depository libraries. The thirteen issues are cumbersome for efficient use. We feel that this consolidation with additions and improvements would contribute to the field.

Other bibliographies on radioecology were compiled by the present authors, at times with others. These are listed in the "Bibliographies" section herein. Through these further compilations and our other activities, we have personally reviewed many more of the references, some especially for this compilation, so that entries are more complete. Many technical errors have been corrected, typographical errors are believed to be fewer, and many more references have been added.

In this compilation we have been somewhat more selective than in original listings. Papers "in press" are not included, although papers presented at symposia, and so on, and issued as an available report are included. In general, entries that deal with nonionizing radiation or nonradioactive nuclides are not included, except where there was evidence brought to our attention that they were directly compared to radionuclides or ionizing radiation. While some ecological studies at various nuclear installations are of interest, references to them are cited only where they are pertinent to the subject of this bibliography. Even so, we have used a broad definition of *radioecology*. Included are references to works that appeared to be estuarine in nature while attempting to exclude marine radioecology. Obviously, the distinction is somewhat arbitrary. On the other hand, many references are included that may not appear to relate to radioecology as some view it. However, an attempt was made to include references of interest to radioecologists.

Emphasis has been placed on studies of wild species of plants and animals in natural environments. Also included are many entries not of this nature that provide information on techiniques and comparative data of interest. Extensive references to environmental radioactivity and radiation from the various natural and artificial sources are included. Some citations not personally reviewed may not be appropriate here. For example, titles and abstracts are often not sufficiently specific as to wave-length of radiation studied. Also, due to the mechanics of compilation, some references were probably omitted, which we regret in spite of the fact that we do not pretend to provide as exhaustive list. Even so, some 20,000 references are listed dating from 1898 to 1979.

Nearly all non-English titles have been translated into English, and re-translations have been made for many entries, providing more readable titles as well as providing our non-English-speaking friends and colleagues with adequate references. Institutional names are entered as they came to us, often in English, often not, but we have attempted to provide the best descriptions available. In this regard, we have listed many references by institution or organization rather than as *Anonymous*.

Authors' names are listed alphabetically. Where possible authors' names are listed as indicated in the publication cited, transliterated in some cases in an accepted manner. An attempt was made to provide orthographic symbols for authors' names, and other material, so far as facilities and original documents permitted, especially for some Slavic and Scandinavian entries. These have affected alphabetical order in some cases, along with some inadvertent misplacements.

References were obtained from a large number of sources. *Nuclear Science Abstracts*, issued from 1948 through 1977, proved to be a major source. Other abstract journals were screened as well, and others were used considerably for confirmation and completion of many references. References to abstracts are shown following entries that may be of use directly or indirectly for availability information.

We considered indexing the list but did not feel that this could be accomplished adequately and efficiently from titles or brief abstracts, nor did we find "key word" lists adequate. Many annotations have been added to supplement title information. It would seem that reviews of the literature based in part on this compilation would be most helpful to the field. Some have been published in the various subfields of radioecology and listed here.

The references are listed in a style readily usable in various ways. Relatively common abbreviations are used for journal titles and other entries, as are commonly used acronyms. A bibliographer's license has been used in hyphenation and other style-related notations.

The "Bibliographies" section at the end of the list may be useful for reference to additional information in a number of subfields. This list also contains references to various organizations' publications, availability information, acronyms and other abbreviations commonly used. In this regard, *Nuclear Science Abstracts* provides similar information in periodic sequence and in special indexes. Such information is provided here.

ABSTRACTS

Abstracts listed as major entries are indicated in titles by "Abstr." Full-length copies available are also indicated under "Other Designations" below. Abstracts available in abstract publications are shown in citations in parentheses. They are abbreviated as follows; abstract numbers or page numbers are given.

AD-	Abstract numbers in Technical Abstract Bulletin, Armed Services Technical Information Agency (ASTIA)
APCA	Air Pollution Control Abstracts
BA	Biological Abstracts
CA	Chemical Abstracts
Diss. Abstr.	Dissertation Abstracts
FA	Forestry Abstracts
NSA	Nuclear Science Abstracts
NSA of Japan	Nuclear Science Abstracts of Japan
PBA	Plant Breeding Abstracts
PHA	Public Health Abstracts
PHEA	Public Health Engineering Abstracts
RDA	Research and Development Abstracts (U.S. AEC)
Ref. zh.	Referativniy Zhurnal (Abstracts of Soviet publications in English; sections)
SFA	Sports Fishery Abstracts
Weed Abstr.	Weed Abstracts
WPA	Water Pollution Abstracts
WR	Wildlife Review

TRANSLATIONS

Translations of papers in English are designated where known and available to the authors. The most common are shown as a report of the U.S. Atomic Energy Commission or one of its laboratories with an acronym followed by "-tr-", "Trans-", "TRANS-", etc. The same is true of others, for example, UK AEA and Australian Atomic Energy Commission. A number of Soviet journals are routinely translated and published with appropriate English titles. Translations into French are shown as French CEA reports. The following indicates our designation of many translations cited.

AD/A-	See AD- above
AEC-tr-	U.S. Atomic Energy Commission (U.S. AEC)
FTD-TT-	U.S. Air Force Wright-Patterson Air Force Base, O.
JPRS-	Joint Publications Research Service, New York
NP-tr-	U.S. AEC file number for miscellaneous translations
NSJ-tr-	Japan Atomic Energy Research Institute
OTS-	U.S. National Technical Information Service order number
TT-	Same (a list of U.S. government-sponsored translations is issued periodically)

OTHER DESIGNATIONS

Various organizations have designated papers, and so on, by a file cryptogram (acronym) number. These papers are generally on file in some form, full-length. Often such designations are incorrectly shown in papers and bibliographies or abstracts as report numbers. We have attempted to distinguish these differences by use of parentheses. Unfortunately, designations are sometimes used by organizations for both report numbers and file numbers. Some of the designations used in this bibliography are listed below.

CONF-	U.S. Atomic Energy Commission (U.S. AEC) and successor organizations (U.S. ERDA, U.S. DOE, U.S. NRC) designation of a conference, symposium, etc.
CEA-CONF-	French Commissariat à l'Energie Atomique (uses a similar system)
AED-	Gmelin Institutes, Max Planck Institutes, Germany, reports and file numbers (uses a similar system)
NP	U.S. AEC file number for non-U.S. AEC papers
OTS-	U.S. National Technical Information Service order number (NTIS)
PB-	U.S. government publication number (Government Printing Office, OTS, or NTIS)
INIS-mf-	International Atomic Energy Agency, Vienna, file number, microfilm copy
Univ. Microfilms	University Microfilms (Dissertations, theses)
DOCKET-	U.S. AEC and U.S. NRC docket numbers for license applications and related documents for licensed nuclear facilities; publicly available at regional and Washington, D.C., Docket Rooms

ACKNOWLEDGMENTS

There are too many people and agencies to thank individually for helping us in our efforts over the years to list the literature on radioecology. We do wish to mention some. Several federal government agencies provided funds through direct salaries and grants to the authors: U.S. Atomic Energy Commission, U.S. Atlantic-Pacific Interoceanic Canal Study Commission, U.S. Department of Health, Education and Welfare, U.S. Energy Research and Development Administration, and U.S. Department of Energy.

This publication was produced without financial support from any agency of government, nor from anyone else except the authors and the publisher and staff. We do wish to give our sincere thanks to the latter, especially to Shirley A. End, Managing Editor.

In addition to our own organizational libraries, we wish to mention the kind assistance provided by librarians and libraries. The National Library of Medicine, Bethesda, Maryland; the main library of the Department of Agriculture, Washington, D.C.; and several major libraries of the Department of Health, Education and Welfare, Rockville, Maryland; Battelle Memorial Institute Slavic Library, Columbus, Ohio; are mentioned, not only for their assistance but as major depositories of literature in the environmental sciences.

We wish to thank the following for their dedicated assistance in organizing and preparing, and technical assistance in contributing to the various issues of the bibliography: Gladys Anderson, Cheryl Anderson, Linda Crane, Earlene Easton, Candess M. Eliason, Wanda Fujimoto, Evelyn S. Hampton, Charles F. Lytle, Patricia L. Schultz, Hugh E. Voress, and Patricia Waldhalm.

Our special thanks are given to Carmen E.

Klement and Taimi Shannon for their assistance in the preparation of this revision and consolidation.

DEDICATION

We have dedicated some issues of the bibliography to some of those who gave us support and encouragement and to some who have been instrumental in the pursuit of the environmental sciences. We list these below, knowing that a number of others would be listed if we could continue this effort. We again dedicate this tome to these people.

R. M. Aleksakhin, Stanley I. Auerbach, D. A. Crossley, Jr., Paul B. Dunaway, Lee L. Eberhardt, Norman R. French, A. Grauby, Wayne C. Hanson, Edward E. Held, Louis A. Krumholz, Roy E. Nakatani, Dan J. Nelson, Eugene P. Odum, Jerry S. Olson, William S. Osburn, Jr., D. E. Reichle, Allyn H. Seymour, Donald G. Watson, F. Ward Whicker, John N. Wolfe, and George M. Goodwell.

Further, we especially dedicate this volume in memory of some of those deceased who made major contributions to the field of radioecology. To the authors they were more: teachers, mentors, scientists, gentlemen, and, above all, friends.

Charles L. Dunham, S. Allan Lough, John R. Olive, and John N. Wolfe.

ALFRED W. KLEMENT, JR.
VINCENT SCHULTZ

Freshwater and Terrestrial Radioecology

Aanyan, V. L. 1966. Effect of fertilizers on yield and radioactivity level of alpine meadow plants. Agrokhimiya (3): 121-126. (In Russian).

Aarkrog, A. and J. Lippert (w/J. Petersen*). 1957-1975. Environmental radioactivity ----. Denmark Atomic Energy Commission, Research Establishment, Risö. (Sea plants, seawater, groundwater, whales, birds, vegetation, animals).

Year Publ.	Place Reptd.	Period Reptd.	Report RISÖ-	No. pp.	NSA Vol. Abstr. No.
1957	Riso	1957-58	3	110	12: 15283
1959	Riso	1958-59	9	50	13: 20003
1960	Riso	1959	14	48	14: 19427
1961	Denmark	1960	23	51	15: 22563
1963*	Denmark	1962	63	147	18: 5479
1963*	Faroes	1962	64	30	18: 5480
1963*	Greenland	1962	65	28	18: 6948
1964	Denmark	1963	85	112	19: 6098
1964	Faroes	1963	86	27	19: 6099
1964	Greenland	1963	22	22	19: 6100
1965	Denmark	1964	107	98	20: 5451
1965	Faroes	1964	108	24	20: 7320
1965	Greenland	1964	109	26	20: 7321
1966	Denmark	1965	130	99	21: 14329
1966	Faroes	1965	131	23	21: 16340
1966	Greenland	1965	132	21	21: 20182
1967	Denmark	1966	154	100	22: 10357
1967	Faroes	1966	155	23	22: 10358
1968	Denmark	1967	180	91	23: 9571
1968	Faroes	1967	181	25	23: 12070
1968	Greenland	1967	182	19	23: 12071
1969	Denmark	1968	201	81	24: 21101
1969	Faroes	1968	202	20	24: 4589
1969	Greenland	1968	203	18	24: 7768
1970	Denmark	1969	220	95	25: 16139
1970	Faroes	1969	221	21	25: 4476
1970	Greenland	1969	222	19	25: 16080

Aarkrog, A. and J. Lippert. (Continued).

Year Publ.	Place Reptd.	Period Reptd.	Report RISÖ-	No. pp.	NSA Vol. Abstr. No.
1971	Denmark	1970	245	95	26: 4743
1971	Faroes	1970	246	21	26: 4716
1971	Greenland	1970	247	18	26: 12277
1972	Denmark	1971	265	103	27: 9556
1972	Faroes	1971	266	20	27: 9557
1972	Greenland	1971	267	18	27: 11964
1973	Denmark	1972	291	99	29: 18323
1973	Faroes	1972	292	22	29: 20861
1973	Greenland	1972	293	16	29: 18324
1974	Denmark	1973	305	99	31: 8610
1974	Faroes	1973	306	20	31: 8611
1974	Greenland	1973	307	20	31: 11250
1975	Faroes	1974	324	25	33: 9261
1975	Greenland	1974	325	20	33: 9262

Aarkrog, A. 1964. Variation in the ratio of strontium-89 to strontium-90 in precipitation samples with sampling height. Nature 204(4960): 771. (NSA 19: 7709).

Aarkrog, A. 1971. A re-examination of plutonium at Thule, Greenland, in 1970. pp. 1213-1218. In: Proc. Intern. Symp. on Radioecology Applied to the Protection of Man and His Environment. (See European Atomic Energy Community, 1971) (NSA 27: 5286).

Aaron-da Cunha, M. I. 1970. Influence of ionizing radiations on crowngall tissues *in vitro*. Bull. Soc. Bot. Fr. 1970: 43-50. (In French) (NSA 28: 47944).

Abak, T. and S. K. Banerjee. 1972. Effects of whole body x-irradiation on the pars intercerebralis cells of the grasshopper, *Acrida gigantea*. Sci. Culture (Calcutta) 38(10): 454-455. (NSA 28: 11073).

Abbatt, J. D., J. R. A. Lakey and D. J. Mathias. 1960. Environmental radiation survey around a nuclear power station. General Electric Co. Atomic Energy Rev. 2: 200-204.

Abdel-Malek, A. 1960. Application of radioactive isotopes in entomological research. Ann. Soc, entomol., Quebec. 6: 80-87.

Abdel-Malek, A. A. 1961. The effect of radioactive phosphorous on the growth and development of Culex pipiens molestus Forsk. (Diptera, Culicidae). Bull. Entomol. Res. 52(4): 701-708.

Abdel-Malek, A. A. 1963. Inhibitory effect of L-methionine-methyl-^{14}C on oviposition by females on the cotton leaf worm, Prodenia litura (F.), induced by radioactive males. Nature 200(4906): 604-605. (NSA 18: 1412).

Abdel-Malek, A. A. 1964. Study of feeding habits of male Anopheles sergenti Theo. at Siwa Oasis using radioisotopes. Bull. Wld. Hlth. Org. 30: 137-139. (^{32}P).

Abdel-Malek, A. A. and M. A. Abdel-Aal. 1966. Study of the dispersion and flight range in Anopheles sergenti Theo. in Siwa Oasis using radioactive isotopes as markers. Bull. Wld. Hlth. Org. 35: 968-973. (NSA 22: 51403).

Abdel-Malek, A. A. and M. F. Abdel-Wahab. 1961. Autoradiography as a technique for radioactive phosphorus, P^{32}, uptake in Culex molestus Forsk. Bull. Soc. Entomol. Eqypte 45: 409-418.

Abdel-Malek, A. A. and M. F. Abdel-Wahab, 1961. Studies on the phosphorus -32 uptake in Schistocerca gregaria (Forsk.) and Anacridium aegyptium (L.). Bull. Soc. Entomol. Egypte 45: 419-425.

Abdel-Malek, A. A. and S. H. Ahmed. 1972. Dominant lethals in the immature states of Culex pipiens molestus induced by gamma irradiation of the parental pupae. Vestn. Cesk. spolecnosti zool. 36(4): 233-236. (NSA 28: 15768).

Abdel-Malek, A. A. and S. H. Ahmed. 1972. Induced dominant lethals in the immature stages of Culex pipiens molestus Forsk. by gamma irradiation. Egypt. J. Genet. Cytol. 1(2): 300-302. (NSA 27: 27986).

Abdel-Malek, A. A. and W. F. Baldwin. 1961. Specificity of plant feeding mosquitoes as determined by radioactive phosphorus. Nature 192(4798): 178-179.

Abdel-Malek, A. A. and D. K. McE. Kevan. 1961. Inhibited oviposition by females of Gryllus assimilis (F.) induced by radioactive males, using L-methionine-methyl-^{14}C. Nature 192(4803): 681-682.

Abdel-Malek, A. A., A. O. Tantawy and A. M. Wakid. 1966-69. Studies on the eradication of Anopheles pharoensis Theobald by the sterile-male technique using ^{60}Co. I. Biological effects of gamma radiation on the different development stages. III. Determination of the sterile dose and its biological effects on different characters related to "fitness" components. VI. Sperm activity in males irradiated with the sterilizing dose. VII. Population dynamics of mosquitoes in Fayoum Area, UAR. J. Econ. Entomol. 59: 672-678; 60: 20-23, 1300-1302; 62(2): 348-351. (NSA 21: 26705, 32818; 22: 10860).

Abdel-Malek, A. A., et al. 1974. Studies on factors influencing the induction of sterility in Anopheles pharoensis Theobald by gamma radiation. pp. 161-174. In: Isotopes in Pesticides and Pest Control. (See Nehme and Hassan, 1974) (NSA 33: 7565).

Abdou, F. M. and S. Larsen. 1964. A radioactive-tracer method for measuring the stability of sparingly soluble phosphates in soil. Soil Science 98 (2): 94-99.

Abdu, R. and G. V. Sumarukov. 1971. Post-radiation recovery of adult insects. Vestn. Mosk. Univ., biol. pochv. 26(5): 112-113. (In Russian) (NSA 29: 768).

Abdu, R. M. and N. F. Shaumar. 1974. Gamma-irradiation of immature stages of the beetle Dermestes frichii Kug. Z. Angew. Entomol. 75(1): 103-106. (NSA 30: 24358).

Abdul Matin, A. S. M. 1966. Susceptibility of adult rice weevil, Sitophilus oryzae (L.) to gamma radiation. pp. 133-135. In: Proc. of the Agricultural Symposium. Pakistan Atomic Energy Center, Dacca. (CONF-660564) (NSA 21: 26637).

Abdul Matin, A. S. M., A. D. Bhuiya and Z. A. Khan. 1966. Attempt to obtain sterility of male rice hispa through gamma radiation. pp. 155-156. In: Proc. of the Agricultural Symposium. Pakistan Atomic Energy Center, Dacca. (CONF-660564) (NSA 21: 26638).

Abdullayev, A. A. 1973. Probe for estimating the characteristics of subsurface or ground water flow. British Patent 1,321,739. 4pp. (NSA 28: 27585).

Abdullayev, A. A. and A. Shutova. 1973. Probe for estimating the characteristics of subsurface or ground water flow. British Patent 1,321,740. 4pp. (NSA 28: 27854).

Abdullayev, A. A., S. Khakimov and A. S. Khasanov. 1972. Accelerated neutron activation determination of strontium in natural waters. Radiokhimiya 14. (Sov. Radiochemistry 14(3): 509-510) (NSA 28: 42).

Abdullayev, A. A., Sh. Khatamov and A. A. Kist. 1972. Sodium and potassium losses during calcination in connection with radioactivation analyses of biological material. Izv. Akad. Nauk Uzbek. SSR (fiz.-mat.) (2): 69-72. (In Russian) (NSA 26: 58506).

Abdullayev, A. A., P. Kh. Nishanov and Ye. B. Shapirov. 1971. Neutron activation method for determining selenium content in natural waters. pp. 125-127. In: Neutron-Activation Analysis. Ye. M. Lobanov, ed. Inst. Yadernoi Fiziki, Akad. Nauk Uzbek. SSR, Tashkent. (In Russian) (NSA 27: 7342).

Abdullayev, A. A., A. Sh. Zakhidov and P. Kh. Nishanov. 1968. Determination of W and Au Content in mine waters by neutron activation analysis. Izv. Akad. Nauk Uzbek. SSR (fiz.-mat.) (5): 60-61. (In Russian) (NSA 23: 28770).

Abdullayev, A. A., Zakhvatayev and V. P. Perelygin. 1968. Determination of the uranium concentration in plants according to traces of uranium fission fragments. Radiobiologiya 8: 765-766. (AEC-tr-7023; pp. 168-170) (NSA 24: 20).

Abdullayev, A. A., et al. 1968. Use of tritium as an indicator for subterranean waters. Z. Angew. Geol. 14: 86-88. (In German) (NSA 23: 51095).

Abdullayev, A. A., et al. 1969. Radiation-based determination of copper in natural waters. Izv. Akad. Nauk Uzbek. SSR (fiz.-mat.) (5): 75-76. (In Russian) (NSA 24: 13794).

Abdullayev, A. A., et al. 1971. Radioactivation determination of Se, Te, and Au in natural waters. Izv. Akad. Nauk Uzbek. SSR (fiz.-mat.) (2): 71-72. (In Russian) (NSA 25: 57370.

Abee, H. H. 1964. Environmental surveys following accidental releases. Nuclear Safety 6(1): 87-89. (NSA 19: 11446).

Abee, H. H. and W. D. Cottrell. 1960. Contamination resulting from the release of radioactive liquid wastes to the Tennessee River system. Am. Indust. Hyg. Assoc. J. 21: 156-161. (NSA 14: 12807).

Abee, H. H., et al. 1970. Environmental monitoring significance of the Clinch River study. pp. 388-393. In: *Environmental Surveillance in the Vicinity of Nuclear Facilities*. (See W. C. Reinig, 1970) (NSA 25: 16180).

Abel, E. and J. Mayer. 1971. Determination of radioactive and stable cesium. Chem. Listy 65: 242-267. (In Slovak) (NSA 25: 31773).

Abel, E., J. Mayer and K. Durcek. 1970. Influence of the secondary effect on the contents of ^{60}Co. in the environment. Radioisotopy 11 (4): 589-597. (In Slovak) (NSA 26: 9539).

Abel, K. H. and L. A. Rancitelli. 1975. Major, minor, and trace element composition of coal and fly ash, as determined by instrumental neutron activation analysis. pp. 118-138. In: *Trace Elements in Fuel*. S. P. Babu, ed. American Chemical Society, Washington. (NSA 32: 21838).

Aberg, B. and F. P. Hungate. (Eds.). 1967. *Radioecological Concentration Processes*. (Proc. Intern. Symposium held in Stockholm, 25-29 April, 1966). Pergamon Press, New York-London. xiv, 1040pp. (CONF-660405) (NSA 21: 36618).

Abood, K. A., J. P. Lawler and K. A. Konrad. 1972. Nuclear plant siting on rivers. vp. In: *Water Quality Considerations: Siting and Operating* OF Nuclear Power Plants. (See Atomic Industrial Forum, 1972) (NSA 28: 29243).

Abood, L. G., K. Kirahasi and M. Perez del Cerro. 1967. Biochemical studies on isolated nerve endings and other particulates of bullfrog brain. Biochim. biophys. Acta 136: 521-532. (NSA 21: 32587) (^{32}P).

Abraham, R. L. 1972. Mortality of mallards exposed to gamma radiation. Radiation Research 49(2): 322-327. (NSA 26: 23103).

Abrahams, J. H., Jr. and R. H. Johnson, Jr. 1966. Soil and sediment analysis: preparation of samples for environmental radiation surveillance. U.S. PHS, Washington, Publ. 999-RH-19. vii, 25pp.

Abrahamson, D. E. 1972. Ecological hazards from nuclear power plants. pp. 795-811. In: *Careless Technology: Ecology and International Development*. M. T. Farvar and J. D. Hilton, eds. Natural History Press, Garden City. (NSA 28: 2568).

Abrahamson, S. and I. H. Herskowitz. 1957. Induced changes in female germ cells of *Drosophila*. II. Oviposition rate and egg mortality in relation to intensity and dosage of x-rays applied to oocytes. Genetics 42: 405-420.

Abram, E., et al. 1975. Neutron activation analysis studies of selected portions of the Mahoning River system using ^{252}Cf. pp. 191-203. In: *Applications of Californium-252*. (Proc. Amer. Nucl. Soc. Nat. Topical Meeting, Austin, Texas, Sept. 11-13, 1972). S. J. Gage, ed. (NSA 32: 16699).

Abramov, I. V. and V. T. Zablotskiy. 1972. Effect of x rays on the virulence of *Theileria annulata*. Veterinariya 10: 91-93. (In Russian) (NSA 29: 8062).

Abro, A. 1964. The effect of x-irradiation on spermatogenesis in *Drosophila melanogaster* Meigen. Arbok Univ. Bergen mat.-nat. (10): 1-21. (NSA 20: 18310.

Abrol, Y. P. and E. E. Conn. 1966. Cyanide metabolism in *Lotus arabicus* and *Lotus tenuis*. Phytochemistry 5(2): 237-242. (^{14}C).

Abu-Samra, A. and J. R. Vogt. 1973. Rapid determination of bromine in water samples by activation analysis. Trans. Amer. Nucl. Soc. 16: 44. (NSA 28: 12615).

Abu-Samra, A., et al. 1974. Determination of molybdenum and selenium in bioenvironmental samples using neutron activation analysis. pp. 206-212. In: *Proc. 2nd Intern. Conf. on Nuclear Methods in Environmental Research*. (See Vogt and Meyer, 1974) (NSA 33: 23330).

Academy of Natural Sciences of Philadelphia. 1951-70. Biological survey of the Savannah River in the vicinity of the Savannah River Project. Savannah River biological survey. South Carolina and Georgia. Radioactivity studies. etc. Summary of studies on the Savannah River, 1951-1970 for E. I. du Pont de Nemours and Co., U.S. AEC report TID-26669 (Pts. 1-3), TID-26671, TID-26673, TID-26674, TID-26675, TID-23709, TID-25393, TID-25634. vp. (NSA 31: 8729, 8730, 8731, 8732, 8734, 8738; 21: 22413; 24: 39471; 25: 18225).

Acharya, N. 1973. Radiotracer investigations on the effect of sulphate and phosphate application on paddy. Indian J. Agric. Chem. 6(1): 23-32. (NSA 32: 22291) (^{35}S).

Acqua, C. 1910. The effects of radium rays on plants. Ann. Bot. (Rome) 8 (2): 223-238.

Acqua, C. 1912. The action of uranium on the plant cell. Arch. farmacol. sper. 14: 81-84. (NP-tr-268).

Ackerman, T. L. and S. A. Bamberg. 1974. Phenological studies in the Mojave Desert at Rock Valley (Nevada Test Site). pp. 215-226. In: Phenology and Seasonality Modeling. H. Lieth, ed. Springer-Verlag, New York.

Acree, F., Jr., F. H. Babers and N. Mitlin. 1956. Preparation of P-32-labeled Bayer L 13/59 and DDVP, and their absorption and translocation in the American cockroach. J. Econ. Entomol. 49(6): 808-812.

Acree, E. H. and F. N. Case. 1969. Method for tagging sand with a gaseous radioactive isotope. U. S. Patent 3,426,205. (NSA 23: 13957).

Adam, G. 1972. Thermoluminescence dating and the sensitization method. Health Phys. 23(4): 579-580. (NSA 26: 58818) (pottery).

Adam, M. and S. L. Iverson. 1968. Occurence and distribution of small mammals in the WNRE controlled area, Pinawa, Manitoba. Whiteshell Nuclear Research Establishment, Atomic Energy of Canada, Ltd., Pinawa, report AECL-3184. 32pp. (NSA 23: 17962).

Adamek, M. and K. Roth. 1970. Biologic properties of molds from uranium mines and their relationship to some diseases in man. Radioisotopy 11(6): 1121-1143. (In Czech) (NSA 26: 2443) (Incl. low-level radiation effect study).

Adamek, M. 1974. Interesting biological phenomenon in uranium mines. Rudy (Prague) 22(8): 232-234. (In Czech) (NSA 32: 14640) (Radionuclides in molds).

Adams, D. R. and P. G. Voillequé. 1971. Effect of stomatal opening on the transfer of $^{131}I_2$ from air to grass. Health Physics 21(6): 771-775. (NSA 26: 9492).

Adams, J. A. S. 1964. Laboratory γ-ray spectrometer for geochemical studies. pp. 485-498. In: The Natural Radiation Environment. (See Adams and Lowder, 1964) (NSA 18: 35888).

Adams, J. A. S. 1974. Real and apparent variations in the terrestrial gamma-ray flux. pp. 59-69. In: Second Workshop on the Natural Radiation Environment. (February 1974). W. M. Lowder, ed. Health and Safety Laboratory, U. S. AEC report HASL-287. (NSA 31: 3321).

Adams, J. A. S. and G. E. Fryer. 1964. Portable γ-ray spectrometer for field determination of thorium, uranium, and potassium. pp. 577-596. In: The Natural Radiation Environment. (See Adams and Lowder, 1964) (NSA 18: 35889).

Adams, J. A. S. and P. Gasparini. 1970. Gamma-Ray Spectrometry of Rocks. Methods in Geochemistry and Geophysics Series, Vol. 10. American Elsevier, New York. 303pp. (NSA 25: 34539).

Adams, J. A. S. and W. M. Lowder. (eds.). 1964. The Natural Radiation Environment. (Proc. Intern. Symposium on the Natural Radiation Environment, held at Houston, Texas, April 10-13, 1964). Univ. of Chicago Press, Chicago. xiv, 1096pp. (NSA 18: 35727).

Adams, J. A. S., W. M. Lowder and T. F. Gesell. (Eds.). 1972. Natural Radiation Environment II. (Proc. Second Intern. Symposium, August 7-11, 1972, Houston, Texas). Vol. 1, 460pp., Vol. 2, 959pp. (CONF-720805-P1 and P2) (NSA 32: 19714, 32: 19722).

Adams, J. A. S., J. K. Osmond and J. J. W. Rogers. 1959. The geochemistry of thorium and uranium. Phys. Chem. Earth 3: 298-348.

Adams, J. R. 1972. Nuclear plant siting on coastal areas in California. vp. In: Water Quality Considerations: Siting and Operating of Nuclear Power Plants. (See Atomic Industrial Forum, 1972) (NSA 28: 29242).

Adams, W. H. and E. B. Fowler. 1970. ^{238}Pu incorporated in fish living in water containing $^{238}PuO_2$. Los Alamos Scientific Lab., U. S. AEC report LA-DC-12899. 8pp. (NSA 29: 18665).

Adams, W. H. and E. B. Fowler. 1974. Studies of the apparent solubility of $^{238}PuO_2$ microspheres in an aquatic environment and the uptake of plutonium from a soil matrix containing $^{238}PuO_2$. pp. 615-645. In: Proc. 2nd AEC Environmental Protection Conf. U. S. AEC report WASH-1332(74).

Adeesan, C., G. W. Rahalkar and A. J. Tamankar. 1971. Radiation-induced behavioral modifications as a measure of radiosensitivity of male khapra beetle (Trogoderma granarum Everts). pp. 373-380. In: Basic Mechanisms in Radiation Biology and Medicine. (See India Dept. of Atomic Energy, 1971) (NSA 26: 28490).

Adkins, T. R., Jr. 1958. Systemic action of two insecticides on arthropod parasites of rabbits and cattle. Dissertation, Auburn Univ., Auburn, Alabama.

Adkins, T. R., Jr. 1961. Use of radioisotopes in entomological research. pp. 168-179. In: Proc. Southeastern Seminar on Atomic Progress in Agriculture. (See Clemson College, 1961).

Adkins, W. C. and L. V. Gibney. 1971. TVA's power plant design: a demonstration in total environmental planning. Proc. Amer. Power Conf. 33: 611-620. (NSA 26: 24815) (Incl. nuclear plant).

Adler, A. 1969. Influence of some environmental factors on the growing and development of silk-worms treated with gamma-irradiation in their embryo-phase. Zooteh. Med. Vet. 12: 145-151. (In Rumanian) (NSA 26: 59114).

Adler, H. I. 1966. Genetic control of radiation sensitivity in the micro-organisms. Adv. Radiat. Biol. 2: 167-191. (Also: Oak Ridge National Lab., U. S. AEC report ORNL-P-1012) (NSA 20: 31041).

Adler, H. I. and M. S. Engel. 1961. Factors influencing the survival of bacteria after exposure to ionizing radiation. J. Cell. Comp. Physiol. 58(Suppl. 1): 95-105.

Adriano, D. C., et al. 1975. Field study to determine plutonium contents of wheat and soil in a warm humid area. Du Pont de Nemours and Co., report DP-MS-75-40. 10pp. (NSA 33: 602).

Aegerter, S. K., H. H. Loosli and H. Oeschger. 1967. Variations in the production of cosmogenic radionuclides. pp. 49-55. In: *Radioactive Dating and Methods of Low-Level Counting*. (See IAEA, 1967) (NSA 22:

Afanas'yev, G. D. (Ed.). 1973. *Geological-Radiological Interpretation of Divergent Values of Age*. Izdatel'stvo Nauka, Moscow. 415pp. (In Russian) (NSA 28: 5583).

Affeltranger, C. E. 1968. Feeding by *Reticulitermes* spp. on radiation-processed wood-plastic combinations. J. Econ. Entomol. 61(2): 398-401.

Afordakos, G., K. Alexopoulos and D. Miliotis. 1974. Using artificial thermoluminescence to reassemble statues from fragments. Nature 250(5461): 47-48. (NSA 30: 26816).

Afrikanova, L. A. 1963. An experimental study of early structural changes of an irradiated, interauricular septum of a frog. (Live and histological observations). Radiobiologiya 3(2): 300-310. (AEC-tr-5435: 230-245 (NSA 17: 30368).

Agafonov, B. M. 1956. Summaries of Hygiene Section reports at the All-Union Conference on Medical Radiology. Publ. House of Medical Literature, Moscow. pp. 26-27. (In Russian).

Agafonov, B. M. 1957. Some experiments in biological de-activation of (radioactive) water. Laboratory tests in small tanks. Trud. Vses. Konf. Med. Radiol., Vopr. Gig. Dozimetr. 79. (In Russian) (WPA 32: 107).

Agafonov, B. M. 1958. Laboratory experiments on the biological deactivation of water in a series of tank-reservoirs. Byull. Ural. otd. Mostov. obshch. isp. prirody 1: 97-102. (In Russian).

Agafonov, B. M. and V. I. Ivanov. 1961. Effect of flow of strontium deactivation of water. Tr. Ural'sk. otd. Moskov. obshch. isp. prirody (3): ____. (In Russian).

Agarwal, A. S. B. R. Singh and Y. Kanehiro. 1972. Immediate effect of gamma-irradiation on the release of mineral nitrogen in fresh tropical soils. Indian J. Agr. Sci. 42(11): 1062-1064. (NSA 29: 549).

Agarwal, H. C. and N. W. H. Houx. 1964. γ-butyobetaine derivative in phospholipids of *Phormia*. Biochim. biophys. Acta 84(3): 353-355. (^{14}C).

Agg, A. R., G. E. Eden and K. V. Melbourne. 1957. A field test for radioactive strontium and barium in water under emergency conditions. Proc. Soc. Water Treatment 6: 142-154.

Agnedal, P.-O. 1967. Calcium and stontium in Swedish waters and fish and accumulation of ^{90}Sr. In: *Radioecological Concentration Processes*. pp. 879-896. (See Aberg and Hungate, 1967) (Also: Sweden Aktiebolaget Atomenergi, Stockholm, report AE-224) (NSA 20: 24836).

Agnedal, P.-O. and S. W. O. Bergstrom. 1966. Recipient capacity of Tvaren, a Baltic bay. Sweden Aktiebolaget Atomenergi, Stockholm, report AE-234 21pp. (NSA 21: 1539.

Agnedal, P.-O., et al. 1958. Biological investigation in the water-recipient at Studsvik, the Research Establishment of the Swedish Atomic Energy Company. Proc. 2nd Intern. Conf. Peaceful uses of Atomic Energy 18: 400-403.

Agosin, M., L. Aravena and A. Neghme. 1965. Enhanced protein synthesis in *Triatoma infestans* treated with DDT. Exptl. Parasitol. 16(3): 318-324. (^{14}C).

Agosin, M., J. Ilivicky and S. Litvak. 1967. The induction of NAD kinase by DDT in *Triatoma infestans*. Can. J. Biochem. 45(5): 619-626. (^{14}C).

Agosin, M., A. Morello and N. Scaramelli. 1964. Partial characterization of the *in vivo* metabolites of DDT-C^{14} in *Triatoma infestans*. J. Econ. Entomol. 57(6): 974-977.

Agosin, M., et al. 1966. The effect of DDT on the incorporation of glucose and glycine into various intermediates in DDT-resistant strains of *Musca domestica* L. Comp. Biochem, Physiol. 19: 339-349. (^{14}C).

Agranat, V. Z. 1958. Certain data concerning cumulation of polonium^{210} by water-living organisms. Med. Radiol. 3(1): 65-69. (In Russian) (NSA 12: 9630).

Agranat, V. Z. 1958. Dynamics of accumulation of Po^{210} in the tissues of fishes. Gigiyena sanitariya 23(10):73-76. (In Russian) (NSA 13: 20073).

Agranat, V. Z. 1962. Artificial sources of radiation. Vol. II, Ch. 2, pp. 26-33. In: *Radiation Hygiene*. (See Marey, 1962).

Agrawal, D. P. and S. Kusumgar. 1975. Radiocarbon dates of neolithic-chalcolithic samples. Curr. Sci. (India) 44(9): 298-300. (NSA 32: 6465).

Agrawal, D. P., S. Kusumgar and D. Lal. 1965. The measurement of radiocarbon activity and some determinations of ages of archeological samples. Curr. Sci. (India) 34: 394-399. (NSA 19: 46617).

Agrawal, D. P., S. Kusumgar and R. P. Sarna. 1964. Radiocarbon dates of archeological samples. Curr. Sci. (India) 33: 402. (NSA 19: 7719).

Agre, A. L. and V. I. Korogodin. 1960. The distribution of radioactive contamination in a stagnant reservoir. Med. radiologiya. 5(1): 67-73: (JPRS-5030; pp. 161-177) (NSA 14: 8633, 14: 21817).

Agre, A. L. and A. P. Raiko. 1964. Accumulation of microquantities of strontium by green and blue-green algae. Fiziol. rastenii 11(1): 135-137. (In Russian) (NSA 18: 41054).

Agre, A. L. and M. M. Telitchenko. 1963. Concerning the accumulation of radioactive strontium by hydrophytes and detritus. Byull. Moskov. obshch. isp. priorody otd. biol. 68(1): 133-137. (JPRS-19937, OTS-63-31160; pp. 29-37) (Also: English transl. Federation Procl. Transl. Suppl. 22: T1218-T1219) (NSA 17: 28-44, 18: 3459).

Agre, A. L., I. V. Molchanova and N. V. Timofeyev-Resovskiy. 1964. Self-purification of cesium-137 from water in weakly running reservoirs at various speeds of flow, volumes of water, and concentrations of cesium. Byull. Moskov. obshch. isp. priorody otd. biol. 69(3): 20-24. (JPRS-26056) (NSA 18: 36928).

Agre, A. L., A. P. Raiko and N. V. Timofeyev-Resovskiy. 1962. The effect of various biomass of water plants on the concentration of microquantities of cesium and strontium in tanks exhibiting a low flow rate. Byull. Moskov. obshch. isp. priorody otd. biol. 67(5): 120-127. (In Russian).

Agricultural National Institute of Agricultural Science. 1956. Investigation of radioactive contamination in river water for irrigation. pp. 785-786. In: Research in the Effects and Influences of the Nuclear Bomb Test Explosions. Vol. II. Japan Society for the Promotion of Science, Ueno, Tokyo.

Ahmad, S. and A. J. Forgashi. 1975. Metabolism of ^{14}C-diazinon by gypsy moth larvae. J. Econ. Entomol. 68(5): 571-576. (NSA 33: 15134).

Ahmed, C. R. 1975. Studies on the mating competition of gamma-arradiated and non-irradiated males of the melon fly (Dacus cucurbitae). Int. J. Appl. Radiat. Isotop. 26(6-7): 407-409. (NSA 32: 22630).

Ahmed, M. I., et al. 1974. Movement of sulfate ion through soils. pp. 159-178. In Isotope and Radiation Techniques in Soil Physics and Irrigation Studies, 1973. (See IAEA, 1974) (NSA 30: 15493) (^{35}S).

Ahmed, M. S. H. 1970. Studies on the filling effect of gamma radiation on Oryzaephilus surinamensis (L.) Coleoptera, Silvanidae in stored dates. Iraq Atomic Energy Commission, Bagdad. 5pp. (NP-18607) (NSA 25: 11244).

Ahmed, S. S. 1969. The effect of x-irradiation and antibiotics on the susceptibility of Aedes togoe and A. aegypti to subperiodic Brugia malayi and B. pahangi. Ann. Trop. Med. Parasitol. 63: 353-357. (NSA 24: 34601).

Ahn, T. H., et al. 1962. Respiration of gamma irradiated Brucella abortus and Mycobacterium tuberculosis. Proc. Soc. Exptl. Biol. Med. 111(3): 771-773. (NSA 17: 7867).

Aidley, D. J. 1965. The effects of strontium and other divalent cations on potassium contracture in a locust leg muscle. J. Physiol. 177: 103-111. (NSA 19: 33725).

Ailam, G. and R. Galun. 1967. Optimum sex ratio for the control of insects by the sterility method. Ann. Entomol. Soc. Am. 60(1): 41-43.

Ailleret, P. 1970. Problems concerning sites for nuclear power stations. Atom. Strom. 16: 149-152. (In German) (NSA 25: 18159).

Ailleret, P. 1971. Location problems for nuclear power stations. Energie Nucléaire (Paris) 13(1): 22-27. (In French) (NSA 25: 36971).

Airey, P. L., et al. 1974. Use of environmental isotopes and artificial tracers to study recharge to groundwater in Burdekin Delta, Queensland. Vol. 1, pp. 115-126. In: Isotope Techniques in Groundwater Hydrology. (See IAEA, 1974) (NSA 31: 19623).

Aitken, M. J. 1969. Thermoluminescent dosimetry of environmental radiation on archeological sites. Archaeometry 11: 109-114. (NSA 24: 29385).

Aitken, M. J. 1970. Physics applied to archaelogy. I. Dating. Rep. Progr. Phys. 33(10): 941-1000. (NSA 26: 194).

Aitken, M. J. 1974. Physics and Archaeology. Claredon Press, Oxford. 291pp. (NSA 32: 17209).

Aitken, M. J. 1974. Radiocarbon dating. pp. 26-84. In: Physics and Archaeology. M. J. Aitken. Caredon Press, Oxford, England. (NSA 32: 17210).

Aitken, M. J. 1974. Thermoluminescent dating. pp. 85-134. In: Physics and Archaeology. M. J. Aitken. Clarendon Press, Oxford, England. (NSA 32: 17211).

Aitken, M. J. and J. C. Alldred. 1972. Assessment of error limits in thermoluminescent dating. Archaeometry 14(2): 257-267. (NSA 26: 50620).

Aitken, M. J. and S. G. E. Bowman. 1973. Thermoluminescent dating: assessment of alpha particle contribution. Archaeometry 17(1): 132-138. (NSA 3509).

Aitken, M. J. and S. J. Fleming. 1972. Thermoluminescence dosimetry in archaeological dating. pp. 1-78. In: Topics in Radiation Dosimetry. F. H. Attix, ed. Academic Press, New York. (NSA 28: 12872).

Aitken, M. J., S. J. Fleming and D. W. Zimmerman. 1967. Thermoluminescence dating of ancient ceramics. pp. 523-530. In: Radioactive Dating and Methods of Low-Level Counting. (See IAEA, 1967) (NSA 22: 6478).

Aitken, M. J. nd. Low-level environmental radiation measurements using natural calcium fluoride. pp. 281-290. In: Proc. 2nd Intern. Conf. on Luminescence Dosimetry. (Gatlinburg, Tennessee, September 23-26, 1968). J. A. Auxier, K. Becker and E. M. Robinson, eds. U. S. AEC report CONF-680920).

Aitken, R. F. 1958. Observations on the uses of atomic energy in entomology. Entomol. Gaz. 9(1): 43-44.

Akaboshi, M., T. Maeda and A. Waki. 1967. Determination of the nucleic acid content of protozoa by activation analysis. Biochim. Biophys. Acta 138: 596-597. (NSA 21: 32579).

Akai, H. and M. Kobayashi. 1965. Sites of fibroin formation in the silk glands in Bombyx mori. Nature 206: 529. (^{3}H).

Akaiwa, H. 1973. Activation analysis of chalcophile elements in natural water. Bunseki Kiki 11(12): 819-826. (In Japanese) (NSA 30: 14787).

Akamsin, A. D. 1960. Sorption of caesium by some sediments from aquatic solutions with different concentrations of sodium, potassium, magnesium and calcium. Sevastopol Biol. Stants. 12: 302-304.

Akamsin, A. D., V. P. Parchevskiy and G. G. Polikarpov. 1960. The radioactivity of some representatives of Black Sea plankton, benthos and nekton. Tr. Sevastopol'sk biol. stantsii Akad. Nauk SSSR 13: 305-308. (JPRS-20630, pp. 26-31) (NSA 17: 35348).

Akamsin, A. D., V. P. Parchevskiy and G. G. Polikarpov. 1960. Algae accumulate radioactivity. Priroda 49(2): 95- ___.

Akatova, N. A., Ye. N. Pavlovskiy and B. K. Shtegman. (Eds.). 1959. Trans. 6th Conf. on the Biology of Inland Waters. (10-19 June 1957). Publ. House Acad. Sci. USSR, Moscow-Leningrad. 734pp. (AEC-tr-2593; TT-68-50492; IPST-Cat. 5136) (CONF-570601) (NSA 24: 2593).

Akerman, B., et al. 1963. Observations on pigeons with prethalmic radiolesions in the nervous pathways from the telencephalon. Acta Physiol. Scand. 56(4): 286-298. (NSA 18: 25099).

Akerman, K. 1972. Application of radionuclides in geology, hydrology, and meteorology. pp. 264-338. In Radiotracer Method in Practical Work. ALFA, Bratislava. (In Slovak) (NSA 33: 29156).

Akhalaya, Ya. G. 1966. Influence of incubation conditions on the radiobiological effects on eggs of Bombyx silk worms. Soobshch. Akad. Nauk Gruz. SSR 44: 727-734. (In Georgian) (NSA 21: 20344).

Akhalaya, Ya. G. 1967. Studies of the genetic nature of radiation effects in the silkworm. Soobshch. Akad. Nauk Gruz. SSR 58: 453-454. (In Russian) (NSA 22: 17014).

Akhalaya, Ya. G. 1968. Studies of the relative role of radiation injuries affecting spermatozoid and oviform cells in radiobiology of the mulberry silkworm. Soobshch. Akad. Nauk Gruz. SSR 59: 689-693. (In Georgian) (NSA 22: 40931).

Akhmetov, Z., et al. 1973. Distribution of radioactive iodine in the organs and tissues of various species of rodents. Biol. Nauki (9): 54-58. (In Russian) (NSA 30: 19034) (Incl. sand eels, jerboa).

Akhromeiko, A. I. and V. A. Shestakova. 1955. The role of microorganisms in the nutrition of trees. In Isotopes in Microbiology. (Trans. Conf. on the Application of Tracer Atoms in Microbiology). A. A. Imshenetzkiy, ed. Publ. House of the Acad. Sci. USSR, Moscow. (AEC-tr-3494; pp. 168-191. (^{32}P).

Akhromeiko, A. I. and V. A. Shestakova. 1958. The influence of rhizospere microorganisms on the uptake and secretion of phosphorus and sulphur by the roots of arboreal seedlings. Proc. 2nd Intern. Conf. on the Peaceful Uses of Atomic Energy 27: 193-199.

Akhromeiko, A. I. and M. V. Zuralev. 1957. The rate of water flow in tree species. Fiziol. rast. 4: 164-170. (FA 19: 78).

Akimov, N. E. (Ed.). 1969. Civil Defense. (Handbooks and Aids for Higher Agricultural Institutions). "Kolos", Moscow. 351pp. (ORNL-tr-2306) (Incl. agricultural aspects).

Akiyama, M. 1966. Soil algal vegetation of gamma irradiated field and natural stronly radioactive district in Japan. Shimane Daigaku Ronshu Shizen Kagaku (16): 126-134. (NSA 22: 8658).

Akov, S. 1966. Retention of theblood-meal in Aedes aegypti following sterilization by chemicals and radiation. Ann. Trop. Med. Parasitol. 60: 482-494. (NSA 22: 2546).

Aksel'rod, F. M. and G. A. Shvartsman. 1967. Determination of the radioactivity of open water reservoirs. Radiokhimiya 9(1): 72-75. (Soviet Radiochemistry 9(1): 72-74) (NSA 21: 28085).

Alabama Power Company. 1970. Joseph M. Farley Nuclear Plant, Units 1 and 2. Birmingham. U. S. AEC DOCKET-50348. 74pp. (NSA 24: 47797) (Incl. environment).

Alamos Cerda, F. 1973. Application of isotopic technique in the hydrological study of the Aconcagua River Basin. Final report for the period 1 November 1979-30 November 1972. International Atomic Energy Agency, Vienna, report IAEA-R-809-F. vp. (In Spanish) (NSA 31: 14453 (^{131}I).

Al-Adil, K. M., W. W. Kilgore and R. R. Painter. 1972. Toxicity and sterilization effectiveness of aflatoxins B_1 and G_1 and distribution of aflatoxin B_1-^{14}C in house flies. J. Econ. Entomol. 65(2): 375-378.

Al-Badri, A. 1975. Groundwater investigation in the Iraqui western desert (Al-Mat area) using nuclear technique. Int. J. Appl. Radiat. Isotop. 26 (4): 207-211. (NSA 32: 3508).

Albert, N. J. 1965. The determination of the LD_{50} with ^{60}Co of certain species of millipeds. Proc. Louisiana Acad. Sci. 28: 121-122. (NSA 20: 18366).

Alberti, G., B. Bettinali and F. Salvetti. 1959. Behavior and determination of Radium-B, Radium-D and Thorium-B in natural waters. Ann. Chim. Roma 49: 193-198.

Albini, A. and C. Triulzi. 1973. CISE activities in radioecology. pp. 47-56. In: Proc. 2nd Conf. on the State of Advancement of Radioecology in Italy. (See Anonymous, 1973) (In Italian).

Albrecht, L. and W. Rounsch. 1971. Radiation protection as a part of environmental protection in generation of electric power from nuclear enery in the German Democratic Republic. Kernenergie 14(7): 227-231. NSA 26: 19811).

Al-Daham, N. K. 1970. Use of chemosterilants, sex hormones, radiation, and hybridization for controlling reproduction in Tilapia species. Thesis, Auburn University, Auburn, Ala. 176pp. (Univ. Microfilms 70-22, 269) (NSA 25: 16326).

Alder, B., H. Oeschger and J. T. Wasson. 1967. ^{26}Al in deep-sea sediments. pp. 189-197. In: Radioactive Dating and Methods of Low-Level Counting. (See IAEA, 1967) (NSA 22: 6459).

Alder, F. 1970. Safety and risk of nuclear power stations. Neue Tech. B12: 201-206. (In German) (NSA 25: 25949).

Alder, F. F. 1972. Nuclear power: world and Australian situation in perspective. Australian Atomic Energy Commission, Coogee, report. 30pp. (INIS-mf-395) (Environmental aspects).

Aleksakhin, R. M. 1973. Achievements and some actual tasks of water radioecology. Vaol. II, pp. 7-19. In: Radioecology of Water Organisms. (See G. P. Andrumaitis, 1973) (In Russian) (NSA 30: 21354).

Aleksakhin, R. M. 1973. Use of the principles and methods of radioecology in solving some current problems of man's effect on the environment. Ekologiya 4(4): 51-56. (Soviet J. Ecol. 4(4): 315-319). (NSA 32: 6446).

Aleksakhin, R. M. and M. A. Bocharova. 1971. Tasks and achievements in the investigations of artificial radionuclides. Migration in forest biogeocenonses. Lesovedeniye (4): 59-70. (In Russian) (NSA 27: 27600).

Aleksakhin, R. M. and M. M. Ravikovich. 1967. On the behavior of alkaline-earth elements - - calcium, magnesium and strontium - - in a forest biogeocoenosis. pp. 443-452. In: Radioecological Concentration Processes. (See Aberg and Hungate, 1967).

Aleksakhin, R. M. 1971. Migration of natural radionuclides and their effects in terrestrial biogeocenoses. Ekologiya 2(6): 107-108. (English transl. pp. 582-583) (Comments on a conference).

Aleksakhin, R. M. and F. A. Tikhomirov. 1971. Radionuclide migration in forest biogeocenoses. Vol. II, pp. 145-164. In: Contemporary Problems of Radiobiology. (See V. M. Klechovskiy, G. G. Polikarpov and R. M. Aleksakhin, 1971).

Aleksakhin, R. M., M. A. Naryshkin and M. A. Bocharova. 1970. Characteristics and quantitative prediction of the cumulative storage of ^{90}Sr in arboreal plants. Dokl. Akad. Nauk SSSR 193: 1192-1194. (In Russian) (NSA 25: 26616).

Aleksakhin, R. M., F. A. Tikhomirov and N. V. Kulikov. 1970. Status and problems of forest radioecology. Ekologiya (1): 27-38. (Sov. J. Ecol. (1): 19-27) (NSA 26: 15439).

Aleksakhin, R. M., et al. 1970. Some aspects of radioactive fission products migration in the forest biogeocenoses and the effects of ionizing radiations on woody plants. pp. 999-1023. In: Proc. Intern. Symp. on Radioecology, Vol. II. (See France Commissariat à l'Énergie Atomique, (1970) (NSA 24: 41788).

Aleksakhin, R. M., et al. 1973. Behavior of radioactive isotopes in soils and their accumulation by wood plants. Pochvovedeniye 5: 51-58. (In Russian) (NSA 30: 306).

Aleksakhin, R. M., et al. 1974. Migration of stable and radioactive isotopes of alkaline and alkali-earth elements in the biological chain soil-plant-milk. Zh. Obshch. Biol. 35(3): 414-422. (In Russian) (NSA 30: 18655).

Aleksandrov, N. N. 1968. Method of determining total radioactivity of atmospheric fallout. Tr. gl. geofiz. observ. (20&): 110-120. (In Russian) (NSA 26: 4429).

Aleksandrov, S. N. and K. F. Galkowskya. 1957. On the relation between the general resistance and radio-resistance in animals. J. Gen. Biol. Moscow 18(1): 52.

Aleksandrov, S. N., D. K. Popov and N. K. Strel'nikova. 1971. Radiational-hygienic significance of natural ^{14}C. Gigiyena sanitariya (3): 63-66. (In Russian) (NSA 25: 39958 (Yeast).

Aleksan'yan, O. M. 1972. Radiochemical analysis of surface water. Gidrokhim. mater. 53: 163-167. (In Russian) (NSA 28: 5193).

Aleksan'yan, O. M. 1973. ^{137}Cs and ^{90}Sr in the Sea of Azov. Vol. II, pp. 225-228. In: Radioecology of Water Organisms. (See Andrumaitis, 1973) (In Russian) (NSA 30: 21120).

Alekseyev, F. A., et al. 1968. Use of isotopes and radiation sources in hydrology. Tr. vses. nauchn.-issled. inst. yadern. geofiz. geokhim. (1): 41-60. (In Russian) (NSA 24: 9605).

Alekseyev, V. A. (ed.). 1968. Nuclear Geophysics. Izdatel'stvo Nedra, Moscow. 288pp. (In Russian) (NSA 22: 51091).

Alescio, T., et al. 1962. The lethal dose for Mus musculus exposed to the γ beam of the Frascati synchrotron. Nuntius Radiol. 28(5): 371-383. (RT/BIO (62)70) (In Italian) (NSA 17: 5863).

Alexander, G. V., R. E. Nusbaum and N. S. MacDonald. 1956. The relative retention of strontium and calcium in bone tissue. J. Biol. Chem. 218(2): 911-918. (Kangaroo rats).

Alexander, J. M. and J. H. Horton. Regional Monitoring Group, Health Physics Section semiannual progress report, January-June 1955. Du Pont de Nemours, U. S. AEC report DPSP-56-25-13(Del.). 41pp. (NSA 30: 9547) (Savannah River Plant).

Alexander, M. E., et al. 1974. Trace element analysis of seawater and fish samples by proton induced x-ray emission spectroscopy. Int. J. Appl. Radiat. Isotopes 25(5): 229-233. (NSA 30: 5936).

Alexander, M. L. 1962. The role of recovery mechanisms and oxygen effects upon changes in radiation sensitivity in sperm treated mature males and fertilized females of Drosophila. Genetics 47(11): 1505-1518. (NSA 17: 7840).

Alexander, P. and K. A. Stacey. 1959. Cross-linking of deoxyribonucleic acid in sperm heads by ionizing radiations. Nature 184(4691): 958-960. (Used sperm of fish).

Alexander, R. E. 1974. Environmental monitoring report for Pantex Plant covering 1973. Mason and Hangar-Silas Co., Amarillo, Texas, U. S. AEC report MHSMP-74-12. 44pp. (NSA 30: 3144).

Alexandre, H. 1974. Effects of x rays on nucleic acid metabolism during cleavage and gastrulation in Pleurodeles waltii eggs. J. Embryol. Exp. Morhpol. 31(1): 147-157. (NSA 32: 1186).

Alexandre, H. and J. Brachet. 1971. Radiosensitivity of the metabolic pathways of incorporation of thymidine-^{3}H and uridine-^{3}H in amphibian embryos. Arch. Intern. Physiol. Biochim. 79(4): 817-818. (NSA 26: 26823).

Alfimov, N. N., P. N. Yagovoy and V. P. Tikhomirov. 1963. On the natural radioactivity of the leaves of some woody species. Botan. zh. 48(5): 724-726. (In Russian).

Algard, F. T., G. B. Friedmann and H. M. McCurdy. 1974. Responses of adult newts (Amphibia: Urodele) to x-rays. Can. J. Zool. 52(6): 665-669. (NSA 30: 24338).

Alibert, J. 1959. Trophallactic exchanges in the yellow-necked termite Calotermes flavicollis Fabr., studies by means of radioactive phosphorus. Comt. rend. 248(7): 1040-1042.

Alicate, J. E. and G. O. Burr. 1949. Preliminary observations on the biological effects of radiation on the life of Trichinella spiralis. Science 109: 595.

Alikhanyan, S. I. and F. S. Klepikova. 1957. Effects of x-radiation on the mutability of Actinomyces subtropicus (The producer of albomycine). Dokl. Akad. Nauk SSSR 115(4): 806-808. (In Russian).

Alimdzhanov, R. A. and R. Khakimova. 1962. Effect of ionizing radiation on Sitona weevils. Dokl. Akad. Nauk Uzbek. SSR 3: 65-66. (In Russian).

Alimdzhanov, R. A. and Z. Sultankhdzhayeva. 1963. Effect of gamm radiation on the development of soil insects. Dokl. Akad. Nauk Uzbek. SSR 1: 59-60. (R. Zh. Biol. No. 8E3).

Aliyev, A. T. and K. P. Kashkin. 1973. Blood plasma esterases of tundra voles (Microtus oeconomus), living under conditions of an increased natural radiation background. Radiobiologiya 13(4): 598-602. (AEC-tr-7499; pp. 146-151) (NSA 30: 7001).

Aliyev, A. T. and P. A. Korzhuyev. 1973. Amino acid composition of blood plasma proteins of root voles (Microtus oeconomus Pall.), living under various ecological conditions. Radiobiolgiya 13(6): 818-823. (AEC-tr-7534; pp. 10-16) (Study on radium and uranium-radium areas).

Aliyev, A. T., P. A. Korzhuyev and K. P. Kashkin. 1972. Blood plasma proteins of the tundra vole Microtus oeconomus, living under conditions of increased natural background radiation. Radiobiologiya 12(1): 63-68. (AEC-tr-7316; pp. 91-99) (NSA 26: 50906).

Al Kital, R. A., L. Chan and E. V. Sayre. 1969. Neutron activation analysis of pottery sherds from Hajar Bin Humeid and related areas. pp. 387-398. In: Hajar Bin Humeid: Investigations at a Pre-Islamic Site in South Arabia. B. W. van Beek ed. Johns Hopkins Press, Baltimore. (NSA 23: 26201).

Al-Kholy, A., et al, 1970. Phosphorous uptake from water by Tilapia zillii (Gervais). Hydrobiologia 36(3-4): 471-478. (SFA 16(2): 13165) (^{32}P).

Allan, H., G. Welch and J. Miquel. 1973. Effects of accelerated particles on the ageing of Drosophila. pp. 689-700. In: Advances in Radiation Research. Vol. 2. Biology and Medicine. J. F. Duplan, ed. Gordon and Breach, New York. (NSA 31: 26188).

Allan, J. D. 1973. Competition and the relative abundances of two cladocerans. Ecology 54(3): 484-498). (^{14}C).

Alldredge, A. W. and F. W. Whicker. 1972. A method for measuring soil erosion and deposition with beta particle attenuation. J. Range Mgmt. 25(5): 393-398.

Alldredge, A. W., J. F. Lipscomb and F. W. Whicker. 1974. Forage intake rates of mule deer estimated with fallout cesium-137. J. Wild. Mgmt. 38 (3): 508-516.

Alldredge, A. W., F. W. Whicker and W. C. Hanson. nd. Environmental impacts associated with Project Rio Blanco. Colorado State University, Fort Collins, U. S. AEC report COO-1156-76. 24pp. (NSA 32: 6360).

Allemand, R., P. Calmels and P. Corompt. 1970. Neutron soil moisture meter for glacial densimetry. French Patent 2,029,938. (In French) (NSA 25: 39775).

Allen, A. L. 1951. Effects of x-irradiation upon embryonic-development in the paradise fish Macropodus opercularis. Thesis, Univ. of California, Berkeley. 168pp.

Allen, A. L. 1958. The effects of x-irradiation upon embryoni-development in the paradise fish, Macropodus opercularis, with some notes on normal states. Brigham Young Univ., U. S. AEC report AECU-3967. 37pp. (NSA 13: 7351).

Allen, A. L. 1958? Cross responses. I. The effects of x-irradiation upon embroyonic development in the paradise fish, Macropodus opercularis. Brigham Young Univ., U. S. AEC report AECU-3978. 46pp. (NSA 13: 4389).

Allen, A. L. and L. M. Mulkay. 1960. X-ray effects on embryos of the paradise fish, with notes on normal stages. Growth 24(2): 131-168. (NSA 15: 27435).

Allen, B. M. 1956. Cell destruction induced by heavy irradiation of recently metamorphosed Bufo. Abstr. Anat. Record 124(2): 387.

Allen, B. M. 1957. The effects of heavy gamma irradiation upon pituitary glands transplanted into hypophysectomied tadpoles of Bufo boreas. J. Exptl. Zool. 136(1): 185-198. (NSA 12: 11243).

Allen, B. M. 1958. Effects of x-irradiation upon the limb-buds of Bufo boreas. Abstr. Anat. Record 130(2): 391-392.

Allen, B. M. 1956. Interpretation of some experiments upon the effects of ionizing irradiation on the tissues of amphibians. Ann. Zool. 2(9): 119-126.

Allen, B. M. and B. L. Bachman. 1956. The effects of critical dosages of gamma irradiation upon the capacity of the pituitary gland to secrete thyrotrophic hormone. J. Exptl. Zool. 132(1): 111-132. (Also: Univ. of Calif., Los Angeles, U. S. AEC report UCLA-324) (NSA 9:2550) (Toads).

Allen, B. M. and M. H. Devick. 1957. Differences in susceptibility to whole-body gamma irradiation in the layers of the retina of Bufo. Biol. Bull. 112(2): 137-143. (Also: Univ. of Calif., Los Angeles, U. S. AEC report UCLA-367) (NSA 10: 3969).

Allen, B. M. and L. M. Ewell. 1960. Resistance to x-irradiation by embryonic cells of the limb-buds of tadpoles. J. Exptl. Zool. 142: 309-335.

Allen, B. M. and O. A. Schjeide. 1951. The influence of anoxia upon destruction and mitoses of the hematopoietic cells of x-irradiated tadpoles. Univ. of Calif., Los Angeles, U. S. AEC report UCLA-149. 13pp. (NSA 5: 4972).

Allen, B. M., O. A. Schjeide and L. B. Hochwald. 1950. Relation of low temperatures to x-ray injury of hematopoietic tissue of tadpoles. Proc. Soc. Exptl. Biol. Med. 73(1): 60-62. (Also: Univ. of Calif., Los Angeles, U. S. AEC report UCLA-50) (NSA 4: 1622).

Allen, B. M., O. A. Schjeide and L. B. Hochwald. 1951. The influence of temperature upon the destruction of x-rayed hematopoietic cells of tadpoles. J. Cell. Comp. Physiol. 38(1): 69-82. (Also: Univ. of Calif., Los Angeles, U. S. AEC report UCLA-93) (NSA 5-286).

Allen, B. M., O. A. Schjeide and R. Piccirillo. 1951. The influence of anoxia upon destruction and mitosis of the hematopoietic cells of x-irradiated tadpoles. Univ. of Calif., Los Angeles, U. S. AEC report UCLA-149. 13pp. (NSA 5: 4972).

Allen, B. M., O. A. Schjeide and R. Piccirillo. 1954. Influence of anoxia upon hematopoietic cells of tadpoles exposed to x-irradiation and colchicine. J. Cell. Comp. Physiol. 44(2): 318-322.

Allen, B. M., et al. 1953. Relation of x-irradiation dosage and elapsed time to the destruction of hematopoietic cells of tadpoles. J. Exptl. Zool. 122(3): 409-422. (Also: Univ. of Calif., Los Angeles, U. S. AEC report UCLS-232) (NSA 7: 475).

Allen, B. M., et al. 1953. The capacity of the anterior pituitary gland to function after heavy irradiation. J. Exptl. Zool. 124(1): 131-146. (Also: Univ. of Calif., Los Angeles, U. S. AEC report UCLA-254) (NSA 7: 3688) (Toads).

Allen, F. and H. J. Grindley. 1957. Radioactive tracers in the Thames estuary. Dock and Harbor Auth. 37(435): 302-306.

Allen, H. E. and R. B. Hahn. 1969. Determination of phosphate in natural waters by activation analysis of tungstophosphoric acid. Environ. Sci. Tech. 3(9): 844- ___.

Allen, J. R. and L. A. Carstens. 1968. Response of Saimiri sciureus monkeys to total body x-irradiation: clinical signs and pathologic changes. Am. J. Veterin. Res. 29: 2179-2188. (NSA 23: 2926).

Allen, J. R. and L. A. Carstens. 1968. Response of Saimiri sciureus monkeys to total body irradiation: electron microscopic changes in the heart. Am. J. Veterin. Res. 29: 2189-2204. (NSA 23: 2925).

Allen, M. B. 1962. Evaluation of carbon-14 technique for measurement of primary production. pp. 477-480. In: Radioecology. (See Schultz and Klement, 1963). (NSA 17: 33599).

Allen, R. E. (Comp.). 1969. Radiation surveillance networks. U. S. AEC report WASH-1148. 140pp. (NSA 24: 20893).

Allen, R. E. 1971. Summary information on accidental releases of radioactive effluent to the atmosphere from underground nuclear detonations designed for containment, August 5, 1963 - June 30, 1971. U. S. AEC report WASH-1183. 28pp. (NSA 25: 48665).

Allen, T. D., M. V. Haigh and A. Howard. 1973. Ultrastructure of giant plastids in a radiation induced mutant of Osmunda regalis. J. Ultrastruct. Res. 42: 491-501. (NSA 28: 21602).

Alley, D. A. and B. G. Hightower. 1966. Mating behavior of the screw-worm fly as affected by differences in strain and size. J. Econ. Entomol. 59: 1499-1502. (NSA 21: 24699).

Allred, D. M. 1962. Mites on grasshopper mice at the Nevada atomic test site. Great Basin Nat. 22(4): 101-104.

Allred, D. M. 1962. Mites on squirrels at the Nevada atomic test site. J. Parasitol. 48(6): 817.

Allred, D. M. 1963. Mites from pocket mice at the Nevada Test Site (Acarina). Proc. Entomol. Soc. Wash. 65(3): 231-233.

Allred, D. M. 1965. Note of phalangids at the Nevada Test Site. Great Basin Nat. 25(1-2): 37-38.

Allred, D. M. 1966. Ecological consideration of nuclear detonations and Acarina related to zoonoses. Brigham Young Univ., U. S. AEC report COO-1355-17. 7pp. (NSA 22: 8495).

Allred, D. M. 1968. Fleas of the National Reactor Testing Station. Great Basin Nat. 28(2): 73-87.

Allred, D. M. 1968. Ticks of the National Reactor Testing Station. Brigham Young Univ. Sci. Bull. (Biol.) 10(1): 1-29. (Also: Brigham Young Univ., U. S. AEC report COO-1559-1) (NSA 23: 6406).

Allred, D. M. 1969. Spiders of the National Reactor Testing Station. Great Basin Nat. 29(2): 105-108.

Allred, D. M. 1970. Mites and lice of the National Reactor Testing Station. Brigham Young Univ. Sci. Bull. (Biol.) 12(1): 1-17.

Allred, D. M. 1971. Mammalian ectoparasite consortism at the National Reactor Testing Station. Great Basin. Nat. 31: 164-168.

Allred, D. M. 1973. Effects of a nuclear detonation on arthropods at the Nevada Test Site. Brigham Young Univ. Sci. Bull. (Biol.) 18(4): 1-20. (NSA 28: 27635).

Allred, D. M. and D E. Beck. 1962. Ecological distribution of mites on lizards at the Nevada atomic test site. Herpetologica 18(1): 47-51.

Allred, D. M. and D E. Beck. 1963. Comparative ecological studies of animals at the Nevada Test Site. pp. 327-331. In: Radioecology. (See Schultz and Klement, 1963) (NSA 17: 33580).

Allred, D. M. and D E. Beck. 1963. Range of movement and dispersal of some rodents at the Nevada atomic test site. J. Mammal. 44(2): 190-200.

Allred, D. M. and D E Beck. 1964. Arthropod associates of plants at the Nevada Test Site. Brigham Young Univ. Sci. Bull. (Biol.) 5(2): 1-16. (Also: Brigham Young Univ., U. S. AEC report TID-21394) (NSA 19: 6901).

Allred, D. M. and M. A. Goates. 1964. Mites from wood rats at the Nevada nuclear test site. J. Parasitol. 50(1): 17.

Allred, D. M. and M. H. Muma. 1971. Solpugids of the National Reactor Testing Station. Great Basin Nat. 31: 164-168.

Allred, D. M. and S. Mulaik. 1965. Two isopods of the Nevada Test Site. Great Basin Nat. 25(1-2): 43-47.

Allred, D. M., D E. Beck and C. D. Jorgensen. 1963. Close-in effects of an underground nuclear detonation on small mammals and selected invertebrates (Project SEDAN). Brigham Young Univ., U. S. AEC report PNE-226P. iv, 19pp. (NSA 17: 19750).

Allred, D. M., D E. Beck and C. D. Jorgensen. 1963. Biotic communities of the Nevada Test Site. Brigham Young Univ. Sci. Bull. (Biol.) 2(2): 1-52; 1 map.

Allred, D. M., D E. Beck and C. D. Jorgensen. 1963. Nevada Test Site study areas and specimen depositories. Brigham Young Univ. Sci. Bull. (Biol.) 2(4): 1-15. (NSA 17: 33505).

Allred, D. M., D E. Beck and C. D. Jorgensen. 1964. Close-in effects of an undergroud nuclear detonation on small mammals and selected invertebrates. Brigham Young Univ., U. S. AEC report PNE-226F. iv, 18pp.

Allred, D. M., D E. Beck and C. D. Jorgensen. 1965. Summary of the ecological effects of nuclear testing on native animals at the Nevada Test Site. Proc. Utah Acad. Sci. 42: 252-260. (NSA 21: 36357).

Allred, D. M., D E. Beck and J. R. Murdock. 1960. Comparative ecological studies of animals exposed to nuclear detonation. Proc. Utah Acad. Sci. (Arts, Ltrs.) 37: 152-153.

Allred, D. M., D. E. Johnson and D E. Beck. 1965. A list of some beeflies of the Nevada Test Site. Great Basin Nat. 25(1-2): 5-11.

d'Almeida, A. 1969. Chemical parameters for medical evaluation of water. Rev. Port. Quim. (Lisbon) 11(1-4): 98-104. (In Portugese) (NSA 24: 46611) (Paramecia).

Alpatov, W. W. and A. W. Shoobina. 1962. The influence of drugs upon the blood serum toxicity acquired under the action of ionizing radiation. Byull. Eksptl. Biol. Med. 53: 28-31. (In Russian) (NSA 16: 23440) (Paramecia).

Alper, T. 1961. Variability in the oxygen effect observed with microorganisms. Part II. Escherichia coli B. Int. J. Radiat. Biol. 3: 369-377. (NSA 15: 25876).

Alstad, J. A., et al. 1959. Determination of radioactive strontium in Norwegian corn and corn products and in animal and human bones. Svensk Kem. Tidskr. 71: 717-720. (In Norwegian, English summary) (NSA 15: 52).

Altmann, H. 1968. Regulation of the synthesis of different nucleic acids within the cell cycle and the influence of irradiation on this system. Austrian Studiengesellschaft für Atomenergie, Seibersdorf, report AGAE-BL-29/1968. 5pp. (For Symposium on Biologically Active Nucleic Acids, Bucharest, Romania, October, 1967) (CONF-671064) (NSA 23: 6353) (Chlorella).

Alvarez Ramis, C. 1973. Natural contamination levels reached by algae due to alpha emitters in the central valley of the Tajo River. Energia Nucl. (Madrid) 17(84): 303-308. (In Spanish) (NSA 30: 725).

Alvarez Ramis, C. and A. Gregorides de los Santos. 1970. Contamination of terrestrial gastrods living in a biological medium with a low level of alpha contamination due to plutonium and uranium. pp. 961-982. In: Proc. Intern. Symp. on Radioecology. Vol. II. (See France Commissariat à l'Énergie Atomique, 1970) (In French) (NSA 24: 41867).

Alvarez Ramis, C. and A. Gregorides de los Santos. 1973. Accumulation of natural alpha emitters in Spanish gastropods in relation with the activity present in soils and plants. Energia Nucl. (Madrid) 17(85): 357-362. (In Spanish) (NSA 30: 3704).

Alvarez Ramis, C. and A. Gregorides de los Santos. 1974. Contamination of terrestrial gastropods in a biotope with low-level α contamination due to plutonium and uranium. pp. 961-982. In: Proc. Intern. Symp. on Radioecology. Vol. II. (See France Commissariat à l'Énergie Atomique, 1970) (LA-tr-7411) (CONF-690918-5) (NSA 30: 18649).

Alverson, D. L. 1972. Bioenvironmental features: an overview. pp. 845-857. In: The Columbia River Estuary and Adjacent Ocean Waters. (See Pruter and Alverson, 1972) (Incl. comments on radionuclides).

Alvinerie, J., et al. 1968. The use of tritium and carbon-14 to define the water situation in the continental environment. pp. 147-153. In: Intern. Congr. on Radioprotection of the Environment. (See Societé Francaise de Radioprotection, 1968) (In French).

Amaldi, U., Jr., et al. 1971. Criteria for selecting the location of nuclear plants. Ann. Ist. Super. Sanita 7: 626-646. (In Italian) (NSA 26: 47574).

Amano, J. 1955. Radiation dosimetry of fish. Shizen 10(3): 43.

Amano, J. and S. Ishidaira. 1969. Land and the characteristic distribution of radioactivity. Doboku Gijutsu Shiryo 11(12): 605-607. (In Japanese) (NSA 26: 4413).

Amano, Y., H. Tozawa and A. Takase. 1956. Studies on fish contaminated by radiation. Nippon Suisan Gakkaishi 21(12): 1261.

Amano, Y., et al. 1955. Studies on fish contaminated with radiation. 1. Radiation of fish tissue. Nippon Suisan Gakkaishi 20(10): 907.

Amato, A. J. 1971. Mathematical analysis of the effects of wind on redistribution of surface contamination. U. S. AEC report WASH-1187. 7pp. (NSA 26: 4428).

Amato, C. G., et al. 1970. New Jersey's development of a comprehensive environmental surveillance program. pp. 129-135. In: Environmental Surveillance in the Vicinity of Nuclear Facilities., (See Reinig, 1970) (NSA 25: 16169; 16209).

Amavis, R. 1970. Introduction to the study of the physico-chemical behavior of radiostrontium in radioactive effluents. pp. 93-123. In: Proc. Intern. Symp. on Radioecology. Vol. II. (See France Commissariat à l'Énergie Atomique, 1970) (In French).

Amavis, R. and J. Smeets. 1968. Radioactive contamination of water: a new approach needed. Euratom Rev. 7: 67-73. (NSA 23: 2417).

Amavis, R., et al. 1973. Development and application of the radiological capacity concept in radiation protection. pp. 583-592. In: Environmental Behavior of Radionuclides Released in the Nuclear Industry. (See IAEA, 1963) (In French) (NSA 29: 29308).

Ambach, W. 1963. The significance of radiation extinction in glacier ice for energy exchange studies. Acta Phys. Austriaca 16: 382-384. (In German) (NSA 17: 37353).

Ambach, W. and H. Eisner. 1965. Radioactivity measurements for determination of the firn layers of an alpine glacier. Naturwissenschaften 52: 154. (In German) (NSA 19: 32441).

Ambach, W. and H. Eisner. 1965. Investigation of the radioactivity of snow layers of an alpine glacier for the establishment of dating boundaries. Acta Phys. Austriaca 20: 58-62. (In German) (NSA 20: 14740).

Ambach, W., H. Eisner and F. A. Prantl. 1968. Vertical distribution of radioactive fission products in accumulation regions of an alpine glacier (remote valley walls and Otztaler Alps.). Acta Phys. Austriaca 27: 46-49. (In German) (NSA 23: 13867).

Ambach, W., H. Eisner and G. Sauzay. 1969. Tritium profiles in two firn cores from alpine glaciers and tritium content in precipitation in the alpine area. Arch. Meteorol. Geophys. Bioklimatol. B17: 93-104. (NSA 23: 47647).

Ambach, W., et al. 1969. The altitude effect on the isotopic composition of precipitation and glacier ice in the Alps. Tellus 20(4): 595-600).

Ambrose, H. W., III. 1969. A comparison of Microtus pennsylvanicus home ranges as determined by isotope and live trap methods. Am. Midl. Nat. 81 (2): 535-555.

Amer, F., S. Mahdi and A. Alradi. 1969. Limitations in isotopic measurements of labile phosphate in soils. J. Soil Sci. 20: 91-100. (NSA 23: 40570).

Amer, N. M. and C. A. Tobias. 1965. Analysis of the combined effect of magnetic fields, temperature and radiation on development. Radiation Research 25: 172-173.

Amer, N. M., J. V. Slater and C. A. Tobias. 1962. Analysis of the combined influence of x-irradiation and elevated temperatures on development. Abstr. E5. Radiation Research 16: 574. (Used Tribolium).

Ameresekere, R. V. W. E. 1969. Circulifer tenellus Baker (Homoptera: Cicadellidae). Control of reproduction by sterilization procedures. Thesis, Univ. of Calif., Riverside. 191pp. (Univ. Microfilms 70-19,221) (NSA 25: 21994).

Ameresekere, R. V. W. E. and G. P. Georghiou. 1971. Sterilization of the beet leafhopper: induction of sterility and evaluation of biotic effects with a model sterilant (OM-53139) and ^{60}Co irradiation. J. Econ. Entomol. 64(5): 1074-1080. (NSA 26: 7128).

Ameresekere, R. V. W. E., G. P. Georghiou and V. Sevacherian. 1971. Histopathological studies of irradiated and chemosterilized beet leafhoppers, Circulifer tenellus. Ann. Entomol. Soc. Am. 64(5): 1025-1031. (NSA 25: 58093).

American Electric Power Service Corp. 1971. Donald C. Cook Nuclear Plant, Units 1 and 2. Benton Harbor Power Plant limnological studies. VII. New York. 90pp. (U. S. AEC DOCKET-50315-41; DOCKET-50316-41) (NSA 25: 44058) (See also Univ. of Michigan, 1971).

American Electric Power Service Corp. 1971. Donald C. Cook Nuclear Plant, Units 1 and 2. Lake Michigan environmental survey. New York. 119pp. (U. S. AEC DOCKET-50315-40; DOCKET-50316-40) (NSA 25: 44057) (See also above and Indiana and Michigan Electric Co., 1971).

American Electric Power Service Corp. 1971. Donald C. Cook Nuclear Plant, Units 1 and 2. Underwater gamma radiation detection system. New York. 54pp. (U. S. AEC DOCKET-50315-39; DOCKET-50316-39) (NSA 25: 44056) (See above).

American Institute of Biological Sciences. 1956. AIBS Connecticut meetings. Science 124(3230): 1017-1018.

American Institute of Chemical Engineers. 1972. Cooling towers. Editors of Chemical Engineering Progress, New York. 145pp.

American Nuclear Society. nd. *Radiation and Isotope Technology in Latin American Development*. (Proc. Amer. Nucl. Soc. Topical Meeting, San Juan, Puerto Rico, May 4-6, 1969). Puerto Rico Nuclear Center, San Juan, U. S. AEC report PRNC-135. 519pp. (In English and Spanish) (CONF-690508) (NSA 24: 39141).

American Nuclear Society. 1971. *Neutron Sources and Applications*. (Proc. Amer. Nucl. Soc. Topical Meeting, April 19-21, 1971, Augusta, Georgia). Savannah River Lab., U. S. AEC report CONF-710402-(Vol. 3). 173pp. (NSA 25: 50615).

American Society for Testing and Materials. 1958. *Symp. on Radioactivity in Industrial Water and Industrial Waste Water*. (61st Annual Meeting, 1958). Philadelphia. ASTM Spec. Tech. Publ. 235. 77pp. (NSA 13: 13437).

American Society for Testing and Materials. 1965. Standard general methods for analysis of radioisotopes. Philadelphia. ASTM E 181-62. 11pp. (NSA 25: 29005).

American Society for Testing and Materials. 1966. Standard method for measurement of alpha particle radioactivity of industrial water and industrial waste water. Philadelphia. ASTM D 1943-66. 7pp. (NSA 25: 29003).

American Society for Testing and Materials. 1966. Standard method for measurement of beta particle radioactivity of industrial water and industrial waste water. Philadelphia. ASTM D 1890-66. 9pp. (NSA 25: 29009).

American Society for Testing and Materials. 1967. Standard method for measurement of gamma radioactivity of industrial water and industrial waste water. Philadelphia. ASTM D 1690-61. 5pp. (NSA 25: 29008).

American Society for Testing and Materials. 1968. Standard method of test for radioactive barium in industrial water and industrial waste water. Philadelphia. ASTM D 2038-68. (NSA 25: 29004).

American Society for Testing and Materials. 1968. Standard methods of test for radioactive iodine in industrial water and industrial waste water. Philadelphia. ASTM D 2334-68. 11pp. (NSA 25: 29007).

American Society for Testing and Materials. 1969. Standard method of test for ^{59}Fe in water. Philadelphia. ASTM D 2461-69. 5pp. (NSA 25: 29010).

American Society for Testing and Materials. 1969. Standard method of test for radioactive cesium in water. Philadelphia. ASTM D 2577-69. 6pp. (NSA 25: 29000).

American Society for Testing and Materials. 1969. Standard method of test for radioactive manganese in water. Philadelphia. ASTM D 2039-69. 6pp. (NSA 25: 29006).

American Society for Testing and Materials. 1970. Standard method of test for radioactive tritium in water. Philadelphia. ASTM D 2476-70. 3pp. (NSA 25: 54447).

American Society for Testing and Materials. 1970. Standard method of test for radionuclides of radium in water. Philadelphia. ASTM D 2460-70. 5pp. (NSA 25: 28999).

Ames, L. L. 1974. Characterization of actinide bearing soils: to sixty centimeters of 216-Z-9 enclosed trench. Battelle Northwest Lab., U. S. AEC report BNWL-1812. 142pp. (NSA 29: 29340).

Amiard, J. C. and M. le Lannou. 1972. Variations in the concentration factor of ^{60}Co in the mussel, *Mytilus edulis* L. CEA Centre de la Hague, France Commissariat à l'Énergie Atomique, Cherbourg, report CEA-CONF-2252. 3pp. (In French) (NSA 29: 10486).

Amman, G. D. and L. A. Rasmussen. 1969. Techniques for radiographing and the accuracy of the x-ray method for identifying and estimating numbers of the mountain pine beetle. J. Econ. Entomol. 62(3): 631-634.

Amman, G. D. and S. Tunnock. 1971. Radiographic detection of *Agathis pumila*, a parasite of the larch casebearer. J. Econ. Entomol. 64(5): 1086-1088.

Ammermann, D. 1971. Morphology and development of the macronuclei of the ciliates *Stylonychia mytilus* and *Euplotes aediculatus*. Chromosoma 33: 209-238. (NSA 25: 58096) (X-rays; autoradiographic experiments).

Ammon, H. and P. Karlson. 1964. On the metabolism of tyrosine in insects. XIV. Radioautographic localization of sclerotizing substances in pupae of *Calliphora erythrocephala*. J. Insect Physiol. 10(3): 525-527. (In German, English summary) (^{14}C).

Ammon, J. 1964. On the question of the mode of action of a radon-containing atmosphere on embryonic states of *Drosophila melanogaster*. Biophysik 2(1): 29-41. (In German) (NSA 18: 27100).

Amoroso, C. and W. Lorenz. 1959. The question "recovery speed or resistance increase" of the x-irradiated skin of frogs (*Rana esculenta*) in small fractionation. Strahlentherapie 108: 84-92. (In German) (NSA 13: 8589).

Amundsen, C. C. 1972. Amchitka Island bioenvironmental program: plant ecology of Amchitka Island, final report. Battelle Columbus Labs., U. S. AEC report BMI-171-139. iii, 22pp. + maps.

Amundsen, C. C. and E. E. C. Clebsch. 1971. Dynamics of the terrestrial ecosystem vegetation of Amchitka Island, Alaska. BioScience 21(12): 619-623. (NSA 25: 49132).

Amy, J.-P. 1971. Study of ^{106}Ru migration in soils. Centre d'Études Nucléaires, France Commissariat à l'Énergie Atomique, Cadarache, report CEA-R-4168. 87pp. (Also: thesis, Nancy Univ.) (In French) (NSA 26: 27990) (Tracer study).

Amy, J.-P., A. Grauby and F. Jacquin. 1970. Environmental mobility of ^{106}Ru during irrigation processes. Compt. rend. D271: 644-647. (In French) (NSA 25: 10692).

Amy, J.-P., A. Grauby and F. Jacquin. 1970. ^{106}Ru migration in irrigated soils from Bas-Rhone. Radioprotection 5: 185-200. (In French) (NSA 25: 21161).

Amy, R. L. 1955. A comparative study of the effect of β-rays, γ-rays, and x-rays on development of *Habrobracon*. Radiation Research 3(2): 166-181.

Amy, R. L. 1958. The changing ultraviolet sensitivity of the developing *Habrobracon* embro. Anat. Record 132(3): 404.

Amy, R. L. 1965. Photorecovery from the effects of ultraviolet radiation in the developing *Bracon* embryo. Radiation Research 25: 674-683. (NSA 19: 42367).

Amy, R. L. and M. L. Ernst. 1958. Sensitivity of the developing *Habrobracon* (I. E. Bracon) embryo to ultraviolet radiation. *Penn. Acad. Sci. Proc.* 32: 254-260.

Ananyan, V. L. and A. Sh. Avetisyan. 1964. Effect of radioactivity of air in soil on nitrification processes in the soil. Soobshch. Lab. Agrokhim. Akad. Nauk Armyan. SSR (5): 19-24. (In Russian) (NSA 21: 9031).

Ananyan, V. L. and A. Sh. Avetisyan. 1971. Contents of radium and thorium in the soils of Armenia. Soobshch. inst. agrokhim. probl. gidroponiki, Akad. Nauk Armyan. SSR (11): 5-13. (In Russian) (NSA 27: 19805).

Ananyan, V. L. and A. Sh. Avetisyan. 1971. Contents of radiostrontium in the main soils of Armenia. Soobshch. inst. agrokhim. probl. gidroponiki, Akad. Nauk Armyan. SSR (11): 19-24. (In Russian) (NSA 27: 19831).

Ananyan, V. L. and A. Sh. Avetisyan. 1971. Accumulation of radium and calcium in plants as influenced by fertilizers. Soobshch. inst. agrokhim. probl. gidroponiki, Akad. Nauk Armyan. SSR (11): 36-41. (In Russian) (NSA 27: 20178).

Ananyan, V. L. and G. M. Martirosyan. 1965. Beta radioactivity of plants in various regions of Armenia. Izv. Akad. Nauk Armyan. SSR (Biol. nauk) 18(1): 63-66. (In Russian) (NSA 21: 12884).

Ananyan, V. L. and B. G. Mnatsakanyan. 1971. Migration of ^{137}Cs on various soils of Armenia. Soobshch. inst. agrokhim. probl. gidroponiki, Akad. Nauk Armyan. SSR (11): 14-18. (In Russian) (NSA 27: 19830).

Ananyan, V. L. and B. G. Mnatsakanyan. 1971. Migration and distribution of ^{54}Mn in the soils of Armenia. Soobshch. inst. agrokhim. probl. gidroponiki, Akad. Nauk Armyan. SSR (11): 25-30. (In Russian) (NSA 27: 19829).

Ananyan, V. L. and B. G. Mnatsakanyan. 1971. Effect of the level of relief on the distribution of radioactive elements. Soobshch. inst. agrokhim. probl. gidroponiki, Akad. Nauk Armyan. SSR (11): 31-35. (In Russian) (NSA 27: 19832) (Sub-alpine meadows).

Ananyan, V. L. and B. G. Mnatsakanyan. 1971. Uptake of K and ^{137}Cs by alpine and subalpine plants influenced by fertilizers. Soobshch. inst. agrokhim. probl. gidroponoki, Akad. Nauk Armyan. SSR (11): 42-60. (In Russian) (NSA 27: 20179).

Ananyan, V. L. and B. G. Mnatsakanyan. 1973. Radioactivity of air found in soils. Biol. Zh. Armyan. SSR 26(11): 66-69. (In Russian) (NSA 30: 6526).

Ananyan, V. L. and G. S. Sarkisyan. 1971. Effect of fertilizers on the accumulation of Ca and radiostrontium in the alpine and subalpine meadow plants. Soobshch. inst. agrokhim. probl. gidroponiki, Akad. Nauk Armyan. SSR (11): 61-73. (In Russian) (NSA 27: 20180).

Ananyan, V. L., B. G. Mnatsakanyan and L. A. Araratyan. 1969. The influence of fertilizers on the distribution of stable and radioactive (^{54}Mn) manganese in alpine field plants. Dokl. Akad. Nauk Armyan. SSR 48: 115-119. (In Russian) (NSA 23: 47938).

Ancel, P. and P. Vintemberger. 1925. Influence of rapidity of development of irradiated *Rana* eggs and length of their survival. Compt. rend. soc. biol. 93: 281-283. (In French).

Ancel, P. and P. Vintemberger. 1927. Influence of temperature on the radiosensitivity of birds' and reptiles' eggs. Compt. rend. soc. biol. 97: 796-799. (In French).

Andelman, J. B. and T. C. Rozzell. 1970. Plutonium in the water environment. I. Characteristics of aqueous plutonium. pp. 118-137. In: *Radionuclides in the Environment*. R. F. Gould, ed. American Chemical Society, Washington.

Anders, G. and H. Ursprung. 1959. Formation of pigmentation in eyes of *Drosophila melanogaster* after experimental damage to image sites. Rev. Suisse Zool. 66(2): 259-265.

Andersen, A., G. Nielsen and H. Sorensen. 1961. Growth chamber for labelling plant material uniformly with radiocarbon. Physiol. Plantarum 14: 378-383. (NSA 19: 10771).

Andersen, B. V. 1959. Hanford environmental monitoring annual report, 1958. Hanford Atomic Products Operation, General Electric Co., U. S. AEC report HW-61676. 18pp. (NSA 14: 7638).

Anderson, A. O. and D. M. Allred. 1964. Kangaroo rat burrows at the Nevada Test Site. Great Basin Nat. 24(3-3): 93-101.

Anderson, C. N., et al. 1969. Aerial radiological measuring surveys (ARMS). Systems and procedures employed through FY69. EG & G, Inc., U. S. AEC report EGG-1183-1457. Tech. Report No. L-967. 49pp. (NSA 26: 28115).

Anderson, E. C., et al. 1958. Barium-140 radioactivity in foods. Science 127(3293): 283-284.

Anderson, G. E., J. B. Gentry and M. H. Smith. 1973. Relationships between levels of radiocesium in the dominant plants and arthropods in a contaminated stream-bed community. Oikos 24(9): 165-170. (Also: Savannah River Ecology Lab., U. S. AEC report SRO-708-2) (NSA 26: 50984; 29: 15647).

Anderson, H. F. and W. E. Sheehan. 1968. Environmental monitoring report: July-December 1967 and 1967 summary. Mound Lab., Miamisburg, O., U. S. AEC report MLM-1464. (NSA 22: 21205).

Anderson, H. F. and W. E. Sheehan. 1968. Environmental monitoring report: January-June 1968. Mound Lab., Miamisburg, O., U. S. AEC report MLM-1522. (NSA 22: 45183).

Anderson, H. F. and W. E. Sheehan. 1969. Environmental monitoring report: July-December 1968 and 1968 summary. Mound Lab., Miamisburg, O., U. S. AEC report MLM-1619. 50pp. (NSA 23: 21823) (Incl. concentration of radionuclides in the Great Miami River).

Anerson, J. B., E. C. Tsivoglou and S. D. Shearer. 1963. Effects of uranium mill wastes on biological fauna of the Animas River (Colorado-New Mexico). pp. 373-383. In: Radioecology. (See Schultz and Klement, 1963) (NSA 17: 33568).

Anderson, K. 1969. Tritium in New York State waters. 1965. Radiological Health Data 10: 93-97. (NSA 24: 2612).

Anderson, K. D. 1972. Environmental radioactivity measurement experience near a fuels reprocessing plant. Trans. Amer. Nucl. Soc. 15(1): 544-545. (NSA 26: 42948).

Anderson, R. V. 1957. Effects of x-rays on the wheat stem rust fungus, *Puccinia graminis* var. *tritici*. Phytopathology 47(9): 515.

Anderson, R. Y. 1955. Biogeochemical reconnaissance of the Annie Laurie Uranium Prospect, Santa Cruz County, Arizona. Econ. Geol. 50: 227-232.

Anderson, R. Y. and E. B. Kurtz. 1954. Factors influencing uranium accumulations in plants. Abstr. Bull. Geol. Soc. Am. 65: 1226.

Anderson, T. D. and J. W. Michel. 1973. Reactor as a source of industrial energy. Oak Ridge Nat. Lab. Rev. 6(3): 19-23. (NSA 29: 28764) (Incl. environmental considerations).

Anderson, T. R. 1970. Temperature adaptation and the phospholipds of membranes in goldfish (*Carassius auratus*). Comp. Biochem. Physiol. 33 (3): 663-687. (SFA 15: 12361) (^{14}C).

Andersson, I. Ö. and I. Nilsson. 1961. Radioactivity in reindeer-breeding Laplanders living in the north of Sweden. Sweden Aktiebolaget Atomenergi, October. (See Sandmeier and Stover, 1962).

Andersson, I. Ö., I. Nilsson and K. Eckerstig. 1962. Investigation of the 800 kev peak in the gamma spectrum of Swedish Laplanders. Sweden Aktiebolaget Atomenergi, Stockholm, report AE-80. (NSA 17: 5763).

Andrawes, N. R. and H. W. Dorough. 1967. Metabolic fate of carbaryl-naphthyl-^{14}C in boll weevils and bollworms. J. Econ. Entomol. 60(2): 453-456.

Andresen, N., C. Chapman-Andresen and H. Holter. 1952. Autoradiographic studies on the amoeba *Chaos chaos* with C^{14}. Compt. rend. trav. lab. Carlsberg chim. 28: 190-220.

Andresen, N., et al. 1950. The distribution of food on amoeba cytoplasm studied by means of auto-radiography. Exptl. Cell Res. 1: 139-142. (NSA 4: 2275).

Andrew, J. T. G., et al. 1965. The use of radioisotopes as ground-water tracers in the Burdekin Delta area of North Queensland, Australia. Research Establishment, Australian Atomic Energy Comm., Lucas Heights, report AAEC/E-137. 44pp. (NSA 19: 42646).

Andrew, R. W. and W. M. Cox. 1965. Sediment properties and their effect on radionuclide retention in the Clinch and Tennessee Rivers. Abstr. Health Physics 11(8): 820.

Andrews, H.L. and S. Warren. 1969. Ion scavenging by the eastern clam and quahog. Health Physics 17: 807-810. (NSA 24: 5979) (^{131}I, ^{85}Sr, ^{59}Fe).

Andrews, J. N. and D. F. Wood. 1972. Mechanism of radon release in rock matrices and entry into groundwaters. Inst. Mining Met. Trans. B81: 198-209. (NSA 27: 11977).

Andrews, R. H. and M. B. Fielding. 1971. Gamma irradiation of raw sewage for sedimentation purposes. Isotopes Radiat. Technol. 8(4): 452-453. (NSA 25: 48537).

Andreyev, A. G., et al. 1969. Estimation of age of uranium-bearing hydromorphous soils on the basis of ionium accumulation. Geokhimiya (6): 724-728. (Geochem. Int. 6: 580-584) (NSA 23: 35335).

Andreyev, P. F., I. V. Andreyeva and E. M. Pogozina. 1962. Salts of uranyl interaction with vegetable tissue components and several of its derivatives. Geokhimiya 4: 313-317. (In Russian) (NSA 16: 20090).

Andreyev, S. V. 1957. Study and formulation of methods of controlling plant pests and diseases by means of radioisotopes and radiation. pp. 621-___. In: Proc. of the All-Union Conference of Workers in Agricultural Science, Moscow, 1957.

Andreyev, S. V., B. K. Martens and V. V. Kablov. 1971. Gamma ray sterilization of pests. Izotopy SSSR (21): 10-14. (In Russian) (NSA 26: 41227).

Andreyev, S. V., B. K. Martens and V. V. Kablov. 1972. Radiation sterilization for pest control purpose in agriculture. Isotopenpraxis 8(2): 68-70. (In German) (NSA 26: 50962).

Andreyev, S. V., B. K. Martens and V. A. Molchanova. 1962. Use of radioisotopes and radiation in the field of plant protection. pp. 23-27. In: _Radioisotopes and Radiation in Entomology_. (See IAEA, 1962) (AEC-tr-5142) (NSA 16: 30190).

Andreyev, S. V., B. K. Martens and V. A. Molchanova. 1969. Development of research in sterilization techniques for agricultural insect pests in the USSR. pp. 57-64. In: _Sterile-Male Technique for Eradication or Control of Harmful Insects_. (See IAEA, 1969).

Andreyev, S. V., V. A. Molchanova and B. K. Martens. 1960. Utilization of radioactive isotopes for marking insects. Zashita Rasteniy 5(2): 45-47. (In Russian).

Andreyev, S. V., Z. I. Samoylova and B. K. Martens. 1964. The potential application of γ radiation for the sterilization of _Chloridea obsoleta_ for the purpose of reducing its population. Radiobiologiya 4(4): 624-626. (AEC-tr-6407; pp. 180-183) (NSA 18: 41190).

Andreyev, S. V., et al. 1958. Some results of the use of tracer techniques in the study of plant protection. Proc. 2nd Intern. Conf. on Peaceful Uses of Atomic Energy 27: 85-92. (Paper A/CONF.15/P/2309) (NSA 13: 6309).

Andreyev, S. V., et al. 1962. Investigation of the effect of the irradiation dose upon mortality and sexual sterilization of the granary weevil (_Calandra granaria_ L.). Radiobiologiya 2(4): 758-762. (AEC-tr-5432; pp. 170-176) (NSA 16: 30190).

Andreyev, S. V., et al. 1962. Possibilities of using gamma radiation for sexual sterilization as a method of controlling pests of agricultural crops. pp. 19-20. In: _Materials of the Symp. on the Use of Biophysics in the Field of Plant Protection_. Leningrad, 1961. (R. Zh. Biol. 1963: 18Zh326).

Andreyev, S. V., et al. 1963. Gamma irradiation in the control of insects. Zashchita Rasteniy 9: 25-26. (Ref. zh. (biol.) 1963: 6E212).

Andreyev, S. V., et al. 1964. Radioactive labeling in the study of insect pests. Vestn. Sel'skokhoz. Nauki, Vses. Akad. Sel'skokhoz. Nauk 9(2): 122-128. (In Russian; English, French, German summaries) (NSA 18: 38724) (_Eurygaster_ and _Hadena_).

Andreyev, S. V., et al. 1967. Using gamma rays for sterilization of harmful insects. Vestn. Sel'skokhoz. Nauki (min. sel'sk.) SSSR (1): 48-52. (In Russian) (NSA 22: 43510).

Andreyev, S. V., et al. 1968. Agricultural pests sterilized by gamma-radiation. Radiobiologiya 8(5): 742-746. (AEC-tr-7023; pp. 122-129; JPRS-47070) (NSA 23: 10128; 24: 935).

Andreyev, S. V., et al. 1971. γ-radiation application in the control of insect pests in agricultural plants. pp. 89-95. In: _Application of Isotopes and Nuclear Radiation_. Atomizdat, Moscow. (In Russian) (NSA 30: 19013).

Andreyev, S. V., et al. 1972. Radio-mutagenesis of phyto- and entomo-pathogenic fungi. Tsitol. genet. 6(5): 395-399. (In Russian) (NSA 27: 27836).

Andreyev, S. V., et al. nd. Use of radioisotopes and radiation in the control of insect pests of plants and animals. 11pp. (JPRS-20425; OTS-63-31422) (NSA 17: 33622).

Andreyeva, Ye. A. 1960. Natural soil radioactivity and total potassium determination by the radiometric method. Pochvovedeniye (5): 21-29. (Soviet Soil Science (5): 477-483).

Andrumaitis (Andrushaitis), G. P. (ed.). 1973. _Radioecology of Water Organisms_. II. _Distribution and Migration of Radionuclides in Freshwater and Sea-Water Biocenoses_. III. _The Effect of Ionizing Radiation on Water Organisms_. Izdatel'stvo Zinatne, Riga. 33, 152pp. (NSA 30: 21082, 24395).

Anellis, A. and R. B. Koch. 1962. Comparative resistance of strains of _Clostridium_ to gamma rays. Appl. Microbiol. 10: 326-330.

Anellis, A., N. Grecz and D. Berkowitz. 1965. Survival of _Clostridium botulinum_ spores. Appl. Microbiol. 13: 397-401. (NSA 19: 27963).

Angeletti, L. 1972. Contamination of pastures by ^{131}I. Part I. Rate of deposition of ^{131}I. Centre d'Études Nucléaires, France Commissariat à l' Énergie Atomique, Fontenay-aux-Roses, report EUR-4872. 26pp. (In French) (NSA 27: 2534).

Angelini, A. M. 1972. Electric power and the environment. Com. Naz. Energ. Nucl. Notiz. 18(8-9): 33-63. (In Italian) (NSA 27: 9551).

Angelovic, J. W. 1965. Some effects of accumulated radium on the productivity of algae. Thesis, Utah State Univ., Logan. 149pp. (NSA 20: 14381).

Angelovic, J. W. and D. W. Engel. 1968. Interaction of gamma irradiation and salinity on respiration of brine shrimp (_Artemia salina_) nauplii. Radiation Research 35(1): 102-108. (NSA 22: 43400).

Angelovic, J. W. and D. W. Engel. 1970. Effects of radiation on estuarine organisms. Marine Pollution Bull. (n.s.) 1: 103-105.

Angelovic, J. W., J. C. White, Jr. and E. M. Davis. 1969. Interactions of ionizing radiation, salinity, and temperature on the estuarine fish, _Fundulus heteroclitis_. pp. 131-141. In: _Symposium on Radioecology_. (See Nelson and Evans, 1969).

Angelovic, J. W., J. C. White, Jr. and D. W. Engel. 1966. Influence of salinity on the response of estuarine animals to ionizing radiation. ASB Bull. 13(2): 29.

Angepova-Latarinskaya, Ye. and A. M. Kuzin. 1971. Dynamics of the excretion of cell metabolites during the post-radiation recovery of yeasts. Radiobiologiya 11(2): 275-278. (AEC-tr-7269; pp. 179-183) (NSA 25: 44973).

Anghileri, L. J. 1957. Decontamination and potability of the waters of the Rio de la Plata after a nuclear explosion. Publs. Com. Nacional de Energia Atomica, Buenos Aires (quim.) 1(12): 141-188. (In Spanish) (NSA 12: 8965; 14: 21828).

Anghileri, L. J. 1958. Decontamination and potabilization of the waters of the Rio Plate following accidental contamination by fission products. Proc. 2nd Intern. Conf. on Peaceful Uses of Atomic Energy 18: 605-619. (Paper A/CONF.15/P/2345) (NSA 13: 6047).

Anghileri, L. J. 1959. Studies of the contamination of _Prochilodus platensis_ (shad) with fission products. Argentina Com. Nacional de Energia Atomica, Buenos Aires, report CNEA-15. 19pp. (In Spanish) (NSA 14: 12446).

Anghileri, L..J. 1960. Studies of the contamination and absorption of ^{90}Sr and ^{137}Cs by _Prochilodus platensis_ (shad). Argentina Com. Nacional de Energia Atomica, Informe No. 21. 14pp. (NP-9368) (HW-tr-41) (15: 28982; 16: 26721).

Angino, E. E., G. K. Billings and N. Anderson. 1966. Observed variations in the strontium concentration in sea water. Chem. Geol. 1: 145-153. (NSA 21: 10485).

Angot, M., M. S. Doty and M. Oguri. 1958. Measure of the primary productivity in water with the carbon-14 technique. Inst. Franc. d'Oceanie (Ocean. Rappt. Sci.) 4: 1-30.

Anguenot, F. and G. Courtois. 1969. Study of natural radioactivity of sediments of the Atlantic Loire. I. Measuring the turbidity by natural radioactivity. II. Studies of sediment deposits. France Dept. of Radioelements, Nantes, report DR/SAR. S/69/FA/GC/JJ. 65pp. (ORNL-tr-2278) (NSA 24: 16673).

Anikeyeva, I. D. and Ye. V. Moskvitin. 1971. Results of combined exposure of _Chlorella_ to vibration and gamma radiation. Kosmich. biol. med. 5(3): 85-87. (JPRS-53801; pp. 133-136) (NSA 25: 49315; 25: 52254).

Anikeyeva, I. D., Ye. N. Vaulina and V. A. Shevchenko. 1964. Influence of UV rays on _Chlorella_. Radiobiologiya 4(6): 883-893. (AEC-tr-6409; pp. 104-118) (NSA 19: 17615).

Anikeyeva, I. D., Ye. N. Vaulina and V. A. Shevchenko. 1966. Effect of UV rays on various states of the cell cycle of _Chlorella_. pp. 16-21. In: _Effect of Ionizing Radiation on Heredity_. (See Dubinin) (NSA 22: 21631).

Anil, Y. D. 1964. The effect of thymidine-^{3}H on the mortality of _Drosophila melanogaster_ larvae. Instab. Univ. Fen. Fak. Mecm. B. 29(3-4): 143-147.

Ankersmit, G. W. 1958. Labelling of insects by radioisotopes and its possible applications. Landbouwk. Tijdschr. Groningen 70(extra no.): 350-356, disc. 356-357. (In Dutch).

Annenkov, B. N., I. K. Dibobesa and R. M. Alexakhin. (Eds.). 1973. _Radiobiology and Radioecology of Domestic Animals_. Atomizdat, Moscow. 224pp. (AEC-tr-7523) (NSA 28: 10882; 29: 29863).

Anno, Y. and H. Oga. 1959. Early reactions on the blood vessels of the frog tongue after irradiation with beta rays. Strahlentherapie 109: 144-149. (In German) (NSA 13: 16731).

Anokhin, A. S. and V. S. Anokhina. 1971. Resistance of seeds of lupine varieties to x radiation. Vestsi Akad. Navuk Belorussk. SSR (biyal.) (3): 56-59. (In Belorussian) (NSA 25: 58149).

Anokhin, V. L. 1974. _Simulation of Migration Processes of Radioisotopes in Dry Land Areas_. Atomizdat, Moscow. 144pp. (NSA 30: 18638).

Anokhin (Anochin), V. L., Yu. (Ju.) M. Svirezhev and A. N. Tyuryukanov (Tjurkukanov). 1967. Mathematical model for migration of radiostrontium in soil. pp. 43-51. In: _Radioecological Concentration Processes_. (See Aberg and Hungate, 1967).

Anonymous.

1915. Radioactivity of the water at Hot Springs, Ark., and Saratoga Springs, New York, and other points in America. Radium 5(1): 10-14.

1953. The Tyler Project. Texas Health Bull. 6(3): 8-15.

1953. The Tyler Project. Texas Health Bull. 6(7): 12-15.

1956. Possible irradiation of birds. Bull. Brit. Ornithol. Club 76(3): 37.

1957. Radiation effects on fish - new study. Atomic 8(2): 38.

1958. Plants and animals accumulate radioactive wastes. Ingeniøren 2 (51): 7.

1959. Water pollution research laboratory. Research (London) 12(5): 241-242. (Incl. radioactivity).

1960 - 1974. Radiological Health Data; Radiological Health Data and Reports; Radiation Data Reports. Reports in semi-journal form were issued mostly on a monthly basis by the U. S. Public Health Service and the U. S. Environmental Protection Agency. Included are routine data on air, water, milk, human diets, and other environmental data. Some articles show authors' names which are listed in this compilation under those names.

1961. Radiostrontium and radiocaesium in drinking water in the United Kingdom. 1961 results. 19pp. (Also: Brit. Waterworks Assoc. J. 44: 415-421) (NSA 16: 24033; 17: 14549).

1961. Radioactive isotope studies. Oceanus 8(1): 20-24. (NSA 16: 26731).

Anonymous. (Continued). (1962, see Intern. Congr. Radiation Research).

1961. Atomic energy used against insect pests. Presse Med. Franc. 69 (7): 326.

1961. Histological investigations on the "premature death" x-irradiated *Drosophila* eggs. Naturwissenschaften 48(8): 310.

1961. Fall-out birds on the fly. New York Herald Tribune, Oct. 2, 1961.

1962. Atomic heat brings down flying birds. Baltimore Sun, Oct. 26, 1962.

1963. Addendum I to National Bureau of Standards Handbook 69. Maximum permissible body burdens and maximum permissible concentrations and radionuclides in air and water for occupational exposure. Health Physics 9(6): 565-568. (NSA 17: 36050).

1963. Hazard of the hot caribou. New Scientist 18(339): 362.

1963. Effect of ionizing radiation on man and his environment. Atomics 16(6): 16-22. (NSA 18: 3516).

1963. Eradicating the melon fly. Agric. Res. (London) 12(2): 5. (Sterile fly technique).

1963. Gamma radiation as a means of insect control in wheat and wheat products. Appl. Atomics 413: 11-12.

1963. Radioisotopes in tropical medicine. World Hlth. Org. Chron. 17: 83-96.

1964. Nature rises from the ashes. Business Week (1832): 116-118. (On Marshall Islands).

1964. Applications of nuclear energy in entomology. Bol. Junta Control Energia At. (Peru) 9(48): 103-106. (In Spanish) (NSA 19: 13631).

1964. Environmental studies. pp. 65-77. In: Annual Report, 1963. Nat. Inst. Radiol. Sci., Chiba, Japan, report NIRS-3. (NSA 19: 31672) (Incl. ^{137}Cs and ^{90}Sr in river water).

1964. Guiding radioactive sewage into surface waters. Part II. Wasser Luft Betrieb 8: 469-473. (In German) (NSA 19: 20234).

1965. Gamma radiation. Emory Magazine 41(1): 5-6. (Radiation facility).

1965. Russian thistle invasion of Sedan crater. NTS News 8(14): 1, 4-5, 7. (Nevada Test Site).

1965. Oriental fruit flies.....gone from Guam. Agric. Res. (Wash.) 13 (11): 12-13. (*Dacus doralis*; sterile male).

1965. How far a fly can fly. Agric. Res. (Wash.) 13(12): 5pp. (*Lucilia*; sterilization by irradiation).

1965. Control of Mediterranean fruit fly in Central America. Plant Prot. Bull. FAO 13(2): 47-48.

1965. Irradiation. A biological control weapon against omnivorous leaf roller and codling moth. Agric. Res. (Wash.) 14(5): 7.

1965. Symposium on the use of radioisotopes in hydrological research. Bull. Amer. Met. Soc. 46: 552-562.

1966. Bulletin of the atmospheric radioactivity, July-September 1966. Taiki Hoshano Kansoku Seiseki (46): 1-60. (NSA 22: 10381).

1966. Irradiation and atherosclerosis in pigeons. Nutrition Rev. 24: 178-179. (NSA 22: 8615).

1966. The radioactive grasshopper. Intern. Atomic Energy Agency Bull. 8 (2): 14-16. (^{192}Ir).

1967. Summaries of ^{137}Cs data from northern Canada. pp. 263-266. In: *Radioecological Concentration Processes*. (See Aberg and Hungate, 1967).

1967. ^{55}Fe levels in fish and in fish-eating populations. Nutrition Rev. 25: 24-26. (NSA 21: 18018).

1967. Mining *in situ* by nature's easy way. Eng. Mining J. 168: 75-80. (NSA 22: 10392).

1967. Radioactivity in drinking water in the United Kingdom: 1966 results. HMSO, London. 20pp. (Also: 1968. Brit. Waterworks Assoc. J. 50: 33-38) (NSA 22: 15017; 40711).

1967. Successes against insects and parasites. Intern. Atomic Energy Agency Bull. 9(5): 26-28. (NSA 22: 31305).

1967. 8th Congress of the International Association of Hydrogeologists. Topics include use of isotopes in groundwater studies, Istanbul, Turkey, September 10-22, 1967. Mem. Ass. Int. Hydrogeol. Reunion Istanbul 8: 1-585. (CONF-670972) (NSA 23: 439).

1968. Flies of Capri. Intern. Atomic Energy Agency Bull. 10: 8-11. (NSA 22: 34004) (Sterile male).

1969. Doing a lake a favor. Cascades 10(2): 23-25. (Concerns tracers in Fern Lake, Washington).

1969. Proving aquatic life and a nuclear plant can exist in harmony. EEI Bull. 37: 146-147. (NSA 23: 40863).

1969. Radioactive fallout monitoring in Italy. Com. Naz. Energ. Nucl. Notiz. 15: 95-97. (In Italian) (NSA 24: 9531).

1969. Radioactivity in drinking water in the United Kingdom: 1967 results. Brit. Water Supply 1: 42-45. (NSA 24: 14482).

1969. Radioisotope directory. Nuclear News 12(12): 1-44. (NSA 24: 7598).

1970. Comparing radioactive emissions. Power 114(4): 31-32. (NSA 24: 20541).

1970. Hunting hot spots in Utah. Nuclear News 13(2): 20-21. (NSA 24: 19096).

1970. Nuclear plants can pollute water with discharges of heat and radioactivity. Power 114(4): 32. (NSA 24: 20540).

1970. Biological irradiation of plants using ^{137}Cs. Tech. Mod. 62: 441-442. (In French) (NSA 25: 21235).

1970. Environmental radioactivity, Moncloa Station (Madrid). Energia Nucleare (Madrid) 14: 67-68. (In Spanish) (NSA 25: 23718).

1970. Proc. 1st Conf. on the State of Advancement of Radioecology in Italy. (Instituto di Zoologica dell'Universita' Degli Studi di Parma e Associanzione Italiana di Fisica Sanitaria e Protezione Controle Radiazioni, Parma, 5-6 Novembre 1970). v, 156pp. (In Italian) (Many references to Italian literature).

1971. Hazards of mercury. Environ. Res. 4(1): 1-69. (NSA 25: 31866) (Neutron activation).

1971. Nuclear power plants in the USA. Atomwirt. Atomtech. 16: 150. (In German) (NSA 25: 25893).

1971. Principles underlying the control of the discharge of radioactive wastes to the environment. Atom (London) 172: 36-38. (NSA 25: 18635).

1971. Radioisotopic sand-tracer study. Isotopes Radiat. Technol. 8(4): 392-395. (NSA 25: 48674).

1971. Tentative method of analysis for tritium content of the atmosphere. Health Lab. Sci. 8(2): 107-110. (NSA 25: 48457).

1971. Radioactive fallout on Greenland and the Faroe Islands in 1966, 1967, 1968, and 1969. Fra Sundhedsstyr. 5: 323-324. (In Danish) (NSA 26: 40838).

1971. Environmental health aspect in the nuclear energy industry. Genshi Doryoku Kenkyukai Teirei Kenkyukai Nenkai Hokokusho (45): 5pp. (NSA 26:2577).

1971. Plutonium-244 on the earth. Nature 234(5325): 123.

1972. 1971 Nuclear Science Symposium and Nuclear Power Symposium, San Francisco, California, November 3-5, 1971. Trans. Nucl. Sci. NS-19(1): 1-909. (NSA 26: 43026).

1972. Peaceful use of nuclear energy. Draft of the 4th atomic program finished. Inform. bild. wiss. (11): 149-150. (In German) (NSA 27: 29604) (Incl. environmental research).

Anonymous. (Continued).

1972. Study group reports on nuclear safety and environment. At. Jap. 16(8): 14-16. (NSA 28: 5552).

1972. Uninhabited islets considered for disposal of radioactive wastes. At. Jap. 16(10): 15-18. (NSA 28: 8837).

1972. Energy and environment. Intern. J. Environ. Stud. 3(4): 301-306. (NSA 30: 18686).

1973. Environmental isotope hydrology. Intern. Atomic Energy Agency Bull. 15(3): 42-48.

1973. Mighty river, nuclear river. Pacific Northwest SEA 6(3): 4-7, 14. (Concerns Columbia River).

1973. Proc. 2nd Conf. on the State of Advancement of Radioecology in Italy. Instituto di Zoologia dell'Universita' Degli Studi di Parma e Associanzione Italiana di Fisica Sanitaria e Protezione Contro le Radiazioni, Parma, 24-25 Maggio 1973. 294pp. (In Italian) (Many references to Italian literature).

1973. Resolution. All-Union Symp. on the Radioecology of Aquatic Organisms. pp. 133-___. In: Radioecology of Water Organisms. III. (See Andrushaitis et al., 1973) (AEC-tr-7592; pp. 86-89).

1973. Environmental consequences: atomic power plants. pp.43-61. In: Notebook on Energy. Trondheim, Norway. (In Norwegian) (NSA 29: 20857).

1973. Environmental radiation under attack. Environ. Sci. Technol. 7 (13): 1104-1105. (NSA 29: 7976).

1973. Use of radioecology in selection or study of electronuclear and industrial sites. (AEC-tr-7476).

1974. Tanzawa pumice found in the western part of Yokahama City and the ^{14}C age of Quaternary formation in Japan. Chikyu Kagaku 28(1): 49-50. (In Japanese) (NSA 30: 31257).

1974. Application of the sterility technique for the eradication of the paddy borer, Tryporyza incertulas. K'un Ch'ung Hsueh Pao 17(2): 135-147. (NSA 31: 33525).

1974. Conclusions. pp. 256-258. In: Chemo-Radioecology of Pelagic and Benthic Zones (Metals and Radionuclides in Aquatic Organisms and Environment). (See Polikarpov, 1974).

1974. Environmental consequences of nuclear and conventional extension of power plants. A statement submitted to the Ministry of Commerce, Shipping, and Industry by the Energy Committee of the Power Plants. Denmark Handelministeriet, Copenhagen. (In Danish) (INIS-mf-1339) (NSA 31: 11243).

Anonymous. (Continued).

1974. Guidelines for measures of environmental monitoring of nuclear power plants with light-water-cooled reactors. Umwelt (Inform. Bundesmin. Inn.) (35): 8-14. (In German) (NSA 32: 3491).

1974. Preliminary study on the reproduction of oncomelaniid snails following ^{60}Co irradiation. Tung Wu Hsueh Pao 20(3): 290. (NSA 31: 33519).

1974. Radioactive effluents from nuclear power stations in the Community. Discharge data radiological aspects. Belgium Directorate for Social Affairs, Commission of the European Communities, Brussels. 44pp. (NP-20826) (NSA 31: 14054).

1974. Sampling of water, food, and other media. Paper 22, vp. In: Environmental Analysis and Environmental Monitoring for Nuclear Power Generation. (See University of California, 1974). (NSA 32: 14312).

1974. Soil moisture, irrigation, and water use efficiency. pp. 269-302. In: Use of Radiations and Radioisotopes in Studies of Plant Productivity. (See India Dept. of Atomic Energy) (NSA 33: 12046) (Techniques).

1975. Nuclear energy: technical, economic and ecological data. Délegation Generale à l'Information, Paris. (Suppl. No. 239, Actualites Service). 119pp. (In French) (NSA 33: 16232).

1975. Nuclear Fuel Cycle. A Survey of the Public Health, Environmental, and National Security Effects of Nuclear Power. MIT Press, Cambridge, Mass. 307pp. (NSA 33: 25560).

nd. Biochemical and chemical studies on Astralagus leaves and roots. 1. Enzymatic. 2. Translocation of ^{75}Se with radioautographs. 3. Absorption and exchange of ^{75}Se in root seedlings. Wyoming Agr. Expt. Sta., Laramie, Bull. 385. 43pp. (NSA 16: 30201).

Anspaugh, L. R. and P. L. Phelps. 1974. Results and data analysis: resuspension element status report. pp. 265-297. In: Dynamics of Plutonium in Desert Environments. (See Dunaway and White, 1974). (NSA 30: 29452).

Anspaugh, L. R., J. J. Koranda and W. L. Robison. 1973. Environmental aspects of natural gas stimulation experiments with nuclear devices. pp. 37-52. In: Radionuclides in Ecosystems. (See D. J. Nelson, 1973). (NSA 26: 4746; 28: 27607).

Anspaugh, L. R., P. L. Phelps and G. W. Huckabay. 1972. Methods for the in situ measurement of radionuclides in soil. pp. 12-38. In: Workshop on Natural Radiation Environment. (See McLaughlin, 1972) (NSA 27: 19826).

Anspaugh, L. R., J. H. Shinn and P. L. Phelps. 1975. Resuspension and redistribution of plutonium in soils. Lawrence Radiation Lab., U. S. AEC report UCRL-76419. 39pp. (NSA 31: 29618).

Anspaugh, L. R., J. H. Shinn and D. W. Wilson. 1974. Evaluation of the resuspension pathway toward protection guidelines for soil contamination with radioactivity. pp. 513-524. In: Population Dose Evaluation and Standards for Man and His Environment. (See IAEA, 1974) (Also: Lawrence Radiation Lab., U. S. AEC report UCRL-75250) (CONF-740529-2, SM-184-13) (NSA 30: 9499; 31: 19555).

Anspaugh, L. R., et al. 1972. In-situ measurement of radionuclides in the environment with ge(Li) spectrometer. pp. 279-303. In: Natural Radiation Environment II. (See Adams, Lowder and Gesell, 1972) (Also: Lawrence Radiation Lab., U. S. AEC report UCRL-73904) (CONF-720805-6) (NSA 27: 22162; 32: 19716).

Anspaugh, L. R., et al. 1973. Wind-driven redistribution of surface deposited radioactivity. pp. 167-184. In: Environmental Behavior of Radionuclides Released in the Nuclear Industry. (See IAEA, 1973) (Also: Lawrence Radiation Lab., U. S. AEC report UCRL-74392) (CONF-730503) (NSA 28: 27259; 29: 29334).

Anspaugh, L. R., et al. 1973. Field spectrometric measurements of radionuclide concentrations and external gamma exposure rates at the Nevada Test Site. Lawrence Radiation Lab., U. S. AEC report UCRL-51412. 19pp. (NSA 28: 21071).

Anthony, D. S. and W. P. Norris. 1946. Apparatus for injection of animals with dangerous amounts of hard beta and gamma emitters. Univ. of Chicago, U. S. AEC report AECD-1965 (CH-3852) (ADC 2; p. 308).

Anthony, L. J. 1966. Sources of Information on Atomic Energy. Vol. 2. International Series of Monographs in Library and Information Science. Pergamon Press, New York. 254pp. (NSA 21: 8186).

Antipov, V. V., et al. 1965. Results of biological experiments carried out under conditions of "Vostok" flights and with the participation of cosmonauts A. G. Nikolayev, P. R. Popovich, and V. F. Bykovskiy. In: Problems of Space Biology. N. M. Sisakyan, ed. Akad. Nauk SSSR, Moscow 4: 239-251. (English transl. NASA report N66-19292).

Antipov, V. V., et al. 1966. Investigation of the complex effect of ionizing radiation and other factors of space flight on the organism. Abstr. 87. p. 23. In: 3rd Intern. Congr. of Radiation Research. (Cortina d' Ampezzo, 26 June - 2 July 1966). G. Silini, ed. North-Holland, Amsterdam.

Antipov, V. V., et al. 1968. Radiobiological experiments on "Cosmos-110" biosatellite. 19th Congress of the International Astronautical Federation, October 1968. Moscos. 38pp. (JPRS-46977).

Anton, H. J. 1967. Measurement of radioactively labeled tissue sections. Histochemie 8: 19-21. (In German) (NSA 21: 30844) (Axolotl).

Antonelli, A., A. Belli and A. Brondi. 1974. Siting problems of nuclear power plants at the sea coast. Ing. Nucl. 15: 11-14. (In Italian) (NSA 32: 16595).

Antonelli, A., et al. 1969. Ecological study of the C. R. N. site at Trisaia. Italy Comitato Nazionale per l'Energia Nucleare, Rome, report RT/PROT-(69)43. 56pp. (In Italian) (NSA 24: 39469).

Antonelli, A., et al. 1970. Ecological site survey at the nuclear research center at Trisaia. Minerva Fisiconucl. 14: 109-119. (In Italian) (CONF-690929-10) (NSA 25: 29479).

Antonelli, A., et al. 1971. Radioecological study of the area around the Eurex plant in Saliggia. pp. 245-285. In: Proc. Intern. Symposium on Radioecology Applied to the Protection of Man and His Environment. (See Commission of the European Communities, 1971) (In Italian) (NSA 27: 5228).

Antonelli, A., et al. 1971. Determinazione di fattori di concentrazione in componenti abiotici e biologici del fiume garigliano. pp. 1305-1320. In: Proc. Intern. Symposium on Radioecology Applied to the Protection of Man and His Environment. (See Commission of the European Communities, 1971) (In Italian) (NSA 27: 5269).

Antonelli, A., et al. 1973. Radioecological investigations around the Brasimone site (Tosco-Emiliano Appennines). Abstr. pp.87-88. In: Proc. 2nd Conf. on the State of Advancement of Radioecology in Italy. (See Anonymous, 1973) (In Italian).

Antonelli, A., et al. 1973. Radioecological survey on the site of Brasimone Lake (Appennini Mountains). G. Fis. Sanit. Prot. Radiza. 17(3-4): 103-112. (In Italian) (NSA 31: 19678).

Antoni, F., et al. 1973. Application of nuclear techniques in environmental protection. Izotoptechnika 16(12): 639-653. (In Hungarian) (NSA 29: 18332).

Antonijevic, M. and D. Lebez. 1962. Influence of time factor on nitrogen excretion by perfused liver after whole-body irradiation. Abstr. In: 2nd Intern. Cong. Radiation Research. (See Anonymous, 1962) (Frogs).

Antonijevic, M. and D. Lebez. 1966. Studies on the nitrogen metabolism in frog liver after whole-body irradiation. IV. Influence of splenectomy, spleen extract and serum of irradiated frogs on the nitrogen excretion by perfused frog liver. Strahlentherapie 130: 280-284. (Also: Yugoslavia Nuklearni Inst., Lujbljana, report NIJS-R-473, 1965) (NSA 20: 18207; 20: 35307).

Antonov, A. S., et al. 1964. Nucleotide composition of rapidly tagged ribonucleic acid in silk gland of *Bombyx mori*. Dokl. Akad. Nauk SSSR 154(1): 216-219. (Doklady Biochem. 154: 1-3) (^{32}P).

Antonucci, L. 1971. Model for preliminary calculation of the environmental receptivity for gaseous radioactive effluents from a boiling water reactor nuclear power station. pp. 531-559. In: Proc. Intern. Symp. Radioecology Applied to the Protection of Man and His Environment. (See Commission of the European Communities, 1971) (In Italian) (NSA 27: 5251).

Antropova, Z. G., et al. 1971. Some peculiarities of the behavior or radioactive and stable strontium in meadow biogeocoenoses. Radiobiologiya 11(2): 301-304. (AEC-tr-7269; pp. 223-227) (NSA 25: 44776).

Antunes, C. M. F., et al. 1971. Study of the effects of gamma-radiation on eggs and miracidia of *Schistosoma mansoni*. Rev. Inst. Med. Trop. São Paulo 13(6): 383-386. (NSA 26: 36449).

Anwar, M. 1968. Some effects of gamma irradiation on *Spodoptera* (*Laphygma*) *exigua* Hb. (Lepidoptera: Noctuidae). pp. 109-121. In: Isotopes and Radiation in Entomology. (See IAEA, 1968) (In French, English summary) (NSA 24: 32166).

Anwar, M. 1968. Sterilization of the European corn borer *Ostrinia nubilalis* by gamma irradiation. pp. 139-148. In: Isotopes and Radiation in Entomology. (See IAEA, 1968) (In French, English summary) (NSA 24: 32318).

Anwar, M. and M. D. Arif. 1971. Sterilization of the spotter bollworm of cotton by radiation. Int. J. Appl. Radiat. Isotopes 22(10): 625-627. (NSA 25: 55368).

Anwar, M. and M. Feron. 1971. Mating behavior of gamma-irradiated and non-irradiated European corn borer, *Ostrinia nubilalis* (Hubner). Int. J. Appl. Radiat. Isotopes 22: 89-93. (NSA 25: 32516).

Anwar, M., et al. 1971. Radiation-sterilization of the Mediterranean fruit fly (Diptera: Tephritidae): comparison of spermatogenesis in flies treated as pupae or adults. Ann. Entomol. Soc. Am. 64(3): 627-633. (NSA 25: 42751).

Anzani, A., et al. 1963. Measure of ambient radioactivity - Ispra 1961. Nuclear Research Center, Ispra, Italy, European Atomic Energy Community, report EUR-233.i. 68pp. (NSA 17: 27591).

Aoki, K. 1963. The effect of whole body x-ray irradiation on the hematopoietic tissue of goldfish. *Carassius auratus*. Zoological (Tokyo) 72 (10): 283-288. (In Japanese) (NSA, Japan 3(5): 01070).

Aoki, K., N. Egami and R. Arai. 1966. Histological changes in interrenal tissue of the goldfish, *Carassius auratus*, following x-irradiation. J. Radiat. Res. (Japan) 7: 83-90. (NSA 21: 13034).

Aoyagi, N. and T. Bojo. 1966. On the radioactivity level of the Tertiary beds in the Tsugawa area, Niigata Prefecture. Cish. Chosaho Geppo 17: 451-458. (In Japanese) (NSA 21: 36168) (Fossils).

Aoyama, I., et al. 1970. Dynamic concentration process of radioactive substances in aquatic organisms. Tentative approach from stochastic theory. Hoken Butsuri 5: 135-140. (In Japanese) (NSA 25: 10998).

Apasheva, L. M., et al. 1972. Investigation of the role of thiols in the determination of the radiosensitivity of various strains of *Chlorella vulgaris* Beijer. Radiobiologiya 12(6): 904-906. (AEC-tr-7421; pp. 137-140) (NSA 28: 2838).

Apasheva, L. M., et al. 1973. Investigation of the role of thiols in the modification of the radiosensitivity of Chlorella during the cell cycle. Radiobiologiya 13(1): 102-105. (AEC-tr-7430; pp. 136-140) (NSA 28: 18442).

Apostol, I. and M. Oncescu. 1964. On the meteorological parameters in the measuring of moisture. Acad. Rep. Populare Romine Inst. Fiz. Atomica, Bucharest, report IFA-MR-21. 12pp. (NSA 19: 7766) (Discusses neutron scattering).

Aquatic Life Advisory Committee. 1960. Acquatic life water quality criteria. J. Water Poll. Control Fed. 32(1): 65-82. (BA 35: 51947; SFA 6: 3915).

Aragao, M. B. 1953. Utilizacao de mosquitos radioactivos na avaliacao do raio de voos. Rev. Bras. Malariol. D. Trop. 5(2): 137-143. (In Portugese).

Aragao, M. B. 1954. Utilizacao de mosquitos radioactivos na avaliacao do raio. Mosquito News 14(1): 41.

Aragao, M. B. 1954. Marcacao de triatomineos com torio. Rev. Bras. Malariol. D. Trop. 6(3): 353-364.

Aragao, M. B., E. F. Pessoa and N. Margem. 1953. A new method for radioactive marking of mosquitoes. Rev. Bras. Malariol. D. Trop. 5(1): 69-73.

Aragao, M. B., E. F. Pessoa and N. Margem. 1956. A new radioactive method for marking mosquitoes and its application. Proc. Intern. Conf. Peaceful Uses of Atomic Energy 10: 525-528. (Paper A/CONF.8/P/140).

Aranda Centurian. B. S. 1971. Application of four different doses of gamma radiation to adult Sitotroga cerealella (Olivier). pp. 37-38. In: Sterility Principle for Insect Control or Eradication. (See IAEA, 1971) (NSA 25: 29618).

Araragi, M. and S. Ishizawa. 1971. Change of microflora especially actinomycete flora in (γ)-field soil. Nogyo Gijutsu Kenkyusho Hokoku B. (22): 59-84. (In Japanese) (NSA 26: 31245).

Arce, A. A. 1968. Possibilities for using the sterile-male technique in torsalo, Dermatobia hominis, and other livestock pests in the Oirsa area. pp. 1-3. In: Control of Livestock Insect Pests by the Sterile-Male Technique. (See IAEA, 1968) (NSA 24: 34806).

Archibald, R. S. 1950. Radiotracers in flow tests. J. Boston Soc. Civ. Engrs. 37(1): 46-116.

Ardashnikov, S. N. 1960. Some properties of peroxides and oxygen effect. pp. 146-152. In: Role of Peroxides and Oxygen in the First Stages of the Radiobiological Effect. Publ. House Acad. Sci. USSR, Moscow. (NP-tr-1479) (NSA 21: 14563) (Tadpoles).

Argentina Comision Nacional de Energia Atomica. 1972. Compilation of the results of monitoring ^{90}Sr and ^{137}Cs due to fallout in the Argentine Republic. Buenos Aires, report RS-34/66. 11pp. (In Spanish) (NP-19357) (NSA 26: 45520).

Arghittu, C. 1960. Studies and experiments in radiomicrobiology. Assimulation of Fe^{59} , I^{131} , and S^{35} by Staphylococcus enterotoxicus. Minerva Nucleare 4: 313-315. (In Italian) (NSA 15: 10663).

Arifov, V. A., et al. 1957. Effect of gamma radiation on live cocoons of the mulberry silkworm. Dokl. Akad. Nauk Uzbek. SSR 4: 9-12.

Arifov, V. A., et al. 1958. The use of radiation to kill silkworm cocoons. Proc. 2nd Intern. Conf. on the Peaceful Uses of Atomic Energy 27: 444-450.

Arifov, V. A., et al. 1961. The use of gamma rays in the primary processing of silkworm cocoons. pp. 432-435. In: Proc. Tashkent Conf. on the Peaceful Uses of Atomic Energy, 1959. Ya. Kh. Turkulov, ed. Publ. House Akad. Sci. Uzbek. SSR, Tashkent. (AEC-tr-6293; pp. 734-738).

Ariizumi, A. and O. Kondo. 1963. Measurement of river-bed variations during a flood. pp. 237-254. In: Radioisotopes in Hydrology. (See IAEA, 1963).

Ariizumi, A. and O. Kondo. 1963. Investigation of the movement of infiltrating acidic hot-spring water in the ground by means of radioisotopes. pp. 365-382. In: Radioisotopes in Hydrology. (See IAEA, 1963) (NSA 18: 1928).

Arkhangel'skiy, M. M. 1958. Application of radioactive isotopes for investigating suspended sediments in rivers. pp.____. In: Proc. All-Union Scientific-Technical Conf. on the Use of Radioactive Isotopes and Radiation in the People's Economy and in Science. Vol. 4. Heat Engineering and Hydrodynamics. M. A. Styrikovich, G. E. Kholodovskiy and M. S. Fomichev, eds. (AEC-tr-4206).

Arkhilov, N. P., et al. 1975. Radionuclide ratios in agricultural and biological chains in the context of the problem of radioactivity protection of the environment. pp. 267-276. In: Isotope Ratios as Pollutant Source and Behavior Indicators. (See IAEA, 1975) (In Russian) (NSA 33: 7617).

Armentano, T. V. and G. M. Woodwell. 1975. Sedimentation rates in a Long Island marsh determined by ^{210}Pb dating. Limnol. Oceanogr. 20(3): 452-456. (NSA 32: 17214).

Armitage, D. B. 1970. Determination and distribution of various trace elements in natural waters by x-ray fluorescence spectroscopy. Thesis, Univ. of Hawaii, Honolulu. 86pp. (Univ. Microfilms 71-4934) (NSA 25: 28919).

Armstrong, N. E. 1968. Numerical solutions of radionuclide transport equations and role of plants in ^{90}Sr transport. Thesis, Univ. of Texas, Austin. 153pp. (NSA 23: 789) (Used Vallisneria americans Michanx).

Armstrong, N. E. and E. F. Gloyna. 1967. Mathematical models for the distribution of radionuclides in aquatic systems. Univ. of Texas, Austin, U. S. AEC report ORO-490-2. 19pp. (NSA 21: 30532).

Armstrong, N. E. and E. F. Gloyna. 1968. Radioactivity transport in water: numerical solution of radionuclide transport equations and role of plants in ^{85}Sr transport. Univ. of Texas, Environmental Health Engineering Research Lab., Austin, Tech. Report 14. 153pp. (NSA 22: 33815).

Armstrong, N. E. and E. F. Gloyna. 1968. Numerical solutions of radionuclide transport equations. Chem. Engrg. Progr. Symp. Ser. 64(90): 47-58. (NSA 23: 13863).

Armstrong, N. E. and E. F. Gloyna. 1969. Mathematical models for the dispersion of radionuclides in aquatic systems. pp. 329-335. In: Symposium on Radioecology. (See Nelson and Evans, 1969).

Armstrong, N. E., L. W. Canter and E. F. Gloyna. 1968. Transport of organic and radioactive wastes. Water Wastes 5: 54-56. (NSA 23: 11653) (in a model river).

Armstrong, R. H. and V. E. Quinn. 1972. Weather predictions and surface radiation estimates for the Cannikin event: final report. Air Resources Lab., U. S. NOAA, U. S. AEC report ARLV-351-26. v, 47pp.

Arnason, A. P., R. A. Fuller and J. W. T. Spinks. 1950. An electronic method of tracing movements of soil-inhabiting insects. Science 111(2871): 5-6.

Arnason, B. and Th. Sigurgeirsson. 1967. Hydrogen isotopes in hydrological studies in Iceland. pp. 35-47. In: Isotopes in Hydrology. (See IAEA, 1967) (NSA 22: 18926).

Arnason, T. J., R. L. B. Irwin and J. W. T. Spinks. 1951. P^{32}-induced lethal mutations in Drosophila. Can. J. Res., D.-Can. J. Zool. 29(3): 234-239.

Arnaud, M. 1966. Continuous exposure of a culture of Chlorella to x-rays. Thesis, Univ. of Grenoble, Grenoble, France. 24pp. (In French) (NP-17125) (NSA 22: 12945).

Arnaudi, C. and G. Venturelli, 1934. The action of radium on plant tumors. Riv. Biol. 16: 61-79. (Rev. Applied Mycol. 13: 497) (CA 29: 878).

Arndt, J. and R. Wolter. 1971. Rapid determination of gamma-ray emitting radionuclides in waters. pp. 311-318. In: Rapid Methods for Measuring Radioactivity in the Environment. (In German) (See IAEA, 1971) (NSA 26: 27817).

Arndt, J., et al. 1974. Natural radioactivity in phosphate fertilizers. F. R. Germany, Bundesgesundheitsamt, Berlin, report WaBoLu-17/74. 16pp. (In German) (NSA 33: 25977).

Arndt, J., et al. 1974. The radium and uranium content of phosphate fertilizers. pp. 555-569. In: Radiation Protection and Environmental Protection. Vol. 2. (In German) (AED-CONF-74-725-000 (Vol. 2) (NSA 33: 25974).

Arnold, C.-G. 1963. Investigation of x-ray mutations of Oenothera berteriana. II. On the reasons for F_1 variations after x irradiation. Biol. Zentr. 82(2): 185-207. (In German) (NSA 18: 11676).

Arnold, C.-G. and M. Kessel. 1965. Increasing the translocation rate in trisomic mutants of Oenothera berteriana. Z. Vererbungsl. 97: 255-271. (In German) (NSA 20: 22770).

Arnold, C.-G. and H. Ranneberg. 1967. Nonreciprocal recombination in Chlamydomonas reinhardi. Ber. Deut. Bot. Ges. 80: 512-516. (In German) (NSA 23: 26760) (Gamma irradiation).

Arnold, C. R. and F. H. Rigler. 1967. A possible source of error in the ^{14}C method of measuring primary productivity. Limnol. Oceanogr. 12(1): 121-124.

Arnold, D. E. 1965. Marking fish with dyes and other chemicals. Thesis, Cornell Univ., Ithaca, N. Y. (Incl. radioisotopes).

Arnold, D. E. 1966. Marking fish with dyes and other chemicals. Tech. Papers, Sport Fish. and Wildlife, No. 10. U. S. Fish and Wildlife Serv., Washington. 44pp. (Discusses radionuclides).

Arnold, G. 1965. An autoradiographic study of RNA synthesis in isolated salivary glands of Drosophila hydei. I. Autoradiographic studies. J. Morphology 116(1): 65-87. (^3H).

Arnold, J. C. and L. C. Walker. 1961. Phosphorus-32 uptake by pine tree roots. Georgia Agr. Res. 3: 11.

Arnold, J. R. and E. C. Anderson. 1957. The distribution of carbon-14 in nature. Tellus 9(1): 28-32.

Arroya Varela, M. and Mellando Brauns. 1964. Effects of the ingestion of radioactive phosphorus on the setting of the fruit fly (Ceratitis capitata) Wied.). Boln. Patol. Veg. Ent. Agric. 27: 183-200.

Arroyo Varela, M., et al. 1965. Application of radioactive isotopes to the investigation of methods for the biological control of pests. I. Tests on marking adults of C. capitata Wied. with ^{32}P. II. Tests on marking larvae of C. capitata with ^{32}P. III. The obtaining of sterile males of C. capitata by irradiation of the pupae with gamma-rays. IV. The effects of gamma-radiation on C. capitata when the dose of irradiation is fractionated. Boln. Patol. Veg. Ent. Agr. 28: 233-292. (In Spanish).

Arslanov, Kh. A., et al. 1968. Radiocarbon dating by the scintillation method. Geokhimiya (2): 198-206. (Geochem. Int. 5(1): 148-157) (NSA 23: 21812).

Artemenkova, L. V. 1971. Methods and equipment used in the USSR for determining environmental radioactivity. pp. 95-107. In: Nuclear Techniques in Environmental Pollution. (See IAEA, 1971) (NSA 25: 32080).

Arthur, B. W. 1962. Metabolism of systemic and other recent insecticides in animals. pp. 65-81. In: Radioisotopes and Radiation in Entomology. (See IAEA, 1962).

Arthur, B. W. and J. E. Casida. 1957. Mode and action of insecticides, metabolism and selectivity of 0,0-dimethyl 2,2,2-trichloro-1-hydroxy-ethyl phosphonate and its acetyl and vinyl derivatives. Agr. and Food Chem. 5 (3): 186-192.

Arthur, B. W. and J. E. Casida. 1959. Biological activity and metabolism of Hercules AC-528 components in rats and cockroaches. J. Econ. Entomol. 52(1): 20-27.

Arthur, C. R. and F. H. Rigler. 1967. A possible source of error on the ^{14}C method of measuring primary productivity. Limnol. Oceanogr. 12(1): 121-124.

Arthur, W. 1959. Radioactive insects? Highlights Agr. Res. (Alabama Sta.) 6(2): 4.

Artom, C., H. B. Lofland and T. B. Clarkson. 1965. Ionizing radiation, atherosclerosis, and lipid metabolism in pigeons. Radiation Research 26: 165-177. (NSA 20: 110).

Aruscavage, P. J. 1972. Neutron activation analysis procedure for the determination of mercury in soil and rock samples. U. S. Geol. Surv. Prof. Paper 800-C: 209-214. (NSA 26: 55682).

Arvy, L., J. A. Boiffard and M. Gabe. The modifications of the hemogram determinations with Rana temporaria L. through irradiation of the hypophysis with x-rays. Compt. rend. Soc. Biol. 141: 455-457. (In French).

Arya, H. C. and A. C. Hildebrandt. 1969. Differential sensitivities to gamma radiation of Phylloxera gall and normal grape stem cells in tissue culture. Can. J. Botany 47: 1623-1628. (NSA 24: 14623).

Asato, Y. 1969. Genetic studies of the blue-green alga, Analystis Midulans Thesis, Univ. of Hawaii, Honolulu. 111pp.

Asato, Y. 1971. Photorecovery of gamma irradiated cultures of blue-green alga, Anacystis nidulans. Radiation Botany 11(4): 313-316. (NSA 25: 49316).

Asakura, K., Y. Kishida and T. Kido. 1969. Effect of x-irradiation on the regeneration of the planarian, Dugesia japonica. Scient. Rept. Kanazawa Univ. (Japan) 14: 253-261. (NSA 24: 46649).

Ashby, W. C., et al. 1967. Ecological dosimetry: radiation levels influenced by plant growth. Science 155: 1430-1432. (NSA 21: 18045).

Ashizawa, A. 1959. Effect if irradiation at lower temperature on the regeneration of the amphibian tail. Nippon Igaku Hoshasen Gakkai Zasshi 19: 1416-1434. (NSA 15: 22100).

Ashley, C. and C. C. Zeigler. 1975. Environmental monitoring at the Savannah River Plant. Annual report, 1974. Du Pont de Nemours and Co., U. S. AEC report DPSPU-75-302. 74pp. (NSA 33: 632).

Ashraf, M., et al. 1974. Reproduction biology and radiation sterilization of the sugar cane top borer, Tryporyza nivella. Entomol. Exp. Appl. 17(1): 61-66. (NSA 31: 815; 31: 33523).

Ashraf, M., et al. 1974. Comparative mating ability of gamma sterilized and normal fruit flies, Dacus zonatus. Radiation Res. 60(3): 541-544. (NSA 31: 8932).

Ashraf, M., et al. 1971. Effects of gamma radiation on the larval midgut of the Indian-meal moth, Plodia interpunctella (Lepidoptera: Phycitidae). Radiation Res. 45: 349-354. (NSA 25: 26916).

Ashrafi, S. H., J. H. Brower and E. W. Tilton. 1972. Gamma radiation effects on testes and mating success of the Indian meal moth, Plodia interpunctella. Ann. Entomol. Soc. Am. 65(5): 1144-1149. (NSA 26: 56749).

Ashrafi, S. H., E. W. Tilton and J. H. Brower. 1972. Inheritance of radiation-induced partial sterility in the Indian meal moth. J. Econ. Entomol. 65(5): 1265-1268. (NSA 26: 59117).

Ashton, M. 1956. The effects of whole-body irradiation on the melanophores of the common frog (Rana temporaria). pp. 419-426. In: Progress in Radiobiology (Proc. 4th Intern. Conf., Cambridge, 14-17 August 1955). J. S. Mitchell, B. E. Holmes and C. L. Smith, eds. Chas. C. Thomas, Springfield, Illinois.

Ashwood-Smith, M. J., J. Copeland and J. Wilcockson. 1968. Response of bacterial spores and Micrococcus radiodurans to ultraviolet irradiation at low temperatures. Nature 217: 337-338. (NSA 22: 33972).

Aslam, K. M. and M. Aslam. 1970. Bacteria-aided water leaching of uranium ores. Nucleus (Karachi) 7: 28-36. (NSA 25: 8337).

Asman, M. 1966. Cytogenetic and developmental effects of gamma-irradiation on Aedes aegypti (L.). Thesis, Univ. of Notre Dame (U. S.). 112pp.

Asman, M. and K. S. Rai. 1965. Radiation-induced chromosomal aberrations in Aedes aegypti. Abstr. 347. Bull. Entomol. Soc. Am. 11(3): 172.

Asman, M. and K. S. Rai. 1965. Radiosensitivity of the yellow-fever mosquito, Aedes aegypti. Proc. Indiana Acad. Sci. 75: 109.

Asman, M. and K. S. Rai. 1967. Gamma radiation and heterosis in Aedes aegypti. Abstr. 310. Bull. Entomol. Soc. Am. 13(3): 201.

Asman, M. and K. S. Rai. 1972. Developmental effects of ionizing radiation in Aedes aegypti. J. Med. Entomol. 9(5): 468-478. (NSA 27: 17677).

Aspinall, A. 1972. Neutron activation analysis of prehistoric flint mine products. Archaeometry 14(1): 41-53. (NSA 26: 25140).

Aspinall, A., et al. 1972. Neutron activation analysis of faience beads. Archaeometry 14(1): 27-40. (NSA 26: 25141).

Asselin-Bagarry, D. 1970. Study of manganese cycle in fresh waters and brine using ^{54}Mn. Thesis, Aix-Marseilles Univ., France. 199pp. (In French) (NSA 24: 48464).

Astapov, S. V. 1958. Experiments in the use of cobalt and iodine radioactive isotopes in land improvement investigations. Gidrotekhnika i Melioratsiya SSSR 10(1): 22-29. (English transl. OTS-60-21151) (NSA 15: 6312).

Astaurov, B. L. 1947. The direct proof of the nuclear nature of the biological effect of x-rays and the independence of the final results of irradiation on primary alterations in the cytoplasm. Zh. obshch. biol. 8: 421-441. (In Russian) (Silkworm).

Astaurov, B. L. 1958. Differential effect of radiation injuries of the nucleus and cytoplasm as a result of their functional specificity. Byul. Mo Mosk. Obshch. Isp. Prirody (biol.) 63: 35-50. (In Russian) (Silkworm).

Astudillo, M. D., F. Sanz and C. Lopez-Zumel. 1966. Effect of γ rays on Na+- K+-ATPase, an enzymatic factor which mediates in active sodium transport through biological membranes. Rev. Espanola fisiol. 22: 115-120. (NSA 21: 24360) (Frog skin).

Astudillo, M. D., F. Sanz and M. C. Lopez-Zumel. 1967. Effect of x rays on the transport of sodium through frog skin. An. real soc. Espanola fis. quim. B63: 827-832. (In Spanish) (NSA 21: 45553).

Atakan, Y. 1972. Atomic weapon tritium hydrology in a ground water body with free groundwater surface. Thesis, Heidelberg Univ., Germany. 101pp. (In German) (NSA 32: 6439).

Atakan, Y., et al. 1974. Sandhausen shallow-groundwater tritium experiment. pp. 21-43. In: Isotope Techniques in Groundwater Hydrology. Vol. 1. (See International Atomic Energy Agency, 1974) (NSA 31: 19618).

Atanesyan, S. G. 1968. Some fallout patterns of nuclear explosion products in the Armenian SSR. State Committee on the Use of Atomic Energy, USSR. Atomizdat, Moscow. UN Document A/AC.82/G/L.1307. (In Russian) (CA 73: 126302a; NSA 24: 9524).

Aten, A. H. W., Jr., J. W. Dalenberg and W. C. M. Bakkum. 1961. Concentration of uranium in sea fish. Health Physics 5(3/4): 225-226. (NSA 15: 23776).

Athalye, V. V. and K. B. Mistry. 1972. Foliar retention, transport, and leaching of ^{210}Po and ^{210}Pb. Radiation Botany 12(4): 287-290. (NSA 26: 25141).

Athanasiu, G. 1926. Radioactivity of various sources of minerals, thermal and fresh waters of Transylvania, Crisana and Banat. Annuar. inst. geol. Romaniei 12(1): ____. (In French).

Athanasiu, G. 1966. Radium determination in the mineral water springs of Romania. Radioactive unbalance of elements in the uranium series. Acad. rep. pop. Romine. Studii cerc. geol. geofiz. geogr. (Geofiz.) (Bucharest) 4: 263. (In French) (NSA 22: 14584).

Athavale, R. N. and D. Lal. 1967. Measurement of tritium activity in natural waters. Part II. Characteristics of global fallout of ^{3}H and ^{90}Sr. Proc. Indian Acad. Sci. A65: 73-103. (NSA 21: 22046).

Atlan, H. 1968. Application of information theory to the study of the stimulating effects of ionizing radiation, thermal energy, and other environmental factors: preliminary ideas for a theory of organization. J. Theor. Biol. 21: 45-70. (NSA 23: 29429; 25: 8742).

Atlan, H., J. Miguel and R. Binnard. 1969. Differences between radiation-induced life shortening and natural aging in Drosophila melanogaster. J. Gerontol. 24: 1-4. (NSA 23: 31889).

Atlan, H., J. Miguel and G. Welch. 1970. Effects of low- and high-LET cyclotron-accelerated α particles on longevity of Drosophila melanogaster. Intern. J. Radiat. Biol. 18: 423-431. (NSA 25: 11248).

Atlavin, A. B., O. E. Veveris and N. A. Dubinskaya. 1970. Neutron activation determination of Mo in biological materials. Latv. PSR Zinat. Akad. Vest. (fiz. teh. zinat. ser.) (6): 3-9. (In Russian) (NSA 25: 29019) (Incl. fish).

Atomic Energy of Canada, Ltd. 1964. Activities of AECL in the fields of biology, health physics and medicine. Chalk River, Ontario, report AECL-2107. 34pp. (NSA 19: 10719) (Incl. ecology).

Atomic Industrial Forum. 1972. Water Quality Considerations: Siting and Operating of Nuclear Power Plants. (Conference held at Washington, D. C., October 1-4, 1972). (CONF-7210105) (NSA 28: 27343).

Atomic Industrial Forum. 1973. Seminar on the Engineering Aspects of Siting and Operation of Power Plants. (February 13-14, 1973, Washington, D. C.). New York. 310pp. (NP-19964) (NSA 30: 6568).

Atrokhin, V. G. 1958. Radioautography in forestry. Lesnoye Khozyayastvo 11(6): ___. (English transl. OTS 60-21868; 9pp.)

Attaway, L. D. 1972. Environmental Protection Agency research programs in determining impact of nuclear plant operation on water quality. vp. In: Water Quality Considerations: Siting and Operating of nuclear Power Plants. (See Atomic Industrial Forum, 1972) (NSA 28: 29240).

Attrep, M., J. A. Enochs and L. D. Broz. 1971. Atmospheric technetium-99. Environ. Sci. Technol. 5(4): 344-345.

Atwal, A. S. and S. L. Sethi. 1964. Effect of gamma radiation on the sterilization of Coryra cephalonica Stanton (L. Pyralididae). pp. 6-7. In: Information Circular on Radiation Techniques and Their Application to Insect Pests, No. 4. International Atomic Energy Agency, Vienna. (WP/31/4).

Atwood, K. C., R. C. von Borstel and A. R. Whiting. 1956. An influence of ploidy on the time of expression of dominant lethal mutations in Habrobracon. Genetics 41(6): 804-813. (NSA 11: 5147).

Atwood, N. D. 1969. Flora of the National Reactor Testing Station, Idaho. Thesis, Brigham Young Univ., Provo, Utah.

Atwood, N. D. 1970. Flora of the National Reactor Testing Station. Brigham Young Univ. Sci. Bull. (Biol.) 11(4): 1-46.

Au, F. H. F., W. F. Beckert and J. C. McFarlane. 1976. Plutonium uptake by a soil microorganism, Aspergillus niger. National Environmental Research Center, Las Vegas, Nevada, report NERC-LV-539-37. 16pp. (NSA 33: 23602).

Aubouin, G., et al. 1968. Destructive and nondestructive analysis of soils and minerals by neutron activation. Centre d'Études Nucléaires, France Commissariat à l'Énergie Atomique, Grenoble, report CEA-CONF-1183. 15pp. (For National Conference on Measurement of Elements Occuring in Traces in Rocks, Nancy, France, December 4, 1968) (In French) (CONF-681205-2) (NSA 23: 13671).

Auclair, J. L. 1963. Aphid feeding and nutrition. Ann. Rev. Entomol. 8: 439-490. (Considers radionuclides).

Auclair, J. L. 1964. Recent advances in the feeding and nutrition of aphids. Can. J. Entomol. 96(1-2): 241-249. (Contains references using radionuclides).

Auerbach, S. I. 1958. The soil ecosystem and radioactive waste disposal to the ground. Ecology 39(3): 522-529.

Auerbach, S. I. 1959. The ecology of White Oak Lake bed - a radioactive waste area, Part I (abstr.). Proc. 9th Intern. Bot. Congr. 2(a): 1.

Auerbach, S. I. 1961. Effect of radiation on an ecological system of the biosphere due to atomic industry operation: introductory statement. International Atomic Energy Agency, Vienna. (pl-30(II)/SR/11).

Auerbach, S. I. 1961. A United States ecological survey limited to the biological effects of radiation. International Atomic Energy Agency, Vienna. (Pl-30(II)/SR/19).

Auerbach, S. I. 1962. Summary of session, First National Symposium on Radioecology 1961, cycling in the terrestrial environmnent. Book 2, pp. 318-347. In: Radioactive Fallout from Nuclear Weapons Tests. (See Klement 1962) (NSA 16: 22454).

Auerbach, S. I. 1962. Analytical requirements for environmental surveys. Nuclear Safety 3(3): 51-54.

Auerbach, S. I. 1963. Behavior of radionuclides in ecosystems. Abstr. Bull. Ecol. Soc. Am. 44(3): 79.

Auerbach, S. I. 1964. Cycling of radioactive materials in natural communities. Abstr. In: 10th Intern. Bot. Congr., Abstr. of Papers. (Edinburgh, August 1964).

Auerbach, S. I. 1965. Radionuclide cycling: current status and future needs. pp. 1355-1361. In: Radiation and Terrestrial Ecosystems. (See Hungate, 1965) (Also: Health Physics 11: 1355-1361; Oak Ridge National Lab., U. S. AEC report ORNL-P-1305) (NSA 20: 12573).

Auerbach, S. I. 1966. Introductory remarks on radioecology and the effects of ionizing radiation on natural populations. pp. 1-2. In: Radiation Effects on Natural Populations. (See Sacher, 1966).

Auerbach, S. I. 1968. Postattack insect problems. pp. 137-142. In: Proceedings of the Symposium on Postattack Recovery from Nuclear War. Off. of Civil Defense, Washington.

Auerbach, S. I. 1969. Introduction. p 1. In: Progress Report in Post-attack Ecology. (See Auerbach, 1969).

Auerbach, S. I. 1969. Progress Report on Post Attack Ecology: Interim Progress Report. Oak Ridge National Lab., U. S. AEC report ORNL-TM-2466. 66pp.

Auerbach, S. I. 1970. Ecological considerations in siting nuclear power plants: the problem of long-term biotic effects. pp. 29-44. In: Nuclear Power and the Public. H. Foreman, ed. Univ. of Minnesota Presss, Minneapolis. (NSA 25: 23206).

Auerbach, S. I. 1970. Ecology around a reactor. Oak Ridge Nat. Lab. Rev. 3(5): 23-28. (NSA 25: 18048).

Auerbach, S. I. 1971. Ecological considerations in siting nuclear power plants: the long-term biotic effects problem. Nuclear Safety 12: 25-34. (NSA 25: 12763).

Auerbach, S. I. 1973. Contributions of radioecology to AEC mission programs. pp. 3-8. In: Radionuclides in Ecosystems. (See Nelson, 1973) (NSA 28: 27230).

Auerbach, S. I. and D. A. Crossley, Jr. 1958. Strontium-90 and cesium-137 uptake by vegetation under natural conditions. Proc. 2nd Intern. Conf. Peaceful Uses of Atomic Energy 18: 494-499. (Paper A/CONF.15/P/401) (NSA 12: 14554).

Auerbach, S. I. and D. A. Crossley, Jr. 1960. A sampling device for soil microarthropods. Acarolgia t. II(3): 279-287.

Auerbach, S. I. and J. S. Olson. 1963. Biological and environmental behavior of ruthenium and rhodium. pp. 509-519. In: Radioecology. (See Schultz and Klement, 1963) (NSA 17: 33603).

Auerbach, S. I. and V. Schultz. 1962. Onsite ecological research of the Division of Biology and Medicine at the Oak Ridge National Laboratory. U. S. AEC report TID-16890. xiii, 146pp. (NSA 17: 13883).

Auerbach, S. I. and S. Warren. 1969. Medical and ecological effects of nuclear weapons. pp. ____. In: Survival and the Bomb: Methods of Civil Defense. E. P. Wigner, ed. Indiana Univ. Press, Bloomington.

Auerbach, S. I., D. A. Crossley, Jr. and M. D. Englemann. 1957. Effects of gamma radiation on Collembola population growth. Science 126(3274): 614. (NSA 11: 13245).

Auerbach, S. I., D. A. Crossley, Jr. and A. F. Shinn. 1968. Postattack insect problems. pp. 137-142. In: Proc. of Symposium on Postattack Recovery from Nuclear War. (Held at Fort Monroe, Virginia, November 6-9, 1967). U. S. Office of Civil Defense and Office of Emergency Planning, Washington.

Auerbach, S. I., J. S. Olson and H. D. Waller. 1963. Landscape studies using radioactive tracers: I. Application and initial behavior of Cs-137 in a tulip poplar forest. Abstr. Bull. Ecol. Soc. Am. 44(3): 73.

Auerbach, S. I., J. S. Olson and H. D. Waller. 1964. Landscape investigations using caesium-137. Nature 201(4921): 761-764. (NSA 18: 15762).

Auerbach, S. I., J. S. Olson and H. D. Waller. 1967. Data on activity transfer to the forest floor. pp. 467-469. In: Radioecological Concentration Processes. (See Aberg and Hungate, 1967).

Auerbach, S. I., et al. 1964. Mineral cycling studies in forest ecosystems using radioactive caesium. Abstr. p. 249. In: 10th Intern. Bot. Congr., Abstracts of Papers, Edinburgh, Scotland, August, 1964.

Auerbach, S. I., et al. 1971. Ecological considerations in reactor power plant siting. pp. 805-820. In: Environmental Aspects of Nuclear Power Stations. (See IAEA, 1971) (CONF-700810-1) (SM146/53) (NSA 24: 45899).

Auerbach, S. I., et al. 1974. Significance of ecological analyses in the interpretation of environmental releases of radionuclides. Trans. Nucl. Sci. MS-21(1): 18-22. (NSA 30: 6486).

Auess, H. E. 1958. The radioactivity of the atmosphere and hydrosphere. Ann. Rev. Nuclear Sci. 8: 243-256.

Aufhammer, G., G. Fischbeck and A. Haisch. 1965. The application of gamma irradiation installations to the treatment of plants. Schrift. Bundesmin. Wiss. Forsch. Radionucl. (3): 1-78. (In German) (NSA 20: 18400).

Aun, P. E., et al. nd. Flow rate measurement of rivers and turbines using radioisotopes. pp. 317-326. In: Radiation and Isotope Technology in Latin American Development. (See Puerto Rico Nuclear Center, nd) (NSA 24: 39042).

Aun, P. E., et al. 1973. Utilization of radioactive tracers in the sedimentary study of the Rio Grande (Brazil) sandbar. Inst. de Pesquisas Radioactivas, Belo Horizonte, Brazil. 44pp. (In Portugese) (INIS-mf-1879) (NSA 33: 29494).

Auraldsson, H., et al. 1971. Further experimental studies on the transfer of radioiodine from pasture to milk. Sweden Forsvarets Forskningsanstalt, Stockholm, report FOA-4-4458. 15pp. (NSA 25: 32394).

Aurand, K. 1963. Question of endangering the environment by the application of radioactive materials in medicine. Strahlentherapie 52: 334-340. (In German) (NSA 20: 36678).

Aurand, K. B. 1965. Methods for monitoring the radioactivity of waters. Thesis, Freie Univ., Berlin. 55pp. (AED-Diss.-65-274) (NSA 20: 43624).

Aurand, K. 1966. Radioactive substances in sewage and drinking water. Röntgenpraxis 19: 266-276. (In German) (NSA 22: 10368).

Aurand, K. 1971. Experience with radioactive effluents as a model for improved monitoring of industrial effluents. pp. 651-667. In: Proc. Intern. Symp. Radioecology Applied to the Protection of Man and His Environment. (See Commission of the European Communities, 1971) (In German) (NSA 27: 5262).

Aurand, K. 1974. Regulation of the release of radioactive waste water from nuclear power plants in the Federal Republic of Germany. pp. 127-137. In: Proc. Seminar on the Monitoring of Radioactive Effluents. (Karlsruhe, Germany, May 14-17, 1974). (See Organization for Economic Co-operation and Development, Paris) (NSA 31: 3342).

Aurand, K. and H. Behrens. 1964. Investigations about the behavior of long living fission products in drinking water. Atompraxis 10(8): 346-351. (In German) (NSA 18: 39536).

Aurand, K. and R. Wolter. 1966. Hazards resulting from discharge of radioactive wastes into municipal sewers. pp. 697-708. In: Disposal of Radioactive Wastes into Seas, Oceans and Surface Waters. (See IAEA, 1966).

Aurand, K., J. Arndt and R. Wolter. 1966. Behavior of radionuclides in sewer systems and purification plants subsequent to discharge of radioactive wastes with concentrations increased over short periods. Inst. für Wasser-, boden- und Lufthygiene, Bundesgesundheitsamt, Berlin, report BMwF-FBK-66-04. 40pp. (In German) (NSA 20: 36533).

Aurand, K., et al. 1970. Control of radioactive waste from nuclear facilities. Inst. für Wasser-, Boden- und Lufthygiene, Bundesgesundheitsamt, Berlin. 26pp. (In German) (NP-18622) (NSA 25: 18094).

Aurand, L., et al. 1970. Control of radioactive waste water from nuclear power plants. Inst. für Wasser-, Boden und Lufthygiene, Bundesgesundheitsamt, Berlin, report WaBoLu-3/70. 13pp. (In German) (NSA 26: 6712).

Aurand, K., et al. 1972. Environmental exposure by radionuclides in gaseous and liquid effluents from nuclear power plants in the Federal Republic of Germany. Kernforschungszentrum, Karlsruhe, Germany, report KFK-1515. 14pp. (In German) (NSA 27: 19124).

Austin, J. H. 1963. Strontium uptake by algae. pp. 231-245. In: Transport of Radionuclides in Fresh Water Systems. (See Kornegay, et al., 1963) (NSA 17: 40531).

Austin, J. H. 1963. Strontium sorption by Chlorella pyrenoidosa in continuous culture. Thesis, Univ. of California, Berkeley. 191pp. (NSA 18: 35299).

Austin, J. H., C. A. Klett and W. J. Kaufman. 1967. Use of continuous cultures for estimating strontium sorption by Chlorella pyrenoidosa. Int. J. Oceanol. Limnol. 1: 1-27. (NSA 22: 31013).

Austin, R. B. and D. C. Longden. 1967. A rapid method for the measurement of rates of photosynthesis using $^{14}CO_2$. Ann. Botany 31(122): 245-253.

Austin, S. A. 1967. Radiosensitive insect. Median lethal dose for whole-body irradiation with ^{60}Co gamma rays of Locusta migratoria migratoriodes R. et F. Brit. J. Radiol. 40: 711-712. (NSA 22: 40868).

Austral Oil Co. 1972. Post-shot activities and programs. Project Rulison. Summary report. U. S. AEC report PNE-R-59. 12pp. (NSA 26: 38389).

Australia Atomic Weapons Tests Safety Committee. 1972. Strontium-90 and caesium-137 in the Australian environment during 1970 and some results for 1971. Victoria, Australia, report AWTSC-4. 45pp. (NSA 28: 18533).

Australian Government Publishing Service. 1972. Biological aspects of fall-out in Australia from French nuclear weapons explosions in the Pacific, June-August, 1971. Canberra. 20pp. (NSA 28: 29868).

Australian Atomic Energy Commission. 1973. Proc. AAEC Symposium on Environmental and Radiological Safety Aspects of the Mining and Processing of Uranium. (Lucas Heights, December 9-10, 1971). Research Establishment. Lucas Heights, report AAEC/E-272. 111pp. (CONF-711227) (NSA 28: 24460).

Avery, J. A. 1963. A study of the haemocytes of the grasshopper, Melanoplus differentialis (Thomas) (Orthoptera: Acrididae). Dissertation Abstr. 24 (4): 1753. (^{14}C).

Avetisyan, G. A., et al. 1965. Experience in the study of the feeding habits of fleas found on the common vole (Microtus arvalis) under alpine conditions in Armenia using radioisotopes. Izv. Akad. Nauk Armyan SSR Nauk 18(9): 102-106. (In Russian) (BA 48: 1967).

Avio, C. M. and L. Lenzerini. 1963. Uptake of ^{137}Cs by Chlorella vulgaris. Pubbl. staz. zool. Napoli 33(2): 69-82. (In Italian) (NSA 18: 29221; 21: 36487).

Avio, C. M. and M. Rossi-Torelli. 1965. Uptake of ^{90}Sr by Chlorella vulgaris. II. Uptake of ^{90}Sr from radioisotopic concentration in the soil. Pubbl. staz. zool. Napoli 34: 346-354. (In Italian) (NSA 21: 36486).

Avio, C. M. and M. Rossi-Torelli. 1966. Accumulation of yttrium and cerium by Chlorella vulgaris (Preliminary communication). pp. 621-622. In: Disposal of Radioactive Wastes into Seas, Oceans and Surface Waters. (See IAEA, 1966).

Avio, C. M., M. Rossi-Torelli and L. Lenzerini. 1963. Uptake of Sr^{90} by Chlorella vulgaris. I. Uptake of Sr^{90} as related to the growth of a population. Pubbl. staz. zool. Napoli 33: 264-273. (NSA 18: 41064; 21: 36487).

Avio, C. M. 1970. Radioactive contamination and biological period. Ann. Med. Nav. 75: 193-242. (In Italian) (NSA 26: 4723) (Incl. marine algae; extensive bibliography).

Avivi, A., R. Galun and M. Warburg. 1968. Rearing and irradiation of the tick, Ornithodoros tholozani. pp. 3-6. In: Radiation, Radioisotopes, and Rearing Methods in the Control of Insect Pests. (See IAEA, 1968) (NSA 24: 32328).

Axelrod, D. I. 1966. Potassium-argon ages of some western Tertiary floras. Am. J. Sci. 264: 497-506. (NSA 21: 176).

Ayala, F. J. 1966. Evolution of fitness. I. Improvement in the productivity and size of irradiated populations of Drosophila serrata and Drosophila birchii. Genetics 53: 883-895. (NSA 20: 31049).

Ayala, F. J. 1967. Evolution of fitness. III. Improvement of fitness in irradiated populations of Drosophila serrata. Proc. Nat. Acad. Sci. (US) 58: 1919-1923. (NSA 22: 31034).

Ayala, F. J. 1969. Evolution of fitness. V. Rate of evolution of irradiated populations of Drosophila. Proc. Nat. Acad. Sci (US) 63: 790-793. (NSA 24: 7809).

Ayres, R. U. 1963-1965. Special aspects of environment resulting from various kinds of nuclear wars. et seq. Hudson Inst., Harmon-on Hudson, N. Y., U. S. Office of Civil Defense reports HI-243-RR, HI-303-RR, HI-518-RR. vp. (NSA 18: 23178; 18: 23821; 18: 18064; 18: 23822; 20: 20930; 20: 20931; 20: 20932).

Ayuzawa, C. and F. Yusa. 1964. On the inactivation of the viruses of the silkworm, Bombyx mori L. by ^{60}Co-gamma ray irradiation. Nippon Sanshigaku Zasshi 33(2): 130-133. (In Japanese).

Azaryan, G. Kh., et al. 1965. Prospective use of radiation to control the hollyhock seed moth (Lepidoptera, Gelechiidae). Entomol. Rev. 44(4): 449.

Babalola, O. and R. Lal. 1974. Neutron moisture meter and beta-gauging techniques to investigate plant-water relations for maize (Zea mays L.) in western Nigeria. pp. 335-349. In: Isotope and Radiation Techniques in Soil Physics and Irrigation Studies, 1973. (See IAEA, 1974) (NSA 30: 32634).

Babenko, L. V. 1960. Radioactive isotopes for the labelling of ticks. Med. parazitol. 29: 320-324.

Babero, B. B. 1952. The effects of x-rays on the life cycle and morphology of Ascardidia gall (Schrank, 1788). Trans. Am. Microscop. Soc. 61: 114-119. (NSA 6: 4367).

Babers, F. H. and C. C. Roan. 1954. Distribution of radioactive phosphorus in susceptible and resistant house flies. J. Econ. Entomol. 47: 973-975.

Babers, F. H., N. Mitlin and T. J. Shortino. 1956. The fate of radiophosphorus ingested by house flies and German cockroaches. J. Econ. Entomol. 49(6): 820-821.

Babers, F. H., C. C. Roan and R. L. Walker. 1954. Tagging boll weevils with radiocobalt. J. Econ. Entomol. 47(5): 928-929.

Babos, L., et al. 1962. Effect of the irradiation (^{60}Co) on pupae of Bombyx mori L. Ann. Biol. Tihany 29:3-9. (In German) (NSA 21: 13021).

Bacastow, R. and C. D. Keeling. 1973. Atmospheric carbon dioxide and radiocarbon in the natural carbon cycle. II. Changes from A. D. 1700 to 2070 as deduced from a geochemical model. pp. 86-134. In: Carbon and the Biosphere. (See Woodwell and Pecan, 1973).

Baccetti, B. 1958. Nuclear energy for control of insect pests. Italia Agr. 95(11): 713-717. (In Italian).

Baccetti, B. 1962-63. Atomic energy in the control of insects. Acad. econ. agrar. Georgofili IX(7): 1-18; 319-334. (In Italian) (NSA 21: 18178).

Baccetti, B. nd. A cytological study of the effects of radiation on the development of the reproductive organs of two species of fruit flies: Dacus oleae and Ceratitis capitata. Inst. Zoologia e Biologia, Siena Univ. (Italy), report No. 7 (Final), December 20, 1962 - December 19, 1965. (IAEA Contract 169/RZ/RB) (NSA 20: 42847).

Baccetti, B. and M. Cappellini. 1961. Investigations on the sterilization of pernicious insects with ionizing radiations. Nuntius radiol. 27: 600-603. (In Italian).

Baccetti, B. and R. de Dominicis. 1963. The effects of gamma radiation on the ovaries of Dacus oleae Gmel. pp. 387-411. In: Radiation and Radioisotopes Applied to Insects of Agricultural Importance. (See International Atomic Energy Agency, 1963).

Baccetti, B. and R. Zocchi. 1962. Tests for control of the pine processionary moth by the use of ionizing radiation. Redia 47: 161-167. (In Italian) (NSA 21: 18199).

Bacchetti, S. and F. Mauro. 1965. Recovery from sublethal x-ray damage in surviving yeast cells. Radiation Research 25(1): 103-114. (NSA 19: 26002).

Bachmann, R. W. 1961. An experimental study of the freshwater zinc cycle. Dissertation, Univ. of Michigan, Ann Arbor. 79pp.

Bachmann, R. W. 1963. Zinc-65 in studies of the freshwater zinc cycle. pp. 485-496. In: Radioecology. (See Schultz and Klement, 1963) (NSA 17: 33601).

Bachmann, R. W., G. W. Saunders and F. B. Trama. 1961. Investigations in lake metabolism-photosynthesis: a modified C^{14} technique for estimations of photosynthesis in large lakes. Abstr. p. 163. In: Proc. 4th Conf. on Great Lakes Research. Great Lakes Research Div., Inst. of Sci. and Tech., Univ. of Michigan, Ann Arbor, Publ. 7.

Bachofer, C. S. and M. E. Gautereaux. 1959. X-ray effects on single nerve fibers. J. Gen. Physiol. 42(4): 723-735. (Used earthworm).

Bachofer, C. S. and S. E. Wittry. 1962. Comparison of electroretinal response to x-rays and to light. Radiation Research 17(1): 1-10. (NSA 16: 21976).

Bachofer, C. S. and S. E. Wittry. 1962. Comparison of stimulus energies required to elicit the ERG in response to x-rays and to light. J. Gen. Physiol. 46(2): 177-187. (NSA 17: 13900) (Rana pipiens).

Bachofer, C. S., M. E. Gautereaux and S. M. Kaack. 1964. Relative sensitivity of isolated nerves to Co^{60} gamma rays. pp. 221-242. In: Response of the Nervous System to Ionizing Radiation. (See Haley and Snider, 1964) (Used frog, turtle, earthworm, Norway rat).

Bachrach, U. and Z. Bachrach. 1974. Radiometric method for the detection of coliform organisms in water. Appl. Microbiol. 28(2): 169-171. (NSA 30: 32634).

Back, A. 1939. A type of lesions produced in Paramecium with x-rays. Compt. rend. soc. biol. 131(19): 1103-1106.

Backman, G. E. and L. W. Roddy. 1965. Radiation control of accidentally contaminated seepage ponds. Abstr. Health Physics 11(8): 821. (Incl. statement on algae).

Bacon, J. P., Jr. 1962. Translocation of radiophosphorus in a stream by bacteria and inorganic ions. Thesis, Michigan State Univ., East Lansing. viii, 124pp.

Bacq, Z. M. 1952. Effect of x-rays, cyanide, and various radiation protective agents on the infusoria. Acta Radiol. 38: 489-505. (In French) (NSA 7: 2968).

Bacq, Z. M. 1962. The influence of certain metabolic factors on radiosensitivity. Laval Med. 34(1): 80-84. (In French) (NSA 17: 27047).

Bada, J. L. and L. Deems. 1975. Accuracy of dates beyond the ^{14}C dating limit using the aspartic acid racemisation reaction. Nature 255(5505): 218-219. (NSA 32: 14325).

Bade, M. L. 1964. Biosynthesis of fatty acids in the roach Eurycotis floridana. J. Insect Physiol. 10(2): 333-341. (^{14}C).

Badgett, C. O. 1971. Radioisotope instruments in pollution measurement and control. Industrial Nucleonics Corp., Columbus, Ohio. (For Conf. on Innovative Applications of Radiation, Dallas, Texas, 22 April 1971) (CONF-710428-3) (NSA 25: 51712).

Badr, F. M. and R. S. Badr. 1971. Effect of X-irradiation on mitotic activity of bone marrow cells of laboratory and wild rats. Folia Biol. (Prague) 17(4): 258-263. (NSA 25: 58116).

Badr, F. M. and R. S. Badr. 1971. Effect of X irradiation on survival of wild and laboratory rats. Radiation Research 48(2): 256-264. (NSA 26: 12471) (Rattus).

Baehr (Bähr), F. W. 1962. The measurement of water flow with radioisotopes. I. and II. Wasserwirtschaft 52: 173-177, 222-226. (In German) (NSA 17: 10885; 10886).

Baerreis, D. A. and R. A. Bryson. 1965. Historical climatology and the southern plains. Abstr. Abstracts Intern. Assoc. Quaternary Research. VII Intern. Congr. Boulder and Denver, Colorado. p. 12. (Carbon dating).

Baertschi, P. and R. Keil. 1973. Influence of varying ^{18}O contents on the refractometric and spectrophotometric D determination in water. Z. Anal. Chem. 266(4): 264-267. (In German) (NSA 29: 2439).

Baetcke, K. P., A. H. Sparrow and C. H. Nauman. 1968. Correlations between nuclear parameters and radiosensitivities of woody herbaceous plants. Abstr. Radiation Research 35(2): 526-527.

Baetsle, L. 1963. Physical chemistry of the migration of the cations into the ground. Euratom Joint Research and Development Program, Special Tech. Report, EURAEC-417. 33pp. (English transl. EUR-141.f) (NSA 17: 29869).

Baetsle, L. H., W. de Laguna and J. Soufiriau. 1968. Remedial actions in case of groundwater contamination of sandy aquifers, 1966. Final report. Belgium Centre d'Étude de l'Énergie Nucléaire, Mol, EURATOM report EUR-4095. 54pp. (NSA 23: 13885).

Baetsle, L. H., W. F. Maes and J. Soufiriau (Souffriau). 1965. Transportation of radioisotopes in the soil. Meded. Landbouwhogesch. Opzoekingssta. Gent 30: 933-940. (In Flemish) (NSA 21: 3181).

Baetsle, L., et al. 1963. Underground radionuclide movement. Final report No. 2. U. S.-Euratom Joint Research and Development Program, Contract 048-61-4, report EURAEC-703. 175pp.

Baetsle, L., et al. 1963. Present status of the study program on the movement of radioelements in the soil at Mol. Euratom report EUR-140.f. 41 pp. (NSA 17: 27602.

Baetsle, L., et al. 1966. Migration of radioelements in soil. Final report. Belgium Centre d'Étude de l'Énergie Nucléaire, Mol, EURATOM report EUR-2481.f. (NSA 20: 20875).

Baez, A. P., L. Galvez and E. Tzontlimatzin. 1975. Comparative measurments of mercury concentrations in the Coatzacoalcos River's estuary using NAA and AAA. Trans. Am. Nucl. Soc. Suppl. 21(2): 55.

Bagge, E. 1962. Isotope determination as an aid to paleontological research. Strahlentherapie 49(Sonderbände): 1-5. (In German) (NSA 19: 615).

Bagheri-Zenous, E. 1971. Influence of the irradiation of the grains of wheat on the development and reproduction of *Sitophilus granarius* L. pp. 23-30. In: *Sterility Principle for Insect Control or Eradication*. (See International Atomic Energy Agency, 1971) .

Bagheri-Zenous, E. 1974. Effect of gamma rays on the larvae of *Plodia interpunctella* (Lepidoptera: Phycitidae). Entomol. Exp. Appl. 17(2): 215-218. (In French) (NSA 31: 29955).

Bagliano, G., P. Chamard and G. F. Clemente. 1972. Experimental comparison between neutron acivation analysis and oscillographic polarography with a double electrolytic cell in determining trace elements in aquatic plant samples. pp. 139-154. In: *Nuclear Activation Techniques in the Life Sciences*. (See International Atomic Energy Agency, 1972) (NSA 27: 22084).

Bagliano, G., et al. 1972. Measurement of ^{137}Cs and ^{90}Sr in soil. Comitato Nazionale per l'Energia Nucleare, Rome, Italy, report RT/PROT-72-11. 27pp. (In Italian) (NSA 26: 45263).

Bahadur, J. 1966. Evaluation of in-place density and moisture content of subsoil by radiation logging. pp. 423-432. In: *Proc. All India Symposium on Radioactivity and Metrology of Radionuclides*. (Bombay, March 14-18, 1966). (NSA 22: 6441).

Bahadur, J. 1967. Study of density and porosity variations of river sand by radiation logging. Irrig. Power 24: 427-431. (NSA 25: 21186).

Bai, A. R. K. and M. K. Manjula. 1974. Effects of tritiated water on the division rate of *Paramecium aurelia*. Indian J. Exp. Biol. 12(5): 410-411. (NSA 32: 1193).

Bai, D. H. 1967. Studies of radiosensitivity. VII. Differences of radiosensitivity by taxonomical relationship in plants. J. Nuclear Sci. (Seoul) 7: 159-168. (In Korean) (NSA 22: 36411).

Bai, D. H. 1969. Radioactive contamination of the atmosphere, soils, and agricultural products in Korea, 1961-1966. pp. 587-596. In: *Symposium on Radioecology*. (See Nelson and Evans, 1969).

Baig, M. M. H., et al. 1972. Gas chromatographic and radiometric study of the behavior of ^{14}C-DDT on mustard plants under tropical conditions. Pak. J. Sci. Inc. Res. 15(3): 220-226. (NSA 28: 756).

Bailey, G. N. A., I. J. Linn and P. J. Walker. 1973. Radioactive marking of small mammals. Mammal. Rev. 3(1): 11-23.

Bailey, P. J. and B. D. Baines. 1964. Irradiation: present and future applications. Nucl. Engrg. 9(97): 213-216. (Incl. entomology).

Bailey, R., et al. 1973. Energy, man, and environment. vp. In: Intern. Symp. Energy-Man Environment. (Rüschlikon, Switzerland, February 3, 1972). (In German) (NSA 28: 29916).

Bailey, R. E. 1953. Radiosurgery and uptake of radioactive iodine by the thyroid of the Oregon junco. Auk 70: 196-199.

Bailey, T. N. 1971. Imobilization of bobcats, coyotes, and badgers with phencyclidine hydrochloride. J. Wildl. Mgt. 35: 847-849. (At National Reactor Testing Station, Idaho).

Bailey, T. N. 1972. Ecology of bobcats with special reference to social organization. Thesis, Univ. of Idaho, Moscow, Idaho. 82pp. (At National Reactor Testing Station, Idaho).

Bailey, T. N. 1972. The elusive bobcat. Natural History 81: 43-49. (At National Reactor Testing Station, Idaho).

Bailey, T. N. 1973. Ecology of bobcats with special reference to social organization. Dissertation, Univ. of Idaho, Moscow, Idaho. 93pp. (At National Reactor Testing Station, Idaho) (WR 150: 46).

Bailey, T. N. 1974. Social organization in a bobcat population. J. Wildl. Mgt. 38(3): 435-446. (At National Reactor Testing Station, Idaho).

Bain, C. A. R. 1972. Radioisotopes in the determination of dispersion coefficients in estuaries for pollution control studies. pp. P11-1-5. In: National Conference on the Technological Application of Nuclear Techniques. (Pelindaba, South Africa, October 12-13, 1972). Atomic Energy Board, Pelindaba, Pretoria, South Africa. (In English and Afrikaans) (CONF-721091) (NSA 28: 21118).

Bain, C. A. R. 1975. Shifting sands. Nucl. Act. (12): 6-10. (NSA 33: 23453) (Radioisotope tracer).

Bair, W. J., J. F. Park and W. J. Clarke. 1967. The radiosensitivity of alligators. Radiation Reserach Soc. 15th Ann. Meeting. Abstr. Academic Press, New York.

Baird, I. E. and R. G. Wetzel. 1968. A method for the determination of zero thickness activity of ^{14}C labeled benthic diatoms in sand. Limnol. Oceanogr. 13(2): 379-382.

Baird, M. B. and A. M. Clark. 1971. X-ray induced life-shortening mutations in Habrobracon: a genetic approach to senescence and duration of life. Exptl. Gerontol. 6(1): 1-8. (NSA 25: 32383; 55185).

Baja, R. 1956. The marking with isotopic tracers in biology. Inform. Sci. (Paris) 1: 1-9. (In French).

Bajaj, Y. P. S. 1970. Effect of gamma-irradiation on growth, RNA, protein and nitrogen contents of bean callus cultures. Ann. Botany 34(138): 1089-1096.

Baker, B. E., B. H. Lauer and E. R. Samuels. 1968. ^{90}Sr and ^{137}Cs levels in the milks of some Arctic species. J. Dairy Sci. 51: 1508-1510. (NSA 22: 49367) (Incl. wild animals).

Baker, B. E., C. H. Neilson and E. R. Samuels. 1970. ^{90}Sr and ^{137}Cs in human and other milks collected in Alaska. J. Dairy Sci. 53: 241-244. (NSA 24: 17019) (Incl. polar bear, dall sheep, harp seal milk).

Baker, B. E., E. R. Samuels and E. Pulliainen. 1972. ^{90}Sr and ^{137}Cs in milk and certain other materials collected in Finland. J. Dairy Sci. 55(5): 633-639. (NSA 26: 38764) (Incl. reindeer milk and meat, lichens).

Baker, C. E. 1970. Relationship of cesium-137 and iron-59 elimination rates to metabolic rates of small rodents. Dissertation, Univ. of Tennessee, Knoxville. xii, 103pp. (Also: C. E. Baker, P. B. Dunaway and S. I. Auerbach, Oak Ridge National Lab., U. S. AEC report ORNL-4568).

Baker, C. E. and P. B. Dunaway. 1969. Retention of ^{134}Cs as an index to metabolism in the cotton rat (Sigmodon hispidus). Health Physics 16(2): 227-230. (NSA 23: 20101).

Baker, C. E., P. B. Dunaway and S. I. Auerbach. 1968. Measurement of metabolism in cotton rats by retention of cesium-134. Oak Ridge National Lab., U. S. AEC report ORNL-TM-2069. v, 46pp. (Also: C. E. Baker, Thesis, Univ. of Tennessee, Knoxville).

Baker, J. F. 1972. Effects of gamma radiation on the spectral sensitivity of the noctuid moth, Heliothis zea. Florida State Univ., U. S. AEC report TID-26078. 14pp. (NSA 26: 45873).

Baker, K. F. 1958. Heterotopic thyroid tissue in fishes. I. The origin and development of heterotopic thyroid tissue in platyfish. J. Morphol. 103: 91-133.

Baker, K. F. 1968. Heterotopic thyroid tissue in fishes. II. The effect of iodine and thiourea upon the development of heterotopic thyroid tissue in platyfish. J. Exptl. Zool. 138(2): 329-353.

Baker, K. F. 1959. Heterotopic thyroid tissue in fishes. III. Extrapharyngial thyroid tissue in montezume swordtails, a guppy and a cherry barb. Zooligica 44(4): 133-141.

Baker, P. S. 1970. Isotopes information center. pp. 399-410. In: Radiation and Isotope Technology in Latin American Development. (See Puerto Rico Nuclear Center, 1970) (NSA 24: 39150).

Baker, V. H., O. Taboada and D. E. Wiant. 1953. Lethal effects of electrons on insects infesting wheat and flour--Part I. Agr. Engr. 34: 755-758.

Baker, V. H., O. Taboada and D. E. Wiant. 1954. Lethal effects of electrons on insects infesting wheat and flour--Part II. Agr. Engr. 35: 407-410.

Baker, V. H., O. Taboada and D. E. Wiant. 1955. Insect control by electron irradiation. Electronics 28(5): 202-208.

Baker-Cohen, K. F. 1959. Renal and other heterotopic thyroid tissue in fishes. pp. 283-301. In: Comparative Endocrinology. A. Gorbman, ed. John Wiley and Sons, New York.

Bakulina, Ye. D., G. L. Pokrovskaya and D. D. Romashov. 1962. The radiosensitivity of the spermatozoa of the loach (Misgurnus fossilis L.). Radiobiologiya 2(1): 92-100. (AEC-tr-5428; pp. 135-147) (NSA 16: 30301).

Bakunov, N. A., A. P. Panarin and L. V. Fedotova. 1973. Accumulation of ^{137}Cs by industrial fish species in the Caspian Sea. pp. 229-234. In: Radioecology of Water Organisms. II. Distribution and Migration of Radionuclides in Freshwater and Sea-Water Biocenoses. (See Andrumaitis, 1973) (In Russian) (NSA 30: 21140).

Bakunov, N. A. 1974. Concentration of cesium-137 in commercial fish of the Atlantic Ocean, Baltic, and Caspian Seas. Radiobiologiya 14(3): 447-449. (AEC-tr-7596; pp. 171-175) (NSA 31: 16538).

Balabukha, V. S. 1958. Acceleration of the elimination of radioactive isotopes from the organism under the effect of complex-forming compounds. In: Accumulation of Radioactive Elements in the Body and Their Elimination. (See Balabukha and Fradkin, 1958) (JPRS-L-1169-N; AEC-tr-5236) (NSA 14: 7205).

Balabukha, V. S. and G. Ye. Fradkin. (eds.). 1958. Accumulation of Radioactive Elements in the Body and Their Elimination. Medgiz, Moscow. (AEC-tr-5236).

Balachandran, S., W. H. Beierwaltes and R. D. Ice. 1975. Tissue distribution of ^{14}C, ^{125}I, and ^{131}I-diphenyl-hydantoin in the toadfish, rat, and human. J. Nucl. Med. 16(8): 775-779. (NSA 32: 22665).

Baldwin, J. P., P. B. Tinker and F. H. C. Marriott. 1971. The measure of length and distribution of onion roots in the field and the laboratory. J. Appl. Ecol. 8(2): 543-554. (^{32}P).

Baldwin, W. F. 1956. Similarities in killing by heat and by x-radiation in the insect Dahlbominus fescipennis (Zett.). Radiation Research 5(1): 46-51. (NSA 10: 9951).

Baldwin, W. F. 1958. Recovery from x-ray induced sensitivity to heat in an insect. Radiation Research 8(1): 17-21.

Baldwin, W. F. 1961. Latent radiation damage and syncronous cell division in the epidermis of an insect. III. Spontaneous reversal of effects leading to delay during mitosis. Radiation Research 14(4): 426-431. (NSA 15: 16848).

Baldwin, W. F. 1965. Visible mutation frequencies in Dahlbominus oogonia produced by acute x-rays and chronic γ-radiation. Mutation Research 2: 55-59. (NSA 19: 38392).

Baldwin, W. F. 1965. Radiosensitivity of female germ cell stages of Dahlbominus. Mutation Research 2: 530-533. (NSA 20: 12691).

Baldwin, W F. 1968. Increased yield of gamma-induced eye color mutations from chronic versus acute exposures in Dahlbominus. pp. 365-375. In: Isotopes and Radiation in Entomology. (See International Atomic Energy Agency, 1968) (Also: 1967, Atomic Energy of Canada, Ltd., Chalk River, report AECL-2791) (NSA 22: 33779).

Baldwin, W. F.1972. Linear dose effect response for mutations in mature oocytes of the wasp (Dahlbominus). Radiation Research 49(1): 190-196. (NSA 26: 23149).

Baldwin, W. F. 1975. Combined effects of heat and radiation on the frequency of eye color mutations and malformations in Dahlbominus. Mutation Research 27(1): 143-145. (NSA 32: 14591).

Baldwin, W. F. and G. Cowper. 1969. The use of radioactive platinum-iridium wire as an internal tag for tracing insects. Can. Entomol. 101(2): 151-152.

Baldwin, W. F. and W. G. Cross. 1966. Effects of fast neutrons on eye color mutations in Dahlbominus. Nature 210: 1396-1397. (NSA 20: 33023).

Baldwin, W. F. and P. A. Knight. 1975. Major differences in the frequencies of radiation-induced malformations and eye color mutants in mature oocytes of the wasp Dahlbominus. Radiation Research 63(2): 320-325. (NSA 32: 22634) (γ radiation).

Baldwin, W. F. and C. A. Narraway. 1957. Interaction of heat and x-rays in killing a chalcid. Nature 179(4567): 971-972.

Baldwin, W. F. and T. N. Salthouse. 1958. Latent radiation damage in the body wall of the insect Rhodnius prolixus. Abstr. Radiation Research 9 (1): 89.

Baldwin, W. F. and T. N. Salthouse. 1959. Oxygen deficiency and radiation damage in the insect Rhodnius. Nature 183(4666): 974. (NSA 13: 11580).

Baldwin, W. F. and T. N. Salthouse. 1959. Latent radiation damage and Syncronous cell division in the epidermis of an insect. I. Non-reversible effects leading to local radiation burns. Radiation Research 10(4): 387-396. II. Reversible effects in burn repair. Radiation Research 10(4): 397-399. (NSA 13: 10809; 10810). 1959.

Baldwin, W. F. and E. L. Shaver. 1963. Radiation-induced sterility in the insect Rhodnius prolixus. Can. J. Zool. 41(4): 637-648. (NSA 18: 3519).

Baldwin, W. F. and E. L. Shaver. 1964. Dose fractionation and recover from x-rays in Rhodnius. Can. J. Zool. 42: 515-518.

Baldwin, W. F. and J. B. Sutherland. 1965. Extreme sensitivity to low-level x-rays in the eye of the cockroach Blaberus. Radiation Research 24(3): 513-518. (NSA 19: 17603).

Baldwin, W. F., J. R. Allen and N. S. Slater. 1966. A practical field method for the recovery of blackflies labelled with phodphorus-32. Nature 212: 959-960.

Baldwin, W. F., H. G. James and H. E. Welch. 1955. A study of predators of mosquito larvae and pupae with a radioactive tracer. Can. Entomol. 87(8): 350-356.

Baldwin, W. F., P. A. Knight and G. D. Chant. 1969. Radiation-induced mutants of Dahlbominus fuscipennis (Hymenoptera: Eulophidae). Can. J. Genet. Cytol. 11: 823-828. (NSA 24: 23161).

Baldwin, W. F., D. F. Riordan and R. W. Smith. 1958. Note on dispersal of radio-active grass-hoppers. Can. Entomol. 90(6): 374-376.

Baldwin, W. F., J. B. Sutherland and J. E. J. Habowsky. 1963. Effects of x-rays on electrical activity in the eye of the cockroach Blaberus giganteus. Nature 199(4893): 616. (NSA 17: 33716).

Baldwin, W. M. 1915. The action of ultraviolet rays upon the frog's egg. Anat. Rec. 9(5): 365-381.

Baldwin, W. M. 1919. The artificial production of monsters conforming to a definite type by means of x-rays. Anat. Rec. 17(3): 135-164. (tadpoles).

Baldwin, W. M. 1920. The artificial production of monsters demonstrating localized defects as the result of injury from x-rays. Am. J. Physiol. 52(2): 296-303. (tadpoles).

Baldwin, W. M. 1921. The artificial production of syringomyelocoele in the tadpole by means of x-rays. Anat. Rec. 22: 305.

Balek, J. 1965. The importance of beta-radioactivity measurements in representative and research basins. pp. 347-350. In: Symposium, Assoc. Intern. Hydrol. Sci., Union Intern. Geodesique et Geophysique. (NSA 21: 45242).

Balek, J. and J. Ralkova. 1965. Communication on the ground water component of river discharge and its relationship to hydrogeology. J. Inst. Water Eng. 19: 486-488. (NSA 21: 41005).

Balesdent-Marquet, M.-L. 1955. Temporary castration by x-rays and determination of temporary sexual character of the isopod crustacean Asellus aquaticus L. Compt. rend. hebd. Séanc. Acad, Sci. (Paris) 240(11): 1275-1277.

Balesdent-Marquet, M.-L. 1955. Localized irradiation with x-rays and determination of the temporary sexual character of the isopod crustacean Asellus aquaticus L. female. Compt. rend. hebd. Séanc. Acad. Sci. (Paris) 241: 609-611.

Balesdent-Marquet, M.-L. and A. Veillet. 1958. Surgical castration, localized irradiation with x-rays and determination of the external sexual characteristics in the female crustacea Isopoda Asellus aquaticus L. Compt. rend. 246: 1753-1756. (In French) (NSA 12: 7659).

Baliga, B. S., P. R. Srinivasan and E. Borek. 1965. Changes in the t-RNA methylating enzymes during insect metamorphosis. Nature 208: 555-557. (^{14}C, Tenebrio molitor).

Balkasheva, L. U., A. Ye. Palchevskaya and S. V. Goncharenko. 1965. The effect of fast neutrons on the culture and morphological characteristics of Bacillus mesentesicus. pp. 84-90. In: Biological Effect of Neutron Radiation. A. A. Gorodetskogo, ed. Publ. House Acad. Sci. Ukr. SSR, Kiev. (JPRS-32,524, TT-65-33003) (NSA 19: 38449).

Ball, E. 1953. Histological effects of absorbed radioisotopes upon the callus of Sequoia sempervirens. Bot. Gaz. 114: 353-363. (BA 28: 14369).

Ball, E. 1953. Modifications in a callus culture of Sequoia sempervirens after growth on P^{32} and S^{35}. Growth 17(4): 215-228.

Ball, E. 1955. Some effects of radioisotopes on the callus culture of Sequoia sempervirens. Abstr. Anat. Rec. 121: 466.

Ball, H. J. 1958. Effect of visible spectrum irradiation on growth and development in several species of insects. J. Econ. Entomol. 51(5): 573-578.

Ball, R. C. and F. F. Hooper. 1963. Translocation of phosphorus in a trout stream ecosystem. pp. 217-228. In: Radioecology. (See Schultz and Klement, 1963).

Ball, R. C. and F. F. Hooper. 1966. Use of ^{74}As-tagged sodium arsenate in a study of effects of a herbicide on pond ecology. pp. 149-163. In: Isotopes in Weed Research. (See International Atomic Energy Agency, 1966).

Ball, R. C., T. A. Wojtalik and F. F. Hooper. 1963. Upstream dispersion of radiophosphorus in a Michigan trout stream. Mich. Acad. Sci. Arts Ltrs. 48: 57-64.

Ballance, W. C. 1974. Radiochemical monitoring of water after the Cannikin Event, Amchitka Island, Alaska, August 1973. Amchitka-42. U. S. Geological Survey report USGS-474-205. 20pp. (NSA 31: 537).

Ballantine, D. S. 1971. Potential role of radiation in wastewater treatment. Isotopes Radiat. Technol. 8(4): 415-420. (NSA 25: 48533).

Ballantine, D. S. and L. A. Miller. 1970. Potential role of radiation in waste water treatment. Chem. Engrg. Progr. Symp. Ser. 66(104): 187-194. (NSA 25: 48523).

Ballantine, D. S., et al. 1969. The practicality of using atomic radiation for wastewater treatment. J. Water Pollut. Control. Fed. 41(3/part 1): 445-458.

Ballard, L. F. and R. P. Gardner. 1965. Compensation of the composition dependence of gamma-scatter gauges for measuring soil density. Trans. Am. Nucl. Soc. 8: 68. (NSA 19: 28567).

Ballard, L. F., et al. 1964. Selected topics on water tracing with radioisotopes. Development of engineering techniques for isotopic water tracing, sample preparation, and radiation measurement. Final Report. (RTI-2513-T9). Research Triangle Inst., Durham, N. C., U. S. AEC report ORO-597. 117pp. (NSA 20: 23321).

Ballardin, E. and P. Metalli. 1962. First results on x-ray-induced genetic damage in Artemia salina Leach. Atti assoc. genet. Ital. 7: 219-231. (NSA 16: 28764).

Ballardin, E. and P. Metalli. 1965. Relation between polyploids and radiosensitivity of oocytes of Artemia salina Leach. Boll. zool. 32: 613-618. (In Italian) (NSA 22: 43277).

Ballardin, E. and P. Metalli. 1966. Effects of x rays on oogenesis and rhythm of molting in Artemia salina Leach. Boll. zool. 33: 195-196. (In Italian) (Also: 1967, Italy Comitato Nazionale per l'Energia Nucleare, Rome, report RT/BIO-(67)26) (NSA 23: 2915).

Ballardin, E. and P. Metalli. 1968. Estimates of some components of fitness in diploid parthenogenetic Artemia salina irradiated over several generations. Atti assoc. genet. Ital. 13: 341-345. (In Italian) (NSA 23: 6561).

Ballardin, E. and P. Metalli. 1972. Transmission of traits in parthenogenetic Artemia salina. Boll. zool. 39(4): 551-564. (In Italian) (NSA 30: 3677) (X-irradiation).

Balloni, W., G. Florenzano and R. Materassi. 1965. Influence of irradiated nitrogen on the nitrogen-fixing activity of blue-green algae. Ann. Inst. Pasteur 109(Suppl.): 62-67. (In French) (NSA 20: 20408).

Ballowe, J. W. 1967. Ben Franklin Dam study. United Nuclear Industries, U. S. AEC report DUN-2349. 61pp. (NSA 30: 9565).

Balock, J. W. 1965. Differential effect of gamma radiation on fruit flies and fruit fly parasites. J. Econ. Entomol. 58(6): 1169.

Balock, J. W., A. K. Burditt, Jr. and L. D. Christenson. 1963. Effects of gamma radiation on various stages of three fruit fly species. Citrus Ind. 44(8): 9-12.

Balock, J. W., A. K. Burditt, Jr. and L. D. Christenson. 1963. Effects of gamma radiation on various stages of three fruit fly species. J. Econ. Entomol. 56(1): 42-46. (NSA 17: 27120).

Balock, J. W., L. D. Christenson and G. O. Burr. 1956. Effect of gamma rays from Co-60 on immature stages of the oriental fruit fly (Dacus dorsalis Hendel) and possible application to commodity treatment problems. Proc. Hawaiian Acad. Sci. 1955-56: 18.

Balock, J. W., et al. 1966. Gamma radiation as a quarantine treatment for Hawaiian fruit flies. J. Econ. Entomol. 59: 202-204. (NSA 20: 39293).

Baltakmens, T. 1974. Profiles of lead-210 and radium-226 in four New Zealand soils. N. Z. J. Sci. 17(4): 435-439. (NSA 33: 29143).

Baltakmens, T. 1975. Simple method for the determination of uranium in soils by two stage ion exchange. Anal. Chem. 47(7): 1147-1149. (NSA 32: 5926).

Baltimore Gas and Electric Co. 1970. Calvert Cliffs Nuclear Power Plant, Units 1 and 2. Environmental report. Baltimore, Md. 214pp. U. S. AEC DOCKET-50317-20; DOCKET-50318-20. (NSA 25:3838).

Balzhiser, R. E. 1973. Role of nuclear energy in preserving environmental values. Publ. Util. Fortn. 91(3): 18-22. (NSA 27: 24202).

Bancher, E., J. Washüttl and H. J. Goller. 1972. Studies on lipids in fractions of spherosomes and mitochondria of γ-irradiated seeds of peanuts and walnuts. Z. Pflanzenphysiol. 67(5): 399-403. (In German) (NSA 27: 27881.

Bancroft, A. R. and R. M. Brown. 1967. A program to survey Canadian Deuterium sources. Atomic Energy of Canada, Ltd., Chalk River, report AECL-2753. (NSA 21: 43047).

Bandirali, G. R., F. A. Paolini and A. C. Pratesi. 1961. Modification of the diffusion of liquids and of reactivity in the skin of animals irradiated during hibernation. Radiobiol. Latina 4: 99-106. (In Italian) (NSA 16: 20213).

Banditsing, C. and P. E. Hunter. 1973. Population development of Acarus siro reared at three temperatures following gamma irradiation. J. Econ. Entomol. 66(1): 82-85. (NSA 27: 17679).

Bando, S., et al. 1969. Choice of activable tracer for spraying agricultural insecticides by helicopter. Radioisotopes (Tokyo): 18: 488-493. (In Japanese) (NSA 24: 2163).

Bandy, L. W. 1972. The bioaccumulation and translocation of ring-labeled chlorine-36 DDT in an old-field ecosystem. Dissertation, Ohio State Univ., Columbus. xi, 241pp.

Bandy, L. W. and T. J. Peterle. 1973. Transfer of chlorine-36-DDT in a meadow. pp. 232-239. In: Radionuclides in Ecosystems. (See D. J. Nelson, 1973) (NSA 25: 41638).

Banegas, A. D. 1968. Effect of gamma radiations on the fertility of torsalo, Dermatobia hominis (Diptera: Cuterebridae). pp. 19-22. In: Control of Livestock Insect Pests by the Sterile-Male Technique. (See International Atomic Energy Agency, 1968) (NSA 24: 34807).

Banegas, A. D. and H. Mourier. 1968. Effects of gamma radiation on the fertility of the torsalo Dermatobia hominis (Diptera: Cuterbridae). Ann. Entomol. Soc. Am. 61(1): 23-25.

Banerjee, M., S. N. Banerjee and R. J. Horsley. 1966. Effects of ionizing radiation on free amino acids in Oedogonium cardiacum. Int. J. Radiat. Biol. 11: 395-397. (NSA 21: 20446).

Banerjee, P. and S. D. Chatterjee. 1966. Radium contents of the Ganges water. Sci. Cult. (Calcutta) 32: 178-180. (NSA 20: 45996).

Banerjee, S. N. and R. J. Horsley. 1968. Radioautographic studies in synchronized cultures of Oedogonium cardiacum. Am. J. Botany 55(4): 514-517.

Banerjee, S. N. and R. J. Horsley. 1973. Effects of metabolic inhibiters on giant cell formation in Oedogonium cardiacum. Radiation Research 54(1): 121-129. (NSA 28: 26965).

Banerji, S. K., S. K. Das and K. C. Banerji. 1973. Radiotracer studies on the mobility and retention of micronutrients in soils. Technology (Sindri) 10(3): 175-181. (NSA 33: 29145).

Baney, G. M. 1970. Effects of gamma irradiation on waste activated sludge. Thesis, Purdue Univ., Lafayette, Indiana. 142pp. (Univ. Microfilms 71-2551) (NSA 25: 24495).

Banham, E. J. 1962. The susceptibility of the confused flour beetle (Tribolium confusum Duv. to gamma radiation. U. K. Atomic Energy Authority, Research Group, Wantage, report AERE-R-3888. 32pp.

Bankes, D. A. 1971. Some effects of localized internode segment and entire shoot X irradiation on the survival and morphology of sunflower plants. Thesis, Ohio State Univ., Columbus. 123pp. (Univ. Microfilms 72-4417) (NSA 26: 23046).

Banks, C. J. 1954. The searching behavior of coccinellid larvae. Brit. J. Anim. Behavior 2(1): 37-38. (Use of radiotracer).

Banks, C. J. 1955. The use of radioactive tantalum in studies of the behavior of small crawling insects on plants. Brit. J. Anim. Behavior 3: 158-159.

Banks, C. J. 1957. The behavior of individual coccillid larvae on plants. Brit. J. Anim. Behavior 5(1): 12-24. (Used ^{182}Ta).

Banks, C. J. and W. G. Duxbury. 1956. The determination of α and β activity in fish. U. K. Atomic Energy Authority, Windscale Works, Sellafield, report IGO-AM/W-36. 3pp. (NSA 10: 4517).

Banks, C. J. and H. L. Nixon. 1958. Effects of the ant, Lasius niger L., on the feeding and excretion of the bean aphid, Aphis fabae Scop. J. Exptl. Biol. 35(4): 703-711.

Banks, C. J. and H. L. Nixon. 1959. The feeding and excretion rates of Aphis fabae Scop. on Vicia faba L. Entomol. Exp. et Appl. 2(2): 77-81.

Banks, C. J., A. Dezfulian and E. S. Brown. 1961. Field studies of the daily activity and feeding behavior of sunn pest, Eurygaster integriceps Put. (Hemiptera: Scutelleridae), on wheat in North Iran. Entomol. Exptl. Appl. 4: 289-300. (Used ^{182}Ta).

Bannister, B. and P. E. Damon. 1972. Dendrochronologically-derived primary standard for radiocarbon dating. pp. H80-H89. In: Proc. 8th Intern. Radiocarbon Dating Conf. Vol. 2. (See T. A. Rafter, 1972) (NSA 31: 8718).

Bannister, B. A. 1970. Ecological life cycle of Euterpe globosa Gaertn. pp. B299-B314. In: A Tropical Rain Forest. (See H. T. Odum and R. F. Pigeon, 1970.

Banterla, G., et al. 1973. Characterization of Samian ware sherds by means of neutron activation analysis. Archaeometry 15(2): 209-220. (NSA 28: 24108).

Banwell, C. J. 1964. Oxygen and hydrogen isotopes in New Zealand thermal areas. pp. 75-138. In: Nuclear Geology on Geothermal Areas. Consiglio Nazionale delle Ricerche, Pisa, Italy. (NSA 20: 45924).

Baptist, J. P. 1963. Savannah Estuary environmental radiological survey (SEERS) project. pp. 11-33. In: Studies of the Fate of Certain Radionuclides in Estuaries and Other Aquatic Environments. (Proc. Symp., (Savannah, Ga., May 25, 1961). J. J. Sabo and P. H. Bedrosian, eds. U. S. Public Health Service Publ. 999-R-3.

Baptist, J. P. 1966. Uptake of mixed fission products by marine fishes. Trans. Amer. Fish. Soc. 95: 145-152. (NSA 21: 14574).

Baptist, J. P. and C. W. Lewis. 1969. Transfer of ^{65}Zn and ^{51}Cr through an estuarine food chain. pp. 420-430. In: Symposium on Radioecology. (See D. J. Nelson and F. C. Evans, 1969)

Baptist, J. P., D. E. Hoss and C. W. Lewis. 1970. Retention of chromium-51, iron-59, cobalt-60, strontium-85, niobium-95, indium-141m, and iodine-131 by the Atlantic croaker. Health Physics 18: 141: 148.

Bara, M. O. 1970. Some aspects of bioaccumulation of cesium-137 in the bobcat (Lynx rufus). Thesis, Univ. of Georgia, Athens. 53pp.

Barabas, B., V. Filep and D. Cojocaru. 1970. Radioactivity of certain mineral waters in the Rodnei Mountains. Rev. Med. (Tirgu-Mures) 16(2): 168-171. (In Rumanian) (NSA 25 51628).

Barabas, B., et al. 1969. Investigations regarding the radioactivity of honey. Rev. Med. (Tirgu-Mures) 15: 214-219. (In Rumanian) (NSA 25: 8665).

Barabas, G. and V. Lupsan. 1967. Investigations on the radioactivity of honey. Igiena (Bucharest) 16: 79-83. (In Rumanian).

Barasch, G. E. and R. H. Beers. 1971. Aerial radiological measuring surveys of the Nuclear Fuel Services Plant, West Valley, New York, 1968 and 1969. E. G. & G., Inc., U. S. AEC report ARMS-68.6.9. (EGG-1183-2235). 39pp. (NSA 26: 17757).

Baranov, V. I. 1939. Assimulation of radioactive elements by plants. Dokl. Akad. Nauk SSSR 24(9): 951-954.

Baranov, V. I. 1968. Development of radiogeology in the USSR. Atomiya energiya 24(5): 419-421. (Soviet At. Energy 24(5): 517-519) (NSA 22: 35933).

Baranov, V. I. 1970. Distribution of Sr^{90} in surface horizons of chernozem soils of the meadow steppe zone. Inst. Geochemistry and Analytical Chem., Moscow. (In Russian).

Baranov, V. I. and N. G. Morozova. 1970. General regularities of natural radioisotope distribution in soils of the USSR. pp. 469-489. In: Proc. Intern. Symp. on Radioecology. (See France Commissariat à l'Énergie Atomique, 1970).

Baranov, V. I. and N. G. Morozova. 1971. Behavior of natural radionuclides in soil. pp. 13-41. In: Contemporary Problems of Radiobiology. (See Klechovskiy, Polikarpov and Aleksakhin, 1971) (In Russian).

Baranov, V. I. and N. A. Titayeva. 1973. Radiogeology. (Textbook for the Geologic Faculties in Universities). Izd. Mosk. Gosud. Univ., Moscow. 242pp. (In Russian) (NSA 31: 22658).

Baranov, V. I. and V. I. Vernadskiy. 1964. Natural radioactivity of soils. Izv. Akad. Nauk SSSR (biol.) 29(1): 159-163. (JPRS-23912) (NSA 18: 14076, 18043).

Baranov, V. I., V. K. Khristianov and B. V. Karasev. 1959. Photoneutronic method of determining the concentration of deuterium in natural water. Dokl. Akad. Nauk SSSR 129(5): 1035-1037. (In Russian).

Baranov, V. I., et al. 1963. Geochemistry of some natural radioactive elements in soils. Pochvovedeniye (8): 11-20. (Soviet Soil Science (8): 733-740) (CA 59: 15071; NSA 19: 30415).

Baranov, V. I., et al. 1965. Distribution of ^{90}Sr over the ground in the Soviet Union from 1959-1960. Atom. energiya 18: 246-250. (English transl. 305-310) (NSA 19: 34706).

Baranov, V. I., et al. 1968. Natural radioelements in surface and soil-subsoil waters. Geokhimiya (3): 334-341. (In Russian) (Abstr., Geochem. Int. GI 68-3-9) (NSA 22: 25509).

Baranov, V. I., et al. 1971. Radioactivity of surface and soil ground waters. pp. 100-109. In: Methods of Radioecological Investigations. (See I. N. Verkhovskaya, 1971) (In Russian).

Baranova, D. D. and G. G. Polikarpov. 1968. Sorption and desorption of radionuclides by shallow-water sediments of the Black Sea. Okeanologiya 8: 427-430. (In Russian) (NSA 23: 11675).

Baranski, L. 1968. Use of radioisotopes in investigations of mixing waste waters with surface bodies of waters. Nukleonika 13: 466-483. (In Polish) (NSA 22: 48926).

Baratta, E. J. and F. E. Knowles, Jr. 1971. Statistical limits of accuracy and precision of gross alpha and beta radioactivity measurements in water. Radiol. Health Data 12(4): 195-198. (NSA 25: 44812).

Barbarin, V. V. and V. A. Kostromskaya. 1963. Qualitative changes in oxidation-reduction processes during gamma irradiation. Sb. Tr. Sverdlovsk Gos. Med. Inst. 39: 11-18. (NSA 19: 8912) (Used hydra).

Barbe, R. 1966. Determination of total strontium in vegetables. Comparison of activation analysis with flame photometry. Thesis, Univ. of Grenoble, Grenoble, France. 22pp. (In French) (NP-17124) (NSA 22: 12938) (Incl. wild plants).

Barber, A. and P. G. Ozanne. 1970. Autoradiographic evidence for the differential effect of four plant species in altering the calcium content of the rhizosphere soil. Proc. Soil Sci. Soc. Am. 34(4): 635.

Barber, D. A. and G. J. Neary. 1958. Effect of radiation on salt uptake in plants. Proc. 2nd Intern. Conf. Peaceful Uses of Atomic Energy 27: 8-12. (Paper A/CONF. 15/P/290) (NSA 13: 6107).

Barber, D. A., G. J. Neary and R. S. Russell. 1957. Radiosensitivity of salt in plants. Nature 180(4585): 556-557. (*Chlorella pyrenoidosa*).

Barber, D. A., J. Sanderson and R. R. Scott. 1968. Influence of microorganisms on the distribution in roots of phosphate labeled with ^{32}P. Nature 217: 644. (NSA 22: 21414).

Barbic, F. and B. Krajincanic. 1971. Effects of microorganisms on the structure of uranium raw materials. pp. 279-281. In: Radiation and Radioisotopes for Industrial Microorganisms. (See International Atomic Energy Agency, 1971) (NSA 26: 22770).

Barbier, G. 1969. Applications of isotopes to the study of soil-plant relationships. Bull. Inform. Sci. Tech. (Paris) (141): 21-30. (In French) (NSA 24: 23089) (Biblographical review).

Barbour, R. W. and M. J. Harvey. 1968. The effect of radioactive tags on the activity of rodents. Am. Midl. Nat. 79(2): 519-522. (NSA 22: 40867).

Barbreau, A., P. Escalier des Orres and R. le Quino. 1971. Nuclear sites and radiological safety studies. Bull. Inform. Sci. Tech. (Paris) (158): 7-13. (In French) (NSA 26: 9255).

Bard, S. T. 1974. Entrainment of stratospheric tritium: another radiological anomaly of the Rocky Mountain West. pp. 73-78. In: 8th Midyear Topical Symposium of the Health Physics Society. (Knoxville, Tennessee). (CONF-741018) (NSA 31: 525).

Bardeen, C. R. 1907. Abnormal development of toad ova fertilized by spermatozoa exposed to Roentgen rays. J. Exptl. Zool. 4(1): 1-44, 5 plates.

Bardeen, C. R. 1909. Variations in susceptibility of amphibian ova to the x-rays at different stages of development. Anat. Record 3(4): 163-166.

Bardeen, C. R. 1911. Further studies on the variation in susceptibility of amphibian ova to x-rays at different stages of development. Am. J. Anat. 11(4): 419-498.

Barden, C. R., et al. 1966. Texas environmental surveillance program. (Paper, 11th Ann. Meeting, Health Physics Soc., Houston, Tex., June 27-30, 1966) (Texas State Department of Health, Austin, 23pp.) (CONF-660621-11) (NSA 20: 46004; Gmelin AED-CONF-66-193-3).

Bardhan, M. 1973. Use of artificial tracers in Indian geo-hydrology and future prospects. Int. Hydrol. Decade Newsl. 22: 13-24. (NSA 33: 29161).

Barendsen, G. W. 1962. Artificial radioactivity in the biosphere and its uptake by man. Ned. Tijdschr. Natuurk. 28(5): 128-138. (In Dutch) (NSA 17: 23564).

Barkai-Golan, R. and R. S. Kahan. 1967. Combined action of diphenyl and gamma radiation of the *in vitro* development of fungi pathogenic to citrus fruits. Phytopathology 57: 696-698. (NSA 22: 594).

Barkai-Golan, R., R. S. Kahan and N. Temkin-Gorodeisli. 1968. Sensitivity of stored melon fruit fungi to gamma irradiation. Int. J. Appl. Radiat. Isotopes 19: 579-583. (NSA 22: 43482).

Barkai-Golan, R., N. Temkin-Gorodeiski and R. S. Kahan. 1966. The effect of gamma radiation on the development of the fungi *Botrytis cinerea* and *Rhizopus nigricans* that cause rot in strawberry fruits. Volcani Inst. of Agricultural Research, Rehovoth, Israel. 6pp. (In English and Hebrew) (NSA 22: 33985).

Barkai-Golan, R., N. Temkin-Gorodeiski and R. Sidney. 1967. Effect of gamma irradiation on development of fungi, *Botrytis cinerea* and *Rhizopus nigricans*, causing rot in strawberry fruits. Food Irradiation 8(1,2): 34-36. (NSA 22: 43486).

Barkai-Golan, R., et al. 1969. Sensitivity to gamma irradiation of fungi pathogenic to pears. Int. J. Appl. Radiat. Isotopes 20: 577-583. (NSA 24: 32266).

Barker, F. B. 1958. Factors affecting the transport of radioactivity by water. J. Am. Waterworks Assoc. 50(5): 603-612.

Barker, H. 1970. Critical assessment of radiocarbon dating. Phil. Trans. Roy. Soc. (London) A269: 37-45. (NSA 25: 32073).

Barker, H., R. Burleigh and N. Meeks. 1969. New method for the combustion of samples for radiocarbon dating. Nature 221: 49-50. (NSA 23: 9546).

Barkhunarov, R. M., A. N. Marey and I. Ya. Novikova. 1973. Factors determining the increased migration of cesium-137 from the environment into the human body. pp. 365-372. In: *Environmental Behavior of Radionuclides Released in the Nuclear Industry*. (See International Atomic Energy Agency, 1973) (In Russian) (NSA 29: 29380).

Barkhunarov, R. M., A. F. Zaitseva and A. N. Marey. 1973. Basic principles for establishing health protection standards with respect to radioisotope concentrations in surface water bodies. pp. 625-632. In: *Environmental Behavior of Radionuclides Released in the Nuclear Industry*. (See International Atomic Energy Agency, 1973) (In Russian) (NSA 29: 29365).

Barloy, J.-J. 1968. How life resumes on irradiated land. Sci. Avenir (261): 906-911. (In French) (NSA 23: 29427).

Barnard, E. A. and J. Marbrook. 1961. Quantitative cytochemistry using directly applied radioactive reagents. Nature 189(4762): 412-413. (NSA 15: 8462).

Barnes, C. A. and M. G. Gross. 1966. Distribution at sea of Columbia River water and its load of radionuclides. pp. 291-302. In: *Disposal of Radioactive Wastes into Seas, Oceans and Surface Waters*. (See International Atomic Energy Agency, 1966).

Barnes, D. W. H., et al. 1955. The Monte Bello rat, November 1953. Research Establishment, U. K. Atomic Energy Authority, Harwell, report AERE-SPAR-1. 22pp.

Barnes, J. and D. Carswell. 1972. Tritium measurements of the Murrumbidgee River at Wagga. J. Geol. Soc. Austral. 19(2): 225-227. (NSA 28: 5580).

Barnes, J. and D. Carswell. 1973. Tritium measurements of the Murrumbidgee River at Wagga. J. Geol. Soc. Austral. 20(2): 175-177. (NSA 29: 18359).

Barnes, M. M. 1959. Radiotracer labeling of a natural tephritid population and flight range of the walnut husk fly (*Rhagoletis completa*). Ann. Entomol. Soc. Am. 52(1): 90-92.

Barnes, R. G. 1970. Nuclear power reactor wastes and our environment. (For 3rd Annual Nat. Pollution Control Conf., San Francisco, Calif., 1 Apr. 1970) (CONF-70040-1) (NSA 25: 34844).

Barnes, R. L. and A. W. Naylor. 1959. Studies on the ornithine cycle in roots and callus tissues of *Pinus serotina* and *Pinus clausa*. Bot. Gaz. 121(2): 63-69. (^{14}C).

Barnes, T. C. 1933. A possible physiological effect of the heavy isotope of H in water. J. Am. Chem. Soc. 55: 4332-4333. (algae, *Planaria*).

Barnes, T. C. 1934. The effect of heavy water of low concentration on *Euglena*. Science 79: 370.

Barnes, T. C. 1934. The controversy concerning the physiological effect of tri-hydrol in liquid water. Science 79: 455-457. (algae, *Planaria*).

Barnes, W. W. 1965. The absorption and metabolism of C^{14}-labeled endosulfan in the housefly. Diss. Abstr. 25(8): 4871.

Barnes, W. W. and G. W. Ware. 1965. The absorption and metabolism of C^{14}-labeled endosulfan in the house fly. J. Econ. Entomol. 58(2): 286-291.

Barnum, A. H. 1964. Orthoptera of the Nevada Test Site. Brigham Young Univ. Univ. Sci. Bull. (Biol.) 4(3): 1-134; xii; 1 map.

Baroughi-Bonab, H. 1965. Study of the post-embryonic development of the ovary in *Ephestia kuhniella* Z. (Lepidoptera, Pyralidae). I. Effect of ionizing radiations. Food Irradiation 6: A13-A14. (NSA 20: 18329).

Baroughi-Bonab, H. 1965. Study of post-embryonic development of the ovary with *Ephestia kuhniella* Z. (Lepidoptera, Phyralidae). Effects of ionizing radiation. Thesis, Univ. of Paris, France. 65pp. (In French).

Barr, R. E. and X. J. Musacchia. 1967. Radiation sensitivity of the hibernating ground squirrel, *Citellus tridecemlineatus*. Proc. Soc. Exptl. Biol. Med. 124: 1204-1207. (NSA 21: 26613).

Barr, R. E. and X. J. Musacchia. 1968. Aspects of radioprotection in the hibernating ground squirrel. Abstr. Radiation Research 35(2): 563.

Barr, R. E. and X. J. Musacchia. 1969. The effect of body temperature and postirradiation cold exposure on the radiation response of the hibernator _Citellus tridecemlineatus_. Radiation Research 38: 437-438. (NSA 24: 9934).

Barr, R. E. and X. J. Musacchia. 1972. Postirradiation hibernation and radiation response of ground squirrels: telemetry surveillance. Radiation Research 51(3): 631-637. (NSA 26: 59090).

Barraclough, J. T. and R. G. Jensen. 1976. Hydrologic data for the Idaho National Engineering Laboratory site, Idaho. Idaho Nat. Engineering Lab. report IDO-22055. 52pp. (NSA 33: 20190).

Barrada, Y. 1965. The neutron moisture meter and its value in water-use efficiency studies. Atom. Energy Rev. 3: 195-196. (NSA 20: 544).

Barrada, Y. 1970. Radiation techniques as means of improving the efficiency of water use. pp. 168-178. In: _Isotopes and Radiation in Investigations of Fertilizer and Water Use Efficiency in Countries of Asia and the Far East_. International Atomic Energy Agency, Vienna, Austria. (CONF-690418-3) (NSA 25: 23711).

Barrata, E. J. 1973. Analysis for tritium in water: intercomparison study of November 1970. pp. 522-530. In: _Tritium_. (See Moghissi and Carter, 1973) (NSA 28: 27021).

Barrata, E. J. and F. E. Knowles, Jr. 1973. Determination of ^{90}Sr in water: collaborative study. J. Assoc. Off. Anal. Chem. 56(1): 208-212. (NSA 27: 19519).

Barrett, J. W. 1964. A cart for a neutron probe and scaler. J. Forestry 62(7): 501-502.

Barrett, M. J., D. Monro and A. R. Agg. 1968. Radiotracer dispersion studies in the vicinity of a sea outfall. Water Research 2: 124-126. (NSA 22: 36003).

Barriera, F. and J. J. Amaral Mendes. 1965. Geographic distribution of ^{90}Sr concentration in sheep bones. Rev. Port. Quin. (Lisbon) 7: 145-153. (In Portugese) (NSA 20: 22540).

Barrington, E. J. W. and L. L. Franchi. 1956. Some cytological characteristics of thyroidal function in the endostyle of the ammocoete larva. Quart. J. Microscop. Sci. 97(3): 393-409. (Used ^{131}I).

Barrington, E. J. W. and B. B. Rawdon. 1967. Influence of thyroxine upon the uptake of S^{35}-labelled sulphate into the branchial arch skeleton of the rainbow trout (_Salmo gairdnerii_). Gen. Comp. Endocrinol. 9(1): 116-128.

Barry, B. D., W. D. Guthrie and E. J. Dollinger. 1967. Evidence of a diffuse centromere in the European corn borer, _Ostrinia nubilalis_ (Lepidoptera: Pyralidae). Ann. Entomol. Soc. Am. 60(2): 487-488.

Barry, P. J. 1965. Exchange rates of radioactive tracers between the atmosphere and snow. Atomic Energy of Canada, Ltd., Chalk River, report AECL-2232. 12pp.

Barry, P. J. 1967. The use of radioactive tracer gases to study the rate of exchange of water vapor between air and natural surfaces. pp. 69-76. In: _Isotope Techniques in the Hydrologic Cycle_. G. E. Stout, ed. Amer. Geophysical Union, Washington.

Barry, P. J. 1967. Perch Lake evaporation study. Report of progress to October 29, 1967. Atomic Energy of Canada, Ltd., Chalk River, report AECL-3022. 27pp. (NSA 22: 35919) (^3H_2O).

Barry, P. J. (Ed.). 1972. Some aspects of the releases of radioactivity and heat to the environment from nuclear reactors in Canada. Atomic Energy of Canada, Ltd., Chalk River, report AECL-4156. 85pp. (NSA 26: 40296).

Barry, P. J. 1973. Estimating dose commitments to populations from radioactive waste disposals into large lakes. pp. 499-506. In: _Environmental Behavior of Radionuclides Released in the Nuclear Industry_. (See International Atomic Energy Agency, 1973) (NSA 29: 29384).

Barry, P. J. and F. Entwistle. 1975. Tritium in lower Perch Lake. Atomic Energy of Canada, Ltd., Chalk River, report AECL-5039. 38pp. (NSA 32: 14306).

Barry, P. J. and W. F. Merritt. 1970. Perch Lake evaporation study. pp. 139-151. In: _Isotope Hydrology 1970_. (See International Atomic Energy Agency, 1970) (Also: Atomic Energy of Canada, Ltd., Chalk River, report AECL-3557) (CONF-700328) (NSA 24: 44090).

Barsdate, R. J. and W. R. Matson. 1967. Trace metals in arctic and sub-arctic lakes with reference to the organic complexes of metals. pp. 711-719. In: _Radioecological Concentration Processes_. (See Aberg and Hungate, 1967).

Barstow, J. M. 1974. Physico-ecological studies on gamma irradiated and non-irradiated _Cladonia sylvatica_ (L.) Hoffm. Thesis, Michigan Technol. Univ., Houghton. 44pp. (Also: U. S. AEC report COO-2085-8) (NSA 30: 9762).

Barsukov, V. S., O. V. Malinovskiy and N. M. Mityushova. 1963. Postradiation recovery of yeast cells irradiated in aerobic and anaerobic conditions. Dokl. Akad. Nauk SSSR 153(5): 1199-1201. (Doklady (biol. sci.): 1487-1488) (NSA 18: 11721).

Barsukov, V. S., O. V. Malinovskiy and N. M. Mityushova. 1965. Oxygen effects on the post-irradiation restoration of yeast cells. Radiobiologiya 5(6): 860-866. (AEC-tr-6603; pp. 124-135) (NSA 20: 14399).

Barsukov, V. S., O. V. Malinovskiy and N. M. Mityushova. 1965. Post-irradiation regeneration of yeast cells during the stationary phase of growth. Dokl. Akad. Nauk SSSR 161(1): 228-229. (Doklady (biophys): 19-20) (NSA 20: 79).

Bartels, O. G. 1972. Estimate of volcanic contributions to the atmosphere and volcanic gases and sublimates as the source of the radioisotopes ^{10}Be, ^{35}S, ^{32}P, and ^{22}Na. Health Physics 22(4): 387-392. (NSA 26: 28066).

Barth, D. S. and R. E. Engels. 1971. Applications of radioecology principles to design of air pollution effects research programs. pp. 603-620. In: Proc. Intern. Symp. Radioecology Applied to the Protection of Man and His Environment. (See Commission of the European Communities, 1971) (NSA 27: 5233).

Barth, J. 1975. Solubility of plutonium from rumen contents of cattle grazing on plutonium-contaminated desert vegetation in in vitro bovine gastrointestinal fluids. pp. 137-149. In: Radioecology of Plutonium and Other Transuranics in Desert Environments. (See White and Dunaway, 1975). (NSA 33:644).

Barth, R. H., Jr., P. P. Bunyard and T. H. Hamilton. 1964. RNA metabolism in pupae of the oak silkworm, *Antheraea pernyi*: the effects of diapause, development, and injury. Proc. Nat. Acad. Sci. (US) 52(6): 1572-1580. (^{3}H).

Bartlett, A. C. 1962. Changes in quantitative traits under selection and irradiation. Thesis, Purdue Univ. 187pp. (Diss. Abstr. 23(10): 3595-3596) (Used *Tribolium castaneum*).

Bartlett, A. C. 1968. Behavior of irradiated boll weevils. II. Reproduction and mortality in cages with untreated boll weevils. J. Econ. Entomol. 61(6): 1680-1684.

Bartlett, A. C. and A. E. Bell. 1961. Radiation effects on reproduction in plateaued and unselected strains of *Tribolium castaneum*. Abstr. Genetics 46(8): 850.

Bartlett, A. C. and A. E. Bell. 1962. The effect of irradiation on reproduction in two strains of *Tribolium castaneum* Herbst. Radiation Research 17: 864-877.

Bartlett, A. C. and A. E. Bell. 1964. The effects of selection and irradiation on quantitative traits in *Tribolium*. Abstr. Genetics 50(2): 234.

Bartlett, A. C. and L. F. Lewis. 1973. Pink bollworm: chromosomal damage and reproduction after gamma irradiation of larvae. J. Econ. Entomol. 66 (3): 731-733. (NSA 29: 13086).

Bartlett, A. C. and N. Mitlin. 1967. Developmental and reproductive effects of heavy water (D_2O) in the diet of the boll weevil. J. Econ. Entomol. 60 (3): 647-655. (BA 48: 124519).

Bartlett, A. C., A. E. Bell and V. L. Anderson. 1966. Changes in quantitative traits of *Tribolium* under irradiation and selection. Genetics 54: 699-713. (NSA 20: 40671).

Bartlett, A. C., P. A. Hooker and D. D. Hardee. 1968. Behavior of irradiated boll weevils. I. Feeding, attraction, mating, and mortality. J. Econ. Entomol. 61(6): 1677-1680.

Bartlett, A. C., R. T. Staten and W. O. Ridgway. 1973. Gamma irradiation of eggs of the pink bollworm. J. Econ. Entomol. 66(2): 475-477. (NSA 28: 30378).

Bartlett, M. R. and R. L. Haedrich. 1966. Techniques in the radiography of fishes. Trans. Amer. Fish. Soc. 95(1): 99-101.

Bartnoff, S. 1973. Siting on rivers and estuaries. New Jersey Power and Light Co., Morristown. (NP-19964, Paper 10). vp. (NSA 30: 6570).

Barton, C. J. and S. V. Kaye. (Comps.). 1975. Evaluation of radiological exposure from Plowshare applications, 1967-1975. Oak Ridge National Lab., U. S. ERDA report ORNL-5071. 33pp. (NSA 32: 27935).

Barton, L. C. and F. W. Stehr. 1970. Normal development of *Anaphes flavipes* in cereal leaf beetle eggs killed with x-radiation and potential field use. J. Econ. Entomol. 63: 128-130. (NSA 24: 23304).

Barton-Browne, L. B. 1964. Water regulation in insects. Ann. Rev. Entomol. 9: 63-78. (Incl. references using radionuclides).

Bartsch, A. F. and D. I. Mount. 1970. Thermal effects on ecological systems. pp. 117-124. In: Northwest Conference on the Role of Nuclear Energy. (See Oregon Governor's Nuclear Development Coord. Comm., 1970) (NSA 25: 23220).

Barwinek, F. 1972. Ecological investigation of the behavior of ^{55}Fe in plants, especially lichen, and in the meat of cattle and game. Thesis, Univ. of Munich, Germany. 146pp. (In German) (NSA 28: 30378).

Bar-Zeev, M and C. H. Schmidt. 1959. Action of a repellent as indicated by a radioactive tracer. J. Econ. Entomol. 52(2): 268-269.

Basham, S. J., Jr. 1973. Nuclear systems: environmental considerations session summary. J. Environ. Sci. 16(5): 17-21. (NSA 29: 24216).

Bashaw, E. C. and B. J. Hoff. 1962. Effects of irradiation on apomicitic common dalligrass. Crop. Sci. 2(6): 501-504. (NSA 17: 7832).

Baslavskaya, S. S. and G. Veber. 1959. Effect of light on conversion of phosphates in plants. Dokl. Akad. Nauk. SSSR 124(1): 227-230. (Doklady (Bot. sci.): 19-22) (Used ^{32}P with *Scenedesmus quadricauda*).

Bassurova, A. N. and V. S. Nikiforov. 1972. Investigations of changes in the luminescence intensity of uranium pearls with time. pp. 238-242. In: Radioecological Investigations in Natural Biogeocenoses. (See I. N. Verkhovskaya, 1972).

Bastin, H. 1915. Radio-active ores and plant life. Scient. American 112 (15): 335.

Bate, L. C. 1971. Determination of mercury in biological samples by neutron activation analysis. pp. 197-200. In: Nuclear Methods in Environmental Research. (See Vogt, et al., 1971) (NSA 26: 17366).

Bate, L. C., W. S. Lyon and E. H. Wollerman. 1974. Gold tagging of elm bark beetles and identification by neutron activation analysis. Radiochem. Radioanal. Lett. 17(1): 77-85. (NSA 30: 2480).

Bate, L. C., S. E. Lindberg and A. W. andren. 1975. Elemental analysis of water and air solids by neutron activation analysis. Trans. Amer. Nucl. Soc. 21(3, Suppl.): 21-22. (NSA 32: 27484).

Bateman, A. J. and W. K. Sinclair. 1950. Mutations induced in Drosophila by ingested phosphorus-32. Nature 165(4186): 117-118.

Batement, W. A. S., et al. 1957. Radioactive fallout in Australia from Operation "Mosaic". Austral. J. Sci. 20(5): 125-135.

Bates, R. G. 1963. Aeroradioactivity survey and areal geology of the National Reactor Testing Station area, Idaho. U. S. Geological Survey, U. S. AEC report CEX-59.4.10. 18pp. (NSA 19: 13622).

Batra, A. and H. C. Arya. 1974. Effect of gamma irradiation on cells of Arachis hypogaea L. and Dolichos lablab L. grown in culture. Indian J. Exp. Biol. 12(1): 67-71. (NSA 30: 27017).

Batsche, H., H. Moser and W. Stichler. 1972. Measurements of deuterium and oxygen-18 contents of karst waters. Geol. Jahrb. Reithe C(2): 275-288. (In German) (NSA 30: 18672).

Batson, W. T. 1955. An ecological study of the land plants and cold-blooded vertebrates of the Savannah River area. I. The flora of the Savannah River Project area. 3. The Myxomycetes (a preliminary listing). 5. The liverworts. 6. The mosses. 6. The mosses (continued). Univ. S. Carolina Publ. (biol. 3) 1(4): 204-206, 209-214, 214-219; 1(5): 267-275.

Batson, W. T. and W. R. Kelley. 1953. Ibid. 1. The monocotyledons. Univ. S. Carolina Publ. (biol. 3) 1(2): 60-89.

Batson, W. T. and W. R. Kelley. 1955. Ibid. VII. Distributional studies of the flora. 1. The sandhills vegetation of Aiken and Barnwell Counties. Univ. S. Carolina Publ. (biol. 3): 1(4): 249-255.

Batson, W. T. and R. N. Tulloch. 1955. Ibid. IV. Succession in fields of the Savannah River Project area. 2. The floristic composition of lowland fields in the third year of abandonment. Univ. S. Carolina Publ. (biol. 3) 1(4): 220-226.

Batson, W. T., et al. 1957. An ecological study of the fauna and flora of the Savannah River Plant area. VII. Distributional studies of the flora. 3. The vegetation of a mature beech-magnolia forest within the Gantt Tract. Univ. S. Carolina Publ. (biol. 3) 2(2): 65-71.

Battelle Memorial Institute. 1956. The development of a gamma radiographic method for the underwater inspection of marine piling. Summary report. Columbus, Ohio.

Battelle Memorial Institute. 1965. Biology at Pacific Northwest Laboratory. Pacific Northwest Lab., Special report, U. S. AEC report BNWC-43. 40pp. (Incl. radioecology program).

Battelle Memorial Institute. 1971. Data for preliminary demonstration phase of the "environmental quality information and planning system (equips)". Pacific Northwest Lab., U. S. AEC report BNWL-B-141. 307pp. (NSA 30: 13299).

Battelle Memorial Institute. 1973. Pacific Northwest Laboratory monthly activities report, November 1973. Pacific Northwest Lab., U. S. AEC report BNWL-1802. 20pp. (NSA 31: 5697) (Incl. uptake of transuranium elements from soil).

Battiston, U. and A. Moauro. 1970. Determination of micro amounts of elements contained in natural water by means of neutron-activation analysis. Italy Comitato per l'Energia Nucleare, Rome, report RT/CHI-(70)33. 36pp. (For Conf. on Analytical and Physical Chemistry of Water, Bormio, Italy, 9 July 1970) (In Italian) (CONF-700717) (NSA 25: 37320).

Battiston, U. and A. Moauro. 1974. Determination of the total content of Ag, As, Au, Co, Cr, Re, Sb, U, and Zn in natural waters by neutro activation analysis. Italy Comitato per l'Energia Nucleare, Rome, report RT/CHI-(74)8. (In Italian) (NSA 31: 2749).

Battistone, G. C., E. Levri and R. Lofberg. 1970. Simplified ultramicrodetermination of copper in biological specimens by neutron activation analysis. Clin. Chim. Acta 30: 429-438.

Baturin, G. N. 1966. Uranium content in Caspian Sea sediments. Atom. energiya 21: 515-516. (Soviet Atomic Energy 21(6): 1228-1230) (NSA 21: 12541).

Baturin, G. N. and A. V. Kochenov. 1969. Migration of uranium in rivers and the duration of its presence in waters of the world's oceans, seas and lakes. Geokhimiya (6): 715-716. (In Russian) (NSA 23: 35332).

Baturin, G. N., A. V. Kochenov and K. M. Shimkus. 1967. Uranium and minor metals in columns of bottom sediments of the Black and Mediterranean seas. Geokhimiya (1): 41-50. (Abstr., Geochem. Int. 4: 29, Am. Geol. Inst. transl. GI-67-1-4) (NSA 21: 12542).

Batygin, N. F. 1971. Irradiation of plants in different periods of ontogenesis. pp. 42-45. In: Use of Isotopes and Nuclear Radiation in Agriculture. Atomizdat, Moscow. (In Russian) (NSA 31: 703).

Batzer, H. O. and N. D. Addy. 1971. Release and recovery of phosphorus-32 tagged jack pine budworm larvae in the field (Lepidoptera: Tortricidae). Ann. Entomol. Soc. Am. 64(5): 1185-1186.

Baudin, G., R. Darras and E. Roth. 1973. Analytical methods for the study of water pollution. Centre d'Études Nucléaires, France Commissariat à l' Énergie Atomique, Saclay, report CEA-CONF-2559. 18pp. (In French) (NSA 30: 2515) (Neutron activation).

Baudin, J. P. 1970. Dynamics of ^{137}Cs absorbed with food in a freshwater fish. Compt. rend. hebd. seances acad. agr.fr. 56: 662-669. (In French) (NSA 25: 21555).

Baudin, J. P. 1973. Experimental study if the fixing and eliminating of ^{65}Zn by *Mytilus galloprovincialis* (Lmk.). Compt. rend. C277(1): 113-116. (In French) (NSA 29: 822).

Baudin, J. P. 1975. An experimental study of zinc cycle in the pond of the Olivier. Vie Milieu 24(1): 59-80. (In French) (NSA 32: 11544) (^{65}Zn).

Baudin, J. P., et al. 1968. Summary of the modality of the experimental contamination of a freshwater fish (*Amiurus nebulosus* Leseuer) by cesium-137. Ann. Fac. Sci. Marseille 41: 81-93. (In French).

Baudouin, M. F. and P. Scoppa. 1974. Accumulation and retention of chromium-51 by freshwater zooplankton. Commission of the European Communities, Ispra, Italy, report EUR-5160e. 23pp. (NSA 30: 32597).

Bauer, H. 1967. Kinetic organization of Lepidoptera chromosomes. Chromosoma 22: 101-125. (NSA 22: 6778) (X-rays).

Bauer, H. 1970. Rearrangements between germ-line limited and somatic chromosomes in *Smittia parthenogetica* (Chironomidae, Diptera). Chromosoma 32: 1-10. (NSA 26: 9519).

Bauer, J. W. and R. E. Monroe. 1969. Metabolism of cholesterol-4-^{14}C in insecticide-resistant and susceptible strains of aseptically reared house flies, *Musca domestica*. Ann. Entomol. Soc. Am. 62(5): 1021-1025.

Bauman, A. 1968. Determination of ^{137}Cs in soil. pp. 335-338. In: *Radiation Protection. Part I.* W. S. Snyder, et al., eds. Pergamon Press, New York. (NSA 23: 11439).

Baumhover, A. H. 1958. Screworm control program. Veterin. Med. 53(4): 216-219. (Irradiation sterility).

Baumhover, A. H. 1963. Influence of aeration during gamma irradiation of scre-worm pupae. J. Econ. Entomol. 56(5): 628-631. (NSA 18: 11865).

Baumhover, A. H. 1965. Sexual aggressiveness of male-screw-worm flies measured by effect on female mortality. J. Econ. Entomol. 58(3): 544-548. (^{60}Co, *Callitroga hominivorax*).

Baumhover, A. H. 1966. Eradication of the screwworm fly: an agent of myiasis. J. Am. Med. Assoc. 196: 240-248. (NSA 21: 6746).

Baumhover, A. H., et al. 1955. Screw-worm control through use of sterilized flies. J. Econ. Entomol. 48(4): 462-466.

Baurmash, L., et al. 1958. Distribution and characterization of fall-out and airborne activity from 10 to 160 miles from GZ, Spring, 1955 (Operation TEAPOT). Univ. of Calif., Los Angeles, U. S. AEC Weapons Test Report WT-1178. 128pp.

van Bavel, C. H. M., E. E. Hood and N. Underwood. 1954. Vertical resolution in the neutron method for measuring soil moisture. Trans. Am. Geophys. Un. 35(4): 595-600.

Baxter, J. P. 1971. Environmental pollution and its control. At. Energy Austral. 14(2): 2-7. (NSA 26: 11373).

Baxter, M. S. and A. Walton. 1970. Theoretical approach to the Suess effect. Proc. Roy. Soc. (London) A318: 213-230. (NSA 24: 46363) (^{14}C).

Baxter, M. S. and A. Walton. 1971. Fluctuations of atmospheric ^{14}C concentrations during the past century. Proc. Roy. Soc. (London) A321: 105-127. (NSA 25: 21203).

Baxter, M. S. and D. D. Harkness. 1975. ^{14}C/^{12}C ratios as tracers of urban pollution. pp. 135-141. In: *Isotope Ratios as Pollutant Source and Behavior Indicators.* (See International Atomic Energy Agency, 1975) (NSA 33: 7319).

Baxter, M. S., et al. 1974. Calibration of the radiocarbon time scale. Nature 249(5452): 93-94. (NSA 30: 12706).

Baxter, P. 1971. Pollution in our environment. Austral. Chem. Proc. Eng. 24(4): 18-23. (NSA 25: 42383) (Concerns nuclear power plants).

Baxter, R. C. and H. A. Blair. 1967. Age of death in *Drosophila* following sublethal exposure to gamma radiation. Radiation Research 31: 287-303. (Also: Univ. of Rochester, U. S. AEC report UR-49-799) (NSA 23: 22359).

Baxter, R. C. and H. A. Blair. 1967. Kinetics of aging as revealed by x-ray dose-lethality relations in *Drosophila*. Radiation Research 30(1): 48-70. (Also: Univ. of Rochester, U. S. AEC report UR-49-735) (NSA 23: 22358).

Baxter, R. C. and H. A. Blair. 1969. Recovery and overrecovery from acute radiation injury as a function of age in *Drosophila*. Radiation Research 39: 345-360. (NSA 23: 41424).

Baxter, R. C. and L. W. Tuttle. 1957. Life span shortening in irradiated *Drosophila melanogaster*. Abstr. Radiation Reserach 7(3): 303).

Bayeva, A. I. and A. B. Akhundova. 1970. Uranium content in various types of plants in the subtropical Lenkoran region of Azerbaidzhan SSR. Dokl. Akad. Nauk Azerb. SSR 26(5): 69-71. (In Russian) (NSA 25: 21574).

Baylor, E. R. and F. E. Smith. 1958. Animal perception of x-rays. Radiation Research 8(6): 466-474. (*Daphnia*).

Beal, J. A. 1958. Status and trends in forest insect research in the United States. Proc. 10th Intern. Congr. Entomol. 4: 323-330.

Beal, J. M. 1951. Negative results following exposure of several kinds of seeds to cosmic rays and other radiations at high altitudes. Bot. Gaz. 112: 533-534. (NSA 5: 6070).

Beam, C. A. 1962. Survival of cells of Hansenula anomala following incorporation of tritiated uridine and deoxythymidine. Abstr. p. 16. In: 2nd Intern. Congr. Radiation Research. (See Anonymous, 1962).

Beam, C. A., et al. 1954. The relationship of radioresistance to budding in Saccharomyces cerevisiae. Arch. Biochem. Biophys. 49: 110-122.

Beament, J. W. L., J. E. Treherne and V. B. Wiggelsworth. (Eds.). 1963. Advances in Insect Physiology. Vol. I. Academic Press, New York. 512pp. (Contains references to radionuclides).

Beard, R. L. 1965. Competition between DDT-resistant and susceptible house flies. J. Am. Med. Assoc. 196: 240-248. (NSA 21: 6746).

Beard, R. L. 1972. Lethal action of UV irradiation on insects. J. Econ. Entomol. 65(3): 650-654.

Beard, R. L. nd. Radiation, insects, and food preservation. Thailand Off. of Atomic Energy for Peace, Bangkok, report THAI. AEC-9. pp. 23-25. (NSA 23: 16452).

Beasley, T. M. 1969. ^{210}Pb production by nuclear devices: 1946-1958. Nature 224(5219): 573. (NSA 24: 4386) (Soil from Bikini, Eniwetok, Rongelap).

Beasley, T. M. and E. E. Held. 1969. Nickel-63 in marine and terrestrial biota, soil, and sediment. Science 164: 1161-1163.

Beasley, T. M. and E. E. Held. 1971. ^{108} Bikini and Eniwetok Atolls. Nature 230: 450-451. (NSA 25: 42377).

Beasley, T. M. and H. E. Palmer. 1966. Lead-210 and polonium-210 in biological samples from Alaska. Science 152(3725): 1062-1063.

Beasley, T. M., E. E. Held and R. M. Conard. 1972. ^{55}Fe in Rongelap people, fish, and soils. Health Physics 22(3): 245-250. (NSA 26: 20355).

Beattie, J. R. 1973. Environmental aspects of nuclear power. Atmos. Environ. 7(12): 1205-1211. (NSA 29: 18322).

Beatley, J. C. 1962. Vascular plants of the U. S. Atomic Energy Commission Nevada Test Site, Nye County, Nevada. Univ. of Calif., Los Angeles, U. S. AEC report UCLA-508. 33pp.

Beatley, J. C. 1963. Annual vegetation of the Northern Mojave Desert. Abstr. Bull. Ecol. Soc. Am. 44(4): 122-123. (Nevada Test Site).

Beatley, J. C. 1963. Vegetation and environments of the Nevada Test Site. Abstr. Bull. Ecol. Soc. Am. 44(4): 123.

Beatley, J. C. 1964. Effects on desert vegetation of a nuclear detonation Nevada Test Site. Abstr. Bull. Ecol. Soc. Am. 45(3): 80.

Beatley, J. C. 1964. Ecological status of introduced species at the Nevada Test Site. Abstr. Bull. Ecol. Soc. Am. 45(3): 78.

Beatley, J. C. 1964. The vascular flora of the Nevada Test Site, Nye County, Nevada. Abstr. Am. J. Bot. 51: 687.

Beatley, J. C. 1965. Ecology of the Nevada Test Site. I. Geographic and ecologic distributions of the vascular flora (annotated checklist). Univ. of Calif., Los Angeles, U. S. AEC report UCLA-12-533. 69pp. (NSA 19: 25817).

Beatley, J. C. 1965. Ecology of the Nevada Test Site. II. Status of introduced species. Univ. of Calif., Los Angeles, U. S. AEC report UCLA-12-554. 39pp.

Beatley, J. C. 1965. Ecology of the Nevada Test Site. III. Survival of winter annuals, 1963-1964. Univ. of Calif., Los Angeles, U. S. AEC report UCLA-12-555. 21pp.

Beatley, J. C. 1965. Ecology of the Nevada Test Site. IV. Effects of the Sedan detonation on desert shrub vegetation in northeastern Yucca Flat, 1962-1965. Univ. of Calif., Los Angeles, U. S. AEC report UCLA-12-571. 55pp.

Beatley, J. C. 1965. Effects of radioactive and non-radioactive dust upon Larrea divaricata Cav., Nevada Test Site. pp. 1621-1625. In: Radiation and Terrestrial Ecosystems. (See Hungate, 1965).

Beatley, J. C. 1965. Survival of winter annuals in the northern Mojave Desert. Abstr. Bull. Ecol. Soc. Am. 46(2): 40. (Nevada Test Site).

Beatley, J. C. 1966. Winter annual vegetation following a nuclear detonation in the northern Mojave Desert (Nevada Test Site). Radiation Botany 6: 69-82. (NSA 20: 18386).

Beatley, J. C. 1967. Survival of winter annuals in the Mojave Desert. Ecology 48(5): 745-750. (Nevada Test Site).

Beatley, J. C. 1968. Rodent densities in relation to winter annuals and precipitation in the northern Mojave Desert. Abstr. Bull. Ecol. Soc. Am. 49(3): 115. (Nevada Test Site).

Beatley, J. C. 1969. Dependence of desert rodents on winter annuals and precipitation. Ecology 50(4): 721-724. (Nevada Test Site).

Beatley, J. C. 1971. Ecologic and geographic distributions of the vascular plants of southern Nye County, and adjacent parts of Clark, Lincoln, and Esmeralda Counties, Nevada. Univ. of Calif., Los Angeles, U. S. AEC report UCLA-12-705(suppl.). 49pp.

Beatley, J. C. 1973. Check list of vascular plants of the Nevada Test Site and central-southern Nevada. Univ. of Cincinnati, U. S. AEC report COO-2307-4. 43pp. (NSA 29: 12745).

Beague, L. A. and R. A. Sjodin. 1968. Transport of cesium in frog muscle. J. Physiol. (London) 194: 105-123. (NSA 22: 45387).

Beavers, J. B., et al. 1971. Some effects of gamma irradiation or the chemosterilant, TEPA, on the citrus red mite and its progeny. J. Econ. Entomol. 64(1): 72-75.

Beavers, P. E., G. W. Esch and R. E. Kuhn. 1971. Some effects of sub-lethal x-irradiation on reproduction and development in two strains of larval Taenia crassiceps. Intern. J. Parasitol. 1: 235-239. (NSA 28: 21785).

Bebbington, W. P. 1973. Nuclear operations and the environment: Savannah River Plant. Savannah River Plant, U. S. AEC report DPSPU-72-30-9. 24pp. (NSA 28: 5527).

Bebbington, W. P. 1973. Environmental effects of a complex nuclear facility. Savannah River Plant, U. S. AEC report DPSPU-72-30-10. 14pp. (NSA 29: 2628).

Bebbington, W. P. 1974. Environmental effect of a complex nuclear facility. Chem. Eng. Progr. 70(3): 85-85. (NSA 29: 29311).

Beck, C. K. 1971. Environmental monitoring for USA licensed power stations. pp. 401-411. In: Environmental Aspects of Nuclear Power Stations. (See IAEA, 1971).

Beck, D E. and D. M. Allred. 1968. Faunistic inventory - BYU ecological studies at the Nevada Test Site. Great Basin Nat. 28(3): 132-141.

Beck, D E., D. M. Allred and E. P. Brinton. 1963. Ticks of the Nevada Test Site. Brigham Young Univ. Sci. Bull. (biol.) 4(1): 1-11, 1 map. (NSA 17: 40539).

Beck, D E., et al. 1964. Nevada Test Site desert ecology. Proc. Utah Acad. Sci. (arts and ltrs.) 41(2): 202-210.

Beck, H. L., B. G. Bennett and T. F. McCraw. 1967. External radiation levels on Bikini Atoll, May 1967. Health and Safety Lab., U. S. AEC report HASL-190. 79pp. (NSA 23: 6217).

Beck, H. L., W. M. Lowder and J. E. McLaughlin. 1971. In situ external environmental gamma-ray measurements utilizing Ge(Li) and NaI(Tl) spectrometry and pressurized ionization chambers. pp. 499-513. In: Rapid Methods for Measuring Radioactivity in the Environment. (See IAEA, 1971) (NSA 26: 28069).

Beck, H. L., J. E. McLaughlin and K. M. Miller. 1976. In situ determinations of environmental plutonium and its related nuclides. IEEE Trans. Nucl. Sci. NS-23(1): 676-682. (NSA 33: 29279).

Beck, H. L., et al. 1971. New perspectives on environmental radiation monitoring around nuclear facilities. Trans. Amer. Nucl. Soc. 14(2): 464. (NSA 26: 2396).

Beck, H. L., et al. 1972. New perspectives on low level environmental radiation monitoring around nuclear facilities. Nucl. Technol. 14(3): 232-239. (NSA 26: 32911).

Beck, J. S. 1961. Effects of radiation upon development of Tribolium confusum. Abstr. 11. Radiation Research 14(4): 449.

Beck, J. S. 1962. Effects of x-irradiation upon cell population and morphogenesis in the wing of Tribolium confusum. Abstr. p. 226. In: 2nd Int. Congr. Radiation Research. (See Anonymous, 1962).

Beck, J. S. 1963. Effects of x-irradiation on cell differentiation and morphogenesis in a developing beetle wing. Radiation Research 19(4): 569-581. (NSA 17: 33730).

Beck, J. S. and T. L. Hayes. 1962. Effects of x-irradiation upon cell population and morphogenesis in the developing beetle wing. Abstr. E4. Radiation Research 16: 573.

Beck, J. S. and T. R. Manney. 1962. Neutron activation analysis for phosphorus in a study of development in a beetle wing. Science 137(3533): 865-866. (Also: Univ. of Calif., U. S. AEC report UCRL-10389) (NSA 16: 30195).

Beck, T. and H. Poschenrieder. 1960. On the eventual identification of radioactive contamination of soil with the use of ecological studies of soil microflora. Mitt. f. Landkult. Mooru. Torfwirt. 8(1): 27-32. (In German).

Beck, T. and R. Roschenthaler. 1960. Radiobiology of micro-organisms. Bayer. landwirtsch. Jahrb. 37(2): 1-82. (In German) (NSA 15: 8535).

Becker, D. A. and P. D. LaFleur. 1972. Determination of trace quantities of uranium in biological materials by neutron activation analysis using a rapid radiochemical separation. Anal. Chem. 44(8): 1508-1511. (NSA 26: 40580).

Becker, D. A. and P. D. LaFleur. 1972. Neutron activation analysis: application to trace element analysis in biological and environmental materials. pp. 447-453. In: Trace Substances in Environmental Health. D. D. Hemphill, ed. Univ. of Missouri, Columbia. (NSA 26: 40589).

Becker, K. 1973. Current studies on integrating environmental monitoring at ORNL. Oak Ridge National Lab., U. S. AEC report CONF-740212-1. 13pp. (NSA 29: 24291).

Becker, K. and A. Moreno y Moreno. 1974. Applications of thermoluminescence measurements in ancient ceramics. Oak Ridge National Lab., U. S. AEC report, ORNL-TM-4572. 35pp. (NSA 30: 18659).

Becker, K. and A. Moreno y Moreno. nd. Some applications of thermoluminescence measurements in ancient ceramics. Oak Ridge National Lab., U. S. AEC report CONF-740833-1. 22pp. (NSA 31: 2653).

Beckman, C. H. and J. E. Kuntz. 1951. Translocation of poisons, dyes, and radioiodine, and its relation to oak wilt. Abstr. Phytopathology 41(1): 2-3.

Beckman, C. H., J. E. Kuntz and A. J. Riker. 1952? The use of radioactive isotopes to ascertain the role of root-grafting in the translocation of water, nutrients, and disease-producing organisms among forest trees. Univ. of Wisconsin, U. S. AEC report AECU-2275. 8pp. (NSA 7: 55).

Beckman, C. H., et al. 1953. Host responses associated with the development of oak wilt. Phytopathology 43: 448-454. (Also: Univ. of Wisconsin, U. S. AEC report AECU-2516) (NSA 7: 3340).

Becknell, D. E., R. H. Marsh and W. Willie, Jr. 1971. Use of anion exchange resin-loaded paper in the determination of trace mercury in water by neutron activation analysis. Anal. Chem. 43(10): 1230-1233. (NSA 25: 44350).

Becquerel, P. 1912. Radioactivity in relation to plant life. Rev. Gen. Sci. 23(15): 576-585.

Bedrosian, P. H., D. G. Easterly and S. L. Cummings. 1970. Radiological survey around power plants using fossil fuel. Eastern Environmental Rad. Lab., Montgomery, Ala., U. S. EPA report EERL-71-3. 17pp. (NSA 26: 196).

Beer, L. P. and B. G. Johnson. 1973. Quality assurance in environmental reports. Power Eng. 77(2): 45-47. (NSA 27: 22484).

Beermann, W. and C. Pelling. 1965. The labelling of individual chromatids of giant chromosomes with ^{3}H-thymidine. Chromosoma 16(1): 1-21. (In German, English summary) (*Chironomus tentans*).

Beetem, W. A. and D. B. Grove. 1966. Radiochemical analysis of groundwater and surface water samples collected after the Salmon Event in the vicinity of Tatum Salt Dome, Lamar County, Mississippi. U. S. Geol. Survey, U. S. AEC report USGS-474-119. 77pp. 1972. *Ibid*. (Dribble-51). 1972. USGS-474-120 (Dribble-53). 1972. USGS-474-124 (Dribble-47 suppl. 1). (NSA 26: 36083).

Beetem, W. A., et al. 1971. Chemical analyses of water samples collected on Amchitka Island, Alaska. Amchitka-29. U. S. Geol. Survey, U. S. AEC report USGS-474-135. 23pp. USGS-474-123; i, 38pp. + 1 map. USGS-135; i, 18pp + 1 map. (NSA 25: 51618; 26: 15233).

Begemann, F. 1957. Tritium assays of natural waters measured in 1956-1957. Univ. of Chicago, U. S. Air Force report AFOSR-TR-58-41. 71pp. (AD-154131) (NSA 12: 8224).

Begemann, F. and W. F. Libby. 1957. Continental water balance, ground-water inventory and storage times, surface ocean mixing rates and world-wide water circulation patterns from cosmic-ray and bomb tririum. Geochim. Cosmochim. Acta 12: 277-296.

Beglaryan, N. P. and A. V. Avetsyan. 1971. Comparative cytological analysis of the effect of x-rays and gibberellic acid on sunflowers. Tsitologiya Genet. (3): 227-230. (In Russian) (NSA 26: 9653).

Behne, D. 1972. Determination of mercury in biological materials by neutron activation analysis. Radiochim. Acta 18(3): 153-156. (In German) (NSA 28: 12635).

Behounek, F. 1961. The effect of nuclear installations on the environment. Jaderna Energie 7: 397-400. (English transl. AECL-1677).

Belanger, G. J., L. M. Babineau and R. H. Despointes. 1970. Radioactivity measurement in biological microsamples. Int. J. Appl. Radiat. Isotopes 21: 437-440. (In French) (NSA 25: 15530).

Belasco, I. J. and W. P. Langsdorf. 1969. Synthesis of C^{14}-labeled Siduron and its fate in soil. J. Agric. Food Chem. 17(5): 1004-1007.

Belgian Royal Institute of Natural Sciences. 1968. *International Symposium on the Application of Neutron Activation Analysis in Oceanography*. (Brussels, 17-22 June 1968). (CONF-680618) (NSA 25: 15535).

Belitsina, N. V., et al. 1963. Informational ribonucleic acid in early development of *Misgurnus fossilis* embryos. Dokl. Akad. Nauk SSSR 153(2): 464-467. (Doklady (biol. sci.) 153: 1406-1408) (^{14}C).

Belitsina, N. V., et al. 1963. The effect of the radiation inactivation of nuclei on the synthesis of imformation ribonucleic acid in the embryos of groundling fish (*Misgurnus fossilis*). Dokl. Akad. Nauk SSSR 153(5): 1204-1206. (Doklady (biol. sci.) 153: 1443-1445) (NSA 18: 11722).

Belitskiy, A. S. 1958. Evaluation of the groundwater sources from territories polluted with radioactive products from uranium fission. Gigiyena sanitariya 23(10): 23-27. (In Russian).

Belitskiy, A. S. 1962. Routes of invasion of radioactive substances into underground waters. Basic factors influencing the migration of radioactive substances in underground waters. Elements of the flow of underground waters which determine the actual speed and time movement of the waters. Movement of radioactive products of uranium fission in underground waters under natural conditions. Vol. II, Ch. 4, pp. 68, 69-70, 71-75, 84-92. In: *Radiation Sanitation*. (See Marey, 1962).

Belitskiy, A. S. 1963. Evaluation of the groundwater sources from territories polluted with radioactive products from uranium fission. Gigiyena sanitariya 28(10): 23-25. (In Russian).

Belitskiy, A. S. and Ye. I. Orlova. 1960. Movement of products of uranium fission in underground waters. Gigiyena sanitariya 25(6): 3-8. (In Russian) (Abstr. Sov. Med. 5(5): 2063).

Belitskiy, A. S. and Ye. I. Orlova. 1968. *Protection of Groundwaters from Radioactive Contamination*. 208pp. Publ. House Medical Literature, Moscow. (In Russian) (NSA 23: 35324).

Belitskiy, A. S., Ye. I. Orlova and A. I. Ryzhov. 1961. Some data on movement of radiostrontium with groundwater current. pp. 306-313. In: *Distribution, Biological Effects, and Migration of Radioactive Isotopes*. A. V. Lebedinskiy and Yu. I. Moskalev, eds. Publ. House Medical Literature, Moscow. (JPRS-15067, pp. 29-40; AEC-tr-7512, pp. 359-367) (NSA 30: 6529).

Belitskiy, A. S., A. D. Turkin and V. M. Kozlov. 1971. Consideration and evaluation of the ICRP standards as applied to the monitoring of radioactive effluents from power reactors. pp. 175-181. In: *Environmental Aspects of Nuclear Power Stations*. (STI/PUB/261) (In Russian) (See IAEA, 1971) (CONF-700810).

Belitskiy, A. S., et al. 1974. Migration of radioactive nuclides under different hydrogeological conditions. pp. 260-265. In: *Proc. 3rd Intern. Congr. Intern. Radiation Protection Assoc.* (In Russian) (CONF-730907-P1) (NSA 30: 26764).

Belk, W. F. and A. E. Gass, Jr. 1969. Relative resistance of gerbils and rats to acute ^{60}Co irradiation. School of Aerospace Medicine, Brooks AFB, Tex., USAF report SAM-TR-69-16. 11pp. (AD-688375) (NSA 23: 41428).

Bell, A. E. 1964. The effects of x-radiation on plateaued populations of *Tribolium castaneum* in regards to reproductive fitness and response to selection. Final report. U. S. AEC report COO-965-21. 8pp. (NSA 18: 27061).

Bell, A. E. 1969. Nature of selection responses in *Tribolium*. Idengaku Zasshi 44(Suppl. 1): 299-309. (NSA 24: 34251).

Bell, C. G., Jr. 1957. Nuclear long-range fallout in surface waters. Proc. Am. Soc. Civil Engrs. 83(SA 5, No. 1400): 1-22.

Bell, C. G., Jr., H. A. Thomas, Jr. and B. L. Rosenthal. 1954. Passage of nuclear detonation debris through water treatment plants. J. Am. Water Works Assoc. 46(10): 973-986.

Bell, C. R. 1970. Seed distribution and germination experiment. Ch. D-6, pp. D177-D182. In: *A Tropical Rain Forest*. (See Odum and Pigeon, 1970).

Bell, G. M. and W. S. Hoar. 1950. Some effects of ultra-violet radiation on sockeye salmon eggs and alevins. Can. J. Res. 28(10): 35-43.

Bell, J. P. 1969. A new design principle for neutron soil moisture cages: the "Wallingford" neutron probe. Soil Science 108(3): 160-164.

Bell, J. P. and C. W. O. Eeles. 1967. Neutron random counting error in terms of soil moisture for nonlinear calibration curves. Soil Science 103(1): 1-__.

Bell, R. K. and F. J. Ward. 1968. Use of liquid scintillation counting for measuring self-absorption of ^{14}C in *Daphnia pulex*. J. Fish. Res. Bd. Can. 25: 2505-2508. (NSA 23: 37911).

Bellamy, A. W., et al. 1949. The 1948 radiological and biological survey of areas in New Mexico affected by the first atomic bomb detonation. Univ. of Calif., Los Angeles, U. S. AEC report UCLA-32. 50pp.

Belle, Yu. S. and O. V. Lebedev. 1972. Experience of determination of low-background gamma-spectrometry. pp. 109-124. In: *Methods of Determination of Radioactivity*. (See Polikarpov and Parchevskiy, 1972).

Bellucci, M. 1959. Modifications in the biological effect of x radiation with respect to the rhythmic fractionation of the radiation beam. Radioterap. radiobiol. e fiz. med. 14(3): 188-204. (In Italian) (NSA 14: 6168).

Belot, Y. and C. Caput. 1973. Effect of vegetation screening on the dispersion and deposition of atmospheric pollutants on the ground. Agriculture (Paris) (366): 175-177. (In French) (NSA 29: 10130).

Belot, Y., et al. 1970. Direct transport of particle pollutants from atmosphere to vegetal communities. Theoretical considerations. pp. 863-881. In: *Proc. Intern. Symp. on Radioecology. Vol. II.* (See France Commissariat à l'Énergie Atomique, 1970) (In French) (NSA 24: 41786).

Belot, Y., et al. 1971. Mineral composition of the particle matter in suspension in the air, in urban areas and in a forest region exposed to industrial pollution. pp. 671-687. In: *Proc. Intern. Symp. Radioecology Applied to the Protection of Man and His Environment.* (See Commission of the European Communities, 1971) (In French) (NSA 27: 5235) (Activation analysis).

Belousova, I. M. 1961. Ch. VI. Radioactivity in the earth's rocks. Ch. VII. Radioactivity in water. Ch. IX. Content of radioactive substances in plants and food products. pp. 61-82, 94-102. In: *Natural Radioactivity*. A. I. Burnazyan, ed. State Publ. House of Medical Literature, Moscow. (NSA 16: 20229).

Belova, E. I. and Z. G. Antropova. 1961. Comparative distribution of strontium-90 and cesium-137 in different types of soils. pp. 72-75. In: *Nuclear Meteorology*. (See Makhon'ko and Malakhov, 1961) (NSA 30: 15429).

Belova, E. I. and Z. G. Antropova. 1971. ^{90}Sr and ^{137}Cs distribution in different types of soils. Tr. Eksp. Meteorol. 21(1): 77-80. (In Russian) (NSA 30: 29403).

Belova, E. I., G. E. Sharonov and A. A. Moiseyev. 1972. Behavior of cesium-137 in sodpodzolic soils of the Ukrainian alluvial plain. State Committee on the Use of Atomic Energy, USSR, Moscow. 15pp. (In Russian) (INIS-mf-1044) (NSA 30: 6525).

Belsky, T. 1966. Chemical evolution and organic geochemistry. Thesis, Univ. of California, Berkeley. 134pp. (NSA 20: 16633).

Belter, W. G. 1962. Radioactive wastes. Intern. Sci. Tech. (12): 42-43. 45-51. (NSA 17: 13873).

Belter, W. G. 1963. Summary discussion. pp. 391-404. In: *Transport of Radionuclides in Fresh Water Systems*. (See Kornegay, et al., 1963).

Belter, W. G., et al. 1969. AEC's position on radioactive waste management. Nuclear News 12(11): 60-65. (NSA 24: 50538).

Beltsev, D. I., et al. 1971. Method of determination of concentrations of ^{90}Sr and ^{137}Cs in tissues of live reindeer. pp. 188-193. In: *Methods of Radioecological Investigations*. (See Verkhovskaya, 1971).

Belyayev, L. I. et al. 1963. Radioactive fallout in the vicinity of Simeiz in 1960-1961. State Committee on the Use of Atomic Energy, Atomizdat, Moscow. 7pp. (UN Document A/AC.82/G/L. 914) (In Russian).

Belyayev, V. A. and V. I. Maslov. 1971. Application of cytogenetic methods in radioecological studies. pp. 176-182. In: Methods of Radioecological Investigations. (See Verkhovskaya, 1971).

Belyayeva, V. N. and G. L. Pokrovskaya. 1957. Arrest of mitosis by x-rays at early developmental stages in loach spawn. Dokl. Akad. Nauk SSSR 119 (1): 361-364. (Or: Compt. rend. (Doklady) Akad. Nauk URSS) (Doklady (biol. sci.) 119: 149-153).

Belyayeva, V. N. and G. L. Pokrovskaya. 1958. Changes in the radiosensitivity of loach spawn during the first embryonic mitoses. Dokl. Akad. Nauk SSSR 125: 632-635. (Doklady (biol. sci.) 125: 192-195) (NSA 13: 12382).

Belyayeva, V. N. and G. L. Pokrovskaya. 1966. Change in radiosensitivity of eggs of loach in process of initial embryonic mitosis. In: 2nd Intern. Oceanographic Congr. Publ. House Acad. Sci. USSR, Moscow.

Benazzi, M. 1962. Action of ionizing radiations on the development of Planaria. Atti accad. naz. lincei rend. (sci. fis. mat. nat.) 32(1): 26-29. (UCLA-Trans-10276) (NSA 17: 5789; 23: 2911).

Benazzi, M. 1962. The action of irradiated sperm in normal fecundation and in pseudogamia in Planaria. Atti accad. naz. lincei rend. (sci. fis. mat. nat.) 32(8): 286-289. (In Italian) (NSA 17: 17180).

Benazzi, M. and V. Scali. 1964. Sensitivity of sperm to x-rays studied in the facultatively parthenogenetic phasmidian Bacillus rossius. Atti accad. naz. lincei rend. (sci. fis. mat. nat.) 36(8): 794-797. (UCRL-Trans-10280) (NSA 19: 35971; 23: 2911) (Used Planaria).

Bende, E. and A. Szabo. 1974. Novel methods in the analysis of cesium. Izotoptechnika 17(8): 413-419. (In Hungarian) (NSA 30: 28976).

Bender, M. E. 1962. The bacterial translocation of radioactive phosphorus through a lotic ecosystem. Thesis, Michigan State Univ., East Lansing. viii, 93pp.

Benecke, P. 1967. Investigations into the behavior of precipitation water in soils by means of ^{131}I. pp. 227-238. In: Isotope and Radiation Techniques in Soil Physics and Irrigation Studies. (See IAEA, 1967) (NSA 22: 23155).

Benes, J. 1974. Natural radioiodine sorbents in the living environment Jaderna Energie 20(9): 316-319. (In Czech) (NSA 33: 641).

Benes, J. 1974. Radioactive contamination of the biosphere. Chesko. Akad. Ved. 232pp. (In Czech) (NSA 31: 19577).

Benes, J. 1975. Investigations of soil-solution systems on the capture of radioiodine in relation to p_H. Coll. Czech. Chem. Commun. 40(70): 2012-2019. (NSA 33: 7326).

Benes, J. nd. Sorption and desorption of radionuclides by Czechoslovak soils. Ceskoslovenska Akad., Prague. pp. 45-52. (In Czech) (INIS-mf-887-3) (NSA 30: 6515).

Benes, J. and Z. Dlouhy. 1972. Retention of radioactive substances in soils: properties of Czechoslovak soils. Jaderna Energie 18(7): 229-232. (In Czech) (NSA 26: 48039).

Benes, J. and Z. Dlouhy. 1973. Behavior of some critical radionuclides in soils: ^{137}Cs, ^{131}I, ^{89}Sr, and ^{90}Sr. pp. 431-433. In: Health Physics Problems of Internal Contamination. (See Bujdoso, 1973) (NSA 28: 29874).

Benes, J. and M. Panyr. 1969. Potassium and ^{137}Cs desorption from soils. Jaderna Energie 15: 299-302. (In Czech) (NSA 24: 4217).

Benes, J. and M. Tomasek, 1970. Dosimetric checking of ^{90}Sr in the biological chain. Radioisotopy 11(5): 823-834. (In Czech) (NSA 25: 57366).

Benes, J., J. Matouskova and M. Panyr. 1971. Dosimetric control of ^{137}Cs in the agricultural product grains of the Czech SSR during 1967 to 1969. Jaderna Energie 17(8): 275-278. (In Czech) (NSA 26: 393).

Benevolinskiy, V. N. and O. M. Meshcheryakova. 1971. Analysis of the dose versus effect curves and postradiation recovery of yeasts irradiated with gamma radiation on the basis of various criteria of survival. Radiobiologiya 11(1): 80-85. (AEC-tr-7233; pp. 117-127) (NSA 25: 44966).

Benevolinskiy, V.N., et al. 1966. The effect of gamma radiation and protons with energies of 127-660 Mev on the radiation injury of yeast cells. pp. 63-64. In: Problems in Aerospace Medicine: Reports of a Conference, May 24-27, 1966. V. V. Parin, ed. (JPRS-38272; TT-66-34698; pp. 77-78) (NSA 21: 16532).

Benfenati, I., et al. 1973. Program of environmental investigations for the installation on the Tosco-Emillian Apennine of a 140 MW fast reactor. G. Fis. Sanit. Prot. Radiaz. 17(3-4): 113-118. (In Italian) (NSA 31: 19677) (Incl. radioecology).

Bengston, M. 1967. ^{32}P labelling of codling moth (Cydia pomenella L.). Queensland J. Agr. Anim. Sci. 24: 113-115. (NSA 22: 52114)

Beninson, D., D. Cancio and C. Menossi. 1969. Étude du transport et de la fixation des produits de fission dans quelques espèces dulcicoles en conditions de non equilibre. pp. 361-394. In: Proc. International Symposium on Radioecology. (See France Commissariat à l'Énergie Atomique, 1969) (In French).

Beninson, D., E. V. Elst and D. Cancio. 1964. Uptake of fission products by aquatic plants. Argentine Comision Nacional de Energia Atomica, Buenos Aires, Informe No. 138. 14pp. (NP-14903) (NSA 25838).

Beninson, D., E. V. Elst and D. Cancio. 1965. Incorporation of fission products by fresh-water plants. pp. 227-232. In: Proc. 5th Inter-American Symp. on the Peaceful Application of Nuclear Energy. Pan American Union, Washington. (In Spanish) (NSA 19: 35939).

Beninson, D., E. V. Elst and D. Cancio. 1966. Biological aspects in the disposal of fission products in surface waters. pp. 337-354. In: Disposal of Radioactive Wastes into Seas, Oceans and Surface Waters. (See IAEA, 1966).

Bennett, B. G. 1973. Fallout tritium in the environment and the dose commitment. pp. 150-170. In: Fallout Program Quarterly Summary Report, September 1, 1972 - December 1, 1972. (See Hardy, 1973) (NSA 27: 14740).

Bennett, B. G. 1974. Environmental tritium and the dose to man. In: Proc. 3rd Intern. Congr. International Radiation Protection Assoc. (CONF-730907-P2) (NSA 30: 27242).

Bennett, G. F. 1963. Use of P^{32} in the study of a population of *Simulium rugglesi* (Diptera: Simuliidae) in Algonquin Park, Ontario. Can. J. Zool. 41(5): 831-840. (NSA 18: 11564).

Bennett, G. F. 1965. The effect of ^{32}P on the fecundity of *Aedes aegypti* (L.) and its use in determining blood meal volumes. Mosquito News 25: 465-469. (NSA 20: 33020).

Bensen, D. W. and A. H. Sparrow. (Eds.). 1971. Survival of Food Crops and Livestock in the Event of Nuclear War. (Proc. Symp. Brookhaven National Lab., Upton, N. Y., September 15-18, 1970). U. S. AEC report CONF-700909. xi, 745pp.

Bentley, R. E., R. P. Parker and M. P. Taylor. 1962. Iodine-131 and other fission products in drinking-waters. Nature 194(4830): 736-737. (NSA 16: 20126).

Benz, G. 1971. Sterilisation of males of the larch bud moth *Zeiraphera diniana* by γ-rays. Experientia 27(9): 1093-1096. (NSA 26: 20622).

Berdjis, C. C., et al. 1961. Virus infection and x radiation: a comparative study of infection with an attenuated Venezuelan equine encephalomyelitis virus and x radiation. J. Infect. Dis. 109(1): 62-70.

Berdnikov, A. I. 1967. The determination of trace amounts of uranium in water. Gigiyena sanitariya 32(10): 57-58. (TT-67-51409/4, pp. 66-68) (NSA 23: 13624).

Berdnikov, A. I. and V. F. Oreshko. 1965. Effect of stable cobalt on absorption of ^{60}Co by pest. Radiokhimiya 7: 494-496. (Soviet Radiochemistry 7: 494-496) (NSA 20: 10756).

Berdnikova, A. V. 1970. Migration of uranium from the environment into a human being. Fak. vnes. sredy Ikh. Znach. Zdorov'ya Naseleniya (2): 135-137. (In Russian) (NSA 27: 7575).

Berend, J. 1957. Study of the vitality of chlamidospores of *Tilletia foetida* under the influenses of external factors. II. The influence of the vitality of *Tilletia foetida* spores through irradiation. Növenyvedelmi Kutato Intezet. Evkön 7: 341-366. (In Hungarian).

Berendes, H. D. 1965. Salivary gland function and chromasomal puffing patterns in *Drosophila hydei*. Chromosoma 17(1): 35-77. (^3H).

Berendyayeva, Ye. L., et al. 1966. Experience of studying contacts within a population of Altai marmots by means of radioactive tagging. Zool. zh. 45(3): 430-435.

Berestovskiy, F. G. 1968. Field experience in concentration of isotopes from water. Gigiyena sanitariya (2): 112-114. (In Russian) (NSA 22: 45199).

Berg, A. 1968. Studies on the metabolism of calcium and strontium in freshwater fish. I. Relative contribution of direct and intestinal absorption. Mem. Ist. Ital. Idrobiol. 23: 161-196. (^{85}Sr).

Berg, A. 1969. Use of ^{85}Sr for the determination of the contribution of direct and intestinal absorption of Ca and Sr in a freshwater fish. pp. 325-337. In: Environmental Contamination by Radioactive Materials. (See IAEA, 1969).

Berg, A. 1972. Studies on the metabolism of calcium and strontium in freshwater fish. IV. Calcium and strontium relationships in fishes of two northern Italian lakes and their general radioecological implications. Mem. Ist. Ital. Idrobiol. 29: 145-167. (SFA 19: 18072).

Berg, A. and N. Verheyden. 1971. The physiological effect of sublethal doses of metallic pollutants and detergents on freshwater fish: modification of the rate of exchange of calcium leaving the water. pp. 759-770. In: Proc. Intern. Symp. Radioecology Applied to the Protection of Man and His Environment. (See Commission of the European Communities, 1971) (In French).

Berg, A., et al. 1961. Accumulation of fission products from fallout in lake biota (Lago Maggiore). Final report, I. A. E. A. Contract No. 59. Comitato Nazionale Energia Nucleare, Rome, report CNI-109. 239pp.

Berg, A., et al. 1973. Radioecological research of the Biological Division of CCR Euratom. pp. 63-76. In: Proc. 2nd Conf. on the State of Radioecology in Italy. (See Anonymous, 1973) (In Italian).

Berg, W., et al. 1966. Mercury content in feathers of Swedish birds from the past 100 years. Oikos 17(1): 71-83. (Activation analysis).

Bergamini, P. G., et al. 1970. Absorption of ^{137}Cs by *Musca domestica* and consequent environmental contamination. Health Physics 18(5): 491-498. (NSA 24: 34234).

Bergeder, H. D. 1958. Potassium loss of cold blooded muscles after x irradiation. Naturwissenschaften 45(3): 61. (In German).

Bergeder, H. D. and D. Böning. 1965. Immediate effects of x-rays on invertebrate animals and frogs. Biophysik 2: 239-246. (In German) (NSA 19: 42271).

Bergeder, H. D. and O. Hockwin. 1960. Alteration of the oxygen consumption of isolated frog muscle after x-irradiation. Naturwissenschaften 47(7): 161. (In German).

Berger, M. 1967. The ability of unicellular synchronized volvcales to divide after x-irradiation. Int. J. Radiat. Biol. 12: 477-486. (In German).

Berger, R. 1972. Tree-ring calibration of radiocarbon dates. pp. A97-A103. In: Proc. 8th Intern. Radiocarbon Dating Conf. Vol. 1. (See Rafter, 1972) (NSA 31: 8684).

Berger, R., W. S. Ting and W. F. Libby. 1965. Coprolites in western caves. pp. 731-744. In: Proc. 6th Intern. Conf. Radiocarbon and Tritium Dating. R. M. Chatters and E. A. Olson, comps. U. S. AEC report CONF-650652).

Bergh, H., et al. 1959. Sampling and analysis of radioactive fall-out in precipitation, peat soil, vegetation, tap water, and milk. Svensk Kem. Tidskr. 71: 695-716. (In Norwegian) (NSA 15: 51).

Bergner, P.-E. E. 1966. Tracer theory: a review. Isotop. Radiat. Technol. 3: 245-262. (NSA 20: 36544).

Bergold, G. H. 1954. Radioactive experiments with silkworms. Can. Dept. A Agr. Forest Biol. Div. Bi-monthly Progr. Rpt. 10(3): 2.

Bergstrøm, R. M. 1961. The effect of 5.3-Mev polonium-210 alpha particles on the electrical activity of α- and β-fibers of the frog sciatic nerve. Ann. Med. Exptl. Biol. Fenn. (Helsinki) 39(3): 211-215. (In English) (NSA 17: 21483).

Bergstrøm, R. M. 1962. The effect of 5.3 Mev ^{210}Po alpha particles on the resting membrane potential and ultrastructure of Nitella flexilis. Ann. Med. Exptl. Biol. Fenn. (Helsinki) 40: 45-53. (NSA 18: 3509).

Bergstrøm, R. M. 1969. Radioactivity from nuclear weapons tests in air and precipitation in Sweden 1967 and 1968. Sweden Forsvarets Forsningsintitut, Stockholm, report FOA-4-4390. 15pp.

Bergstrøm, R. M., R. F. Blafield and M. W. Brenner. 1961. Circumscribing irradiation of surface structures of cylindrical animal and plant cells with polonium-210 α-particles. Nature 191(4794): 1165-1166. (NSA 15: 30482).

Bergstrøm, R. M., R. F. Blafield and M. W. Brenner. 1961. The effect of α particles from a ring-shaped Po^{210} source on the action potential of nerve fibers of the frog. Ann. Acad. Sci. Fenn. A V(82): 10pp. (NSA 17: 23101).

Bergstrøm, R. M., R. F. Blafield and M. W. Brenner. 1962. Selective destruction by heavy nuclear irradiation of the cell membrane in internodal cells of Nitellopsis obtusa. Ann. Med. Exptl. Biol. Fenn. (Helsinki) 40 (Suppl. 1): 44pp. (NSA 17: 23102).

Bergstrøm, R. M., R. F. Blafield and A. Salmi. 1962. Effect of röntgen irradiation on the elasticity of frog skin. Acta Radiol. 58(2): 141-147. (NSA 16: 21921).

Bergstrøm, S. O. W. 1970. Environmental control, a critical review of its means and purposes. pp. 105-117. In: Environmental Surveillance in the Vicinity of Nuclear Facilities. (See Reinig, 1970) (NSA 25: 16167).

Bergstrøm, S. O. W. 1971. Strontium levels in Swedish fish. Health Physics 21(4): 611-612. (NSA 25: 55156).

Berke, B., B. Gyoenoer and Viragh. 1969. Determination of the total beta activity of wastes and fresh waters. Magy. Kem. Lapja 24: 361-365. (In Hungarian) (NSA 23: 47364).

Berlier, Y. 1969. Use of nuclear labels in plant physiology and phytotechnology. Bull. Inform. Sci. Tech. (Paris) (141): 31-43. (In French) (NSA 23: 23090).

Berliner, M. D. 1963. The action of monochromatic ultraviolet radiation on luminescence in Armillaria mellea. Radiation Research 19(3): 392-401. (NSA 17: 27179).

Berlowitz, L. 1965. Correlation of genetic activity, heterochromatization, and RNA metabolism. Proc. Nat. Acad. Sci. (US) 53(1): 68-73. (mealy bug, ^{3}H).

Berlyn, G. P. 1971. Experimental embryogenesis and histogenesis in Pinus lambertiana Dougl. Final report of research activities. Yale Univ., U. S. AEC report TID-26280. 34pp. (NSA 27: 24890).

Bernard, S. R., V. R. Rao Uppulinzi and L. R. Shenton. 1965. Stochastic modes for the distribution of radioactive material in a connected system of compartments. Oak Ridge National Lab., U. S. AEC report ORNL-3809. vi, 45pp.

Bernard, V. V. 1962. Effect of gamma-rays on some groups of soil microflora. Agrobiologiya 4: 610-616. (In Russian) (NSA 21: 441).

Bernardo, B. C. and E. A. Cabalfin. 1963. Ecological survey. A preliminary report. Philippine Atomic Energy Commission, Manilla, report PAEC(D)HP-631. 12pp. (NSA 17: 31791).

Bernhardt, H. 1963. Distribution of fallout radioactivity in a water supply reservoir. Wasserwirtschaft 53: 450-452. (In German) (NSA 21: 178).

Bernhardt, H. and K. Haberer. 1966. The influence of thermal stratification and plankton organisms on the distribution of fallout radioactivity in a water reservoir. Atompraxis 12: 142-147. (In German) (NSA 20: 27215).

Bernhardt, H., H. Hartmann and F. Baumgärtel. nd. Research on the techniques of decontamination of water retained by dams with the installation of rapid filters. France Commissariat à l'Énergie Atomique report CEA-tr-A-1957. 37pp. (Transl. from German to French) (NSA 23: 48029).

Berry, R. J., C. V. Beechey and A. G. Searle. 1973. Cytogenetic radiosensitivity and chiasma frequency in wild and laboratory mice. Mutation Research 19(1): 129-131. (NSA 28: 24843).

Berryman, A. A. 1964. Identification of insect inclusions in x-rays of ponderosa pine bark infested western pine beetle, Dendroctonus brevicornis Le Conte. Can. Entomol. 96(6): 883-888.

Berryman, A. A. 1966. Computer simulation of the sterile male theory. Bull. Entomol. Soc. Am. 12: 263.

Berryman, A. A. 1967. Estimation of Dendroctonus brevicornis (Coleoptera: Scolytidae) mortality caused by insect predators. Can. Entomol. 99(10): 1009-1014. (Radiographically).

Berryman, A. A. 1967. Mathematical description of the sterile male principle. Can. Entomol. 99(8): 858-___.

Berryman, A. A., T. P. Bogyo and L. C. Dickmann. 1973. Computer simulation of population reduction by release of sterile male insects. II. The effects of dynamic survival and multiple mating. pp. 31-43. In: Computer Models and Application of the Sterile-Male Technique. International Atomic Energy Agency, Vienna, Publ. STI/PUB/340.

Berthet, P. L. 1964. Field study of the mobility of Oribatei (Acari), using radioactive tagging. Animal Ecol. 33(3): 443-449.

Bertholf, L. M. 1933. Some physiological effects of ultra-violet radiation on honeybees. J. Agr. Res. 47(7): 375-398.

Berthollet, P. and G. Cavalier. 1971. Analytical methods used by the Geological Section: water. France Commissariat à l'Énergie Atomique, Fontenay-aux-Roses, report CEA-R-3557(2). 189pp. (In French) (NSA 25: 18339).

Berwig, W. 1955. A radioisotope study on cuticular excretion in female ants. Naturwissenschaften 46(21): 610-611. (In German).

Bessis, M. and B. Burte. 1965. Positive and negative chemotaxis as observed after the destruction of a cell by U. V. or laser microbeams. Tex. Rep. Biol. Med. 23(Suppl. 1): 204-212. (NSA 20: 5396).

Bestagno, G, et al. nd. Irradiation of carnation moths. Commission of European Communities, Brussels, report EURISOTOP-ITE-78. 28pp. (In French) (NSA 33: 9587).

Bettis Atomic Power Laboratory. 1966. Environmental radioactivity at the Shippingport Atomic Power Station for the second half of 1965 and calender year 1965. West Mifflin, Pa., U. S. AEC report PNRO-SMD-137. 8pp. (NSA 20: 37026).

Beug, R. J. 1965. Contributions to the postglacial vegetation history of northern Turkey. Abstr. Abstr. Intern. Assoc. Quaternary Research. VII Intern. Congr. Boulder and Denver, Colo. (Carbon dating).

Bevelander, G. 1952. Calcification in mollusks. III. Intake and deposition of ^{45}Ca and ^{32}P in relation to shell information. Biol. Bull. 102(1): 9-15.

Bevilacqua, B. 1965. Changes of the daily rhythm of mitosis in Pinus nigra Arn. caused by gamma rays. Silvae Genet. 14: 81-87. (NSA 20: 42953).

Beyers, R. J. 1964. The microcosm approach to ecosystem ecology. Am. Biol. Teacher 26(7): 491-498. (Discusses radiotracers).

Beyers, R. J., et al. 1971. Standing crops of elements and atomic ratios in a small mammal community. Acta Theriologica 16(14): 203-211. (Savannah River Plant).

Bezzegh, M. M. 1964. Problems of radioactive drinking waters in cisterns and their solutions, Federation Europaischer Gewasserschutz Inform. (10): 69-72. (In German) (NSA 19: 2188).

Bezzegh-Galantai, M. M. 1971. Passage of ^{90}Sr in the air-rain-soil-grass and milk food chain. pp. 1023-1037. In: Proc. Intern. Symp. Radioecology Applied to the Protection of Man and His Environment. (See Commission of the European Communities, 1971) (NSA 27: 5283).

Bezzegh-Galantai and M. Pavoni. 1966. Radioactivity in the plankton of Lake Lucerne; uptake of ^{90}Sr-^{90}Y and ^{89}Sr by plankton. pp. 107-125. In: Disposal of Radioactive Wastes into Seas, Oceans, and Surface Waters. (See IAEA, 1966)

Bhagat, S. K. and E. F. Gloyna. 1963. Environmental radionuclides in municipal wastewater. Water Sewage Works 110(6): 205-209.

Bhagat, S. K. and E. F. Gloyna. 1965. Radioactivity transport in water -- transport of nitrosyylruthenium in an aquatic environment. Environ. Hlth. Engineering Res. Lab., Univ. of Texas, Austin, Tech. Rpt. No. 9. (EHE-11-6502; CRWR-3), U. S. AEC report TID-22870. 159pp. (NSA 20:26606).

Bhagat, S. K., et al. 1971. Trace element analysis of environmental samples by neutron activation method. J. Water Poll. Control Fed. 43(12): 2414-2423. (NSA 26: 8769).

Bhakthan, N. M. G. and K. K. Nair. 1972. Fine structural damage in the somatic tissues of gamma-irradiated house fly. I. Flight muscles. Ann. Entomol. Soc. Am. 65(2): 504-508. (NSA 26: 23152).

Bharatwal, D. S. and G. H. Vaze. 1958. Radiation dose measurements in the monazite areas of Kerala State in India. Proc. 2nd Intern Conf. Peaceful Uses of Atomic Energy 23: 156.

Bhartiya, H. C. and P. N. Srivastava. 1974. Cell migration in the intestine of Indian desert gerbil (Meriones hurrianae Jerdon) and its relationship with the radiosensitivity of the animal. Experientia 30(12): 1397-1398. (NSA 31: 26117).

Bhat, I. S. 1971. Planning and management of environmental pollution control at Tarapur Nuclear Power Station site. pp. 76-108. In: Seminar on Pollution and Human Environment. (New Delhi, India, 26 Aug. 1970) (INIS-mf-160) (CONF-700847-1) (NSA 26: 52543).

Bhat, I. S. 1972. Need for environmental monitoring for nuclear installations. Indian J. Occup. Hlth. 15(11): 8-14. (NSA 28: 29847).

Bhat, I. S. 1973. Daily intake of cobalt by the adult population of Tarapur. Health Physics 24(5): 553-554. (NSA 28: 2598) (Incl. seafood).

Bhat, I. S., P. R. Kamath and A. K. Ganguly. 1972. Dispersal and uptake of radioactive elements in the Tarapur environment. India Bhabha Atomic Research Centre, Bombay, report BARC-644. 255pp. (NSA 28: 21147).

Bhat, I. S., A. A. Kahn and S. Chandramouli. 1971. Radioactivity measurement at Tarapur Nuclear Power Station environment. pp. 493-505. In: Proc. Nat. Symp. on Radiation Physics. (Bhabha Atomic Research Centre, Bombay, India, 24 Nov. 1970) (CONF-701138) (NSA 26: 40828).

Bhat, I. S., A. A. Kahn and P. R. Kamath. 1970. Economic planning of environmental surveys for power reactors. pp. 84-92. In: Environmental Surveillance in the Vicinity of Nuclear Facilities. (See Reinig, 1970) (NSA 25: 16165).

Bhat, I. S., R. S. Lyer and S. Chandramoulis. 1976. Method of separation and measurement of nickel-63 in environmental samples. Anal. Chem. 48 (1): 224. (NSA 33: 16622).

Bhat, I. S., et al. 1971. Radioiodine distribution in the Tarapur Power Reactor environment during commissioning and initial operations. pp. 361-371. In: Health Physics Aspects of Nuclear Facility Siting. (See Voillequé and Baldwin, 1971) (Marine and terrestrial samples).

Bhat, S. G. and S. Krishnaswamy. 1969. Isotopes of uranium and radium in Indian rivers. Proc. Indian Acad. Sci. A(1): 1-17. (NSA 23: 49942).

Bhatia, P. and G. R. Sethi. 1973. Effect of rearing media on the susceptibility of rust red flour beetle, Tribolium castaneum (Herbst) to gamma radiation. Indian Soc. Nucl. Tech. Agr. Biol. Newsltr. 2(1): 30-32. (NSA 31: 3738).

Bhatnagar, P. L., M. Rockstein and M. Dayer. 1965. X-irradiation of pupae of the house fly Musca domestica L. and adult survival. Exptl. Gerontol. 1: 149-159. (NSA 20: 5331).

Bhatt, Y. M. 1966. Uptake of ^{137}Cs by Katelysia marmorata. Current Sci. (India) 35: 512-513. (NSA 21: 8748).

Bhattacharjee, S. B. and N. N. Das Gupta. 1961. Radiation sensitivity of Escherichia coli to β-rays in dry and wet condition. Nature 191(4792): 1015-1017. (NSA 15: 27443).

Bhattacherya, S. 1971. Effect of X rays on the regeneration of the oral structures of Hydra vulgaris. I. Effect of X rays on the rate of regeneration. II. Role of interstitial cells in the process of regeneration. III. Contribution of endoderm cells in wound healing and hypostome differentiation. Z. Biol. 116(6): 480-501. (NSA 26: 23138, 23139 23140).

Bhattacherya, S. 1972. Cell interaction during regeneration of normal and radiated hydras. Indian J. Exptl. Biol. 10(1): 8-12. (NSA 26: 43367).

Bhatti, M. A. and E. Shipp. 1972. Mating competition of irradiated and normal males with normal females of the Queensland fruit fly, Dacus tryoni (Frogg). Nucleus (Karachi) 9(1-2): 107-110. (NSA 27: 12531).

Bhaumik, G. and S. B. Bhattacharjee. 1970. Synergistic killing of bacteria by x-rays, ultraviolet light, and ^{32}P decay. Radiation Research 42: 519-527. (NSA 24: 34776).

Bhodigen, S. J. 1975. Radioactivity in Thailand. Nat. Res. Counc. Thail. 7(1/2/3/4): 24-30. (In Thai) (NSA 33: 12033).

Bhuiya, A. D. and A. S. M. Matin. 1967. The susceptibility of the red flour beetle, Tribolium castaneum (Herbst) to gamma radiation. Pakistan Atomic Energy Centre, Dacca, report AECD-AG-20. 22pp. (NSA 21: 32726).

Bhuiya, A. D., Z. A. Khan and R. Rahman. 1970. Control of insect pests by gamma irradiation in East Pakistan. Pakistan Atomic Energy Centre, Dacca, report AECD-RB-12. 10pp. (NSA 25: 47068).

Bibergal, A. V., et al. 1971. Gamma-unit development for agricultural radiobiology. pp. 96-102. In: Use of Isotopes and Nuclear Radiation in Agriculture. Atomizdat, Moscow. (In Russian) (NSA 31: 661).

Bibor, V. and C. Leray. 1973. Metabolism of iodine and thyroxine in Perca flavescens. J. Fish. Res. Bd. Can. 30(9): 1395-1398. (SFA 19: 17326) (^{125}I).

Bick, Y. A. E. 1970. Comparison of the effects of LSD, heliotrine, and x irradiation on chromosome breakage, and the effects of LSD on the rate of cell division. Nature 226: 1165-1167. (NSA 24: 50832) (Marsupial, Potorous tridactylus).

Bick, Y. A. E. and J. K. Brown. 1969. Variations in radiosensitivity during the cell cycle in a marsupial cell line. Mutation Research 8: 613-622. (NSA 24: 14449).

Bick, Y. A. E. and W. D. Jackson. 1968. Effects of the pyrrolizidine alkaloid heliotrine on cell division and chromosome breakage in cultures of leucocytes from the marsupial Potorous tridactylus. Austral. J. Biol. Sci. 21: 469-481. (NSA 24: 2644) (X-ray).

Bicker, A. E. 1962. Relative effectiveness of 14 Mev neutrons and 200 KVP x-rays for production of lethality in grasshopper embryos. A preliminary study. Thesis, Kansas Univ., Lawrence. 52pp. (NSA 16: 23413).

Biddulph, O. and R. Cory. nd. The relationship between Ca-45, total calcium and fission product radioactivity in plants of Portulaca oleracea growing in the vicinity of the atom bomb test sites on Eniwetok Atoll. Univ. of Washington, U. S. AEC report UWFL-31. 20pp.

Biddulph, S. F. and O. Biddulph. 1953. A description of tumors of Ipomoea tubsa from the A-bomb test sites on Eniwetok. Appendix. In: Radiobiological Survey of Bikini, Eniwetok, and Likiep Atolls - July-August 1949. Univ. of Washington, U. S. AEC report AECD-3446(App.); UWFL-23. 24pp.

Bidwell, K. W. E. and E. E. Foreman. 1957. Distribution of strontium-90 in pond weed and fish. Nature 180(4596): 1195-1196. (NSA 12: 3537).

Bidzilya, N. I. (ed.). 1970. Proc. 1st All-Union Symp. on the Radiobiology of Plants. (Kiev, May 12-16, 1970). Publ. House Acad. Sci. Ukr. SSR, Kiev. 208pp. (In Russian) (NSA 26: 43264).

Bieber, A. M., Jr., et al. 1973. Compositional groupings of some ancient Aegina and eastern Mediterranean pottery. Brookhaven National Lab., U. S. AEC report BNL-18604. 54pp. (NSA 29: 28962).

Biellmann, G. 1961. Effects of local irradiation on Locusta migratoria. Bull. Soc. Zool. France 86(1): 99-106. (In French).

Biebl, R. 1933. Effect of alpha rays on the cells of the moss Bryum capillare. Sitzber. Akad. Wiss. Wien math.-natur. Kl. Abt. IIa 142: 381-386. (In German).

Biebl, R. and K. Hofer. 1966. Diurnal and annual period of radiation resistance of plant cells. Radiation Botany 6: 225-250. (In German) (NSA 20: 31079).

Biebl, R. and W. Uri. 1963. Effects of radiations on the cells of plants. Protoplasma 57: 84-125. (In German) (NSA 22: 4701).

Biemont, J. C. 1973. Effects of x irradiation on the reproductive power of Acanthoscelides obtectus Say (Coleoptera: Bruchidae). Ann. Zool. Ecol. Anim. 5(4): 581-591. (In French) (NSA 33: 3336).

Bien, G. S. and H. E. Suess. 1959. Increase of carbon-14 in the atmoshere from artificial sources measured in a California tree. Z. Physik 154: 172-174. (In English) (NSA 15: 9525).

Bier, K. 1963. Autoradiographic investigations on the function of the egg follicles and the nurse cells during yolk formation and protein synthesis in the fly ovary. Wilhelm Roux Arch. Ent. org. 154(6): 552-575. (In German, English summary) (^{3}H).

Bier, K. 1964. Directed RNA transfer through the cytoplasm. Naturwissenschaften 51(17): 418. (In German) (Musca domestica and Calliphora erythrocephala; ^{3}H).

Bier, K. 1964. The nucleus-plasma relationship and its bearings on the giant growth of the egg cell. Zool. Anz. 27(Suppl.): 84-91. (In German) (Calliphora erythrocephala, Musca domestica, Drosophila, Oncopeltus fasciatus; ^{3}H).

Bier, K. 1965. Study on the transport of macromolecules of the cell across the nuclear membrane. I. RNA synthesis and transfer under conditions of oxygen deficiency or reduced temperature. Chromosoma 16(1): 58-69. (In German, English summary) (Musca domestica; ^{3}H).

Bigger, T. R. L. 1967. γ-radiation and Lepidoptera--a postscript. Entomologist's Gaz. 18: 23-28.

Bigliocca, C., M. de Bortoli and G. Polvani. 1970. Natural cesium in freshwater and fishes. pp. 335-357. In: Proc. Intern. Symp. on Radioecology. Vol. I. (See France Commissariat à l'Énergie Atomique, 1970).

Biles, M. B. and F. E. Coffman. 1974 United States Atomic Energy Commission program for the control, monitoring, and reporting of radioactivity in effluents. pp. 28-49. In: Proc. Seminar on the Monitoring of Radioactive Effluents. (See Organization for Economic Co-operation and Develop., 1974) (NSA 33: 3336).

Biles, M. B. and F. E. Coffman. 1974. The United States Atomic Energy Commission program for the control, monitoring, and reporting radioactivity in effluents. pp. 477-494. In: Proc. 2nd AEC Environmental Protection Conference. U. S. AEC report WASH-1332(74).

Biles, M. B. and F. E. Coffman. 1974. The United States Atomic Energy Commission program for the control, monitoring, and reporting of radioactivity in effluents. (AED-CONF-74-157-002). 6pp. (NSA 30: 26779).

Bileva, D. S. 1965. A study of genetic processes in irradiated populations. I. Numerical dynamics of irradiated populations. Dokl. Akad. Nauk SSSR 164(1): 191-194. (Doklady (biol. sci.) 164: 597-599) (NSA 20: 18324).

Bileva, D. S. 1966. The effects of x irradiation on the numbers and viability of experimental populations of Drosophila melanogaster. pp. 177-187. In: Effects of Ionizing Radiation on Heredity. N. P. Dubinin, ed. Publ. House Acad. Sci. USSR, Moscow. (In Russian) (NSA 22: 21611).

Bileva, D. S. 1967. A study of the genetic processes in irradiated populations of Drosophila melanogaster. III. The fecundity and the concentration of lethal mutations. Genetika 3(4): 66-73. (In Russian) (NSA 21: 28421).

Bilinski, E. 1963. Utilization of lipids by fish. I. Fatty acid oxidation by tissue slices from dark and white muscle of rainbow trout (Salmo gairdnerii). Can. J. Biochem. Physiol. 41(1): 107-112. (^{14}C).

Bilinski, E. and R. E. E. Jonas. 1973. Effects of cadmium and copper on the oxidation of lactate by rainbow trout (Salmo gairdnerii) gills. J. Fish. Res. Bd. Can. 30(10): 1553-1558. (SFA 19: 17337).

Billerbeck, F. W., Jr. 1959. Effect of equally lethal radiation and thermal treatments on bacterial spore germination. Dissert. Abstr. 20: 1515-1516.

Billitteri, A. and G. Gasso. 1958. Metabolic changes in hibernated irradiated animals. I. Changes in serum proteins. Boll. Soc. Med. Chir. Cataniae 26: 80-99.

Billitteri, A. and G. Gasso. 1960. Immunitary variations in animals irradiated during hibernation. V. Behavior of antibodies. Radiobiol. latina 3: 341-346. (NSA 15: 22135).

Bilquez, A. 1955. Action of rays on Crepis zacintha L. BABC: Influence of the intensity of irradiation on the rate of cellular fatality. Compt. rend. 241(3): 327-329. (In French) (NSA 9: 6159).

Bilyk, N. A. and A. V. Dodin. 1972. Radon waters of the Rovno Region, the Ukrainian Soviet Socialist Republic. Geol. zh. (6): 101-104. (In Russian) (NSA 29: 10142).

Binning, A., et al. 1967. The conjugation of phenols with phosphate in grass grubs and flies. Biochemistry J. 103(1): 42-48. (^{14}C).

Bintz, G. L. 1969. Sodium-22 retention as a function of water intake by Citellus lateralis. pp. 45-52. In: Physiological Systems in Semiarid Environments. C. C. Hoff and M. L. Riedesel, eds. Univ. of New Mexico Press, Albuquerque.

Biola, M. T. and R. Le Go. 1966. Use of chromosomal analysis as a biological dosimeter. pp. 33-45. In: Proc. 1st Intern. Congr. Radiation Protection. Vol. I. (English transl. Austral. Atomic Energy Comm., Lucas Heights, report LIB/Trans.-277) (NSA 24: 46609).

Bird, P. M. 1966. Radionuclides in food. Can. Med. Assoc. J. 94: 590-597. (NSA 20: 23390).

Bird, P. M. 1968. Studies of fallout ^{137}Cs in the Canadian North. Arch. Environ. Health 17(4): 631-638.

Birnstiel, M. L., J. L. Sirlin and J. Jacob. 1965. Nucleolus: site of transfer ribonucleic acid synthesis. Biochem. J. 94(1): 10P-11P. (Smittia, ^{14}C).

Biro, T. 1966. The radioactivity of natural waters. Italy Istituto Superiore di Sanita, Rome, report ISS-66/36. 65pp. (NSA 22: 35922).

Biro, T., A. Gristanti and P. Tori. 1970. Measurement of the radioactivity of natural waters. Ann. Ist. Supr. Sanita 6: 95-105. (NSA 25: 13106).

Biro, T., et al. 1974. Determination of natural tritium concentration in environmental samples. Izotoptechnika 17(9-10): 493-497. (In Hungarian) (NSA 31: 10609).

Birron, R., G. Delbrias and J. Labeyrie. 1963. Evolution of the tritium concentration in precipitaion in France. Compt. rend. 256(23): 4951-4954. (In French) (NSA 17: 29144).

Biryukov, D. A. 1957. Mechanism of action of penetrating radiation upon the nervous system. Sechenov. Fiziol. zh. 43(7): 593-597. (In Russian) (Pigeon).

Biscaye, P. 1972. Strontium isotope composition and sediment transport in the Rio de la Plata estuary. Geol. Soc. Am. Mem. (133): 349-357. (NSA 29: 29377)

Bischkopf, R. 1966. Spermatogenesis in drones following irradiation of the young larvae with various doses of x-rays. Deut. Ent. Z. 13(1/3): 237-262. (In German).

Bishop, J. E. and J. B. Bishop. 1968. A technique for selectively marking aquatic insects with ^{32}P. Limnol. Oceanogr. 13(4): 722-724.

Bisir, D. and E. Gaspar. 1967. Determination in situ of soil permeability by punctual dilution method. Hidroteh. Gospod. Apelor Meteorol. Bucharest 12: 305-314. (In Romanian) (NSA 23: 17560).

Bissell, V. C. 1974. Natural gamma spectral peak method for snow measurement from aircraft. pp. 614-623. In: Advanced Concepts and Techniques in the Study of Snow and Ice Resources. (See Santeford and Smith, 1974) (NSA 33: 7310).

Bissell, V. C. and E. L. Peck. 1974. Measurement of snow at a remote site: natural radioactivity technique. pp. 604-613. In: Advanced Concepts and Techniques in the Study of Snow and Ice Resources. (See Santeford and Smith, 1974) (NSA 33: 7309).

Bitakaramire, P. K. 1970. Radioisotope studies on Fasciola gigantica infection of cattle. pp. 35-43. In: Isotopes and Radiation in Parasitology. Part II. (See IAEA, 1970) (NSA 24: 50671).

Bitman, J., et al. 1969. Relationship of beta backscatter counts to egg shell thickness and calcium content. Poultry Sci. 48(6): 2184-2185. ("Japanese quail").

Bittel, R. 1965. Some aspects of radiohydrology. Literature survey. France Commissariat à l'Énergie Atomique, Saclay, report CEA-BIB-59. 46pp. (In French).

Bittel, R. 1971. Use of information from radioecological research relating to metal activation products for the study of the transfer to man of metallic or organometallic chemical pollutants. pp. 745-757. In: Proc. Intern. Symp. Radioecology Applied to the Protection of Man and His Environment. (See Commission of the European Communities, 1971) (In French) (NSA 27: 5278).

Bittel, R. 1974. Pollutions associated with radioactive pollutions in aquatic and irrigated ecosystems. Radioprotection 9(2): 111-130. (In French) (NSA 33: 3008).

Bittel, R. 1975. Consequences of the release of chemical pollutants on the transfers of radioactive products in aquatic systems. France Commissariat à l'Énergie Atomique, Fontenay-aux-Roses, report CEA-CONF-3123. 3pp. (NSA 33: 23321).

Bittel, R. and J. Mancel. 1968. Contribution to the study of maximum levels for liquid radioactive waste disposal into continental and sea water. Treatment of some typical samples. France Commissariat à l'Énergie Atomique, Fontenay-aux-Roses, report CEA-R-3587. 33pp. (EUR-3876) (In French) (NSA 23: 9577).

Bittel, R., et al. 1970. Points of view on the study of the physico-chemical processes of the transfer of radioruthenium in a terrestrial environment. pp. 213-230. In: Proc. Intern. Symp. on Radioecology. (See France Commissariat à l'Énergie Atomique, 1970) (In French).

Bittel, R., et al. 1970. Study of the behavior of radiocobalt through an artificial ecosystem of simple irrigation by submersion. pp. 413-438. In: Proc. Intern. Symp. on Radioecology. (See France Commissariat à l'Énergie Atomique, 1970) (In French).

Bittel, R., et al. 1971. Study of the parameters of radiocobalt transfer in continental aquatic and irrigated ecosystems. France Commissariat à l'Énergie Atomique, Fontenay-aux-Roses, report CEA-R-4157. 36pp. (In French) (NSA 25: 32391).

Bittel, R., et al. 1971. The transfer to man, through his food, of the radioisotopes of ruthenium, cobalt and zinc discharged into continental waters. pp. 869-889. In: Proc. Intern. Symp. Radioecology Applied to the Protection of Man and His Environment. (See Commission of the European Communities, 1971) (NSA 27: 5280).

Bittel, R., et al. 1974. In situ study of a natural aquatic ecosystem contaminated by tritiated liquid effluents with a view to evaluating the sensitivity of the parameters involved in assessing population exposure levels. pp. 613-621. In: Population Dose Evaluation and Standards for Man and His Environment. (See IAEA, 1974) (NSA 31: 19592).

Bizollon, C., P. A. Moret and P. Berger. 1961. Two years of monitoring of the radioactivity of the Rhone and of the ground water table. Extrait du Tome 16(3): 643-664.

Bizollon, C. A., R. Moret and J. P. Molgatini. 1971. Direct measurement method of radium and radon contents in natural waters using γ spectrometry. Ann. Phys. Biol. Med. 5(1): 45-63. (NSA 25: 48610).

Bjegovic, P. 1972. Reproduction of Ooencyrtus kuwanae Howard in the killed gypsy moth eggs with radiation. Zast. Bilja 23: 117-118. (NSA 30: 27205).

Björkman, E. 1960. Monotropa hypoitys L. - an epiparasite on tree roots. Physiol. Plantarum 13(2): 308-327. (^{14}C, ^{32}P) (BA 35: 54912).

Björling, K., D. Lihnell and F. Ossiannilsson. 1951. Marking viruliferous aphids with radioactive phosphorus. Acta Agric. Scand. 1: 301-317.

Björling, T. 1973. Bacterial leaching of uranium shale. Kungliga Tekn. Högskolan, Stockholm, report TRITA-MEL-8. 20pp. (In Swedish) (NSA 29: 9907).

Black, S. C. 1961. Low-level polonium and radiolead analysis. Health Physics 7(1/2): 87-91. (Bats).

Blair, G. J. and F. C. Crofts. 1969. Quantitative and radioactive sulfur (^{35}S) liquid scintillation counting method for determining soil and plant sulfur. Soil Science 107: 277-280. (NSA 23: 21818).

Blair, H. A. and R. C. Baxter. 1970. Radiation life shortening in Drosophila as a function of age and prior exposure. Radiation Research 43: 439-451. (NSA 24: 44588).

Blair, W. F. 1958. Effects of x-irradiation on a natural population of the deer-mouse (Peromyscus maniculatus). Ecology 39(1): 113-118. (NSA 14: 22771).

Blair, W. F. 1959. Effects of radiations on natural populations of vertebrates. Abstr. Proc. 9th Intern. Bot. Congr. 2(A): 3

Blair, W. F. 1960. Radiation-induced genetic damage in the Mexican toad (Bufo valliceps). Texas J. Sci. 12(3): 216-227.

Blair, W. F. 1961. Effects of radiations of natural populations of vertebrates. pp. 1377-1381. In: Recent Advances in Botany. Univ. of Toronto Press, Toronto.

Blair, W. F. 1961. Research on irradiation of natural populations of selected vertebrates. Final Scientific Report for Contract AT-(40-1)-1751. Univ. of Texas, Austin, U. S. AEC report TID-19982. 12pp.

Blair, W. F. and T. E. Kennerly, Jr. 1959. Effects of x-irradiation on a natural population of the woodmouse (Peromyscus leucopus). Tex. J. Sci. 11(2): 137-149. (NSA 14: 27792).

Blake, K. C. H. and L. F. Rapley. 1972. Disposal of radioactivity into a city sewage system. Health Physics 23(6): 861-862. (NSA 27: 2522).

Blake, M. I., A. S. Kaganove and J. J. Katz. 1962. Carbon dioxide uptake studies in algae grown in water and deuterium oxide. J. Pharm. Sci. 51 (5): 375-379. (NSA 16: 23364).

Blakely, E. A. 1975. Growth and viability of Aedes albopictus cell line in vitro after cesium-137 gamma irradiation. Dissertation, Univ. of Illinois, Urbana. 152pp. (Univ. Microfilms No. 75-24) (NSA 33: 29557).

Blanc, D., J. Fontan and A. Birot. 1965. Continuous measurement of the concentration of natural radioactive gases and aerosols. Staub 25: 287-290. (In German) (NSA 20: 7279).

Blanc, G. A. 1970. State role in thermal power plant siting. A report to the State of Washington. Reid, Batchelor and Assoc., San Francisco. 36pp. (NP-18770) (NSA 25: 31525).

Blanchard, R. L. 1965. A rapid method for the determination of ^{210}Pb and ^{210}Po in environmental samples. Abstr. Health Physics 11(8): 831.

Blanchard, R. L. 1967. Relationship between ^{210}Po and ^{210}Pb in man and his environment. pp. 281-296. In: Radioecological Concentration Processes. (See Aberg and Hungate, 1967).

Blanchard, R. L. and B. Kahn. 1971. Pathways for the transfer of radionuclides from nuclear power reactors through the environment to man. pp. 175-189. In: Proc. Intern. Symp. Radioecology Applied to the Protection of Man and His Environment. (See Commission of the European Communities, 1971) (NSA 27: 5226).

Blanchard, R. L. and B. Kahn. 1973. Fate of radionuclides discharged from a PWR nuclear power station into a river. pp. 195-204. In: Environmental Behavior of Radionuclides Released in the Nuclear Industry. (See IAEA, 1973) (NSA 29: 29360).

Blanchard, R. L. and J. W. Kearney. 1967. Natural radioactivity and ^{137}Cs in Alaskan caribou and reindeer samples. Environ. Sci. Technol. 1: 932-939. (NSA 22: 10776).

Blanchard, R. L. and J. B. Moore. 1970. ^{210}Pb and ^{210}Po in tissues of some Alaskan residents as related to consumption of caribou or reindeer meat. Health Physics 18: 127-134. (NSA 24: 19309).

Blanchard, R. L., et al. 1971. Radiological surveillance studies at a BWR nuclear power station: estimated dose rates. pp. 372-384. In: Health Physics Aspects of Nuclear Facility Siting. (See Voillequé and Baldwin, 1971) (NSA 25: 56979).

Bland, D. E. 1963. Lignification in Eucalyptus. Incorporation of phenylalanine, tyrosine, and methionine into Eucalyptus sideroxylon and Eucalyptus Camaldulensis. Biochem. J. 88(3): 523-525. (^{14}C).

Blankenship, J. W. and G. J. Miller. 1968. Incorporation of ^{14}C-labeled acetate and glucose into triglycerides of Vitula edmansae serratilineela larvae. Ann. Entomol. Soc. Am. 61(4): 1036-1037.

Blankenfel'd, Yu. I. and Ye. F. Bogdanovich. 1966. Observations on gas exchange in the soil using ^{14}C. Pochvovedeniye (10): 32-35. (Soviet Soil Sci. (10): 1122-____) (CA 65: 20784e).

Blanusa, M. and M. Kastelan. 1971. Insta-gel scintillant for ^{45}Ca counting in biological samples. Intern. J. Appl. Radiat. Isotopes 22(12): 723-728. (NSA 26: 9351).

Blasius, M. B., et al. 1972. Use of x-ray fluorescence to determine trace metals in water resources. Water Resour. Bull. 8(4): 704-714. (NSA 30: 2589).

Blavoux, B. and P. Olive. 1966. Primary results on the tritium content of waters of Lake Leman. Compt. rend. D262: 2445-2448. (In French) (NSA 20: 35647).

Blavoux, B., et al. 1964. Hydrodynamics and content of tritium in the waters of the Evian Basin (France). Compt. rend. 259: 4323-4326. (In French) (NSA 19: 13636).

Blaylock, B. G. 1963/64. Chromosomal aberrations in a natural population of Chironomus tentans exposed to chronic low-level environmental radiation. (Dissertation Abstr. 24: 3022).

Blaylock, B. G. 1965. Chromasomal aberrations in a natural population of Chironomus tentans exposed to chronic low-level radiation. Evolution 19: 421-429. (NSA 21: 8766).

Blaylock, B. G. 1966. Chromosomal polymorphism in irradiated natural populations of Chironomus. Genetics 53: 131-136. (Also: Oak Ridge National Lab., U. S. AEC report ORNL-P-1560) (NSA 20: 12662).

Blaylock, B. G. 1966. Cytogenetic study of a natural population of Chironomus inhabiting an area contaminated by radioactive waste. pp. 835-846. In: Disposal of Radioactive Wastes into Seas, Oceans and Surface Waters. (See IAEA, 1966) (SM-72/55) (CONF-660507-31) (NSA 20: 40530).

Blaylock, B. G. 1968. The fecundity of a population of Gambusia affinis inhabiting a lake contaminated with radioactive waste. Abstr. Bull. Ecol. Soc. Am. 49(2): 64.

Blaylock, B. G. 1969. The fecundity of Gambusia affinis affinis population exposed to chronic environmental radiation. Radiation Research 37: 108-117.

Blaylock, B. G. 1969. Effects of ionizing radiation on interspecific competition. pp. 61-67. In: Symposium on Radioecology. (See Nelson and Evans, 1969).

Blaylock, B. G. 1971. The production of chromosome aberration in Chironomus riparius (Diptera: Chironomidae) by tritiated water. Can. Entomol. 103(3): 448-453.

Blaylock, B. G. 1974. Effects on aquatic biota of radioactive effluents from nuclear power stations. Trans. Amer. Nucl. Soc. 18: 46-47. (NSA 30: 15372).

Blaylock, B. G. and T. J. Mitchell. 1969. The effect of temperature on the dose response of Gambusia affinis affinis from two natural populations. Radiation Research 40(3): 503-511. (NSA 24: 14596).

Blaylock, B. G. and T. R. Trabalka. 1975. Effectiveness of tritium and ^{239}Pu in producing chromosome aberrations in Chironomus riparius. Oak Ridge National Lab., U. S. AEC report CONF-751126-4. 7pp. (NSA 33: 7605).

Blaylock, B. G. and J. P. Witherspoon, Jr. 1965. Environmental factors affecting glass rod dosimetry. Health Physics 11(6): 549-552.

Blaylock, B. G. and J. P. Witherspoon, Jr. 1975. Dose estimation and prediction of radiation effects on aquatic biota resulting from the nuclear fuel cycle. Oak Ridge National Lab., U. S. AEC report CONF-750662-3. 21pp. (NSA 32: 24338).

Blaylock, B. G., S. I. Auerbach and D. J. Nelson. 1964. Chromosomal aberrations in a natural population of Chironomus tentans exposed to chronic low-level environmental radiation. Oak Ridge National Lab., U. S. AEC report ORNL-3531. vi, 76pp. (Also: Dissertation, Univ. of Tennessee, Knoxville; see Blaylock, 1963/64).

Blazek, J., J. Dockal and J. Kolin. 1964. The possibility of using the ^{60}Co radioisotope against grain storage insects. Mlynsko-Pekarsky Prumysl 10: 636-639. (In Czech) (Calandra granaria).

Bletchly, J. D. 1958. Some laboratory investigations on the eradication of wood-boring insects by gamma radiation. Intern Congr. Entomol. Proc. 10 (4): 385-389.

Bletchly, J. D. 1961. Effect of gamma radiation on some wood-boring insects. Ann. Appl. Biol. 49: 362-370.

Bletchly, J. D. and R. C. Fisher. 1957. Use of gamma radiation for the distribution of wood boring insects. Nature 179(4561): 670. (NSA 11: 6202).

Blincoe, C. 1969. Sequential statistics applied to radiation measurement. Health Physics 17: 679-684. (NSA 24: 4627).

Blincoe, C. and V. R. Bohman. 1969. Fallout concentrations in cattle grazing highly contaminated range. Health Physics 17: 559-565. (NSA 23: 48037).

Blincoe, C. and V. R. Bohman. 1970. Decrease of bovine cesium-137 concentrations following the cessations of atmospheric nuclear weapons testing. pp. 427-435. In: Radionuclides in the Environment. R. F. Gould, ed. American Chemical Society, Washington.

Blincoe, C. and V. R. Bohman. 1971. ^{137}Cs from environmental sources in desert range cattle from 1962 through 1968. Environ. Res. 4(3): 193-200. (NSA 26: 4726).

Blincow, D. W. 1975. Apparatus for measurement of tree core density. U. S. Patent 3,919,551. 4pp. (NSA 33: 14958).

Bliss, W. and L. Dunn. 1971. Measurement of plutonium in soil around the Nevada Test Site. pp. 320-327. In: Proc. Rocky Flats Symp. on Safety in Plutonium Handling Facilities. (See Dow Chemical Co., 1971) (NSA 25: 54755).

Bliss, W. A. and F. M. Jakubowski. 1975. Plutonium distribution in the environs of the Nevada Test Site: status report. pp. 237-250. In: Radioecology of Plutonium and Other Transuranics in Desert Environments. (See White and Dunaway, 1975) (NSA 33: 620).

Bliznik, K. M., et al. 1974. Principles of the formation of radioraces in yeast organisms. 4. Dependence of the yield of saltants on the conditions of postradiation culturing. Radiobiologiya 14(2): 229-236. (AEC-tr-7571; pp. 86-95) (NSA 30: 32396).

Bloch, D. P. and S. D. Brack. 1964. Cytoplasmic synthesis of nuclear histone during spermiogenesis in the grasshopper Chortophaga viridifisciata (de Geer). J. Cell Biol. 22(2): 327-340. (^{3}H).

Block, R. J. 1953. Inorganic sulfur in the synthesis of protein in the rumen from non-protein sources with a note on the synthesis of cystine and methionine in Blattella germanica. Oak Ridge National Lab., U. S. AEC report TID-5115. 492pp. (^{35}S).

Block, W. and H. Schneider. 1967. The problem of the contamination of the Rhine with radioactive nuclides. Gas- Wasserfach. 108: 149-157. (In German) (NSA 22: 23138).

Bloom, S. G. and G. E. Raines. 1967. A note on the origin of exponential functions in models to describe the transfer of radionuclides through an ecosystem. Bioenvironmental and radiological-safety feasibility studies Atlantic-Pacific interoceanic canal. Battelle Columbus Labs., U. S. AEC report BMI-171-009. 10pp. (NSA 21: 39368).

Bloom, S. G. and G. E. Raines. 1970. Kinetic model study of hydrogen flow through the El Verde forest system. Ch. H-7, pp. H123-H128. In: A Tropical Rain Forest. (See Odum and Pigeon, 1970).

Bloom, S. G., et al. 1970. Mathematical methods for evaluating the transport and accumulation of radionuclides. Battelle Columbus Labs., U. S. AEC report, LOCS Memo. BMI-35. 40pp.

Bloom, W. 1959. Cellular responses. Rev. Mod. Phys. 31(1): 21-29. (Involves newt cells and irradiation).

Blume, H. P., U. Zimmermann and K. O. Münnich. 1967. Tritium tagging of so soil moisture: the water balance of forest soils. pp. 315-331. In: Isotope and Radiation Techniques in Soil Physics and Irrigation Studies. (See IAEA, 1967) (NSA 22: 23162).

Blumel, J. 1950. The action of radioactive phosphorus in Drosophila. Science 111(2878): 205-206.

Blumenfeld, S. N. 1966. Nuclear war and soil microflora. RAND Corp., Santa Monica, Calif., U. S. AEC report RM-4827-TAB. 44pp. (NSA 20: 45962).

Blumenthal, R. and M. D. Williams. 1928. Effects of radium on the oxygen consumption of grasshopper eggs. Abstr. Anat. Record 41(1): 45.

Bluzat, R. 1964. Influence of x-rays on the course of embryogenesis of Gryllus domesticus L. (Orthoptera). Bull. soc. zool. Franc. 89(4): 478-495. (In French).

Bluzat, R. 1967. Evidence of radioresistance of germ cells in Leptinotarsa decimliniata (Coleoptera). Compt. rend. D265: 2027-2029. (In French) (NSA 22: 19433).

Bluzat, R. 1968. Destruction of the polar cells of Colorado beetle eggs by irradiation. pp. 103-108. In: Isotopes and Radiation in Entomology. (See IAEA, 1968) (NSA 24: 32165).

Bluzat, R. 1971. Development of radiosensitivity during embryogenesis of Colorado beetle (Leptinotarsa decimlineata Say). Compt. rend. D272(11): 1542-1545. (In French) (NSA 25: 44891).

Boberg, W. W. and D. D. Runnells. 1971. Reconnaisance study of uranium in the South Platte River, Colorado. Econ. Geology 66: 435-450.

Bobovnikova, Ts. (ed.). 1972. Radiochemical Analyses of Objects of the External Medium. (Proc. 3rd Radiochemical Conf., Obninsk, USSR, June 21-24, 1970; 4th Radiochemical Conf., Leningrad, USSR, May 25-28, 1971). Publ. House Hydrometeorological Lit., Moscow. 143pp. (NSA 29: 29044).

Bobovnikova, Ts. I. and K. P. Makhon'ko. 1973. On ^{90}Sr migration in land fresh water. pp. 30-35. In: Radioecology of Water Organisms. II. Distribution and Migration of Radionuclides in Freshwater and Seawater Biocenoses. (See Andrushaitis, et al., 1973) (AEC-tr-7606; pp. 16-19) (NSA 30: 21110).

Bobovnikova, Ts. I and Z. S. Shulenko. nd. ^{90}Sr migration from soils to rivers. (In Russian) (INIS-mf-887-3; pp. 89-94) (NSA 30: 6518).

Boccadoro, B. and G. Veneroni. 1962. Study of pathagenesis of infection with "Salmonella Dublin" tested on rabbits, by the radioactive indicator method. Arch. Vet. Ital. 13(2): 127-146. (In Italian) (Bibl. Agr. 26 (10); WR (108): 19).

Bochenin, V. F. and M. Ya. Chebotina. 1975. Seasonal dynamics of ^{60}Co accumulation by elodea (Elodea canadensis Rich.). Ekologiya 6(5): 80-82. (In Russian).

Bochkarev, V. M., Z. G. Antropova and Ye. I. Belova. 1964. Migration of strontium 90 and cerium 144 in soils of different texture. Pochvovedeniye (9): ___. (Soviet Soil Sci. (9): 936-___).

Bocquet, H. 1959. The "Ecolological pyramids" from the systematic study of the processes of radiocontamination. Acad. Vet. France B32(10): 707-720.

Bocquet, H. 1962. Pollution of the environment by radioactivity. Incidences of radiocontamination of the biological cycle. Cah. Med. Vet. 31 (4): 109-159. (In French) (NSA 18: 33300).

Bocquet, H. 1965. Radioecology and radioactive pollution of the atmosphere. Prod. probl. pharm. 20: 59-80. (In French) (NSA 21: 1728).

Boczek, J. 1965. Radioisotopes in entomological research. Ogrodn. 2: 245-247. (In Polish).

Bodine, J. H. and T. C. Evans. 1934. Respiration and development of individual mud-dauber wasp larvae following x-irradiation. Physiol. Zool. 7(4): 550-555.

Bodine, J. H., L. D. Carson and O. M. May. 1940. Enzymes in ontogenesis (Orthoptera). XII. Some physiological changes in eggs the embryo of which have been destroyed by x-irradiation. Biol. Bull. 78(3): 437-443.

Bodrova, I. M. 1961. Features of the physiological processes and the effect of the nutritive elements on the growth and development of plants infected with gall nematodes. Tr. vses. inst. Zash. Rast. 16: 76-88. (In Russian) (Ref. Zh. biol. No. 13 G108) (^{32}P).

Boedege, R., et al. 1974. Radiological surveillance of the environs at the Kahl, Gundremmingen, Lingen, and Obrigheim nuclear power stations. pp. 285-306. In: Environmental Surveillance Around Nuclear Installations. (See IAEA, 1974) (NSA 31: 14041).

Boeluekbasi, E. K. 1965. The effect of tritiated thymidine and gamma irradiation on the mortality of adult Drosophila melanogaster larvae. II. Cekmece Nuclear Research Center, Atomic Energy Comm., Istanbul, Turkey, report CNAEM-32. 13pp. (NSA 20: 14209).

Boenzi, D., Z. Dlouhy and G. Lenzi. 1965. A study on the sorption properties of the natural tuffs occurring in the Lake Bracciano region (Rome). Comitato Nazionale par l'Energia Nucleare, Rome, report RT/PROT-(65)29. (NSA 20: 14209).

Bogatyrev, I. O. 1962. The distribution of Co^{60} and some radioisotope fission products among the components of experimental bodies of water. Izv. Akad. Nauk SSSR (biol.) (1): 122-126. (JPRS 13,562; pp. 24-31) (NSA 16: 28674).

Bogen, D. C., et al. 1973. Determination of tritium distribution in environmental and biological samples. J. Radioanal. Chem. 13(2): 335-341. (NSA 28: 24157).

Boggie, R. and A. H. Knight. 1962. An improved method for the placement of radioactive isotopes in the study of root systems of plants growing in deep peat. Ecology 50(2): 461-464.

Boggie, R., R. F. Hunter and A. H. Knight. 1958. Studies of the root development of plants in the field using radioactive tracers. Ecology 46: 621-639.

Bogomoletz, V. I. 1961. Effect of x-ray irradiation on the membrane potential and concentration of sodium and potassium in frog muscle fibers. Fiziol. zh. Akad. Nauk Ukr. RSR 7: 214-219. (JPRS-9971) (NSA 15:29004).

Bogomeletz, V. I. and V. A. Maiskiy. 1966. Change in the biophysical parameters of the surface layer of the protoplasm of mollusk neurons under the effect of ionizing radiation. Fiziol. zh. Akad. Nauk Ukr. RSR 12: 63-69. (In Ukrainian) (NSA 20: 22591).

Bogorov, V. A., et al. 1967. On the connection between the physical, chemical, and biological structure of the surface layer of water of the Black Sea. Dokl. Akad. Nauk SSSR 172: 1187-1190. (In Russian) (NSA 22: 8850) (Incl. ^{137}Cs, ^{144}Ce).

Bogoyavlenskaya, M. P. 1955. The possibility of Ca^{45} application in fish labelling. Ryb. khozyaystvo (11): 50-51.

Bogoyavlenskaya, M. P. 1959. A study of calcium metabolism with a view to utilizing Ca^{45} as a label for fish. Vses. Nauchn.-Issled. Inst. Morsk. Ryb. Khoz. Okeanog. 55pp. (English transl. of conclusions Fish. Res. Bd. Can. Transl. ser. No 276).

Bogoyavlenskaya, M. P. 1961. Technique for mass marking of sturgeon young with Ca^{45} by way of the water. Tr. vses. nauchn.-issled. inst. morsk. ryb. khoz. okeanog. 44: 151-155. (Ref. zh. (biol.) 1963: 14111; BA 45: 27219; SFA 10: 7270).

Bogoyavlenskaya, M. P. 1970. On safety engineering in mass labelling of young sturgeons with radioactive calcium. Tr. vses. nauchn.-issled. inst. morsk. ryb. khoz. okeanog. 69(2): 70-75. (NSA 25: 21727).

Bogoyavlenskaya, M. P. and G. S. Karzinkin. 1958. Data regarding calcium metabolism obtained with the aid of radioactive calcium-45. Tr. sovesh. ikhtiol. komissiy Akad. Nauk SSSR 8: 322-326. (Strugeon).

Bogoyavlenskaya, M. P. and G. S. Karzinkin. 1970. Recommendations on safety engineering in mass labelling of young fishes with radioactive ^{45}Ca through water. Tr. vses. nauchn.-issled. inst morsk. ryb. khoz. okeanog. 69(2): 76-78. (NSA 25: 21726).

Bogoyavlenskaya, M. P. and I. A. Shekhanova. 1958. The application of P^{32} and Ca^{45} in the study of some aspects of phosphorus and calcium metabolism in young carp and sturgeon. In: Collection, A Study of Animal Organisms. The Fish Industry and Sea Food Industry. Acad. Sci. USSR, Moscow.

Bogoyavlenskaya, M. P., I. F. Vel'tishcheva and G. S. Karzinkin. 1972. Characteristic features of metabolism and the incorporation of C^{14} into organic compounds in sturgeons reared from large and small eggs. Vopr. Ikhtiol. 12(1): ___. (J. Ichthyol. 12(1): 130-134) (SFA 18: 16590).

Bogoyavlenskaya, M. P., et al. 1971. Application of radioisotopes as tracers in solving some fish industry problems. pp. 245-251. In: Application of Isotopes and Nuclear Radiation in Agriculture. III. Atomizdat, Moscow. (NSA 31: 1020).

Boguleanu, G. 1965. Remarks on the behavior of gamma-irradiation Hyphantria cunea Drury chrysalises. Studdi cerc. biol. (zool.) Bucharest 17: 387-391. (In Romanian) (NSA 21: 36743).

Bohannon, J. R., K. Verghese and J. N. Weaver. 1969. Neutron activation analysis in water resources management for North Carolina. Trans. Amer. Nucl. Soc. 12: 41. (NSA 23: 35038).

Bohart, R. and A. H. Smith. 1956. Radioisotopes for evaluation of a fly nuisance source. Calif. Vector Views 3(2): 5-7.

Bohn, G. 1903. Influence of radium rays on the virility and fertility of eggs, and the the primary stages of development. Compt. rend. 136(18): 1085-1086. (In French) (Strongylocentrotus lividus).

Bohn, G. 1903. Influence of radium rays on development and growth. Compt. rend. 136(17): 1012-1013. (In French) (Bufo vulgaris).

Bohne, F. 1971. Difficulties in measuring the tritium activity of tissue homogenates by the new scintillator instgel. Intern. J. Appl. Radiat. Isotopes 22(6): 384. (NSA 25: 46437).

Böhnert, R. and H. Bonka. 1971. Expected tritium contamination from nuclear reactors to air and water by homogeneous dispersion to the year 2000. Institüt für Reaktorenwicklung, Kernforschungsanlage, Jülich, Germany, report JÜL-763-RG. 95pp. (In German) (NSA 26: 4427).

Boiffard, J. A., M. Gabe and L. Arvy. 1947. Effects of x-rays on the hypophyses of Rana temporaria L. Compt. rend. soc. biol. 141: 340-341.

Boilly, B. 1962. Inhibition of caudal regeneration by x-irradiation in Syllus amica (Annelid polychete). Compt. rend. 255(11): 1414-1416. (In French) (NSA 17: 5808).

Boilly, B. 1968. Origin of regeneration cells in Aricia foetida Clap. (Annelid polychete). Arch. anat. microscop. morphol. exp. 57: 297-308. (In French) (NSA 24: 17130) (X-irradiation).

Boisot, M. and B. Henrion. 1964. Procedure and installation for the destruction of insects with nuclear radiation. Prop. ind. nucl. 7(18): 76. (In French).

Boisseveau, C. H. 1926. The effect of beta rays on bacterial growth. Am. Rev. Tuberc. 14(2): 172-176.

Boiteux, M. 1972. Past and future in the French energy economy. Rev. Fr. Energ. 24(238): 174-186. (In French) (NSA 26: 47964) (Environmental considerations).

Bojovic, P. S. 1968. Problems concerning disposal of radioactive effluents in internation waters. pp. 669-671. In: Radiation Protection, Part 1. W. S. Snyder, et al., eds. Pergamon Press, New York. (In French) (NSA 23: 12106).

Bokowski, D. L. 1974. Liquid scintillation counting for plutonium in environmental samples. Amer. Ind. Hyg. Assoc. J. 35(6): 333-334. (NSA 30: 9090).

Bolch, W. E. and E. F. Gloyna. 1963. Radioactivity transport in water-behavior of ruthenium in algae environments. Environ. Hlth. Engrg. Res. Lab., Univ. of Texas, Austin, Tech. Rept. 4. xi, 55pp. (NSA 18: 1340).

Bolch, W. E., et al. 1971. Ecological approach to environmental surveillance. pp. 252-265. In: Health Physics Aspects of Nuclear Facility Siting. Vol. II. (See Voillequé and Baldwin, 1971) (NSA 25: 55238).

Boldyrev, M. I. 1966. Thomasinia ribis Mar. and its eradication. Izv. Sel'skokhoz. Akad. (4): 151-165. (In Russian).

Bolen, J. J. 1972. Seasonal concentration, turnover, and mode of accumulation of ^{32}P by the juvenile starry flounder Platichthys stellatus (Pallas) in the Columbia River estuary. Oregon State Univ., Corvallis. 152pp. (Univ. Microfilms 72-12,516) (NSA 26: 31382).

Bolin, B. 1958. On the uses of tritium as a tracer for water in nature. Proc. 2nd Intern. Conf. Peaceful Uses of Atomic Energy 18: 336-343.

Bolingbroke, V., E. J. Harris and R. A. Sjodin. 1961. Rubidium and cesium entry, and cation interaction in frog skeletal muscle. J. Physiol. 157 (2): 289-305. (NSA 17: 19797).

Boller, E. F., A. Haisch and R. J. Prokopy. 1971. Sterile-insect release method against Rhagoletis cerasi L. Preparatory ecological and behavioral studies. pp. 77-86. In: Sterility Principle for Insect Control or Eradication. (See IAEA, 1971) (NSA 25: 29742).

Boller, E. F., U. Remund and J. Zehnder. 1975. Sterilization and its influence on the quality of the European cherry fruit fly, Rhagoletis cerasi L. pp. 179-189. In: Sterility Principle for Insect Control, 1974. (See IAEA, 1975) (NSA 32: 11774).

Bolt, G. H. 1968. On the use of radioactive tracers in diffusion experiments. Soil Sci. Soc. Am. Proc. 32(6): 889-891.

Bolton, N. E., et al. 1975. Trace element mass balance around a coal-fired steam plant. pp. 175-187. In: Trace Elements in Fuel. S. P. Babu, ed. Amer. Chem. Soc., Washington. (NSA 32: 21839) (Neutron activation).

Boltwood, B. B. 1905. On the radioactive properties of the waters on the Hot Springs Reservation, Hot Springs, Ark. Am. J. Sci. 20(4): 128-132.

Bolukbasi, E. K. 1965. Effect of tritiated thymidine and irradiation on the mortality of adult Drosophila melanogaster larvae. Turkey Cekmece Nuclear Research Center, Istanbul, report CNAEM-32. 20pp.

Bombace, M. A. and L. Cigna Rossi. 1973. Validity of radioecological methods applied to non-radioactive pollutants. pp. 117-118. In: Proc. 2nd Conf. on the State of Advancement of Radioecology in Italy. (See Anonymous, 1973) (In Italian).

Bond, V. P. 1970. Public and radiation from nuclear power plants. Brookhaven National Lab., U. S. AEC report BNL-14628. 9pp. (NSA 24: 29758).

Bond, V. P. 1971. Evaluation of potential hazards from tritium water. pp. 287-300. In: Environmental Aspects of Nuclear Power Stations. (See IAEA, 1971) (CONF-700810-6) (SM-146/13) (NSA 24: 48540).

Bond, V. P. 1973. Impact of nuclear power on the public; the American experience. Brookhaven National Lab., U. S. AEC report BNL-17897. (NSA 28: 12471).

Bond, V. P. and E. P. Cronkite. 1957. Effects of radiation on mammals. Ann. Rev. Physiol. 19: 299-328.

Bondar, A. I., N. N. Zhdanova and V. D. Pokhodenko. 1971. Survival of some Dematiaceae species collowing ^{60}Co gamma irradiation. Izv. Akad. Nauk SS SSSR (biol.) (1): 80-85. (In Russian) (NSA 25: 52260).

Bondarenko, N. F. and A. M. Globus. 1967. Radioactive tracers in the study of contemporary problems in the physical hydrology of soils. pp. 291-299. In: Isotope and Radiation Techniques in Soil Physics and Irrigation Studies. (See IAEA, 1967) (In Russian) (NSA 22: 23161).

Bonet-Maury, P. 1963. Influence of the kind of radiation on the inactivation of microorganisms. Intern. J. Radiat. Isotopes 14(1): 36-38. (In French) (NSA 17: 24937).

Bonet-Maury, P. 1963. Biological effects of ionizing radiations on the golden carp. Compt. rend. soc. biol. 157(3): 473-477. (In French) (NSA 17: 33655).

Bongiorno, S. 1963. Field populations of small mammals in relation to chronic gamma radiation. Thesis, Rutgers Univ., New Brunswick, N. J.

Bongiorno, S. F. and P. G. Pearson. 1964. Orientation of Peromyscus in relation to chronic gamma radiation and vegetation. Am. Midland Nat. 72(1): 82-92. (WR 115: 21; NSA 21: 26626).

Bonham, K. 1944. Histological effect of x-ray on adult male silver salmon, Oncorhynchus kisutch (Walbaum). Univ. of Washington, U. S. AEC report UWFL-4.

Bonham, K. 1949. Effects of x-rays upon the fresh-water snail, Radix japonica. Univ. of Washington, U. S. AEC report UWFL-21. 31pp. (NSA 3: 505).

Bonham, K. 1950. Some tests with experimental groups of fingerling rainbow trout, Salmo gairdnerii, on uniformity and rate of growth, diet, and photographic size-recording. Trans. Am. Fish. Soc. 79: 94-104.

Bonham, K. 1955. Sensitivity to x-rays of the early cleavage stages of the snail Helisoma subcrenatum. Growth 19(1): 9-18. (Also: Univ. of Washington, U. S. AEC report UWFL-39) (NSA 9: 9845).

Bonham, K. 1958. Radioactivity of invertebrates and other organisms at Eniwetok Atoll during 1954-1955. Univ. of Washington, U. S. AEC report UWFL-53. 52pp.

Bonham, K. 1959. Further contributions on gross beta radioactivity of biological and related samples at the Eniwetok Proving Ground 1952-1958. Univ. of Washington, U. S. AEC report UWFL-63. 41pp.

Bonham, K. and W. H. Bayliff. 1953. Radiography of small fishes for meristic studies. Copeia (3): 150-151.

Bonham, K. and L. R. Donaldson. 1966. Low-level chronic irradiation of salmon eggs and alevins. pp. 869-883. In: Disposal of Radioactive Wastes into Seas, Oceans and Surface Waters. (See IAEA, 1966).

Bonham, K. and L. R. Donaldson. 1972. Sex ratios and retardation of gonadal development in chronically gamma-irradiated Chinook salmon smolts. Trans. Am. Fish. Soc. 101(3): 428-434. (SFA 17: 15576).

Bonham, K. and R. F. Palumbo. 1951. Effects of x-rays on snails, crustacea, and algae. Growth 19(1): 9-18. (Also: Univ. of Washington, U. S. AEC report UWFL-26).

Bonham, K. and A. H. Seymour. 1947. Sections I and II of series of experiments involving the effect of x-ray on fishes: Chinook salmon (Oncorhynchus tschawytscha Walbaum) observed through more than one generation. Univ. of Washington, U. S. AEC report UWFL-6. 41pp.

Bonham, K. and A. D. Welander. 1963. Increase in radioresistance of fish to lethal doses with advancing embryonic development. pp. 353-358. In: Radioecology. (See Schultz and Klement, 1963) (NSA 17: 33584).

Bonham, K., et al. 1947. Lethal effect of x-rays on marine microplankton organisms. Science 106(2750): 245-246. (Reviews lit. on bacteria, yeasts, and Protista).

Bonka, H. 1972. Future global environmental pollution due to tritium and krypton-85 produced in nuclear reactors. ISU Mitt. 2(1): 99-108. (In German) (NSA 29: 18346).

Bonka, H. 1974. Natural radiation load in the FRG. pp. 590-604. In: Radiation Protection and Environmental Protection. Vol. 2. (In German) (AED-CONF-74-725-000(Vol. 2)) (NSA 33: 25982).

Bonka, H. and K. Brüssermann. 1973. Iodine release from nuclear power plan plants and reprocessing plants as well as the estimation of the contamination of the environment in the future. Reaktorentwicklung, Kernforschungsanlage, Jülich, Germany, report JÜL-997-RG. 72pp. (In German) (NSA 29: 18319).

Bonka, H., et al. 1975. Future radioactive environmental pollution in the Federal Republic of Germany by radionuclides from nuclear facilities in normal operation. First report. Kernforschungsanlage, Jülich, Germany, report JÜL-1220. 234pp. (In German) (NSA 33: 25989).

Bonnenberg, H. and J. Meincke. 1971. Nuclear reactors as a contribution to environmental protection. Inst. Reaktorentwicklung, Technische Hochshule, Aachen, Kernforschungsanlage, Jülich, Germany, report JÜL-778-RG. 76pp. (In German) (NSA 26: 3689).

Bonner, F. T., et al. 1961. Chlorine-36 and deuterium study of Great Basin lake waters. Geochim. Cosmochim. Acta 25: 261-266.

Bonner, N. A. and J. A. Mispel. 1965. Radioactivity: distribution from cratering in basalt. Science 150: 489-493. (Also: Lawrence Radiation Lab., Univ. of Calif., Livermore, U. S. AEC report UCRL-14211) (NSA 20: 613).

Bonnet, L. and H. Dalens. 1972. Effect of γ radiation of the resistance and longevity of two oniscoideas species: Porcellio dilatatus and Chaetophiloscia sicula and their mutants. Compt. rend. D274(20): 2794-2797. (In French) (NSA 26: 53504).

Bonnier, G. 1959. Note on the relative sensitivity to irradiation damage in homozygous and heterozygous Drosophila melanogaster with regard to their capacity of producing offspring. Hereditas 40(2): 676-677.

Bonnier, G., U.-B. Jonsson and C. Ramel. 1958. Selection pressure on irradiated populations of Drosophila melanogaster. Hereditas 44(2-3): 378-406.

Bonnyman, J. and J. Molina-Ramos. 1967. ^{90}Sr and ^{137}Cs in some Australian drinking water supplies: 1961-1965. Austral. J. Sci. 30: 171-172. (NSA 22: 8215).

Bonnyman, J., J. C. Duggleby and J. Molina-Ramos. 1972. Lead-210 in the Australian environment, 1964-70. pp. 819-831. In: Natural Radiation Environment II. Vol. 2. (See Adams, Lowder and Gesell, 1972) (NSA 32: 19728).

Bonotto, S. 1965. Study of the morphologic changes at microscopic level produced by Co^{60} gamma irradiation in Fagus silvatica wood. Energia Nucleare 12: 333-336. (In Italian) (NSA 19: 29931).

Bonotto, S. and G. Milic. 1969. Formation of sugars in the wood of Fagus silvatica L. irradiated by ^{60}Co gamma rays. Radiation Botany 9: 375-378. (In French) (NSA 24: 7731).

Bonska, H. 1973. Contamination of the environment by carbon-14 produced in high temperature reactors. Kerntechnik 15(7): 297-300. (In German and English) (NSA 28: 27222).

Book, S. A. 1969. Fallout cesium-137 accumulation in two subpopulations of black-tailed deer (Odocoileus hemionus columbianus). Thesis, Univ. of California, Davis. vi, 61pp. (Also: Univ. of California, Davis, U. S. report UCD-34-P-104-14).

Book, S. A., G. E. Connolly and W. M. Longhurst. 1972. Fallout ^{137}Cs accumulation in two adjacent populations of Northern California deer. Health Physics 22(4): 379-385. (NSA 26: 25595).

Boontiam Disyam. 1966. Radiation effect on red flour beetle (Tribolium castaneum Hbst.). pp. 35-41. In: Insect Eradication by Irradiation. Thailand Off. of Atomic Energy for Peace, Bangkok, report THAI.-AEC-8.

Boormal, J. P. T. 1960. Observations on the feeding habits of the mosquito Aedes (Stegomyia) aegypti (Linnaeus): the loss of fluid after a blood-meal and the amount of blood taken during feeding. Ann. Trop. Med. Parasitol. 54: 8-14. (^{144}Ce, ^{35}S).

Boose, R. B. and L. C. Terriere. 1967. Quantitative aspects of the detoxication of naphthaline by resistant and susceptible house flies. J. Econ. Entomol. 60(2): 580-586. (^{14}C).

Booth, A. H. and E. R. Samuels. 1969. Fall-out strontium-90 levels in Canada. pp. 393-402. In: Environmental Contamination by Radioactive Materials. (See IAEA, 1969).

Booth, R. S. 1975. Compendium of radionuclides found in liquid effluents of nuclear power stations. Oak Ridge National Lab., U. S. AEC report ORNL-TM-3801. 23pp. (NSA 31: 22613).

Booth, R. S. 1975. Systems analysis model for calculating radionuclide transport between receiving waters and bottom sediments. Oak Ridge National Lab., U. S. AEC report ORNL-TM-4751. 37pp. (NSA 32: 3495).

Booth, R. S. and S. V. Kaye. 1971. Preliminary systems analysis model of radioactivity transfer to man from deposition in a terrestrial environment. Oak Ridge National Lab., U. S. AEC report ORNL-TM-3135. 21pp. (NSA 26: 4741).

Booth, R. S., S. V. Kaye and P. S. Rohwer. 1973. A systems analysis methodology for predicting dose to man from a radioactively contaminated terrestrial environment. pp. 877-893. In: Radionuclides in Ecosystems. (See D. J. Nelson, 1973) (NSA 28: 30285).

Boquet, H. 1964. Radioactive pollution of surroundings. Effects of radiocontamination on the biological cycle. Maroc. Med. 43: 156-158. (NSA 19: 18216).

Borak, J. and D. Eisenklamm. 1936. On the influence of röntgen rays on hibernating animals. Wien. Klin. Wochenschr. 49(4): ___. (In German).

Borak, J. and H. Uiberall. 1936. On the influence of röntgen rays on hibernating animals. Wein. Klin. Wochenschr. 49(4): 109-111. (In German).

Bordeen, C. R. 1907. Abnormal development of toad ova fertilized by spermatozoa exposed to the Roentgen rays. J. Exptl. Zool. 4(1): 1-44; 5 pls.

Bordeen, C. R. 1909. Variations in susceptibility of amphibian ova to x-rays at different stages of development. Anat. Record 3(4): 163-165.

Bordeen, C. R. 1911. Further studies in the variation in susceptibility of amphibian ova to x-rays at different stages of development. Am. J. Anat. 11(4): 419-498.

Bordeen, C. R. and F. H. Baetjer. 1904. The inhibitive action of the roentgen rays on regeneration in planarians. J. Exptl. Zool. 1(1): 191-195.

Bores, R. J. and P. J. Bottinot. 1974. Design and dosimetry of a strontium-90-yttrium-90 beta irradiation facility. Health Physics 26(1): 99-101.

Borgeson, D. P. 1959. The movement of radioactive phosphorus through a stream ecosystem. Thesis, Michigan State Univ., East Lansing.

Borghi, C., et al. 1971. Organizing a program for eradication of _Culex pipiens fatigans_ in Recife, Brazil. pp. 373-378. In: _Sterility Principle for Insect Control or Eradication_. (See IAEA, 1971) (Incl. irradiation).

Bormann, F. H. 1961. Intraspecific root grafting and the survival of eastern white pine stumps. Forest Science 7: 247-256.

Bormann, F. H. 1961. Intraspecific root grafts and transport. pp. 1309-1311. In: _Recent Advances in Botany_. Univ. of Toronto Press, Toronto.

Bormann, F. H. and B. F. Graham. 1959. A study of intraspecific root grafts in white pine, _Pinus strobus_ L., utilizing radioisotopes and dyes. Abstr. Proc. 9th Intern. Bot. Congr. 2(A): 4.

Bormann, F. H. and B. F. Graham, Jr. 1959. The occurrence of natural root grafting in eastern white pine, _Pinus strobus_ L., and its ecological implications. Dartmouth Coll., _U. S. AEC_ report AECU-4020. 35pp. (Also: Ecology 40(4): 677-691) (NSA 13: 8489).

Bormann, F. H. and B. F. Graham. 1960. Translocation of silvicides through root grafts. J. Forestry 58: 402-403.

Bormann, F. H. and B. F. Graham. 1961. A study of intraspecific root graft in white pine, _Pinus strobus_ L., by means of radioisotopes and dyes. pp. 1375-1377. In: _Recent Advances in Botany_. Univ. of Toronto Press, Toronto.

Bormann, F. H., P. R. Shafer and D. Mulcahy. 1958. Fallout on the vegetation of New England during the 1957 atom bomb test series. Ecology 39(2): 376-378.

Born, W. 1960. Release of reflexes in snails by x and alpha rays. Strahlentherapie 112: 634-636. (In German) (NSA 14: 22791).

Bornemisza-Pauspertl, P. 1971. Possibility for the nondestructive analysis of plant samples through the fast neutron activation method. Atom. Kutat. Intez. Közl. 13(4): 173-180. (In Hungarian) (NSA 26: 47675).

Bornemisza-Pauspertl, P. 1973. Determination of magnesium in plant materials. J. Radioanal. Chem. 14(1): 189-192. (NSA 19: 27) (Neutron activation).

Boroughs, H. 1962. The use of ionizing radiation for insect control in Latin America. Intern. J. Appl. Radiat. Isotopes 13: 441-443.

Boroughs, H. 1963. The use of ionizing radiation for insect control in Latin America. Turrialba 13(1): 32-34. (In Spanish).

Boroughs, H. and D. Reid. 1957. The role of the blood in the transportation of strontium-90 - yttrium-90 in fish. Anat. Record 128(3): 524. (_Tilapia mossambica_).

Boroughs, H. and D. F. Reid. 1958. The role of the blood in transportation of strontium^{90}-yttrium^{90} in teleost fish. Biol. Bull. 115(1): 64-73.

Boroughs, H., S. J. Townsley and W. Ego. 1958. The accumulation of Y^{90} from equilibrium of Sr^{90}-Y^{90} by _Artemia salina_ L. Limnol. Oceanog. 3(4): 413-417.

Boroughs, H., S. J. Townsley and R. W. Hiatt. 1956. The uptake, accumulation and loss of strontium 89 by fishes. Biol. Bull. 111(3): 336-351.

Borovkov, I. I. 1967. Automatic station for recording gamma-background Gigiyena i sanitariya 32(12): 54-57. (TT-67-51409/4; pp. 376-379) (NSA 23: 13992).

Borowczyk, M. 1964. Determination of physical parameters of soils by isotopic methods in engineering geology. Kwart. geol. 8: 865-872. (In Polish) (NSA 19: 18182).

Bors, J. J. 1972. Trace elements in the sediments of Lake Mendota and Trout Lake. Thesis, Univ. of Wisconsin, Madison. 174pp. (Univ. Microfilms 72-22-077) (NSA 27: 2319).

von Borstel, R. C. 1955. Feulgen-negative nuclear division in _Habrobracon_ eggs after lethal exposure to x-rays or nitrogen mustard. Nature 175(445 (4451): 342-343. (NSA 9: 3024).

von Borstel, R. C. 1960. Population control by release of irradiated males. Science 131(3403): 878-882.

von Borstel, R. C. and A. A. Buzzati-Traverso. 1962. On the role of lethal mutants in the control of populations. pp. 273-278. In: _Radioisotopes and Radiation in Entomology_. (See IAEA, 1962).

von Borstel, R. C. and M. L. Rekemeyer. 1959. Radiation-induced and genetically contrived dominant lethality in _Habrobracon_ and _Drosophila_. Genetics 44(6): 1053-1074.

von Borstel, R. C. and R. W. Rogers. 1958. Alpha-particle bombardment of the of the _Habrobracon_ egg. II. Response of the cytoplasm. Radiation Research 8(3): 248-253. (NSA 12: 7082).

Bose, H. E. 1971. Changes in visual acuity and discrimination learning after proton irradiation of the striate visual area in the squirrel monkey. Thesis, Northern Illinois Univ., De Kalb. 85pp. (Univ. Microfilms 71-18,542) (NSA 25: 49290).

Bose, H. 1968. ^{90}Sr in soil and ^{90}Sr and ^{137}Cs in plants on the territory of the German Democratic Republic - 1966. Staatliche Zentrale für Strahlenschutz, Berlin, GDR, report SZS-12/68. 38pp. 1969. Ibid. -1967. SZS-7/69. 18pp. (In German) (NSA 23: 29475; 24: 31891).

Bose, H. 1970. ^{90}Sr and ^{137}Cs contamination of plants and soils on the territory of the German Democratic Republic during 1966 to 1968. Part II. Kernenergie 13: 156-162. (In German) (NSA 24: 48471).

Boss, M. R. 1972. Plutonium levels in the Rocky Flats environs: CY 1971. pp. I25-I45. In: Fallout Program Quarterly Summary Report, December 1, 1971 - March 1, 1972. (See Hardy, HASL-249) (NSA 26: 28048).

Bostrack, J. M. and A. H. Sparrow. 1969. Effects of chronic gamma irradiation on the anatomy of vegetative tissues of Pinus rigida Mill. Radiation Botany 9(5): 367-374. (NSA 24: 8016).

Bostrack, J. M. and A. H. Sparrow. 1970. Radiosensitivity of gymnosperms. II. On the nature of radiation injury and cause of death of Pinus rigida and P. strobus after chronic gamma irradiation. Radiation Botany 10(2): 131-143. (NSA 24: 48741).

Bostrack, J. M. and A. H. Sparrow. 1970. Radiation injury and cause of death in Pinus rigida and P. strobus after chronic gamma irradiation. Abstr. Radiation Research 43(1): 226-227.

Boswell, J. L. 1966. Some effects of acute ^{60}Co gamma radiation on four developmental stages of the brine shrimp Artemia salina (L.). Thesis, Texas A. and M. Univ., College Station. 88pp. (NSA 21: 34810).

Boswell, R. W. 1973. Atomic energy industry's approach to environmental protection. At. Energy Austral. 16(2): 2-5. (NSA 29: 361).

Botezatu, N. P. 1965. Use of radioactive isotopes in hydrogeological research. Hidrotechnica (Bucharest) 10: 4350440. (In Romanian) (NSA 20: 43555).

Both, A. R. 1962. The influence of ionizing radiations on the periferal frog nerve. Thesis, Universität, Bonn, Germany. 58pp. (NSA 17: 25000).

Bothner, M. H. and R. Carpenter. 1973. Sorption-desorption reactions of mercury with suspended matter in the Columbia River. pp. 73-86. In: Radioactive Contamination of the Marine Environment. (See IAEA, 1973) (NSA 28: 8206)

Botkin, D. B., G. M. Goodwell and N. Temple. 1970. Forest productivity estimated from carbon dioxide uptake. Ecology 51(6): 1057-1060. (At Brookhaven National Lab.).

Bottino, P. J. 1965. Radiosensitivity studies on an interspecific grass hybrid. J. Heredity 56: 225-228. (NSA 20: 12679).

Boudan, M. B., B. Gachot and M. Delpla. 1968. Choice of the measurement parameters for the survey of the water activity in the vicinity of a nuclear power plant. pp. 443-446. In: International Congr. on the Radioprotection of the Environment. (In French) (See Soc. Francais de Radioprotection, 1968).

Boudreaux, H. B. 1960. Further investigations on sex-ratio and fertilization in spider mites. Abstr. Bull. Entomol. Soc. Am. 6(3): 155.

Bougault, H., et al. 1967. Surface deposition of radioisotopes on sands and silts. pp. 233-262. In: Isotopes in Hydrology. (In French) (See IAEA, 1967) (NSA 22: 19063).

Bougnol, M. 1965. Monitoring and control of the radioactivity of waters used for agricultural purposes. Bull. Tech. Genie Rural (71): 1-63. (In French) (NSA 20: 43598).

Boulenger, R., J. Colard and G. Koch. 1973. Relative sensitivity of the measurements of radioactive wastes in disposal in air and water. J. Belge Radiol. 56(5): 439-441. (In French) (NSA 30: 3158).

Boulenger, R., et al. 1964. Radiological monitoring of the surroundings of C. E. N. at Mol. Belgium Centre d'Étude de l'Énergie Nucléaire, Brussels, report R-2298. 88pp. (In French) (NSA 19: 46631) (Incl. surface water).

Boulenger, R., et al. 1971. Study of the determination of the maximum permissible concentration of ^{226}Ra in water in the presence of sulfate. pp. 223-242. In: Proc. Intern. Symp. Radioecology Applied to the Protection of Man and His Environment. (See Commission of the European Communities, 1971) (In French) (NSA 27: 5257).

Bourdeau, P. and G. M. Woodwell. 1964. Field measurements of carbon dioxide exchange of Pinus rigida trees exposed to chronic gamma irradiation Ecology 45(2): 403-406. (Also: Brookhaven National Lab., U. S. AEC report BNL-7054) (NSA 18: 33315).

Bourdeau, P., et al. 1965. Movement of fallout radionuclides in irrigation ecosystems of the Po Valley, Italy. In: Radiation and Terrestrial Ecosystems. Health Physics 11(12): 1429-1444. (See Hungate, 1965).

Bourgin, R. C., R. Krumins and H. Quastler. 1956. Radiation induced delay of pupation in Drosophila. Radiation Research 5(6): 657-673.

Bourke, J. B., M. Faust and L. M. Massey, Jr. 1967. A multiple unit radiometric respirometer for the measurement of the metabolism of ^{14}C labelled substrates. Int. J. Appl. Radiat. Isotopes 18: 619-623. (NSA 21: 39331).

Bourne, N. F. 1959. The determination of carbon transfer from Chlorella vulgaris Beyerinck to Daphnia magna Straus using radioactive carbon (C^{14}) as a tracer. Thesis, Univ. of Toronto, Toronto.

Boush, G. M. and Matsumura. 1967. Insecticidal degradation by Pseudomonas melophthora, the bacterial symbiote of the apple maggot. J. Econ. Entomol. 60(4): 918-920. (^3H, ^{14}C).

Bovard, P. 1964. Technical solution to the continuous monitoring of the radioactivity of liquid wastes and their receiving waters. Fed. Eur. Gewasserschutz Inform. (10): 53-55. (In French) (NSA 19: 3800).

Bovard, P. 1967/68. Methods of studying the dispersal of radioactive effluents in the Rhone Basin. Rev. Inst. Pasteur Lyon 1(5): 787-792. (In French) (NSA 25: 46561).

Bovard, P. 1970. Nuclear pollution. Cah. Ing. Agron. (247): 1-5. (In French) (NSA 25: 42410).

Bovard, P. 1970. Radiocontamination of soils and crops. Prot. Civ. Secur. Ind. (187): 10-13. (In French) (NSA 25: 13402).

Bovard, P. 1970. Socio-psychological factors limiting the emmission of radioactive effluents. France Commissariat à l'Énergie Atomique, Fontenay-aux-Roses, report CEA-CONF-1733. 7pp. (In French) (CONF-701127-1) (NSA 25: 41948).

Bovard, P. 1971. Prospects offered to radioecology in regard to the environment. pp. 639-648. In: Proc. Intern. Symp. Radioecology Applied to the Protection of Man and His Environment. (See Commission of the European Communities, 1971) (NSA 27: 5234).

Bovard, P. 1973. Radioecology and the protection of the environment. Agriculture (Paris) (366): 162-164. (In French) (NSA 29: 7721).

Bovard, P. 1973. Radioecological study of the Rhone Basin: strategy and balance sheet. pp. 507-532. In: Environmental Behavior of Radionuclides Released in the Nuclear Industry. (See IAEA, 1973) (In French) (NSA 29: 29306).

Bovard, P. 1974. Application of radioecological techniques to non-radioactive pollution studies. Application to a specific case. pp. 123-134. In: Comparative Studies of Food and Environmental Contamination. (See IAEA, 1974) (In French) (NSA 30: 15469).

Bovard, P. and L. Foulquier. 1970. Pollution and hydrobiology of fresh water. Bull. Inform. Sci. Tech. (Paris) (151): 51-54. (In French, English summary) (NSA 25: 18937).

Bovard, P. and A. Grauby. 1966. Methods of studying radioactive effluent dispersal in water. pp. 3-10. In: Disposal of Radioactive Wastes into Seas, Oceans and Surface Waters. (See IAEA, 1966).

Bovard, P. and A. Grauby. 1967. The fixation of radionuclides from atmospheric fallout in peat-bog, Sphagnum sp., Polytrichum and Myriophyllum. pp. 533-537. In: Radioecological Concentration Processes. (See Aberg and Hungate, 1967).

Bovard, P. and A. Grauby. 1967/68. Methods of study of the dispersion of radioactive effluent in the Rhodanien Basin. Rev. Inst. Pasteur Lyon 1 (5): 787-792. (In French).

Bovard, P. and A. Grauby. 1975. Cycle of radionuclides released into waters by the nuclear industry. pp. 8.6-1 - 8.6-4. In: Proc. 13th Hydraulics Meeting, Effects of Human Activities on the Hydrometeorological Cycle. (Paris, Sept. 16-18, 1974). Soc. Hydrotechnique de France. (In French) (NSA 33: 29149).

Bovard, P., J. Delmas and A. Grauby. 1972. Effect of agricultural technology on the radioactive contamination of food products. Centre d'Études Nucléaires, France Commissariat à l'Énergie Atomique, Fontenay-aux-Roses, report CEA-CONF-2091. 15pp. (For Intern. Symp. on Chemical Engineering in the Service of Mankind, Paris, Sept. 3, 1972) (In French) (CONF-720960) (NSA 27: 12380).

Bovard, P., A. Grauby and Y. Cattan. 1963. Study of the transfer of a radioactive ion in a slowly-moving stretch of water. Bull. Inform. Sci. Tech. (Paris) (70): 19-26. (In French) (NSA 17: 28659).

Bovard, P., A. Grauby and F. Tchernia. 1970. First results on the effects of gamma irradiation on a Mediterranean-type ecosystem. pp. 1119-1135. In: Proc. Intern. Symp. Radioecology. Vol. II. (In French) (See France Commissariat à l'Énergie Atomique, 1970) (NSA 24: 41793).

Bovard, P., et al. 1975. Effect of the pollutant burden of water on the behavior of radionuclides. pp. 179-190. In: Combined Effects of Radioactive, Chemical, and Thermal Releases to the Environment. (See IAEA, 1975) (In French) (NSA 33: 29175).

Bowen, G. D. and A. D. Rovira. 1969. New techniques to study nutrient relations in plants. At. Energy Austral. 12: 2-7. (NSA 23: 22221).

Bowen, H. J. M. 1956. The biogeochemistry of strontium. Adv. Sci. 12: 585-588.

Bowen, H. J. M. 1960. Biological fractionation of isotopes. Intern. J. Appl. Radiat. Isotopes 7(4): 261-272.

Bowen, H. J. M. 1967. Activation analysis in botany and agriculture. pp. 287-299. In: Nuclear Activation Techniques in the Life Sciences. (See IAEA, 1967) (NSA 23: 4553).

Bowen, H. J. M. 1972. Biochemistry of the trace elements. pp. 393-405. In: Nuclear Activation Techniques in the Life Sciences. (See IAEA, 1972) (NSA 27: 22099) (Cyclical processing and modification by industria pollution).

Bowen, H. J. M. and P. A. Cawse. 1964. Effects of ionizing radiation on soils and subsequent crop growth. Soil Science 97(4): 252-259. (NSA 18: 23318).

Bowen, H. J. M. and P. A. Cawse. 1964. Some effects of gamma radiation on the composition of soil solution and soil organic matter. Soil Science 98(6): 358-361.

Bowen, H. J. M. and J. A. Dymond. 1955. Strontium and barium in plants and soils. U. K. AEA, Harwell, report AERE/SPAR/4. 13pp.

Bowen, H. J. M. and S. R. Smith. 1959. Effects of gamma-radiation on weeds. Nature 183(4665): 907. (NSA 13: 10807).

Bowen, H. J. M. and J. Thick. 1960. Factors from seed extracts that modify radiosensitivity. Radiation Research 13(2): 234-241. (Incl. pine seeds).

Bowen, H. J. M., P. A. Cawse and S. R. Smith. 1962. The effects of low doses of gamma radiation on plant yields. Int. J. Appl. Radiat. Isotopes 13(8): 487-492. (NSA 17: 5827).

Bowen, R. 1969. Progress in isotope hydrology: the combined environmental isotope approach. Sci. Progr. (Oxford) 57: 559-575. (NSA 24: 9509).

Bowen, V. T. 1948. The mineral metabolism of insects. pp. 104-108. In: Brookhaven Conf. in Biology and Medicine. Brookhaven National Lab., U. S. AEC report BNL-C-4.

Bowen, V. T. 1949. Barium metabolism in hornets studied by means of radio-isotopes. Trans. N. Y. Acad. Sci. (2) 2: 68-72. (Also: U. S. AEC report AECU-260) (NSA 3: 592).

Bowen, V. T. 1950. Manganese metabolism of social Vespidae. J. Exptl. Zool. 115: 175-205. (Also: U. S. AEC report AECU-825) (NSA 4: 3646).

Bowen, V. T. 1951. The uptake and distribution of barium-140 and lanthanum-140 in larvae of Drosophila repleta. J. Exptl. Zool. 118: 509-529. (Also: Brookhaven National Lab., U. S. AEC report BNL-1018) (NSA 6: 503).

Bowen, V. T. and A. Carbone. 1951. The uptake of lanthanum by a yeast. Nature 167(4260): 1032.

Bowers, B. and C. M. Williams. 1964. Physiology of insect diapause. XIII. DNA (deoxyribonucleic acid) synthesis during the metamorphosis of the cecropia silkworm. Biol. Bull. 126(2): 205-219. (Hyalophora cecropia, Samia cynthia; ^{3}H).

Bowers, C. Y., A. Segaloff and B. Brown. 1959. Factors affecting the thyroid gland uptake of I^{131} of the Rana catesbeiana tadpole. Endocrinology 65(6): 882-888.

Bowers, W. S. and M. J. Thompson. 1965. Identification of the major constituents of the crystalline powder covering the larval cuticle of Samia cynthia ricini (Jones). J. Insect Physiol. 11(7): 1103-1111. (^{14}C).

Bowes, G. C. and A. P. Baerg. 1971. Tracer method for ^{14}C calibration. Int. J. Appl. Radiat. Isotopes 22(7): 431-434. (NSA 25: 51401).

Bowie, S. H. U., D. Ostle and C. B. Campbell. 1973. Uranium mineralization in northern Scotland, Wales, the Midlands, and southwest England. Inst. Mining Met. Trans. B82: 177-179. (NSA 39: 15558) (Stream sampling).

Bowman, D. H. and K. M. King. 1965. Determination of evapotranspiration using the neutron scattering method. Can. J. Soil Sci. 45(2): 117-126. (NSA 21: 12757).

Bowman, H., et al. 1974. Detailed chemical and radiometric study of geothermal waters and associated rock formations, with environmental implications. pp. 32-35. In: Proc. 2nd Intern. Conf. on Nuclear Methods in Environmental Research. (See Vogt and Meyer, 1974) (NSA 33: 23297).

Bowne, S. W., Jr. and R. R. McGough. 1966. Phototaxis of Euglena with beta and gamma rays. Exptl. Call Res. 42: 193-195. (NSA 20: 24889).

Boxzormenyi, Z., et al. 1962. Methodological problems of photosynthesis measurements made with radioactive tracer C-14 in Lake Balaton. Magy. Tud. Akad. Tihanyi Biol. Kutatoint. Evkonyve 29: 39-63.

Boyer, J. 1967. Effect of fallout on the radioactivity of rain, soil, and certain plants considered as a function of the dates of the experimental nuclear explosions detonated from 1961 to 1965. Bull. Tech. Genie Rural (82): 1-39. (In French) (NSA 22: 25538).

Boyer, J., et al. 1966. Change in soil from effects of gamma irradiation. Bull. Inform. Sci. Tech. (Paris) (106): 1-17. (In French) (NSA 22: 26006).

Boyer, J., et al. 1966. Development of soils under the effect of gamma irradiation. Bull. Inform. Sci. Tech. (Paris) (106): 39-55. (In French) (NSA 21: 11157).

Boyer, J., et al. 1966. Some observations on the action of different soils subjected to gamma irradiation. pp. 920-928. In: French Acad. Agriculture, Verbal Proc. Session 29 June 1966. (In French).

Boyle, W. S. 1968. Comparative radiosensitivity studies in the Gramineae. Final report. Utah State Univ., Logan, U. S. AEC report COO-1404-7. 6pp. (NSA 23: 22251).

Boyle, W. S. 1968. Effect of type and level of polyploidy on radiosensitivity in five species of Triticeae. J. Heredity 59: 151-154. (NSA 22: 45726).

Boyns, P. K. 1973. Aerial measurement of 5% differences in terrestrial background radiation. Trans. Amer. Nucl. Soc. 17: 540-541. (NSA 29: 5114).

Boyns, P. K. and M. D. Sevart. 1973. Aerial radiological measuring survey of the area surrounding the Dresden Nuclear Power Station, Morris, Illinois, September 1968. EG & G, Inc., U. S. AEC report ARMS-68.6.8. 75pp. (NSA 29: 20863).

Bozeman, M. L. 1943. Effects if irradiation of the oöcytes of Sciara. Abstr. Genetics 28(1): 71.

Bozoky, L., J. Kovacs and T. Predmerszky. 1967. Data on the radioactivity of the environment and food in Hungary. Idojaras 71: 281-284. (In Hungarian) (NSA 22: 48946).

Braby, L. A. and D. S. Nachtway. 1973. Lethal and mutagenic responses of Chlamydomonas reinhardi to 1.55 MeV proton irradiation. Battelle Northwest Lab., U. S. AEC report BNWL-SA-4531. 14pp. (For Ann. Meet. Radiation Res. Soc., St. Louis, Mo., April 29, 1973) (CONF-730431) (NSA 28: 5938).

Brace, K. C. 1951. Effects of x-rays on size of yeast cells. Proc. Soc. Soc. Exptl. Biol. Med. 74(4): 751-755.

Brace, K. C. 1952. Histological changes in the tissue of the hibernating marmot following whole-body irradiation. Science 116(3021): 570-571.

Bracha, P. and R. D. O'Brien. 1966. The relation between physical properties and uptake of insecticides by eggs of the large milkweed bug. J. Econ. Entomol. 59(5): 1255-1264. (^{14}C).

Bradley, W. E. and G. E. Stout. 1970. The vertical distribution of tritium in water vapor in the lower troposphere. Tellus 22(6): 699-706.

Bradley, W. E. and M. Van der Westhuizen. 1970. Radioactive nuclear bomb fallout. A relationship between deposition, air concentration and rainfall. Atmos. Environ. 4(3): 321-322.

Bradley, W. F. 1963. Heterogeneous equilibria in sediment under transpor pp. 61-67. In: Transport of Radionuclides in Fresh Water Systems. (See Kornegay, et al., 1963).

Bradley, W. G. and K. S. Moer. 1975. Ecological studies of small vertebrates in Pu-contaminated study areas of NTS and TTR. pp. 151-185. In: Radioecology of Plutonium and Other Transuranics in Desert Environments. (See White and Dunaway, 1975) (NSA 33: 645).

Brady, D. N. 1972. Environmental surveillance sampling results at the Nevada Test Site, July 1969 through June 1970. Reynolds Electrical and Engineering Co., U. S. AEC report NVO-410-11. 58pp. (NSA 27: 24464).

Brady, N. E. (ed.). 1967. Agriculture and the Quality of Our Environment. (A Symp., 133rd Meet. Am. Assoc. Adv. Science, Washington, D. C., Dec. 26-31, 1966). Amer. Assoc. Advancement Sci., Washington, Publ. 85. 472pp. (CONF-661217) (NSA 23: 22298).

Bragin, B. I. and S. A. Matmuratov. 1973. Accumulation of ^{90}Sr and ^{137}Cs by some fish species in freshwater reservoirs. pp. 123-128. In: Radioecology of Water Organisms. II. (See Andrumaitis, 1973) (In Russian) (NSA 30: 21135).

Bragina, A. N., V. N. Gus'kova and V. M. Kupriyanova. 1971. Methods and results of a study of the process of water self-cleaning of waterbodies contaminated with a mixture of radioisotopes. pp. 113-117. In: Methods of Radioecological Investigations. (See Verkovskaya, 1971).

Bramati, L. and M. Pagliari. 1970. ENEL activities in radioecology. pp. 43-46. In: Proc. 1st Congr. on the State of Advancement of Radioecology in Italy. (See Anonymous, 1970) (In Italian).

Bramati, L., G. de Cecco and O. Ilari. 1971. Organizational and economic aspects of environmental investigations and monitoring in the vicinity of nuclear facilities. pp. 105-135. In: Proc. Intern. Symp. Radioecology Applied to the Protection of Man and His Environment. (See Commission of the European Communities, 1971) (NSA 27: 5225).

Bramson, P. E. 1972. Environmental status of the Hanford reservation for 1970. Battelle Northwest Labs., U. S. AEC report BNWL-C-96. 102pp. (NSA 29: 12698).

Bramson, P. E. and J. P. Corley. 1972. Environmental surveillance at Hanford for CY-1971. Et seq., w/similar titles; CY-1972. Battelle Northwest Labs., U. S. AEC reports BNWL-1683, -1683(Add.), 95pp.; -B-228, 85pp.; -B-234, 210pp.; -B-237, 13pp.; -1727, __pp.; -1727(Add.), 112pp. (NSA 27: 2523; 2524; 17256; 28: 8138; 29: 10159; 12697).

Bramson, P. E., J. P. Corley and W. L. Nees. 1973. Ibid. BNWL-B-278. 70pp. (NSA 29: 5139).

Branagan, D. 1971. Convenient method of labelling xodid ticks with radioisotopes for future identification during field studies. Trop. Anim. Health Prod. 3: 101-106. (NSA 32: 17597) (^{144}Ce).

Branca, G. 1965. Technological-hygienic aspects of radioactive waste disposal in Italy. Italy Comitato Nazionale per l'Energia Nucleare, Rome, report RT/PROT(65)-12. 54pp. (For 23rd Congr. Ital. Nat. Assoc. Hygiene, Publ. Hlth., Catania, Italy, October 1963) (CONF-631005-1) (NSA 20: 3604).

Branca, G. and A. Paoletti. 1965. On the concentration of radioactivity on the part of some algae. Igiene sanita pubbl. 20: 360-370. (In Italian) (NSA 22: 47461).

Branca, G., et al. 1971. Principles and methods for the derivation of a formula for the discharge of radioactive liquid wastes in surface waters. Italy Comitato Nazionale per l'Energia Nucleare, Rome, report EUR-4636. 43pp. (EUR F.Fr.6-65) (NSA 25: 34717).

Branca, G., et al. 1973. Applications of a derived formula for the discharge of radioactive liquid wastes. Italy Comitato Nazionale per l' Energia Nucleare, Rome. Commission of the European Communities, Luxembourg, report EUR-4897. 36pp. (NSA 27: 17242).

van den Brande, J. and C. Pelerents. 1963. Some effects of gamma rays on meal worms and various nematodes. Soil & Fert. 26(4): 276.

van den Brande, J. and N. van de Woestijne. 1962. Effect of gamma-radiation on the Mediterranean flour moth Ephestia kuehniella Z. in different stages of development. pp. 563-567. In: 11th Intern. Congr. on Entomology, Vienna, 17-25 August 1960. II. International Congr. for Entomology, Vienna.

Brandenberg, M. K., et al. 1962. Effects of acute gamma radiation on growth and morphology in Pinus monophylla Torr. and Frem. (pinyon pine). Radiation Botany 2(3/4): 251-263. (NSA 17: 15733).

Brandon, S. 1967. Disruption of the circadian rhythm of two species of millipedes by low-level radiation. Proc. La. Acad. Sci. 30: 70-73. (NSA 22: 31218).

Brandt, A. B. and V. A. Krylov. 1971. Stimulation of growth of Chlorella pyrenoidosa by continuous irradiation with ultrviolet of low intensity. Biofizika 16: ____. (Biophysics USSR 16(4): 760-763) (NSA 26: 36347).

Brandt, E.E. and F. J. Finamore. 1963. Protein synthesis in frog eggs. II. Amino acid activation and incorporation by isolated nucleoli. Biochim. Biophys. Acta 68(3): 618-624. (^{14}C).

Brard, E. 1968. Action of x rays on ovulation in chicken. Arch. inst. histol. ambryol. norm. exp. 51: 99-105. (In French) (NSA 24: 23360).

Brard, E. and J. Benoit. 1969. Sterilization of quails by x-rays and interracial grafts. Bull. biol. Fr. Belg. 103: 313-321. (In French) (NSA 25: 10988).

Brasiello, A. R. 1968. Autoradiographic study of ribonucleic acid synthesis during spermatogenesis of Asellus aquaticus (Crust. Isopoda). Exptl. Cell Res. 53: 252-260. (NSA 23: 28430).

Braswell, J. H. and W. L. Hanson. 1969. Bioelimination of tracer ^{65}Zn by the large milkweed bug, Oncopeltus fasciatus in relation to infection with the intestinal flagellate Leptomonas oncopelti. pp. 711-714. In: Symposium on Radioecology. (See Nelson and Evans, 1969).

Brauer, F. P. 1974. Environmental ^{129}I measurements. pp. 328-343. In: Proc. 2nd Intern. Conf. on Nuclear Methods in Environmental Research. (See Vogt and Meyer, 1974) (NSA 33: 23286).

Brauer, F. P. and J. H. Kaye. 1974. Detection systems for the low level radiochemical analysis if iodine-131, iodine-129, and natural iodine in environmental samples. Trans. Nucl. Sci. NS-21(1): 496-502. (NSA 30: 5867).

Brauer, F. P. and N. E. Ballou. 1975. Isotopic ratios of iodine and other radionuclides as nuclear power pollution indicators. pp. 215-209. In: Isotope Ratios as Pollution Source and Behavior Indicators. (See IAEA, 1975) (NSA 33: 7287).

Brauer, F. P. and H. G. Rieck, Jr. 1973. ^{129}I, ^{60}Co, and ^{103}Ru measurements on water samples from the Hanford Project environs. Battelle Northwest Lab., U. S. AEC report BNWL-SA-4478. 41pp. (NSA 32: 27924).

Brauer, F. P., J. H. Kaye and R. E. Connally. 1970. X-ray and β-γ coincidence spectrometry applied to radiochemical analysis of environmental samples. pp. 231-253. In: Radionuclides in the Environment. R. F. Gould, ed. Amer. Chemical Soc., Washington.

Brauer, F. P., et al. 1974. Natural iodine and ^{129}I in mammalian thyroids taken from sites in the USA. pp. 43-46. In: Environmental Surveillance Around Nuclear Installations. Vol. 2. (See IAEA, 1974) (NSA 31: 14031).

Braun, H. J. 1965. Use of isotopes in forstry. Alleg. Forstzeit. 21: 3-6.

Brayton, R. D. 1965. Radiation and fire effects of Gaylussacia baccata and Vaccinium vacillans. Thesis, Rutgers Univ., New Brunswick, N. J. 37pp. (Also: Brookhaven National Lab., U. S. AEC report BNL-8787) (NSA 19: 10872).

Brayton, R. D. and G. M. Goodwell. 1966. Effects of ionizing radiation and fire on Gaylassacia baccata and Vaccinium vacillans. Am. J. Botany 53(8): 816-820.

Brazil Comissão Nacional de Energia Nuclear. 1958. Radioactive products in the soil and in the atmosphere. Inst. Brasileiro de Bibliografia e Documentacão, Rio de Janeiro, report DTC-1. 118pp. (In English).

Brechbill, R. A., R. E. Engel and R. C. Kramp. 1967. Calcium in hock joints of wildlife ruminants in selected areas of the United States. Southwest Radiol. Health Lab., U. S. PHS, Las Vegas, Nev., report SWRHL-34r. 15pp. (NSA 22: 6230).

Brees, D. R. and E. R. Graham. 1969. The effect of soil texture and organic matter on the self-diffusion coefficient of tritiated water in Missouri soils. Soil Science 108(3): 165-169.

Breckenridge, W. J. and J. R. Tester. 1960. Field studies of hibernation of the Manitoba toad (Bufo hemiophrys). Abstr. Bull. Ecol. Soc. Am. 41 (4): 129.

Breckenridge, W. J. and J. R. Tester. 1961. Growth, local movements and hibernation of the Manitoba toad, Bufo hemiophrys. Ecology 42(4): 637-646.

Bregman, A. A. 1968. Formation and size regulation capacity of the serosa in male embryos of the mealy bug, Pseudococcus obscurus Essig. Thesis, Univ. of Rochester, Rochester, N. Y. 84pp. (NSA 23: 6563) (X-irrad.).

Brendakov, V. F., et al. 1967. Levels of radioactive contamination of the ground level atmosperic air and the earth's surface from products of nuclear explosions in 1963-1965 in the vicinity of Moscow. State Comm. on the Use of Atomic Energy, USSR. Atomizdat, Moscow. (AEC-tr-7214; pp. 9-20) (UN Doc. A/AC.82/G/L. 1201) (NSA 22: 21195; 25: 15777).

Brendakov, V. F., et al. 1968. Vertical distribution of fragment elements in some types of soils in the USSR. Atom. energiya 25(4): 331-332 (Sov. At. Energy 25(4): 1134-1136) (NSA 23: 453).

Brendakov, V. F., et al. 1968. Vertical distribution of nuclear explosion products and evaluation of their mobility in some types of soils in the Soviet Union. State Comm. on the Use of Atomic Energy, USSR. Atomizdat, Moscow. (AEC-tr-7030; pp. 152-156) (UN Doc. A/AC.82/G/L. 1258) (NSA 23: 17554).

Brennan, P. A. 1967. Effects different types of radiation on various life stages of cereal leaf beetle, Oulema malaopus (Linnaeus). Thesis, Michigan State Univ., East Lansing. 125pp. (NSA 22: 45681).

Brenner-Holzach, O. and F. Leuthardt. 1965. Origin of the C-2 and C-8 atoms in uric acid in Drosophila melanogaster. Helv. chim. Acta 48(5): 1147-11 1151. (In German) (^{14}C).

Brenner-Holzach, O. and F. Leuthhardt. 1965. Biosynthesis of pterins in Drosophila melanogaster. III. Origin of the sidechain in drosopterin. Helv. chim. Acta 48(7): 1569-1578. (In German) (^{14}C).

Breslavets (Braslavets, Braeslavets, Brayeslavets), L. P. 1946. Plants and X-rays. ___pp. Publ. House Acad. Sci. USSR, Leningrad-Moscow. (English transls.: Am. Inst. Biol. Sci., Washington; also, Brookhaven National Lab., copy may be borrowed from library) (NSA 15: 29065).

Breslavets, L. P. 1951. Alteration in shape and size of plastics in the prothallia of Polypodiaceae and Equisetaceae under the influence of x radiation. Dokl. Akad. Nauk 78(6): 1235-1238. (NSA 5: 6962).

Breslavats, L. P. 1952. Effect of large doses of x rays on prothallium of fern (Dryopteris spinulosum). Dokl. Akad. Nauk SSSR 87: 143-146. (NSA 7: 4271).

Breslavets, L. P. 1954. Comparative-cytological investigation of the effect of low doses of x rays and radium on intussusception in the pericarp of spring rye. Izv. Akad. Nauk SSSR (biol.) (6): 38-51. (NSA 9: 2583).

Breslavets, L. P. and Z. K. Mileshko. 1960. Protection of plants from ionizing radiation. Priroda (5): 50-53. (JPRS-3838).

Bresson, G. 1973. Nuclear energy and the environment. Agriculture (Paris) (366): 160-161. (In French) (NSA 29: 10148).

Bresson, G. and G. Lacourly. 1975. Food contamination and the general investigation of the environment of nuclear sites. Centre d'Études Nucléaires, Fontenay-aux-Roses, France Commissariat à l'Énergie Atomique, report CEA-CONF-3125. 3pp. (NSA 33: 23305).

Bresson, G., G. Lacourly and F. Fitoussi. 1975. Nuclear site selection and environmental protection: the decision-making methods. Centre d'Études Nucléaires, Fontenay-aux-Roses, France Commissariat à l'Énergie Atomique, report CEA-CONF-3124. 6pp. (NSA 33: 23304).

Bresson, G., et al. 1971. Transfer of pollution to man. Introduction to dynamic study of the behavior and transfer of radioactive pollution. Part I. Nuis. Environ. (Dec.): 33-40. (In French).

Breuer, F. and M. de Borotoli. 1973. Behavior of radioiodine in the environment and in man. Comitato Nacionale per l'Energia Nucleare, Rome, Italy, report RT/PROT-(73)13. 81pp. (NSA 29: 12715).

Brewbaker, J. L. and G. C. Emery. 1962. Pollen radiobotany. Radiation Botany 1(2): 101-154.

Brewer, L. W. 1973. Environmental monitoring report for Sandia Laboratories from 1964 through 1972. Sandia Labs., U. S. AEC report SLA-73-339. 55pp. (NSA 28: 366).

Brewer, L. W. 1974. Environmental monitoring report for Sandia Laboratories for 1973. Sandia Labs., U. S. AEC report SLA-74-167. 21pp. (NSA 29: 29343).

Brewer, S. T. and R. A. Livingston. 1972. Application of modeling methods to fuel cycle environmental impact analyses. Trans. Amer. Nucl. Soc. 15 (2): 659-660. (NSA 27: 6963).

Breymeyer, A. and E. P. Odum. 1969. Transfer and bioelimination of tracer ^{65}Zn during predation by spiders on labeled flies. pp. 715-720. In: Symposium on Radioecology. (See Nelson and Evans, 1969).

Brezeanu, A. and H. Titu. 1972. Morphological and ultra-structural modifications induced by ^{60}Co gamma radiations in Festuca pratensis. Rev. Roum. Biol. (bot.) 17(4): 219-226. (NSA 28: 30086).

Brian, M. V., G. Elmes and A. F. Kelly. 1967. Populations of the ant Tetramorium caespitum. J. Animal Ecol. 36(2): 337-342. (Used radionuclides).

Bricteux-Grégoire, S., C. Jeuniaux and M. Florkin. 1964. The biochemistry of the silkworm. XXVII. Biosynthesis of trehalose from pyruvate. Arch. int. physiol. biochim. 72(3): 482-488. (In French) (^{14}C).

Bridges, A. E., J. P. Olivo and V. L. Chandler. 1956. Relative resistances of microorganisms to cathode rays: II. Yeasts and molds. Appl. Microbiol. 4: 147-149.

Bridges, B. A. 1963. Conditions after irradiation affecting survival and recovery of micro-organisms. Int. J. Appl. Radiat. Isotopes 14(1): 48-51. (NSA 17: 24941).

Bridges, R. G., H. D. Crone and J. R. Beard. 1962. A study of the phospholipids of Dieldrin-resistant and susceptible houseflies with particular reference to those of the thoracic ganglion. pp. 145-153. In: Radioisotopes and Radiation in Entomology. (See IAEA, 1962).

Bridges, W. R., B. J. Kallman and A. K. Andrews. 1964. C^{14} DDT in a micro-environment at Denver. U. S. Fish and Wildlife Service Circular 199. Abstr. Fishery Bull. 199: 30-40. (Ampullaria).

Bridgman, J. and R. F. Kimball. 1954. The effect of x-rays on division rate and survival in Tillina magna and Colpoda sp. with an account of delayed death. J. Cell. Comp. Physiol. 44(3): 431-445.

Bridgman, J. and C. King. 1962. A study of the effects of repeated irradiation on the cysts of a ciliate, Tillina magna. Radiation Research 17(2): 203-208. (NSA 16: 26827).

Brien, P. and J. P. van den Eeckhoudt. 1953. Budding and regeneration in Hydra following x-irradiation. Compt. rend. 237(14): 756-758. (In French) NSA 8: 51).

Briggs, A. 1966. The resistances of spores of the genus Bacillus to phenol, heat, and radiation. J. Appl. Bacteriol. 29: 490-504. (NSA 21: 18190).

Bright, W. C. 1971. Rocky Flats environmental research. pp. 316-319. In: Proc. Rocky Flats Symp. on Safety in Plutonium Handling Facilities. (See Dow Chemical Co., 1971). (NSA 25: 54964).

Brill, W. A. 1968. Summary of iron-55 in the environment and levels in humans. Radiol. Hlth. Data (USPHS) 9(4): 195-202.

Brinck, W. L. and B. Kahn. 1970. Radionuclide releases at nuclear power stations. pp. 226-233. In: Environmental Surveillance in the Vicinity of Nuclear Facilities. (See Reinig, 1970) (NSA 25: 16175).

Brinck, W. L., E. D. Harward and R. I. Chissler. 1970. Programs for environmental surveillance around nuclear power plants. pp. 46-54. In: Environmental Surveillance in the Vicinity of Nuclear Facilities. (See Reinig, 1970. (NSA 25: 16160).

Brinkmann, R., K. O. Münnich and J. C. Vogel. 1959. C^{14} age determination of groundwater. Naturwissenschaften 46: 10-12.

Brinkmann, R., K. O. Münnich and J. C. Vogel. 1960. The use of C^{14} methods in soil formation and groundwater circulation. Geol. rundsch. 49: 244-253.

British Standards Institution. 1967. Methods of measurement of liquid flow in open channels. Part II. Dilution methods. 2C. Radioisotope techniques. British Standard 3680. London. 48pp. (NSA 22: 12554).

Brisbin, I. L., Jr., et al. 1973. Accumulation and redistribution of radiocesium by migratory waterfowl inhabiting a reactor cooling reservoir. pp. 373-384. In: Environmental Behavior of Radionuclides Released in the Nuclear Industry. (See IAEA, 1973) (NSA 29: 29304).

Brisbin, I. L., Jr., et al. 1974. Radiocesium concentrations of snakes from contaminated and non-contaminated habitats of the AEC Savannah River Plant. Copeia (2): 501-506. (NSA 32: 9256).

Brisbin, I. L., Jr., et al. 1974. Patterns of radiocesium in the sediments of a stream channel contaminated by production reactor effluents. Health Physics 27(1): 19-27. (NSA 30: 18642).

Broce, A. B., L. J. Goldman and H. L. Cromroy. 1967. Radiosensitivity of cockroaches. Bull. Entomol. Soc. Am. 13(3): 192.

Brock, D. W., et al. 1973. An analysis of mercury residues in Idaho pheasants. pp. 186-198. In: Mercury in the Western Environment. (See Buhler, 1973) (Neutron activation analysis).

Brock, M. A. 1964. Hibernation and temperature effects on the aging of red blood cells. Ann. Acad. Scientiarum Fenn. A(IV)(Biol.) 71: 51-63. (Used radioiron, hampsters).

Brock, T. D. and M. L. Brock. 1969. The fate in nature of photosynthetically assimilated ^{14}C in a blue-green alga. Limnol. Oceanogr. 14($): 604-607.

Brockerhoff, H. and R. J. Hoyle. 1967. Conversion of a dietary triglyceride into depot fat in fish and lobster. Can. J. Biochem. 45: 1365-1370. (1967 Studies, Part 2, Fish. Res. Bd. Can. No. 1186: 439-444) (^{14}C, ^{3}H).

Brockway, A. P. 1955. The effects of x-radiation on larvae and pupae stages of the yellow meal worm Tenebria molitor Linn. Abstr. Biol. Bull. 109 (3): 355.

Broda, E. and T. Schönfeld. 1962. Factors controlling the distribution of fission products in the biosphere. Final report for IAEA Research Contract No. 1. pp. 9-11. In: IAEA Research Contracts, 2nd Ann. Rpt., Ser. 9. Vienna, Austria.

Broda, E. and T. Schoenfield. 1966. The Technical Applications of Radioactivity. Vol. I. Pergamon Press, Oxford. 353pp. (NSA 21: 28175).

Brodo, I. 1964. Field studies of the effects of ionizing radiation on lichens. Bryologist 67: 76-86.

Broecker, W. 1964. Radiocarbon dating: a case against the proposed link between river mollusks and soil humus. Science 143(3606): 596-597. (NSA 18: 12341).

Broecker, W. and A. Kaufman. 1965. Radiocarbon chronology of Lake Lahontan and Lake Bonneville II, Great Basin. Bull. Geol. Soc. Am. 76: 537-566. (NSA 20: 546).

Broecker, W. and E. A. Olson. 1960. Radiocarbon from nuclear tests. II. Science 132(3429): 712-721. (NSA 14: 23897).

Broecker, W. and E. A. Olson. 1961. Lamont radiocarbon measurements. VIII. Radiocarbon 3: 176-204.

Broecker, W. and P. C. Orr. 1958. Radiocarbon chronology of Lake Lahontan and Lake Bonneville. Bull. Geol. Soc. Am. 69: 1009-1032.

Broecker, W. and A. Walton. 1959. The geochemistry of C^{14} in freshwater systems. Geochim. Cosmochim. Acta 16: 15-38.

Broecker, W. and A. Walton. 1959. Radiocarbon from nuclear tests. Science 130(3371): 309-314. (Also: pp. 1987-1999. In: Fallout from Nuclear Weapons Tests. (See U. S. Congress, 1959) (NSA 13: 17741; 14: 9303).

Broertjes, C. 1969. Mutation breeding of Streptocarpus. Euphytica 18: 333-339. (NSA 24: 36739).

Broeshart, H. and D. A. Nethsinghe. 1972. Studies on the pattern of root activity of tree crops using isotope techniques. pp. 453-463. In: Isotopes and Radiation in Soil-Plant Relationships Including Foerstry. (See IAEA, 1972) (NSA 27: 2857).

Broido, A. and A. W. McMasters. 1959. The influence of a fire-induced convection column on radiological fallout patterns. Calif. Forest and Range Expt. Sta., Berkeley, Tech. Paper 32. 52pp. (NSA 15: 25045).

Brookhaven National Laboratory. 1948. Conference on Biological Applications of Nuclear Physics. U. S. AEC report BNL-C-4. 154pp.

Brookhaven National Laboratory. 1953. Abnormal and Pathological Plant Growth. Report of Symposium held August 3-5, 1953, Upton, N. Y. No. 6. U. S. AEC report BNL-258. 310pp. (NSA 8: 5464).

Brookhaven National Laboratory. 1971. Radioecological effects of fallout radiation. Committee 6. pp. 643-644. In: *Survival of Food Crops and Livestock in the Event of Nuclear War*. (See Bensen and Sparrow, 1971).

Brookhaven National Laboratory. 1975. Brookhaven National Laboratory environmental statement. U. S. ERDA report ERDA-1540. vp. (NSA 33: 7291).

Brookhaven National Laboratory. 1975. ERDA/National Laboratories Workshop on Environmental Effects of Energy Held at Rockefeller University, New York, September 4-5, 1975. U. S. ERDA report BNL-20701. 141pp. (NSA 33: 20181).

Brooks, D. B. and J. V. Krutilla. 1969. *Peaceful Use of Nuclear Explosives. Some Economic Aspects*. Johns Hopkins Press, Baltimore. 52pp. (NSA 23: 45435).

Brooks, G. R., Jr. 1962. Resistance of the ground skink, *Lygosoma laterale*, to gamma radiation. Herpetologica 18: 128-129. (NSA 21: 13010).

Brosemer, R. W. 1965. The effect of puromycin and actinomycin D on the development of grasshopper flight muscle glycerolphosphate dehydrogenase. Biochim. Biophys. Acta 99(2): 388-390. (^{14}C).

Brothwell, D. and E. Higgs. (Eds.). 1971. *Science in Archeology, a Survey of Progress and Research*. Thames and Hudson, London. 720pp. (NSA 26: 28097) (Incl. dating).

Brouardel, J. and E. Rinck. 1958. A devise used in the Mediterranean for the study of the production of organic substances. J. Conseil Int. Expl. Mer. 24(1): 10-16. (BA 35: 65642; SFA 6: 3958).

Browder, F. N., P. J. Parsons and H.-v. Phoung. 1969. The control of radioactive waste disposal from both national and international viewpoints. pp. 717-723. In: *Environmental Contamination of Radioactive Materials*. (See IAEA, 1969) (NSA 24: 7919).

Brower, G. R. 1967. Ozonation reactions of selected pesticides for water pollution abatement. Thesis, Washington Univ., St. Louis, Mo. 211pp. (Dissertation Abstr. 28(2): 722-B) (^{14}C).

Brower, J. H. 1963. Changes in insect populations in a natural forest community exposed to chronic gamma radiation stress. Abstr. Bull. Ecol. Soc. Am. 44(3): 69.

Brower, J. H. 1964. Changes in the insect populations of the low shrub synusia in a natural forrest community exposed to chronic gamma radiation stress. Thesis, Univ. of Massachusetts, Amherst.

Brower, J. H. 1965. Ecological changes in the insect community of a pitch pine-oak forest subjected to gamma radiation. Thesis, Univ. of Massachusetts, Amherst. 184pp. (Also: Brookhaven National Lab., U. S. AEC report BNL-9253) (NSA 19: 40307) (*Myzocallis discolor*).

Brower, J. H. 1965. Behavioral changes in an ant colony exposed to chronic gamma irradiation. Abstr. 82. Bull. Entomol. Soc. Am. 11(3): 158. (*Formica integra*).

Brower, J. H. 1966. Behavioral changes in an ant colony exposed to chronic gamma irradiation. Am. Midland Nat. 75: 530-534. (NSA 21: 11022).

Brower, J. H. 1967. A possible radiation avoidance response exhibited by an insect population in nature. Naturwissenschaften 54(2): 39-40.

Brower, J. H. 1972. Combined effects of egg age and radiation dosage on egg hatch of *Tenebrio molitor* (Coleoptera: Tenebrionidae). Can. Entomol. 104(1): 141-144. (NSA 26: 56484).

Brower, J. H. 1972. Interaction of age and radiation dosage on hatch of *Gibbium psylloides* eggs (Coleoptera: Ptinidae). Ann. Entomol. Soc. Am. 65(5): 1237-1238. (NSA 26: 56481).

Brower, J. H. 1973. Gamma radiation sensitivity of malathion-resistant strains of the Indian meal moth. J. Econ. Entomol. 66(2): 461-462. (NSA 28: 2968).

Brower, J. H. 1973. Low-dosage gamma radiation effects on the reproductive capacity of *Callosobruchus maculatus*. Ann. Entomol. Soc. Am. 66(3): 617-619. (NSA 28: 2971).

Brower, J. H. 1973. Depressed flour beetle: sensitivity to gamma irradiation. J. Econ. Entomol. 66(6): 1318-1320. (NSA 29: 13029).

Brower, J. H. 1973. Sensitivity of *Tenebrio molitor* and *T. obscurus* to gamma irradiation. J. Econ. Entomol. 66(5): 1175-1179. (NSA 29: 774).

Brower, J. H. 1974. Inability of populations of *Callosobruchus maculatus* to develop tolerance to exposure of acute gamma irradiation. Ann. Entomol. Soc. Am. 67(2): 287-291. (NSA 29: 27114).

Brower, J. H. 1974. Lack of radioresistance in *Plodia interpunctella* (Lepidoptera: Phycitidae) exposed to sublethal gamma irradiation for 30 generations. Radiation Research 57(1): 73-79. (NSA 29: 18652).

Brower, J. H. 1974. Developmental success of two species of *Ips* (Coleoptera: Scolytidae) in a chronically irradiated forest community. Can. Entomol. 106(3): 233-238. (NSA 32: 11760).

Brower, J. H. 1974. Radioresistance of the red flour beetle, *Tribolium castaneum* (Coleoptera: Tenebrionidae), exposed to sublethal doses of gamma irradiation for 25 generations. Can. Entomol. 106(3): 241-246. (NSA 32: 11761).

Brower, J. H. 1975. Radiosensitivity of the longheaded flour beetle. J. Econ. Entomol. 68(2): 220-222. (NSA 32: 1183).

Brower, J. H. and M. Ashraf. 1972. Histopathology of irradiated yellow mealworm adult and larval midguts. Ann. Entomol. Soc. Am. 65(4): 950-954. (NSA 26: 45875).

Brower, J. H. and H. C. Scott. 1972. Gamma radiation sensitivity of the spider beetle, *Gibbium psylloides* (Coleoptera: Ptinidae). Can. Entomol. 104(10): 1551-1556. (NSA 30: 15979).

Brower, J. H. and E. W. Tilton. 1972. Insect disinfestation of shelled pecans, almonds, and walnuts by gamma radiation. J. Econ. Entomol. 65(1): 222-224. (NSA 26: 23144).

Brower, J. H. and E. W. Tilton. 1972. Gamma-radiation effects on Trogoderma inclusum and T. variable. J. Econ. Entomol. 65(1): 250-254. (NSA 26: 23142).

Brower, J. H., E. W. Tilton and R. R. Cogburn. 1971. Effects of irradiated diets on production of progeny by several successive generations of the Indian meal moth, Plodia interpunctella (Huebner). Radiation Research 48 (2): 283-290. (NSA 26: 12431).

Brower, M. S. 1965. Changes in leaves of Quercus alba exposed to chronic gamma radiation. Thesis. Brookhaven National Lab., U. S. AEC report BNL-9509. 45pp. (NSA 20: 14289).

Brown, A. G. 1969. Use of radioisotopes in forest research. At. Energy Austral. 12(2): 14-17. (NSA 23: 41217).

Brown, F. A., Jr. 1963. An orientational response to weak gamma radiation. Biol. Bull. 125(2): 206-225. (NSA 18: 11673) (Planaria).

Brown, F. A., Jr. and Y. H. Park. 1964. Seasonal variations in sign and strength of gamma-taxis in planarians. Nature 202(4931): 469-471. (^{137}Cs source).

Brown, F. A., Jr. and H. M. Webb. 1968. Some temporal and geographic relations of snail response to very weak gamma radiation. Physiol. Zool. 41: 385-400. (NSA 23: 22361).

Brown, F. A., Jr., H. M. Webb and M. F. Bennett. 1958. Comparisons of some fluctuations in cosmic radiation and in organismic activity during 1954, 1955 and 1956. Am. J. Physiol. 195(1): 237-243.

Brown, F. A., Jr., H. M. Webb and L. G. Johnson. 1962. Orientational responses in organisms effected by very small alterations in gamma (Cs^{137}) radiation. Abstr. Biol. Bull. 123(2): 488-489. (Planaria).

Brown, F. A., Jr., Y. H. Young and J. R. Zeno. 1966. Diurnal variation in organismic response to very weak gamma radiation. Nature 211(5051): 830-833.

Brown, G. A. 1967. Male competiveness in irradiated and non-irradiated Acarus siro L. Bull. Entomol. Soc. 13(3): 194.

Brown, G. A. 1970. Influence of age and radiation dose-rate on some effects of gamma radiation on two stored-product pests. Thesis, Univ. of Georgia, Athens. 93pp. (Univ. Microfilms 71-3716) (NSA 25: 24317).

Brown, G. A. and R. Davis. 1969. Effects of introduced irradiated male grain mites on population development. J. Econ. Entomol. 62(4): 767-771.

Brown, G. A. and R. Davis. 1972. Sensitivity of grain mite eggs to gamma radiation as influenced by dose rate and treatment age. J. Econ. Entomol. 65(6): 1619-1621. (NSA 27: 15090).

Brown, G. A., J. H. Brower and E. W. Tilton. 1972. Gamma radiation effects on Sitophilus zeamais and S. granarius. J. Econ. Entomol. 65(1): 203-205. (NSA 26: 23145).

Brown, G. N. 1964. Cesium in Liriodendron and other woody species: organic bonding sites. Science 143(3604): 368-369. (NSA 18: 9829) (^{137}Cs).

Brown, G. N. 1965. Irradiation-photoperiod interactions in Xanthium. Abstr. Bull. Ecol. Soc. Am. 46(3): 96.

Brown, G. N. and F. G. Taylor, Jr. 1966. Zinc-65 distribution and radiosensitivity of growth and ^{65}Zn uptake in red oak seedlings. Radiation Botany 6(6): 519-524.

Brown, G. N. and F. G. Taylor, Jr. 1967. Interaction of radiation and photoperiodism in Xanthium pensylvanicum: effects of radiation during the light and dark periods of photoinduction. Radiation Botany 7: 67-72.

Brown, K. W. and J. C. McFarlane. 1973. Deposition and retention of ^{131}I on Grayia spinosa following Banberry. Health Physics 24(6): 680-682. (NSA 28: 8524).

Brown, K. W. and B. J. Mason. 1968. Range survey, Area 18, Nevada Test Site. Southwestern Radiological Hlth. Lab, U. S. PHS, Las Vegas, U. S. AEC report SWRHL-52. 42pp. (NSA 23: 26711).

Brown, J. 1965. Determination of the carbon-14 age of Barrow, Alaska. Abstr. p. 45. In: Abstr. Intern. Assoc. Quaternary Research VII Congr., Boulder and Denver, Colo.

Brown, L. C. 1971. Determination of ^{32}P and ^{33}P in aqueous solution. Anal. Chem. 43(10): 1326-1328. (NSA 25: 44365).

Brown, L. J. and C. B. Taylor. 1974. Geohydrology of the Kaikoura Plain, Marlborough, New Zealand. pp. 169-189. In: Isotope Techniques in Groundwater Hydrology. Vol. 1. (See IAEA, 1974) (NSA 31: 19627).

Brown, R. C. 1974. Evidence of seasonal effects on environmental radioactivity. J. Environ. Sci. 17(5): 25-26. (NSA 30: 32080).

Brown, R. E. and J. E. Raymond. 1962. The measurement of Hanford's geohydrologic features affecting waste disposal. pp. 77-98. In: 2nd Conf. on Ground Disposal of Radioactive Wastes. (See Morgan, J. M., et al., 1962) (NSA 16: 25042).

Brown, R. E., K. R. Holtzinger and J. R. Raymond. 1955. Earth Sciences waste disposal monitoring activities, March 1955. General Electric Co., Hanford, U. S. AEC report HW-36411. 13pp. (NSA 29: 23707).

Brown, R. M. 1961. Hydrology of tritium in the Ottawa Valley. Geochim. Cosmochim. Acta 21(3/4): 199-216.

Brown, R. M. and W. E. Grummit. 1956. The determination of tritium in natural waters. Can. J. Chem. 34: 220-226.

Brown, S. L., W. B. Lane and J. L. Mackin. 1970. Beta dosimetry for fallout hazard evaluation. Final report. Stanford Research Inst., Menlo Park, Calif. 57pp. (AD-715017) (NSA 25: 49170).

Brown, S. W. and W. A. Nelso-Rees. 1961. Radiation analysis of a lecanoid genetic system. Genetics 46(8): 983-1007. (NSA 15: 29027).

Brown, V. M. 1962. The accumulation of strontium-90 and yttrium-90 from a continuously flowing natural water by eggs and alevins of the Atlantic salmon and sea trout. UK Atomic Energy Authority report PG-288. 16pp. (NSA 16: 16578; WPA 35: 1874; SFA 8: 5628).

Brown, V. M. and W. L. Templeton. 1964. Resistance of fish embryos to chronic irradiation. Nature 203(4951): 1257-1259. (NSA 18: 43273).

Brownell, J. R. and A. Lauchli. 1969. Measurement of beta emitters in plant material with Cherenkov radiation: correction of color-quenching. Int. J. Appl. Radiat. Isotopes 20: 797-798. (NSA 24: 7706).

Brownell, L. E. and M. Yudelovitch. 1962. Effect of radiation on Mexican fruit-fly eggs and larvae in grapefruit. pp. 193-206. In: Radioisotopes and Radiation in Entomology. (See IAEA, 1962).

Brownell, L. E., J. V. Nehemias and J. J. Bulmer. 1955. The design of a gamma irradiation facility for the control of insect infestation in flour, meal, or grain. U. S. AEC report AECU-3050. 21pp.

Browning, L. S. 1968. Effects of the space environment on radiation-induced damage in the reproductive cells of pupae and adult Drosophila. In: Symposium on the Biosatellite II Experiments. BioScience 18(6): 570-576.

Browning, L. S. and E. Altenberg. 1968. Effects of the space environment on radiation-induced damage in mature reproductive cells of adult Drosophila and in spermatocytes of the immature testis. Abstr. Radiation Research 35(2): 500-501.

Bruant, C., et al. 1973. Study of heavy metals in the grass-milk product chain by means of neutron activation and atomic absorption. Centre d' Études Nucléaires, Grenoble, France Commissariat à l'Énergie Atomique, report CEA-CONF-2496. 18pp. (In French) (NSA 29: 28965)

Brubaker, S. 1972. To Live on Earth: Man and His Environment in Perspective. Johns Hopkins Press, Baltimore. 216pp. (NSA 28: 15477) (Incl. radioactivity).

Bruce, A. K. 1958. Response of potassium retentivity and survival of yeast to far-ultraviolet, near-ultraviolet and visible and x-radiation. J. Gen. Physiol. 41(4): 693-702.

Bruce, A. K. and J. N. Stannard. 1958. The effect of x irradiation on potassium retentivity of yeast. J. Cell. Comp. Physiol. 51(3): 325-340.

Bruce, A. K. and J. N. Stannard. 1958. The effect of temperature on potassium transfer in irradiated yeast. J. Cell. Comp. Physiol. 52(1): 165-175.

Bruce, R. S. 1973. Agricultural aspects of radioactive contamination of the environment by the nuclear power industry. pp. 273-289. In: Environmental Behavior of Radionuclides Released in the Nuclear Industry. (See IAEA, 1973) (NSA 29: 29376).

Bruce-Chwatt, L. J. 1956. Radioisotopes for research on and control of mosquitoes. Bull. World Hlth. Org. 15(3): 491-511.

Bruce-Chwatt, L. J. 1962. Some entomological problems in tropical medicine. pp. 211-229. In: Radioisotopes in Tropical Medicine. (Proc. Symp., Bangkok, 12-16 Dec. 1960). (See IAEA, 1962) (Sterile-male technique, use of radioisotopes).

Bruce-Chwatt, L. J. and J. Hayward. 1956. Uptake of radioactive phosphorus by larvae and adults of mosquitoes. Nature 177(4500): 661-662.

Brucker, W. 1965. On the influence of x rays on the phosphate metabolism of Phycomyces. Ber. Deut. Bot. Ges. 68: 183-192. (In German) (NSA 20: 14264).

Brucker, W. and H. G. Rohde. 1966. On the ribonucleic acid metabolism of datura tissues after x irradiation. Ber. Deut. Bot. Ges. 70: 134-142. (In German) (NSA 21: 24668).

Brunelle, R. L., et al. 1969. Neutron activation and atomic absorption analyses of chemical elements in soil. J. Assoc. Anal. Chem. 52: 911-914. (NSA 23: 47408).

Bruner, H. D. 1973. Distribution of tritium between the hydrosphere and invertebrates. pp. 303-314. In: Tritium. (See Moghissi and Carter, 1973) (NSA 28: 27682).

Brunet, P. C. 1965. The metabolism of aromatic compounds. pp. 49-77. In: Aspects of Insect Biochemistry. T. W. Goodwin, ed. Academic Press, New York. (Mentions radionuclides).

Brungs, W. A., Jr. 1963. The relative distribution of multiple radionuclides in a fresh-water pond. Dissertation, Ohio State Univ., Columbus. 97pp. (NSA 18: 25079).

Brungs, W. A. 1965. Experimental uptake of strontium-85 by freshwater organisms. Health Physics 11(1): 41-46. (NSA 19: 8757).

Brungs, W. A., Jr. 1967. Distribution of cobalt 60, zinc 65, strontium 85, and cesium 137 in a freshwater pond. U. S. PHS Publ. 999-RH-24. vi, 52pp.

Brunner, H. and S. Pretre. (Eds.). 1968. Radiological Protection of the Public in a Nuclear Mass Disaster. (Proc. Symp. Interlaken, Switzerland, 26 May-1 June 1968). Switzerland Fachverband für Strahlenschutz. 688pp. (In several languages) (CONF-680507) (NSA 23: 20178).

Brunner, P. G. 1971. Permissible activity discharge into receiving waters through radioactive waste water from nuclear power plants. Wasser Boden (4): 124-126. (In German) (NSA 26: 25557).

Brunori, A., J. Georgieva and F. d'Amato. 1970. Further observations on the x-irradiation response of G_1 cells in dry seeds. Mutation Research 9: 481-487. (NSA 25: 2555).

Brunovskiy, B. K. 1932. The concentration of radium in various organisms. Trav. lab. biogeochim. acad. sci. (USSR) 2: 9-24.

Brunovskiy, B. K. and K. G. Kunasheva. 1930. The radium content of some plants. Compt. rend. Acad. Sci.USSR 1930A: 537-538.

Brunovskiy, B. K. and K. G. Knasheva. 1935. A contribution on the radium content of plants and water. Trav. lab. biogeochim. acad. sci. (USSR) 3: 31-41.

Bruns, L. E. 1974. Aerial gamma survey by helicopter to measure surficial contamination. Atlantic Richfield Hanford Co., Richland, Wash., U. S. AEC report ARH-SA-176. 18pp. (NSA 31: 8648).

Brunst, V. V. 1944. Investigation of the influence of x-radiation on the development in amphibians. W. Roux. Entw. d. Organ. 142: 706-739.

Brunst, V. V. 1950. Influence of x-rays on limb regeration in urodele amphibians. Quart. Rev. Biol. 25: 1-29. (NSA 4: 3283).

Brunst, V. V. 1954. The effect of local x-ray irradiation upon the development of the anterior part of the head of the axolotl. J. Morphol. 95(2): 373-391.

Brunst, V. V. 1955. The reaction of the skin epithelium of the tails of *Rana catesbiana* tadpoles to local x-irradiation. Radiation Research 2 (6): 556-567. (NSA 9: 6842).

Brunst, V. V. 1955. The influence of roentgen irradiation of the development of the eye of axolotl. Am. J. Roentgen. Rad. Therapy Nucl. Med. 73 (2): 281-293.

Brunst, V. V. 1956. Investigation of the lethal effect if x-irradiation on the young axolotl (*Siredon mexicanum*). Radiation Research 5(3): 267-280. (NSA 10: 11650).

Brunst, V. V. 1957. Histopathology of roentgen death of young axolotl (*Siredon mexicanum*). Am. J. Roentgen. Rad. Therapy Nucl. Med. 78(3): 518-545.

Brunst, V. V. 1958. Effect of irradiation on some portions of head of young axolotl. Proc. Soc. Exptl. Biol. Med. 99(3): 571-574.

Brunst, V. V. 1958. The effect of total-body x-irradiation on the adult axolotl (*Siredon mexicanum*). Radiation Research 8(1): 32-45.

Brunst, V. V. 1958. The effect of different doses of roentgen rays on adult axolotl (*Siredon mexicanum*). Am. J. Roentgen. Rad. Therapy Nucl. Med. 80 (1): 126-142. (NSA 12: 10301).

Brunst, V. V. 1958. The roentgen sensitivity of various portions of the eye of young axolotl (*Siredon mexicanum*). Am. J. Roentgen. Rad. Therapy Nucl. Med. 80(6): 1014-1030. (NSA 13: 1900).

Brunst, V. V. 1959. Further investigation of the effects of roentgen irradiation upon the development of the head of the young axolotl. Am. J. Roentgen. Rad. Therapy Nucl. Med. 82: 708-719. (NSA 13: 20806).

Brunst, V. V. 1959. Roentgen regression and roentgen stimulation in axolotl (*Siredon mexicanum*). Acta Unio Intern. contra Cancrum 15: 568-576. (NSA 14: 7222).

Brunst, V. V. 1960. Reaction of limb regenerates of adult axolotl (*Siredon mexicanum*) to x-irradiation. Radiation Research 12(6): 642-656.

Brunst, V. V. 1963. The successive changes in the skin epithelium of young axolotl (*Siredon mexicanum*). Am. J. Roentgen. Rad. Therapy Nucl. Med. 89 (1): 624-647. (NSA 17: 14000).

Brunst, V. V. 1963. Successive changes in the cornea of young axolotl (*Siredon mexicanum*) after x-irradiation. Radiation Research 20(3): 325-340. (NSA 18: 1418).

Brunst, V. V. 1965. New observations concerning roentgen sensitivity of pigment cells in young axolotls (*Siredon mexicanum*). Am. J. Roentgen. Rad. Therapy Nucl. Med. 93(1): 222-237. (NSA 19: 8812).

Brunst, V. V. 1965. The response of the cornea of the adult axolotl (*Siredon mexicanum*) to x-irradiation. Exptl. Eye Res. 4: 127-134. (NSA 20: 82).

Brunst, V. V. 1965. The response of the olfactory epithelium of the adult axolotl (*Siredon mexicanum*) to roentgen irradiation. Am. J. Roentgen. Rad Therapy Nucl. Med. 94(4): 964-983. (NSA 19: 35965).

Brunst, V. V. 1965. Effects of ionizing radiation on the development of amphibians. Quart. Rev. Biol. 40: 1-67. (NSA 20: 3773).

Brunst, V. V. 1965. Histopathology of the development of the axial skeleton and spinal cord in the irradiated tail of the young axolotl (*Siredon mexicanum*). Am. J. Roentgen. Rad. Therapy Nucl. Med. 95: 992-1012. (NSA 20: 5312).

Brunst, V. V. 1966. Effects of x-radiation on the eye of the one-year-old axolotl (*Siredon mexicanum*). Exptl. Eye Res. 5: 166-172. (NSA 21: 8934).

Brunst, V. V. 1967. The response to the gill epithelium of the adult axolotl (*Siredon mexicanum*) to roentgen radiation. Am. J. Roentgen. Rad. Therapy Nucl. Med. 99: 482-498. (NSA 21: 20353).

Brunst, V. V. 1967. Acute radiation injury to the eye in the adult axolotl (*Siredon mexicanum*). Am. J. Roentgen. Rad. Therapy Nucl. Med. 100: 948-955. (NSA 21: 43521).

Brunst, V. V. 1968. Complete destruction of the eye by strictly local irradiation in adult Mexican salamanders (Siredon mexicanum). Abstr. Radiation Research 35(2): 529-530.

Brunst, V. V. 1968. Lethal effect and visible irradiation damage. Am. J. Roentgen. Rad. Therapy Nucl. Med. 102: 53-70. (NSA 22: 38479) (axolotl).

Brunst, V. V. 1969. Destructive effects of strictly local irradiation of the eye of the adult axolotl (Siredon mexicanum). Radiation Research 39: 26-35. (NSA 23: 35890).

Brunst, V. V. 1969. The effects of partial irradiation on tail regeneration in adult Siredon mexicanum). Am. J. Roentgenol. Rad. Therapy Nucl. Med. 105: 196-206. (NSA 23: 12218).

Brunst, V. V. and F. H. Figge. 1951. The development of secondary tails in young axolotls after local x-ray irradiation. J. Morphol. 89: 111-133. (NSA 5: 6573).

Brunst, V. V., E. A. Sheremetieva-Brunst and F. H. J. Figge. 1953. A comparison of the reactions of the irradiated parts of the bodies of two day old mice and urodele amphibians to roentgen treatment. Am. J. Roentgen. Rad. Therapy Nucl. Med. 70(2): 283-293. (NSA 7: 6345).

Brusse, M. J. 1961. Effects of moisture content of irradiated seeds of Lupinus angustifolius L. New Zeal. J. Agr. Res. 4: 197-202. (NSA 20: 36646).

Brutsärt, P. 1923. Bacteriophage and radium. Compt. rend. soc. biol. 89 (20): 90.

Bruun, P. 1962. Tracing of material movement of sea shores. Shore and Beach (April): 10-15.

Bruun, P. and J. B. Lackey. 1962. Engineering aspects of sediment transport including section on biological aspects. Rev. Engrg. Geol. 1: 39-103. (Use of radionuclides).

Bruynoghe, R. and A. Dubois. 1925. Action of radium on microbes. Compt. rend. soc. biol. 92(3): 211-213.

Bruynoghe, R. and W. Mund. 1925. Irradiated microbes and bacteriophage. Compt. rend. soc. biol. 92: 849-850.

Bryan, G. W. 1963. The accumulation of ^{137}Cs by brackish water invertebrates and its relation to the regulation of potassium and sodium. J. Marine Biol. Assoc. U.K. 43(2): 541-565. (NSA 18: 3463).

Bryan, G. W. 1967. The metabolism of Zn and ^{65}Zn in crabs, lobsters, and freshwater crayfish. pp. 1005-1016. In: Radioecological Concentration Processes. (See Aberg and Hungate, 1967).

Bryan, G. W. and E. Ward. 1962. Potassium metabolism and the accumulation of cesium-137 by decapod crustacea. J. Marine Biol. Assoc. U.K. 42(2): 199-241. (NSA 17: 13937).

Bryan, G. W., A. Preston and W. L. Templeton. 1966. Accumulation of radionuclides by aquatic organisms of economic importance in the United Kingdom. pp. 623-636. In: Disposal of Radioactive Wastes into Seas, Oceans and Surface Wasters. (See IAEA, 1966) (SM-72/39) (CONF-660507) (NSA 20: 40578).

Bryant, F. J., A. Morgan and G. S. Spicer. 1959. The determination of radiostrontium in biological materials. U. K. Atomic Energy Authority, Woolich, report AERE-R-3030. 38pp.

Bryant, P. E. 1968. Survival after fractionated doses of radiation: modification by anoxia of the response of Chlamydomonas. Nature 219: 75-77. (NSA 22: 43483).

Bryant, P. E. 1970. Effect of hypoxia on recovery from sublethal damage in Chlamydomonas. Int. J. Radiat. Biol. 17: 533-539. (NSA 25: 8795).

Bryant, P. E. 1972. Changes in sensitivity of cells during exposure to radiation at low dose rate. Int. J. Radiat. Biol. 22(1): 67-73. (NSA 26: 45754) (Chlamydomonas).

Bryant, P. E. 1973. LET as a determinant of oxygen enhancement ratio and shape of survival curve for Chlamydomonas. Int. J. Radiat. Biol. 23(3): 217-226. (NSA 28: 27682).

Bryant, P. E. 1974. Change in sensitivity of cells after split dose recovery. A further test of the repair hypothesis. Int. J. Radiat. Biol. 26(5): 499-504. (NSA 31: 22934) (Chlamydomonas reinhardii).

Bryant, P. J. 1970. Cell lineage relationships in the imaginal wing disc of Drosophila melanogaster. Develop. Biol. 22: 389-411. (NSA 25: 18967).

Bryant, P. M. 1969. Radioactive wastes. pp. 65-77. In: Technological Injury. (See Rose, 1969) (NSA 24: 25254).

Bryant, P. M. 1971. Derivation and application of limits and reference levels for environmental radioactivity in the United Kingdom. pp. 634-643. In: Health Physics Aspects of Nuclear Facility Siting. III. (See Voillequé and Baldwin, 1971) (NSA 25: 55241).

Bryant, P. M. 1971. Nuclear power industry's need for rapid and accurate estimation of radioactivity in the environment. pp. 33-42. In: Rapid Methods for Measuring Radioactivity in the Environment. (See IAEA, 1971) (NSA 26: 28050).

Bryant, P. M. and J. A. Jones. 1973. Estimation of radiation exposure associated with inert gas radionuclides discharged to the environment by the nuclear industry. pp. 91-107. In: Environmental Behavior of Radionuclides Released in Nuclear Industry. (See IAEA, 1973) (NSA 29: 29202).

Bryant, P. M. and F. Morley. 1971. Disposal of radioactive wastes arising in the United Kingdom from the peaceful uses of atomic energy. U. K. National Radiological Protection Board, Harwell, report NRPB-R-1. 16pp. (NSA 25: 46612) (Incl. food chains).

Bryant, W. C. 1960. Movement of radiophosphorus through the invertebrate community of a trout stream. Thesis, Michigan State Univ., East Lansing.

Bryson, R. A., W. N. Irving and J. A. Larsen. 1965. Radiocarbon and soil evidence of former forest in the southern Canadian tundra. Science 147 (3653): 46-48. (NSA 19: 7721).

Bryson, R. A., J. A. Larsen and W. N. Irving. 1965. Climate and post-glacial forest history in south-western Keewatin, N. W. T.: I. Soils and radiocarbon evidence. Abstr. p. 53. In: Abstr. Intern. Assoc. Quaternary Research. 7th Congr., Boulder and Denver, Colo.

Bryson, R. A., et al. 1969. Radiocarbon isochrones on the disintegration of the Laurentide Ice Sheet. Arctic Alpine Res. 1: 1-14. (NSA 24: 309).

Bubnov, V. D. and I. S. Ivashkov. 1964. Effect of beta rays on larvae of domestic flies. Tr. nauchn.-issled. vet. inst. Minsk 24: 339-___. (In Russian).

Buch, A. 1973. Planning and Site Selection of Power Plants. Concept of the Generation of Electricity in the Federal Republic Taking the Influence of the Environment into Account. Krausskopf, Mainz. 126pp. (NSA 30: 8926).

Buchanan, P. D. 1970. Genetic responses of Habrobracon to radiation in combination with clinostat rotation, vibration, and centrifugation. Thesis, North Carolina State Univ., Raleigh. 84pp. (Univ. Microfilms 71-18,616) (NSA 25: 46957).

Bücher, T. 1965. Formation of the specific structural and enzymic pattern of the insect flight muscle. pp. 15-28. In: Aspects of Insect Biochemistry. T. W. Goodwin, ed. Academic Press, New York. (^{14}C, Locusta).

Buchholtz, C. 1964. Electron microscopic findings in the irradiated supra-esophageal ganglion of Odonata larvae (Calopteryx splendens Haar.). Z. Zellforsh. 63: 1-21. (In German) (NSA 19: 43849).

Buchholtz, C., et al. 1967. Neuroethologic studies on Calopteryx splendens Haar. (Odonata) after x irradiation of the central nervous system. Mikroskop. anat. 82: 282-306. (In German) (NSA 22: 19430).

Buchner, W. 1957. Continuous measurement of radioactive substances in the water. Atompraxis 3(10): 405-409.

Buchold, B., et al. 1971. Some effects of spaceflight on the flour beetle, Tribolium confusum. U. S. NASA report N-72-20053. 17pp. (NSA 26: 50956).

Buchsbaum, R. 1958. Species response to radiation; radioecology. Ch. 5, pp. 124-141. In: Radiation Biology and Medicine. (See Claus, 1958).

Bücker, H. and W. Hanke. 1957. The reaction of the capillaries and of the web tissue of the foot of the frog (Rana tempoaria L.) to UV radiation. Z. Naturforsch. 12b(10): 629-641.

Bücker, H. and W. Hanke. 1958. Comparison of the effect of ultraviolet and x-rays on the capillaries of the frog's web. Nature 182(4646): 1375-1376.

Buckheim, C. E. 1962. Lethal effects in tadpoles of Xenopus laevis (Daudin) after irradiation with high energy electrons. Nature 196(4857): 902-903. (NSA 17: 7861).

Buckhold, B. 1969. Biosatellite II: physiological and somatic effects on insects. pp. 77-83. In: Life Sciences and Space Research VII. W. Vishiac and F. G. Favorite, eds. North-Holland, Amsterdam. (NSA 24: 830).

Buckhold, B. and J. V. Slater. 1969. Effect of temperature and x-irradiation on pupae of the flour beetle, Tribolium confusum. Radiation Research 37: 567-576. (NSA 23: 20335).

Buckley, P. M., et al. 1969. Inactivation of Rhizopus stolonifer sporangiospores by single and combined treatments of heating, chilling, and gamma irradiation. Radiation Research 40: 26-36. (Also: Univ. of California, Davis, U. S. AEC report UCD-34-P-73-24) (NSA 24: 913).

Buckner, M. R. and D. W. Hayes. 1975. Pollutant transport in natural streams. pp. IV123-IV138. In: Proc. Conf. on Computational Methods in Nuclear Engineering (Charleston, S. C., April 15, 1975). U. S. ERDA report CONF-750413-P2. (NSA 32: 6435).

Budakov, N. P. 1958. Calcium accumulation and loss by yearling carp in relation to calcium compound concentration in the surrounding medium. Dokl. Akad. Nauk SSSR 120(3): 661-668. (Doklady (biol. sci.) 120: 349-352).

Budakov, N. P. 1958. Consumption of radioactive and total calcium by yearling carp. Nauch.-tekhn. byull. vses. nauch.-issled. inst. ryb. khoz. (6-7): ____. (In Russian).

Buddemeier, R. W., et al. 1972. Effects of solution exchange on the radiocarbon dating of sediments and natural waters. pp. C73-C86. In: Proc. 8th Intern. Radiocarbon Dating Conf. Vol. 1. (See Rafter, 1972) (NSA 31: 8695).

Budnitz, R. J. 1973. Krypton-85: a review of instrumentation for environmental monitoring. Lawrence Berkeley Lab., Univ. of Calif., U. S. AEC report LBL-1779. 18pp. (For Symp. on Noble Gases) (CONF-730915-1) (NSA 28: 18233).

Budnitz, R. J. 1973. Environmental measurements and monitoring programs around reactors: pre-operational, operational, and for accidents. Lawrence Berkeley Lab., Univ. of Calif., U. S. AEC report LBL-2195. (NSA 29: 5103).

Budnitz, R. J. 1973. Plutonium: a review of measurement techniques for environmental monitoring. Lawrence Berkeley Lab., Univ. of Calif., U. S. AEC report LBL-2039. 9pp. (Also: 1974, Trans. Nucl. Sci. NS-21(1): 430-437. (NSA 29: 7705; 30: 6493).

Budnitz, R. J. 1974. Radon-222 and its daughters. A review of instrumentation for occupational and environmental monitoring. Health Physics 26 (2): 145-163. (NSA 29: 18328).

Budnitz, R. J. 1974. Tritium instrumentation for environmental and occupational monitoring. A review. Health Physics 26(2): 165-178. (NSA 29: 18327).

Budnitz, R. J. 1974. Pre-operational and operational environmental measurements and monitoring programs around power reactors. Lawrence Berkeley Lab., Univ. of Calif., U. S. AEC report LBL-2195 (Rev.). 17pp. (NSA 31: 3325).

Buech, R. R. 1973. Sensitivity of the red-backed vole (*Clethrionomys g. gapperi*) to acute gamma radiation. pp. 1082-1086. In: *Radionuclides in Ecosystems*. (See Nelson, 1973) (NSA 28: 30141).

Buell, M. F. 1970. Time of origin of New Jersey Pine Barrens bog. Bull. Torrey Botan. Club 97(2): 105-108. (^{14}C).

Buenafama, H. D. 1973. Neutron activation analysis of soils. Radioanal. Chem. 18(1-2): 111-118. (NSA 30: 17).

Bugbee, S. L. 1967. A portable vacuum pump for ^{14}C productivity studies. Limnol. Oceanogr. 12(4): 707-708.

Bugher, J. C. and M. Taylor. 1949. Radiophosphorus and radiostrontium in mosquitoes. Science 110(2849): 146-147. (NSA 3: 1194).

Bugher, J. C., J. Coursaget and J. F. Loutit. (Eds.). 1959. *Biological Sciences*. *Progress in Nuclear Energy*. *Vol. 2*. *Series VI*. Pergamon Press, New York. x, 422, XIIpp. (NSA 13: 17757).

Bughio, A. R., Z. A. Qureshi and A. K. Mecci. 1969. Effects of gamma radiation on eggs of *Dacus zonatus* (Saunders). Int. J. Rad. Biol. 16: 183-192. (NSA 24: 6106).

Buhler, D. R. (Ed.). 1973. *Mercury in the Western Environment*. Oregon State Univ., Corvallis. x, 360pp. (Incl. neutron activation).

Bühringer, H. 1973. Radiosensitivity of rainbow trout (*Salmo gairdnerii* Rich.) during embryonal and larval development. Arch. Fischereiwiss. 24 (1/3): 51-63. (In German) (NSA 30: 3634).

Bukatsch, F. 1938. The effects of radon and mineral materials upon photosynthesis of submerged plants. Planta 28: 264-274. (CA 32: 7515).

Büker, H., et al. 1973. Nuclear energy and the environment. Kernforschungsanlage, Jülich, Germany, report KFK-1366. 126pp. (JÜL-929-HT-WT) (In German) (NSA 28: 14598).

Buldakov, L. A. 1961. Distribution of ruthenium-106 in mammals and amphibians. In: *Distribution, Biological Effects, and Migration of Radioactive Isotopes*. A. V. Lebedinskiy and Ya. I. Moskalev, eds. Medgiz, Moscow. (AEC-tr-7512; pp. 93-97) (NSA 30: 7153) (Frog).

Bull, D. L. 1964. Metabolism of di-syston in insects and rats. Abstr. 96. Bull. Entomol. Soc. Am. 10(3): 165. (^{32}P, *Heliothis zea*).

Bull, D. L. 1965. Metabolism of di-syston by insects, isolated cotton leav leaves, and rats. J. Econ. Entomol. 58(2): 249-254. (^{32}P, *Heliothis zea*, *Anthonomus grandis*).

Bull, D. L. 1967. Absorption and metabolism of Trichorfon by insects. Bull. Entomol. Soc. Am. 13(3): 189. (^{32}P).

Bull, D. L. 1972. Metabolism of organophosphorus insecticides in plants. pp. 25-33. In: *Radiotracer Studies of Chemical Residues in Food and Agriculture*. (See IAEA, 1972) (NSA 27: 20186) (^{32}P).

Bull, D. L. and P. L. Adkisson. 1963. Absorption and metabolism of C^{14}-labelled DDT by DDT-susceptible and DDT-resistant pink bollworm adults. J. Econ. Entomol. 56(5): 641-643.

Bull, D. L. and D. A. Lindquist. 1964. Metabolism of 3-hydroxy-N,N-dimethylcrotonamide dimethyl phosphate by cotton plants, insects, and rats. J. Agric. Food Chem. 12: 310-317. (^{32}P, ^{14}C).

Bull, D. L. and D. A. Lindquist. 1965. The effects of chronic doses of organophosphorus inhibitor on chlinesterase activity in boll weevils. Experientia 21(5): 262-263. (^{14}C, *Anthonomus grandis*).

Bull, D. L. and D. A. Lindquist. 1965. A comparative study of insecticide metabolism in photoperiod-entrained and unentrained bollworm larvae *Heliothis zea* (Boddie). Comp. Biochem. Physiol. 16(3): 321-325. (^{32}P).

Bull, D. L. and D. A. Lindquist. 1966. Metabolism of 3-hydroxy-*N*-methyl-cis-crotonamide dimethyl phosphate (Azodrin) by insects and rats. J. Agric. Food Chem. 14(2): 105-109. (^{14}C, ^{32}P).

Bull, J. O. 1960. Insect eradication by sterile male release. An appraisal of the method for control of Australian sheep blowfly. pp. 117-129. In: *The Technological Use of Radiation*. Australia Atomic Energy Commission.

Bull, J. O. and P. B. Cornwell. 1966. The effect of culture environment on the susceptibility of *Sitophilus granarius* (L.) to gamma radiation. pp. 57-69. In: *The Entomology of Radiation Disinfestation of Grain*. (See Cornwell, 1966).

Bull, J. O. and P. B. Cornwell. 1966. A comparison of the susceptibility of the grain weevil *Sitophilus granarius* (L.) to accelerated electrons and ^{60}Co gamma radiation. pp. 157-175. In: *The Entomology of Radiation Disinfestation of Grain*. (See Cornwell, 1966).

Bull, J. O. and T. Wond. 1962. Control of the Mediterranean flour moth *Anagasta kuhniella* Zell. by sterile male release. I. Biological studies related to large scale rearings. U. K. Atomic Energy Authority, Wantage, report AERE-R-3895. 16pp.

Bull, J. O., T. Wond and P. B. Cornwell. 1961. A comparison of the susceptibility of the grain weevil (*Sitophilus granarius* L.) to accelerated electrons and Co^{60} gamma radiation. U. K. Atomic Energy Authority, Wantage, report AERE-R-3890. 45pp.

Bullard, R. W. 1964. Changes in regional blood flow and blood volume, during arousal from hibernation. Ann. Acad. Scient. Fenn. A,IV(biol.) 71: 65-74. (^{86}Rb, ground squirrels).

Bullard, R. W. 1970. Variation of chemical concentration in surface-coated and tableted grain baits. J. Wildl. Mgmt. 34(4): 925-929. (^{14}C).

Bulley, N. R. and E. B. Tregunna. 1970. Sensitivity of the infrared CO_2 gas analyzer to $^{14}CO_2$. Can. J. Bot. 48(6): 1292.

Bulliere, D. 1969. Effects of x rays on regeneration of the tarsus in an insect. Compt. rend. D269: 898-901. (In French) (NSA 24: 7984).

Bullio, P. 1972. Nuclear energy and the environment. Ind. At. Spatiales 16(2): 54-58. (In French) (NSA 26: 58795).

Bulon, V. V. 1972. Use of Sorokin's apparatus to determine the rate of gas exchange in aquatic animals by the radiocarbon method. Gidrobiol. zh. 8 (4): 91-__. (Hydrobiology J. 8(4): 75-81).

Bultynck, H., et al. 1970. Atmospheric dilution factors and calculation of doses in the environment of SCK/CEN Mol for short and long duration stack discharges. Belgium Centre d'Étude de l'Énergie Nucléaire, Mol, report BLG-446(Vol. 1, Vol. 2). 90pp., 113pp. (NSA 25: 48619, 48620).

Bulyginskaya, M. A. 1963. The possibility of utilizing gamma rays for sexual sterilization of the mallow moth (Pectinophora malvella Hb.). Dokl. Akad. Nauk SSSR __: 76-78. (Ref. Zh. (biol.) 1964: 1E180).

Bunce, L. A. and F. W. Sattler. 1966. Radon-222 in natural gas. Radiol. Hlth. Data 7(8): 441-444.

Bunnenberg, C. and W. Kühn. 1974. Movement of water vapor in arid loess soil under conditions of temperature inversions. pp. 89-95. In: Isotope and Radiation Techniques in Soil Physics and Irrigation Studies 1973. (See IAEA, 1974) (NSA 30: 15487) (^{3}H).

Burbank, W. D., R. Grabske and J. R. Comer. 1964. The use of the radioisotope, zinc 65, in a preliminary study of population movements of the estuarine isopod Cyathura polita. Crustaceana 7: __.

Burch, P. R. J., et al. 1964. Studies of the environmental radiation at a particular site with a static γ-ray monitor. pp. 767-780. In: The Natural Radiation Environment. (See Adams and Lowder, 1964) (NSA 18: 35747).

Burch, W. D. 1973. Nuclear waste disposal. pp. 119-127. In: Proc. 4th Annual Symp. on Environmental Pollution. (Edgewood Arsenal, Md., 4-5 April 1973). Edgewood Arsenal, Md. (CONF-730417) (NSA 32: 19685).

Burchell, J. D., J. H. Gordon and A. A. Downs. nd. ^{137}Cs in the Canadian north. pp. 3-13. In: Data from Radiation Protection Programs. Canada Dept. of National Health and Welfare, Ottawa. (NP-16900) (NSA 21: 45614).

Burchill, B. R. 1966. Effects of metabolic inhibitors and radiations on regeneration in the ciliate Stentor coeruleus. Thesis, Western Reserve Univ. 142pp. (NSA 21: 45593).

Burchill, B. R. 1968. Effects of radiation on oral regeneration in Stentor coeruleus. J. Exptl. Zool. 169: 471-480. (NSA 23: 41425).

Burchill, B. R. and R. C. Rustad. 1969. Ultraviolet-microbeam irradiation of regenerating Stentor. J. Protozool. 16(2): 303-306. (NSA 23: 48191).

Burditt, A. K., Jr. and S. T. Seo. 1971. Dose requirements for quarantine treatment of fruit flies with gamma radiation. pp. 33-41. In: Disinfestation of Fruit by Irradiation. (See IAEA, 1971).

Burditt, A. K., Jr., et al. 1975. Application of sterilization techniques to Anastrepha suspensa Loew in Florida, United States of America. pp. 93-101. In: Sterility Principle for Insect Control, 1974. (See IAEA, 1975) (NSA 32: 11767).

Burdon, D. J., et al. 1963. The use of tritium in tracing karst groundwater in Greece. pp. 309-320. In: Radioisotopes in Hydrology. (See IAEA, 1963) (NSA 18: 1925).

Burgess, E. E. 1969. Physical, chemical, and biological control of the alfalfa weevil (Hypera postica: Curculionidae). Thesis, Univ. of Tennessee, Knoxville. 138pp. (Univ. Microfilms 70-17,802) (NSA 25: 26913).

Burgess, E. E. and S. E. Bennett. 1966. Sterilization of the male alfalfa weevil (Hypera postica: Curculionidae) by x radiation. J. Econ. Entomol. 59: 268-270. (NSA 20: 35279).

Burgess, E. E. and S. E. Bennett. 1971. Mortality and abnormalities caused by gamma irradiation of alfalfa weevil larvae. Ann. Entomol. Soc. Am. 64: 88-90. (NSA 25: 16289).

Burgess, E. E. and S. E. Bennet. 1972. Mortality and abnormalities caused by gamma irradiation of alfalfa weevil pupae. Ann. Entomol. Soc. Am. 65 (6): 1331-1333. (NSA 27: 5549).

Burgkhardt, B. and E. Piesch. 1972. Use of CaF_2 thermolumenescent dosimeters for measuring the natural background radiation. Kerntechnik 14(3): 128-134. (In German, English) (NSA 26: 33455).

Burke, G. de P. 1974. Model for estimating variations in terrestrial radiation exposures based on soil-water balance. pp. 140-147. In: 2nd Workshop on the Natural Radiation Environment. (See Lowder, 1974) (NSA 31: 3371).

Burke, G. de P. and A. Shambon. 1972. Investigation of thermoluminescent dosimeters for environmental monitoring. Health and Safety Lab., U. S. AEC report HASL-265. 15pp. (NSA 27: 5327).

Burkholder, P. R. 1963. Radioactivity in some aquatic plants. Nature 198 (4880): 601-603. (NSA 17: 23053).

Burkholder, P. R. and N. H. Giles, Jr. 1947. Induced biochemical mutations in Bacillus subtiles. Am. J. Bot. 34(6): 345-348.

Burkholder, W. E., E. W. Tilton and R. R. Cogburn. 1966. Effects of gamma radiation on the grain mite, Acarus siro. J. Econ. Entomol. 59(4): 976-980.

Burkser, R., M. Shapiro and K. Bronstein. 1929. Radium treatment of foods. Biochem. Z. 211: 323-325. (In German) (Used *Cottus gobio*).

Burlakova, Ye. V., M. L. Kakushkina and N. V. Korneyeva. 1963. X-radiation effects on *Artemia salina* L. eggs. Nauch. Dokl. Vyssh. Shkoly biol. nauk (3): 83-87. (In Russian) (NSA 20: 35291).

Burley, J. W. A., G. E. Gilbert and L. N. Clum. 1962. Preliminary radiological investigations of the vegetation and soils of Neotoma. Neotoma Ecol. and Bioclim. Lab., Ohio State Univ., Columbus, and Ohio Agr. Expt. Sta., Wooster, Special Report No. 10. 39pp.

Burley, W. A. and K. L. Webb. 1963. Dark fixation of carbon-dioxide-14 by salt marsh halophytes. Abstr. Plant Physiol. 38(Suppl.): xl.

Burnett, A. L, R. Baird and F. Diehl. 1962. Method of introducing tritiated thymidine into the tissues of hydra. Science 138(3542): 825-826.

Burnett, A. M., W. H. Mason and S. T. Rhodes. 1969. Reingestion of feces and excretion rates of Zn^{65} in *Popilius disjunctus* versus *Cryptocercus punctulatus*. Ecology 50(6): 1094-1096.

Burnett, W. D., et al. 1961. Radioactive environmental survey at Sandia Corporation. Sandia Corp., Albuquerque, N. Mex., U. S. AEC report SC-4628(M). 34pp. (NSA 15: 29547).

Burnham, J. B., et al. 1969. Nuclear siting survey of the Pacific Northwest and its impact on transmission planning. Proc. Amer. Power Conf. 30: 1169-1180. (NSA 23: 28403).

Burns, R. H. 1972. Radioactive waste management and consideration of the environment. At. Jap. 16(5): 59-80. (NSA 27: 23008).

Burovina, I. V., et al. 1965. Concentration of common caesium in animal and human muscles. Nature 205(4976): 1116-1117. (NSA 19: 25833) (Incl. comments on relationship to ^{137}Cs; fishes, birds, reindeer).

Burov, D. and M. Genkova. 1973. Possibilities of applying radiation sterilization in the control of the cherry fruit fly (*Rhagoletis cerasi* L.). Gradinar. Lozar. Nauka 10: 15-21. (In Bulgarian) (NSA 31: 6342).

Burrage, R. H. and J. G. Saha. 1972. Insecticide residues in pheasant after being fed on wheat seed treated with heptachlor and ^{14}C-lindane. J. Econ. Entomol. 65(4): 1013-1017.

Burris, R. H. 1950. Isotopes as tracers in plants. Botan. Rev. 16: 150-180. (NSA 4: 2976).

Burrows, W. D. and P. A. Krenkel. 1973. Studies on uptake and loss of methylmercury-203 by bluegills (*Lepomis macrochirus* Raf.). Environ. Sci. Technol. 7(13): 1127-1130. (SFA 19: 17234).

Burson, Z. G. 1973. Airborne surveys of terrestrial gamma radiation in environmental research. EG & G, Inc., U. S. AEC report EGG-1183-1621. 16pp. (NSA 30: 32027).

Burson, Z. G. 1974. Airborne surveys of terrestrial gamma radiation in environmental research. Trans. Nucl. Sci. NS-21(1): 558-571. (NSA 30: 6733).

Burson, Z. G. (Comp.). 1974. Research progress in airborne surveys of terrestrial gamma radiation. pp. 70-96. In: *2nd Workshop on the Natural Radiation Environment*. (See Lowder, 1974) (NSA 31: 3322).

Burson, Z. G. and A. E. Fritzsche. 1973. Water equivalent of snow data from airborne gamma radiation surveys: International Field Year for the Great Lakes. EG & G, Inc., U. S. AEC report EGG-1183-1622. 60pp. (NSA 31: 3392).

Burstein, S. 1973. Power plant siting on lakes. Wisconsin Electric Power Co., Milwaukee. vp. (NP-19964, Paper 11) (NSA 30: 6571).

Burton, G. W. 1957. Role of tracers in root development investigations. pp. 71-80. In: *Atomic Energy and Agriculture*. C. L. Comar, ed. Am. Assoc. Adv. Sci., Washington, Publ. 49.

Burton, G. W. 1974. Radiation breeding of warm season forage and turf grasses. pp. 35-39. In: *Polyploidy and Induced Mutations in Plant Breeding*. International Atomic Energy Agency, Vienna, Austria, report STI/PUB-359. (NSA 30: 3556).

Burton, G. W. and J. E. Jackson. 1962. Radiation breeding of apomictic prostrate dallisgrass, *Paspalum dilatatum* var. *paucicilliatum*. Crop Sci. 2 (6): 495-497. (NSA 17: 7831).

Burton, G. W., E. H. deVane and R. L. Carter. 1954. Root penetration, distribution and activity in southern grasses measured by yields, drought symptoms and P-32 uptake. Agronomy J. 46(5): 229-233.

Burton, J. D., W. B. Guenther and H. M. Owen. 1964. Profile of natural radioactivity in Hartsells Loam. Soil Sci. Soc. Am. Proc. 28(4): 500-501. (NSA 18: 35723).

Burton, P. R. 1962. *In vitro* uptake of radioglucose by a frog lung fluke a and correlation with histochemical identification of glycogen. J. Parasitol. 48(6): 874-882. (^{14}C).

Burton, P. R. 1963. The uptake of radiosulphate by *Haematoloechus medioplexus*. J. Parasitol. 49(1): 121-122. (RDA-7: 128) (*Rana pipiens*).

Busdiecker, R. L. and B. W. Maxwell. 1964. Environmental radiation surveillance in the Antarctic. pp. 873-884. In: *The Natural Radiation Environment*. (See Adams and Lowder, 1964) (NSA 18: 35749).

Buscarlet, L. A. 1970. Sterile male technique for insect self destruction. France Commissariat à l'Énergie Atomique, Cadarache, report CEA-R-3980. 32pp. (In French) (NSA 24: 39594).

Buscarlet, L. A. 1970. Biological half-life of radioelements in insects. Ann. Zool. Ecol. Anim. 2(3): 451-470. (In French) (NSA 25: 49075).

Buscarlet, L. A., G. Lasceve and J. Garcin.1974. Use of kinetics of ^{22}Na to estimate the food consumption if Sitophilus granarius in different temperature and feeding conditions. Int. J. Appl. Radiat. Isotop. 25(10): 44 445-453. (In French) (NSA 31: 8978).

Bushland, R. C. 1956. Use of radiation in insect control. pp. 711-715. In: Proc. 10th Congr. Entomology. Vol. 3. (Montreal, Canada, August 17-25, 1956).

Bushland, R. C. 1957. Male sterilization for the control of insects. Adv. Pest Control Res. 3: 1-25. (NSA 19: 33799).

Bushland, R. C. 1959. Insect eradication by release of sterilized males. Abstr. Int. J. Appl. Radiat. Isotop. 7(1): 65.

Bushland, R. C. 1960. Insect eradication by release of sterilized males. Levre T 389. pp. 273-290. In: Large Radiation Sources in Industry. Vol. 2. International Atomic Energy Agency, Vienna, Austria.

Bushland, R. C. 1962. Research to support the southwestern screwworm eradication program. Abstr. Bull. Entomol. Soc. Am. 8(3): 164.

Bushland, R. C. 1966. Radiation sterilization of the black blow fly. Bull. Entomol. Soc. Am. 12: 289.

Bushland, R. C. 1971. Sterility principle for insect control. Historical development and recent inovations. pp. 3-14. In: Sterility Principle for Insect Control or Eradication. (See IAEA, 1971) (NSA 25: 29736).

Bushland, R. C. and D. E. Hopkins. 1951. Experiments with screw worms sterilized with x-rays. J. Econ. Entomol. 44(5): 725-731.

Bushland, R. C. and D. E. Hopkins. 1953. Sterilization of screw worm flies with x-ray and gamma rays. J. Econ. Entomol. 46(4): 648-656. (NSA 8: 50).

Bushland, R. C. and M. Streeter. 1971. Laboratory tests with gamma-irradiated Phormia regina. J. Econ. Entomol. 64(1): 84-87. (NSA 25: 21284).

Bushland, R. C., A. W. Lindquist and E. F. Knipling. 1955. Eradication of screw-worms through release of sterilized males. Science 122(3163): 287-288.

Bustad, L. 1960. Radioisotopes and environmental circumstances: significance of nuclear industry effluents in animal populations. pp. 243-254. In: Radioisotopes in the Biosphere. (See Caldecott and Snyder, 1960).

Butler, C. P. 1962. The light of the atom bomb. Science 138(3539): 483-489. (Birds).

Butler, E. 1935. Studies on limb regeneration in x-rayed Amblystoma larvae. Anat. Record 84(4): 407-413.

Butler, E. E. and J. B. Rowell. 1951/52?. Abnormal spore germination and microthallus development of fungi induced by dilute colloidal polonium. Abstr. Univ. of Minnesota, U. S. AEC report AECU-1813. 2pp. (NSA 6: 1073).

Butler, E. G. 1933. The effects of x-radiation on the regeneration of the forelimb of Amblystoma larvae. J. Exptl. Zool. 65(3): 271-315.

Butler, E. G. 1935. Studies on limb regereration in x-rayed Amblystoma larvae. Anat. Record 62(3): 295-307.

Butler, E. G. and J. P. O'Brien. 1942. Effect of localized x-irradiation on the urodele limb. Anat. Record 84(4): 407-413.

Butler, E. G., H. F. Blum and S. E. Schmidt. 1957. The localized character of ultraviolet effects on the urodele forelimb. J. Cell. Comp. Physiol. 50(3): 381-388.

Butler, F. E. 1962. Sr^{90} monitoring at the Savannah River Plant. Health Physics 8(3): 273-277.

Butler, G. C. and P. J. Barry. 1971. Surveillance of environmental contamination : concepts developed for ionizing radiations. pp. 723-732. In: Proc. Intern. Symp. Radioecology Applied to the Protection of Man and His Environment. (See Commission of the European Communities, 1971) (NSA 27: 5236).

Butler, G. C., et al. 1972. Assessment and control of environmental contamination: experience with artificial radioactivity. Biol. Conserv. 4(3): 177-184. (NSA 26: 45554).

Butler, G. W. 1962. The use of isotopes in soil and plant research in New Zealand. Int. J. Appl. Radiat. Isotopes 13: 343-351. (Statement on natural radioactivity and ^{14}C dating).

Butt, B. A. 1970. Field releases of codling moths sterilized by TEPA or by gamma radiation, 1964-67. J. Econ. Entomol. 63: 912-915. (NSA 24: 34825).

Butt, B. A. 1973. Resume of requirements for a sterile insect release program exclusive of basic laboratory and field cage studies. pp. 145-148. In: Computer Models and Application of the Sterile-Male Technique. (See IAEA, 1973) (NSA 29: 8057).

Butt, B. A. and D. O. Hathaway. 1964. Insect control by the sterile male technique. Wash. State Hort. Assoc. Proc. 60th Ann. Meet. pp. 83-86.

Butt, B. A. and L. F. Steiner. 1969. Methods of estimating the size of populations of codling moths, Carpocapsa pomonella (L.), for sterile-moth release programs. pp. 57-61. In: Insect Ecology and the Sterile-Male Technique. (See IAEA, 1969)

Butt, B. A., et al. 1970. Field releases of codling moths sterilized by TEPA or by gamma irradiation. J. Econ. Entomol. 63(3): 912-915.

von Buttlar, H. and W. F. Libby. 1955. Natural distribution of cosmic-ray produced tritium. II. J. Inorg. Nucl. Chem. 1: 75-91.

von Buttlar, H. and B. Wiik. 1965. Enrichment of tritium by thermal diffusion and measurement of dated Antarctic snow samples. Science 149(3690): 1371-1373. (NSA 19: 46613).

von Buttlar, H., K. Farzine and H.-D. Wohlfahrt. 1965. Tritium concentration of German river waters measured with the proportional-counting technique. Nucl. Instrum. Methods 37: 288-292. (NSA 5654).

Butzel, H. M., Jr. and W. J. van Wagtendonk. 1963. Autoradiographic studies of the differential incorporation of glycine, purine and pyrimidine ribosides by Paramecium aurelia. J. Gen. Microbiol. 30(3): 503-507. (^{14}C).

Buyevich, Ye. A., Ye. M. Karabayev and V. I. Korogodin. 1963. The selection of a model describing the recovery of vitality of yeast cells damaged by gamma irradiation. Radiobiologiya 3(2): 197-203. (AEC-tr-5435; pp. 68-77) (NSA 17: 30360).

Buyske, D. A., et al. 1963. Determination of tritium and C-14 in biological samples by rapid combustion techniques. Vol. I, pp. 185-191. In: Advances in Tracer Methodology. S. Rothchild, ed. Plenum Press, New York.

Buzzati-Traverso, A. A. 1960. Velocity of microevolutionary processes in irradiated populations. Atti Ass. genet. Ital. 5: 183-184. (In Italian, English summary).

Bychkovskaya, I. B. 1959. The problem of the retention of radiation injury in epithelial cells of corneae of frogs irradiated during hibernation. Tsitologiya 1(4): 387-___. (JPRS-2287, pp. 49-58) (NSA 14: 23939).

Bychkovskaya, I. B. 1960. Some data on the mechanism of "conservation" of the radiation effect in the corneal epithelium of the frog under hibernation conditions. Med. radiologiya 5(3): 73. (JPRS-5124, pp. 183-186) (NSA 14: 13619, 25300).

Bychkovskaya, I. B. 1966. Dynamics of lethal radiation effects in various biological species. Radiobiologiya 6(1): 39-45. (AEC-tr-6770; pp. 56-65).

Bychkovskaya, I. B. 1970. Dynamics of Post-Radiation Death in Biological Organisms. Atomizdat, Moscow. 201pp. (In Russian) (NSA 25: 24372).

Bychkovskaya, I. B. and G. K. Ochinskaya. 1964. Study of the "Oxygen Effect" at various radiation dose rates. Radiobiologiya 4(1): 63-66. (AEC-tr-6404; pp. 85-90) (Calandra granaria).

Bychkovskaya, I. B. and G. K. Ochinskaya. 1964. Analysis of the dependence of the protective action of hypoxia on the radiation dose. Radiobiologiya 4(2): 203-209. (AEC-tr-6405; pp. 29-37) (NSA 18: 31296) (Calandra granarius).

Bychkovskaya, I. B. and G. K. Ochinskaya. 1964. Oxygen effect in x- and gamma-irradiation of grain beetles. Radiobiologiya 4(6): 928-929. (AEC-tr-6409; pp. 165-166) (NSA 19: 17622).

Bychkovskaya, I. B. and G. K. Ochinskaya. 1965. Absence of "oxygen aftereffects" in Calandra granaria during hibernation induced with low temperature. Radiobiologiya 5(5): 700-702. (AEC-tr-6602; pp. 101-104) (NSA 20: 5385).

Bychkovskaya, I. B. and G. K. Ochinskaya. 1966. Concerning the system in the reaction of biological objects to radiation. Experiments on granary weevils. Zool. zh. 45: 134-137. (In Russian) (NSA 20: 45466).

Bychkovskaya, I. B. and G. K. Ochinskaya. 1972. Dynamics of post-radiation death in amoebae related to the irradiation dose. Tsitologiya 14(1): 137-140. (In Russian) (NSA 26: 36447).

Bychkovskaya, I. B. and G. K. Ochinskaya. 1972. Viability of progeny of irradiated paramecia during vegetative reproduction and after conjugation. Radiobiologiya 12(5): 767-770. (AEC-tr-7415; pp. 175-178) (NSA 27: 27984).

Bychkovskaya, I. B. and G. K. Ochinskaya. 1973. Viability of the progeny of irradiated cells (experiments on Paramecium caudatum and Amoeba proteus). Radiobiologiya 13(2): 211-215. (AEC-tr-7444; pp. 63-68) (NSA 29: 24844).

Bychkovskaya, I. B. and G. K. Ochinskaya. 1973. The significance of the toxic effect of an irradiated medium in the postradiation death of cells. 2. Experiments on Paramecium caudatum. Radiobiologiya 13(5): 680-685. (AEC-tr-7507, pp. 37-43) (NSA 29: 27113).

Bychkovskaya, I. B. and R. B. Strelkov. 1967. Study of the protective effect of beta-mercaptoethylamine in experiments on frogs (concerning the problem of the method of estimating the effectiveness of radioprotective agents). Radiobiologiya 7(3): 452-454. (AEC-tr-6888; pp. 222-226) (NSA 22: 10816).

Bychkovskaya, I. B., G. K. Ochinskaya and A. I. Ryumtseva. 1971. Significance of the toxic effects of an irradiated medium in the postradiation death of cells. I. Experiments on Amoeba proteus. Radiobiologiya 11(2): 235-238. (AEC-tr-7269; pp. 113-119) (NSA 25: 44942).

Bychkovskaya, I. B., et al. 1972. Stability of radiation injury in cell population of Paramecium caudatum. Dokl. Akad. Nauk SSSR 205(1): 213-215. (In Russian) (NSA 27: 5552).

Bykovskiy, V. A. 1961. Application of radiophosphorus for self-labelling in warm-blooded biocoenosis. pp 32-33. In: Materials of the Symposium on Use of Biophysics in the Field of Plant Protection. Leningrad. (In Russian) (Ref. zh. (biol.) 1963: 31282; BA 45: 27280; WR 116: 7).

Bylinkina, A. A. 1968. Content and distribution of long-lived fission products in the upper Volga reservoirs. pp. 229-239. In: Chemistry of Inland Waters and Factors of Their Pollution and Purification. Publ. House Leningrad Sect. Acad. Sci. USSR, Leningrad. (In Russian) (NSA 24: 25435).

Byrne, A. R., M. Dermelj and L. Kosta. 1971. Neutron activation study of environmental contamination and distribution of mercury in animals and fish. pp. 415-427. In: Nuclear Techniques in Environmental Pollution. (See IAEA, 1971) (NSA 25: 31881).

Cabellero, F., et al. 1967. Use of radioactive isotopes for labelling of Ceratitis capitata. Spain Instituto Nacional de Investigaciones Agronomicas, Madrid. 22pp. (For Symposium on Applications of Radioisotopes, Madrid, June 19-21, 1967) (In Spanish) (CONF-670649-7) (NSA 23: 22355).

Cadahia, D. 1965. Ionizing radiations in applied entomology. Boln Serv. Plagas For. 8(15): 15-27. (In Spanish).

Cadahia, D. 1966. Ionizing radiation in applied entomology. Montes 22: 159-168. (In Spanish) (NSA 22: 10881).

Cadeddu, M. and R. Canazi. 1961. Survey of the radioactivity from fallout in geographic water, rainwater, and in milk samples. Energia Nucleare (Milan) 8: 178-187. (In Italian) (NSA 15: 18345).

Cadet, J. 1966. Distribution of a fission product mixture in a natural environment. Thesis, Univ. of Grenoble, France. 46pp. (In French) (NP-17131) (NSA 22: 12843).

Cadwell, L. L. 1973. Rangeland grasshopper foraging impact. Thesis, Colorado State Univ., Fort Collins. xiii, 109pp. (^{32}P).

Cadwell, L. L. 1973. Colony formation of the western harvester ant in a chronic gamma radiation field. Am. Midl. Naturalist 89(2): 446-488.

Caggiano, M. T. and J. F. Facetti. 1971. Determination of ^{90}Sr and ^{137}Cs in some elements in the Paraguayan diet. Rev. Soc. Cient. Paraguay 12(1-2): 5-10. (In Spanish) (NSA 26: 20358).

Cahu, Y. K. 1971. Determination of trace metals in natural waters. pp. 354-357. In: Identification and Measurement of Environmental Pollutants. B. Westley, ed. National Research Council of Canada, Ottawa. (NSA 29: 2443).

Caillot, A. 1970. Methods for labeling sediments with radioactive indicators. Houille Blanche 25: 661-675. (ORNL-tr-2501) (NSA 25: 41659).

Caillot, A. 1973. Sediment labelling with radioactive isotopes. pp. 169-176. In: Tracer Techniques in Sediment Transport. (See IAEA, 1973) (NSA 28: 31140).

Caillot, A., et al. 1970. Experimental gages for the determination of sediment settling relations and core sample density. Houille Blanche 25(8): 721-724. (In French) (NSA 25: 44712).

Cairns, J., Jr. and G. R. Lanza. 1972. Pollution controlled changes in algal and protozoan communities. pp. 245-272. In: Water Pollution Microbiology. R. Mitchell, ed. John Wiley, New York. (Incl. radioactivity).

Calapaj, G. G. 1970. Research of the Institute of Health of Padova University in environmental radioactivity. pp. 125-128. In: Proc. 1st Conf. on the State of Advancement of Radioecology in Italy. (See Anonymous, 1970) (In Italian).

Calapaj, G. G. 1973. Contamination and decontamination of mussels with ^{90}Sr, ^{137}Cs, mercury and chromium. Abstr. p. 115. In: Proc. 2nd Conf. on the State of Advancement of Radioecology in Italy. (See Anonymous, 1973) (In Italian).

Calapaj, G. G. and D. Ongaro. 1963. Radioactivity of some springs, mud and therapy rooms from Euganean thermal zone. Acta Isotopica 3: 349-366. (In Italian) (NSA 18: 43807).

Calapaj, G. G. and D. Ongaro. 1970. Radioactive contamination of the Venice Lagoon from 1964 to 1969. Minerva Fisiconucl. 14(4): 202-212. (In Italian) (NSA 26: 20356).

Calapaj, G. G. and D. Ongaro. 1972. Environmental radioactivity in the Euganean thermal basin. G. Fis. Sanit. Prot. Radiaz. 16(3): 131-144. (In Italian) (NSA 29: 15635).

Calaprice, J. R. 1970. A preliminary report on x-ray spectrometric analysis and discrimination of salmonids from different geographic areas. Fish. Res. Bd. Can. Tech. Rept. No. 200. 35pp. (SFA 16: 13489).

Calaprice, J. R. 1971. X-ray spectrometric and multivariate analysis of sockeye salmon (Oncorhynchus nerka) from different geographic areas. J. Fish. Res. Bd. Can. 28(3): 369-377. (SFA 16: 13620).

Calaprice, J. R. and F. P. Calaprice. 1970. Marking animals with microtags of chemical elements for identification by x-ray spectroscopy. J. Fish. Res. Bd. Can. 27(2): 317-330.

Calaprice, J. R., H. M. McSheffrey and L. A. Lapi. 1971. Radioisotope x-ray fluorescence spectrometry in aquatic biology: a review. J. Fish. Res. Bd. Can. 28(10): 1583-1594.

Caldarela, P. C. and C. S. Pittendrigh. 1974. Test of the hypothesis that D_2O affects circadian oscillations by diminishing the apparent temperature. Proc. Nat. Acad. Sci. (U.S.) 71(11): 4386-4388. (NSA 31: 8978) (Cockroach).

Caldecott, R. S. and L. A. Snyder. (Eds.). 1960. A Symposium on Radioactive Isotopes in the Biosphere. (Held at Univ. of Minnesota, Minneapolis, October 19-23, 1959). Univ. of Minnesota Print. Dept., Minneapolis. x, 597pp. (NSA 14: 23871).

Calderon, J. K. R. 1969. Ecology of radiations. Salud Ocupac. 14(3-4): 83-98. (In Spanish).

Calderon, M. and M. Gonen. 1971. Effects of gamma radiation on Ephestia cautella (Wlk.) (Lepidoptera, Phycitidae). I. Effects on adults. J. Stor. Prod. Res. 7(2): 85-90. (NSA 25: 55315).

Calderon, M. and M. Gonen. 1974. Sterilization by irradiation of Cadra cautella (WLK.) (Lepidoptera, Pyralidae) males increased by female sex pheronone. Experientia 30(4): 418-419. (NSA 30: 13004).

Caldwell, R. D., R. F. Crosby and M. P. Lockhard. 1970. Radioactivity in coal mine drainage. pp. 438-445. In: Environmental Surveillance in the Vicinity of Nuclear Facilities. (See Reinig, 1970) (NSA 25: 16185).

Calf, G. E. 1969. Catalytic exchange method for low-level tritium measurements in water. Int. J. Appl. Radiat. Isotopes 30: 177-185. (NSA 23: 15813).

Calf, G. E., et al. nd. Nuclear techniques in groundwater hydrology. 8pp. (For Groundwater Symposium, Kensington, Australia, 9 Aug. 1970) (CONF-700840-1) (NSA 25: 32069).

Cali, A. and V. Verga. 1960. Effects of x-rays on the permeability of capillaries (study of the mesentery of *Rana esculenta*). Strahlentherapie 112: 604-607. (In German) (NSA 14: 22787).

California State Resources Agency. 1970. Siting thermal power plants in California. Sacremento. 98pp. (NP-19187) (NSA 26: 24913).

Calkins, J. 1963. Variation of radiation sensitivity of *Paramecium aurelia* as a function of time of irradiation in the interdivision growth cycle. Nature 198(4881): 704. (NSA 17: 23125).

Calkins, J. 1963. Photo-reactivation after division II. Nature 200(4905): 484-485. (NSA 18: 1408) (*Tetrahymena* and *Paramecium*).

Calkins, J. C. 1963. Detailed study of the lethal effects of x rays and ultraviolet light on two ciliate protozoans. Thesis, Univ. of Texas, Austin. 139pp. (NSA 20: 36661).

Calkins, J. 1965. A study of the time course of recovery of *Paramecium aurelia* from the lethal effects of x-rays. Radiation Research 26(1): 124-131. (NSA 19: 45978).

Calkins, J. 1965. Effect of x-ray dose on the growth rate and delay of the first post-irradiation division in a strain of *Paramecium aurelia*. Nature 205(4970): 511-513. (NSA 19: 15137).

Calkins, J. 1966. An alternative multi-target equation useful for fitting survival curves of x-ray irradiated Protozoa. Nature 209: 172-173. (NSA 20: 16273).

Calkins, J. 1967. Unusual form of response in x-irradiated Protozoa and a hypothesis as to its origin. Int. J. Radiat. Biol. 12: 297-301. (NSA 22: 40879).

Calkins, J. and G. Griggs. 1969. U. V. reactivation of Protozoa. Photochem. Photobiol. 10: 61-66. (NSA 24: 7983).

Calkins, J. and G. Griggs. 1969. Photoreactivation of U. V. reactivation in Protozoa. Photochem. Photobiol. 10: 445-449. (NSA 24: 14574).

Callebaut, M. 1968. Sterilization of the left ovary of chicken and quail by the action of thymidine-^{3}H during ovogenesis. Experientia 24: 944-945. (In French) (NSA 23: 6594).

Calow, P. and C. R. Fletcher. 1972. A new radiotracer technique involving ^{14}C and ^{51}Cr, for estimating the assimilation efficiences of aquatic, primary consumers. Oecologia 9(2): 155-170.

Cals-Usciati, J. 1971. Restoration effect of ecdysone after the arresting of post-embryonic development of γ-irradiated *Ceratitis capitata* Wied larvae (Trypetdae Diptera). Compt. rend. D272(26): 3295-3298. (In French) (NSA 25: 58094).

Cals-Usciati, J. 1971. Changes in the pupal morphogenesis in *Ceratitis capitata* Wied (Insecta, Diptera) after γ irradiation in the terminal instar stage. Compt. rend. D273(1): 79-82. (In French) (NSA 26: 12429).

Cals-Usciati, J. 1974. Modifications to the surface ultrastructural characteristics of *Ceratitis capitata* larva cuticles during the puparium formation; effect of γ radiation and β ecdysone on this cuticular morphogenesis. Compt. rend. D278(1): 65-68. (In French) (NSA 30: 19006).

Cambray, R. S., et al. 1970. Methods of collection and analysis of radioactivity from distant nuclear test explosions. U. K. Atomic Energy Authority, Harwell, report AERE-R-5898. 23pp. (NSA 24: 46353).

Cameron, J. F. 1969. Nucleonic soil density and moisture gauges. pp. 81-104. In: *Nuclear Techniques and Mineral Resources*. (Proc. Symp., Buenos Aires, 5-9 Nov. 1968). International Atomic Energy Agency, Vienna, Publ. STI/PUB-198. (NSA 23: 35535).

Cameron, R. J. 1970. Translocation of carbon-14-labelled assimilates in shoots of *Pinus radiata*. J. Exptl. Bot. 21(69): 943-951.

Camp, H. B. and B. W. Arthur. 1967. Absorption and metabolism of carbaryl by several insect species. J. Econ. Entomol. 60(3): 803-807. (^{14}C).

Campbell, B. L. 1964. Sediment research in Australia. At. Energy Austral. 7(4): 13-17. (NSA 19: 7695) (Radiotracers).

Campbell, B. L. and B. W. Seatonberry. 1967. Surface labelling of sand with ^{51}Cr by thermal decomposition of ammonium dichromate ^{51}Cr. pp. 227-232. In: *Isotopes in Hydrology*. (See IAEA, 1967) (NSA 22: 19062).

Campbell, C. A. 1965. Use of naturally occurring ^{14}C to measure the persistence of organic components in soil. Thesis, Univ. of Saskatchewan, Saskatoon, Canada. 163pp. (NSA 21: 14323).

Campbell, F. L. and C. Lukens. 1931. A radioactive indicator method for estimating the solubility of acid lead arsenate within the alimentary tract of the silk worm. J. Econ. Entomol. 24: 88-94.

Campbell, G. M., et al. 1960. Preoperational environmental radiation survey of the site and surrounding area for the Rural Cooperative Power Association nuclear power reactor, Elk River, Minnesota, July 15, 1959 - July 15, 1960. Univ. of Minnesota, Minneapolis, U. S. AEC report TID-12814. 135pp.

Campbell, N. E., et al. 1967. The production of $^{13}N_2$ by 50-MeV protons for use in biological nitrogen fixation. Can. J. Microbiol. 13: 587-600. (NSA 21: 30771).

Campbell, S. D. 1973. Effect of cobalt-60 gamma rays on terrestrial microcosm metabolism. Thesis, Univ. of Michigan, Ann Arbor. 153pp. (Univ. Microfilms 74-3590) (NSA 30: 9792).

Canada Department of National Health and Welfare. 1965. Control of radioactive fallout in water systems. A manual for water engineers. Ottawa. 128pp. (NP-15066) (NSA 19: 30441).

Canada Department of National Health and Welfare. 1963- . Data from radiation protection programs. (Monthly reports are issued; e.g., 5(5): 21pp., 5(6): 24pp., 6(1): 33pp., 6(2): 45pp., 6(3): 30pp., 6(5): 24pp.) (NP-16878; NP-16971; NP-17444; NP-17498; NP-17505; NP-17653) (NSA 21: 39051; 21: 39053; 23: 819; 23: 4747; 23: 4748; 23: 4749) (Incl. concentrations of radionuclides in environmental samples).

Cancio, D., L. Foulquier and A. Grauby. 1968. Ways of uptake for a freshwater bivalve of radiostrontium and its loss (*Anodonta cygnea* (L.)). France Commissariat à l'Énergie Atomique, Cadarache, report CEA-R-3421. 47pp. (In French) (NSA 23: 771).

Candes, P. 1971. Arrangements for the control of radioactive wastes associated with the fast reactor developed in France. pp. 159-173. In: *Environmental Aspects of Nuclear Power Stations*. (See IAEA, 1971) (In French).

Candes, P. 1971. Radiological safety of French reactors and their sites. Centre d'Études Nucléaires, France Commissariat à l'Énergie Atomique, Saclay, report CEA-CONF-1877. 12pp. (ANL-Trans-1030, 25pp.) (CONF-710444-2) (NSA 26: 16981; 33: 16411).

Candes, P. and F. Grandin. 1966. Environmental monitoring in the area of the Cadarache Nuclear Studies Center. pp. 23-29. In: *Manual on Environmental Monitoring in Normal Operation*. International Atomic Energy Agency, Vienna, Safety Ser. No. 16, Publ. STI/PUB/98. (NSA 20: 23405).

Cannizzaro, R. D., T. E. Cullen and R. T. Murphy. 1970. Gas-liquid radiochromatography of C^{14} isopropyl 4, 4'-dibromobenzilate residues in field-weathered soil. J. Agr. Food Chem. 18(4): 728-730.

Cannon, H. L. 1952. Effects of uranium-vanadium deposits on the vegetation of the Colorado Plateau. Am. J. Sci. 250: 735-770.

Cannon, H. L. 1953. Geobotanical reconnaissance near Grants, New Mexico. U. S. Geological Survey, Washington, Cir. 264. 8pp.

Cannon, H. L. 1954. Methods of botanical prospecting for uranium deposits of the Colorado Plateau. U. S. Geological Survey, Denver, Report 442. 103pp.

Cannon, H. L. 1957. Description of indicator plants and methods of botanical prospecting for uranium deposits on the Colorado Plateau. U. S. Geological Survey, Denver, Bull. 1030-M. pp. 399-516.

Cannon, H. L. and F. J. Kleinhampl. 1956. Botanical methods of prospecting for uranium. Proc. Intern. Conf. Peaceful Uses of Atomic Energy 6: 801.

Cantacuzene, A.-M. and C. Ceureau. 1970. Alterations in spermiogenesis of *Locusta migratoria migratorioiedes* after *Corpora allata* implantation and after irradiation. Z. Zellforsch. mikroskop. anat. 103: 351-364. (In French) (NSA 24: 23307).

Cantelow, H. P., et al. 1972. Sampling system for tritium oxide and ^{14}C in environmental air. Health Physics 23(3): 384-385. (NSA 26: 50603).

Cantillon, G. E. 1974. Regulations and practices for the control of radioactive waste in Belgium. pp. 50-60. In: *Proc. Seminar on the Monitoring of Radioactive Effluents*. (See Organ. for Econ. Co-oper. Develop., 1974) (NSA 31: 3337).

Cantillon, G., R. Kirchmann and O. van der Borght. 1970. Fixation of radiomanganese and radiocobalt by sediments, fauna and flora on the Meuse downstream of the central Franco-Belgian Ardennes. pp. 127-160. In: *Proc. Intern. Symp. on Radioecologie. Vol. I.* (See Commissariat à l'Énergie Atomique, 1970) (In French).

Cantwell, G. E. 1967. Inactivation of biological insecticides by irradiation. J. Invert. Pathol. 9: 138-140. (NSA 22: 8665).

Cantwell, G. E. and B. A. Franklin. 1966. Inactivation by irradiation of spores of *Bacillus thuringiensis* var. *thuringiensis*. J. Invert. Pathol. 8(2): 256-258.

Cantwell, G. E. and T. J. Henneberry. 1963. The effects of gamma radiation and apholate on the reproductive tissues of *Drosophila melanogaster* Meigen. J. Insect Pathol. 5(2): 251-264. (NSA 18: 1400).

Capella, J. A. and A. D. Conger. 1967. Radiosensitivity and interphase chromosome volume in the gymnosperms. Radiation Botany 7(2): 137-149.

Capadona, C. 1969. Measurements of movements of solid substances in water by means of stable tracers and activation analysis. pp. 72-75. In: *Modern Trends in Activation Analysis. Vol. I.* (See DeVoe, 1969) (NSA 24: 109).

Caprotti, M., M. Cioci and A. Springhetti. 1965. On the development of oocytes in the *Nauphoeta cinerea* (Blattoidea) during x irradiation of the head. Radiobiol. radioterap. fis. med. 20: 57-74. (In Italian) (NSA 19: 36060).

Caprotti, M., P. G. Rubin and A. Springhetti. 1963. Sterilization of the female adult of *Nauphoeta cinerea* (Blattoidea) by means of x-rays. Genet. agr. 16: 300-306. (In Italian, English summary).

Caputa, A. 1964. Effect of the seasons of the year upon the uptake of I-131 by the thyroid gland in the frog (*Rana ridibunda* Pall.). Biologia 19: 359-365. (NSA 18: 31099).

Carach, J., et al. 1974. Environmental surveillance around the atomic power plant Jaslovske Bohunice, Czechoslavokia. pp. 251-262. In: *Environmental Surveillance Around Nuclear Installations. Vol. 2.* (See IAEA, 1974) (NSA 31: 14039)

Carante, J. P. 1974. Effects of the x ray on the survival period in the complete absence of food on *Ceratitis capitata* Wied. Compt. rend D278(3): 373-376. (In French) (NSA 30: 19005).

Carbon, M. W. and W. A. Houlberg. (Coors.). 1970. Proceedings of a Student Conference on Nuclear Power and the Environment. (Held at Madison, Wis., April 3-4, 1970).

Carcavallo, R. U., E. Rewald and C. A. Carabajal. 1971. Emryological changes in irradiated eggs of *Triatoma infestans* and subsequent development. pp. 503-512. In: *Sterility Principle for Insect Control or Eradication*. (See IAEA, 1971) (In Spanish) (NSA 25: 29620).

Cardon, D. 1974. Study of the scatter of neutron measurement data: application to the measurement of variation in soil moisture storage for two different graminous crops. pp. 71-83. In: *Isotope and Radiation Techniques in Soil Physics and Irrigation Studies, 1973*. (See IAEA, 1974) (In French) (NSA 30: 15486).

Carey, W. E. 1968. Effects of x-ray radiation on survival of the cereal leaf beetle *Oulema melanopus* (L.). Thesis, Ohio State Univ., Columbus. 63pp. (NSA 23: 14245).

Carey, W. M., L. Battist and W. E. Keene. 1974. Measurements of the dispersion of gaseous radioactive effluents from an operating boiling water reactor. pp. 291-295. In: *Proc. 3rd Intern. Congr. Intern. Radiation Protection Assoc.* (CONF-730907-P1) (NSA 30: 26765).

Carfagna, M., A. Solima and A. de Capoa. 1961. Preliminary note on the variation of some fitness factors in periodically irradiated *Drosophila melanogaster* population. Atti assoc. genet. Ital. 6: 277-284. (In Italian) (NSA 17: 10384).

Carfagno, D. G. and B. Robinson. 1975. Annual environmental monitoring report: calender year 1974. Mound Lab., U. S. AEC report MLM-2232. 40pp. (NSA 32: 6416).

Carfagno, D. G. and W. H. Westendorf. 1972. Environmental monitoring report: January-June 1972. Mound Lab., U. S. AEC report MLM-1973. 18pp. 1973. January-June 1973. Report MLM-2085. 21pp. (NSA 27: 2525; 28: 27257).

Carillo, J. L., A. Oriega and J. Rodriguez. 1963/64. Sterilizing effect of gamma-rays and apholate on the Mexican bean beetle, *Epilachna varivestis* Muls. Agricult. téc. Méx. 11(4): 168-175. (In Spanish).

Carlin, F., et al. 1975. Use of environmental isotopes to investigate the inter-connections between the Reno River and groundwater (northern Italy). pp. 179-194. In: *Isotope Ratios as Pollutant Source and Behavior Indicators*. (See IAEA, 1975) (NSA 33: 7322).

Carlson, B. M. 1970. The effect of x-irradiation and beryllium nitrate upon implant-induced supernumerary limb formation in the newt. Oncology 24: 31-47. (NSA 24: 27686).

Carlson, C. A. and M. H. Shealy, Jr. 1972. Marking larval largemouth bass with radiostrontium. J. Fish. Res. Bd. Can. 29(4): 455-458.

Carlson, C. A. and M. H. Shealy, Jr. 1972. Techniques for marking larval largemouth bass with ^{85}Sr for evaluation of predator-prey relationships. Cornell Univ., Water Res. Mar. Sci. Center, Ithaca, N. Y., Tech. Report 49. iii, 21pp.

Carlson, J. G. 1940. Immediate effects of 250r of x-rays on the different stages of mitosis in neuroblasts of the grasshopper, *Chorophaga viridifasciata*. J. Morphol. 66: 11-23.

Carlson, J. G. 1941. Effects of x-radiation on grasshopper chromosomes. Cold Spring Harbor Symposia on Quant. Biol. 9: 104-111.

Carlson, J. G. 1961. The grasshopper neuroblast culture technique and its value in radiobiological studies. Ann. N. Y. Acad. Sci. 95: 932-941. (NSA 16: 26763).

Carlson, J. G. 1969. A detailed analysis of x-ray-induced prophase delay and reversion of grasshopper neuroblasts in culture. Radiation Research 37: 1-14. (NSA 23: 16369).

Carlson, J. G. and M. E. Gaulden. 1964. Grasshopper neuroblast techniques. Methods Cell Physiol. 1: 229-276. (NSA 20: 6844).

Carlson, J. G. and N. G. Harrington. 1955. X-ray induced "stickiness" of the chromosomes of the *Chortophaga* neuroblast in relation to dose and mitotic stage at treatment. Radiation Research 2(1): 84-90.

Carlson, J. G., N. G. Harrington and M. E. Gaulden. 1953. Mitotic effects of prolonged irradiation with low-intensity gamma rays on the *Chortophaga* neuroblast. Biol. Bull. 104(3): 313-322.

Carlson, J. G., M. L. Snyder and A. Hollaender. 1949. Relation of gamma-ray dosage rate to mitotic effect in the grasshopper neuroblast. J. Cell. Comp. Physiol. 33: 365-372.

Carlsson, L. nd. Methods for practical determination of the flow rate of groundwater. Part 3. Field investigations at Gaardsryd. (INIS-mf-271, 229pp.) (In Swedish) (NSA 26: 50618) (^{51}Cr-EDTA tracer).

Carmon, J. L., R. G. Williams and F. B. Golley. 1967. Effects of temperature and radiation on litter size and growth in *Peromyscus polionotus*. Growth 31(1): 79-89.

Carney, G. C. 1959. Differential response of male and female adults of *Trogoderma granarium* Everts towards sterilizing doses of gamma-radiation. Nature 183(4657): 338-339. (NSA 13: 7433).

Carney, G. C. 1964. The effects ionizing radiation on the sarcosomes of housefly flight muscle. Thesis, Univ. of Minnesota, Minneapolis. 98pp. (NSA 19: 42345) (Diss. Abstr. 25: 4917) (*Musca domestica*).

Carney, G. C. 1965. Swelling and shrinkage properties of housefly sarcosomes after *in vivo* exposure to x-rays. Radiation Research 25: 637-645.

Carolina Vaz Carreiro, M. 1969. Absorption of ^{137}Cs by *Spondyllosoma cantharus* L. Lab. de Fisica e Engenharia Nucleares, Sacavem, Portugal Junta de Energia Nucleare, report LFEN-NI-40-a. 14pp. (In Portugese) (NSA 24: 11907).

Carolina Vaz Carreiro, M. 1969. Accumulation and retention of ^{137}Cs by Carassius auratus L. Lab. de Fisica e Engenharia Nucleares, Sacavem Portugal Junta de Energia Nucleares, report LFEN-NI-39-a. 14pp. (In Portugese) (NSA 24: 11906).

Carolina Power and Light Co. 1971. Shearon Harris Nuclear Power Plant, Units 1, 2, 3, and 4. Environmental report. Raleigh, N. C. 104pp. U. S. AEC DOCKET-50400-2, -50401-2, -50402-2, -50403-2. (NSA 25: 51005).

Carosi, G., et al. 1967. Certain aspects of the morphogenesis of the heteroploid cells from abnormal mitosis caused by x rays. Arch. Sci. Med. 124: 592-599. (In Italian) (NSA 23: 31885) (Newts).

Carothers, E. E. 1940. A cytological study of x-rayed grasshopper embryos. J. Morphol. 66(3): 529-553.

Carpenter, B. S. 1969. Determination of trace quantities of uranium in biological materials by the nuclear tract technique. pp. 942-945. In: Modern Trends in Activation Analysis. Vol. II. (See DeVoe, 1969 (NSA 24: 63).

Carpenter, B. S. 1971. Use of the nuclear track technique for boron and uranium analyses in botanical material. Trans. Amer. Nucl. Soc. 14(1): 130-131. (NSA 25: 39436).

Carpenter, B. S. and P. D. LaFleur. 1972. Nitrogen determination in biological samples using the nuclear track technique. Trans. Amer. Nucl. Soc. 15 (1): 118. (NSA 26: 42725).

Carpenter, B. S., D. Samuel and I. Wasserman. 1973. Quantitative applications of ^{17}O tracer. Radiat. Eff. 19(1): 59. (NSA 29: 17).

Carpenter, J. H. 1960. Tracer for circulation and mixing in natural waters. Public Works 91(6): 110-112.

Carr, J. A. and G. Stott. 1974. Monitoring of radioactive gaseous effluents discharged from nuclear installations in the United Kingdom. pp. 73-80. In: Proc. Seminar on the Monitoring of Radioactive Effluents. (See Org. for Econ. Co-op. and Development, 1974) (NSA 31: 3339).

Carrick, R. 1956. Radioiodine as an indicator of free feeding of the rabbit, Oryctolagus cuniculus (L.). C.S.I.R.O. Wildlife Res. (Austral.) 1: 106-113.

Carrigan, P. H., Jr. 1968. Radioactive waste dilution in the Clinch River, Eastern Tennessee. U. S. Geological Survey Prof. Paper No. 433-G. 9pp. (NSA 25: 359).

Carrigan, P. H., Jr. and R. J. Pickering. 1967. Radioactive materials in bottom sediment of Clinch River. Part B. Inventory and vertical distribution of radionuclides in undisturbed cores. Oak Ridge National Lab., U. S. AEC report ORNL-3721(Suppl. 2B). 92pp. (NSA 21: 16029).

Carrigan, P. H., Jr., et al. 1967. Radioactive materials in bottom sediment of Clinch River. Part A. Investigations of radionuclides in upper portion of sediment. Oak Ridge National Lab., U. S. AEC report ORNL-3721(Suppl. 2A). 94pp. (NSA 21: 16028).

Carritt, D. E. and S. H. Goodgal. 1953. The sorptive and zeolitic properties of natural and waterborne silts, with relation to their capacities to remove, transport, concentrate, and regenerate radioactive waste components in natural waters. Johns Hopkins Univ., U. S. AEC report NYO-4591. 97pp.

Carrol, T. E. and A. W. Rudolph. (Comps.). 1969. Dioenvironmental and radiological-safety feasibility studies: Atlantic-Pacific Interoceanic Canal. Possible effects of a sea-level canal on the marine ecology of the American Isthmian region: Abstracts. Battelle Columbus Labs., U. S. AEC report BMI-171-25. 275pp. (NSA 24: 12035).

Carson, H. L. 1959. Effect of irradiation on artificial populations under strong natural selection. Abstr. Genetics 44(4,pt.1): 503.

Carson, H. L. 1964. Population size and genetic load in irradiated populations of Drosophila melanogaster. Genetics 49(3): 521-528. (NSA 18: 13493).

Carstens, L. A. and J. R. Allen. 1969. Response of Saimiri sciureus monkeys to total body x-irradiation: electron microscopic changes in the liver. Am. J. Veterin. Res. 30: 623-635. (NSA 23: 29779).

Carter, M. W., D. T. Wruble and R. E. Jaquish. 1971. Environmental monitoring system for nuclear tests. Western Environmental Research Lab., U. S. EPA, U. S. AEC report SWRHL-501-r. 21pp. (For Intern. Symp. on Rapid Methods for Measuring Radioactivity in the Environment, Munich, Germany, 5 Jul. 1971) (CONF-710705-3) (NSA 25: 44586).

Carvalho, G. and M. P. da Cruz. nd. Differences in sensitivity to gamma radiation in Drosophila of the Willistoni group. Univ. of Brazil, Rio de Janeiro. (AEC-tr-6429) (NSA 18: 29250).

Casa, A. F. 1973. Nuclear techniques in art and archeology. Com. Naz. Energia Nucl. Notiz. 19(1): 15-20. (In Italian) (NSA 28: 2798).

Casarett, G. W. 1956. The effects of ionizing radiations from external sources on gametogenesis and fertility in mammals: a review. Univ. of Rochester, U. S. AEC report UR-441. 45pp.

Casarosa, L. 1968. Effects of total- (partial-) body irradiation on helminthic infections. pp. 105-116. In: Isotopes and Radiation in Parasitology. (See IAEA, 1968).

Casarosa, L. and V. Favati. 1964. Observations on the migration capacity in mice of Neoascaris vitulorum larvae hatched from embryonated eggs subjected to 100,000 R from ^{60}Co. Ann. Fac. Med. Vet. Univ. Pisa 17: 67-74. (In Italian) (NSA 21: 34755).

Casida, J. E. 1956. Mode of action of pesticides, metabolism or organophosphorus insecticides in relation to their antiesterase activity, stability, and residual properties. Agr. Food Chem. 4(9): 772-785. (Radiotracers).

Casida, J. E. 1961. Use of radioisotopes in insecticide studies. pp. 427-438. In: Applications of Radioisotopes and Radiation in the Life Sciences. (See U. S. Congress, 1961) (NSA 15: 27332).

Casida, J. E. 1962. Metabolism of organophosphate insecticides by plants: a review. pp. 49-64. In: Radioisotopes and Radiation in Entomology. (See IAEA, 1962).

Casida, J. E. 1963. Radiotracer approaches to carbamate insecticide toxicology. pp. 223-238. In: Radiation and Radioisotopes Applied to Insects of Agricultural Importance. (See IAEA, 1963) (NSA 17: 21425).

Casida, J. E., et al. 1956. Pesticide residues, residual properties of the systemic insecticide 0,0-dimethyl 1-carbo-methoxy-1-prophen-2-yl phosphate. Agr. Food Chem. 4(3): 236-243.

Casida, J. E., et al. 1964. Progress report on the comparative fate of eight substituted-phenyl N-methylcarbomate-C^{14} insecticides. Parts I and II. Abstrs. 48,49. Bull. Entomol. Soc. Amer. 10(3): 163.

Casida, J. E., et al. 1966. Methylene-C^{14}-dioxyphenyl compounds: metabolism in relation to their synergistic action. Science 153: 1130-1133. (House-flies).

Casolari, A., M. Campanini and G. Cicognani. 1969. Radiosensitivity of heat-pretreated spores. Industria Conserve 44: 199-203. (In Italian) (NSA 24: 44636).

Cassano, G. and G. Lenzi. 1970. General criteria and preliminary experiences for assessing the pollution probabilities of the freatic and superficial waters from radioactive discharges of nuclear reprocessing plants. pp. 145-147. In: Proc. 1st Conf. on the State of Advancement of Radioecology in Italy. (See Anonymous, 1970) (In Italian).

Cassie, R. M. 1962. Microdistribution and other error components of C^{14} primary production estimates. Limnol. Oceanogr. 7(2): 121-130. (NSA 18: 9824) (Marine diatom).

Castagnet, A. C. and H. R. Gomez. nd. Nuclear techniques and equipment in hydrology. pp. 37-52. In: Proc. Conf. on Nuclear Energy and Agricultural Productivity. (Santiago, Chile, January 9-12, 1968). (See Comision Inter-americana de Energia Nuclear, nd.) (In Spanish) (NSA 25: 26353).

Castagnola, D. C. 1969. Tritium anomalies on Amchitka Island, Alaska. Part I. Teledyne-Isotopes, Palo Alto, Calif., U. S. AEC report NVO-1229-113, Part 1. v, 41pp. Part II. Report NVO-1229-113(Pt.2). 23pp. (NSA 26: 42960).

Castellani, E., A. Matta and C. Guerzoni. 1957/58. Pathological effect of beta rays on the plant pathogenic fungus. Turin U. Fac. Sci. Agr. Racc. Mem. 5(134): 8.

Castellani, E., A. Matta and C. Guerzoni. 1958. Pathological effect of beta rays on a phytopathogenic fungus. Minerva Nucleare 2(2): 56-59. (In Italian).

Catanzaro, E. J. 1975. Simple isotope dilution technique for the determination of lead and copper in natural waters. J. Water Poll. Contr. Fed. 47(1): 203-204. (NSA 31: 15965).

Cate, J. L., Jr. and T. O. Hoeger. 1972. Radioisotope monitoring system for sewage effluent. Am. Ind. Hyg. Assoc. J. 33(10): 693-699. (NSA 28: 2671).

Cather, J. N. 1959. The effects of x-radiation on the early cleavage stages of the snail, Ilyanassa obsoleta. Radiation Research 11(5): 720-731. (NSA 14: 1393).

Catrina, I., et al. 1963. The establishment of the process of absorption a and diffusion of systemic insecticides with Populus x euroamericana Dode Guinier <<Robusta>>. pp. 211-220. In: Radiation and Radioisotopes Applied to Insects of Agricultural Importance. (See IAEA, 1963) (In French).

Cattell, W. 1931. The effect of x-rays on the growth of wheat seedlings. Science 73(1898): 531-533.

Causse, R., M. Féron and P. Péreau-Leroy. 1968. Comparative study of the effect of gamma and neutron irradiation on the reproductive organs of the Mediterranean fruit fly Ceratitis capitata Wied. (Diptera: Trypetidae). pp. 355-364. In: Isotopes and Radiation in Entomology. (See IAEA, 1968) (In French, English summary) (NSA 24: 32174).

Cavalcasselle, B., E. de Bellis and A. Liani. 1964. Preliminary investigations on the absorption of parathion-^{32}P by poplar stems. Ente Naz. Cell. Carta. Cent. Sper. Agr. Forest. P. 7: 89-97.

Cavalloro, R. 1960. Uses of ionizing radiation and radioactive isotopes in agricultural entomology. Tobacco 65(700): 1-16. (In Italian) (NSA 22: 47657).

Cavalloro, R. 1961. Studies on the nutrition of Myzus persicae using P^{32}. Tobacco 65: 287-297.

Cavalloro, R. 1961. Study on the nutrition of Myzodes persicae Sulzer by means of ^{32}P. (For 6th Nuclear Congr. Symp. on the Application of Nuclear Energy in Agriculture, Rome, Italy, 15-16 June 1961). 11pp. (In Italian; English and French summaries).

Cavalloro, R. 1963. Study of the trophic relation between aphids and their host plant by means of ^{14}C. pp. 155-156. In: 5th Nat. Congr. on Entomology. (Milan, Italy, September 1963). (In Italian, French summary).

Cavalloro, R. 1967. Data on the accumulation, distribution and elimination of ^{85}Sr, ^{131}I and ^{137}Cs in various insect species. pp. 601-608. In: Radioecological Concentration Processes. (See Aberg and Hungate, 1967).

Cavalloro, R. 1967. Control of harmful insects in agriculture by use of ionizing radiation. Boll. ist. entomol. agr. osserv. fitopatol. Palermo 7(56): 101-112. (In Italian) (NSA 22: 47658).

Cavalloro, R. 1968. Use of ionizing radiations for control of agriculturally harmful insects. pp. 101-111. In: Applications of Nuclear Energy in Agriculture. Italy Comitato Nazionale Energia Nucleare, Rome, Ser. Simp. (In Italian) (CONF-670558) (NSA 24: 32314).

Cavalloro, R. and G. Bonfanti. 1967. Application of ionizing radiation for controlling *Acanthoscelides obsoletus* Say (Bruchidae, Coleoptera) infestations of legume seeds. Boll. zool. agr. Bachicolt. 8: 115-134. (In Italian) (NSA 23: 33917).

Cavalloro, R. and U. Cirio. 1964-5. Accumulation and elimination of radioisotopes in several species of phytophagous and holometabolous insects. Redia 49(2): 239-253. (In Italian, English summary) (^{85}Sr, ^{131}I, ^{137}Cs).

Cavalloro, R. and U. Cirio. 1966. Accumulation and elimination of radioisotopes in some species of phytophagous insects. Euratom Inform. 4(10): 1089. (^{85}Sr, ^{131}I, ^{137}Cs).

Cavalloro, R. and U. Cirio. 1966. The accumulation of radioisotopes in some species of phytophagous homometabolic insects. pp. 78-79. In: *6th Nat. Congr. on Entomology*. (Italy). (In Italian).

Cavalloro, R. and U. Cirio. 1966-67. Research with ionizing radiation on sterilization and radiosensitivity of *Chrysomela decimlineata* Say. Redia 50: 365-377. (In Italian) (NSA 22: 47656).

Cavalloro, R. and U. Cirio. 1967. Distribution of radioactive strontium, cesium, and iodine in the body of some insects. Redia 50: 187-196. (In Italian) (NSA 22: 19262).

Cavalloro, R. and G. Delrio. 1971. Population longevity in γ-irradiated Colembola. Int. J. Appl. Radiat. Isotopes 22: 216. (NSA 25: 29617).

Cavalloro, R. and G. Delrio. 1973. Biological effects of gamma radiation on the life stages of two fruit fly species. pp. 1179-1190. In: *Radionuclides in Ecosystems*. (See Nelson, 1973) (NSA 28: 30240).

Cavalloro, R. and G. Delrio. 1974. Mating behavior and competitiveness of gamma-irradiated olive fruit flies. J. Econ. Entomol. 67(2): 253-255. (NSA 30: 618).

Cavalloro, R. and O. Ravera. 1966. Biological indicator of ^{54}Mn contamination in terrestrial environments. Nature 209: 1259. (NSA 20: 24844).

Cavin,G. E. 1975. Use of the sterile-male technique in plant protection programmes. pp. 3-9. In: *Sterility Principle for Insect Control, 1974*. (See IAEA, 1975) (NSA 32: 11762).

Cavina Pratesi, A. and F. Ghiringhelli. 1964. Behavior of enterochromaffin cells of the intestine of the hedgehog irradiated during hibernation. Radiobiol. radioterap. fis. med. 19: 133-143. (In Italian) (NSA 20: 5377).

Cavina Pratesi, A., G. R. Bandirali and C. Vassallo. 1961. Histological and histochemical study of irradiated hypophysis during hibernation. Radiobiol. Latina 4: 243-256. (In Italian) (NSA 16: 26833).

Cawley, W. A. and W. C. Rutledge. 1966. Application of radioactive tracer techniques. Am. Soc. Civil Engrs., Sanit. Engrg. Div. 92(SA1): 1-_.

Cawse, P. A. 1967. Effects of low sub-sterilizing doses of gamma radiation on carbon, nitrogen, and phosphorus in fresh soils. J. Sci. Food. Agr. 18: 388-391. (NSA 22: 23450).

Cawse, P. A. 1967. Accumulation of nitrite in fresh soils after gamma irradiation. Nature 216: 1142-1143. (NSA 22: 10731).

Cawse, P. A. 1968. Effects of gamma radiation on accumulation of mineral nitrogen in fresh soils. J. Sci. Food Agr. 19: 395-398. (NSA 22: 42865).

Cawse, P. A. 1968. Influence of γ radiation on some of the biochemical properties of soil. Biochem. J. 109: 3pp. (NSA 22: 52038).

Cawse, P. A. 1969. Use of gamma radiation in soil research. U. K. Atomic Energy Authority, Wantage, report AERE-R-6061. 20pp. (NSA 23: 43398).

Cawse, P. A. and A. H. Cornfield. 1969. The reduction of ^{15}N-labelled nitrate to nitrite by fresh soils following treatment with radiation. Soil Biol. Biochem. 1: 267-274. (NSA 24: 27547).

Cawse, P. A. and D. H. Peirson. (Eds.). 1972. An analytical study of trace elements in the atmospheric environment. U. K. Atomic Energy Authority, Harwell, report AERE-R-7134. (NSA 27: 24286) (Neutron activation).

Cawse, P. A. and J. H. Radley. 1970. Response of *Sinapis alba* to irradiation in different growth media. Radiation Botany 10: 491-493. (NSA 25: 604).

Cawse, P. A. and T. White. 1969. Accumulation of nitrate in fresh soils after gamma irradiation. J. Agr. Sci. 72: 331-333. (NSA 23: 26354).

Cawse, P. A. and T. White. 1969. Rapid changes in nitrite after gamma irradiation of fresh soils. J. Agr. Sci. 73: 113-118. (NSA 24: 16694).

Cechacek, Z. and J. Hillova. 1970. DNA synthesis and changes in radiosensitivity during the cell cycle of vegetative culture of *Chlamydomonas reinhardii*. Int. J. Radiat. Biol. 17: 127-134. (NSA 24: 34760).

Cecich, R. A. and J. P. Miksche. 1970. Response of white spruce (*Picea glauca* (Moench) Voss) shoot apices to exposures of chronic gamma radiation. Radiation Botany 10: 457-467. (NSA 25: 603).

Celan, B. and V. Perlea. 1967. Use of radioisotopes when dealing with problems associated with irrigation channels located on loess. pp. 399-403. In: *Isotope and Radiation Techniques in Soil Physics and Irrigation Studies*. (See IAEA, 1967).

CER Nuclear Corp. 1973. Project directors' completion report D + 30 days (detonation related activities). Project Rio Bronco. U. S. AEC report NVO-135. 97pp. (NSA 28: 21035).

CER Nuclear Corp. 1973. Project Rio Blanco. Project directors' preliminary report: D +2 days. U. S. AEC report, PNE-RB-45. 40pp. (NSA 28: 8113).

CER Nuclear Corp. 1973. Project Rio Blanco definition plan. Vol. III. Reentry-related detailed tasks. U. S. AEC report PNE-RB-18(Rev.1). 114pp. (NSA 27: 27535).

CER Nuclear Corp. 1975. Project Rio Blanco radioactivity and the environment. Las Vegas, Nev., U. S. ERDA report PNE-RB-63(Add.). 36pp. (NSA 33: 12028).

Cercek, L., et al. 1968. Development of test systems for use in radiobiolo ology. J. Physiol. (London) 197: 55P-56P. (NSA 23: 14284) (Incl. algae, fern, barley, bean).

Cerda-Olmedo, E. and P. Reau. 1970. Genetic classification of the lethal effects of various agents on heterokaryotic spores of Phycomycetes. Mutation Research 9: 369-384. (NSA 24: 50860).

Cerrai, E. and C. Triulzi. 1965. Investigation of the Seveso River. No. 7. Measurement of the radioactivity and an analysis of the ^{90}Sr and ^{137}Cs concentrations. Acqua Ind. (37): 27-34. (Also: Centro Informazioni Studi Esperienze, Milan, Italy, report CISE-516) (In Italian) (NSA 20: 33562).

Cervigni, T., G. Caserta and C. Bassanelli. 1968. Preliminary investiga tions on thermoluminescence and radiosensitivity in seeds. Radiation Botany 8(1): 71-79. (NSA 22: 31287).

Cervigni, T., et al. 1962. Some aspects of sensitivity of higher plants to chronic gamma irradiation. Abstr. p. 181. In: 2nd Intern. Congr. Radiation Research. (See Anonymous, 1962).

Cervini, A. and S. Giavelli. 1965. Radiosensitivity of different meiotic stages of oocytes in parthenogenetic diploid *Artemia salina* Leach. Mutation Research 2: 452-456. (NSA 20: 5359).

Cervini, A. and S. Giavelli. 1965. Effect of low temperature on radiosensitivity of Artemia oocytes. Progr. Biochem. Pharmacol. 1: 52-58. (NSA 20: 18447).

Chabert, J. E., et al. 1963. Radioactive tracer study of sediment transported in alluvial rivers. pp. 133-142. (See IAEA, 1963) (In French) (NSA 18: 1918).

Chadwick, R. C. and A. C. Chamberlain. 1970. Field loss of radionuclides from grass. Atmos. Environ. 4(1): 51-56.

Chakpitak, S. 1967. The effect of gamma radiation on the hatchability, rate of growth, and reproduction of the fish, *Trichogaster pectoralis* Regan. Thailand Off. of Atomic Energy for Peace, Bangkok, report THAI.-AEC-13. 11pp. (In Thai, English) (NSA 24: 21282).

Chakravarti, D. and R. Eisler. 1959. Strontium-90 and gross beta activity in the fat and non-fat fractions of coconut crab (*Birgus latro*) liver collected at Rongelap Atoll during March 1958. Pacific Sci. 15(1): 155-159. (Also: Univ. of Washington, U. S. AEC report UWFL-59) (NSA 14: 21267).

Chakravarti, D. and E. E. Held. 1960. Potassium and cesium-137 in *Birgus latro* (coconut crab) muscle collected at Rongelap Atoll. Univ. of Washington, U. S. AEC report UWFL-64. 12pp. (Also: J. Mar. Biol. Assoc. India 2(1): 75-81) (NSA 14: 21268).

Chakravarti, D. and E. E. Held. 1963. Chemical and radiochemical composition of the Rongelap diet. J. Food Sci. 28(2): 221-228. (NSA 17: 37371).

Challa, H., P. A. M. Hopmans and J. G. de Swart. 1975. Critical study of β-guaging for water content variation measurement in plant leaves. Newslet. Applic. Nucl. Methods in Biol. Agr. (4): 9-12. (NSA 33: 17270).

Chalov, P. I. and M. A. Isevelev. 1965. On the relative levels of fission product fallout from the stratosphere. Atom. energiya 19(5): 470-472. (Soviet Atomic Energy 19: __) (NSA 20: 14742).

Chalov, P. I., K. I. Merkulova and T. V. Tuzova. 1966. The $^{234}U/^{238}U$ ratio in waters and bottom sediments of the Aral Sea and the absolute age of the basin. Geokhimiya (12): 1431-1438. (Geochem. Int. (6): 1149-1156) (NSA 21: 10452).

Chalov, P. I., K. I. Merkulova and T. V. Tuzova. 1966. Absolute age of the Aral Sea as determined from non-equilibrium uranium. Dokl. Akad. Nauk SSSR 169(1): 89-91. (Doklady (earth sci.) pp. 1-__).

Chalov, P. I., T. V. Tuzova and Ya. A. Musin. 1964. Isotopic ration $^{234}U/^{238}U$ in natural waters and its use for nuclear geochronology. Geokhimiya (5): 404-413. (Geochem. Int. (3): 402-408) (NSA 18: 31762) (Incl. alpine lakes).

Chalov, P. I., T. V. Tuzova and Ya. A. Musin. 1964. Isotopic ratio of $^{234}U/^{238}U$ in natural waters and the use of the ratio in nuclear geochronology. Izv. Akad. Nauk SSSR (geofiz.) 10: 1552-1561. (NSA 19: 6113).

Chalov, P. I., N. A. Svetlichnaya and T. V. Tuzova. 1970. Determination of the absolute age of Lake Balkhash on the basis of nonequilibrium uranium. Dokl. Akad. Nauk SSSR 195: 190-192. (Doklady (earth sci.) 195(1-6): 182-183) (NSA 25: 29210).

Chalov, P. I., N. A. Svetlichnaya and T. V. Tuzova. 1970. Isotopic composition of uranium (^{234}U, ^{238}U) in waters and bottom sediments of Lake Balkhash and the time of the basin existence. Geokhimiya (7): 848-854. (In Russian) (NSA 24: 36403).

Chalov, P. I., et al. 1973. Variations of isotopic composition of uranium (^{234}U,^{238}U) in underground waters after an earthquake. Litol. Polez Iskop. (1): 21-27. (In Russian) (NSA 28: 29889).

Chamberlain, A. C. and R. C. Chadwick. 1953. Experiments on the uptake of iodine 131 vapour by grass. Nucleonics 11(8): 22-24. (Also: U. K. Atomic Energy Authority, Harwell, report AERE-HP/R-993) (NSA 7-513).

Chamberlain, L. L. 1972. Primary productivity in a new and an older California reservoir. Calif. Fish Game 58(4): 254-267. (^{14}C).

Chamberlain, W. F. and E. W. Hamilton. 1964. Absorption, excretion, and metabolism of P^{32}-labeled Metepa by screwworm and stable flies. J. Econ. Entomol. 57(6): 800-803.

Chamberlain, W. M. and F. H. Rigler. 1968. A tracer study of phosphorus circulation in the surface waters of a stratified lake with particular reference to the production and utilization of non-inorganic filtrate phosphorus. Abstr. Bull. Ecol. Soc. Amer. 49(2): 72. (^{32}P).

Chambers, D. L., K. Ohinata and R. T. Cunningham. 1970. Recent research in Hawaii on the Mediterranean fruit fly. pp. 33-42. In: Sterile-Male Technique for Control of Fruit Flies. (See IAEA, 1970) (NSA 25: 24507).

Chambers, D. L., et al. 1970. Sterile-insect technique for eradication or control of the melon fly and oriental fruit fly. Review of current status. pp. 99-109. In: Sterile-Male Technique for Control of Fruit Flies. (See IAEA, 1970) (NSA 25: 24514).

Chamel, A., A. Fourcy and G. Pautou. 1970. Helophyte plants: a bio-indicator of radioactive contamination. pp. 1083-1101. In: Proc. Intern. Symp. on Radioecology. Vol. II. (See Commissariat à l'Énergie Atomique, 1970) (Also: France CEA, Grenoble, report CEA-CONF-1382) (In French) (NSA 23: 50457; 24: 41791).

Champlin, J. B. F. 1967. Transport of ruthenium and sodium through a simulated aquifer. pp. 329-340. In: Proc. Symp. on Groundwater Hydrology. (See Marino, 1967) (NSA 23: 442).

Champlin, J. B. F. 1969. Transport of radioisotopes by fine particulate matter in aquifers. Georgia Inst. Tech., Atlanta. Contract DI-14-01-0001-1629. 200pp. (PB-232179) (NSA 31: 535).

Champlin, J. B. F. 1970. Transport of radioisotopes by fine particulate matter in aqifers. Thesis, Georgia Inst. of Techn., Atlanta. 199pp. (Univ. Microfilms 70-19,264) (NSA 25: 8361).

Chandebois, R. 1965. Variation of the regenerating capacity of irradiated planaria (Dugesia subtentaculata) as a function of the x-ray dose. Compt. rend. 260: 4834-4837. (In French) (NSA 19: 40366).

Chandebois, R. 1968. The respective roles of mitotic activity and of cell differentiation in planarian regeneration. J. Embryol. Exptl. Morphol. 20: 175-188. (NSA 23: 20160) (X-irradiation).

Chandebois, R. 1971. Increase of regeneration power in planaria irradiated with lethal doses by stimulation of the Type I cell system. Compt. rend. D272(10): 1417-1419. (In French) (NSA 25: 44890).

Chandler, C. C., T. G. Storey and C. D. Tangren. 1963. Prediction of fire spread following nuclear explosions. Pacific Southwest Forest and Range Experiment Station, Berkeley, Calif. 110pp. (NP-13147) (NSA 17: 41025).

Chandler, R. P. and D. R. Snavely. 1966. Summary of ^{137}Cs and ^{90}Sr concentrations reported in certain Alaskan populations and foodstuffs, 1961-1966. Radiol. Hlth. Data Repts. 7: 675-689. (NSA 22: 23516).

Chandler, R. P. and S. Wieder. 1963. Radionuclides in the northwestern Alaska foodchain, 1959-1961 - a review. Radiol. Hlth. Data Repts. 4(6): 317-324.

Chandola, A. and J. P. Thapliyal. 1974. Comparison of the effects of chemical, surgical, and radiothyroidectomy in spotted m nia, Lonchura punctata. Gen. Comp. Endocrinol. 22: 128-131. (NSA 30: 27230; 33: 3279).

Chandra, H. G. S. 1962. Cytogenetic studies following high dosage paternal irradiation in the mealy bug, Planococcus citri (Risso). Thesis, Univ. of Calif., Berkeley. 103pp. (NSA 20: 12715) (Diss. Abstrs. 25(4): 2423).

Chandra, H. S. 1963. Cytogenetic studies following high dosage paternal irradiation in the mealy bug, Planococcus citri. I. Cytology of X_1 embryos. II. Cytology of X_1 females and the problem of lecanoid sex determination. Chromosoma 14(3): 310-329, 330-346. (NSA 18: 67, 68, 5030).

Chang, C. and L. Tsuang. 1965. Enhancement of functional activity of pars intermedia following ^{60}Co irradiation in the toad. Shih Yen Sheng Wu Hsueh Pao 10: 33-38. (In Chinese) (NSA 20: 5394).

Chang, C. H. and T. M. Lai. 1964. Studies on methods using radioisotopes for measuring nutrients movement in soil columns. Rep. Taiwan Sugar Exp. Sta. (35): 119-132. (In Chinese) (NSA 20: 33540).

Chang, M. C. and D. M. Hunt. 1959. Lethal effects of gamma radiation in the ferret. Abstr. Fed. Proc. 18(1): 25.

Chang, S. C. 1965. Metabolism of C^{14} labeled TEPA by house flies. Abstr. 41. Bull. Entomol. Soc. Am. 11(3): 156.

Chang, S. C. 1966. Metabolic study of C^{14}-Hempa in male house flies. Abstr. 101. Bull. Entomol. Soc. Am. 12: 286.

Chang, S. C. 1967. Metabolism of C^{14}-hexamethylmelamine (Hemel) in male house flies. Abstr. 101. Bull. Entomol. Soc. Am. 13(3): 192.

Chang, S. C. and A. B. Borkovec. 1969. Comparative metabolism of ^{14}C-labelled Hempa by house flies susceptible or resistant to Isolan. J. Econ. Entomol. 62(6): 1417-1421.

Chang, S. C. and C. W. Kearns. 1964. Metabolism in vivo of ^{14}C-labelled pyrethrin I and cinerin I by house flies with special reference to the synergistic mechanism. J. Econ. Entomol. 57(3): 397-404.

Chang, S. C., A. B. Borkovec and C. W. Woods. 1966. Fate of TEPA uniformly labeled with C^{14} in male house flies. J. Econ. Entomol. 59(4): 937-944.

Chang, S. C., C. W. Woods and A. B. Borkovec. 1970. Metabolism of ^{14}C-labeled N^2,N^2,N^4, N^4-tetramelamine in male house flies. J. Econ. Entomol. 63 (5): 1510-1513.

Chang, S. C., et al. 1968. Metabolism of C^{14}-labeled Hemel in male house flies. J. Econ. Entomol. 61(5): 1357-1365.

Channell, J. K. 1971. Activable rare earth elements as estuarine water tracers. Thesis, Stanford Univ., Stanford, Calif. 205pp. (Univ. Microfilms 71-19,660) (NSA 26: 8774).

Channell, J. K. and P. Kruger. 1969. Post-sampling activation analysis of stable nuclides for estuary water tracing. pp. 81-86. In: Modern Trends in Activation Analysis. Vol. I. (See DeVoe, 1969((NSA 24: 111).

Chant, G. D. and W. F. Baldwin. 1972. Dispersal and longevity of mosquitoes tagged with ^{32}P. Can. Entomol. 104(6): 941-944.

Chanteur, J. and P. Ervet. 1974. French regulations and procedures pertaining to discharges of radioactive effluents. pp. 19-27. In: Proc. Seminar on the Monitoring of Radioactive Effluents. (See Org. Econ. Co-op. and Development, 1974) (In French) (NSA 33: 3335).

Chapman, W. H., H. L. Fisher and M. W. Pratt. 1968. Concentration factors of chemical elements in edible aquatic organisms. Lawrence Radiation Lab., Livermore, U. S. AEC report UCRL-50564. 50pp. (NSA 23: 24507)

Chappell, H. G. 1962. The effect of ionizing radiation on Smilax with special reference to the protection afforded by its production of underground vegetative structures. Thesis, Emory Univ., Atlanta, Ga.

Chappell, H. G. 1963. The effect of ionizing radiation on Smilax with special reference to the protection afforded by its production of underground structures. pp. 289-294. In: Radioecology. (See Schultz and Klement, 1963) (NSA 17: 33573).

Chappell, H. G. and R. B. Platt. 1962. The relation of life form to protection from ionizing radiation in Smilax spp. Abstr. p. 178. In: 2nd Intern. Congr. Radiation Research. (See Anonymous, 1962).

Chapuis, A., et al. 1973. Radiophotoluminescent glasses as dosimeters in radioecological studies. pp. 59-66. In: Dosimetry in Agriculture, Industry, Biology, and Medicine. (See IAEA, 1973) (NSA 28: 15562).

Chararas, C. and K. M'Sadda. 1971. Effect of ^{60}Co radiations on sexual and chemical attraction of Orthotomicus erosus Woll., polygamous coleopter Scolytidae. Compt. rend. D273(20): 1877-1879. (NSA 26: 28487).

Chararas, M. C., C. Boulard and K. M'Sadda. 1974. Blood protein metabolism in the larvae of Phoracantha semipunctata (Coleoptera Cerambycidae Xylophage) and modifications due to cobalt-60 ionizing radiations. Compt. rend. D278(24): 3119-3122. (In French) (NSA 31: 29952).

Charles, D. R., et al. 1958. Genetic effects of chronic x-irradiation exposure in mice. Univ. of Rochester, U. S. AEC report UR-565. 364pp. (NSA 15: 7197).

Charnell, R. L, T. M. Zorich and D. E. Holly. 1969. Hydrologic redistribution of radionuclides around nuclear excavated sea-level canals in Panama and Columbia. Bioenvironmental and Radiological-Safety Feasibility Studies, Atlantic-Pacific Interoceanic Canal. Isotopes-Teledyne, Palo Alto, Calif., Battelle Columbus Labs., U. S. AEC report BMI-171-16. 70pp. (NSA 23: 11632).

Charreau, C. and L. Jacquinot. 1967. Use of tritiated water to study water circulated in a sandy soil in Senegal. pp. 301-314. In: Isotope and Radiation Techniques in Soil Physics and Irrigation Studies. (See IAEA, 1967) (In French) (NSA 22: 23160).

Chase, M. L. (Ed.). 1971. The Fern Lake Studies. College of Fisheries, Univ. of Washington, Seattle, Contribution No. 352. iv, 75pp. (Also: U. S. AEC report RLO-2225-T7-6) (Used radionuclides).

Chatterjee, N. K. and P. N. Bhaduri. 1965. Radiation-induced somatic mutations in Cyathodium. J. Heredity 56(3): 109-111. (NSA 19: 45901).

Chatterjee, S. 1967. Succession of growth phases in irradiated amoeba. Nucleus (Calcutta) 10: 23-27. (NSA 22: 8628).

Chatterjee, S. 1968. X-ray induced changes in the cell body of amoeba. Z. biol. 116: 68-80. (NSA 22: 45682).

Chatterjee, S. 1968. Chemocytology of lethally irradiated amoeba. Sci. & Cult. (Calcutta) 34: 108-110. (NSA 22: 49189).

Chatterjee, S. and S. K. Bhattacharjee. 1975. Effect of near ultraviolet and visible light on amoeba. J. Cell Sci. 19(1): 117-126. (NSA 33: 9591).

Chatterji, S., et al. 1960. Labelling of adults of Bracon brevicornis Wesmael with radioactive phosphorus. Indian J. Entomol. 22(3): 226-230.

Chatterji, S., et al. 1964. Labelling of the adults of Chilo partellus Swinhoe (C. zonellus), the stalk borer of maize and jowar, with radioactive phosphorus (^{32}P). Current Sci. (India) 33: 652-653. (NSA 19: 27927).

Chatters, R. M., R. L. Peterson and R. D. Rice. 1968. Studies on the application of radioisotope techniques in stream pollution problems in the pulp and paper industry. Trans. Amer. Nucl. Soc. 11: 51. (NSA 22: 30779).

Chatters, R. M., R. L. Peterson and R. D. Rice. 1969. Studies on the application of radioisotope techniques in stream pollution problems in the pulp and paper industry, Part I. Paper mill fiber and chip tagging and tracing technique, Part II. Final report. Washington State Univ., Pullman, U. S. AEC report RLO-1951-3. 40pp. (NSA 24: 16674).

Chattopadhyay, A. and R. E. Jervis. 1974. Multielement determination in market-garden soils by instrumental photon activation analysis. Anal. Chem. 46(12): 1630-1639. (NSA 30: 28921).

Chattopadhyay, A., L. G. I. Bennett and R. E. Jervis. 1972. Activation analysis of environmental pollutants. Can. J. Chem. Eng. 50(2): 189-193. (NSA 26: 45198).

Chaudhuri, J. P. 1968. Radioprotective effects of AET, glutathione, and serotonin in an irradiation in air and nitrogen atmospheres of grasshopper chromosomes. Strahlentherapie 35: 250-253. (In German) (NSA 22: 36318).

Chauvin, J. L. and J. Danion. 1963. Measurements of the bed load rate of streams by means of tracers. European Atomic Energy Community report EUR-1837.f. 28pp. (NSA 19: 601).

Chauvin, R. and J. Lecomte. 1964. On the exchanges of the "second degree" among colonies of daughters of Formica polyctena studied by means of radioisotopes. Insectes soc. 11(2): 97-104. (In French, English summary) (^{198}Au).

Chauvin, R. and J. Lecomte. 1965. Evolution of exchanges between different daughter colonies of Formica polyctena measured by means of radioisotopes, ^{198}Au. Insectes soc. 12: 197-200. (In French).

Chauvin, R. and J. Lecomte. 1966. Exchanges between ants of Formica polyctena with the aid of radioisotopes. Insectes soc. 13(1): 1-4.

Chauvin, R., G. Courtois and F. Anguenot. 1962. Ants (Formica rufa) as possible indicators for radioactive fallout. Compt. rend. 256(2): 508-511. (In French) (NSA 17: 15620).

Chavin, W. 1956. Thyroid distribution and function in the goldfish Carassius auratus L. determined by the uptake of tracer doses of radioiodine. Abstr. Anat. Record 124(216): 272.

Chavin, W. and B. N. Bouwman. 1963. Effect of radioiodine on the depression of thyroid iodide uptake in goldfish, Carassius auratus L. Abstr. Am. Zool. 3(4): 539-540. (BA 45: 54046).

Chavin, W. and C. A. Cukrowski. 1968. Effect of ^{131}I on thyroid and pituitary of the goldfish (Carassius auratus L.). Radiation Research 34: 170-199. (NSA 22: 36208).

Chawla, S. S., et al. 1973. Sensitivity of Oryzaephilus surinamensis (Coleoptera: Cuculidae) to cesium-137. I. Irradiation of eggs: the effects on hatching, survival, and fecundity. Can. Entomol. 105(2): 283-288. (NSA 32: 17538).

Chawla, S. S., et al. 1973. Sensitivity of Oryzaephilus surinamensis. II. Irradiation of larvae, pupae, and adults. J. Econ. Entomol. 63(3): 605-608. (NSA 28: 15756).

Chebotina, M. Ya. 1968. Influence of aqueous extraction from tree leaves on the migration of ^{54}Mn in various types of soils. Tr. inst. ekol. rast. zhivotn., Ural. fil. Akad. Nauk SSSR (61): 26-30. (In Russian) (NSA 24: 7773).

Chebotina, M. Ya. 1968. On the properties of desorptive effects of extracts of leaf fall. Tr. inst. ekol. rast. zhivotn., Ural. fil. Akad. Nauk SSSR (61): 12-21. (In Russian) (NSA 24: 7779).

Chefurka, W. 1965. Some comparative aspects of the metabolism of carbohydrates in insects. Ann. Rev. Entomol. 10: 345-382. (Contains references to radionuclides).

Chefurka, W. 1965. Estimation of pathways of carbohydrate metabolism in insects. Proc. Entomol. Soc. Ont. 96: 17-23. ($^{14}CO_2$).

Cheikh, M. 1974. Utilization of radiation sources in the fight against the principal insects. pp. 19-28. In: Isotopes in Pesticides and Pest Control. (See Nehme and Hassan, 1974) (In French) (NSA 33: 7560).

Cheikh, M., et al. 1975. Suppression of the Mediterranean fruit fly in Tunisia with released sterile insects. J. Econ. Entomol. 68(2): 237-243. (NSA 32: 1182).

Chekalov, K. I. and V. P. Illyuviyeva. 1963. Use of C^{14} isotope for the study of the decomposition of organic matter in soil. Pochvovedeniye (5): 40-50. (Sov. Soil Sci. (5): 482-490).

Chen, C. and M. Miyano. 1966. Effect of light and darkness on some organic compounds in Lemna perpusilla. Nippon Nogeikagaku Kaishi 40: 161-164. (In Japanese) (NSA 20: 45255) ($^{14}CO_2$).

Chen, H. C. 1975. Neutron activation analysis of underground water samples taken from blackfoot disease affected area. Nucl. Sci. J. (Taiwan) 12(1): 23-28. (NSA 32: 16709).

Chen, I. C., K. Peng and H. T. Kao-Chen. 1973. Preliminary survey of the condition of genesis of radon in the groundwater in Peking and adjacent regions. Ti Chih Ko Hsueh (2): 167-174. (NSA 30: 18641).

Chen, S. L. 1951. Simultaneous movement of P-32 and C-14 in opposite directions in phloem tissues. Am. J. Botany 38: 203-211.

Chen, T. S., J. N. Beck and P. K. Kuroda. 1970. Daily fluctuations of radioactive fallout particles in groundlevel air. Tellus 22(6): 707-711.

Cheng, W. and D. T. North. 1972. Inhibited sterility in the F_1 progeny of irradiated male pink bollworms. J. Econ. Entomol. 65(5): 1272-1275. (NSA 26: 59118).

Cherdyntsev, V. V. 1972. Basic tasks of biogeochemistry of natural radioelements. pp. 3-7. In: Methods of Determination of Radioactivity. (See Polikarpov and Parchevskiy, 1972) (In Russian).

Cherdyntsev, V. V., T. A. Ibrayev and V. F. Moshalova. 1964. Radioberyllium in fossil bones. Geokhimiya (12): 1240-1243. (Abstr. Geochem. Int. (6): 1148; full transl. Am. Geol. Inst., Washington, doc. GI-64-6-23).

Cherdyntsev, V. V., et al. 1964. Isotopic composition of uranium and thorium in the zone of hypergenesis. Investigation of the peat bog matter. Geokhimiya (5): 399-403. (Geochem. Int. (3): 398-401) (NSA 18: 31761).

Cherdyntsev, V. V., et al. 1965. Radiocarbon dates of the laboratory of the Geological Institute. Geokhimiya (12): 1410-1422. (Abstr. Geochem. Int. (6): 1043; full transl. Am. Geol. Inst., Washington, doc. GI-65-12-3) (NSA 20: 18931).

Chedyntsev, V. V., et al. 1966. Non-equilibrium in uranium carbonate deposits and their age determination. Geokhimiya (2): 139-145. (Abstr. Geochem. Int. 5(1): 92; full transl. Am. Geol. Inst., Washington, doc. GI-66-2-1) (NSA 20: 20887) (Incl. lake waters).

Cherdyntsev, V. V., et al. 1968. ^{239}Pu in nature. Geokhimiya (4): 395-401. (Geochem. Int. 5: 355-356; AEC-tr-6968) (NSA 23: 2411).

Cheremisinoff, P. N. and Y. H. Habib. 1972. Cadmium, chromium, lead, and mercury: plenary account for water pollution. I. Occurrence, toxicity, and detection. Water Sewage Works 119(7): 73-74; 80-86. (NSA 28: 2166).

Cherepanov, A. I. and K. P. Volgina. 1954. Determining the movement of wireworms in the soil by the method of tagged atoms. Dokl. Akad. Nauk SSSR 98(2): 301-302. (In Russian).

Cherezhanova, L. V. and R. M. Aleksakhin. 1971. Cytogenetic effects of long longterm irradiation on natural plant populations. Zh. obshch. biol. 32 (4): 494-500. (In Russian, English summary) (NSA 26: 9520).

Cherezhanova, L. V. and R. M. Aleksakhin. 1975. Biological effect of an increased level of ionizing radiation and the processes of radioadaptation in populations of herbaceaous plants. Zh. obshch. biol. 36(2): 303-311. (In Russian) (NSA 32: 11716).

Cherezhanova, L. V., R. M. Aleksakhin and Ye. G. Smirnov. 1971. Cytogenetic adaptation of plants under a chronic exposure to ionizing radiation. Genetika 7(4): 30-37. (In Russian) (NSA 39957).

Chernilin, Y. F. 1972. Environmental impact of nuclear power development. At. Jap. 16(4Suppl.): 95-114. (NSA 28: 17538).

Cherry, D. S. and W. K. Willard. 1971. Population response of twelve species of insects following exposure to ionizing radiation. Ann. Entomol. Soc. Am. 64: 456-461. (NSA 25: 24313).

Cherry, R. D. 1964. Alpha-radioactivity of plankton. Nature 203(4941): 139-143. (NSA 18: 35274).

Cherry, R. D. 1964. Alpha particle detection techniques applicable to the measurement of samples from the natural radiation environment. In: The Natural Radiation Environment. pp. 407-424. (See Adams and Lowder, 1964) (NSA 18: 35887).

Cherry, R. D., M. M. Skay and L. V. Shannon. 1970. Natural α-radioactivity concentrations in bone and liver from various animal species. Nature 228: 1002-1003. (NSA 25: 21648).

Chertok, R. J. and S. Lake. 1971. Availability in the peccary pig of radionuclides in nuclear debris from Plowshare excavation Buggy. Health Physics 20: 313-316. (NSA 25: 35088).

Chertok, R. J. and S. Lake. 1971. Biological availability of radionuclides produced by the Plowshare event Schooner. II. Retention and excretion rates in peccaries after a single oral dose of debris. III. Accumulation, excretion rates, and body distribution in peccaries after daily ingestion of debris. Health Physics 20: 325-330, 577-584. (NSA 25: 35089; 46690).

Cherubin, L. J. 1956. Potential hazards associated with the release of laboratory liquid waste to the Mohawk River. pp. 406-412. In: Sanitary Engineering Aspects of the Atomic Energy Industry. U. S. AEC report TID-7517 (Pt. Ib).

Cheshire, M. V., C. M. Mundie and H. Shepherd. 1969. Transformation of ^{14}C glucose and starch in soil. Soil Biol. Biochem. 1(2): 117-___.

Chesnutt, M. W., J. C. Drobinski, Jr. and R. H. Gorrie. 1966. Tritium in surface waters, 1964-1965. Radiol. Hlth. Data Repts. 7: 377-380. (NSA 22: 28527).

Chesselet, R. and J.-M. Martin. 1969. Note on the application of gamma spectrometry in situ to the study of solid discharge in the Gironde Estuary. Rev. geogr. phys. geol. dyn. 11: 123-126. (In French) (NSA 24: 2137).

Chesselet, R., D. Nordemann and B. Dussart. 1963. The radioisotopic determination of diffusion coefficients and currents in natural waters. pp. 63-76. In: Radioisotopes in Hydrology. (See IAEA, 1963) (In French) (NSA 18: 1906).

Chesselet, R., D. Nordemann and C. Lalou. 1965. Comparative study of the rate of maritime and continental radioactive fallout. Compt. rend. 260: 2875-2878. (In French) (NSA 19: 24755).

Chesser, A. H. 1963. Anatomical variations in the secondary xylem of Quercus alba (L.) and Liquidambar styraciflua (L.) with respect to in vivo exposure to gamma-neutron radiation. Thesis, Univ. of Georgia, Athens. vi, 47pp.

Chettachai Baditsung. 1966. Study on the life history and the effect of radiation on the rice weevil (Sitophilus oryzae L.). pp. 42-47. In: Insect Eradication by Irradiation. Thailand Office of Atomic Energy for Peace, Bangkok, report THAI.-AEC-8.

Chew, R. M. 1971. The excretion of ^{65}Zn and ^{54}Mn as indices of energy metabolism of Peromyscus polionotus. J. Mammalogy 52(2): 337-350.

Chew, R. M. and J. G. Hemington. 1973. Turnover of zinc-65 as an index of energy metabolism of Perognathus lingimembris, the little pocket mouse. pp. 247-252. In: Radionuclides in Ecosystems. (See Nelson, 1973) (NSA 28: 27673).

Chhabra, A. S. and P. R. Kalmuth. 1964. Environmental studies in some monazite placer areas in southern India. India Atomic Energy Estab., Trombay, Bombay, report AEET/HP/Environ./4. 10pp. (NSA 20: 7294).

Chi, L. W. and F. Boeller. 1968. X radiation of Oncomelania formosana, snail host of Schistosoma japonicum: effect on reproductivity and mortality. Am. J. Trop. Med. Hyg. 17: 900-903. (NSA 23: 29724).

Chiang, J. J. H. 1965. Studies on the inheritable variability of radiation of insects and the biologigical control of (?) sterilized insects. J. Agric. For. (Nanking) 14: 255-269. (In Chinese).

Chiang, S. H. 1963. Study on the biological effects of radiation on insects. I. Biological effects of external x and gamma radiation. J. Agric. For. (Nanking) 12: 114-141. (In Chinese, English summary).

Chiari, E., E. M. Neto and Z. Brener. 1968. Some effects of gamma radiation on Trypanosoma cruzi, culture and blood forms. Rev. Inst. Med. Trop. São Paulo 10: 131-137. (NSA 23: 50377).

Chigarev, G. A. 1963. The present state of the protection of potato against Colorado beetle in the USSR and the problems involved. Tr. vses. inst. zashch. rast. 17: 324-343. (^{60}Co).

Chigusa, S. I. 1968. A study of the effects of x-ray irradiation on a quantitative character, pupal weight, under various genetic backgrounds of Tribolium castaneum. Thesis, Purdue Univ., Lafayette, Ind. 292pp. (NSA 23: 41284).

Chilton, B. S. 1974. Effects of gamma radiation on the ultrastructure and acid phosphatase activity of the midgut of the adult beet armyworm, Spospoptera exigua. Ann. Entomol. Soc. Am. 67(3): 433-441. (NSA 30: 7113).

Ching, T. M. and S. C. Fang. 1963. Utilization of labeled glucose in developing Douglas fir seed cones. Plant Physiol. 38(5): 551-554. (RDA-7: 145) (^{14}C).

Chino, H. and L. I. Gilbert. 1965. Lipid release and transport in insects. Biochim. biophys. Acta 98: 94-110. (^{14}C).

Chino, H. and L. I. Gilbert. 1965. Studies on the interconversion of carbohydrate and fatty acid in Hyalophora cecropia. J. Insect Physiol. 11: 287-295. (^{14}C).

Chira, E. and J. Kantor. 1970. Sensitivity of seeds of some forest species to the action of gamma rays. Acta Univ. Agr. (Brno) C39(4): 315-323. (NSA 26: 15436).

Chirkov, A. M. 1971. Rn^{222} content in the thermal springs of Kamchatka. Dokl. Akad. Nauk SSSR 199: ___. (Doklady (earth sci.) 199(1-6): 196-197).

Chit, K. D. and P. U. Hla. 1968. Radioactive fallout in rain water during monsoon 1966. Un. Burma J. Sci. Tech. 1: 591-593.

Chitani, T., M. Kondo and H. Kotsuro. 1958. Studies of the growth inhibition of Ascaris eggs by Co^{60} gamma rays. pp. 1313-1318. In: Proc. 2nd Japan Conf. on Radioisotopes. (Paper B-2). (See Japan Radioisotopes Assoc., 1958). (AEC-tr-4482) (NSA 15: 25156).

Cho, C. M. 1972. Use of isotopes for ionic transport studies in soil. pp. 145-152. In: Isotopes and Radiation in Soil-Plant Relationships Including Forestry. (See IAEA, 1972) (NSA 27: 2855).

Choban, R. G. 1975. Hyperplasia and associated responses of the ventriculus following cobalt-60 gamma irradiation of Periplaneta americana (L.) (Dictyoptera: Blattidae). Dissertation, Rutgers Univ., New Bruswick, N. J. 95pp. (Univ. Microfilms 75-24,673) (NSA 33: 26295).

Chou, C. C., et al. 1974. Determination of radionuclides in Taiwan research reactor primary cooling water. I. Nucl. Sci. Sci. J. (Taiwan) 11(3): 154-159. (In Chinese) (NSA 31: 10601).

Chow, V. T. 1967. Progress of hydrology. pp. 3-20. In: Isotopes in Hydrology. (See IAEA, 1967) (NSA 22: 18924)

Chowdhury, A. B. and A. K. Bandyopadhay. 1960. Radioactive tagging of the free-living stages of Strongyloides stercoralis with P^{32}. Bull. Calcutta Sch. Trop. Med. 7: 105.

Chowdhury, S. U. and M. M. Hossain. 1967. Lethal doses of gamma radiation on certain spoilage fungi of stored rice. Food Irradiat. 8(1,2): 30-33. (NSA 22: 43487).

Christensen, E. 1954. Root production in plants following localized stem irradiation. Science 119(3082): 127. (Also: Brookhaven National Lab., U. S. AEC report BNL-1596) (NSA 8: 435).

Christensen, E. A. and N. W. Holm. 1964. Inactivation of dried bacteria and bacterial spores by means of ionizing radiation. Acta Pathol. Microbiol. Scand. 60(2): 253-264. (NSA 18: 29253).

Christensen, E. R. 1973. A moving source neutron moisture gauge. Int. J. Appl. Radiat. Isotop. 24(8): 479-481. (NSA 28: 21369).

Christensen, E. R. 1974. Use of the gamma density gauge in combination with the neutron moisture probe. pp. 27-44. In: Isotope and Radiation Techniques in Soil Physics and Irrigation Studies 1973. (See IAEA, 1974) (NSA 30: 15483).

Christenson, L. D. 1956. Recent progress in the development of procedures for eradicating or controlling tropical fruit flies. Proc. 10th Int. Congr. Entomology 3: 11-16.

Christenson, L. D. 1961. Control of insect pests by radiation sterilization techniques. pp. 169-187. In: Applications of Radioisotopes and Radiation in the Life Sciences. (See U. S. Congress, 1961) (NSA 15: 27317).

Christenson, L. D. and R. H. Foote. 1960. Biology of fruit flies. Ann. Rev. Entomol. 5: 171-192.

Chopra, D. P. 1972. Some effects of x-irradiation on the activity of the thyroid and pituitary glands of Xenopus laevis (Amphibia) during metamorphosis. J. Zool. 166(2): 159-162. (NSA 26: 41175).

Chuang, L. A., C. Y. Wong and S. V. Law. 1970. Measurement of absorbed dose rate from terrestrial background radiation in Hong Kong. J. Radiat. Res. (Japan) 11: 53-60. (NSA 25: 19006).

Chuang, T. F., et al. 1968. ^{137}Cs metabolism in Blatta orientalis at various temperatures. Comp. Biochem. Physiol. 24: 997-1002. (NSA 22: 28327).

Chuburkov, Yu. T. and L. M. Lebedev. 1974. Thermal springs and superheavy elements. Radiokhimiya 16(4): 524-529. (Sov. Radiochem. 16(4): 517-521) (NSA 33: 23280).

Chudy, M., et al. 1970. ^{14}C in the atmosphere and biosphere. Radioisotopy 11(5): 935-951. (In Czech) (NSA 25: 57364).

Chueca, A., M. Worwood and D. M. Taylor. 1969. The simultaneous determination of zinc and cadmium in biological materials by neutron activation analysis. Int. J. Appl. Radiat. Isotop. 20: 335-340. (NSA 23: 28762).

Chugunov, V. V. 1968. On a relation of the coefficient of accumulation of radioisotopes in aquatic organisms with the position of the element in the periodic system of Mendeleyev. Gidrobiol. zh. Akad. Nauk Ukr. SSR 4(5): 58-62. (NSA 23: 24588).

Chugunov, V. V., et al. 1971. Investigation of the statistical distribution of ^{144}Ce in the organs and tissue of hydrobionts. Radiobiologiya 11(2): 298-301. (AEC-tr-7269; pp. 217-222) (NSA 25: 44762).

le Chuiton, F. L., L. Tanon and A. Besson. 1961. On the subject of the immersion in oceans and rivers of radioactive wastes from nuclear installations. Bull. Acad. Natl. Med. (Paris) 145: 340-349. (NSA 17: 15526).

Chumichev, V. B. 1963. Accumulation of radioactive strontium by freshwater plants. Comm. on Use of Atomic Energy, USSR, Moscow. 5pp. (AEC-tr-6447) (NSA 18: 35282).

Chumichev, V. C. 1969. Radiochemical method for determining low strontium-90 concentration levels in environmental samples. pp. 335-339. In: Nuclear Meteorology. (See Makhon'ko and Malakhov, 1969) (NSA 30: 14844).

Chun, S. Y. 1971. Determination of bromine, arsenic, mercury, and selenium in plants by neutron activation analysis. Han'guk Sikp'um Kwahakhoe Chi 3(3): 144-150. (NSA 26: 52907).

Chung, K. H. and J. Ryu. 1971. Studies on the rice stem borer control using sterile-male technique. Han'guk Sikmul Poho Hakhoe Chi 10(2): 117-120. (In Chinese) (NSA 27: 27990).

Chung, R. A., D. H. Shaw and E. S. So. 1966. Fatty acid changes in pigeons after lethal-body radiation and bone marrow and thymus administration. Industr. Med. Surg. 35: 607. (NSA 21: 16462).

Chung, R. A., D. H. Shaw and E. S. So. 1967. Fatty acid changes in pigeons after lethal body radiation and bone marrow and thymus administration. Industr. Med. Surg. 36: 202-204. (NSA 21: 32783).

Chung, R. A., E. S. So and D. H. Shaw. 1968. Fatty acid changes in liver fractions of pigeons after lethal-dose radiation and therapy. Poultry Sci. 47: 1127-1130. (NSA 24: 4642).

Chung, S. L. and P. C. Lippold. 1967. Cobalt-60 sterilization studies with the European chafer. II. Bull. Entomol. Soc. Am. 13(3): 192.

Chung, S. L., H. Tashiro and L. M. Massey, Jr. 1971. Gamma irradiation of the European chafer. 3. Longevity, flight behavior, and mating Competitiveness of sterilized males, with observations on reproductive biology. J. Econ. Entomol. 64(4): 883-886. (NSA 25: 49275).

Chung, S. L., et al. 1971. Gamma irradiation of the European chafer. I. Irradiation of third-stadial larvae, with observations on spermatogenesis. J. Econ. Entomol. 64(4): 823-832. (NSA 25: 49279).

Chung, S. L., et al. 1971. Gamma irradiation of the European chafer. 2. Determination of sterilization dose levels for adults, with notes on rearing techniques. J. Econ. Entomol. 64(4): 832-837. (NSA 25: 49276).

Chupka, Sh., M. Petrashova and I. Tsarakh. 1969. Contamination of certain objects of the environment in western Slovakia. Gigiyena sanitariya 5: 75-77. (In Russian) (NSA 25: 46811).

Churayev, N. V. 1960. Study of the water properties, structure and processes of moisture transfer in peat using labelled atoms. Proc. Tashkent Conf. on Peaceful Uses of Atomic Energy, 1959. Ya. Kh. Turakulov, ed. Publ. House Acad. Sci. Uzbek SSR, Tashkent. 2: 243-___. (AEC-tr-6390; pp. 322-338) (NSA 18: 39515).

Churayev, N. V. and N. I. Il'in. 1967. Radiotracer Methods in the Investigation of Underground Water Flow. Atomizdat, Moscow. 124pp. (In Russian) (NSA 21: 30539).

Churayev, N. V. and N. I. Il'in. 1973. Radioindication Methods for Investigating the Motion of Subterranean Waters. Atomizdat, Moscow. 176pp. (In Russian) (NSA 28: 24499).

Churayev, N. V. and L. G. Rode. 1966. Measurement of moisture in organic soils by the neutron method. Pochvovedeniye (1): 96-100. (CA 64: 11800c; NSA 20: 14756).

Church, B. W., E. S. Medling and D. N. Brady. 1975. Different look at Area 13 fidler survey data. pp. 231-236. In: Radioecology of Plutonium and Other Transuranics in Desert Environments. (See White and Dunaway, 1975) (NSA 33: 619).

Church, B. W., et al. 1974. Distribution and inventory element activities on-NTS and off-NTS. pp. 311-319. In: Dynamics of Plutonium in Desert Environments. (Nevada Applied Ecology Group Progress Report as of January 1974). (See Dunaway and White, 1974) (NSA 30: 29453).

Church, K. and D. E. Wimber. 1969. Meiosis in the grasshopper: chiasma frequency after elevated temperature and x-rays. Can. J Genet. Cytol. 11: 209-216. (NSA 25: 8754).

Churchill, M. A., et al. 1965. Concentrations, total stream loads, and mass transport of radionuclides in the Clinch and Tennessee Rivers. Oak Ridge National Lab., U. S. AEC report ORNL-3721(Suppl. 1). 72pp. (NSA 25: 42238).

Churkin, V. M. 1969. Graphoanalytical method for determining cesium-137 concentration in soil samples from the gamma spectrum. pp. 319-334. In: Nuclear Meteorology. (See Makhon'ko and Malakhov, 1969) (NSA 30: 14843).

Chute, J. H., R. A. Clapp and J. P. Quirk. 1970. Alpha-activity of western Australian soils and wheats. J. Roy. Soc. West. Austral. 53: 37-__. (NSA 25: 42427).

Chuvakhin, V. S. 1961. Sexual sterilization as a means of controlling insect pests. Zashchita rasteniy 2: 45-__. (In Russian).

Ciaccio, L. L. (Ed.). 1973. Water and Water Pollution Handbook. Vol. 4. Marcel Dekker, New York. 641pp. (NSA 29: 2445).

Ciftcioglu, O. and D. Taylor. 1972. Moving gamma-source soil density gauges. Soil Science 113(1): 13-17. (NSA 26: 28251).

Cigna, A. A. 1965. Some considerations of radiation contamination from fallout in underground water. pp. 451-464. In: Proc. Conf. on the Problem of Water in Italy. Part I. Underground Waters. T. Milan, ed. (In Italian) (NSA 21: 26389).

Cigna, A. A. 1969. Some aspects of environmental radioactivity research. Arh. Hig. Rada Toksikol. 20: 25-38. (NSA 24: 31902, 46594).

Cigna, A. A. 1972. Conventional and radioactive pollution in the environment. Com. Naz. Energ. Nucl. Notiz. 18(5): 25-17. (In Italian) (NSA 26: 45555).

Cigna, A. A. and L. Cigno-Rossi. 1970. Operative radioecology. pp. 39-42. In: Proc. 1st Conf. on the State of Advancement of Radioecology in Italy. (See Anonymous, 1970) (In Italian).

Cigna, A. A. and C. Polvani. 1973. Radioactive fallout in the Mediterranean region: researches, results, and perspectives. pp. 240-258. In: Proc. Regional Conf. on Radiation. Vol. 1. Y. Feige, ed. Soreq Nucl. Research Centre, Yavne, Israel. (NSA 30: 15387).

Cigna, A. A. and M. Talenti. 1965. The natural radioactivity of some Italalian mineral waters. Minerva Nucl. 9: 248-250. (In Italian) (NSA 20: 9151).

Cigna, A. A., C. Flaoci and C. Polvani. 1974. Health protection and control of environmental radioactive pollution. Com. Naz. Energ. Nucl. Notiz. 20 (2): 25-29. (In Italian) (NSA 30: 6497).

Cigna, A. A., et al. 1963. Composition and deposition in Italy of fallout from U. S. S. R. explosions - fall 1961. Health Physics 9(3): 319-331. (NSA 17: 24875).

Circeo, L. J., Jr. 1969. Nuclear excavation: review and analysis. Eng. Geol. (Amsterdam) 3: 5-59. (NSA 14: 346).

Cirio, U. and I. de Murtas. 1974. Status of Mediterranean fruit fly control by the sterile-male technique on the Island of Procida. pp. 5-16. In: Sterile-Insect Technique and Its Field Applications. (See IAEA, 1974) (NSA 31: 8935).

Ciuffolotti, L. 1967. Symposium on the standardization of radionuclides held by the IAEA (Vienna, October 10-14, 1966). Energia nucleare 14: 300-305. (In Italian) (NSA 21: 30641).

Claassen, H. C. 1970. The determination of low levels of cobalt-60 in environmental waters by liquid scintillation counting. Anal. Chim. Acta 52: 229-235.

Claassen, H. C. 1971. Radiochemical analyses of water samples from selected streams and precipitation collected October 23 - November 9, 1970, in conjunction with the first production-test flaring of gas from test well, Project Rulison. U. S. Geological Survey report USGS-474-108. i, 6pp.

Claassen, H. C. 1971. Radiochemical analyses of water from selected stream and precipitation collected immediately before and after the second production-test flaring, Project Rulison. U. S. Geological Survey report USGS-474-122. i, 6pp. (NSA 26: 17763).

Claassen, H. C. 1973. Water quality and physical characteristics of Nevada Test Site water-supply wells. U. S. Geological Survey, U. S. AEC report USGS-474-158. 152pp. (NSA 28: 397).

Claassen, H. C. and P. T. Voegeli, Sr. 1971. Radiochemical analyses of wat water samples from selected streams and precipitation collected in connection with calibration-test flaring of gas from test well, August 15 - October 13, 1970, Project Rulison. U. S. Geological Survey report USGS-474-107. i, 6pp.

Claeys, R. R. 1973. Introduction to the analytical methods for the determination of mercury compounds. pp. 229-233. In: Mercury in the Western Environment. (See Buhler, 1973) (Incl. neutron activation analysis).

Clajus, P. 1975. Ecology of radium. A literature survey. German Democratic Republic, Staatliches Amt für Atomsicherheit und Strahlenschutz, Berlin, report SAAS-183. 29pp. (In German) (NSA 33: 12009).

Clanton, U. S. 1963. Sorption and release of radionuclides by sediments. pp. 113-125. In: Transport of Radionuclides in Fresh Water Systems. (See Kornegay, et al., 1963) (NSA 17: 40525).

Clanton, U. S., Jr. 1968. Sorption and release of ^{90}Sr and ^{137}Cs by recent sediments of the Guadalupe River of Texas. Thesis, Univ. of Texas, Austin. 96pp. (NSA 23: 11650).

Clanton, U. S. and E. F. Gloyna. 1964. Radioactivity transport in water - sorption and release of radionuclides by sediments of the Guadalupe River. Environ. Hlth. Engrg. Res. Lab., Univ. of Texas, Austin, Tech. Rept. x, 64pp.

Clapper, L. S. 1970. Economic growth and environmental effects of nuclear energy development. pp. 135-142. In: Northwest Conf. on the Role of Nuclear Energy. (See Anonymous, 1970) (NSA 25: 23222).

Clark, A., III and J. R. Hamilton. 1968. Effects of ionizing irradiation on the xylem derivatives of Pinus echinata Mill. West Virginia Univ., Agricultural Expt. Sta. Bull. 567T. 16pp.

Clark, A. M. 1954. The relation of genome number to radiosensitivity in Habrobracon. Brookhaven National Lab., U. S. AEC report BNL-2014. 14pp. (NSA 8: 6623).

Clark, A. M. 1957. The relation of genome number to radiosensitivity in Habrobracon. Am. Naturalist 91(857): 111-119.

Clark, A. M. 1958. Some effects of x rays on longevity in Habrobracon females. Univ. of Deleware and Oak Ridge National Lab., U. S. AEC report TID-6053. 21pp. (NSA 14: 15527).

Clark, A. M. 1960. The relation of genome number to radiosensitivity. Univ. of Deleware, U. S. AEC report TID-6051. 20pp. (NSA 14: 15500).

Clark, A. M. 1961. Some effects of x-irradiation on longevity in Habrobracon females. Radiation Research 15(4): 515-519. (NSA 15: 32041).

Clark, A. M. 1961. The modifications of life span by x-rays for haploids and diploids of the wasp, Habrobracon sp. Abstr. Radiation Research 14 (4): 456, abstr. 30.

Clark, A. M. and K. W. Cole. 1967. The effects of ionizing radiation on the longevity of ploidy types in the wasp Mormoniella vitripennia. Exptl. Gerontol. 2: 89-95. (NSA 21: 41509).

Clark, A. M. and V. J. Cristofalo. 1961. Some effects of oxygen on the insects Anagasta kuehniella and Tenebrio molitor. Physiol. Zool. 34(1): 55-61.

Clark, A. M. and E. B. Herr. 1955. The effect of certain gases on the radiosensitivity of Habrobracon during development. Radiation Research 2(6): 538-543.

Clark, A. M. and E. M. Kelley. 1950. Differential sensitivity of haploid a and diploid prepupae and pupae of Habrobracon. Cancer Research 10: 348-352. (NSA 4: 4432).

Clark, A. M. and C. J. Mitchell. 1951. Radiosensitivity of haploid and diploid Habrobracon during pupal development. J. Exptl. Zool. 117: 489-498. (Also: U. S. AEC report AECU-1444) (NSA 5: 4956).

Clark, A. M. and C. J. Mitchell. 1952. Effects of x-rays upon haploid and diploid embryos of Habrobracon. Biol. Bull. 103(1): 170-177.

Clark, A. M. and D. E. Osmun. 1967. Adult life span in Habrobracon serinopae after x-irradiation during early development. Nature 214: 717-718. (NSA 21: 24566).

Clark, A. M. and M. A. Rubin. 1961. The modification by x-irradiation of the life span of haploids and diploids of the wasp, Habrobracon sp. Radiation Research 15(2): 244-253. (NSA 15: 24709).

Clark, A. M. and R. E. Smith. 1968. The modification of adult life span in Bracon hebetor (Hymenoptera: Braconidae) by irradiation in nitrogen. Ann. Entomol. Soc. Am. 61: 541-542. (NSA 22: 40860).

Clark, A. M, H. A. Bertrand and R. E. Smith. 1963. Life span differences between haploid and diploid males of Habrobracon serinopae after exposure as adults to x rays. Am. Naturalist 97: 203-208. (NSA 18: 8129).

Clark, B. W. and D. C. Coleman. 1970. A comparative study of the effects of acute and chronic gamma irradiation on total soil respiration. Pedobiologia 10: 199-206.

Clark, D. A. 1974. State-of-the-art: uranium mining, milling, and refining industry. National Environmental Res. Cen., Corvallis, Ore., U. S. EPA report EPA-660/2-74-038. 121pp. (NSA 30: 23966) (Environmental effects).

Clark, D. E. and W. C. Cobbin. 1963. Some relationships among particle size, mass level and radiation intensity of fallout from a land surface nuclear detonation. U. S. Naval Radiological Defense Lab., San Francisco, report USNRDL-TR-639. 79pp. (NSA 17: 29138).

Clark, E. W. and P. A. Glick. 1961. Some predators and scavengers feeding upon pink bollworm moths. J. Econ. Entomol. 54(4): 815-816. (^{32}P).

Clark, G. M., D. G. Baker and A. R. Cooke. 1962. The effect of dose and dose-rate of x-rays on the growth of cells with special reference to intermediary metabolism. Abstr. p. 191. In: 2nd Intern. Congr. Radiation Research. (See Anonymous, 1962) (Tetrahymena).

Clark, G. M., et al. 1965. Simulated fallout studies in conifers - a preliminary report. pp. 1627-1636. In: Radiation and Terrestrial Ecosystems. (See Hungate, 1965).

Clark, G. M., et al. 1967. Effects of thermal stress and simulated fallout on conifer seeds. Radiation Botany 7(3): 167-176.

Clark, G. M., et al. 1968. Germination and survival of conifers following chronic gamma irradiation of seed. Radiation Botany 8(1): 59-66. (NSA 22: 33976).

Clark, H. F., et al. 1960. A review of the literature of 1959 on wastewater and water pollution control. J. Water Poll. Contr. Fed. 32: 443-481; 545-593; 681-720.

Clark, J. B. 1966. Final report (on photoprotection against x-ray induced damage), 1966. Univ. of Oklahoma, Norman, U. S. AEC report TID-23347. 32pp. (NSA 22: 45363).

Clark, J. B. and J. Frady. 1959. Variation to radiation resistance in Nocardia corallina. J. Bacteriol. 77(4): 414-416. (NSA 13: 12390).

Clark, J. B. and R. B. Webb. 1957. Effect of successive x irradiation on Nocardia corallina. J. Bacteriol. 74(1): 26-30.

Clark, J. B. and R. B. Webb. 1960. Overlap of site of action of x-ray and ultraviolet irradiation in Nocardia corallina. J. Bacteriol. 80(1): 72-76.

Clark, O. H. 1952. The use of phosphorus-32 in studies of Plasmodium gallinaceum. II. Studies on conditions affecting parasite growth in intact cells and lysates. J. Exptl. Med. 96: 439-450; 451-463.

Clark, R. M. and C. Renfrew. 1973. Tree-ring calibration of radiocarbon dates and the chronology of ancient Egypt. Nature 243(5405): 266-270. (NSA 28: 15491).

Clarke, R. T. 1968. The statistical analysis of a radio-active tracer experiment to determine root activity in potato plants. Biometrics 24(3): 717-722. (Applicable to wild plants).

Clausen, K. E. 1973. Effect of pollen irradiation on reproductive capacity, seedling growth, and variation of Betula nigra. Radiation Botany 13(1): 47-54. (NSA 27: 27879).

Clayton, C. G. and D. B. Smith. 1963. A comparison of radioisotope methods for river flow measurement. pp. 1-24. In: Radioisotopes in Hydrology. (See IAEA, 1963) (NSA 18: 1912).

Clayton, F. E. 1965. Genetic effects from simultaneous irradiation of imm ture and mature Drosophila virilis males. Genetics 52(5): 1081-1092.

Clayton, G. A. and A. Robertson. 1964. The effects of x-rays on quantitative characters. Genetical Res. 5: 410-422. (Drosophila melanogaster).

Clayton, R. B. 1965. The utilization of sterols in insects: a novel approach to the study of cell-membrane structure. Biochem. J. 96(2): 178. (^{14}C, ^{3}H).

Clayton, R. B. and A. M. Edwards. 1962. A novel biological desaturation of the steroid nucleus: conversion of cholestanol to Δ^7-cholestanol in cockroaches. Abstr. 217. Federation Proc. 21: 297. (^{14}C, Eurycotis floridana, Blattella germanica).

Clayton, R. B. and A. M. Edwards. 1963. Conversion of 5 α-cholestan-3β-ol to Δ^7-5α-cholestan-3β-ol in cockroaches. J. Biol. Chem. 238: 1966-1972. (^{14}C, ^{3}H, Eurycotis floridana, Blattella germanica).

Clayton, R. B. and N. L. Lasser. 1964. The intracellular distribution of sterols in the tissues of the cockroach, Eurycotis floridana. Abstr. 1045. Federation Proc. 23(2,pt.1): 275. (^{14}C, ^{3}H).

Clayton, R. K. and P. S. Rudolph. 1962. The lethal effects of alpha irradiation on wild-type and on a high-catalase mutant of Rheudopseudomonas spheroides. Radiation Research 17: 839-846. (NSA 17: 4416).

Clegg, B., J. Koranda and G. Hadley. 1971. System for correlating tritium oxide transport in vegetation with micrometeorological variables. Lawrence Livermore Lab., U. S. AEC report UCRL-73373. 6pp. (For Nuclear Sci. Symp., San Francisco, Calif., 3 Nov. 1971) (CONF-711111-20) (NSA 26: 9486).

Clemente, G. F. 1974. Trace element concentrations in some Italian underground waters, determined by neutron activation analysis. pp. 213-227. In: Comparative Studies of Food and Environmental Contamination. (See IAEA, 1974) (NSA 30: 14764).

Clemente, G. F. 1975. Trace-element pathways from environment to man. Trans. Amer. Nucl. Soc. 21(3 Suppl.): 17. (NSA 32: 27480).

Clemente, G. F. and G. G. Mastinu. 1974. Instrumental method for the determination of trace elements in water samples by neutron activation analysis. J. Radioanal. Chem. 20(2): 707-714. (NSA 30: 20598).

Clementel, E. 1974. Nuclear energy and the environment. Com. Naz. Energ. Nucl. Notiz. 20(7): 25-29. (In Italian) (NSA 30: 31262).

Clements-Merlini, M. 1960. The secretory cycle of iodoproteins in ammocoetes. I. A radioautographic time study of the subpharyngeal gland. J. Morphol. 106(3): 337-342, 7 pls.

Clements-Merlini, M. 1960. The secretory cycle of iodoproteins in ammocoetes. II. A radioautographic study of the transforming larval thyroid gland. J. Morphol. 106(3): 357-359.

Clements-Merlini, M. 1962. Metabolism of I^{131} by the endostyle and notochord of ammocoetes under different conditions of temperature. Gen. Comp. Endocrinol. 2(3): 240-248. (In French) (NSA 18: 9822).

Clemson College. 1961. Proc. of the Southeastern Seminar on Atomic Progress in Agriculture. (February 13-14, 1961, Clemson, S. Car.). xx, 245pp.

Clemson University. 1968. Proc. Conf. on Hydrology in Water Resources Management. (Clemson, S. Car., March 28-29, 1968). Water Resources Res. Inst. report 4. 199pp. (NSA 23: 43397) (Incl. paper on Savannah River Plant).

Cleveland, R. F. and R. A. Deering. 1972. Radiation-induced division delay in Dictyostelium discoideum: relation to survival in resistant and sensitive strains. Int. J. Radiat. Biol. 22(3): 245-256. (NSA 26: 56389).

Clever, U. 1964. Actinomycin and puromycin: effects on sequential gene activation by ecdysone. Science 146: 794-795. (Chironomus tentans, ^{3}H).

Clever, U. 1965. Puffing changes in incubated and in ecdysone treated Chironomus tentans salivary glands. Chromosoma 17(4): 309-322. (^{3}H).

Clifford, C. E. 1964. Effects of the ground on the γ dose from distributed ^{137}Cs sources. Can. J. Phys. 42: 2373-2383. (NSA 19: 7744).

Clifford, H. F. 1959. Response of periphyton to phosphorus introduced into a Michigan trout stream. Thesis, Michigan State Univ., East Lansing.

Cline, J. F. 1967. The effects of substrate conditions on the uptake rate of ^{137}Cs by plants. Battelle Northwest Lab., U. S. AEC report BNWL-SA-1259. 16pp. (Also: pp. 547-552. In: Symposium on Radioecology. (See Nelson and Evans, 1969)) (NSA 21: 28429).

Cline, J. F. and B. Klepper. 1975. Iodine-125 accumulation in plant parts: influence of water use rate and stable iodine content of soil. Health Physics 28(6): 801-804. (NSA 32: 17586).

Cline, J. F. and B. Klepper. nd. Iodine uptake patterns and water use by several crop plants. Battelle Northwest Lab., U. S. AEC report BNWL-SA-5050. 15pp. (CONF-740702-5) (NSA 30: 21437).

Cline, J. F. and W. H. Rickard. 1972. Radioactive strontium and cesium in cultivated and abandoned field plots. Health Physics 23(3): 317-324. (NSA 26: 9486).

Cline, J. F. and W. H. Rickard. 1973. Radioactive strontium and cesium in cultivated and abandoned field plots. pp. 209-211. In: Radionuclides in Ecosystems. (See Nelson, 1973) (NSA 28: 27269).

Cline, J. F., D. O. Wolson and F. P. Hungate. 1965. Effect of physical and biological conditions on deposition and retention of ^{131}I on plants. Health Physics 11(8): 713-717. (Incl. sagebrush).

Cline, J. F., D. W. Uresk and W. H. Rickard. 1975. Characterization of plant communities adjacent to the B-C Cribs controlled area and REDOX pond areas on the 200 Area plateau. Battelle Northwest Lab., U. S. ERDA report BNWL-1916. 59pp. (NSA 33: 664) (At Hanford).

Cline, R. E. and G. W. Pearce. 1963. Unique effects of DDT and other chlorinated hydrocarbons on the metabolism of formate and proline in the housefly. Biochemistry 2: 657-662. (^{14}C).

Cline, R. E. and G. W. Pearce. 1966. Similar effects of DDT and convulsive hydrazides on housefly metabolism. J. Insect Physiol. 12: 153-162. ($^{14}CO_2$).

Cline, R. G. and B. L. Jeffers. 1975. Installation of neutron probe access tubes in stoney and bouldery forest soils. Soil Science 120(1): 71-72. (NSA 32: 19796).

Clopton, J. C. (ed.). 1970. Environmental Radioactivity. (A Symp. held at The Johns Hopkins Univ., January 19-20, 1970). The Johns Hopkins Univ., Baltimore. vii, 272pp.

Close, D. A., J. J. Malanify and C. J. Umbarger. 1974. Detection of the stable isotopes of carbon using a 3-MV Van de Graaff and the application of environmental and biological studies. Trans. Nucl. Sci. NS-21(1): 731-734. (NSA 30: 5878).

Clukey, H. V. 1956. Radioactivity levels of the Columbia River below Richland ---. Hanford Atomic Prod. Oper., U. S. AEC reports HW-45983, 9pp.; HW-46094, 9pp.; HW-47170, 9pp.; HW-49840, 8pp. (NSA 11: 1747, 1748, 3297, 9202).

Clukey, H. V. 1956. Radiation aspects for river navigation through Hanford projects. Hanford Atomic Prod. Oper., U. S. AEC report HW-47152. 8pp. (NSA 33: 7285).

Clukey, H. V. 1957. The Hanford atomic project and Columbia River pollution. Hanford Atomic Prod. Oper., U. S. AEC report HW-54243(rev.). 21pp. (NSA 13: 11038).

Coangelo, S., G. C. Cortellessa and S. Terrani. 1973. Presentation and discussion of the results of the radioactivity measurement in the sewage of some Italian cities. Italy Comitato Nazionale per l'Energia Nucleare, Rome, report RT/PROT-(73)35. 9pp. (In Italian) (CONF-711082-2) (NSA 30: 18639).

Cobian Roa, J. M. 1972. Siting of nuclear power stations. Thesis, Northwestern Univ., Evanston, Ill. 78pp. (Univ. Microfilms 73-10,204) (NSA 28: 4614).

Coffin, C. C., et al. 1949. Exchange of materials in a lake as studied by radioactive phosphorus. Nature 163(4155): 963-964. (NSA 3: 604).

Coffin, C. C., et al. 1949. Exchange of materials in a lake as studied by addition of radiophosphorus. Can. J. Res. 27(C): 207-222. (NSA 3: 1804).

Cogburn, R. R. and H. B. Gillenwater. 1972. Interaction of gamma radiation and fumigation on confused flour beetles. J. Econ. Entomol. 65(1): 245-248. (NSA 26: 23143).

Cogburn, R. R. and P. G. Mahany. 1969. Effect of gamma radiation on the insecticidal efficiency of malathion deposits on wheat and kraft paper. J. Econ. Entomol. 62(4): 829-831.

Cogburn, R. R. and R. D. speirs. 1972. Toxicity of malathion to gamma-irradiated and nonirradiated adult red flour beetles. J. Econ. Entomol. 65(1): 185-188. (NSA 26: 23146).

Cogburn, R. R. and E. W. Tilton. 1966. Gross effects of gamma radiation on the Indian meal moth and the angoumois grain moth. J. Econ. Entomol. 59: 682-685. (NSA 21: 6639).

Cogburn, R. R., J. H. Brower and E. W. Tilton. 1971. Combination of gamma and infrared radiation for control of the angoumois grain moth in wheat. J. Econ. Entomol. 64(4): 923-925. (NSA 25: 49280).

Cogburn, R. R., E. W. Tilton and J. H. Brower. 1972. Bulk-grain gamma irradiation for control of insects infesting wheat. J. Econ. Entomol. 65(3): 818-821.

Cogburn, R. R., E. W. Tilton and J. H. Brower. 1973. Almond moth: gamma radiation effects on the life stages. J. Econ. Entomol. 66(3): 745-751. (NSA 28: 13087).

Coggins, P. B. 1973. Effect if x radiation on spermatogenesis and the fertility of Schistocerca gregaria (Forsk.). J. Embryol. Exp. Morphol. 30 (1): 163-177. (NSA 29: 2941).

Cohen, A. J. and J. N. Smith. 1964. Comparative detoxication. 9. The metabolism of some halogenated compounds by conjugation with glutathione in the locust. Biochem. J. 90(3): 449-456. (^{35}S, Schistocerca gregaria).

Cohen, I. 1965. Insect control by the sterile male technique. Haklauth Israel 10(4): 45-46. (Ceratitis capitata).

Cohen, J. J., L. L. Schwarta and H. A. Tewes. 1974. Economic and environmental evaluation of nuclear waste disposal by underground in situ melting. Trans. Amer. Nucl. Soc. 18: 194-195. (NSA 30: 16494).

Cohen, N. and M. E. Wrenn. 1973. Metabolic characteristics of ^{241}Am in the adult baboon. Radiation Research 55(1): 129-143. (NSA 28: 21826).

Cohen, S. 1963. Carbon dioxide-14 fixation and the accumulation of malonic acid in amphibian hybrids (Rana pipiens females x Rana sylvatica males). Exptl. Cell Res. 29(2): 207-211.

Cohn, S. H. 1958. Persistence of radioactive contamination in animals of Marshall Islands two years after Operation Castle. pp. 211-218. In: The Shorter-Term Biological Hazards of a Fallout Field. (Symp. Dec. 12-14, 1956, Washington, D. C.). U. S. AEC and U. S. DOD, Washington.

Cohn, S. H., J. S. Robertson and H. A. Conard. 1960. Radioisotopes and environmental circumstances: the internal radioactive contamination of a Pacific island community exposed to local fallout. pp. 306-330. In: Radioisotopes in the Biosphere. (See Caldecott and Snyder, 1960.

Cohn, S. H., et al. 1956. Internal radioactive contamination of human beings accidentally exposed to radioactive fallout material. U. S. Naval Radiological Defense Lab., San Francisco, Calif., report USNRDL-TR-86. 22pp. (NSA 10: 6470).

Cohn, S. H., et al. nd. Nature and extent of internal radioactive contamination of human beings, plants, and animals exposed to fallout. U. S. AEC Weapons Test report WT-936. 49pp. (Incl. chickens, ducks, gull eggs).

Cole, M. M., G. C. LaBrecque and G. S. Burden. 1959. Effects of gamma radiation on some insects affecting man. J. Econ. Entmol. 52(3): ___.

Coleman, D. C. 1968. A method for intensity labelling of fungi for ecological studies. Mycologia 60(4): 960-961. (Radionuclides).

Coleman, D. C. 1970. Food webs of small arthropods of a broomsedge field studied with radio-isotope labelled fungi. pp. 203-207. In: Methods of Study in Soil Ecology. UNESCO, Paris. (^{134}Cs, ^{65}Zn).

Coleman, D. C. and B. W. Clark. 1969. Functioning of the litter-soil community when stressed by gamma radiation. Abstr. Bull. Ecol. Soc. Am. 50 (2): 70.

Coleman, D. D. and G. T. Cowley. 1973. Structure and function of hardwood litter and soil subsystems after chronic gamma irradiation. I. Mesofauna, nitrogen, and total soil respiration. pp. 1222-1228. In: Radionuclides in Ecosystems. (See Nelson, 1973) (NSA 28: 30245).

Coleman, D. C. and J. T. McGinnis. 1970. Quantification of fungus-small arthropod food chains in the soil. Oikos 21: 134-137. (^{65}Zn).

Coleman, D. C. and A. MacFadyen. 1966. The recolonization of gamma-irradiated soil by small arthropods. A preliminary study. Oikos 17(1): 62-70.

Coleman, D. C. and C. D. Monk. 1967. The biological half-life of 45-calcium in the fall webworm, Hyphantria cunea (Drury). Abstr. Bull. Ecol. Soc. Am. 48(3): 129.

Coleman, D. C., C. D. Monk and J. T. Callahan. 1969. Effect of food availability on rates of loss of calcium from Hiphantria cunea (Lepidoptera: Arctiidae). Can. Entomol. 101: 328-332. (^{45}Ca).

Coleman, J. R. and R. L. Liderace. 1966. Nuclear power production and estimated krypton-85 levels. Radiological Hlth. Data 7(11): 613-621.

Coleman, R. F. 1974. Comparison of analytical techniques for inorganic pollutants. Anal. Chem. 46(12): 989A-990A; 992A; 994A-996A. NSA 30: 28924) (Neutron activation analysis).

Coles, D. G., J. W. T. Meadows and C. L. Lindenken. 1974. Convenient method for the direct measurement of uranium in soil sample using high-resolution Ge(Li) gamma-ray spectroscopy. Lawrence Livermore Lab., U. S. AEC report UCRL-75619. 21pp. (NSA 30: 23557).

Colette, S. 1960. Evidence of radiosensitive stages during the embryonic development of Gryllus bimaculatus. Compt. rend. 251: 2430-2432. (In French) (NSA 15: 8540).

Collee, R. 1973. Environmental aspects of nuclear energy. Trib. Cebedeau (Cent. Belgé Étude Doc. Eaux) 26(350): 18-28. (In French) (NSA 28: 8137).

Collet, M. 1964. Monitoring of the atmosphere, precipitations, and waters in the Euratom member states. Fed. Europäischer Gewass. Informat. (10): 47-52. (In French) (NSA 19: 2537).

Collet, M., R. Maushart and P. Spaander. 1968. Accumulation of radioactive materials by river silt. pp. 1343-1349. In: Radiation Protection. Part 2. W. S. Snyder, et al., eds. Pergamon Press, New York. (In German) (NSA 23: 13859).

Collier, C. W. and R. Tardif. 1967. Analysis of male gypsy moths for microgram quantities of TEPA. J. Econ. Entomol. 60(1): 28-30. (^{14}C).

Collins, J. C. (Ed.). 1960. Radioactive Wastes, Their Treatment and Disposal. E. & F. N. Spon, London. xxi, 239pp.

Colombo, G. 1958. X-ray induced impairment of fecundity and fertility of Bombyx mori females. Intern. Congr. Genet. Proc. 10(2): 57.

Colombo, G. 1959. The lethal effects of x-rays on oocytes of the silkworm (Bombyx mori L., Lepidoptera). Caryologia 11(3): 273-296. (In Italian).

Colombo, G. 1959. Different sensitivities to x radiation with respect to the stage of development of the embryos of Locusta migratoria migratoriodes (R. and F.). Atti acad. naz. Lincei Rend. (sci. fis., mat. nat.) 26: 583-589. (In Italian) (NSA 14: 3415).

Colombo, G. 1959. First research on the dominant lethals induced with x-rays on the males of Locusta migratoria migratoriodes R. and F. (Orthoptera). Ricerca sci. 29: 2133-2141. (NSA 14: 4252).

Colville, J. E. 1968. The action of x-irradiation on the regeneration field of the adult newt, Diemictylus viridescens. Thesis, Tulane Univ., New Orleans, La. 61pp. (NSA 23: 18096).

Colwell, H. A. and M. S. Thomson. 1926. On some effects of primary and secondary x-rays on the skin of the frog tadpole. Lancet 211(5367): 59-61.

Comar, C. L. 1965. The movement of fallout radionuclides through the biosphere and man. Ann. Rev. Nucl. Sci. 15: 175-206. (NSA 20: 6714).

Comar, C. L. 1966. Early development in the use of radioisotopes in agriculture. Isotopes Radiat. Technol. 4(1): 46-52. (NSA 21: 17791).

Comar, C. L. and J. H. Rust. 1973. Natural radioactivity in the biosphere and foodstuffs. pp. 88-105. In: Toxicants Occurring Naturally in Foods. National Acad. Sci., Washington. (NSA 30: 18627).

Comar, D. 1962. The determination of iodine in biological media using radioactivation analysis. Univ. of Paris, France Commissariat à l'Énergie Atomique, report CEA-2095. 41pp. (In French).

Comar, D. 1969. Biological and pharmacological applications of activation analysis. pp. 91-112. In: Intern. Conf. on Radioactive Isotopes in Pharacology. (See Waser, 1969) (In French) (NSA 25: 23539).

Comar, D. 1971. Use of activation analysis in biology and pharmacology. pp. 218-234. In: Symp. on the Progress of Nuclear Techniques in Pharmacodynamics. (Saclay, France, March 11-13, 1970). Masson, Paris. (In French) (CONF-700343) (NSA 25: 46445).

Comar, D., et al. 1969. The use of neutron capture gamma radiations for the analysis of biological samples. pp. 114-127. In: Modern Trends in Activation Analysis. Vol. I. (See DeVoe, 1969) (NSA 24: 115).

Combescot, C., P. Groussin and F. N'Gore-Kassi. 1967. Radiosensitivity of the male genital gland of the water turtle Emys leprosa. Compt. rend. soc. biol. 151: 157-160. (In French) (NSA 22: 4679).

Come, T. V. 1963. Morphological and reproductive variation under four environmental pressures in an inbred strain of Drosophila melanogaster. Thesis, Pennsylvania State Univ., Univ. Park, Pa. 65pp. (NSA 19: 8941) (Dissertation Abstr. 25: 764).

Comision Interamericana de Energia Nuclear. nd. Proc. Conf. on Nuclear Energy and Agricultural Productivity. (Santiago, Chile, January 9-12, 1968). 245pp. (In English and Spanish) (CONF-680112) (NSA 25: 27127).

Commoner, B. 1965. Conservation of the water resource: the responsibility of the scientist. J. Water Poll. Control Fed. 37: 60-70. (NSA 20: 18) (Mentions radionuclides).

Commonwealth Edison Co. 1969. Dresden Nuclear Power Station, Units 2 and 3. Answers to AEC questions. (Safety Analysis Report), Ammendment 9 for Unit 2 and Ammendment 10 for Unit 3. Chicago. U. S. AEC DOCKET-50249-25; DOCKET-50237-25. 142pp. (NSA 24: 8999) (Incl. environmental monitoring).

Commonwealth Edison Co. 1971. Dresden Nuclear Power Station, Unit 1. Unplanned release of radioactive liquid. Chicago. U. S. AEC DOCKET-50010-95. 2pp. (NSA 25: 56896) (Release to Illinois River).

Commonwealth Edison Co. 1971. Zion Station, Units 1 and 2. Environmental report. Chicago. U. S. AEC DOCKET-50295-26; DOCKET-50304-26. 414pp. (NSA 25: 34289).

Compton, D. M. J. 1970. Destruction of organic substances in wastewater by ionizing radiation. Chem. Engrg. Progr. Symp. Ser. 66(104): 195-204. (NSA 25: 48524).

Compton, D. M. J., S. J. Black and W. L. Whittemore. 1970. Treating wastewater and sewage sludges with radiation: a critical evaluation. Nucl. News 13(9): 58-60. (NSA 24: 46286).

Conard, R. A., et al. 1958. March 1957 medical survey of Rongelap and Utirik people three years after exposure to radioactive fallout. Brookhaven National Lab., U. S. AEC report BNL-501. 29pp. (NSA 12: 12913) (Incl. environmental samples).

Conard, R. A., et al. 1960. Medical survey of Rongelap people five and six years after exposure to fallout (with an addendum on vegetation). Brookhaven National Lab., U. S. AEC report BNL-609(T179). 86pp.

Conard, R. A., et al. 1962. Medical survey of Rongelap people seven years after exposure to fallout. Brookhaven National Lab., U. S. AEC report BNL-727. 88pp. (NSA 16: 20975).

Conaway, C. C. and C. O. Knowles. 1969. Metabolism of diazinon-C^{14} in western corn rootworm beetles. J. Econ. Entomol. 62(2): 286-___.

Conger, A. D. 1970. The relation of interphase chromosome volume (ICV) to an estimated actual chromosome volume. Int. J. Radiat. Biol. 17: 381-384. (NSA 24: 34289).

Conger, A. D. and J. H..Chilton. 1973. Nuclear volumes, DNA contents, and radiosensitivity in whole-body-irradiated amphibians. Radiation Research 54(1): 69-101. (NSA 28: 6030).

Conger, A. D. and H. Cromroy. 1966. Radiosensitivity and nuclear volume in the gymnosperms. Univ. of Florida, Gainesville, U. S. Off. of Civil Def., Washington, Work Unit 3516A. 29pp.

Conger, A. D. and H. Cromroy. 1966. Radiosensitivity and nuclear volume in gymnosperms. Final report. Univ. of Florida, Gainesville. 36pp. (NP-16636) (NSA 21: 22774).

Conger, B. V., C. F. Konzak and J. R. Harle. 1969. Glass manifold vacuum system for controlling atmosphere and water content of seeds for irradiation experiments. Radiation Botany 9: 425-427. (NSA 24: 2345).

Conifer, J. L. 1972. Interrelations among plankton, attached algae, and the phosphorus cycle in artificial open systems. Ecol. Monographs 42(1): 1-23.

Conlan, B., P. Henderson and A. Walton. 1969. A simplified procedure for the assay of picocurie concentrations of ^{226}Ra and its application to a study of the natural radioactivity in surface waters in Scotland. Analyst 94: 15-19. (NSA 23: 19669).

Conley, W. R. 1954. Hanford Atomic Energy Plant - joint discussion. Water Supply. J. Am. Water Works Assoc. 46: 629-633.

Connecticut Light and Power Co., et al. 1969. Millstone Nuclear Power Station, Unit 1. Proposed technical specifications, Rev. 1. (Prelim. Safety Analysis Report). Ammendment 20. Hartford, et al. U. S. AEC DOCKET-50245-20. 160pp. (NSA 24: 9004) (Incl. environmental monitoring).

Connecticut Light and Power Co., et al. 1969. Millstone Nuclear Power Station, Unit 1. Establishment of preoperational environmental survey program. (Millstone Point Co., Wethersfield, Conn.). U. S. AEC DOCKET-50245-64. 1p. (NSA 25: 31491).

Connecticut Yankee Atomic Power Co. 1971. Connecticut Yankee Atomic Power Plant. Environmental radiation survey program. Summary report, Jan.-Dec. 1970. Hartford. U. S. AEC DOCKET-50213-98) (NSA 25: 20617) (see also ICN/Tracerlab for other reports).

Conover, R. J. and V. Francis. 1973. Use of radioactive isotopes to measure the transfer of materials in aquatic food chains. Marine Biology 18(4): 272-283. (SFA 18: 16471; NSA 28: 24487).

Conrad, G. and P. Olive. 1972. Preliminary data on the tritium content of aquifer sheets in northwest and central (Algeria) Sahara; hydrogeological consequences. Compt. rend. D274(14): 2019-2022. (In French) (NSA 26: 45558).

Constantine, D. G., J. A. Jensen and E. S. Tierkel. 1959. The use of radiolabeling in determining prey-predator relationships. J. Mammology 40(2): 240-242. (Bats).

Conte, F. P. 1965. Effects of ionizing radiation on osmoregulation in fish Oncorhunchus kisutch. Comp. Biochem. Physiol. 15(3): 293-302. (BA 46: 106011).

Conte, F. P., H. H. Wagner and T. O. Harris. 1963. Measurement of blood volume in the fish Salmo gairdneri gairdneri. Am. J. Physiol. 205(3): 53 533-540. (BA 45: 1233; SFA 9: 6612) (Used tracers).

Consumers Power Co. 1968. Midland Plant, Units 1 and 2. Preliminary Safety Analysis Report, Vol. 1. Jackson, Mich. U. S. AEC DOCKET-50329-1; DOCKET-50330-1. (NSA 23: 13364) (Incl. environmental study).

Consumers Power Co. 1971. Midland Plant, Units 1 and 2. Environmental report, including appendixes A-S. Jackson, Mich. U. S. AEC DOCKET-50329-27; DOCKET-50330-26. 1411pp. (NSA 25: 59978).

Convair. 1959. Fission products field release test-1. U. S. Air Force Special Weapons Center, Albuquerque, report AFSWC-TR-44; F2K-9-140. 399pp. (Also: Fort Worth, Tex., report NARF-59-32T) (NSA 14: 1715).

Conway, N. F. and A. Dudaitis. 1972. Environmental survey at the A. A. E. C. Research Establishment, Lucas Heights, results for the period January-July 1970. Australian Atomic Energy Commission, Lucas Heights, report AAEC/E-246. 30pp. (NSA 27: 22481) (Woronora Estuary; fish, Zostera, bottom sand; oysters, soil and vegetation).

Cook, C. S. 1968. Basic characteristics of nuclear radiation from fallout. pp. 65-83. In: Radiological Protection of the Public in a Nuclear Mass Disaster. (See Brunner and Pretre, 1968) (NSA 23: 19864).

Cook, L. M. and H. B. D. Kettlewell. 1960. Radioactive labelling of lepidopterous larvae: a method of estimating larval and pupal mortality in the wild. Nature 187(4734): 301-302.

Cooley, J. L. 1969. The effects of chronic gamma irradiation on snails. Abstr. Bull. Ecol. Soc. Am. 50(3): 130.

Cooley, J. L. 1970. Effects of chronic irradiation and temperature on populations of the aquatic snail Physa heterostropha. Thesis, Univ. of Tennessee, Knoxville. ix, 67pp. (Also: Cooley, J. L. and D. J. Nelson, Oak Ridge National Lab., U. S. AEC report ORNL-4612).

Cooley, J. L. 1970. Effects of chronic irradiation and temperature on populations of the aquatic snail, Physa heterostropha. Thesis, Univ. of Tennessee, Knoxville. 96pp. (Univ. Microfilms 71-7619) (NSA 25: 26915).

Cooley, J. L. 1973. Effects of chronic environmental radiation on a natural population of the aquatic snail Physa heterostropha. Radiation Research 54(1): 130-140. (NSA 28: 6071).

Cooley, J. L. 1973. Effects of temperature and chronic irradiation on populations of the aquatic snail Physa heterostropha. pp. 585-590. In: Radionuclides in Ecosystems. (See Nelson, 1973) (NSA 28: 27634).

Cooley, J. L. and F. L. Miller, Jr. 1971. Effects of chronic irradiation on laboratory populations of the aquatic snail Physa heterostropha. Radiation Research 47(3): 716-724. (NSA 25: 55314).

Coons, L. B. and F. E. Guthrie. 1972. High resolution radioautography of ^{3}H-DDT-treated insects. J. Econ. Entomol. 65(4): 1004-1007.

Cooper, E. L. 1963. 1963. The effects of antibiotics and x-irradiation on the survival of scale homografts in Fundulus heteroclitus. Thesis, Brown Univ., Providence, R. I. (Also: 1964. Transplantation 2(1): 2-20) (Dissertation Abstr. 24: 3890; SFA 9: 6737; NSA 18: 19726).

Cooper, E. L. and D. W. Schaefer. 1970. Bone marrow restoration of transplantation immunity in the leopard frog Rana pipiens. Proc. Soc. Exptl. Biol. Med. 135: 406-411. (NSA 25: 8620) (^{60}Co irradiation).

Cooper, J. A., et al. 1975. Feasibility study of in situ sediment analysis by x-ray fluorescence. Nucl. Technol. 26(2): 224-231. (NSA 32: 11172).

Cooper, R. S. 1972. Experimental demonstration of holokinetic chromosomes, and of differential radiosensitivity during oögenesis, in the grass mite Siteroptes gramium (Reuter). J. Exptl. Zool. 182(1): 69-94. (NSA 27: 7784).

Coopey, R. W. 1948. The accumulation of radioactivity as shown by a limnological study of the Columbia River in the vicinity of Hanford Works. Preliminary report. Hanford Atomic Products Operation, General Electric Co., U. S. AEC report HW-11662. 14pp. (NSA 13: 8493).

Coopey, R. W. 1953. A preliminary comparison of the nanoplankton and net plankton of the Columbia River. Hanford Atomic Products Operation, General Electric Co., U. S. AEC report HW-29298. 11pp. (NSA 10: 4121).

Coopey, R. W. 1953. Radioactive plankton from the Columbia River. Trans. Am. Microscop. Soc. 72(4): 315-327. (Also: Hanford Atomic Products Operation, General Electric Co., U. S. AEC report HW-20668) (NSA 6: 513).

Coopey, R. W. 1953. The abundance of the principal crustacea of the Columbia River and the radioactivity they contain. Hanford Atomic Products Operation, General Electric Co., U. S. AEC report HW-25191. 15pp. (NSA 10: 4119).

Cope, J. B., E. Churchwell and K. Koontz. 1960. A method of tagging bats with radioactive gold-198 in homing experiments. Proc. Indiana Acad. Sci. 70: 267-269. (BA 38(2): _____; WR 106: 28).

Cope, O. B. 1965. Agricultural chemicals and fresh-water ecological systems. pp. 115-127. In: Research in Pesticides. (Proc. Conf. on Research Needs and Approaches to the Use of Agricultural Chemicals from a Public Health Viewpoint). C. O. Chester, ed. Academic Press, New York. (^{14}C).

Copeland, B. J. and E. F. Gloyna. 1965. Radioactivity transport in water -- structure and metabolism of a lotic community, Part I (April-July, 1964). Environ. Health Engrg. Res. Lab., Univ. of Texas, Austin, Tech. Rept. -8. ix, 51pp.

Copeland, R. A., R. H. Beethe and W. W. Prater. 1973. Trace element distributions in Lake Michigan fish: a baseline study with calculations of concentration factors and equilibrium radioisotope distributions. Environmental Research Group, Inc., Ann Arbor, Mich., Special Rept. No. 2. 14-143pp. (NSA 29: 411) (Neutron activation analysis).

Coppedge, J. R., et al. 1967. Fate of 2-methyl-2-(Methylthiopropionadehyde) O-(methylcarbamoyl) oxime (Temik) in cotton plants and soil. J. Agr. Food Chem. 15(5): 902-910. (^{14}C, ^{35}S).

Corbean, S. 1970. Uranium determination in crop soils by neutron activation. Stiinta Solului 8(2): 57-69. (In Rumanian) (NSA 25: 26197).

Corbel, J. C. 1965. Remarks on the incorporation of ^{35}S in an intestinal gregarine of an orthopteran insect. Comt. rend. D261(18): 3669-3671. (In French).

Corbella, E. 1930. Influence of Roentgen radiation on the developing embryo of teleosts (Salmo lacustris L., Salmo iridens Gibb., Perca fluviatilis L.). Riv. Biol. (Milan) 12: 93-117. (In Italian).

Corbit, C. D. 1971. Reactors, radiation, and risk. Power Engrg. 75(7): 43-45. (NSA 25: 41362).

Corbridge, A. 1967. Planaria: how they are affected by direct radiation and by irradiated food. Ann. J. Colorado Dent. Assoc. 45: 22-25. (NSA 24: 6107).

Cordiero, A. R., E. K. Marques and A. J. Veiga-Neto. 1973. Radioresistance of a natural population of Drosophila willistoni living in a radioactive environment. Mutation Research 19(3): 325-329. (NSA 29: 2938).

Corey, J. C. 1968. Contribution of tritiated water at various depths in soil to evaporation at the soil surface. Dupont de Nemours, Savannah River Lab., U. S. AEC report DP-MS-67-95. 24pp. (NSA 22: 35921).

Corey, J. C. and J. H. Horton. 1968. Movement of water tagged with ^{2}H, ^{3}H, and ^{18}O through acidic kaolinitic soil. Soil Sci. Soc. Amer. Proc. 32(4): 471-475.

Corey, J. C. and D. R. Nielson. 1973. Some factors influencing the interpretation of soil solute-water interactions. pp. 87-110. In: Soil-Moisture and Irrigation Studies II. (See IAEA, 1973) (NSA 28: 21142).

Corey, J. C., A. R. Boulogne and J. H. Horton. 1970. Determination of soil density and water content by fast neutrons and gamma rays. Water Resources Res. 6: 223-229. (NSA 24: 41428).

Corey, J. C., R. H. Hawkins and R. F. Overman. 1970. Gamma photoneutron method for laboratory studies of soil water. Soil Sci. Soc. Amer. Proc. 34(4): 557-560. (NSA 24: 44104).

Corey, J. C., S. F. Peterson and M. A. Wakat. 1971. Measurement of attenuation of ^{137}Cs and ^{241}Am gamma rays for soil density and water content. Soil Sci. Soc. Amer. Proc. 35: 215-219. (NSA 25: 37494).

Corey, R. A. 1965. Laboratory tests with bidrin insecticide. J. Econ. Entomol. 58(1): 112-114. (Tetranychus telarius, Epilachna varivestis, Erythroneura comes, Chromaphis juglandicola; ^{14}C).

Cork, J. M. 1957. Gamma-radiation and longevity of the flour beetle. Radiation Research 7(6): 551-557. (NSA 12: 2681).

Corley, J. P. 1964. Effects of Hanford operations on Columbia River temperatures. Interim report No. 2. Battelle Northwest Labs., U. S. AEC report HW-80888(Del.). 93pp. (NSA 29: 10165).

Corley, J. P. 1968. Hanford experience with composite environmental dose estimates. Battelle Northwest Labs., U. S. AEC report BNWL-SA-1452-A. 19pp. (NSA 22: 16705).

Corley, J. P. 1971. Routine environmental surveillance program: master schedule, FY 1971. Effective September 1, 1971. Battelle Northwest Labs., U. S. AEC report BNWL-B-136. 28pp. (NSA 26: 4738).

Corley, J. P. 1973. Environmental monitoring at the Pacific Northwest Laboratory by Battelle-Northwest. Battelle Borthwest Labs., U. S. AEC report BNWL-SA-4839. 21pp. (NSA 29: 15571).

Corley, J. P. 1973. Environmental surveillance at Hanford for CY-1970. Battelle Northwest Labs., U. S. AEC report BNWL-1669. 97pp.; BNWL-1669-ADD. 117pp. (NSA 29: 369; 370).

Corley, J. P. 1974. Panel discussion on environmental monitoring and reporting. pp. 677-687. In: Proc. 2nd AEC Environmental Protection Conf. Vol.2. (Held at Albuquerque, New Mexico, April 16-19, 1974). (See U. S. AEC, 1974) (NSA 30: 32037).

Corley, J. P., D. M. Robertson and F. P. Brauer. 1972. Plutonium in surface soil in the Hanford Plant environs. Battelle Northwest Labs., U. S. AEC report BNWL-SA-3967. 11pp. (NSA 29: 24186).

Corley, J. P., et al. 1973. Environmental surveillance for fuel fabrication plants. Battelle Northwest Labs., U. S. AEC report BNWL-1723. 71pp. (NSA 28: 21064).

Cornwell, P. B. 1955. Techniques for labelling trees with radioactive phosphorus. Nature 175(4445): 85-87.

Cornwell, P. B. 1956. Some aspects of mealy bug behaviour in relation to the efficiency of measures for the control of virus diseases of cacao in the Gold Coast. Bull. Entomol. Res. 47: 137-166.

Cornwell, P. B. 1957. Uptake and translocation of ^{32}P in _Theobroma cacao_ L. Trop. Agric. 34: 117-132.

Cornwell, P. B. 1958. The control of pests in grain by gamma irradiation. New Scientist 4(79): 30-33. (NSA 12: 12129).

Cornwell, P. B. 1958. Movement of the vectors of virus diseases of cacao in Ghana. I. Canopy movement in and between trees. Bull. Entomol. Res. 49 (3): 613-630. (^{32}P).

Cornwell, P. B. 1959. The disinfection of foods, particularly grain. Int. J. Appl. Radiat. Isotopes 6: 188-193. (NSA 14: 1310).

Cornwell, P. B. 1960. Insect control by gamma-irradiation: an appraisal of the possibilities and problems involved. J. Sci. Food Agr. 11(12): 754-768.

Cornwell, P. B. 1960. Insect control by ionizing radiations. An appraisal of the potentialities and problems involved. pp. 104-116. In: _The Technological Use of Radiation_. Australian Atomic Energy Commission.

Cornwell, P. B. 1965. Radiation disinfestation of grain -- the last 5 years. Food Irradiat. 6: A2-A5. (NSA 20: 19123).

Cornwell, P. B. (Ed.). 1966. _The Entomology of Radiation Disinfestation of Grain_. Pergamon Press, New York. 256pp.

Cornwell, P. B. 1966. Susceptibility of the grain and rice weevils, _Sitophilus zeamais_ Mots. to gamma radiation. pp. 1-18. In: _The Entomology of Radiation Disinfestation of Grain_. (See Cornwell, 1966 above).

Cornwell, P. B. and D. M. Burson. 1958. Grain weevils, _Calandra granaria_ L. and _C. oryzae_ L. reared on irradiated wheat. Nature 181(4625): 1747-1748.

Cornwell, P. B. and J. A. Morris. 1960. Susceptibility of the grain and rice weevil _Calandra granaria_ L. and _C. oryzae_ L. to gamma radiation. pp. 291-298. In: _Large Radiation Sources in Industry_. _Vol. 2_. International Atomic Energy Agency, Vienna.

Cornwell, P. B., J. O. Bull and J. B. Pendlebury. 1966. Control of weevil populations (_Sitophilus granarius_ (L.)) with sterilising and substerilising doses of gamma radiation. pp. 71-95. In: _The Entomology of Radiation Disinfestation of Grain_. (See Cornwell, 1966)

Cornwell, P. B., L. J. Crook and J. O. Bull. 1957. Lethal and sterilizing effects of gamma radiation on insects infesting cereal communities. Nature 179(4561): 670-672.

Corompt, E. P. and P. Calmels. 1967. Use of information obtainable from a thermalized neutron flux and possible applications to soil- and moisture-density measurements. pp. 49-66. In: _Isotope and Radiation Techniques in Soil Physics and Irrigation Studies_. (See IAEA, 1967) (In French) (NSA 22: 23144).

Corossi, J. C., D. Chanu and P. Couchat. 1971. Equipment for studying soils. French Patent 2,144,086. 10pp. (In French) (NSA 29: 12949)

Corrigan, J. J. 1965. Studies in D-serine and D-2 3-diaminipropionic acid in _Bombyx mori_. Abstr. 93. Bull. Entomol. Soc. Am. 11(3): 159. (^{14}C).

Cortecci, G. 1973. Oxygen-isotope variations in sulfate ions in the water of some Italian lakes. Geochim. Cosmochim. Acta 37(6): 1531-1542. (NSA 28: 5279).

Cosgrove, G. E. 1965. The radiosensitivity of snakes and box turtles. Radiation Research 25(4): 706-712. (NSA 19: 42305).

Cosgrove, G. E. 1966. The radiobiology of snakes and box turtles. pp. 41-43. In: _Radiation Effects on Natural Populations_. (See Sacher, 1966).

Cosgrove, G. E. 1971. Reptilian radiobiology. J. Am. Vet. Med. Assoc. 159 (11): 1678-1684. (NSA 26: 15500).

Cosgrove, G. E. and B. G. Blaylock. 1973. Acute and chronic irradiation effects in mosquito fish at 15 or 25^{0}. pp. 579-584. In: _Radionuclides in Ecosystems_. (See Nelson, 1973) (NSA 28: 27624).

Cosgrove, G. E. and D. J. Bracco. 1973. Determination of minor metallic elements in the water environment. pp. 1315-1356. In: _Water and Water Pollution Handbook_. _Vol. 4_. (See Ciaccio, 1973) (NSA 29: 2446).

Cosgrove, G. E. and T. P. O'Farrell. 1965. Papillomas and other lesions in the stomachs of pine mice. J. Mammal. 46(3): 510-513. (Specimens from Oak Ridge National Lab.).

Cosgrove, G. E., D. Digregorio and P. B. Dunaway. 1974. An early unilateral thymic lymphoma in a postirradiated white-footed mouse. J. Wildl. Diseases 10(1): 32-33.

Cosgrove, G. E., P. B. Dunaway and J. D. Stary. 1969. Malignant tumors associated with subcutaneously implanted ^{60}Co radioactive wires in _Peromyscus maniculatus_. Bull. Wildl. Disease Assoc. 5(3): 311-314.

Cosgrove, G. E., et al. 1964. Infection of laboratory-reared and wild cotton rats with cat tapeworm larvae. J. Tenn. Acad. Sci. 39(3): 140-141. (^{182}Ta labels).

Costa, F., et al. 1967. Influence of irrigation rate and soil type on the vertical migration of iron and manganese in the soils of south-east Spain. pp. 251-271. In: Isotope and Radiation Techniques in Soil Physics and Irrigation Studies. (See IAEA, 1967) (NSA 22: 23157).

Costa, J. J., R. A. Ghelfi and E. O. Brenzoni. 1966. Techniques used in the application of a ^{14}C-labelled herbicide. pp. 9-16. In: Isotopes in Weed Research. (See IAEA, 1966).

Costello, D. P., C. Henley and D. E. Kent. 1957. The effects of P^{32} on mitosis in tail-tips of larval salamanders. J. Exptl. Zool. 136(1): 143-169.

Costes, C., et al. 1964. Studies on the transformation of water in leaves with tritiated water. I. Introduction: bibliographic study of the problem posed by use of labeled water in plant physiology. Ann. physiol. veg. 6: 303-315. (In French) (NSA 21: 24340).

Cotecchia, V. and F. Pantanetti. 1967. Radioisotopes in hydrology and in applied geology. pp. 267-277. In: Scientific Aspects and Techniques of Industrial Applications of Isotopes. (Reports of the 11th Rome Nuclear Congr., Rome, June 23-24, 1966). Italy Comitato Nazionale per l'Energia Nucleare, Rome. (In Italian) (CONF-660661) (NSA 22: 8322).

Cotecchia, V. and E. Pirastru. 1970. Development of nuclear methods for measuring density and natural water content of soils. Italy Comitato Nazionale per l'Energia Nucleare, Rome, report RT/GEO-(70)14. 28pp. (CONF-70106-1) (NSA 26: 17753).

Cotter, D. J. 1962. Effects of radiation on the photoperiodism of Portulaca smalli. Abstr. J. Ala. Acad. Sci. 33: 201.

Cotter, D. J. 1963. Effects of radiation on the photoperiodism of Portulaca smalli. Radiation Botany 3(3): 265-269. (NSA 18: 5084).

Cotter, D. J. 1964. Studies of radiation effects on plant communities. Abstr. J. Ala. Acad. Sci. 35(4): 160.

Cotter, D. J. 1965. Studies on ecological effects of short term exposure of ionizing radiation. J. Ala. Acad. Sci. 36: ___.

Cotter, D. J. and J. T. McGinnis. 1965. Recovery of hardwood stands 3 to 5 years following acute irradiation. pp. 1663-1673. In: Radiation and Terrestrial Ecosystems. (See Hungate, 1965).

Cotter, D. J. and J. T. McGinnis. 1965. Some aspects of recovery of hardwood stands 3 to 5 years following acute radiation. Abstr. ASB Bull. 12(2): 44.

Cottrell, W. B. 1960. Radioactivity in silt of the Clinch and Tennessee rivers. Oak Ridge National Lab., U. S. AEC report ORNL-2847. 47pp. (NSA 14: 4143).

Cottrell, W. B. (Ed.). 1960. Technical progress review. Nuclear Safety 2(1): 119pp. (NSA 15: 7653).

Cottrell, W. B. and J. R. Buchanan. 1970. Summary of environmental information activities of the nuclear information center. Oak Ridge National Lab., U. S. AEC report ORNL-TM-3009. 72pp. (NSA 24: 33365).

Cotty, V. F., S. M. Henry and J. D. Hilchey. 1958. The sulfur metabolism of insects. III. The metabolism of cystine, methionine, taurine, and sulfate by the house fly, Musca domestica L. Contr. Boyce Thompson Inst. 19(2): 379-392. (^{35}S).

Couchat, P. 1967. The neutronic method for measuring soil moisture. France Commissariat à l'Énergie Atomique, Cadarache, report CEA-R-3298. 29pp. (In French) (NSA 22: 4270).

Couchat, P. 1970. Neutron method of soil moisture measurement independent of dry density. Compt. rend. D271: 1264-1267. (In French) (NSA 25: 8570).

Couchat, P. 1973. Recent improvements in development and future prospects of the neutron method of measuring soil moisture. pp. 41-49. In: Soil-Moisture and Irrigation Studies. II. (See IAEA, 1973) (NSA 28: 21136).

Couchat, P. and J. Ramodiharilafy. 1972. Neutron method for measuring the water content of stony soils. Int. J. Appl. Radiat. Isotopes 23(5): 229-235. (In French) (NSA 26: 38425).

Couey, H. M. and W. J. Bramlage. 1965. Effect of spore population and age of infection on the response of Botrytis cinerea to gamma radiation. Phytopathology 55: 1013-1015. (NSA 20: 5371).

Cough, V. L. 1971. Nuclear weapons and ecology. Environ. Contr. Saf. Manage. 141(3): 38-43. (NSA 26: 9496).

Coulamb, R. 1955. Dispersion of the radioelements of the uranium-radium family in natural waters. Bull. soc. franc. min. chryst. 78: 89-96. (CA 49: 15552h).

Coulon, J. and M. Fabries. 1972. Modifications of the reticulo-endoplasmic of cytosomes, and of mitochondria in the bud of the cornelian cherry (Cornus mas L.) gamma irradiated with radiocesium. Compt. rend. D274(11): 1686-1688. (In French) (NSA 26: 48225).

Coulon, M. 1966. Variations in the spectrum of anomalies following x radiation of the embryos of Bombyx mori as a function of hibernation. Compt. rend. D263: 1153-1155. (In French) (NSA 21: 14746).

Coulon, M. 1966. Study of the spectrum of anomalies following aging and with x irradiation of the embryo of Bombyx mori. Thesis, Univ. of Lyon, France. 97pp. (In French).

Coulon, M. 1969. Studies on x-ray damage in the early larva of Bombyx mori. Compt. rend. D268: 959-962. (In French) (NSA 23: 24717).

Coulon, M. 1973. Evolution in the percentage mortality of Bombyx mori after irradiation in the larval stage with x radiation. Compt. rend. D272(16): 1683-1686. (In French) (NSA 29: 29834).

Coulon, R. 1968. Conception and realization of control of radioactive pollution of the food chain. pp. 1035-1038. In: Radiation Protection, Part 2. W. S. Snyder, et al., eds. Pergamon Press, New York. (In French) (NSA 23: 14176).

Coulon, R. 1973. Studies of the effects of siting of nuclear installations. Agriculture (Paris) (366): 184-186. (In French) (NSA 29: 18384).

Coulson, C. L. and A. J. Peel. 1971. The effect of temperature on respiration of ^{14}C-labelled sugars in stems of willow. Ann. Botany 35(139): 9-15.

Coulston, F. and J. H. Wills. 1974. No-effect level and extrapolation as applied to pesticide residues and radioactive substances. pp. 561-571. In: Comparative Studies of Food and Environmental Contamination. (See IAEA, 1974) (NSA 30: 15862).

Courtney, F. E., Jr. 1970. Need for preoperational meteorological programs for nuclear reactor sites for the electric power industry. pp. 140-147. In: Environmental Surveillance in the Vicinity of Nuclear Facilities. (See Reinig, 1970) (NSA 25: 16170).

Courtois, G. 1967. Radioisotopes in sedimentology. pp. 117-165. In: Isotopes in Hydrology. (See IAEA, 1967) (NSA 22: 19058).

Courtois, G. 1970. Sediment dynamics and radioactive tracers: status of work in France. Houille Blanche 25: 617-628. (ORNL-tr-2492) (NSA 25: 34812).

Courtois, G. 1970. Radioactive gaging equipment for sediment measurements. Applications of nuclear instrumentation. Houille Blanche 25(8): 715-720. (In French) (CONF-691135-3) (NSA 25: 44713).

Courtois, G. 1973. Use of artificial radioactive tracers in France. pp. 71-80. In: Trace Techniques in Sediment Transport. (See IAEA, 1973) (NSA 28: 21136).

Courtois, G. and J. Lecompte. 1958. On a procedure for marking bees butineuses average with a radioisotope. Compt. rend. 247(1): 147-149. (In French).

Courtois, G. and J. Lecompte. 1959. On the resistance to gamma radiation of the worker bee. Franc. Inst. Nat. Rech. Argon. Ann. ser. C, Ann. Epiphyt. Ann. l'Abeille 2(4): 285-290.

Courtois, G. and J. Lecompte. 1962. Study of the exchange of food between butineuses of Formica polyctena workers through the different sectors of mais appartenant a la meme ants. Insectes sociaux 9: 323-327. (In French).

Courtois, G. and J. Lecompte. 1963. Some uses of radioelements and radiation in entomology. pp. 5-21. In: Radiation and Radioisotopes Applied to Insects of Agricultural Importance. (See IAEA, 1963) (In French).

Courtois, G. and G. Sauzay. 1966. Count rate balance methods using radioactice tracers for measuring sediment mass flow. Houille Blanche 21: 279-290. (In French) (NSA 21: 16076).

Courvoisier, P. and P. Müller. 1971. Standards for effluents in Switzerland. pp. 183-192. In: Environmental Aspects of Nuclear Power Stations. (See IAEA, 1971).

Courvoisier, P. and E. Nagel. 1968. Contamination of air and the resulting contamination of grass. pp. 1039-1044. In: Radiation Protection, Part 2. W. S. Snyder, et al., eds. Pergamon Press, New York.

Coutts, J. R. H., M. F. Kandil and J. Tinsley. 1968. Use of radioactice ^{59}Fe for tracing soil particle mpvement. II. Laboratory studies of labelling and splash displacement. J. Soil Sci. 19(2): 325-341.

Coutts, J. R. H., et al. 1968. Use of radioactive ^{59}Fe for tracing soil particle movement. I. Field studies of splash erosion. J. Soil Sci. 19 (2): 311-324.

Coutts, P. J. F. 1972. Utility of radio-carbon determinations for dating N. Z. archeological sites. pp. F52-F68. In: Proc. 8th Intern. Radiocarbon Dating Conf. Vol. 2. (See Rafter, 1972) (NSA 31: 8709).

Cowan, F. P. and C. B. Meinhold. 1962. Radiation dosimetry for Co^{60} and Cs^{137} gamma ray field irradiation facilities. Radiation Botany 2(3/4): 241-249. (Source in woods, Brookhaven National Lab.).

Cowan, J. J. 1961. Radiation dosages in the vicinity of an unshielded nuclear reactor. Thesis, Emory Univ., Emory, Georgia.

Cowan, J. J. and R. B. Platt. 1963. Radiation dosages in the vicinity of an unshielded nuclear reactor. pp. 311-318. In: Radioecology. (See Schultz and Klement, 1963) (NSA 17: 33577).

Cowan, R. L., E. W. Hartsook and J. B. Whelan. 1968. Calcium-strontium metabolism in white tailed deer as related to age and antler growth. Proc. Soc. Exptl. Biol. Med. 129(3): 733-737.

Cowan, R. L., et al. 1969. A cage for metabolism and radioisotope studies with deer. J. Wildl. Mgmt. 33(1): 204-208.

Cowart, J. B. 1974. ^{234}U and ^{238}U in the Carrizo Sandstone Aquifer of South Texas. Dissertation, Florida State Univ., Tallahassee. 88pp. (Univ. Microfilms 75-950) (NSA 31: 19541).

Cowart, J. B. and J. K. Osmond. 1974. ^{234}U and ^{238}U in the Carrizo Sandstone Aquifer of South Texas. pp. 131-149. In: Isotope Techniques in Groundwater Hydrology. Vol. 2. (See IAEA, 1974) (NSA 31: 19651).

Cowie, D. B. and E. T. Bolton. 1957. The use of metabolic pools of purine compounds for nucleic acid synthesis in yeasts. Biochim. Biophys. Acta 25: 292.

Cowie, D. B. and R. B. Roberts. 1954. Permeability of microorganisms to inorganic ions, aminoacids and peptides. pp. 1-34. In: Electrolytes in Biological Systems. (Symp., Marine Biol. Lab., Woods Hole, Mass.).

Cowley, G. T. 1970. Fleshy fungi in relation to irradiation and cutting in the Luquillo experimental forest. Ch. F-1, pp. F9-F13. In: A Tropical Rain Forest. (See H. T. Odum and R. F. Pigeon, 1970) (NSA 25: 21946).

Cowley, G. T. 1970. Effect of radiation on the microfungal populations of six litter species in the Luquillo experimental forest. Ch. F-4, pp. F25-F28. In: A Tropical Rain Forest. (See H. T. Odum and R. F. Pigeon, 1970) (NSA 25: 21947).

Cowley, G. T., J. W. Martin and D. C. Coleman. 1973. Structure and function of hardwood litter and soil subsystems after chronic gamma irradiation. II: Microfungi. pp. 1229-1236. In: Radionuclides in Ecosystems. (See Nelson, 1973) (NSA 28: 30043).

Cowper, G. 1969. Sources of inaccuracy in radiation monitoring instruments. pp. 499-511. In: Radiation Protection Monitoring. (See IAEA, 1969) (NSA 23: 47780).

Cowser, K. E. 1964. Current practices in the release and monitoring of I^{131} at NRTS, Hanford, Savannah River and ORNL. Oak Ridge National Lab., U. S. AEC report ORNL-NSIC-3. vii, 103pp. (NSA 18: 37443) (Incl. remarks on vegetation, fish, waterfowl, water, jack rabbits).

Cowser, K. E. and F. L. Parker. 1958. Soil disposal of radioactive wastes at ORNL: criteria and techniques of site selection and monitoring. Health Physics 1(2): 152-163.

Cowser, K. E. and W. S. Snyder. 1966. Safety analysis of radionuclide release to the Clinch River. Final progress report No. 3. Oak Ridge National Lab., U. S. AEC report ORNL-3721(Suppl. 3). (NSA 20: 29452).

Cowser, K. E., W. S. Snyder and M. J. Cook. 1963. Preliminary safety analysis of radionuclide release to the Clinch River. pp. 17-38. In: Transport of Radionuclides in Fresh Water Systems. (See Kornegay, et al., 1963) (NSA 17: 40520).

Cowser, K. E., et al. 1966. Evaluation of radiation dose to man from radionuclides released to the Clinch River. pp. 639-671. In: Disposal of Radioactive Wastes into Seas, Oceans, and Surface Waters. (See IAEA, 1966) (Also: IAEA preprint SM-72/41) (CONF-660507) (Also: Oak Ridge National Lab., U. S. AEC report ORNL-P-2021) (NSA 20: 22514).

Cowser, K. E., et al. 1967. Dose-estimation studies related to proposed construction of an Atlantic-Pacific interoceanic canal with nuclear explosives: Phase I. Oak Ridge National Lab., U. S. AEC report ORNL-4101. x, 208pp. (NSA 21: 26417).

Cowser, K. E., et al. 1973. Methods of estimating dose to man from regional growth of nuclear power. pp. 467-482. In: Environmental Behavior of Radionuclides Released in the Nuclear Industry. (See IAEA, 1973) (NSA 29: 29383).

Cox, H. C., J. R. Young and M. C. Bowman. 1967. Persistence of TEPA in fall armyworm moths. J. Econ. Entomol. 60(4): 1111-1115. (^{14}C).

Cox, S. W. R. and D. E. Filby. 1972. Instrumentation in Agriculture. Hafner, New York. xii, 145pp. (Incl. ionizing radiation).

Craft, T. F. 1964. Effects of nuclear explosions on watersheds. J. Am. Water Works Assoc. 56(7): 846-852. (NSA 18: 33811).

Craft, T. J. 1963. Some effects of x-irradiation on the utilization of C^{14}-labelled L-tyrosine and melanin production in Rana pipiens tadpoles. Thesis, Ohio State Univ., Columbus. 72pp. (NSA 18: 27205).

Crafts, A. S. 1956. Translocation of herbicides. I. The mechanism of translocation: methods of study with C^{14}-labeled 2,4-D. Hilgardia 26: 287-334.

Crafts, A. S. 1959. Further studies on comparative mobility of labeled herbicides. Plant Physiol. 34(6): 613-620.

Crafts, A. S. 1960. Evidence for hydrolysis of esters of 2,4-D during absorption by plants. Weeds 8(1): 19-25.

Crafts, A. S. 1966. Relation between food and herbicide transport. pp. 3-8. In: Isotopes in Weed Research. (See IAEA, 1966).

Crafts, A. S. 1969. Use of radioisotopes in herbicide research. pp. 1010-1019. In: 9th Proc. Japan Conf. on Radioisotopes. (May 13-15, 1969, Tokyo, Japan). Japan Atomic Industrial Forum, Tokyo. (In Japanese) (CONF-690548) (NSA 24: 31823).

Crafts, A. S. and C. L. Foy. 1959. Autoradiography of radioactive dalapon. Down to Earth (Spring): 5pp.

Crafts, A. S. and S. Yamaguchi. 1958. Comparative tests on the uptake and distribution of labeled herbicides by Zebrina pendula and Tradescantia fluminensis. Hilgardia 27(16): 421-454.

Crafts, A. S. and S. Yamaguchi. 1960. Absorption of herbicides by roots. Am. J. Bot. 47(4): 248-255.

Crafts, A. S. and S. Yamaguchi. 1960. Gross autoradiography of solute translocation and distribution in plants. Med. Biol. Illustr. 10(2): 103-109.

Crafts, A. S., H. B. Currier and H. R. Drever. 1958. Some studies on the herbicidal properties of maleic hydrazide. Hilgardia 27(22): 723-757.

Craig, D. K., et al. 1976. SNS source term evaluation. Battelle Northwest Lab., U. S. ERDA report BNWL-1971. 60pp. (NSA 33: 25985).

Craig, D. L., J. S. Kirby-Smith and J. N. Dent. 1956. Mitotic inhibition and chromosome breakage in salamanders of the genus Gyrinophilus. Trans. Microscop. Soc. 75: 234-241. (NSA 10: 9930).

Craig, G. B., Jr. 1964. Applications of genetic technology to mosquito rearing. Bull. Wld. Hlth. Org. 31: 469-473. (NSA 22: 2545) (X-rays).

Craig, G. B., Jr. 1967. Genetic control of Aedes aegypti. Bull. Wld. Hlth. Org. 36: 628-632. (NSA 22: 52108).

Craig, H. 1954. Carbon 13 in plants and the relationships between carbon 13 and carbon 14 variations in nature. J. Geology 62(2): 115-149.

Craig, R. and N. A. Olson. 1951. Rate of circulation of the body fluid in adult Tenebrio molitor Linnaeus, Anasa tristis (de Geer), and Murgantia histrionica (Hahn). Science 113(2945): 648-650. (^{32}P).

Crancher, D. W. 1972. Safety and environmental aspects of nuclear power. Australian Atomic Energy Commission. 32pp. (NP-19420) (NSA 26: 56777).

Crancher, D. W. and D. J. Higson. 1973. Environmental protection in a nuclear power programme. Proc. Roy. Austr. Chem. Inst. 40(11): 321-327. (NSA 29: 18320).

Crancher, D. W., D. J. Higson and L. H. Keher. 1971. Safety and environmental aspects of nuclear energy. Australian Atomic Energy Commission, Coogee. 22pp. (For 43rd Congr. Australian and New Zealand Assoc. for the Advancement of Science, Brisbane, 1971). (CONF-710511-1) (NSA 25: 39857).

Cranmore, G. and F. L. Harrison. 1975. Loss of ^{137}Cs and ^{60}Co from the oyster Crassostrea gigas. Health Physics 28(4): 319-333. (NSA 32: 3831).

Crawford, G. W. (Ed.). 1962. Radiobiologic experiments in Disoverer Satellite XVII. School of Aerospace Medicine, Brooks AFB, U. S. Air Force, Texas, report AF-SAM-62-67. 50pp. (NSA 16: 32723).

Crawford, T. V. and C. E. Bailey. 1975. Environmental sciences and computations: a modular data based systems approach. Savannah River Ecology Lab., U. S. ERDA report DP-MS-75-75. 14pp. (NSA 32: 19733) (Radionuclides).

Creighton, M. 1941. The effect of x-rays on mitotic and meiotic chromosomes and cell division in Chortippus longicotnis (Orthoptera). J. Exptl. Zool. 87(3): 347-370.

Cresp, J. 1970. Action of x-rays on the morphogenesis of buds in Salmacina incrustans (Clap). Compt. rend. D270: 2547-2549. (In French) (NSA 25: 8753).

Crespi, H. L., et al. 1972. Discrimination against tritium by fully deuterated algae. Health Physics 23(1): 111-112. (NSA 26: 41304).

Crews, W. S. and D. D. Tillson. 1969. Analysis of travel time of ^{131}I from the 1301-N crib to the Columbia River during July 1069. Battelle Northwest Lab., U. S. AEC report BNWL-CC-2326. 29pp. (NSA 29: 18356).

Crolius, P. C. 1959. Non-chemical insecticides. Farm Chem. 122(2): 31-32, 34.

Cromroy, H. and A. D. Conger. 1966. Radiosensitivity and nuclear volume in the gymnosperms. Final report, January 28, 1964 - September 30, 1965. Univ. of Florida, Gainesville. 38pp. (AD-648533) (NSA 22: 6923).

Cromroy, H. L. 1969. Mammalian radiosensitivity. Univ. of Florida, Gainesville. 28pp. (AD-702766) (NSA 24: 46650).

Cromroy, H. L. 1971. Successful uses of radiation in agriculture. Univ. of Florida, Gainesville. 30pp. (For Conf. on Innovative Applications of Radiation, Dallas, Tex., 22 April 1971). (CONF-710428-5) (NSA 25: 51968).

Cromroy, H. L., et al. 1967. Uptake of ^{137}Cs from contaminated soil by selected grass crops. Radiol. Hlth. Data Rpts. 8: 421-424. (NSA 22: 28511).

Cromroy, H. L., et al. 1969. The endothelial cell as an indicator of animal radiosensitivity. pp. 127-130. In: Symposium on Radioecology. (See Nelson and Evans, 1969).

Cromroy, H. L., et al. 1971. Prediction of species sensitivity. pp. 419-433. In: Survival of Food Crops and Livestock in the Event of Nuclear War. (See Bensen and Sparrow, 1971).

Crone, H. D. 1964. Phospholipid composition of flight muscle sarcosomes from the housefly, Musca domestica. J. Insect Physiol. 10(3): 499-507. (^{32}P).

Cronkite, E. P., V. P. Bond and C. L. Dunham. 1956. Some effects of ionizing radiation on human beings. U. S. AEC, Washington, report TID-5358. viii, 106pp. (Radioactive fallout in the Marshall Islands, incl. environmental radiation and radioactivity).

Crook, F. M., P. F. Nelson and R. G. Thompson. 1970. Some effects of high energy radiation on the Kraft pulping of eucalypts. Holzforschung 24: 184-190. (NSA 25: 29735).

Crook, L. J. 1962. The susceptibility of the rust-red flour beetle, Tribolium castaneum (Herbst.) to gamma radiation. Research Group, U. K. Atomic Energy Authority, Wantage, report AERE-R-3889. 19pp.

Crooks, R. N., R. G. D. Osmond and T. J. Webber. 1961. Radiostrontium and radiocaesium in drinking water in the United Kingdom, results to December 1960. Reserach Establishment, U. K. Atomic Energy Authority, Harwell, report AERE-R-3552. 20pp. (NSA 15: 29534).

Crooks, R. N., et al. 1959. Radiostrontium and radiocaesium in drinking water in the United Kingdom. Results up to mid 1959. Research Establishment, U. K. Atomic Energy Authority, Harwell, report AERE-R-3127. 9pp. (NSA 11: 11826) (Incl. river and lake water).

Crosby, J. W., III and R. M. Chatters. 1965. New techniques of water sampling for carbon 14 analysis. J. Geophys. Res. 70(12): 2839-2844. (NSA 19: 36681).

Crossley, D. A., Jr. 1962. Radioecological studies of millipedes (Dixidesmus erasus). Abstr. Bull. Ecol. Soc. Am. 43(4): 121.

Crossley, D. A., Jr. 1963. Consumption of vegetation by insects. pp. 427-430. In: Radioecology. (See Schultz and Klement, 1963) (NSA 17: 33591).

Crossley, D. A., Jr. 1963. Movement and accumulation of radiostrontium and radiocesium in insects. pp. 103-105. In: Radioecology. (See Schultz and Klement, 1963) (NSA 17: 33550).

Crossley, D. A., Jr. 1963. Use of radioactive tracers in the study of insect-plant relationships. pp. 43-53. In: Radiation and Radioisotopes Applied to Insects of Agricultural Importance. (See IAEA, 1963) (NSA 17: 21424).

Crossley, D. A., Jr. 1963. Landscape studies using radioactive tracers. III. Interdependence between arthropod food chains in a tulip poplar forest. Abstr. Bull. Ecol. Soc. Am. 44(3): 74.

Crossley, D. A., Jr. 1963. Heterotrophic productivity in the grazing food chain of terrestrial vegetation. Abstr. Proc. Intern. Congr. Zool. 16 (4): 317. (Concerns radiotracers).

Crossley, D. A., Jr. 1964. Biological elimination of radionuclides. Nuclear Safety 5(3): 265-268.

Crossley, D. A., Jr. 1965. Relation of arthropod activities to leaf litter breakdown. Abstr. Bull. Ecol. Soc. Am. 46(3): 129. (Used radiocesium).

Crossley, D. A., Jr. 1967. Radioactive tracer measurements of predation by arthropods. Abstr. Bull. Ecol. Soc. Am. 48(2): 64-64.

Crossley, D. A., Jr. 1969. Comparative movement of ^{106}Ru, ^{60}Co, and ^{137}Cs in arthropod food chains. pp. 687-695. In: Symposium on Radioecology. (See Nelson and Evans, 1969).

Crossley, D. A., Jr. 1974. Radioactive tracer studies of soil and litter arthropod food chains. Summary report, 1971-1974. Univ. of Georgia, Athens, U. S. AEC report SRO-641-14. 64pp. (NSA 30: 26806).

Crossley, D. A., Jr. and K. K. Bohnsack. 1960. Long-term ecological study in the Oak Ridge area. III. The oribatid mite fauna in pine litter. Ecology 41(4): 628-638.

Crossley, D. A., Jr. and H. F. Howden. 1961. Insect-vegetation relationships in an area contaminated by radioactive wastes. Ecology 42(2): 302-317. (NSA 15: 28855).

Crossley, D. A., Jr. and V. A. Merchant. 1971. Feeding by caeculid mites on fungus demonstrated with radioactive tracers. Ann. Entomol. Soc. Am. 64 (4): 760-762. (NSA 25: 42242) (A review).

Crossley, D. A., Jr. and M. E. Pryor. 1960. The uptake and elimination of cesium-137 by a grasshopper - Romalea microptera. Health Physics 4(1): 16-20. (NSA 15: 3831).

Crossley, D. A., Jr. and D. E. Reichle. 1969. Analysis of transient behavior of radioisotopes in insect food chains. BioScience 19(4): 341-343. (^{137}Cs and ^{134}Cs).

Crossley, D. A., Jr. and J. Schnell. 1961. Comparative elimination of radiocesium and radiostrontium by grasshoppers. Ann. Entomol. Soc. Am. 54(2): 459-461. (NSA 15: 21961).

Crossley, D. A., Jr. and R. I. Van Hook, Jr. 1970. Energy assimulation by the house cricket, Acheta domesticus, measured with radioactive ^{51}Cr. Ann. Entomol. Soc. Am. 63(2): 512-515. (NSA 24: 23056).

Crossley, D. A., Jr. and M. Witkamp. 1964. Effects of pesticide on biota and breakdown of forest litter. Trans. 8th Intern. Congr. Soil Science 3: 887-892. (Also: Oak Ridge National Lab., U. S. AEC report ORNL-P-517) (NSA 19: 1960) (^{134}Cs).

Crossley, D. A., Jr. and M. Witkamp. 1964. Forest soil mites and mineral cycling. Acarologia 6: 137-145. (Used radionuclides).

Crossley, D. A., Jr., D. E. Reichle and C. A. Edwards. 1971. Intake and turnover of radioactive cesium by earthworms (Lumbricidae). Pedobiologia 11(1): 71-76.

Crothers, W. C. and A. J. Forgash. 1964. Studies on the rate of penetration of diazinon, dieldrin and malathion in resistant and susceptible M. domestica. Abstr. Bull. Entomol. Soc. Am. 10(3): 163. (^{14}C).

Crouse, H. V. 1950. The differential response to male and female germ cells in Sciara coprophila (Diptera) to irradiation. Am. Naturalist 84: 195-202. (NSA 4: 4433).

Crouse, H. V. 1961. X-ray induced sex-linked recessive lethals and visibles in Sciara coprophila. Am. Naturalist 95: 21-26. (NSA 16: 20138).

Croute, F., J. P. Soleilhavoup and H. Planel. 1972. Effect of autogamy on the sensitivity of cells to natural ionizing radiations. Compt. rend. soc. biol. 166(6): 948-951. (In French) (NSA 28: 8479) (Paramecium).

Croute, F., et al. 1972. Effect of continued radioprotection on successive cycles in Paramecium aurelia. Compt. rend. soc. biol. 166(8): 1170-1173. (In French) (NSA 28: 24964).

Croute, F., et al. 1973. Isolation of stocks of Paramecium aurelia of different radiosensitivities towards natural ionizing radiation. Compt. rend. soc. biol. 167(12): 1984-1989. (In French) (NSA 31: 14300; 26055).

Crouzet, E., et al. 1970. Tritium in investigation of surface hydrology. Experimental determination of coefficient of runoff. J. Hydrol. 11: 217-229. (CSIRO-Trans-10757) (NSA 26: 20337).

Crowson, D. L. 1973. Man-made tritium. pp. 23-27. In: Tritium. (See Moghissi and Carter, 1973) (NSA 28: 27223).

Crowther, J. A. 1926. Action of x-rays on Colpidium colpoda. Proc. Roy. Soc. (London) 100(B704): 390-404.

Crozaz, G. 1967. Dating of glaciers by ^{210}Pb. pp. 385-393. In: Radioactive Dating and Methods of Low-Level Counting. (See IAEA, 1967) (In French) (NSA 22: 6468).

Crozaz, G. 1969. Fission products in Antarctic snow, an additional reference level in January 1965. Earth Planet. Sci. Lett. 6: 6-8. (NSA 23: 33270).

Crozaz, G. and P. Fabri. 1966. Measurement of polonium in the range of 10^{-13} curies, tracing ^{208}Po, and application to the chronology of ice. Earth Planet. Sci. Lett. 1: 446-448. (In French) (NSA 21: 6024).

Crozaz, G. and C. C. Langway, Jr. 1966. Dating Greenland firn-ice cores with ^{210}Pb. Earth Planet. Sci. Lett. 1: 194-196. (NSA 20: 33532).

Crozaz, G., C. C. Langway, Jr. and E. Picciotto. 1966. Artificial radioactivity reference horizons in Greenland firn. Earth Planet. Sci. Lett. 1: 42-48. (Also: U. S. Army Cold Regions Res. and Engrg. Lab., Hanover, N. H., report) (AD-644597) (NSA 20: 25285; 21: 22035).

Crozaz, G., E. Picciotto and W. de Brueck. 1964. Arctic snow chronology with Pb^{210}. J. Geophys. Res. 69(12): 2597-2604. (NSA 18: 25702).

Crystal, M. M. 1965. First efficient chemosterilants against screw-worm flies (Diptera: Colliphoridae). J. Med. Entomol. 2: 317-319. (NSA 20: 42724).

Csiky, I., et al. 1973. Radioactivity of some Hungarian spring-waters and the dose originating from their consumption. pp. 441-444. In: Health Physics Problems of Internal Contamination. (See Bujdoso, 1973) (NSA 28: 29883).

Csovari, M and L. Molnar. 1974. Determination of uranium in biological materials. Magy. Kem. Foly. 80(1): 16-19. (In Hungarian) (NSA 20: 29073).

Csukas-Szatloczky, I., et al. 1968. Fluctuation of radiosensitivity during the ontogenesis of *Drosophila melanogaster* larvae. Acta Biol. Acad. Sci. Hung. 19: 57-64. (NSA 22: 51931).

Csupka, S. 1970. Estimation of future radiostrontium in the system fallout fodder plant, milk. Kernenergie 13(12): 391-394. (In German) (NSA 25: 42378).

Csupka, S. 1971. ^{137}Cs in the soil-plant system in the years 1964-1969. Pol'nohospodarstvo 17(5): 363-370. (In Slovak) (NSA 27: 24605).

Csupka, S. 1972. Radioactivity of waters in the vicinity of Jaslovske Bohunice before startup of the nuclear power plant. Vod. Hospod. B22(7): 183-185. (In Slovak) (NSA 30: 18644).

Csupka, S. 1975. Depth distribution of ^{90}Sr in some soil types. Jaderna Energ. 21(6): 217-220. (In Slovak) (NSA 33: 29142).

Csupka, S. nd. Behavior of ^{90}Sr and ^{137}Cs in the soil-fodder-milk chain. Krajska Hyg. Stanica, Bratislava, Czechoslavakia. (pp. 51-53; INIS-mf-887-2) (In Slovak) (NSA 30: 3178).

Csupka, S., M. Petrasova and J. Carach. 1967. Seasonal variation in the concentration of ^{137}Cs in grass and alfalfa. Nature 213: 1204-1206. (NSA 21: 20114).

Csupka, S., et al. 1970. Forecast calculations of future radiostrontium and radiocesium levels of fodder and cereals. Jaderna Energ. 16: 345-348. (In Czech) (NSA 25: 6813).

Cuany, R. L. and L. E. Guardia. 1962. The radiosensitivity of somatic mutations in tropical plants. Turrialba 12: 146-150. (In Spanish) (NSA 18: 31243).

Cuellar, C. B. 1973. Dynamic aspects of the Pala sterile hybrid field experiment. pp. 113-118. In: Computer Models and Application of the Sterile-Male Techniques. (See IAEA, 1973) (NSA 29: 8053).

Cuellar, C. B. 1973. A view of computer simulation in sterile male experiments. pp. 5-15. In: Computer Models and Application in the Sterile-Male Technique. (See IAEA, 1973) (NSA 29: 8045).

Cuisance, D. and J. Itard. 1971. Techniques for labelling tsetse flies with radioisotopes. pp. 111-130. In: Sterility Principle for Insect Control or Eradication. (See IAEA, 1971) (NSA 25: 29417).

Cuisance, D. and J. Itard. 1975. Releases of *Glossina tachinoides* West. Males at two natural breeding sites: results and prospects. pp. 461-476. In: Sterility Principle for Insect Control, 1974. (See IAEA, 1975) (In French) (NSA 32: 11800) (Sterile males).

Cullen, M. C. and E. T. McGuinness. 1971. Radiochromatographic assay for thiol groups of soluble proteins using mercury-203 labeled methylmercury. Anal. Biochem. 42(2): 455-469. (Environmental use).

Cullen, T. L. 1968. Dosimetric measurements in Brazilian regions of high natural radioactivity leading to cytogenetic studies. pp. 371-378. In: Radiation Protection, Part 1. W. S. Snyder, et al., eds. Pergamon Press, New York. (Incl. "rat" dens).

Culler, F. L., J. O. Blomeke and W. G. Belter. 1972. Current developments in long-term radioactive waste management. Proc. 4th Intern. Conf. on the Peaceful Uses of Atomic Energy 11: 427-443. (See United Nations and IAEA, 1972).

Cumming, B. G. 1967. Correlations between periodicities in germination of *Chenopodium botrys* and variations in solar radio flux. Can. J. Bot. 45: 1105-1113. (NSA 22: 8648).

Cummings, S. L., et al. 1973. Cesium-137 in white-tailed deer as related to vegetation and soils of the Southeastern United States. pp. 123-128. In: Radionuclides in Ecosystems. (See Nelson, 1973) (NSA 28: 27296).

Cummins, D. G. 1962. The effect of gamma and neutron radiation on some chemical and physical properties of soil. Dissertation, Univ. of Georgia, Athens.

Cummins, D. G. and R. A. McCreery. 1964. The effect of gamma and neutron irradiation of soil on the yield and nutrient uptake of rye. Soil Sci. Soc. Am. Proc. 28(3): 390-393. (NSA 18: 29363).

Cummins, J. E. and W. Plant. 1964. The distribution of ^{32}P in ribonucleic acid from nucleate and anucleate half cells of *Amoeba proteus*. Biochim. Biophys. Acta 80(1): 19-30. (NSA 18: 11582).

Cunliffe, F. 1952. Biology of the cockroach parasite Pimeliaphilus podapolipophagus Tragardh., with a discussion of the general Pimeliaphilus and Hirstiella. Proc. Entomol. Soc. Wash. 54(4): 269-280. (NSA 15: 32397).

Cunningham, H. M. 1970. Equipment for the continuous sampling of expired ^{14}C carbon dioxide from animals in a respiration chamber. Can. J. Anim. Sci. 50: 753-754. (NSA 25: 39439).

Cunningham, M. P. 1968. Vaccination of cattle against trypanosomes by injection and treatment. pp. 89-92. In: Isotopes and Radiation in Parasitology. (See IAEA, 1968).

Cunningham, M. P. and E. B. Grainge. 1968. Immune response of cattle infected with Trypanosoma brucei subgroup and the antigenic variation of infecting organisms. pp. 77-88. In: Isotopes and Radiation in Parasitology. (See IAEA, 1968).

Cunningham, M. P., et al. 1973. Some effects of irradiation on the infective stage of Theileria parva harvested from infected ticks. pp. 145-154. In: Isotopes and Radiation in Parasitology III. IAEA, Vienna, report STI/PUB-238. (NSA 28: 15698).

Cupic, Z., M. Miric and A. Damanski. 1971. Effect of x rays on the reproduction and dynamics of some components of the sunflower fruit during its development and ripening. Hrana Ishrana 12(9): 422-428. (In Serbo-Croatian) (NSA 28: 8392).

Curds, C. R. 1963. The flocculation of suspended matter by ciliated protozoa. Abstr. J. Protozool. 10(Suppl.): 30-31. (Used ^{14}C).

Curie, P. 1929. On the study of probability curves relative to the action of x rays on bacillus. Compt. rend. 188(2): 202-204. (In French).

Curlin, J. W. and J. P. Witherspoon. 1968. Components and estimates of dose to roots of potted plants exposed to fast neutron radiation. Health Physics 14(2): 165-168.

Curnow, R. D. 1968. A literature review of zinc-65 in birds with special reference to the mallard. Colorado Coop. Wildlife Res. Unit, Colorado State Univ., Fort Collins. iv, 15pp.

Curnow, R. D. 1968. ^{65}Zn uptake and excretion in mallards. Thesis, Colorado State Univ., Fort Collins.

Curnow, R. D. 1969. Radioactive mallards. Colo. Outdoors 18(6): 12-14.

Curnow, R. D. 1970. Kinetics of 65-zinc in the mallard. Thesis, Colorado State Univ., Fort Collins. 114pp. (Dissertation Abstr. Int. B Sci. Eng. 31(12): 1971).

Curnow, R. D. 1973. Assimilation and retention of zinc-65 by the mallard duck (Anas platyrhynchos). pp. 381-384. In: Radionuclides in Ecosystems. (See Nelson, 1973) (NSA 28: 27676).

Curnow, R. D. and F. A. Glover. 1970. Radiosensitivity of the mallard duck (Anas platyrhynchos). Colorado Coop. Wildlife Res. Unit, Colorado State Univ., Fort Collins, Tech. Paper 13. 9pp.

Curnow, R. D., F. A. Glover and F. W. Whicker. 1967. Uptake, excretion, and distribution of zinc-65 by the mallard. Colorado Coop. Wildlife Res. Unit, Colorado State Univ., Fort Collins, Tech. Paper No. 10. 8pp. (For Ann. Meeting, Ornithologist's Union, August 1967).

Currie, L. A. 1972. Evaluation of radiocarbon measurements and inherent statistical limitations in age resolution. pp. H1-H15. In: Proc. 8th Intern. Radiocarbon Dating Conf. Vol. 2. (See Rafter, 1972) (NSA 31: 8715).

Currie, L. A. 1973. Evaluation of radiocarbon measurements and inherent statistical limitations in age resolution. Final report. National Bureau of Standards, U. S. Dept. of Commerce, Washington, report COM-74-50079. 14pp. (NSA 33: 648).

Currie, L. A. and J. R. DeVoe. 1971. Isotope separator as a tool for low-level radioassay and tract activation analysis. pp. 183-190. In: Nuclear Techniques in Environmental Pollution. (See IAEA, 1971) (NSA 25: 31875).

Curry, LaV. L. 1960. Midge larvae as indicators of radioactive pollution. Engr. Bull. Purdue Univ., Engr. Ext. Ser. No. 106, 45(2): 269-280. (Proc. 15th Industrial Waste Conf., May 3-5, 1960) (NSA 15: 32397).

Curtis, C. F. 1968. Radiation sterilization and the effect of multiple mating of females in Glossina austeni. J. Insect Physiol. 14: 1365-1380. (NSA 23: 20457).

Curtis, C. F. 1968. Theoretical and practical studies on a possible genetic method for tsetse fly control. pp. 233-247. In: Isotopes and Radiation in Entomology. (See IAEA, 1968) (NSA 24: 32320).

Curtis, C. F. 1969. The production of partially sterile mutants in Glossina austeni. Genet. Res. 13: 289-301. (NSA 23: 50973).

Curtis, C. F. 1971. Induced sterility in insects. Adv. Reprod. Physiol. 5: 119-165. (NSA 30: 617).

Curtis, C. F. 1973. The use of simulation in the choice of an optimum radiation dose for control of Glossina morsitans. pp. 177-181. In: Computer Models and Application of the Sterile-Male Technique. (See IAEA, 1973) (NSA 29: 8060).

Curtis, E. J. C. and I. P. Toms. 1972. Techniques for counting carbon-14 and phosphorus-32 labelled samples of polluted natural waters. pp. 167-180. In: Liquid Scintillation Counting, Vol. 2. M. A. Crook, ed. Haydens and Son, London. (NSA 30: 2576).

Curtis, H. J. 1958. Cooperative studies on radiation induced mutations in plants. Proc. 2nd Intern. Conf. on the Peaceful Uses of Atomic Energy 27: 371-373. (Paper A/CONF. 15/P/2530) (Also: Brookhaven National Lab., U. S. AEC report BNL-3883) (See United Nations, 1958) (NSA 13: 5209).

Curtis, H. J. 1961. Radiation effects in dry seeds. pp. 53-55. In: Conf. on Molecular and Radiation Biology. (Gould Hause, Dec. 2-4, 1959, Ardsley-on-Hudson, N. Y.). National Academy of Sciences, Nucl. Sci. Ser. rpt. 31. NAS-NRC-Pub-823. (NSA 15: 17895; 17922).

Curtis, J. T. and R. Dix. 1956. Distribution of alpha-radioactivity in certain forest types. Science 123(3201): 799-800. (NSA 10: 8166).

Curtis, J. T. and R. Dix. 1956. Distribution of alpha radioactivity in native vegetation. Bot. Gaz. 117(3): 233-238. (NSA 12: 13654).

Curtis, W. C. 1936. Effects of x-rays and radium upon regeneration. pp. 411-457. In: Biological Effects of Radiation. Vol. I. B. H. Duggar, ed. McGraw-Hill, New York. (Porifera, Hydrozoa, Turbellaria, Nemertina, Chaetopoda, Tunicata, Amphibia).

Curtis, W. C., J. A. Cameron and K. O. Mills. 1936. Exogastrulation in Amphibia after x-ray exposure. Science 83(2154): 354.

Cushing, C. E. 1967. Concentration and transport of ^{32}P and ^{65}Zn by Columbia River plankton. Limnol. Oceanogr. 12(2): 330-332. (NSA 21: 41426).

Cushing, C. E. 1967. Periphyton productivity and radionuclide accumulation in the Columbia River, Washington, U. S. A. Hydrobiologia 29(1-2): 125-139. (NSA 22: 21591).

Cushing, C. E. 1970. Radiation ecology in freshwater communities. pp. 45-56. In: Man and Aquatic Communities. Water Resources Inst., Oregon State Univ., Corvallis. (Also: Battelle Northwest Lab., U. S. AEC report BNWL-SA-3291) (CONF-700554-1) (NSA 25: 24073).

Cushing, C. E., et al. (Eds.). 1976. Radioecology and Energy Resources. (Proc. 4th National Symp. on Radioecology, May 12-14, 1975, Oregon State Univ., Corvallis, Ore.). The Ecological Soc. of Am. Special Publ. No. 1. Dowden, Hutchinson & Ross, Stroudsberg, Pa. xx, 404pp.

Cushing, C. E. and N. S. Porter. 1969. Radionuclide cycling by periphyton: an apparatus for continuous in situ measurements and initial data on zinc-65 cycling. pp. 285-290. In: Symposium on Radioecology. (See Nelson and Evans, 1969) (Also: Battelle Northwest Lab., U. S. AEC report BNWL-SA-1254).

Cushing, C. E. and L. A. Rancitelli. 1972. Trace element analysis of Columbia River water and phytoplankton. Northwest Sci. 46(2): 115-121. (NSA 26: 38143).

Cushing, C. E. and F. L. Rose. 1970. Cycling of ^{65}Zn by Columbia River periphyton in a closed lotic microcosm. Limnol. Oceanogr. 15: 762-767. (NSA 25: 26593).

Cushing, C. E. and D. G. Watson. 1966. Accumulation and transport of radionuclides by Columbia River biota. pp. 551-570. In: Disposal of Radioactive Wastes into Seas, Oceans, and Surface Waters. (See IAEA, 1966) (NSA 20: 20382).

Cushing, C. E. and D. G. Watson. 1968. Accumulation of phosphorus-32 and zinc-65 by living and killed plankton. Abstr. Bull. Ecol. Soc. Am. 49(2): 65.

Cushing, C. E. and D. G. Watson. 1968. Accumulation of ^{32}P and ^{65}Zn by living and killed plankton. Oikos 19(1): 143-145.

Cushing, C. E. and D. G. Watson. 1973. Cycling of zinc-65 in a simple food web. pp. 318-322. In: Radionuclides in Ecosystems. (See Nelson, 1973) (NSA 25: 39806).

Cushing, C. E. and D. G. Watson. 1973. Decline of radioactivity in biota following shutdown of Hanford reactors. Abstr. Bull. Ecol. Soc. Am. 54 (1): 23-24.

Cushing, C. E. and D. G. Watson. 1974. Aquatic studies of Gable Mountain Pond. Battelle Northwest Lab., U. S. AEC report BNWL-1884. 33pp. (NSA 32: 14278) (^{137}Cs).

Cushing, E. J. 1965. Late-Wisconsin pollen stratigraphy and the glacial sequence in Minnesota. Abstr. Abstrs. Intern. Assoc. Quaternary Res. 7th Intern. Congr., Boulder and Denver, Colo. (Carbon dating).

Cutkomp, L. K. 1968. Uptake of isotopically labelled insecticides in relation to temperature and other physical factors. pp. 179-185. In: Isotopes and Radiation in Entomology. (See IAEA, 1968) (NSA 24: 31794).

Cutshall, N. 1974. Turnover of zinc-65 in oysters. Health Physics 26(4): 327-331. (NSA 29: 29871).

Cutshall, N. and C. Osterberg. 1964. Radioactive particles in sediment from the Columbia River. Science 144(3618): 536-537.

Cwik, T., Z. Gorberg and J. Jaskulski. 1969. Radioactive contamination of the area and the environment of the Institute of Nuclear Research, Zeran Centre, in the period 1961-1967. Poland Inst. Nucl. Res., Warsaw, report INR-114. 29pp. (In Polish) (NSA 24: 23179).

Cwik, T., et al. 1970. Radioactive contamination on-site and in the environs of the Institute of Nuclear Research at Swierk in the period 1958-1967. Poland Inst. Nucl. Res., Warsaw, report INR-1185. 42pp. (In Polish) (NSA 25: 6698).

Czech, H., R. Czech and H. J. Schmermund. 1950. Comparative studies of the effect of particles, electrons and x-rays on the cornea of salamander larvae. Strahlentherapie 81: 187-192. (AEC-tr-3472) (NSA 13: 1892).

Czechoslovakia Commission on Nuclear Energy. 1972. Abstracts. Radioecological Conference: Theoretical and Practical Problems of the Contamination of the Environment with Radioactive Materials. (Stary Smokovec, May 16-19, 1972). (In Russian and Czech) (CONF-7205103-(Absts.)) (NSA 30: 6481).

Czubek, J. A. 1968. Range of nuclear methods for soil moisture and density measurements. Nukleonika 13: 517-533. (In Polish) (NSA 22: 48920).

Dabek, M. and K. Muras. 1968. Some problems of measuring soil density with gamma probes constituting a part of the GGW type assembly. Nukleonika 13: 547-560. (In Polish) (NSA 22: 48925).

Dabrowski, T. E. 1973. Douglas United Nuclear, Inc. reactor and fuel production facilities 1972 environmental release report. Douglas United Nuclear report DUN-8133. 13pp. (NSA 30: 29341).

Dabrowski, Z. T. 1968. Sensitivity of various stages of the grain mite (Acarus siro L.) to gamma radiation. Rocz. Nauk Polon. A93: 613-622. (In Polish) (NSA 22: 42466).

Dafauce, C. 1964. Radioactive labelling of insects. Boln Serv. Plagas for. B7(14): 102-114. (In Spanish, English summary).

Defauce, C., D. Cadahia and F. Astiaso. 1963. Three experiences with Cryptorrhynchus lapathi L. (Curculionidae), by means of radioactive marks. Boln Serv. Plagas for. 6(11): 7-13. (In Spanish, English summary).

Dagg, M. 1967. Neutron-moisture gage in agricultural and hydrological field studies in sub-humid areas. pp. 177-187. In: Isotope and Radiation Techniques in Soil Physics and Irrigation Studies. (See IAEA, 1967) (NSA 22: 23151).

Dahl, B. 1936. The tissue reaction of the larval frog (Rana temporaria) to roentgen radiation. Acta Radiologica 17(3): 275-283. (In German).

Dahl, J. B. 1974. Physics and the environment: pollution studies using isotope techniques. Fra Fys. Verden 36(1): 15-19. (In Norwegian) (NSA 30: 3184).

Dahl, J. B., E. Steinnes and J. Thomassen. 1971. Air pollution studies by means of inactive indium tracer and activation analysis. pp. 283-296. In: Nuclear Techniques in Environmental Pollution. (See IAEA, 1971) (NSA 25: 32089).

Dahl, J. B., et al. 1971. Water pollution studies by means of inactive indium tracer and activation analysis. pp. 663-672. In: Nuclear Techniques in Environmental Pollution. (See IAEA, 1971) (NSA 25: 31886).

Dahl, J. B., et al. 1972. Measurement of groundwater parameters at Korgen Waterworks, Lillehammer, by means of isotope techniques. pp. 67-108. In: Northern Hydrological Conf. 1972. Norway Nordisk Hydrologisk Forening, Oslo. (In Norwegian) (NSA 28: 5579).

Dahlman, R. C. 1971. Prediction of radionuclide contamination of grass from fallout-particle retention and behavior. pp. 492-508. In: Survival of Food Crops and Livestock in the Event of Nuclear War. (See Bensen and Sparrow, 1971).

Dahlman, R. C. and S. I. Auerbach. 1968. Preliminary estimation of erosion and radiocesium redistribution in a fescue meadow. Oak Ridge National Lab., U. S. AEC report ORNL-TM-2343. 24pp. (NSA 23: 16291).

Dahlman, R. C. and C. L. Kucera. 1969. Tagging native grassland vegetation with carbon-14. Ecology 49(6): 1199-1203.

Dahlman, R. C. and C. L. Kucera. 1969. Carbon-14 cycling in the root and soil components of a prairie ecosystem. pp. 652-660. In: Symposium on Radioecology. (See Nelson and Evans, 1969) (NSA 21: 30797).

Dahlman, R. C. and Y. Tanaka. 1973. Ecological-environmental assessments related to the federal repository. Oak Ridge National Lab., U. S. AEC report ORNL-TM-3619. 53pp. (NSA 28: 15494).

Dahlman, R. C. and P. Van Voris. 1975. Cycling of ^{137}Cs in soil and vegetation of a flood plain 30 years after initial contamination. Oak Ridge National Lab., U. S. ERDA report CONF-750503-15. 30pp. (NSA 32: 6429).

Dahlman, R. C., S. I. Auerbach and P. B. Dunaway. 1969. Behavior of ^{137}Cs-tagged particles in a fescue meadow. pp. 153-165. In: Environmental Contamination by Radioactive Materials. (See IAEA, 1969).

Dahlman, R. C., J. J. Beauchamp and Y. Tanaka. 1973. Effects of simulated fallout radiation on reproductive capacity of fescue. pp. 988-998. In: Radionuclides in Ecosystems. (See Nelson, 1973) (NSA 26: 15523).

Dahm, P. A. 1953. Radioactive tracers in insecticide research. III, IV. Soap and Sanitary Chem. 29(9): 136, 139, 161; 29(10): 148-149, 151, 153, 163, 165; 29(11): 141, 143, 145, 147, 165; 29(12): 167-168, 175.

Dahm, P. A. 1953. Studies of insects and insecticides with tracers. pp. 162-199. In: Conf. on the Use of Isotopes in Plant and Animal Research. U. S. AEC report TID-5098. (NSA 8: 82).

Dahm, P. A. 1957. Uses of radioisotopes in pesticide research. pp. 75-124. In: Advances in Pest Control Research. (See Metcalf, 1957).

Dairyland Power Cooperative. 1971. La Crosse Boiling Water Reactor. Environmental monitoring associated with radioactive discharges, January 1, 1969-June 30, 1971. La Crosse, Wis. 46pp. U. S. AEC DOCKET-1155-135. (DPC-851-31) (NSA 25: 56895).

Dalal, V. 1973. Environment, energy, and the need for new technology. Energy Conserv. 13(3): 85-94. (NSA 28: 17686).

Dalcq, A. 1930. Cytological interpretation of the effects on gastrulation of irradiation of the gametes of Rana fusca. Compt. rend. soc. biol. 104: 1055-1058. (In French).

Dalebroux, M. A. 1965. Genetic variability induced by ionizing radiations in quantitative traits of Habrobracon. Thesis, North Carolina State Univ., Raleigh. 95pp. (NSA 20: 36659).

Dalebroux, M. A. and K. Kojima. 1967. An analysis of radiation-induced variation on body weight of Habrobracon juglandis. Genetics 55: 315-328. (NSA 21: 18030).

Dall'Aglio, M. 1971. Study of the circulation of uranium in the supergene environment in the Italian alpine range. Geochim. Cosmochim. Acta 35: 47-59. (NSA 25: 48608).

Dall'Aglio, M. and G. P. Giannotti. 1964. A preliminary hydrogeological and hydrogeochemical survey of the Arrone River in view of a possible immission of radioactive effluents. Com. Naz. Energia Nucl. Notiz. 10(6): 31-45. (In Italian) (NSA 18: 41666).

Dall'Aglio, M. and G. Giannotti. 1967. Distribution and circulation of barium, rubidium, strontium, and caesium in natural waters around the Casaccia Nuclear Centre. pp. 343-366. In: Disposal of Radioactive Wastes into the Ground. (See IAEA, 1967) (NSA 22: 16811).

Dalrymple, A. V. and E. Basler. 1963. Seasonal variation in absorption and translocation of 2,4,5-trichlorophenoxyacetic acid and respiration rates in blackjack oak. Weeds 11(1): 41-45. (^{14}C).

Dalrymple, G. B. 1965. Potassium-argon dating in quarternary correlation. Abstr. Abstrs. Intern. Assoc. Quarternary Res. 7th Intern. Congr., Boulder and Denver, Colo.

Dalrymple, G. B., R. R. Doell and A. Cox. 1965. The transition time for geomagnetic reversals and the precision of potassium-argon dating. Abstr. Abstrs. Intern. Assoc. Quarternary Res. 7th Intern. Congr., Boulder and Denver, Colo.

Dalu, G. 1967. Development of a method for measurement of a natural radioactivity. Ass. Geofis. Ital. Atti Conv. Ann. 16: 211-229. (In Italian) (NSA 24: 36422).

Daly, J. C. and J. J. Gabay. 1969. Diffusion of tritiated moisture from a stream. Trans. Amer. Nucl. Soc. 12: 899-890. (NSA 24: 5749).

Daly, J. C. and J. J. Gabay. 1971. Rapid screening of environmental samples using gamma spectroscopy. Health Physics 20(6): 640-641. (NSA 25: 46830).

Daly, J. C., et al. 1968. Tritiated moisture in the atmosphere surrounding a nuclear fuel reprocessing plant. Radiological Hlth. Data 9(7): 341-346.

Daly, J. C., et al. 1974. Iodine-129 levels in milk and water near a fuel reprocessing plant. Health Physics 26(4): 333-342. (NSA 29: 26835).

Dame, D. A. and R. L. Fye. 1964. Studies on feeding behavior of house flies. J. Econ. Entomol. 57(5): 776-777. (^{32}P).

Dame, D. A. and C. H. Schmidt. 1962. The importance of competitiveness of radiosterilized males in mosquito-control programs. Proc. Ann. Meeting New Jersey Mosquito Exterm. Assoc. 49: 165-168. (NSA 19: 21964).

Dame, D. A. and C. H. Schmidt. 1964. P^{32}-labeled semen for mosquito mating studies. J. Econ. Entomol. 57(5): 669-671.

Dame, D. A. and C. H. Schmidt. 1964. Uptake of metepa and its effect on two species of mosquitos (Anopheles quadrimaculatus, Aedes aegypti) and house flies (Musca domestica). J. Econ. Entomol. 57(1): 77-81. (^{32}P).

Dame, D. A., et al. 1964. Field behavior of sexually sterile Anopheles quadrimaculatus males. Mosquito News 24: 6-14. (NSA 19: 24315).

Dame, D. A., et al. 1965. Central Africa. I. The USDA research team for the chemosterilization of tsetse flies. II. The radio-sterilization of tsetse. Annual Report of the Agricultural Research Council of Central Africa.

Damm, F. C. and A. R. Slonim. 1972. Beryllium uptake by the common guppy using radioisotope Be-7. Preliminary study; tracer mechanics and procedures. Final report. Aerospace Medical Res. Lab., Wright-Patterson AFB, Ohio, report. 17pp. (AD-758814) (NSA 30: 24413).

Dammann, W., M. Elmdust and W. Kühn. 1973. Continuous recording of water vapor in soil with the aid of tritium labelling. Kerntechnik 15(7): 306-311. (In English and German) (NSA 28: 29876).

Damon, P. E. 1965. Radioactive dating of quartenary tephra. Abstr. Abstrs. Intern. Assoc. Quaternary Res. 7th Intern. Congr., Boulder and Denver, Colo.

Damon, P. E., A. Long and D. C. Grey. 1966. Fluctuation of atmospheric C^{14} during the last six millennia. J. Geophys. Res. 71: 1055-1063.

Damon, P. E., A. Long and E. I. Wallick. 1972. Dendrochronologic calibration of the ^{14}C time scale. pp. A28-A43. In: Proc. 8th Intern. Radiocarbon Dating Conf. Vol. 1. (See Rafter, 1972) (NSA 31: 8681.

Dams, R., et al. 1971. Nondestructive multi-element neutron activation anaysis of air pollution particulates. pp. 8-15. In: Nuclear Methods in Environmental Research. (See Vogt, 1971) (NSA 26: 17356).

Dana, S. 1962. Effects of x-irradiation on the testes of the lizard, Uta stansburiana stejnegeri. Thesis, Texas Tech. Coll., Lubbock.

Dana, S. W. and D. W. Tinkle. 1965. Effects of x-radiation on the testes of the lizard, Uta stansburiana stejnegeri. Intern. J. Radiation Biol. 9(1): 67-80. (NSA 19: 36015).

Danali, S. D. 1966. Functional arrangement and organizational problems of a laboratory for monitoring radioactivity in the biosphere. Greece Democritus Nucl. Res. Center, Athens, report 3-4. (UCRL-Trans-1356; 97pp) (NSA 22: 2493).

Dance, J. B. 1973. Radioisotope Experiments in Physics, Chemistry and Biology. Hutchinson Educational, London. 2nd Revised Ed. 239pp. (NSA 29: 5386).

Dancette, C. 1974. Water requirements of the major crops in Senegal. pp. 351-371. In: Isotope and Radiation Techniques in Soil Physics and Irrigation Studies 1973. (See IAEA, 1974) (NSA 30: 15503) (Neutron probe).

Danecka-Palasz, R. and J. Kwapulinski. 1972. Occurrence of ^{90}Sr in the ice of the Limnic Reservoir of the Barrage. Acta Geophys. Pol. 20(3-4): 335-341. (In Polish) (NSA 27: 24610).

Danfors, L. E. and N. Gustafsson. 1967. Tuber formation in potatoes as affected by soil moisture. pp. 379-390. In: Isotope and Radiation Techniques in Soil Physics and Irrigation Studies. (See IAEA, 1967) (NSA 22: 23353).

Daniel, A. P. and B. Highmann. 1951. Some effects of x-irradiation on turtles. J. Exptl. Zool. 118(1): 1-10.

Daniel, C. P. 1960. Irradiation and growth in old-field succession. Abstr. Bull. Ecol. Soc. Am. 41(3): 94.

Daniel, C. P. 1960. Changes in dominance in old field succession under gamma radiation. Abstr. ASB Bull. 7: 26.

Daniel, C. P. 1963. A study of succession in fields irradiated with fast neutron and gamma radiation. pp. 277-282. In: Radioecology. (See Schultz and Klement, 1963) (NSA 17: 33571).

Daniel, C. P. 1964. Dominance areas as a result of ionizing radiation in developing and recovery communities. Abstr. Bull. Ecol. Soc. Am. 45(3): 80.

Daniel, C. P. 1965. Ionizing radiation as an ecological factor in old-field flood plain succession. Thesis, Emory Univ., Atlanta, Ga.

Daniel, C. P. and R. B. Platt. 1968. Direct and indirect effects of short-term ionizing radiation on old-field succession. Ecological Monogr. 38(1): 1-29. (NSA 22: 38501).

Daniel, G. E. and H. D. Park. 1951. The effect of x-ray treated media on hydra tentacles. J. Cell. Comp. Physiol. 42(3): 359-367.

Daniel, G. E. and H. D. Park. 1953. Glutathione and x-ray injury in Hydra and Paramecium. J. Cell. Comp. Physiol. 42(3): 359-367.

Daniel, G. E. and H. D. Park. 1953. The influence of small doses of x- and beta-rays on reproduction of Paramecium. Proc. Soc. Exptl. Biol. Med. 83 (4): 662-665.

Daniel, G. E. and H. D. Park. 1954. A protective effect of inorganic magnesium against roentgen radiation damage to Hydra. Am. J. Roentgenol. Rad. Therapy Nucl. Med. 72(5): 857-866. (NSA 9: 41).

Daniel, J. D., Jr. and R. L. Blain. 1959. Background radiation and endemic faunal range in the San Luis Valley of southern Colorado. Great Basin Nat. 19(1): 37-42.

Daniels, E. W. 1951. Studies on the effect of x-irradiation upon Pelomyxa carolinensis with special reference to nuclear division and plastmotany. J. Exptl. Zool. 117(1): 189-209. (NSA 5: 6058).

Daniels, E. W. 1954. Cell division in the giant amoeba, Pelomyxa carolinensis, following x-irradiation. II. Analysis of therapeutic effects after fusion with nonirradiated cell portions. J. Exptl. Zool. 127(3): 427-461.

Daniels, E. W. 1955. X-irradiation of the giant amoeba, Pelomyxa illinoisensis. I. Survival and cell division following exposure. Therapeutic effects of whole protoplasm. J. Exptl. Zool. 130(2): 183-197.

Daniels, E. W. 1956. Recovery from lethal x-radiation injury in Amoebae after injection of centrifugal portions of nonirradiated cells. Abstr. Radiation Research 5(6): 604-605.

Daniels, E. W. 1958. X-irradiation of the giant amoeba, Pelomyxa illinoisensis. II. Further studies of recovery following supralethal exposure. J. Exptl. Zool. 137(3): 425-442. (NSA 13: 1918).

Daniels, E. W. 1959. Micrurgical studies in irradiated Pelomyxa. Ann. N. Y. Acad. Sci. 78: 662-674. (NSA 13: 19786).

Daniels, E. W. 1962. Limits of transplantation tolerance in large amoebae. I. Microfusion studies using Amoeba proteus, Pelomyxa illinoisensis, and three strains of Pelomyxa carolinensis. J. Protozool. 9(2): 183-187. (NSA 16: 21956).

Daniels, E. W. and E. P. Breyer. 1970. Recovery of supralethally x-irradiated and nitrogen mustard (HN_2)-treated amoebae. Intern. J. Radiat. Biol. 18: 179-187. (NSA 24: 51022).

Daniels, E. W. and E. P. Breyer. 1970. Rescue of supralethally x-irradiated amoeba with nonirradiated cytoplasm. Radiation Research 41: 326-341. (NSA 24: 17127).

Daniels, E. W. and E. P. Breyer. 1971. Supralethally x-irradiated amoebae treated with nonirradiated, homoginized protoplasm. Radiation Research 46: 85-88. (NSA 25: 32521).

Daniels, E. W. and L. E. Roth. 1961. X-irradiation of the giant amoeba, Pelomyxa.illinoisensis. III. Electron microscopy of centrifuged organisms. Radiation Research 14(1): 66-82.

Daniels, E. W. and H. H. Vogel, Jr. 1958. Protective component in nonirradiated protoplasm of amoeba. Proc. 2nd Intern. Conf. on the Peaceful Uses of Atomic Energy 23: 29-33. (Paper A/CONF.15/P/907) (See United Nations, 1958) (NSA 12: 14502).

Daniels, E. W. and H. H. Vogel, Jr. 1959. Effects of nonirradiated protoplasm on recovery of amoebae exposed to supralethal doses of fission neutrons. Radiation Research 10(5): 584-596. (NSA 13: 13197).

Danil'chenko, O. P. 1958. Penetration of strontium-90 from water into the body of fish. Ryb. Khozyaystvo (2): 28-32. (In Russian) (SFA 4: 2701).

Danil'chenko, O. P. 1959. A selection of Sr^{90} (Y^{90}) dosage in fish labelling. Inform. sborn. VNIRO (5): ___. (In Russian).

Danil'chenko, O. P. 1960. Dynamics of Sr^{90} (Y^{90}) metabolism in the sturgeon. Sci. Reports of Higher Schools. Ser. Biologich. Nauki (1): ___. (In Russian).

Danileiko, V. I. 1962. Concerning the question of biologic action of cosmic and solar radiation in mountains. In: Problems in Cosmic Biology. N. M. Sisakyan, ed. (English transl., U. S. Air Force Systems Command, Wright-Patterson AFB, Ohio, report FTD-MT-62-78; pp. 564-575) (NSA 17: 35286).

Danilin, A. I. 1963. The use of radioactive isotopes in certain hydrometeorological measurements and investigations. Tr. vses. nauchn. Meteor. Sov. 4, Prib. met. nabl. 9: 49-60. (JPRS-26742; TT-64-4199, pp 49-60) (NSA 19: 6088).

Danilin, A. I. 1973. Experience with and some features of the use of gamma rays and neutrons in investigating soil moisture content. pp. 51-62. In: *Soil Moisture and Irrigation Studies*. *II*. (See IAEA, 1973) (In Russian, English abstr.) (NSA 28: 21373).

Danneel, I. 1967. Study on food selection of *Aphis fabae* Scop. (Homoptera, Aphididae) in experiments using short time intervals. Z. Angew. Zool. 54: 181-182. (In German).

Danneel, I. 1967. The transmission of ^{32}P-labelled material in the body of *Aphis fabae* Scop. (Homoptera, Aphididae) to its larvae. Z. Angew. Zool. 54(3): 433-444. (In German).

Dansgaard, W., H. B. Clausen and A. Aarkrog. 1966. The ^{32}Si fallout in Scandinavia. A new method for ice dating. Tellus 18: 187-191. (NSA 21: 6061).

Danysz, J. 1903. On the pathogenic action of rays and emanation from radium on different tissues and different organisms. Compt. rend. 136(7): 461-464. (In French) (Rabbit, bacteria, insects).

Dapson, R. W. and L. Kaplan. 1975. Biological half-life and distribution of radiocesium in a contaminated population of green treefrogs *Hyla cinerea*. Oikos 26(1): 39-42. (NSA 32: 9255).

Darab, K. and I. Toeroek. 1971. Factors influencing the movement and fixation of ^{90}Sr in the soil. Agrokem. Talajtan 20(1-2): 147-156. (In Hungarian) (NSA 26: 17733).

Darden, E. B., Jr., E. Maeyens and R. C. Bushland. 1954. A gamma-ray source for sterilizing insects. Nucleonics 12(10): 60-62. (NSA 8: 6917).

Darenskaya, N. G. and G. M. Pravdina. 1966. Some results of the investigation of animal behavior in radiation fields. Abstr. 250. p. 64. In: *3rd Intern. Congr. of Radiation Research*. (Cortina d'Ampezzo, Italy, 26 June-2 July, 1966). G. Silini, ed. North-Holland, Amsterdam.

Darenskaya, N. G., G. M. Pravdina and V. G. Chruschev. 1966. The behavior of living organisms in radiation fields. Radiobiol.-Radiotherap. 7: 81-93. (In German) (NSA 20: 20495).

Darenskaya, N. G., et al. 1974. Seasonal changes of radiosensitivity in mice. Radiobiol.-Radiotherap. 15(3): 399-407. (In German) (NSA 31: 6333).

Darrall, K. G., G. C. M. Hammond and J. F. C. Tyler. 1973. Determination of plutonium-241 in effluents. Analyst 98(1166): 358-363. (NSA 28: 5211).

Darwish, R. O., F. A. Abdel Salam and K. A. Mahmoud. 1958. Some somatic changes observed in *Culex molestus* Forskal 1775. Egypt Atomic Energy Commission, Cairo. Report to the UNSCEAR. 9pp. (NSA 14: 17686).

Das, B. C. 1970. Effects of gamma radiation on germination and seedling development of mulberry. Sci. & Cult. (Calcutta) 36: 60-61. (NSA 25: 21962).

Das, C. C., B. P. Kaufman and H. Gay. 1964. Autoradiographic evidence of synthesis of an arginine-rich histone during spermiogenesis in *Drosophila melanogaster*. Nature 204(4962): 1008-1009. (^{3}H).

Das, H. A. 1969. Examination of amber samples by nondestructive activation analysis. Radiochem. Radioanal. Lett. 1: 289-295. (NSA 23: 40587).

Das, H. A., et al. 1970. Routine procedure for the determination of strontium and barium in rocks and sediments by neutron activation analysis. Radiochem. Radioanal. Lett. 4: 171-180. (NSA 24: 38784).

Das, N. K., E. P. Siegel and M. Alfert. 1965. Synthetic activities during spermatogenesis in the locust. J. Cell Biol. 25(2): 387-395. (^{3}H, *Schistocerca gregaria*).

Dasberg, S. and R. Steinhardt. 1974. Water distribution in an orchard irrigated by sprinkler or trickler irrigation as measured by the neutron method. pp. 467-475. In: *Isotope and Radiation Techniques in Soil Physics and Irrigation Studies 1973*. (See IAEA, 1974) (NSA 30: 15510).

Dasch, E. J. 1969. Strontium isotopes in weathering profiles, deep-sea sediments, and sedimentary rocks. Geochim. Cosmochim. Acta 33: 1521-1552. (NSA 24: 9508).

D'Ascoli, A. and J. C. Gomez-Nuñez. 1966. Note on the means of dispersion of *Rhodnius prolixus* Stål. Acta Cient. Venez. 17(1): 22-25. (In Spanish, English summary) (^{60}Co).

Datta, N. P. and K. B. Shome. 1965. Measurement of ^{32}P in plant material by autoradiographic technique. J. Indian Soc. Soil Sci. 13: 157-160. (NSA 22: 31070).

Daudel, P. 1966. Experimentation with radioisotopes. pp. 955-1005. In: *Applied Radiobiology*, *Vol. I*. J. Loiseleur, ed. Gauthier-Villars, Paris. (In French) (NSA 21: 45533).

Dauer, M., P. L. Bhatnagar and M. Rockstein. 1964. Factors in longevity effects of high-energy radiation on the survival of male house flies. Abstr. Bull. Entomol. Soc. Am. 10(3): 166.

Dauer, M., P. L. Bhatnagar and M. Rockstein. 1965. X-irradiation of pupae of the house fly, *Musca domestica* L., and male survival. J. Gerontol. 20: 219-223. (NSA 19: 33858).

Dauphin, J. 1904. The influence of radium rays on the development and growth of inferior mushrooms. Compt. rend. 138(3): 154-156. (In French).

Dauskurdis, S. I. and Yu. V. Novikov. 1973. Hygienic problems of ^{90}Sr and ^{137}Cs migration along the chain soil-plants-milk and beef. Gigiyena sanitariya (4): 91-94. (In Russian) (NSA 28: 8189).

Davey, D. R. (Ed.). 1975. *Rum Jungle Environmental Studies*. Australia Atomic Energy Commission, Lucas Heights, report AAEC/E-365. 374pp. (NSA 33: 25942) (Uranium mining project).

Davey, D. R. and B. G. O'Brien. 1975. Radioecological aspects. Ch. 9, vp. In: *Rum Jungle Environmental Studies*. (See Davey, 1974) (NSA 33: 25950).

Davey, W. P. 1917. The effects of x-rays on the length of life of Tribolium confusum. J. Exptl. Zool. 22(3): 573-592.

Davey, W. P. 1919. Prolongation of life of Tribolium confusum apparently due to small doses of x-rays. J. Exptl. Zool. 28: 447-458.

Davich, T. B. 1969. Sterile-male technique for control or eradication of the boll weevil, Anthonomus grandis Boh. pp. 65-72. In: Sterile-male Technique for Eradication or Control of Harmful Insects. (See IAEA, 1969) (NSA 24: 32301).

Davich, T. B. and D. A. Lindquist. 1962. Exploratory studies on gamma radiation for the sterilization of the boll weevil. J. Econ. Entomol. 55(2): 164-167. (NSA 16: 32782).

Davich, T. B., et al. 1965. Preliminary field experiments with sterile males for eradication of the boll weevil. J. Econ. Entomol. 58(1): 127-131.

Davids, J. A. G. 1964. Safe disposal of radioactive waste in water and atmosphere. Water-Bodem-Lucht 54(4): 74-84. (NSA 19: 40277).

Davids, J. A. G. 1972. Environmental aspects of electricity production by nuclear power reactors. Atoomenergie Haar Toepass 14: 221-227. (In Dutch) (NSA 27: 2143).

Davidson, H. O. 1959. The effect of radiation on the longevity of the cockroach, Periplaneta americana, as affected by dose, age, sex and food intake. Radiation Research 11(4): 600-615.

Davidson, J. M., J. W. Biggar and D. R. Nielsen. 1963. Gamma-radiation attenuation for measuring bulk density and transient water flow in porous materials. J. Geophys. Res. 68(16): 4777-4783. (NSA 17: 34131).

Davidson, J. M., D. R. Nielsen and E. R. Perrier. 1959. Influence of temperature on soil moisture neutron probes. Soil Sci. Soc. Am. Proc. 23: 251-252.

Davidson, R. S. 1949. The phosphorus metabolism of three species of Helminthosporium as determined by use of radiophosphorus. Abstr. Phytopathology 39: 6. (NSA 2: 1027).

Davidsson, J. 1957. Investigations of sand movements using radio-active sand. Svensk Geogr. Arsb. 33: 107.

Davies, A. G. and J. J. Wilson. 1971. Determination of ^{137}Cs and ^{134}Cs in samples from the Windscale Effluent Treatment Plant. U. K. Atomic Energy Authority, Sellafield, PG-Report-965. 13pp. (NSA 25: 26177).

Davies, A. N., et al. 1959. Exploratory studies on gamma radiation for the sterilization and control of Anopheles quadrimaculatus. J. Econ. Entomol. 52(5): 868-870.

Davies, C. R. 1973. Effects of gamma irradiation on growth and yield of agricultural crops. 3. Root crops, legumes, and grasses. Radiation Botany 13(2): 127-136. (NSA 28: 5954).

Davies, D. R. 1966. Patterns and rates of recovery in synchronous populations of algal cells exposed to gamma radiation. Radiation Research 29: 222-235. (NSA 21: 14770).

Davies, D. R. 1968. Radiation studies on meiotic cells of Chlamydomonas reinhardi. pp. 123-133. In: Effects of Radiation on Meiotic Systems. International Atomic Energy Agency, Vienna, report STI/PUB-173. (CONF-670525) (NSA 24: 21322).

Davies, D. R. 1969. The induction and repair of radiation damage in Chlamydomonas. Curr. Top. Radiat. Res. 5: 75-113. (NSA 23: 38881).

Davies, D. R., P. D. Holt and D. G. Papworth. 1969. The survival curves of haploid and diploid Chlamydomonas reinhardi exposed to radiations of different LET. Int. J. Radiat. Biol. 15(1): 75-87. (NSA 23: 24767).

Davies, D. R. and M. A. P. Thorburn. 1968. The effect of dose rate in relation to hypoxia, and the roll of repair systems in the green alga Chlamydomonas reinhardi. Radiation Research 35: 401-409. (NSA 22: 45638).

Davies, D. R. and E. T. Wall. 1961. The effect of gamma irradiation on the sex ratio of Melandrium species. Heredity 16(1): 131-136. (NSA 15: 27436).

Davies, J. B. 1965. Three techniques for labeling Culicoides (Diptera: Heleidae) with radioactive tracers both in the laboratory and in the field. Mosquito News 25(4): 419-422. (^{32}P).

Davies, S. 1968. Environmental radiation surveillance at a nuclear fuel reprocessing plant. Am. J. Publ. Hlth. 58(12): 2251-2260. (NSA 23: 21794).

Davis, A. N., et al. 1959. Exploratory studies on gamma radiation for the sterilization and control of Anopheles quadrimaculatus. J. Econ. Entomol. 52(5): 868-870.

Davis, G. H., et al. 1967. Seasonal variations in the tritium content of groundwaters of the Vienna Basin, Austria. pp. 451-473. In: Isotopes in Hydrology. (See IAEA, 1967) (NSA 22: 18939).

Davis, G. H., et al. 1970. Geohydrologic interpretations of a volcanic island from environmental isotopes. Water Resources Res. 6: 99-109. (NSA 24: 46345).

Davis, J. J. 1958. The dispersion of radioactive materials by streams: with special reference to the Columbia River. J. Am. Water Works Assoc. 50(11): 1505-1515.

Davis, J. J. 1960. The effects of environmental factors upon the accumulation of radioisotopes by ecological systems. Tex. Engr. Expt. Sta. Misc. Publ. E-72-60: 37-41. (Also: Proc. 2nd Ann. Texas Conf. on the Utilization of Atomic Energy 2: 1-5) (NSA 14: 25209).

Davis, J. J. 1963. Cesium and its relationships to potassium in ecology. pp. 539-556. In: Radioecology. (See Schultz and Klement, 1963) (NSA 17: 33606).

Davis, J. J. 1965. Accumulation of radionuclides by aquatic insects. pp. 211-215. In: Biological Problems in Water Pollution. (See Tarzwell, 1965) (Also: Hanford Labs., U. S. AEC report HW-SA-2848) (NSA 17: 5702).

Davis, J. J. 1971. The Atomic Energy Commission's interest in bioenvironmental research. BioScience 21(12): 600-602. (NSA 25: 49128).

Davis, J. J. and C. L. Cooper. 1951. Effect of Hanford pile effluent upon aquatic invertebrates in the Columbia River. Hanford Atomic Products Operation, General Electric Co., U. S. AEC report HW-20055(Del.). 63pp. (NSA 11: 8260).

Davis, J. J. and R. F. Foster. 1958. Bioaccumulation of radioisotopes through aquatic food chains. Ecology 39(3): 530-535. (NSA 12: 16882).

Davis, J. J., W. C. Hanson and D. G. Watson. 1963. Some effects of environmental factors upon accumulation of world wide fallout in natural populations. pp. 35-38. In: Radioecology. (See Schultz and Klement, 1963) (NSA 17: 33545).

Davis, J. J., D. G. Watson and C. C. Palmiter. 1956. Radiobiological studies of the Columbia River through December 1955. Hanford Atomic Products Operation, General Electric Co., U. S. AEC report HW-36074(Del.). 177pp. (NSA 14: 453).

Davis, J. J., et al. 1958. Radioactive materials in aquatic and terrestrial organisms exposed to reactor effluent water. Proc. 2nd Intern. Conf. on the Peaceful Uses of Atomic Energy 18: 423-428. (Paper A/CONF. 15/P/393) (Also: pp. 1103-1123. In: Industrial Radioactive Waste Disposal, Vol. 2. (See U. S. Congress, 1959)).

Davis, J. M. and R. H. Nagel. 1956. A technique for tagging large numbers of live adult insects with radioisotopes. J. Econ. Entomol. 49(2): 210-211.

Davis, J. P. 1973. Regulations of the environmental effects of nuclear power plants. Nuclear Safety 14(1): 6-13. (NSA 28: 1691).

Davis, J. P. 1974. New Federal Water Pollution Control Act and its impact on nuclear power plants. Nuclear Safety 15(3): 262-275. (NSA 30: 6594).

Davis, J. T. 1959. Biological assay of herbicides for fish toxicity. Weeds 7(4): 397-404. (BA 35: 26021; SFA 5: 3529).

Davis, J. T. and D. Sparks. 1971. Assimilation of $^{14}CO_2$ by catkins of Carya illinoensis and apparent translocation to the pollen. Am. J. Botany 58 (10): 932-938.

Davis, K. A. 1970. Environmental monitoring summary for the Paducah Plant for 1969. Paducah Gaseous Diffusion Plant, Kentucky, U. S. AEC report KY-624. 10pp. (NSA 25: 6700).

Davis, K. A. 1971. Environmental monitoring summary for the Paducah Plant for 1970. Paducah Gaseous Diffusion Plant, Kentucky, U. S. AEC report KY-629. 11pp. (NSA 25: 29502).

Davis, R. J., V. L. Sheldon and S. I. Auerbach. 1956. Lethal effects of gamma radiation upon segments of a natural micobiol population. J. Bacteriology 72(4): 505-510. (NSA 11: 847).

Davis, R. T. and A. L. McDonald. 1962. Perception of oddity by monkeys as a function of aging and radiation. Psychol. Rep. 11: 383-386. (NSA 20: 8695).

Davis, S. N. 1964. Some hydrogeologic factors in water contamination by nuclear cratering. Hazelton-Nuclear Science Corp., U. S. AEC report HNS-1229-49. 32pp. (NSA 19: 6091).

Davis, T. S. 1962. Effect of cobalt-60 gamma radiation on pine seed and one-year-old seedlings. Forest Science 8(4): 411-412.

Davis, W. H., R. W. Barbour and M. D. Hassell. 1968. Colonial behavior of Eptesicus fuscus. J. Mammalogy 49: 44-50. (Radionuclide tag).

Davison, C. and F. Ellinger. 1942. Radiation effects on nervous system and roentgen pigmentation of goldfish (Carassius auretus). Proc. Soc. Exptl. Biol. Med. 49(3): 491-495.

Davy, D. R. 1971. Nuclear power and environmental pollution. Australia Atomic Energy Commission, Lucas Heights. 28pp. (NP-18842) (NSA 25: 37045).

Davy, D. R. 1973. Environmental studies for uranium provinces: aims and methods. pp. IV.1-IV.23. In: Proc. AAEC Symp. on Environmental and Radiological Safety Aspects of the Mining and Processing of Uranium. (See Australia Atomic Energy Commission, 1973) (NSA 28: 24482).

Davy, D. R. 1974. Environmental levels of radon-222. Australia Atomic Energy Commission, Lucas Heights, report AAEC/E-305 report 4. 11pp. (NSA 30: 26778) (Incl. groundwater).

Davy, D. R., M. S. Giles and E. Charash. 1970. Ecological factors in the siting, design and operation of a nuclear power plant. At. Energy Austral. 13(4): 16-28. (Also: Proc. Ecol. Soc. Austral. 5: 153-168) (NSA 25: 36970; 30: 29354).

Davy, D. R., M. S. Giles and N. Conway. 1973. Preoperational assessment of the discharge limits and relative importance of radioactive and other wastes from uranium production in Australia's Northern Territory. pp. 37-54. In: Environmental Behavior of Radionuclides Released in the Nuclear Industry. (See IAEA, 1973) (NSA 29: 29295).

Davydoff-Alibert, S. 1963. Influence of the kind of ambient environment during irradiation on the survival of bacteria. Intern. J. Appl. Radiat. Isotopes 14(1): 41-44. (In French) (NSA 17: 24939).

Davydov, Ye. N. and S. G. Malakhov. 1963. Artificial radioactivity of the ground layer of the atmosphere over Kheysa Island, Franz Josef Land in 1960-1961. State Committee on the Use of Atomic Energy, USSR. Atomizdat, Moscow. 12pp. (In Russian) (UN Doc. A/AC.82/G/L. 912) (NP-14697) (NSA 19: 20157)

Davydov, Ye. N. and M. P. Nekhorosheva. 1964. Determination of annual ^{90}Sr fallout in the polar regions by recording the ^{90}Sr content in the seasonal layers of snow and ice in icebergs of Franz Josef and northern islands. pp. 351-356. In: Radioactive Isotopes in the Atmosphere and Their Use in Meteorology. (Sci. Conf., Obninsk, Feb. 3-6, 1964). I. L. Karol', et al., eds. Atomizdat, Moscow. (CONF-640208) (AEC-tr-6711) (NSA 20: 16703).

Davydov, Ye. N., et al. 1969. Determination of radioisotope fallout from its content in chronological layers of high-altitude glaciers. pp. 29-36. In: Nuclear Meteorology. (See Makhon'ko and Malakhov, 1969) (NSA 30: 15436).

Dawes, B. 1963. Death of Fasciola hepatica L. weakened by x-irradiation. Nature 200(4906): 602-603. (NSA 18: 1411).

Dawson, W. W. and J. C. Smith. 1959. Low-level irradiation and threshold shift in the visual receptor. Science 129(3364): 1670-1671. (Horseshoe crab, Limulus).

Dawson, W. W. and H. Wiederwohl. 1965. Functional alteration of visual receptor units and retinal pigments by x-irradiation. Radiation Research 24(2): 292-304. (NSA 19: 15182) (Limulus polyphemus and Rana catesbiana).

Day, M. F. 1951. Application of isotopic tracers in entomological research. A review. pp. 145-160. In: Proc. Conf. on Applications of Isotopes in Scientific Research. Univ. of Melbourne, Australia.

Day, M. F. and H. Irzykiewicz. 1953. Feeding behavior of aphids Myzus persicae and Brevicoryne brassicae studied with radiophosphorus. Austral. J. Biol. Sci. 6(1): 98-108.

Day, M. F. and H. Irzykiewicz. 1954. Physiological studies on thrips in relation to transmission of tomato spotted wilt virus. Austral. J. Biol. Sci. 7(3): 274-281. (Radiotracer).

Day, M. F. and A. McKinnon. 1951. A study of some aspects of the feeding of the jassid Orosius. Austral. J. Sci. Res. B4(2): 125-135. (^{32}P).

Day, M. F. and I. I. Oster. 1963. 2. Physical injuries F. Radiation. pp. 44-48. In: Insect Pathology, an Advanced Treatise, Vol. I. E. A Steinhaus, ed. Academic Press, New York.

Day, M. F., S. W. Bailey and N. R. Norris. 1961. Uses of radioactivity in entomology and insect-pest control. Austral. Mus. Mag. 13(9): 291-293.

Dayen, A and M. Ahmed. 1969. Susceptibility of the long headed flour beetle, Laetheticus oryzae Waterh. to gamma radiation. Nucl. Sci. Appl. A: 26-30. (NSA 24: 51020).

Dayton, B. R. 1968. Accumulation and transfer of radiostrontium by young loblolly pines. Thesis, Univ. of North Carolina, Chapel Hill. 64pp. (Also: Savannah River Lab., U. S. AEC report TID-24597) (NSA 22: 43274; 23: 43684).

Dayton, B. R. 1970. Slow accumulation and transfer of radiostrontium by young loblolly pines (Pinus taeda L.). Ecology 51(2): 204-216. (NSA 24: 34321).

Deak, J. 1974. Use of environmental isotopes to investigate the connection between surface and subsurface waters in the Nagykunsag Region, Hungary. pp. 157-167. In: Isotope Techniques in Groundwater Hydrology. Vol. 1. (See IAEA, 1974) (NSA 31: 19626).

Deal, L. J. and J. F. Doyle. 1974. Overview of the aerial radiological measurement system (ARMS) program. pp. 543-548. In: Proc. 2nd AEC Environmental Protection Conf. (Albuquerque, N. Mex., April 16-19, 1974). (See U. S. AEC, 1974) (NSA 30: 32034).

Dean, G. J. W. and S. M. Wortham. 1969. Effect of gamma radiation on the tsetse fly Glossina morsitans Westw. Bull. Entomol. Res. 58: 505-519. (NSA 23: 38912).

Dean, G. J. W., D. A. Dame and D. R. Birkenmeyer. 1969. Field cage evaluation of the competitiveness of male Glossina morsitans irientalis Vanderplank sterilized with TEPA or gamma irradiation. Bull. Entomol. Res. 59: 339-344. (NSA 24: 44641).

Dean, G. J. W., R. J. Phelps and B. Williamson. 1968. Sterilization with gamma-rays and field investigations into the breeding of Glossina morsitans. pp. 31-36. In: Control of Livestock Insect Pests by the Sterile-Male Technique. (See IAEA, 1968) (NSA 24: 34808).

Dean, M. H. 1966. A survey of uranium content of vegetation in Great Britain. J. Ecology 54(3): 589-595.

Dean, P. N. 1964. Computer techniques in spectroscopy. pp. 617-626. In: The Natural Radiation Environment. (See Adams and Lowder, 1964) (NSA 18: 35890).

De Angelis, C. 1959. The atomic era and fisheries. G. Pesca 9(13): 3.

Deb, D. L., P. A. Wahid and B. K. Patel. 1974. Isotopically exchangeable phosphorus in major coconut growing soils in Kerala State. J. Indian Soc. Soil Sci. 22(4): 294-299. (NSA 33: 29148) (^{32}P).

DeBach, P. and E. B. White. 1962. Irradiated parasitic wasps: the effect on progeny production and sex ratio. J. Heredity 53(6): 271-276. (NSA 17: 33684).

Debenedetti, A., V. Grossi and A. Lucaroni. 1969. Comparative effects of x-rays and visible light on the reconstruction of the early potentials (ERP) of the isolated frog eye previously abolished with severe light exposition. Boll. Soc. Ital. Biol. Sper. 45: 1371-1373. (In Italian) (NSA 25: 27004).

Debiesse, J. 1960. Natural radioactivity and the effluents at Saclay. Bull. inform. A.T.E.N. 25: 1-7. (In French) (NSA 15: 7185).

Debiesse, J. nd. Natural radioactivity and the wastes of Saclay. Nucleus (Paris) 1: 50-54. (In French) (NSA 15: 24624).

Debolt, J. W. 1973. Preliminary investigation of the effects of gamma irradiation on the fertility, survival, and mating competitiveness of the beet armyworm, Spodaotera exigua (Hübner). pp. 1158-1162. In: Radionuclides in Ecosystems. (See Nelson, 1973) (NSA 28: 30237).

De Boodt, M., R. Hartmann and P. De Meester. 1967. Determination of soil-moisture characteristics for irrigation purposes by neutron-moisture meter and air-purged tensiometers. pp. 147-159. In: Isotope and Radiation Techniques in Soil Physics and Irrigation Studies. (See IAEA, 1967) (NSA 22: 23149).

De Boodt, M, H. Verplancke and A. V. Laksmipathy. 1972. Study on the sensitivity of soil moisture measurements using radiation techniques, nylon resistance units, and a precise lysimeter system. pp. 605-619. In: Isotopes and Radiation in Soil-Plant Relationships Including Forestry. (See IAEA, 1972) (NSA 27: 2605).

de Bortoli, M. C. 1969. Methods of analysis of the radioactivity in the soil. Minerva Fisiconucl. 13: 86-92. (In Italian) (CONF-680535-10) (NSA 25: 34500).

de Bortoli, M. and P. Gaglione (* et al.). 1965-1974. Environmental radioactivity, Ispra, ---. Joint Nuclear Research Center, European Atomic Energy Community (Commission of the European Communities), Ispra, Italy.

Year Publ.	Year Reptd.	Report EUR-	No. pp.	NSA Vol.: Abstr. No.
1965*	1964	2509e	66	20: 2047
1966*	1965	2965e	72	20: 33553
1967*	1966	3554e	--	21: 43056
1968	1967	4088	62	23: 13870
1969	1968	4412	60	24: 23194
1970	1969	4563	50	25: 16131
1972	1970	4805	46	26: 42946
1973	1971	4944	48	28: 5542
1974	1972	5118	63	30: 9523

de Bortoli, M. and P. Gaglione. 1969. Natural and fallout radioactivity in the soil. Health Physics 17: 701-710. (NSA 24: 4379).

de Bortoli, M. and P. Gaglione. 1970. Study of the dispersion of radioactive effluents into a lake. 1. Contamination of the bottom sediments. Minerva Fisiconucl. 14(4): 190-196. (In Italian) (NSA 26: 20319).

de Bortoli, M. and P. Gaglione. 1970. Remarks on the transfer pathways of radionuclides in some components of aquatic and terrestrial environment. pp. 61-67. In: Proc. 1st Conf. on the State of Advancement of Radioecology in Italy. (See Anonymous, 1970) (In Italian).

de Bortoli, M. and P. Gaglione. 1972. ^{226}Ra in environmental materials and foods. Health Physics 22(1): 43-48. (NSA 26: 15537) (Incl. lake waters, fishes).

de Bortoli, M. and P. Gaglione. 1973. The main radioecological works carried out by the Environmental Monitoring Group of C.C.R. Euratom, Ispra in the period 1971-72. pp. 41-46. In: Proc. 2nd Conf. on the State of Advancement of Radioecology in Italy. (See Anonymous, 1973) (In Italian).

de Bortoli, M. and E. van der Stricht. 1969. Some chemical procedures for the radiological monitoring of the environment. European Atomic Energy Community, Ispra, Italy, report EUR-4388. 58pp. (NSA 24:5583).

de Bortoli, M., P. Gaglione and C. Gandino. 1966. Environs monitoring at Ispra. Ingeg. Nucleare 7: 3-11. (In Italian) (NSA 20: 45979).

de Bortoli, M., P. Gaglione and C. Gandino. 1972. JNRC Ispra site characteristics. Joint Nuclear Research Center, European Atomic Energy Community, Ispra, Italy, report EUR-4841. 98pp. (In Italian) (NSA 26: 48067).

de Bortoli, M., P. Gaglione and C. Myttenaere. 1972. Radioiodine transfer in an irrigated grassland ecosystem (marcita). G. Fis. Sanit. Prot. Radiaz. 16(4): 184-190. (NSA 29: 29345).

de Bortoli, M., P. Gaglione and O. Ravera. 1969. Fallout ^{54}Mn and ^{95}Zr in water and fishes of four lakes in northern Italy. Minerva Fisiconucl. 13: 72-77. (CONF-680535-8) (NSA 25: 92239).

de Bortoli, M., et al. 1968. Concentration factors for strontium and cesium in fish of the lakes in the region of Varese (northern Italy). G. Fis. Sanit. Prot. Radiaz. 12: 324-331. (NSA 24: 29707).

De Byle, N. V. 1964. Detection of functional interclonal aspen root connections by tracers and excavation. Forest Science 10(4): 386-396.

de Carvalho, J. P. 1963. Contribution of the radiographic method for the study of Sitotroga cerealella (Oliv.). Agricultura (Lisbon) 19: 22-25. (In Portugese).

de Cecco, G. 1973. Environmental problems in the choice of sites for nuclear power plants. (Discussion). pp. 169-183. In: Proc. 2nd Conf. on the State of Advancement of Radioecology in Italy. (See Anonymous, 1973) (In Italian).

de Cecco, G. 1974. Practices and regulations concerning the control of radioactive effluents in Italy. pp. 61-72. In: Proc. Seminar on the Monitoring of Radioactive Effluents. (See Organization for Economic Co-operation and Development, 1974) (AED-CONF-74-157-003) (NSA 30: 23973; 31: 3338).

de Cesaris, C. L. 1958. Dipeptidase in amphibian embryos, treated with x-rays. Arch. Sci. Biol. (Bologna) 42(3): 205-212.

Dedek, W. 1964. The application of radionuclides in studying the chemistry of compounds for plant protection and pest control. Atompraxis 10(2): 65-70. (In German) (^{32}P, ^{35}S, ^{36}Cl, ^{82}Br, ^{14}C, ^{3}H).

Dedek, W. 1967. Radioactive nuclides in pesticide chemistry. II. Atompraxis 13(3): 126-129. (In German).

Dedek, W. 1970. Radioisotopes in pesticide research. IV. and V. Isotopenpraxis 6: 204-214. (In German) (NSA 25: 92239).

Dedek, W. 1975. Radioisotopes in pesticide research. VI. Review papers, general methods of analysis, insecticides. Isotopenpraxis 11(11): 378-385. (In German) (NSA 33: 29513).

Deering, R. A. 1968. Radiation studies of Blastocladiella emersonii. Radiation Research 34: 87-109. (NSA 22: 31288).

Deering, R. A. 1968. Dictystelium discoideum: a gamma-ray resistant organism. Science 162: 1289-1290. (NSA 23: 10119).

Deering, R. A. and D. S. Jensen. 1973. Nuclear and mitochondrial DNA synthesis in gamma ray-resistant and -sensitive slime mold amebas. Biophys. J. 13(8): 780-794. (NSA 28: 30053).

Deering, R. A., A. C. Adolf and S. M. Silver. 1972. Independence of propagation ability and developmental processes in irradiated cellular slime molds. Intern. J. Radiat. Biol. 21(3): 235-245. (NSA 26: 33658).

Deering, R. A., et al. 1970. Gamma-ray-resistant and -sensitive strains of slime mold (Dictyostelium disoideum). Radiation Research 43: 711-728. (NSA 24: 51102).

Deevey, E. S. 1969. Coaxing history to conduct experiments. BioScience 19(1): 40-43. (Incl. comments on ^{14}C).

Deevey, E. S., Jr. and N. Nakai. 1963. Fractionation of sulfur isotopes in lake waters. pp. 169-178. In: Biogeochemistry of Sulfur Isotopes. M. L. Jensen, ed. Yale Univ. Press, New Haven, Conn.

Deevey, E. S., Jr. and M. Stuiver. 1964. Distribution of natural isotopes of carbon in Linsley Pond and other New England lakes. Limnol. Oceanogr. 9(1): 1-11.

Deevey, E. S., R. F. Flint and I. Rouse. (Eds.). 1964. Radiocarbon 6: 1-376. (NSA 18: 27697).

Deevey, E. S., R. F. Flint and I. Rouse. (Eds.). 1966. Radiocarbon 8: 1-552. (NSA 20: 43560).

Deevey, E. S., M. Stuiver and N. Nakai. 1963. Use of light nuclides in limnology. pp. 471-475. In: Radioecology. (See Schultz and Klement, 1963) (NSA 17: 33598).

Deevey, E. S., Jr., M. Stuiver and N. Nakai. 1964. Isotopes of carbon and sulfur as tracers of lake metabolism. Verh. Intern. Verein. Limnol. 15 (1): 284-288.

Deevey, E. S., et al. 1954. The natural C^{14} contents of materials of hard-water lakes. Proc. Nat. Acad. Sci. (US) 40: 285-288.

de Fernandez, M. L. H. nd. Fruit pests in Bolivia. pp. 129-142. In: Proc. Conf. on Nuclear Energy and Agricultural Productivity. (See Comision Interamericana de Energia Nuclear, nd) (In Spanish) (NSA 25: 27130).

De Franciscis, P., G. De Bella and V. Muto. 1961. Total body x-irradiation in carps. Atompraxis 7(8): 292-294. (In English) (NSA 15: 29013).

De Freitas, J. R. and M. Campos. 1961. Some data on the rate of iron elimination in triatomids measured by radioactive iron (^{59}Fe). Bull. Wrld. Hlth. Org. 25(2): 274-276.

De Freitas, J. R. and M. Campos. 1963. Some data on the elimination of iron in triatomic media with radioactive iron (^{59}Fe). Bol. Ofic. sanit. pan-amer. 54(1): 61-63.

De Freitas, J. R. and A. Guedes da Silveira. 1961. Determination by radioactive iron (^{59}Fe) of the amount of blood injested by insects. Bull. Wrld. Hlth. Org. 25(2): 271-273.

De Freitas, J. R. and A. Guedes da Silveira. 1962. Determination of the amount of blood ingested by insects by the use of radioactive iron (^{59}Fe). Bol. Ofic. sanit. pan-amer. 53(5): 426-429.

de Goeij, J. J. M. 1971. Utilization of activation analysis as a detection technique for mercury. 10pp. (For Meeting on Environmental Pollution, Mol, Belgium, 18 Mar. 1971) (CONF-710325) (NSA 25: 48462) (Hg levels in Dutch birds, river water, river sediments, grass, cows).

de Goeij, J. J. M. and J. P. W. Houtman. 1971. Mercury in the environment. Chem. Weekbl. 67: 13-20. (Also: Netherlands Interuniversitair Reactor Inst., Delft, report IRI-133-71-01) (In Dutch) (NSA 25: 57392) (Neutron activation analysis).

de Goeij, J. J. M., P. Bode and C. Zegers. 1973. Vanadium concentrations in grasslands around Arnhem (Netherlands). Interuniversitair Reactor Inst., Delft, report IRI-133-73-16. 16pp. (In Dutch) (NSA 31: 5544; 32: 13718) (Neutron activation analysis).

de Goeij, J. J. M., et al. 1972. Mercury distribution levels observed in various ecosystems as determined by neutron activation analysis. pp. 295-307. In: Nuclear Activation Techniques in the Life Sciences, 1972. (See IAEA, 1972) (Also: Interuniversitair Reactor Inst., Delft, report IRI-133-72-4) (NSA 26: 47657; 27: 22094) (Water, plants, fish, birds).

Degot, B., et al. 1963. Two uses of bromine-82 in underground hydrodynamics. pp. 321-346. In: Radioisotopes in Hydrology. (See IAEA, 1963) (In French) (NSA 18: 1926).

de Groot, A. J., et al. 1973. Use of activatable tracers. pp. 151-166. In: Trace Techniques in Sediment Transport. (See IAEA, 1973) (NSA 28: 21139).

de Groot, A. J., et al. nd. Activation analysis applied to sediments from various river deltas. Netherlands Reactor Inst., Delft, report RI-133-68-8. 17pp. (For Intern. Conf. on Modern Trends in Activation Analysis Gaithersburg, Md., Oct. 7-11, 1968) (CONF-681003) (NSA 23: 21667).

DeHarpporte, D. R. 1970. Cooling tower site considerations. Power Engrg. 74(8): 49-51. (NSA 24: 47804).

de Irizarry, E. R. and F. K. S. Koo. 1970. Histological damage induced by gamma irradiation in shoot apices of Palicourea riparia. Ch. G-5, pp. G63-G67. In: A Tropical Rain Forest. (See H. T. Odum and R. F. Pigeon, 1970) (NSA 25: 21951)

Deist, J. and W. A. G. Kotze. 1972. Applications of isotope exchange in soil research. pp. R6.bl-R6.b4. In: National Conf. on the Technological Applications of Nuclear Techniques, 1972. (See South Africa Atomic Energy Board, 1972) (NSA 28: 21125).

Dejaegere, R. 1964. Radiobiological effects of tracers used in plant physiology. Ann. Gembloux 70: 12-20. (In French) (NSA 21: 12792).

de la Cruz, A. A. and R. G. Wiegert. 1967. 32-phosphorus tracer studies of a horseweed-aphid-ant food chain. Am. Midland Nat. 77(2): 501-509.

de la Cruz, B. and C. Madriaga. 1970. Effect of orally administered tritiated thymidine on reproductive capacity of field rats. Philipp. Nucl. J. 2(2): 229-231. (NSA 26: 43382).

de la Cruz, B., et al. 1964. A report on the labeling of Schistosoma japonicum cercariae with radioiron. Am. J. Trop. Med. Hyg. 13: 545-547. (NSA 19: 33734).

de Laguna, W. 1968. Tritium and other environmental isotopes in the hydrological cycle. (Book review) (IAEA Tech. Rept. Ser. No. 73) Nucl. Appl. 4: 448-449.

Delahaye, R. P. and E. Tabusse. 1964. Extraterrestrial radiation. I. Biological studies. Rev. Med. Aeron. (Paris) 3: 593-597. (In French) (NSA 19: 36069; 20: 122).

DeLand, F. H. 1971/72. Automated radiometric detection of microorganisms. Isotopes Radiat. Technol. 9(2): 240-243. (NSA 26: 9454) (^{14}C).

de Langguth, E. N. and W. Pohlit. 1968. Effects of x radiation on stationary and exponentially increasing haploid yeast populations. Biophysik 4: 320-328. (In German) (NSA 23: 949).

Delas, J. 1965. Use of labeled elements in agronomy. France Recherches Station d'Agronomie CRASO, Pont-de-la-Maye. 9pp. (For Symp. on Nuclear Science, Bordeaux) (In French) (CONF-650122-1) (NSA 22: 6766).

Delattre, R. 1962. Note on the possibilities of other problems with insecticides. pp. 267-271. In: Radioisotopes and Radiation in Entomology. (See IAEA, 1962) (In French).

Delattre, R. 1963. A preliminary note on the use of radioisotopes in the study of parasites of cotton in Africa. pp. 86-92. In: Radiation and Radioisotopes Applied to Insects of Agricultural Importance. (See IAEA, 1963) (In French).

Delbrück, M. and N. V. Timofeyev-Ressoviy. 1936. Cosmic rays and the origin of species. Nature 137(3461): 358-359.

Delibrias, G. 1970. Validity and interpretation of ^{14}C ages. France Commissariat à l'Énergie Atomique, Saclay, report CEA-CONF-1618. 18pp. (For Meeting on Stratigraphy Methods and Tendencies, Orsay, France, Oct. 12, 1970) (CONF-701008-1) (NSA 24: 50458).

Delibrias, G., M.-T. Guillier and L. Labeysie. 1965. Age measurements using the carbon-14 method. Bull. Inform. Sci. Tech. (Paris) (89): 3-22. (In French) (NSA 19: 40795).

Delinger, D. L. and G. W. Ankersmit. 1973. Studies on the sterile male techniques as a means of control of Adoxophyes orana (Lepidopt., Tortricidae). III. An evaluation of competitiveness of laboratory-reared moths. Neth. J. Plant Pathol. 79(6): 229-235. (NSA 31: 29956).

Della Rosa, R. J., et al. 1969. Maternal-fetal transfer of injected cesium-137 and iodine-131 in deer and sheep. pp. 117-120. In: Radiation Biology of the Fetal and Juvenile Mammal. (See Sikov and Mahlum, 1969).

Delmas, J. and A. Grauby. 1973. Irrigation: transfer channel for radioactive pollution. Agriculture (Paris) (366): 178-180. (In French) (NSA 10150).

Delmas, J., A. Grauby and R. Disdier. 1970. Transfer of radionuclides to plants by irrigation. Bull. Inform. Sci. Tech. (Paris) (151): 67-73. (In French) (NSA 25: 18938).

Delmas, J., A. Grauby and R. Disdier. 1973. Experimental studies on the transfer to crops of certain radionuclides present in nuclear power station effluents. pp. 321-332. In: Environmental Behavior of Radionuclides Released in the Nuclear Industry. (See IAEA, 1973) (In French) (NSA 29: 29379).

Delmas, J., et al. 1971. Experimental radiocontamination of some cultivated species with the aid of effluents of diverse origin. pp. 1103-1119. In: Proc. Intern. Symp. Radioecology Applied to the Protection of Man and His Environment. (See Commission of the European Communities, 1971) (NSA 27: 5284).

Del Monte, G. 1959. The use of ionizing radiation in the preservation of cereal products. Molini d'Italia 10(1): 29-33. (In Italian) (Insects).

Delone, N. L. 1965. Utilization of higher plants as indicators in studying the effect of orbital spaceflight on the living cell. pp. 304-307. In: Problems in Space Biology. H. M. Sisakyan, ed. Acad. Sci. USSR, Moscow. (In Russian) (NSA 20: 18964).

Delone, N. L., et al. 1968. Effect of space flight environments on seeds of higher plants flown on board Cosmos-110. Izv. Akad. Nauk SSSR (biol.) (1): 126-129. (In Russian) (NSA 22: 45488).

De Loose, R. and H. Reyntens. 1963. Some lessons resulting from the application of ionizing radiations in mutagenesis in forage grasses. Bull. Inst. Agron. Sta. Rech. Gembloux 31: 570-575. (In French) (NSA 19: 17527).

Delorenzi, E. and U. Tosco. 1964. Behavior of fungal spores in a confined irradiated environment. Minerva Radiol. Fisioterap. Radiobiol. 9: 582-584. (In Italian) (NSA 19: 19559).

Delorenzi, E. and U. Tosco. 1964. Effects of roentgen irradiation on fungal spores. Minerva Radiol. Fisioterap. Radiobiol. 9: 580-582. (In Italian) (NSA 19: 19558).

Delos, Y. 1970. Radon detection in the sea and the atmosphere. Thesis, Toulouse Univ., France. 104pp. (In French) (NP-18404) (NSA 24: 46472).

Delpla, M., et al. 1968. Some problems of radioactive effluent production at the Chinon nuclear power plant. pp. 469-474. In: International Congr. on Radioprotection of the Environment. (See Societé Francaise de Radioprotection, 1968) (In French).

Delpoux, M. 1970. Comparison of the percentage of radium and uranium of vegetation harvested on uraniferous terrain. pp. 909-917. In: Proc. Intern. Symp. on Radioecology, Vol. II. (See France Commissariat à l'Énergie Atomique, 1970) (In French).

Delrio, G. and R. Cavalloro. 1975. Inherited sterility in Gonocerus acuteangulatus Goeze (Rhynchota, Coreidae). pp. 423-436. In: Sterility Principle for Insect Control. (See IAEA, 1975) (In French) (NSA 32: 11797) (Ionizing radiation).

Delucchi, V. 1969. Potential of the sterile-male technique in integrated insect control programs. pp. 39-44. In: Sterile-Male Technique for Eradication or Control of Harmful Insects. (See IAEA, 1969) (NSA 24: 32297).

Del Vecchio, V., G. Branca and S. U. D. Arca. 1968. Ionizing radiation in purification of waste waters. Ingegnaria Sanit. (1): 3-17. (Also: Italy Comitato Naz. per l'Energia Nucleare, Rome, report RT/PROT (68)15) (In Italian) (NSA 24: 8034).

de Magalhaes Silva, G. 1970. Sterile-male technique for control of the olive fly. Review of work on rearing and radiation. pp. 119-130. In: Sterile-Male Technique for Control of Fruit Flies. (See IAEA, 1970) (NSA 25: 24516).

DeMars, C. J., Jr. 1963. A comparison of radiograph analysis and bark dissection in estimating numbers of the western pine beetle. Can. Entomol. 95(10): 1112-1116. (Correction. 1965. 97(2): 206).

DeMarte, J. A. and R. T. Hartman. 1969. Laboratory studies on uptake and translocation of phosphorus-32 by Myriophyllum exalbescens Fernald. Abstr. Bull. Ecol. Soc. Am. 50(2): 71.

DeMarte, J. A. and R. T. Hartman. 1974. Studies of absorption of ^{32}P, ^{59}Fe, and ^{45}Ca, by water-milfoil (Myriophyllum exalbescens Fernald). Ecology 55 (1): 188-194. (NSA 29: 18672).

de Mazar-Barnett, B. K. 1965. Mutations induced by alkylating agents, x-rays and combined treatments. Abstr. Genetics 52(2,Pt.2): 459.

Dementyev, V. S. and N. G. Syromyatnikov. 1965. Distribution of thorium isotopes between the particles of various degree of dispersion in natural waters. Radiokhimiya 7(6): 710-717. (Sov. Radiochemistry 7(6): 706-712) (NSA 20: 43535).

Dementyev, V. S. and N. G. Syromyarnikov. 1965. On the occurrence form of thorium isotopes in groundwaters. Geokhimiya (2): 211-218. (Geochem. Int. (1): 141-147) (NSA 19: 22542).

Dementyev, V. S. and N. G. Syromyatnikov. 1968. Conditions of formation of a sorption barrier to the migration of uranium in an oxidizing environment. Geokhimiya (4): 459-465. (Geochem. Int. 5: 394-400) (NSA 23: 28999).

Demirchoglyan, G. G., et al. 1962. Results of an investigation of the action of ionizing radiation on the retina and on some other light-sensitive systems. Radiobiologiya 11(3): 442-449. (AEC-tr-5430, pp. 160-170) ("Rain worm").

de Montgareuil, P. G. 1964. French research in the field of nuclear agronomy. Centre d'Études Nucléaires, France Commissariat à l'Énergie Atomique, Cadarache, report CEA-R-2524. 12pp. (In French) (NSA 19: 8720) (Incl. entomology).

Demortier, G. 1974. Determination of light elements in biological samples by prompt nuclear reactions. Radiochem. Radioanal. Lett. 16(6): 329-341. (In French) (NSA 29: 26568).

De Murtas, I. D. 1970. From DDT to ^{60}Co. Papere No. 724. Italy Comitato Naz. per l'Energia Nucleare, Rome, report RT/BIO-(72)55. vp. (In Italian) (NSA 30: 9935).

De Murtas, I. D. 1972. Dispersal of Ceratitis capitata Wied on Procida Island. I. Distribution of sterilized Mediterranean fruit fly on Procida and its relation to irradiation doses and feeding. EPPO Bull. (6): 63-68; 69-76. (Also: Italy Comitato Naz. per l'Energia Nucleare, Rome, report RT/BIO-(72)72; RT/BIO-(72)75) (NSA 30: 9936, 9937) (γ radiation).

De Murtas, I. D. and U. Cirio. 1975. Sterilization of the Mediterranean fruit fly Ceratitis capitata Wied. : an industrial method studied by CNEN for pest autocidal control by the sterile insect technique. pp. 49-55. In: Sterility Principle for Insect Control, 1974. (See IAEA, 1975) (In French) (NSA 32: 11766).

De Murtas, I. D., et al. 1970. Experiment to control the Mediterranean fruit fly on the island of Procida by the sterile-insect technique. pp. 59-70. In: Sterile-Male Technique for Control of Fruit Flies. (See IAEA, 1970) (NSA 25: 24510).

Demyanovsky, S. Ya. and N. S. Rusakova. 1955. The phosphorus metabolism in the oak silk worm Antheraea pernyl. Biokhimiya 20(4): 466-469. (In Russian) (^{32}P).

Demyanovsky, S. Ya. and N. S. Rusakova. 1957. Diapause of oak tree silkworm. Uch. zap. Mosk. gos. ped. inst. im. V. I. Lenina, kaf. org. biol. khim. 98: 49-58. (In Russian) (^{32}P).

Demyanovsky, S. Ya. and N. S. Rusakova. 1957. Phosphorus metabolism in the oak tree silkworm. Uch. zap. Mosk. gos. ped. inst. im. V. I. Lenina, kaf. org. biol. khim. 98: 61-64. (In Russian) (^{32}P).

Demyanovsky, S. Ya. and N. V. Vasil'yeva. 1957. The rate of metabolism in diapausing oak silkworm pupae. Uch. zap. Mosk. gos. ped. inst. im. V. I. Lenina, kaf. org. biol. khim. 98: 65-70. (In Russian) (^{32}P).

Denham, D. H. 1969. Radiological status of the ground water beneath the Hanford Project, July-December 1968. {January-June 1969}. Battelle-Northwest Lab., U. S. AEC report BNWL-1047. 20pp. {BNWL-1233; 20pp.}. (NSA 23: 33261; 24: 7609).

Denham, D. H. 1974. Recommended methodology for environmental surveillance. pp. 531-542. In: *Proc. 2nd AEC Environmental Protection Conf.* (See U. S. AEC, 1974) (NSA 30: 12682).

Denham, D. H., D. A. Waite and J. P. Corley. 1974. Summary of selected AEC contractors' environmental surveillance techniques and capabilities. Battelle-Northwest Lab., U. S. AEC report BNWL-B-384. 44pp. (NSA 31: 14015).

Denis, S. 1972. Some aspects of the environment and electric power generation. Rand Corp., Santa Monica, Calif. 32pp. (NP-19606) (NSA 27: 7509).

De Nettancourt, D. and R. B. Conant. 1966. Comparative study of the effects of chronic gamma irradiation on *Lycopersicum esculentum* Mill. and *L. pimpinellifolium* Dunal. Radiation Botany 6(6): 545.

Denlinger, D. L., G. W. Ankersmit and J. P. W. Noordink. 1973. Studies of the sterile male technique as a means of control of *Adoxophyes orana* (Lepidopt., Tortricidae). III. An evaluation of competitiveness of laboratory reared moths. Neth. J. Plant Pathol. 79(6): 299-235. (NSA 32: 14590).

Dennis, N. M. 1961. The effects of gamma-ray radiation on certain species of stored products insects. J. Econ. Entomol. 54(1): 211-213. (NSA 20: 42904).

Dennister, R. H. 1942. The influence of roentgen ray treatments of the hypophysis on reproducive system of the ground squirrel and rat. J. Exptl. Zool. 91(2): 237-264.

Denny, M. J. S. and T. J. Dawson. 1973. A field technique for studying water metabolism of large marsupials. J. Wildl. Mgmt. 37(4): 574-578. (^{3}H).

Denovan, J. T. 1972. Feasibility of using algae in a process for treating radioactive wastes. Dissertation, Univ. of Michigan, Ann Arbor. 97pp. (Univ. Microfilms 73-11,088) (NSA 28: 5415).

Dent, J. N. 1949. Effects of P^{32} on regenerating amphibian tissues. Abstr. Anat. Record 103: 439. (NSA 3: 21).

Dent, J. N. 1961. Seasonal and sexual variation in the pituitary gland of *Triturus viridescens*. Anat. Record 141(2): 85-95.

Dent, J. N. 1966. Radiation effects. Final report. Univ. of Virginia, Charlottesville, U. S. AEC report ORO-2978-17. 9pp. (NSA 21: 26560) (*Rana*, *Triturus*).

Dent, J. N. and R. L. Amy. 1950. Developmental effects observed in *Habrobracon* after exposure to beta radiation. Growth 14(2): 113-121.

Dent, J. N. and J. S. Kirby-Smith. 1963. Metamorphic physiology and morphology of the cave salamander *Gyrinophilus palleucus*. Copeia (1): 119-130. (Application of ^{131}I).

Dent, J. N. and W. G. Lynn. 1958. A comparison of the effects of giotrogens on thyroid activity in *Triturus viridescens* and *Desmgnathus fuscus*. Biol. Bull. 115(3): 411-420. (RDA-5: 214) (Used ^{131}I).

Denton, T. E. 1966. The effect of a strontium replacement for calcium on carbohydrates and other products of photosynthesis in *Protosiphon botryoides*. Thesis, Univ. of Alabama, University. 62pp. (NSA 21: 34624) (Used radiotracers).

de Nulman, R. M. and G. Burillo. 1970. Design of a preoperational environmental survey for the nuclear center in Mexico. pp. 77-83. In: *Environmental Surveillance in the Vicinity of Nuclear Facilities*. (See Reinig, 1970) (NSA 25: 16164).

de Oliveira, M. L., A. de Moraes Rego and Z. M. de Azevedo. 1969. Effects of gamma rays on eggs of *Culex pipiens fatigans*, Wied, 1928. Bull. Calcutta Sch. Trop. Med. 17: 12-13. (NSA 25: 29614).

de Regt, H. B. and F. C. M. Mattern. 1965. The radioactivity of water, slime, and floating material of the Rhine in the Netherlands. Report concerning measurements issued on behalf of EURATOM in the framework of the investigation of the radioactivity in the Rhine Basin. The Netherlands Rijks Inst. voor Zuivering van Afvalwater, Voorburg and Utrecht, report RA-14. 65pp. (In Dutch) (NSA 28441).

Dermelj, M., V. Ravnik and L. Kosta. 1976. Fast isolation and determination of Cd in some fungi, other biological materials, soil, and Zn metal by NAA. Radiochem. Radioanal. Lett. 24(2): 91-101. (NSA 33: 28295).

de Saint-Blanquat, H. 1968. The fate of ^{14}C. Sciences Avenir (255): 408-415. (In French) (NSA 22: 48936).

Descamps, B. and L. Foulquier. 1974. Experimental contamination by chromium-51 of *Anguilla anguilla*. Centre d'Études Nucléaires, France Commissariat à l'Énergie Atomique, Cadarache, report CEA-R-4548. 100pp. (In French) (NSA 30: 3700).

Descamps, B., et al. 1973. Experimental example on the contamination of *Anguilla anguilla* (L.) by chromium-51 and its excretion. Compt. rend. D276: 1193-1196. (In French).

Deschner, E. and A. H. Sparrow. 1955. Chromosome rejoining capacity with respect to breakage sensitivity to x-rays and thermal neutrons. Genetics 40: 460-475.

Deschreider, A. R. 1961/62. Research in gamma irradiation of wheat flour. Conserva (The Hague) 10: 115-117. (AEC-tr-5773).

DeSelm, H. R. and R. E. Shanks. 1959. Concentrations of radionuclides in the natural vegetation of a contaminated area. Abstr. ASB Bull. 6(2): 24.

DeSelm, H. R. and R. E. Shanks. 1960. Ecological studies on White Oak Lake bed. Abstr. J. Tenn. Acad. Sci. 35(2): 102. (Oak Ridge National Lab.).

DeSelm, H. R. and R. E. Shanks. 1961. Ecology of White Oak Lake bed: a radioactive waste area. pp. 1359-1363. In: Recent Advances in Botany. Univ. of Toronto Press, Toronto, Canada.

DeSelm, H. R. and R. E. Shanks. 1961. Accumulation and cycling of organic matter and chemical constituents during early vegetational succession on a radioactive waste disposal area. Incl. Suppl.: Vegetational history and plant succession on a radioactive waste area. Univ. of Tennessee, Knoxville, U. S. AEC report ORO-580. (NSA 16: 34143).

DeSelm, H. R. and R. E. Shanks. 1962. Biogeochemical cycling in natural ecosystems. II. Five years of accumulation and cycling on a radioactive waste area. Abstr. J. Tenn. Acad. Sci. 37(2): ___.

DeSelm, H. R. and R. E. Shanks. 1962. Basic data from vegetation studies related to the movement of radioactive wastes. 1. Chemical and physical data on soils and plants of White Oak Creek bed, Oak Ridge, Tennessee, 1956-1960. Univ. of Tennessee, Knoxville, U. S. AEC report ORO-585. viii, 88pp. (NSA 17: 4369).

DeSelm, H. R. and R. E. Shanks. 1963. Accumulation and cycling of organic matter and chemical constituents during early vegetational succession on a radioactive waste disposal area. pp. 83-96. In: Radioecology. (See Schultz and Klement, 1963).

DeSelm, H. R. and R. E. Shanks. 1967. Vegetation and floristic changes on a portion of White Oak Lake bed. Ecology 48: 419-425. (NSA 22: 21462).

DeSelm, H. R., P. B. Whitford and J. S. Olson. 1969. The barrens of the Oak Ridge area, Tennessee. Am. Midland Nat. 81(2): 315-330.

de Serres, F. J. 1969. Effects of radiation during space flight on microorganisms and plants on the Biosatellite II and Gemini XI missions. pp. 62-66. In: Life Sciences and Space Research VII. W. Vishniac and G. Favorite, eds. North-Holland, Amsterdam, (NSA 24: 912).

de Serres, F. J. and B. B. Webber. 1968. The combined effect of weightlessness and radiation on inactivation and mutation-induction in Neurospora crassa during the Biosatellite II mission. BioScience 18(6): 590-595.

De Sha, D. L. 1971. Participation of x-rayed cells in a regenerate induced on an x-rayed limb of the salamander, Triturus viridescens. Thesis, Tulane Univ., New Orleans, La. 78pp. (Univ. Microfilms 71-20,218) (NSA 25: 58103).

Deslias, M. D., et al. 1968. Experimental study of the influence of liquid releases from a nuclear center to a pond, on the radioactivity of industrial waters drawn from this reservoir. pp. 161-167. In: Intern. Congr. on Radioprotection of the Environment. (See Societé Francaise de Radioprotection, 1968) (In French).

Desmarais, A. P. and B. T. Helmuth. 1970. Effects of ^{137}Cs radiation on vegetation structure and optical density at El Verde. Ch. D-2, pp. D77-D102. In: A Tropical Rain Forest. (See H. T. Odum and R. F. Pigeon, 1970) (NSA 25: 58103).

De Soete, D., R. Gijbels and J. Hoste. 1972. Neutron Activation Analysis. Chemical Analysis, Vol. 34. 856pp. John Wiley, New York. (NSA 26: 52912).

de Swart, J. G. and P. H. Groenevelt. 1971. Column scanning with 60 keV gamma radiation. Soil Science 112(6): 419-424.

Detroit Edison Co. 1970. Enrico Fermi Atomic Power Plant, Unit 2. Detroit. U. S. AEC DOCKET-50341-18. 61pp. (NSA 24: 47794) (Incl. environment).

Detroit Edison Co. 1971. Enrico Fermi Atomic Power Plant, Unit 2. Environmental report, construction permit stage (revised). Detroit. U. S. AEC DOCKET-50341-34. 176pp. (NSA 25: 59986).

Deutsch, M. and V. E. Engelbert. 1970. Erythropoiesis through clone cell formation in peripheral blood of the white sucker, Catastomus commersoni,

DeVay, J. E., et al. 1965. Effects of thermal neutron irradiation on the frequency of crown gall and bacterial canker resistance in seedlings of Prunus rootstocks. Radiation Botany 5(3): 197-204. (NSA 19: 33887).

Devine, T. and T. J. Peterle. 1968. Possible differentiation of natal areas of North American waterfowl by neutron activation analysis. J. Wildl. Mgmt. 32(2): 274-279.

Devine, T. L., C. E. Venard and W. C. Myser. 1965. Measurement on salivation of Aedes aegypti (L.) feeding on a living host. J. Insect Physiol. 11: 347-353. (^{3}H).

DeVoe, J. R. (Ed.). 1969. Modern Trends in Activation Analysis, Vols. I and II. (Proc. 1968 Intern Conf. held at the National Bureau of Standards, Gaithersburg, Md., Oct. 7-11, 1968). Nat. Bur. of Standards report NBS-Spec. Publ-312 (Vol. 1 and 2). 1362pp. (NSA 24: 104).

Devreux, M. 1963. Effects of chronic gamma radiation on embryogenesis in Capsella bursa-pastoris Moench. pp. 197-217. In: Nuclear Energy in Agriculture. (6th Nuclear Congr., Rome, 15-16 June 1961). Italy Comitato Naz. per l'Energia Nucleare, Rome, Florence, Vallecchi. (In Italian).

de Vries, H. 1959. Measurement and use of radiocarbon. pp. 169-189. In: Researches in Geochemistry. P. H. Abelson, ed. John Wiley, New York.

Dewar, M. A. 1965. Effects of radiation on vegetation. Lockheed Georgia Quart. 2(1): 14-17. (NSA 19: 36038) (Reactor irradiation studies).

Dewey, D. L. 1963. The x-ray sensitivity of Serratia marcescens. Radiation Research 19(1): 64-87. (NSA 17: 24968).

Dewey, D. L. 1969. Survival of Micrococcus radiodurans irradiated at high LET and the effect of acridine orange. Int. J. Radiat. Biol. 16: 583-592. (NSA 24: 34781).

Dey, N. N. and K. C. Pillai. 1971. Radioruthenium in aquatic environment of Trombay. pp. 541-553. In: Proc. Nat. Symp. on Radiation Physics. India Department of Atomic Energy, Trombay, Bombay. (CONF-701138) (NSA 26: 40852).

Dey, N. N., et al. 1973. Environmental survey at Trombay reported for the period January 1, 1970 to December 31, 1972. Bhabha Atomic Research Cen., India Department of Atomic Energy, Bombay, report BARC/I-249. 39pp. (NSA 30: 6480).

de Zeeuw, D. 1965. Nuclear technology in agriculture. Neue Tech. 5: 238-244. (NSA 20: 8527).

Dezsi, Z. 1962. Investigations of the uranium, radium and radium emanation contents of natural waters of high uranium content. Magy. Tud. Akad. Atommag Kut. Int. (Debrecen) Kozl. 4: 93-96. (In Hungarian) (NSA 17: 6293).

Dhar, D. N. and K. S. Singh. 1970. Effect of irradiation on the infective stage larvae of the nematode, Oesophagostomum columbianum and their use as a vaccine. J. Helminthol. 44(1): 11-26. (NSA 25: 26909).

Diabola, J. and L. Taimr. 1965. Some results obtained with the application of the tracer method in insect migration and dispersion studies. Acta ent. bohemosl. 62(6): 413-420. (^{32}P).

Diabola, J. and L. Taimr. 1965. The dispersal flight of Meligethes beetles and spring migration of delphacids with special reference to the application of the tracer method. p. 328. In: Proc. 12th Intern. Congr. of Entomology. P. Freeman, ed. Royal Entomol. Soc., London. (^{32}P).

Dibodes, I. K., et al. 1969. Mechanisms of ^{90}Sr behavior in food chains following atmospheric deposition. Gigiyena sanitariya (1): 41-44. (In Russian) (NSA 25: 24079).

Dick, W. E. 1957. Radioactive materials in the fight against pests. Ch. 4, pp. 74-100. In: Atomic Energy in Agriculture. Butterworths Sci. Publ., London.

Dickerman, R. C. 1967. Fast neutron and x-ray irradiation of Drosophila melanogaster oogonia and oocytes. Genetics 56(3,Pt.2): 555-556.

Dickerman, R. C. 1967. Mutational response of Drosophila melanogaster oocyte and oogonia to x-ray and fast neutron irradiation. Abstr. Eb-7. Radiation Research 31(3): 614.

Dickmann, D. I. and T. T. Kozlowski. 1970. Mobilization and incorporation of photoassimilated ^{14}C by growing vegetative and reproductive tissues of adult Pinus resinosa. Ait. trees. Plant Physiol. 45: 284-288.

Dickson, H. W., et al. 1974. Environmental radiation measurements using in situ and core sampling techniques. pp. 45-52. In: 8th Midyear Topical Symp. of the Health Physics Society. (Knoxville, Tenn.). (CONF-741018) (NSA 31: 524).

Dickstein, H. L. 1974. National environmental hazards and international law. Int. Comp. Law Quart. 23: 426-446. (NSA 30: 24262).

Dieckmann, C., et al. 1950. The radiation death of coliform bacteria produced by fast electrons from a 6 Mev betatron. Naturwissenschaften 37(6): 140-141.

Diehl, F. A. and J. Bouillon. 1966. Observations on the morphogenetic potentialities of irradiated tissues of Cordylophora caspia (Paillas). Bull. Cl. Sci., Acad. Roy. Belg. 5(52): 138-146. (In French) (NSA 20: 45383).

Dienes, G. J. 1970. A kinetic model of radiation response to field irradiation. Ch. C-5, pp. C49-C52. In: A Tropical Rain Forest. (See H. T. Odum and R. F. Pigeon, 1970) (NSA 25: 21927).

Dienhart, A. V. 1972. Nuclear power plant licensing: a worsening predicament. Trans. Amer. Nucl. Soc. 15(1): 72. (NSA 26: 44924).

Dienstbier, Z. and M. Arient. (Eds.). 1966. Review of methods, experimental conditions, and techniques in irradiation. pp. 272-276. In: Experimental Postirradiation Syndrome. Czechoslovakia Statni Zdravotnicke Nakladatelatvi, Prague. (In Czech) (NSA 22: 52014).

Dietz, A. and E. N. Lambremont. 1970. Method of studying food consumption of live honey bee larvae by liquid scintillation counting. Ann. Entomol. Soc. Am. 63: 1340-1342. (NSA 24: 50673) (^{32}P).

Di Ferrante, E. R., E. Gourski and R. R. Boulenger. 1964. Detector for radon-222 measurements at the very low level. pp. 353-358. In: The Natural Radiation Environment. (See Adams and Lowder, 1964) (NSA 17: 32459).

Fi Ferrante, E. R., G. Koch and R. R. Boulenger. 1968. Determination of ^{226}Ra in natural samples. pp. 321-328. In: Radiation Protection, Part 1. W. S. Snyder, et al., eds. Pergamon Press, New York. (CONF-660920) (NSA 23: 12119).

Diffidenti, G. A. 1964. Insects sterilized with atomic energy. Agr. d' Ital. 10(2): 17-29. (In Italian).

Diggle, W. R. and J. Saxon. 1965. An unusually radioactive fossil fish from Thurso, Scotland. Nature 208: 400. (NSA 20: 9152).

DiGrande, F. 1967. Effects of x radiation on gonads and survival of oocytes in embryos of Bufo bufo. Atti. accad. naz. Lincei rend. cl. sci. fis. mat. nat. 43: 393-400. (In Italian) (NSA 22: 40899).

DiGrande, F. 1968. Post-metamorphic development of the gonads and Bidder organs in young Bufo bufo panirradiated with x-rays. Atti accad. naz. Lincei rend. cl. sci. fis. mat. nat. 44: 437 442. (In Italian) (NSA 24: 844).

di Gregorio, D., P. B. Dunaway and G. E. Cosgrove. 1973. Effect of acute gamma irradiation on the reproduction of Peromyscus leucopus. pp. 1076-1081. In: Radionuclides in Ecosystems. (See Nelson, 1973) (NSA 28: 30140).

DiGregorio, D., et al. 1971. Cesium-137 accumulation, dosimetry, and effects on cotton rats. pp. 535-547. In: Survival of Food Crops and Livestock in the Event of Nuclear War. (See Bensen and Sparrow, 1971).

Dillon, W. G. and C. A. Glomski. 1975. Erythrocyte survival in the Mongolian gerbil. J. Nucl. Med. 16(7): 682-684. (NSA 32: 14744) (^{51}Cr and ^{32}P tracers).

Dimaksyan, A. M. and V. P. Dobychin. 1965. Use of neutron scattering method for measuring soil water content. Tr. gos. gidrol. inst. 130: 132-147. (In Russian) (NSA 20: 53558).

Dim(t)chev, T. 1969. Gamma spectrometric investigation of lasting artificial radioactivity in mountainous regions. Izv. Fiz. inst. Aneb, Bulg. Akad. Nauk 19: 199-214. (In Bulgarian) (NSA 24: 16711).

Dim(t)chev, T. and L. Manolov. 1964. Local accumulations of fission products from radioactive fallout in mountain regions. Izv. Fiz. inst. Aneb, Bulg. Akad. Nauk 12: 101-113. (In Russian) (NSA 19: 15800).

Dim(t)chev, T. and Kh. Raichev. 1965. Some local changes of the terrain gamma-field connected with radioactive fallout. Izv. Fiz. inst. Aneb, Bulg. Akad. Nauk 13: 31-42. (In Bulgarian) (NSA 19: 36662).

Dim(t)chev, T. and L. Rajkov. 1972. Intensity and spectral analysis of gamma radiation from natural radionuclides in mountain soils. Kom. Mirno Izpol. Atom. Energ. , Akad. Selskostopanskite Nauki, Sofia, Bulgaria. 13 pp. (In Bulgarian) (INIS-mf-719) (NSA 29: 15627).

Dim(t)chev, T. and L. Rajkov. 1972. Low-background gamma spectrometer for express analysis of radionuclides in soils, plants, and farm produce. Kom. Mirno Izpol. Atom. Energ., Akad. Selskostopanskite Nauki, Sofia, Bulgaria. 13pp. (In Bulgarian) (INIS-mf-720) (NSA 29: 15729).

Dimitroulas, C. and J. Leontiadis. 1971. Investigation under natural conditions of the feasibility of ^{51}Cr-E.D.T.A. as tracer of karst groundwater. Greece Democritus Nuclear Research Center, Athens, report DEMO-71/12. 16pp. (NSA 26: 4408).

Dimitroulas, C. and J. Leontiadis. 1971. Use of radioisotopes in tracing karst groundwater in Greece. II. Investigation on the possible interconnection between Skotini sinkhole and springs of Argos area. Greece Democritus Nuclear Research Center, Athens, report DEMO-71/14. 20pp. (NSA 26: 1969).

Dimitroulas, C., T. Papadimitropoulos and N. Papakis. 1967. Determination of leakages from a large water reservoir using radioisotopic current meter. Mem. Ass. Int. Hydrogeol. 8: 287-293. (NSA 23: 448).

Dimitroulas, C., T. Papadimitropoulos and N. Papakis. 1967. Stratification of radioisotopic effluent in a lake. pp. 341-349. In: Proc. Symp. on Groundwater Hydrology. (See Marino, 1967) (NSA 23: 441).

Dincer, T. 1967. Application of radiotracer methods in streamflow measurements. pp. 93-113. In: Isotopes in Hydrology. (See IAEA, 1967) (NSA 22: 19057).

Dincer, T. 1968. The use of ^{18}O and deuterium concentrations in the water balance of lakes. Water Resources Res. 4: 1289-1305. (NSA 23: 17561).

Dincer, T. and G. H. Davis. 1967. Some considerations on tritium dating and the estimates of tritium input function. Mem. Ass. Int. Hydrogeol. 8: 276-286. (NSA 23: 449).

Dindal, D. L. 1967. Kinetics of Cl-36 DDT in wild waterfowl. Thesis, Ohio State University, Columbus. 214pp.

Dindal, D. L. 1970. Accumulation and excretion of Cl^{36} DDT in mallard and lesser scaup ducks. J. Wildl. Mgmt. 34(1): 74-92.

Dindal, D. L. and T. J. Peterle. 1968. Wing and body tissue relationships of DDT and metabolite residues in mallard and lesser scaup ducks. Bull. Environ. Contam. and Toxicol. 3(1): 37-48. (^{36}Cl).

Dinescu, L. and E. Gaspar. 1970. Considerations on the meteorological parameters in radiohydrological research. Rev. Roum. Phys. 15: 911-922. (NSA 25: 23713).

Dinu, M. 1963. The effect of ionizing radiation on ontogenesis in birds. Comun. Acad. Rep. Populare Romine 13: 891-900. (NSA 18: 5039).

Di Nunno, J. J. 1971. Environmental aspects of nuclear power: some points at issue. 31pp. (For Symp. on Training of Nuclear Facility Personnel; Gatlinburg, Tenn., 19 Apr. 1971) (CONF-710416-19) (NSA 25: 39183).

Di Nunno, J. J., et al. 1970. Survey of environmental aspects of nuclear energy. Nucl. Eng. Int. 15: 607-622. (NSA 25: 2564).

Dios Vidal, R. 1964. Radon-generating activity in different galigian soils in relation to other great world groups and the total lead content. An. Edafol. Agrobiol. (Madrid) 23: 53-63. (In Spanish) (NSA 22: 6826).

Dissanaike, A. S. 1962. The use of radioisotopes on the study of helminth life cycles. pp. 323-340. In: Radioisotopes in Tropical Medicine. (Proc. Symp., Bangkok, 12-16 Dec. 1960). International Atomic Energy Agency, Vienna, Publ. STI/PUB/31.

Dissanaike, A. S., G. A. Dissanaike and W. J. Niles. 1956. Infected mosquito larvae tagged with phosphorus-32. Roy. Soc. Trop. Med. Hyg. Trans. 50(4): 425.

Dissanaike, A. S., G. A. Dissanaike and W. J. Niles. 1957. Production of radioactive infective larvae of Wucheria bancrofti in Culex fatigans. Exptl. Parasitol. 6(1): 52-59.

Dissanaike, A. S., et al. 1957. Further studies on radio-active mosquitoes and filarial larvae using autoradiographic technique. Exptl. Parasitol. 6(3): 261-270.

Dissanaike, G. A., G. M. Jeffery and B. P. Barton. 1958. Radioactive tagging of hookworm larvae (Necator americanus) with P^{32}. Exptl. Parasitol. 7: 249-253.

Dittrich, W., et al. 1949. Studies on the biological effects of high energy electrons and roentgen rays on the pupae of Drosophila. Z. Naturforsch. 4b: 226-229. (AEC-tr-3153) (NSA 12: 5150).

Dittrich, W., et al. 1950. The effects of fast electrons from a 6 Mev Betatron on the eggs of *Drosophila melanogaster*. Strahlentherapie 81: 223-__.

Dix, G. P. and T. J. Dobry. 1972. Critical parameters in plutonium safety evaluations. Health Physics 22(6): 569-574. (NSA 26: 38420).

Dixon, E. B. 1962. Some effects of irradiation on *Cochliomyia hominivorax*. J. Econ. Entomol. 55(6): 826-827. (NSA 17: 17848).

Dixon, H. H. and J. G. Wigham. 1904. Preliminary notes on the action of radiations from radium bromide on some organisms. Sci. Proc. Roy. Dublin Soc. (New Ser.) 10(II, 19): 178-192.

Djurbabic, B., M. Vidakovic and D. Kolbah. 1967. Effect of irradiation on the properties of pollen in Austrian and Scotch pines. Experientia 23: 296-298. (NSA 21: 30928).

Dmitriyev, A. P. and D. M. Grodzinskiy. 1973. Radiosensitization of blue-green alga *Anacystis nidulans* by chemical compounds. Stud. Biophys. 35 (3): 157-163. (In German) (NSA 28: 30087).

Dmitriyev, A. P. and D. M. Grodzinskiy. 1973. Combined effect of chemical compounds and radiation on the cells of blue-green alga *Anacystis nidulans*. Mikrobiologiya 42(2): 307-311. (In Russian) (NSA 30: 12886).

Dmitriyev, A. P. and D. M. Grodzinskiy. 1974. Influence of some radiosensitizers on gamma-ray-induced degradation of DNA in the blue-green alga, *Anacystis nidulans*. Radiation Botany 14(1): 11-15. (NSA 30: 9747).

Dmitriyev, A. P. and D. M. Grodzinskiy. 1974. Influence of a radiosensitizer (monoiodoacetate) on the repair of sublethal damages in cells of the blue-green alga *Anacystis nidulans*. Radiobiologiya 14(3): 382-385. (AEC-tr-7596; pp. 79-83) (NSA 31: 16757).

Dmitriyeva, G. V., Yu. V. Krasnopevtsev and S. G. Malakov. 1969. Distribution of fission products over oceans in the tropical zone and their relation to atmospheric processes. pp. 17-28. In: *Nuclear Meteorology*. (See Makhon'ko and Malakhov, 1969) (NSA 30: 15397).

Dnyansagar, V. R. and A. A. Khatri. 1969. Radiosensitivity of conidiospores to thermal neutrons and isolation of mutants of *Trichoderma glaucum* Abbott. Indian J. Exptl. Biol. 7: 31-33. (NSA 23: 16268).

Doan, V. P. and I. Varga. 1972. Effect of irradiation on the embryonic and postembryonic development of *Amidostomum anseris* (Zeder, 1800). Acta Vet. Acad. Sci. Hung. 22(2): 197-203. (In Hungarian) (NSA 27: 24979).

Dobo, I. 1968. The reliability of soil density measurements using nuclear methods for different soils. Acta Tech. Acad. Sci. Hung. 63: 49-56. (NSA 23: 29237).

Dobosz, E. 1973. Simple method for determination of radioiodine. Radiochem. Radioanal. Lett. 13(5-6): 381-384. (NSA 28: 2215) (Environmental samples).

Dobrokhotov, B. and D. Litvin. 1971. Radioactivity of kestrel pellets ejected after feeding on mice containing P^{32}. Zool. Zh. 50(10): 1591-1592. (In Russian) (WR 147: 97).

Dobrowolski, J. 1967. The influence of x-rays on the development of embryos of tubifex. Folia Biol. (Warsaw) 15: 353-365. (NSA 22: 8610).

Dobrowolski, J. 1969. Radiobiological and radioecological aspects of radioactive pollution of earth atmosphere. Postepy Astronaut. 3: 133-158. (In Polish) (NSA 25: 18941).

Dobson, R. L. 1951. Lethal radiation effects of x-rays, deuterons and alpha particles on the bacterium *Escherichia coli*. Univ. of California, Berkeley, U. S. AEC report UCRL-1140. 64pp. (NSA 5: 3299).

Dobson, R. M. 1962. Marking techniques and their application to the study of small terrestrial animals. pp. 228-239. In: *Progress in Soil Zoology*. (Papers from a Colloquium on Research Methods Organized by the Soil Zoology Committee of the Intern. Soc. of Soil Sci., Rothamsted Experimental Station, Hertfordshire, 10-14 July 1958). P. W. Murphy, ed. Butterworth, London.

Dodakundi, G. B. and S. S. Rodgi. 1974. Liquid scintillation counting of carbon-14 in algae. J. Karnatak Univ. Sci. 19: 37-41. (NSA 33: 29801).

Dodd, A. O. 1956. A survey of some Los Alamos County canyons for radioactive contamination, spring, 1953 to spring, 1955. Los Alamos Scient. Lab., U. S. AEC report LAMS-2038. 17pp. (NSA 13: 11729).

Dodd, J. D. and G. L. Van Amburg. 1970. Distribution of Cs^{134} in *Andropogon scoparius* Michx. clones in two native habitats. Ecology 51(4): 685-689.

Dodd, N. J. F. and M. Ebert. 1969. Free radical reactions modify cellular damage. Nature 221: 1245-1246. (NSA 23: 22438) (*Osmunda regalis*).

Dodd, N. J. F. and M. Ebert. 1970. Effects of ionizing radiation on dried spores of *Osmunda regalis*. I. Electron spin resonance study of spores. II. Modification of survival and its correlation with free radical changes. Int. J. Radiat. Biol. 18: 451-461, 463-473. (NSA 25: 11319, 27069).

Dodd, N. J. F. and M. Ebert. 1972. Post-irradiation temperature effects in dried spores of *Osmunda regalis*. Radiation Botany 12(3): 151-158. (NSA 26: 48227).

Dodd, N. J. F. and M. Ebert. 1973. Correlation of ESR and radiobiological observations of the temperature and oxygen effects in *Osmunda regalis* spores after irradiation with fast electrons. Radiation Botany 13(2): 87-96. (NSA 28: 8372).

Dodd, N. J. F. and M. Ebert. 1973. Effects of storage conditions on survival and free radical populations in dried irradiated spores of *Osmunda regalis*. Biophysik 10(1): 79-88. (NSA 28: 24739).

Dodd, N. J. F. and H. M. Swartz. 1975. Effects of ionizing radiation on dried spores of *Osmunda regalis*. Int. J. Radiat. Biol. 27(3): 205-210. (NSA 32: 17443).

Dodge, J. D. 1965. Effects of ultra-violet light on the survival and nuclear division of dinoflagellate. Protoplasma 59: 485-493. (NSA 19: 31867).

Dodge, J. D. and M. B. E. Godward. 1963. Some effects of x-rays on the nucleus of a dinoflagellate. Radiation Botany 3(2): 99-104. (NSA 17: 33719).

Doederlein, J. M. 1975. Nuclear power and other energy. Forskningsnytt 20 (4): 26-33. (NSA 33: 16234).

Dognon, A. 1925. Temperature coefficient of the effect of x-rays on Ascaris eggs. Compt. rend. soc. biol. 92: 1389-1392. (In French).

Dognon, A. 1926. Effects of temperature on the radiosensitivity of Ascaris eggs. Compt. rend. soc. biol. 94: 466-468. (English transl. SAND-74-6017) (NSA 30: 29766).

Doi, K. and Y. Aoyama. 1973. Uranium content in natural water. Kozan Chishitsu (5): 35-45. (In Japanese) (NSA 32: 19703).

Dojlido, J. 1969. Occurrence of ^{90}Sr in Lake Zegrzynski 1963-1964. Nukleonika 14: 213-220. (In Polish) (NSA 23: 47629).

Dojlido, J. 1969. Radioactive substances in Lake Zegrzynski. Pr. Inst. Gospod. Wodnej 5(4): 111-146. (In Polish) (NSA 28: 8187).

Doke, N. and T. Hirai. 1970. Effects of tobacco mosaic virus infection on photosynthetic CO_2 fixation and $^{14}CO_2$ incorporation into protein in tobacco leaves. Virology 42: 68-77. (NSA 25: 13360).

Dokholyan, V. K., B. P. Kostrov and G. S. Bityutskaya. 1973. Accumulation of ^{90}Sr, ^{90}Y, and the effect on those elements dissolved in water on embryonic development of some fish species in the Caspian Sea. pp. 36-44. In: Radioecology of Water Organisms. III. The Effect of Ionizing Radiation on Water Organisms. (See Andrumaitis, 1973) (In Russian) (NSA 30: 24396).

Dolenc, M. R. 1973. Ecological considerations of Project Wagon Wheel. pp. 215-223. In: Geology and Mineral Resources of the Greater Green River Basin. E. M. Schell, ed. Wyoming Geological Assoc., Casper, Wyo. (NSA 30: 29459).

Dolezal, R., M. Petersen and R. Reinhard. 1971. Measurement of shifting sand on coasts with heavy surf. Exhibit L2, 4th Intern. Conf. on the Peaceful Uses of Atomic Energy, Geneva, 6-16 Sept. 1971. 25pp. (NP-19203) (NSA 26: 20361) (^{46}Sc).

Dolezal, R., et al. 1965. Development and research with a radioactive tracer for measurement of the kinetics, erosion, and transport of sand on a sandy gravel coast. European Atomic Energy Community report EUR-2167.d. 156pp. (In German) (NSA 20: 510).

Dolge, H. 1964. Demarcation of sapwood and heartwood densitiometer measurements on x-ray photographs. Ann. Sci. For. (Nancy) 21: 605-623. (FA 26: 5723).

Dolgov, E. G. 1962. The effect of toad skin gland secretion on the permeability of tissues to vital stains after their exposure to experimental acute radiation. Bull. Exptl. Biol. Med. (USSR) 52(10): 1147-1150. (English transl.) (NSA 16: 20157).

Dolgushina, Z. K. and Yu. I. Onanko. 1973. Radioresistance of some plankton and benthos species of crustacea. pp. 50-56. In: Radioecology of Water Organisms. III. The Effect of Ionizing Radiation on Water Organisms. (See Andrumaitis, 1973) (In Russian) (NSA 30: 24359).

Dolkas, C. B., et al. 1975. Mathematical analysis of the mortality kinetics of Drosophila melanogaster exposed to gamma radiation. Mech. Ageing Dev. 4: 59-69. (NSA 33: 17469).

Domroese, K. A. 1964. Some aspects of lipid catabolism in saturniid moths. Dissertation Abstr. 25(1): 706. (^{14}C, Hyalophora cecropia).

Domroese, K. A. and L. I. Gilbert. 1964. The role of lipid in adult development and flight muscle metabolism in Hyalophora cecropia. J. Exptl. Biol. 41: 573-590. (^{14}C).

Domsch, K. H., J. P. E. Anderson and R. Ahlers. 1973. Method for simultaneous measurement of radioactive and inactive CO_2 evolved from soil samples during incubation with labeled substances. Appl. Microbiol. 25(5): 819-824. (NSA 29: 29393).

Donaldson, D. J. 1972. Effects of x irradiation on lens regeneration in adult Triturus viridescens. Anat. Record 172(1): 45-55. (NSA 26: 25861).

Donaldson, E. M. and U. H. M. Fagerlund. 1968. Changes in the cortisol dynamics of sockeye salmon (Onchorhynchus nerka) resulting from sexual maturation. Gen. Comp. Endochron. 11: 552-561. (SFA 16: 12869).

Donaldson, E. M. and U. H. M. Fagerlund. 1968. Changes resulting from sexual maturation in the half-life ($T_{½}$) and volume distribution (V) of radioactivity after injection of C^{14} cortisol (F) into salmon. Excerta Med. I Int. Congr. Ser. No. 157, Item No. 348. 1968 studies, Part 2. Fish. Res. Bd. Can. (1259): 213. (SFA 16: 13394).

Donaldson, L. R. and K. Bonham. 1964. Effects of low-level chronic irradiation of chinook and coho salmon eggs and alevins. Trans. Am. Fish. Soc. 93(4): 333-341.

Donaldson, L. R. and K. Bonham. 1970. Effects of chronic exposure of chinook salmon eggs and alevins to gamma irradiation. Trans. Am. Fish. Soc. 99(1): 112-119.

Dongmann, G., H. Foerstel and K. Wagener. 1972. ^{18}O-rich oxygen from land photosynthesis. Nature New Biol. 240(99): 127-128. (NSA 27: 7415) (European beech and oak; Spathodea campanulata).

Donini, B. 1967. Effects of chronic gamma-irradiation on Pinus pinea and Pinus halepensis. Radiation Botany 7(3): 183-192. (BA 48: 18636).

Donini, B. 1968. Studies on radiosensitivity of higher plants. Giorn. Bot. Ital. 102: 556-557. (In Italian) (NSA 24: 34763).

Donnellan, J. E., Jr. and H. J. Morowitz. 1957. The irradiation of dry spores of Bacillus subtilis with fast charged particles. Radiation Research 7(1): 71-78.

Donnellan, J. E., Jr. and H. J. Morowitz. 1960. The effect of charged particle irradiation on germination of *Bacillus subtilis*. Radiation Research 12(1): 67-78.

Donnellan, J. F. and B. A. Kilby. 1967. Uric acid metabolism by symbiotic bacteria from the fat body of *Periplaneta americana*. Comp. Biochem. Physiol. 22(1): 235-252. (^{14}C).

Donnelly, G. M. and A. H. John. 1964. Radiosensitivity of the salamander *Desmognathus*. Abstr. Genetics 50(2): 244.

Donnelly, G. M. and A. H. Sparrow. 1963. Large chromosomes: the somatic and meiotic chromosomes of *Amphiuma*. Brookhaven National Lab., U. S. AEC report BNL-7596. 25pp. (NSA 19: 2060) (Mentions radiosensitivity).

Donnelly, G. M., et al. 1968. Radiosensitivity of selected amphibians. Abstr. Radiation Research 35(2): 574.

Donnelly, J. 1965. Possible causes of failure in a field test of the "sterile males" method of control. p. 253. In: *Proc. 12th Intern. Congr. of Entomology*. P. Freeman, ed. Royal Entomol. Soc., London.

Donth, H. H. 1962. Sources of error in monitoring and survey programs due to sampling and their estimation. Bundesministerium für Wissenschaftliche Forschung, Bad Godesberg, W. Germany. 61pp. (In German) (NP-13312) (NSA 18: 4022).

Doran, W. T. 1968. Magnitude of deposited radionuclide problem based on future usage: civilian aspects. Monogr. Nucl. Med. Biol. (2): 471-481. (NSA 23: 24650).

Dorman, F. H. 1968. Some oxygen isotope temperatures and a theory for a 30-million-year world-temperature cycle. J. Geology 76(3): 297-313.

Dorough, H. W. and J. E. Casida. 1964. Nature of certain carbamate metabolites of the insecticide Sevin. J. Agr. Food Chem. 12(4): 294-304. (^{14}C, *Musca domestica*, *Periplaneta americana*).

Dorris, B. J. 1964. Population variability and response to ionizing radiation in *Isoetes melanospora* Engelm. Thesis, Emory Univ., Atlanta, Ga.

Dörschel, E. and W. Stolz. 1970. Uranium determination in water with the help of solid-state track method. Radiochem. Radioanal. Lett. 4: 277-283. (In German) (NSA 24: 46191).

Doshi, G. R., et al. 1969. Ashing procedures for biomaterial. Current Sci. (India) 38: 206-208. (NSA 23: 29305).

Doty, M. S. (Ed.). 1963. *Proc. Conf. on Primary Productivity Measurement, Marine and Freshwater*. (Univ. of Hawaii, August 21-September 6, 1961). Univ. of Hawaii, U. S. AEC report TID-7633. ix, 237pp.

Doty, W. H., D. E. Wood and E. L. Schneider. 1969. Neutron activation analysis of nitrogen in feedstuffs. J. Assoc. Anal. Chem. 52: 953-956. (NSA 23: 47405).

Douault, P. 1966. Study of the external contamination of the bee and its surroundings with a radioisotope introduced in food. Ann. Abeille 9(1): 37-45. (In French, English summary) (BA 48: 20541) (^{32}P).

Douault, P. 1966. Improvements of a research technique in the contamination of the bee labelled with a radioisotope. Ann. Abeille 9(2): 165-169. (In French) (BA 48: 35960) (^{32}P).

Douault, P. 1967. Influence of the brood nest on the exchange of food through workers of bees (*Apis mellifera*). Compt. rend. D264(8): 1092-1095. (In French) (BA 49: 129608) (^{32}P).

Douault, P., J. Louveaux and J. Theurkauff. 1974. Study of the honey storage by bees by means of sugar syrup traced with ^{141}Ce. Apidologie 5(2): 107-126. (In French) (NSA 31: 29785).

Doub, W. O. 1972. Environmental and regulatory aspects of the breeder reactor. Environ. Aff. 2(1): 237-249. (NSA 27: 11270).

Doub, W. O. 1972. Environmental law: a challenge to the legal profession. Publ. Util. Fortn. 90(7): 68-71. (NSA 27: 2226).

Doub, W. O. 1972. Nuclear power is necessary. Catalyst 2(3): 9-__.

Doub, W. O. 1972. U. S. experience in environmental and public acceptance aspects of nuclear power. At. Jap. 16(5 Suppl.): 3-34. (NSA 28: 17539).

Doudoroff, P., et al. 1951. Bio-assay methods for the evaluation of acute toxicity of industrial wastes to fish. Sewage and Industr. Wastes 23(11): 1380-1397.

Douellous, D. 1973. Natural and artificial origins of tritium; its impact on the environment. Rayonn. Ionis. Tech. Mes. Prot. (2): 71-80. (In French) (NSA 28: 27229).

Douglas, G. S. (Ed.). 1967. Radioassay procedures for environmental samples. Envirnmental Health Series (Radiological Health), U. S. Public Health Service, Washington, Publ. No. 999-RH-27. xxi, misc. pp.

Douglas, R. L., S. C. Black and D. S. Barth. 1971. ^{131}I transport through the air-forage-cow-milk system using an aerosol mist (Project Rainout). Southwestern Radiological Health Lab., U. S. EPA, U. S. AEC report SWRHL-43r. iv, 38pp.

Doull, J. and K. P. Du Bois. 1953. Influence of hibernation on survival time and weight loss of x-irradiated ground squirrels. Proc. Soc. Exptl. Biol. Med. 84(2): 367-370. (NSA 8: 1289).

Doull, J., D. F. Petersen and K. P. Du Bois. 1952. Effects of x-irradiation on *Citellus tridecemlineatus*. Abstr. Fed. Proc. 11(1): 340.

Douvoyiannis, J. and H. Mimikos. 1967. Strontium-90 and cesium-137 radioactive contamination of Democrite Center Region (Greece) during 1965 and 1966. Chim. chron. 32A1: 97-103.

Dow, R. P. 1959. Dispersal of adult *Hippelates pusio*, the eye gnat. Ann. Entomol. Soc. Am. 52(4): 372-381.

Dow, R. P. 1967. Radioactive yeast for tagging mosquitoes. Abstr. 331. Bull. Entomol. Soc. Am. 13(3): 202. (^{32}P).

Dow Chemical Company. 1971. Proceedings of the Rocky Flats symposium on safety in plutonium handling facilities, April 13-16, 1971. Golden, Colo., U. S. AEC report CONF-710401. 463pp. (NSA 25: 54937).

Dowell, F. H. 1964. The establishment of Tribolium confusum Duv. populations after stress by parasitism and x-irradiation. Dissertation, Univ. of Tennessee, Knoxville. 92pp. (Dissertation Abstr. 25(5): 3170-3171).

Downes, A. M. and I. W. McDonald. 1964. The chromium-51 complex of ethylenediamine tetraacetic acid as a soluble rumen marker. Brit. J. Nutr. 18: 153-162.

Downs, S. F. 1964. A study of the contribution made to a gene pool of competing groups of radiated and non-radiated Drosophila melanogaster. Thesis, Amherst Coll., Amherst, Mass.

Doyle, J. F., Z. G. Burson and P. K. Boyns. (Eds.). 1973. Aerial radiological measuring surveys of the Oyster Creek Nuclear Power Plant, August 1969. EG&G, Las Vegas, U. S. AEC report ARMS-69.6.7. 22pp. (EGG-1183-1546) (NSA 27: 17255).

Drachev, S. M. and A. A. Bylinkina. 1960. Pollution of reservoir surfaces with radioactive substances. Med. radiologiya 6(6): 54-58. (In Russian) (NSA 14: 22748)

Drachev, S. M. and A. A. Bylinkina. 1960. Contamination of rivers by radioactive fallout from the atmosphere. In: Proc. Conf. on Radiation Hygiene. Minzdrav RSFSR, Leningrad. (In Russian).

Drake, J. L. 1971. Effects of x irradiation on the toxicity of organophosphate insecticides to the house fly (Musca domestica L.). Thesis, Univ. of Arizona, Tucson. 93pp. (Univ. Microfilms 71-23,897) (NSA 26: 15512).

Drake, J. L., L. A. Crowder and G. W. Ware. 1971. Effects of x irradiation on the toxicity of organophosphate insecticides to the house fly. Pest. Biochem. Physiol. (1): 373-380. (NSA 29: 10466).

Draŝković, R., L. Marinkov and R. Raŝković. 1972. Radioactivation analysis and the possibilities of its application at the Boris Kidric Institute of Nuclear Sciences at Vinca. II. Tehnika 26(8): 243-249. (NSA 30: 20611) (Impurities in suspended materials and waters in the Danube River).

Draŝković, R., T. Tasovac and R. Radosavljević. 1971. Neutron activation analysis of the aquatic environment in the Danube. pp. 329-334. In: Nuclear Techniques in Environmental Pollution. (See IAEA, 1971) (NSA 25: 31878).

Draŝković, R., et al. 1971. Interaction between water, trace elements and different components in the Danube River. pp. 1167-1173. In: Proc. Intern. Symp. on Radioecology Applied to the Protection of Man and His Environment. (See Commission of the European Communities, 1971) (NSA 27: 5266).

Draŝković, R., et al. 1972. Neutron activation analysis of the aquatic environment in the Sava River. pp 287-293. In: Nuclear Activation Techniques in the Life Sciences, 1972. (See IAEA, 1972) (NSA 27: 22093).

Dratman, M. B., G. Jacob and C. Dacou. 1969. Radioactive iodocompounds in the neuraxis of Rana pipiens during metamorphosis induced with ^{125}I-thyroxin. Confin. Neurol. 31: 238-246. (NSA 24: 36693).

Dresden, M. H. 1973. Denervation effects on newt limb regeneration: autoradiography with ^{3}H-thymidine. Cell Different. 2: 255-259. (NSA 30: 3760).

Dresser, H. and E. Broda. 1968. Effects of ionizing radiation on endogenous respiration of Bacillus cereus spores. Int. J. Radiat. Biol. 13: 597-600. (In German) (NSA 22: 49235).

Dreusicke, M. 1970. Contaminants in animal products. I. Mathematical formulation of radionuclide metabolism and the significance of concentration factors in the computation of permissible concentrations of contaminants in feedstuffs and food products. Arch. Tierernähr. 20(6): 497-508. (In German) (NSA 26: 58813).

Drew, R. T. and M. Eisenbud. 1965. Feasibility of studying the natural radiation dose to indigenous rodents on the Morro de Ferro, Brazil. Abstr. Health Physics 11(8): 806.

Drew, R. T. and M. Eisenbud. 1966. The natural radiation dose to indigenous rodents on the Morro de Ferro, Brazil. Health Physics 12(9): 1267-1274.

Drew, R. T. and M. Eisenbud. 1970. Pulmonary dose from ^{220}Rn received by indigenous rodents of the Morro de Ferro, Brazil. Radiation Research 42: 270-281. (NSA 24: 31894).

Driggers, J. C., et al. 1951. The transference of radioactive calcium and phosphorus from hen to chick. Poultry Sci. 30(2): 199-204.

Drobek, W. 1958. Necessary means for the protection of water supplies in the development of the atomic industry. Stadtehygiene 9(1): 6-7. (In German).

Drobkov, A. A. 1940. The influence of radioactive elements on yields of plants. Izv. Akad. Nauk SSSR (biol.) 1940: 783-801. (In Russian).

Drobkov, A. A. 1945. Effect of radioactive elements upon development of the root-nodule bacteria and upon the assimilation by them of the molecular nitrogen of the atmosphere. Compt. rend. acad. sci. URSS 49: 224-246. (In Russian) (BA 21: 17817).

Drobkov, A. A. 1947. Influence of radioactive elements on plant life. Nauka i zhizn' 3: 15pp. (NP-331) (AEC-tr-816) (NSA 1: 1770).

Drobkov, A. A. 1949. Radiophotographic determination of natural and artificial radioactive elements in plants. Dokl. Akad. Nauk SSSR 68: 177-180. (In Russian) (CA 44: 710).

Drobkov, A. A. 1951. Part played by natural radioactive elements in biological processes. Usp. sovrem. biol. 31(1): 82-100. (In Russian).

Drobkov, A. A. 1958. Microelements and Natural Radioactive Elements in the Life of Plants and Animals. Acad. Sci. USSR, Moscow. 206pp. (In Russian).

Drobkov, A. A. 1963. Chemical elements and life. Priroda 52(8): 45-52. (JPRS-21,817).

Drobinski, J. C., Jr. 1966. Radiocarbon in the environment. Radiological Hlth. Data 7(1): 10-12.

Drobinski, J. C., Jr., P. J. Magno and A. S. Goldin. 1968. Plutonium, tritium and carbon-14 in man and the biosphere. Abstr. pp. 369-370. In: Radiation Protection, Part 1. W. S. Snyder, et al., eds. Pergamon Press, New York.

Drost, W. 1972. Tracerlog: a new borehole technique for measuring groundwater with radioactive isotopes. pp. 379-401. In: Conf. on Radiation and Isotope Techniques in Civil Engineering. European Atomic Energy Community, Brussels, Belgium, report EURISOTOP-55(Vol. 1). (In German) (NSA 27: 27605).

Drost, W. 1972. Groundwater measurements at the site of a dam in the Degenseemoos Moor near Penzberg (Upper Bavaria). Geol. Jahrb. Reihe C(2): 339-350. (In German) (NSA 30: 21165) (Tracer).

Drost, W. 1973. Application of groundwater measurements by means of radioisotopes on groundwater exploration. pp. 357-369. In: Water for the Human Environment, Vol. I. V. T. Crow, ed. Intern. Water Resources Assoc., Champaign, Ill. (NSA 33: 12056).

Drost, W., et al. 1968. Point dilution methods of investigating ground water flow by means of radioisotopes. Water Resources Res. 4(1): 125-146.

Drost, W., et al. 1972. Isotope methods in groundwater sciences. European Atomic Energy Community, Brussels, Belgium, report EURISOTOP-61. 183pp. (In German) (NSA 27: 19837).

Drouet, F. 1959. Algal flora of the Nevada Test Site. Abstr. J. Colo.-Wyo. Acad. Sci. 4(11): 31.

Drozhzhin, V. M., et al. 1973. Geochemical balance of radioactive elements in the basin of the Black and Azov Seas. III. Balance of radium. Radiokhimiya 15(3): 415-421. (Sov. Radiochem. 15(3): 416-421) (NSA 30: 6536).

Druilhet, A. and J. Fontan. 1974. Application of radon as indicator of the intensity of exchange at the soil (0 to 100 m). Boundary-Layer Meteorology 6(3-4): 387-411. (In French) (NSA 30: 23992).

Drummond, R. O. 1963. Effects of gamma radiation on the fertility of the common cattle grub, Hypoderma lineatum (de Villers). Int. J. Radiat. Biol. 7(5): 491-495. (NSA 18: 15760).

Drummond, R. O., J. G. Medley and O. H. Graham. 1966. Engorgement and reproduction of lone star ticks (Amblyomma amercanum (L.)) treated with gamma radiation. Int. J. Radiat. Biol. 10: 183- 188. (NSA 20: 33009).

Drury, D. E. and J. G. Eales. 1968. The influence of temperature on histological and radiochemical measurements of thyroid activity in the eastern brook trout, Salvelinus fontinalis Mitchell. Can. J. Zool. 46(1): 1-9.

Dryanovska, O. A. and S. I. Glavanakova. 1972. Radiosensitivity of the oocytes in Sitotroga cerealella Oliv. Compt. rend. Acad. Bulg. Sci. 25(2): 253-254. (In Bulgarian) (NSA 26: 53499; 31: 14396).

Dryanovski, P. 1974. On the problem of the irradiation resistance in birds. Vet. Med. Nauki (Sofia) 11(4): 91-97. (In Bulgarian) (NSA 32: 3794).

Drynan, W. R., E. F. Gloyna and D. F. Smallhorst. 1961. Radioactivity of Texas streams. Univ. of Ill. Engrg. Expt. Sta. Cir. 69: 28-33. (See Univ. of Illinois, 1961).

Drzewlocki, J. and T. Yurtsever. 1974. Some conceptual mathematical models and digital simulation approach in the use of tracers in hydrogeological systems. pp. 425-450. In: Isotope Techniques in Groundwater Hydrology Vol. 2. (See IAEA, 1974) (NSA 31: 19666).

D'Souza, T. J., R. Kirchmann and J. J. Lehr. 1972. Distribution of radiostrontium and radiocesium in the organic and mineral fractions of pasture soils and their subsequent transfer to grasses. pp. 595-604. In: Isotopes and Radiation in Soil-Plant Relationships Including Forestry. (See IAEA, 1972) (NSA 27: 2848).

Dubinchuk, V. T., et al. 1974. Conditions for groundwater recharge using natural stable and radioactive isotopes together with analogue simulation. pp. 399-429. In: Isotope Techniques in Groundwater Hydrology, Vol. I. (See IAEA, 1974) (In Russian) (NSA 31: 19639).

Dubinin, N. P. 1966. Radiation and Evolution of Populations. Atomizdat, Moscow. 743pp. (In Russian) (NSA 22: 40655).

Dubinin, N. P. (Ed.). 1966. Effect of Ionizing Radiation on Heredity. Publ. House, Acad. Sci. USSR, Moscow. 320pp. (In Russian) (NSA 22: 21473).

Dubinin, N. P. 1967. Problems of space genetics. Izv. Akad. Nauk SSSR (biol.) (5): 669-681. (In Russian) (NSA 22: 12829).

Dubinin, N. P. (Ed.). 1969. Contemorary Problems in Radiation Genetics. Atomizdat, Moscow. 352pp. (In Russian) (NSA 26: 17957) (Incl. fungi and algae).

Dubinin, N. P., V. A. Shevchenko and M. D. Pomerantseva. 1971. Effect of ionizing radiation on populations (radiation-genetic aspects). pp. 183-228. In: Contemporary Problems of Radiobiology. Vol. 2. Radioecology. (See Klechkovskiy, Polikarpov and Aleksakhin, 1971) (In Russian).

Dubinin, N. P., et al. 1972. Genetic processes in populations subjected to chronic ionizing radiation. Usp. sovrem. genet. (4): 170-205. (In Russian) (NSA 26: 59029) (Natural populations).

Dubinin, N. P., et al. 1975. Effect of ionizing radiations with different LET on survival and mutation in Chlorella. pp. 181-186. In: Life Sciences and Space Research. XIII. Akademie-Verlag GmbH, Berlin. (NSA 33: 29571).

Dubinskaya, N. A. and L. M. Kyun'kye. 1972. Simultaneous neutron activation determination of Cu, Zn, and Mn in plant materials. Latv. PSR Zinat. Akad. vestis fiz. teh. zinat. (2): 3-7. (In Russian) (NSA 27: 9275).

Dubinskaya, N. A. and L. M. Kyun'kye. 1972. Nondestructive neutron activation analysis of Co and some other elements in plant materials. Latv. PSR Zinat. Akad. vestis fiz. teh. zinat. (2): 8-13. (In Russian) (NSA 27: 9276).

Dubis, K. and Z. Srebro. 1970. Morpho-functional study of hypothalmic neurosecretion. VI. The effect of locally administered x-irradiation in Rana temporaria and Rana esculenta. Folia Biol. (Krakow) 18: 247-252. (NSA 25: 8764).

Duboff, H. S. 1957. Factors affecting radiation-induced division delay in Chilomonas paramecium. Physiol. Zool. 30(3): 268-280. (NSA 11: 11861).

Duce, R. A., J. W. Winchester and T. W. Van Nahl. 1966. Iodine, bromine and chlorine in winter aerosols and snow from Barrow, Alaska. Tellus 18: 238-248. (NSA 21: 6040) (Neutron activation).

Duckworth, J. P. 1971. Quarterly report, October 1, 1970 - December 31, 19 1970. Nuclear Fuel Services, Inc., West Valley, N. Y., U. S. AEC report SRO-344-11. 7pp. (NSA 25: 26355) (Monitoring).

Ducoff, H. S. 1965. Recovery from radiation-induced division block in protists. Illinois Univ., U. S. AEC report COO-878-15. 17pp. (NSA 19: 21856).

Ducoff, H. S. 1966. X-ray response of an endoparasitic wasp. Abstr. BC-4. Radiation Research 27: 496.

Ducoff, H. S. 1967. Changes with age in the radiosensitivity of adult flour beetles. Radiation Research 31(3): 612.

Ducoff, H. S. 1972. Causes of death in irradiated adult insects. Biol. Rev. Cambridge Phil. Soc. 47(2): 211-240. (NSA 26: 56480; 33: 15096).

Ducoff, H. S. and G. C. Bosma. 1966. Influence of pupal age on sensitivity to radiation. Biol. Bull. 130: 151-156. (NSA 20: 32975).

Ducoff, H. S. and G. C. Bosma. 1967. Acute lethality after x-irradiation of Tribolium confusum adults. Entomologia Exp. Appl. 10(2): 153-165.

Ducoff, H. S. and H. E. Walburg. 1960. Response of Tribolium larvae to x-irradiation. Abstr. 140. Anat. Record 137: 351.

Ducoff, H. S., B. D. Butler and E. J. Geffon. 1965. The effect of radiations on replication in Brachiomonas submarina Bohlin. Radiation Research 24: 563-571. (NSA 19: 24340).

Ducoff, H. S., A. P. Vaughan and J. L. Crossland. 1971. Dose-fractionation and the sterilization of radiosensitive male confused flour beetles. J. Econ. Entomol. 64: 541-543. (NSA 25: 26355).

Ducoff, H. S., et al. 1971. Acute lethality in irradiated flour beetles: evidence against a role of bacterial infection. J. Invertebr. Pathol. 18 (3): 344-352. (NSA 26: 23150).

Duewer, D. L., B. R. Kowalski and T. F. Schatzki. 1975. Source identification of oil spills by pattern recognition analysis of natural elemental composition. Anal. Chem. 47(9): 1573-1583. (NSA 32: 16729) (Neutron activation).

Dufaure, J. 1965. The precocious destruction of the gonads of the viviparous lizard embryo through irradiation (x-rays) and their effect on the development of the genital tuburcules. Compt. rend. D261: 1735-1737.

Duffey, D. 1972. Environmental impact modeling for breeder reactors. Trans. Amer. Nucl. Soc. 15(2): 646-647. (NSA 27: 7130).

Duffey, J. R. H. 1964. The syntheses of ethane-labelled DDT and o-Cl-DDT and their metabolism in larval Aedes aegypti. Thesis, Fordham Univ., New York, N. Y. (Dissertation Abstr. 25(3): 1562-1563) (^{14}C).

Dugal, L.-P. 1956. The use of isotopes in biology. Ann. Soc. Entomol. Quebec 2: 21-31. (In French).

Dugdale, R. C. and J. J. Goering. 1967. Uptake of new and regenerated forms of nitrogen in primary productivity. Limnol. Oceanogr. 12(2): 196-206.

Dugdale, R. C., et al. 1959. Nitrogen fixation in lakes. Science 130(3379): 859-860. (^{15}N).

Dugdale, V. A. and R. C. Dugdale. 1962. Nitrogen metabolism in lakes. II. Role of nitrogen fixation in Sanctuary Lake, Pennsylvania. Limnol. Oceanogr. 7(2): 170-177. (SFA 7: 5205).

Dugdale, V. A. and R. C. Dugdale. 1965. Nitrogen metabolism in lakes. III. Tracer studies of the assimilation of inorganic nitrogen sources. Limnol. Oceanogr. 10(1): 53-57.

Duggus, J. E. and A. H. Gold. 1967. Relationship of tracer-measured aphid feeding to acquisition of beet western yellows virus and to feeding inhibitors in plant extracts. Phytopathology 57(11): 1237-1241.

Dugle, J. R. 1969. Ecology of the field irradiator-gamma area pre-irradiation studies. I. Plant associations. Whiteshell Nuclear Research Establishment, Atomic Energy of Canada, Ltd., Pinawa, report AECL-3424. 37pp. (NSA 24: 17011).

Dugle, J. R. and M. H. El-Lakany. 1971. Check list revisions of the plants of the Whiteshell area, Manitoba, including a summary of their published radiation sensitivities. Whiteshell Nuclear Research Establishment, Atomic Energy of Canada, Ltd., Pinawa, report AECL-3678. 31pp. (NSA 25: 32616).

Dugle, J. R. and J. E. Guthrie. 1970. Some observations on algae invading a ^{137}Cs contaminated pond. Whiteshell Nuclear Research Establishment, Atomic Energy of Canada, Ltd., Pinawa, report AECL-3463. 12pp. (NSA 24: 27556).

Dugle, J. R. and D. H. Thibault. 1972. Ecology of the field irradiator-gamma area pre-irradiation studies. II. Botanical methods and vegetation sampling procedures. 1974. III. Revisions to botanical methods and vegetation sampling procedures. Whiteshell Nuclear Research Establishment, Atomic Energy of Canada, Ltd., Pinawa, reports AECL-4135; AECL-4668. 32 pp.; 47pp. (NSA 27: 315; 30: 9760).

Dugle, J. R., D. H. Thibault and W. D. Fitzsimmons. 1974. Pre-irradiation study of tree size, age, and growth in the field irradiator-gamma area. Whiteshell Nuclear Research Establishment, Atomic Energy of Canada, Ltd., report AECL-4758. 55pp. (NSA 30: 21168).

Duguid, J. O. 1974. Groundwater transport of radionuclides from buried waste: a case study at Oak Ridge National Laboratory. pp. 511-529. In: Proc. 2nd AEC Environmental Protection Conf., Vol. 2. (See U. S. AEC, 1974) (NSA 30: 3163).

Duhamel, F.-A., M. Menous and M. Candillon. 1958. Disposal of liquid radioactive residues at the CEA, treatment and disposal in a river. Proc. 2nd Intern. Conf. on the Peaceful Uses of Atomic Energy 18: 90-94.

Duke, G. E. 1967. ^{51}Cr in metabolizability and food passage rate studies with the ring-necked pheasant. Thesis, Michigan State Univ., East Lansing. 83pp. (NSA 22: 47493).

Duke, G. E., G. A. Petrides and R. K. Ringer. 1968. Chromium-51 in food metabolizability and passage rate studies with the ring-necked pheasant. Poultry Science 47(4): 1356-1364.

Duke, J. A. 1970. Keys for the identification of seedlings of some prominent woody species in eight forest types in Puerto Rico. Ch. B-15, pp. B239-B274. In: A Tropical Rain Forest. (See H. T. Odum and R. F. Pigeon, 1970).

Duke, J. A. 1970. Postirradiation woody seedlings. Ch. D-8, pp. D189-D192. In: A Tropical Rain Forest. (See H. T. Odum and R. F. Pigeon, 1970) (NSA 25: 21933).

Duke, J. A. (Ed.). 1971. Symposium on Amchitka Island Bioenvironmental Studies. BioScience 21(12): 599-711.

Duke, T. W. 1965. Biogeochemical cycling of radionuclides in the estuarine environment. pp. 315-323. In: Proc. 17th Ann. Conf. Southeastern Assoc. Game Fish Comm.

Duke, T. W. 1967. Possible routes of zinc-65 from an experimental estuarine environment to man. J. Water Poll. Control Fed. 39(4): 536-___.

Duke, T. W. 1968. Salt water radioecology -- a fresh approach for studying the relationships between marine animals, plants and their environment. Wildlife in N. Car. 32(5): 4-6.

Duke, T. W., J. P. Baptist and D. E. Hoss. 1966. Bioaccumulation of radioactive gold used as a sediment tracer in the estuarine environment. Fish. Bull. U. S. Fish Wildl. Serv. 65(2): 427-436.

Duke, T. W., J. N. Willis and D. A. Wolfe. 1968. A technique for studying the exchange of trace elements between estuarine sediments and water. Limnol. Oceanogr. 13(3): 541-545.

Duke, T. W., et al. 1966. Cycling of trace elements in the estuarine environment. I. Movement and distribution of zinc-65 and stable zinc in experimental ponds. Chesapeake Science 7(1): 1-10.

Duke, T. W., et al. 1969. Influence of environmental factors on the concentrations of ^{65}Zn by an experimental community. pp. 355-362. In: Symposium on Radioecology. (See Nelson and Evans, 1969).

Duke Power Company. 1971. Oconee Nuclear Station, Units 1, 2 and 3. Preoperational environmental radioactivity monitoring program. Charlotte, N. Car. U. S. AEC DOCKET-50269-33; DOCKET-50270-31; DOCKET-50287-28. 52pp. (NSA 25: 41333).

Duke Power Company. 1973. Catawba Nuclear Station, Units 1 and 2. USAEC final environmental statement. Charlotte, N. Car. U. S. AEC DOCKET-50413-56. 385pp. (NSA 29: 12290).

Duke University. 1967. A study of the relationship between plant water stress and ionizing radiation. Final report for the period January 1965-May 1966. U. S. AEC report ORO-3094-1. 7pp. (NSA 21: 20424).

Dumasia, M. D. 1967. Water and the effects of radiations on microorganisms, Thesis, Univ. of Saskatchewan, Saskatoon, Canada. 167pp. (NSA 22: 45643).

Dumeste, M., R. Gardel and P. Delafolu. 1959. Contamination of fish by means of radioactivity and decontamination. Study through the physiological repartition of insoluble Ca^{45} in Carassius auratus. Soc. Med. Milit. Franc. 7: 231-235. (In French).

Dunaway, P. B. 1962. Hematology of cotton rats on a radioactive area. Abstr. Bull. Ecol. Soc. Am. 43(4): 125.

Dunaway, P. B. 1962. Litter-size record for eastern harvest mouse. J. Mammal. 43(3): 428-429. (RDA 5: 305) (Observations at Oak Ridge Nat. Lab.).

Dunaway, P. B. 1965. Radiosensitivity of indigenous rodents in eastern Tennessee. Abstr. Bull. Ecol. Soc. Am. 46(3): 121.

Dunaway, P. B. 1966. Problems of evaluating consequences of irradiation on populations of native rodents. pp. 11-18. In: Radiation Effects on Natural Populations. (See Sacher, 1966).

Dunaway, P. B. 1968. Life history and populational aspects of the eastern harvest mouse. Am. Midland Nat. 79: 48-67. (At Oak Ridge Nat. Lab.).

Dunaway, P. B. and S. V. Kaye. 1961. Studies of small-mammal populations on the radioactive White Oak Lake bed. Trans. N. Am. Wildl. Conf. 26: 167-185. (WR 104: 14).

Dunaway, P. B. and S. V. Kaye. 1963. Effects of ionizing radiation on mammal populations on the White Oak Lake bed. pp. 333-338. In: Radioecology. (See Schultz and Klement, 1963) (NSA 17: 33581).

Dunaway, P. B. and S. V. Kaye. 1963. Effects of ionizing radiation on mammalian population dynamics. pp. 141-149. In: Proc. 1st Conf. on Wildlife Disease. (High View, N. Y., June 24-27, 1962).

Dunaway, P. B. and L. L. Lewis. 1965. Taxonomic relation of erythrocyte count, mean corpuscular volume, and body-weight in mammals. Nature 205 (4970): 481-484. (Animals in the Oak Ridge Nat. Lab. area).

Dunaway, P. B. and M. G. White. (Eds.). 1974. Dynamics of Plutonium in Desert Environments. (Nevada Applied Ecology Group Progress Report as of January 1974). Nevada Operations Office, U. S. AEC report NVO-AEIC-74-1; NVO-142. 377pp. (NSA 30: 29439).

Dunaway, P. B., J. D. Story and J. T. Kitchings. 1973. Radiation effects and radionuclide excretion in a natural population of pine voles. pp. 1055-1064. In: Radionuclides in Ecosystems. (See Nelson, 1973) (NSA 27: 12329).

Duncan, D. W. and A. Bruynesteyn. 1971. Microbiological leaching of uranium. pp. 55-61. In: Selected Papers from 1970 Uranium Symp. R. J. Roman, comp. N. Mex. State Bur. Mines and Mineral Resources, Socorro. (NSA 26: 35889).

Dunham, C. L. 1961. Statement. pp. 4-9. In: Applications of Radioisotopes and Radiation in Life Sciences. (See U. S. Congress, 1961).

Dunham, C. L. 1968. The long-range effects session. pp. 405-407. In: Proc. Symp. on Postattack Recovery from Nuclear War. (See U. S. Office of Civil Defense, 1968).

Dunham, R. J. and P. H. Nye. 1974. The influence of soil water content on the uptake of ions by roots. II. Chloride uptake and concentration gradients in soil. J. Appl. Ecol. 11(2): 581-595. (^{36}Cl).

Dunn, D. B. and D. L. Spellman. nd. The revegetation at the New Mexico atomic bomb crater. Univ. of Missouri, U. S. AEC report COO-1296-1. 47pp. (NSA 19: 24236).

Dunn, D. B., M. R. Lewis and S. H. Gordon. 1950. Methods and background values for beta-gamma radioactivity in some plants from New Mexico and California. Univ. of California, Los Angeles, U. S. AEC report UCLA-75. 23pp. (NSA 4: 5566).

Dunning, G. M. (Ed.). 1957. Radioactive Contamination of Certain Areas in the Pacific Ocean from Nuclear Tests. U. S. AEC report. 60pp. (NSA 11: 12638).

Dunning, G. M. 1958. Radiation exposures from nuclear tests at the Nevada Test Site. Health Physics 1(3): 255-267. (Air and water data).

Dunson, W. A. 1969. Concentration of sodium by freshwater turtles. pp. 191-197. In: Symposium on Radioecology. (See Nelson and Evans, 1969).

Dunster, H. J. 1962. A review of procedures for disposal of radioactive waste, 1961. U. K. Atomic Energy Authority, Harwell, report AHSB(RP)-R-29. 37pp. (NSA 17: 15518).

Dunster, H. J. 1969. United Kingdom studies on radioactive materials released in the marine environment. pp. 145-153. In: Biological Imlications of the Nuclear Age. A. M. Golden, ed. U. S. AEC Symposium Ser. 16, report CONF-690303.

Dunster, H. J. 1970. Environmental monitoring in emergencies. pp. 151-161. In: Environmental Surveillance in the Vicinity of Nuclear Facilities. (See Reinig, 1970) (NSA 25: 16171).

Dunster, H. J. 1970. Avoidance of pollution in the dispersal of radioactive wastes. pp. 159-164. In: Northwest Conf. on the Role of Nuclear Energy. (See Governor's Nucl. Develop. Coord. Comm., 1970) (NSA 25: 6708, 23224).

Dunster, H. J. 1970. Objectives of environmental monitoring in the vicinity of nuclear facilities. pp. 5-11. In: Environmental Surveillance in the Vicinity of Nuclear Facilities. (See Reinig, 1970) (NSA 25: 16155).

Dunster, H. J. 1971. Environmental monitoring: British policy and procedures. pp. 427-436. In: Environmental Aspects of Nuclear Power Stations. (See IAEA, 1971).

Dunster, H. J. 1971. Some environmental aspects of nuclear power. U. K. National Radiological Protection Board, Harwell. pp. 25-49. (NP-19476) (NSA 27: 14215).

Dunster, H. J. and T. F. Johns. 1972. Principles of derived working limits and investigation levels in rationalizing the design and interpretation of monitoring programmes. pp. 1743-1754. In: Health Physics Operational Monitoring, Vol. 3. C. A. Willis, ed. Gordon & Reach, New York. (NSA 29: 2914).

Duoll, J. and K. P. DuBois. 1953. Influence of hibernation on survival time and weight loss of x-irradiated ground squirrels. Proc. Soc. Exptl. Biol. Med. 84(2): 367-370. (NSA 8: 1289).

Dupain-Klerkx, L. and G. Germai. 1969. Radiocarbon dating with liquid scintillation. Bull. Soc. Roy. Sci. Liége 38: 50-53. (In French) (NSA 23: 38102).

Duplessy, J. C., L. Chenouard and F. Vila. 1975. Weyl's theory of glaciation supported by isotopic study of Norwegian Core K 11. Science 188 (4194): 1208-1209. (NSA 32: 11549).

Dupuy, P. 1963. Microorganisms resistant to ionizing radiation. Int. J. Appl. Radiat. Isotop. 14(1): 29-35. (In French) (NSA 17: 24936).

Duquesne Light Company. 1962. Environmental radiation monitoring quarterly report, first quarter, Jan.-Mar. 1962. Shippingport, Pa., U. S. AEC report TID-15632. 11pp. (NSA 16: 19185).

Duquesne Light Company. 1971. Beaver Valley Power Station, Unit No. 1. Environmental report, operating license stage. Pittsburgh, Pa., U. S. AEC DOCKET-50334-25. 356pp. (NSA 25: 56962).

Durand, G. L. and B. Guillet. 1966. Dating of the Beillard peat bog by the ^{14}C method. Compt. rend. D262: 28-31. (In French) (NSA 20: 16712).

Duranov, V. S. and I. V. Abramov. 1965. Effects of external and internal irradiation on sheep resistance to causative agents of babesiasis. Veterinariya 42: 41-44. (In Russian) (NSA 22: 49570).

Durham, R. W. 1974. Radionuclide levels in the Great Lakes waters, 1973. Canada Dept. of the Environment, Ottawa, report EC-SS-48. 5pp. (NSA 32: 14307; 33: 29146).

Duron, A. E. 1974. Review of work to combat the Mediterranean fruit fly carried out in Central America and Panama. pp. 17-20. In: Sterile-Insect Technique and Its Field Applications. (See IAEA, 1974) (NSA 31: 8936).

Duryee, W. 1947. The effect of x-rays on chromosomes and nucleoli in ovarian eggs of the salamander, *Triturus pyrrogaster*. Abstr. Biol. Bull. 93 (1): 206-207.

Duryee, W. R. 1949. The nature of radiation injury to amphibian cell nuclei. J. Natl. Cancer Inst. 10: 735-796. (NSA 4: 2516).

Durzan, D. J. and F. C. Steward. 1963. The nitrogen metabolism and seasonal changes in shoots of *Picea glauca* (Moench) Voss. Abstr. Plant Physiol. 38(Suppl.): vi.

Durzan, D. J., A. J. Mia and B. S. P. Wang. 1971. Effects of tritiated water on the metabolism and germination of jack pine seeds. Can. J. Bot. 49(12): 2139-2149.

Dushauskiyene-Duzh, N. F., G. G. Polikarpov and B. I. Styro. 1969. Coefficients of accumulation of ^{90}Sr in certain fish (radioecological investigation). Radiobiologiya 9(1): 113-115. (AEC-tr-7024; pp. 146-149) (NSA 23: 24587).

Dushauskiyene-Duzh, N. F., et al. 1972. Radionuclide uptake of some freshwater hydrobionts. Acad. Sci. Lithuanian SSR 3(59): 201-212. (In Russian, English summary).

Dushauskiyene-Duzh, N. F., et al. 1973. Role of water plants in migration of strontium-90 in some reservoirs of the Lithuanian SSR. pp. 60-72. In: Radioecology of Water Organisms. II. Distribution and Migration of Radionuclides in Freshwater and Sea-Water Biocenoses. (See Andrumaitis, 1973) (In Russian) (NSA 30: 21132).

Dustan, G. G. 1965. Tagging the oriental fruit moth, *Grapholitha molesta* (Busck) with radioactive phosphorus for flight and dispersal studies. Can. Entomol. 97(8): 810-816.

Dustan, G. G. 1966. Effects of tagging amounts of radioactive phosphorus on adults of the oriental fruit moth, *Grapholitha molesta* (Busck) (Lepidoptera: Tortricidae). Can. Entomol. 98: 305-311.

du Toit, G. J. and D. B. Bredenkamp. 1972. Use of radioactive isotopes as groundwater tracers to determine the effective porosity of groundwater aquifers. pp. R24.b1-R24.b9. In: National Conf. on the Technological Applications of Nuclear Techniques, 1972. (See Union of South Africa, 1972) (CONF-721091) (NSA 28: 21126).

Dutton, C. E. 1906. Volcanos and radioactivity. (Paper presented to the National Academy of Sciences). The Englewood Times (N. J.), Apr. 17, 1906. 12pp.

Dutton, J. W. R. 1966. Detection of $^{234}Th/^{234m}Pa$ in sediments from the Ribble Estuary, Lancashire, England. Nature 211: 1395-1396. (NSA 21: 10484).

Dutton, J. W. R. 1968. Gross beta counting of environmental materials. England Fisheries Radiobiological Lab., Lowestoft, report FRL-3. 13pp. (NSA 23: 21936).

Dutton, J. W. R. 1970. Determination of radiocaesium in sea and fresh waters. England Fisheries Radiobiological Lab., Lowestoft, report FRL-6. 9pp. (NSA 25: 31380).

Dutton, J. W. R. and N. T. Mitchell. 1971. Rapid methods for specific radionuclide analysis and their application to aquatic emergency conditions. pp. 745-755. In: Rapid Methods for Measuring Radioactivity in the Enviroment. (See IAEA, 1971) (NSA 26: 27827).

Dutton, J. W. R., et al. 1974. Analytical systems applied to monitoring the aquatic environment in the control of radioactive waste disposal. pp. 155-167. In: Environmental Surveillance Around Nuclear Installations. Vol. 1. (See IAEA, 1974) (NSA 31: 14073).

Duursma, E. K. 1966. Molecular diffusion of radioisotopes in interstitial water of sediments. pp. 355-371. In: Disposal of Radioactive Wastes into Seas, Oceans and Surface Waters. (See IAEA, 1966) (SM-72/20) (CONF-660507) (NSA 20: 45910).

Duval, J. S., Jr. T. Schwarzer and J. A. S. Adams. 1971. Lognormal distribution of trace elements in the environment. pp. 120-131. In: Trace Substances in Environmental Health. IV. (See Hemphill, 1971) (NSA 26: 15183) (^{137}Cs).

Duxbury, R. E. and E. H. Sadun. 1963. Effects of gamma radiation on development of *Dirofilaria uniformis* in *Anopheles quadrimaculatus*. Proc. Helminthol. Soc. 30: 263-265. (NSA 18: 13537).

Dvorak, P., et al. 1972. Radiocesium elimination in the plant-animal-products chain. Czechoslovakia Vyzkumny Ustav Zivocisne Vyroby, report VUZU-1. 103pp. (In Czech) (NSA 30: 9557).

Dyck, W. and A. Y. Smith. 1969. The use of ^{222}Rn in surface waters in geochemical prospecting for uranium. Quart. Colo. Sch. Mines 64: 223-236. (NSA 23:35325).

Dyer, K. F. 1966. Fitness and competitive ability in irradiated populations of *Drosophila melanogaster*. Mutation Research 3: 327-339. (NSA 20: 42922).

Dyer, K. F. 1969. The effect of radiation on small competing populations of *Drosophila melanogaster*. I. The accumulation of genetic damage. II. The changing frequencies of autosomal recessive lethals. III. Changes in competitive ability. Genetics 61: 227-244, 245-274, 275-291. (NSA 23: 33671, 33670, 33669).

Dyer, K. F. 1971. Effects of radiation on populations of *Drosophila melanogaster* with different genetic structures. I. Changes in fitness and productivity. II. Temporal changes in the sex ratio. III. Interaction with population size. Austral. J. Biol. Sci. 24(3): 565-573, 575-583, 585-591. (NSA 25: 49270, 49271, 49155).

Dyer, K. F. 1971. Effect of radiation on small competing populations of *Drosophila melanogaster*. IV. Changes in longevity. Genetics 68(4): 559-569. (NSA 25: 57961).

Dzeyeva, V. P. and S. M. Mashtakov. 1970. Herbicide effects of sodium trichloro-acetate produced by the radiochemical method. Vestsi Akad. Navuk Belorussk. SSR (biol.) (3): 61-67. (In Belorussian) (NSA 24: 50419).

Eakins, J. D. and A. E. Lally. 1970. The simultaneous determination of plutonium alpha activity and ^{241}Pu in biological materials by gel scintillation counting. U. K. Atomic Energy Authority, Harwell, report AERE-R-6640. 17pp. (NSA 25: 20927).

Eales, J. G. 1963. A comparative study of thyroid function in migrant juvenile salmon. Can. J. Zool. 41(5): 811-824. (SFA 8: 6025) (^{131}I).

Eales, J. G. 1964. The influence of temperature on thyroid histology and radioiodine metabolism of yearling steelhead trout, *Salmo gairdneri*. Can. J. Zool. 42: 829-841. (NSA 19: 5609).

Eales, J. G. 1969. *In vivo* uptake of radiothyroxine by the tissues of Atlantic salmon (*Salmo salar* L.) parr, presmolt, and smolt. Can. J. Zool. 47(1): 9-16.

Eales, J. G. 1970. Biliary excretion of radiothyroxine by the brook trout, *Salvelinus fontinalis* (Mitchill). Gen. Comp. Endocrinol. 14: 385-395. (NSA 24: 48378).

Eales, J. G. 1972. Radiothyroxine metabolism in several freshwater teleost fishes. Can. J. Zool. 50(5): 623-631. (SFA 17: 15379).

Eales, J. G., L. A. Welsh and H. H. Chan. 1971. Biliary excretion of 3,5,3'-triiodo-L-thyronine-^{125}I by the brook trout, *Salvelinus fontinalis* (Mitchill). Gen. Comp. Endocrinol. 16(2): 169-175. (SFA 18: 15891).

Ealy, R. P. 1957. Introduction of radioactive solutions into trees and shrubs. Okla. Agr. Expt. Sta., Stillwater, Tech. Publ. 70. 4pp. (NSA 13: 1927).

Eapen, J. and N. Narayanan. 1971. Gross and subcellular distribution of ^{137}Cs in pigeon (*Columba livia*) tissues with special reference to muscles. J. Radiat. Res. 12(2): 51-55. (NSA 26: 18092).

Eapen, J. and C. K. Pushpendran. 1972. Effect of temperature acclimation, tail amputation, and x-irradiation of thyroid gland of the lizard, *Hemidactylus leschenaulti*. Indian J. Exptl. Biol. 10(4): 253-256. (NSA 27: 17659).

Earle, N. W. 1964. {Sterol utilization in the boll weevil}. Abstr. Bull. Entomol. Soc. Am. 10(3): 164. (*Anthonomus grandis*) (Tracer experiment).

Early, A. M. 1970. NAL 1970 archeological survey: the prehistoric occupations of the National Accelerator Laboratory site. National Accelerator Lab., Batavia, Ill., U. S. AEC report NAL-55. 9pp. (NSA 25: 28862).

Easley, C. W. 1969. *Basic Radiation Protection. Principles and Organization.* Gordon & Breach, New York. 137pp. (NSA 24: 25470).

Easter, S. S., Jr. and F. Hutchinson. 1961. Effects of radiations of different LET on *Artemia* eggs. Radiation Research 15(3): 333-340. (Also: School of Aerospace Medicine, Brooks Air Force Base, Tex., report 61-83) (AD-287066).

Eaton, J. L. 1966. Temperature and the action of DDT on the cockroach nervous system. Thesis, Univ. of Illinois, Urbana. 75pp. (Dissertation Abstr. 27(11): 4097-B - 4098-B) (^{14}C).

Eaton, J. L. and J. G. Sternburg. 1967. Uptake of DDT by the American cockroach central nervous system. J. Econ. Entomol. 60(6): 1699-1703. (^{14}C).

Ebeling, G. and P. Seyfried. 1971. Portable multichannel analyser for field measurements of environmental radioactivity. pp. 539-548. In: *Rapid Methods for Measuring Radioactivity in the Environment*. (See IAEA, 1971) (In German) (NSA 26: 28287).

Eberhardt, L. L. 1964. Variability of the strontium-90 and caesium-137 burden of native plants and animals. Nature 204(4955): 238-240. (NSA 19: 1997).

Eberhardt, L. L. 1965. Notes on ecological aspects of the aftermath of nuclear attack. pp. 13-25. In: Problems in Estimating the Biological Consequences of Nuclear War (Three Papers). (See Hollister and Eberhardt, 1965).

Eberhardt, L. L. 1965. Notes on the analysis of natural systems. pp. 27-40. In: Problems in Estimating the Biological Consequences of Nuclear War (Three Papers). (See Hollister and Eberhardt, 1965).

Eberhardt, L. L. 1966. Methodological problems in comparing natural and laboratory populations. pp. 9-10. In: *Radiation Effects on Natural Populations*. (See Sacher, 1966).

Eberhardt, L. L. 1967. Some ecological aspects of nuclear war. Battelle Northwest Lab., U. S. AEC report TID-23939. (NSA 22: 28395).

Eberhardt, L. L. 1972. Sampling for radionuclides and other trace substances. Battelle Northwest Lab., U. S. AEC report BNWL-SA-4362. 24pp. (For: Advanced Institute on Statistical Ecology Conf., University Park, Pa., 15 Jul 1972) (CONF-720715-1) (NSA 56079).

Eberhardt, L. L. 1973. Modeling radionuclides and pesticides in food chains. pp. 894-897. In: *Radionuclides in Ecosystems*. (See Nelson, 1973) (NSA 25: 37747).

Eberhardt, L. L. and R. O. Gilbert. 1972. Statistical analysis of soil plutonium studies, Nevada Test Site. Battelle Northwest Lab., U. S. AEC report BNWL-B-217. 60pp. (NSA 27: 306).

Eberhardt, L. L. and R. O. Gilbert. 1973. Gamma and lognormal distributions as models in studying food-chain kinetics. Battelle Northwest Lab, U. S. AEC report BNWL-1747. 100pp. (NSA 28: 10876) (Incl. radionuclides).

Eberhardt, L. L. and R. O. Gilbert. 1974. General statistical considerations in environmental plutonium studies. pp. 43-49. In: Dynamics of Plutonium in Desert Environments. (See Dunaway and White, 1974) (Also: Battelle Northwest Lab., U. S. AEC report BNWL-SA-4810) (NSA 29: 18347; 30: 29418) (Soil studies).

Eberhardt, L. L. and R. E. Nakatani. 1968. A postulated effect of growth on retention time of metabolites. J. Fish. Res. Bd. Can. 25(3): 591-596. (Radionuclides).

Eberhardt, L. L. and R. E. Nakatani. 1969. Modeling the behavior of radionuclides in some natural systems. pp. 740-750. In: Symposium on Radioecology. (See Nelson and Evans, 1969) (Also: Battelle Northwest Lab., U. S. AEC report BNWL-SA-1258) (NSA 21: 28342).

Eberhardt, L. L., R. L. Meeks and T. J. Peterle. 1971. Food chain model for DDT kinetics in a freshwater marsh. Nature 230(5288): 60-62. (^{36}Cl).

Eberhardt, L. L., et al. 1969. A study of fallout cesium-137 in the Pacific Northwest. J. Wildl. Mgmt. 33(1): 103-112.

Eberle, D. O. and F. A. Gunther. 1965. Chromatographic, spectrometric, and irradiation behavior of five carbamate insecticides. J. Assoc. Off. Agr. Chem. 48: 927-937. (NSA 20: 22593).

Eberline Instrument Company. 1969. Project Gasbuggy: on-site radiological safety during production testing, January 25, 1968 to December 31, 1969. U. S. AEC report PNE-1006. 44pp. (NSA 26: 400).

Eberline Instrument Corporation. 1970. Project Rulison: on-site radiological safety report, April 21 - December 31, 1969. Santa Fe, N. Mex. U. S. AEC report NVO-294-4; PNE-R-35. 17pp. (NSA 24: 48199).

Eberline Instrument Corporation. 1971. H. B. Robinson Unit No. 2. Environmental monitoring report, October 1, 1970 - March 31, 1971. Santa Fe, N. Mex. U. S. AEC DOCKET-50261-61. 31pp. (NSA 25: 36932).

Eberline Instrument Corporation. 1971. Point Beach Nuclear Plant, Unit No. 1. Environmental monitoring program preoperational report. Santa Fe, N. Mex. U. S. AEC DOCKET-50266-41. 61pp. (NSA 25: 15297).

Eberline Instrument Corporation. 1971. Vermont Yankee Nuclear Power Station. Environmental monitoring report ---. Santa Fe, N. Mex.

No.	Period	U. S. AEC DOCKET-	pp.	NSA
1	July - October, 1969	50271-42	39	25: 25920
2	October - December, 1969	50271-43	24	25: 25921
3	January - March, 1970	50271-44	27	25: 25922
4	April - June, 1970	50271-45	25	25: 25923
5	July - September, 1970	50271-46	19	25: 25924
6	October - December, 1970	50271-47	23	25: 25925

Eberline Instrument Corporation. 1972-74. Environmental monitoring report: period covering ----. Santa Fe, N. Mex. El Paso Natural Gas Co. (Incl. deer, moose, elk, antelope, sage grouse; Project Wagon Wheel, vicinity of Pinedale, Wyo.) (See also: El Paso Natural Gas Co.).

Period	U. S. AEC Report	pp.	NSA
Feb. 1 - Apr. 30, 1972	PNE-WW-22	22	27: 7567
May 1 - Jul. 31, 1972	PNE-WW-23	22	27: 7562
Aug. 1 - Oct. 31, 1972	PNE-WW-24	22	27: 5590
Nov. 1 - Jan. 31, 1973	PNE-WW-28	26	28: 10868
Feb. 1 - Apr. 30, 1973	PNE-WW-30	24	28: 21069
May 1 - Jul. 31, 1973	PNE-WW-32	25	29: 15663
1971-1973	PNE-WW-36	16	30: 23983

Eberline Instrument Corporation. 1972-74. Radiological monitoring program: period covering ----. Santa Fe, N. Mex. CER Nuclear Corp. (Project Rio Blanco) (See also: CER Nuclear Corp.).

Period	U. S. AEC Report	pp.	NSA
Oct. 1 - Dec. 31, 1971	PNE-RB-51-1	26	29: 15658
Jan. 1 - Mar. 31, 1972	PNE-RB-51-2	17	29: 15659
Apr. 1 - Jun. 30, 1972	PNE-RB-51-3	22	29: 15660
Jul. 1 - Sep. 30, 1972	PNE-RB-51-4	18	29: 15661
Feb. 15 - May 16, 1973	PNE-RB-51-5	23	29: 15662
May 17 - Jul. 31, 1973	PNE-RB-51-6	--	29: 15593
Nov. 1 - Feb. 4, 1974	PNE-RB-51-8	19	30: 9529
Feb. 4 - Apr. 30, 1974	PNE-RB-51-9	20	30: 29348
May 1 - Jul. 29, 1974	PNE-RB-51-10	16	31: 3328

Ebert, M. and A. Howard. 1961. Modification of radiosensitivity by oxygen, narcotic gases and inert gases. pp. 1039-1043. In: 9th Intern. Congr. of Radiology, Vol. 2. (NSA 15: 24729).

Ebner, H. G. and W. Schwartz. 1974. Geomicrobiological studies. XII. Behavior of microorganisms on uranium-containing rocks. Z. Allg. Mikrobiol. 14(2): 93-102. (In German) (NSA 31: 687).

Echevin, M. 1975. Contribution to the chemistry of antarctic snow: trace element determination by neutron activation. France Centre National de la Recherche. (English transl. AD/A-007106; 84pp.) (NSA 33: 25340).

Echo, J. and D. B. Hawkins. 1966. Algal influence on radionuclides in settling ponds. Nature 209: 1105-1107. (NSA 24815).

Eckart, L. E. and J. H. Leonard. 1972. Method for assessing the relative environmental importance of individual radionuclides. Trans. Amer. Nucl. Soc. 15(1): 81-82. (NSA 26: 42966).

Eckbo, J. K. 1975. Nuclear power plants and environmental protection. Tek. Ukebl. 122(16): 15-16. (In Norwegian) (NSA 32: 8964).

Eckelman, W. R., et al. 1962. Implications of carbon isotopic composition of total organic carbon of some recent sediments and ancient oils. Bull. Am. Assoc. Petrol. Geol. 46: 699-704. (NSA 20: 45917).

Eckert, J. A. and R. B. Evans. 1973. Tritium burdens in two arctic villages. Radiat. Data Rept. 14(5): 273-275. (NSA 28: 10875).

Eckert, J. A., et al. 1970. ^{137}Cs concentrations in Eskimos, spring 1968. Radiol. Hlth. Data 11: 219-225. (NSA 24: 34234).

Eckert, R. E., Jr. and C. R. Blincoe. 1970. Radioisotope uptake by selected range forage and weed species. J. Range Mgmt. 23(5): 367-369.

Eckhoff, N. D., C. J. Pappas and C. W. Deyoe. 1972. Comparison of neutron activation analysis and other analyses procedures for fish samples. Anal. Chem. 44(8): 1506-1508. (NSA 26: 40579).

Economopoulos, A. P. 1972. Sexual competitiveness of x-ray sterilized males of Dacus oleae. Mating frequency of artificially reared and wild females. Environ. Entomol. 1(4): 490-497.

Eden, G. E. and R. Briggs. 1967. Radioisotope techniques developed in water-pollution studies. pp. 191-206. In: Isotopes in Hydrology. (See IAEA, 1967) (NSA 22: 19061).

Edds, G. T., L. E. Swanson and H. Engelbrecht. 1963. Immunization experiments with irradiated lungworm larvae. Am. J. Vet. Res. 24(98): 139-142. (NSA 17: 23100).

Eddy, G. W., et al. 1975. Attractants for the screwworm: irradiation effects on bacteria-inoculated media. Ann. Entomol. Soc. Am. 68(2): 269-270. (NSA 31: 26195).

Eden, G. E., G. H. J. Elkins and G. A. Truesdale. 1954. Removal of radioactive substances from water by biological treatment processes. Atomics 5: 133-142.

Edgar, W. F. 1963. Radium content of drinking water in the central Savannah River area. Savannah River Plant, U. S. AEC report DPSPU-63-30-10. 8pp. (NSA 17: 20239).

Edgington, D. N. and H. F. Lucas, Jr. 1969. A system for the neutron activation analysis of trace elements in samples of biological and environmental interest. Argonne Nat. Lab. Rev. 5: 98-105. (NSA 24: 11485).

Edgington, D. N. and H. F. Lucas, Jr. 1971. Semi-automatic method of neutron activation analysis with application to environmental studies. pp. 311-322. In: Nuclear Techniques in Environmental Pollution. (See IAEA, 1971) (NSA 25: 31876).

Edgington, D. N. and J. A. Robbins. 1974. Application of natural and fallout radionuclides to studies of sedimentation in Lake Michigan. pp. 443-458. In: Proc. 2nd Intern. Conf. on Nuclear Methods in Environmental Research. (See Vogt and Meyer, 1974) (NSA 33: 23278).

Edgington, D. N. and J. A. Robbins. 1975. Behavior of plutonium and other long-lived radionuclides in Lake Michigan. II. Patterns of deposition in the sediments. Argonne National Lab., U. S. AEC report CONF-750662-1. 18pp. (NSA 32: 24343).

Edgington, D. N., J. C. Ritchie and J. A. Robbins. 1975. Comments on the paper: "Use of rivers to predict accumulation in sediment of radionuclides discharged from nuclear power plants" by P. Plato. Health Physics 29(3): 429-431. (NSA 33: 629).

Edgington, D. N., et al. 1975. Plutonium and americium in Lake Michigan sediments. Argonne National Lab., U. S. AEC report CONF-751105-9. 14pp. (NSA 33: 9258).

Edmundson, E. H., Jr. 1971. Effects of acute gamma radiation and temperature on growth and survival of juvenile rainbow trout (Salmo gairdneri). Thesis, Washington State Univ., Pullman. ix, 48pp.

Edney, E. B., W. Allen and J. McFarlane. 1974. Predation by terrestrial isopods. Ecology 55(2): 428-433. (^{32}P).

Edouard, L. 1972. Electrical energy of thermal origin and the environment. Pt. 2. Thermal nuclear power plants. Tech. Mod. 64(10): 31-32. (In French) (NSA 28: 2570).

Edwards, B. A. B. 1961. The fruit fly problem in Australia. Outlook Agr. 3(4): 116-122. (Comments on irradiated males).

Edwards, C. A. 1969. Effects of gamma irradiation on populations of soil invertebrates. pp. 68-77. In: Symposium on Radioecology. (See Nelson and Evans, 1969).

Edwards, C. A., D. E. Reichle and D. A. Crossley, Jr. 1970. The role of soil invertebrates in turnover of organic matter and nutrients. pp. 147-172. In: Analysis of Temperate Forest Ecosystems. D. E. Reichle, ed. Springer-Verlag, New York. (Comments on radiotracers).

Edwards, J. M. and L. E. Holter. 1954. Radio-isotopes for water-input profiles, in water flood operations. Oil Gas J. 53: 60.

Edwards, M. A. 1966. Tabulations of data on announced nuclear detonations by all nations through 1965. Lawrence Radiation Lab., Livermore, U. S. AEC report UCRL-14786. 25pp. (NSA 20: 35670).

Edwards, S. A. and A. T. Wilson. 1965. Carbohydrate metabolism in the sapwood of Populus deltoides. Biochim. Biophys. Acta 97(2): 340-341. (NSA 19: 17463) (Used ^{14}C dating).

Eeche, A. 1970. Test with Drosophila melanogaster on somatic resistance to x-ray irradiation. Hereditas 66: 71-78. (NSA 25: 29613).

Effenberger, E. and J. Haschke. 1967. Measurements of the radioactivity of surface water and sewage in Hamburg. Arch. Hyg. Bakteriol. 151: 740-752. (In German) (NSA 23: 24305).

Egami, N. 1963. Studies on variation in number of anal fin rays in Oryzias latipes. Effect of x-irradiation. Copeia (3): 592-593. (NSA 19: 102).

Egami, N. 1967. Cell population kinetics and aging with special reference to effects of temperature and radiation on cell systems in fish. Jikken Keitaigakushi 21: 366-378. (In Japanese) (NSA 23: 33855).

Egami, N. 1969. Kinetics of recovery from injury after whole-body x-irradiation of the fish Oryzias latipes at different temperatures. Radiation Research 37: 192-201. (NSA 23: 16416).

Egami, N. 1972. Repair potency and aging in organisms. Change in regeneration rate with age in the teleost, Oryzias latipes. Saibo Seibutsugaku Shinpojiumu 23: 203-206. (In Japanese) (NSA 28: 30225) (X-rays).

Egami, N. (Ed.). 1973. Radioactivity and Fishes (Contamination, Injury and Utilization). Koseisha-koseikaku Publ. Co., Tokyo. 398pp. + 2 plates. (In Japanese).

Egami, N. 1973. Heredity effect of external radiation. pp. 174-201. In: Radioactivity and Fishes. (See Egami, 1973) (In Japanese) (NSA 29: 13025).

Egami, N. and K. Eto. 1973. Somatic effect of external irradiation on mature fish (compendium). pp. 65-91. In: Radioactivity and Fishes. (See Egami, 1973) (In Japanese) (NSA 29: 13024).

Egami, N. and A. Hama. 1975. Dose-rate effects on the hatchability of irradiated embryos of the fish, Oryzias latipes. Int. J. Radiat. Biol. 28(3): 273-278. (NSA 33: 20526).

Egami, N. and H. Etoh. 1962. Dose-survival time relationship and protective action of reserpine against x-irradiation in the fish, Oryzias latipes. Annot. Zool. Japan 35(4): 188-198. (BA 45: 4682; SFA 9: 6695).

Egami, N. and Y. Hyodo. 1965. Inhibitory effect of x-irradiation on the development of the ovaries of the fish, Oryzias latipes, in sexually inactive seasons. Annot. Zool. Japan 38: 8-11. (NSA 20: 14307).

Egami, N. and Y. Hyodo-Taguchi. 1969. Hermaphroditic gonads produced in Oryzias latipes by x radiation during embryonic stages. Copeia (1): 195-196. (NSA 23: 45910).

Egami, N. and Y. Hyodo-Taguchi. 1973. Inhibition of formation of spermatozoa of killifishes bred in tritium solution. pp. 6-17. In: Investigation of the Effects of Low Concentrations of Radionuclides on Aquatic Life. Nuclear Safety Research Assoc., Tokyo. (In Japanese) (NSA 31: 6360).

Egami, N. and A. Kazuko. 1966. Effects of x-irradiation of a part of the body of the ovary in the loach, Misgurnus anguillicaudatus. Annot. Zool. Japan 39: 7-15. (NSA 21: 433).

Egami, N., K. Eto and Y. Hyodo-Taguchi. 1973. Effects of external irradiation on the bodies of adult fishes (consideration in detail). pp. 92-173. In: Radioactivity and Fishes. (See Egami, 1973) (In Japanese) (NSA 29: 13023).

Egami, N., Y. Hyodo-Taguchi and K. Konno. 1967. Autoradiographical studies on spermatogenesis in the fish, Oryzias latipes, with special reference to radiation effects on fertility of fish. Gunma Symp. Endocrinol. 4: 147-159. (NSA 23: 913).

Egami, N., T. Ohshima and Y. H. Nakanishi. 1965. Inhibitory effects of x-irradiation on the development of male characteristics in females of the teleost, Oryzias latipes. Jap. J. Zool. 14: 31-43. (NSA 20: 16263).

Egami, N., et al. 1962. Relationship between survival time after exposure to x-rays and dose of the irradiation in the fishes, Carassius auratus and Oryzias latipes. Dobutsugaku Zasshi 71: 313-321. (In Japanese) (NSA 19: 2098).

Egami, N., et al. 1962. Role of the pituitary gland in melanization in the skin of the goldfish, Carassius auratus, induced by x-ray irradiation. Proc. Jap. Acad. 38: 345-347. (In English) (NSA 17: 12167).

Egan, J. T. and G. E. Peterson. 1972. Trace-metal analysis using atomic absorption spectrophotometry. J. Am. Water Works Assoc. 64(10): 669-674. (NSA 27: 70) (Water quality control).

Egeberg, J. 1965. Iodine-concentrating cells in the endostyle of Ammocoetes. Z. Zellforsch. Mikroskop. Anat. 68: 102-115. (NSA 20: 39127).

Eggert, R. C. 1966. Response of x-irradiated limbs of adult urodeles to autografts of normal cartilage. J. Exptl. Zool. 161: 369-389. (NSA 21: 28603).

EG & G, Inc. 1972. Background report and bibliography for a study to forecast nuclear power plant effects on coastal zones. Woods Hole, Mass., report ESTR-1-72. 79pp. (NSA 28: 2610).

EG & G, Inc. 1972. Radiological survey of the Lake Ontario Ordnance Works and vicinity in June, 1972. Technical Report No. L-1076. U. S. AEC report EGG-1183-1554. 22pp. (NSA 27: 9553).

EG & G, Inc. 1972. Radiological survey of the Nevada Test Site. (Survey period: 1970-1971). Technical Report No. L-1064. Las Vegas, Nev., U. S. AEC report EGG-1183-1552. 104pp. (NSA 28: 8149).

EG & G, Inc. 1973. Aerial radiological measuring survey of the Lake Ontario Ordnance Works and vicinity in June 1972. U. S. AEC report ARMS-72.6.1. 24pp. (NSA 28: 21063).

EG & G, Inc. 1973. ARMS: aerial radiological measuring survey of the area surrounding the Point Beach Nuclear Plant, Two Creeks, Wisconsin, August 16 and 17, 1970. U. S. AEC report ARMS-70.6.9. 16pp. (NSA 29: 380).

EG & G, Inc. 1973. ARMS: radiological survey of the area surrounding the Nine Mile Point Nuclear Station, Scriba, New York. U. S. AEC report EGG-1183-1592. 19pp. (NSA 29: 372).

EG & G, Inc. 1973. ARMS: radiological survey of the area surrounding the Peach Bottom Atomic Power Station, Peach Bottom, Pennsylvania. U. S. AEC report EGG-1183-1593. 21pp. (NSA 29: 373).

EG & G, Inc. 1973. Aerial radiological measuring survey of the area surrounding the Enrico Fermi Atomic Power Plant, Monroe, Michigan, September, 1970. U. S. AEC report ARMS-70.6.7. 12pp. (NSA 29: 365).

EG & G, Inc. 1973. Aerial radiological measuring survey of the area surrounding the La Cross Boiling Water Reactor, Genoa, Wisconsin, July 1968. U. S. AEC report EGG-1183-1584; ARMS-68.6.5. 12pp. (NSA 29: 363).

EG & G, Inc. 1973. Aerial radiological measuring survey of the area surrounding the Monticello Nuclear Generating Plant, Monticello, Minnesota, August 1970. U. S. AEC report ARMS-70.6.6. 12pp. (NSA 29: 364).

EG & G, Inc. 1973. Aerial radiological measuring survey of the area surrounding the Pallisades Plant, South Haven, Michigan, July 19, 1970. U. S. AEC report EGG-1183-1585; ARMS-70.6.8. 12pp. (NSA 29: 366).

EG & G, Inc. 1973. Aerial radiological measuring survey of the area surrounding the Vermont Yankee Generating Station and the Yankee Nuclear Power Station, September 18, 1970. U. S. AEC report EGG-1183-1581; ARMS-70.6.11. 24pp. (NSA 29: 367) (Vernon, Vt.).

EG & G, Inc. 1973. Aerial radiological measuring survey of the Fort Calhoun Station, Fort Calhoun, Nebraska, August 9-10, 1972. U. S. AEC report EGG-1183-1596; ARMS-72.6.4. 12pp. (NSA 29: 368).

EG & G, Inc. 1973. Radiological survey of the area surrounding the Elk River Reactor, Elk River, Minnesota. Date of survey: 1,2 August, 24 September 1968. U. S. AEC report EGG-1183-1591. 21pp. (NSA 29: 382).

EG & G, Inc. 1973. Radiological survey of the area surrounding the Robert Emmett Ginna Nuclear Power Plant, Ontario, New York. Date of survey: 8 Sept. 1970. Technical Report No. L-1096. U. S. AEC report ARMS-70.6. 10. 12pp. (NSA 29: 2644).

EG & G, Inc. 1974. Aerial radiological measuring survey of the area surrounding the Hallam Nuclear Power Facility, Hallam, Nebraska, September 20, 1968. U. S. AEC report EGG-1183-1590; ARMS-68.6.7. 16pp. (NSA 30: 3155).

EG & G, Inc. 1974. Aerial radiological measuring survey of the area surrounding the Millstone Nuclear Power Station, Waterford, Connecticut, August 29 and September 11, 1969. U. S. AEC report ARMS-69.6.5. 16pp. (NSA 30: 6499).

EG & G, Inc. 1974. Aerial radiological measuring survey of the Maine Yankee Atomic Power Plant, September 1971. U. S. AEC report EGG-1183-1605; ARMS-71.6.5. 19pp. (NSA 30: 29357) (Wiscasset, Me.).

EG & G, Inc. 1974. Aerial radiological measuring surveys of the Turkey Point Station, April 1972. U. S. AEC report EGG-1183-1602. 19pp. (NSA 30: 29358) (Florida City, Fla.).

EG & G, Inc. 1974. Ground based observations supporting the Shippingport, Pennsylvania, aerial radiation survey of February 1973. U. S. AEC report ARMS-73.6.2. 16pp. (NSA 30: 6512).

EG & G, Inc. 1974. Aerial radiological measuring survey of Cooper Nuclear Station, August, 1972. U. S. AEC report ARMS-72.6.5. 18pp. (NSA 31: 3363) (Brownville, Nebr.).

EG & G, Inc. 1974. Aerial radiological measuring survey of the Fort St. Vrain Nuclear Generating Station, October, 1971. U. S. AEC report ARMS-72.6.9. 15pp. (NSA 31: 3364) (Platteville, Colo.).

EG & G, Inc. 1974. Aerial radiological measuring survey of the Indian Point Station, August, 1969. U. S. AEC report ARMS-69.6.4. 15pp. (NSA 31: 3360) (Indian Point, N. Y.).

EG & G, Inc. 1974. Aerial radiological measuring survey of the Pilgrim Station, September, 1971. U. S. AEC report ARMS-71.6.4. 15pp. (NSA 31: 3362) (Plymouth, Mass.).

EG & G, Inc. 1974. Aerial radiological measuring survey of the Prairie Island Nuclear Generating Plant, September, 1971. U. S. AEC report ARMS-71.6.2. 15pp. (NSA 31: 3361) (Red Wing, Minn.).

EG & G, Inc. 1974. Radiological survey of the area surrounding the Rocky Flats Plant, Golden, Colorado, 3-4 May and 6 October 1972. U. S. AEC report EGG-1183-1641. 20pp. (NSA 33: 603).

Egorova, T. A. 1965. Trehalose in insects. Uch. zap. Mosk. gos. ped. inst. 212: 14-24. (In Russian) (Refers to use of radionuclides).

Ehlermann, D. 1972. Possible identification of an irradiation treatment of fish by means of an electrical (AC) resistance measurement. J. Food Sci. 37(3): 501. (NSA 26: 58965).

Ehrenberg, L. and I. Granhall. 1952. Effects of beta-radiating isotopes in fruit trees. Hereditas 38: 385-419.

Ehrenberg, L. and N. Nybom. 1952. Chemical and biological effects of neutrons and x-rays. Hereditas 38: 482-501.

Ehrenberg, L., et al. 1955. Seed quality and principles of forest genetics. Hereditas 41: 291-366.

Ehrenfeld, D. W. 1970. Biological Conservation. Holt, Rinehart & Winston, New York. xi, 226pp. (See pp. 54-56, 65-66, 148) (Radiobiology).

Ehrhardt, J. P. 1973. Effect of γ irradiation on the respiration of unicellular algae. Compt. rend. soc. biol. 167(11): 1529-1531. (In French) (NSA 30: 18879).

Ehrhardt, J. P. and L. Leguay. 1971. Immediate effects of a subacute gamma irradiation on the photosynthesis of unicellular algae. pp. 141-144. In: Scientific Work, 1971. France Centre de Recherches du Service de Sante des Armeés, Clamart, report FRNC-R-19. (In French) (NSA 28: 30076).

Ehrhardt, P. 1961. Study of food uptake by Megoura viciae Buckt., a phloem sucking aphid (Homoptera, Rhynchota). Experientia 17: 461-462. (In German, English summary).

Ehrhardt, P. 1963. On the problem of food plant selection by aphids. Experientia 19(4): 204-205. (In German) (^{32}P).

Ehrhardt, P. and K. P. Lamb. 1967. Experiments on the uptake and excretion of ^{131}I by Aphis fabae Scop. (Aphidae, Homoptera, Insecta). Int. J. Radiat. Isotopes 18(7): 543-544. (In German) (NSA 21: 32593).

Eiche, A. 1964. During several generations irradiated spermatogonia in Drosophila melanogaster. A population analysis. Abstr. Hereditas 52(2): 243.

Eichelberger, H. H. 1963. Evaluation of the C^{14} method for measuring the primary productivity of the periphyton community in the Ohio River at Louisville, Kentucky. Thesis, Univ. of Louisville.

Eichholz, G. G. and A. N. Galli. 1970. Rapid determination of low-level fission products in water. Radiochem. Radioanal. Lett. 4: 315-323. (NSA 24: 46193).

Eichholz, G. G., T. F. Craft and A. N. Galli. 1967. Trace element fractionation by suspended matter in water. Geochim. Cosmochim. Acta 31: 737-745. (NSA 21: 25985) (Used tracers).

Eichner, F. N. and L. F. Kocher. 1973. Environmental dosimeter of the thermoluminescent type. U. S. Patent 3,789,227. (NSA 29: 24306).

Eidmann, H. H. 1968. Invasion of conifer plantations by radioactively labelled Hylobius abietis L. pp. 75-84. In: Isotopes and Radiation in Entomology. (See IAEA, 1968) (NSA 24: 31899).

Eipper, A. W. 1971. Nuclear power on Cayuga Lake. pp. 112-131. In: Patient Earth. (See Harte and Socolow, 1971) (NSA 26: 53192).

Eipper, A. W. 1971. Nuclear power on Cayuga Lake. Environ. Aff. 1(1): 165-190. (NSA 26: 33365).

Eipper, A. W., C. A. Carlson and L. S. Hamilton. 1970. Impacts of nuclear power plants in the environment. Living Wilderness 34(111): 5-12.

Eisenbud, M. 1962. Sources of radioactive pollution. pp. 153-174. In: Air Pollution. Vol. 2. Academic Press, New York. (NSA 20: 6918).

Eisenbud, M. 1963. Environmental Radioactivity. McGraw-Hill, New York. 435pp. (NSA 17: 21952).

Eisenbud, M. 1968. Radioactivity in the environment: sources of radioactivity in the environment. Pediatrics 41(Suppl.): 174-195. (NSA 22: 43345).

Eisenbud, M. 1968. Radionuclides in the environment. Monogr. Nucl. Med. Biol. (2): 3-18. (NSA 23: 24625).

Eisenbud, M. 1968. Sources of radioactive pollution. pp. 121-147. In: Air Pollution. Vol. 1. A. C. Stern, ed. Academic Press, New York. (1969: 2nd ed.) (NSA 23: 7648, 38117).

Eisenbud, M. 1969. Environmental pollution and its control. Bull. N.Y. Acad. Med. 45: 447-457. (NSA 23: 40869).

Eisenbud, M. 1971. Review of USA power reactor operating experience. pp. 861-878. In: Environmental Aspects of Nuclear Power Stations. (See IAEA, 1971).

Eisenbud, M. 1973. Environmental Radioactivity. Academic Press, New York. 2nd ed. xii, 542pp.

Eisenbud, M., et al. 1964. Naturally occurring radionuclides in foods and waters from the Brazilian areas of high radioactivity. pp. 837-854. In: The Natural Radiation Environment. (See Adams and Lowder, 1964) (NSA 17: 30774).

Eisner, H., W. Dansgaard and J. Meller. 1968. The altitude effect on the isotopic composition of precipitation and glacier ice in the Alps. Tellus 20: 595-600. (NSA 23: 45404).

Eisner, T. and E. O. Wilson. 1958. Radioactive tracer studies of food transmission in ants. Intern. Congr. Entomol. Proc. 10(2): 509-513.

Ekdahl, C. A., Jr. and C. D. Keeling. 1973. Atmospheric carbon dioxide and radiocarbon in the natural carbon cycle. I. Quantitative deductions from the records at Mauna Loa Observatory and at the South Pole. pp. 51-85. (See Woodwell and Pecan, 1973).

Eklund, S. 1972. Nuclear energy and the limitation of environmental pollution. Izotoptechnika 15(12): 597-602. (In Hungarian) (NSA 27: 24574).

Eklund, S. 1974. Nuclear energy, safety, and environment. Rev. Fr. Energ. 25(259): 302-309. (In French) (NSA 30: 20446).

Eklund, S. 1974. Nuclear power, safety, and environment. Int. At. Energ. Agency Bull. 16(1/2): 49-59. (NSA 30: 8921).

Ekman, L. 1967. Mechanisms of uptake and accumulation of radionuclides in terrestrial animals. pp. 547-560. In: Radioecological Concentration Processes. (See Aberg and Hungate, 1967).

Ekman, L. and U. Greitz. 1967. Distribution of radiocesium in reindeer. pp. 655-661. In: Radioecological Concentration Processes. (See Aberg and Hungate, 1967).

Ekman, L., A. Eriksson and U. Greitz. 1973. Transfer of simulated fallout particles from pasture to grazing dairy cattle. II. Experiments with wet deposited 43 to 60 and 100 to 200 μm particles. III. Experiments with 1 μm and 100 to 200 μm particles and dry deposited 100 to 200 μm particles. Sweden Forsvarets Forskningsanstalt, Stockholm, report FOA-4-4528-A-3, FOA-4-4550-A-3. 13, 15pp. (NSA 27: 22816; 28: 27681).

El-Badry, E. A. 1963. The effect of gamma radiation on host-parasitoid relationships. Thesis, Univ. of California, Berkeley. 100pp. Dissertatation Abstr. 24(12, Pt.1): 4929; NSA 18: 35378) (Gnorimoschema operculella, Trichogramma semifumatum, Copidosoma koehleri, Macrocentrus ancylivorus).

El-Badry, E. 1964. Supression of the reproductive potential of the potato tubeworm, Gnorimoschema operculella, by gamma irradiation. J. Econ. Entomol. 57(3): 414-415. (NSA 19: 17573).

El-Badry, E. 1964. The effect of gamma irrradiation on the hemocyte counts of larvae of the potato tubeworm, Gnorimoschema operculella (Zeller). J. Insect Pathol. 6: 327-330.

El-Badry, E. 1965. Some effects of gamma radiation on the potato tubeworm, Gnorimoschema operculella (Lepidoptera: Gelechiidae). Ann. Entomol. Soc. Am. 58(2): 206-209. (NSA 19: 33803).

El-Badry, E. 1965. Some gamma radiation effects on host-parasitoid relationships. Ann. Entomol. Soc. Am. 58(2): 209-212. (NSA 19: 33804).

El-Badry, E. and M. A. Abd Elaal. 1972. Effects of gamma radiation on a spider mite, Tetranychus arabicus. I. Irradiation of eggs. J. Econ. Entomol. 65(4): 947-950. (NSA 26: 50958).

El-Badry, E. A. , et al. 1969. Effects of gamma radiation on the egg stage of the lesser cotton leaf worm, Spodoptera exigua (Hb.). Z. Angew. Entomol. 64: 411-418. (NSA 24: 34603).

El-Badry, E. A., et al. 1971. Studies on the mating competitiveness and restoration of sperm viability in the gamma-irradiated population of Tetranychus arabicus Attiah (Acarina: Tetranychidae). Z. Angew. Entomol. 71 (2): 178-181. (NSA 27: 5548).

El-Badry, E. A., et al. 1971. Sterilization of the cotton leaf worm, Spodoptera littoralis Boisd, by gamma radiation. Isotopes Radiat. Res. 4: 71-76. (NSA 26: 4901).

El Basheir, S. 1967. Cause of resistance to DDT in a Diazinon-selected and a DDT-selected strain of house flies. Entomologia Exp. Appl. 10: 111-126. (^{36}Cl).

El-Bissary, E. A., et al. 1971. Research within the framework of a coordinated programme on the application of radiation techniques to water use efficiency studies. Final report, March 1965-31 July 1970. International Atomic Energy Agency, Vienna, report IAEA-R-310-F. 22pp. (NSA 28: 24636).

Elder, R. A. and S. Vigander. 1967. Capabilities and potential of USAEC's current meter for ultra-low-velocity measurements. pp. 683-695. In: Isotopes in Hydrology. (See IAEA, 1967) (NSA 22: 18941) (^{131}I tracer).

Elder, R. A. and S. Vigander. 1967. Ultra-low velocity measurement in a stratified reservoir by isotopic current meter. pp. 81-84. In: Isotope Techniques in the Hydrologic Cycle. (See Stout, 1967).

Elder, R. L. and W. Moore, Jr. 1965. Comparison of ^{137}Cs binding characteristics in pangola hay and Spanish moss. Radiol. Hlth. Data 6: 586-589. (NSA 20: 32924).

Electric Companies Public Information Program. 1972. Fact book on U. S. nuclear power projects. New York. 11th ed., 176pp. (NSA 28: 31801).

Elfield, M. and R. E. Franklin. 1967. Analysis of gamma-ray spectra from soils and plants. I. Computer program with correction for gain-shift and sample bulk density. Soil Sci. Soc. Am. Proc. 31(1): 39-__.

El-Hadidy, T. T. and M. A. Abdel-Salam. 1971. Effect of gamma irradiation on the survival of Azotobacter chroococcum. Isotopes Radiat. Res. 4: 63-69. (NSA 26: 4886).

Elias, C. A. 1961. The x-ray sensitivity of Escherichia coli in the presence of β-mercaptoethylamine. Radiation Research 15(5): 632-640.

Eliason, J. R. 1967. Field evaluation of ground disposal of reactor effluent: 1301-N crib. Battelle Northwest Lab., U. S. AEC report BNWL-CC-1032. 22pp. (NSA 29: 18318).

Eliassen, R. 1971. Power generation and the environment. Proc. Natl. Acad. Sci. (US) 68(8): 1938-1943. (NSA 25: 54164).

El-Lakany, M. H. H. 1969. Studies on the effects of ionizing radiation on some western coniferous species. Thesis, Univ. of British Columbia, Vancouver. (NSA 25: 8795).

El-Lakany, M. H. 1971. Effects if ionizing radiation on forest trees: a review. Atomic Energy of Canada, Ltd., Pinawa, report AECL-3951. 76pp. (NSA 26: 4705).

El-Lakany, M. H. and O. Sziklai. 1970. Effects of gamma-irradiation on some western conifers. Radiation Botany 10(5): 411-420.

El-Lakany, M. H. and O. Sziklai. 1970. Variation in nuclear characteristics in selected western conifers and its relation to radiosensitivity. Radiation Botany 10(5): 421-427. (NSA 25: 606).

El-Lakany, M. H. and O. Sziklai. 1970. Effects of gamma-irradiation on pollen, on seed, and on seedling characteristics in Douglas-fir. Adv. Frontiers Plant Sci. 25: 41-51.

Ellern, S. J., J. L. Harper and G. R. Sagar. 1970. A comparative study of the distribution of the roots of Avena fatua and A. strigosa in mixed stands using a ^{14}C-labelling technique. J. Ecology 58(3): 865-868.

Ellett, W. H. 1974. EPA water quality criteria and drinking water guidelines. pp. 162-170. In: 6th Ann. National Conf. on Radiation Control: New Challenges. (See U. S. Food and Drug Administration, 1974) (NSA 31: 19547).

Ellinger, F. 1939. Note on the action of x-rays on goldfish (Carassius auratus). Proc. Soc. Exptl. Biol. Med. 41(2): 527-529.

Ellinger, F. 1940. Goldfish as a new biologic test object in experimental radiation therapy. Radiology 35(5): 563-574.

Ellinger, F. 1940. Roentgen-pigmentation in the goldfish. Proc. Soc. Exptl. Biol. Med. 45(1): 148-150.

Ellinger, F. and C. Davison. 1942. Changes in the central nervous system of goldfish, irradiated in the depths of a water phantom. Radiology 39(1): 92-95.

Ellinger, F. and R. Gross. 1941. Depth dosage measurements by means of goldfish. Radiology 37(6): 717-723.

Elliot, A. M. and J. V. Slater. 1951. Actions of radiations on survival of Tetrahymena. Abstr. Proc. Am. Soc. Protozool. 2: abstr. 17.

Elliot, F. C. 1962. The nature of some unusual neutron and x-ray induced translocations in the Gramineae. Abstr. p. 244. In: 2nd Intern. Congr. Radiation Research. (See Anonymous, 1962).

Ellis, C. R. 1972. Susceptibility of two strains of Sitophilus granarius to 1,2-dibromoethane. 2. Strain difference in uptake and metabolism of radiolabeled fumigant. J. Econ. Entomol.65(6): 1573-1577. (NSA 27: 15117) (^{14}C).

Ellis, F. B., E. R. Mercer and G. M. Milbourn. 1968. The contamination of grassland with radioactive strontium. II. Effect of lime and cultivation on the levels of ^{90}Sr in herbage. Radiation Botany 8: 269-284. (NSA 22: 51404).

Ellis, G. 1972. NEPA impact on nuclear power plants. Reactor Technol. 15 (1): 11-13. (NSA 26: 30364).

Ellis, W. C. 1968. Dysprosium as an indigestible marker and its determination by radioactivation analysis. J. Agriculture 16(2): 220-224

Ellis, W. R. 1967. A review of radioisotope methods of stream gaging: Review paper. J. Hydrology 5: 233-257. (NSA 22: 12435).

Ellis, W. R. and G. L. Miles. 1967. Hydrological applications of radioisotopes in Australia. At. Energy Austral. 10: 2-11. (NSA 21: 14320).

Ellis, W. R. and G. L. Miles. 1968. Hydrological applications of radioisotopes in Australia. Isotopes Radiat. Technol. 5: 342-345. (NSA 22: 30713).

El Paso Natural Gas Company. 1972. Technical studies report number two. Project Wagon Wheel. El Paso, Texas. U. S. AEC report PNE-WW-13. 214pp. (NSA 27: 11951).

El-Miniawi, S. F. and M. E. Hassanein. 1964. Radiophosphorus in labelling the desert locust for population estimation. Bull. Soc. Ent. Egypte 48: 27-36.

El-Miniawi, S. F. and M. F. Leheta. 1964. On the effetct of Co^{60} gamma radiation on the desert locust. Bull. Soc. Ent. Egypte 48: 5-10.

El-Sayed, E.-S. I. 1968. Some effects of gamma radiation on the tobacco budworm, Heliothis virescens (Fabricius). Thesis, Louisiana State Univ., Baton Rouge. 79pp. (NSA 23: 900).

El-Sayed, E.-S. I. and J. B. Graves. 1969. Effects of gamma radiation on the tobacco budworm. I. Irradiation of pupae. II. Irradiation of moths. III. Irradiation of eggs and larvae. J. Econ. Entomol. 62: 289-293, 293-296, 296-298. (NSA 23: 24716, 24715, 24714).

El-Sayed, E.-S. I., S. El Basheir and K. A. Lord. 1965. DDT tolerance in diazinon-selected and DDT-selected strains of house flies. Chem. Ind. 37: 1958-1959. (^{36}Cl).

El-Sebay, A. H., R. L. Metcalf and T. R. Fukuto. 1964. Carbamate insecticides: synergism by organothiocyanates. J. Econ. Entomol. 57(4): 478-482. (Musca domestica, ^{14}C).

Elwood, J. W. 1971. Ecological aspects of tritium behavior in the environment. Nuclear Safety 12(4): 326-337. (NSA 25: 48230).

Elwood, J. W. 1973. Tritium behavior in fish from a chronically contaminated lake. pp. 435-439. In: Radionuclides in Ecosystems. (See Nelson, 1973) (NSA 26: 9479) (White Oak Lake).

Elwood, J. W. and D. J. Nelson. 1972. Periphyton production and grazing rates in a stream measured with a ^{32}P material balance method. Oikos 23 (3): 295-303. (NSA 28: 8528).

Emel'yanov, U. M. and V. I. Nekrasov. 1963. Autoradiography of wood from the region of the Tungho meteorite fall. Dokl. Akad. Nauk USSR 148(6): 1418-1421. (In Russian).

Emel'yanov, V. A. 1970. Methods for measuring moisture in saline soils using neutrons. Tr. Vses. Nauch.-Issled. Inst. Gidrotekh. Melior. (48): 44-50. (In Russian) (NSA 29: 7834).

Emel'yanov, V. A. and L. N. Beskin. 1970. Influence of soil density on the reading of neutron moisture meters. Tr. Vses. Nauch.-Issled. Inst. Gidrotekh. Melior. (48): 32-43. (In Russian) (NSA 29: 10282).

Emel'yanov, V. A., L. N. Beskin and VI. I. Isipov. 1963. Neutron method for measurement of moisture content of soils. Future prospects. Pochvovedeniye (7): 109-115. (In Russian) (NSA 19: 7720).

Emerson, D. B., et al. 1974. General environmental guidelines for evaluating and reporting the effects of nuclear power plant site preparation, plant, and transmission facilities construction. Hittman Associates, Baltimore, Md. 134pp. (NP-20377) (NSA 32: 6460).

Emerson, S. and R. Hesslein. 1973. Distribution and uptake of artificially introduced radium-226 in a small lake. J. Fish. Res. Bd. Can. 30(10): 1485-1490. (NSA 29: 18362; SFA 19: 17179).

Emerson, S., W. Broecker and D. W. Schindler. 1973. Gas-exchange rates in a small lake as determined by the radon method. J. Fish. Res. Bd. Can. 30(10): 1475-1484. (SFA 19: 17132).

Emery, J. F. and F. F. Dyer. 1974. Multielement determination in environmental NAA using MONSTR. pp. 123-129. In: Proc. 2nd Intern. Conf. on Nuclear Methods in Environmental Research. (See Vogt and Meyer, 1974) (NSA 33: 23325) (Computer program).

Emery, R. M. and T. R. Garland. 1974. Ecological behavior of plutonium and americium in a freshwater ecosystem. Phase II. Implications of differences in transuranic isotopic ratios. Battelle Northwest Lab., U. S. AEC report BNWL-1879. 29pp. (NSA 31: 25779).

Emery, R. M. and D. C. Klopfer. 1975. Distribution of transuranic elements in a freshwater pond ecosystem. Battelle Northwest Lab., U. S. AEC report BNWL-SA-5424. 23pp. (NSA 33: 598).

Emery, R. M., D. C. Klopfer and W. C. Weimer. 1974. Ecological behavior of plutonium and americium in a freshwater ecosystem. Phase I. Limnological characterization and isotopic distribution. Battelle Northwest Lab., U. S. AEC report BNWL-1867. 82pp. (NSA 31: 5987).

Emery, R. M., et al. 1975. Ecological behavior of plutonium and americium in a freshwater pond. Battelle Northwest Lab., U. S. AEC report BNWL-SA-5346. 38pp. (NSA 32: 8919).

Emler, V. S., et al. 1971. Environmental radioactivity levels and concentrations semiannual summary, first half 1971. Goodyear Atomic Corp., Piketon, O., U. S. AEC report GAT-660. 9pp. (NSA 26: 2386).

Emmerich, H., et al. 1965. On the metabolism of farnesols in the larva of Tenebrio molitor. Z. Naturforschung 20b(3): 211-213. (In German) (^{14}C).

Enders, J. W. 1972. Surface contamination control at a plutonium research facility. pp. 519-537. In: Health Physics Operational Monitoring, Vol. 1. C. A. Willis, ed. Gordon & Breach, New York. (NSA 29: 29312).

Endo, J., T. Kawasaki and Y. Ikuta. 1972. In situ explorations of soil densities and moisture contents using radioisotopes. pp. 195-213. In: Radiation and Isotope Techniques in Civil Engineering. Vol. 1. (Proc. Conf., EURISOTOP Office, Brussels, October 28-30, 1970). A. Proepstl, ed. Commission of the European Communities, Brussels, report EURISOTOP-55(Vol.1). (NSA 27: 27729).

Engel, D. W. 1971. The effects of salinity, temperature, and radiation on the ionic regulation of the blue crab, Callinectes sapidus. Thesis, North Carolina State Univ., Raleigh. 49pp. (Dissertation Abstr. 33(4)).

Engel, D. W. 1973. The radiation sensitivities of three species of fiddler crabs (Uca pugilator, U. pugnax and U. minax). Chesapeake Sci. 14(4): 28 289-291.

Engel, D. W. and J. W. Angelovic. 1968. Effects of radiation and salinity on the respiration of brine shrimp nauplii. Abstr. Radiation Research 35 (2): 563.

Engel, D. W. and D. J. Fluke. 1961. Relation of water content to radiosensitivity for emergence of Artemia larvae from the encysted phase. Abstr. 47. Radiation Research 14(4): 463.

Engel, D. W. and D. J. Fluke. 1962. The effect of water content and postirradiation storage on radiosensitivity of brine shrimp cysts (eggs). Radiation Research 16(2): 173-181. (NSA 16: 17504).

Engel, D. W., et al. 1973. Effect of radiation, salinity, and temperature on the ionic regulation of the blue crab, Callinectes sapidus. pp. 1113-1118. In: Radionuclides in Ecosystems. (See Nelson, 1973) (NSA 28: 30234).

Engel, M. S. and H. I. Adler. 1961. Catalase activity, sensitivity to hydrogen peroxide, and radiation response in the genus Escherichia. Radiation Research 15(3): 269-275. (NSA 15: 20943).

Engel, P. P. 1968. The induction of biochemical and morphological mutants in the moss Physcomitrella patens. Am. J. Bot. 55: 438-466. (NSA 22: 40641) (Used x-rays).

Engel, R. E. and R. A. Brechbill. 1967. Radiation surveillance in wildlife. Congr. Western Assoc. Fish and Game Commissioners, Honolulu, July 16-20, 1967. U. S. Public Health Service, Las Vegas, Nev. (CONF-670724) (NSA 21: 45611).

Engel, R. L. and G. Soli. 1964. Method of monitoring low level radioactivity by the stimulated growth of bacteria. U. S. Patent 3,156,821. (NSA 19: 18231).

Engels, W. and W. Drescher. 1964. Incorporation of H^3 D-glucose during oogenesis in Apis mellifera L. Experientia 20(8): 445-447. (In German, English summary).

Engelhard, H. and T. Houtermans. 1950. Action of alpha rays and of ultraviolet light on Bacterium coli. Z. Naturforschung 5b(5): 264-269.

Engstrand, L. G. and H. G. Östlund. 1962. Stockholm natural radiocarbon measurements. IV. Am. J. Sci. Radiocarbon (Suppl. 4): 115-136.

Engstrom, D. E. 1975. Determination of ^{241}Am in soil using an automated nuclear radiation measurement laboratory. pp. 473-490. In: Radioecology of Plutonium and Other Transuranics in Desert Environments. (See White and Dunaway, 1975) (NSA 33: 649).

Engstrom, D. E. 1975. Environmental sciences information storage and retrieval system. pp. 471-472. In: Radioecology of Plutonium and Other Transuranics in Desert Environments. (See White and Dunaway, 1975) (NSA 33: 667).

Enke, C. G. 1975. Contamination of the waters with environmental chemicals and radioactive materials. Wasser Luft Betr. 19(8): 468-469. (In German) (NSA 33: 14869).

Enkerlin, S. D., R. A. Mancilla Berganza and C. E. Pena Andeliz. 1975. Biological aspects and behavior of Heliothis virescens Fabricius adults from irradiated diapausal pupae. pp. 517-527. In: Sterility Principle for Insect Control, 1974. (See IAEA, 1975) (NSA 32: 11806).

Eno, C. F. 1966. The effects of gamma radiation on soil microorganisms and their metabolic processes. Final report. Univ. of Florida, Gainesville, U. S. AEC report ORO-2754-7. 36pp. (NSA 20: 40637).

Eno, C. F. nd. The effects of gamma radiation on earthworms, algae, and rotifers. Univ. of Florida, Gainesville, U. S. AEC report TID-23280. 11pp. (NSA 22: 31205).

Eno, C. F. and H. Popenoe. 1963. The effect of gamma radiation on the availability of nitrogen and phosphorus in soils. Soil Sci. Soc. Am. Proc. 27(3): 299-301.

Eno, C. F. and H. Popenoe. 1964. Gamma radiation compared with steam and methyl bromide as a soil sterilizing agent. Soil Sci. Soc. Am. Proc. 28 (4): 533-535.

Enomoto, Y. and R. Ichikawa. 1968. ^{85}Sr and ^{45}Ca by rat from brown seaweeds. J. Radiat. Res. (Japan) 9: 26-28. (NSA 22: 45399).

Enseyev, P. K. 1958. Relation of atmospheric precipitations to total radiation, evaporation and absolute humidity. Meteorologiya Gidrologiya USSR (10): 18-23. (English transl. OTS 60-51028).

Ensminger, L. E. and R. W. Peatson. 1957. Uses of macronutrient isotopes in soil fertility research. pp. 19-47. In: Atomic Energy and Agriculture. C. L. Comar, ed. Am. Assoc. Adv. Sci., Washington, Publ. 49.

Entomological Society of Ontario. 1955. Phosphorus-32 labelling the adults of the cabbage maggot, Hylemya brassicae (Bouche) (Diptera: Anthomyidae). Rept. 85: 51-61.

Epshtein, Y. A. and N. F. Lavroskaya. 1959. Effect of ionizing radiation on the protein metabolism of fish. Biokhimiya 24(4): 592-599. (Biochemistry (USSR) 24: 549-555) (NSA 14: 12459).

Eraemetsae, O. and I. Yliruokanen. 1971. Niobium, molybdenum, hafnium, tungsten, thorium, and uranium in lichens and mosses. Suomen Kem. 44(11): 372-374. (NSA 26: 17412).

Erbisch, F. H. and J. J. Kalosis. 1973. Initial observations of the effects of gamma radiation on oxygen consumption, ^{32}P uptake, and phycobiont of Cladonia sylvatica (L.) Hoffm. Radiation Botany 13(6): 361-367. (NSA 29: 27023).

Erdelyszky, Zs. 1968. Application of nuclear techniques in hydrology in the Hungarian People's Republic. Atomtech. Tajek. 11: 647-652. (In Hungarian) (NSA 23: 45406).

Erdelyszky, Zs. and K. Stelczer. 1967. Investigation into the movement of fluvial bed-load with radioisotopes. pp. 263-269. In: Isotopes in Hydrology. (See IAEA, 1967) (NSA 22: 19064).

Erdman, H. E. 1960. Divergence between lethal doses and sterilizing doses of x-rays with progressive development in Habrobracon females. Nature 186(4720): 254-255. (NSA 14: 12487).

Erdman, H. E. 1960. Adult longevity as a sensitive criterion of radiation induced damage when 24 hour Habrobracon (Hymenoptera) embryos are x-rayed. J. Econ. Entomol. 53(5): 971-972. (NSA 15: 4952).

Erdman, H. E. 1960. X rays and the weakest link in Habrobracon ontogeny. Abstr. Radiation Research 12(4): 433.

Erdman, H. E. 1961. Analyses of the differential radiosensitivity of developing reproductive tissues in Habrobracon juglandis (Ashmead) to ionizing radiation. Int. J. Radiat. Biol. 3(2): 183-204. (NSA 15: 14136).

Erdman, H. E. 1962. Comparative x-ray sensitivity of Tribolium confusum and T. castaneum (Coleoptera: Tenebrionidae) at different developmental stages during their lifecycle. Nature 195: 1218. (NSA 16: 31525).

Erdman, H. E. 1962. X-ray effects on single and mixed species populations of Tribolium confusum and T. castaneum (Coleoptera: Tenebrionidae). Abstr. p. 156. In: 2nd Intern. Congr. Radiation Research. (See Anonymous, 1962) (Also: Hanford Labs., U. S. AEC report HW-SA-2686) (NSA 16: 30209).

Erdman, H. E. 1962. Effects of irradiation on the Mediterranean meal moth, Ephestia kuehniella Zeller, cultured on ^{89}Sr-spiked food. Int. J. Radiat. Biol. 5: 331-338. (NSA 17: 76).

Erdman, H. E. 1962. Effects of ingested Pu^{239} on fecundity, fertility and life span of Habrobracon (Hymenoptera: Braconidae). Health Physics 8(4): 635-638.

Erdman, H. E. 1963. The differential sensitivity of flour beetles, Tribolium confusum and T. castaneum to x-ray alteration of reproductive abilities, induced dominant lethals, biomass and survival. J. Exptl. Zool. 153(2): 141-147. (NSA 17: 40546).

Erdman, H. E. 1963. Arrested development in x-rayed larvae of Ephestia kuehniella Zeller (Lepidoptera: Phycitinae). Am. Midland Nat. 69(1): 34-37.

Erdman, H. E. 1964. Differential responses of germ cells in flour beetles, Tribolium castaneum Herbst, due to x-ray dose, hypothermia, sex-exposed and age. Hanford Labs., U. S. AEC report HW-SA-3747. 21pp. (NSA 19: 43715).

Erdman, H. E. 1964. Reproductive performance of x-rayed single-species and mixed-species cultures of Tribolium confusum and T. castaneum reared at different temperatures. Hanford Labs., U. S. AEC report HW-SA-3748. 20pp. (NSA 19: 43716).

Erdman, H. E. 1964. Fast neutron effects on productivity of young and old flour beetles, Tribolium castaneum Herbst and alterations due to temperature and sex-exposed. Hanford Labs., U. S. AEC report HW-SA-3537. 16pp. (NSA 20: 1613).

Erdman, H. E. 1964. Dominant lethal proportions modified by x-radiation, temperatures, and cohabitation in single- and mixed species population of flour beetles, Tribolium confusum and T. castaneum. Abstr. 59. Radiation Research 22(1): 187.

Erdman, H. E. 1965. Dose ratio of x-rays to fast neutrons in producing dominent lethals in flour beetles, Tribolium castaneum. Nature 205(4966): 99-99-100. (NSA 19: 10852).

Erdman, H. E. 1965. Research notes. Tribolium Inform. Bull. 8: 74-75.

Erdman, H. E. 1965. Fast-neutron effects on productivity of young and old flour beetles, *Tribolium castaneum* Herbst, and alterations at different temperatures and after exposure of either or both sexes. Int. J. Radiat. Biol. 9: 305-311. (NSA 20: 84).

Erdman, H. E. 1965. Modifications of productivity in flour beetles, *Tribolium castaneum* Herbst, due to x-ray dose, hypothermia and the sex exposed. Radiation Research 25(2): 341-351. (NSA 19: 28070).

Erdman, H. E. 1965. Effects of dimethyl sulfoxide (DMSO) on productivity of x-rayed flour beetles. Radiation Research 25: 186.

Erdman, H. E. 1966. Effects of dimethylsulfoxide on productivity of x-irradiated flour beetles. Biol. Bull. 130: 157-169. (Also: Battelle Northwest Lab., U. S. AEC report BNWL-307) (NSA 20: 35260).

Erdman, H. E. 1966. X-radiation and temperature modification of reproductive performance of single-species and mixed-species cultures of *Tribolium confusum* and *T. castaneum*. Physiol. Zool. 39: 160-170. (NSA 21: 6618).

Erdman, H. E. 1966. Effects of x-rays on metamorphosis and adult life span of flour beetles. Nature 211: 1427-1428. (NSA 21: 11014).

Erdman, H. E. 1966. The meek shall inherit the earth. Battelle Northwest Lab., U. S. AEC report BNWL-SA-571. 28pp. (NSA 22: 4664) (Insects vs. radiations).

Erdman, H. E. 1966. Metamorphic and adult longevity modifications due to x-irradiation of pupal flour beetles, *Tribolium confusum* (Jacquelin du Val). Bull. Entomol. Soc. Am. 12: 256.

Erdman, H. E. 1966. Modification in productivity and mortality of flour beetle species and strains due to x-ray and the insecticide, DDT. Abstr. 305. p. 78. In: *3rd Intern. Congr. of Radiation Research*. G. Silini, ed. North-Holland, Amsterdam.

Erdman, H. E. 1967. Modification of fitness in species and strains of flour beetles due to x-ray and DDT. Ecology 47(6): 1066.

Erdman, H. E. 1970. Effects of x-radiation and the insecticide DDT on mortality and reproduction of flour beetles, *Tribolium confusum* and *T. castaneum*, with a genetic interpretation for DDT resistance. Ann. Entomol. Soc. Am. 63: 191-197. (NSA 24: 14578).

Erdman, H. E. 1972. Population performance and chronic gamma radiation. Battelle Northwest Lab., U. S. AEC report BNWL-SA-4190. 16pp. (CONF-720203-1) (NSA 26: 23135).

Erdman, H. E. 1973. Comparative species radiation sensitivities of flour beetles to chronic γ radiation. J. Econ. Entomol. 66(1): 251-252. (NSA 27: 17680).

Erdman, H. E. 1973. Effects of chronic ^{60}Co (gamma) radiation, dose rate, temperature, and cohabitation on population performance of flour beetles. Environ. Entomol. 2(1): 41-48. (NSA 28: 6055).

Erdman, H. E. 1974. Productivity modifications to flour beetles (Coleoptera: Tenebrionidae) by gamma radiation (^{60}Co) of one or both sexes and by the addition of radiated males or females to population. Res. Populat. Ecol. (Kyoto) 16(1): 52-58.

Erdelysky, Z. 1974. Measurement of the alluvial deposit transport by means of radioactive tracers. Österr. Wasserwirtsch. 26(3/4): 89-92. (In German) (NSA 33: 7328).

Erdtmann, G., et al. 1974. Determination of polluting elements in soil samples and fly ash by means of activation analysis. pp. 29-41. In: *Workshop on Modern Analytical Methods*. (October 15, 1973). German Federal Republic, Jülich, report JÜL-11(Bd.3). (In German) (NSA 32: 21820).

Eremin, N. N. 1965. Natural radioactivity of the Schirmacher ponds. Inform. Byull. Sov. Antark. Eksped. (English transl. Sov. Antarct. Exped. Inform. Bull. 5(55): 18-19). (NSA 24: 14578).

Ergashev, S. Ye. 1968. Formation of radiochemical compounds in soil waters. Dokl. Akad. Nauk. Uzbek. SSR (7): 3-4. (In Russian) (NSA 23: 4645).

Erickson, D. G. 1964. Anemia in mice exposed to infection with irradiated or non-irradiated *Schistosoma mansoni* cercariae. J. Parasitol. 50: 53. (NSA 19: 126).

Erickson, D. G. 1965. The fate of gamma-irradiated *Schistosoma mansoni* cercariae in mice. Am. J. Trop. Med. Hyg. 14: 574-578. (NSA 20: 5313).

Erickson, D. G. and W. L. Caldwell. 1965. Acquired resistance in mice and rats after exposure to gamma-irradiated cercariae. Am. J. Trop. Med. Hyg. 14: 566-573. (NSA 20: 8591).

Erickson, J. M. 1972. Mark-recapture techniques for population estimates of *Pogonomyrmex* ant colonies: an evaluation of the ^{32}P technique. Ann. Entomol. Soc. Am. 65(1): 57-61.

Erickson, R. C. 1971. Effects of tritiated water and other sources of ionizing radiation on mortality, growth, male characters, courting behavior, and thermal death of *Poecilia reticulata*, the guppy. Thesis, Univ. of Washington, Seattle. 174pp.

Erickson, R. C. 1973. Effects of chronic irradiation by tritiated water on *Poecilia reticulata*, the guppy. pp. 1091-1099. In: *Radionuclides in Ecosystems*. (See Nelson, 1973) (NSA 26: 4837).

Ericson, D. B. and G. Wollin. 1956. Micropaleontological and isotopic determinations of Pleistocene climate. Micropaleontology 2: 257-270.

Eriksson, A. 1970. Radioactive fallout in soil. Grundforbattring 28(Spec. Issue No. 5): 23-30. (In Swedish) (CONF-691139-1) (NSA 26: 7018; 28: 5559).

Eriksson, A. and L. Fredriksson. 1972. Plant uptake of ^{137}Cs from different depths in a homogeneous subsoil layer. Sweden Forsvarets Forskningsanstalt, Stockholm, report FOA-4-4491. 15pp. (NSA 26: 48298).

Eriksson, E. 1961. Natural reservoirs and their characteristics. Geofis. Intern. (Mexico) 1: 27-43. (NSA 21: 25980) (Circulation, radionuclides).

Eriksson, E. 1967. Large-scale utilization of tritium in hydrologic studies. pp. 153-156. In: Isotope Techniques in the Hydrologic Cycle. (See Stout, 1967).

Eriksson, E. 1967. The atmospheric transport of tritium. pp. 56-57. In: Isotope Techniques in the Hydrologic Cycle. (See Stout, 1967).

Eriksson, E. 1967. Isotopes in hydrometeorology. pp. 21-33. In: Isotopes in Hydrology. (See IAEA, 1967) (NSA 22: 18925).

Erlenbach, H. R. 1974. Contribution of factory inspectorate measures to environmental surveillance activities around nuclear installations in the state of North Rhine-Westphalia. pp. 367-378. In: Environmental Surveillance Around Nuclear Installations. Vol. 2. (See IAEA, 1974) (NSA 31: 14046).

Erlenkeuser, H. and H. Willkomm. 1972. Data processing in ^{14}C-measurements. I. Recording and statistical check. II. Age calculation. Atomkernenergie 19(2): 123-126; 1973: 22(2): 90-94. (NSA 26: 28289; 29: 18374).

Ernst, D. and H. Glubrecht. 1966. Investigations on the sterilization of culture soils. pp. 163-167. In: Biophysical Problems of the Actions of Radiation. H. Muth, ed. Georg Thieme Verlag, Stuttgart. (In German) (NSA 21: 22834).

Ertaud, A. 1972. Environment and nuclear reactors. Usine Nouv. (Suppl.): 93-100. (In French) (NSA 28: 31892).

Ervin, P. F. and N. D. Eckhoff. 1973. Measuring mercury in pheasant livers by neutron activation analysis. Trans. Kans. Acad. Sci. 76(1): 1-3. (NSA 31: 2763).

Esch, H., H. Miltenburger and O. Hug. 1964. The influencing of electric potentials of algae cells by means of x radiation. Biophysik 1(4): 380-388. (In German) (NSA 18: 27092).

Esin, T. and M. Alganatay. 1968. Study on the effectiveness of ^{60}Co irradiation against S. granarius and T. confusum Duv. Bitki Koruma Bul. 8(1): 39-60. (In Turkish) (NSA 26: 31360).

Eslyn, W. E. 1959. Radiographical determination of decay in living trees by means of the Thulium x-ray unit. Forest Science 5: 37-48.

Espeland, K. G. 1971. ^{137}Cs in the flesh of important meat producing animals in Norway. Norsk Vet.-Tidsskr. 83(5): 322-328. (In Norwegian) (NSA 17743) (Incl. reindeer).

Espinola, H. N. and J. E. da Silva. 1965. Tagging sandflies with ^{32}P (Diptera, Psychodidae). Rev. Brasil. biol. 25: 295-304. (NSA 22: 4582).

Espinola, H. N., R. Capunay and J. E. da Silva. 1966. Effective half-life of ^{32}P in the sandfly Lutzomyia longipalpis (Clutz and Neivajlaiz) (Diptera, Psychodidae). Rev. Brasil. biol. 26: 175-179. (NSA 22: 4584).

Essac, E. G. and J. E. Casida. 1966. Metabolism and action of methelenedioxyphenyl synergists in the house fly. Abstr. 62. Bull. Entomol. Soc. Am. 12: 266. (^{14}C).

Essig, T. H. (Ed.). 1968. Evaluation of radiological conditions in the vicinity of Hanford, January-June, 1967. Battelle Northwest Lab., U. S. AEC report BNWL-665. 20pp. (NSA 22: 19326).

Essig, T. H. 1969. Radiological status of the ground water beneath the Hanford Project, January-June 1968. Battelle Northwest Lab., U. S. AEC report BNWL-984. 20pp. (NSA 23: 13854).

Essig, T. H. 1970. Radiological impact of Hanford waste disposal on groundwater quality. Battelle Northwest Lab., U. S. AEC report BNWL-SA-3744. 11pp. (For 16th Ann. Health Physics Meeting, New York, N. Y., 12 July 1971). (CONF-710710-2) (NSA 25: 48601).

Essig, T. H. 1970. Radiological status of the groundwater beneath the Hanford Project, January-June 1970. Battelle Northwest Lab., U. S. AEC report BNWL-1539. 23pp. (NSA 25: 23704).

Essig, T. H. 1971. Radiological status of the groundwater beneath the Hanford Project, July-December 1970. Battelle Northwest Lab., U. S. AEC report BNWL-1613. 25pp. (NSA 26: 181).

Essig, T. H. and R. B. Ball. 1966. Environmental status of the Hanford Project. 1965 annual summary. Battelle Northwest Lab., U. S. AEC report BNWL-CC-913. 47pp. (NSA 29: 10122).

Essig, T. H. and R. B. Hall. 1967. Environmental status of the Hanford reservation for December 1966. (Annual summary). Battelle Northwest Lab., U. S. AEC report BNWL-CC-637. 73pp. (NSA 29: 10121).

Essington, E. H. 1971. An interim summary of tritium data for STS "A", Amchitka Island, Alaska July 1, 1970 through June 30, 1971. Teledyne Isotopes, Palo Alto, Calif., U. S. AEC report NVO-1229-171. iii, 66pp. (NSA 26: 42961).

Essington, E. H. 1971. Evaluation and selection of tracers for field measurement of coefficient of dispersion. Teledyne Isotopes, Palo Alto, Calif., U. S. AEC report NVO-1229-176. v, 34pp.

Essington, E. H. 1971. Long-range monitoring network for Amchitka Island, Alaska. Teledyne Isotopes, Palo Alto, Calif., U. S. AEC report NVO-1229-160. 20pp. (NSA 25: 48683).

Essington, E. H., P. R. Fenske and W. E. Nork. 1970. Radioactivity in water: Project Milrow. Isotopes, Inc., Palo Alto, Calif., U. S. AEC report NVO-1229-135. 45pp. (NSA 24: 22770).

Essington, E. H., F. J. Forslow and D. C. Catagnola. 1970. Interim summary of radiochemical data for STS "A", Amchitka Island, Alaska, through June 30, 1969. Teledyne Isotopes, Palo Alto, Calif., U. S. AEC report NVO-1229-121. v, 62pp. (NSA 25: 34720).

Essington, E. H., E. J. Forslow and D. C. Castagnola. 1971. An interim summary of tritium data for STS "A", Amchitka Island, Alaska, July 1, 1969 through June 30, 1970. Teledyne Isotopes, Palo Alto, Calif., U. S. AEC report NVO-1229-157. v, 91pp.

Essington, H. Nishita and A. J. Steen. 1965. Release and movement of radionuclides in soils contaminated with fallout material from an underground thermonuclear detonation. Health Physics 11(8): 689-698. (NSA 19: 36661).

Essington, E. H., et al. 1975. Plutonium, americium, and uranium concentrations in Nevada Test Site soil. Los Alamos Scientific Lab., U. S. AEC report LA-UR-75-1770. 25pp. (NSA 33: 610).

Estlander, A. and A. Kulmala. 1973. Long distance dispersal of radioactive fallout and natural pollen. pp. 61-70. In: Scandinavian Aerobiology. S. Nilsson, ed. NRF, Editorial Services, Stockholm. (NSA 29: 18341).

Etchells, J. L., et al. 1961. Influence of gamma radiation on the microflora of cucumber fruit and blossoms. Appl. Microbiol. 9(1): 45-49. (NSA 17: 17800).

Eto, H. 1963. The effect of γ-irradiation on the physiological responses of the nerve-melanophore system in Carassius carassius. Zool. Mag. (Tokyo) 72(10): 277-282. (In Japanese) (Nucl. Sci. Abstr. Jap. 3: 01075).

Eto, H. 1965. Effect of fractionation of irradiated dose of Co^{60} γ rays the fish, Oryzias latipes. Zool. Mag. (Tokyo) 74(1): 29-30. (In Japanese) (NSA 19: 21904; Nucl. Sci. Abstr. Jap. 4: 01829).

Eto, H. 1968. Changes in incorporation of ^{3}H-thymidine into hematopoietic cells of goldfish following x-irradiation. Dobutsugaku Zasshi 77: 213-219. (In Japanese) (NSA 23: 29781).

Eto, H. 1969. Changes in incorporation of ^{3}H-thymidine into hematopoietic cells of goldfish during recovery period from radiation injury. Annot. Zool. Jap. 42: 159-165. (NSA 24: 14434).

Eto, H. and N. Egami. 1963. Effect of hypophysectomy and adrenalectomy on the length of survival time after x-irradiation in the goldfish, Carassius auratus. Proc. Japan Acad. 39: 503-506. (NSA 19: 10869).

Eto, H. and N. Egami. 1963. Responses to Na^+, K^+, atropine, and adrenaline of the melanophores induced by x-irradiation in the fin of the goldfish, Carassius auratus. Annot. Zool. Jap. 36: 133-139. (NSA 19: 189).

Eto, H. and N. Egami. 1965. Effect of temperature on survival period of the fish Oryzias latipes, following irradiation with different x-ray doses. Annot. Zool. Jap. 38: 114-121. (NSA 20: 16236).

Eto, H. and Y. Taguchi. 1973. Mortality response of goldfish after exposure to fast neutrons and RBE of fast neutrons for inhibition of hatchability in Oryzias embryos. pp. 18-25. In: 1st Rept. on Cancer Therapy with High Energy Particles and the Utilization of Short Lived Radionuclides in Diagnostics. Japan National Institute of Radiological Sciences, Ciba, report NIRS-R-2. (In Japanese) (NSA 32: 6755).

Eto, H., Y. Hyodo-Taguchi and N. Egami. 1971. Mortality of fish under continuous irradiation conditions. pp. 213-218. In: Biological Aspects of Radiation Protection. T. Sugahara, ed. Igoku Shoin, Tokyo. (CONF-691041) (NSA 25: 42634).

Eto, H., Y. Hyodo-Taguchi and T. Maruyama. 1968. Effects of x-irradiation of a part of the body on mortality rate and on histological changes in the goldfish, Carassius auratus. J. Radiat. Res. (Japan) 9: 141-151. (NSA 23: 43864).

Eto, K. 1973. Effect of external irradiation on fish eggs and embryo. pp. 202-233. In: Radioactivity and Fishes. (See Egami, 1973) (In Japanese) (NSA 29: 16005).

Eto, K. 1973. ffect of internal irradiation on fish. pp. 298-316. In: Radioactivity and Fishes. (See Egami, 1973) (In Japanese) (NSA 29: 10474).

Ettenhuber, E. 1970. Accumulation of radionuclides in fish. Germany (GDR) Staatliche Zentral für Strahlenschutz, Berlin, report SZS-7/70. 30pp. (In German) (NSA 24: 48468).

Ettenhuber, E. 1971. Accumulation of radionuclides in fishes. Isotopenpraxis 7(3): 75-84. (In German) (NSA 25: 42249).

Ettenhuber, E. and W. Röhnsch. 1974. Results and conclusions of the environmental monitoring of nuclear installations. pp. 152-158. In: Radiation Protection, Environment, and the Population. (See German Democratic Republic, 1974) (In German) (NSA 31: 8620).

Ettenhuber, E., P. Clajus and W. Röhnsch. 1975. Determination of accumulation factors, fish/water, for radionuclides considered to be significant for radiation exposure of man. Germany (GDR) Staatliches Amt für Atomsicherheit und Strahlenschutz, Berlin, report SAAS-175. 20pp. (In German) (NSA 32: 22284).

Eugster, J. G. 1964. Subradiation experiments concerning the concept of the natural radiation environment. Aerospace Med. 35: 524-526. (NSA 19: 83).

Eugster, J. G., R. Koblet and D. Simons. 1959. Results of experiments on the biological effects of cosmic radiation on seeds of Hordeum bonus 01518/B19 Gustafsson, with special consideration on the effects of heavy primaries. pp. 222-224. In: Proc. 9th Intern. Astronautical Congr. (Amsterdam, 1958). I. F. Hecht, ed. Springer-Verlag, Vienna.

European Atomic Energy Community. 1963. Results of the radioactivity measurements in the Community countries for the years 1961-1962. Brussel, report EUR-461.d,f,i,n. 157pp. (Contains "natural" material).

European Atomic Energy Community. 1964. Results of the radioactivity measurements in the Community countries. Brussels, report EUR-1820.d.f.i.n. 120pp. (NSA 19: 618) (Incl. surface waters).

European Atomic Energy Community. 1965. Tracer Methods for the Measurement of Movement of Solids in Water. Vol. I. Summary of Reports and Report Discussions. Vol. II. Original Papers on the Work Session Protocol of the Work Session of the Eurisotope Bureau. Brussels, Informationsheft des Büro 7 (Ser. Sitzb.-1), 8 (Ser. Sitzb.-2. 161, 479pp. (NSA 22: 42917,8).

European Atomic Energy Community. 1966. Study for the establishment of standards for radioactive wastes in hydrobiological systems. Brussels, report EUR/C-1263/66. (ORNL-tr-1677) (NSA 24: 2648).

European Atomic Energy Community. 1967-1974. Results of environmental radioactivity measurements in the Community countries in 1966. -- 1972. Brussels, Luxembourg, report EUR-3602, -4246, -4605, -4767, -5105. 70, 68, 78, 76, 79pp. (In several languages) (Air, fallout, water, "natural" materials) (NSA 22: 8221; 23: 49951; 25: 37797; 26: 38409; 30: 32043).

European Atomic Energy Community. 1968. General study of the radioactivity of the Rhine Basin. IV. Brussels, report EUR-3741.f. 2 vols., 211, 421 pp. (In French) (NSA 23: 38092, 31416).

European Atomic Energy Community. 1971. Proc. Intern. Symp. on Radioecology Applied to the Protection of Man and His Environment. (Sept. 7-10, 1971, Rome, Italy). Luxembourg, report EUR-4800. 2 vols.; xii, 1421pp. (NSA 27: 5273).

European Atomic Energy Community. 1972. Radioactive effluents from nuclear power stations in the Community. Discharge data radiological aspects. Luxembourg, report. 38pp. (NP-19910) (NSA 29: 14737).

European Atomic Energy Community. 1975. Environment and quality of life: the behavior of chromium in aquatic and terrestrial food chains. Ispra, Italy, report EUR-5375.e. 88pp. (NSA 33: 20578) (^{51}Cr).

European Nuclear Energy Agency. 1971. Radioactive waste management in western Europe. Paris. 146pp. (In French) (NSA 26: 50419).

European Nuclear Energy Agency. 1971. Radioactive waste management practices in western Europe. Paris. 126pp. (In French) (NSA 26: 26141).

Evans. A. G. 1969. New dose estimates from chronic tritium exposures. Health Physics 16: 57-63. (Deer).

Evans, A. G. and J. W. Fenimore. 1960. Radioactivity in soil, vegetation, and ground water. 100-R Area seepage basins. Savannah River Plant, U. S. AEC report DPSPU-60-33. 16pp. (NSA 29: 20859).

Evans, E. A. 1973. Tritium tracers in biology: facts or fiction. pp. 659-675. In: Tritium. (See Moghissi and Carter, 1973) (NSA 28: 27128).

Evans, E. J. and A. J. Dekker. 1969. The fixation and plant recovery of ^{137}Cs. Soil Science 107: 175-180. (NSA 23: 17564).

Evans, F. C. 1967. Symposium on radioecology. BioScience 17(10): 743-744.

Evans, F. C. 1970. Seasonal effects of light, temperature, and radiation on activity patterns of the meadow vole (Microtus pennsylvanicus). Final report. Univ. of Michigan, Ann Arbor, U. S. AEC report COO-1486-5. 20pp. (NSA 24: 21280).

Evans F. C. and M. Katznelson. 1969. Activity patterns in the meadow vole, Microtus pennsylvanicus Ord, under field conditions. Abstr. Bull. Ecol. Soc. Am. 50(2): 106. (^{182}Ta).

Evans, F. R. 1956/57. Preliminary studies on predation of microorganisms as determined with radiotracers. Proc. Utah Acad. Sci. Arts Lett. 34: 35-37.

Evans, K. 1970. Effects of gamma radiation on Heterodera rostochiensis. Nematologica 16: 284-294. (NSA 25: 6767).

Evans, R. B. and D. E. Bernhardt. 1970. Public health evaluation, Project Rulison (production testing). Southwestern Radiological Health Lab., Las Vegas, Nev., U. S. PHS, U. S. AEC report SWRHL-96; PHEP-1; PNE-R-34. 87 pp. (NSA 24: 48544).

Evans, T. C. 1934. Respiration and visible development of individual grasshopper eggs following x-irradiation. Physiol. Zool. 7(4): 556-571.

Evans, T. C. 1935. Variations in suseptibility to x-rays of Melanoplus differentialis eggs during development. Physiol. Zool. 8(4): 521-529.

Evans, T. C. 1936. Qualitative and quantitative changes in radiosensitivity of grasshopper eggs during early development. Physiol. Zool. 9(4): 443-454.

Evans, T. C. 1937. Effects of roentgen radiation on embryonic organization and regulation in Melanoplus differentialis (Orthoptera). Physiol. Zool. 10(3): 258-268.

Evans, T. C., A. J. Levin and N. M. Sulkin. 1941. Inhibition of embryo formation in certain nematodes by roentgen radiation. Proc. Exptl. Biol. Med. 48(3): 624-628.

Evans, T. R. 1967. A new technique for placement of radioactive solutions in root studies and a study of root growth of Paspalum commersonii. Austral. J. Exptl. Agr. Anim. Husb. 7: 447-452. (NSA 23: 6391).

Evans, V. A. 1975. State environmental impact requirements and power facility licensing. Trans. Am. Nucl. Soc. 22: 85-86. (NSA 33: 22647).

Evlakhova, A. A. and B. K. Martens. 1968. Effect of ionizing radiation on the viability and frequency of occurrence of morphological variations of the fungus Beauveria bassiana (Bals.). Vuill. Mikol. Fitopatol. 2: 274-278. (In Russian) (NSA 23: 48258).

Ewert, M. A. 1966. Movements to hibernation sites and hibernation of two species of toads (Bufonidae) in northwestern Minnesota. Abstr. Bull. Ecol. Soc. Am. 47(3): 150. (^{182}Ta).

Ewert, M. A. 1969. Seasonal movements of the toads Bufo americanus and B. cognatus in northwestern Minnesota. Thesis, Univ. of Minnesota, Minneapolis. 217pp. (Dissertation Abstr. 30(6B): __; WR 140: 219) (^{182}Ta).

Ewing, R. A., J. E. Howes and R. B. Price. 1969. Radionuclide and stable-element analyses of environmental samples from Routes 17 and 25. Bioenvironmental and radiological-safety feasibility studies, Atlantic-Pacific interoceanic canal. Battelle Columbus Labs., U. S. AEC report BMI-171-29. 52pp. (NSA 24: 29721).

Extermann, R. C. 1958. Research with Radioisotopes in Plant Biology and Some General Problems. Radioisotopes in Scientific Research. Vol. IV. Pergamon Press, New York. 708pp. (Incl. subjects of an ecological nature).

Eyman, L. D. 1972. Cesium-137 and stable cesium in a hypereuthrophic lake. Thesis, Michigan State Univ., East Lansing. 113pp. (Univ. Microfilms 73-5367) (NSA 27: 27598).

Eyman, L. D. 1975. Changes in ^{137}Cs concentration in fish flesh during preparation for human consumption. Health Physics 28(4): 476-477. (NSA 32: 3500) (Micropterus salmoides).

Eyman, L. D. and N. R. Kevern. 1975. Cesium-137 and stable cesium in a hyper-eutrophic lake. Health Physics 28(5): 549-555. (NSA 32: 11546).

Eyman, L. D. and J. T. Kitchings. 1973. Availability of ^{137}Cs to fishes from ingested clays. Oak Ridge National Lab., U. S. AEC report CONF-740826-1. 16pp. (NSA 30: 24391).

Ezemenari, F. R. C. and J. R. Prescott. 1970. Autumnal peak in ^{109}Cd fall-out in precipitation at Calgary, Canada. J. Geophys. Res. 75(27): 5271-5275.

Faberov, A. I., G. V. Barybina and V. V. Popov. 1972. Effect of x radiation on the pigment epithelium and ganglionic cells of tadpole retina (Rana temporaria). Dokl. Akad. Nauk SSSR 206(4): 1014-1017. (In Russian) (NSA 28: 8463). (Farberov ?; q.v.).

Fabricius, E. 1963. Behavioral effects of radiolesions in the brain of the pigeon (Columba livia). Appendix I. Final technical report. Univ. Stockholm. (AD-635661; NSA 21: 45773).

Fabricius, E. 1964. Behavioral effects of radiolesions in the brain of the pigeon. pp. 708-721. In: Response of the Nervous System to Ionizing Radiation. (See Haley and Snider, 1964).

Fabries, M. 1971. Some ecological and biogeographical aspects of the chronic gamma irradiation of a Mediterranean type ground phytocenose. Centre d'Études Nucléaires, France Commissariat à l'Énergie Atomique, Cadarache, report CEA-R-4300. 100pp. (Also: Thesis, Univ. Toulouse) (In French) (NSA 26: 38635).

Fabries, M. and A. Grauby. 1975. Preliminary data on the effects of low radiation doses on plant life. Centre d'Études Nucléaires, France Commissariat à l'Énergie Atomique, Saclay, report CEA-CONF-3126. 6pp. (In French) (NSA 33: 20491) (Ecosystem).

Fabries, M., A. Grauby and J. L. Trochain. 1972. Study of a Mediterranean type phtyocenose subjected to chronic gamma radiation. Radiation Botany 12(3): 125-135. (NSA 26: 50847).

Fabrikant, I. B. 1966. Influence of prior irradiation on host susceptibility to infection with the attenuated strain of Rickettsia prowazekii. Thesis, Univ. of Maryland, College Park. 153pp. (NSA 21: 18180).

Facugauchi, M. E. 1960. Radioactive measurements in the living fish. Ark. Kemi 15: 375-377. (NSA 16: 20059).

Fähnrich, P. 1965. Inactivation of algae cells with x-radiation. Radiation Botany 5: 507-512. (In German) (NSA 20: 8701).

Failla, P. M. 1959. Some effects of radiation on rotifers and their unirradiated progeny. Abstr. Radiation Research 11(3): 442.

Failla, P. M. and R. C. Rustad. 1970. Protein synthesis and recovery from gamma-ray-induced mitotic delay in fertilized sea urchin eggs. Int. J. Radiat. Biol. 17: 385-388. (NSA 24: 32162).

Fairbairn, D., et al. 1961. Biochemistry of normal and irradiated strains of Hymenolepis diminuta. Exptl. Parasitol. 11(4): 248-263. (NSA 17: 17831).

Fairbanks, L. D. and G. E. Burch. 1966. Turnover of radiosodium in Drosophila melanogaster adults of different ages. J. Insect Physiol. 12: 591-599.

Falk, R. and A. Rahat. 1963. The dependence of viability effects caused by irradiation on the type of mating. Proc. Natl. Acad. Sci. (US) 49(3): 292-298. (NSA 17: 28748) (Drosophila).

Fall, M. W., A. B. Medina and W. B. Jackson. 1971. Feeding patterns of Rattus rattus and Rattus exulans on Eniwetok Atoll, Marshall Islands. J. Mammalogy 52(1): 69-76. (NSA 25: 46771; WR 141: 33).

Faludi, B., I. Csukas-Szatloczky and L. Keresztes. 1962. Incorporation of ^{32}P into ontogenetically different larvae of the Drodophila melanogaster. Ann. Univ. Sci. Budapest (biol.) 5: 87-95. (NSA 22: 23460).

Faludi, B., I. Csukas-Szatloczky and K. Szeplaky. 1960. Transfer to the new generation of the ^{32}P, incorporated by the parents, of the Drosophila melanogaster Maig. Ann. Univ. Sci. Budapest (biol.) 3: 171-178.

Faludi, B., et al. 1961. Transmission of phosphorus-32 incorporated by the parents into descendants of Drosophila melanogaster. Nature 190: 469.

Famelis, S. A. 1973. Study of the radiosensitivity of natural populations of freshwater mollusk Lymnaea stagnalis L. pp. 89-94. In: Radioecology of Water Organisms. III. (See Andrushaitis, 1973) (In Russian) (NSA 30: 24360).

Famelis, S. A. 1975. Radiosensitivity of embryos of Lymnaea stagnalis L. from two natural populations. Ekologiya 6(2): 84-86. (Sov. J. Ecol. 6(2): 156-158) (NSA 33: 29758).

Fanale, F. P. and O. A. Schaeffer. 1965. Helium-uranium ratios for Pleistocene and Tertiary fossil aragonites. Science 149(3681): 312-317. (Also: Brookhaven National Lab., U. S. AEC report BNL-8939) (NSA 19: 38868).

Farberov, A. I., Z. A. Sokolova and V. V. Popov. 1965. X-irradiation influence on frog retina. Radiobiologiya 5(2): 319-320. (AEC-tr-6599; pp. 246-248) (NSA 19: 28087) (Faberov ?; q.v.).

Farcasiu, O. and H. Sahagia. 1969. Measurement of atmosphere and soil radioactivity by aero methods. Use of helicopters for rapid detection of contamination by gamma emitters. Bul. Inst. Politeh. Brasov B11: 57-70. (In Rumaniun) (NSA 25: 32075).

Farges, L. 1970. Automation of a series of radioactive measurements on biological specimens. pp. 707-719. In: Nuclear Techniques in Environmental Pollution. (See IAEA, 1970) (Also: France Commissariat à l'Énergie Atomique, Fontenay-aux-Roses, report CEA-CONF-1647) (In French) (NSA 25: 13229).

Farkas, J. 1964. Investigations into the radiation resistance of *Tribolium confusum* (Duval) and *Tyrophagus dimidiatus* (Hermann). Acta Microbiol. Acad. Sci. Hung. 12: 15-28. (NSA 20: 24874).

Farkas, J. 1965. Study of the irradiation tolerance of some destructive storehouse insects and technical and economic aspects of insect destruction by irradiation. Elelm. Ipar. 19(7): 221-225. (In Hungarian; Russian, German, English summaries).

Farkas, J. 1966. Investigations into the radiation resistance of *Tribolium confusum* (Duval) and *Tyrophagus dimidiatus* (Hermann). Acta Biol. Acad. Sci. Hung. 16: 207-215. (NSA 20: 18302).

Farkas, J., I. Kiss and E. Andrassy. 1967. The survival and recovery of irradiated bacterial spores as affected by population density and some external factors. pp. 343-354. In: Radiosterilization of Medical Products. International Atomic Energy Agency, Vienna, Publ. STI/PUB/157. (CONF-670644) (NSA 22: 52076).

Farmakes, R. (Ed.). 1974. Transactions of the American Nuclear Society Conference on Nuclear Power Plant Siting. (Portland, Ore., August 25-28, 1974). Trans. Amer. Nucl. Soc. 19(1, Suppl.): 1-53. (NSA 30: 25997).

Farmer, J. G. and M. S. Baxter. 1972. Short-term trends in natural radiocarbon. pp. A58-A71. In: Proc. 8th Intern. Radiocarbon Dating Conf. Vol. 1. (See Rafter, 1972) (NSA 31: 8682).

Farmer, J. G. and M. S. Baxter. 1974. Atmospheric carbon dioxide levels as indicated by the stable isotope record of wood. Nature 247(5439): 273-275. (NSA 30: 319).

Farmer, W. J. and C. R. Jensen. 1970. Diffusion and analysis of carbon-14 labeled dieldrin in soils. Soil Sci. Soc. Am. Proc. 34(1): 28-31.

Farnham, A. W., K. A. Lord and R. M. Sawicki. 1965. Study of some of the mechanisms connected with resistance to diazinon and diazoxon in a diazinon-resistant strain of houseflies. J. Insect Physiol. 11(1): 1475-1488. (^{14}C).

Farrar, J. L. 1953. Distribution of radiophosphorus in red pine seedlings. Silv. Leafl. For. Bur. Canada No. 78. 4pp. (FA 15: 3242).

Farris, G. C. 1965. Strontium-90 in antlers and selected bones of Colorado mule deer. Thesis, Colorado State Univ., Fort Collins. vii, 83pp.

Farris, G. C. 1967. Factors influencing the accumulation of strontium-90, stable strontium, and calcium in mule deer. Dissertation, Colorado State Univ., Fort Collins. 189pp.

Farris, G. C., F. W. Whicker and A. H. Dahl. 1967. Effect of age on radioactive and stable strontium accumulation in mule deer bone. pp. 93-102. In: Strontium Metabolism. (See Lenihan, Loutit and Martin, 1967) (NSA 21: 45582).

Farris, G. C., F. W. Whicker and A. H. Dahl. 1969. Strontium-90 levels in mule deer and forage plants. pp. 602-608. In: Symposium on Radioecology. (See Nelson and Evans, 1969) (Also: Colorado State Univ., Ft. Collins, U. S. AEC report COO-1156-23) (NSA 22: 19298).

Farvar, M. A. and H. Cember. 1969. Lethal effects of gamma radiation on the schistosome larva carrier snail (*Australorbis glabratus*). Am. J. Publ. Hlth. 59(11): 2077-2080. (NSA 24: 17125).

Fasolo, G. B., et al. 1974. Environmental radiation surveillance around Italian nuclear centers. pp. 219-237. In: Environmental Surveillance Around Nuclear Installations. Vol. 2. (See IAEA, 1974) (NSA 31: 14037).

Fassuliotis, G. 1961. Observations on the biological effects of ionizing radiations on the life cycle of *Heterodera rostochiensis* Wollenweber, 1923. Thesis, New York Univ. (Univ. Microfilms 61-2593)†

Fassuliotis, G. and A. H. Sparrow. 1955. Preliminary report of x-ray studies on the golden nematode. Plant Disease Rep. 39: 572.

Fatzinger, C. W. and J. C. Dixon. 1965. Use of x-rays to detect southern pine beetles in short-leaf pine bark. J. Forestry 63: 451-455. (*Dendroctonus frontalis*).

Faure, J. 1972. Effect of organic matter in a water course on radioactive pollution. Methods of extracting, fractionating, and identifying the organic constituents and their radionuclide affinities. Centr d'Études Nucléaires, France Commissariat à l'Énergie Atomique, Cadarache, report CEA-R-4335. 63pp. (In French) (NSA 27: 2459).

Faure, J. 1973. Role of organic matter of a waterway with regard to the radioactive pollution. France Commissariat à l'Énergie Atomique report CEA-R-4443. ii, 25pp. + tabs., figs. (ANL-TRANS-943) (NSA 28: 8161).

Faure, J. 1973. Role of the organic materials of a waterway with respect to radioactive pollution.. Centre d'Études Nucléaires, Commissariat à l'Énergie Atomique, Cadarache, report CEA-R-4505. 42pp. (In French) (NSA 29: 7744).

† Dissertation Abstr. 22: 401-402.

Fauth, H. 1973. Hydrogeochemical reconnaissance prospecting. pp. 209-218. In: Uranium Exploration Methods. International Atomic Energy Agency, Vienna, Publ. STI/PUB/334. (NSA 29: 24182).

Faux, F. 1970. Environmental problems in connection with power production. Elektrizitätswirtschaft 69(24): 643-653. (In German) (NSA 27: 29840).

Favino, A., G. F. Besseghini and S. Ghittori. 1973. Radioecological aspects of clinical use of $Na^{131}I$. Abstr. pp. 95-95. In: Proc. 2nd Conf. on the State of Advancement of Radioecology in Italy. (See Anonymous, 1973) (In Italian).

Fawcett, P., D. Green and G. Shaw. 1972. Determination of phosphorus in pollen and sporopollenins using neutron activation techniques. Radiochem. Radioanal. Lett. 9(5-6): 321-327. (NSA 26: 25153).

Fawcett, P., D. Green and G. Shaw. 1973. Application of neutron activation analysis to the determination of trace elements in pollen and sporopollenins. J. Radioanal. Chem. 13(2): 313-318. (NSA 28: 24089).

Fawcett, P., D. Green and G. Shaw. 1974. Determination of phosphorus in pollen grain wall materials using neutron activation analysis. Radiochem. Radioanal. Lett. 17(2): 121-127. (NSA 30: 2476).

Fay, R. W. 1957. The use of radioactive compounds in the study of insecticide metabolism. Proc. Chem. Spec. Mfg. Assoc. 43: 153-155.

Fay, R. W., J. T. Baker and J. A. Jensen. 1959. Radioactive tagging of Culex quinquefasciatus (Say) with P^{32}. Mosquito News 19(3): 139-143.

Fay, R. W., E. M. McCray and J. W. Kilpatrick. 1963. Mass production of sterilized male Aedes aegypti. Mosquito News 23: 210-214. (NSA 20: 35287).

Fay, R. W., J. W. Kilpatrick and J. T. Baker. 1963. Rearing and isotopic labeling of Fanna canicularis. J. Econ. Entomol. 56(1): 69-70.

Fazio, C. 1966. Bees and radioactive ants. Note. Indicators of radioactivity. Agricoltore Ital. 12(2): 3-8. (In Italian).

Federer, B. 1970. Neutron activation determination of the aerosol content of Greenland snow. Pure Appl. Geophys. 79: 120-127. (NSA 24: 48063).

Fedorov, A. F. 1960. The absorption of long-lived beta emitters by plankton. Med. radiologiya 6(6): 51-54. (In Russian) (NSA 14: 22747).

Fedorov, Ye. K. and V. I. Baranov. (Eds.). 1956. Content of natural radioactive substances in the atmosphere and in natural waters in the territory of the USSR. Akad. Nauk USSR, Moscow. (Also: UN Document A/AC.82/G/R. 39) (AEC-tr-3054) (NSA 12: 672).

Fedorov, G. A., et al. 1967. A study of the Cd-109 deposition at several points in the Soviet Union during 1964-1965. Atom. energiya 23(3): 253-255. (Sov. Atom. Energy 23(3): 902-904).

Fedorov, G. A., et al. 1967. Determination of the concentrations of cadmium-109 in the ground level atmospheric air and of fallout at various points in the Soviet Union 1964-1966. State Committee on the Use of Atomic Energy, USSR. Atomizdat, Moscow. UN Document A/AC.82/G/L.1196. 9pp. (AEC-tr-7214; pp. 62-69) (NSA 22: 18946; 25: 15779).

Fedorov, G. A., et al. 1968. Nature of the vertical distribution of cesium-137 in the soil in some areas of the Soviet Union, 1966-1967. State Committee on the Use of Atomic Energy, USSR. Atomizdat, Moscow. UN Document A/AC.82/G/L.1256. 6pp. (AEC-tr-7030; pp. 145-148) (NSA 23: 17553).

Fedorova, G. V. 1963. The effect of water temperature, age, and sex on a fish on the absorption and elimination of C^{14}. Dokl. Akad. Nauk SSSR 150 (1): 168-169. (Doklady (biol. sci.) 150: 669-670) (NSA 17: 24874).

Fedorova, G. V. 1963. Experimental investigation of ^{14}C penetration into developing eggs and larvae of freshwater fishes. Radiobiologiya 3(5): 677-681. (AEC-tr-5438; pp. 73-80) (BA 45: 31719; NSA 18: 23223; SFA 10: 7309).

Fedorova, G. V. 1963. The uptake and release of radioactive carbon (C^{14}) by fish. Nauchn. dokl. vyssh. shkoly (biol.) (4): 84-89. (JPRS-23,604) (NSA 18: 17412).

Fedorova, G. V. 1963. Effect of Sr^{90} on the eggs and larvae of the Ladoga whitefish, Coregonus lavaretus ludoga. Vestn. Leningrad Univ. (biol.) (1): 48-53. (In Russian) (BA 43: 4255; SFA 8: 6132).

Fedorova, G. V. 1964. On the effect of C^{14} on developing ova and larvae of freshwater fish. Vopr. ikhtiol. 4: 723-728. (JPRS-29,452; pp. 62-71) (NSA 19: 24263).

Fedorova, G. V. 1965. Entrance of C-14 into spawn and grubs of autumn spawning fishes. Radiobiologiya 5(5): 690-692. (AEC-tr-6602; pp. 86-90) (NSA 20: 5261).

Fedorova, G. V. 1965. Effects of irradiation with ^{60}Co on embryonic and early post-embryonic development of Coregonus peled. Nauchn. dokl. vyssh. shkoly (biol.) (2): 84-87. (In Russian) (NSA 21: 3447).

Fedorova, G. V. 1968. Effect of ^{14}C on hatching of whitefish larvae from roe. Vestn. Leningrad Univ. (biol.) 4(21): 45-49. (In Russian) (NSA 25: 39907).

Fedorova, G. V. 1972. The biological effect of radiocarbon (^{14}C) on fish in the early development stages. Vopr. ikhtiol. 12(1): 173-177. (In Russian) (SFA 18: 16655).

Fedorova, G. V. 1973. Biological effect of ^{14}C incorporated in fish eggs. Radiobiologiya 12(4): 561-565. (AEC-tr-7397; pp. 109-116) (NSA 27: 17691).

Feely, H. W., et al. 1961. The potential applications of radioisotope techniques to water resource investigations and utilization. Isotopes, Inc., Westwood, N. J., U. S. AEC report NYO-9040. xii, 340pp.

Feher, I. and I. Erdelyvari. 1973. Up-to-date methods in the environmental control of nuclear establishments. Izotoptechnika 16(4): 199-208. (In Hungarian) (NSA 28: 12855).

Feige, Y., B. G. Oltman and J. Kastner. 1968. Production rates of neutrons in soils due to natural radioactivity. J. Geophys. Res. 73(10): 3135-3142.

Feinberg, R. J. 1957. Criteria for pre-startup environmental monitoring programs for nuclear sites. Knolls Atomic Power Lab., U. S. AEC report AECU-4074. 24pp. (NSA 13: 11707).

Feinendegen, L. E. 1967. Tritium-Labeled Molecules in Biology and Medicine. Academic Press, New York. x, 430pp. (NSA 22: 51464).

Feinendegen, L. E. 1968. Problems associated with the use of labelled molecules in biology and medicine. pp. 1-15. In: Biological Effects of Transmutation and Decay of Incorporated Radioisotopes. (Proc. Panel, Vienna, 9-13 October 1967). International Atomic Energy Agency, Vienna, Publ. STI/PUB-183. (CONF-671036) (NSA 24: 31798).

Feinstein, R. N., J. B. Howard and J. T. Faulhaber. 1969. Catalase levels and radiation resistance in three species of wild duck. Int. J. Radiat. Biol. 15: 341-346. (NSA 23: 45645).

Feir, D and E. McClain. 1968. Induced changes in the mitotic activity of hemocytes of the large milkweed bug, Oncopeltus fasciatus. Ann. Entomol. Soc. Am. 61: 416-421. (NSA 22: 23611) (Irradiation).

Feir, D. and G. M. O'Connor. 1965. Mitotic activity in the hemocytes of Oncopeltus fasciatus (Dall). Exptl. Cell Res. 39(2-3): 637-642. (^{3}H).

Feldt, W. 1960. Radioactivity in fish in the Elbe in the vicinity of the Geesthact reactor. Z. Lebensmitt. Untersuch. 111(6): 465-474. (In German) (SFA 9: 6607; WPA 36: 2116).

Feldt, W. 1968. Radiation protection aspects of control of radioactive contamination of fish. pp. 395-400. In: Radiation Protection, Part 1. (See Snyder, et al., 1968) (In German) (NSA 23: 12126).

Feldt, W. 1969. Radiological studies in the Ems below the Lingen Nuclear Power Plant. pp. 235-245. In: Radiation Protection Problems in the Release and Incorporation of Radioactive Materials. (See Hacke and Jacobi, 1969) (In German) (NSA 24: 28934).

Feldt, W. 1969. The concentration and distribution of inactive isotopes as trace pollutants in marine and freshwater environments, and the effect of a calcium injection into a lake. 15pp. (For 5th R. I. S. Symp., Helsinki, Finland, 19 May 1969). (CONF-690527-2) (NSA 25: 37715).

Feldt, W. 1971. Radiation aspects of aquatic radioecology. pp. 445-457. In: Proc. Intern. Symp. on Radioecology Applied to the Protection of Man and His Environment. (See European Atomic Energy Community, 1971) (In German) (NSA 27: 5276).

Feldt, W. 1971. Research on the maximum radioactive burden of some German rivers. pp. 495-506. In: Environmental Aspects of Nuclear Power Stations. (See IAEA, 1971) (CONF-700810-12) (SM 146/29) (NSA 24: 44093).

Feldt, W. 1974. Radioecology of coastal and inland waters. pp. 112-130. In: Contamination and Decontamination of Foods. E. Fischer, ed. AID, Bonn. (In German) (NSA 32: 17205).

Feldt, W. 1974. Radioecology of the ocean and inland waters. pp. 384-402. In: Radiation Protection, Environment, and the Population. Vol. 2. (See German Democratic Republic, 1974) (In German) (AED-CONF-74-725-000(Vol. 2)) (NSA 33: 25968).

Feldt, W. and H. Bühringer. 1970. Damage to fish in the stages of embryonic and gonad development by ionizing radiation. Zentralbl. Veterin. Beih. (11): 42-50. (In German) (NSA 25: 24377).

Feldt, W. and J. Lange. 1962. The ^{90}Sr and ^{137}Cs content of carp from north German inland waters. Z. Lebensmitt. Untersuch. 117: 103-113. (In German) (French transl. Commissariat à l'Énergie Atomique report CEA-tr-A-1880) (NSA 21: 28413, 1710).

Felix Estrada, R. 1972. Theory of somatic mutation in relation to aging. Effect of ionizing irradiation and some antioxidants on the life span of Drosophila melanogaster. Thesis, Univ. Nacional Autonoma de Mexico, Mexico City. 133pp. (In Spanish) (NSA 27: 20149).

Fendley, T. T. 1968. The whole body burden of radioactivity in the cottontail rabbit (Sylvilagus floridanus) of Georgia. Thesis, Univ. of Georgia, Athens. 32pp.

Fendrik, I. and H. Glubrecht. 1967. Investigations of the propagation of plant pollen by an indicator activation method. pp. 325-333. In: Nuclear Activation Techniques in the Life Sciences. (See IAEA, 1967) (NSA 23: 5052).

Fenske, P. R. 1972. Event-related hydrology and radionuclide transport at the Cannikin site, Amchitka Island, Alaska. Desert Research Inst., Reno, Nev., U. S. AEC report NVO-1253-1. iv, 37pp.

Ferber, G. J. and T. R. Palmer. 1970. Weather and fallout studies for interoceanic canal Routes 17 and 25 final report. Environmental Science Services Administration, U. S. Dept. of Commerce, Washington, report ERLTM-ARL-24. vii, 140pp.

Ferber, R. R. (Ed.). 1971. Nuclear Plant Safety. (IEEE Tutorial Course). Inst. Electrical and Electronics Engineers, New York, Course Text 72 CHO 635-3 PWR. 129pp. (NSA 27: 16564) (Incl. environmental effects).

Ferguson, D. L. 1966. Migratory behavior of x-irradiated and nonirradiated larvae of Ascaris suum in pigs. Thesis, Univ. of Nebraska, Lincoln. 142 pp. (NSA 21: 22828).

Ferguson, J. M. 1963. Ground roughness effects for fallout-contaminated terrain: comparison of measurements and calculations. U. S. Naval Radiological Defense Lab., San Francisco, report USNRDL-TR-645. 29pp. (NSA 17: 29139).

Ferguson, T. 1973. Some effects of irradiation on delayed amputation upon regeneration in *Tubularia crocea*. Int. J. Radiat. Biol. 24(4): 421-425. (NSA 29: 18646).

Féron, M. 1962. Study of the winter survival of two species of insects, *Ceratitis capitata*, and *Dacus oleae* (Dipt. Trypetidae), and the factors of pullulation. Ann. Epiphyt. 13(4): 336. (In French) (^{32}P).

Féron, M. 1963. The struggle against insects by autocide methods. Rev. Zool. Agr. 62(4-6): 37-48. (In French) (Discusses radiation).

Féron, M. 1966. Sterilization of the Mediterranean fruit fly, *Ceratitis capitata* Wied., by irradiation of pupae with gamma rays. Ann. Epiphyt. 17(2): 229-239. (In French).

Féron, M. 1969. Studies on the Mediterranean fruit fly in Tunisia. pp. 83-85. In: *Insect Ecology and the Sterile-Male Technique*. (See IAEA, 1969) (In French) (^{32}P).

Féron, M. and L. A. Buscarlet. 1969. Radiation and nuclear tracers in agricultural entomology. Bull. Inform. Sci. Tech. (Paris) (141): 55-62. (In French) (NSA 24: 25630).

Ferraris, M. M. and F. G. Giorcelli. 1965. Networks for the control of environmental radioactivity in Italy. Minerva Nucleare 9: 40-46. (In Italian) (NSA 19: 30467) (Incl. fresh- and seawater).

Ferrell, W. K. and E. E. Hubert. 1952. The use of radioisotopes in forest tree research. Idaho Forester 34: 42-45.

Ferrell, W. K. and F. D. Johnson. 1956. Mobility of calcium-45 after injections into western white pine. Science 124(3117): 364-365.

Ferrer, F. R. 1971. Effects of gamma radiation on the reproduction of the alfalfa weevil *Hypera postica* (Gyllenhal) (Coleoptera: Curculionidae). Thesis, Pennsylvania State Univ., University Park. 160pp. (Univ. Microfilms 71-28,685) (NSA 26: 9607).

Ferrero, B. and S. Orsello. 1967. Effects of continuous exposure to radioactive phosphorus (^{32}P). Dosimetric problems and attempted evaluation of the dose-biological effect relationship in cultures *in vitro*. Cancro 20: 602-617. (In Italian) (NSA 23: 50534).

Ferrill, M. D. 1963. Root extension in a plantation of longleaf pine: investigation of a technique using I-131. Dissertation, Duke Univ., Durham, N. Car. 129pp.

Ferrill, M. D. and F. W. Woods. 1966. Root extension in a longleaf pine plantation. Ecology 47: 97-102. (NSA 20: 36560).

Ferronskiy, V. I. 1971. Neutron techniques in ground water exploration. pp. 48-58. In: *Neutron Sorces and Applications*. (See IAEA, 1971) (NSA 25: 48603).

Ferronskiy, V. I. 1975. *Natural Isotopes of the Hydrosphere*. "Nedra", Moscow. 280pp. (In Russian) (NSA 33: 17175).

Ferronskiy, V. I., et al. 1968. *Radioisotope Investigative Methods in Engineering Geology and Hydrogeology*. Atomizdat, Moscow. 304pp. (AEC-tr-7230) (NSA 25: 48599).

Feteanu, A. and I. Baditoiu. 1964. Radioactive contamination of animal food products. Rev. Sanit. Mil. 60: 423-430. (NSA 19: 640) (Incl. fish).

Feteanu, A., et al. 1973. Experimental study on the protection of chickens and pheasants against syngamosis by use of irradiated larvae. I. Methods for immunization, preparation of immunizing material and identification of serum antibodies by immunoflourescence. pp. 101-111. In: *Isotopes and Radiation in Parasitology. III*. International Atomic Energy Agency, Vienna, Publ. STI-PUB-328. (NSA 29: 15764).

Fetter, F. and H. Altmann. 1968. Investigations on radioinduced changes in nuclei and metabolism of *Chlorella pyrenoidosa*. Bodenkultur 19: 320-335. (In German) (NSA 23: 24530).

Fetterolf, C. M., Jr. and D. J. Seeburger. 1972. Radioactivity in Lake Michigan water and biota, vicinity of Big Rock Point Nuclear Reactor. pp. 211-218. In: *Proc. 14th Conf. on Great Lakes Research*. Intern. Assoc. for Great Lakes Research, Ann Arbor, Michigan. (NSA 26: 45535).

Fiedler, H. J. and H. Reissig. 1965. Contamination of soils and plants with ^{90}Sr as a consequence of nuclear weapons tests during the period 1960 through 1963. Biol. Runsch. 3(4): 149-168. (In German) (NSA 25: 57986).

Field, P. E., R. W. Lichtenheld and R. V. Kilambi. 1964. Effects of various photoperiods and Co^{60} tags upon behavior of steelhead trout. Research in Fisheries, 1963, Contribution No. 166. Univ. of Washington Press, Seattle.

Fields, S. R. 1971. Cost-benefit algorithms for evaluating hydro-environmental effects of nuclear power plants. Hanford Engineering Development Lab., U. S. AEC report HEDL-71-23. 83pp. (NSA 25: 26003).

Figge, H. J. and R. Wichterman. 1959. The protection of radiation sensitized paramecia from lethal doses of roentgen radiation. Abstr. Anat. Record 133: 276.

Figdor, S. K. 1970. Automated calculation of ^{14}C and ^{35}S radioactivity in biological samples by external standardization and a computer program. Comput. Biomed. Res. 3: 201-211. (NSA 26184).

Figura, V. 1961. Present knowledge on the use of radioisotopes and ionizing radiation in entomology. Nuovi Ann. Ig. Microbiol. 12(12): 156-170. (In Italian) (NSA 17: 17688).

Filby, R. H. and K. R. Shah. 1971. Mode of occurrence of trace elements in petroleum and relationship to oil-spill identification methods. pp. 86-96. In: *Nuclear Methods in Environmental Research*. ((See Vogt, et al., 1971) (NSA 26: 17359).

Filby, R. H. and K. R. Shah. 1974. Activation analysis and applications to environmental research. Toxicol. Environ. Chem. Rev. 2(1): 1-44. (NSA 31: 13599).

Filby, R. H., A. I. Davis and K. R. Shah. 1973. Neutron activation analysis for the determination of mercury in environmental samples. pp. 257-264. In: Mercury in the Western Environment. (See Buhler, 1973) (Incl. pheasant, water).

Filby, R. H., K. R. Shah and W. H. Funk. 1974. Role of neutron activation analysis in the study of heavy metal pollution of a lake-river system. pp. 10-23. In: Proc. 2nd Intern. Conf. on Nuclear Methods in Environmental Research. (See Vogt and Meyer, 1974) (NSA 33: 23322).

Filby, R. H., et al. 1970. Determination of mercury in biological and environmental materials by instrumental neutron activation analysis. Mikrochim. Acta (6): 1130-1136. (NSA 25: 34551).

Fillipova, V. N. and I. F. Seitz. 1958. Effect of x-rays on the enzymatic system of coenzyme A in pigeon liver. Biokhimiya 23(1): 119-124.

Fillipovich, Yu. B. 1964. The biosynthisis mechanism of silk fibroin. Usp. sovrem. biol. 57(2): 192-201. (References to radionuclides).

Fimian, W. J., Jr. 1959. Leucocyte infiltration in localized regions of x-rayed amphibian limb tissues. Exptl. Cell Res. 16(2): 418-420.

Fimreite, N., et al. 1971. Mercury in fish and fish-eating birds near sites of industrial contamination in Canada. Can. Field-Naturalist 85(3): 211-220. (Activation analysis).

Findenegg, I. 1964. Types of planktic primary production in the lakes of the eastern Alps as found by the radioactive carbon method. Verh. Intern. Limnol. 15(1): 352-359.

Fine, B. C., M. E. Letelier and M. Agosin. 1966. Effects of various synergists on toxicity and in vivo metabolism of DDT in Triatoma infestans nymphs. Exptl. Parasitol. 19(3): 304-309. (^{14}C).

Fine, B. D., et al. 1967. Resistance to pyrethrins and DDT in a strain of houseflies Musca domestica L. The sorption of a synthetic (^{14}C) pyrethrin I. J. Sci. Food. Agr. 18(5): 220-224.

Fineman, Z. M. and R. McBride. 1962. Variations in jack rabbit bone strontium 90 with age and fallout. Abstr. Bull. Ecol. Soc. Am. 43(4): 124-125.

Fineman, Z. M., R. McBride and J. Detmer. 1963. Use of the jack rabbit as a bio-indicator of environmental strontium-90 contamination. pp. 455-460. In: Radioecology. (See Schultz and Klement, 1963) (NSA 17: 33596).

Fink, B. A. 1957. Radioiodine: a method for measuring thyroid activity. Auk 74(4): 487-493.

Finkelstein, A. 1973. Nuclear energy and the environment: role of the International Atomic Energy Agency. pp. 47-55. In: Nuclear Energy and the Environment. Italy, Comitato Nazionale Energia Nucleare, Rome. (In French) (NSA 29: 15567).

Finland Institute of Radiation Physics. 1967. Studies of environmental radioactivity in Finland 1966. Annual report. Helsinki, report SFL-A8. 58 pp. (NSA 22: 10764).

Fioca, L. and E. A. Pora. 1972. Cellular respiration in ontogeny of the silk gland in Bombyx mori under the influence of microdoses of x and UV irradiations. Studii Cerc. biol. zool. 24(2): 125-137. (In Rumanian) (NSA 28: 2969).

Firsakova, S. K. and V. A. Shirshov. 1971. Degree of ^{90}Sr contamination of pasture fodder in 1965-1967. Agrokhimiya (11): 128-131. (In Russian) (NSA 26: 48059; 27: 22500).

First, M. S. and J. M. Morgan, Jr. (Eds.). 1970. Proc. 11th AEC Air Cleaning Conf. (Held in the Federal Building, Richland, Washington, August 31-September 3, 1970). U. S. AEC report CONF-700816(Vol.2). 411pp. (NSA 25: 21268).

Fischer, E. 1972. Causes and effects of contamination of agricultural products with radionuclides. Ber. Landwirt. 50(2): 267-275. (In German) (NSA 27: 24618).

Fischer, H. A. and E. Schmatolla. 1972. Axonal transport of tritium-labeled putrescine in the embryonic visual system of zebrafish. Science 176(4041 (4041): 1327-1329.

Fisher, H. L., J. R. Coleman and R. O. Grundy. 1966. Strontium-90 concentrations observed in U. S. pasteurized milk compared with strontium-90 levels in precipitation, in total deposition, and in soil, 1960-1963. Radiol. Hlth. Data 7(8): 427-440.

Fishman, M. J. and D. E. Erdman. 1971, 1973. Water analysis. Anal. Chem. 43(5): 356R-388R; 45(5): 361R-403R. (NSA 25: 31787; 28: 2092) (Incl. radionuclides).

Fitch, F. J. and J. A. Miller. 1970. Radioisotope age determination of Lake Rudolf artefact site. Nature 226: 226-228. (NSA 24: 27228).

Fitch, F. W., J. Doull and R. W. Wissler. 1955. Histopathology of the irradiated hibernating ground squirrel. AMA Arch. Pathol. 60(6): 644-650.

Fitzgerald, B. W. 1962. An investigation of the zinc and zinc-65 in shellfish of Fishers Island Sound and its estuaries. Thesis, Univ. of Connecticut, Storrs. 88pp. (NSA 17: 7783).

Fitzgerald, J. J. 1969. Applied Radiation Protection and Control. Gordon & Breach, New York. 2 vols., xii, 1018pp.

Fitzgerald, P. J. 1958. Autoradiography in biology and medicine. Proc. 2nd Intern. Conf. on the Peaceful Uses of Atomic Energy 26: 208-212. (A/CONF. 15/P/406) (NSA 12: 14539).

Fitzner, R. E. and K. R. Price. 1973. Use of Hanford waste ponds by waterfowl and other birds. Battelle Northwest Lab., U. S. AEC report BNWL-1738. 47pp. (NSA 28: 2609).

Fitzner, R. E. and W. H. Rickard. 1975. Avifauna of waste ponds ERDA Hanford Reservation, Benton County, Washington. Battelle Northwest Lab., U. S. Energy Research and Development Admin. report BNWL-1885. 96pp. (NSA 32: 19769).

Fitzpatrick, B. and W. Mulligan. 1968. Variables on the radiation-attenuation of helminth larvae: temperature/oxygen effects. pp. 17-20. In: Isotopes and Radiation in Parasitology. (See IAEA, 1968).

Fitzpatrick, D. E. P., et al. 1966. ^{137}Cs body burdens in Alaskan men, spring, 1966. Radiol. Hlth. Data Rpts. 7: 691-696. (NSA 22: 23517).

Fix, J. J. 1975. Association of long-lived radioactivity with sediment along the Columbia River shoreline, islands, bottom and slough areas. Battelle Northwest Lab., U. S. ERDA report BNWL-SA-5484. 30pp. (NSA 33: 7281).

Fix, J. J. and B. J. Blumer. (Comps.). 1975. Environmental surveillance at Hanford for CY-1974 data. Battelle Northwest Lab., U. S. ERDA report BNWL-1910(Add.). 106pp. (NSA 32: 3479, 3480).

Fix, J. J. and D. E. Robertson. 1976. "Filter-resin" sampling method for measurement of radioactivity in Columbia River water. Battelle Northwest Lab., U. S. Energy Research and Development Admin. report BNWL-SA-5630. 7pp. (NSA 33: 20188).

Flaccus, E., T. V. Armentano and M. Archer. 1974. Effects of chronic gamma radiation on the composition of the herb community of an oak-pine forest. Radiation Botany 14(4): 263-271. (NSA 32: 1084).

Fleck, G. M. 1971. On the generality of first-order rates in isotopic tracer kinetics. J. Theor. Biol. 34(3): 509-___.

Fleming, S. J. 1971. Thermoluminescent dating principles and application. Naturwissenschaften 58(7): 333-338. (NSA 25: 54764) (Pottery).

Fleming, S. J. and D. Stoneham. 1973. Subtraction techniques of thermoluminescent dating. Archeometry 15(2): 229-238. (NSA 28: 24494).

Flemion, F. 1958. Penetration and destruction of plant tissues during feeding by Lygus lineolaris P. de B. Proc. 10th Intern. Congr. Entomol. 3: 475-478. (Used radionuclides).

Flemion, F. nd. Penetration and destruction of plant tissues during feeding by Lygus lineolaris. pp. 1003-1007. In: Report 14th Intern. Horticultural Congr., Netherlands 1955. H. Veenman and Zonen, Wageningen.

Flemion, F. nd. Insect damage as a factor affecting fruit set. In: Proc. campbell Soup Co. Plant Science Symp., 1962.

Flemming, A. S. 1959. Control of radioactive wastes. Publ. Health Rept. 74: 883-884. (SFA 5: 3576; WPA 33: 1126).

Flentje, N. T., H. M. Stretton and A. R. McKenzie. 1967. Mutation in Thanatephorus cucumeris. Austral. J. Biol. Sci. 20: 1173-1180. (NSA 22: 31031).

Flerov, B. A. and V. I. Romanenko. 1970. Comparative study of the elimination of C^{14} tagged antigen in certain cold-blooded and warm-blooded animals. Gidrobiol. zh. 6(2): ___. (Hydrobiol. J. 6(2): 104-106) (NSA 26: 7125).

Fletcher, B. S. and A. Giannakakis. 1973. Sex pheromone production in irradiated males of Dacus (strumeta) tyroni. J. Econ. Entomol. 66(1): 62-64. (NSA 27: 17678).

Fletcher, J. F., et al. 1973. Modelling the regional transport of radionuclides in a major United States river basin. pp. 449-465. In: Environmental Behavior of Radionuclides Released in the Nuclear Industry. (See IAEA, 1973) (Also: Hanford Engineering Development Lab., U. S. AEC report HEDL-SA-507) (NSA 28: 5543; 29: 29305).

Fleyshman, D. G. 1971. Radiometric scintillation methods for samples with small specific gravity. pp. 237-242. In: Methods of Radioecological Investigations. (See Verkhovskaya, 1971).

Fleyshman, D. G. 1971. Accumulation of artificial radionuclides in freshwater fish. pp. 395-421. In: Contemporary Problems of Radiobiology. Vol. 2. Radioecology. (See Klechkovskiy, Polikarpov and Aleksakhin, 1971).

Fleyshman, D. G. and Yu. P. Kanevskiy. 1971. Cosmogenic ^{22}Na in surface waters of dry land. Dokl. Akad. Nauk SSSR 198(6): 1397-1400. (In Russian) (NSA 25: 57568).

Fleyshman, D. G. and Yu. P. Kanevskiy. 1973. Environmental ^{22}Na in freshwater ecosystem studies. pp. 108-115. In: Radioecology of Water Organisms. II. (See Andrushaitis, 1973) (In Russian) (NSA 30: 21134).

Fleyshman, D. G., Yu. P. Kanevskiy and Z. G. Gritchenko. 1975. Determination of the age of natural waters according to cosmogenic and artificial ^{22}Na. Geokhimiya (2): 280-290. (NSA 32: 775).

Fleyshman, D. G., Yu. P. Kanevskiy and A. A. Solyus. 1973. Cs specific activity differences in components of freshwater ecosystems. pp. 160-168. In: Radioecology of Water Organisms. II. (See Andrushaitis, 1973) (In Russian) (NSA 30: 21138).

Fleyshman, D. G., Yu. P. Kanevskiy and A. A. Solyus. nd. Various causes of differences between accumulation coefficients of ^{137}Cs and stable Cs in freshwater organisms under natural conditions. (In Russian) (INIS-mf-887-3; pp. 113-123) (NSA 30: 9549).

Fleyshman, D. G., Ye. S. Romanova and S. M. Baklanova. 1965. Some data on the radioactivity of biospheric objects in the Paratunka River Basin. Vopr. ikhtiol. 5(4): 738-742. (In Russian) (BA 48: 114952).

Flickinger, R. A. 1963. The site of incorporation of C-14-labeled carbon dioxide and glycine in the planarian Dugesia dorotocephala. Exptl. Cell Res. 39(3): 605-607.

Flickinger, R. A. and S. J. Coward. 1962. The induction of cephalic differentiation in regenerating Dugesia dorotocephala in the presence of the normal head and in unwounded tails. Develop. Biol. 5(2): 179-204) (^{14}C).

Flint, H. M. 1964. The effect of cobalt 60 gamma rays on the biology of the eye gnat, Hippelates pusio Loew. Thesis, Univ. of Florida, Gainesville. 67pp. (Dissertation Abstr. 25(7): 3803; NSA 19: 31894).

Flint, H. M. 1965. The effects of γ radiation on the fertility and longevity of Hippelates pusio., J. Econ. Entomol. 58(3): 555-559. (NSA 19: 43803).

Flint, H. M. 1966. Radiation sterilization studies on the tobacco budworm, Heliothis virescens Fab. Abstr. Bull. Entomol. Soc. Am. 12: 286.

Flint, H. M. and E. L. Kressin. 1967. Gamma irradiation of pupae of the tobacco budworm. J. Econ. Entomol. 60(6): 1655-1659.

Flint, H. M. and E. L. Kressin. 1968. Gamma irradiation of the tobacco budworm: sterilization, competitiveness, and observations on reproductive biology. J. Econ. Entomol. 61(2): 477-483. (NSA 22: 36427).

Flint, H. M. and I. C. McDonald. 1972. Genetic loads in two house fly populations treated with sterilizing doses of x-rays for thirty-two generations. Ann. Entomol. Soc. Am. 65(3): 594-597. (NSA 26: 41223).

Flint, H. M. and N. Spencer. 1970. Sterilizing doses of gamma irradiation for the imported cabbageworm, Pieris rapae, and effects on longevity, mating, and fecundity. J. Econ. Entomol. 63: 1008-1009. (NSA 24: 34835).

Flint, H. M., W. R. Bibow and C. K. Lahren. 1966. Radiation studies with the boll weevil: lethal effects on larvae; and adults; male sterility and dose fractionation. J. Econ. Entomol. 59: 1249-1255. (NSA 21: 12971).

Flint, H. M., et al. 1971. Biological effects of irradiation with thermal neutrons on boll weevils fed boric acid containing ^{10}B. J. Econ. Entomol. 64(5): 1002-1008. (NSA 26: 7125).

Flint, H. M., et al. 1973. Gamma-irradiated pink bollworms: attractiveness, mating, and longevity of females. Environ. Entomol. 2(1): 97-100. (NSA 28: 21781).

Floca, L. and E. A. Pora. 1970. Tritiated thymidine fixation in serogenous gland DNA in Bombyx mori under the influence of gamma and UV radiation. Studii. Cerc. biol. zool. 22: 345-351. (In Rumanian) (NSA 25: 32519) (Fioca ?; q.v.).

Flora, D. H. and M. C. Wukasch. 1966. The Galveston environmental health program, 1965. Radiol. Hlth. Data 7(5): 263-267.

Florida Power Corporation. 1971. Crystal River Nuclear Generating Plant, Unit 3. Environmental report. St. Petersburg, Fla. U. S. AEC DOCKET-50302-16. 37pp. (NSA 25: 15308).

Florida Power Corporation. 1971. Crystal River Nuclear Generating Plant, Unit 3. Environmental status report. St. Petersburg, Fla. U. S. AEC DOCKET-50302-23, -24. 80, 84pp. (NSA 25: 23182, 41340).

Florida Power & Light Company. 1966. Turkey Point Nuclear Generating Units No. 3 and No. 4. Preliminary Safety Analysis Report (Ammendment 1), Suppl. 1. U. S. AEC DOCKET-50250-5. 72pp. (NSA 23: 11093) (Incl. oceanographic

Florida Power & Light Company. 1970. Turkey Point Plant. Units No. 3 and No. 4. Environmental report. Miami, Fla. U. S. AEC DOCKET-50250-36, 50251-34. 274pp. (NSA 25: 8009).

Florida Power & Light Company. 1971. Hutchinson Island Plant, Units 1 and 2. Environmental report 1. U. S. AEC DOCKET-50335-19, -50389-7. 300pp. (NSA 25: 34294).

Florida Power & Light Company. 1972. Hutchinson Island Plant, Unit 1. Environmental report, supplement 4. U. S. AEC DOCKET-50335-28. 11pp. (NSA 26: 44904).

Florkin, M. and E. Schaffeniels. 1969. Molecular Approaches to Ecology. Academic Press, New York. x, 203pp. (Contains references to ^{14}C tracer).

Florkowski, T. 1964. The possibility of tritium detection in application to hydrogeology. Kwart. Geol. 8: 887-896. (In Polish) (NSA 19: 18183).

Florkowski, T. and J. Grabszak. 1973. Liquid scintillation measurements of natural tritium in water samples. Poland Inst. Nuclear Techniques, Krakow, report ITJ-36. 43pp. (In Polish) (NSA 28: 17910).

Florkowski, T., B. Holynska and J. Niewodniczanski. 1971. Feasibility of radioisotope x-ray flourescence analysis for measuring water pollution by meats. pp. 335-345. In: Nuclear Techniques in Environmental Pollution. (See IAEA, 1971) (NSA 25: 31823).

Florkowski, T., B. R. Payne and G. Sauzay. 1970. Interlaboratory comparison of analysis of tritium in natural waters. Int. J. Appl. Radiat. Isotopes 21: 453-458. (NSA 24: 46194).

Flory, W. S. and W. R. Singleton. 1959. Initial effects from gamma radiation of several types of nursery stock (conifers, broad-leaved evergreens and deciduous taxa). Abstr. Assoc. South. Agr. Workers Proc. 56: 170.

Flory, W. S. and W. R. Singleton. 1959. A comparison of initial effects from gamma radiation upon some gymnosperms and angiosperms. Abstr. Va. J. Sci. NS10(4): 258-259.

Flowers, H. H. 1963. The effects of x-irradiation on the biological activity of cotton mouth moccasin (Agkistrodon piscivorus) venom. Toxicon 1(3): 131-136. (NSA 18: 33385).

Flowers, H. H. 1963. Effects of x irradiation on the biological activity of cotton mouth moccasin (Agkistrodon piscivorus) venom. U. S. Army Medical Research Lab., Ft. Knox, Ky., report USAMRL-588. 14pp. (AD-413791; NSA 25: 21543).

Flowers, H. H. 1966. Effects of x irradiation on the antigenic character of Agkistrodon piscivorus (cotton mouth moccasin) venum. Toxicon 3: 301-304. (NSA 21: 8727).

Floyd, E. P. and L. Weaver. 1964. Trends of strontium-90 levels in surface waters of the United States, 1959-1963. Radiol. Hlth. Data 5(8): 390-394. (NSA 18: 37386).

Fluke, D. J. 1957. The effect of x-rays on egg hatch and egg laying in Mormoniella (i.e. Nasonia). Abstr. Radiation Research 7(3): 315.

Fogg, G. E. 1964. The significance of measurements of phytoplankton photosynthesis by the radiocarbon method. Abstr. p. 252. In: Abstracts of Papers, 10th Intern. Botan. Congr. (Edinburgh, August, 1964).

Földiak, G. 1973. Isotope and radiation techniques in Hungary. Isotopenpraxis 9(2): 41-47. (In German) (NSA 29: 5263).

Foley, J. M., R. J. Berman and C. E. Smith. 1960. X-ray irradiation of Coccidioides immitis arthrospores: survival curves and avirulent mutants isolated. J. Bacteriology 79(4): 480-487.

Follenius, E. 1952. Effect of x-rays upon the evolution of the testicle and the appearance of secondary sexual characteristics in Lebistes reticulatus. Compt. rend. soc. biol. 146: 1774-1776. (In French) (NSA 7: 4018).

Follenius, E. 1952. Progress of degeneration of the germinal cells in the adult male Lebistes reticulatus after x-irradiation. Compt. rend. soc. biol. 146: 1776-1778. (In French) (NSA 7: 5019).

Follenius, E. 1963. Effect of x rays on the nuclear membrane of spermatocytes of Lebistes reticulatus R. Electron microscope study. Compt. rend. 257(13): 1982-1984. (In French) (NSA 18: 1392).

Folsom, T. R. and G. J. Mohanrao. 1960. Trend of caesium-137 in the effluent of a large city. Nature 188(4755): 979-982. (NSA 15: 7628).

Folsom, T. R., et al. 1963. A study of certain radioactive isotopes in selected wastewater treatment plants. J. Water Poll. Control Fed. 35: 304-333. (NSA 17: 36063)

Folsom, T. R., et al. 1972. Report of analyses of water samples submitted to Scripps, November 1971, for the Lake Michigan intercalibration/intercomparison program (EIIP). Scripps Inst. of Oceanography, U. S. AEC report TID-26199. 7pp. (NSA 27: 2348) (Radionuclides).

Foltin, B. 1973. Zoological gardens next to a nuclear power plant. Harmlessness is to be proven. Handelblatt 28(46): 20. (In German) (NSA 28: 31899).

Fomin, A. N., N. K. Astakhova and S. V. Gvozdeva. 1955. Consumption of organic matter synthesized by microorganisms by the roots of plants. In: Isotopes in Microbiology. (Trans. Conf. on the Application of Tracer Atoms in Microbiology). A. A. Imshenetzkiy, ed. Publ. House Acad. Sci. USSR, Moscow. (In Russian).

Fontaine, M. 1956. Atomic radiations and aquatic life. Semaine Hop. (path. biol.) Ann. Rech. med. : 172-179. (In French).

Fontaine, M. and Y. A. Fontaine. 1956. Determination du pouvois thyreotrope de l'hypophyse et du milieau interieur de teleosteens par mesure de le fixation de ^{131}I par la thyroide de la truite arc en ciel. J. Physiol. (Paris) 48: 881-892. (In French).

Fontaine, M. and J. Leloup. 1957. Specific differences in permeability to radioiodine in the erythrocytes of fishes. J. Physiol. (Paris) 49: 164-169. (In French).

Fontaine, M. and J. Leloup. 1960. The power of concentration of iodides by the thyroid of two salmonoids: the rainbow trout (Salmo gairdnerii) and the young salmon (Salmo salar). Arch. Sci. physiol. 14(1): 15-24. (In French).

Fontaine, Y. A. 1960. Radioactive contamination of aquatic media and organisms. Centre d'Études Nucléaires, France Commissariat à l'Énergie Atomique, Saclay, report CEA-1588. 158pp. (AEC-tr-5358) (NSA 15: 19131).

Fontaine, Y. A. and A. Aeberhardt. 1963. Experimental study of radioactive contamination by cerium-144 of a complex freshwater community studied in the laboratory. Health Physics 9(11): 1047-1056. (In French) (NSA 18: 3448).

Fontes, J. C. 1971. Counting unit for measuring the activity of natural radiocarbon by liquid scintillation. Rev. Geogr. Phys. Geol. Dyn. 13(1): 67-86. (In French) (NSA 26: 4415).

Fontes, J. C., et al. 1967. Tritium and ^{18}O in the Evian Basin. pp. 401-415. In: Isotopes in Hydrology. (See IAEA, 1967) (In French) (NSA 22: 18934).

Food and Agriculture Organization of the United Nations. 1956. Report of the first meeting of the European contact group on the uses of isotopes and radiation in agricultural research. Rome, report FAO/57/3/1637. 41pp. (NP-6353) (Incl. discussion of ^{14}C on wheat bugs).

Food and Agriculture Organization of the United Nations. 1960. Radioactive materials in food and agriculture. Rome. 114pp. (NSA 15: 30353).

Food and Agriculture Organization of the United Nations. 1964. Agricultural and Public Health Aspects of Radioactive Contamination in Normal and Emergency Situations. (Seminar, Scheveningen, the Netherlands, 11-15 December 1961). Rome, FAO Atomic Energy Ser. No. 5. 421pp.

Foote, D. C. 1961. Project Chariot and the Eskimo people of Point Hope, Alaska. 25pp. (NP-14684) (NSA 19: 17413) (Discusses effects of nuclear detonation).

Foote, R. S. 1964. Time variation of terrestrial γ radiation. pp. 757-766. In: The Natural Radiation Environment. (See Adams and Lowder, 1964) (NSA 18: 35746).

Foott, W. H. 1955. Phosphorus-32 labelling of the adults of the cabbage maggot, Hylemya brassicae (Bouche) (Diptera: Anthomyiidae). pp. 56-61. In: 85th Ann. Rept. Entomological Society, Ontario, 1954.

Forage, A. J. and N. E. Gillies. 1969. Enhancement of damage to Escherichia coli Strain B/R after ultrviolet and γ irradiation. J. Gen. Microbiol. 59(3): 329-___.

Forbert, S. and K. Svädstrøm. 1973. Radiometric measuring method for egg shells. Report to the National Environmental Protection Board 1971-1972 (development of the analytical methods for mercury, cadmium). T. Westermark and S. Forberg, eds. Sweden Kungliga Tekniska Holgskolan, Stockholm, report TRITA-KKE-7302. pp. 51-56. (In Swedish) (NSA 29: 10297).

Forcum, D. L. 1966. Behavior of irradiated cottontail rabbits. Thesis, Univ. of Wisconsin, Madison. 39pp.

Forcum, D. L. 1968. Reproduction of normal and whole-body irradiated cottontails. J. Wildl. Mgmt. 32(1): 180-182.

Foreman, E. E. and W. L. Templeton. 1958. The uptake of Zr^{95} and Nb^{95} by Porphyra sp. U. K. Atomic Energy Authority, Risley, report IGR-187 (DB(W)-TN-187). 6pp.

Foreman, E. E. and K. W. E. Bidwell. 1959. A biological investigation of the fate of strontium-90 in a disused filter bed. U. K. Atomic Energy Authority, Risley, report IGR-27 (RD/W). 19pp. (NSA 13: 11535).

Foreman, H. (Ed.). 1970. Nuclear Power and the Public. (Symp. held at Minneapolis, October 10-11, 1969). Univ. of Minnesota Press, Minneapolis. 289pp. (CONF-691054) (NSA 25: 23204).

Foreman, H., M. B. Roberts and E. H. Lilly. 1961. Radioactivity in Cervidae antlers. pp. 61-77. In: Biological and Medical Research Group, H-4, Health Division - Semiannual report July through December 1960. Los Alamos Scientific Lab., U. S. AEC report LAMS-2526. (NSA 15: 23277).

Forman, R. T. T. and G. M. Goodwell. 1967. Effects of chronic gamma irradiation on bryophyte populations of a forest ecosystem. Abstr. Bull. Ecol. Soc. Am. 48: 68.

Forssberg, A. 1941. On reactions of Phycomyces blakesleeanus by x-ray and γ-irradiation. Acta Radiol. 22: 252-259. (AEC-tr-4017) (NSA 14: 10344).

Forssberg, A. 1943. Studies on the biological effects of roentgen and gamma rays specifically on Phycomyces blakesleeanus. Acta Radiol. 27(3/4): 281-293. (In German).

Forssberg, A. 1969. The radiation induced reactions of Phycomyces blakesleeanus: pulsed transfer of cellular particles. Radiation Botany 9: 391-395. (NSA 24: 8017).

Forssberg, A. and R. Novak. 1960. Growth-rate and biochemical changes in Phycomyces at low dose-levels of x-rays. pp. 205-216. In: Proc. Symp., Venice on June 22-26, 1959. A. A. Buzzati-Traverso, ed. Taylor & Francis, Ltd., London. (NSA 15: 3939).

Forssberg, A., R. Brummer and A. Pehap. 1967. Gas exchange, growth and biochemical reactions in Phycomyces blakesleeanus. Radiation Botany 7: 507-512. (NSA 22: 16980).

Forssberg, A., V. Stankovic and A. Pehap. 1964. The radiation response of Phycomyces - structural and biochemical effects. Radiation Botany 4(3): 323-330. (NSA 19: 5692).

Forster, W. O. 1968. Fallout radionuclides (Fe^{55} and Mn^{54}) as indicators of salmon migration in the northeast Pacific relative to neutron activated radionuclide (Zn^{65}). Abstr. Bull. Ecol. Soc. Am. 49(2): 72.

Forsyth, D. J. and T. J. Peterle. 1973. Accumulation of chlorine-36 ring-labeled DDT residues in various tissues of two species of shrew. Arch. Environ. Contam. Toxicol. 1(1): 1-17.

Fort, C. W. and D. T. Wruble. 1972. Off-site radiological safety for the Cannikin event November 6, 1971. National Environmental Research Center, U. S. EPA, U. S. AEC report NERC-LV-539-2. iii, 25pp. (NSA 26: 58789) (Incl. environmental samples).

Fosberg, F. R. 1959. Plants and fallout. Nature 183(4673): 1448. (NSA 15: 13185).

Fosberg, F. R. 1959. Long-term effects of radioactive fallout on plants? Atoll Research Bull. 61, National Academy of Sciences - National Research Council, Washington. 11pp.

Fosberg, F. R. 1963. Effects of nuclear explosions on vegetation. Geographical Rev. 53(Jan.): 135-137.

Foster, J. L. 1970. Sensitization of anoxic fern spores to x-rays by chemicals known to be effective in bacteria. pp. 299-303. In: Radiation Protection and Sensitization. H. Moroson, ed. Barnes & Noble, New York. (NSA 25: 4573).

Foster, R. F. 1946. Some effects of pile area effluent water on young chinook salmon and steelhead trout. Hanford Atomic Products Operation, U. S. AEC report HW-7-4759. 100pp.

Foster, R. F. 1949. Some effects on embryo and young rainbow trout (Salmo gairdnerii Richardson) from exposing the parent fish to x rays. Growth 13(2): 119-142. (Also: Univ. of Washington, Seattle, U. S. AEC report AECD-2079; UWFL-12). (NSA 1: 230).

Foster, R. F. 1955. Aquatic life studies - index of radioactivity control. General Electric Rev. 58(4): 36-39.

Foster, R. F. 1959. The need for biological monitoring of radioactive waste streams. (Preprint V-6, 5th Nuclear Engr. Sci. Conf., Cleveland, O., 1959). (SFA 5: 3577; WPA 33: 1127).

Foster, R. F. 1959. Radioactive tracing of the movement of an essential element through an aquatic community with specific reference to radiophosphorus. Pubbl. staz. zool. Napoli 31(Suppl.): 34-62.

Foster, R. F. 1959. Procedures for the study of the movement of radiophosphorus in an aquarium community. Demonstration. Pubbl. staz. zool. Napoli 31(Suppl.): 63-69.

Foster, R. F. 1960. Biological considerations of radioactive wastes in streams. Hanford Atomic Products Operation, U. S. AEC report HW-SA-2021. 7pp. (NSA 15: 23276).

Foster, R. F. 1960. The significance of the uptake of radioisotopes by fresh-water fishes. pp. 15-20. In: Biological Problems in Water Pollution. (See Tarzwell, 1960).

Foster, R. F. 1962. Evaluation of radiological conditions in the vicinity of Hanford for 1961. Hanford Atomic Products Operation, U. S. AEC report HW-71999. 250pp. (I. C. Nelson, ed.) (NSA 16: 20868).

Foster, R. F. 1962-1965. Radiological status of the Hanford environs for ---. (Summary of radiological data for the month of ---). Hanford Labs., U. S. AEC report HW-72691-1 et seq.; -74307-1 et seq.; -76525-1 et seq.; 80892-1 et seq. 6-25pp. (NSA 29: 12694, 12711, 12712, 15572 et seq.).

Foster, R. F. 1963. Environmental behavior of chromium and neptunium. pp. 569-576. In: Radioecology. (See Schultz and Klement, 1963) (NSA 17: 33609).

Foster, R. F. 1963. Methods used to establish permissible concentrations of radioisotopes in fresh and sea water. Hanford Labs., U. S. AEC report HW-77690. 17pp. (NSA 17: 23546).

Foster, R. F. 1963. Relationships between the concentration of radionuclides in Columbia River water and fish. pp. 269-288. In: Transport of Radionuclides in Fresh Water Systems. (See Kornegay, et al., 1963).

Foster, R. F. 1967. Problems of water pollution associated with thermal energy industries. Battelle Northwest Lab., U. S. AEC report BNWL-SA-1546. 8pp. (NSA 22: 27514).

Foster, R. F. 1970. Effects of Hanford reactors on Columbia River and adjacent land areas. Battelle Northwest Lab., U. S. AEC report BNWL-SA-3679. 48pp. (CONF-701205) (NSA 25: 18224).

Foster, R. F. 1970. Nuclear plants and the environmental "crisis". Battelle Northwest Lab., U. S. AEC report BNWL-SA-3564. 16pp. (For Thermal Power Generation Conf., Pullman, Wash.). (CONF-701021-1) (NSA 25: 10132).

Foster, R. F. 1972. The history of Hanford and its contribution of radionuclides to the Columbia River. pp. 3-18. In: The Columbia River Estuary and Adjacent Ocean Waters. (See Pruter and Alverson, 1972).

Foster, R. F. 1973. Observed versus theorized capacity of the environment for radioactive contaminants. pp. 593-601. In: Environmental Behavior of Radionuclides Released in the Nuclear Industry. (See IAEA, 1973) (NSA 29: 29309).

Foster, R. F. 1974. Nuclear power and pollution. Battelle Northwest Lab., U. S. ERDA report BNWL-SA-5184. 19pp. (NSA 31: 11241).

Foster, R. F. and J. J. Davis. 1956. The accumulation of radioactive substances in aquatic forms. Proc. Intern. Conf. on the Peaceful Uses of Atomic Energy 13: 364-367. (Paper A/CONF. 8/P280) (Also: Int. J. Appl. Radiat. Isotopes 1(1-2): 131-132).

Foster, R. F. and J. F. Honstead. 1967. Accumulation of ^{65}Zn from prolonged consumption of Columbia River fish. Health Physics 13(1): 39-43. (NSA 2 21: 8722).

Foster, R. F. and R. L. Junkins. 1960. Off-project exposure from Hanford reactor effluent. Hanford Atomic Products Operation, U. S. AEC report HW-63654. 47pp. (NSA 14: 11429).

Foster, R. F. and R. E. Rostenbach. 1954. Distribution of radioisotopes in Columbia River. J. Am. Water Works Assoc. 46(7, pt.2): 633-640. (NSA 8: 5795).

Foster, R. F. and J. K. Soldat. 1966. Evaluation of the exposure resulting from the disposal of radioactive wastes into the Columbia River. pp. 683-696. In: Disposal of Radioactive Wastes into Seas, Oceans and Surface Waters. (See IAEA, 1966).

Foster, R. F., R. L. Junkins and C. E. Linderoth. 1960. Waste control at the Hanford plutonium production plant. Hanford Atomic Products Operation, U. S. AEC report HW-SA-1983. 40pp. (NSA 15: 23157).

Foster, S. S. D. 1975. Chalk groundwater tritium anomaly: a possible explanation. J. Hydrology 25(1-2): 159-165. (NSA 33: 17199).

Foti, S., E. Delucchi and V. Akamian. 1972. Determination of picogram amounts of technetium in environmental samples by neutron activation analysis. Anal. Chem. Acta 60(2): 269-276. (Also: Final report 1963-1970. U. S. Naval Ordnance Lab., White Oak, Md., report NOLTR-71-152) (AD-734380; NSA 26: 33034, 47674).

Foulquier, L. and A. Grauby. 1968. Comparative studies of the experimental contamination with Cs-137 of three aquatic vegetal species. pp. 201-208. In: Intern. Congr. on the Radioprotection of the Environment. (See Société Francaise de Radioprotection, 1968) (In French).

Foulquier, L. and A. Grauby. 1970. Influence of age, temperature and luminosity on the experimental contamination of Anodonta cygnea (L.) by cesium-137. pp. 269-316. In: Proc. Intern. Symp. on Radioecology. Vol. I. (See French Commissariat à l'Énrgie Atomique, 1970) (In French).

Foulquier, L. and A. Grauby. 1974. Proposals concerning an international biological radioindicator based on an experimental ecological and economic balance. pp. 67-88. In: Environmental Surveillance Around Nuclear Installations. Vol. 2. (See IAEA, 1974) (In French) (NSA 31: 14077).

Foulquier, L. and A. Lambrechts. 1972. A methodology for the identification of a biological indicator in radioecology (Example: ^{137}Cs and a freshwater mollusc Anodonta cygnea (L.)). France Commissariat à l'Énergie Atomique report CEA-R-4373. iv, 275pp. (In French).

Foulquier, L., D. Asselin and A. Grauby. 1972. Absorption and desorption of manganese with Cyprinus carpio L. Studies with the use of manganese-54. Verh. Int. Verin. Limnol. 18: 1243-1257. (In French).

Foulquier, L., P. Bovard and A. Grauby. 1967. Experimental contamination of Margaritana margaritifera (L.) (Mollusca) with ^{137}Cs. Compt. rend. D265: 1745-1748. (Also: France Commissariat à l'Énergie Atomique, Cadarache, report CEA-R-3054) (In French) (NSA 21: 10721; 22: 19347).

Foulquier, L., G. Marcoux and A. Grauby. 1971. Study of the fixation and desorption of radiostrontium by Anguilla anguilla (L.) in freshwater and salt water and its transfer to man through food. pp. 1177-1210. In: Proc. Intern. Symp. on Radioecology Applied to the Protection of Man and His Environment. (See European Atomic Energy Community, 1971) (In French) (NSA 27: 5566).

Foulquier, L., P. Bovard and A. Grauby. 1973. Results of experiments on zinc-65 fixation by Anodonta cygnea (L.). Malacologia 14: 107-124. (In French, English summary).

Fountain, E. L. 1961. Fallout, its relation to wildlife. pp. 40-45. In: Proc. Interstate Antelope Conf. (^{90}Sr in bighorn sheep).

Fountain, E. L. 1962. Nuclear test program as it is related to the desert bighorn sheep. Desert Bighorn Council Trans. 6: 62-67.

Fountain, E. L. and M. S. Seal. 1967. Strontium-90 in the bones of big game in the western United States. Health Physics 13(7): 1205-1209. (NSA 22: 8559).

Fourcy, A. 1967. Biology and neutron radiation: agronomical applications of radioactivation analysis. Industr. Atomiques 11: 11-12; 39-46. (In French).

Fourcy, A. 1973. Some nuclear methods in environmental studies. Centre d'Études Nucléaires, France Commissariat à l'Énergie Atomique, Grenoble, report CEA-CONF-2755. 20pp. (NSA 31: 22645).

Fourcy, A. and M. Neuburger. 1968. Study of the inorganic components of plants by radioactivation analysis. Application to the physiology of trace elements, to treatment control, to pollution, and to radioecology. Bull. Soc. Chim. Fr. (11): 4681-4686. (In French) (NSA 23: 9327).

Fourcy, A., et al. 1967. Application to radioactivation analysis to radioecological studies. Bull. Inform. Sci. Tech. (Paris) (119): 39-48. (AEC-tr-7041) (NSA 22: 19301; 23: 24111).

Fourcy, A., et al. 1967. Activation analysis in plant biology. Radioecological applications using simple or automated radiochemical separation techniques. France Commissariat à l'Énergie Atomique, Grenoble, report CEA-CONF-1254. 19pp. (CONF-681002) (NSA 23: 21664).

Fourcy, A., et al. 1971. Neutron activation analysis in the detection of pollutant elements retained by plants: application to elements taken up from pesticides and industrial fumes. pp. 507-519. In: Nuclear Techniques in Environmental Pollution. (See IAEA, 1971) (NSA 25: 31885).

Fourcy, A., et al. 1973. Pollution studies by the radioactivation of plants. pp. 81-88. In: 7th National Congr. of the Health Physicist's Assoc. France Centre National de la Recherche Scientifique, Marseilles. (In French) (NSA 30: 2489).

Fourie, H. O. and J. P. Ghijsels. 1969. Radiostrontium in biological material: a precipitation and extraction procedure eliminating the use of fuming nitric acid. Health Physics 17: 685-689. (NSA 24: 7417).

Fowden, L. 1959. Radioactive iodine incorporation into organic compounds of various angiosperms. Physiol. Plant. 12: 657-664.

Fowler, D. B. and K. F. MacQueen. 1972. Effect of low doses of gamma radiation on yield and other agronomic characters of spring wheat (Triticum aestivum). Radiation Botany 12(5): 349-353. (NSA 27: 5475).

Fowler, E. B. (Ed.). 1965. Radioactive Fallout, Soils, Plants, Foods, Man. Elsevier, New York.

Fowler, E. B. and E. H. Essington. 1974. Soils element activities, October 1972-September 1973. pp. 7-16. In: Dynamics of Plutonium in Desert Environments. (See Dunaway and White, 1974) (NSA 30: 29440).

Fowler, E. B. and E. H. Essington. 1975. Soils element activities for the period October 1973-September 1974. pp. 1-26. In: Radioecology of Plutonium and Other Transuranics in Desert Environments. (See White and Dunaway, 1975) (NSA 33: 613).

Fowler, E. B. and V. H. Kollman. 1970. Production of algae containing 93 percent ^{13}C. Los Alamos Scientific Lab., U. S. AEC report LA-4496. 4pp. (NSA 24: 50658).

Fowler, E. B., R. W. Henderson and M. F. Milligan. 1971. Proc. Environmental Plutonium Symp. (Held at LASL, August 4-5, 1971). Los Alamos Scientific Lab., U. S. AEC report LA-4756. 119pp. (CONF-710817) (NSA 26: 15208).

Fowler, E. B., et al. 1968. Soils and plants as indicators of the effectiveness of a gross decontamination procedure. pp. 446-455. In: Radiological Protection of the Public in a Nuclear Mass Disaster. (See Brunner and Pretre, 1968) (NSA 23: 20201).

Fowler, E. E. and W. D. Tucker. 1973. Potential applications of isotopes in pollution control. Brookhaven National Lab., U. S. AEC report BNL-18329. 22pp. (NSA 29: 7722).

Fowler, G. L. 1968. Studies of sperm transfer, sperm storage, sperm motility, and early mating in Drosophila. Thesis, Brown Univ., Providence, R. I. 71pp. (NSA 24: 9920) (X-rays).

Fowler, W. B. 1971-2. Radioisotope monitor for rime accumulation. Isotop. Radiat. Technol. 9(2): 182. (NSA 26: 9406) (High-elevation forests).

Fox, D. P. 1966. The effects of x-rays on the chromosomes of locust embryos. I. The early responses. II. Chromatid interchanges and the organization of the interphase nucleus. (See next entry).

Fox, D. P. 1967. The effects of x-rays on the chromosomes of locust embryos. III. The chromatid abberation types. IV. Dose-response and variation in sensitivity of the cell cycle for the induction of chromatid aberrations. Chromosoma 19: 300-316; 20: 173-194, 386-412, 413-441. (NSA 21: 28607, 18023, 39382, 41385).

Fox, R. L. 1965. The redistribution of radioactive phosphorus (P^{32}) in a closed aquarium system. Am. Biol. Teacher 27(6): 474-479.

Foxon, G. E. H. 1965. The use of x rays as a research instrument: problems in animal biology. Proc. Roy. Soc. Med. 58: 164-165. (NSA 19: 32002) (Comments on studies on frogs and lizards).

Fraga de Azevedo, J. and F. Carvao-Gomez. 1956. The biological action of gamma radiation on freshwater mollusks. Bull. soc. path. exotique 49(5): 917-921. (In French).

Fraga de Azevedo, J., et al. 1956. Study of the metabolism of freshwater mollusks. Metabolism of phosphorus, study with the use of 32P. Bull. soc. path. exotique 49(5): 912-917. (In French).

Fraga de Azevedo, J., et al. 1956. The application of radioisotopes in the study of metabolism of freshwater mollusks. Am. Soc. Belge de Med. Trop. 36(5): 609-613. (In French).

Fraga de Azevedo, J., et al. 1958. Application of radioisotopes in the study of fresh water snails (Gastropoda, Pulmonata). Am. J. Trop. Med. Hyg. 7: 84-89.

Fraizier, A. and A. Vilquin. 1971. An experimental study of ^{137}Cs elimination in the mullet _Mugil chelo_ and the blenny _Blennius pholis_. Marine Biology 10(2): 154-156. (In French).

Fraleigh, P. and R. J. Beyers. 1966. The effect of irradiation on succession in an aquatic microcosm. Abstr. Bull. Ecol. Soc. Am. 47(3): 154.

Fraley, L., Jr. 1971. Response of shortgrass plains vegetation to chronic and seasonally administered gamma radiation. Thesis, Colorado State Univ., Fort Collins. xvi, 170pp.

Fraley, L., Jr. and F. W. Whicker. 1973. Response of a native shortgrass plant stand to ionizing radiation. pp. 999-1006. In: _Radionuclides in Ecosystems_. (See Nelson, 1973) (NSA 26: 71083).

Fraley, L., Jr. and F. W. Whicker. 1973. Response of shortgrass plains vegetation to gamma radiation. I. Chronic irradiation. II. Short-term seasonal irradiation. Radiation Botany 13(6): 331-341, 343-353. (NSA 29: 27025, 29710).

Fram, H., B. E. Proctor and C. G. Dunn. 1950. Effects of x-rays produced at 50 kilovolts on different species of bacteria. J. Bacteriology 60(3): 263-267.

Franchi, C. M. and C. R. Benech. 1969. Apparatus for localized gamma irradiation of small experimental animals. Arch. Soc. Biol. Montevideo 27(Spe (Spec.): 64-68. (NSA 24: 31540).

Francis, C. M. and J. S. Gladstones. 1963. Studies on the use of mutagenic agents in plant breeding. I. The effect of seed moisture content on sensitivity to x-rays in _Lupinus angustifolius_. Austral. J. Agr. Res. 14(1(: 12-19. (NSA 17: 35481).

Francis, C. W. 1973. Plutonium mobility in soil and uptake in plants: review. J. Environ. Qual. 2(1): 67-70. (NSA 28: 384).

Francis, C. W. and T. Tamura. 1973. Cesium-137 soil inventory of a tagged _Liriodendron_ forest 1962-1969. pp. 140-149. In: _Radionuclides in Ecosystems_. (See Nelson, 1973) (NSA 28: 27268).

Francis, G. E., W. Mulligan and A. Wormall. 1959. _Isotopic Tracers_. A theoretical and practical manual for biological students and research workers. Athlone Press, Univ. of London, London. 2nd ed. xiv, 524pp.

Francke, H. C., F. E. Clark and J. W. Strohecker. 1974. Biological denitrification disposal techniques. Oak Ridge National Lab., U. S. AEC report Y-DA-5510. 21pp. (For 2nd AEC Pollution Control Conf., Albuquerque, N. Mex., 16 Apr. 1974). (CONF-740406-4) (NSA 30: 2693).

Francois, B. and M. Limandas. 1971. Counting tritiated water. Int. J. Appl. Radiat. Isotopes 22(10): 632-633. (NSA 25: 57363).

Francois, E., A. Riga and R. Impens. 1968. Ecological study of two aphid vectors of the sugar-beet virus. pp. 53-68. In: _Isotopes and Radiation in Entomology_. (See IAEA, 1968) (In French, English summary) (NSA 24: 31898) (^{32}P).

Frank, A. L. and R. A. Taylor. 1965. Gamma radiation characteristics - angular distribution over a desert terrain fallout field. U. S. Naval Radiological Defense Lab., San Francisco, report USNRDL-TR-856. xii, 76pp.

Frank, M. L. 1971. Effects of acute ionizing radiation on carp (_Cyprinus carpio_ L.) embryos. Thesis, Univ. of Tennessee, Knoxville. vi, 43pp. (Also w/ B. G. Blaylock: Oak Ridge National Lab., U. S. AEC report ORNL-TM-3346).

Frank, M. L. 1973. Radiosensitivity of carp (_Cyprinus carpio_) embryos to acute gamma radiation. pp. 1106-1112. In: _Radionuclides in Ecosystems_. (See Nelson, 1973) (NSA 28: 30143).

Frank, W. H., G. von Hagel and E. Busch. 1964. Studies on the penetration of radioactive substances from fallout into groundwater. Gas-Wasserfach 105: 254-259. (In German) (NSA 21: 190).

Frankenberg, D., G. Heigwer and W. Pohlit. 1970. Irradiation apparatus for microbiological experiments without external hazards. Strahlentherapie 139: 379-384. (In German) (NSA 24: 36473).

Franklin Institute Research Labs. 1971. _Proc. Conf. on Nuclear Energy and the Environment_. (Drexel Univ., June 21-25, 1971). Philadelphia, Pa. 133pp. (CONF-710644) (NSA 26: 49886).

Franklin, R. E., P. L. Gersper and N. Holowaychuk. 1967. Analysis of gamma ray spectra from soils and plants: II. Effect of trees on the distribution of fallout. Soil Sci. Soc. Am. Proc. 31(1): 43-__.

Franquinet, R. and A. Le Moigne. 1975. Uridine ^{3}H-5 and leucine ^{3}H-5 uptake in planarian cells Polycelis tenuis (Iijima) cultivated in vitro. Compt. rend. Hebd. Séances Acad. Sci. D280(19): 2253-2256. (In French) (NSA 33: 12508).

Fransz, H. G. 1969. Effects of x irradiation on adult males of the fruit leaf-roller Adoxophyes reticulana Hb. (Lepidoptera, Tortricidae) with respect to genetic pest control. Neth. J. Zool. 19: 287. (NSA 24: 938).

Franz, E. H. and G. M. Goodwell. 1970. Effects of chronic gamma irradiation on the soil algal community of a forest ecosystem. Abstr. Bull. Ecol. Soc. Am. 51(2): 31.

Franz, E. H. and G. M. Goodwell. 1973. Effects of chronic gamma irradiation on the soil algal community of an oak-pine forest. Radiation Botany 13(6): 323-329. (NSA 29: 27028).

Franz, J. M. and A. Krieg. 1972. Biological Pest Control. Parey, Berlin. 208pp. (In German) (NSA 27: 27988) (Incl. ionizing radiation).

Fraser, D. A. 1956. The translocation of minerals in trees. Canada Dept. of Northern Affairs and National Resources, Forestry Res. Div. Tech. Note 47.

Fraser, D. A. 1958. The translocation of rubidium-86 and calcium-45 in trees. pp. 347-363. In: The Physiology of Forest Trees. (Symp., Harvard Univ., 1957). K. V. Thimann, ed. Ronald Press, New York.

Fraser, D. A. and E. E. Gaertner. 1959. Use of radioisotopes in forestry research. Proc. 9th Intern. Botanical Congr. 2: 122.

Fraser, D. A. and E. E. Gaertner. 1961. Use of radioisotopes in forestry research. pp. 1381-1387. In: Recent Advances in Botany. Univ. of Toronto Press, Toronto.

Fraser, D. A. and E. E. Gaertner. 1962. Use of radioisotopes in forestry research. Tropical Ecology 3: 106-115.

Fraser, D. A. and E. E. Gaertner. 1970. Utilization of radioisotopes in forestry research. Proc. 6th World Forestry Congr. (Madrid) 2: 2264-2274.

Fraser, D. A. and C. A. Mawson. 1953. Movement of radioactive isotopes in yellow birch and white pine as detected with a portable scintillation counter. Can. J. Bot. 31: 324-333.

Freed, S. 1965. Endogenous biochemistry of planarians correlated with learning experiments at Brookhaven National Laboratory. Brookhaven National Lab., U. S. AEC report BNL-981. 21pp. (NSA 20: 45272).

Frederick, B. J. 1963. Labeling lake water with tritium. Int. J. Appl. Radiat. Isotopes 14(8): 401-404. (NSA 17: 40998).

Fredericksen, C. F. and J. H. Lilly. 1955. Measuring wireworm reactions to soil insecticides by tagging with radioactive cobalt. J. Econ. Entomol. 48(4): 438-442.

Fredriksson, L. 1963. Studies on plant accumulation of fission products under Swedish conditions. VI. A new experimental technique for studies on plant absorption of nutrients and fission products under field conditions. Sweden Forsvarets Forskningsanstalt, Stockholm, report FOA-4-4323. 18pp. (NSA 22: 51301).

Fredriksson, L. and A. Eriksson. 1970. Plant uptake of fission products. I. Uptake of ^{90}Sr in pot experiments in relation to uptake under field conditions. Lantbrukshogsk. Ann. 36: 3-18. (NSA 24: 46564) (Incl. hay).

Fredriksson, L., B. Eriksson and A. Eriksson. 1961. Studies on plant accumulation of fission products under Swedish conditions. III. Accumulation of Sr-89 in the aerial parts of different weed species at varying Ca-level in soil. Sweden Forsvarets Forskningsanstalt, Stockholm, report FOA-4-A-4189-4623. 12pp. (NSA 15: 23275).

Fredriksson, L., et al. 1968. Migration and plant availability of fission products in a lysimeter experiment. I. Plant uptake and leaching of ^{137}Cs Sweden Forsvarets Forskningsanstalt, Stockholm, report FOA-4-4357-28. 26 pp. (NSA 22: 38257).

Freeden, F. J., et al. 1953. Mass tagging of black flies (Diptera, Simuliidae) with radiophosphorus. Can. J. Zool. 31(1): 1-15.

Freeman, C. C. 1965. Criteria for radiographic examination of internal insect infestation. J. Assoc. Off. Agr. Chem. 48(6): 1183-1185.

Freeman, F. W., D. P. White and M. J. Bukovac. 1964. Uptake and differential distribution of C^{14}-labeled Simazine in red and white pine seedlings. Forest Science 10(3): 330-334.

Freeman, G. 1964. The role of blood cells in the process of asexual reproduction in the tunicate Perophora viridis. Z. Exptl. Zool. 156: 157-183. (NSA 19: 33939).

Frei, H. 1974. Radiation-induced dominant lethality in oocytes of Drosophila hydei, with emphasis on the differential sensitivity of the immature stages in Class B. Mutation Research 24(2): 125-134. (NSA 30: 27199).

Freidin, A. S. 1958. Effect of radioactive radiation on physical and mechanical properties of wood. Woodworking Industry (USSR) 7(9): 13-15. (English transl. CSIRO, Australia, No. 4417).

Freim, J. O., Jr. 1968. Radiation studies on vegetative Dictyostelium discoideum cells. Thesis, Pennsylvania State Univ., University Park. 131pp. (NSA 23: 50935).

Freitas, J. R. and M. Campos. 1960. Preliminary data on the elimination rate of iron from Triatoma infestans, determined by means of radioactive iron, ^{59}Fe. Cienc. Cult. (Maracaibo) 12(3/4): 164-165. (In Portugese).

Freitas, J. R., W. Mayrink and E. Mansur Neto. 1960. Quantity of blood injested in vivo by Phlebotomus longipalpis (Psychodidae) as determined by means of radioactive iron. Cienc. Cult. (Maracaibo) 12(3/4): 164. (In Portugese).

French, A. L. and R. Hoopingamer. 1966. Absorption and metabolism of insecticides in *Escherichia coli*. Abstr. 65. Bull. Entomol. Soc. Am. 12: 266.

French Commissariat à l'Énergie Atomique. 1970. *Proc. Intern. Symp. on Radioecologie*. (Cadarache, 8 to 12 September 1969). 2 vols., 1192pp. (In French and English) (CONF-690918) (NSA 24: 41784).

French Commissariat à l'Énergie Atomique. 1971. Levels of radioactive contamination of the environment and the food chain. Annual report, 1970. Centre d'Études Nucléaires, Fontenay-aux-Roses, report EUR-4700. 56pp. (In French) (NSA 26: 15525).

French Commissariat à l'Énergie Atomique. 1973. Surveillance for radioactivity in 1973. Centre d'Études Nucléaires, Fontenay-aux-Roses. 71pp. (In French) (NP-20720) (INIS-mf-1974) (NSA 33: 9270, 29120).

French Commissariat à l'Énergie Atomique. 1974. Studies on the levels of radioactive contamination of the food chain and the environment (1961-1971). Centre d'Études Nucléaires, Fontenay-aux-Roses, report EUR-5065f. (In French) (NSA 32: 14311).

French Institute de Droit Compare de Paris. 1969. Notebook of atomic energy law. 2. 263pp. (In French) (NP-18752) (NSA 25: 55252).

French, N. R. 1959. Iodine metabolism in wild jack rabbits. pp. 113-121. In: *Oklahoma Conference - Radioisotopes in Agriculture*. (Oklahoma State Univ., Stillwater, April 2 and 3, 1959). U. S. AEC report TID-7578. (NSA 14: 10290).

French, N. R. 1960. Strontium-90 in Ecuador. Science 131(3417): 1889-1890.

French, N. R. 1961. The occurrence, effect and utilization of specific isotopes in the environment: review and discussion of barium. Univ. of California, Los Angeles, U. S. AEC report UCLA-497. 19pp. (NSA 17: 5755).

French, N. R. 1963. Fallout and natural populations. pp. 152-156. In: *Proc. 1st Intern. Conf. on Wildlife Disease*. (High View, New York, June 24-27, 1962). Wildlife Disease Association.

French, N. R. 1963. The source of ingested radioactivity in desert rodents. Abstr. Bull. Ecol. Soc. Am. 44(3): 74.

French, N. R. 1963. Review and discussion of barium. pp. 557-560. In: *Radioecology*. (See Schultz and Klement, 1963).

French, N. R. 1963. Analysis of reproduction in a black-tailed jack rabbit. Abstr. Proc. XVI Intern. Congr. Zool. 1: 258. (Study at National Reactor Testing Station, Ida.).

French, N. R. 1964. Analysis of dispersal in desert rodents. Abstr. Bull. Ecol. Soc. Am. 45(3): 115. (Nevada Test Site).

French, N. R. 1964. A radiation facility for ecological studies at the AEC Nevada Test Site. Abstr. Bull. Ecol. Soc. Am. 45(3): 79.

French, N. R. 1964. Description of a study of ecolological effects on a desert area from chronic exposure to low level ionizing radiation. Univ. of California, Los Angeles, U. S. AEC report UCLA-12-532. 27pp., 15 figs.

French, N. R. 1965. Mortality and fertility in a desert population of pocket mice. Abstr. Bull. Ecol. Soc. Am. 46(3): 122.

French, N. R. 1965. Radiation and animal populations: problems, progress and projections. In: *Radiation and Terrestrial Ecosystems*. Health Physics 11(12): 1557-1568. (See Hungate, 1965).

French, N. R. 1966. Irradiated desert rodent populations. p.22. In: *Radiation Effects on Natural Populations*. (See Sacher, 1966).

French, N. R. 1967. Comparison of radioisotope assimilation by granivorous and herbivorous mammals. pp. 665-673. In: *Radioecological Concentration Processes*. (See Aberg and Hungate, 1967).

French, N. R. 1969. Radiation sensitivity of rodent species. Nature 222: 1003-1004.

French, N. R. 1970. Chronic low-level gamma irradiation of a desert ecosystem for five years. pp. 1151-1165. In: *Proc. Intern. Symp. on Radioecology. Vol. II*. (See French Commissariat à l'Énergie Atomique, 1970) (NSA 24: 41794).

French, N. R. and H. Bissell. 1963. Strontium-90 in California deer. Abstr. Bull. Ecol. Soc. Am. 44(2): 46.

French, N. R. and H. D. Bissell. 1968. ^{90}Sr in California mule deer. Health Physics 14(5): 489-494. (NSA 22: 36283).

French, N. R. and H. W. Kaaz. 1968. The intrinsic rate of natural increase of irradiated *Peromyscus* in the laboratory. Ecology 49(6): 1172-1179.

French, N. R. and K. H. Larson. 1963. Environmental pathways of radioactive iodine from nuclear tests in arid regions. pp. 77-81. In: *Radioecology*. (See Schultz and Klement, 1963) (Also: Univ. of Calif., Los Angeles, U. S. AEC report UCLA-499) (NSA 17: 6322).

French, N. R. and L. Van Middlesworth. 1958. Biological monitoring of recent air-borne fission products. Proc. 2nd. Intern. Conf. on the Peaceful Uses of Atomic Energy 18: 516-518. (Paper A/CONF.15/P/2497).

French, N. R., P. Hayden and T. Tagami. 1965. The source of ingested radioactivity in desert rodents. Health Physics 11(7): 637-645. (NSA 19: 29868).

French, N. R., B. G. Maza and A. P. Aschwanden. 1965. Duration of life in confined populations of desert rodents. Abstr. Bull. Ecol. Soc. Am. 46 (2): 40. (Studies at Nevada Test Site).

French, N. R., B. G. Maza and A. P. Aschwanden. 1967. Life spans of *Dipodomys* and *Perognathus* in the Mojave Desert. J. Mammal. 48(4): 537-548. (Studies at Nevada Test Site).

French, N. R., B. G. Maza and H. W. Kaaz. 1969. Mortality rates in irradiated rodent populations. pp. 46-52. In: Symposium on Radioecology. (See Nelson and Evans, 1969) (Also: Univ. of California, Los Angeles, U. S. AEC report UCLA-12-634) (NSA 21: 28595).

French, N. R., R. McBride and J. Detmer. 1965. Fertility and population density in the black-tailed jack rabbit. J. Wildl. Mgt. 28(1): 1-13. (At National Reactor Testing Station, Ida.).

French, N. R., T. Y. Tagami and P. Hayden. 1968. Dispersal in a population of desert rodents. J. Mammal. 49(2): 272-280. (At Nevada Test Site).

French, N. R., et al. 1974. A population study of irradiated desert rodents. Ecol. Monogr. 44(1): 45-72. (At Nevada Test Site).

French Service Central de Protection Contre les Rayonnéments Ionisants. 1972. Annual mean contamination of milk by ^{90}Sr as a function of seven French climates. Le Vesinet. 20pp. (In French) (NP-19433) (NSA 26: 56129).

French Service Central de Protection Contre les Rayonnéments Ionisants. 1963/74. {Periodic reports of radioactivity measurements, including environmental samples; reports from Paris, Le Vesinet, and Chatillon Sous Bagneux (Seine); reports SCPRI(RM)-month-year; 12-30pp; in French; most in NSA vols. 17-30; incl. Surveillance of radioactivity in France for 1961 to 1963, SCPRI(T)-98, 22pp.}.

French, R. L. 1964. A comparative study of radioactive source arrangements for simulating fallout gamma radiation fields. Radiation Research Assoc., Inc., Fort Worth, Tex., report AFRRI-CR-65-2. 111pp. (NSA 20: 23345).

Frere, M. H. and H. Roberts, Jr. 1963. The loss of strontium-90 from small cultivated watersheds. Soil Science 27(1): 82-83.

Frere, M. H., et al. 1963. The behavior of radioactive fallout in soils and plants. National Academy of Sciences-National Research Council, Washington, D. C., Publ. 1092. v, 32pp.

Frere, M. H., et al. 1967. Reduction in the plant uptake of ^{90}Sr by soil management treatments. U. S. Dept. of Agriculture Tech. Bull. No. 1378. 38pp. (NSA 24: 19289).

Fretter, V. 1953. Experiments with radioactive strontium ^{90}Sr on certain molluscs and polychaetes. J. Mar. Biol. Assoc. U.K. 32: 367-384.

Frey, D. J. and J. B. Stahl. 1958. Measurements of primary production on Southampton Island in the Canadian Arctic. Limnol. Oceanogr. 3(2): 215-221.

Frey, M. J., H. Jirku and M. Levitz. 1970. Relative conversion of 6,7-^{3}H-estrone to ^{3}H-15α-hydroxyestrogens and ^{3}H-7α-hydroxyestrogens by Glomerella fusarioides. J. Label. Compounds 6: 355-361. (NSA 25: 15613).

Freyden, A. S. 1958. Effect of radioactive radiation on physical and mechanical properties of wood. Woodworking Industry (USSR) 7(9): 13-15. (English transl. CSIRO, Australia, No. 4417).

Freyden, A. S., Y. M. Malinski and D. Y. Karpaw. 1959. The effect of ionizing radiation on wood. Gidrolis. Lesokhimich. Promysl. 12(4): 4-_. (In Russian).

Freytag, H. E. and H. Igel. 1973. Soil respiration and action of minerals. Zentralbl. Bakteriol. Parasitenk. Abt. II 128: 221-231. (In German) (NSA 33: 7645).

Fric, F. and V. Palovcikova. 1975. Automatic liquid scintillation counting of ^{32}P in plant extracts by measuring the Cherenkov radiation in aqueous solutions. Int. J. Appl. Radiat. Isotopes 26(5): 305-311. (NSA 32: 21886).

Fried, M. 1971. Use of isotopes in agronomic soil science. Agrochimica 15 (2-3): 125-137. (NSA 25: 46670).

Friedlander, G. D. 1970. Power, pollution, and the imperiled environment. II. East, Midwest and West Coast: pollution-control plans of some major utilities; role of government in environmental matters; other proposed systems for reducing stack emissions. Spectrum 7(12): 65-75. (NSA 25: 12761).

Friedman, G. B., H. M. McGurdy and F. T. Algard. 1970. Response of newt larvae to x irradiation. Can. J. Zool. 48: 1017-1021. (NSA 25: 16350).

Friedman, M. H., J. T. Tanner and E. Miller. 1972. Rapid instrumental analysis for mercury in biological materials by neutron activation. Trans. Amer. Nucl. Soc. 15(1): 150-151. (NSA 26: 42702).

Friend, A. G. 1963. The aqueous behavior of strontium-85, cesium-137, zinc-65, and cobalt-60 as determined by laboratory type studies. pp. 43-56. In: Transport of Radionuclides in Fresh Water Systems. (See Kornegay, et al., 1963) (NSA 17: 40521).

Friend, A. G. 1963. Research studies on the Clinch River and Mohawk River. pp. 51-73. In: Studies of the Fate of Certain Radionuclides in Estuarine and Other Aquatic Environments. (See Sabo and Bedrosian, 1963).

Friend, A. G. and D. B. Porcella. 1962. Studies on the distribution of radionuclides in the Clinch and Tennessee rivers below Oak Ridge. Preprint Paper No. 75, Engineers Joint Council, New York. 24pp. (NSA 16: 29272).

Friend, A. G., et al. 1965. Behavior of certain radionuclides released into fresh-water environments. Annual report. U. S. Public Health Service Publ. 999-RH-13. ix, 89pp.

Fries, B. A. 1969. Radiotracer studies of water flows in water treatment ponds. Trans. Amer. Nucl. Soc. 12: 501-502. (NSA 24: 5748).

Fries, B. A. and I. M. Whittemore. 1965. Radiotracer measurements of the hydraulic behavior of a large shallow pond. Trans. Amer. Nucl. Soc. 8: 78-79. (NSA 19: 29804).

Fries, N. 1965. The thermosensitizing effect of irradiation with ultraviolet light and x-rays on cells of Ophiostoma and Rhodotorula. Exptl. Cell Res. 39: 693-697. (NSA 20: 1640).

Frigerio, N. A. and W. J. Eisler, Jr. 1968. Low cost, automatic, nest and burrow monitor using radioactive tagging. Ecology 49(4): 788-791.

Frink, C. R. 1971. Fixation of ^{137}Cs by soil clays. Final report. Connecticut Agricultural Experiment Station, U. S. AEC report NYO-2955-25. 18pp. (NSA 26: 182).

Frisancho, P. 1964. Radioactivity in soils, waters, and plants in the area of Lima. Bol. Junta Control Energia At. (Peru) 9(49): 67-76. (In Spanish) (NSA 19: 15795).

Frissel, M. J. and P. Polestra. 1964. A theoretical approach to the movement of strontium through soils. Soil Science 98(4): 274-277.

Frissel, M. J., et al. 1971. Contamination of the soil with mercury. pp. 941-954. In: Proc. Intern. Symp. on Radioecology Applied to the Protection of Man and His Environment. (See Commission of the European Communities, 1971) (Activation analysis, ^{203}Hg).

Frissel, M. J., et al. 1974. Tracing soil moisture migration with ^{36}Cl and tritium. pp. 145-148. In: Isotope and Radiation Techniques in Soil Physics and Irrigation Studies 1973. (See IAEA, 1974) (NSA 30: 15491).

Fritschi, E. W. 1967. A study of the biological and radiological wastes from a primate center and their impact upon the aquatic environment. Thesis, Tulane Univ., New Orleans, La. 127pp. (NSA 22: 47544).

Fritton, D. D. 1969. Resolving tome, mass absorption coefficient, and water content with gamma-ray attenuation. Soil Sci. Soc. Am. Proc. 33(5): 651-655. (NSA 24: 2302).

Frittelli, L. 1971. Rapid environmental monitoring in nuclear emergency situations. pp. 769-777. In: Rapid Methods for Measuring Radioactivity in the Environment. (See IAEA, 1971) (NSA 26: 28059).

Frittelli, L. 1973. On some recent applications of system analysis methods to radioecology. Abstr. p. 135. In: Proc. 2nd Conf. on the State of Advancement of Radioecology in Italy. (See Anonymous, 1973) (In Italian).

Frittelli, L. 1974. Application of system analysis methodology to the determination of the limiting radiological capacity of the area surrounding a nuclear facility. pp. 425-432. In: Environmental Surveillance Around Nuclear Installations. Vol. 1. (See IAEA, 1974) (NSA 31: 14081).

Fritz, T. E., et al. 1968. Interrelations between x-irradiation and the intestinal flora of mice. Monogr. Nucl. Med. Biol. (1): 279-291. (NSA 22: 49548).

Fritze, K. and R. Robertson. 1971. Neutron activation analysis for aluminum in animal tissue. J. Radioanal. Chem. 7(2): 213-220. (NSA 25: 54514).

Fritze, R. and G. Herrmann. 1974. Interpretation of aerosol release measurements at nuclear power plants. pp. 413-423. In: Environmental Surveillance Around Nuclear Installations. Vol. 1. (See IAEA, 1974) (NSA 31: 19763).

Fritz-Niggli, H. 1958. Radiation chimaeras and parabiosis. Strahlentherapie 106(3): 378-390. (In German) (Used Triton alpestres).

Fritz-Niggli, H. and H. R. Schinz. 1961. Biological effectiveness of 30-MeV electrons in dependence on the tissue depth and in comparison with 180-KeV and 31-MeV photons. I. Lethality test on one-hour Drosophila embryos. Strahlentherapie 115: 379-393; 118(4): 503-517 (1962). (In German) (NSA 16: 32822) (Wild Drosophila).

Fritz-Niggli, H., E. Diener and P. Schleuss. 1963. Oxygen effect after conventional x-ray and high-energy electron irradiation in a radioinduced modification (phenocopy). Biophysik 1: 51-59. (In German) (NSA 17: 30307) (Used pre-pupa of Drosophila).

Frizzi, G. 1961. Structures of salivary chromosomes and their contribution to the problem of evolution. Rc. Semin. Fac. Sci. Univ. Cagliari 31: 1-__. (In Italian) (Anopheles atroparvus, irradiated).

Frizzi, G. and M. S. Jolly. 1961. Mutations induced with x rays in Anopheles maculipennis. Atti Assoc. Genet. Ital. 6: 285-290. (In Italian) (NSA 17: 10385).

Fröchlich, K., et al. 1974. Methodical and measuring results using tritium, ^{14}C, and ^{32}Si determinations for hydrogeological problems. Z. Angew. Geol. 20(1): 16-21. (In German) (NSA 33: 661).

Fröhling, W. 1971. Problems of environmental protection; their expansion and importance. Germany Institüt für Reaktorentwicklung, Jülich, report JUL-779-RG. 27pp. (In German) (NSA 26: 4713).

Frome, M. 1967. Wilderness ignored: The Aleutian Islands. Audubon Mag. 69 (1): 28-40. (Concerns nuclear testing on Amchitka Island).

Fromm, P. O. and E. P. Reineke. 1956. Some aspects of thyroid physiology in rainbow trout. J. Cell. and Comp. Physiol. 48(3): 393-403.

Frontali, N. 1964. Brain glutamic acid decarboxylase and synthesis of gamma-aminobutyric acid in vertebrate and invertebrate species. pp. 185-192. In: Comparative Neurochemistry. D. Richter, ed. Pergamon Press, New York. (^{14}C; used different species, incl. Apis mellifera).

Frost, R. 1973. Radioactive methods of geochronology. Vesmir 52(6): 166-169. (In Czech) (NSA 33: 12054).

Frucht, A. H. and F. Kössler. 1961. Important hygienic concepts of radiation from the area of ecology. Z. Ges. Hyg. Ihre Grenzbebiete 7: 567-571. (In German) (NSA 19: 23).

Fry, L. M. 1962. Radium and fission product radioactivity in thermal waters. Nature 195(4839): 375-376. (NSA 16: 26730).

Fry, R. J. M., et al. 1969. Radiation effects and cell renewal in rodent intestine. pp. 255-270. In: Comparative Cellular and Species Radiosensitivity. (See Bond and Sugahara, 1969) (NSA 24: 12292) (Incl. wild species).

Fryer, G. E. and J. A. S. Adams. 1974. Airborne gamma ray spectrometer and its application in nuclear power plant site surveys. Trans. Nucl. Sci. NS-21(1): 572-584. (NSA 30: 6734).

Fu, C.-j., et al. 1964. Influence of the photoperiod on the absorption and accumulation of P^{32} in jute. Zhongshan Daxue Xuebao Acta Sci. Nat. Univ. Sunyatseni (2): 239-247. (In Chinese, English summary) (Ref. Zh. biol. 1965: 5G50) (Abstr. JPRS-32,497, TT-65-32976, p. 10).

Fuchs, A. and F. W. de Vries. 1972. Comparison of methods for the preparation of ^{14}C-labelled plant tissues for liquid scintillation counting. Int. J. Appl. Radiat. Isotopes 23(8): 361-369. (NSA 26: 58552).

Fucugauchi, M. E. 1960. Radioactive measurements in the living fish. Arkiv Kemi 15(4): 375-377. (BA 35: 57886; SFA 6: 3944).

Fuhs, G. W. and E. Canelli. 1970. ^{33}P autoradiography used to measure phosphate uptake by individual algae. Limnol. Oceanogr. 15: 962-967. (NSA 25: 26596).

Fujii, T. and S. Matsumura. 1958. Radiosensitivity in plants. I. Determination of LD-50 in cultivated plants (preliminary report). Idengaku Zasshi 33: 389-397. (NSA 14: 2319).

Fujumoto, Y., M. Watanabe and Y. Anma. 1964. The effects of gamma (^{60}Co) irradiation on the pollen germination of *Pinus desiflora and Pinus thunbergii*. Nippon Ringaku Kaishi 46: 389-394. (In Japanese) (NSA 20: 28980).

Fujinaga, T., et al. 1973. Radioactivation analysis of aluminum, vanadium, copper, molybdenum, zinc, and uranium in natural water samples using organic coprecipitants. Radioanal. Chem. 13(2): 301-311. (NSA 28: 24090).

Fujisaki, K., S. Kitaoka and T. Morii. 1974. Effect of gamma radiation on the feeding, molting, and oviposition of *Haemaphysalis longicornis*. Norinsho Kachiku Eisei Shikenjo Kenkyu Hokoku 68: 12-20. (In Japanese) (NSA 31: 2739).

Fujita, M. 1973. Ecological analysis on low-level radioactive wastes centering around tritium and krypton. pp. 253-286. In: *Siting and Environmental Problems in the Development of Electric Power Station. Its Status and Analysis*. K. Tomioka, ed. Fuji International Co., Tokyo. (In Japanese) (NSA 30: 15461).

Fukai, R. and C. N. Murray. 1973. Environmental behavior of radiocobalt and radiosilver released from nuclear power stations into aquatic systems. pp. 217-242. In: *Environmental Behavior of Radionuclides Released in the Nuclear Industry*. (See IAEA, 1973) (NSA 29: 29362).

Fulford, M. and H. Kersten. 1947. The reaction of *Leucolejeunea clypeata* to variation in dosage of soft x-rays. Bryologist 50: 147-156.

Fuller, R. A., P. W. Riegert and J. W. T. Spinks. 1954. Persistence of radioactivity in grasshoppers (Acrididae) tagged with phosphorus-32. Can. Entomol. 86(5): 201-205.

Fuller, R. A., et al. 1951. Use of radio-active tracers in investigations of soil-inhabiting insects. Rept. Entomol. Soc. Ontario 81: 7-15.

Fuller, R. G. 1969. Pre-event bioenvironmental safety survey and evaluation Project Rulison. Battelle Columbus Labs., U. S. AEC report PNE-R-3. 20pp.

Fuller, R. G. 1971. Amchitka bioenvironmental program. Amchitka biological information summary. Battelle Columbus Labs., U. S. AEC report BMI-171-132. 16pp. (NSA 25: 37743).

Fuller, W. H. and M. F. l'Annunziata. 1969. Movement of algal- and fungal-bound radiostrontium as chelate complexes in a calcareous soil. Soil Science 107: 223-230. (NSA 23: 17563).

Fuller, W. H. and J. E. Hardcastle. 1967. Relative absorption of strontium and calcium by certain algae. Soil Sci. Soc. Am. Proc. 31: 772-774. (NSA 22: 21438).

Funderburk, H. H., Jr. and J. M. Lawrence. 1963. Preliminary studies on the absorption of C^{14} labeled herbicides in fish and aquatic plants. Weeds 11(3): 217-219. (BA 45: 18497; SFA 9: 6864).

Funsch, B. 1971. Estimation of the ^{131}I concentration following a release of fresh fission products. pp. 779-795. In: *Rapid Methods for Measuring Radioactivity in the Environment*. (See IAEA, 1971) (In German) (NSA 26: 27828).

Furnica, Gh. 1968. Study of ^{131}I retention by ionic exchange in the natural suspensions of surface lake waters. Igiena (Bucharest) 17: 135-140. (In Rumanian) (NSA 22: 40219).

Furnica, Gh. 1970. Retention of certain radioactive ions by suspensions of artificial lakes of the area and by certain disturbed soils of Romania. pp. 181-200. In: *Proc. Intern. Symp. on Radioecology. Vol. I.* (See French Commissariat à l'Énergie Atomique, 1970) (In French).

Furr, A. K. and E. Mooney. 1971. Neutron activation studies of the Roanoke River system. Trans. Nucl. Sci. NS-18(1): 79-85. (NSA 25: 57395).

Furtado, V. C., T. J. Kneip and M. Eisenbud. 1970. Measurement of low levels of ^{131}I in reactor atmospheres. Nucl. Appl. Technol. 9: 268-273. (NSA 24: 43967).

Fussell, E. M. 1964. Dispersal studies on radioactive-tagged *Culex quinquefasciatus* Say. Mosquito News 24: 422-426. (NSA 20: 5240).

Fuzeau-Braesch, S. 1957. Fractionation of the oxydo-reduction pigments of the epidermis and the yeux of the cricket. Compt. rend. 245(25): 2401-2404. (In French) (^{14}C).

Fuzeau-Braesch, S. 1959. Research on the origin of cuticular pigments of an insect by the marked element method. Compt. rend. 248(6): 856-858. (In French).

Fuzeau-Braesch, S. 1963. Study of tegumental pigmentation with the aid of radioelements. pp. 289-296. In: Radiation and Radioisotopes Applied to Insects of Agricultural Importance. (See IAEA, 1963) (In French).

Fytizas, E. and A. Bacoyannis. 1968. Action of tepa on the pupal development of Dacus oleae Gmel. (Diptera: Tephritidae). pp. 209-215. In: Isotopes and Radiation in Entomology. (See IAEA, 1968) (In French, English summary) (NSA 24: 32171).

Gaal-Dippold, A. 1966. The radioactivity of some surface waters of Hungary. pp. 131-135. In: Proc. 2nd Symp. on Health Physics. (Pecs, Hungary, September 26-30, 1966, Vol. II). Group Central Research Inst. Physics, Budapest. (NSA 21: 28503).

Gabay, J. J., J. C. Daly and N. I. Sax. 1970. Occurrence of ^{22}Na in medium-sized animals. Health Physics 18: 429-431. (NSA 24: 25438) (Racoon, woodchuck, opossum, flora).

Gabay, J. J., J. A. Dapolito and N. I. Sax. 1966. ^{54}Mn in freshwater clams. Health Physics 12(7): 968-970. (NSA 20: 29462).

Gaborcik, S. and V. Srb. 1967. The germinating capacity of the seeds of some grass varieties of the meadow-grass family (Poaceae) after irradiation with x-rays. Genetica Slechteni 3: 101-108. (In Czech) (NSA 22: 33975).

Gabriel, A. 1968. Effect of the moment of irradiation with x rays on the regeneration and the rate of incorporation of tritiated uridine in the Planaria dugesia genocephala (Turbellaria, Triclidida). Compt. rend. D266: 820-822. (In French) (NSA 22: 40842).

Gabrusewycz-Garcia, N. 1964. Cytological and autoradiographic studies in Sciara coprophila salivary gland chromosomes. Chromosoma 15(3): 312-344. (^{3}H).

Gabuniya, R. I. and B. D. Pavlov. 1971. Use of γ-spectrometer in the study of ^{40}K and ^{137}Cs organ content in laboratory animals. pp. 230-237. In: Methods of Radioecological Investigations. (See Verkhovskaya, 1971).

Gadzhei, E. F. 1963. Fly migration in urban areas. Med. Parazitol. 32(4): 465-468. (^{32}P).

Gaffey, C. T. 1962. Bioelectric effects of high energy irradiation on nerve. pp. 277-296. In: Response of the Nervous System to Ionizing Radiation. T. J. Haley and R. S. Snider, eds. Academic Press, London. (NSA 16: 23486) (Frogs).

Gaffey, C. T. 1970. Electrophysiologic responses of sciatic nerves exposed to 200-kV x-rays and 47.5-MeV protons. Advan. Biol. Med. Phys. 13: 351-371. (NSA 26: 23104) (Frog).

Gaffey, C. T. 1971. The response of maximal and submaximal action potentials from frog sciatic nerve to 200-kV x-rays. Radiation Research 45(2): 311-325. (NSA 25: 26993).

Gage, S. J. 1972. Role of NEPA environmental assessment in nuclear engineering education. Trans. Amer. Nucl. Soc. 15(2): 635-636. (NSA 27: 7128).

Gager, C. S. 1936. The effects of radium rays on plants. (A brief resumé of the more important papers from 1907 to 1932). pp. 987-1013. In: Biological Effects of Radiation. B. M. Duggar, ed. McGraw-Hill Book Co., New York. (Gager, S. C. ?, see below).

Gager, S. C. 1908. Effect of rays of radium on plants. Mem. N.Y. Bot. Gard. 4: VIII, 278.

Gaglione, P. and O. Ravera. 1962. Mn^{54} concentration from fall-out in relation to the age and organs of Unio mancus elongatus (Mollusca; Lamellibranchia). Rend. Sci. Inst. Lombardo Scienze Lett. B96: 157-164. (In Italian, English summary) (Also: Euratom, Ispra, Italy, report EUR-254.i).

Gaglione, P. and O. Ravera. 1964. Manganese-54 concentration in fallout, water, and Unio mussels of Lake Maggiore, 1960-63. Nature 204(4964): 1215-1216. (NSA 19: 10762).

Gaglione, P., F. Möller and C. Polvani. 1970. Sources of variability in the observed concentration factors for fallout ^{90}Sr and ^{137}Cs in freshwater fishes. pp. 439-466. In: Proc. Intern. Symp. on Radioecology. (See French Commissariat à l'Énergie Atomique, 1970).

Gagnaire, J. 1962. Study of the influence of climactic factors on the accumulation of radioelements in plant tissue. Thesis. Centre d'Études Nucléaires, France Commissariat à l'Énergie Atomique, Grenoble, report CEA-2131. 37pp. (In French).

Gagnaire, J. 1963. Absorption of some mineral salts by root system of different woody species and accumulation over a whole vegetative cycle. Centre d'Études Nucléaires, France Commissariat à l'Énergie Atomique, Grenoble, report CEA-2317. 32pp. (In French) (NSA 17: 40514).

Gagnaire, J. 1965. Vegetative rhythms of some woody species. Radicular absorption and transport of alkaline earth (^{45}Ca, ^{140}Ba) salts during a complete vegetative cycle, March 1963-April 1964. Centre d'Études Nucléaires, France Commissariat à l'Énergie Atomique, Grenoble, report CEA-R-2825. 50pp. (In French) (NSA 20: 12590).

Gagnaire, J. 1966. Root uptake and localization of ^{45}Ca in young trees (poplars and thuyas) during a complete growth cycle. pp. 377-389. In: Isotopes in Plant Nutrition and Physiology. (See IAEA, 1966) (In French) (NSA 21: 41344).

Gagnaire, J. and A. Chamel. 1963. The use of plants in the control of radioactive pollution. Bull. Inform. Sci. Tech. (Paris) (70): 39-44. (In French) (NSA 17: 28660).

Gagnaire, J., A. Chamel and B. Lachet. 1967. Study of the interpretation of the variation of the concentration rates of radioelements in black poplar. Bull. Inform. Sci. Tech. (Paris) (119): 29-37. (AEC-tr-7040).

Gagnaire, J, et al. 1967. Attempt to interpret the fluctuations of radioelement concentration in black poplar. Bull. Inform. Sci. Tech. (Paris) (119): 27-37. (In French) (NSA 22: 19304).

Gagnaire, J., et al. 1967. Distribution of a mixture of fission products in a small model simulating ecological conditions of a stream. Bull. Inform. Sci. Tech. (Paris) (119): 9-26. (In French) (NSA 22: 19303).

Gaidamakin, N. A., et al. 1971. Results of studies concerning turtles flown on some space ships. Izv. Akad. Nauk SSSR (biol.) 3: 451-453. (In Russian, English summary) (Environ. Pollut. Bioresearch Today 1(9): 147) (Concerns ionizing radiation).

Gailledreau, C. 1963. Possible utilization nitronitrosyl-ruthenium complexes as tracers in hydrology. pp. 231-235. In: Radioisotopes in Hydrology. (See IAEA, 1963) (In French) (NSA 18: 1922).

Gajewskaja, N. 1923. Influence of roentgen radiation on *Artemia salina*. Verhandl. Intern. Ver theor. angew. limnol. Grund-Vers. Z. Kiel : 359-362. (In German).

Gak, D. Z. 1962. Accumulation of radiophosphorus by various components of water reservoirs. Radiobiologiya 2(6): 938-943. (AEC-tr-5433; pp. 189-197) (NSA 17: 15622).

Gakstatter, J. H. 1966. The uptake from water by several species of freshwater fish of *p*, *p*'-DDT, dieldrin and lindane: their tissue distribution and elimination rate. Thesis, Univ. of North Carolina, Chapel Hill. 146 pp. (Dissertation Abstr. 27: 3820-B) (^{14}C).

Gakstatter, J. H. 1968. Rates of accumulation of ^{14}C-dieldrin residues in tissues of goldfish exposed to a single sublethal dose of ^{14}C-aldrin. J. Fish. Res. Bd. Can. 25(9): 1797-1801.

Gakstatter, J. H. and C. M. Weiss. 1967. The elimination of DDT-C^{14} , dieldrin-C^{14} , and lindane-C^{14} from fish following a single sublethal exposure in aquaria. Am. Fish. Soc. Trans. 96(3): 301-307.

Galegar, W. C. and M. W. DeGeer. 1969. Measuring subsurface spring flow with radiotracers. J. Sanit. Engrg. Div. Am. Soc. Civil Engrs. 95(SA6): 1097-____.

Galkovskaya, K. F. 1951. Recovery of function of denervated limbs of axolotl following x irradiation. Dokl. Akad. Nauk SSSR 81: 945-948. (In Russian) (NSA 6: 3496).

Galkovskaya, K. F. 1952. Stimulating effect of x-rays on the process of regeneration of peripheral nerves. Dokl. Akad. Nauk SSSR 87: 677-679. (In Russian) (NSA 7: 4009).

Gallant, E. M. 1966. Accumulation of beta activity from fallout on leaves. (Paper presented at 75th Ann. Meeting, Ohio Acad. Sci., Columbus, O., Apr. 21-23, 1966). (CONF-660430) (AED-CONF-66-105-2) (NSA 20: 40579).

Gallegos, A. F. 1969. Radiocesium kinetics in the components of a montane lake ecosystem. Thesis, Colorado State Univ., Fort Collins. 342pp. (NSA 24: 27561).

Gallegos, A. F. and F. W. Whicker. 1973. Radiocesium retention by rainbow trout as affected by temperature and weight. pp. 361-371. In: Radionuclides in Ecosystems. (See Nelson, 1973) (NSA 26: 38761).

Gallegos, A. F., F. W. Whicker and T. E. Hakonson. 1971. Accumulation of radiocesium in rainbow trout via a non-foodchain pathway. pp. 477-498. In: Health Physics Aspects of Nuclear Facility Siting. (See Voillequé

Gallien, L. 1969. Provoked chromosome anomalies and mutations in amphibians. Année biol. 8: 319-331. (In French) (NSA 24: 48492).

Gallien, L. and M. Labrousse. 1962. Radiosensitivity to neutrons of fertilized eggs of urodele, *Pleurodeles waltlii* Michah. Compt. rend. 255(2): 371-373. (In French) (NSA 16: 28776).

Gallien, L., M. Labrousse and J.-C. Lacroix. 1963. Chromosomic aberrations associated with the hypomorphoses after γ irradiation of the egg in the amphibian urodele *Pleurodels waltlii* Michah. Compt. rend. 256(25): 5413-5415. (In French) (NSA 17: 33651).

Gallien, L., M. Labrousse and J. Lacroix. 1966. Detection of mitotic chromosomes and lampbrush chromosomes induced by the irradiation of eggs of the urodele amphibian *Pleurodeles waltlii* Michah. Compt. rend. D263: 1984-1987. (In French) (NSA 21: 20357).

Gallien, L., et al. 1965. Experimental modifications of the karyotype in a urodele amphibian (*Pleurodeles waltlii* Michah.) by irradiation of the egg and nuclear grafting. Rev. Suisse Zool. 72: 59-86. (In French) (NSA 21: 6658).

Gallun, R. L. and R. Langston. 1962. Hessian fly feeding studies utilizing radioisotope P-32. Proc. N. Cent. Br. Entomol. Soc. Am. 17: 32.

Gallun, R. L. and R. Langston. 1963. Feeding habits of Hessian fly larvae on P^{32}-labeled resistant and susceptible wheat seedlings. J. Econ. Entomol. 56(5): 702-706).

Galun, R. and M. Warburg. 1968. Irradiation effects on respiration and blood digestion in the tick *Ornithodorus tholozani* and their importance for the sterile-male technique. pp. 249-258. In: Isotopes and Radiation in Entomology. (See IAEA, 1968) (NSA 24: 32172).

Galun, R., M. Warburg and A. Avivi. 1967. Studies on the application of the sterility method in the tick *Ornithodorus tholozani*. Entomologia Exp. Appl. 10(2): 143-152.

Galun, R., M. Warburg and S. Sternberg. 1974. Practical use of the sterility method for the control of soft ticks. pp. 21-25. In: Sterile-Insect Technique and Its Field Applications. (See IAEA, 1974) (NSA 31: 8937).

Gambi, G. 1958. New views in forest research. Monti e Boschi 9(2): 85-86. (In Italian) (Involves ionizing radiation).

Gambino, J. J. 1966. Radiobiological studies on *Perognathus*, a small desert rodetn. pp. 23-27. In: Radiation Effects on Natural Populations. (See Sacher, 1966).

Gambino, J. J. and R. G. Lindberg. 1964. Response of the pocket mouse to ionizing radiation. Radiation Research 22(4): 586-597. (NSA 18: 35362).

Gambino, J. J. and J. W. Towner. 1966. Relationships between cell population kinetics and radiation resistance in pocket mice (Heteromyidae: Perognathus). 3rd Intern. Congr. on Radiation Research, Northrop Space Labs., Hawthorne, Calif., U. S. NASA report NASA-CR-69454. 10pp. (N-68-18954) (CONF-660614-2) (NSA 22: 38424).

Gambino, J. J., B. H. Faulkenberry and P. B. Sunde. 1968. Survival studies on rodents exposed to reactor fast-neutron radiation. Abstr. Radiation Research 35(2): 575-576.

Gambino, J. J., B. H. Faulkenberry and P. B. Sunde. 1968. Survival studies on rodents exposed to reactor fast neutron radiation. Radiation Research 35: 668-680. (NSA 23: 6592).

Gambino, J. J., R. G. Lindberg and P. Hayden. 1965. A search for mechanisms of radiation resistance in pocket mice. Radiaition Research 26: 305-317. (NSA 20: 114).

Gamble, J. F. 1973. A proposed mechanism for the recycling of radiocesium in Florida soil plant systems. pp. 133-139. In: Radionuclides in Ecosystems. (See Nelson, 1973).

Gamble, J. F., et al. 1967/69. Agricultural ecology. Phase I final report. Bioenvironmental and radiological-safety feasibility studies, Atlantic-Pacific Interoceanic Canal. Final report. Univ. of Florida, Gainesville, Battelle Columbus Labs., U. S. AEC report BMI-171-010; BMI-171-20. 40, 212pp. (NSA 21: 39369; 24: 9795).

Gamo, T. and H. Nishiyama. 1956. Some observations of biological influences of radio-active isotopes upon the physiological functions of the silkworm. (1) On the absorption of Ca^{45} administered through the mouth into several tissues of the silkworm and its injurious effects upon the respiratory function. Shinshu Daigaku Sen Igakubu Kenkyu Hokoku 6: 37-41. (In Japanese, English summary).

Gamo, T., H. Nishiyama and S. Midorikawa. 1958. Effect of calcium-45 on the blood cells of silkworm. Dai-2-kai Genshiryoku Shimpojiumu Hobunshu 4: 206-212. (In Japanese).

Gamo, T., H. Nishiyama and S. Midorikawa. 1960. Studies on the influence of radioactive-rays upon the hemocytes of the silkworm, Bombyx mori L. IV. On the sexual differences between the effects of the ingested radioisotopes on the number of hemocytes of the silkworm. Radioisotopes (Tokyo) 9: 17-24. (In Japanese) (NSA 15: 22142).

Gamo, T., et al. 1956. Some observations of biological influences of radioactive isotopes upon physiological functions of silkworms. II. On influence of radioactive phosphorus P^{32} upon development and number of leucocytes of silkworms. pp. 524-527. In: Proc. 1st Japan Conf. on Radioisotopes, 1956. Japan Atomic Industrial Forum, Tokyo. (In Japanese).

Gampel, Z. and J. Toha C. 1969. Aging and repair in Neurospora crassa studied by ultraviolet irradiation. Radiation Research 40(3): 525-533. (NSA 24: 14626).

Gangadharan, P. 1969. Reliability of instruments, effects of climatic conditions and conditions of use. pp. 489-498. In: Radiation Protection Monitoring. (See IAEA, 1969) (NSA 23: 47779).

Gangadharan, S. 1971. Nuclear methods in trace analysis: a review. India Bhabha Atomic Research Centre, Bombay, report BARC-529. 57pp. (NSA 25: 26186).

Gangwere, S. K., W. Chavin and F. C. Evans. 1964. Methods of marking insects, with special reference to Orthoptera (Sens. Lat.). Ann. Entomol. Soc. Am. 57(6): 662-669. (Incl. radionuclides).

Gannutz, T. P. 1968. Biological effects of gamma irradiation on lichens and lichen symbionts. Thesis, Clark Univ., Worchester, Mass. 173pp. (NSA 23: 24766).

Gannutz, T. P. 1970. A study of the lichens of an irradiated rain forest. Ch. D-12, pp. D227-D231. In: A Tropical Rain Forest. (See H. T. Odum and R. F. Pigeon, 1970) (NSA 25: 21937).

Gannutz, T. P. 1972. Effects of gamma radiation on lichens. I. Acute gamma radiation on lichen, algae and fungi. Radiation Botany 12(5): 331-338. (NSA 27: 7744).

Gannutz, T. P. and V. Ahmadjian. 1965. Effects of ionizing radiation on the growth of isolated lichen components. Abstr. Bull. Ecol. Soc. Am. 46(3): 97.

Garby, L., M. Yasuno and Y. Phurivethaya. 1966. Labelling mosquitoes with radioactive iodine, ^{131}I. Trans. Roy. Soc. Trop. Med. Hyg. 60: 136. (NSA 21: 12767).

Garcia, M. N. 1963. Radiation in ecology. Energia Nucl. (Madrid) 7(26): 27-41. (NSA 17: 40581).

Garcia, P. L. 1960. The influence of Larrea divaricata and Atriplex canescens on soil p_H, electrical conductivity and soluble sodium content. Thesis, New Mexico Highlands Univ., Las Vegas, N. Mex. viii, 63pp. (Research at Nevada Test Site).

Gardel, R. 1962. Control of an operation at the EDF Hydroelectric Plant at Lardit. French Service Central de Protection Contre les Rayonnéments Ionisants, report SCPRI(T)-80. 7pp. (In French) (NSA 17: 12499) (See also French SCPRI).

Garder, K. and O. Skulberg. 1964. Sorption phenomena of radionuclides to clay particles in river water. Int. J. Air Water Poll. 8(3/4): 229-241. (NSA 18: 31043).

Garder, K. and O. Skulberg. 1966. An experimental investigation on the accumulation of radioisotopes by freshwater biota. Archiv Hydrobiol. 62 (1): 50-__.

Gardner, W. H. and C. Calissendorff. 1967. Gamma-ray and neutron attenuation in measurement of soil bulk density and water content. pp. 101-112. In: Isotope and Radiation Techniques in Soil Physics and Irrigation Studies. (See IAEA, 1967) (NSA 22: 23145).

Gardner, W. H., G. S. Campbell and C. Calissendorff. 1972. Systematic and random errors in dual gamma energy soil bulk density and water content measurements. Soil Sci. Soc. Am. Proc. 36(3): 393-398. (NSA 26: 42982).

Gardner, W. R. 1967. Water uptake and salt-distribution patterns in saline soils. pp. 335-342. In: Isotope and Radiation Techniques in Soil Physics and Irrigation Studies. (See IAEA, 1967).

Garfinkel, S. B. 1971. Radioactivity standards and the environment. Trans. Amer. Nucl. Soc. 14(1): 124. (CONF-710606) (NSA 25: 39875).

Garg, O. K., B. C. Misra and B. P. Singh. 1972. Effect of pre-sowing exposure of seeds to gamma-radiation on the drought resistance behavior of barley plants (Hordeum vulgare L.). Plant Soil 36(1): 39-45. (NSA 26: 42982).

Garner, J. M., Jr. and O. W. Kochtitzky. 1956. Radioactive sediments in the Tennessee River System. J. Sanit. Engrg. Div. Am. Soc. Civil Engrs. 82 (SA 4): 20pp. Papaer 1051. (NSA 10: 10995).

Garner, R. J. 1971. Transfer of radioactive materials from the terrestrial environment to animals and man. CRC Critical Rev. Environ. Control 2(3): 337-385.

Garnier, A. 1975. Dilution factor of radiactive contamination in the distribution pathway of dietary products. Centre d'Études Nucléaires, France Commissariat à l'Énergie Atomique, Fontenay-aux-Roses, report CEA-CONF-3121. 3pp. (NSA 33: 23275).

Garrec, J.-P. 1968. Nondestructive analysis of major components of plant materials by means of 14-MeV neutrons. France Commissariat à l'Énergie Atomique, Grenoble, report CEA-R-3636. 30pp. (In French) (NSA 23: 17361).

Garrec, J.-P. and A. Fourcy. 1970. Fast analysis by 14-MeV neutron activation of flourine deposits on the vegetation in a polluted area. Centre d'Études Nucléaires, France Commissariat à l'Énergie Atomique, Grenoble, report CEA-CONF-1747. 8pp. (For 3rd European Flourine Chemistry Symp., Aix-en-Provence, France, 14-17 Sept. 1970) (In French) (CONF-700951-2) (NSA 25: 29011).

Garrec, J.-P., et al. 1972. Localization of flourine along polluted Abies alba needles using proton microanalysis. Compt. rend. D274(25): 3467-3471. (In French) (NSA 27: 7345).

Garrec, J.-P., et al. 1972. Exact localization of flourine along polluted Abies alba using microanalysis with protons. Centre d'Études Nucléaires, France Commissariat à l'Énergie Atomique, Grenoble, report CEA-CONF-2123. 9pp. (In French) (CONF-721069) (NSA 27: 16956).

Garrett, A. R., Jr. 1963. An investigation of calcium and Sr-90 in some components of granite outcrop ecosystems of the Georgia Piedmont. Thesis, Univ. of Georgia, Athens.

Garrett, A. R., Jr. 1965. Effects of gamma irradiation on taxonomic diversity within granite outcrop ecosystems. Abstr. Bull. Ecol. Soc. Am. 46 (3): 97.

Garrett, A. R., Jr. 1968. Compositional changes of ecosystem segments during chronic gamma irradiation. Thesis, Emory Univ., Atlanta, Ga. 144pp. (Also: Emory Univ., U. S. AEC report ORO-2412-18) (NSA 23: 14140).

Garrett, A. R., Jr. 1969. Compositional changes of ecosystems during chronic gamma irradiation. pp. 99-109. In: Symposium on Radioecology. (See Nelson and Evans, 1969).

Garry, J. A., F. C. Flack and R. Jakeways. 1965. Radioactive isotopes from fallout in wastewater. J. Water Pollut. Control 37: 871-875. (NSA 21: 1462).

Garsou, J. L. 1960. The ambient radioactivity and its actual increase. Rev. Med. Liége 15: 342-350. (NSA 17: 32276).

Garsou, J. L. 1962. Preliminary results on the effect of high energy photons on lettuce root growth and on Musca domestica. Abstr. p. 5. In: 2nd Intern. Congr. Radiation Research. (See Anonymous, 1962).

Gartner, L. P. 1970. Aging and ionizing radiation: a study of life-span and fine-structural alterations in the Drosophila melanogaster. Thesis, Rutgers Univ., New Brunswick, N. J. 216pp. (Univ. Microfilms 71-12,246) (NSA 25: 42612).

Gartner, L. P. 1973. Long-term effects of ionizing radiation on Drosophila midgut mitochondria. Strahlentherapie 145(6): 725-730. (NSA 28: 30251).

Gartner, L. P. 1973. Ultrastructural examination of ageing and radiation-induced life-span shortening in adult Drosophila melanogaseter. Int. J. Radiat. Biol. 23(1): 23-39. (NSA 27: 24977).

Gartner, L. P. 1973. Radiation-induced life span shortening in Drosophila. Gerontologia 19: 295-302. (NSA 31: 816).

Gartrell, F. E. 1962. Water pollution potential of air pollution control devices. In: Air Pollution. Academic Press, New York. (NSA 20: 7331).

Gartrell, F. E., G. F. Stone and T. A. Wojtalik. 1971. Environmental quality protection: large steam-electric power stations. pp. 453-473. In: Environmental Aspects of Nuclear Power Stations. (See IAEA, 1971) (CONF-700810-14) (Paper SM-146/28) (NSA 24: 45905)

Garver, G. 1972. Energy, Ecology, Economy. W. W. Norton, New York. 235pp. (NSA 27: 29808) (Incl. nuclear power).

Gaspar, E. 1971. Radiological tracers for hydrological works. Romania Institutul de Fizica Atomica, Bucharest, report IFA-MR-36. 24pp. (NSA 26: 40861).

Gaspar, E. 1973. Method with radiotracers and experiments in hydro-karst structures. Romania Institutul de Fizica Atomica, Bucharest, report IFA-MR-39-1973. 25pp. (NSA 28: 21129).

Gaspar, E. and M. Ilie. 1974. Activity of labelled injection sediments for experiments in dynamic sedimentology. Stud. Cercet. Fiz. 26(7): 715-724. (In Romanian) (NSA 31: 19614).

Gaspar, E. and M. Oncescu. 1964. About the measurements of the flowing parameters of underground waters. Acad. Rep. Populare Romine, Institutul de Fizica Atomica, Bucharest, report IFA-MR-23. 13pp. (NSA 19: 42251) (Used tracers).

Gaspar, E., M. Oncescu and D. Bisir. 1967. Field determination of soil permeability by the nuclear methods. Mem. Ass. Int. Hydrogeol. Reunion Istanbul 8: 267-275. (See Anonymous, 1967) (NSA 23: 450).

Gaspar, E., M. Oncescu and O. Serbanescu. 1968. Contribution to the study of movement of sand in the Gulf of Eforia by the method of radioactive tracers. Acad. Rep. Populare Romine, Institutul de Fizica Atomica, Bucharest, report IFA-MR-31. 18pp. (In French) (NSA 23: 28995).

Gass, W. B., et al. 1971. Recovery of residual nitrogen by corn (Zea mays L.) from various soil depths as measured by ^{15}N tracer techniques. Soil Sci. Soc. Am. Proc. 35(2): 290-____.

Gasso, G., A. Billitteri and S. Romeo. 1962. Variation of immutability of irradiated animals in hibernation. Radiobiol. Latina 5(4): 501-505. (In Italian) (NSA 18: 6657).

Gat, J. R. and A. Nir. 1973. Studies of environmental systems by tracer methodology. Weizmann Inst. of Science, Rehovoth, Israel. (INIS-mf-1214) (NSA 30: 12705).

Gatenby, J. B., R. N. Mukerji and S. B. Wigoder. 1929. The effect of x-radiation on the spermatogenesis of Abraxas grossulariata. Proc. Roy. Soc. (London) 105(B739): 446-468.

Gatterdam, P. E., et al. 1964. The absorption, metabolism, and excretion of C^{14}-labelled TDE in certain insects. J. Econ. Entomol. 57(2): 258-264. (Periplaneta americana, Protoparce sexta, Argyrotaenia velutinana).

Gaufin, A. R. 1974. Investigation of the water quality and productivity of Polson Bay, Flathead Lake, Montana. Completion report. Montana State Univ., Boseman. 17pp. (PB-232119) (NSA 31: 11267) (Primary productivity determined by ^{14}C method).

Gaulden, M. E., M. Nix and J. Moshman. 1953. Effects of oxygen concentration on x-ray-induced mitotic inhibition in living Chortophaga neuroblasts. J. Cell. and Comp. Physiol. 41(3): 451-470. (NSA 7: 5002).

Gay, H. 1964. New evidence on chromosome structure and function. Science 146: 425. (Drosophila melanogaster, ^{3}H).

Gazert, V., K. Pötzl and R. Reiter. 1962. Quantitative studies of fission products in samples from the northern Alps and Spitzbergen in 1960. Atomkernenergie 7(3): 106-111. (In German).

Geard, C. R. 1972. Sister chromatid exchanger in chromosomes of the swamp wallaby (Wallabia bicolor) derived from leucocyte cultures. Australian National Univ., Canberra. 11pp. (INIS-mf-289) (NSA 27: 7728) (Used tritiated thymidine).

Geard, C. R. 1975. Chromosomal aberrations in three successive cell cycles of Wallabia bicolor leukocytes after tritiated thymidine incorporation. Radiation Research 61(1): 118-127. (NSA 31: 11684).

Gearing, P. J., C. van Baalen and P. L. Parker. 1973. Response of blue-green algae to technetium. pp. 857-865. In: Radionuclides in Ecosystems. (See Nelson, 1973) (NSA 28: 29831).

Gebhards, S., et al. 1973. Mercury residues in Idaho fishes - 1970. pp. 76-84. In: Mercury in the Western Environment. (See Buhler, 1973) (Neutron activation analysis).

Geckler, R. O. and R. F. Kimball. 1953. Effect of x-rays on micronuclear number in Paramecium aurelia. Science 117: 80-81.

Gedeonov, L. I. and S. P. Rosyanov. 1968. The content of some fission products in forest vegetation. Pochvovedeniye (7): 88-90. (Soviet Soil Sci. (7): 942-944) (NSA 22: 51615).

Gedeonov, L. I. and A. G. Trusow. 1974. Nuclear energy development and environmental protection (a summary). Arch. Energiewirt. 28(8): 386-405. (In German) (NSA 31: 3358).

Gedeonov, L. I., S. P. Rosyanov and V. K. Vinogradova. 1967. Influence of herbaceous vegetation on radioactive contamination in soils. Pochvovedeniye (10): 124-126. (Soviet Soil Sci. (10): 1394-1396) (NSA 22: 45541).

Gedeonov, L. I., S. P. Rosyanov and V. K. Vinogradova. 1969. Distribution and migration of strontium-90 in soils. pp. 76-84. In: Nuclear Meteorology. (See Makhon'ko and Malakhov, 1969) (NSA 30: 15430).

Gedeonov, L. I., S. P. Rosyanov and V. K. Vinogradova. 1971. ^{90}Sr distribution and migration in soils. Tr. Inst. Eksp. Meteorol. 21(1): 80-87. (In Russian) (NSA 30: 29404).

Gedeonov, L. I., et al. 1967. Accumulation of artificial radionuclides on the Earth's surface in the Leningrad District in 1954-1965. State Commitee on the Use of Atomic Energy, USSR. Atomizdat, Moscow. UN Document A/AC.82/G/L.1171. 7pp. (AEC-tr-7214; pp. 21-26) (NSA 22: 21194).

Gedeonov, L. I., et al. 1967. Artificial radioactivity of atmospheric aerosols. USSR, Moscow. UN Document A/AC.82/G/L.1166. (AEC-tr-7214; pp. 51-61) (NSA 25: 15778).

Gedeonov, L. I., et al. 1973. Radiochemical method of determination of uranium, neptunium, plutonium, transplutonium, and rare-earth elements in samples of aerosols, radioactive fallout, and soil. 14pp. (AEC-tr-7566) (NSA 30: 20639).

Gedeonov, L. I., et al. 1974. Collection of samples and determination of tritium in the layer of air near the Earth's surface. pp. 235-241. In: Environmental Surveillance Around Nuclear Installations. (See IAEA, 1974) (In Russian) (NSA 31: 14062).

Gegamyan, G. V. 1974. Behavior of radioisotopes of strontium and cesium in model biogeocenoses. Ekologiya 5(6): 91-93. (Sov. J. Ecol. 5(6): 586-589) (NSA 33: 12043, 12455).

Gegamyan, G. V., N. I. Olesheva and A. N. Tyuryukanov. 1971. Behavior of radioactive isotopes of Sr, Ce and Cs in soils (results of an experimental investigation). pp. 45-49. In: Methods of Radioecological Investigations. (See Verkhovskaya, 1971).

Gehrs, C. W., J. R. Trabalka and E. A. Bardill. 1975. Sensitivity of adult and embryonic calanoid copepods to acute ionizing radiation. Radiation Research 63(2): 382-385. (NSA 32: 22632).

Geier, P. W. 1969. Demographic models of population response to sterile-release procedures for pest control. pp. 33-41. In: Insect Ecology and the Sterile-Male Technique. (See IAEA, 1969).

Geierhaas, B. 1965. Location of radioactive deposits in growing feathers of pigeons after intramuscular Na sulfate (^{35}S) and dl cysteine (^{35}S) injections. Thesis, Univ. of Heidelberg, Germany. 67pp. (AED-Diss. 65-349) (NSA 20: 45317).

Geiger, B., et al. 1974. Power economy and its influence on the environment in the Federal Republic of Germany. Federal Republic of Germany, Inst. für Energiewirtschaft und Kraftwerkstechnik, report DEU-74-01 (Rev. 2). 63pp. (In German) (NSA 33: 17203).

Geiger, R. A., et al. 1972. Distributional patterns of radiocesium in the sediments of a stream channel contaminated by production reactor effluents. Abstr. Bull. Ecol. Soc. Am. 53(4): 9.

Geilmann, W. 1968. Age determination of deer antlers. Z. Anal. Chem. 235: 172-176. (In German) (NSA 22: 33825) (Radionuclides).

Geints, A. Ye. and V. Ye. Garutt. 1964. Determining the absolute age of mammoth and wooly rhinoceros fossil remains from the Siberian permafrost by means of C^{14} dating. Dokl. Akad. Nauk SSSR 154(6): 1367-1370. (Doklady ady (earth sci.) 154(1-6): 168-170) (NSA 18: 23788).

Gelbke, W. 1972. Radioactivity in deer antlers and human strontium-90 uptake. Z. Tierphysiol. Tierernähr. Futterm. 29: 178-195. (In German) (NSA 28: 6073).

Geldiay, R. 1962. Uptake and distribution of the radioactive isotopes Zn^{65} and Co^{60} through aquatic food chains. (Chironomus plumosus, Salmo gairdnerii). Ege Univ. Fac. Sci. Rep. 4. pp. 3-23.

Gellerman, J. L. and H. Schlenk. 1965. Preparation of fatty acids labeled with ^{14}C from Ochromonas danica. J. Protozool. 12: 178-189. (NSA 19: 42251).

Gelskey, D. E., P. Plato and A. P. Jacobson. 1971. In situ measurements of environmental radioactivity. pp. 278-288. In: Health Physics Aspects of Nuclear Facility Siting. Vol. II. (See Voillequé and Baldwin, 1971) (NSA 25: 55239).

Gemrich, E. G., II. 1967. Enzymatic degredation of 2-chloro-4,5-dimethylphenyl N-methylcarbamate by the fat bodies of Blaberus giganteus. J. Agr. Food Chem. 15(4): 617-621. (^{14}C).

Genevois, L. and H. Flavier. 1964-65. Autoradiography of some plant samples from the Landes district taken in 1964. Proces-Verb. sences soc. sci. phys. natur. Bordeaux. pp. 167-172. (In French) (NSA 22: 31066).

Gengozian, N. and J. Batson. 1963. Effects of radiation on a small South American primate, Tamarinus nigricollis. Abstr. Radiation Research 19(1): 223-224.

Gengozian, N. and J. Batson. 1966. Effects of whole-body irradiation on a small South American primate, Saguinus fuscicollis. Int. J. Radiat. Biol. 11: 539-551. (NSA 22: 580).

Gentry, J. B. 1962. The effects of old-field succession on small mammal populations. Thesis, Univ. of Georgia, Athens. 32pp. (At Savannah River Plant).

Gentry, J. B. 1968. Dynamics of an enclosed population of pine mice, Microtus pinetorum. Res. Popul. Ecol. 10(1): 21-30. (At Savannah River Plant).

Gentry, J. B., F. B. Golley and M. H. Smith. 1971. Yearly fluctuations in small mammal populations in a southeastern United States hardwood forest. Acta Theriologica 16(12): 179-190. (At Savannah River Plant).

Gentry, J. B., M. H. Smith and R. J. Beyers. 1971. Use of radioactively tagged bait to study movement patterns in small mammal populations. Ann. Zool. Fennici 8(1): 17-21.

Gentry, J. B., M. H. Smith and R. J. Beyers. 1973. Radioactive isotopes in studies of population dynamics of small mammals. pp. 253-259. In: Radionuclides in Ecosystems. (See Nelson, 1973) (NSA 28: 27326) (^{59}Fe, ^{65}Zn).

Gentry, J. B., M. H. Smith and J. G. Chelton. 1971. An evaluation of the octagon census method for estimating small mammal populations. Acta Theriologica 16(10): 149-159. (At Savannah River Plant).

George, J. A. 1967. Effects of gamma radiation on fertility, mating, and longevity of males of the oriental fruit moth, Grapholitha molesta (Lepidoptera: Tortricidae). Can. Entomol. 99(8): 850-857.

George, J. A. 1967. Effect of mating sequence on egg-hatch from female Aedes aegypti (L.) mated with irradiated and normal males. Mosquito News 27(1): 82-86. (NSA 23: 29948).

George, K. C. and J. Eapen. 1973. Effect of x-irradiation on esterases of tissues of house lizard, Hemidactylus leschenaulti Dum. and Pipr. Indian J. Exp. Biol. 2(2): 76-78. (NSA 29: 18633).

Georgescu, I. I. and H. Skolka. 1970. Radioactivity of the seston and waters of the Black Sea and the alga (Cystoseria barbata) from the Rumanian continental shelf during 1964-1967. Rev. Roum. Phys. 15: 945-958. (In French) (NSA 25: 42361).

Georgescu, I. I., et al. 1963. The radioactivity of the first solid atmospheric precipitations in Bucharest during the winter of 1962-1963. Acad. Rep. Populare Romine Stud. Cercet. Fiz. 14: 805-814. (In Russian) (NSA 18: 12348).

Georgescu, I. I., et al. 1964. On the radioactivity of certain plants from Bucharest region. Rev. Roum. Phys. 9: 867-891. (In French) (NSA 19: 30413) (Wild vegetation).

Georgescu, P. and C. Bratu. 1962. Aspects of the removal and treatment of waste waters with a content of radioactive substances. Meteorol. Hidrol. Gospod. Apelor 7: 205-208. (NSA 17: 29875; PHEA 43(6)).

Georgia Power Company. 1971. Edwin I. Hatch Nuclear Plant, Units 1 and 2. Supplementary information to environmental statement. Atlanta. U. S. AEC DOCKET-50321-23, -50366-12. 4pp. (NSA 25: 46206).

Georgiev, D., Kh. Dilov and S. Avramova. 1972. Investigations on the excretion of ^{14}C by Scenedesmus acutus in nutrient medium depending on the growth phases and cultivation conditions. Bulgaria Komitet za Mirno Izpolzuvane na Atomnata Energiya, Sofia. 10pp. (In Bulgarian) (INIS-mf-709) (NSA 29: 24545).

Georgieva-Todorova, I. 1969. X-rays influence on chromosome abberations in Helianthus annus L. Genet. Selek. 2(6): 469-476. (In Bulgarian) (NSA 25: 42361).

Georgopoulos, S. G., B. Macris and E. Georgiadou. 1966. Measurement of radiation resistance in fruit spoilage fungi by chemicals. Phytopathology 56: 230-234. (NSA 20: 33030).

Gera, F. 1975. Geochemical behavior of long-lived radioactive wastes. Oak Ridge National Lab., U. S. ERDA report ORNL-TM-4481. 99pp. (NSA 32: 14303).

Gera, F. and D. G. Jacobs. 1971. Hazard potential of radioactive waste. pp. 891-907. In: Proc. Intern. Symp. on Radioecology Applied to the Protection of Man and His Environment. (See Commission of European Communities, 1971) (NSA 27: 5483).

Gerard, B. M. 1963. An earthworm labelled with radioactive tantalum. Nature 200(4905): 486-487. (NSA 18: 1335).

Gerardy, J., et al. 1966. Measurement of the radioactivity of the Meuse waters and of the fallout in Belgium during the year 1965. Bull. Soc. Roy. Liége 35(1-2): 106-109. (In French) (NSA 20: 35644).

Gerasimov, I. P. and O. A. Chicgagova. 1971. Some problems in the radiocarbon dating of soil. Pochvovedeniye 5: ___. (Soviet Soil Sci. 5: 519-527).

Gerasimova, G. K. 1963. Ionizing radiation effects on the mechanism of adaptation to rising temperature in the surrounding media. Radiobiologiya 3(5): 711-715. (AEC-tr-5438; pp. 129-135) (NSA 18: 23307) (Frogs).

Gerber, A. 1970. Energy and the environment in electric power generation. pp. 18.1-18.4. In: Energy 70, Proc. Vol. 2. American Nuclear Society, Hinsdale, Ill. (NSA 27: 9499).

Gerber, G. B. 1970. Assessment criteria for biochemical methods used in experimental radiobiology and their influence on interpretation of results. pp. 1-21. In: Biochemical Evidence of Radiation Effects and Their Relationships to Radiation Therapie. G. B. Gerber, ed. Verlag Georg Thieme, Stuttgart. (In German) (NSA 24: 50656).

Gericke, S. 1950. "Radiant plants." Radioactivity as a research aid in botany. Umschau 50: 568-569. (In German) (NSA 5: 33).

Gerish, G. 1971. Problem of locating nuclear power plants. Kernenergie 14 (5): 133-136. (In German) (NSA 25: 60119) (Discusses environment).

German Democratic Republic. 1968. Measurement instructions for determination of the activity of water, wastes, sediments, and biological samples from surface waters. Staatliche Zentrale für Strahlenschutz, Berlin. 15pp. (In German) (NSA 22: 51084).

German Democratic Republic. 1974. Radiation Protection, Environment, and the Population. (1st Seminar, Potsdam, October 16-21, 1972). Staatliche Zentrale für Strahlenschutz, Berlin, report SZS-157. 292pp. (In several languages) (NSA 31: 8873).

German Federal Republic. 1958. Commission on Radiation Protection Data, abstract collection No. IV. Report of the Minister for Atomic Energie and Watersystems (Radiation). Gersbach und Sohn Verlag, Brunswick. 130pp. (In German) (NSA 14: 3715).

German Federal Republic. 1958. First report of the Special Committee on Radioactivity of the German Federal Republic. Georg Thieme Verlag, Stuttgart. 67pp. (In German) (NSA 13: 729).

German Federal Republic. 1961-1968. Environmental radioactivity and radiation burden. Bundesministerium für Atomkernenergie; Bundesministerium für Wissenschaftliche Forschung, Bad Godesberg, report I/61 - IV/67, 1968. 125-285pp. (In German) (NP-18122) (NSA 22: 38290, 51598; 24: 23135).

German Federal Republic. 1964. Symposium on radioactivity and water protection. (Karlsruhe, June 26-28, 1963). Föderat. Europ. Gewässersch. Inform. (10): 1-86. (NSA 19: 2535).

German Federal Republic. 1970. Environmental radioactivity and radiation protection. Measurements in Hamburg 1968 and 1969. Gesundheitsbehörde, Hamburg. 27pp. (INIS-mf-162) (NSA 29: 5101).

German Federal Republic. 1971. Radioactive materials and drinking water supply in a nuclear catastrophe: summary report of the water contamination task group. Bundesministerium des Innern, Bonn. 550pp. (In German) (NP-19938) (NSA 29: 29351, 29252).

German Federal Republic. 1972. Environmental protection. The environmental program of the federal government. Kohlhammer, Stuttgart. 199pp. (In German) (NSA 28: 8178).

German Federal Republic. 1973. Regulations for the measurement and control of radioactive liquid effluents from nuclear power plants with light-water reactors. Bundesgesundheitsblatt 16(15/16): 231. (In German) (NSA 29: 12695).

Gerolt, P. 1965. The fate of dieldrin in insects. J. Econ. Entomol. 58(5): 849-857. (Musca domestica, ^{14}C).

Gerrard, M. 1967. Radioisotope gage for determining suspended-sediment concentration in streams. Isotopes Radiat. Technol. 4: 358-362. (NSA 22: 4273).

Gerrard, M. 1967. Radioisotope program of the International Atomic Energy Agency. Isotopes Radiat. Technol. 4: 430-431.

Gerrard, M. 1968. Measurement of snow-water reserves using natural soil radioactivity. Isotopes Radiat. Technol. 5: 358-362. (NSA 22: 16699).

Gerrard, M. 1969. Winter 1968-1969. Tagging of small animals with radioisotopes for tracking purposes. Isotopes Radiat. Technol. 6(2): 200-204. (NSA 23: 4904).

Gerrard, M. 1969. Tagging of small animals with radioisotopes for tracking purposes: a literature review. Int. J. Appl. Radiat. Isotopes 20: 671-676. (NSA 23: 43656).

Gerrard, M. 1971. Sewage and waste-water processing with isotopic radiation: survey of the literature. Isotopes Radiat. Technol. 8(4): 429-435. (NSA 25: 48534).

Gershenzon, S. M. 1965. Study of mutability in the wasp Mormoniella vitripennis Wlk. Genetika (2): 95-101. (JPRS-35,022; pp. 149-156) (NSA 20: 20464).

Gersper, P. L. 1970. Effect of American beech trees on the gamma radioactivity of soils. Soil Sci. Soc. Am. Proc. 34(2): 318-323. (NSA 24: 31475).

Gerster, R. 1971. Attempt to interpret the kinetics of isotope exchange between $C^{18}O_2$ and the water of a leaf: experiments in the dark. Planta 97 (2): 155-172. (In French) (NSA 25: 46479).

Gertsuskiy, D. F. and L. V. Alekseyenko. 1971. Biological effects of protons and gamma radiation on watercress seeds. Radiobiologiya 11(4): 575-579. (AEC-tr-7304, pp. 121-126) (NSA 26: 36354).

Gertz, S. M. 1973. Fate of a radionuclide in the aquatic environment: a mathematical equation representing radionuclide uptake by algae. Thesis, Drexel Univ., Philadelphia, Pa. 128pp. (Univ. Microfilms 73-18,302) (NSA 29: 838).

Gerwin, R. 1971. Nuclear Power Today and Tomorrow. (Nuclear Research and Nuclear Technique as the Chances of Our Time). Deutsche Verlags-Anstalt GmbH, Stuttgart. 138pp. (In German) (NSA 26: 52321).

Gest, H. and M. D. Kamen. 1948. Studies on the phosphorus metabolism of green algae and purple bacteria in relation to photosynthesis. J. Biol. Chem. 176(1): 299-318.

Getsova, A. B. 1960. Accumulation of various radioisotopes by water insects. Texisy doklada 4 s'yezda Vses. entomol. obshch. Leningrad 1: 37-38. (In Russian).

Getsova, A. B. 1960. The problem of the excretion of radioactive isotopes by various invertebrates. Dokl. Akad. Nauk SSSR 133(2): 459-461. (Doklady (biol.) 133(1-6): 509-511) (NSA 14: 23893).

Getsova, A. B. 1962. On the elimination of radioactive isotopes (desorption) in some aquatic invertebrates. pp. 744-746. In: Proc. 11th Intern. Congr. on Entomology. (Vienna). H. Strouhal and M. Beier, eds. Christoph Reisser's Söhne.

Getsova, A. B. and G. A. Volkova. 1961. The accumulation and elimination of ^{90}Sr by the caddis fly Halesus interpunctatus Zett. Dokl. Akad. Nauk SSSR 139(2): 483-484. (Doklady (biol.) 139: 545-546) (NSA 15: 28960).

Getsova, A. B. and G. A. Volkova. 1962. Accumulation and removal of ruthenium-106, cerium-144 and promethium-147 by the fly Halesus interpunctatus Zett. Dokl. Akad. Nauk SSSR 144(5): 1163-1165. (JPRS-15,088; Doklady (biol.) 144: 502-503) (NSA 16: 28709).

Getsova, A. B. and G. A. Volkova. 1962. Uptake of radioisotopes by some aquatic insects. Entomol. obozreniye 41(1): 109-124. (JPRS-18,423; OTS-63-21923; Entomol. Rev. 41(1); Abstr., Tech. Trans. 10(3): 272).

Getsova, A. B. and G. A. Volkova. 1964. Accumulation of radioactive isotopes of phosphorus, yttrium, iodine and mercury by larvae of aquatic insects. Zool. zh. 43: 1077-1080. Fed. Proc. 24(Transl. Suppl., Pt. II): T683-T684) (NSA 19: 8774, 42247).

Getsova, A. B. and G. A. Volkova. 1967. New data on the accumulation of S^{35} and Ru^{106} by dragonfly nymphs (Odonata). Ecological Monogr. 37(4): 38-__.

Getsova, A. B., Ye. A. Timofeyeva-Resovskaya and N. V. Timofeyev-Resovskiy. 1960. Effect of ethylenediaminetetracetic acid sodium salt on the uptake of various radioisotopes from water by leeches and midges. Dokl. Akad. Nauk SSSR 130(2): 440-442. (Doklady (biol.) 130(1-6): 53-54) (NSA 19: 8829).

Getsova, A. B., et al. 1964. Concentration of chemical elements from aqueous solutions by freshwater organisms. Communication 6. Accumulation of the radioisotopes of eight different elements in tissues of Anodonta cellensis. Nauchn. dokl. vyssh. shkoly (biol.) (4): 82-88.

Getzin, L. W. 1964. The recovery of C^{14}-diazinon and its C^{14} labeled metabolites from soil. Abstr. Bull. Entomol. Soc. Am. 10(3): 176.

Getzin, L. W. 1967. Metabolism of diazinon and zinophos in soils. J. Econ. Entomol. 60(2): 505-508. (^{14}C).

Getzin, L. W. and I. Rosefield. 1966. Persistence of diazinon and zinophos in soils. J. Econ. Entomol. 59: 512-516. (^{14}C).

Geyer-Duszynska, J. 1961. Spindle disappearance and chromosome behavior after partial-embryo irradiation in Cecidomyiidae (Diptera). Chromosoma 12(2): 233-247. (NSA 17: 14023).

Geyer-Duszynska, J. and L. Janota-Bassalik. 1957. Effect of radiations on microorganisms. Postepy Biochem. 3(3-4): 287-307.

Geyh, M. and S. Lorch. 1964. Determining the contribution of γ rays to the natural environmental radiation at ground level. pp. 979-988. In: The Natural Radiation Environment. (See Adams and Lowder, 1964) (NSA 18: 35753).

Geyh, M. and S. Lorch. 1966. Measurement of the terrestrial component of the natural gamma environmental radiation in the Federal Republic of Germany. Germany Bundesministerium für Wissenschaftliche Forschung, Bad Godesberg, report BMwF-FB-K-66-42. 65pp. (In German) (NSA 21: 15993).

Geyh, M. A. 1972. Determination of the initial ^{14}C content in groundwater. pp. D58-D69. In: Proc. 8th Intern. Radiocarbon Dating Conf. Vol. 1. (See Rafter, 1972) (NSA 31: 8700).

Geyh, M. A. 1973. The application of the ^{14}C and ^{3}H methods for environmental studies in hydrology. Z. Deutsch. Geol. Ges. 124: 515-522. (In German) (NSA 33: 14850).

Geyh, M. A. 1974. Experiences with the ^{14}C and tritium methods in applied hydrogeology. Österr. Wasserwirtsch. 26(3/4): 49-54. (In German) (NSA 33: 7330).

Geyh, M. A., J. Merkt and H. Müller. 1970. ^{14}C dating of lacustrine sediments and calibration of the ^{14}C time scale. Naturwissenschaften 57: 564-567. (In German) (NSA 25: 21171).

Ghelelovitch, S. 1960. The determination of the sensitivity to the tumorigenic effect of x rays in Drosopholia melanogaster M. Compt. rend. 250 (7): 1387-1388. (In French).

Ghelelovitch, S. 1963. Physiologic factors determining in the formation of melanotic tumors in drosophila (D. melanogaster). Influence of heat and irradiation. Bull. Assoc. Franc. Étude Cancer 50: 399-411. (In French) (NSA 19: 8829).

Ghelelovitch, S. 1964. The tumorigenic effect of x rays on drosophila (D. melanogaster) and its quantitative development in the course of ontogeny. Bull. Assoc. Franc. Étude Cancer 51: 203-224. (In French) (NSA 27982).

Ghelelovitch, S. 1966. Tumorigenic effect of tritiated leucine, uridine and thymidine in larvae of drosophila. Compt. rend. D262(10): 1173-1176. (In French).

Ghelelovitch, S. 1970. Appearance of sensitivity to the tumor-inducing action of x-rays in the zygone of Drosophila. Int. J. Radiat. Biol. 18: 331-345. (In French) (NSA 25: 11250).

Ghelelovitch, S. 1975. Repair of lesions provoked by x rays in embryos of drosophila (D. melanogaster Meig.). Int. J. Radiat. Biol. 27(6): 513-523. (In French) (NSA 33: 9589).

Gheorghiou, G. P. 1964. Studies on resistance to carbamates in mosquitoes. Abstr. Bull. Entomol. Soc. Am. 10(3): 174. (^{14}C, Culex pipiens quinquefasciatus, Anopheles albimanus).

Gheorghiou, G. P. 1965. Genetic studies on insecticide resistance. Adv. Pest Control Res. 6: 171-230. (Contains references using radionuclides, Musca domestica, Drosophila, Stomoxys calcitrans, Blattella germanica).

Ghosh, A., L. K. Maity and S. K. Banerjee. 1972. Histological changes in testes of the garden lizard, Calotes versicolor (Daudin), following x-irradiation. Indian J. Exptl. Biol. 10(6): 459-461. (NSA 2960).

Ghosh, S. M. and A. K. Hati. 1962. Effect of gamma radiation on the first instar larvae of Culex fatigans. Bull. Calcutta Sch. Trop. Med. Hyg. 10 (3): 119.

Ghosh, S. M., A. K. Hati and S. P. Basu. 1962. Effect of gamma radiation on the fertility of Aedes aegypti. Bull. Calcutta Sch. Trop. Med. Hyg. 9: 111. (NSA 16: 32772).

Ghosh, S. M., A. K. Hati and S. P. Basu. 1962. Effects of gamma radiation on Culex fatigans egg rafts. Bull. Calcutta Sch. Trop. Med. Hyg. 9: 156. (NSA 16: 32773).

Ghys, R., et al. 1967. Comparative radiosensitivity of pathogenic bacteria and viruses. pp. 243-261. In: Radiosterilization of Medical Products. International Atomic Energy Agency, Vienna, Publ. STI/PUB/157. (CONF-670644) (NSA 22: 52083).

Ghysels, J. P. 1972. Determination of plutonium in the environment by extraction and alpha counting with a liquid scintillator. Atomwirt. Atomtech. 17(2): 83-84. (NSA 26: 35836).

Ghysels, J. P. 1972. Determination of Pu in the environment by extraction and liquid scintillation counting. Ind. At. Spatiales 26(1): 33-36. (In French and English) (NSA 26: 45268).

Giacintov, P. and S. Zima. 1970. Concentration of ^{90}Sr in the bones of hares as an indicator of contamination of the biosphere. Zentralbl. Vet. Beih. (11): 142-148. (In German) (NSA 25: 24139).

Giavelli, S. and A. Cervini. 1966. The radiosensitivity of the oocytes of Artemia salina. Atti Assoc. Genet. Ital. 11: 292-300. (In Italian) (NSA 21: 10348).

Giavelli, S. and E. Parazzi. 1963. Effect of low temperature on radiosensitivity of Drosophila pupae. Int. J. Radiat. Biol. 7(5): 465-471. (NSA 18: 31163).

Gibbons, H. L., J. R. Dille and R. G. Cowley. 1965. Inhalant allergy to the screwworm fly: preliminary report. Arch. Environ. Hlth. 10: 424-430. (NSA 19: 43652) (Involves radiosterilized flies).

Gibbs, W., et al. 1964. Comparative study of thyroid and whole-body retention of iodine in box turtles. pp. 20-22. In: 2nd Ann. Oak Ridge Radioisotope Conf. (April 19-22, 1964). U. S. AEC report TID-7689. (NSA 18: 15781).

Gibbs, W. J., et al. 1967. Fallout over Australia from nuclear weapons tests by France in Polynesia during June and July 1967. Austral. J. Sci. 30(6): 217-222.

Gibbs, W. J., et al. 1969. Strontium^{90} in the Australian environment during 1968. Austral. J. Sci. 32(6): 238-243.

Giblin, F. J. and E. J. Massaro. 1974. Mechanism of methyl-mercury transport and transfer to the tissues of the rainbow trout (Salmo gairdneri). pp. 349-355. In: Trace Substances in Environmental Health. VIII. D. D. Hemphill, ed. Univ. of Missouri, Columbia. (NSA 32: 17595) (^{203}Hg).

Gibson, J. A. B. 1964. The gamma radiation background at Grove, Berkshire. U. K. Atomic Energy Authority, Harwell, report AERE-HP-GEN-39. 4pp. (NSA 18: 31751) (Over grassland).

Gibson, J. A. B. and A. K. Burt. 1971. Some techniques for the measurement of tritium. pp. 37-44. In: Protection Against Low Energy or Short Range Radiations and the Biological Effects of Radiation. France Service Central de Protection Contre les Rayonnéments Ionisants, Le Vesinet. (NSA 26: 12302).

Gibson, J. A. B., J. E. Richards and J. Docherty. 1969. Nuclear radiation in the environment: beta and gamma-ray dose rates and air ionization from 1951 to 1968. J. Atmos. Terrest. Phys. 31: 1183-1196. (NSA 24: 322).

Giesecke M., A. A. 1964. Measurements of radioactivity in Peru. Bol. Junta Control Energia At. (Peru) 9(49): 43-52. (In Spanish) (NSA 19: 15794)

Gieskes, W. W. C. and A. J. van Bennekom. 1973. Unreliability of the ^{14}C method for estimating primary productivity in eutrophic Dutch coastal waters. Limnol. Oceanogr. 18(3): 494-495. (NSA 28: 21133).

Giess, M.-C. and H. Planel. 1973. Effect of radioprotection effected at different stages in the development of Drosophila melanogaster on their life times. Compt. rend. D276(6): 1029-1032. (In French) (NSA 28: 24849).

Giess, M.-C., H. Planel and Y. Blanquet. 1970. Influence of very weak ^{60}Co irradiation on the development of Drosophila melanogaster. Compt. rend. D271: 1410-1413. (In French) (NSA 25: 16290).

Giess, M.-C., et al. 1973. Effects of primary and secondary radiation on the development and lifetime of Drosophila melanogaster. Compt. rend. soc. biol. 167(3): 537-541. (In French) (NSA 30: 622).

Gifford, C. A. and R. F. Johnson. 1962. Distribution of calcium in the land crab Cardisoma guanhumi during shell wound recalcification. Comp. Biochem. Physiol. 7(3): 227-231. (RDA-5: 239) (^{45}Ca).

Gifford, C. E. and D. R. Griffin. 1960. Notes on homing and migratory behavior of bats. Ecology 41(2): 378-381. (^{131}I).

Gifford, D. R. 1967. An attempt to use ^{14}C as a tracer in a Scots pine (Pinus silvestris L.) litter decomposition study. pp. 687-693. In: Secondary Productivity in Terrestrial Ecosystems. Vol. II. K. Petrusewicz, ed. Inst. of Ecology, Polish Academy of Sciences, Warsaw.

Gifford, F. A. 1974. Power reactor siting. A summary of United States practice. pp. 46-71. In: Human and Ecologic Effects of Nuclear Power Plants. L. A. Sagan, ed. Charles C. Thomas, Springfield, Ill. (NSA 30: 25991).

Gil, F. P. 1968. Determination of environmental radioactivity. Rev. Ecuator. Hig. Med. Trop. 25: 31-60. (In Spanish) (NSA 25: 11068).

Gilbert, L. I., H. Chino and K. A. Domroese. 1965. Lipolytic activity of insect tissues and its significance in lipid transport. J. Insect Physiol. 11(8): 1057-1070. (Hyalophora cecropia, Periplaneta americana, ^{14}C).

Gilbert, R. O. and L. L. Eberhardt. 1973. Statistical analysis of Pu in soil at NTS: some results (an outline). Battelle Northwest Lab., U. S. AEC report BNWL-SA-4815. 22pp. (NSA 29: 18380).

Gilbert, R. O. and L. L. Eberhardt. 1974. Statistical analysis of Pu in soil at the Nevada Test Site: some results. pp. 51-89. In: Dynamics of Plutonium in Desert Environments. (See Dunaway and White, 1974) (NSA 30: 29444).

Gilbert, R. O. and L. L. Eberhardt. 1975. Evaluation of double sampling for estimating plutonium inventory in surface soil. Battelle Northwest Lab., U. S. ERDA report BNWL-SA-5385. 21pp. (NSA 32: 19713).

Gilbert, R. O., et al. 1975. Statistical design and analysis of environmental studies for plutonium and other transuranics at NAEG "safety shot" sites. Battelle Northwest Lab., U. S. ERDA report BNWL-SA-5505. 13pp. (NSA 33: 2980).

Gilbert, R. O., et al. 1975. Statistical analysis of $^{239-240}$Pu and ^{241}Am contamination of soil and vegetation on NAEG study areas. pp. 339-448. In: Radioecology of Plutonium and Other Transuranics in Desert Environments. (See White and Dunaway, 1975) (NSA 33: 622).

Gilbrat, R. 1973. Nuclear power plants and the environment. Bull. Inform. A. T. E. N. (102): 45-53. (In French) (NSA 30: 2093).

Gilby, A. R. 1965. Lipids and their metabolism in insects. Ann. Rev. Entomol. 10: 141-160. (Mentions radionuclides).

Gile, L. H. and J. W. Hawley. 1968. Age and comparative development of desert soils at the Gardner-Spring radiocarbon site, New Mexico. Soil Sci. Soc. Am. Proc. 32(5): 709-716.

Giles, M. S. 1971. Preliminary ecological studies at Jervis Bay. Australian Atomic Energy Commission Research Establishment, Lucas Heights. 22pp. (CONF-710511) (NSA 25: 42314) (Site of nuclear power plant).

Giles, M. S. 1972. Movement of phosphorus in the Little River Estuary, New South Wales. Australian Atomic Energy Commission Research Establishment, Lucas Heights, report AAEC/TM-616. 21pp. (NSA 27: 22814) (^{32}P, ^{3}H).

Giles, M. S. 1974. Toxicity studies on aquatic organisms and grass-sedge communities in the Magela Creek area. Australian Atomic Energy Commission report AAEC/E-305 (Report 2). pp. 1-15. (NSA 30: 26757) (U mining and processing).

Giles, R. H., Jr. 1964. New tracer techniques for evaluating the effects of an insecticide on the ecology of a forest fauna. Dissertation, Ohio State Univ., Columbus. xvi, 435pp.

Giles, R. H., Jr. 1964. The ecology of a small forested watershed treated with the insecticide malathion S-35. (Dissertation Abstr. 25: 731).

Giles, R. H., Jr. 1970. The ecology of a small forested watershed treated with the insecticide malathion-S^{35}. Wildlife Monogr. No. 24. 81pp.

Giles, R. H., Jr. and T. J. Peterle. 1961/62. New tracer techniques for evaluating the effects of an insecticide (malathion) on the ecology of a forest fauna. Fishery Bull. (U. S. Fish Wildl. Serv.) 167: 100. (^{35}S).

Giles, R. H., Jr. and T. J. Peterle. 1963. Distribution of aerially applied malathion S^{35} in a forest ecosystem. pp. 55-84. In: Radiation and Radioisotopes Applied to Insects of Agricultural Importance. (See IAEA, 1963) (SM-40/29) (Also: Ohio State Univ., U. S. AEC report TID-18606) (NSA 17: 24846).

Giles, R. H., Jr. and T. J. Peterle. 1963-64. New tracer techniques for evaluating the effects of an insecticide (malathion) on the ecology of a forest fauna. p. 100, pp. 119-120. In: Pesticide-Wildlife Studies. U. S. Fish and Wildlife Service, Washington, Circular 167, 199. (^{35}S).

Gilet, R. 1965. Remarks on the extrapolation numbers from survival curves of various populations of Chlorella pyrenoidosa Chick after x irradiation Compt. rend. 261: 5686-5688. (In French) (NSA 20: 16246).

Gilet, R. 1971. Survival curves and cell restoration of gamma-irradiated Chlorella. France Commissariat à l'Énergie Atomique, Grenoble, report CEA-R-4139. 112pp. (In French) (NSA 25: 47031).

Gilet, R. 1972. Potentially lethal damage after gamma irradiation in Chlorella cells (green unicellular alga). Stud. Biophys. 30(1): 37-44. (NSA 27: 5457).

Gilet, R. and P. Ozenda. 1960. Action of x-rays on the growth of cultures of a one-celled chlorophyll organism: Scenesdesmus crassus Chod. (chlorophyll algae). Compt. rend. 250: 1552-1554. (AEC-tr-4601) (NSA 14: 12467; 15: 19160).

Gilet, R. and P. Ozenda. 1961. Analysis of the effect of x rays on the multiplication of a unicellular chlorophyll organism: Scenesdesmus crassus Chod. Compt. rend. 252(24): 3867-3869. (In French) (NSA 15: 25869).

Gilet, R. and J. Terrier. 1967. Application of the Bierman theory to survival curves obtained on Chlorella pyrenoidosa after action of x rays. Compt. rend. D264: 1750-1753. (In French) (NSA 21: 30934).

Gilet, R. and J. Terrier. 1969. Effect of restoration processes make it possible to distinguish acute and subacute irradiations in Chlorella. Compt. rend. D269: 383-385. (In French) (NSA 24: 6142).

Gilet, R. and J. Terrier. 1970. Interpretation of the gamma ray survival curves of Chlorella. Bull. Inform. Sci. Tech. (Paris) (145): 67:72. (In French) (NSA 24: 34766).

Gilet, R., S. Santier and A. Villeroux. 1963. Comparative effect of x rays of medium energy and of soft x rays on the survival rate of Chlorella pyenoidosa Chick. Compt. rend. 257(13): 1985-1988. (In French) (NSA 18: 1393).

Giletti, B. J. 1959. Tritium tracer in arctic problems. Science 129(3353): 901-903.

Gileva, Ye. A. 1960. Accumulation coefficients of seven radioisotopes by three species of plants. Dokl. Akad. Nauk SSSR 132(4): 948-949. (Doklady (biol.) 132-(1-6): 343-344) (NSA 14: 21271) (Cladophora, Spirogyra, Scenedesmus).

Gileva, Ye. A. 1961. Accumulation of twelve radioisotopes from aqueous solution by the alga Scenedesmus quadricauda. In: Summary of Reports of the All-Union Conf. on Cultivation of Single-Cell Algae. Leningrad. (In (Russian).

Gileva, Ye. A. 1963. The accumulation of seven radioisotopes by four different freshwater algae. Dokl. Akad. Nauk SSSR 149(5): 1157-1158. (Doklady (earth sci.) 149(1-6): 178-179) (NSA 17: 28661).

Gileva, Ye. A. 1964. Dynamics of accumulation of chemical elements by the alga Cladophora fracta and the dependence of accumulation on the concentration of elements in the solution. Fiziol. rasteniy 11: 581-586. (In Russian) (NSA 19: 25858).

Gileva, Ye. A. 1965. Effect of the concentration and mode of occurrence of chemical elements in solution on their concentration by Cladophora. pp. 90-91. In: Problems in Hydrobiology. Publ. House Acad. Sci. USSR, Moscow. (In Russian).

Gileva, Ye. A., N. A. Timofeyeva and N. V. Timofeyev-Resovskiy. 1964. The effect of continuous irradiation in a gamma field on the biomass of freshwater algal periphyton. Dokl. Akad. Nauk SSSR 156(2): 455-456. (Doklady (biol.) 156(1-6): 393-394) (NSA 18: 38831).

Gileva, Ye. A., N. A. Timofeyeva and N. V. Timofeyev-Resovskiy. 1965. Influence of ^{60}Co gamma irradiation on the growth of Chlorella cultures. Radiobiologiya 5(5): 732-734. (AEC-tr-6602, pp. 150-154) (NSA 20: 5389).

Gill, E. D. 1965. Radiocarbon dating of past sea levels in south east Australia. Abstr. p. 167. In: 7th Intern. Congr. Abstrs. Intern. Assoc. Quartenary Research, Boulder and Denver, Colo.

Gilles, A. 1959. Comparative study of the effects of different radiations on three species of plants. Bull. Classe Sci. Acad. Roy. Belg. 45(5): 909-912. (In French) (NSA 14: 12463).

Gillet, C. 1962. Immediate effect of x radiation on the cytoplasmic streaming Nitella flexilis. Bull. Classe Sci. Acad. Roy. Belg. 48(5): 1161-1168. (In French) (NSA 17: 24919).

Gillet, C. 1963. Action of x rays on Chara vulgaris L. III. Role of euploidy on the response of Chara to irradiation. IV. The elongation, after irradiation, of the internodal cellules of 2 chromosomic races. Radiation Botany 2(3/4): 195-204; 3(2): 155-161. (In French).

Gillet, C. 1963. Modification of the velocity of cytoplasmic streaming, induced at a distance, by irradiation in Nitella flexilis. Experientia 19 (4): 198-199. (In French) (NSA 17: 24921).

Gillet, C. 1963. Action of x radiation on Chara vulgaris L. V. The modification of the leucoplast after irradiation. Protoplasma 56(4): 637-644. (In French) (NSA 18: 23387).

Gillet, C. 1963. The action of x rays on Chara vulgaris L. I. Abnormal morphogenesis induced by irradiation. II. Details of effects of irradiation on the nuclear development and cytology of the vegetative plant. Rev. Cytol. Biol. Vegetales 21: 59-76, 77-96. (In French) (NSA 19: 36067, 36068).

Gillet, C. 1964. Permeability of cellular membranes and modification of cytoplasmic current of Nitella flexilis after irradiation. Int. J. Radiat. Biol. 8(6): 533-539. (In French) (NSA 19: 25958).

Gillet, C. 1964. Relations between the dose of x-rays applied and the rate of cytoplasmic streaming in Nitella flexilis. Protoplasma 58: 286-293. (In French) (NSA 19: 8900).

Gillet, C. 1965. Restoration of cytoplasmic streaming as soon as possible after irradiation. Protoplasma 60: 24-30. (In French) (NSA 20: 3698).

Gillet, C. 1969. Re-establishment of cytoplasm and motive force during recovery in cyclosis in Nitella cells after irradiation. Protoplasma 67: 269-278. (In French) (NSA 23: 45966).

Gillet, C. and L. Klerckx. 1965. Calcium flux from x-irradiated Nitella cells. Life Sci. 4: 1561-1565. (NSA 19: 43818).

Gillette, R. 1971. AEC's new environmental rules for nuclear plants may open new debate, extend delays, raise plant costs. Science 173(4002): 1112-1113.

Gillette, R. K., et al. 1972. Investigation into the determination of plutonium in soil by a fusion procedure. Mound Lab., U. S. AEC report MLM-1901. 26pp. (NSA 26: 42722).

Gillies, M. T. 1958. A simple autoradiographic method for distinguishing insects labelled with phosphorus-32 and sulphur-35. Nature 182(4650): 1683-1684.

Gillies, M. T. 1961. Studies on the dispersion and survival of Anopheles gambiae Giles in East Africa, by means of marking and release experiments. Bull. Entomol. Res. 52(1): 99-127.

Gillies, M. T. 1961. Marking and release experiments with a tropical mosquito by the use of radioisotopes. pp. 267-279. In: Radioisotopes in Tropical Medicine. (Proc. Symp., Bangkok, 12-16 December 1960). International Atomic Energy Agency, Vienna.

Gillott, C., G. S. Dogra and A. B. Ewen. 1970. A autoradiographic study of endocrine activity following frontal ganglionectomy in virgin females of Melanoplus sanguinipes (Orthoptera: Acrididae). Can. Entomol. 102(9): 1083-1088.

Gilmore, G. R. and G. W. A. Newton. 1975. Radioanalytical chemistry. Radiochemistry (London) 2: 175-238. (NSA 33: 16623).

Gilmour, R. S. and G. H. Dixon. 1972. Biosynthesis of protamine in trout testis. V. In vitro synthesis of protamine in a trout liver cell-free system. J. Biol. Chem. 247(14): 4621-4627. (SFA 17: 15449) (^{14}C).

Gilse, A. C. 1947. Radiations and cell divisions. Quart. Rev. Biol. 22: 253-262.

Gil y Gil, C. 1962. Mean lethal dose for silk worm eggs with radiation of different wave lengths. Abstr. p. 211. In: 10th Intern. Congr. of Radiology. (Montreal, 26 August - 1 September 1962; Book of Abstr. 336(b)).

Gilyarov, M. S. and D. A. Krivolutskiy. 1971. Radioecological studies in soil zoology. Zool. zh. 50(3): 329-342. (In Russian) (NSA 26: 4724).

Gintner, Ye. K., V. I. Ivanov and V. A. Mglinets. 1974. Morphogenetic mosaicism for homoeotic mutation aristapedia in Drosophila melanogaster. Genetika 10(3): 67-75. (In Russian) (NSA 30: 7117).

Gintner, Ye. K., V. I. Ivanov and V. A. Mglinets. 1974. Morphoses of antenna and somatic mosaicism in Drosophila melanogaster (R(1;3)05/y sn; D/ssa. Genetika 10(9): 64-70. (In Russian) (Irradiated).

Gintovt, F. V. and O. G. Mironov. 1963. On the radioactivity of some organisms in the River Neman. Zool. zh. 42(10): 1574-1576. (JPRS-22,714; OTS-64-21275, pp. 64-67).

Gipsh, A. A. and G. G. Kapaturin. 1970. Some results of an investigation of the relationship of natural radioactivity and ash content of coals of the Vorkutsk field. Izv. vyssh. ucheb. zaved. geol. raved. (10): 145-149. (In Russian) (NSA 26: 28076).

Girardi, F. 1973. Use of tracer techniques for the protection of working ambience and in the control of industrial waste. Joint Nuclear Reasearch Center, Commission of the European Communities, Ispra, Italy, report EUR/C-IS-192-73. 12pp. (In Italian) (NSA 30: 6550).

Girardi, F., et al. 1970. Radiotracer technique for the study in vivo of the biological pathway of heavy metals in aquatic organisms. pp. 721-733. In: Nuclear Techniques in Environmental Pollution. (See IAEA, 1970) (NSA 25: 32350).

Girardi, F., et al. 1975. Preparation of ^{203}Pb compounds for studies of pathway and effects of lead pollution. Int. J. Appl. Radiat. Isotopes 26(5): 267-277. (NSA 32: 16909).

Girardie, A. 1964. Action of the pars intercerebralis on the development of *Locusta migratoria* L. J. Insect Physiol. 10(4): 599-609. (In French, English summary) (X-rays).

Giresse, P. and G. Courtois. 1966. An application of the radioactive tracer method: a study of displacements of sediment in the course and at the mouth of the Seine. Rev. geogr. phys. geol. dyn. 8: 193-208. (In French) (NSA 22: 48942).

Girnik, D. V. 1957. The use of isotopes in the investigation of flowering and nectar formation in linden trees. Tr. Fiziol. Rast. Agrokhim. Pochv. 1957: 143-147. (OTS-63-11008) (NSA 18: 25058).

Gist, C. S. 1967. Problems of sampling desert arthropods before and after a thermonuclear cratering test. Great Basin Nat. 27(1): 26-35. (NSA 23: 2805).

Gist, C. S. 1969. Iodine-131 retention in mule deer. Thesis, Colorado State Univ., Fort Collins. 75pp.

Gist, C. S. and F. W. Whicker. 1971. Radioiodine uptake and retention by the mule deer thyroid. J. Wldl. Mgmt. 35(3): 461-468.

Gist, C. S., K. M. Duke and B. C. Patten. 1973. A systems analysis of a model for radioiodine movement in mule deer. pp. 909-914. In: *Radionuclides in Ecosystems*. (See Nelson, 1973) (NSA 28: 30332).

Giudici, G., et al. 1964. On the use of mountain stations observation data for a study of the movements and spreading of "radioactive clouds". Italy Comitato di Studi Nucleare, Comitato Nazionale Energia Nucleare, Casacci, report RT/PROT(64)-1. 27pp. (NSA 18: 37371).

Gjullin, C. M., A. W. Lindquist and J. S. Butts. 1952. Absorption of radioactive DDT by resistant and non-resistant mosquitoes. Mosquito News 12 (3): 201-205.

Gladden, J. B. and D. J. Shure. 1974. Cesium-137 exchanges between a reactor effluent stream and flood plain forest system in South Carolina. Abstr. Bull. Ecol. Soc. Am. 55(2): 42. (At Savannah River Plant).

Glagoleva, T. A. 1962. Photosynthesis in plants of the upper alpine belt in the Eastern Pamirs. Botan. zh. 47: 1567-1581. (In Russian) (^{14}C).

Glagoleva, T. A., Ye. L. Muuk and N. A. Ivanova. 1969. Effect of photophosphorylation inhibitors on ^{14}C metabolism in *Chlorella*. Botan. zh. (12): 1965-1973. (In Russian) (NSA 24: 46574).

Glascock, R. F. 1962. Some examples of the use of radioisotopes in biochemistry. pp. 49-76. In: *Use of Radioisotopes in Animal Biology and the Medical Sciences*. (Proc. Conf. Mexico City, 21 November - 1 December 1962). Academic Press, London. (Insects).

Glaser, R. 1960. On the question of the concentration of radioiodine (I^{131}) in the mussel *Dreissensia polymorpha* Pall. Naturwissenschaften 47(12): 284-287. (In German).

Glaser, R. 1961. On the uptake of radioiodine (^{131}I) from water droppings through the mussel *Dreissensia polymorpha* Pall. Zool. Jb. Physiol. 69(3): 339-378. (In German).

Glaser, R. 1961. On the radioactivity of plankton of the Stechlin and Nehmitz seas near Rheinsberg (Mark Brandenburg). Kernenergie 4(5/61): 398-399. (In German) (NSA 15: 23781).

Glaser, R. 1962. The position of water organisms in radiation problems of nuclear plants. Kernenergie 5(7): 515-533. (AEC-tr-5754) (NSA 17: 1669).

Glaser, R. 1962. On the radioactivity of submerged water plants. Limnologica 1(1): 51-63. (AEC-tr-5753) (NSA 17: 19752).

Glaser, R. 1962. A contribution and question of the dependence on radioactive isotope concentration in water organisms from droppings. Biol. Zentralblatt 81(5): 539-547. (In German) (NSA 18: 3458).

Glaser, R. 1963. One possible use of bioindicators for radiological control of waters. Limnologica 1(4): 281-286. (In German).

Glaser, R. 1963. The increase of radioactivity of water plants of the Stechlin Sea in 1961/1962. Limnologica 1(5): 473-474. (In German) (NSA 18: 5155).

Glaser, R. 1964. Annual investigation of the radioactivity of Stechlin Sea plankton (1960-1962). Limnologica 2(3): 311-320. (In German).

Glaser, R. 1966. Exchange and metabolism of trace elements in freshwater mussels with special reference to accumulation mechanisms of radionuclides. Zool. Jahr. Abt. Allgem. Zool. Physiol. Tiere 72: 1-114. (In German) (NSA 21: 16298).

Glaser, R. and E. Spode. 1960. On the uptake of radioyttrium Y^{91} in some algae. pp. 54-59. In: *Problems and Results in Biophysics and Radiationbiology. II.* (Proc. 4th Biophysics Workshop of the Physical Society, German Democratic Republic). G. Pfennigsdorf and W. Eckart, eds. Akademie-Verlag, Berlin. (In German).

Glass, E. 1964. Nonrandom distribution of x-ray-induced breaks in the chromosomes of *Bellevalia romana*. Port. Acta Biol. A8: 153-176. (In German) (NSA 21: 22788).

Glasstone, S. 1967. *Sourcebook on Atomic Energy*. Van Nostrand, Princeton. 3rd ed. 883pp. (NSA 21: 47055).

Glasstone, S. 1971. Public safety and underground nuclear detonations. U. S. AEC report TID-25708. ix, 276pp.

Glavanakova, S. I. and O. A. Dryanovska. 1970. Induced radiation sterility in the male imago of the grain moth: *Sitotraga cerealella* (Olivier). Compt. rend. Bulg. sci. 23(11): -413-1416. (NSA 25: 46955).

Glavanakova, S. I., et al. 1970. Changes in the ratio of the sexes in F_1 after irradiation with gamma rays of female and male imagos of Sitotraga cerealella Olivier. Compt. rend. Bulg. sci. 23: 1529-1532. (NSA 25: 42610).

Glazunov, V. V., V. P. Parchevskiy and D. G. Fleyshman. 1963. The variation of the fission product content of Cystoseira in the Black Sea. Dokl. Akad. Nauk SSSR 152(5): 1222-1224. (Doklady (earth sci.) 152(1-6): 196-198; JPRS-22,714) (NSA 18: 4979).

Glegg, R. E. and Z. I. Kertesz. 1957. Effect of γ-radiation on cellulose. J. Polymer Sci. 26(114): 289-297. (Incl. wood).

Glembotskiy, Ya. L. 1970. Genetic studies in space. Kosmich. issled. 8: 616-627. (In Russian) (NSA 25: 16292) (Insects).

Glenn, J. L. 1971. Relations among radionuclide content and physical, chemical, and mineral characteristics of Columbia River sediments. U. S. AEC report TID-25786. 125pp. (NSA 25: 57561).

Glenn, N. D. and H. S. Ducoff. 1976. Acute lethality after fast-neutron and x irradiation of Tribolium confusum. Radiation Research 65(1): 120-129. (NSA 33: 26296).

Glocker, R., H. Glocker and M. Langendorff. 1933. On a high Grenz radiation dose for the biological study of roentgen radiation. Naturwissenschaften 21(17): 316. (In German) (Axolotl eggs).

Glocker, R., H. Glocker and A. Reuss. 1933. On the study of roentgen radiation of different wave lengths on biological objects. III. Strahlentherapie 46(2): 517-528. (In German).

Glogowski, I. 1967. The influence of different dosages of roentgen radiation on the mortality and propagation of Rhagoletis cerasi L. Pr. Nauk Inst. Ochr. Rosl. 9: 263-270. (In Polish) (NSA 24: 7981).

Glora, M. A. and B. L. Brown. 1964. The Nevada Test Site environmental surveillance program. Reynolds Elec. & Engineering Co., Las Vegas, Nev., U. S. AEC report NVO-162-15. 7pp. (NSA 19: 7728).

Glover, S. J. 1967. The effect of irradiation upon O_2 uptake in some insects. Abstr. 113. Bull. Entomol. Soc. Am. 13(3): 192.

Glover, S. J. 1968. Effects of gamma radiation on temperature, humidity, and respiration responses in the housefly Musca domestica L. Thesis, Univ. of Georgia, Athens. 56pp. (NSA 23: 31893).

Glojowski, K. 1972. Insect sterilization by irradiation. Wiad. Ekol. 18: 67-77. (In Polish) (NSA 29: 27117).

Gloyna, E. F. 1970. Transport of radionuclides in a model aquatic system. pp. 3-29. In: Proc. Intern. Symp. on Radioecology. Vol. I. (See Commissariat à l'Énergie Atomique, 1970).

Gloyna, E. F. and B. B. Ewing. 1957. Uranium recovery from saline solutions by biological slimes. Nucleonics 15(1): 78-81.

Gloyna, E. F. and W. A. Felsing. 1963. Environmental radionuclides in a water supply. Water Sewage Works 110(7): 245-253.

Gloyna, E. F. and J. O. Ledbetter. 1969. Principles of Radiological Health. Marcel Dekker, New York. 489pp. (NSA 23: 33691).

Gloyna, E. F., E. R. Herman and W. R. Drynan. 1955. Oxidation ponds - waste treatment studies, radioisotope uptake, and algae concentration. Sanitary Engineering Labs., Univ. of Texas, Austin, Tech. Report No. 2, U. S. AEC report AECU-3113. 98pp.

Gloyna, E. F., W. R. Drynan and D. F. Smallhorst. 1961. Radioactivity in the surface waters of Texas. Southwest Water Works J. 43(5): 31-35.

Gloyna, E. F., et al. 1966. Transport of radionuclides in a model river. pp. 11-32. In: Disposal of Radioactive Wastes into Seas, Oceans and Surface Waters. (See IAEA, 1966).

Glubrecht, H. 1959. Radiation biology and isotope utilization in applied botany. Atompraxis 5(7/8): 276-280. (In German) (NSA 14: 2274).

Go, M., Y. Maeda and J. Yamashita. 1964. Radiation effect on the body weight increase of the toad with closed cloaca. Sanfujinka no Shimpo 16: 203-204. (In Japanese) (NSA 21: 28654).

Gochenauer, S. E. and G. M. Goodwell. 1970. Effects of chronic gamma irradiation on the soil microfungi of an oak-pine forest. Abstr. Am. J. Bot. 57: 746-747.

Gockel, A. 1914. The Radioactivity of Soils and Springs. Verlag Friedrich Vieweg u. Sohn, Braunschweig.

Godbold, B. C. and J. K. Jones. (Eds.). 1965. Radiological Monitoring of the Environment. (Proc. Symp., Berkeley, Glouchestershire, October 1963). Pergamon Press, Oxford. 440pp. (CONF-365) (NSA 20: 14810).

Goddyn, H. 1961. The use of radioactive carbon to measure primary productivity. South African Assoc. Marine Biol. Res. Bull. No. 2. pp. 18-23.

Godfrey, G. K. 1953. A technique for finding Microtus nests. J. Mammal. 34(4): 503-505.

Godfrey, G. K. 1954. Tracing field voles (Microtus agrestis) with a Geiger Müller counter. Ecology 35(1): 5-10.

Godfrey, G. K. 1954. Use of radioactive isotopes in small mammal ecology. Nature 174: 951-952.

Godfrey, G. K. 1955. A field study of the activity of the mole (Talpa europaea). Ecology 36(4): 678-685.

Godfroi, E. E. 1958. Hibernation and radioresistance. Proc. 2nd Intern. Conf. on Peaceful Uses of Atomic Energy 23: 76-78.

Godhinho, M. M., et al. 1973. Environmental radioactivity measurements at the Laboratorio de Fisica e Engenharia Nucleares (1966-1970). Portugal Laboratorio de Fisica e Engenharia Nucleares, Junta de Energia Nuclear, Sacavem, report LFEN-158(B). 44pp. (NSA 30: 21073).

Godward, M. B. E. 1954. Irradiation of Spirogyra chromosomes. Heredity 8 (2): 293.

Godward, M. B. E. 1960. Resistance of algae to radiation. Nature 185(4714): 706. (NSA 14: 9349).

Godward, M. B. E. 1962. Invisible radiations. pp. 551-566. In: Physiology and Biochemistry of Algae. R. A. Lewin, ed. Academic Press, New York. (NSA 20: 1681).

Godwin, H. 1960. The Croonian Lecture. Radiocarbon dating and Quaternary history in Britain. Proc. Roy. Soc. (London) B153: 287-320.

Godwin, P. A., H. D. Rule and W. E. Waters. 1964. Some effects of gamma irradiation on the gypsy moth, Portetria dispar. J. Econ. Entomol. 57(6): 986-990.

Goering, J. J. and J. C. Neess. 1964. Nitogen fixation in two Wisconsin lakes. Limnol. Oceanogr. 9(4): 530-539. (^{15}N).

Goering, J. J., R. C. Dugdale and D. W. Menzel. 1964. Cyclic diurnal variations in the uptake of ammonia and nitrate by photosynthetic organisms in the Sargasso Sea. Limnol. Oceanogr. 9(3): 448-451. (^{15}N, ^{14}C).

Gofman, J. W. 1972. Existence of nuclear weapons: a prime environmental threat. Environ. Aff. 1(4): 782-791. (NSA 26: 45492).

Gofman, J. W. 1972. The case against nuclear power. Catalyst 2(3): 18-__.

Goforth, C. D. 1968. Differential effects of radiation on early growth stages of adult Hymenolepis microstoma. Thesis, Univ. of Tennessee, Knoxville. 65pp. (NSA 23: 41426).

Gogate, S. S., T. M. Krishnamoorthy and S. D. Soman. 1975. Mean residence times for tritium in some terrestrial plants. Indian J. Expt. Biol. 13 (3): 264-267. (NSA 33: 7633).

Gogia, U. K. 1970. Gamma irradiation effects on spermatogenesis and reproductive potential of navel orangeworm, Paramyelois transitella (Walker). Thesis, Univ. of California, Berkeley. 99pp. (Univ. Microfilms 71-15,769) (NSA 25: 46959).

Goh, K. M. and B. P. J. Molloy. 1972. Reliability of radiocarbon dates from buried charcoals. pp. G29-G45. In: Proc. 8th Intern. Radiocarbon Dating Conf. Vol. 2. (See Rafter, 1972) (NSA 31: 8712).

Goh, K. M. and J. D. Stout. 1972. Radiocarbon enrichment and the turnover of soil organic matter in a chrono-sequence of soils developed on wind-blown sand in New Zealand. pp. E67-E79. In: Proc. 8th Intern. Radiocarbon Dating Conf. (See Rafter, 1972) (NSA 31: 8650).

Golchert, N. W. and J. Sedlet. 1969. Radiochemical determination of ^{99}Tc in environmental water samples. Anal. Chem. 41: 669-671. (NSA 23: 19671).

Golchert, N. W. and J. Sedlet. 1972. Radiochemical determination of plutonium in environmental water samples. Radiochem. Radioanal. Lett. 12 (4-5): 215-221. (NSA 27: 11721).

Gold, D. D., et al. 1959. Interim report on the Savannah River studies. R. A. Taft Sanitary Engineering Center, U. S. PHS, Cincinnati. j, iv, 257pp.

Gold, R. 1971. Gamma-continuum at the air-land interface. Health Physics 21(1): 79-83. (NSA 25: 48613).

Gold, R. and A. M. Strash. 1971. Preoperational measurement of environmental gamma rays at the Monticello site with a continuum spectrometer. Trans. Amer. Nucl. Soc. 14(2): 41-42. (NSA 25: 51018).

Gold, S., et al. 1964. Measurement of naturally occurring radionuclides in air. pp. 369-382. In: The Natural Radiation Environment. (See Adams and Lowder, 1964) (NSA 18: 36074).

Goldberg, E., et al. 1974. A neutron collimator for animal irradiations. Med. Phys. 1(3): 152-154. (NSA 30: 26706).

Goldberg, E. D. 1963. Geochronology with lead-210. pp. 121-131. In: Radioactive Dating. (See IAEA, 1963) (NSA 17: 30736).

Goldberg, E. D., T. J. Wecker and A. Whishenland. 1951. Phosphate utilization by diatoms. Biol. Bull. 101(3): 274-284.

Goldberg, L. J. and W. R. Leif. 1950. The use of radioactive isotope in determining the retention and initial distribution of airborne bacteria in the mouse. Science 112: 299-300.

Goldberg, M. D. (Ed.). 1972. Energy, Environment and Planning: the Long Island Sound Region. (Proc. Conf., Brookhaven National Lab., October 5-7, 1971). Brookhaven National Lab., U. S. AEC report BNL-50355. 177pp. (NSA 28: 12472).

Golden, G. S. 1960. Gamma-ray spectrometry in environmental monitoring. pp. 4-9. In: 5th Ann. Meeting of the Bio-Assay and Analytical Chemistry Group. (Gatlinburg, Tenn., October 1-2, 1959). U. S. AEC report TID-7591. (NSA 15: 6241).

Golden, J. C., Jr. 1968. Natural background radiation levels in Florida. Sandia Corp., Albuquerque, U. S. AEC report SC-RR-68-196. 32pp. (NSA 22: 40209).

Golden, J. C., Jr. 1971. Pre-operational environmental monitoring programs at nuclear facilities. Thesis, Univ. of Michigan, Ann Arbor. 111pp. (Univ. Microfilms 71-15,574) (NSA 26: 4416).

Golden, J. C., Jr. and G. H. Whipple. 1971. The use of stable element concentrations in pre-operational environmental monitoring. pp. 329-336. In: Health Physics Aspects of Nuclear Facility Siting. (See Voillequé and Baldwin, 1971) (NSA 25: 55240).

Goldhaber, G. and B. Feldman-Muhsam. 1944. Immediate effect of x-rays on the movements of larvae and pupae of mosquitoes. Nature 153: 528.

Goldin, A. S., J. S. Nader and L. R. Setter. 1953. The detectability of low-level radioactivity in water. J. Am. Water Works Assoc. 45(1): 73-80.

Goldin, A. S., R. J. Velten and G. W. Frishkorn. 1959. Determination of radioactive strontium. Anal. Chem. 31(9): 1490-1492.

Goldin, P. B., F. I. Krasnik and L. A. Bishnyakov. 1963. Experimental typhus infection and immunity in irradiated animals. I. Course of typhus infection in cotton rats subjected to x-ray irradiation. Tr. inst. epidemiol. mikrobiol. gig. Pastera 25: 32-41. (In Russian) (NSA 21: 18170).

Goldman, C. R. 1960. Molybdenum as a factor limiting factor limiting primary productivity in Castle Lake, California. Science 132(3433): 1016-1017.

Goldman, C. R. 1961. Primary productivity and limiting factors in Brocks Lake, Alaska. Verh. Intern. Verein. Limnol. 14: 120-124.

Goldman, C. R. 1963. The measurement of primary productivity and limiting factors in freshwater with carbon-14. pp. 103-113. In: Proc. Conf. on Primary Productivity Measurement, Marine and Freshwater. (See Doty, 1963).

Goldman, C. R. 1968. The use of absolute activity for eliminating serious errors in the measurement of primary productivity with C^{14}. J. cons. int. explor. mer 32(2): 172-179.

Goldman, C. R. and R. C. Carter. 1965. An investigation by rapid carbon-14 bioassay of factors affecting the cultural eutrophication of Lake Tahoe, California-Nevada. Water Poll. Control Fed. J. 37: 1044-1059.

Goldman, M., et al. 1965. The comparative metabolism of strontium, calcium, and cesium in deer and sheep. In: Radiation and Terrestrial Ecosystems. Health Physics 11(12): 1415-1422. (See Hungate, 1965).

Goldman, M. I. 1970. Radiological aspects of air and water: discharges from nuclear power installations. pp. 144-158. In: Northwest Conf. on the Role of Nuclear Energy. (See Oregon Governor's Conf., 1970) (NSA 25: 23223).

Goldman, M. I. 1972. Nuclear power and a protected environment. pp. 221-233. In: Proc. Southern Conf. on Environmental Protection from Nuclear Power Plants. (See U. S. EPA, 1972) (NSA 27: 4699).

Goldman, M. I. and A. W. DeAgazio. 1971. New developments in nuclear power plant waste treatment. Trans. Amer. Nucl. Soc. 14(1): 327. (NSA 25: 39589).

Goldsby, R. A. 1961. Tracer studies in hydrogen-adapted Scenedesmus using ^{3}H and ^{14}C. Univ. of California, Berkeley, U. S. AEC report UCRL-9806. 75pp. (NSA 15: 28913).

Goldschmide, E. E., M. Lik-Shing Tsang and J. A. Schiff. 1975. Studies of sulfate utilization by algae. XIII. Adenosine 5'phosphosulfate (APS) as an intermediate in the conversion of adenosine 3'-phosphate 5'-phosphosulfate (PAPS) to acid-volatile radioactivity. Plant Sci. Lett. 4(5): 293-299. (NSA 33: 15139) (^{35}S).

Goldsmith, P. 1967. Diffusiophoretic and thermophoretic effects on particulate matter near condensing and evaporating water surfaces. p. 55. In: Isotope Techniques in the Hydrolic Cycle. (See Stout, 1967).

Goldsmith, W. A. and E. J. Middlebrooks. 1966. Radioactive equilibrium time. Proc. Sanitary Engrg. Div. Am. Soc. Civil Engrs. 92(SA6): 85-__.

Goldsmith, W. A., F. F. Haywood and D. G. Jacobs. 1975. Guidelines for cleanup of uranium tailings from inactive mills. Oak Ridge National Lab., U. S. ERDA report CONF-760202-3. 8pp. (NSA 33: 9266).

Goldsworthy, A. 1966. A simple apparatus for generating an air stream containing a constant concentration of $^{14}CO_2$. J. Exptl. Bot. 17: 147-150. (NSA 20: 35218).

Goldstein, N. P., K. H. Sun and J. L. Gonzales. 1971. Radioactivity in fly ash from a coal burning power plant. Trans. Amer. Nucl. Soc. 14(1): 66-67. (NSA 25: 39605).

Goldstein, R. A. and J. W. Elwood. 1971. Two-compartment, three-parameter model for the absorption and retention of ingested elements by animals. Ecology 52(5): 935-939.

Goldsztein, M. 1957. Geobotanical prospecting for uranium in Esterel. Bull. soc. franc. mineral. crist. 80: 318-324.

Golley, F. B. 1966. The effects of acute gamma radiation on wild mammals. pp. 32-35. In: Radiation Effects on Natural Populations. (See Sacher, 1966).

Golley, F. B. 1967. Methods of measuring secondary productivity in terrestrial vertebrate populations. pp. 99-124. In: Secondary Productivity of Terrestrial Ecosystems. Vol. I. K. Petrusewicz, ed. Inst. of Ecology, Polish Academy of Sciences, Warsaw. (Incl. radionuclides).

Golley, F. B. and J. B. Gentry. 1965. Effect of acute gamma radiation on wild rodents. Abstr. Bull. Ecol. Soc. Am. 46(3): 121.

Golley, F. B. and J. B. Gentry. 1969. Response of rodents to acute gamma radiation under field conditions. pp. 166-172. In: Symposium on Radioecology. (See Nelson and Evans, 1969).

Golley, F. B. and A. H. Loyal. (Eds.). 1965. Onsite ecological research of the Division of Biology and Medicine at the Savannah River Ecology Laboratory. Savannah River Ecology Lab., U. S. AEC report TID-21713. vii, 94pp. (NSA 19: 35887).

Golley, F. B., J. T. McGinnis and R. G. Clements. 1969. Final report, terrestrial ecology. Bioenvironmental and radiological-safety feasibility studies: Atlantic-Pacific Interoceanic Canal. Univ. of Georgia, Athens, Battelle Columbus Labs., U. S. AEC report BMI-171-24. 83pp. (NSA 24: 2594).

Golley, F. B., R. G. Wiegert and R. W. Walter. 1965. Excretion of orally administered zinc-65 by wild small mammals. Health Physics 11(8): 719-722. (NSA 19: 35904).

Golley, F. B., et al. 1965. Response of wild rodents to acute gamma radiation. Radiation Research 24(2): 350-356. (NSA 19: 15186).

Golley, F. B., et al. 1965. Effect of acute gamma radiation on wild opossum, gray fox, racoon and bobcat. In: Radiation and Terrestrial Ecosystems. Health Physics 11(12): 1573-1576. (See Hungate, 1965).

Golley, F. B., et al. 1968. Components of growth in Peromyscus polionotus. Growth 32: 103-111. (NSA 23: 26879) (γ radiation).

Golovinskaya, K. A. and D. D. Ramashov. 1958. Effect of ionizing radiation on the development and reproduction of fishes. Vopr. ikhtiol. 11: 16-38. (In Russian).

Golubenkov, A. M. and A. F. Malenchenko. 1975. Neutron activation method with use of solid-state detectors in the determination of natural uranium in biological materials. Vestsi Akad. Navuk Beleruss. SSR Navuk ser. Fiz. Energ. (1): 42-48. (NSA 32: 16706).

Golubkova, M. G. and D. M. Grodzinskiy. 1972. Intercellular interactions and recovery of a population of the blue-green alga Plectonema boryanum after gamma irradiation. Radiobiologiya 12(1): 144-148. (AEC-tr-7316, pp. 214-219) (NSA 26: 50831).

Golutivina, M. M., et al. 1962. A new rapid method for determining the ^{90}Sr content of bones. Med. radiologiya 7(1): 62-67. (AEC-tr-7232, pp. 104-111).

Gomberg, H. J. 1970. Radioecology at the Puerto Rico Nuclear Center. Ch. A-1, pp. A3-A4. In: A Tropical Rain Forest. (See H. T. Odum and R. F. Pigeon, 1970).

Gomberg, H. J. and S. E. Gould. 1953. Effect of irradiation with cobalt-60 on Trichina larvae. Science 118(3055): 75-77. (NSA 8: 1779).

Gomez, J. C., et al. 1962. The effect of ionizing radiations on the biology and ecology of Rhodnius prolixus, the principal vector of S. cruzi in Venezuela. pp. 185-193. In: 4th Inter-American Symp. on the Peaceful Application of Nuclear Energy. Pan American Union, Washington. (In Spanish).

Gomez, J. C., et al. 1962. The effect of ionizing radiation on the biology and ecology of Rhodnius prolixus, the principal vector of Schizotrypanum (i.e. Trypanosoma) cruzi in Venezuela. Acta cient. Venez. 13(2): 46-52. (In Spanish).

Gomez-Campo, C. and M. Casas-Builla. 1972. Some effects of ionizing radiation on plant phyllotaxis. Radiation Botany 12(3): 165-172. (NSA 26: 50843).

Gomez-Campo, C. and L. Delgado. 1964. Radioresistance in crucifers. Radiation Botany 4(4): 479-483. (NSA 19: 15170).

Gomez-Nuñez, J. C. 1967. Study of the possible control of Rhodnius prolixus by means of ionizing radiations. Acta cient. Venez. 18(Suppl.): 98-105. (In Spanish) (NSA 22: 47653).

Gomez-Nuñez, J. C. 1971. Sterility and Chagas' disease vector control. pp. 157-165. In: Sterility Principle for Insect Control or Eradication. (See IAEA, 1971) (NSA 25: 29745).

Gomez-Nuñez, J. C., A. Gross and C. Machado. 1964. Gamma radiation and the reproductive behavior of male Rhodnius prolixus. Acta cient. Venez. 15: 97-104. (In Spanish) (NSA 19: 15170).

Gomez Ruiz, H. R., G. B. Baro and E. Garcia Agudo. nd. Study of sediment movements in the Plata River with radioactive tracers. pp. 295-306. In: Radiation and Isotope Technology in Latin American Development. (See Puerto Rico Nuclear Center, nd) (In Spanish, English abstr.) (NSA 24: 39040).

Gona, A. G. 1968. Radidiodine studies on prolacting action in tadpoles. Gen. Comp. Endochrinol. 11: 278-283. (NSA 23: 17959).

Gonen, M. and M. Calderon. 1971. Effects of gamma radiation on Ephestia cautella (Wlk.) (Lepidoptera, Phycitidae). II. Effects on progeny of irradiated males. J. Stored Prod. Res. 7(2): 91-96. (NSA 25: 55316).

Gonfiantini, R., et al. 1974. Isotopic study of the continental intercalaire aquifer and its relationship with other aquifers of the northern Sahara. pp. 227-241. In: Isotope Techniques in Groundwater Hydrology. (See IAEA, 1974) (In French) (NSA 31: 19630).

Gong, J. K., et al. 1957. Uptake of fission products and neutron induced radionuclides by the clam. Proc. Soc. Exptl. Biol. Med. 95(3): 451-454.

Gonzales, D. D. and L. E. Wollitz. 1972. Hydrological effects of the Cannikin event. Bull. Seismol. Soc. Am. 62(6): 1527-1542.

Gooding, A. M. and J. G. Ogden. 1965. A radiocarbon dated pollen-sequence from the Wells Mastadon Site near Rochester, Indiana. Ohio J. Sci. 65 (1): 1-11.

Goodspeed, T. H. and F. H. Uber. 1939. Radiation and plant cytogenetics. Botan. Rev. 5:1.

Goodwell, G. M. and B. R. Holt. 1971. Effect of nuclear war on the structure and function of natural communities: an appraisal based on experiments with gamma radiation. pp. 482-491. In: Survival of Food Crops and Livestock in the Event of Nuclear War. (See Bensen and Sparrow, 1971).

Goodyear, C. P., C. C. Coutant and J. R. Trabalka. 1974. Sources of potential biological damage from once-through cooling systems of nuclear power plants. Oak Ridge National Lab., U. S. AEC report ORNL-TM-4180. 46pp. (NSA 30: 26831).

Goodyear Atomic Corporation. 1973. Environmental effects of the construction and operation of a gaseous diffusion plant. Portsmouth, O. U. S. AEC report ORO-725. 33pp. (NSA 29: 24386).

Gopal-Ayengar, A. R. 1966. Atomic energy and life sciences. At. Energy Rev. 4(Comm.): 59-71. (NSA 21: 26364).

Gopal-Ayengar, A. R. and K. B. Mistry. 1962. On the radioactivity of plants from the high radiation areas of the Kerala Coast and adjoining regions. I. Studies on the uptake of gamma-emitting radionuclides. pp. 3-16. In: Radioisotopes in Soil-Plant Nutrition Studies. (See IAEA, 1962) (NSA 16: 3209).

Gopal-Ayengar, A. R., et al. 1970. Biological effects of high background radioactivity: studies on plants growing in the monazite-bearing areas of Kerala Coast and adjoining regions. Indian J. Exptl. Biol. 8: 313-318. (NSA 25: 39949).

Gopalakrishnan, S., et al. 1973. Measurements on airborne and surface fall-out radioactivity in India. Bhabha Atomic Research Center, India Atomic Energy Establishment, Bombay, report BARC-679. 124pp. (NSA 31: 14050).

Goral, I. 1973. Distribution of radioactive products of photosynthesis in Scots pine (Pinus silvestris L.) seedlings during the first vegetation season. Acta Soc. Bot. Pol. 42: 541-553. (NSA 31: 23143).

Goral, I. 1973. Incorporation of radioactive products of photosynthesis into lignin and cellulose of Scots pine (Pinus silvestris L.) seedlings. Acta Soc. Bot. Pol. 42: 555-565. (NSA 31: 23142).

Gorbatyuk, N. V. and N. A. Timofeyeva. 1960. Distribution of disperse elements according to components of bodies of water. III. Stabilization of the activity of a fresh solution of uranium debris with its steady addition to a body of water and uniform dosage inside and outside the body of water. Tr. inst. biol. Akad. Nauk SSSR, Ural'skiy filial 12: 238-277. (In Russian).

Gorbatyuk, N. V. and N. V. Timofeyev-Resovskiy. 1960. Limiting permissible norms of radioactive pollution of water and air. Byull. Ural'skogo otd. Mosk. obshch. ispyt. prirody (biol.) 2: __.

Gorberg, Z. and J. Jaskulski. 1970. Radioactive contamination of the area and the environment of the Institute of Nuclear Research: Centre Zeran area in 1968. ---1969. Poland Institute of Nuclear Research , Warsaw, report INR-1117; INR-1246. 17, 19pp. (In Polish) (NSA 24: 44488; 25: 37731).

Gorberg, Z. and K. Zarnowiecki. 1965. Fallout sampling method. Nukleonika 10(1): 60-61. (In Polish) (NSA 19: 36732).

Gorbman, A. M., M. Clements and R. O'Brien. 1954. Utilization of radio-iodine by invertebrates with special study of several Annelida and Mollusca. J. Exptl. Zool. 127(1): 75-92.

Gorbushina, L. V. and V. G. Tyminskiy. 1974. Radioactive and Stable Isotopes in Geology and Hydrogeology. Atomizdat, Moscow. 104pp. (In Russian) (NSA 30: 18666).

Gordee, R. S. and T. R. Mathews. 1970. Evaluation of systematic antifungal agents in x-irradiated mice. Appl. Microbiol. 20: 624-629. (NSA 25: 16244).

Gordon, B. E. and E. L. Bastin. 1968. The development of a particulate radioactive soil for detergency purposes. Am. Oil Chem. Soc. 45: 754-759. (NSA 23: 19619).

Gordon, B. M. and H. W. Kraner. 1972. Development of a system for trace element analysis in the environment by charged particle x-ray fluorescence. J. Radioanal. Chem. 12(1): 181-188. (Also: Brookhaven National Lab., U. S. AEC report BNL-16182) (CONF-710925) (NSA 26: 4197; 28: 88).

Gordon, D. 1964. Effects of gamma irradiation of cercariae on the host-parasite relationships of Shistosoma mansoni. Thesis, Univ. of Minnesota, Minneapolis. 167pp. (NSA 19: 8944).

Gordon, H. T. 1961. Nutritional factors in insect resistance to chemicals. Role of esterases in detoxication of pyrethrins and carbamates. Ann. Rev. Entomol. 6: 40-42.

Gordon, J. C. and P. R. Larson. 1968. Seasonal course of photosynthesis, respiration, and distribution of ^{14}C in young Pinus resinosa trees as related to wood formation. Plant Physiol. 43: 1617-1624.

Gordon, J. C. and P. R. Larson. 1970. Redistribution of ^{14}C-labeled reserve food in young red pines during shoot elongation. Forest Sci. 16(1): 14-__.

Gordon, S. A. and R. P. Weber. 1950. The effect of x-radiation on IAA and auxin levels in the plant. Abstr. Am. J. Bot. 37(8): 678.

Gordus, A. A. 1970. Neutron activation analysis of archeological artefacts. Phil. Trans. Roy. Soc. (London) A269: 165-174. (NSA 25: 32074).

Goresline, H. E. 1965. Application of ionizing radiation to grain disinfestation. Food Irradiation 6: A10-A12. (NSA 20: 19125).

Gorham, E. 1959. A comparison of lower and higher plants as accumulators of radioactive fall-out. Can. J. Bot. 37: 327-329. (NSA 13: 12377).

Gorham, E. and R. H. Hoffstetter. 1971. Penetration of bog peats and lake sediments by tritium from atmospheric fallout. Ecology 52(5): 898-902.

Gorham, N. 1958. Accumulation of radioactive fallout by plants in the English lake district. Nature 181(4622): 1523-1524.

Goring, C. A. I. and F. E. Clark. 1952. Radioactive phosphorus and the growth and metabolic activities of soil microorganisms. Soil Sci. Soc. Am. Proc. 16: 7-9.

Gorizontov, P. D., et al. 1965. The problem of general radiation sensitivity of organisms. Inst. of Atomic Energy, State Committee on the Use of Atomic Energy, Moscow. 37pp. (In Russian) (NP-16258) (NSA 20: 42849).

Gorman, A. E. 1956. Environmental aspects of the atomic energy industry. Proc. Intern. Conf. on the Peaceful Uses of Atomic Energy 13: 298-304. (Paper A/Conf. 8/P/283).

Gornayeva, G. F. 1969. Roentgenomorphoses of Drosophila melanogaster macrochaeta after irradiation of larvae and pupae at various stages of development. Radiobiologiya 9(1): 151-153. (AEC-tr-7024, pp. 196-199) (NSA 23: 24712).

Gorodilov, Yu. N. 1971. Change in radioresistance of some species of salmon in the early stages of embryonal development. Radiobiologiya 11(6): 930-934. AEC-tr-7307, pp. 159-165) (NSA 26: 41160).

Gorodinskiy, S. M., G. M. Parkhomenko and N. Yu. Tarasenko. (Eds.). 1962. Ch. 3, Influence of radioactive contamination on the sanitary condition of open reservoirs. Ch. 4, Migration of radioactive fission products of uranium in groundwater. Ch. 6, Behavior of radioactive substances in soil and their migration in terrestrial flora and fauna. pp. 34-67, 68-92, 109-126. In: Radiation Hygiene. Vol. II. A. N. Marey, ed. Medgiz, Moscow. (In Russian; Ch. 3, Ch. 6, Table of Contents: JPRS-21,799).

Gorshkov, G. V., V. A. Zyabkin and O. S. Tsvetkov. 1964. On the neutron background at the earth's surface. Atom. Energiya (USSR) 17(6): 492-496. (In Russian) (NSA 19: 7705).

Gorshkov, V. V., L. I. Mekhryusheva and V. I. Gorovenko. 1974. Neutron-activation determination of Cu and Mn in biological materials. Med. Radiologiya 19(10): 56-60. (In Russian) (NSA 30: 31479).

Gortner, W. A. 1962. Radioisotope studies of pesticide metabolism by the pineapple plant. Intern. J. Rad. Isotopes 13: 395-397.

Gosling, A. W., E. A. Jenne and T. T. Chao. 1971. Gold content of natural waters in Colorado. Econ. Geol. 66: 309-313. (NSA 25: 29015) (Neutron activation).

Gösswald, K. 1959. New data on colony life among insects on the basis of radiobiological studies. Natur u. Volk 89(7,8): 209-213; 249-259. (In German).

Gösswald, K. 1962. The ant colony. Math. Naturwiss. Unterricht 12: 208-223. (In German) (Radiotracer).

Gösswald, K. 1962. On the methods of controlling materials for termite resistance with particular consideration of biological data underlying the technique of control. pp. 169-178. In: Intern. Symp. on Termites in the Humid Tropics. (New Delhi, 4-12 October 1960). (^{32}P).

Gösswald, K. and W. Kloft. 1956. Studies on the distribution of radioactively labelled food among ants of a colony of Formica rufopratensis minor. Waldhygiene 1(7): 200-202. (In German).

Gösswald, K. and W. Kloft. 1957. New studies by means of radioisotopes on the social interrelations in the ant. p. 543. In: Proc. 10th Intern. Congr. on Entomol. Vol. 2. (Montreal, August 17-25, 1956). E. C. Becker, ed. Mortimer, Ltd., Ottawa. (In German).

Gösswald, K. and W. Kloft. 1958. Radiobiological investigations on colony-forming insects. Umschau 58(24): 743-745. (In German).

Gösswald, K. and W. Kloft. 1959. Laboratory testing of textile fabrics for termite resistance by means of Calotermes flavicollis Fabr. Entomol. Exptl. et Appl. 2(4): 268-278. (In German) (^{32}P).

Gösswald, K. and W. Kloft. 1960. Radioisotope studies on forest ants. Entomophaga 5(1): 33-41. (In German).

Gösswald, K. and W. Kloft. 1960. Recent studies by means of radioisotopes on the social interrelations among ants. Zool. Beitr. Berl. 5(2/3): 519-566. (In German).

Gösswald, K. and W. Kloft. 1961. Aspects of life in an insect colony, based on radiobiological studies. Imkerfreund 16(1): 7-12. (In German).

Gösswald, K. and W. Kloft. 1963. Study on the elimination of radioactively labelled ants by individuals and groups. pp. 9-17. In: Proc. 4th Congr. of the U. I. E. I. S., Symp. on Genetics and Biology in Italy. Vol. XII. (Pavia, 9-14 September 1961). Tipografia del Libro S. A. S., Pavia. (In German).

Gösswald, K. and W. Kloft. 1963. Tracer experiments on food exchange in ants and termites. pp. 25-40. In: Radiation and Radioisotopes Applied to Insects of Agricultural Importance. (See IAEA, 1963).

Gösswald, K., E.-F. Schulze and W. Kloft. 1963. Problems of application and action of Thiodan studied with S^{35}-labelled insecticide. pp. 241-247. I In: Radiation and Radioisotopes Applied to Insects of Agricultural Importance. (See IAEA, 1963).

Gosz, R., et al. 1969. Distribution of ^{32}P in Ambrosia artemisiifolia; its implications. Abstr. Bull. Ecol. Soc. Am. 50(2): 72.

Got, H. and H. Pauc. 1970. Study of recent dynamic evolution around the Rhone mouth by using radioactive wastes from Marcoule Nuclear Center. Compt. rend. D271: 1956-1959. (In French) (NSA 25: 21165).

Gotchy, R. L. 1972. Project Rulison final operational radioactivity report - production tests. Nevada Operations Off., U. S. AEC report NVO-112. 74pp. (NSA 26: 30977).

Goto, M., T. Okada and H. S. Forrest. 1964. Biosynthesis of drosopterin in the eye pigment of Drosophila melanogaster. Seikagaku Zasshi 56(4): 379. (^{14}C).

Goto, M., T. Okada and H. S. Forrest. 1965. Synthesis of 2-amino-4-hydroxy-6 D-erythrophosphate-10-^{14}C, and its metabolism in Drosophila melanogaster.

Gotecchia, V., G. S. Tazioli and G. Magri. 1974. Isotopic measurements in research on seawater ingression in the carbonate aquifer of the Salentine Peninsula, Southern Italy. pp. 445-463. In: Isotope Techniques in Groundwater Hydrology. Vol. 1. (See IAEA, 1974) (NSA 31: 19641).

Gottikh, R. P., et al. 1968. Isotopic composition of uranium in ground waters. Tr. vses. nauchn.-issled. inst. yadern. geofiz. geokhim. (4): 255-265. (In Russian) (NSA 25: 2356).

Gould, S. E., J. G. Van Dyke and H. J. Gomberg. 1952. Effects of x-ray on trichina larvae. Abstr. Am. J. Pathol. 28(3): 554-555.

Gould, S. E., J. G. Van Dyke and H. J. Gomberg. 1953. Effect of x-rays on trichina larvae. Am. J. Pathol. 29(2): 323-338.

Gould, S. E., et al. 1955, 1957. Studies on Trichinella spiralis. II-VI. Am. J. Pathol. 31(5): 933-963; 33(5): 79-105.

Goulden, A. M. (Ed.). 1969. Biological Implications of the Nuclear Age. U. S. AEC Symposium Ser. 16. ix, 341pp. (CONF-690303).

Gourley, D. R. H. 1952. Failure of P^{32} to exchange with organic phosphorus compounds. Nature 169(4292): 192-193.

Gourlez, P. and G. Sauzay. 1974. Development of a technique for studying mud transfer in an estuary. Centre d'Études Nucléaires de Saclay, France Commissariat à l'Énergie Atomique, 91-Gif-sur-Yvette. (ORNL-tr-2876) (NSA 31: 11249) (^{46}Sc).

Gouvaerts, J., et al. 1967. Measurement of radioactivity of the waters of the Meuse and of fallout in Belgium during the year 1966. Bull. Soc. Roy. Sci. Liége 36: 510-514. (In French) (NSA 22: 21192).

Gowner, J. W. 1965. The effect of radioactive fallout at the Nevada Test Site on the chromosomes of the pocket mouse. Health Physics 11: 1569-1571. (NSA 20: 12666).

Goyffon, M. 1971. Amino-acidemia and whole-body irradiation of scorpions. pp. 132-134. In: Scientific Work, 1971. Centre de Recherches du Service de Sante des Armeés, Clamart, France, report FRNC-R-19. (In French) (NSA 27: 20165).

Goyffon, M., R. Schantz and P. Niaussat. 1969. Effects of whole body irradiation on blood sugar level of the scorpion Androctonus australis L. Compt. rend. soc. biol. 163: 601-605. (In French) (NSA 24: 4563).

Goyffon, M., et al. nd. Spontaneous electric activity of a scorpion nervous system and irradiation by ionizing radiation. pp. 42-43. In: Scientific Works, 1973. Centre de Recherches du Service Sante des Armeés, Clamart, France, report CRSSA-1973. (In French) (NSA 33: 17467).

Graca, M. D. H. and P. V. A. de Andrade. 1974. On environmental pollution. Fomento (Lisbon) 12(2): 133-146. (In Portugese, English abstr.) (NSA 33: 29174).

Grace, J. and H. W. Woolhouse. 1973. A physiological and mathematical study of the growth and productivity of a Calluna-Sphagnum community. III. Distribution of photosynthate in Calluna vulgaris L. Hull. J. Appl. Ecol. 10(1): 77-91. (^{14}C).

Gracheva, L. M. and V. G. Korolev. 1971. Biological effects of the decay of radiophosphorus incorporated into microorganisms. Usp. sovrem. biol. 72 (1): 143-158. (In Russian) (NSA 27: 20165).

Gracheva, L. M., N. Yu. Zheleznyakova and V. G. Korolev. 1970. Effect of radiophosphorus decay on the haploid and diploid cells of yeast Saccharomyces cerevisiae, of different genotypes. Genetika 6(12): 101-106. (In Russian) (NSA 25: 24453).

Graf, P. and H. Zünd. 1969. Treatment and disposal of radioactive waste. Neue Tech. B11: 112-127. (In German) (NSA 24: 46318).

Graf, U. 1972. Radiosensitivity of Drosophila sperm before and after insemination of the egg. Mutation Research 15(3): 287-301. (In German) (NSA 26: 43359).

Graham, B. F., Jr. 1954. A technique for introducing radioactive isotopes into tree stems. Ecology 35(3): 415.

Graham, E. R. 1958. Uptake of waste Sr 90 and Cs 137 by soil and vegetation. Soil Science 86(2): 91-97.

Graham, E. R. 1963. Plants as monitors of radioactive contamination of the environment of Los Alamos, New Mexico. Los Alamos Scientific Lab., U. S. AEC report LAMS-2879. 18pp. (NSA 18: 436).

Graham, E. R. 1963. Factors affecting Sr^{85} and I^{131} removal by runoff water. Water Sewage Works 110(11): 407-410. (NSA 18: 35724).

Graham, H. M., O. T. Robertson and D. F. Martin. 1964. Radiographic detection of pink bollworm larvae in cotton seed. J. Econ. Entomol. 57(3): 419-420.

Graham, H. M., et al. 1972. Dosages of gamma irradiation for full and inherited sterility in adult pink bollworms. J. Econ. Entomol. 65(3): 645-650. (NSA 26: 43350).

Graham, O. H., R. O. Drummond and R. A. Hoffman. 1968. Possibilities of the sterile-male technique for control of livestock insects in the United States of America. pp. 41-44. In: Control of Livestock Insect Pests by the Sterile-Male Technique. (See IAEA, 1968) (NSA 24: 34809).

Graham, W. J. and H. W. Ambrose, III. 1967. A technique for continuously locating small mammals in field enclosures. J. Mammal. 48(4): 639-642.

Graikoski, J. T. and L. L. Kempe. 1962. Effects of field-strength, beam-energy and intermittent irradiation on the lethality of gamma radiation for spores of Clostridium botulinum. Atompraxis 8(12): 467-470. (In English) (NSA 17: 7817).

Grandclement, G. 1963. The use of radioactive tracers in hydrology and hydraulics. Eau L50: 293-300. (In French) (NSA 18: 12370).

Grandin, M. and R. Roux. 1974. Twelve years of environmental monitoring experience at the Cadarache Center for Nuclear Studies. pp. 207-218. In: Environmental Surveillance Around Nuclear Installations. Vol. 2. (See IAEA, 1974) (In French) (Also: France Commissariat à l'Énergie Atomique, Cadarache, report CEA-CONF-2594) (SM-180/48) (CONF-731117-6) (NSA 30: 18615; 31: 14036).

Granhill, I, et al. 1949. X-ray effects in fruit trees. Hereditas 35: 369-379. (NSA 4: 396).

Grant-Taylor, T. L. 1967. Tritium hydrology in New Zealand. pp. 381-400. In: Isotopes in Hydrology. (See IAEA, 1967) (NSA 22: 18935).

Grant-Taylor, T. L. 1972. Extraction and use of plant lipids as a material for radiocarbon dating. pp. E58-E66. In: Proc. 8th Intern. Radiocarbon Dating Conf. Vol. 2. (See Rafter, 1972) (NSA 31: 8706).

Grashchenko, S. M., et al. 1960. Radium concentration in the waters of the Black Sea. Dokl. Akad. Nauk SSSR 132(5): 1171-1172. (Doklady (earth sci.) 132: 458-459) (CA 55: 20663).

Grasso, A., M. Boccacci and M. Quintiliani. 1962. Lehtal effects of x-rays on the housefly, Musca domestica L. Sci. Rep. Ist. Super. Sanita 2(3): 2 292-300.

Grasso, A., M. Boccacci and M. Quintiliani. 1963. Lethal effects of x-rays on Musca domestica L. Intern. J. Rad. Biol. 6(4): 383.

Grasty, R. L. 1973. Snow-water equivalent measurement using natural gamma emission. Nord. Hydrol. 4(1): 1-16. (NSA 29: 7725).

Grasty, R. L., H. S. Loijens and H. L. Ferguson. 1974. Experimental gamma-ray spectrometer snow survey over southern Ontario. pp. 579-593. In: Advanced Concepts and Techniques in the Study of Snow and Ice Resources. (See Santeford and Smith, 1974) (NSA 33: 7307).

Grathwohl, G. 1972. Production and release of tritium by nuclear power plants and reprocessing plants and the expected radiological burden till the year 2000. Inst. für Neutronenphysik und Reaktortechnik, Karlsruhe Univ., Karlsruhe, Germany. 88pp. (NP-19896) (NSA 29: 7702).

Grauby, A. 1970. Continental radioecology. Bull. Inform. Sci. Tech. (Paris) 151: 37-38. (In French).

Grauby, A. 1972. Behavior of radioactivity in the terrestrial environment. pp. 51-68. In: Disposal of Radioactive Waste. (See Organization for Economic Co-operation and Development, 1972) (In French, English summary).

Grauby, A. 1972. Radioecological studies in France. Proc. Jap. Conf. Radioisotopes (10): 302-303. (In Japanese) (NSA 28: 363).

Grauby, A. and P. Raimbault. 1963. The monitoring of the radioactivity of waters. Problems posed by the measurement and the use of the results. Eau L50: 307-313. (In French) (NSA 18: 6981, 12391).

Graves, E., Jr., et al. 1964. Nuclear excavation plan. Annex III, App. 1. Isthmian Canal Studies, 1964. Lawrence Radiation Lab., Livermore, Calif., U. S. AEC report PNE-2000. 67pp. (NSA 20: 1983).

Gray, C. E. (Comp., ed.). 1975. Manual of analytical methods for the Environmental Health Laboratory. Sandia Labs., U. S. ERDA report SAND-75-0014. 180pp. (NSA 32: 16739).

Gray, L. E. and J. W. Gerdemann. 1967. Influence of vesicular-arbuscular mycorrhizas on the uptake of ^{32}P by Liriodendron tulipifera and Liquidambar styraciflua. Nature 213: 106-107. (NSA 21: 8720).

Gray, L. E. and J. W. Gerdemann. 1969. Uptake of phosphorus-32 by vesicullar-arbuscular mycorrhizae. Plant Soil 30(3): 415-422.

Grayevskaya, B. M. 1972. Some results of a study of the radiosensitivity of mammals. Radiobiologiya 12(3): 323-335. (AEC-tr-7402, pp. 1-21) (NSA 27: 17636).

Grayevskiy, Ye. Ya. 1952. Sensitization by beta radiation of bacteria to effects of high and low temperatures and to effects of repeated beta radiation. Dokl. Akad. Nauk SSSR 83: 565-568. (In Russian) (Bacterium coli commune).

Grayevskiy, Ye. Ya. 1963. Theoretical aspects of chemical protection of mammals against ionizing radiation. Zh. obshch. biol. 24(1): 3-22. (JPRS-18,918).

Grayevskiy, Ye. Ya. and A. A. Neyfakh. 1956. Effects of small doses of ionizing radiation on the Paramecium caudatum (contribution on the problem of radiostimulation). Dokl. Akad. Nauk SSSR 111: 1004-1006. (NSA 11: 4790).

Grayevskiy, Ye. Ya. and N. I. Shapiro. 1957. Modern Problems of Radiobiology. A. A. Nichipozovich, ed. Acad. Sci. USSR, Moscow. 95pp.

Grayevskiy, Ye. Ya. and Ye. G. Zinov'yeva. 1956. Radiosensitivity of pituitary glands in total body irradiation of amphibians. Dokl. Akad. Nauk SSSR 110: 379-382. (NSA 11: 1738).

Grayevskiy, Ye. Ya. and Ye. G. Zinov'yeva. 1958. Possibility of changes in radiosensitivity of cells under the influence of fluorchromes. Dokl. Akad. Nauk SSSR 118: 476-478. (Soviet Physics (Doklady) 3: 206-208) (Paramecium caudatum).

Grayevskiy, Ye. Ya. and Ye. G. Zinov'yeva. 1958. An investigation of cell radiosensitivity with repeated exposures to ionizing radiation. Dokl. Akad. Nauk SSSR 121(5): 837-840. (Doklady (biol. sci.) 121(1-6): 538-541) (Paramecia).

Grayevskiy, Ye. Ya., Z. N. Faleyeva and N. K. Flerova. 1966. Radioprotective effect of cysteamine under the action of ionizing radiation on loach sperm. Radiobiologiya 6(6): 886-890. (AEC-tr-6775, pp. 188-195. (NSA 21: 36718).

Grayevskiy, Ye. Ya., I. V. Nekrosova and A. I. Shul'mina. 1962. Study of the radiation-protective action of some protective materials on Protozoa. Radiobiologiya 2(1): 148-155. (AEC-tr-5428, pp. 224-234) (NSA 16: 30309).

Grayum, M. M. 1973. Effects of thermal shock and ionizing radiation on primary productivity. pp. 639-644. In: Radionuclides in Ecosystems. (See Nelson, 1973) (NSA 25: 37756).

Greb, R. J. 1955. Lethal x-ray dose in pheasants. Proc. S. Dak. Acad. Sci. 34(1): 104-105.

Greb, R. J. 1960. Irradiation of pheasant ovaries with gamma rays. Abstr. Proc. S. Dak. Acad. Sci. 39: 170. (BA 36: 43335; WR 103: 60).

Greb, R. J. and W. Morgan. 1961. Treatment of pheasant ovaries with x-rays and gamma rays. Proc. S. Dak. Acad. Sci. 40: 112-119. (WR 107: 71).

Grebinskaya, M. I., et al. 1961. Radiosensitivity of mulberry seeds. pp. 428-432. In: Proc. Tashkent Conf. on Peaceful Uses of Atomic Energy, 1959. Ya. Kh. Turakulov, ed. Publ. House Acad. Sci. Uzbek SSR. (AEC-tr-6293, pp. 728-733).

Grecz, N. 1965. Biophysical aspects of Clostridia. J. Appl. Bacteriol. 28: 17-35. (NSA 21: 22812).

Grecz, N. 1966. Theoretical and applied aspects of radiation D-values for spores of _Clostridium botulinum_. pp. 307-319. In: _Food Irradiation_. (See IAEA, 1966) (Paper SM-73/67) (CONF-660609-5) (NSA 20: 40601).

Grecz, N., J. Upadhyay and T. C. Tang. 1967. Effect of temperature on radiation resistance of spores of _Clostridium botulinum_ 33A. Can. J. Microbiol. 13: 287-293. (NSA 21: 20452).

Green, A. B. 1904. A note on the action of radium on microorganisms. Proc. Roy. Soc. (London) 73(494): 375-381.

Green, B. C. and J. W. T. Spinks. 1955. Automatic plotting of the position of a moving radioactively tagged object. Can. J. Technol. 33: 307-316.

Green, E. L. 1968. Genetic effects of radiation on mammalian populations. Ann. Rev. Genet. 2: 87-120. (NSA 23: 29457).

Green, G. W. 1964. The control of spontaneous locomotor activity in _Phormia regina_ Meigen. -II. Experiments to determine the mechanism involved. J. Insect Physiol. 10(5): 727-752. (Used radio-opaque substances).

Green, G. W., W. F. Baldwin and C. R. Sullivan. 1957. The use of radioactive cobalt in the studies of the dispersal of adult females of the European pine shoot moth, _Rhyacionia buoliana_ (Schiff.). Can. Entomol. 89(8): 379-383.

Green, J. H. and E. T. Pallister. 1965. Radiochemical analyses for ^{40}Sr, ^{137}Cs, ^{226}Ra, and ^{40}K in natural and processed materials in Australia. Austral. J. Sci. 27: 199-202. (NSA 20: 156).

Greenberg, J. and P. Woody-Karrer. 1963. Radiosensitivity in _Escherichia coli_. J. Gen. Microbiol. 33(2): 283-292.

Greene, R. E. 1968. Tracing with activable tracers. (I). Part II. Isotopes Rad. Technol. 5: 269-277; 6: 70-78. (NSA 22: 31008, 48930).

Greenfield, S. M. 1973. Environmental Protection Agency's role in environmental tritium. pp. 12-14. In: _Tritium_. (See Moghissi and Carter, 1973) (NSA 28: 27238).

Greenfield, S. M. 1973. Environment. Subpanel report XIV used in preparing the AEC Chairman's report to the President. U. S. AEC report WASH-1281-14. 402pp. (NSA 29: 26394).

Greenhalgh, G. H. 1973. Nuclear power plants and the environment. Report of Session 4. British Nuclear Forum. 14pp. (NSA 30: 2081).

Greenhous, N. A. and T. F. McCraw. nd. Marshall Islands radiological followup. Brookhaven National Lab., U. S. ERDA report BNL-20767. (NSA 33: 17474).

Greenleigh, S. H. 1973. Federal Water Pollution Control Act Ammendments of 1972 and their impact on AEC operations. pp. 14-16. In: _Proc. AEC Pollution Control Conf._ (Oak Ridge, Tenn., October 25-27, 1972). U. S. AEC report CONF-721030. (NSA 28: 4961).

Greenslade, F. C. 1967. Reduction in DNA and RNA content during the development of haploid _Rana pipiens_ embryos. Experientia 23: 1075-1076. (NSA 22: 25996).

Greenwald, L. 1972. Sodium balance in amphibians from different habitats. Physiol. Zool. 45(3): 229-237.

Greg, T. G. 1959. Genetic studies of irradiated natural populations of _Drosophila_. III. Experimental populations of _Drosophila ananassae_ derived from irradiated natural populations. Univ. of Texas, Austin, Contributions 5914. pp. 207-222.

Gregg, C. T., et al. 1964. Inhibition of oxidative phosphorylation and related reactions in insect mitochondria. Biochem. Biophys. Acta 82: 343-349. (^{32}P).

Gregory, D. J. 1974. Gamma irradiation of the face fly, _Musca autumnalis_ De Geer (Diptera: Muscidae). Thesis, Univ. of Guelph, Ont., Canada. vp. (NSA 30: 32557).

Gregory, L. P. 1972. Simplified separation of strontium, radium, and lead from environmental media by precipitation followed by fractional elution. Anal. Chem. 44(12): 2113-2115. (NSA 26: 55743).

Greifer, B. and J. K. Taylor.1973. Survey of various approaches to the chemical analysis of environmentally important materials. Final report. National Bureau of Standards, Washington, D.C., report COM-74-10469.

Greiff, D. E., et al. 1960. The effects of x-rays and beta rays (tritium) on the growth of _Rickettsia mooseri_ and _Rhickettsia akari_ in embryonic eggs. J. Exptl. Med. 111(6): 841-849.

Grendon, A. 1972. Nuclear power and the environment. Forum (Chicago) 8(1): 70-92. (NSA 27: 29839; 28: 23861).

Grennes-Ravitz, R. A. and R. D. Wilson. 1975. Recreating the pre-Columbian past: the application of nuclear physics to Mesoamerican archeology. Tra Trans. Amer. Nucl. Soc. 21(2, Suppl.): 56-57. (NSA 33: 4951).

Grenon, M. 1971. Calvert Cliffs decision. Rev. Fr. Energ. 23(237): 147-155. (In French) (NSA 26: 58011) (Nuclear power, environmental reports).

Gribkov, A. K. and L. D. Soshin. 1972. Analysis of components of background - a scintillation gamma spectrometer and depression methods. pp. 87-97. In: _Methods of Determination of Radioactivity_. (See Polikarpov and Parchevskiy, 1972).

Gribkov, A. K. and L. D. Soshin. 1972. Methods of treating complicated scintillation gamma-spectra. pp. 184-195. In: _Methods of Determination of Radioactivity_. (See Polikarpov and Parchevskiy, 1972).

Grice, C. F. 1970. Multiple detector submarine radioactivity logging system. U. S. Patent 3,546,456. (NSA 25: 23897).

Griffin, C. S., D. Scott and D. G. Papworth. 1970. Influence of DNA content and nuclear volume on the frequency of radiation-induced chromosome abberations in Bufo species. Chromosoma 30: 228-249. (NSA 25: 6684).

Griffin, D. R. 1949. Radioisotopes as tools for the study of animals under natural conditions. Bull. Ecol. Soc. Am. 30: 58-64. (NSA 4: 1671).

Griffin, D. R. 1952. Radioactive tagging of animals under natural conditions. Ecology 33(3): 329-335.

Griffin, G. F. 1968. Automated determination of magnesium in soil extracts by atomic absorption spectrophotometry. Soil Sci. Soc. Am. Proc. 32(6): 803-805.

Griffiths, E. and R. G. Burns. 1968. Effects of gamma irradiation on soil aggregate stability. Plant Soil 28(1): 167-172. (NSA 22: 40543).

Griffiths, T. D. 1973. Metabolic aspects of dose-fractionation recovery in tenebrionid insects. Thesis, Univ. of Illinois, Urbana. 134pp. (Univ. Microfilms 74-12,028) (NSA 30: 19007).

Griffiths, T. D. and H. S. Ducoff. 1973. Supression by hypoxia of dose-fractionation recovery in tenebrionid beetles. Intern. J. Rad. Biol. 24 (2): 203-206. (NSA 29: 772).

Grigal, D. F. 1973. Calcium cycling: diffusion into a forest soil. pp. 218-224. In: Radionuclides on Ecosystems. (See Nelson, 1973) (NSA 28: 27271) (^{45}Ca).

Grim, L. W., R. K. Meyer and E. M. Donaldson. 1973. Radioimmuno-assay estimates of plasma gonadotropin levels in the spawning pink salmon. Gen. Comp. Endocrinol. 21(1): 69-76. (SFA 19: 17109).

Grisebach, H., R. C. Fuller and M. Calvin. 1957. Metabolism of thiotic acid in algae. Biochim. Biophys. Acta 23: 34-42. (Also: Univ. of Calif., Berkeley, U. S. AEC report UCRL-3360) (NSA 10: 4519).

Grison, G. 1967. Migration of radioelements in **soil**. Mem. Ass. Int. Hydrogeol., Reunion Istanbul 8: 392-414. (See Anonymous, 1967) (In French) (NSA 23: 445).

Gritchenko, Z. G. 1962. Procedure for determining the total activity of fission products in various media. Ch. II, pp. 11-21. In: Radioactive Contamination of the External Environment. (See Shvedov and Shirokhov, 1962) (AEC-tr-6049, pp 11-21) (NSA 17: 8518).

Gritchenko, Z. G. 1963. Investigation of the ruthenium-103 and ruthenium-106 content of a mixture of fission products by the method of beta-gamma coincidences of scintillation spectrometers. State Committee on the Use of Atomic Energy USSR. Atomizdat, Moscow. UN Document A/AC.82/G/L. 915. 20pp. (In Russian).

Grodzinskiy, D. M. 1959. Natural radioactivity of mosses and lichens. Ukr. biokhim. zh. 16(2): 30-38. (In Russian).

Grodzinskiy, D. M. 1965. Natural Radioactivity of Plants and Soils. Publ. House Acad. Sci. Ukr. SSR, Kiev. 216pp. (In Russian) (NSA 20: 29437).

Grodzinskiy, D. M. (Ed.). 1972. Protection Against Radiation and the Post-radiative Rehabilitation of Plants. Publ. House Acad. Sci. Ukr. SSR, Kiev. 168pp. (In Russian) (NSA 29: 24472).

Grodzinskiy, D. M. and I. N. Gudkov. 1969. Methods of protection and recovery of plants damaged with ionizing radiation. Fiz. biokhim. kul't. rast. 1: 147-156. (In Russian) (NSA 25: 21960).

Grodzinskiy, D. M., N. I. Bidzilya and O. P. Golikova. 1965. Factors of radiosensitivity of plants to gamma irradiation. Radiobiologiya 5(4): 596-601. (AEC-tr-6601, pp. 170-179) (NSA 19: 42369) (Incl. quack grass).

Grodzinskiy, D. Ye. 1969. Isotopes in Biology and Medicine. Atomizdat, Moscow. 208pp. (In Russian) (NSA 25: 21538).

Groenevelt, P. H., J. G. de Swart and J. Cisler. 1969. Water content measurement with 60 keV gamma ray attenuation. Bull. Int. Assoc. Sci. Hydrol. 14(2): 67-78. (NSA 23: 47631) (Soil samples).

Gromiec, M. J. and E. F. Gloyna. 1973. Radioactivity transport in water: final report. Univ. of Texas, Austin, U. S. AEC report ORO-23. 80pp. (NSA 29: 5127).

Gronskaya, N. F. 1963. Relationship of the development of radiation damage and repair on the regeneration activity of tissues of Rana temporaria tadpoles. Byull. eksper. biol. med. 55: 108-112. (Bull. Exptl. Biol. Med. USSR 55: 215-218) (NSA 18: 19781).

Gronskaya, N. F. 1965. Reactions of tadpole tissues to x-radiation during starvation conditions. Radiobiologiya 5(6): 850-854. (AEC-tr-6603, pp. 108-115) (NSA 20: 10676).

Groom, D. J. 1972. Siting of nuclear power stations in England and Wales. J. Inst. Nucl. Eng. 13(4): 99-105. (NSA 26: 58310).

Grootes, P. M., et al. 1975. Enrichment of radiocarbon for dating samples up to 75,000 years. Z. Naturforsch. A30(1): 1-14. (NSA 32: 14324).

Gros, C. M. and J. Bloch. 1957. Effect of temperature on the survival of the fish, Carassius carassius (L.) after total irradiation. Compt. rend. soc. biol. 151(3): 602-604. (In French).

Gros, C. M., et al. 1958. Effect of x-rays on the survival of a fish (Carassius carassius) exposed to low temperatures. Compt. rend. soc. biol. 152: 1187-1190. (In French) (NSA 13: 7389; SFA 5: 3571; WPA 33: 946).

Grosch, D. S. 1956. Induced lethargy and the radiation control of insects. J. Econ. Entomol. 49(5): 629-631.

Grosch, D. S. 1956. Lethality induced by feeding radiophosphorus to male Habrobracon. Am. Naturalist 90(852): 200-202.

Grosch, D. S. 1959. Wasp egg production and hatchability after the mothers have been exposed to mixtures of radiations. Atompraxis 5: 290-292. (NSA 14: 2304).

Grosch, D. S. 1961. The influence of antibiotics on the fecundity and fertility of irradiated Habrobracon females. Abstr. 65. Radiation Research 14(4): 470.

Grosch, D. S. 1962. Distribution of zinc-65 in the wasp, Habrobracon, and its effects on reproduction. Nature 195(4839): 356-358. (NSA 16: 26706).

Grosch, D. S. 1962. Entomological aspects of radiation as related to genetics and physiology. Ann. Rev. Entomol. 7: 81-106.

Grosch, D. S. 1964. Localization, persistence and resultant genetic effects in invertebrates of ingested fourth period metals in stable and radioactive forms. Final report. N. Car. Coll., U. S. AEC report TID-21559. 22pp. (NSA 19: 10741).

Grosch, D. S. 1965. Biological Effects of Radiations. Blaisdell Publ. Co., New York. 306pp. (NSA 19: 42346) (Incl. human ecology, pest control).

Grosch, D. S. 1965. Distribution and effective half-life of ^{58}Co in Habrobracon. Nature 208: 906-907. (NSA 20: 8574).

Grosch, D. S. 1965. Biological effects of radiation. Bios 36: 55-62.

Grosch, D. S. 1966. The reproductive capacity of Artemia subjected to successive contaminations with radiophosphorus. Biol. Bull. 131: 261-271. (NSA 21: 12970).

Grosch, D. S. 1968. Habrobracon life span, egg deposit and hatchability after two days of gamma-ray exposure of females during orbit in Biosatellite II. Abstr. Radiation Research 35(2): 546.

Grosch, D. S. 1968. Reproductive performance of female braconids compared after (A) brief and (b) protracted exposures to ionizing radiations. pp. 377-389. In: Isotopes and Radiation in Entomology. (See IAEA, 1968).

Grosch, D. S. 1970. Egg production and embryo lethality for Habrobracon from Biosatellite II and associated postflight vibration experiments. Mutation Research 9: 91-108. (NSA 24: 32177).

Grosch, D. S. 1971. The response of the female arthropod's reproductive system to radiation and chemical agents. pp. 217-227. In: Sterility Principle for Insect Control or Eradication. (See IAEA, 1971) (NSA 25: 29747).

Grosch, D. S. 1975. Combined effects of radiation and chemical agents in altering the fecundity of a braconid wasp. pp. 243-259. In: Sterility Principle for Insect Control, 1974. (See IAEA, 1975) (NSA 32: 11781).

Grosch, D. S. and H. E. Erdman. 1955. X-ray effects on adult Artemia. Biol. Bull. 108(3): 277-282.

Grosch, D. S. and L. E. LaChance. 1956. Fate of radiostrontium fed to Habrobracon females. Science 123(3187): 141-142.

Grosch, D. S. and M. E. Plumb. 1959. A sex difference in radioactivity of Artemia cultured in sea water containing phosphorus-32. Nature 183(4654): 122-123.

Grosch, D. S. and Z. H. Smith. 1957. X-ray experiments with Molgula manhattensis: adult sensitivity and induced zygotic lethality. Biol. Bull. 112 (2): 171-179).

Grosch, D. S. and R. L. Sullivan. 1952. The effect of ingested radiophosphorus on egg production and embryo survival in the wasp, Habrobracon. Biol. Bull. 102(2): 128-140.

Grosch, D. S. and R. L. Sullivan. 1954. Sequelae of rearing Habrobracon on radioactive host larvae. Growth 18(4): 191-205. (NSA 9: 5225).

Grosch, D. S. and R. L. Sullivan. 1954. The quantitative aspects of permanent and temporary sterility induced in female Habrobracon by x-rays and β-radiation. Radiation Research 1(3): 294-319. (NSA 8: 4483).

Grosch, D. S. and R. L. Sullivan. 1953. The fate of radiophosphorus ingested by Habrobracon females. Biol. Bull. 105(2): 296-307.

Grosch, D. S., R. L. Sullivan and L. E. LaChance. 1956. The comparative effectiveness of four beta-emitting isotopes fed to Habrobracon females on production and hatchability of eggs. Radiation Research 5(3): 281-289. (NSA 10: 11651).

Grosch, D. S., R. L. Sullivan and L. E. LaChance. 1957. Biological response to mixed radiations. Nucleonics 15(12): 64, 66. (NSA 12: 2666).

Gross, H. P., W. F. Baldwin and A. S. West. 1972. Introductory studies on the use of radiation in the control of black flies (Diptera: Simuliidae). Can. Entomol. 140(8): 1217-1222. (NSA 30: 15978).

Gross, M. G. 1972. Sediment-associated radionuclides from the Columbia River. pp. 736-754. In: The Columbia River Estuary and Adjacent Ocean Waters. (See Pruter and Alverson, 1972).

Gross, W. J., et al. 1953. The availability of radio-strontium to mammals by way of the food chain. Univ. of Calif., Los Angeles, U. S. AEC report UCLA-259. (NSA 7: 4717).

Grossbard, E. 1966. Potential use of tracer methods, especially autoradiography, in the study of the relation of herbicides to soil biology. pp. 173-189. In: Isotopes in Weed Research. (See IAEA, 1966)

Grossbard, E. 1969. A visual record of the decomposition of ^{14}C-labelled fragments of grasses and rye added to soil. J. Soil Sci. 20(1): 38-51.

Grossbard, E. and D. R. Stranks. 1959. Translocation of ^{137}caesium and ^{60}cobalt in fungi. Abstr. J. Gen. Microbiol. 20(1): iv.

Grossbard, E. and D. R. Stranks. 1959. Translocation of cobalt-60 and cesium-137 by fungi in agar and soil cultures. Nature 184(4683): 301-314.

Grossman, A. 1970. Beta-radioactivity of the tree pin in Sudety. Acta Geophys. Pol. 18: 235-242. (In Polish) (NSA 24: 50869) (Soil, moss, needles of pine, spruce, fir, and dwarf-mount-pine).

Grossman, A. and J. Kwapulinski. 1972. Measurement of the specific radioactivity of water with high salt contents. Przegl. Geofiz. 17(25): 155-164. (In Polish) (NSA 28: 20725).

Grossman, A., J. Kwapulinski and K. Lukasik. 1973. Concentration of ^{226}Ra in the region of Katowice. Nukleonika 18(10): 495-499. (In Polish) (NSA 29: 18325).

Grossman, H. H. 1970. Effect of chronic low level gamma radiation of Polytrichum commune Hedw. Thesis, Pennsylvania State Univ., University Park. 149pp. (Univ. Microfilms 71-6309) (NSA 25: 27075).

Grove, D. B. 1970. Method to describe the flow of radioactive ions in groundwater. Final report. Sandia Corp., Albuquerque, N. Mex., U. S. AEC report SC-CR-70-6139. 41pp. (NSA 25: 21158).

Grove, D. B., et al. 1969. Carbon-14 dates of ground water from a Paleozoic carbonate aquifer, south-central Nevada. pp. C215-C218. In: Geological Survey Research 1969. U. S. Geological Survey Prof. Paper 650C.

Grümm, H. 1972. Production of electricity from nuclear energy: fission product supply and effect on the environment. Österreichische Studiengesellschaft für Atomenergie GmbH, Seibersdorf, report SGAE-2030. 21pp. (In German) (For Conf. of the Electricity Production Assoc., Salzburg, Austria, June 1972) (CONF-720654-1) (NSA 27: 2472).

Grummitt, W. E. 1962. Environmental surveys for nuclear power stations. Atomic Energy of Canada, Ltd., Chalk River, report CRER-1120; AECL-1656. 14pp.

Grummitt, W. E. 1971. Monitoring of radioactive isotopes in environmental materials. pp. 399-403. In: Identification and Measurement of Environmental Pollutants. B. Westley, ed. National Research Council of Canada, Ottawa. (NSA 29: 2631).

Grummitt, W. E. and G. Lahaie. 1973. Method for the determination of strontium-90 and cesium-137 in river sediments and soils. Chalk River Nuclear Labs., Atomic Energy of Canada, Ltd., report AECL-4365. 15pp. (NSA 28: 15232).

Grummitt, W. E. and F. A. Prantl. 1974. Use of thermoluminescent dosimeters for studies of the snow cover. pp. 642-650. In: Advanced Concepts and Techniques in the Study of Snow and Ice Resources. (See Santeford and Smith, 1974) (NSA 33: 7313).

Grune, W. N., H. S. Atlas and G. J. Hamel. 1963-1965. Evaluation of fallout contamination of water supplies. Final report. Merrimack Coll., North Andover, Mass., U. S. Off. of Civil Defense report. 234pp. (NSA 19: 42658).

Grune, W. N., H. S. Atlas and G. J. Hamel. 1966. Evaluation of fallout contamination from surface runoff. Merrimack Coll., North Andover, Mass., U. S. Off. of Civil Defense report. 68pp. (AD-649533) (NSA 21: 41387).

Grune, W. N., D. D. Bartholomew and C. I. Hudson, Jr. 1958. Effects of radioactive materials on anaerobic digestion. I. Radiophosphorus. II. Radioiodine. Sewage and Industr. Wastes 30: 1123-1150, 1399-1410.

Grune, W. N., C. Chueh and C. H. Kaplan. 1963. Effects of C^{14} and Sr^{90} on anaerobic digestion. J. Water Poll. Control Fed. 35: 493-531. (NSA 17: 36064).

Grune, W. N., et al. 1964. Feasibility studies for the establishment of parameters for background radiation measurements. pp. 661-686. In: The Natural Radiation Environment. (See Adams and Lowder, 1964) (NSA 17: 32312).

Grüneberg, H. 1964. Genetical research in an area of high natural radioactivity in South India. Nature 204(4955): 222-224. (NSA 19: 2131) (Incl. wild rats).

Grüneberg, H., et al. 1965. A search for genetic effects of high natural radioactivity in South India. U. K. Medical Research Council, London, Spec. Rept. Ser. No. 307. 63pp. (NSA 20: 22619; 21: 39385).

Grüter, H. 1964. A selective enrichment of the fission product ^{137}Cs in mushrooms. Naturwissenschaften 51(7): 161-162. (In German) (NSA 18: 23183).

Grüter, H. 1967. Behavior of domestic fungus types in relation to the fission product ^{137}Cs. Z. Lebensm.-Untersuch. Forsch. 134: 173-179. (In German) (NSA 23: 29324, 24670).

Grüter, H. 1971. Radioactive fission product ^{137}Cs in mushrooms in West Germany during 1963-1970. Health Physics 20(6): 665-666. (NSA 25: 46688).

Grüter, H. 1973. Radiactive fission product ^{137}Cs in mushrooms in West Germany during 1963-1970. pp. 63-67. In: Chemical and Radionuclide Food Contamination. MSS Information Corp., New York. (NSA 29: 834).

Gruzdev, B. I. 1971. Method of adjoint analysis of some components of biogeocoenoses in the study of migration of radioactive elements. pp. 54-58. In: Methods of Radioecological Investigations. (See Verkhovskaya, 1971).

Gruzdev, B. I. and D. M. Rubtsov. 1972. Accumulation of thorium, uranium, and radium by plants and organogeneous horizons of soils. pp. 112-123. In: Radiological Investigations in Natural Biogeocoenoses. (See Verkhovskaya, 1972) (In Russian) (NSA 26: 56575).

Grzelak, K. and Z. Lassota. 1966. The effect of gamma-rays on the acetylcholine level in insects. Bull. Acad. Pol. Sci. (sci. biol.) 14: 391-395. (NSA 20: 42874).

Grzenda, A. R. and M. L. Brehmer. 1960. A quantitative method for the collection and measurement of stream periphyton. Limnol. Oceanogr. 5(6): 190-194. (SFA 5: 3404).

Grzenda, A. R., D. F. Paris and W. J. Taylor. 1970. The uptake, metabolism, and elimination of chlorinated residues by goldfish (*Carassius auratus*) fed a ^{14}C-DDT contaminated diet. Trans. Amer. Fish. Soc. 99(2): 385-396. (SFA 15: 12198).

Grzenda, A. R., W. J. Taylor and D. R. Paris. 1971. The uptake and distribution of chlorinated residues by goldfish (*Carassius auratus*) fed a ^{14}C-dieldrin contaminated diet. Trans. Amer. Fish. Soc. 100(2): 215-221. (SFA 16: 13543).

Grzbowska, D. 1967. ^{144}Ce + ^{144}Pr, ^{137}Cs and ^{90}Sr content in soil in Poland in 1964. Poland Inst. Nuclear Research, Warsaw, report CLOR-663/D. 26pp. (In Polish) (NSA 21: 36106).

Grzbowska, D. 1972. Uptake of ^{90}Sr and ^{137}Cs by plants from contaminated soils. Nukleonika 17(5-6): 330-340. (In Polish) (NSA 27: 17714).

Grzbowska, D. 1974. Uptake of ^{226}Ra by plants from contaminated soils. Nukleonika 19(1): 71-78. (In Polish) (NSA 30: 7176).

Grzbowska, D. and S. Wlodek. 1971. Contribution to the study of soil-plant transfer factors for ^{90}Sr, ^{137}Cs and ^{226}Ra. pp. 1069-1078. In: *Proc. Intern. Symp. Radioecology Applied to the Protection of Man and His Environment*. (See Commission of the European Communities, 1971) (In French) (NSA 27: 5572).

Gsal, L. 1961. Radioactive pollution of Hungary's surface waters. Hidrol. Kozlony 41(3): 266-268. (JPRS-10595).

Guardabassi, A. 1958. Preliminary observation on the metabolism of Ca^{45} in the larvae of *Bufo bufo bufo*. Monit. Zool. Ital. (Firenze) 66(2/3): 78-83. (In Italian) .

Guardabassi, A., I. Signorelli and P. Clerici. 1968. Oocytes in the testis of ^{131}I thyroidectomized adult toads. Ricerca sci. 38: 976-979. (In Italian) (NSA 23: 50842).

Guardabassi, A., et al. 1966. Uptake of ^{45}Ca on *Bufo bufo* larvae having no endolymphatic sacs. Ricerca sci. 36: 857-861. (In Italian) (NSA 21: 16280).

Gubicza, A. and I. Molnar. 1964. The effect of gamma irradiation on the *varo* race of *Bombyx mori* L. II. Irradiation of eggs in the early and late embryonic stages. Evkonyve 31: 3-13. (Magyar Tud. Akad. Tihanyi Biol. Kutatointez.) (In Hungarian) (NSA 20: 18345).

Guckert, A., T. Chone and F. Jacquin. 1975. Microflora and structural stability of soils. Rev. Ecol. Biol. Sol 12(1): 211-223. (In French) (NSA 26107) (Glucose ^{14}C).

Gude, W. D. 1968. *Autoradiographic Techniques: Localization of Radioisotopes in Biological Material*. Prentice-Hall, Engelwood Cliffs, N. J. xiii, 113pp.

Gudiksen, P. H. 1971. Semiannual report of environmental levels of radioactivity in the vicinity of the Lawrence Livermore Laboratory, January through June 1971. Lawrence Livermore Lab., U. S. AEC report UCRL-51093. 24pp. (NSA 25: 49179).

Gudiksen, P. H. and T. R. Crites. 1975. Comparison of gamma-ray exposure rate measurements at Bikini Atoll. Lawrence Livermore Lab., U. S. ERDA report UCRL-77532. 4pp. (NSA 33: 9352).

Gudiksen, P. H. and O. D. T. Lynch, Jr. 1975. Radioactivity levels in Eniwetok soil. Health Physics 29(1): 17-25. (NSA 32: 27940).

Gudiksen, P. H., et al. 1972. Environmental levels of radioactivity in the vicinity of the Lawrence Livermore Laboratory January through December 1971. Lawrence Livermore Lab., U. S. AEC report UCRL-51242. 47pp. (NSA 26: 45513).

Guedalia, D., et al. 1970. A study on ^{222}Rn emanation from soils. J. Geophys. Res. 75: 357-369. (NSA 24: 9521).

Guegueniat, P. and J. le Gall. 1972. Fixation of ruthenium-106 on various types of sediments. Centre de le Hague, France Commissariat à l'Énergie Atomique, Cherbourg, report CEA-CONF-2223. 2pp. (In French) (NSA 29: 10139).

Guelin, A. and P. Lepine. 1961. Comparative fixation of various radioactive phosphorus samples by bacteria. Ann. Inst. Pasteur 101: 281-284. (In French) (NSA 15: 28944).

Guenet, J.-L. and G. Legeay. 1968. Influence of biological and ecological factors on the radiosensitivity of laboratory animals. France Commissariat à l'Énergie Atomique, Fontenay-aux-Roses, report CEA-R-3409. 35 pp. (In French) (NSA 23: 902).

Guenthner, A. W. and G. W. Ware. 1967. Effects of x-irradiation on toxicity of malathion, heptachlor, and timik to the house fly. J. Econ. Entomol. 60(2): 369-373.

Guenzi, W. D. and W. E. Beard. 1967. Anaerobic biodegredation of DDT to DDD in soil. Science 156: 1116-1117. (^{14}C).

Guerin, P., J. Hugon and R. le Quinio. 1971. Use of activable tracers for studying the diffusion of atmospheric pollutants: industrial applications. pp. 267-282. In: *Nuclear Techniques in Environmental Pollution*. (See IAEA, 1971) (In French) (NSA 25: 32088).

Guerin de Montgareuil, P. 1974. Nuclear indicators in agronomy. Sci. Tech. (Paris) 11: 27-31. (In French) (NSA 30: 21167).

Gueron, J. 1953. Use of radioelements in hydrology. 1st Intern. Congr. of Speleologie 2: 301-306. (In French).

Guerra, A. A. 1972. Sterility induced in tobacco budworms by combination of reserpine and gamma irradiation affected by age and sex of pupae. J. Econ. Entomol. 65(5): 1282-1283. (NSA 26: 59119).

Guerra, A. A. 1975. Sexual sterilization of tobacco budworms with combination of oral chemosterilants and gamma irradiation. J. Econ. Entomol. 68(1): 1-3. (NSA 31: 23042).

Guerra, A. A., R. D. Garcia and H. H. de la Rosa. 1974. Suppression of population of tobacco budworms caged with sterile females. J. Econ. Entomol. 67(3): 333-334. (NSA 30: 15972) (γ irradiation).

Guerra, A. A., M. T. Ouye and H. R. Bullock. 1968. Effect of ultraviolet irradiation on egg hatch, subsequent larval development and adult longevity of the tobacco budworm and the bollworm. J. Econ. Entomol. 61(2): 541-542. (NSA 22: 36347).

Guerra, A. A., D. A. Wolfenbarger and M. J. Lukefair. 1971. Effects of sub-sterilizing doses of reserpine and gamma irradiation on reproduction of the tobacco budworm. J. Econ. Entomol. 64(4): 804-806. (NSA 25: 32088).

Guerra, A. A., et al. 1972. Competitiveness and behavior of tobacco budworms sterilized with reserpine and gamma irradiation. J. Econ. Entomol. 65(4): 966-969. (NSA 26: 50959).

Guha, T. and M. L. De. 1973. Radiation-induced alteration in toad erythrocyte and hemoglobin. Proc. Indian Nat. Sci. Acad. B39(4): 506-511. (NSA 31: 14390).

Guha, T. and M. L. De. 1974. Mortality of the toad <u>Bufo</u> <u>melanostictus</u> after whole-body x-irradiation. Int. J. Radiat. Biol. 26(1): 77-80. (NSA 31: 810).

Guilhamon, J. 1972. Nuclear power plants and the environment. Rev. Fr. Elec. 45(236): 8-13. (In French) (NSA 26: 56076).

Guillet, B. 1975. Iron podsol and iron-humus podsols: correlation between vegetation and vosgian podsols evolution. Rev. Ecol. Biol. Sol 12(1): 405-414. (In French) (NSA 33: 26016) (^{14}C).

Guillot, P. 1972. Some applications of radionuclides in hydrology, mining, and agriculture. pp. P29.p1-P29.p4. In: <u>National Conf. on the Technological Applications of Nuclear Techniques, 1972</u>. (See South Africa Atomic Energy Board, 1972) (CONF-721091) (NSA 28: 21122).

Guillou, R. B. 1964. The aerial radiological measuring surveys (ARMS) program. pp. 705-722. In: <u>The Natural Radiation Environment</u>. (See Adams and Lowder, 1964) (NSA 18: 35742).

Guillou, R. B., J. E. Hand and H. M. Borella. 1961. ARMS II-Las Vegas area aeroradioactivity survey. EG&G, Inc., Las Vegas, Nev., U. S. AEC report CEX-61.6.1. 29pp. (NSA 17: 41018).

Guinn, V. P. 1971. Activation analysis in environmental pollution studies including Hg, Pb, and pesticides. Radioisotopes (Tokyo) 20(11): 612-619. (In Japanese) (NSA 26: 50233).

Guinn, V. P. 1972. Determination of mercury by instrumental neutron activation analysis. pp. 87-97. In: <u>Mercury Contamination in Man and His Environment</u>. (See IAEA, 1972) (NSA 27: 11669) (Incl. fish).

Guinn, V. P. and S. C. Bellanca. 1969. Neutron activation analysis identification of the source of oil pollution of waterways. pp. 93-97. In: <u>Modern Trends in Activation Analysis</u>. <u>Vol</u>. <u>I</u>. (See DeVoe, 1969) (NSA 24: 113).

Guinn, V. P. and R. A. Schmitt. 1963. Determination of pesticide residues by neutron-activation analysis. Residue Rev. 5: 148-174.

Guinn, V. P., D. E. Bryan and H. R. Lukens. 1971. Trace-element characterization of crude oils and fuel oils via instrumental neutron activation analysis. pp. 347-359. In: <u>Nuclear Techniques in Environmental Pollution</u>. (See IAEA, 1971) (NSA 25: 31879).

Guizerix, J. 1973. Applications of tracer techniques to water transfer problems. pp. 27-34. In: <u>Water Pollution, Control, Regulating, Automation, Planning</u>. Assoc. Nationale de la Recherche Technique, Paris. (In French) (NSA 30: 24004).

Guizerix, J., et al. 1963. Apparatus for measuring the relative velocities of groundwaters by the point dilution method. pp. 25-35. In: <u>Radioisotopes in Hydrology</u>. (See IAEA, 1963) (In French) (NSA 18: 1913).

Guizerix, J., et al. 1963. Measures of effecting debits in France with the aid of radioactive tracers by the method of integration. pp. 255-282. In: <u>Radioisotopes in Hydrology</u>. (See IAEA, 1963) (In French).

Guizerix, J., et al. 1967. Tritium and hydrogeology: investigations and measurements carried out at the Grenoble Nuclear Research Center. pp. 433-450. In: <u>Isotopes in Hydrology</u>. (See IAEA, 1967) (In French) (NSA 22: 18937).

Guizerix, J., et al. 1967. Localization of leakages from a large reservoir using radioactive tracers. pp. 587-599. In: <u>Isotopes in Hydrology</u>. (See IAEA, 1967) (In French) (NSA 22: 19071).

Guizerix, J., et al. 1971. Tracer experiments performed in France as part of the study of thermal pollution of streams. pp. 63-79. In: <u>Nuclear Techniques in Environmental Pollution</u>. (See IAEA, 1971) (In French) (NSA 25: 32079).

Guizerix, J., et al. 1974. Analysis of information provided by natural or artificial tracers used to study aquifer systems in hydrogeology. pp. 377-404. In: <u>Isotope Techniques in Groundwater Hydrology</u>. <u>Vol</u>. <u>2</u>. (See IAEA, 1974) (In French) (NSA 31: 19664).

Gulbaba, G. 1974. Determination of heavy metals in trees by neutron activation analysis. Turkey Cekmece Nuclear Research and Training Center, Istanbul, report CNAEM-R-131. 12pp. (NSA 32: 13717).

Gulyakin, I. V. 1963. Influence of systematic fertilization of soils on the accumulation of radioisotopes in crops. Izv. Timiryazev. sel'sko. Akad. (6)::138-150. (In Russian) (NSA 19: 3832).

Gulyakin, I. V. and N. I. Seletkova. 1954. The uptake of radiothorium by plants. In: <u>The Effect of Ionizing Radiation on Biological Objects</u>. Acad. Sci., Moscow. (In Russian).

Gulyakin, I. V. and Ye. V. Yudentseva. 1956. Entry of the radioisotopes of strontium, cesium, ruthenium, zirconium and cerium into plants. Dokl. Akad. Nauk SSSR 111(1): 206-208. (In Russian).

Gulyakin, I. V. and Ye. V. Yudentseva. 1956. Uptake by plants of fission products from soil. In: The Behavior of Radioactive Fission Products in Soils, Their Uptake by Plants and Their Concentration in the Harvest. V. M. Klechkovskiy, ed. Publ. House Acad. Sci. USSR, Moscow. UN Document A/AC.82/G/R.41. (AEC-tr-2867).

Gulyakin, I. V. and Ye. V. Yudentseva. 1956. Supply of radioactive isotopes to the plant by way of the leaves. Dokl. Akad. Nauk SSSR 111: 709-712. (NP-tr-4) (NSA 12: 4066).

Gulyakin, I. V. and Ye. V. Yudentseva. 1958. Uptake of ^{60}Co by plants and its concentration in the harvest. Dokl. Akad. Nauk SSSR 123(2): 368-370. (In Russian).

Gulyakin, I. V. and Ye. V. Yudentseva. 1962. Radioactive Fission Products in Soil and Plants. Gosatomizdat, Moscow. 276pp. (In Russian).

Gulyakin, I. V. and Ye. V. Yudentseva. 1973. Agricultural Radiobiology. Kolos, Moscow. 272pp. (In Russian) (NSA 29: 24470).

Gulyakin, I. V., I. D. Dergunov and V. D. Moroz. 1968. Behavior of radioactive fission products in some soils of the sierozem zone of Uzbekistan. Izv. Timeryaz. sel'sko. Akad. (1): 84-93. (In Russian) (NSA 25: 42319).

Gunar, I. I., E. E. Krastina and A. E. Petrov-Spiridonov. 1958. Periodicity in absorption and excretion by roots. Proc. 2nd U. N. Conf. on the Peaceful Uses of Atomic Energy 27: 74-84.

Gunchev, Kh. and T. Dimchev. 1973. A neutron-activation analysis of lindan and DDT residual concentrations in soil samples. Pochvov. agrokhim. 3(1): 77-86. (In Bulgarian) (NSA 31: 13604).

Gunckel, J. E. and A. H. Sparrow. 1954. Aberrant growth in plants induced by ionizing radiation. pp. 252-279. In: Abnormal and Pathological Plant Growth. (Report of a Symposium, August 3-5, 1953). Brookhaven National Lab. Symp. in Biol. No. 6, U. S. AEC report BNL-258. (NSA 8: 5464).

Gunckel, J. E. and A. H. Sparrow. 1961. Ionizing radiations: biochemical, physiological and morphological aspects of their effects on plants. pp. 556-611. In: Encyclopedia of Plant Physiology. Vol. 16. W. Ruhland, ed. Springer-Verlag, Berlin. (NSA 21: 14764).

Gunckel, J. E., et al. 1953. Variations in the floral morphology of normal and irradiated plants of Tradescantia paludosa. Bull. Torrey Bot. Club 80: 445-456.

Gunckel, J. E., et al. 1953. Vegetative and floral development of irradiated and non-irradiated plants of Tradescantia paludosa. Am. J. Bot. 40(5): 317-332.

Gun-Aazhav, T., et al. 1973. Study of radioactivity of dinosaur bones with a high-resolution gamma spectrometer. Atom. Energiya 35(2): 130-132. (In Russian) (NSA 29: 15553).

Gunner, H. B., et al. 1966. The distribution and persistence of Diazinon applied to plant and soil and its influence on rhizosphere and soil microflora. Plant Soil 25(2): 249-264. (^{14}C).

Gunstrom, G. K. 1973. Gross physical anomalies in F_1 progeny of irradiated and unirradiated coho salmon. Trans. Amer. Fish. Soc. 102(2): 438-442.

Gunter, S. E. and H. I. Kohn. 1956. The effect of x-rays on the survival of bacteria and yeast. I. A comparative study of the dose-survival curves of Azobacter agile, Escherichia coli, Pseudomonas fluorescens, Raodopseudomonas spheroides, and Saccharomyces cerevisiae irradiated in the resting state. J. Bacteriol. 71(5): 571-581. (NSA 10: 9961).

Gunter, S. E., et al. 1956. The relative biological effectiveness of 180-KEVP and 22.5-MEVP x-rays determined from the dose-survival curve of Saccharomyces cerevisiae. Radiation Research 4(4): 326-338.

Günther, J. and O. Hug. 1974. Acute effects of x irradiation on membrane resistance of snail neurones. Radiat. Environ. Biophys. 11(3): 171-187. (In German) (NSA 31: 33524).

Gupta, V. K. and S. D. Soman. 1973. Environmental tritium contamination from nuclear power program. pp. 564-571. In: Tritium. (See Moghissi and Carter, 1973) (NSA 28: 27226).

Guptavanij, P. and C. E. Venard. 1965. A radiographic study of the oesophageal diverticula and stomach of Aedes aegypti (L.). Mosquito News 25 (3): 288-293.

Gurskiy, G. V., V. K. Lukashev and V. A. Tolkach. 1967. Uranium and radium in the zone of active water exchange of the northeast part of the Pripyat Basin. Dokl. Akad. Nauk Belorussk. SSR 11: 1017-1020. (In Russian) (NSA 22: 27905).

Gusarov, I. I. and V. K. Lyapidevskiy. 1963. A method for determining the natural radioactivity of air by products of radon fission. Gigiyena sanitariya 28(10): 10-16. (In Russian) (Fission?).

Gusev, N. G. 1957. Method for rapid determination of specific activity in water in large reservoirs. pp. 124-131. In: Proc. All-Union Conf. on Medical Radiology. Problems of Hygiene and Dosimetry. Medgiz, Moscow. (In Russian).

Gusev, N. G., et al. 1959. Collection of Radiochemical and Dosimetric Methods. Medgiz, Moscow. 460pp. (In Russian).

Guseynov, B. Z. and S. G. Guseynov. 1961. The accumulation and translocation of P^{32} in the white mulberry under the influence of microelements. pp. 262-267. In: Proc. Tashkent Conf. on Peaceful Uses of Atomic Energy, 1959. Ya. Kh. Turakulov, ed. Publ. House Acad. Sci. Uzbek. SSR, Tashkent. (AEC-tr-6293, pp. 451-461).

Guseynov, B. Z. and S. G. Guseynov. 1961. An investigation of the influence of microfertilizers on the intensity of photosynthesis in the white mulberry by the isotopic method. pp. 222-225. In: Proc. Tashkent Conf. on Peaceful Uses of Atomic Energy. Ya. Kh. Turakulov, ed. Publ. House Acad. Sci. Uzbek. SSR, Tashkent. (AEC-tr-6293, pp. 376-383).

Guseynov, D. M. and F. G. Iseyeva. 1962. The effect of radioactive phosphorus on the growth and development of lucerne. Dokl. Akad. Nauk Serb. SSR 18(7): 37-41. (In Russian).

Gus'kova, V. N. and A. N. Bragina. 1958. The hygienic characteristics of natural radioactivity of the soil. Gigiyena i sanitariya 23(10): 32-37. (In Russian).

Gus'kova, V. N. and A. N. Bragina. 1962. The problem of radioactivity of the soil (second communication). Gigiyena i sanitaryia 27(7): 103-105. (JPRS-15,920, pp. 42-46) (NSA 17: 1663, 6288).

Gus'kova, V. N., L. N. Gurfein and A. I. Tikhonova. 1966. Uranium action in a reservoir. Gidrobiol. zh. Akad. Nauk Ukr. SSR 2(6): 53-57. (In Russian) (NSA 23: 6218).

Gus'kova, V. N., et al. 1970. Effect of a mixture of uranium fission products on the sanitary conditions and hydrobionts of weakly-mineralized freshwater Basins. Gidrobiol. zh. Akad. Nauk Ukr. SSR 6(4): 5-11. (Hydrobiol. J. 6(4): 1-6) (NSA 25: 29449).

Gus'kova, V. N., et al. 1971. Dynamics of concentration of strontium-89 in water and carp tissues after a single contamination of a body of water. Gidrobiol. zh. Akad. Nauk Ukr. SSR 7(1): __. (Hydrobiol. J. 7(1): 53-56) (NSA 26: 9495).

Gus'kova, V. N., et al. 1973. Some data on comparative evaluation of radioactive and chemical pollution factors in freshwater reservoirs. pp. 194-199. In: Radioecology of Water Organisms. II. Distribution and Migration of Radionuclides in Freshwater and Seawater Biocenoses. (See Andrumaitis, 1973) (In Russian) (NSA 30: 21139).

Gus'kova, V. N., et al. 1973. Biological effect of ^{106}Ru and ^{131}I on water organisms. pp. 185-193. In: Radioecology of Water Organisms. II. Distribution and Migration of Radionuclides in Freshwater and Seawater Biocenoses. (See Andrumaitis, 1973) (In Russian) (NSA 30: 19010).

Guss, P. L. and T. F. Branson. 1972. The use of ^{75}Se in feeding studies with the corn leaf aphid (Hemiptera: Aphididae). Ann. Entomol. Soc. Am. 65(2): 303-306.

Gustaffson, A. 1954. Swedish mutation work in plants: background and present organization. Acta Agric. Scand. 4: 361-364.

Gustaffson, A. and M. Simak. 1958. Effect of x- and γ-rays on conifer seed. Medd. Stat. Skogsforskinst. 48(5): 20+pp.

Gustafson, P. F. 1965. ^{137}Cs in freshwater fish: human consumption and body burden considerations. Radiol. Hlth. Data 6: 626-629. (NSA 20: 18973).

Gustafson, P. F. 1966. Environmental radiation. Argonne Natl. Lab. Rev. 3: 67-74. (NSA 21: 4626).

Gustafson, P. F. 1967. Comments on radionuclides in aquatic ecosystems. pp. 853-858. In: Radiological Concentration Processes. (See Aberg and Hungate, 1967).

Gustafson, P. F. 1969. Cesium-137 in freshwater fish during 1954-1965. pp. 249-257. In: Symposium on Radioecology. (See Nelson and Evans, 1969)

Gustafson, P. F. 1973. Preparation of environmental impact statements. Trans. Nucl. Sci NS-20(1): 33-35. (NSA 28: 5594).

Gustafson, P. F. and S. S. Brar. 1964. Measurement of γ-emitting radionuclides in soil and calculation of the dose arising therefrom. pp. 499-512. In: The Natural Radiation Environment. (See Adams and Lowder, 1964). (NSA 18: 35740).

Gustafson, P. F., S. S. Brar and S. E. Muniak. 1966. ^{137}Cs in edible freshwater fish. Nature 211: 843-844. (NSA 20: 45507).

Gustafson, P. F., S. S. Brar and D. M. Nelson. 1969. Summary of ^{137}Cs in the Chicago area 1965-1968. pp. 178-180. In: Fallout Program Quarterly Summary Report. (See Hardy and Rivera, 1969

Gustafson, P. F., et al. 1967. Radioactive ^{54}Mn and ^{65}Zn in euryhaline fish. Can. J. Zool. 45(5): 729-735.

Gustafsson, M. 1969. External and internal irradiation of Swedish reindeer breeders 1950-1970. Health Physics 17: 19-34. (NSA 23: 38562).

Gustafsson, M. and K. Lidén. 1969. Total external and internal irradiation of reindeer-breeding Lapps over the 20-year period 1951-1970. Nord. Med. 81: 429-433. (In Swedish) (NSA 24: 4388).

Gutierrez, J. 1973. Comparative study of several strains of the Tetranychus neocaledonicus complex and sterilization tests of males by x rays. Entom. Expt. Appl. 16(1): 123-134. (NSA 30: 19010).

Gutierrez R., C. E. 1974. The use of tracer techniques in water bearing. Mexico Inst. Politecnico Nacional, Mexico City. 76pp. (In Spanish) (INIS-mf-1754) (NSA 33: 26015).

Guthrie, J. E. 1960. Radiostrontium and radiocesium in Ottawa River water (November 1959 to May 1960). Atomic Energy of Canada, Ltd., Chalk River, report CRER-962; AECL-1092. 9pp.

Guthrie, J. E. 1962. Radioactive factors in water pollution. Can. J. Chem. Eng. 40: 273-276. (NSA 17: 6334).

Guthrie, J. E. 1963. Radiostrontium, radiocesium and total-β activities in Ottawa River water. Atomic Energy of Canada, Ltd., Chalk River, report AECL-1765. 34pp. (NSA 17: 27585).

Guthrie, J. E. 1964. Radioactivity in Winnipeg River water, October 1961 to September 1963. (Preliminary results from WNRE pre-operational environmental survey). Atomic Energy of Canada, Ltd., Chalk River, report AECL-1874. 15pp. (NSA 18: 18038).

Guthrie, J. E. 1965. The occurrence of sodium-22 in some species of Canadian biota. Can. J. Zool. 43(5): 889-890. (NSA 20: 4161; WR 120: 6).

Guthrie, J. E. and R. A. Brust. 1969. Elimination rate of radiocesium by an aquatic hemipteran *Lethocerus americanus*. Can. Entomol. 101(8): 856-861.

Guthrie, J. E. and R. A. Brust. 1973. Effects of chronic radiation exposure on mosquitoes (Diptera: Culcidae). I. Effects of rearing in ^{90}Sr + ^{90}Y solutions. pp. 1174-1178. In: *Radionuclides in Ecosystems*. (See Nelson, 1973) (NSA 28: 30239).

Guthrie, J. E. and A. Z. Burzynski. 1972. Accumulation of ^{137}Cs by *Aedes aegypti* larvae and the effect of potassium concentration. Can. Entomol. 104(3): 411-416. (NSA 26: 56567).

Guthrie, J. E. and W. E. Grummitt. 1963. Manual of low-activity environmental analyses. Procedures and methods for preparation and radiochemical analyses of environmental samples at AECL. Atomic Energy of Canada, Ltd., Chalk River, report AECL-1745. v, 112pp. (NSA 17: 32282).

Guthrie, J. E. and A. G. Scott. 1969. Measurement of radiation dose distribution in a pond habitat by lithium fluoride dosimetry. Can. J. Zool. 47(1): 17-20.

Gutknecht, J. 1961. Absorption, accumulation and loss of radioactive scandium by macroalgae. Abstr. Biol. Bull. 121(2): 374-375.

Gwiazdowski, B. 1974. Environmental gamma backgroud radiation measurements in Poland. pp. 570-589. In: *Radiation Protection, Environment, and the Population. Vol. 2*. (See German Democratic Republic, 1974) (AED-CONF-74-725-000(Vol. 2) (NSA 33: 25981).

Gyurdzhian, A. A. 1962. The biological effect of cosmic radiation. Iskusstv. Sputniki Zemli Akad. Nauk SSSR 12: 77-104. (In Russian) (NSA 17: 28735).

Haahr, V. 1963. The use of radioisotopes for soil physical measurements. The neutron scattering method for measuring soil moisture. Grundforbattring 4: 295-315. (In Danish) (NSA 19: 11405).

Haahr, V. 1975. Nuclear methods for detecting root activity. pp. 57-63. In: *Tracer Techniques for Plant Breeding*. International Atomic Energy Agency, Vienna, Publ. STI/PUB-419. (NSA 33: 26330).

Haak, E., A. Eriksson and F. Karlstrøm. 1973. Studies on plant accumulation of fission products under Swedish conditions. XIII. Entry of ^{90}Sr and ^{137}Cs into herbage of contrasting types of pasture. XIV. Uptake of ^{137}Cs by wheat and timothy from six different soils as influenced by rate of K-fertilization and by type and rate of N-fertilization in pot experiments. XV. Uptake of ^{137}Cs by wheat as influenced by type and rate of N-fertilization under field conditions. Sweden Försvarets Forskningsanstalt, Stockholm, report FOA-4-4525-A-3, FOA-4-4557-A-3, FOA-4-4559-A-3. 46, 39, 28pp. (NSA 27: 20174, 29: 10495, 10496).

Habashi, F. and T. Schöfeld. 1960. Determination of "fresh" fission products in the biosphere through gamma spectrometry. Atompraxis 6(10/11): 414-415. (In German) (Rainwater).

Haber, A. H. 1968. Ionizing radiations as research tools. Ann. Rev. Plant Physiol. 19: 463-489. (NSA 25: 10997).

Haber, A. H. 1972. Ionizing radiation effects on higher plants. pp. 231-243. In: *Concepts in Radiation Cell Biology*. G. L. Whitson, ed. Academic Press, New York. (NSA 26: 41091).

Haberer, K. 1962. Criteria for estimating the dangers to drinking water by radioactive fallout. Kerntechnik 4(7): 286-287. (In German) (NSA 16: 28733).

Haberer, K. 1963. Activity losses by evaporation during fallout tests. Atomkernenergie 8(3): 102-113. (In German) (NSA 17: 19802).

Haberer, K. 1969. *Radionuclides in Water. Their Distribution, Application, Measurement and Removal*. Thiemig-Taschenbücher, Band 17. Verlag Karl Thiemig KG, Munich. 319pp. (In German) (NSA 23: 40866).

Haberer, K. and H. Bernhardt. 1965. Studies on the presence of manganese-54 in a drinking water reservoir. Atompraxis 11(1): 27-33. (In German) (NSA 19: 15792).

Haberer, K. and U. Stürzer. 1968. Rapid estimation of radioiodine in water samples. Gas-Wasserfach 109: 1287-1289. (In German) (NSA 23: 17348).

Haberer, K. and U. Stürzer. 1971. Rapid method for the determination of radiostrontium in water. Gas-Wasserfach Wasser Abwasser 112(4): 186-187. (In German) (NSA 28: 2225).

Haberer, K. and U. Stürzer. 1971. Testing and comparison of radiation detection instruments for the monitoring radioactive contamination of water. German F. R. Rheinwasser-Aufbereitungswerk, Wiesbaden-Schierstein. (In German) (NP-19938, Paper 12) (NSA 29: 10496).

Haberer, K. and U. Stürzer. 1972. Rapid determination of radioactive cerium in water samples. Gas-Wasserfach Wasser Abwasser 113(3): 122-123.

Haberer, K. and U. Stürzer. 1974. Rapid analysis of radioactive ruthenium in water samples. Gas-Wasserfach Wasser-Abwasser 115(4): 181-183. (In German) (NSA 31: 5591).

Haberer, K. and W. Weindel. 1966. A rapid method for separation of radiocesium from water samples. Gas-Wasserfach 107: 892-896. (In German) (NSA 23: 40847).

Haccius, B. and H. Reichert. 1964. Restitution phenomena in plant meristems after x irradiation. II. Adventive embryony after seed irradiation of *Eranthis hiemalis*. Planta 62: 355-372. (In German) (NSA 19: 15145).

Hachihama, Y. and S. Takamuku. 1960. The effects of ionizing radiation on lignin and wood. Kogyo Kagaku Zasshi 63(6): 1043-____.

Hacke, J. and W. Jacobi. (Eds.). 1969. Radiation Protection Problems in the Release and Incorporation of Radioactive Materials. (4th Ann. Meeting Assoc. of Specialists for Radiation Protection, Berlin, 28-30 May 1969). Hahn-Meitner Inst. für Kernforschung, Berlin. 638pp. (In German, English) (CONF-690540) (NSA 24: 27593).

Hackman, R. H. and K. N. Saxena. 1964. Tyrosine metabolism in blowfly larvae at pupation. Austral. J. Biol. Sc. 17(3): 803-805. (^{14}C, Lucilia cuprina).

Haddad, H. M., S. P. Lucas and H. Haddad. 1974. Experimental endocrine exophthalmos. I. Effect of TSH on S35 sulfate uptake. Ann. Ophthalmol. 6: 721-725. (NSA 32: 17576) (Goldfish).

Hadley, E. B. and G. M. Woodwell. 1964. The productivity of natural communities. III. Effects of ionizing radiation on rates of metabolism in pine under controlled conditions. Abstr. Bull. Ecol. Soc. Am. 45(3): 74.

Hadley, E. B. and G. M. Woodwell. 1965. Effects of ionizing radiation on rates of CO_2 exchange of pine seedlings. Radiation Research 24(4): 650-656. (Also: Brookhaven National Lab., U. S. AEC report BNL-8414) (NSA 19: 24346).

Hadley, N. F. 1973. Effects of gamma radiation on the respiratory metabolism of adult beet armyworms, Spodoptera exigue (Hübner) (Lepidoptera: Noctuidae). pp. 1163-1168. In: Radionuclides in Ecosystems. (See Nelson, 1973) (NSA 28: 30238).

Haffner, J. W. 1963. Literature survey on world isotope and radiation technology. Illinois Inst. of Technology, Chicago, report IITRI-1194-13. 66 pp. (NSA 17: 37004).

Hagan, M. P. and H. S. Ducoff. 1975. Noncytokinetic injury: sparing effect of dose fractionation for delay of pupariation of the fleshfly, Sarcophaga bullata Parker. Radiation Research 63(2): 378-381. (NSA 32: 22633).

Hagen, K. S., L. Santas and A. Tsecouras. 1963. A technique of culturing the olive fly, Dacus oleae Gmel. on synthetic media under xenic conditions. pp. 333-355. In: Radiation and Radioisotopes Applied to Insects of Agricultural Importance. (See IAEA, 1963).

Häggquist, G. and G. Hevesy. 1958. The combined effect of heavy water and roentgen irradiation on the animal organism. Acta Radiologica 49(5): 321-325. (Used fish).

Haghiri, F. 1970. Fate of ^{90}Sr in soil, plants, and water. Ohio Rep. 55: 74-77. (NSA 25: 16085).

Haghiri, F. 1972. Detailed characterization of soil and vegetation on selected sites to serve as basis for future evaluation of effect of radioactive contamination. Final technical report. Ohio Agricultural Research and Development Center, Wooster, U. S. AEC report COO-414-18. 69pp. (NSA 27: 11976).

Haghiri, F. and R. R. Johnson. (Eds.). 1963. Radionuclides in Foods and Agricultural Products. (Proc. North Central Experiment Stations Workshop, Cincinnati, O., February 19-21, 1963). Ohio Agricultural Experiment Station, Wooster, Special Rept. Ser. No. 1. iv, 145pp.

Haghiri, F. and J. D. Sayre. 1961. ^{90}Sr uptake by plants as influenced by soil types and liming. Soil Sci. Soc. Am. Proc. 25: 120-123. (NSA 20: 6745).

Haglund, D. S., G. M. Friedman and D. S. Miller. 1969. The effect of freshwater on the redistribution of uranium in carbonate sediments. J. Sediment. Petrol. 39: 1283-1296. (NSA 24: 14094).

Hahn, P. F., V. H. Haas and A. Wilcox. 1950. Arrest of development of Plasmodium gallinaceum in mosquitoes (Aedes aegypti) by radiation effect of P^{32}. Science 111(2894): 657-658.

Hahn, R. B. and T. M. Schmitt. 1969. Determination of phosphate in water with ^{185}W. Anal. Chem. 41: 359-360. (NSA 23: 13648).

Hahn, R. B. and C. P. Straub. 1955. Determination of radioactive strontium and barium in water. J. Am. Water Works Assoc. 47: 335-340.

Haigh, M. V. and A. Howard. 1973. Effect of x rays on the growth of the first rhizoid in spore of Osmunda regalis. Radiation Botany 13(1): 37-45. (NSA 27: 27886).

Haigh, M. V. and A. Howard. 1973. Radiation-induced tumorous outgrowths in young gametophytes of Osmunda regalis. Radiation Botany 13(2): 111-119. (NSA 28: 5952).

Hainberger, P. L., et al. 1974. Radioactivity in Brazilian mineral waters. Radiat. Data Repts. 15(8): 483-488. (NSA 30: 29413).

Haines, T. H., S. M. Henry and R. J. Block. 1960. The sulfur metabolism of insects. V. The ability of insects to use sulfate in the synthesis of methionine. Contr. Boyce Thompson Inst. 20(6): 363-364. (^{35}S).

Hairr, L. M. and M. E. Wrenn. 1972. Environmental sampling for river sediments around a nuclear power station. Trans. Amer. Nucl. Soc. 15(1): 82-83. (NSA 26: 42965).

Haisch, A. 1970. Some observations on decreased vitality of irradiated Mediterranean fruit fly. pp. 71-75. In: Sterile-Male Technique for Control of Fruit Flies. (See IAEA, 1970) (NSA 25: 29741).

Haisch, A. 1975. Observations on the flying behavior of the European cherry fruit fly (Rhadoletis cerasi L.). pp. 191-199. In: Sterility Principle for Insect Control, 1974. (See IAEA, 1975) (NSA 32: 11775) (Irradiated, non-irradiated insects).

Haisch, A. and E. F. Boller. 1971. Genetic control of the European cherry fruit fly, Rhagoletis cerasi L. Progress report on rearing and sterilization. pp. 67-76. In: Sterility Principle for Insect Control or Eradication. (See IAEA, 1971) (NSA 29741).

Haisch, A. and A. Süss. 1966. Use of radionuclides and their radiation in plant pest control. Kerntechnik 8: 514-517. (In German) (NSA 21: 4893).

Hajdu, E., L. D. Fairbanks and G. E. Burch. 1965. The effects of injury on the humidity reaction of Drosophila melanogaster. Physiol. Zool. 38(3): 197-201. (^{22}Na).

Hajduković, M., M. Jakovljević and V. Vajgand. 1967. Comparison of three methods of measuring soil moisture. pp. 117-124. In: Isotope and Radiation Techniques in Soil Physics and Irrigation Studies. (See IAEA, 1967) (NSA 22: 23349).

Hajeck, B. F. 1966. Soil survey--Hanford Project in Benton County, Washington. Battelle Northwest Lab., U. S. AEC report BNWL-243. 19pp. (NSA 20: 24818).

Haji-Vassiliou, A. 1970. The association of uranium with naturally occurring organic materials in the Colorado Plateau and other areas. Thesis, Columbia Univ., New York, N. Y. 265pp. (NSA 24: 16712).

Hajnal-Papp, M. and A. Niedetsky. 1967. Effect of ionizing radiation on the center of automatism of frog heart. Acta Biochim. Biophys. (Budapest) 2: 199-204. (NSA 22: 15059).

Haken, J. W. and J. J. Hardon. 1964. Radioactivity control of the Keuringsdienst van Earen, Amsterdam. Deut. Lebensm.-Rundsch. 60: 205-207. (In German) (NSA 20: 20949).

Häkkinen, U. and E. Lakanen. 1968. ^{90}Sr and ^{137}Cs in some Finnish soil profiles. Ann. Agr. Fenn. 7: 123-126. (NSA 23: 26781).

Häkkinen, U. L. and E. Lakanen. 1974. Adsorption and extraction of ^{85}Sr and ^{137}Cs in Finnish soils. pp. 249-258. In: Comparative Studies of Food and Environmental Contamination. (See IAEA, 1974) (NSA 30: 15433).

Hakonson, T. E. 1967. Tissue distribution and excretion of ^{134}Cs in the mule deer. Thesis, Colorado State Univ., Fort Collins. 121pp.

Hakonson, T. E. 1972. Cesium kinetics in a montane lake ecosystem. Thesis, Colorado State Univ., Fort Collins. 154pp. (Neutron activation analysis).

Hakonson, T. E. 1975. Environmental pathways of plutonium into terrestrial plants and animals. Health Physics 29(4): 583-588. (NSA 33: 9289).

Hakonson, T. E. and L. J. Johnson. 1973. Distribution of environmental plutonium in the TRINITY Site ecosystem after 27 years. Los Alamos Scientific Lab., U. S. AEC report LA-UR-73-1291. 6pp. (NSA 29: 374).

Hakonson, T. E. and W. R. Rickard. 1969. Radionuclide deposition and elimination in a darkling beetle. Northwest Sci. 43: 23-28. (NSA 23: 31629).

Hakonson, T. E. and F. W. Whicker. 1969. Uptake and elimination of ^{134}Cs by mule deer. pp. 616-622. In: Symposium on Radioecology. (See Nelson and Evans, 1960) (Also: Colorado State Univ., Fort Collins, U. S. AEC report COO-1156-24) (NSA 22: 19234).

Hakonson, T. E. and F. W. Whicker. 1971. Tissue distribution of radiocesium in the mule deer. Health Physics 21(6): 864-866. (NSA 26: 9462).

Hakonson, T. E., A. F. Gallegos and F. W. Whicker. 1973. Use of cesium-133 and activation analysis for measurement of cesium kinetics in a montane lake. pp. 344-348. In: Radionuclides in Ecosystems. (See Nelson, 1973) (NSA 26: 38129).

Hakonson, T. E., A. F. Gallegos and F. W. Whicker. 1975. Cesium kinetics data for estimating food consumption rates of trout. Health Physics 29 (2): 301-306. (NSA 32: 24536).

Hakonson, T. E., L. J. Johnson and W. D. Purtymun. 1973. Ecodistribution of plutonium in liquid waste disposal areas at Los Alamos. Los Alamos Scientific Lab., U. S. AEC report LA-UR-73-1001. 23pp. (For 3rd. Intern. Congr. Intern. Radiation Protection Assoc.) (NSA 28: 18231).

Hakonson, T. E., L. J. Johnson and W. D. Purtymun. 1973. Distribution of plutonium in liquid waste disposal areas at Los Alamos. Los Alamos Scientific Lab., U. S. AEC report LA-UR-73-1309. 6pp. (NSA 29: 2629).

Hakonson, T. E., J. W. Nyhan and W. D. Purtymun. 1975. Accumulation and transport of soil plutonium in liquid waste discharge areas at Los Alamos. Los Alamos Scientific Lab., U. S. ERDA report LA-UR-75-2061. (NSA 33: 5585).

Hakonson, T. E., et al. 1973. Ecological investigation of radioactive materials in waste discharge areas at Los Alamos for the period July 1, 1972 through March 31, 1973. Los Alamos Scientific Lab., U. S. AEC report LA-5282-MS. 47pp. (Incl. list of biota onsite).

Halbertatädter, L. 1914. Biological action of radiation on trypanosomes. Experimental research. Klin. Woch., Berlin. pp. 252-253. (In German).

Halbertstädter, L. and L. Back. 1942. The effect of x-rays on single colonies of Pandorina . Brit. J. Radiol. 15(172): 124-128.

Halbertstädter, L., G. Goldhaber and O. Hecht. 1943. The effect of x-rays on the development of insects. I. Irradiation in the larval stage. Growth 7(4): 413-425.

Haley, R. J., et al. 1960. Response of the kangaroo rat (Dipodomys merriami Mearns) to single whole-body x-irradiation. Radiation Research 12: 103-111. (Also: Univ. of Calif., Los Angeles, U. S. AEC report UCLA-440) (NSA 14: 6163).

Haley, T. J. and R. S. Snider. (Eds.). 1964. Response of the Nervous System to Ionizing Radiation. (2nd Intern. Symp. held at Univ. of Calif., Los Angeles). Little, Brown & Co., Boston. xx, 749pp.

Halfon, E. 1974. Systems identification: a theoretical method applied to tracer kinetics in aquatic microcosms. pp. 262-265. In: Structure, Functioning, and Management of Ecosystems. The Netherlands, PUDOC, Wageningen. (NSA 33: 14846) (Radionuclide kinetics).

Halkka, O. 1965. X-ray-induced changes in the chromosomes of Limotettix (Homoptera). Chromosoma 16: 185-191. (NSA 19: 27985).

Halkka, O. and L. Heinonen. 1967. Chromosome breakage associated with infection. II. Stained sections. Hereditas 58(1-2): 253-261. (^{3}H).

Hall, R. B. and R. H. Wilson. 1962. Evaluation of discharging radioactive wastes into freshwater streams. Hanford Labs., U. S. AEC report HW-SA-2748. (NSA 17: 2641).

Hall, R. M., Jr. 1970. Thermoluminescent dosimetry. Ch. C-3, pp. C37-C40. In: A Tropical Rain Forest. (See H. T. Odum and R. F. Pigeon, 1970).

Hall, R. M., Jr. and D. I. Ross. 1974. Environmental dose measurements in the vicinity of nuclear facilities. Trans. Nucl. Sci. NS-21(1): 451-455. (NSA 30: 6508).

Hall, W. E. and Y.-P. Sun. 1965. Mechanism of detoxication and synergism of bidrin insecticide in house flies and soil. J. Econ. Entomol. 58(5): 845-849. (^{14}C, Musca domestica).

Hallab, A. H. 1968. Detoxification of pesticidal residues in fish and shell fish. Thesis, Louisiana State Univ., Baton Rouge. 94pp. (NSA 24: 5980).

Hallam, A. and N. B. Price. 1968. Further notes on the strontium contents of unaltered fossil cephalopod shells. Geol. Mag. 105: 52-55. (NSA 23: 8052).

Haller, H. L. 1956. Radioactive materials in insect control. pp. 215-221. In: Radioactive Isotopes in Agriculture. (See U. S. AEC, 1956)

Haller, H. L. 1961. The use of atomic energy to study agricultural chemical residues. pp. 147-160. In: Applications of Radioisotopes and Radiation in Life Sciences. (See U. S. Congress, 1961) (NSA 15: 27315).

Haller, W. A., et al. 1969. The instrumental determination of fifteen elements in plant tissue by neutron activation analysis. pp. 177-183. In: Modern Trends in Activation Analysis. Vol. I. (See Devoe, 1969) (NSA 24: 123).

Hallinan, E. and K. S. Rai. 1973. Radiation sterilization of Aedes aegypti in nitrogen and implications for sterile male technique. Nature 244(5415): 368-369. (NSA 28: 30255).

Halliwell, R. S. and R. Langston. 1965. Effects of gamma irradiation on symptom expression of barley stripe mosaic virus disease and on two viruses in vivo. Phytopathology 55(9): 1039-1040.

Halm, E., W. Herbst and A. Mastrocola. 1962. Measurement of natural radiation gages in Switzerland. Bull. Eidg. Gesundheitsamtes B(6): 1-35. (In German).

Halvas, J. 1963. Study of radioactivity in the Mexican Republic. Program of radiologic protection. Salud Publ. Mex. 5: 903-910. (In Spanish) (NSA 20: 9155).

Halvey, E. 1966. Some health physics implications in the choice of radioactive tracers for hydrological investigations. Health Physics 12(8): 1145-1148. (NSA 20: 39155).

Hamaguchi, H., et al. 1975. Neutron activation analysis for uranium in fossil bones. Anal. Chim. Acta 75(2): 445-448. (NSA 32: 5871).

Hamaguchi, S. and N. Egami. 1975. Post-irradiation changes in oocyte populations in the fry of the fish Oryzias latipes. Int. J. Radiat. Biol. 28(3): 279-284. (NSA 33: 17458).

Hamar, M., G. Suteu and M. Sutova. 1963. "Home range" studies in rodents by marking with ^{32}P. Rev. Biol. 8: 431-446.

Hamar, M., G. H. Suteu and M. Sutova. 1964. Home range and activity study of the mole rat (Spalax leucodon Nordm.) by Co^{60} marking. Rev. Roumaine biol. (zool.) 9: 421-433.

Hamatova, E. 1964. Influence of gamma radiation on the nitrogen fixation activity of Rizobium japonicum in symbiosis with two varieties of soy beans. Acta Microbiol. Polon. 13: 247-254. (NSA 19: 12963).

Hamatova, E. 1969. Influence of chronic gamma-irradiation on nodulating soya. Rostinna Vyroba 15: 197-200. (In Czech) (NSA 24: 51099).

Hamilton, E. W. 1965. Metabolism of ^{36}Cl labeled aldrin and dieldrin in aldrin resistant western corn rootworm. Abstr. 48. Bull. Entomol. Soc. Am. 11(3): 157.

Hamilton, H. E. and G. S. Melville, Jr. 1968. Stress response as a predictor of radiosensitivity in primates. School of Aerospace Medicine, Brooks AFB, Tex., report SAM-TR-68-122. 12pp. (AD-682285; NSA 23: 31900).

Hamilton, J. R. 1963. Characteristics of tracheids produced in a gamma and gamma-neutron environment. Forest Products J. 12(5): 213-227; 13: 62-67. (NSA 18: 36976).

Hamilton, J. R. 1964. Termination report on radiation effects on living trees. Final report. U. S. AEC report. 51pp. (NSA 18: 33282).

Hamilton, J. R. 1964. The influence of ionizing radiation on xylem derivatives of two angiospermous species. Abstr. Bull. Ecol. Soc. Am. 45 (3): 80.

Hamilton, J. R. 1970. Characteristics of xylem tissues exposed to ionizing radiation. Final report. West Virginia Univ., U. S. AEC report ORO-3400-5-a. 8pp. (NSA 24: 44622).

Hamilton, J. R. and A. H. Chesser. 1969. Effects of acute ionizing radiation on the xylem fibers of Quercus alba and Liquidambar styraciflua. Radiation Botany 9(4): 331-339.

Hamilton, J. R. and A. Clark, III. 1970. The influence of ionizing radiation on tracheid morphology and tensile strength in shortleaf pine latewood. Wood Science 3(1): 59-61.

Hamilton, L. 1969. Changes in survival after x-irradiation of Xenopus embryos at different phases of the cell cycle. Radiation Research 37: 173-180. (NSA 23: 16414).

Hamilton, M. A. 1935. Further experiments on the artificial feeding of Myzus persicae (Sulz). Ann. Appl. Biol. 22: 243-258.

Hampl, W., H. Altmann and R. Biebl. 1971. Variation in and effects of repair capacity of Chlorella cells at different stages of the cell cycle. Radiation Botany 11(3): 201-207. (In German) (NSA 25: 47050).

Hampson, M. A. 1967. Uptake of radioactivity by aquatic plants and location in the cells. I. The effect of p_H on the ^{90}Sr and ^{90}Y uptake by the green alga Ulva lactuca and the effect of stable yttrium on ^{90}Y uptake. II. Uptake of ^{144}Ce by the freshwater plant Nitella opaca. J. Exptl. Botany 18: 17-33, 34-53. (NSA 22: 2487, 2488).

Hancock, R. L. 1962. Lethal doses of irradiation for Lumbricus. Life Sci. (11): 625-628. (NSA 17: 35532).

Hancock, R. L. 1965. Irradiation-induced neoplastic and giant cells in earthworms. Experientia 21: 33-34. (NSA 19: 15072).

Hand, J. E. 1964. Instrumentation for aerial surveys of terrestrial γ radiation. pp. 687-704. In: The Natural Radiation Environment. (See Adams and Lowder, 1964) (NSA 18: 35892).

Hand, J. E. and V. F. Weissman. 1968. Humboldt Bay environmental radiation survey (ARMS-II). EG&G, Inc., Las Vegas, Nev., U. S. AEC report CEX-67.6. 2. 39pp. (NSA 24: 29389).

Handley, R. and K. L. Babcock. 1970. Translocation of carrier-free ^{85}Sr, ^{137}Cs, and ^{106} in woody plants. Radiation Botany 10: 577-583. (NSA 25: 10955).

Handley, T. H. and V. A. DeCarlo. 1972. Mercury and cadmium determination in water by prompt (n,γ) method: ^{252}Cf as the neutron source. J. Radioanal. Chem. 11(2): 265-271. (NSA 27: 28).

Handreck, K. and D. S. Riceman. 1969. Cobalt distribution in several pasture species grown in culture solutions. Austral. J. Agr. Res. 20: 213-226. (NSA 23: 38413).

Hanec, W. 1956. A study of the environmental factor affecting the dispersion of house flies (Musca domestica L.) in a dairy community near Fort Whyte, Manitoba. Can. Entomol. 88(6): 270-272.

Haney, W. A. 1960. Chemical effluents technology waste disposal investigations, January, February, March 1960. --- July, August, September 1960. General Electric Co., U. S. AEC report HW-65464-RD, HW-67753-RD. 17, 20p pp. (NSA 29: 12709, 12710).

Haney, W. A. and C. E. Lideroth. 1959. Exploratory field study of a ground waste disposal facility. General Electric Co., U. S. AEC report HW-60115. 22pp. (NSA 29: 18348).

Haney, W. A., D. J. Brown and A. E. Reisenauer. 1962. Fission product tritium in separations wastes and in the ground water. General Electric Co., U. S. AEC report HW-74536. 13pp. (NSA 16: 30156, 29: 12713).

Hanford Atomic Products Operation, General Electric Co. nd. Evaluation of radiological conditions in the vicinity of Hanford for 1960. U. S. AEC report TID-13834. 12pp. (NSA 15: 31029).

Hanford Atomic Products Operation, General Electric Co. 1955. Earth Sciences' waste disposal monitoring activities summary, November 1955. --- December 1955. U. S. AEC report HW-42700, HW-43148. 12, 14pp. (NSA 29: 12708, 12725).

Hanker, I., L. Taimr and A. Kudelova. 1975. Oxygen flask combustion of plant tissues for ^{14}C activity measurements. Radioisotopy 16(2): 181-194. (In Czech) (NSA 32: 5916).

Hannah-Alava, A. 1968. Sex-ratio depression as a brood-pattern criterion of radiation damage in Drosophila melanogaster. Genetica 39: 519-543. (NSA 23: 20338).

Hannerz, L. 1958. Atomic waste and fisheries. Svensk Fisktidskr. 67: 133-135. (In Swedish).

Hannerz, L. 1968. Accumulation, retention and elimination of ^{65}Zn in fresh water organisms studied in pond experiments. Abstr. In: Radiation Protection, Part 1. (See Snyder, et al., 1968).

Hansborough, L. A. and D. Denny. 1951. Distribution of phosphorus^{32} in embryo and larva of the frog. Proc. Soc. Exptl. Biol. Med. 78(2): 437-441.

Hansbury, E., et al. 1961. Contemporary carbon^{14} in the biosphere. Los Alamos Scientific Lab., U. S. AEC report LAMS-2526. pp. 222-244. (NSA 15: 23278).

Hansen, R. L. and G. A. Teter. 1970. Use of radioisotopes in tracing reservoir leakage at Anchor Dam. U. S. Bureau of Reclamation, Denver, Colo., U. S. AEC report TID-25396. 28pp. (NSA 25: 6426).

Hansen, R. O. 1970. Radioactivity of a California terrace soil. Soil Science 110(1): 31-__.

Hansen, R. O. and G. L. Huntington. 1969. Thorium movements and morainal soils of the High Sierra, California. Soil Science 108: 257-265. (NSA 24: 2308).

Hansen, R. O. and P. R. Stout. 1968. Isotopic distributions of uranium and thorium in soils. Soil Science 105(1): 44-50.

Hansen, U.-P. 1969. Decay of the radiation effect on Nitella at repeated irradiation compared with the influence on the mechanism of excitation. Atomkernenergie 14: 207-210. (In German) (NSA 23: 29903).

Hansen, W. R. 1963. Determination of the RBE of 14-Mev neutrons using 14 day old grasshopper embryos (Chortophaga viridifasciata and Encoptplophus sordidus). Thesis, Univ. of Kansas, Lawrence. 65pp. (Also: U. S. AEC report TID-18723) (NSA 17: 24896).

Hansen, W. R. and R. L. Watters. 1970. Plant uptake of ^{210}Po from soil. Radiation Botany 10: 371-375. (NSA 25: 510).

Hansen, W. R. and R. L. Watters. 1971. Unsupported $^{210}PoO_2$ in soil: soil absorption and characterization of soil solution species. Soil Science 112(3): 145-155. (NSA 26: 12039).

Hansen, W. R., R. L. Watters and N. D. Yaney. 1971. $^{210}PoO_2$ movement in a mountain watershed soil. Health Physics 20: 425-429. (NSA 25: 39602).

Hanshaw, B. B. and H. R. Feltz. 1963. Preparation of water sample for carbon-14 dating. U. S. Geological Survey Circ. 480. iii, 3pp.

Hanshaw, B. B., W. Back and M. Rubin. 1965. Radiocarbon determination for estimating groundwater flow velocities in Central Florida. Science 148 (3669): 494-495. (NSA 19: 22580).

Hanshaw, B. B., et al. 1967. Radiocarbon determinations applied to groundwater hydrology. pp. 117-118. In: Isotope Techniques in the Hydrologic Cycle. (See Stout, 1967).

Hanson, G. L., et al. 1973. Input and decayed values of radioactive liquid wastes discharged to the ground in the 200 Areas through 1971. France Centre National de la Recherche Scientifique, Strasbourg, report ARH-2761. 62pp. (NSA 30: 32075).

Hanson, P. J. 1967. Vertical distribution of radioactivity in the Columbia River estuary. Thesis, Oregon State University, Corvallis. 86pp. (NSA 21: 39044, 26: 25586).

Hanson, P. J. 1969. Water tracing with soluble metal chelates and neutron activation analysis: a laboratory field study. Thesis, Oregon State Univ., Corvallis. 100pp. (Also: U. S. AEC report RLO-1750-81) (NSA 25: 15549, 29013).

Hanson, P. J. and W. O. Forster. 1969. Measurement of Columbia River flow time from Hanford to Astoria, Oregon--Summer 1966. Water Resources Res. 5(5): 1129-1131.

Hanson, W. C. 1960. Accumulation of radioisotopes from fallout by terrestrial animals at Hanford, Washington. Northwest Science 34(3): 89-98. (NSA 15: 50).

Hanson, W. C. 1962. Seasonal patterns of stable and radioactive iodine in thyroids of native jack rabbits. pp. 225-232. In: Trans. 27th N. Am. Wildlife and Natural Resources Conf. Wildlife Management Inst., Washington, D. C.

Hanson, W. C. 1963. Iodine in the environment. pp. 581-601. In: Radioecology. (See Schultz and Klement) (NSA 17: 33611).

Hanson, W. C. 1963. Radioactivity in northern Alaskan Eskimos during 1962 and 1963. pp. 1129-1132. In: Fallout, Radiation Standards, and Countermeasures. (See U. S. Congress, 1963).

Hanson, W. C. 1965. Fallout radionuclides in Alaskan food chains. Battelle Northwest Lab., U. S. AEC report BNWL-SA-242. (For 102nd Am. Vet. Med. Assoc. Meeting, Portland, Ore.) (NSA 19: 40780).

Hanson, W. C. 1966. Seasonal cycle of 137-cesium in some Alaskan natives and animals. pp. 6-7. In: Radiation Effects on Natural Populations. (See Sacher, 1966).

Hanson, W. C. 1966. ^{137}Cs body burdens in Alaskan Eskimos during the summer of 1965. Science 153: 525-527. (NSA 20: 45314).

Hanson, W. C. 1966. Fallout radionuclides in Alaskan food chains. Am. J. Veterin. Res. 27: 359-366. (NSA 20: 26609).

Hanson, W. C. 1967. Radioecological concentration processes characterizing arctic ecosystems. pp. 183-191. In: Radiological Concentration Processes. (See Aberg and Hungate, 1967) (Also: Battelle Northwest Lab., U. S. AEC report BNWL-SA-661) (CONF-660405) (NSA 20: 31442).

Hanson, W. C. 1967. ^{137}Cs in Alaskan lichens. Health Physics 13(4): 383-389. (NSA 21: 18055).

Hanson, W. C. 1968. Fallout radionuclides in northern Alaskan ecosystems. Arch. Environ. Hlth. 17: 639-648. (Also: Battelle Northwest Lab., U. S. AEC report BNWL-SA-1327) (NSA 21: 39396).

Hanson, W. C. 1968. Contamination of soft tissues of infants and children with radioactive fallout as exemplified by Cs^{137} and I^{131}. Pediatrics 40(1,II): 240-249. (Alaskan natives).

Hanson, W. C. 1970. ^{210}Po in soils and plants. Thesis, Colorado State Univ., Fort Collins. 221pp. (Univ. Microfilms 71-5336) (NSA 25: 26684).

Hanson, W. C. 1971. ^{137}Cs: seasonal patterns in native residents of three contrasting Alaskan villages. Health Physics 20: 585-591.

Hanson, W. C. 1971. Fallout radionuclide distribution in lichen communities near Thule. Arctic 24(4): 269-276. (NSA 26: 28091).

Hanson, W. C. 1972. Plutonium in lichen communities of the Thule, Greenland region during the summer of 1968. Health Physics 22(1): 39-42. (NSA 26: 15547).

Hanson, W. C. 1974. Ecological considerations of depleted uranium munitions. Los Alamos Scientific Lab., U. S. AEC report LA-5559. (NSA 30: 18684).

Hanson, W. C. 1974. Behavior of plutonium in the environment. pp. 64-71. In: Plutonium Information Meeting. (See U. S. AEC, 1974) (NSA 31: 14018).

Hanson, W. C. 1975. Ecological considerations of the behavior of plutonium in the environment. Health Physics 28(5): 529-537. (NSA 32: 11547).

Hanson, W. C. nd. Biological transfer processes in environmental surveillance. Los Alamos Scientific Lab., U. S. AEC report LA-UR-74-764. 7pp. (NSA 30: 9526).

Hanson, W. C. and R. L. Browning. 1953. Biological monitoring. pp. 180-185. In: Biology Research-Annual Report 1952. Hanford Atomic Products Operation, U. S. AEC report HW-28636. (Upland game and waterfowl).

Hanson, W. C. and A. C. Case. 1963. A method for measuring waterfowl dispersion utilizing phosphorus-32 and zinc-65. pp. 451-453. In: Radioecology. (See Schultz and Klement, 1963) (NSA 17: 33595).

Hanson, W. C. and L. L. Eberhardt. 1969. Effective half-times of radionuclides in Alaskan lichens and Eskimos. pp. 627-634. In: Symposium on Radioecology. (See Nelson and Evans, 1969) (Also: Battelle Northwest Lab., U. S. AEC report BNWL-SA-126).

Hanson, W. C. and H. A. Kornberg. 1956. Radioactivity in terrestrial animals near an atomic site. Proc. Intern. Conf. on the Peaceful Uses of Atomic Energy 13: 385-388. (Paper A/CONF.8/P/281).

Hanson, W. C. and H. E. Palmer. 1964. The accumulation of fallout cesium-137 in northern Alaskan natives. Trans. N. Am. Wildlife Conf. 29: 215-224.

Hanson, W. C. and H. E. Palmer. 1965. Seasonal cycle of ^{137}Cs in some Alaskan natives and animals. In: Radiation and Terrestrial Ecosystems. Health Physics 11(12): 1401-1406.

Hanson, W. C. and L. L. Eberhardt. 1973. Cycling and compartmentalizing of radionuclides in northern Alaskan lichen communities. pp. 71-75. In: Radionuclides in Ecosystems. (See Nelson, 1973) (NSA 25: 37772).

Hanson, W. C., H. E. Palmer and B. I. Griffin. 1964. Radioactivity in northern Alaskan Eskimos and their foods, summer 1962. Health Physics 10(6): 421-429. (NSA 18: 20231).

Hanson, W. C., H. E. Palmer and B. I. Griffin. 1966. Radioactivity in northern Alaskan Eskimos and their foods, summer 1962. pp. 1151-1164. In: Environment of the Cape Thompson Region, Alaska. (See Wilimovsky and Wolfe, 1966).

Hanson, W. C., D. G. Watson and R. W. Perkins. 1967. Concentration and retention of fallout radionuclides in Alaskan Arctic ecosystems. pp. 233-245. In: Radioecological Concentration Processes. (See Aberg and Hungate, 1967) (CONF-660405-3) (NSA 20: 23283).

Hanson, W. C., F. W. Whicker and A. H. Dahl. 1963. Iodine-131 in the thyroids of North American deer and caribou: comparison after nuclear tests. Science 140(3568): 801-802.

Hanson, W. C., et al. 1963. Thyroidal radioiodine concentrations in North American deer following 1961-1963 nuclear weapons tests. Health Physics 9(12): 1235-1239.

Hanusova, J. and B. Havlik. 1974. Determination of uranium in freshwater, sediments, and biological materials. Cesk. Hyg. 19(5): 227-234. (In Czech) (NSA 31: 32500, 33: 19654).

Hanya, T. and T. Araki. 1963. ^{90}Sr content of inland waters in Japan. Nippon Kagaku Zasshi 84: 124-127. (In Japanese) (NSA 17: 21947).

Hanya, T, T. Araki and K. Naka. 1964. Behavior of ^{90}Sr in Lake Haruna, Japan. pp. 335-347. In: Recent Researches in the Fields of Hydrosphere, Atmosphere and Nuclear Geochemistry. Y. Miyake and T. Koyama, eds. Sugawa Vol. Nagoya Univ., Nagoya, Japan. (NSA 20: 24835).

Happ, G. M. and J. Meinwald. 1965. Biosynthesis of arthropod secretions. I. Monoterpene synthesis in an ant (Acanthomyops claviger). J. Am. Chem. Soc. 87(11): 2507-2508. (^{14}C).

Hara, T. and Y. Koshida. 1959/60? Distribution of radiosodium and radiopotassium administered to the adult toad. Dobutsugaku Zasshi 68: 173-176.

Hara, T. and Y. Koshida. 1959/60? Uptake of sodium-24 and potassium-42 by the goldfish. Dobutsugaku Zasshi 68: 177-180. (CA 54: 764a).

Harada, M. 1957. Radioactivity of the granitic soils in the middle district of Tottori Prefecture. Soil Plant Food 2(4): 185-189. (nc. analysis of vegetation).

Harada, Y., T. Ai and S. Kondo. 1959. Radiation effects on germ cells of rainbow trout. Abstr. 3rd Japan Conf. on Radioisotopes, Natl. Inst. Radiological Science. 1pp. (JAIF, JRIA, 59/P-89, A-28).

Haragsim, O., et al. 1965. Comparison of the activity of some geographical races of the honeybee (Apis mellifera L.) on lucerne. Sb. csl. Akad. zemed. Ved 11: 861-872. (In Czech, w/Russian, English, German summaries) (^{32}P, ^{198}Au).

Harbaugh, J. W. 1968. Radiometric age dating. pp. 76-110. In: Stratigraphy and Geologic Time. J. W. Harbaugh, ed. Wm. C. Brown, Dubuque. (NSA 24: 36400).

Harbottle, G. 1970. Neutron activation analysis of potsherds from Knossos and Mycenae. Archeometry 12: 23-34. (Also: 1968: Brookhaven National Lab., U. S. AEC report BNL-1370; 24pp) (NSA 23: 33159, 24: 38798).

Hardcastle, J. E., W. H. Fuller and R. J. Hannapel. 1972. Simultaneous assay of ^{45}Ca and ^{89}Sr in double-isotope biological samples by liquid scintillation counting. Anal. Biochem. 46(2): 534-547. (NSA 26: 33093).

Hardcastle, J. E., W. H. Fuller and R. J. Hannapel. 1973. Assay of ^{89}Sr in plant and soil samples by liquid scintillation counting. J. Radioanal. Chem. 13(1): 141-147. (NSA 28: 12672).

Hardy, E. P., Jr. and S. Klein. 1960. Strontium program quarterly report, September 1 through December 1, 1959. Health and Safety Lab., New York Operations Off., U. S. AEC report HASL-77. x, 222pp. (See next entry. This entry is purposely out of order; listed as the first of the entry below; Hardy as author and senior author of other articles are listed after this and the following entry).

Hardy, E. P., Jr. or with others. (Comps.). 1960-present. Strontium Program Quarterly Report; Fallout Program Quarterly Summary Report; or Environmental Quarterly. Health and Safety Lab., U. S. AEC; U. S. ERDA; Environmental Measurements Lab., U. S. DOE; report HASL- (now EML-); eg., HASL-77 (see above entry) et seq. to HASL-294 (1975) and beyond) (A notable co-compiler was the late J. Rivera) (Reports contain data on all types of environmental samples in systematic order, radiation and radionuclide data, summaries of such data, and specific authored articles, as well as lists of current references on the subject).

Hardy, E. P., Jr. 1970. Sr^{89} fallout from atmospheric nuclear testing. pp. 181-193. In: Fallout Program Quarterly Summary Report. (See Hardy, 1970; report HASL-227).

Hardy, E. P., Jr. and J. H. Harley. (Eds.). 1958. Environmental contamination from weapons tests. Health and Safety Lab., U. S. AEC report HASL-42. xi, 359pp. (Incl. data on caribou and polar bear).

Hardy, E. P., Jr., P. W. Krey and H. L. Volchok. 1973. Global inventory and distribution of fallout plutonium. Nature 241(5390): 444-445. (NSA 27: 27584).

Hardy, E. P., Jr., et al. 1968. Strontium-90 on the earth's surface. Nature 219(5154): 584-587.

Hargrave, B. T. 1970. The utilization of benthic microflora by Hyalella azteca (Amphipoda). J. Animal Ecol. 39(2): 427-437. (^3H, ^{14}C).

Harker, J. E. 1958. Diurnal rhythms in the animal kingdom. Biol. Rev. 33: 1-52. (Incl. x-irradiation).

Harker, J. E. 1964. The Physiology of Diurnal Rhythms. Cambridge Monogr. in Experimental Biology No. 13. Cambridge Univ. Press, London. vii, 114 pp. (Incl. x-irradiation).

Harkness, D. D. and R. Burleigh. 1974. Possible carbon-14 enrichment in high altitude wood. Archeometry 16(2): 121-127. (NSA 31: 3400).

Harkness, D. D. and A. Walton. 1969. Carbon-14 in the biosphere and humans. Nature 223(5212): 1216-1218.

Harkness, D. D. and A. Walton. 1972. Further investigations of the transfer of bomb ^{14}C to man. Nature 240(5379): 302-303. (NSA 27: 9880).

Harley, J. H. 1971. Needs and progress in rapid analysis of environmental samples. pp. 23-31. In: Rapid methods for Measuring Radioactivity in the Environment. (See IAEA, 1971) (NSA 26: 27801).

Harley, J. H. (Ed.). 1972. HASL procedures manual. Health and Safety Lab., U. S. AEC report HASL-300. 602pp. 1974. HASL-300(Suppl. 2). 173pp. (NSA 26: 55644; 30: 28970) (Radioanalytical procedures; looseleaf).

Harley, J. H. 1974. Transuranium elements on land. Trans. Amer. Nucl. Soc. 19: 75-76. (NSA 31: 5973).

Harley, J. H. 1975. Transuranium elements on land. pp. I104-I109. In: Environmental Quarterly. (See Hardy, 1975; report HASL-291) (NSA 31: 33103).

Harley, J. H. and W. M. Lowder. 1971. Natural radioactivity and radiation. pp. I2-I23. In: Fallout Program Quarterly Summary Report. (See Hardy, 1971; report HASL-242) (NSA 25: 32070).

Harley, J. H. et al. 1965. A summary of the panel discussion of approaches to a study of radiological hazards in the environment. pp. 1263-1264. In: Radiation and Terrestrial Ecosystems. (See Hungate, 1965).

Harlimov, V. P. 1964. Some physico-chemical data on interchange of gases in Musca domestica L. and nucleonic DNA composition of their grubs (L_{III}) in the first generation induced by internal ^{32}P beta-irradiation. Radiobiologiya 4: 893-896. (In Russian) (NSA 19: 17616).

Harmon, J. A. and J. M. Leach. 1967. Postattack sanitation, waste disposal, pest and vector control and the effects in waste water and sewer systems. Final report, Phase 2. Engineering-Science, Inc., Arcadia, Calif., U. S. Naval Radiological Defense Lab., San Francisco, report USNRDL-TRC-79. 99pp. (AD-645599; NSA 21: 41055).

Harmut, M., T. Maljkovic and K. Kostial. 1968. Direct counting of ^{47}Ca and ^{85}Sr in biological samples. Arh. Hig. Rada Toksikol. 19: 61-65. (NSA 25: 18997).

Harney, P. J. 1974. Loss of zinc-65 and manganese-54 from the freshwater mollusc Anodonta. Thesis, Oregon State Univ., Corvallis. 116pp. (Also: U. S. AEC report RLO-2227-T12-42) (NSA 30: 21126).

Harremoes, P. 1966. Prediction of pollution from planned wastewater outfalls. J. Water Poll. Control Fed. 38: 1323-1333. (NSA 21: 28118) (Tracer, special boat for assaying).

Harper, F. 1963. Caribou and eskimos. I.U.C.U. Bull. 6(NS):__. (Incl. radionuclides in caribou).

Harrington, N. J. and R. W. Koza. 1951. Effect of x-radiation on the desoxyribonucleic acid and the size of grasshopper embryionic nuclei. Biol. Bull. 102(2): 138-150.

Harris, E. 1957. Radiophosphorus metabolism in zooplankton and microorganisms. Can. J. Zool. 35: 769-782.

Harris, E. K. 1959. Estimating the uptake of radioactivity in streams. Sewage Industr. Wastes 31: 719-728.

Harris, J. B., et al. 1937. Mutations in parthenogenetic Cladocera induced by x-radiation. Physiol. Zool. 10(3): 269-275.

Harris, P. J. 1959. A study of thyroid function in Fundulus heteroclitus. Biol. Bull. 117: 89-99.

Harris, W. B., et al. 1959. Environmental hazards associated with the milling of uranium ore. A.M.A. Arch. Industr. Hlth. 20: 365-382. (NSA 15: 7644).

Harris, W. F. 1970. Ecological aspects of responses of Liriodendron tulipifera L. and Acer saccharinum L. to fast neutron radiation and reduced light. Thesis, Univ. of Tennessee, Knoxville. xvii, 197pp. (Also: Harris, W. F., J. P. Witherspoon and J. S. Olson, Oak Ridge National Lab., U. S. AEC report ORNL-4610) (Univ. Microfilms 71-7642) (NSA 25: 27076).

Harris, W. F. 1972. Morphological responses of Liriodendron tulipifera L. seedlings to acute fast neutron radiation and reduced light. Radiation Botany 12(3): 205-219. (NSA 26: 50846).

Harris, W. F. and J. P. Witherspoon. 1973. Effects of ionizing radiation on processes influencing tolerance of tree seedlings. pp. 961-971. In: Radionuclides in Ecosystems. (See Nelson, 1973) (NSA 28: 30072).

Harrison, F. L. 1967. Concentration factors - their use and abuse. Lawrence Radiation Lab., Livermore, Calif., U. S. AEC report UCRL-50347. 4pp.

Harrison, F. L. 1969. Accumulation and distribution of ^{54}Mn and ^{65}Zn in freshwater clams. pp. 198-220. In: Symposium on Radioecology. (See Nelson and Evans, 1969).

Harrison, F. L. 1969. Radioactive debris from underground nuclear expolosions. 1. Physical and chemical characteristics. 2. Biological availability of the radionuclides to aquatic animanls. Lawrence Radiation Lab., Livermore, Calif., U. S. AEC report UCRL-50596. 28pp. (NSA 23: 41306).

Harrison, F. L. 1970. Some effects of nuclear detonations on aquatic ecosystems. Trans. Amer. Nucl. Soc. 13: 641-642. (NSA 25: 4479).

Harrison, F. L. 1971. Biological implications of nuclear debris in aquatic ecosystems. Nuclear Technol. 11(3): 444-458. (NSA 25: 39808).

Harrison, F. L. 1973. Availability to aquatic animals of short-lived radionuclides from a Plowshare cratering event. Health Physics 24(3): 331-343. (NSA 27: 25067).

Harrison, F. L. and J. J. Koranda. 1973. Tritiation of aquatic animals in an experimental freshwater pool. pp. 425-434. In: Radionuclides in Ecosystems. (See Nelson, 1973) (NSA 26: 9484).

Harrison, F. L. and I. L. Ophel. 1975. Freshwater environment. Lawrence Livermore Lab., U. S. ERDA report UCRL-76552. 11pp. (NSA 31: 29627).

Harrison, F. L. and D. J. Quinn. 1972. Tissue distribution of accumulated radionuclides in freshwater clams. Health Physics 23(4): 509-517. (NSA 26: 56559).

Harrison, F. L., K. M. Wong and R. E. Heft. 1972, 1975. Role of water (solubles) and particulates in radionuclide accumulation in the oyster Crassostrea gigas (in the discharge canal of a nuclear power plant). Lawrence Livermore Lab., U. S. AEC, ERDA, report UCRL-76570(Rev.1,2). 33, 39pp. (NSA 32: 3826, 19759).

Harrison, J. M. and J. G. Harrison. 1956. An Icelandic redshank in freshly moulted summer plummage in November. Bull. Brit. Ornithol. Club 76(2): 23-25. (See next entry).

Harrison, J. M. and J. G. Harrison. 1956. Abnormal seasonal assumption of spring plummage in the redshank (Tringa totanus Linnaeus) in association with possible radioactive contamination. Bull. Brit. Ornithol. Club 76 (4): 60-61.

Harrison, J. M. and J. G. Harrison. 1961. Radioactive contamination in birds. pp. 151-152. In: 12th Ann. Rept. Wildfowl Trust 1959-1960. (WR 104: 49).

Harrison, M. J., R. T. Wright and R. Y. Morita. 1971. Method for measuring mineralization in lake sediments. Appl. Microbiol. 21: 698-702. (NSA 25: 32321) (Used ^{14}C-glucose).

Harrison, M. T., et al. 1965. Nature and availability of iodine in fish. Am. J. Clin. Nutr. 17: 73-77. (NSA 20: 6732).

Harrison, P. R., et al. 1971. Areawide trace metal concentrations measured by multielement neutron activation analysis-- a one day study in northwest Indiana. J. Air Poll. Control Assoc. 21(9): 563-570.

Harrison, S. H. 1975. Neutron activation analysis: a useful tool for evaluating water sampling and analysis techniques. Trans. Amer. Nucl. Soc. 21: 110-111. (NSA 32: 11151).

Harshbarger, J. C. and B. S. Moore. 1966. Some effects of x irradiation on larvae of Galleria mellonella. J. Invertebr. Pathol. 8: 277-279. (NSA 2 21: 11098).

Hart, R. W. 1971. UV- and x-ray response of the desmid, Micrasterias decemdentata. Thesis, Univ. of Illinois, Urbana. (Univ. Microfilms 72-12,195) (NSA 26: 31274).

Hart, S. R. 1966. Current status of radioactive age determination methods. Trans. Am. Geophys. Un. 47: 280-286. (NSA 20: 23325).

Hart, W. G. and S. Ingle. 1966. Influence of radiation on the fecundity of the brown soft scale. Bull. Entomol. Soc. Am. 12: 265.

Harte, J. and R. H. Soclow. 1971. Patient Earth. Holt, Rinehart & Winston, New York. 377pp. (NSA 26: 53191) (Incl. ionizing radiation).

Hartley, P. E. 1965. A method for digital computer processing of data from radioisotope sediment tracing studies. Reserach Establishment, Lucas Heights, Australian Atomic Energy Commission, report AAEC/TM-303. 22pp. (NSA 20: 16632).

Hartley, P. E. 1967. A technique for digital computer processing of data from radioisotope sediment tracing studies. Int. J. Appl. Rad. Isotopes 18: 713-720.

Hartman, A. 1920. On the influence of roentgen radiation on the blood and blood component tissues of Rana temporaria - larvae. Arch. F. Entwicke. d. Org. 47: 131-209. (In German).

Hartman, R. M. 1971. Environmental reports and new power plant schedules. Publ. Util. Fortn. 88(2): 13-22. (NSA 25: 51023).

Hartman, R. T., D. L. Brown and J. R. Moore. 1964. The relationship of undissolved oxygen to photosynthesis in submerged vascular plants. Abstr. Bull. Ecol. Soc. Am. 45(3): 78. (^{14}C).

Hartman, R. T., et al. 1965. Changes in the rate of photosynthesis by submerged vascular hydrophytes during the daylight period. Abstr. Bull. Ecol. Soc. Am. 46(3): 129. (^{14}C).

Hartsook, E. W., et al. 1964. Phosphorus-32 metabolism and antler growth in male white-tailed deer. Abstr. J. Animal Sci. 23(3): 876.

Hartwell, W. V., et al. 1958. Observations of flea transfer between hosts; a mechanism in the spread of bubonic plague. Science 127(3302): 814. (^{144}Ce).

Harvey, H. D., Jr. and M. W. Carter. 1968. Offsite environmental surveillance findings associated with underground nuclear testing; Project Dribble. Radiol. Hlth. Data 9(2): 73-80.

Harvey, H. D., Jr., E. D. Toerber and J. A. Gordon. 1968. Radiological survey of Hampton Roads, Virginia. Southeastern Radiological Health Lab., U. S. Public Health Service, Montgomery, Ala. 23pp. (AD-683208; NSA 23: 31721) (Biota, water samples).

Harvey, J. M. 1967. Excretion of DDT by migratory birds. Can. J. Zool. 45 (5): 629-633. (^{14}C).

Harvey, J. W. 1949. On manganese in sea and fresh water. J. Marine Biol. Assoc. UK 28: 155-164.

Harvey, M. J. 1970. Diel activity patterns of the eastern mole, _Scalopus aquaticus_. Abstr. Bull. Ecol. Soc. Am 51(4): 49. (Radioisotope tag).

Harvey, M. J. and R. W. Barbour. 1964. An improved method for determining home range. A. S. B. Bull. 11: 47. (Radioisotope tag).

Harvey, M. J. and R. W. Barbour. 1965. Home range of _Microtus ochrogaster_ as determined by a modified minimum area method. J. Mammalogy 46: 398-402. (Radioisotope tag).

Harvey, R. A. 1965. Assessment of radiation effects on desert shrubs. Abstr. Bull. Ecol. Soc. Am. 46(2): 50. (At Nevada Test Site).

Harvey, R. S. 1963. Uptake of radionuclides by freshwater algae and fish. Savannah River Plant, U. S. AEC report DPSPU-63-30-3. 11pp. (NSA 17: 23033).

Harvey, R. S. 1964. Savannah River Plant biological monitoring program. Savannah River Plant, U. S. AEC report DPSPU-64-30-4. 7pp. (NSA 19: 11416).

Harvey, R. S. 1964. Uptake of radionuclides by freshwater algae and fish. Health Physics 10(4): 243-248.

Harvey, R. S. 1967. Effects of temperature on the sorption of radionuclides by blue-green algae. Savannah River Plant, U. S. AEC report DP-MS-66-88. 15pp. (NSA 22: 38191).

Harvey, R. S. 1968. Biological indicators of environmental radioactivity. Savannah River Lab., U. S. AEC report DP-MS-67-99. 9pp. (NSA 22: 38253).

Harvey, R. S. 1969. Temperature effects on the sorption of radionuclides by freshwater algae. Abstr. Bull. Ecol. Soc. Am. 50(2): 71.

Harvey, R. S. 1969. Effects of temperature on the sorption of radionuclides by a blue-green alga. pp. 266-269. In: _Symposium on Radioecology_. (See Nelson and Evans, 1969).

Harvey, R. S. 1970. Temperature effects on the sorption of radionuclides by freshwater algae. Health Physics 19: 293-297. (NSA 25: 8598).

Harvey, R. S. 1971. Temperature effects on the maturation of midges (Tendipedidae) and their sorption of radionuclides. Health Physics 20(6): 613-616. (NSA 25: 46687).

Harvey, R. S. 1973. Temperature effects on the sorption of radionuclides by aquatic organisms. Savannah River Lab., U. S. AEC report DP-MS-73-8. 20 pp. (NSA 28: 21159).

Harvey, R. S. 1973. Temperature effects on the sorption of ^{137}Cs, ^{85}Sr and ^{65}Zn by freshwater shrimp. pp. 599-602. In: _Radionuclides in Ecosystems_. (See Nelson, 1973) (NSA 28: 27679).

Harvey, R. S. and S. C. Aiken. 1974. Temperature effects on growth and respiration rates of _Dolania americana_ (Ephemeroptera). Savannah River Lab., U. S. AEC report DP-MS-73-66. 28pp. (NSA 30: 3201).

Harvey, R. S. and R. Patrick. 1966. Concentration of ^{137}Cs, and ^{65}Zn, and ^{85}Sr by fresh-water algae. Savannah River Lab., U. S. AEC report DP-MS-66-12. 15pp. (NSA 21: 17950).

Harvey, W. R. and S. Nedergaard. 1964. Sodium-independent active transport of potassium in the isolated midgut of the cecropia silkworm. Proc. Nat. Acad. Sci. (US) 51: 757-765.

Harvey, W. R., J. A. Haskell and K. Zerahn. 1967. Active transport of potassium and oxygen consumption in the isolated midgut of _Hyalophora cecropia_. J. Exptl. Biol. 46(2): 235-248.

Harwalker, M. R. and K. K. Nair. 1968. Effect of x-irradiation on post-embryonic development of the red cotton bug, _Dysdercus koenigii_. Ann. Entomol. Soc. Am. 61(5): 1107-1111. (NSA 22: 51971).

Harwalker, M. R. and G. W. Rahalkar. 1973. Sperm utilization in the female red cotton bug. J. Econ. Entomol. 66(3): 805-806. (NSA 28: 13088) (X-ray).

Harward, E. D., et al. 1972. Panel discussion: interrelationships of Federal, State, academic, and industrial interests in environmental studies. pp. 194-220. In: _Proc. Southern Conf. on Environmental Protection from Nuclear Power Plants_. (See U. S. EPA, 1972) (NSA 27: 4698).

Häsänen, E. 1970. Determination of mercury in biological material by neutron activation analysis. Suomen Kem. 43B: 251-254. (NSA 25: 10502).

Häsänen, E. 1972. Occurrence of ^{137}Cs in the biosphere evaluated with environmental and metabolic studies. Univ. of Helsinki, Finland, U. S. AEC report TID-26343. 119pp. (Also: U. S. AEC report COO-3011-5) (NSA 28: 12870; 33: 7304) (Incl. aquatic foodchain, reindeer-lichen-man foodchain).

Häsänen, E. 1974. Mercury pollution of the aquatic environment in Finland. pp. 149-157. In: Comparative Studies of Food and Environmental Contamination. (See IAEA, 1974) (NSA 30: 14763).

Häsänen, E. and J. K. Miettinen. 1963. Caesium-137 content of freshwater fish in Finland. Nature 200(4910): 1018-1019.

Häsänen, E. and J. K. Miettinen. 1964. Measurements of caesium-137 in Finnish Lapps in 1962-1964 by a mobile whole-body counter. pp. 193-204. In: Assessment of Radioactivity in Man. Vol. II. International Atomic Energy Agency, Vienna, Publ. STI/PUB/84. (Incl. comments on reindeer).

Häsänen, E. and J. K. Miettinen. 1966. Gamma emitting radionuclides in subarctic vegetation during 1962-1964. Nature 212(5060): 379-382. (NSA 21: 4629).

Häsänen, E., S. Colehmainen and J. K. Miettinen. 1966. Biological half life of ^{137}Cs in perch (Perca fluviatilis L.). Univ. of Helsinki, Finland. 9pp. (NP-16002, paper 2) (NSA 20: 26598).

Häsänen, E., S. Kolehmainen and J. K. Miettinen. 1967. Biological half-time of ^{137}Cs in three species of freshwater fish: perch, roach and rainbow trout. pp. 921-924. In: Radioecological Concentration Processes. (See Aberg and Hungate, 1967) (CONF-660405) (NP-16002, Paper 7) (NSA 20: 26603).

Häsänen, E., S. Kolehmainen and J. K. Miettinen. 1968. Biological half-times of ^{137}Cs and ^{22}Na in different fish species and their temperature dependence. pp. 401-406. In: Radiation Protection, Part 1. (See Snyder, et al., 1968) (NSA 23: 12036).

Haschke, J. 1969. Control of tritium in water and sewage. Arch. Hyg. Bakteriol. 153: 383-389. (In German) (NSA 25: 18360).

Haschke, J. 1972. Determination of precipitation of radioiodine in the radiological examination of contaminated surface waters. Städtehygiene 23(11): 260-265. (In German) (NSA 28: 24156).

Hascoet, M. 1963. Evolution of superficial deposits, diffusion and degradation of two endotherapeutic insecticides: Demeon-S and Endothion, through some marsh plants. pp. 195-210. In: Radiation and Radioisotopes Applied to Insects of Agricultural Importance. (See IAEA, 1963) (In French).

Hasegawa, M., Y. Okada and Y. Anma. 1960. Radiosensitivity of Pinus seedlings and their physiological responses on γ-irradiation. I. Nippon Ringaku Kaishi 42: 183-184. (In Japanese, English summary).

Hashimoto, Y. 1971. Effect of radiation on the germination of spores of Dictostelium discoideum. Nature 231(5301): 316-317. (NSA 25: 47032).

Hashimoto, Y. and K. Yanagisawa. 1970. Effect of radiation on the spore germination of the cellular slime mold Dictostelium discoideum. Radiation Research 44: 649-659. (NSA 25: 27065).

Haskell, C. C. and B. H. Hawkins. 1964. D_2O-Na^{24} method for tracing soil moisture movement in the field. Soil Sci. Soc. Am. Proc. 28(6): 725-728.

Haskell, E. E., Jr., J. S. Leventhal and W. C. Bianchi. 1966. Use of tritium to measure the movement of groundwater toward irrigation wells in western Fresno County, California. J. Geophys. Res. 71: 3849-3859. (NSA 20: 41160).

Hasler, A. D. 1957. Natural and artificially (air-plowing) induced movement of radioactive phosphorus from the muds of lakes. pp. 658-675. In: Intern. Conf. on Radioisotopes in Scientific Research. Vol. 4. UNESCO. Pergamon Press, New York. (UNESCO/NS/RIC/188) (NSA 12: 1206).

Hasler, A. D. and G. E. Likens. 1963. Biological and physical transport of radionuclides in stratified lakes. pp. 463-470. In: Radioecology. (See Schultz and Klement, 1963) (NSA 17: 33597).

Hassan, A., S. M. A. D. Zayed and F. M. Abdel-Hamid. 1965. Metabolism of organo-phosphorus insecticides. V. Mechanism of detoxification of dipterex in Prodenia litura. Biochem. Pharmacol. 14(11): 1577-1584. (^{32}P, ^{14}C).

Hassenteufel, W., R. Jagitsch and F. F. Koczy. 1963. Impregnation of glass surface against sorption of phosphate traces. Limnol. Oceanogr. 8(2): 152-156.

Hasserodt, U. 1972. Current and future environmental problems arising from energy conversion. pp. 28-40. In: Environmental Quality and Safety. Chemistry, Toxicology, and Technology. Vol. I. F. Coulston, ed. Academic Press, New York.

Hassett, C. C. 1950. Radiophosphorus in mosquitoes (Aedes aegypti). Abstr. Fed. Proc. 9(1): 58.

Hassett, C. C. 1954. Uptake of P^{32} by Periplaneta americana; passage through the fore-, mid, and hindgut. Fed. Proc. 13(1): 68.

Hassett, C. C. 1956. Army entomologists find new uses for radioactive isotopes. Armed Forces Chem. J. 10(1): 12.

Hassett, C. C. 1956. Current status of insect control by radiation. Science 124(3230): 1011-1012. (NSA 11: 873).

Hassett, C. C. and D. W. Jenkins. 1949. Production of radioactive mosquitoes. Science 110(2848): 109-110. (NSA 3: 1193).

Hassett, C. C. and D. W. Jenkins. 1952. Use of fission products for insect control. Nucleonics 10(12): 42-46.

Hasuike, J. K. and S. T. Windham. 1973. Construction and operation of an ion exchange cartridge for monitoring radionuclides in the environment. Eastern Environmental Radiation Facility, Montgomery, Ala., report ORP/EERF-73-2. 16pp. (NSA 29: 12726).

Hatanake-Ernst, 1966. Developmental physiological studies on the radio-induced protonema mutant of Funaria hygrometrica Sibith. Z. Pflanzenphysiol. 55: 259-277. (In German) (NSA 21: 24377).

Hatch, F. T., et al. 1968. Studies of liver DNA in tritiated kangaroo rats living at Sedan crater. Lawrence Radiation Lab., Livermore, Calif., report UCRL-50461. 28pp. (NSA 22: 49184).

Hatch, F. T., et al. 1970. Ecology and radiation exposure of kangaroo rats living in a tritiated environment. Radiation Research 44: 649-659. (NSA 25: 27065).

Hatfield, T. W., D. M. Skauen and N. Marshall. 1967. Gross beta radioactivity in Fucus hybrids (rockweed) from Niantic Estuary, Connecticut, 1961-1964. Radiol. Hlth. Data Repts. 8: 481-483. (NSA 22: 40694)

Hathaway, D. O. 1966. Laboratory and field cage studies of the effects of gamma radiation on codling moths. J. Econ. Entomol. 59(1): 35-37.

Hathaway, D. O., B. A. Butt and L. V. Lydin. 1970. Reduction of sexual aggressiveness of male codling moths treated with tepa or gamma irradiation. J. Econ. Entomol. 63: 1881-1883. (NSA 25: 13557).

Hatmosoewarno, S. 1968. Effect of releasing sterile Scirpophaga nivella F. on crop damage and insect population density in sugar cane plantations. pp. 317-329. In: Isotopes and Radiation in Entomology. (See IAEA, 1968) (NSA 24: 32325).

Haug, H. O. 1975. Amount, disposal, and relative toxicity of long-lived fission products and actinides in the radioactive wastes of the nuclear fuel cycle. German F. R. Kernforschungzentrum, Karlsruhe, report KFK-2022. 217pp. (In German) (NSA 33: 23835) (Incl. Western Europe).

Haugberg, M. and K. Steenberg. 1957. A study on nutrient uptake by spruce using radioactive isotopes. Tidsskr. Skogbr. 65(4): 248-258. (In Norwegian) (FA 19: 2698).

Haury, G., W. Schikarski and E. Theone. 1974. Summary of the environmental effects caused in the German Federal Republic by the production and consumption of energy and the legislative and technical measures for their reduction. German F. R. Kernforschungszentrum, Karlsruhe, report KFK-2103UF. 177pp. (In German) (NSA 32: 19736)

Hauschild, A. H. W., C. D. Nelson and G. Krotkov. 1962. The effect of light quality on the products of photosynthesis in green and blue-green algae, and in photosynthetic bacteria. Can. J. Botany 40(12): 1619-1630. (^{14}C).

Haushild, W. L., G. R. Dempster, Jr. and H. H. Stevens, Jr. 1973. Distribution of radionuclides in the Columbia River streambed from nuclear reactors, Hanford Reservation to Longview, Washington. U. S. Geological Survey, U. S. AEC report TID-26531. 78pp. (NSA 29: 24204).

Haushild, W. L., et al. 1971. Radionuclides in transport in the Columbia River from Pasco to Vancouver, Washington. U. S. Geological Survey, U. S. AEC report TID-25894. 77pp. (NSA 26: 22777).

Hausknecht, D. F., D. M. O'Connor and C. F. Smith. 1972. Comparative risks from nuclear and fossil-fueled power plant accidents. Trans. Amer. Nucl. Soc. 15(1): 75-76. (NSA 26: 44735).

Haven, K. F. 1975. Methodology for impact assessment in the estuarine marine environment. Lawrence Livermore Lab., U. S. ERDA report UCRL-51949. 45pp. (NSA 33: 20238).

Haverty, M. I. and G. W. Ware. 1970. Circadian sensitivity and dosage-rate response to x-irradiation in the pink bollworm. J. Econ. Entomol. 63(4): 1296-1300. (NSA 24: 42047).

Havin, E. and P. Oftedal. 1965. Radiosensitivity of Drosophila embryos in air and in N_2, in relation to age. Abstr. 66. Radiation Research 25(1): 196.

Havin, E. and P. Oftedal. 1968. Radiosensitivity in Drosophila embryos in relation to age, in air and in nitrogen. Int. J. Radiat. Biol. 14: 149-160. (NSA 23: 20334).

Havlík, B. 1964. The radioactivity of some Czechoslovak waters. Cesk. Hyg. 9: 551-564. (In Czech) (NSA 19: 32477).

Havlík, B. 1965. Influence of uranium ore mining and processing on surface waters. Cesk. Hyg. 10(10): 577-588. (CA 64: 10916; NSA 20: 45974).

Havlík, B. 1970. Radium-226 content of water and plankton from the Chalk River area. Atomic Energy of Canada, Ltd., Chalk River report AECL-3687. 29pp. (NSA 25: 4473).

Havlík, B. 1971. Radium in aquatic food chains: radium uptake by fresh water algae. Radiation Research 46(3): 490-505. (NSA 25: 44756).

Havlík, B. and E. Robertson. 1973. Radium uptake by freshwater algae. pp. 372-380. In: Radionuclides in Ecosystems. (See Nelson, 1973) (NSA 28: 27690).

Havlík, B., J. Gráfová and B. Nýcová. 1968. Radium-226 liberation from uranium ore processing mill waste solids and uranium rocks into surface streams. I. The effect of different p_H of surface waters. II. The effect of different chemical composition of surface water. Health Physics 14(5): 417-422, 423-430.

Hawkes, C. 1972. The estimation of the dispersal rate of the adult cabbage root fly (Erioischia brassicae (Bouché)) in the presence of a brassica group. J. Appl. Ecol. 9(2): 617-632. (^{32}P).

Hawkes, C. 1974. Dispersal of adult cabbage root fly (Erioischia brassicae (Bouché)) in relation to a brassica crop. J. Appl. Ecol. 11(1): 83-93. (^{32}P).

Hawkins, D. B. and B. L. Schmalz. 1967. Environmental tritium studies at the National Reactor Testing Station. pp. 157-158. In: Isotope Techniques in the Hydrologic Cycle. (See Stout, 1967).

Hawkins, R. H., R. F. Overman and J. C. Corey. 1971. ^{208}Tl and ^{24}Na gamma sources for identifying soil water tagged with deuterium. Soil Sci. Soc. Am. Proc. 35: 199-201. (NSA 25: 37495).

Hawthorne, J. and R. B. Duckworth. 1958. Fall-out radioactivity in a deer's antlers. Nature 182(4645): 1294. (NSA 13: 1924).

Haxel, O. 1962. Carbon-14 in nature. Naturw. Rundschau 15: 133-140. (NSA 19: 32437).

Hay, C. J. and W. C. Myser. 1961. Use of P^{32} as an aid in biological studies of the leafhopper Scaphoideus luteolus. J. Econ. Entomol. 54: 1260-1261. (NSA 17: 1194).

Hayashi, M. and F. Matsumura. 1967. Interactions of DDT with the nervous system of the resistant and suseptible German cockroaches. Nature 215: 1510-1512. (^{14}C).

Hayashi, M. and F. Matsumura. 1967. Insecticide mode of action. Effect of dieldrin on ion movement in the nervous system of Periplaneta americana and Blattella germanica cockroaches. J. Agr. Food Chem. 15(14): 622-627. (^{22}Na).

Hayashi, S. and T. Takayanagi. 1966. Cytological and cytogenetical studies of Paramecium polycaryum. VI. Effects of gamma irradiation especially concerning nuclear division in conjugation. Annot. Zool. Jap. 39: 16-22. (NSA 21: 18034).

Hayes, D. W. and J. E. Hoy. 1973. Chromatographic system for the enrichment and analysis of low-level tritium samples. pp. 127-132. In: Tritium. (See Moghissi and Carter, 1973) (NSA 28: 27019).

Hayes, F. N., E. Hansbury and V. N. Kerr. 1960. Contemporary carbon-14: The p-cymene method. Anal. Chem. 32(6): 617-620.

Hayes, F. N., et al. 1960. Contemporary carbon-14 in lemongrass oil. Z. Physik 158: 374-378.

Hayes, F. R. 1955. The effect of bacteria on the exchange of radiophosphorus at the mud-water interface. Theoret. Angew. Limnol. 12: 111-116.

Hayes, F. R. and N. R. Beckett. 1955. Flow of materials through the thermocline of a lake. Arch. Hydrobiol. 51(3): 391-409.

Hayes, F. R. and C. C. Coffin. 1951. Radioactive phosphorus and exchange of lake nutrients. Endeavour 10(38): 78-81.

Hayes, F. R. and L. H. Jodrey. 1952. Utilization of phosphorus in trout as studied by injection of radioactive phosphorus. Physiol. Zool. 25(2): 134- 144.

Hayes, F. R. and J. E. Phillips. 1958. Lake water and sediment. IV. Radiophosphorus equilibrium with mud, plants, and bacteria under oxidized and reduced conditions. Limnol. Oceanogr.3(4): 459-475.

Hayes, F. R., B. L. Reid and M. L. Cameron. 1958. Lake water sediment. II. Oxidation-reduction relations at the mud-water interface. Limnol. Oceanogr. 3(3): 308-317.

Hayes, F. R., et al. 1952. On the kinetics of phosphorus exchange in lakes. Ecology 40(1): 202-216.

Hayes, J. T. 1969. Studies of a plant-membracid-ant food chain, using ^{32}P as a tracer. Abstr. Bull. Ecol. Soc. Am. 50(3): 131.

Hayes, J. T. 1969. A microscope for radioassay of live insects with automatic counting equipment. Int. J. Appl. Rad. Isotopes 20: 603-604.

Hayes, T. L., R. F. W. Pease and A. S. Camp. 1967. Stereoscopic scanning electron microscopy of living Tribolium confusum. J. Insect Physiol. 13: 1143-1145.

Hayne, D. W. 1969. Implications of radioecology for other research in environmental contamination. pp. 32-38. In: Symposium on Radioecology. (See Nelson and Evans, 1069).

Haynes, C. V. 1967. Bone organic matter and radiocarbon dating. pp. 163-168. In: Radioactive Dating and Methods of Low-Level Counting. (See IAEA, 1967) (NSA 22: 6456).

Hayodom, V., P. Khambanonda and U. Saimongkol. nd. Standard neutron irradiation facility experience in Thailand. International Atomic Energy Agency, Vienna, report IAEA-R-459. 6pp. (NSA 25: 27062).

Hazama, K. 1958. Effect of n-irradiation on mulberry tree. Jap. J. Breeding 8(3): 201.

Hazama, K. 1968. Adaptability of mutant in mulberry tree. Gamma Field Symp. (7): 79-85. (NSA 25: 4486) (Irradiation).

Hazle, A. J. and R. D. Siek. 1971. Environmental sampling. A crucial look at common procedures. pp. 347-351. In: Proc. Rocky Flats Symp. on Safety in Plutonium Handling Facilities. (See Dow Chemical Co., 1971) (NSA 25: 54779).

Hazzaa, I. B. 1969. Determination of the velocity of ground-water flow with radioisotopes. Int. J. Appl. Rad. Isotopes 20: 127-129. (NSA 23: 11651).

Hazzaa, I. B., R. K. Girgis and K. F. Saad. 1965. A method for the determination of aquifer parameters using radioactive isotopes. Int. J. Appl. Rad. Isotopes 16: 487-493. (NSA 19: 42647).

Hazzaa, I. B., A. M. Sharhan and Y. K. Shuhaiber. 1970. Determination of the average velocity of groundwater flow and preliminary estimation of the quantity of groundwater in Shigaya Aquifer in Kuwait using radioisotopes. Isotopes Radiat. Res. 3: 1-7. (NSA 25: 10700).

Hazzaa, I. B., et al. 1972. Tritium content of the Nile Delta zones indicating the movement of the subsoil water. Isotope Radiat. Res. 5: 12-17. (NSA 27: 24611).

Healy, J. W. 1946. Accumulation of radioactive elements in fish immersed in pile effluent water. Hanford Atomic Products Operation, U. S. AEC report HW-3-3442. 9pp.

Healy, J. W. 1956. Reactor effluent monitoring. Hanford Atomic Products Operation, U. S. AEC report HW-45721. 30pp. (NSA 31: 29620).

Healy, J. W. 1972. Problems of plutonium measurement in the environment. Trans. Nucl. Sci. NS-19(1): 219-223. (NSA 26: 42728).

Healy, J. W. 1974. Proposed interim standard for plutonium in soils. Los Alamos Scientific Lab., U. S. AEC report LA-5483-MS. 100pp. (NSA 29: 29342).

Healy, J. W., J. F. Hanstead and W. Y. Matsumoto. 1956. Sanitary engineering aspects of the atomic energy industry. pp. 23-36. In: A seminar, AEC and PHS, Robert A. Taft Sanitary Engineering Center, Cincinnati, O., Dec. 6-9, 1955. U. S. AEC report TID-7517(Pt. II)(Del.). (NSA 13: 13039).

Healy, W. B., W. J. McCabe and G. F. Wilson. 1970. Ingested soil as a sour source of microelements for grazing animals. New Zeal. J. Agr. Res. 13 (3): 503-521. (NSA 26: 15541) (^{60}Co, ^{54}Mn, ^{75}Se, ^{65}Zn).

Hearing, V. J. and J. J. Eppig, Jr. 1969. Electron microscopy of the normal and ^{131}I-treated thyroid gland of the newt Notophthalmus viridescens. J. Morphology 128: 369-386. (NSA 24: 2784).

Hearst, J. R. 1974. Neutron logging in partially saturated media. Lawrence Livermore Lab., U. S. AEC report UCRL-75114. 26pp. (NSA 30: 6764).

Heaslip, M. B. 1959. Effects of seed irradiation on germination and seedling growth of certain deciduous trees. Ecology 40(3): 383-388.

Heaslip, M. B. 1959. Radiation sensitivity of desiduous tree seed. Abstr. Proc. 9th Intern. Bot. Congr. 2(A): 14.

Heaslip, M. B. 1963. Factors affecting tree seed and seedling radiosensitivity. pp. 301-306. In: Radioecology. (See Schultz and Klement, 1963) (NSA 17: 33575).

Heaslip, M. B. 1963. Relative biological effectiveness of neutron and gamma rays on tree seed. Abstr. Bull. Ecol. Soc. Am. 44(3): 69.

Heaslip, M. B. 1964. Radiosensitivity of dormant and physiologically active tree seed to different ratios of neutron and gamma radiations. Abstr. Bull. Ecol. Soc. Am. 45(4): 149.

Heaslip, M. B. 1967. Radiosensitivity of deciduous tree seed to different ratios of fast neutrons and gamma radiations. Radiation Botany 7(5): 415-428.

Heaslip, M. B. 1972. Gamma and fast neutron sensitivity of tree seed and seedlings. Final report. Moorehead State Univ., Ky., U. S. AEC report ORO-2066-16. 49pp. (NSA 27: 12295).

Heaslip, M. B. 1973. Factors affecting the radiosensitivity of Fraxinus americana L. seedlings and seeds. pp. 978-982. In: Radionuclides in Ecosystems. (See Nelson, 1973) (NSA 28: 30074).

Heath, D. F. 1962. Cl^{36}-dieldrin in mice. pp. 83-91. In: Radioisotopes and Radiation in Entomology. (See IAEA, 1962).

Heath, D. F. 1963. Some problems in the determination of residues in plants and mammals. pp. 185-192. In: Radiation and Radioisotopes Applied to Insects of Agricultural Importance. (See IAEA, 1963).

Heatwole, H., et al. 1970. Effects of radiation on a population of the Puerto Rican tree snail, Caracolus caracolla. Ch. E-1, pp. E17-E24. In: A Tropical Rain Forest. (See H. T. Odum and R. F. Pigeon, 1970). (NSA 25: 21814).

Hebel, R. 1973. Contents of ^{137}Cs in different soils as a basis for precisely directed research of the ^{137}Cs contamination of foodstuffs of animal origin. Deut. Tierärztl. Wochenschr. 80(17): 404-408. (In German) (NSA 29: 15634).

Hebrard, J. P. and L. Foulquier. 1975. Study of manganese-54 uptake by Platyhypnidium riparioides (Hedw.). Rev. Bryol. Lichénol. 4(1): 35-54. (In French) (NSA 33: 12042).

Hebrard, J., L. Foulquier and A. Grauby. 1968. Summary of the modalities of the contamination of the moss, Platyhypnidium riparioides (Hedw.) Dix. by cesium-137 and strontium-90. Rev. Bryol. Lichénol. 36(1-2): 119-242. (In French).

Hebrard, J. P., L. Foulquier and A. Grauby. 1972. Use of a terrestrial moss (Grimmia orbicularis (Bruch)) as an indicator of the presence of manganese-54, cesium-137 and strontium-90. Radioprotection 7(3): 159-177. (In French, English summary).

Hebrard, J. P., L. Foulquier and A. Grauby. 1974. Experimental approach to ^{90}Sr transfer processes from rocks to a terrestrial moss: Grimmia orbicularis Bruch. Bull. soc. bot. Fr. 121(7): 235-249. (In French) (NSA 33: 26011).

Heckmann, U. and H. A. Kunkel. 1960. Bone marrow and organ changes in irradiated dormice. Strahlentherapie 112: 92-104. (In German) (NSA 14: 14172).

Heckmann, U. and H. A. Kunkel. 1962. Investigations with cystein and cysteamine on x-irradiated frogs. Strahlentherapie 118: 359-362. (In German) (NSA 16: 31528).

Heddle, J. A., D. Whissell and D. J. Bodycote. 1969. Changes in chromosome structure induced by radiations: a test of the two chief hypotheses. Nature 211: 1158-1160. (NSA 23: 35692) (Rat kangaroo, Potorus tridactylis).

Hedgepath, J. W. 1972. Radioactive contamination in the Antarctic. pp. 97-112. In: Proc. Colloqu. on Conservation Problems in Antarctica. B. C. Parker, ed. Virginia Polytechnic Inst., Blacksburg, Va. (NSA 28: 15482).

Hedgran, A. and B. Lindell. 1970. On the Swedish policy with regard to the limitation of radioactive discharges from nuclear power stations: an interpretation of current international recommendations. Sweden National Inst. of Radiation Protection, Stockholm, SSI-70-27. 25pp. (NSA 25: 12757).

Hedgran, A. and B. Lindell. 1971. Some views on the limitation of releases of radioactive material from nuclear power stations. pp. 125-1[illegible]0. In: 3rd Nordic Radiation Protection Conf. (Copenhagen, August 18-20, 1971). (See Nordic Society, 1971) (In Swedish) (NSA 27: 22477) (Environmental monitoring, radioecological research).

Hedin, P. A., G. Wiygul and N. Mitlin. 1967. Absorption and metabolism of C^{14}-labelled tepa by the boll weevil. J. Econ. Entomol. 60(1): 215-218.

Hedin, P. A., et al. 1967. Sterility induced by tepa in the boll weevil: effective dose and permanency, gonadal changes, and biological turnover of labelled compound. J. Econ. Entomol. 60(1): 209-214. (^{14}C).

Heemstra, R. J., J. W. Watkins and F. E. Armstrong. 1961. Laboratory evaluations of nine water tracers. Nucleonics 19(1): 92-96.

Heffner, R. L., et al. 1973. Effects of power plant operation on Hudson River estuary microbiota. pp. 619-629. In: Radionuclides in Ecosystems. (See Nelson, 1973) (NSA 28: 27357) (Nuclear power plant).

Heidenthal, G. 1945. The occurrence of x-ray dominant lethal mutations in Habrobracon. Genetics 30(2): 197-205.

Heidenthal, G. A. 1959. A comparison of the genetic effects of x-rays and cathode rays on oocytes of Habrobracon. (i.e. Bracon). Intern. Congr. Zool. Proc. 15: 961-963.

Heidenthal, G. 1960. Genetic effects of x-rays and cathode rays on oocytes of Habrobracon. Genetics 45(5): 633-639. (NSA 14: 18781).

Heidenthal, G. 1962. Embryonic lethals induced by x-irradiation of first meiotic metaphase oocytes of Habrobracon. Genetics 47: 685-693. (NSA 16: 30237).

Heidenthal, G., L. B. Clark and J. W. Gowen. 1955. Comparative effectiveness of roentgen rays of 124 KV and 50 MEV on Habrobracon eggs treated in first meiotic metaphase. Am. J. Roentgenol. Rad. Therapy Nucl. Med. 74(4): 677-685. (NSA 9: 7634).

Heidersdorf, S. D. 1965. Alaska radiological health activities. (Paper, Am. Assoc.Advancement of Science 16th Alaska Science Conf., Juneau, Alaska, August 30-September 1, 1965). 12pp. (AED-CONF-65-241-1) (NSA 20: 37042).

Heier, K. S. 1965. Radioactive elements in the continental crust. Nature 208: 479-480. (NSA 20: 9139).

Heier, K. S. and J. L. Carter. 1964. Uranium, thorium, and potassium contents in basic rocks and their bearing on the nature of the upper mantle. pp. 63-86. In: The Natural Radiation Environment. (See Adams and Lowder, 1964) (NSA 18: 35731).

Heine, R. W. 1971. Estimation of conductivity and conducting area of poplar stems using a radioactive tracer. Ann. Botany 34(138): 1019-1024.

Heinemann, H. 1960. Using the radioactive sediment densitometer on Sebetha Lake in Kansas. Soil and Water Conservation Research, U. S. Dept. of Agriculture Reserach Station Reserach Report 343.

Heinonen, J. and E. Häsänen. 1970. Determination of caesium in biological material by neutron activation analysis. Radiochem. Radioanal. Lett. 4: 181-187. (NSA 24: 41230).

Heinonen, J. and O. Suschny. 1972. Reliability of neutron activation analysis for trace elements in biological maretials. pp. 155-173. In: Nuclear Activation Techniques in the Life Sciences, 1972. (See IAEA, 1972) (NSA 27: 22085).

Heinonen, J. and O. Suschny. 1974. Reliability of mercury analysis in environmental materials. J. Radioanal. Chem. 20(2): 499-519. (NSA 30: 20593) (Neutron activation analysis).

Heinonen, J., D. Merten and O. Suschny. 1972. Reliability of mercury analysis in environmental samples. pp. 137-141. In: Mercury Contamination in Man and His Environment. (See IAEA, 1972) (NSA 27: 11671) (Incl. neutron activation analysis).

Heintz, E. 1958. Indirect chemotrophic action of γ rays on the bitterling (Rhodeus amarus B.). Compt. rend. 246: 1309-1312. (In French).

Heitmann, H. G. 1971. Treatment of waste waters in nuclear power stations. Aufbereit.-Tech. (5): 281-284. (In German) (NSA 26: 16929).

Heitzman, M. W. and R. E. Simpson. 1972. Neutron activation analysis of mercury in fish, flour, and standard reference orchard leaves by electrodeposition radiochemistry. J. Assoc. Off. Anal. Chem. 55(5): 960-965. (NSA 26: 58500).

Hekrdle, J. and M. Vydra. 1969. Nomograph for determination of the optimum measuring time of radioactive samples. Jaderna Energie 15: 168-170. (In Czech) (NSA 23: 35471).

Hela, I. 1963. Alternate ways of expressing the concentration factors for radioactive substances in aquatic organisms. Bull. Inst. Oceanogr. Monaco 61(1280): 8pp. (IAEA Radioactivity in the Sea Publ. No. 6) (NSA 21: 22370).

Helbig, W., M. Beer and E. Kuhn. 1967. Distribution measurements for testing new devices of insecticide and fungicide spreading. Isotopenpraxis 3: 334-338. (In German) (NSA 22: 6747) (^{198}Au).

Held, E. E. 1960. Land crabs and radioactive fallout at Eniwetok Atoll. Pacific Science 14(1): 18. (Also: Univ. of Washington, U. S. AEC report UWFL-50) (NSA 12: 5849).

Held, E. E. 1963. Some aspects of the biology of zirconium-95. pp. 577-579. In: Radioecology. (See Schultz and Klement, 1963) (NSA 17: 33610).

Held, E. E. 1963. Qualitative distribution of radionuclides at Rongelap Atoll. pp. 167-169. In: Radioecology. (See Schultz and Klement, 1963).

Held, E. E. 1965. Gamma dose rates at Rongelap Atoll, 1954-1963. Univ. of Washington, U. S. AEC report UWFL-91. iii, 16pp.

Held, E. E. 1971. Amchitka radiobiological program progress report July 1970 to April 1971. Univ. of Washington, U. S. AEC report NVO-269-11. iii, 36pp.

Held, E. E. and A. W. Bellamy. 1950. Irradiation responses in Daphnia magna. I. Increased resistance to x-irradiation under low oxygen tension. Univ. of Calif., Los Angeles, U. S. AEC report UCLA-69. (NSA 4: 3843).

Held, E. E., S. P. Gessel and R. B. Walker. 1965. Atoll soil types in relation to the distribution of fallout radionuclides. Univ. of Washington, U. S. AEC report UWFL-92. vii, 37pp.

Held, E. E. 1971. Radiological resurvey of animals, soils and groundwater at Bikini Atoll, 1969-1970. Univ. of Washington, U. S. AEC report NVO-269-8(Rev.1). v, 44pp. (NSA 26: 41291).

Held, E. E., et al. 1973. Amchitka radiobiological program progress report March 1972 to December 1972. Univ. of Washington, U. S. AEC report NVO-269-19. iv, 90pp. (Incl. mammals, birds, fishes, plants, water).

Helder, R. J. and J. Boerma. 1972. Polar transport of labelled rubidium ions across the leaf of Potamogeton lucens. Acta Bot. Neer. 21(3): 211-__.

Heldern, O. E., Th.-A. Alpen and G. Körn. 1970. Radioactivity in lakes of the GDR, 1968-1969. Kernenergie gesundh. 12(2): 323-326. (In German).

Helf, S. 1967. Environmental radioactivity measurements in the Picatinny Arsenal area, 1963-1966. Feltman Research Labs., Picatinny Arsenal, U. S. Army, Dover, N. J. (AD-651495; NSA 21: 43046).

Helf, S. 1967. Environmental radioactivity in woods soil from Dover, New Jersey during 1964-1966. Radiol. Hlth. Data 8(9): 553-557.

Heller, J. 1961? Biochemistry of insects. In: 5th Intern. Congr. of Bio-Chemistry. (Symp. No. III, Moscow, 10-16 August 1961). (Concerns ^{32}P).

Heller, J. and T. Chojnacki. 1957. Pyrophosphates in the hawkmoth, Celerio euphorbiae. III. Tracer studies with P^{32}. Bull. Acad. Polon. Sci. ser. sci. biol. 5(9): 277-279.

Heller, J., M. J. Peichowska and T. Chojnacki. 1958. Inorganic polyphosphates in the hawkmoth (Celerio euphorbiae L. Acta biochem. Polon. 5(4): 343-354. (^{32}P).

Hellman, A., D. H. Martin and L. J. Wopschall. 1968. The influence of x-irradiation on survival and interferon levels in viral infected mice. Proc. Soc. Exptl. Biol. Med. 128: 455-458.

Hellwig, H. 1963. Effect of roentgen rays on the diatome Nitzschia linearis. Radiation Botany 3(3): 249-257. (NSA 18: 5083).

Helmstetter, C. E. 1961. Radiation sensitivity of synchronous populations of Escherichia coli. Abstr. Radiation Research 14(4): 474.

Helseth, F. 1965. Movement and distribution of radioisotopes on a granitic outcrop in the Georgia Piedmont. Thesis, Univ. of Georgia, Athens.

Hemenway, R. C., Jr. 1971. Effects of gamma radiation on the respiration of Acarus siro L. Thesis, Univ. of Georgia, Athens. 69pp. (Univ. Microfilms 72-2493) (NSA 26: 25880).

Hemenway, R. C., Jr. and R. Davis. 1972. Effects of gamma radiation on the respiration of Acarus siro. J. Ga. Entomol. Soc. 7(2): 140-151. (NSA 26: 45884).

Hemmeter, J. C. and E. Zueblin. 1915. The radioactivity of the mineral waters of Hot Springs, Warm Springs and Healing Springs in Hot Springs, Va. Arch. Intern. Med. 15: 188-203.

Hemphill, D. D. (Ed.). 1971. Trace Substances in Environmental Health. IV. (Proc. 4th Ann. Conf. held at Columbia, Missouri, June 23-25, 1970). Univ. of Missouri, Columbia. 468pp. (CONF-700664) (NSA 26: 15204).

Hemphill, D. D. (Ed.). 1973. Trace Substances in Environmental Health. (Proc. Univ. of Missouri's 7th Ann. Conf., Columbia, Missouri, June 12-14, 1973). Univ. of Missouri, Columbia. 474pp. (CONF-730613) (NSA 30: 5928).

Henderson, B. J., R. C. Baxter and L. W. Tuttle. 1957. The effect of a highly irradiated food medium on life span and development of Drosophila melanogaster. Abstr. Radiation Research 7(3): 321.

Henderson, C. and R. F. Foster. 1957. Studies on small-mouth bass (Micropterus dolomieu) in the Columbia River near Richland, Washington. Trans. Am. Fish. Soc. 86: 112-117.

Henderson, C., G. G. Robeck and R. C. Palange. 1954. Biological aspects of low-level radioactivity in the Columbia River. pp. 153-165. In: Sanitary Engineering Conference. (Baltimore, Md., April 15-16, 1954). U. S. AEC report WASH-275.

Henderson, C., G. G. Robeck and R. C. Palange. 1956. Effects of low-level radioactivity in the Columbia River. Publ. Hlth. Repts. 71(1): 6-14. (NSA 10: 2595).

Henderson, S. A. 1964. RNA synthesis during male meiosis and spermiogenesis. Chromosoma 15(4): 345-366. (Schistocerca gregaria, Cyrtacanthacris tartarica, Chorthippus brunneus; ^{3}H).

Hendricks, D. E. 1974. Tobacco budworm: effects of ^{60}Co irradiation on female attractiveness and mating frequency. J. Econ. Entomol. 67(5): 610-612. (NSA 31: 812).

Hendricks, J. 1963. The effect of low dose x-rays on the auditory threshold in the pigeon. Thesis, Florida State Univ., Tallahassee.

Hendy, C. H., T. A. Rafter and M. W. G. MacIntosh. 1972. Formation of carbonate nodules in the soils of the Darling Downs, Queensland, Australia, and the dating of the Talgai Cranium. pp. D106-D126. In: Proc. 8th Intern. Radiocarbon Dating Conf. Vol. 1. (See Rafter, 1972) (NSA 31: 8703).

Henneberry, T. J. 1963. Effects of gamma radiation on the fertility and longevity of *Drosophila melanogaster*. J. Econ. Entomol. 56(3): 279-281. (NSA 18: 78).

Henneberry, T. J. 1964. Effects of gamma radiation on the fertility of the two-spotted spider mite and its progeny. J. Econ. Entomol. 57(5): 672-673.

Henneberry. T. J. 1967. Susceptibility of various stages of *Drosophila melanogaster* to gamma radiation. J. Econ. Entomol. 60: 1041-1043. (NSA 22: 10864).

Henneberry, T. J. and W. L. McGovern. 1963. Effects of gamma radiation on mating competitiveness and behavior of *Drosophila melanogaster*. J. Econ. Entomol. 56(6): 739-741.

Henneberry, T. J. and W. L. McGovern. 1963. Some effects of gamma radiation on fertility of *Drosophila melanogaster* and variability of sperm. J. Econ. Entomol. 56(6): 819-822.

Henneberry, T. J. and W. N. Sullivan. 1962. The effect of gamma radiation on silkworm eggs. Abstr. 275. Bull. Entomol. Soc. Am. 8(3): 166.

Henneberry, T. J. and W. N. Sullivan. 1963. Effect of gamma-radiation on eggs of the silkworm. Nature 200(4911): 1121-1122. (NSA 18: 8170).

Henneberry, T. J., F. F. Smith and W. L. McGovern. 1964. Some effects of gamma radiation and a chemosterilant on the Mexican bean beetle. J. Econ. Entomol. 57(6): 813-815.

Henning, E. 1959. Study of suction of *Aphis fabae* Scop. with ^{32}P. Naturwissenschaften 46(12): 410-411. (In German).

Henning, E. 1962. Recent investigations on the importance of the so-called probing punctures by aphids. Z. Pfl. Krankh. 69(6): 321-330. (In German) (^{32}P).

Henning, E. 1966. Study on the histology and function of feeding punctures of the black bean aphid, *Aphis fabae* Scop. on *Vicia fabia* plants. J. Insect Physiol. 12: 65-76. (In German, English summary) ($^{32}PO_4$).

Henning, H. 1964. On the determination of the water content of a snow cover by the use of gamma rays at the Fichtelberg meteorological station. Z. Meteorologie 17: 229-233. (In German) (NSA 20: 550).

Henry, S. M. 1962. The significance of microorganisms in the nutrition of insects. Trans. N. Y. Acad. Sci. 24(6): 676-683. (Discusses radionuclides).

Henry, S. M. and R. J. Block. 1960. The sulfur metabolism of insects. IV. The conversion of inorganic sulfate to organic sulfur compounds in cockroaches. The role of intracellular symbionts. Contr. Boyce Thompson Inst. 20(5): 317-329. (^{35}S).

Henry, S. M. and V. F. Cotty. 1957. The rearing of asptic adult house flies for physiological studies. Contr. Boyce Thompson Inst. 19(2): 227-229. (Describes feeding labelled compounds).

Henry, S. M., R. J. Block and T. W. Cook. 1964. Methionine sulfoxide and other combined amino acids in the German cockroach. Adv. Chem. Ser. 44: 85-95. (^{35}S).

Henshaw, P. S. 1935. Changes in sensitivity of *Drosophila* eggs to x-rays. II. Correlation of biological sensitivity and radiosensitivity. Radiology 24(4): 438-443.

Henshaw, P. S. 1940. Further studies on the action of roentgen rays on the gametes of *Arbacia punctata*. IV. Changes in radiosensitivity during the first cleavage cycle. Am. J. Roentgenol. 43: 917-920.

Henshaw, P. S. 1943. Peculiar growth lesions in frogs induced by irradiation of sperm cells with x-rays. J. Natl. Cancer Inst. 3(4): 409-417. (*Rana pipiens*).

Henshaw, P. S. 1963. Radiation effects and peaceful uses of atomic energy in the animal sciences: radiation and biologic capability. pp. 13-17. In: Radioecology. (See Schultz and Klement, 1963).

Henshaw, P. S. and J. M. Golomb. 1940. Responses of *Drosophila* pupae to x-rays. Radiology 34(6): 721-730.

Henshaw, P. S. and C. T. Henshaw. 1933. Changes in susceptibility of *Drosophila* eggs to x-rays. I. A Correlation of changes in radiosensitivity with stages in development. Radiology 21(3): 239-258.

Herbst, W. 1959. Studies on the radioactive contamination of the human environment. I. Soil and plant life. Atompraxis 5: 280-284. (AEC-tr-4003) (NSA 14: 2289, 12437).

Herbst, W. 1961. The situation on worldwide radioactive contamination. pp. 304-316. In: Proc. 4th Aerosol Congr. (Bad Lippspringe, April 20-22, 1961). H. Nuckle, ed. Friedrich-Karl Schattaner-Verlag, Stuttgart. (In German).

Herbst, W. 1962. Radioactive contamination of the environment of man. Vitalstoffe Zivilizationskrankh. 7: 255-260. (In German) (NSA 21: 16362).

Herbst, W. 1962. Barrier and filter effect of vegetation with respect to volatile radioiodine. Strahlentherapie 116: 420-425. (AEC-tr-5381) (NSA 16: 32717).

Herbst, W. 1963. Radiobiological problems of aerosols. Properties and hazards of radioactive aerosols produced by atomic technique. I. Radioiodine (survey). Zentr. biol. Aerosol Forsch. 11(3): 232-273. (AEC-tr-6550).

Herbst, W. 1964. Life tables. 39th continuation. Radiation and atomic fission. Part 6. Radioactive contamination of foodstuffs, air and water through nuclear weapon tests. Vitalstoffe Zivilisationskrankh. 9: 111-115. (In German) (NSA 22: 40699) (Incl. fishes).

Herbst, W. 1964. Investigations of environmental radiation and its variability. pp. 781-796. In: The Natural Radiation Environment. (See Adams and Lowder, 1964) (NSA 17: 31829).

Herbst, W. 1964. The question of radiation-induced stimulation in gymnosperm seedlings. Atompraxis 10(8): 361-367. (In German) (NSA 18: 38807).

Herbst, W. 1965. Filtering and protective effect of forests against radioactive and other impurities of the atmosphere. Forst-Holzwirt 20: 216-220. (In German) (NSA 23: 29430).

Herbst, W. and G. Hübner. 1962. On the variability of the terrestrial components of penetrating external environmental radiation in the open air. Atomkernenergie 7(12): 481-486. (In German) (NSA 17: 7775).

Herbst, W., et al. 1957. Investigation of the radioactivity in vegetation. Atomkernenergie 2(10): 357-367. (In German) (NSA 12: 2199).

Herceg, J. E, 1972-1973. Environmental monitoring in the vicinity of the Los Alamos Scientific Laboratory. Los Alamos Scientific Lab., U. S. AEC report LA-4871, LA-4970, LA-5184. 44, 48, 58pp. (NSA 26: 25561, 48032) (Contains data on wild organisms).

Herceg, J. E. and L. J. Johnson. 1973. Los Alamos Scientific Laboratory's environmental monitoring program: an overview with emphasis on environmental dosimetry. Trans. Amer. Nucl. Soc. 17: 541-542. (NSA 29: 5124).

Herchl, M. and J. Svitel. 1974. Contamination of plants with strontium 90. Voj. Zdravot. Listy 43(1): 29-31. (In Slovak) (NSA 31: 19603).

Hercik, F. (Ed.). 1966. Radiobiological Methods. Acad. Nakladetelstvi Cesk. Akad. Ved., Prague. (In Czech) (NSA 21: 32792).

Herde, K. E. 1946. Studies in the accumulation of radioactive elements in Oncoryhnchus tschawytscha (chinook salmon) exposed to a medium of pile effluent water. Hanford Atomic Products Operation, U. S. AEC report HW-3-5064; M-4130. 21pp. (NSA 13: 9891).

Herde, K. E. 1947. Biological monitoring of the laboratory area by analyses of rodents and other forms of wildlife. Hanford Atomic Products Operation, U. S. AEC report AECD-2888. 12pp. (NSA 4: 5829).

Herde, K. E. 1947. Radioactivity in various species of fish from the Columbia and Yakima Rivers. Hanford Atomic Products Operation, U. S. AEC report M-4129.

Herde, K. E. 1948. A one year study of radioactivity in Columbia River fish. Hanford Atomic Products Operation, U. S. AEC report HW-8668. 4pp.

Herde, K. E. 1948. A second annual check of radioactivity in upland wildfowl near the H. W. project. Hanford Atomic Products Operation, U. S. AEC report HW-11987. 3pp.

Herde, K. E. 1949. Biological monitoring of waste water from 200N storage buildings. Hanford Atomic Products Operation, U. S. AEC report HW-12855. 5pp.

Herde, K. E. 1951. Deposition of radioactivity in rats drinking pile effluent water. Hanford Atomic Products Operation, U. S. AEC report HW-20102. 13pp.

Herde, K. E. 1956. Review of the S. R. P. environmental control program and the influence of nuclear weapons testing. Savannah River Operations Office, U. S. AEC report SRO-29. 22pp. (NSA 13: 9902).

Herde, K. E. and J. F. Cline. 1949. Radioactivity in pekin ducks on the Columbia River. Hanford Atomic Products Operation, U. S. AEC report HW-12079. 11pp.

Herde, K. E., R. L. Browning and W. C. Hanson. 1951. Activity densities in waterfowl of the Hanford Reservation and environs. Hanford Labs., U. S. AEC report HW-18645. 14pp.

Hermanowicz, W., et al. 1960. The examination of the Vistula water within the boundaries of Greater Warsaw. Nukleonika 4: 557-566. (In Polish) (NSA 14: 11372).

Hernadi, F., et al. 1963. Search for antagonistic Actinomycetae in Hungarian soils. VI. The effects of background radiation on streptomycetes. Acta Biol. Avad. Sci. Hung. 14: 111-121. (NSA 18: 19560).

Hernandez, J. A., et al. 1969. Environmental radiometric study in Columbia during 1968. Columbia Instituto de Asuntos Nucleares, Bogata, report IAN-RS-ERA-4. 54pp. (In Spanish) (NSA 24: 31486).

Herrmann, D. B. 1968. New measurement of the radon content of the waters at Bad Brambach. Arch. Phys. Ther. 20: 29-32. (In German) (NSA 24: 16697).

Herrmann, H. 1970. Residual β activity and γ-spectrometric nuclide determination in bees and honeys. Z. Wasser-Abwasser-forsch. 3: 152-153. (In German) (NSA 25: 42374).

Herskowitz, I. H. 1957. Induced changes in female germ cells of Drosophila. III. The effect of maternal dehydration on oviposition rate and mortality of x-rayed oocytes. Genetics 42(3): 289-298.

Herskowitz, I. H. 1963. An influence of maternal nutrition upon the gross chromasomal mutation frequency recovered from x-rayed sperm of Drosophila melanogaster- Genetics 48(5): 703-710. (NSA 17: 27102).

Hertwig, O. 1910. Radium radiation and its effect on the development of an animal eggs. Sitzungsb. K. Pruss. Akad. Wissensch. (24 Feb., 28 Jul.) (In German).

Hertwig, O. 1911. Radium irradiation of unfertilized frog eggs and their development after fertilization with normal sperm. Arch. mikrosk. anat. 77: 165-___. (In German).

Hertwig, O. 1911. The radium sickness of animal germ cells. A contribution to the experimental evidence and the laws of heredity. Arch. mikrosk. anat. 77: 1, 97-__. (In German).

Hertwig, O. 1911. Resulting changes from radium irradiation in the nuclear fission forms in eggs of <u>Ascaris megalocephala</u>. Arch. mikrosk. anat. 76: 301. (In German).

Hertwig, O. 1916. Productive development of half-nucleated triton and fish embryos after radium irradiation. Arch. mikrosk. anat. 87: 63. (In German).

Hervey, R. J. and Z. M. Williams. 1963. Microbial survival in gamma-irradiated soils. Abstr. Texas Rept. Biol. Med. 21(3): 450. (NSA 18: 15885).

Heshamul-Huque. 1963. Some observations on the effect of ionizing radiation against desert locust. Pakistan J. Sci. 15: 267-270. (NSA 21: 43501).

Heslop, J. P. 1964. The estimation of adenosine triphosphate and related compounds in insect tissue. Biochem. J. 91: 183-187. (^{32}P).

Heslop, J. P. 1965. The irreversible nature of glycolysis in the housefly, <u>Musca domestica</u>. Biochem. J. 95: 40. (^{14}C).

Heslop, J. P. and J. W. Ray. 1958. Phosphorus compounds of cockroach nerve and the effect of DDT. Biochem. J. 70(4): 16. (^{32}P).

Hespe, E. D. nd. Use of the critical nuclide critical pathway-critical population group concept in the establishment of water quality criteria. Australian Atomic Energy Commission, Lucas Heights. (For 5th Ann. Australian Convention on Water and Waste Water Assoc., Adelaide, May 1972). 12 pp. (CONF-720597) (NSA 28: 27282).

Hess, B. 1964. Tidal fluctuations of natural radioactivity of the atmosphere. Zentr. Biol. Aerosol Forsch. 11: 495-507. (NSA 19: 18224).

Hess, C. T., et al. 1974. Radioactive isotopic characterization of the environment near Wiscasset, Maine: a preoperational survey in the vicinity of the Maine Yankee Atomic Power Plant. Radiat. Data Rep. 15(2): 39-52. (NSA 29: 26860).

Hess, D. W. 1959. Radiation ecology of the Front Range, report on background gamma radiation of sites in the Rocky Mountain Front Range west of Boulder, Colorado. Inst. of Arctic and Alpine Research, Univ. of Colorado, Boulder. 26pp.

Hess, L. W., D. B. Dunn and G. Leddicotte. 1968. Potential of activation analysis for the comparison of trace minerals in plant species. Trans. Missouri Acad. Sci. 2: 91-99.

Hess, V. F. 1943. On the use of radioactive tracer method in measurement in water. Trans. Am. Geophys. Un. 2: 587-594.

Hetherington, J. A. and D. F. Jefferies. 1974. Distribution of some fission product radionuclides in sea and estuarine sediments. Neth. J. Sea Res. 8(4): 319-338. (NSA 32: 19765).

Heunisch, G. W. 1970. Field method for the estimation of uranium in natural waters. Microchim. Acta (2): 258-261. (NSA 24: 29191).

Heurtebise, M. 1971. Semi-automated determination of iodine in biological fluids by activation analysis. J. Radioanal. Chem. 7(2): 227-233. (NSA 25: 54516).

Hevesey, G. 1945. Rate of renewal of the fish skeleton. Acta Physiol. Scand. 9(3/4): 234-247. (^{32}P).

Hevesey, G. V., D. Lockner and K. Sletten. 1964. On the radiation sensitivity of extramedullary hematopoiesis. Med. Welt (10): 455-460. (In German) (NSA 19: 3937, 29957) (Used fish).

Hewitt, J. E. 1968. Radiation exposures during Biosatellite II flight. BioScience 18(6): 565-569. (Symposium on the Biosatellite II Experiments).

Hewitt, J. E. 1971. Radiation exposures during the Biosatellite II flight. Report N-72-20065. 14pp. (NSA 26: 53365).

Hewlett, J. D., J. E. Douglass and J. L. Clutter. 1964. Instrumental and soil moisture variance using the neutron-scattering method. Soil Science 97: 19-29.

Heydarn, K., J. Lippert and P. Theodorsson. 1957. The radioactivity in the Risö district, measurements up to April 1, 1957. Denmark Atomic Energy Commission, Risö, report RISÖ-1. (U. K. Atomic Energy Authority report AERE-Lib/Trans.-790) (NP-6672) (NSA 13: 16134).

Heyland, K. U., S. Solansky and W. Aufhammer. 1974. Simple laboratory equipment for $^{14}CO_2$-labelling of plants. Angew. Bot. 48(1/2): 63-75. (In German) (NSA 32: 1245).

Hibbert, F. A., V. R. Switzur and R. G. West. 1971. Radiocarbon dating of Flandrian pollen zones at Red Moss, Lancashire. Proc. Roy. Soc. (London) B177: 161-176.

Hibiya, T. and M. Oguri. 1961. Gill absorption and tissue distribution of some radionuclides (Cr-51, Hg-203, Zn-65 and Ag-115m,110), in fish. Bull. Jap. Soc. Sci. Fisheries 27(11): ___. (In English) (NSA,Japan 2: 7).

Hibiya, T. and T. Yagi. 1956. Effects of fission materials upon development of aquatic animals. pp. 1219-1224. In: <u>Research in the Effects and Influences of the Nuclear Bomb Test Expolsions</u>. <u>Vol. II</u>. Japan Soc. for the Promotion of Science, Ueno.

Hickey, J. L. S. 1966. Digest of protection standards and state regulations for radioactivity in water. Radiol. Hlth. Data Repts. 7: 549-554. (NSA 22: 28504).

Hickey, J. L. S. and S. D. Campbell. 1968. ^{226}Ra concentrations in public water supplies. Public Hlth. Repts. 83: 551-557. (NSA 23: 2833).

Hickman, G., J. Gates and E. Edmonds. 1966. Fly tagging: a new approach. J. Environ. Hlth. 28: 277-282. (NSA 21: 8751).

Hickman, G., J. Gates and E. Edmonds. 1967. Radioactive fly tagging. Pest Control 35(4): 20-22. (^{32}P).

Higginbotham, L. B. and J. T. Collins. 1974. U. S. Atomic Energy Commission regulatory program for the control, monitoring and reporting of radioactivity in effluents: an overview. pp. 138-150. In: Proc. Seminar on the Monitoring of Radioactive Effluents. (Karlsruhe, Germany, May 14-17, 1974). (See Organization for Economic Co-operation and Development, 1974) (NSA 31: 3343).

Higgins, E. 1950. Radioactive wastes and their significance in stream ecology. Trans. Am. Fish. Soc. 79: 217-232.

Higgins, E. 1951. Atomic radiation hazards for fish. J. Wildl. Mgt. 15 1-12. (NSA 5: 2348).

Higgins, G. H. 1959. Evaluation of the ground water contamination hazard from underground nuclear explosions. Lawrence Radiation Lab., U. S. AEC report UCRL-5538. 24pp. (NSA 13: 13427).

Higgins, G. H. 1965. Artificial and natural radioactivity in nuclear excavation. Lawrence Radiation Lab., U. S. AEC report UCRL-14191(Rev.1). 19 pp. (NSA 20: 33436).

Higgs, D. A. and J. G. Eales. 1971. Iodide and thyroxine metabolism in the brook trout, Salvelinus fontinalis (Mitchill), during sustained exercise. Can. J. Zool. 49(9): 1255-1269. (^{125}I).

Highnam, K. C. 1962. Neurosecretory control of ovarian development in the desert locust. Mem. Soc. Endocrinol. 12: 370-390. (^{35}S).

Hightower, B. G. 1963. Nocturnal resting places of the screw-worm fly. J. Econ. Entomol. 56(4): 498-500. (Concerns irradiation).

Hightower, B. G. 1969. Population dynamics of the screw-worm fly, Cochliomyia hominivorax (Coquerel), with respect to control by the sterile-male technique. pp. 25-31. In: Insect Ecology and the Sterile-Male Technique. (See IAEA, 1969).

Hightower, B. G. and D. A. Alley. 1965. Observation on the effects of releasing sterile screwworm flies in northwestern Vera Cruz, Mexico. Abstr. 376. Bull. Entomol. Soc. Am. 11(3): 173.

Hightower, B. G. and J. J. Garcia. 1972. Longevity and sexual activity of newly enclosed irradiated screwworm flies held at immobilizing temperatures. J. Econ. Entomol. 65(3): 877-878. (NSA 26: 43355).

Hightower, B. G. and O. H. Graham. 1968. Current status of screw-worm eradication in the southwestern United States of America. pp. 51-54. In: Control of Livestock Insect Pests by the Sterile-Male Technique. (See IA IAEA, 1968) (NSA 24: 34810).

Hightower, B. G., A. L. Adams and D. A. Alley. 1965. Dispersal of released irradiated laboratory-reared screw-worm flies. J. Econ. Entomol. 58: 373-374. (NSA 19: 29844).

Hilchey, J. D. 1957. The action of ionizing radiations on insects. pp. 240-266. In: Radiation Preservation of Food. U. S. Army Quartermaster Corps, U. S. Army Research and Development Service, Washington, report No. 1.

Hilchey, J. D., V. F. Cotty and S. M. Henry. 1957. The sulfur metabolism of insects. II. The metabolism of cystine-S^{35} by house fly, Musca domestica L. Contr. Boyce Thompson Inst 19(2): 189-200.

Hilchey, J. D., et al. 1955. The sulfur metabolism of insects. I. The utilization of cystine and methionine by the German cockroach, Blatella germanica (L.). Contr. Boyce Thompson Inst. 18(2): 109-123. (^{35}S).

Hildreth, P. E. 1967. Developmental effects of x-rays on embryos of Drosophila melanogaster. Proc. Natl. Acad. Sci. (US) 58: 1924-1929. (NSA 22: 31035).

Hill, C. R. 1960. Lead-210 and polonium-210 in grass. Nature 187(4733): 211-212.

Hill, C. R. 1962. An α-spectrometer for normal biological levels of radioactivity. Health Physics 8(1): 91-92.

Hill, C. R. 1967. Rapporteur's synopsis. pp. 359-365. In: Radioecological Concentration Processes. (See Aberg and Hungate, 1967) (Natural radionuclides).

Hill, C. R. 1967. Routes of uptake of ^{210}Po into human tissues. pp. 297-302. In: Radioecological Concentration Processes. (See Aberg and Hungate, 1967).

Hill, C. R. 1968. Natural radiation. pp. 124-127. In: Guidelines to Radiological Health. (See U. S. Public Health Service, 1968) (NSA 23: 50462).

Hill, C. R. and D. S. Woodhead. 1964. Tissue dose due to neutrons of cosmic-ray origin. pp. 449-462. In: The Natural Radiation Environment. (See Adams and Lowder, 1964) (NSA 18: 35405).

Hill, C. R., R. V. Osborne and W. V. Mayneord. 1964. Studies of a radioactivity in relation to man. pp. 395-406. In: The Natural Radiation Environment. (See Adams and Lowder, 1964) (NSA 18: 35271).

Hill, D. K. 1964. The space accessible to albumin with the striated muscle fiber of the toad. J. Physiol. 175: 275-294. (NSA 19: 35918) (^{3}H).

Hill, J. E. 1954. Effects of environment in reducing dose rates produced by radioactive fallout from nuclear explosions. Rand Corp., Santa Monica, Calif., report RM-1285-1. 10pp.

Hill, L. 1965. Incorporation of C^{14}-glycine into the proteins of the fat body of the desert locust during ovarian development. J. Insect Physiol. 11(2): 1605-1615.

Hillel, D. 1971. Application of radiation techniques in water-use efficiency studies. Report to the International Atomic Energy Agency. Israel National and Univ. Inst. of Agriculture, Rehoveth. 87pp. (INIS-mf-529) (NSA 28: 18241).

Hillel, D. and Y. Benyamini. 1974. Experimental comparison of infiltration and drainage methods for determining unsaturated hydraulic conductivity of a soil profile _in situ_. pp. 271-275. In: _Isotope and Radiation Techniques in Soil Physics and Irrigation Studies_. (See IAEA, 1974) (NSA 30: 15499).

Hilliard, T. R. 1965. Water pollution from various sources including uranium mining operations. 9pp. (For Intern. Conf. and Exhibition of the Canadian Nuclear Assoc., Quebec. Toronto, Preprint 65-CNA-219) (CONF-650515-2) (NSA 19: 38290).

Hillova, J. 1966. Recovery of x-ray induced damage in _Chladydomonas reinhardi_. Bull. Boris Kidric Inst. Nucl. Sci. 17(4/A Suppl.): __. (AEC-tr-6646/4A; pp. 11-15) (NSA 23: 12236).

Hillova, J. 1967. Failure of recovery from sub-lethal damage after ultraviolet irradiation in _Chlamydomonas_. Int. J. Radiat. Biol. 13: 383-385. (NSA 22: 36404).

Hillova, J. and V. Drasil. 1967. The inhibitory effect on iodoacetamide on rececery from sub-lethal damage in _Chlamydomonas reinhardti_. Int. J. Radiat. Biol. 12(3): 201-208. (NSA 22: 40963) (X-ray).

Hilton, H. W., N. W. Nomura and S. S. Kameda. 1972. Simplified oxygen bomb combustion procedure for ^{14}C in agricultural materials and soils. Anal. Biochem. 49(1): 285-289. (NSA 26: 55764).

Himes, F. L. and R. Shufelt. 1970. Influence of some organic compounds on the movement of ^{90}Sr in soils. pp. 493-507. In: _Proc. Intern. Symp. on Radioecology_. (See France Commissariat à l'Énergie Atomique, 1970) (NSA 25: 2365).

Hinckley, A. D. 1968. Survival of oak leafminers in gamma-irradiated forest. Abstr. Bull. Ecol. Soc. Am.49(3): 106.

Hinckley, A. D. 1969. Gamma radiation as a tool in population ecology. Abstr. Bull. Ecol. Soc. Am. 50(2): 87.

Hinckley, A. D. 1971. Infestation of irradiated white oaks by the solitary oak leafminer, _Cameraria hamadryadella_ (Lepidoptera Gracilariidae). Ann. Entomol. Soc. Am. 64(5): 1094-1098. (NSA 25: 58092).

Hinckley, A. D. 1973. Concentration of oak aphids, _Myzocallis_ spp., on irradiated white oaks. Environ. Entomol. 66(5): 1071-1073. (NSA 29: 613).

Hinds, W. T. and J. F. Cline. 1975. Techniques for long-term controlled studies of plant uptake of transuranics. Battelle-Northwest Lab., U. S. ERDA report BNWL-SA-5403. 17pp. (NSA 32: 19756).

Hines, B. M., F. A. Harris and N. Mitlin. 1973. _Heliothis virescens_: assimilation and retention of radioactivity in eggs and larvae from ^{32}P-treated adults. J. Econ. Entomol. 66(5): 1071-1073. (NSA 29: 823).

Hines, N. 1962. _Proving Ground_. (An Account of the Radiobiological Studies in the Pacific, 1946-1961). Univ. of Washington Press, Seattle. xvii, 366pp.

Hines, W. J. W. and M. J. H. Smith. 1963. Intermediary metabolism in the desert locust. J. Insect Physiol. 9(4): 463-468. (^{14}C).

Hinrichs, M. A. 1928. Modification of development on the basis of differential susceptibility to radiation. 1. _Fundulus heteroclitus_ and ultraviolet radiation. J. Morphol. Physiol. 41(1): 239-265.

Hinton, H. E. 1954. Radioactive tracers in entomological research. Sci. Progr. 42(166): 292-305.

Hinzpeter, A. 1970. Measuring very low β activities (fallout) in water and ashes after enrichment by selective ion exchange. Atompraxis 16: 414-419. (In German) (NSA 25: 15515).

Hirai, K, T. Miyata and T. Saito. 1973. Penetration of ^{32}P-dimethoate into organophosphate resistant and susceptible citrus red mite, _Panonychus citri_ McGregor (Acarina: Tetranychidae). Appl. Entomol. Zool. 8(3): 183-190. (NSA 33: 3330).

Hirayama, M. 1969. Radioactivity of groundwaters from landslide areas and vicinity. Utsunom. Daigaku Kyoikug. Kiyo (19): 39-43. (In Japanese) (NSA 26: 9252).

Hirth, H. F., et al. 1969. Dispersal of snakes from a hibernaculum in northwestern Utah. Ecology 50(2): 332-339. (^{182}Ta tag).

Hishida, T. 1959. Effect of x-ray irradiation on _Oryzias latipes_ under different conditions. II. Sex difference of radiosensitivity. Nippon Acta Radiol. 19(1): 93-94. (In Japanese).

Hislop, J. S. 1973. Photon activation analysis of biological and environmental samples: a review. pp. 1159-1176. In: _Photonuclear Reactions and Applications_. _Vol. II_. B. L. Berman, ed. (Intern. Conf., Pacific Grove, Calif., March 26-30, 1973) (CONF-730301-P2) (NSA 28: 29584).

Hislop, J. S. and D. R. Williams. 1975. Gamma activation of environmental and biological materials. Trans. Nucl. Sci. NS-21(3): 6-7. (NSA 32: 27471).

Hitchcock, M. and J. N. Smith. 1963. Detoxication mechanisms in the tick _Bophilus decloratus_. Abstr. Biochem. J. 87(1): 34P-35P. (^{14}C).

Hitchcock, M. and J. N. Smith. 1964. Comparative detoxication. XIII. Detoxication of aromatic acids in arachnids; arginine, glutamic acid, and glutamine conjugations. Biochem. J. 93(2): 392-400. (^{14}C, _Bophilus_, _Tegenaria_, _Agelena_, _Epeira_).

Hiyama, Y. (Comp.). 1957. Radiological data in Japan. II. Concentrations of Sr^{90}, Cs^{137}, Pu^{239}, Zn^{65}, Fe^{55}, and Cd^{113}, with Addendum 1. Japan report to the United Nations Scientific Committee on the Effects of Atomic Radiation. 24pp. (NSA 14: 23125).

Hiyama, Y. 1957. Radiological data in Japan. II. Concentrations of Sr^{90}, Cs^{137}, Pu^{239}, and others in various materials on earth's surface. Japan Sci. Monthly 10: 1-17.

Hiyama, Y. (Comp.). 1960. Biogeochemical studies with strontium-90, cesium-137, and others. Nippon Gakuj. Shink. : 79-93. (AEC-tr-4245) (NSA 15: 10668).

Hiyama, Y. 1960. Bio-chemical studies of Sr^{90}, Cs^{137} and similar radionuclides. Annual Report No. 11. Tyokyo. (AEC-tr-4642) (NSA 16: 26688).

Hiyama, Y. 1961. Predation of trout fry traced by radioisotopes. Abstr. A-3. (See Japan Atomic Industrial Forum, 1961).

Hiyama, Y. 1962. Radioisotope marking methods of fish and the public health consideration. Radioisotopes (Tokyo) 11: 389-393. (In Japanese) (NSA 17: 7703).

Hiyama, Y. (Comp.). 1963. Biogeochemical studies with strontium-90, cesium-137, and others. Nippon Gakuj. Snink. : 39-52. (In Japanese) (Incl. fish, deer antlers, wild duck).

Hiyama, Y. and M. Shimizu. 1969. Uptake of radioactive nuclides by aquatic organisms.: the application of the exponential model. pp. 463-475. In: Environmental Contamination by Radioactive Materials. (See IAEA, 1969).

Hiyama, Y., R. Ischikawa and F. Yasuda. 1956. Studies on the process of contamination of fish by the fission products. pp. 1119-1133. In: Research on the Effects and Influences of the Nuclear Bomb Test Explosions. Vol. II. (See Japan Society for the Promotion of Science, 1956).

Hiyama, Y., et al. 1956. State and behavior of fission products in fresh-water, brackish water and marine coastal systems in Japan, 1954-55. pp. 1027-1065. In: Research on the Effects and Influences of the Nuclear Bomb Test Explosions. Vol. II. (See Japan Society for the Promotion of Science, 1956).

Hiyama, Y., et al. 1960. Predation of rainbow trout fry traced by radio-isotopes. Bull. Jap. Soc. Sci. Fish. 26(9): 863-867. (SFA 6: 3852).

Hladky, E., I. Kubik and J. Moravek. nd. General model for evaluating transport of radioactive products released to environment and its consequences. Vyskuny Usav Energeticky, Laslovske Bohunice, Czechoslovakia. (In Czech) (INIS-mf-887, pp. 204-209) (NSA 29: 15590).

Hladky, E., I. Pietrik and J. Moravek. 1974. Use of inorganic sorbents for collecting and monitoring radioiodine in gaseous discharges and in the environs of nuclear power plants. pp. 21-29. In: Environmental Surveillance Around Nuclear Installations. Vol. 2. (See IAEA, 1974) (In Russian) (NSA 31: 14068).

Ho, L. C. and A. J. Peel. 1969. Transport of ^{14}C-labelled assimilates and ^{32}P-labelled phosphate in Salix viminalis in relation to phyllotaxis and leaf age. Ann. Botany 33(132): 743-751.

Ho, L. C. and A. J. Peel. 1969. Investigation of bidirectional movement of tracers in sieve tubes of Salix viminalis. Ann. Botany 33(133): 833-844. (^3H, ^{14}C, ^{32}P).

Hoar, W. S. and J. G. Eales. 1963. The thyroid gland and low-temperature resistance of goldfish. Can. J. Zool. 41(4): 653-669. (SFA 8: 5953) (Used radioiodine).

Hoar, W. S. and J. G. Eales. 1963. Dependence of conversion ratio on the dose of radioiodine in fish. Can. J. Zool. 41(6): 1061-1068. (NSA 18: 11909; SFA 8: 6149) (Goldfish, chum salmon, chub).

Hoather, R. C. and R. F. Rackman. 1963. Some observations on radon in waters and its removal by aeration. J. Inst. Water Eng. 17: 13-22. (NSA 20: 23296).

Hobble, J. E. 1964. Carbon-14 measurements of primary production in two arctic Alaskan lakes. Verh. Intern. Verein. Limnol. 15(1): 360-364.

Hobble, J. E. 1973. Using kinetic analysis of uptake of carbon-14 to measure rates of movement of individual organic compounds in aquatic bacteria. pp. 207-214. In: Modern Methods in the Study of Microbial Ecology. T. Rosswall, ed. NFR, Editorial Services, Stockholm. (NSA 29: 888).

Hober, E. 1968. New site for the eradication of a white grub population (Melolontha vulgaris Fabr. and Melolontha hippocastani Fabr.) by release of irradiated adults. pp. 259-271. In: Isotopes and Radiation in Entomology. (See IAEA, 1968). (See Horber below).

Hobitz, H. 1961. Direct and indirect biological effects of radiation. Arch. Hyg. u. Bakteriol. 145: 561-570. (NSA 16: 20142) (Pollen).

Hochman, L. F., T. E. Jones and B. D. Tan. 1971. Differential sensitivity of Hymenolepis microstoma (Cestoda: Cuclophyllidea) cysticercoid stages to x-rays. J. Parasitol. 57(2): 443-445. (NSA 25: 46960).

Hodge, F. A., W. R. Leif and M. S. Silverman. 1969. Susceptibility to infection with Pasteurella tularensis and the immune response of mice exposed to continuous low dose rate gamma radiation. J. Infect. Dis. 120: 356-365. (NSA 24: 34153).

Hodges, R. and G. Guyer. 1958. The effects of an irradiated wheat diet on the confused flour beetle, granary weevil and the angoumois grain moth. J. Econ. Entomol. 51(5): 674-675.

Hodgson, R. A., C. G. Denny and R. Monaghan. 1962. The sediment density meter. Dresser Industries, Inc., Tulsa, Okla., U. S. AEC report TID-18124. 26pp.

Hodson, S. A. 1968. Comparative responses of two protozoa to two types of radiation. Experientia 24: 59. (NSA 22: 21619).

Hofer, K. and R. Biebl. 1965. Variations of resistance against alpha radiations with the time of day. Protoplasma 59: 506-521. (In German) (NSA 20: 14361).

Hoffert, J. R. and P. L. Fromm. 1959. Estimation of thyroid secretion rate of rainbow trout using radioactive iodine. J. Cell. Comp. Physiol. 54(2): 163-169.

Hoffman, A. C. 1972. Characterization of repair mechanisms operating subsequent to genetic damage induced by ethyl methanesulfonate and gamma radiation in *Bracon hebetor*. Mutation Research 16(2): 175-188. (NSA 26: 59113).

Hoffman, C. M., et al. 1969. Comparison of forensic soil specimens by neutron activation analysis. pp. 251-255. In: *Modern Trends in Activation Analysis. Vol. I.* (See DeVoe, 1969) (NSA 24: 133).

Hoffman, D. A. 1962. The concentration and accumulation of radioactive phosphorus by an aquatic food chain. Thesis, Colorado State Univ., Fort Collins. (Dissertation Abstr. 23: 364-365) (NSA 16: 28732).

Hoffman, D. A. and J. R. Olive. 1961. The use of radiophosphorus in determining food chain relationships in the aquatic environment. Colorado State Univ., Fort Collins, U. S. AEC report TID-13108. 46pp. (NSA 15: 23200; SFA 7: 5168; WPA 35: 270).

Hoffman, D. C., et al. 1971. Detection of ^{244}Pu in nature. Nature 234 (5325): 132-134. (NSA 26: 9241).

Hoffman, G. R. 1972. Accumulation of ^{137}Cs by cryptogams in a *Liriodendron tulipifera* forest. Bot. Gaz. (Chicago) 133(2): 107-119. (NSA 27: 20175).

Hoffman, J. 1940. Uranium determination in springs southeast of the Erzgebirge. Monatsh. Chemie 73: 159-170. (In German).

Hoffman, J. 1941. Experimental determination of uranium in living freshwater algae. Naturwissenschaften 29: 403-404. (AEC-tr-3076) (CA 38: 2360; Chem. Zentr. II: 2096; NSA 12: 1201).

Hoffman, J. 1941. Uranium determination in springs southeast of the Erzgebirge. Monatsh. Chemie 74: 38-52. (In German).

Hoffman, J. 1942. Determination of uranium in living and dead plants. Bod. u. Pflanzenernähr. 26: 318-327. (French transl. CEA-tr-A-580) (NSA 13: 21836).

Hoffman, J. 1942/43. Bioelement uranium in the plant, animal, and humans. Biochem. Z. 313(5/6): 377-387. (In German).

Hoffman, J. 1943. A further contribution to the distribution of U in plant material. Bod. u. Pflanzenernähr. 32: 295-306. (In German) (CA 38: 4653).

Hoffman, J. A. 1971. Effects of selective irradiation of hemopoietic tissue on the hemogram of male imagos of *Locusta migratoria* (Orthoptera). Compt. rend. D273(18): 1604-1607. (In French) (NSA 26: 12437).

Hoffman, J. A. 1972. Modifications of the haemogramme of larval and adult *Locusta migratoria* after selective x-irradiations of the haemocytopoietic tissue. J. Insect Physiol. 18(9): 1639-1652. (NSA 27: 5547).

Hoffman, P. F. and B. M. Zuckerman. 1954. Oak wilt fungus labeled with C-14. Science 120(3107): 106-108.

Hoffman, R. A., A. W. Lindquist and J. S. Butts. 1951. Studies on treatment of flies with radioactive phosphorus. J. Econ. Entomol. 44(4): 471-473.

Hoffman, R. A., C. D. Schmidt and J. J. Matter. 1966. Laboratory and field studies with ^{60}Co sterilized horn flies. Bull. Entomol. Soc. Am. 12: 285.

Hoffman, V. 1922. On the excitation and paralyzation of animal cells with roentgen radiation. I. Experimental investigations on frog eggs and larvae. Strahlentherapie 13: 285-298. (In German).

Hoffmann, J. A. and M. J. Weins. 1974. Proteosynthetic activity of the prothoracic glands and ecdysone titre in permanent larvae of *Locusta migratoria* obtained by selective irradiation of hematopoietic tissue. Experientia 30(7): 821-822. (In French) (NSA 30: 24362). (See Hoffman, J. A. above).

Hoffmeister, H. 1963. Elucidation of the structure of ecdysone. Thesis, Ludwig-Macimilians-Universität, Munich, Germany. 53pp. (In German) (*Calliphora*, ^{3}H).

Hogan, G. R. 1966. Pyrimidines and the x-ray response of *Tribolium confusum*. Thesis, Univ. of Illinois, Urbana. 109pp. (NSA 21: 20417).

Hohn, A. 1955. The effect of x radiation on the larvae of *Xenopus*. Oncologia 8(3): 273-289. (NSA 9: 6577).

Holan, T., et al. 1963. The effect of P^{32} on the early development phases of *Triturus vulgaris* eggs. Fiziol. Normala Pathol. (Bucharest) 9: 169-173. (In Rumanian) (NSA 18: 31161).

Holbrook, F. R. and M. S. Fujimoto. 1970. Mating competitiveness of unirradiated and irradiated Mediterranean fruit flies. J. Econ. Entomol. 63: 1175-1176. (NSA 24: 44649).

Holcomb, C. M., et al. 1973. Occurrence of radionuclides in the exoskeleton of turtles. pp. 385-389. In: *Radionuclides in Ecosystems*. (See Nelson, 1973. (NSA 28: 27677).

Holcomb, H. P., J. H. Horton and E. L. Wilhite. 1976. Actinide-soil interactions in waste management at the Savannah River Plant. Savannah River plant, U. S. ERDA report DP-MS-75-118. 52pp. (NSA 33: 26006).

Holden, A. V. 1962. A study of the absorption of ^{14}C-labelled DDT from water by fish. Ann. Appl. Biol. 50(3): 467-477. (BA 41: 17306; SFA 8: 5769).

Holifield, C. 1969. The scientist's responsibility in the control of man's environment. pp. 1-8. In: *Symposium on Radioecology*. (See Nelson and Evans, 1969).

Hollaender, A. 1961. Initial steps in radiation damage to chromosomes and means of preventing this effect. pp. 201-209. In: The Initial Effects of Ionizing Radiations on Cells. Academic Press, London. (NSA 16: 30277) (Microorganisms).

Hollaender, A. and R. F. Kimball. 1956. Modification of radiation-induced genetic damage. Nature 177(4512): 726-730. (Microorganisms).

Hollaender, A. and G. E. Stapleton. 1956. The influences of chemical pre- and posttreatments on radiosensitivity of bacteria, and their significance for higher organisms. pp. 120-135. In: Ciba Foundation Symp. on Ionizing Radiations and Cell Metabolism. (NSA 11: 5736).

Holland, H. D. and J. L. Kulp. 1954. The transport and distribution of uranium, ionium, and radium in rivers, oceans and ocean sediments. Geochim. Cosmochim. Acta. 5: 197-213.

Holland, J. Z. 1963. The distribution and physical-chemical nature of fallout. Fed. Proc. 22(6): 1390-1397.

Hölldobler, B. 1964. The behavior of male ants during their imaginal life span. Experientia 20(6): 329-330. (In German, English summary) (^{32}P, Camponotus ligniperda, Formica polytena).

Holleman, D. F. and R. A. Dietrich. 1973. Body water content and turnover in several species of rodents as evaluated by the tritiated water method. J. Mammal. 54(2): 456-465.

Holleman, D. F. and J. R. Luick. 1971. Transfer of radiocesium from lichen to reindeer. Health Physics 21(5): 657-666. (NSA 26: 12256).

Holleman, D. F., J. R. Luick and F. W. Whicker. 1971. Transfer of radiocesium from lichen to reindeer. Health Physics 21: 657-666.

Holleman, D. F., J. R. Luick and R. G. White. 1973. Transfer of radium in milk from reindeer cow to calf. pp. 76-80. In: Radionuclides in Ecosystems. (See Nelson, 1973) (NSA 28: 27672).

Holler, J. R. 1968. Effects of irradiation on soil, root, and litter microfungal populations in a Puerto Rican rainforest. Thesis, Univ. of South Carolina, Columbia. 75pp. (NSA 23: 41274).

Holler, J. R. and G. T. Cowley. 1970. Response of soil, root, and litter microfungal populations to radiation. Ch. F-6, pp. F35-39. In: A Tropical Rain Forest. (See H. T. Odum and R. F. Pigeon, 1970) (NSA 25: 21948).

Holley, W. L. 1975. Environmental monitoring report, Sandia Laboratories, 1974. Sandia Labs., U. S. ERDA report SAND-75-0257. 17pp. (NSA 32: 6418).

Holliday, R. 1965. Radiation sensitive mutants of Ustilago maydis. Mutation Research 2: 557-559. (NSA 20: 12694).

Hollingdale, B. and J. S. F. Barker. 1971. Selection for increased abdominal bristle number in Drosophila melanogaster with concurrent irradiation. II. Populations derived from an outbred cage population. Theor. Appl. Genet. 41(6): 263-274. (NSA 26: 18071).

Hollister, H. 1965. Problems in estimating the biological consequences of nuclear war. pp. 1-11. In: Problems in Estimating the Biological Consequences of Nuclear War (Three Papers). (See Hollister and Eberhardt, 1965).

Hollister, H. 1966. Problems in estimating the biological consequences of nuclear war. pp. 52-61. In: Radiation Effects on Natural Populations. (See Sacher, 1966).

Hollister, H. and L. L. Eberhardt. 1965. Problems in Estimating the Biological Consequences of Nuclear War (Three Papers). U. S. AEC report TAB-R-5. vii, 40pp.

Holluta, J. 1958. Radioactivity and the distribution of water from Lake Constance. Schweiz. Ver. Gas-u. Wasserfach. Monatsbull. 38: 102-107, 127-133. (NSA 13: 19105).

Holm, A. W. 1970. Activity measurements of milk, fodder, excrement, water, and soil from farms in Samnanger and Matre, summer and autumn 1970. Norway Forsvarets Forskningsintitut, Kjeller. 32pp. (In Norwegian) (NP-19900) (NSA 29: 7718).

Holm, E. and R. B. R. Persson. 1975. Fallout plutonium in Swedish reindeer. Health Physics 29(1): 43-51. (NSA 32: 27945).

Holmes, J. R. 1962. Applications of radioactive isotopes in veterinary medicine. Vet. Bull. 32: 65-72. (NSA 20: 3859).

Holmes, J. W. 1956. Calibration and field use of the neutron scattering method of measuring soil water content. Austral. J. Appl. Sci. 7: 45-58.

Holmes, R. W. 1961. Size fractionation of C^{14} labelled natural phytoplankton communities. Abstr. Bact. Proc. 1961: 38.

Holmes & Narver, Inc. 1972. Reviews to insure safety of nuclear operations. Los Angeles. U. S. AEC report NVO-48(Rev.2). 15pp. (NSA 26: 48006).

Holmes & Narver, Inc. 1973. California power plant siting study. Executive summary. Dames & Moore, Los Angeles, Calif. Anaheim, Calif., report HN-8145.3(Vol.1). 56pp. (NSA 28: 20537).

Holmquist, C.-E. 1961. Hygienic aspects of tritium formation in heavy-water reactors. Nord. Hyg. Tidskr. 42(11/12): 277-295. (In Swedish) (NSA 18: 454).

Holowaychuk, N., P. L. Gersper and L. P. Wilding. 1969. ^{90}Sr content of soils near Cape Thompson, Alaska. Soil Science 107: 137-144. (NSA 23: 13856).

Holstein, B. 1971. Metabolism of intraperitoneally injected ^{14}C-histamine in the yellow eel (Anguilla anguilla). Comp. Biochem. Physiol. 38(2A): 435-441. (SFA 16: 13775).

Holt, B. R. 1972. Radiosensitivity and distribution of common macrofungi in a chronically irradiated oak-pine forest. Radiation Botany 12(5): 339-342. (NSA 27: 5473).

Holt, B. R. and P. J. Bottino. 1972. Structure and yield of a chronically irradiated winter rye-weed community. Radiation Botany 12(5): 355-359. (NSA 27: 5476).

Holt, G. 1974. Analysis of mercury content in game 1968-1970. Fauna 26(1): 38-58. (NSA 29: 15272) (Neutron activation analysis).

Holt, P. D. 1969. The dosimetry of irradiations of Chlamydomonas cells with high LET radiations from a Van de Graaf accelerator. Phys. Med. Biol. 14: 563-572. (NSA 24: 4623).

Holton, R. L. 1968. Effects of ^{60}Co gamma irradiation on the reproductive performance of the brine shrimp, Artemia. Thesis, Oregon State Univ., Corvallis. 81pp. (NSA 23: 33818).

Holton, R. L., C. L. Osterberg and W. O. Forster. 1973. Effect of gamma irradiation on the reproductive performance of Artemia as determined by individual pair matings. pp. 1191-1197. In: Radionuclides in Ecosystems. (See Nelson, 1973) (NSA 28: 30241).

Holton, R. L., C. L. Osterberg and W. O. Forster. 1973. Effect of gamma irradiation on the maintenance of population size in the brine shrimp, Artemia. pp. 1198-1205. In: Radionuclides in Ecosystems. (See Nelson, 1973) (NSA 28: 30242).

Holtzman, R. B. 1963. The Pb^{210}(RaD) concentrations of some biological materials from Arctic regions. Argonne National Lab., U. S. AEC report ANL-6769. pp. 59-65. (NSA 18: 13400) (Incl. lichens, reindeer and caribou antler, meat and bone).

Holtzman, R. B. 1964. Lead-210 (RaD) and polonium-210 (RaF) in potable waters in Illinois. pp. 227-238. In: The Natural Radiation Environment. (See Adams and Lowder, 1964) (NSA 18: 35738).

Holtzman, R. B. 1966. Natural levels of ^{210}Pb, ^{210}Po and ^{226}Ra in humans and biota of the Arctic. Nature 210: 1094-1097. (NSA 20: 43005).

Holtzman, R. B. 1966. ^{226}Ra and the natural airborne nuclides ^{210}Pb and ^{210}Po in Arctic biota. pp. 51-58. In: Radiological Physics Division Annual Report July 1965 to June 1966. Argonne National Lab., U. S. AEC report ANL-7220. (NSA 21: 26405).

Holtzman, R. B. 1966. ^{226}Ra and the natural nuclides ^{210}Pb and ^{210}Po in Arctic biota. Abstr. Health Physics 12(12): 1807-1808.

Holtzman, R. B. 1968. ^{226}Ra and the natural airborne nuclides ^{210}Pb and ^{210}Po in Arctic biota. pp. 1087-1096. In: Radiation Protection, Part 2. (See Snyder, et al., 1969) (NSA 23: 14185).

Holtzman, R. B. 1969. Concentrations of the naturally occurring radionuclides ^{226}Ra, ^{210}Pb and ^{210} in aquatic fauna. pp. 535-546. In: Symposium on Radioecology. (See Nelson and Evans, 1969).

Holtzman, R. B. 1970. Isotopic composition as a natural tracer of lead in the environment. Environ. Sci. Technol. 4: 314-317. (NSA 24: 22792).

Holy, Z. J. 1973. Comparison of the environmental aspects of nuclear and fossil fuelled power stations. New South Wales Univ., Kensington, Australia. 9pp. (CONF-740555-1) (NSA 30: 26836).

Holzberg, S. and J. H. Schroeder. 1972. Behavioral mutagenesis in the convict cichlid fish, Cichlasoma nigrofasciatum Guenther. I. The reduction of male aggressiveness in the first post-irradiation generation. Mutation Research 16(3): 289-296. (NSA 26: 59083).

Holzer, F. 1970. Plowshare and the environment. Trans. Amer. Nucl. Soc. 13: 632-633. (Also: Lawrence Radiation Lab., U. S. AEC report UCRL-72830) (CONF-701102) (NSA 25: 13430, 4171).

Homsher, P. J. and D. E. Sonenshine. 1972. Spermatogenesis in Dermacentor variabilis (Say) in relation to duration of attachment and the presence of chromosomal aberrations in stocks injected with radioactive glucose (Acarina: Ixodidae). J. Med. Entomol. 9(2): 171-177. (NSA 27: 7790).

Honda, Y., T. Muraoka and T. Koga. 1973. Behavior of ^{60}Co in a non-flowing small scale rice field ecosystem. J. Radiat. Res. (Japan) 14(3): 209-218. (NSA 29: 24205).

Honda, Y., et al. 1971. Uptake and loss of radioactive ruthenium by freshwater organisms. Radioisotopes (Tokyo) 20(8): 376-382. (In Japanese) (NSA 26: 20357).

Honda, Y., et al. 1972. Uptake of ^{106}Ru by eggs and fry of rainbow trout. J. Radiat. Res. (Japan) 13(2): 95-99. (NSA 26: 41306).

Honjo, I., Y. Oki and M. Kato. 1958. Studies of the mechanisms of light sensitizations of the retina by the use of P^{32} -Part 2. pp. 1379-1387. In: Proc. 2nd Japan Conf. on Radioisotopes. (See Japan Radioisotopes Assoc., 1958) (AEC-tr-4482).

Honstead, J. F. 1954. Columbia River survey, 1951, 1952, 1953. Hanford Labs., U. S. AEC report TID-10126. 92pp. (NSA 11: 11666).

Honstead, J. F. 1957. Dispersion of dissolved material in the Columbia River. Hanford Atomic Products Operation, U. S. AEC report HW-49008. 16pp. (NSA 14: 18709).

Honstead, J. F. 1965. Uptake and retention of ^{32}P and ^{65}Zn from the consumption of Columbia River fish. Abstr. Health Physics 11(8): 842.

Honstead, J. F. 1965. Disposal of radioactive wastes into fresh water. Battelle Northwest Lab., U. S. AEC report BNWL-SA-466. 36pp. (NSA 20: 18157).

Honstead, J. F. 1968. Bases for environmental survey design. Battelle Northwest Lab., U. S. AEC report BNWL-SA-1601. (NSA 22: 19327).

Honstead, J. F. 1968. Radionuclide burden-diet relationships near a nuclear facility. Monogr. Nucl. Med. Biol. (2): 19-30. (NSA 23: 24626) (Hanford Plant).

Honstead, J. F. 1969. Recreational use of the Columbia River: evaluation of environmental exposure. Battelle Northwest Lab., U. S. AEC report BNWL-CC-2299. 16pp. (NSA 26: 14594).

Honstead, J. F. 1969. Mechanisms of transfer of radionuclides to man in the vicinity of a nuclear facility. pp. 381-390. In: Environmental Contamination by Radioactive Materials. (See IAEA, 1969).

Honstead, J. F. 1970. Quantitative evaluation of environmental factors affecting population exposure near Hanford. Battelle Northwest Lab., U. S. AEC report BNWL-SA-3203. 12pp.

Honstead, J. F. and D. N. Brady. 1967. The uptake and retention of ^{32}P and ^{65}Zn from the consumption of Columbia River Fish. Health Physics 13(5): 455-463. (Also: Battelle Northwest Lab., U. S. AEC report BNWL-SA-45) (NSA 20: 30969; 21: 22461).

Honstead, J. F. and T. H. Essig. 1972. Evaluation of environmental factors affecting population exposure. pp. 29-57. In: Proc. Southern Conf. on Environmental Protection from Nuclear Power Plants. (See U. S. EPA, 1972) (NSA 27: 2541).

Honstead, J. F. and J. K. Soldat. 1965. Environmental monitoring. Battelle Northwest Lab., U. S. AEC report BNWL-SA-465. 73pp. (NSA 20: 14786).

Hood, D. S. 1967. Probabilistic treatment of a radioisotopic fuel release in shallow water. Trans. Amer. Nucl. Soc. 10: 5-6. (NSA 21: 36176).

Hooper, F. F. 1963. Aspects of biological transport of phosphorus-32 in streams. pp. 247-255. In: Transport of Radionuclides in Fresh Water Systems. (See Kornegay, et al., 1963) (NSA 17: 40532).

Hooper, F. F. and R. C. Ball. 1966. Bacterial transport of phosphorus in a stream ecosystem. pp. 535-549. In: Disposal of Radioactive Wastes into Seas, Oceans and Surface Waters. (See IAEA, 1966) (Preprint SM-72/32) (CONF-660507-21) (NSA 20: 40529).

Hooper, F. F. and D. G. Imes. 1973. Physical and biological dispersion of the hypolimnetic phosphorus of a bog lake system. pp. 401-409. In: Radionuclides in Ecosystems. (See Nelson, 1973) (NSA 28: 27328) (^{32}P).

Hooper, F. F., H. A. Podoliak and S. F. Snieszko. 1961. Use of radioisotopes in hydrobiology and fish culture. Trans. Am. Fish. Soc. 90(1): 49-57. (SFA 6: 4129).

Hooper, G. H. S. 1965. The mechanism of DDT resistance in the spotted root maggot Euxesta notata. J. Econ. Entomol. 58(4): 608-611. (^{14}C).

Hooper, G. H. S. 1970. Sterilization of the Mediterranean fruit fly. A study of laboratory data. pp. 3-12. In: Sterile-Male Technique for Control of Fruit Flies. (See IAEA, 1970) (NSA 25: 24506).

Hooper, G. H. S. 1971. Competitiveness of gamma-sterilized males of the Mediterranean fruit fly: effect of irradiating pupal or adult stage and of irradiating pupae in nitrogen. J. Econ. Entomol. 64(6): 1364-1368.

Hooper, G. H. S. 1971. Gamma sterilization of the Mediterranean fruit fly. pp. 87-95. In: Sterility Principle for Insect Control or Eradication. (See IAEA, 1971) (NSA 25: 29743).

Hooper, G. H. S. 1971. Sterilization and competitiveness of the Mediterranean fruit fly after irradiation of pupae with fast neutrons. J. Econ. Entomol. 64(6): 1369-1372. (NSA 26: 12434).

Hooper, G. H. S. 1972. Sterilization of the Mediterranean fruit fly with gamma radiation: effect on male competitiveness and change in fertility of females alternately mated with irradiated and untreated males. J. Econ. Entomol. 65(1): 1-6. (NSA 26: 23148).

Hooper, G. H. S. 1975. Sterilization of Dacus cucumis French (Diptera: Tephritidae) by γ radiation. I. Effect of dose on fertility, survival and competitiveness. J. Austral. Entomol. Soc. 14: 81-87. (NSA 33: 26293).

Hooper, G. H. S. and K. P. Katiyar. 1971. Competitiveness of gamma-sterilized males of the Mediterranean fruit flv. J. Econ. Entomol. 65(5): 1068-1071. (NSA 26: 7127).

Hooper, G. H. S. and D. J. Nadel. 1970. Pre-release phase of the 1969 Mediterranean fruit fly suppression experiment in Italy. pp. 51-58. In: Sterile-Male Technique for Control of Fruit Flies. (See IAEA, 1970) (NSA 25: 24509).

Hoopingarner, R. A., S. Kumararj and A. L. French. 1965. Gametogenesis and radiation effects in the cereal leaf beetle, Oulema melanopa. Ann. Entomol. Soc. Am. 58(6): 777-781.

Hootman, J. A. and W. S. Nelms. 1931. The radioactivity of Stone Mountain Springs. Am. J. Sci. 3: 37-38.

Hoover, D. L. and E. H. Floyd. 1965. A study of the capability of the rice weevils Sitophilus sasakii and S. oryzae to reproduce parthenogenetically. Ann. Entomol. Soc. Am. 58(4): 565-567.

Hoover, D. L., E. H. Floyd and H. D. Richardson. 1963. Effects of 300-KV x-ray radiation on Sitophilus oryzae. J. Econ. Entomol. 56(5): 584-585. (NSA 18: 11737).

Hoppenheit, M. 1973. Effects on fecundity and fertility of single sublethal x-irradiation of Gammarus duebeni females. pp. 479-486. In: Radioactive Contamination of the Marine Environment. (See IAEA, 1973) (NSA 28: 8486).

Hopke, P. K., et al. 1976. Geochemical profile of Chautaugua Lake sediments. J. Radioanal. Chem. 29(1): 39-59. (NSA 33: 28324) (Neutron activation).

Hopkins, T. L. 1962. Radioisotope techniques and recent research on the metabolism of insecticides in insects. pp. 101-111. In: Radioisotopes and Radiation in Entomology. (See IAEA, 1962)

Hopkins, W. C. and R. P. Gardner. 1973. Monte-Carlo simulation of γ-ray backscatter soil density gauges. Nucl. Eng. Res. 24(3): 332-343. (NSA 28: 2756).

Horber, E. 1963. Eradication of white grub (Melolontha vulgaris F.) by the sterile-male technique. pp. 313-331. In: Radiation and Radioisotopes Applied to Insects of Agricultural Importance. (See IAEA, 1963)

Horber, E. 1963. Cockchafer against cockchafer? An attempt at eradicating larvae of the field cockchafer Melolontha vulgaris Fabr. by means of x-rayed males. Mitt. Schweiz. Landw. 11(10): 145-155. (In German).

Horber, E. 1966. Cockchafer against cockchafer: an experiment for eradication of Melolontha vulgaris Fabr. larvae with x-irradiated males. Waldhygiene 6: 161-170. (In German) (NSA 22: 47654).

Horber, E. 1966. Use of alternative biological control schemes. pp. 167-183. In: Proc. FAO Symp. on Integrated Pest Control. Part 2. (Rome, Italy, 11-15 Oct. 1965). Food and Agriculture Organization, United Nations, Rome.

Horber, E. 1968. New site for the eradication of a white grub population (Melolontha vulgaris Fabr. and Melolontha hippocastani Fabr.) by release of irradiated adults. pp. 259-271. In: Isotopes and Radiation in Entomology. (See IAEA, 1968) (NSA 24: 32321). (Same as Hober above).

Horber, E. 1969. Significance of the sterile-male technique in pest management of the white grub (Melolontha vulgaris F.). pp. 73-85. In: Sterile-Male Technique for the Eradication or Control of Harmful Insects. (See IAEA, 1969) (NSA 24: 32302).

Hori, R. 1959. The uptake of radioactive calcium by eggs of the fish, Oryzias latipes. Abstr. 3rd Japan Conf. on Radioisotopes, JAIF, JRIA, 59/P-202, B-1. (Biological Inst., Toyama Univ., Japan).

Horiguchi, Y. 1974. Activation analysis of extremely small amounts of elements in urban rivers and underground water. Bunseki Kiki 12(4): 224-229. (In Japanese) (NSA 30: 31490).

Horikawa, M. and T. Sugahara. 1960. Studies on the effects of radiation on living cells in tissue culture. I. Radiosensitivity of various imaginal discs and organs in larvae of Drosophila melanogaster. Radiation Research 12(3): 266-275.

Horne, T. and L. E. Brownell. 1962. The use of radiation sources for insect control. pp. 233-254. In: Radioisotopes and Radiation in Entomology. (See IAEA, 1962).

Horneck, G., et al. 1974. Microbial studies in the Biostack experiment of the Apollo 16 mission: germination and outgrowth of single Bacillus subtilis spores hit by cosmic HZE particles. pp. 75-83. In: Life Sciences and Space Research XII. P. H. A. Sneath, ed. Akademie-Verlag, Berlin. (NSA 31: 26058).

Horowitz, M. and N. Hulin. 1971. Effects of gamma radiation on soil and diphenamid. Weed Sci. 19(3): 294-296. (NSA 26: 410).

Horrill, A. D. and G. M. Goodwell. 1973. Structure and cation content of podzolic soil of Long Island, N. Y., seven years after destruction of the vegetation by chronic gamma irradiation. Ecology 54(2): 439-444. (NSA 28: 387).

Horsley, R. J. and L. A. Fucikovsky. 1962. Radiosensitivity of Oedogonium zoospores and short multiple-cell chains at increasing ages. Abstr. In: 2nd Intern. Congr. Radiation Reserach. (See Anonymous, 1962).

Horsley, R. J. and L. A. Fucikovsky. 1963. Variation in radiosensitivity during the cell-cycle of Oedogonium cardiacum. Int. J. Radiat. Biol. 6 (5): 417-429. (NSA 17: 38795).

Horsley, R. J. and L. A. Fucikovsky. 1963. Changes in dose-survival curves during the division cycle of Oedogonium cardiacum. Abstr. Radiation Research 19(1): 219.

Horsley, R. J. and C. M. Pujara. 1969. Study of the inflexion in x-radiation survival curves for synchronized cell populations of the green alga (Oedogonium cardiacum). Radiation Research 40: 440-449. (NSA 24: 9971).

Horsley, R. J., L. A. Fucikovsky and M. Banerjee. 1967. Studies on radiosensitivity during the cell cycle in Oedogonium cardiacum. Radiation Botany 7(3): 241-246.

Hortobagyi, T. and J. Vigassy. 1967. Microorganisms exposed to irradiation in the water circuits of the atomic reactor at Csilleberc. Biol. Kozlem. 14: 113-120. (In Hungarian) (NSA 22: 2557).

Horton, J. H. 1955. Regional Monitoring Group, Health Physics Section, semiannual progress report, July-December 1954. Savannah River Plant, U. S. AEC report DPSP-55-25-34(Del.). 33pp. (NSA 30: 9506).

Horton, J. H. 1961. Radioactive waste management at the Savannah River Plant. Savannah River Lab., U. S. AEC report DP-564.

Horton, J. H. and H. G. Mealing, Jr. 1957. Health physics regional monitoring semiannual report, January-June 1956. Savannah River Plant, U. S. AEC report DPSP-57-25-4(Del.). 38pp. (NSA 30: 9508).

Horton, J. H. and D. I. Ross. 1960. Use of tritium from spent uranium fuel elements as a ground water tracer. Soil Science 90(5): 267-271.

Horton, J. H., J. C. Corey and R. M. Wallace. 1971. Tritium loss from water exposed to the atmosphere. Environ. Sci. Technol. 5: 338-343. (NSA 25: 32076).

Horubala, A. and E. Pijanowski. 1961. Effect of moderate doses of gamma radiation on the keeping quality of bilberries (Vaccinium myrtillus). Bull. Acad. Polon. Sci. (sci. biol.) 9: 167-171. (NSA 16: 28773).

Hory, P., Y. Marque and J. Matutano. 1968. Control of the radioactivity of waters released into the public environment by the Nuclear Center of Saclay. pp. 483-489. In: Intern. Congr. on Radiation Protection of the Environment. (In French) (See Societé Francaise de Radioprotection, 1968).

Hoskins, W. M. nd. The metabolism of DDT in insects. Wld. Rev. Pest Control 3: 85-96. (Contains references to use of radioisotopes).

Hosoda, J., et al. 1963. Ribonucleic acid metabolism in the posterior silk-gland of silkworm, *Bombyx mori*, during the fifth instar. Biochim. Biophys. Acta 72: 544-554. (^{32}P).

Hoss, D. E. 1966. The use of activation analysis methods for tagging post-larval fish. Abstr. Assoc. Southeastern Biol. Bull. 13(2): 37.

Hoss, D. E. 1967. Marking post-larval paralichthid flounders with radioactive elements. Trans. Am. Fish. Soc. 96(2): 151-156.

Hoss, D. E. and J. P. Baptist. 1973. Accumulation of soluble and particulate radionuclides by estuarine fish. pp. 776-782. In: *Radionuclides in Ecosystems*. (See Nelson, 1973) (NSA 28: 29895).

Hossain, M. M. 1963. Effects of massive doses of gamma radiation (120,000r) on the chromosomes of successive generations of *Hymenolepis diminuta*. Thesis, Univ. of Tennessee, Knoxville. 113pp. (NSA 18: 19752).

Hossain, M. M., J. H. Brower and E. W. Tilton. 1972. Radiation sensitivity of successively irradiated generations of the Indian meal moth. J. Econ. Entomol. 65(3): 673-676. (NSA 26: 43351).

Hossain, M. M., J. H. Brower and E. W. Tilton. 1972. Sensitivity to an acute gamma radiation exposure of successively irradiated generations of the cowpea weevil. J. Econ. Entomol. 65(6): 1566-1568. (NSA 27: 15088).

Hossain, M., M. M. Hossain and R. Rahman. 1967. Laboratory rearing and susceptibility of the confused flour beetle (adult) to gamma radiation. Pakistan Atomic Energy Centre, Dacca, report AECD/RB-6. 17pp. (NSA 21: 32727).

Hossain, M. M., M. S. A. Mollah and M. U. Malik. 1967. Studies on an experimental population of *Drosophila melanogaster* raised on irradiated banana diet. Food Irradiation 7(3): 45-49. (NSA 21: 32755; 22: 43508).

Hossain, M. M., et al. 1968. Radiation susceptibility of the flour beetle *Tribolium confusum*. Nucl. Sci. Appl. 4: 7-10. (NSA 23: 6560).

Hosticka, H. E., W. T. Moran and E. T. Oborn. 1952. Studies of herbicidal action on aquatic weeds using radioactive 2,4-D, 5-I. pp. 40-48. In: *Proc. 13th Western Weed Control Conf.*

Hotchin, J., P. Lorenz and C. Hemenway. 1965. Survival of microorganisms in space. Nature 206: 442-445. (NSA 21: 9026).

Houeele, M., C. Mercier and H. Revol. 1973. Radioecological study of the consequences of nuclear wastes in an important river basin. Centre d'Études Nucléaires, Commissariat à l'Énergie Atomique, Saclay, report CEA-CONF-2465. 19pp. (NSA 29: 26829).

Hough, R. A. and R. G. Wetzel. 1972. ^{14}C-assay for photorespiration in aquatic plants. Plant Physiol. 49: 987-990. (NSA 30: 3761).

Hough, W. A. 1963. An investigation of surface root distribution in a long-leaf pine-turkey oak stand using radioiodine. Dissertation, Duke Univ., Durham, N. C.

Hough, W. S. 1963. Effects of gamma radiation on codling moth eggs. J. Econ. Entomol. 56(5): 660-663.

Hours, R. 1955. Radioactive tracers in hydrology. Mem. trav. soc. hydrotechn. 1: 14-24. (In French).

Houtermans, J., H. E. Suess and W. Munk. 1967. Effect of industrial fuel combustion on the ^{14}C level of atmospheric CO_2. pp. 57-68. In: *Radioactive Dating and Methods of Low-Level Counting*. (See IAEA, 1967) (NSA 22: 6493).

Houtermans, J. C., H. E. Suess and H. Oeschger. 1973. Reservoir models and production rate of natural radiocarbon. J. Geophys. Res. 78(12): 1897-1908. (NSA 28: 24480).

Houtman, J. P. W. 1973. Trace element behavior in soil of some Indonesian sawah's and in sludge of an Indonesian river. Delft Progr. Rep. A(1): 5-16. (NSA 30: 5893) (Neutron activation analysis).

Howard, A. 1968. The oxygen requirement for recovery in split-dose experiments with *Oedogonium*. Int. J. Radiat. Biol. 14: 341-350. (NSA 23: 2695 26957).

Howard, A. 1970. Dose-modifying factors and inhibitors of repair. pp. 305-308. In: *Radiation Protection and Sensitization*. H. Moroson, ed. Barnes & Noble, New York. (NSA 25: 4574) (*Oedogonium cardiacum*).

Howard, A. 1970. Rose of oxygen in the repair process. pp. 70-81. In: *Conf. on Time and Dose Relationships in Radiation Biology as Applied to Radiotherapy*. (Carmel, Calif., September 15-18, 1969). (Also: Brookhaven National Lab., U. S. AEC report BNL-50203) (CONF-690915) (NSA 25: 11304) (*Oedogonium cardiacum*).

Howard, A. 1973. Dependence of radiation sensitivity on oxygen tension in *Oedogonium*. I. Extra- and intra-cellular oxygen. Radiation Botany 13(5): 293-296. (NSA 29: 18586).

Howard, A. and M. V. Haigh. 1968. Chloroplast aberrations in irradiated fern spores. Mutation Research 6: 263-280. (NSA 23: 6425).

Howard, A. and M. V. Haigh. 1970. Radiation responses of fern spores during their first cell-cycle. Int. J. Radiat. Biol. 18: 147-158. (NSA 25: 8614).

Howard, A. and R. J. Horsley. 1960. Filamentous green algae for radiobiological study. Int. J. Radiat. Biol. 2: 319-330. (NSA 14: 25331).

Howard, A., N. F. J. Dodd and M. V. Haigh. 1971. Repair of radiation damage in relation to growth processes. pp. 193-195. In: *Radiation Biophysics, Free Radicals. Vol. II.* E. Broda, ed. Wiener Medizinische Akademie, Vienna. (NSA 29: 29711) (Royal fern).

Howard, A., C. W. Gilbert and D. Greene. 1974. Dependence of radiation sensitivity on oxygen tension in *Oedogonium*. II. Effect of temperature and LET. III. Applicability of the Howard-Flanders-Alper formula for 14 MeV neutron irradiation. Radiation Botany 14(2): 101-107; 14(4): 351-353. (NSA 30: 27027; 32: 1060).

Howard-Flanders, P. 1959. Chemical factors in radiation injury to micro-organisms. Abstr. Proc. 9th Intern. Botanical Congr. 2(A): 16.

Howden, H. F. 1957. Investigations on sterility and deformities of Onthophagus (Coleoptera, Scarabaeidae) induced by gamma radiation. Ann. Entomol. Soc. Am. 50(1): 1-9.

Howden, H. F. and S. I. Auerbach. 1958. Some effects of gamma radiation on Trogoderma sternale Jayne. Ann. Entomol. Soc. Am. 51(1): 48-51. (NSA 12: 11233).

Howden, H. F. and D. A. Crossley, Jr. 1961. Insect species on vegetation of the White Oak Lake bed, Oak Ridge, Tennessee. Oak Ridge National Lab., U. S. AEC report ORNL-3094. 38pp. (NSA 15: 15345).

Howe, R. W. and J. E. Currie. 1964. Some laboratory observations on the rates of development, mortality and oviposition of several species of Bruchidae breeding in stored pulses. Bull. Entomol. Res. 55(3): 437-477. (X-rays).

Howell, F. C., et al. 1972. Uranium-series dating of bone from the Isimila prehistoric site, Tanzania. Nature 237(5349): 51-52. (NSA 26: 36106).

Howland, G. P. and E. L. Boyd. 1974. Genetic control of photomorphogenesis. Isolation of nonfilamentous mutants after gamma irradiation of Pteridium aquilinum spores. Radiation Botany 14(4): 281-285. (NSA 32: 1086).

Hoy, J. E. 1971. Environmental radiation monitoring with thermoluminescent dosimeters. Health Physics 21(6): 860-862. (NSA 26: 9385).

Hsiao, S. C. 1963. The radioecology of calcium. pp. 525-532. In: Radioecology. (See Schultz and Klement, 1963) (NSA 17: 33604).

Hsu, F.-T., S. T. Ma and T. C. Chen. 1964. The disturbance by x-rays to the meiosis and spermiogenesis of male Locusta migratoria manilensis Meyen. Acta ent. sin. 13(5): 637-648. (In Chinese, English summary).

Hsu, F.-T., et al. 1964. Influence of ionizing radiation on cells of the bone marrow of mammals. Comparative study of early changes in the cells of the bone marrow of the sternum and of the innominate bone of the baboon after the first whole body x-irradiation. Yuantsengeng (7): 628-634. (In Chinese) (Ref. Zh. (biol.) 1965: 10P506).

Hsu, H. F., S. Y. Li Hsu and J. W. Osborne. 1965. Immunization against Schistosoma japonicum in rhesus monkeys. Nature 206(4991): 1338-1340. (NSA 19: 36045).

Hsu, P., C. Li and H. Chen. 1963. The effect of irradiation with γ-ray: Co^{60} on the development and reproductive power of Trombicula akamushi var. deliensis. Acta ent. sin. 12(5/6): 578-585. (In Chinese).

Huang, C. 1968. Binding of inorganic iodide to the plasma proteins of teleost fishes. J. Fish. Res. Bd. Can. 25: 1651-1666. (NSA 23: 26683) (^{125}I).

Huang, J. F., P. S. Weng and C. M. Tsai. 1974. Tritium concentrations in environmental water samples of north Taiwan reactor sites. Nucl. Sci. J. (Taiwan) 11(3): 122-126. (NSA 31: 10600).

Hubbard, J. E. 1968. ^{137}Cs in an alpine watershed. Thesis, Colorado State Univ., Fort Collins. 107pp. (NSA 23: 40862).

Hubbard, J. S. 1973. Radiorespirometric measurement of metabolic activities in soil. pp. 199-206. In: Modern Methods in the Study of Microbial Ecology. T. Rosswall, ed. NFR, Editorial Services, Stockholm. (NSA 29: 41 419).

Hubbard, J. S., et al. 1970. Measurement of $^{14}CO_2$ assimilation in soils: an experiment for the biological exploration of Mars. Appl. Microbiol. 19: 32-38. (NSA 24: 16949).

Hubbell, D. W. and J. L. Glenn. 1971. Distribution of radionuclides in bottom sediments of the Columbia River estuary. U. S. Geological Survey, Portland, Ore., U. S. AEC report TID-25724. 136pp. (NSA 25: 39596).

Hubbell, S. P., A. Sikora and O. H. Paris. 1965. Radiotracer, gravimetric, and calorimetric studies of ingestion and assimilation rates of an isopod. pp. 1485-1501. (See Hungate, 1965).

Hubbs, C. L., G. S. Bien and H. E. Suess. 1962. La Jolla natural radiocarbon measurements. II. Am. J. Sci. Radiocarbon Suppl. 4: 204-238.

Hüber, K., H. Herrmann and M. Ruf. 1972. Disposal of waste water from the nuclear power station Gundremmingen/Danube: an example for radioecological investigations. Z. Wasser-Abwasserforsch. 5(2): 35-45. (In German) (NSA 27: 4707).

Hüber, K., M. Ruf and H. Herrmann. 1974. Radioecological investigations of the upper reaches of the River Danube. pp. 105-122. In: Comparative Studies of Food and Environmental Contamination. (See IAEA, 1974) (NSA 30: 15373).

Huber, O. 1970. Radioactive substances in the biological environment. Präventivmedizin 15(3): 189-195. (In German) (NSA 26: 31009).

Huber, O., J. Halter and P. Winiger. 1973. Radioactivity of the biosphere: investigations in Switzerland. Freibourg Univ., Switzerland. (INIS-mf-914, pp. 4-15) (NSA 29: 15592).

Huber, P. 1958. Second report of the Federal Commission for the Monitoring of Radioactivity. Atomenenergie 2(3): 13-21. (In French and German) (NSA 13: 156).

Huber, P. 1962. Report of the Swiss Confederation Commission on Surveillance of Radioactivity conducted by the Federal Council for 1961. --1962. Bull. Eidg. Gesundheitsamtes (Switz.) Suppl. B(4): 1-84. (In French and German).

Hubert, P., M. Maybeck and P. Olive. 1970. Limnological dynamics: a study of the dynamics of Lake Geneva using tritium. Compt. rend. D270: 1298-1301. (CSIRO-trans-11258) (NSA 24: 36386; 28: 5021).

Hübner, R. 1974. The radiocarbon clock. Fermelde-Praxis 15(18): 751-761. (In German) (NSA 32: 19766).

Huckabee, J. W., F. O. Cartan and G. S. Kennington. 1972. Environmental influence on trace elements in hair of 15 species of mammals. Oak Ridge National Lab., U. S. AEC report ORNL-TM-3747. 38pp. (NSA 26: 42693) (Neutron activation analysis).

Huckabee, J. W., F. O. Cartan and G. S. Kennington. 1972. Distibution of mercury in pheasant muscle and feathers. J. Wildl. Mgmt. 36(4): 1306-1309. (Neutron activation analysis).

Hudson, J. W. and L. C.-H. Wang. 1969. Thyroid function in desert ground sqirrels. pp. 17-33. In: Physiological Systems in Semiarid Environments. C. F. Hoff and M. L. Riedesel, eds. Univ. of New Mexico Press, Albuquerque. (^{125}I).

Huet, M. 1966. Study of regeneration in Asterina gibbosa Penn. (Echinoderm Asteroid) after irradiation with x rays. Compt. rend. soc. biol. 160: 466-469. (In French) (NSA 21: 3445).

Hufen, T. H., L. S. Lau and R. W. Buddemeier. 1974. Radiocarbon, ^{13}C, and tritium in water samples from basaltic aquifers and carbonate aquifers on the island of Oahu, Hawaii. pp. 111-127. In: Isotope Techniques in Groundwater Hydrology. Vol. 2. (See IAEA, 1974) (NSA 31: 19650).

Huff, D. D. and P. Kruger. 1970. Simulation of the hydrologic transport of radioactive aerosols. pp. 487-505. In: Radionuclides in the Environment. R. F. Gould, ed. Amer. Chemical Soc., Washington.

Huff, F. A. 1965. Radioactive rainout relations on densely gaged sampling networks. Water Resources Res. 1(1): 97-108.

Huffman, J. G., W. D. Kitts and C. R. Krishnamurti. 1971. Effects of alkali treatment and gamma irradiation on the chemical composition and in vitro rumen digestibility of certain species of wood. Can. J. Animal Sci. 51 (2): 457-464. (NSA 26: 25777).

Hug, O. E. 1958. The release of the feeler reflex in snails by x and alpha radiation. Strahlentherapie 106: 155-160. (In German) (NSA 12: 11256).

Hug, O. E. 1960. Reflex-like responses of lower animals and mammalian organs to ionizing radiation. Int. J. Radiat. Biol. (Spec. Suppl.): 217-224. (A. A. Buzzati-Traverso, ed.) (Snails, clams, actiniae, leech, barnacle, sea urchin).

Hug. O. E. 1963. Dose-rate dependence of instantaneous physiological radiation effects. Ann. N.Y. Acad. Sci. 114(1): 169-172. (NSA 18: 19574) (Mimosa pudica).

Hug, O. E. and H. Miltenburger. 1962. Radioinduced turgor movements (radionasty) in mimosas and other sensitive plants. Naturwissenschaften 49(9): 499-500. (In German) (NSA 17: 12166).

Hug, O. E., H. Esch and H. Miltenburger. 1964. Radiation effects on the membranes of plant and animal cells. pp. 213-220. In: Response of the Nervous System to Ionizing Radiation. (See Haley and Snider, 1964) (Concerns primarily Nitella).

Hug, O. E., H. Miltenburger and H. Esch. 1964. Electrophysiological symptoms of radioinduced mothion in Mimosa. Biophysik 1(4): 374-379. (In German) (NSA 18: 27091).

Hughes, A. F. W. and J. A. F. Fozzard. 1961. The effect of irradiation on cell degeneration among developing neurones in Xenopus laevis. Brit. J. Radiol. 34(401): 302-307. (NSA 15: 19162).

Hughes, D. E. 1968. Recovery of microorganisms from the effects of ionizing radiation. pp. 409-423. In: Energetics and Mechanisms in Radiation Biology. G. O. Phillips, ed. Academic Press, London. (NSA 23: 35956).

Hughes, M. W. 1969-70. Tritium content measurements on a variety of water samples. Australia Commonwealth Scientific and Industrial Research Organization, Adelaide, Tech. Mem. 31/1969, __/1970, 25/1970. 10, 7, 8pp. (NP-18638, NP-18639, NP-18844) (NSA 25: 13120, 13103, 37788).

Hughes, M. W. and J. C. Dighton. 1971. Tritium concentration of a variety of water samples. Australia Commonwealth Scientific and Industrial Research Organization, Adelaide. 6pp. (NP-19391) (NSA 26: 48004).

Hughes, M. W. and J. W. Holmes. 1971. Apparatus and method for the measurement of natural abundance of tritium. Australia CSIRO, Adelaide, Tech. Paper No. 3. 15pp. (NP-19005) (NSA 26: 26).

Hugue, H. 1971. Effect of γ radiation on various developmental stages of Corcyra cephalonica STN (Gelleridae Lepidoptera). Int. J. Radiat. Isotopes 22(7): 439-441. (NSA 25: 49273).

Huignard, J. 1971. Study, using gamma irradiation technique of the migration and utilization of spermatozoa in the bean weevil, Acanthoscelides obtectus (Say). pp. 203-216. In: Sterility Principle for Insect Control or Eradication. (See IAEA, 1971) (In French) (NSA 25: 29746).

Hulett, S. H. 1968. Environmental radiation levels and concentrations, first half 1968. Goodyear Atomic Corp., Piketon, O., U. S. AEC report GAT-574. 11pp. (NSA 23: 434).

Hull, A. P. 1971. Recapitulation of effluent releases and related changes in background radiation levels at Brookhaven National Laboratory, and some comparisons with those associated with nuclear power reactor plants. pp. 342-360. In: Health Physics Aspects of Nuclear Facility Siting. Vol. II. (See Voillequé and Baldwin, 1971) (NSA 25: 55200).

Hull, A. P. 1973-74. Environmental monitoring report. Brookhaven National Lab., U. S. AEC report BNL-17874, BNL-18625. 53, 73pp. (NSA 28: 10854; 29: 29390) (1964: with J. A. Ash).

Hull, A. P. 1973. Environmental safety evaluations for nuclear installations. Brookhaven National Lab., U. S. AEC report BNL-18350. 23pp. (NSA 29: 15566).

Hull, A. P. 1973. Average effluent releases from U. S. nuclear power reactors, compared with those from fossil-fueled plants, in terms of currently applicable environmental standards. Brookhaven National Lab., U. S. AEC report BNL-18223. 15pp. (NSA 29: 26177).

Hull, A. P. 1974. Environmental radiation dose criteria and assessment: pathway modeling and surveillance. Trans. Nucl. Sci. NS-21(1): 491-495. (NSA 30: 6491).

Hull, A. P. and J. T. Gilmartin. 1969. 1966 environmental monitoring radiation levels at Brookhaven National Laboratory. Brookhaven National Lab., U. S. AEC report BNL-50196. 32pp. (NSA 25: 4503).

Hull, A. P., J. T. Gilmartin and M. E. Smith. 1970. Evaluation of fission product and activation isotopes in a reactor stack effluent and in the nearby environment. pp. 285-293. In: Environmental Surveillance in the Vicinity of Nuclear Facilities. (See Reinig, 1970) (NSA 25: 16178).

Hull, B. P., Jr. and S. L. Cohen. 1968. Animal investigation program baseline studies: Supplemental Test Site "C". Southwestern Radiological Health Lab., U. S. PHS, U. S. AEC report SWRHL-54-r. 18pp. (NSA 23: 12073) (Incl. 2 mule deer).

Hulubei, H., et al. 1963. Contamination of some forms of vegetation by long-lived fission products from nuclear experiments. Acad. Rep. Populare Romine stud. cerc. fiz. 14: 25-29. (In Rumanian) (NSA 17: 27604).

Huluta, C., et al. 1961. Germination of the seeds and metabolism of ash seedlings (Fraxinus excelsior) after treatment of the seeds with radiophosphorus (P^{32}) solutions. Acad. Rep. Populare Romine stud. cerc. biochim. 4(4): 527-536. (In Rumanian) (NSA 17: 37070).

Hume, D. N. 1972. Analytical methodology for environmental measurements. Final progress report. U. S. AEC report COO-2160-1. 21pp. (NSA 26: 55643).

Hume, D. N. 1973. Pitfalls in the determination of environmental trace metals. pp. 3-16. In: Chemical Analysis of the Environment and Other Modern Techniques. S. Ahuja, ed. Plenum, New York. (NSA 28: 26954) (Incl. neutron activation analysis).

Humphreys, L. L. and R. W. Perkins. 1963. Preliminary measurements of the adsorption of reactor effluent water radionuclides by Columbia River particulate material. Hanford Atomic Products Operation, U. S. AEC report HW-79279. 10pp. (NSA 18: 35761).

Hungate, F. P. (Ed.). 1965. Radiation and Terrestrial Ecosystems. (Proc. Symp. held at Battelle-Pacific Northwest Laboratory, Richland, Wash., May 3-5, 1965). Health Physics 11(12): 1255-1675.

Hungate, F. P. 1965. Conclusions and acknowledgement. p. 1675. In: Radiation and Terrestrial Ecosystems. (See Hungate, 1965, above).

Hungate, F. P. 1965. Symposium on radiation and terrestrial ecosystems. Nuclear Safety 7: 92-99, 102. (Also: Battelle Northwest Lab., U. S. AEC report BNWL-SA-219) (NSA 20: 14237).

Hunkin, G. G. 1975. Environmental impact of solution mining for uranium. Min. Congr. J. 61(10): 24-27. (NSA 33: 17171).

Hunn, J. B. 1963. Uptake, distribution and excretion of the radionuclide I^{131} by rainbow trout (Salmo gairderi). Thesis, Michigan State Univ., East Lansing. 126pp. (NSA 18: 41069).

Hunn, J. B. and P. O. Fromm. 1964. Uptake, turnover and excretion of I-131 by rainbow trout (Salmo gairdneri). Biol. Bull. 126(2): 282-290. (NSA 18: 38745).

Hunn, J. B. and P. O. Fromm. 1966. In vivo uptake of radioiodide by rainbow trout. Water Poll. Control Fed. J. 38(12): 1981-1985.

Hunn, J. B. and E. P. Reineke. 1964. Influence of iodine intake on iodine distribution in trout. Proc. Soc. Exptl. Biol. Med. 115(1): 91-93. (NSA 18: 8094).

Hunt, F. R. 1971. Power station site selection in England and Wales. pp. 647-659. In: Environmental Aspects of Nuclear Power Stations. (See IAEA, 1971).

Hunter, P. E. and V. Krithayakiern. 1971. Effect of gamma radiation upon life expectancy and reproduction in the house cricket, Acheta domesticus (Orthoptera: Gryllidae). Ann. Entomol. Soc. Am. 64: 119-123. (NSA 25: 16288).

Hunter, S. H., et al. 1950. Some approaches to the study of the role of metals in the metabolism of microorganisms. Proc. Am. Philosoph. Soc. 94: 152-170.

Hunter-Jones, P. and P. T. Haskell. 1966. Observations on the effect of radiation on eggs of the desert locust, Schistocerca gregaria (Forsk.). Bull. Entomol. Res. 56: 725-735. (NSA 22: 31243).

Hunzinger, W. 1970. Estimation of tritium discharges into LLW-streams. European Company for the Chemical Processing of Irradiated Fuels, Mol, Belgium, report ETR-267. 17pp. (NP-18737) (NSA 34719).

Huot, L. and W. G. Verly. 1960. The total P^{32} dose to an early larva of Tenebrio molitor L. (Coleoptera: Tenebrionidae). J. Insect Physiol. 4(3): 202-208. (In French).

Huque, H. 1959. Use of radioisotopes in insect physiology. Pakistan J. Sci. 11(6): 307-309.

Huque, H. 1960. Insect tagging with radioisotopes. Pakistan J. Sci. 12(3): 135-136.

Huque, H. 1962. Studies on the persistance, decay and distribution of radiophosphorus in grasshoppers and the Madeira cockroach. pp. 155-162. In: Radioisotopes and Radiation in Entomology. (See IAEA, 1962).

Huque, H. 1962. Role of atomic energy in insect study and control. Agr. Pakistan 13: 77-80. (NSA 21: 34846).

Huque, H. 1963. Some observations on the effect of ionizing radiation against desert locust. Pakistan J. Sci. 15(6): 267-270.

Huque, H. 1963. Preliminary studies on irradiation of some common stored-grain insects in Pakistan. pp. 455-462. In: Radiation and Radioisotopes Applied to Insects of Agricultural Importance. (See IAEA, 1963).

Huque, H. 1971. Effect of gamma radiation on the reproductive organs of Dacus zonatus (Saunders). pp. 31-36. In: Sterility Principle for Insect Control or Eradication. (See IAEA, 1971) (NSA 25: 29617).

Huque, H. 1972. Preliminary report on the residues of carbaryl granules in rice plants. pp. 107-109. In: Radiotracer Studies of Chemical Residues in Food and Agriculture. (See IAEA, 1972) (NSA 27: 20910).

Huque, H. and C. R. Ahmed. 1969. Studies on the control of Dacus ciliatus Loew (Tephritidae: Diptera) by sterile male release technique. Int. J. Appl. Radiat. Isotopes 20: 791-795. (NSA 24: 4706).

Huque, H. and J. Ashraf. 1972. Effects of gamma radiation on the ovaries of desert locust Schistocerca gregaria (Forskal). Int. J. Appl. Radiat. Isotopes 23(12): 557-559. (NSA 27: 15096).

Huque, H. and M. A. Kahn. 1964. Possibilities of controlling Callosobruchis subinnotatus Pic (Bruchidae) by gamma rays. Food Irradiaton 4(3): A2-7. (NSA 18: 23270).

Huque, H. and M. A. Kahn. 1969. Ovicidal effect of gamma radiation on some important lepidopterous pests of food grains and their products. Food Irradiation 9(4): 19-22. (NSA 24: 2839).

Huque, H. and M. A. Khalique. 1972. Some observations on the decay of ^{32}P in the desert locust - Schistocerca gregaria (Forsk.). Pakistan J. Sci. Ind. Res. 15(3): 184-187. (NSA 28: 730).

Huque, H. and Q. R. Malik. 1967. Control of fruit flies Dacus zonatus Saunders by gamma rays. Int. J. Appl. Radiat. Isotopes 18: 658-661. (NSA 21: 45815).

Hurlbert, E. M. 1967. Tracing life processes with radioisotopes. Experiments in biological science. Montgomery Junior College, Takoma Park, Md., U. S. AEC report NYO-3247-4. 93pp. (NSA 22: 16973) (Experiments of interest to ecologists).

Hurley, R. 1966. Experimental infection with Candida albicans in modified hosts. J. Pathol. Bacteriol. 92: 57-67. (NSA 22: 8605).

Hurpin, B. 1968. Research into the application of the sterile-male release technique to the rhinocerus beetle (Coleoptera: Scarabaeidae). pp. 301-315. In: Isotopes and Radiation in Entomology. (See IAEA, 1968) (In French, English summary)) (NSA 24: 32324).

Hurpin, B. 1971. Research on the sterilization of Orycytes rhinoceros L. pp. 489-501. In: Sterility Principle for Insect Control or Eradication. (See IAEA, 1971) (In French) (NSA 25: 29754).

Hursh, J. B. 1954. Radium content of public water supplies. J. Am. Water Works Assoc. 46(1): 43-54. (Also: Univ. of Rochester, U. S. AEC report UR-257) (NSA 7: 4709).

Hursh, J. B. 1957. Natural occurrence of radium in man and in waters and in food. Brit. J. Radiol. (Suppl. 7): 45-53.

Hursh, J. B. and A. A. Gates. 1950. Body radium content of individuals with no known occupational exposure. Nucleonics 7(1): 46-59. (Incl. water).

Hursh, J. B. and A. Lovaas. 1964. A device to measure thorium-228 at natural levels. pp. 345-352. In: The Natural Radiation Environment. (See Adams and Lowder, 1964) (NSA 17: 31984).

Hurst, H. M. and G. H. Wagner. 1969. Decomposition of ^{14}C-labeled cell wall and cytoplasmic fractions from hyaline and melanic fungi. Soil Sci. Soc. Am. Proc. 33(5): 707-___.

Hurwitz, S. 1968. Use of labelled reference substances in studies on intestinal absorption and digestibility. pp. 141-146. In: Isotopes and Radiation in Parasitology. (See IAEA, 1968).

Husseiny, M. M. 1963. Sterilization by gamma radiation for the control of the navel orangeworm Paramyelois transitella. Thesis, Univ. of California, Berkeley. 92pp. (Dissertation Abstr. 24(4): 1758-1759) (Also: 1962: Abstr. 273. Bull. Entomol. Soc. Am. 8(3): 166) (NSA 18: 3573).

Husseiny, M. M. and H. F. Madsen. 1964. Sterilization of the navel orangeworm, Paramyelois transitella (Walker) (Lepidoptera: Phycitidae). Hilgardia 36: 113-137. (NSA 20: 83).

Hussey, R., et al. 1932. Effects of radiation on biological systems. I. Influence of high-frequency x-irradiation on duration of prepupal period of Drosophila. J. Gen. Physiol. 16(2): 207-220.

Hustin, J. E. and W. C. Ellis. 1968. Evaluation of certain properties of radiocerium as an indigestible marker. J. Agric. Food Chem. 16(2): 225-230.

Hutchin, W. H., et al. 1971. Facility for the determination of low-level tritium in biological and environmental samples. 10pp. (For Proc. Tritium Symp., Las Vegas, Nev., August 30, 1971). (CONF-710809-2) (NSA 25: 55003).

Hutchinson, F. and S. S. Easter. 1960. A difference between the biological effects of gamma rays and of heavy ions. Science 132(3436): 1311-1312.

Hutchinson, G. E. 1957. A Treatise on Limnology. Geography, Physics, and Chemistry. Vol. 1. John Wiley & Sons, New York. xiv, 1015pp.

Hutchinson, G. E. and V. T. Bowen. 1947. A direct demonstration of the phosphorus cycle in a small lake. Proc. Natl. Acad. Sci. (US) 33(5): 148-153.

Hutchinson, G. E. and V. T. Bowen. 1950. Limnological study in Connecticut. IX. A quantitative radiochemical study of the phosphorus cycle in Linsley Pond. Ecology 31(2): 194-203. (NSA 14: 3660).

Hutchinson, P. B. and A. W. Osborn. 1971. Irradiation facility for agricultural research. At. Energy Austral. 14: 14-17. (NSA 25: 46582).

Hutnik, R. J. 1964. Functionality of root grafts between scarlet oak trees. Bull. Ecol. Soc. Am. 45(4): 150-151. (^{32}P).

Hutt, R. B. and L. D. White. 1973. Recovery of fertility in male codling moths treated with 38 krad of gamma irradiation. J. Econ. Entomol. 66(2): 388-389. (NSA 28: 712).

Hutter, O. F. and A. E. Warner. 1967. The effect of p_H on the ^{36}Cl efflux from frog skeletal muscle. J. Physiol. 189: 427-443. (NSA 21: 41320).

Hutter, O. F. and A. E. Warner. 1967. Action of some foreign cations and anions on the chloride permeability of frog muscle. J. Physiol. 189: 445-460. (NSA 21: 39360) (Radioisotopes).

Hvinden, T. 1971. Radioactive fallout in Norway. Sivilt Beredskap 14(3): 24-26. (In Norwegian) (NSA 26: 53161) (Incl. reindeer).

Hvinden, T. and A. Lillegraven. 1961. Caesium-137 and strontium-90 in precipitation, soil and animals in Norway. Norway Forsvarets Forskningsinstitutt, Kjeller-Lillestrom, Intern Rapport F-417. 9pp., 3 figs.

Hvinden, T. and A. Lillegraven. 1961. Caesium-137 and strontium-90 in precipitaion, soil and animals. Nature 192(4808): 1144-1146. (Reindeer).

Hyland, K. E., Jr. and J. L. Hamar. 1059. Sternostoma kelloggi, a new species of nasal mite from the catbird (Acarina: Rhinonyssidae). J. Parasitol. 45(2): 24-25. (^{14}C-glycine tracer).

Hyodo, Y. 1964-65. Effect of x-irradiation on the intestinal epithelium of the goldfish, Carassius auratus. I. Histological changes in the intestine of irradiated fish. II. Influence of temperature on the development of histopathological changes in the intestine. Annot. Zool. Jap 37: 104-111. Radiation Research 24(1): 133-141. (NSA 19: 2085, 10880).

Hyodo, Y. 1965. Development of intestinal damage after x-irradiation and ^3H-thymidine incorporation into intestinal epithelial cells of irradiated goldfish, Carassius auratus, at different temperatures. Radiation Research 26: 383-394. (NSA 20: 10555).

Hyodo-Taguchi, Y. 1968. Rate of development of intestinal damage in the goldfish after x-irradiation and mucosal cell kinetics at different temperatures. Monogr. Nucl. Med. Biol. (1): 120-126. (NSA 22: 49540).

Hyodo-Taguchi, Y. 1970. Effect of x-irradiation on DNA synthesis and cell proliferation in the intestinal epithelial cells of goldfish at different temperatures with special reference to recovery process. Radiation Research 41(3): 568-578. (NSA 24: 19034).

Hyodo-Taguchi, Y. and N. Egami. 1969. Change in dose-survival time relationship after x-irradiation during embryonic development in fish, Oryzias latipes. J. Radiation Res. (Japan) 10: 121-125. (NSA 24: 36818).

Hyodo-Taguchi, Y, and N. Egami. 1969. Development of intestinal radiation injury and recovery at different temperatures in fish. pp. 244-254. In: Comparative Cellular and Species Radiosensitivity. (See Bond and Sugahara, 1969) (NSA 24: 12291).

Hyun, J. S. 1969. Present status and potential use of the sterile-male technique for control of rice stem borers. pp. 87-90. In: Sterile-Male Technique for Eradication or Control of Harmful Insects. (See IAEA, 1969) (NSA 24: 32303).

Hyun, J. S., et al. 1972. Study on the mating ability and competitiveness of the radiation irradiated males of rice stem borer. Han'guk Sikmul Poho Hakhoe Chi 11(1): 25-30. (NSA 27: 5545).

Ichikawa, R. 1959. Strontium-calcium discrimination in the rainbow trout. Paper 79. In: 3rd Japan Conf. on Radioisotopes. (JAIF, JRIA, 59/P-79) (Also: 1960: Rec. Oceanogr. Works Japan 5(2): 120-131) (NSA 20: 10555).

Ichikawa, R. 1960. Uptake of potassium and cesium from water by the sockeye salmon. J. Radiation Res. (Japan) 1: 107-110. (NSA 15: 24676).

Ichikawa, R. 1973. Safety from environmental radioactivity. Genshiryoku Kogyo 19(9): 26-30. (In Japanese) (NSA 30: 12678).

Ichikawa, R. and M. Oguri. 1961. Metabolism of radionuclides in fish. I. Strontium-calcium discrimination in gill absorption. Nippon Siusan Gakk. (Bull. Jap. Soc. Sci. Fish.) 27(4): 351-356. (NSA 20: 8576).

Ichikawa, R., Y. Enomoto and F. Sakai. 1964. Strontium and calcium reabsorption in renal tubules of the newt, Triturus pyrrhogaster. Science 144 (3614): 53-54. (NSA 18: 17455) (^{45}Ca, ^{85}Sr).

Ichikawa, R., M. Oguri and N. Fukada. 1961. Metabolism of radionuclides in fish. II. Strontium-calcium discrimination in gill excretion. Bull. Jap. Soc. Sci. Fish. 27(11): 990-995. (In English).

Ichikawa, R., M. Oguri and N. Takada. 1962. Metabolism of radionuclides in fish. III. Comparative behavior of strontium and calcium at gill absorption in environmental water with different amount of calcium. Nippon Suisan Gakk. 28: 1160-1163. (NSA 19: 2035) (Used eel).

Ichikawa, R., M. Oguri and N. Takada. 1963. Uptake of tantalum-182 by fish. Bull. Jap. Soc. Sci. Fish. 29(6): 521-524. (In English).

Ichikawa, T. 1973. Accumulation of radioactive nuclides in fish. pp. 234-297. In: Radioactivity and Fishes. (See Egami, 1973) (In Japanese) (NSA 29: 10490).

Ichikawa, T. 1973. Effects of radioactivity in fish on the human body. pp. 317-351. In: Radioactivity and Fishes. (See Egami, 1973) (In Japanese) (NSA 29: 16061).

Ichikawa, T. 1973. Outline of radiation and biological effects. pp. 1-29. In: Radioactivity and Fishes. (See Egami, 1973) (In Japanese) (NSA 29: 16004)

Ichikawa, T. 1973. Utilization of x rays and radioactive isotopes in ichthyiology and fishery science. pp. 352-378. In: Radioactivity and Fishes. (See Egami, 1973) (In Japanese) (NSA 29: 16098).

Ichikawa, T., K. Eto and N. Egami. 1973. Radiation, ichthyology, and fisheries. pp. 30-64. In: Radioactivity and Fishes. (See Egami, 1973) (In Japanese) (NSA 29: 13027).

Ichimura, S. and Y. Saijo. 1958. Some characteristics of ^{14}C-method to measuring organic matter production in the lake. Botan. Mag. (Tokyo) 71: 174-180.

ICN/Tracerlab. 1970. Millstone Nuclear Power Station, Unit 2. Preoperational environmental radiation survey program. Semiannual report, October 1, 1969-June 30, 1970. Waltham, Mass. U. S. AEC DOCKET-50336-20. 55pp. (NSA 24: 52839).

ICN/Tracerlab. 1971. Connecticut Yankee Atomic Power Plant. Environmental radiation survey program. Semiannual report, January 1-June 30, 1971. Waltham, Mass. U. S. AEC DOCKET-50213-117. 30pp. (NSA 25: 50964).

ICN/Tracerlab. 1971. Millstone Nuclear Power Station, Unit 1. Environmental radiation survey program semiannual report, July 1-December 31, 1970. --- January 1-June 30, 1971. Waltham, Mass. U. S. AEC DOCKET-40245-57, -50245-87. 67, 44pp. (NSA 25: 20626, 50972).

ICN/Tracerlab. 1971. Pilgrim Nuclear Power Station. Preoperational environmental radiation survey program. Summary report through August 31, 1970. --- Quarterly report, January 1-March 31, 1971. --- Quarterly report, April 1-June 30, 1971. Waltham, Mass. U. S. AEC DOCKET-50293-38, -50293-51, -50293-56. 43, 29, 32pp. (NSA 25: 20653, 41338, 59966).

Idaho College. 1974. Proc. National Environmental Research Park Symp., Idaho Falls, Idaho, October 22, 1974. Snake River Regional Studies Center, Caldwell. 62pp. (CONF-741053) (NSA 33: 25938).

Idaho Nuclear Energy Commission. 1970. Transactions of the Isotopic Snow Gage Information Meeting, October 28, 1970, Sun Valley, Idaho. Boise. 66pp. (CONF-701053) (NSA 25: 51925).

Iddings, F. A. 1969. Nuclear methods in air and water pollution analysis. Environ. Sci. Tech. 3(2): 132-140. (NSA 23: 15781).

Igarashi, S. 1966. Temperature-sensitive mutation in Paramecia aurelia. I. Induction and inheritance. II. Modification of mutation frequency by pre- and postirradiation conditions. Mutation Research 3: 13-33. (Also: Oak Ridge National Lab., U. S. AEC report ORNL-P-1355, ORNL-P-1408) (NSA 20: 18348, 18349).

Iirkovskiy, G. G. 1961. Effect of ionizing radiation on the development of various insects. pp. 86-__. In: Papers Presented to the 3rd All-Union Conf. of Young Biologists, Moscow, 1961.

Iizuka, N. 1956. Study on the judgement method of inner rot on the living trees (Todo-fir) by gamma-rays. Bull. Tokyo Univ. Forests 52(143): 143-163.

Ijuin, M., et al. 1973. Determination of the diffusion coefficient of radioelements in the Rhone sediments. Health Physics 24: 665-672.

Ikeda, N., K. Tanaka and R. Seki. 1973. Determination of tritium in hot spring waters. Radioisotopes (Tokyo) 22(11): 623-627. (NSA 30: 307).

Ikeda, N., et al. 1971. New method for radiochemical analysis of iodide and iodate in environmental water. Radioisotopes (Tokyo) 20(7): 336-337. (NSA 25: 54442).

Ikeda, Y., H. Ozaki and Y. Sekizawa. 1974. Absorption, distribution and excretion of $^{3}H\cdot$2-amino-4-phenyl-thiozole, a piscine anesthetic, in carp and crucian carp. Bull. Jap. Soc. Sci. Fish. 40(4): 339-350. (SFA 19: 18002).

Ikenaga, M. and S. Kondo. 1965. Comparative studies of mutation frequencies induced by ^{32}P treatment and gamma-irradiation in the male silkworm. Mutation Research 2: 534-543. (NSA 20: 12692).

Ilari, O. 1973. Introduction. pp. 141-145. In: Proc. 2nd Conf. on the State of Advancement of Radioecology in Italy. (See Anonymous, 1973) (In Italian).

Ilari, O. and C. Polvani. 1972. Development of nuclear energy and environmental safeguard in the Italian licensing system. Technical Meeting No. 9/14. Nuclex, Basel. 10pp. (NSA 29: 26186).

Ilies, C. 1968. Studies of water assimilation and ^{32}P distribution in various parts of cane (Phragmites communis). Celul. Hirtie 17: 97-105. (In Rumanian) (NSA 23: 12026).

Il'in, B. N. 1964. The effect of strontium^{89} on the microflora in biological filters. Gigiyena i sanitariya 29(2): 102-104. (JPRS-23903) (NSA 18: 19451, 27732).

Il'in. B. N. 1969. Role of a biological filter with sedimentation tanks in the decontamination of sewage from radioactive isotopes. Gigiyena i sanitariya (2): 95-97. (In Russian) (NSA 24: 44468).

Il'in, B. N. 1974. Possible introduction into the soil of radiostrontium in soil fertilization with sludge from biological sewage treatment plants. Gigiyena i sanitariya (7): 102-104. (In Russian) (NSA 30: 32081).

Il'in, D. I. and Yu. I. Moskalev. 1961. Distribution, elimination, and coefficients of accumulation of strontium-90, cesium-137, and phosphorus-32 in fish. pp. 322-330. In: Distribution, Biological Effects, and Migration of Radioisotopes. (See Lebedinskiy and Moskalev, 1961) (AEC-tr-7512, pp. 379-387; JPRS-15067, pp. 57-68) (NSA 30: 7171).

Il'in, D. I., Yu. I. Moskalev and A. I. Petrova. 1958. The accumulation of radioelements in certain groups of water organisms. Atom. energiya 5(2): 171-174. (Soviet Atomic Energy 5: 1013-1016; in French: CEA-tr-R-814) (NSA 13: 2885; 14: 22720).

Il'in, D. I., A. I. Petrova and N. Ya. Chepkasova. 1958. Migration of radioactive substances from an open body of water. Atom. energiya 5(1): 75-77. (Soviet Atomic Energy 5: 890-894; in French: CEA-tr-R-814) (NSA 13: 427).

Il'in, L. A., et al. (Eds.). 1972. Radioactive Iodine in the Problem of Radiation Safety. Atomizdat, Moscow. 272pp. (In Russian) (NSA 28: 5574).

Il'ina, G. V. and S. G. Tydkiy. 1964. Effects and aftereffects of radioactive fission products on the yield and quality of grain of certain field crops. Agrokhimiya (3): 80-87. (In Russian) (NSA 19: 35963).

Illinois. 1970. Illinois Environmental Protection Act, 1970. 35pp. (NP-18343) (NSA 24: 46621).

Il'yenko, A. I. 1968. Species differences in accumulation of ^{90}Sr and variability of small mammals caught on a place artificially contaminated with this radionuclide. Zool. zh. 47(11): 1695-1700. (In Russian, English summary) (NSA 25: 21649).

Il'yenko, A. I. 1968. The influence of contamination by strontium-90 on the variability of small-sized mammal populations. Zool. zh. 47(9): 1370-1377. (In Russian, English summary).

Il'yenko, A. I. 1969. Radioecology of freshwater fishes. Vopr. ikhtiol. _: __. (Transl. Am. Fish. Soc. 9(2): 249-260).

Il'yenko, A. I. 1970. Regularities of ^{90}Sr and ^{137}Cs migration in different links of food chains in zoocoenosis. Zh. obshch. biol. 31(6): 698-708. (ORNL-tr-2669) (NSA 25: 24081; 27: 22806).

Il'yenko, A. I. 1970. Concentration of strontium-90 and cesium-137 by freshwater fishes. Vopr. ikhtiol. 10(6): __. (J. Ichthyology 10(6): 860-862).

Il'yenko, A. I. 1971. Infestation of small mammals with gamasid ticks on plots contaminated with ^{90}Sr. Zool. zh. 49(2): 243-246. (In Russian) (NSA 26: 7077).

Il'yenko, A. I. 1971. Some trends in accumulation of artificial radioactive isotopes by birds of the forest biocenose. Zool. zh. 49(12): 1884-1886. (In Russian, English summary) (WR 147: p. 8).

Il'yenko, A. I. 1971. Radioecology of wild animals. pp. 279-316. In: Contemporary Problems of Radiobiology. Vol. II. (See Klechovskiy, Polikarpov and Aleksakhin, 1971) (In Russian).

Il'yenko, A. I. 1972. Some features of cesium-137 concentration in fish populations of a body of freshwater. Vopr. ikhtiol. 12(1): __. (J. Ichthyiology 12(1): 149-153) (SFA 18: 16609).

Il'yenko, A. I. and Ye. A. Federov. 1970. Accumulation of ^{137}Cs in populations of terrestrial vertebrates. Zool. zh. 49: 1370-1376. (In Russian) (NSA 25: 21581).

Il'yenko, A. I. and G. N. Romanov. 1967. Seasonal and age variation in the skeletal radioactivity of Microtus agrestis L. following incorporation of ^{90}Sr under natural conditions. Radiobiologiya 7(1): 76-78. (AEC-tr-6886, pp. 117-121) (NSA 21: 36598).

Il'yinskaya, N. V. and A. S. Troshin. 1954. Labelling flies and mosquitoes with radioactive phosphorus. Zool. zh. 33(4): 841-847. (JPRS-22186; OTS-63-41316) (NSA 18: 3440).

Imamaliyev, G. N. 1963. Influence of instantaneous irradiation of grafts by fast neutrons and x-rays on growth, development, and morphological variability of Vladimirovka cherry trees. Radiobiologiya 3(6): 909-914. (AEC-tr-5439, pp. 198-206).

Imaoka, M. 1964. Biological treatment of radioactive wastewater. Bull. Inst. Publ. Hlth. (Tokyo) 13(2): 90-95. (In Japanese) (NSA Japan 4: 02006; NSA 19: 24196).

Imaoka, M. 1964. Treatment of radioactive wastewater with activated sludge. Bull. Inst. Publ. Hlth. (tokyo) 13(2): 96-104. (In Japanese) (NSA Japan 4: 02008; NSA 19: 24197).

Impens, R., E. Francois and A. Riga. 1964. Utilization of radioisotopes and radiation in applied entomology. Ann. Gembloux 70: 45-63. (In French) (NSA 19: 22828).

Impens, R., E. Francois and A. Riga. 1964. Myzus persicae Sulz and Aphis fabae Scop.; their dispersion followed with the aid of ^{32}P. Rev. Agric. Bruxelles 17(7/8): 905-919. (In French, English summary).

Impens, R., E. Francois and A. Riga. 1965. Radioisotopes in animal ecology: 1. Experience with marking of flea vectors of the yellowing of the sugarbeet. Meded. Landbouwhogesch. Opzoekingssta. Gent 30(2): 1009-1015. (In French) (Myzus; ^{32}P, ^{14}C).

Imre, J. 1968. Testing the air permeability by soils by the radium emanation method. Agrokem. Talaj. 17: 407-424. (In Hungarian) (NSA 24: 22781).

Imshenetskiy, A. A. 1932. Combined effect of dyes and x-rays on yeast. Bull. acad. sci. URSS (sci. math. nat.) (2): 225-240. (BA 8: 15231).

Imshenetskiy, A. A. (Ed.). 1955. Isotopes in Microbiology. (Trans. Conf. on the Application of Tracer Atoms in Microbiology). Akad. Nauk SSSR, Moscow. (AEC-tr-3494) (NSA 13: 8545).

Imshenetskiy, A. A. (Ed.). 1959. Achievements in Soviet Microbiology. Akad. Nauk SSSR, Moscow. (Transl. Consultants Bureau, New York).

India Agricultural Research Institute. 1974. Status report on nuclear techniques for the end of the International Hydrology Decade (1964-1974). Nuclear Research Lab., New Delhi. 22pp. (INIS-mf-1679) (NSA 32: 3503).

India Department of Atomic Energy. 1971. Basic Mechanisms in Radiation Biology and Medicine. (Proc. Symp., All-India Inst. Medical Sciences, New Delhi, February 11-13, 1971). Bombay. 743pp. (CONF-710223) (NSA 26: 28400).

India Department of Atomic Energy. 1971. Radiation and Radioisotopes in Soil Studies and Plant Nutrition. (Proc. Symp., Univ. of Agricultural Sciences, Hebbal, Bangalore, December 21-23, 1970). Bombay. 474pp. (CONF-701227) (NSA 26: 25602).

India Department of Atomic Energy. 1974. Use of Radiations and Radioisotopes in Studies of Plant Productivity. (Proc. Symp., Pantnagar, India, April 12-14, 1974). Bombay. 899pp. (CONF-7404111) (NSA 33: 12311).

India Department of Atomic Energy (Establishment). 1966. Proc. All India Symp. on Radioactivity and Metrology of Radionuclides. (Bombay, March 14-18, 1966). Trombay, Bombay. 534pp. (CONF-660341) (NSA 22: 6544).

Indiana and Michigan Electric Company. 1968. Donald. C. Cook Nuclear Plant. Preliminary Safety Analysis Report, Amendment 7. Fort Wayne, Ind. U. S. AEC DOCKET-50316-11. 115pp. (NSA 23: 4129) (Environmental monitoring program)

Indiana and Michigan Electric Company. 1971. Donald C. Cook Nuclear Power Plant, Units 1 and 2. Environmental report. Fort Wayne, Ind. 132pp. U. S. AEC DOCKET-50315-27, DOCKET-50316-27. (NSA 25: 15316) (See American Electric Power Service Corp. and University of Michigan).

Indruch, I. 1970. Changes of anatomical characters of *Festuca rubra* L., ssp. *genuina* var. *grandiflora* Hack., induced by gamma radiation. Genet. Slechteni 6: 85-93. (In Czech) (NSA 24: 48489).

Ingersoll, J. M. 1964. Special aspects of environment resulting from various kinds of nuclear wars. Part II, Appendix I-2; Part III, Appendix I-2. Historical examples of ecological disaster (II; III). Hudson Institute, Harmon-on-Hudson, N. Y., report HI-303-RR/AI-2. 43, iv, 58pp. (NSA 18: 23179).

Ingerson, E. and F. J. Pearson, Jr. 1964. Estimation of age and rate of motion of groundwater by the ^{14}C method. pp. 263-283. In: Recent Researches in the Fields of Hydrology, Atmosphere and Nuclear Geochemistry. Y. Miyake and T. Koyama, eds. Sugawara Vol. Nagoya Univ., Nagoya, Japan. (NSA 20: 25313).

Inglis, D. R. 1973. Nuclear Energy: Its Physics and Its Social Challenge. Addison-Wesley, Reading, Mass. 406pp. (NSA 27: 24283).

Inglis, J. M., et al. 1968. A device for automatically detecting the presence of small animals carrying radioactive tags. Ecology 49(2): 361-363.

Inman, D. 1968. Evaluation of ^{51}chromium in avian digestive studies. Thesis, Michigan State Univ., East Lansing. 43pp.

Inman, D. L., G. A. Petrides and R. K. Ringer. 1969. Chromium-51 in digestive studies in avian species. Proc. Poultry Sci. Assoc. 48(5): 1826.

Inokukuma, T. 1953. Effect of atomic bomb on trees. Nippon Gakujutsu 1: 264-265. (In Japanese).

Inoue, Y. 1972. ^{85}Kr as an environmental contaminant. Hoken Butsuri 7(3): 157-161. (In Japanese) (NSA 27: 11975).

Inoue, Y. and S. Morisawa. 1970. Safety evaluation of solidified radioactive wastes into the ground. Hoken Butsuri 5(2): 75-80. (In Japanese) (NSA 24: 48251).

Inoue, Y. and S. Morisawa. 1972. Selection of a site for the ground disposal of radioactive wastes. I. An approach to safety analysis. Nippon Genshiryoku Gakk. 14(10): 522-530. (In Japanese) (NSA 27: 19809, 22499) (Ecological transport paths).

Inoue, Y. and S. Morisawa. 1974. Selection of a ground disposal site for radioactive wastes: an approach to its safety evaluation. Health Physics 26(1): 53-63. (NSA 29: 15648).

Inoue, Y. and M. Sakanoue. 1970. Determination of plutonium in soil samples. J. Radiation Res. (Japan) 11: 98-106. (NSA 25: 47).

Inselburg, J. and M. Sussman. 1967. Incorporation of ^{3}H-uridine into RNA during cellular slime mold development. J. Gen. Microbiol. 46(1): 59-64.

Institute of Electrical and Electronics Engineers. 1971. Power plant siting: its impact on system planning and the environment. New York. 28pp. (NSA 29: 20446).

Inter-American Nuclear Energy Commission. 1960. Radioisotopes and Radiation in the Life Sciences. (Proc. 2nd Inter-American Symp. on the Peaceful Application of Nuclear Energy, Buenos Aires, June 1-5, 1959). Pan American Union, Washington. 275pp. (NSA 14: 21227).

International Atomic Energy Agency. 1960. Disposal of Radioactive Wastes. (Conf. Proc., Monaco, 16-21 November 1959). Vienna, Publ. STI/PUB/18. 2 vols., 612, 584pp. (NSA 14: 22652).

International Atomic Energy Agency. 1961. Radioactive Substances in the Biosphere. Vienna, Publ. STI/PUB/28. 43pp. (NSA 15: 14087).

International Atomic Energy Agency. 1961. Radioactive Waste Disposal into the Sea. Vienna, Safety Ser. No. 5, Publ. STI/PUB/14. 174pp.

International Atomic Energy Agency. 1962. Radioisotopes and Radiation in Entomology. (Proc. Symp., Bombay, 5-9 December 1960). Vienna, Publ. STI/PUB/38. xii, 307pp.

International Atomic Energy Agency. 1962. Radioisotopes in Soil-Plant Nutrition Studies. (Proc. Symp., Bombay, 26 February-2 March 1962). Vienna, Publ. STI/PUB/55. 469pp. (NSA 16: 32708).

International Atomic Energy Agency. 1962. IAEA research contracts second annual report. Vienna, Tech. Repts. Ser. No. 9, STI/DOC/10/9. 70pp. (NSA 17: 4).

International Atomic Energy Agency. 1963. Disposal of Radioactive Wastes into Fresh Water. Vienna, Safety Ser. No. 10, STI/PUB/44. 99pp.

International Atomic Energy Agency. 1963. Radiation and Radioisotopes Applied to Insects of Agricultural Importance. (Proc. Symp., Athens, Greece, April 22-26, 1963). Vienna, Publ. STI/PUB/74. 469pp. (NSA 18: 5704).

International Atomic Energy Agency. 1963. Radioisotopes in Hydrology. (Proc. Symp., Tokyo, March 5-9, 1963). Vienna, Publ. STI/PUB/71. 469pp. (NSA 18: 1911).

International Atomic Energy Agency. 1963. Radioactive Dating. (Proc. Symp., Athens, 19-23 November 1962). Vienna, Publ. STI/PUB/68. 443pp. (NSA 17: 30730).

International Atomic Energy Agency. 1963. Information circular on radiation techniques and their application to insect pests. Vienna, report WP/31/2. 30pp. (NSA 18: 3439).

International Atomic Energy Agency. 1964. Isotope Techniques for Hydrology. (Report of the Panel, Vienna, December 17-21, 1962). Vienna, Tech. Rept. Ser. No. 23, STI/DOC/10/23. 38pp. (NSA 18: 18052).

International Atomic Energy Agency. 1964. Laboratory training manual on the use of isotopes and radiation in soil plant relations research. Vienna, Tech. Rept. Ser. No. 29. 165pp.

International Atomic Energy Agency. 1965. A. The IAEA laboratory at Vienna and Seibersdorf. Second report. IV. Agriculture. B. Insect eradication and pest control. 5. Co-ordinated programme on insect eradication and sterile-male technique. Vienna, Tech. Rept. Ser. 41, STI/DOC/10/41).

International Atomic Energy Agency. 1966. Food Irradiation. (Intern. Symp., Karlsruhe, Germany, June 6-10, 1966). Vienna, Publ. STI-PUB-127. (CONF-660609) (NSA 20: 40601).

International Atomic Energy Agency. 1966. Isotopes in Weed Research. (Proc. Symp., Vienna, 25-29 October 1965). Vienna, Publ. STI/PUB/113. 246pp. (CONF-651076) (NSA 20: 42742).

International Atomic Energy Agency. 1966. Guide to the safe handling of radioisotopes in hydrology. Vienna, Safety Ser. No. 20, STI/PUB/131. 38pp. (NSA 21: 14314).

International Atomic Energy Agency. 1966. Isotopes and Radiation in Plant Pathology. Vienna, Tech. Rept. Ser. No. 66, STI-DOC-10/66. 100pp. (NSA 21: 20425).

International Atomic Energy Agency. 1966. Laboratory training manual on the use of isotopes and radiation in animal research. Vienna, Tech. Rept. Ser. No. 60, STI/DOC/10/60. 183pp.

International Atomic Energy Agency. 1966. Laboratory training manual on the use of isotopes and radiation in entomology. Vienna, Tech. Rept. Ser. No. 61, STI/DOC/10/61. (NSA 21: 11008).

International Atomic Energy Agency. 1966. Radioisotopes in the Detection of Pesticide Residues. (Proc. Panel, Vienna, 12-15 April 1965). Vienna, Panel Proc. Ser., Publ. STI/PUB/123. 120pp. (NSA 21: 20029).

International Atomic Energy Agency. 1966. Information circular on radiation techniques and their application to insect pests. Vienna, report WP/31/7. 49pp.

International Atomic Energy Agency. 1966. Disposal of Radioactive Wastes into Seas, Oceans and the Surface Waters. (Proc. Symp., Vienna, 16-20 May 1966). Vienna, Publ. STI/PUB/126. vi, 898pp. (CONF-660507) (NSA 21: 16086).

International Atomic Energy Agency. 1966. Seminar. Current problems in weed research and possible applications of nuclear techniques. pp. 197-228. In: Isotopes in Weed Research. (See IAEA, 1966 above).

International Atomic Energy Agency. 1966. Seminar. Recent advances in weed research resulting from the use of nuclear and other advanced techniques. pp. 87-112. In: Isotopes in Weed Research. (See IAEA, 1966 above).

International Atomic Energy Agency. 1967. Isotopes in Hydrology. (Proc. Symp., Vienna, November 14-18, 1966). Vienna, Publ. STI/PUB/141. (NSA 22: 19048).

International Atomic Energy Agency. 1967. Isotopes in Plant Nutrition and Physiology. (Proc. Symp., Vienna, 5-9 September 1966). Vienna, Publ. STI/PUB/137. (NSA 21: 41316).

International Atomic Energy Agency. 1967. Soil-Moisture and Irrigation Studies. (Proc. Panel, Vienna, 14-18 March 1966). Vienna, Publ. STI/PUB/133. 120pp. (CONF-660338) (NSA 22: 6437).

International Atomic Energy Agency. 1967. Radioactive Dating and Low-Level Counting. (Proc. Symp., Monaco, 2-10 March 1967). Vienna, Publ. STI/PUB/152. 756pp. (CONF-670309) (NSA 22: 6452).

International Atomic Energy Agency. 1967. Tritium and other environmental isotopes in the hydrological cycle. Vienna, Tech. Rept. Ser. No. 73, STI/DOC/10/73. 83pp. (NSA 22: 16773).

International Atomic Energy Agency. 1967. Disposal of Radioactive Wastes into the Ground. (Proc. Symp., Vienna, 29 May-2 June 1967). Vienna, Publ. STI/PUB/156. 67pp. (NSA 22: 16793).

International Atomic Energy Agency. 1967. Isotope and Radiation Techniques in Soil Physics and Irrigation Studies. (Proc. Symp., Istanbul, 12-16 June 1967). Vienna, Publ. STI/PUB/158. 446pp. (CONF-670641) (NSA 22: 23135).

International Atomic Energy Agency. 1967. Coordinated research programme on the application of radiation techniques in studies of water-use efficiency. pp. 97-106. In: Soil-Moisture and Irrigation Studies. (See IAEA, 1967 above).

International Atomic Energy Agency. 1968. Isotopes and Radiation in Entomology. (Proc. Symp., Vienna, 4-8 December 1967). Vienna, Publ. STI/PUB/181. 157pp.

International Atomic Energy Agency. 1968. Economics in managing radioactive wastes. Report resulting from two panels of experts on the economics of radioactive waste management. Vienna, Tech. Rept. Ser. No. 83, STI/DOC-10/83. 46pp. (NSA 24: 7615).

International Atomic Energy Agency. 1968. Guidebook on nuclear techniques in hydrology. Vienna, Tech. Rept. Ser. No. 91, STI/DOC-10/91. 214pp. (NSA 24: 7502).

International Atomic Energy Agency. 1968. Safety considerations in the use of ports and approaches by nuclear merchant ships. Vienna, Safety Ser. No. 27, STI/PUB-206. 20pp. (NSA 24: 7864).

International Atomic Energy Agency. 1968. Application of meteorology to safety at nuclear plants. Vienna, Safety Ser. No. 29, STI/PUB/211. 27pp. (NSA 24: 7522).

International Atomic Energy Agency. 1968. Control of Livestock Insect Pests by the Sterile-Male Technique. (Proc. Panel, Vienna, 23-27 January 1967). Vienna, Publ. STI/PUB/184. 108pp. (CONF-670125) (NSA 23: 34802).

International Atomic Energy Agency. 1968. Elimination of Harmful Organisms from Food and Feed by Irradiation. (Rept. of a panel, Zeist, The Netherlands, 12-16 June 1967). Vienna, Publ. STI/PUB-200. 123pp. (CONF-670671) (NSA 24: 34803).

International Atomic Energy Agency. 1968. Isotopes and Radiation in Entomology. (Proc. Symp., Vienna, 4-8 December 1967). Vienna, Publ. STI/PUB-166. 440pp. (CONF-671209) (NSA 24: 32286).

International Atomic Energy Agency. 1968. Isotopes and Radiation in Soil Organic Matter Studies. (Proc. Symp., Vienna, 15-19 July 1968). Vienna, Publ. STI/PUB-190. 593pp. (NSA 24: 5972).

International Atomic Energy Agency. 1968. Isotope Studies on the Nitrogen Chain. (Proc. Symp. on the Use of Isotopes in Studies of Nitrogen Metabolism in the Soil-Plant-Animal System, 28 August-1 September 1967, Vienna). Vienna, Publ. STI/PUB-161. 350pp. (CONF-670852) (NSA 24: 27486).

International Atomic Energy Agency. 1968. Management of radioactive wastes at nuclear power plants. Vienna, Safety Ser. No. 28, STI/PUB/208. 225pp. (NSA 24: 7616).

International Atomic Energy Agency. 1968. Neutron irradiation of seeds. Part II. Second report of an FAO/IAEA meeting on Co-ordination of Research, Vienna, December 11-15, 1967. Vienna, Tech. Rept. Ser. No. 92, STI/DOC-10/92. 183pp. (CONF-671221) (NSA 24: 17160).

International Atomic Energy Agency. 1968. Radiation, Radioisotopes, and Rearing Methods in the Control of Insect Pests. (Proc. Panel, Tel Aviv, Israel, 17-21 October 1966). Vienna, Publ. STI/PUB-185. 153pp. (CONF-661070) (NSA 24: 32287).

International Atomic Energy Agency. 1968. Radioisotopes in the Pulp and Paper Industry. (Report of a panel, Helsinki, October 9-13, 1967). Vienna, Publ. STI/PUB/202. 117pp. (CONF-671083) (NSA 23: 47713).

International Atomic Energy Agency. 1969. Environmental Contamination by Radioactive Materials. (Proc. Seminar, Vienna, 22-28 March 1969). Vienna, Publ. STI/PUB/226. viii, 746pp.

International Atomic Energy Agency. 1969. Panel recommendations. pp. 139-140. In: Sterile-Male Technique for Eradication or Control of Harmful Insects. (See IAEA, 1969 below) (Summary pp. 3-15).

International Atomic Energy Agency. 1969. Recommendations of the Panel. pp. 95-100. In: Insect Ecology and the Sterile-Male Technique. (See IAEA, 1969 below).

International Atomic Energy Agency. 1969. Induced Mutations in Plants. (Proc. Symp., Pullman, Wash., 14-18 July 1969). Vienna, Publ. STI/PUB/231. 758pp. (CONF-690703) (NSA 24: 41800).

International Atomic Energy Agency. 1969. Environmental Contamination of Radioactive Materials. (Proc. Seminar, Vienna, 24-28 March 1969). Vienna, Publ. STI/PUB/226. 756pp. (CONF-690317) (NSA 24: 7769).

International Atomic Energy Agency. 1969. Insect Ecology and the Sterile-Male Technique. (Proc. Panel, Vienna, 7-11 August 1967). Vienna, Publ. STI/PUB/223. 102pp.

International Atomic Energy Agency. 1969. IAEA laboratory activities. 6th report. Vienna, Publ. STI/DOC/10/98. 142pp.

International Atomic Energy Agency. 1969. Manual on safety aspects of the design and equipment of hot laboratories. Vienna, Safety Ser. No. 30, STI/PUB/169. 102pp. (NSA 24: 7867).

International Atomic Energy Agency. 1969. Quick methods for radiochemical analysis. Vienna, Tech. Rept. Ser. No. 95, STI/DOC-10/95. 63pp. (NSA 23: 48051).

International Atomic Energy Agency. 1969. Radiation Protection Monitoring. (Proc. Regional Seminar for Asia and the Far East, Bombay, 9-13 December 1968). Vienna, Publ. STI/PUB-199. 567pp. (CONF-681203) (NSA 23: 48051).

International Atomic Energy Agency. 1969. Safe operation of nuclear power plants. Vienna, Safety Ser. No. 31, STI/PUB/222. 125pp.

International Atomic Energy Agency. 1969. Sterile-Male Technique for Eradication or Control of Harmful Insects. (Proc. Panel, Vienna, 27-31 May 1968). Vienna, Publ. STI/PUB-224. 151pp. (CONF-680557) (NSA 24: 32288).

International Atomic Energy Agency. 1969. Trace Mineral Studies with Isotopes in Domestic Animals. (Proc. Panel, Vienna, 28-31 October 1968). Vienna, Publ. STI/PUB/218. 151pp.

International Atomic Energy Agency. 1970. Isotope Hydrology 1970. (Proc. Symp., Vienna, 9-13 March 1970). Vienna, Publ. STI/PUB/255. 927pp. (CONF-700328).

International Atomic Energy Agency. 1970. Peaceful Nuclear Explosions. (Phenomenology and Status Report). (Proc. Panel, Vienna, 2-6 March 1970). Vienna, Publ. STI/PUB-273. 464pp. (CONF-700305) (NSA 24: 50478).

International Atomic Energy Agency. 1970. Safeguards Techniques. Vol. I. Vienna, Publ. STI/PUB-260(Vol. 1). 625pp. (CONF-700701) (NSA 25: 21285).

International Atomic Energy Agency. 1970. *Management of Low- and Intermediate-Level Radioactive Wastes*. Vienna, Publ. STI/PUB-264. 825pp. (CONF-700905) (NSA 25: 21287).

International Atomic Energy Agency. 1970. Mediterranean fruit fly: short contributions. Summaries of work done at various institutions. pp. 81-95. In: *Sterile-Male Technique for Control of Fruit Flies*. (See IAEA, 1970 below) (NSA 25: 24513).

International Atomic Energy Agency. 1970. *Sterile-Male Technique for Control of Fruit Flies*. (Proc. Panel, Vienna, 1-5 September 1969). Vienna, Publ. STI/PUB-276. 185pp. (CONF-690969) (NSA 25: 24490).

International Atomic Energy Agency. 1970. *Application of Induced Sterility for Control of Lepidopterous Populations*. (Proc. Panel, Vienna, 1-5 June 1970). Vienna, Publ. STI/PUB-281. 176pp. (CONF-700654) (NSA 25: 47070).

International Atomic Energy Agency. 1970. *Isotopes and Radiation in Parasitology. Part II*. (Proc. Research Co-ordination Meeting, Vienna, 2-6 June 1969). Vienna, Publ. STI/PUB-242. 141pp. (CONF-690642) (NSA 24: 50776).

International Atomic Energy Agency. 1970. *Peaceful Uses of Atomic Energy in Africa*. (Proc. Symp., Kinshasa, Congo, July 28-August 1, 1969). Vienna, Publ. STI/PUB-233. 587pp. (CONF-690711) (NSA 24: 44382).

International Atomic Energy Agency. 1971. *Environmental Aspects of Nuclear Power Stations*. (Proc. Symp., New York, 10-14 August 1970). Vienna, Publ. STI/PUB/261. viii, 970pp. (CONF-700810).

International Atomic Energy Agency. 1971. Disposal of radioactive wastes into rivers, lakes, and estuaries. Vienna, Safety Ser. No. 36, STI/PUB-283. 85pp. (NSA 25: 37564).

International Atomic Energy Agency. 1971. *Nuclear Techniques in Environmental Pollution*. (Proc. Symp., Salzburg, 26-30 October 1970). Vienna, Publ. STI/PUB-268. 824pp. (CONF-701023) (NSA 25: 32371).

International Atomic Energy Agency. 1971. *Safeguards Techniques. Vol. II*. Vienna, Publ. STI/PUB-260(Vol.2). 512pp. (CONF-700701(Vol.2)) (NSA 25: 21286).

International Atomic Energy Agency. 1971. *Sterility Principle for Insect Control or Eradication*. (Proc. Symp., Athens, 14-18 September 1970). Vienna, Publ. STI/PUB-265. 553pp. (CONF-700919) (NSA 25: 29730).

International Atomic Energy Agency. 1971. *Advances in Physical and Biological Radiation Detectors*. (Proc. Symp., Vienna, 23-27 November 1970). Vienna, Publ. STI/PUB-269. 756pp. (CONF-701112) (NSA 25: 57726).

International Atomic Energy Agency. 1971. Environmental isotope data No. 3: world survey of isotope concentration in precipitation (1966-1967). Vienna, Publ. STI/DOC/10/129. 402pp. (NSA 26: 17711).

International Atomic Energy Agency. 1971. Handbook on calibration of radiation protection monitoring instruments. Vienna, Publ. STI/DOC-10/133. 95pp. (NSA 26: 9308).

International Atomic Energy Agency. 1971. *Radiation and Radioisotopes for Industrial Microorganisms*. (Proc. Symp., Vienna, 29 March-1 April 1971). Vienna, Publ. STI/PUB-287. 347pp. (NSA 26: 23012).

International Atomic Energy Agency. 1971. *Rapid Methods for Measuring Radioactivity in the Environment*. (Proc. Intern. Symp., Neuherberg, Germany, July 5-9, 1971). Vienna, Publ. STI/PUB-289. 980pp. (CONF-710705) (NSA 26: 27797).

International Atomic Energy Agency. 1972. *Isotopes and Radiation in Soil-Plant Relationships Including Forestry*. (Proc. Symp., Vienna, 13-17 December 1971). Vienna, Publ. STI/PUB-292. 676pp. (NSA 27: 2851).

International Atomic Energy Agency. 1972. *Mercury Contamination in Man and His Environment*. (Proc. Symp., Amsterdam, May 1967). Vienna, Publ. STI/DOC-10/137. 185pp. (CONF-670560) (NSA 27: 11657).

International Atomic Energy Agency. 1972. *Nuclear Activation Techniques in the Life Science*. (Proc. Symp., Bled, Yugoslavia, April 10-14, 1972). Vienna, Publ. STI/PUB-310. 672pp. (CONF-720425) (NSA 27: 22070).

International Atomic Energy Agency. 1972. *Radiotracer Studies of Chemical Residues in Food and Agriculture*. (Proc. Combined Panel and Research Co-ordination Meeting, Vienna, 25-29 October 1971). Vienna, Publ. STI/PUB-332. 172pp. (NSA 27: 20181).

International Atomic Energy Agency. 1973. *Radioactive Contamination of the Marine Environment*. (Proc. Symp., Seattle, 10-14 July 1972). Vienna, Publ. STI/PUB-313. xii, 786. (NSA 28: 8141).

International Atomic Energy Agency. 1973. *Soil Moisture and Irrigation Studies II*. (Proc. 2nd Panel, Vienna, 2-6 November 1970). Vienna, Publ. STI/PUB-327. 194pp. (CONF-701156) (NSA 28: 21473).

International Atomic Energy Agency. 1973. *Computer Models and Application of the Sterile-Male Technique*. (Proc. Panel, Vienna, 13-17 December 1971). Vienna, Publ. STI/PUB/340. iii, 195pp. (NSA 29: 8045).

International Atomic Energy Agency. 1973. *Environmental Behavior of Radionuclides Released in the Nuclear Industry*. (Proc. Symp., Aix-en-Provence, May 14-18, 1973). Vienna, Publ. STI/PUB-345. 759pp. (NSA 29: 29300).

International Atomic Energy Agency. 1973. *Dosimetry in Agriculture, Industry, Biology, and Medicine*. (Proc. Symp., Vienna, April 17-21, 1972). Vienna, Publ. STI/PUB-311. 695pp. (NSA 28: 15548).

International Atomic Energy Agency. 1973. *Tracer Techniques in Sediment Transport*. (Panel Meeting, Saclay, France, June 21-25, 1971). Vienna, Tech. Rept. Ser. No. 145, STI/DOC-10/145. (CONF-710665) (NSA 28: 21131).

International Atomic Energy Agency. 1974. *Isotope and Radition Techniques in Soil Physics and Irrigation Studies, 1973*. Vienna, Publ. STI/PUB/349. 525pp. (CONF-730817) (NSA 30: 14748).

International Atomic Energy Agency. 1974. Comparative Studies of Food and Environmental Contamination. (Proc. Symp., Otanienii, Finland, 27-31 August 1973). Vienna, Publ. STI/PUB/348. 633pp. (CONF-730817) (NSA 30: 14748).

International Atomic Energy Agency. 1974. Physical Behavior of Radioactive Contaminants in the Atmosphere. (Proc. Symp., Vienna, November 12-16, 1973). Vienna, Publ. STI/PUB-354. 540pp. (NSA 30: 32045).

International Atomic Energy Agency. 1974. Environmental Surveillance Around Nuclear Installations. (Proc. Symp., Warsaw, Poland, November 5-9, 1973) Vienna, Publ. STI/PUB-353. 2 vols., 480, 465pp. (NSA 31: 14019, 14020).

International Atomic Energy Agency. 1974. Isotope Techniques in Groundwater Hydrology. (Proc. Conf., Vienna, March 11-15, 1974). Vienna, Publ. STI/PUB-373. 2 vols., 511, 508pp. (NSA 31: 19609, 19610).

International Atomic Energy Agency. 1974. Peaceful Nuclear Expolosions III. Applications, Characteristics, and Effects. (Proc. Panel, Vienna, November 27-December 1, 1972). Vienna, Publ. STI/PUB-367. 495pp. (NSA 31: 495).

International Atomic Energy Agency. 1974. Population Dose Evaluation and Standards for Man and His Environment. (Proc. Seminar, Portoroz, Yugoslavia, May 20-24, 1974). Vienna, Publ. STI/PUB-375. 657pp. (NSA 31: 19951).

International Atomic Energy Agency. 1974. Sterile-Insect Technique and Its Field Applications. (Proc. Panel, Vienna, November 13-17, 1972). Vienna, Publ. STI/PUB-364. 142pp. (NSA 31: 8929).

International Atomic Energy Agency. 1975. Combined Effects of Radioactive, Chemical, and Thermal Releases to the Environment. (Proc. Symp., Stockholm, 2-5 June 1975). Vienna, Publ. STI/PUB-404. 358pp. (NSA 33: 29123).

International Atomic Energy Agency. 1975. Isotope Ratios as Pollutant Source and Behavior Indicators. (Proc. Symp., Vienna, 18-22 November 1974). Vienna, Publ. STI/PUB-382. 489pp. (NSA 33: 7348).

International Atomic Energy Agency. 1975. Laboratory manual on the use of radiotracer techniques in industry and environmental pollution. Vienna, Publ. STI/DOC-10/161. 120pp. (NSA 31: 19835).

International Atomic Energy Agency. 1975. Objectives and Design of Environmental Monitoring Programmes for Radioactive Contaminants. Vienna, Safety Ser. No. 41, STI/PUB-385. 140pp. (NSA 31: 29607).

International Atomic Energy Agency. 1975. Peaceful Nuclear Explosions IV. (Proc. Tech. Committee Meeting, Vienna, 20-24 January 1975). Vienna, Publ. STI/PUB-414. 488pp. (NSA 33: 9214).

International Atomic Energy Agency. 1975. Radiation for a Clean Environment. (Proc. Intern. Symp., Munich, Germany, March 17-21, 1975). Vienna, Publ. STI/PUB-402. 682pp. (NSA 33: 9099).

International Atomic Energy Agency. 1975. Radiation Techniques for Water-Use Efficiency Studies. (Results of a five-year coordinated research program). Vienna, Publ. STI/DOC-10/168. 231pp. (NSA 32: 22288).

International Atomic Energy Agency. 1975. Siting of Nuclear Facilities. (Proc. Symp., Vienna, 9-13 December 1974). Vienna, Publ. STI/PUB-384. 677pp. (NSA 32: 25383).

International Atomic Energy Agency. 1975. Sterility Principle for Insect Control, 1974. (Proc. Symp., Innsbruck, 22-26 July 1974). Vienna, Publ. STI/PUB-377. 622pp. (NSA 32: 11759).

International Congress of Radiation Research. 1962. 2nd Intern. Congr., Harrogate, Yorkshire, England, August 5-11, 1962. Abstracts of Papers. Silver End Documentary Publ., London. (Some cross ref. to "Anonymous").

International Commission on Radiological Protection. 1965. Radiation Protection. Principles of environmental monitoring related to the handling of radioactive materials. Committee 4, ICRP Publ. 7. Pergamon Press, New York. 15pp.

Iokhel'son, S. V., et al. 1969. Radioactive isotopes of europium produced by nuclear blasts. pp. 37-39. In: Nuclear Meteorology. (See Makhon'ko and Malakhov, 1969) (NSA 30: 15398) (Moss sample).

Iowa State University. 1965. Survey of environmental radioactivity for period January 1-June 30, 1965. U. S. AEC report TID-22173. 16pp. (NSA 19: 44302) (Incl. river water and sediment).

Ireland, R. E., J. E. Guthrie and A. G. Robinson. 1973. Occurrence, radioactivity, and diversity, of Winnepeg River benthic organisms in the vicinity of Whiteshell Nuclear Research Establishment. Atomic Energy of Canada, Ltd., report AECL-4421. 108pp. (NSA 28: 24483).

Irikhimovich, A. I. and A. M. Zelinin. 1961. Distribution of P^{32} in the tissues and organs of mature male carp at different seasons of the year. pp. 64-70. In: Problems of the Hydrobiology and Ichthiology of Molavian Reservoirs. Shtiintsa: Kishniev. (BA 43: 815; Ref. zh. (Biol.) 1962: 18189; SFA 8: 6126).

Irwin, R. L. B., J. W. T. Jenkins and T. J. Arnason. 1950. Deposition of P^{32} in developing Drosophila. Can. J. Res., D.-Can. J. Zool. 28(3): 247.

Isaacson, A. and A. Sandow. 1967. Quinine and caffeine effects on ^{45}Ca movement in frog sartorius muscle. J. Gen Physiol. 50: 2109-2128. (NSA 22: 14970)

Isaacson, R. E. and L. E. Brownell. 1972. Ultimate storage of radioactive wastes in terrestrial environments. Atlantic Richfield Hanford Co., U. S. AEC report ARH-SA-126. 33pp. (For Symp. on the Management of Radioactive Wastes from Fuel Reprocessing, Paris, France, November 27, 1972). (CONF-721107-4) (NSA 27: 9434).

Isaacson, R. E., et al. 1974. Soil moisture transport in arid site vadose zones. Atlantic Richland Hanford Co., U. S. AEC report ARH-SA-169. 23 pp. (NSA 29: 24213) (At Hanford).

Isenberg, H. D., et al. 1964. Calcium^{45} turnover in a mineralizing coccolithophorid protozoan. J. Protozool. 11: 531-534. (NSA 19: 12931) (Hymenomonas).

Ishaque, M., A. H. Cornfield and P. A. Cawse. 1971. Effect of gamma irradiation of an acid tea soil from East Pakistan on nitrogen mineralization and nitrification during subsequent incabation. Plant Soil 35(1): 201-204. (NSA 26: 12573).

Ishibashi, N. 1965. The increase in male adults by gamma-ray irradiation in the root-knot nematode, Meloidogyne incognita Chitwood. Nematologica 11: 361-369. (NSA 21: 24622).

Ishibashi, N. and S. Taguchi. 1965. Studies on the hatching of the root-knot nematode Meloidogyne incognita Chitwood. II. Radiosensitivity of embryos and infectivity of larvae emerging from irradiated egg masses. Annot. Zool. Jap. 38: 12-19. (NSA 21: 22666).

Ishibashi, N. and S. Taguchi. 1965. Effect of ionizing radiations on the propogation of root-knot nematodes, Meloidogyne incognita Chitwood. Annot. Zool. Jap. 38: 182-189. (NSA 20: 28942).

Ishihara, S. and T. Mochizuki. 1969. Chemical characteristics of river water in the eastern part of Shimane Prefecture. With special emphasis on cause of uranium anomaly. Chishitsu Chosasho Geppo 20: 435-448. (In Japanese) (NSA 24: 44134).

Ishihara, T. 1970. Environmental radiation monitoring system. Keisoku To Seigyo 9(12): 981-989. (In Japanese) (NSA 25: 42094).

Ishihara, T. 1973. Disposal of radioactive wastes. Karyoku Gensh. Hatsuden 24(5): 559-566. (In Japanese) (NSA 30: 24810) (Marine and terrestrial).

Ishizuka, K., et al. 1974. Metabolism and accumulation of pesticides in rice plants. pp. 363-372. In: Comparative Studies on Food and Environmental Contamination. (See IAEA, 1974) (NSA 30: 16086) (^{14}C).

Iskra, A. A., V. G. Bakhurov and N. V. Kulikov. 1970. The role of freshwater vegetation in the migration and translocation of natural radioactive elements in a body of water. pp. 251-267. In: Proc. Intern. Symp. on Radioecology. Vol. I. (See France Commissariat à l'Énergie Atomique, 1970).

Iskra, A. A., N. V. Kulikov and V. G. Bakhurov. 1970. Role of freshwater vegetation in processes of migration and distribution of natural radioactive elements in the reservoir. Ekologiya (2): 83-89. (Sov. J. Ecology (2): 157-162) (NSA 26: 15549, 28079).

Isoi, K., K. Ishida and Y. Ikeda. 1970. Study on detection methods of ^{14}C-labelled compounds. Mukogawa Joshi Daigaku Kiyo Shizenk. (17): 357-359. (In Japanese) (NSA 57752).

Isom, B. G. 1973. Colony formation of the western harvester ant in a chronic gamma radiation field. Am. Midland Nat. 89(2): 446-448. (NSA 28: 21775).

Isomaeki, A. M., R. M. Bergström and E. Kivalo. 1962. Ultrstructural changes in the sensory nerve fibers in the skin of the frog (Rana temporaria) after circumscript irradiation with Po^{210} α particles (5.3 Mev). Acta Pathol. Microbiol. Scand. 54: 190-202. (NSA 16: 20134).

Israël, H. 1958. Natural radioactivity in soil, water and air. Beitr. physik Atmosphäre 30: 177-188. (NP-tr-125).

Israël, H. and A. Krebs. 1962. Nuclear Radiation in Geophysics. Academic Press, New York; Springer-Verlag, Berlin. xv, 430pp. (In English, German).

Itagaki, K. and S. Koenuma. 1961. Altitude distribution of fallout contained in rain and snow. J. Geophys. Res. 67: 3927-3933. (NSA 17: 3244).

Italy Comitato Nazionale per l'Energia Nucleare. Data on environmental radioactivity collected in Italy. Rome. (In Italian).

Report Year	Report Period	Report Number	pp.	NSA
1962	Jan.-Jun. 1962	BIO/26/63	363	17: 36024
1963	Jul.-Dec. 1962	BIO/03/63	479	18: 18040
1963	Jan.-Jun. 1963	BIO/32/63	228	19: 42637
1966	Jan.-Jun. 1965	PROT-SAN-02/66	290	21: 39055
1966	Jul.-Dec. 1965	PROT-SAN-10/66	238	21: 39056
1966	Jul.-Dec. 1966	PROT-SAN-12/67	-	22: 23165
1968	Jan.-Jun. 1967	PROT-SAN-6/68	185	24: 20891
1968	Jul.-Dec. 1967	PROT-SAN-12/68	212	24: 20892
1969	Jan.-Jun. 1968	PROT-SAN-8/69	171	25: 29503
1969	Jul.-Dec. 1968	PROT-SAN-13/69	176	25: 29504
1970	Jan.-Dec. 1969	PROT-SAN-10/70	292	27: 11963
1971	Jan.-Dec. 1970	PROT-SAN-11/71	249	28: 5544
1974	Jan.-Dec. 1972	PROT-SAN-03/74	221	32: 3908

Italy Comitato Nazionale per l'Energia Nucleare. 1973. Nuclear Energy and the Environment. (Proc. 17th Nuclear Congr., Rome, March 23-24, 1972). Rome, Ser. Simposi. 344pp. (In several languages) (CONF-720356) (NSA 29: 17687).

Itard, J. 1968. Preliminary results of a gamma-irradiation test on pupae and adult males of Glossina morsitans morsitans. pp. 59-63. In: Control of Livestock Insect Pests by the Sterile-Male Technique. (See IAEA, 1968) (In French) (NSA 24: 34811).

Itard, J. 1968. Sterilization of Glossina tachinoides West. males by gamma ray irradiation. Rev. elevage med. vet. pays trop. 21: 479-491. (In French) (NSA 24: 32309).

Ito, T. and M. Tanaka. 1959. Effect of radiation on the growth of silkworm larvae. J. Sericult. Sci. Japan 28(4): 220-226. (In Italian)

Ito, T. and M. Tanaka. 1963. Effects of γ-irradiation on the activities of digestive enzymes of the silkworm, Bombyx mori L. Nippon Sanshigaku Zass. 32: 330-337. (NSA 19: 8889).

Ivanitskaya, A. F. 1956. Reaction of axolotl cell elements to the total x radiation. Izv. Akad. Nauk SSSR (biol.) (5): 85-97. (In Russian) (NSA 11: 1939).

Ivanitskaya, A. F. 1961. A microcinematography method in the investigation of the effect of x rays on cells. Pathol. biol. (Semaine Hôp.) 9(7/8): 842-843. (NSA 17: 21542) (Axolotl, Ambystoma mexicanum).

Ivanov, K. A. 1958. Radiological testing of water basins. Inform. Byull. Mosk. nauchn,-issled. inst. sanit. gigiyeny 19-20: 44-46.

Ivanov, M. V. 1956. Application of isotopes for study of the intensity of the process of sulfate reduction in Belovod' Lake. Microbiologiya 25(2): 305-309.

Ivanov, M. V. 1967. Study of the intensity of the process of the S^{35} cycle in nature. In: Report of the 6th Conf. on Problems of Biology of Inland Waters. Leningrad. (In Russian).

Ivanov, P. and A. Stanev. 1972. Determination of the isotope-exchangeable phosphorus in certain soil types. Pochovozn. Agrokhim. 7(4): 12-24. (In Bulgarian) (NSA 30: 3162) (^{32}P).

Ivanov, S. N. and Ye. D. Shagalova. 1969. Ratio of ^{90}Sr and calcium in certain mineral and peat-swampy soils. Dokl. Akad. Nauk Belorussk. SSR 13: 269-271. (In Russian) (NSA 23: 35329).

Ivanov, S. N. and Ye. D. Shagalova. 1972. Strontium in the environment. pp. 423-435. In: 2nd Intern. Conf. on Strontium Metabolism. (Glasgow, 16-19 August 1972). (CONF-720818) (NSA 27: 27591).

Ivanov, S. N., Ye. D. Shagalova and S. S. Shifrina. 1970. Certain data concerning the ^{90}Sr content in peat-swampy soils. Dokl. Akad. Nauk Belorussk. SSR 14: 849-852. (In Russian) (NSA 25: 21184).

Ivanov, S. N., Ye. D. Shagalova and S. S. Shifrina. 1970. Effect of biological properties of plant on ^{90}Sr contents in yield. Dokl. Belorussk. SSR 14: 947-950. (In Russian) (NSA 25: 18897).

Ivanov, S. N., Ye. D. Shagalova and S. S. Shifrina. 1970. Effects of the agrochemical properties of soil on the uptake of ^{90}Sr by cereal crops. Dokl. Akad. Nauk Belorussk. SSR 14: 355-358. (In Russian) (NSA 25: 32353).

Ivanov, S. N., Ye. D. Shagalova and S. S. Shifrina. 1970. ^{90}Sr content of pet-bog soils. Dokl. Akad. Nauk Belorussk. SSR 14: 849-852. (In Russian) (NSA 25: 32374).

Ivanov, S. N., Ye. D. Shagalova and S. D. Shifrina. 1974. Certain features in determing the ^{90}Sr discrimination coefficient with respect to calcium in the soil-plant system. Dokl. Akad. Nauk Belorussk. SSR 18(1): 66-69. (In Russian) (NSA 30: 304).

Ivanov, V. A., M. S. Kulish and Ye. I. Yarzev. 1963. A rapid method of intravitam determination of radiostrontium in the animal and human organism. Radiobiologiya 3(2): 322-324. (AEC-tr-5435, pp. 163-264).

Ivanov, V. V., N. I. Gvaita and Ye. N. Yartseva. 1947. Radioactive waters of eastern Transbaikalia. Arkhiv. vses. gos inst. geol. (In Russian).

Ivanova, M. V. 1959. Assessment of sulfur cycle processes in lakes using radioactive sulfur. pp. 160-166. In: Trans. 6th Conf. on the Biology of Inland Waters. Ye. N. Pavlovskiy, et al., eds. Publ. House Acad. Sci. USSR, Moscow-Leningrad. (AEC-tr-6880).

Ives, K. J. 1960. Casesium-137 labelled algae for filtration studies. Int. J. Radiat. Isotopes 9(1): 49-53. (NSA 15: 8462).

Ives, P. T. 1958. The effect of Co^{60} γ radiation on the fecundity of Drosophila males. Abstr. Radiation Research 9(1): 133.

Ives, P. T., R. S. Heilman and H. H. Plough. 1955. The LD_{50} for gamma radiation in Drosophila melanogaster. Abstr. Genetics 40(5): 577.

Iwaide, I. 1953. Reports by forestry team. V. Effect of atomic bomb on fungi. Nippon Gakujutsu Kaigi 1: 266-269.

Iwai, S. and Y. Inoue. 1963. A method for the estimation of groundwater stratification. pp. 297-307. In: Radioisotopes in Hydrology. (See IAEA, 1963) (NSA 18: 1924).

Iwakura, T. 1972. Estimation of present and future levels of excess ^{14}C in the environment from the nuclear tests. 1. Six reservoirs model. Hoken Butsuri 7(1): 7-14. (NSA 26: 48062).

Iwanami, Y. and S. Matsumura. 1963. Effects of irradiation on pollen. I. Irradiation effects of γ-rays on pollen-tube mitosis. Botan. Mag. (Tokyo) 76: 246-255. (NSA 19: 17525).

Iwasaki, I. 1968-69. Distribution of radium (^{226}Ra) in hot springs and the formation (evolution) mechanism of highly radioactive springs. Onsen Kogaku Kaishi 6: 112-114, 165-168; 7(1): 16-24; 7(2): 109-114. (In Japanese) (NSA 23: 26360; 24: 316, 19088; 25: 32374).

Iwasaki, K. 1959. Studies on environmental and human contamination caused by radioactive substances due to A and H-bomb tests. Nippon Igaku Hosh. Gakkai Zasshi (Nippon Acta Radiol.) 19(7): 1447.

Iwasaki, T. 1958. Effects of water content on x-ray sensitivity in Artemia eggs. Abstr. Radiation Research 9(1): 133.

Iwasaki, T. 1964, 1965, 1966, 1971. Sensitivity of Artemia eggs to γ-irradiation. I. Hatchability of encysted dry eggs. II. Effects of water content. IV. After-effect of irradiation in comparison with the effect of heat-treatment. V. Biological after-effect of irradiation in relation to water content and temperature. VII. Relationship between the degree of biological damage and the decay of free radicals in irradiated eggs. VIII. Modification of amount of radiation-induced free radicals by post-irradiation treatments. IX. A correlation between radiation-induced free radicals and hatchability in eggs exposed to β-mercaptoethylamine. J. Radiation Res. (Japan) 5: 69-75, 76-81, 97-101; 6: 11-16. Int. J. Radiat. Biol. 9: 573-580; 11: 153-159; 20: 39-47. (NSA 19: 38383, 38384, 43815; 20: 8602; 21: 20038; 25: 55262).

Iwasaki, T. 1973. The differential radiosensitivity of oogonia and oocytes at different developmental stages of the brine shrimp, Artemia salina. Biol. Bull. 144(1): 151-161. (NSA 28: 18529).

Iwasaki, T. and T. Marutama. 1971. Comparative effects of ^{60}Co γ-rays and 29 MeV electrons on Artemia. Int. J. Radiat. Biol. 20(5): 493-496. (NSA 26: 15509).

Iwasaki, T., Y. Kato and S. Antoku. 1974. Study of the biological effectiveness of high-energy electrons at ultra-high dose rates using dry eggs of Artemia. Radiation Research 57(2): 195-206. (NSA 29: 24510).

Iwasaki, T., T. Maruyama and A. Nichimura. 1973. The relative biological effectiveness on Artemia exposed to fast neutrons. pp. 25-28. In: 1st Report on Cancer Therapy with High Energy Particles and the Utilization of Short Lived Radionuclides in Diagnostics. Japan National Inst. of Radiological Sciences, Ciba, report NIRS-R-2. (In Japanese) (NSA 32: 6786).

Iwasaki, T., et al. 1971. Effects of fast neutrons and ^{60}Co γ-rays on Artemia. Radiation Research 45(2): 288-298. (25: 26917).

Iwashima, K., et al. 1971. Public health responses of local governments to atomic energy facilities with special emphasis on environmental monitoring. Koshu Eiseiin Kenkyu Hokoku 20(4): 212-226. (In Japanese) (NSA 27: 19341).

Iwashita, K. and N. Yamagata. 1966. Environmental contamination with radioruthenium 1961 to 1965. J. Radiation Res. (Japan) 7(2): 91-111. (NSA 21: 36615).

Iyatomi, K. and T. Saito. 1965. Metabolism of parthion in resistant and susceptible rice stem borer. pp. 500-501. In: Proc. 12th Intern. Congr. of Entomology. P. Freeman, ed. Royal Entomological Soc., London. (^{32}P).

Iyengar, M. A. R. and P. M. Markose. 1971. An investigation into the distribution of uranium and daughter in the environment of a uranium ore processing facility. pp. 143-153. In: Radiation and Radioisotopes in Soil Studies and Plant Nutrition. (See India Dept. of Atomic Energy, 1971) (NSA 26: 25601).

Iyengar, M. A. R. and P. M. Markose. 1971. Monitoring of the aquatic environment in the neighborhood of uranium mill at Jaduguda, Bihar. pp. 473-485. In: Proc. National Symp. on Radiation Physics. (Trombay, Bombay, 24 November 1970). India Bhabba Atomic Research Centre, Trombay, Bombay. (CONF-701138) (NSA 26: 40857).

Iyengar, R. and M. Paniegel. 1955. On the excretion of ^{32}P of Drosophila melanogaster and Gryllus domesticus. Compt. rend. 241(13): 823-824. (In French).

Iyer, M. R., D. N. Sharma and A. K. Ganguly. 1974. Fission product nuclear data (FPND) and environmental aspects of nuclear fuel cycle. pp. 3,v.; v.1, pp. 29-50. In: Panel on Fission Product Nuclear Data. IAEA, Vienna, Publ. IAEA-169(Vol.1). (NSA 33: 25988).

Izmerov, N. V. and L. T. Elovskaya. 1975. Hygienic aspects of studies of uranium levels in the environment and its effect on the body. Gigiyena sanitariya (8): 120-121. (In Russian) (NSA 33: 29775).

Izraél', Yu. A. 1964. The influence of uneven terrain on the gamma field over the area contaminated by fallout. Atom. Energiya 17(2): 137-140. (Soviet Atomic Energy 17: 834-848) (NSA 18: 41665). (Izra'yel?).

Izraél', Yu. A. 1967. Mode of origin of radioactive fallout particles and the fractionation of isotopes in an underground nuclear explosion with ejection of soil. Dokl. Akad. Nauk SSSR 169(3): 573-576. (In Russian).

Izraél', Yu. A., et al. 1964. Tritium content in glaciers. Dokl. Akad. Nauk SSSR 156(1): 72-73. (Doklady (earth sci.) 156: 1-2) (NSA 18: 37375).

Izraél', Yu. A., et al. 1969. Radioactive contamination of the environment by underground nuclear explosions, and methods of forecasting it. Moscow. (AEC-tr-7112) (NSA 24: 19093).

Jaag, O. 1963. The control of radioactivity in waters and soil. Oncologia 16(3): 230-244. (In German) (NSA 18: 12393).

Jaakkola, T. 1967. ^{55}Fe and stable iron in some environmental samples in Finland. pp. 247-251. In: Radioecological Concentration Processes. (See Aberg and Hungate, 1967) (CONF-660405-9) (NP-16002, Paper 6) (NSA 20: 26602).

Jaakola, T. 1968. The occurrence of ^{55}Fe in the biosphere and its enrichment in humans, especially in the Lapps. Thesis, Univ. of Helsinki. 95 pp. (In Finnish; English conclusions, summary, tables).

Jaakola, T. 1969. Analysis of ^{55}Fe produced by nuclear tests and its enrichment in Finnish Lapps. Ann. Acad. Sci. Fenn. AII(150): 64pp. (NSA 24: 25434).

Jaakola, T. and P. Kauranen. 1970. Environmental γ-radiation dose rate in Finnish Lapland and southern Finland. Univ. of Helsinki, Finland, U. S. AEC report NYO-3446-14; Paper 28. 13pp. (For 2nd Congr. Intern. Radiation Protection Assoc., Brighton, England). (CONF-700505-16) (NSA 25: 18986).

Jaakola, T., H. Puumala and J. K. Miettinen. 1967. Microelement levels in environmental samoles in Finland. pp. 341-350. In: Radioecological Concentration Processes. (See Aberg and Hungate, 1967) (CONF-660405-7) (NP-16002, Paper 4) (NSA 20: 26600).

Jaakola, T., et al. 1969. ^{137}Cs and ^{90}Sr in blood and urine of Lapps and southern Finns compared with the total body burden of ^{137}Cs and the estimated dietary intake of ^{90}Sr. Radiochim. Acta 11: 214-216. (NSA 23: 35717).

Jacklin, S. W., E. G. Richardson and C. E. Yonce. 1970. Substerilizing doses of gamma irradiation to produce suppression in plum curculio. J. Econ. Entomol. 63(4): 1053-1057. (NSA 24: 44651).

Jacklin, S. W., F. F. Smith and A. L. Boswell. 1965. Egg mortality after gamma irradiation of adults of the omnivorous leaf roller. J. Econ. Entomol. 58: 1168-1169. (NSA 20: 14340).

Jackson, C. G., Jr., C. M. Holcomb and M. M. Jackson. 1972. Strontium-90 in the exoskeletal ossicles of *Dasypus novemcinctus*. J. Mammalogy 53(4): 921-922. (NSA 28: 18238) (Incl. armadillo, jackrabbit, gopher tortoise).

Jackson, C. G., Jr., et al. 1971. Bioaccumulation of radionuclides in turt turtle shells (Reptilia: Testudines). Abstr. ASB Bull. 18(2): 39.

Jackson, N. E., et al. 1967. Gamma irradiation and the microbial population of soils at two water contents. Soil Sci. Soc. Am. Proc. 31: 491-494. (Also: Phillips Petroleum Co., Bartlesville, Okla., U. S. AEC report COO-2169-12) (NSA 21: 45854).

Jackson, W. B. 1967. Rats, bombs, and paradise: the story at Eniwetok. Bowling Green State Univ., Bowling Green, O., U. S. AEC report COO-1435-10. 5pp. (NSA 21: 24543).

Jackson, W. B. 1967. Engebi rats: another chapter. Bowling Green State Univ., Bowling Green, O., U. S. AEC report COO-1485-12. 16pp. (NSA 21: 43445).

Jackson, W. B. 1968. Dragonfly populations at Eniwetok Atoll. BioScience 18(12): 1123.

Jackson, W. B. 1969. Survival of rats at Eniwetok Atoll. Pacific Science 23: 265-275. (NSA 24: 9806).

Jackson, W. B. nd. Field studies of rat populations at Eniwetok Atoll. Final report. Bowling Green State Univ., Bowling Green, O., U. S. AEC report COO-1485-21.

Jackson, W. B. and M. L. Carpenter. 1967. Radioisotope cycling in terrestrial communities at Eniwetok Atoll. Bowling Green State Univ., Bowling Green, O., U. S. AEC report COO-1485-13. (NSA 21: 43318).

Jackson, W. B. and M. L. Carpenter. 1969. Radioisotope cycling in terrestrial communities at Eniwetok Atoll. pp. 644-651. In: *Symposium on Radioecology*. (See Nelson and Evans, 1979).

Jackson, W. B. and P. P. Maier. 1955. Dispersion of marked American cockroaches from sewer manholes in Phoenix, Arizona. Am. J. Trop. Med. Hyg. 4(1): 141-146.

Jacob, J. and J. L. Sirlin. 1958. Labelling of insect spermatozoa by adenine-8-^{14}C. Experientia 14(2): 402-403.

Jacob, J. and J. L. Sirlin. 1964. Electron microscope autoradiography of the nucleolus of insect salivary gland cells. Nature 202: 622-623. (^{3}H, *Smittia*).

Jacob, J. and J. L. Sirlin. 1964. Synthesis of RNA (ribonucleic acid) in *vitro* stimulated in dipteran salivary glands by 1,1,3-tricyano-2-amino-1-propene. Science 144: 1011-1012. (^{3}H, *Smittia parthenogenetica*).

Jacobi, W. 1962. The natural radioactivity of the atmosphere and its significance for the radiation protection of man. Germany F. R. Hahn-Meitner-Institut für Kernforschung, Berlin, report HMI-B-21. 296pp. (NSA 17: 20240).

Jacobi, W. 1971. Transfer of fission products from atmospheric fallout into river water. pp. 1153-1163. In: *Proc. Intern. Symp. Radioecology Applied to the Protection of Man and His Environment*. (See Commission of the European Communities, 1971) (NSA 27: 5265).

Jacobi, W. and N. Stolterfoht. 1969. Contamination of river water by the fallout after extreme reactor accidents. Germany Hahn-Meitner-Institut für Kernforschung, Berlin, report HMI-B-87. 61pp. (In German) (Also ? 1971: NP-19938, Paper 6) (NSA 25: 8660; 29: 29354).

Jacobi, W. and N. Stolterfoht. 1969. Estimation of the surface water's contamination by the near fallout by nuclear explosions. Germany Hahn-Meitner-Institut für Kernforschung, Berlin, report HMI-B-86. 42pp. (In German) (Also ? 1971: NP-19938, Paper 7) (NSA 25: 8354; 29: 29355).

Jacobi, W., J. Eichler and N. Stolterfoht. 1969. Correlation between fallout from nuclear weapons tests and fission product activity in river water. Germany F. R. Hahn-Meitner-Institut für Kernforschung, Berlin, report HMI-B-82. 56pp. (In German) (NSA 23: 31417).

Jacobi, W., N. Stolterfoht and K. Haberer. nd. Evaluation of the contamination of drinking water reservoirs by fallout from nuclear explosions and extreme reactor incidents. Germany F. R. Hahn-Meitner-Institut für Kernforschung, Berlin, report HMI-B-88. 36pp. (Australian Atomic Energy Comm. Research Establishment report LIB/Trans-420) (NP-19938, Paper 8) (NSA 26: 48043; 29: 29356).

Jacobs, D. G. 1963. Movement of radionuclides through the ground. Nuclear Safety 5(1): 109-114. (NSA 18: 422).

Jacobs, D. G. 1968. Sources of tritium and its behavior upon release to the environment. U. S. AEC report TID-24635. vi, 90pp.

Jacobs, D. G. 1971. Survey paper on tritium and nuclear technology. Trans. Amer. Nucl. Soc. 14(1): 160. (CONF-710606) (NSA 25: 41470).

Jacobs, D. G. 1973. Report on the international symposium "Environmental Behavior of Radionuclides Released in the Nuclear Industry" held in Aixen-Provence in May 1973. Atomic Energy Rev. 11(3): 691-697.

Jacobs, D. G. 1974. Impacts in groundwater of effluents arising in the nuclear industry. Paper 16, vp. In: *Environmental Analysis and Environmental Monitoring for Nuclear Power Generation*. (See Univ. of California, 1974) (NSA 32: 14302).

Jacobs, D. G. and T. Tamura. 1960. Retention and movement of radioisotopes in soils. Nuclear Safety 1(3): 65-70.

Jacobson, A. F. and H. Cember. 1971. Effect of temperature upon the uptake of ^{144}Ce by *Chlamydomonas*. pp. 784-795. In: *Health Physics Aspects of Nuclear Facility Siting*. *Vol. III*. (See Voillequé and Baldwin, 1971) (NSA 25: 55162).

Jacobson, A. P. and R. C. Riley. 1974. Role of serotonin in the radioresistance of Mongolian gerbils. Int. J. Radiat. Biol. 26(3): 269-276. (NSA 31: 8926).

Jacobson, A. P., K. A. McDermott and R. C. Riley. 1971. Mortality and bodyweight changes in the Mongolian gerbil after ^{60}Co gamma irradiation. Int. J. Radiat. Biol. 19(3): 247-253. (NSA 25: 46982).

Jacobson, B. S. 1957. Evidence of recovery from x-ray damage in Chlamydomonas. Radiation Research 7(4): 394-406. (NSA 12: 2672).

Jacobson, B. S. 1962. Relationships between cell division and death in x-irradiated Chlamydomonas cultures. Radiation Research 17(1): 82-91. (NSA 16: 21980).

Jacobson, B. S. 1966. Mechanisms of lethal radiation damage and recovery in algal flagellates. Final scientific report. Univ. of Minnesota, Minneapolis, U. S. AEC report COO-1117-10. 37pp. (NSA 21: 13072).

Jacobson, B. S. and T. C. Lee. 1967. Macromolecular synthesis and delayed mitotic death due to radiation and drugs in Chlamydomonas and Chlorella. Radiation Research 31: 368-381. (NSA 21: 34824).

Jacobson, B. S., R. J. Salmon and L. L. Lansky. 1964. Antimitotic effects of chloramphenicol and other inhibitory agents in Chlamydomonas. Exptl. Cell Res. 36: 1-13. (NSA 19: 7075).

Jacobson, G. A., et al. 1965. Michigan State Department of Health Reactor Environmental Surveillance Program - a summary report. Radiol. Hlth. Data 6(1): 49-60.

Jacoby, G. H. and C. P. Hickman, Jr. 1967. A study of the circulating iodocompounds of rainbow trout, Salmo gairdneri, by the method of isotopic equilibrium. Gen. Comp. Endocrinol. 7(2): 245-254. (Na^{125}I, ^{127}I).

Jacques, J. A. and D. A. Karnofsky. 1950. Toxicity and pathological effects of roentgen rays in the chicken. Am. J. Roentgenol. Rad. Therapy 64(2): 289-297.

Jafri, R. H. 1961. Synergistic action of radiation and of Bacillus thuringiensis toxin on protozoan diseases of insects. pp. 510-515. In: Progress in Protozoology. (Proc. 1st Intern. Conf. on Protozoology, Prague, August 22-31, 1961). (NSA 21: 26592).

Jafri, R. H. 1962. Susceptibility of irradiated insects to Bacillus thuringiensis infection. pp. 351-354. In: Colloq. Intern. Pathol. Insects. (Paris, 1962). (NSA 21: 24632).

Jafri, R. H. 1964. Influence of pathogens on the life span of irradiated insects. Rev. path. vég. ent. agric. Fr. 43(1): 37-41. (Tribolium castaneum, Galleria mellonella, Farinocystis tribolii, Bacillus thuringiensis).

Jafri, R. H. 1965. Prospects of integrated radiation and microbial control of harmful insects. pp. 747-748. In: Proc. 12th Intern. Congr. of Entomology. P. Freeman, ed. Royal Entomological Soc., London.

Jafri, R. H. 1965. Influence of pathogens on the life span of irradiated insects. J. Invertebrate Pathol. 7(1): 66-70. (NSA 19: 45908).

Jafri, R. H. 1967. Prospects of integrated microbial and radiation control of harmful insects. J. Invertebrate Pathol. 9: 293-297. (NSA 23: 958).

Jafri, R. H. and M. B. Chaudhry. 1971. Development of Tipula iridescent virus (TIV) in Galleria mellonella larvae exposed to gamma radiation. J. Invertebrate Pathol. 18(1): 46-50. (NSA 26: 12430).

Jafri, R. H. and J. A. Dar. 1975. Cytological study of gamma-irradiated testes of the house fly, Musca domestica L. pp. 387-400. In: Sterility Principle for Insect Control, 1974. (See IAEA, 1975) (NSA 32: 11794).

Jafri, R. H. and A. A. Kahn. 1969. A study of the development of a nuclearpolyhedrosis virus of heliothis in Galleria mellonella larvae exposed to gamma rays. J. Invertebrate Pathol. 14: 104. (NSA 24: 2772).

Jagielski, A. and A. Stanoch. 1962. Contamination of grass by ^{90}Sr in Poland. Nukleonika 7: 357-359. (NSA 17: 4392).

Jagnow, G. and K. Haider. 1972. Evolution of $^{14}CO_2$ from soil incubated with dieldrin-^{14}C and the action of soil bacteria on labelled dieldrin. Soil Biol. Biochem. 4(1): 43-__.

Jahkola, A. 1971. Factors affecting the choice of location for nuclear power plants, particularly in south and west Finland. Sahko 44(5-6): 138-142. (NSA 26: 3752).

Jahn, E., et al. 1966. Study of the dispersal of the nun moth, Lymantria monacha L., by means of rare earths. Anz. Schädlingsk. 39(2): 17-22. (In German; English, French, Russian summaries) (Eu, Dy).

Jaimes, S. and F. B. Lichtenberg. 1965. Host response to eggs of Schistosoma mansoni. IV. Fluorescent antibody titers in mice infected with normal cercariae, gamma-radiated cercariae and with purified eggs. Am. J. Trop. Med. Hyg. 14: 727-735. (NSA 20: 14304).

Jakowska, S., R. F. Nigrelli and A. H. Sparrow. 1954. Local cell responses to induced skin infection in Triturus viridescens after whole-body irradiations. Abstr. Radiation Research 1(6): 569.

Jakowska, S., R. F. Nigrelli and A. H. Sparrow. 1958. Radiobiology of the newt, Diemictylus viridescens. Hematological and histological effects of whole-body x irradiation. Zoologica 43(4): 155-160. (NSA 13: 7460).

Jakowska, S., R. F. Nigrelli and A. H. Sparrow. 1961. Effects of chronic gamma irradiation on adult newts, Diemictylus viridescens. Abstr. Am. Zoologist 1: 361-362.

Jakubick, A. 1972. Studies on the movement of water in partially saturated soils-tritium-tracer-method to determine the charge of groundwater. Thesis, Univ. of Heidelberg, Germany. 53pp. (In German) (NSA 32: 8951).

Jakubowski, M. and J. Sapula. 1968. The use of isotopic methods for determination of physical properties of soils in the construction of the geotechnical cross section. Nukleonika 13: 421-429. (In Polish) (NSA 22: 48922).

Jakutowicz, K. 1973. Application of activation analysis in medical investigation as exemplified by studies in some parasites. Postepy Fiz. Med. 8 (1-2): 109-117. (In Polish) (NSA 30: 31481) (Neutron activation).

James, A. P. and M. M. Werner. 1965. The radiobiology of yeast. Radiation Botany 5: 359-382. (NSA 20: 1663).

James, H. G. 1965. Predators of _Aedes atropalpus_ (Coq.) (Diptera: Culicidae) and other mosquitoes breeding in rock pools in Ontario. Can. J. Zool. 43(1): 155-159. (^{32}P).

James, H. G. 1966. Insect predators of univoltine mosquitoes in woodland pools of the Pre-Cambrian Shield in Ontario. Can. Entomol. 98(5): 550-555. (^{32}P).

James, H. G. 1966. Location of univoltine _Aedes_ eggs in woodland pool areas and experimental exposure to predators. Mosquito News 26: 59-63. (^{32}P).

James, P. E. 1970. Methods and equipment for irradiating insects. Agricultural Research Service, U. S. Dept. of Agriculture, Beltsville, Md. 18pp. (For Meeting, Amer. Soc. Agricultural Engineers, Chicago, Ill.) (CONF-701202-1) (NSA 25: 16285).

James, R. A. and E. H. Fleming, Jr. 1966. Relative significance index of radionuclides for canal studies. Lawrence Radiation Lab., U. S. AEC report UCRL-50050-1. 10pp. (NSA 21: 6496).

Jammet, H. 1964. Mechanisms of radioactive contamination via air and water. pp. 37-42. In: _Agricultural and Public Health Aspects of Radioactive Contamination in Normal and Emergency Situations_. (See Food and Agriculture Organization, 1964).

Janke, A., et al. 1938. The action of radium emanations as well as Gastemer thermal water on microorganisms. Zentr. Bakt. Parasitenk. IIAbt. 98: 97-109. (In German) (CA 32: 5422).

Jansen, H. S. 1962. Depletion of carbon-13 in young kauri tree. Nature 196 (4849): 84-85. (NSA 16: 32704).

Jansen, H. S. 1972. Extending the use of bomb carbon in the life sciences. pp. F1-F26. In: _Proc. 8th Intern. Radiocarbon Dating Conf._ (See Rafter, 1972) (NSA 31: 8707).

Janzer, V. J. 1973. Rapid method for the determination of radioactive cesium isotopes in water. J. Res. USGS 1(1): 113-115. (NSA 27: 24334).

Janzer, V. J., B. P. Robinson and S. J. Rucker. 1965. Radiochemical analysis of water samples collected after the Salmon Event in the vicinity of Tatum Salt Dome, Lamar County, Mississippi. U. S. Geological Survey, U. S. AEC report Dribble 44, USGS-474-115. 33pp. (NSA 26: 20334).

Janzer, V. J., B. P. Robinson and S. J. Rucker. 1971. Radiochemical analyses of ground-water samples collected after the Salmon Event in the vicinity of Tatum Salt Dome, Lamar County, Mississippi. U. S. Geological Survey, U. S. AEC report USGS-474-117. 87pp. (NSA 26: 28078).

Japan Atomic Energy Research Institute. 1970. Environmental monitoring at JAERI, Tokai, in 1969. Tokyo, report JAERI-Memo-4175. 34pp. (NSA 25: 13119).

Japan Atomic Industrial Forum, Inc. 1961. Program and abstracts for the 4th Japan Conf. on Radioisotopes, October 1961. 302pp. (NP-11767) (NSA 16: 23354).

Japan Atomic Industrial Forum, Inc. 1969. Application of radiogaschromatography to agriculture and biology. pp. 907-932. In: _Proc. 9th Japan Conf. on Radioisotopes_. (May 13-15, 1969, Tokyo). (In Japanese) (CONF-690548) (NSA 24: 31248).

Japan Atomic Industrial Forum, Inc. 1969. Research for public hazards, usi using isotope techniques on aquatic organisms. pp. 933-965. In: _Proc. 9th Japan Conf. on Radioisotopes_. (May 13-15, 1969, Tokyo). (In Japanese) (NSA 24: 31897).

Japan Forest Experiment Station. 1962. Studies on absorption and translocation of nutrients for forest trees. (Japan NSA 1: 81).

Japan Forest Experiment Station. 1962. Effects of γ-irradiation on forest trees. (Japan NSA 1: 82).

Japan Meteorological Agency. 1961. Summary of the observations for deposition of radioactivity in the world during IGY-IGC. Tokyo. 145pp. (NP-12006) (NSA 16: 30202).

Japan National Institute of Radiological Sciences. 1962. Research on the radionuclides in the food chain. (Japan NSA 1: 145-146) (Incl. freshwater and marine organisms).

Japan National Institute of Radiological Sciences. 1964. Biochemistry and biophysics; Cytology and morphology; Physiology; Genetics; Environmental studies. pp. 19-56, 65-77. In: Annual Report, 1963. Chiba, report NIRS-3. (NSA 19: 31670, 31672) (Incl. silkworm spermatogonia, _Artemia_ eggs, irradiated _Carassius auratus_, and ^{137}Cs and ^{90}Sr in river water).

Japan National Institute of Radiological Sciences. Radioactivity data in Japan. Chiba. (Incl. environmental data, mentions radioecology).

Year	Report No.	pp.	NSA
1963	1	35	18: 31832
1963	3	65-77	19: 31672
1964	4	31	19: 42639
1968	NIRS-18	31	23: 6437
1968	NIRS-19	18	23: 21824
1968	NIRS-RSD-20	16	24: 23180
1969	NIRS-RSD-22	1-12	24: 31890
1969	NIRS-RSD-23	17	24: 22787
1969	NIRS-RSD-25	11	24: 29392

Japan National Institute of Radiological Sciences. Radioactivity data in Japan. (Cont'd).

Year	Report No.	pp.	NSA
1970	NIRS-RSD-27	17	25: 26392
1970	NIRS-RSD-29	18	25: 37806
1971	NIRS-RSD-31	18	26: 9481
1971	NIRS-RSD-32	21	26: 17709
1972	NIRS-RSD-34	25	26: 56097
1972	NIRS-RSD-37	20	28: 8158

Japan Nuclear Safety Research Association. 1971. Status on some health physics aspects of radio-iodine. Tokyo. 110pp. (In Japanese) (NSA 26: 28092) (Incl. environmental aspects).

Japan Radioisotopes Association. 1958. _Proc. 2nd Japan Conf. on Radioisotopes_. (February 1958). Tokyo. (AEC-tr-4482).

Jaroslow, B. N. 1955. The effect of x irradiation on immunity of the mouse to _Trypanosoma duttoni_. J. Infect. Diseases 96: 242-249.

Jaroslow, B. N. and D. E. Smith. 1961. The immune response in hibernating animals. I. Antigen disappearance in hibernating ground squirrels (_Citellus tridecemlineatus_). pp. 45-47. In: Biological and Medical Research Division summary report, January through December, 1960. Argonne National Lab., U. S. AEC report ANL-6368. (NSA 15: 31987).

Jaroslow, B. N. and D. E. Smith. 1964. Antigen disappearance in ground squirrels in hibernation for 14, 28 and 56 days after injection. Ann. Acad. Scient. Fenn. A(IV biol.) 71: 235-240. (^{131}I).

Jaroslow, B. N. and D. E. Smith. 1964. Effects of hibernation on the latent period in normal and x irradiated ground squirrels. pp. 23-26. In: Biological and Medical Research Division semi-annual report, July through December 1963. Argonne National Lab., U. S. AEC report ANL-6906. (NSA 19: 7013).

Jaroslow, B. N., et al. 1969. Survival of hibernating ground squirrels (_Citellus tridecemlineatus_) after single and fractionated doses of ^{60}Co gamma radiation. Radiation Research 38: 379-388. (NSA 24: 9933).

Jaske, R. T. 1969. An independent view of the thermal and radioactive releases from nuclear power plants. Battelle Northwest Lab., U. S. AEC report BNWL-SA-2279. 30pp. (For Seminar on Hydrologic Problems, Missoula, Mont., Jan. 15, 1969) (CONF-690105) (NSA 23: 17148).

Jaworowski, Z. 1963. Correlation between the concentrations of lead-210 (radium D) and cesium-137 in glacier ice. Nukleonika 8: 853-855. (NSA 18: 22255).

Jaworowski, Z. 1966. Temporal and geographical distribution of radium D (lead-210). Nature 212(5065): 886-889. (Deer antlers, lichen).

Jaworowski, Z. 1969. Radioactive lead in the environment and in the human body. Atom. Energy Rev. 7(1): 3-45. (NSA 23: 33698).

Jaworowski, Z. and J. Bilkiewicz. 1972. Influence of industry on radioactive pollution of environment. Poland Centraine Lab., Ochrony Radiologicznej, Warsaw. (INIS-mf-1236, pp. 3-9) (NSA 33: 589).

Jaworowski, Z., J. Bilkiewicz and E. Zylicz. 1971. ^{226}Ra in contemporary and fossil snow. Health Physics 20: 449-450. (NSA 25: 29209).

Jaworowski, Z., L. Kownecka and E. Wlodek. nd. Effect of traditional industries on environmental contamination with radium-226. Poland Centraine Lab., Ochrony Radiologicznej, Warsaw. (In Russian) (INIS-mf-887, pp. 128-140) (NSA 29: 15610).

Jaworowski, Z., et al. 1972. Artificial sources of natural radionuclides in environment. pp. 809-818. In: _The Natural Radiation Environment II. Vol. 2._ (See Adams, Lowder and Gesell, 1972) (NSA 32: 19743).

Jaworowski, Z., et al. 1974. Artificial sources of natural radioactivity in the environment. pp. 127-135. In: _Radiation Protection, Environment, and the Population_, (See German Democratic Republic, 1974) (NSA 31: 8618).

Jaworoski, Z., et al. 1975. Stable and radioactive pollutants in a Scandinavian glacier. Environ. Pollut. (London) 9(4): 305-316. (NSA 33: 25978).

Jayaraman, S. 1970. Radiation damage and repair after whole-body or partial-body irradiations of _Tenebrio molitor_ larvae. Thesis, Univ. of Illinois, Urbana. 111pp. (Univ. Microfilms 71-5134) (NSA 25: 26914).

Jaylet, A. and C. Bacquier. 1967. Heterozygous chromosomal aberrations in the viable offspring of x-irradiated male newts (_Pleurodeles waltlii_ Michah.). Cytogenetics 6: 390-401. (In French) (NSA 22: 23645).

Jaynes, C. C. 1969. A technique for field use of radioactive phosphorus. J. Range Mgmt. 22(5): 354-___.

Jaynes, H. A. 1957. Some recent developments in white-pine weevil research in the Northwest. Abstr. Bull. Entomol. Soc. Am. 3(3): 40.

Jaynes, H. A. and P. A. Godwin. 1957. Sterilization of white-pine weevils with gamma radiation. J. Econ. Entomol. 50(4): 3930395.

Jayson, G. G. and D. C. Pickering. 1962. The effects of ionizing radiation on the absorption of bivalent ions by green filamentous algae. Abstr. p. 195. In: _2nd Intern. Congr. Radiation Research_. (See Intern. Congr. Radiation Research, 1962).

Jayson, G. G., et al. 1965. Radioprotective action of calcium on _Serratia marcescens_. Naturwissenschaften 52(8): 185. (NSA 19: 36114).

Jeanneau, B. 1967. Study of various processes for marking sediments with ^{198}Au. Centre d'Études Nucléaires, France Commissariat á l'Énergie Atomique, Saclay, report CEA-R-3389. (In French) (NSA 22: 25614).

Jefferies, D. J. 1962. The effects of continuous and fractionated doses of gamma-radiation on the survival and fertility of Sitophilus granarius (Calandra granaria L.). pp. 213-231. In: Radioisotopes and Radiation in Entomology. (See IAEA, 1962).

Jefferies, D. J. 1962. Susceptibility of the saw-toothed grain beetle (Oryzaephilus surinamensis) to gamma radiation. U. K. Atomic Energy Authority, Wantage, report AERE-R-3891. 32pp.

Jefferies, D. J. 1966. Effects of continuous and fractionated doses of gamma radiation on the survival and fertility of Sitophilus granarius (L.). pp. 41-56. In: The Entomology of Radiation Disinfestation of Grain. (See P. B. Cornwell, 1966).

Jefferies, D. J. 1966. Susceptibility of the saw-toothed grain beetle, Oryzaephilus surinamensis (L.) to gamma radiation. pp. 119-130. In: The Entomology of Radiation Disinfestation of Grain. (See Cornwell, 1966).

Jefferies, D. F. 1970. Exposure to radiation from gamma-emitting fission-product radionuclides in estuarine sediments from the northeast Irish Sea. pp. 205-216. In: Environmental Surveillance in the Vicinity of Nuclear Facilities. (See Reinig, 1970) (NSA 25: 16173).

Jefferies, D. J. and E. J. Banham. 1961. The effects of continuous and fractionated doses of gamma radiation on the survival and fertility of Sitophilus granarius (Calandra granaria) L. U. K. Atomic Energy Authority, Harwell, report AERE-R-3503. 48pp.

Jefferies, D. J. and P. B. Cornwell. 1958. Lethal and sterilizing effects of single and fractionated doses of gamma radiations on Calandra granaria L. Nature 182(4632): 402-403.

Jefferson, M. E. 1960. Cobalt-60 irradiation for "sterile male" method of insect eradication. Nucleonics 18(2): 74-76. (Screwworm flies).

Jefferson, S. (Ed.). 1964. Massive Radiation Techniques. John Wiley & Sons, New York; George Newnes, London. 329pp. (Incl. entomology).

Jehanno, C. 1958. New techniques of sampling and of measurement of the radioactivity of air and of atmospheric precipitations at ground level. Centre d'Études Nucléaires, France Commissariat à l'Énergie Atomique, Saclay. 25pp. (NP-14284) (NSA 19: 2531).

Jenkins, C. E. 1969. Radionuclide distribution in Pacific salmon. Health Physics 17: 507-512.

Jenkins, C. E., N. A. Wogman and H. G. Rieck. 1972. Radionuclide distribution in Olympic National Park, Washington. Water Air Soil Poll. 1(2): 181-204.

Jenkins, D. 1965. Determination of primary productivity of turbid waters with carbon-14. J. Water Poll. Control Fed. 37(9): 1281-1288.

Jenkins, D. W. 1949. Use of radioisotopes in ecological research. Bull. Ecol. Soc. Am. 30: 54-57. (NSA 4: 1671).

Jenkins, D. W. 1949. A field method of marking arctic mosquitoes with radiophosphorus. J. Econ. Entomol. 42(6): 944-989. (NSA 4: 2230).

Jenkins, D. W. 1954. Advances in medical entomology using radioisotopes. Exptl. Parasitol. 3(5): 474-490.

Jenkins, D. W. 1956. To what distance a little female fly to fly? Naturaliste Canadien 83(5): 95-102. (In French).

Jenkins, D. W. 1960. Natural control of medically important insects. Abstr. 11th Intern. Congr. Entomology, Vienna.

Jenkins, D. W. 1961. Radioisotopes in entomology and tropical medicine. Science 133(3467): 1836.

Jenkins, D. W. 1963. Use of radionuclides in ecological studies of insects. pp. 431-440. In: Radioecology. (See Schultz and Klement, 1963).

Jenkins, D. W. and C. C. Hassett. 1950. Radioisotopes in entomology. Nucleonics 6(1): 5-14.

Jenkins, D. W. and C. C. Hassett. 1951. Dispersal and flight range of subarctic mosquitoes marked with radiophosphorus. Can. J. Zool. 29(3): 178-187.

Jenkins, D. W. and K. L. Knight. 1950. Ecological survey of the mosquitoes of Great Whale River, Quebec. Proc. Entomol. Soc. Wash. 52(5): 203-223. (^{32}P).

Jenkins, D. W. and A. S. West. 1951. Plant feeding habits of northern mosquitoes studies with radioisotopes. Mosquito News 11(4): 217-219.

Jenkins, J. H. and T. T. Fendley. 1969. The extent of contamination, detection, and health significance of high accumulations of radioactivity in southeastern game populations. pp. 89-95. In: Proc. 22nd Ann. Conf. Southeastern Assoc. Game and Fish Commissions. (WR 138: 6).

Jenkins, J. H. and T. T. Fendley. 1970. Evaluation of the factors involved in bioaccumulation of gamma-emitting radionuclides in white-tailed deer (Odocoileus virginianus). Univ. of Georgia, U. S. AEC report ORO-3905. 35pp. (NSA 25: 18935).

Jenkins, J. H. and T. T. Fendley. 1973. Radionuclide biomagnification in coastal-plain deer. pp. 116-122. In: Radionuclides in Ecosystems. (See Nelson, 1973) (NSA 28: 27295).

Jenkins, J. H., J. R. Moore and F. B. Golley. 1969. Comparison of fallout ^{137}Cs accumulation and excretion in certain southeastern mammals. pp. 623-626. In: Symposium on Radioecology. (See Nelson and Evans, 1969).

Jenkinson, D. S., T. Z. Nawakew and J. D. D. Mitchell. 1972. Growth and uptake of nitrogen by wheat and ryegrass in fumigated and irradiated soil. Plant Soil 36(1): 149-___.

Jenkinson, S. 1971. Studies on the decomposition of C^{14} labelled organic matter in soil. Soil Science 111(1): 64-70.

Jennings, A. R. 1964. Evaluation of selected radioisotopes as groundwater tracers. Thesis, Texas A. & M. Univ., College Station. 57pp. (NSA 19: 13657).

Jennings, A. R. and M. C. Schroeder. 1968. Laboratory evaluation of selected radioisotopes as groundwater tracers. Water Resources Res. 4: 829-838. (NSA 22: 49038).

Jennings, C. D. and C. Osterberg. 1969. Sediment radioactivity in the Columbia River Estuary. pp. 300-306. In: Symposium on Radioecology. (See Nelson and Evans, 1969).

Jennings, D., N. Cutshall and C. Osterberg. 1965. Detection of gamma-ray emission in sediments in situ. Science 148(3672): 948-950.

Jennings, F. W. 1968. Anaemias associated with helminthic infections. pp. 133-140. In: Isotopes and Radiation in Parasitology. (See IAEA, 1968).

Jennings, F. W., W. Mulligan and G. M. Urquhart. 1963. Variables in x-ray inactivation of Nippostrongylus brasiliensis larvae. (Seminar on Parasitic Helminths). Exptl. Parasitol. 13(1): 367-373. (NSA 18: 70).

Jensen, J. A. and R. W. Fay. 1951. Tagging adult house flies and flesh flies with radioactive phosphorus. Am. J. Trop. Med. 31(4): 523-530.

Jensen, J. A., W. F. Durham and G. W. Pearce. 1952. Studies on fate of parathion in rabbits using radioactive isotope techniques. Arch. Industr. Hyg. Occup. Med. 5(4): 326-331.

Jensen, P. A. and E. Somer. 1967. Scintillation techniques in soil-moisture and density measurements. pp. 31-48. In: Isotope and Radiation Techniques in Soil Physics and Irrigation Studies. (See IAEA, 1967) (NSA 22: 23346).

Jermy, T. 1971. Genetic control experiments on the bean weevil, Acanthoscelides obtectus Say. pp. 349-354. In: Sterility Principle for Insect Control or Eradication. (See IAEA, 1971) (NSA 25: 29749).

Jermy, T. and B. Nagy. 1967. Controlling the cockchafer (Melolontha melolontha) by the sterile male technique. Acta Phytopathol. Hung. 2: 211-217. (NSA 23: 2914).

Jermy, T. and B. Nagy. 1968. Investigations of radiation autocide method in Hungary. Atomtech. Tajek. 11: 661-664. (In Hungarian) (NSA 23: 43934) (Insects).

Jermy, T. and B. Nagy. 1969. Sterile-male technique studies in Hungary. pp. 91-95. In: Sterile-Male Technique for Eradication or Control of Harmful Insects. (See IAEA, 1969) (NSA 24: 32304).

Jersey Central Power & Light Co. 1970. Forked River Nuclear Generating Station, Unit 1. Preliminary Safety Analysis Report, Vol. 2. Morristown, N. J. U. S. AEC DOCKET-50363-2. 296pp. (NSA 24: 31026) (Environmental studies).

Jervis, R. E. 1973. Nuclear activation determination of heavy metals in Great Lakes sediments, soils, and vegetation. Univ. of Toronto, Toronto, Canada. 26pp. (AED-CONF-73-562-002) (NSA 33: 671).

Jervis, R. E., B. Tiefenbach and A. Chattopadhyay. 1974. Determination of trace cadmium in biological materials by neutron and photon activation analysis. Can. J. Chem. 52(17): 3008-3020. (NSA 31: 2765) (Incl. soil, vegetation).

Jeschke, W. D., H. Gimmler and W. Simonis. 1967. Incorporation of ^{32}P and ^{14}C into photosynthetic products of Ankistrodesmus braunii as affected by x-rays. Plant Physiol. 42: 380-386. (NSA 21: 24682).

Jimenez, A., et al. 1967. Sterilization of Ceratitis capitata Wied by the use of gamma radiations. 8pp. In: Symp. on Applications of Radioisotopes. (Madrid, Spain). (In Spanish) (CONF-670649-24) (NSA 23: 26974).

Jinks, S. M. and M. Eisenbud. 1972. Concentration factors in the aquatic environment. Radiat. Data Rept. 13(4): 243-247. (NSA 26: 40858).

Job, P. S. 1962. Some effects of irradiating successive generations of Hymenolepis diminuta. Thesis, Univ. of Tennessee, Knoxville. 96pp. (NSA 17: 21575).

Job, P. S., et al. 1963. Decreased variability in the cestode Hymenolepis diminuta following irradiation of successive generations. Evolution 17 (2): 163-169. (NSA 21: 16513).

Jobin, L. J., L. Huot and J. M. Perron. 1970. LD_{50} emergence and β parameters as criteria of sensitivity to gamma radiation in Acheta domesticus. J. Econ. Entomol. 63(1): 114-119. (NSA 24: 23306).

Joergensen, C. B., L. O. Larsen and P. Rosenkilde. 1965. Hormonal dependency of molting in amphibians: effect of radiothyroidectomy in the toad Bufo bufo L. Gen. Comp. Endocrinol. 5: 248-251. (NSA 20: 16252).

Joftes, D. L. 1963. Radioautography in the study of radioisotopically tagged substances in insect control. pp. 155-166. In: Radiation and Radioisotopes Applied to Insects of Agricultural Importance. (See IAEA, 1963).

Joftes, D. L. 1969. Apparatus for administration of gaseous $^{14}CO_2$ to small animals. Int. J. Appl. Radiat. Isotopes 20: 157-165. (NSA 23: 20394).

Johanningsmeier, A. and C. J. Goodnight. 1962. Use of iodine-131 to measure movements of small animals. Science 138(3537): 147-148. (NSA 17: 1198; WR 108: 17).

Johanningsmeier, A. and C. J. Goodnight. 1963. The use of I-131 to measure small mammal movements. Abstr. Bull. Ecol. Soc. Am. 44(3): 68.

Johannsen, K.-H. and W. Lenk. 1969. Sorption of radionuclides on friable soils and other natural absorbents. Germany D. R. Staatliche Zentrale für Strahlenschutz, Berlin, report SZS-6/69. 33pp. (In German) (NSA 24: 9488).

Johansen, I. 1961. The modifying effects of hypothermia and anoxia on the x-ray sensitivity of the 4-day-old chick embryo. Int. J. Radiat. Biol. 3: 411-418. (NSA 15: 25879).

Johansen, I. 1962. Studies on the radiation sensitivity of Drosophila embryos. Abstr. Hereditas 48(3): 543-544. (NSA 18: 25143).

Johansen, O. and E. Steinnes. 1972. Routine determinations of traces of cobalt in soil and plant tissues by instrumental neutron activation analysis. Acta Agric. Scand. 22(2): 103-106. (NSA 26: 52904).

John, J. 1971. Californium based system for the automated detection of heavy metals in inland waters. pp. 72-85. In: Nuclear Methods in Environmental Research. (See Vogt, 1971) (NSA 26: 17358).

Johns, F. B. 1970. Southwestern Radiological Health Laboratory handbook of radiochemical analytical methods, Southwestern Radiological Health Lab., U. S. PHS, Las Vegas, Nev., report SWRHL-11. 122pp. (NSA 24: 36173).

Johns, F. B. and R. E. Jaquish. 1970. Analysis of atmospheric gases. pp. 683-696. In: Proc. 11th AEC Air Cleaning Conf. (See First and Morgan, 1970) (NSA 25: 20928).

Johnsen, S. J., et al. 1972. Oxygen isotope profiles through the Antarctic and Greenland ice sheets. Nature 235(5339): 429-434. (NSA 26: 22815).

Johnson, B., P. Jessen and F. Bergamo. 1971. Investigation of Rulison radioactivity. Kamen Sciences Corp., U. S. AEC report PNE-R-65. 20pp. (NSA 29: 2614).

Johnson, C. D. 1965. Direct x-ray inactivation of the viruses of Teschen and Talfan diseases, foot-and mouth disease and vesicular stomatitis. Nature 207(4992): 37-39. (NSA 19: 36046).

Johnson, C. D. 1968. In-vitro experiments on the radiosensitivity of foot-and-mouth disease virus and other animal viruses to the direct effect of x-irradiation. pp. 65-69. In: Elimination of Harmful Organisms from Food and Feed by Irradiation. (See IAEA, 1968) (NSA 24: 34793).

Johnson, C. H., Jr. and R. S. Foote. 1974. Method and apparatus for aerial radioactivity surveying for surface mineral deposits which compensates for the radioactive decay products in the atmosphere. U. S. Patent 3,825,751. 14pp. (NSA 30: 26761).

Johnson, E. 1926. Effects of x-rays upon growth, development, and oxidizing enzymes of Helianthus annus. Botan. Gaz. 82: 373-402.

Johnson, E. 1928. Growth and germination of sunflowers as influenced by x-rays. Am. J. Bot. 15(1): 65-76.

Johnson, E. 1933. The influence of x-irradiation on Atriplex hortensis L. New Phytol. 32: 297-307.

Johnson, E. 1936. Effects of x-rays upon green plants. pp. 961-985. In: Biological Effects of Radiation. B. M. Duggar, ed. McGraw-Hill Book Co., New York.

Johnson, E. 1936. Susceptibility of seventy species of flowering plants to x-radiation. Plant Physiol. 11(2): 319-342.

Johnson, E. 1939. Floral development of certain species as influenced by x-radiation. Plant Physiol. 14(4): 783-795.

Johnson, E. 1948. Response of Kalanchoe tubiflora to x-radiation. Plant Physiol. 23(4): 544-556.

Johnson, G. T. and G. C. Kyker. 1961. Fission-product and cerium uptake by bacteria, yeasts, and molds. J. Bacteriol. 81(5): 733-740.

Johnson, J. E. 1975. Survey for radioactivity in a swamp. Savannah River Plant, U. S. AEC report DPSPU-75-30-8. 11pp. (NSA 33: 2982) (Below Savannah River Plant).

Johnson, J. E. and A. I. Lovaas. 1970. Retention of near-in fallout by field crops and livestock. Final report. Colorado State Univ., Fort Collins. 53pp. (AD-715423; NSA 25: 49076).

Johnson, J. E., et al. 1965. Fallout determination, separation and recovery of fallouts Cs^{137} from $Zr-Nb^{95}$ in forage samples. J. Agr. Food Chem. 13 (6): 568-___.

Johnson, L. F. and T. S. Osborne. 1954. Survival of fungi in soil exposed to gamma radiation. Can. J. Botany 42(1): 105-113. (NSA 18: 29285).

Johnson, L. J. 1972. Los Alamos land areas environmental radiation survey 1972. Los Alamos Scientific Lab., U. S. AEC report LA-5097. 27pp. (NSA 27: 9554).

Johnson, N. and G. M. Goodwell. 1963. Ecological applications of solid-state dosimetry. Abstr. Bull. Ecol. Soc. Am. 44(3): 68.

Johnson, N. M. and R. L. Blanchard. 1967. Radiation dosimetry from the natural thermoluminescence of fossil shells. Am. Mineral. 52: 1297-1310. (NSA 22: 8219).

Johnson, T. S., R. E. Nakatani and F. B. Conte. 1970. Influence of temperature and x-irradiation on the cellular dynamics of the intestinal epithelium in coho salmon, Oncorhynchus kisutch. Radiation Research 42: 129-140. (NSA 24: 25570).

Johnson, V., N. Cutshall and C. Osterberg. 1967. Retention of ^{65}Zn by Columbia River sediment. Water Resources Res. 3(1): 99-102. (NSA 21: 39048).

Johnson, W. and J. Eakins. 1974. Radiological surveillance around Turkey Point, 1970-1971. Radiat. Data Rept. 15(3): 105-115. (NSA 30: 6495) (Nuclear power plant, Florida).

Johnson, W. and C. L. Nayfield. 1970. Elevated levels of cesium-137 in common mushrooms (Agariaceae) with possible relationship to high levels of cesium-137 in whitetail deer, 1968-1969. Radiol. Hlth. Data 11(10): 527-531.

Johnson, W. S., Jr. 1968. Plutonium contamination of large land areas. pp. 379-384. In: Radiation Protection. Part 1. (See Snyder, et al., 1968).

Johnston, J. W. 1973. Index of adequacy and some other statistical considerations for an environmental monitoring system. Battelle Northwest Lab., U. S. AEC report BNWL-B-250. 75pp. (NSA 28: 364).

Johnstone, C. W. 1964. Detection of natural γ radiation in petroleum exploration boreholes. pp. 115-128. In: The Natural Radiation Environment. (See Adams and Lowder, 1964) (NSA 18: 35733).

Johnstone, G. R. and F. W. Klepinger. 1967. The effects of gamma radiation on germination and seedling development of Yucca brevifolia Engelm. Radiation Botany 7(5): 385-388.

Joiner, R. L. and E. N. Lambremont. 1969. Hydrocarbon metabolism in insects: oxidation of hexadecane-1-^{14}C in the boll weevil and the house fly. Ann. Entomol. Soc. Am. 62(4): 891-894.

Joint Publications Research Service. 1964. Communist Chinese scientific abstracts No. 2. Translation of Abstr. Papers from China Zoological Soc. 30th Anniversary Scientific Conf. (Peking, July 8-18, 1964). New York report JPRS-45744. 500pp. (NSA 22: 51871).

Jokelainen, A. 1965. Diet of the Finnish Lapps and its ^{137}Cs and potassium contents. Acta Agral. Finn. 103: 1-140. (NSA 19: 44305) (Incl. reindeer, fish).

Jokelainen, A. 1967. ^{137}Cs in some Finnish diets in 1962. pp. 87-94. In: Radioecological Concentration Processes. (See Aberg and Hungate, 1967) (Also: NP-16002, Paper 9) (CONF-660405-12) (NSA 20: 26605).

Jollos, V. 1939. Further tests of the role of cosmic radiation in the Drosophila melanogaster. Genetics 24: 113-130.

Joly, P. and G. Biellmann. 1958. Effects of irradiation on Locusta migratoria L. Compt. rend. 247: 243-246. (In French) (NSA 12: 15240).

Jonard, R. 1959. Comparative study of the radiosensitivities of normal and crown-gall tissues of Scorzonera cultivated in vitro. Compt. rend. 248: 2807-2808. (In French) (NSA 13: 13167).

Jonard, R. 1959. Effects of x rays on the development of crown-gall tissues of cultivated Scorzonera in vitro. Compt. rend. 248: 2664-2666. (In French) (NSA 13: 14163).

Jonard, R. 1960. Protective effects of ascorbic acid with respect to irradiation of tissues of the crown-gall of viper's grass by x rays. Compt. rend. 250: 185-187. (In French) (NSA 14: 7242).

Jonard, R. 1960. Effect of indoleacetic acid and of the extract of yeasts on the restoration of the proliferation power of cultures of crown-gall tissues of Scorzonera irradiated with x rays. Compt. rend. 251: 588-590. (In French) (NSA 14: 23955).

Jonard, R. and J. Bayonove. 1972. Effect of irradiation dose rate on the development of crown gall tissue colonies in Scorzonera irradiated with ^{60}Co γ rays. Compt. rend. D275(22): 2489-2492. (In French) (NSA 27: 27869).

Jonard, R. and C. Landre-Ruch. 1966. Radiobiology: effect of x rays on the formation of tumors provoked on tissues of Jerusalem artichoke by Agrobacterium tumefaciens. Compt. rend. D263: 1517-1520. (In French) (NSA 21: 18192).

Jonard, R. and F. Parrot-Miche. 1970. Host-bacteria relationship in the tissues of x-irradiated Jerusalem artichokes. Physiol. veg. 8: 231-234. (In French) (NSA 25: 11306).

Jonard, R. and D. Ravelomanana. 1975. Radiorestoration by S-RNA extracts from yeast and a mixture of their nucleotides on the growth of Scorzonera crown-gall tissue in vitro. Compt. rend. D280(6): 803-805. (In French) (NSA 32: 11712).

Jonas, R. E. E. and E. Bilinski. 1963. Utilization of lipids by fish. III. Fatty acid oxidation by various tissues from sockeye salmon (Oncorhynchus nerka). J. Fish. Res. Bd. Can 21(3): 653-656. (SFA 9: 6817) ^{14}C).

Jonczy, J. 1973. Influence of ionizing radiation on reproduction and mortality of the flour mite (Acarus siro L.). Folia biol. (Krakow) 21(1): 77-84. (NSA 28: 6054; 32: 9193).

Jones, A. R. 1971. Gamma monitor for measuring environmental gamma doses and dose rates. Atomic Energy of Canada, Ltd., Chalk River, report AECL-3989. 17pp. (NSA 25: 57721).

Jones, A. W., et al. 1963. Host-relationships of radiation-induced mutant strains of Hymenoleis diminuta. Ann. N.Y. Acad. Sci. 113(1): 343-359. (NSA 18: 9841).

Jones, E. B., et al. 1973. Areal snowpack water-equivalent determinations using airborne measurements of passive terrestrial gamma radiation, EG&G, Las Vegas, Nev., U. S. AEC report EGG-1183-1615. 11pp. (NSA 29: 5134).

Jones, E. B., et al. 1974. Areal snowpack water-equivalent determinations using airborne measurements of passive terrestrial gamma radiation. pp. 594-603. In: Advanced Concepts and Techniques in the Study of Snow and Ice Resources. (See Santeford and Smith, 1974) (NSA 33: 7308).

Jones, G. B. and G. B. Belling. 1967. The movement of copper, moybdenum and selenium in soils as indicated by radioisotopes. Austral. J. Agr. Res. 18(5): 733-___.

Jones, J. D., et al. 1971. Determination of traces of mercury in biological material by neutron activation. Mikrochim. Acta (3): 399-404. (In French) (NSA 25: 51420).

Jones, J. K., et al. 1973. Experience of the Central Electricity Generating Board in monitoring the environment of its nuclear power stations. Health Physics 24(6): 619-625. (NSA 28: 8182).

Jones, J. M. 1966. The effects of ionizing radiation on the growth of the lichen Parmelia conspersa. Abstr. Assoc. Southeastern Biol. Bull. 13(2): 39.

Jones, J. M. and R. B. Platt. 1969. Effects of ionizing radiation, climate, and nutrition on growth and structure of a lichen Parmelia conspersa (Ach.) Ach. pp. 111-119. In: Symposium on Radioecology. (See Nelson and Evans, 1969).

Jones, L. M. and H. G. Oestlund. 1971. ^{14}C age and tritium content of Lake Vanda, Wright Valley. Antarctic J. (US) 6(5): 200-201. (NSA 26: 9249).

Jones, O. 1956. The determination of radioiodine in grass. Windscale Works, U. K. Atomic Energy Authority, Sellafield, report IGO-AM/W-56. 7pp. (NSA 10: 10314).

Jones, P. H. 1961. Hydrology of radioactive waste disposal at the Idaho chemical processing plant, National Reactor Testing Station, Idaho. U. S. Geological Survey, U. S. AEC report IDO-22041. 16pp.

Jones, P. H. and E. Shuter. 1962. Hydrology of radioactive waste disposal in the MTR-ETR area, National Reactor Testing Station, Idaho. U. S. Geological Survey Prof. Papers No. 450-C, C 113-16. (NSA 17: 1655).

Jones, R. H. 1967. Some irradiation studies and related biological data for Culicoides variipennis (Diptera: Ceratopogonidae). Ann. Entomol. Soc. Am. 60(4): 836-846.

Jones, S. C. and L. Wallace. 1955. Cherry fruit fly dispersion studies. J. Econ. Entomol. 48(5): 616-617.

Jones, T. H. 1967. The uptake of strontium-90 by Chironomus varus. Entomologist 100: 2.

Jones, W. M., C. D. Jennings and N. H. Cutshall. 1975. Cycling of ^{55}Fe and ^{65}Zn in Columbia River carp following reactor shutdown. pp. 309-318. In: Combined Effects of Radioactive, Chemical, and Thermal Releases to the Environment. (See IAEA, 1975) (NSA 33: 29794).

Joosse, J., M. H. Boer and C. J. Cornelisse. 1968. Gametogenesis and oviposition in Lymnaea stagnalis as influenced by gamma-irradiation and hunger. Symp. Zool. Soc. London 22: 213-235.

Jordan, C. F. 1969. Recovery of a tropical rain forest after gamma irradiation. pp. 88-98. In: Symposium on Radioecology. (See Nelson and Evans, 1969).

Jordan, C. F. 1970. Vegetative sprouting following irradiation of a tropical rain forest. Ch. D-19, pp. D305-D308. In: A Tropical Rain Forest. (See H. T. Odum and R. F. Pigeon, 1970) (NSA 25: 21944).

Jordan, C. F. 1970. Flow of soil water in the lower Montane Tropical Rain Forest. Ch. H-18, pp. H199-H200. In: A Tropical Rain Forest. (See H. T. Odum and R. F. Pigeon, 1970).

Jordan, C. F. 1970. Movement of ^{85}Sr and ^{134}Cs by the soil water of a tropical rain forest. Ch. H-19, pp. H201-H204. In: A Tropical Rain Forest. (See H. T. Odum and R. F. Pigeon, 1970) (NSA 25: 21561).

Jordan, C. F. 1970. A progress report on studies of mineral cycles at El Verde. Ch. H-22, pp. H217-H219. In: A Tropical Rain Forest. (See H. T. Odum and R. F. Pigeon, 1970).

Jordan, C. F. 1974. Respiration rates of red pine along a gradient of gamma radiation following eight years of exposure. Radiation Botany 14(4): 337-341. (NSA 32: 1090).

Jordan, C. F. and J. R. Kline. 1972. Mineral cycling: some basic concepts and their application in a tropical rain forest. pp. 33-50. In: Annual Review of Ecology and Systematics. Vol. 3. R. F. Johnston, P. W. Frank and C. D. Michener, eds. Annual Reviews, Inc., Palo Alto, Calif. (Incl. use of radionuclides).

Jordan, C. F., J. R. Kline and D. S. Sasscer. 1973. Simple model of strontium and manganese dynamics in a tropical rain forest. Health Physics 24(5): 477-489. (NSA 28: 2602).

Jordan, C. F., et al. 1969. Tritium movement in a tropical rain forest. Abstr. Bull. Ecol. Soc. Am. 50(2): 72.

Jordan, C. F., et al. 1970. Tritium movement in a tropical ecosystem. BioScience 20(14): 807-812.

Jordan, C. F., et al. 1973. Tritium movement in an old-field ecosystem determined experimentally. pp. 199-203. In: Radionuclides in Ecosystems. (See Nelson, 1973) (NSA 25: 41687).

Jordan, C. F., et al. nd. Experimental test of mathematical model predicting tritium movement through a tropical rain forest. p. 741. In: Radiation and Isotope Technology in Latin American Development. (See Puerto Rico Nuclear Center, nd.) (Also: Puerto Rico Nuclear Center, U. S. AEC report TID-24933) (NSA 23: 44111).

Jordan, H. S. 1974. Release of plutonium to the atmosphere. pp. 16-29. In: Plutonium Information Meeting. (See U. S. AEC, 1974) (NSA 31: 14017).

Jordan, S. and W. Schikarski. 1973. Evaluation of radioactive and non-radioactive trace constituents emitted from fossil-fuel and nuclear power plants. pp. 525-535. In: Environmental Behavior of Radionuclides Released in the Nuclear Industry. (See IAEA, 1973) (NSA 29: 29335).

Jordan, W. H. 1971. Nuclear power: social needs and benefits. Trans. Nucl. Sci. NS-18(1): 28-30. (NSA 25: 60002) (Compares effects of fossil and nuclear plants).

Jorgenson, C. D. 1968. Home range as a measure of probable interactions among populations of small mammals. J. Mammalogy 49(1): 104-112. (At Nevada Test Site).

Jorgenson, C. D. 1968. Reaction of native animals living in the vicinity of an underground nuclear detonation. Final report. Brigham Young Univ., U. S. AEC report COO-1336-6. 115pp. (NSA 22: 33776) (Nevada Test Site).

Jorgenson, C. D. 1970. Free living mites of the Nevada Test Site. Final report. Brigham Young Univ., U. S. AEC report COO-1731-4. 47pp. (NSA 28: 27231) (Incl. effects of nuclear weapons).

Jorgenson, C. D. and C. L. Hayward. 1965. Mammals of the Nevada Test Site. Brigham Young Univ. Sci. Bull. (biol.) 6(3): 1-81. (NSA 19: 33702).

Jorgenson, C. D. and W. W. Tanner. 1963. The application of the density probability function to determine the home ranges of *Uta stansburiana* and *Cnemidophorus tigris*. Herpetologica 19(2): 105-115. (Nevada Test Site).

Jorgenson, C. D., D. M. Allred and D E. Beck. 1963. Some effects of an underground nuclear detonation on biotic communities at the Nevada Test Site. Proc. Utah Acad. Sci. Arts Lett. 40(1): 49-61. (NSA 17: 33507).

Joshi, L. U., C. Rangarajan and S. S. Gopalakrishnan. 1971. Investigations on ^{210}Pb concentrations in varius regions of India. Health Physics 20(6): 655-668. (NSA 25: 42387).

Joshima, H. and O. Matsuoka. 1970. Liquid scintillation counting of plutonium in biological samples. pp. 53-56. In: Research Report "On Internal Exposures to Plutonium" in April 1969-March 1970. Japan National Inst. of Radiological Sciences, Chiba, report NIRS-Pu-7. (In Japanese) (NSA 26: 55741).

Jost, E. and J. D. Amirkhanian. 1971. Comparative studies of the effect of x-rays and 1,3-propanedioldimethanesulfonate on the mosquito *Culex pipiens* L. Mutation Research 13(1): 49-57. (NSA 26: 7007).

Jost, E. and H. Laven. 1971. Meiosis in translocation heterozygotes in the mosquito *Culex pipiens*. Chromosoma 35: 184-205. (NSA 26: 43363).

Joubert, J., et al. 1965. Investigations on the deposition of microorganisms in ionizing fields. Staub 25: 295-299. (In German) (NSA 20: 6710).

Jouzel, J. 1974. Measurement of natural concentrations of tritium present in small quantities of water. Centre d'Études Nucléaires, France Commissariat à l'Énergie Atomique, Saclay, report CEA-N-1685. 75pp. (In French) (NSA 30: 84).

Jouzel, J. and L. Merlivat. 1973. Tritium in the natural water cycle. Bull. inform. sci. tech. (Paris) (178): 55-57. (In French) (NSA 28: 24486).

Javanovic, M. 1971. Isotopes in animal production. Agrochimica 15(2-3): 150-171. (NSA 25: 46669).

Joyner, T. and R. Eisler. 1961. Retention and translocation of radioactive zinc by salmon fingerlings. Growth 25(2): 151-156. (NSA 15: 28965).

Judson, S. and J. K. Osmond. 1955. Radioactivity in ground and surface water. Am. J. Sci. 253(2): 104-116.

Juguet, B. 1969. Monitoring of the radioactivity of air and water. Tech. Sci. Munic. 64: 123-130. (In French) (NSA 24: 320).

Juguet, B., et al. 1966. Importance of mud and suspended matter in the transport and elimination of radionuclides contained in surface waters; a study of the Seine Basin. pp. 127-137. In: *Disposal of Radioactive Wastes into Seas, Oceans and Surface Waters*. (See IAEA, 1966) (In French) (CONF-660507-7) (SM-72/7) (NSA 20: 43544).

Julian, E. C. 1968. Gross radioactivity of the solids in selected surface waters of the United States, 1961-1965. Radiol. Hlth. Data 9(1): 1-12.

Jumar, A. and K. Sieber. 1967. Residue studies in rape seed oil and honey with toxaphene-^{36}Cl. Z. Lebensmitt.-Forsch. 133(6): 357-364. (In German).

Jun, J. S. 1969. Some practical considerations on counting efficiency for environmental samples and the gammagraph calibration. New Phys. (Korea) 9: 75-82. (NSA 25: 2448).

Junkins, R. L. 1960. Removal of radionuclides from the Pasco supply by conventional treatment. J. Am. Water Works Assoc. 52(7): 834-840.

Junkins, R. L. 1963. Arsenic and its isotopes in the environs. pp. 615-618. In: *Radioecology*. (See Schultz and Klement, 1963).

Junkins, R. L. 1969. Reactor releases of radionuclides. pp. 133-142. In: *Biological Implications of the Nuclear Age*. (See Goulden, 1969).

Junkins, R. L., et al. 1960. Evaluation of radiological conditions in the vicinity of hanford for 1959. Hanford Atomic Product Operations, U. S. AEC report HW-65534. 50pp. (Incl. Jan.-Mar. 1960) (NSA 14: 21808).

Juo, A. S. R. and S. A. Barber. 1970. Retention of strontium by soils as influenced by p_H, organic matter, and saturation cations. Soil Science 109: 143-148. (NSA 24: 25113).

Jurain, G. 1960. Methods and results of a study of radioactivity from radon in natural water. Geochim. Cosmochim. Acta 20: 51-82. (AEC-tr-7042) NSA 15: 392; 23: 26351).

Jurain, G. 1965. Solubility of radon in water and lack of equilibrium. Compt. rend. congr. natl. soc. Savantes sci. 90(Pt.2): 231-234. (In French) (CA 64: 252e) (Incl. aquatic organisms).

Jurasek, L. 1958. Sensitivity of a wood-annoying fungus to radioactive gamma radiation. Drevo 13(11): 341-342. (In German).

Jurasek, L. and M. Lehky. 1961. Radioisotopes in Wood Technology. Prague. 147pp. (In Czech) (Incl. 85 refs. to articles published 1946-1960 on the effects of radiation on the chemical, physical, mechanical, and biological properties of wood and the use of radiotracers in wood research and for control of woodworking machinery).

Kurkiewicz, L. 1964. Solution of geological-engineering and hydro-geological problems by means of isotopic methods. Kwart. geol. 8: 854-864. (In Polish) (NSA 19: 18181).

Justus, L. 1965. Garden under attack. Atlantic J. & Constitution Mag. (March 21, 1965): 6,7,22. (On Platt's ecological research, Emory Univ.).

Kaack, B. 1970. Effect of beta irradiation on oxygen consumption in the frog sciatic nerve. Radiation Research 42: 405-412. (NSA 24: 31845).

Kadokawa, M. 1975. Studies on the methods of environmental radiation monitoring. Japan Atomic Energy Research Inst., Tokyo, report JAERI-M-5957. 189pp. (In Japanese) (NSA 33: 12024).

Kaddah, M. T. 1968. Cesium uptake by dusangrass seedlings from four soils in southern California. Soil Science 105: 369-375. (NSA 22: 33745).

Kaddou, I. K., D. L. Bull and D. A. Lindquist. 1964. Joint action of DDT and dieldrin on the boll weevil. Abstr. Bull. Entomol. Soc. Am. 10(3): 163. (Tracer).

Kadish, K. M., W. van Willis and K. E. Arnold. 1974. Mercury in an estuarine mussel population: quantitative results obtained by activation, spectrophotometric, and electronanalytical techniques. Trans. Nucl. Sci. NS-21(1): 602-604. (NSA 30: 5877).

Kadota, H., H. Miyoshi and H. Shibata. 1959. Effects of some post-irradiation treatments on the survival of bacteria irradiated with (60) Co gamma rays. I. Bull. Jap. Soc. Sci. Fish. 24: 1001-1007.

Kadyrov, N. B., et al. 1969. Isotopic composition of uranium and actinium in minerals and natural waters. Geokhimiya (4): 492-494. (Geochem. Int. 5: 418-419) (NSA 23: 93).

Kadyrov, N. B., et al. 1970. Isotopic composition of uranium and the content of actinium in minerals and natural waters. Z. Angew. geol. 16: 251-252. (In German) (NSA 25: 34738).

Kafiani, K. A., et al. 1966. Effect of roentgen-ray irradiation on synthesis of information RNA in the early embryogenisis of the loach. Biokhimiya 31: 365-371. (Biochemistry (USSR) 31(4): 319-324) (NSA 21: 22724).

Kafkafi, U. 1965. A noncontaminating probe for injecting radioactive solution into soil. Soil Sci. Soc. Am. Proc. 29: 324-325. (NSA 19: 36685).

Kagan, M. S. 1967. Content of radon, radium, uranium, and mesothorium in mineral waters of the Rostov Province. Geokhimiya (7): 842-849. (Abstr. Geochem. Int. 4: 698; Am. Geol. Inst. transl. GI 67-7-11) (NSA 21: 36114).

Kahlos, H. 1969. ^{222}Rn concentration of some water samples determined with gamma-spectrometric measurements. Finland Inst. of Radiation Physics, Helsinki, report SFL-A-14. 34pp. (NSA 23: 49733).

Kahlos, H. and L. Blomqvist. 1971. In situ spectroscopy as a method for environmental surveillance. pp. 203-214. In: 3rd Nordic Radiation Protection Conf. (Copenhagen, August 18-20, 1971). (See Nordic Soc. for Radiation Protection, 1971) (In Swedish) (NSA 27: 22482).

Kahn, B. 1973. Determination of radioactive nuclides in water. pp. 1357-1388. In: Water and Water Pollution Handbook. Vol. 4. (See Ciaccio, 1973) (NSA 29: 2462).

Kahn, B. 1973. Environmental radiation monitoring. Trans. Amer. Nucl. Soc. 17: 536. (NSA 29: 5110).

Kahn, B. 1973. Radionuclides in the environment at nuclear power stations. pp. 30-36. In: Radionuclides in Ecosystems. (See Nelson, 1973) (NSA 28: 27293).

Kahn, B., H. L. Krieger and H. E. Kolde. 1971. Radionuclide release at a PWR nuclear power station. Trans. Amer. Nucl. Soc. 14(1): 326-327. (CONF-710606) (NSA 25: 41361).

Kahn, B., et al. 1971. Radiological surveillance studies at a pressurized water nuclear power reactor. Nat. Environmental Research Center, Cincinnati, O., U. S. EPA. 107pp. (NSA 28: 29885).

Kahn, B., et al. 1971. Radiological surveillance studies at a boiling water nuclear power reactor. Bur. of Radiological Health, U. S. PHS report BRH/DER-70-1. 117pp. (NSA 24: 26708).

Kahn, B., et al. 1971. Radiological surveillance studies at a boiling water power reactor. pp. 535-548. In: Environmental Aspects of Nuclear Power Stations. (See IAEA, 1971).

Kahols, H. and M. Asikainen. 1973. Natural radioactivity of ground water in the Helsinki area. Finland Inst. Radiation Physics, Helsinki, report SFL-A-19. 59pp. (NSA 30: 9551) (Kahlos?).

Kaindl, K., C. Chwala and R. Kirschner. 1962. The influence of γ-radiation from a ^{60}Co-source on the seeds of Solanum nigrum with reference to the physiological and morphological qualities of the plants. Abstr. In: 2nd Intern. Congr. Radiation Research. (See Anonymous, 1962).

Kajanne, P. and A. Hollming. 1958. Moisture determination in wood chips using gamma-ray scattering. Paperi ja Puu 40(4a): 153.

Kajita, S. 1953. Reports by forest research team. VII. Effect of atomic bomb on physical property of trees (strength tests of trees exposed to the atomic bomb). Nippon Gakujutsu Kaigi 1: 273-277. (In Japanese).

Kakhmanyuk, F. S. and M. F. Petryna. 1967. Effect of x and gamma radiation on the development of unmatched gypsy moths (Ocneria dispar L., Lymantridae, Lepidoptera). Vestsi Akad. Navuk Belorussk. SSR (Biol.) (2): 98-104. (In Belorussian) (NSA 22: 8642).

Kalamarz, E. 1963. Study of the Mallophaga. I. Use of the isotope ^{59}Fe for the determination of the nature of food of the hen. Zess. Nauk Wyz. Szkol. Rolnicz. Olszt. 15(2): 247-251. (In Polish).

Kalandiya, A. A., M. A. Vashakidze and Z. S. Shonvadze. 1968. Efficiency of water decontamination from radioactive substances by means of soil filtration. Gigiyena sanitariya (3): 48-50. (In Russian) (NSA 22: 30696).

Kalina, Z. K. 1973. Ratio $^{90}Sr/Sr$ in plankton of Latvian lakes of various types. pp. 81-87. In: Radioecology of Water Organisms. II. Distribution and Migration of Radionuclides in Freshwater and Seawater Biocenoses. (See Andrumaitis, 1973) (In Russian) (NSA 30: 21113).

Kallay, N. and A. Tigyi-Sebes. 1967. Localization of $^{32}PO_4^{3-}$ in isolated muscle fibril. Biochem. Biophys. Acta 2(2): 221-224. (Insect).

Kalmon, B. 1965. Environmental radiation levels and concentrations, first half 1965. Goodyear Atomic Corp., Piketon, O., U. S. AEC report GAT-505. 10pp. (NSA 19: 42656) (Incl. surface water).

Kalmon. B. and S. H. Hulett. 1970-71. Environmental radiation levels and concentrations, first half 1970. --- second half and annual summaries, 1970. Goodyear Atomic Corp., Piketon, O., U. S. AEC report GAT-628, GAT-629. 10, 14pp. (NSA 25: 6697, 32370).

Kalmon, B. and F. A. Koehler. 1974. Portsmouth gaseous diffusion plant environmental monitoring report, 1973. Goodyear Atomic Corp., Piketon, O., U. S. AEC report GAT-781. 27pp. (NSA 30: 29346).

Kalmykov, P. G. 1963. Changes of oxidation processes in flies following γ-irradiation. pp. 55-56. In: Problems in Parasitology. (Proc. 4th Conf. Inst. Zoology). Akad. Nauk Ukr. SSR, Lvov. (In Russian).

Kalmykov, P. G. 1965. Changes of oxidation processes in insects under the influence of γ rays. Radiobiologiya 5: 505-510. (AEC-tr-6601, pp. 25-32) (NSA 19: 42309).

Kalmykov, P. G. 1970. Effects of Ionizing Radiations on Insects. Atomizdat, Moscow. 144pp. (In Russian) (NSA 25: 24316).

Kalmykov, P. G. and N. V. Kuftyreva. 1963. The insecticidal properties of radioactive DDT and their influence on adult flies and larvae. pp. 56-59. In: Problems in Parasitology. (Proc. 4th Conf. Inst. Zoology). Akad. Nauk Ukr. SSR, Lvov. (In Russian) (^{14}C, Protophormia tenae-novae).

Kalninya, Z. K. and S. A. Osipenko. 1969. Coefficients of accumulation of strontium and ^{90}Sr in lake plants and plankton. Radiobiologiya 9(1): 111-112. (AEC-tr-7024, pp. 143-145) (NSA 23: 24585; 25: 46692).

Kalninya, Z. K. and G. G. Polikarpov. 1968. ^{90}Sr accumulation factor in plankton of different bodies of water. Radiobiologiya 8(1): 160-161. (AEC-tr-6914, pp. 84-89) (NSA 22: 47500).

Kalninya, Z. K. and G. G. Polikarpov. 1968. The concentration coefficients of strontium-90 in plankton in various kinds of reservoirs. pp. 3-5. In: Biological Migration of Radionuclides in Freshwater and Saltwater Reservoirs. State Committee on the Use of Atomic Energy, USSR. Atomizdat, Moscow. UN Document A/AC.82/G/L.1243. (AEC-tr-7128, pp. 2-5) (NSA 23: 246 24604; 24: 12024).

Kalninya, Z. K. and G. G. Polikarpov. 1969. Strontium-90 concentration factors of lake plankton, macrophytes, and substrates. Science 164(3887): 1517-1519.

Kalninya, Z. K., G. G. Polikarpov and N. S. Risik. 1970. Concentration factors of ^{90}Sr and uranium of hydrobionts living in lakes of different types. pp. 233-249. In Proc. Intern. Symp. on Radioecology. (See France Commissariat à l'Énergie Atomique, 1970) (NSA 25: 2546).

Kal'niy, V. S. 1967. Radioresistance of Amoeba proteus in relation to irradiation conditions and mitotic states. Tsitologiya 9: 1535-1542. (In Russian) (NSA 22: 36351).

Kal'niy, V. S. 1969. A comparative radiosensitivity of two strains of Amoeba proteus. Tsitologiya 11(1): 68-71. (In Russian) (NSA 23: 35857).

Kamada, H., M. Hukawa and M. Saiki. 1973. Removal of fallout ^{90}Sr from land to river. Radioisotopes (Tokyo) 22(6): 279-285. (In Japanese) (NSA 28: 29885).

Kamada, H., M. Mita and M. Saiki. 1971. Measurements of ^{89}Sr and ^{90}Sr in environmental samples of low background β-ray spectrometer. Radioisotopes (Tokyo) 20(5): 217-223. (In Japanese) (NSA 25: 37314).

Kamada, H., M. Mita and M. Saiki. 1971. Measurements of ^{89}Sr and ^{90}Sr in environmental samples by a low-background beta-ray spectrometer. pp. 427-439. In: Rapid Methods for Measuring Radioactivity in the Environment. (See IAEA, 1971) (NSA 26: 27826).

Kamada, H., M. Yukawa and M. Saiki. 1974. Studies on the removal of ^{90}Sr, ^{106}Ru, ^{137}Cs, and ^{144}Ce on land and in freshwater. pp. 137-146. In: Environmental Surveillance Around Nuclear Installations. Vol. 2. (See IAEA, 1974) (NSA 31: 14034).

Kamal, T. H. and A. E. Abdelaal. 1972. Seasonal changes in ^{32}P and ^{45}Ca metabolism in friesians and water buffaloes. pp. 111-120. In: Isotope Studies on the Physiology of Domestic Animals. International Atomic Energy Agency, Vienna, Publ. STI/PUB-309. (CONF-720345) (NSA 27: 28068).

Kamalova, G. V. and A. I. Kandov. 1969. Effects of gamma-radiation on pollen metabolism in some mallow type plants. Dokl. Akad. Nauk Uzbek. SSR (9): 49-50. (In Russian) (NSA 24: 25597).

Kamath, P. R. 1969. Recent guidelines for developing environmental monitoring programs (land and water) near nuclear installations. pp. 295-308. In: Environmental Contamination by Radioactive Materials. (See IAEA, 1969).

Kamath, P. R. 1970. Environmental radiation surveillance laboratory. A guide to design, layout, staff, and equipment requirements. World Health Organization, Geneva. 44pp. (NSA 25: 10763) (Kamuth ?, q.v.).

Kamath, P. R. and I. S. Bhat. 1969. Field monitoring and investigations for the evaluation of public health hazards through contamination of water sources. pp. 435-445. In: Environmental Contamination by Radioactive Materials. (See IAEA, 1969).

Kamath, P. R., I. S. Bhat and A. K. Ganguly. 1971. Environmental behavior of discharged radioactive effluents at Tarapur Atomic Power Station. pp. 475-494. In: Environmental Aspects of Nuclear Power Stations. (See IAEA, 1971) (CONF-700810-22) (SM-146/58) (NSA 24: 45909).

Kamath, P. R., M. A. R. Iyengar and I. S. Bhat. 1972. Changing radiation environment in India. pp. 917-928. In: The Natural Radiation Environment. Vol. 2. (See Adams, Lowder and Gesell, 1972) (NSA 32: 19730).

Kamchatova, I. Ye., et al. 1975. Study of the lethal action of ionizing radiation on Chlorella cells with various contents of intracellular cysteine. Radiobiologiya 15(2): 218-221. (ERDA-tr-51, pp. 52-56) (NSA 33: 5699).

Kameda, K. 1959. Radioactivation analysis of rare-earth elements in fish. In: 3rd Japan Conf. on Radioisotopes. (JAIF, JRIA 59/P-105).

Kamen, M. D. and S. Spiegelman. 1948. Studies on the phosphate metabolism of some unicellular organisms. Cold Spring Harbor Symp. on Quantitative Biology 13: 151-163.

Kamenkova, K. V. and V. A. Molchanova. 1962. Determination of facts on supplementary feeding of insects with the aid of radioactive phosphorus and its effects on the maturing eggs of Meniscus agnatus, a parasite of Hadena basilinea Sch. pp. 29-30. In: Materials on the Symp. on the Use of Biophysics in the Field of Plant Protection. (Leningrad, 1961). Leningrad. (In Russian) (Ref. zh. (biol.) 1961: 18Zh374).

Kametani, K., et al. 1970. Chemical and hygienic studies on natural radioactive spa. 2. Natural radioactive nuclides in hot spring water at Misasa spa area. 3. Natural radioactive nuclides in hot spring water at Masutomi spa area. Eisei Kagaku 16(1): 9-18. (In Japanese) (NSA 25: 26743, 13470).

Kamon, E. 1965. Distribution of scorpion venom in locusts. J. Insect Physiol. 11(7): 933-945. (^{131}I).

Kamra, O. P. and S. K. Kamra. 1962. Factors influencing x-ray tolerance of Haplopappus gracilis seeds. Can. J. Genet. Cytol. 4(3): 255-262. (NSA 17: 7825).

Kamra, S. K. 1967. Detection of mechanical damage and internal insects in seed by x-ray radiography. Svensk. Bot. Tidskr. 61(1): 43-48.

Kamuth, P. R., et al. 1964. Environmental radioactivity measurements at Trombay Establishment. pp. 957-978. In: The Natural Radiation Environment. (See Adams and Lowder, 1964) (NSA 18: 35752) (Kamath ?, q.v.).

Kanada, S. 1974. Fundamental policy on the environmental problems of nuclear power plant. Karyoku Hatsuden 25(4): 401-411. (In Japanese) (NSA 32: 17184).

Kananishi, N. and K. Yanagisawa. 1970. Radiation resistance and ploidy in the cellular slime mold Dictyostelium discoideum. Gamma Field Symp. (9): 19-31. (NSA 26: 31252).

Kanevskiy, Yu. P. 1973. K, Rb, and Cs exchange dynamics in fish and lamprey migrating from sea to freshwater (before spawning). pp. 129-135. In: Radioecology of Water Organisms. II. Distribution and Migration of Radionuclides in Freshwater and Seawater Biocenoses. (See Andrumaitis, 1973) (In Russian) (NSA 30: 21136).

Kanevskiy, Yu. P. and D. G. Fleyshman. 1971. Accumulation of rubidium, cesium, and potassium in roach and bass under natural conditions. Ekologiya 2(2): 10-14. (Sov. J. Ecology 2: 97-100) (Isotope dilution).

Kanevskiy, Yu. P. and D. G. Fleyshman. 1971. Investigation of food chains in an ichthyocoenosis of Lake Dal'nyi (Kamchatka) according to the concentration of rubidium and cesium in hydrobionts. Ekologiya 2(3): 5-8. (Sov. J. Ecology 2: 191-193) (Isotope dilution).

Kang, Y. 1968. Study of the variation of tritium and ^{14}C concentrations in the environment. New Phys. (Korea) 8: 1-10. (In Korean) (NSA 23: 13880).

Kang, Y. S., et al. 1964. On the rate of laying of eggs and development in D. melanogaster induced by x-rays. J. Nucl. Sci. (Seoul) 4: 64-68. (In Korean, English summary).

Kannan, S. 1971. Plasmalemma: the seat of dual mechanisms of ion absorption in Chlorella pyrenoidosa. Science 173(4000): 927-929. (^{86}Rb).

Kannowski, P. B. 1959. The use of radioactive phosphorus in the study of colony distribution of the ant Lasius minutus. Ecology 40(1): 162-165.

Kansche, G. A. and H. Haardick. 1948. On the intake of radioactive phosphorus $_{15}P^{32}$ by colibacteria. Z. Naturforsch. 3b: 433-437. (In German) (NSA 4: 372).

Kansu, A. 1961. Radioisotopes in research on insect physiology and toxicology. Bitki Koruma Bult. 2(9): 27-49. (In Turkish).

Kansu, A. 1961. The use of radioisotopes for research in insect ecology and biology. Bitki Koruma Bult. 2(7): 24-45. (In Turkish).

Kansu, A. 1962. A new method to control insects: use of radiation. Bitki Koruma Bult. 2: 40-60. (In Turkish) (NSA 18: 31150).

Kansu, A. 1962. Preliminary experiments on the sterilization of the pupae of the khapra beetle by irradiation with gamma rays. Z. angew. entomol. 49(2): 224-228. (In German) (NSA 18: 31250).

Kansu, A. 1963. Labelling methods of insects with radioisotopes. pp. 11-12. Yearbook 1962, Faculty of Agriculture, Univ. of Ankara, Turkey.

Kansu, A. 1963. Use of radioisotopes in entomology. Yuks. Zir. Enstit. Derg. : 47-72. (In Turkish).

Kansu, A. 1964. Radioisotopes in entomology. Yuks. Zir. Enstit. Derg.: 7-28. (In Turkish).

Kanter, M. 1967. Low level tritium measuring system. Thesis, Technion-Israel Inst. of Tech., Haifa. 58pp. (In Hebrew, English) (NP-18661) (NSA 25: 15886).

Kao, C. W., E. R. Graham and R. W. Blanchard. 1971. Determination of sulphate in soils as the $^{133}BaSO_4$ precipitate. Soil Science 112(4): 221-224. (NSA 26: 8736).

Kao, H. 1965. Autoradiographic research on the incorporation of $Na_2{}^{35}SO_4$ in the regenerating limb of the axolotl (Ambystoma mexicanum) after x-ray irradiation. Shih Yen Sheng Wu Hsueh Pao 10: 285-297. (In Chinese) (NSA 21: 6429).

Kapil, R. 1964. Effects of non-lethal doses of x rays on the exploring and feeding behavior of the larvae of Drosophila melanogaster Meig. Thesis, Lyon Univ., France. 48pp. (NP-15559) (NSA 20: 5295).

Kaplan, I. S., et al. 1972. Effect of chronic irradiation with ^{90}Sr and external γ-irradiation on cytogenetic changes of some agricultural crops. Genetika 8(8): 52-59. (In Russian, English summary).

Kaplan, W. D. and P. Oftedal. 1969. Genetic effects of tritiated thymidine and evidence for its incorporation into a cytoplasmic component of the adult testis of Drosophila melanogaster. Mutation Research 8: 127-138. (NSA 23: 31712).

Kaplan, W. D. and S. R. Pelc. 1956. Autoradiographic studies of Drosophila gonads following the feeding of carbon-14-labeled formaldehyde. Z. indukt. Abstamm.-Vererblehre 87(3): 356-364.

Kaplan, W. D. and J. E. Sisken. 1960. Genetic and autoradiographic studies of tritiated thymidine in testes of Drosophila melanogaster. Experientia 16: 67-69.

Kaplan, W. D., H. D. Gugler and K. K. Kidd. 1965. The distribution of sex-linked recessive lethals induced in Drosophila males by tritiated deoxycytidine. Abstr. Genetics 52(2): 451.

Kaplan, W. D., et al. 1964. Non-random distribution of lethals induced by tritiated thymidine in Drosophila melanogaster. Genetics 49(4): 701-714.

Kaplanis, J. N., et al. 1965. The utilization and fate of β-sitosterol in the larva of the housefly, Musca domestica L. J. Insect Physiol. 11: 251-258. (^3H).

Kapoor, I. P., et al. 1970. Comparative metabolism of methoxychlor, methiochlor, and DDT in mouse, insects, and in a model ecosystem. J. Agric. Food Chem. 18(6): 1145-1152. (^3H, ^{14}C).

Kapp, R. O. and A. M. Gooding. 1962. A radiocarbon-dated pollen profile from Sunbeam Prairie Bog, Darke Co., Ohio. Abstr. Bull. Ecol. Soc. Am. 43(4): 118.

Kapp, R. O. and A. M. Gooding. 1964. A radiocarbon dated pollen profile from Sunbeam Prairie Bog, Darke County, Ohio. Am. J. Sci. 262(2): 259-266.

Kapul'tsevich, Yu. G. and K. M. Bliznik. 1967. Role of recovery in modification of yeast survival curves during cell growth in media with different nutrient contents. Radiobiologiya 7(4): 572-575. (AEC-tr-6889, pp. 136-142) (NSA 22: 19359).

Kapul'tsevich, Yu. G., et al. 1972. Analysis of the radiobiological responses of yeast cells. 2. Forms of inactivation and post irradiation modification of radiation lesion. Radiobiologiya 12(3): 408-415. (AEC-tr-7402, pp. 129-140) (NSA 27: 17536).

Karaban, R. T. and F. A. Tikhomirov. 1967. Influence of radioactive contamination of the soil on pine and spruce seedlings. Radiobiologiya 7(2): 275-280. (AEC-tr-6887, pp. 188-197) (NSA 21: 34817).

Karaban, R. T. and F. A. Tikhomirov. 1971. Action of radiation of ^{90}Sr and ^{90}Y on seeds and seedlings of conifers in radioactive contaminated soils. pp. 66-79. In: Methods of Radioecological Investigations. (See Verkhovskaya, 1971) (In Russian).

Karaki, H., M. Ikeda and N. Urakawa. 1967. Increase in ^{45}Ca uptake during barium-induced contracture in guinea pig Taenia coli. Jap. J. Pharmacol. 17: 496-497. (NSA 22: 49207).

Karavayeva, E. N. 1968. Effects of soil moisture on the distribution of ^{35}S in the basic groups of chemical compounds of pea and pine plants. Tr. inst. ekol. rast. zhivotn. Ural. fil. Akad. Nauk SSSR (61): 61-64. (In Russian) (NSA 24: 7715).

Karavayeva, E. N. and I. V. Molchanova. 1971. Uptake of chemical analogs (Sr^{90}-Ca and Cs^{137}-K) by plants as a function of soil moistness. Ekologiya 2(5): 86-98. (Ecology USSR 2: 470-471).

Karawya, M. S., E. M. Abdelkader and A. Hifny Saber. 1962. The effect of γ radiations on Datura innoxia Miller growing in Egypt. Bull. Fac. Pharm. Cairo Univ. 2: 137-147. (NSA 19: 35980).

Karawya, M. S., S. I. Balbaa and A. Hifny Saber. 1963. Pre-sowing γ irradiation of Datura stramonium L., Datura stramonium var inermis Safford, Datura tatula L. and Datura ferox L. and its effect on growth and alkaloids. Bull. Fac. Pharm. Cairo Univ. 2: 65-80. (NSA 19: 35981).

Karches, G. J., et al. 1971. Field determination of dose from ^{133}Xe in the plume from a pressurized water reactor. pp. 515-523. In: Rapid Methods for Measuring Radioactivity in the Environment. (See IAEA, 1971) (NSA 26: 28144).

Kardashev, A. V. 1966. Research in the USSR on the conservation of fish and fish products by gamma irradiation. pp. 65-72. In: Application of Food Irradiation in Developing Countries. International Atomic Energy Agency, Vienna, Tech. Rept. Ser. No. 54, STI/DOC-10/54. (NSA 20: 43776).

Karimov, A. D. and O. V. Malinovskiy. 1966. On the role of gamma treatment of oocytes in the realization of radiation injury of Drosophila sperm cells. Genetika (6): 140-143. (In Russian) (NSA 20: 36625).

Karin, M., S. Rahman and M. Rahman. 1971. Rb^{86} as tracer for potassium: I. Uptake of Rb and K by rice plant in nutrient solution. Plant Soil 35(1): 179-181.

Karlicek, V., et al. 1969. Determination of 10 trace elements in biological materials by the method of neutron activation analysis. Cas. Lek. Cesk. 108: 1017-1021. (In Czech) (NSA 24: 2153).

Karlson, P. 1965. Biochemical studies of ecdysone control of chromosomal activity. J. Cell. Comp. Physiol. 66(2/II): 69-76. (Chironomous, Calliphora; radionuclides).

Karlson, P. and P. Herrlich. 1965. Tyrosine metabolism in insects. XVI. Tyrosine metabolism in the locust Schistocerca gregaria Forsk. J. Insect Physiol. 11(1): 79-89. (In German, English summary) (^{14}C).

Karlson, P. and G. Peters. 1965. The mechanism of action of hormones. IV. The effect of ecdysone on nucleic acid metabolism in Calliphora larvae. Gen. Comp. Endocrinol. 5(2): 252-259. (In German, English abstr.) (^{32}P).

Karlson, P., D. Mergenhagen and C. E. Sekeris. 1964. Tyrosine metabolism in insects. XV. The o-diphenol oxidase system in Calliphora erythrocephala. Hoppe-Seyler's Z. physiol. chem. 338(1/2): 42-50. (In German, English summary) (^{14}C).

Karlson, P., C. E. Sekeris and R. Maurer. 1964. The mechanism of action of hormones. I. Distribution of tritium labelled ecdysone in larvae of Calliphora erythrocephala. Hoppe-Seyler's Z. physiol. chem. 336(1/3): 100-106. (In German, English summary) (^{3}H).

Karlstrom, E. L. 1957. The use of Co^{60} as a tag for recovering amphibians in the field. Ecology 38(2): 187-195.

Karnaukh, I. M., P. S. Mazur and A. Yu. Ratmanskiy. 1971. Local measurements of γ-radiation with an automatic compensation for γ-background. pp. 246-252. In: Methods of Radioecological Investigations. (See Verkhovskaya, 1971) (In Russian).

Karnaukh, I. M., et al. 1971. Specific device for measuring radioactive contamination due to β- and γ-radiation in field conditions. pp. 242-246. In: Methods of Radioecological Investigations. (See Verkhovskaya, 1971) (In Russian).

Karol', I. L. 1968. Stratospheric fallout of radioactive nuclear explosion products over the continents and oceans in Temperate latitudes of the Northern Hemisphere. State Committee on the Use of Atomic Energy. Atomizdat, Moscow. UN Document A/AC.82/G/L.1249. 27pp. (AEC-tr-7030, pp. 75-105) (NSA 23: 17626).

Karol', I. L. and S. G. Malakhov. (Eds.). 1962. Problems in Nuclear Meteorology. Gosatomizdat, Moscow. (In Russian) (NSA 18: 10603).

Karol', I. L., S. G. Malakhov and K. P. Makhon'ko. 1970. Works of the Soviet scientists in nuclear meteorology. Izv. Akad. Nauk SSSR (fiz. atmos. okeana) 6: 352-364. (In Russian) (NSA 24: 39081).

Karol', I. L., et al. 1965. Comparative analysis of global fallout of nuclear explosion products in the continents and oceans. State Committee on the Use of Atomic Energy, USSR. Moscow. 24pp. (In Russian) (NP-16517) (NSA 21: 34654).

Karol', I. L., et al. 1966. Comparative analysis of global distribution of nuclear test explosion fallout over continents and oceans. Tr. inst. okeanol. 82: 56-71. (In Russian) (NSA 20: 31421).

Karol', I. L., et al. nd. Global contamination of the atmosphere by ^{85}Kr from worldwide nuclear power plants and the radiation danger. USSR Inst. Experimental Meteorology, Obninsk. 17pp. (JPRS-53174) (NSA 25: 54781).

Karolczyk, J., J. Kwapulinski and H. Majchrzyk. 1973. Studies on the radioactivity of the carcass of carp bred in southern districts of the Katowice Province. Acta Hydrobiol. 15(4): 413-425. (NSA 31: 22644).

Karpenko, C. P. and D. T. North. 1973. Ovipositional response elicited by normal, irradiated, F_1 male progeny, or castrated male Trichoplusia ni (Lepidoptera: Noctuidae). Ann. Entomol. Soc. Am. 66(6): 1278-1280. (NSA 29: 2943).

Karpov, A. Ye. 1959. On polyhedral disease in the silkworm induced by x-rays. Dopovidi Akad. Nauk Ukr. RSR (9): 1015-1018. (In Ukrainian).

Karpov, A. Ye. 1963. Experimental activation of latent polyhedral virus at the embryonic stages of the silkworm. Dopovidi Akad. Nauk Ukr. RSR 15: 661-664. (In Ukrainian).

Karpov, V. G. 1962. P^{32}-tests for investigating the competition between tree roots and regrowth in southern Taiga forests. Dokl. Akad. Nauk SSSR 146: 717-719. (In Russian) (NSA 17: 2806) (Birch, fir).

Karpov, V. L. and V. F. Filippov. 1965. Effects of ^{14}C β particles in Chlorella growing in a medium containing ^{14}C-carbonates. Radiobiologiya 5(4): 580-583. (AEC-tr-6601, pp. 145-150) (NSA 19: 42253).

Karsten, J. H. M. and F. J. Haasbroek. 1973. A new method of calibrating a neutron moisture meter and a correction for the effect of soil bulk density. Agrochemophysica 5: 35-40. (NSA 31: 14215).

Kartman, L., et al. 1958. New knowledge on the ecology of sylvatic plague. Ann. N. Y. Acad. Sci. 70(3): 668-711. (Labelling with ^{144}Ce).

Karulin, B. Ye. 1970. Technique for using isotopes for the study of mobility and activity of small mammals. Zool. sh. 49(3): 444-450. (In Russian, English summary).

Karzinkin, G. S. 1955. Wider inculcation of the radioisotope technique in fish cultivation science and practice. Ryb. khozyaystvo 31(11): 48-50. (English transl. U. S. Fish and Wildlife Service).

Karzinkin, G. S. 1961. Results and future plans in the development of physiological research in the fish industry. Tr. vses. nauchn.-issled. inst. morsk. ryb. okeanogr. 44: ___. (In Russian).

Karzinkin, G. S. 1962. The Use of Radioisotopes in the Fishing Industry. Pishchepromizdat, Moscow. 73pp. (JPRS-21,109) (NSA 17: 37057).

Karzinkin, G. S. 1962. Radioactive isotopes in pisciculture. Rybovod. rybolov. 1: 8-9. (In Russian) (BA 41: 17315; SFA 8: 5771).

Karzinkin, G. S., A. V. Kardashov and L. L. Lagunov. 1959. Use of radioisotopes and radiation in the fish industry. In: Radioisotopes and Nuclear Radiation in the National Economy of the USSR. Vol. 2. (Coll. of Reports, Conf.). U. S. S. R. Ministry of Heavy Industry.

Karzinkin, G. S., Ye. V. Soldatova and I. A. Shekhanova. 1959. Some results of mass marking of "non-standard" sturgeon fingerlings by means of radiophosphorus. pp. 27-40. In: Collection on Animal Migration. Vol. 1. (Fish. Res. Bd. Can. transl. ser. 279; NRC C-3391) (SFA 6: 3833).

Karzinkin, G. S., Ye. V. Soldatova and I. A. Shekhanova. 1959. Results of mass labelling by radiophosphorus of sturgeon fish fry at the Kurinsk commercial experimental plant. Tr. vses. nauchn.-issled. inst. morsk. ryb. okeanogr. 44: __. (In Russian).

Karzinkin, G. S., I. F. Vel'tishcheva and M. P. Boboyavlenskaya. 1970. Study of the rate of incorporation of C^{14} in the organic matter of sturgeon (Acipenser guldenstadti (Brandt) eggs and young. Vopr. ikhtiol. 10(1): 84-88. (In Russian) (SFA 16: 13020).

Kasai, A. 1970. Environmental monitoring outside atomic energy installations. 2. Monitoring by measurement of activities in environmental materials. Hoken Butsuri 5(2): 89-95. (In Japanese) (NSA 25: 15790).

Kasatkina, S. V. 1971. Effect of ^{90}Sr+^{90}Y on the development and functioning of the hatching glands in Atlantic salmon. Tr. polyar. nauchn.-issled. Proekt. inst. morsk. ryb. okeanogr. (29): 55-61. (AEC-tr-7418, pp. 66-74) (NSA 27: 28002).

Kasatkina, S. V., et al. 1973. Chronic effect of ^{144}Ce and ^{137}Cs radioisotopes on embryonic development of the Atlantic salmon. pp. 25-35. In: Radioecology of Water Organisms. III. Effect of Ionizing Radiation on Water Organisms. (See Andrumaitis, 1973) (In Russian) (NSA 30: 24374).

Kashchenko, L. A. 1971. Radiation reaction of the adrenohypophysis-gonad system in cold-blooded animals. Tr. polyar. nauchn.-issled. Proekt. inst. morsk. ryb. okeanogr. (29): 135-158. (AEC-tr-7418, pp. 166-194) (NSA 27: 27949) (Grass frog).

Kashchenko, L. K. 1962. Radiation damage to ovulation mechanism in the frog. Radiobiologiya 2(6): 868-872. (AEC-tr-5433, pp. 92-98) (NSA 17: 15743) (Kashchenko, L. A. ?).

Kashchenko, L. K. 1963. Changes in the anterior hypophysis in irradiated frogs. Radiobiologiya 3(1): 76-80. (AEC-tr-5434, pp. 102-107) (NSA 17: 30341).

Kashiwagi, M. and H. Etoh. 1970. Histological and autoradiographical observations of the effects of x-irradiation on gill epithelium of the goldfish, Carassius auratus. Annot. Zool. Jap. 43(2): 93-101. (NSA 24: 51033).

Kashkina, G. B. and Yu. D. Abaturov. 1967. The survival of soil microorganisms on gamma irradiation of the soil. Radiobiologiya 7(2): 263-266. (AEC-tr-6887, pp. 170-174) (NSA 21: 34830).

Kashkina, G. B. and Yu. D. Abaturov. 1968. Postradiation changes in a community of soil microorganisms. Radiobiologiya 8(6): 908-912. (AEC-tr-7013, pp. 172-178) (NSA 23: 18138).

Kashkina, G. B. and Yu. D. Abaturov. 1969. Influence of chronic irradiation on the microflora of the soil of a gamma field. Radiobiologiya 9(3): 460-462. (AEC-tr-7026, pp. 185-188) (NSA 24: 9804).

Kasida, Y., et al. 1962. Carbon-14 in a biological environment. I. Annual variation of carbon-14 concentration in thymol. Bull. Chem. Soc. Japan 35(11): 1811-1813. (Plants).

Kasinov, V. B. 1966. Rate of reproduction and survival of duckweed Lemna minor after gamma irradiation. Radiobiologiya 6(4): 605-612. (AEC-tr-6773, pp. 206-218) (NSA 21: 22787).

Kasinov, V. B. 1966. Reversed development order of corymbs in irradiated colonies of Lemna minor. Dokl. Akad. Nauk SSSR 167: 201-204. (In Russian) (NSA 21: 9017).

Kasinov, V. B. and G. V. Kasinova. 1971. Dynamics of radiation damage to the plant organism with the lesser duckweed as an example. Radiobiologiya 11(4): 580-586. (AEC-tr-7304, pp. 127-136. (NSA 26: 36355).

Kasinov, V. B. and G. V. Kasinova. 1971. Radiation-induced deformities in vegetatively reproduced duckweed. Dokl. Akad. Nauk SSSR 199(2): 460-463. (In Russian) (NSA 26: 15429).

Kasinov, V. B. and G. V. Kasinova. 1971. Permanently polymorphic clones of duckweed as a model for study of some problems of developmental biology. Ontogenez 2(6): 555-564. (Sov. J. Develop. Biol. 2(6): 441-448) (NSA 28: 5961) (X-irradiation).

Kasinova, G. V. 1963. Effect of x-rays on the dynamic quantity of green and colorless strains of Euglena gracilis. Tsitologiya 5(6): 692-693. (In Russian).

Kasinova, G. V. 1964. Influence of different culture conditions in green and colorless cells of Euglena gracilis on their survival after x-ray irradiations. Acta Protozool. 2: 279-285. (In Russian) (NSA 19: 24267).

Kasinova, G. V. 1964. X-ray sensitivity of the green and colorless cells of Euglena gracilis Klebs. Radiobiologiya 4(4): 603-606. (AEC-tr-6407, pp. 156-160) (NSA 19: 41187).

Kasperek, K. 1974. Analytical techniques for the determination of 20 trace elements in biological samples by means of instrumental thermal neutron activation analysis. pp. 154-167. In: Proc. 2nd Intern. Conf. on Nuclear Methods in Environmental Research. (See Vogt and Meyer, 1974) (NSA 33: 23327).

Kassai, T., B. Fitzpatrick and W. Mulligan. 1966. Variables in the radiation attenuation of helminth larvae: the effect of differences in quality of radiation. Parasitology 56: 651-656. (NSA 21: 16456).

Kasting, R. and A. J. McGinnis. 1962. Nutrition of the pale western cutworm. IV. Amino acid requirements determined with glucose-U-C^{14}. J. Insect Physiol. 8(1): 97-103.

Kasting, R. and A. J. McGinnis. 1966. Radioisotopes and the determination of nutrient requirements. Ann. N.Y. Acad Sci. 139: 98-107. (^{14}C).

Kasting, R., G. R. F. Davis and A. J. McGinnis. 1962. Nutritionally essential and nonessential amino acids for the prairie grain wireworm determined with glucose-U-C^{14}. J. Insect Physiol. 8(Nov./Dec.): 589-596.

Kasting, R. and A. J. McGinnis. 1964. Amino acid requirements for the wheat stem sawfly determined with glucose-U-C^{14} after vacuum-infiltration. Can. Entomol. 96(8): 1133-1137. (Cephus cinctus).

Kasting, R. and A. J. McGinnis. 1965. Measuring consumption of food by an insect with carbon 14-labelled compounds. J. Insect Physiol. 11(9): 1253-1260. (Agrotis orthogonia).

Kastner, J., B. G. Oltman and L. D. Marinelli. 1962. Progress report on flux and spectrum measurements of the cosmic-ray neutron background. pp. 441-448. In: The Natural Radiation Environment. (See Adams and Lowder, 1964) (NSA 18: 36220).

Kastner, J., F. R. Shonka and J. E. Rose. 1964. A muscle-equivalent environmental radiation meter of extreme-sensitivity. pp. 655-660. In: The Natural Radiation Environment. (See Adams and Lowder, 1964) (NSA 17: 27706).

Kasturi Bai, A. R. and L. Dilli. 1974. Effects of radiations on the rate of regeneration in Homalozoon vermiculare (Stokes) and Urostyla. Indian J. Exptl. Biol. 12(4): 343-345. (NSA 31: 14400).

Kasturi Bai, A. R. and M. E. Manjula. 1974. Effects of tritiated water on the division rate of Spirostomum ambiguum minor. Stimul. Newsl. (6): 34-41. (NSA 32: 28223).

Kas'yanov, V. L. and N. A. Lukina. 1971. Effect of x irradiation on the incorporation of glycine-^{14}C in the tissue of Atlantic salmon larvae. Tr. polyar. nauchn.-issled. Proekt. inst. morsk. ryb. okeanogr. (29): 110-115. (AEC-tr-7418, pp. 134-141) (NSA 27: 27948).

Katagiri, K. 1970. Differences in mutation rate and mutation spectrum after acute gamma irradiation of mulberry saplings. Nippon Sanshigaku Zasshi 39: 194-200. (In Japanese) (NSA 25: 24090).

Katayama, T. 1953. Cytogenetical studies on the sterile wild senna (Cassia tora L.) produced by atomic bomb explosion. II. Crossing experiments between normal and asynaptic plants, and inheritance of the asynaptic plant. III. Cytological observations of asynaptic plant. IV. On the gigas mutants segregated from the asynaptic wild senna (Cassia tora L.). Sci. Bull. Fac. Agr. Kyushu Univ. 14: 195-215; J. Fac. Agr. Kyushu Univ. 10: 119-132. (In Japanese, English summary) (PBA 24: 3249, 3250, 2354).

Kater, R. and K. Fröhlich. 1975. Measuring equipment for the determination of natural ^{32}Si activities. Isotopenpraxis 11(3): 103-105. (In German) (NSA 32: 22351) (Atmosphere and ground waters).

Kathren, R. L. 1972. Instruments in the field: use, abuse, and misuse. pp. 811-823. In: Health Physics Operational Monitoring. Vol. 2. C. A. Willis, ed. Gordon & Breach, New York. (NSA 29: 29478).

Kathren, R. L. 1973. Environmental impact statements for nuclear power plan plants. J. Environ. Sci. 16(5): 11-16. (NSA 29: 24217).

Katiyar, K. P. 1962. Possibilities of eradication of the Mediterranean fruit fly Ceratitis capitata Wied. from Central America by gamma irradiated males. pp. 211-217. In: 4th Inter-American Symp. on the Peaceful Application of Nuclear Energy. Vol. 2. Pan American Union, Washington.

Katiyar, K. P. 1964. Control of insects by sterilization by gamma radiation. Bol. Junta Control Energia At. (Peru) 9(48): 25-28. (In Spanish) (NSA 19: 3910).

Katiyar, K. P. and F. T. Ferrer. 1967. Effects of high-temperature on irradiated medfly pupae. Turrialba 17: 31-34. (In Spanish) (NSA 21:39619).

Katiyar, K. P. and F. T. Ferrer. 1968. Rearing technique, biology and sterilization of the coffee leaf miner, Leucoptera coffeella Guer. (Lepidoptera: Lyonetiidae). pp. 165-175. In: Isotopes and Radiation in Entomology. (See IAEA, 1968) (NSA 24: 32319).

Katiyar, K. P. and E. Ramirez. 1970. Mating frequency and fertility of Mediterranean fruit fly females alternately mated with normal and irradiated males. J. Econ. Entomol. 63: 1247-1250. (NSA 24: 44589).

Katiyar, K. P. and E. Ramirez. 1972. Suppression of the reproductive potential of a wild strain Mediterranean fruit fly by gamma irradiated males in caged coffee trees. Turrialba 22(2): 156-159. (NSA 28: 8477).

Katiyar, K. P. and S. J. Valerio. 1963. Studies on the dispersion and length of life of the Mediterranean fruit-fly C. capitata labelled with P^{32}. Turrialba 13(3): 181-184. (In Spanish).

Katiyar, K. P. and S. J. Valerio. 1964. Effect of introducing males sterilized by γ-radiation into a normal population of Mediterranean fruit fly (Ceratitis capitata). Turrialba 14(4): 211-212. (In Spanish, English summary).

Katiyar, K. P. and S. J. Valerio. 1965. Further studies on the possible use of sterile-male release technique in controlling or eradicating the Mediterranean fruit fly Ceratitis capitata Wied. from Central America. pp. 197-202. In: Proc. 5th Inter-American Symp. on the Peaceful Application of Nuclear Energy. Pan American Union, Washington. (NSA 19: 43825).

Katkevich, Yu. Yu., P. N. Odintsov and M. K. Pendere. 1970. Gamma radiation-induced changes in the anatomic structure and chemical composition of spruce wood. Dokl. Akad. Nauk SSSR 199(2): 460-463. (In Russian) (NSA 26: 45774).

Kato, M. 1963. Application of radioisotopes in hydrology. Problems and perspective. Genshiryoku Kogyo 9(8): 10-12. (In Japanese) (NSA 18: 10291).

Kato, M., Y. Kato and T. Oishi. 1967. Radioluminous paints as activator of photoreceptor systems studied with swallow-tail butterfly and quail. Proc. Jap. Acad. 43: 220-223. (NSA 22: 25758).

Kato, M., et al. 1961. Field experiment on measurement of flow rate of river stream using radioactive tracer. Abstr. E-29. In: 4th Japan Conf. on Radioisotopes. (See Japan Atomic Industrial Forum, 1961).

Kato, M., et al. 1963. Radiotracer experiments on littoral drift in Japan. pp. 143-174. In: Radioisotopes in Hydrology. (See IAEA, 1963) (NSA 18: 1919).

Kato, M., et al. 1963. A study of river engineering based on the results of field measurements of flow velocities with radioisotopes in the Sorachi River, Japan. pp. 89-110. In: Radioisotopes in Hydrology. (See IAEA, 1963) (NSA 18: 1916).

Katona, E. 1966. On the behavior of irradiated "snail enzyme". Radiol. Clin. Biol. 35: 305-310. (NSA 21: 26359).

Katov, A. Ye., et al. 1971. Kinetics of polonium-210 in the tissues of fish at $0^{o}C$ and in water of different salinities. Tr. Atl. nauchn.-issled. inst. ryb. khoz. okeanogr. 44: 60-65. (In Russian) (NSA 30: 3710).

Katsura, K. 1970. Radioactive springs in the Kinki District and their radon sources. Onsen Kagaku Kaish. 21(3-4): 121-133. (In Japanese) (NSA 26: 48047).

Katsura, K. and J. Hatsuda. 1968. Distribution of radon content of gas associated with spring waters at Awara Hot Springs. Onsen Kagaku Kaish. 19: 105-110. (In Japanese) (NSA 23: 31421).

Katsurayama, K. 1968. Accumulation of radioactivity in rice fields. Ann. Rep. Res. Reactor Inst. Kyoto Univ. 1: 293-297. (In French) (NSA 24: 2611).

Katz, F. F. 1960. The effect of irradiation on reproduction by the heterogenetic generation of Strogyloides papillosus. I. Irradiation of males and females. J. Parasitiol. 46: 383-391. (NSA 15: 2465).

Katz, J. 1951. Plutonium metabolism: a literature review. Hanford Works, U. S. AEC report HW-21868. 24pp. (NSA 33: 9624).

Katz, J. J. and H. L. Crespi. 1970. Isotope effects in biological systems. pp. 286-363. In: Isotope Effects in Chemical Reactions. C. J. Collins, ed. Van Nostrand Reinhold, New York. (NSA 25: 37359) (Incl. algae).

Katz, S. A. 1974. Application of neutron activation analysis with californium-252 to the measurement of silver of ancient coins. Lab. Pract. 23(3): 112-113. (NSA 30: 9042).

Katz, S. A. 1974. Determination of silver in ancient and rare coins by neutron activation analysis with 252-californium. Trans. Nucl. Sci. NS-21(1): 633-634. (NSA 30: 5866).

Katznelson, H. and J. A. Robb. 1962. The use of gamma radiation from cobalt-60 in the control of diseases of the honey-bee and the strilization of honey. Can. J. Microbiol. 8(2): 175-179. (BA 39: 16702; NSA 16: 32664).

Kaufman, A. and W. Broecker. 1965. Comparison of ^{230}Th and ^{14}C ages for carbonate materials from lakes Lahontan and Bonneville. J. Geophys. Res. 70: 4039-4054. (NSA 19: 44271).

Kaufman, B. P. and H. Gay. 1963. Cytological evaluation of differential radiosensitivity in spermatogeneous cells of Drosophila. pp. 375-408. In: Repair from Genetic Radiation Damage. F. H. Sobels, ed. Macmillan Co., New York. (NSA 18: 11841).

Kaufman, G. and M. Wasserman. 1957. Effects of irradiation on the screw-worm, Callitroga hominivorax (Coq.). pp. 246-259. In: Studies in the Genetics of Drosophila. Univ. of Texas, Austin, Publ. 5721.

Kaufman, G. A. and R. J. Beyers. 1973. Effects of chronic gamma irradiation on the fish, Oryzias latipes. pp. 1119-1124. In: Radionuclides in Ecosystems. (See Nelson, 1973) (NSA 28: 30144).

Kaufman, G. E. and M. W. Miller. 1970. A statistical evaluation of Sparrow et al.'s relationship of D_o to chromosome volume. Radiation Research 42 (1): 181-187. (NSA 24: 25593).

Kaufman, S. and W. F. Libby. 1954. The natural distribution of tritium. Physical Rev. 93(6): 1337-1344.

Kaufman, W. J. and P. C. Klingerman. 1963. Nature and disposition of sediment sorbed radionuclides entering San Francisco Bay. pp. 149-157. In: Transport of Radionuclides in Fresh Water Systems. (See Kornegay, et al., 1963) (NSA 17: 40527).

Kaufman, W. J. and G. T. Orlob. 1956. An evaluation of ground water tracers. Trans. Am. Geophys. Un. 37(3): 297-306.

Kaufman, W. J. and G. T. Orlob. 1956. Measuring ground water movement with radio-active and chemical tracers. J. Am. Water Works Assoc. 48(5): 559-572.

Kauranen, P. 1969. ^{210}Po and ^{210}Pb concentrations of some water and fish samples from Finland. Univ. of Helsinki, Finland, U. S. AEC report NYO-3446-13 (Paper 23). 6pp. (For 5th R. I. S. Symp., Helsinki, May 19-20, 1969) (NSA 23: 41304).

Kauranen, P. and J. K. Miettinen. 1967. ^{210}Po and ^{210}Pb in environmental samples in Finland. pp. 275-280. In: Radioecological Concentration Processes. (See Aberg and Hungate, 1967) (Also: Univ. of Helsinki, Finland, NP-16002, Paper 5) (NSA 20: 26601).

Kauranen, P. and J. K. Miettinen. 1969. ^{210}Po and ^{210}Pb in the arctic food chain and the natural radiation exposure of Lapps. Health Physics 16: 287-295.

Kauranen, P. and J. K. Miettinen. 1970. Polonium and radiolead in some aqueous ecosystems in Finland. Univ. of Helsinki, Finland, U. S. AEC report NYO-3446-14, Paper 27. 18pp. (For Meeting on the Biology and Ecology of Polonium and Radiolead, Sutton, England) (CONF-700412-2) (NSA 25: 18930).

Karaunen, P., J. K. Miettinen and E. Pullianen. 1970. Polonium and radiolead in some land animals in Finland. Univ. of Helsinki, Finland, U. S. AEC report NYO-3446-14, Paper 26. 10pp. (For Meeting on the Biology and Ecology of Polonium and Radiolead, Sutton, England) (CONF-700412-1) (NSA 25: 18929).

Kaushanskiy, D. A. 1971. MRKh-γ-25M type isotopic γ-ray source for microbiological radiation-chemical investigations. Atom. energiya 30: 83-85. (In Russian) (NSA 25: 32139).

Kauss, H. 1962. Influence of the lack of thiamine and biotin on the metabolism of Ochromonas malbamensis. Vort. Gesamtg. Botan. Ges. (1): 129-132. (^{14}C).

Kautsky, H. 1966. Possible accumulation of discrete radioactive elements in river mouths. pp. 163-175. In: Disposal of Radioactive Wastes into Seas, Oceans and Surface Waters. (See IAEA, 1966) (SM-72/9) (CONF-660507) (NSA 20: 45225).

Kawabata, T. 1954. Contamination of fish by radiation. Koshu Eisei 16(6): 16.

Kawabata, T. 1967. Studies on the sorption and release of radionuclides by river sediments. J. Radiat. Res. (Japan) 8: 20-31. (NSA 22: 70216).

Kawabata, T. and E. E. Held. 1959. A method for the determination of strontium-90 in biological materials. Univ. of Washington, Seattle, U. S. AEC report UWFL-52. 9pp.

Kawakami, H. 1969. Determination of microgram amounts of mercury in hot spring water by neutron activation analysis. Onken Kiyo 20(3): 226-233. (In Japanese) (NSA 23: 40602).

Kawakami, H. and T. Oda. 1969. Application of activation analysis in hot spring chemistry and sanitary chemistry. Onken Kiyo 21(1): 47-56. (In Japanese) (NSA 25: 26195).

Kawamura, H. and H. Kawakami. 1971. Radiochemical studies on effect of bath bath. 1. Determination of mercury in spring mud by neutron activation analysis. Eisei Kagaku 17(4): 284-286. (In Japanese) (NSA 27: 17).

Kawano, M. and S. Nakatani. 1964. Some properties of natural radioactivity in the atmosphere. pp. 291-314. In: The Natural Radiation Environment. (See Adams and Lowder, 1964) (NSA 17: 32280).

Kawano, M., et al. 1969. Some properties of ionizing radiations and radioactive ions in the natural environment. Mem. Fac. Eng. Nagoya Univ. 21: 250-272. (NSA 24: 36416).

Kawase, K. 1972. Studies on radioactive fallout in soils of various types. Proc. Jap. Conf. Radioisotopes (10): 562-567. (In Japanese) (NSA 28: 378).

Kawase, K. and E. Yokohama. 1973. Strontium-90 and cesium-137 levels in soils of various types at Niigata Prefecture, Japan. J. Radiat. Res. (Japan) 14(1): 40-48. (NSA 28: 21110).

Kawashima, K. 1972. Siting and construction of atomic power plants. Characteristics of civil engineering works for atomic power plants. Doboku Gakkaishi 57(2): 13-21. (NSA 27: 4710) (Incl. ecological studies).

Kawatani, T. and S. Hemmi. 1962. Effects of radiation on the germination and growth of Rauwolfia serpentina Benth. Eisei Shikenjo Hokoku 80: 166-167. (In Japanese) (NSA 19: 27999).

Kawatani, T. and T. Ohno. 1962. Influence of x-ray radiation on the germination and early growth of Artemisia kurramensis Qazilbash. Eisei Shikenjo Hokoku 80: 163-164. (In Japanese) (NSA 19: 25948).

Kay, K. 1968. Look at the future of hazardous contamination of the circumpolar environment. Arch. Environ. Hlth. 17: 653-661. (NSA 23: 29478).

Kaye, N. W. and E. E. le Bourhis. 1958. Uptake and turnover of a single injected dose of I^{131} in tadpoles of Rana clamitans. Zoologica (N.Y.) 43(3): 73-76.

Kaye, S. V. 1960. Gold-198 wires used to study movements of small mammals. Science 131(3404): 824.

Kaye, S. V. 1961. Movements of harvest mice tagged with gold-198. J. Mammal. 42(3): 323-337. (WR 104: 22).

Kaye, S. V. 1962. Miniature glass rod dosimeters in radiation ecology. Abstr. Bull. Ecol. Soc. Am. 43(4): 125.

Kaye, S. V. 1963. Use of glass rod dosimetry in the ORNL ecology program. Abstr. Health Physics 9(8): 888.

Kaye, S. V. 1965. Use of miniature glass rod dosimeters in radiation ecology. Ecology 46(1,2): 201-206.

Kaye, S. V. 1968. Distribution and retention of orally administered radiotungsten in the rat. Health Physics 15(5): 399-417.

Kaye, S. V. 1973. Assessing potential radiological impacts to aquatic biota in response to the National Environmental Policy Act (NEPA) of 1969. Oak Ridge National Lab., U. S. AEC report. 14pp. (For Symp. on Environmental Behavior of Radionuclides in the Nuclear Industry) (CONF-730503-2) (NSA 28: 5539).

Kaye, S. V. and S. J. Ball. 1969. Systems analysis of a coupled compartment model for radionuclide transfer in a tropical environment. pp. 731-739. In: Symposium on Radioecology. (See Nelson and Evans, 1969).

Kaye, S. V. and D. A. Crossley. 1968. Radiotungsten retention by two insect species. Health Physics 14(2): 162-165. (NSA 22: 30998).

Kaye, S. V. and P. B. Dunaway. 1962. Bioaccumulation of radioactive isotopes by herbivorous small mammals. Health Physics 7(3/4): 205-217. (WR 106: 24).

Kaye, S. V. and P. B. Dunaway. 1963. Estimation of dose rate and equilibrium state from bioaccumulation of radionuclides by mammals. pp. 107-111. In: Radioecology. (See Schultz and Klement, 1963) (NSA 17: 33551).

Kaye, S. V. and D. J. Nelson. 1968. Analysis of specific-activity concept as related to environmental concentration of radionuclides. Nuclear Safety 9(1): 53-58.

Kaye, S. V. and P. S. Rohwer. 1969. Methods of estimating population exposures from Plowshare applications. pp. 550-584. In: Proc. Symp. on Public Health Aspects of Peaceful Uses of Nuclear Explosives. U. S. Public Health Service, Las Vegas, Nev., report SWRHL-82.

Kaye, S. V. and P. S. Rohwer. 1970. Dose-estimation studies related to proposed construction of an Atlantic-Pacific interoceanic canal with nuclear explosives: Phase III. Oak Ridge National Lab., U. S. AEC report ORNL-4579. ix, 135pp. (Also: Battelle Columbus Labs., U. S. AEC report BMI-171-50).

Kayser, P. 1962. Contribution to the study of the level of natural radionuclides of mineral waters of Mondorf-les-Bains and Plombieres. Thesis, Faculté de Pharmacie, Univ. Nancy, France. (In French) (NSA 19: 18168).

Kayser, P. 1966. Radioactivity of water, suspended materials and sediments in the Rhine Basin. Abstr. Health Physics 12(12): 1821.

Kayser, P. and A. Meunier. 1964. Natural radioelement content (U^{nat}, Th^{nat}, ^{40}K, ^{226}Ra, ^{222}Rn, and ^{210}Pb) of the thermal waters of Plombieres-les-Bains. Ann. Pharm. Franc. 22: 365-372. (In French) (NSA 19: 18150).

Kayuha, H. J., I. Aoki and D. L. Wireman. 1974. REECo field activities and sample logistics in support of the Nevada Applied Ecology Group. pp. 17-19. In: Dynamics of Plutonium in Desert Environments. (See Dunaway and White, 1974) (NSA 30: 29441).

Kaznacheyev, V. P., Z. N. Kashafutdinova and Ye. V. Baibikov. 1973. Deuterium and free radicals in the radon water of Belokurikh Health Resort. Vopr. Kurortol. Fiz. Lech. Fiz. Kul't. 38: 275-276. (In Russian) (NSA 31: 5573).

Keck, G. 1967. Long-lived radioactivity in high mountains from Jan. 1, 1965 to Dec. 1965. Wien. Tierärztl. Mschr. 54: 237-270. (In German) (NSA 23: 24319).

Keck, G. and M. Küster. 1967. The long-lived radioactivity in the air following the Chinese nuclear weapons test of October 28, 1966. Wien. Tierärztl. Mschr. 54: 25-26. (In German) (NSA 23: 24320) (Incl. pasture grass).

Keck, G., et al. 1967. Long lasting radioactivity in high mountains (January 1st 1965 to December 31, 1966. Wien. Tierärztl. Mschr. 54: 237-270. (In German).

Keefer, D. A. and M. Dauer. 1970. Natural environmental radioactivity in South Florida sands and soils, February-June 1968. Radiological Health Data 11(9): 441-448. (NSA 25: 13107).

Keil, R. 1968. On the occurrence of ^{90}Sr in biological materials. Nahrung 12: 399-405. (In German) (NSA 24: 31974) (Carp).

Keil, R. A. 1968. Investigation of fluctuation ranges in natural environmental radiation. pp. 220-229. In: Molecular Structure and Radiation Effects. (Ann. Meeting of Molecular Structure and Radiation Effects, Hanover, Germany, June 2-3, 1966). H. Glubrecht, ed. Georg Thieme Verlag, Stuttgart. (In German) (CONF-660671) (NSA 23: 14210).

Keil, R., F. Randow and H.-A. Schulze. 1966. ^{90}Sr in Mecklenburg hay. Arch. Tierernähr. 16: 125-130. (In German) (NSA 20: 36671).

Keil, R., et al. 1970. Radioactive contaminated lichens and the peculiar importance for radiologic hygiene. Z. Gesam. Hyg. Ihre Grenzgeb. 16: 898-902. (In German) (NSA 26: 25599).

Keiser, I. and E. L. Schneider. 1969. Need for immediate sugar and ability to withstand thirst by newly emerged oriental fruit flies, melon flies, and Mediterranean fruit flies untreated or sexually sterilized with gamma radiation. J. Econ. Entomol. 62: 539-540. (NSA 23: 35979).

Keiser, I. and E. L. Schneider. 1969. Longevity; resistance to deprivation of food and water, and susceptibility to malathion and DDT of oriental fruit flies, melon flies, and Mediterranean fruit flies sexually sterilized with tepa or radiation. J. Econ. Entomol. 62(3): 663-667.

Keiser, I., L. F. Steiner and H. Kamasaki. 1965. Effect of chemosterilants against the oriental fruit fly, melon fly and Mediterranean fruit fly. J. Econ. Entomol. 58(4): 682-685. (Dacus cucurbitae, Ceratitis capitata; incl. γ-rays).

Keith, L. B. and E. C. Meslow. 1968. Trap response by snow hares. J. Wildl. Mgmt. 32(4): 795-841. (^{45}Ca).

Keith, M. L. and G. M. Anderson. 1963. Radiocarbon dating: fictitious results with mollusk shells. Science 141(3581): 634-637. (NSA 17: 34135).

Keith, M. L. and J. N. Weber. 1964. Carbon and oxygen isotopic composition of selected limestones and fossils. Geochim. Cosmochim. Acta 28: 1787-1816. (NSA 19: 4364).

Keith, M. L., G. M. Anderson and R. Eichler. 1964. Carbon and oxygen isotopic composition of mollusk shells from marine and freshwater environments. Geochim. Cosmochim. Acta 28: 1757-1786. (NSA 19: 4363).

Kelemen, L., S. Barfai and J. Meszner. 1973. Proposition for the international control of radioactive contamination of the Danube, in accordance with the atomic energy development programs of Danube countries. Izotoptechnika 16(12): 654-662. (In Hungarian) (NSA 29: 18366).

Kelkar, D. N. and P. V. Joshi. 1969. Rapid method for estimating radium and radon in water. Health Physics 17: 253-257. (NSA 23: 35979).

Kelldorf, W. F. N. 1970. Radioactive tracer surveying: a comprehensive report. J. Petrol. Technol. 22: 661-669. (NSA 24: 46348).

Kelleher, W. J. 1969. Environmental surveillance around a nuclear fuel reprocessing installation, 1965-1967. Radiol. Hlth. Data 10: 329-339. (NSA 24: 2322) (Deer, fish, soil, plants).

Keller, E. C., Jr. 1970. Xanthine dehydrogenase activity in parental and F_1 _Drosophila_ and _Habrobracon_ under conditions of hypogravity. BioScience 20(19): 1045-1049. (Incl. ionizing radiation).

Keller, T. 1971. Air pollution effects on vegetation. Städtehygiene 22(6): 130-136. (In German) (NSA 27: 5558).

Kelley, H. L. and C. R. Phillips. 1973. Review of tritium monitoring devices. pp. 133-143. In: _Tritium_. (See Moghissi and Carter, 1973) (NSA 28: 27411).

Kellner, A., et al. 1955. Effect of radiation on bacteria. Bacteriol. Rev. 19: 22-44.

Kellogg, F. E. 1965. Determination of the cobalt-60 $LD_{50(30)}$ for the wild-caught cotton rat (_Sigmodon hispidus_). Thesis, Univ. of Georgia, Athens. v, 30pp.

Kellogg, F. E., E. E. Provost and M. R. Pelton. 1968. Determination of a ^{60}Co $LD_{50(30)}$ for the wild-caught cotton rat (_Sigmodon hispidus_). Radiation Research 35: 458-464. (NSA 22: 49518).

Kelly, M. J., S. S. Kirslis and R. G. West. nd. Environmental impact of HGTR power stations. Oak Ridge National Lab., U. S. AEC report. 9pp. (CONF-74051) (NSA 29: 28635).

Kelsall, J. P. and J. R. Calaprice. 1972. Chemical content of waterfowl plumage as a potential diagnosic tool. J. Wildl. Mgmt. 36(4): 1088-1097. (X-ray flourescence using ^{241}Am).

Kemp, N. E. 1962. Alterations in frog skin after heavy gamma irradiation. p. vv-3. In: _Electron Microscopy_. _Vol. 2_. Academic Press, New York. (NSA 16: 31419).

Kemp, W. G. 1970. Effect of gamma radiation on carnation mottle virus in crude _Dianthus barbatus_ Sap. Can. J. Microbiol. 16: 1374-1375. (NSA 25: 21985).

Kempton, J. P. 1966. Radiocarbon dates from Altonian and Twocreekan deposits at Sycamore, Illinois. Trans. Ill. State Acad. Sci. 59(1): 39-42. (NSA 22: 23141).

Kenady, R. M. 1962. The soils of Rongelap Atoll, Marshall Islands. Thesis, Univ. of Washington, Seattle. 99pp. (Also: U. S. AEC report TID-21432) (NSA 19: 9363).

Kenaga, D. J. and E. B. Cowling. 1959. Effect of gamma radiation on ponderosa pine: hygroscpicity, swelling and decay susceptibility. Forest Products J. 9(3): 112-116.

Kenaga, E. E., et al. 1965. Laboratory tests with dursban insecticide. J. Econ. Entomol. 58(6): 1043-1050. (^{36}Cl).

Kenagy, G. J. and C. B. Smith. 1973. Radioisotopic measurement of depth and and determination of temperatures in burrows of heteromyid rodents. pp. 265-273. In: _Radionuclides in Ecosystems_. (See Nelson, 1973) (NSA 28: 27327) (^{60}Co, ^{24}Na, ^{64}Cu, ^{198}Au).

Kendall, K. and D. S. Nachtwey. 1968. Inhibition of regeneration of planaria by midlethal exposures to x rays. U. S. Naval Radiological Defense Lab., San Francisco, report USNRDL-TR-68-29. 24pp. (AD-668330; NSA 23: 16374).

Kennedy, E. J., et al. 1971. Environmental studies of mercury and other elements in coal and lake sediments as determined by neiutron activation analysis. pp. 205-215. In: _Nuclear Methods in Environmental Research_. (See Vogt, et al., 1971) (NSA 26: 17368).

Kennedy, N. C. and H. G. Booth. 1974. Measurements of meteorological parameters. pp. 235-239. In: _Dynamics of Plutonium in Desert Environments_. Nevada Applied Ecology Group Progress Report as of January 1974. (See Dunaway and White, 1974) (NSA 30: 29451).

Kennedy, W. R. and W. D. Purtymun. 1970. Plutonium and strontium in soil in the Los Alamos, Espanola, and Sante Fe, New Mexico areas. Los Alamos Scientific Lab., U. S. AEC report LA-4562. 4pp. (NSA 25: 15763).

Kennington, G. S. and D. R. Percival. 1971. Radiation doses as a function of density and biological half-life in populations of _Tribolium confusum_ Duval. Physiol. Zool. 44(1): 48-57. (NSA 26: 7022).

Kenny, A. W. 1957. Radioactive discharge to sewers and rivers. Inst. Sewage Purif., J. Proc. Pt.4: 383-390.

Kenny, A. W. 1966. The degree of treatment required for low and intermediate level radioactive wastes for the prevention of hazardous pollution of the environment. Paper SM-71/64. In: _Practices in the Treatment of Low- and Intermediate-Level Radioactive Wastes_. (Proc. Symp., Vienna, 6-10 December 1965). International Atomic Energy Agency, Vienna, Publ. STI-PUB-116. (CONF-651202) (NSA 20: 26559).

Kenny, A. W. 1969. Radioactive contamination arising from the peaceful uses of atomic energy: past, present and foreseeable future. pp. 15-26. In: _Environmental Contamination by Radioactive Materials_. (See IAEA, 1969).

Kenny, A. W., R. N. Crooks and J. R. W. Kerr. 1966. Radium, radon, and daughter products in certain drinking waters in Great Britain. J. Inst. Water Engr. 20(2): 123-124. (CA 64: 19183; NSA 20: 37031).

Kent, C. E., S. Levy and J. M. Smith. 1971. Effluent control for boiling water reactors. pp. 339-356. In: Environmental Aspects of Nuclear Power Stations. (See IAEA, 1971).

Kent, E. 1964. The effect of tritiated thymidine and gamma irradiation on the mortality of Drosophila melanogaster larvae. Cekmece Nuclear Research Center, Turkey Atomic Energy Commission, Istanbul, report CNAEM-16. 7pp. (NSA 18: 36947).

Kent, E. 1965. The effect of tritiated thymidine and gamma irradiation on the mortality of adult Drosophila melanogaster. Cekmece Nuclear Research Center, Turkey Atomic Energy Commission, Istanbul, report CNAEM-32. 11pp. (NSA 20: 10602).

Kenworthy, W. 1956. The effect of oxygen concentration on the dose-action survival curves obtained for Habrobracon eggs irradiated during meiotic prophase and metaphase. Am. Naturalist 90: 119-126. (NSA 11: 5173).

Kenworthy, W. 1962. Radiation effects on the cytoplasm of Habrobracon eggs. U. S. AEC report TID-16471. 22pp. (NSA 16: 28751).

Keough, R. F. and G. J. Powers. 1970. Determination of plutonium in biological materials by extraction and liquid scintillation counting. Anal. Chem. 42: 419-421. (NSA 24: 16450).

Keppel, H. 1969. Morphological, physiological and environmental conditions affecting the accumulation of long-lived radionuclides from fission products by plants. pp. 99-111. In: Environmental Contamination by Radioactive Materials. (See IAEA, 1969).

Kepp, R., K. Öhlert and G. Öhlert. 1957. Studies of the biological dose equivalent for hard x-rays (180 KV) and beta rays from strontium-90 on eggs of Ascaris megalocephala. Strahlentherapie 103(3): 392-397. (In German).

Kerpen, W. and H. W. Scharpenseel. 1967. Movement of ions and colloids in undistrubed soil and parent rock material columns. pp. 213-226. In: Isotope and Radiation Techniques in Soil Physics and Irrigation Studies. (See IAEA, 1967) (NSA 22: 23154).

Kerr, G. W. 1961. Statistical treatment of environmental radio-assay data. Health Physics 6(3/4): 153-162.

Kerr, G. W. 1962. Use of statistical methodology in environmental monitoring. Am. Industr. Hygiene Assoc. J. 23(1): 75-82.

Kersten, H. and G. G. Snider. 1935. Daphnia magna as a biological dosimeter for soft x-rays. Radiology 25: 285-288.

Kerstetter, T. H., L. B. Kirschner and D. D. Rafuse. 1970. On the mechanisms of sodium transport by the irrigated gills of rainbow trout (Salmo gairdneri). J. Gen. Physiol. 56(3): 342-359. (^{22}Na).

Kerth, K. 1973. Radula replacement and cell proliferation in the x-irradiated radular gland of the slug Limas flavus L. On the function of the radular gland epithelia. Arch. Entwicklungsmech. Organ. 172(4): 317-348. (In German) (NSA 29: 27118) (Kerth, L. ?).

Kerth, L. and G. Krause. 1969. Investigations into the x-ray influenced radula-replacements in the naked snail Limax flavus. Arch. Entwicklungsmech. Organ. 164(1): 48-82. (In German) (NSA 24: 27669) (Kerth, K. ?).

Kessler, R. and W. Luther. 1957. The action of x-rays on the testes and secondary sex characteristics of Lebistes reticulatus Peters. Z. Verg. Physiol. 40(5): 492-528. (In German).

Ketcheson, J. W. and J. J. Onderdonk. 1974. Use of radioisotopes to determine surface movement of phosphorus fertilizer on sloping land. pp. 195-201. In: Isotopes and Radiation Techniques in Soil Physics and Irrigation Studies 1973. (See IAEA, 1974) (NSA 30: 15495).

Kettlewell, H. B. D. 1952. Use of radioactive tracer in the study of insect populations (Lepidoptera). Nature 170(4327): 584-585.

Kettlewell, H. B. D. 1955. Labelling locusts with radioactive isotopes. Nature 175(4462): 821-822.

Kettlewell, H. B. D. 1956. Labelling of locusts with radioactive isotopes. Proc. Intern. Conf. on the Peaceful Uses of Atomic Energy 12: 214-215. (A/CONF.8/P/1079)

Kettlewell, H. B. D. and M. J. Heard. 1961. Accidental radioactive labelling of a migratory moth. Nature 189(4765): 676-677. (NSA 15: 12689).

Kevern, N. R. 1964. Strontium and calcium uptake by the green alga, Oocystis eremosphaeria. Science 145(3639): 1445-1446. (NSA 18: 38735).

Kevern, N. R. 1965. The feeding rate and energy requirement of carp (Cyprinus carpio Linnaeus). Abstr. Bull. Ecol. Soc. Am. 46(3): 133. (^{137}Cs).

Kevern, N. R. 1966. Feeding rate of carp established by a radioisotopic method. Trans. Am. Fish. Soc. 94: 363-371.

Kevern, N. R. 1970. A measurement of low level cesium isotope concentration in a fresh water lake. Suppl. to final report. Michigan State Univ., East Lansing, U. S. AEC report COO-1795-7. 71pp. (NSA 25: 39433).

Kevern, N. R. 1970. Effects of biogeochemical factors on the accumulation of ^{137}Cs fallout by largemouth bass (Micropterus salmoides). Suppl. to final report. Michigan State Univ., East Lansing, U. S. AEC report COO-1795-6. 107pp. (NSA 25: 37750).

Kevern, N. R. 1971. Effects of limnological factors on uptake of ^{137}Cs fallout by fish. Final report, June 1968-June 1971. Michigan State Univ., East Lansing, U. S. AEC report COO-1795-3. 6pp. (NSA 25: 37749).

Kevern, N. R. and S. A. Spigarelli. 1973. Effects of selected limnological factors on the accumulation of cesium-137 fallout by largemouth bass (Micropterus salmoides). pp. 354-360. In: Radionuclides in Ecosystems. (See Nelson, 1973) (NSA 25: 46801).

Keyl, H.-G. 1958. Investigations on the karyotype of Chironomus thummi II. Structural changes in the salivary gland chromosomes following roentgen irradiation of embryos and larvae. Chromosoma 9(5): 441-483.

Khademi, B. and A. Mesghali. 1971. Investigation and measurement of radium in Ramsar mineral water. Health Physics 21(3): 464-465. (NSA 26: 1980).

Khademi, B. and A. Tahsili. 1972. Environmental radioactivity in certain parts of Iran. pp. 369-374. In: Natural Radiation Environment II. Vol. 1. (See Adams, Lowder and Gesell, 1972) (NSA 32: 19718).

Khakimova, R. K. 1964. Effect of ionizing radiation on the survival rate of Agrotis segetem Schiff, and its ability to breed. pp. 255-257. In: Problems of Biology and Regional Medicine, 1963. Akad. Nauk Uzbek. SSR, Tashkent. (In Russian).

Khalil, G. M. 1971. Biochemical and physiological studies of certain ticks (Ixodoidea). Incorporation of tritiated tyrosine in the digestive system of nymphal Argas (Persicargas) arboreus (Argasidae). ---in the excretory system ---. Ann. Entomol. Soc. Am. 64(5): 1149-1159.

Khan, K. U. and Y. Hiyama. 1964. Mutual effect of Sr-Ca upon their uptake by fish and freshwater plants. Rec. Oceanog. Works Japan 7(2): 107-122.

Khan, Z. A. and A. S. M. Abdul Matin. 1967. The susceptibility of the different stages of rice weevil, Sitophilus oryzae L. to gamma radiation. Pakistan Atomic Energy Centre, Dacca, report AECD/AG-22. 15pp. (NSA 22: 10822).

Khan, Z. A. and A. S. M. Abdul Matin. 1968. Susceptibility of the different developmental stages of rice weevil, Sitophilus oryzae L. to gamma radiation. Nucl. Sci. Appl. 4: 11-18. (NSA 23: 6628).

Kharchuk, A. I. 1973. Methods of determination of the accumulation of radionuclides by plants from aquatic environment. pp. 128-132. In: Radioecology of Water Organisms. III. (See Andrumaitis, 1973) (In Russian) (NSA 30: 24412).

Kharlamov, V. P. 1965. A change in the activity of feeding and mobility of the fleas Xenopsylla cheopis marked with radioactive phosphorus ^{32}P. Zool. zh. 44(4): 547-551.

Kharlamova, L. A. 1972. Investigation of the radiobiological reactions in Amoeba proteus. 1. Survival, delay of cell division, and forms of death after gamma irradiation of radiosensitive and radioresistant strains. 3. DNA content influence of modifying factors on the survival of amoebas after irradiation. Radiobiologiya 12(3): 401-407; 13(2): 260-264. (AEC-tr-7402, pp. 120-128; AEC-tr-7444, pp. 125-130) (NSA 27: 17535; 28: 24845).

Kharmalova, L. A. 1973. Investigation of the radiobiological responses in Amoeba proteus. Radiobiologiya 12(6): 934-937. (AEC-tr-7421, pp. 181-184) (NSA 28: 2840).

Kharlampovich, S. I. and V. A. Gulyayev. 1967. Electron microscopic changes in the thyroid gland in local irradiation. Med. radiologiya 12(4): 33-36. (In Russian) (NSA 21: 28610) (Frog).

Kheissina, V. Y. 1956. Effect of x-rays on the early period of embryogenesis. Dokl. Akad. Nauk SSSR 110: 57-60. (In Russian).

Khera, A. K., E. Steinnes and A. Øien. 1970. Comparative study of the aluminum determination in ammonium acetate soil extracts by spectrophotometry, activation analysis, and atomic absorption. Acta Agr. Scand. 20(1): 33-34. (NSA 25: 29020).

Khmeleva, N. N. 1957. Experiment in the applications of radioactive phosphorus for study of mineral fertilization in ponds. In: Proc. 6th Conf. on Problems in Biology of Inland Waters. (10-19 June 1957). B. S. Kuzin and S. I. Kuznetsov, eds. Publ. House Acad. Sci. USSR, Leningrad-Moscow. (AEC-tr-6880; TT-50492, pp. 41-48).

Khmeleva, N. N. 1961. Comparative determination of the primary production of plankton in ponds of the Latvian SSR by the oxygen and radiocarbon methods. pp. 209-213. In: Primary Production of the Seas and Inland Waters. G. G. Vinberg, ed. Ministry of Special Technical and Voacational Training, Beloruss. SSR, Minsk. (Australia CSIRO/No. 6775; TT-64-15424) (BA 43: 4550; Ref. zh. (biol.) 1962: 20Zh451; SFA 8: 6139).

Khmeleva, N. N. 1962. Accumulation of radioactive isotopes of calcium, yttrium and cerium by Amphipoda. Radiobiologiya 2(6): 944-946. (AEC-tr-5433, pp. 198-202) (NSA 17: 15623).

Khmeleva, N. N. and G. M. Laurent'yeva. 1966. Accumulation of radioactive strontium, yttrium, and cerium isotopes by filamentous algae in the presence of EDTA and dissolved organic compounds. Radiobiologiya 6(1): 117-121. (AEC-tr-6670, pp. 184-191) (NSA 20: 38570).

Kholodov, Yu. A. (Ed.). 1973. Effect of certain cosmic and geophysical factors on the earth's biosphere. Probl. Kosm. Biol. 18. 210pp. (In Russian) (NSA 30: 9690).

Khoury, A. T. and R. A. Deering. 1973. Sedimentation of DNA of Dictyostelium discoideum lysed on alkaline sucrose gradients. Role of single strand breaks in gamma ray lethality of sensitive and resistant strains. J. Mol. Biol. 79(2): 267-284. (NSA 29: 517).

Khristianov, V. K. and B. N. Korchuganov. 1971. On radon content in the water of the upper Volga. Geokhimiya (4): 492-496. (In Russian) (NSA 25: 41929).

Khristianova, L. A., V. V. Anikiyev and N. N. Vinogradova. 1973. Distribution of radioactive isotopes in water reservoirs. State Committee on the Use of Atomic Energy, Moscow. (In Russian) (INIS-mf-1041) (NSA 30: 6533).

Khropova, V. I., K. V. Kvitko and I. A. Kakharov. 1964. Comparative study of mutagenic effects of various types of radiation and ethylenimine on Chlorella. Issled. genet. 2: 69-76. (In Russian) (NSA 22: 51496).

Khrushchev, V. G., G. M. Pravdina and N. G. Darenskaya. 1962. Behavior of the fruit fly (Drosophila melanogaster) during irradiation. Radiobiologiya 2(2): 272-279. (AEC-tr-5429, pp. 150-160) (NSA 16: 28814).

Khudadov, G. D. 1959. A method for tagging insects by giving them radioactive isotopes with food. Byull. Mosk. obshch. ispyt. prirody (biol.) 64 (3): 34-45. (JPRS-2,453; Tech. Transl. 3(11): 700) (NSA 14: 13548).

Khudadov, G. D. 1959. Radioautographic method of finding insects and ticks marked with radioisotopes. Med. parazitol. 23(1): 60-64. (In Russian).

Khudadov, G. D. 1959. On the use of radiophosphorus to study the movement of flies from inhabited premises. Trans. Central Inst. Sci. Res. on Disinfestation (11): 319-324. (In Russian).

Khudadov, G. D. 1959. Use of radioactive isotopes for the labelling of epidemiologically important arthropods. Dokl. Akad. Nauk SSSR. (In Russian) (Ref. zh. (biol.) 1959: 50899D).

Khudadov, G. D. 1961. An apparatus for labelling Arthropoda by means of spraying them with radioactive isotopes. Byull. Mosk. obshch. isypt. prirody (biol.) 66(3): 40-50. (In Russian).

Khusnullin, M. Kh., F. M. Sattarova and V. I. Zaytsev. 1974. Enrichment of stratified layer water with radium isotopes. Nefteprom. Delo (12): 18-21. (NSA 33: 17192).

Khvatova, L. 1965. Sterilization of parasites of the ash tree. Zashchita rasteniy (12): 51-52. (In Russian).

Kidd, C. C., Sr. 1970. *Pontoporeia affinis* (Crustacea, Amphipoda) as a monitor of radioactivity in Lake Michigan. Thesis, Univ. of Michigan, Ann Arbor. 81pp. (Univ. Microfilms 71-15,197) (NSA 25: 49126).

Kiefer, H. and R. Maushart. 1958. The natural radioactivity in water. Nukleonik 1(1): 23-27. (In German)

Kiefer, H. and R. Maushart. 1963. Method and apparatus for the continuous measurement of the tritium content of water. German Patent 1, 498, 832. (In German) (NSA 28: 18319).

Kiefer, H. and R. Maushart. 1972. *Radiation Protection Measurements*. R. Friese, transl. Pergamon Press, Oxford. 577pp. (NSA 26: 45808).

Kiel, G. R. 1972. Radioactive waste management at Hanford. Atlantic Richfield Hanford Co., U. S. AEC report ARH-SA-113. 12pp. (CONF-720347-1) (NSA 28: 360).

Kieschnick, E. 1971. Decontamination of radioactive effluents. Sulzer Tech. Rev. 53(4): 227-235. (NSA 26: 27963).

Kigoshi, K. 1962. Krypton-85 in the atmosphere. Bull. Chem Soc. Japan 35 (6): 1014-1016.

Kigoshi, K. 1966. Secular variation of atmospheric radiocarbon concentration. Nippon Kagaku Zasshi 87: 209-220. (NSJ-tr-104) (NSA 20: 27220).

Kigoshi, K. and K. Endo. 1961. Variation of the atmospheric carbon-14 in recent years in Tokyo. Bull. Chem. Soc. Japan 34(11): 1740.

Kigoshi, K. and Y. Tomikura. 1960. Variation of radiocarbon concentration in modern wood. Bull. Chem. Soc. Japan 33: 1576-1580. (NSA 15: 10622).

Kigoshi, K. and Y. Tomikura. 1961. Tritium and carbon-14 in tree rings. Bull. Chem. Soc. Japan 34(11): 1739.

Kigoshi, K., et al. 1964. Introduction of systematic studies of the variation of C^{14}-contents in the atmosphere for the past thousand years. pp. 466-469. In: *Intern. Conf. on Cosmic Rays*. India Tata Inst. of Fundamental Research, Bombay. (NSA 19: 34376).

Kihira, H. 1960. Effects of x-rays on a single nerve fibre. Nippon Igaku Hoshasen Gakkai Zasshi 20: 46-60. (NSA 15: 25896).

Kikuchi, C. 1974. Environmental impacts on nuclear power plants. Brazil Instituto de Energia Atomica, São Paulo, report IEA-31. 126pp. (NSA 33: 29367).

Kilcawley, E. J., et al. 1953. The fate of fission products deposited in the reservoirs of the Troy, New York area following nuclear detonations during the spring tests of 1953. J. Am. Water Works Assoc. 40(11): 1101-1111. (Also: U. S. AEC report NYO-4569) (NSA 10: 9723).

Kilgore, W. W. and R. R. Painter. 1966. Insect chemosterilants: incorporation of 5-fluorouracil into house fly eggs. J. Econ. Entomol. 59: 746-747. (^{14}C).

Kilgore, W. W. and R. R. Painter. 1966. Persistance of tepa ingested by fall army-worm moths. Abstr. Bull. Entomol. Soc. Am. 12: 288. (^{14}C).

Killebrew, R. and L. Brou. 1965. Effects of ^{60}Co (gamma radiation) on the melanophores of *Gambusia affinis* and *Mollienisia latipinna*. J. Miss. Acad. Sci. 11: 197-204. (NSA 20: 42905).

Killebrew, R. and J. A. Winkler, III. 1965. Effects of ^{60}Co irradiation of *Euglena gracilis* (K_1). J. Miss. Acad. Sci. 11: 205-208. (NSA 20: 42906).

Kilpatrick, J. W., R. W. Fay and J. T. Baker. 1957. The rearing and radioactive tagging of *Fannia canicularis*. Abstr. Bull. Entomol. Soc. Am. 3 (3): 26-27.

Kim, C. M., et al. 1970. Studies on the radiation breeding of perennial plants. I. Radiosensitivities of some economic and ornamental tree species in Korea. J. Nucl. Sci. (Seoul) 9(2, Pt.2): 75-87. (In Korean) (NSA 25: 4724).

Kim, J. H. 1968. Studies on the cellular metabolism in microorganisms as influenced by gamma irradiation. IV. On the carbohydrate metabolism of yeast irradiated by γ rays. Korean J. Microbiol. 6(2): 41-53. (In Korean) (NSA 25: 47040).

Kim, J. H., et al. 1968. Studies on the cellular metabolism in microorganisms as influenced by gamma irradiation. V. On the membrane permeability changes and leakage of cellular constituents of irradiated yeast cells. Korean J. Microbiol. 6(2): 54-62. (In Korean) (NSA 25: 44968).

Kim, S. M. 1969. Precision of natural radiocarbon measurement. Radiochem. Radioanal. Lett. 2: 187-193. (NSA 24: 7515).

Kimball, R. F. 1949. The induction of mutations in Paramecium aurelia by beta radiation. Genetics 34(2): 210-222.

Kimball, R. F. 1950. The effect of radiations on genetic mechanisms of Paramecium aurelia. J. Cell. Comp. Physiol. 35(Suppl.1): 157-169.

Kimball, R. F. 1952. Genetic effects of radiation. Ann. Rev. Nucl. Sci. 1: 479-494. (Microorganism).

Kimball, R. F. 1955. The effects of radiation on protozoa and the eggs of invertebrates other than insects. pp. 285-331. In: Radiation Biology. Vol. 2. A. Hollaender, ed. McGraw-Hill Book Co., New York.

Kimball, R. F. 1955. The role of oxygen and peroxide in the production of radiation damage in Paramecium. Ann. N.Y. Acad. Sci. 59: 638-647.

Kimball, R. F. 1957. Modification of the genetic effects of x rays by treatment after irradiation. pp. 252-255. In: Proc. Intern. Genetics Symp. (Tokyo, 1956). Cytologia (Suppl.). (Microorganism).

Kimball, R. F. 1957. Nongenetic effects of radiation on microorganisms. Ann. Rev. Microbiol. 11: 199-220.

Kimball, R. F. 1958. Experiments with Stentor coeruleus on the nature of the radiation-induced delay in fission in ciliates. J. Protozool. 5(2): 151-155.

Kimball, R. F. 1961. Postirradiation processes in the induction of recessive lethals by ionizing radiation. J. Cell. Comp. Physiol. 58(Suppl.1): 163-170. (Microorganism).

Kimball, R. F. 1962. Fractionation of dose and radiation intensity in the induction of recessive lethal mutations by x-rays in Paramecium aurelia. Abstr. p. 6. In: 2nd Intern. Congr. Radiation Research. (See Anonymous, 1962).

Kimball, R. F. 1963. X-ray dose rate and dose fractionation studies on mutation in Paramecium. Genetics 48(4): 581-595. (NSA 17: 21515).

Kimball, R. F. 1963. The relation of repair differntial radiosensitivity in the production of mutations in Paramecium. pp. 167-178. In: Repair from Genetic Radiation. Pergamon Press, New York. (NSA 17: 38854).

Kimball, R. F. 1963. Repair and differential sensitivity to mutation induction: summary and synthesis. pp. 427-441. In: Proc. Symp. in Repair from Genetic Radiation Damage and Differential Radiosensitivity of Germ Cells. (Leidén, 1962). F. H. Sobels, ed. Pergamon Press, New York. (NSA 18: 1446) (Paramecium).

Kimball, R. F. 1964. Studies on radiation mutagenesis in microorganisms. pp. 227-234. In: Genetics Today. Pergamon Press, New York. (NSA 19: 21981).

Kimball, R. F. 1964. The distribution of x-ray induced mutations to chromosomal strands in Paramecium aurelia. Mutation Research 1: 129-138. (NSA 19: 7114).

Kimball, R. F. 1966. Repair of premutational damage. Adv. Radiat. Biol. 2: 135-166. (Also: Oak Ridge National Lab., U. S. AEC report ORNL-P-983) (NSA 20: 31040).

Kimball, R. F. 1969. Studies on mutations induced by ultraviolet radiation in Paramecium aurelia with special emphasis on photoreversal. Mutation Research 8: 79-89. (NSA 23: 31709).

Kimball, R.F. and N. Gaither. 1951. The influence of light upon the action of ultraviolet on Paramecium aurelia. J. Cell. Comp. Physiol. 37(2): 211-233.

Kimball, R. F. and N. Gaither. 1952. The role of externally produced hydrogen peroxide in damage to Paramecium aurelia by x rays. Proc. Soc. Exptl. Biol. Med. 80(3): 525-529

Kimball, R. F. and N. Gaither. 1953. Influence of oxygen upon genetic and nongenetic effects of ionizing radiation on Paramecia aurelia. Proc. Soc. Soc. Exptl. Biol. Med. 82(3): 471-477.

Kimball, R. F. and N. Gaither. 1956. Behavior of nuclei at conjugation in Paramecium aurelia. II. The effects of x-rays on diploid and haploid clones, with a discussion of dominant lethals. Genetics 41(6): 715-728. (NSA 12: 2698).

Kimball, R. F. and J. Z. Hearon. 1955. Tests for role of H_2O_2 in x-ray mutagenesis. I. Estimations of the concentration of H_2O_2 inside the nucleus. Radiation Research 3(3): 283-294. (Microorganism).

Kimball, R. F. and S. W. Perdue. 1962. Quantitative cytochemical studies on Paramecium. V. Autoradiographic studies of nucleic acid syntheses. Exptl. Cell Res. 27(3): 405-415. (RDA-5: 245) (^3H).

Kimball, R. F. and S. W. Perdue. 1962. Studies on the refractory period for the induction of recessive lethal mutation by x-rays in Paramecium. Genetics 47(11): 1595-1607. (NSA 17: 7842).

Kimball, R. F. and S. W. Perdue. 1967. Comparison of mutagenesis of x-rays and triethylene melamine in Paramecium, with emphasis on the role of mitosis. Mutation Research 4: 37-50. (NSA 21: 16324).

Kimball, R. F. and L. Vogt-Köhne. 1963. Effects of radiation on cell and nuclear growth in Paramecium aurelia. Exptl. Cell Res. 28: 228-238. (NSA 17: 10408).

Kimball, R. F., N. Gaither and S. W. Perdue. 1961. Metabolic repair of premutational damage in Paramecium. Int. J. Radiat. Biol. 3: 133-147. (NSA 15: 14133).

Kimball, R. F., N. Gaither and S. M. Wilson. 1957. Postirradiation modification of mutagenesis in Paramecium by streptomycin. Genetics 42(5): 661-669. (NSA 12: 2698).

Kimball, R. F., N. Gaither and S. M. Wilson. 1959. Reduction of mutation by post irradiation treatment after ultraviolet and various kinds of ionizing radiations. Radiation Research 10(4): 490-497. (Microorganism).

Kimball, R. F., N. Gaither and S. M. Wilson. 1959. Recovery of stationary-phase paramecia from radiation effects leading to mutation. Proc. Natl. Acad. Sci. (US) 45(6): 833-839.

Kimball, R. F., J. Z. Hearon and N. Gaither. 1955. Tests for a role of H_2O_2 in x-ray mutagenesis. II. Attempts to induce mutations by peroxide. Radiation Reserach 3(4): 435-443. (Microorganism).

Kimball, R. F., R. P. Geckler and N. Gaither. 1952. Division delay by radiation and nitrogen mustard in *Paramecium*. J. Cell. Comp. Physiol. 40(3): 427-459.

Kimber, G. and A. Booth. 1958. Radioactivity of plant material collected after the accident at Windscale No. 1 Pile. Nature 181(4620): 1391-1392.

Kimbrough, R. D., Jr. and T. B. Gaines. 1971. γ irradiation of DDT: radiation products and their toxicity. J. Agr. Food Chem. 19(5): 1037-1038. (NSA 25: 44968).

Kimeldorf, D. J. 1967. Prompt reactions of organisms to ionizing radiations. Abstr. Bull. Ecol. Soc. Am. 48(3): 145. (Mentions insects).

Kimeldorf, D. J. and E. L. Hunt. 1965. *Ionizing Radiation: Neural Function and Behavior*. Academic Press, New York. xii, 331pp. (NSA 20: 10689) (Incl. data on insects, frogs, planaria, daphnia, snails, fish, crab, turtle).

Kimmins, J. P. 1970. The relative contributions of leaching litter-fall, and insect defoliation to the removal of cesium-134 from red pine trees. Abstr. Bull Ecol. Soc. Am. 51(2): 24.

Kimura, K. and T. Nakai. 1953. Radium content of the soil and well water in the vicinity of hypocenter at Nagasaki and Hiroshima. Nippon Gakujutsu Kaigai 1: 61-66.

Kimura, K. and Y. Uemura. 1953. Artificial radioactice elements found in soil collected at Takasu, Hiroshima. Nippon Gakujutsu Kagai 1: 54-58.

Kimura, K., K. Noguchi and S. Ueno. 1953. Contamination of underground water due to the atomic bomb in Hiroshima. Nippon Gakujutsu Kagai 1: 66-69.

Kimura, K., K. Noguchi and S. Ueno. 1953. Contamination of underground water due to the atomic bomb in Nagasaki. Nippon Gakujutsu Kaigai 1: 69-72.

Kimura, K., et al. 1953. Artificial radioactive elements found in soil collected at Nishiyama. Nippon Gakujutsu Kagai 1: 53-54.

Kimura, T. J., R. Duffy and A. W. A. Brown. 1965. Dehydrochlorination and DDT resistance in *Culex* mosquitoes. Bull. Wld. Hlth. Org. 32(4): 557-561. (^{14}C).

Kimura, T., H. L. Keegan and T. Haberkorn. 1967. Dehydrochlorination of DDT by Asian bloodsucking leeches. Am. J. Trop. Med. Hyg. 16(5): 688-690. (^{14}C).

Kimura, Y., et al. 1968. Studies on the behavior and distribution of radioactive substances in coastal and estuarine waters. Kinki Daigaku Genshir. Kenky. Nenpo 7: 21-31. (In Japanese) (NSA 25: 34760).

Kind, N. V. 1965. Absolute chronology of the main stages of last glaciation and postglacial period in Siberia (according to radiocarbon data). The radiocarbon chronology of the last glaciation and of the post-glacial interval in Siberia. Abstr. pp. 267, 269. In: Abstrs. Intern. Assoc. Quaternary Research, 7th Intern. Congr., Boulder and Denver, Colo.

King. L. G. 1967. Gamma-ray attenuation for soil-water-content measurements using ^{241}Am. pp. 17-29. In: *Isotope and Radiation Techniques in Soil Physics and Irrigation Studies*. (See IAEA, 1967) (NSA 22: 23143).

King, R. C. 1950. Genetic changes accompanying the uptake of P^{32} by *Drosophila melanogaster*. Genetics 35(1): 118-119.

King, R. C. 1952. Calculation of radiation doses to P^{32}-labelled male *Drosophila melanogaster*. Nucleonics 10(11): 88-89.

King, R. C. 1952. The mutagenic effectiveness of radioactive phosphorus in *Drosophila melanogaster*. Abstr. Genetics 37: 595.

King, R. C. 1953-54. Studies with radiophosphorus in *Drosophila*. I. The mutagenic effectiveness of radioactive phosphorus in adult male *Drosophila melanogaster*. II. The turnover and distribution of phosphorus in adult *Drosophila*. III. The effect of P^{32}-treatment upon developing flies. J. Exptl. Zool. 122(3): 541-575; 125: 331-352; 126(2): 323-336.

King, R. C. 1953. Studies with radiophosphorus in *Drosophila*. 2. The effect of P^{32} treatment upon development. 3. The turnover of phosphorus in adult *D. melanogaster* and *D. simulans*. Brookhaven National Lab., U. S. AEC report BNL-1406, BNL-1405. 13, 11pp.

King, R. C. 1954. The effect of yeast upon the incorporation of phosphorus into the tissue of adult fruit flies. Evolution 9(1): 93-94.

King, R. C. 1954. The effect of yeast on phosphorus uptake by *Drosophila*. Am. Naturalist 88(840): 115-158.

King, R. C. 1957. Oogenesis in adult *Drosophila melanogaster*. III. Radiation-induced ovarian tumors. Growth 21(3): 129-135.

King, R. C. and A. C. Rubinson. 1957. Distribution of calcium in adult *Drosophila melanogaster*. Science 125(3247): 546. (^{45}Ca).

King, R. C. and L. P. Wilson. 1954-55. Studies with radiophosphorus in *Drosophila*. IV. Experiments on flies homogeneously labeled with P^{32}. V. Quantitative studies on the phosphorus balance of adult female flies feeding upon yeast. J. Exptl. Zool. 126(3): 401-407; 130(1): 71-82.

King, R. C. and P. Wilson. 1955. Studies of the radiation syndrome in Drosophila melanogaster. Radiation Research 2(6): 544-555.

King, S. F. 1964. Uptake and transfer of cesium-137 by Chlamydomonas, Daphnia, and bluegill fingerlings. Ecology 45(4): 852-859.

King, V. 1969. A study of the mechanism of water transfer across frog skin by a comparison of the permeability of the skin to deuterated and tritiated water. J. Physiol. (London) 200: 529-538. (NSA 23: 35102)

Kinkade, M. L. and H. E. Erdman. 1973. Confused flour beetles: x-ray and/or malathion modification of population fitness. J. Econ. Entomol. 66(4): 9 967-972. (NSA 28: 21778).

Kinnamon, K. E. 1963. Some independent and combined effects of copper, molybdenum, and zinc on the placental transfer of zinc-65 in the rat. J. Nutrition 81(4): 312-320. (NSA 18: 15794) (Discusses ecology).

Kinsman, S. 1957. Radiological Health Handbook. R. A. Taft Sanitary Engineering Center, U. S. PHS, Cincinnati, O. 358pp. (Revised ed., 1960) (PB-121784, PB-121784R) (NSA 11: 6212).

Kipiani, R. and T. V. Tsetskhladze. 1957. The killing of and the preservation of silkworm cocoons by gamma rays. Soobshch. Akad. Nauk Gruz. SSR 1957: 657-662. (In Russian).

Kipiani, R. Ya., E. V. Tvalchrelidze and G. V. Kvachadze. 1969. Levels of strontium-90, stable strontium, and calcium in Aurantiaceae of the Georgian SSR in 1965-1967. Tr. inst. zashch. rast. Tiflis 21: 355-357. (In Russian) (NSA 28: 8526).

Kipp, K. L., Jr. 1973. Radiological status of the groundwater beneath the Hanford Project, January-June 1972. Battelle Northwest Lab., U. S. AEC report BNWL-1737. 31pp. (NSA 28: 10872).

Kips, R. H. and J. van den Brande. 1958/1959. Radioactive rays and radioactive isotopes in applied entomology and phytopharmacy. Land-en Tuinbouw Jaarb. 13(A58/59): 143-148. (In Flemish).

Kirby, H. V. 1971. Effect of temperature on the feeding behavior of Daphnia rosea. Limnol. Oceanogr. 16(3): 580-581. (^{32}P).

Kirby, H. W. 1974. Geochemistry of the naturally occurring radioactive series. Mound Lab., U. S. AEC report MLM-2111. 82pp. (NSA 30: 31996).

Kirby, J. A., et al. 1976. Detector system for in situ spectrometric analysis of ^{241}Am and Pu in soil. Trans. Nucl. Sci. NS-23(1): 683-689. (NSA 33: 29367).

Kirchberg, E. 1956. On the application of radioisotopes to pest control. Prakt. Schädlingsbekämpfer 8(11): 129-132. (In German).

Kirchen, R. V. 1963. Radiopathology and dosimetry in the thyroid region of goitrous killifish fed I^{131} under euryhaline conditions. Abstr. Health Physics 9(12): 1341.

Kirchmann, R. 1964. Contaminations of the biosphere by artificial radioelements. Ann. Gembloux 70: 215-223. (In French) (NSA 20: 43596).

Kirchmann, R. 1971. Radioactive contamination of the soil and the associated radiological risks. Ann. Gembloux 77: 207-223. (In French) (NSA 27: 14750, 17269).

Kirchmann, R. and G. Cantillon. 1971. The behavior in a river of ^{60}Co and ^{54}Mn from effluents of a PWR power station. pp. 425-441. In: Proc. Intern. Symp. on Radioecology Applied to the Protection of Man and His Environment. (See Commission of the European Communities, 1971) (In French) (NSA 27: 5260).

Kirchmann, R. and T. J. D'Souza. 1972. Behavior of ruthenium in an established pasture soil and its uptake by grasses. pp. 587-593. In: Isotopes and Radiation in Soil-Plant Relationships Including Forestry. (See IAEA, 1972) (NSA 27: 2847).

Kirchmann, R. and J. Lambinon. 1973. Vegetable bioindicators of the contamination of a water flow by the wastes from a pressurized water nuclear power plant. J. Bull. Soc. Roy. Bot. Belg. 106: 187-201. (ORNL-tr-2950) (NSA 30: 3166; 33: 623).

Kirchmann, R. and J. van den Hoek. 1972. Tritium on the biosphere and its transfer through the food chain. J. Belg. Radiol. 55(2): 233-242. (In French) (NSA 26: 48301).

Kirchmann, R., S. Bonotto and G. Cantillon. 1974. Contamination of living organisms in an estuary by the ^{226}Ra present in liquid effluents of industrial origin. pp. 167-183. In: Environmental Surveillance Around Nuclear Installations. Vol. 2. (See IAEA, 1974) (In French) (NSA 31: 14033).

Kirchmann, R., et al. 1971. The transfer of radium-226 from industrial effluents discharged into watercourses in the human food chain. pp. 203-220. In: Proc. Intern. Symp. on Radioecology Applied to the Protection of Man and His Environment. (See Commission of the European Communities, 1971) (In French) (NSA 27: 5274).

Kirchmann, R., et al. 1972. Comparison between the rate of transfer of ^{226}Ra from potable water on hay to cow milk. Compt. rend. soc. biol. 166(11): 1557-1562. (In French) (NSA 29: 2956).

Kirchmann, R., et al. 1973. Study of the biological cycle of radioactive nuclei. Belgium Centre d'Étude de l'Énergie Nucléaire, Brussels, report BLG-477 (Vol.1, 2). 42, 125pp. (In French) (NSA 29: 400, 401).

Kirchmann, R., et al. 1973. Studies on the food chain contamination by tritium. pp. 341-348. In: Tritium. (See Moghissi and Carter, 1973) (NSA 28: 27317).

Kirchmann, R., et al. 1974. Use of biological indicators for environmental monitoring in the vicinity of nuclear facilities. pp. 105-118. In: Environmental Surveillance Around Nuclear Installations. Vol. 2. (See IAEA, 1974) (In French) (NSA 31: 14033).

Kirk, W. P. 1972. ^{85}Kr: a review of the literature and an analysis of radiation hazards. Eastern Environmental Radiation Lab., U. S. EPA report. 60pp. (NP-19251) (NSA 26: 31381).

Kirkegaard, P. 1969. Monte Carlo study of neutrons in soil for surface moisture gauges. Denmark Atomic Energy Commission, Research Establishment, Risö, report RISÖ-M-896. 54pp. (NSA 23: 35534).

Kirkham, D. 1968. Use of nuclear energy in water studies for increasing agricultural productivity. Iowa State Univ., Ames. (For Conf. on the Application of Nuclear Energy to Agricultural Productivity, Santiago, Chile) (CONF-680112-1) (NSA 22: 23131).

Kirkham, D. and J. C. Corey. 1973. Recent progress in the design of radiation equipment and its practical implications. pp. 17-39. In: Soil-Moisture and Irrigation Studies II. (See IAEA, 1973) (NSA 28: 21379).

Kirkham, D., E. Rolston and D. D. Fritton. 1967. Gamma-radiation detection of water content in two dimensional evaporation prevention experiments. pp. 3-16. In: Isotope and Radiation Techniques in Soil Physics and Irrigation Studies. (See IAEA, 1967) (Also: Iowa State Univ., Ames, U. S. AEC report COO-1269-15) (NSA 22: 6796).

Kirkov, P., et al. 1974. Determination of the origin of water in springs from the simultaneous application of natural and artificial isotopes. Some aspects of the origin of water in the Rasche Spring. pp. 465-479. In: Isotope Techniques in Groundwater Hydrology. Vol. 1. (See IAEA, 1974) (NSA 31: 19642).

Kirkpatrick, R. L. and R. A. Wilbur. 1965. The development and habits of the granary weevil Sitophilus granarius within the kernel of wheat. J. Econ. Entomol. 58(5): 979-985. (Radiographs).

Kirkwood, J. B. 1969. Amchitka bioenvironmental program: bienvironmental safety studies, Amchitka, Alaska. Milrow: D + 7 days report. Battelle Columbus Labs., U. S. AEC report BMI-171-122(Rev.). 19pp. (NSA 24: 12211).

Kirkwood, J. B. 1970. Bioenvironmental safety studies, Amchitka Island, al Alaska. Milrow D+2 report. Battelle Columbus Labs., U. S. AEC report BMI-171-126. 34+pp. (NSA 24: 23130).

Kirkwood, J. B. 1971. Bioenvironmental studies at Amchitka Island, Alaska. BioScience 21(12): 602-606. (NSA 25: 49129).

Kirkwood, J. B. 1971. Summary of ecological effects of Milrow. BioScience 21(12): 707-711. (NSA 25: 49148).

Kirkwood, J. B. 1974. Amchitka bioenvironmental program. Bioenvironmental safety studies, Amchitka Island, Alaska. Spring, 1974 Task Force report. Battelle Columbus Labs., U. S. AEC report BMI-171-157. 43pp. (NSA 30: 26822).

Kirkwood, J. B. (Comp.). 1975. Bioenvironmental and hydrologic studies, Amchitka Island, Alaska. Fall, 1974 Task Force report. Battelle Columbus Labs., U. S. AEC report BMI-171-158. 48pp. (NSA 31: 25791).

Kirkwood, J. B. and R. G. Fuller. 1971. Amchitka bioenvironmental program: bioenvironmental-effects predictions for the proposed Cannikin underground nuclear detonation at Amchitka Island, Alaska. Battelle Columbus Labs., U. S. AEC report BMI-171-141. 28pp.

Kirkwood, J. B. and R. G. Fuller. (Comps.). 1972. Bioenvironmental safety studies, Amchitka Island, Alaska. Cannikin D+2 months report. Battelle Columbus Labs., U. S. AEC report BMI-171-147. ix, 62pp. + App. (NSA 26: 56039).

Kirkwood, R. C., M. M. Robertson and J. E. Smith. 1966. Differential absorption as a factor influencing the selective toxicity of MCPA and MCPB. pp. 47-60. In: Isotopes in Weed Research. (See IAEA, 1966).

Kirpichnikov, V. S. 1961. Growth and viability of carp marked with calcium-45 and influence of ^{45}Ca marking on heredity. Izv. gos. nauchn.-issled. inst. ozern. rechn. ryb. khoz. 51: 255-265. (BA 45: 27236; NSA 19: 36024; Ref. zh. (biol.) 1962: 9118; SFA 19: 7278).

Kirpichnikov, V. S., A. N. Svetovidov and A. S. Troshin. 1956. The labelling of carp with the radioactive isotopes of phosphorus and calcium. Dokl. Akad. Nauk SSSR 111(1): 221-224. (In Russian).

Kirpichnikov, V. S., A. N. Svetovidov and A. S. Troshin. 1956. Absorption and elimination of radioactive calcium by Daphnia, Cyclops and Lebistes reticulatus (Peters). Dokl. Akad. Nauk SSSR 110(6): 1122-1125. (In Russian).

Kirpichnikov, V. S., A. N. Svetovidov and A. S. Troshin. 1958. Tagging fish with radioactive isotopes of phosphorus and calcium. pp. 307-321. In: Proc. Sessions of the Ichthyological Commission. Session on Physiology of Fish, 1956. No. 8. Akad. Nauk SSSR. (Fish. Res. Bd. Can. Transl. Ser. No. 313).

Kirschner, H. 1970. The effect of copper on the radiosensitivity of bacteria. Int. J. Radiat. Biol. 17: 81-85. (NSA 24: 34797).

Kisner, R. L. 1961. Meiosis, mitosis, and the induction of chromosomal abnormalities in Hymenolepis minuta by gamma rays. Univ. of Tennessee, Knoxville. (Dissertation Abstr. 22: 3304; NSA 16: 26791).

Kiss, B., M. Liszonyi and J. Kovacs. 1974. Investigation of radioactive contamination in foodstuffs of animal origin. pp. 145-150. In: Radiation Protection, Environment, and the Population. (See German Democratic Republic, 1974) (NSA 31: 8664) (Incl. fish).

Kitihara, K. and T. Akayama. 1963. The radioactivity in rain water, settled dust, and activated sludge observed in the Kagoshima District. 11. Med. J. Kagoshima Univ. 15: 331-339. (In Japanese) (NSA 20: 534).

Kitaoka, S., et al. 1967. Effects of gamma radiation and chemosterilants on the cattle tick, Boophilus microplus. Jap. J. sant. zool. 18(2-3): 126-129.

Kitchin, R. M. 1969. Radiation analysis of a Comstockiella chromosome system (Homoptera: Coccoidea). Thesis, Univ. of California, Berkeley. 130pp. (NSA 24: 48482).

Kitchings, J. T., III and P. B. Dunaway. 1970. Blood changes in irradiated cotton rats and rice rats. Radiation Research 42: 331-352. (NSA 24: 32186).

Kitchings, J. T., III, et al. 1968. Use of radioiron (^{59}Fe) as an index to hemopoietic damage caused by ionizing radiation. J. Tenn. Acad. Sci. 43 (3): 85-87. (7 species of wild rodents).

Kitchings, J. T., III, et al. 1969. Uptake and excretion of ^{134}Cs from fall-out simulant and vegitation by cotton rats. pp. 18-39. In: Progress Report in Post Attack Ecology. (See Auerbach, 1969) (Also: Health Physics 17: 265-277).

Kitchings, T., D. DiGregorio and P. van Voris. 1975. Review of the ecological parameters of radionuclide turnover in vertebrate food chains. Oak Ridge National Lab., U. S. ERDA report. (CONF-750503-16) (NSA 32: 6440).

Kitzmiller, J. B. 1958. X-ray induced mutation in the mosquito, *Culex fatigans*. Exptl. Parasitol. 7(4): 439-462.

Kival, J., M. J. Barrett and L. Battist. 1972. Issues in power plant cost-benefit analysis. Trans. Amer. Nucl. Soc. 15(1): 74. (NSA 26: 44925).

Kivi, E. I. and A. P. James. 1962. The influence of environment on radiation-induced mutations affecting growth. Heredtas 48(2): 247-263. (NSA 17: 17842).

Kiyoku, M. and R. Tsukuda. 1968. Effects of ^{137}Cs on the sterility of *Callobruvhus chinensis* L. 2. Study on the ecology of insects which were sterilized artificially (gamma radiation). Okayama Daigaku Nogabuku Gaku. Hok. (32): 15-23. (In Japanese) (NSA 23: 41527).

Kiyoku, M. and R. Tsukuda. 1969. Studies on the ecology of insects sterilized artificially (gamma radiation. 3. Effects of ^{137}Cs on the sterility of *Spodoptera littoralis* (Boisd.). Nippon Oyo Dobutsu Konch. Zasshi 13(2): 61-69. (In Japanese) (NSA 24: 9983).

Kiyoku, M. and R. Tsukuda. 1970. Studies on the ecology of insects sterilized artificially (gamma radiation). 5. Development and reproduction of the progeny from irradiated pupae, *Spodoptera littoralis* Boisd. by gamma radiation. Okayama Daigaku Nogabuku Gaju. Hok. (36): 19-26. (In Japanese) (NSA 26: 409).

Kiyoku, M., R. Tsukuda and T. Wada. 1969. Studies on the ecology of insects which were strilized artificially (gamma radiation). 4. Sexual behavior of *Spodoptera littoralis* (Boisd.) sterilized as pupae by gamma radiation. Okayama Daigaku Nogabuku Gaku. Hok. (34): 15-21. (In Japanese) (NSA 25: 11338).

Klainer, A. S., S. Gorbach and L. Weinstein. 1967. Studies of intestinal microflora. VII. Effect of the diet and fecal microbial flora on survival of animals exposed to x irradiation. J. Bacteriol. 94: 383-387. (NSA 22: 557).

Klassen, W. and J. F. Creech. 1973. Population suppression with dominant and conditional lethal mutations: some important considerations and approaches. pp. 65-79. In: Computer Models and Application of the Sterile-Male Technique. (See IAEA, 1973) (NSA 29: 8050).

Klassen, W. and J. F. Norland. 1967. Sterilization of boll weevils with combinations of chemosterilants and radiation. Abstr. 358. Bull. Entomol. Soc. Am. 13(3): 203.

Klassen, W., J. F. Norland and R. W. Briggs. 1969. Sterilization of boll weevils with combinations of chemosterilants and x-rays, gamma rays, thermal neutrons, or fast neutrons. J. Econ. Entomol. 62: 1204-1216. (NSA 24: 4707).

Klechkovskiy, V. M. (Ed.). 1956. The Behavior of Radioactive Fission Products in Soils, Their Uptake by Plants and Their Concentration in the Harvest. Publ. House Acad. Sci. USSR, Moscow. UN Document A/AC.82/G/R.41. 176pp. (AEC-tr-2867).

Klechkovskiy, V. M. and I. V. Gulyakin. 1958. Behavior of microquantities of strontium, cesium, ruthenium, and zirconium in soils and in plants. Pochvovedeniye (3): 1-15. (CA 52: 17587e).

Klechkovskiy, V. M. and G. N. Tselishcheva. 1956. Behavior of radioactive fission products in soils. In: The Behavior of Radioactive Fission Products in Soils, Their Uptake by Plants and Their Concentration in the Harvest. (See Klechkovskiy, 1956).

Klechkovskiy, V. M., G. G. Polikarpov and R. M. Aleksakhin. (Eds.). 1971. Contemorary Problems of Radiobilogy. Vol. 2. Radioecology. Atomizdat, Moscow. 422pp. (English transl. John Wiley & Sons, New York, 1973, xii, 371pp.).

Klechkovskiy, V. M., G. G. Polikarpov and R. M. Aleksakhin. 1971. Introduction. pp. 5-12. In: Contemporary Problems of Radiobiology. Vol. 2. Radioecology. (See above).

Klechkovskiy, V. M., A. V. Sokolov and V. M. Zezyulinskiy. (Eds.). 1955. Tagged Atoms in the Study of Plant Nutrition and Use of Fertilizers: Conf. Trans. Publ. House Acad. Sci. USSR, Moscow. (AEC-tr-3376) (NSA 13: 1053).

Klechkovskiy, V. M., L. M. Sokolova and G. N. Tselischeva. 1958. Absorption of trace amounts of strontium and cesium in soils. Proc. 2nd Intern. Conf. on the Peaceful Uses of Atomic Energy 18: 486-493. (Paper A/CONF.15/P/2310) (NSA 13: 6944).

Klein, D. T. and R. M. Klein. 1965. Photoprotection against genetic damage in *Neurospora* caused by x-rays. Photochem. Photobiol. 4: 623-624. (NSA 20: 14358).

Klein, J. R. 1961. Protoporphyrin and heme formation by erythrocytes from the x-irradiated and bled duck. Am. J. Physiol. 201: 663-667. (Also: Brookhaven National Lab., U. S. AEC report BNL-5527) (NSA 15: 32026).

Klein, R. M. and D. T. Klein. 1971. Post-irradiation modulation of ionizing radiation damage to plants. Botan. Rev. 37(4): 397-436. (NSA 26: 23050).

Klein, W. L. and R. A. Brechbill. 1972. Preliminary radiation surveillance of an aquatic system near the Nevada Test Site June-July, 1967. Western Environmental Research Lab., U. S. EPA, U. S. AEC report SWRHL-65r. iii, 23pp.

Kleinbergs-Krisans, S. and R. H. Catlett. 1975. Radiosensitivity of the arenicolous lizard, Uma notata. Radiation Research 61(1): 110-117. (NSA 31: 11649).

Klement, A. W., Jr. 1959. A review of potential radionuclides produced in weapons detonation. U. S. AEC report WASH-1024. iii, 102pp. (NSA 14: 1341).

Klement, A. W., Jr. 1960. Radioactive fallout data 1959, collected from ten AEC installations. U. S. AEC report TID-11851. iii, 110pp.

Klement, A. W., Jr. 1960. Radioactive fallout from nuclear detonations of February and April 1960. U. S. AEC report TID-6235. 13pp. (French tests).

Klement, A. W., Jr. (Ed.). 1962. Radioactive Fallout from Nuclear Weapons Tests. (Proc. Conf., Germantown, Md. November 15-17, 1961). U. S. AEC report TID-7632. 2 books, viii, 541pp. (NSA 16: 22433).

Klement, A. W., Jr. 1963. Sources of radioactivity in foods. pp. 4-8. In: Proc. North Central Experiment Station Workshop on Radionuclides in Foods and Agricultural Products. (Cincinnati, O., February 19-21, 1963). F. Haghiri and R. R. Johnson, eds. Ohio Agricultural Experiment Station, Wooster, Special Report No. 1.

Klement, A. W., Jr. 1963. Current research on radioactive contamination of soils and plants. pp. 59-63. In: Proc. North Central Experiment Station Workshop on Radionuclides in Foods and Agricultural Products. (Cincinnati, O., February 19-21, 1963). F. Haghiri and R. R. Johnson, eds. Ohio Agricultural Experiment Station, Wooster, Special Report No. 1.

Klement, A. W., Jr. 1965. Natural radionuclides in foods and food source materials. pp. 113-155. In: Fallout - Soils, Plants, Foods, Man. E. B. Fowler, ed. Elsevier Press, Amsterdam. (Incl. wild organisms).

Klement, A. W., Jr. 1965. Radioactive fallout and mechanisms. pp. 1267-1274. In: Radiation and Terrestrial Ecosystems. (See Hungate, 1965).

Klement, A. W., Jr. (Ed.). 1965. Radioactive Fallout from Nuclear Weapons Tests. (Proc. 2nd Conf., Germantown, Md., November 3-7, 1964). U. S. AEC Symp. Ser. No. 5, report CONF-765. xi, 953pp.

Klement, A. W., Jr. 1969. Radiological safety research for nuclear excavation projects. pp. 652-653. In: Proc. Symp. on the Public Health Aspects of Peaceful Uses of Nuclear Explosives. (Held in Las Vegas, Nev., April 7-11, 1969). Southwestern Radiological Health Lab., U. S. PHS, Las Vegas, Nev., report SWRHL-82.

Klement, A. W., Jr. (Ed.). 1970. Nuclear Cratering Explosion Effects for Interoceanic Canal Feasibility Studies. Nevada Operations Office, U. S. AEC, Las Vegas, report NVO-67. viii, 154pp.

Klement, A. W., Jr., et al. 1972. Estimates of Ionizing Radiation Doses in the United States 1960-2000. Special Studies Group, Criteria and Standards Div., Off. of Radiation Programs, U. S. EPA, Washington, report ORP/CSD 72-1. ix, 171pp. (An interagency study; U. S. DOD, AEC, HEW, EPA; begun as a U. S. Federal Radiation Council study in 1970).

Klement, A. W., Jr. 1978. Some data on natural radiation. App. 3.7, pp. 611-626. In: Handbook of Radiation Measurement and Protection. A. Brodsky, ed. CRC Press, Boca Raton, Fla. (The volume contains many articles of interest to ecologists as do volumes in press to be available in 1980, especially the volume: Protection of the Environment; and Health Physics Methods in Industrial, Academic, and Research Facilities).

Klemm, M. and W. Klemm. 1964. The application of radioisotopes for continuing studies of the progress of rapid transpiration by trees. Flora 154: 89-93. (In German).

Klemm, W. 1959. The use of radioactive isotopes in forestry. Archiv Forstwesen 8: 726-742. (In German).

Klemm, W. 1966. The elaboration of the transmission method for studying the water mechanisms in trees. Isotopenpraxis 2: 262-267. (In German) (NSA 20: 45267).

Klimov, A. G. nd. Radioecology problems of nuclear power reactor development. International Atomic Energy Agency, Vienna. (In Russian) (INIS-mf-887, pp. 14-22) (NSA 29: 17677).

Klimovskaya, L. D. 1962. Effect of ionizing radiation upon the function of ciliated epithelium of frog esophagus. Radiobiologiya 2(2): 222-227. (AEC-tr-5429) (NSA 16: 28809).

Klimpinya, A. Ye. 1965. The effect of gamma-radiation on Pegomyia hyposcyami. Zashch. rast. vredit. bolez. 1965: 41-43. (In Russian).

Klimpinya, A. Ye. 1966. Use of gamma rays to reduce numbers of beet flies. Latv. PSR Kinat. Akad. Vest. (11): 144-146. (In Russian).

Klimpinya, A. Ye. 1967. The effect of gamma radiation om Pegomyia hyoscyami. Entomophaga 12: 1294.

Klimpinya, A. Ye. 1968. Effect of gamma rays on the mangold fly. Zashch. rasteniy 12(1): 20-21. (In Russian) (NSA 24: 32317).

Kline, A. B., Jr. and M. S. Terpilak. (Eds.). 1972. The environmental and ecological forum 1970-71. U. S. AEC report TID-25857. vii, 186pp.

Kline, J. R. 1964. An investigation of copper and other selected elements in soils utilizing gamma radiations induced by thermal neutron activation. Thesis, Univ. of Minnesota, Minneapolis. 199pp. (NSA 20: 7041).

Kline, J. R. 1966. Radionuclide behavior and distribution in tropical forests. Bioenvironmental and radiological-safety feasibility studies, Atlantic-Pacific Interoceanic Canal. Puerto Rico Nuclear Center, Battelle Columbus Labs., U. S. AEC report BMI-171-001. 20pp. (NSA 21: 41010).

Kline, J. R. 1970. Retention of fallout radionuclides by tropical forest vegetation. In: Ch. H-17, pp. H191-H197. In: A Tropical Rain Forest (See H. T. Odum and R. F. Pigeon, 1970) (NSA 25: 21550).

Kline, J. R. 1973. Mathematical simulation of soil-plant relationships and soil genesis. Soil Science 115(3): 240-249. (Incl. ^{3}H, ^{90}Sr).

Kline, J. R. and S. S. Brar. 1969. Instrumental analysis of neutron irradiated soils. Soil Sci. Soc. Am. Proc. 33(2): 234-238.

Kline, J. R. and C. F. Jordan. 1970. Tritium movement in soil of a tropical rain forest. Ch. H-8, pp. H129-H131. In: A Tropical Rain Forest. (See H. T. Odum and R. F. Pigeon, 1970) (NSA 25: 21637).

Kline, J. R. and N. Mercado. 1970. Preliminary studies of radionuclide cycling in understory plants in the rain forest. Ch. H-20, pp. H205-H209. In: A Tropical Rain Forest. (See H. T. Odum and R. F. Pigeon, 1970) (NSA 25: 21552).

Kline, J. R. and H. T. Odum. 1970. Comparisons of amounts of fallout radionuclides in tropical forests. Ch. H-15, pp. H181-H186. In: A Tropical Rain Forest. (See H. T. Odum and R. F. Pigeon, 1970) (NSA 25: 21549).

Kline, J. R. and M. L. Stewart. 1974. Tritium uptake and loss in grass vegetation which has been exposed to an atmospheric source of tritiated water. Health Physics 26(6): 567-575. (NSA 30: 10054).

Kline, J. R., F. A. Colon and S. S. Brar. 1973. Distribution of ^{137}Cs in soils and vegetation on the island of Purto Rico. Health Physics 24(5): 469-475. (NSA 28: 2589).

Kline, J. R., J. E. Foss and S. S. Brar. 1969. Lanthanum and scandium distribution in three glacial soils of western Wisconsin. Soil Sci. Soc. Am. Proc. 33: 287-291. (NSA 23: 28749) (Activation analysis).

Kline, J. R., C. F. Jordan and P. Kovac. 1972. Use of tritiated water for determination of plant transpiration and biomass under field conditions. pp. 419-437. In: Isotopes and Radiation in Soil-Plant Relationships Including Forestry. (See IAEA, 1972).

Kline, J. R., C. F. Jordan and R. C. Rose. 1972. Transpiration measurement in pines using tritiated water as a tracer. Isotopes Radiat. Techn. 9(3): 348-350.

Kline, J. R., C. F. Jordan and R. C. Rose. 1973. Transpiration measurements in pines using tritiated water as a tracer. pp. 190-198. In: Radionuclides in Ecosystems. (See Nelson, 1973) (NSA 25: 41688).

Kline, J. R., H. T. Odum and J. C. Bugher. 1970. Effect of gamma radiation on leaching of ^{137}Cs and ^{54}Mn from tropical forest foliage and litter. Ch. H-16, pp. H187-189. In: A Tropical Rain Forest. (See H. T. Odum and R. F. Pigeon, 1970) (NSA 25: 21638).

Kline, J. R., M. L. Stewart and C. F. Jordan. 1972. Estimation of biomass and transpiration in coniferous forests using tritiated water. pp. 159 166. In: Proc. - Research on Coniferous Forest Ecosystems - a Symp. J. F. Franklin, L. J. Dempster and R. H. Waring, eds. Pacific Northwest Forest and Range Expt. Sta., U. S. Forest Service, Portland, Ore.

Kline, J. R., et al. 1970. Measurement of transpiration in tropical trees with tritiated water. Ecology 51: 1068-1071. (NSA 25: 15997).

Klingmuller, W. 1961. Modification of radiation damage in seeds by post treatment. Nature 189(4762): 413-414. (NSA 15: 8556).

Kljajic, R. 1960. Determination of lethal doses for some phytopathogenic fungi radiated with Co-60 gamma rays. Arh. za Poljoprivredne Nauke 13 (39); 96-103. (In Serbo-Croatian).

Klobe, W. D. 1972. Tritiated water movement in semiconfined growth systems. J. Environ. Qual. 1(2): 189-192.

Kloft, W. 1956/57. Studies on plant-sucking insects and the reactions of the plant host-tissue. Ber. Phys.-Med. Ges. Wurzburg 68: 64-72. (In German).

Kloft, W. 1960. Exchange reactions between plant-sucking insects and host-tissues of the plant. Parts I and II. Z. angew. entomol. 45(4): 337-381; 46(1): 42-70. (In German).

Kloft, W. 1962. Technical problems of radioisotope measurement in insect metabolism. pp. 163-172. In: Radioisotopes and Radiation in Entomology. (See IAEA, 1962).

Kloft, W. 1966. Insects annihilate themselves. Naturwiss. Medizin 3(13): 29-37. (In German).

Kloft, W. 1967. Radioecological investigation of forest Formicidae. Wiss. Z. Tech. Hochsch. Dresden 16(2): 582-583. (In German).

Kloft, W. and P. Ehrhardt. 1959. Studies on the sucking activity and damaging effects of the Sitka spruce louse Liosomaphis abietina (Walk) (Neomyzaphis abietina Walk.). Phytopathol. Z. 35(4): 401-404. (In German).

Kloft, W. and P. Ehrhardt. 1959. The problem of saliva injection during the suction process of Thrips tabaci Lind. (Thysanoptera, Terebrantia). Naturwissenschaften 46(16): 586-587. (In German).

Kloft, W. and P. Ehrhardt. 1962. Studies on the assimilation and excretion of labelled phosphate in aphids. pp. 181-190. In: Radioisotopes and Radiation in Entomology. (See IAEA, 1962).

Kloft, W. and K. Gosswald. 1964. Studies by means of the tracer method of the significance of wood-destroying carpenter ants to forestry. Anz. Schädlingst. 37(11): 163-169. (In German) (^{131}I).

Kloft, W. and B. Hölldobler. 1964. The wood-destroying carpenter ant and its importance in forestry, studied by tracer technique. Anz. Schädlingst. 37: 163-169. (In German, English summary) (^{131}I, Camponotus) (Error in cited pp. ?; see above entry).

Kloft, W. and H. Kunkel. 1962. Aspects of saliva release by Myzus ascalonicus Donc., elucidated by means of the tracer technique. pp. 746-749. In: Proc. 11th Intern. Congr. on Entomology. Vol. 1. (Vienna, 17-25 Aug. 1960). H. Strouhal and M. Beier, eds. Christoph Reisser's Söhne, Vienna. (In German).

Kloft, W. and E. F. Schulze. 1967. Use of tritiated water (THO) for continuous registration of insect transpiration. Atompraxis 13: 265-268. (In German) (NSA 21: 45555).

Kloft, W., P. Ehrhardt and H. Kunkel. 1968. Radioisotopes in the investigation of interrelationships between aphids and host-plants. pp. 23-30. In: Isotopes and Radiation in Entomology. (See IAEA, 1968) (NSA 24: 31827).

Kloft, W., B. Hölldobler and A. Haisch. 1965. Tracer studies through separation of actual wood destruction by carpenter ants (Camponotus herculeanus L. and C. ligniperda Latr.). Entomologia exp. appl. 8(1): 20-26. (In German, English summary) (^{131}I).

Kloft, W., J. E. Lloyd and A. P. Bhatkar. 1975. Demonstration of reflex-breeding in fireflies (Coleoptera) by help of radiotracers. Experient-31(4): 450-451. (In German) (NSA 32: 1247).

Kloke, A. and K. Riebartsch. 1966. Weed seed labelling with rare earth elements. pp. 165-172. In: Isotopes in Weed Research. (See IAEA, 1966)

Klotz, D. and H. Moser. 1969. Using radioactive isotopes in hydrology. VII. Measuring small filter velocities by the point dilution method. Atomkernenergie 14: 423-430. (In German) (NSA 2307).

Klun, J. A. and T. A. Brindley. 1966. Role of 6-methoxybenzoxazolinone in inbred resistance of host plant (maize) to first-brood larvae of European corn borer. J. Econ. Entomol. 59: 711-718. (^{14}C).

Knälmann, M. 1970. Effect of concentration supplied on the iodide sorption by various soils. Atomkernenergie 16: 247-252. (In German) (NSA 25: 10694, 15772) (^{131}I).

Knälmann, M. 1972. Dependence of iodide sorption on various soil components. Atomkernenergie 19(4): 329-332. (In German) (NSA 26: 45528) (^{131}I).

Knapp, S. E., et al. 1956. A method for labeling the lone star tick with a radioactive indicator (P^{32}). J. Econ. Entomol. 49(3): 393-395.

Knauss, H. J. and J. W. Porter. 1954. The absorption of inorganic ions by by Chlorella pyrenoidosa. Plant Physiol. 29(3): 229-234. (Also: Hanford Atomic Products Operation, U. S. AEC report HW-29993) (NSA 8: 1300).

Knauss, R. W. and W. H. Thomas. 1954. The growth and inorganic nutrition of Scendesmus obliquis in mass culture. Plant Physiol. 29(3): 205-214.

Kneebone, L. R. 1959. High yielding mushroom from irradiated spores. Sci. Farmer (Penn. Station) 6(4): 14.

Kneip, T. J. and G. J. Lauer. 1973. Trace metal concentration factors in aquatic ecosystems. pp. 43-62. In: Chemical Analysis of the Environment and Other Modern Techniques. S. Ahuja, ed. Plenum, New York. (NSA 28: 27321).

Kneitz, G. 1963. Tracer studies of food distribution in forest ants. pp. 38-50. In: Symp. on Genetics and Biology in Italy. (Proc. 4th Congr., Pavia, 9-14 September 1961). Tipografia del Libro S. A. S., Pavia. Vol. XII. (In German).

Knievel, D. P. and D. A. Schmer. 1971. Preliminary results of growth characteristics of buffalograss, blue gramma, and western wheatgrass, and methodology for translocation studies using ^{14}C as a tracer. U. S. International Biological Program, Grassland Biome, Colorado State Univ., Fort Collins, Tech. Rept. 86. 28pp.

Knight, A. 1961. Translocation of radiophosphorus through an aquatic ecosystem. Thesis, Michigan State Univ., East Lansing.

Knight, A., R. C. Ball and F. F. Hooper. 1962. Some estimates of primary production rates in Michigan ponds. Mich. Acad. Sci. Arts Lett. 47: 219-233.

Knight, A. H., R. Boggie and H. Shepherd. 1972. The effect of ground water level on water movement in peat: a study using tritiated water. J. Appl. Ecol. 9(2): 633-641. (Also ^{131}I).

Knight, F. B. and W. Albertin. 1965. Portable x-ray equipment in forestry research. J. Forestry 63(7): 543-544.

Knipling, E. F. 1955. The possibilities of insect control or eradication through use of sexually strilized males. J. Econ. Entomol. 48(4): 459-462.

Knipling, E. F. 1957. Control of screw-worm fly by atomic radiation. Scientific Monthly 85(4): 195-202.

Knipling, E. F. 1959. Sterile-male method of populations control. Science 130(3380): 902-904.

Knipling, E. F. 1959. Screwworm eradication. Concepts and research leading to the sterile-male method. Smithsonian Inst. Ann. Report 1957/58: 409-418.

Knipling, E. F. 1960. Control of screwworm fly by atomic radiation. pp. 169-182. In: Am. Assoc. Adv. Sci., Washington, Publ. 61.

Knipling, E. F. 1960. Use of insects for their own destruction. J. Econ. Entomol. 53(3): 415-420.

Knipling, E. F. 1960. The eradication of the screw-worm fly. Scient. Amer. 203(4): 54-61.

Knipling, E. F. 1961. The use and limitation of isotopes and radiation sterility in meeting insect problems. p. 16. In: 10th Pacific Science Congr., Honolulu, Hawaii.

Knipling, E. F. 1961. Control of insects with radiation. pp. 152-167. In: Proc. Southeastern Seminar on Atomic Progress in Agriculture. (See Clemson College, 1961).

Knipling, E. F. 1962. The use and limitations of isotopes and radiation sterility in meeting insect problems. Int. J. Appl. Radiat. Isotopes 13 (7): 417-426. (NSA 17: 15559).

Knipling, E. F. 1963. The autocidal method in pest control. Umschau 63(20): 632-636. (In German).

Knipling, E. F. 1964. The potential role of the sterility method for insect population control with special reference to combining this method with conventional methods. Agricultural Research Service, U. S. Department of Agriculture, report ARS-33-98. 54pp.

Knipling, E. F. 1965. The sterility principle of insect population control. pp. 251-252. In: Proc. 12th Intern. Congr. of Entomology. P. K. Freeman, ed. Royal Entomological Soc., London.

Knipling, E. F. 1965. Opportunities for developing alternate ways to control activities. Am. J. Publ. Hlth. 55(7 Suppl.): 20-26. (A Survey of Pesticide Problems and Control Activities) (Discusses irradiation).

Knipling, E. F. 1965. The sterility method of pest population control. pp. 233-249. In: Research in Pesticides. C. O. Chichester, ed. Academic Press, New York.

Knipling, E. F. 1966. Knipling on pest control. Agric. Chem. 21(4): 34-36.

Knipling, E. F. 1966. Some basic principles in insect population suppression. Bull. Entomol. Soc. Am. 12(1): 7-15.

Knipling, E. F. 1967. Sterile technique - principles involved, current application, limitations, and future application. pp. 587-616. In: Genetics of Insect Vectors of Disease. J. W. Wright and R. Pal, eds. Elsevier, Amsterdam.

Knipling, E. F. 1969. Concept and value of eradication or continuous suppression of insect populations. pp. 19-32. In: Sterile-Male Technique for Eradication or Control of Harmful Insects. (See IAEA, 1969) (NSA 24: 32295).

Knipling, E. F. 1970. Suppression of pest Lepidoptera by releasing partially sterile males. BioScience 20(8): 465-470.

Knizhnikov, V. A., et al. 1965. Dietary intake of the products of nuclear tests in the Soviet Union in 1963-64. State Committee on the Use of Atomic Energy USSR, Moscow. UN Document A/AC.82/G/L.1083. 22pp., (AEC-tr-6797).

Knizhnikov, V. A., et al. 1967. Intake of strontium-90 and cesium-137 with food by the population of the Soviet Union in 1965-1966 as a result of stratospheric fallout. State Committee on the Use of Atomic Energy USSR. Atomizdat, Moscow. pp. 3-12. UN Document A/AC.82/G/L.1186. (In Russian) (NSA 22: 33802).

Knizhnikov, V. A., et al. 1969. Dietary intake of strontium-90 by the population of the USSR over the period 1959-1967 as a result of stratospheric fallout. State Committee on the Use of Atomic Energy USSR. Atomizdat, Moscow. UN Document A/AC.82/G/L.1328. (AEC-tr-7232, pp. 45-56) (NSA 25: 39853).

Knizley, J. B., G. B. Morgan and J. B. Lackey. 1962. Comparative uptake of radioisotopes by various microorganisms. Abstr. Bull. Ecol. Soc. Am. 43 (4): 121-122.

Knobf, V. I. 1951. Preparation of samples and correction of data. Oak Ridge National Lab., U. S. AEC report ORNL-1048. 26pp.

Knobf, V. I. 1951. Studies on radioactivity in fish from White Oak Lake and the Clinch River. Oak Ridge National Lab., U. S. AEC reoprt ORNL-1031. 38pp. (NSA 5: 4942).

Knoerr, K. R. 1959. Summer evapotranspiration measurements using a radioactive soil moisture probe. J. Geophys. Res. 64(6): 691.

Knoll, K. C. 1969. Effect of detergents on adsorption of radionuclides by soil. Battelle Northwest Lab., U. S. AEC report BNWL-859. 6pp. (NSA 23: 13852).

Knoop, H.-U. 1971. Investigation of the primary reactions in the irradiation of single-celled algae with thermal neutrons through substitution of ^{14}N and ^{10}B. Germany Kernforschungsanlage Institut für Botanik u. Mikrobiol., Jülich, report JÜL-727-BO. 96pp. (In German) (NSA 25: 52255).

Knoop, H.-U. 1972. Biological effectiveness of the primary reactions of thermal neutrons. 1. Damage rates and effectiveness of the capture reactions of thermal neutrons in Chlorella pyrenoidosa. Radiation Botany 12(4): 261-269. (In German) (NSA 26: 59015).

Knoop, H.-U. and M. Paschke. 1973. RBE of the $^{10}B(n,\alpha)Li$ capture reaction of thermal neutrons inside the green alga Chlorella pyrenoidosa. Germany Kernforschungsanlage, Jülich. (Also: 1974. pp. 215-219. In: Biological Effects of Neutron Irradiation. International Atomic Energy Agency, Vienna, Publ. STI/PUB-352. (NSA 29: 15875; 30: 3528).

Knowles, F. E., Jr. and E. J. Baratta. 1971. Determination of accuracy and precision limits for tritium in water. Radiol. Hlth. Data 12(8): 405-409. (NSA 26: 31).

Knox, J. B. 1971. Airborne radiation from the nuclear power industry. Nuclear News 14(2): 27-32. (NSA 25: 21205).

Knox, J. B. and T. V. Crawford. 1974. Atmospheric release advisory capability. Research and progress. Lawrence Livermore Lab., U. S. AEC report UCRL-75633(Rev.1). 33pp. (CONF-740406-13) (NSA 30: 9531).

Knox, J. B., et al. 1971. Comparison of U.S. and U.S.S.R. methods of calculating transport, diffusion, and deposition of radioactivity. Lawrence Livermore Lab., U. S. AEC report UCRL-51054. 52pp. (NSA 25: 54782).

Knutson, C. F. 1973. Project Rio Blanco: evaluation of possible radioactivity transport in ground water. CER Geonuclear Corp., U. S. AEC report PNE-RB-46. 105pp. (NSA 28: 21036).

Knutson, C. F. 1973. Rio Blanco. Hydrologic testing through January 30, 1973. CER Geonuclear Corp., U. S. AEC report PNE-RB-39. 106pp. (NSA 28: 27536).

Knutson, C. F. 1973. Project Rio Blanco: results from the initial alluvial water quality program. CER Geonuclear Corp., U. S. AEC report PNE-RB-59. 39pp. (NSA 30: 29312).

Knutson, C. F. 1974. Project Rio Blanco: comparative postdetonation hydrologic testing. CER Geonuclear Corp., U. S. AEC report PNE-RB-60. 53pp (NSA 30: 29313).

Knutson, G. nd. Studies of ground-water flow in various aquifers using tracers. Lund Univ., Sweden. (INIS-mf-272) (NSA 26: 50619) (^{51}Cr-EDTA, HTO, NH_4Br^{82}).

Knutsson, G., K. Ljunggren and H. G. Forsberg. 1963. Field and laboratory tests of chromium-51-EDTA and tritium water as a double tracer for ground-water flow. pp. 347-363. In: Radioisotopes in Hydrology. (See IAEA, 1963) (NSA 18: 1927).

Kobal, I. and K. Kristan. 1972. Determination of ^{222}Rn concentration in hot spring waters using a scintillation chamber. Radiochem. Radioanal. Lett. 10(5): 291-295. (NSA 26: 42726).

Kobal, I. and J. Kristan. 1973. Extension of Raghavayya's method for the determination of radon in water. Mikrochim. Acta (2): 219-224. (NSA 28: 2228).

Kobal, I., et al. 1974. Sorption-emanation method for soluble radium in waters. Health Physics 27(4): 381-384. (NSA 31: 5597).

Kobayashi, H. 1964. Dew-point hygrometry below 0°C by the use of a radioactive source. Sci. Papers Inst. Phys. Chem. Res. (Tokyo) 58(3): 106-110.

Kobayashi, H., A. Gorbman and A. W. Wolfson. 1960. Thyroidal utilization of radioiodine in the white throated sparrow and weaver finch. Endochrinology 67: 153-161.

Kobayashi, J. and T. Yamamoto. 1971. Effects of irradiation on spermatogenesis in the guppy, Lebistes reticulatus. Tokushima J. Exptl. Med. 18: 21-27. (NSA 27: 534).

Kobayashi, K. and H. Akitake. 1975. Studies on the metabolism of chlorophenols in fish. 1. Absorption and excretion of PCP by goldfish. 2. Turnover of absorbed PCP in goldfish. Nippon Suisan Gakkaishi 41(1): 87-99. (NSA 33: 3338, 3339) (^{14}C).

Kobayashi, K. and T. Tomiyama. 1959. Studies on absorption and metabolism of P^{32} in aquatic organisms. II. Bull. Jap. Soc. Fish. 25: 576-580.

Kobayashi, M. 1963. The chemistry and physiology of the brain hormone. pp. 226-233. In: Proc. 16th Intern. Congr. of Zoology. Vol. 4. J. A. Moore, ed. Washington. (^{14}C, Cephus cinctus).

Kobayashi, M. 1972. Problems in tracer applications. Proc. Jap. Conf. on Radioisotopes (10): 264-266. (NSA 28: 2399) (Incl. environmental samples).

Kobayashi, M. 1973. Techniques of pollution control utilizing radiation. Possibility of purifying sewage with radiation energy. Kogyo Rea Metaru (51): 53-57. (In Japanese) (NSA 30: 12337).

Kobayashi, M. and A. Nukazawa. 1963. Measurement of river dilution and flow by means of radioactive tracers and activation analysis. pp. 433-449. In: Radioisotopes in Hydrology. (See IAEA, 1963) (NSA 18: 1933).

Kobayashi, S. and H. Hirata. 1957. Effects of x-irradiation upon rainbow trout (Salmo irideus). I. Influence on the feeding activity in rainbow trout fry. Bull. Fac. Fish. Hokkaido Univ. 9(2): 89-94.

Kobayashi, S. and M. Mogami. 1958. Effects of x-irradiation upon rainbow trout (Salmo irideus). III. Ovary growth in the stages of fry and fingerling. Bull. Fac. Fish. Hokkaido Univ. 9: 89-94. (pp. ?).

Kobayashi, T., K. Matsumoto and G. Yoshii. 1972. Effects of post-irradiation incubation on the fibroin synthetic activity of silkgland fibosomes. J. Radiat. Res. (Tokyo) 13(4): 214-220. (NSA 29: 10469).

Kobayashi, T., et al. 1974. Windowless Si radiation detector and biomedical telemetry system for studying the function of organs of free-ranging animals. Trans. Nucl. Sci. NS-21(1): 735-740. (NSA 30: 6675).

Kobayashi, U. 1969. Distribution and transfer mechanism of the radioactive materials in snow layers. Niigata Daigaku Kogak. Kenkyu Hokoku (18): 1-14. (NSA 23: 48041).

Kobelak, A., et al. 1967. Beta radioactivity of woodrings of spruce in lower Silesia area. Nukleonika 11: 511-515. (In Polish) (NSA 21: 8736).

Koch, A., D. Klotz and H. Moser. 1967. The use of radioactive isotopes in hydrology. VI. The influence of the filter tube on the determination of filtration velocity by the point dilution meyhod. Atomkernenergie 12: 361-370. (In German) (NSA 21: 41006).

Koch, A. L. 1966. On the difference between the lethal effects of ^{3}H and ^{32}P in bacteria. Radiation Research 19: 18-32. (NSA 20: 45421).

Kocharov, G. Ye. 1970. Variations in radioactive carbon content in earth atmosphere. Atom. energiya 28(5): 444-446. (In Russian).

Kochenov, A. V. and G. N. Baturin. 1967. The distribution of uranium in the bottom sediments of the Aral Sea. Okeanologiya 7: 623-627. (Oceanology (USSR) 7: 484-487) (NSA 22: 25548; 23: 19859).

Kochenov, A. V., V. T. Dubinchuk and E. V. Germogenova. 1973. Uranium occurrence in phosphatic remains of fossil fish. Sov. Geol. (3): 69-77. (In Russian) (NSA 29: 2438).

Kochenov, A. V., et al. 1965. Uranium and organic matter in sediments of the Black and Mediterranean seas. Geokhimiya (3): 302-313. (English abstr. Geochemistry (2): 212; full transl. Am. Geol. Inst. GI 65-3-6) (NSA 18: 22585).

Koczy, F. F. 1963. The natural radioactive series in organic material. pp. 611-613. In: Radioecology. (See Schultz and Klement, 1963) (NSA 17: 33613).

Kodaira, K., M. Kato and M. Ishikawa. 1973. Radiochemical analysis of ^{210}Pb, as well as distribution of ^{90}Sr and ^{137}Cs in soil profiles. 1. Radioisotopes (Tokyo) 22(7): 331-337. (NSA 29: 9852; 31: 5984).

Kodrich, W. R. and C. A. Tryon. 1973. Effect of season of thyroid release of iodine-131 in free-ranging eastern chipmunks (Tamias striatus). pp. 260-264. In: Radionuclides in Ecosystems. (See Nelson, 1973) (NSA 26: 19889).

Koehn, R. K. and B. B. Knowles. 1967. Radiation levels in the barrel cactus, Ferocactus wislizeni, and their relation to some aspects of cacti ecology. Southwest Nat. 12: 119-126. (NSA 22: 40688).

Koga, A. and K. Kikkawa. 1970. Tritium content of Beppu Hot Spring water. Oita-ken Onsen Chosa Kenk. Hokoku (21): 1-4. (In Japanese) (NSA 25: 34732).

Koga, T, S. Tsukada and Y. Honda. nd. Behavior and distribution of radioactive ruthenium in the freshwater environment. Role of food chains. Kinki Daigaku Gensh. Kenk. Nenpo 10: 5-16. (In Japanese) (NSA 31: 8668).

Kogai, Ye. S., M. K. Kadyrova and Z. P. Shcherban. 1966. Biological effects of ionizing ^{60}Co radiation on the preimaginal stages of mosquito Culex pipiens pipiens Linn. Dokl. Akad. Nauk Uzbek. SSR 8: 44-46. (In Russian).

Kogan, R. M. and Sh. D. Fridman. 1967. Effect of soil moisture on the structure of the gamma field in the atmosphere near the earth. Izv. Akad. Nauk SSSR (fiz. zemli) (3): 82-85. (In Russian) (NSA 21: 25977).

Kogan, R. M., I. M. Nazarov and Sh. D. Fridman. 1969. Gamma Spectrometry of Natural Environments and Formations. (TT-70-50092; xiv, 337pp.).

Kogan, R. M., et al. 1965. The determination of water supplies in snow deposits by aerial gamma exposure. Meteorologiya gidrol. (4): 51-55. (In Russian) (NSA 20: 549).

Kogotkov, A. Ya. and Ye. I. Belova. 1972. Natural leaching of ^{90}Sr from soils. Gigiyena sanitariya (5): 111-113. (In Russian) (NSA 26: 56114).

Kogotkov, A. Ya. and R. F. Popova. 1962. Sorption of long-lived fission products by soils and argillaceous minerals. III. Selectivity of soils and clays toward Sr^{90} under various conditions. Radiokhimiya 4(3): 328-334. (Soviet Radiochemistry 4(3): 292-296).

Kogure, M. and M. Nakajima. 1956. The effect of P^{32} on the silkworm, administered to the body with mulberry leaves. J. sericul. sci. (Tokyo) 25: 238.

Kogure, M., et al. 1951. Studies on P^{32} in the silkworm. Dobstsugaku zasshi 60: 24. (In Japanese).

Kogure, S. and M. Nakashima. 1958. Stages of developmental variations in radiation susceptibility of the silkworm seminal gland - especially low hatching rate and its histocytological and biochemical proof. In: Proc. 2nd Japan Conf. on Radioisotopes. (Paper B-1) (See Japan Radioisotope Association, 1958) (AEC-tr-4482, pp. 1295-1312).

Koh, W. Y., C. I. Morehouse and V. L. Chandler. 1956. Relative resistance of microorganisms to cathode rays: I. Nonsporeforming bacteria. Appl. Microbiol. 4: 143-146.

Kohman, L. and N. Saito. 1954. Radioactivity in geology and cosmology. pp. 401-402. In: Annual Review of Nuclear Science. Vol. 4. J. G. Beckerley, M. D. Kamen and L. I. Schiff, eds. Annual Reviews, Inc., Stanford, Calif.

Kohn, H. I., G. C. McLeod and K. A. Wright. 1967. Inhibition of chlorophyll A and B synthesis in Chlamydomonas reinhardi mutant strain y-2 by ionizing radiation. Radiation Botany 7: 123-128. (NSA 21: 22280).

Koide, M., K. W. Bruland and E. D. Goldberg. 1973. $^{228}Th/^{232}Th$ and ^{210}Pb geochronologies in marine and lake sediments. Geochim. Cosmochim. Acta 37(5): 1171-1187. (NSA 28: 2605).

Koide, M., J. J. Griffin and E. D. Goldberg. 1975. Records of plutonium fallout in marine and terrestrial samples. J. Geophys. Res. 80(30): 4153-4162. (NSA 33: 14847).

Koidsumi, K. 1930. Quantitative studies on the lethal action of x-rays upon certain insects. J. Soc. Trop. Agr. 2: 243-262.

Koiwai, S. 1957. Meiotic abnormalities induced by the exposure to radium in grasshopper (Podisma sapporense) spermatocytes. Jap. J. Genet. 32(5): 165-167.

Kok, B. and L. H. Bongers. 1961. Radiation tolerances in photosynthesis and consequences of excesses. pp. 299-322. In: Medical and Biological Aspects of the Energies of Space. Columbia Univ. Press, New York. (NSA 16: 20238) (Algae).

Kokke, R. 1972. Radioisotopes applied to environmental toxicity research with microbes. pp. 15-23. In: Radiotracer Studies of Chemical Residues in Food and Agriculture. (See IAEA, 1972) (NSA 28: 20185).

Kokke, R., J. T. van Zuilekom and T. O. Wiken. 1969. Detection and isolation of radionuclide-accumulating bacteria by autoradiography. Antonie van Leeuwenhoek J. Microbiol. Serol. 35: 121-128. (NSA 24: 559).

Kokke, R., et al. 1970. Accumulation of radionuclides by bacteria and yeasts. pp. 637-650. In: Proc. Intern. Symp. on Radioecology. Vol. I. (See France Commissariat à l'Énergie Atomique, 1970).

Kolaczkowski, A. M. and W. A. Jester. 1972. Activation analysis of heavy metals in surface waters using ion exchange filter paper and cyanide compplexing. Trans. Amer. Nucl. Soc. 15(2): 643-644. (NSA 27: 4934).

Kolb, W. 1971. ^{181}W and other short lived fission products in ground-level air in north Germany and north Norway. Nature 232(5312): 552-553. (NSA 25: 54785).

Kolb, W. and U. Lauterbach. 1971. Dose rate meter for the measurement of environmental radiation. pp. 565-572. In: Rapid Methods for Measuring Radioactivity in the Environment. (See IAEA, 1971) (In German) (NSA 26: 28145).

Kolehmainen, S. E. 1969. The balances of ^{137}Cs, stable cesium, and the feeding rates of bluegill (*Lepomis marochirus* Raf.) in White Oak Lake. Dissertation, Univ. of Tennessee, Knoxville. (Also: w/D. J. Nelson, Oak Ridge National Lab., U. S. AEC report ORNL-4445, xiii, 109pp.).

Kolehmainen, S. E. 1970. Daily feeding rates of bluegill determined by a radioisotopic method. Proc. Ann. Conf. Southeast. Assoc. Game Fish Comms. 23: 422-430. (^{137}Cs).

Kolehmainen, S. E. 1972. Balances of ^{137}Cs, stable cesium, and potassium of bluegill (*Lepomis macrochirus* Raf.) and other fish in White Oak Lake. Health Physics 23(3): 301-315. (NSA 26: 53535).

Kolehmainen, S. E. 1974. Daily feeding rates of bluegill (*Lepomis macrochirus*) determined by a refined radiisotope method. J. Fish. Res. Bd. Can. 31(1): 67-74. (SFA 19: 17583) (^{137}Cs).

Kolehmainen, S. E. 1975. Radioactivity of aquatic environment. Puerto Rico Nuclear Center, U. S. AEC report TID-26722. 28pp. (In Spanish) (NSA 31: 29961).

Kolehmainen, S. E., E. Häsänen and J. K. Miettinen. 1966. ^{137}Cs levels in fish of different limnological types of lakes in Finland during 1963. Health Physics 12(7): 917-922.

Kolehmainen, S. E., E. Häsänen and J. K. Miettinen. 1967. ^{137}Cs in fish, plankton, and plants in Finnish lakes during 1964-1965. pp. 913-919. In: Radioecological Concentration Processes. (See Aberg and Hungate, 1967) (CONF-660405-11) (NP-16002, Paper 8) (NSA 20: 26604).

Kolehmainen, S. E., E. Häsänen and J. K. Miettinen. 1968. ^{137}Cs in the plants, plankton, and fish of the Finnish lakes and factors affecting its accumulation. pp. 407-415. In: Radiation Protection. Part 1. (See Snyder, et al., 1968) (NSA 23: 12109).

Kolehmainen, S. E., E. Häsänen and J. K. Miettinen. 1968. Biological half-times of ^{137}Cs and ^{22}Na in different fish species and their temperature dependence. pp. 401-406. In: Radiation Protection. Part 1. (See Snyder, et al., 1968).

Kolehmainen, S. E., S. Takatalo and J. K. Miettinen. 1969. A tracer experiment with I-131 in an oligotrophic lake. pp. 278-284. In: Symposium on Radioecology. (See Nelson and Evans, 1969).

Kolehmainen, S., et al. 1968. Pollution experiments with Cs-137 in lakes of two limnologically different types. pp. 181-188. In: Intern. Congr. on Radioprotection of the Environment. (See Societé Francaise de Radioprotection, 1968).

Kolesnikov, N. V., I. A. Gorshkova and Y. F. Biryulin. 1970. Variations in concentration of ^{14}C in the atmosphere from 1957 to 1968 (by dendrological data). Izv. Akad. Nauk SSSR (fiz. atmos. okeana) 6(6): 647-649. (In Russian) (CA 73: 11694r; NSA 24: 41437).

Kollar, J., D. Chorvat and L. Klemova. 1970. Estimation of radioactivity of biological samples containing a mixture of ^{144}Ce, ^{144}Pr, and ^{59}Fe. Atompraxis 16: 247-249. (NSA 25: 11044).

Kollar, J., D. Chorvat and M. Tatara. 1970. Radiometric problems related to studies of tritium and ^{14}C in health physics practice. Pracovni Lekarstvi 22: 18-24. (In Slovak) (NSA 24: 36176) (Literature review).

Kollar, J., et al. 1974. Determination of natural tritium concentrations in water samples. pp. 232-238. In: Radiation Protection, Environment, and the Population. (See German Democratic Republic, 1974) (NSA 31: 8192; 33: 17191).

Kollar, J., et al. nd. Significance for radiation hygiene of monitoring the environmental contamination with tritium. Czechoslovakia Vyskumny Ustav Hygieny Prace a Chorob z Povolania, Bratislava. (In Slovak) (INIS-mf-887, pp. 196-203) (NSA 29: 24190).

Kollmann, R. 1965. Localization of the functioning sieve cells in secondary phloem tissue of *Metasequoia glyptostroboides*. Planta 65(2): 173-179. (In German) ($^{14}CO_2$).

Kolodny, Y. and I. R. Kaplan. 1970. Deposition of uranium in the sediment and interstitial water of an anoxic fjord. 43pp. (For Intern. Symp. on Hydrogeochemistry and Biochemistry, Tokyo, Japan, 10 Sep. 1970) (CONF-700965-1) (NSA 25: 48672).

Komala, Z. 1968. The observations on the influence of a small dose of x-rays on the conjugating paramecia. Folia biol. (Krakow) 16: 299-306. (NSA 23: 33819).

Komala, Z. 1970. Observations on the behavior of *Paramecium jenningsi* under the influence of a sublethal dose of x-rays. Folia biol. (Krakow) 18: 1-8. (NSA 24: 51021).

Komala, Z. 1975. Effect of x-rays on strains and syngens of *Paramecium aurelia* differing in genetic affinity. Folia biol. (Krakow) 23(4): 341-423. (NSA 33: 29757).

Kommendahl, T., et al. 1953. The physiology and genetics of plant pathogenic microorganisms when grown in the presence of various radioisotopes. Ohio Agricultural Experiment Station, U. S. AEC report AECU-2705. 16pp. (NSA 8: 28).

Kondo, S. 1964. Variation in mutagenicity and radiation resistance with genome complexity and evolution. Idengaku Zasshi 39: 176-198. (NSA 19: 219-21922) (Bacteria, fungi, viruses, insects, plants, mice).

Kondo, S. 1965. RBE of fast neutrons to γ-rays for mutations in relation to repair mechanisms. Idengaku Zasshi 40(Suppl.): 97-106. (NSA 19: 43786) (Silkworms).

Kondrat'yev, V. P. 1961. Application of radioactive isotopes in commercial fishing. Ryb. khozyaystvo 37(5): 62-64. (JPRS-21,989; TT-63-41204) (BA 411; NSA 18: 5698; SFA 8: 5558).

Konermann, G. 1970. Capability for learning and memorizing of trained quails following head irradiation. Strahlentherapie 140: 757-764. (In German) (NSA 25: 29687).

König, L. A. 1970. Contribution of poor exchange weather conditions to the radiation burden of the environment of atomic energy facilities. Germany Kernforschungszentrum, Karlsruhe, report KFK-1336. 14pp. (English transl. Australian AEC report LIB/Trans-382) (NSA 25: 32084; 26: 197).

König, L. A. 1972. Environmental aspects of ^{129}I. Germany Kernforschungszentrum, Karlsruhe, report KFK-1543. 7pp. (In German) (NSA 26: 31007).

König, L. A. 1972. Environmental burden of tritium. Germany Kernforschungszentrum, Karlsruhe, report KFK-1560. 26pp. (In German) (NSA 26: 45509).

König, L. A. and M. Winter. 1971. Environmental contamination by tritium. pp. 287-301. In: Proc. Intern. Symp. on Radioecology Applied to the Protection of Man and His Environment. (See Commission of the European Communities, 1971) (In German) (NSA 27: 5229).

König, L. A. and M. Winter. 1974. Investigation of the tritium level in the environment of the Karlsruhe Nuclear Research Center. pp. 400-405. In: Proc. 3rd Intern. Congr. Intern. Radiation Protection Association. (CONF-730907-P1) (NSA 30: 26782) (Surface water).

König, L. A. and M. Winter. 1974. Environmental exposure to tritium. KFK Nachr. 6(3): 33-38. (In German) (NSA 32: 768).

König, L. A., J. G. Wilhelm and H. G. Dillmann. 1973. Continuous sampling for detecting tritiated water. Atomwirt. Atomtech. 18(12): 582-583. (In German) (NSA 30: 212).

König, V. 1975. Nuclear physics as a device in archaeology. Phys. Bl. 31 (3): 104-114. (In German) (NSA 32: 27952).

Konno, K. and N. Egami. 1966. Notes on effects of x-irradiation on the fertility of the male of Oryzias latipes (Teleostei, Cyprinodontidae). Annot. Zool. Jap. 39: 63-70. (NSA 21: 39509).

Konopac, J. 1966. Isotopic composition of mineral waters. Fysiat. Reumatol. Vestn. 44: 157-163. (In Czech) (NSA 21: 32272).

Konrad, J. G., D. E. Armstrong and G. Chesters. 1967. Soil degradation of diazinon, a phosphorothioate insecticide. Agron. J. 59(6): 591-594. (^{14}C).

Konstantinov, I. Ye. and G. A. Federov. 1972. Some methods of measurement of small activity with the assistance of scintillation γ-spectrometers. pp. 153-162. In: Methods of Determination of Radioactivity. (See Polikarpov and Parchevskiy, 1972) (In Russian).

Konstantinov, V. M. 1965. Use of tracer elements in the evaluation of radioactive anomalies in arid climate regions. Atom. energiya 18(6): 657-659. (Soviet Atomic Energy 18(6): 834-836) (NSA 19: 36692).

Konstantinov, V. P. and A. M. Gorodnitskiy. 1968. Radioactivity in the surface layer of water in the Black Sea and its relationship to some physical characteristics. Tr. nauchn.-issled. inst. geol. Arkitiki (5): 236-239. (In Russian) (NSA 24: 31519).

Konzak, C. F., et al. 1968. Irradiation of seeds. A review of procedures employed at Washington State University. pp. 83-96. In: Neutron Irradition of Seeds. Part II. (See IAEA, 1968) (NSA 24: 17026).

Koo, F. K. S. and E. R. de Irizarry. 1970. Nuclear volume and radiosensitivity of plant species at El Verde. Ch. G-1, pp. G15-G20. In: A Tropical Rain Forest. (See H. T. Odum and R. F. Pigeon, 1970) (NSA 25: 21949).

Koo, F. K. S. and R. C. Venator. 1970. Radiation effect on survival and growth of bromeliads. Ch. D-18, pp. D299-D303. In: A Tropical Rain Forest. (See H. T. Odum and R. F. Pigeon, 1970) (NSA 25: 21943).

Koo, F. K. S., H. T. Odum and E. R. de Irizarry. 1970. Effects of gamma radiation on chlorophyll A content in bromeliads. Ch. D-17, pp. D295-D298. In: A Tropical Rain Forest. (See H. T. Odum and R. F. Pigeon, 1970) (NSA 25: 21942).

Kopec, J. and J. Benes. 1970. Radiation safety of nuclear reactors. Radioisotopy 11(4): 559-577. (In Czech) (NSA 26: 9537).

Kopelman, M., P. Markakis and B. S. Schweigert. 1967. Effect of ionizing radiations on resting conidia of Aspergillus flavus. J. Food Sci. 32: 694-696. (NSA 22: 52044).

Kopp, J. F. and R. C. Kroner. 1967. Tracing water pollution with an emission spectrograph. J. Water Poll. Control Fed. 39: 1659-1668. (NSA 23: 24078).

Koranda, J. J. 1965. Preliminary studies of the persistence of tritium and ^{14}C in the Pacific Proving Ground. pp. 1445-1457. In: Radiation and Terrestrial Ecosystems. (See Hungate, 1965).

Koranda, J. J. 1967. Residual tritium at Sedan crater. Lawrence Radiation Lab., U. S. AEC report UCRL-70292. 36pp. (NSA 21: 28446).

Koranda, J. J. 1969. Residual tritium at Sedan crater. pp. 696-708. In: Symposium on Radioecology. (See Nelson and Evans, 1969).

Koranda, J. J. and J. R. Martin. 1969. Persistence of radionuclides at sites of nuclear detonations. pp. 159-187. In: Biological Implications of the Nuclear Age. (See Goulden, 1969) (Incl. wild plants and animals).

Koranda, J. J. and J. R. Martin. 1972. Movement of tritium in ecological systems. Lawrence Livermore Lab., U. S. AEC report UCRL-73178(Rev.1). 47pp. (For Tritium Symp., Las Vegas, Nev., 30 August 1971) (CONF-71809-18) (NSA 26: 50973).

Koranda, J. J. and J. R. Martin. 1973. Gamma-emitting radionuclides in Alaskan environments 1967-1970. pp. 81-107. In: Radionuclides in Ecosystems. (See Nelson, 1973) (NSA 26: 4744; 28: 27294).

Koranda, J. J., J. R. Martin and L. R. Anspaugh. 1971. Significance of tritium releases to the environment. Lawrence Livermore Lab., U. S. AEC report UCRL-73546. 18pp. (For Nuclear Science Symp., San Francisco, Calif., 3 November 1971) (CONF-711111-23) (NSA 26: 9487).

Koranda, J. J., J. R. Martin and R. Wikkerink. 1967. Radioecological studies at Amchitka Island, Alaska. I. Tritium measurements. Lawrence Radiation Lab., U. S. AEC report UCRL-50367. 9pp.

Koranda, J. J., J. R. Martin and R. W. Wikkerink. 1970. Leaching of radionuclides at Sedan crater. pp. 97-117. In: Radionuclides in the Environment. R. F. Gould, ed. Am. Chem. Soc., Washington.

Koranda, J. J., et al. 1969. Radioecological studies of Amchitka Island, Aleutian Islands, Alaska. II. Gamma-emitting radionuclides in the terrestrial environment. Lawrence Radiation Lab., U. S. AEC report UCRL-50786. 44pp. (NSA 24: 27562).

Koranda, J. J., et al. 1970. Postshot distribution and movement of radionuclides in nuclear crater ejecta. pp. 400-421. In: Symp. on Engineering with Nuclear Expolosives. Vol. 1. (See Anonymous, 1970) (NSA 24: 41490).

Koranda, J. J., et al. 1971. Sampling and analytical systems for measurement of environmental radioactivity. Lawrence Livermore Lab., U. S. AEC report UCRL-72987. 41pp. (For Intern. Symp. on Rapid Methods for Measuring Radioactivity in the Environment, Munich, Germany, 5 July 1971) (CONF-710705-2) (NSA 25: 46622).

Korkisch, J. and D. Dimitriadis. 1973. Use of ion exchange for determining trace elements in natural waters. III. Thorium. Talanta 20(12): 1303-1308. (In German) (NSA 30: 2561).

Korkisch, J. and L. Gödl. 1974. Determination of uranium in natural waters after anion-exchange separation. Anal. Chim. Acta 71: 113-121. (NSA 30: 12232).

Korkisch, J. and L. Gödl. 1974. Use of ion exchange methods for the determination of trace elements in natural waters. IV. Uranium, cobalt, and cadmium. Talanta 21(10): 1035-1046. (In German) (NSA 31: 8187).

Korkisch, J., P. Antal and F. Hecht. 1959. Determination of uranium and thorium in natural waters after pre-concentration with Amberlite IRA-400 and Dowex-50. Mikrochimica Acta 5: 694-705.

Kormondy, E. J. 1965. Uptake and loss of zinc-65 in the dragonfly Plathemis lydia. Limnol. Oceanogr. 10(3): 427-433. (NSA 19: 45789).

Kormondy, E. J. 1969. Concepts of Ecology. Prentice-Hall, Englewood Cliffs, N. J. xiii, 209pp. (Contains section on ionizing radiation).

Kormondy, E. J. 1969. Studies on zinc in freshwater ecosystems. Final report. Oberlin Coll., O., U. S. AEC report COO-1499-3. 32pp. (NSA 23: 20143).

Kornberg, H. A. 1959. Biological implications of radioactive wastes. pp. 385-412. In: 6th Intern. Congr. and Exhibition of Electronics and Atomic Energy. (Rome, June 1959). Italy Comitato Nazionale Ricerche Nucleare, Rome. (Also: U. S. AEC report TID-7579, pp. 1-16) (NSA 13: 23064; 14: 9289).

Kornberg, H. A. 1965. Introduction. p. 1255. In: Radiation and Terrestrial Ecosystems. (See Hungate, 1965).

Kornberg, H. A. and J. J. Davis. 1966. Food chains and freshwater. pp. 383-418. In: Radioactivity and Human Diet. (See Russell, 1966) (NSA 21: 43355).

Kornegay, B. H., et al. (Eds.). 1963. Transport of Radionuclides in Fresh Water Systems. (A Working Conf., Austin, Tex., January 30-February 1, 1963). U. S. AEC report TID-7664. 404pp.

Körner, W. J. Murglat and M. Hannemann. 1974. Radiation protection criteria for the evaluation of geohydrologic conditions of nuclear plant sites. German Democratic Republic Staatliche Zentrale für Strahlenschutz, Berlin, report SZS-156. 23pp. (In German) (NSA 30: 21178).

Korneyev, G. A. 1967. Ectoparasitic contacts between some mammalian species in great gerbil colonies. Parazitologiya 1(3): 233-237. (In Russian).

Korneyev, G. A. 1967. On the utilization character of Rhombomys opimes colonies by Meriones merideonalis according to results of radioactive labeling of rodents. Nauchn. dokl. vyssh. shkoly (biol.) (1): 26-30. (In Russian, English summary).

Korneyev, N. A., et al. 1970. Migration of ^{90}Sr in the biological chain and possibilities of its reduction in the latter. The results of observations on cultivated plants and animals. Zh. obshch. biol. 31: 690-697. (In Russian) (NSA 25: 24026).

Korneyev, N. A., et al. 1971. Radioecology of farm animals. pp. 316-325. In: Contemporary Problems of Radiobiology. Vol. 2. (See Klechovskiy, Polikarpov and Aleksakhin, 1971).

Körnicke, M. 1904. On the study of roentgen and radium radiation on plant tissues and cells. Ber. Deutsch. Bot. Ges. 23: 148, 155. (In German).

Korogodin, V. I. 1957. Certain irregularities in the growth of macrocolonies after irradiation of yeast cells with gamma rays from radiocobalt. Biofizika 2(2): __. (Biophysics (USSR) 2: 180-188).

Korogodin, V. I. 1957. Some regularities in the retardation of the first division in yeast cells acted on by ^{60}Co gamma rays. Biofizika 2(5): 536-580. (Biophysics (USSR) 2: 566-570) (NSA 12: 2642).

Korogodin, V. I. 1958. Forms of inactivation of yeast cells by ionizing radiations. Biofizika 3(2): 206-213. (Biophysics (USSR) 3: 189-197).

Korogodin, V. I. 1958. Some regularities in the post-irradiation changes in resting yeast cells. Biofizika 3(6): 703-710. (Biophysics (USSR) 3: 663-668).

Korogodin, V. I. 1962. Genetics. pp. 45-100. In: *The Contribution of Radiology to the Growth of Medico-Biological Disciplines*. P. V. Petrov, et al., eds. Ministry of Special Technological and Vocational Training, Beloruss. SSR, Minsk.

Korogodin, V. I. and K. M. Bliznik. 1972. Principles of the formation of radioraces in yeasts. Radiobiologiya 12(2): 163-170. (AEC-tr-7362, pp. 1-15) (NSA 27: 2727).

Korogodin, V. I. and G. G. Polikarpov. 1957. Primary processes in radiation injury. On the problem of the mechanisms of intensification of radiobiological effect. Usp. sovrem. biol. 44(1(4)): 93-102.

Korogodin, V. I., et al. 1962. Reactivation of yeast cells of various ploidy irradiated with alpha particles. Abstr. p. 18. In: *2nd Intern. Congr. of Radiation Research*. (See Anonymous, 1962).

Korogodina, Yu. V. and V. I. Korogodin. 1971. Radiobiological reactions of amoebae with few chromosomes. Radiobiologiya 11(6): 865-870. (AEC-tr-7307, pp. 68-75) (NSA 26: 41217).

Korogodina, Yu. V., et al. 1973. Nucleic acid contents in some species of amoebae differing in their radiosensitivity. Tsitologiya 15(5): 613-616. (In Russian) (NSA 28: 21578).

Korol, I. L., et al. 1972. Worldwide contamination of the atmosphere with ^{85}Kr from peaceful uses of nuclear energy. Tr. inst. eksp. meteorol. (1): 27-40. (In Russian) (NSA 29: 15618).

Korolev, M. B. and A. A. Neyfakh. 1965. Radiation investigation of morphogenetic function of the nuclei in tissue differentiation in early development of fish embryo (loach, *Misgurnus fossilis*). Zh. obshch. biol. 26: 352-357. (In Russian) (NSA 21: 1799).

Korte, F. 1966. Metabolism of chlorinated insecticides. pp. 38-48. In: *Radioisotopes in the Detection of Pesticide Residues*. (Proc. Panel, Vienna, 12-15 April 1965). International Atomic Energy Agency, Vienna, Publ. STI/PUB/123. (^{14}C).

Korte, F. 1974. Global inputs and burdens of chemical residues in the biosphere: the problem and control measures. pp. 3-22. In: *Comparative Studies of Food and Environmental Contamination*. (See IAEA, 1974) (NSA 30: 15384) (Incl. radionuclides).

Korte, F. and M. Stiasni. 1964. Insecticides in metabolism. VI. Conversion of telodrin-^{14}C by microorganisms and mosquito larvae. Justus Liebigs Annln. Chem. 673: 146-152. (In German).

Kortshak, H. P. 1962. Mutagenic effect of x-rays on moths. Nature 196(4853 (4853): 490. (NSA 17: 7858).

Korver, J. A., C. R. Boardman and D. E. Rawson. 1964. Shoal post-shot environment. Lawrence Livermore Lab., U. S. AEC report UCID-15771. 14pp. (NSA 30: 29317).

Koshcheyeva, I. Ya., G. M. Varshal and T. K. Velyukhanova. 1973. Method of chemiluminescence used for research of interaction of ruthenium submicroquantities with organic material dissolved in natural waters. pp. 116-122. In: *Radioecology of Water Organisms. III*. (See Andrushaitis, 1973) (In Russian) (NSA 30: 23593).

Kosheleva, V. V. 1971. Accumulation of radioactive isotopes by the developing eggs of the Atlantic salmon. Tr. Polyar. nauchn.-issled. Proekt. inst. morsk. ryb. khoz. okeanogr. (29): 8-15. (AEC-tr-7418, pp. 7-15) (NSA 27: 28046).

Kosheleva, V. V. 1971. Change in peripheral blood in the embryos and larvae of Atlantic salmon under the influence of radioactive contamination of the water and x-rays. Tr. Polyar. nauchn.-issled. Proekt. inst. morsk. ryb. khoz. okeanogr. (29): 62-73. (AEC-tr-7418, pp. 75-88) (NSA 27: 27946).

Koshi, P. T. 1966. Soil-moisture measurement by the neutron method in rocky wildland soils. Soil Sci. Soc. Am. Proc. 30: 282-284. (NSA 20: 31435).

Koshida, Y. 1965. Behavior of radiosilver administered to amphibian larvae. Zool. Mag. (Tokyo) 74(1): 6-10. (In Japanese) (NSA Japan 4: 01826; NSA 19: 21826).

Koshlakova, K. G., et al. 1965. Effect of gamma irradiation on the storage of sunflower seeds. Prikl. biokhim. mikrobiol. (4): 471-473. (In Russian) (NSA 20: 45416).

Koshy, T. and K. R. Chandra. 1972. New approaches to the control of harmful insects. J. Sci. Ind. Res. 31(10): 515-523. (NSA 29: 769) (Radiotracer; sterile male technique).

Koshy, T. and S. P. Singh. 1970. Laboratory studies on radiation-induced dominant lethality in sperms in population control of the mosquito *Culex pipiens fatigans* Wied. Int. J. Radiat. Biol. 18: 521-530. (NSA 25: 16293).

Kosmowski, A. 1964. The radioactivity of water and its control. Wasser Luft Betrieb 8: 719-721. (NSA 20: 11021).

Kössler, F. 1961. Important concepts of radiation hygiene in the field of ecology. Z. ges. hyg. Grenzgebiete 7(8): 567-571. (In German, English summary).

Kössler, F. 1961. On the accumulation of radioactive materials in the biosphere. Z. ges. hyg. Grenzgebiete 7(8): 571-580. (In German, English summary) (NSA 17: 35442).

Kössler, F. 1964. Respiration and photosynthesis of *Chlorella* during gamma irradiation. Thesis, Humboldt-Universität, Berlin. 76pp. (In German) (NSA 19: 29992).

Kössler, F. 1964. Breathing depression after chronic γ irradiation and photoreactivation in *Chlorella*. Naturwissenschaften 51(12): 289. (In German) (NSA 18: 31193).

Kössler, F. 1965. Respiration and photosynthesis of Chlorella under chronic γ irradiation. Radiation Botany 5(2): 115-128. (In German) (NSA 19: 24333).

Kosta, L. and A. R. Byrne. 1969. Activation analysis for mercury in biological samples at anaogram level. Talanta 16: 1297-1303. (NSA 23: 40604).

Kosta, L., V. Ravnik and J. Dumanović. 1969. Determination of nitrogen in plant seeds by fast neutron activation analysis. pp. 161-168. In: New Approaches to Breeding for Improved Plant Protein. (Proc. Panel, Röstanga, 17-21 June 1968). International Atomic Energy Agency, Vienna, Publ. STI/PUB/212. (NSA 23: 31307).

Kostolanska, A. 1962. Effects of low temperatures on post-radiation changes in eggs of Bombyx mori L. Biologia 17(8): 626-628. (In Czech) (NSA 17: 17816).

Kostromskaya, V. A. 1972. Change in the intensity of oxidative phosphorylation and the activity of catalase during irradiation of various species of invertebrates. Radiobiologiya 12(3): 437-441. (AEC-tr-7402, pp. 173-178) (NSA 27: 17675).

Kotelev, V. V. 1955. Penetration of P^{32} into cells of microorganisms and transfer to the host plant. In: Isotopes in Microbiology: Trans. Conf. on the Application of Tracer Atoms in Microbiology. A. A. Imshenetskiy, ed. Akad. Nauk SSSR, Moscow. (AEC-tr-3494, pp. 229-237).

Kotel'nikov, V. N., V. N. Markelov and B. A. Nelepo. 1965. New data on the atmospheric radioactivity and fallout density near the Black Sea. Atom. energiya 19(5): 469-470. (Soviet Atomic Energy 19: 1447-1449) (NSA 21: 14741).

Kotov, M. M. 1969. The variation of the oak (Quercus robur L.) seedlings grown from x-irradiated acorns. Genetika 5(6): 32-39. (In Russian) (NSA 24: 2830).

Kovac, J. and L. Taimr. Use of radioisotopes in research on insecticides and the toxicology of insects. Cesk. Akad. Zemedel. Ved. Sborn. Rost. Vyroba. 32(3): 447-456.

Kovacic, A. 1972. Theory of radiobiological effects of intensive ^{60}Co γ-radiation in Helianthus annuus L. Tsitol. genet. 6(6): 529-540. (In Russian) (NSA 27: 24903).

Kovacic, A. 1973. Theory of radiobiological effect of intensive ^{60}Co γ radiation in Helianthus annuus L. Tsitol. genet. 7(1): 25-32. (In Russian) (NSA 30: 3546).

Kovacic, A. and R. Skalska. 1968. Influence of moisture in sunflower seeds on the reaction to acute gamma radiation (^{60}Co). Genet. Slechteni 4: 109-114. (In Czech) (NSA 24: 27717).

Kovacic, A. and R. Skalska. 1969. Effect of exposure from the acute gamma radiation ^{60}Co to the sowing of achenes on the initial development and growth of the sunflower. Genet. Slechteni 5: 115-119. (In Czech) (NSA 24: 25595).

Kovacic, A. and R. Skalska. 1969. Preliminary tests on the effect of acute gamma radiation of ^{60}Co on sunflower plants. Genet. Slechteni 5: 23-28. (In Czech) (NSA 24: 27716).

Kovacic, A. and R. Skalska. 1971. Modification of the effect of the acute gamma irradiation by ^{60}Co in Helianthus annuus L. Genet. Slechteni 7(4): 245-258. (In Czech) (NSA 26: 23036).

Kovacic, A., R. Skalska and R. Kryzanek. 1970. The effetcs of ^{60}Co gamma radiation on the development, growth, and biochemical process dynamics in sunflower. Genet. Slechteni 6: 11-24. (In Czech) (NSA 24: 34769).

Kovacs, J. and L. Simon. 1964. Investigation of the radoactive contamination in domestic and imported fishes. Magy. Allatorv. Lapja 19: 424-426). (NSA 19: 22546).

Kovacs, M., et al. 1972. Gamma spectrometric data processing for environmental sample analysis. Magyar Tudomanyos Akad. Kozponti Fizikai Kutato Intezete, Budapest, report KFKI-72-51. 15pp. (NSA 27: 7363).

Kovacs, M., et al. 1974. Gamma spectrometric data processing for environmental analysis. pp. 247-257. In: Radiation Protection, Environment and the Population. (See German Democratic Republic, 1974) (NSA 31: 8193).

Kovalev, N. M. 1938. Investigation of snow radioactivity. Meteorologiya gidrologiya (11-12): 35-__.

Kovaleva, N. Ye. 1948. Investigations of the effect of x-rays on Paramecium caudatum. (Thesis, Leningrad, 1947). Uch. zap. Leningrad. gos. ped. inst. kaf. zool. darVin. 70: 75-144. (In Russian).

Kovaleva, N. Ye. 1962. Effect of cultivation temperature on sensitivity of infusoria to injurious action of x-rays. Tsitologiya 4(3): 306-317. (Fed. Proc. 22(Transl. Suppl., Pt.II): T507-T513) (NSA 17: 30318).

Kovaleva, N. Ye. 1964. The influence of temperature on the content of cytoplasmic ribonucleic acid on Paramecium caudatum irradiated with x-rays. Tsitologiya 6: 709-717. (In Russian) (NSA 21: 9033).

Kovaleva, N. Ye. 1966. The influence of ionizing radiation on some pathways of metabolic processes in paramecia. Tsitologiya 8: 629-638. (In Russian) (NSA 21: 24365).

Kovaleva, N. Ye. and M. A. Frenkel. 1967. The investigation of the activity of some dehydrogenases in infusoria after x-irradiation. Tsitologiya 9: 1550-1558. (In Russian) (NSA 22: 36184).

Kovaleva, N. Ye. and M. A. Frenkel. 1969. The sensitivity to x-irradiation of some dehydrogenases in paramecia adapted to low temperature. Tsitologiya 11: 1266-1274. (In Russian) (NSA 24: 7980).

Kovaleva, N. Ye. and G. V. Selivanova. 1966. The influence of temperature variation in the medium on the DNA content of paramecia after ionizing irradiation. pp. 36-45. In: Effects of Ionizing Radiation on Heredity. N. P. Dubinin, ed. Publ. House Acad. Sci., Moscow. (In Russian) (NSA 22: 21608).

Kovaleva, N. Ye. and A. V. Yankovskiy. 1965. The effect of ionizing radiation on nuclear reorganization processes in Paramecium putrinum. I. Nuclear processes in irradiated conjugants. Zh. obshch. biol. 26: 176-189. (In Russian) (NSA 20: 29007).

Kovaleva, N. Ye. and A. V. Yankovskiy. 1966. Nuclear processes at conjugation of paramecia following x-irradiation. Tsitologiya 8: 80-89. (In Russian) (NSA 20: 33047).

Kovalevskiy, A. L. 1962. Natural radioactive elements in plants. Izv. Sibirsk. otd. Akad. Nauk SSSR (4): 108-114. (JPRS-15,274) (NSA 16: 26725).

Kovalevskiy, A. L. 1963. On certain regularities on the uptake of the second-group elements of the Mendeleyev table. Izv. Sibirsk. otd. Akad. Nauk SSSR (biol. med.) (4): 53-61. (In Russian) (Incl. Ra).

Kovalevskiy, A. L. 1965. Absorption of radium by various kinds of plants. Izv. Sibirsk. otd. Akad. Nauk SSSR (biol. med.) (1): 43-47. (Higher natural plants).

Koval'skiy, V. V. 1958. Geochemical ecology. Priroda 47(9): 100-101. (In Russian).

Koval'skiy, V. V. and S. V. Letunova. 1961. Role of phyto- and zooplankton of bodies of water in the migration of cobalt. Zool. zh. 40(6): 809-817. (In Russian).

Koval'skiy, V. V. and S. V. Letunova. 1970. Role of microorganisms in the biogenic migration of uranium. Biol. nauki 18: 521-530. (NSA 25: 16293).

Koval'skiy, V. V. and I. Ye. Vorotnitskaya. 1965. Biogenic migration of uranium in the Lake Tssyk-Kul. Geokhimiya (6): 724-732. (Geochemistry Int. (3): 544-552)

Koval'skiy, V. V. and I. Ye. Vorotnitskaya. 1966. Regularities in the biogenic migration of uranium. Ukr. biokhim. zh. 38(4): 419-424. (RFP-TRANS-141) (NSA 29: 10146).

Koval'skiy, V. V., V. I. Baranov and V. Ye. Sinel'nikov. 1963. Problems of biogeochemistry in radiation hygiene. Gigiyena sanitariya 28(12): 81-86. (JPRS-23,224) (NSA 18: 13459).

Koval'skiy, V. V., et al. 1967. Biogeochemical food chains of uranium in aquatic and terraneous organisms. pp. 329-332. In: Radioecological Concentration Processes. (See Aberg and Hungate, 1967).

Kovarik, A. F. and L. W. McKeehan. 1925. Radioactivity. Bull. Nat. Res. Council 10(Pt.1, 51): 203pp.

Kowal, N. E. and D. A. Crossley, Jr. 1971. The ingestion rates of microarthropods in pine mor, estimated with radioactive calcium. Ecology 52(3): 444-452.

Kowalski, B. R., T. F. Schatzki and F. H. Stross. 1972. Classification of archeological artifacts by applying pattern recognition to trace element data. Anal. Chem. 44(13): 2176-2180. (NSA 26: 58547) (X-ray fluorescence).

Kozlov, A. A. 1971. Effect of background doses of gamma irradiation on the rate of division of infusorians. Radiobiologiya 11(6): 935-937. (AEC-tr-7307, pp. 166-169) (NSA 26: 41219).

Kozlov, A. A. 1974. Influence of preliminary irradiation of the nutrient medium and infusoria on the division rate. Soobshch. Akad. Nauk Gruz. SSR 74(1): 213-215. (In Russian) (NSA 30: 21351).

Kozlova, M. V. and V. V. Lisunova. 1969. On the method of determining ^{137}Cs in environmental objects. Gigiyena sanitariya (2): 52-55. (In Russian) (NSA 23: 31303).

Kozlova, M. V., et al. 1971. Study of ^{210}Po content in the environment and the human body. Gigiyena sanitariya (4): 44-47. (In Russian) (NSA 25: 42429).

Krabbenhoft, K. L. 1965. Significance of pigments in radiation resistance of Micrococcus radiodurans. Thesis, Oregon State Univ., Corvallis. 130pp. (NSA 20: 29016) (Food irradiation).

Krabbenhoft, K. L., A. W. Anderson and P. R. Elliker. 1965. Ecology of Micrococcus radiodurans. Appl. Microbiol. 13: 1030-1037. (NSA 20: 6805).

Kraemer, H. J. and B. Neidhart. 1974. Accumulation of mercury in fish. pp. 213-223. In: Proc. 2nd Intern. Conf. on Nuclear Methods in Environmental Research. (See Vogt and Meyer, 1974) (NSA 33: 2331) (^{203}Hg tracer).

Krafft, B. 1965. On a possible exchange of substances through individuals in the social spider Agelena consociata Denis. Compt. rend. D260(20): 5376-5378. (^{32}P).

Kramer, H. H., W. J. Gemmill and E. R. Johnson. 1971. Translocation of yttrium-90. Carolina Tips 34(7): 25-26. (Describes technique using domesticated plants).

Kramer, H. P., D. W. Moeller and D. A. Pecsok. 1958. Radioactivity in surface water. J. Am. Water Works Assoc. 50: 1516-1522. (NSA 13: 19101).

Kramer, H. P., D. W. Moeller and D. A. Pecsok. 1959. Preprint V-22. For: 5th Nuclear Engineering Science Conf., Cleveland, O. 8pp.

Kramer, P. J. 1951. Effects of respiration inhibitors on accumulation of radioactive phosphorus by roots of loblolly pine. Plant Physiol. 26(1): 30-36. (NSA 5: 3335).

Kramer, P. J. and H. H. Wiebe. 1952. Longitudinal gradients of P-32 absorption in roots. Plant Physiol. 27: 661-674.

Kramer, P. J. and K. M. Wilbur. 1949. Absorption of radioactive phosphorus by mycorrhizal roots of pine. Science 110(2844): 8-9. (NSA 3: 597).

Kramp, R. C. 1965. Preliminary report on ^{90}Sr levels in hock joints of desert bighorn sheep and mule deer. Desert Bighorn Council Trans. 9: 87-92.

Kraner, H. W., G. L. Schroeder and R. D. Evans. 1964. Measurements of the effetcs of atmospheric variables of radon-222 flux and soil-gas concentrations. pp. 191-215. In: The Natural Radiation Environment. (See Adams and Lowder, 1964) (NSA 18: 35737).

Kranz, L. and L. Weise. 1961. An example for the measurement of river water low activities with methane flow counters. Mitteilungsbl. Strahlungsmess. (6): 8-9. (In German) (NSA 18: 14095).

Krapu, G. L., G. A. Swanson and H. K. Nelson. 1973. Mercury residues in pintails breeding in North Dakota. J. Wildl. Mgmt. 37(3): 395-399. (Neutron activation analysis).

Krasil'nikov, N. A. 1938. The sensitivity of the Actinomycete to radon. Mikrobiologiya 7: 721-731. (In Russian).

Krasil'nikov, N. A. 1958. Absorption of natural radioactive elements by soil organisms. Priroda 47(9): 97-99. (NIH Transl. MF #9088620). (NSA 13: 1077).

Krasil'nikov, N. A. and A. A. Drobkov. 1965. Accumulation of natural radioactive elements by azotobacter, algae, and protozoa. Dokl. Akad. Nauk SSSR 163(2): 486-487. (Doklady (biol. sci.) 163: 483-486) (NSA 20: 6738).

Krasil'nikov, N. A., A. A. Drobkov and L. G. Shirokov. 1958. The accumulation of naturally radioactive elements by soil microorganisms. Dokl. Akad. Nauk SSSR 120(5): 1136-1137. (Doklady (biol. sci.) 120: 453-455) (CA 52: 18644).

Krasnik, F. I. and P. B. Goldin. 1963. Experimental typhus infection and immunity in irradiated animals. II. Effects of total irradiation on immunity from typhus fever. Tr. inst. epidemiol. mikrobiol. gig. Pastera 25: 42-49. (In Russian) (NSA 21: 18169).

Kratz, F. L. 1975. Radioresistance in natural populations of *Drosophila nebulosa* from a Brazilian area of high background radiation. Mutation Research 27(3): 347-355. (NSA 32: 28307).

Kraus, M. P. 1969. Resistance of blue-green algae to ^{60}Co gamma radiation. Radiation Botany 9(6): 481-489. (NSA 24: 12331) (23 species).

Kraus, O. 1957. A genetic effect in nature from the radioactivity in uranium deposits. Natur u. Volk 87: 399-401. (AEC-tr-3194) (NSA 12: 9576).

Kraus, W. and R. Friemert. 1970. Quantitative alpha-autoradiography: a documentation. German Democratic Republic Staatliche Zentrale für Strahlenschutz, Berlin, report SZS-15/70. 21pp. (In German) (NSA 25: 8584).

Krause, H. 1967. The growth of *Microsporum gypseum* following the effect of beta radiations on fungus and culture medium. Hautarzt 18: 351-355. (In German) (NSA 23: 3888).

Kraybill, H. F. 1959. The effect of ionizing radiation on insects. Int. J. Appl. Radiat. Isotopes 5(3): 187. (NSA 14: 1309).

Krebs, A. T. 1961. Immediate reaction of the red-eared turtle (*Pseudemys scripta elegans*) to x-irradiation. Radiation Research 15(3): 372-377. (Also: U. S. Army Medical Research Lab., Ft. Knox, Ky., report AMRL-445) (NSA 15: 2426).

Krebs, A. T. 1965. Radiation-induced premature aging in leaves and autumnal events in nature. Beitr. biol pflanzen 41: 157-174. (NSA 20: 36609).

Krebs, A. T. and B. W. Benson. 1966. Effects of ^{60}Co gamma radiation on natural digging and tunneling behavior of the ant *Pogonomyrmex californicus*. Naturwissenschaften 53: 131. (NSA 20: 24924).

Krebs, A. T. and M. M. McLaughlin. 1968. Behavior changes in gamma-irradiated ants. Abstr. Radiation Research 35(2): 575.

Kreft, A. and A. Zuber. 1972. Radioisotope measurements of flow rates of streams for determining leakages to a mine. Nukleonika 17(5-6): 305-312. (In Polish) (NSA 27: 14757).

Kreger, D. R. 1957. X-ray interferences of barium sulphate in fungi and algae. Nature 180(4591): 867-868.

Kreitsberg, Z. N., V. N. Sergeyeva and Ya. K. Grabovskiy. 1968. Effect of ionizing radiation on the contents of ortho-oxyphenol in poplar shoots. Khim. Drev. (2): 13-15. (In Russian) (NSA 25: 13598).

Krepelka, J. nd. Effect of radioactivity of the waste waters from the nucear research institutes at Rez on the environment. Czechoslovakia Ustav Jaderneho Vyzkumu, Reg. (In Czech) (INIS-mf-887-3, pp. 169-186) (NSA 30: 6577).

Krepelka, J. and J. Chysky. 1972. Radioactivity in waste waters in nuclear research institute, Rez, and environment. Jaderna Energie 18(5): 146-150. (In Czech) (NSA 26: 45565).

Kreuzer, W. 1971. Bioindicators for determining control of ecosystems through their ^{137}Cs contamination. pp. 405-423. In: Proc. Intern. Symp. on Radioecology Applied to the Protection of Man and His Environment. (See Commission of the European Communities, 1971) (In German) (NSA 27: 5275).

Kreuzer, W. 1973. Radioactive contamination of the environment and its effects on livestock and food of animal origin. Arch. Lebensmittelhyg. 24 (5): 104-108. (In German) (NSA 29: 29389).

Krey, P. W. 1974. ^{239}Pu contamination in the Denver area. Health Physics 26(1): 117-120. (NSA 29: 15599).

Krey, P. W. and E. P. Hardy. 1970. Plutonium in soil around the Rocky Flats plant. Health and Safety Lab., U. S. AEC report HASL-235. 44pp. (NSA 24: 46355).

Krey, P. W. and B. T. Krajewski. 1972. Plutonium isotopic ratios at Rocky Flats. pp. 167-194. In: Fallout Program Quarterly Summary Report, December 1, 1971-March 1, 1972. (See Hardy; HASL-249) (NSA 26: 28075).

Kriebig, A.† Sarcoma through x-irradiated fungus. Zentralbl. Chir. : 1897.

Kriegel, H. 1960. Incorporation and decorporation of radioisotopes. Med. Klin. (Munich) 55: 1833-1839. (NSA 17: 41).

Krieger, H. L. and F. J. Burmann. 1969. Effective half-times of ^{85}Sr and ^{134}Cs for a contaminated pasture. Health Physics 17: 811-824. (NSA 24: 6014).

Krieger, H. L. and G. W. Frishkorn. 1971. Evaluation of ion-exchange surveillance sampler for analyzing radioactive liquid effluents. Health Physics 21(4): 591-595. (NSA 15: 57351).

Krieger, H. L., S. Gold and B. Kahn. 1973. Tritium releases from nuclear power stations. pp. 557-564. In: Tritium. (See Moghissi and Carter, 1973) (NSA 28: 27225).

Kriger, I. A. and O. M. Zorina. 1959. The effect of x-rays and γ-rays on the unilateral permeability of frog skin. Biofizika 4: __. (Biophysics (USSR) 4: 88-94) (NSA 14: 9338).

Kriks, H.-J., W. Rasp and V. Siegel. 1971. Rapid determination of fission products in the surroundings of reactors. pp. 479-486. In: Rapid Methods for Measuring Radioactivity in the Environment. (See IAEA, 1971) (In German) (NSA 26: 28232).

Krishnamoorthy, P. N. 1971. Methods of effluent control in nuclear plants. pp. 317-324. In: Environmental Aspects of Nuclear Power Stations. (See IAEA, 1971) (NSA 25: 28623).

Krishnamurthy, B. S., S. N. Ray and G. C. Joshi. 1962. A note on preliminary field studies of the use of irradiated males for reduction of C. fatigans Wied. populations. Indian J. Malariol. 16: 365-373. (NSA 19: 43796).

Krishnamurthy, K. and S. M. Rao. 1972. Theory for the quantitative estimation of bed load transport using radioisotopes. Isotopenpraxis 8(11-12): 455-457. (Also: India Bhabba Atomic Research Centre, Bombay, report BARC-624) (NSA 27: 17282, 27607).

Krishnamurthy, K. and S. A. Rao. 1973. Comparison of radioactive glasses of scandium and iridium as tracers in sediment-transport studies. J. Hydrol. 19(3): 189-204. (NSA 30: 9564) (Estuaries and rivers).

Krishnaswami, S. 1974. Man-made plutonium in freshwater and marine environments. Proc. Indian Acad. Sci. A80(3): 116-123. (NSA 32: 22282).

† 1931.

Kristensen, H. 1972. Isolation of radiation resistant microorganisms from the environments of cobalt-60 sources; preliminary results. pp. 43-49. In: Experiences in Radiation Sterilization of Medical Products. International Atomic Energy Agency, Vienna, Publ. IAEA-159. (NSA 31: 11548).

Kristensen, K. J. 1973. Depth intervals and topsoil moisture measurement with the neutron depth probe. Nord. Hydrol. 4(2): 77-85. (NSA 29: 12738).

Krisyuk, E. M., et al. 1971. Rapid method of measuring environmental radioactivity used in the USSR. In: Rapid Methods for Measuring Radioactivity in the Environment. (See IAEA, 1971) (CONF-710705-7) (NSA 26: 56151).

Kritsky, G. A. 1958. Effect of x-irradiation on hypoxanthine biosynthesis and glycine metabolism in pigeon liver. Biokhimiya 23(1): 87-91. (Biochemistry (USSR) 23: 80-86).

Krivolutskiy, D. A. and M. N. Fedorova. 1973. Effect of soil heating by ^{239}Pu on field fauna. Zool. zh. 52(4): 601-603. (ORNL-tr-2824) (NSA 30: 15369).

Krivolutskiy, D. A. and T. L. Kozhevnikova. 1972. Effect of radioactive contamination of meadow soil microfauna. Ekologiya. (Sov. J. Ecol. 3(2): 148-151) (NSA 30: 16009).

Krivolutskiy, D. A. and A. D.Pokarzhevskiy. 1974. Role of soil animals in the biogenic migration of Ca and ^{90}Sr. Zh. obshch. biol. 35(2): 263-269. (In Russian) (NSA 30: 12702).

Krivolutskiy, D. A., A. V. Smurov and M. A. Snetkov. 1972. Effect of the radioactive pollution of the soil with ^{90}Sr on variability in some organisms. Zh. obshch. biol. 33(5): 587-591. (In Russian) (NSA 27: 12364).

Krivolutskiy, D. A., A. L. Tikhomirova and V. A. Turchaninova. 1972. Changes in structure of animal populations (terrestrial and soil invertebrates) due to soil contamination with ^{90}Sr. Pedobiologia 12: 374-380. (ORNL-tr-2823) (NSA 30: 9978).

Kröbel, W. and G. Krohm. 1959. The effects of low radiation doses on the signal production and propagation properties in frog nerves. Atomkernenergie 4(7/8): 280-286. (In German).

Kröbel, W. and K. Vanselow. 1962. Investigation of the influence of ion dose rates on the radiation effects on the ischiadicus of frogs. Atomkernenergie 7(12): 464-470. (In German) (NSA 17: 7816).

Kröger, H. 1957. An analysis of x-ray-induced modified wings in the flour beetle Ephestia kühniella. Wilhelm Roux' Arch. f. Entwicklmech. der Org. 150(1): 77-104. (In German).

Kronborg, O. J. and E. Steinnes. 1975. Simultaneous determination of arsenic and selenium in soil ny neutron-activation analysis. Analyst 100(1196): 835-837. (NSA 33: 28291).

Kronfeld, J. 1971. ^{234}U/^{238}U disequilibrium in ground waters of central Texas: a progress report on research completed. pp. 303-310. In: Nuclear Techniques in Geology and Geochemistry. (Final report). Rice Univ., U. S. report GJO-935-1(Pt.2). (NSA 26: 31018).

Krotowicz, J. 1967. Isotopic indicators in the investigation of underground water flow. Przegl. Geol. 15: 30-35. (In Polish) (NSA 21: 22031).

Krouse, H. R. 1967. General comment on microbiological activity and its effect on isotopic studies of water. p. 196. In: Isotope Techniques in the Hydrologic Cycle. (See Stout, 1967).

Krueger, H. R., R. D. O'Brien and W. C. Dauterman. 1960. Relationship be tween metabolism and differential toxicity in insects and mice of Diazinon, Dimethoate, Parathion, and Acethion. J. Econ. Entomol. 53(1): 25-31.

Krueger, H. W. 1965(?). The preservation and dating of collagen in ancient bones. In: Radiocarbon and Tritium Dating. (Proc. 6th Intern. Conf., Washington State Univ., Pullman, June 7-11, 1965). R. M. Chatters and E. A. Olson, comps. (CONF-650652-16) (AED-CONF-65-158-19; NSA 20: 7290).

Kruger, P. 1966. Nuclear civil engineering. Outline syllabus for the course CE 277 given winter quarter, 1966. Stanford Univ., U. S. AEC report TID-23748. iv, 330pp.

Kruger, P. 1969. ^{210}Pb in surface air along slopes of Mauna Loa Volcano, Hawaii. Final report. Dept. of Civil Engineering, Stanford Univ., Tech. Rept. No. 117, U. S. AEC report SU-326-PA-16-3. 24pp. (NSA 24: 16704).

Kruger, P. 1970. Hydrologic transport of radionuclides from nuclear craters and quarries. Stanford Univ., U. S. Army Engineer Nuclear Cratering Group, Livermore, Calif., report NCG-TR-30. 131pp. (NSA 26: 22793).

Kruglov, V. T. and N. M. Vikhrenko. 1969. ^{90}Sr uptake by agricultural plants and animals. Veterinariya 46: 93-95. (In Russian) (NSA 25: 11050).

Krumholz, L. A. 1954. Summary of the findings of the ecological survey of White Oak Creek, Roane County, Tennessee, 1950-1953. Oak Ridge National Lab., U. S. AEC report ORO-132. 54pp. (NSA 8: 2757).

Krumholz, L. A. 1954. An ecological survey of White Oak Creek, 1950-1953. Oak Ridge National Lab., U. S. AEC report ORO-587. 3 vols. vi,158, 180, 208pp. (NSA 17: 18332, 21376, 21932).

Krumholz, L. A. 1962. Radionuclides in the freshwater environment. pp. 360-371. In: Radioactive Fallout from Nuclear Weapons Tests. (See Klement, 1962) (NSA 16: 22456).

Krumholz, L. A. 1963. Biological transfer of natural and fallout radionuclides in streams. pp. 261-265. In: Transport of Radionuclides in Fresh Water Systems. (See Kornegay, et al., 1963) (NSA 17: 40533).

Krumholz, L. A. 1965. A radioecological study of the biota of Doe Run, Meade County, Kentucky. Final report. Univ. of Louisville, Ky., U. S. AEC report. 71pp.

Krumholz, L. A. 1967. Accumulation of radioactive fallout materials in the biota of Doe Run, Meade County, Kentucky, 1959-1963. pp. 791-818. In: Radioecological Concentration Processes. (See Aberg and Hungate, 1967).

Krumholz, L. A. and A. H. Emmons. 1953. Preparation of fish tissues for gross beta radioassay. J. Wildl. Mgmt. 17(4): 456-459.

Krumholz, L. A. and R. F. Foster. 1957. Accumulation and retention of radioactivity from fission products and other radiomaterials by fresh-water organisms. pp. 88-95. In: The Effects of Atomic Radiation on Oceanography and Fisheries. (See National Academy of Sciences, 1957).

Krumholtz, L. A. and J. H. Rust. 1954. Osteogenic sarcoma in a muskrat from an area of high environmental radiostrontium. A.M.A. Arch. Pathol. 57: 270-278.

Krumholz, L. A. and C. S. Shoup. 1953. Summary of papers on ecological effects of radioactive effluents. pp. 481-483. In: The Role of Atomic Energy in Agricultural Research. (Proc. Oak Ridge Summer Symp., 4th Ann. Symp., Oak Ridge, Tenn., 1952). U. S. AEC report TID-5115. (NSA 8: 2757).

Krumholz, L. A., E. D. Goldberg and H. Boroughs. 1957. Ecological factors involved in the uptake, accumulation, and loss of radionuclides by aquatic organisms. pp. 69-79. In: The Effects of Atomic Radiation on Oceanography and Fisheries. (See National Academy of Sciences, 1957).

Krzywicka, A. M. and G. H. Wagner. 1975. Laboratory culture for ^{14}C-labeling of Chlorella and Oedogonium. Int. J. Appl. Radiat. Isotopes 26(9): 515-518. (NSA 33: 9697).

Kubelka, V. and P. Strohal. 1967. Determination pf the underground water-flow rate in a borehole in the karst by means of the radioactive method. Mem. Ass. Int. Hydrogeol. Reunion Istanbul 8: 294-297. (See Anonymous, 1967) (NSA 23: 447).

Kubica, J., B. Bukowska and Z. Cichowlas. 1962. Changes in intestinal micro-flora in x-irradiated rats. Acta Microbiol Polon. 11(1/2): 111-118. (In Polish, English summary) (NSA 17: 27074).

Kudinov, M. A. 1968. Increase in the soil germination of seeds of woody plants under the influence of gamma radiation. Radiobiologiya 8(5): 763-764. (AEC-tr-7023, pp. 165-167) (NSA 24: 921).

Kudinov, M. A. 1971. Resistance of carrageen and indigo bush to gamma rays. Radiobiologiya 11(2): 292-294. (AEC-tr-7269, pp. 208-211) (NSA 25: 44976).

Kudinov, M. A. B. and B. Issled. 1967. Intensity of the photosynthesis of Oregon ash grown from irradiated seeds. Bot. issled. Belorussk. otd. vses. bot. obshch. (9): 134-136. (In Russian) (NSA 23: 8085) (Should be Kudinov, M. A. 1967. ?).

Kudo, A. 1969. Radioactivity transport in water: interaction between flowing water and bed sediment. Thesis, Univ. of Texas, Austin. 168pp. (NSA 24: 9519).

Kudo, A. and E. F. Gloyna. 1971. Transport of ^{137}Cs. II. Interaction with bed sediments. Water Research 5: 71-79. (NSA 25: 29204).

Kudritskiy, Yu. K. (Ed.). 1973. All-Union Symp. on Theoretical and Practical Aspects of the Effects of Small Doses of Ionizing Radiation. (Syktyvkar, USSR, September 4-6, 1973). Syktyvkar Inst. Biol., Akad. Nauk SSSR. 207pp. (In Russian) (CONF-730996-(Absts.) (NSA 30: 3173).

Kudzinau, M. A. 1966. Gamma radiation effects on the growth and development of seedlings of a variety of maples and birches. Vestsi Akad. Navuk Belorussk. SSR (biol.) (3): 54-63. (In Ukrainian) (NSA 21: 22784).

Kuehnert, C. C. and J. P. Miksche. 1964. Application of the 22.5 Mev deuteron microbeam to the study of morphogenic problems within the shoot apex of Osmunda calytoniana. Am. J. Botany 51: 743-747. (NSA 19: 8807).

Kuhlman, H. H. 1961. Studies on the morphology and growth rate of the cestode Hymenolepis microstoma after gamma irradiation of the cysticercoids and eggs. Dissertation, Univ. of Tennessee, Knoxville.

Kuhn, C. W. and H. J. Teas. 1968. Increased resistance to tobacco mosaic virus in gamma irradiated tobacco leaves. Radiation Botany 8: 317-323. (NSA 22: 52027).

Kühn, W. K. G. 1967. Determination by indicator activation analysis of the sorption conditions, the movement, and the de-enrichment factor of anions at a site on the Danube. pp. 271-278. In: Isotope and Radiation Techniques in Soil Physics and Irrigation Studies. (See IAEA, 1967) (NSA 22: 23158).

Kühn, W. and J. Handl. 1971. Moisture measurement in thin soil layers by absorption of gamma-rays in a capillary porous body. Kerntech. Atompraxis. 13: 15-17. (In German) (NSA 25: 23709).

Kühn, W. and H. P. Schätzler. 1973. Moisture measurement in thin soil layers by absorption of gamma rays in a capillary porous body. pp. 67-76. In: Soil-Moisture and Irrigation Studies. II. (See IAEA, 1973) (NSA 28: 21374).

Kühn, W., J. Handl and H. P. Schätzler. 1973. Transport of ^{131}I, ^{137}Cs, ^{106}Ru, ^{144}Ce, and ^{54}Mn in an undisturbed soil under natural conditions. pp. 347-362. In: Environmental Behavior of Radionuclides Released in the Nuclear Industry. (See IAEA, 1973) (NSA 29: 29344).

Kujala, N. F., I. L. Larsen and C. L. Osterberg. 1969. Radioisotope measurements of the viscera of Pacific salmon. pp. 440-449. In: Symposium on Radioecology. (See Nelson and Evans, 1969)

Kukita, Y. and N. Egami. 1969. Effects of x irradiation on rejection of transplanted scale in goldfish. Dobutsugaku Zasshi 78: 112-113. (In Japanese) (NSA 23: 41448).

Kulebakina, L. G. 1969. Dependence of the coefficients of accumulation of ^{90}Sr in hydrobionts on the ash content. Radiobiologiya 9(5): 776-777. (AEC-tr-7136, pp. 182-185) (NSA 24: 11912).

Kulik, I. L. 1967. Use of radioactive isotopes in ecological and biocoenotic studies of terrestrial vertebrates. Zool. zh. 46(8): 1234-1246. (In Russian, English summary) (BA 49: 1968).

Kulikov, N. V. 1957. Effect of uranium debris on biomass and on the structure of experimental phytocoenosis. Bot. zh. 42: 377-394. (In Russian).

Kulikov, N. V. 1957. Studies in experimental biogeocoenology. II. The effect of soaking seeds in a mixture of beta-emitters on the biomass and structure of an experimental phytocoenosis. Tr. inst. biol., Akad. Nauk SSSR, Ural'skiy filial (9). Sbornik rabot. lab. biofizika 1: 252-291. (In Russian).

Kulikov, N. V. 1960. Accumulation of Co^{60} during the fall-winter season and its distribution in birch seedlings. Bot. zh. 45: 1549-1552. (In Russian).

Kulikov, N. V. 1970. Radiosensitivity of tench (Tinca tinca L.) embryos at early stages of development. Radiobiologiya 10(1): 127-130. (AEC-tr-7154, pp. 201-206) (NSA 24: 27693).

Kulikov, N. V. 1970. Radiosensitivity of roe of pike (Esox lucius L.) during fertilization and early cleavage. Radiobiologiya 10: 768-770. (AEC-tr-7215, pp. 201-204) (NSA 25: 29659).

Kulikov, N. V. 1971. Radioecology of freshwater plants and animals. pp. 367-384. In: Contemporary Problems of Radiobiology. Vol. 2. Radioecology. (See Klechkovskiy, Polikarpov and Aleksakhin, 1971) (Transl. pp. 323-337).

Kulikov, N. V. 1973. Radionuclides in the system water-freshwater plants. pp. 44-55. In: Radioecology of Water Organisms. II. (See Andrumaitis, et al., 1973) (AEC-tr-7606, pp. 25031) (NSA 30: 21130).

Kulikov, N. V. and N. M. Korobitsyn. 1968. Effects of the vegetable cover on the vertical migration of ^{144}Ce in the soil. Tr. inst. ekol. rast. zhivot., Ural. fil. Akad. Nauk SSSR (61): 47-54. (In Russian) (NSA 24: 7776).

Kulikov, N. V. and I. V. Molchanova. 1974. Reduction of ^{90}Sr content of soil-plant cover of tundra. Ekologiya 5(2): 69-70. (Sov. J. Ecol. 5(2): 150-151) (NSA 32: 8923).

Kulikov, N. V. and I. V. Molchanova. 1975. Continental Radioecology (Soil and Freshwater Ecosystems). "Nauka", Moscow. 184pp. (In Russian).

Kulikov, N. V. and I. V. Molchanova. nd. Radioisotopes in the soil-solution system. (INIS-mf-887-3, pp. 2-13) (In Russian).

Kulikov, N. V. and L. I. Piskunov. 1970. Role of plants in the vertical migration of ^{90}Sr and ^{137}Cs in soil. Agrokhimiya 7(7): 115-121. (In Russian) (NSA 25: 39810).

Kulikov, N. V. and L. I. Piskunov. 1971. Quantitative estimation of the action of vegetation on the vertical migration of radioactive isotopes in soil. pp. 58-66. In: Methods of Radioecological Investigations. (See Verkhovskaya, 1971) (In Russian).

Kulivov, N. V. and L. N. Ozhegov. 1975. Accumulation of ^{60}Co by developing eggs pike, perch and rig (Coregonus). Ekologiya 6(1): 100-102. (In Russian).

Kulikov, N. V., V. A. Bezel' and L. N. Ozhegov. 1970. Accumulation of radioisotopes by developing roe of tench (*Tinca tinca* L.) and perch (*Perca fluviatilis* L.). Ekologiya 1(5): 73-77. (Sov. J. Ecol. 1(5): 425-428) (NSA 26: 48061).

Kulikov, N. V., V. G. Kulokova and S. A. Lyubimova. 1971. The movement of strontium-90 and cesium-137 along with fish eggs during spawning season. Ekologiya 2(4): 12-16. (Sov. J. Ecol. 2(4): 296-299).

Kulikov, N. V., S. A. Lyubimova and D. G. Fleyshman. 1971. Study of the accumulation of ^{137}Cs by freshwater plants. pp. 94-100. In: *Methods of Radioecological Investigations*. (See Verkhovskaya, 1971) (In Russian).

Kulikov, N. V., S. A. Lyubimova and N. A. Timofeyeva. 1970. Role of freshwater plants in processes of coprecipitation of ^{90}Sr with calcium carbonate. Ekologiya 1(4): 55-58. (Sov. J. Ecol. 1(4): 315-317) (NSA 26: 15551).

Kulikov, N. V., I. V. Molchanova and Ye. N. Karavayeva. 1973. Influence of the condition of soil moisture on the passage of strontium-90, cesium-137, and cerium-144 from the soil into solution. Ekologiya 4(4): 57-62. (Sov. J. Ecol. 4(4): 320-323) (NSA 32: 6433).

Kulikov, N. V., I. V. Molchanova and L. I. Piskunov. 1975. On the migration of ^{90}Sr in soil-plant cover of subarctic tundra. Ecologiya 6(3): 88-90. (In Russian).

Kulikov, N. V., N. A. Timofeyeva and L. K. Al'shits. 1969. Influence of ionizing radiations on the early stages of development of *Limnaea stagnalis* L. Radiobiologiya 6(6): 908-912. (AEC-tr-6775, pp. 228-236) (NSA 21: 36717).

Kulikov, N. V., N. A. Timofeyeva and L. K. Shishenkova. 1968. Radiosensitivity of developing tench embryos (*Tinca tinca* L.). Radiobiologiya 8(3): 391-395. (AEC-tr-6916, 75-81) (NSA 22: 51954).

Kulikov, N. V., et al. 1971. Change in the radiosensitivity of plants as a result of preliminary radiation influence. Radiobiologiya 11(4): 630-632. (AEC-tr-7304, pp. 201-204) (NSA 26: 36358) (Scotch pine seeds).

Kulivova, G. M. 1960. The feasibility of using confervoid algae for the purification of radioactively contaminated water. Dokl. Akad. Nauk SSSR 135(4): 978-980. (Doklady (biol. sci.) 135(1-6): 874-876).

Kulikova, V. G. and N. V. Kulikov. 1973. Migration of ^{90}Sr from fish organisms into progeny. pp. 169-176. In: *Radioecology of Water Organisms. II*. (See Andrumaitis, et al., 1973) (Transl. pp. 101-104) (NSA 30: 21432).

Kulikova, V. G., Ye. L. Beregovaya and N. V. Kulikov. 1972. Transfer of strontium-90 from the fish organism to progeny. Ekologiya 3(4): 48-52. (Sov. J. Ecol. 3(4): 328-331) (NSA 30: 16058).

Kulkarni, P. V. and I. L. Preiss. 1975. Use of radioisotope induced by x-ray fluorescence for environmental studies. J. Radioanal. Chem. 24(2): 423-432. (NSA 32: 2996).

Kulkarni, V. V. 1972. Distribution of natural radioactivity and trace elements in the soils and sands from the high radiation coastal belt of India. Thesis, Bombay Univ., India. 128pp. (NSA 30: 305).

Kulkarni, V. V., T. N. V. Pillai and A. K. Ganguly. 1973. Distribution of natural radioactivity and trace elements in the soils and sands from the high radiation coastal belt of India. India Bhabba Atomic Research Centre, Bombay, report BARC-702. 104pp. (NSA 31: 5982).

Kullyev, P. 1965. Content of nucleic acids in silk glands of silkworm varieties differing by their productivity. I. Akad. Nauk Turkmen. SSR (biol.) 1: 70-74. (In Russian) (^{3}H, *Bombyx mori*).

Kullyev, P., et al. 1964. RNA (ribonucleic acid) biosynthesis in the silk gland of the silkworm. Biokhimiya 29(3): 470-476. (Biochemistry (USSR) 29(3): 407-412) (^{14}C, ^{32}P).

Kulp, J. L. and A. R. Schulert. 1961-62. Strontium-90 in man. Vol. I. Summary. Vol. II. Analytical data. Lamont Geological Observatory, Columbia Univ., U. S. AEC report NYO-9934. vp. (NSA 16: 14448, 20870) (Foods).

Kulp, J. L. and A. R. Schulert. 1962. Strontium-90 in man V. Science 136 (3516): 619-632. (Incl. foods).

Kumagai, M. 1967. Influence of post-treatment humidity on the irradiated rice weevil adult, *Sitophilus zeamais* Motschulsky (Coleoptera: Curculioniade). Appl. Entomol. Zool. 2(1): 51-57.

Kumagai, M. 1969. Effects of irradiation on the eggs of adzuki-bean weevil, *Callosobruchus chinensis* L. (Coleoptera: Bruchidae). Appl. Entomol. Zool. 4: 9-15. (NSA 23: 48195).

Kumagai, Y. 1961. On the effects of ionizing radiation emitted from ^{60}Co upon the development of ascaris eggs. Report II. Studies on the relationship between irradiation dose and its inhibitory effects upon the development of eggs. Nippon Univ. J. Med. 3: 301-320. (NSA 18: 27139).

Kumagai, Y., R. Kumagai and N. Mori. 1974. On the effects of ionizing radiation upon the development of ascaris eggs. 3. Nippon Univ. J. Med. 16 (3): 229-240. (NSA 32: 19994).

Kumar, A. and P. N. Srivastava. 1971. Histopathological study of spleen after 600 r ^{60}Co external irradiation in Indian desert gerbil (*Meriones hurrianae* Jerdon). Strahlentherapie 142(6): 721-725. (NSA 26: 31356).

Kumar, S. S. and E. Hodgson. 1964. Metabolism of choline in *Periplaneta americana*. Abstr. Bull. Entomol. Soc. Am. 10(3): 164. (^{14}C).

Kumpf, W., B. Dieterich and J. Kumpf. 1956. Radioactive substances and water: a documentation. R. Oldenbourg Verlag, Munich. 140pp. (In German) (NSA 12: 11267).

Kunasheva, K. G. 1944. The amount of radium in plants and animals. Tr. biogeokhim. lab. Akad. Nauk SSSR 7: 98-105. (In Russian).

Kunasheva, K. G. and B. K. Brunovskiy. 1935. The influence of soil sand on the determination of radium in plants. Tr. biogeokhim. lab. Akad. Nauk SSSR 3: 45-48. (In Russian).

Kunc, F. 1974. Respiration and metabolic processes of bacteria in autoclaved and gamma-irradiated soil. Bull. Astron. Inst. Czech. 25(3): 218-228. (NSA 30: 24204.

Kunert, J. 1972. Radiation protection problems of discharge of low radioactivity waste waters from laboratories into sewer systems and surface wat ers. German Democratic Republic Staatliche Zentrale für Strahlenschutz, Berlin, report. (In German) (INIS-mf-1236, pp. 138-149) (NSA 33: 590).

Kunert, J. 1973. Environmental monitoring and special surveillance of the contamination of flowing waters with long-lived radionuclides. German Democratic Republic Staatliche Zentrale für Strahlenschutz, Berlin, report SZS-148. pp. 130-133. (For: German-Soviet Working Meeting on the Question of Radiation Protection, Berlin, 8 June 1972) (In German) (NSA 30: 18640).

Kunert, J. and F. Moldenhawer. 1974. Results and conclusions of fallout monitoring. pp. 116-126. In: Radiation Protection, Environment, and the Population. (See German Democratic Republic, 1974) (In German) (NSA 31: 8617).

Kunert, J. and D. Schneider. 1969. Fallout radioactivity surveillance of waters in the German Democratic Republic, 1961-1968. German Democratic Republic Staatliche Zentrale für Strahlenschutz, Berlin, report SZS-19/69. 41pp. (In German) (NSA 24: 27221).

Kunert, J. and H. Zindler. 1973-74. Results of fallout monitoring in the German Democratic Republic (in 1971?). --- in 1972. German Democratic Republic Staatliche Zentrale für Strahlenschutz, Berlin, report SZS-152, SZS-161. (In German) (NSA 30: 6485; 31: 527).

Kunert, J., et al. 1972. Results of fallout monitoring in the German Democratic Republic 1970. German Democratic Republic Staatliche Zentrale für Strahlenschutz, Berlin, report SZS-128. 47pp. (In German) (NSA 26: 53156).

Künkel, H. A. 1959. Effects of higher radiation doses on the dormouse. Z. Naturforsch. 14b: 757-759. (In German) (NSA 14: 7268).

Künkel, H. A. 1959. Incorporation of radioactive iron into erythrocytes of irradiated dormaouse. Naturwissenschaften 46(22): 627-628. (In German).

Künkel, H. A. 1961. Investigations of radiation biology on hibernating dormouse. Chemotherapie 3(3): 200-224. (In German) (BA 37(5): __; WR 106: 35).

Künkel, H. A. 1965. Dependence of the biological effect on radiation frequency. pp. 175-178. In: Symposium on High-Energy Electrons. (Montreux, Switzerland, 7-11 September 1964). A. Zuppinger, ed. Springer-Verlag, New York. (In German) (NSA 22: 51758) (Drosophila).

Künkel, H. A. and F. Oberheuser. 1962. Ultrafractionation and biological efficiency of fast electrons. Investigations on "eggs" of Drodophila melanogaster of different stages of development. Abstr. p. 6. In: 2nd Intern. Congr. Radiation Research. (See Anonymous, 1962).

Künkel, H. A., G. Höhne and H. Maass. 1956. Radiobiological investigations on the hibernating lair (Glis glis). pp. 176-180. In: Advances in Radiobiology. G. C. de Hevesy, A. G. Forssberg and J. S. Abbatt, eds. Oliver and Boyd, Edinburgh-London.

Künkel, H. A., G. Höhne and H. Maass. 1957. The effect of cysteine and hibernation on the food intake of the irradiated dormouse (Glis glis). Z. Naturforsch. 12b(3): 144-147. (In German).

Kuntz, J. E. and A. J. Riker. 1956. The use of radioactive isotope to ascertain the role of root grafting in the translocation of water, nutrients, and disease-inducing organisms among forest trees. Proc. Intern, Conf. on the Peaceful Uses of Atomic Energy 12: 144-148. (Paper A/CONF.8/P/105).

Kunz, S. E. and J. L. Eschle. 1971. Possible use of the sterile-male technique for eradication of the horn fly. pp. 145-156. In: Sterility Principle for Insect Control or Eradication. (See IAEA, 1971) (NSA 25: 29744).

Kunz, W. F. 1972. Response of the alga Chlorella sorokiniana to ^{60}Co gamma radiation. Nature 236(5343): 178-179. (NSA 26: 33656).

Kuper, S. W. A and S. R. Pelc. 1952. Radioactive labelling experiments with Plasmodium. Parasitology 42(1/2): 269-271.

Kupriyanova, V. M. and Ye. S. Romanova. 1958. Preliminary data on the characteristics of radioactivity of surface waters and possibilities of their pollution with sewage. Gigiyena sanitariya 23(10): 71-72. (Also: Int. J. Appl. Radiat. Isotopes 5(3): 1-14(1959)) (NSA 13: 20072).

Kuptsov, V. M. and V. V. Cherdyntsev. 1969. Decay products of uranium and thorium in active volcanism in the USSR. Geokhimiya (6): 643-658. (Geochem. Int. 6: 532-545) (NSA 23: 31473).

Kuptsov, V. M., V. V. Cherdyntsev and V. L. Zverev. 1969. Radioisotopes in water of neovolcanic regions of the Caucasus. Geokhimiya (9): 1153-1155. (Geochem. Int. 6: 919-922) (NSA 24: 36406).

Kurata, K. 1968. Incidence of virus diseases in asceptically-grown silkworm, Bombyx mori L. 2. Effect of feeding of chemicals and irradiation of gamma-rays. Norinsho Sanshi Shikenjo Hokoku 23: 149-171. (In Japanese) (NSA 23: 40868).

Kurbatov, L. M. 1936. Radioactivity of bottom deposits of seas and lakes in the USSR. Arktika 4: 95. (In Russian).

Kurchatov, B. V., et al. 1959. Investigation of the radioactive strontium content of the atmosphere, soil, food products, and human bones. In: Soviet Scientists on the Danger of Atomic Testing. A. V. Lebedinskiy, ed. Atomizdat, Moscow. (UCRL-Trans-549).

Kurczne, C. I., et al. 1972. Radioactivity of some Hungarian springwaters and the dose originating from their consumption. Izotoptechnika 15(6): 354-357. (In Hungarian) (NSA 27: 7787).

Kurdyumov, G. V., M. B. Neyman and G. M. Frank. 1957. The use of radioactive isotopes in the USSR. Atom. energiya 3(11): 465-478. (Sov. Atomic Energy 3(11): 1345-1360) (NSA 12: 10283).

Kurihara, K. 1970. Present status of recovery of uranium. Recovery of uranium by bacterial leaching. Genshiryoku Kogyo 16(12): 23-27. (In Japanese) (NSA 25: 37482).

Kurihara, T. 1963. Radioactive tagging of _Culex pipiens_ with P^{32}. Jap. J. Sanit. Zool. 14(1): 16-19.

Kurir, A. 1963. Possibility of control of termites with atomic waste. Holzforsch. Holzverwert. 15: 67-70. (In German) (NSA 20: 10644).

Kuroda, T. 1968. Distribution of naturally occurring radionuclides belong ing to the uranium decay series. 5. ^{210}Po and ^{226}Ra concentrations in drinking water and air. Igaku to Seibutsugaku 76: 34-37. (In Japanese) (NSA 23: 40868).

Kurokawa, Y. 1971. Data on natural radioactivity in Japan. Koken Butsuri 6(4): 192-193. (In Japanese) (NSA 27: 310).

Kursanov, A. L. 1963. Metabolism and the transport of organic substances in phloem. pp. 209-278. In: _Advances in Botanical Research_. _Vol. I_. R. D. Preston, ed.

Kursanov, A. 1971. Isotopes in plant physiology and biochemistry. Agrochimica 15(2-3): 101-124.

Kurz, J. C. and R. L. Marchinton. 1972. Radiotelemetry studies of feral hogs in South Carolina. J. Wildl. Mgmt. 36(4): 1240-1248. (At Savannah River Plant).

Kusaka, Y. and T. Ozaki. 1972. Zinc analysis in land water samples by radioisotope dilution method. Jap. Anal. 21(10): 1352-1358. (In Japanese) (NSA 27: 11685).

Kusakari, S. I. and Y. Tagaki. 1973. Fungistatic activity of soil sterilized by gamma radiation. Can. J. Microbiol. 19(10): 1333-1334. (NSA 29: 26994).

Kushnuruk, V. A. 1962. The radiosensitivity of birds. Probl. lab. radiobiol. L'vovsk. gos. univ. (1): 81-83. (JPRS-23169; OTS-64-215311) (NSA 17: 30416).

Kusumgar, S., D. Lal and R. P. Sarna. 1963. Radiocarbon dating: results. Proc. Indian Acad. Sci. A58: 141-152.

Kusumgar, S., D. Lal and V. K. Sharma. 1963. Radiocarbon dating: techniques. Proc. Indian Acad. Sci. A58: 125-137.

Kutlakhmedov, Ye. A. and D. M. Grodzinskiy. 1973. Phenomonological peculiarities of the photomodification of radiation damage in the blue-green alga _Plectonema boryanum_. Radiobiologiya 13(3): 448-451. (AEC-tr-7471, pp. 159-163) (NSA 30: 6586).

Kuusi, J. 1973. Application of nuclear techniques in industry and environmental studies. Finland Technical Reseach Centre, Otaniemi. (NP-19936-P1-2; Paper 16, vp) (INIS-mf-1168/2) (NSA 29: 20739; 30: 15742).

Kuusi, J. and R. Kuoppamaeki. 1974. Development of methods and instruments for environmental tracer studies. pp. 430-442. In: _Proc. 2nd Intern. Conf. on Nuclear Methods in Environmental Research_. (See Vogt and Meyer, 1974) (NSA 33: 23300).

Kuwatsuka, S. and J. E. Casida. 1965. Synthesis of methylene-C^{14}-doxyphenyl compounds: radioactive safrole, dihydrosafrole, myristicin, piperonyl butoxide, and diasteroisomers of sulfoxide. J. Agric. Food Chem. 13(6): 528-533. (^{14}C, _Musca domestica_).

Kuzin, A. M. 1954. Marked atoms in research on agriculture. Akad. Nauk SSSR, Moscow. (In Russian).

Kuzin, A. M. 1959. _The Threat of Nuclear Explosions to Mankind_. Akad. Nauk SSSR, Moscow. 131pp. (In Russian).

Kuzin, A. M. 1960. The application of radioisotopes in biology. Developments in the Peaceful Application of Nuclear Energy No. 7. Review Series. International Atomic Energy Agency, Vienna, Publ. STI/PUB/15/7). (In Russian and English) (NSA 15: 10645).

Kuzin, A. M. 1960. The danger of increasing ^{14}C concentration in the atmosphere. Vest. Akad. Nauk SSSR (9): 48-51. (Also: UN Document A/AC.82/G/L. 413) (AEC-tr-4532).

Kuzin, A. M. 1964. Radiation ecology. Ch. 11, pp. 360-378. In: _Radiation Biology_. A. M. Kuzin and N. I. Shapiro, eds. Publ. House Acad. Sci., USSR, Moscow. (In Russian).

Kuzin, A. 1966. Radiobiology and space investigations. Aviat. Kosmonavt. (8): 25-41. (In Russian) (Incl. microorganisms, plants).

Kuzin, A. M. and Z. Angelova-Patarinska. 1970. Stimulating effect of small doses of x radiation of the simultaneous gemmation of yeast cells. Compt. rend. bulg. sci. 23(11): 1421-1424. (NSA 25: 47038).

Kuzin, A. M. and A. K. Kasymov. 1963. Appearance pf substances which depress growth and development in irradiated potato tubers. Radiobiologiya 3(3): 472-476. (AEC-tr-5436, pp. 203-209).

Kuzin, A. M. and P. F. Minayev. 1966. Postirradiative modification of radiation damage in frogs (_R. ridibunda_). Radiobiologiya 6(5): 690-692. (AEC-tr-6774, pp. 72-76) (NSA 21: 10933).

Kuzin, A. M. and A. A. Peredel'skiy. 1956. Natural conservation and several problems of radioactive-ecological connections. Okhrana prirody zap. delo SSSR Byull. 1: 65-78. (In Russian).

Kuzin, A. M. and V. I. Tokarskaya. 1959. Complete labelling of the organic substances of plants by radioactive carbon as a method for studying metabolic disturbances. Biokhimiya 24(1): 80-86. (Biochemistry (USSR) 24(1): 71-77) (NSA 12: 21851).

Kuzin, A. M. and M. P. Silayev. 1963. A new possible way of using ionizing radiation in the meat industry. Radiobiologiya 3(4): 545-548. (AEC-tr-5437, pp. 80-85).

Kuzin, A. M. and N. I. Yusfov. 1967. Radiation sensitivity of different stages of development of _Ephestia ruchniella_ L. in the light of the concept of DNA transinduction. _Radiobiologiya 7(1)_: 3-6. (AEC-tr-6886, pp. 1-7) (NSA 21: 36711).

Kuzin, A. M., A. K. Kasymov and L. M. Kryukova. 1964. On the mechanism of stimulating and supressing effect of irradiation of potato tubers. Radiobiologiya 4(1): 144-149. (AEC-tr-6404, pp. 201-208).

Kuzin, A. M., L. V. Slozhenikina and V. L. Migushina. 1971. Study of radiosensitivity of thyroxine induction of urea-forming enzymes in _Rana temporaria_ tadpoles. Radiobiologiya 11(5): 754-757. (AEC-tr-7306, pp. 169-174) (NSA 26: 38676).

Kuzin, A. M., Ye. K. Uzorin and B. I. Chirkovsky. 1963. Investigation of remote after-effects of gamma-irradiation of seeds on some _Nicotiana_ plants. Radiobiologiya 3(6): 903-909. (AEC-tr-5439, pp. 188-197)

Kuzin, A. M., et al. 1964. On the mutagenic effectiveness of incorporated ^{14}C. Radiobiologiya 4(6): 804-809. (AEC-tr-6409, pp. 5-12) (NSA 19: 17606).

Kuznetsov, S. I. 1955. The use of radiocarbon dioxide (C^{14}) for the study of comparative values of photo- and chemosynthesis in lakes of various types. In: _Isotopes in Microbiology_. (Trans. Conf. on the Application of Tracer Atoms in Microbiology). A. A. Imshenetskiy, ed. Publ. House Acad. Sci. USSR, Moscow. (AEC-tr-3494, pp. 132-143).

Kuznetsov, S. I. 1956. Application of radioactive isotopes to the study of processes of photosynthesis and chemosynthesis in lakes. Proc. Intern. Conf. on the Peaceful Uses of Atomic Energy 12: 368-376. (Paper A/CONF.8/P/700).

Kuznetsov, S. I. 1956. The problem of the possibility of "radiosynthesis". Mikrobiologiya 25(2): 195-199. (In Russian).

Kuznetsov, S. I. (Ed.). 1968. _Chemistry of Inland Waters and Factors of Their Pollution and Purification_. Publ. House Leningrad Sect. Acad. Sci. USSR, Leningrad. 255pp. (In Russian) (NSA 24: 25436) (Incl. radioactivity).

Kvaliashvili, A. A., et al. 1958. Effect of hibernation upon the manifestation of radiation sickness in the Transcaucasian hamster. Soobshch. Akad. Nauk Gruz. SSR 20(5): 601-607. (In Russian).

Kvaratskheliya, N. T. and G. G. Glonty. 1965. Migration of ^{90}Sr in the soils of Georgia. Pochvovedeniye (10): 64-__. (Sov. Soil Science (10): 1187-1193(1966)).

Kvaratskhelya, N. T., et al. 1967. The influence of some natural factors on the behavior of radioactive strontium in soils. pp. 19-24. In: _Radio-ecological Concentration Processes_. (See Aberg and Hungate, 1967).

Kvelland, I. 1962. Radiosensitivity in different stages of spermatogenesis in _Drosophila melanogaster_. Hereditas 48(2): 221-242. (NSA 17: 17841).

Kvelland, I. 1965. Some observations on the mating activity and fertility of _Drosophila melanogaster_ males. Hereditas 53: 281-306. (NSA 20: 10644).

Kwapulinski, J. 1968. Distribution of beta activity in the bottom material from Kozlov Mountain water basin. Nukleonika 13: 239-249. (In Polish) (NSA 22: 27900).

Kwapulinski, J. 1969. Influence of some factors on the radioactivity of water in the shallow reservoir of the Barrage. Nukeonika 14: 107-122. (In Polish) (NSA 23: 40871).

Kwapulinski, J. 1972. Statistical analysis of β radioactivity of the periphyton in the shallow reservoir of the Barrage. Nukleonika 17(1): 529-535. (In Polish) (NSA 28: 382).

Kwapulinski, J. 1972. β-radioactivity of fungi in Poland. Acta Geophys. Pol. 20: 217-222. (In Polish) (NSA 29: 20739).

Kwapulinski, J. 1972. Beta radioactivity of periphyton in certain dam reservoirs. Acta Hydrobiol. 14(3): 273-286. (NSA 30: 15460).

Kwapulinski, J. 1973. Influence of some factors on the radioactivity of bottom sedimentation in a limnic reservoir. Nukeonika 18(5): 219-228. (In Polish) (NSA 30: 15424).

Kwapulinski, J. 1973. Characteristic of the radioactive contamination of the Vistula River. Gaz, Woda Tech. Sanit. 47(9): 302-305. (In Polish) (NSA 33: 29106).

Kwapulinski, J. 1973. Cumulation of radioactive substance in dam reservoirs. Acta Hydrobiol. 15(2): 215-225. (NSA 31: 19594).

Kwapulinski, J. 1974. Influence of mineralization of rainfall and snowfall on the specific radioactivity. Nukleonika 19(4): 347-353. (In Polish) (NSA 30: 15424).

Kwapulinski, J. and R. Danecka-Palasz. 1973. Influence of rain and of dustiness on the concentration of ^{90}Sr in different reservoirs. Acta Geophys. Pol. 21(2): 139-151. (In Polish) (NSA 28: 24481).

Kwapulinski, J. and T. Trzcionka. 1970. Problem of the co-relation between the specific radioactivity of surface water and the residue after heating. Woda Tech. Sanit. 44: 203-205. (In Polish) (NSA 26: 190).

Kwapulinski, J. and T. Zubik. 1972. Influence of some hydrological parameters on changes in the radioactivity of the waters of rivers Czarna Przemsza and Przemsza. Acta Hydrobiol. 14(3): 257-271. (NSA 30: 18643).

Kwiatkowski, T. and W. Surewicz. 1971. Effect of ionizing radiation on the chemical components of spruce wood irradiated in different media. Nukleonika 16(1-2): 23-29. (In Polish) (NSA 25: 46500).

Kwon, S. H. and K. H. Jeong. 1973. Use of radiation for pest control. J. Korean Nucl. Soc. 5(1): 65-71. (In Korean) (NSA 28: 8476).

Kylin, A. 1966. Uptake and loss of Na^+, Rb^+, and Cs^+ in relation to an active mechanism for extrusion of Na^+ in Scenedesmus. Plant Physiol. 41: 579-584.

Kyrs, M., L. Holeckova and L. Neumann. 1962. Cs-137 concentrations and isolation from supply and river water by the extraction of cesium polyiodides with nitrobenzene. Jaderna Energia 8: 429-431. (In Czech) (NSA 17: 8455).

Kyun'ke, L. M. and N. A. Dubinskaya. 1974. Determination of molybdenum in plants by neutron activation analysis. Latv. PSR Zinat. Akad. Vestis. fiz. teh. Zinat. ser. (2): 12-17. (In Russian) (NSA 30: 14795).

Kyuregyan, T. N. 1970. Radon waters in the Caucasus. Isv. Akad. Nauk Armyan. SSR (zemle) 23(4): 92-93. (In Russian) (NSA 26: 22798).

L., S. 1957. Absorption of radio-active substances by molluscs. Atom. energiya 3(9): 275-276. (Sov. Atomic Energy 3(9): 1079).

Labaw, L. W., V. M. Mosley and R. W. G. Wyckoff. 1950. Radioactive studies of the phosphorus metabolism of Escherichia coli. J. Bacteriol. 59(2): 251-262.

Labeyrie, J., C. Lalou and G. Delibrias. 1967. Study of marine transgressions on a Pacific atoll by means of ^{14}C and the $^{234}U/^{230}Th$ ratio. pp. 349-358. In: Radioactive Dating and Methods of Low-Level Counting. (See IAEA, 1967) (In French).

Labeyrie, J., et al. 1967. Study of temperatures prevailing in ancient times by measurement of the ^{18}O, ^{13}C, and ^{14}C content of concretions in caves. pp. 153-160. In: Radioactive Dating and Methods of Low-Level Counting. (See IAEA, 1967) (NSA 22: 6455).

Labrador, J. R. 1963. Application of radioisotopes in entomology. Ingenieria Agron. 11: 19-25. (In Spanish).

LaBrecque, G. C. and J. C. Keller. (Eds.). 1965. Advances in insect population control by the sterile-male technique. Panel report. International Atomic Energy Agency, Vienna, Tech. Rept. Ser. No. 44. 81pp. (Dacus olea, Ceratitis capitata, et al.).

LaBrecque, G. C., D. W. Meifert and D. E. Weidhaas. 1975. Potential of the sterile-male technique for the control or eradication of stable flies, Stomoxys calcitrans Linnaeus. pp. 449-460. In: Sterility Principle for Insect Control, 1974. (See IAEA, 1975) (NSA 32: 11799).

Lacaze, J. C., C. Hallopeau and M. Voight. 1969. Preliminary laboratory research in experimental brackish ecosystems. Bull. Mus. Nat. Hist. 41(5): 1278-1289. (EPA-TR-58-75) (NSA 33: 12008) (Incl. radioactivity).

LaChance, L. E. 1955. Effects of delayed oviposition on x-ray induced sterility. Nucleonics 13(4): 49-50. (NSA 9: 3736).

LaChance, L. E. 1959. The effect of chlation and x-rays on fecundity and induced dominant lethals in Habrobracon. Radiation Research 11(2): 218-228. (NSA 13: 18843).

LaChance, L. E. 1960. An independence of intensity for embryo dominant lethals induced by x-rays in first meiotic metaphase eggs of Habrobracon. Genetics 45(6): 665-668. (NSA 14: 20094).

LaChance, L. E. 1961. Postirradiative effects of nitrogen and carbon monoxide on hatchability of Habrobracon eggs treated in first meiotic metaphase. Int. J. Radiat. Biol. 4(1): 15-20. (NSA 16: 23452).

LaChance, L. E. 1961. Radiosensitivity of the various stages of oogenesis in Callitroga hominivorax. Abstr. Genetics 46(8): 877.

LaChance, L. E. 1962. Effects of radiations on insects. Proc. N. Cen. Br. Entomol. Soc. Am. 17: 11-14.

LaChance, L. E. 1963. Enhancement of radiation-induced sterility in insects by pretreatment in CO_2 + air. Int. J. Radiat. Biol. 7(4): 321-331. (NSA 18: 23276).

LaChance, L. E. 1967. The induction of dominant lethal mutations in insects by ionizing radiation and chemicals - as related to the sterile-male technique of insect control. pp. 617-650. In: Genetics of Insect Vectors of Disease. J. W. Wright and R. Pal, eds. Elsevier, Amsterdam.

LaChance, L. E. 1973. Radiation induced sterility in insects for the control and/or eradication of agricultural pests. pp. 1463-1470. In: Advances in Radiation Research. Biology and Medicine. Vol. 3. J. F. Duplan, ed. Gordon & Breach, New York. (NSA 31: 26191).

LaChance, L. E. 1974. Status of the sterile-insect release method in the world. pp. 55-62. In: Sterile-Insect Technique and Its Field Applications. (See IAEA, 1974) (NSA 31: 8940).

LaChance, L. E. 1974. Dominant lethal mutations in insects with holokinetic chromosomes. II. Irradiation of sperm of cabbage looper. Ann. Entomol. Soc. Am. 67(1): 35-39. (NSA 29: 16043).

LaChance, L. E. 1975. Induced sterility in irradiated Diptera and Lepidoptera. pp. 401-411. In: Sterility Principle for Insect Control, 1974. (See IAEA, 1975) (NSA 32: 11795).

LaChance, L. E. and S. B. Bruns. 1963. Oogenesis and radiosensitivity in Cochiliomyia hominovorax (Diptera: Calliphoridae). Biol. Bull. 124(1): 65-83. (NSA 18: 11671).

LaChance, L. E. and M. M. Crystal. 1965. Induction of dominant lethal mutations in insect oocytes and sperm by gamma rays and an alkylating agent: dose-response and joint action studies. Genetics 51: 699-708. (NSA 19: 24289).

LaChance, L. E. and M. Degrugillier. 1969. Chromosomal fragments transmitted through three generations in Oncopeltus (Homoptera). Science 166: 235-236. (NSA 24: 7810) (X-rays).

LaChance, L. E. and D. E. Hopkins. 1962. Mutations in the screw-worm fly. J. Econ. Entomol. 55(5): 733-737. (NSA 21: 8752).

LaChance, L. E. and A. P. Leverich. 1962. Radiosensitivity of developing reproductive cells in female Cochliomyia hominovorax. Genetics 47(6): 721-735. (NSA 16: 30239).

LaChance, L. E. and D. T. North. 1967. Genetic consideration of radioresistance and inherited sterility in lepidopteran species. Abstr. 106. Bull. Entomol. Soc. Am. 13(3): 192.

LaChance, L. E. and R. D. Richard. 1973. Irradiation of sperm and oocytes in Oncopeltus fasciatus (Hemiptera: Lygaeidae): sex ratio, fertility, and chromosome aberrations in the F_1 progeny. Can. J. Genet. Cytol. 15(4): 713-721. (NSA 30: 15981).

LaChance, L. E. and R. D. Richard. 1974. Effect of nitrogen on the competitiveness of irradiated Musca domestica. J. Econ. Entomol. 67(4): 562-563. (NSA 30: 24355).

LaChance, L. E. and J. G. Riemann. 1964. Cytogenetic investigations on radiation and chemically induced dominant lethal mutations in oocytes and sperm of the screw-worm fly. Mutation Research 1(3): 318-333. (NSA 19: 45950).

LaChance, L. E. and J. G. Riemann. 1973. Dominant lethal mutations in insects with holokinetic chromosomes. I. Irradiation of Oncopeltus (Hemiptera: Lygaeidae) sperm and oocytes. Ann. Entomol. Soc. Am. 66(4): 813-819. (NSA 28: 18528).

LaChance, L. E., M. Degrugillier and A. P. Leverich. 1970. Comparative effects of chemosterilants on spermatogenetic stages in the house fly. II. Recovery of fertility and sperm transfer in successive matings after sterilization with 1,3-propanediol dimethanesulfonate or x-rays. Ann. Entomol. Soc. Am. 63: 422-428. (NSA 24: 23419).

LaChance, L. E., C. H. Schmidt and R. C. Bushland. 1967. Radiation-induced sterilization. pp. 147-196. In: Pest Control. Biological, Physical, and Selected Chemical Methods. W. Kilgore and R. L. Doutt, eds. Academic Press, New York.

Lachet, B. 1972. Laboratory studies of the purification kinetics of river water (Isere) by sedimentary suspensions. Case of ^{51}Cr, ^{59}Fe, ^{60}Co, ^{85}Sr, ^{137}Cs, ^{141}Ce, and ^{203}Hg. Radioprotection 7(3): 143-157. (In French) (NSA 27: 17022).

Lachet, B. and R. Magnaval. 1973. Empiracal equation for the purification of surface waters. pp. 185-192. In: Environmental Behavior of Radionuclides Released in the Nuclear Industry. (See IAEA, 1973) (In French) (NSA 29: 29359).

Lachet, B., J. Gagnaire and R. Plebin. 1972. ^{106}Ru distribution in a reduced model simulating river-banks; some hydrodynamic and kinetic aspects of its adsorption on sediments. Centre d'Études Nucléaires, France Commissariat à l'Énergie Atomique, Grenoble, report CEA-R-4304. 28pp. (In French) (NSA 26: 36076).

Lachiver, F. 1957. Relation between the amount of I^{127} and the relationship monoiodotyrosine-diiodotyrosine on the thyroid. Its influence on the speed of hormone secretion; study of the fact with the hibernating lerot (Eliomys quercinus) with the aid of I^{131}. Compt. rend. soc. biol. 151(4): 649-652. (In French).

Lachiver, F. 1964. Thyroid activity in the garden dormouse (Eliomys quercinus) studies from June to November. Ann. Acad. Scient. Fenn. A IV(Biol.) 71: 283-294. (^{131}I).

Lackey, J. B. 1957. The suspended microbiota of the Clinch River and adjacent waters in relation to radioactivity in the summer of 1956. Oak Ridge National Lab., U. S. AEC report ORNL-2410. 36pp. (Also: Tech. Paper 145, Florida Engrg. Industr. Expt. Sta., Gainesville) (NSA 12: 1188).

Lackey, J. B. 1957. Microscopy in radioactive waste treatment by dilution. pp. 32-35. In: Blazing Trails with the Peaceful Atom. Florida Engrg. Industr. Expt. Sta. Bull. Ser. No. 88.

Lackey, J. B., G. B. Morgan and O. H. Hert. 1959. Turbidity effects in natural waters in relation to organisms and the uptake of radioisotopes. Engrg. Progr. Univ. Fla. 13(8): 9pp. (Tech. Paper 167, Florida Engrg. Industr. Expt. Sta., Gainesville) (Also: Trans. Industr. Waste Conf., Honey Harbor, Ontario, Canada, June 15, 1959).

La Cour, L. F. 1953. The Luzula system analyzed by x-rays. Heredity 6 (Suppl.): 77-81.

Lacourly, G. 1971. Problems raised by the evaluation of the limiting radiological capacity of a receiving medium - general principles of approach. pp. 459-470. In: Proc. Intern. Symp. on Radioecology Applied to the Protection of Man and His Environment. (See Commission of the European Communities, 1971) (In French) (NSA 27: 5230).

Lacourly, G. 1973. Radiological capacity of the environment as a basis for future radiation protection studies. pp. 603-612. In: Environmental Behavior of Radionuclides Released in the Nuclear Industry. (See IAEA, 1973) (In French) (NSA 29: 29310).

Lacroix, J. C. and M. T. Loones. 1971. X-ray induced fragmentation of the organizer of a lampbrush chromosome differentiation in Pleurodeles waltlii. Chromosoma 36: 112-118. (NSA 26: 25852).

La Croix, J. D. 1964. Radiosensitivity of jack pine seed to cobalt-60. Forest Science 10(3): 293-295.

Lacroix, M. 1971. Isotope effects in biology. Part II. Biological observations and applications. Bull. Soc. Roy. Sci. Liége 40(3-4): 179-198. (In French) (NSA 26: 20623) (Bacteria, yeasts, algae, molluscs, echinoderms).

Lacy, W. J. and M. J. Stangler. 1962. The postattack water-contamination problem. Health Physics 8: 423-427. (NSA 16: 29263).

Ladd, T. L., Jr. 1973. Sterilization of Japanese beetles by gamma radiation. J. Econ. Entomol. 66(5): 1047-1048. (NSA 29: 775).

Ladefoged, K. 1960. A method of measuring the water consumption of larger intact trees. Physiol. Plant. 13: 648-658.

Ladygin, V. G., G. A. Semenova and S. V. Tageyeva. 1972. Plastid ultrstructure of the white and yellow mutants of Chlamydomonas reinhardi. Tsitologiya 14(12): 723-729. (In Russian) (NSA 27: 20024) (γ irradiation).

Lafuma, J. 1975. Biological behavior of plutonium 239. Centre d'Études Nucléaires, France Commissariat à l'Énergie Atomique, Fontenay-aux-Roses, report CEA-CONF-3153. 12pp. (In French) (NSA 33: 26311) (Man, environment).

Lag, J. and E. Steinnes. 1971. Study of mercury and iodine distribution in Norwegian forest soils by neutron activation analysis. pp. 429-438. In: Nuclear Techniques in Environmental Pollution. (See IAEA, 1971) (NSA 25: 31882).

Lag, J. and E. Steinnes. 1972. Distribution of chlorine, bromine, and iodine in Norwegian forest soils studied by neutron activation analysis. pp. 383-395. In: Isotopes and Radiation in Soil-Plant Relationships Including Forestry. (See IAEA, 1972) (NSA 27: 2313).

Lagoni, H. L. 1961. Radioactive irradiation in agriculture and food production. Verbraucherdienst 6: 210-211. (In German).

Lagrange, E. 1963. Effect of x-rays on metacercaria in Fasciola hepatica. Compt. rend. soc. biol. 157: 1310-1311, 1826. (In French) (NSA 18: 5037, 25128).

Laguarda-Figueras, A. 1969. Effect of x, gamma, and UV radiations on populations of Dugesia dorotocephela (Platyhelminthes: Turbellaria) under acttion of the serotonin-sulfate complex of creatinine. Thesis, Univ. Nac. Autonoma de Mexico, Mexico City, D. F. 98pp. (In Spanish) (NSA 24: 39519).

Laguarda-Figueras, A. and R. Villalobos-Pietrini. 1967. Protection by serotonin-creatinine sulfate complex of the planarian Dugesia tigrina against the lethal effects of x-rays. Proc. Soc. Exptl. Biol. Med. 126: 667-669. (NSA 22: 23597).

Lagunov, L. L. 1934. Radioactivity of the rocks in the Far East and its influence on forests. Bull. Far East Branch Acad. Sci. USSR 9: 105-109. (In Russian) (CA 29: 3910).

Lahiri, S. K., et al. 1967. Ionic distribution studies in haemolymph from different parts of Periplaneta americana by radioactive tracers. Indian J. Exptl. Biol. 5(1): 27-28. (NSA 23: 9852).

Lahoud, J. A. and D. L. Orphal. 1971. Plowshare and nuclear power plant siting. Trans. Amer. Nucl. Soc. 14(2): 713-714. (NSA 26: 3749).

Lakanen, E. and A. Paasikallio. 1968. Effects of soil factors on the uptake of radiostrontium by plants. Part I. Part II. Ann. Agr. Fenn. ser. Agrogeol.-Chim.-Phys. 7(35): 89-94; 9(46): 133-138. (NSA 23: 47942; 25: 21645).

Lal, D. and H. E. Suess. 1968. Radioactivity of the atmosphere and hydrosphere. Ann. Rev. Nucl. Sci. 18: 407-434. (NSA 23: 9574).

Lal, H., S. Ginocchio and E. J. Hawrylewicz. 1963. Procedure for bioassaying mosquito repellents in laboratory animals. Proc. Soc. Exptl. Biol. Med. 113: 770-772. (^{131}I, ^{14}C).

Lallemand, A. and G. Grison. 1967. Laboratory studies of four tracers, ^{51}Cr EDTA-^{59}Fe EDTA-^{45}Ca EDTA-$(CN)_2$ ^{110}Ag. Comparisons of the results. Mem. Ass. Int. Hydrogeol. Reunion Istanbul 8: 383-391. (See Anonymous, 1967) (In French) (NSA 23: 446).

LaMarche, V. C., Jr. and T. P. Harlan. 1973. Accuracy of tree ring dating of bristlecone pine for calibration of the radiocarbon time scale. J. Geophys. Res. 78(36): 8849-8858. (NSA 30: 3182).

Lamb, K. P., P. Ehrhardt and V. Moericke. 1967. Labelling of aphid saliva with rubidium-86. Nature 214: 602-605.

Lamb, K. P., E. Hassan and D. R. Scotter. 1971. Dispersal of scandium 46 labeled Partorhytes weevils in Papual cacao plantations. Ecology 52(1): 178-182.

Lamb, M. J. 1964. The effects of radiation on the longevity of female Drosophila subobscura. J. Insect Physiol. 10: 487-497. (NSA 19: 122).

Lamb, M. J. 1965. The effects of x-irradiation on the longevity of triploid and diploid female Drosophila melanogaster. Exptl. Gerontol. 1: 181-187.

Lamb, M. J. 1966. Relationship between age at irradiation and life-shortening in adult Drosophila. pp. 163-174. In: Radiation and Aging. P. J. Lindop, ed. (NSA 22: 43407).

Lamb, M. J. and R. E. Monroe. 1968. Lipid synthesis from acetate-1-C^{14} by the cereal leaf beetle, Oulema melanopus. Ann. Entomol. Soc. Am. 61(5): 1164-1166.

Lamb, M. J. and J. M. Smith. 1964. Radiation and ageing in insects. Exptl. Gerontol. 1(1): 11-20.

Lambert, G. and G. Polian. 1963. On the diffusion of lower altitude continental and oceanic air masses in each hemisphere. Compt. rend. 256: 4265-4267.

Lambert, G., J.-C. Le Roulley and J. Sanak. 1971. Correlations between the ^{210}Pb deposit in the Antarctic and fluctuations in the earth's magnetic field. Compt. rend. B272: 177-180. (In French) (NSA 25: 29211).

Lambert, G., C. Lorius and B. Ardouin. 1968. Dating of snow samples off the border of Antarctica by measurement of the natural and artificial radioactivity. Earth Planet. Sci. Lett. 3: 429-433. (In French) (NSA 22: 18916).

Lambert, G., et al. 1971. Balance of ^{90}Sr over Antarctica: existence of protected area. Earth Planet. Sci. Lett. 11(4): 317-323. (NSA 25: 54788) (Firn cores).

Lambert, G., et al. 1972. Natural radioactivity balance in the atmosphere of the Southern Hemisphere. pp. 787-807. In: The Natural Radiation Environment II. Vol. 2. (See Adams, Lowder and Gesell, 1972) (NSA 32: 19727). 19727).

Lambert, J. P. F., et al. 1969. Neutron acivation analysis of selenium in biological samples. J. Assoc. Anal. Chem. 52: 915-917. (NSA 23: 47407).

Lambet, M. S. and D. S. Nikolayev. 1962. The forms of uranium in the waters of the Sea of Azov and some estuaries and rivers of the Azov-Black Sea Basin. Dokl. Akad. Nauk SSSR 142(3): 681-682. (Doklady (earth sci.) 142: 155-156).

Lambina, V. A. 1962. The effect of radioactive phosphorus-32 on photosynthesizing purple sulfur bacteria. Radiobiologiya 2(3): 492-496. (AEC-tr-5430, pp. 237-243) (NSA 16: 32868).

Lambremont, E. N. 1964. Distribution of newly synthesized fatty acids to the the principal lipid fractions of the boll weevil. Abstr. 99. Bull. Entomol. Soc. Am. 10(3): 164. (^{14}C).

Lambremont, E. N. 1965. The metabolism of glucose and its conversion into long-chain fatty acids by the boll weevil. Abstr. 89. Bull. Entomol. Soc. Am. 11(3): 159. (^{14}C).

Lambremont, E. N. 1965. Biosynthesis of fatty acids in aseptically reared insects. Comp. Biochem. Physiol. 14(3): 419-424. (^{14}C, Anthonomus grandis).

Lambremont, E. N. and C. I. Stein. 1964. Metabolic interconversions of dietary fatty acids in the boll weevil. Abstr. Bull. Entomol. Soc. Am. 10 (3): 164. (^{14}C).

Lambremont, E. N. and C. I. Stein. 1965. $C^{14}O_2$ production in the boll weevil, Anthonomus grandis, after injection of C^{14}-1-acetate. Ann. Entomol. Soc. Am. 58(5): 765-766.

Lambremont, E. N., C. I. Stein and A. F. Bennett. 1965. Synthesis and metabolic conversion of fatty acids by the larval boll weevil. Comp. Biochem. Physiol. 16(3): 289-302. (^{14}C, Anthonomus grandis).

Lambrev, Zh. and N. Balabanov. 1963. Natural radioactivity of organisms. I. Radioactivity of certain vertebrates. Radiobiologiya 3(6): 875-876. (AEC-tr-5439, pp. 143-146) (NSA 18: 19517).

Lambrev, Zh. and T. Dushev. 1965. Effect of ionizing radiation x-rays on the silkworm (Bombyx mori). Nauchn. trud. vyssh. sel. inst. Vasil Kolarov 14(2): 179-182. (In Bulgarian, German summary).

Lammers, W. T. 1967. Biophysical limnology: the assessment of low-level contaminants in natural water. Oak Ridge Gaseous Diffusion Plant, U. S. AEC report K-L-6131. 24pp. (NSA 21: 22409).

Lammers, W. T. 1968. The distribution of ^{60}Co, ^{106}Ru and ^{137} among suspended and dissolved particles in White Oak Lake. Oak Ridge Gaseous Diffusion Plant, U. S. AEC report K-1758. 20pp. (NSA 22: 48915).

Lamy, R. and H. J. Muller. 1939. Evidence of the non-genetic nature of the lethal effect of radiation on Drosophila embryos. pp. 180-181. In: Proc. 7th Intern. Congr. Genetics (Suppl.).

Land, J. D. 1963/4. Absorption and distribution of DDT in the boll weevil and the influence of synergists on toxicity. (Dissertation Abstr. 24: 3029) (^{14}C).

Landa, V. and B. Rezábová. 1966. Problems in insect sterilization. Agrochémia (Bratislava) 5: 150-152. (In Czech).

Landis, J. W. 1971. Nuclear radiation molehill. pp. 95-113. In: Inst. of Nuclear Materials Management 12th Ann. Meeting. (June 29-July 1, 1971). H. L. Toy, Columbus, O. (CONF-710617) (NSA 26: 3749).

Landis, J. W. 1972. Environmental aspects of nuclear energy. At. Jap. 16 (4, Suppl.): 5-56. (CONF-720341-1) (NSA 27: 24204).

Landreth, H. F., P. B. Dunaway and G. E. Cosgrove. 1974. Effects of whole-body gamma irradiation on various life stages of the toad, Bufo woodhousei fowleri. Radiation Research 58(3): 432-438. (NSA 30: 9887).

Landstrom, O. and C. G. Wenner. 1965. Trace elements in Quaternary research and ground water geology. Abstr. p. 279. In: Abstr. Intern. Assoc. Quaternary Research 7th Intern. Congr., Boulder and Denver, Colo.

Landstrom, O., K. Samsahl and C. G. Wenner. 1969. An investigation of trace elements in marine and lacustrine deposits by means of a neutron activation method. pp. 353-366. In: Modern Trends in Activation Analysis. Vol. I. (See DeVoe, 1969) (NSA 24: 246).

Lane, R. H. 1967. Use of a tracer element in studying the movements of some small mammals. Thesis, Bowling Green State Univ., Bowling Green, O., U. S. AEC report COO-1485-14. (NSA 21: 41312).

Lane, R. H., R. M. Grodner and J. L. Graves. 1976. Irradiation studies of mallard duck eggs material containing MIREX. J. Agric. Food Chem. 24(1): 192-193. (NSA 33: 17457).

Lane, W. B. 1970. Fallout simulant development: leaching of fission products from Nevada fallout and properties of iodine-tagged simulant. Final report. Stanford Research Inst., Menlo Park, Calif. 56pp. (AD-717281; NSA 25: 48614).

Lang, C. A. 1963. The accumulation of zinc by the mosquito (*Aedes aegypti* (Linn.), *Culex pipiens* (Linn.), *Culex molestus* (Forskal)). J. Genet. Physiol. 46: 617-627. (^{65}Zn).

Lange, C. S. 1968. Studies on the cellular basis of radiation lethality. I. The pattern of mortality in the whole-body irradiated planarian (Tricladida, Paludicola). Int. J. Radiat. Biol. 13: 511-530. (NSA 22: 49517).

Lange, C. S. 1968. An outline of studies on the cellular basis of planarian radiation lethality. J. Physiol. (London) 197: 54P-55P. (NSA 23: 2913).

Lange, C. S. 1968-69. Studies on the cellular basis of radiation lethality. II. Survival-curve parameters for standardized planarian-populations. III. (w/C. W. Gilbert). The measurement of stem-cell repopulation probability. V. A survival curve for the reproductive integrity of the planarian neoblast and the effect of polyploidy on the radiation response. Int. J. Radiat. Biol. 14: 119-132, 373-388; 15(1): 51-64. (NSA 23: 20331, 33817, 22360) (*Dugesia lugubris*).

Lange, R. 1966. Food distribution among workers of a forest ant colony. Habilitätionsschrift, Freiburg Univ. (In German).

Lange, R. 1967. Food distribution in a forest ant colony. Wiss. Z. Tech. Hochsch. Dresden 16(2): 594-596. (In German) (^{32}P).

Langenauer, H. D. 1972. Effects of acute x-irradiation upon growth of *Parthenocissus tricuspidata* axillary buds. Thesis, Univ. of Massachusetts, Amherst. 130pp. (Univ. Microfilms 73-6691) (NSA 27: 27877).

Langenauer, H. D., T. S. Osborne and D. A. Haskell. 1972. Effects of acute x-irradiation upon growth of *Parthenocissus tricuspidata* axillary buds. 1. Morphological damage and recovery. Radiation Botany 12(4): 297-306. (NSA 26: 56396).

Langendorf, H., M. Langendorf and H. J. Melching. 1965. On the influence of external factors upon the efficiency of radioprotective substances. Progr. Biochem. Pharmacol. 1: 418-426. (NSA 20: 18472).

Langer, H. 1964. Phosphate metabolism of the compound eye, in darkness and in light. Helgoländer wiss. Meersunters. 9(1/4): 251-260. (^{32}P, *Calliphora erythrocephala*).

Langford, J. C. 1969. Procedure for the determination of ^{210}Pb and total lead in biological samples. Anal. Chem. 41: 1716. (NSA 23: 47394).

Langham, W. H. 1965. Considerations of biospheric contamination by radioactive fallout. pp. 3-18. In: *Radioactive Fallout - Soils, Plants, Foods, Man*. (See Fowler, 1965).

Langham, W. H. 1968. The problem of large-area plutonium contamination. Paper No. 002. For U. S. PHS Seminar, Los Alamos Scientific Lab, N. Mex. 15pp. (NP-18208) (NSA 24: 29738).

Langley, P. A. and H. Maly. 1971. Control of the Mediterranean fruit fly (*Ceratitis capitata*) using sterile males. Effects of nitrogen and chilling during gamma-irradiation of puparia. Entomologia Exp. Appl. 14(2): 137-146. (NSA 26: 59115).

Langley, P. A., C. F. Curtis and J. Brady. 1974. Viability, fertility, and behavior of tsetse flies (*Glossina morsitans*) sterilized by irradiation under various conditions. Entomologia Exp. Appl. 17(1): 97-111. (NSA 31: 814).

Langley, T. M. and W. L. Marter. 1973. Savannah River Plant site. Savannah River Lab., U. S. AEC report DP-1323. 160pp. (NSA 28: 380).

Langway, C. C., Jr., et al. 1965. Sampling polar ice for radiocarbon dating. Nature 206(4983): 500-501. (NSA 19: 28565).

L'Annunziata, M. F. and W. H. Fuller. 1968. The chelation and movement of Sr^{89}-Sr^{90}(Y^{90}) in a calcareous soil. Soil Science 105(5): 311-319.

Lanzola, E. 1970. Past and current research of the Intituto di Igiene of Pavia University. pp. 55-60. In: *Proc. 1st Conf. on the State of Advancement of Radioecology in Italy*. (See Anonymous, 1970) (In Italian).

Lapage, R. (Ed.). 1974. Peaceful uses of nuclear explosives. Review paper No. 1. Radioactivity and PNE. Atomic Weapons Research Establishment, U. K. Atomic Energy Authority, Aldermaston, report AWRE-0-74. 114pp. (NSA 29: 29262).

Lappenbusch, W. L. 1970. Effect of circadian rhythm on the radiosensitivity of the rough-skinned newt (*Taricha granulosa*). J. Radiat. Res. (Japan) 11: 134-137. (NSA 25: 32570).

Lappenbusch, W. L. 1972. Effect of circadian rhythm on the radiation response of the Chinese hamster (*Cricetulus griseus*). Radiation Research 50(3): 600-610. (NSA 26: 41213).

Lappenbusch, W. L. and D. L. Willis. 1969. The effect of dimethyl sulfoxide (DMSO) on the radiosensitivity of the rough-skinned newt (*Taricha granulosa*). Abstr. Radiation Research 39(2): 553.

Lappenbusch, W. L. and D. L. Willis. 1970. The effect of dimethyl sulfoxide on the radiation response of the rough-skinned newt (*Taricha granulosa*). Int. J. Radiat. Biol. 18(3): 217-233.

Lappenbusch, W. L., D. G. Watson and W. L. Templeton. 1971. *In situ* measurement of radiation dose in the Columbia River. Health Physics 21(2): 247-251. (NSA 25: 46841).

Lappenbusch, W. L., et al. 1973. Effect of 2450-MHz microwaves on the radiation response of x-irradiated Chinese hamsters. Radiation Research 54 (2): 294-303. (NSA 28: 8455).

Lara, F. J. S., C. R. Lopes and L. Foresti. 1965. Sedimentation characteristics of rapidly labeled RNA from the salivary gland cells of *Rhynchosciara angelae*. Life Sci. 4(5): 579-582. (^{3}H).

LaRoche, G. and C. P. Leblond. 1954. Destruction of thyroid of Atlantic salmon (Salmo salar) by means of radio-iodine. Proc. Soc. Exptl. Biol. Med. 87: 273-276. (NSA 9: 828).

LaRoche, G., C. L. Johnson and A. N. Woodall. 1965-66. Thyroid function in rainbow trout (Salmo gairdneri Rich.). I. Biochemical and histological evidence of radiothyroidectomy. II. (w/J. E. Halver). Effects of thyroidectomy on the development of young fish. Gen. Comp. Endocrinol. 5: 145-159; 6(2): 249-266. (NSA 19: 43773).

LaRoche, C. L. Johnson and A. N. Woodall. 1966. Iodine metabolism in young chinook salmon (Oncorhynchus tschaytscha, Walbaum): I. Thyroidal impairment with the use of I^{131}. Gen. Comp. Endocrinol. 7(3): 512-524.

LaRoche, G., et al. 1963. Iodine metabolism in adult king salmon (Oncorhychus tschawyscha, Walbaum). p. 154. In: Proc. 16th Intern. Congr. Zoology. Vol. 2. (SFA 9: 6407) (^{125}I, ^{127}I, ^{131}I).

Larsen, I. L., W. C. Renfro and N. H. Cutshall. 1973. Zinc-65 specific activity in Mytilus californianus tissues. pp. 747-751. In: Radionuclides in Ecosystems. (See Nelson, 1973) (NSA 28: 29893).

Larsen, P. R. and J. C. Gordon. 1969. Leaf development, photosynthesis, and C^{14} distribution in Populus deltoides seedlings. Am. J. Bot. 56(9): 1058-1066.

Larsen, W. 1961. A study of the embryology of the cockroach, Blaberus craniifer Burmeister. (Dissertation Abstr. 21: 3565) (Used x-rays).

Larsen, W. 1963. Survival of isolated insect tissues following radiation. Ann. Entomol. Soc. Am. 56(5): 720.

Larsen, W. 1963. Some effects of x-irradiation on embryos of the cockroach Blaberus craniifer. Ann. Entomol. Soc. Am. 56(4): 442-448.

Larsen, W. 1964. Cell proliferation in an insect tissue culture. Life Sci. 3(2): 103-106. (Blaberus craniifer).

Larsen, W. 1964. Survival of isolated embryonic heart fragments following treatment with some radiation protecting chemicals. Life Sci. 3: 539-543. (NSA 19: 13103) (Blaberus craniifer).

Larson, C. E. 1973. Not in retrospect. pp. xiii-xvi. In: Radionuclides in Ecosystems. (See Nelson, 1973).

Larson, C. E. and H. C. Brown, Jr. 1972. The energy-environment equation. Proc. 4th Intern. Conf. on the Peaceful Uses of Atomic Energy 11: 529-545.

Larson, J. D. and W. A. Beetem. 1970. Chemical and radiochemical analyses of water from streams, reservoirs, wells and springs in the Rulison Project area, Garfield and Mesa counties, Colorado. U. S. Geological Survey, Denver, report USGS-47-67. 16pp. (NSA 24: 14111, 16413).

Larson, K. H. 1963. Continental close-in fallout: its history, characteristics and measurement. pp. 19-25. In: Radioecology. (See Schultz and Klement, 1963) (NSA 17: 33543).

Larson, K. H. and J. W. Neel. 1959. Findings related to the testing program at Nevada Test Site. pp. 2000-2020. In: Fallout from Nuclear Weapons Tests. (See U. S. Congress, 1959) (Also: Univ. of California at Los Angeles, U. S. AEC report UCLA-438. (NSA 14: 9307).

Larson, K. H., H. A. Hawthorne and J. H. Olafson. 1962. Nevada Test Site fallout: some characteristics, its apparent environmental equilibrium and biological availability. pp. 4-24. In: Radioactive Fallout from Nuclear Weapons Tests. Book 1. (See Klement, 1962) (NSA 16: 22434).

Larson, K. H., et al. 1951. The 1949 and 1950 radiological soil survey of fission product contamination and some soil-plant interrelationships of areas in New Mexico affected by the first atomic bomb detonation. Univ. of California at Los Angeles, U. S. AEC report UCLA-140. 83pp. (NSA 13: 11069).

Larson, K. H., et al. 1951. Alpha activity due to the 1945 atomic bomb detonation at Trinity, Alamogordo, New Mexico. Univ. of California at Los Angeles, U. S. AEC report UCLA-108. 44pp. (NSA 13: 11746).

Larson, K. H., et al. 1951. Alpha activity due to the 1945 atomic bomb detonation at Trinity, Alamogordo, New Mexico. Univ. of California at Los Angeles, U. S. AEC report UCLA-111. 30pp.

Larson, K. H., et al. 1952. Field observations and preliminary field data obtained by the UCLA survey group, Operation JANGLE, Nov. 1951. Univ. of California at Los Angeles, U. S. AEC report UCLA-182. 29pp. (NSA 13: 11747).

Larson, K. H., et al. 1966. Distribution, characteristics, and biotic availability of fallout, Operation PLUMBBOB. Projects 37.1, 37.2, 37.2a, 37.3 and 37.6. Univ. of California at Los Angeles, U. S. AEC report WT-1488. 281pp. (NSA 21: 1461).

Larson, P. R. 1969. Incorporation of ^{14}C in the developing walls of Pinus resinosa tracheids (earlywood and latewood). Holzforschung 23: 17-26.

Larson, T. E. and R. L. Weatherford. 1960. Natural radioactivity in Illinois water resources. J. Am. Water Works Assoc. 52: 769-778. (Also: Illinois State Water Survey, Urbana, Cir. No. 81) (NSA 15: 1638).

Larson, T. E. and R. L. Weatherford. 1961. Natural radioactivity in Illinois water resources. Engrg. Expt. Sta., Univ. of Illinois, Champaign, Cir. 69: 42-49. (See Univ. of Illinois).

Larsson, S. 1972. Low level tritium and ^{14}C determination. Anal. Biochem. 50(1): 245-254. (NSA 27: 4984).

La Sala, A. M., Jr. and G. C. Doty. 1971. Preliminary evaluation of hydrologic factors related to radioactive waste storage in basaltic rocks at the Hanford Reservation, Washington. U. S. Geological Survey, Richland, Wash., U. S. AEC report TID-25764. 72pp. (NSA 25: 48606).

Lasceve, G. 1974. Radioactivation analysis of the mineral content of a honey. Apidologie 5(3): 201-223. (In French) (NSA 32: 11139).

Lasceve, G., L. A. Buscarlet and A. Bossy. 1972. Radioactive analysis of codling moths labelled with ^{191}Ir. Int. J. Appl. Radiat. Isotopes 23(6): 265-270. (In French) (NSA 26: 43407).

Laschi, R. and A. Abbati. 1960. Early vascular changes following x-irradiation. Rass. Intern. Elettron. Nucl. 7 (5th Congr. Nucl., Rome, 1960) 3: 445-475. (NSA 17: 37123) (Comments on frog).

Lasher, R. and R. Rugh. 1962. The Hertwig effect in teleost development. Biol. Bull. 123(3): 582-588. (NSA 17: 23107).

Lasser, N. L., R. B. Clayton and A. M. Edwards. 1963. The dynamic state of sterols in the cockroach. Abstr. Fed. Proc. 22(2): 590. (^{14}C).

Lassota, Z. 1963. Action of γ rays on eggs, larvae, and pupae of Bombyx mori. Acta Biochim. Pol. 10(4): 379-386.

Lassota, Z. 1965. Uric acid in normal and gamma-irradiated eggs of Bombyx mori. Acta Biochim. Pol. 12: 271-278. (NSA 20: 8643).

Lassota, Z. 1965. Oxygen uptake and ammonia release in normal and gamma-irradiated eggs of Bombyx mori. Acta Biochim. Pol. 12: 369-377. (NSA 20: 18301).

Lassota, Z. 1966. Intestinal damage and water imbalance in gamma-irradiated larvae of Bombyx mori. Bull. Acad. Polon. Sci. (sci. biol.) 14: 293-296. (NSA 21: 3436).

Lassota, Z. 1966. Nucleic acids in normal and γ-ray irradiated eggs of Bombyx mori. Acta Biochim. Pol. 14: 7-20. (NSA 22: 28373).

Latarjet, R. 1944. Comparative primary actions of x-rays and ultraviolet on the yeast Saccharomyces ellipsoideneuses. Ann. Inst. Pasteur 70: 270-.

Latarjet, R. and L. R. Caldas. 1952. Restoration induced by catalase in irradiated microorganisms. J. Gen. Physiol. 35(3): 455-470.

Laudani, H. and E. W. Tilton. 1966. USDA research program and facilities for the use of gamma irradiation in the control of stored-products insects. Stored-Products Insects Research and Development Lab., U. S. Dept. of Agriculture, Savannah, Ga. Reprint No. 34C. Amer. Inst. Chemical Engineers, New York. 19pp. (NSA 21: 28711).

Laudani, H., E. W. Tilton and J. H. Brower. 1965. USDA research program and facilities for the use of gamma irradiation in the control of stored-product insects. Food Irradiation 6: A6-A9. (NSA 20: 19124).

Lauderdale, R. A. and R. Eliassen. 1956. The removal of radioactive fall-out from water by municipal and industrial water treatment plants. Massachusetts Inst. Tech., U. S. AEC report NYO-4441. 79pp. (NSA 11: 1402).

Laven, H. and E. Jost. 1971. Inherited semisterility for control of harmful insects. I. Productions of semisterility due to translocations in the mosquito, Culex pipiens L., by x-rays. Experientia 27(4): 471-473. (NSA 25: 42335).

Laviolette, P. and P. Cuir. 1959. Action of x-rays on the gonads of Arion rufus L. (Gastropod mollusc). Arch. Anat. Micro. Morphol. Exptl. 48(1): 25047. (In French).

Laviolette, P. and P. Nardon. 1963. Influence of irradiation on the adults of Sitophilus sasakii takahashi (Curculionidae) and their descendents. pp. 431-440. In: Radiation and Radioisotopes Applied to Insects of Agricultural Importance. (See IAEA, 1963) (In French).

Laviolette, P. and P. Nardon. 1963. The action of γ-rays of cobalt-60 on mortality and fertility of the adults of a rice weevil. Bull. Biol. 97: 305-333. (In French).

Lavrov, M. T. and V. A. Bogomaz. 1958. The effect of radiation on the cockchafer Melolontha vulgaris. Zashchita rasteniy 6: 53. (In Russian).

Law, Y. M. C. and J. G. Eales. 1973. Deiodination of radiothyroxine by tissue homogenates of brook trout, Salvelinus fontinalis (Mitchill). Comp. Biochem. Physiol. 44(4B): 1175-1183. (SFA 18: 16473).

Lawless, G. P., N. A. MacGillivray and P. R. Nixon. 1963. Soil moisture interface effects upon readings of neutron moisture probes. Soil Sci. Soc. Am. Proc. 27(5): 502-507.

Lawrence, C. H. and F. W. Gilcreas. 1965. Removal of low-level radioisotopes from wastewater by acrobic treatment. J. Water Poll. Control Fed. 37(9): 1289-1301.

Lawrence, C. W. 1962. Radiosensitivity in repeatedly irradiated line of Arabidopsis thaliana. Abstr. p. 157. In: 2nd Intern. Congr. Radiation Research. (See Anonymous, 1962).

Lawrence, C. W. 1965. Influence of non-lethal doses of radiation on recombination in Chlamydomonas reinhardi. Nature 206(4986): 789-799. (NSA 19: 31848).

Lawrence, C. W. 1965. The effect of dose duration in the influence of irradiation on recombination in Chlamydomonas. Mutation Research 2(6): 487-493. (NSA 20: 12688).

Lawrence, C. W. 1967. Influence of nonlethal doses of radiation on allelic recombination in Chlamydomonas reinhardi. Genetical Res. 9: 123-127. (NSA 21: 22383).

Lawrence, C. W. 1968. Effect of fractionated doses of radiation on recombination. Heredity 23: 143-146. (NSA 22: 49310).

Lawrence, C. W. 1968. Radiation effects on genetic recombination in Chlamydomonas reinhardi. pp. 135-144. In: Effects of Radiation on Meiotic Systems. International Atomic Energy Agency, Vienna, Publ. STI/PUB-173. (CONF-670525) (NSA 24: 21130).

Lawrence, C. W. 1970. Dose dependence for radiation-induced allelic recombination in Chlamydomonas reinhardi. Mutation Research 10: 557-566. (NSA 25: 24122).

Lawrence, C. W. and P. D. Holt. 1970. Effect of gamma radiation and alpha particles on gene recombination in Chlamydomonas reinhardi. Mutation Research 10: 545-555. (NSA 25: 24123).

Lawrence, F. S. 1959. Estimates of standing crop of plankton in Clear Lake, Iowa. Iowa State College, Ames, U. S. AEC report TID-11579. 37pp. (NSA 15: 7147).

Lawrence, W. H. and J. H. Rediske. 1960. Radiotracer technique employing scandium-46 as a radiotracer for Douglas fir seed. pp. 99-101. In: Proc. Soc. Amer. Foresters Meeting 1959. (WR 98: 15).

Lawrence, W. H. and J. H. Rediske. 1962. Fate of sown Douglas-fir seed. Forest Science 8(3): 210-218. (WR 108: 4) (Used ^{46}Sc as tag).

Lawson, F. R. 1967. Theory of control of insect populations by sexually sterile males. Ann. Entomol. Soc. Am. 60(4): 713-722. (BA 48: 114017).

Lawson, F. R., G. B. Lucas and N. S. Hall. 1954. Translocation of radioactive phosphorus injected by the green peach aphid into tobacco plants. J. Econ. Entomol. 47(5): 749-752.

Laycock, G. 1967. The beautiful, sad face of Amchitka. Audubon 70(6): 8-12, 14, 16, 18-20, 22-25. (Concerns use of nuclear devices).

Layne, G. W. 1951. Studies on the movements of the plum cuculio, Conotrachelus nenuphar Herbst., in a peach orchard with radiophosphorus (P^{32}) as a tracer. Thesis, Ohio State Univ., Columbus.

Lazard, L. 1959. Influence of "neutral" grafts on the regeneration of extremities of Amblystoma. Compt. rend. 249(18): 1819-1820. (Used radiation).

Lazard, L. 1965. Regeneration of irradiated members in axolotl, induced by grafts of testicle. Compt. rend. 260: 1277-1279. (In French) (NSA 19: 21899).

Lazard, L. 1967. Restoration of regenerative capacity in irradiated axolotl limbs by grafts of different heterotropic tissues. J. Embryol. Exptl. Morphol. 18: 321-342. (In French) (NSA 22: 52010).

Lazard, L. 1973. Development of transplanted gonads in axolotl limbs. Effects of irradiation and amputation of the host limb. Compt. rend. D276(1): (10): 1601-1604. (In French) (NSA 28: 21750).

Lazarev, K. F., D. S. Nikolayev and S. M. Grashchenko. 1961. The concentration of thorium isotopes in sea water. Radiokhimiya 3(5): 623-634. (Sov. Radiochemistry 3(5): 112-124) (CA 56: 5768) (Sea of Azov).

Lazarev, K. F., et al. 1965. Radium concentrations in waters of the Azov Sea and of certain lagoons. Dokl. Akad. Nauk SSSR 164(5): 1151-1153. (Doklady (earth sci.) 164: 202-204) (CA 64: 439d; NSA 20: 18905).

Lazarev, K. F., et al. 1966. Geochemistry of radium isotopes in the Black and Azov seas and some estuaries. pp. 166-172. In: Chemical Processes in Seas and Oceans. S. V. Bruyevich, ed. Publ. House Acad. Sci., Moscow. (In Russian) (NSA 21: 4638).

Lazda, E. E. and P. N. Odintsov. 1967. Effect of gamma-rays on pine resin channel anatomical structure and physiologic condition. Latv. PSR Zinat. Akad. Vest. (1): 36-44. (In Russian) (NSA 21: 28677).

Lazda, E. E. and P. N. Odintsov. 1969. Physiological peculiarities of live pinewood cells after irradiation of cambium. Latv. PSR Zinat. Akad. Vest. (fiz. tehn.) (9): 61-71. (In Russian) (NSA 24: 12333).

Lea, D. E., R. B. Haines and E. Bretscher. 1941. The bactericidal action of x-rays, neutrons and radioactive radiations. J. Hygiene 41(1): 1-16.

Lea, D. E., R. B. Haines and C. A. Coulson. 1936. Mechanism of the bactericidal action of radioactive radiations. Proc. Roy. Soc. (London) B120 (816): 47-75.

Leach, B. J., T. R. Bauman and † . 1962. Thyroxine secretion rate of Missouri Valley mole, Scalopus aquaticus. Proc. Soc. Biol. Med. 110: 681-682.

Leach, P. A. 1970. Gene and chromosomal behavior in a parasitic wasp. Thesis, University of Maine, Orono. 102pp. (Univ. Microfilms 71-20,270) (NSA 25: 57962) (X-rays).

Leach, W. M. 1964. Retention of tritiated thymidine in grasshopper neuroblasts. Exptl. Cell. Res. 35(1): 201-204. (Chortophaga viridifasciata).

Leach, W. M. 1965. Some factors influencing the utilization of tritiated thymidine in grasshopper embryos. (Dissertation Abstr. 26: 1232) (Chortophaga viridifasciata).

Leal, A. R. and F. A. Roig. 1965. Preliminary investigations of the possible influence of radiation on the natural vegetation of uraniferous deposits of the Cuyo region. Rev. Fac. Cienc. Agr., Univ. Nac. Cuyo (Mendoza) 12(2): 99-149. (AEC-tr-6945).

Leamy, M. L. and T. A. Rafter. 1972. Isotope ratios preserved in pedogenic carbonate and their application in paleopedology. pp. D42-D57. In: Proc. 8th Intern. Radiocarbon Dating Conf. Vol. 1. (See Rafter, 1972) (NSA 31: 8699).

Lean, D. R. S. 1973. Movements of phosphorus between its biologically important forms in lake water. J. Fish. Res. Bd. Can. 30(10): 1525-1536. (SFA 19: 17310) (^{32}P).

Lean, G. H. and M. J. Crickmore. 1963. Methods for measuring sand transport using radioactive tracers. pp. 111-131. In: Radioisotopes in Hydrology. (See IAEA, 1963) (NSA 18: 1917).

Leavitt, V. D. 1970. Soil survey of Area 18, Nevada Test Site. Southwestern Radiological Health Lab., U. S. AEC report SWRHL-74-r. 119pp. (NSA 24: 44096).

Leavitt, V. D. 1974. Soil surveys of five plutonium contaminated areas on the test range complex in Nevada. Nat. Environmental Research Center, U. S. EPA, Las Vegas, Nev., U. S. AEC report NERC-LV-539-23. 84pp. (NSA 30: 3194).

† C. W. Turner.

Leavitt, V. D. 1974. Soil surveys of five plutonium-contaminated areas on the test range complex in Nevada. pp. 21-27. In: Dynamics of Plutonium in Desert Environments. (Nevada Applied Ecology Group Progress Report as of January 1974). (See Dunaway and White, 1974) (NSA 30: 29442).

Leavitt, V. D. and B. J. Mason. 1971. Soil survey of Area 15 Nevada Test Site. Western Environmental Research Lab., U. S. EPA, U. S. AEC report SWRHL-106-r. 46pp. (NSA 25: 46560).

Lebedev, D. V. 1950. Tagged atoms in investigations of movement of soil insects. Priroda 39(12): 56-57. (In Russian).

Lebedeva, G. D. 1957. Plankton as an indicator of pollution of freshwater reservoirs by radioactive substances. Med. radiologiya 2(6): 65-69. (JPRS-NY-L-441) (BA 35: 28636; NSA 12: 5182; PHEA 39(9): 31; SFA 5: 3545).

Lebedeva, G. D. 1959. The accumulation of cerium-144 by year-old carp and water thyme. Med. radiologiya 4(4): 72-77. (JPRS-2592) (NSA 14: 16503; (SFA 6: 4434; WPA 34(5): 949).

Lebedeva, G. D. 1961. Concentration of strontium-90 by young Leucaspius and carp. pp. 319-322. In: Distribution, Biological Effects and Migration of Radioisotopes. (See Lebedinskiy and Moskalev, 1961) (JPRS-15067, pp. 51-56; AEC-tr-7512, pp. 375-378) (BA 43: 4277; NSA 30: 7147).

Lebedeva, G. D. 1962. Principle points of entry of strontium-89 into benthos-eating fishes in freshwater. Radiobiologiya 2(1): 43-49. (AEC-tr-5428, pp. 62-71) (NSA 16: 30204).

Lebedeva, G. D. 1963. Experimental study of the effect of stable phosphorus upon accumulation of P^{32} in fishes. Radiobiologiya 3(3): 377-382. (AEC-tr-5436, pp. 70-77).

Lebedeva, G. D. 1964. Strontium-90 and phosphorus-32 accumulation by freshwater fish at different salinities. pp. 52-58. In: Radioisotopes in Hydrobiology and Methods of Sanitary Hydrobiology. Publ. House Acad. Sci. USSR, Moscow-Leningrad. (In Russian).

Lebedeva, G. D. 1966. Effect of various saline compositions of water on the buildup of ^{90}Sr and ^{137}Cs in freshwater fish. pp. 176-181. In: Distribution and Biological Effects of Radioactive Isotopes. (See Moskalev, 1966) (AEC-tr-6944(rev.), pp. 227-234. (NSA 23: 22248).

Lebedeva, G. D. 1966. Influence of different salt composition of water on ^{137}Cs accumulation and excretion in freshwater fish. Radiobiologiya 6(4): 556-559. (AEC-tr-6733, pp. 118-124) (NSA 21: 22312).

Lebedeva, G. D. 1968. The influence of stable and radioactive strontium on freshwater organisms. Gidrobiol. zh. Akad. Nauk Ukr. SSR 4(4): 3-10. (In Russian) (NSA 23: 38518).

Lebedeva, G. D. and G. A. Kuznetsova. 1969. Uptake of ^{65}Zn by freshwater fish (experimental investigation). Gigiyena sanitariya (1): 119-120. (In Russian) (NSA 25: 24009).

Lebedeva, G. D. and A. N. Marey. 1962. Role of aquatic flora and fauna in changes of the sanitary condition of a water system -- aquatic organisms. pp. 41-49. In: Radiation Hygiene. Vol. 2. (See Marey, 1962) (JPRS-21,700, pp. 13-27).

Lebedeva, G. D. and M. M. Saurov. 1958. Aquatic organisms as indicators for detection of small amounts of radioactive strontium in bodies of water. pp. __. In: All-Union Methodological Conf. on Determination of Small Amounts of Radioactive Substances in the External Environment. Akad. Nauk SSSR, Leningrad. (In Russian).

Lebedeva, G. D. and S. G. Sinevid. 1958. The effect of radioactive strontium upon survival and propagation of Daphnia magna. Dokl. Akad. Nauk SSSR 122: 586-588. (In Russian) (NSA 13: 5657).

Lebedeva, L. I., T. D. Osetrova and Yu. Ya. Kerkis. 1963. Comparative radiosensitivity of the hereditary structures of mammalian cells in vitro. Dokl. Akad. Nauk SSSR 152(5): 1225-1226. (Doklady (biol. sci.) 152: 1283-1284).

Lebedinets, L. N. 1971. Growth and morphogenesis of pecan under the effect of γ-rays of ^{60}Co. Tsitologiya 5(1): 45-48. (In Russian) (NSA 26: 17973).

Lebedinskiy, A. V. 1957. The consequences of strontium-90 fallout. Med. radiologiya 2(5): 22-23. (JPRS-210, pp. 11-26) (NSA 12: 15223).

Lebedinskiy, A. V. (Ed.). 1959. Soviet Scientists on the Hazards of Nuclear Arms Testing. Atomizdat, Moscow. (UCRL-Trans-549) (NSA 15: 22567).

Lebedinskiy, A. V. and Yu. I. Moskalev. (Eds.). 1961. Distribution, Biological Effects, and Migration of Radioactive Isotopes. Medgiz, Moscow. 342pp. (JPRS-15,067) (NSA 18: 13396).

Lebez, D. 1970. Labeling of snake venoms with radioactive isotopes. pp. 75-86. In: Radiation Sensitivity of Toxins and Animal Poisons. (See IAEA, 1970) (NSA 24: 50669).

Lebez, D., M. Antonijevic and M. Furlan. 1965. Changes in the composition of frog serum proteins after whole-body irradiation. Yugoslavia Nulearni Inst. "Jozef Stefan", Ljubljana. 3pp. (NSA 20: 18243).

Lebez, D., M. Antonijevic and M. Furlan. 1966. Studies on the nitrogen metabolism in frog liver after whole-body irradiation. V. Changes in the composition of frog serum proteins after whole-body irradiation. Strahlentherapie 130: 437-439. (NSA 20: 42787).

Lebez, D., Z. Maretić and J. Kristan. 1965. Studies on labeled animal poisons. I. Distribution of ^{32}P-labeled Latrodectus tredecimguttatus venom. Toxicon 2(4): 251-253.

Leboeuf, M. B. 1956. Analysis of vegetation for I^{131}. Hanford Atomic Products Operation, U. S. AEC report HW-15743. (NSA 11: 2240).

Lech, J. J. 1973. Isolation and identification of 3-trifluoromethyl-4-nitophenyl glucuornide from bile of rainbow trout exposed to 3-trifluoromethyl-4-nitrophenol. Toxicol. Appl. Pharmacol. 24(1): 114-124. (SFA 18: 16244) (^{14}C).

Lecis, A. 1963. Effects of x-rays on the larvae at the fourth stage of Anopheles maculipennis atroparvus. Atti Ass. Genet. Ital. 8: 236-241. (In Italian, English summary).

Lecis, A. 1966. Effects of radiation from a ^{60}Co source on mature ovary cycles of A. mac. atroparvus. Atti Ass. Genet. Ital. 11: 267-275. (In Italian) (NSA 21: 13047).

Lecis, A. R. and G. Orru. 1974. Radioprotective effects and toxicity of dimethylsulfoxide on panirradiated adult males of Anopheles mac. atroparvus (Diptera: Nematocera). Riv. Biol. 67(3): 319-329. (NSA 33: 26292).

Leclipteur, J. 1963. Introduction of chemical errors in the radiocarbon age calculation. Bull. Soc. Roy. Sci. Liége 32: 680-687. (In French) (NSA 18: 5477).

Lecomte, J. d? Techniques for the study of populations of insect pollinators. Ann. Abeille 5(3): 201-213. (In French) (Considers radionuclides).

Lecomte, J. 1962. Methods for the study of dynamics of insect populations. Methods for the study of populations of insect pollinator. (Res.) Ann. Epiphyt. 13(4): 333. (In French).

Lecomte, J. 1963. Study of the exchange of food in a colony of hummers with radioisotopes. Compt. rend. 257(23): 3664-3665. (In French).

Lecomte, J. 1965. Study of nutritional interchanges in the ant Formica polyctena by means of radioisotopes. Mede. Landb. Hoogsch. OpzoeksSta. Gent 30(2): 1029-1034. (In French).

Lecomte, J. and A. Pouvreau. 1968. Use of radioactive gold, ^{198}Au, to study the radius of action of colonies of bumble-bees (Bombus sp.) with a view to pollination of cultivated plants. pp. 17-21. In: Isotopes and Radiation in Entomology. (See IAEA, 1968) (In French) (NSA 24: 31793).

Lecoq, J. J., et al. 1958. Prospecting for uranium and thorium materials in the desert countries and in the equatorial forest regions of the French Union. Proc. 2nd Intern. Conf. on the Peaceful Uses of Atomic Energy 2: 744-773. (Paper A/CONF.15/P/1245).

Leddicotte, G. W. 1969. Specific activation analysis techniques and methods for the assay of trace substances in aquatic and terrestrial environments. pp. 76-80. In: Modern Trends in Activation Analysis. Vol. I. (See DeVoe, 1969) (NSA 24: 110).

Leddicotte, G. W. 1971. Uses of neutron sources in pollution control. pp. III.39-46. In: Neutron Sources and Applications. (Proc. Amer. Nuclear Soc. Nat. Topical Meeting, Augusta, Ga.) (CONF-710402-Vol. 1) (NSA 26: 11543).

Leddicotte, G. W. 1971. Activation analysis of the biological trace elements. Meth. Biochem. Anal. 19: 345-434. (NSA 26: 17384).

Lederer, C. M. and J. M. Hollander. 1973. Survey of nuclear data use in applied fields. Lawrence Berkeley Lab., U. S. AEC report LBL-1262. 16 pp. (For Symp. on Applications of Nuclear Data in Science and Technology, Paris, France, March 12, 1973) (CONF-730302-8) (NSA 28: 1955) (Environmental studies and oceanography).

Lee, C., C. W. Ji and K. S. Park. 1970. Determination of dissolved uranium contents in hot spring waters of Korea. J. Nucl. Sci. (Seoul) 9 (2,Pt.1): 17-19. (In Korean) (NSA 24: 41228).

Lee, C., N. B. Kim and E. B. Park. 1973. Simultaneous determination of mercury, bromine, arsenic, and cadmium in biological materials by neutron activation analysis. J. Korean Nucl. Soc. 5(4): 265-285. (In Korean) (NSA 30: 14800).

Lee, C. C. and K. H. Chung. 1970. Studies on insect pest control using the sterile-male technique. 2. On the mating competition between sterile males and Ri labeleling in adult moths. J. Nucl. Sci. (Seoul) 10(2): 55-61. (In Korean) (NSA 25: 29731).

Lee, C. C., et al. 1970. Studies on the insect pest control using the sterile male technique. I. On the radiosensitivity of fall webworm (Hyphantria cunea Drury). J. Nucl. Sci. (Seoul) 9(2,Pt.2): 89-94. (In Korean) (NSA 25: 4735).

Lee, C. K. 1974. Characteristics of replication and radiation response of Aedes albopictus cell line in vitro. Dissertation, Univ. of Illinois, Urbana. 88pp. (Univ. Microfilms 75-11, 806) (NSA 32: 28305).

Lee, J. J., et al. 1973. Effects of environmental stress on the community structure, productivity, energy flow, and mineral cycling in salt marsh epiphytic communities. pp. 816-828. In: Radionuclides in Ecosystems. (See Nelson, 1973) (NSA 25: 37753) (^{14}C, ^{35}S, ^{32}P).

Lee, K. B., J. K. Hwang and K. S. Lee. 1968. Exploratory studies on the eradication of the Korean pine caterpillar Dendrolimus spectabilis Butler by means of radiation. pp. 273-286. In: Isotopes and Radiation in Entomology. (See IAEA, 1968) (NSA 24: 32322).

Lee, K. B., J. K. Hwang and K. S. Lee. 1968. Exploratory studies on the eradication of the Korean pine moth Dendrolimus spectabilis Butler by using gamma irradiation. pp. 79-85. In: Radiation, Radioisotopes, and Rearing Methods in the Control of Insect Pests. (See IAEA, 1968) (NSA 24: 32329).

Lee, P. K. and S. I. Auerbach. 1959. Determination and evaluation of the radiation field above White Oak Lake bed. Oak Ridge National Lab., U. S. AEC report ORNL-2755. vi, 63pp. (NSA 13: 21833).

Lee, W. R. 1958. The dosage response curve for radiation induced dominant lethal mutations in the honey bee. Genetics 43(4): 480-492. (NSA 13: 1913).

Lee, W. R. 1961. The nonrandom distribution of foraging honey bees between apiaries. J. Econ. Entomol. 54(5): 928-933. (^{32}P).

Lee, W. R. 1964. Partial-body radiation of queen honey bees. Apicult. Res. 3: 113-116. (NSA 20: 38631).

Lee, W. R. 1965. Relation of distance to foraging intensity of honey bees (Apis mellifera) on natural food sources. Ann. Entomol. Soc. Am. 58(1): 94-100. (^{32}P, ^{131}I).

Lee, W. R. 1965. Feeding radioactive isotopes to specific larval stages of Drosophila melanogaster. Drosophila Infor. Serv. 40: 101. (^{32}P, ^{3}H).

Lee, W. R., et al. 1965. Mutagenic effect on mature Drosophila spermatozoa of P^{32} incorporated into DNA. Abstr. Genetics 52(2): 455.

Lees, D. H. and F. J. Francis. 1972. Effect of gamma radiation on anthocyanin and flavonol pigments in cranberries (Vaccinium macrocarpon Ait). J. Am. Soc. Hort. Sci. 97(1): 128-132. (NSA 26: 28394).

Le Faucheux, O. 1971. Pollution and protection of seas and beaches. pp. 81-122. In: Pollution and Protection of Seas and Beaches. M. Avargues, ed. Guy le Prat, Paris. (In French) (NSA 30: 12680).

LeFebvre, E. A. 1962. Energy metabolism in the pigeon (Columba livia) at rest and in flight. Thesis, Univ. of Minnesota, Minneapolis. 85pp. ($D_2{}^{15}O$).

LeFevre, G., Jr. 1965. The mutability of mature sperm of Drosophila melanogaster irradiated in the female and in the male. Genetics 51: (NSA 19: 15085).

LeFevre, G., Jr. and U.-B. Jonsson. 1964. The effect of mating intensity on mutation frequency patterns detected after irradiation of Drosophila melanogaster males. Genetics 50: 879-890. (NSA 19: 109).

Lefevre, G., Jr. and U.-B. Jonsson. 1964. X-ray indiced mutability in male germ cells of Drosophila melanogaster. Mutation Research 1: 231-246. (NSA 19: 10840).

Legan, R. W. and R. T. Jaske. 1970. Cost analysis of improved regulation of the Columbia River to increase minimum flow in the Hanford Plant vicinity. Battelle Northwest Lab., U. S. AEC report BNWL-1343. 76pp. (NSA 24: 27929).

Legay, J.-M. and P. Teulade. 1962. Sensitivity to x rays of the development time of Bombyx mori eggs when the irradiation is done several hours after the start of incubation. Compt. rend. 255(15): 1784-1785. (In French) (NSA 17: 7829).

Léger, C., M. T. Perquis and G. Delibrias. 1959. Modification of the natural radioactivity caused by carbon-14. Bull. Inform. sci. tech. (Paris) 30: 20-22. (In French) (NSA 13: 19097).

Leggo, D., et al. 1964. Post-harvest sterilization of oranges against Queensland fruit fly. Food Preserv. Quar. 24(1): 15-19.

LeGrand, H. E. 1962. Graphic evaluation of hydrologic factors in management of radioactive wastes. pp. 67-76. In: 2nd Ground Disposal of Radioactive Wastes Conf. U. S. AEC report TID-7628. (See Morgan, et al., 1962) (NSA 16: 25041).

Lehman, J. W. 1973. Tritium cycling in a Lake Erie marsh ecosystem. Proc. Conf. on Great Lakes Res. 16: 65-75. (NSA 30: 6543).

Lehr, J. 1972. Indirect radioactive contamination of the food chain, determination of the factors of transfer soil/agricultural produce and soil/milk in the European Community. Centre d'Études Nucléaire, France Commissariat à l'Énergie Atomique, Fontenay-aux-Roses, report EUR-4901. 42pp. (NSA 27: 14755).

Lehr, J. and C. Myttenaere. 1971. Contamination by rain (^{90}Sr, ^{137}Cs) of parts above ground of crops. pp. 989-1013. In: Proc. Intern. Symp. on Radioecology Applied to the Protection of Man and His Environment. (See Commission of the European Communities, 1971) (In French) (NSA 27: 5282).

Leighty, R. D. 1966. Instruments and methods: snow density profiling by nuclear means. J. Glaciol. 6: 171-176. (NSA 20: 39130).

Leinerte, M. P. 1968. Experiments on the irradiation of duckweed (Lemna minor L.) with γ-rays and fast neutrons. Radiobiologiya 8(1): 156-157. (AEC-tr-6914, pp. 273-276) (NSA 22: 47636).

Leinerte, M. P. 1969. Growth of higher freshwater plants after irradiation with gamma-rays and fast neutrons. Latv. PSR Zinat. Akad. Vestis (11): 68-72. (In Russian) (NSA 24: 36852). (See also Leynerte).

Leinerte, M. P. 1969. Influence of external gamma irradiation of higher aquatic plants on their accumulation of ^{90}Sr, ^{137}Cs, and ^{144}Ce. Radiobiologiya 9(3): 427-432. (AEC-tr-7026, 131-137) (NSA 24: 9986).

Leinerte, M. P., et al. 1973. Study of the ratio Sr^{90}/Sr in a Latvian seaside lake. pp. 73-80. In: Radioecology of Water Organisms. II. (See Andrumaitis, et al., 1973) (AEC-tr-7606, pp. 43-48) (NSA 30: 21112).

Leitch, J. L. 1951. Summary of the radiological findings in animals from the biological surveys of 1947, 1948, 1949, and 1950. Univ. of California at Los Angeles, U. S. AEC report UCLA-111. 30pp. (NSA 11: 12315).

Leitch, J. L. and N. S. Hagen. 1965. A FORTRAN IV program for the evaluation of post-irradiation survival times. Univ. of California at Los Angeles, U. S. AEC report UCLA-12-573. 39pp. (NSA 20: 14744).

Lekven, J. 1972. ^{55}Fe in human blood in Norway, 1967-1969. Nature 235 (5336): 284-286. (NSA 26: 18083) (Food web).

Le Louarn, H. and B. Dassonville. 1972. First results of radioactive labeling of the field voles Microtus arvalis and Pitymys multiplex. Mammalia 36(1): 107-111. (In French, English summary).

Lemaitre, C. 1973. Modifications produced by x irradiation on the weight and water content of Ceratitis capitata. Compt. rend. D276(16): 2419-2422. (In French) (NSA 28: 21786).

Lemaitre-Boyeaux, C. 1972. Modifications in the thermogenesis of Ceratitis capitata during the puparium state produced by x-rays. Compt. rend. D275 (5): 687-690. (In French) (NSA 27: 17683).

Le Masne, G. and A. Bonavita-Cougourdan. 1972. Preliminary results of a prolonged irradiation with cesium of the populations of ant hills, in Haute-Provence. Ekologie Polska 20(14): 129-144. (In French).

Le Moigne, A. 1965. Effects of irradiations with x rays on the embryonic development and the regeneration capacity for hatching of Polycelis nigra (turbullarian, triclaidan). Compt. rend. 260: 4627-4629. (In French) (NSA 19: 31826).

Le Moigne, A. 1966. Study of embryonic development and regeneration in the embryo of the planarian Polycelis nigra (turbullarian, tricladidan). J. Embryol. Exptl. Morphol. 15: 39-60. (In French) (NSA 21: 28648).

Le Moigne, A. 1968. Electron microscopic study of the development of x-irradiated embryonic planarian structures. J. Embryol. Exptl. Morphol. 19: 181-192. (In French) (NSA 22: 40928).

Lenaers, W. M. 1971. Photochemical degradation of sediment organic matter: effect on ^{65}Zn release. Thesis, Oregon State Univ., Corvallis. 64pp. (NSA 26: 48038).

Lenarz, H. H. 1972. A contribution to a revision of the ^{14}C-method for measuring primary production in inland waters. Univ. Freiburg i. B. 83pp. (In German((NSA 32: 8952).

Lender, T. and A. Gabriel. 1965. Neoblasts labeled by tritiated uridine migrate and build up the regeneration blastema of fresh water planarians. Compt. rend. 260: 4095-4097. (In French) (NSA 19: 35987).

Lengerová, A. 1955. A method for evaluating the effects of ionizing radiation on microorganisms. Folia Biol. (Prague) 1: 54-61. (In Russian) (NSA 13: 19788).

Lenihan, J. M. A. and S. J. Thomson. (Eds.). 1965. Activation Analysis. Academic Press, New York. 211pp.

Lenihan, J. M. A., J. F. Loutit and J. H. Martin. (Eds.). 1967. Strontium Metabolism. (Proc. Intern. Symp., Chapelcross, Glasgow and Strontian, May 5-7, 1966). Academic Press, New York. 354pp. (CONF-660561) (NSA 21: 45558).

Lentsch, J. W., et al. 1971. Manmade radionuclides in the Hudson River Estuary. pp. 499-528. In: Health Physics Aspects of Nuclear Facility Siting. Vol. II. (See Voillequé and Baldwin, 1971) (NSA 25: 55160).

Lentsch, J. W. 1972. Aquatic radiological assessments at nuclear power plants. vp. In: Water Quality Considerations: Siting and Operating of Nuclear Power Plants. (See Atomic Industrial Forum, 1972) (NSA 28: 2722 27228).

Lenz, H. 1973. Electronic data processing and radiometric age determinations. Geol. Jahrb. C(1): 75-80. (In German) (NSA 33: 14849).

Lenzi, G. and G. Cassano. 1972. Possibility of pollution of underground and surface waters from radioactive discharge from a nuclear fuel reprocessing plant. Ing. sanit. (1): 7-18. (In Italian) (NSA 28: 6472, 8159).

Leonard, O. A. and A. S. Crafts. 1956. Translocation of herbicides. III. Uptake and distribution of radioactive 2,4-D by brush species. Hilgardia 26: 366-413.

Leonard, O. A. and R. J. Hull. 1966. Translocation of ^{14}C-labelled substances and $^{32}PO_4$ in mistletoe-infected and uninfected conifers and dicotyledonous trees. pp.31-46. In: Isotopes in Weed Research. (See IAEA, 1966).

Leonardos, O. 1958. On the radioactivity of the brazilnut. An. Acad. Brasil. cien. 30(4): 51-52. (In Portugese).

Leontiadis, J. and C. Dimitroulas. 1971. Use of radioisotopes in tracing karst groundwater in Greece. I. Further investigation on the possible interconnection between Parthini sinkhole and various springs of the near area. Greece Democritus Nuclear Research Center, Athens, report DEMO-71/10. 13pp. (NSA 26: 1967).

Lepp, N. W. and G. J. Dollard. 1974. Studies on lateral movement of ^{210}Pb in woody stems. Oecologia 16(2): 175-184. (NSA 32: 1214).

Leppard, L. P. 1965. MPC objectives for drinking water contaminated by certain uranium-thorium-daughter mixtures. Preprint No. 65-CNA-209. For: Intern. Conf. and Exhibition of the Canadian Nuclear Association, Quebec. Canadian Nuclear Assoc., Toronto. (CONF-650515) (NSA 19: 36743).

Lerman, A. 1972. ^{90}Sr in the Great Lakes: concentration-time model. J. Geophys. Res. 77(18): 3256-3264. (NSA 26: 45532).

Lerman, A. 1973. Transport of radionuclides in sediments. pp. 936-944. In: Radionuclides in Ecosystems. (See Nelson, 1973) (NSA 28: 29881).

Lerman, A. and T. A. Lietzke. 1975. Uptake and migration of tracers in lake sediments. Limnol. Oceanogr. 20(4): 497-510. (NSA 33: 2988, 12076) (^{90}Sr, ^{137}Cs as tracers).

Lerman, A. and H. Taniguchi. 1973. Strontium-90 and cesium-137 in water and deep sediments of the Great Lakes. pp. 460-464. In: Radionuclides in Ecosystems. (See Nelson, 1973) (NSA 28: 27280).

Lerman, J. C. 1972. Carbon 14 dating: origin and correction of isotope fractionation errors in terrestrial living matter. pp. H16-H28. In: Proc. 8th Intern. Radiocarbon Dating Conf. Vol. 2. (See Rafter, 1972) (NSA 31: 8716).

Lerman, J. C., W. G. Monk and J. C. Vogel. 1967. Effect of the Tunguska meteor and sun-spots on radiocarbon in tree rings. Nature 216: 990-991. (NSA 22: 6485).

Le Roux, E. J. 1954. The absorption, distribution and site of action of DDT in DDT-resistant and DDT-susceptible house flies using carbon^{14} labeled DDT. J. Econ. Entomol. 47(6): 1058-1066.

Leski, R. 1963. Studies of the biology and ecology of populations of the cherry fruitfly Rhagoletis cerasi (L.). Polskie Pismo ent. B(3-4): 153-240. (In Polish; English, Russian summaries).

Leski, R. 1969. Population studies of the cherry fruit fly, Rhagoletis cerasi L. pp. 1-7. In: Insect Ecology and the Sterile-Male Technique. (See IAEA, 1969).

Lessler, M. A. 1959. Low-level x-ray damage to amphibian erythrocytes. Science 129(3362): 1551-1553. (NSA 13: 14211).

Lessler, M. A. 1964. Nuclear-cytoplasmic interaction with ionizing radiation. Int. Rev. Cytol. 16: 155-172. (NSA 19: 17566) (Injuries in amphibian eggs).

Lessler, M. A. and F. M. Herrara. 1961. Electron microscope studies of x-ray damage to frog blood cells. Abstr. Radiation Research 14(4): 482.

Lessler, M. A. and F. M. Herrara. 1962. Electron-microscope studies of x-ray damage to frog blood cells. Radiation Research 17(2): 111-117. (NSA 16: 26821).

Lessler, R. M. and F. W. Guy. 1965. Gamma dose rates and integrated doses from neutron-induced residual radioactivity in soil. Lawrence Radiation Lab., U. S. AEC report UCRL-12339. 2 vols, 52, 295pp. (NSA 19: 24750).

Lessler, R. M. and F. W. Guy. 1965. Neutron-induced activity in earth and sea-water from buried and surface neutron sources. Lawrence Radiation Lab., Lab., U. S. AEC report UCRL-12407. 62pp. (NSA 19: 32422).

Letavet, A. A. 1957. Hygienic problems of radiology. Med. radiologiya 2(1): 11-21. (In Russian).

Letavet, A. A. and Ye. B. Kurlyandskaya. (Eds.). 1962. The Toxicology of Radioactive Substances. Vol. 1. Strontium, Cesium, Ruthenium, Radium. Vol. 2. Cobalt, Sodium, Plutonium and Gold. Medgiz, Moscow. (English transl., R. E. Travers, transl.; E. Lloyd, ed.; Pergamon, New York, Oxford; Macmillan Co. 236, 196pp.) (NSA 18: 11613, 4994).

Letavet, A. A. and Ye. B. Kurlyandskaya. (Eds.). 1970. Toxicology of Radioactive Substances. Vol. 5. ^{65}Zn. Pergamon, Elmsford, New York. 191pp. (NSA 25: 21871).

Lettinga, G. 1972. Radioactivity and water supplies. Thesis, Technische Hogeschool, Delft, The Netherlands. 197pp. (NP-19637) (NSA 27: 14752).

Letunova, S. V., V. V. Koval'skiy and S. N. Romanova. 1970. Geochemical ecology of microorganisms under conditions of different uranium content in mud deposits. Zh. obshch. biol. 31: 111-120. (In Russian) (NSA 24: 34232).

Leuthold, U. 1966. The feeding of radiation-sensitizing substances to Drosophila. Thesis, Eidgenössische Technische Hochschule, Zurich, Switzerland. (In German).

Levenbook, L. 1958. Intracellular water of larval tissues of the southern armyworm (Prodenia eridania) as determined by the use of C^{14} carboxyl-inulin. J. Cell. Comp. Physiol. 52(2): 329-339.

Levenbook, L. 1959. Insect biochemistry. A report on Symposium XII. Proc. 4th Intern. Congr. on Biochemistry. (Vienna). L. Levenbook, ed. Pergamon, London. 12: 239-252. (Discusses use of radionuclides).

Levenbook, L. 1965. Meetings: Insect biochemistry. Science 150(3696): 643-644. (Review of various papers using radionuclides).

Levenbook, L. and M. L. Dinamarca. 1965. A simplified ionization chamber procedure for the continuous measurement of respiratory $C^{14}O_2$ of insects. Anal. Biochem. 11(2): 391-394.

Leventhal, J. S. and W. F. Libby. 1970. Tritium fallout in the Pacific United States. J. Geophys. Res. 75: 7628-7633. (NSA 25: 10702).

Leventhal, J. S. and W. F. Libby. 1973. Tritium hydrology studies in California. pp. 724-736. In: Tritium. (See Moghissi and Carter, 1973) (NSA 28: 27330).

Leveque, P. C. 1971. Evolution of tritium activities in superficial waters from Austria to the Aquitaine shore between 1968 and 1970. Compt. rend. D273(1): 19-22. (In French) (NSA 26: 1976).

Leveque, P. and M. Vigneaux. 1965. Results of the tritium activity measurement of some waters of the Southwest. Compt. rend. 261: 4168-4171. (In French) (NSA 20: 14744).

Leveque, P., M. Vigneaux and J. Alvinerie. 1967. Evolution of tritium activity in the southwest of France. pp. 417-432. In: Isotopes in Hydrology. (See IAEA, 1967) (In French) (NSA 22: 18936).

Leveque, P. C., et al. 1974. Use of isotope techniques to solve hydrological problems in civil engineering: a study of three specific cases. pp. 223-239. In: Isotope Techniques in Groundwater Hydrology. Vol. 2. (See IAEA, 1974) (In French) (NSA 31: 19656).

LeVert, F. E. and E. L. Helminski. 1973. Literature review and commercial source evaluation of americium-241. Tuskegee Inst., U. S. AEC report ORO-4333-1. v, 415pp.

Levi, E. 1966. Handling plants for macro-autoradiography. pp. 189-197. In: Isotopes in Weed Research. (See IAEA, 1966).

Levin, A. A., et al. 1970. Estimation of potential radionuclide ingestion by native populations of eastern Panama and northwestern Colombia. Battelle Columbus Labs., U. S. AEC report BMI-171-45. 73pp. (Concerns interoceanic canal feasibility studies).

Levin, G. V., et al. 1959. Rapid radioactive test for coliform organisms. J. Am. Water Works Assoc. 51(1): 101-104.

Levin, G. V. 1963. Rapid microbiological determinations with radioisotopes. Adv. Appl. Microbiol. 5: 95-133. (NSA 20: 3667).

Levin, M. D. 1960. A comparison of two methods of mass-marking foraging honey bees. J. Econ. Entomol. 53(4): 696-697.

Levin, S. G. and R. K. Stoms. 1969. Natural background radiation dose rate measurements in Michigan, Colorado and Minnesota. Am. J. Publ. Hlth. 59 (1): 102-109.

Levine, H., W. P. Kirk and H. J. L. Rechen. 1967. Plutonium and strontium-90 in precipitation, August 1966 through March 1967. Radiol. Hlth. Data 8(10): 574-576.

Levitt, J. 1972. Responses of Plants to Environmental Stresses. Academic Press, New York. 732pp. (Incl. ionizing radiation).

Levy, H. B. 1972. Evaluating the hazards of groundwater contamination by radioactivity from an underground nuclear explosion. Lawrence Livermore Lab., U. S. AEC report UCRL-51278. 25pp. (NSA 27: 14753).

Levy, L. W. and R. A. Munoz. 1959. Biosynthesis of radioactive rotenone. pp. 225-228. In: Radioisotopes and Radiation in the Life Sciences. (See Inter-American Nuclear Energy Commission, 1959) (NSA 14: 21249).

Levy, O. 1956. Microscopic examination in experiments on the effect of radium radiation on the embryonal and regenerative development after survival of the preparation of Prof. Alfred Schaper. Roux. Arch. Entwicklungsmech. Org. 21: 130. (In German).

Levy, R. 1971. Relationship between insect mineral content and radiation sensitivity. Thesis, Univ. of Florida, Gainesville. 155pp. (Univ. Microfims 72-21078) (NSA 26: 56478).

Levy, R., H. L. Cromroy and J. A. Cornell. 1973. Major and trace elements as bioindicators of acute insect radiosensitivity. Radiation Research 56(1): 130-139. (NSA 28: 30257).

Lewin, V. H. 1965. Radioactivity and sewage treatment. J. Proc. Inst. Sewage Purif. Part 2: 166-170. (NSA 19: 36724).

Lewis, C. T. 1963. Some applications of radioisotopes to the study of the contamination of insects by insecticide solutions. pp. 135-142. In: Radiation and Radioisotopes Applied to Insects of Agricultural Importance. (See IAEA, 1963).

Lewis, C. T. and N. Waloff. 1964. The use of radioactive tracers in the study of dispersion of Orthotylus virescens (Douglas and Scott) (Miridae, Heteroptera). Entomologia exptl. appl. 7(1): 15-24. (^{32}P).

Lewis, D. C. and R. H. Burgy. 1964. The relationship between oak tree roots and groundwater in fractured rock as determined by tritium tracing. J. Geophys. Res. 69: 2579-2588.

Lewis, L. 1972. Terrestrial radiological monitoring programs at Duke Power Company's Oconee and McGuire Nuclear Stations. pp. 131-160. In: Proc. Southern Conf. on Environmental Protection from Nuclear Power Plants. (See U. S. EPA, 1972) (NSA 27: 2528).

Lewis, L. F. and G. W. Eddy. 1964. Some effects of gamma radiation on the horn fly. J. Econ. Entomol. 57(2): 275-276. (NSA 18: 29313).

Lewis, L. L., et al. 1969. Radiation effects in the Soricidae, Cricetidae, and Muridae. pp. 173-184. In: Symposium on Radioecology. (See Nelson and Evans, 1969).

Lewis, N. F. 1971. Studies on a radio-resistant coccus isolated from Bombay duck (Harpodon nehereus). J. Gen. Microbiol. 66(1): 29-35. (NSA 25: 47065).

Lexow, S., E. P. P. Maneschi and A. M. Sa. 1948. Preliminary note on the presence of uranium in living things in Argentina. An. Asoc. quim. Arg. 36: 203-207.

Ley, F. J. 1963. The influence of dose rate on the inactivation of microorganisms. Int. J. Appl. Radiat. Isotopes 14(1): 38-41. (NSA 17: 24938).

Leynerte, M. P., et al. 1974. Study of the isotopic equilibrium $^{90}Sr/Sr$ in an ultraeutrophic lake of Latvia. Ecologiya 5(5): 77-79. (Sov. J. Ecology 5(5): 462-464) (NSA 32: 27931). (Leinerte, q.v.).

Leyton, L. 1974. Actual and potential uses of isotopes and radiation techniques in forest water relation research. pp. 3-10. In: Isotope and Radiation Techniques in Soil Physics and Irrigation Studies 1973. (See IAEA, 1974) (NSA 30: 15482).

Lezzi, M. 1961. Behavior of the puff pattern of the salivary-gland chromosomes in Chironomus under the influence of Drosophila egg contents. Thesis, Eidgenössische Technische Hochschule, Zurich, Switzerland. 30pp. (In German) (X-rays).

Lezzi, M. and H. Kröger. 1966. Uptake of ^{22}Na in the cell nuclei of the salivary glands of Chironomus thummi. Z. Naturforsch. 21b: 274-277. (In German) (NSA 20: 24850).

Li, Y., T. Hu and T. Fang. 1964. Sensitivity of Bunodactis asiaticus Uchid to ^{60}Co gamma radiation. pp. 113-114. In: Chung Kuo Hai Yang Hu Chao Hsueh Hui 1963 Nien Hsueh Shu Nein Hui Lun Wen Chai Yao Hui Pien. Peking. (In Chinese) (NSA 20: 45385).

Libby, L. M. (Comp.). 1970. Fifty environmental problems of timely importance. Rand Corp., Santa Monica, Calif., report P-4415(RAND). 67pp. (NSA 25: 24076) (Incl. ionizing radiation).

Libby, L. M. and W. F. Libby. 1972. Vulcanism and radiocarbon dates. pp. A72-A75. In: Proc. 8th Intern. Radiocarbon Dating Conf. Vol. 1. (See Rafter, 1972) (NSA 31: 8683).

Libby, L. M. and H. R. Lukens. 1973. Production of radiocarbon in tree rings by lightning bolts. J. Geophys. Res. 78(26): 5902-5903. (NSA 29: 26852).

Libby, W. F. 1955. Tritium in nature. J. Wash. Acad. Sci. 45(10): 301-314.

Libby, W. F. 1955. Radiocarbon Dating. Univ. of Chicago Press, Chicago. 2nd ed. 186pp. (NSA 20: 7288).

Libby, W. F. 1956. Radiocarbon dating. Am. Scientist 44(1): 98-112.

Libby, W. F. 1963. Accuracy of radiocarbon dates. Science 140(3564): 278-280. (NSA 17: 19740).

Libby, W. F. 1963. Radiocarbon dating. Ann. Intern. Med. 59(4): 566-578.

Libby, W. F. 1966. Isotope chronology and recorded history. Intern. Atomic Energy Agency Bull. 8(2): 28-37. (NSA 20: 31429).

Libby, W. F. 1967. The state of the art in ^{14}C dating. Trans. Amer. Nucl. Soc. 10: 20. (NSA 21: 36112).

Libby, W. F. 1967. History of radiocarbon dating. pp. 3-25. In: Radioactive Dating and Methods of Low-Level Counting. (See IAEA, 1967) (NSA 22: 6453).

Libby, W. F. 1969. Radiocarbon dating. Chem. Brit. 5: 548-552. (NSA 24: 9520).

Libby, W. F. 1973. History of tritium. pp. 3-11. In: Tririum. (See Moghissi and Carter, 1973) (NSA 28: 27125).

Licciardello, J. J. and J. T. R. Nickerson. 1962. Effect of radiation environment on the thermal resistance of irradiated spores of *Clostridium sporogenes* P. A. 3679. J. Food Sci. 27(3): 211-218. (NSA 16: 32783).

Licciardello, J. J. and J. T. R. Nickerson. 1963. Effect of radiation environment on the thermal resistance of irradiated spores of *Bacillis subtilis*. Appl. Microbiol. 11(3): 216-219. (NSA 17: 35468; 18: 8131).

Licciardello, J.J., C. A. Ribich and S. A. Goldblith. 1969. Effect of irradiation temperature on inactivation of *Clostridium botulinum* toxin type E by gamma rays. J. Appl. Bacteriol. 32: 476-480. (NSA 24: 27543).

Lichtenstein, E. P. 1972. Increase in toxicity of ^{14}C-insecticides with polychlorinated biphenyl compounds. pp. 137-141. In: Radiotracer Studies of Chemical Residues in Food and Agriculture. (See IAEA, 1972) (NSA 27: 2521).

Lidén, K. 1961. Cs-137 burdens in Swedish Laplanders and reindeer. Acta Radiologica 56(3): 237-240. (NSA 15: 32389).

Lidén, K. 1964. Increasing ^{137}Cs level in Lapps and other people in northern Sweden. (For 4th Symp. on Radioactivity in Scandanavia, Risö, Denmark, October 506, 1964). Univ. of Lund, Sweden. (CONF-747-8) (NSA 20: 42736).

Lidén, K. 1966. Accumulation of radionuclides in lichens and mosses in the food chain: lichen, reindeer, and man. Suomen Ken. A39(2): 18-28. (In Swedish) (NSA 20: 35705).

Lidén, K. and I. Ö. Andersson. 1962. Caesium-134 in man. Nature 195(4846): 1040-1043. (Incl. reindeer data).

Lidén, K. and G. Bengtsson. 1967. Comment on the metabolism of cesium in reindeer. p. 663. In: Radioecological Concentration Processes. (See Aberg and Hungate, 1967).

Lidén, K. and M. Gustafsson. 1967. Relationships and seasonal variation of ^{137}Cs in lichen, reindeer and man in northern Sweden 1961 to 1965. In: Radioecological Concentration Processes. (See Aberg and Hungate, 1967).

Lidén, K. and Y. Naversten. 1964. The enhanced radiocaesium levels of people in northern Sweden. pp. 107-182. In: Assessment of Radioactivity in Man. Vol. II. (See IAEA, 1964) (Incl. comments on reindeer).

Lieberman, J. A. 1971. Ecological standards, energy needs, and emission characteristics. Trans Amer. Nucl. Soc. 14(1): 73-74. (CONF-710606) (NSA 25: 39811).

Lieberman, J. A. 1972. Man and his envioronmental responsibilities. pp. 1-14. In: Proc. Southern Conf. on Environmental Protection from Nuclear Power Plants. (See U. S. EPA, 1972) (NSA 27: 2521).

Lieberman, J. A. 1972. EPA, environmental legislation and energy. Trans. Amer. Nucl. Soc. 15(1): 73. (NSA 26: 42934).

Lieberman, J. A. and J. R. Janis. 1970. Environmental problems and nuclear chemical engineering. Chem. Engrg. Progr. Symp. Ser. 66(104): 142-147. (NSA 25: 49123).

Lieberman, J. A., E. D. Harward and C. L. Weaver. 1970. Environmental monitoring and surveillance. pp. 60-72. In: Nuclear Power and the Public. H. Foreman, ed. Univ. of Minnesota Press, Minneapolis. (NSA 25: 23209).

Lieberman, J. A., E. D. Harward and C. L. Weaver. 1970. Environmental surveillance around nuclear power reactors. Radiol. Hlth. Data 11: 325-332. (NSA 24: 44514).

Lieberman, K. W. and H. H. Kramer. 1970. Cadmium determination in biological tissue by neutron activation analysis. Anal. Chem. 42: 266-267. (NSA 24: 11477).

Liebscher, K. and T. Schönfeld. 1960. Measurement of radioactivity of surface waters with the gamma spectrometer. Mitt. Österr. Sanitätsverw. 61: 329-331. (In German).

Liebscher, K., F. Habashi and T. Schönfeld. 1961. Observations on the behavior of fission products in surface waters. Atompraxis 7(3): 94-100. (SLA TR-G-inf. ser. 48, pp. 96-182) (NSA 15: 18344).

Lieser, K. H. 1972. Tasks of the nuclear and radiochemistry in environmental research and technology. Radiochim. Acta 18(3): 115-120. (In German) (NSA 28: 12606, 20664).

Lieser, K. H. 1973. Application of radiochemical methods in environmental research. Z. Anal. Chem. 263(4): 304-307. (In German) (NSA 28: 20675).

Lievens, F. 1962. Recent progress on the application of nuclear energy to agriculture. Belgium Centre d'Étude de l'Énergie Nucléaire, Brussels, report BLG-167. 26pp. (NSA 17: 19748).

Liggins, J. A. and H. Singh. 1971. Effects of fractionated dose of gamma irradiation on fecundity, mortality and longevity of the lesser grain borer. J. Econ. Entomol. 64(5): 1306-1307.

Light, A. E. 1935. A histologic study of the effects of x-rays on frog skin. Radiology 25(6): 734-738.

Lighthart, B. and P. E. Tiegs. 1972. Exploring the aquatic carbon web. pp. 289-300. In: _Proc. - Research on Coniferous Forest Ecosystems - a Symp._ Pacific Northwest Forest and Range Expt. Sta., U. S. Forest Service, Portland, Oregon. (^{14}C).

Lih, K. H. 1956. A note on the distribution of radioactive substances in some Chinese river and well waters. Report to the U. N. Scientific Commitee. Republic of China, Taiwan, report No. UN RC 56-1.

Likens, G. E. 1962. Transport of radioisotopes in lakes. Thesis, Univ. of Wisconsin, Madison. 191pp. (Dissertation Abstr. 23(4); NSA 17: 4715).

Likens, G. E. 1965. Vertical water motions in a small ice-covered lake. J. Geophys. Res. 70(10): 2333-2344. (Used radionuclides).

Likens, G. E. and A. D. Hasler. 1960. Movement of radiosodium in a chemically stratified lake. Science 131(3414): 1676-1677.

Likens, G. E. and A. D. Hasler. 1964. Transport of radionuclides in an ice-covered lake. Abstr. Verh. Intern. Verein. Limnol. 15(1): 258.

Likens, G. E. and P. L. Johnson. 1969. Measurements of background radiation in aquatic habitats in Alaska. pp. 319-328. In: _Symposium on Radioecology_. (See Nelson and Evans, 1969).

Likins, R. C., E. G. Berry and A. S. Posner. 1963. Comparative fixation of calcium and strontium by snail shell. Ann. N.Y. Acad. Sci. 109(1): 269-277. (NSA 18: 4974).

Lim, H. C., D. K. Kevan and V. R. Vickery. 1972. Effects of gamma radiation on the chromosomes of Gryllidae (Orthoptera). Can. J. Genet. Cytol. 14 (4): 753-761. (NSA 29: 8064).

Limpel, L. E. and J. E. Casida. 1957. Iodine metabolism in insects. I. _In vivo_ metabolism of radioiodide. II. _In vivo_ distribution and metabolism of iodoamino acids and related studies with _Periplaneta_. J. Exptl. Zool. 135(1): 19-27; 136(3): 595-613.

Limpert, F. A. and J. L. Smith. 1974. Utility of isotope profiling snow gage for water management. pp. 624-631. In: _Advanced Concepts and Techniques in the Study of Snow and Ice Resources_. (See Santeford and Smith, 1974) (NSA 33: 7311).

Lin, K., E. Pirie and D. Taylor. 1969. Moving gamma-source method of measuring soil densities. Nucl. Instrum. Methods 72: 325-328. (NSA 23: 38341).

Lin, K., D. Taylor and E. Pirie. 1969. Some design criteria for gamma back-scatter soil density gauges. Soil Science 108(4): 300-302. (NSA 24: 2305).

Lin, T. P. and E. D. Goldberg. 1951. Accumulation of radioiodine in the thyroid gland of elasmobranchs. Endocrinology 48(4): 485-488.

Lind, O. T. and R. S. Campbell. 1969. Comments on the use of liquid scintillation for routine determination of ^{14}C activity in production studies. Limnol. Oceanogr. 14(5): 787-789.

Lindberg, R. G. 1959. Factors influencing the biological consequences of environmental contamination by nuclear debris. pp. 42-59. In: _Proc. 2nd Plowshare Symp. Part II._ (May 13-15, 1959, San Francisco, Calif.). Lawrence Radiation Lab., Livermore, U. S. AEC report UCRL-5675. (NSA 14: 22688).

Lindberg, R. G. and K. H. Larson. 1956. The short term biological fate and persistence of radioactive fallout as measured at various locations within fallout patterns. pp. 197-204. In: _The Shorter-Term Biological Hazards of a Fallout Field_. G. M. Dunning and J. A. Hilchen, eds. U. S. AEC - U. S. DOD, Washington.

Lindberg, R. G., et al. 1954. Environmental and biological fate of fall-out from nuclear detonations in areas adjacent to the Nevada Proving Grounds, Operation UPSHOT-KNOTHOLE. Univ. of California at Los Angeles, U. S. AEC report WT-812. 49pp. (NSA 13: 11758).

Lindberg, R. G., et al. 1959. The factors influencing the biological fate and persistence of fallout (Operation TEAPOT). Univ. of California at Los Angeles, U. S. AEC report WT-1177. 78pp. (Interim report, 1955, ITR-1177) (NSA 13: 11720; 14: 23892).

Lindberg, R. G., et al. 1963. Investigation of _Perognathus_ as an experimental organism for research in space biology. Final report NASr 91. Northrop Space Lab., Hawthorne, Calif., report NSL 62-125-5. vii, 126pp. (Incl. radiation effects).

Lindeborg, R. G. 1964. A stem well method of introducing radioisotopes into plants. Ecology 45(2): 406-410.

Lindeken, C. L., D. G. Coles and J. W. Meadows. 1975. Ge(Li) spectrometer measurements of isotopic uranium ratios in soils treated with apatite-derived fertilizers. pp. 255-265. In: _Isotope Ratios as Pollutant Source and Behavior Indicators_. (See IAEA, 1975) (NSA 33: 7323).

Lindeken, C. L., D. J. Hodgkins and T. O. Hoeger. 1971. Estimating natural terrestrial background in the 17- and 60-keV energy regions from the Compton continuum peak. Lawrence Radiation Laboratory, Livermore, U. S. AEC report UCRL-72927. 12pp. (For Symp. on Safety in Plutonium-Handling Facilities, Rocky Flats, Colo., 13 Apr 1971) (NSA 25: 34727).

Lindeken, C. L., et al. 1972. Geographical variation in environmental radiation background in the United States. pp. 317-322. In: _Natural Radiation Environment. II. Vol. 1._ (See Adams, Lowder and Gesell, 1972) (Also: Lawrence Radiation Lab., Livermore, U. S. AEC report UCRL-73822) (CONF-720805-1) (NSA 26: 48035; 32: 19717).

Lindell, B. 1974. Establishment of standards and operative limits for radioactive contaminants. pp. 391-398. In: Environmental Surveillance Around Nuclear Installations. Vol. 2. (See IAEA, 1974) (NSA 31: 14048).

Lindell, B. and S. Løfveberg. 1972. Nuclear Power, Man, and Safety. Allmänna Fölaget, Stockholm. 446pp. (NSA 27: 29848) (Incl. behavior in ecological systems).

Lindell, B. and A. Magi. 1967. Observed levels of ^{137}Cs in Swedish reindeer meat. pp. 217-219. In: Radioecological Concentration Processes. (See Aberg and Hungate, 1967)

Lindgren, D., D. Nikolic and M. Simak. 1972. Germination of γ-irradiated and non-irradiated seeds of Pinus silvestris under various temperature conditions. 23pp. (NP-19439) (NSA 26: 56371).

Lindgren, W. A., et al. 1971. Automated primary productivity measurements in water bodies. Phase II-Design, fabrication and test of demonstration instrument. Biospherics, Inc., Rockville, Md., U. S. AEC report NYO-3993-2. xx, 254pp. (^{14}C).

Lindquist. A. W. 1952. Radioactive materials in entomological research. J. Econ. Entomol. 45(2): 264-270.

Lindquist, A. W. 1955. The use of gamma radiation for control or eradication of the screw-worm. J. Econ. Entomol. 48(4): 467-469. (NSA 6: 4723).

Lindquist, A. W. 1957. Studies of insects and insecticides with radioactive materials. pp. 527-535. In: Proc. Inter-American Symp. on the Peaceful Application of Nuclear Energy. (See U. S. AEC, 1957) (NSA 13: 8537).

Lindquist, A. W. 1958. Entomological uses of radioisotopes. Ch. 7, pp. 688-712. In: Radiation Biology and Medicine. (See Claus, 1958).

Lindquist, A. W. 1958. Radioactive tracer methods. pp. 76-82. In: Methods of Testing Chemicals in Insects. H. H. Shepard, ed. Burgess, Minneapolis.

Lindquist, A. W. 1959. Use of sexually sterile males for eradication of screw-worms. pp. 229-235. In: Radioisotopes and Radiation in the Life Sciences. (See Inter-American Nuclear Energy Commission, 1959) (NSA 14: 21353).

Lindquist, A. W. 1960. New aspects of biological control. pp. 104-111. In: Conf. on Biological Control of Insects of Medical Importance. (Washington, D. C., February 3-4, 1960). Am. Inst. of Biological Sciences, Washington, Tech. Rept.

Lindquist, A. W. 1961. New ways to control insects. Pest Control 29(6): 9-19, 36-40. (NSA 19: 45960).

Lindquist, A. W. (Ed.). 1962. Insect Population Control by the Sterile-Male Technique. (Comprehensive Report of a Panel, Vienna, October 16-19, 1962). International Atomic Energy Agency, Vienna, Tech. Rept. Ser. No. 21, Publ. STI/DOC/10/21. 59pp. (NSA 18: 3580).

Lindquist, A. W. 1969. Biological information needed in the sterile-male method of insect control. pp. 33-37. In: Sterile-Male Technique for Eradication or Control of Harmful Insects. (See IAEA, 1969) (NSA 24: 32296).

Lindquist, A. W., et al. 1951. Use of radioactive tracers in studies of penetration and metabolism of DDT in house flies. J. Econ. Entomol. 44 (3): 167-172.

Lindquist, A. W., et al. 1951. Studies of the flight habits of three species of flies tagged with radioactive phosphorus. J. Econ. Entomol. 44 (3): 397-400.

Lindquist, A. W., et al. 1967. Dispersion studies of Culex pipiens fatigans tagged with ^{32}P in the Kemmendine area of Rangoon, Burma. 1967. Bull. Wld. Hlth. Org. 36: 21-37. (NSA 22: 17003).

Lindquist, D. A. and P. A. Dahm. 1956. Metabolism of radioactive DDT by the Madeira roach and European corn borer. J. Econ. Entomol. 49(5): 579-584.

Lindquist, D. A., D. L. Bull and R. L. Ridgway. 1965. Systemic activity of bidrin in the cotton plant. J. Econ. Entomol. 58(2): 2--203. (^{32}P, Aphis gossypii).

Lindsay, E. and R. Craig. 1942. The distribution of radiophosphorus in the waxmoth, mealworm, cockroach and firebrat. Ann. Entomol. Soc. Am. 35(1): 50-56.

Lindsay, R. H., et al. 1966. Accumulation of I^{131} by channel catfish (Ictalurus punctatus) ovaries in vivo and in vitro. Gen. Comp. Endocrinol. 6 (2): 231-238. (SFA 11: 8908).

Lindstedt, K. D. and P. Kruger. 1969. Neutron activation analysis of vanadium in natural waters. pp. 87-92. In: Modern Trends in Activation Analysis. Vol. 1. (See DeVoe, 1969) (NSA 24: 112).

Lindstedt, K. D. and P. Kruger. 1969. Vanadium concentrations in Colorado River Basin waters. J. Am. Water Works Assoc. 61: 85-88. (NSA 23: 13685) (Neutron activation analysis).

Lindstedt, K. D. and P. Kruger. 1970. Determination of vanadium in natural waters by neutron activation analysis. Anal. Chem. 42: 113-115. (NSA 24: 9263).

Lindsten, D. C., S. K. Hasuike and A. G. Friend. 1965. Removal of radioactive contaminants from a seminatural water source with U. S. Army water purification equipment. Health Physics 11(8): 723-730. (NSA 19: 36717).

Ling, V. and G. H. Dixon. 1970. The biosynthesis of protamine in trout testis. II. Polysome patterns and protein synthetic activities during testis maturation. J. Biol. Chem. 245(12): 3035-3042.

Linn, D. U. 1962. The effects of radium on certain goldfish blood constituents and organs. Dissertation, Utah State Univ., Logan. 105pp.

Linn, I. and J. Shillito. 1960. Rings for marking very small mammals. Proc. Zool. Soc. (London) 134(3): 489-495.

Lion, P. 1969. Process for decontaminating radioactive waters in view of returning them back to the natural environment or rendering them drinkable. French Patent 2,031,748. (In French) (NSA 25: 42384).

Liotti, F. S. 1959. Influence of cobalt salts on the mortality caused by x-rays in the larvae of Bufo vulgaris. Spermentale 109(1): 87-99. (In Italian).

Lipke, E. J., Jr. 1971. Effects of environmental parameters on the uptake of radioisotopes in freshwater fish. Thesis, Univ. of Michigan, Ann Arbor. 130pp. (Univ. Microfilms 71-23,807) (NSA 26: 9488).

Lipke, H., B. Graves and S. Leto. 1965. Polysaccaride and glycoprotein formation in the cockroach. II. Incorporation of D-glucose-^{14}C into bound carbohydrate. J. Biol. Chem. 240(2): 601-608.

Lipke, H., S. Leto and B. Graves. 1965. Carbohydrate-amino acid conversions during cuticle synthesis in Periplaneta americana. J. Insect Physiol. 11 (9): 1225-1232. (^{14}C).

Lippold, P. C. 1967. Plastic enclosures for in vivo radioassay of insects. Int. J. Appl. Radiat. Isotopes 18: 656-658. (NSA 21: 41352).

Lippold, P. C. and F. L. Gambrell. 1965. Effects of ionizing radiation on eggs and adult insects. Abstr. 79. Bull. Entomol. Soc. Am. 11(3): 158. (^{60}Co).

Lippold, P. C. and F. L. Gambrell. 1966. Cobalt-60 radiation studies on the European chafer. Abstr. Bull. Entomol. Soc. Am. 12: 286.

Lippold, P. C., F. L. Gambrell and L. M. Massey, Jr. 1968. Effects of ionizing radiation on the European chafer, the plum curculio and the large milk-weed bug. Ann. Entomol. Soc. Am. 61(1): 151-157.

Lippold, P. C., et al. 1969. Degradation of insecticides by ^{60}Co gamma radiation. J. Econ. Entomol. 62: 1509-1510. (NSA 24: 34101).

Liptakova, H. and T. Sabova. nd. Current problems of soil and plant radio-ecology. Safarika Univ., Kosice, Czechoslovakia. (In Russian) (INIS-mf-887-2, pp. 1-42) (NSA 30: 3177).

Lisachenko, Ye. P. 1972. Some practical aspects of low background gamma-ray spectroscopy. pp. 53-65. In: Methods of Determination of Radioactivity. (See Polikarpov and Parchevskiy, 1972).

Lischi, G. and A. Pacciardi. 1952. Action of x-rays on the cardian function of Bufo vulgaris. II. Electrocardiographic observations. Radioterapia radiobiol. fis. med. 4: 308-314. (In Italian) (NSA 6: 3166).

Lisitsin, A. K. 1975. Uranium-carrying capacity of underground water. pp. 216-223. In: Radioactive Elements in Rocks. V. A. Kuznetsov, R. S. Zhuravlev and D. K. Osipov, eds. "Nauka", Novosibirsk. (In Russian) (NSA 33: 11993).

Lisk, D. J. 1966. Detection and measurement of pesticide residues. Science 154: 93-98. (Radionuclides).

Lisk, D. J. 1971. Ecological aspects of metals. N.Y. State J. Med. 71(21): 2541-2555. (NSA 26: 9490) (Incl. radiocuclides).

Little, H. F. 1967. Midgut epithelium of adult Chelisoches morio (Dermaptera: Chelisochidae) and Ceratitis capitata (Diptera: Tephritidae) following ionizing radiation. Ann. Entomol. Soc. Am. 60: 412-414. (NSA 22: 34757).

Little, J. B. and R. S. Chang. 1968. Effect of ionizing radiation on cell division in an amoeboid cell. Radiation Research 35: 132-146. (NSA 22: 40914).

Litver, B. Ya., A. A. Moiseyev and P. B. Ramzayev. 1967. Strontium-90 and polonium-210 in the bones of the inhabitants of the Far North in 1965. State Committee on the Use of Atomic Energy. Atomizdat, Moscow. UN Document A/AC.82/G/L.1190. 8pp. (AEC-tr-7214, pp. 108-114) (NSA 22: 33803; 25: 16067).

Litver, B. Ya., et al. 1967. Natural radioactivity in the lichen-reindeer-man chain. pp. 105-106. In: Science-Engineering Conf. on Radiation Hygiene. Leningrad. (In Russian).

Litver, B. Ya., et al. 1969. Lead-210 and polonium-210 in Arctic regions of the Northeast. State Committe on the Use of Atomic Energy. Atomizdat, Moscow. UN Document A/AC.82/G/L.1293. (AEC-tr-7128, pp. 180-193) (NSA 24: 12032).

Litvin, V. Yu. 1967. Optimal doses of ^{32}P for isotope labeling of the common vole. Zool. zh. 46: 1088-1093. (In Russian) (NSA 23: 16230).

Litvin, V. Yu. and Ye. V. Karaseva. 1968. A study of the daily migrations of Microtus oeconomus Pall. voles by radioactive tagging. Zool. zh. 47 (11): 1701-1706. (In Russian, English summary).

Litvin, V. Yu., et al. 1973. Repeated catches and radioisotope marking in studying the utilization of territory by rodents (common vole taken as an example). Zool. zh. 52(6): 931-938. (In Russian, English summary).

Litvinova, I. B. 1959. Effects of ionizing radiation on Paramecium cultivation at varius temperatures. Dokl. Akad. Nauk SSSR 124(2): 448-451. (Doklady (biol. sci.) 124: 22-24).

Litvinova, I. B. 1959. The effect of postirradiation temperature on the reaction of paramecia to different doses of ionizing radiation. Dokl. Akad. Nauk SSSR 124(3): 678-680. (Doklady (biol. sci.) 124: 25-27).

Litynski, T and A. Domnicz. 1967. ^{90}Sr-contamination of Polish soils. II. The ^{90}Sr-contamination of soils of southern and central Poland in 1963. Rocz. Glebozn. 18: 43-55. (In Polish) (NSA 23: 26532).

Livingston, G. K. 1971. Experimental studies on the induction of haploid parthenogenesis in Douglas fir and the effects of radiation on the germination and growth of Douglas fir pollen. Thesis, Univ. of Washington, Seattle. 212pp. (Univ. Microfilms 72-7383) (NSA 26: 25770).

Livingstone, D. A. and A. H. Estes. 1967. A carbon-dated pollen diagram from the Cape Breton plateau, Nova Scotia. Can. J. Bot. 45(3): 339-359.

Livnat, R. 1971. Laboratory measurement of sand density gradient by gamma radiation. Israel Negev Inst. for Arid Zone Research, Beersheba, report NIAZR-118. 20pp. (In Hebrew) (NSA 28: 21368).

Ljunggren, K. 1967. A review of the use of radioisotope tracers for evaluating parameters pertaining to the flow of materials in plant and natural systems. Isotopes Radiat. Technol. 5(1): 3-24.

Ljunggren, K. 1968. Nuclear techniques for water pollution control in Sweden. Trans. Amer. Nucl. Soc. 11: 50. (NSA 22: 30777).

Ljunggren, K., et al. 1959. Tracing water flow by means of radioactive isotopes and scintillation counters. Int. J. Appl. Radiat. Isotopes 5(3): 204-212.

Ljunggren, K., et al. 1971. Activation analysis of mercury and other environmental pollutants in water and aquatic ecosystems. pp. 373-405. In: Nuclear Techniques in Environmental Pollution. (See IAEA, 1971) (NSA 15: 31880).

Ljunggren, P. 1960. Fallout radioactivity in southern Sweden. Nature 186 (4725): 655-658. (NSA 14: 14648).

Ljunggren, P. 1960. The atomic bomb detonation and the increased radioactive contamination of forest trees. Skogen (Stockholm) 47(1): 16. (In Swedish).

Lloyd, D. and A. G. Callely. 1965. The assimilation of acetate and propionate by *Prototheca zopfii*. Biochem. J. 97: 176-179. (^{14}C).

Loaharanu, S., et al. 1975. Mating competitiveness of the radiosterilized male armyworm, *Spodoptera exigua* Hb., and the male mosquito, *Aedes aegypti* L., in field cages. pp. 317-323. In: Sterility Principle for Insect Control. (See IAEA, 1975) (NSA 32: 11786).

Löbbecke, E.-A. and O. Oltmanns. 1961. On a change in the spectrum of somatic mutations in *Ephestia kuehniella* Z. by temperature treatment before irradiation. Z. Vererbungslehre 92(3): 246-251. (In German) (NSA 17: 23206).

Lobo, P. F..S., et al. 1974. Determination of the time of residence of humic fractions of two ferallitic soils using natural and thermonuclear radiocarbon. Cah. ORSTOM (pedol.) 12(1): 115-123. (In French) (NSA 32: 17218).

Lochmueller, C. H., J. W. Galbraith and R. L. Walter. 1974. Trace metal analysis in water by proton-induced x-ray emission analysis of ion-exchange membranes. Anal. Chem. 46(3): 440-442. (NSA 29: 20635).

Locke, M., W. V. Condoulis and L. F. Hushman. 1965. Molt and intermolt activities in the epidermal cells of an insect. Science 149: 437-438. (^{3}H, *Calpodes*).

Lockner, D., A. Pettersson and U. Ericson. 1972. Radiosensitivity of haemopoiesis in fish. II. Studies at 18° and 10°C. Int. J. Radiat. Biol. 21 (6): 559-572. (NSA 26: 41209).

Lodhi, M. A. and A. Skaukat. 1967. Labelling sand with radioisotopes. Progress report, March 1963. Pakistan Atomic Energy Centre, Lahore, report PAEC/Chem/Misc-1. (NSA 22: 28018).

Loe, J. W. 1967. Investigation of effluent control standards and practices. Douglas United Nuclear, Inc., Richland, Wash., U. S. AEC report DUN-3155. 69pp. (NSA 30: 26774) (Columbia River).

Loeffel, R. E and W. O. Forster. 1970. Determination of movement and identity of stocks of coho salmon in the ocean using the radionuclide ^{65}Zn. Res. Rep. Fish. Comm. Oregon 2(1): 1-13. (NSA 26: 28087).

Loew, O. and K. J. Stiller. 1973. Studies on quantitative morphology. IV. Relationship between follicle size and ^{131}I uptake in the avian thyroid gland. Exp. Pathol. 8(5-6): 371-373. (NSA 31: 11714) (*Melopsittacus undulatus*).

Loewenstein, H. 1965. An injection probe for rapid placement of radioisotopes in soil. Soil Sci. Soc. Am. Proc. 29: 328-329. (NSA 19: 38870).

Loewenthal, K. 1953. Effect of roentgen rays on *Microsporum audouini*. An *in vitro* study. Arch. Dermatol. Syphilol. 67: 191-193. (NSA 7: 2964).

Lofts, B., A. J. Marshall and J. Rotblat. 1960. Effects of whole-body irrariation on the breeding plumage of the weaver finch, *Quelea quelea*. Nature 187(4737): 615-616. (NSA 14: 21323).

Loginov, L. I. 1963. The feasibility of utilizing sewage water containing natural radioelements for the irrigation of fields. Gigiyena sanitariya 18(6): 20-23. (JPRS-20871; OTS-63-31657).

Logson, J. E. 1972. Radioactive waste discharges to the environment from nuclear power facilities. Radiat. Data Rept. 13(3): 117-129. (NSA 26: 30365).

Logson, J. E. and R. I. Chissler. 1970. Radioactive waste discharges to the environment from nuclear power facilities. Bureau of Radiological Health, U. S. PHS, report BRH/DER-70-2. 77pp. (NSA 24: 28933).

Logson, J. E. and M. A. Culliton. 1972. Environmental radioactivity surveillance around nuclear power facilities. Nuclex, Basel, Tech. Meeting No. 9/11. 14pp. (NSA 29: 25980).

Logson, J. E. and T. L. Robinson. 1971. Radioactive waste discharged to the environment from nuclear power facilities. U. S. EPA report BRH/DER-70-2 (Add.1), ORP/SID-71-1. 33pp. (NSA 26: 20318).

Loh, C. K. 1968. Potassium exchange kinetics in amphibian atrial myocardium. Thesis, Univ. of Minnesota, Minneapolis. 169pp. (NSA 23: 24173) (^{42}K).

Lohse, G. E., D. W. Rhodes and B. R. Wheeler. 1970. Preventin activity release at the Idaho Chemical Processing Plant. pp. 411-418. In: Management of Low- and Intermediate-Level Radioactive Wastes. International Atomic Energy Agency, Vienna, Publ. STI/PUB-264. (CONF-700905) (NSA 25: 21123).

Loiseleur, J. (Ed.). 1966. Applied Radiobiology. Gauthier-Villars, Paris. 2 vols. (In French) (NSA 21: 45766).

Loken, A. 1963. The use of radiography in forest tree seed analyses and experiments. X-Ray Focus 4: 1-5.

Lollar, A. Q., D. C. Coleman and C. E. Boyd. 1971. Carnivorous pathway of phosphorus uptake by Ultricularia inflata. Arch. Hydrobiol. 69(3): 400-404. (^{32}P).

Lombard, L. S., C. F. Decker and A. M. Brues. 1961. The $LD_{100/35}$ for Sr^{90} in the Everglades rat snake (Elaphe obseleta rossaleni). pp. 78-81. In: Biological and Medical Research Division Summary Report, January through December, 1960. Argonne National Lab., U. S. AEC report ANL-6368.

Lombard, S. M. 1974. Theoretical, analytical, and statistical interpretation of environmental data. Los Alamos Scientific Lab., U. S. AEC report LA-UR-75-872. 16pp. (NSA 32: 19737).

Lomenick, T. F. 1963. Movement of ruthenium in the bed of White Oak Lake. Health Physics 9(8): 835-845. (NSA 17: 39275).

Lomenick, T. F. 1963. Movement of ruthenium in the bed of White Oak Lake. pp. 217-227. In: Colloq. Intern. Retention and Migration of Ions in Radioactive Sols. Centre d'Études Nucléaires, Saclay; and Press Universitaire, Paris. (NSA 18: 17348).

Lomenick, T. F. and T. Tamura. 1965. Naturally occurring fixation of ^{137}Cs on sediments of lacustrine origin. Soil Sci. Soc. Am. Proc. 29(4): 383-387. (NSA 19: 46681).

Lonati, R. D. and F. Tonolini. 1960. Study of alpha radioactivity in the soil and in biological material by spectrum determination. Energia Nucleare 7: 107-110. (In Italian).

Long, D. B. 1955. Field observations on adults of the wheat bulb fly (Leptohylemyia coarctata (Fall.)). Bull Entomol. Res. 49(1): 77-94. (^{32}P).

Long, I. F. and B. K. French. 1967. Measurement of soil moisture in the field by neutron moderation. J. Soil Sci. 18: 149-166. (NSA 21: 36120).

Long, M. 1960. How does radiation affect our environment? The Emory Alumnus 36(2): 8-15.

Long, R. L. and G. L. Liebler. 1972. Environmental problems associated with the repair of a nuclear power reactor primary coolant system. pp. 388-392. In: Inst. of Environmental Sciences 1972 Proc. Inst. of Environmental Sciences, Mt. Prospect, Ill. (NSA 26: 49849).

Long, W. H. and J. H. Lilly. 1958. Wireworm behavior in response to chemical seed treatments. J. Econ. Entomol. 51(3): 291-295. (^{60}Co).

Longhurst, W. M. and R. E. Kepner. 1969. Strontium-90 and cesium-137 food chain relationships to deer and sheep. Summary report May 1, 1963-September 30, 1969. Univ. of California, Davis, U. S. AEC report AT(11-1)-34P. 104. 25pp.

Longhurst, W. M, M. Goldman and R. J. Della Rosa. 1967. Comparison of the environmental and biological factors affecting the accumulation of ^{90}Sr and ^{137}Cs in deer and sheep. pp. 635-648. In: Radioecological Concentration Processes. (See Aberg and Hungate, 1967) (CONF-660405) (NSA 20: 20388).

Longhurst, W. M., et al. 1970. Total water and water turnover in sheep and deer. Am. J. Vet. Res. 31(4): 673-677. (Tritium oxide).

Longin, R. 1971. New method of collagen extraction for radiocarbon dating. Nature 230: 241-242. (NSA 25: 37312).

Look, M. and L. R. White. 1970. Simplified synthesis for ^{14}C-labeling of aryl dimethylurea herbicides. J. Agr. Food. Chem. 18(4): 745.

Loos, W. E. 1962. Effect of gamma radiation on the toughness of wood. Forest Prod. J. 12(6): 261-264.

Loosli, H. H., H. Oeschger and W. Wiest. 1970. ^{37}Ar, ^{39}Ar and ^{81}Kr in the atmosphere and tracer studies on these isotopes. J. Geophys. Res. 75: 2895-2900. (NSA 24: 31492) (Water reservoirs).

Lopatkina, A. P. 1964. Patterns of uranium migration in natural waters of humid areas and their use in the determination of hydrochemical background of uranium. Geokhimiya (8): 802-810. (Geochem. Int. (4): 788-795) (NSA 18: 39497).

Lopatkina, A. P. 1967. Conditions of accumulation of uranium in peat. Geokhimiya ( ): __. (Geochem. Int. 4: 545-550. (NSA 22: 40214).

Lopatkina, A. P., et al. 1970. Concentration of uranium by living and dead peat-forming plants. Geokhimiya (3): 372-377. (Geochem. Int. 7: 277-282) (NSA 25: 13116).

Lopez, D. F. 1970. Sterile-male technique for eradication of the Mexican and Caribbean fruit flies. Review of current status. pp. 111-117. In: Sterile-Male Technique for Control of Fruit Flies. (See IAEA, 1970) (NSA 25: 24515).

Lopez Perez, B. 1974. Environmental contamination by the nuclear industry. Energia Nuclear (Madrid) 19(87): 5-14. (In Spanish).

Loprieno, N. and A. Nannarone. 1958. X ray sensitivity of conidia of Collectortrichum coccodes (Wallr.) Hughes. Atti Ass. Genet. Ital. 8: 42-44. (In Italian) (NSA 17: 28708).

Lorch, S. 1971. Automatic recording of the moisture density and the water content variation in soil by means of γ-rays. Z. Pflanzenernähr. Bodenk. 130(2): 136-151. (In German) (NSA 28: 2751).

Lorenz, K. P. and P. S. Henshaw. 1941. Radiobiologic action and the killing effects of x-rays on Achromobacter Fischer. Radiology 36(4): 471-481.

Lorenz, P. R., et al. 1968. Survival of micro-organisms in space. Results of Gemini-IX-A, Gemini-XII, and Agena-VIII satellite-borne exposure and collection experiments. Space Life Sci. 1: 118-130. (NSA 22: 33988).

Lorenzola, J. B. and M. P. Anselmetti. 1967. Determination of the beta and gamma activity in samples of foods in common use, in drinking water and in air taken from the province of Vercelli, and relative methods. Boll. Lab. Chim. Prov. (Bologna) 18: 501-513. (In Italian) (NSA 22: 51588).

Lorz, H. W. 1971. Effects of x irradiation, diethylstibestrol, and size at time of release on the early sexual maturation of coho salmon Oncorhynchus kisutch. Thesis, Oregon State Univ., Corvallis. 108pp. (Univ. Microfilms 71-25,068) (NSA 26: 15499).

Loshak, A. Ya. 1962. Hygienic evaluation of radioactive iodine accumulation in freshwater fish. Gigiyena sanitariya 27(6): 93-95. (DA A-1365; OTS or SLA 62-33563) (NSA 17: 5757).

Loucks, O. L. 1972. Systems methods in environmental court actions. pp. 419-473. In: Systems Analysis and Simulation in Ecology. Vol. II. B. C. Patten, ed. Academic Press, New York. (Incl. Rulison event).

Louisiana Power & Light Co. 1971. Waterford Steam Electric Station, Units 3 and 4. Applicant's environmental report, construction permit stage. New Orleans, La. U. S. AEC DOCKET-50382-6; DOCKET-50383-6. 96pp. (NSA 25: 15334).

Louloudes, S. J., et al. 1961. Lipogenesis from C^{14}-acetate by the American cockroach. Ann. Entomol. Soc. Am 54(1): 99-103.

Lourie, I. M. and T. E. Shea. 1961. Protection against ionizing radiation. Outlines, description, and considerations of the establishment of programs of this type in the Americas. Bol. Ofic. Sanit. Panama 51: 113-118. (NSA 18: 5150) (Discusses ecology and screwworm control).

Loutit, J. F. 1966. Radiation in man's environment. pp. 3-26. In: Radioactivity and Human Diet. (See Russell, 1966) (NSA 21: 43404).

Loutit, J. F. and R. S. Russell. (Eds.). 1959. Operation BUFFALO, Part 5: the entry of fission products into food chains. Atomic Weapons Research Establishment, U. K. Atomic Energy Authority, Aldermaston, report AWRE-T-57/58. 216pp.

Love, S. K. 1951. Natural radioactivity in water. Industr. Engr. Chem. 43(7): 1541-1544.

Lovelace, F. E., and H. A. Podoliak. 1952. Absorption of radioactive calcium by brook trout. Progr. Fish-Culturist 14: 154-158.

Lovelius, N. V. 1974. Possibility of estimating the effect of supernova explosions on growth of tree species. Bot. zh. 59(7): 992-995. (In Russian) (NSA 32: 17447).

Lovellette, E. and F. J. Ratty. 1962. Comparisons of inbred and random bred larval survival to 1200 r. Drosophila Inform. Serv. 36: 87-88.

Lovric, G. and P. Strohal. 1972. Radioecological investigations in a thermal spring region. Limnol. Oceanogr. 17(3): 441-444. (NSA 26: 58815).

Lowder, W. M. 1972. Time variations of environmental radiation. pp. 1-7. In: Workshop on Natural Radiation Environment. (See McLaughlin, 1972) (NSA 27: 19819).

Lowder, W. M. (Ed.). 1974. 2nd Workshop on the Natural Radiation Environment. (February, 1974). Health and Safety Lab., U. S. AEC report HASL-287. 168pp. (NSA 31: 3320).

Lowder, W. M. and L. R. Solon. 1956. Background radiation, a literature search. Health and Safety Lab., U. S. AEC report NYO-4712. iii, 43pp.

Lowder, W. M., H. L. Beck and W. J. Condon. 1964. Spectrometric determination of dose rates from natural and fall-out gamma-radiation in the United States, 1962-63. Nature 202(4934): 745-749. (NSA 18: 25694).

Lowder, W. M., H. L. Beck and J. E. McLaughlin. 1970. Assessment of radioactive contamination in the environment by field gamma spectrometry. pp. 419-425. In: Environmental Surveillance in the Vicinity of Nuclear Facilities. (See Reinig, 1970) (NSA 25: 16184).

Lowder, W. M., W. J. Condon and H. L. Beck. 1964. Field spectrometric investigations of environmental radiation in the U. S. A. pp. 597-616. In: The Natural Radiation Environment. (See Adams and Lowder, 1964. (NSA 18: 35741).

Lowder, W. M., A. Segall and W. J. Condon. 1964. Environmental radiation survey in northern New England. pp. 907-918. In: The Natural Radiation Environment. (See Adams and Lowder, 1964) (NSA 18: 35751).

Lowe, C. H., Jr. and K. S. Norris. 1950. Study of the reinvasion of the crater by reptiles. pp. 65-66. In: Quarterly progress report for period ending June 30, 1950. Univ. of California at Los Angeles, U. S. AEC report UCLA-76, Part II.

Lowe, H. N., et al. 1956. Destruction of microorganisms in water, sewage, and sewage sludge by ionizing radiations. J. Am. Water Works Assoc. 48: 1663-1667. (NSA 11: 4236).

Lowe, H. N., et al. 1957. Destruction of microorganisms in water, sewage and sewage sludge by ionizing radiation (Salty Dog VIII). U. S. Army Engineer Res. and Develop. Labs., Ft. Belvoir, Va., Research Report. 34pp.

Lowenhaupt, B. 1956. The transport of calcium and other cations in submerged aquatic plants. Biol. Revs. Cambridge Phil. Soc. 31: 371-395. (Also: Lawrence Radiation Lab., Berkeley, U. S. AEC report UCRL-3247) (NSA 10: 3164).

Lowenhaupt, B. 1958. Active cation transport in submerged aquatic plants. I. Effect of light upon the absorption and excretion of calcium by Potamogeton crispus (L) leaves. J. Cell. Comp. Physiol. 51: 199-208. (Also: Lawrence Radiation Lab., Berkekey, U. S. AEC report UCRL-3460) (NSA 10: 9104).

Lowman, F. G. 1963. Iron and cobalt in ecology. pp. 561-567. In: Radioecology. (See Schultz and Klement, 1963) (NSA 17: 33608).

Lowman, F. G. 1969. Radionuclides of interest in the specific activity approach. BioScience 19(11): 993-999, 1005.

Lowman, F. G. 1969. The effects of the marine biosphere and hydrosphere upon the specific activity of contaminant radionuclides. pp. 436-457. In: Proc. Symp. on Public Health Aspects of Peaceful Uses of Nuclear Explosives. Southwestern Radiological Health Lab., U. S. PHS, Las Vegas, Nev., report SWRHL-82.

Lowman, F. G., R. F. Palumbo and D. J. Smith. 1957. The occurrence and distribution of radioactive non-fission products in plants and animals on the Pacific Proving Ground. Univ. of Washington, U. S. AEC report UWFL-51. 61pp. (NSA 12: 10397).

Lowman, F. G., et al. 1970. Estuarine and marine ecology. Vol. I. Final report and Appendix A. Vol. II. Appendixes B through E. Vol. III. Appendix F. Vol. IV. Appendixes G through N. Bioenvironmental and radiological- safety feasibility studies Atlantic-Pacific interoceanic canal. Puerto Rico Nuclear Center, Mayaguez; Battelle Columbus Labs., U. S. AEC report BMI-171-32(vols.1-4). 143, 240, 233, 283pp. (NSA 26: 40869, 40870, 40871, 40872).

Lowson, R. T. 1975. Bacterial leaching of uranium. A review. Australia Atomic Energy Commission, Lucas Heights, report AAEC/E-356. 52pp. (NSA 33: 14293).

Lowton, R. J., J. H. Martin and J. W. Talbot. 1966. Dilution, dispersion and sedimentation in some British estuaries. pp. 189-206. In: Disposal of Radioactive Wastes into Seas, Oceans and Surface Waters. (See IAEA, 1966)

Lozina-Lozinskiy, L. K. 1961. The resistance of paramecia adapted to life in warm radioactive springs to different external factors. Tsitologiya 3(2): 154-156. (JPRS-) (BA 38: 16469; Ref. zh. (biol.) 1961: 20D8).

Lu, C.-H., et al. 1964. Metabolism of ribonucleic acid (RNA) in the silk gland of Samia ricini. K'o Hsueh T'ung Pao (Scientia) 4: 340-342. (In Chinese) (^{14}C, ^{32}P).

Lu, C.-H., et al. 1964. Ribonucleic acid in the silk gland of silkworm (Attacus ricini). Shi Yen Sheng Wu Hsueh Pao 9(3): 300-310. (In Chinese) (14C, ^{32}P).

Lu, C.-P. 1956. Effects of atomic energy on silkworm and mulberry tree. J. Agric. Assoc. China (Taipei) 15: 99-103. (In Chinese, English summary) (^{14}C, ^{32}P).

Lu, H.-s., S.-m. Chang and T.-h. Chuang. 1964. The use of P^{32} for the determination of ribonucleic acid of the silk worm on the multiplication of a nuclear-polyhedrosis virus. Yuan Chia-neng (7): 635-638. (In Chinese) (Ref. zh. (biol.) 1965: 13B83).

Lu, S.-C., Y. Tai and Y.-C. Hsu. 1964. Radiocarbon concentration in Taiwan wood. Chinese J. Phys. (Taiwan) 2: 1-5. (NSA 18: 31759).

Lubin, M. and P. B. Schneider. 1957. The exchange of potassium for caesium and rubidium in frog muscle. J. Physiol. (London) 138(1): 140-155.

Lubyanov, I. P. 1962. Study of radioactivity of benthonic animals in fresh-water reservoirs. Radiobiologiya 2(2): 255-258. (AEC-tr-5429, pp. 122-128).

Lucas, A. C. 1965. A miniature thermoluminescent dosimeter and its application to radioecology. EG&G, Santa Barbara, Calif., Tech. Rept. EGG 1183-2033, o. S-285-R. 16pp.

Lucas, A. C. and N. R. French. 1966. A miniature thermoluminescent dosimeter and its application in radioecology. pp. 402-411. In: Luminescence Dosimetry. (Proc. Conf., June 1965, Stanford Univ., Stanford, Calif.) F. H. Attix, ed. U. S. AEC Symposium Ser. No. 8.

Lucas, A. C. and B. M. Shoffner. 1976. Stabilized scintillation chrystal system for monitoring gamma ray exposure in the environment. Trans. Nucl. Sci. NS-23(1): 699-703. (NSA 33: 29281).

Lucas, A. C., Z. G. Burson and R. E. Lagerquist. 1966. Design of a shielded source for the irradiation of natural animal populations. EG&G, Santa Barbara, Calif., U. S. AEC report CEX-63.10. 44pp. (NSA 20: 40518).

Lucas, H. F., Jr. 1964. A fast and accurate survey technique for both radon-222 and radium-226. pp. 315-330. In: The Natural Radiation Environment. (See Adams and Lowder, 1964) (NSA 18: 35487).

Lucas, H. F., Jr. and F. H. Ilcewicz. 1958. Natural radium-226 content of Illinois water supplies. J. Am. Water Works Assoc. 50: 1523-1532.

Lucas, H. F., Jr. and D. P. Krause. 1960. Preliminary survey of radium 226 and radium 228 (MsThI) contents of drinking water. Radiology 74(1): 114. (NSA 14: 6568).

Lucas, H. F., Jr., D. N. Edgington.and P. J. Colby. 1970. Concentrations of trace elements in Great Lake fishes. J. Fish. Res. Bd. Can. 27: 677-684. (NSA 25: 21644).

Lucas, H. F., Jr., et al. 1962. Correlation of the ^{228}Ra and ^{226}Ra content of man with that of the food and water of his environment. Abstr. p. 102. In: 2nd Intern. Congr. Radiation Research. (See Anonymous, 1062).

Lucas, J. W. and D. C. Pickering. 1958. Direct absorption of dissolved strontium-90 and yttrium-90 by tadpoles of Rana temporaria. Nature 182 (4644): 1242-1243. (NSA 13: 1072).

Luce, C. F. 1972. Nuclear energy least damaging. Catal. Environ. Qual. 2 (3): 15-17. (NSA 27: 14227).

Lück, H. and F. Dell. 1963. Action of radiation on cellulose and plant structural substance. Chimia (Asrau) 17: 1-8. (In German) (NSA 17: 27088).

Lüdicke, M. 1952. On the uptake of radioactivity secondary sodium phosphate by Lucanus cervus L. Z. vergl. Physiol. 34(5): 508-524. (In German).

Lüdicke, M. 1953. Studies on the radioactivity of insect wings after feeding with P^{32} labelled disodium hydrogen phosphate. Zool. Anz. Suppl. 18, Verh. dtsch. zool. Ges. pp. 412-417. (In German).

Lüdicke, M. 1954. On the distribution in the emerging imago of Vanesa io L. of P^{32}-labelled $Na_2HP^{32}O_4$ solution, taken up at the larval stage. Z. vergl. Physiol. 36(5): 508-530. (In German).

Lüdicke, M. and H. Peterhänsel. 1965. Labelling of certain areas of pigmentation on butterfly wings with ^{35}S-sodium sulphate and ^{35}S-DL-cystine solutions. Naturwissenschaften 52(5): 113-114. (In German).

Lüdicke, M. and D. Plesse. 1970. Description of pigment change in the wings of seasonal dimorphous forms of Araschnia levanaprorsa L. (Nymphalidae) after oral application of ^{35}S-sodium sulphate in the larval phase. Z. Naturforschung 25b: 399-406. (In German) (NSA 24: 27496).

Ludlam, S. D. and J. G. McGowan. 1971. Ecological-engineering study of an agricultural or silvacultural system utilizing heated water from power plants. pp. 711-721. In: 1971 Intersociety Energy Conversion Engineering Conf. (Boston, Mass., 3 August 1971). Society of Automotive Engineers, Inc. (CONF-710830) (NSA 26: 56154).

Ludwieg, F. 1962. Radioactive Isotopes in Fodder and Feed. Karl Thiemig KG, Munich. (In German).

Luebs, R. E., M. J. Brown and A. F. Laag. 1968. Determining water content of different soils by the neutron method. Soil Science 106(3): 207-212.

Lugo, A. 1970. Photosynthetic studies on four species of rain forest seedlings. Ch. I-7, pp. I81-I102. In: A Tropical Rain Forest. (See H. T. Odum and R. F. Pigeon, 1970) (NSA 25: 21952).

Lugo, A. E., S. C. Snedaker and J. F. Gamble. 1973. Models of matter flow in a southern mixed hardwood forest in Florida: preliminary results. pp. 929-935. In: Radionuclides in Ecosystems. (See Nelson, 1973) (NSA 28: 30376) (Discusses ^{137}Cs).

Lukens, H. R., D. E. Bryan and M. A. Hiatt. 1971. Oil-slick identification certainties with the method of neutron activation analysis. pp. 62-71. In: Nuclear Methods in Environmental Research. (See Vogt, 1971) (NSA 26: 17357).

Lukacsovics, F. and A. Gubicza. 1963. The effect of gamma rays on "vara" race of Bombyx mori L. (Lepidoptera). I. Study of the diapause of eggs. Ann. inst. biol. Tihany 30: 67-72. (In Hungarian, German summary).

Lukiewica, S. 1957. Physical factors determining the lethal action of electric fields on tadpoles of Xenopus laevis Daud. Folia Biol. (Krakow) 5(4): 267-282. (In Polish).

Lumb, C. 1969. Water pollution. pp. 29-51. In: Technological Injury. (See Rose, 1969) (NSA 24: 25432) (Mentions radioactivity).

Lunan, K. D. and H. K. Mitchell. 1964. Tyrosine metabolism and tyrosine-O-phosphate in Drosophila. Abstr. Fed. Proc. 23(2,Pt.1): 370. (^{14}C).

Lund, L., et al. 1962. A study of Sr-90 and Cs-137 in Norway 1957-1959. Norway Defense Research Establishment, Kjeller, report K-253. 129pp. (NSA 17: 37054).

Lüning, K. G., et al. 1957. The recovery phenomenon after irradiation in Drosophila melanogaster. I. Recovery or differential sensitivity to x-rays. II. Recovery of recessive lethals. III. The inactivation dose of the recovery process. Hereditas 43(3-4): 549-562, 563-570, 571-577.

Lur'ye, A. A. 1975. Radioisotope labelling of mosquitos. Usp. sovrem. biol. 79(1): 64-77. (In Russian) (NSA 32: 26120).

Luse, R. A. 1970. The phosphorus cycle in a tropical rain forest. Ch. H-12, pp. H161-H166. In: A Tropical Rain Forest. (See H. T. Odum and R. F. Pigeon, 1970).

Lustick, S., T. Voss and T. J. Peterle. 1972. Translocation and bioaccumulation of Cl-36 in a terrestrial system (effects of DDT on steroid metabolism and energetics in bobwhite quail (Colinus virginianus)). Ohio State Univ. Research Foundation, Columbus, Final Rept. RF Project 2464, Rept. No. 2. 38pp.+ 3 figs. (WR 146: 11).

Lustick, S., T. Voss and T. J. Peterle. 1972. Effects of DDT on steroid metabolism and energetics in bobwhite quail (Colinus virginianus). pp. 213-233. In: Proc. 1st National Bobwhite Quail Symp. (Oklahoma State Univ., Stillwater, Okla., April 23-26, 1972). J. A. Morrison and J. C. Lewis, eds. (^{131}I).

Luther, W. 1938. Radiosensitivity and cell division. I. Experiments and frog eggs. Strahlentherapie 62(3): 436-449. (In German).

Luther, W. 1939. The action of radiation on amphibian skin before and after metamorphosis. Naturwissenschaften 27(43): 713-720. (In German).

Luther, W. 1940. Further studies on the supression of cell division in the skin of irradiated salamander larvae. Strahlentherapie 68(2): 185-205. (In German).

Luttge, U. 1964. Microautoradiographic evidence of the uptake of $^{35}SO_4$ in the roothairs of maple seedlings. Naturwissenschaften 51: 296-297. (In German).

Lutz, G. A., D. W. Hubbell and H. H. Stevens, Jr. 1973. Discharge and flow distribution, Columbia River Estuary. Open-file report. U. S. AEC report TID-26445. 80pp. (NSA 29: 2641) (Incl. radionuclides).

Lutz, G. J. 1971. Determination of lead in environmental samples by photon activation analysis. pp. 144-149. In: Nuclear Methods in Environmental Research. (See Vogt, 1971) (NSA 26: 17362).

Lutz, G. J. 1974. Analysis of biological and environmental samples for lead by photon activation. J. Radioanal. Chem. 19(2): 239-244. (NSA 30: 2463).

Luyanus, V. Yu., R. Yu. Yasyulenis and B. I. Styro. 1972. Use of multi-dimensional spectrometers in measuring atmospheric radioactivity. pp. 145-153. In: Methods of Determination of Radioactivity. (See Polikarpov and Parchevskiy, 1972) (In Russian).

Lyalikova, N. N. 1957. Study of the assimilation of free CO_2 by purple sulfur bacteria in Lake Beloved'. Mikrobiologiya 26(1): 97-104. (Microbiology 26(1): 92-99).

Lyapin, Ye. N., et al. 1973. Dependence of the accumulation of radionuclides in the tissues of fish on temperature and saline conditions in the environment. Tr. Atl. nauchn.-issled. inst. ryb. khoz. okeanogr. (44): 45-59. (In Russian) (NSA 30: 3708).

Lyles, D. I. 1969. Effects of x irradiation on embryogenesis, infectivity, and migratory behavior of the larvae of Toxocara canis (Werner, 1782) in white mice. Thesis, Ohio State Univ., Columbus. 162pp. (NSA 24: 48670).

Lynch, D. E. 1950. Soil and water uranium and radium survey. Health and Safety Lab., U. S. AEC report NYO-1521. 45pp. (NSA 10: 2248).

Lynch, E. J. 1964. Transport of radionuclides by ground water -- some theoretical aspects. Hazelton-Nuclear Science Corp., Palo Alto, Calif., U. S. AEC report HNS-1229-61. 56pp.

Lynch, E. J., R. R. Fielder and R. C. McFarlane. 1965. Flow of groundwater around nuclear craters. Bradbury and Assoc., Inc., Los Altos, Calif., U. S. AEC report CBA-1229-F4. 29pp. (NSA 22: 168).

Lynch, O. D. T., Jr., et al. 1975. Radiological resurvey of food, soil, air, and groundwater at Bikini Atoll, 1972. U. S. ERDA report ERDA-34. 33pp. (NSA 32: 6458).

Lynch, V. H. and M. Calvin. 1952. Carbon dioxide fixation by microorganisms. J. Bacteriol. 63(4): 525-531. (Also: Lawrence Radiation Lab., Berkeley, U. S. AEC report UCRL-1411) (NSA 5: 5036).

Lynn, M. 1967. Ionizing radiations in forests and forestry (excluding the use of radioactive tracers). Forest. Abstr. 28: 1-18. (NSA 22: 47499).

Lynn, R. L., et al. 1970. Gamma radiation spectra in the vicinities of Projects Shoal and Faultless. EG&G, Inc., Santa Barbara, Calif., U. S. AEC report EGG-1183-2255. iv, 15pp. (Incl. desert rodents).

Lynn, W. G. and J. N. Dent. 1060. The action of various goitrogens in inhibiting localization of radioiodine in the thyroid and thymus glands of larval tree toads. Biol. Bull. 118: 430-438.

Lynn, W. G. and J. N. Dent. 1961. A comparison of the responses of Triturus and Desmognathus to thyroid-stimulating hormone administration. Biol. Bull. 120(1): 54-61.

Lyon, G. L., P. J. Peterson and R. R. Brooks. 1969. ^{51}Cr transport in the xylem sap of Leptospermum scoparium (Manuka). New Zeal. J. Sci. 12: 541-545. (NSA 24: 31834).

Lyon, W. S., L. C. Bate and J. F. Emery. 1972. Environmental pollution: use of neutron activation analysis to determine the fate of trace elements from fossil fuel combustion in the ecological cycle. pp. 253-262. In: Nuclear Activation Techniques in the Life Sciences. (See IAEA, 1972) (NSA 27: 22090).

Lyubimova, S. A. 1973. Sorption of ^{90}Sr and ^{137}Cs by bottom deposits of a freshwater lake. pp. 96-101. In: Radioecology of Water Organisms. II. (See Andrumaitis, 1973) (In Russian) (NSA 30: 21115).

McAllan, J. W and W. T. A. Neilson. 1965. Labeling the apple maggot with strontium-89. J. Econ. Entomol. 58(1): 168. (NSA 19: 25863).

McBain, J. B. and J. J. Menn. 1965. Fate of ^{14}C {N-(mercaptomethylphthalimide S-(O,O-dimethylphosphorodithioate)} (Imidan) in rats and insects. Abstr. 46. Bull. Entomol. Soc. Am. 11(3): 157.

McBain, J. B., J. J. Menn and J. E. Casida. 1968. Metabolism of carbonyl-C^{14} labelled Imidan N-(Mercaptomethyl)phthalimide-S-(0,0-dimethylphosphorodithionate) in rats and cockroaches. J. Agric. Food Chem. 16(5): 813-820.

McBrayer, J. F. and D. E. Reichle. 1971. Trophic structure and feeding rates of forest soil invertebrate populations. Oikos 22(3): 381-388.

McBride, J. A. 1970. Regulatory controls on radioactive effluents. Chem. Engrg. Progr. 66(4): 74-77. (NSA 24: 22975).

McBride, J. R. and D. Hill. 1969. Offsite radiological surveillance for Project Gasbuggy, June 1967 - July 1968. Radiol. Hlth. Data 10: 535-546. (NSA (NSA 24: 31893).

McBride, J. R. and D. T. Wruble. 1974. Environmental radiological surveillance around United States nuclear explosive testing sites. pp. 307-326. In: Environmental Surveiilance Around Nuclear Installations. Vol. 2. (See IAEA, 1974) (NSA 31: 14042).

McBride, R. 1963. Radioiodine uptake in native animals at the NRTS and environs. Health Physics 9(12): 1227-1230.

McCabe, R. A. 1961. A direct evaluation of productivity in animal populations through development of a new tracer technique for ecological research (progress report). U. S. AEC report TID-13564. 13pp. (NSA 15: 28909).

McCabe, R. A. and G. A. LePage. 1958. Identifying progeny from pheasant hens given radioactive calcium (Ca^{45}). J. Wildl. Mgmt. 22(2): 134-141.

McCallan, S. E. A. 1954. Use of isotopes in fungicidal research. Agric. Ch Chem. 9(2): 52-53, 119, 121.

McCarley, W. H. 1959. A study of the dynamics of a population of _Peromyscus gossypinus_ and _P. nuttalli_ subjected to the effects of x-irradiation. Am. Midland Nat. 61(2): 447-469.

McCarter, J. A., et al. 1952. Movement of materials in the hypolimnion of a lake as studied by the addition of radioactive phosphorus. Can. J. Zool. 30: 128-133.

McCartney, W. C. 1970. Arboreal behavior of the Polynesian rat (_Rattus exulans_). BioScience 20(19): 1061-1062. (Eniwetok Atoll).

McCauley, G. N. and J. F. Stone. 1972. Source-detector geometry effect on neutron probe calibration. Soil Sci. Soc. Am. Proc. 36(2): 246-___. (Soil moisture measurement).

McClanahan, R. J. 1965. Sterilization of the carrot rust fly by irradiation with ^{137}Cs. Can. J. Entomol. 97: 1042-1045. (NSA 20: 14317).

McClanahan, R. J. and H. S. Simmons. 1966. Sterilization of onion maggots by irradiation with ^{137}Cs. Can. J. Entomol. 98: 931-935. (NSA 22: 4725).

McClearen, H. A. 1974. Panel discussion on environmental monitoring and reporting: Savannah River Plant. pp. 689-692. In: _Proc. 2nd AEC Environmental Protection Conf._ (See U. S. AEC, 1974) (NSA 30: 32038).

McClearen, H. A. 1974. Plutonium in soil at the Savannah River Plant. pp. 495-510. In: _Proc. 2nd AEC Environmental Protection Conf._ (See U. S. AEC, 1974) (NSA 30: 32127).

McClendon, L. T. 1974. Selective determination of chromium in biological and environmental matrices. pp. 255-257. In: _Trace Substances in Environmental Health. VIII._ D. D. Hemphill, ed. Univ. of Missouri, Columbia. (NSA 32: 16711) (Neutron activation analysis).

McColloch, R. J., J. W. Hamilton and S. K. Brown. 1963. An apparent seleniferous leaf wax from _Stanleya bipinnata_. Biochem. Biophys. Res. Comm. 11: 7-13. (NSA 17: 17657) (Used ^{75}Se).

McConiga, M. W. 1960. MPC's for ingestion of rare earths in Columbia River water. Hanford Atomic Products Operation, U. S. AEC report HW-66026. 7pp. (NSA 26: 38433).

McConiga, M. W., J. M. Selby and J. K. Soldat. 1958. Radioactive contamination in the Hanford environs for the period July, August, September 1957. Hanford Atomic Products Operation, U. S. AEC report HW-54841. 40pp. (NSA 13: 12566).

McConnell, J. V., M. L. Clay and R. B. Kauffman. 1961. Evaluation of LD_{50} level in decapitated _Planaria_. Univ. of Michigan, Ann Arbor, U. S. AEC report TID-15984. 7pp. (NSA 16: 23414).

McConnon, D. 1962. Dose-rate measurements on beaches and islands on the Columbia River between Ringold and Richland. Hanford Atomic Products Operation, U. S. AEC report HW-72229(Rev.). 22pp. (NSA 19: 36703).

McCorkell, R., E. L. Fireman and C. C. Langway, Jr. 1967. ^{26}Al and ^{10}Be in Greenland ice. Science 158: 1690-1692. (NSA 22: 10371).

McCormack, M. L. 1963. A study of techniques for determining root extension using radioactive tracers. Dissertation, Duke Univ., Durham, N. C. 193pp.

McCormack, M. L. and F. W. Woods. 1961. Determining root extension with radioactive tracers. Abstr. J. Elisha Mitchell Sci. Soc. 77(2): 106.

McCormick, D. J., F. L. Glesius and J. C. Kroon. 1976. Microprocessor based environmental radiation monitoring system. Trans. Nucl. Sci. NS-23(1): 710-714. (NSA 33: 29283).

McCormick, J. F. 1959. Ecological analysis of selected granite outcrop communities and their response to chronic gamma radiation. Thesis, Emory Univ., Atlanta, Ga.

McCormick, J. F. 1960. Responses of selected granite outcrop communities to chronic low-level gamma irradiation. Abstr. Bull. Ecol. Soc. Am. 41: 94.

McCormick, J. F. 1961. Changes in a granite outcrop community one year following chronic exposure to low-level ionizing radiation gradients. Abstr. Bull. Ga. Acad. Sci. 19(1-2): 2.

McCormick, J. F. 1962. Effects of ionizing radiation on a natural plant community. Thesis, Emory Univ., Atlanta, Ga. 117pp. (NSA 17: 15801).

McCormick, J. F. 1963. Changes in a herbaceous plant community during a three-year period following exposure to ionizing radiation gradients. pp. 271-276. In: _Radioecology_. (See Schultz and Klement, 1963) (NSA 17: 33570).

McCormick, J. F. 1964. Interactions of gamma radiation and other environmental stresses upon plant populations and ecosystems. Abstr. Bull. Ecol. Soc. Am. 45(3): 75.

McCormick, J. F. 1964. Interactions of gamma radiation and drought upon _Pinus elliotti_ and _Pinus palustris_. BioScience 14(7): 43-44.

McCormick, J. F. 1964. Effects of ionizing radiation and other environmental stresses upon _Pinus elliotti_. Assoc. Southeastern Biol. Bull. 11:49. (NSA 21: 11125).

McCormick, J. F. 1965. Ecological studies in a tropical rain forest exposed to ionizing radiation. Abstr. Bull. Ecol. Soc. Am. 46(3): 96.

McCormick, J. F. 1967. Effects of ionizing radiation on seeding diversity and abundance in a tropical forest. Univ. of North Carolina, Chapel Hill, U. S. AEC report ORO-3299-7. 27pp. (NSA 21: 39372).

McCormick, J. F. 1969. Effects of ionizing radiation on a pine forest. pp. 78-87. In: _Symposium on Radioecology_. (See Nelson and Evans, 1969) (Also: Univ. of North Carolina, Chapel Hill, U. S. AEC report ORO-3299-6) (NSA 21: 36747).

McCormick, J. F. 1970. Patterns of radiation exposure in the tropical rain forest. Ch. C-4, pp. C41-C47. In: A Tropical Rain Forest. (See H. T. Odum and R. F. Pigeon, 1970) (NSA 25: 21688).

McCormick, J. F. 1970. Growth and survival of the sierra palm under radiation stress in natural and simulated environments. Ch. D-9, pp. D193-D199. In: A Tropical Rain Forest. (See H. T. Odum and R. F. Pigeon, 1970) (NSA 25: 21934).

McCormick, J. F. 1970. Direct and indirect effects of gamma radiation on seedling diversity and abundance in a tropical forest. Ch. D-10, pp. D201-D211. In: A Tropical Rain Forest. (See H. T. Odum and R. F. Pigeon, 1970) (NSA 25: 21935).

McCormick, J. F. and D. J. Cotter. 1964. Radioactivity on southern granite outcrops. Bull. Ga. Acad. Sci. 22(1): 20-28.

McCormick, J. F. and F. B. Golley. 1964. A portable radiation facility for ecological studies. Abstr. Bull. Ecol. Soc. Am. 45(3): 74.

McCormick, J. F. and F. B. Golley. 1966. Irradiation of natural vegetation. An experimental facility, procedures and dosimetry. Health Physics 12(10): 1467-1474. (NSA 20: 45266).

McCormick, J. F. and R. B. Platt. 1962. Effects of ionizing radiation on a natural plant community. Radiation Botany 2(3/4): 161-188. (NSA 17: 15727).

McCormick, J. F. and W. N. Rushing. 1963. Effects of ionizing radiation on cytogenetic races of Sedum pulchellum Michx. Abstr. Bull. Ecol. Soc. Am. 44(3): 70.

McCoy, O. R., V. F. Downing and S. N. van Voorhis. 1941. The penetration of radioactive phosphorus into encysted Trichinella larvae. J. Parasitol. 27: 53-58.

McCrackin, F. A. 1971. Broad view of the environment. A national crisis: progress, pollution, and people. pp. I.1-68. In: Environment. (See Southern California Edison Co., 1971) (NSA 26:

McCraw, T. F. nd. Levels of environmental radioactivity in Bikini Atoll. U. S. AEC report WASH-1289. 13pp. (NSA 31: 3412).

McCray, E. M., Jr., J. A. Jensen and H. F. Schoof. 1961. Cobalt-60 sterilization studies with Aedes aegypti (L.). pp. 110-115. In: Proc. 48th Ann. Meeting New Jersey Mosquito Ext. Assoc. (NSA 16: 26843; PHEA 42: 8).

McCreedy, H. H., V. F. Harrison and W. A. Gow. 1969. A proposed method, using bacteria, for the continuous leaching of a uranium ore. Can. Mining Met. Bull. 62: 135-140. (Also: Canada Dept. of Energy, Mines and Resources, Ottawa, Internal Rept. EMA68-5; for 70th Ann. Meeting of Mining and Metallurgy, Vancouver, B. C.) (CONF-680426-2) (NSA 23: 339, 17533).

McCurdy, H. M., F. T. Algard and G. B. Friedmann. 1974. Responses of metamorphosing Tarichi torosa to x-rays. Can. J. Zool. 52(6): 671-676. (NSA 30: 24317).

McDonald, D. J. 1961. X-irradiation of the developing male germ cells, T. confusum. Genetics 46(11): 1511-1517.

McDonald, D. J. and H. C. Long. 1961. The effect of neutron irradiation on the fertility of T. confusum. Amer. Nat. 95(881): 124-126.

McDowell, L. L., J. C. Stephens and E. H. Stewart. 1969. Radiocarbon chronology of the Florida Everglades peat. Soil Sci. Soc. Am. Proc. 33(5): 743-745. (NSA 24: 2303).

McDowell, W. J. 1974. High-resolution liquid scintillation method for the analytical determination of alpha-emitters in environmental samples. Oak Ridge National Lab., U. S. AEC report CONF-741212-4. 20pp.

McEnroe, W. D. and A. J. Forgash. 1957. The in vivo incorporation of carbon-14 formate in the ureide groups of uric acid by Periplaneta americana. Ann. Entomol. Soc. Am. 50(5): 429-431.

McEnroe, W. D. and A. J. Forgash. 1958. Formate metabolism in the American cockroach, Periplaneta americana (L.). Ann. Entomol. Soc. Am. 51(2): 126-129.

McErlean, A. J. and H. J. Brinkley. 1971. Temperature tolerance and thyroid activity of the white perch Roccus (=Morone) americanus. J. Fish. Biol. 3(1): 97-114. (Radionuclide used).

McEwan, E. H. and P. E. Whitehead. 1971. Measurement of the milk intake of reindeer and caribou calves using tritiated water. Can. J. Zool. 49(4): 443-447.

McFarlane, J. C. and B. J. Mason. 1970. Plant radioiodine relationships: a review. Southwestern Radiological Health Lab., Las Vegas, Nev., U. S. PHS, U. S. AEC report SWRHL-90-r. 53pp. (NSA 25: 13395).

McFarlane, J. C., K. W. Brown and R. R. Kinnison. 1975. In situ soil gamma analyzer: a new approach to determine the movement of pollutants in soils. Soil Science 120(2): 140-146. (NSA 33: 3016).

McGaughey, R. W. and M. C. Chang. 1970. X-irradiation of pregnant Mongolian gerbils. Anat. Record 167: 37-53. (NSA 24: 44604).

McGinnis, A. J. and R. Kasting. 1962. Nutrition of the pale western cutworm. V. Utilization of protein labeled with carbon-14. J. Nutrition 76(4): 333-340.

McGinnis, J. T. 1960. Effects of ionizing radiation on tree growth. Abstr. ASB Bull. 7: 33.

McGinnis, J. T. 1962. Ionizing radiation effects on litter production of white oaks. Abstr. ASB Bull. 9(2): 32.

McGinnis, J. T. 1963. Effects of ionizing radiation on the litter production of selected tree species. Abstr. Bull. Ecol. Soc. Am. 44(3): 69.

McGinnis, J. T. 1963. Effects of radiation from an air-shielded reactor on forest litter production. pp. 283-287. In: Radioecology. (See Schultz and Klement, 1963) (NSA 17: 33572).

McGinnis, J. T. 1963. Some environmental changes in forest stands following exposures to radiation. Abstr. ASB Bull. 10(2): 33.

McGinnis, J. T. 1964. Effects of mixed neutron-gamma radiation on forest litter production. Thesis, Emory Univ., Atlanta, Ga. 118pp. (NSA 19: 13082).

McGinnis, J. T., et al. 1967. Terrestrial ecology. Phase I final report. Bioenvironmental and radiological-safety feasibility studies Atlantic-Pacific interoceanic canal. Univ. of Georgia, Athens; Battelle Columbus Labs., U. S. AEC report BMI-171-011. ii, 114pp. (NSA 21: 39370).

McGivern, J. J. and K. S. Rai. 1972. Radiation-induced paracentric inversion in *Aedes aegypti*. Cytogenetic and interchromosomal effects. J. Heredity 63(5): 247-255. (NSA 27: 17682).

McGovern, W. L., et al. 1975. Irradiated boll weevils: pheromone production determined by GLC analysis. J. Econ. Entomol. 68(4): 521-523. (NSA 33: 7572).

McGowan, M. 1974. Depths of water extraction by roots: application to soil-water balance studies. pp. 435-445. In: *Isotope and Radiation Techniques in Soil Physics and Irrigation Studies 1973*. (See IAEA, 1974) (NSA 30: 15508) (Neutron probe).

McGrath, R. A. 1962. X-ray induced incorporation of tritiated thymidine into grasshopper neuroblast chromosomes. Thesis, Univ. of Tennessee, Knoxville. 62pp. (NSA 17: 21576).

McGrath, R. A., W. M. Leach and J. G. Carlson. 1965. Cell stages refractory to thymidine incorporation induced by x-rays. Exptl. Cell Res. 37(1): 39-44. (^{3}H, *Chortophaga viridifasciata*).

McGrath, R. A., R. W. Williams and R. B. Setlow. 1964. Increased ^{3}H-thymidine incorporation into DNA of irradiated slime mold. Int. J. Radiat. Biol. 8(4): 373-380. (NSA 19: 17554).

McGregor, D. L. and R. C. Wetzel. 1968. Self-absorption of C^{14} radiation in freshwater ostracods. Ecology 49(2): 352-355.

McGregor, J. F. and H. B. Newcombe. 1968. Major malformations in trout embryos irradiated prior to active organogenesis. Radiation Research 35: 282-300. (NSA 22: 45685).

McGregor, J. F. and H. B. Newcombe. 1972. Dose-response relationships for yields of major eye malformations following low doses of radiation to trout sperm. Radiation Research 49(1): 155-169. (NSA 26: 23116).

McGregor, J. F. and H. B. Newcombe. 1972. Decreased risk of embryo mortality following low doses of radiation to trout sperm. Radiation Research 52(3): 536-544. (NSA 27: 20103).

McGregor, J. H. 1908. Abnormal development of frog embryos as a result of treatment of ova and sperm with roentgen rays. Science 27(690): 445-446.

McHenry, J. R. 1962. Determination of densities of reservoir sediments *in situ* with a gamma probe. Agricultural Research Service, U. S. Dept. of Agriculture ser. ARS-41-53. 10pp. (NSA 21: 1663).

McHenry, J. R. 1969. Use of tracer technique in soil erosion research. Isotopes Radiat. Techn. 6: 280-287. (NSA 23: 15969).

McHenry, J. R. and F. E. Dendy. 1964. Measurement of sediment density by attenuation of transmitted gamma rays. Soil Sci. Soc. Am. Proc. 28(6): 817-822.

McHenry, J. R. and A. C. Gill. 1967. The influence of bulk density, slow neutron absorbers, and time on the calibration of neutron-moisture probes. pp. 83-100. In: *Isotope and Radiation Techniques in Soil Physics and Irrigation Studies*. (See IAEA, 1967) (NSA 22: 23348).

McHenry, J. R. and P. H. Hawks. nd. Sedimentation survey of lake Guaybal, Puerto Rico using a nuclear sediment density probe. pp. 269-286. In: *Radiation and Isotope Technology in Latin American Development*. (See Puerto Rico Nuclear Center, nd.) (NSA 24: 39039).

McIntyre, G. I. 1962. Studies on the translocation in *Agropyron repens* of ^{14}C-labelled 2,2-dichloropropionic acid. Weed Research 2(3): 165-176.

McKim, J. M., et al. 1975. Effects of pollution on freshwater fish. J. Water Pollut. Control. Fed. 47(6): 1711-168. (NSA 33: 20533) (Incl. radioactivity).

McKinney, C. O. and F. B. Turner. 1971. Genetic variation in irradiated and nonirradiated populations of the lizard, *Uta stansburiana*. Radiation Research 47(2): 530-536. (NSA 25: 55320).

McKown, D., et al. 1971. Multi-element surveys of bioenvironmental samples using neutron activation analysis. pp. 150-157. In: *Nuclear Methods in Environmental Research*. (See Vogt, 1971) (NSA 26: 17369).

McLachlan, K. D. and M. Crawford. 1970. Determination of sulfur in plant material by x-ray emission spectrometry. J. Sci. Food Agr. 21: 408-410.

McLain, M. E., Jr. and G. W. Leddicotte. 1971. Critical look at the current problems in mercury analysis. pp. 216-225. In: *Nuclear Mehtods in Environmental Research*. (See Vogt, 1971) (NSA 26: 17369).

McLaren, A. D. 1969. Radiation as a technique in soil biology and biochemistry. Soil Biol. Biochem. 1(1): 63-73. (NSA 24: 14638).

McLaren, A. D., R. A. Luse and J. J. Skujins. 1962. Sterilization of soil by irradiation and some further observations on soil enzyme activity. Soil Sci. Soc. Am. Proc. 26: 371-377.

McLaren, A. D., L. Reshetko and W. Huber. 1957. Sterilization of soil by irradiation with an electron beam, and some observations on enzyme activity. Soil Science 83: 497-502.

McLaughlin, J. E. (Ed.). 1972. *Workshop on Natural Radiation Environment (March 1972)*. Health and Safety Lab., U. S. AEC report HASL-269. 80pp. (NSA 27: 19818).

McLaughlin, J. E. and H. L. Beck. 1973. Environmental radiation dosimetry for nuclear facilities and problems. Trans. Nucl. Sci. NS-20(1): 36-42. (NSA 28: 5553).

McLean, E. O., C. Lakshmanan and F. P. Miller. 1969. Relative absorption and desorption of strontium and calcium to and from soils and soil clays: column sturation-displacement and acid displacement. Soil Science 107: 206-212. (NSA 23: 17569).

McLendon, H. R. 1975. Soil monitoring for plutonium at the Savannah River Plant. Health Physics 28(4): 347-354. (NSA 32: 3493).

McLendon, H. R., et al. 1975. Relationships among plutonium contents of soil, vegetation, and animals collected on and adjacent to an integrated nuclear complex in the humid southeastern United States. Savannah River Plant, U. S. AEC report DP-MS-75-30. 23pp. (NSA 33: 12004).

McLeod, J. and J. Donnelly. 1957. Individual and group marking methods for fly-population studies. Bull. Entomol. Res. 48(3): 585-592.

McMahan, E. A. 1963. A study of termite feeding relationships, using radio-isotopes. Ann. Entomol. Soc. Am. 56(1): 74-82.

McMahan, E. A. 1963. New radioactive tests show how termites feed. Pest Control 31(2): 32-34.

McMahan, E. A. 1966. Food transmission within _Cryptotermes brevis_ colony (Isoptera: Kalotermitidae). Ann. Entomol. Soc. Am. 59(6): 1131-1137. (^{57}Co).

McMahan, E. A. 1969. Effects of ionizing radiation on three-neotropical termite species. Ann. Entomol. Soc. Am. 62(1): 120-125.

McMahan, E. A. 1970. Radiation and the termites at El Verde. Ch. E-8, pp. E105-E122. In: _A Tropical Rain Forest_. (See H. T. Odum and R. F. Pigeon, 1970) (NSA 25: 21817).

McMahan. E. A. and N. F. Sollins. 1970. Diversity of microarthropods after irradiation. Ch. E-11, pp. E151-E158. In: _A Tropical Rain Forest_. (See H. T. Odum and R. F. Pigeon, 1970) (NSA 25: 21818).

McMahan, J. W. and C. N. Wright. 1974. Field measurement of cesium-137 in deer. Trans. Nucl. Sci. NS-21(1): 513-516. (NSA 30: 6676).

McMahon, J. W. 1964. The dispersion of tritium in Perch Lake. Atomic Energy of Canada, Ltd., Chalk River, report AECL-1889. i, 9pp. + 6 figs. (NSA 18: 18039).

McMahon, J. W. 1973. Membrane filter retention: a source of error in the ^{14}C method of measuring primary production. Limnol. Oceanogr. 18(2): 319-324. (NSA 28: 11119).

McMahon, J. W. and A. E. Docherty. 1975. Temporal variations in the phosphorus content of a nothern dystrophic lake. Atomic Energy of Canada, Ltd., Chalk River, report AECL-5103. 17pp. (NSA 32: 17208) (^{32}P tracer).

McMahon, J. W. and F. H. Rigler. 1965. Feeding rate of _Daphnia magna_ Straus in different foods labelled with radioactive phosphorus. Limnol. Oceanogr. 10(1): 105-113.

McMillan, G. R., S. D. Falkenberg and W. D. Thompson. 1972. Early behavioral effects of ionizing radiation on four species of subhuman primates. Trans. Amer. Nucl. Soc. 15(Suppl.1): 7-8. (NSA 26: 20607).

McMurray, W. R. and L. O. Stander. 1963. Survey of background radioactivity in the Hartbeespoort Dam. pp. 147-158. In: _National Conf. on Nuclear Energy; Application of Isotopes and Radiation_. South Africa Atomic Energy Board, Pelindaba. (NSA 19: 11459).

McNabb, R. A. and G. E. Pickford. 1970. Thyroid function in male killifish, _Fundulus heteroclitus_, adapted to high and low temperatures and to fresh water and sea water. Comp. Biochem. Physiol. 33(4): 783-792. (SFA 15: 12360) (^{131}I).

McNelis, D. N., S. C. Black and E. L. Whittaker. 1971. Radioiodine field studies with synthetic aerosols. Southwestern Radiological Health Lab., U. S. EPA, Las Vegas, Nev., U. S. AEC report SWRHL-103r. iii, 23pp.

McPeak, D. W., G. M. Lodde and W. H. Parr. 1965. Changes in ^{137}Cs retention and routes of elimination as effected by external factors. U. S. Army Medical Research Lab., Ft. Knox, Ky., report USAMRL-617. (AD-464714; NSA 25: 21544) (Cold and x-irradiation).

MacDonald, D. R. 1963. Biological interactions with ecosystems. Abstr. Bull. Ecol. Soc. Am. 44(3): 82.

Macdonald, H. F. 1971. Assessment of environmental hazards following nuclear reactor accidents. pp. 43-54. In: _Rapid Methods for Measuring Radioactivity in the Environment_. (See IAEA, 1971) (NSA 26: 28051).

Macdonald, H. F. and R. H. Clarke. 1972. Prediction of environmental releases from nuclear reactors under normal and accident conditions. pp. 817-830. In: _New Trends in Radiation Protection_. (6th Intern. Congr., Bordeaux). Societé Francaise de Radioprotection, Montrouge. (CONF-730328) (NSA 27: 17240).

Macdonald, H. F., P. J. Darley and R. B. Pepper. 1974. Behavior of radionuclides in the environment: impressions of two recent international symposia. U. K. Central Electricity Generating Board, Berkeley Nuclear Labs., Berkeley, report RD/B/M-2893. 38pp. (NSA 30: 297).

Macek, K. J., et al. 1970. The uptake, distribution and elimination of dietary ^{14}C-DDT and ^{14}C-dieldrin in rainbow trout. Trans. Am. Fish. Soc. 99 (4): 689-695. (SFA 16: 12790).

Macfarlane, J. J. 1966. Control of the Queensland fruit fly by gamma irradiation. J. Econ. Entomol. 59: 884-889. (NSA 21: 20384).

Macfarlane, W. V., B. Howard and B. F. Good. 1974. Tracers in field measurements of water, milk, and thyroxine metabolism of tropical ruminants. pp. 1-23. In: _Tracer Techniques in Tropical Animal Production_. International Atomic Energy Agency, Vienna, Publ. STI/PUB-360. (NSA 29: 29951) (Incl. banteng, eland, wildebeest, oryx, domestic ruminants).

MacGregor, R. A. 1969. Bacterial leaching of uranium. Nucl. Appl. 6: 68-72. (NSA 23: 7611).

Machado, J. O., et al. 1970. Preliminary data on the activity of ionizing radiation on fertile eggs of triatomids Rhodnius prolixus. Hospital (Rio de Janeiro) 77(1): 203-207. (In Portugese) (NSA 25: 46963).

Mache, R. and R. Patouraux. 1973. Radiosensitivity of the liverwort Marchantia polymorpha L. to internal radioisotopes. Nature 241(5391): 547-548. (NSA 27: 28020).

Machi, S. 1971. Application of radiation to the treatment of waste water. Genshiryoku Kogyo 17(2): 45-50. (In Japanese) (NSA 25: 44380).

Machi, S. 1973. Environmental protection by using radiation. Genshiryoku Kogyo 19(5): 16-20. (NSA 29: 24405).

MacIntosh, N. W. G. 1972. Radiocarbon dating as a pointer in time to the arrival and history of man in Australia and islands to the north west. pp. XLIV-LVI. In: Proc. 8th Intern. Radiocarbon Dating Conf. Vol. 1. (See Rafter, 1972) (NSA 31: 8679).

Machta, L. 1968. Radioiodine fallout over the Midwest in May. Science 160 (3823): 64-66.

Machta. L. 1973. Prediction of CO_2 in the atmosphere. pp. 21-30. In: Carbon and the Biosphere. (See Goodwell and Pecan, 1973) (Nuclear test ^{14}C).

Machta, L., K. Telegadas and D. L. Harris. 1970. ^{90}Sr fallout over Lake Michigan. J. Geophys. Res. 75: 1092-1096. (NSA 24: 16695).

Mackereth, F. J. 1953. Phosphorus utilization by Asterionella formosa Hass. J. Exptl. Bot. 4: 296-313.

Mackethun, K. M. 1969. Radioactive wastes. Ch. 9, pp. 131-135. In: The Practice of Water Pollution Biology. U. S. Federal Water Pollution Control Administration, Washington.

Macko, V and J. Jasic. 1961. The use of γ-radiation in applied entomology. Polnohopodarstvo 8(2): 139-147. (In Polish).

Macksoud, S. W. 1971. Reserach within the framework of a coordinated programme on the application of radiation techniques to water use efficiency studies. Final report, 1 March 1965-31 December 1970. International Atomic Energy Agency, Vienna, report IAEA-R-311-F. 21pp. (NSA 28: 24637).

MacLean, J., et al. 1968. Pathophysiology of Fasciola hepatica infections. pp. 117-124. In: Isotopes and Radiation in Parasitology. (See IAEA, 1968).

MacLeod, J. 1959. Atomic war against an insect; British trials of a new insect control technique using gamma rays to sterilise male flies. Chem. & Drug. 172(4151): 4-6. (McLeod ?, q.v.).

MacLeod, J. and J. Donnelly. 1956. Methods for study of blowfly populations. II. The use of laboratory-bred material. Ann. Appl. Biol. 44(4): 643-648. (Radiotracer).

MacLeod, J. and J. Donnelly. 1961. Failure to reduce an isolated blowfly population by the sterile males method. Entomol. Exptl. App. 4(2): 101-118.

MacRae, I. C., K. Raghu and T. F. Castrol. 1967. Persistence and biodegradation of four common isomers of benezene hexachloride in submerged soils. J. Agric. Food Chem. 15(5): 911-914. ($^{14}CO_2$).

Maddox, D. M. and M. E. Resnik. 1968. Radioisotopes - a potential means of evaluating the host specificity of phytophagous insects. J. Econ. Entomol. 61(6): 1499-1502.

Madison, D. M. and C. R. Shoop. 1970. Homing behavior, orientation, and home range of salamanders tagged with tantalum-182. Science 168(3939): 1484-1487. (WR 140: 224).

Madsen, H. F. and R. S. Downing. 1967. Codling moth control. Pesticide Progr. 5(3): 61-62.

Maestas, S. and A. E. Gallegos. 1974. Radiochemical analysis for Pb-210 and Po-210 in environmental samples. Technical completion report. New Mexico State Univ., University Park. 65pp. (PB-237907) (NSA 32: 38).

Maestrini, A. A., et al. nd. Radioactive tracers applied to the studies of sediment transportation. pp. 307-316. In: Radiation and Isotope Technology in Latin American Development. (See Puerto Rico Nuclear Center, nd) (NSA 24: 39041).

Maetz, J. 1972. Interaction of salt and ammonia transport in aquatic organisms. pp. 105-154. In: Nitrogen Metabolism and the Environment. J. W. Campbell and L. Goldstein, eds. Academic Press, New York. (^{24}Na, goldfish).

Magalhaes, L. E., et al. 1965. On lethals and their supressors in experimental populations of Drosophila willistoni. Mutation Research 2: 45-54. (NSA 19: 38391).

Magata, N. Ya. 1963. Use of tritium, ^{14}C, ^{90}Sr, and ^{137}Cs. Genshiryoku Kogyo 9(8): 13-16. (In Japanese) (NSA 19: 30425) (Water studies).

Magi, A. and B. Lindell. 1967. ^{222}Rn in milk. In: Radioecological Concentration Processes. (See Aberg and Hungate, 1967).

Magi, A., J. O. Snihs and G. A. Swedjemark. 1970. Some measurements of radioactivity in Sweden caused by nuclear test explosions. Radiol. Hlth. Data 11(10): 487-509. (Incl. fish, reindeer).

Magin, G. B., Jr. and O. M. Bizzell. 1964-65. Some applications of radioisotope technology to water resources investigations and utilization. Isotopes Radiat. Techn. 2(2): 124-133. (NSA 19: 13830).

Magni, G. E. 1956. Problems concerning the radiosensitivity of yeast cells. Compt. rend. Lab. Carlsberg 26: 273-284.

Magno, P. J. 1973. Studies of dose pathways from a nuclear fuel reprocessing plant. pp. 537-551. In: Environmental Behavior of Radionuclides Released in the Nuclear Industry. (See IAEA, 1973) (NSA 29: 29385).

Magno, P., T. Reavey and J. Apidianakis. 1970. Liquid waste effluents from a nuclear fuel reprocessing plant. Northeastern Radiological Health Lab., U. S. PHS, Winchester, Mass., report BRH/HERHL 70-2. xii, 59pp.

Magno, P., T. Reavey and J. Apidianakis. 1971. Liquid waste effluents from a nuclear fuel reprocessing plant. pp. 208-220. In: Health Physics Aspects of Nuclear Facility Siting. (See Voillequé and Baldwin, 1971) (NSA 25: 54753).

Magno, P. J., T. C. Reavey and J. C. Apidianakis. 1972. ^{129}I in the environment around a nuclear fuel reprocessing plant. U. S. EPA report ORP/SID-72-5. 31pp. (NSA 27: 22483).

Maguire, B., Jr. 1970. Aquatic communities in bromeliad leaf axils and the influence of radiation. Ch. E-6, pp. E95-E101. In: A Tropical Rain Forest. (See H. T. Odum and R. F. Pigeon, 1970) (NSA 25: 21835).

Maguire, B., Jr. and W. E. Neill. 1971. Species and individual productivity in phytoplankton communities. Ecology 52(5): 903-907. (NSA 26: 9491) (^{14}C).

Mahar, H. 1974. Combined effects of cobalt-60 irradiation and selected herbicides on Spirodela polyrrhiza. Dissertation, Univ. of Michigan, Ann Arbor. 121pp. (Univ. Microfilms 74-25,259) (NSA 31: 8850).

Mahdavi, A. 1964. The thorium, uranium, and potassium contents of Atlantic and Gulf Coast beach sands. pp. 87-114. (See Adams and Lowder, 1964) (NSA 18: 35732).

Mahmoud, K. A., et al. 1968. Affinity of Ismailia Canal bottom sediments for certain radionuclides. Isotopes Radiat. Res. 1: 9-18. (NSA 24: 31482).

Mahmoud, K. A., et al. 1973. Transfer mechanism and concentrating processes of certain radionuclides in the aquatic biology of the Ismailia Canal. pp. 205-215. In: Environmental Behavior of Radionuclides Released in the Nuclear Industry. (See IAEA, 1973) (NSA 29: 29361).

Mahrous, M. A. and Z. M. F. Rostom. 1962. Effect of x-rays on the duration of life of the rice weevil, Sitophilus oryzae L. Int. J. Radiat. Biol. 5: 191-193. (NSA 17: 69).

Maier, P. 1973. Irradiation of mature Drosophila sperm: dependence of the rates of mutation on the genotype and age of the tested female. Arch. Genet. 46(3): 173-195. (In German) (NSA 33: 26274).

Maine Yankee Atomic Power Co. 1970. Maine Yankee Power Station. Environmental report. Wiscasset, Me. U. S. AEC DOCKET-50309-25. 46pp. (NSA 25: 3837).

Maine Yankee Atomic Power Co. 1971. Maine Yankee Power Station. Environmental report, addenda. Westboro, Mass. U. S. AEC DOCKET-50309-29. 18pp. (NSA 25: 20667).

Mainx, F. and E. Doschek. 1967. Genetic crossing over in both sexes of Megasellia schlaris. Molec. Gen. Genetics 99: 203-218. (In German) (NSA 22: 2495) (X-rays).

Mairhofer, J. 1967. Groundwater flow and direction measurement by means of radioisotopes in a single well. pp. 119-129. In: Isotope Techniques in the Hydrologic Cycle. (See Stout, 1967).

Majerczyk, J. and A. Zuber. 1966. The influence of composition and density of soils on readings of neutron moisture meters. In: Radioisotope Instruments in Industry and Geophysics. (Proc. Symp., Warsaw, Poland, October 18-22, 1965). International Atomic Energy Agency, Vienna, Publ. STI/PUB-112. (Preprint SM-68/39) (CONF-651023) (NSA 20: 14771).

Major, W. J., et al. 1974. Determination of ^{239}Pu and ^{241}Am in large NAEG vegetations samples. pp. 107-118. In: Dynamics of Plutonium in Desert Environments. (See Dunaway and White, 1974) (NSA 30: 28971).

Makarov, P. O. and Ye. S. Ismailov. 1967. The influence of pulsed microwaves on Opalina ranarum. Nerv. Sist. 8: 140-143. (In Russian) (NSA 23: 50827).

Makarov, P. V. 1965. Cytochemical investigation of silk glands in Antheraea pernyi. Tsitologiya 7(5): 616-621. (In Russian) (^{3}H, ^{14}C).

Makarov, P. V. 1965. Autoradiographic study of protein-producing cells. Arkh. anat. cistol. embriol. 48(4): 3-16. (In Russian) (^{14}C, Antheraea pernyi, Rana temporaria).

Makhonina, G. I. 1965. On the occurrence of ^{56}Fe, ^{60}Co, ^{65}Zn, ^{90}Sr, and ^{137}Cs in the pine, larch, aspen, and birch. Byull. Mosk. obshch. ispyt. prirody (biol.) 70(2): 120-123. (In Russian) (NSA 19: 45773).

Makhonina, G. I. 1968. Distribution of various radioisotopes in the pine seedlings. Tr. inst. ekol. rast. zhivotn., Ural. fil. Akad. Nauk SSSR (61): 58-60. (In Russian) (NSA 24: 14412).

Makhonina, G. I. and Ye. A. Gileva. 1968. Uptake of ^{65}Zn, ^{115}Cd and ^{203}Hg by freshwater plants and effects of EDTA on the concentration of these isotopes. Tr. inst. ekol. rast. zhivotn., Ural. fil. Akad. Nauk SSSR (61): 72. (In Russian) (NSA 24: 7713).

Makhonina, G. I. and I. V. Molchanova. 1961. A study of the behavior of microquantities of iron and zinc in soils. Nauchn. dokl. vyssh. shkoly (biol.) (4): 218-225. (In Russian).

Makhonina, G. I., et al. 1960. An attempt to analyze the distribution of radioisotopes in natural biogeocoenoses. Dokl. Akad. Nauk SSSR 133: 484-487. (Doklady (bot. sci.) 133: 141-143) (BA 37: 16943; NSA 14: 23894; WR 106: 4).

Makhonina, G. I., et al. 1961. Distribution of strontium-90 and cesium-137 in components of biogeocoenosis. Dokl. Akad. Nauk SSSR 140(5): 1209-1212. (In Russian).

Makhonina, G. I., et al. 1963. Distribution of Sr-90 and Ru-106 in the basic organs of pine. Dokl. Akad. Nauk SSSR 151(6); 1456-1457. (JPRS-22,340, pp. 8-11; Doklady (biol. sci.) 151: 1085-1086).

Makhon'ko, K. P. 1968. Deposition of radioactive debris and its elimination from the atmosphere by precipitation. State Committee on the Use of Atomic Energy, USSR. Atomizdat, Moscow. U. N. Document A/AC.82/G/L.1262. 42pp.

Makhon'ko, K. P. nd. Migration of global radioisotopes in various spheres. (INIS-mf-887, pp. 235-251) (In Russian) (NSA 29: 24196).

Makhon'ko, K. P. and V. B. Chumichev. 1966. Vertical migration of basic fission products in soils. Inform. byull. Akad. Nauk SSSR (biol.) (9): 52-55. (In Russian).

Makhon'ko, K. P. and S. G. Malakhov. (Eds.). 1969. Nuclear Meteorology. (Proc. All-Union Conf., Obninsk, 23-28 June 1969). 380pp. (TT-74-50011) (CONF-691076) (NSA 30: 15383).

Makky, A.-M. M. 1966. Radioactive nicotine: biosynthesis, toxicity and metabolism. Thesis, Univ. of Minnesota, Minneapolis. 115pp. (Dissertation Abstr. 27(3): 724-B) (^{3}H, *Aedes aegypti*).

Makovaki, E. 1967. Radiometric method of seepage control in dams and their environs. pp. 601-616. In: Isotopes in Hydrology. (See IAEA, 1967) (In Russian) (NSA 22: 19072).

Makowski, J. 1972. Choice of radioisotopes and their activity for hydrogeological research purposes. Isotopenpraxis 8(9): 327-332. (In German) (NSA 27: 19843).

Makowski, J. 1972. Appropriate choice of radioisotopes and their activity for hydrogeological investigation according to model measurements. Geol. Jahrb. Reihe C(2): 329-337. (In German) (NSA 30: 21164; 31: 11269).

Maksimovic, M. 1971. Effect of ^{60}Co irradiation of male pupae of the gypsy moth, *Lymantria dispar* L., on biological functions of male moths. pp. 15-22. In: Sterility Principle for Insect Control or Eradication. (See IAEA, 1971) (NSA 25: 29616).

Maksimovic, M. 1972. Two releases of sterile gypsy moth males (*Lymantria dispar* L.) in the park of Jelsa an the island of Hvar. Zast. Bilja 23: 117-118. (NSA 30: 27206) (Irradiated males).

Maksymov, J. K. 1965. New perspectives in insect control. Schweiz. Z. Forst. 116(3): 235-242. (In German, French summary).

Malakhov, S. G. and K. P. Makhon'ko. 1963. Statistics of the occurrence of "hot" particles in the ground layers of the air of the Moscow area in 1961-1962. State Committee on the Use of Atomic Energy, USSR. Moscow. UN Document A/AC.82/G/L.917. 7pp.

Malakhov, S. G., Ye. N. Davydov and N. A. Karagod. 1963. The fallout of ^{141}Ce, ^{144}Ce, ^{95}Zr+^{95}Nb and ^{90}Sr over the territory of the USSR in the second half of 1962. State Committee on the Use of Atomic Energy, USSR. Atomizdat, Moscow. UN Document A/AC.82/G/L.902. 9pp.

Malavolta, E., O. J. Crocomo and C. C. Delwiche. 1965. Fixation of carbon dioxide by nitrifying bacteria. pp. 285-299. In: Proc. 5th Inter-American Symp. on the Peaceful Applications of Nuclear Energy. Pan American Union, Washington. (In Spanish) (NSA 19: 35941).

Maldiney and Thouvenin. 1898. The influence of x-rays on germination. Rev. gen. bot. 10(111): 81-86. (In French).

Malekghassemi, B. 1969. Biological studies on culture and elimination of the beet fly (*Pegomyia betae* Curtis) by means of the self destruction method (sterile male technique). Part 2. Bayer. Landwirt. Jahrb. 46: 958-980. (In German) (NSA 24: 44638).

Malhotra, N. and P. N. Srivastava. 1975. The hematological effects of irradiation on the Indian desert gerbil (*Meriones hurrinae* Jerdon). Strahlentherapie 150(3): 344-349. (In German) (NSA 33: 26274).

Malich, C. W. 1965. Thin window holder for particle irradiation of *Drosophila*. Drosophila Inform. Serv. 40: 99.

Malik, M. A., M. Mohyuddin and N. Osman. 1972. Effects of radiation on fungi. Part I. Ionizing radiations. Nucleus (Karachi) 9(1-2): 57-60. (NSA 27: 12287).

Malik, S. R., S. A. Durrani and J. H. Fremlin. 1973. Comparative study of the spatial distribution of uranium and of TL-producing minerals in archeological materials. Archeometry 15(2): 249-253. (NSA 28: 24495).

Malla, D. S., J. F. Diehl and D. K. Salunkhe. 1967. *In vitro* susceptibility of strains of *Penicillium viridicatum* and *Aspergillus flavus* to β-irradiation. Experientia 23: 492-493. (NSA 21: 34836).

Mallik, M. U. 1975. Studies on the incorporation of ^{14}C in tissues of *Galleria mellonella*. Int. J. Appl. Radiat. Isotopes 26(9): 567-569. (NSA 33: 9627).

Mallik, M. U., M. M. Hossain and S. A. Mollah. 1968. Preliminary study on the stimulating effect of low dose gamma radiation on the larvae of silk worm (*Bombyx mori*). Nucl. Sci. Appl. 4: 19-20. (NSA 23: 6559).

Mallory, E. C., Jr., J. O. Johnson and R. C. Scott. 1961? Water load of uranium, radium, and gross beta activity at selected gaging stations, water year 1960-1961. U. S. Geological Survey, Washington. 31pp. (NP-18582) (NSA 25: 6424).

Malone, C. R. and D. J. Nelson. 1969. Feeding rates of freshwater snails (*Goniobasis clavaeformis*) determined with ^{60}Co. Ecology 50(4): 728-730. (NSA 24: 7708).

Malone, C. R. and D. E. Reichle. 1973. Chemical manipulation of soil biota in a fescue meadow. Soil Biol. Biochem. 5: 629-639. (NSA 29: 15637).

Maloney, M. A. and F. R. Mraz. 1969. The effect of whole body gamma irradiation on survivors' egg production in the white leghorn, *Coturnix coturnix japonica* and bobwhite quail. Poultry Sci. 48: 1939-1944. (NSA 24: 25564).

Malov, G. A., V. F. Oreshko and M. A. Pinigin. 1958. The radioactivity of water and bottom deposits of city ponds. Gigiyena sanitariya 23(10): 28-32. (In Russian) (NSA 13: 20063).

Malovitskaya, L. M. and Yu. I. Sorokin. 1961. An experimental study of the feeding of *Diaptomus* (Crustacea, Copepoda) using C^{14}. Tr. inst. biol. vodokhran. Akad. Nauk SSSR 4: 262-273. (In Russian).

Mal'skaya, R. V. 1965. Radioactivity of the underground waters in the west Ukrainian region. Geokhimiya (12): 1487-1490. (Geochem. Int. (6): 1088-1091) (NSA 20: 23289).

Maltz, J. 1958. Sodium exchange in lower vertebrates studied by means of ^{24}Na - its incorporation as a "clearance" of radioactive water. Proc. 2nd Intern. Conf. on the Peaceful Uses of Atomic Energy 24: 138-142. (Paper A/CONF.15/P/1224) (NSA 13: 6156).

Malueg, K. W. 1966. An ecological study of *Chaoborus*. Thesis, Univ. of Wisconsin, Madison. xix, 231pp. (^{131}I).

Malvicini, A., et al. 1963. Significant increase of manganese-54 in radioactive fallout. Comit. Naz. Energia Nucl. Notiziario 9(8-9): 44-51. (In Italian) (NSA 17: 41055).

Malvano, R. and P. Grosso. 1968. Iodine activation analysis in materials of biological interest. J. Nucl. Biol. Med. 12: 86-97. (NSA 23: 17382).

Malykhin, V. M. 1972. Optimization of conditions and criteria of quality of radiometry of biological samples with ultra small levels of activity. pp. 28-49. In: *Methods of Determination of Radioactivity*. (See Polikarpov and Parchevskiy, 1972) (In Russian).

Malysowa, E., J. Bors and L. Szerszen. 1967. ^{90}Sr in the profiles of some arable soils of Lower Silesia. Rocz. Glebozn. 18: 33-41. (In Polish) (NSA 23: 26353).

Mamedov, K. and R. Yu. Rakhimkulov. 1971. Review of the literature on the effect of ionizing radiation and chemical mutagens on plants. III. Izv. Akad. Nauk Turkmen. SSR (biol.) (2): 86-92. (In Russian) (NSA 26: 7112).

Mamuro, T. 1971. Activation analysis of atmospheric pollutants. Genshiryoku Kogyo 17(8): 19-25. (In Japanese) (NSA 26: 11547).

Mamuro, T. 1974. Activation analysis of environmental pollutants. Radioisotopes (Tokyo) 23(6): 355-363. (In Japanese) (NSA 31: 32482).

Mamuro, T., Y. Matsuda and A. Mizohata. 1972. Identification of air pollution source by instrumental neutron activation analysis. Radioisotopes (Tokyo) 21(3): 183-185. (In Japanese) (Vicinity of a factory).

Mamuro, T., T. Matsunami and T. Ishiyama. 1971. Detection of radiotungsten in rainwater. Health Physics 20: 86-89. (NSA 25: 15794).

Mamuro, T., et al. 1961, 1962, 1969, 1970. Observations radioactive air-borne dusts. Observations of fallout around our laboratory. Activation analysis of river water. Ann. Rept. Radiation Center, Ozaka Prefecture, Japan 2: 1-9, 10-19; 3: 1-5; 10: 21-23; 11: 14-28. (NSA 16: 20121, 20122; 17: 33531; 24: 38786; 26: 6502).

Mamuro, T., et al. 1971. Activation analysis of polluted river water. Radioisotopes (Tpkyo) 20: 111-116. (In Japanese) (NSA 25: 31852)

Mamuro, T., et al. 1972. Instrumental neutron activation analysis of environmental pollution. Proc. Jap. Conf. Radioisotopes (10): 493-495. (In Japanese) (NSA 28: 2138).

Mandl, A. M. 1964. The radiosensitivity of germ cells. Rev. Camb. Phil. Soc. Biol. 39(3): 288-371. (Incl. insects).

Mandler, J. W., J. H. Reed and R. B. Moler. 1971. Oil slick identification utilizing charged particle activation techniques. pp. 111-116. In: *Nuclear Methods in Environmental Research*. (See Vogt, 1971) (NSA 26: 17361).

Mangeno, J. J. and M. E. Miles. 1970. Disposal of radioactive wastes from U. S. Naval nuclear-powered shops and their support facilities, 1969. Radiol. Hlth. Data 11(8): 373-378.

Mangold, O. and T. Peters. 1956. On the action of similar roentgen doses on different stages of development of *Triton alpestris*. Beitr. path. anat. 116: 480-498. (In German).

Mani, A., et al. 1967. Distribution of global and net radiation over the Indian Ocean and its environments. Arch. Meteorol. Geophys. Bioklimatol. B15: 82-98. (NSA 22: 40265).

Maniago, D. P. and A. S. Crandon. 1973. Near-shore and riverine siting of floating nuclear plants. Trans. Amer. Nucl. Soc. 17: 74. (NSA 29: 7123).

Manikas, G. and W. F. Baldwin. 1969. Radiation induced sterility in the control of the olive fruit fly, *Dacus oleae* (Gmel.). pp. 21-23. In: *Insect Ecology and the Sterile-Male Technique*. (See IAEA, 1969).

Manil, P. and P. Demalsy. 1965. Effects on *Medicago sativa* L. of acute doses of gamma radiation. Radiation Botany 5(2): 109-114. (In French) (NSA 19: 24332).

Mann, L. A. 1971. Biological-gamma-radiation system for sewage processing. Isotopes Radiat. Techn. 8(4): 439-444. (NSA 25: 48535).

Manna, G. K. and S. C. Mazumder. 1968. Induced x-chromosomal aberrations in the study of intra- and interspecific radiosensitivity of grasshoppers. Nucleus (Calcutta) (Suppl.): 197-207. (NSA 24: 25464).

Manolov, L. S. 1968. Peculiarities of snow cover pollution from dry and wet radioactive fallout in mountainous places. pp. 43-55. In: *Investigations of Atmospheric Self-Scavanging Processes from Radioactive Isotopes*. (See Styro, et al., 1968) (In Russian) (NSA 23: 17590).

Manoto, E. C. 1973. Some biological and histopathological effects of gamma radiation on the gonads of the oriental fruit fly, *Dacus dorsalis* Hendel. Thesis, Univ. of Hawaii, Honolulu. 138pp. (Univ. Microfilms 74-17,213) (NSA 30: 27204).

Manoto, E. C. 1975. Gamma radiation: its effects on the male reproductive organ of the oriental fruit fly (*Dacus dorsalis* Hendel). Nucleus (Quezon City) 13(5): 4-5. (NSA 33: 20538).

Manoto, E. C., G. B. Fernandez and G. B. Viado. 1970. Some effects of gamma radiation on the striped stemborer of rice, *Chilo suppressalis* Walker. Philippine Nucl. J. 2(2): 177-191. (NSA 26: 43368).

Manowitz, B. 1966. Status of gamma irradiator design. Brookhaven National Lab., U. S. AEC report BNL-10257. 39pp. (For Am. Inst. Chem. Engineers, Atlantic City, N. J., Sept. 18-21, 1966) (CONF-660903-1) (NSA 20: 35606).

Manowitz, B., M. Smith and L. Newman. 1971. Simultaneous use of isotope radiotracer and sulphur hexachloride as a dual-tracer system for sulphur oxides in the atmosphere. pp. 223-239. In: Nuclear Techniques in Environmental Pollution. (See IAEA, 1971) (NSA 25: 32086).

Manowitz, B., et al. nd. Experimental determination of dose distribution in the proposed fir gamma irradiator. Chem. Eng. Progr. 55 (Symp. Ser. 22): 127-137. (Listed in error, not ecology).

Manowitz, B., et al. 1971. Isotope ratio tracer method: applications in atmospheric sulfur pollution studies. pp. 993-999. In: Proc. 2nd Intern. Clean Air Congr. H. M. Englund, ed. Academic Press, New York. (NSA 26: 45568).

Manoylov, S. Ye. 1959. Importance of the ionization of iron-containing compounds during x-irradiation of an organism. pp. 189-191. In: Proc. 1st All-Union Conf. on Radiation Chemistry. Part 4. (Moscow, 1957). (In Russian) (NSA 13: 22118) (Used frog heart).

Manoylov, S. Ye. and K. P. Ivanov. 1955. Some characteristics of the biological activity of penetrating radiation. Dokl. Akad. Nauk SSSR 105(1): 180-181. (In Russian) (Ref, zh. (biol.) 1955: 3724) (Used frog heart).

Manoylov, S. Ye., A. V. Lazovskaya and V. A. Orlov. 1956. Effects of x-rays emitted from various anodes upon the work of isolated frog heart. Dokl. Akad. Nauk SSSR 110: 125-128. (In Russian) (NSA 11: 1737).

Mansell, R. S., L. C. Hammond and R. M. McCurdy. 1973. Coincidence and interference corrections for dual energy gamma ray measurements of soil density and water content. Soil Sci. Soc. Am. Proc. 37(4): 500-504. (NSA 28: 21370).

Mansfeld, A. and E. Hanslik. nd. Radioactive contamination of waters: the effect of radioactive raw material exploitation and processing. Czechoslovakia Vyzkumny Ustav Vodohospodarsky, Prague. (In Czech) (INIS-mf-887-3, pp. 53-62) (NSA 30: 6531).

Mansfeld, A., J. Ruzicka and E. Hanslik. 1970. Contribution to the study of natural radionuclides in waste and surface waters. Radioisotopy 11(5): 715-723. (In Czech) (NSA 25: 54774).

Mansingh, A. 1964. The effect of malathion on the catabolism of labelled glucose in Blattella germanica. Indian J. Entomol. 26(4): 423-433. (^{14}C).

Mansingh, A. 1965. Glycine catabolism in Blattella germanica (L.). J. Insect Physiol. 11(7): 1031-1037. (^{14}C).

Mansingh, A. 1965. The effect of malathion on the metabolism of amino acids in the German cockroach Blattella germanica. J. Insect Physiol. 11(10): 1389-1400. (^{14}C).

Manunta, G. 1963. Radioactivity in Turdus musicus, Turdus merula merula, and Scolopax rusticola. Boll. Soc. Ital. Biol. Sper. 39: 527-529. (In Italian) (NSA 18: 17429).

Manuilova, N. A. 1972. Radioautographic study of RNA syntheses in the retina of birds during development. Biol. Nauki 15(7): 33-38. (In Russian) (BA 57(1):).

Maqsood, M., E. P. Reinike and P. O. Fromm. 1960. In vitro uptake of iodine by tissues of rainbow trout. Nature 188(4755): 1030-1031. (NSA 15: 7191).

Maqsood, M., E. P. Reineke and P. O. Fromm. 1961. In vitro uptake by tissues of rainbow trout. Physiol. Zool. 34(1): 34-38.

Maragoudakis, M. E., T. E. King and V. H. Cheldelin. 1964. Acidic amino acids from α-hydr xy-γ-oxoglutarate-^{14}C in rat liver and larvae of the blowfly, Sarcophaga bullata. Biochim. Biophys. Acta 93(3): 646-649.

Marais, P. G. 1963. The use of radioisotopes and ionizing radiation in agricultural research. S. Afr. J. Sci. 59: 319-324. (In Afrikaans).

Marais, P. G., et al. 1964. ^{60}Co irradiation unit for agricultural research. S. Afr. J. Agr. Sci. 7: 837-862. (NSA 20: 697).

Maramorosch, K. 1963. Arthropod transmission of plant viruses. Ann. Rev. Entomol. 8: 369-414. (Mentions use of radionuclides).

Marathe, S. G., et al. 1970. Gamma-ray spectrometric analysis applied to estimation of radionuclides in reactor coolant water. Radiochim. Acta 14: 141-143. (NSA 25: 12956).

Marble, G. 1972. Pollution, aggression, and radioprotection. Radioprotection 7(4): 197-205. (In French) (NSA 28: 24458).

Marcenko, E. 1965. Restoration of irradiated algae after a period of darkness. Nature 207(4996): 542-543. (NSA 19: 38410).

Marcenko, E. 1966. The effects of gamma rays on algae (Desmidiacean). Protoplasma 62: 156-183. (In German) (NSA 21: 36760).

Marcesse, J. 1967. In situ determination of soil-retention capacity using a neutron-moisture meter. pp. 137-144. In: Isotope and Radiation Techniques in Soil Physics and Irrigation Studies. (See IAEA, 1967) (In French) (NSA 22: 23148).

Marcesse, J. 1969. Nuclear techniques in problems connected with the water supply of soil and plants. Bull. Inform. Sci. Tech. (Paris) (141): 11-20. (In French) (NSA 24: 9506).

Marcesse, J. and P. Couchat. 1974. Hydrodynamic soil studies involving the use of an automatic neutron moisture gage. pp. 277-293. In: Isotope and Radiation Techniques in Soil Physics and Irrigation Studies 1973. (See IAEA, 1974) (In French) (NSA 30: 15500).

March, R. B., et al. 1955. Metabolism of systox in the white mouse and American cockroach. J. Econ. Entomol. 48(4): 355-363.

March, R. B., et al. 1956. Fate of P^{32} labeled malathion in the laying hen, white mouse, and American cockroach. J. Econ. Entomol. 49(2): 185-195.

March, R. B., et al. 1956. Fate of P^{32} labeled malathion sprayed on Jersey heifer calves. J. Econ. Entomol. 49(5): 679-682.

Marchart, H. 1968. Radiotracer study of the predators on *Distantiella theobroma* (Distant) (Hemiptera: Miridae). pp. 3-15. In: *Isotopes and Radiation in Entomology*. (See IAEA, 1968) (NSA 24: 31792).

Marchyulenene, D. P. and V. B. Nyanishkene. 1972. Accumulation and distribution of ^{90}Sr, ^{137}Cs, ^{144}Ce, ^{106}Ru in *Dreissena polymorpha* Pall and *Lymnea stagnalis* L. Lietvos TSR Mokslu Akad. Darbei C(2): 141-145. (In Russian) (NSA 28: 8521).

Marchyulenene, D. P. and G. G. Polikarpov. 1969. Pathways of penetration of radionuclides into freshwater mollusks and insect larvae. Radiobiologiya 9(2): 286-289. (AEC-tr-7025, pp. 146-150) (NSA 23: 29481).

Marchyulenene, D. P., N. K. Stryupkuvene and G. G. Polikarpov. 1973. Uptake of ^{90}Sr by organisms of *Gammarus* sp. and *Chironumus* f. l. *plumosus* L. pp. 152-159. In: *Radioecology of Water Organisms. II*. (See Andrushaitus, et al., 1973) (AEC-tr-7606, pp. 92-95) (NSA 30: 21341).

Marckwordt, U. and J. Lehr. 1971. Factors of transfer of ^{137}Cs from soils to crops. pp. 1057-1066. In: *Proc. Intern. Symp. on Radioecology Applied to the Protection of Man and His Environment*. (See Commission of the European Communities, 1971) (NSA 27: 5571).

Marckwordt, U. and J. M. Mousny. 1970. Availability of radiocesium in relation to ammonium fertilization and potassium status of two soils from Allonville. pp. 571-587. In: *Proc. Intern. Symp. on Radioecology*. (See France Commissariat à l'Énergie Atomique, 1970) (In French) (NSA 25: 2529).

Marcoux, G., et al. 1971. Preliminary experimental data on the dynamics of ^{85}Sr in *Anguilla anguilla* (L.). Centre d'Études Nucléaires, France Commissariat à l'Énergie Atomique, Cadarache, report CEA-N-1451. 14pp. (In French) (NSA 26: 4676).

Marcovich, H. 1965. Discussion on the genetic control of the sensitivity of microorganisms to the lethal action of radiations. Biol. Med. (Paris) 54: 477-484. (In French) (NSA 20: 35264).

Marcovich, H. 1966. Some theoretical and experimental data on the utilization of microorganisms in radiology. pp. 229-236. In: *Applied Radiobiology. Vol. I*. J. Loiseleur, ed. Gauthier-Villars, Paris. (In French) (NSA 21: 45850).

Marcuzzi, G. and P. Degasperi. 1964. Application of autoradiographic techniques to the study of the excretion in the coleopterous insect *Tenebrio molitor* L. pp. 127-129. In: *Preparation and Biomedical Application of Labelled Molecules*. (Proc. Symp., Venice, 23-29 August 1964). European Atomic Energy Community report EUR-2200.e. (CONF-774) (NSA 19: 31762, 32260) (^{14}C).

Marcuzzi, G. and V. Santoro. 1959. Studies on the water exchange of *Tenebrio molitor* by tritiated water. Ricerca scient. 12: 2576-2581. (In Italian).

Marek, J. 1961. Effects on plant cells of aphid punctures. Entomol. exptl. appl. 4: 20-34. (In German) (^{32}P).

Marenco, A. and J. Fontan. 1972. Sources of polonium-210 within the troposphere. Tellus 24(1): 38-46.

Marey, A. N. 1955. Water organisms as a sanitary index of contamination of water bodies with radioactive agents. Gigiyena sanitariya 20(8): 3-9. (In Russian).

Marey, A. N. 1956. Radioactive wastes and problems of health. Med. radiologiya 1(4): 3-7. (In Russian) (NSA 11: 903).

Marey, A. N. 1956. Several problems of sanitation protection of bodies of water from pollution by radioactive substances. Gigiyena sanitariya 21(9): 7-11. (In Russian) (Ref. zh. (med.) 4(2): 44).

Marey, A. N. 1957. Hygienic evaluation of radioactive strontium as a factor in contamination of the environment. Med. radiologiya 2(5): 89-95. (JPRS-210, pp. 108-115) (NSA 12: 15231).

Marey, A. N. 1958. Sanitary protection of open bodies of water from pollution by radioactive substances. pp. 32-90. In: *Sanitary Protection of Open Bodies of Water from Pollution by Radioactive Substances*. Medgiz, Moscow. (JPRS-L-514-N) (NSA 13: 19080).

Marey, A. N. 1958. Radioactive waste disposal. Priroda 47(12): 47-50. (In Russian).

Marey, A. N. 1961. Radiostrontium as a factor in environmental contamination. In: *Distribution, Biological Effects, and Migration of Radioactive Isotopes*. (See Lebedinskiy and Moskalev, 1961) (AEC-tr-7512, pp. 349-358) (NSA 30: 6539).

Marey, A. N. (Ed.). 1962. *Radiation Hygiene. Vol. 2. Communal Hygiene*. Medgiz, Moscow. 216pp. (In Russian; Table of Contents, Chs. 3 and 6, JPRS-21,799).

Marey, A. N. 1962. Natural sources of radiation. Distribution of radioactive substances in nature. Ch. 2, pp. 9-10. In: *Radiation Hygiene. Vol. 2*. (See Marey, ed., 1962) (In Russian).

Marey, A. N. 1962. Factors determining the behavior of radioactive substances in water systems. Significance of the bottom as a regulator of the specific radioactivity of the water of a water system. Significance of hydrobionts as a factor reducing the concentration of radioactive materials in a water system. Dispersion of radioactive substances falling into a water system. Indicators characterizing the influence of radioactive substances on the sanitary condition of water systems. Ch. 3, pp. 34-35, 36-40, 59-60, 61-64, 65-67. In: *Radiation Hygiene. Vol. 2*. (See Marey, ed., 1962) (JPRS-21,799, pp. 1-5, 5-13, 39-41, 41-49, 49-55).

Marey, A. N. 1971. Environmental radioactive pollution and man. pp. 164-182. In: Contemorary Problems of Radiobiology. Vol. 2. Radioecology. (See Klechkovskiy, Polikarpov and Aleksakhin, 1971) (English transl. pp. 141-156) (NSA 29: 12733).

Marey, A. N. 1973. Environmental monitoring. pp. 124-129. In: Radiation Protection. (German-Soviet Working Meeting, Berlin, 8 June 1972). German Democratic Republic, Staatliche Zentrale für Strahlenschutz, Berlin, report SZS-148. (In Russian) (NSA 30: 18912).

Marey, A. N. and R. M. Barkhudarov. 1974. Principles of substantiating permissible concentrations of radioactive substances in freshwater bodies. Gigiyena sanitariya (1): 50-54. (In Russian) (NSA 31: 19593).

Marey, A. N. and B. K. Borisov. 1967. Strontium-90 in the bone tissue of the population of the Soviet Union. State Committee on the Use of Atomic Energy, USSR. Atomizdat, Moscow. 10pp. UN Document A/AC.82/G/L.1184. (In Russian) (NSA 22: 33801).

Marey, A. N. and M. M. Saurov. 1967. Material for assessment of the role played by food chains in ^{90}Sr migration from freshwater reservoirs into the diet of man. pp. 135-141. In: Radioecological Concentration Processes. (See Aberg and Hungate, 1967).

Marey, A. N., R. M. Barkhudarov and N. Ya. Novikova. 1974. Radiation-hygienic evaluation of geochemical provinces with increased cesium-137 migration. pp. 254-259. In: Proc. 3rd Intern. Congr. Radiation Protection. (In Russian) (CONF-730907-P1) (NSA 30: 26801).

Marey, A. N., V. A. Knizhnikov and A. N. Karmayeva. 1967. The effect of calcium in drinking water on the accumulation of ^{226}Ra and ^{90}Sr in the human body. pp. 333-336. In: Radioecological Concentration Processes. (See Aberg and Hungate, 1967).

Marey, A. N., Ye. I. Yartsev and Ye. I. Moiseyenko. 1965. On problems of distribution of strontium-90 in skeletons of adults. State Committee on the Use of Atomic Energy, USSR. Moscow. UN Document A/AC.82/G/L.1086. 10pp. (AEC-tr-6797).

Marey, A. N., A. F. Zaitseva and R. M. Barkhudarov. 1974. Sanitary standardization of radioactive isotopes in freshwater reservoirs (pertinent principles, criteria, and calculation methods). pp. 28-32. In: Radiation Protection, Environment and Population. (See German Democratic Republic, 1974) (In Russian) (NSA 31: 8615).

Marey, A. N., et al. 1965. Strontium-90 in human bone in the USSR (data for 1964). State Committee on the Use of Atomic Energy, USSR. Moscow. UN Document A/AC.82/G/L.1084. 8pp. (AEC-tr-6797).

Marey, A. N., et al. 1967. Cesium-137 in the organisms of the inhabitants of Moscow. State Committee on the Use of Atomic Energy, USSR. UN Document A/AC.82/G/L.1185. 5pp. (In Russian) (NSA 20: 19323).

Marey, A. N., et al. 1972. Effect of natural factors on ^{137}Cs accumulation in the bodies of residents in some geographical regions. Health Physics 22(1): 9-15. (NSA 26: 15529).

Margalef, R. 1964. Experimental models of phytoplankton populations: further observations on pigments and fixation of inorganic carbon. Investigation Pesquera 26: 195-203. (^{14}C).

Margocsy, N., et al. 1973. Study of the absorption of natural U and ^{210}Pb by population in relation to radioactivity of the environment in an area of Romania with increased natural background. pp. 401-406. In: Health Physics Problems of Internal Contamination. (See Bujdoso, 1973) (NSA 28: 29846).

Marichal, M. 1963. The control of the surrounding radioactive pollution. Bull. Inform. Sci. Tech. (Paris) (72-73): 65-72. (In French) (NSA 17: 42424).

Marichal, M. 1972. Control of radioactive pollution around the Marcoule Center. pp. 853-870. In: New Trends in Radiation Protection. (6th Intern. Congr., Bordeaux, France). France Societé de Radioprotection, Montrouge. (In French) (CONF-720328) (NSA 27: 17241).

Marichal, M. and H. Peyresblanques. 1973. Safety measures in a nuclear center in view of protecting the environment. pp. 19-34. In: 7th National Congr. of the Health Physicist's Assoc. France Centre de la Recherche Scientifique, Marseilles. (In French) (NSA 30: 6488).

Maramuthu, K. M., et al. 1972. Effects of space flight (Biosatellite II) and radiation on female gametophyte development in Tradescantia. Am. J. Bot. 59(4): 359-366.

Marin, A., et al. 1960. Considerations on the changes in the blood capillaries after ionizing radiations. II. Histologic and histochemical study of the blood capillaries of the frog tongue after single exposure to x rays. Riv. Anat. Patol. Oncol. 18(5): 606-622. (In Italian) (NSA 17: 23194).

Marine, I. W. 1967. The use of a tracer test to verify an estimate of the groundwater velocity in fractured crystalline rock at the Savannah River Plant near Aiken, South Carolina. pp. 171-179. In: Isotope Techniques in the Hydrologic Cycle. (See Stout, 1967).

Marino, M. A. (Ed.). 1967. Proc. Symp. on a Groundwater Hydrology. (San Francisco, California, November 6, 7, and 8, 1967). Amer. Water Works Resources Assoc., Urbana, Ill., Proc. Ser. No. 4. 365pp. (NSA 23: 444).

Marinov, V., D. Apostolov and M. Lazarova. 1974. Instrumental neutron-activation analysis of plants after elimination of ^{24}Na and ^{42}K. Byl. Akad. Nauk. Sofia Univ. 8pp. (In Bulgarian) (INIS-mf-1485) (NSA 32: 27460).

Mark, H. B., et al. 1969. Application of semipermeable ion exchange membranes to in situ determination of trace metal ions in aqueous systems by electrochemical and neutron activation techniques. Critical evaluation of techniques. Environ. Sci. Technol. 3: 165-168. (NSA 23: 15819).

Markee, E. H., Jr. 1968. Turbulent transfer characteristics of radioiodine effluents from air to grass. pp. 589-601. In: Proc. U. S. AEC Meteorological Information Meeting. (Chalk River Nuclear Labs., September 11-14, 1967). C. A. Mawson, ed. Atomic Energy of Canada, Ltd., Chalk River, report AECL-2787. (NSA 22: 40252).

Markee, E. H., Jr. 1971. Studies of the transfer of radioiodine gas to and from natural surfaces. A summary of controlled environmental release test analyses (1963-1969). Air Resources Lab., U. S. NOAA report NOAA-TM-ERL-ARL-29. 28pp. (NSA 26: 22800).

Markham, O. D. 1971. Effects of ^{60}Co radiation and social interaction on the pika (Ochotona princeps). Thesis, Colorado State Univ., Fort Collins. 62pp. (Univ. Microfilms 72-16,117) (NSA 26: 41178).

Markham, O. D. 1974. Environmental and radiological monitoring at the National Reactor Testing Station during FY-1973 (July 1972-June 1973). Radiat. Data Rept. 15(5): 227-246. (NSA 30: 18621).

Markham, O. D. and F. W. Whicker. 1970. Radiation $LD_{50(30)}$ of pikas (Ochotona princeps) in the natural environment and captivity. Am. Midland Nat. 84(1): 248-252. (Also: Colorado State Univ., U. S. AEC report COO-1156-32) (NSA 24: 32179).

Markham, O. D. and F. W. Whicker. 1973. Intraspecific competition and response of pikas (Ochotona princeps) to radiation. pp. 1070-1075. In: Radionuclides in Ecosystems. (See Nelson, 1973) (NSA 26: 41162; 28: 30139).

Markham, O. D., F. W. Whicker and R. M. Hansen. 1970. Radiation $LD_{50(30)}$ of Richardson ground squirrels. Health Physics 18: 731-732. (Also: Colorado State Univ., U. S. AEC report COO-1156-33) (NSA 24: 32180).

Marko, A. M. 1970. Biological significance of radioactive releases to the environment. (For Proc. 10th Ann. Intern. Conf. of the Canadian Nuclear Assoc., Toronto, Canada, May 24, 1970). 12pp. (CONF-700564) (70-CNA-641) (NSA 25: 55194).

Marko, A. M., et al. 1972. Nuclear power and the environment. Proc. 4th Intern. Conf. on the Peaceful Uses of Atomic Energy 11: 475-485.

Markus, B. and E. Sticinsky. 1961. Experiments on the effect of the energy spectrum of fast electrons on biological reactions. Strahlentherapie 115: 394-403. (In German) (Used Drosophila eggs).

Markus, B. and E. Stricinsky. 1962. The action of 14-MeV electrons, 14-MVp and 200-KVp x-rays and the influence of the energy spectrum of 14-MeV electrons on eggs of Drosophila. Abstr. p. 244. In: 2nd Intern. Congr. Radiation Research. (See Anonymous, 1962).

Markus, B. and E. Stricinsky. 1963. The effect of the energy spectrum from 14 MeV electrons, and the relative effects of 14 MeV electrons and 14 MV and 200 kV x-rays on Drosophila eggs. Strahlentherapie 120(2): 262-268. (In German).

Markussen, E. K. 1970. Radiochemical procedures for the determination of plutonium in environmental samples. Denmark Atomic Energy Commission, Risö, report RISÖ-M-1242. 5pp. (NSA 24: 31241).

Marley, W. G. 1971. Evaluation of environmental monitoring data for nuclear facilities. pp. 799-808. In: Rapid Methods for Measuring Radioactivity in the Environment. (See IAEA, 1971) (NSA 26: 28060).

Marlow, W. F. and P. D. Lafleur. 1971. Standard reference materials for the analysis of environmental samples. pp. 91-94. In: Nuclear Techniques in Environmental Pollution. (See IAEA, 1971) (NSA 25: 31822).

Marois, H. and R. J. Wilson. 1974. Environmental monitoring and control of radioactive materials released from Quebec and Ontario hydro nuclear power stations. pp. 15-___. In: Canadian Nuclear Assoc. 14th Ann. Intern. Conf. Vol. 6. Nuclear Energy and the Public. (74-CNA-600) (NSA 31: 19545).

Marouzzi, G. 1959. Total water and interstitial liquid in insects preliminary note). Boll. Zool. 23(2): 609-620. (In Italian) (Used ^{35}S).

Marples, T. G. 1965. A radionuclide tracer study of arthropod food chains in a Spartina salt marsh ecosystem. Thesis, Univ. of Georgia, Athens. (Dissertation Abstr. 26: 2392).

Marples, T. G. 1965. The distribution of forest floor fauna along a gamma radiation gradient. Abstr. Bull. Ecol. Soc. Am 46(3): 97.

Marples, T. G. 1966. A radionuclide tracer study of arthropod food chains in a Spartina salt marsh ecosystem. Ecology 47(2): 270-277.

Marples, T. G. and E. P. Odum. 1964. A radionuclide tracer study of arthropod food chains in a Spartina alterniflora salt marsh. Abstr. Bull. Ecol. Soc. Am. 45(3): 81.

Marquardt, W. 1968. The influence of fog on radioactive contamination in the biosphere. Z. Meteorol. 20: 37-42. (In German) (NSA 23: 40883).

Marquardt, W., R. Höhle and U. Schuh. 1970. Radioactive emissions by coal-fed power stations. Z. Gesam. Hyg. Ihre Grenzgeb. 16: 188-191. (In German) (NSA 26: 33334).

Marques, B. E. and M. R. Sales Grade. 1970. Radioactive pollution in natural water. Rev. Port. Quim. (Lisbon) 12(3): 140-147. (In Portugese) (NSA 26: 58551).

Marques, B. E. and M. R. Sales Grade. 1972. Radioactive contamination in the Tejo Estuary. pp. 887-920. In: New Trends in Radiation Protection. (6th Intern. Congr., Bordeaux, France). Societé Francaise de Radioprotection, Montrouge. (In french) (CONF-720328) (NSA 27: 17270).

Marques, E. K. 1966. Genetic effects of gamma radiation on Drosophila populations under interspecific competition. Abstr. 590. p. 149. In: 3rd Intern. Congr. Radiation Research. (Cortina d'Ampezzo, Italy, 26 June-2 July 1966) G. Silini, ed. North-Holland, Amsterdam.

Marques, E. K. 1973. Development of radioresistance in irradiated Drosophila nebulosa populations. Mutation Research 17(1): 59-72. (NSA 27: 9877).

Marques, E. K. and C. M. P. Maciel. 1961. Some components of adaptive values of heterozygous Drosophila willistoni from irradiated natural populations. Experientia 17: 404-405.

Marsden, E. 1959. Radioactivity of soils, plant ashes and animal bones. Nature 183(4666): 924-925.

Marsden, E. 1964. Radioactivity of some rocks, soils, plants and bones. pp. 807-824. In: The Natural Radiation Environment. (See Adams and Lowder, 1964) (NSA 18: 35382; 22: 2215).

Marsh, K. V., et al. 1975. Radiological and chemical studies of the ground water at Enewetak Atoll. 1. Sampling, field measurements, and analytical methods. Lawrence Livermore Lab., U. S. ERDA report UCRL-51913(Pt.1). 28pp. (NSA 33: 7300).

Marshall, J. 1963. Can we eradicate the codling moth? West. Fruit Gr. 17 (4): 19-20. (Concerns use of sterile males).

Marshall, J. S. 1961. The effects of continuous gamma radiation on the intrinsic rate of natural increase of *Daphnia pulex*. Industry Program, Coll. of Engineering, Univ. of Michigan, Ann Arbor, report IP-491. vi, 68pp. (Also: 1962, Ecology 43(4): 598-607).

Marshall, J. S. 1963. The effects of continuous, sublethal gamma radiation on the intrinsic rate of natural increase and other population attributes of *Daphnia pulex*. pp. 363-366. In: Radioecology. (See Schultz and Klement, 1963).

Marshall, J. S. 1966. Population dynamics of *Daphnia pulex* as modified by chronic radiation stress. Ecology 74(4): 561-572. (NSA 21: 16496).

Marshall, J. S. 1967. Radiation stress in exploited *Daphnia* populations. Limnol. Oceanogr. 12(1): 154-158.

Marshall, J. S. 1967. Iron, manganese, cobalt, and zinc cycles in a southeastern reservoir lake. Savannah River Lab., U. S. AEC report DP-MS-66-81. 12pp. (NSA 22: 37875).

Marshall, J. S. 1969. Competition between *Daphnia pulex* and *D. magna* as modified by radiation stress. Abstr. Bull. Ecol. Soc. Am. 50(2): 70.

Marshall, J. S. and J. H. LeRoy. 1973. Iron, manganese, cobalt, and zinc cycles in a South Carolina reservoir. pp. 465-473. In: Radionuclides in Ecosystems. (See Nelson, 1973) (NSA 28: 27281) (Savannah River Plant).

Marshall, J. S. and P. A. Nessle. 1962. Role of Cladocera in the freshwater strontium cycle and influence of dissolved salts. Abstr. p. 174. In: Proc. 5th Conf. on Great Lakes Research, Toronto, Canada, April 9-10, 1962. Great Lakes Research Div., Univ. of Michigan, Ann Arbor, Publ. 9.

Marshall, J. S., A. M. Beeton and D. C. Chandler. 1964. Role of zooplankton in the freshwater strontium cycle and influence of dissolved salts. Verh. Intern. Verein. Limnol. 15(2): 665-672.

Marshall, J. S., B. J. Waller and E. M. Yaguchi. 1974. Plutonium in the Laurentian Great Lakes: food-chain relationships. Argonne National Lab., U. S. AEC report. 15pp. (CONF-740826-3) (NSA 30: 29415).

Marston, R. B. 1958. Use of a nuclear soil moisture meter in studies of evapotranspiration on mountain soils. Proc. Soc. Am. Foresters.

Marteka, V. 1965. Radiation and plant life. Current Science 51(11): 87. (At Lockheed reactor site, Georgia).

Martel, Y. A. and E. A. Paul. 1974. Use of radiocarbon dating of organic matter in the study of soil genesis. Soil Sci. Soc. Am. Proc. 38(3): 501-506. (NSA 30: 21158).

Martell, E. A. 1955 (Rev. 1957). Strontium-90 concentration data from biological materials, soils, waters, and air filters. Enrico Fermi Inst. of Nuclear Studies, Univ. of Chicago, Project Sunshine Bull. 11, Bull. 12, U. S. AEC report AECD-3763, AECU-3297(Rev.). 67, 70pp. (The latter also: pp. 617-650. In: The Nature of Fallout and Its Effects on Man. (See U. S. Congress, 1957) (NSA 11: 6612, 6613).

Martell, E. A. 1970. Transport patterns and resident times for atmospheric trace constituents vs. altitude. pp. 138-157. In: Radionuclides in the Environment. R. F. Gould, ed. Amer. Chem. Soc., Washington.

Marter, W. L. 1963. Radioiodine release incident at the Savannah River Plant. Health Physics 9(12): 1105-1109. (Incl. grass).

Marter, W. L. 1970. Savannah River water quality studies, 1965-1969. Savannah River Lab., U. S. AEC report DPST-70-445. 27pp. (NSA 26: 20315).

Marter, W. L. 1970. Radioactivity in the environs of Steel Creek. Savannah River Plant, U. S. AEC report DPST-70-425. 9pp. (NSA 29: 20860).

Marter, W. L. 1974. Environmental effects of a tritium gas release from the Savannah River Plant on May 2, 1974. Savannah River Plant, U. S. AEC report DP-1369. 28pp. (NSA 31: 14096).

Marter, W. L. 1974. Radioactivity from SRP operations in a down-stream Savannah River swamp. Savannah River Plant, U. S. AEC report DP-1370. 51pp. (NSA 30: 32086).

Marter, W. L. and C. M. Patterson. 1971. Monitoring of tritium in gases, liquids and in the environment. Trans. Amer. Nucl. Soc. 14(1): 162-163. (NSA 25: 39437).

Martha, G. 1969. Tagging of small animals with radioisotopes for tracking purposes: a literature review. Int. J. Appl. Radiat. Isotopes 20: 671-676. (NSA 23: 43656).

Martin, A. and R. L. Blanchard. 1969. The thermal volitilization of ^{137}Cs, ^{210}Po, and ^{210}Pb from *in vivo* labeled samples. Analyst 94: 441-446. (NSA 24: 565) (Caribou bone, reindeer muscle).

Martin, A., Jr., J. N. Dent and L. Joseph. 1948. Effects of beta rays on *Habrobracon juglandis*. Abstr. Genetics 33(6): 619-620.

Martin, A. O. 1965. Studies on the rate of spermatogenesis in *Drosophila*: effects of x-rays and streptonigrin. Z. Verebungsl. 96: 28-35.

Martin, C. N. 1954. Accumulative effects on the global surface caused by thermonuclear explosions. Compt. rend. 239(20): 1287-1289. (In French).

Martin, D. E. 1973. Eventual Great Lakes radiological impact due to nuclear power generation. Trans. Amer. Nucl. Soc. 17: 62-63. (NSA 30: 5109).

Martin, J. A., Jr. 1973. Calculations of environmental radiation exposures and population doses due to effluents from a nuclear fuel reprocessing plant. Radiat. Data Rept. 14(2): 59-76. (NSA 27: 27581) (Fish, deer; water pathways were calculated).

Martin, J. E., E. D. Harward and C. L. Weaver. 1971. Public Health Service role in providing information relative to the nuclear power controversy. pp. 724-735. In: Health Physics Aspects of Nuclear Facility Siting. Vol. III. (See Voillequé and Baldwin, 1971) (NSA 25: 56981).

Martin, J. E., et al. 1971. Radioactivity from fossil fuel and nuclear power plants. pp. 325-336. In: Environmental Aspects of Nuclear Power Stations. (See IAEA, 1971) (Paper SM-146/19) (CONF-700810-24) (NSA 24: 45911).

Martin, J. H. 1966. Environmental monitoring around the nuclear power station at Chapelcross, Annan, Scotland. pp. 45-54. In: Manual on Environmental Monitoring in Normal Operation. International Atomic Energy Agency, Vienna, Safety Ser. No. 16, Publ. STI-PUB-98. (NSA 20: 23407).

Martin, J. J. and P. Beau. 1972. Effluents from nuclear power plants and environmental monitoring. pp. 871-886. In: New Trends in Radiation Protection. (6th Intern. Congr., Bordeaux, France). Societé Francaise de Radioprotection, Montrouge. (In French) (CONF-720328) (NSA 27: 17602).

Martin, J.-M. 1969. Use of natural radioelements for sediment measurement. Houille Blanche 25(8): 757-764. (In French) (NSA 26: 1975).

Martin, J.-M. 1970. Seasonal variations in the radioactivity of matter suspended in rivers. Compt. rend. D271: 1934-1937. (In French) (NSA 25: 21164).

Martin, J. R. and J. J. Koranda. 1970. Distribution, residence time, and inventory of tritium in Sedan crater ejecta. Lawrence Radiation Lab., U. S. AEC report UCRL-72572. 22pp. (For Meeting, Amer. Nucl. Soc., Washington, 15 Nov 1970) (CONF-701102-18) (NSA 25: 29244, 4181, 37512).

Martin, J. R. and J. J. Koranda. 1972. Biological half-life studies of tritium in chronically exposed kangaroo rats. Radiation Research 50(2): 426-440. (NSA 26: 38762).

Martin, J. R. and J. J. Koranda. 1973. Recent measurements of cesium-137 resident time in Alaskan vegetation. pp. 108-115. In: Radionuclides in Ecosystems. (See Nelson, 1973) (NSA 25: 49122).

Martin, J. R., et al. 1970. Radioecological studies of tritium movement in a tropical rain forest. pp. 422-438. In: Symp. on Engineering with Nuclear Explosives. Vol. 1. (See U. S. AEC, 1970) (NSA 24: 41491).

Martin, J. W. 1971. Some effects of ionizing irradiation on the microfungal populations of Quercus rubrum litter. Thesis, Univ. of South Carolina, Columbia. 85pp. (Univ. Microfilms 71-21,847) (NSA 25: 58150).

Martin, R. E., S. I. Auerbach and D. J. Nelson. 1964. Growth and movement of smallmouth buffalo, Octiobus bubalus (Rafinesque), in Watts Bar Reservoir, Tennessee. Oak Ridge National Lab., U. S. AEC report ORNL-3530. viii, 98pp. (Also: Thesis, Univ. of Tennessee, Knoxville; R. E. Martin) (Contains data on radionuclides in fish tissues and scales).

Martin, S. S., W. T. Helm and W. F. Sigler. 1969. Accumulation of ^{226}Ra in two aquatic ecosystems. pp. 307-318. In: Symposium on Radioecology. (See Nelson and Evans, 1969).

Martin, V. J., et al. 1962. The effect of culture environment on the susceptibility of the grain weevil (Sitophilus granarius L.) to gamma radiation. U. K. Atomic Energy Authority, Wantage, report AERE-R-3893. 26pp.

Martin, W. E. 1962. Applications of fundamental biology to the needs of man. 4. Radioecology and the study of environmental radiation. Univ. of California at Los Angeles, U. S. AEC report TID-16060. 70pp. (NSA 16: 21891).

Martin, W. E. 1962. Immediate effects of a nuclear detonation on desert vegetation. Abstr. Bull. Ecol. Soc. Am. 43(4): 123-124.

Martin, W. E. 1963. Close-in effects of an underground nuclear detonation on vegetation. I. Immediate effects of cratering, throw-out, and blast (Project Sedan). Univ. of California at Los Angeles, U. S. AEC report PNE-228P. 41pp. (NSA 17: 21398).

Martin, W. E. 1963. Loss of I^{131} from fallout-contaminated vegetation. Abstr. Bull. Ecol. Soc. Am. 44(3): 76.

Martin, W. E. 1963. Loss of I^{131} from fallout-contaminated vegetation. Health Physics 9(12): 1141-1148. (Also: Univ. of California at Los Angeles, U. S. AEC report TID-18759) (NSA 17: 25483) (Sagebrush, shadscale, rabbits; Nevada Test Site and vicinity).

Martin, W. E. 1963. Notes on the deposition of fallout in relation to topography and local meteorological conditions. Univ. of California at Los Angeles, U. S. AEC report UCLA-513. 16pp. (NSA 17: 29128).

Martin, W. E. 1964. Radioecology and the study of environmental radiation. Bull. Torrey Botan. Club 91(4): 283-323.

Martin, W. E. 1964. Losses of Sr^{90}, Sr^{89}, and I^{131} from fallout-contaminated plants. Radiation Botany 4(3): 275-284. (NSA 19: 6104).

Martin, W. E. 1965. Early food-chain kinetics of radionuclides following close-in fallout from a single nuclear detonation. pp. 758-782. In: Radioactive Fallout from Nuclear Weapons Tests. (See Klement, 1965) (Native plants and animals at Sedan site).

Martin, W. E. 1965. Radiation facilities and objectives. Abstr. Bull. Ecol. Soc. Am. 46(2): 49. (At Nevada Test Site).

Martin, W. E. 1965. Interception and retention of fallout by desert shrubs. In: Radiation and Terrestrial Ecosystems. Health Physics 11(12): 1341-1354. (See Hungate, 1965) (NSA 20: 12572).

Martin, W. E. 1965. Interception and retention of radioactive fallout by desert shrubs in the Sedan fallout field. Univ. of California at Los Angeles, U. S. AEC report PNE-238F. 40pp. (NSA 20: 7318).

Martin, W. E. 1969. Bioenvironmental studies of the radiological-safety feasibility of nuclear excavation. BioScience 19(2): 135-137.

Martin, W. E. 1969. Radioecology and the feasibility of nuclear canal excavation. pp. 9-22. In: Symposium on Radioecology. (See Nelson and Evans, 1969).

Martin, W. E. 1971. Mathematical simulation of radionuclide transport in tropical ecosystems. pp. 107-119. In: Trace Substances in Environmental Health. IV. (See Hemphill, 1971) (NSA 26: 15214).

Martin, W. E. and G. E. Raines. 1969. Ecological transfer mechanisms--terrestrial. pp. 401-433. In: Proc. Symp. on Public Health Aspects of Peaceful Uses of Nuclear Explosives. Southwestern Radiological Health Lab. Lab., Las Vegas, Nev., report SWRHL-82.

Martin, W. E. and F. B. Turner. 1963. Increased environmental radiation in southern Nevada, October-December, 1961. Univ. of California at Los Angeles, U. S. AEC report UCLA-518. 45pp. (Incl. data on soil, plants, and cotton-tail and jack rabbits).

Martin, W. E. and F. B. Turner. 1965. Food-chain relationships of radiostrontium in the Sedan fallout field. Univ. of California at Los Angeles, U. S. AEC report PNE-237F. 74pp. (NSA 19: 32421).

Martin, W. E. and F. B. Turner. 1966. Transfer of Sr^{89} from plants to rabbits in a fallout field. Health Physics 12(5): 621-631.

Martin, W. E. 1966. Close-in effects of nuclear excavation and radiation on desert vegetation. pp. 7-9. In: Radiation Effects on Natural Populations. (See Sacher, 1966).

Martin, W. E., S. G. Bloom and R. J. Yorde, Jr. 1974. NAEG plutonium study modeling program: plutonium transport and dose estimation model. pp. 331-360. In: Dynamics of Plutonium in Desert Environments. (See Dunaway and White, 1974) (NSA 30: 29455).

Martin, W. E., et al. 1971. Mathematical models of radionuclides in ecosystems. pp. 39-99. In: Nuclear Cratering Explosion Effects for Interoceanic Canal Feasibility Studies. (See Klement, 1971).

Martin, W. S., R. C. Grubbs and M. A. Lessler. 1956. Effects of low-level x-irradiation of oxygen consumption of bullfrog erythrocytes. Am. J. Physiol. 187(3): 505-508. (NSA 11: 4231).

Martinec, J. 1975. Subsurface flow from snowmelt traced by tritium. Water Resources Res. 11(3): 496-498. (NSA 32: 14323).

Martinec, J., et al. 1974. New insights into the run-off mechanism by environmental isotopes. pp. 129-143. In: Isotope Techniques in Groundwater Hydrology. Vol. 1. (See IAEA, 1974) (NSA 31: 19624).

Martinec, M. 1973. Effect of uranium ore mining on ^{226}Ra content in surface water, reed, and freshwater fish. Cesk. Hyg. 18(1): 475-478. (In Czech) (NSA 31: 11265; 32: 19755).

Martinelli, P. 1970. Radiation analysis techniques. Tech. Moderne 62: 424-427. (In French) (NSA 25: 26164).

Martinez-Silva, R., et al. 1969. Trypanosoma cruzi: effects of gamma radiation on growth and infectivity. Exptl. Parasitol. 25: 162-170. (NSA 24: 29846).

Martinsen, D. L. and D. J. Kimeldorf. 1972. Prompt detection of ionizing radiations by carpenter ants. Biol. Bull. 143(2): 403-419. (NSA 27: 22758).

Martoja, R. 1962. Assimilation of the cellulose and microorganisms by the intestines of Gryllus bimaculatus de Geer (Insecta, Orthoptera, Grylloidea). Note. Compt. rend. 254(16): 3040-3042. (In French).

Martoja, R. 1964. Biochemical and histochemical data on the incorporation of radioactive sodium sulfate with Gryllus bimaculatus de Geer (Insecta, Orthoptera). Compt. rend. hebd. Séanc. 258(13): 3550-3553. (In French).

Martz, F. A., et al. 1971. Activation analysis of ^{197}Au for measuring rate of ingesta passage in ruminants. J. Dairy Sci. 54(6): 926-928. (NSA 25: 37333).

Maruyama, T., E. Tanaka and T. Hiramoto. 1971. Role of low-background beta-ray spectrometers for rapid assay of environmental samples. pp. 357-371. In: Rapid Methods for Measuring Radioactivity in the Environment. (See IAEA, 1971) (NSA 26: 28230).

Marzocchi, A., et al. 1971. Disposal of the radioactive wastes at the Trisaia Nuclear Research Center. Minerva Fisiconucl. 15(1): 38-45. (In Italian) (NSA 26: 26140).

Mascia, C. 1967. Research on the presence and distribution of ^{90}Sr in bones of birds. Minerva Fisiconucl. 11: 172-177. (In Italian) (NSA 24: 2551).

Mashiko, K. and K. Jozuka. 1962. Studies on the calcium uptake by teleost fishes. I. ^{45}Ca uptake by the crucian carp. Sci. Rept. Kanazawa Univ. (Japan) 8: 107-126. (NSA 20: 20405).

Mashiko, K. and K. Jozuka. 1964. Absorption and excretion of calcium by teleost fishes with special reference to routes followed. Annot. Zool. Jap. 37: 41-50. (NSA 20: 42750).

Mashneva, N. I. and S. Ya. Sukal'skaya. 1971. Kinetics of accumulation and biological activity of iodine-131 during the development of whitefish and carp. Gidrobiol. zh. 7(3): __. (Hydrobiol. J. 7(3): 29-34).

Mashneva, N. I. and S. Ya. Sukal'skaya. 1973. Experimental study of the injuring effect of a mixture of fission-product radionuclides on the embryonic development of freshwater fishes according to the dose received. pp. 45-49. In: Radioecology of Water Organisms. III. (See Andrumaitis, 1973) (In Russian) (NSA 30: 24376).

Maslov, V. I. 1965. Specific features of population inhabitance of wild animals in the areas of increased uranium and radium soil content. In: Abstr. Communications Delivered at the All-Union Inter-College Conf. on the Problem of Microelements and Natural Radioactivity. Vol. 3. Petrozavodsk. (In Russian).

Maslov, V. I. 1971. Animal ecology study methods in regions of elevated radiation. pp. 128-140. In: Methods of Radioecological Investigations. (See Verkhovskaya, 1971) (In Russian).

Maslov, V. I. 1972. Conducting complex radioecological investigations in biogeocenoses with a high radioactive background. pp. 9-21. In: Radioecological Studies in Natural Biogeocenoses. (See Verkhovskaya, 1972) (In Russian) (NSA 26: 56094).

Maslov, V. I. 1972. Introduction. pp. 7-_. In: Radioecological Studies in Natural Biogeocenoses. (See Verkhovskaya, 1972) (In Russian).

Maslov, V. I. 1972. Radiation conditions in the homes and refuges of muriform rodents in biogeocenoses with a high natural radiation background. pp. 216-226. In: Radioecological Studies in Natural Biogeocenoses. (See Verkhovskaya, 1972) (In Russian) (NSA 26: 56455).

Maslov, V. I. 1972. Radioecology of grouse in biogeocenoses of thorium regions. pp. 191-215. In: Radioecological Studies in Natural Biogeocenoses. (See Verkhovskaya, 1972) (In Russian) (NSA 26: 26554).

Maslov, V. I. and K. I. Maslova. 1972. Radioecological groups of mammals and birds in biogeocenoses of regions with a high natural radioactive background. pp. 161-173. In: Radioecological Studies in Natural Biogeocenoses. (See Verkhovskaya, 1972) (In Russian) (NSA 26: 56552).

Maslov, V. I. and K. I. Maslova. 1972. Some problems of the radioecology of the otter in biogeocenoses of thorium regions. pp. 173-191. In: Radioecological Studies in Natural Biogeocenoses. (See Verkhovskaya, 1972) (In Russian) (NSA 26: 56553).

Maslov, V. I., K. I. Maslova and I. N. Verkhovskaya. 1967. Characteristics of the radioecological groups of mammals and birds of biogeocoenoses with high natural radiation. pp. 561-571. In: Radioecological Concentration Processes. (See Aberg and Hungate, 1967).

Mason, B. J., H. W. Hop and C. L. Miller. 1973. Transfer of tritium from methane to vegetation. pp. 455-461. In: Tritium. (See Moghissi and Carter, 1973) (NSA 28: 27318).

Mason, B. J., W. J. Wipper and V. D. Leavitt. 1969. Tritium uptake following a thermonuclear test. pp. 167-172. In: Environmental Contamination by Radioactive Materials. (See IAEA, 1969).

Mason, B. J., et al. 1973. Desert vegetation uptake of tritium from Project Gasbuggy effluent. pp. 177-182. In: Radionuclides in Ecosystems. (See Nelson, 1973) (NSA 28: 27256).

Mason, B. J., et al. nd. Desert vegetation uptake of tritium from Project Gasbuggy effluent. Western Environmental Research Lab., U. S. EPA, U. S. AEC report PNE-G-78. 15pp. (NSA 26: 4742).

Mason, H. C., M. C. Papageorgiou and S. W. Jacklin. 1965. Mating preference and reproduction of irradiated Drosophila melanogaster. Abstr. 83. Bull. Entomol. Soc. Am. 11(3): 159.

Mason, H. C., M. C. Papageorgiou and S. W. Jacklin. 1969. Effect of irradiation on mating preference and reproduction suppression of Drosophila melanogaster. J. Econ. Entomol. 62(4): 818-821. (NSA 23: 43930).

Mason, W. H. and E. P. Odum. 1969. The effect of coprophagy on retention and bioelimination of radionuclides by detritus-feeding animals. pp. 721-724. In: Symposium on Radioecology. (See Nelson and Evans, 1969).

Massachusetts Institute of Technology. 1970. Man's Impact on the Global Environment. Assessment and Recommendations for Action. MIT Press, Cambridge, Mass. 341pp. (NSA 30: 9500) (Incl. radioactive waste).

Massaux, F. and C. Misse. 1974. Study of radicular absorption of the cacao tree using a radioactive tracer under South Cameroon ecological conditions. Cafe, Cacao, The 18(4): 225-236. (In French) (NSA 32: 11883) (^{32}P).

Mastinu, G. G. and G. P. Santaroni. 1973. About the problem of radioactivity content in drinking mineral waters. Abstr. p. 97. In: Proc. 2nd Conf. on the State of Advancement of Radioecology in Italy. (See Anonymous, 1973) (In Italian).

Mastinu, G. G. and G. P. Santoroni. 1974. Natural radioactivity of mineral waters. G. Fis. Sanit. Prot. Radiaz. 18(1-2): 9-13. (In Italian) (NSA 31: 19586).

Masubuchi, M. 1974. Early replicating DNA in heterochromatin of Plagiochila ovalifolia (liverworts). Shokubutsugaku Zasshi 87(10007): 229-235. (NSA 33: 3334) (^{3}H-thymidine autoradiography).

Masui, K., et al. 1953. Bacteria colonies in soil of the area of explosion. Nippon Gakujutsu 1: 241-247.

Mater, J. 1957. Chemical effects of high energy irradiation of wood. Forest Prod. J. 7(6): 208-209.

Mather, J. D. 1968. Literature survey of the use of radioisotopes in groundwater studies. Water Supply Papers, Inst. Geol. Sci., National Environmental Research Council, London. 33pp. (NSA 25: 37333).

Mathew, R. and B. K. Gaur. 1973. Radiation stimulation of germination in dormant cocklebur (Xanthium stramonium) seeds. Stimulation Newsletter (5): 43-47. (NSA 30: 32427).

Mathies, J. B. 1972. Annual consumption of cesium-137 and cobalt-60 labeled pine seeds by small mammals in an oak-hickory forest. Thesis, Michigan State Univ., East Lansing. xviii, 213pp. (Univ. Microfilms 72-8740) (Also: Mathies, J. B., et al., Oak Ridge National Lab., U. S. AEC report ORNL-TM-3912) (NSA 26: 28529; 27: 9203).

Mathur, R. P., P. Khanna and I. P. Saraswat. 1968. Radio-isotope (^{32}P) and water quality. J. Inst. Eng. (India), Publ. Hlth. Eng. Div. 48: 206-212. (NSA 23: 50048).

Mathur, R. S. 1963. Cytochemistry of the normal and x-irradiated spermatogenesis of *Stenobothrus viridulus* (Fischer). Cytologia (Tokyo) 28(2): 113-117. (NSA 18: 19615).

Matin, A. S. M. A. 1975. Possible development of radioresistance in *Sitophilus oryzae* (L.) following exposures of successive generations to ionizing radiation. Int. J. Appl. Radiat. Isotopes 26(10): 625-626. (NSA 33: 9590).

Matlack, G. M., et al. 1975. Plutonium release rates from ^{238}Pu fuels exposed to aquatic environments. Trans. Amer. Nucl. Soc. 22: 100-101. (NSA 33: 20468).

Matousek, J., F. Mertl and S. Patocka. 1962. Some data from the measurement of artificial radioactivity of external environment. Part II. Plzensky Lekarsk. Sb. (19): 43-53. (In Czech) (NSA 17: 37359).

Matousek, J., F. Mertl and S. Patocka. 1962. Some observations dealing with artificial radioactivity of the external environment. Plzensky Lekarsk. Sb. (18): 57-63. (In Czech) (NSA 17: 37358).

Matsubaya, O., et al. 1971. Isotopic study of hot springs in Japan. 1. Techniques for oxygen isotopic analysis of spring waters. Okayama Daig. Onsen Kenkyu. Hokoku (40): 33-40. (In Japanese) (NSA 26: 17406).

Matsue, Y. and R. Hirano. 1956. Accumulation of radioactivity in plankton cultured or kept in water infusions of the Bikini ashes. pp. 1099-1104. In: *Researches in the Effects and Influences of the Nuclear Bomb Test Explosion*. (See Japan Soc. for the Promotion of Science, 1956).

Matsumoto, S. and Y. Kagami-Ishi. 1970. Temperature dependence of mortality rate of radiosensitive strains of *E. coli* and *S. cerevisae*. Jap. J. Genet. 45: 153-160. (In Japanese) (NSA 24: 48749).

Matsumoto, T., et al. 1959. The lethal effect of ^{60}Co γ-rays on growing hen's egg. Radioisotopes (Tokyo) 8: 52-55. (In Japanese) (NSA 15: 22139).

Matsumura, F. and G. M. Boush. 1966. Malathion degradation by *Trichoderma viride* and a *Pseudomonas* species. Science 153: 1278-1280. (^{14}C).

Matsumura, F. and M. Hayashi. 1965. Absorption of C^{14} dieldrin to the nerve components of the German cockroach. Abstr. 53. Bull. Entomol. Soc. Am. 11(3): 157.

Matsumura, F. and M. Hayashi. 1966. Dieldrin: interaction with nerve components of cockroaches. Science 153: 757-759. (^{14}C).

Matsumura, F. and M. Hayashi. 1966. Interaction of dieldrin with the subcellular components of both resistant and susceptible straines of *Aedes aegypti*. Mosquito News 26(2): 190-194. (^{14}C).

Matsumura, F. and C. J. Hogendijk. 1964. The enzymatic degradation of malathion in organophosphate resistant and susceptible strains of *Musca domestica*. Entomologia exptl. appl. 7: 179-193. (^{32}P, ^{14}C).

Matsumura, F. and C. J. Hogendijk. 1964. The enzymatic degradation of parathion in organophosphate-susceptible and -resistant houseflies. J. Agric. Food. Chem. 12(5): 447-453. (^{32}P).

Matsumura, F. and R. D. O'Brien. 1966. Absorption and binding of DDT by the central nervous system of the American cockroach. J. Agric. Food Chem. 14(1): 36-39. (^{14}C).

Matsumura, F. and R. D. O'Brien. 1966. Interactions of DDT with components of American cockroach nerve. J. Agric. Food Chem. 14(1): 39-43. (^{14}C).

Matsumura, F. and G. Voss. 1964. Mechanism of malathion and parathion resistance in the two-spotted spider mite, *Tetranychus urticae* . J. Econ. Entomol. 57(6): 911-917. (^{14}C, ^{3}H).

Matsumura, F. and G. Voss. 1965. Properties of partially purified malathion carboxyesterase of the two-spotted spider mite. J. Insect Physiol. 11(2): 147-160. (^{14}C, ^{3}H, *Tetranychus urticae*).

Matsumura, S., T. Fujii and S. Kondo. 1960. Co^{60}-irradiator and its use for plant genetics. pp. 179-193. In: *Large Radiation Sources in Industry*. (Conf. Proc., Warsaw, Poland, 8-12 September 1959). International Atomic Energy Agency, Vienna, Publ. STI/PUB/12. (NSA 15: 83).

Matsunami, T., A. Mizohata and T. Mamuro. 1974. Some attempts to interpret the variations with time of monthly depositions of ^{210}Pb, ^{7}Be, ^{137}Cs and ^{144}Ce. Ann. Rept. Radiat. Cen. Osaka Prefect. 15: 24-29. (NSA 32: 19746) (Incl. soil samples).

Matsuoka, S. 1963. Structure of the chlorophyll-less mutant of *Euglena* cell by neutron irradiation. Acta Scholae Med. Gifu 10: 95-99. (NSA 19: 8805).

Matsuoka, S. 1964. Effects of radiation on the growth of *Euglena* cells. Bull. Inst. Chem. Res. Kyoto Univ. 42(1): 1-9. (NSA 18: 17501; 21: 11067).

Matsuoka, S. 1965. Effect of gamma-irradiation on the RNA content of *Euglena* cells. Bull. Inst. Chem. Res. Kyoto Univ. 43: 17-21. (NSA 19: 33826).

Matsuoka, S. 1966. Diffusion of some nucleotides into outer medium from non-irradiated and gamma irradiated *Euglena* cells. Bull. Inst. Chem. Res. Kyoto Univ. 44: 57-65. (NSA 20: 28904).

Matsuoka, S. 1967. Effect of gamma irradiation on uracil-^{14}C incorporation into *Euglena* cells. Bull. Inst. Chem. Res. Kyoto Univ. 45: 21-29. (NSA 21: 22823).

Matsuoka, S. 1967. Lethal effect of gamma irradiation on Euglena cells in two different growth phases. Bull. Inst. Chem. Res. Kyoto Univ. 45: 12-20. (NSA 21: 22822).

Matsuoka, S. 1967. Effects of gamma irradiation on the protein content of Euglena cells and on the alanine-^{14}C incorporation into the cells. Bull. Inst. Chem. Res. Kyoyo Univ. 45: 30-42. (NSA 21: 22824).

Matsuoka, S. 1972. Response of some nucleotides in vivo to gamma radiation on Euglena cells in two different growth phases. Bull. Inst. Chem. Res. Kyoto Univ. 50(1): 38-44. (NSA 26: 33597).

Matsusaka, N., Y. Nishimura and R. Ichikawa. 1972. Incorporation of radioactive cobalt into the eggs of domestic fowls. J. Radiat. Res. (Japan) 13(3): 156-162. (NSA 26: 56550) (Incl. Coturnix).

Matsuura, H., S. Tanifuji and M. Iwanuchi. 1961. Effects of past temperature. Treatments on the x-ray induced chromosome aberrations in the PMCS of Trillium (preliminary report). Jap. J. Genet. 36: 419-422. (NSA 17: 1303).

Matsuyama, A., Y. Okazawa and M. Namiki. 1959. Microbial growth and radiations. Hakko Kyokaishi 17: 346-357; 561-572.

Matsuyama, A., M. J. Thornley and M. Ingram. 1964. The effect of freezing on the radiation sensitivity of vegetative bacteria. J. Appl. Bacteriol. 27(1): 110-124. (NSA 18: 33326).

Matsuyama, A., et al. 1973. Isolation of radioresistant vegetative bacteria from the high background radioactivity area. Sci. Pap. Inst. Phys. Chem. Res. (Japan) 67(1): 35-40. (NSA 29: 5294).

Matter, B. E. 1967. Hydrogen sulfide-radiomimetic effects and radioprotective effects on Drosophila eggs. Radiol. Clin. Biol. 36: 299-307. (In German) (NSA 22: 28604).

Mattern, F. C. M. and L. Strackee. 1966. Investigation of radioactive materials in surface waters and precipitation. Measured results in 1965. Netherlands Rijks Inst. Volksgezondheid, Utrecht. 22pp. (In Dutch) (NSA 21: 25997).

Matthews, P. O. 1964. A study of two-hit aberrations in Haplopappus as scored at metaphase. Thesis, Univ. of Kansas, Lawrence. 66pp. (NSA 18: 29246).

Mattick, W. 1973. Nuclear Energy and Environment. Brown, Boveri, Mannheim. vp. (In German) (NSA 28: 32104).

Mattoni, R. H. T. 1968. Space-flight effects and gamma radiation interaction on growth and induction of lysogenic bacteria, a preliminary report. In: Symp. on the Biosatellite II Experiments. BioScience 18(6): 602-608. (NSA 22: 47645).

Mattoni, R. H. T., et al. 1971. Induction of lysogenic bacteria in the space environment. U. S. NASA (?) report N-72-20063. 16pp. (NSA 26: 53364).

Mattsson, L. J. S. 1972. ^{137}Cs in the reindeer lichen Cladonia alpestris; deposition, retention and internal distribution, 1961-1970. Radiofysiska Inst., Lund Univ., Sweden, report LURI-1972-8. 50pp. (NSA 28: 30377).

Mattsson, L. J. S. 1973. ^{155}Eu, ^{144}Ce, ^{125}Sb, ^{106}Ru,^{95}Zr, ^{54}Mn, and ^{7}Be in the reindeer lichen Cladonia alpestris: deposition, retention and internal distribution, 1961-1970. Radiofysiska Inst., Lund Univ., Sweden, report LURI-1972-9. 50pp. (NSA 29: 403).

Mattsson, L. J. S. 1975. ^{137}Cs in the reindeer lichen Cladonia alpestris: deposition, retention and internal distribution, 1961-1970. Health Physics 28(3): 233-248. (NSA 32: 3835).

Mattsson, L. J. S. 1975. Deposition, retention, and internal distribution of ^{155}Eu, ^{144}Ce, ^{125}Sb, ^{106}Ru, ^{95}Zr, ^{54}Mn, and ^{7}Be in the reindeer lichen Cladonia alpestris, 1961-1970. Health Physics 29(1): 27-41. (NSA 32: 28327).

Mattsson, L. J. S. and R. B. R. Persson. 1971. Radioecology of ^{22}Na, ^{210}Po, and ^{210}Pb in the unique food-chain: lichen-reindeer-man. pp. 1135-1151. In: Proc. Intern. Symp. on Radioecology Applied to the Protection of Man and His Environment. (See Commission of the European Communities, 1971) (NSA 27: 5285).

Matulyavichene, N. I. 1970. Coefficients of accumulation of the radioisotope ^{59}Fe by certain species of freshwater plants. Radiobiologiya 10: 461-463. (AEC-tr-7193, pp. 220-225) (NSA 24: 50697).

Matulyavichene, N. I. and V. P. Matulyavichyus. 1969. Peculiarities of the water basin model used in investigating the accumulation of radioactive substances with the help of ecosystem components. Lietvos TSR Mokslu Akad. Darbai B(3): 193-198. (In Russian) (NSA 23: 26714).

Matulyavichene, N. I., V. P. Matulyavichyus and B. I. Styro. 1968. About some methods which may be used in investigating the dynamics of accumulation of radioactive isotopes in a freshwater ecosystem. Lietvos TSR Mokslu Akad. Darbai B(3): 187-191. (In Russian) (NSA 23: 26713).

Matutano, J., P. Hory and L. Grivaud. nd. Continuous measurement of α and β radioactivity of water in view of maximum permissible concentrations for the public. Centre d'Études Nucléaires, France Commissariat à l'Énergie Atomique, Saclay, report CEA-CONF-2007. 11pp. (For 6th Intern. Congr. on New Trends in Radiation Protection, Bordeaux, France, 27 March 1972) (In French) (CONF-720328-1) (NSA 26: 48031).

Matuszek, J. M., et al. 1974. Environmental levels of ^{129}I. pp. 3-20. In: Environmental Surveillance Around Nuclear Installations. Vol. 2. (See IAEA, 1974) (NSA 31: 14030).

Matuszek, J. M., et al. 1974. Permanent gas measurements as part of an environmental surveillance program. pp. 225-233. In: Environmental Surveillance Around Nuclear Installations. Vol. 1. (See IAEA, 1974) (NSA 31: 14061).

Mauchline, J., A. M. Taylor and E. B. Ritson. 1964. The radioecology of a beach. Limnol. Oceanogr. 9(2): 187-194. (NSA 18: 21762; SFA 9: 6651).

Maurin, V. and J. Zötl. 1967. Specialists' conference on the tracing of subterranean waters held in Graz, March 28 -April 1, 1966. Steirische Beiträge zur Hydrogeologie Jahrgang 1966/67. Leykam AG, Graz. (NSA 22: 27915).

Maurav'yev, M. N. and A. B. Tsypin. 1967. Action of gamma rays of ^{60}Co on the automatic activity of the lymphatic hearts of the frog. Dokl. Akad. Nauk SSSR 177: 479-482. (Doklady (biol. sci.) 177(1-6): 799-802) (NSA 22: 31132).

Mauro, J. J. 1973. Accumulation of ^{137}Cs by Fundulus heteroclitus in the Hudson River Estuary. 131pp. Thesis, New York Univ., New York. (Univ. Microfilms 74-16,845) (NSA 30: 26808).

Mauro, J. J. and M. E. Wrenn. 1974. Investigation into the reasons for a lack of a trophic level effect for ^{137}Cs in Fundulus heteroclitus and its food in the Hudson River Estuary. pp. 285-290. In: Proc. 3rd Intern. Congr. Intern. Radiation Protection Assoc. (CONF-730907) (NSA 30: 27263).

Maurus, P., et al. 1970. Studies on the varying ^{137}Cs contamination of wild game. Zentralbl. Veterinärmed. Beih. (11): 135-141. (In German) (NSA 25: 24138).

Mautz, W. W. 1967. Determination of the usefulness of chromium-51 in digestive studies of the white-tailed deer. Thesis, Michigan State Univ., East Lansing.

Mautz, W. W. 1970. Investigation of some digestive parameters of the white-tailed deer using the radioisotope ^{51}chromium. Thesis, Michigan State Univ., East Lansing. 81pp. (BA 30(12B); WR 140: 102).

Mautz, W. W. 1971. Comparison of the $^{51}CrCl_3$ ratio and total collection techniques in digestibility studies with a wild ruminant, the white-tailed deer. J. Animal Sci. 32(5): 999-1002.

Mautz, W. W. 1971. Confinement effects on dry-matter digestibility coefficients by deer. J. Wildl. Mgmt. 35(2): 366-368. (Used ^{51}Cr).

Mautz, W. W. and G. A. Petrides. 1967. The usefulness of chromium-51 in digestive studies of the white-tailed deer. Trans. N. Amer. Wildl. & Nat. Resources Conf. 32: 420-429.

Mautz, W. W. and G. A. Petrides. 1971. Food passage rate in the white-tailed deer. J. Wildl. Mgmt. 35(4): 723-731. (Used ^{51}Cr).

Mavashev, B. Z. 1967. Anomalies in the radioactivity of the Tashkent mineral waters during 1965 and 1966. Dokl. Akad. Nauk Uzbek. SSR (5): 38-39. (In Russian) (NSA 22: 8211).

Mavljanova, M. I. 1966. On the reaction of infusoria of the genus Colpoda to ^{60}Co gamma radiations. Pedobiologia 6: 129-139. (In German) (NSA 21: 16543).

Mawson, C. A. 1956. Report on waste disposal system at the Chalk River Plant Plant of Atomic Energy of Canada Limited. Atomic Energy of Canada, Ltd., Chalk River, Ontario, report CRB-658. 15pp.

Mawson, C. A. 1961. Processing of radioactive wastes. Review Ser. - Collection Monogr. No. 18. Atomic Energy of Canada, Ltd., Chalk River, Ontario. 44pp.

Mawson, C. A. 1964. Survey of hydrological application of tracers in Canada. Atomic Energy of Canada, Ltd., Chalk River, Ontario, report AECL-2005. 14pp. (NSA 19: 7715).

Mawson, C. A. 1966. Quantitative aspects of environmental monitoring in Canada. pp. 261-273. In: Proc. All India Symp. on Radioactivity and Metrology of Radionuclides. (See India Atomic Energy Establishment, 1966).

Mawson, C. A. 1968. Environmental contamination from nuclear industry. pp. 121-126. In: Guidelines to Radiological Health. S. C. Skoryna and D. Waldron-Edwards, eds. McGill Univ., Montreal.

Mawson, C. A. 1970. Objectives of environmental surveillance: a Canadian viewpoint. pp. 12-13. In: Environmental Surveillance in the Vicinity of Nuclear Facilities. (See Reinig, 1970) (NSA 25: 16156).

Mawson, C. A. and A. E. Russell. 1960. Facilities for waste management at Chalk River, Canada. pp. 361-371. In: Disposal of Radioactive Wastes. Vol. 1. (See IAEA, 1960) (NSA 14: 22668).

Maxia, V., S. Meloni and N. Genova. 1970. Some radiochemical and activation analysis applications to ecological problems. pp. 139-143. In: Proc. 1st Conf. on the State of Advancement of Radioecology in Italy. (See Anonymous, 1970) (In Italian).

Maxwell, J. 1938. Inactivation of sperm by x-radiation in Habrobracon. Biol. Bull. 74(2): 253-255.

May, H. and L. D. Marinelli. 1964. Cosmic-ray contribution to the background of low-level scintillation spectrometers. pp. 463-480. In: The Natural Radiation Environment. (See Adams and Lowder, 1964) (NSA 18: 4081).

May, L. H. 1962. Review of use of radioisotopes in Australia in research on growth regulators and herbicides. Int. J. Appl. Radiat. Isotopes 13: 403-406.

May, J. T. and H. G. Posey. 1958. The effect of radiation by cobalt-60 gamma rays on the germination of slash pine seed. J. Forestry 56: 854-855.

May, M. J. and I. F. Stuart. 1970. Comparison of calculated and measured long term gamma doses from a stack effluent of radioactive gases. pp. 234-248. In: Environmental Surveillance in the Vicinity of Nuclear Facilities. (See Reinig, 1970) (NSA 25: 16176).

May, P. F., A. R. Till and M. J. Cumming. 1972. Systems analysis of ^{35}sulfur kinetics in pastures grazed by sheep. J. Appl. Ecol. 9(1): 25-49.

May, P. F., A. R. Till and M. J. Cumming. 1973. Systems analysis of the effect of application methods of the entry of sulfur into pastures grazed by sheep. J. Appl. Ecol. 10(2): 607-626. (^{35}S).

May, R. and H. Schneider. 1960. Artificial radioactivity in spring and running water. Atomkernenergie 5(9): 333-335. (In German) (NSA 15: 387).

May, R.-M., J. Guimard and J. P. Denefle. 1964. Comparison between the action of x-rays and detoxified venom from the snake Naja naja on adrenal cortical mitoses in the mouse. Arch. Anat. Microscop. Morphol. Exptl. 53: 57-68. (In French) (NSA 19: 8814).

May, R.-M., M. Maria and J. Guimard. 1964. Differential action of x and ultraviolet rays on the tardigrade Macrobiotes areolatus in the active and desiccated state. Bull. biol. Fr. Belg. 98(2): 349-367. (In French) (NSA 19: 38373).

Mayas, I. 1974. Effects of fractionation of the sterilizing dose of radiation on emergence, fertility, and competitivity of the Mediterranean fruit fly, Ceratitis capitata Wied. pp. 146-152. In: Isotopes and Pesticides and Pest Control. (See Nehme and Hassan, 1974) (In French) (NSA 33: 7563).

Mayas, I. A. 1975. Effects of fractionating the gamma-ray sterilization dose on the eclosion, fertility and competitivity of the Mediterranean fruit fly, Ceratitis capitata Wied. pp. 229-236. In: Sterility Principle for Insect Control, 1974. (See IAEA, 1974) (In French) (NSA 32: 11779).

Mayer, M. S. 1963. Biological and histopathological effects of gamma radiation on three life stages of Anthonomus grandis Boheman. Thesis, Texas A. &. M. Univ., College Station. 154pp. (Dissertation Abstr. 25: 43; NSA 18: 41216).

Mayer, M. S. and J. R. Brazzel. 1961. Certain biological effects produced in the boll weevil by tagging it with P^{32}. J. Econ. Entomol. 54: 1197-1203. (NSA 17: 1296).

Mayer, M. S. and J. R. Brazzel. 1966. Laboratory studies to sterilize the boll weevil with radiation. Ann. Entomol. Soc. Am. 59(2): 284-290.

Mayer, R. A., et al. 1973. Ecological and environmental thesaurus. Battelle Columbus Labs., U. S. AEC report BMI-X-649. 935pp. (NSA 28: 15478).

Maynard-Smith, J. 1962. The cause of ageing. Proc. Roy. Soc. (London) B157: 115-127. (Habrobracon).

Mayneord, W. V. 1959. Spectroscopic identification of alpha-emitting nuclides in biological material. Nature 184(4687): 667-669. (NSA 14: 1322).

Mayneord, W. V. 1964. Background radiations and their origins. Advanc. Sci. 21: 127-131. (NSA 19: 32423; 20: 5409) (Soils, air, water, foods, man).

Mayneord, W. V. 1965. Man's natural radiation environment. pp. 1-12. In: Radioactive Isotopes in the Hospital and Research. Urban u. Schwarzenberg, Munich-Berlin. Band VI. (NSA 19: 46688).

Mayneord, W. V. and C. R. Hill. 1969. Natural and man-made background radiation. pp. 401-451. In: Radiation Dosimetry. Vol. III. F. H. Attix, ed. Academic Press, New York. (NSA 25: 24162).

Mayneord, W. V., R. C. Turner and J. M. Radley. 1960. The alpha activity of certain botanical materials. Nature 187(4733): 208-211. (NSA 14: 18766).

Mayo, J., et al. 1966. Multiplication of foot-and-mouth disease virus in irradiated rodents. Rev. Invest. Agropecuar. (ser. 4) 3(6): 58-69. (In Spanish) (NSA 22: 51429).

Maza, B. G., N. R. French and A. P. Aschwanden. 1973. Home range dynamics in a population of heteromyid rodents. J. Mammal. 54(2): 405-425. (Nevada Test Site).

Mazor, E., et al. 1974. Kalahari groundwaters: their hydrogen, carbon, and oxygen isotopes. pp. 203-225. In: Isotope Techniques in Groundwater Hydrology. Vol. 1. (See IAEA, 1974) (NSA 31: 19629).

Mazzotti, L., et al. 1961. Irradiation of scorpions of the species Centruroides limpidus with gamma rays. Rev. Inst. Salubridad Enferm. Trop. (Mex.) 21: 125-127. (In Spanish) (NSA 17: 38829).

Mead, C. G. 1964. A deoxyribonucleic acid-associated ribonucleic acid from Drosophila melanogaster. J. Biol. Chem. 239(2): 550-554. (^{32}P).

Mealing, H. G., Jr. (w/J. H. Horton; R. S. Harvey). 1957, 1958. Health physics regional monitoring semiannual report, January-June 1957. ---July-December 1957. Savannah River Plant, U. S. AEC report DPSP-57-25-43(Del.), DPSP-58-25-17(Del.). 43, 50pp. (NSA 30: 9509, 9510).

Medica, P. A., F. B. Turner and D. D. Smith. 1973. Effects of radiation on a fenced population of horned lizards (Phrynosoma platyrhinos) in southern Nevada. J. Herpetol. 7(2): 79-85. (Also: 1972, Univ. of California at Los Angeles, U. S. AEC report UCLA-12-886) (NSA 27: 522).

Medica, P. A., F. B. Turner and D. D. Smith. 1973. Hormonal induction of color change in female leopard lizards, Crotaphytus wislizenii. Copeia (4): 658-661. (Nevada Test Site).

Medina, F. I. S. 1975. Effects of radiation on field rats. Nucleus (Diliman, Quezon City) 13(4): 11, 17-18. (NSA 33: 17483) (Laboratory study).

Medina, F. I. S., et al. 1973. Observations on the induction of mutation by ionizing radiation in Rattus rattus ssp. (Journal not given in abstr.) 8(3): 28-30. (NSA 31: 14415) ("Field rats").

Medler, J. T. and R. O. Wagner. 1964. Isotopes to estimate colony size of Formica cinerea Mayr (Hymenoptera: Formicidae). J. N.Y. Entomol. Soc. 72 (3): 151-159. (^{198}Au).

Medvedev, Zh. A. 1958. Rapid determination of the radioactivity of proteins and of the relative proportions of various fractions of radioactive substances present in leaves. Biokhimiya 23(6): 801-808. (Biochemistry USSR 23: 759-766) (NSA 13: 16682).

Medvedev, Zh. A. and I. A. Chaban. 1968. The effect of ionizing radiation on the RNA synthesis in the nucleated erythrocytes of tadpoles and frogs. Tsitologiya 10(12): 1569-1579. (In Russian) (NSA 23: 12024).

Meek, B. D., A. J. MacKenzie and K. R. Stockinger. 1964. Evaluation of a radioactive tracer method for measuring water intake of soils. Soil Sci. Soc. Am. Proc. 28(2): 153-156.

Meeks, R. L. 1968. The accumulation of ^{36}Cl ring-labeled DDT in a freshwater marsh. J. Wildl. Mgmt. 32(2): 376-398.

Meeks, R. L. and T. J. Peterle. 1964. The cycling of Cl^{36} labeled DDT in a marsh ecosystem. p. 122. In: Pesticide-Wildlife Studies, 1963: a Review of Fish and Wildlife Service Investigations During the Calender Year. U. S. Fish and Wildlife Service, Washington, Cir. 199.

Meeks, R. L. and T. J. Peterle. 1967. The cycling of ^{36}Cl labeled DDT in a marsh ecosystem. Thesis, Ohio State Univ., Columbus. 234pp. (NSA 21: 34626).

Megumi, K. and T. Mamuro. 1972. Emanation of radon and thoron gases from soil. Ann. Rep. Radiat. Center Osaka Prefect. 13: 35-37. (NSA 28: 29879).

Megumi, K. and T. Mamuro. 1973. Radon and thoron exhalation from the ground. J. Geophys. Res. 78(11): 1804-1808. (NSA 28: 29878).

Mehas, T. C. nd. Columbia River continuous monitoring system. Hanford Atomic Products Operation, U. S. AEC report HW-SA-2171. 12pp. (NSA 15: 22557).

Mehas, T. C. 1961. An automatic monitoring station for the Columbia River. Hanford Labs., U. S. AEC report HW-SA-2126. 9pp. (NSA 16: 30673).

Mehendale, H. M. and E. Hodgson. 1965. Metabolism of carnitine in the blow fly, *Phormia regina*. Abstr. 100. Bull. Entomol. Soc. Am. 11(3): 159. (^{14}C).

Mehl, J. 1974. Objectives of radioactive effluent monitoring. pp. 10-15. In: *Proc. Seminar on the Monitoring of Radioactive Effluents*. (See Organization for Economic Co-operation and Development, 1974) (NSA 31: 3334).

Mehra, P., N. Palta and K. Harinder. 1969. Radiobiological investigations on *Pteris vittala* L. I. X-ray induced plastid variations in the gametophytes. Radiation Botany 9: 93-103. (NSA 23: 35691).

Mehrotra, K. N. and G. R. Sethi. 1966. Advances in insect population control by the sterile-male technique. J. Sci. India Res. 25: 539-543. (NSA 21: 22720).

Mehrotra, K. N., G. R. Sethi and Y. Prasad. 1968. Metabolism of malathion in the desert locust *Schistocerca gregaria* Forskal: distribution and degradation of ^{32}P-labelled malathion *in vivo*. pp. 187-198. In: *Isotopes and Radiation in Entomology*. (See IAEA, 1968).

Meier, H., et al. 1973. Anomalies of the natural isotopic abundance of ^{239}Pu. From 24th IUPAC Congr., Hamburg, Germany, 2 September 1973. vp. (In German) (NSA 29: 15559).

Meikle, R. W. 1964. The fate of sulfuryl fluoride in wheat flour. J. Agric. Food Chem. 12(5): 464-467. (^{35}S).

Meinertz, T. 1963. An investigation of cell division activity in *Axolotl* larvae and larvae of *Triton punctatus*, and also experiments with irradiation by ultraviolet light and constant electric light. Arch. Entwicklunsmech. Organ. 154: 513-525. (In German) (NSA 19: 17511).

Meinertz, T. 1963. An investigation of the cell division activity of x irradiated larve of *Triton punctatus*. Arch. Entwicklungsmech. Organ. 154: 195-215. (In German) (NSA 19: 15044).

Meinhof, W. and A. Wiskemann. 1965. Investigations on the effect of beta rays on dermatophytes. Arch. Klin. Exp. Dermatol. 223: 199-216. (In German) (NSA 21: 14774).

Meinke, W. W. 1973. Is radiochemistry the ultimate in trace analysis? Pure Appl. Chem. 34(1): 93-104. (NSA 28: 5131).

Meisner, H. M. 1964. Influence of tonicity on mitotic recovery in x-irradiated grasshopper neuroblasts. Thesis, Emory Univ., Atlanta, Ga. 77pp. (Dissertation Abstr. 25: 2182-2183; NSA 19: 13119).

Meisner, H. M. 1965. Influence of tonicity on mitotic recovery in x-irradiated grasshopper neuroblasts. Physiol. Zool. 38: 276-288.

Mejdahl, V. 1969. Thermoluminescence dating of ancient Danish ceramics. Archeometry 11: 99-104. (NSA 23: 35691).

Mejdahl, V. 1970. Measurement of environmental radiation at archeological excavation sites. Archeometry 12: 147-159. (NSA 25: 2352).

Mejdahl, V. 1971. Dating of pre-historical pottery by thermoluminescence. Kosmos (Stockholm) 48: 57-60. (In Danish) (NSA 28: 21132).

Mejdahl, V. 1971. Progress in TL dating at Risö. pp. 930-947. In: *3rd Intern. Conf. on Luminescent Dosimetry*. Denmark Atomic Energy Commission, Risö, report RISÖ-249(Pt.3). (NSA 27: 17285) (Archeological).

Mejdahl, V. 1972. Dosimetry techniques in thermoluminescence dating. Denmark Atomic Energy Commission, Risö, report RISÖ-261. 46pp. (NSA 27: 7629) (Archeological).

Mejstrik, V. 1970. Uptake of ^{32}P by different kinds of ectotrophic Mycorrhiza of *Pinus*. New Phytol. 69: 295-298. (NSA 25: 8603).

Melin, E. and H. Nilsson. 1952. Transport of labelled nitrogen from an ammonium source to pine seedlings through mycorrhizal mycelium. Sv. Bot. Tids. 46: 281-285.

Melin, E. and H. Nilsson. 1954. Transport of labelled phosphorus to pine seedlings through the mycelium of *Cortinarius glaucopus* (Shaeff, Ex. Fr.). Sv. Bot. Tids. 48: 555-558.

Melin, E. and H. Nilsson. 1955. Ca-45 used as indicator of transport of cations to pine seedlings by means of mycorrhizal mycelium. Sv. Bot. Tids. 49: 119-122.

Melin, E. and H. Nilsson. 1957. Transport of C-14 labelled photosynthate to the fungal associate of pine Mycorrhizae. Sv. Bot. Tids. 51: 166-186.

Melis, A. 1962. The protection of the olive tree at the present time. Redia 47: 1-79. (Dacus oleae).

Melis, A. 1963. Nuclear energy control of Dacus oleae. Agricoltore Ital. 9(4): 11-18.

Melis, A. and B. Baccetti. 1960. Old and new experimental methods against the principle phytophagous pests of the olive in Tuscany in 1960. Redia 45: 193-217. (In Italian) (Concerns sterile-male technique).

Mellado, L., M. Arroyo and P. Ros. 1974. Control of Ceratitis capitata Wied. by the sterile-male technique in Spain. pp. 63-73. In: Sterile-Insect Technique and Its Field Applications. (See IAEA, 1974) (NSA 31: 8941).

Mellanby, K. 1967. Pesticides and Pollution. Collins, London. (NSA 21: 41431) (Incl. radiation).

Mellen, W. J. 1968. ^{131}I conversion ratio in domestic birds. Gen. Comp. Endocrinol. 10: 315-329. (NSA 22: 51451) (Incl. quail).

Mellinger, P. J. 1972. The comparative metabolism of cadmium, mercury and zinc as environmental contaminants in the freshwater mussel, Margaritifera margaritifera. Dissertation, Oregon State Univ., Corvallis. (Tracers).

Mellinger, P. J. 1973. The comparative metabolism of two mercury compounds as environmental contaminants in the freshwater mussel, Margaritifera margaritifera. pp. 173-180. In: Trace Substances in Environmental Health - VI. (See Hemphill, 1973).

Mellinger, P. J. and V. N. Smith. 1973. An instrumental neutron activation method for the determination of mercury: with comparison to atomic absorption spectrophotometry. pp. 250-256. In: Mercury in the Western Environment. (See Buhler, 1973) (Incl. sea lions).

Mel'nikova, M. K. 1971. Radioactive isotope application for investigation into migration of moisture and solid particles in soils, mobility of mineral nutrients and their accessibility to plants. pp. 160-167. In: Application of Isotopes and Nuclear Radiation in Agriculture. II. Atomizdat, Moscow. (In Russian) (NSA 30: 21161).

Mel'nikova, M. K. and S. V. Kovenya. 1971. Use of radioactive tracers for simulating the lessivage process. Pochvovedeniye 5: __. (Sov. Soil Science 5: 611-618.

Melville, C. 1958. An apparent beneficial effect of gamma-radiation on the flour mite. Nature 181(4620): 1403-1404. (NSA 12: 10311).

Mendel, J. E. 1973. Review of leaching test methods and the leachability of various solid media containing radioactive wastes. Battelle Northwest Lab., U. S. AEC report BNWL-1765. 50pp. (NSA 28: 29758).

Mendelson, D. 1973. Effect of caffeine on a repair system in Drosophila melanogaster. Genen Phaenen 16(3): __. (NSA 29: 18648) (X-radiation).

Mendgen, K. 1975. Tracer techniques for the study of host-parasite relations. pp. 73-77. In: Tracer Techniques for Plant Breeding. International Atomic Energy Agency, Vienna, Publ. STI/PUB-419. (NSA 33: 26383).

Mengle, D. C. and L. L. Lewallen. 1964. Metabolism of P^{32} labeled parathion in susceptible and phosphate-resistant strains of the mosquito Aedes nigromaculis (Ludlow). Abstr. Bull. Entomol. Soc. Am. 10(3): 165.

Mengle, D. C. and L. L. Lewallen. 1966. Biochemical-radiological determinations of parathion resistance in Aedes nigromaculis. J. Econ. Entomol. 59: 743-744. (^{32}P).

Menn, J. J. and J. B. McBain. 1964. Metabolism of phthalimidimethyl O,O-dimethylphosphorodithioate (Imidan) in cotton plants. J. Agric. Food Chem. 12(2): 162-166. (^{14}C).

Menn, J. J., J. B. McBain and I. M. Ford. 1965. Metabolism of phthalimidimethyl-O,O-dimethylphosphorodithioate (Imidan) in water, insects and the rat. p. 486. In: Proc. 12th Intern. Congr. of Entomology. P. Freeman, ed. Royal Entomological Soc., London. (^{14}C, Blattella germanica).

Menon, M. and K. K. Nair. 1972. Sex pheromone productive behavior in gamma-irradiated Tenebrio molitor. J. Insect Physiol. 18: 1323-1331. (NSA 26: 48274).

Menoux, A., P. Bovard and A. Doury. 1963. Arrangements for the monitoring of environmental radioactivity in conjunction with safety surveys of nuclear sites. pp. 79-88. In: Siting of Reactors and Nuclear Research Centres. International Atomic Energy Agency, Vienna. (In French) (NSA 17: 41083).

Mentzer, C. 1957. On a method of administration of radioactive solutions to des végéteaux ligneux. Compt. rend. 245(25): 2354-2355. (In French).

Menzel, D. B., et al. 1961. The metabolism of C^{14}-labeled DDT in the larvae, pupae and adults of Drosophila melanogaster. J. Econ. Entomol. 54(1): 9-12.

Menzel, R. G. 1965. Soil-plant relationships of radioactive elements. In: Radiation and Terrestrial Ecosystems. Health Physics 11(12): 1325-1332. (See Hungate, 1965).

Menzel, R. G. 1967. Airborne radionuclides and plants. pp. 57-75. In: Agriculture and the Quality of the Environment. (See Brady, 1967) (NSA 23: 22301).

Menzel, R. G. 1968. Uranium, radium, and thorium content in phosphate rocks and their possible radiation hazard. J. Agric. Food Chem. 16(2): 231-234. (NSA 22: 21193).

Menzel, R. G. 1974. Land surface erosion and rainfall as sources of ^{90}Sr in streams. J. Environ. Qual. 3(3): 219-223. (NSA 31: 16536).

Menzel, R. G. and P. E. James. 1971. Treatments for farmland contaminated with radioactive material. U. S. Agricultural Research Service, Beltsville, Md., Agricultural Handbook No. 395, U. S. AEC report TID-25845. 18pp. (NSA 26: 48274).

Menzer, R. E. and J. E. Casida. 1965. Nature of toxic metabolites formed in mammals, insects, and plants from 3-(dimethoxyphosphinyloxy)-N-N-dimethy cis-crotonamide and its N-methyl analog. J. Agric. Food Chem. 13(2): 102-112. (^{32}P, ^{14}C).

Mergen, F. 1963. Evaluation of spontaneous, chemical and radiation-induced mutations in the Pinaceae. Brookhaven National Lab., U. S. AEC report BNL-6896. 34pp. (NSA 17: 19804).

Mergen, F. 1964. Evaluation of spontaneous, chemical and radiation-induced mutations in Pinaceae. World Consultation Forest Genet. Tree Improvement Proc., Stockholm, Aug. 23-30, 1963. Food and Agric. Organ., United Nations, Rome. 14pp. (NSA 21: 6725).

Mergen, F. 1966. Effect of ionizing radiation on the reproductive capacity of a pine-oak forest. Brookhaven National Lab., U. S. AEC report BNL-9888. 13pp., 9 figs.

Mergen, F. and J. Cummings. 1965. Germination of *Pinus rigida* seeds after gamma irradiation. Effects of post-irradiation treatments, moisture content and individual trees. Radiation Botany 5(1): 39-51. (NSA 19: 15177).

Mergen, F. and T. S. Johansen. 1963. Effect of ionizing radiation on microsporogenesis in *Pinus rigida* Mill. Radiation Botany 3(4): 321-331. (Also: Brookhaven National Lab., U. S. AEC report BNL-6991).

Mergen, F. and T. S. Johansen. 1964. Effect of ionizing radiation on seed germination and seedling growth of *Pinus rigida* (Mill). Radiation Botany 4(4): 417-427. (NSA 19: 15165).

Mergen, F. and B. A. Simpson. 1967. Effect of ionizing radiation on megasporogenesis on *Pinus rigida* Mill. Radiation Botany 7(3): 247-252. (BA 48: 118642).

Mergen, F. and G. R. Stairs. 1961. Progeny test from a pitch pine-oak forest, damaged by low level chronic gamma radiation. Brookhaven National Lab., U. S. AEC report BNL-6466. (NSA 17: 2819).

Mergen, F. and G. R. Stairs. 1962. Comulative radiation effects on sexual reproduction in pine and oak. So. Forest Tree Improvement Comm. Publ. 22: 79-87. (Also: Brookhaven National Lab., U. S. AEC report BNL-6467) (NSA 17: 2820).

Mergen, F. and G. R. Stairs. 1962. Low level chronic gamma irradiation of a pitch pine-oak forest - its physiological and genetical effects on sexual reproduction. Radiation Botany 2(3/4): 205-216. (Also: Brookhaven National Lab., U. S. AEC report BNL-6368) (NSA 17: 15730).

Mergen, F. and G. R. Stairs. 1963. Progeny from a pitch pine-oak forest damaged by low level chronic gamma radiation. pp. 3-8. In: *Proc. 10th* Conf. Northeast Forest Tree Improvement.

Mergen, F. and G. R. Stairs. 1970. The effect of ionizing radiation on the reproductive capacity of forest trees. Health Physics 19: 37-45. (NSA 24: 44627).

Mergen, F. and B. A. Thielges. 1965. Effects of differential chronic exposures to Co^{60} radiation on survival, height growth, productivity, and cytology of *Pinus rigida* seedlings. Brookhaven National Lab., U. S. AEC report BNL-9579. 18pp.

Mergen, F. and B. A. Thielges. 1966. Effects of chronic exposures to Co^{60} radiation on *Pinus rigida* seedlings. Radiation Botany 6(3): 203-210.

Mergen, F. and B. A. Thielges. 1967. Radiation-induced damage in vegetative tissues of *Pinus rigida* Mill. Cytologia 32: 248-254. (NSA 22: 40955).

Mericle, L. W. and R. P. Mericle. 1965. Reassessing the biological role of background terrestrial radiation as a constituent of the natural environment. In: *Radiation and Terrestrial Ecosystems*. Health Physics 11(12): 1607-1620. (See Hungate, 1965).

Mericle, L. W. anf R. P. Mericle. 1965. Biological discrimination of differences in natural background radiation level. Radiation Botany 5: 475-492. (Also: Michigan State Univ., East Lansing, U. S. AEC report COO-1400-4) (NSA 20: 8698).

Mericle, L. W., R. P. Mericle and W. S. Osburn. 1963. Is mutation rate significantly altered by five-fold differences in natural background radiation? Abstr. Radiation Research 19(1): 185.

Mericle, L. W., R. P. Mericle and W. S. Osburn. 1964. Factors associated with differences in radiation level discrimination. Abstr. Genetics 50 (2): 268.

Mericle, L. W., R. P. Mericle and W. S. Osburn. 1964. Somatic mutation rate as a biological discriminator of natural background radiation. Abstr. Radiation Research 22(1): 214-215.

Mericle, L. W., R. P. Mericle and A. H. Sparrow. 1962. Cumulative radiation damage in oak trees. Radiation Botany 2(3/4): 265-271. (Also: Brookhaven National Lab., U. S. AEC report BNL-6374) (NSA 17: 15734).

Merlini, M. 1967. The freshwater clam as a biological indicator of radiomanganese. pp. 977-982. In: *Radioecological Concentration Processes*. (See Aberg and Hungate, 1967).

Merlini, M. 1968. Trace element analysis of aquatic organisms. pp. 75-101. In: *Intern. Symp. on the Application of Neutron Activation in Oceanography*. (See Anonymous, 1968) (NSA 25: 15540).

Merlini, M. and R. Bittel. 1970. Uptake, loss, and distribution of radiocobalt in a freshwater fish, *Lepomis gibbosus* L. pp. 399-412. In: *Proc. Intern. Symp. on Radioecology*. (See French Commissariat à l'Énergie Atomique, 1970) (NSA 25: 2526).

Merlini, M., F. Girardi and G. Pozzi. 1967. Activation analysis in studies of an aquatic ecosystem. pp. 615-629. In: Nuclear Activation Techniques in the Life Sciences. (See IAEA, 1967) (NSA 23: 4576).

Merlini, M., O. Ravera and C. Bigliocca. 1969. Nondestructive determination of elements in specific freshwater microplankton by activation analysis. pp. 475-481. In: Modern Trends in Activation Analysis. Vol. I. (See DeVoe, 1969) (NSA 24: 162).

Merlini, M., et al. 1965. The stable manganese content of molluscs from Lake Maggiore determined by activation analysis. Limnol. Oceaogr. 10(3): 371-378. (NSA 19: 45788).

Merlini, M., et al. 1971. Trends in the concentration of heavy metals in organisms of a mesotrophic lake as determined by activation analysis. pp. 447-458. In: Nuclear Techniques in Environmental Pollution. (See IAEA, 1971) (NSA 25: 31883).

Merlini, M., et al. 1971. The effects of sublethal amounts of cadmium and mercury on the metabolism of zinc-65 by freshwater fish. pp. 1327-1344. In: Proc. Intern. Symp. on Radioecology Applied to the Protection of Man and His Environment. (See Commission of the European Communities, 1971) (NSA 27: 5568).

Merlini, M., et al. 1973. The biological pathway of zinc (^{65}Zn) in freshwater fish and its alteration by heavy metals. pp. 285-306. In: Radionuclides in Ecosystems. (See Nelson, 1973) (NSA 28: 27674).

Merlivat, L., et al. 1973. Tritium and deuterium content of the snow in Greenland. Earth Planet. Sci. Lett. 19(2): 235-240. (NSA 28: 27266).

Merriam, R. A. 1959. Nuclear probe compared with other soil moisture measurement methods. Forest Science, Forest Res. Notes 146, U. S. Dept. of Agriculture.

Merriam, R. A. 1960. Moisture sampling with a neutron probe. Iowa State J. Sci. 34(4): 641-648.

Merriam, R. A. and O. L. Copeland, Jr. 1963. Brief concluding comments about neutron probe sensitivity. Note. Soil Science 27(5): 604.

Merriman, J. 1973. Soil tests with nuclear density-moisture equipment: laboratory and field tests of probe-equipped surface-type nuclear density-moisture equipment for evaluation toward future use in earthwork control. U. S. Bureau of Reclamation, Denver, Colo., report REC-ERC-73-21. 36pp. (PB-231646) (NSA 31: 621).

Merritt, W. F. 1962. Routine measurements of ground-water velocity using S^{35}. Health Physics 8(2): 185-190.

Merritt, W. F. 1965. A study of dilution in the Winnipeg River near WNRE. Atomic Energy of Canada, Ltd., Chalk River, report AECL-2295. (NSA 20: 7293).

Merritt, W. F. 1967. Techniques of groundwater tracing using radionuclides. pp. 169-170. In: Isotope Techniques in the Hydrologic Cycle. (See Stout, 1967).

Merritt, W. F. 1968. Further studies of dilution in Winnipeg River near WNRE. Atomic Energy of Canada, Ltd., Chalk River, report AECL-3256. 5pp. (NSA 23: 21921).

Merritt, W. F. 1971. Identification and measurement of trace elements in freshwater by neutron activation. pp. 358-362. In: Identification and Measurement of Environmental Pollutants. B. Westley, ed. National Research Council of Canada, Ottawa. (NSA 29: 2426).

Merton, A. 1971. Environmental effect of nuclear energy: social responsibility of scientists. Inst. für Reaktorsicherheit der Technischen Uberwachsungs-Vereine e. V., German Federal Republic, Köln, report IRS-S-4. 6pp. (In German) (NSA 30: 12676).

Mery, J. and R. Roux. 1973. Disposal of liquid effluents from the CEN Cadarache. Measures taken to avoid environmental pollution. pp. 35-45. In: Control of Ionizing Radiation Pollution, Dosimetry, Applications. France Centre National de la Recherche Scientifique, Marseilles. (In French) (NSA 30: 6537).

Meserve, F. G. and M. J. Kenney. 1934. The effects of x-rays on Planaria doratocephala. Science 79(2053): 408-409.

Meshitsuka, G., et al. 1963. Effects of growth conditions on the radiosensitivity of yeasts. Annual Rept., Tokyo Metropolitan Isotope Centre 2: 129-136. (In Japanese).

Meslow, E. C. and L. B. Keith. 1968. Demographic parameters of a snowshoe hare population. J. Wildl. Mgmt. 32(4): 812-834. (^{45}Ca).

Mestres, G. and P. Laviolette. 1968. Action of x rays on the osmotic pressure and the p_H of the homolymph of Galleria mellonella (Lepidoptera In secta). Compt. rend. D266: 2093-2096. (In French) (NSA 22: 43462).

Meszena, G. 1960. Calculations by Bruns-series for the statistical evaluation of uranium tracers occurring in natural waters. Magyar Tudom. Aka. Atommag Kutato Int. (Debrecen) Kozl. 2: 99-107. (In Hungarian) (NSA 15: 6222).

Metalli, P. and E. Ballardin. 1962. First results on x-ray-induced genetic damage in Artemia salina Leach. Atti assoc. genet. Ital. 7: 219-231. (NSA 16: 26771).

Metalli, P. and E. Ballardin. 1962. Artemia salina Leach, a new material for radiobiological research. Strahlentherapie 51(Sonderbande): 126-128. (In German) (NSA 18: 1437).

Metalli, P. and E. Ballardin. 1965. Effects of x radiation on the life span of various strains of Artemia salina Leach. Boll. zool. 32: 587-602. (In Italian) (NSA 22: 43470).

Metalli, P. and E. Ballardin. 1965. On the relationship between radiation resistance and recuperation ability of Artemia salina oocytes. (For Ann. Meeting European Soc. for Radiation Biology (Genetical Effects), Utrecht, Sept. 16-18, 1965). 10pp. (Also: Italy Comitato Nazionale per l'Energia Nucleare, Rome, report RT-BIO-(65)-31) (CONF-650968-1) (NSA 20: 22618).

Metalli, P. and E. Ballardin. 1972. Radiobiology of Artemia: radiation effects and ploidy. Curr. Topics Rad. Res. Quart. 7(2): 181-240. (NSA 26: 23155).

Metcalf, R. L. (Ed.). 1957, 1958, 1960. Advances in Pest Control Research. Vols. I-III. Interscience, New York. vii, 514; vii, 426; vii, 448pp.

Metcalf, R. L. 1960. Radioisotopes in the study of the mechanisms of action of insecticides. pp. 237-252. In: Radioisotopes and Radiation in the Life Sciences. (2nd Inter-American Symposium on the Peaceful Application of Nuclear Energy, Buenos Aires). U. S. AEC report TID-7554.

Metcalf, R. L. 1967. Mode of action of insecticide synergists. Ann. Rev. Entomol. 12: 229-256. (Radionuclides).

Metcalf, R. L. 1974. Laboratory model ecosystem for evaluating the chemical and biological behavior of radiolabelled micropollutants. pp. 49-63. In: Comparative Studies of Food and Environmental Contamination. (See IAEA, 1974) (NSA 30: 15539).

Metcalf, R. L. 1974. Screening compounds for early warnings about environmental pollution. pp. 213-217. In: Trace Substances in Environmental Health. VIII. (See Hemphill, 1974) (NSA 32: 17212) (Radiotracer techniques).

Metcalf, R. L. and T. R. Fukuto. 1965. Effects of chemical structure on intoxication and detoxication of phenyl N-methylcarbamates in insects. J. Agric. Food. Chem. 13(3): 220-231. (^{14}C).

Metcalf, R. L., R. B. March and T. R. Fukuto. 1954. Radiotracers in study of systemic insecticides. Agr. Chem. 9: 33-35.

Metcalf, R. L., M. F. Osman and T. R. Fukuto. 1967. Metabolism of C^{14}-labeled carbamate insecticides to $C^{14}O_2$ in the house fly. J. Econ. Entomol. 60(2): 445-450.

Metcalf, R. L., G. K. Sangha and I. P. Kapoor. 1971. Model ecosystems for the evaluation of pesticide biodegradability and ecological magnification. Environ. Sci. Technol. 5(8): 709-713. (^{3}H, ^{14}C).

Metcalf, R. L., et al. 1955. Pesticide residues, Schradan residues in cotton and cottonseed products. Agr. & Food Chem. 3(12): 1011-1013.

Metcalf, R. L., et al. 1966. Metabolism of 2-methyl-2-(methylthio)propionaldehyde O-(methylcarbomoyl)-oxime in plant and insect. J. Agric. Food Chem. 14(6): 579-584. (^{14}C).

Metivier, H., et al. 1972. Neoplasms induced in baboon apes (Papio papio) by the inhalation of PuO_2. Compt. rend. D275(25): 3069-3071. (In French) (NSA 28: 2979).

Metropolitan Edison Co. 1970. Three Mile Island Nuclear Station, Unit 1 and Unit 2. Environmental report: operating license stage. Reading, Pa. U. S. AEC DOCKET-50289-26. 38pp. (NSA 24: 50032).

Metropolitan Edison Co. 1970. Three Mile Island Nuclear Station, Units 1 and 2. Environmental report. Reading, Pa. U. S. AEC DOCKET-20289-27, DOCKET-50320-18. 91pp. (NSA 24: 52834).

Metropolitan Edison Co. 1970. Three Mile Island Nuclear Station, Units 1 and 2. Environmental report: operating license stage, erratum. Reading, Pa. U. S. AEC DOCKET-50289-28, DOCKET-50320-19. 2pp. (NSA 24: 52835).

Mettern, F. C. M. and L. Strackee. 1966. Investigation of radioactive materials in surface waters and precipitation. Measurement results in 1965. The Netherlands Rijks Inst. voor de Volksgezondheid, Utrecht, report RA-24. 22pp. (NSA 21: 25997).

Mewissen, D. J., J. Dambloz and Z. M. Bacq. 1959. Comparative sensitivity to radiation of seeds from a wild plant grown on uraniferous and non-uraniferous soils. Nature 183(4673): 1449. (NSA 13: 13186).

Meyer, G. L. 1969. Use of environmental isotopes in hydrology: results from WMO/IAEA Network. Wrld. Meteorol. Organ. Bull. 18: 81-85. (NSA 23: 31424) (Meteorology, oceanography, hydrology).

Meyer, G. L. 1971. Applications of neutron sources to ground-water hydrology. pp. III.21-III.38. In: Neutron Sources and Applications. Vol. 1. (Proc. Amer. Nuclear Soc. National Topical Meeting, Augusta, Ga., April 19-21, 1971). (CONF-710402-Vol.1) (NSA 26: 12028).

Meyer, M. W. and L. T. Alexander. 1965. Strontium 90 sources in soils of arid regions: rainfall, dry fallout, and irrigation water. Soil Science 99(4): 234-235. (NSA 19: 26549).

Meyer, M. W., et al. 1968. ^{90}Sr on the earth's surface. IV. Summary and interpretation of a world-wide soil sampling program: 1961-1967 results. Health and Safety Lab., U. S. AEC report TID-24341. (NSA 22: 27917).

Meyer, S. and E. Schweidler. 1927. Radioactivity. Teubner, Berlin. 2nd ed. 722pp. (In German).

Meyer, W. T. 1969. Uranium in lake water from the Kaipokok region, Labrador. Quart. Colo. Sch. Mines 64: 377-394. (NSA 23: 40865).

Meyer-Arendt, J. R. 1961. Theory and application of autoradiography. Acta Histochem. 13(1/4): 47-61. (NSA 16: 20054).

Meyers, P., L. van Zeist and E. V. Sayre. 1973. Determination of major components and trace elements in ancient silver by thermal neutron activation analysis. J. Radioanal. Chem. 16(1): 67-78. (NSA 30: 50).

Meysel', M. N. 1955. The application of isotopes in microbiology. pp. __. In: Isotopes in Microbiology: Trans. Conf. on the Application of Tracer Atoms in Microbiology. A. A. Imshenetskiy, ed. Publ. House Acad. Sci. USSR, Moscow. (AEC-tr-3494, pp. 3-16).

Meysel', M. N. 1956. The peculiarities of inactivation of microorganisms by means of radiation sterilization. pp. 106-126. In: Session Acad. Sci. USSR on the Peaceful Uses of Atomic Energy. (1-5 July 1955). (AEC-tr-2435, Pt.4, pp. 79-88; Consultants Bureau, New York, vp.) (NSA 9:7617).

Meysel', M. N. and N. D. Chernyayev. 1956. Scientific and practical problems connected with sterilization and pasteurization by radiation. Vestn. Akad. Nauk SSSR 26: 38-45. (In Russian).

Meysel', M. N. and V. M. Manteyfel'. 1963. The nature of micronecrotic foci of hemopoietic organs of animals exposed to radiation. Izv. Akad. Nauk SSSR (biol.) (6): 871-878. (JPRS-22,915).

Meysel', M. N., et al. 1964. Cytophysiological and biochemical investigations of microorganisms during their restoration following radiation injury. Izv. Akad. Nauk SSSR (biol.) 29(6): 827-850. (JPRS-28,405, pp. 1-32) (NSA 19: 17496).

Mezzadri, M. G. and L. Tassi Pelati. 1973. Research on the radioactive pollution of Lake Santo Parmense. Abstr. p. 103. In: Proc. 2nd. Conf. on the State of Advancement of Radioecology in Italy. (See Anonymous, 1973) (In Italian).

Mezzadri, M. G. and L. Tassi Pelati. 1974. Research on the radioactive pollution of Lake Santo Parmese. G. Fis. Sanit. Prot. Radiaz. 18(1-2): 26-39. (In Italian) (NSA 31: 19589).

Mia, M. M. 1965. Atomic energy for agriculture. Nucleus (Lahore) 2(1): 30-34. (NSA 19: 33726).

Miceli, R. and G. Prodi. 1957. The effects of roentgen rays on the metachromatic substance of the skin of the axolotl. Radioter. Radiobiol. Fis. Med. 12(1): 36-42.

Michael, H. N. and E. K. Ralph. (Eds.). 1972. Dating Techniques for the Archaelogist. MIT Press, Cambridge, Mass. 208pp.

Michaels, G. S. and B. Blumberg. 1965. Study on RNA from 80S ribosomes from Drosophila virilis larvae. Abstr. 127. J. Cell Biol. 27(2): 65A. (^{32}P).

Michaels, R. M. 1969. Mating of Schistosoma mansoni in vitro. Exptl. Parasitol. 25: 58-71. (NSA 24: 34599) (X-irradiation).

Michalski, T. 1963. Radiation-induced mutations in Digitalis lanata Ehrh. Biul. Inst. Roslin Leczniczych 9B: 45-51. (In Polish) (NSA 19: 43746).

Michels, D. E. 1972. Plutonium resuspension from soil: how measurable? Rocky Flats Div., Dow Chemical Co., U. S. AEC report RFP-1927. 24pp. (For National Industr. Hygiene Assoc. Conf., Denver, Colo., 5 October 197-1972) (CONF-721076) (NSA 27: 24594).

Michels, D. E. 1974. Analysis of paired data sequential in space or time and the relationship to sampling continuous cyclic distributions. Dow Chemical Co., Golden, Colo., U. S. AEC report RFP-2165. 9pp. (NSA 30: 23993) (Example of ^{137}Cs in soil samples).

Michels, D. E. 1974. Cesium-137 fallout in Red Desert Basin of Wyoming, USA. Rocky Flats Div., Dow Chemical Co., Golden, Colo., U. S. AEC report RFP-2164. iv, 7pp. (NSA 30: 21079) (Soil).

Michielsen, N. C. 1966. Intraspecific and interspecific competition in the shrews Sorex araneus L. and S. minutus L. Arch. Néerl. Zool. 17: 73-174. (Radionuclide tag).

Micholet-Cote, C. M., et al. 1973. Study of radiocontamination of fish in the Meuse River. pp. 413-427. In: Environmental Behavior of Radionuclides Released in the Nuclear Industry. (See IAEA, 1973) (In French) (NSA 29: 29381).

Michon, G. 1964. Radioactive pollution. Biological evolution. Maroc Med. 43: 167-176. (In French) (NSA 19: 31685).

Michon, G. 1965. Radioecology and radiological protection. Energie Nucleaire (Paris) 7: 333-339. (In French) (NSA 20: 11039).

Michon, M. G. 1959. Biological cycles of radioactive contaminants. Rev. Path. Gen. Physiol. Clin. (707): 505-514. (Also: France Commissariat à l'Énergie Atomique, Saclay, report CEA-1273) (In French, English summary).

Micklin, P. P. 1974. Environmental hazards of nuclear wastes. Bull. At. Sci. 30(4): 36-42. (NSA 30: 16496).

Micle, F., et al. 1971. Some data on ^{65}Zn and ^{55}Fe uptake in the seeds of some Caryophyllaceae. Stud. Univ. Babes-Bolyai (biol.) 16(1): 95-100. (NSA 26: 18098).

Middendorf, P. A. and P. G. Pearson. 1970. P^{32} movement through the soil fauna of a ragweed dominated field. Bull. Ecol. Soc. Am. 51(2): 28. (Title only).

Midura, T. F. 1964. The resistance of Clostridium perfringens type A spores to gamma radiation. Thesis, Univ. of Michigan, Ann Arbor. 157pp. (NSA 20: 1679).

Miettinen, J. K. 1963. A preliminary review of a Finnish intercomparison programme between seven laboratories studying fallout accumulation in reindeer and their environment. (Panel on Standardization of Low Level Activity Measurements, Vienna, October 28-31, 1963). International Atomic Energy Agency, Vienna, Publ. PL-86. 6pp. (NSA 18: 12347).

Miettinen, J. K. 1965. Radioactive food chains in subarctic regions. Naringsforskning (4): 97-106. (In Dutch) (NSA 21: 36592).

Miettinen, J. K. 1966. Enrichment of radionuclides by foodstuffs and man in arctic, nutrient-deficient regions. (For 7th Intern. Congr. on Nutrition, Hamburg, Germany, August 3-10, 1966) (CONF-660827) (NSA 21: 41378).

Miettinen, J. K. 1967. Radioactive food chains in subarctic regions. Bibl. Nutr. Dieta. 9: 43-58. (NSA 22: 10768).

Miettinen, J. K. 1967. Concentration of ^{137}Cs and ^{55}Fe through food chains in arctic and subarctic regions. pp. 267-274. In: Radioecological Concentration Processes. (See Aberg and Hungate, 1967) (CONF-660505-6) (NSA 20: 27194).

Miettinen, J. K. 1968. Enrichment of radioactivity by Lapps and reindeer. pp. 333-338. In: Intern. Congr. on Radioprotection and the Environment. (See Societé Francaise de Radioprotection, 1968) (Also: Univ. of Helsinki, Finland, report) (CONF-670313-2) (NSA 21: 36570).

Miettinen, J. K. 1969. Enrichment of radioactivity by arctic ecosystems in Finnish Lapland. pp. 23-31. In: Symposium on Radioactivity. (See Nelson and Evans, 1969) (Also: Univ. of Helsinki, Finland, report) (CONF-670503-18) (NSA 21: 36571).

Miettinen, J. K. 1969. The present situation and its late development regarding the accumulation of ^{137}Cs, ^{90}Sr and ^{55}Fe in arctic foodchains. pp. 145-150. In: Environmental Contamination by Radioactive Materials. (See IAEA, 1969) (Also: Univ. of Helsinki, Finland, U. S. AEC report NYO-3446, Paper 22) (CONF-690317) (NSA 23: 41303).

Miettinen, J. K. 1970. Chemical and radioactive environmental problems in Finland. Univ. of Helsinki, Finland, U. S. AEC report NYO-3446-14, Paper 30. 24pp. (NSA 25: 18931).

Miettinen, J. K. and E. Häsänen. 1965. Results of Project Lapland at the beginning of 1966. Univ. of Helsinki, Finland. 8pp. (CONF-660405-15) (Paper not in Proc.) (NSA 20: 31444).

Miettinen, J. K. and E. Häsänen. 1967. ^{137}Cs in Finnish Lapps and other Finns in 1962-1965. pp. 221-231. In: Radioecological Concentration Processes. (See Aberg and Hungate, 1967) (CONF-660405-14) (NSA 20: 27194).

Miettinen, J. K., P. Roine and K. Lidén. Relationships of radiocesium and potassium in the diet and body of Finnish Lapps. pp. 511-516. In: Radioactivity in Man. R. Meneely and S. M. Linde, eds. Charles C. Thomas, Springfield, Ill. (NSA 20: 26599).

Miettinen, J. K., et al. 1964. Cesium-137 and potassium in Finnish Lapps and their diet. Radiol. Hlth. Data 5(2): 83-97. (NSA 18: 14149) (Incl. reindeer, fish).

Miettinen, J. K., et al. 1963. Cs^{137} and potassium in people and diet - a study of Finnish Lapps. Ann. Acad. Scient. Fenn. II. Chem. ser. A120: 5-46.

Miettinen, J. K., et al. 1969. Distribution and excretion rate of phenyl- and methylmercury nitrate in fish, mussels, molluscs and crayfish. Univ. of Helsinki, Finland. 11pp. (For 9th Japan Conf. on Radioisotopes, Tokyo, May 13-15, 1969; Paper B/(11-17).

Migalovskaya, V. N. 1971. Chronic effect of ^{90}Sr+^{90}Y on the frequency of chromosomal aberrations in the embryic cells of the Atlantic salmon. Tr. polyar. nauch.-issled. Proekt. inst. morsk. ryb. khoz. okeanogr. (29): 74-81. (AEC-tr-7418, pp. 89-99) (NSA 27: 28003).

Migalovskaya, V. N. 1971. Effect of x irradiation on the gametes and embryonic cells of the Atlantic salmon. Tr. polyar. nauch.-issled. Proekt. inst. morsk. ryb. khoz. okeanogr. (29): 82-90. (AEC-tr-7418, pp. 100-112) (NSA 27: 27947.

Migalovskiy, I. P. 1971. Development of Atlantic salmon eggs under conditions of radioactive contamination of water by ^{90}Sr-^{90}Y and ^{144}Ce. Tr. polyar. nauch.-issled. Proekt. inst. morsk. ryb. khoz. okeanogr. (29): 16-27. (AEC-tr-7418, pp. 16-30) (NSA 27: 27999).

Migalovskiy, I. P. 1971. Development of fish eggs and the early period of gametogenesis in the embryos and larvae of the Atlantic salmon under conditions of radioactive contamination of water. Tr. polyar. nauch.-issled. Proekt. inst. morsk. ryb. khoz. okeanogr. (29): 32-44. (AEC-tr-7418, pp. 36-52) (NSA 27: 28001).

Migita, N. 1953. Reports by forest research team. IX. A-bomb effects on the chemical properties of trees. Nippon Gakujutsu Kaigi 1: 278-279.

Mihailov, M. G., et al. nd. Natural radioactivity in food products in a region with increased natural radioactive background. Roentgenol. Radiol. 6(4): 236-239. (In Bulgarian) (NSA 25: 21680).

Mikai, K. and T. Hayakawa. 1970. Radioactive springs in Tono District in Gifu Prefecture. Onsen Kagaku 21(3-4): 104-120. (In Japanese) (NSA 26: 58811).

Mikami, Y., H. Watanabe and K. Takano. 1956. The influence of radioactive rain-water on the growth and differentiation of a tropical fish "Zebra Danio". pp. 1225-1229. In: Research in the Effects and Influences of the Nuclear Bomb Test Explosions. Japan Soc. for the Promotion of Science, Ueno, Tokyo.

Mikami, Y., et al. 1956. Effect of radioactive rainwater upon the development of *Bufo vulgaris* eggs. Mie Med. J. (Tsu) 6(1/2): 27-39.

Mikami, Y., et al. 1965. Effects of rainwater radioactivity on organisms. Mie Kaibo Gyosekishu 16: 1-27. (In Japanese) (NSA 22: 28671) (*Zebra danio*).

Miki, M. 1970. Effect of x and γ irradiation on embryo in silkworm: *Bombyx mori* L. 1. The effect on larval growth. 2. Effect on larval growth observed after irradiation at different developmental stages. Nippon Sansgigaku Zasshi 39(4): 293-297; 40(6): 445-451. (In Japanese) (NSA 25: 46956; 27: 22761).

Miki, T., et al. 1970. Radiation hazard control system. Kinki Daigaku Genshiryoku Kenky. Nenpo 8: 29-37. (In Japanese) (NSA 26: 58796) (Environmental monitoring).

Miksche, J. P. and T. D. Rudolph. 1968. Use of nuclear variables to investigate radiosensitivity of gymnosperm seed. Radiation Botany 8(2): 187-192. (NSA 22: 45725).

Miksche, J. P., A. H. Sparrow and A. F. Rogers. 1961. Effects of chronic gamma irradiation on the apical meristems of *Pinus strobus* and *Taxus media*. Abstr. Am. J. Bot. 48(6,Pt.II): 529.

Miksche, J. P., A. H. Sparrow and A. F. Rogers. 1962. The effects of chronic gamma irradiation on the apical meristem and bud formation of *Taxus media*. Brookhaven National Lab., U. S. AEC report BNL-6203. 21pp. (NSA 17: 12122).

Milani, R. 1971. Genetics of factors affecting fertility and of sex-ratio distortions in the house fly. pp. 381-397. In: Sterility Principle for Insect Control or Eradication. (See IAEA, 1971) (NSA 25: 29751).

Milbourn, G. M. and R. Taylor. 1965. The contamination of grassland with radioactive strontium. I. Initial retention and loss. Radiation Botany 5(4): 337-347. (NSA 19: 38332).

Milde, G., Ch. Klinger and K. Fröchlich. 1974. Hydrogeological possibilities of applying short-term tritium studies on groundwater from jointed porous and jointed cavernous groundwater aquifer types. Z. Angew. Geol. 20(4): 159-163. (In German) (NSA 33: 659).

Miles, M. E. and J. J. Mangeno. 1967, 1969. Disposal of radioactive wastes from U. S. Naval nuclear-powered ships and their facilities. Radiol. Hlth. Data 8(12): 699-701; 10: 135-138. (NSA 23: 31744).

Milham, R. C. 1971. Concentrating radionuclides in environmental water samples with a centrifugal contractor. Trans. Amer. Nucl. Soc. 14(2): 463-464. (NSA 26: 1979).

Milham, R. C. 1973. Determination of phosphorus-32 in aqueous environmental samples. Savannah River Plant, U. S. AEC report DP-MS-73-35. 13pp. (NSA 29: 2451).

Milham, R. C. 1974. Determination of promethium-147 in aqueous environmental samoles. Savannah River Plant, U. S. AEC report DP-MS-73-77. 11pp. (NSA 30: 28969).

Mill, A. J., et al. 1973. Effect of hypoxia on radiation life-shortening in Drosophila melanogaster. Int. J. Radiat. Biol. 24(3): 297-305. (NSA 29: 2939).

Miller, C. E. 1968. ^{154}Eu retention patterns in small animals. Abstr. Radiation Research 35(2): 507. (Incl. one chinchilla).

Miller, C. F. 1963. Fallout nuclide solubility, foliage contamination, and plant part uptake contour ratios. Stanford Research Inst., Menlo Park, Calif. 33pp. (NP-13115) (NSA 17: 38732).

Miller, C. F. 1963. Types of catastrophes and their physical proportions. Abstr. Bull. Ecol. Soc. Am. 44(3): 79.

Miller, C. F. 1964. Biological and radiological effects of fallout from nuclear explosions. Ch. 1, 2. Stanford Research Inst., Menlo Park, Calif., U. S. Office of Civil Defense, report SRI Proj. No. IMU-4536. vii, 82pp.

Miller, C. F. 1965. Physical damage from nuclear explosions. pp. 1-10. In: Ecological Effects of Nuclear War. (See Woodwell, 1965).

Miller, C. F. 1968. Nature and behavior of local fallout. pp. 49-64. In: Radiological Protection of the Public in a Nuclear Mass Disaster. (See Brunner and Pretre, 1968).

Miller, C. R. and M. Inclan-Suarez. 1971. Critical review of the assumptions involved in establishing the MPD and MPC's in air and water. pp. 667-675. In: Health Physics Aspects of Nuclear Facility Siting. Vol. III. (See Voillequé and Baldwin, 1971) (NSA 25: 55254).

Miller, C. W., B. M. Zuckerman and A. J. Charig. 1966. Water translocation of diazinon-C^{14} and parathion-S^{35} off a model cranberry bog and subsequent occurrence in fish and mussels. Trans. Am. Fish. Soc. 95(4): 345-349.

Miller, D. R. 1970. Theoretical survival curves for radiation damage in bacteria. J. Theor. Biol. 26(3): 383-398.

Miller, G. L. 1968. Influence of season on the radiation sensitivity of an old field community. Thesis, Univ. of North Carolina, Chapel Hill. 262pp. (NSA 23: 43719) (At Savannah River Plant).

Miller, J. J., V. V. Kingsley and C. Ramirez. 1965. Comparative sensitivity to gamma radiation of yeast mitosis and meiosis. Radiation Botany 5: 383-390. (NSA 20: 1664).

Miller, J. M. and D. Ostle. 1973. Radon measurement in uranium prospecting. pp. 237-247. In: Uranium Exploration Methods. International Atomic Energy Agency, Publ. STI/PUB-334. (NSA 29: 24184).

Miller, J. N. 1961. The atom taps the forest's secrets. Popular Mechanics 116(3): 144-147, 260, 262.

Miller, K. R. and T. F. Allsop. 1970. Oxygen flask combustion method for the radioactive analysis of ^{35}S and ^{75}Se in biological samples. New Zeal. J. Sci. 13: 149-159. (NSA 25: 51393).

Miller, L. and F. W. Woods. 1965. Root-grafting in loblolly pine. Botan. Gaz. 126(4): 252-255. (Used ^{32}P).

Miller, L. N. 1965. Changes in radiosensitivity of pine seedlings subjected to water stress during chronic gamma irradiation. Health Physics 11(12): 1653-1662. In: Radiation and Terrestrial Ecosystems. (See Hungate, 1965).

Miller, L. P. and S. E. A. McCallan. 1955. Some examples of the use of radioisotopes in studies of pesticide action. pp. 546-555. In: Proc. 14th Intern. Hort. Congr. Vol. 1. (The Hague-Scheveningen).

Miller, L. S. 1957. Tracing vole movements by radioactive excretory products. Ecology 38(1): 132-136.

Miller, M. W. 1966. Relation between extrapolation number and apical cell number in gemmae of Marchantia polymorpha L. Nature 210: 748-749. (Also: Brookhaven National Lab., U. S. AEC report BNL-9135) (NSA 20: 31065).

Miller, M. W. 1968. The radiosensitivity of gemmae of Lunularia cruciata to acute gamma irradiation. Abstr. Radiaition Research 35(2): 727.

Miller, M. W. 1968. Radiosensitivity of gemmae of Lunularia cruciata to acute gamma irradiation. Bryologist 71(2): 73-81. (NSA 23: 16441).

Miller, M. W. and A. H. Sparrow. 1964. Relationship between nuclear volume and radiosensitivity of different cell types in gemmae of *Marchantia polymorpha* L. Nature 204(4958): 596-597. (Also: Brookhaven National Lab., U. S. AEC report BNL-8444) (NSA 19: 7119).

Miller, M. W. and A. H. Sparrow. 1965. The radiosensitivity of *Marchantia* thalli as related to nuclear and interphase chromosome volumes. Brookhaven National Lab., U. S. AEC report BNL-9611. 6pp.

Miller, M. W. and A. H. Sparrow. 1965. The radiosensitivity of thalli of *Marchantia polymorpha* L. to acute gamma irradiation. Radiation Botany 5: 567-580. (Also: Brookhaven National Lab., U. S. AEC report BNL-9211) (NSA 20: 8706).

Miller, M. W., E. D. Garber and P. D. Voth. 1962. Nutritionally deficient mutants of *Marchantia polymorpha* induced by x-rays. Botan. Gaz. 124(2): 94-102. (NSA 18: 9847).

Miller, M. W., A. H. Sparrow and A. F. Rogers. 1965. The radiosensitivity of gemmae of *Marchantia polymorpha* L. to acute gamma irradiation. Bryologist 68(1): 31-47.

Miller, S. and A. S. Perry. 1964. Separation and purification of DDT-degrading enzymes from the human body louse. J. Agric. Food Chem. 12(2): 167-169. (^{14}C).

Miller, T. A. 1964. Effect of x-irradiation upon the infective larvae of *Ancylostoma caninum* and the immunogenic effect in dogs of a single infection with 40 kR-irradiated larvae. J. Parasitol. 50: 735-742. (NSA 19: 33861).

Miller, T. A. 1965. Studies on canine ancylostomiasis: double vaccination with x-irradiated *Ancylostoma caninum* larvae. J. Am. Vet. Med. Assoc. 146: 41-44. (NSA 19: 36025).

Miller, T. A. 1966. Blood loss during hookworm infection, determined by erythrocyte labeling with radioactive ^{51}Cr. I. Infection of dogs with normal and with x-irradiated *Ancylostoma caninum*. J. Parasitol. 52: 844-855. (NSA 21: 22321).

Miller, T. A. 1068. Development of an x-irradiated vaccine for ancylostomiasis: current progress. pp. 13-16. In: *Isotopes and Radiation in Parasitology*. (See IAEA, 1968).

Miller, W. R. 1969. Simple statistical test for significance of environmental radioactivity data. Health Physics 17: 736-738. (NSA 24: 2663).

Millikan, C. R., B. D. Hanger and E. N. Bjarnason. 1968. Effect of phosphorus and zinc levels in the substrate on ^{65}Zn distribution in subterranean clover and flax. Austral. J. Biol. Sci. 21(4): 619-___.

Millikan, C. R., A. W. Kellock and T. Woodcock. 1971. Effect of recent red leaf virus infection on the distribution of ^{65}Zn, ^{45}Ca, and ^{54}Mn in subterranean clover. Austral. J. Agr. Res. 22(6): 909-916.

Millo, A. and A Debenedetti. 1964. Changes of the permeability of the isolated frog heart irradiated with x-rays. Protective action of glucose, of cysteamine, and of sodium thiosulphate. Riv. Biol. 57(Suppl.2): 151-167. (In Italian) (NSA 20: 3794).

Mills, H. L. and L. M. Shields. 1961. Root absorption of fission products by *Bromus rubens* L. from the AEC Nevada Test Site soil contaminated by an underground nuclear explosion. Radiation Botany 1(1): 84-91.

Mills, R. B. 1965. Laboratory studies of the biology and behavior of the angoumois grain moth, *Sitophaga cerealella* (Oliv.). Thesis, Kansas State Univ., Manhattan. 148pp. (Dissertation Abstr. 25: 1424-1425) (Radiography).

Mills, R. B. and D. A. Wilbur. 1967. Radiographic studies of angoimois grain moth development in wheat, corn, and sorghum kernels. J. Econ. Entomol. 60(3): 671-677.

Mills, W. A. 1974. EPA role in nuclear environmental standards. pp. 145-149. In: *6th Ann. National Conf. on Radiation Control: New Challenges*. (See U. S. Food and Drug Administration, 1974) (NSA 31: 19546).

Miltenburger, H., H. Esch and O. Hug. 1965. Electrophysiological potentials in plant cells under x irradiation. Bayer. Landwirtsch. Jahrb. 42 (1): 63-68. (In German) (NSA 19: 25933) (*Mimosa*, *Nitella*).

Milyutina, G. A. and V. I. Ivanov. 1961. Accumulation of chemical elements from aqueous solutions by freshwater organisms. Accumulation of radioisotopes of 12 different elements by various species of freshwater plants. Tr. Ural'skogo otd. Moskov. obshch. Isp. Prirody (Sverdlovsk) (3): ___. (In Russian).

Minayev, P. F. and G. M. Musagaleva. 1969. Change in the content of adrenalin and noradrenalin in various tissues of irradiated animals. Radiobiologiya 9(4): 625-628. (AEC-tr-7109, pp. 209-215) (NSA 24: 7994).

Minckley, W. L., J. E. Craddock and L. A. Krumholz. 1963. Natural radioactivity in the food web of the banded sculpin *Cottus carolinae* (Gill). pp. 229-236. In: *Radioecology*. (See Schultz and Klement, 1963) (NSA 17: 33567).

Mingo-Perez, E., F. F. Soprunov and L. P. Riskina. 1958. Radioisotope marking of flies. Med. Parazitol. Moskova. 27(6): 688-693. (In Russian).

Min'ko, I. F. 1971. ^{32}P transmission from plant to plant. Vestsi Akad. Navuk Belorussk. SSR (biyal.) (3): 20-24. (In Belorussian) (NSA 25: 57950).

Minnesota Atomic Development Problems Committee. nd. Basic data regarding atomic development problems in Minnesota. 121pp.

Minnesota Dept. of Health. 1969. Survey of environmental radioactivity, July-December 1968. Rural Cooperative Power Association, Elk River, U. S. AEC report COO-651-72. 38pp. (NSA 24: 31888).

Miquel, J., K. G. Bensch and D. E. Philpott. 1972. Viruslike particles in the tissues of normal and γ irradiated *Drosophila melanogaster*. J. Invert. Pathol. 19(1): 156-159. (NSA 26: 28493).

Miquel, J., et al. 1972. Natural aging and radiation-induced life shortening in *Drosophila melanogaster*. Mech. Ageing Develop. 1(1): 71-97. (NSA 28: 30249).

Miraldi, F. D. and E. J. Morgan. 1971. Gamma radiation source and method for the treatment of sewage. U. S. Patent 3,603,788. (NSA 26: 95).

Mirchink, T. G., G. B. Kashkina and Yu. D. Abaturov. 1968. Resistance of the dark-colored fungi *Stemphylium botryosum* Wallr. and *Cladosporium cladosporioides* (Fries) de Vries to gamma radiation. Mikrobiologiya 37(5): 865-869. (Microbiology (USSR) 37: 724-727) (NSA 23: 43909).

Mirchink, T. G., G. B. Kashkina and Yu. D. Abaturov. 1972. Resistance of fungi containing various pigments to gamma-irradiation. Mikrobiologiya 41(1): 83-86. (In Russian) (NSA 27: 20019).

Mirinov, B. A. 1971. Method of study of the influence of soil moisture on migration of radiaoactive isotopes in an artificial biogeoceoenosis. pp. 39-45. In: *Methods of Radioecological Investiagations*. (See Verkhovskaya, 1971).

Mironov, B. A. and Ye. N. Karavayeva. 1968. Influence of soil moisture on the behavior of ^{35}S, ^{60}Co, ^{90}Sr in the soil-plant system. Tr. inst. ekol. rast. zhivotn. Ural. fil. Akad. Nauk SSSR (61): 35-46. (In Russian) (NSA 7775).

Mironov, O. G. 1961. The radioactivity of a number of the higher algae in the region of Theodosia. Dokl. Akad. Nauk SSSR 138: 686-687. (Doklady (biol. sci.) 138(1-6): 362-363) (NSA 15: 25804).

Mironov, O. G. 1965. Contamination of vegetation by radioactive fallout. Atom. energiya 18(6): 659-661. (Sov. Atomic Energy 18: 837-839) (NSA 19: 35888).

Miserex, A. 1968. Radioactivity in foods in 1967. Mitt. Geb. Lebensmit. Hyg. 59: 156-192. (In French) (NSA 25: 16087) (Incl. water).

Mishev, I. T., M. A. Radicheva and T. M. Troshev. 1964. Investigation of the radioactive contamination around the nuclear reactor IRT-1000. Atom. energiya 16(4): 344-349. (In Russian) (NSA 18: 31772).

Mishima, J. and E. P. Odum. 1963. Ecretion rate of Zn^{65} by *Littorina irrorata* in relation to temperature and body size. Limnol. Oceanogr. 8(1): 39-44. (Salt marsh snail).

Mishra, N. C. and S. F. H. Threlkeld. 1968. Genetic studies in *Eudorina*. Genetical Res. 11: 21-31. (NSA 22: 38241) (Irradiation).

Mishra, U. C. 1972. Natural and fallout gamma nuclides in Indian soils. pp. 333-345. In: *Natural Radiation Environment. II*. (See Adams, Lowder and Gesell, 1972) (NSA 32: 19748).

Mishra, U. C. and S. Sadasivan. 1971. Gamma spectrometric measurement of soil radioactivity. Int. J. Appl. Radiat. Isotopes 22: 256-257. (NSA 25: 57567).

Mishra, U. C. and S. Sadasivan. 1971. Natural radioactivity levels in Indian soils. J. Sci. Ind. Res. 30(2): 59-62. (NSA 26: 4725).

Mishra, U. C. and S. Sadasivan. 1971. Rapid and accurate method of estimating soil concentrations of gamma emitting isotopes. pp. 39-48. In: *Radiation and Radioisotopes in Soil Studies and Plant Nutrition*. (See India Dept. of Atomic Energy, 1971) (NSA 26: 25583).

Mishra, U. C. and S. Sadasivan. 1972. Fallout radioactivity in Indian soils. Health Physics 23(1): 55-62. (NSA 26: 40848).

Miskel, J. A. 1971. Production of tritium by nuclear weapons. Lawrence Radiation Lab., U. S. AEC report UCRL-73270. (For Tritium Symposium, Las Vegas, Nev., 30 August 1971) (CONF-710809-3) (NSA 25: 48664).

Miskus, R. P., D. P. Blair and J. E. Casida. 1964. Conversion of DDT to DDD by bovine rumen fluid, lake water and reduced porphyrins. J. Agric. Food Chem. 13(5): 481-483. (^{14}C).

Misra, U. K., et al. 1973. Activation analysis of aluminum from water extract of soil. Soil Sci. Soc. Am. Proc. 37(2): 193-196. (NSA 28: 2124).

Mistry, K. B., B. M. Bhujbal and T. J. D'Souza. 1973. Influence of agronomic practices on uptake of fission products by crops from soils of regions adjoining nuclear installations in India. pp. 303-320. In: *Environmental Behavior of Radionuclides Released in the Nuclear Industry*. (See IAEA, 1973) (NSA 29: 29378).

Mistry, K. B., A. R. Gopal-Ayengar and K. G. Bharathan. 1965. On the radioactivity of plants from the high radiation areas of the Kerala Coast and adjoining regions. II. Studies on the uptake of alpha and gamma emitters. In: *Radiation and Terrestrial Ecosystems*. Health Physics 11(12): 1459-1470. (See Hungate, 1965).

Mistry, K. B., V. B. Yadav and B. M. Bhujbal. 1974. Movement of polyphosphate fertilizers and radiostrontium in contrasting Indian soil types. pp. 179-193. In: *Isotope and Radiation Techniques in Soil Physics and Irrigation Studies*. (See IAEA, 1974) (NSA 30: 15494). (^{32}P).

Mitchell, G. J., S. Hontzeas and W. Cameron. 1972. Accumulation of ^{90}Sr in pronghorn antelope (*Antilocapra americana americana*) females in Alberta. Can. J. Zool. 50(2): 213-216. (NSA 26: 28522).

Mitchell, H. C., et al. 1973. Boll weevils: tagging for hibernation and field studies. J. Econ. Entomol. 66(2): 563-564. (NSA 28: 741) (^{65}Zn, ^{32}P).

Mitchell, H. H. 1961. Ecological problems and postwar recuperation. A preliminary survey from the civil defense viewpoint. Rand Corp., Santa Monica, Calif., report RM-2801-PR. 40pp. (NSA 19: 19504).

Mitchell, H. H. 1965. Floods and the post attack biology problem. A preliminary survey. Rand Corp., Santa Monica, Calif., report RM-4238-TAB. 30pp. (NSA 19: 11421).

Mitchell, H. H. 1966. Plague in the United States: as assessment of its significance as a problem following a thermonuclear war. Rand Corp., Santa Monica, Calif., U. S. AEC report RM-4968-TAB. 54pp. (NSA 20: 43587).

Mitchell, H. H. 1966. Survey of the infectious disease problem as it relates to the post attack environment. Rand Corp., Santa Monica, Calif., U. S. AEC report RM-5090-TAB. 89pp. (NSA 20: 43588).

Mitchell, J. E. 1972. An analysis of the beta-attenuation technique for estimating standing crop of prairie range. J. Range Mgmt. 25(4): 300-304.

Mitchell, J. W. and P. J. Linder. 1950. Some methods used in tracing radioactive growth-regulation substances in plants. Botan. Gaz. 112: 126-129.

Mitchell, N. T. 1968-1973. Radioactivity in surface and coastal waters of the British Isles, 1967. ---1971. Fisheries Radiobiological Lab., U. K. Ministry of Agriculture, Fisheries and Food, Lowestoft, report FRL-2, FRL-5, FRL-7, FRL-8, FRL-9. 41, 39, 33, 38, 39pp. (NSA 23: 2433; 24: 21099; 25: 26397; 26: 25585; 29: 20877).

Mitchell, N. T. 1969. Monitoring of the aquatic environment of the United Kingdom and its application to hazard assessment. pp. 449-461. In: *Environmental Contamination by Radioactive Materials*. (See IAEA, 1969)

Mitchell, N. T. 1974. Roles of effluent and environmental monitoring in surveillance of radioactive wastes released from nuclear installations. pp. 399-411. In: *Environmental Surveillance Around Nuclear Installations. Vol. 2*. (See IAEA, 1974) (NSA 31: 14049).

Mitchell, N. T. 1974. Transfer of radionuclides to man through environmental tal pathways. pp. 485-500. In: *Population Dose Evaluation and Standards for Man and His Environment*. (See IAEA, 1974) (NSA 31: 19604).

Mitchell, N. T. 1974. Safe disposal of liquid radioactive wastes from the United Kingdom nuclear power programme. pp. 301-307. In: *Proc. 3rd Intern. Congr. Intern. Radiation Protection Assoc*. (CONF-730907) (NSA 30: 26767).

Mitchell, N. T. and G. E. Eden. 1962. Radioactive strontium in the River Thames. J. Inst. Water Engrs. 16(2): 175-187. (NSA 17: 35440).

Mitchell, N. T. and D. F. Jefferies. 1973. Control of low-level liquid radioactive waste disposal: UK experience in the derivation and use of environmental capacity. pp. 633-647. In: *Environmental Behavior of Radionuclides Released in the Nuclear Industry*. (See IAEA, 1973) (NSA 29: 30193).

Mitchell, N. T., K. R. Brendish and G. E. Eden. 1962. The behavior of radioactive isotopes in a sewer. J. Proc. Inst. Sewage Purif. London Pt.6: 586-591. (NSA 17: 36061).

Mitlin, N. and P. H. Babers. 1956. Action of radiostrontium in the housefly and German cockroach. J. Econ. Entomol. 49(5): 714-715.

Mitlin, N., A. C. Bartlett and I. C. Keller. 1965. Elimination rate and effect on reproduction of ingested radiophosphorus in the boll weevil. J. Econ. Entomol. 58(1): 119-121. (NSA 19: 25970).

Mitra, J. 1969. Relative radio-sensitivity of different cell-types in hydra. Gegenbaurs Morphol. Jahrb. 114: 258-268. (NSA 25: 42613).

Mitra, J. 1970. Differential x-ray sensitivity of hydras in relation to age. Indian J. Exptl. Biol. 8: 302-303. (NSA 25: 39906).

Mitsui, E., et al. 1970. Radiation disinfestation of cereal insects. 2. Sensitivity of *Sitophilus zeamais* and *S. oryzae*. Nippon Shokuhin Kogyo Gakkaishi 17: 379-384. (In Japanese) (NSA 25: 21822).

Mituo, I. and S. Kondo. 1965. Comparative studies of mutation frequencies induced by ^{32}P treatment and γ-irradiation in the male silkworm. Mutation Research 2: 534-543.

Miura, I. 1953. Report by forestry research team. VIII. Anatomical survey of trees damaged by A-bomb. Nippon Gakujutsu Kaigi 1: 277-278.

Miura, K, et al. 1965. Biosynthesis of unsaturated and saturated fatty acids in aseptically reared larvae of *Calliphora erythrocephala*. Arch. int. physiol. biochem. 73(1): 65-72. (^{14}C).

Miura, Y., et al. 1964. Protein synthesis in silk glands. V. Relation of ribosomes to endoplasmic reticulum during fibroin synthesis. Seikagaku Zasshi 55(6): 623-628. (^{14}C).

Miura, Y., et al. 1965. Protein synthesis in silk glands. VI. RNA metabolism in fifth instar larvae. Seikagaku Zasshi 58(3): 293-299. (^{14}C, *Attacus ricni*, *Bombyx mori*).

Mix, M. C. 1972. Chronic tissue degeneration in the Pacific oyster: *Crassostrea gigas*, following acute γ-irradiation. Radiation Research 49(1): 176-189. (NSA 26: 23141).

Mix, M. C. and A. K. Sparks. 1971. Repair of digestive tubule tissue of the Pacific oyster, *Crassostrea gigas*, damaged by ionizing radiation. J. Invertebr. Pathol. 17(2): 172-177. (NSA 26: 43361).

Mix, T. 1934. Radium transfer through *Phthirius pubis* in a bovine-equine environment. Esp. Bull. forn. Ann. 30(3): 117-119.

Miyake, H. and M. Michijima. 1972. Inequilibrium of thorium series radionuclides in the sediments around a river mouth. Radioisotopes (Tokyo) 21 (2): 74-79. (In Japanese) (NSA 26: 58542).

Miyake, N. 1963. 00832 a radiation-using, snow-fall telemeter. Electrical Power (Kyoto) 47(10): 32-35, 40. (In Japanese).

Miyake, Y. and H. Tsubota. 1963. Estimation of the direct contribution of meteoric water to river waters by means of fallout radiocesium and radiostrontium. pp. 425-432. In: Radioisotopes in Hydrology. (See IAEA, 1963) (NSA 18: 1932).

Miyake, Y., Y. Katsuragi and Y. Sugimura. 1968. Deposition of plutonium in Tokyo through the end of 1966. Papers Meteorol. Geophys. 19(2): ___.

Miyake, Y., Y. Katsuragi and Y. Sugimura. 1970. Study on plutonium fallout. J. Geophys. Res. 75: 2329-2330. (NSA 24: 25142).

Miyake, Y., Y. Sugimura and H. Tsubota. 1964. Content of uranium, radium, and thorium in river waters in Japan. pp. 219-226. In: The Natural Radiation Environment. (See Adams and Lowder, 1964) (NSA 17: 30790).

Miyanaga, I., A. Kasai and K. Imai. 1973. Preliminary experiment on the deposition of gaseous radioiodine onto environmental materials. pp. 157-165. In: Environmental Behavior of Radionuclides Released in the Nuclear Industry. (See IAEA, 1973) (NSA 29: 29296).

Miyazaki, S. and G. M. Boush. 1967. Microbial degradation of Chloropropylate (isopropyl f,f'-dichlorobenzilate). Bull. Entomol. Soc. Am. 13(3): 192. (^{14}C).

Miyoshi, T.,1953. Reports by forestry research team. VI. Effect of atomic bomb on physical property of trees. Nippon Gakujutsu Kaigi 1: 269-273.

Mochida, O. and Y. Miyahara. 1974. Effect of gamma radiation on male pupae and adults of *Spodoptera litura* (F.) (Lepidoptera: Noctuidae) following pupal exposure. Appl. Entomol. Zool. 9(1): 41-47. (NSA 33: 3270).

Mochizuki, T. and S. Nakao. 1972. Uranium contents of suspended matters in the lakes Shinji and Naka-Umi, southwestern Japan. Chishitsu Chosasho Geppo 23(7): 393-398. (In Japanese) (NSA 30: 308).

Moden, D. D., J. W. Mullins and R. E. Jaquish. 1973. Laboratory system for tritium analysis of large numbers of environmental samples. pp. 512-516. In: Tritium. (See Moghissi and Carter, 1973) (NSA 28: 27020).

Moeken, H. H. Ph. and J. J. H. Alderhout. 1961. The radioactivity in air and rain in relation to the radioactivity in grass. European Co. for the Chemical Processing of Irradiated Fuels, Mol, Belgium, report ETR-105. 21pp. (NP-12548) (NSA 18: 37370).

Moeken, H. H. and J. J. H. Alderhout. 1963. The relationship between the concentration of world-wide fall-out in air and in rain. Air Water Poll. 7(2/3): 91-93. (NSA 17: 24873).

Moeller, D. W. 1957. Radionuclides in reactor cooling water-identification, source and control. Oak Ridge National Lab., U. S. AEC report ORNL-2311. 161pp.

Moeller, D. W. and A. S. Goldin. 1969. Environmental protection for nuclear applications. Proc. Am. Soc. Civil Engrg. 95(SA3): 373-385.

Moeller, D. W., A. S. Goldin and R. Pavanello. 1970. Environmental radiation surveillance; a comparison of the recommendations of four international agencies. pp. 29-39. In: Environmental Surveillance in the Vicinity of Nuclear Facilities. (See Reinig, 1970) (NSA 25: 16159).

Moeller, P. and K. Wagener. 1967. Dating soil layers by ^{10}Be. pp. 177-188. In: Radioactive Dating and Methods of Low-Level Counting. (See IAEA, 1967).

Moerner, N. A. 1975. Climatic cycles. J. Interdisc. Cycle Res. 6(1): 15-24. (NSA 32: 14320) (Radiocarbon).

Moffitt, H. R. and L. D. White. 1972. Susceptibility of gamma-irradiated and unirradiated codling moths to azinphosmethyl, carbonyl, and DDT. J. Econ. Entomol. 65(4): 1008-1010. (NSA 26: 50960).

Moghissi, A. A. 1970. Analysis and public health aspects of environmental tritium. pp. 419-426. In: Radionuclides in the Environment. R. F. Gould, ed. Amer. Chem. Soc., Adv. Chem. Ser. 93, Washington.

Moghissi, A. A. and M. W. Carter. (Eds.). 1973. Tritium. (Conf., Las Vegas, Nevada, August 28-September 3, 1971). Messenger Graphics, Phoenix. 815pp. (NSA 28: 27237).

Moghissi, A. A. and M. W. Carter. 1973. Environmental tritium. pp. 399-405. In: Tritium. (See above entry) (NSA 28: 27240).

Moghissi, A. A. and R. Lieberman. 1972. Comparison between tritium concentration in drinking water and body water. Health Physics 22(5): 509-511. (NSA 26: 31380).

Moghissi, A. A. and B. Moghissi. 1973. Tritium: history of discovery, application and hazards to the environment. Phys. Bl. 29(12): 543-547. (In German) (NSA 29: 29297).

Moghissi, A. A. and C. R. Porter. 1968. Tritium in surface waters of the United States, 1966. Radiol. Hlth. Data 9: 337-339. (NSA 23: 12092).

Moghissi, A. A., E. W. Bretthauer and M. W. Carter. 1971. Monitoring of tritium under emergency conditions. pp. 263-266. In: Rapid Methods for Measuring Radioactivity in the Environment. (See IAEA, 1971) (NSA 26: 27815).

Moghissi, A. A., E. W. Bretthauer and E. H. Compton. 1973. Separation of water from biological and environmental samples for tritium analysis. Anal. Chem. 45(8): 1565-1566. (NSA 28: 10672).

Moghissi, A. A., R. G. Patzer and M. W. Carter. 1973. Biokinetics of environmental tritium. pp. 314-322. In: Tritium. (See Moghissi and Carter above) (NSA 28: 27646).

Moghissi, A. A., et al. 1971. Rapid liquid scintillation technique for monitoring ^{14}C in air. pp. 391-394. In: Rapid Methods for Measuring Radioactivity in the Environment. (See IAEA, 1971) (NSA 26: 27822).

Moghissi, A. A., et al. 1975. Oxygen bomb combustion of environmental and biological samples for tritium analysis. Int. J. Appl. Radiat. Isotopes 26(6-7): 339-342. (NSA 32: 21882).

Moh, C. C. 1963. The application of nuclear energy to agriculture (annual report). Inter-American Inst. of Agricultural Sciences, Turrialba, Costa Rica, U. S. AEC report TID-18936. 62pp. (NSA 17: 28678) (Pines).

Moh, C. C. 1963. Radiosensitivity of tropical plant species: Carica papaya, Manihot dulcis, and Swietenia humilis. Turrialba 13: 180-181. (In Spanish) (NSA 18: 33380).

Mohammad Rabbani, G. 1974. X-ray induced chromosomal translocations with reference to genetic control of the mosquito Anopheles albimanus Wiedmann. Dissertation, Univ. of Illinois, Urbana. 69pp. (Univ. Microfilms 75-408) (NSA 31: 20069).

Mohindra, V. K. 1967. ^{137}Cs burdens in the Canadian North. Acta Radiol. Ther. Phys. Biol. 6: 481-490. (NSA 22: 28491).

Mohindra, V. K. and J. H. Gordon. 1965. ^{137}Cs in the Canadian North, January-June, 1965. Data from Radiation Protection Programs (Ottawa) 3(10): 3-12. (NP-15763) (NSA 20: 9175).

Mohn, S. Fr. and J. Teige. 1968. ^{137}Cs concentrations in Norwegian lynx and wolverine. Nytt Mag. Zool. (Oslo) 16: 50-52. (NSA 23: 45726).

Mohyuddin, M. and W. P. Skoropad. 1970. Effect of ^{60}Co gamma irradiation on the survival of some fungi in single samples of each of three different grades of wheat. Can. J. Bot. 48(2): 217-219. (NSA 24: 34759).

Mohyuddin, M., N. Osman and W. P. Skoropad. 1972. Inactivation of conidiospores and mycelia of Aspergillus flavus by γ-radiation. Radiation Botany 12(6): 427-431. (NSA 27: 17563).

Moiseyenko, Ye. I. 1959. Experimental application of the biogeochemical survey to the search for uranium in swampy areas. Geokhimiya (1): 96-98. (In Russian).

Moiseyenko, Ye. V. 1960. Nucleic acid content and radiosensitivity of Escherichia coli. Biofizika 5(2): 176-179. (Biophysics (USSR) 5: 203-207).

Moiseyenko, Ye. V. 1964. Action of x rays on desoxyribonucleoprotein of Misgurnus fossilis sperm. Radiobiologiya 4(1): 52-59. (AEC-tr-6404, pp. 70-80) (NSA 18: 17574).

Moiseyenko, Ye. V. 1966. Primary damage and enzymatic decomposition of loach sperm and E. coli deoxyribonucleoprotein under the action of radiation. Dissertation, Moscow. (In Russian).

Moiseyev, A. A. 1966. Cesium-137 in humans. Gigiyena sanitariya 31(1): 82-88. (In Russian).

Moiseyev, A. A. 1967. Body burdens of cesium-137 in various groups of native populations of the Nenets National District in 1965. State Committee on the Use of Atomic Energy, USSR. Atomizdat, Moscow. UN Document A/AC.82/G/L.1188. (AEC-tr-7214, pp. 1-8) (NSA 25: 16063).

Moiseyev, A. A. 1967. Forecast of radiation levels in the case of the indigenous population of the Far North resulting from incorporated global cesium-137. State Committee on the Use of Atomic Energy, USSR. Atomizdat, Moscow. UN Document A/AC.82/G/L.1198. 9pp. (AEC-tr-7214, pp. 27-36) (NSA 25: 16127).

Moiseyev, A. A. and V. I. Ivanov. 1964. Short Handbook of Radiation Protection and Dosimetry. Atomizdat, Moscow. vp. (JPRS-27,535; TT-64-51784) (NSA 19: 13122) (Incl. natural radiation).

Moiseyev, A. A. and P. V. Ramzayev. 1975. Cesium-137 in the Biosphere. Atomizdat, Moscow. 184pp. (In Russian) (NSA 33: 25998).

Moiseyev, A. A., et al. 1967. Distribution of gamma-radiation tissue doses of ^{137}Cs incorporated into the body of a reindeer (Rangifera tarandus). pp. 649-654. In: Radioecological Concentration Processes. (See Aberg and Hungate, 1967).

Moiseyev, A. A., et al. 1967. Characteristics of the migration of global cesium-137 from Podzolic turf soils to the human organism through food chains. State Committee on the Use of Atomic Energy, USSR. Atomizdat, Moscow. UN Document A/AC.82/G/L.1187. 22pp. (AEC-tr-7030, pp. 1-13) (NSA 22: 25799; 23: 17999).

Moiseyev, P. A. 1957. The effect of an atomic explosion on fishing. Ryb. khozyaystvo (5): 81-84. (In Russian).

Moiseyev, P. A. 1958. New data on the effect of thermonuclear explosions on aquatic organisms. Ryb. khozyaystvo (7): 22-24. (In Russian).

Mokady, R. S. and D. Zaslavsky. 1967. On the use of radioactive tracers in diffusion experiments. Soil Sci. Soc. Am. Proc. 31(5): 604-___.

Mokinari, J. 1973. Tracer techniques in the study of pollution problems. Nusances Environ. : 191-206. (In French) (NSA 30: 18654).

Molchanov, A. A., et al. 1968. Distribution of major radioactive fission products deposited from the atmosphere in global fallout in forests of various areas of the USSR during the period 1964-1965. State Committee on the Use of Atomic Energy, USSR. UN Document A/AC.82/G/L.1302. 10pp. (In Russian) (NSA 24: 25421).

Molchanova, I. V. 1968. Behavior of ^{144}Ce in various types of soil. Tr. inst. ekol. rast. zhivotn. Ural. fil. Akad. Nauk SSSR (61): 4-11. (In Russian) (NSA 24: 7772).

Molchanova, I. V. and M. Ya. Chebotina. 1975. Role of fallen leaves in the vertical migration of ^{90}Sr in soil. Ekologiya 6(2): 78-80. (Sov. J. Ecol. 6(2): 149-151) (NSA 33: 29099).

Molchanova, I. V. and M. Ya. Karavayeva. 1971. The distribution of radioisotopes in a soil-soil solution-plant system as a function of soil moisture tension. Ekologiya 2(1): 107-109. (Sov. J. Ecol. 2(1): 86-87).

Molchanova, I. V. and N. V. Kulikov. 1972. Radioactive Isotopes in Soil-Plant Systems. Atomizdat, Moscow. 86pp. (In Russian) (NSA 33: 9282).

Molchanova, I. V. and N. V. Kulikov. 1970. Distribution of ^{90}Sr and ^{137}Cs in the soil and vegetation cover of the tundra. Dokl. Akad. Nauk SSSR 195: 959-961. (In Russian) (NSA 25: 37497).

Molchanova, I. V., Ye. N. Karavayeva and N. V. Kulikov. 1972. Influence of soil moisture content on strontium-90 uptake by plants. Ekologiya 3(3): 78-80. (Sov. J. Ecol. 3(3): 257-259) (NSA 30: 16066).

Molchanova, V. I., N. V. Kulikov and L. I. Piskunov. 1973. Experimental study of strontium-90 migration in soils of forest biogeocoenoses. Ekologiya 4(5): 61-65. (Sov. J. Ecol. 4(5): 416-419) (NSA 32: 6432, 22276).

Molchanova, V. I., B. A. Mironov and N. V. Kulikov. 1971. Experimental study methods of ^{90}Sr migration in a natural forest biogeocoenosis. pp. 34-39. In: Methods of Radioecological Investigations. (See Verkhovskaya, 1971) (In Russian).

Moldenhawer, F. and U. Schulze-Wollgast. 1971. Fallout radioactivity in the biosphere in the GDR during 1969. Kernenergie 14(7-8): 261-268. (In German) (NSA 26: 20351).

Moldenhawer, F., H. Bose and B. Korsetz. (Comps.). 1970. Conference of the German-Polish Work Group for the Surveillance of the Radioactive Contamination of the Environment. Part II, 1. (Berlin, June 16-21, 1969). German Democratic Republic Staatliche Zentrale für Strahlenschutz, Berlin, report SZS-10/70. 46pp. (In German) (NSA 24: 301).

Mole, R. H. 1963. Dose rate and time factor. pp. 18-33. In: Radiation Effects in Physics, Chemistry and Biology. North-Holland, Amsterdam. (NSA 17: 25019) (Drosophila eggs, mice, birds, Paramecium, silkworm larvae).

Mole, R. H. 1965. Dose response relationships, particularly in mammalian radiobiology. Ann. Rev. Nucl. Sci. 15: 207-240. (NSA 20: 6803).

Molev, Ye. V. 1958. Culture of midges of the genus Culicoides under laboratory conditions. Zool. zh. 37(10): 1563-1568. (JPRS-22,187) (Use of radionuclides).

Molinari, J. 1968. Contribution of radioactive tracers to the determination of discharges of watercourses and polluted effluents. Tech. sci. munic. 63: 133-146. (In French) (NSA 24: 301) (Mokinari ?, qv.).

Molinari, V. 1964. On resistance to physical and chemical agents of Endamoeba histolytica in relation to its presence in drinking water and waste water. Nuovi Ann. Ig. Microbiol. 15: 455-462. (In Italian) (NSA 20: 5366).

Molla, S. A., M. M. Hossain and M. U. Mallik. 1968. Mutagenic effect of gamma rays on the immature and mature sperms of silkworm, Bombyx mori. Pakistan Atomic Energy Centre, Dacca, report AECD-RB-8. 6pp. (NSA 23: 6407).

Möller, F. 1973. Compartment systems as stochastic processes. Abstr. p. 133. In: Proc. 2nd Conf. on the State of Advancement of Radioecology in Italy. (See Anonymous, 1973) (In Italian).

Molnar, G. and I. Tolg. 1960. Roentgenological investigation of the duration of gastric digestion in the pike-perch (Lucioperca lucioperca L.). Acta Biologica 9(2): 103-108.

Molnar, I., et al. 1962. The killing of cocoons by irradiation from radioactive sources. Magyar Text.-Tech. 15: 196-199. (In Hungarian).

Molnar, I., et al. 1964. A study of silk cocoons from the eggs of Bombyx mori L. irradiated with Co^{60}. Magyar Text.-Tech. 16: 449-451. (In Hungarian) (NSA 19: 24302).

Monakov, A. V. and Yu. I. Sorokin. 1959. An attempt to investigate the predatory behavior of Cyclops with the air of the isotopic method. Dokl. Akad. Nauk SSSR 125(1): 201-204. (Doklady (biol. sci.) 125: 319-321).

Monakov, A. V. and Yu. I. Sorokin. 1960. An experimental investigation of Daphnia nutrition using C^{14}. Dokl. Akad. Nauk. SSSR 135: 1516-1518. Doklady (biol. sci.) 135(1-6): 925-926).

Monash, S. and F. J. Roth, Jr. 1959. The effect of x-rays, grenz rays, thorium-X, and radioactive phosphorus on superficial mycotic agents. A.M.A. Arch. Dermatol. 80(3): 288-293.

Monib, M., M. A. A. Salam and T. T. El-Hadidy. 1971. Cumulative effect of gamma irradiation on the microbial soil population and its activity. Folia Microbiol. (Prague) 16(6): 457-464. (NSA 26: 17952).

Monk, C. D. 1966. Effects of short-term gamma irradiation on an old field. Radiation Botany 6(4): 329-335. (NSA 20: 35201).

Monk, C. D. 1967. Radioisotope tagging through seed soaking. Bull. Georgia Acad. Sci. 25(1): 13-17.

Monk, C. D. 1967. Effects of gamma irradiation on growth in slash pine seedlings when different parts are exposed. Bull. Georgia Acad. Sci. 25(4): 197-201. (NSA 22: 28710).

Monk, C. D. 1967. Effects of gamma irradiation on prickly pear cactus. Quart. J. Fla. Acad. Sci. 30(3): 168-172.

Monk, C. D. 1971. Leaf decomposition and loss of ^{45}Ca from diciduous and evergreen trees. Am. Midland Nat. 86(2): 379-384.

Monro, J. 1960. The possible use of sterile males in controlling or eradicating Queensland fruitfly. pp. 130-137. In: The Technological Use of Radiation. Australia Atomic Energy Commission.

Monro, J. 1963. Population control in animals by overloading resources with sterile animals. Science 140: 496-497.

Monro, J. 1966. Population flushing with sexually sterile insects. Science 151: 1536-1538. (NSA 20: 22742).

Monro, J. 1968. Marking insects with dietary manganese for detection by neutron activation. Ecology 49(4): 774-776.

Monro, J. and P. T. Bailey. 1965. Influence of radiation on avarian maturation and histolysis of pupal fat body in Diptera. Nature 207(4995): 437-438. (NSA 19: 38406).

Monroe, J. R. 1966. The whole body burden of radioactivity in selected predator and prey species. Thesis, Univ. of Georgia, Athens. 36pp.

Monroe, R. E. 1964. Metabolism and utilization of cholestorol-4-^{14}C for growth and reproduction of aseptically reared houseflies, *Musca domestica* L. (Dissertation Abstr. 25: 1539).

Monson, A. M. 1960. Movement of radioisotopes in *Rhizoctonia solani*. Abstr. Phytopatology 50(9): 646.

Monson, A. M. and T. W. Sudia. 1963. Translocation in *Rhizoctonia solani* Botan. Gaz. 12(6): 440-443. (Used ^{65}Zn, ^{35}S, ^{89}Sr, ^{32}P).

Montganer, H. 1964. Étude du compartement alimenaire et des relations trophallactiques des males an sein isotope. Insectes soc. 11(4): 301-316. (In French) (^{198}Au).

Montagner, H. 1964. On the characterization of intestinal transit with social wasps. The biological consequences. Compt. rend. hebd. séanc. acad. sci. 259(12): 2016-2019. (In French) (*Paravespula germanica*, *P. vulgaris*, ^{198}Au).

Montagner, H. 1966. The mechanisms and consequences of the nutritional requirements in the wasps of the genus *Vespa*. Thesis, Univ. of Nancy, France. (In French) (^{198}Au).

Montoloy, F., M. Heurtebise and J. Lubkowitz. 1972. Simple method for the activation analysis of mercury in fish by using iodinated resins. J. Radioanal. Chem. 11(2): 167-173. (NSA 27: 33).

Mook, L. J. and D. M. Davies. 1966. The European praying mantis (*Mantis regligiosa* L.) as a predator of the red-legged grasshopper (*Melanopus femurrubrum* De Geer). Can. Entomol. 98(9): 913-918. (^{32}P).

Mook, W. G. 1972. Application of natural isotopes in groundwater hydrology. Geol. Mijnbouw 51(1): 131-136. (NSA 26: 31035).

Mook, W. G. 1972. Reconstruction of the initial ^{14}C content of groundwater from the chemical and isotopic composition. pp. D31-D41. In: *Proc. 8th Intern. Radiocarbon Dating Conf. Vol. 1*. (See Rafter, 1972) (NSA 31: 8698).

Mook, W. G. 1975. Absolute dating with carbon-14. Ned. Tijdschr. Natuurk. 41(2): 17-22. (In Dutch) (NSA 32: 3510).

Mook, W. G., A. V. Munaut and H. T. Waterbolk. 1972. Determination of age and duration of stratified prehistoric bog settlements. pp. F27-F40. In: *Proc. 8th Intern. Radiocarbon Dating Conf*. (See Rafter, 1972) (NSA 31: 8708).

Mookerjee, A. 1959. Effect of x-radiation on the survival of yeast in dry and wet conditions. Nature 184(Suppl.19): 1502-1503.

Mookerjee, S. and A. Aditya. 1966. Regeneration time in irradiated hydras. Indian J. Exptl. Biol. 4: 201-205. (NSA 21: 20368).

Mookerjee, S. and S. Chatterjee. 1970. Formation of supernumary surface layers on irradiated amoebae. Current Sci. (India) 39: 468-469. (NSA 25: 11249).

Mooney, E. J. 1972. Neutron activation studies of the Roanoke River system from its headwaters to Niagara Dam. Virginia Polytechnic Inst., Blacksburg, Va. 104pp. (Univ. Microfilms 72-23,841) (NSA 27: 2317).

Moor, K. S. and W. G. Bradley. 1974. Ecological studies of vertebrates in plutonium-contaminated areas of the Nevada Test Site. pp. 187-212. In: *Dynamics of Plutonium in Desert Environments*. (See Dunaway and White, 1974) (NSA 30: 29449).

Moore, A. K. and M. L. Riedesel. 1974. Metabolism of ^{14}C-urea in relation to starvation, water deptivation and hibernation in the rat and golden mantled ground squirrel. Abstr. Bull. Ecol. Soc. Am. 55(2): 16.

Moore, D. and T. H. Essig. (Eds. and comps.). 1966. Evaluation of radiological conditions in the vicinity of Hanford for 1965. Appendices. Battelle Northwest Labs., U. S. AEC report BNWL-316(App.). 146pp. (NSA 21: 34649).

Moore, J. D. 1968. Environmental monitoring. Semiannual report, July 1-December 31, 1968 and annual report 1968. Atomics International, Canoga Park, Calif., U. S. AEC report TID-25018. 28pp. (NSA 23: 29507) (Soil, vegetation, water, air).

Moore, J. D. 1972. Environmental monitoring. Annual report, 1972. Atomics International, Canoga Park, Calif., U. S. AEC report TID-26442. 27pp. (NSA 28: 27234).

Moore, J. L. 1973. Comparative accumulation of fallout radionuclides on contrasting plant surfaces. Thesis, Univ. of Utah, Salt Lake City. 187pp. (Univ. Microfilms 73-21,374) (NSA 29: 378).

Moore, J. R. and R. T. Hartman. 1964. Productivity and community structure in a small pond. Abstr. Bull Ecol. Soc. Am. 45(3): 78. (Used ^{14}C).

Moore, R. and G. Gregory. 1963. Chromosome aberrations in *Protemnodon bicolor* after irradiation. I. Irradiation in the premitotic period. Int. J. Radiat. Biol. 7(6): 549-560. (NSA 18: 41120).

Moore, R. V. and O. W. Propheter. 1973. Neutron activation analysis of bottom sediments. Southeast Environmental Research Lab., U. S. EPA report EPA-R2-73-009, 16020 GHQ.

Moore, S. T., M. F. Schuster and F. A. Harris. 1974. Radioisotope technique for estimating lady beetle consumption of tobacco budworm eggs and larvae. J. Econ. Entomol. 67(6): 703-705. (NSA 31: 9047) (^{32}P).

Moore, W. G. 1935. The effects of x-rays on fertility in *Drosophila melanogaster*. Biol. Bull. 62(3): 294-305.

Moore, W. S. 1967. Amazon and Mississippi River concentrations of uranium, thorium, and radium isotopes. Earth Planet. Sci. Lett. 2: 231-234. (NSA 21: 32270).

Moore, W. S. and D. F. Reid. 1973. Extraction of radium from natural waters using manganese-impregnated acrylic fibers. J. Geophys. Res. 78(36): 8880-8886. (NSA 30: 2602).

Moraczewski, S. and F. Kelsey. 1948. Distribution and rate of metabolism of phosphorus compounds in Trypansoma equiperdum. J. Infect. Dis. 82: 45-51.

Moreira, L. and C. Lalou. 1972. Experimental study of the ratio $^{234}U/^{238}U$ in natural waters having passed through different types of rocks. An. Acad. Brasil. Cienc. 44(1): 13-18. (In French) (NSA 27: 24561).

Morekas, S. 1973. National disposal sites for hazardous wastes. pp. 111-114. In: Proc. 4th Ann. Symp. on Environmental Pollution. (Edgewood Arsenal, Maryland, 4-5 April 1973). A. E. Hilsmeier, comp. U. S. Army Edgewood Arsenal, Md., report. (CONF-730417) (NSA 32: 19684).

Moreland, D. E. 1950. A study of the translocation of radioactive phosphorus in loblolly pine (Pinus taeda L.). J. Elisha Mitchell Sci. Soc. 66(2): 175-181.

Morello, A. 1964. Role of DDT-hydroxylation in resistance. Nature 203: 785-786. (^{14}C, Triatoma infestans).

Morello, A. 1965. Induction of DDT-metabolizing enzymes in microsomes of rat liver after administration of DDT. Can. J. Biochem. 43(8): 1289-1293. (^{14}C).

Moreng, R. E., et al. 1960. Response of the domestic fowl to x-ray irradiation at one day of age. Radiation Research 12(4): abstr. 110.

Moreshet, S., G. Stanhill and D. Koller. 1968. A radioactive tracer technique for the direct measurement of the diffusion resistance of stomata. J. Exptl. Bot. 19: 460-467. (NSA 23: 38386).

Morette, A. and J.-P. Divin. 1968. Influence of a tensioactive substance on the behavior of radioactive water in an artificial earth. Eau 55: 221-224. (In French) (NSA 23: 4950).

Morgan, A. and G. M. Arkell. 1961. The fate of radioactive effluent discharge from A.E.R.E. Harwell into the River Thames. I. A preliminary survey. Research Establishment, U. K. Atomic Energy Authority, Harwell, report AERE-R-3555. 26pp. (NSA 15: 29535).

Morgan, A. and D. G. Stanbury. 1961. The contamination of rivers with fission products from fallout. Health Physics 5(3/4): 101-107.

Morgan, A., A. Holmes and C. Gold. 1972. Studies on the solubility of constituents of chrysotile asbestos in vivo radioactive tracer techniques. Environ. Res. 5: 558-___.

Morgan, A. V. 1974. Land use assessment at Chalk River Nuclear Laboratories: Perch Lake and environs. Final report. Waterloo Univ., Ont., report. 17 pp. (INIS-mf-1694) (NSA 32: 24357) (Waste disposal).

Morgan, B. H. and J. B. Reed. 1954. Resistance of bacterial spores to gamma irradiation. Food Research 19: 357-366.

Morgan, F. 1964. The uptake of radioactivity by fish and shellfish. I. ^{134}cesium by whole animals. J. Marine Biol. Assoc. U.K. 44(1): 259-271. (NSA 18: 29216) (SFA 9: 6512).

Morgan, G. B. 1961. The absorption of radioisotopes by certain microorganisms. Quart. J. Fla. Acad. Sci. 24(2): 94-100.

Morgan, K. Z. 1971. Health physics and the environment. pp. 3-21. In: Rapid Methods for Measuring Radioactivity in the Environment. (See IAEA, 1971) (NSA 26: 28046) (Incl. comments on ecology program at ORNL).

Morgan, K. Z. and E. G. Struxness. 1971. Criteria for control of radioactive effluents. Int. J. Environ. Stud. 2: 111-119. (NSA 26: 15213).

Morgan, J. M., Jr. 1959. Stream survey in the uranium mining and milling area of the Colorado Plateau. Thesis, Johns Hopkins Univ., Baltimore, Md. 354pp. (Univ. Microfilms 59-3894) (Dissertation Abstr. 20: 1313).

Morgan, J. M., Jr., D. K. Jamison and J. D. Stevenson. (Eds.). 1962. 2nd Conf. on Ground Disposal of Radioactive Wastes. (Chalk River, Canada, September 26-29, 1961). U. S. AEC report TID-7628. viii, 635pp. (NSA 16: 25037).

Morgan, P. B. and G. C. LaBrecque. 1964. Studies of the effect of meat exposed to gamma radiation or chemosterilants on the reproductive capacity of a blow fly Phaenicia cuprina (Wiedmann). Florida Entomol. 47(1): 31-33. (^{60}Co).

Morgan, T. J. 1973. Uptake and elimination of tritiated water by fish. Thesis, Purdue Univ., Lafayette, Ind. 97pp. (Univ. Microfilms 73-28,115) (NSA 29: 16071).

Mori, N. 1956. Radioactive contamination of the mulberry field soil in suburbs of Tokyo. p. 731. In: Reserach in the Effects and Influences of the Nuclear Bomb Test Explosions. Japan Soc. for the Promotion of Science, Ueno, Tokyo.

Mori, N. 1963. On the effects of gamma rays upon the development of Ascaris eggs. Sogo-Kaiyokagaku 5: 1-12. (In Japanese) (NSA 18: 27163).

Mori, N. 1963. The effects of ionizing radiation upon yeast cells. Nippon Univ. J. Radiat. Med. Biol. 1: 1-12. (NSA 19: 7127).

Mori, N., Y. Kunagai and Y. Soeno. 1962. The effects of ionizing radiation emitted from ^{60}Co upon the development of Ascaris eggs. Abstr. p. 241. In: 2nd Intern. Congr. Radiation Reserach. (See Anonymous, 1962).

Mori, T. and M. Saiki. 1956. Studies on the distribution of administered radioactive zinc in the tissues of fish. pp. 1205-1208. In: Research in the Effects and Influences of the Nuclear Bomb Test Explosions. Vol. II. Japan Soc. for the Promotion of Science, Ueno, Tokyo.

Morikawa, O. and T. Saito. 1966. Degradations of Vamidothion and Dimethoate in plants, insects, and mammals. Botyo-kagaku 31(3): 130-135. (^{32}P).

Morisawa, S. 1973. Selection of a site for the ground disposal of radioactive wastes. Nippon Genshiryoku Gakkaishi 15(7): 503-508. (In Japanese) (NSA 29: 15633).

Morisawa, S. and Y. Inoue. 1974. Selection of a ground disposal site by sensitivity analysis. Health Physics 26(3): 251-261. (NSA 29: 19027).

Morishima, H., et al. 1973. Determination of uranium isotopic ratio ^{235}U/^{238}U in environmental samples by neutron activation analysis. Hoken Butsuri 8(3): 153-160. (In Japanese) (NSA 29: 9797).

Moriuchi, S., K. Imai and I. Miyanaga. 1974. New low level environmental radiation monitoring around nuclear facilities; discriminating from natural radiation. pp. 382-387. In: Proc. 3rd Intern. Congr. Intern. Radiation Protection Assoc. (CONF-730907-P1) (NSA 30: 27021).

Moriwaki, D., N. Yoskiko and H. Ikeda. 1967. A shift of sex ratio in the progeny from irradiated males in Drosophila melanogaster. Nippon Idengaku Zasshi 42: 23-38. (NSA 22: 15102).

Moriwaki, D., et al. 1958. Survival rates of progeny of irradiated Drosophila melanogaster. Abstr. Radiation Research 9(1): 155.

Morlan, H. B., et al. 1962. Field tests with sexually sterile males for control of Aedes aegypti. Mosquito News 22(3): 295-300. (NSA 18: 25047).

Morley, F. 1973. Radiological limits and environmental capacities. pp. 573-581. In: Environmental Behavior of Radionuclides Released in the Nuclear Industry. (See IAEA, 1973) (NSA 29: 29307).

Morley, F. and P. M. Bryant. 1969. Basic and derived radiological protection standards for the evaluation of environmental contamination. pp. 255-272. In: Environmental Contamination by Radioactive Materials. (See IAEA, 1969).

Moroni, A. 1970. From natural to human ecosystem. pp. 27-35. In: Proc. 1st Conf. on the State of Advancement of Radioecology in Italy. (See Anonymous, 1970) (In Italian).

Moroni, A. 1973. Quality promotion of the human environment. (Discussion). pp. 147-168. In: Proc. 2nd Conf. on the State of Advancement of Radioecology in Italy. (See Anonymous, 1973) (In Italian).

Morozova, N. G. 1973. Natural radioelements in contemporary bottom sediment of freshwater lakes of Estonia. pp. 116-122. In: Radioecology of Water Organisms. II. (See Andrumaitis, 1973) (NSA 30: 21116).

Morozova, V. P. 1963. Immunization of irradiated Citellus suslica Güldenstädt with penicillin and Penicillinalum leptospira vaccine. Zh. mikrobiol. epidemiol. immunobiol. (7): 64-68. (In Russian) (NSA 19: 33948).

Morozova, V. P. 1963. Search for methods of stimulating leptospirosis immunity in irradiated animals. Zh. mikrobiol. epidemiol. immunobiol. (10): 75-80. (In Russian) (NSA 18: 9908).

Morris, D. A. 1967. Use of chemical and radioactive tracers at the National Reactor Testing Station, Idaho. pp. 130-142. In: Isotope Techniques in the Hydrologic Cycle. (See Stout, 1967).

Morris, R. L. 1958. The effect of radioactive rains on surface water supplies. Water Sewage Works 105: 92-94. (NSA 13: 9920).

Morrison, D. L. 1973. General environmental siting criteria: a review of current approaches. Trans. Amer. Nucl. Soc. 17: 70. (NSA 29: 7118).

Morrison, D. L., et al. 1972. Environmental benefit-cost analysis for nuclear power generation. Nuclear News 15(6): 50-57. (NSA 26: 47386).

Morrison, F. O. and W. F. Oliver. 1949. The distribution of radioactive arsenic in the organs of insect larvae. Can. J. Research 27: 265-269. (NSA 4: 367).

Morrison, H. G. and D. V. Cormack. 1965. Variation of the x-ray sensitivity of Serratia marcescens with relative humidity. Nature 208: 91-93. (NSA 20: 99).

Morrison, R. B. 1965. Radiocarbon chronologies of Lakes Lahontan and Bonneville: a stratigraphic evaluation. Abstr. Abstr. Intern. Assoc. Quatenary Res. (7th Intern. Congr., Boulder and Denver, Colo.).

Mortensen, J. L., E. C. Marcusiu and N. Holowaychuk. 1963. Strontium exchange characteristics of soils from the Ogotoruk Creek watershed in Alaska. Ohio J. Sci. 63(5): 225-231. (NSA 18: 33777).

Mortimer, R. K. 1958. Radiobiological studies on yeast (discussion). Lawrence Radiation Lab., Berkeley, U. S. AEC report UCRL-84-75. 9pp. (NSA 13: 5229).

Mortimer, R. K. and R. C. von Borstel. 1963. Radiation induced dominant lethality in haploid and diploid sperm of the wasp Mormoniella. Genetics 48(11): 1545-1549. (NSA 18: 73).

Morton, R. J. (Ed.). 1961-1966. Status report No. 1 on Clinch River Study. ---No. 3, 4, 5, 6. Oak Ridge National Lab., AEC report ORNL-3119, -3370, -3409, -3721, -3941. vii, 75; x, 117; xi, 110; xiii, 142; 84pp. (NSA 17: 2810, 37055; 21: 3174).

Morton, S. D. and P. H. Derse. 1968. Use of gamma radiation to control algae. Environ. Sci. Technol. 2(11): 1041-1043. (NSA 23: 5057).

Mortreuil, M. 1959. Lethal effects of x-rays on Calandra granaria L. Compt. rend. soc. biol. 153: 393-394. (In French) (NSA 13: 17682).

Mortreuil, M. 1959. Radioactive marking of the puceron of the aphid (Aphis leguminosae) and its parthenogenetic progeny. Compt. rend. soc. biol. 153(5): 755-757. (In French).

Mortreuil, M. 1959. Sterilizing action of x-rays on the weevil. Compt. rend. soc. biol. 153: 1165-1166. (In French).

Mortreuil, M. and I. M. Brader. 1962. Radioactive marking of ants through les plantations d'ananas. pp. 39-45. In: Radioisotopes and Radiation in Entomology. (See IAEA, 1962) (In French).

Mortreuil-Langlois, M. 1960. Effect of x-rays on the middle intestine of Blabera fusca Br. Compt. rend. soc. biol. 154: 1769-1770. (In French) (NSA 15: 17940).

Mortreuil-Langlois, M. 1963. Histopathological study of the midgut of Blabera fusca Br. (Orthoptera) over an extended post-irradiation period. Bull. soc. zool. Fr. 98: 539-546. (In French).

Moscicki, J. 1972. Soil density measuring by the γ-γ method range 0.5 to 1.4 g/cm^3. Poland Inst. of Nuclear Research, Krakow, report ITJ-28. 33 pp. (In Polish) (NSA 28: 2747).

Moseley, B. E. B. 1967. The isolation and some properties of radiation-sensitive mutants of Micrococcus radiodurans. J. Gen. Microbiol. 49(2): 293-300. (NSA 22: 49309).

Moser, H. and F. Neumaier. 1972. Nuclear methods in hydrology for construction engineering. pp. 335-378. In: Conf. on Radiation and Isotope Techniques in Civil Engineering. Commission of the European Communities, Brussels, report EURISOTOP-55(Vol.1). (In German) (NSA 27: 27604).

Moser, H. and F. neumaier. 1973. Groundwater: experiments with isotopes. Umsch. Wiss. Tech. 73(16): 501-502. (In German) (NSA 29: 10154).

Moser, H., F. Neumaier and W. Rauert. 1963. New experiences with the use of radioactive isotopes in hydrology. pp. 283-295. In: Radioisotopes in Hydrology. (See IAEA, 1963) (NSA 18: 1923).

Moser, H., F. Neumaier and W. Rauert. 1973. Application of isotope methods for the investigation of groundwater pollution. Z. Deutsch. Geol. Ges. 124: 501-514. (In German) (NSA 30: 32113).

Moser, H., F. Neumaier and W. Rauert. 1973. Application of isotope methods for the investigation of groundwater pollution. Gesellschaft für Strahlen und Umweltforschung mbH, F. R. Germany, Nürnberg. (For Symp. on Groundwater and Environment, Essen, Germany, 2 May 1973). vp. (In German) (NSA 28: 29903).

Mosetti, F. 1965. A new interpretation of a radioactive tracer experiment in the Timavo River. Boll. Geofis. Teor. Appl. 7: 218-243. (In Italian) (NSA 20: 35656).

Moshagen, D. 1973. Analysis of isotopic exchange in isolated frog skin. (For 41st Meeting of the German Physiology Society, Düsseldorf, Germany, 27 March 1973). vp. (In German) (NSA 28: 21832) (^{24}Na).

Moskalev, Yu. I. 1966. Distribution and Biological Effects of Radioactive Isotopes. Atomizdat, Moscow. 718pp. (AEC-tr-6944(Rev.)) (NSA 24: 39418).

Moskalev, Yu. I. 1968. Tritium Oxides. Atomizdat, Moscow. 396pp. (In Russian) (NSA 23: 4881).

Moskalev, Yu. I. (Ed.). 1970. Radioisotopes in the External Environment and in Organisms. Atomizdat, Moscow. 264pp. (In Russian) (NSA 25: 24010).

Moskalev, Yu. I. and V. G. Kulikova. 1961. Distribution of cerium-144 in animals. pp. 53-78. In: Distribution, Biological Effects, and Migration of Radioactive Isotopes. (See Lebedinskiy and Moskalev, 1961) (AEC-tr-7512) (NSA 28: 13098) (Incl. frogs).

Moss, C. E. and E. J. Sternglass. 1973. Environmental aspects of radioisotopes excreted in nuclear medicine procedures. p. 268. In: Health Physics in the Healing Arts. Food and Drug Admin., U. S. Dept. of Health, Education, and Welfare, Washington, report FDA-73-8029. (NSA 29: 29373).

Mosse, B. and D. S. Hayman. 1973. Effects of mycorrhiza on phosphate uptake of some plant species from Meathop Wood soil. Abstr. Bull. Brit. Ecol. Soc. 4(3): 6. (^{32}P).

Mosselmans, G. and J. Vaccarezza. 1973. Radioactive pollution: its elimination. Part II. Radioactive wastes processing plants. European Atomic Energy Community, Joint Nuclear Research Center, Ispra, Italy, report EUR-4981. 27pp. (In French) (NSA 28: 18062).

Mossige, J. 1958. Translocations and recessive sex-linked lethals induced in Drosophila melanogaster by ingested radiophosphorus. Abstr. p. 197. In: Proc. 10th Intern. Congr. on Genetics. Vol. 2. (Montreal). Univ. of Toronto Press, Toronto.

Mossige, J. and P. Oftedal. 1958. The mutagenic efficiency of P^{32} contained in Drosophila melanogaster males. Abstr. Radiation Research 9(1): 156.

Mostafa, I. Y., I. M. I. Fakhr and Y. A. El-Zawahry. 1974. Metabolism of organophosphorus insecticides. XV. Translocation and degradation of ^{32}P-malathion in bean and cotton plants. pp. 385-392. In: Comparative Studies of Food and Environmental Contamination. (See IAEA, 1974) (NSA 30: 16077).

Mostova, R. S. 1959. The effect of bactericidal irradiation on the virulence of microorganisms. pp. 150-157. In: Collected Papers, Institute of Radiation Hygiene. (JPRS-8,727) (NSA 15: 28996).

Mosulishili, L. M., N. E. Kuchava and E. N. Ginturi. 1970. Neutron activation analysis of some mineral waters of Georgia. Soobshch. Akad. Nauk Gruz. SSR 59: 93-96. (In Russian) (NSA 24: 48055).

Motomura, I. and M. Yamamoto. 1968. Biological effect of electron beam irradiation of small creatures. I. Res. Rep. Lab. Nucl. Sci., Tohoku Univ. 1(1): 92-93. (In Japanese) (NSA 24: 21278) (Newts, Echinus spawns).

Mott, W. E. 1971. Isotopic techniques in the study and control of environmental pollution. pp. 3-46. In: Nuclear Techniques in Environmental Pollution. (See IAEA, 1971) (NSA 25: 32143).

Moule, J. W. and P. F. Nace. 1964. Radioiodide metabolism in Amphiuma after its intraperitoneal administration. Can. J. Physiol. Pharmacol. 42(5): 623-626. (NSA 18: 38748).

Mound Laboratory. 1970, 1972. Environmental monitoring report. January-June 1970. --- January-June 1972. Miamisburg, O., U. S. AEC report MLM-1784, MLM-1880. 36, 41pp. (NSA 25: 16077; 26: 15185).

Mounib, M. S. and J. S. Eisan. 1969. Metabolism of pyruvate and glyoxylate by eggs of salmon (Salmo salar). Comp. Biochem. Physiol. 29(1): 259-364. (^{14}C).

Mount, M. S. and A. H. Ellingboe. 1969. ^{32}P and ^{35}S transfer from susceptible wheat to Erysiphe graminis f. sp. tritici during primary infection. Phytopathology 59(2): 235.

Mourikis, P. A. and E. Fytizas. 1970. Review of olive-fly ecology in relation to the sterile-male technique. pp. 131-139. In: Sterile-Male Technique for Control of Fruit Flies. (See IAEA, 1970) (NSA 25: 24517).

Moustacchi, E. 1962. X-ray sensitivity of ^{32}P-labeled haploid Saccharomyces cerevisiae. Abstr. p. 19. In: 2nd Intern. Congr. Radiation Research. (See Anonymous, 1962).

Mouton, R. F. and O. Tremeau. 1970. Microbiological evolution under radiation: a phenotype of a radioresistant mutant of Escherichia coli K_{12} induced and selected by successive exposures to gamma radiation from ^{60}Co. Int. J. Radiat. Biol. 17: 237-248. (In French) (NSA 25: 11019).

Moutschen, J. 1957. Action of x-rays on the growth of internodal cells of the alga Chara vulgaris L. Experientia 13(6): 240-244. (In French) (NSA 11: 9199).

Moutschen, J. 1958. Growth modifications due to x-rays. Proc. 2nd. Intern. Conf. on the Peaceful Uses of Atomic Energy 27: 217-222.

Moutschen, J. 1959. X-rays and polyploidy in mosses. Radiobiol. Latina 2: 105-111. (In French) (NSA 14: 8372).

Moutschen, J. 1959. Action of x-rays on a peroxide of the barley seed and some cryptogams. Radiobiol. Latina 1(1): 15-36. (In French).

Moutschen, J. 1960. Combined action of myleran and x-rays and its genetic importance. Hereditas 46: 471-480. (In French) (NSA 15: 5909).

Movsesijan, M., et al. 1967. Effect of irradiation of metacercariae on the subsequent morphogenesis of Fasciola hepatica in albino mice. Exptl. Parasitol. 29: 249-254. (NSA 22: 25985).

Moxham, R. M. 1964. Some aerial observations on the terrestrial component of environmental γ radiation. pp. 737-746. In: The Natural Radiation Environment. (See Adams and Lowder, 1964) (NSA 18: 35744).

Mraz, F. R. 1966. Embryonic response to ^{131}I. Poultry Sci. 45: 96-99. (NSA 20: 35296).

Mraz, F. R. 1969. Calcium and strontium metabolism in Japanes and bob-white quail. Poultry Sci. 48: 737-739. (NSA 23: 47937) (^{45}Ca, ^{89}Sr).

Muckenthaler, F. A. 1964. Autoradiographic study of nucleic acid synthesis during spermatogenesis in the grasshopper, Melanoplus differentialis. Exptl. Cell Res. 35(3): 531-547. (Dissertation Abstr. 25: 2673-2674) (^{3}H).

Mudherjea, A. K., et al. 1960. Radioiodine uptake by Entamoeba histolytica in culture. Bull. Calcutta Sch. Trop. Med. 7: 103-105.

Mugera, G. M., P. K. Bitakaramire and W. K. Munyua. 1973. Preliminary studies on the immunization of cattle against east coast fever using irradiated ticks Rhipicephalus appendiculatus infected with Theileria parva. pp. 129-137. In: Isotopes and Radiation in Parasitology III. International Atomic Energy Agency, Vienna. (NSA 28: 15765).

Mugmon, H. M. and R. C. De Vall. (Comps.). 163. Thyroid metabolism in salmon. pp. 35-38. In: Progress in Sport Fishery Research 1962. U. S. Fish and Wildlife Serv. Cir. 160.

Muhlmann, M. M. 1963. Contamination of the environment by ionizing radiation arising from radioactive rocks. Semana Med. (Buenos Aires) 123: 709-718. (In Spanish) (NSA 18: 31768).

Mukai, T. and I. Yoshikawa. 1964. Heterozygous effects of radiation-induced mutations on viability of homozygous and heterozygous genetic background in Drosophila melanogaster (preliminary report). Nippon Idengaku Zasshi 38: 282-287.

Mukai, T., I. Yoshikawa and K. Sano. 1966. The genetic structure of natural populations of Drosophila melanogaster. IV. Heterozygous effects of radiation-induced backgrounds. Genetics 53: 513-527. (NSA 20: 18333).

Mukherjee, A. B. 1965. Pattern of DNA synthesis in mosquito chromosomes. Abstr. 102. Bull. Entomol. Soc. Am. 11(3): 159. (^{3}H, Aedes cataphylla).

Mukherjee, A. S. and W. Beermann. 1965. Synthesis of ribonucleic acid by the x-chromosomes of Drosophila melanogaster and the problem of dosage compensation. Nature 207: 785-786. (^{3}H).

Mukhin, I. Ye. 1961. Examination of water for radioactivity in selecting underground sources for drinking water supply. Gigiyena sanitariya 26(11): 9-14. (NSA 16: 20889).

Mukhin, I. Ye. and L. I. Nagovitsina. 1967. Strontium-90 content in total fallout in the territory of the Ukranian SSR in 1963-1966. State Committee on the Use of Atomic Enery, USSR. Atomizdat, Moscow. UN Document A/AC.82/G/L.1199. 17pp. (In Russian).

Mukhin, I. Ye., F. M. Levinina and R. I. Khil'chevskaya. 1960. Radioactivity of the water of the Dnieper River and some underground sources and their hygienic estimation. In: Proc. Conf. on Radiation Hygiene. Minzdrav RSFSR, Leningrad. (In Russian).

Mukhin, I. Ye. 1973. Radioactivity of the environment, food products, and man in the Ukraine in the period 1962-1969. State Committee on the Use of Atomic Energy, USSR. Atomizdat, Moscow. 23pp. (In Russian) (INIS-mf-35) (NSA 30: 6540).

Mukhin, I. Ye., N. M. Borovikova and L. I. Nagovitsina. 1974. Potentiation of strontium-90 biological action on an organism by agricultural chemicals. pp. 136-140. In: Proc. 3rd Intern. Congr. Intern. Radiation Protection Assoc. (CONF-730907-P1) (NSA 30: 27221).

Mukhin, I. Ye., et al. 1973. Global ^{137}Cs and ^{90}Sr migration in food chains of the population of Ukrainian woody boggy areas. State Committee on the Use of Atomic Energy, USSR. Atomizdat, Moscow. 18pp. (In Russian) (INIS-mf-1135) (NSA 30: 12700).

Mulla, M. S. and R. B. March. 1959. Flight range dispersal patterns and population density of the eye gnat, Hippelates collusor. Ann. Entomol. Soc. Am. 52(6): 641-646. (Used ^{32}P).

Mullen, R. K. 1974. Internal dosimetry in small vertebrates resident in a fallout field. Trans. Amer. Nucl. Soc. 18: 391. (NSA 30: 16005).

Müller, A. 1970. Safety measures for nuclear power stations with regard to protection of water, soil, and air. Bundesgesundheitsblatt 13(3): 25-30. (In German) (NSA 25: 23321).

Muller, H. D. 1961. The avian embryo and x-irradiation. Abstr. Radiation Research 14(4): 488.

Muller, H. D. and R. E. Moreng. 1960. Growth and fecundity of chickens hatched from embryos surviving x-ray irradiation. Abstr. Poultry Sci. 39(5): 1278.

Muller, H. D., R. E. Moreng and P. A. Thornton. 1959. Response of the avian embryo to x-ray irradiation. Abstr. Poultry Sci. 38(5): 1232.

Muller, H. J. 1938. Comparative effects of 200kv x-rays and gamma rays on the pupae of Drosophila melanogaster. Am. J. Cancer 32: 565-___.

Muller, H. J. 1963. Mechanisms of life-span shortening. pp. 235-245. In: Cellular Basis and Aetiology of Late Somatic Effects of Ionizing Radiation. Academic Press, New York. (NSA 17: 40647) (Comments on Drosophila, Habrobracon).

Müller, H. P. 1968. Soil sterilization with ionizing radiation. Atompraxis 14: 480-484. (In German) (NSA 23: 38906).

Müller, I., E. A. Löbbecke and O. Oltmanns. 1962. Dose effect curve of somatic mutations in Ephestia kühniella Z. for low dose range (0-200r). Nature 194(4830): 783-784. (NSA 16: 20281).

Müller, I., S. Pentz and C. Harte. 1966. On the relation between the somatic mutation spectrum of Ephestia kühniella and the developmental stage of the irradiated animal. Z. Vererbungsl. 97: 353-360. (In German) (NSA 21: 10844).

Müller, P. 1958. Experiments on current flow and ionic movements in single myelinated nerve fibers. Exptl. Cell Res. (Suppl.5): 118-152. (Frog).

Müller, P. W. and P. H. Payot. 1966. Fate of ^{14}C-labelled triazine herbicides in plants. pp. 61-71. In: Isotopes in Weed Research. (See IAEA, 1966).

Muller-Olsen, C. and M. Simak. 1954. X-ray photography employed in germination analysis of Scots pine (Pinus silvestris L.) Meddel. Fran Statens. Skogsforskninginstitut 44(6): 1-19.

Muller-Olsen, C., M. Simak and A. Gustafsson. 1956. Germination analysis by the x-ray method: Picea abies (L.). Karst. Medd. Stat. Skogforskinst. 46(1): 12pp.

Mulligan, W. 1963. Application of radiobiology to the study of helminth disease. World Vet. Congr. Proc. (Hanover) 1: 831-832. (NSA 21: 18147).

Mulligan, W. and D. F. Stewart. 1961. The use of ionizing radiation in the study of immunity to parasitic nematodes. pp. 245-249. In: Radiobiology. P. L. T. Ilbery, ed. Butterworths Sci. Publ., London. (NSA 13: 30520).

Mullins, J. W. and J. L. Stein. 1972. Evaluation of tritium in ground and surface waters of the western United States, April 1968-December 1969. Radiat. Data Rept. 13(2): 59-67. (NSA 26: 25589).

Mullins, J. W., V. E. Andrews and L. M. Dunn. 1972. Techniques for monitoring and analysis of environmental tritium. Trans. Nucl. Sci. NS-19(1): 177-180. (NSA 26: 42956).

Mulvey, P. F., Jr. 1960. The uptake and fate of radioactive elements in Trypanosoma equiperdum. J. Infect. Dis. 107: 155.

Muminov, M. 1964. Radioactivity of the most common grass-like plants. Izv. Akad. Nauk Uzbek. SSR (fiz.-mat.) (5): 68-70. (In Russian) (NSA 19: 28587).

Muminov, M. and Yu. Maksudov. 1969. The radioactivity of grain: leguminous plants. Izv. Akad. Nauk Uzbek. SSR (fiz.-mat.) (2): 75-77. (In Russian) (NSA 24: 6018).

Mumma, R. O., et al. 1966. Dieldrin: extraction of accumulations by root uptake. Science 152: 530-531. (^{36}Cl).

Münchberg, P. 1964. Demonstrating the haemolymph circulation in dragonfly wings by means of injections of $Na_2{}^{35}SO_2$ and $Na_2H^{32}PO_4$ solutions. Z. Naturforschung 19b(7): 634-640. (In German).

Münnich, H. 1973. Transport mechanisms of radioactive substances in the water cycle. pp. 43-55. In: Papers Presented at the Meeting at Göttingen. Germany Bundesministerium des Innern, Bonn. (In German) (AED-CONF-73-697-000) (NSA 32: 19751).

Münnich, K. O. 1963. Tritium of atomic bombs as an indicator in hydrology. Physik. Bl. 19: 418-421. (In German) (French transl. CEA-tr-A-1729) (NSA 19: 20174).

Münnich, K. O. and W. Röther. 1963. A comparison of carbon-14 and tritium ages of groundwater. pp. 397-406. In: Radioisotopes in Hydrology. (See IAEA, 1963) (NSA 18: 1930).

Münnich, K. O. and J. C. Vogel. 1958. Radioactive carbon in the atmosphere produced by atomic explosions. Naturwissenschaften 45: 327-329. (In German) (NSA 12: 16146).

Münnich, K. O. and J. C. Vogel. 1959. C^{14}-age determinations of freshwater of calcareous deposits. Naturwissenschaften 46: 168-169. (In German).

Münnich, K. O. and J. C. Vogel. 1959. The recent increase in the C^{14} content of the atmosphere, the biosphere and the ocean. Germany Zweitas Physikalisches Inst. der Univ. Heidelberg, Federal Ministry of Nuclear Energy and Water Economy, Bad Godesberg. UN Document A/AC.82/G/L.279. (NSA 14: 15846).

Münnich, K. O. and J. C. Vogel. 1960. C^{14} age determination of deep groundwater. I.A.S.H. Comm. of Subterranean Waters, Publ. 52: 537-541.

Münnich, K. O., W. Töther and L. Thilo. 1967. Dating of groundwater with tritium and ^{14}C. pp. 305-320. In: Isotopes in Hydrology. (See IAEA, 1967) (NSA 22: 18929).

Munteanu, G., et al. 1964. Contributions to the study of changes in the β activity of the vegetation as a function of the place of collection. Acad. Rep. Populare Romine Studii cerc. fiz. 15: 591-600. (In Rumanian) (NSA 19: 30404, 42663) (Incl. pine needles, moss, grass).

Muntiu, N., et al. 1961. Influence of ionizing radiation on the evolution and neosalvarsan treatment of experimental trypanosomiasis in the rat. Arch. Roum. pathol. exptl. microbiol. 20: 76-85. (NSA 16: 20256).

Muntzing, L. M. 1972. AEC's response to environmental legislation. Trans. Amer. Nucl. Soc. 15(1): 73. (NSA 26: 42933).

Münzner, R. 1969. Some factors affecting the radiosensitivity of molds. Arch. Mikrobiol. 64: 349-356. (In German) (NSA 23: 35942).

Münzner, R. 1973. Identification of irradiated mushrooms. Germany Bundesforschungsanstalt für Lebensmittelfrischhaltung, Karlsruhe. vp. (In German) (NSA 29: 15815).

Murachi, K. 1944. The influence of radiation upon fish eggs. (1) The influence of KCl upon the heart of an embryo which has been kept under radiation of x-rays. Zool. Mag. (Tokyo) 56(8): 5-7. (In Japanese).

Murakami, A. 1966. Relative biological effectiveness of 14 Mev neutrons to gamma-rays for inducing mutations in mature sperm of the silkworm. Nippon Idengaku Zasshi 41: 17-26. (NSA 20: 38626).

Murakami, A. 1970. Strain sensitivity of silkworm eggs to killing and mutation by ultraviolet light. Radiation Research 44: 146-153. (NSA 25: 6766).

Murakami, A. 1971. Comparison of the stage sensitivity to x-rays during meiosis in the egg of the silkworm, Bombyx mori. Int. J. Radiat. Biol. 19: 167-176. (NSA 25: 42609).

Murakami, A. 1971. Relative biological effectiveness of 14 MeV neutrons for inducing dominant lethals in mature sperm of the silkworm: a comparison of the RBE for dominant lethals and specific-locus mutations. Jap. J. Genet. 46(2): 67-74. (NSA 26: 7012).

Murakami, A. 1972. Effects of ultraviolet light on the early embryos of the silkworm and its biological significance. Dobutsugaku Zasshi 81(1): 41-48. (In Japanese) (NSA 28: 6053).

Murakami, A. 1973. Comparison of mutation induction by 14 MeV fast neutrons and ^{137}Cs γ rays in meiotic spermatocytes in the silkworm. Mutation Research 17(1): 73-80. (NSA 27: 9878).

Murakami, A. and S. Kondo. 1963. Relative biological effectiveness of 14.1 Mev neutrons in killing dormant silkworm eggs. Ann. Rep. Nat. Inst. Genet. (14): 106-107. (NSA 20: 72).

Murakami, A. and S. Kondo. 1964. Relative biological effectiveness of 14 Mev neutrons to γ-rays for inducing mutations in silkworm gonia. Nippon Idengaku Zasshi 39: 102-114. (NSA 19: 21915).

Murakami, A. and M. Miki. 1972. Age-dependent changes of radiosensitivity in embryo of Bombyx mori. J. Radiat. Res. (Japan) 13(4): 183-192. (NSA 27: 12349).

Murakami, A. and Y. Tazima. 1962. Modification of x-ray induced mutation rate in the silkworm by pre- or post-irradiation treatment with halogenated bases analogues. Rep. Nat. Inst. Genet. Misima 13: 89-91.

Murakami, A. and Y. Tazima. 1967. Studies on strain differences in radiosensitivity in the silkworm. I. Screening of sensitive and resistant strains to embryonic radiation killing. II. Relation between sensitivity to embryonic killing and mutability. Ann. Rep. Nat. Inst. Genet. Misima (17): 98-100, 100-102. (NSA 23: 43853, 43735).

Murakami, A., S. Kondo and Y. Tazima. 1965. Comparison of fission neutrons and γ-rays in respect to their efficiency in inducing mutations in silkworm gonia. Nippon Idengaku Zasshi 40: 113-124. (NSA 19: 45883; 21: 43359).

Murakami, Y., et al. 1958. Chemical prospecting of uranium deposits in Japan. Proc. 2nd Intern. Conf. on the Peaceful Uses of Atomic Energy 2: 131-139. (Paper A/CONF.15/P/1357).

Muraki, K. and D. Morikawa. 1956. Genetic effects induced on plants. p. 615. In: Research in the Effects and Influences of the Nuclear Bomb Test Explosions. Vol. II. Japan Soc. for the Promotion of Science, Ueno, Tokyo.

Murav'yev, M. N. 1963. A study of certain manifestations of automatism in frog intestine after whole-body gamma-irradiation. Radiobiologiya 3(4): 540-544. (AEC-tr-5437, pp. 73-79) (NSA 18: 6669).

Murayama, N. 1965. Highly active single particles in radioactive fallout. Kisocho Kenkyu Jiho 17: 308-316. (In Japanese) (NSA 20: 41138).

Murdie, G. 1973. Development of a population model of cotton red bollworm as a basis for testing pest control strategies. pp. 119-128. In: Computer Models and Application of the Sterile-Male Technique. (See IAEA, 1973) (NSA 29: 8054).

Murphy, C. E., Jr. and J. C. Corey. 1974. Absorption of tritiated water vapor from the atmosphere by the needles of pine trees. Savannah River Plant, U. S. AEC report DP-MS-74-63. 14pp. (NSA 31: 29985).

Murphy, K. J. 1974. Effective disinfection of sewage water by gamma radiation. Ind. Miljoe 5(8): 5-9. (In Norwegian) (NSA 33: 7331).

Murphy, P. G. 1970. Ecological effects of acute beta irradiation from simulated fallout particles upon a natural plant community. Dissertation, Univ. of North Carolina, Chapel Hill. xii, 97pp.

Murphy, P. G. 1970. Tree growth at El Verde and the effects of ionizing radiation. Ch. D-4, pp. D141-D171. In: A Tropical Rain Forest. (See H. T. Odum and R. F. Pigeon, 1970) (NSA 25: 21930).

Murphy, P. G. 1975. Response of a forest ecotone to ionizing radiation. Summary report, 15 Oct. 1972-14 Jul. 1975. Michigan State Univ., East Lansing, U. S. AEC report COO-2283-6. 26pp. (NSA 32: 19910).

Murphy, P. G. and J. F. McCormick. 1971. Ecological effects of acute beta irradiation from simulated-fallout particles on a natural plant community. pp. 454-581. In: Survival of Food Crops and Livestock in the Event of Nuclear War. (See Bensen and Sparrow, 1971).

Murphy, P. G. and J. F. McCormick. 1973. Effects of beta radiation from simulated fallout on the reproductive potential of a dominant plant species in a natural community. pp. 983-987. In: Radionuclides in Ecosystems. (See Nelson, 1973) (NSA 25: 42718).

Murphy, P. W. 1958. Radioisotope method for determination of rate of disappearance of leaf litter in woodland. pp. 1-6. In: Colloquim on Research Methods in Soil Zoology. Int. Soc. Soil Sci., Rothamsted, England, No. 9.

Murray, C. H. and L. Murray. 1973. Absorption-desorption equilibria of some radionuclides in sediment-fresh-water and sediment-seawater systems. pp. 105-122. In: Radioactive Contamination of the Marine Environment. (See IAEA, 1973) (Also: Radioact. Sea (36): 105-124. (NSA 28: 8165; 33: 29112).

Murrmann, R. P., R. W. Winters and T. G. Martin. 1971. Determination of trace elements in soils and clay minerals by resonance neutron activational analysis. Soil Sci. Soc. Am. Proc. 35(4): 647-652. (NSA 25: 57375).

Murty, G. S., H. C. Moser and J. A. Hobbs. 1963. A tracer technique for rapid determination of nutrient movement in soil. Soil Sci. Soc. Am. Proc. 27(4): 477-478.

Musacchia, X. J. and R. E. Barr. 1968. Survival of whole-body-irradiated hibernating and active ground squirrels: Citellus tridecemlineatus. Radiation Research 33: 348-356. (NSA 22: 19367).

Musacchia, X. J. and R. E. Barr. 1969. Comparative aspects of radioresistance with depressed metabolism. pp. 569-607. In: Depressed Metabolism. X. J. Musacchia, ed. American Elsevier, New York. (NSA 24: 27702) (Ground squirrel, hamster).

Musacchia, X. J. and D. D. Westhoff. 1964. Absorption of D-glucose by segments of intestine from active and hibernating, irradiated and non-irradiated ground squirrels, Citellus tridecemlineatus. Ann. Acad. Sci. Fenn. ser. A, IV (biol.) (71): 345-359. (NSA 19: 45841).

Musagalyeva, G. M., A. B. Uteshev and K. A. Ulmanova. 1969. Radiosensitivity of different cell fractions of adrenal medulla. Tr. Kazak. nauchn.-issled. inst. onkol. radiol. 6: 22-24. (In Russian) (NSA 24: 34130).

Muth, H., et al. 1960. The normal radium content and the Ra^{226}/Ca ration of various foods, drinking water and different organs and tissues of the human body. Health Physics 2(3): 239-245.

Muzheyev, _. and _. Shitova. 1956. Influence of radon irradiation on the oxygen consumption of an isolated frog muscle. pp. 69-77. In: Problems of Radiobiology. M. N. Pobedenskiy and P. N. Kiseleva, eds. Medgiz, Leningrad. (In Russian).

Muzzarelli, R. 1973. Chemical behavior of radionuclides in natural waters and their interaction with chelating agents. Final report, 15 March 1970-14 May 1973. International Atomic Energy Agency, Vienna, report IAEA-R-895-F. vp. (INIS) (NSA 30: 9108).

Myers, D. A., et al. 1975. Evaluation of the use of sludge containing plutonium as a soil conditioner for food crops. Lawrence Livermore Lab., U. S. ERDA report UCRL-77318. 19pp. (NSA 33: 2987).

Myers, D. S., J. F. Tinney and P. H. Gudiksen. 1971. Health physics aspects of LRL tritium release. Lawrence Radiation Lab., U. S. AEC report UCRL-73310. 18pp. (NSA 25: 46870).

Myers, D. S., J. F. Tinney and P. H. Gudiksen. 1973. Health physics aspects of a large accidental tritium release. pp. 611-622. In: Tritium. (See Moghissi and Carter, 1973) (NSA 28: 27227).

Myers, W. G. and F. A. White. 1971. Mass spectrometry and fission track analysis in nuclear environmental measurements. pp. 38-43. In: Nuclear Methods in Environmental Research. (See Vogt, 1971) (NSA 26: 17430).

Myburgh, A. C. 1963. Lethal and sterilizing effects of cobalt-60 gamma rays on Argyoploce leucotreta. pp. 514-525. In: National Conf. on Nuclear Energy. Application of Isotopes and Radiation. South Africa Atomic Energy Board, Pelindaba.

Myhill, J. 1969. Investigation of the effect of data error in the analysis of biological tracer data from three compartment systems. J. Theoret. Biol. 23: 218-231. (NSA 23: 50260).

Myllymäki, A. 1969/1970. Trapping experiments on the water vole, Arvicola terrestris (L.), with the aid of the isotope technique. pp. 39-55. In: Energy Flow Through Small Mammal Populations. K. Petrusewicz and L. Ryszkowski, eds. PWN-Polish Scientific Publishers, Warsaw. (^{32}P).

Myllymäki, A., A. Paasikallio and U. Häkkimen. 1971. Analysis of a 'standard trapping' of Microtus agrestis (L.) with triple isotope marking outside the quadrat. Ann. Zool. Fenn. 8(1): 22-34. (^{32}P, ^{131}I, ^{51}Cr).

Myllymäki, A., et al. 1971. Removal experiments on small quadrats as a means of rapid assessment of the abundance of small mammals. Ann. Zool. Fenn. 8(1): 177-185. (^{32}P, ^{131}I).

Myser, W. C. and W. E. Carey. 1969. X-ray radiation effects on survival and sterility of adults of the cereal leaf beetle. J. Econ. Entomol. 62 (3): 543-550.

Myttenaere, C. and P. Bourdeau. 1968. Some aspects of the radioactive contamination of irrigated cultures. pp. 289-296. In: Intern. Congr. on Radioprotection of the Environment. (See Societé Francaise de Radioprotection, 1968) (In French).

Myttenaere, C. and U. Marckwordt. 1973. Influence of water regime on the indirect absorption of radiocesium, radiostrontium, and radiocobalt by lowland rice. pp. 212-217. In: Radionuclides in Ecosystems. (See Nelson, 1973) (NSA 28: 27270).

Myttenaere, C. and M. Masset. 1966. Effect of irrigation on absorption and localization of caesium and strontium in Oryza sativa L. pp. 499-517. In: Disposal of Radioactive Wastes into Seas, Oceans and Surface Waters. (See IAEA, 1966).

Myttenaere, C. and J. Mousny. 1971. Influence of the quality of the irrigation water on the transfer of radiocaesium and radiostrontium in irrigated rice-fields. pp. 1255-1275. In: Proc. Intern. Symp. on Radioecology Applied to the Protection of Man and His Environment. (See Commission of the European Communities, 1971) (In French) (NSA 27: 5574).

Myttenaere, C., P. Bourdeau and M. Masset. 1969. Relative importance of soil and water in the indirect contamination of flooded rice with radiocaesium. Health Physics 16: 701-707. (NSA 23: 35613).

Nachtigall, W. 1965. Investigations of the electroretinogram on the effect of ionizing radiation on the complex eyes of insects. Biophisik 2: 145-165. (In German) (NSA 19: 21882).

Nachtwey, D. S. 1968. Differential lethal effects of mixed reactor radiations on logarithmic-phase and synchronized Tetrahymena pyriformis. U. S. Naval Radiological Defense Lab., San Francisco, report USNRDL-TR-68-53. 18pp. (AD-671943; NSA 23: 6558).

Nadareishvili, K. Sh. and O. I. Saneblidze. 1966. Ionizing radiation effects on contractile functions of frog striated muscles. Soobshch. Akad. Nauk Gruz. SSR 44: 741-748. (In Russian) (NSA 21: 20048).

Nadareishvili, K. Sh. and O. I. Saneblidze. 1967. The mechanism of radiation-induced contracture of striated muscles. Radiobiologiya 7(2): 214-220. (AEC-tr-6887, pp. 91-101) (NSA 21: 34784).

Nadareishvili, K. Sh., B. S. Gugushvili and E. D. Kakhiani. 1967. Local ionizing radiation effects on the heart activity in frogs. Soobshch. Akad. Nauk. Gruz. SSR 47: 451-456. (In Russian) (NSA 22: 12905).

Nadel, D. J. and G. Guerreiri. 1969. Experiments on Mediterranean fruit fly control with the sterile-male technique. pp. 97-105. In: Sterile-Male Technique for Eradication or Control of Harmful Insects. (See IAEA, 1969) (NSA 24: 32305).

Nadkarni, R. A. 1974. Role of neutron activation analysis in biomedical and environmental studies. Radiochem. Radioanal. Lett. 19(1): 17-24. (NSA 31: 5).

Nadkarni, V. B. and A. Gorbman. 1966. Structure of the corpuscle of Stannius in noermal and radiothyroidectomized chinook fingerlings and spawning Pacific salmon. Acta Zoologica (Stockholm) 47(1-8): 61-66. (BA 47: 106053; SFA 11: 8946).

Nagagawa, K. 1970. Radiation sterilization of laboratory animal diets. Genshiryoku Kogyo 16(12): 42-46. (In Japanese) (NSA 25: 47073).

Nagai, M. and S. Ikeda. 1973. Carbohydrate metabolism in fish. IV. Effect of dietary composition on metabolism of acetate-U-^{14}C and L-alanine-U-^{14}C in carp. Bull. Jap. Soc. Sci. Fish. 39(6): 633-634. (In English) (SFA 19: 17165).

Nagamatsu, T. 1949. Cytogenetics of the asyaptic sterile hedionda (Cassia occidentalis L.) collected in Nakasaki City after atomic bomb explosion. Nippon Idengaku Zasshi 24: 46-48.

Nagamatsu, T. 1952. The heritability of aberrant derivations discovered among the progeny of the seeds collected in the exploded area in the city of Nakasaki. J. Hiroshima Med. Assoc. 5: 45. (In Japanese).

Nagamatsu, T. and T. Katayama. 1956. Cytogenetical studies on the sterile wild senna (Cassia tora L.) produced by the atomic bomb explosion in Nagasaki. pp. 617-628. In: Research in the Effects and Influences of the Nuclear Bomb Test Explosions. Vol. II. Japan Soc. for the Promotion of Science, Ueno, Tokyo.

Nagaraj, G., C. Sriramamurty and N. C. Gopalachari. 1970. Improved method for the determination of ^{32}P in plant material. Int. J. Appl. Radiat. Isotopes 21: 563-564. (NSA 24: 46197).

Nagasawa, K., K. Kametani and Y. Kido. 1959. Analysis of Cs^{137} in fresh water. Hisei Shikenjo Hokoku (77): 1-4.

Nagasawa, K., et al. 1956. Studies on radioactive contamination due to A- and H-bomb tests. (4) Radioactive contamination of drinking water, vegetables and fruits. (5) Contamination and radiochemical analysis of fish. (7) Radiation contamination of fish and radio-chemical analysis. Eisei Shikenjo Hokoku 74: 209, 213; 75: 393.

Nagatsuka, S., H. Suzuki and K. Nakajima. 1971. Activation analysis of lanthanide elements in natural water. Radioisotopes (Tokyo) 20(7): 305-309. (NSA 25: 54504).

Nagatsuka, S., Y. Tanizaki and Y. Okano. 1974. Neutron activation analysis of trace elements in water samples. Radioisotopes (Tokyo) 23(12): 693-699. (In Japanese) (NSA 32: 19250).

Nagayama, S. and T. Tamaki. 1963. Leaching of radionuclides into river waters and variation of radionuclide concentrations in tap waters. Bull. Univ. Osaka Pref. A12(1): 127-138. (NSA of Japan 4(3): 01919).

Nagayama, S., et al. 1973. Radioecological study of strontium-90 in Osaka District, Japan. Bull. Univ. Osaka Pref. A21(2): 331-338. (NSA 29: 15632).

Nagel, E. 1971. First measures for ascertaining radiation danger in the neighborhood of the Federal Institute for Reactor Research (EIR) after an accident. pp. 695-703. In: Rapid Methods for Measuring Radioactivity in the Environment. (See IAEA, 1971) (In German) (NSA 26: 28057).

Nagel, R. H. 1959. Influence of nematodes and a tagging solution of the flight performance of the Engelmann spruce beetle. Abstr. Bull. Entomol. Soc. Am. 5(13): 131.

Nagel, R. H. and J. M. Davis. 1956. Results from studies on the Engelmann spruce beetle (Dendroctomus engelmanni). Entomol. Soc. Am. North Central States Branch Proc. 11: 20-21.

Nagler, M. 1964. Importance of the use of radioactive substances and ionizing radiations in phytopathology. Natura (biol.) 16(6): 17-25. (In Rumanian) (NSA 21: 438).

Nagpal, M. K., K. K. Nagpal and A. K. Bhan. 1974. Uranium content in plants a preliminary report. Current Sci. (India) 43(1): 8-9. (NSA 29: 16088).

Nagy, A. 1968. New achievements in the neutron moisture measurements. Atomtech. Tajek. 11: 421-433. (In Hungarian) (NSA 23: 43615).

Nagy, A. Z. and T. Tazga. 1966. Radioisotopic combined moisture-density meter. J. Sci. Instrum. 43: 383-387. (NSA 20: 33537).

Nagy, B. 1967. Possibilities of using ionizing radiation in the control of insects in Hungary. Atomtech. Tajek. 10: 1-6. (In Hungarian) (NSA 22: 12949).

Nagy, B., B. Pozsár and L. Szalay-Marzsó. 1965. The use of ionizing radiations for the control of harmful animals. Izotóp Intézet. Országos Atomenergia Bizottság, Budapest. iii, 83pp. (In Hungarian).

Nagy, I. 1967. Nomograms for use in research with radioactive isotopes. Fiziol. Norm. Pathol. (Bucharest) 13: 153-159. (In Rumanian) (NSA 23: 31302).

Nagy, J. 1973. Determination of microquantities of ^{137}Cs in liquids. Izotoptechnika 16(12): 674-678. (In Hungarian) (NSA 29: 18072).

Nagy, K. A. 1974. Isotopically labeled water used to measure energy and material balance in animals. Trans. Amer. Nucl. Soc. 19: 71-72. (NSA 31: 5552).

Naharin, A. 1970. Effects of ionizing radiation on the susceptibility of store pests. Technion-Israel Inst. Tech., Haifa. 140pp. (In Hebrew) (INIS-mf-51) (NSA 26: 4898).

Naharin, A., M. Calderon and Y. Kott. 1971. Effect of dose rate on the radiation sensitivity of the warehouse pest Callosobruchus maculatus F. Israel Atomic Energy Commission, Tel Aviv, report IA-1237. 11pp. (NSA 26: 15507).

Naharin, A., M. Calderon and Y. Kott. 1972. Effect of dose rate on the radiation sensitivity of the warehouse pest Callosobruchus maculatus F. pp. 42-45. In: Proc. 6th National Conf. Israel Health Physics Society. T. Schlesinger, ed. (In Hebrew) (NSA 29: 16047).

Naidu, J. R. and A. H. Seymour. 1969. Accumulation of zinc by oysters in Willapa Bay, Washington. pp. 463-474. In: Proc. Symp. on Mollusca - Part II. Bangalore Press, Bangalore.

Nair, G. G. 1961. Floral study of the sterile Crotolaria striata L. observed in the radioactive monazite sand. Trans. Bose Res. Inst. 24: 67-72.

Nair, K. K. 1962. Preliminary studies on the effects of gamma-radiation on housefly pupae with special reference to the critical periods in relation to the mechanism of emergence. pp. 207-211. In: Radioisotopes and Radiation in Entomology. (See IAEA, 1962).

Nair, K. K. and N. M. G. Bhakthan. 1969. Preliminary studies on the ultrastructural damage in the flight muscles of gamma-irradiated housefly. Int. J. Radiat. Biol. 16: 397-399. (NSA 24: 34600).

Nair, K. K. and G. W. Rahalkar. 1963. Studies on the effects of gamma radiation on the different developmental stages of the Khapra beetle, Trogoderma granarium Everts. pp. 465-477. In: Radiation and Radioisotopes Applied to Insects of Agricultural Importance. (See IAEA, 1963).

Nair, K. K. and G. Subramanyam. 1963. Effects of variable dose-rates on radiation damage in the rust-red flour beetle, Tribolium castaneum Herbst. pp. 425-429. In: Radiation and Radioisotopes Applied to Insects of Agricultural Importance. (See IAEA, 1963).

Nair, K. K., G. Bhaskaran and P. Sivasubramanian. 1967. Effect on adult emergence of whole and partial x-irradiation of pupae of the housefly, Musca domestica nebulo. Can. Entomol. 99(6): 597-598.

Nair, K. S. S. and J. C. George. 1965. Effects of gamma radiation on the postembryonic stages of a dermestid beetle, *Anthrenus vorax* Waterhouse. Entomologia exper. app. 8(4): 241-248. (NSA 22: 6915).

Nair, P. V. R. 1966. Standardization of ^{14}C stock solutions using scintillation and biological methods in primary production studies. pp. 15-20. In: *Proc. All India Symp. on Radioactivity and Metrology of Radionuclides.* (Bombay, March 14-18, 1966). (CONF-660341) (NSA 22: 6724).

Najdenov, M., et al. 1972. Determining certain macro- and microelements in soils with the aid of neutron activation analysis. (For 2nd National Conf. on Uses of Radioisotopes and Ionizing Radiation in Biology and Agriculture, Sofia, Bulgaria, 28 Nov. 1972). 9pp. (In Bulgarian) (INIS-mf-717) (NSA 29: 15230).

Nakagawa, S., G. J. Farias and L. F. Steiner. 1966. The attractiveness of male lures to sperm-deficient female Mediterranean fruit flies. Bull. Entomol. Soc. Am. 12: 250.

Nakahara, H. and G. Bevelander. 1969. Electron microscope study of ingestion of thorotrast by amebocytes of *Pinctada radiata*. Tex. Rep. Biol. Med. 27: 101-110. (NSA 23: 35605).

Nakai, T. 1938. The minor constituents of the mineral springs of Japan. V. The concentration of uranium from the water of mineral springs. J. Chem. Soc. Japan 59: 1187-1192.

Kakajima, M. 1972. Radiation effects on silkworm, *Bombyx mori* L., with special reference to the changes in the frequencies of induced somatic mutations during different stages of larval development. Nippon Saikingaku Zasshi 41(5): 359-364. (In Japanese) (NSA 29: 27115).

Nakajima, S, K. Shimizu and Y. Nakajima. 1963. Studies on the biological influence on the termites exposed to Co^{60} gamma source. I. Influences on the adult of Formosan termite and its offspring. Miyazaki Daigaku Nogakubu Kenkuy Jiho 8: 340-346. (In Japanese) (NSA 18: 31186).

Nakamura, K. 1953. Reports by forestry research team. I. Effect of atomic bomb on trees. Nippon Gakujutsu Kaigi 1: 263-264.

Nakanishi, Y. H. and S. Makino. 1964. Phase cinematography studies on the effects of radiation and of some chemicals on cells and chromosomes. Int. Soc. Cell Biol. Symp. 3: 47-62. (*Bombyx mori*, *Podisma sapporense*).

Nakanishi, Y. H., T. Iwasaki and H. Kato. 1965. Cytological studies on the radiosensitivity of spermatogonia of the silkworm. Nippon Idengaku Zasshi 40(Suppl.): 49-67. (NSA 19: 43783).

Nakanishi, Y. H., Y. Ohnuki and S. Makino. 1961. Phase cinematography studies on the effects of radiation on the cell, with special regard to the behavior of the chromosomes in grasshopper spermatocytes in response to x- and beta-irradiations. In: *Proc. Symp. on Genetic Effect of Radiation.* (Misima, Nov. 7-8, 1960). Jap. J. Genetics 36(Suppl.): 115-135.

Nakao, Y. 1956. Action of x-rays on amphibian egg development. II. The effect of oxygen upon the relative sensitivity of various stages during the first cleavage in the toad egg. pp. 363-369. In: *Progress in Radiobiology.* (Proc. 4th Intern. Conf., Cambridge, 14-17 August 1955). J. S. Mitchell, B. E. Holmes and C. L. Smith, eds. Chas. C. Thomas, Springfield, Ill.

Nakao, Y. 1965. Comparisons between mutagenic and cell-killing effects induced by radiation and radiomimetic substances. Nippon Idenhaku Zasshi 40(Suppl.): 107-118. (NSA 19: 43787) (Used silkworms).

Nakao, Y. and S. Taguchi. 1962. The protective action of DNP (dinitrophenol) against the lethal effects of x-irradiation in toad embryos. Abstr. p. 96. In: *2nd Intern. Congr. Radiation Research.* (See Anonymous, 1962).

Nakashima, Y. and T. Aoki. 1974. Comparison of dosimetric methods used in environment. J. Nucl. Sci. Technol. 11(1): 15-21. (In Japanese) (NSA 30: 6714).

Nakatani, R. E. 1962. A method for force-feeding radioisotopes to yearling trout. Prog. Fish Cult. 24(2): 56-59. (Also: Hanford Labs., U. S. AEC report HW-SA-2303) (NSA 17: 5738; SFA 7: 5051).

Nakatani, R. E. 1966. Biological response of rainbow trout (*Salmo gairdneri*) ingesting zinc-65. pp. 809-823. In: *Disposal of Radioactive Wastes into Seas, Oceans and Surface Waters.* (See IAEA, 1966).

Nakatani, R. E. 1970. Nuclear power plants and water quality criteria. Battelle Northwest Lab., U. S. AEC report BNWL-SA-3282. 21pp. (For Colloquium on Environmental Quality, Richland, Wash.) (CONF-700425-1) (NSA 25: 7978).

Nakatani, R. E. and R. F. Foster. 1963. Effect of chronic feeding of Sr^{90}-Y^{90} on rainbow trout. pp. 359-362. In: *Radioecology.* (See Schultz and Klement, 1963) (Also: Hanford Labs., U. S. AEC report HW-SA-2255) (NSA 17: 5701).

Nakatani, T. 1968. On the effects of ionizing radiation emitted from ^{60}Co upon the germination and the growth of timothy grass (*Phleum pratense* L.). Report 2. Studies on dose-effects, dose-rate effects, and summation effects. Nichidai Igaku Zasshi 27: 797-810. (In Japanese) (NSA 23: 2984).

Nakatsugawa, T., N. M. Tolman and R. A. Dahm. 1967. Metabolism of ^{35}S-parathion in insects. Abstr. Bull. Entomol. Soc. Am. 13(3): 191.

Nakatsugawa, T., N. M. Tolman and R. A. Dahm. 1969. Metabolism of ^{35}S-parathion in the housefly. J. Econ. Entomol. 62(2): 408-411.

Nakayama, F. A. and W. L. Ehrler. 1964. Beta ray gauging technique for measuring leaf water content changes and moisture status of plants. Plant Physiol. 39: 95-98.

Nakayama. H. 1953. Biological investigation of atomic bomb disaster. 2. Effect on plants. Ch. 4. Radiation effects mainly on plants. Nippon Gakujutsu Kaigi 1: 239-241.

Nakazawa, K. 1960. Effect of x-ray irradiation upon the affinity of bullfrog's sciatic nerve tissue for protoporphyrin. Mie Igaku 4: 293-314. (NSA 15: 25885).

Nakhla, S. M. and G. Delibrias. 1967. Use of ^{14}C of thermonuclear origin to study the dynamics of carbon in the soil. pp. 169-176. In: Radioactive Dating and Methods of Low-Level Counting. (See IAEA, 1967) (In French) (NSA 22: 6457).

Nalborczyk, E., B. Zelawska and M. Kolakawska. 1971. Effects of x-rays on seed germination and growth of Scots pine (*Pinus silvestris* L.) seedlings of different provenence. Acta Soc. Bot. Pol. 40: 403-412. (NSA 27: 15028).

Namenwirth, M. R. M. 1969. Inheritance of cell differentiation during limb regeneration in the axolotl. Thesis, Indiana Univ., Bloomington. 118pp. (Univ. Microfilms 70-7484) (NSA 24: 51066) (X-irradiation).

Narang, N. and S. Narang. 1974. Chromosomal studies during spermatogenesis and the radiation induced chromosomal aberrations in *Biomphalaria glabrata*. Rev. Brasil. Pesquis. Med. Biol. 7(4): 419-425. (NSA 32: 22635).

Narayanan, E. S. 1957. The phenomena of insect parasitism and their practical utilisation in the biological control of insect pests. Proc. Indian Soc. Congr. 44(2): 167-189. (Mentions radiotracer).

Narayanan, E. S., S. S. Chawla and S. Ghai. 1963. Effects of gamma rays on mortality, longevity and fecundity of some species of fruit flies. Proc. Natl. Inst. Sci. India B29: 613-621. (NSA 19: 5682).

Narayanan, E. S., et al. 1959. Labelling of adults of an insect parasite *Bracon gelechiae* Ashmead with radioactive phosphorus (P^{32}). Proc. Indian Acad. Sci. B49(3): 149-155.

Narayanan, E. S., et al. 1959. Studies on the effect of beta radiations on insects. I. The effect of beta radiations (irradiation of eggs and 1st instar larvae) on the life-history of *Corcyra cephalonica* Stainton. Proc. Indian Acad. Sci. 50(2): 82-87.

Narayanan, N. and J. Eapen. 1971. ^{137}Cs in pigeon tissues with special reference to muscles. pp. 609-619. In: Basic Mechanisms in Radiation Biology and Medicine. (See India Dept. of Atomic Energy, 1971) (NSA 26: 15507).

Narayanan, N. and J. Eapen. 1971. Effect of malonate, succinate, and glucose, *in vivo*, on the uptake of ^{137}Cs by pigeon (*Columba livia*) muscles. J. Radiation Res. 12(2): 56-60. (NSA 26: 18091).

Narayanan, N. and J. Eapen. 1971. ^{137}Cs in tissues of the lizard *Hemidactylus leschenaulti*. J. Anim. Morphol. Physiol. 18(2): 171-175. (NSA 27: 12372).

Narayanaswamy, S. 1963. Effect of x-ray on rhizogenetic phenomena in Jerusalem artichoke tissue cultivated *in vitro*. Compt. rend. 257(2): 508-510. (In French) (NSA 17: 35495).

Nardon, P. 1962. Repercussion of the effect of γ rays on descendents of *Sitophilus saskii takashashi* (Coleopt. Curcul.). Compt. rend. 254(13): 2454-2456. (In French) (NSA 16: 20159).

Nardon, P. 1963. Possibilities of use of radioations in the control of insects. Phytoma 15: 7-12. (In French) (NSA 20: 38671).

Naryshkin, M. A., R. M. Aleksakhin and N. N. Mishenkov. 1973. On the migration of ^{90}Sr, ^{106}Ru, ^{137}Cs and ^{144}Ce of the global fallout in a forest biogeocoenosis. Dokl. Akad. Nauk SSSR (3): 10-17. (In Russian).

Nasyrov, Yu. S., et al. 1962. The use of ^{14}C for studying the photosynthetic balance of plant communities. Botan. zh. 47: 96-99. (In Russian).

Nascimento Filho, V. F. and F. M. Wiendl. 1972. Possibility of replacement of Geiger-Mueller detector by a liquid scintillator in measuring the radioactivity of the African bee *Apis mellifera adansoni* (L.) labelled with ^{32}P. Brazil Centro de Energia Nuclear na Agricultura, Sao Paulo, report CENA-BT-1. 38pp. (NSA 29: 16093).

Nash, D. and W. Plaut. 1965. On the presence of DNA in larval salivary gland nucleoli in *Drosophila melanogaster*. J. Cell Biol. 27(3): 682-686. (^{3}H).

Nasim, M. 1971. Environmental aspects of Karachi nuclear power plant. pp. 781-789. In: Environmental Aspects of Nuclear Power Stations. (See IAEA, 1971).

Natarajan, P. R. 1969. Chemical methods of analysis of plutonium: a literature survey. India Bhabba Atomic Research Centre, Bombay, report BARC-422. 13pp. (NSA 24: 11421).

Nathans, M. W. and W. D. Holland. 1971. Analysis of ^{239}Pu particles collected near the Rocky Flats facility. Final report. Trapelo, U. S. AEC report TLW-6105. 29pp. (NSA 28: 2584).

National Academy of Sciences. 1970. Source material for radiochemistry. Nuclear Science Series No. 42. Washington. 117pp. (NSA 25: 15718).

National Academy of Sciences-National Research Council. 1956. Report of the Committee on the Effects of Atomic Radiation on Oceanography and Fisheries. Report of the Committee on Disposal and Dispersal of Atomic Wastes. Biological Effects of Atomic Radiation: Summary Reports. Washington. pp. 71-84, 99-108. (NP-6017) (NSA 10: 7460, 7428).

National Academy of Sciences-National Research Council. 1957. The Effects of Atomic Radiation on Oceanography and Fisheries. Washington, Publ. 551. ix, 137pp. (NSA 12: 5848).

National Academy of Sciences-National Research Council. 1963. Damage to Livestock from Radioactive Fallout in the Event of Nuclear War. Washington, Publ. 1078. vi, 93pp.

National Academy of Sciences-National Research Council. 1964. Postattack resources. pp. 21-27. In: Civil Defense, Project Harbor Summary Report. (Report of a Summer Study Group at Woods Hole, Mass.). Washington, Publ. 1237.

National Academy of Sciences-National Research Council. 1965. Geochronology of North America. Washington, Publ. 1276. v, 315pp.

National Academy of Sciences-National Research Council. 1971. Radioactivity of the Marine Environment. Washington. ix, 272pp.

National Academy of Sciences-National Research Council. 1972. Effects on Populations of Exposure to Low Levels of Ionizing Radiation. (Report of the Advisory Committee on the Biological Effects of Ionizing Radiations). Washington. 228pp. (NP-19638) (NSA 27: 17578; 28: 2908).

National Academy of Sciences-National Research Council. 1972. Environmental transport and effects of radionuclides. pp. 21-40. In: Effects on Populations of Exposure to Low Levels of Ionizing Radiations. (See above entry) (NSA 27: 17580).

National Academy of Sciences-National Research Council. 1973. Radionuclides in Foods. Washington. 95pp. (COD-3167-58) (NSA 28: 12871) (Incl. water).

National Council on Radiation Protection and Measurements. 1952. Radiological monitoring methods and instruments. U. S. National Bureau of Standards, Washington, report NCRP 10 1952. 33pp. (NSA 25: 34894).

National Council on Radiation Protection and Measurements. 1971. Basic radiation protection criteria. Washington, report NCRP-39. 135pp. (NSA 25: 19007).

National Council on Radiation Protection and Measurements. 1971. Protection against neutron radiation. Washington, report NCRP-38. 153pp. (NSA 25: 19007).

National Council on Radiation Protection and Measurements. 1975. Krypton-85 in the atmosphere: accumulation, biological significance, and control technology. Washington, report NCRP-44. 86pp. (NSA 32: 25903).

National Council on Radiation Protection and Measurements. 1975. Natural background radiation in the United States. Washington, report NCRP-45. 163pp. (NSA 33: 29611).

National Lead Company. 1974. Feed Materials Production Center environmental monitoring annual report for 1973. U. S. AEC report NLCO-1109. 28pp. (NSA 30: 29347).

National Wildlife Federation. 1971. Nuclear power plants and you: a citizens handbook. Washington. 27pp.

Naulleau, G. 1967. Technique for labeling serpents with radioisotopes. Compt. rend. D265: 1976-1979. (In French) (NSA 22: 19432).

Naulleau, G. and G. Courtois. 1965. Utilization of ^{60}Co for the marking of serpents. Compt. rend. 260: 6219-6222. (In French) (NSA 19: 42239).

Naumov, R. L. and A. A. Lur'ye. 1972. Marking Ochotona alpina with Co^{60}. Zool. zh. 50(11): 1728-1731. (In Russian) (WR 147: 30).

Naversten, Y. and K. Lidén. 1965. High ^{137}Cs body burdens of Lapps and other people in northern Scandinavia and the food chains involved. pp. 517-526. In: Radioactivity in Man. R. Menelly and S. M. Linde, eds. Chas. C. Thomas, Springfield, Ill. (NSA 20: 7334).

Nay, U. 1967. The absorption of fallout ^{90}Sr at the surface of different grass species. pp. 489-500. In: Radioecological Concentration Processes. (See Aberg and Hungate, 1967).

Nayar, G. G. 1961. Floral study of the sterile Crotalaria striata L. observed in the radioactive monazite sand at Manavalkurichy in the Madras state. Trans. Bose Res. Inst. (Calcutta) 24: 68-73.

Nayar, G. G., K. P. George and A. R. Gopal-Ayengar. 1970. On the biological effects of high background radioactivity: studies on Tradescantia grown in radioactive monazite sand. Radiation Botany 10(2): 287-292. (NSA 24: 41857).

Nayar, G. G., K. T. Jacob and S. Bose. 1960. A preliminary note on the comparative study of the monazite sand at Manavalkurichy in the Madras state. Sci. and Cult. 26(1): 42-43.

Nayfield, C. L., W. Johnson and B. P. Prewitt. 1971. Preoperational radiological surveillance of the Florida Power Corporation's Chrystal River power reactor site. Radiol. Hlth. Data 12(9): 441-449. (NSA 26: 17729).

Nazyrov, N. N. and O. A. Gogenko. 1966. Effect of radiostimulation of wild pasture grasses growing in various ecological conditions. Uzbek. biol. zh. (6): 3-7. (In Russian) (NSA 21: 26670).

Neal, E. C., B. C. Patten and C. E. DePoe. 1967. Periphyton growth on artificial substrates in a radioactively contaminated lake. Ecology 48(6): 918-924.

Neal, W. T. L. 1964. Sources and nature of environmental radioactive contamination. pp. 19-32. In: Agricultural and Public Health Aspects of Radioactive Contamination in Normal and Emergency Situations. (See United Nations, 1964).

Neal, W. T. L. 1969. An historical review of local monitoring for the contamination of land through disposal to the atmosphere. pp. 339-344. In: Environmental Contamination by Radioactive Materials. (See IAEA, 1969).

Neal, W. T. L. 1969. The United Kingdom approach to environmental monitoring in relation to the discharge of radioactive wastes. pp. 345-356. In: Environmental Contamination by Radioactive Materials. (See IAEA, 1969).

Nease, F. R. 1953. Contamination by fission products and its effects on plants growing in the White Oak Creek area, Tennessee. Dissertation, Duke Univ., Durham, N. C.

Nebbia, G. 1971. Man and His Environment: an International Survey. (Meeting of Intern. Union for Conservation of Nature, Milan, Italy, April, 1970). Tamburini Editore s.p.a., Milan. 335pp. (NSA 26: 22780) (Incl. nuclear energy).

Nebesar, C. V. 1967. Maternal transfer and fetal absorption and retention of injected radionuclides in deer and sheep. Thesis, Univ. of California, Davis. iv, 44pp.

Nebraska Department of Health. 1970. Hallam Nuclear Power Facility. Semi-annual report of the environmental surveillance. Lincoln. U. S. AEC DOCKET-1153-17. 3pp. (NSA 25: 10134).

Nebraska Public Power District. 1971. Cooper Nuclear Station. Environmental report, operating license stage. Columbus. U. S. AEC DOCKET-50298-22. 343pp. (NSA 25: 56955).

Nechayev, L. N., et al. 1972. Kinetics of metabolism of some radioisotopes in fish tissue as related to temperature and salinity. pp. 348-352. In: Biological Effects of Radiation from External and Internal Sources. Yu. I. Moskalev and V. S. Kalistatova, eds. "Meditsina", Moscow. (AEC-tr-7457, pp. 510-515).

Neel, J. W. and K. H. Larson. 1963. Biological availability of Sr^{90} to small native animals in fallout patterns from the Nevada Test Site. pp. 45-49. In: Radioecology. (See Schultz and Klement, 1963) (Also: 1961. Univ. of California at Los Angeles, U. S. AEC report UCLA-500) (NSA 16: 10309).

Nees, W. L. and J. P. Corley. 1974. Environmental surveillance at Hanford for CY-1973. Battelle Northwest Lab., U. S. AEC report BNWL-1811, BNWL-1811(Add.). 85, 132pp. (NSA 30: 9503; 31: 3319).

Nees, W. L. and J. P. Corley. 1975. Environmental status of the Hanford Reservation for CY-1973. Battelle Northwest Lab., U. S. AEC report BNWL-B-336. 71pp. (NSA 31: 29602).

Neess, J. C., et al. 1963. Use of nitrogen-15 for measurement of rates in the nitrogen cycle. pp. 481-484. In: Radioecology. (See Schultz and Klement, 1963) (NSA 17: 33600).

Nefedova, A. I. 1972. Accumulation and ways of entry of ^{35}S from sodium sulfate and methionine in the organism of fasting loach (Misgurnus fossilis L.). pp. 154-160. In: Radioecological Studies in Natural Biogeocenoses. (See Verkhovskaya, 1972) (In Russian) (NSA 26: 56551).

Nefedova, A. I. and Ye. I. Popova. 1965. On distribution of ^{22}Na in a reservoir. Radiobiologiya 5(3): 451-456. (AEC-tr-6600, pp. 170-177; JPRS-32,524; TT-65-33003, p. 47) (NSA 19: 35926).

Nefedova, A. I. and I. N. Verkhovskaya. 1968. Growth and survival of Pteris longifolia L. as a function of the dose of x-irradiation and time of sowing of spores after irradiation. Radiobiologiya 8(1): 124-130. (AEC-tr-6914, pp. 208-217) (NSA 22: 47638).

Neharin, A., M. Calderon and O. Yacobi. 1964. Susceptibility of Callosobruchus maculatus to high-dose-rate gamma irradiation. Israel Atomic Energy Commission report IA-1010. 11pp. (NSA 19: 21857).

Nehme, M. and A. Hassan. (Eds.). 1974. Use of Isotopes in Pesticides and Pest Control. (Proc. Symp., Beirut, Lebanon, March 1974). Middle Eastern Regional Radioisotope Centre for the Arab Countries, Cairo, Egypt. 267pp. (In French, English) (CONF-740379) (NSA 33: 7559).

Neidinger, J. W., R. A. Baratz and W. S. Moos. 1965. Oxygen uptake of three life stages of Tenebrio molitor prior and subsequent to massive doses of low energy x-rays. Atompraxis 11: 564-568. (NSA 20: 5314).

Neilson, J. A., Jr. 1963. The use of autoradiographs in the study of root systems of herbaceous plants growing in close association. Abstr. Bull. Ecol. Soc. Am. 44(2): 48.

Neilson, J. A., Jr. 1964. Autoradiography for studying individual root systems in mixed herbaceous stands. Ecology 45(3): 644-646. (Used ^{14}C).

Nelkin, D. 1971. Nuclear Power and Its Critics. The Cayuga Lake Controversy. Cornell Univ. Press, Ithaca, N. Y. 138pp. (NSA 25: 34317).

Nellis, D. W. 1967. Mammalian feces marking. Thesis, Univ. of Georgia, Athens. 20pp. (Incl. radionuclides).

Nellis, D. W., J. H. Jenkins and A. D. Marshall. 1967. Radioactive zinc as a feces tag in rabbits, foxes, and bobcats. pp. 205-207. In: Proc. 21st Ann. Conf. Southeastern Game and Fish Commissions.

Nelp, W. B., H. N. Wagner, Jr. and F. J. Schwartz. 1963. Total body radioactivity and the lack of excretion of Co^{60} vitamin B_{12} by the aglomerular oyster toadfish Opsanus tau. Chesapeake Sci. 4(4): 192-194. (SFA 9: 6632).

Nelson, C. D. 1964. The production and translocation of photosynthate-C^{14} in conifers. pp. 243-257. In: The Formation of Wood in Forest Trees. M. H. Zimmerman, ed. Academic Press, New York.

Nelson, C. T. and R. L. Koontz. 1957. Automatic water sampling device. pp. 159-164. In: Proc. Health Physics Society.

Nelson, D. J. 1960. Biological concentration of radioisotopes from aquatic environments. Nuclear Safety 2(1): 81-83.

Nelson, D. J. 1962. Strontium, strontium-90, and calcium analyses of Clinch and Tennessee River clams. Oak Ridge National Lab., U. S. AEC report ORNL-TM-270. 20pp. (NSA 16: 23410).

Nelson, D. J. 1962. Clams as indicators of strontium-90. Science 137: 38-39. (NSA 16: 21889).

Nelson, D. J. 1962. Use of specific activities for interpretation of ecological tracer studies. Abstr. Bull. Ecol. Soc. Am. 43(4): 119-120.

Nelson, D. J. 1963. The strontium and calcium relationships in Clinch and Tennessee River mollusks. pp. 203-211. In: Radioecology. (See Schultz and Klement, 1963) (NSA 17: 33566).

Nelson, D. J. 1963. Role of bottom organisms. pp. 193-201. In: Transport of Radionuclides in Fresh Water Systems. (See Kornegay, et al., 1963) (NSA 18: 40529).

Nelson, D. J. 1964. Biological vectors and reservoirs of strontium-90. Nature 203(4943): 420. (NSA 18: 36918).

Nelson, D. J. 1964. Interpretation of radionuclide uptake from aquatic environments. Nuclear Safety 5(2): 196-199. (NSA 18: 13434).

Nelson, D. J. 1965. Ecological aspects of radionuclides in streams. Abstr. Bull. Ecol. Soc. Am. 46(2): 39.

Nelson, D. J. 1965. Prediction of ^{90}Sr uptake in fish. Abstr. Bull. Ecol. Soc. Am. 46(3): 101.

Nelson, D. J. 1967. The prediction of ^{90}Sr uptake in fish using data on specific activities and biological half-lives. pp. 843-851. In: Radioecological Concentration Processes. (See Aberg and Hungate, 1967).

Nelson, D. J. 1967. Ecological behavior of radionuclides in the Clinch and Tennessee Rivers. pp. 169-187. In: Reservoir Fishery Resources Symposium. Southern Div. Amer. Fish. Soc.

Nelson, D. J. 1969. Cesium, cesium-137, and potassium concentrations in white crappie and other Clinch River fish. pp. 240-248. In: Symposium on Radioecology. (See Nelson and Evans, 1969).

Nelson, D. J. (Ed.). 1973. Radionuclides in Ecosystems. (Proc. 3rd National Symp. on Radioecology, May 10-12, 1971, Oak Ridge, tennessee). U. S. AEC report CONF-710501-P1, -P2. xvi, pp. 1-678; xvi, pp. 679-1268. (NSA 28: 27291, 29839).

Neldon, D. J. and B. G. Blaylock. 1962. The preliminary investigation of salivary gland chromosomes of Chironomus tentans Fabr. from the Clinch River. pp. 367-372. In: Radioecology. (See Schultz and Klement, 1963) (NSA 17: 33585).

Nelson, D. J. and F. C. Evans. (Eds.). 1969. Symposium on Radioecology. (Proc. 2nd National Symp., Ann Arbor, Michigan, May 15-17, 1967). U. S. AEC report CONF-670503. xii, 774pp.

Nelson, D. J. and S. V. Kaye. 1971. Specific activity concept applied to aquatic ecosystems. pp. 735-746. In: Nuclear Techniques in Environmental Pollution. (See IAEA, 1971) (NSA 25: 32376).

Nelson, D. J., S. V. Kaye and R. S. Booth. 1972. Radionuclides in river systems. pp. 367-387. In: River Ecology and Man. R. T. Oglesby, C. A. Carlson and J. A. McCann, eds. Academic Press, New York. (NSA 27: 7557).

Nelson, D. J., et al. 1969. Estimates of periphyton mass and stream bottom area using ^{32}P. Water Research 3(5): 367-373. (NSA 24: 545).

Nelson, D. J., Jr. and D. L. Snow. nd. Environmental radiation surveillance. U. S. PHS Publ. 1224. 11pp. (NP-17401) (NSA 22: 51570) (Incl. caribou, reindeer).

Nelson, D. M., P. F. Gustafson and J. Sedlet. 1971. Fallout radionuclides as tracers of lake mixing. Proc. Conf. Great Lakes Res., Int. Assoc. Great Lakes Res. 13(1): 490-494.

Nelson, D. M., G. P. Romberg and W. Prepejchal. 1971. Radionuclide concentrations near the Big Rock Point Nuclear Power Station. Proc. Conf. Great Lakes Res., Int. Assoc. Great Lakes Res. 14: 268-276. (NSA 26: 45536).

Nelson, D. M., et al. 1972. Effects of routine radioactive release on aquatic radionuclide concentrations. Proc. Conf. Great Lakes Res., Int. Assoc. Great Lakes Res. 15: 260-269.

Nelson, D. R., H. M. Flint and T. S. Adams. 1970. Effect of partial or whole-body irradiation of pupal or adult female house-flies on ovarian maturation and retention of pupal fat body. Int. J. Radiat. Biol. 18: 71-79. (NSA 24: 51019).

Nelson, F. R. S. 1968. Comparative sterility effects of gamma radiation and a chemosterilant on Drosophila melanogaster Meigen. Thesis, Univ. of Massachusetts, Amherst. 86pp. (NSA 23: 43852).

Nelson, F. R. S. 1973. Drosophila melanogaster: effects of γ radiation on fecundity and longevity. J. Econ. Entomol. 66(1): 257-258. (NSA 27: 17681).

Nelson, I. C. (Ed.). 1961. Evaluation of radiological conditions in the vicinity of Hanford, April-June 1961. Hanford Atomic Products Operation, U. S. AEC report HW-70552. 25pp. (NSA 15: 30994).

Nelson, J. L. and W. L. Haushild. 1970. Accumulation of radionuclides in bed sediments of the Columbia River between Hanford reactors and McNary Dam. Water Resources Res. 6(1): 130-137. (NSA 24: 46108).

Nelson, J. L., R. W. Perkins and W. L. Haushild. 1966. Determination of Columbia River flow times downstream from Pasco, Washington, using radioactive tracers introduced by the Hanford reactors. Water Resources Res. 2: 31-39. (Also: Battelle Northwest Lab., U. S. AEC report BNWL-SA-176) (NSA 20: 20916).

Nelson, J. L., R. W. Perkins and J. M. Nielsen. 1964. Progress in studies of radionuclides in Columbia River sediments. A summary of Hanford achievements in this program under General Electric, 1963-1964. Hanford Atomic Products Operation, U. S. AEC report HW-83614. 26pp. (NSA 19: 15784).

Nelson, J. L., et al. 1966. Reactions of radionuclides from the Hanford reactors with Columbia River sediments. pp. 139-161. In: Disposal of Radioactive Wastes into Seas, Oceans and Surface Waters. (See IAEA, 1966.

Nelson, J. M. 1969. Radiobiology of the Mongolian gerbil (Meriones unguiculatus). Thesis, Univ. of Michigan, Ann Arbor. 152pp. (NSA 24: 34654).

Nelson, J. M. 1971. Survival time response of the Mongolian gerbil after total-body irradiation. Radiation Research 48(1): 189-198. (NSA 26: 7091).

Nelson, J. M. 1972. Influence of intestinal mucosal renewal rate on development of gastrointestinal injury survival time of irradiated gerbils. Radiation Research 49(2): 367-380. (NSA 26: 23111).

Nelson, L., Jr. 1970. Effects of sublethal, cerebral x-irradiation on movement and home-range patterns of black-tailed jackrabbits. Thesis, Utah State Univ., Logan. viii, 62pp.

Nelson, N. N., L. M. Shields and R. B. Platt. 1969. Incidence of pine needle gall fly infestation in pine species two years postirradiation. Botan. Gaz. 130(4): 226-230. (NSA 24: 25591).

Nelson, N. S. and J. H. Rust. 1965. Use of ionizing radiation for measuring biological phenomena. pp. 59-170. In: Methods of Animal Experimentation. Vol. 2. W. J. Gay, ed. Academic Press, New York. (NSA 20: 14386).

Nelson, R. D. 1968. Effects of gamma radiation on the biology and population suppression of the two-spotted spider mite, *Tetranychus urticae* Koch. Thesis, Univ. of California, Davis. 114pp. (Also: w/M. Stafford, Hilgardia 41(12): 299-342(1972)) (NSA 24: 14637; 28: 15767).

Nelson, R. W. 1964. Analysis of waste released by seepage to the Columbia River from the 1301-N crib. Hanford Labs., U. S. AEC report HW-81306. 34pp. (NSA 29: 7700).

Nelson, R. W. 1965. A sequence for predicting waste transport by ground water. Battelle Northwest Lab., U. S. AEC report BNWL-63. 12pp. (NSA 19: 40268).

Nelson, R. W. and A. E. Reisenauer. 1963. Application of radioactive tracers in scientific groundwater hydrology. pp. 207-230. In: Radioisotopes in Hydrology. (See IAEA, 1963).

Nelson, S. O. 1965. Towards better insect control. Crops Soils 18(1): 8-9. (NSA 21: 26704).

Nelson, S. O. 1966. New ways to control insects. Farm, Ranch, Home Quart. (132): 15-16. (NSA 22: 17068).

Nelson, T. E. 1966. The effects of gamma radiation upon various stages of *Fannia canicularis* (L.). Bull. Entomol. Soc. Am. 12: 278.

Nelson, W. C. 1961. Experiments on the uptake of radioactive phosphorus from the water and from a single oral dose by brook and rainbow trout. Colorado Dept. of Game and Fish, U. S. AEC report TID-13109. 22pp. (NSA 15: 2991, 23201; SFA 7: 5171; WPA 35(2): 392).

Nelson, W. C. and F. W. Whicker. 1969. Cesium-137 in some Colorado game fish, 1956-66. pp. 258-265. In: Symposium on Radioecology. (See Nelson and Evans, 1969) (Also: Colorado State Univ., Ft. Collins, U. S. AEC report COO-1156-21) (NSA 21: 22234).

Nelson-Rees, W. A. 1962. The effects of radiation damaged heterochromatic chromosomes on male fertility in the mealy bug, *Planococcus citri* (Risso). Genetics 47(6): 661-683. (NSA 16: 30236).

Nemtseva, L. S. 1971. Protracted mutagenesis induced by ionizing radiation and its dependence on the water content of a biosystem. Genetika 7 (8): 19-30. (In Russian) (NSA 25: 55167) (Seeds).

Nerpina, S. B. (Ed.). 1967. Use of Isotopes and Irradiation in Studies in Agriculture. (Proc. Sci. Res. Conf. of the Members of the Council on Mutual Assistance in Agriculture, 1965). Izdatel'stvo Kolos, Moscow. 264pp. (In Russian) (NP-17449) (NSA 22: 51445).

Netherlands Instituut voor Toepassing van Atoomenergie in de Landbouw. 1972. Application of atomic energy in agriculture. Commission, Wageningen, report EUR-4868. 90pp. (NSA 28: 24634).

Netherlands Normalisatie- Instituut. 1962. Determination of beta-radioactivity in water. 'sGravenhage, report NEN-2395 1962. 4pp. (In Dutch) (NSA 26: 20014).

Netherlands Rijks Instituut voor de Volksgezondheit. nd. Radioactivity in surface water and precipitation. Measurements during 1964. Utrecht, report RA-23. 24pp. (NSA 21: 28442).

Nethsinghe, D. A. and H. Broeshart. 1975. Root activity patterns of some tree crops. (Results of a five-year coordinated research program of the Joint FAO/IAEA Division of Atomic Energy in Food and Agriculture). International Atomic Energy Agency, Vienna, Publ. STI/DOC-10/170. 152pp. (NSA 33: 15112) (^{32}P solution soil injection technique).

Neuhaus, G. M. 1965. Discharge of radioactive waste waters to public waters. Gas-Wasserfach 106: 1035-1039. (NSA 22: 23250).

Neuhold, J. M. and R. K. Sharma. 1969. Interacting effects of gamma radiation and sodium halide concentrations on rainbow trout. pp. 142-151. In: Symposium on Radioecology. (See Nelson and Evans, 1969) (Also: Utah State Univ., Logan, U. S. AEC report COO-1421-4) (NSA 21: 43498).

Neumaier, F. 1973. Isotope techniques in hydrology. pp. 228-238. In: Worldwide Radiation from Nuclear and Conventional Energy Production. H. Braun, ed. Georg Thieme Verlag, Stuttgart. (In German) (NSA 31: 3398).

Neuschel, S. K. 1963. Aeroradioactivity survey and areal geology of the District of Columbia and parts of Maryland, Virginia and West Virginia (ARMS-I). U. S. Geological Survey, Washington, U. S. AEC report CEX-59.4.17. 48pp. (NSA 20: 20928).

Neustroyev, G. V. and V. N. Podymakhin. 1966. Respiration of *Salmo salar* L. spawn in radioactively contaminated waters. Radiobiologiya 6(1): 115-116. (AEC-tr-6770, pp. 181-183) (BA 47: 117874; NSA 20: 42820).

Neustroyev, G. V. and V. N. Podymakhin. 1966. Developmental rate of Atlantic salmon (*Salmo salar*) eggs when there is radioactive contamination of the water by Sr^{90}-Y^{90}. Radiobiologiya 6(2): 321-323. (AEC-tr-6771, pp. 230-233) (BA 48: 13718; SFA 11: 9000).

Neustroyev, G. V., V. G. Kondratenko and A. M. Kuzin. 1967. Effect of gamma radiation and radiotoxins on the synthesis of DNA in early embryogenesis of the groundling Misgurnus fossilis. Izv. Sibirsk. otd. Akad. Nauk SSSR (biol. med.) (5): 95-99. (In Russian) (NSA 22: 23423).

Neustroyev, G. V., V. G. Kondratenko and A. M. Kuzin. 1968. Influence of AET on loach embryos, Misgurnus fossilis, under the action of gamma rays and radiotoxins. Radiobiologiya 8: 588-593. (AEC-tr-7014, pp. 137-144) (NSA 23: 861).

Nevstruyeva, M. A., et al. 1964. Characteristics of radioactive pollution of the biological chain moss-reindeer-man in the Far North of the USSR from 1961 to 1964. State Committee on the Use of Atomic Energy, USSR. Atomizdat, Moscow. 7pp. (In Russian).

Nevstrueyeva, M. A., et al. 1967. The nature of ^{137}Cs and ^{90}Sr transport over the lichen-reindeer-man food chain. pp. 209-215. In: Radioecological Concentration Processes. (See Aberg and Hungate, 1967).

Neuwirth, R. 1957. Occasional measurement of radioactive substances in water. Atompraxis 3(10): 402-404. (In German).

Nevin, R. L. 1964. Redistribution of radiocontaminants by surface water flows. Preliminary report. Hazelton-Nuclear Science Corp., Palo Alto, Calif., U. S. AEC report HNS-1229-46. 36pp. (NSA 18: 37368).

Newbould, P. and R. L. Lucas. 1959. Effect of the level of microbial population on isotopically exchangeable phosphate in soil. Nature 184(Suppl. 5): 1155-1156.

Newbould, P., E. R. Mercer and P. M. Lay. 1968. Estimation of changes in the water status of soils under field conditions from the attenuation of β radiation by water held in absorbent nylon pads. Exptl. Agr. 4: 167-177. (NSA 22: 48938).

Newcombe, C. L. 1960. The possible use of meromictic lakes for the disposal of radioactive wastes. p. 173. In: Proc. 9th Pacific Science Congr., Thailand, 1957.

Newcombe, H. B. and J. F. McGregor. 1967. Major congenital malformations from irradiations of sperm and eggs. Mutation Research 4: 663-673. (NSA 22: 15072) (Rainbow trout).

Newcombe, H. B. and J. F. McGregor. 1972. Increased embryo production following low doses of radiation to trout spermatozoa. Radiation Research 51(2): 402-409. (NSA 26: 50946).

New England Nuclear Corporation. 1957. Proc. Symp. on Tritium in Tracer Applications. Boston. 43pp. (NSA 13: 7462).

Newman, J. S., J. Hacskaylo and M. E. Bloodworth. 1965. An improved technique for subsoil placement of radioisotopes. Agron. J. 57: 315. (NSA 20: 8557).

New York State Electric and Gas Corporation. 1969. Bell Station. Preliminary Safety Analysis Report, Ammendment 3. Binghamton. U. S. AEC DOCKET-50319-7. 356pp. (NSA 23: 15542) (Incl. environmental monitoring).

New Zealand National Radiation Laboratory. 1971. Environmental radioactivity in New Zealand. Quarterly report, April-June 1971 and Pacific area monitoring, 4 June-31 August 1971. Christchurch, report NRL-F-45. 28pp. (NSA 26: 12276).

New Zealand National Radiation Laboratory. 1974, 1975. Environmental radioactivity annual report, 1973. ---1974. Christchurch, report NRL-F-52, NRL-F-54. 39, 43pp. (NSA 31: 3327; 33: 17186).

Neyfakh, A. A. 1956. The changes of radiosensitivity in the course of fertilization in the loach Misgurnus fossilis. Dokl. Akad. Nauk SSSR 109(5): 943-946. (In Russian).

Neyfakh, A. A. 1956. Effects of ionizing radiation on the sex cells of Misgurnus fossilis. Dokl. Akad. Nauk SSSR 111(3): 585-588. (In Russian) (NSA 11: 3657).

Neyfakh, A. A. 1959. The effect of ionizing radiation on the early development of fishes. Tr. inst. morfol. zhivot. (24): 135-159. (In Russian).

Neyfakh, A. A. 1959. X-ray inactivation of nuclei as a method for studying their function in the early development of fishes. J. Embryol. Exptl. Morphol. 7(2): 173-192).

Neyfakh, A. A. 1959. Using the method of radiation inactivation of nuclei for investigation of their functions in the early development of fish. Zh. obshch. biol. 20(3): 202-213. (In Russian) (Ref. zh. (biol) 1961: 13A174).

Neyfakh, A. A. 1960. Radiation inactivation of cell nuclei as a method of study of their role in the development of respiration in fish embryos. Biokhimiya 25(4): 669-674. (Biochemistry (USSR) 25: 503-511) (NSA 15: 12735).

Neyfakh, A. A. 1960. An investigation of the functions of the nuclei in the development of Strongylocentrotus dröbachiensis by the method of radiation inactivation. Dokl. Akad. Nauk SSSR 132(6): 1458-1461. (Doklady (biol. sci.) 132: 376-379).

Neyfakh, A. A. 1961. Radiation studies of the morphogenetic functions of the nuclei in early development of tailless apmphibians. Dokl. Akad. Nauk SSSR 136(5): 1248-1251. (Doklady (biol. sci.) 136: 41-44).

Neyfakh, A. A. 1961. A comparative study of the morphogenetic function of the nuclei in the development of animals. Zh. obshch. biol. 22(1): 42-57. (In Russian).

Neyfakh, A. A. 1962. Embryology. pp. 101-123. In: The Contribution of Radiology to the Growth of the Medico-Biological Disciplines. P. V. Petrov, et al., eds. Ministry of Special Technical and Vocational Training, Belorussian SSR, Minsk. (In Russian).

Neyfakh, A. A. and V. V. Radayevskaya. 1967. On the morphogenetic function of nuclei in the goldfish x loach (Carassius auratus auratus x Misgurnus fossilis) hybrids. Genetika (12): 80-88. (In Russian) (NSA 22: 21584).

Neyfakh, A. A. and N. N. Rott. 1958. A study of the radiation injuries in fish and in early development stages. Dokl. Akad. Nauk SSSR 119(2): 261-264. (Doklady (biol. sci.) 119: 134-137 (NSA 12: 11240) (*Misgurnus fossilis*, *Acipenser stellatus*).

Neyfakh, A. A. and N. N. Rott. 1959. Synchronization of cell division in early embryos of the loach *Misgurnus fossilis* by the action of a lower temperature. Dokl. Akad. Nauk SSSR 125(2): 432-535. (In Russian).

Neyfakh, A. A. and I. M. Shapiro. 1962. A radiation study of the function of oocytes in the process of ovulation in the loach. Dokl. Akad. Nauk SSSR 144(4): 942-944. (Doklady (biol. sci.) 144: 527-528) (NSA 16: 26795).

Neystadt, M. I., et al. 1964. Radiocarbon and pollen dating of Holocene strata. Dokl. Akad. Nauk SSSR 144: ___. (Doklady (biol. sci.) 144: 115-117).

Nezami, M. 1966. Contribution to the study of geophysical behavior of ^{210}Pb by application of alpha spectrometry. Thesis, Univ. of Paris. Centre d'Études Nucléaires, France Commissariat à l'Énergie Atomique, Saclay, report CEA-R-2886. 95pp. (In French) (NSA 20: 23284).

Nezami, M. 1966. Separation of ^{210}Po for an accurate measurement, by alpha spectrometry, of the rate of accumulation of polar snow. J. Chim. Phys. 63: 603-606. (In French) (NSA 20: 27151).

Ng, Y. C. 1965. Neutron activation of the terrestrial environment as a result of underground nuclear explosions. Lawrence Radiation Lab., Livermore, U. S. AEC report UCRL-14249. 38pp.

Ng, Y. C., W. L. Robinson and D. W. Wilson. 1973. Modeling radiation exposures to populations from radioactivity released to the environment. Lawrence Livermore Lab., U. S. AEC report UCRL-74389. 14pp. (For Symp. on Environmental Behavior of Radionuclides Released in the Nuclear Industry; Aix-en-Provence, France, 14 May 1973) (CONF-730503) (NSA 28: 27315).

Nguyen, B. C. 1971. A study using radioactive tracers of the exchange between various zones of the atmosphere over the continents and oceans. Thesis, Univ. of Paris. 175pp. (Also: Commissariat à l'Énergie Atomique, report CEA-R-3977) (In French) (NSA 25: 21200).

Nguyen, Huu Van and C. Lalou. 1969. Simultaneous determination of the isotope ratio ^{234}U/^{238}U and of the ^{238}U content of natural waters. Radiochim. Acta 12: 156-160. (In French) (NSA 24: 2143).

Niagara Mohawk Power Corporation. 1964. Nine Mile Point Nuclear Station. Preliminary Hazards Summary Report, Vol. 1, Exhibit D. ---Exhibit D (Suppl.). Syracuse, N. Y. U. S. AEC DOCKET-50220-14, -50220-16. 276, 226pp. (NSA 23: 15533, 15535) (Incl. site and environment).

Niagara Mohawk Power Corporation. 1967. Nine Mile Point Nuclear Station. Appendixes. Final Safety Analysis Report, Ammendment 2, Vol. 2, Exhibit D-2. Syracuse, N. Y. U. S. AEC DOCKET-50220-10. 369pp. (NSA 23: 17151) (Environmental monitoring).

Niaussat, P. 1968. Study of radioresistance in various animals. Perspectives thus opened in radioprotection. Rev. Corps Sante Armées 9: 243-246. (In French) (NSA 24: 48671).

Niaussat, P., et al. 1963. Contingent relation between the resistance to γ radiation of certain arthropods of the Sahara and the nucleic acid content of their tissues. Bull. Soc. Med. Mil. Fr. 57: 295-299. (In French) (NSA 18: 36965).

Niaussat, P., et al. 1963. Behavior of the Saharan scorpion *Androctonus amoreuxi* (Aud. and Sav.) after exposure to high doses of radiation and ailments following this irradiation. Compt. rend. 257(5): 1177-1180. (In French) (NSA 17: 40578).

Nichev, L. 1956. Insect marking with radioisotopes. Priroda (Sofia) 5(4): 61-62. (In Bulgarian).

Nichols, B. L. 1972. Balancing environmental impact and economic progress: the impact of the courts. Nuclear Safety 13(3): 209-215. (NSA 26: 32912).

Nickel, E. 1965. Effects of 2,4-dinitrophenol on the radioinduced dominant lethal factors in *Drosophila melanogaster*. Strahlentherapie 126: 263-268. (In German).

Nicklas, R. B. and R. A. Jaqua. 1965. X-chromosome DNA replication, developmental shift from synchrony to asynchrony. Science 147(3661): 1041-1043. (^{3}H, *Melanoplus differentialis*).

Nicolas, C. and R. Kirchmann. 1974. Contamination of freshwater by ^{54}Mn and ^{60}Co. European Atomic Energy Community, Brussels, report EUR-5167e. (NSA 31: 3387).

Nicoletti, B., A. Olivieri and G. Olivieri. 1964. Further data on the interaction between ultrviolet and ionizing radiations on the production of recessive lethals in *Drosophila melanogaster*. Atti Assoc. Genet. Ital. 9: 172-184. (In Italian).

Niedergerke, R. and S. Page. 1969. A new method for the determination of calcium fluxes in the frog heart by means of high precision measurement of ^{45}Ca concentrations. Pflügers Arch. Europ. J. Physiol. 306: 354-356. (NSA 23: 33610).

Niedetzky, A. 1967. Effect of radiation of isolated frog hearts stopped as a result of Ca deficiency. Acta Biochim. Biophys. (Budapest) 2: 205-210. (NSA 22: 15068).

Niedetzky, A. and Cs. Lajtai. 1969. Effect of radioactive radiations on the lifetime of sperms. Acta Biochim. Biophys. 4: 211-216. (NSA 24: 34602).

Nielsen, D. R. and J. W. Biggar. 1967. Radioisotopes and labelled salts in soil-water movement. pp. 61-76. In: *Soil-Moisture and Irrigation Studies*. (See IAEA, 1967) (NSA 22: 6448).

Nielsen, D. R. and J. W. Biggar. 1973. Analyzing soil water and solute movement under field conditions. pp. 111-127. In: *Soil-Moisture and Irrigation Studies*. *II*. (See IAEA, 1973) (NSA 28: 21143).

Nielsen, D. R., et al. 1974. Soil-water and solute movement studies. pp. 117-133. In: Isotope and Radiation Techniques in Soil Physics and Irrigation Studies 1973. (See IAEA, 1974) (NSA 30: 15489) (Tracers).

Nielsen, J. M. 1963. Behavior of radionuclides in the Columbia River. pp. 91-105. In: Transport of Radionuclides in Fresh Water Systems. (See Kornegay, et al., 1963) (NSA 17: 40524).

Nielsen, J. M. and R. W. Perkins. 1962. The removal of radioisotopes from the Columbia River by natural processes. Hanford Labs., U. S. AEC report HW-SA-2411. 13pp. (NSA 17: 6310).

Nielsen, J. M. and N. A. Wogman. 1971. Some nuclear techniques for atmospheric dilution and dispersion studies. pp. 257-265. In: Nuclear Techniques in Environmental Pollution. (See IAEA, 1971) (NSA 25: 32087).

Nielsen, K. 1968. Hypoproteinaemias associated with helminthic infections, with special reference to gastro-intestinal parasites. pp. 125-132. In: Isotopes and Radiation in Parasitology. (See IAEA, 1968) (Radionuclide tags).

Nielsen, R. A. 1971. Radiation biology of the greater wax moth (Galleria mellonella L.): effects on developmental biology, bionomics, mating-comppetitiveness and F1 sterility. Thesis, Utah State Univ., Logan. 154pp. (Univ. Microfilms 73,908) (NSA 27: 20147).

Niemann, E. G., W. Kuehn and A. K. Kaul. 1975. Nuclear techniques for determining biomass production, evaporation and transpiration, root development, and nutritional value. pp. 79-83. In: Tracer Techniques for Plant Breeding. International Atomic Energy Agency, Vienna, Publ. STI/PUB-419. (NSA 33: 26384).

Niemeyer, L. E., R. A. McCormick and J. H. Ludwig. 1971. Environmental aspects of nuclear power plants. pp. 711-721. In: Environmental Aspects of Nuclear Power Stations. (See IAEA, 1971) (Paper SM-146/42) (CONF-700810-25) (NSA 24: 46138).

Niethammer, G. and D. Sauerbeck. 1961. Radioactivity of northern winter birds. Bonn. Zool. Beitr. 12(3/4): 316-323. (Wild birds).

Nieto Garcia, M., M. J. Bengoechea Pere and J. G. Maganto Fernadez. 1975. Effect of acclimation temperature on proliferation of erythropoietic precursors in Carassius auratus and on the uptake of tritiated thymidine in haematopoietic and intestinal tissue. pp. 319-327. In: Combined Effects of Radioactive, Chemical, and Thermal Releases to the Environment. (See IAEA, 1975) (In Spanish) (NSA 33: 29795).

Nietzche, K. 1961. Testing of decay in wooden poles by means of radioisotopes. Abstr. Isotopentechnik 1(5/6): 137. (In German).

Nifontov, B. I., et al. 1965. Underground Nuclear Explosions. Problems of Industrial Nuclear Explosions. Atomizdat, Moscow. (AEC-tr-6777; PNE-3004).

Nifontova, M. G. 1976. Accumulation of ^{90}Sr and ^{137}Cs by lichens under experimental conditions. Ekologiya 7(1): 89-92. (In Russian).

Nigon, V. and S. Gillot. 1964. Incorporation of thymidine in the course of ovogenesis and embryonic development in Drosophila. Exptl. Cell Res. 33: 29-38. (^{14}C).

Nigon, V., et al. 1968. DNA synthesis in the giant nuclei of insects - control machinery and structures observed in the silk-producing gland of Bombyx mori. pp. 407-416. In: Isotopes and Radiation in Entomology. (See IAEA, 1968) (In French, English summary).

Nijampurkar, V. N., et al. 1966. "Dating" ground waters of ages younger than 1,000-1,500 years using natural ^{32}Si. Nature 210: 478-480. (NSA 20: 31432).

Nikiforov, V. S. 1965. Algorithm for the description of the radiation environment. pp. 46-49. In: Materials of a Scientific Conf. of the Komi Republic Youth. (Abstracts of Reports). Syktyvar. (In Russian).

Nikiforov, V. S. and V. Ya. Ovchenkov. 1972. Determination of radon concentrations in the burrows and refuges of some burrowing animals. pp. 230-237. In: Radiological Studies in Natural Biogeocenoses. (See Verkhovskaya, 1972) (In Russian) (NSA 26: 56136).

Nikiforov, V. S. and I. N. Verkhovskaya. 1971. Comments on the estimation of the radiation situation. pp. 121-128. In: Methods of Radioecological Investigations. (See Verkhoskaya, 1971) (In Russian).

Nikiforov, V. S., et al. 1971. Calculation of doses to a bioobject placed in compound natural γ-fields. pp. 141-148. In: Methods of Radioecological Investigations. (See Verkhovskaya, 1971) (In Russian).

Nikolayev, D. S. 1957. Geochemistry of uranium and thorium. pp. __. In: Radiometric Methods in Prospecting for Uranium Ore. Gostekhizdat, Moscow. (In Russian).

Nikolayev, D. S., K. F. Lazarev and S. M. Grashchenko. 1961. The concentration of thorium isotopes in the waters of the Azov Sea. Dokl. Akad. Nauk SSSR 138(3): 674-676. (Doklady (earth sci.) 138: 489-490) (CA 56: 8474).

Nikolayev, D. S., et al. 1965. Isotopic composition of uranium in the waters and sediments of the Black and Azov Seas. Dokl. Akad. Nauk SSSR 165: 187-189. (In Russian) (NSA 20: 33515).

Nikolayev, D. S., et al. 1966. Geochemical balance of radioactive elements in the Black and Azov Seas. Radiokhimiya 8(4): 469-476. (Sov. Radiochem. 8: 433-439) (NSA 21: 4635).

Nikolayev, D. S., et al. 1969. Geochemical balance of radioactive elements in the basin of the Black and Azov Seas. II. Reserves of thorium isotopes. Radiokhimiya 11(6): 688-698. (Sov. Radiochem. 11(6): 674-682) (NSA 24: 19117).

Nikolyuk, V. F. and M. I. Mavlyanova. 1963. Significance of presowing treat-of cotton seeds with Protista irradiated with gamma-rays of Co^{60}. Biol. zh. 7(4): 59-61. JPRS-21966; OTS-63-41211) (NSA 18: 3502).

Nilubol, M. L., A. P. Sungtong and D. Chamnirokasarnt. nd. Soil analysis by neutron activation method. Thailand Office of Atomic Energy for Peace, Bangkok, report THAI-AEC-36. 8pp. (NSA 25: 34545).

Nimlos, T. J., W. P. Van Meter and L. A. Daniels. 1968. Rooting patterns of forest understory species as determined by radioiodine absorption. Ecology 49(6): 1146-1151.

Nininger, R. D. 1956. Minerals for Atomic Energy. D. Van Nostrand, New York. 2nd ed., xvi, 399pp.

Nir, A. 1967. Development of isotope methods applied to groundwater hydrology. pp. __. In: Isotope Techniques in the Hydrologic Cycle. (See Stout, 1967) (CONF-651108-5) (AED-CONF-65-347-10) (NSA 20: 39142).

Nir, A., et al. 1966. Natural tritium. Rev. Geophys. 4: 441-456. (NSA 21: 3179).

Nish, G. and P. Phelps. 1968. Large area R. F. linked, gamma radiation monitoring system with centralized data recording. Univ. of California, Davis, U. S. AEC report UCID-15381. 16pp. (NSA 30: 26857).

Nishi, Y. 1959. Studies on the character and prevention of the virus disease of garden crops. II. Studies on the mechanism of aphid transmission of mosaic disease of Japanese radish, using radioactive phosphorus. Proc. Ass. Plant Prot. Kyushu 5: 24-25. (In Japanese).

Nishida, K. and S. Hibino. 1952. Studies in absorption and mechanism of transport of P^{32} in the leaf-itcher of Sarracenia falava L. Ann. Rept. Research Committee on the Application of Artificial Radioactive Isotopes in Japan 2: 42-47.

Nishida, T., et al. 1967. Radiosensitivity and induction of somatic mutations in woody perennials under chronic gamma ray irradiation. Gamma Field Symp. (6): 19-45. (CONF-670739) (NSA 24: 31928).

Nishimura, K. 1967. On the formation of visceral inversion caused by radioactive iron and calcium. Mie Med. J. 16: 253-262. (NSA 22: 10857) (Bufo).

Nishimura, K. 1967. Influence of ^{137}Cs and ^{90}Sr on the growth of early stage amphibian eggs. Mie Med. J. 16: 245-251. (NSA 22: 10875).

Nishimura, K. 1967. Abnormal formation of visual organs of amphibian larvae induced by radioactive rainwater. Mie Med. J. 16: 263-267. (NSA 22: 15049).

Nishimura, K., K. Omoto and Y. Koseki. 1972. Tohoku University radiocarbon measurements. I. Sci. Rep. Tohoku Univ. Ser. 7 22(2): 271-274. (NSA 30: 3183).

Nishimura, M. and N. Urakawa. 1972. Effect of estrogenization on the distribution of radiomercury in quail. Jap. J. Pharmacol. 22(4): 535-543. (NSA 30: 16059).

Nishimura, M., N. Urakawa and M. Ikeda. 1971. An autoradiographic study on the distribution of mercury and its transfer to the egg in the laying quail. Jap. J. Pharmacol. 21(5): 651-659. (BA 55: 1973).

Nishimura, S. 1971. Fission track dating of archeological materials from Japan. Nature 230(5291): 242-243. (NSA 25: 32078).

Nishimura, S. 1973. Radioactivity in earth's crust. Seibutsu Kagaku 25 (2): 66-77. (In Japanese) (NSA 30: 23970).

Nishina, J., H. Nakamura and H. Nakayama. 1940. Effect of neutrons on photosynthesis. Bull. Inst. Phys. Chem. Res. 19(10): 1343-1347. (In Japanese) (Chlorella, Scenedesmus).

Nishio, K., et al. 1962. The radioactive contamination of rabbits fed with clover contaminated by fallout. II. Ann. Rep. Radiation Center Osaka Prefect. 3: 137-140. (NSA 17: 33534).

Nishita, H. and M. Hamilton. 1969. Heating effects on the thermoluminescence of gamma irradiated soils. Soil Science 108: 1-10. (NSA 23: 40860).

Nishita, H. and M. Hamilton. 1971. Influence of temperature during the gamma irradiation period on the subsequent thermoluminescence of soils and LiF and CaF_2 dosimeters. Soil Science 111(6): 393-398. (NSA 25: 39599).

Nishita, H. and M. Hamilton. 1972. Soil thermoluminescence in relation to radiation exposure under field conditions. Soil Science 114(6): 430-439. (NSA 27: 17291).

Nishita, H. and K. H. Larson. 1957. Summary of certain trends in soil-plant relationship studies of the biological availability of fallout debris. Univ. of California at Los Angeles, U. S. AEC report UCLA-401. 68pp.

Nishita, H. and W. A. Rhoads. 1970. Ecological and environmental effects from local fallout from Schooner. 1. Soil thermoluminescence in relation to radiation exposure under field conditions. Univ. of California at Los Angeles, U. S. AEC report PNE-526. 21pp. (NSA 24: 19121).

Nishita, H., D. Dixon and K. H. Larson. 1960. A comparison of uptake of strontium 90 and cesium 137 by bryophytes and vascular plants. Univ. of California at Los Angeles, U. S. AEC report UCLA-468. 9pp. (NSA 15: 12706).

Nishita, H., E. M. Romney and K. H. Larson. 1961. Uptake of radioactive fission products by crop plants. Agr. Food Chem. (2): 101-106. (Also: Univ. of California at Los Angeles, U. S. AEC report UCLA-459).

Nishiwaki, Y. 1973. Activities of the IAEA related to environmental protection. pp. 279-284. In: Proc. Regional Conf. on Radiation Protection. Vol. 1. Y. Feige, ed. Soreq Nuclear Research Center, Yavne, Israel. (NSA 30: 18037).

Nishiwaki, Y. and H. Kawai. 1962. Measurement of the tritium concentration in natural water. Mass Spectroscopy (Tokyo) 10(21): 83-90. (In Japanese) (NSA of Japan 2(3): 00020).

Nishiwaki, Y. and H. Kawai. 1965. Measurement of tritium in natural waters in Japan. Abstr. Health Physics 11(8): 803.

Nishiwaki, Y., et al. 1973. Behaviour and distribution of radioactive substances in coastal and estuarine waters. pp. 177-192. In: Radioactive Contamination of the Marine Environment. (See IAEA, 1973) (NSA 28: 8167).

Nishiyama, I., S. Ichikawa and E. Amano. 1964. Radiobiological studies in plants. X. Mutation rate induced by ionizing radiations at AL locus of sand oats. Radiation Botany 4(4): 503-516. (NSA 19: 15172).

Nishizawa, T., Y. Nishi and T. Kimura. 1959. Studies on the mechanism of aphid transmission of mosaic disease of Japanese radish using radioactive phosphorus 32. Virus 9(2): 130-133. (In Japanese).

Nishizawa, T., Y. Nishi and H. Yoshii. 1958. Studies on the varietal resistance of garden crops to the virus disease. V. On the course of aphids transmission of the mosaic disease of Japanese radish determined by P^{32} or S^{35} (1). Bull, Kyushu Agric. Expt. Sta. 5: 35-40. (In Japanese).

Nitschke, A. 1932. On the influence of hibernation through irradiated Ergosterin. Z. Ges. Exptl. Med. 82(1/2): 227-235. (In German).

Nixon, H. L. and C. R. Ribbands. 1952. Food transmission within the honey bee community. Proc. Roy. Soc. (London) B140: 43-50.

Nizam, J. 1968. Effects of cosmic rays on plants. Nucleus (Calcutta) (Suppl.): 224. (Also: In: Intern. Seminar on Chromosome - Its Structure and Function. (Calcutta, August 11-13, 1968; Univ. of Calcutta, India) (CONF-680846) (NSA 24: 25608).

Nishnikov, A. I., et al. 1969. Cesium-137 in lichen-reindeer-man in the Far North of the USSR 1962-1968. State Committee on the Use of Atomic Energy, USSR. Atomizdat, Moscow. 16pp. UN Document A/AC.82/G/L.1327. (AEC-tr-7232, pp. 27-44) (NP-18381) (NSA 25: 11037, 39825).

Nishnikov, A. I., et al. 1973. ^{210}Pb, ^{210}Po, ^{226}Ra, ^{228}Th, and ^{239}Pu in the lichen-reindeer-man chain in the USSR Actic. State Committee on the Use of Atomic Energy, USSR. Atomizdat, Moscow. 10pp. (In Russian) (INIS-mf-1132). (NSA 30: 12699).

Nizza, P. and F. Daburon. 1973. Radiocontamination of human food chains. Experimental research with large domestic animals. Agriculture (Paris) (366): 170-174. (In French) (NSA 29: 10149).

Noack, D. 1965. Testing of the interaction between microorganisms of a pupulation by statistical methods. Biophysik 2: 361-367. (In German) (NSA 20: 6706).

Noack, D. and W. Kleuters. 1960. The determination of moisture content of wood by means of radioactive isotopes. Wood as Raw and Construction Material 18(8): 304-308.

Noakes, J. E., K. A. Schneider and B. L. Brandau. 1972. Mobile archeological laboratory: microsample extraction and radiocarbon dating. pp. B94-B100. In: Proc. 8th Intern. Radiocarbon Dating Conf. Vol. 1. (See Rafter, 1972) (NSA 31: 8693).

Noakes, J. E., I. R. Supernaw and L. K. Akers. 1967. Anomalies in the $^{230}Th/^{232}$ activity ratio in some Mississippi River sediments. J. Geophys. Res. 72: 2679-2682. (NSA 21: 24109).

Noble-Nesbitt, J. 1963. A site of water and ionic exchange with the medium in Podura aquatica. J. Exptl. Biol. 40: 701-711. (Used ^{24}Na).

Noda, K. and N. Egami. 1975. Regeneration of transplanted intact cell populations in lethally irradiated hydra. Radiation Research 63(1): 174-184. (NSA 33: 7567).

Noethel, H. 1974. Investigations on radiosensitivity and radioresistant populations of Drosophila melanogaster. V. Relations between the relative radioresistance and some recombination properties in the irradiated population. ROE I. Mutation Research 25(3): 325-336. (NSA 31: 29953).

Noguchi, T., A. Shigematsu and Y. Igarashi. 1961. Studies on the mode of ascaricidal action, tracing S-35 and P-32 elements incorporated into organisms in vivo. Abstr. B-9. (See Japan Atomic Industrial Forum, 1961).

Nokes, M. A. and M. Simic. 1975. X-ray sensitivity of photosynthesis and reproductive systems in Chlorella. Photochem. Photobiol. 21(4): 265-268. (NSA 32: 28206).

Nokukuma, T. 1953. Effect of atomic bomb on trees. Nippon Gakujutsu 1: 264-265.

Noller, H. G. and H. Breitner. 1959. The dependence of the radiation effect on the body temperature. Strahlentherapie 109: 97-100. (In German) (NSA 13: 16728).

Nomi, M. 1971. Treatment, disposal, and discharge to the environment of radioactive wastes. FAPIG (Tokyo) (63): 136-142. (In Japanese) (NSA 26: 50586).

Noordink, J. P. W. 1965. Some applications of radioactive isotopes in ecological research. Ent. Ber. 25(7): 130. (In Dutch).

Noordink, J. P. W. 1971. Irradiation, competitiveness, and the use of radioisotopes in sterile-male studies with the onion fly, Hylemya antique Meigen. pp. 323-328. In: Sterility Principle for Insect Control or Eradication. (See IAEA, 1971) (NSA 25: 29748).

Noordink, J. P. W., H. P. Maas Geesteranus and C. A. van der Anker. 1967. Exudates of germinating spores of Bacillus thuringeiensis. pp. 249-250. In: Insect Pathology and Microbial Control. P. A. van der Laan, ed. North-Holland, Amsterdam.

Norberg, E. R., et al. 1973. Pheasant feeding study using seeds treated with methyl and phenylmercury. pp. 212-218. In: Mercury in the Western Environment. (See Buhler, 1973) (Neutron activation analysis).

Nordau, G. 1962. The bee and radioactivity. Inds. atomiques 6: 75-84. (In French).

Nordby, A. and K. Steenberg. 1959. Application of radioactive plant protection chemicals illustration and quantitative evaluation of spray deposits. J. Agr. Engr. Res. 4(2): 174-180. (NSA 14: 17668).

Nordenskiold, H. 1962. Studies of meiosis in Luzula purpurea. Hereditas 48(3): 503-519. (NSA 18: 75).

Nordenskiold, H. 1963. A study of meiosis in the progeny of x-irradiated Luzula purpurea. Hereditas 49(1-2): 33-47. (NSA 18: 76).

Nordenskiold, H. 1964. The effect of x-irradiation on diploid and polyploid Luzula. Hereditas 51: 344-374. (NSA 19: 3929).

Nordic Society for Radiation Protection. 1971. Proc. 3rd Nordic Radiation Protection Conf. (Copenhagen, August 18-20, 1971). Helsinki, Finland. 394pp. (In several languages) (CONF-710847) (NSA 27: 22693).

Nork, W. E. 1969. Radioactivity in the hydrologic environment: Project Rulison. Final pre-shot report. Isotopes, Inc., Palo Alto, Calif., U. S. AEC report NVO-1229-108. 13pp. (NSA 24: 5769).

Nork, W. E. and P. R. Fenske. 1970. Radioactivity in water: Project Rulison. Isotopes, Inc., Palo Alto, Calif., U. S. AEC report PNE-R-22. 20pp. (NSA 24: 22769).

Nork, W. E., E. J. Forslow and E. H. Essington. 1971. Radioactivity in water. Teledyne Isotopes, Palo Alto, Calif., U. S. AEC report NVO-1229-175. 84pp. (NSA 26: 56118).

Normand, M. 1974. Method of calibrating a neutron soil-moisture gauge on the basis of density measurements from an associated gamma density gauge. pp. 53-69. In: Isotope and Radiation Techniques in Soil Physics and Irrigation Studies 1973. (See IAEA, 1974) (NSA 30: 15485).

Norris, D. O. 1966. Radiothyroidectomy in the salmonid fishes Salmo gairdnerii Richardson and Oncorhynchus tshawytscha Walbaum. Thesis, Univ. of Washington, Seattle. 133pp. (NSA 21: 36744).

Norris, D. O. 1969. Depression of growth following radiothyroidectomy of larval chinook salmon and steelhead trout. Trans. Am. Fish. Soc. 98(1): 104-106.

Norris, J. S., D. O. Norris and J. T. Windell. 1973. Effect of simulated meal size on gastric acid and pepsin secretory rates in bluegill (Lepomis macrochirus). J. Fish. Res. Bd. Can. 39(2): 201-204. (SFA 18: 16256) (^{14}C).

Norris, K. R. 1965. The bionomics of blow flies. Ann. Rev. Entomol. 10: 47-68. (Refers to radionuclides).

Norris, L., R. E. Norris and M. Calvin. 1954. A survey of the rates and products of photosynthesis in plants of 9 phyla. Univ. of California, Berkeley, U. S. AEC report UCRL-2563. 28pp. (NSA 8: 5118).

Norris, R. A. 1957. Breeding bird census, two AEC Savannah River Plant, South Carolina, censuses. Audubon Field Notes 11(5): 2-3.

Norris, R. A. 1958. Some effects of x-irradiation on the breeding biology of eastern bluebirds. Auk 75(4): 444-455.

North, D. T. 1967. Radiation-induced male sterility exhibited in the P_1 and F_1 generation in lepidoptera. Abstr. Eb-9. Radiation Research 31(3): 615.

North, D. T. and G. G. Holt. 1967. The cytogenetic basis of radioresistance in lepidoptera species, Trichoplusia ni. Genetics 56(3,Pt.2): 580.

North, D. T. and G. G. Holt. 1968. Genetic and cytogenetic basis of radiation-induced sterility in the adult male cabbage looper Trichoplusia ni. pp. 391-403. In: Isotopes and Radiation in Entomology. (See IAEA, 1968) (NSA 24: 32327).

North, D. T. and G. G. Holt. 1968. Inherited sterility in progeny of irradiated male cabbage loopers. J. Econ. Entomol. 61(4): 928-931. (NSA 22: 43464).

North, D. T., et al. 1975. Corn earworms: quality of sperm in sterile males released for population supression on St. Croix Island. J. Econ. Entomol. 68(5): 595-598. (NSA 32: 15095).

North Carolina State Board of Health. 1962. Background radioactivity in surface water supplies of North Carolina 1958-1962. Annual report 1962. Raleigh.

Northern Indiana Public Service Company. 1971. Bailey Generating Station, Nuclear I. Environmental report. Hammond, Ind. 304pp. U. S. AEC DOCKET-50367-11. (NSA 25: 23197).

Northern States Power Company. 1967. Monticello Nuclear Generating Plant, Monticello, Minnesota, Unit 1. (Facility Description and Safety Analysis Report). Ammendment 4. ---Ammendment 8. Minneapolis, Minn. 215, 8pp. U. S. AEC DOCKET-50263-11, -13. (NSA 23: 4118, 4121) (Environmental safety; Mississippi River flow and pollution discussed).

Northern States Power Company. 1971. Monticello Nuclear Generating Plant, Unit 1. Environmental monitoring and ecological studies 1970 annual report. ---errata and addenda. Minneapolis, Minn. 330, 3pp. U. S. AEC DOCKET-50263-77, -80. (NSA 25: 46196, 50981).

Northern States Power Company. 1971. Prairie Island Nuclear Generating Plant, Units 1 and 2. Environmental Report, operating license stage. Minneapolis, Minn. U. S. AEC DOCKET-50282-33; DOCKET-50306-33. (NSA 25: 34286).

Norton, H. T. 1957. The turbulent diffusion of river contamination. Hanford Labs., U. S. AEC report HW-49195. 13pp. (NSA 11: 7160).

Norton, L. B. and R. Hansberry. 1941. Radioactive tracer methods for determination of the disposition of arsenic in the silk worm. J. Econ. Entomol. 34(3): 431-436.

Norwood, W. D. 1972. Atomic power safety: a common sense approach. J. Am. Med. Assoc. 221(12): 1392-1396. (NSA 26: 58311).

Nosaki, H., I. Furuta and K. Makino. 1955. Radiation found in livestock given feed contaminated with Bikini ashes (4) radiation in chickens fed Zn^{65}. Nippon Chikusangakkai Kaiho 26(1): 8.

Nosaki, H., et al. 1956. Bikini ashes appearing in hens eggs. pp. 749-761. In: Research in the Effects and Influences of the Nuclear Bomb Test Explosions. Vol. II. Japan Soc. for the Promotion of Science, Ueno, Tokyo.

Noshkin, V. E. 1972. Ecological aspects of plutonium dissemination in aquatic environments. Health Physics 22(6): 537-549. (NSA 26: 4715).

Noshkin, V. E., et al. 1975. Plutonium radionuclides in the ground waters at Eniwetok Atoll. Lawrence Livermore Lab., U. S. ERDA report UCRL-76725. 35pp. (NSA 33: 626).

Nöthel, H. 1963. The influence of x-rays on the vitality of Drosophila melanogaster. Studies of mortality, fecundity, and fertility after various doses. Thesis, Univ. of Berlin, Germany. 80pp. (In German) (NSA 18: 19750). (See also Noethel).

Nöthel, H. 1963. The influence of x-rays on longevity, fecundity and fertility of Drodophila melanogaster. Int. J. Radiat. Biol. 6(4): 381-382.

Nöthel, H. 1965. The influence of x radiation on the fitness characteristics of Drosophila melanogaster. Part I. Investigations of the lifespan. Strahlentherapie 126: 269-282. (In German) (NSA 19: 17648).

Nöthel, H. 1967, 1968. The influence of x radiation on the fitness characteristics of Drosophila melanogaster. II. Investigations on fecundity. IV. Investigations on fertility. Strahlentherapie 134: 609-624; 135: 493-499. (In German) (NSA 22: 23618, 33967).

Nöthel, H. 1968. Correlations, interactions, and differences between radiation effects on longevity and natural aging. pp. 87-102. In: Isotopes and Radiation in Entomology. (See IAEA, 1968) (NSA 24: 32164) (Insects).

Nöthel, H. 1972. Investigations of radiosensitive and radioresistant populations of Drosophila melanogaster. II. Restriction of the decrease in radiosensitivity of the irradiated population ROE I to Stage-7 oocytes. Mutation Research 15(3): 277-286. (NSA 26: 43358).

Nöthiger, R. and Dübendorfer. 1971. Somatic crossing-over in the housefly. Mol. Gen. Genet. 112(1): 9-13. (NSA 26: 12267) (X-rays).

Nourteva, P. and L. Reinius. 1953. Incorporation and spread of C^{14} labelled oral secretions of wheat bugs in wheat kernels. Ann. Entomol. Fenn. 19(3): 95-104.

Novick, S. 1965. Cesium-137 in Alaskans. Science 147(3665): 1598. (Comments on paper by Palmer, et al., 1965, q.v.).

Novikov, E. and G. D. Hanks. 1961. Analysis of irradiated Drosophila populations for meiotic drive. Nature 190(4780): 989-900.

Novikov, Yu. V. 1974. Health Problems in Studying Uranium Content in the Environment and Its Effects on the Organism. "Meditsina", Moscow. 232pp. (In Russian) (NSA 31: 554).

Novikov, Yu. V. I. I. Rezanov. 1962. Hygienic issues on the study of the effect of augmented natural radioactivity background environment on the population's health. Gigiena sanitariya 27(3): 78-84. (Open waters, air, Ra in water, humans).

Novikov, Yu. V., V. A. Liperovskiy and A. A. Polynkova. 1962. The fallout of radioactive substances during precipitation (snow). Atom. energiya 13(4): 385-387. (English transl., Consultants Bureau, New York, pp. 998-1000) (NSA 17: 1664).

Novikova, N. Ya. and T. A. Kazakova. 1970. Determination of ^{137}Cs in food products. Vopr. Pitaniya 29: 74-76. (In Russian).

Novikova, R. Ye. 1966. Techniques employed in irradiating experimental animals. Med. radiologiya 11(1): 88-93. (In Russian) (NSA 20: 35282).

Novikova, S. K. 1968. Distribution of $strontium^{90}$ in a meadow. Pochvovediniye (12): 46-52. (Sov. Soil Science (12): 1642-1647).

Novoreshchenova, N. S., et al. 1961. Use of radioactive carbon for the labeling of fleas. Med. parazitol. 30(1): 72-76. (In Russian).

Novoselova, G. P., T. F. Voronina and M. M. Golutvina. 1972. Determination of ^{239}Pu in bone. Gigiena sanitariya (1): 73-75. (In Russian) (NSA 26: 33096).

Noyce, J. R., et al. 1971. Temporal distributions of radioactivity and $^{89}Sr/^{90}Sr$ ratios during rainstorms. J. Geophys. Res. 76: 646-656. (NSA 15791).

Nozaki, T., et al. 1970. Neutron activation analysis of uranium in human bone, drinking water, and daily diet. J. Radioanal. Chem. 6(1): 33-40. (NSA 25: 46457).

Nozaki, Y. and H. Yagi. 1969. Determination of metals in life tissue by atomic absorption method. 1. Method for determination of calcium and strontium. Nagasaki Daigaku Suisan. Kenkyu Hokoku (27): 65-71. (In Japanese) (NSA 24: 16428).

Nriagu, J. O. 1973. Sulfur isotope abundances in Great Lakes waters: a preliminary report. Proc. Conf. Great Lakes Res. 16: 1038-1043. (NSA 30: 6591).

Nuclear Energy Agency. 1972. Disposal of Radioactive Waste. (Proc. Inform. Meeting, 12-14 April 1972). Organization for Economic Co-operation and Development, Paris. 123pp.

Nuclear Fuel Services, Inc. 1968. West Valley Reprocessing Plant. Environmental monitoring survey. West Valley, N. Y. 112pp. U. S. AEC DOCKET-50201-27. (NSA 24: 7496).

Nuclear Fuels Services, Inc. 1970. West Valley Reporcessing Plant. Low-level liquid waste treatment plant. Final Safety Analysis Report. Rockville, Md. 45pp. U. S. AEC DOCKET-50201-58. (NSA 25: 10660) (Concerns stream waste disposal).

Nuclear Fuels Services, Inc. 1971, 1973. West Valley Reprocessing Plant. Quarterly report. ---Quarterly report, January 1, 1973-March 31, 1973. West Valley, N. Y. vp, 8pp. U. S. AEC DOCKET-50201-21, -141. (NSA 25: 44412; 28: 5541) (Environmental monitoring).

Nussenzweig, R. S., et al. 1967. Protective immunity produced by the injection of x-irradiated sporozoites of Plasmodium berghei. Nature 216 (5111): 160-162. (NSA 21: 45849).

Nussig, N. and R. Perez del Viso. 1965. Radiochemical method of determination of chloride (^{36}Cl) absorption of mycelium of Aspergillus phoenicis (Group A. niger). pp. 33-38. In: 1st Interamerican Conf. on Radiochemistry. Pan American Union, Washington. (In Spanish) (NSA 20: 35282).

Nuzhdin, N. I. and R. L. Dozortseva. 1972. Genetic changes induced by space flight factors in the barley seeds taken aboard the spaceships Soyuz-5 and Soyuz-9. Zh. obshch. biol. 33(3): 336-346. (In Russian) (NSA 26: 53384).

Nuzhdin, N. I. and G. V. Nizhnik. 1960. The effect of gamma-rays from cobalt on early stages of embryogenesis in the rabbit. Tr. iqst. genet. Akad. Nauk SSSR 27: 348-358. (In Russian) (BA 42: 8583).

Nuzhdin, N. I. and N. A. Pastushenko-Stretlets. 1967. Influence of ecological conditions of plant growing on seed radiosensitivity. pp. 37-49. In: Experimental Data on the Effects of Ionizing Radiation on Organisms. N. I. Nuzhdin, ed. Publ. House Acad. Sci. USSR, Moscow. (In Russian) (NSA 22: 17053).

Nuzhdin, N. I., R. L. Dozortseva and N. S. Samokhvalova. 1963. Influence of cysteine on cytological damage caused by Co^{60} gamma rays and protons during the irradiation of barley seed. Zh. obshch. biol. 24(4): 261-275. (JPRS-22,147).

Nuzhdin, N. I., R. L. Dozortseva and N. S. Somokhvalova. 1963. Frequency of chromosome aberrations as a function of irradiation intensity and the physiological status of barley seeds. Izv. Akad. Nauk SSSR (biol.) (4): 514-525. (In Russian).

Nuzhdin, N. I., N. A. Pastushenko-Stretlets and G. N. Shangin-Berezovskiy. 1965. The effect of ecological cultivation conditions and the physiological conditions of seeds (degree of maturity) on radiosensitivity and the frequency and character of hereditary changes in gamma irradiated barley. Tr. inst. genet. Akad. Nauk SSSR (32): 18-68. (In Russian) (NSA 22: 17053).

Nyanishkene, V. M. B. and G. G. Polikarpov. 1970. Quantitative characteristics of the means of penetration of ^{90}Sr into the bodies of gastropod mollusks. Radiobiologiya 10: 928-930. (AEC-tr-7225, pp. 198-202) (NSA 25: 29440).

Nybom, N. 1953. Some experiences from mutation experiments in Chlamydomonas Hereditas 39(4): 317-324. (X- and UV-irradiation).

Nydal, R. 1963. Increase in radiocarbon from the most recent series of thermonuclear tests. Nature 200(4903): 212-214.

Nydal, R. 1966. Variation in ^{14}C concentration in the atmosphere during the last several years. Tellus 18: 271-279. (NSA 21: 6062).

Nydal, R. 1967. Transfer of radiocarbon in nature. pp. 119-129. In: Radioactive Dating and Methods of Low-Level Counting. (See IAEA, 1967) (NSA 22: 6497).

Nydal, R. 1968. Further investigations on the transfer of radiocarbon in nature. J. Geophys. Res. 73: 3617-3635. (NSA 22: 33573).

Nydal, R. and K. Loevseth. 1970. Prospective decrease in atmospheric radiocarbon. J. Geophys. Res. 75: 2271-2278. (NSA 24: 25139).

Nygaard, G. 1955. On the productivity of five Danish waters. Proc. Intern. Assoc. Limnol. 12: 123-133.

Nygaard, O. F. 1962. Insensitivity of ionizing radiation of DNA-polymerase in a slime mould. Abstr. p. 26. In: 2nd Intern. Congr. Radiation Research. (See Anonymous, 1962).

Nygaard, O. F. and S. Guttes. 1962. Effects of ionizing radiation on a slime mould with synchronous mitosis. Int. J. Radiat. Biol. 5: 33-44. (NSA 17: 57).

Nygaard, O. F., et al. 1973. Correlation between sensitivity to ionizing radiation and DNA replication in Physarum polycephalum. pp. 989-995. In: Advances in Radiation Research, Biology and Medicine. Vol. 2. J. F. Duplan, ed. Gordon & Breach, New York. (NSA 31: 26057).

Nygaard, P. 1970. Studies on the germination of pine pollen (Pinus mugo) in vitro. II. Effects of different ions. Physiol. Plant. 23: 372-384. (NSA 25: 489).

Nyhan, J. W. and T. E. Hakonson. 1976. Summary of recent studies of soil plutonium in the Los Alamos and Trinity site environs. Los Alamos Scientific Lab., U. S. ERDA report LA-UR-76-318. 56pp. (NSA 33: 25976).

Nyire Nya, E. U. 1964. The influence of whole body irradiation on the system hypothalmus-posterior pituitary lobe of the loir (Glis glis). Studies during hibernation and application of cysteine after irradiation. Thesis, Univ. of Hamburg, Germany. (AED-Diss. 64-486) (NSA 20: 3703).

Oakley, D. T. 1972. Natural radiation exposure in the United States. Thesis, Harvard Univ., Boston, Mass. 75pp. (Also: U. S. EPA report ORP-SID-72-1. (NSA 27: 9841).

Oakberg, E. F. 1968. Mammalian gametogenesis and species comparisons in radiation response of the gonads. pp. 3-15. In: Effects of Radiation on Meiotic Systems. International Atomic Energy Agency, Vienna.

Oak Ridge Associated Universities. 1965. Oak Ridge Associated Universities Health Physics Manual. U. S. AEC report TID-25375. 210pp. (NSA 24: 50928).

Oak Ridge National Laboratories. nd. Environs monitoring. pp. 2-35. In: Health Physics and Safety Annual Report for 1966. U. S. AEC report ORNL-4146. (NSA 21: 45249).

Oba, K. (w/T. Furukawa, M. Masafumi, M. Murai). 1966, 1967. Studies on radiosensitivity of forest trees. 1. ^{14}C fixation in radiation induced bud sports of Cryptomeria japonica D. Don and Chamaecyparis obtusa Endl. under cut-branch condition. 2. Growth and somatic mutations of Alnus, Populus, and Betula species under chronic gamma irradiation in a gamma field. 4. Mineral contents in leaves of trees growing under chronic gamma irradiation in a gamma field. 5. Difference in radiosensitivity of self- and cross-pollinated seeds, and of seeds with a marker gene in sugi, Cryptomeria japonica D. Don. Nippon Ringaku Kaishi 48(1): 153-157, 12-19, 163-175; 49(1): 45-52. (In Japanese) (NSA 21: 12811, 22796, 24664; 22: 25999). (See Ohba)

Oberhausen, E. and H. Muth. 1963. Natural and artificial radioisotopes in the biosphere and the human organism. Schrift. Bundesmin. Wiss. Forsch. Strahlensch. (Munich) (25): 123-143. (In German) (NSA 19: 3834).

Oberhauser, F. and H. A. Künkel. 1963. Ultrafractionation and relative biological effectiveness of fast electrons. Investigation on Drosophila embryos of different development stages. Biophysik 1: 11-19. (In German) (NSA 17: 30304).

Oberlander, H., et al. 1965. RNA and DNA synthesis during activation and secretion of the prothoracic glands of saturniid moths. J. Exptl. Zool. 159: 15-31. (^{3}H, Samia cynthia ricni, Antheraea polyphemus).

Oberländer, H. E. and K. Roth. 1973. Transformation of ^{14}C-labelled green compost in soil. Three years' outdoor pot experiment. Landwirt. Forsch. 25(4): 295-318. (NSA 28: 30403).

Oberpriller, J. C. 1968. The action of x irradiation on the regeneration field of the forelimb of the adult newt, Diemictylus viridescens. J. Exptl. Zool. 168: 403-422. (NSA 23: 20346).

Obo, F. 1954. Radioactive rains and fishes in the Kagoshima area. Igaku Seibutsugaku 33: 19-23. (In Japanese).

Obo, F., et al. 1955. Radioactivity of fish. II. Igaku Seibutsugaku 255-258. (CA 52: 2296e).

Oborn, E. T., E. T. Lyons and L. O. Timblin, Jr. 1954. Studies of herbicidal action on aquatic weeds using radioactive 2,4-D-1-C^{14}. U. S. Bureau of Reclamation, Denver, Colo., Lab. Rept. SI-4. 17pp., 16 plates.

Obradovich, J. D. 1965. Isotopic ages related to pleistocene events. Abstr. Abstracts Intern. Assoc. Quaternary Reserach. 7th Intern. Congr., Boulder and Denver, Colo.

Obreshkove, V. 1934. The quantitative biological effect of x-rays of different effective wave-lengths. Physiol. Zool. 7(3): 330-339. (Cladocera).

Obreshkove, V. and A. J. King. 1932. A change in the morphology of a cladoceran effected by x-ray treatment. Physiol. Zool. 5(3): 472-478. (Simocephalus vetulus).

Obreshkove, V. and A. J. King. 1932. The effect of x-rays on the gaseous metabolism of a cladoceran. Physiol. Zool. 5(3): 457-471.

O'Brien, J. P. and W. L. Gojmerac. 1956. Radiosensitivity of larval and adult amphibia in relation to temperature during and subsequent to irradiation. Proc. Soc. Exptl. Biol. Med. 92(1): 13-16. (NSA 10: 9939).

O'Brien, R. D. 1964. Reaction of carbamates with cholinesterase. Abstr. 46. Bull. Entomol. Soc. Am. 10(3): 163. (^{14}C).

O'Brien, R. D. and F. Matsumura. 1964. DDT a new hypothesis of its mode of action. Science 146: 657-658. (^{14}C).

O'Brien, R. D. and L. S. Wolfe. 1964. Radiation, radiosensitivity, and insects. Monograph Ser. on Radiation Biology. Academic Press, New York. 221pp. (NSA 18: 15756).

O'Brien, R. D., E. C. Kimmel and P. R. Sferra. 1965. Toxicity and metabolism of a famphur in insects and mice. J. Agric. Food Chem. 13(4): 366-369. (^{3}H).

Ochiahi, Kimura and Takahashi. 1958. Measurement of underground water by use of Na^{24}. Paper A-20, pp. 1701-1717. In: Proc. 2nd Japan Conf. on Radioisotopes. (See Japan Radioisotopes Assoc., 1958) (AEC-tr-4482) (NSA 15: 25802).

Ochiai, T. 1964. Investigation of undergroud streams by measuring the background radioactivity. Genshiryoku Kogyo 10(1): 73-77. (In Japanese) (NSA 19: 13653).

Ochiai, T. 1971. Investigation of groundwater by measurement of natural radioactivity. Radioisotopes (Tokyo) 20(5): 245-251. (In Japanese) (NSA 26: 9242).

Ochiai, T. 1971. New method for exploring groundwater by measuring natural radioactivity. Chikasui Ido Ponpu 14(1): 2-12. (In Japanese) (NSA 26: 48065).

Ochiai, T. 1972. Prospecting for underground water by the surface detection of natural radioactivity. Chigaku Zasshi 81(5): 307-318. (In Japanese) (NSA 30: 3169).

Ochiai, T. 1974. Application of radioisotopes and radiation in the field of agricultural engineering. Izotoptechnika 17(3): 140-146. (In Hungarian) (NSA 30: 9682).

Ochiai, T. and H. Kawasaki. 1969. Evaluation of groundwater resources by tritium dating. pp. 253-254. In: Proc. 9th Japan Conf. on Radioisotopes. (May 13-15, 1969, Tokyo, Japan). Japan Atomic Industrial Forum, Tokyo. (In Japanese) (CONF-690548) (NSA 24: 31479).

Ochinskaya, G. K. and I. B. Bychkovskaya. 1965. On the existence of system in the response of biological objects to irradiation (analysis of data obtainable on Paramecium caudatum). Dokl. Akad. Nauk SSSR 160(6): 461-463. (Doklady (biophys.) 160: 4-6) (NSA 19: 27997).

O'Connor, J. T. 1968. Fate of zinc in natural surface waters. Final report. Civil Engineering Studies, Sanitary Engrg. Ser. No. 49, Univ. of Illinois, Urbana, U. S. AEC report COO-1264-5. 164pp. (NSA 23: 11646).

O'Connor, J. T. and C. E. Renn. 1963. Evaluation of procedures for the determination of zinc. J. Am. Water Works Assoc. 55(5): 631-638.

Oda, T. 1968. Determination of rare earth elements in hot spring spring waters. 3. Activation method in combination with ion exchange separation method. Onken Kiyo 20(3): 205-213. (In Japanese) (NSA 23: 40601).

Oda, T. 1969. Determination of rare earth elements in hot spring water by radioactivation method. Radioisotopes (Tokyo) 18: 39-43. (In Japanese) (NSA 23: 17383).

Oda, T. 1970. Determination of rare earth elements in hot spring waters by activation method in combination with ion exchange separation method. Onsen Kagaku Kaishi 21(1): 7-16. (In Japanese) (NSA 26: 17373).

Oda, T. 1970. Determination of rare earth elements in hot springs by activation analysis, using Ge(Li) detector, and geochemical studies on these elements. Onsen Kagaku Kaishi 21(1): 17-25. (In Japanese) (NSA 26: 17372).

Oda, T. 1971. Measurement of ^{14}C in Beppu Hot Spring. Oita-ken Onsen Chosa Kenkyu-kai Hokoku (22): 80-83. (In Japanese) (NSA 29: 15644).

Oda, T. 1971. Natural radiocarbon in Beppu Hot Spring waters. Nippon Kagaku Zasshi 92(11): 1025-1026. (In Japanese) (NSA 26: 17727).

Oda, T. and H. Kawakami. 1969. Determination of the $^{35}Cl/^{37}Cl$ ratio in Beppu Hot Spring water by fast neutron activation analysis. Oita-ken Onsen Chosa Kenkyu-kai Hokoku (20): 63-65. (In Japanese) (NSA 24: 16456).

Oden, S. 1964. The occurrence of ^{14}C and tritium throughout Scandinavia, with applications in soil science and hydrology. Grundforbattring 17: 122-142. (In Swedish) (NSA 21: 179).

Odintsov, P. and Yu. Katkevich. 1967. Submicroscopic structure of spruce tracheids formed under the influence of gamma radiation. Latv. PSR Zinat. Akad. Vestis (7): 70-75. (In Russian) (NSA 22: 10737).

Odum, E. P. 1956. Consideration of the total environment in power reactor waste disposal. Proc. Intern. Conf. on the Peaceful Uses of Atomic Energy 13: 350-353. (Paper A/CONF.8/P 480).

Odum, E. P. 1956. Ecological aspects of waste disposal. pp. 95-103. In: Conf. on Radioactive Isotopes in Agriculture. (See U. S. AEC, 1956) (NSA 11: 4804).

Odum, E. P. 1957. Ecology and the atomic age. ASB Bull. 4(2): 27-29.

Odum, E. P. (Ed.). 1957. Environmental aspects of nuclear radiation. Bull. Ga. Acad. Sci. 15: 127-136.

Odum, E. P. 1959. Radiation ecology. Ch. 14, pp. 452-486. In: Fundamentals of Ecology. W. B. Saunders, Philadelphia.

Odum, E. P. 1961. Excretion rate of radio-isotopes as indices of metabolic rates in nature: biological half-life of zinc-65 in relation to temperature, food consumption, growth and reproduction in arthropods. Biol. Bull. 121(2): 371-372.

Odum, E. P. 1962. Relationships between structure and function in the ecosystem. Jap. J. Ecology 12(3): 108-118.

Odum, E. P. 1963. Ecology. Holt, Rinehart & Winston, New York. vii, 152pp. (Contains comments on radionuclides and ionizing radiation).

Odum, E. P. (Moderator). 1963. Panel discussion on education and research training. pp. 643-645. In: Radioecology. (See Schultz and Klement, 1963).

Odum, E. P. 1965. Feedback between radioecology and general ecology. In: Radiation and Terrestrial Ecosystems. Health Physics 11(12): 1257-1262. (See Hungate, 1963).

Odum, E. P. 1965. Summary. pp. 69-72. In: Ecological Effects of Nuclear War. (See Woodwell, 1965).

Odum, E. P. 1971. Radiation ecology. Ch. 17, pp. 451-467. In: Fundamentals of Ecology. W. B. Saunders, Philadelphia. 3rd ed.

Odum, E. P. and F. B. Golley. 1963. Laboratory of Radiation Ecology, AEC Savannah River Plant. Am. Zoologist 3(3): 357-358.

Odum, E. P. and F. B. Golley. 1963. Radioactive tracers as an aid to the measurement of energy flow at the population level in nature. pp. 403-410. In: Radioecology. (See Schultz and Klement, 1963) (NSA 17: 33588).

Odum, E. P. and E. J. Kuenzler. 1963. Experimental isolation of food chains in an old-field ecosystem with the use of phosphorus-32. pp. 113-120. In: Radioecology. (See Schultz and Klement, 1963) (NSA 17: 33562).

Odum, E. P. and T. G. Marples. 1963. Bioelimination of Zn^{65} and Fe^{59} in relation to kind of food eaten. Abstr. Bull. Ecol. Soc. Am. 44(3): 74-75.

Odum, E. P. and W. H. Mason. 1965. The radionuclide zinc-65 as an ecological tool. Abstr. Bull. Ecol. Soc. Am. 46(3): 100.

Odum, E. P. and A. J. Pontin. 1961. Population density of the underground ant, Lasius flavus, as determined by tagging with P^{32}. Ecology 42(1): 186-188.

Odum, E. P. and W. K. Willard. 1959. The fate of radioisotopes introduced into pioneer vegetation. Abstr. Proc. 9th Intern. Botan. Congr. 2(A): 25.

Odum, E. P., C. E. Connell and L. B. Davenport. 1962. Population energy flow of three primary consumer components of old-field ecosystems. Ecology 43(1): 88-96. (At Savannah River Plant).

Odum, E. P., E. J. Kuenzler and M. X. Blunt. 1958. Uptake of P^{32} and primary productivity in marine benthic algae. Limnol. Oceanogr. 3(3): 340-348.

Odum, E. P., R. P. Martin and B. C. Loughman. 1962. Scanning systems for the rapid determination of radioactivity in ecological materials. Ecology 43 (1): 171-173. (WR 106: 10).

Odum, H. T. 1957. Preliminary measurements in eleven Florida springs and a marine turtle grass community. Limnol. Oceanogr. 2: 85-97.

Odum, H. T. 1963. The element ratio method for predicting biogeochemical movements from metabolic measurements in ecosystems. pp. 209-224. In: Transport of Radionuclides in Fresh Water Systems. (See Kornegay, et al., 1963) (NSA 17: 40530).

Odum, H. T. 1967. Hydrogen budget and compartments in the rain forest at El Verde, Puerto Rico, pertinent to consideration of tritium metabolism. Bioenvironmental and radiological safety feasibility studies, Atlantic-Pacific interoceanic canal. Puerto Rico Nuclear Center, Battelle Columbus Labs., U. S. AEC report BMI-171-002. 36pp. (NSA 21: 39321).

Odum, H. T. 1970. In: A Tropical Rain Forest. (See H. T. Odum and R. F. Pigeon, 1970, below).

Preface. pp. ix-xi.

The rain forest and man: an introduction. Ch. A-2, pp. A5-A11.

The El Verde Study Area and the rain forest system of Puerto Rico. Ch. B1, pp. B3-B32.

The AEC rain forest program. Ch. C-1, pp. C3-C22.

Rain forest structure and mineral-cycling homeostasis. Ch. H-1, pp. H3-H-52.

Summary, an emerging view of the ecological system at El Verde. Ch. I-10, pp. I191-I289.

w/G. Cintron. Forest chlorophyll and radiation. Ch. I-2, pp. I21-I33.

w/G. Drewry. The cesium source at El Verde. Ch. C-2, pp. C23-C36.

w/C. F. Jordan. Metabolism and evapotranspiration of the lower forest in a giant plastic. Ch. I-9, pp. I165-I189.

w/A. Lugo. Metabolism of forest-floor microcosms. Ch. I-3, pp. I35-I56.

w/G. A. Briscoe and C. B. Briscoe. Fallout radioactivity and epiphytes. Ch. H-13, pp H167-176.

Odum, H. T. 1970. (with others, continued).

w/G. Drewry and J. R. Kline. Climate at El Verde, 1963-1966. Ch. B-22, pp. B347-B418.

w/A. M. Moore and L. A. Burns. Hydrogen budget and compartments in the rain forest. Ch. H-6, pp. H105-H122.

et al. Metabolism and evapotranspiration of some rain forest plants and soil. Ch. I-8, pp. I103-I164.

et al. Estimates of chlorophyll and biomass of the Tabonuco Forest of Puerto Rico. Ch. I-1, pp. I3-I19.

et al. Effects of gamma radiation on the forest at El Verde. Ch. D-1, pp. D3-D75.

Odum, H. T. and R. F. Pigeon. (Eds.). 1970. A Tropical Rain Forest. (A Study of Irradiation and Ecology at El Verde, Puerto Rico). Puerto Rico Nuclear Center, U. S. AEC report TID-24270; PRNC-138. xxiii, 1644pp. (NSA 25: 21633) (Incl. many articles on radioecology incl. herein as well as other articles on ecological studies of tropical rain forests).

Oechel, W. C., B. R. Strain and W. R. Odening. 1972. Tissue water potential, photosynthesis, ^{14}C-labeled photosynthate utilization, and growth in the desert shrub Larrea divaricata Cav. Ecol. Monogr. 42(2): 127-141.

Oertle, E., R. B. Emerson and H. E. Wheeler. 1953. Transfer of radioactivity from worker to drone honey bees after ingestion of radioactive sucrose. Ann. Entomol. Soc. Am. 46(4): 596-698. (Also: U. S. Dept. of Agriculture, U. S. AEC report AECU-2630) (NSA 7: 5700).

Oestmann, M. J. 1970. Tritium sources and release to the environment. Trans. Amer. Nucl. Soc. 13: 486-487. (CONF-701102) (NSA 25: 5941).

O'Farrell, T. P. 1969. Effects of acute ionizing radiation in selected Pacific Northwest rodents. pp. 157-165. In: Symposium on Radioecology. (See Nelson and Evans, 1969) (Also: Battelle Northwest Lab., U. S. AEC report BNWL-SA-1260) (NSA 21: 28587).

O'Farrell, T. P. 1973. Effects of acute radiation in captive and free-ranging pocket mice, Perognathus parvus. pp. 1065-1069. In: Radionuclides in Ecosystems.. (See Nelson, 1973) (NSA 25: 37868).

O'Farrell, T. P. and J. V. Dilley. 1975. Comparison of radiation response, cyanide, toxicity, and sulfur transferase activity in native North American rodents. Comp. Biochem. Physiol. 50B: 443-447. (NSA 33: 9586).

O'Farrell, T. P. and P. B. Dunaway. 1967. Incorporation and tissue distribution of a thymidine analog in the cotton rat Sigmodon hispidus. Comp. Biochem. Physiol. 22: 435-450.

O'Farrell, T. P. and P. B. Dunaway. 1969. Effects of acute ionizing radiation on incorporation of ^{131}IUDR by cotton rats, Sigmodon hispidus. Radiation Research 38(1): 109-124.

O'Farrell, T. P. and R. O. Gilbert. 1975. Transport of radioactive mater ials by jackrabbits on the Hanford Reservation. Health Physics 29(1): 9-15. (NSA 32: 27948).

O'Farrell, T. P., S. I. Auerbach and P. B. Dunaway. 1966. Incorporation of a thymidine analog by indigenous rodents. (Thesis). Oak Ridge National Lab., U. S. AEC report ORNL-3907. xx, 172pp. (Also: T. P. O'Farrell, thesis, Univ. of Tennessee, Knoxville).

O'Farrell, T. P., et al. 1972. Effects of ionizing radiation on survival, longevity, and reproduction in free-ranging pocket mice, Perognathus parvus. Radiation Research 49(3): 611-623. (NSA 26: 28481).

O'Farrell, T. P., et al. 1973. Radiation effects in free-ranging pocket mice, Perognathus parvus, during the breeding season. Science 179(4070): 289-291. (NSA 27: 12347).

Offori, E. D. 1969. Effects of gamma radiation on the reproduction and longevity of the stable fly, Stomoxys calcitrans. Thesis, North Dakota State Univ., Fargo. 94pp. (Univ. Microfilms 70-11,517) (NSA 25: 8822).

Offori, E. D. 1970. Gamma irradiation of Stomoxys calcitrans. J. Econ. Entomol. 63(2): 574-___.

Offori, E. D. 1970. Cytology of gamma-irradiated gonads of Stomoxys calcitrans (Diptera: Muscidae). Ann. Entomol. Soc. Am. 63: 706-712. (NSA 24: 34596).

Offori, E. D. 1972. Neutron irradiation and its possible application for insect control. pp. 233-237. In: Radiobiological Applications of Neutron Irradiation. International Atomic Energy Agency, Vienna, Publ. STI/PUB-325. (NSA 27: 20150).

Offori, E. D. and K. H. Czock. 1975. Use of fast neutrons and gamma radiation to sterilize the tsetse fly Glossina tachinoides Westw. Int. J. Appl. Radiat. Isotopes 26(5): 257-260. (NSA 32: 17539).

Offutt, M. S. 1962. Effect of gamma ray and thermal neutron irradiation on seed of Korean lespdeza Lespedeza stipulacea maxim. Crop Science 2: 325-330. (NSA 16: 31488).

Offutt, M. S. and R. D. Riggs. 1970. Radiation-induced resistance to root-knot nematodes in Korean lespedeza. Crop Science 10: 49-50. (NSA 24: 34284).

Oftedal, P. 1958. The prediction of mutation patterns after exposure to chronic irradiation. Proc. 2nd Intern. Conf. on the peaceful Uses of Atomic Energy 22: 298-302. (Paper A/CONF.15/P/588).

Oftedal, P. 1959. Study of the retention and the mutagenic mode of action of radioactive phosphorus in Drosophila melanogaster. Hereditas 45(2/3): 245-331.

Oftedal, P. 1961. Studies with radioyttrium in flies. I. Retention and distribution in Drosophila after injection. II. Retention and distribution in Drosophila and Musca after ingestion. Int. J. Radiat. Biol. 3(2): 211-221, 222-230.

Oftedal, P. 1962. Radiosensitivity of Drosophila spermatogonia. Abstr. Hereditas 48(3): 543.

Oftedal, P. 1963. Induction of mutations and killing of cells in irradiated spermatogonia of Drosophila. Nature 199(4900): 1301-1302. (NSA 17: 38822).

Oftedal, P. 1964. The radiosensitivity of Drosophila spermatogonia. I. Acute doses. Genetics 49(2): 181-193. (NSA 18: 9856; 21: 18124).

Oftedal, P. 1964. Radiosensitivity of Drosophila spermatogonia. III. Comparison of acute and protracted irradiation efficiences in relation to cell killing. Mutation Research 1: 63-76. (NSA 19: 3940).

Oftedal, P. and J. C. Mossige. 1957. Incorporation and mutagenicity of P^{32} in Drosophila sperm. pp. 457-462. In: Advances in Radiobiology. (Proc. 5th Intern. Conf. on Radiobiology, Stockholm). G. C. Hevesy, A. G. Forssberg and J. D. Abbatt, eds. Oliver & Boyd, London.

Oftedal, P. and J. Mossige. 1957. Incorporation of phosphorus-32 and mutagenesis during spermatogenesis in Drosophila. Nature 179(4550): 104-105.

Oga, H. 1959. Effects of P^{32} irradiation on skin and blood vessels. Yonago Med. J. 10(4): 850-865. (Frog).

Ogaki, M. and E. Nakashima-Tanaka. 1966. Inheritance of radioresistance in Drosophila. I. Mutation Research 3: 438-443. (NSA 21: 4787).

Ogaki, M. and E. Tanaka. 1963. Inheritance of tolerance to gamma radiation in Drosophila melanogaster. Abstr. Genetics 48(7): 904-905.

Ogaki, M., E. Nakashima-Tanaka and S. Matsunaga. 1969. Genetic control of radiosensitivity in Drosophila melanogaster. Gamma Field Symp. (8): 32-52. (CONF-690731) (NSA 25: 52026).

Oganesyan, S. A. 1971. Histogenesis and functioning of the hypophysis and thyroid gland in the larvae of Atlantic salmon exposed to ionizing radiation. Tr. polyar. nauchn.-issled. Proekt. inst. morsk. ryb. khoz. okeanogr. (29): 45-54. (AEC-tr-7418, pp. 53-65) (NSA 27: 27945).

Ogawa, Y. 1960. Synthesis of contractile proteins in the x-irradiated embryo. Nature 186(4718): 77-78. (NSA 14: 12483).

Ogawa, Y. 1968. Growth, differentiation and regeneration. 44. Effects of repeated x-irradiation on the formation of the early Triturus embryo. Igaku Seibutsugaku 76: 81-84. (In Japanese) (NSA 23: 48229).

Ogden, J. G. 1963. The Squibnocket Cliff peat: radiocarbon dates and pollen-stratigraphy. Am. J. Sci. 261(4): 344-353.

Ogden, J. G. 1964. Forest history research at OWU. Ohio Wesleyan Mag. 41 (3): 3-8. (^{14}C dating incl.).

Ogden, J. G. 1965. Radiocarbon and pollen evidence for a sudden change in climate in the Great Lakes region approximately 10,000 years ago. Abstr. Abstracts Intern. Assoc. Quaternary Research. 7th Intern. Congr., Boulder and Denver, Colo.

Ogden, J. G. and R. J. Hay. 1964. Ohio Wesleyan University natural radiocarbon measurements. I. Radiocarbon 6: 340-348.

Oglesby, R. T., C. A. Carlson and J. A. McCann. (Eds.). 1972. River Ecology and Man. Academic Press, New York. 492pp. (Incl. radionuclides).

Oglobin, K. F. and Ch. M. Khalifa-Zade. 1974. Distribution of uranium in shells of recent and fossil mollusks. Geokhimiya (2): 315-320. (In Russian) (NSA 30: 6476).

Ogoshi, S. 1965. Fibroin synthesis in posterior silk glands. VI. Metabolic changes in RNA during the first instar. Seikagaku Zasshi 57: 209-216. (^{14}C).

Ogura, H. and T. Sato. 1963. Mutants of Pellicularia filamentosa (Pat.) Rogers induced by irradiation. Nippon Shokubustu Byori Gakkaiho 28: 280-282. (In Japanese) (NSA 20: 35293).

Oguri, M., N. Takada and R. Ichikawa. 1965. Metabolism of radionuclides in fish. IV. Strontium-calcium discrimination in the renal excretion of fish. Nippon Suisan Gakkaishi 31: 435-438. (NSA 20: 6741).

Ohba, K. 1961. Radiation sensitivity of pine seeds of different water content. Hereditas 47(2): 283-294. (Oba ? q.v.).

Ohba, K. 1964. Studies on radiosensitivity of somatic mutations in forest trees. In: Mutations in Quantitatve Traits. (Rept. of Symp., Ohmiya-machi, Ibarakiken, Japan, July 28-29, 1964). Gamma Field Symp. (3): 111-141. (NSA 20: 18421).

Ohba, K. 1971. Radiation breeding of forest trees. Hoshasen Ikushujo Kenkyu Hokoku (2): 1-102. (NSA 29: 21142).

Ohba, K. 1971. Studies on the radiation breeding of forest trees. Acta Radiobot. Genet. (2): 1-98. (NSA 26: 33668).

Ohba, K. and M. Murai. 1969. Cross-fertility of pine pollen, Pinus densiflora, exposed to acute gamma irradiation. Nippon Ringaku Kaishi 51(4): 79-84. (In Japanese) (NSA 23: 41499).

Ohba, K. and M. Murai. 1971. Recessive genes producing albino and light green seedlings in sugi, Cryptomeria japonica D. Don. Nippon Rin Gakkaishi 53(6): 177-180. (In Japanese) (NSA 26: 45771).

Ohba, K. and M. Simak. 1961. Effect of x-rays on seeds of Scots pine from different provenances (Pinus silvestris L.). Silvae Genet. 10: 84-90). (NSA 21: 11121).

Ohinata, K., et al. 1971. Sterilization of the Mediterranean fruit fly by irradiation: comparative mating effectiveness of treated pupae and adults. J. Econ. Entomol. 64(4): 781-784. (NSA 25: 49278).

Ohkubo, F. 1954. Research report on radioactive rain and radioactive fish in Kagoshima area. Igaku Seibutsugaku 33(1): 19.

Öhlkers, F. and H. Berdfeld-Gärtner, 1962. Induction of mutations by radioisotopes. Z. Vererbungslehre 93(2): 264-279. (In German) (NSA 17: 23207) (Oenothera flavens).

Ohlsen, H. 1969. Determination of the mean population burden from natural external radiation in the Soviet Zone of Germany. German Democratic Republic Staatliche Zentrale für Strahlenschutz, Berlin, report SZS-14/69. 70 pp. (AEC-tr-7216).

Ohmomo, Y., H. Suzuki and M. Saiki. 1972. Application of isotope techniques for studies on mechanisms of land contamination and water pollution. Proc. Jap. Conf. Radioisotopes (10): 499-503. (In Japanese) (NSA 28:2574).

Ohmomo, Y., et al. 1969. Studies on the distribution of ^{203}Hg-labelled methyl mercury and phenyl mercury in pike. Univ. of Helsinki, U. S. AEC report NYO-3446-13. 15pp. (For 5th R.I.S. Symp., Helsinki, Finland, May 19-20, 1969) (CONF-690527-4) (NSA 25: 37716).

Ohno, S. 1971. Determination of iodine and bromine in biological materials by neutron-activation analysis. Analyst 96(1143): 423-426. (NSA 25: 44344).

Ohno, S., M. Hiki and M. Yatazawa. 1972. Determination of some trace elements in soil by neutron activation analysis. Radioisotopes (Tokyo) 21(7): 411-414. (NSA 26: 47660).

Ohno, S., M. Suzuki and M. Yatazawa. 1970. Simultaneous determination of copper and manganese in plants by neutron-activation analysis. Analyst 95: 995-999. (NSA 25: 10509).

Ohno, S., et al. 1971. Determination of europium in biological materials by neutron activation analysis. Radioisotopes (Tokyo) 29(5): 224-227. (NSA 25: 37332).

Ohse, T., et al. 1956. Investigation on contamination of migratory birds due to ashes of death. Fukuiken Eiseikenkyusho Gyoseki Shoroku 1: 15.

Ohta, N., K. Tomura and M. Omori. 1974. Aluminum, titanium, and vanadium in recent and fossil shells. V. Nippon Kagaku Kaishi (4): 703-711. (In Japanese) (NSA 30: 20587).

Ojima, T. and H. Toratani. 1965. Behavior of water in soil aggregates. II. Under unsaturated condition. Nippon Dojo Hiryogaku Zasshi 36: 339-342. (In Japanese) (NSA 21: 22030) (^3H_2O tracer).

Okabayashi, H. and M. Suzuki. 1970. Study on bioassay of plutonium. pp. 78-79. In: Research Report "On Internal Exposure to Plutonium" in April 1969-March 1970. Japan National Institute of Radiological Sciences, Chiba, report NIRS-Pu-7. (In Japanese) (NSA 26: 55742).

Okabe, K. and K. Ueda. 1959. Studies on the effects of radiation on chick embryos. I. Lethal effects using a single large dose irradiation of x-rays. Radioisotopes (Tokyo) 8: 177-184. (In Japanese) (NSA 15: 23335).

Okada, I., et al. 1956. On the influence of γ-ray radiation on aquatic animals. On the influence in the early development of goldfish. (*Carassius auratus* L.). pp. 1211-1218. In: *Research in the Effects and Influences of the Nuclear Bomb Test Explosions. Vol. II.* (See Japan Soc. for the Promotion of Science, 1956).

Okada, K. 1953. Biological investigation of atomic bomb disasters. I. Effect on animals. Nippon Gakujutsu Kaigi 1: 217-225.

Okada, T. and M. Goto. 1965. Synthesis of 2-amino 4-hydroxy-6-(hydroxymethyl) pteridine-10-^{14}C in *Drosophila melanogaster*. Seikagaku Zasshi 58 (5): 458-462).

Okawachi, A., et al. 1960. Natural radioactivities around our laboratory. Ann. Rept. Rad. Center Osake Prefect. (Japan) 1: 1-9.

Okazawa, Y. 1971. Isolation of molds from Austrian apple juice and their radiosensitivity. Rikagaku Kenkyusho Hokoku 47(5): 205-210. (In Japanese) (NSA 26: 23033).

Okazawa, Y. 1971. Synergistic effect between radiation and heat on molds. Rikagaku Kenkyusho Hokoku 47(5): 211-216. (In Japanese) (NSA 26: 23032).

Okazawa, Y., M. Namiki and A. Matsuyama. 1960. The mechanism of radiation enhancement of the lethal effect of sodium chloride on microorganisms. Bull. Agr. Chem. Soc. (Japan) 24: 435-443. (NSA 15: 27414).

Okazawa, Y., et al. 1960. Enhancement of the over-all lethal effect of ionizing irradiation on microorganisms by sodium chloride. Bull. Agr. Chem. Soc. (Japan) 24: 235-242. (NSA 15: 27413).

O'Kelley, J. C. and W. R. Herndon. 1959. Effect of strontium replacement for calcium on production of mobile cells in *Protosiphion*. Science 130 (3377): 718.

Oketay, M. and A. Senguen. 1963. Incorporation of iron-59 into the cells of different tissues of *Chironomus plumosus*. Nature 197(4867): 613-614. (NSA 17: 13945).

Oki, I. and S. Yoshida. 1964. Investigation of environmental fallout contamination by the measurement of the total beta rays. Osaka-furitsu Koshueisei Kenkyusho Kenkyo Hokoku (1): 29-36. (NSA 20: 7274).

Okonski, J., F. W. Lengemann and C. L. Comar. 1961. Incorporation of I^{131} into chicken eggs. Health Physics 6(1/2): 27-31. (NSA 15: 30358).

Oksbjerg, E. B. 1958. Investigations of the distribution of roots and root competition for phosphate in stands of *Picea excelsa* and *Abies alba*. Oikos 9(1): 57-76. (FA 20: 257).

Okuda, A. and S. Ida. 1961. Metabolism of glucose and acetate in *Chlorella* cells in relation to the active transport of mineral nutrients. Abstr. A-14. In: *Program and Abstracts for the 4th Japan Conf. on Radioisotopes.* (See Japan Atomic Industrial Forum, 1961).

Okun, D. A. 1973. Potential for the use of cobalt-60 irradiation in the treatment of wastewaters. North Carolina Univ., Chapel Hill, U. S. AEC report TID-26341. 33pp. (NSA 28: 10970).

Okun, P. A., et al. 1962. A review of the literature of 1961 on wastewater and water pollution control. J. Water Poll. Control Fed. 34(7): 629-703. (SFA 8: 5882) (Incl. remarks on radioactive wastes).

Okunewick, J. P. 1966. Effects of acute doses of gamma radiation on pine trees. RAND Corp., Santa Monica, Calif., Memo. RM-4904-TAB. vii, 30pp.

Okunewick, J. P. and S. E. Herrick. 1967. Response of pines irradiated with lethal and sublethal doses of gamma radiation. Nature 214: 514-515. (NSA 21: 22776).

Okunewick, J. P., S. Herrick and E. N. Carlsen. 1964. Early response of *Pinus thunbergii* to acute γ-irradiation. Nature 204(4956): 394-395. (NSA 19: 3944).

Okura, K. 1953. Biological investigation of atomic bomb disaster. 2. Effect on plants. Ch. I. Outline of survey. Nippon Gakujutsu Kaigi 1: 225-227.

Olafson, J. H. and K. H. Larson. 1963. Plutonium, its biology and envioronmental persistence. pp. 633-639. In: *Radioecology*. (See Schultz and Klement, 1963) (Also: Univ. of California at Los Angeles, U. S. AEC report UCLA-501) (NSA 16: 9812).

Olafson, J. H., H. Nishita and K. H. Larson. 1957. The distribution of plutonium in the soils of central and northeastern New Mexico as a result of the atomic bomb test of July 16, 1945. Univ. of California at Los Angeles, U. S. AEC report UCLA-406. 25pp. (NSA 11: 13247).

Olafson, J. H., et al. 1953. Preliminary study of off-site air-borne radioactive materials, Nevada Proving Grounds. I. Fallout originating from SNAPPER 6, 7, and 8 at distances of ten to fifty miles from Ground Zero. Univ. of California at Los Angeles, U. S. AEC report UCLA-243. 123pp. (NSA 13: 11748).

Olenov, Y. M., K. F. Galkovskaya and A. D. Pushnitsyna. 1959. Materials for describing the effect of ionizing radiation on individual development. Tsitologiya 1(3): 60-77. (JPRS-3398) (*Drosophila melanogaster*).

Ølgaard, P. L. and V. Haahr. 1968. On the sensitivity of subsurface neutron moisture gauges to variations in bulk density. Soil Science 105(1): 62-64.

Olive, J. H. and J. H. Morrison. 1967. Variations in distribution of ^{14}C in cell extracts of phytoplankton living under natural conditions. Limnol. Oceanogr. 12(3): 383-___.

Olive, J. H., D. M. Benton and J. Kishler. 1969. Distribution of C^{14} in products of photosynthesis and its relationship to phytoplankton composition and rate of photosynthesis. Ecology 50(3): 380-386.

Oliver, G. D., et al. 1964. A compilation of Camp Mercury environmental monitoring data from June 28, 1962 to October 9, 1963. U. S. Army Nuclear Defense Lab., Edgewood Aersenal, Md., report NDL-TR-63. 41pp. (NSA 18: 31787) (In Greenland).

Oliver, J. H., R. L. Osburn and J. R. Roberts, Jr. 1972. Cytogenetics of ticks (Acari: Ixodoidea). 9. Chromosomes of Rhipicephalus sanguineus (Latreille) and effects of gamma radiation on spermatogenesis. J. Parasitol. 58(4): 824-827. (NSA 27: 2823).

Olivereau, M. 1957. Radiothyroidectomy of an eel (Anguilla anguilla L.). Arch. Anat. Microscop. Morphol. Exptl. 46: 39-60. (In French).

Olivereau, M. 1957, Presence of bound iodine in organ tissues of certain endostyle cells and in the thyroid. Compt. rend. bull. assoc. anat. 92: 1113-1132. (In French) (Lampreys).

Olivereau, M. 1963. Effects of radiothyroidectomy on the hypophysis of the eel. Discussion on the para distals of teleosts. Gen. Comp. Endocrinol. 3(3): 312-332. (In French) (NSA 18: 11727).

Olivereau, M. 1963. Effect of radiothyroidectomy on eel interrenal tissue. J. Physiol. 55: 311-312. (In French) (NSA 18: 21795).

Olivereau, M. and G. La Roche. 1965. Effects of low iodine intake or destructive doses of ^{131}I on pituitary histochemistry of young chinook salmon (Oncorhynchus tshawytscha, Walbaum). Am. Zool. 5: 234. (NSA 19: 35900).

Olivereau, M., G. La Roche and A. N. Woodall. 1964. Cytological modifications of trout pituitary as a result of an iodine deficiency and a radiothyroidectomy. Ann. Endocrinol. (Paris) 25: 481-490. (In French) (NSA 19: 33745).

Olivieri, F. 1929. On the action of roentgen rays on Bufo vulgaris. Monit. Zool. Ital. 40(11/12): 442-444. (Riv. radiol. fisica med. 1: 527) (In Italian).

Olivieri, G. and A. Olivieri. 1965. Autoradiographic study of nucleic acid synthesis during spermatogenesis in Drosophila melanogaster. Mutation Research 2(4): 366-380. (^{3}H).

Olivieri, G. and A. Olivieri. 1965. The mutagenic effect of tritiated uridine in Drosophila spermatocytes. Mutation Research 2(4): 381-384.

Olsen, S. 1950. Phosphate absorption and isotopic exchange in lake muds. Verh. int. ver. limnol. 13: 915-923.

Olsen, S. 1964. Phosphate equilibrium between reduced sediments and water. Verh. int. ver. limnol. 15(1): 333-341.

Olson, A. R. and G. N. Lewis. 1928. Natural radioactivity and the origin of species. Nature 12(3052): 673-674.

Olson, E. A. and W. S. Broecker. 1958. Sample contamination and reliability of radiocarbon dates. Trans. N.Y. Acad. Sci. (III)21: 593-604.

Olson, J. S. 1961. Ecological sampling and meteorological calculations of fallout on forests near Oak Ridge. Oak Ridge National Lab., U. S. AEC report ORNL-3181. v, 57pp.

Olson, J. S. 1962. Surveys of environmental radioactivity. Nuclear Safety 4(1): 88-95.

Olson, J. S. 1963. Analog computer models for movement of nuclides through ecosystems. pp. 121-125. In: Radioecology. (See Schultz and Klement, 1963) (NSA 17: 33563).

Olson, J. S. 1964. Models for simulating ecological systems with the aid of analog and digital computers. Abstr. Biometrics 20(2): 392. (Incl. concentration of radioisotopes).

Olson, J. S. 1964. Advances in radiation ecology. Nuclear Safety 6(1): 78-81. (NSA 19: 10733).

Olson, J. S. 1965. Equations for cesium transfer in Liriodendron forest. In: Radiation and Terrestrial Ecosystems. Health Physics 11(12): 1385-1392. (See Hungate, 1965).

Olson, J. S. 1965. Use of tracer techniques for the study of biogeochemical cycles. Oak Ridge National Lab., U. S. AEC report ORNL-P-1820. 42pp. (NSA 20: 8534).

Olson, J. S. 1966. Progress in radiation ecology: radionuclide movement in major environments. Nuclear Safety 8: 53-68.

Olson, J. S. 1968. Use of tracer techniques for the study of biogeochemical cycles. pp. 271-288. In: Functioning of Terrestrial Ecosystems at the Primary Production Level. (Proc. Copenhagen Symp.). UNESCO, Paris.

Olson, J. S. and S. I. Auerbach. 1960. Biological contamination and dispersal of radioactive wastes. Nuclear Safety 1(3): 62-65.

Olson, J. S. and D. A. Crossley, Jr. 1963. Tracer studies of the break-down of forest litter. pp. 411-416. In: Radioecology. (See Schultz and Klement, 1963) (NSA 17: 33589).

Olson, J. S. and H. D. Waller. 1965. Two-year redistribution of cesium 137 in vegetation of a tagged Liriodendron forest. Abstr. Bull. Ecol. Soc. Am. 46(3): 100.

Olson, J. S., et al. 1960. Distribution of radioactive isotopes in plants, insects and soils. Nuclear Safety 1(3): 62-68.

Olson, K. R., H. L. Bergman and P. O. Fromm. 1973. Uptake of methyl mercuric chloride by trout: a study of uptake pathways into the whole animal and uptake by erythrocytes in vitro. J. Fish. Res. Bd. Can. 30(9): 1293-1299. (SFA 19: 17169) (^{203}Hg).

Olson, P. A., Jr. 1948. Some effects of pile area effluent water on young silver salmon. Hanford Atomic Products Operation, U. S. AEC report HW-8944. 21pp. (NSA 13: 8492).

Olson, P. A., Jr. and R. F. Foster. 1952. Accumulation of radioactivity in Columbia River fish in the vicinity of Hanford Works. Hanford Atomic Products Operation, U. S. AEC report HW-23093. 59pp. (NSA 11: 13569).

Olson, P. R. 1964. Accumulation of calcium-45 in developing coho salmon eggs and fry reared in varying concentrations of stable calcium. Thesis, Univ. of Washington, Seattle.

Olson, P. R. 1973. Accumulation of calcium-45 in developing coho salmon eggs and fry reared in varying concentrations of stable calcium. pp. 874. In: Radionuclides in Ecosystems. (See Nelson, 1973) (NSA 28: 30331).

Olsson, I. U. 1965. C^{14}/C^{12} ratio during the last thousand years and the reliability of C^{14}-dates. Abstr. Abstracts Intern. Assoc. for Quaternary Research. 7th Intern. Congr., Boulder and Denver, Colo.

Olsson, I. U. 1968. Modern aspects of radiocarbon datings. Earth-Sci. Rev. 4: 203-218. (NSA 23: 40867).

Olsson, I. U. (Ed.). 1970. Radiocarbon Variations and Absolute Chronology. John Wiley, New York. 655pp. (NSA 25: 21194).

Olsson, I. U. 1974. Eighth International Conference on Radiocarbon Dating. Geol. Foren. Stockholm Forh. 96(556): 37-44. (NSA 31: 3402).

Olsson, I. U., I. Karlen and A. Stenberg. 1966. Radiocarbon variations in the atmosphere. Tellus 18: 293-297. (NSA 21: 6066).

Olteanu, V., O. D. Kuznetsova and I. P. Arman. 1970. Recovery phenomena in yeast. I. The lethality and mitotic recombination induced by gamma-rays and chemicals in a UV-supersensitive mutant of yeast. Genetika 6(2): 123-129. (In Russian) (NSA 24: 34768).

Oltman, B. G., et al. 1973. A radiological envrionmental survey at EBR-II. pp. 278-288. In: Radiological and Environmental Research Division Annual Report, Center for Human Radiobiology, July 1972-June 1973. Argonne National Lab., U. S. AEC report ANL-8069, Part II. (At National Reactor Testing Station, Idaho).

Omanwar, P. K. and J. A. Robertson. 1971. A micro and rapid method for measuring P^{32} in plant samples. Plant Soil 34(1): 225-227.

Oncescu, M. and O. Serbanescu. 1963. The β radioactivity of the zooplankton of the Black Sea. Acad. Rep. Populare Romine Studii Cerc. Fiz. 14: 133-141. (In Russian) (NSA 17: 31843).

Oncescu, M., E. Gaspar and L. Dinescu. 1973. Determination of some constants of the tracer dynamics in filed investigations. Rev. Roum. Phys. 18(8): 1023-1033. (NSA 29: 17169).

Ondrej, M. 1965. Influence of interaction between EDTA and x-rays on mutation and cross-over frequencies in Drosophila melanogaster. pp. 26-29. In: Induction Mutations Mutation Process. (Proc. Symp., Prague, 1963).

O'Neill, R. V. 1973. Error analysis of ecological models. pp. 898-908. In: Radionuclides in Ecosystems. (See Nelson, 1973) (NSA 25: 57947).

O'Neill, R. V. and C. E. Styron. 1970. Applications of compartment modeling techniques to Collembola population studies. Am. Midland Nat. 83(2): (Incl. ionizing radiation).

Ono, Y., A. Tsuya and M. Kawakami. 1969. Effects of head irradiation on paradoxical sleep (EEG after-reaction) in the rabbit. Radiation Research 38: 365-378. (NSA 24: 9937).

Onodera, K. and T. Komano. 1964. Glycine-specific transfer ribonucleic acid (t-RNA) in the posterior silk gland of Bombyx mori. Biochem. Biophys. Acta 87(2): 338-340. (^{14}C).

Onodera, K., et al. 1965. Presence of a different sequence in the terminal region of glycine-specific (transfer RNA). Agric. Biol. Chem. (Tokyo) 29 (7): 693-695. (^{14}C, Bombyx mori).

Onuf, C. P. and W. W. Murdoch. 1974. Mediterranean as a system. II. Small ecosystems. Int. J. Environ. Stud. 6(1): 29-34. (NSA 33: 12041) (Incl. radioactivity).

Ooi, H. 1961. Studies on the effect of gamma-ray irradiation at embryonic stages upon the larval and cocoon characters in the silkworm. Nippon Sanshigaku Zasshi 30(5): 401-404.

Oonnithan, E. S. and R. Miskus. 1964. Metabolism of C^{14}-dieldrin by dieldrin-resistant Culex pipiens quinquefasciatus mosquitoes. J. Econ. Entomol. 57(4): 425-426.

Ophel, I. L. 1957. The disposal of radioactive wastes. Atomic Energy of Canada, Ltd., Chalk River, report AECL-450. 5pp. (NSA 14: 10245).

Ophel, I. L. 1960. Investigations of the effects of radioactive material on aquatic life in Canada. Abstr. p. 21. In: Biological Problems in Water Pollution. (See Tarzwell, 1960).

Ophel, I. L. 1963. The fate of radiostrontium in a freshwater community. pp. 213-216. In: Radioecology. (See Schultz and Klement, 1963) (NSA 17: 7781).

Ophel, I. L. 1973. Environmental capacity of freshwaters for waste radionuclides. pp. 613-624. In: Environmental Behavior of Radionuclides Released in the Nuclear Industry. (See IAEA, 1973) (NSA 29: 29364).

Ophel, I. L. and C. D. Fraser. 1957. The Chalk River liquid disposal area, 1956. Atomic Energy of Canada, Ltd., Chalk River, report CRHP-709. 42pp.

Ophel, I. L. and C. D. Fraser. 1970. Calcium and strontium discrimination by aquatic plants. Ecology 51(2): 324-327.

Ophel, I. L. and C. D. Fraser. 1973. The fate of cobalt-60 in a natural freshwater ecosystem. pp. 323-327. In: Radionuclides in Ecosystems. (See Nelson, 1973) (NSA 28: 27302).

Ophel, I. L. and J. Judd. 1962. Absorption of radiostrontium by the gills of freshwater fish. Nature 194(4834): 1187-1188. (NSA 16: 20106).

Ophel, I. L. and J. Judd. 1966. Effects of internally deposited radionuclides on the thermal tolerance of fish. pp. 825-833. In: Disposal of Radioactive Wastes into Seas, Oceans and Surface Waters. (See IAEA, 1966) (Also: Atomic Energy of Canada, Ltd., Chalk River, report AECL-2409) (NSA 20: 31028).

Ophel, I. L. and J. M. Judd. 1967. Experimental studies of radiostrontium accumulation by freshwater fish from food and water. pp. 859-865. In: Radioecological Concentration Processes. (See Aberg and Hungate, 1967).

Ophel, I. L. and J. M. Judd. 1967. Skeletal distribution of strontium and calcium and strontium/calcium ratios in several species of fish. pp. 103-109. In: Strontium Metabolism. J. M. A. Lenihan, J. F. Loutit and J. H. Martin, eds. Academic Press, New York. (NSA 21: 45578).

Ophel, I. L. and J. M. Judd. 1968. Marking fish with stable strontium. J. Fish. Res. Bd. Can. 25(7): 1333-1337.

Ophel, I. L. and J. M. Judd. 1969. Strontium-calcium relationships in aquatic food chains. pp. 221-225. In: Symposium on Radioecology. (See Nelson and Evans, 1969).

Ophel, I. L., C. D. Fraser and J. M. Judd. 1971. Strontium concentration factors in biota and bottom sediments of a freshwater lake. pp. 509-527. In: Proc. Intern. Symp. on Radioecology Applied to the Protection of Man and His Environment. (See Commission of the European Communities, 1971) (NSA 27: 5255).

Oppelt, J. 1965. Radioisotope flowmeter. Jaderna Energie 11: 382-384. (In Czech) (NSA 20: 4227).

Oppenheim, R. W., J. R. Jones and G. Gottlieb. 1970. Embryonic motility and posthatching perception in birds after prenatal gamma irradiation. J. Comp. Physiol. Psychol. 71(1): 6-21. (NSA 24: 41970) (Ducks, chickens).

Oppenoorth, F. J. 1967. Two types of sesamex-suppressible resistance in the housefly. Entomologia Exp. Appl. 10: 75-86. (^{14}C).

Oppermann, K. 1913. The development of trout eggs after fertilization with radium irradiated spermatozoa. Arch. microsc. anat. 83(2): 141-189, pls. 5-7. (In German).

Orbe Murillo, F. 1969. Radioactive contamination of the environment. Atmosphere, water and food products. Politecnica (Quito) 1(3): 75-112. (In Spanish) (NSA 27: 19814).

Ord, M. J. and J. F. Danielli. 1956. The site of damage on amoebae exposed to x-rays. Quart. J. Microscop. Sci. 97: 29-37.

Oregon Governor's Nuclear Development Coordinating Committee. 1970. Northwest Conference on the Role of Nuclear Energy. (2nd Ann. Governor's Conf. on Conservation, Portland, Oregon, December 4-5, 1969). Portland. 215pp. (CONF-691223) (NSA 25: 23111).

Oren, O. H. 1965. Study of the physico-chemical-biological processes affecting the dispersion of radionuclides. Final report, December 1-1961-December 31, 1064. Technion-Israel Inst. of Tech., Haifa. 48pp. (NP-15861) (NSA 20: 14731).

Orenski, S. W., J. R. Murray and K. Maramorosch. 1965. Further studies on the feeding habits of aster yellows virus-carrying corn leafhoppers. Contr. Boyce Thompson Inst. Plant Res. 23(3): 47-50. (^{32}P, Callistophus chinensis, Dalbulus maidis).

Organization for Economic Co-operation and Development. 1972. Disposal of Radioactive Waste. (Proc. Information Meeting, Paris, April 12-14, 1972). Paris. 294pp. (In French and English) (CONF-720453) (NSA 28: 6463).

Organization for Economic Co-operation and Development. 1974. Panel: conclusions of the seminar. pp. 399-429. In: Proc. Seminar on the Monitoring of Radioactive Effluents. (See below) (NSA 31: 3357).

Organization for Economic Co-operation and Development. 1974. Proc. Seminar on the Monitoring of Radioactive Effluents. (Karlsruhe, Germany, May 14-17, 1974). Paris. 446pp. (In French and English) (CONF-740556) (NSA 31: 3333).

Orlov, A. J. 1962. Accumulation of phosphorus by seedlings of several arboreal plants. Dokl. Akad. Nauk SSSR 147(1): 233-236. (In Russian).

Orlova, Ye. I. 1962. Sorption of radioactive products of uranium fission by rock strata. Ch. 4, pp. 74-83. In: Radiation Hygiene. Vol. 2. (See Marey, 1962) (In Russian).

Orlova, Ye. I. 1965. Migration of ^{90}Sr in the ground waters of various salt compositions. Gigiyena sanitariya 30(7): 46-50. (In Russian) (NSA 19: 46602).

Orlova, Ye. I., V. A. Smirennaya and R. A. Chelysheva. 1974. Migration of ^{60}Co in losse rocks contaminated with effluents from an atomic electric power station (AES). Gigiyena sanitariya (3): 50-53. (In Russian) (NSA 30: 9545).

Orphanidis, P. S. and P. E. Kalmoukos. 1971. Possibilities of integrated control of Dacus oleae (Gmelin) by autocidal and chemical methods. pp. 55-66. In: Sterility Principle for Insect control or Eradication. (See IAEA, 1971) (In French) (NSA 25: 29740).

Orphanidis, P. S., C. D. Soultanopoulos and M. G. Karandeinos. 1963. Preliminary report on the dispersion of adult Dacus oleae with ^{32}P. In: Radiation and Radioisotopes Applied to Insects of Agricultural Importance. (See IAEA, 1963) (In French).

Orr, H. 1967. Excretion of orally administered zinc-65 by the cotton rat in the laboratory and field. Health Physics 13(1): 15-20. (NSA 21: 6413).

Orr, P. C. and R. Berger. 1965. Radiocarbon age of a Nevada mummy. Science 148(3676): 1466-1467. (NSA 19: 30428).

Orrego, H., J. de Dios Vial and I. Mena. 1965. Gastric absorption of fat in the toad (Bufo spinolosus) and in rats. pp. 97-102. In: Proc. 5th Inter-American Symp. on the Peaceful Uses of Atomic Energy. Pan American Union, Washington. (In Spanish) (NSA 19: 35938).

Orsini, A. and E. Pirastru. 1970. Some significant results of measurement of density and natural water content of soils with gamma-ray and neutron probes in geotechnology. Italy Comitato Nazionale per l'Energia Nucleare, Rome, report RT/GEO-(70)15. 23pp. (For 10th Conf. on Geological Techniques, Bari, Italy, 26 Oct. 1970) (In Italian) (CONF-701061-2) (NSA 26: 9237).

Ortins de Bettencourt, A., A. Severo and M. M. Ribau Teixeira. 1974. Environmental surveillance at the Nuclear Physics and Engineering Laboratory (SACAVEM): experience acquired over twelve years. 1974. pp. 239-250. In: Environmental Surveillance Around Nuclear Installations. Vol. 2. (See IAEA, 1974) (In French) (NSA 31: 14038).

Ortiz Magana, J. R., et al. 1974. Transport of tritium in an agricultural environment in Mexico: study program of contamination with tritium in an agricultural area. pp. 147-166. In: Environmental Surveillance Around Nuclear Installations. (See IAEA, 1974) (In Spanish) (NSA 31: 14035).

Ortega, R. F. and B. K. Lee. 1970. Neutron activation of ancient pigments from murals of Cholula and Teotihuacan. Archeometry 12: 197-203. (NSA 25: 2133).

Orvini, E., T. E. Gills and P. D. LaFleur. 1972. Nuclear activation analysis of Se, As, Zn, Cd, and Hg in environmental matrices. Trans. Amer. Nucl. Soc. 15(2): 642-643. (NSA 27: 4933).

Orvini, E., T. E. Gills and P. D. LaFleur. 1974. Method for determination of selenium, arsenic, zinc, cadmium, and mercury in environmental matrices by neutron activation analysis. Anal. Chem. 46(9): 1294-1297. (NSA 30: 18177).

Osborn, A. W., E. Shipp and P. B. Hutchinson. 1966. Biology and radiation sterilization of sugar cane leafhoppers. At. Energy Austral. 9(4): 11-19. (NSA 21: 20463).

Osborn, G. K. and D. J. Kimeldorf. 1957. Some radiation responses of two species of bats exposed to warm and cold temperatures. J. Exptl. Zool. 134(1): 159-170. (Also: U. S. Naval Radiological Defense Lab., San Francisco, report USNRDL-TR-76) (NSA 10: 7399).

Osborne, T. S. and J. A. Bacon. 1960. Radiosensitivity of seeds. I. Reduction of stimulation of seedling growth as a function of gamma-ray dose. Radiation Research 13(5): 686-690.

Osborne, T. S. and M. J. Constantin. 1966. Table 41. Sensitivity to ionizing radiation: dormant seeds. pp. 183-190. In: Environmental Biology. P. L. Altman and D. S. Dittmer, eds. Am. Soc. Exptl. Biol., Bethesda, Md. (Incl. wild plants).

Osburn, W. S., Jr. 1960. Phytogram comparisons of Penstemon virens growing in a relatively radioactive site and four other habitats. Abstr. J. Colo.-Wyo. Acad. Sci. 5(1): 47.

Osburn, W. S., Jr. 1961. Variation in clones of Penstemon growing in natural areas of differing radioactivity. Science 134(3475): 342-343. (NSA 15: 27474).

Osburn, W. S., Jr. 1962. Modifying influences of four mountain environments on pea plants grown from irradiated seeds. Abstr. p. 179. In: 2nd Intern. Congr. Radiation Research. (See Anonymous, 1962).

Osburn, W. S., Jr. 1963. Influence of four Rocky Mountain regional environments on pea plants grown from irradiated seeds. pp. 319-324. In: Radioecology. (See Schultz and Klement, 1963) (NSA 17: 33578).

Osburn, W. S., Jr. 1963. The dynamics of fallout distribution in a Colorado alpine tundra snow accumulation ecosystem. pp. 51-71. In: Radioecology. (See Schultz and Klement, 1963) (NSA 17: 33547).

Osburn, W. S., Jr. 1964. Radioecology of the Colorado Front Range. Terminal report. Univ. of Colorado, Boulder, U. S. AEC report COO-1191-10. 173pp. (NSA 20: 20379).

Osburn, W. S., Jr. 1965. Primordial radionuclides: their distribution, movement, and possible effect within terrestrial ecosystems. In: Radiation and Terrestrial Ecosystems. Health Physics 11(12): 1274-1295. (See Hungate, 1965).

Osburn, W. S., Jr. 1967. Radioecology of the Colorado Front Range. Terminal report. Colorado State Univ., Fort Collins, U. S. AEC report COO-1445-5. 60pp. (NSA 21: 26387).

Osburn, W. S., Jr. 1967. Ecological concentration of nuclear fallout in a Colorado mountain watershed. pp. 675-709. In: Radioecological Concentration Processes. (See Aberg and Hungate, 1967).

Osburn, W. S., Jr. 1968. Forecasting long-range recovery from nuclear attack. pp. 107-135. In: Proc. Symp. on Postattack Recovery from Nuclear War. (See U. S. Office of Civil Defense, 1968).

Osburn, W. S., Jr. 1969. Primula parryi: correlation of morphological variability with background radioactivity. pp. 185-190. In: Symposium on Radioecology. (See Nelson and Evans, 1969).

Osburn, W. S., Jr. 1969. Accountability of nuclear fallout deposited in a Colorado high mountain bog. pp. 578-586. In: Symposium on Radioecology. (See Nelson and Evans, 1969).

Öschger, H., et al. 1972. ^{14}C and other isotope studies on natural ice. pp. D70-D92. In: Proc. 8th Intern. Radiocarbon Dating Conf. Vol. 1. (See Rafter, 1972) (NSA 31: 8701).

Öschger, H., et al. 1974. ^{39}Ar dating of groundwater. pp. 179-190. In: Isotope Techniques in Groundwater Hydrology. Vol. 2. (See IAEA, 1974) (NSA 31: 19653).

Osloond, J. H. and D. L. Newcomb. 1968. Radioactive waste disposal data for the National Reactor Testing Station, Idaho. Idaho Operations Office, U. S. AEC report IDO-12040(Suppl. 3). 14pp. (NSA 23: 591).

Osloond, J. H., et al. 1973. Tritium air sampling method for environmental and nuclear plant monitoring. pp. 531-533. In: Tritium. (See Moghissi and Carter, 1973) (NSA 28: 27262).

Osmond, J. K. 1964. The distribution of the heavy radioelements in rocks and waters of Florida. pp. 153-160. In: The Natural Radiation Environ-Environment. (See Adams and Lowder, 1964) (NSA 18: 35735).

Osmond, J. K. 1965. Pleistocene dating by uranium-ionium in fossil carbonates. Abstr. Abstracts Intern. Assoc. Quaternary Research. 7th Intern. Congr., Boulder and Denver, Colo.

Osmond, J. K., M. I. Kaufman and J. B. Cowart. 1974. Mixing volume calculations, sources and aging trends of Floridan aquifer water by uranium isotopic methods. Geochim. Cosmochim. Acta 38(7): 1083-1100. (NSA 30: 21163).

Osmond, R. G. D., J. R. W. Kerr and P. Metson. 1964. Iodine-131 in water supplies after nuclear attack. J. Inst. Water Engrs. 18: 459-464. (NSA 19: 26596).

Ostensson, L. 1964. Sterilization of insects. Mjolkpropagandan 4(5): 139pp. (In Swedish) (BA 29: 52644).

Oster, I. I. 1957. Modification of x-ray mutagenesis in Drosophila. Relative sensitivity of spermatids and mature spermatozoa. pp. 475-480. In: Proc. 5th Intern. Conf. on Radiobiology. (Stockholm, Aug. 15-19, 1957). G. C. de Hevesy, A. G. Forssberg and J. D. Abbott, eds. Oliver & Boyd, London.

Oster, I. I. 1964. Modification of genetic damage produced by ionizing radiation. Abstr. p. 135. In: Research and Development in Progress. Biology and Medicine. No. 3. U. S. AEC report TID-4203.

Oster, I. I. 1965. Mutagenic effects of chronically and intensely delivered radiation in Drosophila. Nippon Idengaku Zasshi 40(Suppl.): 83-96.

Oster, I. I. 1968. Effects of weightlessness on radiation-induced somatic damage. In: Symp. on the Biosatellite II Experiments. BioScience 18(6): 576-582.

Oster, I. I. 1971. Genetic implications of space-flight. U. S. NASA report N-72-20051. 14pp. (NSA 26: 50955) (Drosophila).

Oster, I. I. and A. Z. Cicak. 1958. Mortality of irradiated preimaginal stages of Drosophila. Drosophila Inform. Serv. 32: 143-144.

Oster, I. I. and D. Good. 1968. Irradiation of Drosophila under space-flight conditions. Abstr. Radiation Research 35(2): 500.

Osterberg, C. 1962. Fallout radionuclides in Euphausids. Science 138: 529-530. (NSA 17: 47).

Osterberg, C. 1963. The effluent of the Columbia River as a source of radioactivity. Abstr. Trans. Am. Geophys. Un. 44(1): 212-213.

Osterberg, C. 1965. Radioactivity from the Columbia River. pp. 968-979. In: Ocean Science and Ocean Engineering 1965. Vol. 2. (Trans. Joint Conf. and Exhibit - 14-17 June 1965, Washington, D. C.). Marine Technology Soc., Washington.

Osterberg, C., et al. 1966. Some non-biological aspects of Columbia River radioactivity. pp. 321-335. In: Disposal of Radioactive Wastes into Seas, Oceans and Surface Waters. (See IAEA, 1966) (Also: Oregon State Univ., Vorvallis, U. S. AEC report RLO-1750-4) (CONF-660507-3) (NSA 20: 25282).

Osterberg, C., J. Pattullo and W. Pearcy. 1964. Zinc 65 in euphausiids as related to Columbia River water off the Oregon coast. Limnol. Oceanogr. 9(2): 249-257.

Ostlind, D. A. 1966. Effects of x rays on the developmental biology of Heterakis gallinrum and on the pathogenicity of Histomonas meleagridis. Thesis, Kansas State Univ., Manhattan. 276pp. (NSA 21: 43535).

Östlund, G. and E. Berry. 1970. Modification of atmospheric tritium and water vapor by lake Tahoe. Tellus 22(4): 463-465. (NSA 25: 41938).

Östlund, G. and L. B. Lundgren. 1964. Stockholm natural tritium measurements. Tellus 16(3): 118-130. (NSA 18: 41659).

Östlund, H. G. 1967. Hurricane tritium I: preliminary results on Hilda 1964 and Betsy 1965. pp. 58-60. In: Isotope Techniques in the Hydrologic Cycle. (See Stout, 1967).

Otgonsuren, O., V. P. Pereygin and D. Chultem. 1970. Deposition of uranium in the bones of animals. Atom. energiya 29(4): 301-302. (Sov. Atomic Energy 29(4): 1035-1036) (NSA 25: 26744) (Dinosaurus, giant turtle shells).

Otiefa, B. A., Y. Barrada and D. M. el Gindi. 1958. An approach for using labelled radioactive phosphorus (P^{32}) in physio-pathological studies of plant-nematode diseases. Proc. 2nd Intern. Conf. on the peaceful Uses of Atomic Energy 27: 48-50. (Paper A/CONF.15/P149) (NSA 13: 6209).

Otlet, R. L. 1968. Low-level tritium measurements for hydrological applications. pp. 10-17. In: Nucleonic Instrumentation. Inst. of Electrical Engineers, London. (NSA 23: 21816).

O'Toole, J. J., et al. 1971. Environmental trace element survey at a heavy metals refining site. pp. 172-185. In: Nuclear Methods in Environmental Research. (See Vogt, 1971) (NSA 26: 17364).

Ottar, B. 1962. Fallout activity in the outlet from Groset-Tjern as calculated from precipitation in the Mosvatn area. Norway Defense Research Establishment, Kjeller, report K-278. 13pp.

Ottar, B. and T. Wik. 1962. Distribution of strontium-90 and cesium-137 in Norway spruce (Picea abies). Norway Defense Research Establishment, Kjeller, report K-251. 37pp. (NSA 17: 37053).

Ottaway, B. 1973. Dispersion diagrams. A new approach to the display of ^{14}C dates. Archeometry 15(1): 5-12. (NSA 28: 388).

Ottaway, B. and J. H. Ottaway. 1973. Suess calibration curve and archeological dating. Nature 239(5374): 512-513. (NSA 27: 9575).

Ottaway, B. and J. H. Ottaway. 1974. Irregularities in dendrochronological calibration. Nature 250(5465): 407-408. (NSA 30: 29430).

Otte, R. 1964. The influence of soft x rays on in vitro cultures of pathogenic fungi. Thesis, Univ. of Giessen, Germany. 82pp. (AED-Diss.-64-204) (In German) (NSA 19: 17663).

Ouchi, S, et al. 1970. Uranium contamination of the Shinkawa River. pp. 91-98. In: Semi-annual Progress Report of Power Reactor and Nuclear Fuel Development Corp., Tokai Works, Japan, report PNC-N-831-70-2, January-June 1970. (NSA 26: 33342).

Oughton, J. 1951. "Tagging" root maggot flies (Hylemya spp., Anthomyidae) by means of radioactive phosphorus. Rept. Entomol. Soc. Ontario 81: 91-92.

Ousheva, N. and K. Popov. 1970. Radiometric method of serial measurement of photosynthesis in unicellular algae. Comt. rend. acad. Bulg. sci. 23: 1151-1154. (NSA 25: 23945).

Outram, I. 1967. Factors affecting the resistance of insect eggs to sulphuryl fluoride. I. The uptake of sulphuryl-^{35}S fluoride in insect eggs. II. The distribution of sulphuryl-^{35}S fluoride in insect eggs after fumigation. J. Stored Prod. Res. 3(3): 255-260; 3(4): 353-358.

Outram, I. and F. Call. 1967. Apparatus for the synthesis and application of radioactive sulphuryl fluoride to insect eggs. J. Stored Prod. Res. 3(1): 71-74.

Ouye, M. T., R. S. Garcia and D. F. Martin. 1964. Determination of the optimum sterilizing dosage for pink bollworms treated as pupae with gamma radiation. J. Econ. Entomol. 57(3): 387-389.

Ovanesyan, V. O., et al. 1963. Beta-radioactivity of some samples of animal bones, grass and soil. Vopr. Radiobiol. Akad. Nauk Armyan. SSR sek. radiobiol., Sb. Tr. 3-4: 197-200. (In Russian) (NSA 18: 27030).

Ovchenkov, V. Ya. 1972. Radium migration from sites of its concentration under natural conditions. pp. 147-153. In: Radioecological Studies in Natural Biogeocenoses. (See Verkhovskaya, 1972) (In Russian) (NSA 26: 56112).

Ovchenkov, V. Ya. and V. S. Nikoforov. 1965. Vacuum method of determination of radon in the burrows of digging animals. In: Materials of a Scientific Conf. of the Komi Republic Youth. (Abstr. of Reports). Syktyvav. (In Russian).

Ovchenkov, V. Ya. and V. A. Lodygin. 1971. Concentrating ultra small quantities of U and Th from weak mineralized natural waters. pp. 109-113. In: Methods of Radioecological Investigation. (See Verkhovskaya, 1971) (In Russian).

Ovchenkov, V. Ya. and D. M. Rubtsov. 1972. Content of radioactive isotopes of the uranium and thorium series in soils. pp. 86-94. In: Radioecological Studies in Natural Biogeocenoses. (See Verkhovskaya, 1972) (In Russian) (NSA 26: 56111).

Overman, R. F., J. C. Corey and R. H. Hawkins. 1968. Use of gamma-neutron reaction for determining D_2O in H_2O and in tracing soil moisture. Trans. Amer. Nucl. Soc. 11: 74. (NSA 22: 30693).

Overrein, L. N. 1968. Lysimeter studies on tracer nitrogen in forest soils: l. Nitrogen losses by leaching and volitalization after addition of urea-N^{15}. Soil Science 106(4): 280-290.

Overrein, L. N. 1970. Immobilization and mineralization of tracer nitrogen in forest raw humus. II. Effect of temperature and incubation time on the interchange of urea-, ammonium-, and nitrate N^{15} under waterlogged conditions. Plant Soil 32(1): 207-220.

Ovington, J. D. and D. B. Lawrence. 1964. Strontium 90 in maize field, cattail marsh, and oakwood ecosystems. J. Appl. Ecol. 1(1): 175-181. (NSA 21: 26442).

Ovington, J. D. and J. S. Olson. 1970. Biomass and chemical content of El Verde lower montane rain forest plants. Ch. H-2, pp. H53-H77. In: A Tropical Rain Forest. (See H. T. Odum and R. F. Pigeon, 1970).

Owens, M., N. S. Thom and G. E. Eden. 1961. The uptake and release of radioactive strontium by freshwater plants. Proc. Soc. Water Treatment Exam. 10(1): 53-65. (NSA 16: 17434).

Owers, M. J. and A. W. Powell. 1974. Deposition velocity of sulfur dioxide on land and water surfaces using a ^{35}S tracer method. Atmos. Environ. 8 (1): 63-67. (NSA 30: 18669).

Owston, P. W., J. L. Smith and H. G. Halverson. 1970. Development of some radioisotope procedures for measuring water movement in trees. Isotopes Radiat. Technol. 7(4): 396-401. (NSA 24: 29704).

Oyen, L. C. 1970. Radiation surveillance: monitoring the environs of a nuclear station. Mech. Engr. 92(8): 32-37. (NSA 24: 48556).

Ozanne, P. G. and C. H. E. Werkhoven. 1971. Calcium uptake by subterranean clover bur shown by using ^{45}Ca. At. Energy Austral. 14: 16-18. (NSA 25: 26614).

Ozbek, N., T. Aksoy and G. Celebi. 1967. Neutron-moisture meter in studies of the effect of fallout on water conservation in arid regions. pp. 161-176. In: Isotope and Radiation Techniques in Soil Physics and Irrigation Studies. (See IAEA, 1967) (NSA 22: 23150).

Ozenda, P. 1965. Some aspects of the radiobiology of chlorophyllic microorganisms. pp. 341-366. In: Internal Works at Lucien Plantefol. Masson et Cie, Paris. (In French) (NSA 21: 20438).

Ozenda, P. 1967. Role of plants in continental radioecology. The case of the Grenoble site. Bull. Inform. Sci. Techn. (Paris) (119): 3-8. (AEC-tr-7038) (NSA 22: 1932; 23: 29417).

Ozenda, P. 1970. Cartographical representation of natural biological systems: its applications in radioecology. pp. 1105-1117. In: Proc. Intern. Symp. on Radioecology. Vol. II. (See French Commissariat à l'Énergie Atomique, 1970) (In French) (NSA 24: 41792).

Ozenda, P. 1970. Chlorella, a radiobiological material. Bull. Inform. Sci. Techn. (Paris) (145): 49-52. (In French) (NSA 24: 34757).

Ozerskiy, M. I. 1969. Effects of x radiation on the permeability of the alga Nitella flexilis to ^{22}Na ions. Radiobiologiya 9(1): 120-122. (AEC-tr-7024, pp. 156-159) (NSA 23: 24765).

Ozol, A. A. 1973. Induced mutagenesis in strains of Chlorella vulgaris differing in their radiosensitivity. Genetika 9(4): 39-48. (In Russian) (NSA 28: 10994).

Ozol, A. M. 1965. Ionizing Radiation in Biology. Isdatel'stvo Zinatne, Riga. 155pp. (NSA 20: 45257).

Paakkola, O. 1964. Decontamination methods for radioactive waters. Föd. Europ. Gewassershutz Inform. (10): 66-68. (In German) (NSA 19: 2187).

Paakkola, O. 1966. Radiostrontium in milk, grass and some other biological samples in Finland. Finland Inst. of Radiation Physics, Helsinki. 34pp. (SFL-A3) (NSA 20: 41134).

Paakkola, O. 1970. Monitoring the radioactivity of animals and animal products. Zentralbl. Veterinärmed. Beih. (11): 88-94. (In German) (NSA 25: 24016).

Paakkola, O. and J. K. Miettinen. 1963. Strontium-90 and cesium-137 in plants and animals in Finnish Lapland during 1960. Ann. Acad. Scient. Fenn. AII(Chem.) (125): 1-15.

Paakkola, O. and A. Voipio. 1965. Strontium-90 in the Baltic Sea. Suomen. Kemist. 38: B11-B17. (NSA 19: 38901).

Paasikallio, A., U. Häkkinen and E. Lakanen. 1971. Effects of soil factors on the uptake of radiostrontium by plants. Part III. Ann. Agr. Fenn. 10(54): 125-130. (NSA 26: 18099).

Pacific Gas and Electric Company. 1961. Humboldt Bay Power Plant, Unit No. 3. Final hazards summary report, ammendment 11. San Francisco, Calif. 496pp. U. S. AEC DOCKET-50133-1. (NSA 23: 15530) (Incl. site and environment).

Pacific Gas and Electric Company. 1971. Diablo Canyon Nuclear Power Plant, Units 1 and 2. Environmental report. San Francisco, Calif. 637pp. U. S. AEC DOCKET-50278-22; -50275-15. (NSA 25: 48187).

Pacific Gas and Electric Company. 1971. Mendocino Power Plant, Units 1 and 2. Environmental report, construction permit stage. San Francisco, Calif. 120pp. U. S. AEC DOCKET-50398-8. (NSA 25: 56973).

Packard, C. 1914. The effects of radium radiations on the fertilization of Nereis. J. Exptl. Zool. 16(1): 85-129.

Packard, C. 1918. The effect of radium radiation on the development of Chaetopterus. Biol. Bull. 35(1): 50-70.

Packard, C. 1934. Biological dosimeters in radiology. Cold Spring Harbor Symp.on Quantitative Biol. 2: 264-273. (Incl. Drosophila).

Packard, C. 1935. The relation between age and radiosensitivity of Drosophila eggs. Radiology 25(2): 223-230.

Packard, C. 1939. The biological measurement of radiations. Am. J. Roentg. Rad. Therapy 41(3): 441-443. (Data on Drosophila survival).

Padilla, G. M. and P. A. Van Dreal. 1963. Radiation-induced division delay in exponentially and synchronously dividing Astasia longa. Abstr. J. Cell Biol. 19(2): 54A.

Padilla, G. M., P. A. Van Dreal and N. G. Anderson. 1966. Studies on synchronized cells: radiation-induced division delay in the flagellate Astasia longa. Radiation Research 28: 157-165. (Also: Oak Ridge National Lab., U. S. AEC report ORNL-P-118) ((NSA 20: 31085).

Padzik, A. and J. Niemcyzynowicz. 1966. Laboratory investigations of the determination of the nonactive carrier influence on the measurement of water filtration horizontal velocity by means of radioactive ^{131}I. Nukleonika 11: 651-658. (In Polish) (NSA 21: 10459).

Paez, A. S. 1961. Migratory birds with a radiotracer. Novedades, December 4, 1961. Veracruz, Ver., Mexico. (In Spanish).

Paglia, D. E. 1968. Hematopathologic surveys of kangaroo rats (Dipodomys microps) populating plutonium contaminated regions of the Nevada Test Site. Health Physics 15: 493-498. (NSA 23: 7975).

Pagliari, M. 1973. And now, poor ecology? pp. 121-128. In: Proc. 2nd Conf. on the State of Advancement of Radioecology in Italy. (See Anonymous, 1973) (In Italian).

Pahl, G. 1962. Effect of irradiation temperature on hatching of brine shrimp eggs. Abstr. p. 241. In: 2nd Intern. Congr. Radiation Research. (See Anonymous, 1962).

Pahl, G. 1969. Radioactive and stable strontium analysis of upper Mississippi River clamshells. pp. 234-239. In: Symposium on Radioecology. (See Nelson and Evans, 1969).

Pahl, L. G. 1961. Irradiation temperature and Ascaria eggs. Abstr. Radiation Research 14(4): 490.

Pahlke, H. 1972. Isotope techniques for the continuous and automatic determination of density and transport concentration in hydraulic transport. pp. 285-331. In: Conf. on Radiation and Isotope Techniques in Civil Engineering. Commission of European Communities, Brussels, report EURISOTOP-55(Vol.1). (In German) (NSA 27: 22730).

Pahlke, H. 1973. Some applications using artificial radioactive tracers in Germany. pp. 81-96. In: Trace Techniques in Sediment Transport. (See IAEA, 1973) (NSA 28: 21137).

Pahlke, H. and W. Schob. 1968. Investigations on the density determination of water-sand mixtures by isotope techniques. Germany Bundesministerium für Wissenschaftliche Forschung report BMwF-FBK-68-23. (ORNL-tr-2048) (NSA 23: 11723).

Pain, J., B. Roger and P. Douault. 1975. Study of the seasonal variation in the exchanges of food labelled with ^{198}Au from a worker bee to a queen (Apis mellifera ligustica S.). Compt. rend. hebd. Seances Acad. Sci. D280 (23): 2677-2680. (In French) (NSA 33: 23613).

Paine, D. J., E. Johnson and R. L. Watters. 1971. Plutonium movement in aquatic systems: a review. pp. 407-410. In: Proc. Rocky Flats Symp. on Safety in Plutonium Handling Facilities. (See Dow Chemical Co., 1971) (NSA 25: 55150).

Painter, R. R. and W. W. Kilgore. 1965. Incorporation of glycine 2-C-14 into untreated and chemosterilant treated house flies. Abstr. 479. Bull. Entomol. Soc. Am. 11(3): 178.

Painter, R. R. and W. W. Kilgore. 1965. Chemosterilant effect of 5-fluororotic acid on house flies. J. Econ. Entomol. 58(5): 888-891. (^{14}C).

Painter, R. R. and W. W. Kilgore. 1967. The effect of Apholate and Thiotepa on nucleic acid synthesis and nucleotide ratios in housefly eggs. J. Insect Physiol. 13(7): 1105-1118. (^{14}C).

Pajni, H. R. 1975. Irradiation-induced sterility in the males and females of Callosobruchus analis (F.) (Bruchidae: Coleoptera). Current Science (India) 44(18): 669-670. (NSA 33: 3267).

Pakendorf, K. W. 1970. Studies on the use of mutagenic agents in Lupinus. I. The effect of gamma-ray dose on plant survival in L. angustifolius, L. albus, L. luteus, and L. mutabilis. Z. Pflantzensücht. 63: 102-113. (In German) (NSA 24: 51096).

Pakulo, A. G. 1971. Passage of ^{137}Cs from a fish to the ultimate serving during culinary preparation. Gigiyena sanitariya (10): 49-51. (In Russian) (NSA 26: 33819).

Pakulo, A. G. 1973. Role of water and food in uptake of ^{137}Cs in fish. pp. 136-139. In: Radioecology of Water Organisms. II. (See Andrumaitis, 1973) (In Russian) (NSA 30: 21137).

Pakulo, A. G. 1974. Effect of certain chemical elements on the accumulation of ^{137}Cs in freshwater fish. Gigiyena sanitariya (6): 111-113. (In Russian) (NSA 30: 24400).

Pal, R. and B. S. Krishnamurthy. 1959. Induced mutations of x-ray irradiations in Culex fatigans Wied (1828). Nature 184(Suppl.9): 658. (NSA 14: 1380).

Palais, C. J. 1970. Applications of the liquid scintillation spectrometer to radioecology, emphasis is placed on the detection of Cerenkov radiation of beta-emitting nuclides. pp. 201-211. In: Proc. Intern. Symp. on Radioecology. (See French Commissariat à l'Énergie Atomique, 1970).

Palamarchuk, A. S. 1960. Stimulation of the growth of the roots of plants of the Chenopodiaceae, Criciferae, and Umbelliferae in soils of peat and minerals by gamma rays of radium and cobalt, and by x-rays. Ukr. Bot. zh. 17(3): 3-18. (French transl. French Commissariat à l'Énergie Atomique report CEA-tr-R-1678) (NSA 17: 27068).

Palamarchuk, A. S. and V. Ye. Bondarenko. 1975. Seasonal accumulation of strontium-90 and cesium-137 in certain kinds of medicinal plants of the legume family. Dokl. Akad. Nauk Belorusk. SSR 19(2): 184-185. (In Russian) (NSA 32: 1216).

Palange, R. C., G. G. Robeck and C. Henderson. 1956. Radioactivity in stream pollution. Industr. Engrg. Chem. 48: 1847-1850.

Palayer, P. 1958. Fixation of radioactive phosphorus through different parts of the brain, notably the epiphysis and through tissues of the rainbow trout (Salmo gairdnerii R.). Compt. rend. soc. biol. 152: 305. (In French).

Palliy, V. F. and G. G. Iirkovskiy. 1972. Studies on the effect of ionizing radiation on the development of insects. Vsesoyuz. Ent. obshch Kirgiz. otd. sborn. ent. rabot. 1: 70-79. (In Russian).

Pallister, E. T. and J. H. Green. 1964. Radium-226 content of natural waters and coal in New South Wales. Nature 201(4921): 813-814. (NSA 18: 16211).

Palmer, H. E. and T. M. Beasley. 1965. ^{55}Fe in humans and their foods. (For Health Physics Soc. Ann. Meeting, Los Angeles, Calif., June 14-17, 1965). 9pp. (CONF-650616) (NSA 20: 9171).

Palmer, H. E. and R. W. Perkins. 1963. Cesium-134 in Alaskan eskimos and in fallout. Science 142(3588): 66-67.

Palmer, H. E., W. C. Hanson and B. I. Griffin. 1962. Cs-137 excretion and body burdens in Alaskan eskimos. Hanford Labs., U. S. AEC report HW-SA-2764. 20pp. (NSA 17: 2798).

Palmer, H. E., et al. 1962. Radioactivity in Alaskan eskimos. Hanford Labs., U. S. AEC report HW-SA-2718. 20pp. (For 2nd Symp. on Radioactivity in Man, Applications of Whole Body Counting; Northwestern Univ. Med. School, Chicago, September 5-7, 1962).

Palmer, H. E., et al. 1963. Cesium-137 in Alaskan eskimos. Science 142 (3588): 64-66. (Also: Hanford Labs., U. S. AEC report HW-SA-2869) (NSA 17: 41066).

Palmer, H. E., et al. 1964. Whole-body counting studies of Cs-137 in Alaskan eskimos. Hanford Atomic Products Operation, U. S. AEC report HW-SA-3689. (NSA 19: 623).

Palmer, H. E., et al. 1964. Radioactivity measurements in Alaskan eskimos in 1963. Science 144(3620): 859-860. (NSA 18: 23823).

Palmer, H. E., et al. 1964. Measurements of radioactivity in Alaskan natives in 1964. Hanford Works, U. S. AEC report HW-SA-3750. 7pp. (NSA 19: 44289).

Palmer, H. E., et al. 1968. Levels of iron-55 in humans, animals, and food, 1964-1967. Radiol. Hlth. Data 9(8): 387-390.

Palmieri, G. G. 1961. The problem of hibernation in radiobiology. 9th Intern. Congr. Radiology 2: 1013-1017. (In German) (NSA 15: 24724).

Palms, J. M., B. K. Tanner and R. E. Wood. 1974. Ground-level air concentrations of natural and man-made radionuclidies. pp. 128-139. In: 2nd Workshop on the Natural Radiation Environment. (See Lowder, 1974) (NSA 3370).

Palms, J. M., et al. 1974. Baseline ground-level airborne radionuclide survey for nuclear fuel reprocessing plant. Trans. Nucl. Sci. NS-21(1): 585-595. (NSA 30: 6506).

Paltineau, I. C. 1974. Possibilities of using the neutron method for water application efficiency studies in sprinkler and furrow irrigation. pp. 477-506. In: Isotope and Radiation Techniques in Soil Physics and Irrigation Studies 1973. (See IAEA, 1974) (NSA 30: 15511).

Palumbo, R. F. 1955. Uptake of iodine-131 by the red alga Asparagopsis taxiformis. Univ. of Washington, U. S. AEC report UWFL-44. 11pp. (NSA 11: 4261).

Palumbo, R. F. 1959. The difference in uptake of radioisotopes by marine and terrestrial organisms. Abstr. Proc. 9th Intern. Bot. Congr. 2(A): 26-27.

Palumbo, R. F. 1961. The differences in uptake of radioisotopes by marine and terrestrial organisms. pp. 1367-1372. In: Recent Advances in Botany. Univ. of Toronto Press, Toronto. (NSA 17: 2807).

Palumbo, R. F. 1962. Recovery of the land plants at Eniwetok Atoll following a nuclear detonation. Radiation Botany 1(2): 182-189.

Palumbo, R. F. 1962. Radioactivity of the biota at islands in the Central Pacific 1954-1958. Univ. of Washington, U. S. AEC report UWFL-79. 63pp.

Palumbo, R. F. 1963. Factors controlling the distribution of the rare earths in the environment and in living organisms. pp. 533-537. In: Radioecology. (See Schultz and Klement, 1963) (NSA 17: 33605).

Palumbo, R. F. 1963. Radionuclide content of foodstuffs collected at Christmas Island and at other islands of the Central Pacific during Operation DOMINIC, 1962. Univ. of Washington, U. S. AEC report UWFL-87. 37pp.

Palumbo, R. F. and F. G. Lowman. 1958. The occurrence of antimony-125, europium-155, iron-55, and other radionuclides in Rongelap Atoll soil. Univ. of Washington, U. S. AEC report UWFL-56. 27pp.

Palumbo, R. F., A. H. Seymour and A. D. Welander. 1966. Radionuclides in foods from the Central Pacific, 1962. Nature 209: 1190-1192. (NSA 20: 27156).

Palumbo, R. F., et al. 1963. Uptake of I^{131} by fresh-water organisms. Abstr. Health Physics 9(12): 1213.

Panama Canal Company. 1964. Construction of a sea-level canal by nuclear methods. Annex III. Isthmian Canal Studies, 1964. ---App. 1, Incl. A. ---App. 1, App. 2, Incl. A. Balboa Heights, Canal Zone, U. S. AEC report PNE-1999, PNE-2005, PNE-2006. 22, 77, 25pp. (NSA 20: 471, 472, 473).

Panikovskaya, L. I. and N. A. Troitskiy. 1968. Genetic effects of intermediate neutrons. I. The deletion frequency and fertility in Drosophila melonogaster. Genetika (4): 15-20. (In Russian) (NSA 22: 19315).

Panitz, R. 1964. Hormone-controlled gene activity in the giant chromosomes of Acricotopus lucidus. Biol. Zbl. 83: 197-230. (In German, English summary).

Pankova, N. V. 1965. Chromosome breakage in some cell generations in irradiated Misgurnus fetuses. Radiobiologiya 5(2): 248-252. (AEC-tr-6599, pp. 120-126) (NSA 19: 28079).

Pankova, N. V. 1974. Investigation of radiation-induced chromatid abberations in amphibian lymphocytes in vitro at the G_2 phase upon the varying doses of x-rays. Genetika 10(5): 52-56. (In Russian) (NSA 30: 18869).

Pankova, N. V. nd. Variations in chromosomal damage in plants and animals. Priroda (4): 78-82. (In Russian) (NSA 32: 26030).

Panov, D. A. and Yu. I. Sorokin. 1967. Determination of threshold food concentrations for fish larvae using radioactive carbon. Vopr. ikhtiol. 7(1): 129-141. (French transl. French Commissariat à l'Énergie Atomique report CEA-tr-R-1947) (NSA 23: 35593).

Panteleyeva, A. P. 1972. Some regularities of the interaction of alginic acid with cations of metals. pp. 112-115. In: Radiation and Chemical Ecology of Aquatic Organisms. (See Polikarpov, 1972) (In Russian).

Pany, J. 1959. Simple apparatus for labeling of biological substances with gaseous tritium. Naturwissenschaften 46: 515. (In German) (NSA 13: 21880).

Paoletti, A. 1971. Lessons from radioecology in the field of chemical contamination of the environment. pp. 779-795. In: Proc. Intern. Symp. Radioecology Applied to the Protection of Man and His Environment. (See Commission of the European Communities, 1971) (In French) (NSA 27: 5237).

Papadopoulos, J. and C. A. Ziegler. 1966. Radioisotope technique for monitoring sediment concentration in rivers and streams. In: Radioisotope Instruments in Industry and Geophysics. (Proc. Symp., Warsaw, 18-22 October 1965). International Atomic Energy Agency, Vienna, Publ. STI-PUB-112. (Paper SM-68/26) (CONF-651023) (NSA 20: 23280).

Papadopoulos, J., et al. 1964. Sediment gauge for rivers and streams. p. 142. In: 2nd Ann. Oak Ridge Radioisotope Conf. (April 19-22, 1964). U. S. AEC report TID-7689. (NSA 18: 16195).

Papadopoulou, C. P. 1963. Disinfestation of dried figs by gamma radiation. pp. 485-490. In: Radiation and Radioisotopes Applied to Insects of Agricultural Importance. (See IAEA, 1963) (NSA 18: 5120).

Pape, G. and O. Bobleter. 1970. Small growth chamber for labelling plants with ^{14}C. Atomkernenergie 15: 301-302. (In German) (NSA 24: 44389).

Parchevskaya, D. S. 1972. Some considerations of experimental planning in radiobiological investigations. pp. 166-171. In: Methods of Determination of Radioactivity. (See Polikarpov and Parchevskiy, 1972).

Parchevskaya, D. S. and I. S. Zaburunova. 1967. Strontium-90 in algae, flowering plants, mollusks, crustaceans, and fishes of the Black Sea (1965-1967). State Committee on the Use of Atomic Energy, USSR. Atomizdat, Moscow. (In Russian).

Parchevskiy, V. P. 1964. The radioactivity of certain organisms of the Black Sea. In: Radioactive Contamination of the Seas and Oceans. V. I. Baranov and L. M. Khitrov, eds. Oceanographic Commission, Acad. Sci., USSR. Publ. House Acad. Sci. USSR, Moscow. (AEC-tr-6641, pp. 128-144) (NSA 21: 8810).

Parchevskiy, V. P. and I. A. Sokolova. 1971. Content of ^{90}Sr in some organisms in the Black Sea (1965-1968). Dokl. Akad, Nauk SSSR 199(3): 705-707. (In Russian) (NSA 26: 22796).

Parfenov, G. P. 1963. Causes of gamete lethality in Drosophila following flights on the Vostok-3 and Vostok-4 satellites. Kosmich. Issled. 2: 335-340. (In Russian) (NSA 18: 38860).

Parfenov, G. P. 1964. Occurrence of crossing over in Drosophila males under the influence of vibration, acceleration and γ-irradiation. Kosmich. Issled. 2(4): 648-653. (FTD-TT-64892; Cosmic Res. 2: 232-241).

Parfenov, G. P. 1965. The development of dominant lethality in fruit flies exposed to vibration, acceleration, and gamma radiation. Kosmich. Issled. 3: 643-651. (In Russian) (NSA 20: 14343).

Parfenov, G. P. 1967. Genetic studies in the cosmos. Kosmich. Issled. 5: 140-155. (In Russian) (NSA 21: 39386).

Parfenov, Yu. D. 1974. Polonium-210 in the environment and in humans. Atom. Energy Rev. 12(1): 75-143. (NSA 30: 6548).

Paris, O. H. 1963. The dispersal of radioactive isopods. Abstr. Bull. Ecol. Soc. Am. 44(3): 75.

Paris, O. H. 1965. Vagility of P^{32}-labelled isopods in grassland. Ecology 46(5): 635-648.

Paris, O. H. and A. Sikora. 1965. Radiotracer demonstration of isopod herbivory. Ecology 46(5): 729-734.

Paris, O. H. and A. Sikora. 1967. Radiotracer analysis of the trophic dynamics of natural isopod populations. pp. 741-771. In: Secondary Productivity of Terrestrial Ecosystems. K. Petrusewicz, ed. Inst. of Ecology, Polish Acad. Sci., Warsaw.

Parish, J. C. and B. W. Arthur. 1965. Mammalian and insect metabolism of chemosterilant thiotepa. J. Econ. Entomol. 58(5): 976-979. (^{32}P, Blattella germanica, Musca domestica, Stomoxys calcitrans, Anthonomus grandis).

Parisi, V. 1973. Ecological aspects of the Po River. pp. 185-197. In: Proc. 2nd Conf. on the State of Advancement of Radioecology in Italy. (See Anonymous, 1973) (In Italian).

Park, H. D. 1958. Sensitivity of hydra tissues to x-rays. Physiol. Zool. 31(3): 188-193.

Park, K., et al. 1965. Strontium-90 and cesium-137 in Columbia River plume, July 1965. Nature 208(5015): 1084-1085.

Park, K. C. 1968. Chemical study of the hot spring waters in Korea. III. Determination of radon. Onsen Kagaku 19: 49-53. (In Japanese) (NSA 23: 26356).

Park, K. E. and J. S. Hyun. 1968. Preliminary study on the biological effects of gamma ray on the silkworm, Bombyx mori L. J. Nucl. Sci. (Seoul) 8(2): 19-25. (In Korean) (NSA 23: 5014).

Park, T., P. P. H. DeBruyn and J. A. Bond. 1958. The relation of x-irradiation to the fecundity and fertility of two species of flour beetles. Physiol. Zool. 31(2): 151-170.

Parkash, O. 1965. Induction of sex-linked recessive lethals and visible mutations by feeding x-irradiated DNA to Drosophila melanogaster. Nature 205: 312-313.

Parkash, O. 1969. On the radiomimetic effect of irradiated desoxyribonucleic acid (DNA_m) on Drosophila melanogaster. Naturwissenschaften 52: 142.

Parke, D. V. 1965. Radioisotopes in the study of the metabolism of foreign compounds. Isotopes Exptl. Pharmac. : 315-342.

Parker, F. L. 1958. Radioactive tracers in hydrologic studies. Trans. Am. Geophys. Un. 39: 434-439.

Parker, F. L. 1963. Clinch River studies. pp. 161-180. In: Transport of Radionuclides in Fresh Water Systems. (See Kornegay, et al., 1963) (NSA 17: 40528).

Parker, F. L. 1964. Disposal of radioactive wastes in fresh water. Nuclear Safety 6(1): 89-97. (NSA 19: 12858).

Parker, F. L. 1969. Status of radioactive waste disposal in U.S.A. Proc. Amer. Soc. Civil Engrs. 95(SA3): 439-467.

Parker, F. L., B. J. Frederick and P. H. Carrigan. 1964. Movement of radionuclides in the Clinch River. Oak Ridge National Lab., U. S. AEC report ORNL-P-605. 14pp. (NSA 19: 1946).

Parker, F. L., et al. 1961. Dispersion of radioactive contamination in an estuary. Health Physics 6(1/2): 66-85. (NSA 15: 30360).

Parker, H. E. and P. J. West. 1974. Management of wastes from nuclear fuel reprocessing for the protection of man and his environment. (For 9th World Energy Conf., Detroit, Michigan, September 22-27, 1974; Div. 4: Energy Conversion; Paper 4.2-15). 18pp. (NSA 33: 16701).

Parker, H. M. 1948. Speculations on long-range waste disposal hazards. Hanford Works, U. S. AEC report HW-8674. 11pp. (NSA 29: 12706).

Parker, H. M. 1949. Hazards of waste storage at Hanford Works. Hanford Works, U. S. AEC report HW-14058. 60pp. (NSA 29: 20216).

Parker, H. M. 1956. Radiation exposure from environmental hazards. Proc. Intern. Conf. on the Peaceful Uses of Atomic Energy 13: 305-310. (Paper A/CONF.8/P/279).

Parker, H. M. 1959. Statement. pp. 2356-2372. In: Industrial Radioactive Waste Disposal. (See U. S. Congress, 1959).

Parker, H. M., et al. 1965. North American experience in the release of low-level wastes to rivers and lakes. Proc. 3rd Intern. Conf. on the Peaceful Uses of Atomic Energy 14: 62-68.

Parker, M. 1966. Studies on the distribution of cobalt in lakes. Thesis, Univ. of Wisconsin, Madison. 74pp. (Used ^{57}Co).

Parker, R. A. and R. Helm. 1965. The relationship between inorganic phosphate uptake and oxygen consumption by Daphnia schødleri Sars. Ecology 46(6): 881-882. (Used ^{32}P).

Parkinson, T. F. 1975. Environmental assessments: why, who, and how. Trans. Amer. Nucl. Soc. 22: 81. (NSA 33: 22642).

Parouty, G. and C. Chaisemartin. 1974. Experimental analysis of zinc salt transfer to final consumer in a freshwater ecosystem. Compt. rend. soc. biol. 168(1): 78-82. (In French) (NSA 31: 29628) (^{65}Zn).

Parr, V. H., et al. 1974. Atomic energy. pp. 86-117. In: Report from the Norwegian Nature Consevancy Association. (Prep. by the Assoc. Energy Group). H. Parr, ed. Norwegian Naturvernforbund, Oslo. (In Norwegian) (NSA 30: 29343) (Environmental aspects of nuclear power).

Parrish, W. B. 1958. Attenuation of gamma-rays by wood. Thesis, Ohio State Univ., Columbus.

Parrish, W. B. 1961. Detecting defects in wood by the attenuation of gamma rays. Forest Science 7: 136-143.

Parrish, W. B., W. C. Myser and M. L. Pool. 1957. Radioisotopes - a possible means of finding defects in logs and trees. J. Forestry 55(11): 851-852.

Parry, W. H. and J. B. Ford. 1967. The artificial feeding of phosphamidon to Myzus perisicae: I. Intraspecific differences exhibited by this aphid on feeding through a parafilm membrane. Entomologia Wxp. Appl. 10(3-4): 437-452. (^{32}P).

Parsons, P. A. 1969. A correlation between the ability to withstand high temperatures and radioresistance in Drosophila melanogaster. Experientia 25: 1000-1001. (NSA 24: 2713).

Parsons, P. A. 1975. Differences between Drosophila melanogaster and its sibling species D. simulans to radioresistance using isofemale strains from natural populations. Int. J. Radiat. Biol. 27(3): 297-300. (NSA 32: 19993).

Parsons, P. A. 1975. Quantitative variation in natural populations. Genetics 79(Suppl.): 127-136. (NSA 32: 17431) (^{60}Co irradiation, Drosophila).

Parsons, P. J. 1960. Movement of radioactive wastes through soil. 1. Soil and ground-water investigations in lower Perch Lake Basin. Atomic Energy of Canada, Ltd., Chalk River, report AECL-1038; CRER-932. 15pp. (NSA 14: 25103).

Parsons, P. J. 1962. Movement of radioactive waste through soil. 4. Migration from a single source of liquid waste deposited in porous media. Atomic Energy of Canada, Ltd., Chalk River, report CRER-1077; AECL-1485. 22pp.

Parsons, P. J. 1963. The movement of tritium from the Chalk River liquid disposal area. Atomic Energy of Canada, Ltd., Chalk River, report CRER-1146. 16pp. (NSA 17: 29866).

Parsons, P. J. 1967. Symposium on waste disposal into surface waters. Nuclear Safety 8: 249-257. (NSA 21: 24187).

Parsont, M. A. 1967. The distribution of ^{226}Ra in an aquatic environment. Thesis, Colorado State University, Fort Collins. 133pp. (NSA 22: 47498).

Partanen, C. R. 1958. Quantitative technique for analysis of radiation-induced tumorization in fern prothalli. Science 128(3330): 1006-1007. (NSA 13: 1274).

Partanen, C. R. 1958. The quantification of a tumorous response to ionizing radiations in fern prothalli. Intern. Congr. Genet. Proc. 10(2): 214.

Partanen, C. R. 1960. Suppression of radiation-induced tumorization in fern prothalli. Science 131(3404): 926-927. (NSA 14: 10369).

Partanen, C. R. and E. M. Gifford. 1959. Application of autoradiographic techniques to studies of shoot apices. Nature 182: 1747-1748.

Partridge, J. E., et al. 1973. Comparison of CaF_2: Mn and LiF thermoluminescent dosimeters for environmental radiation monitoring. Eastern Environmental Radiation Facility, Montgomery, Ala., U. S. EPA report EPA-520/5-73-006. 14pp. (NSA 33: 17230).

Partskhaladze, N. N. 1970. Study of the survival of frog embryos subjected to x-irradiation at the stage of the zygote. Soobshch. Akad. Nauk Gruz. SSR 60: 685-687. (In Russian) (NSA 25: 21900).

Parvinen, M., E. Soini and P. Tuohimaa. 1973. Automatic gas counter for quantitative micro determinations of tritium in biological material. Anal. Biochem. 55(1): 193-200. (NSA 28: 29621).

Pascaud, X. and P. Niaussat. 1963. Nucleic acid concentration and radiosensitivity of the scorpion Androctonus amoreuxi Aud and Sav. Compt. rend. soc. biol. 157(3): 503-506. (In French) (NSA 17: 33656).

Paschinger, H. and T. Vanicek. 1974. Effects of gamma irradiation on the two mechanisms of Rb(K) uptake by Chlorella. Radiation Botany 14(4): 301-307. (NSA 32: 1059).

Pasechnik, A. M. 1960. Variation of microorganisms under the influence of ionizing radiation. Mikrobiol. zh. Akad. Nauk Ukr. RSR 22(3): 69-74. (JPRS-7263) (NSA 15: 4935).

Pasechnik, A. M., M. D. Matveyeva and G. G. Tulina. 1963. The effect of x-ray and nuclear irradiation on *Bac. perfringens*. I. The change in the morphology and cultural characteristics of *Bac. perfringens* types A and B induced by the action of gamma-neutron radiation. Mikrobiol. zh. Akad. Nauk Ukr. RSR 25: 15. (In Russian) (NSA 17: 35537).

Pasechnik, P. P. 1962. Nuclear explosions are detectable regardless of their point of occurrence on the globe: science has proved it. Priroda (7): 3-13. (In Russian).

Pashneva, G. Ye., T. P. Slavnina and V. V. Serebrennikov. 1965. Rare earth and thorium content of soils of Tomsk region. Izv. Sibirsk. otd. Akad. Nauk SSSR (1): 48-52. (In Russian) (CA 63: 14586h; NSA 20: 2065).

Pasic, H and Z. Petrovic. 1969. Effect of meteorological factors and the enclosed Sarajevo Valley on the distribution of radioactive aerosols. Meteorologie (10): 241-249. (In French) (NSA 28: 8155).

Passera, L. 1971. Effect of gamma radiations on the ant *Plagiolepis pygmaea* Latr. (Formicidae Hym.) and two of its social parasites: comparison of their life spans. Compt. rend. D273(17): 1502-1505. (In French) (NSA 26: 12436).

Passonneau, J. V. 1954. The effect of x-irradiation on the metabolism of phosphorus-containing compounds in *Melanoplus differentialis* eggs. Physiol. Zool. 27: 119-128. (Also: Argonne National Lab., U. S. AEC report AECU-2304; UAC-663. (NSA 7: 470).

Pate, D. A. and H. H. Funderburk, Jr. 1966. Absorption, translocation and metabolism of ^{14}C-labelled dichlobenil. pp. 17-26. In: *Isotopes in Weed Research*. (See IAEA, 1966).

Patel, B. and G. R. Doshi. 1971. *Lamellidens marginalis* (Lamarck) as indicator of ^{54}Mn. Arch. Oceanogr. Limnol. 17: 27-42.

Patel, B. and A. K. Ganguly. 1969. Concept of acute and chronic tissue concentration of elements in radioecology. Proc. Symp. on Mollusca 2: 446-455. (Aquatic).

Patel, B. and A. K. Ganguly. 1973. Occurrence of ^{75}Se and ^{113}Sn in oysters. Health Physics 24(5): 559-562. (NSA 28: 2599).

Patel, R. J. 1970. Effects of irradiation (high energy β rays) on Cladophorales. I. *Chaetomorpha melogonium*. Nucleus (Calcutta) 13(2): 111-117. (NSA 25: 58153).

Patetta Queirolo, M. A. 1964. Skin absorption of the beta rays emitted from a $^{90}Sr/^{90}Y$ source. Acta Cient. Venezolana 15: 26-28. (In Spanish) (NSA 19: 12915, 33791) (Used *Bufo marinus* L.).

Patetta Queirolo, M. A. nd. Action of beta radiation upon leaves of various lactiferous plants (Fam. Moraceae). pp. 109-112. In: *Radiation and Isotope Technology in Latin American Development*. (See Puerto Rico Nuclear Center, nd.) (In Spanish, English abstr.) (NSA 24: 39589).

Patin, S. A., V. L. Pachkurenkov and I. A. Shekhanova. 1971. Kinetics and mechanism of accumulation of plutonium by *Misgurnus fossilis* spawn. Radiobiologiya 11(5): 742-746. (AEC-tr-7306, pp. 153-159) (NSA 26: 38759).

Paton, J. M. and R. F. S. Hearmon. 1957. Effect of exposure to gamma-rays on the hygroscopicity of Sitka spruce wood. Nature 180(4587): 651. (NSA 12: 1174).

Patrick, H., R. E. Emerson and H. M. Hyre. 1965. Design and usage of a ^{137}Cs gamma irradiator for breeding and physiological studies. W. Va. Univ. Agr. Exp. Sta. Bull. No. 518T. 11pp. (NSA 21: 34409).

Patrick, K. H. 1963. The effect of irradiation on sporogenesis, gametophyte development and seed set in yellow-anthered dallisgrass. Thesis, Texas A. & M. Coll., College Station. 105pp. (NSA 18: 8193).

Patrick, R. 1967. Concentration of ^{137}Cs, ^{65}Zn, and ^{85}Sr by fresh-water algae. Biotechnol. Bioeng. 9: 449-456. (NSA 22: 25754).

Patrick, R., J. Cairns, Jr. and S. S. Roback. 1966. An ecosystematic study of the fauna and flora of the Savannah River. Proc. Acad. Nat. Sci. Phila. 118: 109-407. (NSA 22: 6775).

Patrusheva, O. I. and I. A. Tarchevskiy. 1965. β-irradiation influence of frequency of systoles in mollusk embryo. Radiobiologiya 5(2): 314. (AEC-tr-6599, pp. 234-235) (NSA 19: 26025).

Patt, H. and M. Swift. 1948. Influence of temperature on radiation to frogs. Am. J. Physiol. 155(3): 388-393.

Patten, B. C. 1964. The systems approach in radiation ecology. Oak Ridge National Lab., U. S. AEC report ORNL-TM-1008. 17pp. (NSA 19: 12905).

Patten, B. C. 1965. Community organization and energy relationships in plankton. Oak Ridge National Lab., U. S. AEC report ORNL-3634. 65pp. (NSA 19: 14916) (Used ^{14}C).

Patten, B. C. and R. L. Iverson. 1966. Photosynthesis and uptake of ^{85}Sr in freshwater plankton. Nature 211: 96-97. (Also: Oak Ridge National Lab., U. S. AEC report ORNL-P-1622) (NSA 20: 36568).

Patten, B. C. and M. Witcamp. 1967. Systems analysis of ^{134}cesium kinetics in terrestrial microcosms. Ecology 48(5): 813-___.

Patterson, C. C. 1963. Dispersion measurements. pp. 295-306. In: *Transport of Radionuclides in Fresh Water Systems*. (See Kornegay, et al., 1963) (NSA 17: 40534).

Patterson, C. C. 1963. The dispersion of radionuclides in open channel flow. Thesis, Univ. of Texas, Austin. 107pp. (NSA 18: 35725).

Patterson, C. C. and E. F. Gloyna. 1963. Radioactivity transport in water - the dispersion of radionuclides in open channel flow. Los Alamos Scientific Lab and Univ. of Texas, U. S. AEC report ORO-619. xvi, 87pp.

Patterson, J. T. 1929. The effects of x-rays in producing mutations in somatic cells of *Drosophila melanogaster*. J. Exptl. Zool. 53(3): 327-372.

Patterson, J. T. 1930. The production of mutations in somatic cells of *Drosophila melanogaster* by means of x-rays. Science 68(1750): 41-43.

Patterson, R. S., B. J. Smittle and R. T. DeNeve. 1969. Feeding habits of male southern house mosquitoes on ^{32}P-labeled plants. J. Econ. Entomol. 62(6): 1455-1458.

Patterson, R. S., B. J. Smittle and C. S. Lofgren. 1967. Distribution of ^{32}P in the male reproductive system of *Culex pipiens quinquefasciatus* (Say). Bull. Entomol. Soc. Am. 13(3): 201.

Patterson, R. S., B. J. Smittle and C. S. Lofgren. 1968. Accumulation and transfer of ^{32}P by male southern house mosquitoes. J. Econ. Entomol. 61 (6): 1546-1548.

Patzer, R. G., A. A. Moghissi and D. N. McNelis. 1973. Accumulation of tritium in various species of fish reared in tritiated water. pp. 403-412. In: *Environmental Behavior of Radionuclides Released in the Nuclear Industry*. (See IAEA, 1973) (NSA 29: 29867).

Paul, G. I., J. E. Guthrie and O. E. Acres. 1974. Is there a place for precipitation collection in routine radiation surveillance. pp. 263-271. In: *Environmental Surveillance Around Nuclear Installations*. *Vol. 2*. (See IAEA, 1974) (NSA 31: 14069).

Pauley, G. B. and R. E. Nakatani. 1968. Metabolism of the radioisotope ^{65}Zn in the freshwater mussel *Anodonta californiensis*. J. Fish. Res. Bd. Can. 25(12): 2691-2694.

Paulov, S. 1961. Effect of low temperature on the development of radiation injury in eggs of the silkworm *Bombyx mori* L. Folia biol. (Prague) 7: 281-284. (NSA 15: 30467).

Pavan, C. 1960. Effect of radiation on puffs of polytenic chromosomes of *Rhynchosciara angelae*. pp. 42-52. In: *2nd Inter-American Symp. on the Peaceful Application of Nuclear Energy*. (See U. S. AEC, 1960).

Pavan, C. 1965. Nucleic acid metabolism in polytene chromosomes and the problem of differentiation. pp. 222-241. In: *Genetic Control of Differentiation*. (Report of Symp., June 7-9, 1965). Brookhaven National Lab., U. S. AEC report BNL-931. (CONF-650665) (^{3}H, *Rhynchosciara angelae*).

Pavan, C. and R. Basile. 1963. Effects of radiation on the synthesis of nucleic acids in polytene chromosomes. Natl. Cancer Inst. Monogr. 14: 199-204. (NSA 19: 8883) (*Rhynchosciara angelae*).

Pavdernik, J. 1962. Measurement of radioactivity at the Czechoslovak Water Works Laboratories compared to foreign measurement technology. Isotopen Technik 2(11): 341-343. (JPRS-18391).

Pavlicsek, I., A. Veres and S. O. Rakvacsne. 1972. Measurement of the tritium concentration of natural waters. Izotoptechnika 15(12): 619-625. (In Hungarian) (NSA 27: 24609).

Pavlotskaya, F. I. 1960. Radiochemical methods of analyzing artificial radioisotopes in various natural objects. Acad. Sci. USSR, Moscow. 38pp. UN Document A/AC.82/G/L.549. (In Russian).

Pavlotskaya, F. I. 1969. Correlation between rates of entry of ^{90}Sr from the atmosphere, its content in soil cover and quantity of precipitation. State Committee on the Use of Atomic Energy, USSR. Atomizdat, Moscow. UN Document A/AC.82/G/L.1300. (AEC-tr-7128, pp. 214-234. (NSA 24: 11690).

Pavlotskaya, F. I. 1971. Arrival and distribution of radioactive products of nuclear explosions on the earth's surface. pp. 41-81. In: *Contemporary Problems of Radiobiology*. *Vol. 2*. (See Klechovskiy, Polikarpov and Aleksakhin, 1971) (Transl. pp. 30-68) (NSA 29: 12717).

Pavlotskaya, F. I. 1973. Relative mobility, state and forms of existence of strontium-90, stable strontium, and calcium in soils. State Committee on the Use of Atomic Energy, USSR. Atomizdat, Moscow. 39pp. (In Russian) (INIS-mf-1040) (NSA 30: 6523).

Pavlotskaya, F. I. 1974.. *Migration of Radioactive Fallout Products in Soils*. Atomizdat, Moscow. 216pp. (In Russian) (NSA 31: 536).

Pavlotskaya, F. I. and Ye. V. Babicheva. 1973. Long-lived artificial and natural radioisotopes in seeds of crops in Moscow environs. State Committee on the Use of Atomic Energy, USSR. Atomizdat, Moscow. 31pp. (In Russian) (INIS-mf-1137) (NSA 30: 15380).

Pavlotskaya, F. I. and G. A. Fedoseyev. 1967. Using total beta-activity data for the assessment of radioactive element in soils and plants. Pochvovedeniye (10): 119-124. (Sov. Soil Science (10): 1389-1393) (NSA 22: 45540).

Pavlotskaya, F. I. and L. N. Satsepina. 1966. Studying deposit on the ground of certain fission products. Atom. energiya 20(4): 333-337. (Sov. Atom. Energy 20(4): 373-377).

Pavlotskaya, F. I. and Ye. B. Tyuryukanova. 1967. The influence of natural conditions on the content and distribution of radioactive strontium in the soil cover. Atom. energiya 23(3): 229-233. (Sov. Atom. Energy 23(3): 954-958) (Also: State Committee on the Use of Atomic Energy, USSR. Atomizdat, Moscow. 16pp. UN Document A/AC.82/G/L.1175) (NSA 22: 23128).

Pavlotskaya, F. I. and L. N. Zatsepina. 1965. Problems in studies of the form of uptake of various fission products on the earth's surface. State Committee on the Use of Atomic Energy, USSR. Atomizdat, Moscow. (In Russian).

Pavlotskaya, F. I., M. I. Blokhina and G. N. Arnautov. nd. Distribution of ^{90}Sr onto groups and fractions in organic matter of soils. (In Russian) (INIS-mf-887-3, pp. 77-88) (NSA 30: 6517).

Pavlotskaya, F. I., Ye. B. Tyuryukanova and V. I. Baranov. 1967. On the mobility of strontium and some other components of global fallout in soils and their accumulation in plants. pp. 33-38. In: *Radioecological Concentration Processes*. (See Aberg and Hungate, 1967).

Pavlotskaya, F. I., E. B. Tyruyukanova and V. I. Baranov. 1970. Global distribution of radioactive strontium over the earth's surface. Publ. House Acad. Sci. USSR, Moscow. 160pp. (In Russian) (NSA 24: 44128).

Pavlotskaya, F. I., et al. 1964. A method for the determination of strontium-90, stable strontium and calcium in soil-vegetation cover samples. Pochvovedeniye (2): 105-112. (Sov. Soil Science (2): 197-204) (NSA 19: 35890).

Pavlotskaya, F. I., et al. 1966. Problems in studies on detection of the form of uptake of certain fission products in soils. Inform. Byull. Akad. Nauk SSSR (biol.) (9): 3-9. (In Russian).

Pavlotskaya, F. I., et al. 1967. Distribution of radioactive and stable isotopes of alkaline and alkali-earth elements in the above-ground organs of agricultural plants. State Committee on the Use of Atomic Energy, USSR. Atomizdat, Moscow. 17pp. (In Russian) UN Document A/AC.82/G/L.1174.

Pavlov, V. A., I. Ye. Konstantinov and G. A. Fedorov. 1968. An exponential source as a model for radioactive contamination of the soil. State Committee on the Use of Atomic Energy, USSR. Atomizdat, Moscow. 13pp. UN Document A/AC.82/G/L.1250. (AEC-tr-7030, pp. 106-119) (NSA 23: 17551).

Pavlov, V. A., I. Ye. Konstantinov and G. A. Fedorov. 1968. Evaluation of gamma-radiation attenuation by snow cover. Atom. energiya 25(1): ___. (Sov. Atomic Energy 15(1): 814-816) (NSA 22: 51104).

Pavlov, V. A., I. Ye. Konstantinov and G. A. Fedorov. 1971. Gamma emission in soil as a result of radioactive fallout. Atom. energiya 31(6): 639-640. (In Russian) (NSA 26: 12038).

Payez, J. F., R. A. Deering and J. O. Freim, Jr. 1972. Lethal effect of alkylating agents or ultraviolet light on the gamma-ray resistant parent strain and a gamma-ray sensitive strain of the cellular slime mold, *Dictyostelium discoideum*. Mutation Research 15(1): 82-85. (NSA 26: 31242).

Payne, B. R. 1967. Contribution of isotope techniques to the study of some hydrological problems. pp. 62-68. In: *Isotope Techniques in the Hydrologic Cycle*. (See Stout, 1967).

Payne, B. R. 1972. Isotope hydrology in Latin America. Intern. Atomic Energy Agency Bull. 14(4): 17-20. (NSA 26: 53168).

Payne, B. R. and T. Dincer. 1965. Isotope survey of the Karst region of southern Turkey. In: *Radiocarbon and Tritium Dating*. (Proc. 6th Intern. Conf., Washington State Univ., Pullman, Wash., June 7-11, 1965). R. M. Chatters and E. A. Olsen, comps. (CONF-650652-6) (AED-CONF-65-158-8) (NSA 20: 538).

Payne, B. R. and T. Dincer. 1969. Nuclear techniques in hydrology. pp. 126-148. In: *Progress in Hydrology*. *Vol. I*. (See Univ. of Illinois, 1969) (NSA 25: 2359).

Payne, B. R. and Y. Yurtsever. 1974. Environmental isotopes as a hydrogeological tool in Nicaragua. pp. 193-202. In: *Isotope Techniques in Groundwater Hydrology*. *Vol. 1*. (See IAEA, 1974) (NSA 31: 19628).

Payne, B. R., et al. 1967. The measurement of hydrogen and oxygen isotopes in natural waters and its application to meteorology and hydrology. Final report. International Atomic Energy Agency, Vienna, report IAEA-3162-1. 119pp. (NSA 21: 36146).

Payne, D. W. and W. A. L. Evans. 1964. Transglycosylation in the desert locust, *Schistocerca gregaria* Forsk. J. Insect Physiol. 10(5): 675-688. (^{14}C).

Payne, F. 1913. A study of the effect of radium upon the eggs of *Ascaris megaloc*. Arch. Entwicklungsmech. 36: 287-___.

Paz-Castillo, F. and P. Kruger. 1972. Design of a waterway connecting the Orinoco and Rio Negro rivers in the Federal Territory of Amazonas, Venezuela. International Atomic Energy Agency, Vienna. (Also: Dept. of Civil Engineering, Stanford Univ., Tech. Rept. No. 153) (UCRL-Trans.-10701) (NP-19361) (NSA 30: 3118) (Concerns nuclear and conventional excavation).

Pazdernik, J. 1971. Production of radioactive waste waters in Czechoslovakia. Vod. Hospod. B. 21(5): 132-134. (In Czech) (NSA 26: 56078).

Peakall, D. B. 1962. Rate of incorporation of amino acids into the web proteins of the spider *Araneus diadematus* Cl. p. 13. In: *Proc. 16th Intern. Congr. of Zoology*. *Vol. 1*. J. A. Moore, ed. Washington. (^{3}H, ^{14}C).

Peakall. D. B. 1964. Effects of cholinergic and anticholinergic drugs on the synthesis of silk fibroins of spiders. Comp. Biochem. Physiol. 12(4): 465-470. (^{14}C, *Araneus sericatus*).

Peakall, D. B. 1965. Regulation of the synthesis of silk fibroins of spiders at the glandular level. Comp. Biochem. Physiol. 15(4): 509-515. (^{14}C, *Araneus sericatus*).

Peakall, D. B. 1967. Pesticide-induced enzyme breakdown of steroids in birds. Nature 216: 505-506. (^{14}C).

Pearson, E. A. 1959. Tracer methodology and pollutional analysis of estuaries. pp. 556-568. In: *Proc. 1st Intern. Conf. on Waste Disposal in the Marine Environment*. Pergamon Press, New York.

Pearson, F. J., Jr. 1966. Ground-water ages and flow rates by the ^{14}C method. Thesis, Univ. of Texas, Austin. 105pp. (NSA 21: 32273).

Pearson, F. J., Jr. and W. V. Swarzenki. 1974. ^{14}C evidence for the origin of arid region groundwater, Northeastern Province of Kenya. pp. 95-109. In: *Isotope Techniques in Groundwater Hydrology*. *Vol. 2*. (See IAEA, 1974) (NSA 31: 19649).

Pearson, F. J., Jr., M. S. Bedinger and B. F. Jones. 1972. ^{14}C ages of water from the Arkansas hot springs. pp. D19-D30. In: *Proc. 8th Intern. Radiocarbon Dating Conf*. *Vol. 1*. (See Rafter, 1972) (NSA 31: 8697).

Pearson, I. A., et al. 1948. Studies on the behavior of fungi in the presence of radioactive isotopes. J. Bacteriol. 56(3): 397-402. (NSA 2: 37).

Pearson, J. D. 1966. Use of ^{51}-Cr-labelled haemoglobin and ^{47}Sc as inert faecal markers. Int. J. Appl. Radiat. Isotopes 17: 13-16. (NSA 20: 12612).

Pearson, J. E. 1967. Natural environmental radioactivity from radon 222. U. S. PHS Publ. No. 999-RH-26. vi, 31pp.

Pearson, J. E. and G. E. Jones. 1965. Emanation of ^{222}Rn from soils and its use as a tracer. J. Geophys. Res. 70: 5279-5290. (NSA 20: 528).

Pearson, J. E. and G. E. Jones. 1966. Soil concentrations of emanating ^{226}Ra and the emanation of ^{222}Rn from soils and plants. Tellus 18: 655-662. (NSA 21: 6078).

Perason, J. E., D. H. Rimbey and G. E. Jones. 1965. A soil-gas emanation measurement system used for radon-222. J. Appl. Meteorology 4(3): 349-356. (NSA 19: 36664).

Pecci, J., et al. 1965. Radiochemical procedures for selected radionuclides in environmental samples. U. S. Air Force Cambridge Research Labs., report AFCRL-65-410 (Instrumentation Papers No. 68). vi, 64pp.

Pechen, A. M., I. C. Bonici and N. G. Bazan. 1974. Distributional changes of ^{32}P-labelled acid-soluble phosphates and phospholipids among subcellular fractions during early vertebrate embryonic development. Biochim. Acta 372(2): 388-399. (NSA 32: 17631) (Bufo).

Pechenko, B. F. 1922. The effect of radium rays on amoeba. Vestn. rentgenol. radiol. 2(1): 39-48.

Pechkurenkov, V. L. 1970. Rise of chromosomal aberrations of Misgurnus fossilis L. larvae developing in the ^{90}Sr-^{90}Y solutions of various activity. Genetika 6(10): 67-78. (In Russian) (NSA 25: 16093).

Pechkurenkov, V. L., I. A. Shekhanova and I. G. Telysheva. 1972. The effect of chronic small dose irradiation on the embryonic developemnt of fishes and the validity of various assessment methods. Vopr. ikhtiol. 12: ___. (J. Ichthyol. 12(1): 71-79) (SFA 18: 16481).

Peck, E. L., et al. 1971. Evaluation of snow water equivalent by airborne measurement of passive terrestrial gamma radiation. Water Resources Res. 7(5): 1151-1159. (NSA 26: 12045).

Pedigo, R. A. 1959. Effects of radiation on the ecology of Pinus taeda. Abstr. ASB Bull. 7: 37.

Pedigo, R. A. 1960. The effects of ionizing radiation on the ecology of Pinus Taeda. Abstr. Bull. Ecol. Soc. Am. 41: 94.

Pedigo, R. A. 1962. The effects of ionizing radiation on the ecology of Pinus taeda (L.). Dissertation, Emory Univ., Atlanta, Ga. 132pp. (NSA 17: 15802).

Pedigo, R. A. 1963. Effects of ionizing radiation of Pinus taeda L. pp. 295-299. In: Radioecology. (See Schultz and Klement, 1963) (NSA 17: 33574).

Pedigo, R. A. and R. B. Platt. 1959. Studies in radiation ecology: II. Plant communities of a 10,000 acre experimental area in north Georgia. Bull. Ga. Acad. Sci. 17: 44-60.

Peeva, S. 1972. Uranium content of drinking waters from the central water supply system in Gabrovo and Russe districts and its bearing on the chemical composition of the water. Letop. Khig. Epidemiol. Inst. 6(37): 109-114. (In Bulgarian) (NSA 30: 32091).

Pegel'man, S. G. and Yu. I. Vakher. 1965. Influence of sub-lethal gamma irradiation on growth and development of chicks. Radiobiologiya 5: 382-389) (In Russian) (NSA 19: 33913).

Pekerman, S. M. 1963. Depression of antibody production in monkeys after gamma-irradiation in high doses. Med. radiologiya 8(9): 61-65. (JPRS-22121) (NSA 18: 1404).

Pelczar, T. 1966. Studies on beta radioactivity of mineral components of medicinal herbs. Acta Polon. Pharm. 23: 132-136. (NSA 22: 2490).

Peleg, B. A. and S. Gothilf. 1964. Labeling of eggs of the carob moth, Ecromyelois ceratoniae (Zeller), with P^{32} for ecological studies. Izrael J. Agr. Res. (Ktavim) 14(2): 75-76.

Peleg, B. A. and D. J. Nadel. 1966. A method of P^{32} labelling of the armored and soft scale predator Chilocorus bipustulatus L. Izrael J. Agr. Res. Ktavim 16: 2.

Pelekassis, C. E. D., P. A. Mourikos and D. N. Bantzios. 1963. Preliminary studies of the field movement of the olive fruit fly (Dacus oleae Gmel.) by labelling a natural population with radioactive phosphorus (P^{32}). pp. 105-113. In: Radiation and Radioisotopes Applied to Insects of Agricultural Importance. (See IAEA, 1963).

Pelerents, C. 1962. Effect of gamma radiation on wintering bees. Med. Landb. Hogesch. Opzoeklstus. Gent 27: 896-907. (ORNL-tr-1899) (NSA 23: 5010).

Pelerents, C. and J. van den Brande. 1961. The effect of γ-rays on eggs of Ephestia kuehniella Zell. Med. Landb. Hogesch. Opzoeklstus. Gent 26(3): 1401-1408. (In Flemish).

Pelerents, C. and J. van den Brande. 1963. Some supplementary data concerning the influence of γ rays on worker bees. Bull. Inst. Agron. Sta. Rech. Gembloux 31: 576-583. (In French) (NSA 19: 17528).

Peling, C. 1964. RNA synthesis by the giant chromosomes of Chironomus tentans. Chromosoma 15(1): 71-122. (In German, English summary) (^{3}H).

Pellerin, P. 1971. Environmental control at nuclear power stations. pp. 387-398. In: Environmental Aspects of Nuclear Power Stations. (See IAEA, 1971).

Pellerin, P. 1972. Nuclear power stations: environmental surveillance of radioactivity. Int. At. Energy Agency Bull. 14(3): 21-31. (NSA 27: 12316).

Pellerin, P. and J. P. Moroni. 1973. Principles of environmental monitoring at nuclear power stations. pp. 111-125. In: Environmental Behavior of Radionuclides Released in the Nuclear Industry. (See IAEA, 1973) (In French) (NSA 29: 29301).

Pellerin, P. and J. P. Moroni. 1974. Nuclear plants and environmental protection. Ann. Mines (1): 61-70. (In French) (NSA 30: 18620).

Pellerin, P., et al. 1959. Low-temperature autoradiography for the detection of tritium in tissue, with removal of luminescence induced by tritium. Nature 184(4696): 1385-1386. (NSA 14: 3370).

Pelletier, C. A. 1970. Performance design of an environmental survey. pp. 55-65. In: Environmental Surveillance in the Vicinity of Nuclear Facilities. (See Reinig, 1970) (NSA 25: 16161).

Pelletier, C. A. and J. D. Zimbrick. 1970. Kinetics of environmental radioiodine transport through the milk-food chain. pp. 257-267. In: Environmental Surveillance in the Vicinity of Nuclear Facilities. (See Reinig, 1970) (NSA 25: 16082).

Pelton, M. R. 1966. The effects of radiation on survival and reproduction. of wild cotton rats (Sigmodon hispidus) in enclosed areas of natural habitat. Thesis, Univ. of Georgia, Athens. 44pp. (At Savannah River Plant).

Pelton, M. R. and E. E. Provost. 1969. Effects of radiation on survival of wild cotton rats (Sigmodon hispidus) in enclosed areas of natural habitat. pp. 39-45. In: Symposium on Radioecology. (See Nelson and Evans, 1969).

Pelton, M. R. and E. E. Provost. 1973. Effects of radiation on reproduction of irradiated cotton rats (Sigmodon hispidus) trapped from enclosed areas of natural habitat. pp. 1048-1054. In: Radionuclides in Ecosystems. (See Nelson, 1973) (NSA 28: 30138).

Pelz, F. M. 1939, 1940. Anz. Akad. Wiss. Wien, math.-nat. Klasse 1939: 117-119. Chem. zentr. 1940: II: 73-74. (Natural radiation, wood, plants).

Pendlebury, J. B. 1962. Some effects of gamma radiation on the lesser grain borer, tropical warehouse moth, Indian meal moth, and cigarette beetle. UK Atomic Energy Research Establishment, Wantage, report AERE-R-4003. 23pp.

Pendelbury, J. B. 1966. The influence of temperature upon the radiation susceptibility of Sitophilus granarius (L.). pp. 27-40. In: Entomology of Radiation Disinfestation of Grain. P. B. Cornwell, ed. Pergamon Press, Oxford.

Pendlebury, J. B. and E. J. Banham. 1962. The influence of temperature on the radiation sensitivity of Sitophilus granarius L. UK Atomic Energy Research Establishment, Wantage, report AERE-R-3641. 38pp.

Pendlebury, J. B., et al. 1966. Some effects of gamma radiation on Rhizopertha dominica (F.), Cadra cautella (Wlk.), Plodia intepunctella (Hübn.) and Lasioderma serricorne (F.). pp. 143-156. In: Entomology of Radiation Disinfestation of Grain. P. B. Cornwell, ed. Pergamon Press, Oxford.

Pendleton, R. C. 1956. Labeling animals with radioisotopes. Ecology 37(4): 686-689.

Pendleton, R. C. 1956. Bioaccumulation of radioelements from KAPL waste in aquatic organisms. Hanford Atomic Products Operation, U. S. AEC report HW-43112. (NSA 22: 21497).

Pendleton, R. C. 1964. A trophic level effect on the accumulation of Cs^{137}. Abstr. Bull. Ecol. Soc. Am. 45(4): 150.

Pendleton, R. C. 1964. Environmental effects on the content of Cs^{137} and Sr^{90} in milk, forage plants, and soils. Abstr. Bull. Ecol. Soc. Am. 45 (4): 150.

Pendleton, R. C. 1965. Accumulation of cesium-137 through the aquatic food web. pp. 355-363. In: Biological Problems in Water Pollution. (See Tarzwell, 1965).

Pendleton, R. C. and A. W. Grundmann. 1954. Use of P^{32} in tracing some insect-plant relationships of the thistle, Cirsium undulatum. Ecology 35 (2): 187-191.

Pendleton, R. C. and W. C. Hanson. 1958. Absorption of cesium-137 by components of an aquatic community. Proc. 2nd Intern. Conf. on the Peaceful Use of Atomic Energy 18: 419-422. (Paper A/CONF.15/P/392) (NSA 12: 14553).

Pendleton, R. C. and R. D. Lloyd. 1970. Environmental levels of radioacticity in Utah following Operation Pristine. Radiol. Hlth. Data 11: 65-67. (NSA 24: 34198).

Pendleton, R. C. and R. D. Lloyd. 1970. Zinc-65 in Utah vegetation produced during the 1966 growing season. Radiol. Hlth. Data 11(6): 277-282.

Pendleton, R. C. and E. W. Smart. 1954. A study of the food relations of the least chub, Lotichthys phlegethontis (Cope), using radioactive phosphorus. J. Wildl. Mgmt. 18(2): 226-228.

Pendleton, R. C. and R. L. Uhler. 1960. Accumulation of cesium-137 by plants grown in simulated pond, wet meadow and irrigated field environments. Nature 185(4714): 707-708. (NSA 14: 9284).

Pendleton, R. C., et al. 1962. Accumulation of Cs^{137}, Sr^{90} in forage, humans and milk from ecologically dissimilar area. Abstr. Bull. Ecol. Soc. Am. 43(4): 24.

Pendleton, R. C., et al. 1964. Fallout plutonium-239 and zirconium-95 in the lungs of deer. Nature 202(4933): 715-716. (NSA 18: 25693).

Pendleton, R. C., et al. 1964. Trophic level effect on the accumulation of cesium-137 in cougars feeding on mule deer. Nature 204(4959): 708-709. (NSA 19: 5621).

Pendleton, R. C., et al. 1964. Radium contamination at the Jordan River delta of the Great Salt Lake. J. Water Poll. Control Fed. 36(7): 925-927. (NSA 19: 20220).

Pendleton, R. C., et al. 1965. A trophic level effect on ^{137}Cs concentration. In: Radiation and Terrestrial Ecosystems. Health Physics 11(12): 1503-1510. (See Hungate, 1965).

Pendleton, R. C., et al. 1973. Radioecological studies in Utah subsequent to the Baneberry event. pp. 150-169. In: Radionuclides in Ecosystems. (See Nelson, 1973) (NSA 28: 30138).

Peng, C. T. 1973. Tritium-labeling. A review. pp. 676-704. In: Tritium. (See Moghissi and Carter, 1973) (NSA 28: 27129).

Penna-Franca, E. 1960. Analysis of radioactive contaminants in biological materials. pp. 65-75. In: Radioisotopes and Radiation in the Life Sciences. (See Inter-American Nuclear Energy Commission, 1960).

Penna-Franca, E. 1968. Radioactivity in the diet of the inhabitants of the Brazilian regions of elevated natural radiation. Thesis, Univ. Fed. do Rio de Janeiro, Brazil. 163pp. (In Portugese) (NSA 25: 18927).

Penna-Franca, E. and C. Chagas. 1958. Preliminary examination of radioactivity in brazilnuts, almonds and peanuts. An. Acad. Brasil. Cien. 30(4): 58.

Penna-Franca, E. and O. Gomes de Freitas. 1963. Radioactivity of biological materials from Brazilian areas rich in thorium compounds. Nature 197 (4872): 1062-1065.

Penna-Franca, E., et al. 1965. Survey of radioactive content of food grown on Brazilian areas of high natural radioactivity. In: Radiation and Terrestrial Ecosystems. Health Physics 11(12): 1471-1484. (See Hungate, 1965) (Also: Univ. of Brazil, U. S. AEC report NYO-3273-5) (CONF-650523-2) (CA 64: 3937; NSA 19: 36708; 20: 12980).

Penna-Franca, E., et al. 1965. Status of investigations in the Brazilian areas of high natural radioactivity. Health Physics 11(8): 699-712. (NSA 19: 36004) (Discusses ecological studies).

Penna-Franca, E., et al. 1972. Natural radioactivity in Brazil: a comprehensive review with a model for dose-effect studies. pp. 929-940. In: Natural Radiation Environment. II. Vol. 2. (See Adams, Lowder and Gesell, 1972) (NSA 32: 19731).

Pennak, R. W. 1963. Ecological and radiocarbon correlations in some Colorado mountain lake and bog deposits. Ecology 44(1): 1-15.

Pennington, W., et al. 1973. Observations on lake sediments using fallout ^{137}Cs as a tracer. Nature 242(5396): 324-326. (NSA 28: 5558).

Pennington, W. H. 1973. Environmental impact statements: new AEC guidelines. pp. 17-28. In: Proc. AEC Pollution Control Conf. (See U. S. AEC, 1973) (NSA 28: 4962).

Pennington, W. H. 1974. AEC-response to recent developments in NEPA implementation. pp. 29-44. In: Proc. 2nd AEC Environmental Protection Conf. U. S. AEC report WASH-1332(74).

Pennsylvania Department of Education. 1975. Environmental impact of power generation: nuclear and fossil. Harrisburg. 240pp. (ERDA-69) (NSA 33: 20578).

Pennsylvania Power and Light Company. 1971. Susquehanna Steam Electric Station, Units 1 and 2. Environmental report, construction permit stage. Allentown. 100pp. U. S. AEC DOCKET-50387-8, -50388-8. (NSA 25: 25947).

Pennsylvania State University. 1974. Lethal and mutagenic effects in microorganisms grown in $^{3}H_2O$. University Park, U. S. AEC report COO-3423-4. 3pp. (NSA 29: 21247).

Penot, M. 1959. Rate of phosphorus-32 transport in wood and phloem of Pelargonium. Compt. rend. 249(3): 450-452. (In French).

Pensko, J. 1968. Environmental radiation measurements in Poland by means of scintillation methods. pp. 499-506. In: Intern. Congr. on Radioprotection of the Environment. (See Societé Francaise de Radioprotection, 1968).

Pensko, J. 1972. Fast methods of environmental radioactivity measurement. Postepy Fiz. Med. 7(1): 33-42. (In Polish) (NSA 29: 18331).

Pensko, J. 1973. Terrestrial γ radiation field in the lower atmosphere and method of its investigation. Poland Inst. of Nuclear Research, Warsaw, report INR-1473/19/D/B. 239pp. (In Polish) (NSA 33: 14827).

Pensko, J. 1974. Means of evaluation of population dose from natural background radiation for control of exposure in the environment of nuclear power plants. Nukleonika 19(1): 47-56. (In Polish) (NSA 30: 6498).

Pensko, J., J. Jagielak and K. Mamont. 1970. Spectrometric methods of measurement of the terrestrial gamma background radiation and concentration of radionuclides in the soil. Nukleonika 15: 573-583. (In Polish) (NSA 25: 21162).

Pensko, J., T. Wardaszko and M. Wochna. 1972. Influence of some geophysical factors on gamma background and ^{222}Rn concentration in soil and atmosphere. G. Fis. Sanit. Prot. Radiaz. 16(4): 157-167. (NSA 29: 29339).

Pensko, J., et al. 1971. Ionizing background radiation in the Kowary Basin. Nukleonika 16(5-6): 293-300. (In Polish) (NSA 26: 12319).

Pensko, J., et al. 1971. Combined environmental radioactivity measurements for rapid estimation of the gamma-radiation field. pp. 443-458. In: Rapid Methods for Measuring Radioactivity in the Environment. (See IAEA, 1971) (NSA 26: 28053).

Pentreath, R. J. 1973. The accumulation from water of ^{65}Zn, ^{54}Mn, ^{60}Co and ^{59}Fe by the mussel, Mytilus edulis. J. Marine Biol. Assoc. UK 53(1): 127-143.

Pentz, S. and G. Krause. 1968. Clusters of aberrant scales after x irradiation of the wing hereditary factors in female diapause and nondiapause caterpillars of Plodia (Lepidoptera). Arch. Entwicklungsmech. Organ. 160: 167-186. (In German) (NSA 22: 51531).

Peperstraete, H., et al. 1973. Physico-chemical characteristics of five ruthenium salts during fresh-water to marine transition. pp. 219-221. In: Radioactive Contamination of the Marine Environment. (See IAEA, 1973) (NSA 28: 8169).

Pepper, R. B. 1974. Experience in complying with authorization conditions for the discharge of liquid and gaseous radioactive effluents from CEGB nuclear power stations. pp. 180-190. In: Proc. Seminar on the Monitoring of Radioactive Effluents. (See Organization for Economic Co-operation and Development, 1974) (NSA 31: 3346).

Peray, R., G. Juste and A. Descamps. 1974. Continuous monitoring of radioactive contamination in suspicious effluents. pp. 328-341. In: Proc. Seminar on the Monitoring of Radioactive Effluents. (See Organization for Economic Co-operation and Development, 1974) (In French) (NSA 31: 3355).

Perea, E. P., E. I. Gonzalez and D. L. Saenz. 1962. Treatment and handling of radioactive wastes in the Andujar uranium mill. Spain Junta de Energia Nuclear, Madrid, report JEN-108-DMa/1-15. 37pp. (NSA 17: 13871).

Peredel'skiy, A. A. 1957. The principles and problems of radioecology. Zh. obshch. biol. 18(1): 17-30. (French transl. France Commissariat à l'Énergie Atomique report CEA-tr-R-1369) (NSA 16: 5150).

Peredel'skiy, A. A. 1957. Ecological studies of the effect of ionizing radiation. pp. 379-392. In: Biological Science. 1. Radiobiology. A. M. Kuzin, ed. Acad. Sci. USSR, Moscow. (JPRS-859D).

Peredel'skiy, A. A. 1958. Questions on radioecology. Priroda (8): 27-32.

Peredel'skiy, A. A. 1962. Radioecology. Vopr. ekologii 4: 66-67. (JPRS-19496) (Ref. zh. (biol.) 1963: 3D13).

Peredel'skiy, A. A. 1964. Penetrating radiation and radioecology. pp. 449-470. In: The Earth in the Universe. V. V. Fedynskiy, ed. Publ. House Acad. Sci. USSR, Moscow. (In Russian).

Peredel'skiy, A. A. and I. O. Bogatyrev. 1959. The scattering of radioactive isotopes by aquatic insects. Byull. Mosk. obshch. ispyt. prirody (biol.) 64(2): 150. (In Russian).

Peredel'skiy, A. A. and I. O. Bogatyrev. 1959. Radioactive contamination of land areas by insects emerging from contaminated water bodies. Izv. Akad. Nauk SSSR (biol.) (2): 186-192. (JPRS-17812).

Peredel'skiy, A. A. and I. O. Bogatyrev. 1961. Differences in the nature of Ca^{45} and Sr^{90} accumulation in plant leaves and roots. Zh. obshch. biol. 22(1): 74-75. (In Russian).

Peredel'skiy, A. A., et al. 1957. The role of earthworms in the purification of soil from pollution with radioactive isotopes. Dokl. Akad. Nauk SSSR 115(4): 809-812. (Doklady (biol. sci.) 115: 638-641).

Peredel'skiy, A. A., et al. 1957. Use if ionizing radiation against pests of stored grain. Biofizika 2(2): 209-214. (Biophysics USSR 2: 209-213) (NSA 12: 2644).

Peredel'skiy, A. A., et al. 1960. The influence of earthworms (Lumbricidae) and of wireworms (Elateridae) on the incorporation of radioactive isotope Ca^{45} and Sr^{90} by plants from the soil. Dokl. Akad. Nauk SSSR 134(6): 1450-1452. (Doklady (biol. sci.) 134: 704-705) (NSA 15: 5892).

Perdel'skiy, A. A., et al. 1960. Dispersion of radioactive isotopes in the soil by earthworms. Dokl. Akad. Nauk SSSR 135(1): 185-188. (Doklady (biol. sci.) 135: 868-870) (NSA 15: 12707).

Perelygin, V. P., et al. 1972. Development and use of dielectric detectors for measuring concentration of fission elements in samples. pp. 78-86. In: Methods of Determination of Radioactivity. (See Polikarpov and Parchevskiy, 1972) (In Russian).

Perez Luina, A. 1971. Nuclear energy: risks and benefits. Energ. Nucl. (Madrid) 15(74): 491-501. (In Spanish) (NSA 26: 42867).

Perkins, E. J. and B. R. H. Williams. 1963. The biology of the Solway Firth in relation to the movement and accumulation of radioactive materials. I. General introduction. UK Atomic Energy Authority, Annon, Scotland, report PG-Report-500. 16pp. (NSA 18: 4890).

Perkins, E. J., B. R. H. Williams and J. Gorman. 1966. The biology of the Solway Firth in relation to the movement and accumulation of radioactive materials. V. Radioactivity in the fauna. IX. General discussion. UK Atomic Energy Authority, Annon, Scotland, report PG-Report-752, -753. 22, 7pp. (NSA 21: 12803, 8747).

Perkins, R. W. 1961. A large well chrystal for the direct measurement of trace amounts of radioisotopes in environmental samples. Health Physics 7(1/2): 81-86.

Perkins, R. W. 1965. The direct measurement of radionuclides in bioassay samples by multidirectional gamma-ray spectrometry. pp. 179-200. In: Proc. 10th Ann. Bioassay and Analytical Chemistry Meeting. (Cincinnati, Ohio, October 8-9, 1964). U. S. AEC report CONF-727. (Incl. water, fish data).

Perkins, R. W. and J. M. Nielsen. 1959. Zinc-65 in foods and people. Science 129(3341): 94-95. (Concerns Columbia River).

Perkins, R. W. and J. M. Nielsen. 1965. Cosmic-ray produced radionuclides in the environment. In: Radiation and Terrestrial Ecosystems. Health Physics 11(12): 1297-1304. (See Hungate, 1965).

Perkins, R. W. and J. M. Nielsen. 1968. Deposition of a wide spectrum of radionuclides in the food chain by nuclear production and testing operations. pp. 553-564. In: Intern. Congr. on Radioprotection of the Environment. (See Societé Francaise de Radioprotection, 1968).

Perkins, R. W. and L. A. Rancitelli. 1971. Nuclear techniques for trace element and radionuclide measurements in natural waters. Battelle Northwest Lab., U. S. AEC report BNWL-SA-3993. 32pp. (For Meeting on Nuclear Methods in Environmental Research) (CONF-710818-1) (NSA 26: 210).

Perkins, R. W., W. A. Haller and H. G. Rieck. 1971. ^{252}Cf neutron activation for terrestrial, sea floor, and possible planetary surface analysis. Battelle Northwest Lab., U. S. AEC report BNWL-SA-3803. 15pp. (For Meeting on Neutron Sources and Applications, Augusta, Ga., 18 Apr. 1971) (CONF-710402-2) (NSA 25: 26188).

Perkins, R. W., J. L. Nelson and W. L. Haushild. 1966. Behavior and transport of radionuclides in the Columbia River between Hanford and Vancouver, Washington. Limnol. Oceanogr. 11: 235-248. (NSA 20: 25290).

Perkins, R. W., et al. 1971. ^{252}Cf neutron activation for terrestrial, sea floor, and possible planetary surface analysis. pp. III.20-III.29. In: Neutron Sources and Applications. (See U. S. AEC, 1971) (NSA 25: 48469).

Perlowagora-Szumlewicz, A. 1964. Survival, growth, and fecundity of *Australorbis glabratus* snails developed from eggs exposed to ionizing radiation. Exptl. Parasitol. 15(3): 232-241. (NSA 18: 38836).

Perlowagora-Szumlewicz, A. 1964. Effects of radiation on the population kinetics of the snail *Australorbis glabratus*: age at exposure and immediate and late effects of x-rays. Radiation Research 23(3): 377-391. (NSA 19: 2138).

Perlowagora-Szumlewicz, A. 1964. Effect of ionizing radiation on the population kinetics of the snail *Australorbis glabratus*: age at exposure and the effects on reproduction. Radiaiton Research 23(3): 392-404. (NSA 19: 2138).

Perlowagora-Szumlewicz, A. P. 1964. Schistomiasis: age of snails and susceptibility to x-irradiation. Science 144: 302-304.

Perlowagora-Szumlewicz, A. 1966. Studies on acquired resistance to *Schistosoma mansoni* in mice exposed to x-irradiated cercariae of one sex. Rev. inst. med. trop. São Paulo 8: 203-218. (NSA 21: 22662).

Perlowagora-Szumlewicz, A. and E. G. Berry. 1964. Effects of ionizing radiation on *Australorbis glabrata* eggs. Exptl. Parasitol. 15(3): 226-231. (NSA 18: 38835).

Perlowagora-Szumlewicz, A., E. G. Berry and H. L. Andrews. 1962. Effects of ionizing radiation on the snail, *Australorbis glabratus*, an immediate host of *Schistosoma mansoni*. Abstr. p. 100. In: 2nd Intern. Congr. Radiation Reserach. (See Anonymous, 1962).

Permikina, Z. A. and Z. P. Yakovleva. 1969. Detection of ^{90}Sr in the winter and summer feed of dairy cattle. Veterinariya 46: 95-97. (In Russian) (NSA 25: 24000).

Perova, A. A., et al. 1969. The dynamics of fission-product accumulation in vegetables. State Committee on the Use of Atomic Energy USSR. Atomizdat, Moscow. 9pp. UN Document A/AC.82/G/L.1294. (In Russian).

Perri, T. 1950. The biologic action of x-rays on embryos of amphibians. Experiments of transplantation. Riv. biol. 42: 119-154. (In Italian) (NSA 4: 5348).

Perri, T. 1959. The effect of x rays on the larvae of the *Anfibi anuri*. Grafts of joints and administration of thyroxine. Atti accad. naz. lincei. rend. Classe. sci. fis. mat. nat. 27(8): 258-263. (In Italian) (NSA 14: 14677) (Amphibians Anurans ?).

Perrin, F. 1963. Experimental study of the radioactive contamination of cultures, especially by irrigation water. Compt. rend. acad. agr. Fr. 49: 611-620. (In French) (NSA 18: 33239).

Perry, A. S., D. J. Hennessey and J. W. Miles. 1967. Comparative toxicity and metabolism of p, p'-DDT and various substituted DDT-derivatives by susceptible and resistant house flies. J. Econ. Entomol. 60(2): 568-573. (^{3}H-DDT).

Perry, A. S., J. A. Jensen and G. W. Pierce. 1955. Resistance to insecticides, colorimetric and radiometric determinations of DDT and its metabolites in resistant house flies. Agr. Food Chem. 3(12): 1008-1011.

Perry, A. S., G. W. Pearce and A. J. Buckner. 1964. The absorption, distribution, and fate of C^{14}-aldrin and C^{14}-dieldrin by susceptible and resistant house flies. J. Econ. Entomol. 57(6): 867-872.

Perry, J. J. 1970. The survival of actinomycetes in a radiation field. Ch. F-10, pp. F67-F-68. In: A Tropical Rain Forest. (See H. T. Odum and R. F. Pigeon, 1970) (NSA 25: 21971).

Perry, R. P. 1964. Quantitative autoradiography. pp. 304-326. In: Methods in Cell Physiology. Vol. 1. D. Prescott, ed. Academic Press, New York. (NSA 20: 12616).

Pershad, G. D., et al. 1961. Distribution of cytological aberrations after gamma irradiation. Radiation Research 14(2): 184-191. (NSA 15: 10744).

Persiani, C. 1971. Detection of selected air pollutants using a laboratory neutron generator. Am. Industr. Hyg. Assoc. J. 32(9): 573-578. (NSA 26: 17375).

Person, F. and J. Brun. 1974. Physical and chemical mutagenesis on mycophagous nematode *Aphelenchoides composticola* (M. T. Franklin, 1957). Ann. Zool. Ecol. Anim. 6(1): 111-130. (In French) (NSA 33: 12432).

Persson, B. 1967. ^{55}Fe from fallout in lichen, reindeer and Lapps. pp. 253-258. In: Radioecological Concentration Processes. (See Aberg and Hungate, 1967).

Persson, B. R. 1970. ^{210}Pb-atmospheric deposition in lichen-carpets in northern Sweden during 1961-1969. Tellus 22: 564-571. (NSA 25: 23708).

Persson, B. R. 1971. ^{90}Sr in northern Sweden: relationships and annual variations from 1961-1969 in lichen, reindeer, and man. Health Physics 20: 393-402. (NSA 25: 29480).

Persson, B. R. 1972. Lead-210, polonium-210, and stable lead in the food-chain lichen, reindeer and man. pp. 347-367. In: Natural Radiation Environment II. Vol. 1. (See Adams, Lowder and Gesell, 1972) (NSA 32: 19757).

Persson, G. 1973. Tritium data from the ground water of the Kristianstad Plain 1963-1972. Preliminary report. Vannet Norden 6(3): 3-17. (NSA 29: 15642).

Persson, G. N. 1974. Tritium data from groundwater in the Kristianstad Plain, southern Sweden. pp. 45-56. In: Isotope Techniques in Groundwater Hydrology. Vol. 1. (See IAEA, 1974) (NSA 31: 19619).

Persson, R. B. R. 1968. $^{134}Cs/^{137}Cs$ activity ratio in the biosphere from 1956 until 1966. Health Physics 14: 241-250. (NSA 22: 28482).

Persson, R. B. R. 1969. Iron-55 in northern Sweden; relationships and annual variation from 1956 until 1967 in lichen and reindeer as well as uptake and metabolism in man. Health Physics 16(1): 69-78.

Persson, R. B. R. 1974. Stable lead and ^{210}Pb in the food chain lichen-reindeer-man. pp. 231-247. In: Comparative Studies of Food and Environmental Contamination. (See IAEA, 1974) (NSA 30: 15450).

Pertsov, L. A. 1964. Natural Radioactivity of the Biosphere. Atomizdat, Moscow. 315pp. (AEC-tr-6714) (NSA 19: 24).

Pertsov, L. A. 1964. The concentration of natural radioisotope fragments in the biosphere. Radiobiologiya 4(4): 619-623. (AEC-tr-6407, pp. 175-179) (NSA 18: 41658).

Pertsov, L. A. 1973. Ionizing Radiation in the Biosphere. Atomizdat, Moscow. 288pp. (In Russian) (NSA 29: 2637).

Pertsov, L. A., et al. 1967. Radioactivity of tissues of inhabitants of various regions of the Soviet Union. State Committee on the Use of Atomic Energy USSR. Atomizdat, Moscow. 17pp. UN Document A/AC.82/G/L.1182. In: (In Russian) (NSA 22: 33800).

Pertsov, L. A., et al. 1971. Method of estimation of functional significance of radioactive isotope contamination of biogeocoenoses. pp. 161-167. In: Methods of Radioecological Investigations. (See Verkhovskaya, 1971) (In Russian).

Perutik, R. 1966. An x-ray device for the defectoscopy of seeds. To determine healthy, infested and barren seeds of oats and winter barley. Ochr. Rost. Min. Zem. Les. Hosp. UVTI Sb. 39(1): 37-42. (In Czech).

Pesson, P. 1962. Research works on the use of isotopes and nuclear radiations in applied entomology in France and associated nations. pp. 297-300. In: Radioisotopes and Radiation in Entomology. (See IAEA, 1962).

Pesson, P. 1963. Some experimental data on cobalt-60 radiation doses capable of arresting insect infestation of cereals and flour. Food Irradiation 3(4): A18-A21.

Pesson, P. and G. K. Girish. 1968. Radiosensitivity of Sitophilus zeamais Mots (S. oryzae L.) at various stages of development: a study of the larval stages by means of radiography and "actographic" recording. pp. 123-137. In: Isotopes and Radiation in Entomology. (See IAEA, 1968) (NSA 24: 32167).

Pesson, P. and M. Ozer. 1965. Application of an actographic recording method to the study of the effects of irradiation on Calandra granaria larvae. Food Irradiation 6(1-2): A15.

Peter, C. W. 1974. Instrumentation and environmental radiation assessment systems. Trans. Nucl. Sci. NS-21(1): 470-477. (NSA 30: 6673).

Peterfi, S., T. Pavel and F. Nagh-Toth. 1966. Uptake of ^{90}Sr and ^{134}Cs in Scenedesmus acutus Meyen - green algae. Rev. Roum. Biol. (bot.) 11: 327-331. (In German) (NSA 22: 4705).

Peterle, T. J. 1966. The use of isotopes to study pesticide translocation in natural environments. J. Appl. Ecol. 3(Suppl.): 181-191. (WR 124: 8).

Peterle, T. J. 1967. Translocation and bioaccumulation of Cl-36 DDT in a freshwater marsh. pp. 297-307. In: 7th Intern. Congr. Game Biol. (WR 140: 17).

Peterle, T. J. 1969. Pyramiding damage. Environment 11(6): 34-40. (Used ^{36}Cl in study of DDT cycling in marsh).

Peterle, T. J. 1969. Radioisotopes and their use in wildlife research. pp. 109-118. In: Wildlife Management Techniques. R. H. Giles, Jr., ed. The Wildlife Society, Washington.

Peterle, T. J. and R. H. Giles, Jr. 1964. New tracer techniques for evaluating the effects of an insecticide on the ecology of a forest fauna. Ohio State Univ., Columbus, U. S. AEC report TID-20355. 444pp. (^{35}S).

Peters, L. N. and J. P. Witherspoon. 1972. Retention of 44 to 88μ similated fallout particles by grasses. Health Physics 22(3): 261-266. (NSA 26: 20354).

Peters, L. N., J. S. Olson and R. M. Anderson. 1969. Eight-year summary of plant-soil data on ^{137}Cs in a tagged Appalachian forest dominated by Liriodendron tuilipifera. Oak Ridge National Lab., U. S. AEC report ORNL-TM-2737. 235pp. (NSA 24: 9802).

Peters, T. 1955. Development of an unsegmented salamander embryo after treatment with x radiation in combination with oxygen deficiency, cold, and hydrocyanic acid. Strahlentherapie 98: 628-639. (In German) (NSA 10: 2591).

Peters, T. 1959. Growth damages at variuos doses in salamander larvae after irradiation in incipient food absorption stage. Strahlentherapie 108: 383-402. (In German) (NSA 13: 11590).

Peters, T. 1960. The effects of x rays on fertilized egg cells of Triton alpestris with special emphasis on the effects of small doses and less damage. Strahlentherapie 112(4): 525-542. (In German).

Petersen, H. I. 1966. Translocation of ^{14}C-labelled 2,4-D in cereals. pp. 27-31. In: Isotopes in Weed Research. (See IAEA, 1966).

Petersen, J. 1962. Environmental radioactivity at Copenhagen, February 1961 -August 1962. Denmark Atomic Energy Commission, Risö, report RISÖ-51. 25pp. (NSA 17: 12498).

Peterson, D. E. 1971. Modeling of water transport of radionuclides. Trans. Amer. Nucl. Soc. 14(2): 458-459. (NSA 26: 2382).

Peterson, H. T., Jr. 1971. Stable element and radionuclide reconcentration in aquatic ecosystems. I. Stable element effects in the equilibrium situation. pp. 289-328. In: Health Physics Aspects of Nuclear Facility Siting. Vol. II. (See Voillequé and Baldwin, 1971) (NSA 25: 55158).

Peterson, H. T., Jr. and J. M. Smith. 1970. Guides for predicting thyroid dose from environmental measurements following radioiodine releases. pp. 172-188. In: Environmental Surveillance in the Vicinity of Nuclear Facilities. (See Reinig, 1970) (NSA 25: 16172).

Peterson, H. T., Jr., et al. 1969. Environmental tritium contamination from increasing utilization of nuclear energy sources. pp. 35-59. In: Environmental Contamination by Radioactive Materials. (See IAEA, 1969).

Petin, V. G. 1969. Comparative study of the principles of the inactivation and recovery of yeast cells after alpha and gamma-irradiation. Radiobiologiya 9(3): 421-426. (AEC-tr-7026, pp. 122-130) (NSA 24: 9965).

Petin, V. G. and P. G. Polit. 1969. Influence of the radiation dose rate on the survival and recovery of yeast cells. Radiobiologiya 9(4): 492-499. (AEC-tr-7109, pp. 15-24) (NSA 24: 8021).

Petit, D. 1974. ^{210}Pb and stable lead isotopes in lake sediments. Earth Planet. Sci. Lett. 23(2): 199-205. (In French) (NSA 31: 8606).

Petri, V. N. (Ed.). 1962. Compilation of laboratory works on biophysics. IV. Radioactive contamination of the biosphere and protective measures against it. Tr. inst. biol. Ural. fil. Akad. Nauk SSSR (22): 1-91. (In Russian) (NSA 17: 41074).

Petri, V. N. (Ed.). 1970. Effects of ionizing radiation on hydrobionts and land plants. Tr. inst. ekol. rast. zhivot. (74): 1-98.

Petrides, G. A. 1968. Chromium-51 as an indicator of energy flow characteristics in small mammals. Abstr. Small Mammal News Letters 2(7): 139.

Petrides, G. A. 1968. The use of chromium-51 in the determination of energy-flow and other digestive characteristics in animals. pp. 25-31. In: Proc. Symp. on Recent Advances in Tropical Ecology. Vol. I. R. Misra and B. Gopal, eds. Intern. Soc. for Tropical Ecology, Banares Hindu Univ., Varanasi, India.

Petrides, G. A. and P. G. Stewart. 1969/70. Determination of energy flow in small mammals using ^{51}chromium. pp. 131-142. In: Energy Flow Through Small Mammal Populations. K. Petrusewicz and L. Ryszkowski, eds. PWN-Polish Scient. Publ., Warsaw.

Petrov, D. and G. Gecheva. 1974. The effect of gamma rays on the eggs of Rhipicephalus bursa. Vet. med. nauki (Sofia) 11(2): 21-25. (In Bulgarian) (NSA 32: 3803).

Petrov, P. 1972. Measuring soil moisture with the aid of isotopes. Pochvo Agrokhim. 7(3): 3-8. (In Bulgarian) (NSA 30: 426).

Petrov, R. V. and V. F. Sosova. 1963. Relativity of radiation induced immunological tolerance. Radiobiologiya 3(1): 99-103. (AEC-tr-5434, pp. 138-144) (NSA 17: 30345).

Petrov, R. V., et al. 1963. Radioactive Fallout. (Physics, Biological Effects and Protective Measures). A. I. Burnazyan, ed. Publ. House Med. Lit. Acad. Sci. USSR, Moscow. (AEC-tr-6634).

Petrova, J. 1963. The "direct" and "indirect" effects of ionizing radiation on the alga Zygnema. Folia Biol. (Prague) 9(1): 51-59. (NSA 17: 17837).

Petrovich, I. K. and Yu. I. Moskalev. 1963. On the biological action of fast neutrons. Radiobiologiya 3(6): 785-794. (AEC-tr-5439, pp. 1-14) (NSA 18: 19677).

Petrucci, D. 1963. The pentose phosphate cycle in the respiration of amphibian embryos. Ric. Sci. Rend. Sez. B3(1): 61-68. (In Italian)

Petrushova, N. I. 1967. Sexual sterilization in the fight against the codling moth in apples. Zashchita Rasteniy (10): 26-27. (In Russian).

Petrusheyskaya, M. G. 1962. The role of the skeleton growth in Radiolaria for their systematics. Zool. zh. 41(3): 331-341. (In Russian).

Pettit, R. D. and C. C. Jaynes. 1971. Use of radiophosphorus and soil-block techniques to measure root development. J. Range Mgmt. 24(1): 63-65.

Petukhova, Ye. V. and V. A. Knizhnikov. 1968. Uptake of strontium-90 and cesium-137 in the diet of the population of the Soviet Union on 1966-1967 as a result of stratospheric fallout. State Committe on the Use of Atomic Energy USSR. Atomizdat, Moscow. UN Document A/AC.82/G/L.1245. (AEC-tr-7030, pp. 40-48) (NSA 23: 18001) (Incl. fish, water).

Peyremorte, P., G. Philippeau and J. Marcesse. 1972. Optimization of sampling for determination of the water balance in cultivated land by means of a neutron moisture gaga. pp. 631-647. In: Symp. on the Use of Isotopes and Radiation in Research on Soil-Plant Relationships Including Applications in Forestry. (In French) (NSA 26: 50617).

Pezard, A. G. 1958. Effect of maturity and x-ray castration on the color centers of the retina in Chelonians. Compt. rend soc. biol. 152: 414-416. (In French).

Pezard, A. G. 1958. Effect of sexual maturity and of castration induced by x-rays on turtles. J. Physiol. (Paris) 50(2): 450-451.

Pfau, A. 1957. General considerations on the measurement and observation of radioactivity in water. Atompraxis 3: 389-398. (AEC-tr-3980) (NSA 14: 8625).

Pfister, H. and H. Pauly. 1974. The content of natural radioactive materials in phosphate fertilizers and their contribution to the radiation exposure of the human population. pp. 537-554. In: Radiation Protection, Environment, and the Population. Vol. 2. German Democratic Republic, Staatliche Zentrale für Strahlenschutz, Berlin, report SZS-157. 292pp. (In several languages) (NSA 31: 8873).

Phair, G. and D. Gottfried. 1964. The Colorado Front Range, Colorado, U. S. A., as a uranium and thorium province. pp. 7-38. In: The Natural Radiation Environment. (See Adams and Lowder, 1964) (NSA 18: 35728).

Phelps, D. K. 1967. Partitioning of the stable elements Fe, Zn, Sc, and Sm within a benthic community, Anasco Bay, Puerto Rico. pp. 721-734. In: Radioecological Concentration Processes. (See Aberg and Hungate, 1967).

Phelps, P. L. and L. R. Anspaugh. 1974, 1975. Resuspension element status report. pp. 221-233; pp. 197-203. In: Dynamics of Plutonium (and Other Transuranics) in Desert Environments. (See Dunaway and White, 1974, 1975) (NSA 30: 29450; 33: 616).

Phelps, P. L., et al. 1970. A ge(Li) gamma-ray spectrometer of high sensitivity and resolution for biological and environmental counting. pp. 202-230. In: Radionuclides in the Environment. R. F. Gould, ed. Amer. Chemical Soc., Washington.

Philadelphia Electric Company. 1970. Limerick Generating Station, Units 1 and 2. Applicant's Environmental Report, construction permit stage. Philadelphia, Pa. 173pp. U. S. AEC DOCKET-50352-8, -50353-8. (NSA 25: 1838).

Philadelphia Electric Company. 1971. Limerick Generating Station, Units 1 and 2. Responses to comments made by Department of Interior, Fish and Wildlife Service. Philadelphia, Pa. 12pp. U. S. AEC DOCKET-50352-21, -50353-21. (NSA 25: 59989).

Philadelphia Electric Company. 1971. Peach Bottom Atomic Power Station, Units 1 and 2. Environmental Report, operating license stage. Philadelphia, Pa. 358pp. U. S. AEC DOCKET-50277-20, -50278-20. (NSA 25: 36937).

Philbin, T. W. 1973. Environmental report: its impact on the licensing process. Proc. Amer. Power Conf. 35: 300-309. (NSA 30: 2195).

Philbin, T. W. and J. V. Halvorsen. (Eds.). nd. Preparation of environmental reports for nuclear power plants. Atomic Industrial Forum, New York. 34pp. (NP-20164) (NSA 30: 34301).

Philbin, T. W. and G. H. Whipple. 1970. Environmental survey for the Nine Mile Point Power Reactor. pp. 93-104. In: Environmental Surveillance in the Vicinity of Nuclear Facilities. (See Reinig, 1970) (NSA 25: 16166).

Philippine Atomic Energy Commission. 1966. Radioisotopes in soils and nutrition. Manila, report PAEC(A)-IN-65. 114pp. (NSA 21: 24298).

Philip, H. G. 1975. Effect of gamma radiation on DDT resistance and DDT-dehydrochlorinase levels in Musca domestica. J. Econ. Entomol. 68(2): 135-139. (NSA 32: 1184).

Phillips, A. M., Jr., et al. 1957, 1958, 1959. The nutrition of trout. N. Y. Conservation Dept., Courtland Hatchery Rept. 26, 27, 28; Fish. Res. Bull. 21, 22, 23. 93, 87, 83pp.

Phillips, F. T. 1963. The application and measurement of labelled residual insecticides in some physicochemical studies. pp. 147-152. In: Radiation and Radioisotopes Applied to Insects of Agricultural Importance. (See IAEA, 1963).

Phillips, G., A. B. Leonard and C. A. Leane. 1951-52? The effect of a large single dose of phosphorus-32 in the bone marrow of the cotton rat, Sigmodon hispidus texianus. Univ. of Kansas, U. S. AEC report AECU-2157. 29pp. (NSA 6: 5260).

Phillips, S. J. and J. R. Raymond. 1975. Monitoring and characterization of radionuclide transport in the hydrogeologic system. Battelle Northwest Lab., U. S. AEC report BNWL-SA-5494(Pt.1, Pt.2). 23, 84pp. (NSA 33: 7282, 7283).

Phuc, D. V. and I. Varga. 1972. Effect of irradiation of the embryonic and postembryonic development of Amidostomum anseris (Zeder, 1800). Acta Vet. Acad. Sci. Hung. 22(2): 197-203. (NSA 27: 22757).

Piatt, J. 1962. The effect of temperature upon the development of x-irradiated ear grafts in Ambystoma. J. Exptl. Zool. 149(1): 59-73. (NSA 17: 33681).

Piatt, J. 1962. Comparative radiosensitivity of forelimb and hindlimb primordia in Ambystoma punctatum. J. Exptl. Zcol. 150(2): 143-149. (NSA 17: 19851).

Piatt, J. 1964. Development of the ear in x-irradiated parabionts of Ambystoma. J. Exptl. Zool. 155(1): 117-128. (NSA 18: 25148).

Piatt, J. 1964. A summary study of direct and indirect radiation damage upon the developing ear in Ambystoma. J. Exptl. Zool. 156: 331-343. (NSA 19: 8864).

Piatt, J. 1966. X-irradiation on lateral-line primordia in Ambystoma. J. Exptl. Zool. 162(2): 193-207. (NSA 21: 13037).

Piatt, J. 1968. Dessociability of developmental processes in the morphogenesis of the ear in Ambystoma, as revealed by x-ray damage. Anat. Record 160(1): 143-151. (NSA 22: 40921).

Piatt, J. and A. Raventoss. 1953. Transplantation of x-irradiated tissues in embryonic Ambystoma. J. Exptl. Zool. 124(1): 167-197.

Picat, P. and A. Morisset. 1970. Study of river bed pollution. The development of equipment simulating natural flow. Bull. Inform. Sci. Tech. (Paris) (151): 39-49. (In French) (NSA 25: 18940).

Picat, P., P. Bovard and A. Grauby. 1968. Radionuclides released in waterways. pp. 155-158. In: Intern. Congr. on Radioprotection of the Environment. (See Societé Francaise de Radioprotection, 1968).

Picat, P., et al. 1970. Development of the radiocontamination of irrigation from the Rhone. pp. 33-76. In: Proc. Intern. Symp. on Radioecology. Vol. I. (See French Commissariat à l'Énergie Atomique, 1970) (In French).

Picat, P., et al. 1971. Review of the dispersion of radionuclides in river basin prospects in relation to the nuclear activities: processing plants and power reactors. pp. 139-172. In: Proc. Intern. Symp. on Radioecology Applied to the Protection of Man and His Environment. (See Commission of the European Communities, 1971) (In French) (NSA 27: 5256).

Picciotto, E. and S. Wilgain. 1963. Fission products in Antarctic snow, a reference level for measuring accumulation. J. Geophys. Res. 68(20): 5965-5972. (NSA 17: 41008).

Picciotto, E., G. Crozaz and W. de Breuck. 1964. Rate of accumulation of snow at the South Pole as determined by radioactivity measurements. Nature 203(4943): 393-394. (NSA 18: 37379).

Picciotto, E., S. Deutsch and L. Aldaz. 1966. The summer 1957-1958 at the South Pole. An example of an unusual meteorological event recorded by the oxygen isotope ratios in the firn. Earth Planet. Sci. Lett. 1: 202-204. (NSA 20: 33518).

Picciotto, E., et al. 1967. ^{210}Pb and ^{90}Sr in an alpine glacier. Earth Planet. Sci. Lett. 3: 237-242. (NSA 22: 12436).

Picciotto, E., et al. 1968. Determination of the rate of snow accumulation at the Pole of relative inaccessibility, Eastern Antarctica: a comparison of glaciological and isotopic methods. J. Glaciology 7: 273-287. (NSA 23: 2420).

Picer, M. 1968. Measurement of total beta-radioactivity of natural waters by a new evaporation method. Arh. Hig. Rada Toksikol. 19: 225-232. (In Serbo-Croatian) (NSA 23: 37885).

Pick, J. 1965. The fine structure of sympathetic neurons in x-irradiated frogs. J. Cell. Biol. 26: 335-351. (NSA 19: 42285).

Pickering, R. J. 1970. Composition of water in Clinch River, Tennessee River, and White-Oak Creek as related to disposal of low-level radioactive liquid wastes. U. S. Geological Survey Prof. Paper 433-J: 1-15. (NSA 25: 21163).

Pickering, R. J. 1966. Radioactivity in bottom sediment of the Clinch and Tennessee rivers. pp. 57-88. In: Disposal of Radioactive Wastes into Seas, Oceans and Surface Waters. (See IAEA, 1966) (SM-72/4) (CONF-660507-6) (NSA 20: 45223).

Pickersgill, B. 1967. The use of tritiated water to investigate plants in pit heaps. Abstr. J. Animal Ecol. 4(2): 56.

Pickett, C. and W. G. Friend. 1965. Nutritionally essential amino acids of Rhodnius prolixus (Stal) determined with glucose-U-^{14}C. J. Insect Physiol. 11(12): 1617-1623.

Pickford, R. 1971. Studies on the reproductive biology of the migratory grasshopper Melanoplus sanguinipes (Fabricius) (Orthoptera: Acrididae) with particular reference to the role of the male. Thesis, Univ. of Saskatchewan, Saskatoon. 169pp. (Univ. Microfilms 71-26,236) (NSA 26: 25875).

Pieau, C. 1966. Effects of x-rays on the young embryo of the blindworm (Anguis fragilis L.). Compt. rend. soc. biol. 160: 59-62. (In French) (NSA 20: 18410).

Pieau, C. and J. Vasse. 1970. General effects of x rays on green lizard embryos. Compt. rend. D271: 2023-2026. (In French) (NSA 25: 21850).

Piechowska, M. J. 1965. Effect of ionizing radiation on the endocrine system in insects. Bull Acad. Polon. Sci. ser. Sci. Biol. 13: 139-144. (NSA 19: 45861).

Piechowska, M. J. 1966. The effect of 2600 R of gamma radiation dose on the endocrinal processes in Celerio euphorbiae L. butterfly. Polske Towarzystwo Biochemiczne Monogr. Biochem. 13. Panstwowe Wydawnictwo Nauk., Warsaw. 79pp. (NSA 22: 31269).

Piek, T. 1962. Synthesis of wax in the honeybee, Apis mellifera L. Thesis, Utrecht Rijkuniv., The Netherlands. 65pp. (In Dutch, English summary) (Also: J. Insect Physiol. 10(4): 563-572. (^{14}C).

Piekos, B. 1969. Effect of x irradiation on tooth primordia in the golden hamster in early stages of development. Folia Med. Cracov. 11(4): 471-514. (In Polish) (NSA 25: 42649).

Pierpoint, G. 1966. Measuring surface moisture with the neutron depth probe and a surface shield. Soil Science 101: 189-192.

Pietsch, P. and A. K. Bruce. 1965. The significance of post-irradiation growth in regenerating limb and tail blastema of Ambystoma larvae. J. Cell. Comp. Physiol. 66: 243-249. (NSA 20: 6837).

Piggot, C. S. 1931. Radium in rocks II. Granites of eastern North America from Georgia to Greenland. Am. J. Sci. 3: 28-36.

Pijck, J. and A. Defalque. 1963. The rapid detection of coliform microorganisms by a radiometric method. J. Pharm. Belg. 18: 180-190. (In French) (NSA 18: 29181).

Pijnacker, J. P. 1969. The influence of temperature, x irradiation, and centrifugation on the sex of the stick insect Carausius morosus Br. Neth. J. Zool. 19: 282-283. (NSA 24: 833).

Pikulev, A. T. and L. S. Tal'kovskaya. 1968. Content of ammonia and glutamine in the brain upon x-ray irradiation during inhibition of the function of the adrenal cortex. Dokl. Akad. Nauk Belorussk. SSR 12: 1048-1049. (In Russian) (NSA 24: 2575).

Pilcher, J. R. 1973. Pollen analysis and radiocarbon dating of a peat on Slieve Gallion. Co., N. Ireland. New Phytol. 72(3): 681-689. (NSA 29: 413).

Pilgrim, D. H. and K. K. Watson. 1964. The application of radioisotope techniques to water engineering investigations. J. Inst. Eng. Austral. 37: 175-184. (NSA 20: 548).

Pillai, K. C. 1971. Aquatic pollution control systematics for discharge of radioactive effluents. From: Seminar on Pollution and Human Environment; New Delhi, India, 26 August 1970. (INIS-mf-160, pp. 164-191) (NSA 26: 50610).

Pillai, K. C. and P. R. Kamath. 1971. Control of radioactive pollution in India. Marine Poll. Bull. 2(5): 74-75. (NSA 26: 9285).

Pillai, M. M. K. and A. W. A. Brown. 1965. Physiological and genetical studies on resistance to DDT substitutes in Aedes aegypti. J. Econ. Entomol. 58(2): 255-266. (^{14}C).

Pillai, T. N. V. and P. R. Kamath. 1966. Radioactivity measurements in environmental samples in monazite areas. pp. 274-286. In: Proc. All-India Symp. on Radioactivity and Metrology of Radionuclides. (See India Atomic Energy Establishment, 1966) (NSA 22: 6440).

Pillay, K. K. S. 1975. Activation analysis and dendrochronology for estimating pollution histories. Trans. Amer. Nucl. Soc. 21(Suppl., 3): 22-23. (NSA 32: 27486).

Pillay, K. K. S. and C. C. Thomas, Jr. 1971. Determination of the trace element levels in atmospheric pollutants by neutron activation analysis. J. Radioanal. Chem. 7(1): 107-118. (NSA 25: 48484).

Pillay, K. K. S., C. C. Thomas, Jr. and J. A. Sondel. 1971. Activation analysis of airborne selenium as a possible indicator of atmospheric sulfur pollutants. Environ. Sci. Technol. 5(1): 74-77.

Pillay, K. K. S., C. C. Thomas, Jr. and C. M. Hyche. 1973. Neutron activation analysis of some of the biologically active trace elements in fish. J. Radioanal. Chem. 20(2): 597-606. (NSA 30: 20602).

Pillay, K. K. S., C. C. Thomas, Jr. and G. F. Mahoney. 1972. Examination of evidence materials for environmental contamination using activation an analysis. Paper 4, pp. 1-11. In: 2nd Intern. Conf. on Forensic Activation Analysis. (Glasgow, 27-29 September 1972; Text of Papers). (NSA 29: 9733).

Pillay, K. K. S., et al. 1971. Determination of mercury in biological and environmental samples by neutron activation analysis. Anal. Chem. 43(11): 1419-1425. (NSA 25: 54508).

Piltingsrud, H. V. and J. R. Stencel. 1973. Direct method for evaluating plutonium-239 and americium-241 content in large soil samples. Health Physics 25(4): 448-450. (NSA 28: 24169).

Pimentel, D. and R. W. Fay. 1955. Dispersion of radioactively tagged Drosophila from pit privies. J. Econ. Entomol. 48(1): 19-22.

Pimentel, K. D. 1975. Toward a mathematical theory of environmental monitoring: the infrequent sampling problem. Thesis, Univ. of California. 404pp. (NSA 32: 22300).

Pinder, J. E., et al. 1975. Field study to obtain plutonium contents of old field vegetation and soil under humid climatic conditions. Savannah River Ecology Lab., U. S. ERDA report DP-MS-74-65. 7pp. (NSA 32: 17195).

Pinet, J.-M. and M. H. B. Hamouda. 1971. Exceptional molting of Rhodnius prolixus caused by irradiations applied at various times during postembryonic development. Compt. rend. D273(15): 1314-1317. (In French) (NSA 26: 15511).

Pinkas, J. and J. Pluhar. 1965. Tracing radioactive grassland contamination in the area of CSSR. Vet. Med. (Prague) 10: 705-710. (NSA 20: 27165).

Pinkerton, C., et al. 1964. Background radioactivity monitoring of a pilot study community in Washington County, Maryland. pp. 919-956. In: The Natural Radiation Environment. (See Adams and Lowder, 1964) (NSA 17: 41135).

Piotrowski, F. and T. Rudnicki. 1965. Retention of radioactive phosphorus in the body of the human louse (Pediculus humanus L., Anoplura). Acta Physiol. Polon. 16: 435-439. (In Polish) (NSA 19: 43684).

Piovella, C., G. F. Mazzoleni and A. de Silvestri. 1960. Hepatic, microvascular modification induced by irradiation with gamma rays from radium and observed in vivo by transillumination. Minerva nucleare 4: 202-207. (In Italian) (NSA 25334).

Piper, A. M. 1966. Potential effects of Project Chariot on local water supplies, northwestern Alaska. U. S. Geological Survey, Washington, Prof. Paper 539. 47pp. (NSA 23: 2457).

Piper, N. R. 1970. Manual of procedures for the rapid analysis of radioactivity, radionuclides, and beryllium in environmental samples. Australian Atomic Energy Commission Reserach Establishment, Lucas Heights, report AAEC/M-85. 41pp. (NSA 26: 8726).

Piperingos, G. N., W. A. Chisolm and R. C. Kepferle. 1965. Geology and uranium deposits in the Cave Hills area, Harding County, South Dakota. U. S. Geological Survey, Washington, Prof. Paper 476-A. 64pp. (CA 63: 17707) (Incl. surface and well waters).

Pirastru, E., F. Carlin and G. Dai Pra. 1967. First application of neutron probe for the detection of a stable tracer (Cd) in hydrogeology. pp. 279-297. In: *Scientific Aspects and Techniques of Industrial Applications of Isotopes*. (Report of the 11th Rome Nuclear Congr., Rome, June 23-24, 1966). Italy Comitato Nazionale Energia, Rome. (In Italian) (CONF-660661) (NSA 22: 8214).

Pirie, A. 1961. Difference in reaction to x-irradiation between chicken and rabbit lens. Radiation Research 15(2): 211-219. (NSA 15: 24708).

Pirie, E., K. Lin and D. Taylor. 1968. Soil density measurements with gamma-rays. Soil Science 106: 411-414. (NSA 23: 6221).

Pisarev, V. V. and I. S. Tsybizov. 1973. Effect of some natural factors on transition of ^{90}Sr from soil surface to open waters. pp. 36-43. In: *Radioecology of Water Organisms. II*. (See Andrushaitis, et al., 1973) (AEC-tr-7606, pp. 20-24) (NSA 30: 21111).

Pisarev, V. V., et al. 1972. Leaching of strontium-90 from soil by surface water. Pochvovediniye 4: ___. (Sov. Soil Scince 4(2): 193-201) (NSA 28: 12866).

Piskunov, L. I. 1969. An incorrect relationship between the buildup of ^{90}Sr and ^{144}Ce in freshwater plants. Dokl. Akad. Nauk SSSR 188: 917-919. (Doklady (biol. sci.) 188(1-6): 696-698) (NSA 24: 21062).

Piskunov, L. I. 1970. Statistical study of the accumulation of radioisotopes in aquatic organisms. Gidrobiol. zh. 6(5): 70-72. (Hydrobiol. J. 6(5): 58-60) (NSA 26: 45926).

Piskunov, L. I. 1971. Quantitative aspects of the accumulation of radioisotopes by freshwater organisms. Gidrobiol. zh. 7(1): 64-88. (Hydrobiol. J. 7(1): 49-52) (NSA 26: 9494).

Piskunov, L. I. and S. I. Treyger. 1970. Discrimination of ^{90}Sr with respect to calcium in freshwater plants. Radiobiologiya 10: 788-790. (AEC-tr-7215, 234-238) (NSA 25: 29452).

Piskunov, L. I. and B. V. Popov. 1974. Probability and statistical characteristics of concentration of radionuclides in freshwater plants. Ekologiya 5(3): 90-93. (Sov. J. Ecology 5(3): 276-278) (NSA 32: 8924; 33: 647).

Piskunov, L. I., et al. 1975. Results of yearly observations on radiation levels in the Belojarskaya AES by I. V. Kurchatov. Gigiyena sanitariya (11): 111-113. (In Russian).

Piskunov, L. I., et al. 1975. Use of radioecological investigations in dosimetric control of radioactivity of the environment. pp. 212-214. In: *Radiation Hygiene V*. Scientific Research Inst. of Radiation Hygiene, Leningrad. (In Russian).

Pitkin, J. A., S. K. Neuschel and R. G. Bates. 1964. Aeroradioactivity surveys and geologic mapping. pp. 723-736. In: *The Natural Radiation Environment*. (See Adams and Lowder, 1964) (NSA 18: 35743).

Pitkyanen, G. B. and N. G. Safronova. 1973. Distribution and migration of ^{90}Sr in experimental still-water reservoir. pp. 140-146. In: *Radioecology of Water Organisms. II*. (See Andrushaitis, et al., 1973) (AEC-tr-7606, pp. 85-88) (NSA 30: 21117).

Pitkyanen, G. B. and Yu. A. Zaitsev. 1974. The nature of radiation conditions in the development of eggs of freshwater fish belonging to various ecological groups. Ekologiya 5(6): 73-75. (Sov. J. Ecology 5(6): 566-568) (NSA 33: 12454) (*Esox lucius* L.).

Pittilio, J. D. 1965. Fallout in woody plants on and off granitic outcrops. Dissertation, Univ. of Georgia, Athens.

Pittilio, J. D. and G. L. Plummer. 1965. Accumulation of radioisotopes in *Juniperus virginiana*. Bull. Ga. Acad. Sci. 23(3-4): 95-105.

Pittilio, J. D. and G. L. Plummer. 1965. Bioaccumulation of fallout by *Juniperus virginiana* on and off granitic outcrops. Abstr. ASB Bull. 12: 49.

Pittilio, J. D., G. L. Plummer and H. F. Perkins. 1966. Accumulation of radioisotopes within *Pinus taeda*, *Juniperus virginiana*, and *Quercus georgiana* in two ecosystems. Abstr. ASB Bull. 13(2): 44.

Pizon, P. 1962. The pollution of the waters of rivers and oceans by radioactive wastes. Presse Med. 70(41): 1901-1904. (In French) (NSA 18: 14145).

Plamondon, J. E. and J. A. Bassham. 1966. Glycolic acid labeling during photosynthesis with $^{14}CO_2$ and tritiated water. Plant Physiol. 41: 1272-1275. (NSA 21: 22266) (*Chlorella*).

Planel, H., J. P. Soleilhavoup and F. Croute. 1975. Effects of space balloon flights on reproductive activity in *Paramecium aurelia*. pp. 173-180. In: *Life Sciences and Space Research XIII*. Akademie-Verlag, Berlin. (NSA 33: 29762).

Planel, H., J.-P. Soleilhavoup and M.-C. Geiss. 1969. Influence of the medium and the stock on the role of natural radiation in *Drosophila melanogaster*. Compt. rend. D268: 584-587. (In French) (NSA 23: 29908).

Planel, H., J.-P. Soleilhavoup and R. Tixador. 1965. Investigations on the action of natural ionizing radiations on the growth of one-cell animals. Compt. rend. 260: 3770-3773. (In French) (NSA 19: 27988).

Planel, H., J.-P. Soleilhavoup and R. Tixador. 1967. Effects of low doses of ionizing radiation on multiplication of *Paramecium aurelia*. Compt. rend. D264: 2945-2948. (In French) (NSA 22: 19435).

Planel, H., et al. 1966. Test of experimental demonstration of the biological activity of natural radiation. Compt. rend. D262: 2767-2770. (In French) (NSA 20: 42886).

Planel, H., et al. 1967. Action of natural ionizing radiations on the reproduction of metazoans: studies on *Drosophila melanogaster*. Compt. rend. D264: 755-758. (In French) (NSA 21: 26594).

Planel, H., et al. 1967. Studies on the action of natural ionizing radiation in *Drosophila melanogaster*. Compt. rend. soc. biol. 161: 467-470. (In French) (NSA 22: 17045).

Planel, H., et al. 1967. Some evidence of retardation of development in *Drosophila melanogaster* with reduction of the natural ambient radioactivity. Compt. rend. D264(6): 865-868. (In French).

Planel, H., et al. 1968. Evidence of action of natural ionizing radiation on the reproduction of *Drosophila melanogaster*. Compt. rend. assoc. anatom. (139): 1031-1035. (In French) (NSA 22: 47591).

Planel, H., et al. 1968. Estimate of biological effects from very low level irradiations. pp. 647-654. In: *Proc. Intern. Congr. on Radioprotection of the Environment*. (See Societé Francaise de Radioprotection, 1968) (In French).

Planel, H., et al. 1968. Effect of protection against natural irradiation in *Paramecium aurelia* after autogamy. Compt. rend. soc. biol. 162: 990-995. (In French).

Planel, H., et al. 1968. Influence of radioactivity of the environment in the study of cellular growth. Compt. rend. soc. biol. 162: 1215-1219. (In French) (NSA 23: 38410) (*Paramecium*).

Planel, H., et al. 1969. Research on the biological action of natural radiation: study of the growth of *Paramecium aurelia* and *Paramecium caudatum* in subterranean laboratories. Compt. rend. D269: 1697-1700. (In French) (NSA 24: 12228).

Planel, H., et al. 1970. Study of the development of *Paramecium aurelia* in an underground laboratory: influence of natural ionizing radiation. Compt. rend. soc. biol. 164: 654-658. (In French) (NSA 25: 13556).

Planel, H., et al. 1971. Detection of a biological response to very weak radiations. pp. 336-339. In: *Protection Against Low Energy or Short Range Radiations and the Biological Effects of Radiation*. France Service Central de Protection Contre les Rayonnéments Ionisants, le Visinet. (In French) (NSA 26: 12432) (*Paramecium*).

Planel, H., et al. 1974. Study of cosmic ray effects on *Artemia salina* eggs during Apollo 16 and 17 flights. pp. 85-89. In: *Life Sciences and Space Research XII*. P. H. A. Sneath, ed. Akademie-Verlag, Berlin. (NSA 31: 26194).

Plapp, F. W., Jr. and D. A. Lindquist. 1963. Radioisotopes in the study of the fate of insecticides applied to animals and plants. pp. 171-183. In: *Radiation and Radioisotopes Applied to Insects of Agricultural Importance*. (See IAEA, 1963).

Plapp, F. W., Jr., G. A. Chapman and J. W. Morgan. 1965. DDT resistance in *Culex tarsalis* Coquillett: cross resistance to related compounds and metabolic fate of a ^{14}C labeled DDT analog. J. Econ. Entomol. 58(6): 1064-1069.

Plata Bedmar, A., W. Sanchez and C. Szulak. 1973. Studies of aquifers of the Gurgueia and Fidalgo river basins in the State of Piaui using environmental isotopes. Brazil Inst. de Energia Atomica, São Paulo, report IEA-315. 46pp. (In Portugese) (NSA 30: 32108).

Plata Bednar, A., W. Sanchez and C. Szulak. 1973. - Radioisotopic techniques applied to the study of the aquifers of the basins of the Gurgueia and Fidalgo rivers, Piaui. Brazil Inst. de Energia Atomica, São Paulo, report IEA-313. 56pp. (In Portugese) (NSA 30: 32107).

Plato, P. A. 1968. Predicting the movement of a radionuclide through soil. Thesis, Iowa State Univ., Iowa City. 154pp. (NSA 23: 45407).

Plato, P. A. 1972. Distribution of ^{137}Cs and naturally occurring radionuclides in sediments of Lake Michigan. Radiat. Data Rept. 13(4): 181-187. (NSA 26: 36085).

Plato, P. A. 1975. ^{137}Cs in Lake Michigan sediments. Health Physics 28(5): 635-636. (NSA 32: 11534).

Plato, P. A. and J. T. Denovan. 1974. Influence of potassium on the removal of ^{137}Cs by live *Chlorella* from low radioactive wastes. Radiation Botany 14(1): 37-41. (NSA 30: 6496).

Plato, P. A. and D. E. Gelskey. 1971. Use of field instruments in the operation of environmental surveys around nuclear power plants. Instr. Power Ind. 14: 1-5. (NSA 26: 40380).

Plato, P. A. and D. E. Gelskey. 1974. Use of field instruments in the operation of environmental surveys around nuclear power plants. pp. 176-183. In: *Water Quality Instrumentation*. *Vol. 1*. J. S. Scales, ed. Instrument Soc. of Amer., Pittsburgh. (NSA 33: 12039).

Plato, P. A. and G. C. Goldman. 1972. Use of fallout ^{137}Cs as a tracer to define the recent deltaic facies of a river. Radiat. Data Rept. 13(12): 653-657. (NSA 27: 17294).

Platova, T. P. 1965. Revealing of intermediate stages of protein synthesis in the nuclei of *Drosophila melanogaster* salivary gland cells. Tsitologiya 7(3): 318-325. (In Russian) (^{14}C).

Platt, J. 1968. Dessociability of developmental processes in the morphogenesis of the ear in *Ambystoma*, as revealed by x-ray damage. Anat. Record 160: 143-152. (NSA 22: 40921).

Platt, R. B. 1959. Radiation ecology studies at Emory University. Abstr. Proc. 9th Intern. Botan. Congr. 2(A): 28.

Platt, R. B. 1959. Studies in radiation ecology. Abstr. Proc. 9th Botan. Congr. 2(A): 28.

Platt, R. B. 1960. Radiation biology: what it means to you. The Emory Univ. Quart. 16: 65-76.

Platt, R. B. 1960. Mixed gamma neutron radiation effects on flood plain and oak-hickory-pine forests. Abstr. Bull. Ecol. Soc. Am. 41: 94.

Platt, R. B. 1962. Recovery of Pinus echinata from exposure to ionizing radiation two preceding summers. Bull. Ecol. Soc. Am. 43(3): 81-82.

Platt, R. B. 1962. Radiation and plant life. Discovery 23(11): 42-47.

Platt, R. B. 1963. Ecological effects of ionizing radiation on organisms, communities and ecosystems. pp. 243-255. In: Radioecology. (See Schultz and Klement, 1963) (NSA 17: 33569).

Platt, R. B. 1963. Homeostasis and succession in disturbed ecosystems. Abstr. Bull. Ecol. Soc. Am. 44(3): 82.

Platt, R. B. 1964. Man's pollution of his environment. The Emory Univ. Quart. 20(4): 220-225. (Incl. comments on nuclear aspects).

Platt, R. B. 1965. Ionizing radiation and homeostasis of ecosystems. pp. 39-60. In: Ecological Effects of Nuclear War. (See Goodwell, 1965).

Platt, R. B. 1965. Radiation effects on plant populations and communities: research status and potential. In: Radiation and Terrestrial Ecosystems. Health Physics 11(12): 1601-1606. (See Hungate, 1965).

Platt, R. B. and J. F. McCormick. 1964. Manipulatable terrestrial ecosystems. tems. Ecology 45(3): 649-650.

Platt, R. B. and J. A. Mohrbacher. 1959. Studies in radiation ecology: I. The program of study. Bull. Ga. Acad. Sci. 17: 1-30.

Platt, R. B., J. T. McGinnis and J. J. Cowan. 1964. An Automatically controlled outdoor gamma irradiation facility at Emory University. Bull. Ga. Acad. Sci. 22(3/4): 75-80.

Platt, R. B., H. L. Ragsdale and D. J. Shure. 1973. An ecological approach to the evaluation of radioactivity within the man-environment ecosystem. pp. 9-17. In: Radionuclides in Ecosystems. (See Nelson, 1973).

Platt, R. B., et al. 1964. The importance of environment to life. BioScience 14(7): 25-29. (Incl. comments on nuclear aspects).

Platt, R. B., et al. 1973. Empirical benefits derived from an ecosystem approach to environmental monitoring of a nuclear fuel reprocessing plant. pp. 673-700. In: Environmental Behavior of Radionuclides Released in the Nuclear Industry. (See IAEA, 1973) (NSA 29: 29203).

Platt, R. B., et al. 1974. Environmental control in nuclear fuel reprocessing. Chem. Engrg. Progr. 70(4): 87-88. (NSA 30: 6496).

Plaut, W. and D. Nash. 1964. Localized DNA synthesis in polytene chromosomes and its implications. pp. 113-135. In: The Role of Chromosomes in Develoment. M. Locke, ed. Academic Press, New York. (^{3}H, Drosophila melanogaster).

Plecas, M., J. Bittner and V. Voinescu. 1969. Effect of gamma-irradiation on spores of some species of the Clostridium genus. Acta Biol. Med. Germ. (DDR) 23: 921-923. (NSA 24: 34788).

Plesset, M. S. and D. J. Dugas. 1966. Decay of a disturbance in the natural distribution of carbon-14. RAND Corp., Santa Monica, Calif., Memo. RM-5100-TAB. x, 42pp.

Plough, H. H. 1952. Radiation tolerances and genetic effects. Nucleonics 10(8): 16-20.

Plough, H. H. and C. W. Sheppard. (Eds.). 1954. Genetic effects of fast neutrons from nuclear detonations. U. S. AEC, Washington. 59pp. (Drosophila, Mormoniella).

Pluchennik, H. 1965. Mutation process of Chlorella induced by incorporation of radioactive carbon dioxide. Genetika (5): 19-25. (In Russian) (NSA 20: 22674).

Pluman, I. I. 1971. Uranium content in black argillites of the Volga stage of the West Siberian platform as criterion of geochemical conditions of sedimentation. Geokhimiya (9): 1138-1143. (In Russian) (NSA 26: 1998).

Plummer, G. L. 1958. Certain aspects of the effect of gamma radiation on the growth and competition exhibited by five domestic grasses. Agron. Abs. 1958: 57.

Plummer, G. L. 1960. Biometric analysis of a growth response of two plant species in a radioactive waste area. Oak Ridge National Lab., U. S. AEC report ORNL-2903. iii, 10pp. (NSA 14: 11442).

Plummer, G. L. 1962. Growth responses of Helianthus annuus to external calcium-45. Botan. Gaz. 123: 272-278. (NSA 17: 2840).

Plummer, G. L. 1963. Movement of radionuclides through granite rock outcrop ecosystems. Abstr. Bull. Ecol. Soc. Am. 44(3): 75.

Plummer, G. L. 1963. Radioactive fallout and vegetation in the Georgia Piedmont. Abstr. ASB Bull. 10: 36.

Plummer, G. L. 1964. Radioactive fallout in Georgia Piedmont. Abstr. Bull. Ga. Acad. Sci. 22: 2.

Plummer, G. L. 1966. Bioaccumulation of radionuclides on granitic outcrops in the Georgia Piedmont. Final report. Univ. of Georgia, Athens. 68pp. (NP-16464) (NSA 21: 19759).

Plummer, G. L. 1969. Fallout radioisotopes in Georgia lichens. pp 571-577. In: Symposium on Radioecology. (See Nelson and Evans, 1969).

Plummer, G. L. and F. Helseth. 1965. Movement and distribution of radionuclides on the granitic outcrops within the Georgia Piedmont. In: Radiation and Terrestrial Ecosystems. Health Physics 11(12): 1423-1428. (See Hungate, 1965).

Plummer, G. L. and J. B. Kethley. 1964. Foliar absorption of amino acids, peptides, and other nutrients by the pitcher plant, Sarracenia flava. Botan. Gaz. 125(4): 245-260. (Incl. ants, ^{32}P, ^{35}S).

Plummer, G. L. and J. B. Moncrief, Jr. 1964. Lichen growth on granitic flat rocks in Georgia. Bull. Ga. Acad. Sci. 22(3-4): 58-64. (Comment on ^{137}Cs)

Plummer, G. L., D. A. Crossley, Jr. and D. A. Gardiner. 1965. Growth response of two sedges inhabiting a radioactive waste disposal area. Ecology 46(3): 346-348. (NSA 19: 38331).

Plummer, G. L., T. M. Pullen, Jr. and E. E. Provost. 1969. Cesium-137 and a population of Georgia white-tailed deer. pp. 609-615. In: Symposium on Radioecology. (See Nelson and Evans, 1969).

Po-Chedley, D. S. 1964. Effect of x-rays on the meal worm, *Tenebrio molitor*, embryo. I. Changes in the protein and free α-amino nitrogen fraction. Final report for Contract AT(30-1)-2814. D'Youville Coll., U. S. AEC report NYO-2814-1. 7pp. (NSA 19: 73).

Po-Chedley, D. S. 1965. The effects of x-rays on the meal worm, *Tenebrio molitor*, embryo. p. 180. In: Proc. 12th Intern. Congr. Entomology. (London, 8-16 July 1964). P. Freeman, ed. Royal Entomological Soc., London.

Po-Chedley, D. S. 1966. The effect of x-rays and visible light upon the one day old meal worm, *Tenebrio molitor*, embryo. Abstr. 710. In: 3rd Intern. Congr. of Radiation Research. (Cortina d'Ampezzo, Italy, 26 June-2 July 1966). G. Silini, ed. North-Holland, Amsterdam. (AED-Conf-226-121).

Po-Chedley, D. S. 1969. Radio-sensitivity and water content for yellow mealworm embryos. J. Econ. Entomol. 62: 1505-1506. (NSA 24: 14576).

Podder, N. G., et al. 1970. Effect of x-irradiation on hydra in the presence of arginine and citrulline and study of the radiolytic products. Indian J. Exptl. Biol. 8: 37-38. (NSA 25: 2588).

Podolyan, V. Ya. 1957. The biological effect of radioactive cobalt (Co^{60}) on the preimaginal stages of the blue carrion fly *Protophorma terrae novae* R. D. In: Proc. 3rd Conf. All-Union Entomological Soc., USSR. (In Russian).

Podolyan, V. Ya. 1957. The biological effect of radioactive cobalt (Co^{60}) on the preimaginal stages of the carrion fly *Protophormia terrae novae* R. D. Tr. voyenno-med. Akad. RKKA 105: __. (In Russian).

Podolyan, V. Ya. 1958. The effect of radioactive cobalt emanation on different states of flies. In: Proc. 1st Zoological Conf. of the Belorussian SSR. (In Russian).

Podolyan, V. Ya. 1959. The biological effect of radioactive cobalt (Co^{60}) on the preimaginal stages of the spring carrion fly *Protophormia terrae novae* R. D. Tr. voyenno-med. ord. Len. Akad. S. M. Korov 105: __. (In Russian) (See 1957 above; same?).

Podolyan, V. Ya. 1963. Pathogenic effect of low and medium doses of gamma-rays on the progeny of irradiated insects. Tr. inst. zool. Akad. Nauk Kazakh. SSR 19: 220-225. (JPRS-22107; OTS-63-41271) (NSA 18: 5010).

Poelstra, P. and M. J. Frissel. 1967. Migration of water and ions in undisturbed soil columns and its description by simulation models. pp. 203-212. In: Isotope and Radiation Techniques in Soil Physics and Irrigation Studies. (See IAEA, 1967) (NSA 22: 23153).

Poelestra, P. and M. J. Frissel. 1971. The Sr-90 contamination of grass, based on some ecological parameters. pp. 1039-1055. In: Proc. Intern. Symp. Radioecology Applied to the Protection of Man and His Environment. (See Commission of the European Communities, 1971) (NSA 27: 5240).

Poelestra, P., et al. 1974. Behavior of mercury compounds in soils: accumulation and evaporation. pp. 281-292. In: Comparative Studies of Food and Environmental Contamination. (See IAEA, 1974) (NSA 30: 15471) (^{203}Hg).

Poet, S. E. and E. A. Martell. 1972. ^{239}Pu and ^{241}Am contamination in the Denver area. Health Physics 23(4): 537-548. (NSA 26: 56113).

Poet, S. E. and E. A. Martell. 1974. Reply to plutonium-239 contamination in the Denver area by P. W. Krey. Health Physics 26(1): 120-122. (NSA 29: 15598).

Pogosyan, R. I. 1962. The sensitivity of skin chromatophores to the ionizing radiation. Byull. Eksptl. biol. med. 53: 82-84. (In Russian) (NSA 16: 23444) (Frog).

Poglazova, M. N. 1958. Biological features of inactivation of microorganisms by radiation. The problem of sterilization by radiation. Inst. Mikrobiol. Akad. Nauk SSSR. (BA 44: 21846; Ref. zh. (biol.) 1959: 91766D).

Pohley, H. J. and H. Esser. 1958. On the state of mutant scales of the posterior wings of the meal moth *Ephestia kühniella* following irradiation of the pupae. Z. Vereblehre 89(5): 707-714. (In German).

Pohl-Rüling, J. and F. Scheminsky. 1972. Natural radiation environments of Badgastein, Austria, and its biological effects. pp. 393-419. In: Natural Radiation Environment II. Vol. 1. (See Adams, Lowder and Gesell, 1972) (NSA 32: 19720).

Pointel, J.-G. and M. Pham Van Sam. 1969. Effect of ionizing radiation on development of *Dermestes maculatus*. J. Stored Prod. Res. 5: 95-109. (In French) (NSA 24: 9978).

Poisson, J., B. Cahagnier and A. Guilbot. 1971. Gamma sensitivity of the main corn moulds conidia. Incidence on the radurization of corn. Mycopathol. Mycol. Appl. 45(3-4): 193-209. (In French) (NSA 26: 28348).

Poivilliers de la Queriere, F. 1960. Study of the content of iodine-127 of the thyroid and the plasma of 2 species of Columbae, the ringdove pigeon (*Columba palumbus*) and the domestic pigeon (*Columba livia* var. *domestica*) Compt. rend. 250(17): 2933-2935. (In French).

Pokidin, V. K., et al. 1972. Radioactivity and rate of sediment formation in the Caspian Sea. Geokhimiya (7): 834-843. (In Russian) (NSA 26: 58794).

Polach, H. A., et al. 1972. Correlation of ^{14}C activity of NBS oxalic with Arizona 1850 wood and ANU-sucrose radiocarbon dating standards: a preliminary report of investigations and results. pp. H90-H120. In: Proc. 8th Intern. Radiocarbon Dating Conf. (See Rafter, 1972) (NSA 31: 8719).

Poland Bureau of the Atomic Energy Ministry. 1959. First National Symp. on Use of Radioisotopes in Biology, Medicine, and Agriculture. (Warsaw, April 2-3, 1959). Warsaw. (In Polish) (NSA 15: 12692).

Poland Centrainy Osrodek Pomiarow Skazen Promieniotworoczych. 1969. Reports of the Conf. of Polish-German Working Group on Territorial Radioactive Fallout. (Warsaw, June 17, 1968). Warsaw. 146pp. (In Several languages) (CONF-680664) (NSA 24: 25127).

Polatnick, J. and H. L. Backrach. 1968. Ionizing radiation of foot-and-mouth disease virus and its ribonucleic acid. Arch. Gesamt. Virusforsch. 23: 96-104. (NSA 22: 52045).

Polezhayev, L. V. 1966. Mechanism of recovery of the regenerative capacity inhibited by x irradiation. Izv. Akad. Nauk SSSR (biol.) (2): 254-265. (In Russian) (NSA 20: 29038).

Polezhayev, L. V. 1966. Restoration of regererative capacity suppressed by x irradiation. Izv. Akad. Nauk SSSR (biol.) (1): 37-58. (In Russian) (NSA 20: 18337).

Polezhayev, L. V. and S. Ya. Tuchkova. 1968. Transplantation of heterologous tissue as a means of restoring axolotl limb regenerability after depression by x-irradiation. Dokl. Akad. Nauk SSSR 180: 754-757. (Doklady (biol. sci.) 180(1-6): 277-280) (NSA 22: 45688).

Polezhayev, L. V., N. A. Teplits and S. Ya. Tuchkova. 1962. Regeneration of axolotl extremities depressed by x-rays and prevention of x-ray induced ulcers. pp. 210-216. In: Recovery Processes in Radiation Injuries. (Meeting of the Acad. Med. Sci. USSR). N. A. Krayevskiy and A. V. Lebedinskiy, eds. Atomizdat, Moscow. (In Russian) (NSA 18: 38969).

Polezhayev, L. V., N. A. Teplits and S. Ya. Tuchkova. 1963. Factors affecting the restoration of regenerative capacity of the extremities of axolotls (larval salamanders), which had been suppressed by x-ray irradiation. Dokl. Akad. Nauk SSSR 150(3): 694-697. (Doklady (biol. sci.) 150(1-6): 690-693) (NSA 17: 33665).

Polezhayev, L. V., N. A. Teplits and S. Ya. Tuchkova. 1964. Regeneration ability restored with the aid of nucleic acid in axolotl extremities, inhibited by x-ray irradiation. Dokl. Akad. Nauk SSSR 159(3): 682-685. (Doklady (biol. sci.) 159(1-6): 798-801) (NSA 19: 26118).

Polezhayev, V. G., et al. 1962. Study of rodent migrations by means of tagging with radioactive isotopes. Tr. tsentr. nauch.-issled. dezinfektsio inst. 15: 295-299. (In Russian) (BA 45: 40842; Ref. zh. (biol.) 1963: 191161; WR 117: 25).

Polikarpov, G. G. 1957. Features of the reactions of radiation aftereffect in Pelmatohydra oligactis. pp. 34-35. In: Biochemical and Physico-Chemical Basis of Radiation.Effects. (February 23-28, 1957; Abstr. of Reports). Moscow State Univ., Moscow. (In Russian).

Polikarpov, G. G. 1957. On the problem of the development of the reactions of radiation aftereffects. Biofizika 2(2): 174-177. (Biophysics USSR 2: 176-179) (NSA 12: 2641).

Polikarpov, G. G. 1958. The accumulation of cerium radioisotopes by freshwater mollusks. Priroda (5): 86-87. (In Russian) (NSA 12: 13694).

Polikarpov, G. G. 1964. Radioecology of Marine Organisms. V. P. Shvedov, ed. Atomizdat, Moscow. {Radioecology of Aquatic Organisms. (English Translation and Revised Edition, 1966. V. Schultz and A. W. Klement, Jr., eds. North-Holland, Amsterdam; Reinhold, New York. xxviii, 314pp.} (NSA 21: 20115).

Polikarpov, G. G. 1965. The radioactivity of the hydrosphere and problems of the radioecology of aquatic organisms. pp. 344-345. In: Problems of Hydrobiology. Publ. House Acad. Sci. USSR, Moscow. (In Russian).

Polikarpov, G. G. 1967. Regularities of uptake and accumulation of radionuclides in aquatic organisms. pp. 819-826. In: Radioecological Concentration Processes. (See Aberg and Hungate, 1967).

Polikarpov, G. G. 1970. Unsolved problems of aquatic radioecology. Radiobiologiya 10(2): 242-252. (AEC-tr-7171, pp. 121-137) (NSA 24: 39473).

Polikarpov, G. G. 1971. Radioecology of hydrobionts. Priroda (10): 47-55. (In Russian) (NSA 26: 17742).

Polikarpov, G. G. (Ed.). 1972. Radiation and Chemical Ecology of Aquatic Organisms. "Naukova Dumka", Kiev. 118pp. (In Russian) (NSA 26: 53532).

Polikarpov, G. G. and D. S. Parchevskaya. 1967. All-union symposium on radioecology of fish eggs. Vopr. ikhtiol. 7(1): __. (In Russian).

Polikarpov, G. G. and V. P. Parchevskiy. (Eds.). 1972. Methods of Determination of Radioactivity. "Naukova Dumka", Kiev. 196pp. (In Russian) (NSA 29: 2461).

Polikarpov, G. G., V. I. Timoshchuk and V. P. Parchevskiy. 1967. Strontium-90 in the Danube River and in the adjacent international zone of the Black Sea. Gidrobiol. zh. 6: 68. (In Russian).

Polikarpov, G. G., A. Ya. Zesenko and A. A. Lyubimov. 1972. Dynamics of physico-chemical transformation of radionuclide polyvalent elements in the environment and their accumulation by aquartic organisms. pp. 5-42. In: Radiation and Chemical Ecology of Aquatic Organisms. (See Polikarpov, 1972) (In Russian).

Polikarpov, G. G., et al. 1967. Strontium-90 in reservoirs of the saltwater and freshwater type (1966). State Committee on the Use of Atomic Energy, USSR. Atomizdat, Moscow. 23pp. UN Document A/AC.82/G/L.1178. (AEC-tr-7214, pp. 161-186) (NSA 25: 15760).

Polikarpov, G. G., et al. 1967. Strontium-90 in algae, aquatic plants, mollusks, crustaceans and fish of the Black Sea (1965-1966). State Committee on the Use of Atomic Energy, USSR. Atomizdat, Moscow. 29pp. UN Document A/AC.82/G/L.1194. (In Russian) (NSA 22: 33805).

Polikarpov, G. G., et al. 1968. Biological migration of radionuclides in freshwater and saltwater reservoirs. State Committee on the Use of Atomic Energy, USSR. Atomizdat, Moscow. UN Document A/AC.82/G/L.1243. (AEC-tr-7128, pp. 6-13) (NSA 23: 24574; 24: 12025).

Polikarpov, G. G., et al. 1971. Radioecology and chemoecology in the service of the protection of nature. pp. 773-777. In: Proc. Intern. Symp. on Radioecology Applied to the Protection of Man and His Environment. (See Commission of the European Communities, 1971) (NSA 27: 5279).

Pollack, S. S., H. Lentz and W. Zeichman. 1971. X-ray diffraction study of humic acids. Soil Science 112(5): 318-324.

Polles, S. G. and S. B. Vinson. 1972. Penetration, distribution, and metabolism of ^{14}C-endrin in resistant and susceptible tobacco budworm larvae. J. Agric. Food Chem. 20(1): 38-41.

Polozhentsev, P. A. and L. P. Khvatova. 1972. Influence of ionizing irradiation on the sexual sterilization of the leopard moth. Radiobiologiya 12 (2): 313-315. (AEC-tr-7362, pp. 235-237) (NSA 27: 2818).

Polozhensev, P. A. and D. I. Tavolzhanskiy. 1969. Experimental sexual sterilization of cockchafer males by ionizing radiations. Radiobiologiya 9(6): 897-900. (AEC-tr-7145, pp. 155-160) (NSA 24: 17123).

Polyakov, I. M. and A. N. Dmiriyeva. 1955. New ways of investigating the processes of fertilization in higher plants with the aid of radioactive isotopes. J. Obcej. biologii 16: 3-16.

Polyakov, Yu. A. 1956. Problem of contamination of soil and agricultural products by radioactive disintegration. Pochvovedeniye (8): __. (In Russian).

Polyakov, Yu. A. 1959. The sorption of strontium and calcium ions by soils. Kolloid zh. 21: 221-225. (In Russian).

Polyakov, Yu. A., N. I. Anichkova and L. K. Leont'yev. 1964. Beta-activity of water, ooze, and vegetation in the Rybinsk water reservoir. In: Radioactive Contamination of the Seas and Oceans. V. I. Baranov and L. M. Khirov, eds. Oceanographic Commission, Acad. Sci. USSR. Publ. House Acad. Sci., Moscow. (AEC-tr-6641, pp. 64-71) (NSA 21: 8475).

Polyakov, Yu. A., V. F. Gol'tsov and V. G. Grakovskiy. 1968. Diffusion of strontium-90 in soils. State Committee on the Use of Atomic Energy, USSR. Atomizdat, Moscow. 15pp. UN Document A/AC.82/G/L.1242. (AEC-tr-7030, pp. 14-25) (NSA 23: 17550).

Polyakov, Yu. A., G. M. Kader and V. V. Krinitskiy. 1971. Behavior of strontium-90 and cesium-137 in soils. pp. 90-118. In: Contemporary Problems of Radiobiology. Vol. 2. Radioecology. (See Klechovskiy, Polikarpov and Aleksakhin, 1971) (English transl. pp. 78-102) (NSA 29: 12723).

Polyakov, Yu. A., L. N. Kalishina and L. F. Nazarova. 1967. Distribution of ^{90}Sr in the soil profile in the moderately northern latitudes of the U. S. S. R. pp. 39-42. In: Radioecological Concentration Processes. (See Aberg and Hungate, 1967).

Polyakov, Yu. A., A. M. Leont'yev and L. K. Mel'nikov. 1962. Certain aspects of Sr^{90} fallout in the middle latitudes of the USSR. Pochvovedeniye (11): 45-50. (Sov. Soil Science (11): 1256-1259) (Bryophytes, lichens).

Polymakhin, V. N. 1971. Natural radiation loads on the eggs of marine and freshwater organisms. Tr. polyar. nauch.-issled. proekt. inst. morsk. ryb. khoz. okeanogr. (29): 116-127. (AEC-tr-7418, pp. 142-156) (NSA 27: 28005).

Polymakhin, V. N. 1971. Radiometric and dosimetric characteristics of experiments for determining the influence of radioactivity of a water medium on the development of eggs of the Atlantic salmon. Tr. polyar. nauch.-issled. proekt. inst. morsk. ryb. khoz. okeanogr. (29): 91-103. (AEC-tr-7418, pp. 113-133) (NSA 27: 28004).

Polymakhin, V. N. 1971. Data on the dependence of dose effect for eggs of Atlantic salmon. Tr. polyar. nauch.-issled. proekt. inst. morsk. ryb. khoz. okeanogr. (29): 28-31. (AEC-tr-7418, pp. 31-35) (NSA 27: 28000).

Polzer, W. L. 1971. Solubility of plutonium in soil/water environments. A theoretical study. pp. 411-429. In: Proc. Rocky Flats Symp. on Safety in Plutonium Handling Facilities. (See Dow Chemical Co., 1971) (NSA 25: 54756).

Pomerantseva, M. D. 1957. Effect of anesthesia and natural hibernation on radiosensitivity in animals. Zh. obshch. biol. 18(3): 194-207. (AEC-tr-3611) (NSA 12: 9646; 13: 10794).

Pomeroy, D. L. 1971. Releases of tritium from nuclear facilities. Trans. Amer. Nucl. Soc. 14(1): 160. (CONF-710606) (NSA 25: 41469).

Pomeroy, L. R. 1960. Residence time of dissolved phosphate in natural waters. waters. Science 131(3415): 1731-1732.

Pomeroy, L. R. 1963. Isotopic and other techniques for measuring benthic primary production. pp. 97-102. In: Proc. Conf. on Primary Productivity Measurement, Marine and Freshwater. (See Doty, 1963).

Pomeroy, L. R. 1970. The strategy of mineral cycling. pp. 171-190. In: Annual Review of Ecology and Systematics. R. F. Johnston, P. W. Frank and C. D. Michener, eds. Annual Reviews, Palo Alto. (Incl. radionuclides).

Pomeroy, L. R. and H. H. Haskin. 1954. The uptake and utilization of phosphate ions from sea water by the American oyster Grassostrea virginica (Gmel.). Biol. Bull. 107(3): 123-129.

Pomeroy, L. R., E. E. Smith and C. M. Grant. 1965. The exchange of phosphate between estuarine water and sediments. Limnol. Oceanogr. 10(2): 167-172.

Pomeroy, L. R., et al. 1966. Flux of ^{32}P and ^{65}Zn though a salt-marsh ecosystem. pp. 177-188. In: Disposal of Radioactive Wastes into Seas, Oceans and Surface Waters. (See IAEA, 1966) (SM-72/10) (CONF-660507) (NSA 20: 40573).

Pomeroy, L. R., et al. 1969. The phosphorus and zinc cycles and productivity of a salt marsh. pp. 412-419. In: Symposium on Radioecology. (See Nelson and Evans, 1969).

Pomoshchnikova, N. A. 1955. Study of phosphorus exchange in the organisms of the yeasts. In: Isotopes in Microbiology. (See Imshenetzkiy, 1955) (AEC-tr-3494. pp. 95-105) (^{32}P).

Poonawalla, Z. T. and M. S. Quraishi. 1970. Dose dependence of the rate of penetration of topically applied DDT-C^{14} into the house fly. Can. Entomol. 102(9): 1136-1138.

Popa, A. and Gh. Mihalache. 1965. On the radioactive isotope utilization for forest protection. Rev. Padurilor (Bucharest) 80(2): 59-62. (In Rumanian) (NSA 21: 6472).

Popenoe, H. and C. F. Eno. 1962. The effect of gamma radiation on the microbial population of the soil. Soil Sci. Soc. Am. Proc. 26(2): 164-167. (NSA 18: 38899).

Popham, T. W. and S. J. Ursic. 1969. Computer program for converting neutron probe readings to soil water equivalents. Soil Science 107: 302. (NSA 23: 21819).

Popov, A. S. 1963. The influence of urea peroxide on the post-irradiation recovery of infusoria. Radiobiologiya 3(1): 121-125. (AEC-tr-5434, pp. 170-176) (NSA 17: 30418).

Popov, A. S. 1965. Combined action of H_2O_2 and nucleic compounds on infusoria. Radiobiologiya 5(5): 652-655. (AEC-tr-6602, pp. 28-33) (NSA 20: 6709).

Popov, A. S. and Yu. P. Vinetskiy. 1967. Effect of irradiated glucose solution on phage and infusoria. Radiobiologiya 7(4): 630-631. (AEC-tr-6889, pp. 235-240) (NSA 22: 15122).

Popov, D. V. 1969. Protective effect of cystamine on the induction capacity of the irradiated eye. Vestn. Mosk. Univ. (biol., pochv.) (1): 110-111. (NSA 23: 48203) (Rana temporaria tadpoles).

Popov, V. V. and A. I. Faberov. 1960. The effect of irradiation on the capacity of the eye for cornea induction. Dokl. Akad. Nauk SSSR 132(3): 719-722. (Doklady (biol. sci.) 132: 395-397) (NSA 14: 20093).

Popov, V. V. and V. A. Golichenkov. 1964. Resistance of triton lens to radiation and trauma. Nauchn. dokl. vyssh. shkoly (biol.) (3): 82-86. (JPRS-28782, pp. 36-41) (NSA 19: 21866).

Popov, V. V. and V. V. Ul'yanova. 1967. Stimulating and depressing effect of x-rays on the functional state of the retina and on its ability to induce cornea formation. Dokl. Akad. Nauk SSSR 177: 728-731. (Doklady (biol. sci.) 177(1-6): 753-756) (NSA 22: 33926).

Popov, V. V., O. P. Melikhova and Yu. P. Kozlov. 1965. The relation between certain physical and chemical changes in irradiated skin and its ability of rapid transformation into cornea. Dokl. Akad. Nauk SSSR 165(1): 241-244. (Doklady (biol. sci.) 165(1-6): 724-727) (NSA 20: 32986).

Popova, O. N. and R. P. Kodaneva. 1965. Regarding some modes of participation of radium and uranium in the mineral exchange of substances in plants. pp. 44-46. In: Materials of a Scientific Conf. of the Komi Republic Youth. (Abstr. of Reports). Syktyvkar. (In Russian).

Popova, O. N. and A. N. Kyrchanova. 1971. Sap analysis in investigations of ^{226}Ra movement between soil and plant. Ekologiya 2(6): 33-37. (Sov. J. Ecology 2(6): 511-513).

Popova, O. N. and M. B. Sokolov. 1971. Statistical models and estimation criteria of results of experiments on growing of plants on plots with elevated natural radioactivity. pp. 215-223. In: Methods of Radioecological Investigations. (See Verkhovskaya, 1971) (In Russian).

Popova, O. N., R. P. Kodaneva and P. P. Vavilov. 1964. On the distribution in plants of radium absorbed from the soil. Zh. fiziol. rasteniy SSSR 11(3): 436-441. (In Russian).

Popova, O. N., A. I. Taskayev and B. V. Testov. 1974. Aerial uptake of ^{210}Po by plants on plots with elevated radium content. Ekologiya 5(6): 22-28. (Sov. J. Ecology 5(6): 522-526) (NSA 33: 12466).

Popova, Ye. I. 1960. Accumulation of radioactive phosphorus by several freshwater mollusks. Tr. Komi fil. Akad. Nauk SSSR (9): __. (In Russian).

Popova, Ye. I. 1965. Concentration of radium by various water plants when its content in natural waters is above normal. pp. 349-350. In: Problems of Hydrobiology. Publ. House Acad. Sci. USSR, Moscow. (In Russian).

Pora, A. E. and A. I. Vertunov. 1958. Effect of the external salt concentration on the penetration of $P^{32}O_4HNa_2$ through the branchiae of Rutilus rutilus. Proc. 2nd Intern Conf. on the Peaceful Uses of Atomic Energy 24: 112-113. (Paper A/CONF.15/P/1290) (NSA 13: 6169).

Pora, E. A. and St. Manciulea. 1972. Action of very small doses of raditions on certain enzymes in Rana esculenta. Acad. rep. pop. Romine fil. Cluj Studii cerc. biol. 24(2): 139-142. (In Rumanian) (NSA 28: 2959).

Pora, E. A., St. Manciulea and A. Suteu-Rusdea. 1968. The action of very low gamma-ray doses on frog liver tissue respiration during spring recovery. Rev. Roum. biol. (zool.) 13: 193-195. (In French) (NSA 23: 43875).

Pora, E., et al. 1965. Investigation of ^{32}P-uptake and clearance in some brackish organisms of the Agigea and the Eforie lakes. Acad. rep. pop. Romine fil. Cluj Studii cerc. biol. (zool.) 17(4): 329-337. (In Rumanian).

Porcella, D. B. and A. G. Friend. 1966. Field studies of specific radionuclides in fresh water. J. Water Poll. Control Fed. 38(1): 102-110. (NSA 20: 33581).

Porcella, D. B., C. E. Rixford and J. V. Slater. 1967. Factors influencing radiostrontium accumulation in Daphnia. Health Physics 13: 391-399. (NSA 21: 17985).

Porte, A., J. P. Zahnd and A. Batzenschlager. 1963. The ultrstructure aspects of the liver in Lacerta stirpium and Accara latifrons after intraperitoneal injection of thorium dioxide (Thorotrast). Compt. rend. soc. biol. 157: 1791-1795. (In French) (NSA 18: 25066) (Lizard, fish).

Portela, A., et al. 1971. Radiation effects on water permeability and distribution in frog muscle cells. Radiation Research 47(3): 704-715. (NSA 25: 55323).

Porter, C. A. and E. G. Jaworski. 1965. Biosynthesis of chitin during various stages in the metamorphosis of Prodenia eridania. J. Insect Physiol. 11(9): 1151-1160. (^{14}C).

Porter, C. R., et al. 1965. Procedures for determination of stable elements and radionuclides in environmental samples. U. S. PHS, Washington, Publ. 999-RH-10. vii, 83pp.

Porter, J. W. and H. J. Knauss. 1954. Inhibition of growth of Chlorella pyrenoidosa by β-emitting radioisotopes. Plant Physiol. 29: 60-63. (Also: Hanford Atomic Products Operation, U. S. AEC report HW-29666) (NSA 8: 975).

Porter, J. W. and H. J. Knauss. 1954. The effect of tritium oxide on some synthetic processes of Chlorella pyrenoidosa. Radiation Research 1(3): 253-261. (Also: Hanford Atomic Products Operation, U. S. AEC report HW-30252) (NSA 8: 2086).

Porter, J. W. and M. S. Watson. 1954. Gross effects of growth-inhibiting levels of tritium oxide on Chlorella pyrenoidosa. Am. J. Bot. 41(7): 550-555. (Also: Hanford Atomic Products Operation, U. S. AEC report HW-30056) (NSA 8: 1498).

Porter, N. S., N. C. Hoitink and C. E. Cushing. 1966. Measurement of radionuclide uptake by algae. p. 159. In: Proc. 19th Ann. Conf. Engineering in Medicine and Biology. Vol. 8. (San Francisco, Calif., November 14-17, 1966). G. G. Vurek, ed. Inst. Electrical and Electronics Engineers, New York, and Instrument Soc. of Amer., Pittsburgh. (NSA 22: 43275).

Porter, R. P., M. Vuksanovic and M. Dauer. 1966. Radiosensitivity of the tamarin to whole-body irradiation. Radiology 87: 74. (NSA 20: 33040).

Porter, S. W., Jr. 1973. Monitoring of tritium in the aquatic environment of power reactors. pp. 517-521. In: Tritium. (See Moghissi and Carter, 1973) (NSA 28: 27283).

Porter, S. W., Jr. 1975. Radiological environmental surveillance philosophy and implementation for US nuclear power stations. pp. 183-185. In: Radiation Protection Measurement: Philosophy and Implementation. (See Recht and Lakey, 1975) (NSA 33: 26194).

Portland General Electric Company. 1974. Terrestrial ecology baseline study. Trojan Nuclear Plant, Oregon. Final report 1972-1973. Portland, Ore., report. 230pp. (PB-233322) (NSA 31: 3407).

Porz, F. 1973. Production and release of radioiodines by nuclear power plants and reprocessing plants and the expected radiological burden till the year 2000. 82pp. (In German) (NSA 30: 9498).

Posner, H. B. 1965. Effects of gamma irradiation on growth, colony-forming ability and some cellular constituents of Chlorella pyrenoidosa. Radiation Botany 5(2): 129-141. (NSA 19: 24334).

Posner, H. B. and A. H. Sparrow. 1964. Survival of Chlorella and Chlamydomonas after acute and chronic gamma radiation Radiation Botany 4(3): 253-257. (Also: Brookhaven National Lab., U. S. AEC report BNL-7426) (NSA 19: 5685).

Posselt, H. S. 1971. Environmental chemistry of cadmium in aqueous systems. Thesis, Univ. of Michigan, Ann Arbor. 189pp. (Univ. Microfilms 72-7495) (NSA 26: 22766) (^{115m}Cd).

Postlethwait, J. H. and K. R. Girton. 1974. Development in genetic mosaics of aristapedia. Genetics 76(4): 767-774. (NSA 30: 9934) (X-ray).

Postlethwait, J. H. and H. A. Schneiderman. 1973. Pattern formation in imaginal discs of Drosophila melanogaster after irradiation of embryos young larvae. Develop. Biol. 32(2): 345-360. (NSA 28: 21779).

Postlethwait, S. N. and B. Rogers. 1958. Tracing the path of the transpiration stream in trees by the use of radioactive isotopes. Am. J. Bot. 45(10): 753-757.

Poswillo, D. E., W. J. Hamilton and D. Sopher. 1972. Marmoset as an animal model for teratological research. Nature 239(5373): 460-462. (NSA 27: 2805) (Incl. x-irradiation).

Potter, J. L. and R. D. O'Brien. 1964. Parathion activation by livers of aquatic and terrestrial vertebrates. Science 144: 55-57. (^{3}H).

Potter, L. D. and M. Marr. 1969. Cesium-137 concentrations in Alaskan arctic tundra vegetation, 1967. Arctic Alpine Res. 1(3): 147-154.

Potts, W. H. 1957. The effect of gamma rays on reproductive organs of female Glossina morsitans (Westw.). Trans. Roy. Soc. Trop. Med. Hyg. 51: 292.

Potts, W. H. 1957/58. Some effects of gamma rays on tsetse pupae (Glossina morsitans Westw.). Roy. Entomol. Soc. (London) Proc. ser. C. J. Mtg. 22(1): 2.

Potts, W. H. 1958. Sterilization of tse-tse flies (Glossina) by gamma radiation. Ann. Trop. Med. Parasit. 52(4): 484-499.

Potts, W. H. 1958. Irradiation and the control of insect pests. Times Sci. Rev. 28: 13-14, 17.

Potts, W. H. 1965. The gamma irradiation of Glossina puparial stages and control. pp. 254-255. In: Proc. 12th Intern. Congr. of Entomology. (London, 8-16 July 1964). P. Freeman, ed. Royal Entomological Soc., London.

Potts, W. T. W. and P. P. Rudy, Jr. 1969. Water balance in the eggs of the Atlantic salmon Salmo salar. J. Exptl. Biol. 50(1): 223-237. (Used tritiated water).

Potzl, K. and R. Reiter. 1960. Quantitative determinations of nuclear fission products deposited in northern Alpine regions. Atomkernenergie 5(7/8): 285-287. (In German) (NSA 14: 21819).

Poulsen, D. F. and V. T. Bowen. 1952. Organization and function of the inorganic constituents of nuclei. Exptl. Cell. Res. (Suppl. 2)::161-180. (Review; Drosophila; ^{59}Fe, ^{64}Cu).

Poulsen, D. F. and V. T. Bowen. 1951. The copper metabolism of Drosophila. Science 114(2967): 486. (^{64}Cu).

Poulsen, D. F., et al. 1952. The copper metabolism of Drosophila. Proc. Natl. Acad. Sci. (US) 38: 912-921. (Also: Brookhaven National Lab., U. S. AEC report BNL-1027) (NSA 6: 1114).

Povelyagina, Z. S. and M. M. Telitchenko. 1959. Concentration of the radioactive isotopes of phosphorus and strontium by various species of freshwater mollusks. Byull. Mosk. obshch. ispyt. prirody (biol.) 64(2): 79-83. (In Russian).

Povinec, P. nd. Contamination with tritium of nuclear power reactor environments. Komenskeho Univ., Bratislava. (In Slovak) (INIS-mf-887, pp. 221-234) (NSA 29: 15611).

Povinec, P. 1973. Tritium in nature. I. Origin, distribution, and methods of counting. Acta Fac. Rerum Natur. Univ. Comen. Phys. 13: 251-275. (NSA 31: 3359, 33125).

Povinec, P., M. Chudy and S. Saro. 1973. Tritium in nature. II. Experimental equipment for low tritium activity measurements. Acta Fac. Rerum Natur. Univ. Comen. Phys. 13: 277-288. (NSA 31: 32517).

Power Authority of the State of New York. 1968. James A. Fitzpatrick Nuclear Power Plant. Preliminary Safety Analysis Report, Vol. 3. 599pp. U. S. AEC DOCKET-50333-3. (NSA 23: 11109) (Incl. site environmental studies).

Power Authority of the State of New York. 1971. James A. Fitzpatrick Nuclear Plant. Environmental report, operating license stage. 161pp. U. S. AEC DOCKET-50333-36. (NSA 25: 36943).

Powers, E. L., Jr. 1948. Death after autogamy in Paramecium aurelia following exposure in solution to the radioactive isotope P^{32} and $Sr^{89,90}$, Y^{90}. Abstr. Genetics 33(1): 120-121.

Powers, E. L. 1955. Radiation effects in Paramecium. Ann. N.Y. Acad. Sci. 59: 619-637. (NSA 9: 3014).

Powers, E. L. 1965. Some recent advances in the radiation biology of bacteria. Radiol. Clin. N. Am. 3: 197-208.. (NSA 20: 3776).

Powers, E. L. and D. Shefner. 1950. Effects of high dosages of x-rays in Paramecium aurelia. Genetics 35: 131.

Pradhan, S. et al. 1971. Feasibility of controlling stored grain pests by the sterile-male technique. pp. 365-371. In: Sterility Principle for Insect Control or Eradication. (See IAEA, 1971) (NSA 25: 29750).

Prantl, F. A., R. M. Koerner and E. Robertson. 1974. Nuclear techniques for snow and ice studies in Canadian subpolar regions (Devon Island). pp. 632-641. In: Advanced Concepts and Techniques in the Study of Snow and Ice Resources. (See Santeford and Smith, 1974) (NSA 33: 7312).

Prasad, A. B. 1968. Histological changes induced in shoot apical meristems of canary grass after irradiation. Nucleus (Calcutta) 11: 75-78. (NSA 23: 33906).

Prasad, B. N. and M. B. E. Godward. 1968. Cytology of irradiated Mougeotia sp. Nucleus (Calcutta) 11: 43-49. (NSA 23: 33907).

Prasad, K. N. 1968. Necturus maculosus as an experimental animal to study cellular injuries of the irradiated gut in vivo. Monogr. Nucl. Med. Biol. (1): 377-384. (NSA 22: 49555).

Prasad, K. N. and H. A. Johnson. 1968. Differential radiosensitivity of the hepatic melanocytes of Amphiuma tridactylum. Radiation Research 33(2): 403-416. (Also: Brookhaven National Lab., U. S. AEC report BNL-12119) (NSA 22: 21580).

Praslicka, M. and M. Hill. 1959. The effect of a low temperature on the development of radiation injury in frogs. Folia Biol. (Prague) 5: 348-354. (In Russian) (NSA 14: 4234).

Praslicka, M., et al. 1955. The influence of the simultaneous effect of x-ray irradiation on the mortality of tadpoles. Folia Biol. (Prague) 1: 375-380. (In Russian) (NSA 13: 19789).

Pratesi, A. C. and C. Vassalo. 1963. Hepatic lesions produced by irradiation of the animal in hibernation. Histological and histochemical study. Ann. Radiol. Sem. hop. 6(11/12): R906-R917. (In French) (NSA 18: 13472).

Pravdina, G. M. 1965. Behavior of fish in the area of radiation sources. Byull. eksper. biol. med. 60(11): 88-91. (Bull. Exptl. Biol. Med. 60: 1305-1308) (NSA 20: 40656).

Pravdina, G. M. 1966. Behavior of fish in a radiation zone. pp. 97-103. In: Problems of General Radiobiology. M. P. Domshlak, ed. Atomizdat, Moscow. (In Russian) (NSA 22: 23663) (Cyprinus carpio).

Pravdina, G. M. and I. B. Zagorskaya. 1965. Seasonal variations of radiation sensitivity of rabbits. USSR Inst. of Atomic Energy, State Committee on the Use of Atomic Energy, Moscow. 9pp. (NSA 20: 42848).

Preece, A. W. 1969. Gamma irradiation of Stentor polymorphus. J. Protozool. 16: 289-294. (NSA 23: 48190).

Preiss, K. 1967. Design of a neutron scattering water content gage for soils. pp. __. In: Isotope Techniques in the Hydrologic Cycle. (See Stout, 1967) (CONF-651108-4) (AED-CONF-65-347-5) (NSA 20: 39141).

Preiss, K. 1970. Relation between energy and error due to nuclear statistics for density measurement by gamma ray transmission. Soil Science 110 (3): 151-___.

Premvati. 1965. Some observations on the effect of gamma radiations on the eggs of _Hoemonchus contortus_ (Rudolphi, 1803). Current Science (India) 34: 380. (NSA 19: 42281).

Preobrazhenskaya, Ye. I. 1967. Radiation resistance of plants and phylogeny. Dokl. Akad. Nauk SSSR 177: 243-245. (In Russian) (NSA 22: 23682).

Preobrazhenskaya, Ye. I. 1967. The relationship of radiosensitivity of plants with phylogenetic age. Izv. Sibirsk. otd. Akad. Nauk SSSR (biol.-med.) (5): 79-84. (In Russian) (NSA 22: 23675).

Preobrazhenskaya, Ye. I. 1971. Radioresistance of higher plants and phylogenetic relations. pp. 52-58. In: _Radioactive Isotopes and Nuclear Radiation in Agricultural Research_. Atomizdat, Moscow. (In Russian) (NSA 30: 18900).

Preobrazhenskaya, Ye. I. 1971. _Radiation Sensitivity of Plant Seeds_. Atomizdat, Moscow. 232pp. (In Russian) (NSA 26: 7113).

Preobrazhenskaya, Ye. I. 1971. Significance of eco-geographical factors for plant radiosensitivity. pp. 79-83. In: _Methods of Radioecological Investigations_. (See Verkhoskaya, 1971) (In Russian).

Preobrazhenskaya, Ye. I. and N. V. Timofeyev-Resovskiy. 1962. The correlation between germination and survival in different species of cultivated plants after irradiation of the seed with various doses of Co^{60} γ-rays. Dokl. Akad. Nauk SSSR 143: 448-451. (In Russian) (NSA 16: 20167).

Preobrazhenskaya, Ye. I. and N. V. Timofeyev-Resovskiy. 1962. A possible connection between radiation resistance and the phylogenetic system in cultivated plants. Dokl. Akad. Nauk SSSR 143(5): 1219-1221. (In Russian) (NSA 16: 21169).

Press, E. 1972. Nuclear and fossil-fueled plants generating electricity. Their environmental health hazards. J. Am. Med. Assoc. 222(10): 1281-1283. (NSA 27: 11974).

Pressman, M. and P. B. Pruett. 1961. The SM-1 environmental radiological monitoring program, November 1954-December 1960. U. S. Army Engineer Research and Development Labs., Ft. Belvoir, Va., report 1691-TR. 36pp.

Preston, A. 1959. Leukocytosis in response to bacteraemia as a feature of the acute radiation syndrome in the plaice. Nature 183(4664): 832-833.

Preston, A. 1966. Site evaluations and the discharge of aqueous radioactive wastes from civil nuclear power stations in England and Wales. pp. 725-737. In: _Disposal of Radioactive Wastes into Seas, Oceans and Surface Waters_. (See IAEA, 1966) (SM-72/146) (CONF-660507-28) (NSA 20: 47090).

Preston, A. 1968. Control of radioactive pollution in a North Sea oyster fishery. Helgolaender Wiss. Meersunters. 17: 269-279. (NSA 25: 26742).

Preston, A. 1969. Aquatic monitoring programmes. pp. 309-323. In: _Environmental Contamination by Radioactive Materials_. (See IAEA, 1969).

Preston, A. 1971. The United Kingdom approach to the application of ICRP standards to control disposal of radioactive waste resulting from nuclear power programs. pp. 147-156. In: _Environmental Aspects of Nuclear Power Stations_. (See IAEA, 1971).

Preston, A. 1972. Critical path analysis applied to the control of radioactive waste disposal to aquatic environments. pp. 121-134. In: _Disposal of Radioactive Waste_. (See Nuclear Energy Agency, 1972).

Preston, A. 1974. Application of critical path analysis techniques to the assessment of environmental capacity and the control of environmental waste disposal. pp. 573-583. In: _Comparative Studies of Food and Environmental Contamination_. (See IAEA, 1974) (NSA 30: 15540).

Preston, A. and D. F. Jefferies. 1969. Aquatic aspects in chronic and acute contamination situations. pp. 183-209. In: _Environmental Contamination by Radioactive Materials_. (See IAEA, 1969).

Preston, A. and D. F. Jefferies. 1969. The I. C. R. P. critical group concept in relation to the Windscale sea discharges. Health Physics 16: 33-46. (NSA 23: 6018).

Preston, A., J. W. Dutton and B. R. Harvey. 1968. Detection, estimation and radiological significance of silver-110m in oysters in the Irish Sea and the Black-Water Estuary. Nature 218(5142): 689-690.

Preston, A., D. F. Jefferies and J. W. R. Dutton. 1967. The concentration of ^{137}Cs and ^{90}Sr in the flesh of brown trout taken from rivers and lakes in the British Isles between 1961 and 1966: the variables determining the concentrations and their use in radiological assessments. Water Research 1 (17): 475-496. (NSA 22: 10773, 28460).

Preston, E. M. 1971. The importance of ingestion in chromium-51 accumulation by _Crassostrea virginica_ (Gmelin). J. Exptl. Mar. Biol. Ecol. 6(1): 47-54.

Preston, R. J., J. G. Brewen and N. Gengozian. 1974. Persistence of radiation-induced chromosome aberrations in marmoset and man. Radiation Research 60(3): 516-524. (NSA 31: 8917).

Pretorius, R. and R. E. Wainerdi. 1970. Nuclear activation analysis for ultrasensitive determinations of environmental pollutants. S. Afr. Med. J. 44: 169-171. (NSA 24: 38793).

Priadceneu, A. 1964. Utilization of ionizing radiation and radioactive isotopes in agriculture and forestry investigations in Rumania, 1958-1964. pp. 83-90. In: _Peaceful Uses of Radioisotopes in the Rumanian People's Republic, 1958-1964_. Rumanian Committee for Nuclear Energy, Bucharest. (NP-15543) (NSA 20: 5235).

Pribyl, J. and J. Napravnik. nd. Radioiodine removal from waste water after a reactor accident. Ceskoslovenska Akad. Ved, Rez, Ustav Jaderneho Vyzkumu, report UJV-2360.Ch. 11pp. (In Czech) (NSA 25: 23654).

Price, G. M. 1963. The effects of anoxia on metabolism in the adult housefly, _Musca domestica_. Biochem. J. 86: 372-378. (^{14}C).

Price, G. M. 1965. Aspects of amino acid and nucleic acid metabolism in the larva of the blowfly Calliphora erythrocephala. Abstr. Biochem. J. 95 (3): 39P-40P. (^{14}C).

Price, G. M. 1965. The incorporation in vitro of l-valine into the fat body protein of the larva of the blowfly, Calliphora erythrocephala. Abstr. Biochem. J. 97(3): 33P-34P. (^{14}C).

Price, H. L. 1970. Regulatory program of the U. S. Atomic Energy Commission. pp. 105-116. In: Northwest Conf. on the Role of Nuclear Energy. (See Governor's Nuclear Development Coordinating Committee, Oregon, 1970) (NSA 25: 23219).

Price, K. R. 1965. A field method for studying root systems. In: Radiation and Terrestrial Ecosystems. Health Physics 11(12): 1521-1525. (See Hungate, 1965).

Price, K. R. 1965. Uptake of ^{131}I from the soil as related to root distribution and phenology in sagebrush vegetation. Thesis, Washington State Univ., Pullman. 92pp. (NSA 20: 28883).

Price, K. R. 1971. Critical review of biological accumulation, discrimination, and uptake of radionuclides important to waste management practices, 1943-1971. Battelle Northwest Lab., U. S. AEC report BNWL-B-148. 67pp. (NSA 26: 23416).

Price, K. R. 1972. Uptake of ^{237}Np, ^{239}Pu, ^{241}Am, and ^{244}Cm from soil by tumbleweed and cheatgrass. Battelle Northwest Lab., U. S. AEC report BNWL 1688. (NSA 27: 5570).

Price, K. R. 1973. Tumbleweed and cheatgrass uptake of transuranium elements applied to soil as organic acid complexes. Battelle Northwest Lab., U. S. AEC report BNWL-1755. (NSA 28: 21108).

Price, K. R. 1973. A review of transuranic elements in soils, plants, and animals. J. Environ. Qual. 2(1): 62-66. (NSA 30: 12703).

Price, K. R. and W. H. Rickard. 1973. Vascular plants of waste storage sites in the 200 areas of the Hanford Reservation. Battelle Northwest Lab., U. S. AEC report BNWL-1796. 8pp. (NSA 29: 18379).

Price, M. A. and W. H. Newton. 1962. The screwworm and its control. Southwest Vet. 16(1): 43-49.

Price, T. J. 1965. Accumulation of radionuclides and the effects of radiation on mollusks. pp. 202-210. In: Biological Problems in Water Pollution. (See Tarzwell, 1965).

Prikhod'ko, L. S. and L. K. Klyshev. 1971. Effect of x-radiation on some aspects of plant metabolism under conditions of substrate salinization. Tr. inst. bot. Akad. Nauk Kaz. SSR 29: 86-95. (In Russian) (NSA 29: 614).

Pristavko, V. P. and G. S. Orgel. 1970. Sensitivity of male and female codling moths to x and gamma radiation with various dose rates. Radiobiologiya 10: 878-881. (AEC-tr-7225, pp. 110-116) (NSA 25: 29612).

Pristavko, V. P. and G. S. Orgel. 1972. Influence of x irradiation on the hemolymph and phagocytosis in larvae of the codling moth Laspeyresia pomonella. Radiobiologiya 12(2): 307-310. (AEC-tr-7362, pp. 225-230) (NSA 27: 2817).

Pristavko, V. P. and N. Ye. Sem'yanova. 1973. Circadian rhythym of the radiosensitivity of the codling moth Laspeyresia pomonella L. (Lepidoptera, Tortricidae). Radiobiologiya 13(2): 286-289. (AEC-tr-7444, pp. 159-162) (NSA 28: 24846).

Pristavko, V. P., L. V. Sherman and V. N. Chaika. 1974. Influence of x-irradiation on olfaction in the codling moth Laspeyresia pomonella L. Radiobiologiya 14(1): 136-139. (AEC-tr-7548, pp. 173-176).

Prister, B. S. 1969. Behavior of uranium in the biological chain State Committee on the Use of Atomic Energy USSR. Atomizdat, Moscow. UN Document A/AC.82/G/L.1298. (AEC-tr-7128, pp. 194-207) (NSA 24: 12033; 25: 52010).

Prister, B. S. and S. S. Zubach. 1968. Use of arsenazo-III in U determination in soils and biological media. Radiokhimiya 10(6): 743-746. (Sov. Radiochemistry 10(6): 733-736) (NSA 23: 13646).

Pritchard, D. W. 1958. Factors affecting the dispersal of fission products in estuarine and inshore environmnets. Proc. 2nd Intern. Conf. on the Peaceful Uses of Atomic Energy 18: 410-413. (Paper A/CONF.15/P/1835) (NSA 12: 14536).

Pritchard, D. W. 1960. Problems related to disposal of radioactive wastes in estuarine and coastal waters. pp. 22-32. In: Biological Problems in Water Pollution. (See Tarzwell, 1960).

Pritchard, D. W. and J. H. Carpenter. 1960. Measurements of turbulent diffusion in estuarine and inshore waters. Int. Assoc. Sci. Hydrol. Publ. 51: 361.

Pritchard, D. W. and A. B. Joseph. 1963. Disposal of radioactive wastes: its history, status and possible impact on the environment. pp. 27-31. In: Radioecology. (See Schultz and Klement, 1963) (NSA 17: 33544).

Pritchett, W. L. and W. K. Robertson. 1960. Problems relating to research in forest fertilization with southern pines. Soil Sci. Soc. Am. Proc. 24(6): 510-512. (Used radionuclides).

Privalov, G. F. 1963. Experimental mutations of the vegetative organs of arboreal plants. Dokl. Akad. Nauk SSSR 150(3): 661-664. (Doklady (biol. sci.) 150: 718-720).

Privalov, G. F. 1963. Radiosensitivity of seeds of some species of woody plants. Radiobiologiya 3(5): 770-772. (AEC-tr-5438, pp. 227-232) (NSA 18: 23312).

Privalov, G. F. 1964. Morphological changes in pine induced by treatment of the seed with gamma rays and colchicine. Izv. Sibirsk. otd. Akad. Nauk SSSR (8): biol. med. (2): 98-102. (In Russian) (NSA 19: 10829).

Privalov, G. F. 1968. Investigations of experimental mutagenesis in arboreous plants. Genetika 4(6): 144-157. (In Russian) (NSA 22: 45499).

Privalov, G. F. 1968. Mutability in box-elder (*Acer negundo* L.) induced by gamma and x-rays. Genetika 4(5): 39-42. (In Russian) (NSA 22: 40624).

Privalov, G. F. 1972. Effect of combined treatment of seeds with mutagens and growth regulators on viability and growth of *Acer negundo* L. seedlings. Genetika 8(2): 36-45. (In Russian) (NSA 26: 33664).

Prochazka, H. and R. Jilek. 1971. Determination of environmental radioactivity by liquid-scintillation spectrometry. pp. 373-380. In: *Rapid Methods for Measuring Radioactivity in the Environment*. (See IAEA, 1971) (NSA 26: 27820).

Prochazka, H., et al. nd. Mathematical modelling of radionuclide migration in certain regions of the biosphere. (In Czech) (INIS-mf-887-1, pp. 46-73) (NSA 30: 3156).

Procházka, Z. and W. Mulligan. 1965. Immunological studies on *Nippostrongylus brasiliensis* infection in the rat: experiments with irradiated larvae. *Exptl. Parasitol.* 17: 51-56. (NSA 21: 16485).

Procházka, Z. and J. Tománek. 1968. Some practical aspects on the production of radiation vaccines against helminthic diseases. pp. 21-34. In: *Isotopes and Radiation in Parasitology*. (See IAEA, 1968).

Procter, B. W., et al. 1954. The use of ionizing radiations in the eradication of insects in packaged military rations. Food. Technol. 8: 536-540.

Proctor, C. M., E. Papadopulos and R. H. Firminhae. 1962. Gamma scintillation probe for field use and measurements of radiation background in Puget Sound. Limnol. Oceanogr. 7(3): 280-286.

Proctor, C. M., et al. 1962. Response of gamma scintillation detectors for field survey use. Limnol. Oceanogr. 7(3): 273-279.

Prokhorov, V. M. 1966. Diffusion of various radioactive fission oroducts in soils. Infor. Byull. Akad. Nauk SSSR (biol.) (9): 46-51. (In Russian).

Prokhorov, V. M. 1969. Kinetics of absorption of ^{90}Sr on the bottom of confined water basins. Radiokhimiya 11(3): 317-325. (Sov. Radiochem. 11(3): 310-315) (NSA 24: 308).

Prokhorov, V. M. 1969. Calculation of the reduction of strontium-90 content in lake and pond water as a result of bottom absorption. State Committee on the Use of Atomic Energy, USSR. Atomizdat, Moscow. UN Document A/AC. 82/G/L.1304. (AEC-tr-7128, pp. 254-262) (NSA 24: 11691).

Prokhorov, V. M. 1970. Information on the International Symposium of Radio-Ecology in Cadarache (France). Radiobiologiya 10(2): 304-305. (AEC-tr-7171, pp. 215-217).

Prokhorov, V. M. 1971. Ion diffusion in soils and its role in radionuclide migration. pp. 118-145. In: *Contemporary Problems of Radiobiology. Vol. 2. Radioecology*. (See Klechkovskiy, Polikarpov and Aleksakhin, 1971) (English transl. pp. 103-125) (NSA 29: 12724).

Prokhorov, V. M. nd. Mathematical modelling of radionuclide migration in certain regions of the biosphere. (In Russian) (INIS-mf-887, pp. 252-265) (NSA 29: 15629).

Prokhorov, V. M. and Chai Tien-ying. 1963. Diffusion of cesium-137 in soil. Radiokhimiya 5(5): 639-642. (Sov. Radiochem. 5(5): 598-600) (NSA 18: 14089).

Prokhorov, V. M. and Chai Tien-ying. 1964. Diffusion of cerium-144 in the soil. Pochvovedeniye (7): 107-108. (Sov. Soil Science (7): 706-707) (NSA 18: 35722).

Prokhorov, V. M. and A. S. Frid. 1972. Effect of humus content and composition of strontium-90 mobility in soil. Pochvovedeniye 4(3): __. (Sov. Soil Science 4(3): 333-340) (NSA 28: 12867).

Prokhorov, V. M. and L. R. Ginzburg. 1971. Modelling the process of migration of radionuclides in forest ecosystems and description of the model. Ekologiya 2(5): 11-19. (Sov. J. Ecol. 2(5): 396-402).

Prokhorov, V. M. and S. V. Kovenya. 1974. Estimation of the rate of extraction of radionuclides from fused mineral perticles. Radiokhimiya 16(4): 529-535. (Sov. Radiochem. 16(4): 522-526) (NSA 33: 23281).

Prokhorov, V. M., A. S. Frid and M. V. Ryzhinskiy. 1970. Effect of liming of soil on the diffusion rate of ^{90}Sr. Agrokhimiya 7(2): 40-48. (In Russian) (NSA 25: 39598).

Prokhorov, V. M, M. V. Ryzhinskiy and A. M. Margolin. 1973. Numerical method of calculating the dynamics of sorption of a substance in a medium with variable transfer coefficients. Radiokhimiya 15(3): 298-304. (Sov. Radiochem. 15(3): 302-306) (NSA 30: 6527).

Prokhorycheva, N. P. 1973. Radioactive pollution of industrial water organisms of western basin. pp. 248-256. In: *Radioecology of Water Organisms. II*. (See Andrushaitis, et al., 1973) (AEC-tr-7606, pp. 148-152) (NSA 30: 21143).

Prokof'yev, O. N. and G. V. Pokhvalinskaya. 1972. Sensitivity of the anion exchange method of determining iodine-131 in milk and water in the β-radiometry of the resin of various counters. Radiokhimiya 14(6): 909-911. (Sov. Radiochem. 14(6): 949-950) (NSA 28: 17917).

Prokof'yev, O. N., et al. 1969. Cesium-137 from global fallout in foodstuffs and humans. State Committee on the Use of Atomic Energy, USSR. 21pp. UN Document A/AC.82/G/L.1326. (In Russian) (NP-18382) (NSA 25: 13340).

Prokof'yeva-Bel'govskaya, A. A. and N. V. Pankova. 1965. Differential radiation injury of parent sets of chromosomes. Dokl. Akad. Nauk SSSR 163(4): 1001-1002. (Doklady (biol. sci.) 163(1-6): 146-148) (BA 47: 70290; NSA 20: 6819).

Prokof'yeva-Bel'govskaya, A. A., O. N. Kapitonova and Z. N. Shamina. 1962. Changes in actinomycete cells caused by radiation. Tsitologiya 4(1): 18-26. (BA 42: 8589).

Proshold, F. I. and J. A. Bartell. 1970. Inherited sterility in progeny of irradiated male tobacco budworms: effects on reproduction, developmental time and sex ratio. J. Econ. Entomol. 63: 280-285. (NSA 24: 23415).

Proshold, F. I. and J. A. Bartell. 1972. Postembryonic growth and development of F_1 and F_2 tobacco budworms (Lepidoptera: Noctuidae) from partially sterile males. Can. Entomol.104(2): 165-172. (NSA 26: 56486).

Proshold, F. I. and J. A. Bartell. 1972. Inherited sterility and post-embryonic survival of two generations of tobacco budworms, Heliothis virescens (Lepidoptera: Noctuidae), from partially sterile males. Can. Entomol. 104(2): 221-230. (NSA 26: 56485).

Proshold, F. I. and J. A. Bartell. 1972. Difference in radiosensitivity of two colonies of tobacco budworms, Heliothis virescens (Lepidoptera: Noctuidae). Can. Entomol. 104(7): 99 -1002. (NSA 30: 15980).

Proshold, F. I. and J. A. Bartell. 1973. Fertility and survival of tobacco budworms, Heliothis virescens (Lepidoptera: Noctuidae). Can Entomol. 105 (3): 377-382. (NSA 32: 11816).

Prospero, J. M. and T. N. Carlson. 1970. ^{222}Rn in the North Atlantic trade winds: its relationship to dust transport from Africa. Science 167: 974-977. (NSA 24: 16705).

Prosser, C. L., et al. 1946. Accumulation and distribution of radioactive strontium, barium-lanthanum, fission mixture and sodium in goldfish. Univ. of Chicago, U. S. AEC report MDDC-496; CH-3233. 42pp. (ADC 1: 298).

Proverbs, M. D. 1961. Progress on the use of induced sexual sterility for the control of the codling moth, Carpocapsa pomonella (L.) (Lepidoptera: Olethreutidae). Proc. Entomol. Soc. Ontario 92: 5-11.

Proverbs, M. D. 1962. Sterilization of the codling moth by gamma-irradiation. Nature 194(4835): 1297. (NSA 16: 21960).

Proverbs, M. D. 1964. The sterile male technique and its possible use for codling moth eradication. Can. Entomol. 96(1-2): 143.

Proverbs, M. D. 1969. Induced sterilization and control of insects. Ann. Rev. Entomol. 14: 81-102. (NSA 23: 20455).

Proverbs, M. D. 1974. Codling moth control by the sterility principle in British Columbia: estimated cost and some biological observations related to cost. pp. 81-88. In: Sterile-Insect Technique and Its Field Applications. (See IAEA, 1974) (NSA 31: 8943).

Proverbs, M. D. and J. R. Newton. 1960. Could this be death to the codling moth? Brit. Columb. Orchardist 1(4): __. (Also: pp. 13-16. In: Gamma Irradiation in Canada. Atomic Energy of Canada, Ltd., Ottawa, report AECL-1329(Vol. 2)).

Proverbs, M. D. and J. R. Newton. 1962. Some effects of gamma radiation on the reproductive potential of the codling moth, Carpocapsa pomonella (L.) (Lepidoptera: Olethreutidae). Can. Entomol. 94(11): 1162-____.

Proverbs, M. D. and J. R. Newton. 1962. Suppression of the reproductive potential of the codling moth by gamma irradiated males in caged orchard trees. J. Econ. Entomol. 55(6): 934-936.

Proverbs, M. D. and J. R. Newton. 1962. Influence of gamma radiation on the development and fertility of the codling moth, Carpocapsa pomenella (L.) (Lepidoptera: Olethreutidae). Can. J. Zool. 40: 401-420.

Proverbs, M. D., D. M. Logan and B. E. Carty. 1973. Some biological observations related to codling moth control by the sterility principle. pp. 149-163. In: Computer Models and Application of the Sterile-Male Technique. (See IAEA, 1973) (NSA 29: 8058).

Proverbs, M. D., J. R. Newton and D. M. Logan. 1966. Orchard assessment of the sterile male technique for control of the codling moth, Carpocapsa pomenella (L.) (Lepidoptera: Olethreutidae). Can. Entomol. 98(1): 90-95.

Proverbs, M. D., J. R. Newton and D. M. Logan. 1967. Autocidal control of the codling moth by release of males and females sterilized as adults by gamma radiation. J. Econ. Entomol. 60: 1302-1306. (NSA 22: 10899).

Proverbs, M. D., J. R. Newton and D. M. Logan. 1969. Codling moth control by release of radiation-sterilized moths in a commercial apple orchard. J. Econ. Entomol. 62(6): 1331-1334. (NSA 24: 14635).

Proverbs, M. D., et al. 1975. Codling moth control by the release of radiation-sterilized moths in a pome fruit orchard and observations of other pests. J. Econ. Entomol. 68(4): 555-560. (NSA 33: 7574).

Provost, E. E. 1964. Resistance of feral small mammals to acute gamma irradiation. Abstr. Bull. S. Car. Acad. Sci. 26: 28.

Provost, E. E. 1965. Acute gamma irradiation effects on feral rodents. Abstr. Bull. Ga. Acad. Sci. 23(2): 49.

Provost, E. E., M. R. Pelton and F. E. Kellogg. 1966. The effect of acute gamma irradiation on survival in feral rodents. pp. 35-39. In: Radiation Effects on Natural Populations. (See Sacher, 1966).

Provost, M. W. 1952, 1957, 1960. The dispersal of Aedes taeniorhynchus. 1. Preliminary studies. 2. The second experiment. 3. Study method for migratory studies. Mosquito News 12(3): 174-190; 17(4): 233-247; 20(2): 148-161. (Used ^{32}P).

Proz, E. 1974. Production and release of radioiodines by nuclear power plants and reprocessing plants and the expected radiological burden till the year 2000. German F. R. Kernforschungszentrum, Karlsruhe, report KZK-1912. 79pp. (In German) (NSA 31: 16523).

Pruitt, W. O., Jr. 1962. A new "caribou problem." Beaver Outfit 293: 24-25. (On radioactive fallout).

Pruitt, W. O. 1963. Lichen, caribou and high radiation in eskimos. Audubon Mag. 65(5): 284-287.

Pruter, A. T. and D. L. Alverson. 1972. The Columbia River Estuary and Adjacent Ocean Waters: Bioenvironmental Studies. Univ. of Washington Press, Seattle. xiii, 868pp.

Ptydz, S. and J. E. Mikkelborg. 1975. Thermoluminescent labelling: useful for the study of sedimentation in rivers. Int. J. Appl. Radiat. Isotopes 26(2): 92-93. (NSA 31: 29641).

Pshenichnov, R. and Ye. G. Noskova. 1963. The effect of x-rays on R. prowazeki cultivated in the carrier's organism. pp. 47-52. In: Rickettsial and Viral Infections. Perm'. (In Russian) (Ref. zh. (biol.) 1963: 19B 144).

Pshenin, L. P. 1960. The accumulation of U-UX_1 by nitorgen-fixing microorganisms of the Black Sea. Dokl. Akad. Nauk SSSR 133(6): 1448-1450. (Doklady (biol. sci.) 133: 589-591).

Public Service Company of Colorado. 1970. Fort St. Vrain Nuclear Generating Station, Unit 1. Applicant's environmental report: operating license stage. Denver. 124pp. U. S. AEC DOCKET-50267-25. (NSA 25: 12737).

Public Service Company of Colorado. 1971. Fort St. Vrain Nuclear Generating Station, Unit 1. Preliminary safety analysis report, amendment 22 (includes additions and revisions to final safety analysis report). Plans for coordination with Bureau of Sport Fisheries and Wildlife and Colorado Game, Fish and Parks Department. Denver. 41, 3pp. U. S. AEC DOCKET-50267-32, -30. (NSA 25: 36935, 44047).

Public Service Electric & Gas Company. 1971. Newbold Island Nuclear Generating Station, Units 1 and 2. Environmental report; construction permit stage. ---amendment 1. Newark, N. J. 70, 139pp. U. S. AEC DOCKET-50354-11, -13; -50355-11, -13. (NSA 25: 20676, 25937).

Public Service Electric & Gas Company. 1971. Salem Nuclear Generating Station, Units 1 and 2. Environmental report, operating license stage. --- Final safety analysis report, vol. 1. ---vol. 5. Newark, N. J. 314, 386, 379pp. U. S. AEC DOCKET-50272-11, -14, -18; -50311-11, -14, -18. (NSA 25: 41334, 50984, 50988).

Public Service Electric & Gas Company. 1973. Salem Nuclear Generating Station, Units 1 and 2. Final environmental statement. Newark, N. J. 270pp. U. S. AEC DOCKET-50272-57. (NSA 27: 29648).

Pubtsov, D. M. 1971. Investigation of soils of separate biogeocoenoses with elevated levels of natural radioactive elements. pp. 24-34. In: Methods of Radioecological Investigations. (See Verkhovskaya, 1971) (In Russian).

Puckett, W. 1936. The effect of x-radiation on limb development and regeneration in Ambystoma. J. Morphol. 59(1): 173-213.

Puckett, W. 1937. X-irradiation and thyroid induced metamorphosis in anuran larvae. J. Exptl. Zool. 76(2): 303-323.

Pudovkina, I. B. 1967. Comparison of results of measurements of atmospheric fallout of strontium-90 in various countries. State Committee on the Use of Atomic Energy, USSR. Atomizdat, Moscow. 26pp. UN Document A/AC.82/G/L.1202. (AEC-tr-7214, pp. 267-293) (NSA 22: 21196; 25: 15780).

Puerto Rico Nuclear Center. nd. Radiation and Isotope Technology in Latin American Development. (Proc. American Nuclear Society Topical Meeting, San Juan, Puerto Rico, May 4-6, 1969). San Juan, P. R., U. S. AEC report PRNC-135. 519pp. (In English and Spanish) (CONF-690508) (NSA 24: 39141).

Puerto Rico Water Resources Authority. 1971. Aguirre Nuclear Plant, Unit No. 1. Environmental report. San Juan, P. R. 293pp. U. S. AEC DOCKET-50376-14. (NSA 25: 34305).

Puerto Rico Water Resources Authority 1971. Bonus Nuclear Power Plant. Contract for radioactivity surveillance. San Juan, P. R. 14pp. U. S. AEC DOCKET-1154-5. (NSA 25: 46195).

Pugh, P. R. 1973. Evaluation of liquid scintillation counting techniques for use in aquatic primary production studies. Limnol. Oceanogr. 18(2): 310-319. (NSA 28: 11120).

Pugleise, A. 1961. Concentration of radioactivity present in the principal species of edible fish from Lake Maggiore. Rend ist. Lombardo Sci. Lett. B95: 216-231. (In Italian) (NSA 17: 10349).

Pugliese, A. and L. Vido. 1962. Concentration of radioactivity present in some species of edible fish of the Lago Maggiore (1959). Italy Comitato Nazionale per l'Energia Nucleare, Ispra, report CNI-106. 19pp. (NSA 16: 33307).

Puishes, A. 1961. Applied radiation - a radioactive control for pigeon population. Nucleonics 19(7): 100.

Pujara, C. M., R. J. Horsley and S. N. Banerjee. 1970. Radiation-induced division and mitotic delay studies in green algae (Oedogonium cardiacum). Radiation Research 44: 413-420. (NSA 25: 8644).

Pullen, T. M., Jr., G. L. Plummer and E. E. Provost. 1967. Strontium-90, cesium-137 and other fallout radionuclides in the food chain of a white-tailed deer population in the Georgia Piedmont. Bull. Ga. Acad. Sci. 25 (1): 26-42.

Pullum, P. A. and F. H. Erbisch. 1972. Effects of gamma radiation on the lichen Cladonia verticillata (Hoffm.)Schaer. Bryologist 75(1): 48-53.

Punning, Ya. M., T. Kakum and R. Rayamyae. 1975. List of samples dated by the radiocarbon method at the Institute of Geology of the Academy of Science of the Estonian SSR. II. Izv. Akad. Nauk Est. SSR Khim. Geol. 4(2): 160-165. (In Russian) (NSA 32: 17216).

Punt, A. and P. J. van Nieuwenhoven. 1957. The use of radioactive bands in tracing hibernating bats. Experientia 13: 51-54.

Purdom, C. E. 1965. Genetic effect of incorporated ^{14}C in Drosophila melanogaster. Mutation Research 2(2): 156-167.

Purdom, C. E. 1966. Radiation and mutation in fish. pp. 861-867. In: Disposal of Radioactive Wastes into Seas, Oceans and Surface Waters. (See IAEA, 1966).

Purdom, C. E. and B. A. Bridges. 1964. Effect of sulphydryl reagents on the frequency of x-ray-induced autosomal recessive lethal mutations in Drosophila melanogaster. Nature 203: 419.

Purdue, L. J., J. Bryant and B. Ortiz de Montellano. 1969. Analysis of pesticide residues by a dual-column, dual-electron-capture detector method. J. Agric. Food Chem. 17(2): 264-266.

Purica, I. I. 1972. Pollution sources in nuclear power generation. Progr. Stiintei 8(2): 166-168. (In Rumanian) (NSA 26: 56077).

Purnell, R. E., M. A. Ledger and J. B. Oatre. 1972. Some effects of irradiation on Theileria parva in the salivary glands of the tick Rhipicephalus appendiculatus. Parasitology 65(1): 23-26. (NSA 27: 567).

Purnell, R. E., et al. 1972. Some effects of irradiation on the tick Rhipicephalus appendiculatus. Parasitology 64(3): 429-440. (NSA 29: 770).

Purnell, R. E., et al. 1973. ^{60}Co-irradiation of the tick Rhipicephalus appendiculatus. pp. 139-144. In: Isotopes and Radiation in Parasitology III. (See IAEA, 1973) (NSA 28: 15766).

Purnell, R. E., et al. 1974. East Coast fever: ^{60}Co-irradiation of Theileria parva in its tick vector, Rhipicephalus appendiculatus. Int. J. Parasitol. 4(5): 507-511. (NSA 32: 17421).

Pursglove, D. L., et al. 1971. Effects of x-irradiation on the genetically-determined melanoma system of xiphophorin fish. Experientia 27(6): 695-697. (NSA 25: 49300).

Purtymun, W. D. 1970. Plutonium in stream channel alluvium in the Los Alamos area, New Mexico. Los Alamos Scientific Lab., U. S. AEC report LA-4561. 7pp. (NSA 25: 15762).

Purtymun, W. D. 1973. Regional survey of tritium in surface and ground water in the Los Alamos area, New Mexico, August 1966-May 1969. Los Alamos Scientific Lab., U. S. AEC report LA-5234-MS. 9pp. (NSA 28: 2562).

Purtymun, W. D. 1973. Underground movement of tritium from solid-waste storage shafts. Los Alamos Scientific Lab., U. S. AEC report LA-5286-MS. 7pp. (NSA 28: 12865).

Purtymun, W. D. 1974. Dispersion and movement of tritium in a shallow aquifer in Mortandad Canyon at the Los Alamos Scientific Laboratory. Los Alamos Scientific Lab., U. S. AEC report LA-5716-MS. 10pp. (NSA 30: 32077).

Purtymun, W. D. 1974. Storm runoff and transport of radionuclides in DP Canyon, Los Alamos County, New Mexico. Los Alamos Scientific Lab, U. S. AEC report LA-5744. 9pp. (NSA 31: 3312).

Purushothaman, K. 1971. Radionuclide transport in an aquatic model system. pp. 174-185. In: Trace Substances in Environmental Health. IV. (See Hemphill, 1971) (NSA 26: 15206).

Purvis, A. E. and R. S. Foote. 1964. Atmospheric attenuation of γ radiation pp. 747-756. In: The Natural Radiation Environment. (See Adams and Lowder, 1964) (NSA 18: 35745).

Puschmann, H. 1970. Research on sorption of the radionuclides ^{85}Sr, ^{137}Cs, and ^{131}I by unconsolidated sediments. Sand from the low-terraces of the rivers Weser, Danube, Elbe/Germany. Deut. Gewässerk. Mitt. 14(3): 63-72. (In German) (NSA 25: 34737).

Pushkareva, M. I. 1958. Diffusible toxic compounds in radiation damage in plants. Biofizika 3(4): 447-453. (Biophysics (USSR) 3(4): 430-437) (NSA 13: 11550).

Pushpendran, C. K. and J. Eapen. 1971. Effect of temperature acclimation, x irradiation, and tail amputation on incorporation of leucine-^{14}C into hepatic proteins of the house lizard (Hemidactylus leschenaulti). Comp. Biochem. Physiol. 40B(3): 651-656.

Pushpendran, C. K. and J. Eapen. 1972. Electrophoretic studies on proteins of tail of house lizard. J. Anim. Morph. Physiol. 19(2): 191-195. (NSA 31: 8914) (X-irradiation).

Putnam, J. L. and S. Jefferson. 1956. Application of radioisotopes to leakage and hydraulic problems. Proc. Intern. Conf. on the Peaceful Uses of Atomic Energy 15: 147-150. (Paper A/CONF.8/P/462).

Putnam, J. L., et al. 1954. Thames siltation investigation - preliminary experiments on the use of radioactive tracers for indicating mud movements. UK Hydraulics Research Station, Wallingford, Berks., England. 24pp.

Pyatkovskaya, N. Ya. and M. Yu. Dolmatova. 1972. Study of the properties of some acid hydrolysis products of alginates. pp. 90-94. In: Radiation and Chemical Ecology of Aquatic Organisms. (See Polikarpov, 1972) (In Russian).

Qayyum, H. A. 1962. The scope of the use of radioisotopes and radiation sources in entomology in Pakistan. pp. 281-285. In: Radioisotopes and Radiation in Entomology. (See IAEA, 1962).

Quam, W. M. and W. O. Wilde. 1969. Experimental study of a large-area ecological irradiator. EG&G, Inc., Santa Barbara, Calif., U. S. AEC report EGG-1183-2203. 26pp. (NSA 23: 26404).

Quan, S. F., et al. 1957. Cesium 144 as a tag for arthropods of medical importance. Trans. Roy. Soc. Trop. Med. Hyg. 51(1): 87-88.

Quarterman, K. D., J. W. Kilpatrick and W. Mathis. 1954. Fly dispersal in a rural area near Savannah, Georgia. J. Econ. Entomol. 47(3): 413-419.

Quarterman, K. D., W. Mathis and J. W. Kilpatrick. 1954. Urban fly dispersal in the area of Savannah, Georgia. J. Econ. Entomol. 47(3): 405-412.

Quarterman, K. D., et al. 1955. Flight dispersal of rice field mosquitoes in Arkansas. J. Econ. Entomol. 48(1): 30-32.

Quastler, H. and M. Baer. 1950. Inhibition of plant growth by irradiation. III. Successive radiation effects; homologous responses. V. Radiation effects of initiation and completion of growth. J. Cell. Comp. Physiol. 35(1): 75-94; Cancer Research 10(10): 604-612.

Quastler, H., A. M. Schertige and W. N. Stewart. 1952. Inhibition of plant growth by irradiation. IV. Growth arrest vs. effects on mitotic activity. J. Cell. Comp. Physiol. 39(3): 357-369.

Quednau, W. 1960. Radioactive labelling of certain hymenoptera. Atompraxis 6(10/11): 427-431. (In German).

Queirazza, G. and E. Smedile. 1973. The role of freshwater periphyton in the accumulation of heavy metals and radioactive elements. Abstr. pp. 101-102. In: Proc. 2nd Conf. on the State of Advancement of Radioecology in Italy. (See Anonymous, 1973) (In Italian).

Queirazza, G., E. Smedile and E. Tibaldi. 1969. Transfer of some radionuclides traversing the links of a river food chain. Atti. accad. naz. lincei rend. cl. sci. fis. mat. nat. 46: 81-90. (In Italian) (NSA 24: 4591).

Quell, P. 1973. Activity release from nuclear power plants. pp. 158-170. In: Worldwide Radiation from Nuclear and Conventional Energy Production. H. Braun, ed. Georg Thieme Verlag, Stuttgart. (In German) (NSA 31: 3313).

Quick, H. F. 1965. Small mammal populations and ecological variations in an alpine watershed with a preliminary estimate of biomass cycle. Abstr. Bull. Ecol. Soc. Am. 46(3): 120. (Mentions radioactive fallout).

Quick, T. F., H. G. Abbott and W. J. Mellen. 1970. Locating tree seed caches of small mammals with a radioisotope. Forest Science 16(2): 147-148.

Quimby, G. H., J. G. Denison and J. E. Stein. 1965. Liquid scintillation counting. Water Sewage Works 112: 72-73. (NSA 20: 30997) (Primary production).

Quraishi, M. S. 1963. Use of isotopes for investigating the behavior and ecology of insect pests in some recent studies. pp. 93-98. In: Radiation and Radioisotopes Applied to Insects of Agricultural Importance. (See IAEA, 1963).

Quraishi, M. S. 1963. Water and food relationship of the eggs and first instar nymph of Eurygaster integriceps with the aid of P^{32}. J. Econ. Entomol. 56(5): 666-668.

Quraishi, M. S. 1964. Study of the flight range and gonotrophic cycles in Anopheles stephensi using P^{32}. pp. 426-428. In: The Role of Science in the Development of Natural Resources with Particular Reference to Pakistan, Iran and Turkey. Pergamon Press, New York.

Quraishi, M. S. 1965. Swarming, mating and density in nature of Anopheles stephensi mysorensis. J. Econ. Entomol. 58(5): 821-824. (^{32}P).

Quraishi, M. S. 1968. Uptake of ^{32}P by yellow-fever mosquito larvae of equal age, and those pupating synchronously: its loss and transfer during development, mating, and oviposition. J. Econ. Entomol. 61(2): 530-533. (NSA 22: 36165).

Quraishi, M. S. 1970. Volatilization of DDT-^{14}C and lindane-^{14}C from treated surfaces and metabolism of DDT-^{14}C by house fly pupae. Can. Entomol. 102 (9): 1189-1195.

Quraishi, M. S. and W. E. Kershaw. 1966. The effect of x-irradiation on the survival of Liponyssus bacoti and on its susceptibility to infection with Litomosoides carinii. Ann. Trop. Med. Parasitol. 60(1): 57-62.

Quraishi, M. S. and M. Metin. 1963. Radioactivity of various stages of Callosobruchus chinensis L. pp. 479-484. In: Radiation and Radioisotopes Applied to Insects of Agricultural Importance. (See IAEA, 1963).

Quraishi, M. S. and Z. T. Poonawalla. 1969. Radioautographic study of the diffusion of topically applied DDT-^{14}C into the house fly and its distribution in internal organs. J. Econ. Entomol. 62(5): 988-994.

Quraishi, M. S., R. A. Brust and L. P. Lefkovitch. 1966. Uptake, transfer and loss of P^{33} during metamorphosis, mating and oviposition in Aedes vexans. J. Econ. Entomol. 59(6): 1331.

Quraishi, M. S., M. A. Faghih and N. Esghi. 1966. Flight range, lengths of gonotrophic cycles, and longevity of ^{32}P-labeled Anopheles stephensi mysorensis. J. Econ. Entomol. 59: 50-55. (NSA 20: 35195).

Quraishi, M. S., M. A. Faghih and N. Esghi. 1966. Flight range, lengths of gonotrophic cycles, and longevity of ^{32}P-labelled Anopheles stephensi mysorensis. Arch. Environ. Hlth. 12: 50-55.

Quraishi, M. S., R. F. Lamsachi and C. Ergul. 1963. Laboratory studies on tagging of Anopheles stephensi. J. Econ. Entomol. 56(5): 672-674.

Qureshi, Z. A. 1966. Effects of sub-lethal gamma radiations on the biology and behavior of the angoumois grain moth Sitotroga cerealella Olivier. Thesis, Kansas State Univ., Manhattan. 138pp. (NSA 20: 38705).

Qureshi, Z. A. 1969. The ability of sterilized irradiated males of Dacus cucurbitae Coq. and Dacus zonatus (Saunders) to compete with normal males. pp. 43-46. In: Insect Ecology and the Sterile-Male Technique. (See IAEA, 1969).

Qureshi, Z. A. and A. R. Bughio. 1969. Sterilization and competitive ability of gamma sterilized males to normal males of Dacus zonatus (Saunders). Int. J. Appl. Radiat. Isotopes 20: 473-476. (NSA 24: 932).

Qureshi, Z. A. and D. A. Wilbur. 1966. Effect of sub-lethal gamma on eggs, early, intermediate, and last instar larvae of the angoumois grain moth, Sitotroga cerealella Oliv. pp. 112-124. In: Proc. Agricultural Symp. Pakistan Atomic Energy Center, Dacca. (NSA 21: 26636).

Qureshi, Z. A., R. B. Mills and D. A. Wilbur. 1964. Sensitivity of various stages of angoumois grain moth to gamma radiation. Abstr. Bull. Entomol. Soc. Am. 10(3): 163.

Qureshi, Z. A., S. Mohiuddin and N. Akram. 1975. Effects of gamma radiation and Hempa on the midgut of the desert locust Schistocerca gregaria. Folia Biol. (Krakow) 23(1): 123-127. (NSA 32: 11815).

Qureshi, Z. A., D. A. Wilbur and R. B. Mills. 1965. Effects of gamma radiation on prepupae and pupae of angoumois grain moth. Bull. Entomol. Soc. Am. 11(3): 158.

Qureshi, Z. A., D. A. Wilbur and R. B. Mills. 1968. Sub-lethal gamma radiation effects on prepupae, pupae, and adults of angoumois grain moth. J. Econ. Entomol. 61(6): 1699-1705.

Qureshi. Z. A., D. A. Wilbur and R. B. Mills. 1970. Irradiation of early instars of the angoumois grain moth. J. Econ. Entomol. 63: 1241-1247. (NSA 24: 44648).

Qureshi, Z. A., et al. 1974. Rearing, reproductive behavior, and gamma sterilization of fruit fly, Dacus sonatus (Diptera: Tephrilidae). Entomol. Exp. Appl. 17(4): 504-510. (NSA 32: 14592).

Raaen, H. P. 1972. Environments and isotopes. Oak Ridge National Lab., U. S. AEC report ORNL-IIC-39. 118pp. (NSA 26: 48180).

Raats, P. A. C. 1967. Kinematics of soil water. pp. 191-200. In: Isotope and Radiation Techniques in Soil Physics and Irrigation Studies. (See IAEA, 1967) (NSA 22: 23152) (Tracer).

Rabinowitz, D. C., C. R. Holmes and G. W. Gross. 1973. Forced exchange of tritiated water with clays. pp. 471-485. In: Tritium. (See Moghissi and Carter, 1973) (NSA 28: 27127).

Rabinowitz, J. L. and R. M. Myerson. 1966. Exposure of tropical fish to dimethyl sulfoxside (DMSO) with special reference to toxicity and effects on the uptake of radioactive dyes. Proc. Soc. Exptl. Biol. Med. 121(4): 1065-1067. (BA 47: 107230; SFA 11: 8945).

Rabon, E. W. 1968. Some seasonal and physiological effects on ^{137}Cs and $^{89,90}Sr$ content of the white-tailed deer, Odocoileus virginianus. Health Physics 15: 37-42. (NSA 22: 49194).

Rabon, E. W. 1969. Comparison of ^{137}Cs, $^{89,90}Sr$, and calcium content in pregnant white-tailed deer (Odocoileus virginianus) and related fetuses. Abstr. Bull. Ecol. Soc. Am. 50(2): 73-74.

Rabon, E. W. 1972. Radiological monitoring results for the 1971 deer hunts pp. 15-19. In: Status of Savannah River Plant Deer Herd. Savannah River Plant, U. S. AEC report SRO-154. (NSA 26: 36461).

Rabon, E. W. and J. E. Johnson. 1973. Rapid field-monitoring of cesium-137 in white-tailed deer. Health Physics 25(5): 515-516. (NSA 29: 826).

Rabson, R. 1969. Activities of the United States Atomic Energy Commission related to radiation techniques used in control of insects. pp. 45-49. In: Sterile-Male Technique for Eradication or Control of Harmful Insects. (See IAEA, 1969) (NSA 24: 32298).

Rachinskiy, V. V. 1974. Course on the Principles of Atomic Techniques in Agriculture. Atomizdat, Moscow. 392pp. (In Russian) (NSA 32: 1257).

Rachkov, L. D. 1961. A study of the interaction of plants in the natural community, using C^{14}. pp. 218-222. In: Proc. Tashkent Conf. on Peaceful Uses of Atomic Energy. Ya. Kh. Turakulov, ed. Publ. House Acad. Sci. Uzbek. SSR, Tashkent. (AEC-tr-6293, pp. 370-375).

Rachkov, L. D. and N. V. Timofeyev-Resovskiy. 1962. A study of migration of cobalt-60 in experimental phytocoenoses. Dokl. Akad. Nauk Tadzhiksk. SSR 5(4): 43-47. (In Russian) (BA 45: 80800).

Racz, G. J., D. A. Rennie and W. L. Hutcheon. 1964. The P^{32} injection method for studying the root system of wheat. Can. J. Soil. Sci. 44: 100-108.

Radeleff, R. D., R. C. Bushland and D. E. Hopkins. 1952. Phosphorus-32 labeling of the screw-worm fly. J. Econ. Entomol. 45(3): 509-514. (NSA 6: 5296).

Radeleff, R. D., R. C. Bushland and D. E. Hopkins. 1956. Phosphorus-32 labeling of the screw-worm fly. J. Econ. Entomol. 49(5): 714-715.

Radicheva, M. A. and I. T. Mishev. 1966. Study of the deposition of radioactive fallout in certain places in Bulgaria. Izv. fiz. inst. Aneb. Bulg. Akad. Nauk 15: 5-10. (In Russian) (NSA 21: 16007).

Radiochemical Centre. 1964. Labelled insecticides. Amersham, Buckinghamshire, England, Tech. Bull. 64/19. 4pp.

Radosavljević, R. and T. Tasovac. 1970. Capture of cobalt-60 by suspended particles in the Danube. pp. 83-91. In: Proc. Intern. Symp. on Radioecology. (See French Commissariat à l'Énergie Atomique, 1970).

Radosavljević, R., et al. 1965. The movement of radioactivity emanating from ^{40}K in the precipitation and waters of the rerritory of Vinca. Tehnika (Belgrade) (6): 1053-1056. (JPRS-31469) (NSA 19: 46590).

Radovanović, R., et al. 1970. Natural activity and fission-product contamination of the environment and of the food chain on Serbian territory from 1961 to 1968. pp. 1065-1082. In: Proc. Intern. Symp. on Radioecology. (See French Commissariat à l'Énergie Atomique, 1970) (In French) (NSA 24: 41790).

Radriquez, J. G. 1948. The effect of a 3-month exposure of mites to radiophosphorus. Ohio State Univ. Eng. Expt. Sta. News 22: 1-3.

Radu, V. G., et al. 1971. Radioisotope researches concerning the feeding and the assimilation of ^{45}Ca in terrestrial isopods. Pedobiologia 11: 296-303.

Radwan, M., et al. 1965. Indirect method determining the contamination of ground with fall-out. Nukleonika 10(2): 123-129. (In Polish) (NSA 19: 28596).

Radwan, M. A. 1967. Translocation and metabolism of C^{14}-labeled tetramine by Douglas-fir, orchard grass, and blackberry. Forest Science 13(3): 265-.

Radwan, M. A. and W. E. Dodge. 1970. Fate of radioactive tetramine in small mammals and its possible use as a seedling protectant. Northwest Sci. 44 (1): 25-30.

Rafter, T. A. 19?? Recent developments in the interpretation and the reporting of carbon-14 activity measurements from New Zealand. pp. 33-41. In: Radioactive Tracers in Oceanography. I.U.G.G. Monograph No. 20. (NSA 19: 20238) (Incl. trees).

Rafter, T. A. 1965. Nuclear studies in chemistry and biology. J. New Zeal. Inst. Chem. 29: 64-73. (NSA 20: 18909).

Rafter, T. A. 1965. ^{14}C variations in nature. Part 1. Technique of ^{14}C preparation, counting, and reporting of results. New Zeal. J. Sci. 8: 451-471. (NSA 20: 31433).

Rafter, T. A. (Comp.). 1972. Proc. 8th Intern. Radiocarbon Dating Conf. (Held at Lower Hutt, New Zealand, October 18-25, 1972). Royal Soc. of New Zealand, Wellington. 2 vols., vp. (NSA 31: 8677, 8704).

Rafter, T. A. and G. J. Fergusson. 1957. Recent increase in the C^{14} content of the atmosphere, biosphere, and surface waters of the oceans. New Zeal. J. Sci. Tech. B38(8): 871-883.

Rafter, T. A. and G. J. Fergusson. 1957. Atomic bomb effect. Recent increase of carbon-14 content of the atmosphere and biosphere. Science 126 (3273): 557-558.

Rafter, T. A. and S. H. Wilson. 1963. The examination of sulphur isotopic ratios in the geothermal and volcanic environment. New Zealand Inst. of Nuclear Sciences, Lower Hutt, report INS-R-25. 59pp. (NSA 18: 43800).

Rafter, T. A. and S. H. Wilson. 1964. Examination of sulfur isotopic ratios in the geothermal and volcanic environment. pp. 139-172. In: Nuclear Geology on Geothermal Areas. Italy Consiglio Nazionale delle Ricerche, Pisa. (NSA 20: 31433).

Rafter, T. A., et al. 1972. New Zealand radiocarbon reference standards. pp. H29-H79. In: Proc. 8th Intern. Radiocarbon Dating Conf. Vol. 2. (See Rafter, 1972) (NSA 31: 8717).

Raghavan, K. G., G. B. Nadkarni and A. Sreenivasan. 1971. Effect of gamma irradiation on development of rice moth (Corcyra cephalonica). pp. 151-158. In: Basic Mechanisms in Radiation Biology and Medicine. (See India Dept. of Atomic Energy, 1971) (NSA 26: 28489).

Ragland, P. C. 1964. Autoradiographic investigations of naturally occurring materials. pp. 129-152. In: The Natural Radiation Environment. (See Adams and Lowder, 1964) (NSA 18: 35734).

Ragotzkie, R. A. and G. E. Likens. 1964. The heat balance of two Antarctic lakes. Limnol. Oceanogr. 9(3): 412-415. (Used ^{131}I).

Ragsdale, H. L. 1968. The effects of biotic complexity and fast neutron radiation on cesium-137 and cobalt-60 kinetics in aquatic microcosms. Thesis, Univ. of Tennessee, Knoxville. xv, 171pp. (Also: w/ J. P. Witherspoon and D. J. Nelson, Oak Ridge National Lab., U. S. AEC report ORNL-4318) (NSA 23: 12043).

Ragsdale, H. L. and W. A. Rhoads. 1974. Four-year post-exposure assay of vegetation surrounding Project Pin Stripe: demonstration of the utility of delayed damage appraisals. Radiation Botany 14(4): 229-236. (NSA 32: 1078).

Ragsdale, H. L. and D. J. Shure. 1973. Flood plain transfer and accumulation of ^{137}Cs from a reactor effluent stream. pp. 243-253. In: Environmental Behavior of Radionuclides Released in the Nuclear Industry. (See IAEA, 1973) (NSA 29: 29363).

Ragsdale, W. H. L. and D. J. Shure. 1972. Redistribution of reactor effluent ^{137}Cs between stream and flood plain ecosystems in South Carolina. Abstr. Bull Ecol. Soc. Am. 53(2): 32.

Rahalkar, G. W. and R. L. Doutt. 1965. A comparison of procedures for marking adult endoparasitic wasps with ^{32}P. J. Econ. Entomol. 58: 278-281. (NSA 19: 35913).

Rahalkar, G. W. and K. K. Nair. 1966. Influence of diapause on the radiosensitivity of kapra beetle larvae. Proc. Entomol. Soc. Ontario 97: 126.

Rahalkar, G. W. and K. K. Nair. 1968. Influence of diapause on the radiosensitivity of khapra beetle larvae. pp. 149-154. In: Isotopes and Radiation in Entomology. (See IAEA, 1968) (NSA 24: 32168).

Rahalkar, G. W. and V. Ramakrishnan. 1968. Effects of irradiation on insect host-parasite relations. pp. 69-74. In: Isotopes and Radiation in Entomology. (See IAEA, 1968) (NSA 24: 32163).

Rahalkar, G. W. and A. P. D'Souza. 1971. Temperature induced modification of radiosensitivity in khapra beetle (Trogoderma granarium Everts) larvae. pp. 381-391. (In: Basic Mechanisms in Radiation Biology and Medicine. (See India Dept. of Atomic Energy, 1971) (NSA 26: 28491).

Rahalkar, G. W., et al. 1971. Labeling adults of red palm weevil (Rhynchophorus ferrugineus) with cerium for detection by neutron activation. Ecology 52(1): 186-188. (NSA 25: 23536).

Rahalkar, G. W., et al. 1972. Laboratory studies on radiation sterilization of the red palm weevil (Rhynchophorus ferrugineus Oliv.) males. pp. 141-146. In: Proc. 1st National Symp. on Plantation Crops. N. M. Nayar, ed. Indian Soc. for Plantation Crops, Kasargod. (NSA 33: 971).

Rahalkar, G. W., et al. 1975. Laboratory studies on sterilization of the male red palm weevil, Rhynchophorus ferrugineus Oliv. pp. 261-267. In: Sterility Principle for Insect Control, 1974. (See IAEA, 1975) (NSA 32: 11782).

Rahimov, Z. R. 1973. Distribution of total strontium-90 in the skeleton and teeth of adult wild boar in the Karakalpak SSR. Sb. rab. Leningrad Vet. Inst. (33): 275-277. (In Russian) (NSA 31: 33554).

Rahman, W. M. 1972. Importance of micronutrient studies in soil-plant relationship by activation analysis: a review. Nucl. Sci. Appl. A6: 41-45. (NSA 28: 20688).

Rahmouni, O. 1971. Determination of the water profile of a soil column using the attenuation of a γ ray beam. Mes. Reg. Automat. 36(8-9): 102-106. (In French) (NSA 26: 6795).

Rahn, O. 1934. The physio-chemical basis of biological radiations. Cold Spring Harbor Symp. Quant. Biol. 2: 226-240. (Incl. yeast, sea urchin eggs, amphibians).

Rahola, T. and J. K. Miettinen. 1970. Body burden of ^{137}Cs and the radiation dose of a population group in Finnish Lapland. Univ. of Helsinki, Finland, U. S. AEC report NYO-3446-14. 12pp. (For 2nd Congr. Intern. Radiation Protection Assoc.) (CONF-700505-17) (NSA 25: 18987).

Rahola, T. and J. K. Miettinen. 1970. Radionuclides in plants and reindeer meat in Lapland. Univ. of Helsinki, Finland, U. S. AEC report NYO-3446-14, Paper 31. 6pp. (NSA 25: 18932).

Rahola, T. and J. K. Miettinen. 1971. Body burden of ^{137}Cs and radiation dose in Finnish Lapps. pp. 299-308. In: 3rd Nordic Radiation Protection Conf. (See Nordic Soc. for Radiation Protection, 1971) (NSA 27: 22793) (Lichen, reindeer).

Rahola, T. and J. K. Miettinen. 1973. Accumulation of ^{137}Cs in Finnish lapps. Arch. Envirn. Hlth. 26(2): 67-69. (NSA 27: 20158) (Incl. lichen, reindeer, water, fish).

Rahola, T., T. Hattula and J. K. Miettinen. 1969. Body burdens of ^{137}Cs of Finnish Lapps in spring 1969. Univ. of Helsinki, Finland, U. S. AEC report NYO-3446-13, Paper 21. 5pp. (For 5th R.I.S. Symp., Helsinki, May 19-20, 1969) (CONF-690527-2) (NSA 23: 41302).

Rahola, T., T. Hattula and J. K. Miettinen. 1970. Present situation regarding the accumulation of ^{137}Cs in Finnish Lapps. pp. 1039-1056. In: Proc. Intern. Symp. on Radioecology. Vol. II. (See French Commissariat à l'Énergie Atomique, 1970) (NSA 24: 41789).

Rai, K. S. 1967. Cytogenetics of *Aedes aegypti*. Bull. Wld. Hlth. Org. 36: 363-365. (NSA 22: 49319) (X-rays).

Rai, K. S. 1968. Techniques for studying the effects of radiation on meiosis and related processes in mosquitoes with particular reference to *Aedes aegypti*. pp. 185-200. In: Effects of Radiation on Meiotic Systems. International Atomic Energy Agency, Vienna, Publ. STI/PUB-173. (CONF-670525) (NSA 24: 21133).

Rai, K. S. 1969. Status of the sterile-male technique for mosquito control. pp. 107-114. In: Sterile-Male Technique for Eradication or Control of Harmful Insects. (See IAEA, 1969) (Also: Univ. of Notre Dame, South Bend, Ind., U. S. AEC report COO-38-613) (NSA 24: 32306).

Rai, K. S. and P. T. McDonald. 1971. Chromosomal translocation and genetic control of *Aedes aegypti*. pp. 437-452. In: Sterility Principle for Insect Control or Eradication. (See IAEA, 1971) (NSA 25: 29753).

Rai, K. S. and P. T. McDonald. 1972. Application of radiation-induced translocations for genetic control of *Aedes aegypti*. pp. 77-94. In: Genetics and Our Health. Indian Council of Medical Research, New Delhi. (NSA 29: 10463).

Raikov, L. 1965. Natural radioactivity of Bulgarian soils. Rasteniyev'dni Nauk (Bulgaria) 2: 31-36. (In Russian) (NSA 22: 40211).

Raikov, L. 1972. Tests for limiting the penetration of ^{90}Sr from the soil into plants. Bulgaria Komitet za Mirno Izpolzuvane na Atomnata Energiya, Akad. Selsk. Nauki, Sofia. 22pp. (In Bulgarian) (INIS-mf-715) (NSA 29: 15625).

Raikov, L., M. Najdenov and St. Darkova. 1972. Pollution of mountain soils with radioactive strontium and cesium. Bulgaria Komitet za Mirno Izpolzuvane na Atomnata Energiya, Akad. Selsk. Nauki, Sofia. 14pp. (In Bulgarian) (INIS-mf-716) (NSA 29: 15626).

Raikov, L., et al. 1970. Total β-radioactivity and ^{40}K content in brown forest soils of the central Balkans and the northwestern Rhodopes. Gorskostop. Nauka 7(2): 27-35. (In Bulgarian) (NSA 26: 48040).

Raimundo, A. C. and L. Roda Santos. 1965. Insect control. I. Labelling the fruit fly *Ceratitis capitata* Wied. with radioactive phosphorus for the sterile male technique. Estudos agron. 6(3): 105-112. (In Portugese).

Raimundo, A. C. and L. Roda Santos. 1965. Insect control. I. Labelling the fruit fly (*C. capitata*) with radioactive phosphorus for the sterile-male method. Garcia de Orta 13(3): 351-358. (In Portugese, English and French summaries).

Rainbolt, C. 1973. New thermoluminescent dosimeters for x-ray machine surveys and environmental monitoring. pp. 425-429. In: Health Physics in the Healing Arts. U. S. Food and Drug Administration report FDA-73-8029. (NSA 29: 29432).

Raines, G. E. and S. G. Bloom. 1969. Simulation studies as related to the ecological effects of underground testing of nuclear devices on Amchitka Island. Battelle Columbus Labs., U. S. AEC report BMI-171-118. 50pp. (NSA 25: 16074).

Raines, G. E., S. G. Bloom and A. A. Levin. 1969. Ecological models applied to radionuclide transfer in tropical ecosystems. BioScience 19(12): 1086-1091.

Raines, G. E., et al. 1971. Mathematical simulation of sea otter population dynamics, Amchitka Island, Alaska. BioScience 21(12): 686-691. (NSA 25: 49146).

Rainey, C. T., et al. 1954. Distribution and characteristics of fallout at distances greater than ten miles from Ground Zero, March and April, 1953. (Operation UPSHOT/KNOTHOLE). Univ. of California at Los Angeles, U. S. AEC report WT-811. 98pp. (NSA 13: 11757).

Rajagopalan, P. K., M. Yasuno and G. C. Labrecque. 1973. Dispersal and survival in the field of chemosterilized, irradiated, and cytoplasmically incompatible male Culex pipeins fatigans. Bull. Wrld. Hlth. Org. 48: 631-635. (NSA 30: 13005).

Rajasekhara, K. 1970. Effects of gamma radiation on the beet armyworm, Spodoptera exigua (Hubner) (Lepidoptera: Noctuidae). Thesis, Univ. of California, Berkeley. 85pp. (Univ. Microfilms 71-15871) (NSA 25: 49274).

Rajewsky, B. and A. Redhardt. 1964. On the kinetics of radical reactions in x-ray irradiated feather quills. Biophysik 1(4): 299-302. (In German) (NSA 18: 27085).

Rajput, M. A. and A. H. Khan. 1971. Radiosensitivity studies in sunflower (Helianthus annus). Nucleus (Karachi) 8(1-2): 84-87. (NSA 25: 58148).

Rakhteyenko, I. N. 1957. Seasonal rhythm of absorption and excretion of phosphorus by roots of woody plants. Lesn. khoz. 10(3): 31-34. (In Russian).

Rakitin, A. A. 1963. The use of radioisotopes in the marking of Eurygaster inegriceps Put. Entomol. obozreniye 42(1): __. (Entomol. Rev. 42(1): 20-24).

Rakitin, Y. V. and A. V. Krylov. 1958. A study of the translocation, distribution and transformation of some physiologically active substances in plants. Proc. 2nd Intern. Conf. on the Peaceful Use of Atomic Energy 27: 68.

Rakoczi, L. 1973. Critical review of current suspended sediment gauges. pp. 217-225. In: Tracer Techniques in Sediment Transport. (See IAEA, 1973) (NSA 28: 21378).

Rakovic, M. 1970. Activation Analysis. Academia, Prague. 339pp. (NSA 24: 29234).

Ralph, E. K. 1973. Cyclic solution for the relationship between magnetic and atmospheric ^{14}C changes. pp. A76-A84. In: Proc. 8th Intern. Radiocarbon Dating Conf. Vol. 1. (See Rafter, 1973) (NSA 31: 8627).

Rama, M. Koide and E. D. Goldberg. 1961. Lead-210 in natural waters. Science 134(3472): 98-99.

Ramakrishnan, C. V. 1971. Possible use of gamma radiation in waste water treatment. J. Sci. Ind. Res. 30(5): 228-234. (NSA 26: 22971).

Ramakrishnan, C. V. 1973. Effect of irradiation on the micro-ecosystem of sewage. Part of a coordinated programme on the study and improvement of biospheric resources. Final report, 1 April 1970-31 May 1972. International Atomic Energy Agency, Vienna, report IAEA-R-795-F. 5pp. (NSA 30: 27022).

Ramakrisnan, S. P., B. S. Krishnamurthy and S. N. Ray. 1962. Laboratory studies on the use of irradiated sterile males to reduce C. fatigans Wied. populations. Indian J. Malariol. 16: 357-364. (NSA 20: 24904).

Ramamurty, P. S. 1963. On the origin of the RNA in the growing egg cells of Panorpa communis (Insecta, Mecoptera). Naturwissenschaften 50(10): 383-384. (In German) (^{3}H).

Ramamurty, P. S. 1964. On the contribution of the follicle epithelium to the deposition of yolk in the oocyte of Panorpa communis (Mecoptera). Exptl. Cell Res. 33: 601-605.

Rambelli, A. 1962. Preliminary research on the behavior of some physiological groups of microorganisms in soil irradiated with cobalt-60. Publ. Centro Sper. Agr. Forest. 6: 119-129. (In Italian) (NSA 17: 30334).

Rambelli, A. 1963. Further research on the effects of gamma radiation from ^{60}Co on soil microflora. Publ. Centro Sper. Agr. Forest. 7: 129-146. (In Italian) (NSA 22: 45730).

Ramel, C., G. Bonnier and U. B. Jonsson. 1958. Selection pressure on irradiated populations of Drosophila melanogaster. Abstr. Radiation Research 9(1): 170.

Ramenskaya, G. P. 1965. The incorporation of uridine-^{3}H into cellular structures of the silk gland and the effect on the process of actinomycin D. Tsitologiya 7(1): 88-90. (In Russian).

Ramirez-Martinez, J. R. and A. D. McLaren. 1966. Some factors influencing the determination of phosphatase activity in natural soils and in soils sterilized by irradiation. Enzymologia 31: 23-38. (NSA 21: 12774).

Ramodiharilafy, J. and Ph. Couchat. 1974. Tritium and oxygen-18 labelling as applied to the study of saturated and non-saturated soil dispersion effects. pp. 135-144. In: Isotope and Radiation Techniques in Soil Physics and Irrigation Studies 1973. (See IAEA, 1974) (In French) (NSA 30: 15490).

Ramos, E. and M. Celma. 1964. Environmental radioactivity surveys in Spain. pp. 989-1007. In: The Natural Radiation Environment. (See Adams and Lowder, 1964) (NSA 18: 35754).

Ramos, V. P. 1964. The use of sterilized male insects in pest control. Gaz. Agric. Mocamb. 16(177): 47-53. (In Portugese).

Ramzayev, P. V., et al. 1965. Indirect assessment of total body burden of ^{137}Cs in humans. Med. radiologiya 6: 22-28. (In Russian) (NSA 19: 35922).

Ramzayev, P. V., et al. 1967. Basic results of radiation hygiene investigations into the migration of total fallout in the arctic regions of the USSR in 1959-1966. State Committee on the Use of Atomic Energy, USSR. Atomizdat, Moscow. 13pp. UN Document A/AC.82/G/L.1189. (In Russian).

Ramzayev, P. V., et al. 1967. Metabolism of cesium-137 of global origin among inhabitants of arctic regions. State Committee on the Use of Atomic Energy, USSR. Atomizdat, Moscow. 19pp. UN Document A/AC.82/G/L.1183. (In Russian).

Ramzeyev, P. V., et al. 1967. Radiation-hygienic conditions in the Far North (1959-1965). Tr. radiats. gig. Leningrad. nauchn.-issled. inst. rad. gig. (3): 251-260. (AEC-tr-7000) (NSA 23: 14150) (Lichen-reindeer-man food chains).

Ramzayev, P. V., et al. 1969. Results of investigations of nuclear fallout over the territory of the USSR. Atom. energiya 26(1): 62-63. (Sov. Atom. Energy 26(1): 73-75) (NSA 23: 49954) (Incl. lichens, venison, grass, river fish).

Ramzayev, P. V., et al. 1971. Study methods of radioactive isotope migration in lichen-reindeer-man. pp. 167-175. In: Methods of Radioecological Investigations. (See Verkhovskaya, 1971) (In Russian).

Ramzayev, P. V., et al. 1973. ^{137}Cs, ^{90}Sr, ^{210}Pb, and ^{210}Po in the food chain of lichen-reindeer-man. pp. 417-420. In: Health Physics Problems of Internal Contamination. E. Bujdose, ed. (CONF-720503) (NSA 28: 29898).

Rancon, D. 1967. Changes in the moisture of a soil continuously irrigated over a small surface. pp. 125-135. In: Isotope and Radiation Techniques in Soil Physics and Irrigation Studies. (See IAEA, 1967) (In French) (NSA 22: 23147) (Neutron-moisture meter).

Rancon, D. 1971. Study and demarcation of pollution in non-saturated soils. pp. 551-563. In: Nuclear Techniques in Environmental Pollution. (See IAEA, 1971) (In French) (NSA 25: 32081).

Rancon, D. 1972. Practical use of the distribution coefficient in measuring the radioactive contamination of minerals, rocks, soil and underground water. Centre d'Études Nucléaires, France Commissariat à l'Énergie Atomique, Cadarache, report CEA-R-4274. 35pp. (In French) (NSA 26: 31015).

Rancon, D. 1973. Soil contamination. 2. Behavior of radioactive ions in unsaturated soils after water disposal into a limited surface. Centre d'Études Nucléaires, France Commissariat à l'Énergie Atomique, Cadarache, report CEA-R-3635(2). 50pp. (BNWL-tr-152) (NSA 28: 376; 33: 2981).

Randolph, N. 1959. With the help of radiactive cobalt, science attacks the screwworm. Nature Mag. 52(9): 460-463.

Randolph, P. D. 1970. Engineering the isotopic snow gage. pp. 36-47. In: Trans. Isotopic Snow Gage Information Meeting. (See Idaho Nuclear Energy Commission, 1970) (NSA 25: 51927).

Randow, F. and H.-A. Schulze. 1971. Studies on the artificial radioactivity of fish in 1967 to 1968. Nahrung 15(1): 81-89. (In German) (NSA 25: 35096).

Rangnekar, P. V. and D. F. Forward. 1969. Foliar nutrition and growth in red pine: the fate of photoassimilated carbon in a seedling tree. Can. J. Bot. 47: 897-906. (^{14}C).

Ranzi, S. 1970. Some freshwater radioecological problems. pp. 21-25. In: Proc. 1st Conf. on the State of Advancement of Radioecology in Italy. (See Anonymous, 1970) (In Italian).

Rao, A. R. and P. N. Srivastava. 1967. Ovarian changes induced by chronic gamma radiation emitted by sealed ^{60}Co source placed inside the abdomen in the Indian desert gerbil. Strahlentherapie 133: 594-601. (NSA 21: 45812).

Rao, N. J., R. K. Iyer and M. S. Das. 1967. Investigation of trace element distribution in potsherds by activation analysis. pp. 498-504. In: Proc. of the Nuclear and Radiation Chemistry Symp. (Poona, March 6-9, 1967). India Dept. of Atomic Energy, Bombay. (CONF-670335) (NSA 23: 19).

Rao, P. J. and K. N. Mehrotra. 1972. Effects of ionizing radiation on consumption, digestion, and utilization of food by the desert locust, Schistocerca gregaria Forskal. Indian J. Exptl. Biol. 10(5): 350-356. (NSA 28: 21782).

Rao, S. R., M. D. Borkar and S. N. Shah. 1971. Distribution of ^{226}Ra in soil and water. pp. 507-512. In: Proc. National Symp. on Radiation Physics. India Bhabha Atomic Research Centre, Trombay, Bombay. (CONF-701138) (NSA 26: 40849).

Rasmussen, N. C. 1974. Nuclear power station safety. ERA 47(12): 238-242. (NSA 33: 8780).

Rast, D., A. G. McInnes and A. C. Neish. 1963. Synthesis of raffinose in spruce twigs. Can. J. Bot. 41(12): 1681-1686. (Used ^{14}C).

Rastogi, R. K., Y. Hyodo-Taguchi and N. Egami. 1969. Effects of fractionated whole-body x irradiation on cell proliferation in the intestine of the goldfish, Carassius auratus. Annot. Zool. Jap. 42: 193-202. (NSA 24: 32159).

Rasulov, F. K. 1963. Effect of gamma rays on the sterility of the cotton moth (Laphygma exigua). Khlopkodstvo (7): 41-42. (In Russian) (NSA 18: 23283).

Rasulov, F. K. and S. A. Anastasov. 1963. The control of warehouse pests by gamma radiation. Vestn. Sel'skokhoz. Nauki 9: 34-35. (In Russian).

Ratcliffe, D. A. 1970. Changes attributable to pesticides in egg breakage frequency and eggshell thickness in some British birds. J. Appl. Ecol. 7(1): 67-107. (Considers ionizing radiation).

Rathi, S. K. 1969. Response of testes to TP and irradiation in the Indian catfish, Heteropneustes fossilis (Bloch). Indian J. Zootomy 10(3): 175-190. (NSA 30: 3684).

Rathi, S. K. 1970. Radiosensitizing action of androgen on the ovaries of Heteropneustes fossilis (Bloch). pp. 415-418. In: Radiation Protection and Sensitization. H. Moroson, ed. Barnes & Noble, New York. (NSA 25: 4591).

Rathi, S. K. 1973. Radiation-induced zoning in the cytoplasm of early oocytes of Heteropneustes fossilis Bloch. Biologia (Bratislava) 28(12): 997-1001. (NSA 31: 6362).

Rathsack, R. and W. Brucker. 1962. On the induced increased resistance of bacteria against conventional roentgen irradiation. Acta Biol. Med. Ger. 8: 547-558. (In German) (NSA 17: 1245).

Rattan Lal, et al. 1960. Studies on the effect of beta-radiations on insects. II. The effect of beta radiations (Irradiation of full-grown caterpillars and freshly-formed pupae) on the life-history of Corcyra cephalonica Stainton. Indian J. Entomol. 22(1): 15-21.

Rattazzi, M. 1964. Free amino acids in the embryos of Rana esculenta and Bufo vulgaris. Quantitative determination in normal ova and in ova fertilized with irradiated sperm. Atti accad. Naz. lincei rend. cl. sci. fis. mat. nat. 36: 75-85. (In Italian) (NSA 19: 12978).

Ratty, F. J. and E. J. Loveliette. 1967. Chromosome substitution and radiosensitivity in Drosophila melanogaster. Radiation Research 30: 825-831.

Rauert, W. and W. Stichler. 1974. Groundwater investigations with environmental isotopes. pp. 431-443. In: Isotope Techniques in Groundwater Hydrology. Vol. 1. (See IAEA, 1974) (NSA 31: 19640).

Raude, D., et al. 1974. Energy and the environment in Baden-Württemburg. German F. R. Kernforschungszentrum, Karlsruhe, report KFK-1966-UF. 183pp. (In German) (NSA 30: 12714).

Raudszuz, O. 1958. Marking the queens with a radioactive substance. Gleanings Bee Cult. 86(1): 18-19.

Raun, E. S., et al. 1967. Gamma irradiation of European corn borer larvae. J. Econ. Entomol. 60(6): 1724-1730.

Rausch, L. 1973. Actual problems of environmental protection in energy production with consideration of the relation between environmental protection and radiation protection. pp. 1-17. In: Worldwide Radiation from Nuclear and Conventional Energy Production. H. Braun, ed. Georg Thieme Verlag, Suttgart. (In German) (NSA 31: 3330).

Rausch, L. (Ed.). 1973. Podium discussion: environmental protection in nuclear and conventional energy production. pp. 187-221. In: Worldwide Radiation from Nuclear and Conventional Energy Production. H. Braun, ed. Georg Thieme Verlag, Stuttgart. (In German) (NSA 31: 3314).

Ravera, O. 1961. Influence of the concentration of stable manganese on assumption of radiomanganese (Mn^{54}) by Unio pictorum, L. Italy Servicis Sanitario CCR-EURATOM, Ispra (Varese). Final report to IAEA (Contract No. 59).

Ravera, O. 1964. Distribution of Mn^{54} from fall-out in population of freshwater lamellibranchs (Unio mancus elongatus Pfeiffer). Verh. Int. Verein. Limnol. 15(2): 885-892.

Ravera, O. 1964. The radioactivity of Viviparus ater, Crist. and Jan, freshwater mollusk, in relation to that of the sediment. Foder. Europ. Gewassersch. Inform. (10): 61-65. (NSA 19: 1996).

Ravera, O. 1965. Radioactivity in freshwater organisms of some lakes of northern Italy. pp. 195-201. In: Biological Problems in Water Pollution. (See Tarzwell, 1965).

Ravera, O. 1966. Effects of x-irradiation on various stages of the life cycle of Physa acuta Draparnaud, a fresh-water gastropod. pp. 799-808. In: Disposal of Radioactive Waste into Seas, Oceans and Surface Waters. (See IAEA, 1966) (SM-72/52) (CONF-660507-30) (NSA 20: 40700).

Ravera, O. 1966.. The effect of x rays on demographic characteristics of a freshwater gastropod, Physa acuta Drap. Malacologia 5: 51. (NSA 21: 34747).

Ravera, O. 1967. The effect of x rays on the demographic characteristics of Physa acuta (Gastropoda: Basommatophora). Malacologia 5: 95-109. (NSA 22: 6885).

Ravera, O. 1970. Activities of the Biological Division of C.C.R EURATOM in radioecology. pp. 47-54. In: Proc. 1st Conf. on the State of Advancement of Radioecology in Italy. (See Anonymous, 1970) (In Italian).

Ravera, O. 1971. Sr-85 and Cs-134 uptake by freshwater seston. Rev. Int. Oceanogr. Med. 21: 109-124.

Ravera, O. 1972. Water, water everywhere, or any drop to drink... . Euro-Spectra 11(2): 34-46. (NSA 27: 2529) (Incl. program at Ispra, Italy).

Ravera, O. and R. de Bernardi. 1970. Effects of gamma and x irradiation on the variety normal and albino of Biomphalaria glabrata, Say (Gastropoda, Prosobranchiata). pp. 1137-1150. In: Proc. Intern. Symp. on Radioecology. (See French Commissariat à l'Énergie Atomique, 1970) (NSA 24: 41963).

Ravera, O. and P. Gaglione. 1962. Concentration of Mn^{54} in fallout in relation to age and organ of Unio mancus elongatus (Mollusk, Lamellibranch). Ist. Lombardo (rend. sci.) B96: 157-164. (In Italian).

Ravera, O. and M. Merlini. 1961. Different biometrics and ponderali of populations of Unio mancus var. elongatus, Pfeiffer (Mollusk, Lamellibranch) of Lake Maggiore. Boll. Zoologia 28(2): 457-468. (In Italian, English summary) (Discusses ^{54}Mn in population).

Ravera, O. and M. Merlini. 1960. A research program for the study of radioactivity in mollusks as a possible index to the contamination of a lake. Italy Comitato Nazionale per l'Energia Nuclear, Ispra, report CNI-58. 13 pp. (NSA 16: 22428).

Ravera, O. and G. Premazzi. 1971. A method to study the history of any persistent pollution in a lake by the concentration of ^{137}Cs from fallout. pp. 703-718. In: Proc. Intern. Symp. on Radioecology Applied to the Protection of Man and His Environment. (See Commission of the European Communities, 1971) (NSA 27: 5292).

Ravera, O. and L. Vido. 1961. Measurement of ^{54}Mn in populations of Unio pictorum L. (Mollusca, Lamellibranchia) of Lake Maggiore. Mem. ist. Ital. Idrobiol. Dott. Marco de Marchi 13: 75-84. (In Italian) (NSA 20: 16185).

Ravera, O., M. Merlini and S. Musso. 1969. Uptake and excretion of radiocobalt by a freshwater mollusc (Unio mancus var. elongatus). Minerva fisiconucl. 13: 301-305.. (In Italian) (CONF-680535-29) (NSA 25: 29441).

Ravier, R. H. and G. Schneebeli. 1955. Use of tracers in hydrology. Assoc. Franc. Étude Eaux (40): 1-2. (In French).

Ravoire, J., et al. 1970. Tritium content in a firn core from Antarctica. J. Geophys. Res. 75: 2331-2335. (NSA 24: 25134).

Ray, B. J. and D. L. Johnson. 1972. Method for the neutron activation analysis of natural waters for arsenic. Anal. Chim. Acta 61(1): 196-199. (NSA 27: 43).

Ray, D. T. 1957. X-ray sensitivity of meiotic stages of *Mormoniella* eggs. Abstr. Genetics 42(4): 391.

Ray, D. T. 1963. Fecundity studies on x-rayed *Mormoniella vitripennis*. Abstr. Genetics 48(7): 906.

Ray, E. L. and S. E. Hammond. 1960. Use of alpha pulse-height analyzer in environmental monitoring. pp. 10-17. In: *Proc. 5th Ann. Meeting Bio-Assay and Analytical Chemistry Group*. (Gatlinburg, Tennessee, October 1-2, 1959). U. S. AEC report TID-7591.

Ray, J. 1957. Will atomic radiation control Africa's tsetse fly? Foreign Agr. 21(6): 15-16.

Ray, O. M. 1938. Effects of roentgen rays on the activation and production of the enzyme tyrosinase in the insect egg (Orthoptera). Radiology 31(4): 428-437.

Ray-Chaudhuri, S. P. and I. Sarkar. 1952. Frequency of dicentric bridges in meiosis in the grasshopper *Gesonia punctifrons*, produced by different dosages of x-rays. Science 116(3018): 479-482. (NSA 7: 1041).

Ray-Chaudhuri, S. P. and I. Sarkar. 1958. The effect of P^{32} on the meiotic chromosomes of the grasshopper, *Gesonula punctifrons*. Nucleus 1(1): 123-130.

Raynaud, A. 1960. Test of destruction by localized irradiation with x-rays of the hypophysial roots of the embryo of the blindworm (*Anguis fragilis* L.). Compt. rend. 251: 2416-2418. (In French) (NSA 15: 8541).

Raynaud, A. and M. Frilley. 1965. The development of external genital organs of the embryo of the blindworm (*Anguis fragilis* L.) in the absence of genital glands. Compt. rend. 261: 5641-5644. (In French) (NSA 20: 24880).

Read, J. 1963. Restoration of supralethally irradiated amoebae. Int. J. Radiat. Biol. 7(1): 101-102. (NSA 18: 19637).

Read, P. A. 1967. Neutron-induced activity on a beach: a method for calculating the dose rate. U. S. Naval Radiological Defense Lab., San Francisco, report USNRDL-TR-67-133. 25pp. (AD-663103; NSA 22: 35938).

Rebiffe, J. 1973. Tritium: physical, chemical, and radiological data. Rayonn. Ionis. Tech. nes. prot. (2): 63-70. (In French) (NSA 28: 27110).

Rebuck, A. L. 1964. Changes in the structure of an oak-pine forest exposed for two years to chronic gamma irradiation. Dissertation, Duke Univ., Durham, N. C. x, 88pp. (Also: Brookhaven National Lab., U. S. AEC report BNL-8675).

Rebuck, A. L. and G. M. Woodwell. 1964. Patterns of mortality in an oak-pine forest exposed to chronic gamma irradiation. Abstr. Bull. Ecol. Soc. Am. 45(3): 73.

Rebuck, A. L. and G. M. Woodwell. 1965. The relationship between mortality and size of trees in a forest exposed to chronic gamma irradiation. Abstr. Bull. Ecol. Soc. Am. 46(3): 98.

Rechen, H. J. L., et al. 1968. ^{137}Cs concentrations in Alaskans during the spring of 1967. Radiol. Hlth. Data 9: 705-717. (NSA 23: 31781).

Recher, H. F. 1970. Population density and seasonal changes of the avifauna in a tropical forest before and after gamma irradiation. pp. E69-E93. In: *A Tropical Rain Forest*. (See H. T. Odum and R. F. Pigeon, 1970) (NSA 25: 21841).

Recht, P. 1971. Health protection of the environment in industrialized countries. Health Physics 21(2): 145-161. (In French) (NSA 25: 46881).

Recht, P. 1972. Ecological aspects of radiation protection. Int. Atom. Energy Agency Bull. 14(2): 26-32. (NSA 26: 50596).

Recht, P. 1973. Comparison of the influence on the environment of nuclear and other industries. Assoc. des Ingenieurs Electriciens, Liége, Belgium. 25pp. (In French) (INIS-mf-579) (NSA 29: 5102).

Recht, P. and M. Collet. 1966. Systematic international study of the radioactivity in a river basin (program and organization). Abstr. Health Physics 12(12): 1819-1820.

Recht, P. and M. Collet. 1968. Systematic study of the radioactivity of a drainage waterway organized on an international plan. pp. 1337-1341. In: *Radiation Protection. Part 2*. W. S. Snyder, et al., eds. Pergamon Press, New York. (In French) (NSA 23: 13858).

Recht, P. and J. R. A. Lakey. (Eds.). 1975. *Radiation Protection Measurement: Philosophy and Implementation*. (Selected Papers of the Intern. Symp. Held at Aviemore, Scotland, 2-6 June 1974). Commission of the European Communities, Luxembourg, report EUR-5397e. 297pp. (In French and English) (NSA 33: 26179).

Reda, A. and G. V. Sumarukov. 1970. Minimum sterilizing doses of x-ray irradiation for the first and second generations of the rice and granary weevil. Vestn. Mosk. Univ. (biol. yochv.) 25(5): 8-10. (In Russian) (NSA 25: 58161).

Reddi, O. S. 1963. The genetical response of *Drosophila* spermatozoa to different types of radiation treatment. Int. J. Radiat. Biol. 7(2): 301-305. (NSA 18: 19647).

Reddi, O. S. and M. S. Rao. 1964. Genetic effects of cosmic radiation in Drosophila melanogaster. Nature 201: 96-97.

Reddi, O. S., et al. 1965. Lack of mutagenic effect of irradiated Drosophila medium. Nature 208: 702. (NSA 20: 6850).

Rédei, G. P. 1962. Supervital mutants in Arabidopsis. Genetics 47(4): 443-460. (NSA 16: 21943).

Redfield, A. C. and E. M. Bright. 1918. Temperature coefficient of the action of beta rays upon the egg of Nereis. J. Gen. Physiol. 1(2): 255-260.

Redfield, A. C. and E. M. Bright. 1921. The physiological changes produced by radium rays and ultrviolet light in the egg of Nereis. J. Physiol. 55 (1/2): 61-85.

Redies, M. and I. Vimpany. 1966. Determination of total potassium in soils using the gamma emission of ^{40}K. Nature 210: 1078-1079. (NSA 20: 39280).

Rediske, J. H. and A. A. Selders. 1953. The absorption and translocation of strontium by plants. Plant Physiology 28(4): 594-605.

Rediske, J. H. and A. A. Selders. 1954. The uptake and translocation of yttrium by higher plants. Am. J. Bot. 41: 238-242. (Also: Hanford Atomic Products Operation, U. S. AEC report HW-29091) (NSA 8: 996).

Rediske, J. H., J. F. Cline and A. A. Selders. 1955. The absorption of fission products by plants. Hanford Atomic Products Operation, U. S. AEC report HW-36734. 17pp. 17pp. (NSA 9: 5837).

Rediske, J. H., I. Gauditz and N. E. Johnson. 1970. Distribution of dimethoate-P^{32} in Douglas-fir following stem injection. Forest Science 16 (1): 106-___.

Reed, J. R. 1971. Uptake and excretion of ^{60}Co by black bullheads Ictalurus melas (Rafinesque). Health Physics 21(6): 835-844. (NSA 26: 9464).

Reed, J. R. and B. A. Martinedes. 1973. Uptake and retention of tungsten-181 by crayfish (Cambarus longulus longerostris Ort.). pp. 390-393. In: Radionuclides in Ecosystems. (See Nelson, 1973) (NSA 28: 27678).

Reed, J. R. and D. J. Nelson. 1969. Radiostrontium uptake in blood and flesh in bluegills (Lepomis macrochirus). pp. 226-233. In: Symposium on Radioecology. (See Nelson and Evans, 1969) (Also: Oak Ridge National Lab., Lab., U. S. AEC report ORNL-P-3160) (NSA 21: 34656).

Reeder, H. O. and L. L. Thatcher. 1963. Tritium used as a groundwater tracer between Lake McMillan and major Johnson Springs, Eddy County, New Mexico. U. S. Geological Survey, U. S. AEC report TEI-839. 135pp. (NSA 18: 25684).

Rees, G. H. 1962. Effects of gamma radiation on two decapod crustaceans, Palaemonetes pugio and Uca pugnax. Chesapeake Sci. 3(1): 29-34. (NSA 16: 21931).

Reeves, M. and J. O. Duguid. 1975. Application of soil and ground-water pollutant-transport model. Oak Ridge National Lab., U. S. AEC report CONF-750530-1. 15pp. (NSA 32: 6430).

Reginato, R. J. 1973. Swelling and shrinking of soil in situ as determined by a dual gamma-ray transmission technique. Thesis, Univ. of California, Riverside. 80pp. (Univ. Microfilms 73-23837) (NSA 29: 415).

Reginato, R. J. 1974. Gamma radiation measurements of bulk density changes in a soil pedon following irrigation. Soil Sci. Soc. Am. Proc. 38(1): 24-29. (NSA 29: 24210).

Reginato, R. J. and C. H. M. van Bavel. 1964. Soil water measurement with gamma attenuation. Soil Sci. Soc. Am. Proc. 28(6): 721-724.

de Regt, H. J. B. 1960. Radioactive wastes. V. Radioactivity investigations of water. Atoomenergie 2: 53-57. (In Dutch) (NSA 14: 14609).

Rehak, W. and B. Feddersen/G. Ubl. 1970. Radiochemical determination of ^{137}Cs in soil. Radiochemical determination of ^{144}Ce and ^{141}Ce in environmental samples. Rapid radiochemical determination of ^{137}Cs in water. German D. R. Staatliche Zentrale für Strahlenschutz, Berlin, report SZS-1/70. pp. 8-18, 19-30, 1-7. (In German) (NSA 24: 29213, 29214, 29212).

Rehak, W. and B. Feddersen. 1973. Radiochemical determination of strontium-90 in samples of environmental and personal monitoring. pp. 17-28. In: Radiation Monitoring, Radiation Chemistry, and Radioactive Wastes. German D. R. Staatliche Zentrale für Strahlenschutz, Berlin, report SZS-149. (In German) (NSA 29: 18076).

Rehak, W. and F. Moldenhawer. 1974. Probe analytics for radiation protection measurements in the Federal Center for Radiation Protection of the DDR. pp. 258-263. In: Radiation Protection, Environment, and the Population. (See German Democratic Republic, 1974) (In German) (NSA 31: 8623).

Reichardt, K. and P. L. Libardi. 1974. New equation to estimate soil-water diffusivity. pp. 45-51. In: Isotope and Radiation Techniques in Soil Physics and Irrigation Studies 1973. (See IAEA, 1974) (NSA 30: 15484).

Reichert, S. O. 1962. Radionuclides in ground water at the Savannah River Plant waste disposal facilities. J. Geophys. Res. 67: 4363-4374. (NSA 17: 4700).

Reichert, S. O. and J. W. Fenimore. 1962. Lithology and hydrology of radioactive waste disposal sites at the Savannah River Plant. Savannah River Plant, U. S. AEC report DPSPU-62-30-21B. 37pp. (NSA 17: 2639).

Reichert, W. L. and D. C. Malins. 1974. Interaction of mercurials with salmon serum lipoproteins. Nature 247(5442): 569-570. (SFA 19: 17396) (^{203}Hg).

Reichle, D. E. 1965. ^{134}Cesium behavior and energetics in terrestrial isopods. Abstr. Bull. Ecol. Soc. Am. 46(3): 133.

Reichle, D. E. 1967. Measurement of elemental assimilation in arthropods. Abstr. Bull. Ecol. Soc. Am. 48(2): 64. (Incl. radionuclides).

Reichle, D. E. 1967. Radiocesium turnover rates and energy flow in terrestrial isopod populations. Ecology 48(3): 351-366.

Reichle, D. E. 1968. Relation of body size to food intake, oxygen consumption, and trace element. Ecology 49(3): 538-542. (Radiocesium).

Reichle, D. E. 1969. Lethal effects of gamma radiation on eggs and larvae of the bagworm, Thyridopteryx ephemeraeformis. Ann. Entomol. Soc. Am. 62(5): 1113-1117.

Reichle, D. E. 1969. Measurement of elemental assimilation by animals from radioisotope retention patterns. Ecology 50(6): 1102-1104.

Reichle, D. E. and D. A. Crossley, Jr. 1965. Radiocesium dispersion in a cryptozoan food web. In: Radiation and Terrestrial Ecosystems. Health Physics 11(12): 1375-1384. (See Hungate, 1965).

Reichle, D. E. and D. A. Crossley, Jr. 1967. Investigation on heterotrophic productivity in forest insect communities. pp. 563-587. In: Secondary Productivity of Terrestrial Ecosystems. Vol. II. K. Petrusewicz, ed. Inst. of Ecology, Polish Acad. Sci., Warsaw. (^{137}Cs).

Reichle, D. E. and D. A. Crossley, Jr. 1969. Trophic level concentrations of cesium-137, sodium, and potassium in forest arthropods. pp. 678-686. In: Symposium on Radioecology. (See Nelson and Evans, 1969) (Also: Oak Ridge National Lab., U. S. AEC report ORNL-P-3065) (NSA 21: 32603).

Reichle, D. E. and R. I. Van Hook, Jr. 1970. Radionuclide dynamics in insect food chains. Manitoba Entomol. 4: 22-32. (NSA 29: 29388).

Reichle, D. E. and R. I. Van Hook, Jr. 1971. Effect of temperature and radiation stresses on the survivorship of isopods: Armadillidium vulagre and Cylisticus convexus. Radiation Research 46(3): 588-600.

Reichle, D. E., P. B. Dunaway and D. J. Nelson. 1970. Turnover and concentration of radionuclides in food chains. Nuclear Safety 11: 43-55. (NSA 24: 12038).

Reichle, D. E., M. H. Shanks and D. A. Crossley, Jr. 1969. Calcium, potassium, and sodium content of forest floor arthropods. Ann. Entomol. Soc. Am. 62(1): 57-62.

Reichle, D. E., R. I. Van Hook, Jr. and E. A. Bardill. 1969. Effect of temperature and radiation stresses on the survivorship of isopods: Armadillidium vulgare (Latrielle) and Cyclisticus convexus (DeGeer). Abstr. Bull. Ecol. Soc. Am. 50(2): 68-69.

Reichle, D. E., et al. 1971. Effects of beta-gamma radiation of earthworms under simulated-fallout conditions. pp. 527-534. In: Survival of Food Crops and Livestock in the Event of Nuclear War. (See Bensen and Sparrow, 1971).

Reichle, D. E., et al. 1973. Organic matter and ^{137}Cs turnover in forest soil by earthworm populations: application of bioenergetic models to radionuclide transport. pp. 240-246. In: Radionuclides in Ecosystems. (See Nelson, 1973) (NSA 28: 29892).

Reid, C. P. P. and F. W. Woods. 1967. Atmospheric transfer of ^{14}C: a problem in fungus translocation studies. Science 157: 712-713. (NSA 21: 36532).

Reid, C. P. P. and F. W. Woods. 1969. Translocation of C^{14}-labeled compounds in mycorrhizae and its implications in interplant nutrient cycling. Ecology 50(2): 179-187.

Reid, D. F., W. T. Ego and S. J. Townsley. 1959. Ion exchange through epithelia of fresh and sea water adapted teleost studies with radioactive isotopes. Anat. Record 134(2): 628.

Reid, D. F., S. J. Townsley and W. T. Ego. 1959. Uptake of Sr^{85} and Ca^{45} through epithelia of freshwater and sea-water adapted Tilapia mossambica. 1p. In: Proc. Hawaii Acad. Sci. 34th Ann. Meeting, 1958-1959. (NSA 15: 2424).

Reid, G. W. 1971. Radioisotope technique for the measurement and treatment process modelling of biological slimes used in waste treatment. pp. 749-754. In: Nuclear Techniques in Environmental Pollution. (See IAEA, 1971) (NSA 25: 32323).

Reid, G. W. 1971-72. Radioisotope technique for studying biological slime-treatment processes. Isotopes Radiat. Technol. 9(2): 244-245. (NSA 26: 9455) (^{32}P).

Reidar, N. and K. Loevseth. 1965. Distribution of radiocarbon from nuclear tests. Nature 206(4988): 1029-1031. (NSA 19: 32483) (Incl. fish).

Reinecke, L. H., W. Klassen and J. F. Norland. 1969. Damage to testes and recovery of fertility in boll weevils fed chemosterilants. Ann. Entomol. Soc. Am. 62: 511-525. (NSA 24: 4676) (Incl. x-rays).

Reiners, W. A. 1963. Changes in evapotranspiration rates in a forest ecosystem damaged by ionizing radiation. Abstr. Bull. Ecol. Soc. Am. 44(3): 69.

Reiners, W. A. 1964. Ecology of a shrub synusia in an irradiated ecosystem. Dissertation, Rutgers Univ., New Brunswick, N. J. (Abstr. Bull. Ecol. Soc. Am. 45(3): 75).

Reiners, W. A. and G. M. Goodwell. 1966. Water budget of a forest damaged by ionizing radiation. Ecology 47: 303-306. (Also: Brookhaven National Lab., U. S. AEC report BNL-8466) (NSA 20: 3653; 21: 8749).

Reinhardt, K. 1967. Problems of irradiation investigations during hypothermia and hibernation. Strahlentherapie 133: 396-405. (In German) (NSA 21: 45818) (Amphibians).

Reinholz, E. 1959. Effecting morphogenesis of embryonic organs by ionizing radiation. I. Embryonic anomalies by x-irradiation of Arabidopsis thaliana embryos in various development stages. Strahlentherapie 109: 537-553. (In German) (NSA 13: 19799).

Reinig, W. C. 1963. The 1951 preoperational environmental survey for the Savannah River Plant - in retrospect. Health Physics 9(1): 83-85.

Reinig, W. C. (Ed.). 1970. Environmental Surveillance in the Vicinity of Nuclear Facilities. (Proc. Symp. Held in Augusta, Georgia, January 24-26, 1968). Chas. C. Thomas, Springfield, Ill. 477pp. (CONF-680108) (NSA 25: 16154).

Reinig, W. C. 1970. Objectives of environmental surveillance: panel discussion. pp. 21-25. In: Environmental Surveillance in the Vicinity of Nuclear Facilities. (See Reinig, 1970 above) (NSA 25: 16158).

Reinig, W. C. and E. L. Albensius. 1963. Control of tritium health hazards at the Savannah River Plant. Am. Industr. Hyg. Assoc. J. 24(3): 276-283. (APCA 9: 5399).

Reinig, W. C., et al. 1954. A ecological study of the land plants and cold blooded vertebrates of the Savannah River Project area. Natural radioactive contents of the environs of the Savannah River Project. Univ. of S. Car. Publ. (biol. ser. 3) 1(3): 180-186.

Reiniger, P, E. Levi and M. Coppola. 1974. Measurement of surface water storage on ryegrass and clover by nuclear methods. pp. 327-334. In: Isotope and Radiation Techniques in Soil Physics and Irrigation Studies 1973. (See IAEA, 1974) (NSA 30: 15501).

Reininger, P., M. J. Frissel and P. Poelstra. 1970. Calculation by digital computer of the migration rate of ^{90}Sr in the soil. pp. 589-609. In: Proc. Intern. Symp. on Radioecology. (See French Commissariat à l'Énergie Atomique, 1970) (In French) (NSA 25: 2366).

Reiquam, H. 1972. Establishing priorities among environmental stresses. pp. 71-82. In: Indicators of Environmental Quality. W. A. Thomas, ed. Plenum Press, New York. (Incl. ionizing radiation).

Reissig, H. 1963. Sr^{90} contents of soils and plants from the territory of the DDR in 1961. ---1962. Kernenergie 6(5): 231-238; 7(2): 117-127. (In German) (NSA 17: 30284; 18: 15909).

Reissig, H. 1966. Investigations on ^{90}Sr washout in soils. Chem. Erde 25: 204-229. (In German) (NSA 22: 10370).

Reitemeier, R. F. 1968. Soil-plant relationships of radioactive nuclides in fallout. pp. 143-148. In: Proc. Symp. on Postattack Recovery from Nuclear War. U. S. Office of Civil Defense, Washington.

Reitemeier, R. F., H. Hollister and L. T. Alexander. 1967. The extent and significance of soil contamination with radionuclides. pp. 269-282. In: Agriculture and the Quality of Our Environment. (See Brady, 1967).

Reiter, R. 1958. Natural and artificial air radioactivity in the Alps at different heights. Ber. Deut. Wetterdienstes 54: 38-53. (In German) (NSA 16: 26724).

Reiter, R. 1959. Measurements of the radioactivity in air, precipitation, and surface water during catastrophic high water in the northern Alps. Atomkernenrgie 4: 490-492. (In German) (NSA 14: 6559).

Reiter, R. 1961. Fission-product radioactivity in precipitation on glacier surfaces, in grasses and animals from measurements at different altitudes in the northern Alps in the years 1957 to 1960. Archiv. Meteorol. Geophys. Bioklimatol. A12(2): 222-261. (In German, English summary) (NSA 16: 28731).

Reiter, R. 1961. Natural and Artificial Radioactivity in Mountains. Friedrich-Karl Schattauer Verlag, Stuttgart. 78pp. (In German).

Reiter, R. 1962. On the current situation of atmospheric contamination through fission products. Naturwissenschaften 49(4): 83-84. (In German).

Reiter, R. and R. Sladkovic. 1970. Influx of fresh nuclear fission prodcts from China beginning January 1969. Meteorol. Rundsch. 23: 45-47. (In German) (NSA 24: 44133).

Rejesus, R. S. and E. B. Lapis. 1972. Disinfestation of stored products by gamma irradiation. I. Defining lethal doses for the red flour beetle, Tribolium castaneum Herbst, and saw-toothed grain beetle, Oryzaephilus surinamensis (Lin.). II. Survival pattern and reproductive potential of the red flour beetle, Tribolium castaneum Herbst, exposed to sub-lethal doses of gamma radiation. pp. 49-56, 74-80. In: Proc. Radioisotope Soc. Philippines. 1972. 1973. (CONF-720486) (NSA 30: 3675; 31: 29950).

Rekers, P. E. 1951. The effect of x-irradiation on rats with and without Bartinella muris. J. Infect. Diseases 88(3): 224-229.

Rekharskaya, V. M. and L. N. Polupanova. 1973. Distribution of uranium in fossil bone detritus. Izv. Akad. Nauk SSSR (geol.) (3): 122-129. (In Russian) (NSA 29: 2440).

Remezova, T. S. 1957. Effect of γ-neutron radiation on microorganisms. pp. 77-83. In: Summary of Reports of the All-Union Conf. on Application of Isotopes and Irradiation in National Economy and Science. Acad. Sci. USSR, Moscow. (In Russian).

Remmenga, E. E. and F. W. Whicker. 1967. Sampling variability in radionuclide concentrations in plants native to the Colorado Front Range. Health Physics 13(9): 977-983. (NSA 21: 41414).

Remy, M. L. and P. Pellerin. 1968. Natural radioactivity of 250 French mineral water springs. Bull. inst. nat. sante rech. med. 23: 23-61. (In French) (NSA 22: 48932).

Renaud, B. C. 1973. Calibration and operation of a soil volumetric moisture and weight meter which uses radioactive sources. Thesis, Univ. Nacional Autonoma de Mexico, Mexico City. 110pp. (In Spanish) (NSA 30: 6769).

Renfrew, C. 1971. Carbon 14 and the prehistory of Europe. Scient. Amer. 225(4): 63-70, 72.

Renfrew, C. and R. M. Clark. 1974. Problems of the radiocarbon calender and its calibration. Archeometry 17(1): 5-18. (NSA 29: 26850).

Renfro, W. C. 1968. Dynamics of ^{65}Zn in biota of the Columbia River Estuary. Abstr. Bull. Ecol. Soc. Am. 49(2): 72.

Renfro, W. C. 1968. Radioecology of ^{65}Zn in an arm of the Columbia River Estuary. Thesis, Oregon State Univ, Corvallis. 94pp. (NSA 22: 45469).

Renfro, W. C. 1972. Radioecology of zinc-65 in Alder Slough, an arm of the Columbia River Estuary. pp. 755-776. In: The Columbia River Estuary and Adjacent Ocean Waters. (See Pruter and Alverson, 1972).

Renfro, W. C. 1973. Seasonal radionuclide inventories in Alder Slough, an ecosystem in the Columbia River Estuary. pp. 738-746. In: Radionuclides in Ecosystems. (See Nelson, 1973) (NSA 28: 29892).

Renfro, W. C. and C. Osterberg. 1969. Radiozinc decline in starry flounders after temporary shutdown of Hanford reactors. pp. 372-379. In: Symposium on Radioecology. (See Nelson and Evans, 1969).

Renfro, W. C., W. O. Forster and C. L. Osterberg. 1972. Seasonal and areal distributions of radionuclides in the biota of the Columbia River Estuary. pp. 777-798. In: The Columbia River Estuary and Adjacent Ocean Waters. (See Pruter and Alverson, 1972).

Renfro, W. C., R. V. Phelps and T. F. Guthrie. 1969. Influence of preservation and storage of radiozinc and total zinc concentrations in fish. Limnol. Oceanogr. 14(1): 168-170. (NSA 23: 26780).

Renk, H. and J. Chrapan. 1969. Measurement of extremely low ^{14}C activities with GM counters. Acta Fac. Rerum Nat. Univ. Comen. Phys. (9): 61-75. (In Slovak) (NSA 25: 21406).

Renn, C. E. 1964. Adsorption of zinc (and silver) on silt and water handling surfaces. Final report. Johns Hopkins Univ., Baltimore. Md., U. S. AEC report NYO-10544(Rev.). 27pp. (NSA 18: 45176).

Rensing, L. 1969. A circadian rhythm of sensitivity of x irradiation in Drosophila. Z. Vergl. Physiol. 62: 214-220. (In German) (NSA 23: 43854).

Rerábek, J. and A. Bubeník. 1956. Studies on the mineral metabolism of antlered deer by means of radioactive isotopes. Z. Jagdwissensch. 2(3): 119-123. (AEC-tr-4942).

Rerábek, J. and A. Bubeník. 1963. The metabolism of phosphorus and iodine in deer. Unpublished report. (AEC-tr-5631) (^{131}I and ^{32}P) (NSA 17: 35408).

Retief, J. V. and P. Kruger. 1971. Use of nuclear explosives for water resources development in arid regions. Stanford Univ., U. S. AEC report SU-326-P-31-2. 108pp. (NSA 25: 48685).

Retnakaran, A. 1971. Method for determining sperm precedence in insects. J. Econ. Entomol. 64(3): 578-580. (NSA 25: 42259) (^{3}H-, ^{14}C-sperm).

Reuter, F. W., III. 1975. Numerical matrix correction technique for the measurement of trace elements in plant materials by x-ray fluorescence spectroscopy. Anal. Chem. 47(11): 1763-1766. (NSA 32: 21876).

Reverberi, G., et al. 1960. An analysis of the ascidian development by the use of radioisotopes. Acta Embryol. Morphol. Exp. 2: 202-212.

Reynier, B., L. Foulquier and A. Grauby. 1970. Experimental contamination of Anguilla anguilla (L.) by ^{137}Cs and decontamination. Ann. Limnol. 6(1): 101-129. (In French) (NSA 25: 24141).

Reynolds, S. A. and T. G. Scott. 1975. Determination of plutonium in environmental samples. I. Development of methods. Radiochem. Radioanal. Lett. 23(4): 269-274.

Reynolds, S. L., et al. 1968. Effects of x-irradiation on the immune response of guinea pigs to Q fever vaccine. Radiation Research 36: 98-106. (NSA 22: 51469).

Reynolds, T. D. 1963. Sorption and release of radionuclides by sediments. pp. 127-144. In: Transport of Radionuclides in Fresh Water Systems. (See Kornegay, et al., 1963) (NSA 17: 40526).

Reynolds, T. D. 1963. Transport of strontium and cesium by stream and estuarine sediments. Thesis, Univ. of Texas, Austin. 135pp. (NSA 18: 35726).

Reynolds, T. D. and E. F. Gloyna. 1963. Radioactivity transport in water - transport of strontium and cesium by stream and estuarine sediments. Environ. Hlth. Enrg. Res. Lab., Inst. Marine Sci., Univ. of Texas, Austin, Tech. Rept. -1. xvii, 117pp.

Reynolds Electrical and Engineering Company. 1968, 1969, 1971. Environmental radioactivity at the Nevada Test Site, July 1966-June 1967. --- July 1967-June 1968. --- July 1968-June 1969. Las Vegas, Nev., U. S. AEC report NVO-410-15, -22, -16. 70, 89, 64pp. (NSA 28: 24465, 24466, 29: 29325).

Reyss-Brion, M. 1962. Modifications of the competence of the ectoderm of the gastrula of batrachians after irradiation with x-rays. Compt. rend. 225: 1013-1014. (In French) (NSA 17: 5807).

Reyss-Brion, M. 1963. On the structures differentiated from x-irradiated amphibian ectoderm. J. Embryol. Exptl. Morphol. 11(4): 649-657. (In French) (NSA 18: 19649).

Reyss-Brion, M. 1964. The effect of x rays on the respective potentialities of competent ectoderm and its natural inducer in the early amphibian gastrula. Arch. Anat. Microscop. Morphol. Exptl. 53: 397-465. (In French) (NSA 19: 33809).

Rezio, A. 1969. Various applications of radioactive isotopes in hydrology. Rev. Port. Quim. (Lisbon) 11(1-4): 142-152. (In Portugese) (NSA 24: 29462).

Rhoads, W. A. 1974. Analysis of vegetation cover in certain Pu-contaminated areas using aerial photography. pp. 119-133. In: Dynamics of Plutonium in Desert Environments. (See Dunaway and White, 1974) (NSA 30: 29446).

Rhoads, W. A. and L. A. Franks. 1975. Radiation doses and possible radiation effects of low-level, chronic radiation in vegetation. pp. 89-95. In: Radioecology of Plutonium and Other Transuranics in Desert Environments. (See White and Dunaway, 1975) (NSA 33: 615).

Rhoads, W. A. and R. B. Platt. 1968. Local fallout radiation effects on vegetation from the nuclear excavation experiment Palanquin. Abstr. Radiation Research 35(2): 495-496.

Rhoads, W. A. and R. B. Platt. 1971. Beta radiation damage to vegetation from close-in fallout from two nuclear detonations. BioScience 21(22): 1121-1125.

Rhoads, W. A. and H. L. Ragsdale. 1973. *Artemisia* shrub size and radiation damages to *Artemisia* from local fallout from Project Schooner. pp. 953-960. In: Radionuclides in Ecosystems. (See Nelson, 1973) (NSA 25: 42717).

Rhoads, W. A., A. D. Kantz and H. L. Ragsdale. 1972. Ecological and environmental effects from local fallout from Project Schooner. 2. The beta and gamma radiation effects from close-in fallout. EG&G, Inc., Santa Barbara, Calif., U. S. AEC report PNE-529. vii, 58pp. (NSA 26: 45542).

Rhoads, W. A., R. B. Platt and R. A. Harvey. 1967. Radiosensitivity of certain perennial shrub species based on a study of the nuclear excavation experiment, Palanquin, with other observations of effects on the vegetation. EG&G, Inc., Santa Barbara, Calif., U. S. AEC report CEX-68.4. 42 pp. (NSA 23: 33904).

Rhoads, W. A., et al. 1968. Ecological and environmental effects from local fallout from Cabriolet. I. Radiation doses and short-term effects on the vegetation from close-in fallout. U. S. AEC report PNE-956. viii, 62pp.

Rhoads, W. A., et al. 1971. Radiation doses to vegetation from close-in fallout at Project Schooner. pp. 352-369. In: Survival of Food Crops and Livestock in the Event of Nuclear War. (See Benson and Sparrow, 1971).

Rhode, C. J. 1957. Studies on the biology of the mite *Caloglyphus mycophagus* Megn., 1874 (Acarina: Acaridae) including the effects of gamma radiation upon certain developmental stages. Entomol. Soc. Am. B. 3(3): 30-31.

Rhode, R. H. 1969. Problems in large-scale sterile-male technique programs in developing countries. pp. 51-56. In: Sterile-Male Technique for Eradication or Control of Harmful Insects. (See IAEA, 1969) (NSA 24: 32299).

Rhode, R. H. 1970. Application of the sterile-male technique in Mediterranean fruit fly suppression. A follow-up experiment in Nicaragua. pp. 43-50. In: Sterile-Male Technique for Control of Fruit Flies. (See IAEA, 1970) (NSA 25: 24508).

Rhode, R. H., et al. 1961. Effect of gamma radiation on the reproductive potential of the Mexican fruit fly. J. Econ. Entomol. 54(1): 202-203. (NSA 20: 3745).

Rhode, R. H., et al. 1971. Application of the sterile-insect-release technique in Mediterranean fruit fly suppression. J. Econ. Entomol. 64(3): 708-713. (NSA 25: 42750).

Rhodes, D. W. 1952. Preliminary studies of plutonium absorption in Hanford soil. Hanford Labs., U. S. AEC report HW-24548. 22pp. (NSA 29: 12693).

Rhodes, S. T. and W. H. Mason. 1971. ^{65}Zn excretion as an index to metabolic rate in the wood-feeding cockroach, *Cryptocercus punctulatus*. Ann. Entomol. Soc. Am. 64(2): 450-452.

Ricci, E. 1975. Methodology for high-flux absolute multielement neutron activation analysis: environmental baselines by analysis of tree rings. Anal. Chem. Acta 79: 109-124. (NSA 33: 16595).

Ricci, E., T. H. Handley and F. F. Dyer. 1973. Environmental analysis at ORNL's high-flux neutron activation laboratory. Trans. Amer. Nucl. Soc. 16: 43-44. (NSA 28: 12614).

Ricci, E., W. F. Harris and D. E. Reichle. 1974. Environmental baselines by NAA of tree-ring samples. pp. 144-153. In: Proc. 2nd Intern. Conf. on Nuclear Methods in Environmental Research. (See Vogt and Meyer, 1974) (NSA 33: 23284).

Ricciuti, E. R. 1967. Animals in atomic research. U. S. AEC, Washington. 54pp.

Rice, T. R. 1963. Accumulation of radionuclides by aquatic organisms. pp. 35-50. In: Studies of the Fate of Certain Radionuclides in Estuarine and Other Aquatic Environments. (See Sabo and Bedrosian, 1963).

Rice, T. R. 1963. Review of zinc in ecology. pp. 619-631. In: Radioecology. (See Schultz and Klement, 1963) (NSA 17: 33615).

Rice, T. R. 1965. Radioisotope techniques in fishery research. Trans. Am. Wildl. Conf. 30: 66-75.

Rice, T. R. 1967. A review of "Radioecology of Aquatic Organisms," by G. G. Polikarpov. Limnol. Oceanogr. 12(2): 364-365.

Rice, T. R. and J. P. Baptist. 1974. Ecologic effects of radioactive emissions from nuclear power plants. pp. 373-439. In: Human and Ecologic Effects of Nuclear Power Plants. L. A. Sagan, ed. Chas. C. Thomas, Springfield, Ill. (NSA 30: 23979).

Rice, T. R. and R. J. Smith. 1958. Filtering rates of the hard clam (*Venus mercenaria*) determined with radioactive phytoplankton. U. S. Fish and Wildlife Service 58: Fishery Bull. 129: 73-82.

Rice, T. R., et al. 1970. Potential hazards from radioactive pollution of the estuary. FAO Techn. Conf. on Marine Pollution and Its Effects on Living Resources and Fishing, Rome, Italy, 9-18 December 1970. UN Food and Agriculture Organ., Rome, reportFIR:MP/70/E-34. 9pp.

Rich, E. R. and L. E. Waldman. 1973. A radionuclide technique for the assay of physiological activity of algae. pp. 652-659. In: Radionuclides in Ecosystems. (See Nelson, 1973).

Richards, A. 1915. Recent studies on the biological effects of radioactivity. Science 42(1079): 287-300.

Richardson, B. L. 1971. Effects of x-rays on the free amino acid concentration of the hemolymph during the last larval instar of the wax moth, Galleria mellonella. Thesis, Ohio State Univ., Columbus. 153pp. (Univ. Microfilms 72-4621) (NSA 26: 25878)

Richardson, B. L. and W. C. Myser. 1973. Radiation effects on the hemolymph free amino acid pool of the honeybee prepupa (Apis mellifera) and the wax-moth larvae (Galleria mellonella). Radiation Research 54(2): 274-283. (NSA 28: 8483).

Richardson, C. J. 1972. The use of stomatal-resistance, photopigments, nitrogen, water potential, and radiation to estimate net photosynthesis in Liriodendron tulipifera L. - a physiological index. Thesis, Univ. of Tennessee, Knoxville.

Richardson, J. A. 1974. Safety and environmental protection requirements for light water nuclear power reactors to meet current regulations in the USA. J. Brit. Nucl. Energy Soc. 13(2): 149-171. (NSA 30: 8807).

Richardson, K. A. 1964. Thorium, uranium, and potassium in the Conway Granite, New Hampshire, U.S.A. pp. 39-50. In: The Natural Radiation Environment. (See Adams and Lowder, 1964) (NSA 18: 35729).

Richmond, C. R., W. H. Langham and T. T. Trujillo. 1962. Comparative metabolism of tritiated water by mammals. J. Cell. Comp. Physiol. 59(1): 45-53. (NSA 16: 17420) (Mouse, rat, kangaroo rat, rabbit, dog, horse, man).

Richmond, C. R., T. T. Trujillo and D. W. Martin. 1960. Volume and turnover of body water in Dipodomys deserti with tritiated water. Proc. Soc. Exptl. Biol. Med. 104(1): 9-11. (NSA 14: 14639).

Richter, D. 1965. On the localization of ^{90}Sr fallout in soils. Chem. Erde 24: 67-76. (In German) (NSA 21: 3188).

Richter, D. 1972. Regulatory aspects of environmental protection with regard to radioactive wastes and effluents produced by nuclear installations in the GDR. pp. 1001-1006. In: New Trends in Radiation Protection. (6th Intern. Congr., Bordeaux, France). Societé Francaise de Radioprotection, Montrouge. (CONF-720328) (NSA 27: 17946).

Richter, H. G. and A. S. Gillespie. 1964. A radio-relaease technique for tracing stream flows. Final report. Reserach Triangle Inst., Durham, N. C., N. C., U. S. AEC report TID-21881. v, 26pp.

Richter, J. and D. Schirrmeister. 1965. New results in the determination of dose distribution by means of computers. Radiobiol. Radiotherap. 6: 371-375. (In German) (NSA 19: 36735).

Richter, W. 1968. The effect of x radiation on the regenerative events in the brain of Ambystoma mexicanum. Z. Mikrosk.-Anat. Forsch. 79: 316-341. (In German) (NSA 23: 48193).

Rickard, W. H. 1963. Vegetational analyses in a creosote bush community and their radioecologic implications. pp. 39-44. In: Radioecology. (See Schultz and Klement, 1963) (NSA 17: 33546; 18: 4961).

Rickard, W. H. 1964. Spring precipitation and the strontium-90 contamination of wheat in the semiarid regions of Idaho and Montana. Nature 210 (4916): 309-310. (NSA 18: 10286).

Rickard, W. H. 1965. Field observations on fallout accumulation by plants in natural habitats. J. Range Mgmt. 18(3): 112-114.

Rickard, W. H. 1965. Sodium and potassium accumulation by greasewood and hopsage leaves. Botan. Gaz. 126(2): 116-119. (At Hanford, mentions radionuclides).

Rickard, W. H. 1965. ^{137}Cs in Bromus tectorum L. in relation to precipitation regimes and harvest yields. J. Range Mgmt. 18: 343. (NSA 20: 43527).

Rickard, W. H. 1966. ^{137}Cs in litter and understory vegetation. Northwest Sci. 40: 25-30. (Also: Battelle Northwest Lab., U. S. AEC report BNWL-SA-215) (NSA 20: 33585).

Rickard, W. H. 1967. Accumulation of ^{137}Cs in litter and understory plants of forest stands from various climatic zones of Washington. pp. 527-531. In: Radioecological Concentration Processes. (See Aberg and Hungate, 1967) (Also: Battelle Northwest Lab., U. S. AEC report BNWL-SA-530) (NSA 20: 20879).

Rickard, W. H. 1969. Cesium-137 in Cascade mountain vegetation. pp. 556-560. In: Symposium on Radioecology. (See Nelson and Evans, 1969) (Also: Battelle Northwest Lab., U. S. AEC report BNWL-SA-1212) (NSA 21: 26406).

Rickard, W. H. 1971. Radiocesium fallout in the forest floor. J. Forestry 69(3): 158-160.

Rickard, W. H. and J. F. Cline. 1965. Mineral transfer in a greasewood community and ion uptake by grasses. In: Radiation and Terrestrial Ecosystems. Health Physics 11(12): 1371-1374. (See Hungate, 1965).

Rickard, W. H. and L. L. Eberhardt. 1973. Fallout radiocesium in sedges and trout of a Cascade mountain bog. pp. 349-353. In: Radionuclides in Ecosystems. (See Nelson, 1973) (NSA 25: 37773).

Rickard, W. H. and H. A. Sweany. 1975. Radionuclides in Canada goose eggs. Battelle Northwest, U. S. ERDA report BNWL-SA-5451. 12pp. (NSA 33: 7280).

Rickard, W. H., A. D. Higgins and J. K. Fremstad. 1963. Caesium-137 in dried milk products in relation to phytoclimatic zones. Nature 197(4863): 197-198. (Also: Hanford Labs., U. S. AEC report HW-SA-2337) (NSA 17: 5752).

Rickard, W. H., et al. 1965. Gamma-emitting radionuclides in Alaskan tundra vegetation 1959, 1960, 1961. Ecology 46(3): 352-356. (Also: Hanford Atomic Products Operation, U. S. AEC report HW-SA-3316) (NSA 19: 9366; 20: 525).

Rickards, J. and J. F. Ziegler. 1975. Trace-element analysis of water samples using 700-keV protons. Appl. Phys. Lett. 27(12): 707-709. (NSA 33: 8856).

Ridgway, R. L. and D. A. Lindquist. 1966. Systematic activity of Shell SD-9129 in cotton plants. J. Econ. Entomol. 59(4): 961-964. (^{32}P).

Ridgway, R. L., et al. 1967. Fertilizers impregnated with systematic insecticides for control of cotton insects. J. Econ. Entomol. 60(2): 592-594. (^{32}P).

Ridgway, S. H. and D. D. Dailey. 1972. Cerebral and cerebellar involvement of trematode parasites in dolphins and their possible role in stranding. J. Wildl. Diseases 8(1): 33-43. (^{35}S).

Riedel, M. 1967. The biology, breeding, and sterilization of the cabbage fly _Phorbia brassicae_ Bouche, with special consideration of its occurrence in black radish cultivation. Research on sterilization by radioactive irradiation. Bayer. Landwirt. Jahrb. 44: 414-429. (In German) (NSA 22: 40981).

Rieder, J. B. 1971. Measurement with radioactive sources of soil moisture and soil density in differently treated pastures. Bayer. Landwirt. Jahrb. 48 (3): 259-305. (In German) (NSA 26: 183).

Riedesel, M. L., J. J. Young and G. H. Bryan. 1967. Cesium-137 metabolism in active and hibernating _Citellus lateralis_. pp. 219-234. In: _Mammalian Hibernation III_. K. C. Fisher, et al., eds. American Elsevier, New York.

Riekerk, H. 1965. Mineral cycling in a Douglas fir forest stand. In: _Radiation and Terrestrial Ecosystems_. Health Physics 11(12): 1363-1369. (See Hungate, 1965).

Riekerk, H. 1967. Movement of phosphorus, potassium, and calcium in a Douglas fir forest ecosystem. Dissertation, Univ. of Washington, Seattle. 143pp. (^{32}P, ^{86}Rb, ^{45}Ca).

Riel, G. K. 1967. Concentration of radioactive isotopes in environmental water measured by underwater gamma spectrometry. Thesis, Univ. of Maryland, College Park. 206pp. (Also: U. S. Naval Ordnance Lab., White Oak, Md., report) (AD-648810; NSA 21: 40992; 22: 10373).

Riel, G. K. 1969. Addition of automatic temperature and pressure sensors to underwater gamma spectrometer. U. S. Naval Ordnance Lab., White Oak, Md., U. S. AEC report TID-25484. 20pp. (NSA 24: 46527).

Riel, G. K. 1970. Radioactive cesium in estuaries. Radiol. Hlth. Data 11: 659-665. (NSA 25: 21642).

Riel, G. K. 1972. Distribution of fallout cesium-137 in the Chesapeake Bay. pp. 883-896. In: _Natural Radiation Environment II_. _Vol. 2_. (See Adams, Lowder and Gesell, 1972) (NSA 32: 19729).

Riel, G. K. and D. Duffey. 1968. Monitoring of radioisotopes in environmental water. Trans. Amer. Nucl. Soc. 11: 52-53. (NSA 22: 30708).

Riel, G. K., J. P. Dolfis and D. G. Simons. 1960. Spectrum of radiation background under water. Nature 188(4747): 266-268.

Riemann, J. G. 1964. Cytological effects of radiation on testes of the screw-worm fly, _Cochliomyia hominivorax_. Abstr. Genetics 50(2): 279.

Riemann, J. G. 1967. A cytological study of radiation effects on testes of the screw-worm fly, _Cochliomyia hominivorax_ (Diptera: Callophoridae). Ann. Entomol. Soc. Am. 60: 308-320. (NSA 21: 34760).

Riemann, J. G. 1973. Ultrastructure of sperm in F_1 progeny of irradiated males of the Mediterranean flour moth, _Anagasta kuehniella_. Ann. Entomol. Soc. Am. 66(1): 147-153. (NSA 27: 15093).

Riemann, J. G. and H. M. Flint. 1967. Irradiation effects on midguts and testes of the adult boll weevil, _Anthonomus grandis_, determined by histological and shielding studies. Ann. Entomol. Soc. Am. 60(2): 298-320.

Riemann, J. G. and B. J. Thorson. 1969. Comparison of effects of irradiation on the primary spermatogonia and mature sperm of three species of Diptera. Ann. Entomol. Soc. Am. 62(3): 613-617.

Riemann, J. G. and B. J. Thorson. 1974. Viability and use of sperm after irradiation of the large milkweed bug. Ann. Entomol. Soc. Am. 67(6): 871-876. (NSA 31: 6343).

Rigler, F. H. 1964. The phosphorus fractions and the turnover time of inorganic phosphorus in different types of lakes. Limnol. Oceanogr. 9(4): 511-518. (Used ^{32}P).

Rinehart, R. R. 1962. Some effects of nitric oxide and oxygen on dominant lethal production in x-irradiated _Drosophila viridis_ males. Thesis, Univ. of Texas, Austin. 73pp. (NSA 17: 7885).

Rinehart, R. R. 1964. Influence of oxygen, helium, and metabolic inhibition of x-ray induced dominant lethality in Stage 7 and Stage 14 oocytes of _Drosophila melanogaster_. Genetics 49: 855-863.

Rinehart, R. R. and F. J. Ratty. 1965. Mutation in _Drosophila melanogaster_ cultured on irradiated food. Genetics 52(6): 1119-1126.

Rinehart, R. R. and F. J. Ratty. 1967. Mutation in _Drosophila melanogaster_ cultured on irradiated whole food or food components. Int. J. Radiat. Biol. 12(4): 347-354.

Rinehart, R. R., R. M. Valencia and J. I. Valencia. 1963. Comparative lethal rates from x-irradiated _D. melanogaster_ males and females mated together or to unirradiated partners. Drosophila Inform. Serv. 38: 71-72.

Rings, R. W. 1953. Fruit insect behavior studies aided by radioisotopes. Ohio Agr. Expt. Sta. Bull. (725): 59-60.

Riordan, D. F. 1964. Effect of continuous low-level gamma-irradiation on successive generations of Dahlbominus fuscipennis (Zett.) (Hymenoptera: Eulophidae). Can. J. Zool. 42: 685-688. (NSA 19: 3918).

Riordan, D. F. 1964. Effects of low-level gamma-radiation on the immature stages of Dahlbominus fuscipennis (Zett.) (Hymenoptera. Eulphidae). Can. J. Zool. 42: 675-684. (NSA 19: 95).

Riordan, D. F. 1965. Emergence and fertility of Aedes aegypti (L.) reared in various concentrations of P^{32}. Can. J. Zool. 43: 497-501. (NSA 19: 25939).

Riordan, D. F. and D. P. Peschken. 1970. Method for obtaining ^{32}P-labelled eggs of the flea beetle Altica carduorum (Coleoptera. Chrysomelidae). Can. Entomol. 102: 1613-1616. (NSA 25: 39792).

Ripper, W. E., R. M. Greenslade and G. S. Hartley. 1951. Selective insecticides and biological control. J. Econ. Entomol. 44(4): 448-459. (Uses of tracers discussed).

Risley, C., Jr. 1963. Radioactivity in Lake Michigan: a preliminary report. pp. 124-132. In: Great Lakes Research Div., Univ. of Michigan, Ann Arbor, Publ. 10.

Risley, C. 1965. Radioactivity in Lake Michigan and its tribularies. pp. 160-167. In: Proc. 8th Conf. on Great Lakes Research. Great Lakes Research Div., Univ. of Michigan, Ann Arbor, Publ. 13.

Ritchie, J. C. 1962. Distribution of fallout cesium-137 in litter, humus, and surface soil layers under natural vegetation in the Great Smoky Mountains. Thesis, Univ. of Tennessee, Knoxville, v, 47pp.

Ritchie, J. C. 1962. Distribution of fallout cesium-137 in the Great Smoky Mountains. Abstr. ASB Bull. 9(2): 31.

Ritchie, J. C. 1967. Natural gamma radiation in northeastern and east-central Georgia. Thesis, Univ. of Georgia, Athens. 98pp. (NSA 22: 48939).

Ritchie, J. C. 1969. Absolute pollen frequencies and carbon-14 age of a section of Holocene lake sediment from the Riding Mountain area of Manitoba. Can. J. Botany 47(9): 1345-1349.

Ritchie, J. C. and J. R. McHenry. 1973. Determination of fallout ^{137}Cs and naturally occurring gamma-ray emitters in sediments. Int. J. Appl. Radiat. Isotopes 24(10): 575-578. (NSA 28: 29618).

Ritchie, J. C. and J. R. McHenry. 1973. Vertical distribution of fallout cesium-137 in cultivated soils. Radiat. Data Rept. 14(12): 727-728. (NSA 29: 18352).

Ritchie, J. C. and G. L. Plummer. 1969. Natural gamma radiation in northeast and east-central Georgia. Bull. Ga. Acad. Sci. 27: 173-194. (NSA 24: 17040).

Ritchie, J. C. and G. L. Plummer. 1966. Thorium, uranium, and potassium in certain Georgia soils and their biological implications. Abstr. ASB Bull. 13(2): 45.

Ritchie, J. C., E. E. C. Clebsch and W. K. Rudolph. 1970. Distribution of fallout and natural gamma radionuclides in litter, humus, and surface mineral soil layers under natural vegetation in the Great Smoky Mountains, North Carolina - Tennessee. Health Physics 18(5): 479-489. (NSA 24: 36387).

Ritchie, J. C., J. R. McHenry and A. C. Gill. 1973. Dating recent reservoir sediments. Limnol. Oceanogr. 18(2): 254-263. (NSA 28: 10885).

Ritchie, J. C., J. R. McHenry and A. C. Gill. 1973. The use of fallout ^{137}Cs to determine sediment movement in small watersheds and trap efficiencies of small reservoirs. Abstr. Bull. Ecol. Soc. Am. 54(1): 20.

Ritchie, J. C., J. A. Spraberry and J. R. McHenry. 1974. Estimating soil erosion from the redistribution of fallout ^{137}Cs. Soil Sci. Soc. Am. Proc. 38(1): 137-139. (NSA 29: 24211).

Ritchie, J. C., et al. 1972. Fallout ^{137}Cs in reservoir sediment. Health Physics 22(1): 96-98. (NSA 26: 12032).

Ritchie, J. C., et al. 1973. Distribution of cesium-137 in a small watershed in northern Mississippi. pp. 129-132. In: Radionuclides in Ecosystems. (See Nelson, 1973) (NSA 28: 27267).

Ritter, G. and H. Lyr. 1965. The significance of mycorrhizal fungi for the utilization of different sources of phosphate by Pinus silvestris L. pp. 277-182. In: Plant Microbes Relationships. Czech. Acad. Sci., Prague

Ritossa, F. M. 1964. Behavior of RNA and DNA synthesis at the puff level in salivary gland chromosomes of Drosophila. Exptl. Cell Res. 36(3): 515-523. (^{3}H).

Ritossa, F. M. 1964. Experimental activation of specific loci in polytene chromosomes of Drosophila. Exptl. Cell Res. 35(3): 601-607. (^{3}H).

Ritossa, F. M. and S. Spiegelman. 1965. Localization of DNA complementary to ribosomal RNA in the nucleolus organizer region of Drosophila melanogaster. Proc. Natl. Acad. Sci. (US) 53(4): 737-745. (^{3}H, ^{32}P.

Ritossa, F. M., R. C. von Borstel and H. Swift. 1964. The action of ribonuclease on salivary glands cells of Drosophila. Genetics 50: 279. (^{3}H.

Ritossa, F. M., et al. 1965. On the action of ribonuclease in salivary gland cells of Drosophila. Chromosoma 16(2): 144-151. (3H).

Ritzerfeld, W., E. Schnepper and D. H. Winterhoff. 1964. The influence of cobalt-60 gamma radiation on aerobic bacterial respiration. Arch. Hyg. Bakteriol. 148(3): 206-216. (In German) (NSA 18: 29266).

Riviere, R., et al. 1972. Iodine activation analysis in biological media. pp. 349-366. In: Colloquium of INSERM on Experimental Methodology in Thyroidism Physiology and Physiopathology. France Inst. National de la Sante et de la Recherche Medicale, Boulogne-Billancourt. (In French) (NSA 26: 42707).

Rivosecchi, L. 1962. An experiment in the field with irradiated male Musca domestica in a rural zone in the Province of Latium. Riv. Parassit. 23 (1): 71-74. (In Italian).

Riza-Zade, R. R. and N. L. Suleymanova. 1970. Dynamics of the appearance of radiation stimulation. Radiobiologiya 10(1): 98-102. (AEC-tr-7154, pp. 150-157) (NSA 24: 27719) (Sunflower).

Riza-Zade, R. R. and N. L. Suleymanova. 1970. Dependence of the radiobiological effect on the time between irradiation and planting of the sunflower. Radiobiologiya 10: 772-775. (AEC-tr-7215, pp. 209-214) (NSA 25: 29717).

Riza-Zade, R. R. and N. L. Suleymanova. 1970. Dependence of the radiobiological effect on the level of the mineral nutrition of the sunflower. Radiobiologiya 10: 907-909. (AEC-tr-7225, pp. 161-165) (NSA 25: 29714).

Riza-Zade, R. R. and N. L. Suleymanova. 1971. Effect of repeated irradiation of seeds after storage. Radiobiologiya 11(2): 306-307. (AEC-tr-7269, pp. 232-235) (NSA 25: 44971) (Sunflower).

Riza-Zade, R. R. and N. L. Suleymanova. 1971. Dynamics of biological radiation effect manufestation. pp. 70-77. In: Radioactive Isotopes and Nuclear Radiation in Agricultural Research. I. Atomizdat, Moscow. (In Russian) (NSA 31: 704) (Sunflowers).

Riza-Zade, R. R., N. L. Suleymanova and A. T. Tagiyev. 1970. Irradiation of sunflower and barley seeds before sowing. Izotopy SSSR (18): 50. (In Russian) (NSA 25: 21956).

Roberge, M. R. 1970. Behavior of urease added to unsterilized, steam-sterilized, and gamma radiation-sterilized black spruce humus. Can. J. Microbiol. 16: 865-870. (NSA 26: 4902).

Roberge, M. R. 1971. Respiration of a black spruce humus sterilized by heat of irradiation. Soil Science 111(2): 124-128. (NSA 25: 24528).

Roberge, M. R. and R. Knowles. 1967. Sterilization by gamma radiation of a humus in the black spruce (Picea mariana). Natur. Can. 94: 221-227. (In French) (NSA 23: 38907).

Roberge, M. R. and R. Knowles. 1968. Urease activity in a black spruce humus sterilized by gamma radiation. Soil Sci. Soc. Am. Proc. 32: 518-521. (NSA 22: 49674).

Röbbelen, G. 1958. Instable gene appearance following x irradiation of Arabidopsis thaliana (L.) Heynh. Z. Naturforschung 13b: 14-16. (In German) (NSA 12: 12942).

Röbbelen, G. 1962. Comparative effect between ethyl methane sulfonate and x-radiation in mutation research with Arabidopsis thaliana. Naturwissenschaften 49(3): 65. (In German) (NSA 16: 20208).

Robbins, J. A. and D. N. Edgington. 1975. Determination of recent sedimentation rates in Lake Michigan using ^{210}Pb and ^{137}Cs. Geochim. Cosmochim. Acta 39(3): 285-304. (NSA 31: 22651).

Robbins, W. E. 1963. Studies on the utilization, metabolism and function of sterols in the house-fly, Musca domestica. pp. 269-279. In: Radiation and Radioisotopes Applied to Insects of Agricultural Importance. (See IAEA, 1963).

Robbins, W. E. and P. A. Dahm. 1955. Absorption and excretion, distribution, and metabolism of carbon-14-labeled DDT by the American cockroach. J. Agric. Food Chem. 3(6): 500-508.

Robert, M. and H. Kroeger. 1965. Localization of the additional RNA synthesis in trypsin-treated giant chromosomes of Chironomus thummi. Experientia 21(6): 326-327. (In German, English summary).

Roberts, B. R. 1964. Effects of water stress on the translocation of photosynthetically assimilated carbon-14 in yellow poplar. pp. 273-288. In: The Formation of Wood in Forest Trees. M. H. Zimmermann, ed. Academic Press, New York.

Roberts, E. F. and D. P. Snyder. 1973. Use of iodine-125 for identifying mother-offspring relationships in the eastern chipmunk, Tamias striatus pp. 274-281. In: Radionuclides in Ecosystems. (See Nelson, 1973) (NSA 26: 22499).

Roberts, T. A. 1968. Resistance of spores of Clostridium welchii. pp. 95-100. In: Elimination of Harmful Organisms from Food and Feed by Irradiation. (See IAEA, 1968) (NSA 24: 34794).

Roberts, T. A. and M. Ingram. 1965. The resistance of spores of Clostridium botulinum type E to heat and radiation. J. Appl. Bacteriol. 28: 125-138. (NSA 19: 33848).

Roberts, T. A. and M. Ingram. 1965. Radiation resistance of spores of Clostridium species in aqueous suspension. J. Food Sci. 30: 879-885. (NSA 20: 18340).

Roberts, T. A., P. J. Ditchett and M. Ingram. 1965. The effect of sodium chloride on radiation resistance and recovery of irradiated anaerobic spores. J. Appl. Bacteriol. 28: 336-348. (NSA 20: 6834).

Roberts, W. A. 1968. Preparation of liquid scintillation mixtures for the measurement of ^{14}C and ^{3}H samples. Lab. Pract. 17: 703-706. (NSA 23: 21659).

Robertson, D. E. and R. Carpenter. 1974. Neutron activation techniques for the measurement of trace metals in environmental samples. National Academy of Sciences-National Research Council, Washington, report NAS-NRC-3114. 84pp. (NSA 29: 20586).

Robertson, D. E. and R. W. Perkins. 1975. Radioisotope ratios in characterizing the movement of different physical and chemical species through natural soils. pp. 123-131. In: Isotope Ratios as Pollutant Source and Behavior Indicators. (See IAEA, 1975) (NSA 33: 7299).

Robertson, D. E., et al. 1972. Transport and depletion of radionuclides in the Columbia River. Trans. Amer. Nucl. Soc. 15(1): 66. (NSA 26: 42963).

Robertson, D. E., et al. 1973. Transport and depletion of radionuclides in the Columbia River. pp. 141-155. In: Radioactive Contamination of the Marine Environment. (See IAEA, 1973) (NSA 28: 8166).

Robertson, G. 1971. Marking red-scale with ^{14}C and its effect on reproduction and survival. Entomologia Exp. Appl. 14(4): 449-456. (NSA 26: 23169).

Robertson, G. 1973. Sensitivity of scale insects to ^{14}C β-particles and to ^{60}Co γ-rays. Int. J. Radiat. Biol. 23(4): 313-323. (NSA 28: 27665).

Robertson, G. 1974. Effects of carbon-14 on the survival and reproduction of red scale males. Entomologia Exp. Appl. 17(1): 31-35. (NSA 31: 813).

Robertson, J. B. 1974. Digital modeling of radioactive and chemical waste transport in the Snake River Plain aquifer at the National Reactor Testing Station, Idaho. U. S. Geological Survey, U. S. AEC report IDO-22054. 47 pp. (NSA 30: 18636).

Robertson, J. B. 1974. Application of digital modelling to the prediction of radioisotope migration on groundwater. pp. 451-478. In: Isotope Techniques in Groundwater Hydrology. Vol. 2. (See IAEA, 1974) (NSA 31: 19667).

Robertson, J. B., R. Shoen and J. T. Barraclough. 1974. Influence of liquid waste disposal on the geochemistry of water at the National Reactor Testing Station, Idaho: 1952-1970. U. S. Geological Survey, U. S. AEC report IDO-22053. 231pp. (NSA 29: 30186).

Robertson, J. S. 1957. Theory and use of tracers in determining transfer rates in biological systems. Physiol. Revs. 37: 133-154.

Robertson, J. S. 1973. Tritium turnover rates in mammals. pp. 322-327. In: Tritium. (See Moghissi and Carter, 1973) (NSA 28: 27665).

Robertson, T. J. M. 1975. Neutron radiography as a non-destructive tool in archaeology. Non-Destruct. Test. 8(1): 17-20. (NSA 31: 19517).

Robiller, F. and A. Unverricht. 1972. Radiation dose and radioprotective effect in the examination of the thyroid gland of birds. Radiobiol.-Radiother. 13(3): 393-396. (In German) (NSA 26: 53515) (^{131}I).

Robinson, A. S. 1973. Increase in fertility, with repeated mating, of gamma irradiated male codling moths, Laspeyresia pomonella (L.) (Lepidoptera: Olethreutidae). Can. J. Zool. 51(4): 427-430. (NSA 29: 2940).

Robinson, A. S. 1974. Gamma radiation and insemination in the codling moth, Laspeyresia pomonella (Lepidoptera: Olethreutidae). Entomol. Exp. Appl. 17(3): 425-432. (NSA 31: 29957).

Robinson, A. S. 1975. Influence of anoxia during gamma irradiation on the fertility and competitiveness of the adult male codling moth, Laspeyresia pomonella (L.). Radiation Research 61(3): 526-534. (NSA 31: 29951).

Robinson, A. S. and C. van Heemert. 1975. Preliminary radiobiological studies on Hylemya antiqua Meigen and data on three radiation-induced (0.5 krad) chromosomal rearrangements. pp. 375-385. In: Sterility Principle for Insect Control, 1974. (See IAEA, 1975) (NSA 32: 11793).

Robinson, A. V., et al. 1975. Microbial transformation of plutonium. Battelle Northwest Lab., U. S. AEC report BNWL-SA-5531. 12pp. (NSA 33: 7634).

Robinson, E. S. and A. W. Jones. 1971. Moniliformis dubius: x-irradiation and temperature effects on morphogenesis in Periplaneta americana. Exptl. Parasitol. 29(2): 292-301. (NSA 25: 46962).

Robinson, F. E. 1963. Results obtained with lightweight rate meter for neutron soil moisture measurements. Soil Science 96(3): 218-219.

Robinson, G. A. and K. H. Tam. 1974. Distribution of ^{99m}Tc and ^{131}I among tissues of the Japanese quail. Poultry Sci. 53(6): 2209-2215. (NSA 33: 20576).

Robinson, J. B., C. T. Corke and D. A. Jones. 1970. Survival of bacteriophage in soil sterilized with gamma irradiation. Soil Sci. Soc. Am. Proc. 34: 703-704. (NSA 24: 44650).

Robinson, J. R. 1969. ^{33}P: a superior radiotracer for phosphorus? Int. J. Appl. Radiat. Isotopes 20: 531-540. (NSA 23: 33599).

Robinson, J. R. and W. Chefurka. 1964. Continuous measurement of $C^{14}O_2$ respired by insects. An ionization chamber method. Anal. Biochem. 9: 197-203. (NSA 19: 38).

Robinson, L. R., Jr. 1965. Monitoring streams for radioactive wastes. Water Sewage Works (112): R152-R160.

Robinson, T. 1963. Biochemical mutants of higher plants. Final report. Univ. of Massachusetts, U. S. AEC report NYO-10616. 4pp. (NSA 17: 37063).

Robredo, F. 1964. Use of radioactive isotopes and ionizing radiation in entomology. Bol. Serv. Plagas Forest 7(13): 42-51. (In Spanish) (NSA 21: 32770).

Roche, J., S. Andre and G. Salvatore. 1962. Metabolism of iodine (^{131}I) and formation of iodinated scleroprotein (Gorgonine) of the corneous skeleton of Eunicella verrucosa (Pallas). Comp. Biochem. Physiol. 1(4): 286-292. (In French) (NSA 16: 32697).

Roche, J., G. Salvatore and I. Covelli. 1962. Metabolism of I^{131} and thyroidal function of the larvae (ammocoete) of a cyclostome Petromyzon planeri Bl. Comp. Biochem. Physiol. 2(2): 90-99. (In French) (NSA 16: 32698).

Rockstein, M. 1966. Differential gerontomimetic effects of x-irradiation to pupae of male and female house flies. pp. 175-178. In: Radiation and Aging. P. J. Lindop, ed. (NSA 22: 43408).

Rockstein, M. 1972. Effects and mechanisms of action of ionizing radiation on the house fly, *Musca domestica* L. Final report, November 1, 1968-October 31, 1972. Miami Univ., U. S. AEC report ORO-3862-2. 19pp. (NSA 27: 24974).

Rockstein, M. and G. T. Baker. 1974. Effects of x-irradiation of pupae on aging of the thoracic flight muscle of the adult house fly, *Musca domestica* L. Mech. Ageing Dev. 3(5-6): 271-278. (NSA 33: 29763).

Rockstein, M. and P. L. Bhatnagar. 1966. X-irradiation and wing retention in the common house fly, *Musca domestica* L. Naturwissenschaften 53: 702-703. (NSA 21: 30897).

Rockstein, M., P. L. Bhatnagar and M. Dauer. 1965. The effects of high level x-irradiation on adult house fly flight ability and longevity. Abstr. 78. Bull. Entomol. Soc. Am. 11(3): 158.

Rockstein, M., M. Dauer and P. L. Bhatnagar. 1965. Adult emergency of the house fly, *Musca domestica*, from x-irradiated pupae. Ann Entomol. Soc. Am. 58: 375-379. (NSA 20: 14306).

Rockstein, M., M. Dauer and P. L. Bhatnagar. 1967. Further studies on the effect of x-irradiation on the house fly, *Musca domestica* L. Radiation Research 31: 840-845. (NSA 22: 555).

Rocznik, K. 1969. Total radioactivity and the weather. Meteorol. Rundsch. 22: 144-146. (In German) (NSA 24: 46366).

Rode, L. G. and O. M. Nachayev. 1970. Operation of a moisture meter with an epithermal neutron detector. Tr. vses. nauchn.-issled. inst. gidrotech. melior. (48): 70-76. (In Russian) (NSA 29: 7835).

Rodgers, C. A. and D. L. Stalling. 1970. 2,4-D butoxythanol ester uptake, distribution and elimination in organs of rainbow trout, channel catfish and bluegills. Abstr. p. 97. In: Abstr. Meeting, Weed Soc. Am. (SFA 18: 16044; Weed Abstr. 21: 252) ^{14}C).

Rodhe, W. 1958. The primary production in lakes: some results and restrictions of the C^{14} method. Rapp. Proc.-Verb. Cons. Int. Explor. Mer 144: 122-128.

Rodier, J. 1970. Water and the nuclear industry. Meded. Vlaam. Chem. Ver. 32: 26-44. (In French) (NSA 25: 8326).

Rodier, J., M. Marichal and R. Benoit. 1966. Experimental study of the fixation on plants and fish of radioisotopes originating from plutonium production centre effluent. pp. 519-533. In: *Disposal of Radioactive Wastes into Seas, Oceans and Surface Waters*. (See IAEA, 1966).

Rodier, J., et al. 1963. The discharge into the Rhone of liquid radioactive effluents from the Marcoule Centre. Bull. Inform. Sci. Tech. (Paris) (70): 7-18. (In French) (NSA 17: 29871).

Rodier, J., et al. 1966. Return of waste water to the environment; the experience of seven years' operation of the Marcoule Plutonium Production Centre. pp. 709-723. In: *Disposal of Radioactive Wastes into Seas, Oceans and Surface Waters*. (See IAEA, 1966).

Rodina, A. G. 1957. Possibility of application of the marked atom method for resolution of the problem of food selectivity of aquatic animals. Zool. zh. 36(3): 337-343. (In Russian).

Rodina, A. G. and A. S. Troshin. 1954. Application of marked atoms in the study of the feeding of aquatic animals. Dokl. Akad. Nauk SSSR 98(2): 297-300. (In Russian).

Rodina, A. G. and A. S. Troshin. 1954. The path of phosphorus introduced into pond water with plant fertilizer. Dokl. Akad. Nauk SSSR 98(4): 665-668. (In Russian).

Rodionov, V. M. 1962. The biochemical mechanisms of radiation damage. Vestn. Akad. Med. Nauk SSSR 17(9): 48-58. (In Russian) (BA 42: 4421).

Rodionova, L. F. and S. Ya. Sukal'skaya. 1969. Accumulation of ^{140}Ba and ^{140}La by freshwater plankton. Gidrobiol. zh. 5(6): 87-91. (In Russian) (NSA 25: 24012).

Rodionova, L. F. and S. Ya. Sukal'skaya. 1971. Methods and results of a study of accumulation of a mixture of radioactive isotopes by planktonic organisms from water bodies. pp. 118-121. In: *Methods of Radioecological Investigations*. (See Verkhovskaya, 1971) (In Russian).

Rodionova, L. Z. 1958. Change in the feeding activity of *Calandra granaria* L. mites irradiated by x-rays. Tr. inst. zerna prod. ego pererabotki 35: 58-61. (In Russian).

Rodnan, G. P., G. F. Ebaugh, Jr. and M. R. S. Fox. 1957. The life span of the red blood cell and the red blood cell volume in the chicken, pigeon and duck as estimated by use of $Na_2Cr^{51}O_4$. With observations on red cell turnover rate in the mammal, bird and reptile. Blood 12(4): 355-366.

Rodriguez, A. 1976. Behavioral and electrophysiological evidence for the mechanisms involved in the detection of ionizing radiations by the crayfish *Pacifastacus trowbridgii*. Dissertation, Oregon State Univ., Corvallis. 104pp. (Univ. Microfilms 75-29987) (NSA 33: 29759).

Rodriguez, J. G. 1948. The effects of a three-month exposure of mites to radiophosphorus. Ohio State Univ. Engr. Expt. Sta. News 20(5): 54-56.

Rodriguez, J. G. 1954. Radiophosphorus in metabolism studies in the two-spotted spider mite. J. Econ. Entomol. 47(3): 514-517.

Rodriguez, J. G., R. Hampton and M. Chaplin. 1964. Utilization of amino acids by the two-spotted mite, *Tetranychus urticae* Koch. Abstr. Bull. Entomol. Soc. Am. 10(3): 166.

Rodriguez, J. G., et al. 1967. Ingestion in the two-spotted spider mite, *Tetranychus urticae* Koch, as influenced by wave-length of light. J. Insect Physiol. 13(6): 925-932. (^{32}P).

Rodriguez, P. H. 1970. Gamma radiation-induced effects on the genetic fitness of the cestode, *Hymenolepis nana*, and of the flour beetle vector, *Tribolium confusum*. Thesis, Univ. of Rhode Island, Kingston. 103pp. (Univ. Microfilms 71-10496) (NSA 25: 39921).

Rodriguez Diaz, A. J. 1972. National program for the use of peaceful nuclear explosives in the national economy of the Republic of Venezuela. International Atomic Energy Agency, Vienna. (UCRL-Trans-10701, pp. 234-239) (NSA 30: 3119).

Roessler, C. E. 1967. ^{137}Cs and other gamma radioactivity in the Florida environment: a study of selected media. Thesis, Univ. of Florida, Gainesville. 181pp. (NSA 23: 6220).

Roessler, C. E., et al. 1972. Utility sponsored environmental surveillance and radioecological research program for a coastal nuclear power plant. Am. J. Publ. Hlth. 62(10): 1378-1386. (NSA 28: 367).

Rogalya, A. M. 1971. History of the development of physical methods for investigating the radioactivity in the atmosphere. Vopr. isotor. Estestvozn. Tekh. 1(34): 31-41. (In Russian) (NSA 29: 18337).

Rogatykh, N. P. 1971. Radiostimulation of the growth of Chlorella at various stages of the cellular cycle. Radiobiologiya 11(1): 92-95. (AEC-tr-7233, pp. 137-143) (NSA 25: 44967).

Rogers, A. H., et al. 1967. The use of radioactivity for the assessment of dose in bioassays involving bacterial insect pathogens. pp. 296-302. In: Insect Pathology and Microbial Control. P. A. van der Laan, ed. North Holland, Amsterdam.

Rogers, D. R. 1975. Mound Laboratory environmental plutonium study, 1974. Mound Lab., U. S. AEC report MLM-2249. 143pp. (NSA 32: 27956).

Rogers, J. J. W. 1964. Statistical tests of the homogeneity of the radioactive components of granitic rocks. pp. 51-62. In: The Natural Radiation Environment. (See Adams and Lowder, 1964) (NSA 18: 35730).

Rogers, L. R. and C. C. Gamertsfelder. 1971. USA regulations for control of releases of radioactivity into the environment in effluents from nuclear facilities. pp. 127-144. In: Environmental Aspects of Nuclear Power Stations. (See IAEA, 1971).

Rogers, L. R. and C. L. Henderson. 1973. AEC implementation of the National Environmental Policy Act in its licensing and regulation of nuclear facilities. pp. 63-67. In: Radionuclides in Ecosystems. (See Nelson, 1973) (NSA 28: 29435).

Rogers, R. W. 1955. α-particle dosimetry and the inhibition of mitoses in the grasshopper neuroblast by low dosage x-irradiation. Radiation Research 3(1): 18-37. (NSA 9: 7237).

Rogers, R. W. and R. C. von Borstel. 1957. Alpha-particle bombardment of the Habrobracon egg. I. Sensitivity of the nucleus. Radiation Research 7(5): 484-490. (NSA 12: 2676).

Rogers, W. I. and J. D. Hilchey. 1957. The effect of beta radiation on the feeding activity of Tribolium spp. Abstr. Bull. Entomol. Soc. Am. 3(3): 25.

Rogers, W. I. and J. D. Hilchey. 1960. Studies on the postirradiation feeding activity of Tribolium castaneum (Tenebrionidae: Coleoptera). Ann. Entomol. Soc. Am. 53: 584-590.

Rogowski, A. S. and T. Tamura. 1965. Movement of ^{137}Cs by runoff, erosion, and infiltration on the alluvial captina silt loam. In: Radiation and Terrestrial Ecosystems. Health Physics 11(12): 1333-1340. (See Hungate, 1965).

Rogowski, A. S. and T. Tamura. 1970. Environmental mobility of ^{137}Cs. Radiation Botany 10(1): 35-45. (NSA 24: 27563).

Rogowski, A. S. and T. Tamura. 1970. Erosional behavior of ^{137}Cs. Health Physics 18(5): 467-477. (NSA 24: 29383).

Rohde, C. J. 1959. Studies on the biologies of two mite species, predator and prey, including some effects of gamma radiation and selected developmental states. Ecology 40(4): 572-579. (NSA 14: 9343).

Rohde, H. 1975. Rapid method for the determination of mercury traces by neutron activation analysis and mercury content of biological materials of different geographic and time origins. German F. R. Kernforschungszentrum, Karlsruhe, report KFK-2201. 83pp. (In German) (NSA 33: 22707).

Rohleder, K. 1967. Radioactive decontamination of edible mushrooms by blanching. Ind. Obst-Gemuseverwert. 52: 64-66. (In German) (NSA 23: 22315).

Röhnsch, W. and K. Kunert. 1968. Environmental monitoring of nuclear establishments. Kernenergie 11: 209-214. (In German) (NSA 23: 465).

Rohrbough, S. F., J. G. Ballinger and L. E. Koehler. 1967. The baloonborne hygrometer system for the measurement of atmospheric water vapor. pp. 47-54. In: Isotope Techniques in the Hydrologic Cycle. (See Stout, 1967).

Rohwer, P. S. 1972. Responding to public concern over man-made radioactivity. in the environment. Trans. Amer. Nucl. Soc. 15(1): 545-546. (NSA 26: 42943).

Rohwer, P. S., M. J. Kelly and R. S. Booth. 1973. Guidance for limiting environmental releases of tritium. pp. 422-430. In: Tritium. (See Moghissi and Carter, 1973) (NSA 28: 27241).

Rokitskiy, P. F. 1959. A study of the process of artificial selection under conditions of natural variability and x-ray induced variability. Byull. Mosk. obshch. ispyt. prirody (biol.) 64(4): 75-97. (In Russian) (Used Drosophila melanogaster).

Rollason, G. S. 1949. X-radiation of eggs of Rana pipiens at various maturation stages. Biol. Bull. 97(2): 169-186. (NSA 4: 33).

Rom, F. E. 1973. Summary of the study of disposal of nuclear waste into space. U. S. NASA, Cleveland, O., report N-73-24834. 14pp. (NSA 29: 5612).

Romanenko, V. I. 1967. Comparison of the oxygen and radiocarbon methods for determining the intensity of phytoplanktonic photosynthesis. Tr. inst. biol. vnutren. vod. Akad. Nauk SSSR 15(18): 54-60. (In Russian).

Romanoff, A. L. and A. J. Romanoff. 1972. Ch. V. Ionizing radiation. pp. 107-149. In: Pathogenesis of the Avian Embryo. A. L. Romanoff and A. J. Romanoff. Wiley-Interscience, New York.

Romanov, G. N. and V. A. Ukhanova. 1969. Some quantitative characteristics of the direct contamination of the above-ground part of plants by global radioactive fallout. State Committee on the Use of Atomic Energy, USSR. Atomizdat, Moscow. UN Document A/AC.82/G/L.1291. (AEC-tr-7128, pp. 152-162) (NSA 24: 12062).

Romanov, V. V. and V. N. Soifer. 1968. Use of natural tritium in hydrogeological studies. Tr. vses. nauchn.-issled. inst. yadern. geofiz. geokhim. (4): 289-304. (In Russian) (NSA 25: 15765).

Romantsev, Ye. F. and S. K. Khalikov. 1968. Formation of formyl derivatives of folic acid in the hyaloplasm of pigeon liver after irradiation and administration of protector. Radiobilogiya 8(1): 107-110. (AEC-tr-6914) (NSA 22: 47627).

Romantsev, Ye. F., et al. 1966. Radioprotective action of Stellin. pp. 244-248. In: Protection Against and Repair of Radiation Injuries. Ye. Ya. Grayevskiy, V. I. Ivanov and V. I. Korogodin, eds. Publ. House Acad. Sci. USSR, Moscow. (In Russian) (NSA 22: 25854) (From sturgeon milt).

Romashov, D. D. and V. N. Belyayeva. 1964. The cytology of the radiation gynogenesis and androgenesis in Misgurnus fossils L. Dokl. Akad. Nauk SSSR 157(4): 964-967. (Doklady (biol. sci.) 157: 503-506) (NSA 19: 2102).

Romashov, D. D. and V. N. Belyayeva. 1965. Radiosensitivity of loach (Misgurnus fossilis L.) eggs. Genetika (5): 101-109. (In Russian) (NSA 20: 24894).

Romashov, D. D. and V. N. Belyayeva. 1966. Constancy of radiation-damaged chromosomes in fish embryogenesis. Genetika (4): 4-15. (In Russian) (NSA 20: 35270).

Romashov, D. D. and K. A. Golovinskaya. 1960. Radiation biology and the genetics of fishes. pp. 9-49. In: Ionizing Radiation and Heredity. D. P. Dubinin, ed. Progress in Science Ser. No. 3, Acad. Sci. USSR, Moscow. (In Russian).

Romashov, D. D., et al. 1960. Radiation damage in fish. Acad. Sci. USSR, Moscow. 18pp. UN Document A/AC.82/G/L.426. (In Russian) (NSA 15: 15407).

Romashov, D. D., et al. 1960. Radiation induced diploid gynogenesis in fish. Biofizika 5(4): 461-467. (Biophysics (USSR) 5(4): 524-532) (NSA 15: 15430).

Romashov, D. D., et al. 1963. Radiation injuries in fish. In: Radiation Genetics (Collected Works). Acad. Sci. USSR. (English Abstr. Sov. Block and Mainland China Tech. Ser. 6 (bio-sci.) 62-11, 145(26): np; SFA 10: 7338).

Romashov, D. D., et al. 1963. Possibilities of producing diploid radiation-induced gynogenesis in sturgeons. Radiobiologiya 3(1): 104-110. (AEC-tr-5434, pp. 144-154) (NSA 17: 37100).

Romberg, G. P. and W. C. Renfro. 1973. Radioactivity in juvenile Columbia River salmon: a model to distinguish differences in movement and feeding habits. Trans. Amer. Fish. Soc. 102(2): 317-322. (NSA 28: 13103).

Romberg, G. P., et al. 1974. Migratory behavior of fish tagged at a nuclear power plant discharge into Lake Michigan. Argonne National Lab., U. S. AEC report CONF-740813-2. 31pp. (NSA 30: 29462).

Romney, E. M. and J. D. Childress. 1963. Reactions of tungsten in soils and its uptake by plants. pp. 521-524. In: Radioecology. (See Schultz and Klement, 1963).

Romney, E. M. and J. J. Davis. 1972. Ecological aspects of plutonium dissemination in terrestrial environments. Health Physics 22(6): 551-557. (Also: Univ. of Calif. at Los Angeles, U. S. AEC report UCLA-12-848) (NSA 26: 9483, 38419).

Romney, E. M., H. M. Mork and K. H. Larson. 1970. Persistence of plutonium in soil, plants, and small mammals. Health Physics 19: 487-491. (NSA 25: 4478).

Romney, E. M., A. Wallace and J. D. Childress. 1973. Revegetation problems following nuclear testing activities at the Nevada Test Site. pp. 1015-1022. In: Radionuclides in Ecosystems. (See Nelson, 1973) (NSA 25: 37757).

Romney, E. M., et al. 1963. Contamination of plant foliage with radioactive fallout. Ecology 44(2): 343-349. (Also: Univ. of Calif. at Los Angeles, U. S. AEC report TID-15870) (NSA 17: 7774).

Romney, E. M., et al. 1967. Concentration of radionuclides by plants grown on ejecta from the Sedan thermonuclear cratering detonation. pp. 391-398. In: Radioecological Concentration Processes. (See Aberg and Hungate, 1967).

Romney, E. M., et al. 1973. Persistence of radionuclides in soil, plants, and small mammals in areas contaminated with radioactive fallout. pp. 170-176. In: Radionuclides in Ecosystems. (See Nelson, 1973) (NSA 25: 37792).

Romney, E. M., et al. 1975. $^{239-240}Pu$ and ^{241}Am contamination of vegetation in aged plutonium fallout areas. pp. 43-87. In: Radioecology of Plutonium and Other Transuranics in Desert Environments. (See White and Dunaway, 1975) (NSA 31: 29626; 33: 614, 625).

Ronca, L. B. and E. J. Zeller. 1965. Thermoluminescence as a function of climate and temperature. Am. J. Sci. 263: 416-428. (NSA 20: 29398).

Roncucci, R. R. and R. J. Kirchmann. 1964. Effect of p_H on the foliar retention of iodine-131, in the chemical forms I_2 and I^-, in Helodea canadensis Rich. Compt. rend. soc. biol. 158: 2212-2216. (In French) (NSA 19: 29863) (Aquatic plant).

Roncucci, R. R., et al. 1968. Simultaneous liquid scintillation determination of tritium and ^{35}S in biological low-level samples using the oxygen flask method. Anal. Biochem. 26: 118-136. (NSA 23: 9321).

Rondanelli, E. G., et el. 1965. Pseudo-amitoses provoked by x rays in erythropoietic cells in vitro. Nouvelle Rev. Franc. Hematol. 5: 213-220. (In French) (NSA 19: 43826) (Involves the triton Molge vulgaris).

Rondanelli, E. G., et al. 1967. Morphogenesis of the cells at several nuclei by the action of x radiation. Study in contrast of phase on the erythropoietic cells to the living state. Exptl. Cell Res. 47: 222-228. (In French) (NSA 21: 43509) (Triton spleen).

Rone, E. and W. D. Urry. 1952. Radium and uranium contents of ocean and river waters. Am. J. Sci. 250(3): 241-262. (CA 46: 6497g).

Rongstad, O. J. 1965. Calcium-45 labeling of mammals for use in population studies. Thesis, Univ. of Wisconsin, Madison.

Rongstad, O. J. 1965. Calcium-45 labeling of mammals for use in population studies. In: Radiation and Terrestrial Ecosystems. Health Physics 11(12): 1543-1556. (See Hungate, 1965).

Ronkin, R. R. 1950. The uptake of radioactive phosphate by the excised gill of the mussel Mytilus edulis. J. Cell. Comp. Physiol. 35(2):;241-260.

Ronzani, C. A. 1968. Evolution of organic matter in the soil: its determination with natural ^{14}C. Acta Cient. Venez. 19: 180-183. (In Spanish) (NSA 24: 16692).

Rook, D. A. 1971. Effect of undercutting and wrenching on growth of Pinus radiata D. Don seedlings. J. Appl. Ecol. 8(2): 477-490. (^{14}C).

Rook, H. L., T. E. Gills and P. D. LaFleur. 1972. Method for determination of mercury in biological materials by neutron activation analysis. Anal. Chem. 44(7): 1114-1117. (NSA 26: 33059).

Ros, J. P. 1975. Genetic control of Ceratitis capitata Wied. by the sterile-male technique: activities carried out in Spain (1969-73). pp. 47-92. In: Sterility Principle for Insect Control, 1974. (See IAEA, 1975) (In Spanish) (NSA 32: 11765).

Rose, D. J. 1970. Hazards of tritium from controlled fusion. Health Physics 18: 439-440. (NSA 24: 27565).

Rose, F. C. and S. M. Rose. 1965. The role of normal epidermis in recovery of regenerative ability in x-rayed limbs of Triturus. Growth 29: 361-393. (NSA 20: 20465).

Rose, F. C. and S. M. Rose. 1967. Nerve penetration and regeneration after x-rayed limbs of Triturus viridescens were covered with unx-rayed epidermis. Growth 31: 375-380. (NSA 22: 43451).

Rose, F. C. and S. M. Rose. 1974. Regeneration of aneurogenic limbs of salamander larvae after x irradiation. Growth 38(1): 97-108. (NSA 29: 29818).

Rose, F. C., H. Quastler and S. M. Rose. 1955. Regeneration of x-rayed salamander limbs provided with normal epidermis. Science 122(3178): 1018-1019. (NSA 10: 521).

Rose, F. L. and C. E. Cushing. 1970. Periphyton: autoradiography of zinc-65 adsorption. Science 168: 576-577.

Rose, G. B. 1970. Accumulation and transfer of ^{45}Ca by the biota of a tagged cornfield. Abstr. Bull. Ecol. Soc. Am. 51(2): 27.

Rose, G. B., C. D. Monk and R. G. Wiegert. 1969. Accumulation and transfer of ^{45}Ca by the biota of a tagged cornfield. pp. 672-677. In: Symposium on Radioecology. (See Nelson and Evans, 1969).

Rose, J. (Ed.). 1969. Technological Injury. The Effect of Technological Advances on Environment, Life, and Society. Gordon & Breach, New York. 240pp. (NSA 24: 25433).

Rosen, A. A. 1964. Natural carbon-14 activity of organic substances in streams. Science 143(3611): 116 -1164. (SFA 9: 4975).

Rosen, A. S. and M. Rubin. 1965. Discriminating between natural and industrial pollution through carbon dating. J. Water Pollut. Control Fed. 37(9): 1302-1307. (NSA 21: 19765).

Rosen, D. and K. B. Dawson. 1960. Search for immediate effects of x-radiation on frog nerve-muscle preparations. Radiation Research 12(4): 357-370.

Rosen, D. and K. B. Dawson. 1962. Fall in threshold to electrical excitation of x-irradiated isolated frog nerve. Int. J. Radiat. Biol. 5: 535-541. (NSA 17: 10417).

Rosen, J. C., N. Cohen and M. E. Wrenn. 1972. Short-term metabolism of ^{241}Am in the adult baboon. Health Physics 22(6): 621-626. (NSA 26: 38773).

Rosenstein, M. and A. S. Goldin. 1965. Statistical technics for quality control of environmental radioassay. Health Lab. Sci. 2: 93-102. (NSA 19: 36716).

Rosenthal, G. M., Jr. and M. L. Stewart. 1973. Tritium incorporation in algae and transfer in simple aquatic food chains. pp. 440-444. In: Radionuclides in Ecosystems. (See Nelson, 1973) (NSA 25: 41689).

Rosenthal, G. M., Jr., D. J. Nelson and D. A. Gardiner. 1965. Deposition of strontium and calcium in snail shell. Nature 206(4979): 51-54. (Also: Oak Ridge National Lab., U. S. AEC report ORNL-P-608) (NSA 19: 35923).

Rosenthal, H. L. 1956. Uptake and turnover of calcium-45 by the guppy. Science 124(3222): 571-574. (NSA 10: 11682).

Rosenthal, H. L. 1957. Uptake of calcium-45 and strontium-90 from water by freshwater fishes. Science 126(3276): 699-700. (NSA 12: 2728).

Rosenthal, H. L. 1957. The metabolism of strontium-90 and calcium-45 by Lebistes. Biol. Bull. 113(3): 442-450. (WPA 31: 431).

Rosenthal, H. L. 1958. Comparative uptake of calcium-45 and strontium-90 by wild and lordotic guppies. Proc. Soc. Exptl. Biol. Med. 97(3): 624-627. (SFA 4: 19).

Rosenthal, H. L. 1960. Accumulation of strontium-90 and calcium-45 by freshwater fishes. Proc. Soc. Exptl. Biol. Med. 104(1): 88-91. (NSA 14: 14650).

Rosenthal, H. L. 1961. The uptake and turnover of S^{35} sulfate by *Lebistes*. Biol. Bull. 120(2): 183-191. (NSA 15: 23255).

Rosenthal, H. L. 1963. Uptake, turnover, and transport of bone seeking elements in fishes. Ann. N.Y. Acad. Sci. 109(1): 278-293. (BA 45: 953; NSA 18: 4975; SFA 9: 6611).

Roser, F. X. 1958. On the assimilation of natural radioactive isotopes by plants. Acad. Bras. Cien. An. 30(3): 24. (In Portugese, English summary).

Roser, F. X. and T. L. Cullen. 1958. On the intensity levels of natural radioactivity in certain selected areas of Brazil. Brazil Comissão Nacional de Energia Nuclear, Rio de Janeiro, report DTC-1. pp. 5-52.

Roser, F. X. and T. L. Cullen. 1962. Radiation levels in selected regions of Brazil. An. acad. Brasil. cien. 34: 23-35. (NSA 16: 26722).

Roser, F. X. and T. L. Cullen. 1964. External radiation levels in high-background regions of Brazil. pp. 825-836. In: *The Natural Radiation Environment*. (See Adams and Lowder, 1964) (NSA 18: 35748).

Roser, F. X., G. Kegel and T. L. Cullen. 1964. Radiogeology of some high-background regions of Brazil. pp. 855-872. In: *The Natural Radiation Environment*. (See Adams and Lowder, 1964) (NSA 18: 35713).

Rosholt, J. N. 1967. Open system model for uranium-series dating of Pleistocene samples. pp. 299-311. In: *Radioactive Dating and Methods of Low-Level Counting*. (See IAEA, 1967) (NSA 22: 6463).

Ross, A. 1966. Estimation of growth patterns of protozoa in radiation experiments. Johns Hopkins Univ., Baltimore, Md. 9pp. (For: Eastern Regions Joint Meeting, Biometric Soc. Inst. Mathematical Statistics, Upton, N. Y., Apr. 27-29, 1966) (CONF-660441-1) (AED-CONF-66-118-1) (NSA 20: 45472).

Ross, M. H. and D. G. Cochran. 1963. Some early effects of ionizing radiation on the German cockroach, *Blatella germanica*. Ann. Entomol. Soc. Am. 56(3): 256-260.

Rossi, G. 1973. Gamma ray instruments for the measurement of the water equivalent of a snow pack. Italy Ente Nazionale per l'Energia Elettrics, Rome. 16pp. (In Italian) (INIS-mf-995) (NSA 30: 6762).

Rossi, L. C. and A. A. Cigna. 1965. Biological investigations concerning the environmental radioactivity. I. Plan of study. Minerva Nucleare 9: 138-148. (In Italian) (NSA 20: 9202).

Rossi, L. C. and A. A. Cigna. 1968. Biological studies of ambient radioactivity. II. The ecology of radionuclides. G. Fis. Sanit. Prot. Radiaz. 12: 261-271. (In Italian) (NSA 24: 29727).

Rossi, L. C., et al. 1970. Applied ecological studies on the environment of the Casaccia Nuclear Center, Rome Italy. pp. 71-76. In: *Environmental Surveillance in the Vicinity of Nuclear Facilities*. (See Reinig, 1970) (NSA 25: 16163).

Rossmoore, H. W. and E. A. Hoffman. 1971. Effect of ^{137}Cs radiation on growth, development, and mortality of the larvae of the tussock moth, *Hemerocampa leucostigma*. J. Invertebr. Pathol. 17(2): 277-281. (NSA 26: 45885).

Rossouw, S. F. (Comp.). 1966. Radioisotopes and radiation in helminthology. South Africa Atomic Energy Board, Pelindaba, Pretoria. 29pp. (NSA 21: 20323).

Rost, T. L. and J. Van't Hof. 1973. Radiosensitivity, RNA, and protein metabolism of leaky and arrested cells in sunflower root meristems (*Helianthus annuus*). Am J. Bot. 60(2): 172-181. (NSA 28: 18438).

Rostand, J. 1957. Monstrous frogs and radioactivity. Compt. rend. 245(14): 1175-1176. (In French).

Rostenbach, R. E. 1956. Radioactivity levels of the Columbia River below Richland, Washington for the period July, August, September 1955. --- October, November, December 1955. Hanford Atomic Products Operation, U. S. AEC report HW-42181, -42946. 9, 9pp. (NSA 10: 3979, 6491).

Rostenbach, R. E. 1959. Radioactivity levels and temperature variations of the Columbia River. Chem. Engr. Progr. 55(Symp. ser. 22): 37-43. (NSA 14: 4144).

Roth, A. R. and G. W. Eddy. 1957. Tests with DOW ET-57 against cattle grubs. in Oregon. J. Econ. Entomol. 50(3): 244-246.

Roth, A. R. and R. A. Hoffman. 1952. A new method of tagging insects with P^{32}. J. Econ. Entomol. 45(6): 1091.

Roth, J. S. 1962. Biochemical studies on irradiated protozoa. I. Effect of metabolites on the respiration of x-irradiated *Tetrahymena pyriformis*. J. Protozool. 9(2): 142-146.

Roth, J. S. and G. Buccino. 1965. Biochemical studies on irradiated protozoa. III. Catalase activity in *Tetrahymena pyriformis* W. J. Protozool. 12: 432-437. (NSA 20: 22598).

Roth, R. 1943. On the cultivation of *Paramecium caudatum*, the ashes of brood clones modified, the entire ashes, under the influence of roentgen radiation. Strahlentherapie 74(1): 169-181. (In German).

Roth, S. J. and G. Huckabay. 1974. Comparison of the *in-situ* measurement of terrestrial americium with a Ge(Li) spectrometer and fidler. Trans. Nucl. Sci. NS-21(1): 438-443. (Also: Lawrence Livermore Lab., U. S. AEC report UCRL-75204) (NSA 30: 6762; 31: 2836).

Roth, T. F. and K. R. Porter. 1964. Yolk protein uptake in the oocyte of the mosquito *Aedes aegypti* L. J. Cell Biol. 20(2): 313-332. (^3H).

Rothschild, S. (Ed.). 1966. Advances in Tracer Methodolgy. Vol. 3. (A Collection of Papers Presented at the 9th and 10th Symp. on Tracer Methodology). Plenum Press, New York. 341pp. (NSA 21: 30794).

Rothstein, M. and E. M. Saffran. 1963. Lysine biosynthesis in algae. Arch. Biochem. Piophys. 101(3): 373-377. (Used ^{14}C).

Roti, J. L. and S. V. Kaye. 1967. Calibration of $^{90}Sr+^{90}Y$ plaques for irradiating insects. Oak Ridge National Lab., U. S. AEC report ORNL-TM-1921. 22pp. (NSA 21: 39403).

Rott, N. N. 1959. The effect of colchicine upon irradiated Misgurnus fossilis embryos. Dokl. Akad. Nauk SSSR 125: 917-920. (In Russian) (NSA 13: 13170).

Rottschafer, J. M., J. D. Jones and H. B. Matk, Jr. 1971. A simple rapid method for determining trace mercury in fish via neutron activation analysis Environ. Sci. Technol. 5(4): 336-338. (NSA 25: 31850; SFA 16: 13721).

Rouhanizadeh, N. 1974. Environmental protection and radioecology. Centre d'Études Nucléaires, France Commissariat à l'Énergie Atomique, Fontenayaux-Roses, report CEA-R-4570. 25pp. (In French) (NSA 30: 3143).

Roushdy, H. M. and H. S. El-Kashef. 1970. Fluctuations in sugar levels in blood and hepatopancreatic tissue of gamma-irradiated Nile fish Tilapia nilotica. Isotop. Radiat. Res. 3: 49-55. (NSA 25: 21562).

Roushdy, H. M. and A. E. Imam. 1971. Growth rate measurements in gamma-irradiated catfish. Z. Naturforschung 26b(9): 962-967. (NSA 26: 7078).

Roushdy, H. M., H. S. El-Kashef and G. M. Haggag. 1969. Deactivation of some hydrolytic enzymes in fish after radiation exposure. Isotop. Radiat. Res. 2: 27-32. (NSA 24: 44427).

Roussel, J., P. Schoumacher and M. Pernot. 1958. Sensitivity of yeasts to various ionizing radiations. J. Radiol. Electrol. 39(3): 202-203.

Rouston, W. J., C. A. Schuntner and H. J. Schnitzerling. 1966. Metabolism of Coumaphos in larvae of the cattle tick Boophilus microplus. Austral. J. Biol. Sci. 19(4): 619-633. (^{32}P).

Routson, R. C. 1973. Review of studies on soil-waste relationships on the Hanford Reservation from 1944-1967. Battelle Northwest Lab., U. S. AEC report BNWL-1464. 63pp. (NSA 28: 2559).

Routson, R. C. 1975. Effect of soil concentration on the tumbleweed uptake of ^{90}Sr and ^{137}Cs from a Burbank sand. Battelle Northwest Lab., U. S. AEC report BNWL-1905. 14pp. (NSA 32: 8918).

Roux, F., M. Chauvet and J. P. Bisset. 1973. Hydrological surveillance of the Cadarache CEN site. pp. 47-51. In: Control of Ionizing Radiation: Pollution, Dosimetry, Applications. France, Centre National de la Recherche Scientifique, Marseilles. (In French) (NSA 30: 6479).

Roux, J.-C. 1966. Radiosensitivity of Chlorella after medium energy accelerated electron irradiation. Centre d'Études Nucléaires, France Commissariat à l'Énergie Atomique, Grenoble, report CEA-R-2984. 72pp. (In French) (NSA 20: 45353).

Roux, J.-C. 1970. Radiosensitivity of Chlorella to alpha particles. Bull. Inform. Sci. Tech. (Paris) (145): 61-66. (In French) (NSA 24: 34758).

Rovinskiy, F. Ya. 1962. Methods of extracting radioactive elements from soil. Pochvovediniye (6): 99-101. (Sov. Soil Science (6): 654-___).

Rovinskiy, F. Ya. 1965. A method for calculating radioactive impurities in water and the noncirculating residue layer in water basins. Atom energiya 18(4): 379-383. (Sov. Atomic Energy 18(4): 480-485) (NSA 19: 30408).

Rovinskiy, F. Ya. 1967. Migration of ^{90}Sr in stagnant waters. Radiokhimiya 9(1): 80-84. (Sov. Radiochemistry 9(1): 76-79) (NSA 21: 28086).

Rovinskiy, F. Ya. and A. L. Agre. 1966. Predicting ^{90}Sr uptake in fish. Gigiyena sanitariya (8): 117-119. (In Russian) (NSA 21: 4780).

Rovinskiy, F. Ya., et al. 1973. Migration possibility of radionuclides and their transition into water when applying atomic energy for peaceful purposes. pp. 20-29. In: Radioecology of Water Organisms. II. (See Andrushaitis, et al., 1973) (AEC-tr-7606, pp. 160-162) (NSA 30: 21146).

Rowe, D. R. 1965. The transport of ^{65}Zn in an aqueous environment. Thesis, Univ. of Texas, Austin. 116pp. (NSA 20: 6715).

Rowe, D. R. and E. F. Gloyna. 1964. Radioactivity transport in water - the transport of Zn^{65} in an aqueous environment. Environmental Health Engrg. Labs., Univ. of Texas, Austin, Tech Rept. -5. xiii, 101pp.

Rowe, W. D. 1974. Radiological environmental monitoring: the EPA approach. Trans. Nucl. Sci. NS-21(1): 416-422. (NSA 30: 6490).

Rowell, J. B. and T. W. Sudia. 1960. Ten years of research show effects of radioactivity on microorganisms. Minn. Farm Home Sci. 17(3): 19-22. (Involves plant pathogens).

Rowinski, P. 1962. The biological effects of variation of radioactivity of potassium in the perfused isolated heart. Final report for IAEA Contract 52. 35pp. (NP-14664) (NSA 19: 12953) (Rana esculenta).

Roy, R. M., B. Donini and A. Brunori. 1972. Biochemical and cytological studies on developing cotyledons of Pinus pinea following x-irradiation of dry seeds. Radiation Botany 12(4): 240-260. (NSA 26: 59014).

Rozhanskaya, L. I. 1968. Concentration factors of manganese, copper and zinc in some hydrobionts of the Azov Sea. pp. 21-25. In: Biological Migration of Radionuclides in Freshwater and Saltwater Reservoirs. State Committee on the Use of Atomic Energy, USSR. Atomizdat, Moscow. UN Document A/AC.82/G/L.1243. (AEC-tr-7128, pp. 30-38) (NSA 23: 24576; 24: 12027).

Rozhanskaya, L. I. 1968. The role of some marine organisms in the migration of manganese, copper and zinc in the Azov Sea. pp. 11-20. In: Biological Migration of Radionuclides in Freshwater and Saltwater Reservoirs. State Committee on the Use of Atomic Energy, USSR. Atomizdat, Moscow. UN Document A/AC.82/G/L.1243. (AEC-tr-7128, pp. 14-29) (NSA 23: 24575; 24: 12026).

Rozhanskaya, L. I. 1970. Determination of Mn, Cu, Zn and Co in hydrobionts and water. pp. 24-25. In: Marine Radioecology. G. G. Polikarpov, ed. Publ. House Acad. Sci. Ukr. SSR, Kiev. (In Russian).

Rozhanskaya, L. I. 1970. Mn, Cu and Zn in water and organisms of the Azov Sea. pp. 182-207. In: Marine Radioecology. G. G. Polikarpov, ed. Publ. House Acad. Sci. Ukr. SSR, Kiev. (In Russian).

Rozhdestvenskaya, L. F., et al. 1959. Influence of gamma-rays from cobalt-60 on nutritional properties of mulberry leaves. Dokl. Akad. Nauk Uzbek. SSR (8): 11-13. (In Russian).

Rozhdestvenskaya, L. F., et al. 1961. Influence of the gamma rays of Co^{60} on the nutritional properties of mulberry leaves. pp. 435-438. In: Proc. Tashkent Conf. on the Peaceful Uses of Atomic Energy, 1959. Publ. House Acad. Sci. Uzbek. SSR, Tashkent. (In Russian).

Rozkowska, A. and A. Rozkowski. 1973. Unbalanced ratio of activity of uranium isotopes as a natural hydrogeological marker. Tech. Poszukiwan 13(47-48): 40-43. (In Polish) (NSA 31: 22652).

Rozyanov, S. P., V. K. Vinogradova and L. I. Gustova. 1967. Distribution of strontium-90 and cesium-137 in the prfile of soils under natural conditions. State Committee on the Use of Atomic Energy, USSR. Atomizdat, Moscow. UN Document A/AC.82/G/L.1169. (AEC-tr-7214, pp 82-86) (NSA 25: 15758).

Ruan, E. S., et al. 1967. Gamma irradiation of European corn borer larvae. J. Ecom. Entomol. 60(6): 1724-1730.

Ruban, Ye. L. and S. P. Lyakh. 1968. The stability of Nadsoniella nigra to the action of short-wave electromagnetc irradiation. Izv. Akad. Nauk SSSR (biol.) (3): 402-407. (In Russian) (NSA 23: 8089).

Rubin, M. and E. Spiker. 1972. ^{14}C activity as an indicator of course of dissolved organic carbon in rivers and lakes. pp. D13-D18. In: Proc. 8th Intern. Radiocarbon Dating Conf. Vol. 1. (See Rafter, 1972) (NSA 31: 8696).

Rubin, M. and D. W. Taylor. 1963. Radiocarbon activity of shells from living clams and snails. Science 141(3581): 637. (NSA 17: 34136).

Rubin, M., R. C. Likens and E. G. Berry. 1963. On the validity of radiocarbon dates from snail shells. J. Geology 71(1): 84-88.

Rubini, P. G. and M. Caprotti. 1966. Observations on the lethal effects of x rays of different hardness on the pupa of Musca domestica L. Symp. Genet. Biol. Ital. 13: 387-392. (In Italian) (NSA 23: 20320).

Rubinstein, D. 1965. Gamma ray degradation of chloroplast structure and photoreduction. Thesis, Univ. of California, Davis. 116pp. (NSA 20: 12627).

Rublevskiy, V. P., A. S. Zykova and A. D. Turkin. 1974. Atomic electric power stations as sources of carbon-14 discharge. pp. 296-300. In: Proc. 3rd Intern. Congr. Intern. Radiation Protection Assoc. Inst. Biophysics, USSR, Moscow. (In Russian) (CONF-730907-P1) (NSA 30: 21146).

Rubtsov, D. M. 1966. Distribution of thorium in various soils. Pochvovedinye (3): 55-67. (In Russian) (NSA 20: 23333).

Rubtsov, D. M. 1972. Distribution of uranium and radium in the podzolic mountain soils of thin forests. pp. 42-53. In: Radioecological Studies in Natural Biogeocenoses. (See Verkhovskaya, 1972) (In Russian) (NSA 26: 56108).

Rubtsov, D. M. 1972. Thorium and radium content in the silt fraction of the podzolic mountain soils of thin forests. pp. 53-66. In: Radioecological Studies in Natural Biogeocenoses. (See Verkhovskaya, 1972) (In Russian) (NSA 26: 56109).

Rubtsov, D. M. and Ye. I. Pravdina. 1972. Content and distribution of uranium, radium, and thorium in mountain tundra soils. pp. 67-86. In: Radioecological Studies in Natural Biogeocenoses. (See Verkhovskaya, 1972) (In Russian) (NSA 26: 56110).

Rubtsov, I. A. 1964. Mode and range of black fly (Diptera: Simulidae) larval migrations. Entomol. Oboz. 43(1): 52-66. (Entomol. Rev. (USSR) 43(1): 27-33) (Tracer).

Ruchhoft, C. C. and L. R. Setter. 1953. Application of biological methods in the treatment of radioactive wastes. Sewage Industr. Wastes 25: 48-60.

Rudakov, N. P. 1958. Comparative absorption of calcium-45 and strontium-90 by young salmon. Ryb. Khozaystvo (4): 19-20. (In Russian) (SFA 4: 2710).

Rudakov, N. P. 1958. Marking of fish by radioactive cerium (Ce^{144}-Pr^{144}). Ryb. Khozaystvo 34: 29-31. (In Russian)

Rudakov, N. P. 1958. The use of radioisotopes for studying the uptake and release of certain mineral ions in fishes and their labeling. (Author's summary, thesis, All-Union Research Inst. for Lake and River Fishing. 14pp. (In Russian).

Rudakov, N. P. 1961. Characteristics of Ca^{45}, Sr^{90}, Ce^{144}, Cs^{137} in the body of a fish and methods of marking young fish with radioactive isotopes. Izv. gos. nauchn.-issled. inst. ozern. rechn. ryb. khoz. 51: 165-254, 271-271. (BA 45: 27258; NSA 19: 35910; Ref. zh. (biol.) 1963: 14156; SFA 10: 7286) (In Russian).

Rudakov, N. P. 1961. The isotope effect in the determination of primary production of organic matter in waters by the radioisotope method. pp. 214-218. In: Primary Production of the Seas and Inland Waters. G. G. Vinberg, ed. Ministry of Higher and Secondary Trade and Professional Education, Beloruss. SSR, Minsk. (In Russian) (BA 43: 4424; Ref. zh. (biol.) 1962: 20Zh385; SFA 8: 6135).

Rudenchik, Yu. V. 1963. Application of radioactive tracers for the study of intrapopulational connections as an epizootological factor in the settlement of Rhombomys opimus. Zool. zh. 47(12): 1849-1856. (In Russian).

Rudenchik, Yu. V., et al. 1967. A quantitative estimate of the possibilities of a territorial advance of an epizootic plague in a population of great gerbils (northern Kyzyl-Kum). Zool. zh. 46(1): 117-123. (In Russian) (Radionuclides).

Rudin, S. and H. Hart. 1971. Monitor for ^{133}Xe contamination of air. J. Nucl. Med. 12: 145-146. (NSA 29: 21403).

Rudnicki, T. 1961. The relative biological effect of P^{32} β radiation. I. Comparison of P^{32} β radiation and x-rays as to their biological effects on Drosophila melanogaster. Acta Phys. Polon. 12(1): 154-157. (In Polish).

Rudolph, A. W., et al. (Comps.). 1970. Index. Bioenvironmental and radiological-safety feasibility studies, Atlantic-Pacific Interoceanic Canal. Battelle Columbus Labs., U. S. AEC report BMI-171-43. 269pp. (NSA 24: 48455).

Rudolph, T. D. 1967. Effects of x-irradiation of seed on X_1 and X_2 generations in Pinus banksiana Lambert. Radiation Botany 7(4): 303-312. (BA 48: 123802).

Rudolph, T. D. 1969. Seed yield and quality in white spruce self-pollinated with gamma-irradiated pollen. pp. 207-217. In: Induced Mutations in Plants. (See IAEA, 1969) (NSA 24: 42027).

Rudolph, T. D. 1971. Gymnosperm seedling sensitivity to gamma radiation: its relation to seed radiosensitivity and nuclear variables. Radiation Botany 11: 45-51. (NSA 25: 35250).

Rudolph, T. D. 1972. Effect of acute gamma irradiation at various stages of seasonal growth in Pinus banksiana Lambert. pp. 289-300. In: Isotopes and Radiation in Soil-Plant Relationships Including Forestry. (See IAEA, 1972) (NSA 27: 2739).

Rudolph, T. D. 1973. Seed yield and quality in white spruce cross-pollinated with gamma-irradiated pollen. pp. 972-977. In: Radionuclides in Ecosystems. (See Nelson, 1973) (NSA 28: 30073).

Rudolph, T. D. (Ed.). 1974. Enterprise, Wisconsin, radiation forest: preirradiation ecological studies. U. S. Forest Service, U. S. AEC report TID-26113. 58pp. (NSA 30: 9772).

Rudolph, T. D. and J. P. Miksche. 1970. The relative sensitivity of the soaked seeds of nine gymnosperm species to gamma radiation. Radiation Botany 10(5): 401-409.

Rudolph, W. K. 1963. Concentrations of gamma-emitting fallout radionuclides from Picea rubens and Rhododendron maximum of the Great Smoky Mountains. Thesis, Univ. of Tennessee, Knoxville. (NSA 18: 419).

Ruf, M. 1965. Radioactivity in freshwater fish. Zentralbl. Veternarmed. Reihe B12(7): 605-612. (BA 47: 95613; SFA 8925).

Ruf, M. 1966. Accumulation of fission products by fresh water fish. Bull. Off. Int. Epizoot. 65: 707-713. (NSA 22: 31076).

Ruf, M. 1968. Contamination of fresh water with radioactive material with special attention to sludge of river dams. Wasserwirtschaft 58: 16-22. (In German) (NSA 23: 9547).

Ruf, M. 1970. Disposal into human environment of radioactive waste from atomic reactors, with special reference to water supply. Zentralbl. Veterinarmed. Beih. (11): 167-180. (In German) (NSA 25: 26009).

Ruf, M. 1971. Radioecological analysis of the Upper Danube. pp. 477-506. In: Proc. Intern. Symp. on Radioecology Applied to Man and His Environment. (See Commission of the European Communities, 1971) (In German) (NSA 27: 5261).

Ruf, M. 1974. Radioecology of the Danube. pp. 403-431. In: Radiation Protection, Environment, and the Population., Vol. 2. (See German Democratic Republic, 1974) (In German) (AED-CONF-74-725-000(Vol.2) (NSA 33: 25969).

Ruf, M. and H. Rohde. 1973. Determination of mercury contents in diverse samples of fish and other biological materials by neutron activation analysis. Z. Anal. Chem. 263(2): 116-120. (NSA 28: 2143).

Ruff, M. D. and M. F. Hansen. 1967. Elimination of male Ascaridia galli (Nematoda) by x irradiation of embryonated eggs. J. Parasitol. 53: 1285-1287. (NSA 22: 40898).

Ruff, M. D., M. F. Hansen and D. A. Ostlind. 1965. Effects of x-irradiation on the development of Ascaridia galli, intestinal worm of chickens. Trans. Kan. Acad. Sci. 68: 498-507. (NSA 20: 42968).

Rugh, R. 1939. Developmental effects resulting from exposure to x-rays. I. Effect on the embryo of irradiation of frog sperm. II. (w/F. Exner, 1940) Development of leopard frog eggs activated by bullfrog sperm. Proc. Am. Philos. Soc. 81(3): 447-471; 83(5): 607-619.

Rugh, R. 1949. Some prenatal effects of Ambystoma opacum larvae exposed to 25,000 r x-radiation. Abstr. Anal. Record 103(3): 500-501.

Rugh, R. 1949. The immediate and delayed morphological effects of x-radiation on meiotic chromosomes. Columbia Univ., U. S. AEC report AECU-649 (CUF-3). 20pp. (NSA 4: 383) (Triturus pyrrhogaster).

Rugh, R. 1953. Vertebrate radiobiology: embryology. Ann. Rev. Nuclear Sci. 3: 271-302. (NSA 8: 978).

Rugh, R. 1954. The effect of ionizing radiations on amphibian development. J. Cell. and Comp. Physiol. 43(Suppl.1): 39-67. (NSA 8: 5095).

Rugh, R. 1954. X-irradiation of the frog embryo: gross effects. Columbia Univ., U. S. AEC report NYO-474. 25pp.

Rugh, R. 1954. The effect of various levels of x-irradiation on the gametes and early embryos of Fundulus heteroclitus. Abstr. Biol. Bull. 107(2): 319-320.

Rugh, R. 1959. Vertebrate radiobilogy (embryology). Ann. Rev. Nuclear Sci. 9: 493-522.

Rugh, R. and H. Clugston. 1955. Effects of various levels of x-irradiation on the gametes and early embryos of *Fundulus heteroclitus*. Biol. Bull. 108(3): 318-325.

Rugh, R. and E. Grupp. 1958. The effect of x-irradiation on the early fish embryo. Abstr. Biol. Bull. 115(2): 362.

Rugh, R. and R. H. Van Dyke. 1962. Reactions of the x-irradiated embryonic eye in a normal field and *vice versa*. J. Exptl. Zool. 149: 217-231. (NSA 17: 7848) (*Ambystoma punctatum*).

Rugh, R. and R. H. Van Dyke. 1964. The fate and effect of x-irradiated neuroblasts (presumptive neural ectoderm) in a normal environment. J. Exptl. Zool. 157: 197-215. (NSA 19: 28026) (*Ambystoma punctatum*).

Rukavishnikov, B. I. 1963. Possible uses of ionizing radiation for sterilizing and destroying insect pests. Rasten. sel'skoye khoz. Rubezom. 3: 53-58. (In Russian).

Rukavishnikov, B. I. 1968. Genetic methods of insect pest control. Genetika 4(6): 128-143. (In Russian) (NSA 22: 45744).

Rule, H. D., P. A. Godwin and W. E. Waters. 1965. Irradiation effects on spermatogenesis in the gypsy moth, *Porthetria dispar* (L.). J. Insect Physiol. 11: 369-378. (NSA 19: 36030).

Rumiantsev, P. D. and V. F. Ratanova. 1958. Effect of ionizing radiation on grain mites. Tr. inst. zerna prod. ego pererabotki 35: 55-57. (In Russian).

Runner, G. A. 1916. Effect of roentgen rays on the tobacco or cigarette beetle. J. Agr. Res. 6(11): 383-388.

Rupp, R. S. and S. E. DeRoche. 1965. Standing crops of fishes in three small lakes compared with C-14 estimates of net primary productivity. Trans. Am. Fish. Soc. 94(1): 9-25.

Ruppert, D. F., et al. 1974. Arsenic concentrations and distribution in Chautauqua Lake sediments. J. Radioanal. Chem. 28(1-2): 159-169. (NSA 31: 21993) (Neutron activation analysis).

Rush, R. E. and G. W. Ware. 1969. Toxicity of DDT, Carbaryl, and Azinphos-methyl to gamma-radiated pink bollworm moths. J. Econ. Entomol. 62(5): 1 1124-1125.

Rushing, D. E. 1967. Determination of dissolved radium-226 in water. J. Am. Water Works Assoc. 59(5): 593-___.

Rushton, W. 1916. Biological notes, some experiments with fry. Salmon Trout Mag. (82-85): 57-67. (Used x-rays).

Russanova, G. V. 1962. On the problem of studying the leaching processes and migration of radium in soils. Pochvovedeniye (9): 85-87. (In Russian) (CA 58: 5411).

Russanova, G. V. 1964. Behavior of radium and calcium in the soil-plant system. Pochvovedeniye (3): 63-70. (Sov. Soil Science (3): 275-280) (NSA 19: 54).

Russanova, G. V. 1965. Properties of soils in a district with elevated natural radiation. Tr. Komi fol. Akad. Nauk SSSR (14): 3-11. (CA 64: 10964).

Russanova, G. V. 1972. Radium content of soils in regions with a high radiation background and the effect of fertilizers on radium mobilty. pp. 22-33. In: *Radioecological Studies in Natural Biogeocenoses*. (See Verkhovskaya, 1972) (In Russian).

Russanova, G. V. and I. N. Verkhovskaya. 1972. The distribution and migration of calcium as a representative of alkaline-earth elements in soils of a podzolic type. pp. 33-41. In: *Radioecological Studies in Natural Biogeocenoses*. (See Verkhovskaya, 1972) (In Russian).

Russell, I. J. 1971. Radial distribution of fallout nuclides and radium isotopes in a 107 year oak trunk. Boston College, Chestnut Hill, Mass., U. S. AEC report NYO-3756-7. 22pp. (NSA 25: 46808).

Russell, J. L. and P. B. Hahn. 1971. Public health aspects of ^{129}I from the nuclear power industry. Radiol. Hlth. Data 12(4): 189-194. (NSA 25: 44811) (Incl. deer thyroids).

Russell, J. L. and P. B. Hahn. 1971. Public health aspects of I-129 from the nuclear power industry. pp. 241-251. In: *Health Physics Aspects of Nuclear Facility Siting*. (See Voilliequé and Baldwin, 1971) (Incl. deer).

Russell, R. S. 1965. Interception and retention of airborne material on plants. In: *Radiation and Terrestrial Ecosystems*. Health Physics 11(12): 1305-1315. (See Hungate, 1965).

Russell, R. S. 1966. Nature of food chains and the nuclides of major concern. pp. 45-62. In: *Radioactivity and Human Diet*. (See Russell, 1966 below) (NSA 21: 43406).

Russell, R. S. (Ed.). 1966. *Radioactivity and Human Diet*. Pergamon Press, Oxford. xi, 552pp.

Russell, R. S. 1969. Contamination of the biosphere with radioactivity. Biol. Conserv. 2(1): 2-9.

Russell, R. S. and R. S. Bruce. 1969. Environmental contamination with fallout from nuclear weapons: a review. pp. 3-12. In: *Environmental Contami- by Radioactive Materials*. (See IAEA, 1969).

Russell, R. S. and F. B. Ellis. 1968. Estimation of the distribution of plant roots in soil. Nature 217: 582-583. (NSA 22: 23471).

Russell, R. S. and L. J. Middleton. 1956. Use of radioactive isotopes in the study of certain agricultural problems. pp. 75-86. In: *Progress in Nuclear Energy. Biological Sciences*. Ser. 6. J. C. Bugher, J. Coursaget and J. F. Loutit, eds. Pergamon Press, London.

Russell, R. S. and M. G. T. Shone. 1963. Movement of strontium and caesium in soils and its significance in studies on the contamination of food chains. pp. 151-159. In: Intern. Colloq. on Retention and Migration of Radioactive Ions in Soils. France Commissariat à l'Énergie Atomique, Saclay. (NSA 18: 16239).

Russell, R. S., H. M. Squire and R. P. Martin. 1955. The effects of Operation HURRICANE on plants and soils. UK Atomic Energy Authority, Harwell, report AERE/SPAR/3. 106pp. (NSA 10: 5474).

Russo, P. F. and R. C. Fix. 1970. Comparison of the threshold, absolute contribution, and graded approaches to environmental radioactivity site surveys. pp. 66-70. In: Environmental Surveillance in the Vicinity of Nuclear Facilities. (See Reinig, 1970) (NSA 25: 16162).

Russo-Caia, S. 1964. Recent autoradiographic observations on the incorporation of labeled precursors of nucleic acids and proteins in the fatty bodies of Musca domestica. Atti accad. naz. lincei rend. cl. sci. fis. mat. nat. 37(6): 518-520. (In Italian) (^{14}C).

Rustamov, R., et al. 1972. Radioactivation analysis of the iron, cobalt, zinc, scandium, cesium, and antimony in soils and plants. Atom. energiya 33(6): 975-977. (In Russian) (NSA 27: 14494).

Rustamov, R., et al. 1973. Activation method of determining Mo, Al, Ca, Cl, Mn, Na, and K in soil-plant samples. Atom. energiya 34(6): 476. (In Russian) (NSA 29: 7489).

Rustamov, R., et al. 1974. Multielement, nondestructive activation analysis of biological soil samples. Izv. Akad. Nauk Uzbek. SSR ser. fiz.-mat. Nauk (3): 63-65. (In Russian) (NSA 30: 31484).

Russu, I. G., et al. 1961. The influence of radioactive phosphorus (P^{32}) on the growth and differentiation of the tadpoles of Rana temporaria. Acta Anat. 46(3): 233-239. (NSA 17: 27072).

Rüther, W., et al. 1974. Preliminary results on the action of cosmic heavy ions on the development of eggs of Artemia saldna. pp. 69-74. In: Life Sciences and Space Research XII. P. H. A. Sneath, ed. Akademie-Verlag, Berlin. (NSA 31: 26193).

Ruzicka, J. and C. G. Lamm. 1969. Automated determinations of traces of mercury in biological materials by substoichiometric radioisotope dilution. Talanta 16: 157-168. (NSA 23: 13650).

Ryabchenko, V. A. and N. M. Yelmanov. 1971. Effects of radiation dose on disinfection of water by γ radiation. Gigiyena sanitariya (5): 99-100. (In Russian) (NSA 26: 28353).

Rydell, H. S. 1969. The implications of uranium isotope distribution associated with the Floridan aquifer of North Florida. Thesis, Florida State Univ., Tallahassee. 127pp. (NSA 24: 14064).

Ryder, M. L. 1956. The localization of radioisotopes in tissue sections by autoradiography. Microscope 11(3): 49-53.

Rygg, B. 1974. Release of radioactive materials from nuclear power plants. Report No. 2. Dispersion mechanisms, transport paths, and concentration factors for radionuclides in the cooling water recipient. Norsk Inst. Vannforskning, Oslo. 36pp. (In Norwegian) (NP-20251) (NSA 34: 11262).

Ryle, G. J. A. 1970. Partition of assimilates in an annual and a perennial grass. J. Appl. Ecol. 7(2): 217-227. (^{14}C).

Ryle, G. J. A. 1972. A quantitative analysis of the uptake of carbon and of the supply of ^{14}C-labelled assimilates to areas of meristematic growth in Lolium temulentum. Ann. Botany 36(146): 497-512.

Ryman, R. G., C. E. Powell and G. J. A. Ryle. 1975. Comparison of two methods of measuring $^{14}CO_2$ respired by plants. J. Appl. Radiat. Isotopes 26 (8): 483-484. (NSA 33: 7643).

Rymaszewska, M. 1969. Ratio of chloride ion concentration to radon in mineral waters used as hydroradiochemical tracer. Acta Geophys. Pol. 17: 197-202. (In Polish) (NSA 23: 45403).

Rymaszewska, M. and I. Zejmo. 1969. A note on the long-lived alpha radioactivity of mineral waters in Swieradow. Acta Geophys. Pol. 17: 39-43. (In Polish) (NSA 23: 40870). (Szczawno ?).

Rymaszewska, M. and I. Zejmo. 1970. Note on the long-lived alpha and beta radioactivity of mineral waters in Szczawno. Acta Geophys. Pol. 18(1): 99-102. (In Polish) (NSA 24: 46349).

Ryther, J. H. 1954. Inhibitory effects of phytoplankton upon the feeding of Daphnia magna with reference to growth, reproduction and survival. Ecology 35(4): 522-533.

Ryther, J. H. 1956. The measurement of primary production. Limnol. Oceanogr. 1(1): 72-84.

Ryther, J. H. and D. W. Menzel. 1965. Comparison of the ^{14}C-technique with direct measurement of photosynthetic carbon fixation. Limnol. Oceanogr. 10(3): 490-492.

Ryzhkov, L. P. 1957. Gas metabolism variability in young fish when acted upon by radioactive phosphorus. Izv. Akad. Nauk. Armyan. SSR 10(3): 101-103. (In Russian).

Ryzhkov, V. L. 1955. Isotopes in the study of phytopathogenic viruses. pp. 17-26. In: Trans. Conf. on the Application of Tracer Atoms in Microbiology. A. A. Imshenetskiy, ed. Publ. House Acad. Sci., USSR, Moscow. (AEC-tr-3494, pp. 17-26).

Ryznar, L. and V. Drasil. 1967. Photoreactivation in Chlamydomonos reinhardi Dangaard. Folia Microbiol. (Prague) 12: 524-528. (NSA 22: 31286).

Saakyan, A. G. 1959. Effects of radon mineral water on isolated frog heart subjected to block of tissue sulfhydryl groups. Farmakol. Toksikol. 22: 178-182.

Saas, A. 1973. Contamination of soils by industrial radioactive wastes. Mechanisms and factors of migration and evolution. Sci. Sol Bull. A.F.E.S. (4): 255-264. (In French, English summary) (NSA 31: 14070).

Saas, A, and A. Grouby. nd. Application of emission spectroscopy to the determination of 19 trace and major elements in the soil, the various granular fractions and soil extracts. Centre d'Études Nucléaires, France Commissariat à l'Énergie Atomique, Cadarache, report CEA-CONF-1182. 24pp. (For: National Conf. on Measurement of Elements Occurring in Traces in Rocks, Nancy, Dec. 4, 1968) (In French) (CONF-681205) (NSA 23: 13619).

Saas, A. and A. Grouby. 1968. Influence in the organic matter in soils on the fixation and the migration of cesium-137, strontium-90, manganese-54, cerium-144, ruthenium-106, and ruthenium-103. pp. 255-265. In: Intern. Congr. on Radioprotection of the Environment. (See Societé Francaise de Radioprotection, 1968).

Saas, A. and A. Grauby. 1970. Experimental study of the liberation of radionuclides and mineral elements starting with two types of litter. pp. 509-567. In: Proc. Intern. Symp. on Radioecology. Vol. I. (See French Commissariat à l'Énergie Atomique, 1970) (In French).

Saas, A. and A. Grauby. 1970. Fixation and migration in soils of the principal products of fission and activation. Bull. Inform. Sci. Tech. (Paris) (151): 55-65. (In French) (NSA 25: 18528).

Saas, A. and A. Grauby. 1973. Mechanisms for the transfer to cultivated soils of radionuclides discharged by nuclear power stations into the system: river-irrigated soil-groundwater. pp. 255-269. In: Environmental Behavior of Radionuclides Released in the Nuclear Industry. (See IAEA, 1973) (Also: Centre d'Études Nucléaires, France Commissariat à l'Énergie Atomique, Cadarache, report) (SM-172-57) (LIB/TRANS-507) (NSA 29: 15630, 29303).

Saas, A. and A. Grauby. 1973. Transfer mechanisms in cultivated soils of waste radionuclides from electronuclear power plants in the system river-irrigated soil-underground water level. Centre d'Études Nucléaires, France Commissariat à l'Énergie Atomique, Cadarache, report CEA-CONF-2443. 20pp. (In French) (NSA 29: 10160).

Saas, A. and A. Grauby. 1974. Radionuclide migration as a function of the quality of irrigation water. Compt. rend. Hebd. Seances Acad. agr. Fr. 60(5): 1181-1192. (In French) (NSA 33: 12037).

Saas, A. and A. Grauby. 1975. Techniques for rapid determination of effects of synergy between radionuclides and pollutants. pp. 145-154. In: Combined Effects of Radioactive, Chemical, and Thermal Releases to the Environment. (See IAEA, 1975) (In French) (NSA 33: 29102).

Saas, A. and M. Matteoli. 1972. Investigation of the organic-mineral complexes of iron in the biological cycle of plants. Kerntechnik 14(6): 269-271. (^{59}Fe).

Saas, A., P. Bovard and A. Grauby. 1971. Biologic cycle of manganese 54 and cobalt 60 in soil. Kerntech. Isotop. Chem. 13(5): 230-232.

Saas, A., P. Bovard and A. Grauby. 1975. Effect of chronic gamma irradiation on decay of oak (Quercus pubescens Willd) and dogwood (Cornus mas L.) leaves and subjacent litter. Radiation Botany 15(2): 141-151. (In French) (NSA 33: 888).

Saas, A., A. Grauby and J. Andres. 1971. Cycle of some radioelements in natural soils. Main data for the control of a site. pp. 381-393. In: Proc. Intern. Symp. on Radioecology Applied to the Protection of Man and His Environment. (See Commission of the European Communities, 1971) (NSA 27: 5254).

Sabo, J. J. and P. H. Bedrosian. 1963. Studies of the Fate of Certain Radionuclides in Estuarine and Other Aquatic Environments. U. S. PHS, Publ. 999-R-3. xiii, 73pp. (NSA 17: 29151).

Sabo, J. J., J. P. Baptist and F. G. Rueter. 1963. Savannah estuary environmental radiological survey (SEERS) project. pp. 11-33. In: Studies of the Fate of Certain Radionuclides in Estuarine and Other Aquatic Environments. (See Sabo and Bedrosian, 1963 above).

Sacca, G. 1961. Studies on houseflies, sterilized with x rays. Rend. ist. super. sanita 24(1): 5-12. (In Italian, English summary) (NSA 17: 15751).

Sacca, G. 1961. Studies on houseflies, sterilized with x-rays. Atti accad. naz. Ital. entomol. rend. 8: 91-98. (In Italian) (NSA 21: 28709).

Sacchi, L., A. Grigolo and M. Caprotti. 1973. Influence of temperature on x-ray dose fractionation effect in Piophila casei L. Genet. agr. 27(4): 410-414. (In Italian) (NSA 30: 24361).

Sachdev, M. S. and P. Chhabra. 1974. Transformation of S35-labelled sulphate in aerobic and flooded soil conditions. Plant Soil 41(2): 335-341. (NSA 31: 29638).

Sacher, G. A. 1961. Effects of x-rays on the survival of Drosophila imagoes. Argonne National Lab., U. S. AEC report TID-14624. 41pp. (NSA 17: 7791).

Sacher, G. A. 1963. Effects of x-rays on the survival of Drosophila imagoes. Physiol. Zool. 36(4): 295-311.

Sacher, G. A. 1966. Comparative studies on aging and acute and chronic radiation effects in Rodentia. I. Motivations and methods. pp. 27-28. In: Radiation Effects on Natural Populations. (See Sacher, 1966 below).

Sacher, G. A. (Ed.). 1966. Radiation Effects on Natural Populations. (Colloq. Radiation Research Soc. Meeting, Philadelphia, Penn., May 23, 1965). Argonne National Lab., Lemont, Ill. iii, 61pp.

Sacher, G. A. 1968. Mathematical analysis of the division delay produced by ionizing radiation. Radiation Research 33: 644-658. (NSA 22: 31240) (Incl. Paramecium).

Sacher, G. A. and E. Trucco. 1962. A theory of the improved performance and survival produced by small doses of radiations and other poisons. pp. 244-251. In: Biological Aspects of Aging. N. W. Schock, ed. Columbia Univ. Press, New York. (Contains remarks on insects).

Sacher, G. A., et al. 1962. Effect of x-rays on the fission rate of Paramecium aurelia. Argonne National Lab., U. S. AEC report TID-15521. 37pp. (NSA 17: 5778).

Sackett, W. M. 1966. Manganese nodules: ^{230}Th:^{231}Pa ratios. Science 154: 646-647. (NSA 21: 3199).

Sackett, W. M. and G. Cook. 1969. Uranium geochemistry of the Gulf of Mexico. Trans. Gulf Coast Ass. Geol. Soc. 19: 233-238. (NSA 25: 57587) (Incl. rivers).

Sacremento Municipal Utility District, California. 1971. Rancho Seco Nuclear Generating Station, Unit 1. Environmental report. 335pp. U. S. AEC DOCKET-50312-25. (NSA 25: 34292).

Sado, T. 1961. Spermatogenesis of the silkworm and its bearing on the radiation induced sterility. In: Proc. Symp. on Genetic Effect of Radiation. (Misima, Nov. 7-8, 1960). Nippon Idengaku Zasshi 36(Suppl.): 136-151. (NSA 19: 10859).

Sado, T. 1961. Cytological estimation of LD_{50} x ray dose for secondary spermatogonia of the silkworm. Rep. Natl. Inst. Genet. Misima 13: 91-92.

Sado, T. 1963. Recovery of irradiated spermatogonia of the silkworm and the stage limit for the production of functional sperm. Rep. Natl. Inst. Genet. Misima 10: 119.

Sado, T. 1963. Spermatogenesis of the silkworm and its bearing on radiation induced sterility. I. --- II. J. Fac. Agr. Kyushu Univ. 12: 359-385, 387-404. (NSA 19: 8865, 10831).

Sado, T. 1966. Cytological evaluation of dose-effects of radiation on mutation frequency of silkworm gonia. I. Kinetics of proliferation and killing of spermatogonia during chronic irradiation. II. (w/K. Oishi) Evidence of G_2 accumulation of primary spermatogonia after irradiation. Mutation Research 3(6): 510-521, 522-536. (NSA 21: 20135).

Sadun, E. H. 1968. Immune mechanisms in mice immunized with irradiated Plasmodium berghei. pp. 75-76. In: Isotopes and Radiation in Parasitology. (See IAEA, 1968).

Sadun, E. H., et al. 1950. The effect of x-ray upon the growth and infectivity of Endoamoeba histolytica. Am. J. Trop. Med. 30: 635-640. (NSA 5 5:12).

Sadun, E. H., et al. 1968. Relative role of eggs and adult worms in development of liver damage in mice infected with Schistosoma mansoni. pp. 43-58. In: Isotopes and Radiation in Parasitology. (See IAEA, 1968).

Saeki, A. and K. Sano. 1956. Absorption and metabolism of fission products in goldfish. pp. 1105-1117. In: Reserach in the Effects and Influences of the Nuclear Bomb Test Explosions. Vol. II. (See Japan Soc. for the Promotion of Science, 1956).

Saeki, M. and T. Koyanagi. 1965. Movement of radioisotopes in the drain system. Nuclear Eng. (Tokyo) 11: 55-58. (In Japanese) (NSA 20: 27161).

Saeki, S., S. Okano and J. Moritaka. 1957. Radioactive substance in radiation contaminated fish. Nippon Suisan Gakkai Shi 20(10): 902. (Also: Bunseki Kagaku 6(7): 443).

Saeki, S., T. Yasumato and K. Mori. 1958. Studies on nuclear species in contaminated fish and their movement in the body of the fish. pp. 1560-1597. In: Proc. 2nd Japan Conf. on Radioisotopes. (See Japan Radioisotopes Assoc., 1958) (Paper A-9) (AEC-tr-4482) (NSA 15: 25793; SFA 7: 5286; WPA 35: 570).

Safiyazov, Zh. and L. H. Eidus. 1965. Radiation aftereffects on Bac. megaterium spores. Radiobiologiya 5(1): 32-34. (AEC-tr-6598, pp. 52-56) (NSA 19: 17624).

Safronova, N. G. 1973. Kinetics of the self-purification of a body of water containing strontium-90 as a result of absorption of the radionuclide by the bottom deposits. Ekologiya 4(2): 12-18. (Sov. J. Ecology 4(2): 101-107) (NSA 30: 29414).

Sagan, L. A. 1972. Human costs of nuclear power. Science 177(4048): 487-493. (NSA 26: 55959).

Sagan, L. A. (Ed.). 1974. Human and Ecologic Effects of Nuclear Power Plants. Chas. C. Thomas, Springfield, Ill. 552pp. (NSA 30: 25990; 32: 30348).

Sagane, R., et al. 1953. Characteristics of artificial radioactive elements isolated from soil collected in the Nishiyama area, Nagasaki (high velocity portion of β spectrum of $_{59}Pr^{144}$). Nippon Gakujutsu Kaigi 1: 58-61.

Sager, R. and Z. Ramanis. 1974. Mutations that alter the transmission of chloroplast genes in Chlamydomonas. Proc. Nat. Acad. Sci. (US) 71(12): 4698-4702. (NSA 31: 11577).

Saha, J. G., et al. 1969. Factors affecting extraction of dieldrin-^{14}C from soil. J. Agric. Food Chem. 17(4): 877-882.

Saha, M., A. K. Sumner and J. G. Saha. 1973. Comparison of three extraction and cleanup methods for determining ^{14}C-labeled residues from wheat plants grown in soil treated with Dyfonatering-^{14}C. J. Assoc. Off. Anal. Chem. 56(1): 45-48. (NSA 27: 19517).

Said, E. E.-D. and M. A.-K. Mahrous. 1960. On the uptake of radioactive phosphorus by the organs of the scorpion Buthus quinquestriatus H.E. Proc. Egypt. Acad Sci. 25: 29-32. (NSA 17: 35431).

Saijo, H. and T. Ichimura. 1958. Application of C^{14} to measurement of the productivity of lakes and swamps. In: Proc. 2nd Japan Conf. on Radioisotopes. (See Japan Radioisotopes Assoc., 1958) (Paper S-6) (AEC-tr-4482, pp. 451-461).

Saijo, Y. and S.-E. Ichimura. 1963. A review of recent development of techniques measurinf primary production. pp. 91-96. In: Primary Productivity Measurement, Marine and Freshwater. (See Doty, 1963).

Saiki, M. 1970. Safe control of environmental radioactivity around atomic energy plants. Genshiryoku Kogyo 16(6): 28-32. (In Japanese) (NSA 25: 13471).

Saiki, M. and T. Mori. 1955. Studies on the distribution of administered radioactive zinc in the tissues of fish. Bull. Jap. Soc. Sci. Fish. 21 (8): 945-949. (In Japanese, English summary).

Saiki, M., et al. 1971. Environmental considerations and public acceptance aspects of nuclear plant site selection in coastal area of Japan. pp. 821-846. In: Environmental Aspects of Nuclear Power Stations. (See IAEA, 1971) (SM-146/41) (CONF-700810-42) (NSA 24: 45923).

Sailor, V. L. 1972. Cost and benefit of nuclear power. Brookhaven National Lab., U. S. AEC report BNL-17266. 39pp. (NSA 27: 4457).

Sailor, V. L. 1972. United States energy crisis: supply and demand. Brookhaven National Lab., U. S. AEC report BNL-17280. 43pp. (For: Intern. Convocation: 1972 Meeting, New York, N. Y., October 14, 1972) (CONF-721050-1) (NSA 27: 9445) (Discusses environmental effects).

St. Amand, W. 1956. Mitotic inhibition and chromosome breakage induced in grasshopper neuroblasts by x-irradiation at known mitotic syages. Radiation Research 5(1): 65-78. (NSA 10: 9952).

Saito, N. 1970. Radioactive precipitates in Tamagawa and Misasa hot springs. Chigaku Kenkyu 21(11-12): 387-392. (In Japanese) (NSA 26: 6794).

Saito, N. and K. Kimura. 1972. Studies on the measurement of soil moisture content and density using the nuclear gages. Obayashi-Gumi Gijutsu Kenk. (6): 133-137. (In Japanese) (NSA 30: 3406).

Saito, S. 1959. Studies on the dehydrogenase system of Bacterium tularense IV. Effect of x irradiation and animal passage. Fukushima. J. Med. Sci. 6(1): 45-47. (NSA 14: 17726).

Saito, T. 1960. Distribution of P^{32} labeled Schradan in various insects. Botyu Kagaku 25(2): 64-71.

Saito, T. 1963. Studies on the selectivity of Schradan. pp. 255-265. In: Radiation and Radioisotopes Applied to Insects of Agricultural Importance. (See IAEA, 1963).

Saito, T. 1964. Toxicological aspects of insecticide resistance. Jap. J. Med. Sci. Biol. 17(1): 44-46. (^{32}P, Chilo suppressalis).

Sakagishi, S. 1963. Investigation of the movement of mobile bed material in the estuary and on the coast by means of radioactive material. J. Oceanog. Soc. Jap. 19: 27-36. (NSA 18: 4010).

Sakagishi, S. 1970. Safety measures associated with the development of atomic energy. Environmental safety. Genshiryoku Kogyo 16(4): 129-132. (In Japanese) (NSA 25: 23227).

Saikai, R. K., et al. 1972. Sterility enhancement in the mosquito, Culex tritaeniorhynchus. Can. J. Genet. Cytol. 14(2): 215-220. (NSA 26: 56483).

Sakamoto, S., et al. 1964. Studies on measurement of moisture content by neutron scattering. Radioisotopes (Tokyo) 13(5): 368-376. (In Japanese).

Sakanoue, M. and T. Hashimoto. 1964. A study of $^{234}U/^{238}U$ ratio in natural samples. Nippon Kagaku Zasshi 85(10): 622-627. (In Japanese) (NSA 19: 6118; NSA of Japan 4: 01879) (Mineral hot springs).

Sakanoue, M. and T. Tsuji. 1971. Plutonium content of soil at Nagasaki. Nature 234(5324): 92-93. (NSA 26: 6799).

Sakanoue, M., M. Nakaura and T. Imai. 1971. Determination of plutonium in environmental samples. pp. 171-181. In: Rapid Methods for Measuring Radioactivity in the Environment. (See IAEA, 1971) (NSA 26: 27807).

Sakka, M. 1965. Radiation induced shortening of life of Sarcophaga peregrina and its modification by dose fractionation. Nippon Act. Radiol. 24 (11): 1185-1188. (In Japanese) (NSA of Japan 4: 01835).

Sakka, M. 1968. Age-dependent radiosensitivity in adult Sarcophaga peregrina. Nippon Igaku Hoshasen Gakkai Zasshi 28: 569-572. (In Japanese) (NSA 23: 898).

Sakka, M. 1968. Independency of inhibition of imago formation in Sarcophaga pupae upon dose-rate. Nippon Igaku Hoshasen Gakkai Zasshi 28: 1176-1179. (In Japanese) (NSA 23: 29725).

Sakka, M. 1970. Recovery and accumulation of radiation effects at low dose-rate in Sarcophaga pupae. Nippon Igaku Hoshasen Gakkai Zasshi 29: 1405-1407. (In Japanese) (NSA 25: 16297).

Sakka, M. 1970. Effects of anoxic conditions on radiation-induced life-shortening in Sarcophaga adults. Nippon Igaku Hoshasen Gakkai Zasshi 30 (9): 849-852. (In Japanese) (NSA 26: 18072).

Sakka, M. 1973. Cause of life shortening in adult flesh fly, Sarcophaga perigrina. Nippon Igaku Hoshasen Gakkai Zasshi 33(2): 130-133. (In Japanese) (NSA 28: 24848) (X-ray).

Sakki (Sakka?), M. 1965. X-ray induced shortening of life span of adult Sarcophaga peregrina and its modification by dose fractionation. Tohoku J. Exp. Med. 86: 325-333. (NSA 20: 16291).

Sakovich, I. S. 1971. Characteristics of a Chlorella culture in the transitional period of increase in optical density after irradiation. Radiobiologiya 11(4): 633-635. (AEC-tr-7304, pp. 205-209) (NSA 26: 36340).

Sakovich, I. S. 1973. Influence of fractional irradiation on the survival and mutability of Chlorella. Radiobiologiya 13(5): 758-761. (AEC-tr-7507, pp. 142-146) (NSA 29: 26989).

Sakovich, I. S. and O. F. Ovchinnikova. 1971. Variation of the survival of Chlorella as a function of the density of the irradiated suspension. Radiobiologiya 11(3): 472-473. (AEC-tr-7303, pp. 216-218) (NSA 26: 33622).

Sakovich, I. S. and L. K. Vekshina. 1969. Survival of Chlorella upon continuous cultivation after a single gamma-irradiation. Kosm. biol. med. 3: 24-27. (In Russian) (NSA 24: 23399).

Sakovich, I. S. and L. K. Vekshina. 1970. Survival of Chlorella irradiated with MeV protons. Kosm. biol. med. 4(3): 86-87. (In Russian) (NSA 25: 52257).

Sakovich, I. S. and L. K. Vekshina. 1975. Dependence of the viability and mutability of Chlorella on post-radiation storage conditions. Kosm. biol. aviakosm. med. 9(2): 84-85. (JPRS-64929, pp. 143-145) (NSA 32: 14471).

Sakovich, I. S., et al. 1967. A mathematical model for Chlorella-cultivation parameters under conditions of ionizing irradiation. Kosm. biol. med. 1: 44-47. (In Russian) (NSA 22: 52058).

Saks, N. M. and J. J. Lee. 1973. The differential sensitivity of various species of salt marsh epiphytic algae to ionizing radiation and thermal stress. pp. 565-570. In: Radioactive Contamination of the Marine Environment. (See IAEA, 1973).

Saks, N. M. and J. J. Lee. 1973. Radiosensitivity and light quality responses of salt marsh epiphytic algae. pp. 1023-1028. In: Radionuclides in Ecosystems. (See Nelson, 1973).

Saksena, R. S., S. Chandra and B. P. Singh. 1974. Gamma transmission method for the determination of moisture content in soils. J. Hydrol. 23(3/4): 341-352. (NSA 32: 11644).

Salati, E., J. Menezes Leal and M. Mendes Campos. 1974. Environmental isotopes used in a hydrogeological study of northeastern Brazil. pp. 259-283. In: Isotope Techniques in Groundwater Hydrology. Vol. 1. (See IAEA, 1974) (NSA 31: 19632).

Salbu, B., E. Steinnes and A. C. Pappas. 1975. Multielement neutron activation analysis of fresh water using Ge(Li) gamma spectrometry. Anal. Chem. 47(7): 1011-1016. (NSA 32: 5860).

Salceda, V. M. 1967. Recessive lethals in second chromosomes of Drosophila melanogaster with radiation histories. Genetics 57: 691-699. (NSA 22: 6790).

Saleh, T. M. 1966. Radioautography of plant tissues using tritium labeled lignin precursors. Univ. of Washington, Seattle. 100pp. (NSA 21: 30795) (Thesis ?).

Salmon, L. and M. G. Creevy. 1971. On-line computer system for instrumental activation analysis of air, water, and soil. pp. 47-61. In: Nuclear Techniques in Environmental Pollution. (See IAEA, 1971) (NSA 25: 31780).

Salmon, L. and A. E. R. Toureau. 1971. Rapid evaluation of radioactive samples by gamma-ray spectrometry. pp. 339-355. In: Rapid Methods for Measuring Radioactivity in the Environment. (See IAEA, 1971) (NSA 26: 27819).

Salo, A. 1966. ^{90}Sr and ^{137}Cs in water in Finland. Finland Inst. Radiation Physics, Helsinki, report SFL-A4. 14pp. (NSA 20: 33510).

Salo, A. and J. K. Miettinen. 1964. Strontium-90 and caesium-137 in arctic vegetation during 1961. Nature 201(4925): 1177-1179. (NSA 18: 18047).

Salo, A. and R. Saxen. 1974. Role of humic substances in the transport of radionuclides. Finland Inst. Radiation Physics, Helsinki, report SFL-A-20. 32pp. (NSA 32: 6437).

Salo, A. and A. Voipio. 1973. Transport of radionuclides in lake and river systems flowing through areas characterized by Precambrian bedrock and peat-bogs. pp. 195-217. In: Radioactive Contamination of the Marine Environment. (See IAEA, 1973) (NSA 28: 8168).

Salo, E. O. 1969. Some environmental factors to be considered in the design of thermal power plants in the Northwest. Trend. Eng. Univ. Wash. 21: 3-7, 32. (NSA 24: 18547).

Salo, E. O. and W. L. Leet. 1969. The concentration of ^{65}Zn by oysters maintained in the discharge canal of a nuclear power plant. pp. 363-371. In: Symposium on Radioecology. (See Nelson and Evans, 1969).

Salonen, L. and H. A. Vaajakorpi. 1974. Bioaccumulation of ^{14}C-DDT in a small pond. pp. 201-211. In: Comparative Studies of Food and Environmental Contamination. (See IAEA, 1974) (NSA 30: 15470).

Salonius, P. O., J. B. Robinson and F. E. Chase. 1967. A comparison of autoclaved and gamma-irradiated soils as media for microbial colonization experiments. Plant Soil 27: 239-248. (NSA 22: 26021).

Salpeter, M. M. and A. O'Connor. 1965. Acetylcholinesterase in motor endplates evaluated by electron microscope autoradiography. Abstr. J. Cell Biol. 27(2): 93A. (^{3}H).

Saltman, P. and H. Boroughs. 1960. The accumulation of the zinc by fish liver slices. Arch. Biochem. Biophys. 86(2): 169-174.

Salvato, J. A., Jr. 1972. Environmental Engineering and Sanitation. John Wiley & Sons, New York. 2nd ed., 936pp. (NSA 27: 9559) (Incl. ionizing radiation).

Salvetti, C. 1973. Nuclear power plants and the environment. Com. Naz. Energ. Nucl. Notiz. 19(12): 52-63. (In Italian) (NSA 29: 31370).

Sambuichi, H. 1964. Effects of radiation on frog eggs. I. The development of eggs transplanted with irradiated nuclei or cytoplasm. Idengaku Zasshi 39: 259-267. (NSA 19: 36006).

Sambuichi, H. 1966. Effects of radiation on frog eggs. II. Diploid and haploid eggs with heavily γ-irradiated cytoplasm. Embryologia (Nagoya) 9: 196-204. (NSA 21: 24623).

Samburenko, I. Z. 1964. Contribution to the problem of studying the radioactivity of inland waters. Soviet Hydrology (5): 499-510. (NSA 20: 5386).

Samokhvalova, G. V. 1935. The effect of x-rays on the sexual gland and secondary sexual characteristics of Lebistes reticulatus. pp 213-229. In: Works on Growth Dynamics. Vol. 10. Acad. Sci. USSR, Moscow. (In Russian, English summary and conclusions).

Samostrel'skiy, A. I., et al. 1966. Experiment on the use of radioactive isotopes to mark Ornithodoros papillipes ticks. Tr. inst. epidem. pastera 29: 86-89. (In Russian).

Samperio, J. G. nd. Cooperative program on the investigation of the med fly in Central America. pp. 143-150. In: Proc. Conf. on Nuclear Energy and Agricultural Productivity. (See Comision Interamericana de Energia Nuclelear, nd.) (In Spanish) (NSA 25: 27131).

Samsahl, K. 1972. Accuracy and precision studies of radiochemical multi-element method for activation analysis in the field of life sciences. Sweden Aktiebolaget Atomenergi, Studsvik, report AE-458. 18pp. (NSA 26: 58495).

Samuels, E. R. and V. K. Mohindra. 1965. Reactor environment monitoring, 1964. ---Canadian Fallout Data for March 1965. ---for October 1965. pp. 3-22, 22-29, 21-33. Canada Dept. of National Health and Welfare, Ottawa. (NP-15519, -15641, -15853) (NSA 19: 44294; 20: 7306, 14738) (Incl. rabbits, fish).

Samuels, L. D. 1964. A study of environmental exposure to radium in drinking water. pp. 239-252. In: The Natural Radiation Environment. (See Adams and Lowder, 1964) (NSA 17: 32314).

Sanchez, W., G. Guidicini and R. F. da Silva. 1975. Water percolation conditions at the entrance of the canal locks of the Parana River using radioisotope techniques. Brazil Inst. de Energia Atomica, São Paulo, report IEA-389. 21pp. (In Portugese) (NSA 33: 23448) (^{131}I tracer).

Sand, M. D., P. A. LaRock and R. E. Hodson. 1975. Radioisotope assay for the quantification of sulfate-reducing bacteria in sediment and water. Appl. Microbiol. 29(5): 626-634. (NSA 32: 8950).

Sandberg, G. R., J. S. Olson and E. E. C. Clebsch. 1969. Internal distribution and loss from roots by leaching in cesium-137 inoculated into Liriodendron tulipifera L. seedlings grown in sand culture. Oak Ridge National Lab., U. S. AEC report ORNL-TM-2660. ix, 63pp. (Also: Thesis, Univ. of Tennessee, Knoxville).

Sandermann, W., W. Schweers and H. Hoheisel. 1965. Analysis of pine-tree extracts by means of beta radiation. Chim. Ind. (Paris) 90: 58-60. (In French) (NSA 20: 23561).

Sanders, C. J. and W. F. Baldwin. 1969. Iridium-192 as a tag for carpenter ants of the genus Camponotus (Hymenoptera: Formicaridae). Can. J. Entomol. 101(4): 416-418.

Sanders, C. L., Jr. 1962. Diverse physiological effects of gamma irradiation stress in the cotton rat (genus Sigmodon). Thesis, Texas A&M Coll., College Station.

Sanderson, G. C. 1966. The study of mammal movements - a review. J. Wildl. Mgmt. 30(1): 215-235. (Incl. radionuclide tags).

Saneblidze, O. I. 1967. Radioinduced contraction of skeletal muscles in frogs. Soobshch. Akad. Nauk Gruz. SSR 47: 457-462. (In Russian) (NSA 22: 12904).

Saneblidze, O. I., K. Sh. Nadareishvili and Yu. Ya. Shafir. 1967. Radiation effects on electrical and mechanical activity of skeletal muscles in frogs. Soobshch. Akad. Nauk Gruz. SSR 45: 499-506. (In Russian) (NSA 21: 24616).

Sándi, E. 1962. Radioactivity in snail shells due to fallout. Nature 193 (4812): 290.

San Diego State College. 1969. Radiation resistance of Drosophila larvae. Three year comprehensive report. California, U. S. AEC report SAN-529-12. 59pp. (NSA 23: 48015).

Sandmeier, H. A. and J. H. Stover. 1962. High Cs^{137} body burden measured on Laplanders. European Scientific Notes 16-1, pp. 16-17. Office of Naval Research, U. S. Embassy, London, England.

Sandquist, G. M. and G. A. Whan. 1973. Environmental aspects of nuclear stimulation. pp. 293-313. In: Geothermal Energy. P. Kruger, ed. Stanford Univ. Press, Stanford, Calif. (NSA 28: 17669).

Sandström, A. E., M. A. Pomerantz and B.-O. Grönkvist. 1963. Sea level cosmic ray intensity and threshhold rigidity. Tellus 15(2): 184-___.

Sanides, F. 1956. The lethal and teratogene operation of roentgen radiation of endangered embryos and gastrula of Triton alpestris in different environments. Biol. Zbl. 75: 149-177. (In German).

Sankaranarayanan, K. 1963. Fitness of irradiated populations of Drosophila melanogaster. Abstr. 5.58. p. 75. In: Genetics Today. Vol. I. (Proc. 11th Intern Conf. of Genetics, The Hague, September 1963). S. J. Geerts, ed. Pergamon Press, Oxford.

Sankaranarayanan, K. 1964. Genetic loads in irradiated experimental populations of Drosophila melanogaster. Genetics 50(1): 131-150. (NSA 18: 29295) (Incl. survival, hatchability).

Sankaranarayanan, K. 1965. Genetic loads in irradiated populations of Drosophila melanogaster. (Dissertation Abstr. 25(6): 3766).

Sankaranarayanan, K. 1965. Relative magnitude of genetic loads in experimental populations of Drosophila melanogaster relaxed from x-irradiation. Abstr. Genetics 52(2, Pt.2): 471-472.

Sankaranarayanan, K. 1965. Further data on the genetic loads in irradiated experimental populations of Drosophila melanogaster. Genetics 52(1): 153-164. (NSA 19: 36003).

Sankaranarayanan, K. 1965. Inbreeding and genetic loads in irradiated experimental populations of Drosophila melanogaster. Nature 207(5002): 1216-1217. (NSA 19: 45952).

Sankaranarayanan, K. 1966. Some components of the genetic loads in irradiated experimental populations of Drosophila melanogaster. Genetics 54: 121-130. (NSA 20: 36618).

Sankaranarayanan, K. 1967. Influence of selection on the viability of irradiated experimental populations of Drosophila melanogaster. Genetics 57 (3): 687-690.

Sankaranarayanan, K. 1969. The effects of oxygen and nitrogen post-treatments on the mortality of Drosophila eggs irradiated as stage-7 oocytes. Mutuation Research 7: 357-368. (NSA 23: 38537).

Sankaranarayanan, K. 1973. Ionizing radiations, genetic loads, and population fitness: a review. Arch. Genet. 46(3): 137-139. (NSA 33: 7552).

Sankaranarayanan, K. 1974. Search for radioresistance in experimental populations of Drosophila with radiation histories. Mutation Research 24(2): 213-217. (NSA 30: 27198).

Sansoni, B. 1971. Rapid methods for radioactive environmental monitoring in the Federal Republic of Germany. pp. 655-675. In: Rapid Methods for Measuring Radioactivity in the Environment. (See IAEA, 1971) (In German) (NSA 26: 28055).

Sansoni, B. 1971. Development of rapid radiochemical methods for monitoring environmental alpha and beta contamination in biological material and foodstuff by using a new method of wet-ashing with OH radicals in aqueous solution. Final report for the period July 1, 1969-January 31, 1971. International Atomic Energy Agency, Vienna, report IAEA-R-770-F. 31pp. (NSA 27: 14509) (Environmental samples).

Sansoni, B. 1972. Analytical methods in environmental control. Germany Inst. fur Strahlenschutz, Neuherberg bei Munich, report GSF-S-188. 25pp. (In German) (NSA 28: 17838) (Trace analysis of radionuclides).

Sansoni, B. 1973. Analytical methods of environmental monitoring. pp. 238-258. In: Worldwide Radiation from Nuclear and Conventional Energy Production. H. Braun, ed. Georg Thieme Verlag, Stuttgart. (In German) (NSA 31: 3331).

Sansoni, B. and W. Kracke. 1971. Rapid determination of low-level alpha- and beta activities in biological material using wet ashing by -OH radicals. pp. 217-231. In: Rapid Methods for Measuring Radioactivity in the Environment. (See IAEA, 1971) (In German) (NSA 26: 27811).

Sansoni, B., W. Kracke and R. Winkler. 1969. Rapid assay of environmental radioactive contamination with special reference to a new method of wet ashing. pp. 487-500. In: Environmental Contamination by Radioactive Materials. (See IAEA, 1969) (NSA 24: 7416).

Santeford, H. S. and J. L. Smith. (Comps.). 1974. Advanced Concepts and Techniques in the Study of Snow and Ice Resources. (An Interdisciplinary Symp., Monterey, California, December 2-6, 1973). National Academy of Sciences, Washington. 789pp. (NSA 31: 22797; 33: 7306).

Santos, I. S. 1964. Reduction of sensitivity to ^{60}Co gamma rays in Phaseolus mungo L. through pre- or post-irradiation heat treatment of the seed. Sci. Rev. 5: 25-31.

Santroch, J. nd. Ten years of radioactive contamination monitoring by the Hydrometeorological Institute. Czechoslovakia Hydrometeorologicky Ustav, Prague. (In Czech) (INIS-mf-887, pp. 210-219) (NSA 29: 15591).

São Paulo University (Brazil). 1967. Comparative analysis of the behavior of natural and of radiation induced lethals in natural and laboratory populations. Final report. U. S. AEC report NYO-2733-16. 23pp. (NSA 23: 26718).

Sapankevic, P. V. 1964. Development of seedlings obtained from Fraxinus excelsior samaras treated with x-rays. Lesn. Z. Arhangel'sk 7: 8-11. (FA 26: 5052).

Saparlyev, Ch., L. N. Zatsepina and L. N. Kalishina. 1969. Express-method for determining ^{90}Sr in plants. Izv. Akad. Nauk Turkman. SSR (biol. nauk) (6): 19-22. (In Russian) (NSA 25: 4039).

Sarkisyan, S. M. and P. N. Mashadyan. 1969. Effect of irradiation on egg formation and egg fertility of the mulberry silkworm. Biol. zh. Armenii 22(9): 8-15. (In Armenian) (NSA 24: 34595).

Saro, S. 1973. Statistics of low level counting. Low level counting of ^{3}H and ^{14}C. Acta Fac. Rerum Nat. Univ. Comenianae phys. (13): 131-143. (NSA 31: 2854, 32510).

Sarosiek, J. and H. Wozakowska-Natkaniec. 1967. Response of Marchantia polymorpha L. to chronic gamma radiation under natural conditions. Acta Soc. Bot. Pol. 36: 187-197. (NSA 23: 50405).

Sarosiek, J. and H. Wozakowska-Natkaniec. 1968. The effect of low-level acute gamma radiation on the growth of Marchantia polymorpha L. originating from various ecological populations. Acta Soc. Bot. Pol. 37: 413-426. (NSA 23: 38879).

Sarosiek, J., W. Kola and A. Orda. 1973. The radioecology of Pogonatum urnigerum (L.) Pal. Beauv. Acta Soc. Bot. Pol. 42: 241-252. (NSA 30: 29691; 31: 22951) (Moss).

Sartory, A. and R. Sartory. 1922. Combined action of thorium sulfates and shakings on the growth of Phycomyces splendens Banier. Compt. rend. soc. biol. 88: 743-746.

Sarty, K. S., R. R. Murty and P. S. Sarma. 1958. Zinc toxicity in the larvae of the rice moth, Corcyra cephalonica. Biochem. J. 69(3): 425-428. (Used ^{65}Zn).

Sasaki, R. 1956. Some observations on the biological influences of radioactive isotopes upon physiological functions. Proc. Intern. Conf. on the Peaceful Uses of Atomic Energy 12: 330-333. (Paper A/CONF.8/P/1067) (^{45}Ca with silkworm larvae).

Sasaki, S. 1966. Effects of x-irradiation on postembryonic development of Sarcophaga peregrina Robineau-Desviody. I. Dobutsugaku Zasshi 75: 207-214. (In Japanese) (NSA 21: 22752).

Sasaki, T. (S. ?, qv.). 1968. Effect of x-irradiation on post-embryonic development of *Sarcophaga peregrina* Robineau-Desvoidy. 2. Change of radiosensitivity during prepupal and early pupal stages. Dobutsugaku Zasshi 77: 109-116. (In Japanese) (NSA 23: 20333).

Sasaki, T., et al. 1964. Neutron irradiation effect on cellulose. Chem. High Polymers (Tokyo) 21(234): 637-641. (In Japanese) (Used wood).

Sasscer, D. S., C. F. Jordan and J. R. Kline. 1973. Dynamic model of water movement in soil under various climatological conditions. pp. 485-495. In: *Tritium*. (See Moghissi and Carter, 1973) (NSA 28: 27319).

Sasscer, D. S., C. F. Jordan and J. R. Kline. 1973. Mathematical model of tritiated and stable water movement in an old-field ecosystem. pp. 915-923. In: *Radionuclides in Ecosystems*. (See Nelson, 1973).

Sastri, C. S. and V. Sivaramakrishnan. 1968. Study of the radioactivity of separated minerals of beach sands of Manavalakurichi, Madras State. Current Sci. (India) 37: 550. (NSA 23: 474).

Sastry, B. S., et al. 1970. Laboratory determination of soil density using gamma ray attenuation. Technology (Sindri) 7(3): 115-119. (NSA 26: 186).

Sathyanesan, A. G. and W. Chavin. 1966. Effects of whole-body irradiation on the adrenal of the goldfish *Carassius auratus* L. Radiation Research 28(2): 352-364. (BA 47: 100532; NSA 20: 31094; SFA 11: 8937).

Sathyanesan, A. G. and W. Chavin. 1966. Effect of whole-body irradiation on the preoptico-hypophyseal neurosecretory system and the hypophysis of the goldfish *Carassius auratus* L. Radiation Research 29(1): 100-113. (BA 48: 10902; NSA 20: 45428; SFA 11: 8980).

Sathyanesan, A. G. and S. S. Srivastava. 1970. Radioiodine induced thyroid tumours in the Indian freshwater cat-fish *Mystus vittatus* (Bloch). Current Sci. (India) 39: 350-351. (NSA 24: 46569).

Sato, C. 1966. Effects of x-irradiation on the development of the respiratory system and survival time in pupae of *Sarcophagus peregrina*. Tohoku J. Exptl. Med. 90: 1-8. (NSA 21: 12977).

Sato, C., et al. 1966. Effects of x-rays on respiration of *Sarcophaga* pupae. Nippon Igaku Hoshasen Gakkai Zasshi 26(9): 1180-1183. (In Japanese).

Sato, K. 1953. Reports by forestry research team. III. Effect of atomic bomb on forests. Nippon Gakujutsu Kaigi 1: 265-266.

Sato, K. and K. Susuki. 1969. Significance of the Bursa of Fabricius after whole-body x irradiation. Poultry Science 48: 1195-1198. (NSA 24: 32233).

Sato, Y. and J. Miyamoto. 1967. An aspect concerning the movement of Dimethoate from its granular formation in soil. Nayaku Seisan Gijutsu 16: 23-28. (^{32}P).

Satter, L. P. and J. H. Harley. 1965. "Trinitite": cobalt-60, and europium-152. Science 148(3672): 954-955. (Nuclear test, Alamagordo, N. Mex.).

Sauer, J. 1968. The effect of radiation on the viability of worm eggs. Z. Parasitenk. 31: 15. (In German) (NSA 23: 38761).

Sauer, J. H., J. H. Frick and J. A. Hair. 1974. Control of ^{36}Cl by isolated salivary glands of the lone star tick. J. Insect Physiol. 20(9): 1771-1778. (NSA 32: 3827).

Sauer, M. E. 1960. Radiation injury resulting from nuclear labeling with thymidine-H^3 in the chick embryo. Abstr. Anat. Record 136: 272.

Sauer, M. E. and B. E. Walker. 1961. Radiation injury resulting from nuclear labeling with tritiated thymidine in the chick embryo. Radiation Research 14(5): 633-642.

Sauerbeck, D. 1960. Labeling of plants with C^{14}. Atompraxis 6: 221-225. (AEC-tr-4353) (NSA 14: 22724).

Sauerwein, K. 1967. Flow of a stream through a reservoir. pp. 697-719. In: *Isotopes in Hydrology*. (See IAEA, 1967) (NSA 22: 18940) (Tracer).

Saunders, G. W. 1961. Investigations in lake metabolism-bacteria: distribution and activities. Abstr. Great Lakes Res. Div., Univ. of Michigan, Ann Arbor, Publ. 7. (Proc. 4th Conf. on Great Lakes Research, p. 162).

Saunders, G. W., F. B. Trama and R. W. Bachmann. 1962. Evaluation of a modified C-14 technique for shipboard estimation of photosynthesis in large lakes. Great Lakes Res. Div., Univ. of Michigan, Publ. 8. 61pp.

Saunders, J. F. 1968. The effects of gamma radiation combined with weightlessness on biologic systems in Biosatellite II. Abstr. Radiation Research 35(2): 498-499.

Saur, W. and J. Hoigne. 1970. Disinfestation of insect-infested wheat by means of ionizing radiation. Mühle 107(8): 104-106. (In German) (NSA 25: 52288).

Saurov, M. M. 1957. Radioactive contamination of fish in water containing strontium. pp. 66-73. In: *Proc. Conf. on Medical Radiology. Problems of Hygiene and Dosimetry*. Medgiz, Moscow. (AEC-tr-3746) (NSA 13: 19079; WPA 32: 107).

Saurov, M. M. 1962. Absorption and migration of radioactive substances in soils. pp. 109-113. In: *Radiation Hygiene. Vol. 2*. (See Marey, 1962) (JPRS-21799, pp. 56-62).

Saurov, M. M. 1962. Accumulation of radioactive substances in terrestrial vegetation. pp. 114-123. In: *Radiation Hygiene. Vol. 2*. (See Marey, 1962) (JPRS-21799, pp. 62-72).

Saurov, M. M. 1962. Some data on the accumulation and expulsion of radioactive substances in animals. pp. 124-126. In: *Radiation Hygiene. Vol. 2*. (See Marey, 1962) (JPRS-21799, pp. 72-84).

Saurov, M. M., et al. 1973. Role of freshwater as possible supplier of ^{32}P to population under conditions of pollution of natural basin by this radionuclide. pp. 177-184. In: *Radioecology of Water Organisms. II*. (See Andrumaitis, 1973) (In Russian) (NSA 30: 21422).

Sauzay, G. 1968. Count rate balance method of measuring sediment transport of sand beds by radioactive tracers. Thesis, Univ. of Toulouse, France. France Commissariat à l'Énergie Atomique, Saclay, report CEA-R-3431. 168 pp. (In French) (NSA 23: 15964).

Sauzay, G. 1973. Principles of tracer methods. pp. 9-12. In: Trace Techniques in Sediment Transport. (See IAEA, 1973) (NSA 28: 21135).

Sauzay, G. 1974. Sampling of lysimeters for environmental isotopes of water. pp. 61-68. In: Isotope Techniques in Groundwater Hydrology. Vol. 2. (See IAEA, 1974) (NSA 31: 19647).

Sauzay, G., et al. 1974. Study of sediment transport on the sea bed and on the beach at the site of the Hadera power plant. Report to the Israel Electric Corporation. Israel Atomic Energy Commission, Tel Aviv. 59pp. (INIS-mf-1687) (NSA 32: 8927) (Radionuclide and fluorescent tracers).

Savannah River Plant. 1959-1974. Health physics regional monitoring semiannual report. Effect of Savannah River Plant on environmental radioactivity. Environmental monitoring at the Savannah River Plant. U. S. AEC report DPST- or DPSPU-.

Publ. Year	Period	Report No.	pp.	NSA
1959	Jul.-Dec. 1958	-59-11-23	46	30: 9516
1959	Jan.-Jun. 1959	-59-11-30	41	30: 9517
1960	Jul.-Dec. 1959	-60-11-9	44	30: 9518
1960	Jan.-Jun. 1960	-60-25-2(Del.)	56	30: 9511
1961	Jul.-Dec. 1960	-61-25-4(Del.)	56	30: 9512
1962	Jan.-Jun. 1961	-62-25-2(Del.)	68	30: 9513
1962	Jul.-Dec. 1961	-62-25-9(Del.)	67	30: 9514
1963	Jan.-Jun. 1962	-63-25-3(Del.)	50	30: 9515
1964	1963	-64-11-12	39	30: 9519
1965	-	-65-30-2	16	19: 44288
1966	Jul.-Dec. 1966	-67-30-1	20	21: 16336
1967	-	-67-30-2	--	21: 43055
1967	Jul.-Dec. 1967	-68-30-1	--	22: 18965
1968	Jan.-Jun. 1968	-68-30-2	18	22: 45178
1969	Jul.-Dec. 1968	-69-30-1	--	23: 29468
1969	Jan.-Jun. 1969	-69-30-2	18	24: 2654
1969	Jul.-Dec. 1969	-70-30-1	19	24: 25473
1970	Jan.-Jun. 1970	-70-30-2	18	24: 50897
1971	Jul.-Dec. 1970	-71-30-1	18	25: 24151
1971	Jan.-Jun. 1971	-71-30-16	18	25: 57996
1972	-	-72-30-2	29	30: 9520
1973	1972	-73-30-1	38	28: 12877
1973	-	-73-30-2	33	30: 9521
1973	-	-74-30-1	31	30: 9522
1974	1974	-75-30-1	38	32: 6411

Savannah River Plant. 1965. Sorption of Sr and Cs by stream beds. U. S. AEC report TID-26519. 4pp. (NSA 29: 18358) (At Savannah River Plant).

Savannah River Plant. 1966. Tritium and cesium in Par Pond. U. S. AEC report TID-26515. 5pp. (NSA 29: 18357) (At Savannah River Plant).

Savannah River Plant. 1967. Excerpts from SRL monthly reports on waste management. U. S. AEC report TID-26518. 6pp. (NSA 29: 24203).

Savannah River Plant. 1973. Environmental activities and programs at the Savannah River Plant. U. S. AEC report DPST-73-436. 23pp. (NSA 28: 29840).

Savin, V. N. and O. G. Stepanenko. 1968. Effect of ^{60}Co gamma radiation on the sunflower. Sel'skokhoz. biol. 3: 921-922. (In Russian) (NSA 24: 44630).

Savin, V. N. and O. G. Stepanenko. 1969. Change in growth relationships in sunflower plants exposed to ^{60}Co rays. Sb. Tr. agron. fiz. (17): 93-101. (In Russian) (NSA 24: 46682).

Savin, V. N. and O. G. Stepanenko. 1970. Mechanism of the intensified growth of lateral shoots of irradiated plants. Dokl. Akad. Nauk SSSR 193: 929-931. (In Russian) (NSA 25: 21967).

Savin, V. N. and O. G. Stepanenko. 1971. Effects of ^{60}Co γ radiation on fertility in sunflowers following irradiation of vegetative plants. pp. 45-50. In: Radioactive Isotopes and Ionizing Radiation in Agricultural Research. I. Atomizdat, Moscow. (In Russian) (NSA 30: 18899).

Sax, K. 1955. The effect of ionizing radiation on plant growth. Am. J. Bot. 42(4): 360-364.

Sax, K. and L. A. Schairer. 1963. The effect of chronic irradiation on apical dominance of trees. Radiation Botany 3(3): 283-285. (Also: Brookhaven National Lab., U. S. AEC report BNL-6858) (NSA 18: 3475).

Sax, K. and C. P. Swanson. 1941. Differential sensitivity of cells to x-rays. Am. J. Bot. 28(1): 52-59.

Sax, N. I. and J. J. Gaby. 1970. Surveillance technique for tritiated moisture in the atmosphere. pp. 382-387. In: Environmental Surveillance in the Vicinity of Nuclear Facilities. (See Reinig, 1970) (NSA 25: 16179).

Sax, N. I., et al. 1969. Radioecological surveillance of the waterways around a nuclear fuels processing plant. Radiol. Hlth. Data 10: 289-296. (NSA 23: 45684).

Saxena, G. S., R. E. Franklin and G. S. Taylor. 1970. Importance of pulse-height analysis in the gamma-ray attenuation technique for measurement of soil moisture. Soil Science 110(4): 283-___.

Saygili, N., T. Altan and F. Pamuk. 1974. Source and mineral bath waters of Turkey with respect to their U content and radioactivity. Tech. J. Ankara Nucl. Res. Cen. 1(3): 125-135. (In French) (NSA 31: 5572).

Sayles, C. D., J. D. Procunier and L. W. Browder. 1973. Radiolabelling of *Drosophila* embryos. Nature New Biol. 241(111): 215-216. (NSA 27: 22812).

Sayre, E. V. 1972. Activation analysis applications in art and archeology. Advan. Actv. Anal. 2: 155-184. (NSA 26: 30659).

Sayre, E. V. and R. W. Smith. 1973. Analytical studies of ancient Egyptian glasses. Brookhaven National Lab., U. S. AEC report BNL-18562. 36pp. (NSA 29: 20600).

Sayre, E. V., L.-H. Chan and J. A. Sabloff. 1968. High resolution gamma ray spectroscopic analyses of Mayan fine orange pottery with brief comments on their archaeological background and implications. Brookhaven National Lab., U. S. AEC report BNL-13133. 64pp. (CONF-680918-6) (NSA 23: 21662).

Sayre, W. W. and D. W. Hubbell. 1963. Transport of radionuclides in fresh water: dispersal of bed sediments. pp. 327-350. In: Transport of Radionuclides in Fresh Water Systems. (See Kornegay, et al., 1963) (NSA 17: 40536).

Sayre, W. W., H. P. Guy and A. R. Chamberlain. 1963. Uptake and transport of radionuclides by stream sediments. U. S. Geological Survey Prof. Paper 433-A. iii, 35pp.

Scalf, M. R., J. L. Witherow and C. P. Priesing. 1968. Iron-59 as solids tracer in aqueous suspensions. J. Sanit. Engrg. Div. Am. Soc. Civil Engrs. 94(6): 1195-1211.

Scalf, M. R., et al. 1968. Fate of DDT and nitrate in ground water. U. S. Federal Water Control Admin., Ada, Okla., and Southwestern Great Plains Research Center, Bushland, Tex. ix, 46pp. (Tritiated water).

Scandolios, J. G. 1964. Some effects of x rays on root primordia in the poplar. Radiation Botany 4(4): 355-359. (NSA 19: 15158).

Scarascia, G. T. 1959. Observations on the radiodiagnostics of seeds. Nuovo giorn. botan. Ital. NS66: 289-295. (In Italian) (NSA 14: 15509).

Scarascia-Mugnozza, G. T. 1971. Application of nuclear energy in agriculture: biological aspects. Agrochimica 15(2-3): 87-100. (In Italian) (NSA 25: 47047).

Scarbakova, M. A. 1964. Determining conifer seed quality by radiography. Lesn. Hoz. 17: 52-56. (FA 26: 5056).

Scarborough, B. B. and R. G. Addison. 1962. Conditioning in fish: effects of x-irradiation. Science 136(3517): 712-713.

Schabinski, G. 1959. Experimental contribution to the question of sterilitization with ^{60}Co. Deut. Gesundheitsw. 14: 859-861. (NSA 21: 16539).

Schaefer, C. H. 1964. Free amino acids of the Virginia pine sawfly, Neodiprion pratti Dyar: their chromatographic determination and biosynthesis. J. Insect Physiol. 10(2): 363-369. (^{14}C).

Schaefer, C. H. 1965. Fatty acids of the Virginia pine sawfly, Neodiprion pratti Dyar. Can. J. Entomol. 97(9): 941-944. (^{14}C).

Schaefer, C. H. and U. P. Sun. 1967. A study of dieldrin in the house fly central nervous system in relation to dieldrin resistance. J. Econ. Entomol. 60(6): 1580-1583. (^{14}C).

Schaefer, C. H., J. N. Kaplanis and W. E. Robbins. 1965. The relationship of the sterols of the Virginia pine sawfly, Neodiprion pratti Dyar, to those of two host plants, Pinus virginiana Mill and Pinus rigida Mill. J. Insect Physiol. 11(7): 1013-1021. (^{14}C).

Schaefer, M. B. 1960. New research required in support of radioactive waste disposal. pp. 265-282. In: Disposal of Radioactive Wastes. (See IAEA, 1960).

Schaeffer, O. A. 1967. Direct dating of fossils by the helium-uranium method. pp. 395-402. In: Radioactive Dating and Methods of Low-Level Counting. (See IAEA, 1967) (Also: State Univ. of New York, Stony Brook, U. S. AEC report NYO-3629-7) (NSA 21: 39043; 22: 6469).

Schaeffer, O. A., G. Megrue and S. O. Thompson. 1964. Experiments to test the presence of cosmogenic nuclides in ocean sediments. Ann. N.Y. Acad. Sci. 119: 347-350. (NSA 19: 4358).

Schaeffer, R. 1974. Effects of radioecological reconcentration on maximum permissible concentrations for river water. pp. 525-538. In: Population Dose Evaluation and Standards for Man and His Environment. (See IAEA, 1974) (In French) (NSA 31: 19605).

Schäfer, H. 1972. Energy technology and the environment. Brennst.-Wärme-Kraft 24(7): 280-282. (In German) (NSA 26: 50632).

Schafer, K. L., Jr. 1970. Visual effects of ionizing radiation in the noctuid moth, Heleiothis zea. Thesis, Florida State Univ., Tallahassee. 94 pp. (Univ. Microfilms 71-7097) (NSA 25: 26910).

Schales, F, R. Keil and T. Franke. 1973. Sources of terrestrial radiation. Natural radionuclides in the soil. Germany Gesellschaft für Strahlen und Umweltforschung mbH, Frankfurt/Main. (For: Symp. on Natural Radiation Exposure of Man: Basis for Judging the Radiation Risk, Frankfurt am Main, 18 July 1973). (In German) (NSA 28: 29875).

Schaper, A. 1904. Experimental studies on the effect of radium-radiation and radium-emanation on the embryos and regenerative organ development. Anat. Anz. 25(12/13): 298-320; 25(14/15): 326-337. (In German).

Scharpenseel, H. W. 1967. Studies on tagged clay migration due to water movement. pp. 279-287. In: Isotope and Radiation Techniques in Soil Physics and Irrigation Studies. (See IAEA, 1967) (NSA 22: 6469).

Scharpenseel, H. W. 1971. Radiocarbon dating of soils. Pochvovedeniye (1): 34-42. (Sov. Soil Science (1): 76-84) (CA 74: 89816).

Scharpenseel, H. W., F. Penningsfeld and K. Duquin. 1965. Automatic control of irrigation from above and below by a neutron probe. pp. 147-153. In: Isotopes and Radiation in Soil-Plant Nutrition. (Proc. Symp., Ankara, Turkey, June 28-July 2, 1965). International Atomic Energy Agency, Vienna, Publ. STI-PUB-108. (In French) (CONF-650614-3) (NSA 20: 633).

Schaudinn, F. 1899. On the effect of roentgen radiation on protozoans. Pflügers Arch. ges. Physiol. 77(1/2): 29-43. (In German).

Schauer, T., W. Kreuzer and P. Maurus. 1971. Sampling for rapid estimation of ^{137}Cs contamination of meat from herbivores. pp. 621-626. In: Rapid Methods for Measuring Radioactivity in the Environment. (See IAEA, 1971) (In German) (NSA 26: 28090).

Schauer, T., W. Kreuzer and H. Schmidt. 1970. Activity and penetration of ^{137}Cs in various soils. Zentralbl. Veterinärmed. Beih. (11): 149-153. (In German) (NSA 25: 24080).

Scheele, W. 1963. Use of radioisotopes, especially in the form of open preparations, in wood technological problems. II. Holztechnologie 4: 99-104. (In German) (NSA 20: 711).

Scheffer, F. and F. Ludwieg. 1963. Demonstration of Mn^{54} and Co^{57} in plants as a result of Russian nuclear tests. Atompraxis 9(3): 75-77. (In German) (NSA 17: 17739).

Scheffer, T. C. 1963. Effect of gamma radiation on decay resistance of wood. Forest Products J. 13: 308.

Scheidhauer, J. 1970. Surveillance of the radioactive waste disposal site at La Hague (Manche Centre). pp. 104-116. In: Conf. on Health Implications of the Storage of Radioactive Substances on and in the Ground. (Cherbourg-La Hague, France, 14 April 1970). EURATOM, report EUR-4736. (NSA 26: 17761).

Scheidhauer, J., G. Cassabois and R. Gandon. 1970. Accidental release of ^{131}I at the UP 2 Plant on October 2nd and 3rd 1968. France Commissariat à l'Énergie Atomique, La Hague, report CEA-R-4008. 45pp. (In French) (NSA 25: 8657).

Scheidhauer, J., L. Messainguiral and A. M. Meiraneisio. 1963. Plutonium separation and the measurement of its concentration in natural waters. Centre de Marcoule, France Commissariat à l'Énergie Atomique, Marcoule, report CEA-235. i, 6pp. (In French).

Scheitanov, G. and G. Russev. 1963. Effect of contact of roentgen-irradiated salamander embryos on non-irradiated salamanders. Radiobiol. Radiotherap. 4(2): 159-165. (In German) (NSA 17: 33743).

Schelenz, R. and J. F. Diehl. 1973. Application of neutron activation analysis to the determination of mercury in foodstuffs. Z. Anal. Chem. 265(2): 93-97. (In German) (NSA 28: 20684).

Schell, W. R. 1970. An internal gas proportional counter for measuring low level environmental radionuclides. pp. 173-201. In: Radionuclides in the Environment. R. F. Gould, ed. Amer. Chemical Soc., Washington.

Schell, W. R. 1974. Sedimentation rates and mean residence times of stable Pb and ^{210}Pb in Lake Washington, Puget Sound estuaries, and a coastal region. Univ. of Washington, Seattle, U. S. AEC report RLO-2225-T14-6. 44pp. (NSA 31: 22647).

Schell, W. R. and R. L. Watters. 1975. Plutonium in aqueous systems. Health Physics 29(4): 589-597. (NSA 31: 22632; 33: 9287).

Schell, W. R. and A. I. C. Yang. 1973. Long-lived radionuclides produced at Bikini and Eniwetok atolls (I). Univ. of Washington, Seatte, U. S. AEC report RLO-22250T018-3. 29pp. (NSA 28: 29616).

Schelske, C. L. 1964. Ecological implications of radioactivity accumulated by molluscs. Abstr. Bull. Ecol. Soc, Am. 45(4): 149-150.

Schelske, C. L. 1966. The fate of fallout radioactivity in an estuary. Am. Biol. Teacher 28(5): 373-380.

Schelske, C. L. 1973. Fallout ^{54}Mn accumulated by bay scallops Argopecten irradians (Lamarck) near Beaufort, North Carolina. pp. 331-345. In: Radioactive Contamination of the Marine Environment. (See IAEA, 1973).

Schelske, C. L., D. A. Wolfe and D. E. Hoss. 1973. Ecological implications of fallout radioactivity accumulated by estuarine fishes and mollusks. pp. 791-806. In: Radionuclides in Ecosystems. (See Nelson, 1973) (NSA 27: 15166).

Scherf, E. and G. Meszéna. 1960. Mathematical-statistical studies of the physical conditions for uranium enrichment of natural waters. Magyar Tudomanyos Akad. Atom. Kutató Int. (Debrecen) Közl. 2: 109-143. (In Hungarian) (NSA 15: 6223).

Scherfke, R. 1971. Exactness in determining soil moisture by neutron probe. Pr. Zakresu Les. 32: 163-178. (In Polish) (NSA 31: 624). (Schefke ?).

Scherney, F. and A. Haisch. 1967. Study on mass breeding and sterilization of the Mediterranean fruit fly Ceratitis capitata Wied. A contribution to the autocidal technique. Bayer. Landwirtsch. Jb. 44(6): 748-756. (In German).

Scheufler, H. 1963. The effects of a single whole-body irradiation on the growth and osteogenesis of the house mouse (Mus musculus domesticus Rutty 1772). Naturwissenschaften 50(11): 403-404. (In German) (NSA 17: 28774).

Schiagar, K. J., et al. 1971. Biological/ecological considerations for Project Rio Blanco. Vol. I. Pre-event inventory and potential impact analysis. Vol. II. Appendices containing environmental inventory data from various sources. Colorado State Univ., Ft. Collins, CER Geonuclear Corp., U. S. AEC report PNE-RB-11(Vol.1), -12(Vol.2). xiii, 207; 348pp. (NSA 26: 36029, 36030).

Schier, G. A. 1970. Seasonal pathways of ^{14}C-photosynthate in red pine labeled in May, July, and October. Forest Science 16(1): 2-__.

Schiff, S. O. 1965. Ribonucleic acid synthesis in neuroblasts of Chortophaga viridifiasciata. Thesis. (Dissertation Abstr. 25: 5459) (^{3}H).

Schiff, S. O. 1965. Ribonucleic acid synthesis in neuroblasts of Chortophaga viridifiasciata (De Geer), as determined by observations of individual cells in the mitotic cycle. Exptl. Cell Res. 40(2): 264-276. (^{3}H).

Schiffman, R. H. 1961. The uptake of strontium from diet and water by rainbow trout. Hanford Labs., U. S. AEC report HW-SA-1997; HW-72107. 8pp. (NSA 16: 1539, 11951, 28698; SFA 8: 5627; WPA 35: 1873).

Schiffman, R. H. 1961. A perfusion study of the movement of strontium across the gills of rainbow trout (*Salmo gairdnerii*). Biol. Bull. 120(1): 110-117. (Also: Hanford Atomic Products Operation, U. S. AEC report HW-SA-1978) (NSA 15: 15356; SFA 6: 4059).

Schiffman, R. H. 1965. Strontium-calcium transport across the gill of rainbow trout, (*Salmo gairdnerii*). J. Cell. Comp. Physiol. 65: 1-5.

Schikarski, W. and H. Wild. 1971. Environmental exposure from a sodium-cooled fast breeder power plant. pp. 502-505. In: *Reactor Conference*. (Bonn, Germany, 30 March 1971). Deutsches Atomforum, Bonn. (In German) (CONF-710346) (NSA 26: 42546).

Schildmacher, H. 1964. Radioactive substances in the feathers of migratory birds. Kernenergie 7(1): 59. (In German) (NSA 18: 13405).

Schiller, E. L. 1957. Investigations on use of x-irradiation as a mechanism for facilitating the study of morphological variation in *Hymenolepis nana*. J. Parasitol. 43(Suppl.): 43.

Schiller, E. L. 1959. Experimental studies on morphological variation in the cestode *Hymenolepis*. III. X-irradiation as a mechanism for facilitating analyses in *H. nana*. Exptl. Parasitol. 8(5): 427-470.

Schiller, P. and G. B. Cook. 1971. Determination of trace amounts of gold in natural sweet waters by nondestructive activation analysis after pre-concentration. Anal. Chim. Acta 54: 364-368. (NSA 25: 29014).

Schiller, P. and G. B. Cook. 1971. Determination of trace amounts of gold in natural waters by non-destructive activation analysis using preconcentration before irradiation. pp. 137-142. In: *Nuclear Techniques for Mineral Exploration and Exploitation*. (Proc. Panel, Crakow, 8-12 December 1969). International Atomic Energy Agency, Vienna, Publ. STI/PUB-279. (CONF-691229) (NSA 25: 41771).

Schiller, P., G. B. Cook and C. K. Beswick. 1971. Contribution to geobotanical and geochemical prospecting for gold by activation analysis. pp. 129-135. In: *Nuclear Techniques for Mineral Exploration and Exploitation*. (Proc. Panel, Crakow, 8-12 December 1969). International Atomic Energy Agency, Vienna, Publ. STI/PUB-279. (CONF-691229) (NSA 25: 41770).

Schiller, P., et al. 1972. Determination of trace (parts per 10^9) amounts of gold in plants by non-destructive activation analysis. 3. Society for Analytical Chemistry Conf. Analyst 97(1157): 601-604. (NSA 26: 50230) (Incl. grass).

Schindler, D. W. 1966. A liquid scintillation method for measuring ^{14}C uptake in photosynthesis. Nature 211: 844-845. (NSA 20: 45311).

Schindler, D. W. 1968. Feeding, assimulation and respiration rates of *Daphnia magna* under various environmental conditions and their relation to production estimates. J. Animal Ecol. 37(2): 369-385. (^{14}C).

Schindler, D. W. and S. K. Holmgren. 1971. Primary production and phytoplankton in the experimental lakes area, northwestern Ontario, and other low-carbonate waters, and a liquid scintillation method for determining ^{14}C activity in photosynthesis. J. Fish. Res. Bd. Can. 28(2): 189-201. (SFA 16: 13588).

Schindler, D. W., R. V. Schmidt and R. A. Reid. 1972. Acidification and bubbling as an alternative to filtration in determining phytoplankton production by the ^{14}C method. J. Fish. Res. Bd. Can. 29(11): 1627-1631. (SFA 18: 15949).

Schindler, J. E. 1971. Food quality and zooplankton nutrition. J. Animal Ecol. 40(3): 589-595. (^{14}C tag).

Sching, H. R. and H. Fritz-Niggli. 1954. Malformations of the adult claw frog (*Xenopus laevis* Daud) after x-radiation in the embryonic stages. Strahlentherapie 94: 147-151. (In German) (NSA 8: 5093).

Schink, D. R. 1969. Natural ^{32}Si for age-dating ground waters: a neglected potential. Trans. Amer. Nucl. Soc. 12: 500. (NSA 24: 5747).

Schinz, H. R. and H. Cocchi. 1944. The radiation sensitivity of the testes in comparison to egg stock, description of the hybrid gland of *Arion empiricorum* Fér. Radiol. Clinc. 13: 301-___. (In German) (Snail eggs).

Schinz, H. R. and H. Fritz-Niggli. 1957. Radiation sensitivity of early embryonic stages of *Drosophila melanogaster*. Strahlentherapie 104(3): 328-337. (In German).

Schinz, H. R., H. Fritz-Niggli and E. Frey. 1952. Effect of ultra-hard radiation (31 Mev Betatron) on the eggs of *Drosophila melanogaster*. Experientia 8(1): 16-18.

Schjeide, O. A. and B. M. Allen. 1951. The relation of mitosis to the manifestation of x-ray damage in hematopoietic cells of tadpoles. J. Cell. Comp. Physiol. 38(1): 51-67. (Also: Univ. of California at Los Angeles, U. S. AEC report UCLA-94) (NSA 5: 287) (*Rana catesbiana*).

Schjeide, O. A., M. J. Millard and B. M. Allen. 1953. The effects of x-irradiation on the circulating blood cells of the tadpole of *Rana catesbiana*. Univ. of California at Los Angeles, U. S. AEC report UCLA-241. (NSA 7: 1872).

Schlagbauer, A. 1967. A rational method of measurement and evaluation for the determination of the biological half-life of radioisotopes and its application in entomology. Int. J. Appl. Radiat. Isotopes 18: 595-603. (In German) (NSA 21: 39350).

Schlagbauer, A. 1967. A method for computing the effective and biological half-life from the measurements on surviving organisms. Glas-Instrum.-Tech. Fachztg. Lab. (12): 1073-1080. (In German) (Tracers in entomology).

Schlagbauer, A. W. J. 1968. Further results of determination of biological half-lives of radioactive tracers for species of insects with different nutritional physiology (Heteroptera: Insecta). Int. J. Appl. Radiat. Isotopes 19: 757-763. (In German) (NSA 23: 2784).

Schlegel, D. E. and M. Hudson. 1969. Increasing sensitivity of immunodiffusion reactions with radioautography. Phytopathol. 59(6): 886.

Schliep, H.-J. 1962. The early radiation response of annelid nerve and muscle. Ärztl. Forsch. 16(5): 263-265. (In German) (NSA 17: 33626).

Schlossberger-Racke, I. and P. Karlson. 1964. Tyrosine metabolism of insects. XIII. Autoradiographic localization of tyrosine metabolites in the cuticle of Schistocerca gregaria. J. Insect Physiol. 10(2): 261-266. (In German, English summary) (^{14}C).

Schlosser, K. and F. J. P. Oszuszky. 1971. Coverage of the environmental influence as a result of the first Austrian nuclear power plant. Österreichische Studiengesellschaft für Atomenergie, Seibersdorf, report SGAE-RT-40/1971. 76pp. (In German) (NSA 25: 51012).

Schlumberger, H. G. 1958. Effect of radiothyroidectomy in the parakeet. A.M.A. Arch. Pathol. 66(6): 747-753.

Schlumberger, H. G. and U. K. Henschke. 1956. Effect of total body x-irradiation on the parakeet. Proc. Soc. Exptl. Biol. Med. 92(2): 313-315. (NSA 10: 9942).

Schlundt, H. 1910. The radioactivity of some spring waters at Madison, Wisconsin. Trans. Wis. Acad. Sci. 16(2): 1245-1251.

Schmalz, B. L. 1972. Radionuclide distribution in soil mantle of the lithosphere as a consequence of waste disposal at the National Reactor Testing Station. Idaho Operations Office, U. S. AEC report IDO-10049. 80pp. (NSA 27: 9569).

Schmalz, B. L. and W. L. Polzer. 1969. Tritiated water distribution in unsaturated soil. Soil Science 108(1): 43-47.

Schmelzer, L. L. 1969. Control of radiation hazards. Role of the environmental health scientist. J. Occup. Med. 11: 33-36. (NSA 23: 18033).

Schmid, P. 1967. Effect of x-radiation on the growth and division process of Tetrahymena pyriformis. Biochem. Biophys. Res. Commun. 26: 615-620. (NSA 21: 18025).

Schmid, R. 1969. Biosatellite II. Naturwiss. Rundschau 22: 24-25. (In German) (NSA 24: 21119) (^{85}Sr, Drosophila, Tribolium, Habrobracon).

Schmidt, C. H. and P. A. Dahm. 1956. The synthesis of C^{14} labeled piperonyl butoxide and its fate in the Madeira roach. J. Econ. Entomol. 49(6): 729-735.

Schmidt, C. H. and G. C. LaBrecque. 1959. Acceptability of toxicity of poisoned baits to house flies resistant to organophosphorus insecticides. J. Econ. Entomol. 52(2): 345-346.

Schmidt, C. H. and C. N. Smith. 1963. The role of radionuclides in insect behavior studies. pp. 441-443. In: Radioecology. (See Schultz and Klement, 1963) (NSA 17: 33593).

Schmidt, C. H. and D. E Wiedhaas. 1959. Effect of varying conditions in a laboratory testing technique on the mortality of mosquito larvae. J. Econ. Entomol. 52(5): 977-979.

Schmidt, C. H., D. A. Dame and D. E. Wiedhaas. 1964. Radiosterilization vs. chemosterilization in house flies and mosquitoes. J. Econ. Entomol. 57(5): 753-756.

Schmidt, G. 1968. Determination of some trace elements in Rhine water by neutron activation analysis. Germany Kernforschungszentrum, Karlsruhe, report KFK-863. 70pp. (In German) (NSA 23: 37914).

Schmidt, H. 1961. The effect of x-rays on wood-destroying insects. Holzforschung 13(1): 8-11. (In German).

Schmidt, R. G. 1962. Aeroradioactivity survey and areal geology of the Savannah River Plant area, South Carolina and Georgia (ARMS-1). U. S. Geological Survey, U. S. AEC report CEX-58.4.2. 41pp.

Schmidt, R. G. 1962. Aeroradioactivity survey and areal geology of the Hanford Plant area, Washington and Oregon (ARMS-1). U. S. Geological Survey, U. S. AEC report CEX-59.4.11. 25pp.

Schmitt, E. D. 1963. Preparation of radioactive samples for measurement. I. Preparation of water. Glas-Instr.-Tech. 7(3): 91-98. (NSA 19: 26588).

Schmitz, J., J. Schneider and H. Vogg. 1972. Practical experience with neutron activation analysis of water. Gas-Wasserfach Wasser-Abwasser 113(7): 318-324. (In German) (NSA 28: 40, 2149).

Schmotzer, J. K., et al. 1972. Sources of water pollution established by using a neutron activatable tracer. Trans. Amer. Nucl. Soc. 15(2): 641-642. (NSA 27: 5272).

Schneider, E. 1926. The biological effect of roentgen radiation on single celled organisms through investigations of Paramecium. Strahlentherapie 22: 92-106. (In German).

Schneider, E., et al. 1969. Studies on alpha radiation sensitivity of embryonic stages of Drosophila melanogaster. Strahlentherapie 137: 119-121. (In German) (NSA 23: 31888).

Schneider, G., and C. Waldenmaier. 1969. Distribution of radiation induced breaks in the sex chromosomes of Microtus agrestis. Humangenetik 8: 230-241. (In German) (NSA 24: 14481).

Schneider, H. 1968 (w/W. Block), 1969, 1970. On the question of the loading capacity of the Rhine for radioactive nuclides. Sorption of radionuclides by sediments of the Rhine. 4. Sorption of radionuclides by the plankton fraction of suspended substances in the Rhine. 6. Sorption of radionuclides on selected materials. Gas-Wasserfach 109: 1410-1415; 110: 647-652; 111(1): 21-26. (In German) (NSA 23: 31420; 24: 14481; 25: 2350).

Schneider, K. J. 1971. Radioactive wastes from nuclear power in the U.S.A. Battelle Northwest Lab., U. S. AEC report BNWL-SA-3850. 16pp. (For: Environmental Quality Conf., Washington, D. C., 7 June 1971) (CONF-710662-1) (NSA 27: 24589).

Schneider, L. 1961. Electron microscope studies of the effect of radiations on the cytoplasm. I. The early effects of x rays on the cytoplasm of *Paramecium*. II. The late effects of x rays on the cytoplasm of *Paramecium*. Protoplasma 53: 530-553, 554-574. (In German) (NSA 21: 24680, 24681).

Schneiderman, H. A. and L. I. Gilbert. 1964. Control of growth and development in insects. Science 143: 325-333. (^{3}H, *Hyalophora cecropia*).

Schneider-Minder, A. 1962. Cytological investigations to interpret the differential radiosensitivity observed in *Drosophila* eggs of different ages. Arch. Julius Klaus-Stift Vereb-Forsch. 37: 38. (In German).

Schneider-Minder, A. 1966. Cytological study of embryological development of *Drosophila melanogaster* with x irradiation in early development. Int. J. Radiat. Biol. 11: 1-20. (In German) (NSA 21: 20145).

Schnell, J. H. 1963. The effect of neutron-gamma radiation on free-living small mammals at the Lockheed Reactor Site. pp. 339-344. In: *Radioecology*. (See Schultz and Klement, 1963) (NSA 17: 33593).

Schnell, J. H. 1964. An experimental study of carrying capacity based on the disappearance rates of cotton rats (*Sigmodon hispidus komareki*) introduced into enclosed areas of natural habitat. Thesis, Univ. of Georgia, Athens. 27pp. (At Savannah River Plant).

Schnell, J. H. 1964. Some effects of neutron-gamma radiation on late summer bird populations. Auk 81(4): 528-533.

Schnell, J. H. 1968. The limiting effects of natural predation on experimental cotton rat populations. J. Wildl. Mgmt. 32(4): 698-711. (^{60}Co, ^{65}Zn tags).

Schnell (Schell ?), W. R. 1963. Measurements of carbon-14 in known age samples and their geophysical implications. Thesis, Univ. of Washington, Seattle. 115pp. (NSA 19: 6116) (Used tree rings).

Schoeller, H. 1959. Types of tracer, microcirculation of water in aquifers radioactive tracers. Ch. 5, pp. 85-115. In: Arid Zone Research No. 12. UNESCO.

Schoen, A. 1973. AEC posture regarding federal, state, and local pollution control regulations. pp. 8-13. In: *Proc. AEC Pollution Control Conf.* (Oak Ridge, Tennessee, October 25-27, 1972). U. S. AEC report CONF-721030) (NSA 28: 4960).

Schoenborn, H. W. 1954. Mutations in *Astasia longa* induced by radiation. J. Protozool. 1(3): 170-173.

Schoettger, R. A. 1966. The toxicology of thiodan in several fish and aquatic invertebrates. Thesis, Colorado State Univ., Ft. Collins. 138pp. (Dissertation Abstr. 27(12,Pt.1): 4609-B-4610-B) (^{14}C).

Schofield, J. C. 1967. Origin of radioactivity at Niue Island. New Zeal. J. Geol. Geophys. 10: 1362-1371. (NSA 22: 37878).

Scholander, P. F., et al. 1961. Composition of gas bubbles in Greenland ice. J. Glaciol. 3: 813-824. (^{14}C).

Schönfeld, T. and K. Liebscher. 1959. Radioactive fission products in water from streams and rivers in the Vienna area. Tech. Rept. to IAEA, Vienna.

Schönfeld, T. and K. Liebscher. 1961. Investigations on the state of fission products in rain and in surface waters. Austria Inst. Physical Chemistry and Inorganic Chemistry, Univ. of Vienna. 37pp.

Schönwälder, H. 1961. ^{60}Co irradiation of two actinomycin-producing *Streptomyces* strains. Arch. Mikrobiol. 39(1): 13-21. (In German) (NSA 18: 38803).

Schoof, H. F. 1959. How far do flies fly? And what effect does flight pattern have on their control? Pest Control 27(4): 16-24. (Used ^{32}P).

Schoof, H. F. and G. A. Mail. 1953. Dispersal habits of *Phormia regina* in Charleston, W. Va. J. Econ. Entomol. 46(2): 258-262.

Schoof, H. F. and R. E. Siverly. 1954. Multiple release studies on the dispersion of *Musca domestica* at Phoenix, Arizona. J. Econ. Entomol. 47(5): 830-838.

Schoof, H. F. and R. E. Siverly. 1954. The occurrence and movement of *Periplaneta americana* (L.) within an urban sewerage system. Am. J. Trop. Med. Hyg. 3(2): 367-371.

Schoof, H. F. and R. E. Siverly. 1954. Urban fly dispersal studies with sp special reference to movement pattern of *Musca domestica* L. Am. J. Trop. Med. Hyg. 3(3): 539-547.

Schoof, H. F., R. E. Siverly and J. A. Jenson. 1952. Housefly dispersion studies in metropolitan areas. J. Econ. Entomol. 45(5): 675-683.

Schoonees, R., J. N. de Klerk and G. P. Murphy. 1969. Correlation of prostatic blood flow with ^{65}Zn activity in intact, castrated, and testosterone-treated baboons. Invest. Urology 6: 476-484. (NSA 23: 47925).

Schramel, P. and K. Samsahl. 1973. Activation analysis for environmental protection. pp. 296-299. In: *Worldwide Radiation from Nuclear and Conventional Energy Production*. H. Braun, ed. Georg Thieme Verlag, Stuttgart. (In German) (NSA 31: 2754).

Schreckhise, R. G. and F. W. Whicker. 1975. Model for predicting strontium-90 levels in mule deer. Battelle Northwest Lab., U. S. AEC report BNWL-SA-5367. 35pp. (NSA 32: 6829).

Schroder, L. J. 1971. Determination of tritium in water. U. S. Geological Survey, Denver, Colo., U. S. AEC report USGS-474-134. i, 18pp. (NSA 26: 8729).

Schroder, L. J. and W. C. Balance. 1972-73. Radiochemical monitoring of water after the Cannikin event, Amchitka Island, Alaska, January 13-April 5, 1972. --- July 1972. --- October 1972. U. S. Geological Survey, U. S. AEC report USGS-474-159, -166, -169. i, 13; i, 11; i, 11pp. (NSA 26: 50611).

Schroder, L. J. and W. C. Balance. 1973. Summary of chemical and radiochemical monitoring of water for the Cannikin event, Amchitka Island, Alaska, fiscal year 1972. U. S. Geological Survey, U. S. AEC report USGS-474-167. ii, 35pp.

Schroder, L. J., W. A. Beetem and E. Villasana. 1971. Radiochemical analyses of water from wells, springs, and streams in central Nevada. U. S. Geological Survey, U. S. AEC report USGS-474-97. i, 20pp. + 1 map.

Schroeder, G. L., H. W. Kraner and R. D. Evans. 1965. Diffusion of radon in several naturally occurring soil types. J. Geophys. Res. 70(2): 471-474. (NSA 19: 13639).

Schroeder, J. H. 1969. Inheritance of radiation-induced spinal curvatures in the guppy _Lebistes reticulatus_. Can. J. Genet. Cytol. 11: 937-947. (NSA 24: 23158).

Schroeder, J. H. 1969. Radiation-induced spermatogonial exchange between the X and Y chromosomes in the guppy. Can. J. Genet. Cytol. 11: 948-954. (NSA 24: 23160).

Schroeder, J. H. 1969. X-ray induced mutations in the poeciliid fish, _Lebistes reticulatus_ Peters. Mutation Reserach 7: 75-90. (NSA 23: 22259).

Schroeder, J. H. 1969. Quantitative changes in breeding groups of _Lebistes_ after irradiation. Can. J. Genet. Cytol. 11: 955-960. (NSA 25: 21656).

Schroeder, J. H. and S. Holzberg. 1972. Population genetics of _Lebistes_ (_Poecilia_) _reticulatus_ Peters (Poeciliidae, Pisces). I. Effects of radiation-induced mutations on the segregation ratio in postirradiation F_2. Genetics 70(4): 621-630. (NSA 26: 38696).

Schroeder, K. J. P. and W. Roether. 1975. Releases of krypton-85 and tritium to the environment and tritium to krypton-85 ratios as source indicators. pp. 231-253. In: _Isotope Ratios as Pollutant Source and Behavior Indicators_. (See IAEA, 1975) (NSA 33: 7297).

Schroeder, W. J., D. L. Chambers and R. Y. Miyabara. 1973. Reproduction of the melon fly: mating activity and mating compatibility of flies treated to function in sterile-release programs. J. Econ. Entomol. 66(3): 661-663. (NSA 28: 13085) (γ-irradiation).

Schubert, G., H. A. Künkel and H. Mass. 1957. Electrophoretic studies of serum of roentgen-irradiated and hibernation of the dormouse, _Glis glis_. Strahlentherapie 103(3): 368-375.

Schubert, J. 1967. Inadequacies in radiobiological experiments on medium effects. Int. J. Radiat. Biol. 13: 297-300. (NSA 22: 36420) (Microorganisms).

van der Schueren and J. Bonte. 1958. Disturbance of the ion transport across the irradiated skin of amphibians. Proc. 2nd Intern. Conf. on the Peaceful Uses of Atomic Energy 22: 191-200. (Papaer A/CONF.15/P/111) (NSA 13: 6069).

Schulert, A. 1961. Strontium-90 in Alaska. pp. 232-247. In: _Strontium-90 in Man and His Environment_. _Vol_. _3_. (See Kulp and Schulert, 1961).

Schulert, A. 1962. Strontium-90 in Alaska. Science 136(3511): 146-148.

Schulte, H. F. 1968. Monitoring airborne radioactivity. pp. 393-424. In: _Air Pollution_. _Vol_. _II_. _Analysis, Monitoring, and Surveying_. A. C. Stern, ed. Academic Press, New York. 2nd ed. (NSA 23: 38119).

Schults, D. W. and K. W. Malueg. 1973. Uptake of radiophosphorus by rooted aquatic plants. pp. 417-424. In: _Radionuclides in Ecosystems_. (See Nelson, 1973) (NSA 28: 27688).

Schultz, J. 1933. X-ray effects on _Drosophila pseudoobscura_. Genetics 18 (3): 284-291.

Schultz, V. (Ed.). 1961, 1963, 1965. Offsite ecological research of the Division of Biology and Medicine - terrestrial and freshwater. U. S. AEC report TID-13358, -(Rev.1), -(Rev.2). vii, 138; v, 122; v, 129pp. (NSA 15: 28907).

Schultz, V. 1964. Sampling white-tailed deer antlers for strontium-90. J. Wildl. Mgmt. 28(1): 45-49.

Schultz, V. 1965. Comparison of strontium-90 levels between antler and mandible of white-tailed deer. J. Wildl. Mgmt. 29(1): 33-38.

Schultz, V. 1966. Activities of the Environmental Sciences Branch, U. S. AEC. pp. 3-6. In: _Radiation Effects on Natural Populations_. (See Sacher, 1966).

Schultz, V. and V. Flyger. 1965. Relationship of sex and age to strontium-90 accumulation in white-tailed deer mandibles. J. Wildl. Mgmt. 29(1): 39-43.

Schultz, V. and A. W. Klement, Jr. 1963. _Radioecology_. (Proc. 1st National Symposium on Radioecology, held at Fort Collins, Colorado, September 10-15, 1961). Reinhold Publ. Corp., New York, and American Institute of Biological Sciences, Washington. xvii, 746pp. (NSA 17: 33539).

Schultz, V. and A. W. Klement, Jr. (Eds.). 1966. _Radioecology of Aquatic Organisms_. (By G. G. Polikarpov, qv.). (English transl. and updating of _Radioecology of Marine Organisms_, 1964). North-Holland, Amsterdam. xxviii, 314pp.

Schultz, V. and W. M. Longhurst. 1963. Accumulation of strontium-90 in yearling Columbia black-tailed deer, 1950-1960. pp. 73-76. In: _Radioecology_. (See Schultz and Klement, 1963) (NSA 17: 33548).

Schultz, V. and F. W. Whicker. 1972. Ecological aspects of the nuclear age: selected readings in radiation ecology. Washington State University, Pullman, and Colorado State University, Fort Collins, U. S. AEC report TID-25978. 588pp. (NSA 26: 36092).

Schulz, R. K. 1965. Soil chemistry of radionuclides. In: Radiation and Terrestrial Ecosystems. Health Physics 11(12): 1317-1324. (See Hungate, 1965).

Schulz, R. K. and H. H. Riedel. 1961. Effect of aging on fixation of ^{90}Sr by soils. Soil Science 91: 262-264. (NSA 21: 6023).

Schulz, R. K., G. A. Tompkins and K. L. Babcock. 1975. Uptake of plutonium and americium by plants from soils. Univ. of California, Berkeley, report UCB-34P211-2. 38pp. (NSA 32: 24538) (Used Nevada Test Site soil).

Schulze, H. A. and F. Randow. 1970. Artificial radioactivity from the fall-out and food in the northern districts of GDR in the year 1968. Z. Gesam. Hyg. Ihre Grenzgeb. 16: 68-71. (In German) (NSA 27: 317) (Incl. fish).

Schulze-Rettmer, R. and K. H. Rattay. 1965. Water economy and waste water treatment of the Jülich Nuclear Research Installation in the years 1960-1962. Germany Kernforschugsanlage, Jüluch, report JÜL-261-DE. 30pp. (In German) (NSA 19: 48078) (Incl. monitoing lakes and rivers).

Schumacher, E. and I. Welker. 1961. Radioactive contamination of drinking water; its control, evaluation, and removal. Vierteljahresschr. Schweiz. Sanität. 38(1): 57-66. (In German) (NSA 17: 21950).

Schumacher, G. J. and L. A. Whitford. 1965. Respiration and P^{32} uptake in various species of fresh-water algae as affected by a current. J. Phycol. 1: 78-80.

Schumann, G. 1958. The International Geophysical Year, XV. Radioactivity of the atmosphere, the earth and the oceans. Umschau 58(5): 141-143. (In German).

Schuntner, C. A. and P. G. Thompson. 1966. Detection and estimation of (^{14}C) Diazinon in vivo by countercurrent distribution and inhibition assay. Biochim. Biophys. Acta 115(1): 225-227. (Tick).

Schurr, J. M. and M. N. Stamper. 1962. Model for the accumulation of strontium and calcium by recently molted crayfish (Cambarus longulus longirostris Ort.). Limnol. Oceanogr. 7(4): 474-477. (NSA 17: 13941).

Schurr, K. M. 1959. A survey of radioactivity in naturally occurring organisms. Thesis, Bowling Green State Univ., Bowling Green, O.

Schurr, S. H. 1971. Energy Research Needs. Resources for the Future, Washington. 890pp. (PB-207516) (NSA 28: 1948) (Incl. environmental impact).

Schüssler, R. 1962. Investigations on the question of silicosis and natural radioactivity of the earth's crust. Fortschr. Gebiete Röntgenstrahlen Nuklearmed. 97: 196-200. (In German) (NSA 16: 33328).

Schuster, G., et al. 1972. Nuclear power generation in Western Europe: safety and environmental implications. Trans. Amer. Nucl. Soc. 15(2): 645-666. (NSA 27: 5296).

Schuster-Woldan, E. 1936. X-ray studies of the intestine of Cyprinus carpio and Timea vulgaris as a contribution to the problem of the significance of smell and sight in fish in the search for food. Z. Fischerei 34: 245.

Schüttelkopf, H. 1971. Determination of plutonium in environmental surveillance samples. pp. 183-197. In: Rapid Methods for Measuring Radioactivity in the Environment. (See IAEA, 1971) (In German) (NSA 26: 27808).

Schwalm, F. E. 1965. Cell and mitosis pattern of normal and x-irradiated regulating embryonic rudiments in Gryllus domesticus. Z. Morphol. Okol. Tiere 55: 915-1023. (In German) (NSA 23: 784).

Schwartz, D. 1954. An interesting phenomenon associated with irradiation of dry maize seeds. Science 119(3079): 45-46.

Schwartz, L. L., et al. 1975. Potential pathways to man for radionuclides from excavation applications of peaceful nuclear explosives. pp. 321-342. In: Peaceful Nuclear Explosions. IV. (See IAEA, 1975) (NSA 33: 9285).

Schwarz, J., et al. 1968. Mobile radioactive tracer laboratory for hydrological studies. Israel Atomic Energy Commission, Tel Aviv, report IA 1127. 42pp. (NSA 23: 435).

Schwarzer, W. 1971. Nuclear power plants: a risk for the environment. At. Strom. 17(7-9): 118-121. (In German) (NSA 26: 1597).

Schwarzer, W. 1972. Environmental problems in the F. R. Germany. Limiting the discharge of airborne radioactive effluents in regions of high nuclear power plant density. At. Jap. 16(4 Suppl.): 71-94. (NSA 28: 17537).

Schwien, W. G., et al. 19__. Carbon-14 fixation in pollen of yellow lupine (Lupinus luteus Linn.). Kansas Agr. Expt. Sta., Manhatten, U. S. AEC report TID-15982. 12pp. (NSA 16: 21890).

Schweizer, H. 1961. Increase of the antibiotic formation of strains of Streptomyces by roentgen irradiation. Arch. Mikrobiol. 39(3): 230-241. (In German) (NSA 18: 38804).

Schweizer, P. 1969. Effects of x radiation on the embryonic development of male gonads in Drosophila melanogaster. Strahlentherapie 137: 749-759. (In German) (NSA 23: 38763).

Schwibach, J. 1969. Experience from large discharges of radioactive liquid waste. Atomkernenergie 14: 65-68. (In German) (NSA 23: 7636) (Concerns Columbia and Clinch rivers).

Schwibach, J. 1971. Emergency planning at nuclear installations as a basis for rapid monitoring programs. pp. 679-693. In: Rapid Methods for Measuring Radioactivity in the Environment. (See IAEA, 1971) (NSA 26: 28056).

Schwibach, J. 1972. Methods of disposal of radioactive wastes into water and their control. pp. 145-156. In: Disposal of Radioactive Waste. (See Organization for Economic Co-operation and Development, 1972).

Schwibach, J. 1972. Discharge of radioactive liquid effluents in water and their control. Bundesgesundheitsblatt 15(18): 260-263. (In German) (NSA 28: 11362).

Schwibach, J. 1972. Radiation protection standards for the licensing of discharge of radioactive substances from nuclear installations. pp. 19-29. In: Institute for Reactor Safety Technical Meeting. (Cologne, Germany, November 8, 1971). Report IRS-T-23. (In German) (NSA 27: 15325) (Into a river).

Schwibach, J. 1973. Safety and environmental protection problems in the handling of radioactive wastes. Atomwirt. Atomtech. 18(5): 231-234. (In German) (NSA 28: 15476).

Schwibach, J. and I. Winkelmann. 1974. Monitoring of radioactive gaseous effluents from nuclear power plants in the Federal Republic of Germany. pp. 81-109. In: Proc. Seminar on the Monitoring of Radioactive Effluents. (Karlsruhe, Germany, May 14-17, 1974) (See Organization for Economic Co-operation and Development, 1974) (NSA 31: 3340).

Schwieger, R. G. 1971. Survey shows nuclear plants will limit polluting effluents. Power 115(1): 49-53. (NSA 25: 18055).

Schwinghamer, E. A. 1959. The relation between radiation dose and the frequency of mutations for pathogenicity in Melamspora lini. Phytopathology 49: 260-269. (NSA 13: 13189).

Schwoerbel, J. 1970. Methods on Hydrobiology. Pergamon, Elmsford, N. Y. ix, 200pp. (Incl. radionuclide techniques).

Sciesinska, E. 1970. Accuracy of the neutron method of moisture measurement. Nukleonika 15: 593-612. (NSA 25: 42177).

Scott, A. G. 1972. Radiation dose to mosquito larvae reared in radioactive media. Atomic Energy of Canada, Ltd., Pinawa, report AECL-3789. 23pp. (NSA 27: 24972).

Scott, A. G. and J. E. Guthrie. 1972. Accumulation of radioactive cesium, phosphorus, strontium, and yttrium by mosquito larvae and its relevance to radiation dose. Can. J. Zool. 50(6): 767-770. (NSA 26: 56560).

Scott, C. M. 1934. Action of x-rays on eggs of Calliphora. Proc. Roy. Soc. (London) B115(792): 100-119.

Scott, D. P. 1961. Radioactive iron as a fish mark. J. Fish. Res. Bd. Can. 18(3): 383-391. (SFA 6: 4313).

Scott, D. P. 1962. Radioactive cesium as a fish and lamprey mark. J. Fish. Res. Bd. Can. 19: 149-151. (NSA 16: 20069).

Scott, M. R. 1968. Thorium and uranium concentrations and isotope ratios in river sediments. Earth Planet. Sci. Lett. 4: 245-252. (NSA 22: 33557).

Scott, R. C. 1963. Radium in natural waters in the United States. pp. 237-240. In: Radioecology. (See Schultz and Klement, 1963) (NSA 17: 33568).

Scott, R. C. and F. B. Barker. 1962. Ground-water sources containing high concentrations of radium. Article 414, pp. 357-359. In: Geological Survey, research 1961. U. S. Geological Survey Prof. Paper 424D.

Scott, R. C. and F. B. Barker. 1962. Data on uranium and radium in ground water supplies in the United States 1954-1957. U. S. Geological Survey Prof. Paper 426. 116pp. (CA 58: 1237).

Scott, R. M., D. Seiz and H. J. Shaughnessy. 1964. Rapid carbon-14 test for coliform bacteria in water. Am. J. Publ. Hlth. 54(5): 827-833. (NSA 18: 39534).

Scott, T. G. and S. A. Reynolds. 1975. Determination of plutonium in environmental samples. II. Procedures. Radiochem. Radioanal. Lett. 23(4): 275-281. (NSA 33: 22732).

Scully, N. J. 1962. Nature of x-irradiation induced growth stimulation and arrest in Xanthium. Abstr. p. 6. In: 2nd Intern. Congr. Radiation Research. (See International Congress of Radiation Research,11962).

Seabloom, R. W. 1975. Field irradiator γ: preirradiation occurrence of breeding birds in three boreal habitats. Atomic Energy of Canada, Ltd., Pinawa, report AECL-5115. 17pp. (NSA 33: 9532).

Seaborg, G. T. 1970. Peaceful uses of nuclear energy: a collection of speeches. U. S. AEC, Washington. 159pp. (NSA 25: 34817).

Seaborg, G. T. 1971. Environment: a global problem, an international challenge. pp. 3-8. In: Environmental Aspects of Nuclear Power Stations. (See IAEA, 1971) (CONF-700810-37) (NSA 24: 45921).

Seaborg, G. T. 1972. Energy and environment. Int. J. Environ. Stud. 3(4): 301-306. (NSA 26: 55967).

Seaborg, G. T. 1972. Nuclear energy and the environment. Int. Atom. En. Agency Bull. 14(2): 2-11. (NSA 26: 50595).

Seaborg, G. T. and W. R. Corliss. 1971. Man and Atom. E. P. Dutton, New York. 411pp. (NSA 26: 27427).

Seaborg, G. T., J. L. Bloom and E. L. Nelson. 1973. Public policy toward environment 1973: a review and appraisal. Ann. N.Y. Acad. Sci. 216: 79-88. (NSA 29: 28763).

Sealander, J. A. 1970. Short term effects of acute sublethal gamma radiation on populations of the old-field mouse, Peromyscus polionotus. Health Physics 19: 299-306. (NSA 25: 8760).

Searle, A. G., R. J. Berry and C. V. Beechey. 1970. Cytogenetic radiosensitivity and chiasma frequency in wild-living male mice. Mutation Research 9: 137-140. (NSA 24: 31950).

Seay, T. N. and J. G. Rodriguez. 1965. The effect of colored lights on ingestion by immature two-spotted spider mites. Abstr. Bull. Entomol. Soc. Am. 11(3): 160. (^{32}P, Tetranychus urticae).

Sechkina, Ye. R. 1973. Isolation of radiation-sensitive strains of Drosophila melanogaster from natural populations. Genetika 9(10): 160-162. (In Russian) (NSA 29: 16044).

Sedell, J. F. 1973. Feeding rates and food utilization of stream caddisfly larvae of the genus Neophylax (Trichoptera: Limnephilidae) using cobalt-60 and carbon-14. pp. 486-491. In: Radionuclides in Ecosystems. (See Nelson, 1973) (NSA 28: 27693).

Sedel'nikov, G. S. and I. M. Karatayeva. 1966. Rubidium content of Gulf of Kara-Bogaz brine and Caspian Sea water. Dokl. Akad. Nauk SSSR 170(4): 939-940. (Doklady (earth sci.) 170(1-6): 214-___).

Sedivy, J., et al. 1966. Evaluation of activity of honeybee colonies moved to a lucerne seed field. Acta ent. bohemosl. 63(1): 1-9. (^{32}P, ^{198}Au).

Sedlet, J. (w/A. F. Stehney; F. S. Iwami; N. W. Golchert and T. L. Duffy). Environmental radioactivity (monitoring) at Argonne National Laboratory. Argonne National Lab., U. S. AEC report:

Publ. Year	Year Reported	Report ANL-	pp.	NSA
1957	1953	5289	35	11: 7399
1958	1957	5934	42	13: 5490
1959	1958	6047	43	14: 2287
1963	--	6736	61	17: 36023
1965	1964	7104	45	20: 12998
1973	1972	8007	71	28: 359
1974	1973	8078	88	29: 29315
1975	1974	75-18	78	32: 751

Seeley, B. A., J. B. Peterson and M. H. Smoler. 1966. Relative biological effectiveness of x-ray and gamma radiation. Drosophila Inform. Serv. 41: 178.

Seelye, J. G. 1975. Predictive capabilities of the specific activity hypothesis for Cs and Zn in freshwater systems. Dissertation, Michigan State Univ., East Lansing. 131pp. (Univ. Microfilms 75-20888) (NSA 33: 9263).

Segebade, C. 1973. Quantitative determination of bromide in a pesticide by activation analysis with 14-MeV neutrons. Radiochem. Radioanal. Lett. 15 (4-5): 251-260. (In German) (NSA 29: 7511).

Segner, W. F. and C. F. Schmidt. 1966. Radiation resistance of spores of Clostridium botulinum Type E. In: Food Irradiation. (Proc. Intern Symp., Karlsruhe, Germany, June 6-10, 1966). International Atomic Energy Agency, Vienna, Publ. STI-PUB-127. (SM-73/28) (CONF-660609-19) (NSA 20: 40699).

Sehe, C. T. 1965. Comparative studies on the ultimobranchial body in reptiles and birds. Gen. Comp. Endocrinol. 5(1): 45-59. (Used ^{131}I and ^{35}S).

Sehmel, G. A. and F. D. Lloyd. 1974. Particle resuspension rates. Battelle Northwest Lab., U. S. AEC report BNWL-SA-5124. 14pp. (NSA 31: 3368).

Sehmel, G. A. and F. D. Lloyd. 1974. Resuspension of plutonium at Rocky Flats. Battelle Northwest Lab., U. S. AEC report BNWL-SA-5085. 39pp. (NSA 31: 3367).

Seifriz, W. 1936. Reactions of protoplasm to radium radiation. Protoplasma 25(2): 196-200.

Seitz, K. A. 1966. Normal development of the embryo of the arachnid Cupiennius salei Keyserling and its regulatory capacity after x irradiation. Zool. Jahrb. Abt. Anat. Ontog. Tiere 83: 327-447. (In German) (NSA 21: 34773).

Seitz, K. A. 1967. Studies on the dynamic processes in the ovum of the spider Cupiennius salei (Ctenidae) by means of x-ray barriers. Zool. Jahrb. Abt. Anat. Ontog. Tiere 84: 343-374. (In German) (NSA 22: 28637).

Seitz, M. G. and R. E. Taylor. 1974. Uranium variations in a dated fossil bone series from Olduvai Gorge, Tanzania. Archeometry 16(2): 129-135. (NSA 31: 3401).

Sekeris, C. E. 1963. Tyrosine metabolism in the blowfly, Calliphora erythrocephala. pp. 281-287. In: Radiation and Radioisotopes Applied to Insects of Agricultural Importance. (See IAEA, 1963).

Sekeris, C. E. 1965. Action of ecdysone on RNA and protein metabolism in the blowfly, Calliphora erythrocephala. pp. 149-167. In: Mechanisms of Hormone Action. P. Karlson, ed. Academic Press, New York. (^{14}C).

Sekeris, C. E. and N. Lang. 1964. Induction of DOPA-decarboxylase activity by insect messenger RNA in an in vitro amino acid incorporating systems from rat liver. Life Sci. 3: 625-632. (^{14}C, Calliphora erythrocephala).

Sekeris, C. E. and N. Lang. 1965. Biosynthesis of proteins in insects. In vitro incorporation of amino acids in proteins through fractions from the epidermal cells of the bluebottle fly, Calliphora erythrocephala. Hoppe-Seyler's Z. physiol. chem. 342(1-3): 161-165. (In German, English summary) (^{14}C).

Sekeris, C. E., N. Lang and P. Karlson. 1965. The mechanism of action of hormones. V. The effect of ecdysone on RNA metabolism in the epidermis of the blowfly, Calliphora erythrocephala. Hoppe-Seyler's Z. physiol. chem. 341(1-3): 36-43. (In German, English summary) (^{32}P).

Seki, M., K. Itaoka and T. Ookura. 1951. Radiation dosimetry of atomic bombed areas in Hiroshima and experimental studies on action upon animals in the soil. J. Jap. Balneo-climatol. Assoc. 16(1): 22.

Selby, J. M., et al. 1961. A proportional continuous river and stream water sampler at Georgia Nuclear Laboratories. Health Physics 6(3/4): 149-151.

Selby, J. M., et al. 1961. Radiation effects of a 10-MW reactor on environs of the Georgia Nuclear Laboratories. Health Physics 6(3/4): 126-135.

Selders, A. A. 1950. The absorption and translocation of several fission elements by Russian thistles. Hanford Atomic Products Operation, U. S. AEC report HW-18034. 25pp. (NSA 10: 4118).

Selders, A. A. 1968. Autoradiograph of pine trunk cross section. Forestry 66(2): 137-___.

Selders, A. A., J. H. Rediske and R. F. Palmer. 1953. The absorption and translocation by plants of radioactive elements from "Jangle" soil. Hanford Atomic Products Operation, U. S. AEC report HW-27620. (NSA 7: 4326) (Operation JANGLE, Nevada Test Site).

Selecki, A. and S. Filipek. 1966. Single-well method for determining the direction of flow of ground water using a neutron source. Int. J. Radiat. Isotopes 17: 351-352. (NSA 20: 31430).

Self, L. S. 1965. Adaptation of insects to tobacco. Thesis, N. Carolina State Univ., Raleigh. 66pp. (Dissertation Abstr. 26: 1133) (Protoparce sexta, Heliothis virescens; $^{14}CO_2$, ^{14}C-nicotine).

Self, L. S., F. E. Guthrie and E. Hodgson. 1964. Adaptation of tobacco hornworms to the ingestion of nicotine. J. Insect Physiol. 10(6): 907-914. (^{14}C, Protoparse sexta).

Sellers, B., C. A. Ziegler and J. Papadopoulos. 1967. Development of a radioisotope gage for monitoring sediment concentration in rivers and streams. pp. 3-10. In: Isotope Techniques in the Hydrologic Cycle. (See Stout, 1967).

Sellers, L. G. and F. E. Guthrie. 1972. Distribution and metabolism of ^{14}C-dieldrin in the resistant and susceptible house fly. J. Econ. Entomol. 65(2): 378-385.

Sellstedt, H., L. Engstrand and N.-G. Gejvall. 1966. New applications of radiocarbon dating to collagen residue in bones. Nature 212: 573-574. (NSA 21: 4618).

Semenov, D. I. 1957. Effect of complexons on the behavior of metals and radioactive substances in the body. I. Mode of action of complexons. II. (w/I. P. Tregubenko) Ethylenediaminetetraacetic acid sodium salt. Sb. rabot lab. biofiz., Tr. inst. biol. Akad. Nauk SSSR, Ural. fil. 1: 4-19, 20-56. (In Russian).

Semenov, D. I. 1958. Effect of complexons on the behavior of radioactive substances in the body and the mode of action of complexons. Byull. Ural. otd. Mosk. obshch. ispyt. prirody 1: 19-38. (In Russian).

Semenova, L. M. 1957. Study of the permeability of the integument of soil insects to salt by the method of tagged atoms, for example the larvae of the crane fly, Tipula paludosa Mtig. (Diptera, Tipulidae). Zool. zh. 36 (12): 1826-1830. (In Russian).

Semerdzhyan, S. P., N. G. Nor-Arevyan and Sh. G. Megroyan. 1964. Effect of high oxygen pressure on the radiation effect on silkworm granules. Izv. Akad. Nauk Armyan. SSR 16(11): 91-94. (In Armenian).

Semerdzhyan, S. P., et al. 1971. Role of sulfhydryl groups in radiosensitivity of silkworm eggs. Biol. zh. Armyan SSR 24(10): 83-84. (In Russian) (NSA 26: 56482).

Sen, S. and N. Ghosh. 1968. Delayed germination in gamma-rayed dry seeds of jute (Corchorus sp.). Sci. Culture (Calcutta) 34: 346-347. (NSA 23: 38880).

Sen'chuk, Ye. Z. 1956. Natural radioactivity in plants. Student Nauk. Pratzi Kiev Derzhav. Univ. im. T. G. Shevchenka biol. zbirnik (2): 75-78. (In Russian) (CA 51: 18145g).

Senegacnik, M. and S. Paljk. 1969. Fallout monitoring of waters. I. Radiochemical determination of ^{137}Cs and $^{89,90}Sr$ by ion-exchange. II. Radiochemical determination of ^{144}Ce, ^{137}Cs, $^{89,90}Sr$, and ^{140}Ba by ion-exchange. Z. Anal. Chem. 244: 306-309, 375-379. (NSA 23: 26187, 26179).

Sengel, C. 1959. Is the caudal region of planarian capable of inducing the regeneration of the pharynx? J. Embryol. Exptl. Morphol. 7(1): 73-85. (Involves irradiation).

Senger, C. M. 1964. Effect of x-ray irradiation on oxygen uptake by planarians. Worm Runner's Dig. 6: 46-51. (NSA 18: 31248).

Sen Gupta, A. K. and C. O. Knowles. 1970. Rate of formetanate-^{14}C acaracide in the rat. J. Econ. Entomol. 63(1): 10-__.

Sen'ko, Ye. Ye. and B. G. Starikov. 1968. Evaluation of thoron exhalation. Izv. Akad. Nauk SSSR (fiz. atmos. okeana) 4: 367-368. (In Russian) (NSA 23139).

Sen-Sarma, P. D. and W. Kloft. 1965. Trophallaxis in pseudoworkers of Kalotermes flavicollis (Fabricius) (Insecta: Isoptera: Kalotermitidae) using radioactive I^{131}. Proc. Zool. Soc. Calcutta 18: 41-46.

Senvar, C. B. 1965. Dating of nuclear explosions by measuring gross beta activity in precipitation. Atompraxis 11(3): 144-146. (In German) (NSA 19: 22606) (Snow).

Senvar, C. B. and U. Yucelik. 1963. Radioactive fallout and purification of the water contaminated by the fallout. Turkey Atomic Energy Commission, Ankara, Ser. No. 2. 9pp. (In Turkish).

Senvar, C. B. and U. Yucelik. 1963. Removal of fallout radionuclides from water. Atompraxis 9(6): 222-223. (In German) (NSA 17: 30765).

Senvar, C. B., et al. 1966. Turkish Atomic Energy Commission third activity report of the Radioactive Fallout Laboratory, August 1965-July 1966. Turkey Atomic Energy Commission, Ankara, report TAEC-D-5. 12pp. (NSA 22: 181).

Seo, S. T., et al. 1973. Hawaiian fruit flies in papaya, bell pepper, and eggplant: quarantine treatment with gamma irradiation. J. Econ. Entomol. 66(4): 937-939. (NSA 28: 21777).

Seo, S. T., et al. 1974. Mango weevil: cobalt-60 γ irradiation of packaged mangoes. J. Econ. Entomol. 67(4): 504-505. (NSA 30: 24354).

Serbanescu, O. and P. Jitaru. 1965. On ^{32}P uptake in the body of some Black Sea fishes. An. Stiint. Univ. Al. I. Cuza, Iasi Sect. IIa 11(1): 7-12. (In Romanian) (NSA 21: 39357).

Serebrennikov, V. S. 1959. Seasonal variation in the uranium content of groundwater. pp. 218-223. In: Data on Geological Mine Deposits, Petrography, Mineralogy and Geochemistry. Acad. Sci. USSR, Moscow. (In Russian).

Serebrovskiy, A. S. 1969. On the possibility of a new method for the control of insect pests. pp. 123-137. In: Sterile-Male Technique for Eradication or Control of Harmful Insects. (See IAEA, 1969) (NSA 24: 32308).

Sereda, G. A. 1961. Pollution of the seas and oceans by artificial radioactive materials. Atomizdat, Moscow. UN Document A/AC.82/G/L.704. 17pp. (In Russian).

Sereda, G. A. 1962. Contamination of the seas and oceans by artificial radioactive substances. pp. 259-271. In: Problems of Nuclear Meteorology. (See Karol' and Malakhov, 1962) (AEC-tr-6128, pp. 312-328) (NSA 18: 10278).

Sereda, G. A. and Ts. I. Bobovnikova. 1962. Radiochemical methods for bulk control of strontium-90 content in freshwater. pp. 223-234. In: Warsaw Symp. Tech. Zagadnien Ochrony Promieniowaniem. (NSA 18: 39509).

Sereda, G. A. and Ts. I. Bobovnikova. 1963. A radiochemical method for the strontium-90 content of freshwater reservoirs. State Committee on the Use of Atomic Energy USSR. Atomizdat, Moscow. 12pp. UN Document A/AC.82/G/L. 903. (AEC-tr-6457).

Sereda, G. A. and Ts. I. Bobovnikova. 1963. Strontium-90 in freshwater reservoirs in the Soviet Union in 1961 and 1962. State Committee on the Use of Atomic Energy USSR. Atomizdat, Moscow. 28pp. UN Document A/AC.82/G/L. 896. (In Russian) (NP-14728) (NSA 19: 20160).

Sereda, G. A. and Ts. I. Bobovnikaova. 1964. ^{90}Sr in the freshwater reservoirs of the Soviet Union in 1961-1962. pp. 361-372. In: Radioactive Isotopes in the Atmosphere and Their Use in Meteorology. (Scientific Conf. on Nuclear Meteorology, Obninsk, Feb. 3-6, 1964). I. L. Karol', et al., eds. Atomizdat, Moscow. (AEC-tr-6711) (CONF-640208) (NSA 20: 16705).

Sereda, G. A. and F. Ya. Robinskiy. 1963. Shifting of the radioactive equilibrium between strontium-90 and yttrium-90 in lake waters. Atom. energiya 14(3): 326-327. (Sov. Atomic Energy 14(3): 327-328; Mound Lab., U. S. AEC report MLM-1470(tr)) (NSA 17: 21940).

Sergiou, C. 1975. Sterile-male technique for control of the Mediterranean fruit fly, Ceratitis capitata Wied., in the Mediterranean Basin. pp. 11-28. In: Sterility Principle for Insect Control, 1974. (See IAEA, 1975) (NSA 32: 11763).

Sergiou, C., J. Zyngas and A. Krambias. 1973. Preliminary studies of Mediterranean fruit fly, Ceratitis capitata Wied. populations in Cyprus. pp. 165-176. In: Computer Models and Application of the Sterile-Male Technique. (See IAEA, 1973) (NSA 29: 8059).

de Serres, F. J. 1961. Some aspects of the influence of environment of the radiosensitivity of micro-organisms. In: Microbial Reaction to Environment. Symposia of the Soc. Gen. Microbiol. No. XI.

Setter, L. R. 1962. Radioactive contamination of food. J. Am. Dietetic Assoc. 38: 561-566. (NSA 16: 29264).

Setter, L. R. and A. S. Goldin. 1956. Measurements of low-level radioactivity in water. J. Am. Water Works Assoc. 48(11): 1373-1379. (NSA 14: 7279).

Setter, L. R. and A. S. Goldin. 1956. Radioactive fallout in surface waters. Industr. Engr. Chem. 48: 251-255. (NSA 10: 4509).

Setter, L. R. and H. H. Russell. 1959. The radioactive contamination of the environment in the Cincinnati area. J. Am. Water Works Assoc. 51(4): 447-462.

Setter, L. R. and C. P. Straub. 1958. The distribution of radioactivity from rain. Trans. A. Geophys. Un. 39: 451-458.

Setter, L. R., G. R. Hagee and C. P. Straub. 1958. Analysis of radioactivity in surface waters - practical laboratory methods. Am. Soc. Testing Materials Spec. Publ. (227): 35-40; (235): 56-66.

Setter, L. R., J. E. Regnier and E. A. Diephaus. 1959. Radioactivity of surface waters in the United States. J. Am. Water Works Assoc. 51(11): 1377-1392.

Seuberling, H.-B. 1971. High resistance to x rays of the activity and light induced activation of NADP-dependent glyceraldehyde phosphate dehydrogenase in Ankistrodesmus braunii. Planta 98: 270-273. (In German) (NSA 25: 55353).

Seufert, R. 1965. The radiation sensitivity of different development stages of tetraploid Oenothera berteriana. Radiation Botany 5: 153-169. (In German) (NSA 19: 24335).

Seymour, A. H. 1958. The use of radioisotopes as a tag for fish. pp. 118-124. In: Proc. Gulf and Caribbean Fish. Inst. 10th Ann. Session.

Seymour, A. H. 1961. The use of radioisotopes and radiation to study plant and animal life in fresh and marine waters. pp. 266-276. In: Applications of Radioisotopes and Radiation in the Life Sciences. (See U. S. Congress, 1961) (NSA 15: 27323).

Seymour, A. H. 1964. Contributions of radionuclides to our understanding of aquatic ecosystems. Verh. Int. Verein. Limnol. 15(1): 227-236.

Seymour, A. H. 1969. Review of: Symposium on Radioecology. (D. J. Nelson and F. C. Evans, eds.). Trans. Am. Fish. Soc. 98(4): 728.

Seymour, A. H. 1970. Radiological aspects of nuclear power and the aquatic environment. In: Proc. of a Student Conf. on Nuclear Power and the Environment. (See Carbon and Houlberg, 1970) (Also: Univ. of Washington, Seattle, U. S. AEC report RLO-2225-T-1-1) (CONF-700415-1) (NSA 24: 39467).

Seymour, A. H. 1973. The rate of loss of mercury by Pacific oysters. pp. 85-90. In: Mercury in the Western Environment. (See Buhler, 1973) (^{203}Hg).

Seymour, A. H. 1974. Impact of radiological effluents in surface waters. Paper 13, vp. In: Environmental Analysis and Environmental Monitoring for Nuclear Power Generation. (See Univ. of California at Berkeley, 1974) (NSA 32: 14275).

Seymour, A. H. and G. B. Lewis. 1964. Radionuclides of Columbia River origin in marine organisms, sediments, and water collected from the coastal and offshore waters of Washington and Oregon, 1961-1963. Univ. of Washington, Seattle, U. S. AEC report UWFL-86. 78pp. (NSA 19: 27910).

Seymour, A. H. and V. A. Nelson. 1973. Biological half-lives for zinc and mercury in the Pacific oyster, Crassostrea gigas. pp. 849-856. In: Radionuclides in Ecosystems. (See Nelson, 1973) (NSA 28: 37218).

Seymour, R. and K. B. Dawson. 1967. Variation in the response and threshold to electrical excitation of x-irradiated isolated frog nerve with dose and dose-rate. Int. J. Radiat. Biol. 12: 1-11. (NSA 21: 36715).

Seymour, R. and K. B. Dawson. 1967. Effects of x-rays combined with other agents on the excitability of frog sciatic nerve. Int. J. Radiat. Biol. 13: 171-178. (NSA 22: 21568).

Shabadash, A. L. 1960. Cytochemical signs of radiobiological reactivity of the nervous system and their functional significance. Vopr. Tsitol. obshch. fiziol. : 368-381. (JPRS-12921; CSO: 6905-N) (Used frogs and rats).

Shabadash, A. L., S. O. Yenenko and L. V. Orlova. 1963. Cytochemical studies of glycogen in the central nervous system of the frog after gamma-irradiation. Arkh. Anat. Gistol. Embriol. 44(5): 26-36. (JPRS-21170(del.); OTS-63-31802(Del.)) (NSA 17: 37069).

Shabanova, Ye. A. 1972. Influence of chronic gamma irradiation on Chlorella. Radiobiologiya 12(1): 140-144. (AEC-tr-7316, pp. 207-213) (NSA 26: 50830).

Shackelford, B. W. 1973. Siting on coastal areas. Pacific Gas & Electric Co., San Francisco, Calif. vp. (NP-19964, Paper 9) (NSA 30: 8627).

Shacklette, H. T. 1965. Element content of bryophytes. U. S. Geological Survey Bull. 1198-D. 21pp. (CA 64: 5450f) (Incl. K, Nd, Pb, V, Sm, Gd, Hf, Re, U).

Shacklette, H. T. and M. Rubin. 1969. Radiocarbon dating of ash deposits on Amchitka Island, Alaska. U. S. Geological Survey Prof. Paper 650-B, pp. 81-83. (NSA 23: 31428).

Shafrir, N. H., J. Laichter and E. Gilat. 1971. Fast qualitative and quantitative determination of environmental radioactivity by Ge(Li) gamma spectrometry. pp. 291-300. In: Rapid Methods for Measuring Radioactivity in the Environment. (See IAEA, 1971) (NSA 26: 28229).

Shahjahan, A. K. M. and H. U. Ahmed. 1972. Effect of radiation on the growth behavior of Collectotrichum falcatum Went. Nucl. Sci. Appl. A6: 54-56. (NSA 28: 21579).

Shahjahan, M. and M. A. Hamid Miah. 1972. Comparative resistance of some radiation induced jute plant mutants to jute weevil, Apion corchori Marsh. Nucl. Sci. Appl. A6: 59-63. (NSA 28: 21614).

Shaikh, H. 1965. Effects of irradiation on biology and immunology of infective larvae of Nippostrongylus brasiliensis (Travassos, 1914) Lane, 1923, in rats. Thesis, Texas A&M Univ., College Station. (NSA 20: 3786).

Shaikh, U. 1969. Radioisotopes as tracers in water resource investigations and sanitary engineering problems. Pakistan Inst. Nuclear Science and Technology, Islamabad, report PINSTECH/HP-9. 69pp. (NSA 23: 45400).

Shakhov, A. A. 1962. The effect of cosmic radiation on plant activity. Zh. obshch. biol. 23(2): 81-89. (FTD-TT-62-771) (AD-287707; NSA 17: 5781).

Shalayeva, T. V. and A. T. Miller. 1970. Accumulation of radioactive ^{90}Sr by meadow plants under soil conditions of Latvia. Latv. PSR Zinat. Akad. Vest. (5): 20-23. (In Russian) (NSA 24: 50805).

Shalayeva, T. V. and G. S. Sabardina. 1971. Radioecological characteristic of the main components of vegetation cover of the Latvian SSR meadows and pastures. Latv. PSR Zinat. Akad. Vest. (11): 41-46. (In Russian) (NSA 26: 25593).

Shalayeva, T. V., G. S. Sabardina and A. T. Miller. 1971. Effect of copper on on ^{90}Sr content in timothy. Latv. PSR Zinat. Akad. Vest. (7): 23-25. (In Russian) (NSA 26: 22803).

Shalhevet, J. 1973. Effect of mineral type and soil moisture content on plant uptake of ^{137}Cs. Radiation Botany 13(3): 165-171. (NSA 29: 10497).

Shalhevet, J. and B. Yaron. 1967. Ion distribution, moisture content and density of soil columns measured with gamma radiation. Soil Sci. Soc. Am. Proc. 31(2): 153-___.

Shalmon, E. and J. Chanteur. 1972. Health implications of the disposal of radioactive wastes and wastes from other sources. pp. 205-223. In: Disposal of Radioactive Waste. Organization for Economic Co-operation and Development, Paris. (In French and English) (CONF-720453) (NSA 28: 5535) (Incl. environment).

Shamos, M. H., A. Liboff and J. Siderowitz. 1964. Thin-wall ion chambers for low-level radiation environment. pp. 627-642. In: The Natural Radiation Environment. (See Adams and Lowder, 1964) (NSA 17: 25604).

Shank, B. B. 1974. Results of radiobiological experiments on satellites. pp. 313-351. In: Space Radiation Biology and Related Topics. C. A. Tobias, ed. Academic Press, New York. (NSA 30: 15780).

Shanks, R. E. and H. R. DeSelm. 1958. Vegetation studies in an area contaminated with radioactive waste. Abstr. J. Tenn. Acad. Sci. 32(2): 150.

Shanks, R. E. and H. R. DeSelm. 1963. Factors related to concentration of radiocesium in plants growing on a radioactive waste disposal area. pp. 97-101. In: Radioecology. (See Schultz and Klement, 1963) (NSA 17: 33549).

Shannon, L. V. and M. J. Orren. 1970. Rapid method of analysis for plutonium content of atmospheric particulate matter. Health Lab. Sci. 7: 141-145. (NSA 25: 2112).

Shapiro, M. A. 1962. A limnological study of the upper Ohio River. Final report covering hydromechanics and radiation phases of the study. U. S. AEC contract AT(30-1)-2411. iii, 27pp. + 32 figs., 5 tables.

Shapiro, M. A., et al. 1964. Some factors of importance in evaluating sites for nuclear industry as determined by a limnological study of the upper Ohio River. Verh. Int. Verein. Limnol. 15(1): 299-306.

Shapiro, S. nd. Final report on morphogenetic effects of radiation. Oregon State Univ., Eugene, U. S. AEC report RLO-1783-13. 18pp. (NSA 22: 4700) (*Populus nigra*, x-rays).

Sharif, M. 1971. Research within the framework of a coordinated programme on the application of radiation techniques to water use efficiency studies. Final report, 15 December 1965-14 December 1970. International Atomic Energy Agency, Vienna, report IAEA-R-379-F. 59pp. (NSA 28: 24638).

Sharif, M., M. Tahir and F. M. Chaudhry. 1974. Studies in water-use efficiency for the maize crop. pp. 389-399. In: *Isotope and Radiation Techniques in Soil Physics and Irrigation Studies 1973*. (See IAEA, 1974) (NSA 30: 15505) (Neutron probe).

Sharifi, S. 1972. Radiographic studies of the parasite *Choctospila elegans* on the maize weevil, *Sitophilus zeamaize*. Ann. Entomol. Soc. Am. 65(4): 852-856.

Sharipov, Ye. B. and Yu. Khudaiberganov. 1970. Radioactive method for determining indium and gallium in water. Izv. Akad. Nauk Uzbek. SSR (fiz.-mat.) (6): 55-56. (In Russian) (NSA 25: 44346).

Sharipov, Ye. B., et al. 1972. Activation method for simultaneous determination of Fe, Zn, and Se in mineralized waters. Izv. Akad. Nauk Uzbek. SSR (fiz.-mat.) (5): 62. (In Russian) (NSA 27: 16977).

Sharitz, R. R., et al. 1975. Uptake of radiocesium from contaminated floodplain sediments by herbaceous plants. Health Physics 28(1): 23-28. (NSA 31: 23091).

Sharma, M. C. 1971. Radioactive dust fallout in Bhavnagar area in May 1966. Sci. Culture (Calcutta) 37(1): 59-60. (NSA 26: 200).

Sharma, M. J. and J. Tunny. 1972. Measuring soil moisture near the surface with the neutron subsurface probe. J. Austral. Inst. Agr. Sci. 38(3): 221-223. (NSA 28: 2750).

Sharma, M. P. and W. H. Vanden Born. 1971. Effect of picloram on $^{14}CO_2$-fixation and translocation of ^{14}C-assimilates in Canada thistle, soybean, and corn. Can. J. Bot. 49(1): 69-74.

Sharma, R. K. 1968. Effects of gamma radiation and its interactions with sodium halides on rainbow trout. Thesis, Utah State Univ., Logan. 107pp. (NSA 23: 24730).

Sharma, V. P. 1972. Effectiveness of chemosterilants and gamma radiation in the sterilization of the tropical house-mosquito *Culex fatigans*. pp. 129-134. In: *Genetics and Our Health*. Indian Council of Medical Research, New Delhi. (NSA 29: 10465).

Sharman, G. B. 1959. Some effects of x-rays on dividing cells in the testis and bone marrow of the marsupial *Potorus tridactylus*. Int. J. Radiat. Biol. 1: 115-130. (NSA 13: 14170).

Sharp, A. J. 1946. Preliminary report on some botanogeographical studies in Mexico and Guatemala. Rev. Soc. Mex. Hist. Nat. 7(1-4): 35-39. (In Spanish. (Contains comment on natural radiation).

Sharp, J. L. 1972. Effects of increasing dosages of gamma irradiation on wingbeat frequencies of *Dacus dorsalis* Hendel males and females at different age levels. Proc. Hawaii. Entomol. Soc. 21(2): 257-262. (NSA 29: 18649).

Sharp, J. V. A. and J. Westphal. 1969. Environmental tritium in conjunctive use studies of continental river systems. Trans. Amer. Nucl. Soc. 12: 501. (NSA 24: 5746).

Shaughnessy, J. 1973. Groundwater and Project Wagonwheel. El Paso Natural Gas Co., U. S. AEC report PNE-WW-37. 22pp. (CONF-730595-1) (NSA 30: 18578).

Shaver, E. L. 1975. Effects of γ rays on the developing embryos of *Calotes virsicolor*. Experientia 31(10): 1213-1215. (NSA 33: 5797).

Shavlovskiy, G. N. 1955. The part of the microorganisms of the rhizosperes in vitamin and amino acid mourishment of plants. In: *Isotopes in Microbiology*. (See Imshenetzkiy, 1955) (AEC-tr-3494, pp. 201-214) (^{32}P).

Shaw, J. 1972. Nuclear siting controversy in the United States of America. J. Inst. Nucl. Eng. 13(3): 85-90. (NSA 26: 58309).

Shaw, J. 1975. Review of the I.A.E.A. symposium on the siting of nuclear facilities. Ann. Nucl. Energy 2(7/8): 579-587. (NSA 33: 19578).

Shchegoleva, G. I. 1966. Disinfestation of dried fruits. Zashchita rasten. (Pt.9): 14-15. (In Russian) (Insects and mites, ^{60}Co source, *Ephestia elutella*, *Plodia interpunctella*, et al.).

Shchepoteva, Ye. S., et al. 1961. The specificity of manifestation of oxygen effect under the action of α-rays. Izv. Akad. Nauk SSSR (biol.) 4: 642-649. (In Russian) (NSA 15: 30480).

Shealy, H. C., A. J. Hazle and T. M. Gerusky. 1974. Is environmental monitoring necessary? pp. 259-264. In: *6th Ann. National Conf. on Radiation Control: New Challenges*. (See U. S. Food and Drug Administration, 1974) (NSA 31: 19564).

Shearer, D. S., Jr. 1962. The leachability of radium-226 from uranium mill waste solids and river sediments. Dissertation, Univ. of Wisconsin, Madison. 176pp. (Dissertation Abstr. 23(4):__; NSA 17: 5653).

Shearer, S. D. (D.S.?). 1964. Leachability of radium-226 from uranium mill solids and river sediments. Health Physics 10(4): 217-228.

Shearin, R. L. and C. R. Phillips. 1971. Contingency response for rapid evaluation of radiological pollution from a space vehicle launch accident. pp. 757-767. In: Rapid Methods for Measuring Radioactivity in the Environment. (See IAEA, 1971) (NSA 26: 28058).

Shearin, R. L., C. R. Porter and S. L. Cummings. 1971. Study of the feasibility of measuring ^{85}Kr through a national surveillance system. pp. 649-653. In: Rapid Methods for Measuring Radioactivity in the Environment. (See IAEA, 1971) (NSA 26: 28070).

Shechmeister, I. L., et al. 1962. The effect of x-irradiation on goldfish. I. The effect of x-irradiation on survival and susceptibility of the goldfish, Carassius auratus, to infection by Aeromonas salmonicida and Gyrodactylus spp. Radiation Research 16(1): 89-97. (NSA 16: 17498).

Shedd, J. C. 1912-13(?). Radioactivity of the mineral springs of Manitou. Proc. Col. Sci. Soc. 10: 233-263.

Sheets, B. H., et al. 1972. Summary of environmental radiation surveillance activities in Arkansas. Radiat. Data Rept. 13(1): 19-26. (NSA 26: 25568).

Sheibley, D. W. 1975. Trace elements by instrumental neutron activation analysis for pollution monitoring, pp. 98-117. In: Trace Elements in Fuel. S. P. Babu, ed. Amer. Chemical Soc., Washington. (NSA 32: 21837).

Shekhanova, I. A. 1955. Use of P^{32} for marking sturgeon fry. Ryb. Khozyaystvo 11: 51-53. (In Russian).

Shekhanova, I. A. 1956. The possibility of assimilation of inorganic phosphorus from the water by fish. Dokl. Akad. Nauk SSSR 106(1): 161-164. (In Russian) (SFA 3: 146).

Shekhanova, I. A. 1959. The Study of Phosphorus Metabolism in Carp and Sturgeon Fry Using Radioactive Phosphorus. Fisheries Press, Moscow. 78pp. (In Russian).

Shekhanova, I. A. 1961. Some problems concerning phosphorus metabolism in fish. Tr. vses. nauchn.-issled. inst. morsk. ryb. khoz. okeanogr. 44: 60-77. (In Russian) (BA 45: 22991; Ref. zh. (biol.) 1963: 13151; SFA 10: 7248) (^{32}P).

Shekhanova, I. A. 1966. Uptake of biogenic elements by developing fish eggs. pp. 84-85. In: Summaries of Reports. All-Union Conf. on the Ecology and Physiology of Fishes. Moscow. (In Russian) (Radionuclides).

Shekhanova, I. A. 1968. Accumulation of ^{90}Sr and ^{90}Y dissolved in the water and its effect on the embryonic development of the loach. Vopr. ikhtiol. 8(4): 689-701. (In Russian) (NSA 25: 46693).

Shekhanova, I. A. 1970. Possibility of using ^{89}Sr for a mass labelling of of young food-fishes. Tr. vses. nauchn.-issled. inst. morsk. ryb. khoz. okeanogr. 69(2): 58: 69. (NSA 25: 21565).

Shekhanova, I. A. 1970. Accumulation and distribution of water soluble radioisotopes in fish eggs during embryogenesis. Tr. vses. nauchn.-issled. inst. morsk. ryb. khoz. okeanogr. 69(2): 19-27. (In Russian) (NSA 25: 21576).

Shekhanova, I. A. and V. L. Pechkurenkov. 1968. Accumulation of strontium-90, dissolved in the water and its effect on the embryonic development of the loach. Vopr. ikhtiol. 8(4):__. (Am. Fish. Soc. transl. 8(4): 551-561).

Shekhanova, I. A. and Ye. A. Voronina. 1971. Effect of radioactive contamination of the water on reproductive capacities of fish. Tr. vses. nauchn.-issled. inst. morsk. ryb. khoz. okeanogr. 79: 95-101. (In Russian; English, French summaries).

Shekhanova, I. A., V. S. Bel'makov and V. I. Lapin. 1969. Effect of strontium-90--yttrium-90 on breeding loach kept in solutions of various activities. II. Vopr. ikhtiol. 9(5): 627-635. (In Russian) (SFA 15: 12198).

Shekhanova, I. A., et al. 1970. Effect of dissolved-in-water radioisotopes on the fish roe during embryonic development. Tr. vses. nauchn.-issled. inst. morsk. ryb. khoz. okeanogr. 69(2): 34-48. (In Russian) (NSA 25: 21563).

Shekhanova, I. A., et al. 1973. Effect of ionizing radiation on the biological condition of fish. pp. 7-24. In: Radioecology of Water Organisms. III. (See Andrushaitis, et al., 1973) (AEC-tr-7592, pp. 2-12) (NSA 30: 24305; 31: 11667).

Shekhtman, Ya. L., I. Breyksh and V. A. Struchkov. 1966. Relative biological efficiency of fast neutrons and gamma rays on loach sperm and DNA of irradiated sperm. Radiobiologiya 6(4): 495-499. (AEC-tr-6773, pp. 12-21) (NSA 21: 22755).

Shelby, C. A. 1963. Radioactive fallout in Sargassum drift on Texas Gulf Coast beaches. Publ. Inst. Marine Sci. (Univ. of Texas) 9: 33-36. (NSA 18: 37382).

Sheldon, G. L. and R. K. Stoms. 1969. Natural background gamma radiation dose rate measurements in Michigan, Colorado and Minnesota. Am. J. Publ. Hlth. 59(1): 102-109.

Sheldon, M. G., et al. 1963. Laboratory studies of pesticide using carbon-14 labeled compounds. Canada geese: feeding test with C-14 tagged 2,4-D. p. 46. In: Pesticide-Wildlife Studies. A Review of Fish and Wildlife Service Investigations during 1961 and 1962. U. S. Fish and Wildlife Service, Washington, Cir. 167.

Sheldon, W. H. and H. Bauer. 1962. The role of predisposing factors in experimental fungus infections. Lab. Invest. 11: 1184-1191. (NSA 17: 38809) (Incl. ionizing radiation).

Shellabarger, C. J., et al. 1956. Some quantitative and qualitative aspects of I^{131} metabolism in turtles. Endocrinology 59: 331-339. (NSA 11: 899).

Shelly, W. J. 1974. Environmental concerns in uranium mining, milling, and conversion. Trans. Amer. Nucl. Soc. 19: 211. (NSA 31: 6354).

Shemakhanova, N. M. 1955. The use of tracer atoms in the study of the mycotrophy of trees. In: Isotopes in Microbiology. (See Imshenetskiy, 1955) (AEC-tr-3494, pp. 215-228).

Shemanchuk, J. A., F. J. H. Fredeen and A. M. Kristjanson. 1955. Studies of flight range and dispersal habits of Aedes flavescens (Muller) (Diptera: Culicidae) tagged with radiophosphorus. Can. Entomol. 87(9): 376-379.

Shemanchuk, J. A., J. W. T. Spinks and F. J. H. Fredeen. 1953. A method of tagging prairie mosquitoes (Diptera: Culicidae) with radiophosphorus. Can. Entomol. 85(7): 269-272.

Shem'i-zade, A. E. 1967. Correlation of the concentration of radioactive natural and fallout aerosols in the lower atmosphere. Atom. energiya 22(1): 58-59. (Sov. Atomic Energy 22(1): 67-68) (NSA 21: 16020).

Shen, C. C., et al. 1967. Metabolism of hempa uniformly labeled with C^{14} in male house flies. J. Econ. Entomol. 60(6): 1623-1630.

Shen, S. K. and A. A. Berryman. 1967. The male reproductive system and spermatogenesis of the European pine shoot moth, Rhyacionia buoliana (Lepidoptera: Olethreutidae), with observations on the effects of gamma irradiation. Ann. Entomol. Soc. Am.60(4): 767-774. (BA 48: 114452).

Shen, S. P., S. A. Korff and H. A. C. Neuberg. 1963. Tritium content of Antarctic snow. Nature 199(4888): 60-61. (NSA 17: 28614).

Shen-Miller, J. 1970. Notes after a conference on Great Lakes research. BioScience 20(24): 1294-1296. (Incl. comments on nuclear power reactors).

Shenberg, C. and M. Boazi. 1975. Rapid qualitative determination of main components in archeological samples by radioisotope-excited x-ray fluorescence analysis. J. Radioanal. Chem. 27(2): 457-463. (NSA 33: 11488).

Shepard, D. C., A. C. Giese and C. L. Brandt. 1956. Action of x-rays on Didinium nasutum. Radiation Research 4(2): 154-157. (NSA 10: 5459).

Shepard, F. P. and J. R. Curray. 1965. Carbon-14 determinations of sea level changes in stable areas. Abstr. Abstr. Intern. Assoc. Quaternary Research. 7th Intern. Congr., Boulder and Denver, Colo.

Shepard, M., R. Noblet and P. E. Hunter. 1972. Biology and flight ability of the house fly following irradiation of the pupae. J. Econ. Entomol. 65(6): 1721-1725. (NSA 27: 15091).

Sherman, W. F., et al. 1971. Dechlorination of DDT in solution by ionizing radiation. Nature 232(5306): 118-119. (NSA 25: 48538).

Sherwood, J. D., D. T. Moore and P. A. Kuroda. 1973. Radioactive cerium isotopes. The fallout from recent French and Chinese nuclear weapons tests. Health Physics 24(5): 491-495. (NSA 28: 2587) (Rain and snow).

Shestakov, S. V., V. M. Glazer and V. D. Zhevner. 1973. UV-radiation-induced incorporation of thymine and thymidine into the DNA of blue-green algae. Biol. Nauki (5): 107-111. (In Russian) (NSA 30: 18896).

Shevchenko, V. A. 1964. Regularity of distribution of Crepis capillaris chromosome breakage in different types of aberrations under the influence of x-rays and ethylenamine. Radiobiologiya 4(6): 870-878. (AEC-tr-6409, pp. 88-97) (NSA 19: 17614).

Shevchenko, V. A. 1965. X-ray influence on survival and mutagenic process in Chlorella. Radiobiologiya 5(2): 253-259. (AEC-tr-6599, pp. 127-133) (NSA 19: 28080).

Shevchenko, V. A. 1966. Analysis of the nature of radiation damage to Chlorella cells. pp. 8-15. In: Effect of Ionizing Radiation on Heredity. D. P. Dubinin, ed. Publ. House Acad. Sci. USSR, Moscow. (In Russian) (NSA 22: 21474).

Shevchenko, V. A. 1967. Natural and induced mutation in Chlorella. Usp. sovrem. genet. (1): 246-278. (NSA 23: 26754).

Shevchenko, V. A., et al. 1967. The vital activity of Chlorella under space flight conditions. Kosm. biol. med. 1: 25-28. (NSA 22: 17057).

Shevchenko, V. A., et al. 1969. Investigation of mutagenesis induced by acute and chronic ionizing irradiation in populations of Chlorella and Chlamydomonas. Genetika 5(9): 61-73. (In Russian) (NSA 24: 675).

Shevchenko, V. A., et al. 1973. Effect of fission products of ^{235}U on the mutation process in Chlorella cultures. pp. 102-108. In: Radioecology of Water Organisms. III. (See Andrushaitis, et al.1973) (AEC-tr-7592, pp. 65-69) (NSA 30: 24377; 31: 11685).

Shevchuk, A. I. 1968. Accumulation and distribution of radioactive elements in natural phytocoenoses of the Carpathians. Ukr. bot. zh. 26(4): 18-24. (In Ukrainian) (NSA 26: 45551).

Shibamoto, T. 1953. Reports by forestry research team. IV. Effect of atomic bomb on red pine and organic matter in soil. Nippon Gakujutsu Kaigi 1: 266.

Shibuya, M. 1969. Application to research and its problems: agriculture. Genshiryoku Kogyo 15(5): 25-30. (In Japanese) (NSA 24: 9737) (Radionuclides).

Shibuya, M. 1970. Trends in the utilization of radiation and radioisotopes. Applications in agriculture. Genshiryoku Kogyo 16(4): 104-106. (In Japanese) (NSA 25: 21658).

Shibuya, M. and Y. Kamemoto. 1963. Application of neutron activation analysis to studies of agricultural damages by water pollution. Mizu-Syori-Gijutsu (Osaka) 4(7): 17-25. (NSA of Japan 3: 00592).

Shidei, T., S. Okada and K. Yoshikawa. 1957. On the contamination of seedlings at nursery by some radiation substances contained in rainfall. Jap. Forestry Soc. J. 39(5): 183-184.

Shieh, H. S. and B. H. Reddy. 1972. Incorporation of $\{^{14}C\}$ amino-acids by a fish pathogen, Aeromonas salmonicida. Int. J. Biochem. 3(14): 167-170. (SFA 18: 15880).

Shields, L. M. 1959. An appraisal of radiation effects on vegetation within the Nevada Test Site. Abstr. Proc. 9th Intern. Bot. Congr. 2(A): 33.

Shields, L. M. 1959. Recovery of vegetation in the vicinity of ground zero sites. Abstr. J. Colo.-Wyo. Acad. Sci. 4(11): 30-31.

Shields, L. M. and W. H. Rickard. 1960. Recovery of vegetation in the vicinity of ground zeros at the Nevada Test Site. Abstr. Bull. Ecol. Soc. Am. 41(4): 119.

Shields, L. M. and W. H. Rickard. 1961. A preliminary evaluation of radiation effects at the Nevada Test Site. pp. 1387-1390. In: Recent Advances in Botany. Univ. of Toronto Press, Toronto.

Shields, L. M. and P. V. Wells. 1962. Effects of nuclear testing on desert vegetation. Science 135(3497): 38-40.

Shields, L. M. and P. V. Wells. 1963. Recovery of vegetation on atomic target areas at the Nevada Test Site. pp. 307-310. In: Radioecology. (See Schultz and Klement, 1963) (NSA 17: 33576).

Shields, L. M., L. W. Durrell and A. H. Sparrow. 1961. Preliminary observations of radiosensitivity of algae and fungi in soils from the Nevada Test Site. Ecology 42(2): 440-441.

Shields, L. M., P. V. Wells and W. H. Rickard. 1963. Vegetational recovery on atomic target areas in Nevada. Ecology 44(4): 697-705. (NSA 20: 8661).

Shigematsu, H. and H. Takeshita. 1968. Effect of gamma-irradiation of late embryo on the growth of silkglands and on silk production in the silkworm, Bombyx mori. Nippon Sanshigaku Zasshi 37(2): 88-94. (In Japanese) (NSA 23: 12188).

Shigematsu, H. and H. Takeshita. 1968. Effects of gamma-irradiations on development of silkglands and silk formation of the silkworm, Bombyx mori L. Norin-sho Sanshi Shikenjo Hokoku 23: 121-148. (In Japanese) (NSA 24: 21277).

Shigematsu, H. and H. Takeshita. 1968. Formation of silk proteins by the silkworm, Bombyx mori, after gamma-ray irradiation in the embryonic stage. J. Insect Physiol. 14: 1013-1024. (NSA 23: 6388).

Shih, C. S. and E. F. Gloyna. 1966. Radioactivity transport in water-transport of ^{85}Sr and ^{137}Cs in an aquatic environment. Univ. of Texas, Austin, report EHE-01-6602; CRWR-10, U. S. AEC report TID-22877. 110pp. (NSA 20: 24833).

Shih, C. S. and E. F. Gloyna. 1967. Radioactivity transport in water: mathematical model for the transport of radionuclides. Univ. of Texas, Austin, report EHE-04-6702, CRWR-18, U. S. AEC report ORO-490-12. 179pp. (NSA 22: 10356).

Shih, C. S. and E. F. Gloyna. 1968. Influence of sediments on the transport of soultes. Univ. of Texas, Austin, U. S. AEC report ORO-490-12a. 41pp. (NSA 22: 37876).

Shih, C. S. and E. F. Gloyna. 1970. Mathematical model for the transport of radionuclides in stream system. pp. 320-341. In: Environmental Surveillance in the Vicinity of Nuclear Facilities. (See Reinig, 1970) (NSA 25: 16083).

Shim, J. W. and J. S. Hyun. 1965. Some effects of x-ray irradiation on the rice stem borer. J. Nucl. Sci. (Seoul) 6: 193-199.

Shim, J. W., S. Y. Choi and J. S. Hyun. 1967. The effects of γ-ray irradiation on the pupal mortality and hatchability of the rice stem borer (Chilo suppressalis Walker). J. Nucl. Sci. (Seoul) 7: 50-55. (In Korean) (NSA 22: 36397).

Shimizu, M. 1973. Biological concentration of radioactive materials in the environment. Gamma Field Symp. 22(11): 662-673. (NSA 30: 15459).

Shimizu, M., et al. 1971. Uptake of ^{58}Co by mussel, Mytilus edulis. J. Radiat. Res. (Japan) 12(1): 17-28. (NSA 26: 4727).

Shimokata, K. and H. Ishihara. 1967. The content of thorium series elements of hot spring waters in Japan. 12. On hot springs in the Kanto and Chubu districts. Onsen Kagaku 17(3,4): 134-139. (In Japanese) (NSA 22: 6447).

Shimotomae, N., et al. 1953. Biological investigation of atomic bomb disaster. 2. Effect on plants. Ch. 2. On the spot investigation in Hiroshima. Nippon Gakujutsu Kaigi 1: 227-235.

Shindel'man, Yu. M. 1969. Acidity of pasture soil and accumulation of strontium-90 in the grass, skeleton of cattle, and in their milk. Latv. Lauksaimn. Akad. Raksti (22): 182-184. (In Russian) (NSA 28: 8188).

Shinholster, D. L. 1974. Effects of x-irradiation and chemotherapy on the host-parasite relationship between Tribolium castaneum (Herbst) and two protozoan parasites: Nosema whitei Weiser and Adelina tribolii Hesse. Dissertation, Cornell Univ., Ithaca, N. Y. 156pp. (Univ. Microfilms 75-4265) (NSA 31: 23041).

Shinn, A. F. 1964. Radioecology of mud-dauber wasps. Abstr. Bull. Entomol. Soc. Am. 10(3): 172. (Sceliphron).

Shinn, A. F. 1967. Freshwater fish in a nuclear disaster. pp. 51-55. In: Proc. Commercial Fish Farming Conf. (Texas A&M Univ., February 1-2, 1967, College Station, Texas). (Also: Oak Ridge National Lab., U. S. AEC report ORNL-P-2954) (CONF-670208) (NSA 21: 24413; 22: 12861).

Shinn, A. F. 1969. Honey bee irradiation studies. pp. 47-49. In: Progress Report in Postattack Ecology. (See Auerbach, 1969).

Shinn, J. H. and L. R. Anspaugh. 1975. Resuspension: new results in predicting the vertical dust flux. pp. 207-215. In: Radioecology of Plutonium and Other Transuranics in Desert Environments. (See White and Dunaway, 1975) (NSA 33: 617).

Shipler, D. B. and R. F. Foster. 1973. National Environmental Policy Act and nuclear power plant licensing. Battelle Northwest Lab., U. S. AEC report BNWL-SA-4670. 26pp. (CONF-730447-1) (NSA 28: 20427).

Shipp, E. 1966. The effect of rearing medium on the susceptibility of Tribolium confusum Duv. and Sitophilus granarius (L.) to gamma radiation. pp. 97-105. In: The Entomology of Radiation Disinfestation of Grain. P. B. Cornwell, ed. Pergamon Press, Oxford.

Shipp, E. 1966. Susceptibility of Australian strains of Sitophilus and Tribolium species to gamma radiation. pp. 131-141. In: The Entomology of Radiation Disinfestation of Grain. P. B. Cornwell, ed. Pergamon Press, Oxford.

Shipp, E. and A. W. Osborn. 1968. Irradiation of Queensland fruit fly pupae to meet quarantine requirements. J. Econ. Entomol. 61(6): 1721-1726.

Shipp, E., A. W. Osborn and P. B. Hutchinson. 1966. Radiation sterilization of sugar cane leafhoppers of the family Delphacidae. Nature 211: 98-99. (NSA 20: 37146).

Shipp, O. E. and J. R. Brazzel. 1968. Distribution of C^{14}-labelled dieldrin in dieldrin-resistant and susceptible bollweevils, Anthonomus grandis. J. Econ. Entomol. 57(1): 174-175.

Shirai, K. and K. Mori. 1958. Studies on the detection of Cd^{113m} and Cd^{115m} in radioactive contaminated fish and their movement in the body of the fish. Paper A-8, pp. 1541-1559. In: Proc. 2nd Japan Conf. on Radioisotopes. (See Japan Radioisotopes Assoc., 1958) (AEC-tr-4482) (NSA 15: 25792).

Shirai, K. and K. Mori. 1959. On the combination of cadmium and protein in fish blood. In: 3rd Japan Conf. on Radioisotopes. Japan Atomic Industrial Forum, Japan Radioisotopes Assoc., 59/P-78.

Shiraishi, Y. and H. Okabayashi. 1967. Radioactive contamination of fruit trees by fission products. 1. Distribution of fallout ^{137}Cs in the orchard (in 1965). Soil Sci. Plant Nutr. (Tokyo) 38: 333-338. (In Japanese) (NSA 23: 16295).

Shirshova, R. A. 1961. The effect and aftereffect of the radioactive isotope strontium-90 on plants. Dokl. Akad. Nauk SSSR 138(4): 373-375. (Am. Inst. Biol. Sci. transl.).

Shishido, T. and J. Fukami. 1963. Studies on the selective toxicities of organic phosphorus insecticides. II. The degradation of ethyl Parathion, methyl Parathion, methyl Paraoxon and Sumithion in mammal, insect and plant. Botyu-kagaku 28(3): 69-76. (In Japanese) (^{32}P).

Shivaji, S. 1975. X-ray induced changes in the levels of glucose, fructose, and glycogen in Periplaneta americana. Current Sci. (India) 43(23): 757-758. (NSA 31: 8930).

Shivaji, S. and S. C. Rastogi. 1974. X-ray induced changes in the activities of acid and alkaline phosphatase in Periplaneta americana. Current Sci. (India) 43(23): 757-758. (NSA 31: 8930).

Shiyan, P. N. and A. I. Merezhko. 1972. Effect of hydrogen ion concentration on photosynthesis and radiocarbon metabolism in water plants. Gidrobiol. zh. 8(2): 41-__. (Hydrobiol. J. 8(2): 23-29).

Shleien, B. 1970. An estimate of radiation doses received by individuals living in the vicinity of a nuclear fuel reprocessing plant in 1968. U. S. PHS report BRH/NERHL 70-1. x, 19pp. (Incl. fish and deer).

Shleien, B., J. A. Cochran and P. J. Magno. 1970. Strontium-90, strontium-89, plutonium-239, and plutonium-238 concentrations in ground level air, 1964-1969. Environ. Sci. Tech. 4(7): 598-602.

Shlyk, A. A. 1955. Experimental peculiarities of the method of tracer atoms. In: Isotopes in Microbiology. (See Imshenetzkiy, 1955) (AEC-tr-3494, pp. 254-259).

Shlyumberger, V. 1966. Determination of the salt tolerance of pike (Esox lucius) by means of Na^{22}. pp. 161-167. In: Ecology of Aquatic Organisms. Acad. Sci. USSR, Moscow. (In Russian) (Ref. zh. (biol.) 1967: 11118K).

Shmalganzen, I. I. 1958. The ways in which radiation damage appears in early development in fish. Dokl. Akad. Nauk 119(1): __. (Doklady (biol. sci.) 119: 134-137).

Shmelev, I. P. 1959. Methods and apparatus radiotopography. Biofizika 4(1): 108-116. (Biophysics USSR 4(1): 108-118) (NSA 14: 2275).

Shmuklarsky, M. J. 1972. Environmental radioactivity in Illinois, 1970. Radiat. Data Rept. 13(11): 589-618. (NSA 27: 17251).

Shoji, S., Y. Ando and J. Masui. 1974. ^{14}C-age of peats in Tanemori, Akita Prefecture. ^{14}C-age of the Quaternary deposits in Japan (101). Chikyu Kagaku 28(5): 197. (In Japanese) (NSA 33: 3014).

Shoji, S., I. Yamada and J. Masui. 1974. ^{14}C-age of humus layer in volcanic ash soil in Mamiana, Utsunomiya City and Shimokusawa City. ^{14}C-age of Japanese Quaternary strata (99). Chikyu Kagaku 28(4): 149-150. (In Japanese) (NSA 32: 6450).

Shoop, C. R. 1971. A method for short-term marking of amphibians with 24-sodium. Copeia (2): 371.

Shoop, C. R. 1975. Effects of ionizing radiation on salamander orientation. Final report. Univ. of Rhode Island, U. S. AEC report COO-3486-3. 36pp. (NSA 33: 20525).

Shoop, C. R. and T. L. Doty. 1972. Migratory orientation by marbled salamanders (Ambystoma opacum) near a breeding area. Behavioral Biol. 7(1): 131-136. (^{182}Ta).

Shoop, R. C. (C.R.?). 1970. Effects of radiation on migratory orientation of salamanders. Final report. Wellesley Coll., U. S. AEC report NYO-3554-8. 17pp. (NSA 25: 4475).

Shore, B. W. 1969. The fate and importance of radionuclides produced in nuclear events. pp. 595-651. In: Proc. Symp. on Public Health Aspects of Peaceful Uses of Nuclear Explosives. (See U. S. PHS, 1969).

Shorey, H. H., et al. 1958. The residual effects of insecticides applied to meadow and pasture sod for control of the European chafer. J. Econ. Entomol. 51(6): 765-767.

Short, D. J. and D. P. Woodnott. (Eds.). 1969. The Institute of Animal Technicians (I.A.T.) Manual of Laboratory Animal Practice and Techniques. Crosby Lockwood, London. 2nd ed., 471pp. (NSA 24: 2532) (Incl. Ch. on radioisotopes).

Short, Z. F., et al. 1969. The uptake of iodine-131 by the biota of Fern Lake, Washington, in a laboratory and a field experiment. Ecology 50(6): 979-989. (NSA 24: 23155).

Short, Z. F., et al. 1973. Uptake of molybdenum, marked with ^{99}Mo, by the biota of Fern Lake, Washington, in a laboratory and a field experiment. pp. 474-485. In: Radionuclides in Ecosystems. (See Nelson, 1973) (NSA 25: 44287).

Shotton, F. W. 1972. Example of hand-water error in radiocarbon dating of vegetable matter. Nature 240(5382): 460-461. (NSA 27: 17014).

Shoukry, A., A. A. Abdel-Malek and N. F. Shehata. 1974. Anatomical effects of gamma irradiation on the female reproductive organs of the Mediterranean fruit fly Ceratitis capitata Wied. pp. 153-160. In: Isotopes in Pesticides and Pest Control. (See Nehme and Hassan, 1974) (NSA 33: 7564).

Shreve, D. C. and A. J. Brown. 1974. Development and field testing of a remote radioisotopic snow gage. pp. 661-675. In: Advanced Concepts and Techniques in the Study of Snow and Ice Resources. (See Santeford and Smith, 1974) (NSA 33: 7314).

Shrivastava, S. C. and A. G. Richards. 1965. An autoradiographic study of the relation between hemocytes and connective tissue in the wax moth Galleria mellonella L. Biol. Bull. Mar. Biol. Lab. (Woods Hole) 128: 337-345. (^{3}H).

Shrivastava, S. P., M. Tsukamoto and J. E. Casida. 1969. Oxidative metabolism of C^{14}-labeled Baygon by living house flies and by house fly enzyme preparations. J. Econ. Entomol. 62(2): 483-498.

Shubko, V. M. and B. V. Kurchatov. 1961. Long-lived fission products fallout o the territory of the USSR 1959-1960. State Committee on the Use of Atomic Energy USSR. Atomizdat, Moscow. 9pp. UN Document A/AC.82/G/L.706. (In Russian).

Shubko, V. M. and A. M. Yeremicheva. 1963. Strontium-90 fallout over the USSR territory in the fourth quarter of 1961 and the first half of 1962. State Committee on the Use of Atomic Energy USSR. Atomizdat, Moscow. 6pp. UN Document A/AC.82/G/L.916. 6pp. (In Russian).

Shull, R. D. 1968. Radioactive transport in water: simulation of sustained releases to selected river environments. Thesis, Univ. of Texas, Austin. 254pp. (NSA 23: 11649).

Shull, R. D. and E. F. Gloyna. 1968. Radioactivity transport in water: simulation of sustained releases to selected river environments. Univ. of Texas, Austin, report EHE-04-6801, CRWR-25, U. S. AEC report ORO-490-15. 252pp. (NSA 22: 35924).

Shum, Y. S., W. D. Loveland and E. W. Hewson. 1974. Use of artificial activable trace elements to monitor pollutant source strengths and dispersal patterns. Oregon State Univ., Corvallis, U. S. ERDA report RLO-2227-T7-28. 25pp. (NSA 31: 19608).

Shumakov, Ye. M. 1968. Difficulties and successes in the mass rearing of insects in the laboratory, and the possibility of autocidal control of some harmful species. pp. 219-232. In: Isotopes and Radiation in Entomology. (See IAEA, 1968) (In Russian, English summary).

Shuping, R. E., C. R. Phillips and A. A. Moghissi. 1970. ^{85}Kr levels in the environment determined from dated krypton gas samples. Radiol. Hlth. Data 11: 671-672. (NSA 25: 21641).

Shura-Bura, B. L. 1952. Experimental study of housefly migration by means of radioactive tracers. Zool. zh. 31(3): 410-412. (In Russian).

Shura-Bura, B. L. 1955. Use of tagged atoms in studying fly migration from a dump. Gigiyena sanitariya 20(9): 12-15. (In Russian).

Shura-Bura, B. L. 1957. Use of radioisotopes for the study of the role of flies in epidemiology. J. Hyg. Epidem. Microbiol. (Prague) 1(3): 249-255. (In French).

Shura-Bura, B. L. 1957. Use of tagged atoms in entomology. Usp. sovrem. biol. 44(1): 103-120. (In Russian).

Shura-Bura, B. L. and V. P. Kharlamov. 1960. Autoradiography as a method of identifying labelled rodents and their ectoparasites when studying migration problems. Zool. zh. 40(2): 258-263. (In Russian).

Shura-Bura, B. L., O. I. Sukhomlinova and B. I. Isarova. 1962. Use of radioactive tracers in studying the ability of synanthropic flies to fly across water barriers. Entomol. obozren. 41(1): 99-108. (JPRS-16616).

Shura-Bura, B. L., R. A. Tararin and N. S. Klyuchnik. 1961. Experiment on the radiomarking of gray rats for a study of their migration in a port. p. 34. In: Materials of the Symp. on the Use of Biophysics in the Field of Plant Protection. Leningrad. (In Russian) (BA 43: 4646; Ref. zh. (biol.) 1962: 18I266; WR 113: 28).

Shura-Bura, B. L., R. A. Tararin and B. K. Melnikov. 1960. On the methods of radioactive labelling the rats Rattus norvegicus for the purpose of studying migration problems. Zool. zh. 39(11): 1700-1706. (In Russian).

Shura-Bura, B. L., et al. 1956. Routes of fly dispersion from places of mass breeding in Leningrad. Entomol. obozren. 35(2): 334-336. (In Russian).

Shura-Bura, B. L., et al. 1958. On the character of the dispersion of synanthropic flies beginning from their release. Entomol. obozren. 37(2): 336-346. (Entomol. Rev. USSR 37: 282-290).

Shurashkevich, O. N. 1966. Effect of ionizing radiation on the metabolic processes of Chlorella. Ukr. botan. zh. 23: 79-81. (In Russian) (NSA 21: 45847).

Shure, D. J. 1968. Radionuclide analysis of the trophic transfer from ragweed (Ambrosia artemisiifolia). Abstr. Bull. Ecol. Soc. Am. 49(3): 106. (^{32}P).

Shure, D. J. 1969. Insecticides effects on trophic relationships in an old field ecosystem. Abstr. Bull. Ecol. Soc. Am. 50(2): 67. (^{32}P).

Shure, D. J. 1970. Radionuclide analysis of the trophic transfer from wild radish (Raphanus raphanistrum). Abstr. Bull. Ecol. Soc. Am. 51(2): 28.

Shure, D. J. and M. R. Gottschalk. 1975. Cesium-137 dynamics within a reactor effluent stream in South Carolina. Emory Univ., Atlanta, Ga., U. S. AEC report ORO-2412-64. 27pp. (NSA 32: 27928).

Shure, D. J. and P. G. Pearson. 1969. Distribution of ^{32}P in Ambrosia artemisiifolia; its implications for trophic transfer studies. Ecology 50(4): 724-726. (Abstr. Bull. Ecol. Soc. Am. 50(2): 72-73).

Shure, D. J. and H. L. Ragsdale. 1973. Cesium-137 dynamics within a reactor effluent stream in South Carolina. Abstr. Bull. Ecol. Soc. Am. 54(1): 24.

Shuvalova, V. S. 1957. Experiment with radioactive calcium (^{45}Ca) in the study of inorganic fertilization of ponds. pp. 33-40. In: Trans. 6th Conf. on Problems in the Biology of Inland Waters. Ye. N. Pavlovskiy, et al., eds. Publ. House Acad. Sci. USSR, Moscow-Leningrad. (AEC-tr-6880; TT-50492; pp. 33-40) (BA 37: 21387; SFA 7: 5137) (Ref. has been listed as V. S. Shuvalov, 1959; Ref. zh. (biol.) 1961: 12D676).

Shvarts, S. S. (Ed.). 1971. Problems of the Radioecology of Aquatic Organisms. (Symp., Miass, USSR, July 10-15, 1968). Tr. inst. ekol. rast. zhivot. (78): 1-240. (In Russian).

Shvedov, V. P. (Ed.). 1959. Evaluation of the Contamination of the Biosphere by Products of Nuclear Tests. (Collected Papers). Acad. Sci. USSR, Moscow. 253pp. UN Document A/AC.82/G/L.323. (AEC-tr-4599).

Shvedov, V. P. and L. I. Gedeonov. 1962. Contamination of the biosphere in the environs of Leningrad by nuclear explosion products. pp. 45-66. In: What Russian Scientists Say about Fallout. A. V. Lebedinskiy, ed. Collier Books, New York.

Shvedov, V. P. and A. M. Maksimova. 1962. Content of certain long-lived fission products in surface waters. pp. 246-250. In: Radioactive Contamination of the External Environment. (See Shvedov and Shirokov, 1962) (AEC-tr-6049).

Shvedov, V. P. and S. A. Patin. 1968. Radioactivity in the Oceans and Seas. Atomizdat, Moscow. 287pp. (In Russian) (NSA 22: 51122).

Shvedov, V. P. and S. I. Shirokov. (Eds.). 1962. Radioactive Contamination of the External Environment. Gosatomizdat, Moscow. 275pp. (AEC-tr-6049) (NSA 17: 12516; 18: 8518).

Shvedov, V. P. and V. K. Vinogradova. 1962. The content, distribution, and migration of long-lived fission products in soils. pp. 225-232. In: Radioactive Contamination of the External Environment. (See Shvedov and Shirokov, 1962) (AEC-tr-6049) (NSA 18: 8524).

Shvedov, V. P., Ye. P. Ankudinov and B. G. Bunin. 1962. Total activity of fission products in surface waters. pp. 233-246. In: Radioactive Contamination of the External Environment. (See Shvedov and Shirokov, 1962) (AEC-tr-6049).

Shvedov, V. P., Z. G. Gritchenko and L. I. Gedeonov. 1962. The concentration of Be^7 in the ground layer of air and in atmospheric precipitation. Atom. energiya 12(1): 64-66. (Sov. Atomic Energy 12(1): 68-69) (NSA 16: 10280).

Shvedov, V. P., G. G. Polikarpov and I. A. Sokolova. 1973. Accumulation of Sr^{90}, Ca and Sr by fish in seas and oceans. pp. 263-268. In: Radioecology of Water Organisms. II. (See Andrushaitis, et al., 1973) (AEC-tr-7606, pp. 157-159) (NSA 30: 21145).

Shvedov, V. P., S. P. Rosyanov and Z. G. Gritchenko. 1963. The contamination of vegetation by strontium-90 and cesium-137 in 1961. State Committee on the Use of Atomic Energy USSR. Atomizdat, Moscow. 7pp. UN Document A/AC.82/G/L.913. (AEC-tr-6448) (NSA 18: 33235).

Shvedov, V. P., G. V. Yakaleva and L. M. Ivanova. 1959. Measurements of the content of radioactive products of artificial origin in the water of the Neva River during 1955-1958. pp. 62-70. In: Evaluation of the Contamination of the Biosphere by Products of Nuclear Tests. (Collected Papers). (See Shvedov, 1959) (AEC-tr-4599).

Shvedov, V. P., et al. 1958. Radioactive fallout in the Leningrad area. Atom. energiya 5(5): 577-582. (Sov. Atomic Energy 5(5): 1500-1507).

Shvedov, V. P., et al. 1959. Determination of radioactive strontium in water samples. Radiokhimiya 1(5): 616-618.

Shvedov, V. P., et al. 1959. Strontium-90 fallout from the atmosphere in the Leningrad area. Atom. energiya 7(5): 479. (Sov. Atomic Energy 7(5): 948-949).

Shvedov, V. P., et al. 1961. Investigation of radioactive pollution in the Black Sea area in 1959. Moscow. 28pp. UN Document A/AC.82/G/L.703. (In Russian) (NSA 17: 1631).

Shvedov, V. P., et al. 1963. Study of the distribution of the isotopes of cerium, ruthenium, zirconium and niobium in samples in the breakdown of mineral acids of radioactive fallout and atmospheric precipitation. State Committee on the Use of Atomic Energy USSR. Atomizdat, Moscow. 44pp. UN Document A/AC.82/G/L.898. (AEC-tr-6446) (NSA 18: 37372).

Shvedov, V. P., et al. 1963. The strontium content of the Black Sea in 1959-1961. State Committee on the Use of Atomic Energy USSR. Atomizdat, Moscow. 8pp. UN Document A/AC.82/G/L.909. (In Russian) (NP-14796) (NSA 19: 13630).

Shvedov, V. P., et al. 1970. State of ^{137}Cs and ^{90}Sr in sewage solutions in the presence of synthetic detergents and soaps. Radiokhimiya 12(5): 784-786. (Sov. Radiochem. 12(5): 754-756) (NSA 26: 1840).

Shveikina, R. V. 1967. Absorbtiometric determination of iodine in water. Izv. vyssh. ucheb. zaved. khim. techn. 10: 1288-1289. (In Russian) (NSA 23: 24080).

Shvydkiy, V. V., I. M. Parkhomenko and Ye. V. Burlakova. 1974. Radioprotective action of natural substances contained in the diet of radioresistant rodents. Radiobiologiya 14(2): 247-250. (AEC-tr-7571, pp. 111-115) (NSA 30: 32375) (Wild gerbils).

Shvydko, N. S., et al. 1969. Measurements of concentrations of strontium-90 and cesium-137 in various food products during the period 1957 to 1967. State Committee on the Use of Atomic Energy USSR. Atomizdat, Moscow. (In Russian).

Shyamala, M. B. and J. V. Bhat. 1965. Essential amino acid requirements of the silkworm, Bombyx mori. Indian J. Biochem. 2(3): 210-202. (^{14}C).

Sidorenko, G. I., M. A. Pinigin and V. T. Savenkova. 1963. The use of the combined effect of gamma-radiation and chlorine for sterilization of water. Gigiyena sanitariya 28(9): 97-98. (JPRS-21975, pp. 10-12) (NSA 18: 5697).

Sidorov, B. N. and N. N. Sokolov. 1964. Radiation analysis of the structure and reproduction of chromosomes. Radiobiologiya 4(6): 828-835. (AEC-tr-6409, pp. 35-44) (NSA 19: 17610) (Crepis capillaris).

Sidorov, V. Ye. and I. M. Grokhovskaya. 1964. The effect of x-rays on adult Hyalomma asiaticum. I. Med. parazitol. parazitarn. bolezni (5): 560-563. (In Russian) (NSA 19: 36039).

Sidorova, V. F. 1951. Regeneration of x-ray-irradiated extremity of axolotl on transplantation of unspecific tissues. Dokl. Akad. Nauk SSSR 81: 297-299. (In Russian) (NSA 6: 5943).

Sidorova, V. F. 1957. On the causes of regeneration ability loss in the axolotl extremity after x-rays. Byull. eksp. biol. med. 43(2): 215-220. (Bull. Exptl. Biol. Med. 43: 84-89) (NSA 12: 9582).

Sidorova, V. R. 1962. Effect of Co^{60} gamma rays on the small cutworm. pp. 154-158. In: Problems of Biology and Regional Medicine. Biol. Sci. Sect., Acad. Sci., Uzbek. SSR, Tashkent. (In Russian) (Ref. zh. (biol.) 1962: 13E182).

Sidwell, R. W., B. D. Thorpe and L. P. Gebhardt. 1964. Studies of latent Q fever infections. I. Effects of whole-body x-irradiation upon latently infected guinea pigs, white mice, and deer mice. Am. J. Hyg. 79: 113-124. (NSA 19: 24269).

Sievert, R. M. and A. Forssberg. 1931, 1936. The time factor in the biological action of x-rays. I. Investigations on Drosophila eggs. II. Investigations at very short irradiation times. Acta Radiologica 16(6): 535-551; 17(3): 290-298.

Sigler, W. F. 1965. Some effects of accumulated radium on the productivity of algae. Final progress report. Utah State Univ., U. S. AEC report TID-22322. 31pp. (NSA 20: 26).

Sigler, W. F. and D. W. Linn. 1967. The effects of radium on goldfish blood. Proc. Utah Acad. Sci. Arts Lett. 44(2): 600-608.

Sikov, M. R. and D. D. Mahlum. (Eds.). 1969. Radiation Biology of the Fetal and Juvenile Mammal. (Proc. 9th Ann. Symposium, Richland, Wash., May 5-8, 1969). U. S. AEC Symp. Ser. No. 17, report CONF-690501. 1026pp. (NSA 24: 23310).

Silini, G., P. Metalli and G. Vulpis. 1972. Radiotoxicity of tritium in mammals. Commission of European Communities, Brussels, Belgium, report EUR-5033. 36pp. (NSA 33: 29773).

Silker, W. B. 1957. A continuous river sampling system and methods of radiochemical analyses for strontium-89, -90, barium-140, and phosphorus-32. Hanford Atomic Products Operation, U. S. AEC report HW-47578. 10pp. (NSA 12: 12145).

Silker, W. B. 1964. Variations in elemental concentrations in the Columbia River. Limnol. Oceanogr. 9(4): 540-545. (Incl. radionuclides).

Silker, W. B., R. W. Perkins and H. G. Rieck. 1971. Sampler for concentrating radionuclides from natural waters. Ocean Engrg. 2: 49-55. (NSA 25: 34704).

Sill, C. W. 1975. Problems in measuring plutonium in the environment. Health Physics 29(4): 619-626. (NSA 33: 9280).

Sill, C. W. and J. K. Flygare, Jr. 1960. Iodine monitoring at the National Reactor Testing Station. Health Physics 2(3): 261-268.

Sill, C. W. and R. L. Williams. 1969. Radiochemical determination of uranium and the transuranium elements in process solutions and environmental samples. Anal. Chem. 41: 1624-1632. (NSA 23: 47396).

Sill, C. W. and R. L. Williams. 1971. Rapid identification and determination of alpha emitters in environmental samples. pp. 201-211. In: Rapid Methods for Measuring Radioactivity in the Environment. (See IAEA, 1971) (NSA 26: 27809).

Sill, C. W., K. W. Puphal and F. D. Hindman. 1974. Simultaneous determination of alpha-emitting nuclides of radium through californium in soil. Anal. Chem. 46(1): 113-118. (NSA 29: 15636).

Silow, R. A. 1957. The application of nuclear energy to agriculture and biology. pp. 401-411. In: *Proc. Scientific Congress. Vol. 1. Nuclear Session*. (NSA 13: 5268).

Silver, G. L. 1971. Plutonium in natural waters. Mound Lab., Miamisburg, O., U. S. AEC report MLM-1870. 15pp. (NSA 26: 8786).

Silver, W. J., et al. 1974. Environmental monitoring at the Lawrence Livermore Laboratory. 1974 annual report. Lawrence Livermore Lab., U. S. AEC report UCRL-50027-74. 53pp. (NSA 32: 758).

Silver, W. J., et al. 1974. Environmental levels of radioactivity in the vicinity of the Lawrence Livermore Laboratory. Lawrence Livermore Lab., U. S. AEC report UCRL-51547. 49pp. (NSA 29: 29332).

Silverman, M. S., et al. 1969. Cellular response of mice to infection with *Pasturella tularensis* (live vaccine strain) following continuous exposure to low-dose gamma radiation. J. Infect. Dis. 120: 366-371. (NSA 24: 19307).

Silvestri, M. 1970. Nuclear energy in the present and in the future. Energia nucleare 17: 523-532. (In Italian) (NSA 24: 49999).

Simak, M. and A. Gustafsson. 1953. X-ray photography and sensitivity in forest tree species. Hereditas 39: 458-468. (PBA 24: 655).

Simak, M. and A. Gustafsson. 1954. X-ray photography and sensitivity in forest tree species. pp. 518-525. In: *11th Congr. Intern. Union of Forest Research Organizations*. (Rome, 1953).

Simak, M., K. Ohba and B. Suszka. 1961. Effect of x-irradiation on seeds of different weight from individual trees of Scots pine (*Pinus silvestris*). Botan. Notiser 114(3): 300-312.

Simakov, Yu. G., L. M. Poluyektova and V. V. Popov. 1969. Influence of large doses of radiation on the lipid content in the chrystalline lens of the grass frog. Radiobiologiya 9(5): 784-785. (AEC-tr-7136, pp. 197-200) (NSA 24: 11949).

Simavonyan, V. G. and E. I. Dzhikidze. 1968. Roentgenological changes in the skeleton of sacred baboons in prolonged gamma irradiation. Med. radiologiya 13(7): 36-40. (In Russian) (NSA 23: 2942).

Simeon, C. and J. Bonnefoy-Claudet. 1968. Organization of the agricultural ecological control in the Rhone Valley in the vicinity of the Pierrelatte Plant. pp. 477-482. In: *International Congress of Radioprotection of the Environment*. (See Societé Francaise de Radioprotection, 1968) (In French)

Simeon, C. and J. Bonnefoy-Claudet. 1968. Control of liquid effluent releases and diagnosis of chronic pollution by means of hydrobiological methods. pp. 609-615. In: *Intern. Congr. of Radioprotection of the Environment*. (See Societé Francaise de Radioprotection, 1968) (In French).

Simmons, G. M., Jr. and J. R. Reed, Jr. 1973. Some ecological considerations in locating a nuclear-powered electrical generating facility on the North Anna River, Virginia. pp. 497-510. In: *Radionuclides in Ecosystems*. (See Nelson, 1973) (NSA 28: 27333).

Simmons, M. J. 1974. X-ray induced heterozygous viability loads in *Drosophila melanogaster*. Dissertation, Univ. of Wisconsin, Madison. 84pp. (Univ. Microfilms 74-19) (NSA 31: 8933).

Simmons, M. L. and J. Franklin. 1967. Acute radiation mortality in mice inoculated with pigmented or non-pigmented *Pseudomonas*. Res. Vet. Sci. 8: 71-73. (NSA 21: 22742).

Simón F., J. E. 1968. Large-scale breeding of *Dysdercus peruvianus* G. and its sterilization by gamma rays. pp. 287-300. In: *Isotopes and Radiation in Entomology*. (See IAEA, 1968) (In Spanish) (NSA 24: 32323).

Simón F., J. E. 1969. Present stage of research into the eradication of the Mediterranean and South American fruit flies and the cotton stainer in Peru by the sterile-male technique. pp. 115-121. In: *Sterile-Male Technique for Eradication or Control of Harmful Insects*. (See IAEA, 1969) (NSA 24: 32307).

Simón F., J. E. 1974. Present status of the Peruvian project on Mediterranean fruit fly control by sterile male technique. pp. 95-101. In: *Sterile-Insect Technique and Its Field Applications*. (See IAEA, 1974) (NSA 31: 8944).

Simón F., J. E. and B. G. García. 1963. Effects of gamma-radiation on the cotton stainer. Boln. Soc. Ent. Agr. Peru 3.

Simonis, W. and H. B. Seuberling. 1973. Sensitivity of chlorophyll formation and of increase of NADP-dependent GPD-activity in greening *Euglena gracilis* to x-rays and to other inhibitors of protein synthesis. Atmos. Envir. 13(5): 297-300. (NSA 29: 18575).

Simonis, W. and D. Urbach. 1966. On the action of x rays on photosynthesis and phosphate metabolism of unicellular green algae (*Ankistrodesmus braunii*). Z. Pflanzenphysiol. 54: 321-332. (In German) (NSA 21: 3388).

Simpson, E. S., W. A. Beetem and F. H. Ruggles. 1958. Radiotracer experiments in the Mohawk River, New York, to study sewage path and dilution. Trans. Am. Geophys. Un. 39: 427-433.

Simpson, H. J. 1970. Tritium in Crater Lake, Oregon. J. Geophys. Res. 75 (27): 5195-5207.

Sims, M. A. 1975. Electron microscope autoradiographic investigation of the accumulation of zinc-65 by a species of *Eutreptia*. Experientia 31(4): 426-427. (NSA 32: 1217).

Sinclair, W. K., S. E. Gunter and A. Cole. 1959. The relative biological effectiveness of 200 KVP x-rays, cobalt-60 gamma-rays, and 22-MEVP x-rays, determined from dose-survival curve of *Saccharomyces cerevisiae*. Radiation Research 10: 418-432.

Singer, G., J. M. Rozental and D. M. Norris. 1974. ^{35}S incorporated in neural membrane protein involved in Periplaneta americana chemoreception. Experientia 30(12): 1384-1385. (NSA 31: 26208).

Singh, B. R. and Y. Kanehiro. 1970. Effects of gamma irradiation on the available nitrogen status of soils. J. Sci. Food Agr. 21(2): 61-64.

Singh, B. R., A. S. Agarwal and Y. Kanehiro. 1969. Recovery of added nitrogen from gamma-irradiated soils. Soil Science 108(2): 85-88.

Singh, H. 1971. Acute low-level gamma ray effects on the reproduction and adult survival of Rhyzopertha dominica (Coleoptera: Bostrichidae). J. Ga. Entomol. Soc. 6(4): 258-264. (NSA 26: 25876).

Singh, H. and J. N. Liles. 1972. Effects of gamma rays on the acute lethality and reproductive potential of lesser grain borer adults. J. Econ. Entomol. 65(3): 656-659. (NSA 26: 43352).

Singh, H. and J. N. Liles. 1972. Sex differences in the radiosensitivity of Rhyzopertha dominica adults. J. Econ. Entomol. 65(4): 1213. (NSA 26: 50961).

Singh, H. and J. N. Liles. 1972. Effects of irradiated food on the adult survival and reproduction of Rhyzopertha dominica (F.) (Coleoptera, Bosthrichidae). J. Stored Prod. Res. 8(2): 155-157. (NSA 26: 56382).

Singh, H. and J. N. Liles. 1972. Temperature modifications of the post irradiation survival of the lesser grain borer adults. Environ. Entomol. 1(4): 395-___.

Singh, J. and B. S. Gill. 1975. Effect of gamma irradiation on oocysts of Eimeria necatrix. Parasitology 71(1): 117-124. (NSA 33: 9592).

Singh, J. S. and D. C. Coleman. 1973. Carbon-14 transfers and evaluation of functional root biomass in a shortgrass prairie ecosystem. Abstr. Bull. Ecol. Soc. Am. 54(1): 17.

Singh, M. S. and R. Lawrence. 1972. Production and release of ^{13}N and ^{18}F from power reactors. Trans. Amer. Nucl. Soc. 15(1): 80-81. (NSA 26: 42964).

Singh, T. P. 1968. Seasonal changes in radioiodine uptake and epithelial cell height of the thyroid gland in the freshwater teleosts Esomus danricus (Ham.) and Mystus vittatus (Bloch) under varying conditions of illumination. Z. Zellforsch. Mikroskop. Anat. 87: 422-428. (NSA 22: 43248).

Singh, T. P. 1968. Thyroidal ^{131}I uptake and TSH potency of the pituitary in response to graded doses of methyltestosterone in Mystus vittatus (Bloch). Gen. Comp. Endocrinol. 11: 1-4. (NSA 23: 24564).

Singh, T. P. 1969. Thyroidal radioiodine uptake and thyrotropic potency of the pituitary in a freshwater catfish, Mystus vittatus (Bloch), in response to L-thyroxine, antithyroid drugs, and heavy doses of ^{131}I. Z. Zellforsch. Mikroskop. Anat. 94: 172-180. (NSA 23: 50288).

Singh, T. P. 1970. X-irradiation and thyroid activity in a teleost, Mystus vittatus (Bloch). Experientia 26: 1261-1262. (NSA 25: 16332).

Singh, V. P. 1974. Effect of gamma and UV radiation and KSCN and LiCl on the rhizoid differentiation in the golden fern, Onychium auratum Kaulf. Indian J. Exp. Biol. 12(1): 63-66. (NSA 30: 27052).

Sinha, M. K. 1972. Organic matter transformations in soil. II. Nature of carbohydrates in soils incubated with C^{14}-labelled oat roots under aerobic and anaerobic conditions. Plant Soil 36(2): 295-299.

Sinha, S. 1969. After-effects of irradiation. pp. 32-37. In: Proc. Symp. on Radiation and Radiomimetric Substances in Mutation Breeding. (Bombay, September 26-29, 1969). India Dept. of Atomic Energy, Bombay. (CONF-690974) (NSA 25: 39909) (Drosophila melanogaster).

Sipple, A. and E. Heim. 1962. The distribution of life spans in biological collections. A method for the evaluation of studies in which the life span of biological collections was affected by irradiation or the introduction of drugs. Z. Naturforsch. 17b: 406-414. (In German) (NSA 16: 26842).

Sisigina, T. I. 1969. Assessment of radon emanation from the surface of extensive territories. pp. 239-244. In: Nuclear Meteorology. (See Makhon'ko and Malakhov, 1969) (TT-74-50011) (NSA 30: 15383).

Sissakian, N. M. 1967. Some problems of ecophysiology. pp. 49-60. In: Proc. 2nd Intern. Symp. on Basic Environmental Problems of Man in Space. (June 14-18, 1965). H. Bjurstedt, ed. Springer Verlag, Vienna. (NSA 21: 43531).

Sissons, J. B. and C. L. Brooks. 1971. Dating of early postglacial land and sea level changes in the western Forth Valley. Nature Phys. Sci. 1(4): 395-___.

Sittkus, A. and E. Welte. 1959. Dependence of the strontium-90 content in various grass samples on the strontium-90 content of their respective soils in the federal territory. Naturwissenschaften 46: 399-400. (In German).

Sivak, B. and M. Shaw. 1970. Microradioautography of potato leaves inoculated with ^{3}H-labeled conidia of Phytophora infestans. Can. J. Bot. 48(3): 613-615.

Sivasubramanian, P. 1973. Puparium formation in the fleshfly, Sarcophagus bullata P. Thesis, Iniv. of Illinois, Urbana. 121pp. (Univ. Microfilms 74-12191) (NSA 30: 19008) (X-irradiation).

Sivasubramanian, P., G. Bhaskaran and K. K. Nair. 1970. Differentiation of imaginal muscles in x-irradiated house fly pupae. Ann. Entomol. Soc. Am. 63: 1019-1022. (NSA 24: 41955).

Sivasubramanian, P., G. Bhaskaran and K. K. Nair. 1970. Effects of x-rays on morphogenesis in the housefly. J. Insect Physiol. 16: 89-97. (NSA 24: 17124; 26: 31367).

Sivasubramanian, P., H. S. Ducoff and G. Fraenkel. 1974. Effect of x irradiation on the formation of the puparium in the fleshfly, *Sarcophagus bulata*. J. Insect Physiol. 20(7): 1303-1317. (NSA 32: 3799).

Skakicky, J. and R. Karasek. nd. Contamination of the Svratka River Basin with waste waters from uranium industry and the hazards for a critical population group. Czechoslovakia Ustav Hygieny Prace v Uranovem Prumyslu, Pribram. (In Czech) (INIS-mf-887-3, pp. 131-151) (NSA 30: 6532).

Skalka, M. and M. Hill. 1959. The effect of x-irradiation on pigeons. Folia Biol. (Prague) 5: 60-70.

Skalon, I. S. 1955. Influence of carbon dioxide upon the development of heterotrophic microorganisms in relation to conditions in the medium. In: *Isotopes in Microbiology*. (See Imshenetzkiy, 1955) (AEC-tr-3494, pp. 157-167) (Used radiotracer).

Skauen, D. M. 1963. Tritium in ecology - a review. pp. 603-609. In: *Radioecology*. (See Schultz and Klement, 1963) (NSA 17: 33612).

Skauen, D. M. 1964. The effects of tritium oxide on aquatic organisms. Final report. Univ. of Connecticut, U. S. AEC report NYO-3039-1. 17pp. (NSA 19: 33).

Skauen, D. M. 1973. Analysis of tritium oxide from selected areas of the Connecticut River. Completion report. Univ. of Connecticut, Storrs. 9pp. (PB-223499) (NSA 32: 37).

Skauen, D. M., N. Marshall and R. J. Fragala. 1971. A liquid scintillation method for assaying ^{14}C-labelled benthic microflora. J. Fish. Res. Bd. Can. 28(5): 769-770.

Skauen, D. M., et al. 1961. Gross beta activity in the Thames River (Conn.). Nature 191(4792): 1017.

Skeen, J. N. and G. L. Plummer. 1966. Morphological heterogeneity in *Andropogon* growing on areas of high and low natural background radioactivity. Bull. Ga. Acad. Sci. 24(2-3): 100-116. (NSA 21: 14768).

Skinner, D. M. 1963. Incorporation of labeled valine into the proteins of the *Cecropia* silkworm. Biol. Bull. 125(1): 165-176. (Used ^{14}C).

Sklaroff, D. M. 1958. The uptake of radioactive rubidium (Rb-86) by yeast. Am. J. Roentgenol. 79(6): 994-998.

Skok, J., W. Chorney and W. S. Broecker. 1962. Uptake of CO_2 by roots of *Xanthium* plants. Botan. Gaz. 124: 118-120.

Skoog, F. 1935. The effect of x-irradiation on auxin and plant growth. J. Cell. Comp. Physiol. 7(2): 227-270.

Skou, J. P. 1962. Studies on microflora of soil under chronic irradiation. J. Gen. Microbiol. 28: 521-530. (NSA 17: 1298).

Skou, J. P. 1964. *Aureobasidium pullilans* (De By.) Arnaud - a common and very radio-resistant fungus on fresh fruits and vegetables. IVA Medd. 138: 63-70. (NSA 19: 13009).

Skou, J. P., K. Bech and K. Lundsten. 1974. Effects of ionizing radiation on mushrooms as influenced by physiological and environmental conditions. Radiation Botany 14(4): 287-299. (NSA 32: 1062).

Skowron, S. and H. Roguski. 1958. Regeneration from implanted dissociated cells. I. Regenerative potentialities of limb and tail cells. Folia Biol. (Krakow) 6: 163-173. (Used axolotl).

Skreb, Y. 1968. The effect of certain agents on radiosensitivity of amoeba and on recovery from radiation damage. Arh. Hig. Rada Toksikol. 19(Suppl. 1): 117-123. (NSA 24: 7985).

Skreb, Y. and D. Horvat. 1966. Influence of gamma-radiation in the incorporation of ^{14}C-adenine and ^{14}C-phenylalanine into whole amoebae and nucleate and anucleate amoeba fragments. Exptl. Cell Res. 43: 639-644. (NSA 21: 10797).

Skreb, Y. and D. Horvat. 1971. Modification of the effects of radiation on amoebas. I. Sensitizing action of actimomycin D before gamma irradiation. Int. J. Radiat. Biol. 20(2): 129-135. (In French) (NSA 26: 4692).

Skreb, Y., M. Eger and D. Horvat. 1965.., Effects of gamma rays on the ribonucleic acid and protein content of amoeba. Biochim. Biophys. Acta 103(1): 180-182. (ANL-Trans-475) (NSA 19: 38371; 21: 28686).

Skujins, J. J. 1963. Enzymatic and microbial activity in radiation sterilized soil. Thesis, Univ. of California, Berkeley. 121pp. (NSA 18: 120).

Skulberg, A. 1965. The resistance of *Clostridium botulinum* Type E toxin to radiation. J. Appl. Bacteriol. 28: 139-141. (NSA 19: 33849).

Skuratovich, A. A. and L. B. Koznova. 1966. Biological effect of ultra-fractionated irradiation. pp. 281-294. In: *Problems of General Radiobiology*. M. P. Domshlak, ed. Atomizdat, Moscow. (In Russian) (NSA 22: 23667) (Incl. various organisms).

Skuratovich, A. A. and G. M. Pravdina. 1965. Features peculiar to the biological action of ultrafractional pulse radiation. Med. radiologiya 10(2): 28-35. (In Russian) (NSA 19: 24308) (*Drosophila*).

Slade, D. H. 1971. Meteorologic questions concerning plant siting. pp. 56-82. In: *Health Physics Aspects of Nuclear Facility Siting*. *Vol. 1*. (See Voillequé and Baldwin, 1971) (NSA 25: 54793).

Slade, P. 1966. Possible appearance of degradation products of paraquat in crops. pp. 113-134. In: *Isotopes in Weed Research*. (See IAEA, 1966) (^{14}C).

Sladen, B. K. 1969. Ecology of animal communities. pp. 87-100. In: *Biology of Populations*. B. K. Sladen and F. B. Bang, eds. American Elsevier, New York. (NSA 25::11006) (Discusses radioactivity in animal communities)

Sladkovic, R. 1969. Investigation of the transport of fallout from the seventh Chinese nuclear explosion into the Alps. Arch. Meteorol. Geophys. Bioklimatol. A18: 87-110. (In German) (NSA 24:321).

Slagle, K. A. 1971. Water quality. pp. III.1-III.27. In: Environment. (See Southern California Edison Co., 1971) (NSA 26: 58827).

Slankis, V., V. C. Runeckles and G. Krotkov. 1964. Metabolites liberated by roots of white pine (*Pinus strobus* L.) seedlings. Physiol. Plant. 17: 301-313.

Slansky, C. M. 1971. Radioactive waste management in relation to the environment. Monitoring and pretreatment analysis: radioactive waste management. Principles and practices in the disposal of radioactive waste to fresh waters. Allied Chemical Corp., Idaho Falls, Ida. 15, 13, 14pp. (For: Regional Training Course in Radioactive Waste Management Conf., Tokyo, Japan, 4 Nov. 1971) (CONF-711113-2, -3, -4) (NSA 26: 6775, 6776, 9233).

Slater, J. V. 1957. Radioactive cobalt accumulation in *Tetrahymena*. Biol. Bull. 112(3): 390-399.

Slater, J. V. 1961. Comparative accumulation of radioactive zinc in young rainbow, cutthroat and brook trout. Copeia 2: 158-161.

Slater, J. V., N. M. Amer and C. A. Tobias. 1962. Modification of radiation response during embryonic development by the use of elevated temperatures. Abstr. p. 241. In: 2nd Intern. Congr. of Radiation Research. (See International Congress of Radiation Research, 1962) (*Tribolium*).

Slater, J. V., B. Buckhold and C. A. Tobias. 1968. Effect on a flour beetle of irradiation during space flight. In: Symp. on the Biosatellite II Experiments. BioScience 18(6): 595-597.

Slater, J. V. B. M. Buckhold and C. A. Tobias. 1968. Synergism of x-irradiation and space flight factors in the flour beetle, *Tribolium confusum*. Abstr. Radiation Research 35(2): 501.

Slater, J. V., B. Buckhold and C. A. Tobias. 1969. Space-flight enhancement of irradiation effects in the flour beetle, *Tribolium confusum*. Radiation Research 39: 68-81. (NSA 23: 35694).

Slater, J. V., M. E. Yu and C. A. Tobias. 1964. Oxygen dependence for radiation sensitivity during development of insects. Abstr. Radiation Research 22(1): 236. (*Tribolium confusum*).

Slater, J. V., et al. 1961. Comparative influence of accelerated heavy nuclei on anomalous development in *Tribolium*. Abstr. Radiation Research 14(4): 503.

Slater, J. V., et al. 1961. Yttrium localization in the desert pocket mouse *Perognathus penicillatus*. Proc. Soc. Exptl. Biol. Med. 108(1): 198-199.

Slater, J. V., et al. 1963. Temperature dependence of wing abnormality in *Tribolium confusum*. Science 140: 408-409. (Compares results with x-irradiation).

Slater, J. V., et al. 1964. Heavy ion localization of sensitive embryonic sites in *Tribolium*. Radiation Research 21(4): 541-549. (NSA 18: 19668).

Slavic, I., et al. 1973. Computer program for determination of concentrations of trace elements in components of water systems by non-destructive activation analysis. Isotopenpraxis 9(3): 87-91. (In German) (NSA 20: 2424).

Slesinski, R. S. and A. H. Ellingboe. 1971. Transfer of ^{35}S from wheat to the powdery mildew fungus with compatible and incompatible parasite/host genotypes. Can. J. Bot. 49(2): 303-310.

Sletten, K., D. Lockner and G. Hevesy. 1964. Radiosensitivity of haemopoiesis in fish. I. Studies at 18^0C. Int. J. Radiat. Biol. 8(4): 317-328. (NSA 19: 17549).

Slew, Y. C. 1965. The endocrine control of adult reproductive diapause in the chrysomelid beetle, *Galeruca tanaceti* (Linn.). Thesis, Univ. of London, London, England. 190pp. (^{35}S).

Slife, F. W., et al. 1962. Penetration, translocation, and metabolism of 2,4-D and 2,4,5-T in wild and cultivated cucumber plants. Weeds 10(1): 29-35. (^{14}C).

Slipka, J. 1952. Intake of I^{131} through cuticle of larva of *Tipula maxima*. Biol. Listy (Prague) 32: 316-322.

Slizewicz, P. 1963. The discharge of radioactive waste into the sea. Bull. Inform. Sci. Tech. (Paris) (70): 31-37. (In French) (NSA 17: 29873).

Slizewicz, P. 1964. Tritium used as a tracer in geophysics. pp. 107-111. In: Proc. Seminar on Protection Against Tritium Hazards. (Le Vesinet, France, April 16-18, 1964). France Commissariat à l'Énergie Atomique, Saclay. (In French) (CONF-640413) (NSA 20: 23320).

Sloka, J. J., et al. 1973. ^{90}Sr in water, sediments, and water organisms in Latvian lakes of various types. pp. 56-59. In: Radioecology of Water Organisms. II. (See Andrushaitis, 1973) (In Russian) (NSA 30: 21131).

Slonim, A. R. and F. C. Damm. 1972. Beryllium uptake by the common guppy using radioisotope Be-7. II. Beryllium concentration in fish. Final report. Aerospace Medical Research Lab., Wright-Patterson Air Force Base, O. 25pp. (AD-758815; NSA 30: 24414).

Small, L. F. 1959. Estimates of standing crop of plankton in Clear Lake, Iowa. Iowa State Coll., U. S. AEC report TID-11579. (NSA 15: 7147).

Small, L. F. 1962. Use of radioisotopes in laboratory energy assimilation investigations with *Daphnia*. Nature 196: 787-788.

Small, L. F. 1967. On the standardization of C^{14} for primary production estimates in aquatic environments. Iowa State J. Sci. 42(1): 63-71.

Small, L. F. 1969. Experimental studies on the transfer of ^{65}Zn in high concentration by euphausiids. J. Exptl. Mar. Biol. Ecol. 3: 106-123. (NSA 23: 33614).

Smedile, E. and E. Tibaldi. 1973. Ecological and radioecological investigations on the trophic structure in a stretch of the Po River. Abstr. pp. 99-100. In: Proc. 2nd Conf. of the State of Advancement in Radioecology in Italy. (See Anonymous, 1973) (In Italian).

Smedile, E. and E. Tibaldi. 1974. Ecological and radioecological researches on the food-web in a stretch of the Po River. G. Fis. Sanit. Prot. Radiaz. 18(1-2): 14-16. (In Italian) (NSA 31: 19600).

Smedile, E. and C. Triulzi. 1973. Evolution of the artificial radioactivity in river sediments. Abstr. p. 91. In: Proc. 2nd Conf. on the State of Advancement of Radioecology in Italy. (See Anonymous, 1973) (In Italian).

Smedile, E. and C. Triulzi. 1973. Evolution of artificial radioactivity in fluvial sediments. G. Fis. Sanit. Prot. Radiaz. 17(3-4): 119-124. (In Italian) (NSA 31: 19588).

Smeets, J. 1969. Radioactive contamination of surface waters in the European Community. pp. 49-62. In: Water for Tomorrow. Italy Federazione delle Associazioni Scientifiche e Techniche, Milan. (CONF-680475) (NSA 25: 46856).

Smeets, J. G. P. M. 1962. The behavior of radionuclides in the food chain. pp. 75-97. In: Symp. on the Health Aspects of Radioactive Substances in Foods. N. V. Uitgeverij, ed. W. P. van Stockum and Zn, The Hague. (In Dutch) (NSA 19: 24791).

Smelova, I. V. 1961. The penetration of various compounds of S-35 from the water into bodies of fishes. Tr. vses. nauchn.-issled. inst. morsk. ryb. khoz. okeanogr. 44: 37-47. (In Russian).

Smelova, I. V. 1961. The distribution in the body of fish of sulfur-35, administered in live foods. Vopr. ikhtiol. 17: 110-114. (In Russian) (BA 42: 4434; Ref. zh. (biol.) 1962: 5I36).

Smierzchalska, K. 1972. Strontium-90 and caesium-137 as dangerous components of radioactive contamination in the food chain (soil-plant-animal-man) man). Postepy Nauk Roln. 24(5): 91-107. (In Polish) (NSA 33: 29153).

Smierzchalska, K. 1972. Influence of potassium, ammonium and calcium ions on the sorptionpdesorption patterns of Cs-137 in various soils. Pol. J. Soil Sci. (5): 117-125. (NSA 31: 14071).

Smil, V. J. 1972. Energy and the environment: a long-range forecasting study. Thesis, Pennsylvania State Univ., University Park. 272pp. (Univ. Microfilms 72-33204) (NSA 27: 24203).

Smirnoff, W. A. 1967. Effect of gamma radiation on the larvae and the nuclear-polyhedrosis virus of the eastern tent caterpillar Malacosoma americanum. J. Invert. Pathol. 9: 264-266. (NSA 22: 6906).

Smirnoff, W. A. and M. Cantin. 1967. Effect of gamma irradiation on the growth rate of species of Bacillus cereus group. J. Invert. Pathol. 9(3): 357-363. (NSA 23: 6621).

Smirvov, Ye. G. and F. A. Tikhomirov. 1975. Accumulation of ^{90}Sr by herbaceous plants in a birch forest. Ekologiya 6(2): 16-22. (Sov. J. Ecol. 6 (2): 102-106) (NSA 33: 29800).

Smirnov, V. N., et al. 1964. Participation of ribosomes in the biosynthesis of silk fibroin. Dokl. Akad. Nauk SSSR 156(5): 1221-1224. (Doklady Biochem. 156: 189-192) (^{14}C).

Smirnov, V. N., et al. 1964. Ribonucleic acid synthesis of silk secreting gland of Bombyx mori. Dokl. Akad. Nauk SSSR 155(4): 957-960. (In Russian) (^{32}P).

Smirnova, I. B. 1960. The effect fo temperature on the response of paramecia to irradiation. Dokl. Akad. Nauk SSSR 130(4): 901-903. (Doklady (biol. sci.) 130: 28-30) (NSA 14: 14693).

Smirnova, I. B. and R. V. Latsis. 1970. Change of radiosensitivity and of content of SH groups in early embryogenesis of the loach (Misgurnus fossilis). Dokl. Akad. Nauk SSSR 192: 915-916. (In Russian) (NSA 25: 24318).

Smirnova, I. B. and R. V. Latsis. 1973. Correlation of the amounts of protein and nonprotein thiols and the level of radiosensitivity in loach eggs during cleavage. Radiobiologiya 13(2): 277-280. (AEC-tr-7444, pp. 146-150) (NSA 28: 24802).

Smirnova, I. B., R. V. Latsis and Ye. V. Frolova. 1973. Radioactivity of loach eggs during cleavage under conditions of colchicine-induced depression of mitosis. Radiobiologiya 13(3): 442-445. (AEC-tr-7471, pp. 150-154) (NSA 30: 6974).

Smith, A. E. and W. E. Moore. 1972. Report of the radiological clean-up of Bikini Atoll. Western Environmental Research Lab., U. S. EPA, U. S. AEC report SWRHL-111-r. 45pp. (NSA 26: 15212).

Smith, A. H. and R. M. Bohart. 1956. Evaluating fly nuisance source. Flies tagged with radioisotopes used as a field test to establish migratory tendencies of a normal fly population. Calif. Agric. 10(7): 11-12. (^{32}P).

Smith, A. R. and H. A. Wollenberg. 1972. Geology and natural terrestrial dose rates. pp. 8-11. In: Workshop on Natural Radiation Environment. (March 1972). (See McLaughlin, 1972) (NSA 27: 19825).

Smith, A. R., et al. 1975. Investigation of radon-222 in subsurface waters as an earthquake predictor. Lawrence Berkeley Lab., U. S. ERDA report LBL-4445. 15pp. (NSA 33: 20202).

Smith, B. M., et al. 1961. Natural radioactivity in ground water supplies in Maine and New Hampshire. J. Am. Water Works Assoc. 53(1): 75-88. (NSA 20: 37006).

Smith, C. D. and J. G. Eales. 1971. Influence of acclimation temperature on rate constants for thyroid uptake of iodide by brook trout, Salvelinus fontinalis (Mitchell). Can. J. Zool. 49(5): 783-786. (SFA 16: 13732).

Smith, C. N. 1963. Prospects for vector control through sterilization procedures. Bull. Wrld. Hlth. Org. 29(Suppl.): 99-106. (NSA 18: 17502).

Smith, C. N. 1967. Possible use of the sterile-male technique for control of Aedes aegypti. Bull. Wrld. Hlth. Org. 36: 633-635. (NSA 22: 52110).

Smith, C. N. 1972. Genetic control of mosquitoes: programme of the WHO/ICMR collaborative research unit. pp. 120-122. In: Genetics and Our Health. Indian Council of Medical Research, New Delhi. (NSA 29: 10464).

Smith, D. B. 1969. Growth potential of isotopes. S. Afr. Mining Eng. J. 80: 628-636. (NSA 24: 2312) (Hydrology, etc.).

Smith, D. B. 1973. Use of artificial radioactive tracers in the United Kingdom. pp. 97-102. In: Trace Techniques in Sediment Transport. (See IAEA, 1973) (NSA 28: 21138).

Smith, D. B. and W. E. Clark. 1963. An investigation of the hydraulics of an underground pool with iodine-131. pp. 77-88. In: Radioisotopes in Hydrology. (See IAEA, 1963) (NSA 18: 1915).

Smith, D. C., J. A. Bassham and M. Kirk. 1961. Dynamics of the photosynthesis of carbon compounds. II. Amino acid synthesis. Biochim. Biophys. Acta 48(2): 299-313. (Also: Univ. of California, Berkeley, U. S. AEC report UCRL-9323) (RDA-3(1): 40) (Chlorella).

Smith, D. D. 1972. Radiation surveillance of ruminants on and about the Nevada Test Site. National Environmental Research Center, Las Vegas, Nev., U. S. EPA, U. S. AEC report NERC-LV-539-18. i, 10pp. (Incl. deer and bighorn sheep).

Smith, D. D. and K. R. Giles. 1974, 1975. Animal investigation program 1970 annual report. ---Annual report. National Environmental Research Center, Las Vegas, Nev., U. S. EPA, U. S. AEC report NERC-LV-539-16, -20. 56, 39 pp. (NSA 32: 27957) (Incl. cattle, water, wild animals on or near the Nevada Test Site).

Smith, D. D., et al. 1976. Tissue burdens of selected radionuclides in beef cattle on and around the Nevada Test Site. National Environmental Research Center, Las Vegas, Nev., U. S. AEP, U. S. ERDA report NERC-LV-539-29. 90 pp. (NSA 33: 20204).

Smith, D. E. 1958. The influence of x-irradiation upon dormancy in vertebrates. J. Exptl. Zool. 139(1): 85-94. (Bat, ground squirrel, frog).

Smith, D. E. 1959. Protection of irradiated ground squirrel by cysteine. Radiation Research 10(3): 335-338.

Smith, D. E. 1960. The effects of ionizing radiation in hibernation. Bull. Mus. Comp. Zool. Harvard Univ. 124: 493-506. (NSA 15: 12737).

Smith, D. E. 1960. Failure of cysteine given post-irradiation to protect the hibernating ground squirrel. Radiation Research 12(1): 79-80.

Smith, D. E. and J. F. Thomson. 1959. Physiological and biochemical studies on various species exposed to massive x-irradiation. Radiation Research 11(2): 198-205. (NSA 13: 18841).

Smith, D. E., D. R. Russ and E. M. Jackson. 1955. Response of the bat (Myotis lucifugus) to x-irradiation. Radiation Research 2(4): 330-338. (NSA 9: 5215).

Smith, D. E., G. Svihla and H. M. Patt. 1951. The effects of x radiation on circulation in the wing of the bat. Physiol. Zool. 24(3): 249-257. (Myotis lucifugus).

Smith, D. G. 1970. Investigation of a nuclear fuel reprocessing plant upon its environment. pp. 832-836. In: Proc. 11th AEC Air Cleaning Conf. (See First and Morgan, 1970) (NSA 25: 21686).

Smith, D. M. and R. Y. Mixer. 1959. The effects of lignin on the degredation of wood by gamma irradiation. Radiation Research 11(6): 776-780. (NSA 14: 3375).

Smith, D. W., C. B. Fliermans and T. D. Brock. 1973. Isotopic technique for measuring the autotropic activity of soil microorganisms in situ. pp. 243-246. In: Modern Methods in the Study of Microbial Ecology. T. Rosswall, ed. NFR Editorial Services, Stockholm. (NSA 29: 889).

Smith, E. C. 1936. The effects of radiation on fungi. Ch. XXVII, pp. 889-918. In: Biological Effects of Radiation. Vol. II. B. M. Duggar, ed. McGraw-Hill, New York. (Incl. x-rays).

Smith, E. R. 1934. The isotopic fractionation of water by physiological processes -- an addition. Science 79: 454-455. (Willow).

Smith, F. and M. M. Grenan. 1951. Effect of hibernation upon survival time following whole body irradiation in the marmot (Marmota monax). Science 113(2946): 686-688. (NSA 5: 5483).

Smith, F. and M. M. Grenan. 1951. Circulating blood cells following radiation in the hibernating woodchuck. Abstr. Fed. Proc. 10(1): 128.

Smith, F. A. 1973. Preliminary studies of mercury tissue levels from game birds and fish in Utah. pp. 140-145. In: Mercury in the Western Environment. (See D. R. Buhler, 1973) (Neutron activation analysis).

Smith, G. F. and H. Kersten. 1942. Auxin and calines in seedlings from x-rayed seed. Am. J. Bot. 20: 785-791.

Smith, G. M. 1932. Eruptions of corial melanophores and general cutaneous melanosis in the goldfish (Carassius auratus) following exposure to x-ray. Am. J. Cancer 16(4): 863-870.

Smith, G. M. 1932. Melanophores induced by x-ray compared with those existing in patterns seen in Carassius auratus. Biol. Bull. 63(3): 484-491.

Smith, G. N., B. S. Watson and F. S. Fischer. 1966. The metabolism of (^{14}C) O,O-diethyl O-(3,5,6-trichloro-2-pyridyl) phosphorothioate (Dursban) in fish. J. Econ. Entomol. 59(6): 1464-1475.

Smith, G. N., et al. 1964. Simple apparatus for combustion of samples containing ^{14}C-labeled pesticides for residue analysis. J. Agric. Food Chem. 12(2): 172-175.

Smith, H. G., R. L. Goulding and J. L. Priano. 1970. A dispenser for simultaneous introduction of ion exchange resin into preparations for radioisotopic determination of cholinesterase activity. J. Econ. Entomol. 63(2): 683-.

Smith, H. H. 1972. Environmental modulation of plant response to neutron irradiation. Radiation Botany 12(4): 229-237. (NSA 26: 59012).

Smith, H. L. and W. L. Heilman. 1974. Numerical rating of environmental impacts for power plant siting. Trans. Amer. Nucl. Soc. 19(Suppl.)(1): 11-12. (NSA 30: 26054).

Smith, J. C. and D. J. Kimeldorf. 1963. The bioelectrical response of the insect eye to beta radiation. U. S. Naval Radiological Defense Lab., San Francisco, Calif., report USNRDL-TR-698. 30pp. (AD-429190; NSA 18: 15814).

Smith, J. C. and D. J. Kimeldorf. 1964. The bioelectrical response of the insect eye to beta radiation. J. Insect Physiol. 10(5): 838-847. (*Agrotis ypsilon*, *Pseudaletia unipuncta*).

Smith, J. C., D. J. Kimeldorf and E. L. Hunt. 1963. Motor responses of moths to low-intensity x-ray exposure. Science 140(3568): 805-806). (Also: U. S. Naval Radiological Defense Lab., San Francisco, Calif., report USNRDL-TR-622) (NSA 17: 19808).

Smith, J. C., et al. 1964. Immediate response in the pigeon to brief x-ray exposure. Abstr. Radiation Research 22(1): 237.

Smith, J. L. and D. W. Willen. 1969. Gage device for measurement of density profiles of snowpack. U. S. Patent 3,432,656. (NSA 23: 16170).

Smith, J. L., D. W. Willen and M. S. Owens. 1965. Measurement of snowpack profiles with radioactive isotopes. Weatherwise 18: 247-251. (NSA 21: 1449).

Smith, J. L., D. W. Willens and M. S. Owens. 1967. Isotope snow gages for determining hydrologic characteristics of snowpacks. pp. 11-21. In: *Isotope Techniques in the Hydrologic Cycle*. (See Stout, 1967).

Smith, J. L., H. G. Halverson and R. Jones. 1969. Profiling snow gage and its utility in water supply forecasting. Trans. Amer. Nucl. Soc. 12: 500-501. (NSA 24: 5828).

Smith, J. M., Jr. 1963. Locating nuclear-power plants with reference to environmental safety. pp. 133-145. In: *Siting of Reactors and Nuclear Research Centres*. International Atomic Energy Agency, Vienna. (NSA 17: 41086).

Smith, J. W. and H. B. Boudreaux. 1972. An autoradiographic search for the site of fertilization in spider mites. Ann. Entomol. Soc. Am. 65(1): 69-74.

Smith, K. A. and E. R. Mercer. 1970. Determination of ^{226}Ra and ^{228}Ra in so soils and plants, using ^{225}Ra as a yield tracer. J. Radioanal. Chem. 5(2): 303-312. (NSA 26: 4203).

Smith, L. D. 1972. Evolution of opposition to the peaceful uses of nuclear energy. Nucl. Eng. Int. 17(193): 461-468. (NSA 26: 40826).

Smith, L. L. 1961. Algae: radioactivity uptake irradiation effects, a literature search. Savannah River Lab., U. S. AEC report TID-11749. (NSA 15: 8504).

Smith, L. S., et al. 1971. Responses to teleost fish to environmental stress. U. S. EPA Water Pollution Control Research Ser. 18050EBK02171. xvii, 114pp. (Incl. use of ^{14}C and ^{3}H).

Smith, L. W. 1969. Combustion and liquid scintillation determination of ^{14}C in biological samples. Anal. Biochem. 29: 223-229. (NSA 23: 28732).

Smith, M. H., J. L. Carmon and F. B. Golley. 1969. Effects of temperature and gamma radiation on reproduction in the old-field mouse. Abstr. Bull. Ecol. Soc. Am. 50(3): 130.

Smith, R. C. and R. C. von Borstel. 1972. Genetic control of insect populations. Science 178(4006): 1164-1174. (NSA 27: 15092).

Smith, R. F. 1967. The vegetation structure of a Puerto Rican rain forest before and after short-term gamma irradiation. Thesis, Univ. of Georgia, Athens. 184pp. (NSA 22: 49283).

Smith, R. F. 1970. The vegetation structure of a Puerto Rican rain forest before and after short-term gamma irradiation. Ch. D-3, pp. D103-D140. In: *A Tropical Rain Forest*. (See H. T. Odum and R. F. Pigeon, 1970) (NSA 25: 21650).

Smith, R. F. and W. H. Fuller. 1961. Identification and mode of action of a component of positively-ionized air causing enhanced growth in plants. Plant Physiol. 36: 747-751. (NSA 21: 24667).

Smith, R. J. 1958. Filtering efficiency of hard clams in mixed suspensions of radioactive phytoplankton. Proc. Natl. Shell-fisheries Assoc. 48: 115-124.

Smith, R. L. 1966. *Ecology and Field Biology*. Harper & Row, New York. xiv, 686pp. (Contains sections on ionizing radiation).

Smith, T. E. and R. T. Taylor. 1969. Incorporation of tritium from tritiated water into carbohydrates, lipids, and nucleic acids. Lawrence Radiation Lab., Livermore, U. S. AEC report UCRL-50781. 24pp. (NSA 24: 21060).

Smith, T. L. 1965. A serviceable irradiatior-shield device. Ann. Entomol. Soc. Am. 58(2): 245-247.

Smithers, S. R. 1962. Immunizing effect of irradiated cercariae of *Shistosoma mansoni* in rhesus monkeys. Nature 194(4834): 1146-1147. (NSA 16: 20202).

Smithsonian Institute Center for Short-Lived Phenomena. 1970. National and international environmental monitoring activities. Cambridge, Mass. 306 pp. (NSA 25: 37496).

Smittle, B. J. 1967. Effect of aeration on gamma irradiation of house fly pupae. J. Econ. Entomol. 60(6): 1594-1595.

Smittle, B. J. 1968. Effect of gamma irradiation on female Culex pipiens quinquefasciatus. Florida Entomol. 51: 59-61. (NSA 23: 29726).

Smittle, B. J., S. O. Hill and F. M. Philips. 1967. Migration and dispersal patterns of Fe^{59}-labeled lone star ticks. J. Econ. Entomol. 60(4): 1029-1031. (BA 48: 124498).

Smittle, B. J., G. C. LaBrecque and E. E. Carroll. 1971. Comparative effectiveness of fast neutrons and gamma rays in producing sterility in house flies. J. Econ. Entomol. 64(5): 1030-1032. (NSA 26: 7126).

Smittle, B. J., R. S. Patterson and C. S. Lofgren. 1969. P^{32}-labeling of common malaria mosquitoes: assimilation by larvae and retention and transfer during mating. J. Econ. Entomol. 62(4): 851-853.

Smythe, L. E. and P. S. Davis. 1960. Results of strontium ninety determinations on samples from Lucas Heights and Richmond, N.S.W. 1959-1960. Australia Atomic Energy Commission, Lucas Heights. 25pp. (NSA 15: 5234).

Snavely, D. R. and W. A. Bill. 1968. Comparison of environmental radioactivity levels after foreign nuclear weapons tests of May 9, 1966 and December 27, 1966. Radiol. Hlth. Data 9(2): 63-69.

Snelling, R. N. 1970. Environmental survey of uranium mill tailings pile, Monument Valley, Arizona. Radiol. Hlth. Data 11: 511-517. (NSA 25: 11045).

Snelling, R. N. 1971. Data acquisition and analysis system for emergency environmental surveillance. Western Environmental Research Lab., U. S. EPA, U. S. AEC report SWRHL-502-r. 30pp. (For: Intern. Symp. on Rapid Methods for Measuring Radioactivity in the Environment, Munich, Germany, 5 Jul 1971) (CONF-710705) (NSA 25: 44730).

Snelling, R. N. 1971. Environmental survey of uranium mill tailings pile, Mexican Hat, Utah. Radiol. Hlth. Data 12: 17-28. (NSA 25: 32409).

Snelling, R. N. and S. D. Shearer, Jr. 1969. Environmental survey of uran ium mill tailings pile, Tuba City, Arizona. Radiol. Hlth. Data 10: 475-487. (NSA 24: 31895).

Snider, G. and H. Kersten. 1935. The action of soft x-rays on Cladocera (Daphnia magna). Physiol. Zool. 8(4): 530-538.

Snider, G. and H. Kersten. 1936. Susceptibility to soft x-rays of Daphnia magna during its development from eggs to young in the brood pouch. J. Exptl. Zool. 74(1): 1-6.

Snihs, J. O. 1973. Content of some natural radioactive elements, especially ^{222}Rn in some potable waters in Sweden. Nord. Hydrol. 4(4): 256-274.

Snipes, W. C. and W. Gordy. 1963. Radiation damage to Artemia cysts: effects of water vapor. Science 142(3591): 503-504. (NSA 17: 40622).

Snow, J. A. and A. W. Wolfe. 1964. Radioactivity in arctic peoples. Scient. Citizen 6(9-10): 26-33. (NSA 18: 43329).

Snow, J. W., W. J. Lewis and A. N. Sparks. 1970. Mating of normal and sterilized colonized and native corn earworms with emphasis on failure of pairs to separate after copulation. J. Econ. Entomol. 63: 1873-1878. (NSA 25: 16296).

Snow, J. W., et al. 1972. Effects of irradiation on ability of adult male corn earworms to transfer sperm, and field attractiveness of females mated to irradiated males. J. Econ. Entomol. 65(3): 906-908. (NSA 26: 43356).

Snook, G. W., et al. 1964. Sterilization of anthrax-contaminated mink pelts by radiation. Proc. U.S. Livestock Sanit. Assoc. 68: 238-240. (NSA 21: 34847).

Snyder, E. H. C. Grigsby and J. V. Hidalgo. 1961. X-radiation of southern pine seed at various moisture contents. Silvae Genet. 10: 125-160.

Snyder, J. M. and R. B. Platt. 1973. Effects of chronic gamma irradiation on the growth of a crustose lichen, Trapelia ornata (Sommerfelt) Hertel. Radiation Botany 13(5): 269-271. (NSA 29: 18588).

Snyder, W. S. 1972. Environmental pathways from reactor to man. Oak Ridge national Lab., U. S. AEC report. 26pp. (For: 2nd Intern. Summer School on Radiation Protection, Herceg Novi, Yugoslavia, 21 August 1973) (CONF-730814) (NSA 28: 27304).

Snyder, W. S. (Ed.). 1974. Proc. 3rd Intern. Congr. of the International Radiation Protection Association. (Washington, D. C., September 9-14, 1973). International Radiation Protection Assoc., Washington. 2 Pts., 798, 677pp. (In several languages) (CONF-730907-P1, -P2) (NSA 30: 27056, 27094).

Snyder, W. S., et al. (Eds.). 1968. Proc. 1st Intern. Congr. of Radiation Protection. Pergamon Press, New York. 2 Pts.; xix, 800pp.; xviii, 821pp.

Soares de Gouvein, A. J. and M. I. Moreira. 1962. Use of x-rays to evaluate the effectiveness of insecticide treatments. Garcia de Orto 10(4): 661-665. (In Portugese) (Phaseolus vulgaris, Zabrotes subfasciatus, Acanthoscelides obtectus).

Sobels, F. H. 1964. Radiosensitivity and repair in different germ cell stages of Drosophila. pp. 235-255. In: Proc. 11th Intern. Congr. on Genetics. (The Hague, Sept. 1963). Pergamon Press, Oxford. (NSA 19: 21989).

Sobels, F. H. 1964. The contrasting effects of O_2 and N_2 in determining initial sensitivity and post-radiation recovery in Drosophila sperm and spermatids. Int. J. Radiat. Biol. 7(5): 505.

Sobels, F. H. 1969. A study of the causes underlying the differences in radiosensitivity between mature spermatozoa and late spermatids in Drosophila. Mutation Research 8: 111-125. (NSA 23: 31894).

Sobatka, A. 1969. Influence of chronic effects of gamma radiation on the formation of short roots of Scots pine. Lesnictvi 15: 429-436. (In Czech) (NSA 24: 44624).

Societé Francaise de Radioprotection. 1968. Intern. Congr. on Radioprotection of the Environment. (Related to the Development of the Use of Peaceful Uses of Nuclear Energy, Toulouse, 14, 15, 16 March 1967). Paris. 661 pp. (In French or English).

Sodd, V. J. and K. L. Scholz. 1969. Analysis of tritium in water: a collaborative study. J. Assoc. Off. Anal. Chem. 52: 90-93. (NSA 23: 9338).

Sodd, V. J. and R. J. Velten. 1959. Estuarine and marine waters and biota. Fate of radioactive contaminants in water. U. S. Dept. of Health, Education, and Welfare, Washington, Progr. Rept. No. 1, Ch. III, Sec. TR-R 59-2, pp. 17-22.

Södergren, A. 1968. Uptake and accumulation of C^{14}-DDT by Chlorella sp. (Chlorophyceae). Oikos 19(1): 126-138.

Soeno, Y. 1961. On the effects of ionizing radiation emitted from ^{60}Co upon the development of Ascaris eggs. I. Studies on the relation between dose rate and its inhibitory effects upon the development of eggs. Nippon Univ. J. Med. 3: 283-300. (NSA 18: 27138).

Soeren Mattsson, L. J. 1972. ^{22}Na in the food-chain: lichen-reindeer-man. Health Physics 23(2): 223-230. (NSA 26: 45909).

Sofer, Z. and J. R. Gat. 1975. Isotope composition of evaporating brines: effect of the isotopic activity ratio in saline solutions. Earth Planet. Sci. Lett. 26(2): 179-186. (NSA 32: 24355).

Sokol, D. 1970. Ground water safety evaluation. Project Gasbuggy. U. S. AEC report PNE-1009; PNE-G-74. 38pp. (NSA 25: 26406).

Sokolic, A. 1968. Immunological studies on Fasciola hepatica. Clinical and immunological response in mice to infection with irradiated and normal fluke cysts. pp. 93-104. In: Isotopes and Radiation in Parasitology. (See IAEA, 1968).

Sokoloff, A. 1961. Irradiation experiments with Tribolium. Tribolium Inform. Bull. 4: 28-33. (NSA 21: 24608).

Sokolov, A. N. 1935. The radioactivity of some Russian mineral waters, muds and soils. Zh. Russ. Fiz.-khim. 37(17:___. (In Russian).

Sokolov, N. N., et al. 1967. Effects of ionizing radiation on the chromosomes of plants and animals. Izv. Sibirsk. otd. Akad. Nauk SSSR (biol.-med.) (5): 124-131. (In Russian).

Sokolov, V. Ye. and A. I. Il'yenko. 1969. Radioecology of land vertebrates. Usp. sovrem. biol. 67(2): 235-255. (In Russian) (BA 51: ___; WR 140: 18).

Sokolov, Yu. L. and A. V. Gurskiy. 1962. Experimental studies of cosmic radiation effects on higher plants. pp. 164-169. In: Problems of Space Biology. Vol. 2. N. M. Sisakyan and V. I. Yazdovskiy, eds. Publ. House Acad. Sci., Moscow. (In Russian) (NSA 17: 27201).

Sokolova, G. A. and Yu. I. Sorokin. 1958. Determination of the rate of bacterial reduction of sulfates in bottom deposits of the Gorky Resrvoir using S^{35}. Dokl. Akad. Nauk SSSR 118(1): 404-406. (Doklady (biol. sci.) 118: 81-83).

Sokolova, I. A., et al. 1968. Methods and materials for the investigation of the contamination of the hydrobiosphere by ^{90}Sr. Okeanologiya 8: 1092. (In Russian) (NSA 23: 35013).

Sokolova, Z. A., A. A. Gusovskaya and V. V. Popov. 1969. Effects of x irradiation on the retina of grass frog tadpoles. Radiobiologiya 9(6): 863-867. (AEC-tr-7145, pp. 103-109) (NSA 24: 17139).

Sokurova, Ye. N. 1956. On certain regularities in ionizing radiation effects on microorganisms. Izv. Akad. Nauk SSSR (Biol.) (6): 35-53. (UK AEA re port AERE-lib.-Trans.-772) (NSA 11: 3279).

Sokurova, Ye. N. 1957. Effetct of beta emitters on rootnodule bacteria. Mikrobiologiya 26(4): ___. (Microbiology 26: 444-449).

Sokurova, Ye. N. 1957. The influence of beta radiation on the development of Azotobacter and its physiological activity. Mikrobiologiya 26(5): 519-525. (Microbiology 26: 514-519).

Sokurova, Ye. N. 1974. Forms of inactivation of yeast organisms of various taxonomic groups after irradiation with the gamma rays of ^{60}Co. Radiobiologiya 14(2): 224-228. (AEC-tr-7571, pp. 80-85) (NSA 30: 32395).

Sokurova, Ye. N. 1974. Radiobiological characteristics and ability for recovery from radiation damages of yeast organisms of various taxonomic groups. Radiobiologiya 14(1): 53-57. (AEC-tr-7548, pp. 64-70) (NSA 30: 12880).

Sokurova, Ye. N. and M. N. Meysel'. 1957. The combined effect of ultraviolet and x-rays on Bac. anthrocoides spores. Biofizika 2(4): 483-486. (Biophysics (USSR) 2(4): 474-477).

Sokurova, Ye. N. 1958. A comparison of the effects of alpha and beta-rays on microorganisms. Biofizika 3(4): 474-478. (Biophysics (USSR) 3(4): 453-457) (NSA 13: 11554).

Solberg, A. N. 1938. The susceptibility of the germ cells of Oryzias latipes to x-radiation and recovery after treatment. J. Exptl. Zool. 78(4): 417-439.

Solberg, A. N. 1938. The susceptibility of Fundulus heteroclitus embryos to x-radiation. J. Exptl. Zool. 78(4): 441-469.

Soldat, J. K. 1961. Some radioactive materials measured in various waters in the United States. A literature search. Hanford Atomic Products Operation, U. S. AEC report HW-70706. 19pp. (NSA 15: 27893).

Soldat, J. K. 1962. Compilation of basic data relating to the Columbia River. Section 8. Dispersion of reactor effluent in the Columbia River. Hanford Atomic Products Operation, U. S. AEC report HW-69369. 241pp. (NSA 29: 10126).

Soldat, J. K. 1963. The relationship between I^{131} concentrations in various environmental samples. Health Physics 9(12): 1167-1171. (At Hanford Labs).

Soldat, J. K. 1968. A statistical study of habits of fishermen utilizing the Columbia River below Hanford. Battelle Northwest Lab., U. S. AEC report BNWL-SA-1451. 16pp. (NSA 22: 16999).

Soldat, J. K. 1971. Environmental problems in production of electricity from nuclear plants. Battelle Northwest Lab., U. S. AEC report BNWL-SA-3912. 35pp. (For: Environmental Symp., Klamuth, Oregon, 20 May 1971) (CONF-710521-1) (NSA 25: 48284).

Soldat, J. K. 1971. Modeling of environmental pathways and radiation doses from nuclear facilities. Trans. Amer. Nucl. Soc. 14(2): 459-460. (Also: Battelle Northwest Lab., U. S. AEC report BNWL-SA 3939) (CONF-711008) (NSA 26: 2395, 4751).

Soldat, J. K. 1974. Environmental behavior and radiation doses from iodine-129. Battelle Northwest Lab., U. S. AEC report BNWL-SA-4879. 27pp. (CONF-740702-6) (NSA 30: 21405).

Soldat, J. K. 1974. Radiological impact of a nuclear center on the environment. Battelle Northwest Lab., U. S. AEC report BNWL-B-333. 19pp. (NSA 30: 26763).

Soldat, J. K. and C. M. Unruh. 1971. Environmental surveillance of a nuclear facility. Battelle Northwest Lab., U. S. AEC report BNWL-SA-3786. 51pp. (For: 16th Ann. Health Physics Soc. Meeting, New York, N. Y., 12 Jul 1971) (CONF-710710-5) (NSA 25: 46816).

Soldat, J. K., N. M. Robinson and D. A. Baker. 1974. Models and computer codes for evaluating environmental radiation doses. Battelle Northwest Lab., U. S. AEC report BNWL-1754. 170pp. (NSA 29: 29346).

Soldat, J. K., et al. 1973. The radioecology of iodine-129: an interim report BNWL-1783. iv, 74pp.

Soliman, M. H. 1972. Age at pupation and population effects on survival of adults of the flour beetle _Tribolium_ after x-irradiation. Int. J. Radiat. Biol. 22(5): 425-430. (NSA 27: 2822).

Soliman, M. H. 1972. Variable effects of x-irradiation on _Tribolium castaneum_ and _T. confusum_ due to strain and age at pupation. Dissertation, Univ. of Alberta, Edmonton. 91pp. (NSA 31: 33520).

Soliman, M. H. 1972. Correlated response in the induction of split wing after selection for developmental tome and x-irradiation in _Tribolium castaneum_. Abstr. Genetics 71: S61.

Soliman, M. H. 1973. "Geographic" variation in sensitivity to x-irradiation in _Tribolium_. I. Differences in adult emergence due to species, strain, and age at pupation. II. Incidence of developmental abnormalities. Radiation Research 54(3): 510-520, 521-530. (NSA 28: 15755, 15757).

Soliman, M. H. 1973. Lethality: dependence on developmental rate in x irradiated pupae of _Tribolium_. Experientia 29(5): 550-551. (NSA 28: 21774).

Soliman, M. H. 1974. Repair of x-irradiation damage in _Tribolium castaneum_ strains selected for differences in developmental rates. Genetics 78(3): 897-904. (NSA 31: 20068).

Solodikhina, L. D. 1959. On the natural radioactivity of precipitation. Izv. Akad. Nauk SSSR (geofiz.) (2): 276-283. (In Russian).

Sologub, M. I. 1962. Effects of different doses of x-ray irradiation on changes of intra-cellular resting potentials of muscular fibers. Vestn. Leningrad Univ. (biol.) (15): 138-145. (In Russian) (NSA 19: 24362) (Used _Rana temporaria_).

Solon, L. R., et al. 1959. Further investigations of natural environmental radiation. Health and Safety Lab., New York, U. S. AEC report HASL-73.

Soloway, S., J. H. Welsh and A. K. Solomon. 1953. Studies of a Ca^{45} transport in crayfish nerve. J. Cell. Comp. Physiol. 42: 471-485. (Also: Harvard Univ., U. S. AEC report AECU-2414) (NSA 7: 1905).

Solyus, A. A. 1973. Ratio between ^{137}Cs and K concentration factors in freshwater organisms. pp. 102-107. In: _Radioecology of Water Organisms. II._ (See Andrushaitis, et al, 1973) (AEC-tr-7606, pp. 62-65) (NSA 30: 21133).

Solyus, A. A., D. G. Fleyshman and V. G. Leont'yev. 1970. Cs^{137} concentration by the fishes of freshwater lakes. Vopr. ikhtiol. 10(6): 827-832. (Transl. pp. ?).

Soma, S. D., et al. 1975. Tritium behavior pattern in some soil-plant systems in a tropical environment. pp. 195-210. In: _Isotope Ratios as Pollutant Source and Behavior Indicators._ (See IAEA, 1975) (NSA 33: 7632).

Somayajulu, B. L. K. 1969. Cosmic-ray-produced ^{32}Si in near-coastal waters. Proc. Indian Acad. Sci. A69: 338-346. (NSA 23: 47661).

Somko, V. A. 1973. Determination of radon in water. Gigiyena sanitariya (5): 94-96. (In Russian) (NSA 28: 20722).

Sommer, N. F., et al. 1963. Recovery of gamma irradiated _Rhizopus stolonifer_ sporangiospores during auto-inhibition of germination. J. Cell. Comp. Physiol. 61(1): 93-98. (NSA 17: 27118).

Sommer, N. F., et al. 1964. An oxygen-dependent postirradiation restoration of _Rhizopus stolonifer_ sporangiospores. Radiation Research 22(1): 21-28. (NSA 18: 23288).

Sommer, N. F., et al. 1968. Heat sensitization of control of grey mold of strawberry fruits by gamma irradiation. Radiation Botany 8: 441-448. (NSA 32008).

Sommer, N. F., et al. 1972. Comparative sensitivity to gamma radiation of conidia, mycelia, and sclerotia of Botrytis cinerea. Radiation Botany 12 (2): 99-103. (NSA 26: 41108).

Sondhaus, C. A. 1964. Effect of high-energy protons and alpha particles on small mammals. pp. 97-103. In: 2nd Symp. on Protection Against Radiation in Space. (Gatlinburg, Tennessee, October 12-14, 1964). A. Reetz, Jr., ed. U. S. NASA, Washington, report NASA-SP-71. (NSA 20: 1614).

Sonehara, S. 1933. Studies of the effects of x-rays upon the development of a pond snail Lymnaea (Radix) japonica. J. Sci. Hiroshima Univ. B1: 151-169.

Sonenshine, D. E. 1968. Radioisotopes in studies on the ecology of tick vectors of disease. pp. 31-52. In: Isotopes and Radiation in Entomology. (See IAEA, 1968) (NSA 24: 31828).

Sonenshine, D. E. 1970. Fertility and fecundity of Dermacentor variabilis reared from radioisotope-tagged larvae. J. Econ. Entomol. 63: 1675-1676. (NSA 25: 16295).

Sonenshine, D. E. 1970. Use if radioisotopes for studies on the ecology of tick vectors of disease. Final progress report, April 1, 1966-December 31, 1970. Old Dominion Coll., Norfolk, Va., U. S. AEC report ORO-3514-10. 129pp. (NSA 25: 18934).

Sonenshine, D. E. 1971. Mass rearing of radioisotope-tagged larval ticks for ecological investigations. J. Econ. Entomol. 64(6): 1423-1429.

Sonenshine, D. E. and G. M. Clark. 1968. Field trials on radioisotope tagging of ticks. J. Med. Entomol. 5(2): 229-235.

Sonenshine, D. E. and G. Esch. 1970. Radiolabeling Dermacentor variabilis with strontium-90 and strontium-90 plus carbon-14 mixtures. J. Econ. Entomol. 63(5): 1635-1638.

Sonenshine, D. E. and G. F. Levy. 1971. The ecology of the lone star tick Amblyomma americanum (L.) in two contrasting habitats in Virginia Acarina: Ixodidae). J. Med. Entomol. 8(6): 623-635. (Radionuclide-tagged larvae).

Sonenshine, D. E. and C. E. Yunker. 1968. Radio-labeling of tick progeny by inoculation of procreant females. J. Econ. Entomol. 61(6): 1612-1617.

Sonnenblick, B. P. and L. P. Gartner. 1967. Lifespan studies with strains of gamma-irradiated Drosophila adults. Radiation Research 31(3): 612.

Sonnenblick, B. P. and J. Grodis. 1963. Can a dose of 4-5000 r of x rays double the longevity of D. melanogaster? Drosophila Inform. Serv. 37: 130. (NSA 17: 35501).

Sonnenblick, B. P. and D. Rockford. 1964. Mortality and life span of D. melanogaster (a wild-type and a mutant strain with a ring-x chromosome) exposed as young adults to x-rays or to high doses of cobalt-60 gamma rays. Drosophila Inform. Serv. 39: 113-114.

Sonnischen, J. C., Jr., et al. 1970. Dispersion characteristics of the Columbia River between Miles 383 to 355. Battelle Northwest Lab., U. S. AEC report BNWL-1477. (NSA 25: 6422).

Sonoda, H. 1972. Effect of gamma irradiation on fertility and mating competiveness of the mosquito, Culex pipiens molestus F. (Diptera: Culicidae). Appl. Entomol. Zool. 7(3): 103-108. (NSA 29: 18647).

Sons, E. 1957. The use of radioisotopes in tracing sewage introduced into the sea and coastal waters. Atompraxis 3: 443-444. (NSA 11: 11906).

Sons, E. 1957. Use of radioactive isotopes by tracing introduction of polluted water in lakes and coastal waters. Atompraxis 3(8/9): 306. (In German).

Sood, S. K. and E. O. Knutson. 1972. Scavenging study of snow and ice chrystals. Final technical report, April 1, 1971-March 31, 1972. IIT Research Inst., U. S. AEC report IITRI-C-6105-19. 75pp. (NSA 26: 50601).

Sorenson, J. C., C. B. Smith and V. Francisco. 1971. Dispersion studies using stable and radioactive tracers in fresh water streams. pp. 187-197. In: Proc. Institute of Environmental Sciences 17th Ann. Tech. Meeting and Equipment Exposition. (April 26-30, 1971, Los Angeles, California). Inst. of Environmental Sciences, Mt. Prospect, Ill. (CONF-710409) (NSA 25: 29213).

Soria, F. 1963. Study of populations and dispersion of Ceratitis capitata Wied. (Dipt. Trypetidae) in Tunis with the use of radioisotopes. pp. 357-362. In: Radiation and Radioisotopes Applied to Insects of Agricultural Importance. (See IAEA, 1963) (In French).

Sorokin, B. P. (Ed.). 1971. Effect of ionizing radiation on the organism. The problem of the effect of radioactive water pollution on the reproduction of commercial fishes. Tr. polyar. nauchn.-issled. proekt. inst. mor morsk. ryb. khoz. okeanogr. (29): 1-182. (AEC-tr-7418) (NSA 27: 27944).

Sorokin, Yu. I. 1955. Determination of the importance of chemosynthesis in water of Rubinsk Reservoir using ^{14}C. Dokl. Akad. Nauk SSSR 105(6): 1343-1345. (In Russian).

Sorokin, Yu. I. 1955. A study of the exchange of phosphorus in the chemosynthesis of sulfate-reducing bacteria with the aid of radiophosphorus. In: Isotopes in Microbiology. (See Imshenetskiy, 1955) (AEC-tr-3494, pp. 144-156).

Sorokin, Yu. I. 1956. Processes of bacterial synthesis in silt deposits. Mikrobiologiya 24(4): 393-399. (In Russian) (Ref. zh. (biol.) 1956: 43583).

Sorokin, Yu. I. 1956. On the use of the radioactive carbon^{14} for the study of productivity in a water basin. Tr. vses. gidrobiol. obshch. 7: 271-286. (U. S. Navy Hydrographic Office, Washington, report NHO-TRANS-147).

Sorokin, Yu. I. 1957. Determination of the photosynthetic yield of phytoplankton with the aid of C^{14}. pp. 564-576. In: All-Union Conf. on Photosensitivity: Report of Problems of Photosensitivity. Vol. 2. (English transl. U. S. Dept. of Agriculture, Washington).

Sorokin, Yu. I. 1958. Application of radioactive carbon (C^{14}) for the investigation of the cycle of organic substances in water basins. pp. 188-193. In: Studies of the Animal Organism. Acad. Sci. USSR, Moscow. (In Russian).

Sorokin, Yu. I. 1958. Results and prospects of using isotopic carbon for the investigation of carbon cycle in water basins. pp. 663-684. In: Proc. Intern. Conf. on Radioisotopes in Scientific Research. Vol. 4. R. C. Exterman, ed. UNESCO, Paris.

Sorokin, Yu. I. 1959. Determination of the photosynthetic productivity of phytoplankton in water using C^{14}. Fiziol. rasteniy 6(1): 125-133. (In Russian).

Sorokin, Yu. I. 1960. Determination of the isotope effect during labeled carbon dioxide assimilation in photosynthesis and chemosynthesis. Mikrobiologiya 29(2): 204-208. (Microbiology 29: 153-156).

Sorokin, Yu. I. 1962. Determination of correction coefficients for self-absorption of C^{14} emission in estimation of photosynthetic and chemosynthetic production in waters. Mikrobiologiya 31(1): 121-128. (Microbiology 29: 96-102) (BA 42: 20478).

Sorokin, Yu. I. 1963. Radioactive carbon in hydrobiological research. Vest. Akad. Nauk SSSR (6): 69-71. (JPRS-20614).

Sorokin, Yu. I. 1966. Carbon-14 method in the study of the nutrition of aquatic animals. Int. rev. ges. hydrobiol. hydrogr. 51: 209-224.

Sorokin, Yu. I. and A. N. Meshkov. 1958. The application of radioactive C^{14} to determine the assimilation of protococcoid algae by larvae of the midge Tendipes plumosus. Dokl. Akad. Nauk SSSR 118(1): 205. (Doklady (biol. sci.) 118: 122-125).

Sorokin, Yu. I. and A. N. Meshkov. 1959. Use of radioactive carbon in studying the nutrition of aquatic invertebrates. Tr. inst. biol. vod. Akad. Nauk SSSR 2(5): 7-14. (In Russian).

Sorokin, Yu. I. and Ye. D. Mordukhai-Boltovskaya. 1962. The study on the feeding of rotifers Asplanchna by means of C^{14}. Byull. inst. biol. vod. Akad. Nauk SSSR (12): 17-20. (In Russian).

Sorokin, Yu. I. and D. A. Panov. 1967. The use of C^{14} for the quantitative study of the nutrition of fish larvae. Int. rev. gesamten hydrobiol. 51 (5): 743-756. (BA 48: 73764).

Sorokin, Yu. I. and D. A. Panov. 1968. Experimental determination of the food consumption of hatchlings and fry of silver carp with the aid of ^{14}C. Dokl. Akad. Nauk SSSR 182(1): 208-210. (In Russian) (SFA 17: 14744).

Sorokina, O. N., I. D. Anikeyeva and Ye. L. Iofa. 1964. The protective effect of metabolites of radioresistant plants. Radiobiologiya 4(2): 279-283. (AEC-tr-6405, pp. 144-151 (NSA 18: 31308) (Incl. moss).

Sorokina, Z. O. 1971. Kinetics of environmental ^{42}K exchange for potassium ions of frog skeletal muscle fibers. Fiziol. zh. Akad. Nauk Ukr. RSR 17 (6): 765-768. (In Ukrainian) (NSA 26: 36466).

Sorsa, V. 1970. Fern cytology and the radiation field. Ch. G-3, pp. G39-G50. In: A Tropical Rain Forest. (See H. T. Odum and R. F. Pigeon, 1970) (NSA 25: 21651).

Sorvachev, K. F. and O. V. Belokopytova. 1960. Fish consumption of inorganic carbon from the surrounding medium and its participation in metabolism. Biokhimiya 25(3): 459-464. (Biochemistry USSR 25: 351-355).

Sosna, M. 1957. Cosmic rays as a factor in the cytological process in Hordeum. Folia biol. (Prague) 3(5): 308-313.

Sosna, M. 1957. Mitotic indices in cells of the root meristem under the influence of cosmic radiation. Csl. biol. 6(3): 222-226.

Sosna, M. 1967. Effect of secondary cosmic rays on spontaneous mutability Biol. plant acad. sci. Bohemoslov 9: 473-475. (NSA 22: 25813) (Equisetum).

Sotobayashi, T., T. Suzuki and A. Furusawe. 1969. Inter-hemispheric transfer of fresh debris from French tests in 1968. Nature 224: 1096-1097. (NSA 24: 9529).

Sousselier, Y. 1969. Selection criteria for siting reprocessing plants in Europe. France Commissariat à l'Énergie Atomique, Paris, report CEA-CONF-1528. 14pp. (For: 17th Conf. on Remote Systems Technology, San Francisco, Calif., Nov. 30-Dec. 4, 1969) (CONF-691102-24) (NSA 24: 44092).

Sousselier, Y. 1972. Present situation and outlook concerning radioactive waste disposal. Centre d'Études Nucléaires, France Commissariat à l'Énergie Atomique, Fontenay-aux-Roses, report CEA-CONF-2017. 8pp. (For: Meeting on Radioactive Waste Disposal, Paris, France, 12 Apr. 1972) (In French) (CONF-720452) (NSA 26: 59354).

South Africa Atomic Energy Board. 1972. National Conf. on the Technological Applications of Nuclear Techniques. (Pelindaba, South Africa, October 12-13, 1972. Pelindaba, Pretoria. 242pp. (In English, Africaans) (CONF-721091) (NSA 28: 21460).

South Carolina Electric and Gas Company. 1971. Virgil C. Summer Nuclear Station, Unit 1. Environmental report. Columbia. 186pp. U. S. AEC DOCKET-50395-2. (NSA 25: 44072).

Southern California Edison Company. 1971. Environment. Los Angeles. 200pp. (NSA 26: 58780).

Southwood, T. R. E. 1966. Radioactive isotopes. Ch. 3, pp. 63-71. In: Ecological Methods. Methuen, London.

Sovershayev, P. F. 1971. Effect of γ irradiation on some cytophysiological and biochemical processes in germinated pine, fir, and Siberian laerch seeds. Radiobiologiya 11(5): 737-741. (AEC-tr-7306, pp. 146-152) (NSA 26: 38632).

Spaander, J. and L. Strackee. 1971. Radioecology as an aid to the general environmental pollution problem. pp. 587-599. In: Proc. Intern. Symp. on Radioecology Applied to the Protection of Man and His Environment. (See Commission of the European Communities, 1971) (NSA 27: 5232).

Spalding, R. F. and W. M. Sackett. 1972. Uranium in runoff from the Gulf of Mexico distributive province: anomalous concentrations. Science 175 (4022): 629-631. (NSA 26: 20341).

Spangler, M. B. 1974. Environmental and social issues of site choice for nuclear power plants. Energy Policy 2(1): 18-32. (NSA 30: 2095).

Sparr, M. C. 1960. I. Quantitative chemical analyses of lake water using ion exchange resins. II. Investigations of the nutrient status of bog lakes using the growth of *Chlorella pyrenoidosa* as an index. Dissertation, Univ. of Wisconsin, Madison.

Sparrow, A. H. 1948. Relative x-ray sensitivity of metaphase anterphase chromosomes. Nature 162(412): 651-652.

Sparrow, A. H. 1951. Radiation sensitivity of cells during mitotic and meiotic cycles with emphasis on possible cytochemical changes. Ann. N.Y. Acad. Sci. 51: 1508-1540.

Sparrow, A. H. 1951. Some cytological and morphological changes induced in plants by ionizing radiation. Abstr. Science 114(2967): 488.

Sparrow, A. H. 1954. Somatic mutations induced in plants by treatment with x and γ radiation. In: Proc. Intern. Congr. of Genetics, 9th Congr. (Bellagio, Italy, 1953). Catyologia 6(Suppl. Pt.2): 1105-1106. (PBA 27: 1479).

Sparrow, A. H. 1956. Cytological changes induced by ionizing radiations and their possible relation to the production of useful mutations in plants. pp. 76-113. In: Work Conf. on Radiation Induced Mutations. Brookhaven National Lab., Upton, L. I., New York.

Sparrow, A. H. 1960. Uses of large sources of ionizing radiation in botanical research and some possible practical applications. pp. 195-219. In: Large Radiation Sources in Industry. (Conf. Proc., Warsaw, 8-12 September 1959). International Atomic Energy Agency, Vienna, Publ. STI/PUB/12. (NSA 15: 84).

Sparrow, A. H. 1961. Comments on the conference with respect to advances in the fundamental aspects of plant and seed radiobiology. pp. 641-644. In: Symp. on the Effects of Ionizing Radiation on Seeds and Its Significance for Crop Improvement. (Karlsruhe, Germany). International Atomic Energy Agency, Vienna.

Sparrow, A. H. 1962. Predictions of radiosensitivities of plant species. Abstr. p. 178. In: 2nd Intern. Congr. Radiation Reserach. (See International Congress of Radiation Research, 1962).

Sparrow, A. H. 1962. The role of the cell nucleus in determining radiosensitivity. Brookhaven Lecture Ser. No. 17. Brookhaven National Lab., U. S. AEC report BNL-766(T-287). ii, 88pp. (NSA 17: 38739).

Sparrow, A. H. 1963. The tolerance of plants to ionizing radiation: variations, modifications and predictions. In: Thermonuclear War: Some Effects on People and Nature. T. Blaisdell, ed. Stonier, New York. (Also: Brookhaven National Lab., U. S. AEC report BNL-7435).

Sparrow, A. H. 1963. Summary of comments of plant growth stimulation by ionizing radiation. Brookhaven National Lab., U. S. AEC report BNL-8491. 4pp. (NSA 19: 2065).

Sparrow, A. H. 1964. Variations in the sensitivity of plant taxa to ionizing radiation. Abstr. p. 319. In: 10th Intern. Botanical Congr. (Abstr. of Papers; Edinburgh, August, 1964).

Sparrow, A. H. 1965. Comparisons of the tolerances of higher plant species to acute and chronic exposures of ionizing radiation. Idengaku Zasshi 40 (Suppl.): 12-37. (Also: Brookhaven National Lab., U. S. AEC report BNL-8539) (NSA 19: 43782).

Sparrow, A. H. 1965. Research uses of the gamma field and related radiation facilities at Brookhaven National Laboratory. Brookhaven National Lab., U. S. AEC report BNL-8892. (NSA 19: 21855).

Sparrow, A. H. 1966. Research uses of the gamma field and related radiation facilities at Brookhaven National Laboratory. Radiation Botany 6: 377-405.

Sparrow, A. H. 1966. Variations in the radiation sensitivity of chromosomes associated with their size and DNA content. Abstr. p. 208. In: 3rd Intern. Congr. Radiation Research. (Cortina, Italy, June 26-July 2, 1966).

Sparrow, A. H. and E. Christensen. 1953. Tolerance of certain higher plants to chronic exposure to gamma radiation from cobalt-60. Science 118(3075): 697-698. (Also: Brookhaven National Lab., U. S. AEC report BNL-1588) (NSA 8: 434).

Sparrow, A. H. and R. L. Cuany. 1959. Radiation-induced somatic mutations in plants. Brookhaven National Lab., U. S. AEC report BNL-4296. 10pp. (See also: U. S. AEC, 1959, TID-7578, pp. 153-156) (NSA 14: 29).

Sparrow, A. H. and H. J. Evans. 1961. Nuclear factors affecting radiosensitivity. I. The influence of nuclear size and structure, chromosome complement and DNA content. Brookhaven Symposia in Biology 14: 76-100.

Sparrow, A. H. and F. Forro. 1953. Cellular radiobiology. Ann. Rev. Nucl. Sci. 3: 339-368.

Sparrow, A. H. and J. E. Gunckel. 1956. The effects on plants of chronic exposure to gamma radiation from radiocobalt. Proc. Intern. Conf. on the Peaceful Uses of Atomic Energy 12: 52-59

Sparrow, A. H. and C. F. Konzak. 1958. The use of ionizing radiation in plant breeding. Accomplishments and prospects. pp. 425-452. In: Camellia Culture. E. C. Tourje, ed. MacMillan Co., New York.

Sparrow, A. H. and M. Maldawer. 1950. Differential rejoining as a factor in apparent sensitivity of chromosomes to x-ray breakage. Proc. Natl. Acad. Sci (US) 36(11): 636-643.

Sparrow, A. H. and J. P. Miksche. 1961. Correlation of nuclear volume and DNA content with higher plant tolerance to chronic radiation. Science 134(3474): 282-283. (NSA 15: 24710).

Sparrow, A. H. and B. A. Rubin. 1952. Effects of radiations on biological systems. Surv. Biol. Progr. 2: 1-52.

Sparrow, A. H. and L. A. Schairer. 1958. Some factors influencing radioresistance and tumor induction in plants. Proc. 2nd Intern. Conf. on the Peaceful Uses of Atomic Energy 27: 335-340. (Paper A/CONF.15/P/2172) (NSA 12: 14509).

Sparrow, A. H. and L. A. Schairer. 1962. Plant radiosensitivities and potential damage from fallout radiation. Abstr. Radiation Research 16(4): 584.

Sparrow, A. H. and W. R. Singleton. 1953. The use of radiocobalt as a source of gamma rays and some effects of chronic irradiation on growing plants. Am. Naturalist 87: 29-48.

Sparrow, A. H. and R. C. Sparrow. 1954. The formation of polyploid microspores in Trillium erectum following x-ray treatment at first meiotic metaphase. Abstr. Radiation Research 1(2): 229.

Sparrow, A. H. and G. M. Woodwell. 1962. Prediction of the sensitivity of plants to chronic gamma irradiation. Radiation Botany 2: 9-26. (Also: Brookhaven National Lab., U. S. AEC report BNL-5932) (NSA 16: 30247).

Sparrow, A. H. and G. M. Woodwell. 1963. Prediction of the sensitivity of plants to chronic gamma irradiation. pp. 257-270. In: Radioecology. (See Schultz and Klement, 1963).

Sparrow, A. H., M. Furuya and S. S. Schwemmer. 1968. Effects of x- and gamma radiation on anthocyanin content in leaves of Rumex and other plant genera. Radiation Botany 8(1): 7-16.

Sparrow, A. H., M. J. Moses and R. Dubow. 1952. Relationships between ionizing radiation, chromosome breakage and certain other nuclear disturbances. Exptl. Cell Res. (Suppl. 2): 245-267.

Sparrow, A. H., M. J. Moses and R. Steele. 1952. A cytological and cytochemical approach to an understanding of radiation damage in dividing cells. Brit. J. Radiol. 25(292): 182-188.

Sparrow, A. H., A. F. Rogers and S. S. Schwemmer. 1968. Acute gamma radiosensitivity of 28 woody-plant species. Abstr. Radiation Research 35(2): 575.

Sparrow, A. H., A. F. Rogers and S. S. Schwemmer. 1968. Radiosensitivity studies with woody plants. I. Acute gamma irradiation survival data for 28 species and prediction for 190 species. Radiation Botany 8(2): 149-186. (NSA 22: 45482).

Sparrow, A. H., L. A. Schairer and K. M. Marimuthu. 1968. Genetic and cytologic studies of Tradescantia irradiated during orbital flight. In: Symp. on the Biosatellite II Experiments. BioScience 18(6): 582-590.

Sparrow, A. H., L. A. Schairer and K. M. Marimuthu. 1971. Radiologic studies of Tradescantia plants orbited in Biosatellite 2. Brookhaven National Lab., U. S. NASA report N-72-20054. 24pp. (NSA 26: 53382).

Sparrow, A. H., L. A. Schairer and J. P. Miksche. 1961. Tolerance of vegetation to chronic exposures of ionizing radiations. Abstr. Bull. Ecol. Soc. Am. 42: 95.

Sparrow, A. H., L. A. Schairer and R. C. Sparrow. 1963. Relationship between nuclear volumes, chromosome numbers, and relative radiosensitivities. Science 141(3576): 163-166. (Also: Brookhaven National Lab., U. S. AEC report BNL-6999) (NSA 17: 31939).

Sparrow, A. H., L. A. Schairer and G. M. Woodwell. 1965. Tolerance of Pinus rigida trees to a ten-year exposure to chronic gamma irradiation from cobalt-60. Radiation Botany 5(1): 7-22. (Also: Brookhaven National Lab., U. S. AEC report BNL-8404) (NSA 19: 15174).

Sparrow, A. H., S. S. Schwemmer and P. J. Bottino. 1971. Effects of external gamma radiation from radioactive fallout on plants, with special reference to crop production. Radiation Botany 11(2): 85-118. (NSA 25: 47035).

Sparrow, A. H., S. S. Schwemmer and P. J. Bottino. 1973. Influence of dose, environmental conditions, and nuclear volume on survival times in several gamma-irradiated plant species. Int. J. Radiat. Biol. 24(4): 377-388.

Sparrow, A. H., S. S. Schwemmer and K. H. Thompson. 1976. Radiosensitivity studies with woody plants. III. Predictions of limits of probable acute and chronic LD_{50} values from lognormal distributions of interphase chromosome volume in gymnosperms. Radiation Research 65(2): 315-326. (NSA 33: 29594).

Sparrow, A. H., R. C. Sparrow and L. A. Schairer. 1960. The use of x-rays to induce somatic mutations in Saintpaulia. African Violet Mag. 13: 32-37. (Also: Brookhaven National Lab., U. S. AEC report BNL-4667) (NSA 15: 7201).

Sparrow, A. H., A. G. Underbrink and R. C. Sparrow. 1967. Chromosomes and cellular radiosensitivity. 1. The relationship of D_o to chromosome volume and complexity in seventy-nine different organisms. Radiation Research 32(4): 915-945. (NSA 22: 10854).

Sparrow, A. H., et al. 1950. Sensitivity to chromosomes to breakage by x-rays and its relationship to the nucleic acid cycle in dividing cells. Abstr. Cancer Research 10(4): 241-242.

Sparrow, A. H., et al. 1956. Tumor formation and other porphogenic responses in an amphidiploid tobacco hybrid exposed to chronic gamma irradiation. Am. J. Bot. 43(5): 377-388.

Sparrow, A. H., et al. 1961. Some factors affecting the responses of plants to acute and chronic radiation exposures. Radiation Botany 1(1): 10-34. (Also: Brookhaven National Lab., U. S. AEC report BNL-4912) (NSA 15: 30491).

Sparrow, A. H., et al. 1963. The radiosensitivity of gymnosperms. I. The effect of dormancy on the response of Pinus strobus seedlings to acute irradiation. Radiation Botany 3(2): 169-173. (NSA 17: 33723).

Sparrow, A. H., et al. 1970. Woody plants: changes in survival in response to long-term (8 years) chronic gamma irradiation. Science 169: 1082-1084. (NSA 24: 48738).

Sparrow, A. H., et al. 1970. Radiosensitivities of selected amphibians in relation to their nuclear and chromosome volumes. Radiation Research 42: 353-371. (NSA 24: 32195).

Sparrow, A. H., et al. 1970. Radiosensitivity studies with woody plants. II. Survival data for 13 species irradiated chronically for up to 8 years. Radiation Research 44(1): 154-177. (NSA 25: 6816).

Sparrow, A. H., et al. 1965. The use of nuclear and chromosomal variables in determining and predicting radiosensitivities. pp. 101-131. In: The Use of Induced Mutations in Plant Breeding. Pergamon Press, New York.

Sparrow, A. H., et al. 1975. Chromosomes and cellular radiosensitivity. IV. New radiotaxon represented by the ferns. Int. J. Radiat. Biol. 27(4): 343-354. (NSA 32: 17445).

Sparrow, R. C. and A. H. Sparrow. 1965. Relative radiosensitivities of woody and herbaceous spermatophytes. Science 147(3664): 1449-1451. (Also: Brookhaven National Lab., U. S. AEC report BNL-8470) (NSA 19: 17640).

Sparrow, S., Jr. 1973. Chronic radiation stress on soil microorganisms of a grassland. Thesis, Colorado State Univ., Fort Collins. xi, 103pp.

Spaulding, J. F. and P. McWilliams. 1965. Radiosensitivity, circadian rhythmicity, and radiobiology. Health Physics 11: 647-651. (Also: Los Alamos Scientific Lab., U. S. AEC report LA-DC-6404(Rev.)) (NSA 20: 5336).

Spear, W. G. and M. O. Rankin. 1955. River probe. Hanford Atomic Products Operation, U. S. AEC report HW-38884. 5pp.

Speas, C. J. 1965. Method of treating grass seed. U. S. Patent 3,197,640. (NSA 19: 46004) (Concerns radiation effects on root systems and growth).

Spector, W. S. (Ed.). 1956. Handbook of Biological Data. W. B. Saunders Philadelphia. (Incl. information on ionizing and UV effects).

Speer, D. R. and D. A. Waite. 1975. Statistical distributions as applied to environment surveillance data. Battelle Northwest Lab., U. S. ERDA report BNWL-SA-5482. 30pp. (NSA 33: 7290).

Speers, C. F. 1956. Radioisotopes in forest insect studies. Proc. Assoc. south. Agr. Workers 53: 130. (Used ^{192}Ir).

Speicher, B. R., K. G. Speicher and F. L. Roberts. 1965. Genetic segregation in the unisexual wasp Devorgilla. Genetics 52: 1035-1041. (Also: Oak Ridge National Lab., U. S. AEC report ORNL-P-1330) (NSA 20: 1643).

Spencer, J. L. and A. F. Blakeslee. 1954. Induced pollen lethals from seeds of Datura stramonium exposed to a nuclear detonation. Proc. Natl. Acad. Sci. (US) 40(6): 441-446. (NSA 8: 4863).

Spencer, J. L. and A. F. Blakeslee. 1955. Induced pollen lethals from seeds of Datura stramonium exposed to different radiations. Proc. Natl. Acad. Sci. (US) 41(5): 307-312. (NSA 9: 4671).

Spencer, N. R. 1971. Sterilization of insect diet by gamma irradiation. J. Econ. Entomol. 64(3): 753-754. (NSA 25: 42749).

Sperlich, D. 1967. Population genetics. Part 1. Drosophila. Fortschr. Zool. 18: 223-278. (n German) (NSA 22: 8509).

Spiers, F. W., M. J. McHugh and D. B. Appleby. 1964. Environmental γ-ray dose to populations: surveys made with a portable meter. pp. 885-906. In: The Natural Radiation Environment. (See Adams and Lowder, 1964) (NSA 18: 35750).

Spigarelli, S. A. 1973. Ecological factors affecting the accumulation of cesium-137 fallout by a natural population of largemouth bass (Micropterus salmoides). pp. 328-333. In: Radionuclides in Ecosystems. (See Nelson, 1973) (NSA 25: 46802).

Spikes, J. D. 1963. Radiation effects and peaceful uses of atomic energy in the plant and soil sciences. pp. 5-11. In: Radioecology. (See Schultz and Klement, 1963).

Spikes, J. D., B. C. Mayne and R. W. Van Norman. 1958. Gamma radiation effects on Chlorella metabolism. Plant Physiol. 33(Suppl.): xxvii.

Spinks, J. W. T. 1956. Studies of special problems in agriculture and silviculture by the use of radioisotopes. Proc. Intern. Conf. on the Peaceful Uses of Atomic Energy 12: 75-86. (Paper A/CONF.8/P/10).

Spinks, J. W. T. 1958. Nuclear energy in agriculture and biology. In: 38th Ann. Meeting and Convention of the Agricultural Institute of Canada. (Acadia Univ., Wolfville, N. S., June 24, 1958).

Spinks, J. W. T., M. R. Berlie and J. B. O'Neil. 1949. Determination of the fate of calcium in the laying hen by means of radiocalcium (Ca^{45}). Science 110(2857): 332-333.

Spinks, J. W. T., et al. 1948. Lethal effects of absorbed radioisotopes on plants. Can. J. Research 26C: 249-262. (NSA 1: 901).

Spirin, V. D., Yu. S. Belle and V. F. Chimina. 1965. Measurement of the radon concentration in water by gamma radiation. Med. radiologiya 10(12): 11-13. (In Russian) (NSA 20: 20959).

Spitsyn, E. Ya. 1965. Treatment and Disposal of Radioactive Wastes. Atomizdat, Moscow. 107pp. (AEC-tr-6881; TT-68-50326; IPST-Cat.-5084) (NSA 23: 542).

Spitsyn, V. I., et al. 1958. A study of the migration of radioelements in soils. Proc. 2nd Intern. Conf. on the Peaceful Uses of Atomic Energy : Paper A/CONF.15/P/2207. (NSA 13: 7073).

Splittstoesser, D. F. and L. M. Massey., Jr. 1967. Irradiation of composts used in propagation of Agaricus bisporus. Appl. Microbiol. 15: 646-649. (NSA 21: 28704).

Spofford, W. O., Jr. and H. A. Thomas, Jr. 1963. A least cost evaluation of low level liquid atomic waste treatment-simplified deterministic model. pp. 353-386. In: Transport of Radionuclides in Fresh Water Systems. (See Kornegay, et al., 1963).

Spragg, W. T. and R. E. Fox. 1974. Use of a radioactive tracer to study the nesting system of Mastotermes darwiniensis Froggat. Australia Atomic Energy Commission, Lucas Heights, report AAEC/E-303. 10pp. (NSA 30: 726).

Spring, E. 1969. Increased radiosensitivity following simultaneous ultrasonic and gamma-ray irradiation. Radiology 93: 175-176. (NSA 23: 45948) (Lolium italicum).

Springell, P. H. 1963. Irradiation. A sterilization weapon against the corn borer. Agric. Res. 12(3): 6.

Spronk, N. 1965. Copper in Lymnaea stagnalis L. pp. 643-657. In: Radioisotopes in Animal Nutrition and Physiology. (Proc. Symp., Prague, 23-27 November 1964). International Atomic Energy Agency, Vienna, Publ. STI/PUB/90. (Neutron activation analysis).

Spronk, N. 1967. Nuclear activation in the animal sciences. pp. 335-352. In: Nuclear Activation Techniques in the Life Sciences. (See IAEA, 1967) (NSA 23: 4557).

Spurny, K. and E. Holy. 1970. Radioactive fallout in Prague, 1969. Zdrav. Tech. Vzduchotech. 13(3): 141-143. (In Czech) (NSA 26: 199).

Spyrou, N. M. 1973. Nuclear data and neutron activation analysis of biological samples. pp. 335-349. In: Nuclear Data in Science and Technology. Vol. I. International Atomic Energy Agency, Vienna, Publ. STI/PUB-343. (NSA 29: 15256).

Spyrou, N. M., K. Ingle and F. Ozek. 1974. Cyclic activation in trace element analysis of environmental samples. pp. 106-122. In: Proc. 2nd Intern. Conf. on Nuclear Methods in Environmental Research. (See Vogt and Meyer, 1974) (NSA 33: 23324).

Spyrou, N. M., et al. 1974. Realistic detection limits for neutron activation analysis of biological samples. pp. 471-487. In: Comparative Studies of Food and Environmental Contamination. (See IAEA, 1974) (NSA 30: 14768).

Squire, H. M. 1960. Changes with time in the availability of strontium-90 in soil. Nature 188(4749): 518-519. (NSA 15: 7188).

Squire, R. D. 1969. An analysis of fitness components in the brine shrimp Artemia salina, with reference to gamma irradiation and inbreeding depression. Thesis, North Carolina State Univ., Raleigh. 129pp. (NSA 24: 14575).

Squire, R. D. 1970, 1973. The effects of acute gamma irradiation on the brine shrimp, Artemia. I. (w/D. S. Grosch) Life spans and male reproductive performance. II. Female reproductive performance. III. Male F_1 reproductive performance following paternal irradiation of mature sperm. Biol. Bull. 139(2): 363-374, 375-385; 144(1): 192-199. (NSA 25: 24309, 21285; 28: 1830, 21776).

Srapenyants, R. A. 1971. The results of work on developing and putting into operation the radiation technology methods and facilities. pp. 178-182. In: Isotope and Nuclear Radiation Application in Agriculture. 2. Isotope Application in Crop Science and Agriculture. Atomizdat, Moscow. (In Russian) (NSA 31: 5997).

Srebro, Z. 1970. Mammalian and frog pars tuberalis in some experimental conditions including x-irradiation. Folia Biol. (Krakow) 18: 335-341. (NSA 25: 24365).

Sridhara, S. and L. V. Bhat. 1965. Incorporation of mevalonate-2-^{14}C into the lipids of the silkworm Bombyx mori. Life Sci. 4(2): 167-171.

Srinath, E. G. and S. C. Pillai. 1967. Removal of radioactive phosphorus from sewage by activated sludge. Water Waste Treat. 11: 410-414. (NSA 22: 10534).

Srinivasan, N. G., J. J. Corrigan and A. Meister. 1965. Biosynthesis of D-serine in the silkworm, Bombyx mori. J. Biol. Chem. 240(2): 796-800. (^{14}C).

Srisan Ruangopas. 1966. A preliminary study on the use of gamma radiation in the control of cigarette beetle (Lasioderma serricorne F.). pp. 31-34. In: Insect Eradication by Irradiation. Thailand Office of Atomic Energy for Peace, Bangkok, report THAI.-AEC-8.

Srisan Ruangopas. 1966. Studies on the use of gamma radiation in the control of pea weevil (Callosobruchus chinensis L.). pp. 18-30. In: Insect Eradication by Irradiation. Thailand Office of Atomic Energy for Peace, Bangkok, report THAI.-AEC-8.

Srivastava, M. D. L. and K. S. Bachher. 1966. Effect of x rays on the chromosomes of Chrotogonus incertus Bolivar (Acrididae). Caryologia 19: 513-521. (NSA 21: 32622).

Srivastava, P. N. 1960. Thyroidal control of radiophosphorus metabolism in yearling salmon, Salmo salar L. Can. J. Zool. 38(2): 249-256. (SFA 5: 3397).

Srivastava, P. N. 1960. Influence of thyroid on radiophosphorus metabolism in fish. Nature 118(4749): 512-513. (SFA 6: 3829).

Srivastava, P. N. 1960. Effect of bacteria on radiophosphorus metabolism in yearling salmon, Salmo salar. Can. J. Zool. 38: 525-532.

Srivastava, P. N. 1963. Effect of hypophysectomy and thiourea treatment of radiophosphorus metabolism in goldfish, Carassius auratus L. Naturwissenschaften 50(15): 528-529. (NSA 17: 37049).

Srivastava, P. N. 1966. Effect of ionizing radiation on the ovaries of Japanese medaka, Orizias latipes (T. and S.). Acta Anat. 63: 434-440. (NSA 21: 16488).

Srivastava, P. N. and P. U. Devi. 1970. Role of pituitary ACTH in the regulation of radiocalcium and radiophosphorus utilization in the Indian catfish, Heteropneustes fossilis Bloch. Strahlentherapie 139: 236-242. (NSA 24: 34057).

Srivastava, P. N. and G. H. Nair. 1970. Effect of thyroid stimulating hormone and radiophosphorus irradiation on the blood components of Heteropneustes fossilis Bloch. Strahlentherapie 139: 565-570. (NSA 25: 590).

Srivastava, P. N. and S. Ram. 1966. Effect of ^{60}Co internal irradiation on the liver of Heteropneustes fossilis Bloch. pp. 360-372. In: Proc. All India Symp. on Radioactivity and Metrology of Radionuclides. (See India Atomic Energy Establishment, 1966) (NSA 22: 6874).

Srivastava, P. N. and S. Ram. 1967. Pattern of cell damage due to ionizing radiation in the epithelial lining of the intestine in Heteropneustes fossilis Bloch. Experientia 23: 358-359. (NSA 21: 30905).

Srivastava, P. N. and S. Ram. 1968. Effect of internally injected radioisotopes on the connective tissue in the Indian catfish, Heteropneustes fossilis Bloch. Strahlentherapie 135: 479-485. (NSA 22: 33971).

Srivastava, P. N. and S. Ram. 1969. Effect of ^{60}Co internal irradiation on the liver of the Indian catfish, Heteropneustes fossilis Bloch. Strahlentherapie 137: 499-507. (NSA 23: 38758).

Srivastava, P. N. and A. Ramesharao. 1968. Effect of beta irradiation on the function of the follicle stimulating hormone in the female Indian desert gerbil, Meriones hurrianae Jerdon. Bull. Nat. Inst. Sci. India 36: 68-74. (NSA 26: 20616).

Srivastava, P. N. and A. R. Rao. 1966. ^{60}Co-induced radiation changes in the unilaterally ovarectomized Indian desert gerbil Meriones hurrianae Jerdon. pp. 373-380. In: Proc. All India Symp. on Radioactivity and Metrology of Radionuclides. (See India Atomic Energy Establishment, 1966) (NSA 22: 6873).

Srivastava, P. N. and S. K. Rathi. 1965. Effect of internal radiophosphorus irradiation of the ovary of Puntius sophore Hamilton. Experientia 21(3): 133-134. (NSA 19: 21908) (Fish).

Srivastava, P. N. and S. K. Rathi. 1967. Radiation-induced castration in the female teleostean fish, Heteropneustes fossils Bloch. Experientia 23: 229-230. (NSA 21: 22698).

Srivastava, P. N. and S. K. Rathi. 1967. Action of exogenous fish and internal ^{32}P irradiation on the ovary of the Indian catfish, Heteropneustes fossilis Bloch. Arch. Biol. (Liége) 78: 633-643. (NSA 22: 45654).

Srivastava, P. N. and S. K. Rathi. 1970. Effect of radiation on the reproductive system of the Indian catfish, Heteropneustes fossilis Bloch. I. Annual cycle in the development of ovarian eggs. Acta Anat. 75: 114-125. (NSA 25: 21858).

Srivastava, P. N. and C. Tachi. 1964. Sodium excretion as a factor causing death in fish due to irradiation. Nature 202(4938): 1234-1235. (NSA 18: 29330) (Goldfish).

Srivastava, P. V. 1963. Further studies on the thyroidal control of radiophosphorus metabolism as determined by the effects of thyroxine and pituitary extract on thiourea and radioiodine treated goldfish, Carassius auratus L. Endocrinol. Japon. 10: 190-198. (NSA 19: 25855).

Srivastava, P. V., et al. 1964. Thyroid activity and radiosensitivity relationship in goldfish, Carassius auratus L. Strahlentherapie 125: 305-308. (NSA 19: 155).

Ssmokhvalova, G. V. 1935. The influence of x-rays on the sex glands and the secondary sexual characters in Lebistes reticulatus. Tr. Dinamike Razvitiya 10: 213-229. (In Russian).

Ssmokhalova, G. V. 1938. Effect of x-rays on fishes (Lebistes reticulatus, Ziphophorus helleri and Carassius vulgaris). Biol. zh. 7: 1023-1024. (In Russian).

Staatz, M. H., C. M. Bunker and C. A. Bush. 1971. Geochemical prospecting for thorium veins by steam-sediment sampling, Lemhi Pass Quadrangle, Idaho and Montana. U. S. Geological Survey Prof. Paper 750-C: 136-140. (NSA 26: 22771).

Stadnyk, L., R. S. Campbell and B. T. Johnson. 1971. Pesticide effect on growth and ^{14}C assimilation in a freshwater alga. Bull. Environ. Contam. Toxicol. 6(1): 1-8. (SFA 17: 15299).

Staebler, G. R. and J. H. Rediske. 1959. Progress in developing a radioactive tracer technique for mapping roots of Douglas fir. Proc. Soc. Am. Forest 1958: 164-166.

Staffeldt, E. and G. A. Sacher. 1966. Comparative studies on aging and acute and chronic radiation effects in Rodentia. II. Considerations in the selection of experimental species. pp. 28-30. In: Radiation Effects on Natural Populations. (See Sacher, 1966).

Stahl, W., H. Aust and A. Dounas. 1974. Origin of artesian and thermal waters determined by oxygen, hydrogen, and carbon isotope analyses of water samples from the Sperkhios Valley, Greece. pp. 317-339. In: Isotope Techniques in Groundwater Hydrology. Vol. 1. (See IAEA, 1974) (NSA 31: 19635).

Stahler, N. 1971. Changes in survival and tolerance levels in a radiation-resistant strain of Aedes aegypti (Diptera::Culicidae) during 90 generations. Ann. Entomol. Soc. Am. 64(6): 1247-1249. (NSA 26: 15513).

Stahler, N. and L. A. Terzian. 1963. The response of blood-fed Aedes aegypti to gamma radiation. J. Econ. Entomol. 56(3): 416-417. (NSA 18: 79).

Stahler, N. and L. A. Terzian. 1966. Reproductive and immunological characteristics of a strain of Aedes aegypti selected for resistance to gamma radiation. Ann. Entomol. Soc. Am. 59: 763-765. (NSA 21: 13052).

Stahler, N. and L. A. Terzian. 1968. Effect of blood-feeding on the longevity of radiation-resistant yellow-fever mosquitoes. J. Econ. Entomol. 61: 1447. (NSA 23: 2912).

Stairs, G. R. 1963. Acute gamma irradiation of Quercus seed - its effect on germination and seedling growth. Brookhaven National Lab. U. S. AEC report BNL-6892. 37pp.

Stairs, G. R. 1963. Chronic and acute radiation of microsporogenesis and mature pollen in Quercus. Brookhaven National Lab., U. S. AEC report BNL-6894. 40pp.

Stairs, G. R. 1963. Effects of chronic gamma radiation on seed production in an oak forest. Brookhaven National Lab., U. S. AEC report BNL-6893. 33pp.

Stairs, G. R. 1963. Microsporogenesis and embryogenesis in Quercus. Brookhaven National Lab., U. S. AEC report BNL-6895. 20pp.

Stairs, G. R. 1963. Acute gamma irradiation of Quercus seed. Silvae Genet. 12(6): 192-200.

Stairs, G. R. 1964. Effects of chronic and acute gamma irradiation of male flower buds and mature pollen in Quercus. Forest Science 10(4): 397-409.

Stairs, G. R. 1964. Effetcs of chronic gamma radiation on seed production in an oak forest. Adv. Frontiers Plant Sci. 9: 207-216. (NSA 21: 246660).

Stairs, G. R. and D. B. Houston. 1969. Nuclear and cytoplasmic radiosensitivity of conifer pollen. pp. 219-230. In: Induced Mutations in Plants. (See IAEA, 1969) (NSA 24: 42028).

Stairs, G. R. and R. M. Jeffers. 1967. Effects of gamma radiation on juvenile and mature cuttings of quaking aspen. Forest Science 13(3): 240-245. (Also: State Univ. of New York, Syracuse, U. S. AEC report NYO-3571-3. (NSA 23: 22436).

Stalling, D. L. and J. N. Huckins. 1971. Gas-liquid chromatography - mass spectrometry characterization of polychlorinated biphenyls and ^{36}Cl-labeling of aroclors 1248 and 1254. J. Assoc. Off. Anal. Chem. 54: 801-807. (Aquatic).

Stanek, Z. 1973. Radionuclides in water works reservoirs and their influence on living environments of man. pp. 271-277. In: Proc. 6th Conf. on Radiation Hygiene. II. T. Trnovec and J. Stojkovic, eds. Hradec Kralove, Czechoslovakia. (NSA 32: 14308).

Stange, E. 1969. Determination of the beta radioactivity in water with commercially available detectors (VA-Z-310, VA-Z-530, VA-T-84) from the viewpoint of radiation protection. Deutsches Demokratische Republic, Staatliche Zentrale für Strahlenschutz, Berlin, report SZS-1/69. 34pp. (In German) (NSA 23: 21655).

Stange, L., et al. 1962. $^{14}CO_2$ incorporation into nucleic acids of synchronously growing Chlorella cells. Biochim. Biophys. Acta 61: 681-695.

Stankovic, D. 1962. Present status and the future development of radiotechnology. Tehnika (Belgrade) 17: 137-145. (Incl. entomology).

Stanley, D. R. 1955. Sand filtration studied with radiotracers. Am. Soc. Civil Engrs. 81: 592.

Stanley, P. E. and D. J. D. Nicholas. 1969. Some uses of isotopes in metabolic studies with soil microorganisms. At. Energ. Austral. 12: 11-20. (NSA 24: 36694).

Stanley, R. E. and S. R. Lloyd. 1972. Technic for implanting a radioactive particle in a rodent lung. Lab. Anim. Sci. 22(3): 424-427. (NSA 27: 28019).

Stannard, J. N. 1973. Toxicology of radionuclides. Ann. Rev. Pharmacol. 13: 325-357. (NSA 28: 24878) (Incl. discussion of ecosystem and lower organisms).

Stanovick, R., J. Giddens and R. A. McCreery. 1961. Effect of ionizing radiation on soil microorganisms. Soil Science 92(3): 183-187. (NSA 15: 29060).

Stanwix, P. and J. Hawthorn. 1961. Strontium-90 from radioactive fallout in central Scotland. Chem. Industr. : 524-529.

Stapleton, G. B. and R. H. Thomas. 1972. Estimation of the induced radioactivity of the ground water system in the neighborhood of a proposed 300 GeV high energy accelerator situated on a chalk site. Health Physics 23 (5): 689-699. (NSA 26: 58805).

Stapleton, G. B. and R. H. Thomas. 1973. Effect of sorption on the migration of ^{7}Be from a high energy accelerator constructed on a chalk site. Water Research 7(9): 1259-1268. (NSA 28: 29877).

Stapleton, G. E., A. Sbarra and A. Hollaender. 1955. Some nutritional aspects of bacterial recovery from ionizing radiations. J. Bacteriol. 70: 7-14.

Stara, J. F., et al. 1971. Comparative metabolism of radionuclides in mammals: a review. Health Physics 20: 113-137. (NSA 25: 32326).

Starik, I. Ye. (Ed.). 1957. Trans. V. G. Khlopin Radium Inst. Vol. VI. 176pp. (AEC-tr-4208) (NSA 15: 5859).

Starik, I. Ye., et al. 1958. Uranium content in natural waters of the USSR. Trans. V. G. Khlopin Radium Inst. Vol. VIII. (Chemistry and Geochemistry). pp. 250-261. I. Ye. Starik, ed. Publ. House Acad. Sci. USSR, Moscow-Leningrad. (AEC-tr-4474, pp. 303-318) (NSA 15: 22197).

Starik, I. Ye., et al. 1961. The ratio of radioactivities in deposits of the Sea of Azov and the Black Sea. Dokl. Akad. Nauk SSSR 139(2): 456-459. (Doklady (earth sci.) 139: 687-689) (CA 56: 6955).

Starik, I. Ye., et al. 1964. Natural radioactivity of the waters and sedim ments of the Black Sea and the Sea of Azov. In: Radioactive Contamination of the Seas and Oceans. (See Baranov and Khitrov, 1964) (AEC-tr-6641, pp. 10-13) (NSA 21: 8474).

Stark, N. 1973. Radiotracer studies of nutrient cycling pathways. pp. 225-231. In: Radionuclides in Ecosystems. (See Nelson, 1973) (NSA 28: 27300) (^{32}P, ^{59}Fe, ^{75}Se, ^{65}Zn).

Starzycki, S. and J. Pala. 1963 The influence of small doses of gamma rays at various dose rates on the seed of three speciec of grass. Hodowla Roslin, Akim, Nasienn. 7: 593-604. (In Polish) (NSA 20: 10646).

Starzycki, S., J. Pala and E. Konopko. 1967. The effect of low doses of gamma rays upon the rate of infection of plants with grass mildew (Erysiphe graminis DC). Hodowla Roslin Akim. Nasienn. 1: 103-108. (In Polish) (NSA 24: 6141).

Staton, M. A., I. L. Brisbin, Jr. and R. A. Geiger. 1974. Some aspects of radiocesium retention in naturally contaminated captive snakes. Herpetologica 30(2): 204-211. (NSA 32: 9257).

Statler, M. W. 1970. Effects of gamma radiation on the ability of the adult female gypsy moth to attract males. J. Econ. Entomol. 63(1): 163-164. (NSA 24: 23421).

Stauber, E. 1971. How safe are nuclear power plants? Atomwirt. Atomtech. 16(11): 570-573. (In German) (NSA 26: 6228).

Stauber, E. 1973. Environmental protection and nuclear power plants in the Federal Republic of Germany. pp. 173-178. In: Nuclear Energy and the Environment. (See Italy Comitato Nazionale Energia Nucleare, 1973) (NSA 29: 17688).

Stearner, S. P. 1947, 1948. I. Hematology of the frog (Rana pipiens). II. Effect of 1500 r upon the peripheral blood of Rana Pipiens. Effects of various doses of x-rays on frogs. I. Survival. II. Hematology. III. Histology. I. Acute changes following 900 r x-irradiation in the peripheral blood of the frog (Rana pipiens). pp. 42-50, 36-39, 38-47. In: Biology and Medicine Division, Argonne National Lab., U. S. AEC report ANL-4108, -4147, -4163.

Stearner, S. P. 1950. The effects of x-irradiation on Rana pipiens (leopard frog), with special reference to survival and to the response of the peripheral blood. J. Exptl. Zool. 115: 251-262. (NSA 5: 917).

Stearner, S. P. 1951. The effects of variation in dosage rate of roentgen rays on survival in young birds. Am. J. Roentgenol. Rad. Therapy 65(2): 265-271.

Stearner, S. P. and S. A. Tyler. 1957. An analysis of the role of dose and dosage rate in the early radiation mortality of the chick. Radiation Research 7(3): 253-266.

Stearner, S. P. and S. A. Tyler. 1962. Comparative aspects of the acute radiation response. Abstr. p. 7. In: 2nd Intern. Congr. Radiation Research. (See International Congress of Radiation Research, 1962) (Birds).

Stearner, S. P. and S. A. Tyler. 1962. Comparative aspects of acute radiation mortality in birds. Int. J. Radiat. Biol. 5: 205-215. (Also: Argonne National Lab., U. S. AEC report TID-14627) (NSA 17: 71).

Stearner, S. P., M. H. Sanderson and E. J. Christian. 1957. The initial post-irradiation period in the pigeon. pp. 56-67. In: Biological and Medical Research Division Semiannual Report. Argonne National Lab., U. S. AEC report ANL-5732. (NSA 11: 12642)

Stearner, S. P., et al. 1955. Role of hypotension in the initial response of x-irradiated chicks. Am. J. Physiol. 182: 204-210.

Stearner, S. P., et al. 1956. Initial radiation syndrome in the adult chicken. Am. J. Physiol. 184(1): 134-140.

Stearner, S. P., et al. 1958. Evidence of circulatory failure in the early radiation response of the chicken. Am. J. Physiol. 192(3): 620-624.

Stearner, S. P., et al. 1960. Modes of radiation death in the chick embryo. I. Acute syndromes. Radiation Research 12(3): 286-300.

Stearner, S. P., et al. 1960. Factors influencing the initial lethal response of the irradiated chick. Abstr. Radiation Research 12(4): 475-476.

Stearner, S. P., et al. 1961. Acute radiation mortality in the parakeet. pp. 5-9. In: Biological and Medical Research Division Summary Report. Argonne National Lab., U. S. AEC report ANL-6368. (NSA 15: 31979).

Stearner, S. P., et al. 1961. Mechanisms of resistance and reversal in the initial radiation response in the chick. Radiation Research 14(6): 732-747.

Steen, L. 1963. Effects of telencephalic radiolesions on learning and retention in the pigeon (Columba livia). Appendix II. Final Tech. Report. Stockholm Univ., Sweden. 22pp.

Steere, W. C. 1970. Bryophyte studies on the irradiated and control sites in the rain forest at El Verde. Ch. D-11, pp. D213-D225. In: A Tropical Rain Forest. (See H. T. Odum and R. F. Pigeon, 1970) (NSA 25: 21936).

Stefanov, G. and L. Daieva. 1972. Non-destructive determination of micro-quantities of iridium in soil samples through activation analysis. Isotopenpraxis 8(4): 146-148. (In German) (NSA 26: 47671).

Steffan, A. W. and W. J. Kloft. 1973. The possibilities of using the sterile male technique for aphid control - a theoretical discussion. pp. 129-136. In: Computer Models and Application of the Sterile-Male Technique. (See IAEA, 1973) (NSA 29: 8055).

Steger, R. 1956. Embryo mortality among the progeny of male Drosophila flies following irradiation with 180 KEV and 31 MEV. Oncologia 9: 12-32. (AEC-tr-3340).

Steel, E. W. and E. F. Gloyna. 1954. Oxidation ponds - radioactivity uptake and algae concentration. Univ. of Texas, Austin, U. S. AEC report AECU-2837. (NSA 8: 3227).

Steel, E. W. and E. F. Gloyna. 1955. Concentration of radioactivity in oxidation ponds. Sewage Industr. Waste 27(8): 941-956.

Steele, J. H. 1957. A comparison of plant production estimates using C^{14} and phosphate data. J. Mar. Biol. Assn. UK 37: 233-241.

Steele, V. E. 1975. Characterization of an organ-specific differentiator substance. Thesis, Univ. of Rochester, Rochester, N. Y. 150pp. (NSA 32: 19992) (Planaria, x-irradiation).

Steeman-Nielsen, E. 1951. The interaction of photosynthesis and respiration and its importance for the determination of carbon-14. Nature 167(4252): 684-685.

Steeman-Nielsen, E. 1955. The interaction of photosynthesis and respiration and its importance for the determination of C^{14} discrimination in photosynthesis. Physiol. Plant 8: 945-957.

Steeman-Nielsen, E. 1957. The chlorophyll content and the light utilization in communities of plankton algae and terrestrial higher plants. Physiol. Plant 10: 1009-1021.

Steeman-Nielsen, E. and A. A. Al Kholy. 1956. Use of the C^{14}-technique in measuring photosynthesis of phosphorus or nitrogen deficient algae. Physiol. Plant 9: 144-153.

Steffensen, D. and L. E. LaChance. 1960. Radioisotopes and the genetic mechanism: cytology and genetics of divalent metals in nucleic acids and chromosomes. pp. 132-145. In: Symp. on Radioisotopes in the Biosphere. R. S. Caldecott and L. A. Snyder, eds. Univ. Of Minnesota Press, Minneapolis. (Mentions Habrobracon).

Stehney, A. F. 1955. Radium and thorium X in some potable waters. Acta Radiologica 43(1): 43-51.

Stein, C. H., T. Alper and E. E. Anderssen. 1952. The movement of a radiophosphorus-labeled insecticide in groundnut plants. J. Sci. Food Agric. 1955-56: 17-18.

Stein, J. D. 1968. Labeling larvae of the elm sawfly with P^{32}. J. Econ. Entomol. 61(2): 508-570. (NSA 22: 36179).

Stein, O. L., A. H. Sparrow and L. A. Schairer. 1959. Leaf tumors induced in Graptopetalum paraguavense by gamma radiation. Cancer Research 19: 746-748.

Steinberg, M. and J. Pruzansky. 1970. Radiation treatment of mine waste waters. U. S. Patent 3,737,966. (NSA 25: 13036).

Steinberg, M., et al. 1968. Removal of iron from acid mine drainage waste with the aid of high energy radiation. Part III. Brookhaven National Lab. U. S. AEC report BNL-12115.. 17pp. (For: Coal Industry Advisory Comm. Meeting to ORSANCO, Pittsburgh, Pa., May 14-15, 1968) (CONF-680536-1) (NSA 23: 9416).

Steiner, L. F. 1969. Mediterranean fruit fly research in Hawaii for the sterile fly realease program. pp. 73-82. In: Insect Ecology and the Sterile-Male Technique. (See IAEA, 1969).

Steiner, L. F. 1969. Methods of estimating the size of populations of sterile pest Tephritidae in release programs. pp. 63-72. In: Insect Ecology and the Sterile-Male Technique. (See IAEA, 1969).

Steiner, L. F. and L. D. Christenson. 1956. Potential usefulness of the sterile fly release method in fruit fly eradication programs. Proc. Hawaii Acad. Sci. 1955-56: 17-18.

Steiner, L. F., W. C. Mitchell and A. H. Baumhover. 1962. Progress of fruit-fly control by irradiation sterilization in Hawaii and the Marianas Islands. Int. J. Appl. Radiat. Isotopes 13(7): 427-434. (NSA 17: 15560).

Steiner, L. F., et al. 1965. Melon fly eradication by overflooding with sterile flies. J. Econ. Entomol. 58(3): 519-522. (Dacus curcurbitae).

Steiner, L. F., et al. 1970. Eradication of the oriental fruit fly from the Mariana Islands by the methods of male annihilation and sterile insect release. J. Econ. Entomol. 63(1): 131-___.

Steinmuller, D. and A. G. Motulsky. 1967. Treatment of hereditary spherocytosis in Peromyscus by radiation and allogeneic bone marrow transplantation. Blood 29: 320-330. (NSA 21: 22758).

Steinnes, E. 1973. Rapid and sensitive neutron activation method for uranium in water. Radiochem. Radioanal. Lett. 16(1): 25-29. (NSA 29: 15236).

Stem, V. M. and E. I. Schlinger. 1960. A study of the dispersal habits of Trichogramma fasciatum Perkins using radio-active P^{32}. Abstr. Bull. Entomol. Soc. Am. 5(3): 117.

Stenersen, J. H. V. 1965. DDT-metabolism in resistant and susceptible stable-flies and in bacteria. Nature 207: 660-661. (^{14}C, Stomoxys calcitrans).

Stepanenko, B. N. and I. Z. Sergiyenko. 1958. Study of the effect of adrenalin on the fixation of carbon dioxide in the living organism and the perfection of the methods of synthesis of glycogen and glucose, marked in the 3rd and 4th positions. Studies of Animals, Fish Industry, Sea Food Industry, Acad. Sci. USSR. (In Russian).

Stepanov, P. S. 1963. Development and course of syphilis in irradiated organism. Med. radiologiya 8(7): 56-58. (NSA 17: 37083).

Stepanov, Yu. S. and R. M. Barkhudarov. 1968. Some patterns of contamination of items of the external environment by strontium-90 during a period of stratospheric fallout. State Committee on the Use of Atomic Energy USSR. Atomizdat, Moscow. 5pp. UN Document A/AC.82/G/L.1252. (AEC-tr-7030, pp. 131-133) (NSA 23: 18002).

Stepanov, V. A. 1957. Desorption of some split radioisotopes from the bottom sediments. Gigiyena sanitariya 22(11): 3-7. (In Russian).

Stephens, N. H., E. H. Acree and F. N. Case. 1968. Evaluation of four methods using ^{198}Au for surface labeling sand and a new technique for simulating mass labeling. Oak Ridge National Lab., U. S. AEC report ORNL-4338. 9pp. (NSA 23: 7697).

Stephens, R. E. 1970. Action of actinomycin D on mitotic activity and on x-ray induced prophase reversion in neuroblasts of _Chortophaga viridifasciata_ (De Geer). Thesis, Univ. of Tennessee, Knoxville. 57pp. (Also: U. S. AEC report ORO-2575-51) (NSA 26: 23136).

Stepin, A. A. 1973. Alimentary tract in accumulation of ^{137}Cs by gudgeon. pp. 147-151. In: _Radioecology of Water Organisms. II._ (See Andrushaitis, et al., 1973) (In Russian) (NSA 30: 21430).

Stern, V. M. 1969. Insect pests of major food crops, their reinvasion potential and the effects of radiation on arthropods. Final report. Univ. of California, Riverside, U. S. Off. of Civil Defense report TRC-68-70. vii, 199pp.

Stern, V. M. 1971. Insect-induced agroecological imbalances as an analog to fallout effects. pp. 434-453. In: _Survival of Food Crops and Livestock in the Event of Nuclear War._ (See Benson and Sparrow, 1971).

Stern, V. M., E. I. Schlinger and W. R. Bowen. 1965. Dispersal studies of _Trichogramma semifumatum_ (Hymenoptera: Trichogrammatidae) tagged with radiophosphorus. Ann. Entomol. Soc. Am. 58(2): 234-240.

Sternau, R., et al. 1967. Radioisotope tracers in large-scale recharge studies of groundwater. pp. 489-505. In: _Isotopes in Hydrology._ (See IAEA, 1967) (NSA 22: 19067).

Sternberg, S., B. A. Peleg and R. Galun. 1973. Effect of irradiation on mating competitiveness of the male tick _Argas persicus_ (Oken). J. Med. Entomol. 10(2): 137-142. (NSA 28: 21780).

Stettler, R. F. 1968. Irradiated _Mentor_ pollen: its use in remote hybridization of black cottonwood. Nature 219: 746-747. (NSA 22: 49307).

Stettler, R. F., K. S. Bawa and G. K. Livingston. 1969. Experimental induction of haploid parthenogenesis in forest trees. pp. 611-619. In: _Induced Mutations in Plants._ (See IAEA, 1969) (NSA 24: 41847).

Stevart, M. D. 1973. Aerial radiological measuring survey of the area surrounding Big Rock Point Nuclear Plant, Big Rock Point, Michigan. EG&G, Inc., Las Vegas, Nev., U. S. AEC report ARMS-68.6.3. 36pp. (NSA 29: 5137).

Stevens, J. and M. E. Gaulden. 1959. The effects of radiation on mitotic rate of individual stages of mitosis in the grasshopper neuroblast. ASB Bull. 6(2): 33.

Stevenson, G. 1958. Use of ^{15}N in the study of fixation of atmospheric nitrogen by non-nodulated seed plants. Proc. 2nd Intern. Conf. on the Peaceful Uses of Atomic Energy 27: 51-57. (Paper A/CONF.15/P/1507) (NSA 13: 6217).

Stevenson, J. R. and R. Buchsbaum. 1961. Aging in irradiated and nonirradiated hydras. Science 134(3475): 332-333. (NSA 15: 27473).

Stewart, B. and M. C. Wukasch. 1967. The Galveston environmental surveillance program, 1966. Radiol. Hlth. Data 8(8): 484-486.

Stewart, G. L. 1967. Fractionation of tritium and deuterium in soil water. pp. 159-168. In: _Isotope Techniques in the Hydrologic Cycle._ (See Stout, 1967).

Stewart, G. L. 1972. Clay-water interaction, the behavior of ^{3}H and ^{2}H in adsorbed water, and the isotope effect. Soil Sci. Soc. Am. Proc. 36(3): 421-___.

Stewart, G. L. 1973. Behavior of tritium in the soil. pp. 462-470. In: _Tritium._ (See Moghissi and Carter, 1973) (NSA 28: 27272).

Stewart, G. L. and R. K. Farnsworth. 1968. United States tritium rainout and its hydrologic implications. Water Resources Res. 4(2): 273-289. (NSA 22: 37888).

Stewart, G. L. and T. A. Wyerman. 1970. Tritium rainout in the United States during 1966, 1967, and 1968. Water Resources Res. 6(1): 77-87. (NSA 24: 44132).

Stewart, G. L., et al. 1972. Tritium in pine trees from selected locations in the United States, including areas near nuclear facilities. U. S. Geological Survey Prof. Paper 800-B: 265-271. (NSA 26: 38418).

Stewart, M. L. 1970. Distribution and movement of radioactive nuclides, both natural and artificial, in soils, vegetation, and wildlife. Thesis, Univ. of Utah, Salt Lake City. 181pp. (Univ. Microfilms 70-14444) (NSA 25: 21643).

Stewart, M. L., J. R. Kline and C. F. Jordan. 1972. A tritiated water recovery system. Int. J. Appl. Radiat. Isotopes 23(8): 387-388.

Stewart, M. L., R. C. Pendleton and J. L. Lords. 1972. Flocculant filter ion-exchange system for the concentration of ^{137}Cs in water. Int. J. Appl. Radiat. Isotopes 23(7): 345-347. (NSA 26: 50290).

Stewart, M. L., G. M. Rosenthal and J. R. Kline. 1973. Tritium: discrimination and concentration in freshwater microcosms. pp. 452-459. In: _Radionuclides in Ecosystems._ (See Nelson, 1973) (NSA 25: 41691).

Stewart, W. D. P. 1967. Nitrogen turnover in marine and brackish habitats. II. Use of ^{15}N in measuring nitrogen fixation in the field. Anns. Botany 31(122): 385-407.

Stich, H. and M. Grell. 1955. Incorporation of phosphorus-32 into the melphighian tubes during the metamorphosis of _Culex pipiens._ Nature 176 (4489): 930-931.

Stigall, G. E. 1967. Effects of nuclear power station liquid waste dischrges on Lake Michigan. NUS Corp., Washington, D. C., Commonwealth Edison Co., Chicago, Ill. report NUS-TM-S-60. 17pp. (NSA 24: 20969).

Stigall, G. E., T. W. Fowler and H. L. Kreiger. 1971. ^{131}I discharge from an operating boiling water reactor nuclear power station. Health Physics 20(6): 593-599. (NSA 25: 42388).

Stiller, K. J. and F. Robiller. 1974. Radiation dosage in metabolic investigations on avian thyroid glands. Exptl. Pathol. 9(5/6): 374-376. (NSA 32: 20007) (Budgerigar).

Stiller, M., I. Carmi and K. O. Muennich. 1975. Water transport through Lake Kinneret sediment traced by tritium. Earth Planet. Sci. Lett. 25(3): 297-304. (NSA 32: 11543).

Stimmann, M. W., W. W. Wolf and H. H. Toba. 1972. Pheronome response of gamma-irradiated cabbage loopers in the field. J. Econ. Entomol. 65(5): 1496-1498. (NSA 26: 59116).

Stinson, B. D. 1958. Radioprotection of regeneration capacity in urodele larvae. Anat. Record. 132(3): 510-512.

Stinson, B. D. 1963, 1964. The response of x-irradiated limbs of adult urodeles to normal tissue grafts. I. Effects of autographs of sixty-day forearm regenerates. III. Comparative effects of autografts of complete forearm regenerates and longitudinal half forearm regenerates. J. Exptl. Zool. 153(1): 37-52; 156(1): 1-18. (NSA 18: 13499; 19: 13013) (Used _Triturus viridescens_).

Stinson, R. H. 1972. Boron autoradiography of botanical specimens. Can. J. Bot. 50(1): 245-246. (NSA 26: 25946).

Stobbart, R. H. 1965. The effect of some anions and cations upon the fluxes and net uptake of sodium in the larva of _Aedes aegypti_ (L.). J. Exptl. Biol. 42: 29-43. (^{22}Na).

Stobbart, R. H. 1967. The effect of some anions and cations upon the fluxes and net uptake of chloride in the larva of _Aedes aegypti_ (L.), and the nature of the uptake mechanisms for sodium and chloride. J. Exptl. Biol. 47 (1): 35-57.

Stockner, J. G. 1968. Algal growth and primary productivity in a thermal stream. J. Fish. Res. Bd. Can. 25(10): 2037-2058. (^{14}C).

Stoddart, D. M. 1970. Individual range, dispersion and dispersal in a population of water voles (_Arvicola terrestris_ (L.)). J. Animal Ecol. 39(2): 403-425. (^{32}P).

Stoklasa, J. 1914. The function of radioactivity in the physiology of plants. Centr. Bakt. Paraistenk Abt.II 40(2): 266-280. (In German) (CA 8: 2739).

Stoklasa, J. 1925. The effect of radioactivity of energy exchange and metabolism of animal and plant cells. Deut. med. wochschr. 51: 1057-1060. (In German).

Stoklasa, J. 1925. Significance of radioactivity in physiology. Strahlentherapie 4: 1-18. (Also: Zentr. Biochem. Biophys. 16: 890) (In German) (Bacteria).

Stoklasa, J. 1928. Influence of radioactivity of water on the biological action of the cells of lower and higher organisms. Strahlentherapie 29: 324-332. (Also: Wasser Abwasser 26: 31-32) (In German) (Algae, Bacteria).

Stoklasa, J. 1930. The physiological action of α-, β- and γ-rays on the protoplasm of the plant cell. Letter. Naturwissenschaften 18(24): 567. (In German) (Incl. letter, comments of N. Feichtinger).

Stoklasa, J. and J. Penkava. 1924. The radioactivity from the gas eruption of Vesuvius and the Solfataren of Campanien and their effect on the development of bacteria and higher plants. Compt. rend. 179: 819-822.

Stoklasa, J. and J. Penkava. 1932. _Biology of the Radiums and the Radioactive Elements. Vol. 1._ Paul Parey, Berlin. xiv, 958pp. (In German) (An extensive review of naturally radioactive substances, their accumulation in biological systems and the biological effects of radiation).

Stoklasa, J. and V. Zdobnicky. 1913. Influence of radiation on vegetatation. Compt. rend. 157: 1082-1084. (In French) (CA 8: 1808).

Stoklasa, J., J. Cakir and J. Penkava. 1930. Assimilation of carbon dioxide under the influence of radioactivity in the process of vegetative development of chlorophyll-containing cells. Biochem. Z. 224(1-3): 102-134. (In German) (_Juglans regia_).

Stoklasa, J., et al. 1928. The biology of uranium. Biochem. Z. 194: 15-77. (CA 22: 2581).

Stone, B. F. and A. W. A. Brown. 1967. Mechanisms of resistance to Fenthion in _Culex fatigans_. Abstr. 42. Bull. Entomol. Soc. Am. 13(3): 189. (^{32}P).

Stone, G. F. 1972. Aquatic radiological monitoring, Browns Ferry nuclear plant. pp. 161-176. In: _Proc. Southern Conf. on Environmental Protection from Nuclear Power Plants_. (See U. S. EPA, 1972) (NSA 27: 2536).

Stone, J. F., R. H. Shaw and D. Kirkham. 1960. Statistical parameters and reproducibility of the neutron method of measuring soil moisture. Soil Sci. Soc. Am. Proc. 24: 435-438.

Stone, K. J. 1967. Relative biological effectiveness, comparison of hard x-ray, soft x-ray and beta radiation. Abstr. 108. Bull. Entomol. Soc. Am. 13(3): 192.

Stone, W. E. 1963. Effects of ionizing radiation on insects and other arthropods. pp. 301-309. In: _Radiation and Radioisotopes Applied to Insects of Agricultural Importance_. (See IAEA, 1963).

Stone, W. S., et al. 1957, 1958, 1959, 1962. Genetic studies of irradiated natural populations of _Drosophila_. --- II. 1957 tests. IV. 1958 tests. V. Summary and discussion of tests of populations collected in the Pacific Proving Ground from 1955 through 1959. Univ. of Texas, Austin, Publ. 5721: 260-316; 5914: 223-233; 6205: 1-54; Proc. Natl. Acad. Sci. (US) 44: 565-575.

Stonier, T. 1956. Labeling crown gall bacteria with P^{32} for radioautography. J. Bacteriol. 72(2): 259-268. (NSA 10: 11038).

Storebø, P. 1971. Siting of nuclear power plants. Fossekallen 18(6): 16-18. (In Norwegian) (NSA 26: 60736).

Storer, J. B. and D. Grahn. 1960. Vertebrate radiobiology: late effects. Ann. Rev. Nucl. Sci. 10: 561-582. (NSA 15: 4938).

Storms, H. A., D. C. Carlson and F. F. Hunter. 1974. Determination of plutonium in soils by mass spectrometry. pp. 276-285. In: Proc. 2nd Intern. Conf. on Nuclear Methods in Environmental Research. (See Vogt and Meyer, 1974) (NSA 33: 23277).

Storteir, S. and A. Palmgren. 1971. Long-term recording of incubating rhythm and feeding frequency with the aid of radioactive tagging. Ornis Fenn. 48(1): 33-35. (In Swedish, English summary) (BA 53(10): __; WR 146: 89).

Story, A. H. and E. F. Gloyna. 1963. Radioactivity transport in water - environmental behavior of nitrosylruthenium. Environmental Health Engrg. Lab., Univ. of Texas, Austin, Tech. Rept. -3, U. S. AEC report ORO-620. xiii, 89pp.

Stott, G. and J. A. Hetherington. 1974. Monitoring of radioactive liquid effluent discharged into the environment from nuclear installations in the United Kingdom. pp. 110-126. In: Proc. Seminar on the Monitoring of Radioactive Effluents. (See Organization for Economic Co-operation, 1974) (AED-CONF-74-157-014) (CONF-741053-_) (NSA 30: 21059; 31: 3341).

Stotzky, G. and J. L. Mortenson. 1957. Effect of gamma radiations on some metabolic activities and growth of microorganisms in an organic soil. Agron. Abstrs. 49: 17.

Stotzky, G. and J. L. Mortenson. 1959. Effect of gamma radiation on growth and metabolism of microorganisms in an organic soil. Soil Sci. Soc. Am. Proc. 23: 125-127.

Stotzky, G., et al. 1964. Some effects of gamma irradiation on seeds and rhizomes of Musa. Am. J. Bot. 51(7): 724-729.

Stout, G. E. (Ed.). 1967. Isotope Techniques in the Hydrologic Cycle. Am. Geophysical Union, Washington. (NSA 21: 36126).

Stout, J. D. and B. J. O'Brien. 1972. Factors affecting radiocarbon enrichment in soil and turnover of soil organic matter. pp. E13-E26. In: Proc. 8th Intern. Radiocarbon Dating Conf. (See Rafter, 1972) (NSA 31: 8705).

Stout, P. R. and T. C. Broyer. 1957. Use of radioisotopes in plant nutrition studies. pp. 487-503. In: Proc. Inter-American Symp. on the Peaceful Application of Nuclear Energy. (Brookhaven National Lab., May 13-17, 1957). U. S. AEC report TID-7554. (NSA 13: 8535).

Stout, P. R. and D. R. Hoagland. 1939. Upward and lateral movement of salt in certain plants as indicated by radioactive isotopes of potassium, sodium, and phosphorus absorbed by roots. Am. J. Bot. 26: 320-324.

Stout, V. F. 1963. Effects of ionizing radiation on lipids of fish. U. S. Fish and Wildlife Service, U. S. AEC report TID-18657. iii, 18pp. (NSA 17: 37064).

Stoute, J. R. D. 1974. Monitoring of gaseous effluents from several installations at the Petten Research Centre. pp. 153-165. In: Proc. Seminar on the Monitoring of Radioactive Effluents. (See Organization for Economic CO-operation and Development, 1974) (NSA 31: 3344).

Stover, B. J. and W. S. S. Jee. (Eds.). 1972. Radiobiology of Plutonium. J. W. Press, Salt Lake City. 583pp. (NSA 26: 41253).

Stpiczynska, Z. and J. Jaskulski. 1972. Radioactive contaminations of the Intitute of Nuclear Research, Centre Zeran site and environs in 1970. Pol- and Inst. of Nuclear Research, Warsaw, report INR-1392. 17pp. (In Polish) (NSA 26: 40847).

Stradomskiy, V. B. 1947. On the problem of distribution of long-lived fragmental nuclides in surface waters on land. Gidrokhim. mater. 43: __. (In Russian).

Stradomskiy, V. B. 1963. Strontium-90 in the water of the River Don Basin. State Committee on the Use of Atomic Energy USSR. Atomizdat, Moscow. 5 pp. UN Document A/AC.82/G/L.918. (AEC-tr-6458) (NSA 35776).

Stradomskiy, V. B. 1970. Form of occurrence of long-lived radionuclides in the water of the River Don. Gidrokhim. mater. 54: 91-94. (In Russian) (NSA 28: 29888).

Strand, J. A., W. L. Templeton and P. A. Olson. 1975. Fixation and long-term accumulation of tritium from tritiated water in an experimental aquatic environment. Battelle Northwest Lab., U. S. ERDA report BNWL-SA-5416. 24pp. (NSA 33: 7303).

Strand, J. A., W. L. Templeton and E. G. Tangen. 1973. Accumulation and retention of tritium (tritiated water) in embryonic and larval fish, and radiation effect. pp. 445-451. In: Radionuclides in Ecosystems. (See Nelson, 1973) (NSA 25: 37714).

Strand, J. A., et al. 1973. Suppression of Chondrococcus columnaris immune response in rainbow trout sublethally exposed to tritiated water during embryogenesis. pp. 543-549. In: Radioactive Contamination of the Marine Environment. (See IAEA, 1973) (NSA 28: 8462).

Straney, D. O., et al. 1975. Radiocesium in birds of the Savannah River Plant. Health Physics 28(4): 341-345. (NSA 32: 3501).

Straub, C. P. 1952. Removal of waste from water. Nucleonics 10(1): 40-44.

Straub, C. P. 1953. Treatment and disposal of liquid radioactive wastes. pp. 450-469. In: Industrial Wastes. Reinhold, New York.

Straub, C. P. 1955. Limitations of water treatment methods for removing radioactive contaminants. Publ. Hlth. Repts. 70(9): 897-904.

Straub, C. P. 1957. Effects of and on receiving waters and environment (flora and fauna) and effects of water treatment processes and personnel by radioactivity in the water supply. R. A. Taft Sanitary Engrg. Center, Cincinnati, U. S. PHS report SEC No. 931. Mimeo. 9pp.

Straub, C. P. 1959. Statement. pp. 963-1060. In: Fallout from Nuclear Weapons Tests. (See U. S. Congress, 1959) (NSA 14: 7276).

Straub, C. P. 1960. Pollution problems created by power reactors and other uses of atomic energy. pp. 33-39. In: Biological Problems in Water Pollution. (See Tarzwell, 1960).

Straub, C. P. 1960. Research in the radioactive contamination of the environment. A.M.A. Arch. Environ. Hlth. 1: 302-310.

Straub, C. P. 1961. Significance of radioactivity data. J. Am. Water Works Assoc. 53(6): 704-713. (NSA 15: 22580).

Straub, C. P. 1961. Water supply implications and methods of removal. Univ. of Illinois Engr. Expt. Sta. Cir. 69: 34-41. (See Univ. of Illinois, 1961) (NSA 15: 23792).

Straub, C. P. 1964. Low-Level Radioactive Wastes. U. S. AEC report, Washington. xvi, 430pp. (Contains ecological information).

Straub, C. P. 1964. Types and objectives of general and local monitoring programs: air and water. pp. 191-227. In: FAO Atomic Energy Series, Rome, No. 5. (See Food and Agriculture Organization, 1964).

Straub, C. P. 1967, 1968. Review of the literature of 1966 on waste water and water-pollution control. Industrial wastes - radioactive. ---1967. J. Water Poll. Control Fed. 39(6): 915-945; 40(6): 1199-1219. (NSA 23: 2389; 24: 39016).

Straub, C. P. 1970. Public health implications of radioactive waste releases. World Health Organization, Geneva. 61pp. (NSA 25: 21679).

Straub, C. P. 1975. Radioactive wastes. J. Water Poll. Control Fed. 47(6): 1498-1513. (NSA 33: 20192).

Straub, C. P. and J. H. Fooks. 1963. Cooperative field studies on environmental factors influencing I^{131} levels in milk. Health Physics 9(12): 1187-1195. (Incl. grass).

Straub, C. P., M. W. Carter and D. W. Moeller. 1964. Environmental behavior of nuclear debris. J. Sanit. Eng. Div. Am. Soc. Civil Engrs. 90: 25-40. (NSA 19: 38857).

Straub, C. P., A. S. Goldin and A. G. Friend. 1960. Environmental implications of radioactive waste disposal as related to stream environments. pp. 407-419. In: Disposal of Radioactive Wastes. Vol. 1. (See IAEA, 1960).

Straub, C. P., R. J. Morton and O. R. Placak. 1951. Studies on the removal of radioactive contaminants from water. J. Am. Water Works Assoc. 43: 773-792.

Straub, C. P., et al. 1960. Strontium-90 in surface water in the United States. J. Am. Water Works Assoc. 52(6): 756-768. (NSA 15: 1637).

Strauch, S. 1973. The year 2000 - nuclear power and man. pp. 53-62. In: Radionuclides in Ecosystems. (See Nelson, 1973) (NSA 28: 27608).

Strauch, S., et al. 1973. Study of the potential radiological impact of an expanding nuclear power industry on the Tennessee Valley region. pp. 431-447. In: Environmental Behavior of Radionuclides Released in the Nuclear Industry. (See IAEA, 1973) (NSA 29: 29382).

Strauss, B. S. 1959. The effect of the distribution of sulfur on the lethal action of incorporated $S^{35}O_4$. Radiation Research 11(3): 345-356. (Used Neurospora).

Strautman, F. I. and V. A. Kushniruk. 1961. A study of radiosensitivity among certain vertebrate species. In: Conf. on Problems "Use of Atomic Energy"; Transactions. Acad. Sci. Ukr. SSR, Kiev. (In Russian) (Incl. birds).

Strawinski, K. 1958. The application of radioactive isotopes in entomology. Pol. Pismo Ent. 1-2(9-10): 41-45. (In Polish).

Strayer, J. R. 1965. Effect of radiophosphorus ^{32}P on light response of Aedes aegypti (L.) larvae. Fla. Entomol. 48: 81-84. (NSA 22: 6916).

Strehler, B. L. 1962. Further studies on the thermally induced aging of Drosophila melanogaster. J. Gerontol. 17: 347-352.

Strehler, B. L. 1964. Studies on the comparative physiology of aging. III. Effects of x-radiation dosage on age-specific mortality rates of Drosophila melanogaster and Campanularia flexosa. J. Gerontol. 19(1): 83-87. (NSA 18: 27119).

Strickberger, M. W. 1965. Experimental control over the evolution of fitness in laboratory populations of Drosophila pseudoobscura. Genetics 51: 795-800. (NSA 19: 24292).

Strickberger, M. W. 1969. Factors determining rates of the evolution of fitness in laboratory populations of Drosophila pseudoobscura. Genetics 62: 639-651. (NSA 24: 4678) (X-rays).

Strigini, P., C. Rossi and G. Sermonti. 1963. Effects of disintegration of incorporated ^{32}P in Aspergillus nidulans. J. Mol. Biol. 7: 683-699. (NSA 19: 28035).

Strike, T. A. 1965. The lethal effects of fission radiations and 250 KVP x-rays in the chinchilla. U. S. Armed Forces Radiobiology Research Inst., Bethesda, Md. 15pp. (For: 13th Ann. Meeting Radiation Research Soc., Philadelphia, Pa., May 23-26, 1965) (AED-CONF-65-157-21) (CONF-650560-11) (NSA 20: 6900).

Strike, T. A. and L. J. Seigneur. 1969. Acute mortality of chinchillas exposed to mixed gamma-neutron radiations or 250-kVp x-rays. Radiation Research 38: 414-424. (NSA 24: 9936).

Stroganov, N. S. and M. M. Telitchenko. 1958. Accumulation and distribution of several radioactive substances in hydrobionts. Byull. Mosk. obshch. ispyt. prirody (biol.) 63(4): 155-156. (In Russian).

Stroganov, N. S. and M. M. Telitchenko. 1958. The chronic effect of small doses of U-238 on bleak gonads. (Read in November 1957). Byull. Mosk. obshch. ispyt. prirody (biol.) 63(4): 154. (In Russian) (BA 45: 54124; Ref. zh. (biol.) 1959: 44599; SFA 10: 7517).

Strogonov, A. A. and Ye. P. Lisachenko. 1969. Accumulation of ^{137}Cs by Black Sea hyponeuston. Radiobiologiya 9(4): 604-607. (AEC-tr-7109, pp. 176-180) (NSA 24: 7796).

Strohal, P. 1973. Present and future activities of the radioecological group in the "Rudjer Boskovic" Institute. pp. 51-61. In: Proc. 2nd Conf. on the State of Advancement of Radioecology in Italy. (See Anonymous, 1973) (In Italian).

Strohal, P., S. Lulic and O. Jelisavcic. 1969. The loss of cerium, cobalt, manganese, protactinium, ruthenium, and zinc during dry ashing of biological material. Analyst 94: 678-680. (NSA 23: 47388) (Used radionuclides and *Mytillus galloprovincialis*).

Ströhl, G. 1960. Reactions between radioactive pollution and suspended matter in streams. Deut. gewässerkundl. Sonderheft. : 36-39. (In German) (NSA 16: 20038).

Stromberg, W. L. and C. J. Goodnight. 1973. Relationships between phosphorus-32 accumulation in algae, bacteria, and tubificids. pp. 410-416. In: Radionuclides in Ecosystems. (See Nelson, 1973) (NSA 28: 27691).

Strømnaes, Ø. 1968. Radiation sensitivity and repair of chromosomal damage. pp. 341-354. In: Isotopes and Radiation in Entomology. (See IAEA, 1968) (NSA 24: 32173).

Stroosnijder, L. 1973. Collimation of a 60-keV column scanner. pp. 77-84. In: Soil-Moisture and Irrigation Studies. II. (See IAEA, 1973) (NSA 28: 21375) (Soil column scanner).

Strong, F. E. 1963. Fatty acids: *in vivo* synthesis by the green peach aphid. Science 140(3570): 983-984. (Used ^{14}C).

Strunnikov, V. A. 1960. The relative effect of primary radiation injuries to the nucleus and cytoplasm of sex cells of the silkworm. Tsitologiya 2(5): 573-580. (JPRS-7884, pp. 114-126) (NSA 19: 16784).

Strunnikov, V. A. 1961. The relative sensitivity to x-ray radiation of the nucleus and cytoplasm of the germ cells of the mulberry silkworm. pp. 109-115. In: Proc. Tashkent Conf. on Peaceful Uses of Atomic Energy, 1959. Publ. House Acad. Sci. Uzbek. SSR, Tashkent. (AEC-tr-6293, pp. 181-189).

Strunnikov, V. A. and L. M. Gulamova. 1957. Developing a mulberry silkworm breed by using radiation methods in breeding. Nestn. sel'skokhoz. nauki 2(8): 143-147. (In Russian).

Struxness, E. G., et al. 1967. Comprehensive report on the Clinch River study. Oak Ridge National Lab., U. S. AEC report ORNL-4035. iv, 115pp. (NSA 21: 22117).

Stuart, T. P., et al. 1973. Aerial radiological survey of Eniwetok Atoll. Trans. Amer. Nucl. Soc. 17: 540. (NSA 29: 5113).

Stuckenrath, R., Jr. 1965. ^{14}C and the unwary archeologist. In: Radiocarbon and Tritium Dating. (Proc. 6th Intern. Conf., Washington State Univ. Univ., Pullman, June 7-11, 1965). R. M. Chatters and E. A. Olsen, comps. (CONF-650652-9) (AED-CONF-65-158-14) (NSA 20: 4147).

Stuiver, M. 1964. Carbon isotopic distribution and correlated chronology of Searles Lake sediments. Am. J. Sci. 262: 377-392. (NSA 21: 24106).

Stuiver, M. 1965. Carbon-14 content of 18th- and 19th-century wood: variations correlated with sunspot activity. Science 149(3683): 533-536. (NSA 19: 36675).

Stuiver, M. 1967. Origin and extent of atmospheric ^{14}C variations during the past 10,000 years. pp. 27-40. In: Radioactive Dating and Methods of Low-Level Counting. (See IAEA, 1967) (NSA 22: 6490).

Stuiver, M. 1970. Tree ring, varve, and ^{14}C chronologies. Nature 228: 454-455. (NSA 25: 15771).

Stuiver, M. 1970. Oxygen and carbon isotope ratios of fresh-water carbonates as climatic indicators. J. Geophys. Res. 75(27): 5247-5257.

Stuiver, M. and E. S. Deevey. 1961, 1962, 1963. Yale natural radiocarbon measurements. VI. VII. VIII.(w/I. Rouse). Radiocarbon 3: 126-140; 4: 250-262; 5: 312-341. (NSA 17: 6298, 6299).

Sturbaum, V., A. L. Brooks and R. O. McClellan. 1970. Tissue distribution and dosimetry of ^{144}Ce in Chinese hamsters. Radiation Research 44: 359-367. (NSA 25: 24036).

Sturges, F. W. 1968. Radiosensitivity of song sparrows and slate-colored juncos. Wilson Bull. 80(1): 108-109.

Sturges, K. L. and R. E. Sundin. 1968. Gross alpha and beta radiation in waters at a Wyoming mountain bog. Water Resources Res. 4: 159-162. (NSA 22: 25833).

Sturges, M. M. and I. Levin. 1921. Influence of radium and x-rays on the frogs leucocytes. Proc. Exptl. Biol. Med. 18(8): 295-297.

Stuy, J. H. 1961. Studies on the radiation inactivation of microorganisms. VII. Nature of the x-ray-induced break-down of deoxyribonucleic acid in *Haemophilus influenzae*. Radiation Research 14(1): 56-65.

Styro, B. I. 1935. On the nature of the radioactivity of atmospheric precipitation. Nauchn. soobshch. inst. geol. geogr. Akad. Nauk Litov. SSR 2: 63-72. (In Russian).

Styro, B. I. 1959. Problems of Nuclear Meteorology. Publ. House Acad. Sci. Lithuanian SSR, Vil'nyus. 414pp. (In Russian).

Styro, B. I. 1959. On the quality of the activity of atmospheric precipitation. Tr. Akad. Nauk Litov. SSR B1: ___. (In Russian).

Styro, B. I. 1959. On the question of the exchange of natural radioactive materials betwenn the earth's surface and the atmosphere. Tr. Akad. Nauk Litov. SSR (Lietuvos TSR Mokslu Akad. geol. geogr. inst. Moksliniai Pran.) B10(1): 55-61. (In Russian).

Styro, B. I., Ch. A. Garbalyauskasa V. Yu. Luyanasa. (Eds.). 1968. Investigations of Atmospheric Self-Scavenging Processes from Radioactive Isotopes. (Collected Reports on the Conf. on Palanga, Lithuanian SSR, July 7-9, 1966). Izdatel'stvo Mintis, Vil'nyus. 352pp. (In Russian) (CONF-660728) (NSA 23: 17586).

Styro, B. I., et al. 1972. Use of alpha-radiography in the study of weak radioactivity of atmospheric matter.

Styron, C. E. 1964. Observations on the ecology of a granite outcrop isopod, Lirceus fontinalis Raf., with emphasis on ionizing radiation effects. Abstr. ASB Bull. 11(2): 56.

Styron, C. E. 1964. The ecology of a granite outcrop population of Lirceus fontinalis Raf., an aquatic isopod, with emphasis on the effects of ionizing radiation. Thesis, Emory Univ., Atlanta, Ga.

Styron, C. E., Jr. 1965. An ecological study of granite outcrop populations of an aquatic isopod, Lirceus fontinalis Raf., with emphasis on the effects of ionizing radiation. Abstr. Bull. Ecol. Soc. Am. 46(3): 97.

Styron, C. E. 1968. Effects of beta and gamma radiation on Pseudosinella (Collembola). Abstr. Bull. Ecol. Soc. Am. 49(3): 106.

Styron, C. E. 1969. Effects of beta and gamma radiation of Sinella curviseta (Collembola). pp. 40-46. In: Progress Report in Post Attack Ecology. (See Auerbach, 1969).

Styron, C. E. 1969. Ecology of two populations of an aquatic isopod (Lirceus fontinalis Raf.), with emphasis on ionizing radiation effects. pp. 53-60. In: Symposium on Radioecology. (See Nelson and Evans, 1969).

Styron, C. E. 1971. Effects of beta and gamma radiation on a population of springtails, Sinella curviseta (Collembola). Radiation Research 48(1): 53-62. (NSA 26: 7076).

Styron, C. E. 1973. Effects of chronic beta radiation from simulated fallout on springtails, Folsomia sp. (Collembola). Health Physics 25(3): 281-284. (NSA 29: 780).

Styron, C. E. and W. D. Burbanck. 1967. Ecology of an aquatic isopod, Lirceus fontinalis Raf., emphasizing radiation effetcs. Am. Midland Nat. 78 (2): 389-415.

Styron, C. E. and G. J. Dodson. 1971. Responses of some grassland arthropods to ionizing radiation. pp. 509-520. In: Survival of Food Crops and Livestock in the Event of Nuclear War. (See Benson and Sparrow, 1971).

Styron, C. E. and G. J. Dodson. 1973. Responses of a grassland arthropod community to simulated radioactive fallout. Radiation Research 55(3): 487-500. (NSA 28: 30252).

Stzhizhovskiy. A. D. 1961. Radiation response of mechanisms regulating mitotic activity. Radiobiologiya 1(4): 469-475. (AEC-tr-5425, pp. 7-19).

Subbiah, B. V. and J. C. Katyal. 1970. Recent soil, fertilizer and physiological studies with ^{32}P on high yielding varieties of paddy. pp. 61-81. In: Isotopes and Radiation in Investigations of Fertilizer and Water Use Efficiency in Countries of Asia and the Far East. (See IAEA, 1970) (NSA 25: 32314).

Subbotina, Ye. N. and N. V. Timofeyev-Resovskiy. 1961. Coefficients of accumulation of several disperse elements from aqueous solutions by lichen. Bot. zh. 46(2): 212-221. (In Russian).

Subrahmanyam, U. 1967. Some effects of gamma irradiation on navel orangeworm, Paramyelois transitella (Walker). Abstr. 356. Bull. Entomol. Soc. Am. 13(3): 203.

Sudia, T. W. and A. J. Linck. 1963. Methods for introducing radionuclides into plants. pp. 417-425. In: Radioecology. (See Schultz and Klement, 1963) (NSA 17: 33590).

Suess, H. E. 1955. Radiocarbon concentrations in modern wood. Science 122 (3168): 415-417.

Suess, H. E. 1958. The radioactivity of the atmosphere and hydrosphere. Ann. Rev. Nucl. Sci. 8: 243-256.

Suess, H. E. 1967. Bristlecone pine calibration of the radiocarbon time scale from 4100 to 1500 B.C. pp. 143-151. In: Radioactive Dating and Methods of Low-Level Counting. (See IAEA, 1967) (NSA 22: 6454).

Suess, H. E. 1967. Chronology of ancient Egypt. Z. Physik 202: 1-7. (In German) (NSA 21: 30550) (^{14}C).

Suess, H. E. 1973. Natural radiocarbon. Endeavor 32(115): 34-38.

Sueyhiro, Y., et al. 1956. Transmission and metabolism of strontium-90 in aquatic animals. pp. 1135-1142. In: Research on the Effects and Influences of the Nuclear Bomb Test Explosions. Vol. II. (See Japan Soc. for the Promotion of Science, 1956).

Sugahara, T. and M. Horikawa. 1958. Tissue culture analysis of delayed lethal irradiation effects in D. melanogaster. Genet. Ann. Rpt. 8: 75-76.

Sugai, E. 1965. Effect of temperature on radiation induced male sterility in the silkworm, Bombyx mori L. Nippon Oyo Dob. Konch. Zasshi 9: 266-270. (English transl. Japan Atomic Energy Res. Inst., Tokyo, report NSJ-tr-89) (NSA 23: 6627).

Sugihara, T. 1968. On the contents of radium C and radon in radioactive mineral spring waters in Japan. Yamanashi Daigaku Gaku. Kenk. Hohoku (19): 272-279. (NSA 23: 50364).

Sugihara, T. 1971. ^{226}Ra and ^{224}Ra(ThX) contents in hot spring water. Yamanashi Daigaku Kyol. Kenk. Hokoku 21: 182-189. (NSA 30: 3171).

Sugihara, T. 1972. Radioactive mineral springs in Yamanashi Prefecture. Onsen Kagaku 23(2): 84-93. (In Japanese) (NSA 30: 309).

Sugimoto, T. 1951. Cytogenetical studies on the sterile wild senna (Cassia tora L.) produced by atomic bomb explosion. I. Morphological observations on the asynaptic sterile plant. Sci. Bull. Fac. Agr. Kyushu Univ. 13: 14-19. (In Japanese) (NSA 6: 6268).

Sugino, H., Y. Okuno and J. Yoshinobu. 1971. Effect of transplanted pieces from non-x-irradiated worms on irradiated ones in Dugesia japonica. Osaka Kyoiku Daigaku Kiyo 3(19): 63-71. (NSA 29: 21218).

Suita, N. 1953. Biological investigation of atomic bomb disaster. 2. Effect on plants. Ch. 3. On unusual plants. Nippon Gakujutsu Kaigi 1: 235-239.

Sukhija, B. S. and F. A. S. Rama. 1973. Evaluation of groundwater recharge in semi-arid region of India using environmental tritium. Proc. Indian Acad. Sci. A77(6): 279-292. (NSA 28: 24459).

Sukhoverkhov, F. M., P. V. Krymova and V. G. Farberov. 1963. The influence of cobalt on growth and the hematological characteristics of carp. Ryb. khozyaystvo (8): 35-38. (In Russian).

Sulerzhitskiy, L. D. and V. S. Forova. 1966. Radiocarbon in wood from regions of contemporary volcanicity. Dokl. Akad. Nauk SSSR 171(6): 1421-1423. (Doklady (earth Sci.) 171: 232-237).

Sullivan, C. R. 1953. Use of radioactive cobalt in tracing the movements of white-pine weevil, Pissodes strobi Peck. (Coleoptera: Curculionidae). Can. Entomol. 85: 273-276.

Sullivan, C. R. 1961. The survival of adults of the white pine weevil, Pissodes strobi (Peck), labelled with radioactive cobalt. Can. Entomol. 93(1): 78-79.

Sullivan, R. R., et al. 1971. Inactivation of thirty viruses by gamma radiation. Appl. Microbiol. 22(1): 61-65. (NSA 25: 47064).

Sullivan, R. L. and D. S. Grosch. 1953. The radiation tolerance of the adult wasp. Nucleonics 11(3): 21-23.

Sullivan, W. D. 1959. The effect of ultraviolet radiation and the interaction with sulfur compounds on growth and division of Tetrahymena. Trans. Am. Microscop. Soc. 78(2): 181-193.

Sullivan, W. D. and J. V. Boyle. 1961. The effects of x-radiation on the succinic dehydrogenase activity of the dividing stages of Tetrahymena pyriformis Gl. Brotéria sér. cienc. nat. 30: 77-85. (NSA 16: 28772).

Sumarukov, G. V. 1958. The dynamic radiation injury in Calandra granaria under various irradiation conditions. Biofizika 3(3): 374-376. (Biophysics (USSR) 3(3): 359-361).

Sumarukov, G. V. 1962. Correlation between the redox potential of the lymph of crickets during irradiation and radiosensitivity. Abstr. p. 95: In: 2nd Intern. Congr. Radiation Research. (See Anonymous, 1972).

Sumarukov, G. V. 1962. Correlation between the magnitude of the oxidation-reduction potential of the hemolymph of the cricket, and the radiosensitivity of the cricket. Radiobiologiya 2(3): 374-377. (AEC-tr-5430).

Sumarukov, G. V. 1966. Mechanism of the oxygen effect. pp. 163-170. In: Protection Against and Repair of Radiation Injuries. Ye. Ya. Grayevskiy, V. I. Ivanov and V. I. Korogodin, eds. Publ. House Acad. Sci. USSR, Moscow. (In Russian) (NSA 22: 25869) (Crickets).

Sumiya, H. 1973. Environmental problems in the promotion of nuclear power generation and the countermeasures. pp. 145-180. In: Siting and Environmental Problems in the Development of Electric Power Stations. (Its Status and Analysis). K. Tomioka, ed. Fuji International Co., Tokyo. (In Japanese) (NSA 30: 14604).

Sumrall, C. L., III and E. J. Middlebrooks. 1968. Removal of radioisotopes from water by slurrying with yazoo and zilpha clays. J. Am. Water Works Assoc. 60(4): 485-494.

Sun, S. and W. Lee. 1965. Histopathological studies on the effects of gamma irradiation on the ovarian tissues of Ostrinia nubilalis (Hübner). Acta ent. sin. 14(5): 431-432. (In Chinese, English summary).

Sun, Y. P., C. H. Schaefer and E. R. Johnson. 1966. Effects of application methods on the toxicity and distribution of Dieldrin and resistance in house flies. Bull. Entomol. Soc. Am. 12: 248. (^{14}C).

Sun, Y. P., C. H. Schaefer and E. R. Johnson. 1967. Effects of application methods on the toxicity and distribution of Dieldrin in house flies. J. Econ. Entomol. 60(4): 1033-1037. (^{14}C).

Sunada, T. 1969. Treatment of polluted waters with radiation. Soda to Enso (Tokyo) 20: 123-134. (In Japanese) (NSA 23: 45408).

Sundstrøm, G. and C. A. Wachtmeister. 1973. Synthesis of ^{14}C-labelled and unlabelled PCB compounds. pp. 73-86. In: PCB Conf. 2. E. Ahnland, ed. Allmaenna Fölaget, Stockholm. (NSA 28: 24347).

Sunquvist, B., et al. 1974. Fast method for nitrogen determination in single seeds. Int. J. Appl. Radiat. Isotopes 25(6): 277-281. (NSA 30: 14794).

Suomalaine, E., O. Turpeinen and R. Niini. 1956. Mutations in Drosophila melanogaster grown on media containing carbon-14 labelled sugars. Nature 178(4529): 357-358.

Suomela, J. 1974. Method for determination of longlived radon daughters in water and biological samples. Sweden Statens Straalskyddsinstitut, Stockholm, report SSI-1974-007. 17pp. (In Swedish) (NSA 30: 31536).

Surabyan, A. S. 1969. The influence of gamma irradiation upon the population dynamics and sex ratio in model populations of *Drosophila melanogaster*. Zh. obshch. biol. 30: 374-376. (In Russian) (NSA 23: 35860).

Surganova, N. A., et al. 1971. Extraction and radiochemical methods in the determination of uranium isotopes in ground waters. Izv. vyssh. uch. zaved. geol. razved. 14(1): 100-105. (In Russian) (NSA 26: 33361).

Susorova, N. A. 1962. Methods of sampling and preparation for radiochemical analysis. pp. 233-246. In: *Radioactive Contamination of the External Environment*. (See Shvedov and Shirokov, 1962) (AEC-tr-6049, pp. 1-10) (NSA 18: 8519).

Süss, A. 1973. Application of radiotracers in studying the persistence, sorption, and degradation of herbicides in soil. Part of a coordinated programme on isotope-tracer aided studies of the origin and fate of foreign chemical residues in the agricultural environment. Final report, 1 April 1971-31 March 1973. International Atomic Energy Agency, Vienna, report IAEA-R-1003-F. 3pp. (NSA 30: 21155).

Süss, A., G. Schurmann and G. Pfann. 1966. Behavior of ^{14}C-labelled diesel oil in different soils. pp. 125-134. In: *Isotopes in Weed Research*. (See IAEA, 1966).

Süss, A., et al. 1972. Reactions of herbicides with soil and plants. pp. 111-121. In: *Radiotracer Studies of Chemical Residues in Food and Agriculture*. (See IAEA, 1972) (NSA 27: 20191).

Süss, A., et al. 1974. Fate of a urea herbicide in the soil and in the plants. pp. 373-384. In: *Comparative Studies of Food and Environmental Contamination*. (See IAEA, 1974) (NSA 30: 15473) (^{14}C).

Süss, A., et al. 1974. Experimental irradiation facility for the sterilization of sewage sludge. Kerntecknik 16(2): 65-70. (In German and English) (NSA 30: 3171).

Suszka, B., K. Ohba and M. Simak. 1960. On the growth of pine seedlings from x-irradiated seeds. Medd. Stat. Skogforskinst. 49(9): 18pp. (In German).

Sutcliffe, D. W. 1961, 1962. Studies on salt and water balance in caddis larvae (Trichoptera). I. Osmotic and ionic regulation of body fluids in *Limnephilus affinis* Curtis. II. Osmotic and ionic regulation of body fluids in *Limnephilus stigma* Curtis and *Anabolia nervoso* Leach. III. Drinking and excretion. J. Exptl. Biol. 38: 501-509, 521-530; 39: 141-160. (^{24}Na).

Sutton, W. W. 1965. Effects of ionizing radiations on peritrichs. Thesis, Howard Univ., Washington, D. C. 95pp. (NSA 20: 16295).

Suzuki, A. 1967. Analysis of the effects of neutron radiation of developing *Triturus* embryos. J. Exptl. Zool. 165(1): 111-115. (NSA 22: 10877).

Suzuki, A. 1968. Studies on the blood types of fish with radioactive antibodies - I. Preparation of ^{131}I-marked antibodies and dried blood samples. Bull. Tokai Fish. Res. Lab. Tokyo (56): 117-124. (In Japanese, English summary).

Suzuki, K. (Ed.). 1971. *Selected Papers in Biochemistry. Vol. 4. Radiation Biology of Microorganisms*. Univ. Park Press, Baltimore. xx, 312pp.

Suzuki, K. and T. Yamamoto. 1959. Uptake of yttrium by microorganisms. Paper JRIA 59/P-203, B-2. In: *3rd. Japan Conf. on Radioisotopes*. Japan Atomic Industrial Forum, Tokyo. (English abstract).

Suzuki, M. 1969. Fission track dating and uranium contents of obsidian. Dai-4-Ki Kenkyu 8: 123-130. (In Japanese) (NSA 25: 13112) (Archaeology).

Suzuki, T. and O. Tanabe. 1963. Studies on the effects of β-rays irradiation on the mold. Rept. Ferment. Res. Inst. (Chiba) 24: 57-67. (In Japanese) (NSA of Japan 3(5): 01088).

Suzuki, Y., R. Nakamura and T. Ueda. 1972. Accumulation of strontium and calcium in freshwater fishes of Japan. J. Radiat. Res. 13(4): 199-207. (NSA 27: 12375; 28: 15817).

Suzuki, Y., et al. 1974. ^{137}Cs content in Japanese diet from 1963 to 1971. J. Radiat. Res. 15(4): 181-188. (NSA 32: 19762).

Svensson, G. K. 1967. The increasing ^{137}Cs level in forest moss in relation to the total ^{137}Cs fallout from 1961 through 1965. pp. 539-546. In: *Radioecological Concentration Processes*. (See Aberg and Hungate, 1967).

Svensson, G. K. and K. Lidén. 1965. The quantitative accumulation of ^{95}Zr+^{95}Nb and ^{140}Ba+^{140}La in carpets of forest moss. A field study. Health Physics 11: 1033-1042. (NSA 19: 42249).

Svensson, G. K. and K. Lidén. 1965. The transport of ^{137}Cs from lichen to animal to man. In: *Radiation and Terrestrial Ecosystems*. Health Physics 11(12): 1393-1400. (See Hungate, 1965).

Svirezhev, Yu. M., V. L. Anokhin and A. N. Tyuryukanov. 1971. Quantitative models in radiation biocoenology. pp. 207-215. In: *Methods of Radioecological Investigations*. (See Verkhovskaya, 1971) (In Russian).

Sviridov, G. G. 1963. The use of radioactive isotopes in the study of some problems of flea ecology. 2. The contact of animals and intensity of the exchange of ectoparasites in *Rhombomys opimus* population. Zool. zh. 42(6): 947-949. (JPRS-20864; OTS-63-31650) (NSA 17: 35438).

Sviridov, G. G., et al. 1963. Use of radioactive isotopes for the study of certain problems of flea ecology. I. Alimentary relations of fleas of the *Xenopsylla* with *Rhombomys opimus* Pall. under natural conditions. Zool. zh. 42(4): 546-550. (JPRS-20207).

Svoboda, J. A. 1965. Dealkylation of H^{3}-γ-sitosterol in American cockroaches and tobacco hornworms. Abstr. 97. Bull. Entomol. Soc. Am. 11(3): 159.

Swaminathan, M. S., et al. 1963. Mutations: incidence in *Drosophila melaonogaster* reared on irradiated medium. Science 141(3581): 637-638. (NSA 17: 33747).

Swanson, C. A. 1957. Translocation of trees. Ohio J. Sci. 57: 331-336.

Swanson, C. R. 1966. Recent research on the fate of herbicides in crop plants. pp. 135-148. In: Isotopes in Weed Research. (See IAEA, 1966).

Swanson, J. L. 1973. Nature of actinide species retained by sediments at Hanford. Interim progress report. Battelle Northwest Lab., U. S. AEC report BNWL-B-296. 25pp. (NSA 29: 12692).

Swartz, W. 1973. Use of microorganisms for leaching processes. Metall. 27 (12): 1202-1206. (In German) (NSA 30: 2694).

Sweden Radiofysiska Institutionen. nd. Physical aspects of the radioactive fallout over Scandanavia, especially during the period October 1958-October 1959. (See Sandmeier and Stover, nd.).

Swedjemark, G. A. 1971. Viewpoints on "non-action levels" for foodstuffs in case of radioactive contamination. pp. 131-138. In: 3rd Nordic Radiation Protection Conf. (See Nordic Society for Radiation Protection, 1971) (In Swedish) (NSA 27: 22493) (^{137}Cs in reindeer meat).

Switsur, V. R., et al. 1974. New sample combustion bomb for radiocarbon dating. Int. J. Appl. Radiat. Isotopes 25(3): 113-117. (NSA 29: 29394).

Switzerland Eidgenössisches Dept. des Innern. 1968. Water protection in connection with coolant uptake and discharge in conventional and nuclear power plants. Bern. 116pp. (In German) (NP-19736) (NSA 29: 26183).

Switzerland Gottlieb Duttweiler Institute. 1972. Intern. Symp. on Energy, Man, and the Environment. (Zurich, February 3-5, 1972). 193pp. Zurich. (CONF-720234) (NSA 29: 31502).

Switzerland Organisation Mondiale de la Sante. 1969. Surveillance of radionuclides in the air and water at the present time. Geneva. 72pp. (In French) (NSA 25: 1845).

Szabó, Á. 1964. I. F. A. - Cluj contributions to the study of environmental radioactivity in the last 15 years. Acad. rep. populare Romine studii cerc. fiz. 16(9): 1091-1116. (In Rumanian) (NSA 20: 23410) (Mineral and fresh waters, gases, minerals).

Szabó, Á. 1968. An experimental method for determining the total beta-gamma soil radioactivity. Stiint. Solului 6(4): 23-27. (In Rumanian) (NSA 24: 25117).

Szabó, Á. nd. Research into radioisotope contamination of the living environment in Rumania. (INIS-mf-887-1) (In French) (NSA 29: 29299).

Szabó, Á. and I. Chereji. 1967. Researches on mineral water radioisotopes in Crisana District. Acd. rep. populare Romine studii cerc. geol. geofiz. geogr. (geofiz) 5: 281-288. (In Rumanian) (NSA 23: 6071).

Szabó, Á. and I. Chereji. 1967. Abundance research of radioisotopes in mineral waters of Crisana District. Inst. Fiz. Atomica, Acad. Rep. Pop. Romine, Bucharest, report IFA-IS-44. (NSA 22: 27896).

Szabó, Á. and O. Serbanescu. 1963. Radioactivity of mineral waters RPR. XII. The natural radioactivity of the water, sediments, and certain aquatic organisms of the Black Sea and Techirghiol Lake. An. Stiint. Univ. Iasi II9(1): 157-162. (In Rumanian) (CA 60: 10385; NSA 20: 10512).

Szabo, B. J. 1971. Comment concerning the applicability of the open system model to dating of fossil bones from San Joaquin soil, California. Earth Plant. Sci. Lett. 10: 252. (NSA 25: 21170).

Szabo, B. J. 1972. Uranium-series systematics in natural materials from the Newport area, Oregon. U. S. Geological Survey Prof. Paper 800-C: 199-201. (NSA 26: 56146).

Szabo, B. J. and J. N. Rosholt. 1969. Uranium-series dating of Pleistocene molluscan shells from southern California. J. Geophys. Res. 74: 3253-3260. (NSA 23: 33265).

Szabo, B. J., H. E. Malde and C. Irwin-Williams. 1969. Dilemma posed by uranium-series dates on archaeologically significant bones from Valsequilli, Puebla, Mexico. Earth Plant. Sci. Lett. 6: 237-244. (NSA 23: 49939).

Szabo, B. J., et al. 1970. Distribution of uranium in uranium-series dated fossil shells and bones show by fission tracks. U. S. Geological Survey Prof. Paper 700-B: 90-92. (NSA 24: 46346).

Szabo, E. 1972. Determination of uranium in the soil. Igiena (Bucharest) 21(5): 309-312. (NSA 29: 7515).

Szafránski, P., J. Lutowicz and L. puźyńska. 1964. The ribonucleic acid from the silk gland of the silkworm and the amino acid code. Acta biochim. Polon. 11(1): 71-81. (^{32}P).

Szalay, S. and E. Scherf. 1960. Uranium trace content of waters deriving from the Tokaj Mountains and their forelands. Magyar Tudomanyos Akad. Atommag Kuatató Int. (Debrecen) Közlemenyek 2: 71-98. (In Hungarian) (NSA 15: 6221).

Szentesi, A. 1975. Effect of substerile irradiation doses on the progeny of treated bean weevil adults (Acanthoscelides obtectus Say, Col., Bruchidae). pp. 269-278. In: Sterility Principle for Insect Control, 1974. (See IAEA, 1975) (NSA 32: 11783).

Szepke, R. 1961. An estimation of radiation exposure from fallout. Poland Central Lab. for Radiation Protection, Warsaw, report CLOR-7. 24pp. (NSA 15: 29118).

Szepke, R. 1961. Plants as monitors of radioactive pollution of the environment. Poland Central Lab. for Radiation Protection, Warsaw, report CLOR-9. 88pp.

Szepke, R. 1963. An estimate of decontamination factors for the drainage area of the Vistula River. Atompraxis 9(1): 1-2. (NSA 17: 15617).

Szepke, R. 1963. Pine needles as a bioindicator of radioactivity contamination of the environment. Atompraxis 9(6): 218-219. (NSA 17: 30763).

Szepke, R. 1965. Graphic statistical analysis applied to investigation on radioactivity of some surface waters in Poland. Ekol. Polska B11: 83-89. (In Polish) (NSA 19: 15799).

Szepke, R. and D. Grzybowska. 1964. Strontium-90 in tap water and in some rivers in Poland, 1962. Poland Central Lab. for Radiation Protection, Warsaw, report CLOR-32. 11pp. (NSA 19: 24701).

Szepke, R., Z. Gorberg and T. Deszczak. 1960. Measurement of radioactivity in Warsaw, Poland, during the year 1959. Inst. of Nuclear Research, Poland Acad. of Sci., Warsaw, report 140/doz. 30pp. (NP-8676) (NSA 14: 13952).

Szepke, R., T. Wardaszko, and J. Pensko. 1961. Practical method of the radioactive air contamination determinations in the self-luminous times establishments. Poland Central Lab. for Radiation Protection, Warsaw, report CLOR-6. 22pp. (NSA 15: 29118).

Szepke, R., et al. 1961, 1962. Water radioactivity in Poland 1960. ---1961. Poland Central Lab. for Radiation Protection, Warsaw, report CLOR-10, -13. 50, 31pp. (NSA 16: 22400).

Szepke, R., et al. 1961. Strontium-90 in soil, Poland, 1960. Poland Central Lab. for Radiation Protection, Warsaw, report CLOR-12. 19pp.

Szilagyi-Gyoeri, E. and P. Fodor-Csanyi. 1967. The measurement of beta-emitting isotopes with liquid scintillator in samples of biological origin. Ann. Univ. Sci. Budapest. Rolando Eotvos Non. (chim.) 9: 41-47. (NSA 23: 17360).

Szilvinyi, A. V. and G. Kernmayer. 1964. Giant cells by β radiation. Naturwissenschaften 51(10): 241-242. (In German) (NSA 18: 29331).

Szmuk, P. and S. Watanabe. 1971. Dating of Brazilian Indian pottery by thermoluminescent dosimetry. Brazil Inst. de Energia Atomica, São Paulo, report IEA-231. 7pp. (NSA 26: 22808).

Szulc, M. 1963. Natural radioactivity of potassium in some food products of animal origin. Med. Weterynar. (Poland) 19(4): 196-199. (In Polish) (NSA 17: 29148).

Szulc, M. and A. Bruniak. 1971. Content of cesium in muscles of broilers, hens, and partridges. Med. Weternyar. (Poland) 27: 297-299. (In Polish) (NSA 26: 45548).

Szumie, M. 1961. Preliminary measurements of radiation intensity in ponds. Acta Hydrobiol. 3(2/3): 133-142.

Szumlewicz, A. P. 1964. Schistosomiasis: age of snails and susceptibility to x-irradiation. Science 144(3616): 302-303. (NSA 18: 19698).

Szy, D. and A. Toth. 1970. Determination of $^{234}U/^{238}U$ ratios in samples of natural water. Mikrochim. Acta (5): 960-964. (In German) (NSA 25: 15519).

Szyszko, E., J. Wozniak and J. Malesa. 1962. Effects of ionizing radiation on pests in cereals. II. Effects of cobalt-60 radiation on *Tyroglyphus farinae* mites. Roczn. pánst. Zakl. Hig. 13(5): 493-501. (In Polish).

Tabert, G. G. and D. B. Lackman. 1965. The radioisotope precipitation test for study of Q fever antibodies in human and animal sera. J. Immunol. 94: 959-965. (NSA 18: 45739).

Tadmor, J. 1971. Consideration of stable elements in environment in the evaluation of the maximum permissible concentrations for radionuclides. pp. 397-404. In: *Proc. Intern. Symp. on Radioecology Applied to the Protection of Man and His Environment*. (See Commission of the European Communities, 1971) (NSA 27: 5250).

Tadmor, J. 1971. The importance of deposition of pollutants in the evaluation of environmental pollution. pp. 561-583. In: *Proc. Intern. Symp. on Radioecology Applied to the Protection of Man and His Environment*. (See Commission of the European Communities, 1971) (NSA 27: 5231).

Tadmor, J. 1972. Environmental behavior of radioactive sodium: transport, uptake, and population dose. Harvard Univ., U. S. AEC report COO-3019-2. 117pp. (NSA 27: 17685).

Tadmore, J. 1972. Atomic energy in the environment. At. Jap. 16(5): 35-47. (NSA 27: 27574).

Taege, M. 1963. Investigations on the spermatogenesis and embryonic development following irradiation of *Calliphora erythrocephala* Meig. Diptera, Calliph. males. Dtsch. Entomol. Z. 10: 335-377. (In German).

Taga, N. 1956. Preliminary studies on removal of radioactive substances from polluted water by the biological method. On the capacity of water bacteria for absorption of fission products. pp. 1245-1250. In: *Research in the Effects and Influences of the Nuclear Test Explosions*. Japan Soc. for the Promotion of Science, Ueno, Tokyo.

Tagliati, S. and O. Ilari. 1966. Environmental monitoring around the Nuclear Research Centre at La Casaccia, Italy. pp. 30-44. In: *Manual on Environmental Monitoring in Normal Operation*. International Atomic Energy Agency, Vienna, Safety Ser. No. 16, Publ. STI-PUB-98. (NSA 20: 23406).

Tahmisian, T. N. 1949. The effect of x-radiation on the metabolic processes of the resting cell. J. Exptl. Zool. 112: 449-463. (Grasshopper).

Tahmisian, T. N. 1961. Cytological effects. pp. 333-352. In: *Mechanisms in Radiobiology. Vol. I. General Principles*. M. Errera and A. Forssberg, eds. Academic Press, New York. (Concerns grasshopper embryos).

Tahmisian, T. N. 1965. Biochemical and developmental response of grasshopper embryos and spermatogonia to x-irradiation. Argonne Natl. Lab. Rev. 2(3): 52-53, 59. (NSA 19: 33813).

Tahmisian, T. N. and D. M. Adamson. 1950. Oxidase increase in *Melanoplus differentialis* eggs caused by x-irradiation. J. Exptl Zool. 115(2): 379-398.

Tahmisian, T. N. and R. L. Devine. 1955. Repression and enhancement of irradiation effects on grasshopper cells by metabolic poisons and oxygen. Radiation Research 3(2): 182-190. (NSA 10: 45).

Tahmisian, T. N. and R. L. Devine. 1961. The influence of x rays on organelle induction and differentiation in grasshopper spermatogenesis. J. Biophys. Biochem. Cytol. 9(1): 29-45. (NSA 17: 14040).

Tahmisian, T. N. and J. Gasvada. 1951. Morphological and respiratory changes induced in *Melanoplus differentialis* by β irradiation. Abstr. Argonne National Lab., U. S. AEC report AECU-1258; UAC-350. 1p. (NSA 5: 3296).

Tahmisian, T. N. and H. H. Vogel. 1953. Relative biological effectiveness of fast neutrons, gamma rays, x-rays on grasshopper nymph ovarioles (*Melanoplus differentialis*). Proc. Soc. Exptl. Biol. Med. 84(3): 538-543. (NSA 8: 1501).

Tahmisian, T. N. and B. J. Wright. 1957. Teratogenesis in grasshopper (*Melanoplus differentialis*) embryos as a function of the dose rate of x-irradiation. Anat. Record 128(3): 632.

Tahmisian, T. N. and B. J. Wright. 1971. Morphological changes in cellular organelles of x-irradiated bacteria-free grasshopper diapause embryos. Radiation Research 46: 165-185. (NSA 25: 32520).

Tahmisian, T. N., J. V. Passonneau and D. M. Adamson. 1954. The effect of x irradiation and lowered metabolic rate on the morphogenesis of developing *Melanoplus differentialis* embryos. J. Natl. Cancer Inst. 14: 941-953. (NSA 8: 2305).

Taimr, L. 1958. Radioisotopes in the scientific research on insecticides. Bull. Radiobiol. Komise 3: 241. (In Czech).

Taimr, L. and J. Diribek. 1967. Definition of the range of host plants of the frit fly (*Oscinella frit* L.) by means of ^{32}P in the feeds of the insects. Ved. Pr. vyzk. Ust rostl. Vyroby Prace-Ruzni 12: 29-36. (In Czech with English, German and Russian summaries).

Taimr, L. and J. Dlabola. 1963. Radioisotopes as tracers used for migration studies of the leafhopper species *Caligypona pellucida* F. Acta agron. Hung. 12(3/4): 321-334. (In Hungarian).

Taimr, L. and J. Dlabola. 1965. Application of radiophosphorus for the mating of leafhoppers in the study of their migration. Ochr. Rost. 3: 75-83. (In Czech with Russian and English summaries).

Taimr, L., et al. 1967. Further experience obtained in studies on dispersal flights of *Meligethes aeneus* F., marked with ^{32}P (Coleoptera). Acta entomol. Bohemoslov. 64: 325-332. (NSA 22: 8467).

Taira, T. and T. Morita. 1958. Melanotic tumors induced in *Drosophila melanogaster* by ionizing irradiation. Mishima. Natl. Inst. Genet. Ann. Rpt. 8: 82-84.

Taira, T. and S. Nawa. 1957. The effects of x-irradiation on nitrogen metabolism in *Drosophila melanogaster*. Mishima. Natl. Inst. Genet. Ann. Rpt. 7: 83-84.

Tajima, E. 1972. Atomic energy in the environment. At. Jap. 16(5): 35-47. (NSA 27: 27574).

Takahashi, J. 1956. Vitamin B_{12} in silkworm. III. Nippon Nogei-Kagaku Kaishi 30: 563-565. (Used ^{60}Co).

Takahashi, K. 1970. Environmental contamination associated with the development of nuclear power. Doboku Gakkaishi 55(2): 41-42. (In Japanese) (NSA 25: 13437).

Takahashi, T. 1964. Measurement of tritium at natural levels. Genshiryoku Kogyo 10(2): 65-68. (NSJ-tr-49) (NSA 20: 31398) (Tap water, river water).

Takahashi, T., T. Hamada and F. Yamasaki. 1965. Measurement of tritium in rain, city water, ground water and hot spring. Sci. Papers Inst. Phys. Chem. Res. (Tokyo) 59: 49-51. (NSA 19: 30414).

Takahashi, T., M. Nishida and T. Hamada. 1969. Improvement in enrichment procedure for analysis of tritium in natural water. II. Radioisotopes (Tokyo) 18: 559. (NSA 24: 9251).

Takahashi, T., et al. 1968. Chemical and isotopic balances for a meromictic lake. Limnol. Oceanogr. 13: 272-292. (NSA 22: 45182).

Takahashi, T., et al. 1969. Tritium concentration in wine, rain, and ground water. Radioisotopes (Tokyo) 18: 560-563. (NSA 24: 9252).

Takanaka, J., et al. 1964. Determination of the hardness of water by the isotope dilution method. J. Water and Waste (Tokyo) 6(5): 389-392. (In Japanese).

Takano, H. and K. Umeda. 1974. Radiation disinfestation of cereal insects. III. Radiosensitivity of *Plodia interpunctella*. Nippon Shokuhin Kogyo Gak. 21(6): 267-272. (In Japanese) (NSA 31: 3740).

Takase, A. 1972. Application of radioisotopes for agricultural and biological studies on environmental pollution. Proc. Jap. Conf. Radioisotop. (10): 504-506. (In Japanese) (NSA 27: 28101).

Takase, A. 1972. Pollution of environment and radioactive tracer techniques. Nippon Koshu Eisei Zasshi 19(5): 245-249. (In Japanese) (NSA 30: 334).

Takase, A. and K. Yamada. 1955. Distribution of radioactivity in various tissues of fish, and group separation of radioactive elements in them. Bull. Inst. Publ. Hlth. (Tokyo) 4(3): 17-21. (CA 49: 15096d).

Takase, A., T. Bito and A. Ishihara. 1959. The utilization of radio-isotope-labelled bacteria in the food sanitation study-II, Y-90-labelled *E. coli* and bacterial contamination of living fish or shellfish. Paper. In: *3rd Japan Conf. on Radioisotopes*. Japan Atomic Industrial Forum, Yokyo. Tokyo. (JAIF, JRAI-59/P-81, A-20) (English abstr.).

Takase, A., A. Kawamura and Y. Izumo. 1969. A method for the measurement of digestibility in fish using radio-chromic oxide. pp. 468-470. In: *Proc. 9th Japan Conf. Radioisotopes*. (May 13-15, 1969, Tokyo, Japan). Japan Atomic Industrial Forum, Tokyo. (CONF-690548) (NSA 24: 31782).

Takase, A., et al. 1969. A bioassay method for the evaluation of river water pollution using shellfish's ability on the uptake of radioactive particles. pp. 471-473. In: Proc. 9th Japan Conf. on Radioisotopes. (May-13-15, 1969, Tokyo, Japan). Japan Atomic Industrial Forum, Tokyo. (In Japanese) (CONF-690548) (NSA 24: 31783).

Takatalo, S. and J. K. Miettinen. 1969. Excretion rate of strontium in three species of fresh water fish: cruvian carp (Cyprinus carassius), perch (Perca fluviatilis) and rainbow trout (Salmo gairdneri). Univ. of Helsinki, Finland, U. S. AEC report NYO-3446-13. 6pp. (For: 5th R.I.S. Symp., Helsinki, May 19-20, 1969) (CONF-690527).

Takatalo, S. and J. K. Miettinen. 1969. Total balance experiments with ^{32}P in eutrophic and oligotrophic lakes. Univ. of Helsinki, Finland, U. S. AEC report NYO-3446-13. 28pp. (For: 5th R.I.S. Symp., Helsinki, May 19-20, 1969) (CONF-690527).

Takatalo, S., S. Kolehmainen and J. K. Miettinen. 1970. Excretion of iodine by perch (Perca fluviatilis). Transl. Am. Fish. Soc. 99(3): 515-517. (SFA 15: 12403) (^{131}I).

Takeda, H. and K. Tanaka. 1963. Study on the influence of x-ray upon the hematocytes of the silkworm larvae Bombyx mori L. Proc. Japan Conf. Radioisotopes 5(4): 234-236. (In Japanese) (NSA 17: 30221).

Takeda, H., T. Tanaka and K. Asano. 1959. Studies on the aftereffects of x-rays on the larval development and the cocoons during the embryonic stages. Res. Rpts. Fac. Text. Sericult. Shinshu Univ. 9: 60-73. (Silkworm).

Takesita, I. and S. Kakuda. 1957. Radioactivity in the river water, brackish water, sea water and the layer at the river mouth, derived from radioactive rain water. J. Fac. Fish. Animal Husb. (Hiroshima Univ.) 1(3): 303-310.

Takeshita, K. 1962. Measurements of radioactivity in Nishiyama District, Nagasaki, Japan. J. Radiat. Res. (Japan) 3: 177-181. (NSA 18: 14082) (Incl. mud, water).

Takeuchi, T. and H. Okutsu. 1974. ^{14}C age of alluvium in Yokokura, Tsunoda City, Miyagi Prefecture. Chikyu Kagaku 28(4): 150-151. (In Japanese) (NSA 31: 33159).

Takhomirov, F. A. 1971. Tasks of dosimetry in radioecology. pp. 261-279. In: Contemporary Problems of Radiobiology. Vol. 2. (See Klechovskiy, Polikarpov and Aleksakhin, 1971) (In Russian).

Takhomirov, F. A. and R. M. Aleksakhin. 1971. Effect of ionizing radiation on forest biogeocenoses. pp. 228-260. In: Contemporary Problems of Radiobiology. Vol. 2. (See Klechkovskiy, Polikarpov and Aleksakhin, 1971) (In Russian).

Takasawa, Y. 1968. Environmental disturbances due to radiation fallout. Naika 21: 859-863. (In Japanese) (NSA 24: 2601) (Mentions radioecology).

Takasawa, Y. 1968. Radioecological studies of environmental radioactivity to man. Niigata Igakkai Zasshi 82: 565-568. (NSA 23: 14139).

Talburt, D. E. 1965. Some effects of rare earths and yttrium on the growth and metabolism of Aspergillus niger. Thesis, Univ. of Arkansas, Fayetteville. 93pp. (NSA 20: 16191).

Talburt, D. E. and G. T. Johnson. 1967. Some effects of rare earth elements and yttrium on microbial growth. Mycologia 59: 493-503. (NSA 21: 43348).

Talenti, M. and A. Cigna. 1964. Problem of radioactive mineral waters from the hygienic point of view. Nuovi Ann. Ig. Microbiol. 15: 568-571. (In Italian) (NSA 21: 6523).

Talibudeen, O. 1964. Natural radioactivity in soils. Soils Fertilizer 27: 347-359. (NSA 21: 3182).

Tallentire, A. and C. O. Chiori. 1963. Heat and gamma-radiation resistance of Bacillus megaterium spores. J. Pharm. Pharmacol. 15(Suppl.): 148T-149T. (NSA 18: 13504).

Tallentire, A. and D. J. G. Davies. 1961. A post-irradiation oxygen effect in bacterial spores and its dependence on water content. Exptl. Cell Res. 24: 148-150. (NSA 15: 29021).

Tallentire, A. and E. L. Powers. 1963. Modification of sensitivity to x-irradiation by water in Bacillus megaterium. Radiation Reserach 20(2): 270-287. (Also: Argonne National Lab., U. S. AEC report TID-18378; UAC-7155) (NSA 17: 40614).

Tallentire, A., N. A. Dickenson and J. H. Collett. 1963. A dependence of water content of bacterial efficiency of gamma-radiation. J. Pharm. Pharmacol. 15(Suppl.): 180T-181T. (NSA 18: 13505).

Talmage, R. V., S. B. Stephen and C. W. Yates. 1962. The effect of temperature on the uptake of radioiodine by the thyroid gland of the frog, Rana pipiens. Gen. Comp. Endocrinol. 2: 266-272.

Talvite, N. A. 1971. Radiochemical determination of plutonium in environmental and biological samples by ion exchange. Anal. Chem. 43(13): 1827-1830. (NSA 26: 1798).

Tamada, T., et al. 1973. Genetic analysis of developmental stages of the silkworm egg. Nippon Sanshigaku Zasshi 41(1): 21-27. (In Japanese) (NSA 28: 6052).

Tamers, M. A. 1967. Radiocarbon ages of groundwater in an arid zone unconfirmed aquifer. pp. 143-152. In: Isotope Techniques in the Hydrologic Cycle. (See Stout, 1967).

Tamers, M. A. 1967. Surface-water infiltration and ground-water movement in arid zones of Venezuela. pp. 339-353. In: Isotopes in Hydrology. (See IAEA, 1967) (NSA 22: 18931).

Tamers, M. A. 1967. Filtration of surface water and movement underground water in arid zones. Acta Cient. Venez. 18(Suppl.3): 358-371. (In Spanish) (NSA 22: 42983).

Tamers, M. A. 1969. Radiocarbon dating of recent events. Atompraxis 15: 271-276. (NSA 24: 315).

Tamers, M. A., C. Ronzani and H. W. Scharpenseel. 1969. Observation of naturally occurring ^{36}Cl. Atompraxis 15: 433-437. (In German) (NSA 24: 11969).

Tamers, M. A., J. J. Stipp and R. Weiner. 1975. Radiocarbon ages of ground water as a basis for the determination of safe limits of aquifer exploitation. Environ. Res. 9(3): 250-264. (NSA 33: 7324).

Tami, T. M., T. A. Gibson, Jr. and A. L. Prindle. 1970. Far-out collection program. Project Schooner. Lawrence Livermore Lab., U. S. AEC report PNE-527. 17pp. (NSA 25: 26404).

Tamm, C. O. 1951. Seasonal variation in composition of birch leaves. Physiol. Plant. 4: 461-469.

Tamm, C. O. and T. G. Ostlund. 1960. Radiocarbon dating of soil humus. Nature 185: 706-707.

Tamplin, A. R. 1970. Regulation of man-made radiation in the biosphere. pp. 45-51. In: Nuclear Power and the Public. H. Foreman, ed. Univ. of Minnesota Press, Minneapolis. (NSA 25: 23207).

Tamura, T. 1968. Movement of cesium-137 by runoff, erosion, and infiltration from a soil under different cover conditions. pp. 140-167. In: Proc. Symp. on Postattack Recovery from Nuclear War. U. S. Off. of Civil Defense, Washington.

Tamura, T. 1974. Distribution and characterization of plutonium in soils from Nevada Test Site. pp. 29-42. In: Dynamics of Plutonium in Desert Environments. (See Dunaway and White, 1974) (NSA 30: 29443).

Tamura, T. 1975. Characterization of plutonium in surface soils from Area 13 of the Nevada Test Site. pp. 27-41. In: Radioecology of Plutonium and Other Transuranics in Desert Environments. (See White and Dunaway, 1975) (NSA 33: 665).

Tamura, T. 1975. Distribution and characterization of plutonium in soils from Nevada Test Site. J. Environ. Qual. 4(3): 350-354. (NSA 33: 5587).

Tamura, T. 1975. Physical and chemical characteristics of plutonium in existing contaminated soils and sediments. Oak Ridge National Lab., U. S. ERDA report. 19pp. (CONF-751105-4) (NSA 33: 7284).

Tan, B. D. (D. B.?). 1965. X-ray-induced abnormalities and recovery in Hymenolepis microstoma. Thesis, Univ. of Tennessee, Knoxville. 73pp. (NSA 20: 36662).

Tan, D. B. (B. D.?) and A. W. Jones. 1966. X-ray induced abnormalities and recovery in Hymenolepis microstoma. Exptl. Parasitol. 18: 355-373. (NSA 21: 24638).

Tanaka, K. 1970. Quality control of soils in banking using the radioisotope apparatus. Kajima Ken. Gijutsu Kenky. Nempo 18: 107-114. (In Japanese) (NSA 26: 189).

Tanaka, K. 1974. Estimation of the rate of precipitation recharge to the groundwater by tritium measurements. Chikyu Kagaku 7-8(1): 9-16. (In Japanese) (NSA 32: 19767).

Tanaka, N. 1958. Cytological effect of radioactive rainfall. Intern. Congr. Genet. Proc. 10(2): 287-288.

Tanaka, N. 1964. Effects of x-ray irradiation on secondary gill formation and tail regeration of Branchiura sowerbyi. I. Nippon Igaku Hoshasen Gak. Zasshi 23(11): 1314-1325. (In Japanese) (NSA 19: 21951).

Tanaka, N. 1966. Effect of fractionated x-rays and γ-rays irradiation on tadpole of Amphibia. I. Rana nigromaculata. II. On the Rhacophorus schlegelii arborea. Nippon Igaku Hoshasen Gak. Zasshi 26: 910-913, 913-921. (In Japanese) (NSA 21: 22753; 22: 28663).

Tanaka, N. 1966. The protective effects of some narcotic substances against x irradiation in aquatic animals. I. Effects of chlorobutanol in larvae of Rhacophorus arboreus. II. Protective effects of ethyl alcohol. III. Effects of urethane and changes in concentration of chlorobutanol solution against irradiation in larvae of Rhacophorus arboreus. Nippon Igaku Hoshasen Gak. Zasshi 26: 1001-1011, 1012-1015, 1016-1021. (NSA 21: 41464, 36667; 22: 28552) (III in English).

Tanaka, N. 1966. Effects of x-rays on the secondary gill formation and the tail regeneration in Branchiura sowerbyi. III. The protective effect of chlorobutanol against x-irradiation on tail regeneration in Branchiura sowerbyi. Nippon Igaku Hoshasen Gak. Zasshi 26: 922-926. (In Japanese) (NSA 21: 34732).

Tanaka, N. 1967. The effects of x-irradiation on the treated amphibian larvae in 2,4-dinitophenol and azaid. Nippon Igaku Hoshasen Gak. Zasshi 26: 1384-1387. (In Japanese) (NSA 21: 36644).

Tanaka, N., et al. 1958. Cytological effect of radioactive rain on mitotic cells of Tradescantia paludosa. Jap. J. Genet. 33(9): 327. (In Japanese).

Tanaka, R. 1942. Effect of x-ray irradiation on the hatching of the eggs of Oncorhynchus keta Walbaum. Bull. Jap. Soc. Sci. Fish. 11(1): 39-40. (In Japanese).

Tanaka, Y. and F. Goto. 1966. Experimental studies on the characteristics of various genes in the silkworm. 1. Sanshi Kagaku Kenky. Iho (14): 20-25. (In Japanese) (NSA 21: 39520).

Tang, C.-w. and C. J. Constantine. 1970. Elimination of ^{24}Na and ^{42}K interferences in activation analysis of biological samples. Science 167: 52-54. (NSA 24: 92611).

Taniguchi, H. 1963. Strontium-90 in Canadian soils, 1960-1961. Radiol. Hlth. Data 4(7): 383-385.

Tanner, A. B. 1964. Radon migration in the ground: a review. pp. 161-190. In: The Natural Radiation Environment. (See Adams and Lowder, 1964) (NSA 18: 35739).

Tanner, A. B. 1964. Physical and chemical controls on distribution of radium-226 and radon-222 in ground water near Great Salt Lake, Utah. pp. 253-276. In: The Natural Radiation Environment. (See Adams and Lowder, 1964) (NSA 18: 35739).

Tanner, T. M., L. A. Rancitelli and W. A. Haller. 1972. Multielement analysis of natural waters by neutron activation, group chemical separations, and gamma-ray spectrometric techniques. Water Air Soil Poll. 1(2): 132-__.

Tanner, W. W. 1963. Natural radiation effects of vertebrate animals inhabiting the uranium areas of southeastern Utah. pp. 325-326. In: Radioecology. (See Schultz and Klement, 1963) (NSA 17: 33579).

Tanner, W. W. 1965. A comparative population study of small vertebrates in the uranium areas of the upper Colorado Basin. Brigham Young Univ. Sci. Bull. (Biol. ser.) 7(1): 31pp. + 3 plates. (NSA 19: 38330)

Tansley, K., L. H. Gray and F. G. Spear. 1948. A preliminary note on some biological effects of alpha radiation on the frog tadpole. Brit. J. Radiol. 21(251): 567-570. (NSA 2: 33).

Tantawy, A. O. 1963. Developmental homeostasis in x-rayed populations of Drosophila pseudoobscura. Genetica 33: 222-243. (NSA 18: 72).

Tantawy, A. O. 1963. Effects of temperature and x-ray irradiation on intrinsic growth rate in populations of Drosophila pseudoobscura. Genetica 34(1): 34-45. (NSA 18: 6633).

Tantawy, A. O., A. A. Abdel-Malik and A. M. Wakid. 1966,1967. Studies on the eradication of Anopheles pharoensis by the sterile-male technique using ^{60}Co. II. Induced dominant lethals in the immature stages. IV. Mating behavior and its frequency in the sterilized mosquitoes. V. Mating competiveness in radiosterilized males. J. Econ. Entomol. 59(6): 1392-1394; 60 (1): 23-26; 60(3): 696-699. (NSA 21: 34777).

Tantawy, A. O., G. E. Nasrat and F. F. Saad. 1969. Studies on natural populations of Drosophila. VII. Egg production in natural and x-irradiated populations of Drosophila melanogaster in relation to geographical distribution. Isotopes Radiat. Res. 2: 39-47. (NSA 24: 44585).

Tape, G. F. 1970. National policy on peaceful uses of nuclear explosives. pp. 3-9. In: Education for Peaceful Uses of Nuclear Explosives. (See Weaver, 1970) (NSA 24: 27261).

Taplin, G. V., O. M. Meredith, Jr. and H. Kade. 1958. Evaluation of the acute inhalation hazard from radioactive fall-out materials by analysis of results from field operations and controlled inhalation studies in the laboratory. Univ. of California at Los Angeles, U. S. AEC report WT-1172. 26pp. (NSA 12: 5850).

Tarasov, B. N. 1954. Principles of biological action of radioactive radiation. pp. 3-138. In: Principles of the Biological Effect of Radioactive Radiations. Medgiz, Moscow. (PB-11230) (NSA 13: 14992).

Tarasov, B. N. 1962. Initial Processes of Radiation Injury. Gosatomizdat, Moscow. 95pp. (In Russian).

Tarasov, S. A. 1967. Determination of field vole age by the Roentgen technique. Zool. zh. 45(8): 1247-1250. (In Russian, English summary).

Tarchanoff and Moldenhauer. 1905. Inductive and natural radioactivity in plants. Intern. akad. Blatt. 1905: 728-734. (Krakow).

Tardieux, P., et al. 1971-72. Preliminary study of effects of gamma irradiation on growth of fungi. Agrochimica 16(1-2): 83-98. (In French) (NSA 26: 56369).

Tarpley, W., et al. 1953. Radiation sterilization I. J. Bacteriol. 65(3): 305-309.

Tarr, H. L. A. 1964. Incorporation of radioactive purine and pyrimidine bases bases and of thymidine into deoxyribonucleic acid of salmon milts. Can. J. Biochem. 42(1): 51-57. (BA 45: 50209).

Tarr, H. L. A. and J. Roy. 1966. Incorporation of radioactive compounds into purine and pyrimidine bases and 2-deoxyribose of salmon testes deoxyribonucleic acid. Can. J. Biochem. 44: 1435-1446. (1967 Studies, Part 1, Fish. Res. Bd. Can. Studies No. 1112: 15-26).

Tarrant, R. M., Jr. 1971. Seasonal variation in the accumulation and loss of ^{131}I by tissues of adult female channel catfish, Ictalurus punctatus (Rafinesque). Trans. Am. Fish. Soc. 100(2): 237-246. (SFA 16: 13544).

Tarzwell, C. M. (Comp.). 1960, 1965. Biological Problems in Water Pollution. (Trans. 2nd {3rd} Seminar, April 20-24, 1959 {August 13-17, 1962}, Cincinnati, Ohio). R. A. Taft Sanitary Engineering Center, U. S. PHS, Cincinnati, O., report W60-3, 999-WP-25. xv, 285pp.; ix, 424pp.

Tashiro, F. 1972. Effects of irradiation ^{60}Co gamma-ray on the maturation of rainbow trout. Bull. Jap. Soc. Sci. Fish. 38(8): 793-797. (In Japanese, English summary) (NSA 29: 27079; SFA 18: 17012).

Tasovac, T. and R. Radosavljevíc. 1969. Investigation of the radioactivity of the Danube. pp. 427-432. In: Environmental Contamination by radioactive Materials. (See IAEA, 1969).

Tasovac, T. and R. Radosavljevíc. 1970. Uptake of fallout radioactivity by plants on the territiory of the Boris Kidric Institute of Nuclear Studies. pp. 983-995. In: Proc. Intern. Symp. of Radioecology. Vol. II. (See French Commissariat à l'Énergie Atomique, 1970).

Tasovac, T., R. Radosavljevíc and M. Zaric. 1974. Environmental characteristics of the Danube River system and the problems of radiological safety standards. pp. 427-432. In: Population Dose Evaluation and Standards for Man and His Environment. (See IAEA, 1974) (NSA 31: 19574).

Tatchell, R. J. 1967. Salivary excretion in the cattle tick as a means of water elimination. Nature 213: 940-941. (^{3}H).

Tate, K. R. 1972. Radiocarbon dating in studies of soil organic matter-vegetation relationships. pp. E27-E38. In: Proc. 8th Intern. Radiocarbon Conf. Vol. 2. (See Rafter, 1972) (NSA 31: 8649).

Tatsumi, S. 1964. Calcium metabolism in the trees II. The effects of the temperature on the ascension of calcium in poplar cuttings. J. Japan. Forestry Soc. 46(10): 325-335. (In Japanese) (Used tracer).

Tauber, H. 1967. Copenhagen radiocarbon measurements. VIII. Geographic variations in atmospheric ^{14}C activity. Radiocarbon 9: 246-256. (NSA 21: 36160).

Taylor, A. M., J. Mauchline and E. P. Ritson. 1962. Notes on the radioecology of Sellafield Beach. UK Atomic Energy Authority, Windscale, report PG-353(W). 33pp.

Taylor, A. W. 1968. Strontium retention in acid soils of the North Carolina coastal plain. Soil Science 106: 440-447. (NSA 23: 6222).

Taylor, A. W. 1969. The selectivity coefficient of slowly exchangeable calcium and strontium. Soil Science 107: 58-62. (NSA 23: 21814).

Taylor, C. B. 1964. Tritium content of Antarctic snow. Nature 201(4915): 146-147. (NSA 18: 12353).

Taylor, C. B. 1969. A comparison of tritium and strontium-90 fallout in the Southern Hemisphere. Tellus 20(4): 559-576.

Taylor, C. B. 1971. Influence of the 1968 French thermonuclear tests on tritium fallout in the Southern Hemisphere. Earth Planet. Sci. Lett. 10: 196-198. (NSA 25: 21202).

Taylor, C. B., H. A. Polach and T. A. Rafter. 1963. Tritium measurements in nature. A review of the work of the Tritium Laboratory, Institute of Nuclear Sciences, New Zealand. New Zealand Inst. of Nuclear Sciences, Lower Hutt, report INS-R-24. 64pp. (NSA 19: 6093).

Taylor, C. E. and W. J. Taylor. 1974. Multielement analysis of environmental samples by spark source mass spectrometry. Environmental Research Lab., U. S. EPA, Athens, Ga., report EPA-660/2-74-001. 30pp. (NSA 30: 23520) (Water samples).

Taylor, C. V., J. O. Thomas and M. G. Brown. 1933. Studies on Protozoa. IV. Lethal effects of the x-radiation of a sterile culture medium of Colpidium camphylum. Physiol. Zool. 6(4): 467-492.

Taylor, E. A. nd. Review of method of sterile insect release to control and eradicate insect populations. pp. 123-128. In: Proc. Conf. on Nuclear Energy and Agricultural Productivity. (See Comision Interamericana de Energia Nuclear, nd.) (NSA 25: 27129).

Taylor, E. A. 1971. Mediterranean fruit fly suppression using the sterility principle: prospects and programs. pp. 41-48. In: Sterility Principle for Insect Control or Eradication. (See IAEA, 1971) (NSA 25: 29738).

Taylor, E. W. 1962. Radiactivity in water. Chem. Ind. (London): 1229. (NSA 17: 6337).

Taylor, E. W. 1964. Radiation in water. pp. 23-41.. In: Radiation Levels. Air, Water and Food. Royal Society of Health, London.

Taylor, F. G., Jr. 1965. Nuclear volume changes in southeastern tree species before spring growth. Radiation Botany 5(1): 61-66. (Also: Oak Ridge National Lab., U. S. AEC report ORNL-P-337) (NSA 19: 1496) (At Oak Ridge).

Taylor, F. G., Jr. 1966. Predicted seasonal radiosensitivity of southeastern tree species. Radiation Botany 6: 307-311. (Also: Oak Ridge National Lab., U. S. AEC report ORNL-P-1870) (NSA 20: 36551).

Taylor, F. G., Jr. 1968. Some effects od acute gamma radiation in giant sequoia seedlings. Radiation Botany 8(1): 67-70. (NSA 22: 28706).

Taylor, F. G., Jr. and J. P. Witherspoon. 1972. Retention of simulated fallout particles by lichens and mosses. Health Physics 23(6): 867-869. (NSA 27: 2849).

Taylor, J. F. and E. Hodgson. 1965. The origin of phospholipid ethanolamine in the blowfly, Phormia regina (Meig.). J. Insect Physiol. 11: 281-285. (^{14}C).

Taylor, J. H. 1964. The arrangement of chromosomes in the mature sperm of the grasshopper. J. Cell Biol. 21(2): 286-289. (^{3}H, Romalea microptera).

Taylor, J. H. 1965. Distribution of tritium labeled DNA among chromosomes during meiosis. I. Spermatogenesis in the grasshopper. J. Cell Biol. 25 (2, Pt.2): 57-67. (^{3}H, Romalea microptera).

Taylor, K. B., A. A. Berryman and J. L. Blackwell. 1966. Progress in the sterilization of the European pine shoot moth. Abstr. 98. Bull. Entomol. Soc. Am. 12: 286.

Taylor, L. S. 1971. Radiation protection standards. --- Part II. CRC press Crit. Rev. Envirn. Control 2(1): 81-124; 2(2): 147-205.

Taylor, R. E. and W. T. Oakley. 1968. Water-well investigations before and after the Sterling Event in Tatum Salt Dome, Lamar County, Mississippi. U. S. Geological Survey, U. S. AEC report USGS-474-121. 10pp. (NSA 26: 30979).

Taylor, R. W. and A. R. Boulogne. 1967. Rapid determination of ^{32}P. Health Physics 13: 657-659. (NSA 21: 25791) (Incl. algae, fish).

Taylor, W. R. 1960. Some results of studies of the uptake of radioactive waste materials by marine and estuarine phytoplankton organisms using continuous culturing techniques. Chesapeake Bay Inst., Johns Hopkins Univ. Tech. Rept. 21. 49pp.

Tazima, Y. 1957. Comparative studies on the differential sensitivity to radiation of silkworm germ-cells at different stages of the gametogenesis. I. Jap. J. Genet. 32(8): 262-263. (In Japanese).

Tazima, Y. 1957. Effect of external irradiation with β-rays upon the germ-cells of the silkworm. Mishima Natl. Inst. Genet. Ann. Rept. 7: 76-78.

Tazima, Y. 1957. Genic control of x-ray sensitivity of silkworm germ-cells. Mishima Natl. Inst. Genet. Ann. Rept. 7: 78-79.

Tazima, Y. 1957. Differences in sensitivity of germ-cells and chromosomes to radiation among some mutant strains of the silkworm. pp. 280-286. In: Proc. Intern. Genetics Symp. 1956. (Tokyo and Kyoto, September, 1956). Science Council of Japan, Tokyo.

Tazima, Y. 1958. Changes in sensitivity of silkworm germ cells to x-rays with development. Abstr. Radiation Research 9(1): 193.

Tazima, Y. 1958. Mutation response pattern of silkworm germ cells to x-rays. Intern. Congr. Genet. Proc. 10(2): 291-292.

Tazima, Y. 1961. Considerations on the changes in observed mutation rates in the silkworm after irradiation of various stages of gametogenesis. In: Proc. Symp. on Genetic Effect of Radiation. (Mishima, Nov. 7-8, 1960). Jap. J. Genet. 36(Suppl.): 50-64.

Tazima, Y. 1959. Some observations relevant to the possible cause of radiation induced sterility in the male silkworm. Mishima Natl. Inst. Genet. Ann. Rept. 10: 118.

Tazima, Y. 1967. Changes in the mutation response of post-meiotic silkworm germ cells to gamma rays with progressing spermiogenesis. Ann. Rep. Nat. Inst. Genet. (Japan) (17): 97-98. (NSA 23: 43739).

Tazima, Y. 1969. Analysis of radiation sensitivity of silkworm spermatogonia and its implications in the study of mutation. pp. 164-172. In: Comparative Cellular and Species Radiosensitivity. (See Bond and Sugahara, 1969) (NSA 24: 12223).

Tazima, Y. and S. Kondo. 1962. Further studies on two types of dose-rate dependence of radiation-induced mutation rates in spermatogonia and oogonia of the silkworm. Abstr. p. 7. In: 2nd Intern. Congr. Radiation Research. (See International Congress of Radiation Research, 1962).

Tazima, Y. and S. Kondo. 1963. Differential radiation-sensitivity of germ cells as a possible interpretation of sex difference in dose-rate dependence of induced mutation rates in the silkworm. pp. 237-252. In: Repair from Genetic Radiation Damage. F. H. Sobels, ed. McMillan Co., New York. (NSA 18: 11839).

Tazima, Y. and A. Murakami. 1962. The increase in induced mutation frequency after fractionated irradiation of gonial cells of the silkworm. Ann. Rep. Nat. Inst. Genet. (Japan) (13): 87-89. (NSA 18: 31143).

Tazima, Y. and A. Murakami. 1969. Analysis of strain differences in radiosensitivity of the silkworm. Gamma Field Symp. (8): 53-68. (CONF-690731) (NSA 25: 52027).

Tazima, Y. and K. Onimaru. 1958. Radiation mutagenesis in the silkworm. I-III. Mishima Natl. Inst. Genet. Ann. Rept. 8: 71-75.

Tazima, Y. and K. Onimaru. 1959. Nature of radiation induced mutations detected in the silkworm by the method of specific loci. Mishima Natl. Inst. Genet. Ann. Rept. 10: 115-116.

Tazima, Y. and K. Onimaru. 1959. X-ray induced dominant lethals from irradiation of different stages of oogenesis in the silkworm. Mishima Natl. Inst. Genet. Ann. Rept. 10: 117.

Tazima, Y. and K. Onimaru. 1959. Some observations relevant to the possible cause of radiation induced sterility in the male silkworm. Mishima Natl. Inst. Genet. Ann. Rept. 10: 118.

Tazima, Y. and K. Onimaru. 1967. Mechanisms of mutation induction by mitomycin-C in the silkworm. Ann Rep. Nat. Inst. Genet. (Japan) (17): 94-95. (NSA 23: 43738).

Tazima, Y. and K. Onimaru. 1967. Modification of gamma ray-induced mutation frequencies in the silkworm by post-treatment of spermatids and spermatozoa with nitrogen gas. Ann. Rep. Nat. Inst. Genet. (Japan) (17): 92-94. (NSA 23: 43737).

Tazima, Y. and K. Onimaru. 1967. Repair of radiation-induced premutational damages revealed by fractionated irradiation of silkworm spermatids. Ann. Rep. Nat. Inst. Genet. (Japan) (17): 91-92. (NSA 23: 45736).

Tazima, Y. and T. Sado. 1967. Postirradiation modification and mechanism of reverse dose-rate effect on mutation induction of silkworm spermatids. Ann. Rep. Nat. Inst. Genet. (Japan) (17): 90. (NSA 23: 45706).

Teas, H. W., H. J. Sax and K. Sax. 1965. Cycasin: radiomimetic effect. Science 149(3683): 541-542. (NSA 19: 35929).

Tector, M. A. and E. Kaefer. 1962. Radiation-induced chromosomal aberrations and lethals in Aspergillus nidulans. Science 136(3521): 1056-1075. (NSA 16: 21983).

Tedford, C. F. 1968. Magnitude of deposited radionuclide problem based on future usage: defense aspects. Monogr. Nucl. Med. Biol. (2): 482-486. (NSA 23: 24651).

Telford, J. N. and F. Matsumura. 1971. Electron microscopic and autoradiographic studies on distribution of dieldrin in the intact nerve tissues of German cockroaches. J. Econ. Entomol. 64(1): 230-238.

Telitchenko, M. M. 1958. Accumulation and distribution of uranium, thorium and strontium with Daphnia and fish. Byull. Mosk. obshch. ispyt. prirody (biol.) 63(3): 57-63. (In Russian).

Telitchenko, M. M. 1958. Chronic effects of small doses of ^{238}U, ^{232}Th, and $^{89+90}Sr$ on a number of generations of Daphnia magna Straus. Nauchn. dokl. vyssh. shkoly (biol.) (1): 114-118. (In Russian).

Telitchenko, M. M. 1961. Accumulation of radioisotopes of strontium-(89+90) and yttrium-90 by mirror carp. Ryb. khozyaystvo (5): 40-43. (JPRS-21989, pp. 5-10) (NSA 18: 5148).

(NOTE: Telitchenko is listed here consistently, but may appear in some abstracts with different spellings).

Telitchenko, M. M. 1962. The continuing effect of elevated concentrations of uranium on the fertility and survival of guppies (Lebistes reticulatus). Byull. Mosk. obshch. ispyt. prirody (biol.) 67(3): 156-157. (In Russian) (BA 42: 358).

Telitchenko, M. M. 1962. The role of metabolism in the accumulation of radioisotopes by fishes. Nauchn. dokl. vyssh. shkoly (biol.) (3): 90-93. (JPRS-15871) (NSA 17: 5708).

Telitchenko, M. M. and T. I. Chistyakova. 1965. Changes in concentration of nucleic acids in the gonads of Leucaspius delineatus after exposure to uranium. Dokl. Akad. Nauk SSSR 161(4): 962-963. (Doklady (biophys) 161: 37-38) (NSA 19: 43769).

Telitchenko, M. M. and Ye. N. Levitova. 1959. Histologic study of the gonads of bleaks (Alburnus L.) exposed to small chronic doses of U^{238}. Vestn. Mosk. Univ. (biol. pochv. geol. geogr.) (1): 45-48. (AEC-tr-12410).

Telitchenko, M. M. and L. A. Timofeyeva. 1969. Edentate Anodonta complanata as indicators of water contamination by cerium, ruthenium, and cesium radioisotopes. Biol. nauki (6): 58-61. (In Russian) (NSA 24: 4592).

Telushkina, Ye. L. 1957. Accumulation of radioactive cesium in some water organisms. pp. 74-78. In: Proc. All-Union Conf. on Medical Radiology. (Pro-lems of Hygiene and Dosimetry). Medgiz, Moscow. (CA 52: 17555d).

Telysheva, I. G. and Yu. A. Shulyakovskiy. 1973. Radioresistance of eggs of some industrial fish. pp. 78-88. In: Radioecology of Water Organisms. III. (See Andrushaitis, et al., 1973) (In Russian) (NSA 30: 24375).

Tempelis, C. H. and M. G. Lysenko. 1965. Effect of x-irradiation on Tympanosoma lewisi infection in the rat. Exptl. Parasitol. 16: 174-181. (NSA 19: 33835).

Templeton, W. L. 1959. Fission products and aquatic organisms. pp. 125-140. In: The Effects of Pollution on Living Material. (Inst. of Biology, London September 1958) (WPA 32: 773; NSA 16: 19210).

Templeton, W. L. 1964. Transfer of radionuclides from the environment through aquatic food products to man. pp. 43-64. In: FAO Atomic Energy Ser. No. 5. (See Food and Agriculture Organization, 1964).

Templeton, W. L. 1966. Resistance of fish eggs to acute and chronic irradiation. pp. 847-860. In: Disposal of Radioactive Wastes into Seas, Oceans and Surface Waters. (See IAEA, 1966).

Templeton, W. L. 1968. Book review: Radioecology of Aquatic Organisms by G. G. Polikarpov. J. Wildl. Mgmt. 32(2): 432.

Templeton, W. L. 1971. Disposal of liquid wastes into coastal waters. Effluent Water Treat. J. 11(5): 251-258. (NSA 25: 44577).

Templeton, W. L. 1972. Review of effects of radiation on aquatic populations. Battelle Northwest Lab., U. S. AEC report BNWL-SA-4435. 2pp. (For: Panel on Radiation Effects on Population Dynamics in Ecosystems, Reykjavik, Iceland, October 2, 1972). (NSA 27: 9874).

Templeton, W. L. and V. M. Brown. 1964. The relationship between the concentrations of calcium, strontium and strontium-90 in wild brown trout, Salmo trutta L. and the concentrations of the stable elements in some waters in the United Kingdom, and the implications in radiological health studies. Int. J. Air Water Poll. 8(1): 49-75. (NSA 18: 16243).

Templeton, W. L. and A. Preston. 1966. Transport and distribution of radioactive effluents in coastal and estuarine waters of the United Kingdom. pp. 267-289. In: Disposal of Radioactive Wastes into Seas, Oceans and Surface Waters. (See IAEA, 1966).

Templeton, W. L., et al. 1967, 1969. Freshwater ecology. Phase I final report. --- Final report. Bioenvironmental and radiological-safety feasibility studies Atlantic-Pacific interoceanic canal. Battelle Northwest Lab., Battelle Columbus Labs., U. S. AEC report BMI-171-008, -22. vi, 131; 131pp. (NSA 21: 39367; 24: 14442).

Templeton, W. L., et al. 1969. Freshwater ecological studies in Panama and Columbia. BioScience 19(9): 804-808. (Contains comments on potential intake of radionuclides).

Tencer, R. 1961. Incorporation of tritium-labeled thymidine in Bufo ♀ x Rana temporaria hybrid embryos. Nature 191(4770): 100.

Teneva, M. 1972. Dependence of natural radioactivity of the surface layer on soil conditions. Khidrol. Meteorol. 21(5): 35-39. (NSA 30: 32082).

Tennessee Valley Authority. 1968. Sequoya Nuclear Plant (Units 1 and 2). Preliminary Safety Analysis Report, Vol. 1. Knoxville, Tenn. 446pp. U. S. AEC DOCKET-50327-1. (NSA 23: 4134) (Environment).

Tennessee Valley Authority. 1970. Browns Ferry Nuclear Plant, Units 1, 2, and 3. Sequoya Nuclear Plant, Units 1 and 2. Environmental Statements, proposed procedure. Chattanooga, Tenn. 2pp. U. S. AEC DOCKET-50259-19. (NSA 24: 50027).

Tennessee Valley Authority. 1971. Browns Ferry Nuclear Plant, Units 1 and 2. --- Units 1, 2, and 3. Environmental radioactivity levels. Semiannual report, January- June 1970. --- July-December 1970. Chattanooga, Tenn. 32, 32pp. U. S. AEC DOCKET-50259-21, -24; -50260-20, -23; -50296-20. (NSA 25: 12735, 46195).

Tennessee Valley Authority. 1971. Watts Bar Nuclear Plant, Units 1 and 2. Draft Environmental Statement. Knoxville, Tenn. 124pp. U. S. AEC DOCKET-5050390-6, -50391-6. (NSA 25: 36969).

Tennessee Valley Authority. 1972. Watts Bar Nuclear Plant, Units 1 and 2. Draft Environmental Statement, supplements and additions. Chatanooga, Tenn. U. S. AEC DOCKET-50390-12. 183pp. (NSA 26: 27520).

Tennessee Valley Authority. 1972. Browns Ferry Nuclear Power Station. Units 1, 2, and 3. Final Environmental Statement, Vol. 1. Chattanooga, Tenn. 432pp. U. S. AEC DOCKET-50259-225. (NSA 29: 12249).

Tennessee Valley Authority. 1974. Quality of water in Wilson and Pickwick reservoirs. Chattanooga, Tenn. 97pp. (NSA 33: 14842) (Incl. radioactivity).

Tenore, K. R., D. B. Horton and T. W. Duke. 1968. Effects of bottom substrate on the brackish water bivalve Rangia cuneata. Chesapeake Sci. 9(4): 238-248. (^{65}Zn, ^{32}P).

Teplits, N. A. 1963. Changes in the nucleic acid content in the limbs of the axolotl after irradiation with roentgen rays and restoration of the power of regeneration. Dokl. Akad. Nauk SSSR 147(1): 244-247. (Doklady (biol. sci.) 147: 1312-1314) (NSA 17: 7834).

Teplits, N. A. 1964. Variation in the content of nucleic acids in the tissues of axolotl extremities following their x-ray treatment as against normal regeneration. Dokl. Akad. Nauk SSSR 159(2): 442-445. (Doklady (biophys.) 159: 169-172) (NSA 19: 25942).

Teplykh, L. A. 1973. Radioecological concentration processes in line waterfish. pp. 257-262. In: Radioecology of Water Organisms. II. (See Andrushaitis, 1973) (In Russian) (NSA 30: 21144).

Terada, K., S. Yamanaka and T. Kiba. 1974. Neutron activation analysis of rhenium in rocks and sediments by extraction chromotography on a pyridine-kel-F column. J. Radioanal. Chem. 20(1): 27-39. (NSA 30: 9045).

Teravskiy, I. K. 1966. Action of ionizing radiation on argasid mites. Zool. zh. 45: 371-374. (In Russian) (NSA 20: 45467).

Teresi, J. D. and C. L. Newcomb. 1966. An estimate of the effects of fallout beta radiation on insects and associated invertebrates. U. S. Naval Radiological Defense Lab., San Francisco, Calif., report USNRDL-TR-982. 88pp. (AD-633024; NSA 20: 28925).

Teriu, M. and Y. Harada. 1969. Gamma irradiation for control of diseases in stored apples. Hirosaki Daigaku Nogakubu Gak. Hokoku (15): 21-31. (In Japanese) (NSA 23: 48308)) (Studied 6 fungi).

Ternovoy, V. I. 1961. A study of the migration of Wolfartia flies by the method of radiolabelling. p. 31. In: Materials of the Symp. on the Use of Biophysics in the Field of Plant Protection. Acad. Sci. USSR, Leningrad. (In Russian).

Ternovskaya, Yu. G. and A. N. Vorsin. 1959. Experimental use of radioactive tracers for marking the water rat (Arvicola amphibius). pp. 210-216. In: The Water Rat and Its Control in Western Siberia. Novosibirsk. (In Russian) (BA 41: 17428; Ref. zh. (biol.) 1961: 1D644).

Terrani, M. and S. Villani. 1973. Characterization of archaeological findings by means of activation analysis. Com. Naz. Engr. Nucl. Notiz. 19(5): 77-79. (In Italian) (NSA 28: 24101).

Terranova, A. C. and C. H. Schmidt. 1966. Metabolism of hempa in the cabbage looper. Abstr. 102. Bull. Entomol. Soc. Am. 12: 288. (^{14}C).

Terriere, L. C. and R. D. Schonbrod. 1955. The excretion of a radioactive metabolite by houseflies treated with carbon-14 labeled DDT. J. Econ. Entomol. 48(6): 736-739.

Terrill, J. G., Jr. 1971. Limitations and needs for nuclear techniques in environmental control. pp. 81-89. In: Nuclear Techniques in Environmental Pollution. (See IAEA, 1971) (NSA 25: 32375).

Terrill, J. G., Jr., et al. 1969. Public health factors in reactor site selection. Proc. Am. Soc. Civil Engr. 95(SA3): 545-563.

Terry, L. L. 1968. Dimensions of the problem of fallout. Pediatrics 41 (Suppl.): 170-173. (NSA 22: 25834).

Teruo, N. 1963. Some results of studies carried out at the Institute of Radiation Breeding in 1962. pp. 79-92. In: Cell Differentiation and Somatic Mutation. Inst. of Plant Breeding, Japan Ministry of Agriculture and Forestry, Ohmiya. (NSA 19: 33955).

Terwilliger, R. C. and C. K. Levy. 1964. Behavioral response of the fiddler crab (Uca pugilator) to light and ionizing radiation. Abstr. Physiologist 7(3): 270.

Terzi, M. 1961. Comparative analysis of inactivating efficiency of radiation on different organisms. Nature 191(4787): 461-463. (NSA 15: 25887).

Terzi, M. 1965. Radiosensitivity and genetic complexity. J. Theor. Biol. 8: 233-243. (NSA 19: 21934) (Viruses, bacteria, fungi, yeasts).

Terzian, L. A. 1961. Morphological changes produced by gamma radiation on the sporogonous cycle of Plasmodium gallinaceum. Exptl. Parasitol. 11: 102-110. (NSA 16: 28782).

Terzian, L. A. and N. Stahler. 1958. A study of some effects of gamma radiation on the adults and eggs of Aedes aegypti. U. S. Naval Medical Research Inst., Bethesda, Md., report NM-52-01-00.05.01. 13pp. (NSA 13: 14145).

Terzian, L. A. and N. Stahler. 1962. A study of some effects of gamma radiation on the adults and eggs of Aedes aegypti. Biol. Bull. 115(3): 536-550.

Terzian, L. A. and N. Stahler. 1966. A selected strain of Aedes aegypti resistant to gamma radiation. Radiation Research 28: 643-646. (NSA 20: 33032).

Teshirogi, W. 1963. Fusion of different body segments using x-ray-irradiated animals and non-irradiated ones in a freshwater planarian Bdellocephala brunnea. Sci. Rpt. Fac. Lit. Sci. Hirosaki Univ. 10(1): __. (NSA 18: 31232).

Tesić, Z. P. and D. I. Marjanović. 1971. Bacterial leaching of uranium in Yugoslavia. pp. 221-231. In: Recovery of Uranium. (Proc. Symp., São Paulo, Brazil, 17-21 August 1970). International Atomic Energy Agency, Vienna, Publ. STI/PUB-262. (CONF-700836) (NSA 25: 23674).

Tester, J. R. 1963. Techniques for studying movements of vertebrates in the field. pp. 445-450. In: Radioecology. (See Schultz and Klement, 1963) (NSA 17: 33594).

Tester, J. R. 1963. Radiation and other pollutants: introductory remarks. p. 140. In: Proc. 1st Intern. Conf. on Wildlife Disease. (High View, New York, June 24-27, 1962). Wildlife Disease Assoc.

Tester, J. R. and W. J. Breckenridge. 1964. Winter behavior patterns of the Manitoba toad, Bufo hemiophrys, in Northwestern Minnesota. Ann. Acad. Scientarium Fenn. AIV(Biol.) 71: 421-430. (^{182}Ta tag).

Tester, J. R., M. A. Ewert and D. B. Siniff. 1970. Effects of ionizing radiation on natural and laboratory populations of Manitoba toads, Bufo hemiophrys. Radiation Research 44: 379-389. (NSA 25: 8628).

Tester, J. R., F. McKinney and D. B. Siniff. 1968. Mortality of three species of ducks (Anas discors, A. crecca, and A. clypeata) exposed to ionizing radiation. Radiation Reserach 33: 74-81. (NSA 22: 19420).

Tetteh, G. K. 1967. Variations in the x-ray sensitivity of Serratia marcescens. Thesis, Univ. of Saskatchewan, Saskatoon. 146pp. (NSA 22: 49435).

Tetteh, G. K. and D. V. Cormack. 1968. Inositol, oxygen, and water effects on the x-ray sensitivity of Serratia marcescens. Radiation Research 34: 532-543. (NSA 22: 43492).

Teulade, P. 1964. Variations of the sensitivity to x rays on the embryonic development and duration of hibernation of the eggs of Bombyx mori. Thesis, Troisiéme Cycle Faculté, Lyon, France. 103pp. (In French).

Teulade, P. 1965. Continuous irradiation at low dose rate on eggs of Bombyx mori and the concept of effective dose rate. J. Insect Physiol. 11: 379-386. (In French) (NSA 19: 45904).

Teulade, P. 1966. Recent experiences in radiobiology with Bombyx mori. Atom. Energy Rev. 4(2): 3-57. (In French).

Teulon, F. 1968. Choice of biological detectors for the fluor and uranium contamination of waters. pp. 597-602. In: Intern. Congr. on Radioprotection of the Environment. (See Societé Francaise de Radioprotecion, 1968).

Tewes, H. A. and L. L. Schwartz. 1975. Potential pathways to man for radionuclides from contained applications of peaceful nuclear explosives. pp. 297-320. In: Peaceful Nuclear Explosives. IV. (See IAEA, 1975) (NSA 33: 9284).

Thailand Office of Atomic Energy for Peace. 1966. Certain papers presented at the seminar on insect eradication by irradiation, 28-29 June 1966. Bangkok, report THAI.-AEC-8. 63pp. (NSA 22: 40979).

Thailand Office of Atomic Energy for Peace. 1968. Neutron activation analysis of gold in teak (Tectona grandis). Bangkok, report THAI.-AEC-17. 4pp. (NSA 23: 28745).

Thaine, R. and M. C. Walters. 1955. Experiments on the application of autoradiographic techniques to the study of problems in plant physiology. Austral. J. Biol. Sci. 8: 354-368.

Thatcher, L. L. 1961. Radioisotopes in the study of the hydrologic cycle. pp. 252-265. In: Applications of Radioisotopes and Radiation in the Life Sciences. (See U. S. Congress, 1961) (NSA 15: 27322).

Thatcher, L. L. 1967. Water tracing in the hydrologic cycle. pp. 97-108. In: Isotope Techniques in the Hydrologic Cycle. (See Stout, 1967).

Thatcher, L. L. 1969. Principles of the application of nuclear techniques to hydrologic investigations. pp. 149-193. In: Progress in Hydrology. Vol. I. New Developments in Hydrology. (See Univ. of Illinois, 1969) (NSA 25: 2360).

Thatcher, L. L. and C. M. Hoffman. 1963. Tritium fallout over North America from the Soviet tests in 1961. J. Geophys. Res. 68(20): 5899-5901. (NSA 17: 41059).

Thatcher, L. L. and J. O. Johnson. 1971. Comprehensive program in neutron activation analysis in water quality. pp. 323-328. In: Nuclear Techniques in Environmental Pollution. (See IAEA, 1971) (NSA 25: 31877).

Thatcher, L. L. and J. O. Johnson. 1973. Determination of trace elements in water and aquatic biota by neutron activation analysis. pp. 277-298. In: Bioassay Techniques and Environmental Chemistry. G. E. Glass, ed. Ann Arbor Science Publ., Ann Arbor, Mich. (NSA 31: 15917).

Theiss-Seuberling, H. B. 1973. Effect of x rays and inhibitors of protein synthesis on the synthesis of chlorophyll and NADP-linked glyceraldehyde 3-phosphate dehydrogenase in greening Euglena gracilis. Arch. Mikrobiol. 92(4): 331-344. (In German) (NSA 30: 487).

Theodorescu, M. and V. J. A. Novak. 1970. Effect of x irradiation of the head on neurosecretory cells in Gryllus domesticus L. Acta Entomol. Bohemoslov. 67(4): 207-210. (In French) (NSA 25: 32549, 58095).

Theodorsson, P. 1967. Natural tritium in groundwater studies. pp. 371-380. In: Isotopes in Hydrology. (See IAEA, 1967) (NSA 22: 18933).

Theodorsson, P. 1968. Tritium in groundwater and glaciers in Iceland. Joekull (Reykjavik) 18: 350-358. (In Icelandic) (NSA 26: 6796).

Theodorsson, P. 1975. Study of ^{14}C penetration into filters in primary productivity measurements using double side counting. Limnol. Oceanogr. 20 (2): 288-291. (NSA 32: 6916).

Theodorsson, P. and B. Arnason. 1964. Environmental radioactivity in Iceland. Univ. of Iceland, Reykjavik. 15pp. (For: 4th Symp. on Radiactivity in Scandinavia, Risø, Denmark, Oct. 5-6, 1964) (CONF-747-10) (NSA 20: 45908).

Thesis, C. V. 1963. Hydrologic phenomena affecting the use of tracers in timing groundwater flow. pp. 193-206. In: Radioisotopes in Hydrology. (See IAEA, 1963) (NSA 18: 1921).

Theunissen, J. 1971. Radiation pathology in Hylemya antiqua (Meigen): outlines of research. pp. 329-339. In: Sterility Principle for Insect Control or Eradication. (See IAEA, 1971) (NSA 25: 29619).

Theunissen, J. 1974. Possibility of application of the sterile-male technique for the control of the onionfly, Hylemya antiqua (Meig.). Histopathological and cytopathological consequences of sterilizing doses of radiation. Final report for the period 1 November 1969 - 30 Noveber 1974. International Atomic Energy Agency, Vienna, report IAEA-R-676F. 19pp. (NSA 33: 29753).

Thiebaux, M. L., Jr. and H. J. Thiebaux. 1974. Radionuclide decay in steady reservoirs. Water Research 8(7): 447-453. (NSA 30: 18647).

Thimann, K. V. 1968. Biosatellite II experiments: preliminary results. Proc. Natl. Acad Sci. (US) 60: 347-361. (NSA 22: 47503).

Thobbi, V. V. and T. Srihari. 1967. Growth response of Prodenia litura Fabr. in relation to the castor leaves obtained from irradiated plants. Indian J. Entomol. 29(2): 154-156.

Thode, H. G. and J. Monster. 1964. Sulfur isotope abundances in evaporites and in the ancient oceans. pp. 589-600. In: Chemistry of the Earth's Surface. (Memorial to V. J. Vernadskiy). Vol. 2. Moscow. (In Russian) (NSA 20: 41135).

Thode,H. G. and J. Monster. 1965. Sulfur-isotope geochemistry of petroleum evaporites, and ancient seas. pp. 367-377. In: Fluids in Subsurfaces Environments: a Symposium. Amer. Assoc. Petroleum Geologists Memoir No. 4. (NSA 21: 24107).

Thomas, C. W., et al. 1970. The measurement and behavior of airborne radionuclides since 1962. pp. 158-172. In: Radionuclides in the Environment. R. F. Gould, ed. Amer. Chemical Soc., Washington.

Thomas, H. A., et al. 1953. Radioactive fallout in Massachusetts surface waters. J. Am. Water Works Assoc. 45(6): 562-568.

Thomas, J. M. and L. L. Eberhardt. 1973. Similarity ratios and patterns in retention of trace substances. pp. 924-928. In: Radionuclides in Ecosystems. (See Nelson, 1973) (NSA 25: 37748).

Thomas, J. M. and L. L. Eberhardt. 1975. Ecological impact assessment. Battelle Northwest Lab., U. S. ERDA report BNWL-SA-5458. 17pp. (NSA 32: 24356).

Thomas, J. M. and L. L. Eberhardt. 1975. Extrapolation of animal radionuclide retention data to man: the use of similarity ratios. Battelle Northwest Lab., U. S. ERDA report BNWL-SA-5366. 13pp. (NSA 33: 12000).

Thomas, J. M., et al. 1976. Radionuclide cycling in a Utah dairy farm. Battelle Northwest Lab., U. S. ERDA report BNWL-1937. vp. (NSA 33: 23274).

Thomas, J. W., J. G. Terrill and R. A. Gilmore. 1959. Radioactive pollutants. Am. Soc. Civil Engrs. Proc. 85(SA1): 75-85.

Thomas, L. J. and H. Quastler. 1950. Preliminary report of x-ray effects on the nematode Rhabditis strongyloides. Science 112(2909): 356-357. (NSA 6265).

Thomas, P. and G. W. Rahalkar. 1975. Disinfestation of fruit flies in mango by gamma irradiation. Current Sci. (India) 44(21): 775-776. (NSA 33: 17470).

Thomas, W. A. 1966. Cycling of ^{45}Ca by dogwood trees. Oak Ridge National Lab., U. S. AEC report ORNL-P-3067. 14pp. (NSA 21: 32562).

Thomas, W. A. 1967. Dye and calcium ascent in dogwood trees. Plant Physiol. 42(12): 1800-1802. (^{47}Ca).

Thomas, W. A. 1968. Calcium distribution in leaves: autoradiography versus chemical analysis. Int. J. Appl. Radiat. Isotopes 19: 544-545. (NSA 22: 43215).

Thomas, W. A. 1969. Accumulation and cycling of calcium by dogwood trees. Ecol. Monogr. 39(2): 101-120. (^{45}Ca).

Thomas, W. A. 1969. Cycling of ^{45}Ca by dogwood trees. pp. 661-664. In: Symposium on Radioecology. (See Nelson and Evans, 1969).

Thomas, W. A. 1970. Decomposition of irradiated persimmon leaves. Radiation Botany 10(1): 7-10. (NSA 24: 25603).

Thomas, W. A. 1970. Energy turnover by dogwood Cornus florida L. trees. Oikos 21: 71-75. (At Oak Ridge National Lab.).

Thomas, W. A. 1970. Retention of calcium-45 by dogwood trees. Plant Physiol. 45: 510-511.

Thomas, W. A. 1970. Review: Environmental Contamination by Radioactive Materials. Nuclear Safety 11(6): 492.

Thomas, W. A. 1973. Accumulation of ^{144}Ce by hickory and ^{60}Co by black gum seedlings. pp. 187-189. In: Radionuclides in Ecosystems. (See Nelson, 1973) (NSA 28: 27298).

Thomas, W. A. 1975. Accumulation of rare earths and circulation of cerium by mockernut hickory trees. Can. J. Bot. 53(12): 1159-1165. (NSA 33: 676) (^{144}Ce tracer).

Thomas, W. A. 1975. Cobalt accumulation and circulation by blackgum trees. Forest Science 21(3): 222-226. (NSA 33: 17509).

Thomas, W. D. E. 1957. An to work on radioactive tracers. Sci. Hort. 12(1955/56): 104-109. (Insecticide research).

Thomas, W. H. 1963. Physiological factors affecting the interpretation of phytoplankton production measurements. pp. 147-162. In: Proc. Conf. on Primary Productivity Measurement, Marine and Freshwater. (See Doty, 1963).

Thomas, W. H. 1964. An experimental evaluation of the C^{14} method for measuring phytoplankton production, using cultures of Dunaliella primolecta Butcher. U. S. Fish and Wildlife Service, Fish. Bull. 63(2): 273-292.

Thomasson, W. N. 1972. Uptake and translocation of ^{85}Sr, ^{59}Fe, ^{185}W and ^{134}Cs by banana plants and coconut plants following foliar application. Thesis, Univ. of Florida, Gainesville. 154pp. (Univ. Microfilms 73-971) (NSA 27: 20176).

Thomou, H. 1963. Sterilization of Dacus oleae by gamma radiation. pp. 413-424. In: Radiation and Radioisotopes Applied to Insects of Agricultural Importance. (See IAEA, 1963) (NSA 18: 5119).

Thompson, I. M. G. 1975. Environmental monitoring. J. Inst. Nucl. Eng. 16 (3): 67-71. (NSA 33: 7298).

Thompson, J. B. 1974. Studies on 31 Cyanophyta species. I. Amino acid analysis. II. Effects of irradiation with ^{60}Co. Dissertation, Univ. of Houston, Houston, Tex. 170pp. (Univ. Microfilms 74-21400) (NSA 31: 708).

Thompson, R. C. 1954. Biological application of tritium. Nucleonics 12(9): 31-35. (NSA 8: 6399).

Thompson, R. C. 1960. Vertebrate radiobiology: metabolism of internal emitters. Ann. Rev. Nucl. Sci. 10: 531-560. (Also: Hanford Atomic Products Operation, U. S. AEC report HW-SA-1900) (NSA 15: 4904).

Thompson, R. C. 1971. A review of laboratory animal experiments related to the radiobiology of tritium. Battelle Northwest Lab., U. S. AEC report BNWL-SA-3739. 19pp. (For: 19th Ann. Meeting Radiation Research Soc., Boston, Mass., 9 May 1971) (CONF-710508-3) (NSA 25: 46673).

Thompson, R. C. and W. J. Bair. (Eds.). 1972. Biological Implications of the Transuranium Elements. (Proc. 11th Hanford Biology Symp., held at Richland, Washington, 27-29 September 1971). Health Physics 22(6): 533-954. (NSA 26: 38766).

Thompson, S. E. 1965. Effective half-life of fallout radionuclides on plants with special emphasis on iodine-131. Lawrence Radiation Lab., Livermore, U. S. AEC report UCRL-12388. 17pp. (NSA 19: 24752).

Thompson, T. 1963. Some observations of variations of the natural background radiation. Tellus 15(3): 313-318. (NSA 18: 10290) (Correlation with precipitation).

Thompson, T. J. 1972. Nuclear power plants: present, past, and future. pp. 57-71. In: Environmental and Ecological Forum, 1970-1971. U. S. AEC report TID-25857. (NSA 26: 52480).

Thomson, J. F. and E. W. Daniels. 1961. Cytochrome oxidase in radiosensitive and radioresistant amoebae. Proc. Soc. Exptl. Biol. Med. 107(4): 916-919. (NSA 15: 30488).

Thornley, M. J. 1963. Radiation resistance among bacteria. J. Appl. Bacteriol. 26(3): 334-345. (NSA 18: 17527).

Thornton, D. G. and J. M. Davis. 1956. A method of sampling for the drop size of aerial spray deposits. J. Econ. Entomol. 49(1): 80-83.

Thresh, J. C., J. F. Beale and E. V. Suckling. 1958. The Examination of Waters and Water Supplies. Taylor & Churchill, London. 850pp. (Incl. radioactivity).

Thurber, D. L., W. S. Broecker and A. Kaufman. 1965. The comparison of radiocarbon ages of carbonates with uranium series ages. In: Radiocarbon and Tritium Dating. (Proc. 6th Intern Conf., Washington State University, Pullman, Wash., June 7-11, 1965). R. M. Chatters and E. A. Olson, comps. (AED-CONF-65-158-17; CONF-650652) (NSA 20: 7289).

Thurman, D. C., Jr. and R. C. Husbands. 1951. Preliminary report on mosquito flight dispersal studies with radioisotopes in California, 1950. Commun. Disease Cen. Bull. 10(4): 1-9.

Thygeson, T. 1966. Radioactive labelling of insects. Ugeskr. Land. 32: 531-537. (In Danish, English summary).

Ticheler, J. and Ph. W. Noordink. 1968. Application of the sterile-male tecynique on the onionfly, Hylemya antiqua Meig., in the Netherlands. Progress report. pp. 111-115. In: Radiation, Radioisotopes, and Rearing Methods in the Control of Insect Pests. (See IAEA, 1968) (NSA 24: 32330).

Ticknor, R. L. 1953. The entry of nutrients through the bark and leaves of diciduous fruit as indicated by radioactive isotopes. Michigan State Coll., U. S. AEC report AECU-2753. (NSA 8: 722).

Tietz, A. 1963. Fat synthesis in cell-free preparations of locust fat-body. pp. 85-89. In: Biosynthesis of Lipids. G. Popjak, ed. Pergamon Press, New York. (^{14}C, Locusta migratoria).

Tigyi, J. 1963. The effect of 15-Mev electron rays on the activity and excitability of the isolated frog heart. Acta Physiol. Acad. Sci. Hung. 24: 129-135. (NSA 18: 25098).

Tigyi, J. 1971. Effect of gamma-irradiation on the Ca content of striated muscles. pp. 403-407. In: Radiation Biophysics, Free Radicals. Vol. II. E. Broda, ed. Wiener Med. Akad., Vienna. (NSA 30: 588) (Frog).

Tikhomirov, F. A. 1971. Tasks of dosimetry in radioecology. pp. 261-279. In: Contemporary Problems of Radiobiology. Vol. 2. Radioecology. (See Klechovskiy, Polikarpov and Aleksakhin, 1971) (English transl. pp. 225-240).

Tikhomirov, F. A. 1972. Effect of Ionizing Radiation on Ecological Systems. Atomizdat, Moscow. (In Russian) (NSA 33: 7288).

Tikhomirov, F. A. 1973. Effects of ionizing radiation on a gypsy moth population. Ekologiya 4(5): 15-21. (Sov. J. Ecology 4(5): 380-385) (NSA 32: 6788).

Tikhomirov, F. A. and R. M. Aleksakhin. 1971. Effect of ionizing radiation on forest biogeocenoses. pp. 228-260. In: Contemporary Problems of Radiobiology. Vol. 2. Radioecology. (See Klechovskiy, Polikarpov and Aleksakhin, 1971) (English transl. pp. 197-224).

Tikhomirov, F. A. and R. T. Karaban'. 1971. β-radiation dosimetry of radioactive contaminated soil and plants. pp. 252-255. In: Methods of Radioecological Investigations. (See Verkhovskaya, 1971) (In Russian).

Tikhomirova, M. M. 1966. Comparative study of radiosensitivity of different Drosophila strains. Genetika (3): 93-97. (In Russian) (NSA 20: 32994).

Tikhomirova, M. M. 1967. Comparative study of the radiosensitivity at different stages of oogenesis. Genetika (6): 24-32. (In Russian) (NSA 21: 32756).

Tokhomirova, M. M., A. A. Dzhimelin and O. Ya. Belyatskaya. 1967. Genetical after-effects of x-rays depending on the temperature regime. Genetika 3 (5): 95-104. (In Russian).

Till, A. R. and P. F. May. 1970. Nutrient cycling in grazed pastures. II. Further observations with ^{35}S, gypsum. Austral. J. Agr. Res. 21(2): 253-_.

Tilly, L. J. and S. C. Aiken. 1974. Periphyton colonization and productivity in the reactor cooling reservoir: PAR Pond. Savannah River Plant, U. S. AEC report DP-MS-73-67. 18pp. (NSA 30: 3202).

Tilton, E. W. and J. H. Brower. 1973. Utilization of irradiated sperm from successive matings by the female cowpea weevil. J. Econ. Entomol. 66(2): 545-546. (NSA 28: 715).

Tilton, E. W., J. H. Brower and R. R. Cogburn. nd. Irradiation studies with insects infesting bulkgrain and packaged commodities. pp. 241-244. In: 8th Ann. AEC Food Irradiation Contractors' Meeting. (October 16-17, 1968). (NSA 23: 18176).

Tilton, E. W., J. H. Brower and R. R. Cogburn. 1974. Gamma irradiation for control of insects in wheat flour. J. Econ. Entomol. 67(3): 430-432. (NSA 30: 15971).

Tilton, E. W., J. H. Brower and R. R. Cogburn. 1974. Insect control in wheat flour with gamma irradiation. Int. J. Appl. Radiat. Isotopes 25(7): 301-305. (NSA 30: 27007).

Tilton, E. W., W. E. Burkholser and R. R. Cogburn. 1964. Sterilization effects of gamma radiation on eight insect and one mite species that infest stored products. Abstr. Bull. Entomol. Soc. Am. 10(3): 163.

Tilton, E. W., W. E. Burkholder and R. R. Cogburn. 1965. The effects of confining confused flour beetles in gelatin capsules before, during, and after gamma irradiation. J. Econ. Entomol. 58(1): 175-176. (NSA 19: 28024).

Tilton, E. W., W. E. Burkholder and R. R. Cogburn. 1965. Notes on the effect of preconditioning confused flour beetles with temperature variations or carbon dioxide prior to gamma irradiation. J. Econ. Entomol. 58(1): 179-180.

Tilton, E. W., W. E. Burkholder and R. R. Cogburn. 1966. Mating competition of gamma-irradiated and non-irradiated male Trogoderma glabrum Herbst. J. Econ. Entomol. 59: 168-169. (NSA 20: 40673).

Tilton, E. W., W. E. Burkholder and R. R. Cogburn. 1966. Effect of gamma radiation on Trogoderma glabrum and Attagenus piceus. J. Econ. Entomol. 59(4): 944-948.

Tilton, E. W., W. E. Burkholder and R. R. Cogburn. 1966. Effects of gamma radiation on Rhyzopertha dominica, Sitophilus oryzae, Tribolium confusum, and Lasioderma serricorne. J. Econ. Entomol. 59(6): 1363-1368.

Tilton, E. W., et al. 1972. Combination of gamma and microwave radiation for control of the angoumois grain moth in wheat. J. Econ. Entomol. 65 (2): 531-533. (NSA 26: 28496).

Timblin, L. O., Jr. and A. J. Peterka. 1963. Use of radioisotopes for open-channel flow measurements. pp. 37-61. In: Radioisotopes in Hydrology. (See IAEA, 1963) (NSA 18: 5119).

Timofeyev, B. N. and Yu. Nesytov. 1969. Predictions of Radioactive Contamination. Voyen. Izdatel'stvo Min. Oborony SSSR, Moscow. 102pp. (AEC-tr-7137; TT-70-50151; IPST-Cat-5842) (NSA 23: 21827; 25: 48615).

Timofeyeva, N. A. 1960. On the migration of radioactive strontium in biogeocenoses. Dokl. Akad. Nauk SSSR 133(2): 488-491. (Doklady (Biol. sci.) 133: 144-146) (NSA 14: 23895).

Timofeyeva, N. A. 1961. Dependence of the accumulation of strontium-90 by freshwater plants upon calcium concentration in the water. Tr. Ural'-skogo otd. Mosk. obshch. ispyt. prirody Akad. Nauk SSSR (3): ___. (In Russian).

Timofeyeva, N. A. 1965. Features of radiostrontium concentration by freshwater organisms. pp. 416-417. In: Problems in Hydrobiology. Publ. House Acad. Sci. USSR, Moscow. (In Russian).

Timofeyeva, N. A. and A. L. Agre. 1965. Coefficients of ^{90}Sr accumulation by freshwater plants from solutions of different specific activity. Radiobiologiya 5(3): 457-458. (AEC-tr-6600, pp. 178-180; JPRS-32524; TT-65-3303, p. 69).

Timofeyeva, N. A. and N. V. Kulikov. 1967. The role of freshwater plants in accumulation of ^{90}Sr and its distribution over the components of a reservoir. pp. 835-841. In: Radioecological Concentration Processes. (See Aberg and Hungate, 1967).

Timofeyeva, N. A. and N. V. Kulikov. 1968. Role of freshwater plants on uptake and distribution of ^{90}Sr in waterbasin components. Tr. inst. ekol. rast. zhivotn. Ural. fil. Akad. Nauk SSSR (61): 65-71. (In Russian) (NSA 24: 7778).

Timofeyeva, N. A. and A. A. Titlyanova. 1959. Sorption of strontium-90 by soils. Izv. Akad. Nauk SSSR (biol.) (1): 111. (In Russian).

Timofeyeva, N. A. and A. A. Titlyanova. 1959. Sorption of microquantities of chemical elements by the soil. 3. Absorption of radioactive isotopes (cobalt-60, strontium-90, yttrium-90, ruthenium-106, cesium-137 and cerium-144) by the soil. Tr. Ural'skogo otd. Mosk. obshch. ispyt. prirody Akad. Nauk SSSR (2): ___. (In Russian).

Timofeyev-Resovskiy, N. V. 1956. Biophysical interpretation of effects of radio-stimulation of plants. Biofizika 1(7): 616-627. (In Russian).

Timofeyev-Resovskiy, N. V. 1957. Use of radiation and radioactive substances in experimental biogeocenology. Botan. zh. 42(2): 162-194. (In Russian).

Timofeyev-Resovskiy, N. V. 1960. Distribution of trace elements in the components of water bodies. 1. Some considerations. (Collected Papers of the Biophysics Laboratory). Tr. inst. biol. Ural. fil. Akad. Nauk SSSR (2) 2: 189-193. (In Russian) (BA 42: 463).

Timofeyev-Resovskiy, N. V. 1962. Some problems of radiation biogeocenology. Dissertation, Inst. of Biology, Urals Affiliate, Acad. Sci. USSR, Sverdlovsk. (In Russian).

Timofeyev-Resovskiy, N. V. 1962. On the radioactive contamination of the biosphere and measures combating these contaminations. Tr. inst. biol. Ural. fil. Akad. Nauk SSSR (22): 7-16. (FTD-TT-63-664) (NSA 17: 37056).

Timofeyev-Resovskiy, N. V. 1964. Some problems of radiation biogeocenology. Probl. Kibernetiki (12): 210-231. (In Russian) (NSA 19: 36051).

Timofeyev-Resovskiy, N. V. and Ye. A. Timofeyeva-Resovskaya. 1959. Distribution of radioactive substances in bodies of water. pp. 1-21. In: A Conf. on Aspects of the Exploitation of the Kamsk Reservoir. Urals Affiliate, Acad. Sci. USSR, Perm'. (In Russian).

Timofeyev-Resovskiy, N. V. and Ye. A. Timofeyeva-Resovskaya. 1960. Effect of ethylenediaminetetraacetate (EDTA) on the coefficient of accumulation of various radioisotopes from aqueous solution by leeches. Byull. Ural'skogo otd. Mosk. obshch. ispyt. prirody (biol.) (3): ___. (In Russian).

Timofeyev-Resovskiy, N. V., et al. 1957. The radiostimulation of plants. I. The effect of small doses of ionizing radiation on the growth and development of plants. Tr. inst. biol. Ural. fil. Akad. Nauk SSSR 9: 129-201. (In Russian) (BA 45: 30160).

Timofeyev-Resovskiy, N. V., et al. 1960. Rate of uptake of 16 different radioisotopes by freshwater organisms and the effect of EDTA on some of them. Dokl. Akad. Nauk SSSR 132(5): 1191-1194. (Doklady (biol. sci.) 132: 369-372; French transl., CEA-tr-R-1243) (NSA 14: 21272; 15: 19134).

Timofeyeva-Resovskaya, Ye. A. 1957. Soil-biological deactivation of water in ponds and sedimentation tanks. Byull. Mosk. obshch. ispyt. prirody (biol.) 62(1): 37-41. (In Russian).

Timofeyeva-Resovskaya, Ye. A. 1958. Rate of underwater overgrowth (periphyton formation) in the presence of weak concentrations of radioactive substances. Byull. Ural'skogo otd. Mosk. obshch. ispyt. prirody (biol.) (1): 87-96. (In Russian).

Timofeyeva-Resovskaya, Ye. A. 1958. Effect of ethyldiaminetetraacetic acid sodium salt on strontium, ruthenium, cerium, cobalt, zinc and cesium concentration factors. pp. 159-163. In: Complexing Agents. (Their Synthesis, Properties and Uses in Biology and Medicine). Urals Affiliate, Acad. Sci. USSR, Sverdlovsk. (In Russian).

Timofeyeva-Resovskaya, Ye. A. 1963. Distribution of radioisotopes by basic components in bodies of freshwater. Tr. inst. biol. Akad. Nauk SSSR Ural. fil. 30: 1-78. (JPRS-21816) (BA 45: 36049; NSA 18: 1342; Ref. zh. (biol.) 1963: 11D312; SFA 10: 7337).

Timofeyeva-Resovskaya, Ye. A., et al. 1958, 1959. Concentration of chemical elements from aqueous solutions by freshwater organisms. 1. Concentration of radioisotopes of phosphorus, zinc, strontium, ruthenium, cesium, cerium by different species of freshwater mollusks. 2. Rate of uptake of different radioisotopes by the pond snail (Limnea stagnalis). 3. The concentration factors of various radioisotopes by three species of water plants. Byull. Mosk. obshch. ispyt. prirody (biol.) 63(3): 65-78; 63(5): 123-131; 64(5): 117-131. (In Russian).

Timofeyeva-Resovskaya, Ye. A. and N. V. Timofeyev-Resovskiy. 1960. Distribution of trace elements in the components of water bodies. 2. Soil-biological deactivation of water in ponds and sedimentation tanks. Tr. inst. biol. Akad. Nauk SSSR Ural. fil. (12): Coll. Papers Biophysics Lab. 3: 194-223. (In Russian).

Timofeyeva-Resovskaya, Ye. A. and N. V. Timofeyev-Resovskiy. 1960. On the effect of ethylenediaminetetraacetate (EDTA) on the coefficients of different radioisotopes from aqueous solution by freshwater plants. Dokl. Akad. Nauk SSSR 130(1): 210-213. (Doklady (biol. sci.) 130: 9-11).

Timofeyeva-Resovskaya, Ye. A. and N. V. Timofeyev-Resovskiy. 1966. Influence of various doses of ^{60}Co gamma-rays on duckweed and water thyme. pp. 235-237. In: Effect of Ionizing Radiation on Heredity. (See Dubinin, 1966) (In Russian) (NSA 22: 21632).

Timofeyeva-Resovskaya, Ye. A., B. B. Agafonov and N. V. Timofeyev-Resovskiy. 1960. Soil-biological deactivation of water. Tr. inst. biol. Akad. Nauk SSSR Ural. fil. (13): Coll. Papers Biophysics Lab. 3: 35-48. (In Russian).

Timofeyeva-Resovskaya, Ye. A., N. V. Timofeyev-Resovskiy and Ye. A. Gileva. 1961. On specific accumulators of separate radioactive isotopes among freshwater organisms. Dokl. Akad. Nauk SSSR 140(6): 1437-1440. (Doklady (biol. sci.) 140: 780-783) (NSA 16: 5137).

Timofeyeva-Resovskaya, Ye. A., et al. 1960. The coefficient of accumulation for radioactive strontium, ruthenium, cesium, and cerium by freshwater organisms. Zool. zh. 39(10): 1449-1453. (In Russian).

Timoshchuk, V. I. and I. A. Sokolova. 1970. Radioecological investigations of the Danube and adjacent part of the Black Sea. pp. 142-150. In: Marine Radioecology. (See Polikarpov, 1970) (In Russian).

Ting, P. and R. L. Litle. 1973. Continuous monitoring of aqueous tritium activity. pp. 170-183. In: Tritium. (See Moghissi and Carter, 1973) (NSA 28: 27414).

Tinkle, D. W. 1965. Effects of radiation on the natality, density and breeding structure of a natural population of lizards, *Uta stansburiana*. In: *Radiation and Terrestrial Ecosystems*. Health Physics 11(12): 1995-1999. (See Hungate, 1965) (NSA 20: 12585).

Tinkle, D. W. 1966. Effects of radiation on the natality, density and breeding structure of a natural population of lizards, *Uta stansburiana*. p. 45. In: *Radiation Effects on Natural Populations*. (See Sacher, 1966).

Tinney, J. F., J. J. Koch and C. T. Schmidt. 1972. Plutonium survey with an x-ray sensitive detector. pp. 851-875. In: *Health Physics Operational Monitoring*. *Vol. 2*. C. A. Willis, ed. Gordon & Breach, New York. (NSA 29: 29480).

Tinney, J. F., et al. 1971. Environmental sampling program following an accidental tritium release. Lawrence Livermore Lab., U. S. AEC report UCRL-51029. 18pp. (NSA 25: 46869).

Tipton, S. R. and G. S. Amand. The effect of x-rays on the respiratory metabolism of eggs and embryos of the grasshopper *Chortophaga viridifasciata*. Physiol. Zool. 27(4): 311-317.

Titayeva, N. A. 1966. Possibility of absolute dating of organic sediments by the ionium method. Geokhimiya 3: __. (Geochem. Int. 3: 941-950) (NSA 22: 10366).

Titayeva, N. A. 1967. Association of radium and uranium with peat. Geokhimiya (12): 1493-1499. (Geochem. Int. 4: 1168-1174) (NSA 22: 8209; 23: 17525).

Titayeva, N. A. and Ye. V. Viasova. 1968. Form of occurrence of radium in peat. Geokhimiya (3): 384-387. (In Russian).

Titayeva, N. A., et al. 1973. Behavior of uranium and thorium isotopes in the system crystalline rocks-surface waters under conditions of a cold humid climate. Geokhimiya (10): 1522-1528. (In Russian) (NSA 30: 29353).

Tite, M. S. 1970. Impact of the natural sciences on archaeology. Contemp. Phys. 11: 523-539. (NSA 25: 2354).

Tite, M. S. 1975. Physical methods in archaeology. Phys. Bull. (London) 26: 112-114. (NSA 32: 3507).

Titlyanova, A. A. 1960. Sorption of microquantities of chemical elements by soil. V. Behavior of radioactive cesium in soil. Tr. Ural'skogo otd. Mosk. obshch. ispyt. prirody (3): __. (In Russian).

Titlyanova, A. A. 1962. Behavior of cesium and rubidium in soils. Pochvovedeniye (3): 53-61. (In Russian).

Titlyanova, A. A. 1963. The influence of cesium concentration on its accumulation by plants. Dokl. Akad. Nauk SSSR 152(2): 441-442. (Doklady (earth sci.) 152: 183-184; JPRS-22656) (NSA 18: 8064).

Titlyanova, A. A. and V. I. Ivanov. 1961. Concentration of cesium by three species of freshwater plants from aqueous solutions of different concentrations. Dokl. Akad. Nauk SSSR 136(3): 721-722. (In Russian) (NSA 15: 16798).

Titlyanová, A. A. and N. A. Timofeyeva. 1962. Sorption of radioactive isotopes by the soil. Tr. inst. biol. Akad. Nauk Ural. fil. (22): ___. (In Russian).

Titlyanova, A. A., A. N. Tyuryukanov and G. I. Makhonina. 1959. Desorbant effect of natural extracts. Dokl. Akad. Nauk SSSR 126(6): 1346-1350. (In Russian).

Titterton, E. W. 1973. Radiation in man's environment. Aust. J. Forensic Sci. 6(2): 96-100. (NSA 30: 29353).

Tixador, R., et al. 1969. Generation time is lengthened in *Paramecium aurelia* cultivated under radioprotection. Compt. rend. soc. biol. 163: 1429-1432. (In French) (NSA 24: 51012).

Tkacheva, G. A. 1968. Ontogenetic dependence of the interrenal gland reaction on irradiation. Byull. eksper. biol. med. 65(5): 109-113. (In Russian) (NSA 22: 45695).

Tkalich, S. M. 1956. The biogeochemical method of prospecting for ore deposits. In: *Materials of the Soviet Geologists of Eastern Siberia and the Far East on Methods of Geological Surveys and Prospecting Operations*. Chita State Geological Technical Publ. House, USSR. (In Russian).

Toba, H. H., A. N. Kishaba and D. T. North. 1972. Reduction of populations of caged cabbage loopers by release of irradiated males. J. Econ. Entomol. 65(2): 408-411. (NSA 26: 48495).

Tobia, S. K. and E. V. Sayre. 1973. Analytical comparison of various Egyptian soils, clays, shales, and some ancient pottery by neutron activation. pp. 99-128. In: *Recent Advances in Science and Technology of Materials*. *Vol. 3*. A. Bishay, ed. Plenum, New York. (NSA 31: 32470).

Tobias, C. A. 1974. Radiation and molecular and biological evolution. pp. 197-225. In: *Space Radiation Biology and Related Topics*. C. A. Tobias, ed. Academic Press, New York. (NSA 30: 15757).

Tobias, C., et al. 1967. Interaction of radiation and other environmental stresses in biological systems. pp. 141-158. In: *Proc. 2nd Intern. Symp. on Basic Environmental Problems of Man in Space*. (Paris, June 14-18, 1965). H. Bjurstedt, ed. Springer Verlag, Vienna. (NSA 21: 43533).

Todd, D. K. and D. E. McNulty. 1974. Polluted ground water: a review of the significant literature. General Electric Co., Santa Barbara, Calif., Environmental Monitoring Ser., U. S. EPA report EPA-600/4-74-001. 224pp. (PB-235556) (NSA 31: 22630).

Todd, R. and R. Logan. 1966. ^{237}Pu labeling: a new technique for plutonium metabolic studies in animals. Int. J. Appl. Radiat. Isotopes 17: 253-255. (NSA 20: 18221).

Todisco, S., et al. 1964. Determination of ^{32}P in biological samples in relation to the maximum permissible dose for internal irradiation in man. Minerva Nucleare 8: 278-281. (In Italian) (NSA 19: 4016) (Incl. fish).

Todorovic, P., V. Vukmirovic and R. Vukotic. 1967. Contribution to the kinetic theory of bed-material discharge. pp. 271-290. In: Isotopes in Hydrology. (See IAEA, 1967) (NSA 22: 18996) (Tracers).

Todorović, Z. and A. Filip. 1967. Movement of complex compounds through different soil types. pp. 241-250. In: Isotope and Radiation Techniques in Soil Physics and Irrigation Studies. (See IAEA, 1967) (NSA 22: 23156).

Toelgyessy, J. and S. Varga. 1972. Nuclear Analytical Chemistry. II. Radioactive Indicators in Chemical Analysis. University Park Press, Baltimore. 436pp. (NSA 28: 29625).

Török, G. and J. Farkas. 1960. The importance of exterminationg insects in warehouses from the viewpoint of national economy and a new way of insect control: ionizing radiation. Élelmezesi Ipar 14(7): 199-206. (In Hungarian).

Toeroek, I. 1968. Measurement of soil humidity with neutron-scattering humidity measuring instrument. Atomtech. Tajek. 11: 580-584. (In Hungarian) (NSA 23: 43614).

Toeroek, I. 1969. Soil moisture determination by a neutron scattering moisture meter. Agrokem. Talaj. 13: 107-116. (In Hungarian) (NSA 24: 11698).

Toeroek, I. 1971. Determination of ^{137}Cs in the soil with a Ge(Li) detector. Agrokem. Talaj. 20(1-2): 203-204. (In Hungarian) (NSA 26: 20015).

Toeroek, I. 1972. ^{90}Sr contamination of some soils in Hungary. Agrokem. Talaj. 21(3-4): 315-320. (In Hungarian) (NSA 27: 24606).

Toeroek, I. and J. Kovacs. 1973. Distribution of radioactive soil contamination in Hungary. Izotopthechnika 16(1): 21-31. (In Hungarian) (NSA 27: 19827).

Togasaki, R. K. and M. Gibbs. 1963. Pre-illumination experiments with Anacystis and Chlorella. Abstr. Plant Physiol. 38(Suppl.): xi. (Used ^{14}C).

Tohver, V. and E. Lokk. 1966. Prototrophic properties and x-ray resistance of some nodule bacteria. Izv. Akad. Nauk Eston. SSR (biol.) 3: 347-356. (In Russian) (NSA 23: 24771).

Tojyo, I. 1963. Effects of immersion of wintering bud of mulberry in the ^{32}P solution on the pollen mother cells. Nippon Sanshi. Zasshi 32: 108-112. (In Japanese) (NSA 18: 37137).

Tojyo, I, 1969. Influence of gamma-irradiation on the pollen formation of the Moraceae. Sanshi Kenkyu (70): 22-23. (NST-tr-158) (NSA 25: 8794).

Tokarev, A. N. and A. V. Shcherbakov. 1956. Radiohydrogeology. State Publ. House of Sci.-Tech. Lit. on Geology, etc., USSR, Moscow. (AEC-tr-4100, 346pp.) (NSA 14: 24356).

Tolbert, N. E. and P. B. Pearson. 1952. Atomic energy and the plant science nces. Adv. in Agronomy 4: 279-303.

Tomanek, J. and Z. Prochaska. 1968. Effects of the length of storage upon the invasiveness of x-irradiated Dictyocaulus viviparus larvae. Veterin. (Czech.) 7: 261-265. (NSA 24: 32176).

Tomes, G. A. R. and M. V. Brian. 1946. An electronic method of tracing the movements of beetles in the field. Nature 158(4016): 551.

Tominaga, H. 1971. Radioisotope instruments in environmental pollution control. Genshiryoku Kogyo 17(8): 32-36. (In Japanese) (NSA 26: 9412).

Tomiyama, T. and K. Kobayashi. 1958. Direct uptake of radioisotopes by fish. p. 16. In: Proc. 9th Pacific Sci. Congr. Oceanography. Thailand Dept. of Science, Bangkok.

Tomiyama, T. and K. Kobayashi. 1958. Direct absorption of radioisotopes in water by salt water and fresh water fish. In: Proc. 2nd Japan Conf. on Radioisotopes. (See Japan Radioisotopes Assoc., 1958) (AEC-tr-4482, pp. 1488-1529) (NSA 15: 25790).

Tomiyama, T., S. Ishio and K. Kobayashi/ Tomiyama, T., K. Kobayashi and S. Ishio. 1956. In: Research in the Effects and Influences of the Nuclear Bomb Test Explosions. Japan Society for the Promotion of Science, Ueno, Tokyo. Vol. II. (Japanese and English editions).

Absorption of dissolved ^{45}Ca by Carassius auratus. pp. 1151-1156.

Absorption of dissolved ^{45}Ca by goldfish. pp. 1163-1167.

Absorption of dissolved ^{45}Ca by Rhizodrilus limasus. pp. 1173-1176.

Excretion of absorbed ^{45}Ca by goldfish. pp. 1169-1172.

Absorption of $^{90}Sr(^{90}Y)$ by carp. pp. 1181-1187.

Distribution and excretion of intramuscularly administered $^{90}Sr(^{90}Y)$ in carp. pp. 1189-1193.

Absorption by Carassius auratus of ^{45}Ca contained in Rhizodrilus limasus. pp. 1157-1162.

Absorption of $^{32}PO_4$ ion by carp. pp. 1195-1200.

Distribution and excretion of intramucularly administered $^{32}PO_4$ by carp. pp. 1201-1203.

Tomizawa, C. and J. Fukami. 1956. Biochemical studies on the action of insecticides. III. Phosphorus metabolism of insects and the influence of insecticides. Botyu Kagaku 21(4): 133-139. (In Japanese) (Used ^{32}P).

Tomizawa, C. and Y. Koshimizu. 1966. Studies on the residue analysis of pesticides in plant materials. II. Neutron activation analysis of mercury in rice grains. Shokuhin Eiseigaku Zasshi 7: 33-38. (In Japanese) (NSA 21: 28401).

Tomizawa, C., Y. Uesugi and T. Murai. 1972. Fate and significance of fungicides in rice. pp. 103-105. In: Radiotracer Studies of Chemical Residues in Food and Agriculture. (See IAEA, 1972) (NSA 27: 20189).

Tomkins, D. J. and W. F. Grant. 1972. Comparative cytological effects of the pesticides menazon, metrobromuron and tetrachloroisophthalonitrile in Hordeum and Tradescantia. Can. J. Genet. Cytol. 14(2): 245-256. (NSA 26: 56394) (X-irradiation).

Tomlinson, R. E., et al. 1970. Management of the Hanford water table and waste management implications. Atlantic Richfield Hanford Co., U. S. AEC report ARH-1-37. 43pp. (NSA 29: 10155).

Tong, W. and I. L. Chaikoff. 1961. ^{131}I utilization by the aquarium snail and the cockroach. Biochim. Biophys. Acta 48(2): 347-351.

Tonges, S. D. 1971. ^{65}Zn uptake by a bacterium isolated from Alder Slough, Columbia River Estuary. Thesis, Oregon State Univ., Corvallis. 38pp. (NSA 26: 25944).

Toombs, G. L. 1965. Radiological survey of the lower Columbia River in Oregon, August 1963-July 1964. Radiol. Hlth. Data 6: 563-568. (NSA 20: 33590).

Toombs, G. L. and J. G. Bailey. 1964. Radiological survey of the lower Columbia River in Oregon, January 1962-July 1963. Radiol. Hlth. Data 5(8): 395-400. (NSA 18: 37387).

Topp, G. C. 1970. Soil water content from gamma ray attenuation: a comparison of ionization chamber and scintillation detectors. Can. J. Soil Sci. 50: 439-447. (NSA 25: 13110).

Toribara, T. Y., C. Predmore and P. A. Hargrave. 1962. The separation and determination of plutonium in diverse biological samples. Univ. of Rochester, U. S. AEC report UR-606. 11pp.

Torossian, C. 1970. Effects of x and γ radiations on the resistance and longevity of orphan workers of the ant Dolichoderus quadrupunctatus. Compt. rend. D271: 987-990. (In French) (NSA 25: 11245).

Torossian, C. 1971. Study of changes in the fertility of orphan workers of the ant Dolichoderus quadripunctatus irradiated with x and γ rays. Compt. rend. D272(7): 966-969. (In French) (NSA 44889).

Torossian, C. and R. Causse. 1967. Experimental study of the suspension of inhibition of egglaying of workers of the ant Dolichoderus quadripunctatus raised in the presence of queens having sustained heavy irradiation. Compt. rend. D264: 2789. (In French).

Torossian, C. and R. Causse. 1968. The effect of gamma radiation on the fertility and longevity of Dolichoderus quadripunctatus (Hymenoptera: Formicoidea Dolichoderidae). pp. 155-164. In: Isotopes and Radiation in Entomology. (See IAEA, 1968) (In French) (NSA 24: 32169).

Torissian, C. and R. Causse. 1971. Modifications induced by gamma irradiation in the oviposition of workers and females of the ant Dolichoderus quadripunctatus. Ann. Zool.-Ecol. Anim. 3(4): 427-441. (In French) (NSA 26: 45877).

Torre, C. 1963. The energy distribution of radioisotopes in groundwater flow. Atomkernenergie 8(6): 226-231. (In German) (NSA 17: 30685).

Torres de Arauz, R. 1967, 1970. Human ecology studies, Panama. Phase I Final report. --- Human ecology studies of Route 17 (Sasardi-Morti) region, Darien, Panama. Bioenvironmental and radiological-safety feasibility studies, Atlantic-Pacific interoceanic canal. Battelle Columbus Labs., U. S. AEC report BMI-171-006, -39. 131, 199pp. (NSA 21: 39365; 24: 50797).

Torroja, E. 1964. Genetic loads in irradiated experimental populations of Drosophila pseudoobscurus. Genetics 50: 1289-1298. (NSA 19: 8849) (Concerns viability).

Toth, A., T. Zsoldos and A. Urban. 1966. Investigation of the gamma activities and gamma spectra of snow samples. Acta Phys. Acad. Sci. Hung. 20: 305-322. (NSA 20: 25283).

Toth, G. 1974. Determination of the mercury content in natural waters by activation analysis. Period. Polytech. Chem. Eng. 18(1): 3-7. (NSA 31: 13602).

Totter, J. R. 1972. Research programs of the Atomic Energy Commission's Division of Biology and Medicine relevant to problems of health and pollution. In: Proc. 6th Berkeley Symp. on Mathematical Statistics and Probability. Vol. VI. L. M. le Cam, M. Neyman and E. L. Scott, eds. Univ. of California Press, Berkeley. (NSA 28: 684).

Touchberry, R. W. and J. C. DeFries. 1964. Genetic effects of gamma-irradiation on egg production and adult emergence of Drosophila melanogaster. Genetics 49(3): 387-400. (NSA 18: 13489).

Touhill, C. J., et al. 1969. The effects of radiation on Chicago Metropolitan Sanitary District municipal and industrial wastewaters. J. Water Poll. Control Fed. 41(Suppl.): R44-R60. (NSA 23: 50960).

Towheed, S. M. and S. M. H. Kabir. 1970. Effects of ^{60}Co gamma radiation on the egg and larval stage of the rice weevil, Sitophilus oryzae (L.). Nucleus (Karachi) 7: 59-64. (NSA 25: 13558).

Towner, J. W. 1965. The effect of radioactive fallout at the Nevada Test Site on the chromosomes of native populations of the pocket mouse. In: Radiation and Terrestrial Ecosystems. Health Physics 11(12): 1569-1571. (See Hungate, 1965).

Towner, J. W. and R. G. Lindberg. 1966. Persistent radiation induced chromosomal changes in native pocket mouse populations of the Nevada Test Site. Final report. Northrup Space Labs., Hawthorne, Calif., U. S. AEC report SAN-510-2. 81pp. (NSA 21: 11007).

Toy, A. J. and C. L. Lindeken. 1975. Implications of sampling from a log-normal population. Lawrence Livermore Lab., U. S. AEC report UCRL-76936. 19pp. (NSA 33: 2996) (Radioactivity).

Trabalka, J. R. 1971. Radioecology of $Ce^{144}Pr$ in a littoral fresh-water microcosm. Thesis, Univ. of Michigan, Ann Arbor. 225pp. (Univ. Microfilms 72-15021) (NSA 26: 45553).

Tracey, K. M., S. Jakowska and V. M. Fodor. 1959. The effects of x-irradiation on the life cycle of Tenebrio molitor. Abstr. Radiation Research 11(3): 473.

Trach, H. C. 1970. Study of fallout of artificial ^{55}Fe. Application to the evaluation of the fallout of natural iron of stratospheric origin. Thesis, Univ. of Paris. France Commissariat a l'Energie Atomique, Saclay, report CEA-R-3918. 102pp. (In French) (NSA 24: 25126).

Train, R. E. 1970. Balanced environment: the impact of thermal electric power generation on the environment. pp. 55-60. In: Northwest Conf. on the Role of Nuclear Energy. (See Governor's Nuclear Development Coordinating Committee, 1970) (NSA 25: 23218).

Tralau, H. W. 1957. Contribution to information on the variability of spruce. II. The effects of γ-radiation on Picea abies. Bot. Notizer 110(4): 442-454. (In German).

Trampusch, H. A. L. 1958. Action of x-rays on morphogenetic field. I. Heterotopic grafts on irradiated limbs. Proc. kon. ned. Akad. Wet. 61: 417-420. (Used axolotl).

Trampusch, H. A. L. 1959. New experiments on the effect of x-ray on the morphogenetic field. Ned. T. Geneesk. 103(13): 695-696. (Used amphibian).

Trapelo/West. 1970. Fort Calhoun Reactor. Pre-operational environmental survey. 8th report, 3rd quarter 1970. Richmond, Calif. 25pp. U. S. AEC DOCKET-50285-40. (NSA 25: 18085).

Trapelo/West. 1971. Fort Calhoun Reactor. Pre-operational background radiological survey. Summary report, October 1968-October 1970. Richmond, Calif. 41pp. U. S. AEC DOCKET-50285-41. (NSA 25: 18086).

Traut, H. and P. Schmidt. 1967. Repair of dominant lethal damage induced by x-rays in immature oocytes of Drosophila melanogaster. Int. J. Radiat. Biol. 13: 405-415. (NSA 22: 45673).

Trautmann, J. 1961. Basic data for experiments on embryonic damage by ionizing radiations. Strahlentherapie 114: 147-152. (In German) (NSA 15: 8573).

Travis, J. R. 1973. Mathematical model for the transport of environmental plutonium. Los Alamos Scientific Lab., U. S. AEC report LA-UR-73-954. 7pp. (NSA 29: 385).

Travis, J. R. 1974. Model for predicting the redistribution of particulate contaminants from soil surfaces. Los Alamos Scientific Lab., U. S. AEC report LA-UR-74-1340; LA-6035-MS. 59, 63pp. (NSA 31: 3372; 33: 609).

Trebichavsky, I. and J. Nedvidek. 1971. Incorporation of radiophosphorus into nucleic acids of frog embryo. Folia Biol. (Prague) 17(1): 54-58. (NSA 25: 26617).

Treherne, J. E. 1954. The exchange of labelled sodium in the larva of Aedes aegypti L. J. Exptl. Biol. 31(3): 386-401.

Treherne, J. E. 1962. Radioisotopes and the insect central nervous system. pp. 137-144. In: Radioisotopes and Radiation in Entomology. (See IAEA, 1962).

Treherne, J. E. 1965. Active transport in insects. pp. 1-13. In: Aspects of Insect Biochemistry. T. Goodwin, ed. Academic Press, New York. (References to use of radionuclides).

Treherne, J. E. 1965. The distribution and exchange of inorganic ions in the central nervous system of the stick insect Carausius morosus. J. Exptl. Biol. 42: 7-27. (^{14}C).

Treherne, J. E. and D. S. Smith. 1965. The penetration of acetylcholine into the intact central nervous tissues of an insect (Periplaneta americana L.). J. Exptl. Biol. 43: 13-21. (^{14}C).

Treherne, J. E. and D. S. Smith. 1965. The metabolism of acetylcholine in the intact central nervous system of an insect (Periplaneta americana). J. Exptl. Biol. 43(3): 441-454. (^{3}H).

Treip, C. S. and G. Divita. 1966. Histopathology of the cerebellum, medulla, and labyrinth of Xenopus laevis after high doses of total-body acute irradiation. J. Neuropathol. Exptl. Neurol. 25: 296-310. (NSA 21: 20094).

Trent, D. S. 1975. Mathematical modeling of transport processes in aquatic systems. pp. IV.119-IV.121. In: Proc. Conf. on Computational Methods in Nuclear Engineering. Vol. 2. (Charleston, South Carolina, April 15-17, 1975). U. S. ERDA report CONF-750413-P2. (NSA 32: 6434).

Trevino, G. 1972. On the natural redistribution of surface contamination. U. S. AEC report WASH-1225. 8pp. (NSA 28: 377).

Tritremmel, C. 1971. Methods of measurement for radiation monitoring of the environment in Austria. pp. 615-620. In: Rapid Methods for Measuring Radioactivity in the Environment. (See IAEA, 1971) (In German) (NSA 26: 28054).

Tritremmel, C., et al. 1966. Behavior of radioisotopes released to a stream. pp. 89-105. In: Disposal of Radioactive Wastes into Seas, Oceans and Surface Waters. (See IAEA, 1966).

Triulzi, C. 1970. The CISE contribution to radioecology. pp. 119-124. In: Proc. 1st Conf. on the State of Advancement of Radioecology in Italy. (See Anonymous, 1970) (In Italian).

Trivelloni, J. C., A. O. Donoso and G. E. Bur. 1959. Thyroidectomy by radioactive iodine in the toad. Rev. soc. arg. biol. 35: 238-241.

Troitskaya, M. N., et al. 1969. Cesium-137, strontium-90, and lead-210 in the far north regions. pp. 67-71. In: Nuclear Meteorology. (See Makhon'ko and Malakhov, 1969) (In Russian) (NSA 30: 15428).

Troitskaya, M. N., et al. 1971. ^{137}Cs, ^{90}Sr, and ^{210}Pb in soils of the arctic regions. Tr. inst. eksp. meteorol. 21(1): 72-76. (In Russian) (NSA 30: 29405).

Troitskaya, M. N., et al. 1971. Radioecology of landscapes of the Far North. pp. 325-353. In: Contemporary Problems of Radiobiology. Vol. 2. Radioecology. (See Klechovskiy, Polikarpov and Aleksakhin, 1971) (In Russian).

Troitskaya, M. N., et al. 1974. Analysis of geographical differences in the rate of purification of the chain lichens-reindeer-man from cesium-137. pp. 266-271. In: Proc. 3rd Intern. Congr. of the International Radiation Protection Association. (CONF-730907-P1) (In Russian) (NSA 30: 26802).

Troitskiy, N. A. and G. V. Khromenko. 1966. Use of gamma rays in radiative selection of Actinomyces aureofaciens. Vest. Akad. Navuk Belorussk. SSR (biyal.) (1): 59-61. (In Russian) (NSA 20: 24935).

Troitskiy, V. and Z. Pershina. nd. Radiobiology. In: Great Medical Encyclopedia 27: 770-789. (JPRS-18414).

Troshin, A. S. 1956. Radioactive isotope technique and its application in hydrobiology. Zhizn' Presnykh vod SSSR 4(1): __. (In Russian).

Troshin, A. S. and V. I. Zhadin. 1957. Radiomarking of vimba and bleak for disclosing the operational efficiency of the vimba-bleak nursery. Tr. probl. temot. soveshch. zool. inst. Akad. Nauk SSSR 7: 57-61. (In Russian).

Trott, K.-R. 1966. Radiation effect of the potential and resistance of frog skin. pp. 125-129. In: Biophysical Problems of the Actions of Radiation. H. Muth, ed. Georg Thieme Verlag, Stuttgart. (In German) (NSA 21: 22361).

Trout, W. E., III. 1965. The so-called recovery phenomenon after irradiation in Drosophila melanogaster. Thesis, Indiana Univ., Bloomington. 108pp. (NSA 21: 20155).

Truckenbrodt, W. 1965. The sensitivity of undifferentiated and differentiated nuclei following x-irradiation in the early stage of development of the egg of the termite Kalotermes flavicollis Fabr. Arch. Entwicklungsmech. Organ. 156(2): 101-126. (In German).

Truckenbrodt, W. 1968. Mitosis in the cleavage and germ disc stages of termite eggs after total x radiation with a dose of 5000 R. Strahlentherapie 136: 632-638. (In German) (NSA 23: 20330).

Truesdale, G. A. 1954. Use of radioactive isotopes in tracing sewage flow. Atomics Atomic Tech. 5: 304-312.

Trump, J. G. and R. J. van de Graaff. 1948. Irradiation of biological materials by high energy Roentgen rays and cathode rays. J. Appl. Phys. 19: 599-604.

Tryfiates, G. P. 1959. Algal photosynthesis as measured by absorption of radioactive carcon from water. Thesis, Bowling Green State Univ., Bowling Green, O. (Also: 1960 Ohio J. Sci. 60(2): 77-82).

Tryon, C. A., Jr. 1963. Radiation and other pollutants: discussion. pp. 150-151. In: Proc. 1st Intern. Conf. on Wildlife Disease.

Tryon, C. A., Jr. 1966. Designs for the experimental radiation of populations of free-ranging animals. pp. 19-21. In: Radiation Effects on Natural Populations. (See Sacher, 1966).

Tryon, C. A. Jr. and D. P. Snyder. 1973. The effect of exposure to 200 and 400 R of ionizing radiation on the survivorship curves of the eastern chipmunk (Tamias striatus) under natural conditions. pp. 1037-1041. In: Radionuclides in Ecosystems. (See Nelson, 1973) (NSA 26: 23088).

Tsai, C. M. and P. S. Weng. 1973. ^{226}Ra concentrations in Taiwan hot spring and river waters. Health Physics 24(4): 429-430. (NSA 27: 27589).

Tschurlovits, M. and E. Unfried. 1975. Physical methods for environmental monitoring in the vicinity of nuclear facilities. Atomkernenergie 25(1): 1-18. (In German) (NSA 32: 11535).

Tsevelev, M. A., P. I. Chalov and K. P. Makhonko. 1970. Study of the irregularities in the washing out of radioactive fission products in mountains and foothills. Tr. inst eksp. meteorol. (5): 102-115. (English transl. FTD-HT-23-927-74; AD-777682) (NSA 30: 32026).

Tsvetkov, D., et al. 1972. Sterilizing doses of gamma irradiation for control of certain storehouse pests on stored wheat grain. Bulgaria Komitet za Mirno Izpolzuvane na Atomnata Energiya, Sofia. 14pp. (In Hungarian) (NSA 29: 29: 24508).

Tsiropoulos, G. J. and M. E. Tzanakakis. 1970. Mating frequency and inseminating capacity of radiation-sterilized and normal males of the olive fruit fly. Ann. Entomol. Soc. Am. 63: 1007-1010. (NSA 24: 42045).

Tsitsipis, J. A. and M. E. Tzanakakis. 1968. Recent radiation work with the olive fruit fly at the Democritus Nuclear Research Center, Greece. pp. 117-121. In: Radiation, Radioisotopes, and Rearing Methods in the control of Insect Pests. (See IAEA, 1968) (NSA 24: 32331).

Tsivoglou, E. C. 1961. Method for grass radioactivity analysis of environmental samples. R. A. Taft Sanitary Engineering Center, Cincinnati, O., U. S. PHS. 15pp. (Mimeo).

Tsivoglou, E. C. 1963. Research for the control of radioactive pollutants. J. Water Poll. Control Fed. 35(2): 242-259. (NSA 18: 9816).

Tsivoglou, E. C. 1966. Environmental monitoring around a uranium mill in the United States of America. pp. 63-70. In: Manual on Environmental Monitoring in Normal Operation. International Atomic Energy Agency, Vienna, Safety Ser. No. 16, Publ. STI-PUB-98. (NSA 20: 23409).

Tsivoglou, E. C. 1968. ^{85}Kr measurement of stream reaeration rates. Trans. Amer. Nucl. Soc. 11: 51-52. (NSA 22: 30707).

Tsivoglou, E. C. 1971. Nuclear power: risks and social concerns. Trans. Nucl. Sci. NS-18(1): 19-27. (NSA 25: 60001).

Tsivoglou, E. C. and G. W. Burke, Jr. 1963. Radioactivity standards for water. J. Am. Water Works Assoc. 55(6): 668-676. (NSA 17: 27613).

Tsivoglou, E. C. and R. L. O'Connell. 1962. Waste guide for the uranium milling industry. R. A. Taft Sanitary Engineering Center, Cincinnati, O., U. S. PHS, Tech. Rept. W-62-12. 78pp. (NSA 17: 21383) (Remarks on field studies).

Tsivoglou, E. C. and W. W. Towne. 1957. Sources and control of radioactive water pollutants. Sewage Industr. Wastes 29(2): 143-156.

Tsivoglou, E. C., E. D. Harward and W. M. Ingram. 1957. Stream surveys for radioactive waste control. J. Am. Water Works Assoc. 49(6): 750-766.

Tsivoglou, E. C., E. D. Harward and W. M. Ingram. 1957. Stream surveys for radioactive waste control. pp. 388-398. In: 2nd Nuclear Engr. and Sci. Conf. (Philadelphia).

Tsivoglou, E. C., M. Stein and W. W. Towne. 1960. Control of radioactive pollution of the Animas River. J. Water Poll. Control Fed. 32(3): 262-287. (BA 35: 55003; SFA 6: 3937).

Tsivoglou, E. C., et al. 1956. Report of survey of contamination of surface waters by uranium recovery plants. R. A. Taft Sanitary Engineering Center, Cincinnati, O., U. S. PHS.

Tsivoglou, E. C., et al. 1958. Effects of uranium ore refinery wastes on receiving waters. Sewage Industr. Wastes 30(8): 1012-1027.

Tsivoglou, E. C., et al. 1959, 1960. Survey of interstate pollution of the Animas River (Colorado-New Mexico). --- Part II. R. A. Taft Sanitary Engineering Center, Cincinnati, O., U. S. PHS. v, 116; iii, 53pp.

Tsivoglou, E. C., et al. 1968. Tracer measurement of stream reaeration. II. Field studies. J. Water Poll. Control Fed. 40: 285-305. (NSA 22: 48937) (^{85}Kr).

Tsugawa, Y. and K. Yamamoto. 1960, 1962. On Lebistes reticulatus' perception of x-radiation. I. II. Kanazawa Med. Sci. Ser. 57; 66. (In Japanese).

Tsugo, T. 1971. Environmental contamination and human food and bodies. pp. 111-188. In: Studies on Radiation Effects. Y. Hiyama, ed. Tokyo Daigaku Shuppankai, Tokyo. (In Japanese) (NSA 27: 27589).

Tsujii, T. 1963. The change of pearl colors by the irradiation with γ-ray or neutron ray. J. Radiation Res. (Tokyo) 2(2-3-4): 120-125.

Tsutsui, T. (Comp.). nd. Studies of the utilization of radioisotopic tracers in the field of sanitary engineering. Research Reactor Inst., Kyoto Univ., Japan, report KURRI-TR-7. 11pp. (In Japanese) (NSA 23: 16040).

Tsutsui, T. and K. Nishimaki. 1972. Movement of radionuclides through saturated sand zones. Doboku Gak. Ronbun Hokokushu (200): 25-29. (In Japanese) (NSA 28: 2591).

Tsypin, A. B. 1964. Conditioned reflex during ^{60}Co gamma irradiation of fish and rabbits. pp. 46-49. In: Pathogenesis, Experimental Prophylaxis and Therapy of Radiation Injuries. A. V. Lebedinskiy, ed. Izdatel'stvo Meditsina, Moscow. (In Russian) (NSA 20: 38655).

Tsypin, A. B. and Yu. A. Kholodov. 1964. Development of a conditioned reflex to ionizing radiation in fish and rabbits. Radiobiologiya 4: 402-408. (In Russian) (NSA 18: 33366) (Used carp).

Tsypin, A. B. and N. M. Irunova. 1964. Reactions of the retina to ionizing radiation. In: 10th Congr. All-Union Pavlov Physiological Soc. Vol. II. No. 2. Izdatel'stvo Nauka. (FTD-TT-65-48, pp. 590-591) (NSA 19: 40359) (Incl use of frog).

Tsytsugina, V. G., N. S. Pisik and G. Ye. Lazorenko. 1973. Artificial and Natural Radionuclides in the Life of Aquatic Organisms. "Naukova Dumka, Kiev. 152pp. (In Russian) (NSA 28: 15818).

Tuchkova, S. Ya. 1964. Causes underlying the loss of regeneration capacity of salamander extremities, after exposure to x-rays. Dokl. Akad. Nauk SSSR 158(6): 1420-1423. (Doklady (biophys.) 158: 159-162) (NSA 19: 19544).

Tuchkova, S. Ya. 1964. Variation in the content of nucleic acids, accompanying suppressed regeneration of extremities in salamanders following x-ray treatment. Dokl. Akad. Nauk SSSR 159(1): 215-218. (Doklady (Biol. sci.) 159: 791-793) (NSA 19: 19546).

Tuchkova, S. Ya. 1965. Histological and histochemical investigations of the restoration of the ability to regenerate limbs of axolotl, suppressed by x-ray irradiation. Radiobiologiya 5(2): 207-210. (AEC-tr-6599, pp. 65-71).

Tuchkova, S. Ya. 1966. Role of trauma in the restoration of regeneration capacity in axolotl extremities inhibited by x-ray irradiation. Dokl. Akad. Nauk SSSR 168: 473-476. (Doklady (biol. sci.) 168(1-6): 349-352) (NSA 21: 8986).

Tuchkova, S. Ya. 1967. Variation in mitotic activity in extremity tissues of axolotl in the case of restored regeneration ability inhibited by x-ray treatment. Dokl. Akad. Nauk SSSR 172: 222-224. (In Russian) (NSA 21: 34749).

Tuchkova, S. Ya. 1969. Autoradiographic study of DNA synthesis from the inclusion of thymidine-^{3}H during restoration of regenerative ability in axolotl extremities, suppressed by x-ray irradiation. Dokl. Akad. Nauk SSSR 186: 724-727. (Doklady (Biol.) 186(1-6): 452-456) (NSA 23: 50824).

Tuchkova, S. Ya. 1973. Autoradiographic data on the participation of irradiated tissues in the process of axolotl extremity regeneration suppressed by x-ray irradiation. Dokl. Akad. Nauk SSSR 210(4): 986-988. (In Russian) (NSA 28: 20178).

Tueller, P. T., A. D. Bruner and J. B. Davis. 1972. Ecology of Hot Creek Valley - vegetation and soil response to underground detonation. Univ. of Navada, Reno, U. S. AEC report NVO-409-1. ii, 131pp.

Tugarinov, V. V. and V. V. Kuznetzov. 1966. Combined effects of x-rays and streptomycin on the mutability of Chlorella. Genetika (2): 57-63. (In Russian) (NSA 20: 32993).

Tugarinov, V. V. and V. V. Kuznetzov. 1966. Combined effects of x-rays and streptomycin on the mutability of Chlorella algae. Tr. Mosk. obshch. ispyt. prirody (biol.) 22: 163-169. (In Russian) (NSA 22: 51511).

Tuite, P. T. 1970. Disposing of nuclear plant solid wastes. Power 114(11): 78-79. (NSA 25: 2425).

Tul'tseva, N. M. and B. L. Astaurov. 1958. Increased radioresistance of Bombyx mori L. polyploids and the general theory of the biological action of ionizing radiation. Biofizika 3(2): 197-215. (Biophysics (USSR) 3 (2): 183-189).

Tumanishvili, G. D. 1962. Regeneration of tissues under the action of extracts irradiated by γ- and x-rays. Tr. inst fiz. Akad. Nauk Gruz. SSR 8: 109-114. (In Russian) (NSA 17: 12117).

Tumanishvili, G. D., V. P. Manzhgaladze and Kh. N. Dzhanelidze. 1964. Effect of ionizing radiation on the stimulation properties of tissue extracts. Soobshch. Akad. Nauk Gruz. SSR 33: 549-556. (NSA 19: 8917) (Used Rana ribibunda).

Tukey, H. B., Jr. 1970. Leaching of metabolites from foliage and its implication in the tropical rain forest. Ch. H-11, pp. H155-H160. In: A Tropical Rain Forest. (See H. T. Odum and R. F. Pigeon, 1970).

Tukey, H. B., et al. 1956. Utilization of radioactive isotopes in resolving the effectiveness of foliar absorption of plant nutrients. Proc. Intern. Conf. on the Peaceful Uses of Atomic Energy 12: 138-143. (Papaer A/CONF.8/ P106).

Tumanyan, M. A. and Z. G. Pershina. 1971. Effects of ionizing radiation on microorganisms and radiation sterilization (survey). Zh. mikrobiol epidemiol. immunobiol. (5): 75-80. (In Russian) (NSA 26: 20563).

Tung, H. H. N. 1965. The effects of ionizing radiation on the ecology of Viguiera porteri (A. Gray) Blake. Thesis, Emory Univ., Atlanta, Ga.

Tung, P. S. 1970. Microspectrometric and cytological studies of the effect of ^{60}Co irradiation on spermatogenesis and DNA synthesis in germ cells of the face fly. Thesis, Pennsylvania State Univ., University Park. 114pp. (Also w/R. B. Mitchell and A. Anthony. 1971: J. Insect Physiol. 17(2): 277-284) (Univ. Microfilms 71-16681) (NSA 25: 32524, 46958).

Tuominen, Y. 1967. Studies of the strontium uptake of the Cladonia alpestris thallus. Ann. Bot. Fenn. 4: 1-28. (NSA 22: 15112).

Tur, J. 1904. Embryonic malformations through the action of radium on the eggs of the ponle. Compt. rend. soc. biol. 57: 236-238. (In French).

Turchaninova, V. A. and F. A. Tikhomirov. 1973. Effect of contamination of soils with radioactive isotopes strontium-90 and yttrium-90 on soil-dwelling invertebrates. Ekologiya 4(5): 101-102. (Sov. J. Ecology 4(5): 453-454. (NSA 32: 6820).

Turekian, K. K., et al. 1970. An evaluation of the uranium-helium method of dating fossil bones. Earth Planet. Sci. Lett. 7: 420-424. (NSA 24: 25114).

Turk, A., et al. 1974. Environmental Science. W. B. Sanders, Philadelphia. 576pp. (NSA 30: 6487) (Incl. radioactivity)

Turkey Atomic Energy Commission. 1970. Sixth activity report of the Radioactive Fallout Laboratory, August 1968-August 1969. Ankara. 17pp. (In Turkish and English) (NP-18379) (NSA 24: 44125).

Turner, F. B. 1962. Some influences of an underground nuclear detonation on close-in populations of whiptailed lizards (Cnemidophorus tigris) at the Nevada Test Site. Abstr. Bull. Ecol. Soc. Am. 43(4): 124.

Turner, F. B. 1963. Influence of a cratering device on close-in populations of lizards (Project Sedan). Univ. of California at Los Angeles, U. S. AEC report PNE-224F. vi, 35pp.

Turner, F. B. 1963. A mathematical model for time-specific relationships of I^{131} between food consumed and the thyroid burden. Abstr. Bull. Ecol. Soc. Am. 44(3): 75.

Turner, F. B. 1963. Biotic communities of the Nevada Test Site. A review. Ecology 44(3): 633-634.

Turner, F. B. 1963. Quantitative relationships between fallout radioiodine on native vegetation and in thyroids of herbivores. Health Physics 9(12): 1246.

Turner, F. B. 1965. Uptake of fallout radionuclides by mammals and a stochastic simulation of the process. pp. 800-820. In: Radioactive Fallout from Nuclear Weapons Tests. (See Klement, 1965) (Jackrabbit data from the Sedan site, Nevada Test Site).

Turner, F. B. 1965. Radiation effects on lizards in a Puerto Rican forest. Univ. of California at Los Angeles, U. S. AEC report UCLA-12-567. 11pp. (NSA 20: 38601).

Turner, F. B. 1966. Growth rate of lizards (Uta stansburiana) exposed to chronic gamma radiation. pp. 46-52. In: Radiation Effects on Natural Populations. (See Sacher, 1966).

Turner, F. B. 1975. Effects of continuous irradiation on animal population. Adv. Radiat. Biol. 5: 83-144. (NSA 32: 28299).

Turner, F. B. and C. S. Gist. 1965. Influences of a thermonuclear cratering test on close-in populations of lizards. Ecology 46: 845-852. (Also: Univ. of California at Los Angeles, U. S. AEC report UCLA-12-522) (NSA 18: 23181; 20: 42732) (At Nevada Test Site).

Turner, F. B. and C. S. Gist. 1970. Observations of lizards and tree frogs in an irradiated Puerto Rican forest. Ch. E-2, pp. E25-E49. In: A Tropical Rain Forest. (See H. T. Odum and R. F. Pigeon, 1970) (NSA 25: 21839).

Turner, F. B. and R. I. Jennrich. 1965. The uptake of fallout radioisotopes by mammals and a stochastic simulation of the process. Abstr. Biometrics 21(3): 773. (Jackrabbits ?).

Turner, F. B. and R. I. Jennrich. 1967. The concentration of ^{131}I in the thyroids of herbivores and a theoretical consideration of the expected frequency distribution of thyroidal ^{131}I in a large consumer population. pp. 175-182. In: Radioecological Concentration Processes. (See Aberg and Hungate, 1967).

Turner, F. B. and J. R. Lannom, Jr. 1968. Radiation doses sustained by lizards in a continuously irradiated natural enclosure. Ecology 49(3): 548-551. (NSA 22: 45470).

Turner, F. B. and W. E. Martin. 1963. Food-chain reletionships of iodine-131 following two nuclear tests in Nevada (Project Sedan). Univ. of California at Los Angeles, U. S. AEC report PNE-236P. 70pp. (NSA 17: 40516).

Turner, F. B. and W. E. Martin. 1964. Food-chain relationships of iodine-131 in Nevada following the Sedan test of July 1962. Univ. of California at Los Angeles, U. S. AEC report PNE-936F. 54pp. (NSA 19: 34).

Turner, F. B. and W. A. Rhoads. 1966. Book review - Proving Ground: An Account of the Radiobiological Studies in the Pacific, 1946-1961. By N. O. Hines. Am. Midland Nat. 76(1): 248-250.

Turner, F. B., G. A. Hoddenbach and J. R. Lannom, Jr. 1965. Growth of lizards in natural populations exposed to gamma irradiation. In: Radiation and Terrestrial Ecosystems. Health Physics 11(12): 1585-1593. (See Hungate, 1965) (NSA 20: 12669).

Turner, F. B., G. A. Hoddenbach and J. A. Lannom. 1965. Growth of Uta Stansburiana in four experimental areas at the Nevada Test Site. Abstr. Bull. Ecol. Soc. Am. 46(2): 51.

Turner, F. B., G. A. Hoddenbach and J. A. Lannom, Jr. 1967. A comparison of some demographic attributes of irradiated and non-irradiated populations of Uta stansburiana in Nevada. Abstr. Bull. Ecol. Soc. Am. 48(3): 137.

Turner, F. B., R. H. Rowland and R. A. White. 1966. Nuclear engineering and wildlife: radioactivity in jackrabbits after the Sedan test. J. Wildl. Mgmt. 30(2): 433-443.

Turner, F. B., R. H. Rowland and R. A. Wood. 1965. Radioactivity in jack rabbits after the Sedan Test of 1962. University of California at Los Angeles, U. S. AEC report UCLA-12-572. 26pp.

Turner, F. B., et al. 1962. Uptake of radioactive materials from a nuclear reactor by small mammals at the Nevada Test Site. Univ. of California at Los Angeles, U. S. AEC report TID-16835. (NSA 16: 32693).

Turner, F. B., et al. 1963. Uptake of radioactive materials from a nuclear reactor by small mammals at the Nevada Test Site. Univ. of California at Los Angeles, U. S. AEC report TID-18864. 13pp. (NSA 17: 27054).

Turner, F. B., et al. 1969. A demographic analysis of fenced populations of the whiptail lizard Cnemidophorus tigris at southern Nevada. Southwestern Nat. 14(2): 189-201. (Nevada Test Site).

Turner, F. B., et al. 1970. The demography of the lizard, Uta stansburiana Baird and Girard, in southern Nevada. J. Animal Ecol. 39(2): 505-519. (Nevada Test Site, gamma radiation).

Turner, F. B., et al. 1973. Radiation-induced sterility in natural populations of lizards (Crotaphytus wislizenii and Cnemidophorus tigris). pp. 1131-1143. In: Radionuclides in Ecosystems. (See Nelson, 1973) (NSA 28: 30145).

Turner, N. C. and L. D. Incoll. 1971. The vertical distribution of photosynthesis in crops of tobacco and sorghum. J. Appl. Ecol. 8(2): 581-591. (^{14}C).

Turner, W. K. and L. F. Charity. 1971. Determining response of insects to radiation by continuous monitoring of their carbon dioxide output. Ann. Entomol. Soc. Am. 64(2): 419-424. (NSA 25: 24312).

Turpin, R. A. and R. A. Schlising. 1971. A new method for studying pollen dispersal using iodine-131. Radiation Botany 11(1): 75-78.

Turska, R. 1971. Pathologic changes in the ultrastructure of hepatocytes in the axolotl (Ambystoma tigrinum Green) under the influence of roentgen rays. Acta Med. Pol. 12(2): 231-233. (NSA 25: 58112).

Turska, R. 1971. Pathological changes of ultrastructures of hepatocytes of axolotls after x-irradiation. Zool. Pol. 21(1): 91-99. (NSA 26: 48269).

Tutarova, A. B. 1971. Background radioactivity of the ground waters of the Ukraine. Geol. zh. 31(1): 91-97. (In Russian) (NSA 26: 20335).

Twigg, G. I. and H. Miller. 1963. The use of calcium^{45} as an agent for labeling rat populations. J. Mammalogy 44(3): 335-337. (WR 111: 20).

Tyler, S. A. and S. P. Stearner. 1960. Modes of radiation death in the chick embryo. II. A model of lethal mechanisms. Radiation Research 12(3): 301-316.

Tyuryukanov, A. N., Yu. D. Abaturov and A. N. Letova. 1971. Experimental investigation of the turnover of radioactive isotopes in plant-solution system. pp. 84-86. In: Methods of Radioecological Investigations. (See Verkhovskaya, 1971) (In Russian).

Tyuryukanova, Ye. B. 1968. Procedure for investigating the behavior of radioactive strontium in soils of different geochemical regions. State Committee on the Use of Atomic Energy USSR. Atomizdat, Moscow. 15pp. UN Document A/AC.82/G/L.1253. (AEC-tr-7030, pp. 134-144) (NSA 23: 17552).

Tyuryukanova, Ye. B. 1969. Accumulation of strontium-90 in soils. pp. 61-66. In: Nuclear Meteorology. See Makhon'ko and Malakhov, 1969) (In Russian) (NSA 30: 15427).

Tyuryukanova, Ye. B. 1971. Landscape-geochemical aspects of strontium-90 migration. pp. 81-90. In: Contemporary Problems of Radiobiology. Vol. 2. Radioecology. (See Klechkovskiy, Polikarpov and Aleksakhin, 1971) (In Russian).

Tyuryukanova, Ye. B. 1971. Investigation of behavior of radioactive strontium in soils of different geochemical landscapes. pp. 16-24. In: Methods of Radioecological Investigations. (See Verkhovskaya, 1971) (In Russian).

Tyuryukanova, Ye. B. 1972. Migration of strontium-90 and cesium-137 in soils. State Committee on the Use of Atomic Energy USSR, Moscow. 11pp. (In Russian) (INIS-mf-1036) (NSA 30: 6521).

Tyuryukanova, Ye. B. and V. A. Kalugina. 1971. The behavior of thorium in soils. Ekologiya 2(5): 93-95. (Sov. J. Ecology 2(5): 467-469).

Tyuryukanova, Ye. B., N. I. Konova and S. Saboiyev. 1971. Radioactive strontium in the pasture soils of East Pamir. Ekologiya 2(1): 45-48. (Sov. J. Ecology 2(1): 31-34).

Tyuryukanova, Ye. B., N. I. Konova and V. V. Yeme'yanov. 1973. Aqueous migration of strontium-90. State Committee on the Use of Atomic Energy USSR, Moscow. 15pp. (In Russian) (INIS-mf-1037) (NSA 30: 6522).

Tyuryukanova, Ye. B., F. I. Pavlotskaya and V. I. Baranov. 1965. Specific features of strontium-90 in different soils in the European region of the USSR in 1961. State Committee on the Use of Atomic Energy USSR, Moscow. 19pp. (In Russian) (NP-16527) (NSA 21: 30531).

Tyuryukanova, Ye. B., F. I. Pavlotskaya and V. I. Baranov. 1967. Distribution of radioactive strontium in the soils of different natural zones. State Committee on the Use of Atomic Energy USSR. Atomizdat, Moscow. 16 pp. UN Document A/AC.82/G/L.1176. (AEC-tr-7214, pp. 143-160. (NSA 22: 23128; 25: 15759).

Tyuryukanova, Ye. B., et al. 1964. Distribution of strontium-90 in the top horizons of soils in relation to the soil group and landscape. Pochvovedeniye (8): 88-95. (Sov. Soil Science (8): 853-859) (NSA 19: 22548).

Tyuruykanova, Ye. B., et al. 1964. Migration and distribution of strontium-90 and cerium-144 in soils of the Mosow region. Pochvovedeniye (10): 66-73. (In Russian) (NSA 19: 22549).

Tyuryukanova, Ye. B., et al. 1973. Landscape-geochemical aspects of ^{90}Sr behavior in the forest and flood-plain ecosystems of wooded boggy areas. State Committee on the Use of Atomic Energy USSR, Moscow. 40pp. (In Russian) (INIS-mf-1136) (NSA 30: 12696).

Tyruyukanova, Ye. B., et al. nd. Geochemistry of ^{90}Sr and ^{137}Cs in soils of the USSR. (In Russian) (INIS-mf-887-3, pp. 63-76) (NSA 30: 6516).

Tzanakakis, M. E. 1974. Current status and prospects of applying the sterile-insect release method against Dacus oleae. pp. 109-115. In: Sterile-Insect Techniques and Its Field Applications. (See IAEA, 1974) (NSA 31: 8946).

Tzanakakis, M. E., et al. 1966. Gamma radiation-induced dominant lethality to the sperm of the olive fruit fly. J. Econ. Entomol. 59: 214-216. (NSA 20: 40674).

Ubell, K. 1967. Investigations into groundwater balance by applying radioisotope tracers. pp. 521-530. In: Isotopes in Hydrology. (See IAEA, 1967) (NSA 22: 19069).

Ucer, E. 1965. The mortality effect of radiation on Drosophila melanogaster larvae of different ages. Istanbul Univ. Fen. Fak. Mecm. B30: 171-175. (NSA 22: 2549).

Uchida, H. 1973. Safety, environment, and licensing problems of nuclear power plant on Japan. Nucl. Eng. Int. 18(206): 560-562. (NSA 28: 17643).

Uchida, T., H. S. Rahmati and R. D. O'Brien. 1965. The penetration and metabolism of H^3-dimethoate in insects. J. Econ. Entomol. 58(5): 831-835.

Uchida, T., J. Zschintsch and R. D. O'Brien. 1966. Relation between synergism and metabolism of dimethoate in mammals and insects. Toxicol. Appl. Pharmacol. 8: 259-265. (^{3}H).

Udaltsova, N. I. and L. L. Leonova. 1970. Uranium, thorium, and rare-earth elements in thermal waters of the Kamchatka Peninsula. Geokhimiya (12): 1504-1510. (In Russian) (NSA 25: 10697).

Ueda, K. and K. Okabe. 1960. Modes of radiation death in the chick embryo. II. The lethal effect of internal β-irradiation of ^{32}P. Radioisotopes (Tokyo) 9: 25-32. (In Japanese) (NSA 15: 22143).

Ueda, K., K. Okabe and S. Yamamoto. 1960. Retinal anomaly of the chick embryo produced by x-ray irradiation. J. Radiat. Res. (Japan) 1: 124-130. (NSA 15: 24696).

Ueno, A. M. 1974. Incorporation of tritium from tritiated water into nucleic acids of Oryzias latipes eggs. Radiation Research 59(3): 629-637. (NSA 30: 27279).

Ueno, K. 1960. Determination of radioactive fallout from nuclear weapons tests, (I) gamma-ray spectrometric determination of small amounts of Zr^{95}-Nb^{95} absorbed on pine leaves. J. Atom. Energy Soc. Japan 2(2): 67-72. (In Japanese, English abstr.) (NSA 14: 12811).

Ueno, M., K. Yoshihara and T. Okada. 1967. Living root system distinguished by the use of ^{14}C. Nature 213: 530-532. (NSA 21: 12778).

Ueno, T., et al. 1969. Survey of radioactivity of Osaka Bay. Ann. Rep. Radiat. Cen. Osaka Pref. 10: 86-89. (NSA 24: 48552).

Ujihara, A., S. Mori and S. Hori. 1965. Genetic studies on radioresistant strain of Chlorella ellipsoidea. I. Radiosensitivity and radioresistant strain induced by repeated gamma-irradiation in Chlorella ellipsoidea. Ann. Rep. Eadiat. Cen. Osaka Pref. 6: 90-92. (NSA 20: 38605).

Ulrich, H. and F. E. Würgler. 1964. Genetic and non-genetic radiation effects on Drosophila eggs. Zool. Anz. 27(Suppl.): 602-606. (In German).

Ulrich, M. I. 1965. Factors concerning post-irradiation bacteremia in mice. Thesis, Stanford Univ., Stanford, Calif. 121pp. (NSA 20: 8721).

Ulrikson, G. U. 1973. Radiation effects on serum proteins, hematocrits, electrophoretic patterns and protein components in the bluegill (Lepomis macrochirus). pp. 1100-1105. In: Radionuclides in Ecosystems. (See Nelson, 1973) (NSA 28: 30142).

Ulubekova, M. V. and L. A. Kuz'mina. 1953. Application of C^{14} in the study of the photoreduction and chemosynthesis of green algae. Dokl. Akad. Nauk SSSR 93(5): 915-918. (In Russian).

Umanskiy, S. R. 1968. Influence of irradiation on the composition of messenger RNA during early loach embryogenesis. Radiobiologiya 8(3): 347-353. (AEC-tr-6916, pp. 12-22) (NSA 22: 51388).

Underbrink, A. G. and A. H. Sparrow. 1968. Chromosomes and cellular radiosensitivity. II. (w/V. Pond). Use of interrelationships among chromosome volume, nucleotide content and D_o of 120 diverse organisms in predicting radiosensitivity. III. Note on the inability to demonstrate a correlation between nucleic acid base ratios and radiation response at D_o in several diverse organisms. Radiation Botany 8(3): 205-238; Radiation Research 35: 311-317. (NSA 22: 47478).

Underbrink, A. G., A. H. Sparrow and R. A. Owens. 1969. The fine structure of the alga Brachiomonas submarina Bohlin after x- and γ-irradiation. Radiation Botany 9(4): 241-250.

Underdal, B. 1968. Measurement of ^{137}Cs in meat from different species, 1965-1967. Nord. Veterinarmed 20: 344-353. (NSA 23: 50468) (Incl. Lynx, wolverine, reindeer).

Underdal, B. and S. Hauge. 1968. Measurements of radioactive nuclides from fallout in Norway. Nord. Veterinarmed 20: 71-74. (In Norwegian) (NSA 23: 20241) (Incl. reindeer).

Unger, A. 1959. Radioactivity and water. Progress report of the state radioactivity control organization in the State Bacteriological Institute in Munich. Öff. Gesund.-Dienst 21(1): 16-25. (In German).

Union Carbide Corporation. 1974. Environmental monitoring report, calender year 1973. Paducah Gaseous Diffusion Plant, U. S. AEC report UCC-ND-279. 28pp. (NSA 30: 29351).

Union Carbide Corporation. 1974. Environmental monitoring report: United States Atomic Energy Commission, Oak Ridge facilities. Calender year 1973. U. S. AEC report UCC-ND-280. 46pp. (NSA 30: 29352).

Union of South Africa Atomic Energy Board. 1972. National Conf. on the Technological Applications of Nuclear Techniques. (Pelindaba, South Africa, October 12-13, 1972). Pelindaba, Pretoria. 242pp. (In English and Afrikaans) (CONF-721091) (NSA 28: 21460).

Union of Soviet Socialist Republics. 1956. Conf. on the Peaceful Uses of Atomic Energy. (July 1-5, 1955, Session of the Division of Biological Science, Acad. Sci. 198pp. (English transl. Consultants Bureau, N. Y.).

Union of Soviet Socialist Republics. nd. Radiobiology in the USSR. Transl. of abstracts of articles from selected Soviet periodicals. Joint Publications Reserach Service, New York, report JPRS(NY)-899. 19pp. (NSA 14: 1277).

USSR Institute of Oceanology. 1960. Technical Handbook on Determination of the Primary Production of Organic Matter in Resrvoirs by the Radiocarbon Method. Acad. Sci., Minsk. (English transl. U. S. Fish and Wildl. Serv., FWSS(?), Fishery Lit., Misc. Ser. No. 36).

United Kingdom Atomic Energy Authority.

1956. The determination of alpha and beta activities in vegetation, roots, seaweed, soil, seabed sludge, seashore sand, silt and river mud. Industrial Group, Risley, report IGO-AM/W-35. 4pp. (NSA 10: 4786).

1956. The determination of alpha and beta activities in mussels and other invertebrates. Industrial Group, Risley, report IGO-AM/W-34. 2pp. (NSA 10: 4785).

1959. The radiochemical determination of strontium 89 and strontium 90 in vegetation, fish and seaweed ash. Industrial Group, Risley, report IGO-AM/W-185. 16pp.

1960. Analytical methods for the determination of caesium-137, strontium-89 and -90 in rain, lake and sea water. Production Group, Windscale, report PG-155(W)(Rev. ed.). 19pp.

1961. Analytical method for the determination of cerium-144 in vegetation, seaweed, fish flesh, and natural waters. Production Group, Windscale, report PG-203(W). 10pp. (NSA 15: 23451).

UK Department of Energy. 1975. Nuclear safety and the environment. Evidence by the Department of Energy to the Royal Commission on Environmental Pollution in connection with its study of radiological safety. London. 38pp. (NSA 33: 14836).

UK Department of Scientific Industrial Research. 1956. Radioactive tracers in the Thames Estuary. Hydraulic Research Station, London, Res. Paper No. 20.

UK Institute of Cancer Research. 1970. Biology and ecology of polonium and radiolead. Abstrs. Proc. Meeting held on 30th April and 1st May 1970. Sutton, England. 38pp. (CONF-700412-(absts.)) (NSA 24: 48377).

UK Ministry of Agriculture, Fisheries, and Food. 1968. Annual report of the Director of Fishery Research. Lowestoft. 135pp. (NSA 22: 45545) (Incl. radionuclides).

United Kingdom Ministry of Housing and Local Government (London); Secretary of State for Scotland, Scottish Development Department (Edinburgh); and Ministry of Health and Local Government (Belfast). 1963. Radioactivity in drinking water in the United Kingdom, 1962 results. 24pp. (NP-13388) (NSA 18: 4023).

United Nations. 1956. *Proc. Intern. Conf. on the Peaceful Uses of Atomic Energy*. Geneva. (Pertinent papers are listed individually by authors; applies as well to 2nd, 3rd, and 4th conferences).

United Nations. 1958, 1962, 1964, 1966, 1969, 1972. Report of the United Nations Scientific Committee on the Effects of Atomic Radiation. New York. vp. (A/3838, A/5216, A/5814, A/6314, A/7613, E.72.IX.17, 18) (Reports contain extensive data on environmental radiation and radioactivity, dosimetry, radioanalytical considerations, health physics, biological transport of radionuclides, radiation effects, etc.).

U. S. Air Force. 1962. Nuclear radiation guide. Aerospace Medical Div., 6570th Aerospace Medical Research Lab., Wright-Patterson Air Force Base, O., report MRL-TDR-62-61. 233pp. (NSA 17: 29120) (Incl. items related to ecology).

U. S. Army Engineer Nuclear Cratering Group. 1964. Excavation with nuclear explosions. Isthmian Canal Studies, 1964. Annex III, App.1, Incl. A. Livermore, Calif., U. S. AEC report PNE-2001. 119pp. (NSA 20: 1984).

U. S. Atomic Energy Commission.

1953. *A Conf. on the Use of Isotopes in Plant and Animal Research*. (Kansas State Coll., July 12, 13, qnd 14, 1952). Report TID-5098. iii, 272pp. (NSA 8-71).

1956. *A Conf. on Radioactive Isotopes in Agriculture*. (Michigan State Univ., January 12, 13, and 14, 1956). Report TID-7512. 420pp. (NSA 13: 4804).

1957. *Proc. Inter-American Symposium on the Peaceful Applications of Nuclear Energy*. (Brookhaven National Lab., May 13-17, 1957). Report TID-7554. 2 books, iii, 363; iv, 250pp. (NSA 13: 8500)

1961, 1962, 1964. Atmospheric radioactivity and fallout research. Report TID-12616, -12616(Rev.1), -12616(Rev.2). 76, 98, 132pp. (NSA 15: 23768; 19: 13631).

1961. Atomic energy research in the life and physical sciences. v, 175pp.

1962. Research on radiation in the environment. pp. 311-352. In: *Fundamental Nuclear Energy Research*. Report TID-18468. (NSA 17: 19746).

1963. Survey of environmental radioactivity. Report No. 1. Ames Lab., Ames, Ia., report TID-20369. (NSA 18: 16205) (At reactor site).

1964. General construction plan. Isthmian Canal Studies, 1964. Annex III, App. 1, Incl. D. Nevada Operations Office, report PNE-2004. 55pp. (NSA 20: 1986).

U. S. Atomic Energy Commission. (Continued).

1965. Environmental concentrations of radioactive materials near the Paducah Plant. Report for the first half of 1965. Report TID-22172. 9pp. (NSA 19: 44301).

1969. Nuclear power and the environment. 30pp.

1969. Terrestrial ecology. pp. 141-164. In: *Fundamental Nuclear Energy* Research - 1968.

1969. *Biological Implications of the Nuclear Age*. (Proc. Symp., Livermore, Calif., March 5-7, 1969). Symp. Ser. No. 16, report CONF-690303. 334pp. (NSA 24: 9796).

1969. Operating history of U. S. nuclear power reactors. Report TID-25537. 88pp. (NSA 49995).

1969. Questions and answers to the Colorado Committee for Environmental Information. Nevada Operations Office, report PNE-G-48. 80pp. (NSA 23: 47663).

1969. Technical discussions of offsite safety programs for underground nuclear detonations. Nevada Operations Office, report NVO-40(Rev.2). 311 pp. (NSA 23: 45426).

1969. Underground nuclear testing. Report TID-25180. 59pp. (NSA 24: 4397).

1970. Summaries of USAEC environmental research and development. Report TID-4065. v, 277pp. (NSA 24: 46593).

1970. *Symp. on Engineering with Nuclear Explosives*. (Las Vegas, Nevada, January 14-16, 1970, Proc.). Report CONF-70010. 2 vols., 1785pp. (NSA 24: 41468).

1970. Dresden Nuclear Power Station, Unit 3. Detailed statement on the environmental considerations. Division of Reactor Licensing. DOCKET-50249-52. 184pp. (NSA 25: 10146).

1970. Indian Point Nuclear Generating Unit No. 2. Detailed statement on environmental considerations. Division of Reactor Licensing. DOCKET-50247-45, -50251-34. 212pp. (NSA 25: 8008).

1970. North Anna Power Station, Units 1 and 2. Detailed statement on the environmental considerations. Division of Reactor Licensing. DOCKET-50338-26, -50339-25. 43pp. (NSA 25: 3841).

1970. Safeguards glossary. Report WASH-1162. 22pp. (NSA 25: 18626).

1970. Trojan Nuclear Plant. Detailed statement of the environmental considerations. Division of Reactor Licensing. DOCKET-50344-18. 106pp. (NSA 25: 3843).

1970. Project Wagon Wheel: preliminary site evaluation report. Report NVO-55 (PNE-WW-2). 31pp. (NSA 26: 19187).

U. S. Atomic Energy Commission. (Continued).

1971. Calvert Cliffs Nuclear Power Plant, Units 1 and 2. Draft detailed statement on environmental considerations. Division or Reactor Licensing. DOCKET-50317-24, -50318-24. 434pp. (NSA 25: 25933).

1971. Enrico Fermi Atomic Power Plant, Unit 2. Final detailed statement on the environmental considerations. Division of Radiological and Environmental Protection. DOCKET-50341-30. 74pp. (NSA 25: 41347).

1971. Forecast of growth of nuclear power. Report WASH-1139. 187pp. (NSA 25: 43994).

1971. Maine Yankee Atomic Power Station. Draft detailed statement on the environmental considerations. Division of Reactor Licensing. DOCKET-50309-32. 22pp. (NSA 25: 25931).

1971. Palisades Plant. Detailed statement on the environmental considerations. Division of Reactor Licensing. DOCKET-50255-38. 189pp. (NSA 25: 50978).

1971. Point Beach Nuclear Power Plant, Unit. No. 1. Detailed statement on the environmental considerations. Division of Radiological and Environmental Protection. DOCKET-50301-32. 444pp. (NSA 25: 34290).

1971. Quad Cities Station, Units 1 and 2. Final detailed statement on environmental considerations. Division of Radiological and Environmental Protection. DOCKET-50254-42, -50265-38. 108pp. (NSA 25: 44040).

1971. Reviews insure safety of nuclear operations. Nevada Operations Office, report NVO-48(Rev.1). 3pp. + 11 encl.

1971. Science information available from the Atomic Energy Commission. Report TID-4550. vi, 28pp.

1971. Supplement to the environmental statement: radioactive waste repository, Lyons, Kansas. Report TID-25727(Suppl.). 103pp. (NSA 25: 48605).

1971. Surry Power Station, Units 1 and 2. Draft detailed statement on environmental considerations. Division of Reactor Licensing. DOCKET-50280-37, -50281-37. 101pp. (NSA 25: 25928).

1971. Vermont Yankee Nuclear Power Station. Draft detailed statement on the environmental considerations. Division of Reactor Licensing. DOCKET-50271-39. 205pp. (NSA 25: 20645).

1971. Wm. H. Zimmer Nuclear Power Station, Unit 1. Draft detailed statement on environmental considerations. Division of Radiological and Environmental Protection. DOCKET-50358-21. 25pp. (NSA 25: 46214).

1971. Elk River reactor dismantling. Environmental statement. Report WASH-1516(Dr.). 43pp. (NSA 26: 17078).

1971. Growth of nuclear power, 1972-1985. Report WASH-1139(Rev.1). 18pp. (NSA 26: 55267).

U. S. Atomic Energy Commission. (Continued).

1971. Report on releases of radioactivity from power reactors in effluents during 1970. --- 1971. Report WASH-1198, -1198(1971). 9, 12pp. (NSA 26: 14448; 27: 16646).

1971. Underground nuclear test programs. Nevada Test Site (tests of one megaton or less). Environmental statement, Report WASH-1501, -1504. 88, 89pp. (NSA 26: 17765; 27: 17275).

1971. Project Cannikin. Report TID-26311. 42pp. (NSA 28: 5504).

1972. Project Cannikin D+30 day report. Nevada Operations Office, report NVO-108. iii, 28pp.

1972. Elk River Reactor dismantling, Elk River, Minnesota. Environmental statement. Report WASH-1516. 123pp. (NSA 26: 35548).

1972. Environmental and Ecological Forum, 1970-1971. Report TID-25857. 194pp. (NSA 26: 52545).

1972. Fort Calhoun Station, Unit 1. Final environmental statement. DOCKET-50285-71. 272pp. (NSA 26: 55301).

1972. Guide for submission of information on costs and benefits of environmentally related alternative designs for defined classes of completed nuclear facilities. Report TID-26167. 47pp. (NSA 26: 60734).

1972. Indian Point Nuclear Generating Unit 2. Final environmental statement, vol. 2. DOCKET-50247-83. 291pp. (NSA 26: 58205).

1972. Land acquisition, Rocky Flats Plant, Colorado. Environmental statement. Report WASH-1518(Draft), -1518. 23, 49pp. (NSA 26: 25607, 31041).

1972. Leasing of AEC controlled uranium bearing lands, Colorado, Utah, New Mexico. Environmental statement. Report WASH-1523. 130pp. (NSA 26: 45572).

1972. Maine Yankee Atomic Power Station. Final environmental statement. DOCKET-50309-61. 231pp. (NSA 26: 47473).

1972. Oconee Nuclear Station, Units 1, 2, and 3. Environmental statement. DOCKET-50269-50. 379pp. (NSA 26: 27481).

1972. Palisades Plant. Final environmental statement. DOCKET-50255-92. 470pp. (NSA 26: 44850).

1972. Pilgrim Nuclear Power Station. Final environmental report. DOCKET-50293-70. 186pp. (NSA 26: 32866).

1972. Plutonium recovery facility, Rocky Flats Plant, Colorado. Environmental statement. Report WASH-1507. 127pp. (NSA 26: 28104).

1972. Point Beach Nuclear Plant, Units 1 and 2. Final environmental statement. DOCKET-50266-67. 188pp. (NSA 26: 35515).

U. S. Atomic Energy Commission. (Continued).

1972. Quad Cities Station, Units 1 and 2. Final environmental statement. DOCKET-50254-120. 351pp. (NSA 26: 55285).

1972. Rio Blanco gas stimulation project, Rio Blanco County, Colorado. Environmental statement. Report WASH-1519(Draft), PNE-RB-1; -1519, PNE-RB-25. 170, 259pp. (NSA 26: 17768, 40874).

1972. Shoreham Nuclear Power Station, Unit 1. Final environmental statement. DOCKET-50322-30. 555pp. (NSA 26: 60699).

1972. Status of Savannah River Plant deer herd. Savannah River Operations Office, report SRO-154. 19pp. (NSA 26: 36460).

1972. Surry Power Station, Unit 2. Final environmental statement. DOCKET-50281-73. 292pp. (NSA 26: 44870).

1972. Turkey Point Plant, Units 3 and 4. Final environmental statement. DOCKET-50250-84. (NSA 26: 47413).

1972. Wagon Wheel gas stimulation project, Sublette County, Wyoming. Environmental statement. Report WASH-1524(Draft), PNE-WW-3; -1524. 109, 208pp. (NSA 26: 17769, 33363).

1972. Edwin I. Hatch Nuclear Plant, Units 1 and 2. Final environmental statement. DOCKET-50321-54. (NSA 27: 2114).

1972. Environmental survey of the nuclear fuel cycle. Report WASH-1237. 339pp. (NSA 27: 9558).

1972. Environment survey of transportation of radioactive materials to and from nuclear power plants. Report WASH-1238. 124pp. (NSA 27: 23019).

1972. Fort St. Vrain Nuclear Generating Station. Final environmental statement. DOCKET-50267-51. 279pp. (NSA 27: 16678).

1972. Hanford No. 2 Nuclear Power Plant. Final environmental statement. DOCKET-50397-33. 279pp. (NSA 27: 9108).

1972. Joseph M. Farley Nuclear Plant, Units 1 and 2. Final environmental statement. DOCKET-50348-50. 260pp. (NSA 27: 19217).

1972. Kewaunee Nuclear Power Plant. Final environmental statement. DOCKET-50305-61. 357pp. (NSA 27: 11450).

1972. Midland Plant, Units 1 and 2. Final environmental statement. DOCKET-50329-34. 379pp. (NSA 27: 19195).

1972. Monticello Nuclear Generating Plant, Unit 1. Final environmental statement. DOCKET-50263-161. 294pp. (NSA 27: 4577).

1972. Research and development in progress: biomedical and environmental research program. Report TID-4060(2nd ed.). 369pp. (NSA 27: 9809).

1972. Research contracts in the life sciences. Report WASH-1200(72). 83 pp. (NSA 28: 2816).

U. S. Atomic Energy Commission. (Continued.

1972. S8G prototype, Kesserling Site, West Milton, New York. Environmental statement. Report WASH-1525. 265pp. (NSA 27: 19255).

1972. Three Mile Island Nuclear Station, Units 1 and 2. Final environmental statement. DOCKET-50289-80. 250pp. (NSA 27: 9078).

1973. Effects evaluation and documentation plan for Project Rio Blanco. Nevada Operations Office, report NVO-126. vii, 92pp.+ App.

1973. Environmental statement. Underground nuclear testing program, Nevada Test Site. Report WASH-1526. iv, 60pp.+ App.

1973. The potential radiological implications of nuclear facilities in the Upper Mississippi River Basin in the year 2000. Report WASH-1209. vp. (NSA 28: 8136).

1973. AEC Project Director's operation plans for Project Rio Blanco. Nevada Operations Office, report NVO-127; PNE-RB-44. 141pp. (NSA 28: 8110).

1973. Arkansas Nuclear One, Unit 1. Final environmental statement. DOCKET-50313-64. 369pp. (NSA 27: 21784).

1973. Bailly Generating Station, Nuclear I. Final environmental statement. DOCKET-50367-34. 336pp. (NSA 27: 21811).

1973. Beaver Valley Power Station, Unit 1. --- Unit 2. Final environmental statement. DOCKET-5050334-80, -50412-25. 301, 312pp. (NSA 28: 20477, 20486).

1973. Calvert Cliffs Nuclear Power Plant, Units 1 and 2. Final environ mental statement. DOCKET-50317-87. 322pp. (NSA 27: 29695).

1973. Cleanup summary report: Tatum Dome Test Site, Mississippi. Nevada Operations Office, report NVO-129. 53pp. (NSA 28: 2611).

1973. Continuing environmental challenge. pp. 45-64. In: _1972 Atomic Energy Programs: Operating and Developmental Functions_. Report TID-26277. (NSA 28: 7622).

1973. Chrystal River Nuclear Generating Plant, Unit 3. Final environmental statement. DOCKET-50302-75. 404pp. (NSA 28: 10397).

1973. Davis-Besse Nuclear Power Station. Final environmental statement. DOCKET-50346-41. 284pp. (NSA 27: 21806).

1973. Diablo Canyon Nuclear Power Plant, Units 1 and 2. Final environmental statement. DOCKET-50275-44. 508pp. (NSA 28: 10365).

1973. Donald C. Cook Nuclear Plant, Units 1 and 2. Final environmental statement. DOCKET-50315-129. 429pp. (NSA 28: 20471).

1973. Duane Arnold Energy Center. Final environmental statement. DOCKET-50331-87. 333pp. (NSA 27: 21795).

U. S. Atomic Energy Commission. (Continued).

1973. Environmental monitoring at major U. S. Atomic Energy Commission Sites. Report WASH-1259. 1195pp. (NSA 28: 27236).

1973. Forked River Nuclear Generating Station Unit 1. Final environmental statement. DOCKET-50363-47. 252pp. (NSA 27: 21809).

1973. Grand Gulf Nuclear Station, Units 1 and 2. Final environmental statement. DOCKET-50416-42. 481pp. (NSA 28: 26738).

1973. Haddam Neck Plant, Unit 1. Final environmental statement. DOCKET-50213-228. 332pp. (NSA 28: 29285).

1973. Highland Uranium Mill. Final environmental statement. DOCKET-408102-7. 120pp. (NSA 27: 27369).

1973. Measuring and monitoring of radioactivity in the environs of nuclear power plants. Regulation REG/G-4.1(1-73). 4pp. (NSA 27: 19813).

1973. Nine Mile Point Nuclear Station, Unit 2. Final environmental statement. DOCKET-50410-35. 966pp. (NSA 28: 12446).

1973. North Anna Power Station, Units 1, 2, 3, and 4. Final environmental statement. DOCKET-50338-60. 471pp. (NSA 27: 29715).

1973. Peach Bottom Atomic Power Station, Units 2 and 3. Final environmental statement. DOCKET-50277-83. 419pp. (NSA 27: 29655).

1973. Prairie Island Nuclear Generating Plant, Units 1 and 2. Final environmental statement. DOCKET-50282-118. 434pp. (NSA 28: 12329).

1973. Proc. AEC Pollution Control Conf. (Oak Ridge, Tennessee, October 25-27, 1972). Report CONF-721030. 557pp. (NSA 28: 2563).

1973. Project Rio Blanco. Project Director's preliminary report: D + 2 days. Nevada Operations Office, report PNE-RB-45. 40pp. (NSA 28: 8113).

1973. Rio Blanco gas stimulation project, Rio Blanco County, Colorado. Environmental statement. Report WASH-1519(Add.), PNE-RB-25. 173pp. (NSA 28: 2615).

1973. St. Lucie Plant, Unit 1. Final environmental statement. DOCKET-50335-63. 284pp. (NSA 28: 12404).

1973. San Onofre Nuclear Generating Station, Units 2 and 3. Final environmental statement. DOCKET-50361-37. 459pp. (NSA 27: 27011).

1973. Summaries of USAEC environmental research and development. Report TID-4065-R-1. 276pp. (NSA 28: 362).

1973. Susquehanna Steam Electric Station, Units 1 and 2. Final environmental statement. DOCKET-50387-40. 322pp. (NSA 28: 12433).

1973. Test and research reactor fuel element fabrication plant. Environmental statement. DOCKET-701319-4. 135pp. (NSA 28: 27335).

U. S. Atomic Energy Commission. (Continued).

1973. Trojan Nuclear Plant. Final environmental report. DOCKET-50344-66. 379pp. (NSA 28: 20482).

1973. Underground nuclear testing program, Nevada Test Site. Environmental statement. Report WASH-1526. 135pp. (NSA 28: 5531).

1973. Virgil C. Summer Nuclear Station, Unit 1. Final environmental statement. DOCKET-50395-36. 302pp. (NSA 27: 19224).

1973. Enewetak radiological survey. Nevada Operations Office, report NVO-140(vol.1), (vol.2), (vol.3). 747, 617, 687pp. (NSA 30: 24007, 24008, 24009). (Eniwetok Proving Grounds vicinity).

1973. Environmental impact of electrical power generation: nuclear and fossil. A minicourse for secondary schools and adult education. Report WASH-1261. 96pp. (NSA 29: 17911).

1973. General environmental siting criteria for nuclear power plants: topics and bases. Report WASH-1294(Draft). 135pp. (NSA 29: 20895).

1973. Government reactors: monitoring. pp. 273-309. In: Tritium Control Technology. Report WASH-1269. (NSA 30: 8795).

1973. Operating history of U. S. nuclear power reactors. Report WASH-1203-73. 83pp. (NSA 29: 31153).

1973. Preparation of environmental reports for nuclear power plants. Regulatory Guide 4.2. Report TID-26473. 76pp. (NSA 29: 9609).

1973. Summary report on the assessment of environmental radioactivity in the vicinity of the Shippingport Power Station. Report WASH-1256. 80pp. (NSA 29: 5108).

1973. Teacher's guide for the environmental impact of electrical power generation: nuclear and fossil. A minicourse for secondary schools and adult education. Report WASH-1262. 165pp. (NSA 29: 17912).

1974. Environmental monitoring report, 1973. Report TID-26578. 25pp. (NSA 30: 9530) (Pinellas Plant, Texas).

1974. Environmental survey of the uranium fuel cycle. Report WASH-1248. 277pp. (NSA 30: 23748).

1974. High-level radioactive waste management alternatives. Report WASH-1297. 94pp. (NSA 29: 30190) (Incl. seabeds and ice caps).

1974. Nuclear power growth 1974-2000. Report WASH-1139-74. 78pp. (NSA 30: 14676).

1974. Proc. 2nd AEC Environmental Protection Conf. (Albuquerque, New Mexico, April 16-19, 1974). Report WASH-1332(74)(Vol.1, Vol.2). 567, 590pp. (NSA 30: 32018, 32035).

U. S. Atomic Energy Commission. (Continued).
1974. Title listing of power reactor DOCKET information. Report PRDI-74-1, -2, -10. 50, 49, 59pp. (NSA 29: 17882, 20385; 30: 34438).

1974. USSR reports on natural and fallout radioactivity. (Transl. of Russian articles). Report AEC-tr-7570. (NSA 30: 21063).

1974. Plutonium information meeting, Los Alamos, New Mexico, January 4-5, 1974. Advisory Committee on Reactor Safeguards, Washingotn. Report CONF-740115. 108pp. (NSA 31: 3357).

1975. Summary of radioactivity released in effluents from nuclear power plants during 1973. 53pp. (PB-239191) (NSA 33: 14830).

U. S. atomic Energy Commission and U. S. Public Health Service. 1956. *Sanitary Engineering Aspects of the Atomic Energy Industry*. (A Seminar held at the R. A. Taft Sanitary Engineering Center, Cincinnati, Ohio, December 6-9, 1955). U. S. AEC report TID-7517(Pts. Ia, Ib, II(del.)). vp. (NSA 11: 1026; 13: 13037).

U. S. Congress. 1957. *The Nature of Radioactive Fallout and Its Effects on Man*. (Hearings before the Special Subcommittee on Radiation, May 27-June 4, 1957). Joint Committee on Atomic Energy, Washington. 2 vols. + index + summary, xiv, 2235pp.

U. S. Congress. 1959. *Fallout from Nuclear Weapons Tests*. (Hearings before the Special Subcommittee on Radiation, May 5-8, 1959). Joint Committee on Atomic Energy. 4 vols. + summary-analysis, xxi, 2735pp. (NSA 14: 6109, 7275, 9300).

U. S. Congress. 1959. *Biological and Environmental Effects of Nuclear Warfare*. (Hearings before the Special Subcommittee, June 22-26, 1959). Joint Committee on Atomic Energy. vi, 1014pp. (Incl. separate summary-analysis).

U. S. Congress. 1961. *Applications of Radioisotopes and Radiation in the Life Sciences*. (Hearings before the Special Subcommittee on Research, Development and Radiation, March 27-30, 1961). Joint Committee on Atomic Energy. ix, 535pp. (Incl. separate summary-analysis). (NSA 15: 27309).

U. S. Congress. 1963. *Fallout, Radiation Standards, and Countermeasures*. (Hearings before the Special Subcommittee on Research, Development, and Radiation, August 20-27, 1963). Joint Committee on Atomic Energy. 2 parts, xi, 1298pp.

U. S. Congress. 1970. *Environmental Effects of Producing Electric Power*. (Hearings before the Joint Committee on Atomic Energy). Part 2, 2 vols., vii, pp. 1109-2708.

U. S. Congress. 1971. *AEC Licensing Procedure and Related Legislation*. (Hearings before the Subcommittee on Legislation, June 22, 23; July 13, 14, 1971). Joint Committee on Atomic Energy. 4 pts., 2131pp. (NSA 26: 8537, 8538, 8539, 8540).

U. S. Congress. 1971. *Nuclear Power and Related Energy Problems, 1968 Through 1970*. (Report of the Joint Committee on Atomic Energy, December 1971). 1120pp. (NSA 26: 22776).

U. S. Congress. 1972. *Interim Nuclear Licensing*. (Hearings before the Subcommittee on Merchant Marine and Fisheries, House of Representatives, March 22, 24, 27, 1972). Joint Committee on Atomic Energy. 365pp. (Related to H.R. 13752; Ser. No. 92-18) (NSA 26: 32910).

U. S. Congress. 1972. *Selected Materials on the Calvert Cliffs Decision: Its Origin, and Aftermath*. Joint Committee on Atomic Energy. 681pp. (NSA 26: 27528).

U. S. Congress. 1972. *Background Report on Power Plant Siting*. (For Use of the Committee on Commerce, Senate). Joint Committee on Atomic Energy. 80pp. (NSA 27: 19241).

U. S. Council on Environmental Quality. 1973. *Energy and the Environment: Electric Power*. Washington. 62pp. (NSA 28: 32108).

U. S. Delegation to the United Nations. 1959. United States at Geneva, 1958. (2nd Intern. Conf. on the Peaceful Uses of Atomic Energy). U. S. Atomic Energy Commission, Washington. 194pp. (NSA 14: 1191).

U. S. Department of Agriculture. 1962. Protection of food and agriculture against nuclear attack. (A guide for agricultural leaders). Agricultural Research Service, Washington, Agr. Handbook No. 234. 41pp.

U. S. Department of Agriculture. 1962. Questions and answers about screw-worm eradication. Agricultural Research Service, Beltsville, Md., report ARS 91-38.

U. S. Department of Agriculture. 1963. USDA radiological monitoring handbook. Agricultural Research Service, Beltsville, Md., Agr. Handbook No. 246. 47 pp. (NSA 18: 14109) (Describes procedures for monitoring radioactive fall-out on agricultural and forest lands and in water).

U. S. Department of Agriculture. 1964. The potential role of the sterility method for insect population control with special reference to combining this method with conventional methods. Agricultural Research Service, report ARS-33-98. 54pp.

U. S. Department of the Interior. 1972. United States energy. (A summary review). Washington. 50pp. (Incl. environmental protection).

U. S. Department of the Interior. 1975. Final environmental statement of the approval of the Department of the Interior of a lease of the Ute Mountain Ute Tribal Lands for uranium exploration and possible mining. Bureau of Indian Affairs, Albuquerque, N. Mex., report FES-75-94. 164pp. (NSA 33: 9248).

U. S. Energy Research and Development Administration. 1975. Environmental monitoring at major U. S. Energy Research and Development Administration contractor sites: calender year 1974. Washington, report ERDA-54(Vol.1)(74), -54(Vol.2)(74). 797, 577pp. (NSA 33: 604, 25897).

U. S. Energy Research and Development Administration. 1975. Final environmental statement, waste management operations, Hanford Reservation, Richland, Washington. Washington, report ERDA-1538(Vol.1), -1538(Vol.2). 593, 428pp. (NSA 33: 17893, 17894).

U. S. Energy Research and Development Administration. 1975. Oak Ridge Reservation land-use plan. Oak Ridge Operations Office, report ORO-748. 70pp. (NSA 33: 9245).

U. S. Energy Research and Development Administration. 1976. Environmental monitoring at major U. S. ERDA sites. Calender year 1974. Supplement. Report ERDA-54(Suppl.)(74). 102pp. (NSA 33: 23336).

U. S. Energy Research and Development Administration. 1977. Nuclear reactors built, being built, or planned in the United States as of June 30, 1977. Washington, report TID-8200-R36. 48pp. (Reports revised semiannually).

U. S. Environmental Protection Agency.
1971. Environmental surveillance for Peewee 1, experimental plans II and III, November 21 and December 4, 1968. Final report. Western Environmental Research Lab., Las Vegas, Nev., U. S. AEC report SWRHL-87-r. 33pp. (NSA 25: 60196).

1971. Final report of off-site surveillance for the NRX-A6 reactor test series. Western Environmnetal Research Lab., Las Vegas, Nev., U. S. AEC report SWRHL-78-r. 46pp. (NSA 25: 60195).

1971. Off-site surveillance for Project Schooner, December 8, 1968. Final report. Western Environmental Research Lab., Las Vegas, Nev., U. S. AEC report PNE-524. 91pp. (NSA 25: 57574).

1972. Final report of off-site surveillance for the Baneberry event, December 18, 1970. Western Environmental Research Lab., Las Vegas, Nev., U. S. AEC report SWRHL-107-r. 171pp. (NSA 26: 55301).

1972. Off-site surveillance activities of the Southwestern Radiological Research Laboratory from January through June 1968. ---Janauary-June 1970. Western Radiological Research Lab., Las Vegas, Nev., U. S. AEC report SWRHL-81-r, -108-r. 107, 68pp. (NSA 26: 25564, 56091).

1972. Environmental monitoring report for the Nevada Test Site, January-December 1972. National Environmental Research Center, Las Vegas, Nev., U. S. AEC report NERC-LV-539-1. 91pp. (NSA 27: 292).

1972. Environmental radioactivity surveillance guide. Washington, report ORP/SID-72-2. 31pp. (NSA 27: 17244).

1972. Off-site environmental surveillance for the XE Prime Nuclear Rocket Engine Test Series. National Environmental Research Center, Las Vegas, Nev Nev., U. S. AEC report NERC-LV-539-13. 44pp. (NSA 27: 5241).

1972. Proc. South Conf. on Environmental Radiation Protection from Nuclear Power Plants. (April 21-22, 1971). Washington, report ORP/SID-72-4. 240 pp. (NSA 27: 2540).

U. S. Environmental Protection Agency. (Continued).
1973. Environmental monitoring report for the Nevada Test Site and other test areas used for underground nuclear detonations, January-December 1972. National Environmental Research Center, Las Vegas, Nev., U. S. AEC report NERC-LV-539-23. 143pp. (NSA 28: 10858).

1973. Assessment of environmental radioactivity in the vicinity of Shippingport Atomic Power Station. Report EPA-520/5-73-5. 35pp. (NSA 30: 18616).

1973. Environmental analysis of the uranium fuel cycle. --- Part II. Nuclear power reactors. Office of Radiation Programs, Washington, report EPA-520-9-73-3-D, -C. 150, 169pp. (NSA 30: 223, 31354).

1973. Reviewing environmental impact statements: power plant cooling systems, engineering aspects. Washington, report EPA-660-2-73-16. 93pp. (NSA 29: 31355).

1974. Environmental monitoring report for the Nevada Test Site and other test areas used for underground nuclear detonations, January through December 1973. National Environmental Research Center, Las Vegas, Nev., U. S. AEC report NERC-LV-539-31. 110pp. (NSA 30: 3145).

1974. Environmental radioactivity laboratory intercomparison studies programs, 1973-1974. National Environmental Research Center, Las Vegas, Nev., report EPA-680-4-73-1-b. 27pp. (NSA 30: 6483, 33: 22792).

1974. Environmental radiation dose committment: an application to the nuclear power industry. Washington, report EPA-520-4-73-2. 88pp. (NSA 30: 3175).

1974. Final report of the radiological surveillance program for the Project Gasbuggy production test, May 15, 1973 to November 6, 1973. National Environmental Research Center, Las Vegas, Nev., U. S. AEC report NERC-LV-539-30. 63pp. (NSA 30: 23982) (Vegetation).

1974. Off-site surveillance activities of the National Environmental Research Center from July through December 1970. U. S. AEC report NERC-LV-539-17. 153pp. (NSA 29: 29322).

1975. Environmental monitoring report for the Nevada Test Site and other test areas used for underground nuclear detonations, January through December 1974. National Environmental Research Center, Las Vegas, Nev., U. S. AEC report NERC-LV-539-39. 129pp. (NSA 32: 6459).

1975. Proceedings of public hearings: plutonium and other transuranium elements. Vol. 2. (Proc. Hearings in Denver, Colorado, January 10, 1975). Washington, report OSD/CSD-75-1(Vol.2). 550pp. (NSA 33: 2993).

U. S. Federal Radiation Council. Washington. Staff Reports.
1960, 1961. Background material for the development of radiation protection standards. No. 1, No. 2. iii, 39; iii, 19pp.

1962. Health implications of fallout from nuclear weapons testing through 1961. No. 3.

U. S. Federal Radiation Council. Washington. Staff Reports. (Continued). 1963. Estimates and evaluation of fallout in the United States from nuclear weapons testing conducted through 1962. No. 4. iii, 31pp. (Incl. glossary of terms; Staff Report No. 3 is appended).

1964. Background material for the development of radiation protection standards. No. 5. iii, 29pp.

1964. Revised fallout estimates for 1964-1965 and verification of 1963 predictions. No. 6. 29pp.

1965. Background material for the development of radiation protection standards, protection action guides for strontium-89, strontium-90 and cesium-137. No. 7. iv, 44pp. + errata.

1967. Guidance for the control of radiation hazards in uranium mining. No. 8. iv, 60pp.

U. S. Food and Drug Administration, Department of Health Education and Welfare. 1974. _6th Annual National Conf. Control: New Challenges_. (April 28-May 2, 1974, San Antonio, Texas). Bureau of Radiological Health, Rockville, Md., report DHEW/FDA-75-8010. 299pp. (NSA 31: 19561).

U. S. Geological Survey, Department of the Interior. Geological survey research, 1970. Prof. Paper 700-A. 426pp. (NSA 25: 34741) (Incl. radionuclides).

U. S. National Aeronautics and Space Administration. 1961. First planning conference on biomedical experiments in extra-terrestrial environments. (Washington, D. C., June 20, 1960). Washington, report NASA-TN-D-781. 86pp. (NSA 15: 8432).

U. S. National Field Investigations Center-Denver. 1973. Reconnaissance study of radiochemical pollution from phosphate rock mining and milling. Denver, Colo. 106pp. (PB-241242) (NSA 33: 23287).

U. S. Nuclear Regulatory Commission. 1975. Environmental survey of transportation of radioactive materials to and from nuclear power plants. Suppl. I. Environmental survey. Washington, report NUREG-75/038. 16pp. (NSA 33: 23875).

U. S. Nuclear Regulatory Commission. 1975. General site suitability criteria for nuclear power stations. Washington, Regulatory Guide REG/G-4.7(Rev.1). 32pp. (NSA 33: 17894).

U. S. Nuclear Regulatory Commission. 1975. Preparation of environmental reports for commercial uranium enrichment facilities. Washington, Regulatory Guide REG/G-4.9(Rev.1)(10-75). 32pp. (NSA 33: 14831).

U. S. Office of Civil Defense. 1968. _Proc. Symp. on Postattack Recovery from Nuclear War_. (Monroe, Virginia, November 6-9, 1967). Washington. vii, 434pp.

U. S. Office of Science and Technology. 1968. Considerations affecting steam power plant site selection. U. S. AEC report TID-24936. 145pp. (NSA 23: 15658).

U. S. Office of Science and Technology. 1970. Electric power and the environment. Washington. xi, 71pp. (NP-19170) (NSA 26: 16920).

U. S. National Committee of the World Energy Conference. 1971. _Proc. National Energy Forum_. (Washington, D. C., September 23-24, 1971). 229pp. (CONF-710982) (NSA 28: 29407).

U. S. Atlantic-Pacific Interoceanic Canal Study Commission. 1970. _Interoceanic Canal Studies 1970_. Washington. vp. 129pp. + 5 annexes, 17 appendices. (Concerns feasibility study for nuclear, conventional, combination excavation of a sea-level canal in Central America, especially in Panama and Colombia; extensive environmental data and estimates, incl. radiological considerations and other ecological considerations).

U. S. Public Health Service, Department of Health, Education, and Welfare. 1957. Determination of gross radioactivity in water, silt, and biological materials. R. A. Taft Sanitary Engrg. Cen., Cincinnati, O.

1958, 1959, 1961. National Water Quality Network annual compilation of data, October 1, 1957-September 30, 1958 (w/Suppl. 1, Statistical summary of selected data). --- 1958-1959. --- 1959-1960. R. A. Taft Sanitary Engineering Center, Cincinnati, O., Publ. 663. iv, 239; iv 164(suppl.1): x, 323; ix, 424pp.

1959. National Water Quality Control Network. pp. 167-168. In: _Fallout from Nuclear Weapons Tests_. (See U. S. Congress, 1959).

1959. Course manual. Radioactive pollutants in water. R. A. Taft Sanitary Engineering Center, Cincinnati, O. vp.

1959. Course manual. Sanitary engineering aspects of nuclear energy. R. A. Taft Sanitary Engineering Center, Cincinnati, O. vp.

1962. Process and waste characteristics at selected uranium mills. R. A. Taft Sanitary Engeneering Center, Cincinnati, O., Tech. Rept. W-62-17. v, 94pp. (Comments on environment).

1962. Radionuclide analysis of environmental samples. R. A. Taft Sanitary Engineering Center, Cincinnati, O., Tech. Rept. R-59-6.

1964. Radioactive pollutants in the environment. R. A. Taft Sanitary Engineering Center, Cincinnati, O. 325pp. (NP-14254) (NSA 18: 41648).

1964. Analysis of radionuclides in water. R. A. Taft Sanitary Engineering Center, Cincinnati, O. 193pp. (NP-14255) (NSA 18: 39531).

1964. Measurement of airborne radioactivity. R. A. Taft Sanitary Engineering Center, Cincinnati, O. 102pp. (NP-14258) (NSA 18: 39532).

1966. Routine surveillance of radioactivity around nuclear facilities. Interlaboratory Technical Advisory Committee Report No. 1. Washington. 28pp. (NSA 21: 20176).

1967. Guide for environmental surveillance around nuclear facilities. National Center for Radiological Health, Rockville, Md. (NSA 25: 15335).

U. S. Public Health Service. (Continued).

1968. Environmental contamination by radioactive substances. Environmental Radiation Exposure Advisory Committee. Washington. 24pp. (NSA 22: 45184).

1968. Final report of off-site surveillance for the NRX-A5 test series. Southwestern Radiological Health Lab., Las Vegas, Nev., U. S. AEC report SWRHL-32r. v, 45pp.

1968. Guidelines to radiological health. (Lectures presented at the International Conf. at McGill University, Montreal, Canada, August 17-21, 1967 1967). Bureau of Radiological Health, Rockville, Md., report PHS-999-RH-33. 173pp. (CONF-670848) (NSA 23: 50452).

1968. Nimbus-B/SNAP-19 launch: May 18, 1968, off-site radiological surveillance. Southwestern Radiological Health Lab., Las Vegas, Nev., U. S. AEC report SWRHL-50-r. 37pp. (NSA 23: 13872).

1968. Off-site surveillance for Project Longshot. Final report. Southwestern Radiological Health Lab., U. S. AEC report SWRHL-6-r. 8pp. (NSA 23: 13896).

1968. Off-site surveillance activities of the Southwestern Radiological Health Laboratory from July-December 1964. Southwestern Radiological Health Lab., U. S. AEC report SWRHL-22-r. 43pp. (NSA 23: 15975).

1969. Evaluation of ^{222}Rn near uranium tailings piles. Final report. Bureau of Radiological Health, report DER-69-1. 74pp. (NSA 24: 31961).

1969. Off-site radiological surveillance program for Project Gasbuggy. Southwestern Radiological Health Lab., Las Vegas, Nev., U. S. AEC report PNE-G-46. 26pp. (NSA 23: 40878).

1969. Evaluation of radon 222 near uranium tailings piles. Bureau of Radiological Health, Rockville, Md. 67pp. (PB-188691) (NSA 30: 9538).

1970. Environmental surveillance for Project Gasbuggy production test phase. Southwestern Radiological Health Lab., Las Vegas, Nev., U. S. AEC report SWRHL-100-r; PNE-G-63. 39pp. (NSA 24: 41433).

1970. Final report of off-site surveillance for Project Palanquin. Southwestern Radioligical Health Lab., Las Vegas, Nev., U. S. AEC report PNE-910-F. 128pp. (NSA 24: 36445).

1970. Off-site surveillance activities of the Southwestern Radiological Health Laboratory from January through June 1967. Las Vegas, Nev., U. S. AEC report SWRHL-47-r. 121pp. (NSA 24: 50034).

1970. Project Gasbuggy, off-site radiological safety report. GB-2R Phase 1 program. Southwestern Radiological Health Lab., Las Vegas, Nev., U. S. AEC report SWRHL-104-r; PNE-G-65. 20pp. (NSA 24: 44127).

1971. Off-site surveillance activities of the Southwestern Radiological Health Laboratory from July-December 1968. --- July-December 1969. Las Vegas, Nev., U. S. AEC report SWRHL-83-r, -98-r. 121, 67pp. (NSA 25: 46826, 24113).

U. S. Public Health Service. (Continued).

1971. Report of off-site surveillance for Project Cabriolet. Southwestern Radiological Health Lab., Las Vegas, Nev., U. S. AEC report PNE-959. 70pp. (NSA 26: 9262).

nd. SWRHL 1969. (Annual report). Southwestern Radiological Health Lab., Las Vegas, Nev., U. S. AEC report. 53pp. (Contains data on ^{90}Sr in mule deer and bighorn sheep).

U. S. President's Science Advisory Committee. 1965. Restoring the quality of our environment. Environmental Pollution Panel. Washington. xii, 317 pp. (Contains statements on radioactivity).

U. S. Senate. 1960. Radiation Research in the Life Sciences. (Current Projects in the United States and Throughout the World; Information on the Magnitude, Organization, and Distribution of Programs). 176pp. (NSA 15: 8470).

University of California. 1972, 1973. Instrumentation for environmental monitoring: radiation. Lawrence Berkeley Lab., U. S. AEC report LBL-1. 4 vols. w/adds., vp. (NSA 27: 12005, 24630; 29: 20911, 26934; 30: 6780).

University of California. 1974. Environmental analysis and environmental monitoring for nuclear power generation. Berkeley. vp. (NP-20456) (NSA 32: 14274).

University of Illinois. 1961. Proc. 2nd Sanitary Engineering Conf. on the Radiological Aspects of Water Supplies. Champaign, Cir. 58(83); Engrg. Expt. Sta. Cir. 69. 80pp.

University of Illinois. 1969. Progress in Hydrology. Vol. I. New Developments in Hydrology. (Proc. 1st Intern. Seminar for Hydrology Professors held at Urbana, Illinois, July 13-25, 1969). 511pp. (CONF-690726-Vol.1) (NSA 25: 2358).

University of Michigan. nd. Biological and psychological effects of radiation in the planarian. Ann Arbor, Tech. Report No. 1, U. S. AEC report TID-18949. 22pp. (NSA 17: 28679).

University of Michigan. 1971. Donald C. Cook Nuclear Plant, Units 1 and 2. Benton Harbor Power Plant limnological studies. Part VIII. Winter operations, 1970-1971. Great Lakes Res. Div., Ann Arbor. 37pp. U. S. AEC DOCKET-50315-45, -50316-45. (NSA 25: 50995).

University of North Carolina. 1974. Proc. Symp. on Power Generation and Air Quality. (May 9-10, 1974, Raleigh, North Carolina). Chapel Hill. 119pp. (NSA 33: 25986) (Incl. ecological aspects).

University of Texas. 1958. Effects of ionizing radiations on growth responses, metabolism and electrical pattern of Avena coleptiles and earthworms. Austin, U. S. AEC report ORO-179. 11pp. (NSA 13: 1045).

University of Texas. 1968. Radioactivity transport in water. Interaction between flowing water and bed sediments. Environmental Health Engineering Research Lab., Austin, report EHE-69-3; CRWR-36; U. S. AEC report ORO-490-17. 165pp. (NSA 23: 24301).

University of Washington. Applied Fisheries Lab. (Lab. Radiation Biology). 1947. Radiobiological resurvey of Bikini Atoll during the summer of 1947. Seattle, U. S. AEC report UWFL-7. 62pp. (NSA 13: 11751).

1949. Bikini radiobiological resurvey of 1948. Seattle, U. S. AEC report UWFL-16. 51pp. (NSA 13: 11074).

1949. Eniwetok radiological resurvey July 1948. Seattle, U. S. AEC report UWFL-19. 38pp.

1950. Radiobiological survey of Bikini, Eniwetok, and Likiep Atolls; July-August 1949. Seattle, U. S. AEC report UWFL-23; AECD-3446. 145pp. (NSA 6: 5934).

1955. A radiological study of Rongelap Atoll, Marshall Islands, during 1954-1955. Seattle, U. S. AEC report UWFL-42. 75pp. (NSA 12: 18).

1955. Radiobiological resurvey of Rongelap and Ailinginae Atolls, Marshall Islands, October-November 1955. Seattle, U. S. AEC report UWFL-43. 91pp. (NSA 11: 9192).

1956. Program of the Applied Fisheries Laboratory, University of Washington, for the 1956 test series at Bikini and Eniwetok Atolls, Marshall Islands. Seattle, U. S. AEC report UWFL-45. 27pp. (NSA 11: 4250).

University of Wisconsin. 1970. Point Beach Nuclear Plant, Units 1 and 2. Environmental studies. Dept. of Botany, Milwaukee. 7pp. U. S. AEC DOCKET-50266-31, -50301-25. (NSA 25: 1832).

Unlu. M. Y. 1970. Uptake of mercury (^{203}Hg) by _Tapes decussatus_ L. Turkey Cekmece Nuclear Research Center, Istanbul. 27pp. (In Turkish) (NP-18429) (NSA 25: 8625).

Unlu, M. Y. 1974. Radioecology of the Kucuk Cekmece Lagoon. Final report, 1 November 1970-31 October 1973. International Atomic Energy Agency report IAEA-R-960-F. vp. (NSA 30: 24000).

Unruh, C. M. 1970. Nuclear power as a source of contamination on a global scale. Battelle Northwest Lab., U. S. AEC report BNWL-SA-3665. 16pp. (CONF-701205) (NSA 25: 18080).

Unruh, C. M. 1970. Present and projected sources of environmental radiation (the givens). Battelle Northwest Lab., U. S. AEC report BNWL-SA-3354. 13 pp. (For: 15th Ann. Meeting of the Health Physics Society, Chicago, Ill.). (CONF-700609-4) (NSA 27: 17684).

Unruh, C. M. 1972. Radiological safety aspects of the nuclear power program. Battelle Northwest Lab., U. S. AEC report BNWL-SA-4460. 46pp. (NSA 27: 19815) (Incl. environmental monitoring).

Unruh, C. M. 1974. Nuclear radiation: sources and impact. Battelle Northwest Lab., U. S. AEC report BNWL-SA-4962. 29pp. (NSA 30: 3138).

Unruh, C. M. 1975. Preparation of environmental impact statements: radiological considerations. Battelle Northwest Lab., U. S. AEC report BNWL-SA-5497. 44pp. (NSA 33: 23874).

Upadhya, M. D. and J. L. Brewbaker. 1966. Irradiation of mangoes for control of the mango seed weevil. Hawaii Farm Sci. 15: 6-7. (NSA 22: 52120).

Updegraff, D. M. 1969. Relationship of microorganisms to uranium deposits. Final report. Denver Research Inst., report DRI-2498, U. S. AEC report GJO-932-2. 73pp. (NSA 24: 7712).

Upton, A. C. 1969. _Radiation Injury: Effects, Principles, and Perspectives_. Univ. of Chicago Press, Chicago. 350pp.

Upton, A. C. 1974. Presentation of U.S.A. National Academy of Sciences report on the effects on populations of exposure to low levels of ionizing radiation. Part III. Somatic effects. pp. 43-50. In: _Proc. 3rd Intern. Congr. International Radiation Protection Association_. (CONF-730907-P1) (NSA 30: 27061).

Urbain, P. 1954. On the actual state of hydrological research through the radioactive tracer method. pp. 238-247. In: _Meeting of the Intern. Assoc. of Scientific Hydrology_. _Vol. 2_. (Rome). (Publ. 37).

Urbani, E. and S. Russo-Caia. 1964. Cytochemical and autoradiographic observation of nucleic acid metabolism in the oogenesis of _Dytiscus marginalis_. Rend. 1st sci. univ. Camerino 5(1): 19-50. (In Italian, English summary) (^{3}H, ^{14}C).

Uribe, R. L. 1965. Effect of gamma radiation on the longevity and morphollogy of _Sacodes pyralis_ Dyar. Colombia Inst. de Asuntos Nucleares, Bogota, report IAN-A-3. (In Spanish) (NSA 21: 1797).

Urquhart, G. M. 1968. Immune response of young animals to infection with x-irradiated and normal nematode larvae. pp. 35-42. In: _Isotopes and Radiation in Parasitology_. (See IAEA, 1968).

Urrutia, J. L. F. 1969. Aspects of industrial pollution. Rev. sanid. hig. publica. 43: 663-713. (In Spanish) (NSA 24: 46595).

Ursino, D. J. 1973. Effects of chronic internal β-radiation from photoassimilated $^{14}CO_2$ on the retention and distribution of ^{14}C in young white pine plants. Plant Physiol. 51: 954-959. (NSA 28: 21804).

Ursprung, H., J. Leone and L. Stein. 1968. Blastular arrest and chromosome abnormalities produced by x rays in two amphibians: _Rana pipiens_ and _Xenopus laevis_. J. Exptl. Zool. 168: 379-386. (NSA 23: 24745).

Usacev, S. nd. Contamination of the environment with ^{90}Sr and ^{137}Cs fallout. Komenskeho Univ., Bratislava, Czechoslovakia. (In Slovak) (INIS-mf-887, pp. 114-127) (NSA 29: 24189).

Usacev, S. 1973. Special methods of radioactive fallout measurement in Iraq. Acta fac. rerum natur. univ. Comeniannae phys. 13: 111-115. (NSA 31: 3377).

Usacev, S. and Z. Dobiosova. 1970. Calculation of ^{90}Sr content changes in atmospheric fallout and on the ground in west Slovakia up to the year 2000. Jaderna Energie 16: 420-421. (In Czech) (NSA 25: 21204).

Ushakova, A. P., et al. 1973. Evaluation of the internal population dose from global cesium-137 in a number of nationalities of the Far North of the Soviet Union. 8pp. (In Russian) (INIS-mf-1042) (NSA 30: 9555).

Ushioda, T. and T. Krose. 1956. Mulberry field contamination with the radioactivity caused by the explosion of the hydrogen bomb. pp. 791-793. In: Research in the Effects and Influences of the Nuclear Bomb Test Explosions. Vol. II. Japan Soc. for the Promotion of Science, Ueno, Tokyo.

Utah International, Inc. 1972, 1975. Shirley Basin Uranium Mill. Environmental statements, supplemental information. --- Environmental report to accompany source material license application. San Francisco, Calif. 104, 158pp. U. S. AEC DOCKET-406622-4, -7. (NSA 26: 50317; 33: 17172).

Uthe, J. F., F. A. J. Armstrong and M. P. Stainton. 1970. Mercury determination in fish samples by wet digestion and flameless atomic absorption spectrometry. F. Fish. Res. Bd. Can. 27(4): 805-811. (SFA 15: 12160).

Vacca, G. 1964. The principle natural and artificial radioactive substances encountered in the surroundings. Maroc Med. 43: 159-166. (NSA 19: 18217).

Vachaud, G., et al. 1974. Automatic in situ checking of water transfer in the unsaturated zone. pp. 251-270. In: Isotope and Radiation Techniques in Soil Physics and Irrigation Studies 1973. (See IAEA, 1974) (In French) (NSA 30: 15498) (Neutron probe).

Vachon, M., et al. 1964. Comparative radiosensitivity, with respect to γ radiation of certain species of scorpion. Compt. rend. 259(19): 3389-3391. (In French) (NSA 19: 8840).

Vadzis, D. R. and Z. K. Kalnina. 1973. ^{90}Sr distribution on water and bottom sediments of Latvian lakes. pp. 88-95. In: Radioecology of Water Organisms. II. (See Andrumaitis, et al, 1973) (In Russian) (NSA 30: 21114).

Vagn, M. nd. Dosimetry problems related to the thermoluminescence dating of ancient ceramics. pp. 868-882. In: Proc. 2nd Intern. Conf. on Luminescence Dosimetry. (Gatlinburg, Tennessee, September 23-26, 1968). J. A. Auxier, K. Becker and E. M. Robinson, eds. U. S. AEC report CONF-680920) (NSA 23: 7860).

Vago, C. 1963. 11. Predispositions and interrelations in insect diseases. 3. Irradiation injury. p. 351. In: Insect Pathology. An Advanced Treatise. Vol. I. E. A. Steinhaus, ed. Academic Press, New York.

Vaidya, V. G., et al. 1974. Effect of gamma irradiation on the protein metabolism in Drosophilidae. Radiation Research 57(3): 545-550. (NSA 29: 27116).

Vajdic, A. H. 1971. Gamma irradiation of waters and waste waters for disinfection purposes. Isotopes Radiat. Technol. 8(4): 451. (NSA 25: 48536).

Vajdic, A. H. 1971. Gamma-ray treatment of tastes and odors. J. Am. Water Works Assoc. 63(7): 459-461. (NSA 25: 49325).

Val'dshtein, Ye. A. and V. D. Zhestyanikov. 1966. Postirradiation recovery in bacteria. pp. 5-19. In: Protection Against and Repair of Radiation Injuries. Ye. Ya. Grayevskiy, V. I. Ivanov and V. I. Korogodin, eds. Publ. House Acad. Sci. USSR, Moscow. (In Russian) (NSA 22: 26010) (Lit. review).

Valencia, R. M. and J. I. Valencia. 1962. Studies of the genetic effects of irradiation during the perifertilization stages of Drosophila melanogaster- pp. 179-184. In: 4th Inter-American Symp. on the Peaceful Application of Nuclear Energy. Vol. 2. Pan American Union, Washington.

Valencia, R. M. and J. I. Valencia. 1964. The radiosensitivity of mature germ cells and fertilized eggs in Drosophila melanogaster. pp. 345-360. In: Mammalian Cytogenetics and Related Problems in Radiobiology. Pergamon Press, New York.

Valett, B. 1970. Uptake, retention, and distribution of ^{60}Co, ^{65}Zn, ^{85}Sr, and ^{131}I in the rough-skinned newt (Taricha granulosa): a radioecological study. Thesis, Oregon State Univ., Corvallis. 110pp. (NSA 24: 48470).

Valodzin, U. G., M. A. Kartel and Z. I. Lisouskaya. 1972. Criteria of plant radiosensitivity. Vest. Akad. Navuk Belorussk. SSR (biol.) (1): 65-68. (In Belorussian) (NSA 26: 43274).

Van, S. S. 1964. The speed of the ascending stream in forest trees. Dokl. Akad. Nauk SSSR 156: 706-707. (In Russian).

van As, et al. Environmental radioactivity at the National Nuclear Research Centre, Pelindaba. South Africa Atomic Energy Board, Pelindaba, Pretoria.

Year Publ.	Year Reptd.	Report No.	pp.	NSA	(Incl.
1965	1964	PEL-89	75	22: 180	gamma,
1967	1966	PEL-160	51	22: 10763	^{40}K;
1967	1967	PEL-181	20	23: 6450	soil, fish,
1969	1968	PEL-192	26	24: 23195	water, air,
1970	1969	PEL-205	33	25: 32401	milk, farm
1971	1970	PEL-216	21	26: 31008	products,
1972	1971	PEL-208	21	28: 10867	vegetation,
1973	1972	PEL-229	20	30: 21078	plankton,
1974	1973	PEL-240	20	31: 22617	sediments)
1975	1974	PEL-246	19	33: 636	

van As, D. 1968. Fallout over South Africa from nuclear weapons tested by France in 1966. Health Physics 14(4): 307-310.

van As, D. 1972. Activable tracers in the control of pollution and the study of environmental transport processes. pp. P34.p1-P34.p3. In: National Conf. on the Technological Applications of Nuclear Techniques, 1972. (See Union of South Africa, 1972) (CONF-721091) (NSA 28: 21123).

van As, D. and J. K. Basson. 1969. Study of radioactivity in the environment of the National Nuclear Research Centre. S. Afr. J. Sci. 65(1): ___.

Van Bavel, C. H. M. 1963. A further note on the efficiency of soil moisture probes. Soil Sci. Soc. Am. Proc. 27(5): 604.

Van Bavel, C. H. M. and G. B. Stirk. 1967. Soil water measurement with a ^{241}Am-Be neutron source and an application to evaporimetry. J. Hydrology 5: 40-46. (NSA 22: 25511).

Van Bavel, C. H. M., N. Underwood and R. W. Swanson. 1956. Soil moisture measurement by neutron moderation. Soil Science 82: 29-41.

van Caneghem, P. 1964. Cause of death after irradiation of lower extremities of frogs. Strahlentherapie 125: 441-448. (In German) (NSA 19: 7150).

van Dalen, A., et al. 1975. Neutron activation analysis on water samples from the environment. Trans. Amer. Nucl. Soc. 21(3 Suppl.): 20. (NSA 32: 27483).

Vandehey, R. C. 1964. Genetic variability in Aedes aegypti (Diptera: Culicidae). III. Plasticity in laboratory populations, Ann. Entomol. Soc. Am. 57: 488-496. (NSA 19: 88) (Incl. some irradiation data).

van den Bruel, W. E. and D. Bollaerts. 1960. Resistance of different stages of development of Sitophilus granarius (L.) and S. oryzae L. irradiated with gamma rays (Co^{60}). Bull. inst. agron. Gemblouxhors 2: 883-905. (In French).

van den Hende, A. 1968. New contribution concerning the use of radioisotopes in plant physiology. Atompraxis 14: 294-296. (In German) (NSA 22: 49190).

van den Hoek, J. 1963. Absorption, secretion, and excretion of fallout radionuclides in farm animals. Tijdschr. Diergeneesk. 88: 627-652. (In Dutch) (NSA 19: 25886) (Incl. comments on soil-chemical and botanical aspects).

van den Hoek, J. and J. A. Eygenraam. 1965. Radioactive contamination in a wild bird population. Ardea 53(1/2): 108.

van den Hoek, J. and R. J. Kirchmann. 1970. Transfer coefficients of strontium and cesium from grass to milk in dairy cattle after a single contamination of pasture. Zentralbl. Veterinarmed. Beih. 11: 95-100. (NSA 25: 21579).

van den Hoek, J., J. A. Eygenraam and W. J. D. van Troostwijk. 1964. Research on radioactive contamination in geese. pp. 245-249. In: Proc. 1st European Wildfowl Conserv. (St. Andrews, Scotland, 16-18 October 1963). (In English, French summary).

van den Hoek, J., et al. 1969. Importance of some methods of pasture feeding or pasture type, and of seasonal factors on ^{85}Sr and ^{134}Cs transfer from grass to milk. Health Physics 17: 691-700. (NSA 24: 4612).

van den Winkel, P. 1970. Determination of trace elements in biological materials by neutron activation analysis. Verh. Kon. Vlaam. Acad. Wetensch. Belg. Kl. Wetensch. 32(114): 189pp. (In Flemish) (NSA 25: 48492).

Vanderbeek, J. W. (Ed.). 1962. Evaluation of radiological conditions in the vicinity of Hanford, July-September 1962. Hanford Labs., U. S. AEC report HW-75431. 31pp. (NSA 17: 8427).

Vandenberg, J. P. 1964. Synthesis and transfer of DNA, RNA and protein during vitellogenesis in Rhodinus prolixus (Hemiptera). Biol. Bull. 125: 556-575. (^{3}H).

van der Borght, O. 1963. In- and outfluxes of Ca ions in freshwater gastropods. Arch. Intern. Physiol. Biochem. 71: 46-50. (NSA 17: 35416).

van der Borght, O. 1963. Accumulation of radium-226 by the freshwater gastropod Lymnaea stagnalis L. Nature 197(4855): 612-613. (NSA 17: 13944).

van der Borght, O. 1972. Experimental study of the biological availability of a heavy metal, ruthenium, in laboratory conditions. J. Belgé Radiol. 55(2): 243-252. (In Dutch) (NSA 27: 17713) (Freshwater and marine animals).

van der Borght, O. and S. van Puymbroeck. 1964. Active transport of alkaline earth ions as physiological base of the accumulation of some radionuclides in freshwater molluscs. Nature 204(4958): 533-534. (NSA 19: 6971).

van der Borght, O. and S. van Puymbroeck. 1966. Calcium metabolism in a freshwater mollusk: quantitative importance of water and food as supply for calcium during growth. Nature 210: 791-793. (NSA 20: 30991).

van der Borght, O. and S. van Puymbroeck. 1967. Ionic regulation and radiocontamination: "Active Transport" of alkaline earth ions in freshwater gastropods (Mollusca). pp. 925-930. In: Radioecological Concentration Processes. (See Aberg and Hungate, 1967).

van der Borght, O. and S. van Puymbroeck. 1970. Initial uptake, distribution, and loss of soluble ^{106}Ru in marine and freshwater organisms in laboratory conditions. Health Physics 19: 801-811. (NSA 25: 26587).

van der Borght, O. and S. van Puymbroeck. 1971. Kinetics of the direct uptake of Ca and Sr ions from water by freshwater gastropods. Environ. Physiol. 1(2): 83-95. (NSA 26: 48291) (^{45}Ca, ^{85}Sr).

van der Borght, O., et al. 1974. Long-term laboratory experiments on the fixation of various ^{106}Ru compounds by freshwater organisms. pp. 31-41. In: Environmental Surveillance Around Nuclear Installations. Vol. 2. (See IAEA, 1974) (NSA 31: 14441).

van der Brande, J. and C. Pelerents. 1962. Some effects of gamma radiation on the farina moth and on nematodes. Compt. rend. rech. 1(28): 11-47. (In French) (NSA 18: 25125).

van der Kloot, W. G. 1966. Exchange of radioactive cations by somatic and cardiac muscles of the crayfish. Comp. Biochem. Physiol. 17: 1019-1043. (NSA 21: 24318).

van der Mark, W. 1973. Determination of fluorine in plant materials. J. Radioanal. Chem. 16(2): 637-641. (NSA 30: 5854) (Neutron activation).

Vanderplank, F. L. 1958. Studies on the coconut pest, Pseudotheraptus wayi Brown (Coreidae), in Zanzibar. I. Method of assessing the damage caused by the insect. Bull. Entomol. Res. 49(3): 559-584. (^{32}P, ^{35}S).

Vanderploeg, H. A., R. S. Booth and F. H. Clark. 1975. Specific activity and concentration model applied to cesium-137 movement in a eutrophic lake. U. S. NOAA, Oak Ridge, Tenn. 31pp. (CONF-750503-7) (NSA 32: 6436).

Vanderploeg, H. A., et al. 1975. Bioaccumulation factors for radionuclides in freshwater biota. Oak Ridge National Lab., U. S. ERDA report ORNL-5002. 235pp. (NSA 33: 5586).

Van der Sloot, H. A. and H. A. Das. 1974. Determination of mercury in water samples from the environment by neutron activation analysis. Anal. Chim. Acta 73(2): 235-244. (NSA 31: 28979).

Van der Sloot, H. A., R. Massee and H. A. Das. 1975. Determination of uranium in sea and surface waters by neutron activation analysis after adsorption on charcoal. J. Radioanal. Chem. 25(1): 99-105. (NSA 32: 5876).

van der Westhuizen, M. 1969. Radioactive nuclear bomb fallout. A relationship between deposition, air concentration and rainfall. Atmosph. Environ. 3(3): 241-___.

van der Westhuizen, M. 1972. Use of isotopes in soil physical studies. pp. R25.b1-R25.b5. In: National Conf. on the Technilogical Applications of Nuclear Techniques. (See Union of South Africa, 1972) (NSA 28: 21127).

van der Westhuizen, M. and M. J. Smith. 1971. Change in concentration of tritium in water during evaporation and the possibility of using it to determine evaporation from water, soil, and plant surface. South Africa Atomic Energy Board, Pelindaba, Pretoria, report PEL-214. 15pp. (NSA 26: 25562).

van der Westhuizen, M. and H. A. J. van Rensburg. 1973. Movement of some nuclear fission and activation products in Duinefontein soil. South Africa Atomic Energy Board, Pelindaba, Pretoria, report PEL-226. 15pp. (NSA 29: 2647).

van der Westhuizen, M., D. J. van der Bank and A. Streutker. 1973. The use of ^{51}Cr-EDTA for the determination of the hydraulic conductivity of soil and the rate of flow of ground water at the Vaalharts Government Irrigation Scheme. Agrochemophysica 5: 67-70. (In Afrikaans) (NSA 33: 662).

Van de Vloed, A. 1963. Transportation of radioactive pollution by water to great distances. Tech. Sci. Munic. (Netherlands) 58(2): 55-68. (NSA 17: 29876).

Van de Woestijne, N. and J. van den Brande. 1960. Insect resistance to ionizing radiation: some results with the flour moth, Ephestia kuehniella. Z. Gembloux. Inst. Agron. B. Inst. Agron. Sta. Rech. Gembloux (Extrord.) 2: 872-882.

Van Die, J., P. Leeuwangh and S. M. R. Hoekstra. 1970. Translocation of assimilates in Fritillaria imperialis L. I. The secretion of ^{14}C-labeled sugars by the nectaries in relation to phyllotaxis. Acta Botan Neerl. 19 (1): 16-23.

Van Dyne, G. M. 1965. Application of some operations research techniques to food chain analysis problems. In: Radiation and Terrestrial Ecosystems. Health Physics 11(12): 1511-1519. (See Hungate, 1965).

Van Dyne, G. M. 1966. Ecosystems, systems ecology, and systems ecologists. Oak Ridge National Lab., U. S. AEC report ORNL-3957. 33pp. (NSA 20: 40488).

Van Groenewoud, H. 1958. Humus sterilization by gamma irradiation. Can. Phytopath. Soc. Proc. 25: 12.

van Haaren, F. W. J. 1964. The limiting of radioactive isotope in soil (underground water). Föderation Europ. Gewassersch. Inform. (10): 56-61. (In German) (NSA 19: 2538).

van Heemert, C. 1975. General discussion. pp. 68-76. In: Chromosomal Rearrangements in the Onion Fly Hylemya antiqua (Meigen), Induced and Isolated for Genetic Insect Control Purposes. C. van Heemert, ed. Wageningen, The Netherlands. (NSA 32: 14584) (X-ray).

Van Hook, R. I. Jr. 1970. Energy and nutrient dynamics of predator and prey arthropod populations in a grassland ecosystem. Thesis, Clemson Univ., Clemson, S. C. xi, 102pp. (Also w/D. E. Reichle and S. I. Auerbach: Oak Ridge National Lab., U. S. AEC report ORNL-4509) (Used radioisotopes).

Van Hook, R. I., Jr. 1971. Survival of crickets, Acheta domesticus (L.) as affected by variations in gamma dose rate. pp. 521-526. In: Survival of Food Crops and Livestock in the Event of Nuclear War. (See Bensen and Sparrow, 1971).

Van Hook, R. I., Jr. and D. A. Crossley, Jr. 1969. Assimilation and biological turnover of cesium-134, iodine-131, and chromium-51 in brown crickets, Acheta domesticus (L.). Health Physics 16(4): 463-467.

Van Hook, R. I., Jr. and S. L. Deal. 1973. Tritium uptake and elimination by tissue-bound and body-water components in crickets (Acheta domesticus). J. Insect Physiol. 19(3): 681-687.

Van Hook, R. I., Jr. and E. T. Herbert. 1971. Effects of gamma irradiation and temperature on the biological assimilation and retention of ^{137}Cs by Acheta domesticus (L.). Health Physics 21(6): 845-850. (NSA 26: 9608).

Van Hook, R. I., Jr. and D. E. Reichle. 1970. Nutrient dynamics in arthropod components of a grassland ecosystem. Abstr. Bull. Ecol. Soc. Am. 51(2): 27. (Used radioisotopes).

Van Hook, R. I., Jr. and A. J. Yates. 1975. Transient behavior of cadmium in a grassland arthropod food chain. Environ. Res. 9(1): 76-83. (NSA 33: 20194) (^{109}Cd).

Van Hook, R. I., Jr, et al. 1973. Radioisotope techniques in delineation of the environmental behavior of cadmium. Oak Ridge National Lab., U. S. AEC report. 28pp. (For: Symp. on Nuclear Techniques in Comparative Studies of Food and Environmental Contamination) (IAEA, SM-175/22) (CONF-730817-2) (NSA 28: 20610).

Vaninskaya, Yu. M. and P. A. Pinskaya. 1967. Biological resistance of wood modified by the method of radiative chemistry. Dokl. Akad. Nauk Belorussk. SSR 11: 1024-1026. (In Russian).

Van Mameren, J. and H. Houwing. 1968. Effect of irradiation of *Anisakis* larvae in salted herring. pp. 73-80. In: *Elimination of Harmful Organisms from Food and Feed by Irradiation*. (See IAEA, 1968) (NSA 24: 34830).

Van Middlesworth, L. 1975. Radioiodine ^{131}I in animal thyroids during nuclear tests in both hemispheres. Health Physics 29(6): 861-963. (NSA 33: 26321).

Vanselow, K. 1967. Studies on short-term recovery processes in the stimulation threshold of periferal nerve fibres altered by x-irradiation. Atompraxis 13: 34-37. (In German) (NSA 21: 18123) (Frog).

van Steenis, H. 1971. Differences in radiosensitivity of the chromosomes of rat kangaroo (*Potorous tridactylis*) cells *in vitro*. Mutation Research 11 (4): 445-447. (NSA 25: 42348).

van Steenis, H., R. Tuscany and B. Leigh. 1974. Distribution of x-ray induced chromosomal abnormalities in kangaroo rat (*Potorous tridactylis*) cells *in vitro*. Mutation Research 23(2): 223-228. (NSA 33: 17455).

Van Weers, A. W. 1973. Uptake and loss of ^{65}Zn and ^{60}Co by the mussel *Mytilus edulis* L. pp. 385-400. In: *Radioactive Contamination of the Marine Environment*. (See IAEA, 1973) (NSA 28: 8518).

Van Zon, A. Q. and W. P. J. Overmeer. 1972. Induction of chromosome mutations by x-irradiation in *Tetranychus urticae* (Tetranychidae) with respect to a possible method of genetic control. Entomologia Exp. Appl. 15(2): 195-202. (NSA 26: 48275).

Varga, I. 1964. Immunization experiments with irradiated larvae. I. Studies on the effect of x-rays on eggs and larvae of *Ascaridia galli*. II. The resistance-inducing effect of treatment with irradiated larvae of *Ascaridia galli* in chickens. III. The effect of x-rays upon eggs and larvae of *Syngamus trachea*. Acta Vet. Acad. Sci. Hung. 14(1): 95-103, 399-409, 411-418. (NSA 18: 38793; 19: 35957; 20: 70).

Varga, I. 1965. Experiments to immunize chicks with irradiated larvae of *Syngamus trachea*. Allotorv. Lapja 20: 205-210. (In Hungarian) (NSA 21: 6650).

Varga, I. 1968. Immunization experiments with irradiated larvae of *Syngamus trachea* in chickens. pp. 1-12. In: *Isotopes and Radiation in Parasitology*. (See IAEA, 1968).

Varga, I. 1973. Effect of ionizing radiation on animal parasites. pp. 81-86. In: *Manual on Radiation Sterilization of Medical and Biological Materials*. International Atomic Energy Agency, Vienna. (NSA 29: 16046).

Vargas, L. 1964. The use of labeled mosquitoes for studying population problems. Boln. inst estud. méd. biol. 22(3): 443-451. (In Spanish).

Various Authors. 1954. Symposium on effects of radiation and other deleterious agents on embryonic development. Oak Ridge, Tennessee, April 20, 21, 1953. J. Cell. Comp. Physiol. 43(Suppl.1): 333pp.

Various Authors. 1960. Radioactivity in water supply and waste water systems: peacetime detection and control. Bureau of Yards and Docks, U. S. Navy report NAVDOCKS-TP-TE-6. 92pp. (PB-17109) (NSA 15: 27895).

Various Authors. 1960, 1961. Annual report no. 11 (on genetics). --- no. 12. Japan Inst. of Genetics, Mishima. 118, 135pp. (NP-12940, -12941) (NSA 17: 28674, 28675) (On *Drosophila*, silkworm, *Triturus*).

Various Authors. 1961. Additions to Bioenvironmental Committee's Project Chariot report of March 1961. U. S. AEC, Washington. 17pp.

Various Authors. 1961. Proc. 15th Industrial Waste Conf. May 3, 4, and 5, 1960. Engineering Bull. 45(2): 624pp.

Various Authors. 1962. *Mechanisms in Radiobiology*. *Vol. 2. Multicellular Organisms*. Gosatomizdat, Moscow. 315pp. (In Russian).

Various Authors. 1962. The fallout of Sr^{90}, Cs^{137}, Ce^{144}, and other radioactive isotopes onto the earth's surface and their content in the atmosphere. Distribution of radioactive fission products in the atmosphere and global fallout. Ch. X, pp. 202-224; Ch. IX, pp. 161-210. In: *Radioactive Contamination of the External Environment*. (See Shvedov and Shirokov, 1962).

Various Authors. 1963. Disposal of radioactive wastes into fresh water. International Atomic Energy Agency, Vienna, Safety Ser. No. 10. 102pp. (NSA 17: 24738).

Various Authors. 1964. A review of the Dutch elm disease. Canada Dept. of Forestry, For. Entomol. Pathol. Br. Bi-monthly Progr. Rept. 20(4): 1-8. (Discusses study of irradiated seeds).

Various Authors. 1970. Project Crested Ice. USAF Nuclear Safety 65(2, spec.): 97pp. (Contains section on radiation ecology).

Various Authors. 1973. *Chemical and Radionuclide Food Contamination*. (Reprints). MSS Information Corp., New York. 188pp.

Various Authors. 1973. *Freshwater Pollution. II*. (Reprints). MSS Information Corp., New York. 228pp. (NSA 30: 12681).

Various Authors. 1973. Round table discussion: interregional aspect of environmental conservation concerning nuclear pollution. pp. 209-294. In: *Proc. 2nd Conf. on the State of Advancement of Radioecology in Italy*. (See Anonymous, 1973) (In Italian).

Varma, M. M. and G. W. Reid. 1963. The relationship between adsorbed P^{32} and p_H of the substrate. Water Sewage Works 110(11): 364-365.

Varma, M. M. and D. H. Stonefield. 1966. Uptake of ^{32}P and ^{45}Ca by pure and mixed cultures of bacteria. Purdue Univ. Eng. Bull. 50: Ext. ser. (121): 103-116. (NSA 21: 39338).

Varney, D. M. and W. A. Rhoads. 1977. Chronic radiation induced chromosomal aberrations in native shrubs at Nevada Test Site. pp. 351-378. In: Transuranics in Natural Environments. (See White and Dunaway, 1977).

Varshal, G. M., I. Ya. Koshcheyeva and R. P. Morozova. 1972. Possible forms of ruthenium migration in surface waters and soils. Geokhimiya (8): 971-976. (In Russian) (NSA 26: 58806).

Vartazarov, S. Ya. 1967. Uses of Radioactive Isotopes in Hydraulic and Hydrological Studies. Atomizdat, Moscow. 185pp. (In Russian) (NSA 21: 28093).

Varzandeh, M. and W. S. Moos. 1963. The effect of x-radiation on longevity, emergence, and DDT-susceptibility of the house fly. Proc. Entomol. Soc. Am. No. Central Br. 18: 55-59.

Vasconcellos, M. B. A. 1973. Applications of activation analysis to solve problems in archaeology and art. Brazil Inst. de Energia Atomica, São Paulo, report IEA-71-24. 18pp. (In Portugese) (NSA 30: 23501).

Vashkov, V. I. and V. D. Poleshchuk. 1966. Effect of x-irradiation on the insect sensitivity to insecticides. Zh. mikrobiol. epidemiol. immunobiol. 43(8): 9-11. (In Russian) (NSA 21: 1824).

Vashkov, V. I., G. D. Khudadov and V. I. Zakolodina. 1965. The rate of penetration and accumulation of chlorophos-^{32}P in various organs and tissues of house flies. Zh. mikrobiol. epidemiol. immonobiol. 42(8): 3-6. (In Russian).

Vasil'yev, G. and G. Filev. 1974. Integral dosimeter study of gamma-ray dose in the vicinity of a nuclear reactor. pp. 285-288. In: Environmental Surveillance Around Nuclear Installations. (See IAEA, 1974) (NSA 31: 14063).

Vasil'yev, I. B., V. I. Lyvchiyev and R. G. Ivanova. 1972. Studies on the reaction of the housefly (Musca domestica L.) to the presence of radiation of gamma-rays. Compt. rend. Acad. Bulg. Sci. 25(9): 1269-1271. (NSA 31: 14394).

Vasil'yev, I. M. 1962. Effects of Ionizing Radiations on Plants; Radiophysiological Investigations. Publ. House Acad. Sci. USSR, Moscow. (AEC-tr-5836) (NSA 18: 33284).

Vasil'yev, O. P. and A. I. Sherstnev. 1965. Effect of strontium-90 on the formal elements of fish blood. pp. 56-57. In: Problems of Hydrobiology. Publ. House Acad. Sci. USSR, Moscow. (In Russian).

Vasil'yev, V. 1973. A method for reading of the residual beta-activity in plants in the radiometric determination of ^{40}K. Vet. Med. Nauki (Sofia) 10(5): 61-65. (In Bulgarian)) (NSA 31: 13645).

Vasil'yeva, Ye. N., L. A. Okuneva and Yu. P. Kukel'. 1960. A hygienic study of grain irradiated with radioactive cobalt. Vopr. pitanya 19(5): 59-61. (JPRS-7169).

Vasiljevic, L. J. 1970. Gypsy moth strilization with ^{60}Co γ radiation. Zast. Bilja. 21(109): 201-206. (NSA 27: 27989).

Vasilyan, V. V. 1961. The effect of ionizing radiation upon the development of the mallow moth. Entomol. Obozreniye 39(3): 599-604. (Entomol. Rev. 39: 425-429).

Vasse, J. 1970. Action of x rays on the unformed limbs of green lizard embryos. Compt. rend. D271: 2193-2196. (In French) (NSA 25: 21849).

Vassilaki, M., L. Salmon and J. A. B. Gibson. 1965. Measurement of radioactivity in soil. UK Atomic Energy Authority, Harwell, report AERE-R-4996. ii, 11pp.

Vassileva-Dryanovska, O. and R. Belcheva. 1965. Radiation gynogenesis in Salmo irideus Gibb. Compt. rend. acad. Bulg. sci. 18: 359-362. (NSA 21: 22686).

Vasudeva, R. S., et al. 1959. Stimulating effect of ionizing radiation on certain microorganisms. Indian Phytopath. 12(1): 19-24.

Vates, W. W., et al. 1951. Treatment of mosquito larvae and adults with radioactive phosphorus. J. Econ. Entomol. 44(1): 34-37.

Vaughn, J. C. 1961. Monitoring the radioactivity in Chicago water supply. pp. 72-78. In: Radiological Aspects of Water Supplies. (See Univ. of Illinois, 1961).

Vaughn, J. C., et al. 1958. Studies of radioactivity in the Chicago water supply. J. Am. Water Works Assoc. 50(5): 581. (NSA 12: 10336).

Vaughn, J. W. and M. E. Miles. 1966. Disposal of radioactive wastes from U. S. Naval nuclear-powered ships and their support facilities. Radiol. Hlth. Data 7(5): 257-262.

Vavilov, P. P., O. N. Popova and R. P. Kodaneva. 1964. The behavior of radium in plants. Dokl. Akad. Nauk SSSR 157(4): 992-994. (Doklady (biol. sci.) 157: 134-136) (NSA 19: 2016).

Vavilov, P. P., et al. 1971. Use of stereo uranium-radium irradiators in experimental radioecological investigations. pp. 49-53. In: Methods of Radioecological Investigations. (See Verkhovskaya, 1971) (In Russian).

Vavilov, P. P., et al. 1972. Investigation of meiosis in irradiated plants and in plants grown in soils with a high uranium and radium content. pp. 103-112. In: Radioecological Studies in Natural Biogeocenoses. (See Verkhovskaya, 1972) (In Russian) (NSA 26: 56386).

Vavilov, P. P., et al. 1972. Conditions of radium accumulation by plants from the soil. pp. 95-103. In: Radioecological Studies of Natural Biogeocenoses. (See Verkhovskaya, 1972) (In Russian) (NSA 26: 56574) (Sunflower).

Vaynson, A. A., A. S. Merkulov and A. M. Kuzin. 1969. Effects of radiotoxins on the development of sea urchin embryos. Radiobiologiya 9(3): 457-459. (AEC-tr-7026, pp. 180-184) (NSA 24: 9917).

Vekshina, L. K., I. S. Sakovich and V. A. Shevchenko. 1974. Influence of densley ionizing radiation on Chlorella. Radiobiologiya 14(3): 386-389. (AEC-tr-7596, pp. 84-88) (NSA 31: 16758).

Vekshina, L. K., et al. 1969. Repair of visible mutations and lethal injuries in Chlorella. Biol. Nauki (10): 105-107. (In Russian) (NSA 24: 27576).

Vekshina, L. K., et al. 1969. Effect of alpha-irradiation on Chlorella survival and mutation. Kosm. biol. med. 3(5): 34-38. (In Russian) (NSA 24: 6025).

Vekshina, L. K., et al. 1970. Relative biological effectiveness of multicharged ions during single irradiation of Chlorella. Kosm. biol. med. 4 (5): 39-42. (JPRS-52121, pp. 58-62) (NSA 25: 16375).

Veksler, V., A. Savvaitov and V. Stelle. 1974. Work of the Radiocarbon Laboratory of the All-Union Research Institute of Marine Geology and Geophysics (Riga) on the reliability of measurements of natural radiocarbon activity. Izv. Akad. Nauk Est. SSR khim. geol. 23(1): 80-82. (In Russian) (NSA 30: 6564).

Vela, G. R. 1964. Radiation resistance of soil bacteria in their natural state. Thesis, Univ. of Texas, Austin. 125pp. (NSA 18: 41215).

Vela, G. R. 1969. The effects of ionizing radiation on nitrification in the soil. Tex. J. Sci. 20: 315-322. (NSA 23: 45976).

Vela, G. R. and O. Wyss. 1965. Radiation resistance of soil Azotobacter. J. Bacteriol. 89: 1280-1285. (NSA 19: 28108).

Velichko, A. A., et al. 1964. The first determinations of absolute age in loess soils of the Russian Steppe. Dokl. Akad. Nauk SSSR 155(3): 555-558. (In Russian) (NSA 18: 25685).

Velkov, D. 1972. Gamma radiation sensitivity of Scots pine seeds. Gorskostop. Nauka 9(4): 11-22. (In Bulgarian) (NSA 30: 15844).

Vel'tishcheva, I. F. 1961. Penetration of carbonate carbon (C^{14}) from water, and its distribution in the fish body. Tr. vses. nauchn.-issled. inst. morsk. ryb. khoz. okeanogr. 44: 23-36. (In Russian) (BA 45: 23000; Ref. zh. (biol.

Venator, R. and F. K. S. Koo. 1970. Inherent and radio-induced cytological abnormalities in Palicourea reaparia Benth. Ch. G-2. pp. G21-G31. In: A Tropical Rain Forest. (See H. T. Odum and R. F. Pigeon, 1970) (NSA 25: 21950).

Vende, G. V. and V. P. Parchevskiy. 1964. Liquid 4-pi scintillation counter for the absolute measurement of beta-activity of preparations of film and methods of preparing them. Radiobiologiya 4(3): 465-466. (AEC-tr-6406, pp. 203-206).

Vende, G. V. and V. P. Parchevskiy. 1966. Radioactive contamination of organisms of the Black Sea. Results and procedure for determining strontium-90 in Cytoseira. In: Radioactive Contamination of the Seas and Oceans. (See Baranov and Khitrov, 1966) (AEC-tr-6641, pp. 121-127; JPRS-26002, pp. 77-86) (NSA 21: 8809).

Venge, O. 1963. A note on induced sterility in mink. J. Reprod. Fert. 5(1): 83-86. (NSA 18: 83).

Verain, A. 1957. Plasmodium berghei and x-rays. Compt. rend. soc. biol. 151(6): 1164-1166. (In French).

Verduin, J. 1961. Measurement of environmental radioactivity with basic counting equipment. Ohio J. Sci. 61(1): 1-6. (NSA 15: 14116).

Verduin, J. 1961. A large discrepancy between C-14-based and p_H-based photosynthesis measurements. Abstr. p. 164. In: Proc. 4th Conf. on Great Lakes Research Techniques, Great Lakes Res. Div., Inst. Sci. Tech., Univ. of Michigan Publ. 7.

Verduin, J. 1963. Radioactivity of suspensoids in aquatic environments of northwestern Ohio. Ohio J. Sci. 63(1): 39-43.

Verigo, A. B. 1938. Determination of the radium emanation content of atmospheric air at arctic stations. Gidrometeozdat, Leningrad. (In Russian).

Vereecke, A. 1969. Reproducing capacity of gamma-irradiated adult males of the confused flour beetle. J. Econ. Entomol. 62: 357-359. (NSA 23: 24713).

Vereecke, A. and C. Pelerents. 1965. The effect of γ-irradiation on fecundity and fertility of Tribolium confusum Duval. Meded. Landb. Opzoek. Gent 30(2): 1017-1027. (In Flemish; English, French, German summaries).

Vereecke, A. and C. Pelerents. 1965. The influence of gamma rays on the longevity of Tribolium confusum Duval. Meded. Landb. Opzoek. Gent 30(3): 1824-1835. (In Flemish; English, French, German summaries).

Vereecke, A. and C. Pelerents. 1969. The influence of dose-rate of gamma-irradiation on the reproducing capacity of adult females of Tribolium confusum Duval. Meded. Landb. Opzoek. Gent 32(3/4): 890-903. (In Flemish; English, French, German summaries).

Verhagen, B. T. and J. P. F. Sellschop. 1965. Tritium measurements of subterranean waters in some South African sites. In: Radiocarbon and Tritium Dating. (Proc. 6th Intern. Conf., Washington State University, Pullman, Wash., June 7-11, 1965). R. M. Chatters and E. A. Olson, comps. (CONF-650652-20) (AED-CONF-66-176-2) (NSA 20: 41167).

Verhagen, B. T. and J. P. F. Sellschop. 1972. Environmental radioisotopes in water studies. pp. P26.p1-P26.p11. In: National Conf. on the Technological Applications of Nuclear Techniques. (See Union of South Africa, 1972) (CONF-721091) (NSA 28: 21120) (^{14}C, ^{3}H).

Verhagen, B. T. and J. P. F. Sellschop. 1974. Radiocarbon and tritium evidence for direct rain recharge to ground waters in the northern Kalahari. Nature 249(5458): 643-644. (NSA 30: 18661).

Verho, S., P. M. Rissanen and E. Spring. 1973. Temperature dependence of the radiosensitivity of grass seeds. Strahlentherapie 146(4): 469-472. (NSA 29: 24471).

Verkhovskaya, I. N. 1962. Bromine in the Animal Organisms and the Mechanism of Its Action. Acad. Sci., USSR, Moscow. In Russian).

Verkhovskaya, I. N. (Ed.). 1971. Methods of Radioecological Investigations. Atomizdat, Moscow. 260pp. (In Russian) (NSA 26: 12255).

Verkhovskaya, I. N. 1971. Tasks, methods and perspectives of radioecological investigations in different biogeocoenoses. pp. 3-15. In: Methods of Radioecological Investigations. (See Verkhovskaya, 1971 above) (In Russian).

Verkhovskaya, I. N. (Ed.). 1972. Radioecological Studies in Natural Biogeocenoses. "Nauka", Moscow. 267pp. (In Russian, English table of contents and summaries) (NSA 26: 56134).

Verkhovskaya, I. N. and A. I. Nefedova. 1966. Influence of x rays on germination of spores and development of sprouts of fern (Pteris longifolia L.). Radiobiologiya 6(5): 744-748. (AEC-tr-6774, pp. 153-159) (NSA 21: 11114).

Verkhovskaya, I. N. and V. S. Nikiforov. 1972. General problems of compiling descriptions (algorhythms) of radiation peculiarities in radiological experiments and observations. pp. 227-230. In: Radioecological Studies in Natural Biogeocenoses. (See Verkhovskaya, 1972 above) (In Russian).

Verkhovskaya, I. N., V. I. Maslov and K. I. Maslova. 1965. The action of minute doses of radiation and incorporated natural radioactive elements on spermatogenesis in Microtus oeconomus under natural conditions. Radiobiologiya 5(5): 720-729. (AEC-tr-6602, pp. 133-146).

Verkhovskaya, I. N., V. I. Maslov and K. I. Maslova. 1965. The effects of increased natural radiation and incorporated natural radioactive elements on spermatogenesis in Microtus oeconomus. Abstr. Communications, All-Union Inter-College Conf. on the Problem of Microelements and Natural Radioactivity. Vol. 3. Petrozavodsk. (In Russian).

Verkhovskaya, I. N., P. P. Vavilov and V. I. Maslov. 1967. The migration of natural and radioactive elements under natural conditions and their distribution according to biotic and abiotic environmental components. pp. 313-328. In: Radioecological Concentration Processes. (See Aberg and Hungate, 1967).

Verkhovskaya, I. N., P. P. Vavilov and V. I. Maslov. 1972. Distribution and redistribution of uranium, radium, and thorium in natural biogeocenoses. pp. 243-256. In: Radioecological Studies in Natural Biogeocenoses. (See Verkhovskaya, 1972 above) (RFP-trans-162) (NSA 26: 56137; 30: 21080).

Verkhovskaya, I. N., et al. 1972. Distribution of natural radioactive elements in the basic components of natural aquatic microbiogeocenoses. pp. 124-147. In: Radioecological Studies in Natural Biogeocenoses. (See Verkhovskaya, 1972 above) (NSA 26: 56135).

Verly, W. G. 1959. Tritium in biology. In: Marine Biological Applications of Radioisotope Research Techniques. Pubbl. stazi. zool. Napoli 31(Suppl.): 70-94. (See Zoological Station of Naples, 1959).

Vermeulen, C. A. and G. Venema. 1971. Autoradiographic estimation of competence and the relationship between competence and transformability in cultures of Bacillus subtilis. J. Gen. Microbiol. 69(2): 239-252.

Vermont Yankee Nuclear Power Corporation. 1969, 1970, 1972. Vermont Yankee Nuclear Power Station. Environmental surveillance program. ---Environmental report. ---Final environmental statement. Rutland, Vt. 70, 77, 357pp. U. S. AEC DOCKET-50271-13, -27, -88. (NSA 24: 9014, 45951; 26: 47435).

Vernadskiy, V. I. 1926. The Biosphere. F. Alcan, Paris. 232pp. (In French) (Incl. radioactivity).

Vernadskiy, V. I. 1926. Isotopes in living matter. Compt. rend. acad. sci. URSS 1926: 215-218. (In Russian).

Vernadskiy, V. I. 1930. The concentration of radium by plant organisms. Compt. rend acad. sci. URSS 1930A: 539-542. (In Russian) (CA 25: 4911).

Vernadskiy, V. I. 1930. Radium in aquatic organisms. Compt. rend. 191(1): 421-423. (In French).

Vernadskiy, V. I. 1931. Isotopes and living organisms. Compt. rend. 192(1): 131-133. (In French) (CA 25: 1873).

Vernadskiy, V. I. 1933. Oceanography and geochemistry. Mineral. petrogr. mitt. 44: 168-192. (In Russian).

Vernadskiy, V. I. 1934. Outlines of Geochemistry. Gorgeoneftizdat, Moscow-Leningrad. (In Russian) (German transl. E. Kordes Geochemie in Ausgewählten Kapiteln. Akad. Verlag, Leipzig., 370pp.).

Vernadskiy, V. I. 1938. Some fundamental problems in biogeochemistry. Izv. Akad. Nauk SSSR (geol.) 18: 19-34. (In Russian).

Vernadskiy, V. I. 1940. Biogeochemical Outlines. 1922-1932. Acad. Sci. USSR, Moscow-Leningrad. 249pp. (In Russian).

Vernadskiy, V. I. and B. K. Brunovskiy. 1929. Radium concentration by organisms. Compt. rend. acad. sci. URSS 1929: 33-34. (In Russian) (CA 24: 5773; Chem. Zentr. 1930 I: 98).

Vernadskiy, V. I. and A. P. Vinogradov. 1931. The elementary chemical composition of some species of the genus Lemna as a species characteristic. Compt. rend. acad. sci. URSS 1931A(6): 148-152. (In Russian).

Vernadskiy, V. I., B. K. Brunovskiy and K. G. Kunasheva. 1933. Mesothorium I in Lemna. Compt. rend. 197: 1556-1557. (In French) (BA 9: 11122).

Vernadskiy, V. I., B. K. Brunovskiy and K. G. Kunasheva. 1937. Concentration of mesothorium I by duckweed (Lemna). Nature 140: 317-318.

de Vessey, E. and G. Czerny. 1957. A study of underground flow of radioactive isotopes and tracing ions. Hidrologiai Kolony 37: 44-56.

Viado, G. B. 1960. Radioisotopes in the biological sciences. Philippine Atomic Energy Commission, Manila. 255pp. (NSA 15: 28928).

Viado, G. B. 1962. The future of radioisotopes in insect-control investigations in the Philippines. pp. 287-295. In: Radioisotopes and Radiation in Entomology. (See IAEA, 1962).

Viado, G. B. and E. C. Manoto. 1963. Effects of gamma radiation on three species of Philippine insect pests. pp. 443-453. In: Radiation and Radioisotopes Applied to Insects of Agricultural Importance. (See IAEA, 1963) (NSA 17: 19918).

Viado, G. B. and E. C. Manoto. 1964. Effects of gamma rays on three species of Philippine insect pests. Scient. Rev. 5(3): 11-15.

Viado, G. B., et al. 1972. Some factors affecting the biological half-life of ^{51}Cr and ^{32}P in the oriental migratory locust. pp. 39-48. In: 1972 Proc. Radioisotope Soc. of the Philippines and the Philippine Soc. of Nuclear Medicine, Joint Conv. (Quezon City, April 7, 1972). (CONF-720486) (NSA 30: 3701).

Vichuda Krithayakiern. 1966. Preliminary study of the effect of gamma radiation in controlling oriental fruit fly (Dacus dorsalis Hendel) in banana. pp. 53-60. In: Insect Eradication by Irradiation. Thailand Office of Atomic Energy for Peace, Bangkok, report THAI.-AEC-8.

Vickers, D. H. and C. E. Boyd. 1973. Effects of organic insecticides upon carbon-14 uptake by freshwater phytoplankton. pp. 492-496. In: Radionuclides in Ecosystems. (See Nelson, 1973) (NSA 28: 27694).

Vidakovic, M. 1967. Growth of Austrain and Scots pine plants after gamma-irradiation of pollen. Radiation Botany 7(6): 529-542.

Vielkind, J., et al. 1971. Uptake of bacterial H^3-DNA into fish embryos. Experientia 27(3): 347-348. (SFA 16: 13704).

Viereck, L. A. 1964. Radioactivity in the arctic food chain. Alaska Cons. Soc. News Bull. 5(4): 2-4.

Vijayan, E. and A. G. Sathyanesan. 1971. Effect of varied doses of x irradiation on the thyroid histopathology of the Indian palm squirrel Funambulus pennanti (Wroughton). Strahlentherapie 141(2): 235-243. (NSA 25: 39935).

Vijayan, E. and A. G. Sathyanesan. 1971. Occurrence of ectopic thyroid in the normal and experimental Indian palm squirrel Funambulus pennanti (Wroughton). Strahlentherapie 141(4): 485-490. (NSA 25: 44933).

Vijayan, E. and A. G. Sathyanesan. 1971. Histopathological changes in the thyroid gland of the Indian palm squirrel Funambulus pennanti (Wroughton) in response to varied doses of radioiodine. Endokrinologie 58(1): 37-54. (NSA 26: 1797).

Viktor, J. 1932. On the influence of radium radiations on the coloration, development and consumption of oxygen of the nymphs of Tenebrio molitor. Main. Sec. May Sci. Beheme (Prague) (Paper 9): 1-26. (In French).

Viktor, J. 1934. On the study of radium radiation and ultraviolet light on the coloration, metamorphoses and oxygen consumption of pupae of Tenebrio molitor L. Biol. Generalis 10(2): 483-520. (In German).

Viktor, J. and V. Kocian. 1933. On the oxygen consumption of the pupae of Tenebrio molitor L. during various metamorphoses. Temperature, after stigma agglutination, bisection of trachea, and after injection of Thyrozin and Germaninlesungen. Zool. Jahrb. Abt. Allg. Zool. Physiol. 52(4): 561-590. (In German).

Vilenskiy, V. D. 1969. ^{210}Pb and ^{226}Ra distribution in some soils. Geokhimiya (12): 1507-1510. (English transl. Am. Geol. Inst. GI 12-11-69; ORNL-tr-2989) (Abstr. Geochem. Int. 6: 1196-1197; NSA 24: 9496; 32: 14304).

Vilenskiy, V. D. and V. V. Anikiyev. 1974. ^{210}Pb and ^{226}Ra in bottom deposits of freshwater basins. Geokhimiya (8): 1259-1263. (In Russian) (NSA 31: 3385).

Vilenskiy, V. D. and Z. V. Yeremenko. 1970. Determination of ^{226}Ra in soils by an α-scintillation method. Radiokhimiya 12(6): 864-870. (Sov. Radiochem. 12(6): 834-839) (NSA 26: 1797).

Vilenskiy, V. D., et al. 1972. Use of isotopic methods for the determination of the present rate of snow accumulation on the Antarctic continent. Geokhimiya 9: 1071-1082. (In Russian) (NSA 27: 329).

Villee, C. A. 1946. Some effects of x-rays on development in Drosophila. J. Exptl. Zool. 101(2): 261-280.

Villegas, L. 1965. A study of the chloride movement in the gastric mucosa by radioactive tracers. pp. 85-91. In: Proc. 5th Inter-American Symp. on the Peaceful Application of Nuclear Energy. Pan American Union, Washington. (In Spanish) (NSA 19: 35937).

Villegas, R. and G. M. Villegas. 1959. Pathway for water diffusion in the giant nerve fiber of the squid. pp. 77-81. In: Radioisotopes and Radiation in the Life Sciences. (See Inter-American Nuclear Energy Commission, 1959).

Villella, J. B., S. E. Gould and H. J. Gomberg. 1958. Effect of cobalt-60 and x-ray on infectivity of Ascaris eggs. J. Parasitol. 44(1): 85-92.

Villella, J. B., S. E. Gould and H. J. Gomberg. 1960. Effect of cobalt-60 and x-ray on infectivity of cysticercoids of Hymenolepis diminuta. J. Parasitol. 46: 165-169. (NSA 15: 3894).

Vinberg, G. G. 1961. Primary production of plankton in lake studies using three methods: oxygen, radioactive carbon and chlorophyll. Verh. intern. Ver. theor. angew. limnol. 14: 109-112. (SFA 8: 5595; WPA 35: 862).

Vinberg, G. G. and V. I. Gaponenko. 1958. Development of several aquatic organisms in cultures with different concentrations of P^{32} radioactivity. Tr. biol. stants. oz. Naroch' (1): 193-196. (In Russian).

Vinberg, G. G., T. N. Godney and V. I. Gaponenko. 1955. Experience in the application of phosphorus radioisotopes in studying the fertilization of ponds. Dokl. Akad. Nauk SSSR 100(3): 575-578. (In Russian).

Vinberg, G. G., et al. 1963. The primary production of bodies of water. (AEC-tr-5692(Books 1 and 2, 601pp.).

Vinck, W. F., H. A. Maurer and L. Leonardini. 1971. Engineering safety factors and their influence on siting practice for nuclear power plants in the European Communities. pp. 661-685. In: Environmental Aspects of Nuclear Power Stations. (See IAEA, 1971).

Vinogradov, A. P. 1956. The biological role of K^{40}. Dokl. Akad. Nauk SSSR 110(3): 375-378. (In Russian).

Vinogradov, A. P. 1957. Biological role of potassium-40. Nature 179(4223): 308-309. (Aspergillus niger).

Vinogradov, A. P. 1957. Biological role of potassium-40. Nature 180(4584): 507-508. (Aspergillus niger).

Vinogradov, A. P. 1957. The Geochemistry of Rare and Trace Elements in Soils. Acad. Sci. USSR, Moscow. 2nd ed. (Consultants Bureau, New York, 1959, 209 pp.).

Vinogradov, A. P. 1957. The isotope ^{40}K and its biological role. Biokhimiya 22(1-2): 14-20. (Biochemistry (USSR) 22: 12-18).

Vinogradov, A. P. 1964. ^{14}C dating of the upper anthropogene in the USSR. Radioact. Sea (IAEA, Monaco) (9): 1-14. (NSA 19: 46621).

Vinogradov, A. P. and V. V. Koval'skiy. 1957. Radioactivity of K^{40} in animals. Dokl. Akad. Nauk SSSR 113(2): 315-318. (Doklady (biol. sci.) 113: 212-215) (BA 36: 43388; WR 103: 9).

Vinogradov, A. P., A. L. Diverts and Ye. I. Dobkina. 1961. The increase in content of radioactive carbon as a result of nuclear explosions. Dokl. Akad. Nauk SSSR 137: 688-691. (Doklady (earth sci.) 137: 246-248) (NSA 15: 17915).

Vinogradov, A. P., A. L. Diverts and Ye. I. Dobkina. 1968. The current tritium contents of natural waters. Geokhimiya (10): 1147-1162. (Geochem. Int. 5: 952-966) (NSA 23: 49938).

Vinogradov, A. P., V. M. Kutorskiy and I. K. Zadorzhniy. 1959. Fractionation of the atmospheric carbon dioxide isotope. Geokhimiya (3): 195-205. (Geochemistry (3): 241-253).

Vinogradov, A. P., et al. 1963. Determination of the boundary between the late and the post-glacial time from carbon-14 and spore-pollen analyses. Geokhimiya (11): 971-981. (Geochem. Int. (11): 1009-1019).

Vinogradov, A. P., et al. 1969. New datings by radiocarbon of late-Quaternary deposits. Report 6. Geokhimiya (10): 1171-1178. (Transl. Am. Geol. Inst. GI 69-9-7) (Abstr. Geochem. Int. 6: 1011; NSA 24: 2300).

Vinogradova, I. D. and Ya. L. Shekhtman. 1968. Effects of certain protectors on the survival of embryos and the viscous properties of DNA in the case of irradiation of loach sperm. Radiobiologiya 8(8): 811-815. (AEC-tr-7013, pp. 37-43) (NSA 23: 18084).

Vinogradova, I. D., Z. N. Faleyeva and Ya. L. Shekhtman. 1968. Question of the permeability and connection of ^{35}S-labeled cysteamine with components of the sperm cell of Misgurnus fossilis. Radiobiologiya 8(1): 34-39. (AEC-tr-6914, pp. 54-62) (NSA 22: 47488).

Vinogradova, V. K., L. I. Gedeonov and S. P. Rosyanov. 1973. Method of ^{90}Sr determination of bottom samples. pp. 123-127. In: Radioecology of Water Organisms. III. (See Andrushaitis, et al., 1973) (AEC-tr-7592, pp. 79-81) (NSA 30: 23565; 31: 10562).

Vinson, S. B. 1966. The penetration, distribution, and metabolism of ^{14}C labelled DDT in the larvae of the tobacco budworm, Heliothis virescens. Thesis, Mississippi State Univ. (Dissertation Abstr. 26: 5611).

Vinson, S. B. 1967. Effect of several nutrients on DDT resistance in tobacco budworm. J. Econ. Entomol. 60(2): 565-568. (^{14}C).

Vinson, S. B. and J. R. Brazzel. 1966. The penetration and metabolism of ^{14}C-labelled DDT in resistant and susceptible tobacco budworm larvae Heliothis virescens (F.). J. Econ. Entomol. 59: 600-604.

Vinson, S. B., R. L. Londono and A. C. Bartlett. 1969. Effect of gamma radiation on tissues of the tobacco budworm, Heliothis virescens. Ann. Entomol. Soc. Am. 62: 1340-1347. (NSA 24: 4677).

Vintemberger, P. 1928. On the variation of radiosensitivity during early mitoses of segmentation on the eggs of Rana fusca. Compt. rend. soc. biol. 98: 532-535. (In French).

Vintemberger, P. 1928. On the effects of application of x rays to localized protoplasm or the nuclear region of the cell. Compt. rend. soc. biol. 99: 1968-1971. (Frog eggs).

Vintemberger, P. 1928. On the use of x rays in embryology as a localized destructive agent. A new technique for the study of two early blastomeres of frog eggs. Compt. rend. soc. biol. 99(33): 1590-1592. (In French).

Vintemberger, P. 1930. Experimental studies on observed mitosis as a factor of radiosensitivity. Arch. anat. 12: 229-___. (In French).

Vintemberger, P. 1930. On the results of application of three heavy doses of x rays to diverse regions of the eggs of the frog. Compt. rend. soc. biol. 102: 1055-1057. (In French).

Vinter, V. and B. Vechet. 1964. Spores of microorganisms. XV. The alteration of heat sensitivity and its relation to radiation resistance of bacterial spores. Folia Microbiol. (Prague) 9: 238-248. (NSA 19: 8844).

Virginia Electric and Power Company. 1970. Surry Power Station, Units 1 and 2. Applicant's environmental report: operating license stage. --Preoperational environmental radiation surveillance program, May 1, 1968-June 30, 1970. Preliminary safety analysis report, ammendment 16, Pt. C. Richmond, Va. 169, 68pp. U. S. AEC DOCKET-50280-32, -50281-32; -50280-29, 50281-29. (NSA 25: 12739, 5928).

Virk, D. S. 1959. Genetical studies on the effects of x-rays in the silkworm Bombyx mori L. I. Jap. J. Genet. 34(9): 285-292.

Virkki, N. 1965. Insect gametogenesis as a target. Agr. Sci. Rev. 3(3): 24-37.

Virkki, N. 1970. Karyotype rearrangements and malformation of gonads in the walkingsticks (Phasmatoptera) of the El Verde Radiation Center. Ch. G-4, pp. G51-G62. In: A Tropical Rain Forest. (See H. T. Odum and R. F. Pigeon, 1970) (NSA 25: 21652).

Virkkunen, J. and A. P. U. Vuorinen. 1962. Strontium-90 in reindeer bone. Suomen Kemistilexti B35: 35-38. (NSA 16: 23402).

Vis, R. D. and H. Verheul. 1975. Analysis of phosphorus and chlorine in natural waters by alpha-particle activation. J. Radioanal. Chem. 25(2): 293-297. (NSA 32: 21848).

Viscrian, I. 1961. Carbon-14 in nature and its detection. Acad. rep. populare Romine, fil. Iasi studii cerc. stiint. fiz. stiinte Tehn. 12: 111-120. (In Rumanian) (NSA 17: 30682).

Viste, A. 1974. Iodine contents in drinking water and precipitation from various places in Norway, spring and summer 1974. Norway Forsvarets Forskningsinstitut, Kjeller, report FFIF-F-605. 17pp. (NSA 31: 29619) (^{131}I).

Vivien, J. H. 1950. Masculinization of Xiphophores females by x-irradiation of the gonad. Compt. rend. 231(21): 1166-1168. (In French) (NSA 15: 2339).

Vivien, J. 1953. Total sterilization of the gonads after treatment with radioactive phosphorus, in the cyprinodontids: Lebistes and Xiphophores. Compt. rend. 236: 2172-2174. (In French) (NSA 7: 4713).

Vivien, J. 1953. Effects of radiophosphorus P^{32} on the gonads of viviparous cyprinodontids. Compt. rend. 236: 535-538. (In French) (NSA 7-1877).

Vivyurka, A. J. 1975. Rehabilitation of uranium mines tailings. Can. Min. J. 96(6): 44-45. (NSA 33: 2978).

Vlasyuk, P. A. 1955. Effects of Nuclear Radiations on Plants. Publ. House Acad. Sci. USSR, Moscow. 339pp. (In Russian) (NSA 9: 7618).

Vlasyuk. P. A. 1960. Trace elements and natural radioactivity of soils of the Ukrainian SSR. Biol. stiint. agric. Bucaresti 1960: 759-767. (Soil and Fertilizer 24(6): 2886).

Vlasyuk, P. A. (Ed.). 1964. Physiological-Biochemical Effects of Nuclear Radiation on Plants. Publ. House Acad. Sci. Ukr. SSR, Kiev. 210pp. (In Russian) (NSA 19: 17659).

Vlasyuk, P. A. and N. I. Bidzilya. 1958. The influence of beta-particles of radioactive isotopes on size variation on chloroplasts of Elodea canadensis. Dokl. Akad. Nauk SSSR 119(1): 65-67. (Doklady (biol. sci.) 119: 130-133) (NSA 12: 10306).

Vlasyuk, P. A. and D. M. Grodzinskiy. 1956. On the effect of nuclear radiation on plants. Dokl. Akad. Nauk SSSR 106: 562-564. (In Russian) (Saccharomyces cerevisiae, Uromyces onobrychides).

Vlasyuk, P. A. and G. V. Porutskiy. 1958. The combined action of herbicides and radioactive substances. Nauchni tr. Ukr. inst. fiziol. rast. 15: 61-70. (In Russian).

Vodovozova, I. G. 1969. Horizontal water migration of strontium-90. pp. 85-87. In: Nuclear Meteorology. (See Makhon'ko and Malakhov, 1969) (In Russian) (NSA 30: 15437).

Vodovozova, I. G. 1971. Horizontal water migration of ^{90}Sr. Tr. inst. eksp. meteorol. 21(1): 87-90. (In Russian) (NSA 30: 29407).

Vodovozova, I. G., S. Ya. Zaidman and Z. G. Antropova. 1973. Interaction of radioactive isotopes with organic soil material. State Committee on the Use of Atomic Energy USSR, Moscow. 15pp. (In Russian) (INIS-mf-1932) (NSA 30: 6520).

Vodyanitskiy, V. A. 1958. Discussions on the atomic industrial waste disposal into the Black Sea. Priroda 47(2): 46-52. (AEC-tr-3296).

Voegeli, P. T., Sr. and H. C. Claassen. 1971. Radiochemical analyses of water from selected streams, wells, springs, and precipitation collected prior to reentry drilling, Project Rulison. U. S. Geological Survey, U. S. AEC report USGS-474-83, -101; PNE-R-48. i, 16; 14pp. (NSA 25: 34728).

Voets, J. P. and M. Dedeken. 1965. The behavior of some amino acids in gamma irradiated soils. Naturwissenschaften 52(16): 476. (NSA 19: 45941).

Voets, J. P. and M. Dedeken. 1965. Influence of high frequency and gamma irradiation on the microflora and the soil enzymes. Meded. Landb. Opzoek. Gent 30: 2037-2049. (In Dutch) (NSA 21: 26693).

Vogel, H. H. and E. Staffeldt. 1966. Acute gamma-irradiation mortality in five species of wild rodents. pp. 31-32. In: Radiation Effects on Natural Populations. (See Sacher, 1966).

Vogel, H. H., J. W. Clark and D. L. Jordon. 1954. The relative biological effectiveness of fast neutrons and Co^{60} gamma rays. Abstr. Radiation Research 1(2): 233. (Grasshoppers).

Vogel, J. C. 1967. Investigation of groundwater flow with radiocarbon. pp. 355-369. In: Isotopes in Hydrology. (See IAEA, 1967) (NSA 22: 18932).

Vogel, J. C. and M. van Dijken. 1972. Determination of groundwater recharge with tritium. pp. P28.p1-P28.p7. In: National Conf. on the Technological Applications of Nuclear Techniques, 1972. (See Union of South Africa, 1972) (NSA 28: 21121).

Vogel, J. C. and D. Ehhalt. 1963. The use of carbon isotopes in groundwater studies. pp. 383-396. In: Radioisotopes in Hydrology. (See IAEA, 1963) (NSA 18: 1929).

Vogel, J. C. and I. Uhlitzsch. 1975. Carbon-14 as an indicator of CO_2 pollution in cities. pp. 143-152. In: Isotope Ratios as Pollutant Source and Behavior Indicators. (See IAEA, 1975) (NSA 33: 7320).

Vogel, J. C., W. A. Casparie and A. V. Munaut. 1969. ^{14}C in subfossil pine stubs. Science 166: 1143-1145. (NSA 24: 2315).

Vogel, J. C, D. Ehhalt and W. Roether. 1963. A survey of the natural isotopes of water in South Africa. pp. 407-415. In: Radioisotopes in Hydrology. (See IAEA, 1963) (NSA 18: 1907).

Vogl, M. 1964. The migration of marked phosphorus in poplar wood. Flora 154: 94-98.

Vogt, H.-H. 1969. Canal construction in Central America: a biological catastrophe? Naturwiss. Rundschau 22: 209-210. (In German) (NSA 24: 9545).

Vogt, J. R. and W. Meyer. (Eds.). 1974. Proc. 2nd Intern. Conf. on Nuclear Methods in Environmental Research. (Univ. of Missouri, Columbia, July 29, 30, and 31, 1974). Coll. of Engineering, Univ. of Missouri, Columbia. 479pp. (CONF-740701) (NSA 33: 7320).

Vogt, J. R., J. E. Howes and R. A. Ewing. 1968. Radionuclide and stable element analyses of environmental samples from Amchitka Island, Alaska. Annual progress report. Battelle Columbus Labs., U. S. AEC report BMI-171-110. 50pp. (NSA 25: 16072).

Vogt, J. R., T. F. Parkinson and R. L. Carter. (Eds.). 1971. Nuclear Methods in Environmental Research. (Proc. Amer. Nuclear Soc. Topical Meeting, Columbia, Missouri, Aug. 23-24, 1971). Univ. of Missouri, Columbia. 252 pp. (NSA 26: 17340).

Vogt, K. J. 1969. Criteria for the release of radioactive materials from nuclear plants with the atmosphere. pp. 22-31. In: Radiation Protection Problems in the Relaease and Incorporation of Radioactive Materials. (See Hacke and Jacobi, 1969) (In German) (NSA 24: 27234).

Vogt, K. J., et al. Environmental radioactivity at the Nuclear Research Establishment, Jülich. Annual report. Germany Kernforschungsanlage, Jülich.

Year Publ.	Year Reptd.	Report No.	pp.	NSA
1962	1961	JÜL-55-ST	87	17: 307
1963	1962	-	110	18: 1952
1965	-	-	79	20: 4178
1966	-	-379-ST	88	21: 1446

Vogt, K. J., et al. (Continued).

Year Publ.	Year Reptd.	Report No.	pp.	NSA
1967	1966	JÜL-485-ST	78	22: 2193
1969	1968	-603-ST	7-84	24: 27234

Vogt, K. J. 1971. Environmental contamination and exposure to radiation due to radioactive emissions from the Jülich Nuclear Research Centre in 1970. pp. 395-396. In: Proc. Intern. Symp. on Radioecology Applied to the Protection of Man and His Environment. (See Commission of the European Commuities, 1971).

Vohra, K. G., et al. 1966. Estimates of biospheric contamination and radiation dose from fallout for all the pre-treaty tests of nuclear weapons. Health Physics 12: 1087-1098. (NSA 20: 37004).

Voilleque, P. G. and B. R. Baldwin. (Comps.). 1971. Health Physics Aspects of Nuclear Facility Siting. (Proc. 5th Ann. Health Physics Soc. Midyear Topical Symp., Idaho Falls, Ida., November 3-6, 1970). 3 vols., vi, 795pp. (CONF-701106) (NSA 25: 55193).

Voilleque, P. G., D. R. Adams and J. B. Echo. 1970. Transfer of ^{85}Kr from air to grass. Health Physics 19: 835. (NSA 25: 13124).

Voilleque, P. G., D. G. Hill and D. N. Tillitt. 1973. Automatic data handling system for low level environmental radiation monitoring. Trans. Amer. Nucl. Soc. 17: 543-544. (NSA 29: 5126).

Voino-Yasenetskiy, A. V. 1955. Effect of Co^{60} gamma radiation on the process of ovulation, fertilization, and embryonic development in the frog. Dokl. Akad. Nauk SSSR 100: 389-391. (In Russian) (NSA 9: 2581).

Vojvodic, V. 1967. Working principles of three-dimension gamma-ray contamination simulator for radioactive fallout. Nukl. Energ. 4(6): 28-31. (In Serbo-Croatian) (NSA 25: 15788).

Volarovich, M. P. and N. V. Churayev. 1960. Use of the radioactive tracer method to study problems of the translocation of water in the peat stratum during drainage. pp. 230-___. In: Proc. Tashkent Conf. on Peaceful Uses of Atomic Energy, 1959. Vol. 2. Ya. Kh. Turakulov, ed. Publ. House Acad. Sci. Uzbek. SSR, Tashkent. (AEC-tr-6390, pp. 304-321) (NSA 18: 39514).

Volchok, H. L. 1966. The global ^{90}Sr budget. J. Geophys. Res. 71: 1515-1518. (NSA 20: 23291).

Volchok, H. L. 1968, 1969, 1972. Worldwide deposition of ^{90}Sr through 1967. --- through 1968. --- through 1971. pp. I-2 - I-13, I-2 - I-12, I.18-I.32. In: Fallout Program Quarterly Summary Report. (See Hardy; HASL-200, -et al.).

Volchok, H. L. 1972. High fallout in the western United States. pp. I.2-I.32. In: Fallout Program Quarterly Summary Report (March 1, 1972 - June 30, 1972). (See Hardy) (NSA 26: 45517).

Volchok, H. L. 1973. Worldwide fallout: its effects on science. pp. 227-239. In: Proc. Regional Conf. on Radiation Protection. Vol. 1. Y. Feige, ed. Israel Soreq Nuclear Research Centre. (NSA 30: 15386).

Volger, C. 1964. Contribution to the study of translocation of aminotriazol in Pteridium aquilinum (L.) Kuhn. Meded. Landb. bent. 29: 663-664. (^{14}C).

Volker, A. R. 1965. Determination of the density of rocks by the scintillation counter by means of the method of γ absorption of a small radiation beam. Geophys. geol. 7: 84-95. (In German) (NSA 22: 25513) (Incl. soils).

Volkova, A. P. 1963. The effect of gamma irradiation on the survival of Bacerium coli in an external environment. Gigiyena sanitariya 28(3): 101-103. (In Russian) (NSA 17: 35510).

Volkova, G. A. 1963. On the coefficient of accumulation of radioisotopes of some chemical elements by aquatic insects. Entomol. obozreniye 42(3): 516-519. (JPRS-22034, pp. 280-282) (NSA 18: 3456; 19: 25857).

Volkova, G. A. 1963. Contribution on the accumulation and excretion of radioactive isotopes of seven chemical elements by dragonfly larvae. Zool. zh. 42(1): 138-139. (JPRS-18548(Excerpt)) (NSA 17: 17738).

Volkova, M. Ya., G. I. Makhonina and A. A. Titlyanova. 1964. Effect of natural extracts upon the sorption of some radioisotopes by the soil. Pochvovedeniye (3): 52-57. (Sov. Soil Science (3): 275-270) (NSA 19: 53).

Vollenweider, R. A. 1969. A Manual on Methods for Measuring Primary Production on Aquatic Environments. IBP Habdbook. F. A. Davis, Philadelphia. 213pp. (^{14}C).

Vollenweider, R. A. and A. Nauwerck. 1961. Some observations on the C-14 method for measuring primary production. Verh. inern. Ver. theor. angew. Limnol. 14: 134-139. (SFA 8: 5598; WPA 35: 865).

Volodkin, B. B. 1970. Relationship between sulfur and phosphorus in the nutrition of Microcystis aeruginosa Kütz. Gidrobiol. zh. 6(3): ___. (Hydrobiol. J. 6(3): 47-54) (^{35}S, ^{32}P).

von Aufsess, G. 1967. Somatic mutations in wing scales of the dried fruit moth Plodia interpunctella Huebner after x irradiation of the caterpillars. Arch. Entwicklungsmech. Org. 158: 124-146. (In German) (NSA 22: 51505).

von Borstel, R. C. 1963. Effects of radiation on germ cells of insects: dominant lethals, gametic inactivation and gonial-cell killing. pp. 367-383. In: Radiation and Radioisotopes Applied to Insects of Agricultural Importance. (See IAEA, 1963) (NSA 17: 21572).

von Borstel, R. C. 1968. Control of adult pests by the irradiation-of-male method. pp. 331-337. In: Isotopes and Radiation in Entomology. (See IAEA, 1968) (NSA 24: 32326).

von Borstel, R. C. and W. St. Amand. 1963. Stage sensitivity to x-radiation during meiosis and mitosis in the egg of the wasp Habrobracon. pp. 87-100. In: Repair from Genetic Radiation. Pergamon Press, New York. (NSA 18: 3568).

von Borstel, R. C., et al. 1968. The Habrobracon experiment in the Biosatellite II spacecraft. Abstr. Radiation Research 35(2): 501.

von Borstel, R. C., et al. 1968. Mutational response of Habrobracon in the Biosatellite II experiment. In: Symp. on the Biosatellite II Experiments. BioScience 18(6): 598-601.

von Borstel, R. C., et al. 1969. Mutational responses of insects in the Biosatellite II Experiment . pp. 70-76. In: Life Sciences and Space Research VII. W. Vishniac and F. G. Favorite, eds. North-Holland, Amsterdam. (NSA 24: 829).

von Borstel, R. C., et al. 1970. Biological responses of Habrobracon to spaceflight. Life Sci. Space Res. 8: 6-11. (NSA 26: 41226).

von Borstel, R. C., et al. 1973. Habrobracon in the Biosatellite Experiment. pp. 701-705. In: Advances in Radiation Research. Vol. 2. Biology and Medicine. J. F. Duplan, ed. Gordon & Breach, New York. (NSA 31: 26189).

von Buttlar, H. 1967. Radioactivity and earth history. Bild Wiss. 4: 902-911. (In German) (NSA 22: 25522).

von Engelhardt, W., et al. 1975. Energy, water and urea metabolism in the llama. pp. 111-122. In: Tracer Studies on the Non-Protein Nitrogen for Ruminants II. International Atomic Energy Agency, Vienna, Publ. STI/PUB/389. (NSA 33: 29825).

von Hevesy, G. 1966. Radioactive tracers and their applications. Isotopes Radiat. Technol. 4(1): 9-12. (NSA 21: 17784).

von Hollen, H. J. 1972. PWR nuclear power plant systems for reducing radioactive releases. pp. 87-101. In: Proc. Southern Conf. on Environmental Protection from Nuclear Power Plants. (See U. S. EPA, 1972) (NSA 27: 4447).

Vorobotsova, I. Ye. 1964. A comparative investigation of the radiosensitivity of different strains of Drosophila melanogaster. Dokl. Akad. Nauk SSSR 153(4): 943-946. (Doklady (biol. sci.) 153(1-6): 1476-1479).

Vorob'yev, V. A., et al. 1970. Method for separating the radiation dose rate of artificial and natural radioactive isotopes in the ground. Atom. energiya 29(2): 133-135. (Sov. Atom. Energy 29(2): 840-842) (NSA 25: 8360).

Voronin, M. V. 1956. Apparatus for application of radioactive isotopes on external covering of insects. Abstr. Vses. nauchn.-issled. inst. vet. sanit. ektoparazitol. (nauch.-tekh. inform.) Bl: 26-27. (In Russian).

Voronitskaya, I. Ye. 1965. Biogenic migration in Lake Issyk-kul'. Dissertation. (In Russian).

Vos, J. 1973. Anionic electrophoretic pattern of five ruthenium salts in fresh and sea water: effects of ageing and dilution. Anal. Chim. Acta 67: 480-485. (NSA 29: 20851).

Voss, M. D. Survey of environmental radioactivity ---. Ames Lab., Iowa State Univ., Ames, Ia., U. S. AEC report.

Year Publ.	Year Reptd.	Report No.	pp.	NSA
1968	1967	IS-1776	45	22: 25532
1970	1969	-2260	41	24: 21149
1971	1970	-2541	37	25: 34743
1971	1971(½)	-2658	20	25: 55162
1972	1971(½)	-2926	22	26: 50590
1972	1971	-2791	39	26: 22778
1973	1972	-3048	44	27: 19817
1974	1973	-3313	56	30: 9525

Vostokova, T. A. and Ye. B. Tyurukanova. 1972. Migration of strontium-90 and cesium-137 in soil-plant cover of moraine regions. State Committee on the Use of Atomic Energy USSR, Moscow. 16pp. (In Russian) (INIS-mf-1043) (NSA 30: 6524).

Vovk, I. F. 1970. Method to determine the absorbed dose of natural radioactivity in underground waters. Dopovidi Akad. Nauk Ukr. RSR B(10): 881-883. (In Ukrainian) (NSA 25: 15773).

Vovk, I. F. and A. V. Litovchak. 1975. Effect of radioactive radiation on the interaction of rocks and minerals with water. pp. 171-177. In: Radio-Elements in Rocks. V. A. Kuznetsov, R. S. Zhuravlev and D. K. Osipov, eds. "Nauka", Novosibirsk. (In Russian) (NSA 33: 11545).

Vredenburg, R. T. and W. O. Willis. 1970. Effects of snow on soil water measurement by neutron moderation. Can. J. Soil Sci. 50: 463-464. (NSA 25: 15769).

deVries, H. 1958. Variation of concentration of radiocarbon with time and location on earth. Proc. Koninkl. Ned. Akad. Vet. B61: 94-102.

deVries, H. 1958. Atomic bomb effect: variation of radiocarbon in plants, shells and snails in the past 4 years. Science 128(3318): 250-251. (NSA 12: 13657).

Vroman, H. E., J. N. Kaplanis and W. E. Robbins. 1964. Chloresterol turnover in the American cockroach, Periplaneta americana. J. Lipid Res. 5(3): 418-421. (^{14}C).

Vroman, H. E., J. N. Kaplanis and W. E. Robbins. 1965. Effect of allatectomy on lipid biosynthesis and turnover in the female American cockroach, Periplaneta americana (L.). J. Insect Physiol. 11(7): 897-904. (^{14}C).

Vukmirovic, V. 1963, 1965. The application of radioactive tracers in the study of sand movements in rivers. --- Report No. 4(Final). Yugoslavia Inst. za Vodoprivredu "Jaroslav Cerni", Belgrade, IAEA contract 207/Rl/RB. 114, 93pp. (NP-14667) (NSA 19: 13618; 20: 43549).

Vukmirovic, Z. M. and M. D. Matic. 1968. Distribution of hydrogen and oxygen isotopes in some natural sites in Yugoslavia. Bull. Soc. Chim. Beograd. 33: 265-270. (In Serbo-Croatian) (NSA 24: 25122).

Vukmirovic, V, et al. 1964. Radioactive tracers in studying bed load movements. Saopstenja Inst. Vodoprivredu "Jaroslav Cerni" 9(31): 31-38. (English transl. OTS 64-11459/2, pp. 31-38).

Wachholz, B. W. (Comp.). 1974. Plutonium and Other Transuranium Elements: Sources, Environmental Distribution and Biomedical Effects. U. S. AEC report WASH-1359. 327pp. (NSA 31: 16524).

Wachsmann, F. 1950. Differences in the biological effect of conventional and ultrahard radiation. Strahlentherapie 81: 273-280. (AEC-tr-3442) (NSA 13: 1057).

Wachsmann, F. and J. Schwibach. 1971. Considerations for siting nuclear power plants in areas with high population density in the Federal Republic of Germany. pp. 791-802. In: Environmental Aspects of Nuclear Power Stations. (See IAEA, 1971).

Wack, B. F. 1965. Apparatus for measuring the humidity and mass of soils. U. S. Patent 3,428,806. (NSA 23: 16171).

Wack, B. 1967. Experimental study of the water content measurements of soil using intermediate neutron transmissions. Compt. rend. AB264: 1098-1101. (AEC-tr-6977) (NSA 23: 433).

Wack, B. 1969. Radioactivity measurement in soil mechanics. Bull dir. étud. rech. A(2): 1-135. (In French) (NSA 24: 16688).

Wada, S. 1966. Analysis of the head-neck region of Tachycines (Saltoria) by morphological units. II. Experimental-teratological findings in the cranial skeleton with reference to the central nervous system. Zool. Jahrb. Abt. Anat. Ontog. Tiere 83: 235-326. (In German) (NSA 21: 18141).

Waddington, C. H. 1942. Some developmental effects of x-rays in Drosophila. J. Exptl. Biol. 19(2): 101-117.

Waddington, C. H. 1959. Autoradiography and some of its uses. pp. 115-122. In: Marine Biological Applications of Radioisotope Research Techniques. Pubbl. stazi. zool. Napoli 31(Suppl.). (See Italy Zoological Station of Naples, 1959).

Wadley, G. W. 1971. Effect of x-irradiation under various temperature conditions and embryonic developmental stages on the growth and mortality of chinook salmon. Thesis, Univ. of Washington, Seattle. 125pp. (Univ. Micro-Microfilms 71-28488) (NSA 26: 4838, 23112).

Wadley, G. W. and A. D. Welander. 1971. X-ray and temperature: combined effects on mortality and growth of salmon embryos. Trans. Am. Fish. Soc. 100 (2): 267-275. (NSA 25: 44925) (SFA 16: 13545).

Wagner, C. K. 1968. Relationship between oxygen consumption, ambient temperature and excretion of 32-phosphorus in laboratory and field populations of cotton rats. Thesis, Univ. of Georgia, Athens. 37pp. (At Savannah River Plant).

Wagner, C. K. 1970. Oxygen consumption, ambient temperature, and excretion of ^{32}P in cotton rats. Ecology 51(2): 311-317. (NSA 24: 34230).

Wagner, G. A. and M. Reimer. 1972. Dating error. Nature 237(5349): 57. (NSA 26: 53181).

Wagner, G. H. and V. K. Mutatkar. 1968. Amino components of soil organic matter formed during humification of ^{14}C glucose. Soil Sci. Soc. Am. Proc. 32(5): 683-686.

Wagner, J. F. and M. Parker. 1973. Primary production and limiting nutrients in a small subalpine Wyoming lake. Trans. Am. Fish. Soc. 102(4): 698-706. (SFA 19: 17075) (^{14}C).

Wagner, P. T. 1958. Development of a procedure for estimating the amount of radioactivity in brackish water. Mare Island Naval Shipyard, U. S. Navy, Mare Island, Calif., report 6051-57. 19pp. (AD-202977; NSA 14: 6537).

Wagner, R. H. 1964. The autumnal seed rain in an irradiated forest. Abstr. Bull. Ecol. Soc. Am. 45(3): 73.

Wagner, R. H. 1965. Radiation and successional pattern. Abstr. Bull. Ecol. Soc. Am. 46(3): 98.

Wagner, R. H. 1965. The annual seed rain of adventive herbs in a radiation damaged forest. Ecology 46(4): 517-520.

Wagner, R. H. 1966. Herbaceous recolonization of the Brookhaven irradiated forest: the first three years. Radiation Botany 6(6): 561-565.

Wagner, R. H. 1971. Padiation. Ch. 11, pp. 196-221. In: _Environment and Man_. W. W. Norton, New York.

Wagner, R. H. and T. G. Marples. 1966. The breeding success of various passerine birds under chronic radiation stress. Auk 83(3): 437-440.

Wagner, W. W. 1973. Cesium-137 accumulation and distribution in plants and soils of an alpine community of Utah. Thesis, Univ. of Utah, Salt Lake City. 286pp. (Univ. Microfilms 73-21508) (NSA 29: 377).

Wagoner, D. E. 1969. Linkage group-karyotype correlation in the house fly, _Musca domestica_ L., confirmed by cytological analysis of x-ray induced y-autosomal translocations. Genetics 62: 115-121. (NSA 24: 659).

Wagoner, D. E., C. A. Nickel and O. A. Johnson. 1969. Chromosomal translocation heterozygotes in the house fly. Egg hatch, sex ratios and transmission to progeny of 193 translocations. J. Heredity 60: 301-304. (NSA 24: 36738) (X-rays).

Wahlgren, M. A. and J. S. Marshall. 1974. Distribution studies of plutonium in the Great Lakes. pp. 267-275. In: _Proc. 2nd Intern. Conf. on Nuclear Methods in Environmental Research_. (See Vogt and Meyer, 1974) (NSA 33: 23276).

Wahlgren, M. A. and J. S. Marshall. 1975. Behavior of plutonium and other long-lived radionuclides in Lake Michigan. I. Biological transport, seasonal cycling, and residence times in the water column. Argonne National Lab. 21pp. (CONF-750662-2) (NSA 32: 24344).

Wahlgren, M. A. and D. M. Nelson. 1974. Studies of plutonium cycling and sedimentation in Lake Michigan. pp. 212-218. In: _Proc. 17th Conf. on Great Lakes Research_. (NSA 33: 25979).

Wahlgren, M. A., et al. 1974. Spatial distribution of radionuclides in Lake Michigan biota near the Big Rock nuclear plant. Argonne National Lab. 15 pp. (CONF-750503-17) (NSA 32: 6441).

Wahlstrøm, G. 1969. Effects of radiothyroidectomy by ^{131}I on the self-selected circadian rhythm and activity in the canary. Acta Soc. Med. Upsal. 74: 161-178. (NSA 24: 32239).

Wai, N. and L. Ou. 1960. Co^{60} radiation-induced phenotypic mutation of soil bacteria. Bull. Inst. Chem. Acad. Sinica (2): 1-28. (NSA 15: 23949).

Waid, J. S., K. J. Preston and P. J. Harris. 1973. Autoradiographic techniques to detect active microbial-cells in natural habitats. pp. 317-322. In: _Modern Methods in the Study of Microbial Ecology_. T. Rosswall, ed. NFR Editorial Services, Stockholm. (NSA 29: 890).

Wainerdi, R. R., L. E. Fite and W. E. Kuykendall. 1971. Nuclear analytical methods and systems for the measurement of trace constituents in the environment. pp. 459-480. In: _Nuclear Techniques in Environmental Pollution_. (See IAEA, 1971) (NSA 25: 31884).

Waisal, Y. and A. Fahn. 1965. A radiological method for the determination of cambial activity. Physiologia 18: 44-46. (NSA 20: 12614) (^{14}C).

Waisal, Y., N. Liphschitz and T. Arzee. 1967. Phellogen activity in _Robinia pseudacacia_ L. New Phytol. 66: 331-335. (^{14}C).

Waisal, Y., I. Noah and A. Fahn. 1966. Cambial activity in _Eucalyptus camaldulensis_ Dehn. II. The production of Phloem and xylem elements. New Phytol. 65: 319-324. (^{14}C).

Waite, D. A. 1973. Analytical technique for distributing air ampling locations around nuclear facilities. Battelle Northwest Lab., U. S. AEC report BNWL-SA-4534. 12pp. (CONF-730603) (NSA 28: 12861).

Waite, D. A. 1974. Impact of radiation on man and his environment. Battelle Northwest Lab., U. S. AEC report BNWL-SA-5013. 6pp. (NSA 30: 12912).

Waite, D. A. and P. E. Bramson. 1975. Interpretation of near-background environmental surveillance data by distribution analysis. Battelle Northwest Lab., U. S. ERDA report BNWL-SA-5355. 23pp. (NSA 33: 7289).

Waite, D. A. and D. H. Denham. 1974. Use of critical pathway models and log-normal frequency distributions for siting nuclear facilities. pp. 407-415. In: _Siting of Nuclear Facilities_. International Atomic Energy Agency, Vienna, Publ. STI/PUB-384. (NSA 31: 11278; 32: 24444).

Waite, D. A. and D. H. Denham. 1975. Log-normal statistics applied to the planning, collection, and analysis of preoperational environmental surveillance data. pp. 150-163. In: _Radiation Protection Measurement: Philosophy and Implementation_. (See Recht and Lakey, 1975) (Also: Battelle Northwest Lab., U. S. AEC report BNWL-SA 4840) (NSA 30: 3141; 33: 26191).

Wakid, A. F. M. 1973. Effect of nitrogen during γ irradiation of purpuria and adults of the Mediterranean fruit fly on emergence, sterility, longevity and competitiveness. Environ. Entomol. 2(1): 37-40. (NSA 29: 10467).

Wakid, A. M. and M. Y. Y. Ahmed. 1972. Effects of gamma radiation on the adult stage of the lesser cotton leaf worm, Spodoptera exigua (Hb.). Z. angew. entomol. 71(4): 355-359. (NSA 28: 6057).

Wakid, A. F. M. and J. M. Hayo. 1974. Inherited sterility in progeny of irradiated male cotton leaf worm, Spodoptera littoralis Boisd. pp. 125-130. In: Isotopes in Pesticides and Pest Control. (See Nehme and Hassan, 1974) (NSA 33: 7561).

Wakid, A. M., A. A. Abdel-Malek and A. O. Tantawy. 1969, 72, 75. Studies on the eradication of Anopheles pharoensis Theobald by the sterile male technique using ^{60}Co. VIII. Uptake, distribution, persistence and decay of radiophosphorus in adults of Anopheles pharoensis. X. Biological effects of larval labelling with ^{32}P on different characters related to fitness components. XI. Release-capture experiments for flight range and dispersion. Isotopes Radiat. Res. 2: 49-54; 5: 27-30. pp. 507-513. In: Sterility Principle for Insect Control, 1974. (See IAEA, 1975) (NSA 24: 44408; 27: 24980; 32: 11805).

Wakid, A. M., E. A. Elbadry and M. A. Elaal. 1972. Effects of gamma radiation on the fertility of the spider mite Tetranychus arabicus Attiah. --- II. Irradiation of larvae. Ann. Zool.-Ecol. Anim. 4(3): 375-378, 379-383. (NSA 27: 24975, 24976; 28: 8480, 8481).

Wakid, A. M., et al. 1970. Biological effects of gamma radiation on different characters related to fitness components of the cotton leaf worm Spodoptera littoralis Boisd. Isotopes Radiat. Res. 3: 35-40. (NSA 25: 21827).

Wakita, H. and K. Kigoshi. 1964. Activation analysis of thorium in shells. J. Chem. Soc. Japan (Pure chem.) 85(8): 476-480. (In Japanese).

Wakshal, E. and F. Yaron. 1974. ^{234}U/^{238}U disequilibrium in waters of the Judea Group (Cenomanian-Turonian) Aquifer in Galilee, northern Israel. pp. 151-177. In: Isotope Techniques in Groundwater Hydrology. Vol. 2. (See IAEA, 1974) (NSA 31: 19652).

Walburg, H. E., Jr. 1973. Role of the microbial environment in radiation-induced life shortening and its relation to host-defense systems. pp. 1367-1380. In: Advances in Radiation Research. Vol. 3. Biology and Medicine. J. F. Duplan, ed. Gordon & Breach, New York. (NSA 31: 26177) (Laboratory mice).

Walburg, H. E., Jr. 1975. Radiation-induced life-shortening and premature aging. Adv. Radiat. Biol. 5: 145-179. (NSA 32: 28300).

Walder, J. M. M. and F. M. Wiendl. 1974. Effects of high dose of gamma radiation on the longevity of adults of Callosobruchus maculatus (Fabr. 1792) (Col. Bruchidae). Brazil Centro de Energia Nuclear na Agriculture, Piracicaba, report CENA-BC-013. 11pp. (In Portugese) (NSA 33: 29751).

Walden, S. J. 1973. Effects of tritiated water on the embryonic development of the three-spine stickleback, Gasterosteus aculeatus, Linnaeus. pp. 1087-1090. In: Radionuclides in Ecosystems. (See Nelson, 1973) (NSA 26: 2415).

Waldichuk, M. 1971. A summary report of: Radioactivity in the Marine Environment. U. S. AEC report WASH-1185. vi, 28pp.

Waldrop, R. H., R. Peel and H. F. Schoof. 1958. Fly movement study shows benefits of control program. Mod. Sanit. Build. Maint. 10(9): 43-45. (^{32}P).

Walker, C. H., G. A. El-Zorgani and J. Kenny. 1974. Radiotracer-aided studies on the comparative metabolism of dieldrin analogues by vertebrates. pp. 65-66. In: Isotope Tracer Studies of Chemical Residues in Food and the Agriculture Environment. International Atomic Energy Agency, Vienna, Publ. STI/PUB-363. (NSA 30: 10200).

Walker, D., R. Harwood and K. Groves. 1956. Determining insect populations in empty grain bins by the use of P^{32}. Abstr. Bull. Entomol. Soc. Am. 2(3): 25.

Walker, D. W. 1968. Potential for control of the sugar-cane borer through radio-induced sterility. pp. 131-140. In: Radiation, Radioisotopes, and Rearing Methods in the Control of Insect Pests. (See IAEA, 1968) (NSA 24: 32332).

Walker, D. W. 1973. Insect sterility program. Technical report No. 7. Puerto Rico Nuclear Center, U. S. AEC report PRNC-161. 41pp. (NSA 29: 5357).

Walker, D. W. and V. Quintana-Muñiz. 1966. Mortality staging of dominant lethals induced in the sugarcane borer Diatraea saccaralis (F.) (Crambidae, Lepidoptera). Bull. Entomol. Soc. Am. 12: 250.

Walker, D. W. and V. Quintana. 1967. Gamma-induced sterility in the sugarcane moth (Diatraea saccharalis (Fab.)) (Lepidoptera: Crambidae). Abstr. Eb-1. Radiation Research 31(2): 612.

Walker, D. W. and V. Quintana. 1967. Inherited sterility in survivors of irradiated Diatraea saccharalis (Fab.) (Lepidoptera: Crambidae). Abstr. 107. Bull. Entomol. Soc. Am. 13(3): 192.

Walker, D. W. and V. Quintana. 1968. Inherited partial sterility among survivors from irradiation-eradication experiment. J. Econ. Entomol. 61: 318-319. (NSA 22: 23700) (Insect).

Walker, D. W. and V. Quintana-Muñiz. 1968. Mortality staging of dominant lethals induced in the F_1 generation of the sugarcane borer; Diatraea saccharalis (F.). Radiation Research 36: 138-143. (NSA 22: 52124).

Walker, D. W., A. Alemany and M. Figueroa. 1964. Induced sterility of adult Diatraea saccharalis (Fab.) by gamma irradiation. Abstr. Bull. Entomol. Soc. Am. 10(3): 167.

Walker, D. W., V. Quintana-Muñiz and F. Padovani. 1971. Effect of gamma irradiation of immature sugarcane borers. pp. 513-524. In: Sterility Principle for Insect Control or Eradication. (See IAEA, 1971) (NSA 25: 29755)

Walker, D. W., V. Quintana and J. Torres. 1971. Genetic collapse of insect populations. I. Extinction of inbred and outbred lines in laboratory populations of the sugarcane borer. J. Econ. Entomol. 64(3): 660-667. (NSA 25: 42611) (γ irradiation).

Walker, D. W., H. Singh and K. P. MacKay. 1975. Gamma-induced sterility of the greater wax moth, Galleria mellonella L., Pyralidae, Lepidoptera. pp. 585-592. In: Sterility Principle for Insect Control, 1974. (See IAEA, 1975) (NSA 32: 11813).

Walker, J. R. 1962. Evaluation of control of European corn borer, Ostrinia nubilalis (Huebner), by x-ray induced sterility. Dissertation, Iowa State Univ., Ames. (Dissertation Abstr. 23: 761).

Walker, J. R. 1966. Reproductive potential of the sweet potato weevil after exposure to ionizing radiations. J. Econ. Entomol. 59: 1206-1208. (NSA 21: 24695).

Walker, J. R. and T. A. Brindley. 1963. Effect of x-ray exposure on the European corn borer. J. Econ. Entomol. 56(4): 522-525. (NSA 18: 5052).

Walker, L. C. 1959. Isotope tracer methods in tree nutrition. pp. 25-30. In: Proc. 1st North American Forest Soils Conf. (East Lansing, Mich., 1958).

Walker, L. C., E. Beck and E. Dumbroff. 1959. Absorption rates of 1,4,5-T, tagged with carbon-14. Forest Science 5: 128-137.

Walker, P. T. 1965. Phorate for insect control in maize. I. Control of frit fly. II. Uptake and distribution of radioactive phorate. Tropical Pesticides Research Unit, Porton Down, Salisbury, England, report No. 293. (^{32}P).

Walker, R. B., E. E. Held and S. P. Gessel. 1961. Radiocesium in plants grown on Rongelap Atoll soils. pp. 1363-1367. In: Recent Advances in Botany. Univ. of Toronto Press, Toronto. (NSA 17: 2893).

Walker, R. L. 1975. Uptake, distribution, and incorporation of ^{59}Fe in tissue and blood of rainbow trout (Salmo gairdneri). Dissertation, Michigan State Univ., East Lansing. (Univ. Microfilms 75-20000) (NSA 33: 9629).

Wall, J. P. 1967. Analysis of trace elements and of radioactive backgrounds in lakes. Final report. Univ. of Wisconsin, Madison, U. S. AEC report COO-1281-9. (NSA 22: 16697).

Wall, M. A. 1964. Application of statistical techniques to environmental radiation surveillance. Publ. Hlth. Rept. 79: 1057-1064.

Wallace, A. and R. T. Mueller. 1969. Effect of chelating agents on the availability to plants of carrier-free ^{59}Fe and ^{65}Zn added to soils to simulate contamination from fallout. Soil Sci. Soc. Am. Proc. 33: 912-914. (NSA 24: 34037).

Wallace, A. and E. M. Romney. 1971. Estimation of some aspects of nitrogen and phosphorus cycling in the northern Mojave and southern Great Basin deserts. Abstr. Bull. Ecol. Soc. Am. 52(3): 19. (At Nevada Test Site).

Wallace, A. and E. M. Romney. 1972. Radioecology and ecophysiology of desert plants at the Nevada Test Site. Univ. of California at Los Angeles, U. S. AEC report TID-25954. viii, 439pp. (NSA 45755).

Wallace, A. and E. M. Romney. 1975. Feasibility and alternate procedures for decontamination and post-treatment management of Pu-contaminated areas in Nevada. pp. 251-337. In: Radioecology of Plutonium and Other Transuranics in Desert Environments. (See White and Dunaway, 1975) (NSA 31: 3390; 33: 621).

Wallace, A., W. A. Rhoads and E. F. Frolich. 1968. Germination behavior of Salsola as influenced by temperature, moisture, depth of planting, and gamma irradiation. Agron. J. 60: 76-78. (NSA 22: 40957).

Wallace, A., E. M. Romney and R. A. Wood. 1973. Cycling of stable cesium in a desert ecosystem. pp. 183-186. In: Radionuclides in Ecosystems. (See Nelson, 1973) (NSA 26: 2415).

Wallace, H., S. Wessels and H. Conn. 1971. Radioresistance of nerves in amphibian limb regeneration. Arch. Entwicklungsmech. Org. 166: 219-225. (NSA 25: 32525).

Wallen, D. G. and G. H. Geen. 1968. Loss of radioactivity during storage of ^{14}C-labelled phytoplankton on membrane filters. J. Fish. Res. Bd. Can. 25: 2219-2224. (NSA 24: 2539).

Waller, H. D. and J. S. Olson. 1967. Prompt transfers of ^{137}Cs to the soils of a tagged Liriodendron forest. Ecology 48: 15-25.

Walls, N. W. nd. Physiological studies on Clostridium botulinum, Type F. Three year summary report of progress, March 1, 1965-November 30, 1967. Georgia Inst. Technology, Atlanta, U. S. AEC report ORO-3347-5. 17pp. (NSA 22: 33986).

Walter-Levy, L. and R. Strauss. 1968. The absorption of strontium by Chara fragilis Desvaux. Compt. rend. D266: 1486-1489. (In French) (NSA 22: 40520).

Walterscheid, E. C. 1966. Nuclear excavation. Ind. Res. : 54-59. (NSA 20: 31336).

Walton, A. and M. S. Baxter. 1968. Calibration of the radiocarbon time scale. Nature 220: 475-476. (NSA 23: 2416).

Walton, A., M. Ergin and D. D. Harkness. 1970. ^{14}C concentrations in the atmosphere and carbon dioxide exchange rates. J. Geophys. Res. 75: 3089-3098. (NSA 24: 31502).

Walton, A., et al. 1967. ^{14}C concentrations in environmental materials and their temporal fluctuations. pp. 41-47. In: Radioactive Dating and Methods of Low-Level Counting. (See IAEA, 1967).

Wambersie, A. and A. Dutreix. 1966. Sources of radiation. pp. 85-162. In: Applied Radiobiology. Vol. I. J. Loiseleur, ed. Gauthier-Villars, Paris. (In French) (NSA 21: 45528).

Wanderer, E. 1969. Accumulation of ^{235}U fission products on polymicron river sediments. Naturwissenschaften 56: 281. (In German) (NSA 23: 31423).

Wanderer, E. 1970. Studies on the selfcleaning capacity on radionuclides of a small river. pp. 165-178. In: Proc. Intern. Symp. on Radioecology. Vol. I. (See Societé Francaise de Radioprotection, 1970).

Wanderer, E. 1970. Radioprotective survey of the main drainage channel of the Reactor Center Seibersdorf. Isotop. Ind. Landwirt. 7(1): 14-19. (In German) (NSA 25: 21682).

Wang, A.-c. and C.-y. Wang. 1960. The effect of x-irradiation of the parent on the embryonic development of goldfish. Tung Wu Hsueh Pao 12: 127-130.

Wang, C. H. and D. L. Willis. 1965. Radiotracer Methodology in Biological Science. Prentice-Hall, Englewood Cliffs, N. J. xvii, 382pp. (Incl. experiments using freshwater plants and crayfish).

Wang, P. 1963. Ionizing-radiation-induced death. Wei Sheng Wu Hsueh Pao 201-202. (NSA 20: 26697).

Wang, S. and T. Wang. 1962. Study of the effect of ionizing radiation on the physiological activity of Azotobacter. Wei Sheng Wu Hsueh Pao 8: 446-454. (In Chinese) (NSA 22: 21428).

Wang, Y. (Ed.). 1969. Handbook of Radioactive Nuclides. CRC Press, Cleveland. (NSA 24: 50918).

Wangersky, P. J. 1963. Manganese in ecology. pp. 499-508. In: Radioecology. (See Schultz and Klement, 1963) (NSA 17: 33602).

Warburton, J. A. 1967. The tracing of radioisotopes released into the lower troposphere. p. 61. In: Isotope Techniques in the Hydrologic Cycle. (See Stout, 1967).

Ward, A. A. 1957. Outline of the use of radio-isotopes at the West African Cocoa Research Institute. pp. 89-91. In: South Africa Scientific Council of Africa South of the Sahara, Pretoria, CSA Publ. No. 28.

Ward, B. C., et al. 1972. Radiation mortality in the Chinese hamster, Cricetulus griseus, in relation to age. Radiation Research 51(3): 599-607. (NSA 26: 59092).

Ward, F. J. and M. Nakanishi. 1971. A comparison of Geiger-Mueller and liquid scintillation counting methods in estimating primary productivity. Limnol. Oceanogr. 16(3): 560-563.

Ward, F. J. and M. Nakanishi. 1973. A comparison of liquid scintillation and Geiger-Müller estimates of primary productivity in an in situ experiment. J. Fish. Res. Bd. Can. 30(5): 708-711.

Ward, F. J., B. Wong and G. G. C. Robinson. 1970. A liquid scintillation procedure for determining the effect of size on self-absorption of ^{14}C in Daphnia pulex. Limnol. Oceanogr. 15(4): 648-652.

Ward, P. R. B. 1967. Continuous recording of bird nesting visits using radioactive tagging. Nature 216(5115): 592-593.

Ward, P. R. B. 1968. Radioactive labeling of bird nest. Umsch. Wiss. Tech. 68(7): 217.

Ward, P. R. B. and P. Wurzel. 1968. The measurement of riverflow with radioactive isotopes with particular reference to the method and time of sampling. Int. Assoc. Sci. Hydrol. Bull. 13: 40-48. (NSA 22: 48952).

Ware, G. W. 1968. DDT-C^{14} translocation in alfalfa. J. Econ. Entomol. 61 (5): 1451.

Ware, G. W. and A. W. Guenthner. 1966. Effects of x-irradiation on toxicity of insecticides. Abstr. 33. Bull. Entomol. Soc. Am. 12: 252.

Ware, G. W. and C. K. Irwin. 1969. Conversion of a therapy x-ray unit for insect irradiations. J. Econ. Entomol. 62(3): 721-722.

Ware, G. W. and G. K. Whitacre. 1970. Similar effects of x-irradiation and piperonyl butoxide on toxicity of malathion and malaoxon to house flies. J. Econ. Entomol. 63(5): 1621-1622.

Ware, G. W., B. J. Estesen and W. P. Cahill. 1970. Uptake of C^{14}-DDT from soil by alfalfa. Bull. Environ. Cont. Toxicol. 5(1): 85-86.

Warembourg, F. R. 1975. Isotopic techniques for measuring the biological activity in plant rhizosphere. Rev. Ecol. Biol. Sol 12(1): 261-272. (In French) (NSA 33: 26368).

Warner, R. M. 1959. Radioactive tagging for tracing movements of Drosophila. Calif. Fig Inst. Proc. Ann. Res. Conf. 13: 35-37.

Warren, S. (Chairman). 1968. Session II: Biosatellite II. Radiation-weightlessness experiments. In: Symp. on the Biosatellite II Experiments. BioScience 18(6): 570.

Warren, S. E. 1973. Geometrical factors in the neutron activation analysis of archaeological specimens. Archaeometry 15(1): 115-122. (NSA 28: 34).

Washburn, E. W. and E. R. Smith. 1934. The isotopic fractionation of water by physiological processes. Science 79: 188-189. (Willow).

Washburn, G. E. 1952. Progress report of mosquito studies in California. Mosquito News 12(4): 235-238. (Used ^{32}P).

Washington Public Power Supply System. 1971. Hanford No. 2 Nuclear Power Plant. Environmental report. Kennewick, Wash. 314pp. U. S. AEC DOCKET-50397-2. (NSA 25: 48199).

Wasik, S. P. and W. Tsang. 1970. Determination of trace amounts of contaminants in water by isotope dilution gas chromatography. Anal. Chem 42: 1649-1651. (NSA 25: 2107).

Watanabe, N. and M. Suzuki. 1969. Fission track dating of archaeological glass materials from Japan. Nature 222: 1057-1058. (NSA 23: 33267).

Watanabe, W. 1960. Effect of γ-rays upon food microorganisms. V. Studies on the survival of Esch. coli irradiated with γ-rays. Bull. Agr. Chem. Soc. Jap. 24: 84-92. (NSA 15: 27412).

Waterhouse, D. F. 1962. Insect control by radiation sterilization in Australia. Int. J. Appl. Radiat. Isotopes 13: 435-439. (NSA 17: 15561).

Waters, T. F. 1961. Standing crop and drift of stream bottom organisms. Ecology 42(3): 532-537.

Watkins, M. J. 1964. Evidence for a lampbrush type structure in grasshopper spermatocyte chromosomes. Exptl. Cell Res. 36(1): 14-18. (^{3}H, Melanoplus differentialis).

Watson, D. G. 1975. Iodine-129 in aquatic organisms near nuclear fuels processing plants. Battelle Northwest Lab., U. S. ERDA report BNWL-SA-5411. 14pp. (NSA 32: 8920).

Watson, D. G. and J. J. Davis. 1957. Concentration of radioisotopes in Columbia River whitefish in the vicinity of the Hanford Atomic Products Operation. Hanford Atomic Products Operation, U. S. AEC report HW-48523(Del.). 136pp. (NSA 13: 1266).

Watson, D. G. and W. L. Templeton. 1973. Thermoluminescent dosimetry of aquatic organisms. pp. 1125-1130. In: Radionuclides in Ecosystems. (See Nelson, 1973) (NSA 25: 37795).

Watson, D. G., J. J. Davis and W. C. Hanson. 1961. Zinc-65 in marine organisms along the Oregon and Washington coasts. Science 133(3467): 1826-1828. (NSA 15: 19135).

Watson, D. G., J. J. Davis and W. C. Hanson. 1963. Interspecies differences in accumulation of gamma emitters by marine organisms near the Columbia River Mouth. Limnol. Oceanogr. 8(2): 305-309.

Watson, D. G., W. C. Hanson and J. J. Davis. 1964. Strontium-90 in plants and animals of Arctic Alaska, 1959-1961. Science 144(3621): 1005-1009.

Watson, D. G., et al. nd. Fallout radionuclides in the biota of the Cape Thompson region, Alaska - a listing of measurements. Project Chariot. Hanford Atomic Products Operation, U. S. AEC report PNE-480. 135pp.

Watson, D. G., et al. 1966. Radionuclides in terrestrial and freshwater biota. pp. 1165-1200. In: Environment of the Cape Thompson Region, Alaska. (See Wilimovsky and Wolfe, 1966) (NSA 20: 31428).

Watson, D. G., et al. 1968. Effect of Hanford reactor shutdown on Columbia River biota. Battelle Northwest Lab., U. S. AEC report BNWL-SA-1248. 21 pp. (Also: pp. 291-299. In: Symposium on Radioecology. (See Nelson and Evans, 1969) (NSA 21: 26407).

Watson, D. G., et al. 1970. Radioecological studies on the Columbia River. Part I. Part II. Battelle Northwest Lab., U. S. AEC report BNWL-1377. 46, vp. (NSA 24: 36726).

Watson, D. G., et al. 1971. Cycling of radionuclides in Columbia River biota. pp. 144-157. In: Trace Substances in Environmental Health. IV. (See Hemphill, 1971) (NSA 26: 15205).

Watson, G. M. 1972. Environmental hazards of fossil and nuclear power production. pp. 7.1-7.20. In: Symp. on the Role of Nuclear Energy in Australia's Development. (Canberra, Australia, 1 June 1972). Australia Atomic Energy Commission, Sydney. (Also: Lucas Heights, report AAEC-TM-627) (CONF-720627) (NSA 26: 45510, 56073).

Watson, G. M. 1973. Philosophy of environmental protection and control. pp. II.1-1 -II.1-5. In: Proc. of the AAEC Symp. on Environmental and Radiological Safety Aspects of the Mining and Processing of Uranium. (Lucas Heights, December 9-10, 1971). See Australia Atomic Energy Commission, 1973) (NSA 28: 24851).

Watson, H. 1970. Branch elongation and the radiation field. Ch. D-16, pp. D287-D294. In: A Tropical Rain Forest. (See H. T. Odum and R. F. Pigeon, 1970) (NSA 25: 21941).

Watson, J. E. 1973. Opening remarks for panel discussion on power plant siting. Tennessee Valley Authority, Chatanooga, Tenn. (NP-19964, Paper 8) (NSA 30: 6569).

Watson, J. P. 1971. Comparison of chromium ^{51}-versenate and tritiated water movement in a termite mound and soil. Soil Science 111(3): 188-191.

Watson, M. A. and H. L. Nixon. 1953. Studies on the feeding of Myzus persicae (Sulz.) on radioactive plants. Ann. Appl. Biol. 40(3): 537-545.

Watters, F. L. 1965. Physical methods of insect control. Proc. Entomol. Soc. Manitoba 21: 18-27.

Watters, F. L. 1966. Effects of gamma irradiation on the longevity and fertility of five species of stored-product insects. Bull. Entomol. Soc. Am. 12: 265.

Watts, J. R. and R. S. Harvey. 1963. Uptake and retention of Cs^{137} by a blue-green alga in continuous flow and batch culture systems. Limnol. Oceanogr. 8(1): 45-49. (NSA 17: 19798).

Wave, H. E. and T. J. Henneberry. 1967. Sterile-male releases control populations of Drosophila melanogaster in cage tests. J. Econ. Entomol. 60 (6): 1758-1759.

Weatherford, R. L. 1961. Preparation of samples for radioactivity counting. pp. 56-57. In: Radiological Aspects of Water Supplies. (See Univ. of Illinois, 1961).

Weatherley, P. E. and B. T. Watson. 1969. Some low-temperature effects on sieve tube translocation in Salix viminalis. Ann. Botany 33(133): 845-853. (^{14}C).

Weaver, C. L. 1967. Proposed radioactivity concentration guide for shellfish. Radiol. Hlth. Data Repts. 8: 491-494. (NSA 22: 23513).

Weaver, C. L., E. D. Harward and H. T. Peterson, Jr. 1969. Tritium in the environment from nuclear power plants. Publ. Hlth. Repts. 84: 363-371. (NSA 23: 38101).

Weaver, J. N. 1974. Rapid, instrumental neutron activation analysis for the determination of uranium in environmental matrices. Anal. Chem. 46(9): 1292-1294. (NSA 30: 18176).

Weaver, L. 1966. Stream surveillance programs. J. Water Poll. Control Fed. 38: 1334-1344. (NSA 21: 22416).

Weaver, L., A. W. Hoadley and S. Baker. 1963. Radioactivity in surface waters of the United States, 1957-1962. Radiol. Hlth. Data 4(6): 306-316. (NSA 18: 8572).

Weaver, L. E. (Ed.). 1970. Education for Peaceful Uses of Nuclear Energy. (Symp. held at Tucson, Arizona, March 31-April 2, 1969). Univ. of Arizona Press, Tucson. 356pp. (NSA 24: 27260).

Weaver, N. and J. H. Law. 1960. Heterogeneity of fatty acids from royal jelly. Nature 188: 938-939. (^{14}C).

Webb, K. L. and J. W. A. Burley. 1965. Dark fixation of $C^{14}O_2$ by obligate and facultative salt marsh halophytes. Can J. Bot. 43(2): 281-285.

Webb, K. L. and R. H. Hodgson. 1960. Some effects of ionizing radiation on translocation in plants. Science 132(3441): 1762-1763. (NSA 15: 7226).

Webb, J. V., et al. 1976. Temperature dependence of repair of sublethal radiation damage in Drosophila. Radiation Research 65(1): 187-190. (NSA 33: 26297).

Webb, S. J., D. V. Cormack and H. G. Morrison. 1964. Relative humidity, inositol and the effect of radiations on air-dried microorganisms. Nature 210(4584): 1103-1105. (NSA 18: 17551).

Webbe, G. and W. W. C. Read. 1962. The use of cobalt-60 for labeling snails in the study of field populations. Ann. Trop. Med. Parasitol. 56(2): 206-209. (NSA 17: 21406).

Weber, F. 1922. Forced static plants through roentgen radiation. Biochem. Zeitsch. 128: 495-507. (In German).

Webster, E. J. 1967. A autoradiographic study of invertebrate uptake of DDT-Cl^{36}. Ohio J. Sci. 67(5): 300-307.

Wegorek, W., K. Glogowski and E. Czaplicki. 1963. Possibility of contamination of insectivorous game birds during ecological investigations using insects labeled with ^{60}Co. Prace Nauk. Inst. Ochrony Rosl. 5(2): 19-28. (In Polish) (NSA 21: 6471).

Wegorek, W., K. Glogowski and E. Czaplicki. 1965. Investigation of hibernation of Perillus bioculatus Fabr. tagged with ^{60}Co. Ekol. Pol. 13A(23): 451-462.

Wehrhahn, C. F. 1973. An approach to modelling spatially heterogeneous populations and the simulation of populations subject to sterile insect re lease programs. pp. 45-64. In: Computer Models and Application of the Sterile-Male Technique. (See IAEA, 1973) (NSA 29: 8049).

Weichold, K. E. (Ed.). 1962. Bioenvironmental features of the Ogotoruk Creek area, Cape Thompson, Alaska: including predictions and assessments of nuclear detonations proposed for an excavation experiment. A second summary compiled by the Committee on Environmental Studies for Project Chariot. U. S. AEC, Washington. xi, 183pp. (NSA 17: 6273).

Weidhaas, D. E. 1965. Research on radioisotopes and radiation in insect control related to the development of the sterile-male technique. pp. 189-195. In: Proc. 5th Inter-American Symp. on the Peaceful Application of Nuclear Energy. Pan American Union, Washington. (NSA 19: 36098) (Screw-worm fly).

Weidhaas, D. E. 1973. Field studies on insect sterilization with mosquitoes, house flies and stable flies. pp. 137-139. In: Computer Models and Application of the Sterile-Male Technique. (See IAEA, 1973) (NSA 29: 8056).

Weidhaas, D. E. 1973. Principles of the sterile male technique with emphasis on ecology. pp. 17-22. In: Computer Models and Application of the Sterile-Male Technique. (See IAEA, 1973) (NSA 29: 8046).

Weidhaas, D. E. and C. H. Schmidt. 1962. Field studies on the release of sterile males for the control of Anopheles quadrimaculatus. Mosquito News 22(3): 283-291. (NSA 18: 19654).

Weidhaas, D. E., C. H. Schmidt and M. C. Bowman. 1960. Effects of heterogeneous distribution and codistillation on the results of tests with DDT against mosquito larvae. J. Econ. Entomol. 53(1): 121-125.

Weidhaas, D. E., C. H. Schmidt and W. F. Chamberlain. 1962. Metabolism of radio-labeled systemic insecticides in animals. pp. 93-98. In: Radioisotopes and Radiation in Entomology. (See IAEA, 1962).

Weidhaas, D. E., C. H. Schmidt and W. F. Chamberlain. 1962. Research on radiation in insect control. pp. 257-265. In: Radioisotopes and Radiation in Entomology. (See IAEA, 1962).

Weidinger, N. and E. Jahn. 1964, 1965. Insect labelling with rare earths for studying forest pests. Abstr. Anz. Schlädlingsk. 38: 16; 39: 77.

Weimer, W. C., M. G. Rodel and D. E. Armstrong. 1975. Loss of ^{32}P-phosphate and ^{14}C-carbonate activity during liquid scintillation counting of aqueous samples. Environ. Sci. Technol. 9(10): 966-972. (NSA 33: 19673).

Weinberg, A. M. 1973. Technology and ecology: is there a need for confrontation? BioScience 23(1): 41-45. (NSA 27: 14745) (Nuclear power).

Weinberger, D. and J. W. Porter. 1954. Metabolism of hydrogen isotopes by rapidly growing Chlorella pyrenoidosa cells. Arch. Biochem. Biophys. 50: 160-168. (NSA 8: 4207).

Weinbren, M. P. and B. M. Weinbren. 1970. Observations on the mosquito population in the irradiated forest at El Verde. Ch. E-12, pp. E159-E167. In: *A Tropical Rain Forest*. (See H. T. Odum and R. F. Pigeon, 1970) (NSA 25: 21819).

Weinbren, M. P. and B. M. Weinbren. 1970. A study of the natural virus cycles before, during, and after irradiation. Ch. F-9, pp. F57-F65. In: *A Tropical Rain Forest*. (See H. T. Odum and R. F. Pigeon, 1970) (NSA 25: 21970).

Weinmann, H. P. 1964. Investigations with labelled amino acids of the protein metabolism of normal and lethal genotypes of *Drosophila melanogaster*. Z. vergl. physiol. 48(4): 429-461. (In German, English summary) (^{14}C).

Weinreich, R. 1971. Determination of nuclides in radioactive fallout employing Ge(Li) detector. pp. 405-416. In: *Rapid Methods for Measuring Radioactivity in the Environment*. (See IAEA, 1971) (In German) (NSA 26: 27824).

Weir, R. N. 1964. An island of 1,000 inhabitants suitable for comparative studies on high background radiation. Cent. Afr. J. Med. 10: 169-173. (NSA 19: 31823) (Comments on fish, vegetation, and wildlife studies being of interest).

Weirich, G. and P. Karlson. 1969. Distribution of tritiated ecdysone in salivary glands and other tissues of *Rhynchosciara* and *Chironomus* larvae. An autoradiographic study. Arch. Entwicklungsmech. Org. 164: 170-181. (NSA 24: 27487).

Weisbart, M. 1973. The distribution and tissue retention of mercury-203 in the goldfish (*Carassius auratus*). Can. J. Zool. 51(2): 143-150. (SFA 18: 16404).

Weisbrot, D. R. 1967. Competitive interactions among irradiated strains of *Drosophila melanogaster*. Genetics 56(3,Pt.2): 596.

Weish, P. and E. Bruber. 1975. *Radioactivity and Environment*. Fischer, Stuttgart. 159pp. (NSA 32: 28241).

Weishaas, D. E. and C. H. Schmidt. 1963. Mating ability of male mosquitoes, *Aedes aegypti* L., sterilized chemically or by gamma radiation. Mosquito News 23(1): 32-34. (NSA 17: 28743).

Weiss, B. H., et al. 1975. Detailed measurement of ^{131}I in air, vegetation and milk around three operating reactor sites. U. S. NRC, Washington, report NUREG-75/021. 81pp. (NSA 33: 17187).

Weiss, C. M., et al. 1968. A review of the literature of 1967 on wastewater and water pollution control. J. Water Poll. Control Fed. 40: 897-1222. (NSA 23: 40837).

Weiss, G. B. 1966. The effect of potassium on nicotine-induced contracture and ^{45}Ca movements in frog sartorius muscle. J. Pharmacol. Exptl. Ther. 154: 595-604. (NSA 22: 6737).

Weiss, G. B. 1966. The effect of p_H on nicotine-induced contracture and ^{45}Ca movements in frog sartorius muscle. J. Pharmacol. Exptl. Ther. 154: 605-612. (NSA 22: 6738).

Weiss, G. B. 1972. Uptake and distribution of nicotine-^{14}C in frog rectus abdominis muscle. Arch. Int. Pharmacodyn. 195: 99-108. (NSA 28: 734).

Weiss, G. B. and C. P. Bianchi. 1965. The effect of potassium concentration on Ca^{45} uptake in frog sartorius muscle. J. Cell. Comp. Physiol. 65: 385-392. (NSA 19: 35912).

Weiss, H. V. 1956. Residual contamination of plants, animals, soils, and water of the Marshall Islands two years following Operation CASTLE fallout. pp. 205-210. In: *The Shorter-Term Biological Hazards of a Fallout Field*. G. M. Dunning and J. A. Hilchen, eds. U. S. AEC and U. S. DOD, Washington.

Weiss, H. V. 1971. Determination of selenium in glacial ice by radioactivation. Anal. Chim. Acta 56(1): 136-139. (NSA 25: 57394).

Weiss, H. V. and K. K. Bertine. 1973. Simultaneous determination of Mn, Cu, As, Ca, Sb, and Hg in glacial ice by radioactivation. Anal. Chim. Acta 65(2): 253-259. (NSA 28: 12638).

Weiss, H. V. and W. L. Shipman. 1957. Biological concentration by killer clams of cobalt-60 from radioactive fallout. Science 125(3250): 695.

Weiss, H. V., et al. 1956. Residual contamination of plants, animals, soil, and water of the Marshall Islands two years following Operation CASTLE fallout. U. S. Naval Radiological Defense Lab., San Francisco, Calif., report USNRDL-455. iv, 52pp.

Weissbuch, H., et al. 1973. Behavior of ^{137}Cs, ^{85}Sr, and ^{65}Zn radionuclides released to water basins. A study of an experimental laboratory model. pp. 421-430. In: *Health Physics Problems in Internal Contamination*. (See E. Bujdoso, 1973) (NSA 28: 29882).

Welander, A. D. 1945. Studies of the effects of roentgen rays on the growth and development of the embryos and larvae of the chinook salmon (*Oncorhynchus tschawytscha*). Univ. of Washington, Seattle, U. S. AEC report UWFL-2. 131pp.

Welander, A. D. 1954. Some effects of x-irradiation of different embryonic stages of the trout (*Salmo gairdnerii*). Growth 18(4): 227-255. (Also: Univ. of Washington, Seattle, U. S. AEC report UWFL-38) (NSA 9: 5844).

Welander, A. D. 1969. Distribution of radionuclides in the environment of Eniwetok and Bikini attols, August 1954. pp. 346-354. In: *Symposium on Radioecology*. (See Nelson and Evans, 1969).

Welander, A. D. 1971. Effects of irradiation and temperature on young salmonids. Final report. Univ. of Washington, Seattle, U. S. AEC report RLO-2225-T-8-5. 16pp. (NSA 26: 4839).

Welander, A. D., G. W. Wadley and D. K. Dysart. 1971. Growth and fecundity of rainbow trout (Salmo gairneri) exposed to single sublethal doses of x-rays during the eyed embryo stage. J. Fish. Res. Bd. Can. 28(8): 1181-1184. (SFA 17: 14295).

Welander, A. D., et al. 1948. The effects of roentgen rays on the embryos and larvae of the chinook salmon. Growth 12(3): 203-242. (Also: Univ. of Washington, U. S. AEC report UWFL-8; MDDC-1689) (NSA 2: 27).

Welander, A. D., et al. 1949. The effects of roentgen rays on adult rainbow trout. Univ. of Washington, Seattle, U. S. AEC report UWFL-17; AECU-188. 20pp. (NSA 2: 1888).

Welander, A. D., et al. 1966, 1967. Bikini-Eniwetok studies, 1964. Part. I. Ecological observations. Part II. Radiobiological studies. Univ. of Washington, Seattle, U. S. AEC report UWFL-93(Pt.1, Pt.2). 277, 233pp. (NSA 23: 9909, 9944).

Welander, A. D., et al. 1971. Growth and fecundity of rainbow trout (Salmo gairdneri) exposed to single sublethal doses of x rays during the eyed embryo stage. J. Fish. Res. Bd. Can. 28(8): 1181-1184. (NSA 26: 9634).

Welch, G. P. 1968. Some effects of chronic irradiation on a steady-state yeast population. Radiation Research 36: 274-286. (NSA 23: 6624).

Welch, H. E. 1960. Two applications of a method of determining the error of population estimates of mosquito larvae by the mark and recapture technique. Ecology 41(1): 228-229.

Welch, N. H. and P. B. Allen. 1973. Field calibration and evaluation of a nuclear sediment gage. Water Resources Res. 9(1): 154-158. (NSA 28: 12961).

Wellde, B. T. and E. H. Sadun. 1968. Resistance produced in rats and mice by exposure to irradiated Plasmodium berghei. pp. 59-74. In: Isotopes and Radiation in Parasitology. (See IAEA, 1968).

Weller, R. I. 1964. Low-level radioactive contamination. pp. 567-576. In: The Natural Radiation Environment. (See Adams and Lowder, 1964) (NSA 18: 35441).

Wells, C. 1960. The response of Tetrahymena pyriformis to ionizing radiation strain specific radiosensitivities. J. Cell. Comp. Physiol. 55(2): 207-219.

Wells, C. 1961. Evidence for micronuclear function during vegetative growth and reproduction of the ciliate, Tetrahymena pyriformis. J. Protozool. 8 (3): 284-290.

Wells, M. M. 1962. Nuclear differentiation in Tetrahymena pyriformis as affected by near infrared and x-radiation. (Dissertatation Abstr. 22: 3824-3825; NSA 16: 21938).

Wells, M. R. and J. D. Yarbrough. 1972. Retention of ^{14}C-DDT in cellular fractions of vertebrate insecticide-resistant and susceptible fish. Toxicol. Appl. Pharmacol. 22(3): 409-414. (SFA 17: 15401) (Gambusia).

Wells, M. R. and J. D. Yarbrough. 1973. In vivo and in vitro retention of (^{14}C) aldrin and (^{14}C) dieldrin in cellular fractions from brain and liver tissues of insecticide-resistant and susceptible Gambusia. Toxicol. Appl. Pharmacol. 24(2): 190-196. (SFA 18: 16205).

Welte, E. and K. Mechsner. 1959. Biological action of potassium-40 in naturally occurring potash salts. Nature 184(Suppl. 25): 1958-1959. (Used Torula utilis).

Welty, C. G., Jr. and M. B. Biles. 1973. U. S. Atomic Energy Commission program for monitoring the behavior of radionuclides released to the environment. pp. 139-156. In: Environmental Behavior of Radionuclides Released in the Nuclear Industry. (See IAEA, 1973) (NSA 29: 29302).

Wendt, I, et al. 1967. Model experiments for ^{14}C water-age determinations. pp. 321-337. In: Isotopes in Hydrology. (See IAEA, 1967) (NSA 22: 1893).

Weng, P. S. and A. P. L. Liu. 1971. ^{90}Sr and ^{137}Cs in canned seafood, Taiwan, Republic of China, 1970-1971. Radiol. Hlth. Data 12(12): 615-616. (NSA 26: 22804).

Weng, P. S., T. C. Chu and C. M. Tsai. 1973. Environmental radioactivity on north Taiwan. Health Physics 25(2): 123-128. (NSA 28: 21077).

Weng, P. S., C. M. Tsai and T. C. Chu. 1972. Environmental radioactivity and radiation measurements in Taiwan, Republic of China. pp. 375-392. In: Natural Radiation Environment. (See Adams, Lowder and Gesell, 1972) (NSA 32: 19719).

Weng, P. S., et al. 1971. Background activity of Taiwan. Hoken Butsuri 6 (3): 131-136. (NSA 26: 48041).

Wensel, A. 1964. Measurement of environmental radiation with plastic scintillators. pp. 643-654. In: The Natural Radiation Environment. (See Adams and Lowder, 1964) (NSA 18: 35891).

Went, D. F. and F. E. Wuergler. 1972. Sterilization of paedogenetic Heteropeza larvae with x-rays. Experientia 28(1): 100-101. (NSA 26: 25874).

Wenzel, D. R. 1970. Environmental monitoring of fission product release from the loss-of-fluid test. pp. 118-128. In: Environmental Surveillance in the Vicinity of Nuclear Facilities. (See Reinig, 1970) (NSA 25: 16168).

Werkema, G. J. and M. A. Thompson. 1974. Annual environmental monitoring report. Rocky Flats Plant, January through December 1973. Dow Chemical Co., Rocky Flats, Plant, U. S. AEC report RFP-ENV-73. 37pp. (NSA 30: 6503).

Wermuth, J. F. 1968. The effects of x-irradiation on stolon growth and hydranth production in Campanularia flexuosa: glutathione as a radio-protective agent. Thesis, Indiana Univ., Bloomington. 81pp. (NSA 23: 31991) (Hydroid).

Werner, L. B. 1969. Radioactivity in the hydrologic environment. pp. 280-305. In: *Proc. Symp. on Public Health Aspects of Peaceful Uses of Nuclear Explosives*. (Las Vegas, Nev., April 7-11, 1969). Southwestern Radiological Health Lab., U. S. PHS report SWRHL-82. (NSA 24: 22772).

West, A. S. 1966. Tagging and recovery of black flies. Bull. Entomol. Soc. Am. 12: 289.

West, A. S. and D. W. Jenkins. 1951. Plant feeding habits of northern mosquitoes studies with radioisotopes. Mosquito News 11(4): 217-219.

West, F. K., P. W. West and T. V. Ramakrishna. 1967. Stabilization and determination of traces of silver in waters. Environ. Sci. Technol. 1: 717-720. (NSA 21: 45238).

West African Cacao Research Institute. 1953, 1957. Annual report, April 1952-March 1953. --- 1955-56. 39, 20pp. (Entomological use of ^{32}P).

Westerdahl, H. E. 1973. Particulate organic ^{32}P in a simulated sediment-water system. Thesis, Univ. of Oklahoma, Norman. 201pp. (Univ. Microfilms 74-12326) (NSA 30: 18657).

Westerman, J. M. and P. A. Parsons. 1972. Radioresistance and longevity of inbred lines of *Drosophila melanogaster*. Int. J. Radiat. Biol. 21(2): 145-152. (NSA 26: 24873).

Westerman, J. M. and P. A. Parsons. 1973. Asymmetrical responses to directional selection for radiation resistance and sensitivity in Drosophila. Experientia 29(6): 722-724. (NSA 28: 21787).

Westerman, J. M. and P. A. Parsons. 1973. Variations in genetic architecture at different doses of gamma radiation as measured by longevity in *Drosophila melanogaster*. Can. J. Genet. Cytol. 15(2): 289-298. (NSA 29: 8063).

Westerman, M. 1967. The effect of x-irradiation on male meiosis in *Schistocerca gregaria* (Forskal.). I. Chiasma frequency response. Chromosoma 22 (4): 401-416.

Westerman, M. 1972. Dose-response of chiasma frequency to x-radiation in *Chorthippus brunneus*. Mutation Research 15(1): 55-65. (NSA 26: 31364).

Westermark, T. 1972. Activation analysis of mercury in environmental studies. Adv. Activ. Anal. 2: 57-88. (NSA 26: 30663).

Westinghouse Electric Corporation. 1958. Pre-operational radiation survey of the Shippingport Atomic Power Station site and surrounding area. Bettis Plant, Pittsburgh, Pa., U. S. AEC report WAPD-CTA(IH)-208. 88pp. (NSA 12: 11259).

Westley, B. (Ed.). 1971. *Identification and Measuerement of Environmental Pollutants*. (Proc. Symp., Ottawa, Ontario, Canada, June 14, 15, 16, 17, 1971). Canada National Research Council, Ottawa. 458pp. (NSA 29: 2630).

Wetch, J. R. 1970. Air pollution and power generation. Proc. Am. Power Conf. 32: 504-511. (CONF-700426) (NSA 25: 18045).

Wetherell, D. F. and R. W. Krauss. 1957. X-ray-induced mutations in *Chlamydomonas engametos*. Am. J. Botany 44(7): 609-619.

Wettstein, R. and J. R. Sotelo. 1969. Electron microscope study on the action of ionizing radiation on the testis of *Gryllus argentinus*. Arch. Soc. Biol. Montevideo 27(Spec.): 140-145. (NSA 24: 29847).

Wetzel, R. G. 1964. A comparative study of the primary productivity of higher aquatic plants, periphyton, and phytoplankton in a large, shallow lake. Int. Rev. ges. Hydrobiol. 49(1): 1-61. (^{14}C).

Wetzel, R. G. 1964. Primary productivity of aquatic macrophytes. Verh. int. verein. limnol. 15: 426-436. (^{14}C).

Wetzel, R. G. 1965. Necessity for decontamination of filters in C^{14} measured rates of photosynthesis in fresh waters. Ecology 46(4): 540-542.

Wetzel, R. G. 1965. Techniques and problems of primary productivity measurements in higher plants and periphyton. In: *Proc. I. B. P. Symp. on Primary Productivity in Aquatic Environments*. (Pallanza, Italy, April 1965). Men. ist. Ital. Idrobiol. 18(Suppl.): 249-267. (Also: In: *Primary Productivity in Aquatic Environments*. C. R. Coleman, ed. Univ. of California, Berkeley. 1966) (^{14}C).

Wey, M.-T. 1964. Determination of contents of manganese and copper during the life cycle of *Chlorella* cells by neutron activation analysis. Chemistry (Taipei) (4): 165-169. (In Chinese) (NSA 19: 45775).

Whaley, D. L. 1966. Conditioned punishers established by pairing with x-irradiation and their effects on two-bar responding with the tamarin. Thesis, Florida State Univ., Tallahassee. 83pp. (NSA 21: 43536).

Wharton, D. R. A. 1957. Action of ionizing radiations on Protozoa. pp. 225-234. In: *Radiation Preservation of Foods*. U. S. Army Research and Development Ser. No. 1.

Wharton, D. R. A. and M. L. Wharton. 1959. The effect of radiation on the longevity of the cockroach, *Periplaneta americana*; as affected by dose, age, sex, and food intake. Radiation Research 11(4): 600-613.

Wharton, D. R. A. and M. L. Wharton. 1961. Effects of radiation on nitrogen excretion by the cockroach, *Periplaneta americana* L. Radiation Research 14(4): 432-443. (NSA 15: 16849).

Wharton, D. R. A., M. L. Wharton and J. E. Lola. 1965. Blood volume and water content of the male American cockroach, *Periplaneta americana* L. - methods and the influence of age and starvation. J. Insect Physiol. 11(4): 391-404. (^{14}C).

Wharton, R. H., et al. 1962. House fly populations and their dispersion in Malaya with particular reference to the fly problem in the Cameron Highlands. Med. J. Malaya 17(2): 115-131. (Used ^{32}P).

Whatley, A. T. 1972. Environmental decisions in nuclear energy. Trans. Nucl. Sci. NS-19(1): 168-171. (NSA 26: 42950).

Wheaton, E. and G. B. Pratt. 1962. Radiation survival. Curves of *Clostridium botulinum* spores. F. Food Sci. 27: 327-334. (NSA 17: 7849).

Wheeler, B. 1950. Halogen metabolism of *Drosophila gibberosa*. I. Iodine metabolism studies by means of ^{131}I. *J. Exptl. Zool.* 115(1): 83-107.

Wheeler, F. 1964. Growing plants under a nuclear reactor. New Scientist 24: 174-175. (NSA 19: 5678).

Wheeler, H. E. 1952. The use of radiocarbon for tagging fungi. Phytopathology 42: 431-435. (Also: Louisiana State Univ., U. S. AEC report ORO-71) (NSA 6: 4387).

Whelan, J. B., et al. 1967. Uptake and distribution of ^{32}P in male white-tailed deer. Pennsylvania Coop. Wildl. Res. Unit, Pennsylvania State Univ., University Park, Paper No. 125. 21pp.

Whicker, F. W. 1962. Distribution and behavior of radionuclides in a mule deer environment. Colorado Coop. Wildl. Res. Unit, Colorado State Univ., Fort Collins, Quart. Rept., January-March, 1962, April-June 1962. 15(3): 32-36; 15(4): 20-25.

Whicker, F. W. 1965. Factors influencing the accumulation of fallout cesium-137 in mule deer. Dissertation, Colorado State Univ., Fort Collins xvii, 220pp.

Whicker, F. W. and A. H. Dahl. 1963. Accumulation of fallout radionuclides in Colorado mule deer. Abstr. P/90. Health Physics 9(8): 892. (NSA 17: 34147).

Whicker, F. W. and L. Fraley, Jr. 1974. Effects of ionizing radiation on terrestrial plant communities. Adv. Radiat. Biol. 4: 317-366. (NSA 32: 22512).

Whicker, F. W. and C. M. Loveless. 1968. Relationships of physiography and microclimate to fallout deposition. Ecology 49(2): 363-366. (NSA 22: 40269).

Whicker, F. W., G. C. Farris and A. H. Dahl. 1966. Radioiodine in Colorado deer and elk thyroids during 1964-65. J. Wildl. Mgmt. 30: 781-785. (NSA 21: 30776).

Whicker, F. W., G. C. Farris and A. H. Dahl. 1967. Concentration patterns of ^{90}Sr, ^{137}Cs, and ^{131}I in a wild deer population and environment. pp. 621-633. In: *Radioecological Concentration Processes*. (See Aberg and Hungate, 1967).

Whicker, F. W., G. C. Farris and A. H. Dahl. 1968. Wild deer as a source of radionuclide intake by humans and as indicators of fallout hazards. pp. 1105-1110. In: *Proc. 1st Intern. Congr. of Radiation Protection*. W. S. Snyder, ed. Pergamon Press, New York.

Whicker, F. W., C. A. Little and T. F. Winsor. 1974. Plutonium behavior in the terrestrial environs of the Rocky Flats installation. pp. 89-103. In: *Environmental Surveillance Around Nuclear Installations. Vol. 2*. (See IAEA, 1974) (NSA 31: 14032).

Whicker, F. W., W. C. Nelson and A. F. Gallegos. 1972. Fallout ^{137}Cs and ^{90}Sr in trout from mountain lakes in Colorado. Health Physics 23(4): 519-527. (NSA 26: 56555).

Whicker, F. W., E. E. Remmenga and A. H. Dahl. 1965. Factors influencing the accumulation of I-131 in Colorado deer thyroids following 1961-1962 nuclear weapons tests. Health Physics 11: 293-296.

Whicker, F. W., R. A. Walters and A. H. Dahl. 1967. Fall-out radionuclides in Colorado deer liver. Nature 214: 511-513. (NSA 21: 22414).

Whicker, F. W., et al. 1965. Factors influencing the accumulation of fallout ^{137}Cs in Colorado mule deer. In: *Radiation and the Terrestrial Environment*. Health Physics 11(12): 1407-1414. (See Hungate, 1965) (NSA 20: 12984).

Whipple, G. H. 1957. Pre-operational environmental survey for a power reactor. Am. Ind. Hyg. Assoc. Quart. 18(4): 315-318.

Whipple, G. H. 1961. The problems of monitoring. pp. 50-53. In: *Radiological Aspects of Water Supplies*. (See Univ. of Illinois, 1961).

Whipple, G. H. 1968. Environmental planning: a health physicist's view. J. Publ. Hlth. 58: 1410-1413. (NSA 23: 20259).

Whipple, G. H. 1970. Objectives of environmental surveillance: the American viewpoint. pp. 14-18. In: *Environmental Surveillance in the Vicinity of Nuclear Facilities*. (See Reinig, 1970) (NSA 25: 16157).

Whipple, R. T. P. 1962. Considerations on the siting of outfalls for the area disposal of radioactive effluent in tidal waters. Preprint Paper No. 46. Pergamon Press, Oxford. 24pp. (NSA 16: 32623).

Whitacre, G. K. and G. W. Ware. 1970. Effects of gamma- and x-irradiation on the toxicity of malathion to house flies. J. Econ. Entomol. 63(2): 424-426. (NSA 24: 25555).

White, J. C., Jr. 1964. Fractionated doses of x-radiation: a preliminary study of effects on telost embryos. Int. J. Radiat. Biol. 8(1): 85-91. (NSA 19: 5662).

White, J. C., Jr. and J. W. Angelovic. 1973. Interactions of chronic gamma radiation, salinity, and temperature on the morphology of postlarval pinfish, *Lagodon rhomboides*. pp. 199-217. In: *Proc. Workshop on Egg, Larval and Juvenile Stages of Fish in Atlantic Coast Estuaries*. U. S. NOAA Tech. Publ. No. 1. (SFA 19: 17688).

White, J. E. 1969. An in-progress report on the effects of temperature on the uptake and distribution of cesium-137 in *Cnemidophorus neomexicanus* (New Mexico whiptail lizard). pp. 149-151. In: *Physiological Systems in Semiarid Environments*. C. C. Hoff and M. L. Riedesel, eds. Univ. of New Mexico Press, Albuquerque.

White, L. D. and D. M. Allred. 1961. Range of kangaroo rats in areas affected by atomic detonations. Proc. Utah Acad. Sci. Arts Lett. 38: 101-110. (NSA 21: 34628).

White, L. D. and R. B. Hutt. 1970. Effects of gamma irradiation on longevity and oviposition of the codling moth. J. Econ. Entomol. 63: 866-869. (NSA 24: 32161).

White, L. D. and R. B. Butt. 1971. Codling moth catches in sex and light traps after exposure to 0, 25, or 40 krad of gamma irradiation. J. Econ. Entomol. 64(5): 1249-1250. (NSA 26: 7074).

White, L. D. and R. B. Hutt. 1972. Effects of treating adult codling moth with sterilizing and substerilizing doses of gamma irradiation in a low temperature environment. J. Econ. Entomol. 65(1): 140-143. (NSA 26: 23147).

White, L. D. and R. B. Hutt. 1975. Codling moth: catches of irradiated and untreated laboratory-reared and native males in synthetic sex attractant traps. J. Econ. Entomol. 68(4): 449-450. (NSA 33: 7571).

White, L. D., R. B. Hutt and B. A. Butt. 1969. Releases of unsexed gamma-irradiated codling moths for population suppression. J. Econ. Entomol. 62: 795-798. (NSA 23: 43929).

White, L. D., et al. 1972. Longevity and reproduction of codling moths irr radiated with ^{60}Co or ^{137}Cs. J. Econ. Entomol. 65(3): 692-697. (NSA 26: 43353).

White, M. G. and P. B. Dunaway. 1974. Nevada Applied Ecology Group environmental studies of plutonium. pp. 931-947. In: Proc. 2nd AEC Environmental Protection Conf. Vol. 2. (See U. S. AEC, 1974) (NSA 30: 32078).

White, M. G. and P. B. Dunaway. (Eds.). 1975. Radioecology of Plutonium and Other Transuranics in Desert Environments. (Nevada Applied Ecology Group progress report as of January 1975). Nevada Operations Office, Las Vegas, U. S. ERDA report NVO-153. 600pp. (NSA 33: 612).

White, M. G. and P. B. Dunaway. (Eds.). 1977. Transuranics in Natural Environments. (Proc. Symp., Gatlinburg, Tennessee, October 1976). U. S. ERDA report NVO-178. vi, 710pp.

White, M. G., P. B. Dunaway and D. L. Wireman. (Eds.). 1978. Transuranics in Desert Ecosystems. (Nevada Applied Ecology Group November 1977). Nevada Operations Office, Las Vegas, U. S. DOE report NVO-181. v, 469pp.

White, M. J. D. 1937. The effect of x-rays on the first meiotic division in three species of Orthoptera. Proc. Roy. Soc. (London) B124: 183-195.

White, R. P. and B. G. Ellis. 1968. Routine counting of ^{32}P in colored solutions from dry ashed plant samples utilizing Cerenkov radiation. Soil Sci. Soc. Am. Proc. 32(5): 740.

White, S. A. 1971. Effect of ionizing radiation on the stable fly, Stomoxys calcitrans (Linnaeus). Thesis, Univ. of Florida, Gainesville. 147pp. (Univ. Microfilms 72-16670) (NSA 26: 45883).

White, W. B. 1970. Radiography to facilitate bagworm egg counts. J. Econ. Entomol. 63(3): 910-911.

Whitehead, N. E. and R. R. Brooks. 1969. Radioecological observations on plants of the lower Buller Gorge region of New Zealand and their significance for biogeochemical prospecting. J. Appl. Ecol. 6(2): 301-310.

Whitehead, N. E., R. R. Brooks and G. E. Coote. 1971. Gamma radiation of some plants and soils from a uraniferous area in New Zealand. N. Zeal. J. Sci. 14(1): 66-76. (NSA 26: 45529).

Whitehead, N. E., R. R. Brooks and P. J. Peterson. 1971. Nature of uranium occurrence in the leaves of Coprosma australis (A. Rich.) Robinson. Austral. J. Biol. Sci. 24: 67-73. (NSA 25: 32405).

Whitfield, P. 1974. Determination of free chlorine in natural waters. Oklahoma State Univ., Stillwater, U. S. AEC report ORO-4254-17. 12pp. (NSA 30: 9035) (Proton and neutron activation analysis).

Whitford, L. A. and G. J. Schumacher. 1961. Effect of current on mineral uptake and respiration by freshwater alga. Limnol. Oceanogr. 6(4): 423-425.

Whitford, L. A. and G. J. Schumacher. 1964. Effect of current on respiration and mineral uptake in Spirogyra and Oedogonium. Ecology 45(1): 168-170. (Used ^{32}P).

Whiting, A. R. 1939. Sensitivity to x-rays of stages of oogenesis of Habrobracon. Abstr. Genetics 24(Jan.): 89-90.

Whiting, A. R. 1940. Sensitivity to x-rays of different meiotic stages in unlaid eggs of Habrobracon. J. Exptl. Zool. 83(2): 249-269.

Whiting, A. R. 1945. Dominant lethality and correlated chromosome effects in Habrobracon eggs x-rayed in diplotene and in late metaphase I. Biol. Bull. 89(1): 61-71.

Whiting, A. R. 1945. Effects of x-rays on hatchability and on chromosomes of Habrobracon eggs treated in first meiotic prophase and metaphase. Am. Naturalist 79: 193-227.

Whiting, A. R. 1948. Incidence and origin of androgenetic males in x-rayed Habrobracon eggs. Biol. Bull. 95(3): 354-360. (NSA 2: 704).

Whiting, A. R. 1949. Androgenesis, a differentiator of cytoplasmic injury induced by x-rays in Habrobracon eggs. Biol. Bull. 97(2): 210-220.

Whiting, A. R. 1950. Failure of pupation of Ephestia larvae following exposure to x-rays. Abstr. Anat. Record 108: 609.

Whiting, A. R. 1954. The effects of oxygen on the frequency of x-ray-induced mutations in Habrobracon eggs. Genetics 39(6): 851-858. (NSA 9: 2592).

Whiting, A. R. 1961. Temperature effects on lethal mutation rates of Habrobracon oocytes x-irradiated in first meiotic metaphase. Genetics 46(7): 811-816. (NSA 15: 27434).

Whiting, A. R. 1963. X-ray induced visible mutations in Habrobracon oocytes. Genetics 48(4): 491-495. (NSA 17: 21513).

Whiting, A. R. and C. H. Bostian. 1931. The effects of x-radiation on larvae in Habrobracon. Genetics 16(Nov.): 659-680.

Whiting, A. R. and W. E. Murphy. 1956. Differences in response of irradiated eggs and spermatozoa of Habrobracon to anoxia. Genetics 54: 297-303. (NSA 11: 5719).

Whiting, A. R., R. H. Smith and R. C. von Borstel. 1968. Methods for radiation studies during oogenesis in Habrobracon juglandis (Ashmead). pp. 201-208. In: Effects of Radiation on Meiotic Systems. International Atomic Energy Agency, Vienna.

Whiting, A. R., et al. 1958. Stages of death of x-ray-induced embryo lethals in haploids and in heterozygotes of Habrobracon. Radiation Research 8(2): 195-202.

Whiting, P. W. 1929. X-rays and parasitic wasp. J. Heredity 20(6): 268-276.

Whiting, P. W. 1938. Decrease in biparental males by x-raying sperm in Habrobracon. Pacific Acad. Sci. Proc. 12: 74-76.

Whitley, J. E. 1972. Nuclear techniwues for environmental studies. Trans. Soc. Occup. Med. 22(4): 126-129. (NSA 29: 9724).

Whitley, J. E. and C. Muir. 1971. Determination of sodium in newt egg cell nuclei by activation analysis. Radiochem. Radioanal. Lett. 8(1): 53-57.

Whitley, J. E. and C. Muir. 1972. Determination of sodium and potassium in the nuclei of single newt oocytes. Scottish Research Reactor Centre, East Kilbride, Scotland. 8pp. (For: Int. Colloqu. on the Activation Analysis of Microquantities of Elements in Very High Purity; Inorganic and Organic Substances and in Biological Media; Saclay, France, October 2, 1972) (CONF-721069-22) (NSA 28: 20668).

Whitley, J. E. and C. Muir. 1974. Determination of sodium and potassium in the nuclei of single newt oocytes. J. Radioanal. Chem. 19(2): 257-262. (NSA 30: 2461) (Neutron activation analysis).

Whitmore, F. W. and R. Zahner. 1967. Evidence for a direct effect of water stress in the metabolism of cell wall in Pinus. Forest Science 13: 397-400. (^{14}C).

Whitson, G. L. 1972. Radiation-induced biochemical changes in Protozoa. pp. 123-151. In: Concepts in Radiation Cell Biology. G. L. Whitson, ed. Academic Press, New York. (NSA 26: 41224).

Whittaker, R. H. 1961. Experiments with radiophosphorus tracer in aquarium microcosms. Ecol. Monogr. 31(2): 157-188.

Whittaker, R. H. and G. M. Woodwell. 1967. Surface area relations of woody plants and forest communities. Am. J. Botany 54(8): 931-939. (At Brookhaven National Lab.).

Whittaker, R. H. and G. M. Woodwell. 1968. Dimension and production relations of trees and shrubs in the Brookhaven Forest, New York. J. Ecology 56(1): 1-25.

Whittaker, R. H. and G. M. Woodwell. 1969. Structure, production and diversity of the oak-pine forest at Brookhaven, New York. J. Ecology 57(1): 155-174.

Whittaker, R. H., N. Cohen and J. S. Olson. 1963. Net production relations of three tree species at Oak Ridge, Tennessee. Ecology 44(4): 806-810.

Whittemore, W. L., D. M. J. Compton and S. J. Black. 1970. Ionizing radiation for the treatment of municipal waste waters. Final report, February 26, 1969-February 28, 1970. Gulf General Atomic, Inc., San Diego, Calif., U. S. AEC report GA-9924. 165pp. (NSA 25: 6820).

Whitten, B. K. and C. J. Goodnight. 1965. The accumulation of ^{89}Sr in tubificid worms. Abstr. Bull. Ecol. Soc. Am. 46(3): 100.

Whitten, B. K. and C. J. Goodnight. 1967. The accumulation of ^{89}Sr and ^{45}Ca by an aquatic oligochaete. Physiol. Zool. 41: 371-385. (NSA 23: 31620).

Whitten, B. K. and C. J. Goodnight. 1969. The role of tubificid worms in the transfer of radioactive phosphorus in an aquatic ecosystem. pp. 270-277. In: Symposium on Radioecology. (See Nelson and Evans, 1969).

Whitten, M. J. and W. C. Taylor. 1970. Role for sterile females in insect control. J. Econ. Entomol. 63: 269-272. (NSA 24: 23416) (γ-rays).

Whittingham, C. P., M. Bermingham and R. G. Hiller. 1963. The photometabolism of glucose in Chlorella. Z. Naturforschung 18b(11): 701-706. (^{14}C).

Whittington, W. J. (Ed.). 1969. Root Growth. Butterworth, London. (Discusses radionuclides).

Wichtermann, R. 1948. The biological effects of x-rays on mating types and conjugation of Paramecium bursaria. Biol. Bull. 94(2): 113-127.

Wichtermann, R. 1948. Mating types and conjugation of four different races of Paramecium calkinsi and the effects of x-rays on the mating reaction. Abstr. Biol. Bull. 95(2): 271-272.

Wichtermann, R. 1951. Effects of high dosage irradiation with x-rays upon dividing and non-dividing cells and conjugant stages: Paramecium. Abstr. Anat. Record 111(3): 527.

Wichtermann, R. 1951. The effects of x-rays upon conjugant and vegetative stages of Paramecium calkinsi. Abstr. Biol. Bull. 101(2): 232-233.

Wichtermann, R. 1953. Studies on the effects of x-radiation upon Paramecium: variations among and within species. Abstr. Biol. Bull. 105(2): 387.

Wichtermann, R. 1953. The Biology of Paramecium. McGraw-Hill, New York. 527pp.

Wichtermann, R. 1954. Variation in x-radiation sensitivity of different stages of growth in Paramecium. Abstr. Biol. Bull. 107(2): 322.

Wichtermann, R. 1954. Paramecia and high dosage radiation. Reserach Rev. (O.N.R.) October: 7-14.

Wichtermann, R. 1955. The usefulness of the one-celled animal, *Paramecium*, in studying the biological effects of high-dosage x-radiation. Proc. Penn. Acad. Sci. 29: 78-93.

Wichtermann, R. 1955. Survival and other effects following x-radiation of the flagellate, *Euglena gracillis*. Abstr. Biol. Bull 109(3): 371.

Wichtermann, R. 1957. X-irradiation of the giant amoeba, *Chaos chaos*. Abstr. Biol. Bull. 113(2): 359.

Wichtermann, R. 1957. Biological effects of radiations on Protozoa. Bios 28: 3-20.

Wichtermann, R. 1959. Mutation in the protozoan *Paramecium multimicronnucleatum* as a result of x-irradiation. Science 129(3343): 207-208. (NSA 13: 4424).

Wichtermann, R. 1960. Production of viable races of *Paramecium caudatum* after micronuclear elimination with x-rays. Abstr. Biol. Bull. 119(2): 348.

Wichtermann, R. 1961. Survival and reproduction of *Paramecium* after x-irradiation. J. Protozool. 8(2): 158-162.

Wichtermann, R. 1972. Biological effects of ionizing radiations on Protozoa: some discoveries and unsolved problems. BioScience 22(5): 281-289.

Wichtermann, R. and F. H. J. Figge. 1952. The effects of hematoporphyrin and other substances upon x-rays in *Paramecium*. Abstr. Biol. Bull. 103 (2): 302.

Wichtermann, R. and F. H. J. Figge. 1952. Environmental factors concerned with the lethality of x-rays in *Paramecium*. Abstr. Biol. Bull. 103(2): 311.

Wichtermann, R. and F. H. J. Figge. 1953. The influence of hematoporphyrin, sodium nitrate, and nembutal on radiation sensitivity in *Paramecium*. Proc. Am. Assoc. Cancer Res. 1: 16.

Wichtermann, R. and F. H. J. Figge. 1954. Lethality and the biological effects of x-rays in *Paramecium*: radiation resistance and its variability. Biol. Bull. 106(2): 253-263.

Wichtermann, R. and C. M. Honegger. 1958. Action of x-rays on the two common amoebas, *Chaos diffluens* and *Chaos chaos*. Proc. Penn. Acad. Sci. 32: 240-253.

Wichtermann, R., H. Solomon and F. H. J. Figge. 1958. The influence of protoporphyrin-nitroresorcinol and other phenols on x-radiation sensitivity of *Paramecium caudatum*. Abstr. Biol. Bull. 115(2): 369-370.

Wickman, B. E. 1964. A comparidon of radiographic and dissection methods for measurement siricid populations in wood. Can. Entomol. 96(3): 508-510. (*Sirex longicauda*).

Wickman, B. E. 1966. Use of radiography to detect mortality of California flatheaded borers in pine bark. J. Econ. Entomol. 59(4): 1028-____.

Wiecek, C. S. and J. Skok. 1968. Effects of brief withholding of essential elements on radiosensitivity of sunflower plants. Radiation Botany 8: 245-250. (NSA 22: 43478).

Wiechen, A. 1971. Determination of uranium in biological materials by the fission-track method. pp. 213-216. In: *Rapid Methods for Measuring Radioactivity in the Environment*. (See IAEA, 1971) (In German) (NSA 26: 27810).

Wiechen, A. 1974. Results with the fission-track method for determination of uranium in soil samples. Radiochem. Radioanal. Lett. 17(5-6): 389-392. (In German) (NSA 30: 14788).

Wiechen, A. 1974. Determination of tritium in milk and surface water from the immediate surroundings of a nuclear power plant in the north of Germany. Milchwissenschaft 29(6): 346-348. (In German) (NSA 31: 29622).

Wiegert, R. G. 1964. Detritus production and turnover rate on three southeastern oil fields. Abstr. Bull. Ecol. Soc. Am. 45(3): 87. (At Savannah River Plant).

Wiegert, R. G. 1968. Radioecology. pp. 328-332. In: *McGraw-Hill Yearbook of Science and Technology*. D. I. Eggenberger, ed. McGraw-Hill, New York.

Wiegert, R. G. 1970. Effects of ionizing radiation on leaf fall, decomposition, and the litter microarthropods of a montane rain. Ch. H-4, pp. H89-H100. In: *A Tropical Rain Forest*. (See H. T. Odum and R. F. Pigeon, 1970) (NSA 25: 21636).

Wiegert, R. G. and R. G. Lindeborg. 1964. A "stem well" method of introdicing radioisotopes into plants to study food chains. Ecology 45(2): 406-409.

Wiegert, R. G. and E. P. Odum. 1964. A radionuclide tracer study of arthropod food chains in an experimental one-year fallow field. Abstr. Bull. Ecol. Soc. Am. 45(3): 81.

Wiegert, R. G. and E. P. Odum. 1969. Radionuclide tracer measurement of food web diversity in nature. pp. 709-710. In: *Symposium on Radioecology*. (See Nelson and Evans, 1969).

Wiegert, R. G., D. C. Coleman and E. P. Odum. 1970. Energetics of the litter-soil subsystem. pp. 93-98. In: *Methods of Study of Soil Ecology*. UNESCO, Paris. (Mentions tracers and activation analysis).

Wiendl, F. M. 1971. Some gamma-irradiation effects on survival, longevity, and reproduction of *Zabrotes subfasciatus* (Boh.). pp. 525-530. In: *Sterility Principle for Insect Control or Eradication*. (See IAEA, 1971) (NSA 25: 29756).

Wiendl, F. M., V. L. Tornisielo and R. E. Domarco. 1974. Influence of gamma irradiation on the mortality of Tribolium castaneum (Herbst). Brazil Centro de Energia Nuclear na Agriculture, Piracicaba, report CENA-BC-024. 26pp. (In Portugese) (NSA 33: 29752).

Wiendl, F. M., et al. 1975. Method of determining the gamma radiation doses for the sterilization of stored product insects. pp. 289-315. In: Sterility Principle in Insect Control, 1974. (See IAEA, 1975) (NSA 32: 11785).

Wiendl, F. M., et al. 1974. Mortality and reproduction in Sitophilus zeamais Mots. on irradiated rice. São Paulo Univ., Piracicaba, Brazil Centro de Energia Nuclear na Agriculture, report CENA-BC-015. 19pp. (In Portugese) (NSA 33: 7558).

Wiens, A. W. and L. I. Gilbert. 1965. Regulation of cockroach fat-body metabolism by the corpus cardiacum in vitro. Science 150: 615-616. (^{14}C, Leucophaea maderae).

Wiersum, L. K. 1967. Root-system development. pp. 83-96. In: Soil-Moisture and Irrigation Studies. (See IAEA, 1967) (NSA 22: 6449).

Wieser, P. H. and F. Wörner. 1970. Artificial radioactivity of ground-level air, precipitation, and dust in Stuttgart-Hohenheim in the years 1963-1968. Meteorol. Rundsch. 23: 99-101. (In German) (NSA 25: 23717).

Wiesner, R. 1939. Determination of the radium content of algae. Sitzber Akad. Wiss. Wien Mattr.-naturw. Kl. Abt. IIA 147: 521-528.

Wiggins, A. D. 1960. On a multicompartment migration model with chronic feeding. Biometrics 16(4): 642-658. (Radionuclide cycling).

Wightman, J. A. 1975. Improved technique for measuring assimilation efficiency by the ^{51}Cr-^{14}C twin tracer method. Oecologia 19(4): 273-284. (NSA 33: 15135) (Used 3rd instar larvae of a Kenyan cetoniid).

Wijker, H. 1972. Some guidelines on handling radioactive materials in Holland and the disposal of liquid radioactive wastes in rivers. pp. A.4.1-A.4.5. In: Safety Measure in Nuclear Research. G. Fieuw, ed. Belgium Centre d'Étude de l'Énergie Nucléaire, Mol. (NSA 30: 1000).

Wijnands-Staeb, K. J. A. and M. J. Frissel. 1973. Computer simulation for genetic control of the onion fly Hylemya antiqua (Meigen). pp. 95-111. In: Computer Models and Application of the Sterile-Male Technique. (See IAEA, 1973) (NSA 29: 8052).

Wijnands-Staeb, K. J. A. and C. van Heemert. 1974. Radiation induced semi-sterility for genetic control purposes in the onion fly Hylemya antiqua (Meigen). Theor. Appl. Genet. 44(3): 111-119. (NSA 30: 24357).

Wiklander, L. 1964. Uptake, adsorption, and leaching of radiostrontium in a lysimeter experiment. Soil Science 97(3): 168-172.

Wilbur, K. M. and L. H. Jodrey. 1952. Studies on shell formation: I. Measurement of the rate of shell formation using Ca^{45}. Biol. Bull. 103(2): 269-276.

Wilbur, K. M. and L. H. Jodrey. 1955. Studies on shell formation: V. The inhibition of shell formation by carbonic anhydrase inhibitors. Biol. Bull. 108(3): 359-365.

Wild, A. 1963. A red eye color mutation in Culex pipiens after x-irradiation. Nature 200(4909): 917-918. (NSA 18: 5067).

Wild, A. and K. Egle. 1966. On the aerophytic growth of a mutant Chlorella. Beitr. Biol. Pflanz. 42: 471-477. (In German) (NSA 22: 43291) (^{32}P, ^{35}S).

Wildman, B. E. 1966. The United States Atomic Energy Commission's Columbia River program. pp. 673-682. In: Disposal of Radioactive Wastes into Seas, Oceans and Surface Waters. (See IAEA, 1966).

Wildung, R. E., T. R. Garland and D. A. Cataldo. 1975. Accumulation of technetium from soil by plants: a potential mechanism for uptake and toxicity. Battelle Northwest Lab., U. S. ERDA report BNWL-SA-5393. 13pp. (NSA 33: 597).

Wilgain, S., E. Picciotto and W. de Brueck. 1965. ^{90}Sr fallout in Antarctica. J. Geophys. Res. 70: 6023-6032. (NSA 20: 9145).

Wilhelmi, M. 1974. Literature study on the diffusion mechanism of released plutonium. German F. R. Kernforschungszentrum, Karlsruhe, report KFK-1974. 52pp. (In German) (NSA 30: 32009).

Wilhm, J. L. 1970. Transfer of radioisotopes between detritus and benthic macroinvertebrates in laboratory microecosystems. Health Physics 18: 277-284. (NSA 24: 21105).

Wilimovsky, N. J. 1961. An application of radioactive isotopes in fish tagging. Proc. Alask. Sci. Conf. 11: 93-99.

Wilimovsky, N. J. and J. N. Wolfe. (Eds.). 1966. Environment of the Cape Thompson Region, Alaska. U. S. AEC report PNE-481. 1265pp. (NSA 20: 32885) (Concerns Project Chariot).

Wilkening, M. H. 1964. Radon-daughter ions in the atmosphere. pp. 359-368. In: The Natural Radiation Environment. (See Adams and Lowder, 1964) (NSA 17: 30773).

Wilkening, M. H. 1974. Radon-222 from the Island of Hawaii: deep soils are more important than lava fields or volcanoes. Science 183(4123): 413-415. (NSA 29: 18340).

Wilkening, M. H., W. E. Clements and D. Stanley. 1972. Radon 222 flux measurements in widely separated regions. pp. 717-730. In: Natural Radiation Environment II. Vol. 2. (See Adams, Lowder and Gesell, 1972) (NSA 32: 19723).

Wilkinson, J. E. 1972. Comparative study of the early effects of neutron gamma radiation on the Sidman avoidance performance of rhesus monkeys and baboons. Thesis, Baylor Univ., Waco, Tex. 105pp. (Univ. Microfilms 73-7326) (NSA 27: 27962).

Wilkinson, P. N. and F. E. Hoecker. 1953. Selective placental transmission of radioactive alkaline earths and plutonium. Trans. Kansas Acad. Sci. 56: 341-363. (NSA 15: 7182).

Wilkniss, P. E. and D. J. Bressan. 1971. Use of radionuclides to evaluate the transport of pollutants from the hydrosphere into the atmosphere. pp. 297-307. In: Nuclear Techniques in Environmental Pollution. (See IAEA, 1971) (NSA 25: 32144).

Willard, B. E. 1973. Environmental considerations of nuclear stimulation of natural gas. Trans. Amer. Nucl. Soc. 17: 14. (NSA 29: 5067).

Willard, W. K. 1960. Avian uptake from an area contaminated by low level atomic wastes. Thesis, Univ. of Georgia, Athens.

Willard, W. K. 1960. Avian uptake of fission products from an area contaminated by low-level atomic wastes. Science 132(3420): 148-150. (NSA 14: 17683).

Willard, W. K. 1963. Relative sensitivity of nestlings of wild passerine birds to gamma radiation. pp. 345-349. In: Radioecology. (See Schultz and Klement, 1963) (NSA 17: 33583).

Willard, W. K. 1965. Long term effects of acute lowlevel x-rays on the population dynamics of the yellow fever mosquito, Aedes aegypti. Thesis, Univ. of Tennessee, Knoxville. 50pp. (Also: In: Radiation and the Terrestrial Environment. Health Physics 11(12): 1577-1583. (See Hungate, 1965)) (NSA 20: 10692).

Willard, W. K. 1969. Radiosensitivity and regulation of population density in the yellow fever mosquito (Aedes aegypti). Radiation Research 37: 493-498. (NSA 23: 20459).

Willard, W. K. 1973. Dynamics of Conotrachelus nenuphar (Coleoptera: Curculionidae) populations following exposure to ionizing radiation. pp. 1144-1149. In: Radionuclides in Ecosystems. (See Nelson, 1973) (NSA 28: 30235).

Willard, W. K. 1975. Radiocobalt cycling in a small mammal food web. Oak Ridge National Lab., U. S. ERDA report. 14pp. (CONF-750503-8) (NSA 32: 3497).

Willard, W. K. and D. S. Cherry. 1975. Comparative radiosensitivity in the Class Insecta. J. Theor. Biol. 52(1): 149-158. (NSA 33: 12433).

Willcock, E. G. 1904. The action of the rays from radium upon some simple forms of animal life. J. Physiol. (London) 30(5/6): 449-454.

Williams, C. and G. Brown. 1971. Uranium content of peaty soils rich in molybdenum and selenium from Co. Limerick, Eire. Geoderma 6(3): 223-225. (NSA 26: 22772).

Williams, C. E. and G. R. Luetkehans. 1973. Project Director's completion report D+30 days (detonation related activities). U. S. AEC report NVO-135. v, 88pp. (Project Rio Blanco).

Williams, D. B. 1958. Effects of x-rays on fission in the predaceous holotrich , Spathidium spathula. J. Protozool. 5(Suppl.): 25.

Williams, D. B. 1962. Sensitivity of Spathidium spathula tp low doses of x-radiation. J. Protozool. 9(2): 119-122. (NSA 16: 21954).

Williams, D. B. 1966. Effects of x rays on cell lethality and micronuclear number in the ciliate Spathidium spathula. J. Protozool. 13: 272-277. (NSA 21: 8977).

Willaims, D. B. 1968. Effects of radiation, antibiotics and alkylating agents on cell division and growth in the ciliate Spathidium spathula. J. Protozool. 15(4): 811-816. (NSA 23: 38755).

Williams, E. C., Jr. and D. E. Reichle. 1968. Radioactive tracers in the study of energy turnover by a grazing insect (Chrysochus auratus Fab.; Coleoptera Chrysomelidae). Oikos 19(1): 10-18.

Williams, H. 1965. Bikini Atoll - 1964. Alumnus (Univ. Wash.) 55(2): 28-33. (Radioecology).

Williams, H. 1967. Bikini nine years later. Science J. (April): 2-7.

Williams, J. E. 1963. Harvest and infra-red gas analysis measurements of primary net productivity under natural conditions and a comparison of methods. Thesis, Univ. of Georgia, Athens. 62pp. (At Savannah River Plant).

Williams, J. E. 1968. Effects of napthalene on decomposition soil-litter organisms and dispersion of zinc 65 from litter in a coastal plain broomsedge (Andropogon) community. Thesis, Univ. of Georgia, Athens. 81pp. (At Savannah River Plant).

Williams, J. E. 1969. Biological transport of ^{65}Zn in a South Carolina broomsedge field. pp. 665-671. In: Symposium on Radioecology. (See Nelson and Evans, 1969).

Williams, J. F. and C. W. Cristina. 1972. Influence of ionizing irradiation on infectivity of eggs of Echinococcus granulosus in laboratory rodents. J. Parasitol. 58(3): 427-430. (NSA 26: 59121).

Williams, L. G. 1960. Uptake of cesium-137 by cells and detritus of Euglena and Chlorella. Limnol. Oceanogr. 5(3): 301-311.

Williams, L. G. 1964. Relative strontium and calcium uptake by green algae. Science 146(3650): 1488. (^{90}Sr, Oocystis eremosphaeria).

Williams, L. G. and Q. Pickering. 1961. Direct and food-chain uptake of cesium^{137} and strontium^{85} in bluegill fingerlings. Ecology 42(1): 205-206. (NSA 15: 24675).

Williams, L. G. and H. D. Swanson. 1958. Concentration of cesium-137 by algae. Science 127(3291): 187-188. (NSA 12: 6424).

Williams, L. G., M. Howell and C. P. Straub. 1960. Packaged organic materials as monitoring tools for radionuclides. Science 132(3439): 1554-1555. (NSA 15: 5896).

Williams, M. 1923. Observations on the action of x-rays on plant cells. Ann. Bot. 37(146): 217-223. (Saxafraga umbosa).

Williams, N. W. and N. R. Benson. 1966. Transfer of C^{14} components from Psylla pyricola (Foer.) to pear seedlings. J. Insect Physiol. 12(2): 251-254.

Williams, R. B. and M. B. Murdoch. 1969. The potential importance of Spartina alterniflora in conveying zinc, manganese, and iron into estuarine food chains. pp. 431-439. In: Symposium on Radioecology. (See Nelson and Evans, 1969).

Williams, R. G. and J. L. Carmon. 1966. Effect of radiation on variability of weights in Peromyscus polionotus. Bull. Ga. Acad. Sci. 24: 93-99. (NSA 21: 14719).

Williams, R. G., J. L. Carmon and M. H. Smith. 1968. Influence of temperature on the susceptibility of the old-field mouse (Peromyscus polionotus) to acute radiation. Radiation Research 35(3): 709-713. (NSA 22: 52015).

Williams, R. B. and M. B. Murdoch. 1973. The effects of continuous low-level gamma radiation on estuarine microcosms. pp. 1213-1221. In: Radionuclides in Ecosystems. (See Nelson, 1973) (NSA 28: 30244).

Williamson, L. L. 1969. A study of potential game animal utilization as an emergency food supply during nuclear disaster recovery. Thesis, Univ. of Georgia, Athens. iv, 29pp.

Williamson, P. 1972, 1973. Environment of nuclear stations in England and Wales. Central Electricity Generating Board, London, England. 13, 16pp. (INIS-mf-285, -1731) (NSA 28: 1677; 32: 25373).

Williams-Walls, N. J. 1969. Effects of growth and toxin production of exposure of spores of Clostridium botulinum type F to sublethal doses of gamma irradiation. Appl. Microbiol. 17: 128-134. (NSA 23: 29915).

Willis, C. P., D. G. Olson and C. W. Sill. 1970. Radiochemical determination of ^{35}S in large samples of vegetation. Anal. Chem. 42: 124-126. (NSA 24: 9248).

Willis, D. L. 1974. Radiosensitivity of the rough-skinned newt. Oregon State Univ., Corvallis, U. S. AEC report RLO-2227-T8-6. 44pp. (NSA 32: 3746).

Willis, D. L. and M. Mahlon. 1968. Radiosensitivity of the western fence lizard (Sceloporus occidentalis) and the desert crested lizard (Dipsosaurus dorsalis). Abstr. Radiation Research 35(2): 574.

Willis, D. L. and B. B. Valett. 1973. Metabolism of iodine-131 in two amphibian species (Taricha granulosa and Rana pipiens). pp. 394-400. In: Radionuclides in Ecosystems. (See Nelson, 1973) (NSA 25: 44288).

Willkomm, H. and H. Erlenkeuser. 1972. ^{14}C measurements on water, plants, and sediments of lakes. pp. D1-D12. In: Proc. 8th Intern. Radiocarbon Dating Conf. Vol. 1. (See Rafter, 1972) (NSA 31: 8630).

Wilson, A. T. 1958. Techniques for the use of tritium as a tracer in vivo biological systems. Proc. 2nd Intern. Conf. on the Peaceful Uses of Atomic Energy 25: 213-214. (Paper A/CONF.15/1506) (NSA 13: 6216).

Wilson, A. T. 1961. Carbon-14 from nuclear explosions as a short-term dating system: use to determine the origin of heartwood. Nature 191(4789): 714. (NSA 15: 28926).

Wilson, A. T. and J. M. Gumbley. 1963. Carbohydrate metabolism in the sapwood of Pinus radiata. Nature 197(4868): 711-712. (^{14}C).

Wilson, A. T., J. M. Gumbley and D. J. Spedding. 1963. Resin metabolism in the sapwood of Pinus radiata. Nature 198(4839): 500. (NSA 17: 23057) (Used ^{12}C, ^{13}C, ^{14}C).

Wilson, C. B. (Ed.). 1970. Evaluation of radiological conditions in the vicinity of Hanford for 1968. ---(w/T. H. Essig). 1969. ---Appendices. ---(w/T. H. Essig). January-June 1969. Battelle Northwest Lab., U. S. AEC report BNWL-1341, -1341(App.), -1505, -1505(App.), -CC-2478. ?, 136, 73, 120, 94pp. (NSA 25: 521, 16075, 16129; 29: 12702).

Wilson, D. W. 1968. A mathematical model of the transport of ^{137}Cs from fallout to milk. Thesis, Colorado State Univ., Fort Collins. 152pp. (NSA 23: 41273).

Wilson, D. W. and G. M. Ward. 1968. A prediction model for the transport of fallout ^{137}Cs from the atmosphere to plants. pp. 275-280. In: Intern. Congr. on the Radioprotection of the Environment. (See Societé Fransaise de Radioprotection, 1968).

Wilson, D. W., Y. C. Ng and W. L. Robinson. 1975. Evaluation of plutonium at Eniwetok Atoll. Health Physics 29(4): 599-611. (NSA 31: 19676; 33: 9288).

Wilson, D. W., G. M. Ward and J. E. Johnson. 1965. Deposition of fallout ^{137}Cs at Fort Collins, Colorado, 1962-1964. Radiol. Hlth. Data 6: 675-676. (NSA 20: 18919).

Wilson, E. D., W. D. Gibbs and C. C. Lushbaugh. 1964. Thyroid uptake and whole-body retention of ^{131}I in box turtles. Texas Rept. Biol. Med. 22 (1): 211-212. (NSA 18: 33254).

Wilson, E. O. and T. Eisner. 1957. Quantitative studies of liquid food transmission in ants. Insectes Sociaux 4(2): 157-166.

Wilson, G. B. and A. H. Sparrow. 1960. Configurations resulting from iso-chromatid and iso-subchromatid unions after meiotic and mitotic prophase irradiation. Chromosoma 11: 229-244.

Wilson, G. B., A. H. Sparrow and V. Pond. 1958. Radiation-induced subchromatid breakage and rejoining during meiosis of Trillium erectum L. Abstr. pp. 317-318. In: Proc. 10th Intern. Congr. Genetics. Vol. 2.

Wilson, G. B., A. H. Sparrow and V. Pond. 1959. Subchromatid rearrangements in Trillium erectum L. I. Origin and nature of configurations induced by ionizing radiation. Am. J. Botany 46(4): 309-316.

Wilson, J. D. 1966. Changes in x-ray sensitivity during germination of Penicillium atrovenetum conidiospores. Thesis, Univ. of Illinois, Urbana. 148pp. (NSA 21: 34837).

Wilson, J. D. and E. L. Powers. 1970. X-ray sensitivity and modifying effects of water in conidia of Aspergillus nidulans. Radiation Research 43: 698-710. (NSA 24: 46684).

Wilson, O. E. and G. L. Hunt, Jr. 1966. Habitat selection by the queens of two field-dwelling species of ants. Ecology 47(3): 485-489. (^{131}I).

Wilson, R. H. 1962. The Columbia River environmental monitoring program. Hanford Labs., U. S. AEC report HW-75076. 32pp.

Wilson, R. H. (Ed.). 1963, 1965. Evaluation of radiological conditions in the vicinity of Hanford for 1962. ---January-March, 1963. ---April-June, 1963. ---. Hanford Labs., U. S. AEC report HW-76526, -77533, -78395; Battelle Northwest Lab., U. S. AEC report BNWL-165. 183, 28, 39, 29pp. (NSA 17: 30750; 18: 8545; 20: 7297) (Incl. waterfowl, fish, marine organisms).

Wilson, W. A. and W. H. A. Wilde. 1971. Some effects of gamma radiation on earwigs in Ontario, Canada (Dermaptera: Forficulidae). Can. Entomol. 103 (8): 1201-1206. (NSA 26: 20621).

Wilson, W. L. 1957. The treatment, conveyance and disposal of radioactive wastes. pp. 253-272. In: The Treatment of Trade-Waste Waters and the Prevention of River Pollution. P. C. G. Isaac, ed. Dept. of Civil Engineering, Univ. of Durham, Durham, N. C., Bull. 10.

Wimer, L. T. 1963/64. Biochemical changes associated with growth and development of the larval blowfly, Phormia regina (Meigen). (Dissertation Abstr. 24: 2631-2632) (^{14}C).

Windham, S. T. and C. R. Phillips. 1973. Radiological survey of New London Harbor, Thames River, Conn., and environs. Radiat. Data Rep. 14(11): 659-666. (NSA 29: 18360).

Winograd, I. J. and G. M. Farlekas. 1974. Problem in ^{14}C dating of water from aquifers of deltaic origin. An example from the New Jersey coastal plain. pp. 69-93. In: Isotope Techniques in Groundwater Hydrology. Vol. 2. (See IAEA, 1974) (NSA 31: 19848).

Winsor, T. F. 1973. Effects of acute gamma radiation on Japanese quail (Coturnix coturnix japonica) in relation to functional levels in a thermoneutral environment, and thermoregulatory responses to abrupt increase in ambient air temperature. Thesis, Washington State Univ., Pullman. xxxii, 259pp.

Winter, J. 1972. Radiocarbon dating by thermoluminescent dosimetry. Archeometry 14(2): 281-286. (NSA 26: 50621).

Winter, M. 1974. Monitoring of radioactive liquid effluents at the Karlsruhe Nuclear Research Center. pp. 342-359. In: Proc. Seminar on the Monitoring of Radioactive Effluents. (Karlsruhe, Germany, May 14-17, 1974). (See (Organization for Economic Co-operation and Development, 1974) (NSA 31: 3356).

Winterhoff, D. H. 1965. The effect of Co-60 gamma radiation on respiration of aerobic bacteria. Thesis, Univ. of Münster, Germany. 59pp. (In German) (AED-Diss.-65-26) (NSA 20: 3699).

Winteringham, F. P. W. 1952. Some aspects of insecticide biochemistry. Endeavor 11(41): 22-28. (Discusses use of radionuclides).

Winteringham, F. P. W. 1956. Labelled metabolic pools for studying quantitatively the biochemistry of toxic action. Int. J. Appl. Radiat. Isotopes 1(1/2): 57-65; Analyst 81(964): 440-441.

Winteringham, F. P. W. 1960. The labeled pool technique with particular reference to pesticide research. pp. 75-124. In: Advances on Pest Control Research. Vol. III. (See Metcalf, 1960).

Winteringham, F. P. W. 1960. Phosphorylated compounds in the head and thoracic tissues of the adult housefly, Musca domestica L., during flight, rest, anoxia and starvation. Biochem. J. 75(1): 38-45. (Used ^{32}P).

Winteringham, F. P. W. 1962. Radioactive tracer techniques in insect biochemistry. pp. 113-134. In: Radioisotopes and Radiation in Entomology. (See IAEA, 1962).

Winteringham, F. P. W. 1965. Some distinctive features of insect metabolism. pp. 29-37. In: Aspects of Insect Biochemistry. T. W. Goodwin, ed. Academic Press, New York. (Mentions studies using radionuclides).

Winteringham, F. P. W. 1975. Food and environmental quality protection. Int. At. Energy Agency Bull. 17(5): 14-25. (NSA 33: 12529).

Winteringham, F. P. W. and A. Harrison. 1956. Study of anticholinesterase action in insects by labelled pool technique. Nature 178(4524): 81-83.

Winteringham, F. P. W., P. M. Bridges and G. C. Hellyer. 1955. Mode of insecticidal action studied with labelled systems. Phosphorylated compounds in the muscle of the adult housefly, Musca domestica L. Biochem. J. 59(1): 13-21. (^{32}P).

Winteringham, F. P. W., A. Harrison and R. G. Bridges. 1952. Radioactive tracer-paper chromatography techniques. Analyst 77: 19-28.

Winteringham, F. P. W., A. Harrison and R. G. Bridges. 1952. Radioactive-tracer techniques in paper chromitography. Nucleonics 10(3): 52-57.

Winteringham, F. P. W., P. M. Loveday and G. C. Hellyer. 1953. Phosphorus metabolism in the housefly Musca domestica. Abstr. Biochem. J. 55(5): xxxiii-xxxiv. (Used ^{32}P).

Winteringham, F. P. W., H. Marchart and J. I. Munn. 1974. Food and environmental contamination: a scientific approach to the problems as a whole. pp. 603-612. In: Comparative Studies of Food and Environmental Contamination. (See IAEA, 1974) (NSA 30: 15453) (Incl. radionuclides).

Wireman, D. L. and C. E. Rosenberry, Jr. 1975. REECo activities and sample logistics in support of the Nevada Applied Ecology Group. pp. 465-469. In: Radioecology of Plutonium and Other Transuranics in Desert Environments. (See White and Dunaway, 1975) (NSA 33: 666).

Wirjosimin, S. 1961. Environmental radioactivity in Bandung area. Madjalah Inst. Tekn. Bandung Proc. 1(2): 52-64. (NSA 17: 37372).

Wisconsin Department of Health and Social Services. 1970. Point Beach Nuclar Plant, Units 1 and 2. Kewaunee Nuclear Power Plant. Environmental radiation survey annual report, July 1969-June 1970. Madison. 27pp. U. S. AEC DOCKET-50266-42, -50301-27, -50305-18. (NSA 25: 20641).

Wisconsin Electric Power Company. 1971. Point Beach Nuclear Plant, Unit 1. Environmental studies, report No. PBR-2. Milwaukee, Wis. 90pp. U. S. AEC DOCKET-50301-40. (NSA 25: 56947).

Wisconsin Michigan Power Company. 1970. Point Beach Nuclear Plant, Unit No. 2. Appleton, Wis. 359pp. U. S. AEC DOCKET-50301-24. (NSA 24: 47778) (Incl. environment).

Wisconsin Public Service Corporation. 1971. Kewaunee Nuclear Power Plant. Environmental report, operating license stage. ---supplementrary information. Green Bay, Wis. 68, 39pp. U. S. AEC DOCKET-5050305-12, -21. (NSA 25: 15309, 41341).

Wise, J. S., Jr. 1968. The whole body burden of cesium-137 in the gray squirrel in seven physiographic provinces. Thesis, Univ. of Georgia, Athens. v, 24pp.

Wise, K. R. and B. W. Bentley. 1971. Modeling of radionuclide release from nuclear power and reprocessing plants. Trans. Amer. Nucl. Soc. 14(2): 457-458. (NSA 26: 3747).

Wiser, C. W. and D. J. Nelson. 1964. Uptake and elimination of cobalt-60 by crayfish. Am. Midland Nat. 72(1): 181-202.

Witherspoon, J. P. 1962. Cycling of cesium-134 in white oak trees. Abstr. Bull. Ecol. Soc. Am. 43(4): 120-121.

Witherspoon, J. P., Jr. 1963. Cycling of cesium-134 in white oak trees on sites of contrasting soil type and moisture. I. 1960 growing season. pp. 127-132. In: Radioecology. (See Schultz and Klement, 1963) (NSA 17: 33540).

Witherspoon, J. P. 1965. Radiation damage to forest surrounding an unshielded fast reactor. In: Radiation and Terrestrial Ecosystems. Health Physics 11(12): 1637-1642. (See Hungate, 1965).

Witherspoon, J. P. 1965. Responses of Pinus strobus to acute gamma irradiation of roots and shoots. Abstr. Bull. Ecol. Soc. Am. 46(3): 96.

Witherspoon, J. P. 1968. Effects of internal ^{137}Cs radiation on seeds of Liriodendron tulipifera. Radiation Botany 8(1): 45-48. (NSA 22: 28702).

Witherspoon, J. P. 1969. Radiosensitivity of forest tree species to acute fast neutron radiation. pp. 120-126. In: Symposium on Radioecology. (See Nelson and Evans, 1969).

Witherspoon, J. P. 1969. Ecological recovery. pp. 27-32. In: A Report to the Atomic Energy Commission by a Committee of the National Academy of Sciences. U. S. AEC report TID-24690. (NSA 23: 14156).

Witherspoon, J. P. 1970. Interception and retention of radioactive aerosols by vegetation. Atmos. Environ. 4: 591. (NSA 25: 29444).

Witherspoon, J. P. 1971. Field studies of fallout retention by plants. pp. 396-404. In: Survival of Food Crops and Livestock in the Event of Nuclear War. (See Benson and Sparrow, 1971).

Witherspoon, J. P. 1973. Radioecology. pp. 250-257. In: Medical Radiation Biology. G. V. Dalrymple, et al., eds. W. B. Saunders. Philadelphia.

Witherspoon, J. P. and and G. N. Brown. 1965. Translocation of cesium-137 from parent trees to seedlings of Liriodendron tulipifera. Botan. Gaz. 126(3): 181-185.

Witherspoon, J. P. and F. G. Taylor, Jr. 1969. Retention of a fallout simulant containing ^{134}Cs by pine and oak trees. Health Physics 17: 825-829. (NSA 24: 6017).

Witherspoon, J. P. and F. G. Taylor, Jr. 1969. Radiation-induced anatomical modifications in forest trees. J. Tenn. Acad. Sci. 44: 118-124. (NSA 24: 17016).

Witherspoon, J. P. and F. G. Taylor, Jr. 1969. Effects of external beta radiation on higher plants. pp. 2-17. In: Progress Report in Post Attack Ecology. (See Auerbach, 1969).

Witherspoon, J. P. and F. G. Taylor, Jr. 1970. Interception and retention of a simulated fallout by agricultural plants. Health Physics 19: 493-499. (NSA 25: 512).

Witherspoon, J. P., S. I. Auerbach and J. S. Olson. 1962. Cycling of cesium-134 in white oak trees on sites of contrasting soil type and moisture. Dissertation, Univ. of Tennessee, Knoxville. xv, 139pp. (Also: Oak Ridge National Lab., U. S. AEC report ORNL-3328) (NSA 17: 1230).

Witkamp, M. 1960. Biological uptake of radionuclides. Nuclear Safety 2(2): 65-69.

Witkamp, M. 1961. Influence of radiation from an unshielded reactor on a natural microflora (preliminary report). Oak Ridge National Lab., U. S. AEC report ORNL-TM-29. 9pp. (NSA 15: 31996).

Witkamp. M. 1962. Influence of environmental factors on microbial immobilization of certain radionuclides in forest litter. Abstr. Bull. Ecol. Soc. Am. 43(4): 121.

Witkamp, M. 1964. Environmental influences on microbial populations and activity of the forest floor. Oak Ridge National Lab., U. S. AEC report ORNL-P-586. 19pp. (NSA 19: 1961).

Witkamp, M. 1966. Biological concentrations of ^{137}Cs and ^{90}Sr in arctic food chains. Nuclear Safety 8: 58-62. (NSA 21: 12798).

Witkamp, M. 1967. Assessment of microbial immobilization of mineral elements. Soil Biology (July): 36-37. (Tracer).

Witkamp, M. 1968. Accumulation of ^{137}Cs by Trichoderma viride relative to ^{137}Cs in soil organic matter and soil solution. Soil Science 106(4): 309-311. (Also: Oak Ridge National Lab., U. S. AEC report ORNL-P-3594) (NSA 23: 2831).

Witkamp, M. 1968. External factors influencing mineralization and immobilization of some radionuclides from tree litter. pp. 231-240. In: Isotopes and Radiation in Soil Organic-Matter Studies. (See IAEA, 1968) (NSA 24: 6016).

Witkamp, M. 1969. Environmental effects on microbial turnover of some mineral elements. Part I--Abiotic factors. ---Part II--Biotic factors. Soil Biol. Biochem. 1(3): 167-176, 177-184. (Radionuclides).

Witkamp, M. 1970. Aspects of soil microflora in a gamma-irradiated rain forest. Ch. F-5, pp. F29-F33. In: A Tropical Rain Forest. (See H. T. Odum and R. F. Pigeon, 1970) (NSA 25: 21969).

Witkamp, M. 1970. Mineral retention by epiphyllic organisms. Ch. H-14, pp. H177-H179. In: A Tropical Rain Forest. (See H. T. Odum and R. F. Pigeon, 1970).

Witkamp, M. 1972. Transfer of ^{137}Cs from detritus to primary producer. pp. 341-347. In: Isotopes and Radiation in Soil-Plant Relationships Including Forestry. (See IAEA, 1972).

Witcamp, M. and B. Barzansky. 1968. Microbial immobilization of ^{137}Cs in forest litter. Oikos 19: 392-395. (Also: Oak Ridge National Lab., U. S. AEC report ORNL-P-3610) (NSA 23: 38520).

Witkamp, M. and D. A. Crossley, Jr. 1966. The role of arthropods and microflora in breakdown of white oak litter. Pedobiologia 6: 293-303.

Witkamp, M. and M. L. Frank. 1964. First year of movement, distribution and availability of Cs^{137} in the forest floor under tagged tulip poplars. Radiation Botany 4(4): 485-495. (NSA 19: 14951).

Witkamp, M. and M. L. Frank. 1967. Retention and loss of ^{137}Cs by components of the groundcover in a pine (Pinus virginiana L.) stand. Health Physics 13(9): 985-990. (NSA 21: 41335).

Witkamp, M. and M. L. Frank. 1969. Cesium-137 kinetics in terrestrial microcosms. pp. 635-643. In: Symposium on Radioecology. (See Nelson and Evans, 1969).

Witkamp, M. and M. L. Frank. 1969. Loss of weight, ^{60}Co and ^{137}Cs from tree litter in three subsystems of a watershed. Environ. Sci. Technol. 3: 1195-1198. (NSA 24: 6019).

Witkamp, M. and M. L. Frank. 1970. Effects of temperature, rainfall, and fauna on transfer of ^{137}Cs, K, Mg, and mass in consumer-decomposer microcosms. Ecology 51(3): 465-474.

Witkamp, M. and V. A. Merchant. 1973. Effects of light, temperature, and soil fertility on distribution of manganese-54 and cesium-137 in producer-consumer microcosms. pp. 204-208. In: Radionuclides in Ecosystems. (See Nelson, 1973) (NSA 28: 27299).

Witkamp, M. and H. D. Waller. 1963. Landscape studies using radioactive tracers: II. Behavior of Cs-137 in the floor of a tulip poplar forest. Abstr. Bull. Ecol. Soc. Am. 44(3): 73.

Witt, J. M. 1965. The metabolism of ^{14}C-labelled DDT by the large milkweed bug, the house fly, and the American cockroach. Thesis, Univ. of California, Berkeley. 224pp. (Dissertation Abstr. 26: 7020).

Wittmann, W. 1964. A cytological study of the "free nuclear divisions" in Macrozamia communis (Johns.) carried out on γ-irradiated material. Radiation Botany 4(3): 331-340. (NSA 19: 5693) (Cycad).

Wittwer, S. H., H. G. Tukey and M. J. Bukovac. 1959. Absorption of radionuclides applied to above ground plant parts. Michigan State Univ., East Lansing, U. S. AEC report AECU-4597. 20pp. (NSA 14: 5010).

Wixson, B. G. and W. J. Clark. 1967. Gamma radiation effects on nymphs of the damselfly Argia translata. Ann. Entomol. Soc. Am. 60(2): 485-486.

Wlodawer, P. 1961. Incorporation of P^{32} into phosphorus compounds of the wax moth larvae. Acta biochem. polon. 8: 321-335.

Wlodek, S. 1964. Radioactivity of a series of aquatic and terrestrial organisms in relation to radioactivity in their ambient medium. Ekol. Pol. A12: 415-427. (Also: Poland Centralne Lab. Ochrony Radiologicznej, Warsaw, report CLOR-2b) (In French) (NSA 18: 31753; 20: 23392).

Wlodek, S. 1965. Preliminary research on the migration of ^{137}Cs in the food chain: water-plant-mollusk-fish. France Commissariat à l'Énergie Atomique, Grenoble. 17pp. (In French) (NP-16594) (NSA 21: 22367).

Wlodek, S. 1965. Behavior of ^{137}Cs in a reservoir of fresh water, taking account of the reactions of the phytocoenoses to contamination. Note Bio-V/9. France Commissariat à l'Énergie Atomique, Grenoble. 72pp. (In French) (NP-16590) (NSA 21: 24412).

Wlodek, S. 1965. Radioactivity of the plants of the tributaries of the Isere River in 1964-1965. Note Bio-V/10. France Commissariat à l'Énergie Atomique, Grenoble. 29pp. (In French) (NP-16593) (NSA 21: 20174)

Wlodek, S. 1967. The behavior of ^{137}Cs in freshwater reservoirs, with particular reference to the reactions of plant communities to contamination. pp. 897-912. In: Radioecological Concentration Processes. (See Aberg and Hungate, 1967).

Wlodek, S. 1968. Existence and the role of cesium in the natural environment, particularly aquatic environments. II. Results of research on the contamination of freshwater ecosystems by radiocesium. Ekol. Pol. B14: 199-220. (In Polish) (NSA 25: 32377).

Wlodek, S. nd. Effect of human activities on ^{226}Ra migration in environment. Poland Centralne Lab. Ochrony Radiologiczrej, Warsaw. (In Russian) (INIS-mf-887, pp. 280-293) (NSA 29: 24202).

Wlodek, S., M. Bysiek and D. Grzybowska. 1970. Autochtonic cesium and behavior of allochtonic cesium through a fish culture pond. pp. 319-331. In: Proc. Intern. Symp. on Radioecology. Vol. I. (See French Commissariat à l'Énergie Atomique, 1970) (In French).

Wlodek, S., Grzybowska and M. Bysiek. 1970. Influence of human activity upon ^{226}Ra migration in the environment. Nukleonika 15: 675-681. (In Polish) (NSA 25: 32373).

Wobrauschek, P. 1974. Effects on the environment of nuclear power stations. Elektotech. Maschinenbau 91(6): 324-328. (In German) (NSA 31: 19551).

Wohlgemuth, R. 1973. Comparative investigations on the effect of x rays or γ rays (^{60}Co) on the grain weevil (Silophilus granarius). Z. angew. geol. 74(1): 7-24. (In German) (NSA 30: 7118).

Wojtalik, T. A. 1963. The transfer of radiophosphorus within a stream and its relationship to light intensity and insect migration. Thesis, Michigan State Univ., East Lansing. xi, 208pp.

Wojtalik, T. A. and C. W. Voigtlander. 1971. The elements of a monitoring program to assess isotope accumulation and thermal effects of a nuclear power plant in a TVA reservoir. pp. 469-480. In: Reservoir Fisheries and Limnology. G. E. Hall, ed. Amer. Fisheries Soc., Washington.

Wojtczak, L. and A. B. Wojtczak. 1960. Uncoupling of oxidative phosphorylation and inhibition of adenosine triphosphate-orthophosphate exchange by a substance from insect mitochondria. Biochim. Biophys. Acta 39: 277-286. (^{32}P).

Wolbert, P. 1969. The ecdysial rate of the greater wax moth, Galleria mellonella L. (Lepidoptera, Pyralidae), after x-irradiation wounding, and combined injury in the pupal instar. Arch. Entwicklungsmech. Org. 164: 134-169. (In German) (NSA 24: 36806).

Wolf, L. E. 1951. Diet experiments with trout. Prog. Fish. Cult. 13: 17-24.

Wolf, M. 1973. Synthesis of benzene over acetic acid for the low level measurement of tritium counts. Int. J. Appl. Radiat. Isotopes 24(5): 299-301. (In German) (NSA 28: 12679).

Wolfe, D. A. 1967. Accumulation of fallout cesium-137 by an estuarine clam. Abstr. Bull. Ecol. Soc. Am. 48(3): 128.

Wolfe, D. A. 1967. Seasonal variations of ^{137}Cs from fallout in a clam, Rangia cuneata Gray. Nature 215: 1270-1271. (NSA 21: 41351).

Wolfe, D. A. 1967. Molluscs in radiobiology and radioecology. Bull. N. Car. Shell Club (4): 20-23.

Wolfe, D. A. 1970. Levels of stable Zn and ^{65}Zn in Crassostrea virginica from North Carolina. J. Fish. Res. Bd. Can. 27(1): 47-57.

Wolfe, D. A. 1971. Fallout cesium-137 in clams (Rangia cuneata) from the Neuse River Estuary, North Carolina. Limnol. Oceanogr. 16(5): 797-805. (NSA 26: 15544).

Wolfe, D. A. and C. B. Coburn, Jr. 1970. Influence of salinity and temperature on the accumulation of cesium-137 by an estuarine clam under laboratory conditions. Health Physics 18: 499-505.

Wolfe, D. A. and C. D. Jennings. 1973. Iron-55 and ruthenium-103 and -106 in the brackish-water clam Rangia cuneata. pp. 783-790. In: Radionuclides in Ecosystems. (See Nelson, 1973) (NSA 28: 29896).

Wolfe, D. A. and T. R. Rice. 1968. Safe levels of radioactivity in aquatic environments. Scientia (Milan) 103: 469-487. (23: 22317).

Wolfe, D. A. and T. R. Rice. 1972. Cycling of elements in estuaries. U. S. NOAA Fishery Bull. 70(3): 959-972. (NSA 31: 33100; SFA 17: 15496) (Incl. radionuclides).

Wolfe, D. A. and C. L. Schelske. 1966. Counting efficiences for carbon-14 in marine productivity measurements. Abstr. Bull. Ecol. Soc. Am. 47(3): 133-134.

Wolfe, D. A. and C. L. Schelske. 1969. Accumulation of fallout radioisotopes by bivalve molluscs from the lower Trent and Neuse rivers. pp. 493-504. In: Symposium on Radioecology. (See Nelson and Evans, 1969).

Wolfe, D. A., et al. 1975. Trace element interactions in the estuarine zone of the Anasco River, Puerto Rico. pp. 155-178. In: Combined Effects of Radioactive, Chemical, and Thermal Releases to the Environment. (See IAEA, 1975) (NSA 33: 29103) (Neutron activation, radiotracers).

Wolfe, J. A. 1962. Analytical procedures for plant and soil samples used in vegetation studies related to the movement of radioactive wastes. Dept. of Botany, Univ. of Tennessee, Knoxville. vi, 58pp. Mimeo.

Wolfe, J. N. 1959. The ecological aspects of Project Chariot. pp. 60-66. In: Proc. 2nd Plowshare Symp. Part II. (May 13-15, 1959, San Francisco, Calif.). Lawrence Radiation Lab., Livermore, Calif., Plowshare Ser. Rept. No. 2, U. S. AEC report UCRL-5676. (NSA 14: 22745).

Wolfe, J. N. 1959. Ecological aspects. App. H. to statement of C. L. Dunham. pp. 144-146. In: Fallout from Nuclear Weapons Tests. Vol. 1. (See U. S. Congress, 1959).

Wolfe, J. N. 1959. Long-time ecological effects of nuclear war. pp. 832-842. In: Biological and Environmental Effects of Nuclear War. (See U. S. Congress, 1959) (Also: U. S. AEC report TID-5561) (NSA 13: 18005).

Wolfe, J. N. 1963. Impact of atomic energy on the environment and environmental science. pp. 1-2. In: Radioecology. (See Schultz and Klement, 1963) (NSA 17: 33540).

Wolfe, J. N. 1965. New perspectives in ecology. pp. 215-218. In: Proc. 5th Inter-American Symp. on the Peaceful Application of Nuclear Energy. Pan American Union, Washington. (NSA 19: 35891).

Wolfe, J. N. 1969. Radioecology: retrospection and future. pp. xi-xii. In: Symposium on Radioecology. (See Nelson and Evans, 1969).

Wolfenbarger, D. A. and J. G. Darroch. 1965. A sequential sampling plan for determining the status of corn earworm control in sweet corn. J. Econ. Entomol. 58(4): 651-654.

Wolfenbarger, D. A. and H. M. Graham. 1970. Toxicity of insecticides to irradiated and unirradiated pink bollworm moths. J. Econ. Entomol. 63: 1692-1693. (NSA 25: 16294).

Wolfenbarger, D. A. and A. A. Guerra. 1971. Response of strains and sexes of the tobacco budworm to gamma irradiation. J. Econ. Entomol. 64(6): 1412-1415. (NSA 26: 12435).

Wolfenbarger, D. A. and A. A. Guerra. 1972. Sexual competitiveness of ^{60}Co-irradiated tobacco budworm. J. Econ. Entomol. 65(6): 1596-1600. (NSA 27: 15089).

Wolfenbarger, D. A. and C. L. Mangum. 1972. Fertility of moths of pink bollworm reared from irradiated eggs. J. Econ. Entomol. 65(3): 919-921. (NSA 26: 43357).

Wolfenbarger, D. A., A. A. Guerra and S. H. Robinson. 1973. Tobacco budworms: sterilization, sexual competitiveness and sperm transfer of moths reared from eggs irradiated with ^{60}Co. J. Econ. Entomol. 66(5): 1067-1070. (NSA 29: 776).

Wolfenden, R. N. and F. W. Forbes-Ross. 1898. Action of roentgen rays upon growth and activity of bacteria and microorganisms. Lancet 1898 I (3904): 1752-1753.

Wolff, E. and M. Kieny. 1958. Results of localized irradiations of the tips of the anterior limbs in the bird. Compt. rend. soc. biol. 152: 712-714. (In French) (NSA 13: 1908).

Wolff, E. and T. Lender. 1962. The neoblast and phenomena of induction and inhibition in regeneration in planaria. Annee Biol. 38: 499-529. (In French) (NSA 18: 29263).

Wolff, E. and M. Reyss-Brion. 1957. Effect of x-rays on gastrula of a batrachian urodele: Pleurodeles waltii. Compt. rend. 244(26): 3172-3175. (In French).

Wolff, S., et al. 1974. Uniformity of normalized radiation-induced mutation rates among different species. Genetics 78(1): 133-134. (NSA 31: 16762) (Bacteria, fungi, plants, insects, mammals).

Wollenberg, H. A. 1974. Radioactivity of Nevada hot spring systems. Geophys. Res. Lett. 1(8): 359-362. (Also: Lawrence Berkeley Lab., U. S. AEC report LBL-2482) (NSA 29: 20878; 31: 29621).

Wollenberg, H. A. and A. R. Smith. 1964. Studies of terrestrial γ radiation. pp. 513-566. In: The Natural Radiation Environment. (See Adams and Lowder, 1964) (NSA 17: 32264).

Wollenberg, H. A. and A. R. Smith. 1972. Geological factors controlling terrestrial gamma-ray dose rates. pp. 457-481. In: Natural Radiation Environment II. Vol. 2. (See Adams, Lowder and Gesell, 1972) (NSA 32: 19691).

Wollenberg, H. A., et al. 1969. Natural and fallout radioactivity in the San Francisco Bay area. Health Physics 17: 313-321. (NSA 23: 38100).

Wolman, A. 1969. The nature and extent of radioactive waste disposal problems. pp. 84-92. In: Water, Health, and Society. (Selected Papers by Abel Wolman). G. F. White, ed. Indiana Univ. Press, Bloomington. (NSA 23: 40819).

Wolman, A. and A. E. Gorman. 1950. Waste materials in the United States Atomic Energy Program. U. S. AEC report WASH-8. 20pp. (NSA 5: 924).

Wolter, R. and J. Arndt. 1975. Radioactive waste waters from nuclear power stations in the Federal Republic of Germany for 1974. Bundesgesundheitsblatt 18(7): 113-115. (NSA 32: 23966).

Wong, A.-c. and C.-y. Wang. 1960. The effect of embryonic development of goldfish from exposing the parent fish to x-rays. Acta zool. sinica 12 (1): 127-130.

Wong, P. T. 1970. Effects of x irradiation on DNA synthesis in salivary gland cells of Drosophila hydei. Thesis, Oregon State Univ., Corvallis. 87pp. (Univ. Microfilms 70-14144) (NSA 25: 11246).

Wood, D. L. and R. W. Stark. 1966. Effects of gamma radiation on the biology and behavior of adult Ips confusus (LeConte) (Coleoptera: Scolytidae). Can. Entomol. 98: 1-10. (NSA 20: 42883).

Wood, E. M. 1958. Fixation of radioiodide in the eggs and alevens of some salmonids (Salmo fairo L., Salmo iridus G.I.B., Salmo salar L.). Ann. Sta. Centr. Hydrobiol. Appl. 7: 285-299. (In French).

Wood, F. C. and J. B. Goodey. 1957. Effects of gamma-ray irradiation on nematodes infesting cultivated mushroom beds. Nature 180(4589): 760-761.

Wood, J. M. 1974. Biological cycles for toxic elements in the environment. Science 183(4129): 1049-1052.

Wood, K. G. 1971. Self-absorption corrections fot the ^{14}C method with $BaCO_3$ for measurement of primary productivity. Ecology 52(3): 491-498.

Wood, R. A., et al. 1976. Determination of ^{238}Pu and ^{242}Cm in aqueous solid extracts. Univ. of California at Los Angeles, U. S. ERDA report UCLA-12-1047. 10pp. (NSA 33: 25386) (Soil).

Wood, R. E. and J. M. Palms. 1974. Gamma-ray spectrum analysis technique for low-level environmental radionuclides. Trans. Nucl. Soc. NS-21(1): 536-542. (NSA 30: 6730).

Wood, V. G. 1964. The effect of x radiation on longevity in Drosophila melanogaster. Final report, Contract AT-(11-1)-885. U. S. AEC report TID-20550. 83pp. (NSA 18: 19554).

Woodbridge, D. D., L. A. Mann and W. R. Garrett. 1970. Application of gamma radiation to sewage treatment. Nuclear News 13(9): 60-64. (NSA 24: 46283).

Woodhead, D. S. 1970. The assessment of the radiation dose to developing fish embryos due to the accumulation of radioactivity by the egg. Radiation Research 43(3): 582-597. (NSA 24: 50923).

Woodley, R. 1967. Effect of dose on the energy dependence of LIF thermoluminescent radiation dosimeters. Radiation Botany 7(1): 1-5.

Woods, F. W. 1965. A study of root distribution in two hardwood stands in the Duke forest. Determination of the extent of root distribution by the use of radiotracer techniques. Final report. Duke Univ., U. S. AEC report TID-21545, -21544. 116, 12pp. (NSA 19: 14950, 14949).

Woods, F. W. 1970. Interspecific transfer of inorganic materials by root systems of woody plants. J. Appl. Ecol. 7(3): 481-486.

Woods, F. W. and K. Brock. 1964. Interspecific transfer of Ca^{45} and P^{32} by root systems. pp. 62-65. In: 2nd Ann. Oak Ridge Radioisotope Conf. (April 19-22, 1964). U. S. AEC report TID-7689. (Also: Ecology 45(4): 886-889) (NSA 18: 15757).

Woods, F. W. and M. L. McCormack. 1963. Two scintillation probes for radioecological investigations. pp. 205-209. In: Advancing Frontiers of Plant Sciences. Vol. 2. R. Vira, ed. India Inst. for the Advancement of Science and Culture, New Delhi.

Woods, F. W. and D. O'Neal. 1965. Tritiated water as a tool for ecological field studies. Science 147(3654): 148-149.

Woods, F. W., M. D. Ferrill and M. L. McCormack. 1962. Methyl bromide for increasing I^{131} uptake in pine trees. Radiation Botany 2(3/4): 273-277. (NSA 17: 15735).

Woods, F. W., et al. 1962. Studies of root extension by the use of I-131. Abstr. Bull. Ecol. Soc. Am. 43(4): 120.

Woodson, R. D. 1972. Logical approaches to power supply and environment. J. Power Div., Amer. Soc. Civil Engrs. 98(PO2): 181-186. (NSA 27: 5291).

Woodward, R. N. 1964. Strontium-90 and caesium-137 in antarctic snows. Nature 204(4965): 1291. (NSA 19: 11400).

Woodwell, G. M. 1962. Effects of ionizing radiation on terrestrial ecosystems. Science 138(3540): 572-577. (Also: Brookhaven National Lab., U. S. AEC report BNL-6408) (NSA 17: 1332, 35436).

Woodwell, G. M. 1963. Design of the Brookhaven experiment on the effects of ionizing radiation on a terrestrial ecosystem. Radiation Botany 3(2): 125-133. (Also: Brookhaven National Lab., U. S. AEC report BNL-6672) (NSA 17: 33537).

Woodwell, G. M. (Pres.). 1963. Symposium: Some Approaches to the Effects of Nuclear Catastrophies on Ecological Systems. Bull. Ecol. Soc. Am. 44(3): 78-79, 82-83. (Individual abstrs. listed: see Miller, C. F.; Auerbach, S. I.; Broido, A.; Woodwell, G. M. and A. H. Sparrow; Platt, R. B; McDonald, D. R.).

Woodwell, G. M. 1963. The ecological effects of radiation. Sci. American 298(6): 40-49. (NSA 17: 23195).

Woodwell, G. M. 1965. Radiation and the patterns of nature. Brookhaven National Lab., U. S. AEC report BNL-924(T-381). iii, 15pp.

Woodwell, G. M. (Ed.). 1965. Ecological Effects of Nuclear War. (Proc. Symp. Amherst, Mass., August 1963). Brookhaven National Lab., U. S. AEC report BNL-917(C-43). v, 72pp.

Woodwell, G. M. 1966. Sensitivity to ionizing radiation: major ecosystems and dominant plant species. Table 40, pp. 181-182. In: Environmental Biology. P. L. Altman and D. S. Dittmer, eds. Am. Soc. Exptl. Biol., Berhesda, Md.

Woodwell, G. M. 1967. Toxic substance and ecological cycles. Sci. American 216(3): 24-31. (NSA 21: 18016) (Incl. radionuclides).

Woodwell, G. M. 1967. Radiation and the patterns of nature. Science 156: 461-470. (NSA 21: 24373).

Woodwell, G. M. 1967. Radioactivity and fallout: the model pollution. Brookhaven National Lab., U. S. AEC report BNL-12401. (NSA 22: 27916).

Woodwell, G. M. 1968. Forest ecology. pp. 185-187. In: McGraw-Hill Yearbook of Science and Technology. D. I. Eggenberger, ed. McGraw-Hill, New York. (Discusses ionizing radiation).

Woodwell, G. M. 1969. Radioactivity and fallout: the model pollution. BioScience 19(10): 884-887.

Woodwell, G. M. 1970. Effects of pollution on the structure and physiology of ecosystems. Science 168: 429-433. (NSA 24: 34203) (Incl. ionizing radiation).

Woodwell, G. M. 1970. The energy cycle of the biosphere. Sci. American 223 (3): 64-74. (Incl. ionizing radiation).

Woodwell, G. M. 1970. Radioactivity and fallout: the model pollution. pp. 159-169. In: Challenge for Survival, Land, Air, and Water for Man in Metropolis. P. Dansereau, ed. Columbia Univ. Press, New York. (NSA 24: 25431).

Woodwell, G. M. 1974. Threshold problems in ecosystems. Brookhaven National Lab., U. S. AEC report BNL-19244. 14pp. (NSA 31: 8868).

Woodwell, G. M. nd. Ecologist's perspective on electrical power. Brookhaven National Lab., U. S. AEC report BNL-16979. 10pp. (For: Conf. on Power, Johnson, Vermont, January 1972) (CONF-720124-1) (NSA 27: 2470).

Woodwell, G. M. nd. Effects of ionizing radiation on natural ecosystems. Brookhaven National Lab., U. S. AEC report BNL-7200. 11pp. (NSA 18: 35306).

Woodwell, G. M. and F. R. Becherer. 1963. Field measurement of CO_2 exchange rates of a forest ecosystem exposed to chronic gamma irradiation. Abstr. Bull. Ecol. Soc. Am. 44(3): 69.

Woodwell, G. M. and F. H. Borman. 1964. The distribution of radiosensitivity among natural populations of plants: an hypothesis. Abstr. Bull. Ecol. Soc. Am. 45(3): 73.

Woodwell, G. M. and P. F. Bourdeau. 1962. Measurement of dry matter production of the plant cover. Brookhaven National Lab., U. S. AEC report BNL-6152. 26pp. (Data from an irradiated oak-pine stand).

Woodwell, G. M. and J. H. Brower. 1967. Aphid population explosion induced by chronic gamma irradiation of a forest. Ecology 48(4): 680-683. (NSA 22: 8500).

Woodwell, G. M. and T. P. Gannutz. 1967. Effects of chronic gamma irradiation on lichen communities of a forest. Am. J. Bot. 54(10): 1210-1214.

Woodwell, G. M. and E. B. Hadley. 1964. The productivity of natural communities. II. Rates of metabolism of an oak-pine forest damaged by chronic gamma irradiation. Abstr. Bull. Ecol. Soc. Am. 45(3): 74.

Woodwell, G. M. and E. C. Hammond. 1962. A descriptive technique for study of the effects of chronic ionizing radiation on a forest ecological system. Brookhaven National Lab., U. S. AEC report BNL-715. 15pp. (NSA 16: 18758).

Woodwell, G. M. and B. R. Holt. 1971. Effect of nuclear war on the structure and function of natural communities: an appraisal based on experiments with gamma radiation. pp. 482-491. In: Survival of Food Crops and Livestock in the Event of Nuclear War. (See Benson and Sparrow, 1971).

Woodwell, G. M. and T. G. Marples. 1968. The influence of chronic gamma irradiation on production and decay of litter and humus in an oak-pine forest. Ecology 49(3): 456-465. (NSA 22: 45471).

Woodwell, G. M. and L. N. Miller. 1963. Chronic gamma radiation effects the distribution of radial increment in Pinus rigida stems. Science 139(3551): 222-223. (Also: Brookhaven National Lab., U. S. AEC report BNL-6458) (NSA 17: 7878).

Woodwell, G. M. and J. K. Oosting. 1965. Effects of chronic gamma irradiation on the development of old field plant communities. Radiation Botany 5(3): 205-222. (Also: Brookhaven National Lab., U. S. AEC report BNL-8398) (NSA 19: 33735).

Woodwell, G. M. and E. V. Pecan. (Eds.). 1973. Carbon and the Biosphere. (Proc. 24th Brookhaven Symp. in Biology, Upton, New York, May 16-18, 1972). U. S. AEC Symp. Ser. 30, report CONF-720510. vii, 392pp.

Woodwell, G. M. and A. L. Rebuck. 1967. Effects of chronic gamma radiation on the structure and diversity of an oak-pine forest. Ecol. Monographs 37(1): 53-69.

Woodwell, G. M. and A. H. Sparrow. 1963. Predicted and observed effects of chronic gamma radiation on a near-climax forest ecosystem. Radiation Botany 3(3): 231-237. (NSA 18: 5081).

Woodwell, G. M. and A. H. Sparrow. 1965. Effects of ionizing radiation on ecological systems. pp. 26-27. In: Ecological Effects of Nuclear War. (See Woodwell, 1965) (NSA 20: 1588).

Woodwell, G. M. and R. H. Whittaker. 1968. Effects of chronic gamma irradiation on plant communities. Quart. Rev. Biol. 43(1): 42-55. (NSA 22: 51487).

Woodwell, G. M. and R. H. Whittaker. 1968. Promary production in terrestrial ecosystems. Am. Zool. 8: 19-30. (Forest, Brookhaven National Lab.).

Woodwell, G., F. R. Becherer and E. B. Hadley. 1964. The productivity of natural communities. I. An automatic data recording system for monitoring metabolism and microclimate in the field. Brookhaven National Lab., U. S. AEC report BNL-7901. 30pp. (NSA 19: 17397).

Wooldridge, C. B. (Ed.). 1968. Environmental surveillance in the vicinity of Hanford for November-December 1967. ---1967. ---January-June 1968. ---July-December 1968. ---October 1968. Battelle Northwest Lab., U. S. AEC report BNWL-CC-1197-6, -983, -1135, -CC-2026, -778-10. 84, 87, 67, 84, 17 pp. (NSA 23: 20175, 29464, 43747; 29: 10123, 12701).

Wooldridge, D. D. 1965. Tracing soil particle movement with Fe-59. Soil Sci. Soc. Am. Proc. 29(4): 469-471.

Worgul, B. V. 1974. Investigation of radiation cataract by hormonal manipulation of lense growth. Dissertation, Univ. of Vermont, Burlington. 277 pp. (Univ. Microfilms 74-22035) (NSA 31: 811) (Frog).

World Health Organization. 1968. Routine surveillance for radionuclides in air and water. United Nations, Geneva. 64pp. (NSA 23: 2829).

World Health Organization. 1972. Protection against ionizing radiations. A survey of current world legislation. United Nations, Geneva. 329pp. (NSA 27: 7239).

Worseck, M. 1961. Effect of ionizing radiation on bacteria. A literature review. Isotopentechnik 1: 77-82. (In German) (NSA 15: 14138).

Wortley, G. 1970. Toxicity of pesticides labelled with radionuclides. pp. 57-65. In: Nuclear Techniques for Studying Pesticide Residue Problems. (See IAEA, 1970) (NSA 25: 8600).

Woskressensky, N. M. 1928. On the effect of roentgen radiation on embryo growth. Arch. Entwicklungsmech. Org. 113: 447. (In German).

Wozniak, J. 1973. Calibration of neutron gauges for measuring of soil moisture content in the 4π geometry. I. Two group diffusion theory, case of an empty hole. Poland Inst. of Nuclear Techniques, Krakow, report ITJ-27. 49pp. (NSA 28: 12957).

Wray, E. T. 1969, 1970. Environmental monitoring associated with discharges of radioactive waste during 1968 from UKAEA establishments. ---1969. UK Atomic Energy Authority, Harwell, report AHSB(RP)R-97, -105. 22, 22pp. (NSA 24: 29720; 25: 16128).

Wray, F. J. and J. A. Richardson. 1964. Paths of water transport in higher plants. Nature 202(4930): 415-416. (Used tritium to study water movement in willows).

Wrenn, M. E., et al. 1972. Natural radioactivity in Hudson River Estuary samples and their influence on the detection limits for gamma emitting radionuclides using NaI gamma spectrometry. pp. 897-916. In: Natural Radiation Environment II. Vol. 2. (See Adams, Lowder and Gesell, 1972) (NSA 32: 19752).

Wrenn, M. E., et al. 1973. Radiocesium distribution in water, sediment, and biota in the Hudson River Estuary from 1964 through 1970. pp. 334-343. In: Radionuclides in Ecosystems. (See Nelson, 1973) (NSA 28: 27303).

Wrenn, M. E., et al. 1974. ^{137}Cs and ^{134}Cs distribution in sediment, water and biota of the lower Hudson River and their dosimetric implications for man. pp. 279-284. In: Proc. 3rd Intern. Congr. International Radiation Protection Association. (CONF-730907-P1) (NSA 30: 26803).

Wright, C. N. and W. F. Splichal, Jr. 1973. Portable cesium counter for field use. Health Physics 25(5): 516-517. (NSA 28: 29940).

Wright, J. H. 1970. Power and the environment. Proc. Amer. Power Conf. 32: 487-498. (CONF-700426) (NSA 25: 18044).

Wright, J. H. 1974. Environmental aspects of the Clinch River Breeder Reactor Project. Nucl. Eng. Int. 19(221): 851-854. (NSA 31: 10329).

Wright, J. H. 1975. Environment and nuclear energy. IV. Use of energy and its relation with the environment. Energia nuclear (Madrid) 19(94): 111-114. (In Spanish) (NSA 32: 11536).

Wright, J. H., J. B. F. Champlin and O. H. Davis. 1971. The impact of environmental radiation and discharge heat from nuclear power plants. pp. 549-559. In: Environmental Aspects of Nuclear Power Stations. (See IAEA, 1971).

Wright, R. T. 1964. Dynamics of a phytoplankton community in an ice-covered lake. Limnol. Oceanogr. 9(2): 163-178. (SFA 9: 6649) (Used ^{14}C).

Wright, S. J. L. and E. C. Hill. 1968. The development of radiation-resistant cultures of Escherichia coli I by a process of growth-irradiation cycles. J. Gen. Microbiol. 51(1): 97-106.

Wruble, D. T., V. E. Andrews and G. A. Boysen. 1973. Environmental tritium surveillance for Project Rulison. pp. 592-602. In: Tritium. (See Moghissi and Carter, 1973) (NSA 28: 27242).

Wruble, D. T., et al. 1964. Radioactivity in waters and sediments of the Colorado River Basin, 1950-1963. Radiol. Hlth. Data 5(11): 557-567.

Wu, Y. Y., et al. 1974. Neutron activation analysis of arsenic, cadmium, and mercury in polluted river water. Nucl. Sci. J. (Taiwan) 11(2): 105-110. (NSA 31: 9).

Wukasch, M. C. and L. M. Cook. 1972. Environmental surveillance in south Texas. pp. 845-862. In: Natural Radiation Environment II. Vol. 2. (See Adams, Lowder and Gesell, 1972) (NSA 32: 19699) (Surface and groundwater).

Wukasch, M. C. and L. M. Cook. 1974. High radioactivity in drinking water and ground water in south Texas. pp. 272-278., In: Proc. 3rd Intern. Congr. International Radiation Protection Association. CONF-730907-P1) (NSA 30: 26797).

Würgler, F. E. 1962. Experimental analysis of a linear dose effect curve resulting from x-irradiation of Drosophila eggs. Abstr. p. 199. In: 2nd Intern. Congr. Radiation Research. (See Intern. Congr. Rad. Res.).

Würgler, F. E. 1963. The radiosensitivity of early developmental stages of Drosophila. Abstr. Int. J. Radiat. Biol. 6(4): 378-379.

Würgler, F. E. 1964. Studies on the radiosensitivity of early embryonic stages of Drosophila melanogaster. Thesis, Univ. of Zurich, Switzerland. 151pp. (In German).

Würgler, F. E. 1968. Electronic computers and survival-curve analysis. Int. J. Radiat. Biol. 14: 193-196. (NSA 23: 8053) (Drosophila eggs and embryos).

Würgler, F. E., H. Ulrich and A. Schneider-Minder. 1963. Variation of radiosensitivity during meiosis and early cleavage in newly laid eggs of Drosophila melanogaster. pp. 101-106. In: Repair from Genetic Radiation Damage. F. H. Sobels, ed. Macmillan Co., New York. (NSA 18: 11833).

Würgler, F. E., et al. 1972. Radiosensitivity of oocytes of Drosophila. 3. Sensitivity of Class-B oocytes of XX/O and X/X females. Int. J. Radiat. Biol. 22(4): 367-377. (NSA 27: 568).

Wurzel, P. 1972. Nuclear techniques in Rhodesian hydrology. pp. P23.p1-P23.p9. In: National Conf. on the Technological Applications of Nuclear Techniques. (See Union of South Africa, 1972) (NSA 28: 21119).

Wurzel, P. and R. R. B. Ward. 1969. Groundwater studies in the Sabi Valley, Rhodesia, using natural tritium measurements. J. Hydrology 8: 48-58. (NSA 24: 2306).

Wyatt, G. R. 1959. Phosphorus compompounds in insect development. pp. 161-178. In: Proc. 4th Intern. Congr. on Biochemistry. (Vienna 1958). L. Levenbock, ed. Pergamon Press, London.

Wyatt, G. R. 1963. Metabolic regulation in the development of insects. pp. 179-188. In: Control Mechanisms in Respiration and Fermentation. B. Wright, ed. Ronald Press, New York. (^{14}C).

Wyatt, G. R. and B. Linzen. 1965. The metabolism of ribonucleic acid in cecropia silkmoth pupae in diapause, during development and after injury. Biochim. Biophys. Acta 103(4): 588-600. (^{32}P).

Wyerman, T. A., R. K. Farnesworth and G. L. Stewart. 1970. Tritium in streams. in the United States, 1961-1968. Radiol. Hlth. Data 11(9): 421-439.

Wykes, E. R. 1966. Cesium-137 and strontium-90 in a lowland environment. Radiol. Hlth. Data 7(9): 545-547. (Incl. swamps, woods, uncultivated land).

Wyttenbach, A. 1966. Determination of some trace elements in water by activation analyses. Chimia 20: 119-121. (In German) (NSA 20: 24990).

Wyttenbach, A. and P. A. Schubiger. 1973. Trace element content of Roman lead by neutron activation analysis. Archeometry 15(2): 199-207. (NSA 28: 24107).

Yaalon, D. H. and E. Ganor. 1968. Chemical composition of dew and dry fallout in Jerusalem, Israel. Nature 217: 1139-1140. (NSA 22: 23168).

Yager, C. M. 1964. The uptake of radioactive zinc, cadmium, and copper by the freshwater snail Taphius glabratus. Malacologia 1: 339-353. (NSA 22: 4581).

Yager, C. M. 1966. Uptake of heavy metal ions by Taphius glabratus, a snail host of Schistosoma mansoni. Exptl. Parasitol. 19: 174-182. (NSA 22: 4585).

Yager, C. M. and H. W. Harry. 1962. Studies on the uptake of Zn^{65} by the snail, Taphius glabratus. Abstr. Bull. Ga. Acad. Sci. 20: 2.

Yagi, N. 1958. Ca^{45} and P^{32} tracing in the body of Pieris rapae crucivora and its parasites. New Entomol. 7(1): 1-2.

Yagi, N. 1958. Sexual-mark variation in male and female Pieris rapae pterin-protein complex by feeding with Ca^{45} and radium γ-ray irradiation. Dai-2-kai Gensh. Shimpojiuma Hobun. 4: 218-221.

Yagovoy, P. N. 1967. Capacity of leaves on trees and bushes to retain global radioactive fallout. Gigiyena sanitariya (2): 32-37. (In Russian) (NSA 21: 20202).

Yagovoy, P. N., et al. 1971. Characterizing the natural β-activity of some kinds of mineral water. Vopr. pitaniya 30(3): 62-65. (In Russian) (NSA 26: 4759).

Yaguchi, E. M., B. J. Waller and J. S. Marshall. 1974. Plutonium distribution in Lake Michigan biota. Argonne National Lab., U. S. AEC report. 24 pp. (CONF-740813-1) (NSA 30: 23998).

Yakushev, B. I. 1966. Intensity of the absorption of ^{32}P by pines and some aspects of other soil covering plants. Dokl. Akad. Nauk. Belorussk. SSR 10: 698-700. (In Russian) (NSA 21: 22369).

Yamada, J., et al. 1962. On improving the efficiency of feed for fish culture--II. Digestibilities of feeding stuff for rainbow trout and some trials on the improvement. Bull. Jap. Soc. Sci. Fish. 28(9): 905-908. (SFA 8: 5446) (^{32}P label).

Yamada, K. and K. Amano. 1965. Reduction of coliform counts in the shucked baby clam, Vererupis semidecussata, by gamma irradiation. Tokaiku Suisan Kenkyu. Kenkyo Hokoku (43): 91-96. (In Japanese) (NSA 20: 22753).

Yamada, K. and A. Takase. 1955. Radiation distribution of disposed fish. Nippon Koshu Eisei Zasshi 2(2 Zokango): 309.

Yamada, K., et al. 1955, 1956. Studies on radiation contaminated fish. (2) Results of qualitative analysis of radiation. (3) Confirmation of ^{65}Zn. Nippon Suisan Gakkaishi 20(10): 916, 921.

Yamada, M. and T. Morimoto. 1973. Nuclear energy environmental problems and ARGs. FAPIG (Tokyo) (70): 81-87. (In Japanese) (NSA 30: 20442).

Yamada, O., H. Wada and T. Sameshima. 1972. ^{14}C dating by liquid scintillation counting on synthesized methanol and its application on wooden remnants in ejecta of the Fuji Volcano. Chishitsugaku Zasshi 78(5): 235-239. (In Japanese) (NSA 29: 10153).

Yamada, T. and M. E. Roesel. 1964. Effects of actinomycin D on the lens regenerating system. J. Embryol. Exptl. Morphol. 12(4): 713-725. (NSA 19: 45738) (Triturus).

Yamada, Y., S. Tamai and T. Miyaguchi. 1958. The measurement of the thickness of leaves using S-35. pp. 624-627. In: Proc. 2nd Japan Conf. on Radioisotopes. (See Japan Radioisotopes Association, 1958).

Yamafuji, K. and H. Omura. 1954. Study of polyhedral virus formation with radioactive phosphorus. Enzymologia 17(1): 28-30. (Used silkworm larvae).

Yamagata, H. 1969. The 2000 Curie ^{60}Co gamma-ray irradiation facility for biological studies. Bull. Inst. Chem. Res. Kyoto Univ. 47: 67-76. (NSA 23: 35367).

Yamagata, H., Y. Kowyama and K. Syakudo. 1969. Radiosensitivity and polyploidy in some non-tuber bearing Solanum species. Radiation Botany 9: 509-521. (NSA 24: 12329).

Yamagata, N. and K. Iwashima. 1962. Balance of zinc in Japanese people and diet with reference to environmental contamination with radioactive zinc-65. Koshu Eiseiin Kenkyu Hokoku 11: 131-138. (NSA 19: 13681) (Incl. marine fishes, nuts).

Yamagata, N., and K. Iwashima. 1965. Environmental contamination with ^{95}Zr-^{95}Nb in Japan. Koshu Eiseiin Kenkyu Hokoku 14: 137-147. (NSA 20: 43615).

Yamagata, N., M. Chiba and H. Kobayashi. 1969. Total ground deposition of ^{90}Sr in Japan. J. Radiat. Res. (Japan) 10(2): 62-72. (NSA 23: 47650) (Incl. statistical analysis of soil samples).

Yamagata, N., T. Hanya and S. Nagayama. 1961. The concentration of ^{90}Sr and ^{137}Cs in land waters in Japan. Japan report to the UN Scientific Committee on the Effects of Atomic Radiation. 7pp.

Yamagata, N., S. Matsuda and M. Chiba. 1969. Radiecology of ^{137}Cs and ^{90}Sr in a forest. J. Radiat. Res. (Japan) 10(3-4): 107-112. (NSA 24: 36684).

Yamagata, N., S. Matsuda and K. Kodaira. 1963. Run-off of caesium-137 and strontium-90 from rivers. Nature 200(4907): 6961).

Yamagata, N., T. Okita and K. Kodaira. 1963. The runoff of fall-out radio-strontium from a river and its applications in hydrology. pp. 417-424. In: Radioisotopes in Hydrology. (See IAEA, 1963) (NSA 18: 1931).

Yamagata, N., T. Yamagata and S. Matsuda. 1959. The different distribution of rubidium and cesium in natural plants. Bull. Chem. Soc. Japan 32: 407-414. (NSA 13: 19787).

Yamaguchi, M. and J. D. Campbell. 1973. Gamma-irradiation of mushrooms and its effect on active and latent forms of O-diphenol oxidase. Radiation Botany 13(1): 55-58. (NSA 27: 27848).

Yamaguchi, S. and A. S. Crafts. 1958. Autoradiographic method for studying absorption and translocation of herbicides using C^{14}-labeled compounds. Hilgardia 28(6): 161-191.

Yamaguchi, S. and A. S. Crafts. 1959. Comparative studies with labeled herbicides on woody plants. Hilgardia 29(4): 171-204.

Yamaguchi, Y. 1963. Possible use of Ca^{41} in nuclear dating. Progr. Theoret. Phys. (Kyoto) 30: 567. (NSA 18: 6961).

Yamamoto, M. and I. Motomura. 1968. Biological effect of electron beam irradiation on small creatures. 2. Change in the digestive canal of newts due to 60-MeV electron beam. Res. Rep. Lab. Nucl. Sci. Tohoku Univ. 1(2): 143-147. (In Japanese) (NSA 24: 21279).

Yamamoto, T., et al. 1973. Environmental contamination by uranium. 3. Effects of carbonate ion on uranium adsorption to and desorption from soil. J. Radiat. Res. (Japan) 14(3): 219-224. (NSA 29: 24199).

Yamashina, K. and S. Ogasawara. 1957. On the movement and distribution of P^{32} in akamatsu (Pinus densiflora Sieb et Zucc.) and kuromatsu (P. thunbergii Parl.). Jap. Forestry Soc. J. 39(8): 293-297. (In Japanese, English summary).

Yamashita, T. and T. Momoi. 1968. Radiactive nuclides in atmosphere and soil air near the ground surface. Kumamoto Daigaku Kyoikug. Kiyo (16): 3-10. (In Japanese) (NSA 22: 51101).

Yamazaki, T. 1957. The effect of x-ray on the survival of yeast. Jap. J. Genet. 32(8): 256-266. (In Japanese).

Yamazaki, T., et al. 1964. The breeding of medicinal plants exposed to gamma-ray irradiation. I. The appearance of variations in the X_1 generation of several medicinal plants. Syoyakugaku Zasshi 18: 20-25. (In Japanese).

Yanders, A. F. 1959. The effect of x-rays on sperm activity in Drosophila Genetics 44(4, Pt.1): 545-546.

Yanders, A. F. 1964. The effects of x rays on insemination and sperm retention in Drosophila. Genetics 49(2): 309-317. (NSA 18: 9859).

Yanders, A. F. 1965. A relationship between sex ratio and paternal age in Drosophila. Genetics 51: 481-486. (NSA 19: 15088).

Yang, A. I. C. and A. W. Fairhall. 1972. Variations of natural radiocarbon during the last 11 millenia and geophysical mechanism for producing them. pp. A44-A57. In: Proc. 8th Intern. Radiocarbon Dating Conf. Vol. 1. (See Rafter, 1972) (NSA 31: 8626).

Yang, C. 1967. Response and recovery studies on x-irradiated Tribolium confusum larvae. Thesis, Univ. of Illinois, Urbana. 147pp. (NSA 22: 45672).

Yang, C. H. and C. A. Tobias. 1974. Interaction between radiation effects, gravity, and other environmental factors in Tribolium confusum. pp. 21-30. In: Life Sciences and Space Research XII. P. H. A. Smith, ed. Akademie-Verlag, Berlin. (NSA 31: 26192).

Yang, C. H., I. L. Silver and C. A. Tobias. 1975. Effects of weightlessness, gravity compensation and radiation on the flour beetle, Tribolium confusum. Univ. of California, Berkeley, U. S. ERDA report LBL-4702. 70pp. (NSA 33: 26291).

Yang, K. R. and C. K. Pak. 1973. Environmental radioactivity at KO-rI nuclear power plant site December 1970-November 1972. J. Korean Nucl. Soc. 5(3): 240-248. (NSA 29: 12034).

Yang, T. C.-H. 1968. X-ray response and recovery in Tribolium confusum larvae. Abstr. Radiation Research 35(2): 516.

Yang, T. C.-H. and H. S. Ducoff. 1971. Recovery studies of x-irradiated Tribolium castaneum (flour beetle) larvae. Radiation Research 46(2): 290-300. (NSA 25: 37862).

Yankov, A., A. Jolov and N. Petkov. 1973. Combined effect of infrared and ultraviolet rays on some silk moth (Bombyx mori L.). Zhivotnovud. Nauki 10(6): 121-126. In Bulgarian) (NSA 31: 14398).

Yankov, A., N. Petkov and A. Jolov. 1973. Concerning the influence of x-rays on the silk moth (Bombyx mori L.). II. Change in the technological qualities of cocoons and raw silk. Zhivotnovud. Nauki 10(10): 129-133. (In Bulgarian) (NSA 31: 14397).

Yankov, A., N. Petkov and A. Jolov. 1973. Some biological and economic qualities of the silk moth (*Bombyx mori* L.) as affected by x-rays. Zhivotnovud. Nauki 10(4): 135-140. (In Bulgarian) (NSA 31: 8931).

Yankyavichyus, K. K., et al. 1972. Development of planktonic organisms of a eutrophic reservoir and their interaction with radioactive environment. Lietvos TSR Mokslu Akad. Darbei C3(59): 3-22. (In Russian) (NSA 27: 15124).

Yannacone, V. J., Jr. 1974. Environment and the law. Forum (Chicago) 9(5): vp. (NSA 30: 26118).

Yano, K., et al. 1975. Surveys of radiation-resistant asporogenic bacteria in natural environments. pp. 85-98. In: *Radiation for a Clean Environment*. (See IAEA, 1975) (NSA 32: 26020; 33: 9465).

Yao, K. T. S. 1971. Haploid cells in the kangaroo rat corneal endothelium cultures and x-ray-induced chromosome abberations. Genetics 67(3): 399-409. (NSA 25: 42347).

Yapp, W. B. (Ed.). 1959. *Effects of Pollution on Living Material*. Inst. of Biology, London. (Incl. radionuclides).

Yarborough, J. D. 1964. Liquid scintillation determination of ^{45}Ca and ^{89}Sr in *Chlamydomonas reinhardti* Dangeard and *Protosiphon botryoides* Klebs. Thesis, Univ. of Alabama, University. 60pp. (NSA 20: 1603).

Yaron, B., A. R. Swoboda and G. W. Thomas. 1967. Aldrin adsorption by soils and clays. J. Agr. Food. Chem. 15(4): 671-675.

Yashchenko, M. L. and Ye. S. Barshvskaya. 1971. *Short Survey of Applications of Strontium and Lead Isotopes in Geology*. Publ. House Acad. Sci. USSR, Leningrad. 140pp. (In Russian) (NSA 26: 42968).

Yastrebov, M. T. 1970. Gamma spectrometric determination of radioisotopes in soils. Vestn. Mosk. Univ. VI (biol. pochv.) 25(5): 75-80. (In Russian) (NSA 26: 29).

Yastrebov, M. T. 1971. Natural radioactivity of some black soils in the European areas of the USSR. Vestn. Mosk. Univ. (biol. pochv.) 26(5): 93-97. (In Russian) (NSA 28: 29825).

Yataro, T. 1965. Mechanisms controlling two types of dose-rate dependence of radiation-induced mutation frequencies in silkworm gonia. Idengaku Zasshi 40(Suppl.): 68-82. (NSA 19: 43784).

Yatazawa, M. and T. Ishihara. 1955. The absorption by plants of unseparated fission products derived from the hydrogen bomb detonated in the spring of 1954 at Bikini Atoll. Nippon Nogei-Kagaku Kaishi 29: 229-234. (In Japanese) (Also: Soil Plant Food 1: 21-24).

Yates, C. M. Gjullin and A. W. Lindquist. 1951. Treatment of mosquito larvae and adults with radioactive phosphorus (summary). Mosquito News 11(2): 111.

Yates, W. W., A. W. Lindquist and J. S. Butts. 1952. Further studies of dispersion of flies tagged with radioactive phosphoric acid. J. Econ. Entomol. 45(3): 547-548.

Yates, W. W., et al. 1951. Treatment of mosquito larvae and adults with radioactive phosphorus. J. Econ. Entomol. 44(1): 34-37. (NSA 5: 3338; 17: 10423).

Yeates, B. D., A. S. Goldin and D. W. Moeller. 1970. Radiation from natural sources in the urban environment. Harvard School of Public Health, Boston, report HSPH/EHS-70-2. 81pp. (NSA 26: 56165) (Incl. extensive bibliogr.).

Yegorova, T. A. 1965. Trehalose in insects. Uch. zap. Mosk. gos. ped. inst. 212: 14-24. (In Russian) (Contains references on use of radionuclides).

Yeh, C. and C. Sun. 1965. Effects of gamma radiation from ^{60}Co on the survival, reproduction and rate of infection of watersnails. Yuan Tzu Neng. (3): 272-279. (NSA 20: 31116).

Yemel'yanov, V. A., L. I. Beskin and V. I. Osipov. 1964. Neutron method for measuring soil moisture content and its future. Pochvoedeniye (7): 109-115. (Sov. Soil Science (7): 708-714) (NSA 19: 7720).

Yemel'yanov, V. A., et al. 1969. Neutron measurements of moisture in arable layer. Atom. energiya 26(1): 10-14. (Sov. Atom. Energy 26(1): 12-15) (NSA 23: 14059).

Yemel'yanov, V. S. (Ed.). 1959. Radioecology of agricultural plants and animals. p. 363. In: *Short Encyclopedia of Atomic Energy*. Great Soviet Encyclopedia, Moscow. (In Russian).

Yemel'yanov, V. S. 1962. A further note on the problem of radioactive waste. Novoye vremya (39): 16-18. (In Russian).

Yemel'yanov, V. S. 1964. *Atoms and Peace*. Atomizdat, Moscow. 275pp. (In Russian).

Yeremin, N. N. 1965. Natural radioactivity of the Schirmacher ponds. Sov. Antarct. Exped. Inform. Bull. 5(55): 18-19. (In Russian) (NSA 22: 10365).

Yerington, A. P. and R. M. Warner. 1961. Flight distances of *Drosophila* determined with radioactive phosphorus. J. Econ. Entomol. 54(3): 425-428.

Yermoleyeva-Makovskaya, A. P., L. A. Persov and D. K. Popov. 1968. Polonium-210 in the body and environment. State Committee on the Use of Atomic Energy USSR. Atomizdat, Moscow. 10pp. UN Document A/AC.82/G/L.1260. (AEC-tr-7030, pp. 163-170) (NSA 23: 18004).

Yermoleyeva-Makovskaya, A. P., L. A. Pertsov and D. K. Popov. 1971. Method of determination and content of ^{210}Po in man. pp. 182-188. In: *Methods of Radioecological Investigations*. (See Verkhovskaya, 1971) (In Russian).

Yliruokanen, I. 1975. A chemical study on the occurence of rare earths in plants. Ann. Acad. Sci. Fenn AII(176): 1-28. (NSA 33: 29802) (Incl. comprehensive literature review).

Yli-Vaakburi, P. 1953. Study on the organic connecting roots between trees in a Scots pine stand. Acta Forestalia Fennica 60: 1-117. (In German).

Yoshida, I., K. Sugimura and M. Hayashi. 1962. Killing effect of ^{60}Co gamma rays and aquired resistance of _Bacillus anthracis_ by successive irradiation. Natl. Inst. Animal Hlth. Quart. 2: 207-215. (NSA 17: 15718).

Yoshida, S. and G. H. N. Towers. 1963. Synthesis of shikimic acid-C^{14} in _Pinus resinosa_ Ait. from some C^{14} compounds. Can. J. Biochem. Physiol. 41(3): 579-586.

Yoshida, T., et al. 1974. Control of harmful insects in timbers by irradiation. Doses required for sterilization and inhibition of emergence of the minute pine bark beetle. Nippon Oyo Dobutsu Konchu Gak.-Shi 18(2): 52-58. (In Japanese) (NSA 33: 3269).

Yoshida, T., F. Takashima and T. Watanabe. 1973. Distribution of ^{14}C PCBs in carp. Ambio 2(4): 111-113. (NSA 29: 27135).

Yoshida, Y., et al. 1957. Induction of _situs inversus cordis et viscerum_ by culturing in radioactive Ca and Fe solution. Mie Med. J. 7(1): 79-93. (Used toad).

Yoshii, G., N. Watanabe and Y. Okada. 1956. Biological decontamination of fission products. Science 124(3216): 320-321.

Yoshii, G., N. Watanabe and Y. Okada. 1956. Studies on the uptake of P^{32}, S^{35}, Ca^{45}, and Sr^{89} in the tissues of fish. Bull. Jap. Soc. Sci. Fish. 22(4): 240-243. (In Japanese) (Used goldfish, _Spirogyra_).

Yoshii, G., N. Watanabe and Y. Okada. 1956/1957. The uptake of phosphorus-32, sulfur-35, calcium-45, and strontium-89. Nippon Suisangaku Kaishi 22: 240-243. (Algae and goldfish) (Same as above?).

Yoshimeki, M., L. D. Newson and J. L. Roddy. 1967. Effect of host eggs derived from irradiated pupae of the sugarcane borer on Trichogramma parasitism. Abstr. 357. Bull. Entomol. Soc. Am. 13(3): 203.

Yoshimura, N. and H. Etoh. 1969. Notes on the effects of beta-rays from ^{90}Sr- ^{90}Y on spermatogenesis in the teleost, _Oryzias latipes_. Annot. Zool. Jap. 42(2): 75-79. (NSA 23: 50877).

Yoshioka, T. 1973. Nuclear power in Japan. J. Brit. Nucl. Energy Soc. 12(2): 135-138. (NSA 28: 4666) (Incl. environmental safety).

Yoshioka, T., T. Itakura and T. Hashimoto. 1971. Management of effluents from JAPC's nuclear power stations. pp. 439-451. In: _Environmental aspects of Nuclear Power Stations_. (See IAEA, 1971).

Yost, H. T., Jr. and A. F. Blakeslee. 1954. The effect of fast-neutron radiation from a nuclear detonation on chromosome abberation in _Datura_. Proc. Natl. Acad. Sci. (US) 40: 447-451. (NSA 8: 4864).

Yotsukura, N. 1963. Turbulent dispersion of miscible materials in open channels. pp. 311-323. In: _Transport of Radionuclides in Fresh Water Systems_. (See Kornegay, et al., 1963) (NSA 17: 40535).

Young, D. R. 1970. Distribution of cesium, rubidium, and potassium in the quasi-marine ecosystem of the Salton Sea. Thesis, Univ. of California, San Diego. 234pp. (Univ. Microfilms 71-17678) (NSA 25: 49125).

Young, J. A., et al. 1970. Short lived cosmic ray-produced radionuclides as tracers of atmospheric processes. pp. 506-521. In: _Radionuclides in the Environment_. R. F. Gould, ed. Amer. Chemical Soc., Washington.

Young, R. G. 1963. The biosynthesis of beeswax. Life Sci. 2(9): 676-679. (Used ^{14}C).

Young, R. S. 1968. Biological experiments in space. Space Sci. Rev. 8: 665-689. (NSA 23: 29717) (Incl. frog eggs, Amoebae, larval and adult _Drosophila_, _Tribolium confusum_, _Habrobracon_).

Young, S. E., J. D. Dodd and E. R. Ibert. 1970. Tritium collection and extraction techniques for plant-water relationship studies. Ecology 51(3): 535-537.

Young, S. Y., III. 1967. Radiosynthesis and metabolism of methylthio aryl phosphorothionates in calves and blood sucking arthropods. Thesis, Auburn Univ., Auburn, Ala. 115pp. (NSA 22: 33741).

Younis, S. E. A. and A. M. M. Ali. 1969. Genetic effects of gamma-irradiation on _Fusarium moniliforme_. Isotopes Radiat. Res. 2: 65-70. (NSA 24: 44625).

Yousef, Y. A. 1965. The transport of ^{58}Co in an aqueous environment. Thesis, Univ. of Texas, Austin. (NSA 20: 6716).

Yousef, Y. A. and E. F. Gloyna. 1964. Radioactivity transport in water - the transport of Co^{58} in an aqueous environment. Envirnmental Health Engrg. Res. Lab., Inst. Marine Sci., Univ. of Texas, Austin, Tech Rept. - 7. xiv, 121pp.

Yousef, Y. A., T. J. Padden and E. F. Gloyna. 1975. Diurnal changes in radionuclides uptake by phytoplankton in small-scale ecosystems. Water Res. 9(2): 181-187. (NSA 32: 1218, 6878).

Ytterborn, K. H. 1962. X-ray sensitivity of spermatogonia and spermatozoa in _Drosophila melanogaster_. Nature 194: 797-798.

Yubuki, I. 1965. Study on the effects of radiation produced by means of an atomic reactor upon _Ascaris_ eggs. Nichidai Igaku Zasshi 24: 1007-1017. (In Japanese) (NSA 20: 33026).

Yudin, L. A. 1961. Salivary gland function studied by radioactive isotopes. Med. radiologiya 6(10): 78-85. (In Russian).

Yule, W. N., M. Chiba and H. V. Morley. 1967. Fate of insecticide residues. Decomposition of lindane in soil. J. Agr. Food Chem. 15(6): 1000-1004.

Yupares Permswarng. 1966. Effect of gamma irradiation on the development and sterility of angoumois grain moth (Sitotroga cerealella Oliv.). pp. 48-52. In: Insect Eradication by Irradiation. Thailand Office of Atomic Energy for Peace, Bangkok, report THAI.-AEC-8.

Yurkiewicz, W. J. 1965. Flight exhaustion studies on the blow fly Phaenicia sericata using C^{14}-labeled sucrose. Ann. Entomol. Soc. Am. 58(5): 766.

Yushkov, P. I. 1962. Physiology of resin formation in pine trees. pp. 50-52. In: Report 2nd Scientific and Technical Conf. of Young Specialists in Forestry Production of the Urals in Accordance with the Overall Task. Sverdlovsk. (Used ^{14}C).

Yusifov, N. I., I. K. Kolomiitseva and A. M. Kuzin. 1968. Investigation of the causes of radiation disruption of the pupation of Ephestia kuehniella larvae. Radiobiologiya 8(4): 510-513. (AEC-tr-7014, pp. 36-40) (NSA 23: 5012).

Yu-sun, C. C. C. 1964. Biochemical and morphological mutants of Ascobolus immersus. Genetics 50: 987-998. (NSA 19: 111).

Yuthamanop, A. nd. Environmental isotope investigation in the Totes Gebirge Styria, Austria. Thailand Office of Atomic Energy for Peace, Bangkok, report THAI.-AEC-35(pp. 9-20). (NSA 25: 34726) (Used ^{3}H and ^{14}C).

Zabka, H. and K. Möbert. 1971. Program for determining environmental safety of nuclear power plants. Kernenergie 14(1): 21-25. (In German) (NSA 25: 37049).

Zabyalendzik, S. F. 1971. Phosphorus feeding of buckwheat and yellow lupine in pure and mixed plants. Vestsi Akad. Navuk Belorussk. SSR (biyal.) (3): 25-29. (In Belorussian) (NSA 25: 57949) (^{32}P).

Zaduban, M. 1972. Nuclear power development and its effect on living environment. pp. 181-208. In: Living Environment and Civilization. ALFA, Bratislava. (In Slovak) (NSA 33: 25279).

Zaduban, M. nd. Necessity of solving radioecology problems on nuclear power development. Univ. P. J. Safarika Kosice Czech. (In Russian) (INIS-mf-887, pp. 33-82) (NSA 29: 17678).

Zagruzniy, P. L. 1968. Genetic effect of natural radioactivity. Tsitol. genet. 2(1): 72-74. (In Russian) (NSA 27: 480) (Wheat and barley seeds soaked in radioactive mineral water).

Zaiman, H., R. G. Howard and C. J. Miller. 1961. Immune response in rats effected with Trinchinella spiralis larvae subjected to roentgen radiation. Am. J. Trop. Med. Hyg. 10(2): 215-219. (NSA 16: 20137).

Zakharov, I. A. and T. N. Kozhina. 1967. A yeast mutant, supersensitive to ultraviolet radiation. Dokl. Akad. Nauk SSSR 176: 1417-1418. (Doklady (biol.) 176(1-6): 605-606) (NSA 22: 47519).

Zakhorov, I. A. and A. S. Kriviskiy. 1972. Radiation Genetics of Microorganisms. Atomizdat, Moscow. (In Russian) (NSA 28: 8358).

Zakhorov, I. A. and V. V. Tugarinov. 1964. Radiosensitivity of the unicellular alga Chlorella vulgaris. Radiobiologiya 4(1): 92-95. (AEC-tr-6404, 126-130) (NSA 18: 17581).

Zaki, M. M., et al. 1969. Effects of gamma radiation on egg and larval development of the cotton leaf worm Spodoptera littoralis Boisd. Isotopes Radiat. Res. 2: 33-38. (NSA 24: 44586).

Zaki, M. M., et al. 1970. Effects of gamma radiation on the larval stage of the lesser cotton leaf worm, Spodoptera exigua Hb. Z. angew. Entomol. 66: 54-59. (NSA 25: 11247).

Zakladnoy, G. A. 1966. Influence of gamma irradiation dose rate on the lifetime of granary weevils. Radiobiologiya 6(3): 478-479. (AEC-tr-6772, pp. 199-201) (NSA 21: 6672).

Zakladnoy, G. A. 1967. Radiosensitivity of the granary weevil as a function of temperature. Vestn. sel'skokhoz. nauki (3): 138-141. (In Russian, French summary).

Zalenskiy, O. V., O. A. Semikhatova and V. L. Voznesenskiy. 1955. Methods in the Use of Carbon C^{14} in the Study of Photosynthesis. Publ. House Acad. Sci. USSR, Moscow. (AEC-tr-3432) (NSA 13: 2644).

Zambruni, L. 1963. The response of the brown spots character of Drosophila melanogaster to treatment with x-rays. Rend. ist. Lombardo sci. lett. B97: 3-18. (In Italian) (NSA 20: 42945).

Zanetti, M. and F. Cutrufelli. 1961. Levels of radioactivity found in some specimens of non-migratory and migratory birds. Nuovi Ann. Igiene Microbiol. 12(5): 405-411. (Wild birds).

Zanetti, M. and F. Cutrufelli. 1962. Radioactivity levels found in some specimens of non-migratory and migratory birds. pp. 7-11. In: 3rd Intern. Congr. on Hygiene and Preventive Medicine. (Vienna, May 27-30, 1962). (NSA 17: 38734).

Zanetti, M. and F. Cutrufelli. 1963. Basic research on the levels of radioactivity in migratory birds. Nuovi Ann. Igiene Microbiol. 14(5): 339-341. (Wild birds).

Zapol'skaya, N. A. and Ye. D. Pavutskaya. 1971. A general method approach to the study of kinetics of radiation damage and restoration by incorporating radioisotopes. pp. 148-154. In: Methods of Radioecological Investigations. (See Verkhovskaya, 1971) (In Russian).

Zapol'skaya, N. A., Ye. D. Pavlitskaya and A. V. Fedorova. nd. Problems of migration and metabolis of carbon-14. Leningradskiy Nauchno-Issledovatel. Inst. Radiatsionnoy Gigieny MZ PSFSR, USSR. (In Russian) (INIS-mf-887, pp. 189-195) (NSA 29: 24538).

Zaslavsky, D. 1968. More on the diffusion of radioisotopes. Soil Sci. Soc. Am. Proc. 32(6): 891-892.

Zatsepina, L. N., Ch. Saparlyev and F. I. Pavlotskaya. 1966. On the method of determining ^{90}Sr in plants. Gigiyena sanitariya (10): 68-70. (In Russian) (NSA 21: 30820).

Zavadsky, M. and J. Cechova. nd. Radioactive contamination of waters in the vicinity of uranium ore mining and processing plants. Czechoslovakia Ustov Hygieny Prace Uranoveho Prumyslu, Kamenna. (In Slovak) (INIS-mf-887-3, pp. 124-134) (NSA 30: 6471).

Zavarzina, N. B. 1955. Development of methods for growing cultures of Scenedesmus quadricauda and Daphnia longispina labeled with radioactive phosphorus. Mikrobiologiya 24: 31-35. (In Russian) (NSA 9: 3358).

Zavel'skiy, F. S. 1964. Mass coefficients for the absorption of γ radiation in soils and the error involved in measurements using the γ radiation method. Atom. energiya 16(3): 266-268. (In Russian) (NSA 18: 25700).

Zavitkovski, J. and T. D. Rudolph. 1973. Predicted effects of chronic gamma irradiation on northern forest communities. pp. 1007-1014. In: Radionuclides in Ecosystems. (See Nelson, 1973) (NSA 28: 30075).

Zayed, S. M. A. D. and A. Hassan. 1965. Metabolism of organophosphorus insecticides. I. Distribution and metabolism of dipterex in adult larvae of the cotton leaf worm (Prodenia litura F.). VII. (w/I. Y. Mostafa) Transformation of ^{32}P-labeled dipterex through microorganisms. Can. J. Biochem. 43(8): 1257-1262; Arch. Mikrobiol. 51(2): 118-121.

Zayed, S. M. A. D., A. Hassan and T. M. Hussein. 1965. Studies on organophosphorus insecticides. II. Distribution and metabolism of ^{32}P-O-O-dimethyl-2,2,2-trichloro-1-methoxyethyl phosphonate in the adult larvae of "Prodenia litura F." Z. Naturforschung 20b(6): 587-591).

Zayed, S. M. A. D., A. Hassan and T. M. Hussein. 1966. Metabolism of carbamate drugs. II. Degradation of 1-naphthyl-N-methyl carbamate (Sevin) in adult larva of the cotton leaf worm (Prodenia litura F.). Biochem. Pharmacol. 15(12): 2057-2064. (^{14}C).

Zaytsev, Yu. P. and G. G. Polikarpov. 1964. Problems of the radioecology of the hyponeuston. Okeanologiya 4(3): 423-430. (JPRS-25966, pp. 28-41) (NSA 19: 22).

Zaytsev, Yu. P. and G. G. Polikarpov. 1965. The hyponeuston and aspects of its radioecology. p. 161. In: Problems of Hydrology. Publ. House Acad. Sci. USSR, Moscow. (In Russian).

Zeid, M. M., et al. 1953. Studies with radioactive pyrethrins. I. Tissue distribution, excretion of $C^{14}O_2$, and degradation of radioactive pyrethrins administered to the American cockroach, Periplaneta americana (L.). J. Econ. Entomol. 46: 324-336. (Also: Kansas Agr. Expt. Sta., U. S. AEC report AECU-1411) (NSA 7: 1904).

Zelawski, W. and E. Nalborczyk. 1971. Productivity of photosynthesis in Scots pine (Pinus silvestris L.) seedlings grown from seed irradiated by x-rays. Acta Soc. Bot. Pol. 40: 413-421. (NSA 27: 486).

Zeleny, V. nd. Estimating the consequences of environmental contamination with artificial radioactivity: experience of the U. N. Radiation Committee. Czechoslovakia Akad. Ved., Prague. (In Czech) (INIS-mf-887, pp. 174-188) (NSA 29: 24477).

Zel'venskiy, Ya. D., et al. 1965. Concentration of water samples for determining tritium content. Atom. energiya 18(4): 367-372. (Sov. Atom. Energy 18(4): 466-472) (NSA 19: 30407).

Zentgraf, E. and J. Barner. 1955. The uptake of radioactive phosphorus by fir roots. Allg. Forst. Jagdztg. 126(1): 23-24. (FA 18: 84).

Zhadin, V. I. 1958. Application of radioactive isotopes in hydrobiology and pisciculture. Priroda (6): 58-62. (In Russian).

Zhadin, V. I. 1959. Results of the North Caucasus hydrobiological expedition. Tr. inst. zool. Akad. Nauk SSSR 26: __. (In Russian).

Zhadin, V. I. 1960. Biological purification of water basins. Vestn. Akad. Nauk SSSR 9: 61-64. (In Russian).

Zhadin, V. I. 1964. Radioactive Isotopes in Hydrobiology and Methods of Sanitary Hydrobiology. Publ. House Acad. Sci. USSR, Mosow-Leningrad. 192pp. (In Russian) (NSA 19: 3829).

Zhadin, V. I. 1965. Perspectives of developing radiohydrobiology of freshwaters. Gidrobiol. zh. 1: 20-24. (In Russian) (NSA 21: 8763).

Zhadin, V. I., S. I. Kuznetsov and N. V. Timofeyev-Resovskiy. 1958. Isotopes in solving hydrobiology problems. Proc. 2nd Intern. Conf. on the Peaceful Uses of Atomic Energy 27: 200-207. (Paper A/CONF.15/P/2317) (Also: Papers by Soviet Scientists. Atomizdat, Moscow 6: 335-346) (NSA 13: 7316; Ref. zh. (biol.) 1961: 6D448).

Zhadin, V. I., A. G. Rodina and A. S. Troshin. 1957. Study of the biological productivity of bodies of water with the aid of radioisotopes. pp. 194-198. In: Summary of Reports at the All-Union Conf. on Application of Isotopes and Irradiation in the Public Economy and Science. Acad. Sci. USSR, Moscow. (In Russian) (BA 44: 9328; Ref. zh. (biol.) 1959: 61048; SFA 9: 1).

Zhadin, V. I., et al. 1955. Problems and methods of tagging insects and fishes with radioactive isotopes. pp. 276-284. In: Proc. Conf. on the Achievements and Tasks of Soviet Biophysics in Agriculture. (See Kuzin, 1955) (In Russian).

Zharova, T. V. 1962. Accumulation of radioactive isotopes of strontium, ruthenium, cesium and cerium by some bacteria. Mikrobiologiya 30(5): 871-876. (Microbiology 30: 713-716) (NSA 16: 26705).

Zharova, T. V. 1965 Filamentous bacteria such as Sphaerotilus in water bodies and their role in the accumulation of radioisotopes. pp. 152-153. In: Problems of Hydrobiology. Publ. House Acad. Sci. USSR, Moscow. (In Russian).

Zhdanova, N. N. and G. N. P'yankov. 1967. Effect of the dose rate of ^{60}Co gamma-rays on the survival of some Dematiaceae species. Izv. Akad. Nauk SSSR (biol.) (3): 438-442. (In Russian) (NSA 21: 28689).

Zhdanova, N. N., G. N. P'yanakov and A. I. Bondar. 1968. On the direct action of ^{60}Co gamma-radiation effect on conidia of Dematiaceae. Izv. Akad. Nauk SSSR (biol.) (4): 584-590. (In Russian) (NSA 23: 8087).

Zhestyanikov, V. D. 1964. Formation of elongated forms of *E. coli* during culturing under conditions of continuous gamma radiation. Radiobiologiya 4(1): 96-101. (AEC-tr-6404, pp. 131-139).

Zhilkina, M. I., N. V. Vasil'yeva and L. I. Gedeonov. 1973. Radiochemical determination of the radioisotopes ^{89}Sr and ^{90}Sr insamples from the environment. Radiokhimiya 15(4): 584-590. (Sov. Radiochem. 15(4): 585-589) (NSA 30: 12250).

Zhezhel, N. G. 1956. Effect of natural radioactive substances on the yield of agricultural crops. pp. 149-161. In: *Session of the Acad. Sci. USSR on the Peaceful Uses of Atomic Energy*. (1-5 July 1955). Publ. House Acad. Sci. USSR, Moscow. (English transl. Consultants Bureau, New York) (NSA 9: 7619).

Zhogova, V. M. 1957. Some data on the effect of radioactive strontium on the processes of self-decontamination of water reservoirs. Med. radiologiya 2(6): 69-72. (In Russian).

Zhogova, V. M. 1958. The effect of various concentrations of isotopes on a suspension of *Escherichia coli* in a physiological saline solution. Gigiyena sanitariya 23(11): 80. (In Russian).

Zhogova, V. M. 1961. Accumulation of strontium-90 by bacterial cells. Gigiyena sanitariya 26(4): 5-9. (In Russian).

Zhogova, V. M. 1961. The effect of radioactive strontium on water microflora and processes of mineralization of organic substances. pp. 314-319. In: *Distribution, Biological Effects and Migration of Radioisotopes*. (See Lebedinskiy and Moskalev, 1962) (JPRS-15067, pp. 41-50; AEC-tr-7512, pp. 368-374) (NSA 30: 7177).

Zhogova, V. M. 1962. Role of aquatic flora and fauna in changes of the sanitary condition of a water system. Bacterial flora. Ch. 3, pp. 50-59. In: *Radiation Hygiene*. (See Marey, 1962) (JPRS-21799, pp. 27-39).

Zhogova, V. M. 1964. *Problems of Radiation Microbiology in the Sanitary Protection of Reservoirs*. Medgiz, Moscow. 158pp. (In Russian) (NSA 18: 29911).

Zhogova, V. M. 1973. Survival and gas formation of denitrification bacilli in the presence of radioactive agents in a water medium. pp. 57-61. In: *Radioecology of Water Organisms. III*. (See Andrumaitis, 1973) (In Russian) (NSA 30: 24378).

Zhukov, B. G., D. A. Kaushanskiy and V. P. Averkiyev. 1968. Soviet gamma apparatus for radiation investigations in biology and medicine. Med. radiologiya 13(7): 47-52. (NSA 22: 49039).

Zhukov-Verezhnikov, N. N., et al. 1966. Effects of cosmic radiation and other space flight factors on lysogenic bacteria and human cells in culture. Izv. Akad. Nauk SSSR (biol.) (4): 592-593. (In Russian) (NSA 21: 18022).

Ziad Al-ahmadi, A. 1967. Effects of gamma radiation on development of *Dermacentor occidentalis* (Ixodidae). Bull. Entomol. Soc. Am. 13(3): 200.

Zicker, E. I., K. C. Berge and A. D. Hasler. 1956. Phosphorus release from bog lake beds. Limnol. Oceanogr. 1(4): 296-303.

Ziemer, R. R. 1971. Translocation of ^{14}C in ponderosa pine seedlings. Can. J. Bot. 49(1): 167-171.

Zilka, L., M. Matucha and K. Svihel. 1971. Radio-gas chromatography of ^{14}C-labeled higher fatty acids of the radioactive cultivated alga *Chlorella vulgaris*. Radioisotopy 12(5): 767-778. (NSA 26: 25196).

Zill, L. P. and N. E. Tolbert. 1958. The effect of ionizing and ultraviolet radiations on photosynthesis. Arch. Biochem. Biophys. 76: 196-203. (NSA 12: 12124).

Zillich, J. A. 1974. Biological impacts of AEC Rocky Flats Plant wastes discharged to surface waters. pp. 193-260. In: *Proc. 2nd AEC Environmental Protection Conf. Vol. 1*. (See U. S. AEC, 1974) (NSA 30: 32102).

Zilversmit, D. B., C. Enteman and M. C. Fishler. 1943. On the calculation of "turnover time" and "turnover rate" from experiments involving the use of labeling agents. J. Gen. Physiol. 26: 325-331.

Zimerman, A. M. and L. Silberman. 1967. Studies on the incorporation of ^{3}H-thymidine in *Arbacia* eggs under hydrostatic pressure. Exptl. Cell Res. 46: 469-476. (NSA 21: 434).

Zimmering, S. and G. Fowler. 1966. X-irradiation of the *Drosophila* male and its effects on the number of sperm transferred to the female. Z. Vererbungel 98: 150-151. (NSA 21: 26650).

Zimmerman, D. W. 1971. Thermoluminescent dating using fine grains from pottery. Archaeometry 13: 29-52. (NSA 25: 29212).

Zimmerman, D. W. 1971. Uranium distributions in archologic ceramics: dating of radioactive inclusions. Science 174(4011): 818-819. (NSA 26: 9248).

Zimmerman, D. W. nd. Dating of ancient pottery by thermoluminescence. pp. 858-867. In: *Proc. 2nd Intern. Conf. on Luminescence Dosimetry*. (Gatlinburg, Tennessee, September 23-26, 1968). J. A. Auxier, K. Becker and E. M. Robinson, eds. U. S. AEC report CONF-680920. (NSA 23: 7859).

Zimmerman, D. W. and J. Huxtable. 1969. Recent applications and developments in thermoluminescent dating. Archeometry 11: 105-108. (NSA 24: 29386).

Zimmerman, M. H. 1960. Absorption and translocation: transport in the phloem. Ann. Rev. Plant Physiol. 11: 167-190.

Zimmermann, M. 1965. Effects of x-rays on excitation conduction of nerves of cold-blooded animals. Thesis, Technische Hochschule, Karlsruhe, Germany. 73pp. (In German) (AED-Diss. 65-256) (NSA 20: 29021).

Zimmermann, U., D. Ehhalt and K. O. Münnich. 1967. Soil-water movement and evapotranspiration: changes in the isotopic composition of the water. pp. 567-585. In: Isotopes in Hydrology. (See IAEA, 1967) (NSA 22: 18939).

Zimmermann, U., K. O. Münnich and W. Roether. 1967. Downward movement of soil moisture traced by hydrogen isotopes. pp. 28-36. In: Isotope Techniques in the Hydrologic Cycle. (See Stout, 1967).

Zindler, H. 1972. On the determination of the permissible release of radioactive materials into the atmosphere via the exhaust air of nuclear power plants. pp. 29-32. In: Radioactive Aerosol Control and Release. German Democratic Rep. Staatliche Zentrale für Strahlenschutz, Berlin, report SZS-142. (In German) (NSA 27: 17237).

Zinov'yeva, Ye. G. 1958. Causes of death in infusorians under the influence of x-rays. Dokl. Akad. Nauk SSSR 121: 80-83. (Doklady (Biol. sci.) 121: 521-524) (NSA 12: 16880).

Zinov'yeva, Ye. G. 1958. Dependence of the survival of Paramecium caudatum on their concentration on irradiation by x-rays. Dokl. Akad. Nauk SSSR 118: 694-697. (Sov. Physics (Doklady) 3: 209-212) (NSA 12: 9585; 13: 4426).

Zinov'yeva, Ye. G. 1962. Paramecium injury caused by the radiolytic products of water at the moment of radiation interaction. Dokl. Akad. Nauk SSSR 145(6): 1389-1392. (Doklady (biol. sci.) 145: 722-724) (NSA 17: 52).

Zinov'yeva, Ye. G. 1966. Investigation of the radioresistance of Paramecium caudatum during nuclear regeneration. pp. 28-35. In: Effect of Ionizing Radiation on Heredity. (See Dubinin, 1966) (In Russian) (NSA 22: 21606).

Zirkle, R. E. 1934. Modification of x-ray injury by carbon dioxide and ammonia. Abstr. Am. J. Physiol. 109(1): 115-116. (Fern, Pteris longifolia).

Zirkle, R. E. P. C. Aebersol and E. R. Demjsster. 1937. The relative biological effectiveness of fast neutrons and x-rays upon different organisms. Am. J. Cancer 29(3): 556-562.

Zlotowski, I. and M. Rynaszewska. 1963. A note of the radioactivity of Polish mineral waters. Part II. Acta Geophys. Polon. 11: 59-65. (In Polish) (NSA 17: 30684).

Zmijewska, W. and J. Minczewski. 1969. Determination of molybdenum in soil extracts by neutron activation analysis. Chem. Anal. (Warsaw) 14: 23-27. (NSA 23: 19684).

Zolotov, A. V. 1961. Data from the study of soil and plant samples in the region of the Tungus meteor crash of 1908. Dokl. Akad. Nauk SSSR 140: 103-106. (In Russian).

Zoltan, D. 1963. Studies on the determination of the radioactive materials content of natural waters. Fiz. Szemle 13: 298-304. (In Hungarian) (NSA 18: 12378).

Zitikov, A. A. 1960. Effect of ionizing radiation on the luminous bacteria B. issatchenkoi. Radiological problems of the time factor and of recovery. Biofizika 5(2): 170-175. (Biophysics (USSR) 5(2): 197-202) (NSA 15: 1257).

Zotimov, N. V. 1965. Use of ground radioactivity method for measuring snow water reserve. Tr. gos. gidrol. inst. 130: 148-162. (In Russian) (NSA 20: 43559).

Zotov, V. G. 1971. Radioactivity of materials of veterinary interest. Veterinariya 10: 31-32. (In Russian) (NSA 29: 412).

Zschintzsch, J., R. D. O'Brien and E. H. Smith. 1965. The relation between uptake and toxicity of organophosphates for eggs of the large milkweed bug. J. Econ. Entomol. 58(4): 614-621. (^{14}C, Oncopeltus fasciatus).

Zubairi, M. Y. 1964. Metabolism of the carbamate insecticide dimethylan-C^{14} 1-dimethylcarbamoyl-5-methyl-3-pyrazolyl dimethylcarbamate-(carbonyl-C^{14}). Abstr. 47. Bull. Entomol. Soc. Am. 10(3): 163.

Zubairi, M. Y. 1965. Detoxication of dimetilan in cockroaches and houseflies. Thesis, Univ. of Wisconsin, Madison. 105pp. (Dissertation Abstr. 26: 3620) (^{14}C).

Zubairi, M. Y. and J. E. Casida. 1965. Detoxication of dimetilan in cockroaches and houseflies. J. Econ. Entomol. 58(3): 403-409. (^{14}C, Blattella germanica, Periplaneta americana, Musca domestica).

Zuber, A. 1971. A method and device for quick, point measurements of leakage rates through the bottom of reservoirs, canals, rivers. International Atomic Energy Agency, Vienna, report IAEA-R-580. 14pp. (NSA 25: 21155).

Zuccaro Labellarte, G. 1974. Distribution model for radioactive and other persistent pollutants in the environment and in the food chain. pp. 585-594. In: Comparative Studies of Food and Environmental Contamination. (See IAEA, 1974) (NSA 30: 15452).

Zullini, A., et al. 1970. Preliminary data concerning the use of artificial substrata for the ecological study of periphyton and for the identification of environmental radioactivity in a running-water ecosystem. Minerva Fisiconucl. 14: 124-131. (In Italian) (CONF-690929-12) (NSA 25: 29478).

Zuppi, G. M., J. C. Fontes and R. Letolle. 1974. Environmental isotopes and movements of sulphurated waters in Latium, Italy. pp. 341-361. In: Isotope Techniques on Groundwater Hydrology. Vol. 1. (See IAEA, 1974) (In French) (NSA 31: 19636).

Zuppinger, A. 1928. Radiobiological studies of Ascasis eggs. Strahlentherapie 28: 639-758.

Zurabyan, A. S. 1967. Gamma radiation effects on model populations of *Drosophila melanogaster*. Biol. zh. Armyan. SSR 20(7): 25-31. (In Russian) (NSA 22: 36238).

Zureshi, Z. A. 1969. The ability of sterilized irradiated males of *Dacus cucurbitae* Coo. and *Dacus zonatus* (Saunders) to compete with normal males. pp. 43-46. In: *Insect Ecology and the Sterile-Male Technique*. (See IAEA, 1969) (NSA 25: 27154).

Zutshi, M. K. 1968. Effect of continuous and fractional doses of gamma radiation on *Calandra oryzae* L. Bull. Grain Technol. 6(3): 164-165. (NSA 26: 12572).

Zverev, V. L. 1972. Methods of investigation of heavy radioelements in the hydrosphere. pp. 8-14. In: *Methods of Determination of Radioactivity*. (See Polikarpov and Parchevskiy, 1972) (In Russian).

Zverev, V. L. and A. I. Spiridonov. 1974. Uranium isotopes in meteoritic waters of Central Asia. Geokhimiya (8): 1256-1259. (NSA 31: 3307).

Zvol'skiy, S. T. and P. K. Vovk. 1971. Gamma-ray properties of dispersive soils. Dopovidi Akad. Nauk Ukr. RSR B(2): 133-135. (In Ukranian).

Zykova, A. S., et al. 1968. Strontium-90 and cesium-137 content in some items of the external environment and in the human body, 1958-1967. State Committee on the Use of Atomic Energy USSR. Atomizdat, Moscow. 15pp. UN Document A/AC.82/G/L.1248. (AEC-tr-7030, pp. 65-74) (NSA 23: 17571).

Bibliographies

Anonymous, 1958. A classified bibliography on the application of isotopes in pure and applied research. I. Agriculture. Int. J. Appl. Radiat. Isotopes 3(1): 46-47. (NSA 12: 5873).

Anthony, L. J. 1960. Sources of information in atomic energy. U. K. Atomic Energy Authority, Harwell, report AERE-Lib/L-1. 3rd ed., 44pp. (NSA 14: 10255).

Atomic Bomb Casualty Commission. nd., 1961, 1962, 1970, 1971. Bibliography of publications concerning effect of nuclear explosions 1945-1960. ---II. 1945-1961. ---etc. National Academy of Science-National Research Council, Washington, U. S. AEC report TID-18968; ABCC-00-62; ABCC, Hiroshima, Nagasaki, Japan, report ABCC-TR-0-71. 187, 22, 11, 6, 8pp.

Ballentine, D. S. and L. E. Kopp. 1965. Bibliography of technical publications and reports of the Radiation Division, Brookhaven National Laboratory, January 1963--June 1965 inclusive. Brookhaven National Lab., U. S. AEC report BNL-943. 5pp. (NSA 20: 28859).

Becker, C. D. 1973. Aquatic bioenvironmental studies in the Columbia River at Hanford, 1945-1971. A bibliography with abstarcts. Battelle Northwest Lab., U. S. AEC report BNWL-1734. 155pp.

Becker, K. 1968, 1971, 1972. Bibliographies in nuclear science and technology. Ser. C, Sect. 21. Dosimetry. Germany Zentralstelle für Atomkernenergie-Dokumentation, Frankfurt/Main, report AED-C-21-7, -8, -9, 10. 58, 147, 199, 154pp. (In several languages) (NSA 26: 7800, 344, 48286, 58861) (See also: Beising, Stahlhofen).

Becker, K. 1969. Radioluminescence dosimetry bibliography. II. Health Physics 17: 631-636. (NSA 23: 45559).

Beising, K. (Comp.). 1970. Bibliographies in nuclear science and technology, No. 10, April 1970. Contamination and decontamination of food. Germany Zentralstelle für Atomkernenergie-Dokumentation, Frankfurt/Main, report AED-C-02-10. 72pp. (NSA 24: 46596).

Berton, M. 1971, 1972. Peaceful uses of nuclear expolosions: geology, seismology, behavior of the ground and rocks. ---mines, chemistry, and gas and oil extraction. France Commissariat à l'Énergie Atomique, Bruyeres-le-Chatel, report CEA-BIB-152(Add.1), -129(Add.1). 203, 285pp. (In French and English) (NSA 26: 17680, 58758).

Benggeli, M. 1963, 1965, 1967, 1969. Radioisotopes and ionizing radiation in entomology. ---(1961-1963). ---(Vol. III (1964-1965)). ---(Vol. IV (1966-1967)). International Atomic Energy Agency, Vienna, Biblio. Ser. No. 9, 15, 24, 36; Publ. STI/PUB/__, /__, /21/24, /21/36. 414, 564, 454, 821pp. (NSA 17: 28768; 22: 12896).

Bittel, R. 1968. Bibliographic discussion of the physico-chemical behavior and the radioecology of ruthenium in hydrobiological systems. ---cobalt---. ---cerium and other lanthanides. France Commissariat à l'Énergie Atomique, Fontenay-aux-Roses, report CEA-Bib-123, -130, -138(EUR-4107). 66, 72, 88pp. (In French) (NSA 22: 51486; 23: 12041, 43715).

Blakely, J. P. and A. Klein. 1969. Index to nuclear safety. A technical progress review by chronology, permuted title, and author, Vol. 1, No. 1 through Vol. 9, No. 6. Oak Ridge National Lab., U. S. AEC report ORNL-NSIC-60. 94pp. (NSA 23: 23943).

Bock-Werthmann, W., H. Niebuhr and B. Sanson. 1969, 1971. Bibliographies in nuclear science and technology. Ser. C, Sect. 14. Activation analysis. Germany Zentralstelle für Atomkernenergie-Dokumentation, Frankfurt/Main, report AED-C-14-4, -5. vp, 276pp. (In German, English, other languages) (NSA 24: 11463; 25: 57368).

Bost, W. E. (w/H. E. Voress, S. F. Lanier, N. K. Smelcer). 1959, 1961, 1965. Radioactive fallout, a literature search. ---a bibliography of the world's literature. U. S. AEC report TID-3528, -3086, -3086(Suppl.2). 41, 194, 264pp. (NSA 19: 20164).

Brigham Young University. 1972. Publications dealing with animals collected as part of the Brigham Young University ecology projects at the Nevada Test Site. Provo, Utah. 6pp.

Brown, R. W. 1959. Bibliography of the geology of south central Washington and pertinent adjacent areas. Hanford Atomic Products Operation, U. S. AEC report HW-51030(Rev.1). 24pp. (NSA 29: 10161) (Incl. Hanford area).

Buskirk, B. A. and A. R. Spiewak. 1975. Lists of environmental statements projects. Oak Ridge National Lab. report ORNL-TM-4985. 44pp. (NSA 32: 25378

Button, J. C. 1971. Radiological health and safety aspects of uranium mining and ore treatment. Safety section bibliography. Australia Atomic Energy Commission, Luca Heights, (INIS-mf-238) (NSA 26: 33335).

Caja, J. L. 1964. The use of radioisotopes in studies of foliar absorption. Bol. Junta Control Energie At. (Peru) 9(49): 89-104. (In Spanish) (NSA 19: 12921) (A bibliographic review).

Charamathieu, A. 1968. Tritium in health physics. France Commissariat à l' Énergie Atomique, Fontenay-aux Roses, report CEA-Bib-131. 104pp. (In French) (NSA 23(5): 7727).

Chilean Nitrate Educational Bureau. 1948. Bibliography of the literature on the minor elements and their relation to plant and animal nutrition. Vol. 1. 4th ed., 1037pp.

Copenhaver, E. D. and B. K. Wilkinson. 1974. Transport of hazardous substances through soil processes. Vol. 1. Arsenic, beryllium, cadmium, chromium, copper, cyanide, lead, mercury, selenium, zinc, and others. Oak Ridge National Lab. U. S. AEC report ORNL-EIS-74-70(Pt. 1). 153pp. (NSA 31: 11310).

Copenhaver, E. D., et al. 1973. Cadmium in the environment: an annotated bibliography. Oak Ridge National Lab., U. S. AEC report ORNL-EIS-73-17. 444pp. (NSA 28: 2642).

Corrill, L. S., E. D. Copenhaver and H. V. Leland. 1974. Heavy metals and other trace elements in aquatic environments. Oak Ridge National Lab., U. S. AEC report ORNL-EIS-74-69. (NSA 30: 32089).

Crockett, P. W. 1975. Radioactivity from nuclear power plants. A bibliography with abstracts. Period covered: 1964-March 1975. U. S. National Technical Information Service, Springfield, Va., Publ. NTIS/PS-75/429. 277pp. (NSA 32: 27927).

Curlin, J. W. and J. D. Jones. 1972. Environmental law abstracts. Volume II. Oak Ridge National Lab., U. S. AEC report ORNL-EIS-72-9. 205pp. (NSA 27: 7237).

Curnow, R. D. 1968. A literature review of zinc-65 in birds with special reference to the mallard. Colorado Coop. Wildlife Res. Unit, Colorado State Univ., Ft. Collins, report COO-1744-1. iv, 15pp.

Dacquisto, M. P. 1959. Acquired radioresistance - a review of the literature and report of a confirmatory experiment. Radiation Research 10: 118 129.

David, L. R. 1954. Biological effects of neutrons, gamma radiation, and neutron-gamma combinations. A selctive bibliography of unclassified literature. U. S. AEC report TID-3052. 38pp. (NSA 8: 3948)

Davis, T. F. (w/H. E. Voress). 1960, 1964, 1966. Radioactive waste processing and disposal. A literature search. ---An annotated bibliography of selected literature. U. S. AEC report TID-3555, -3311 (Suppl. 1), -3311(Suppl. 2). 98; ix, 184, 182pp. (NSA 14; 20042; 21: 8567).

Dedek, W. 1967. Radioactive nuclides and pesticide chemistry. III. Atompraxis 13(4/5): 202-208. (In German) (Bibliography, 505 ref.).

Demine, O. 1965. Bibliography concerning transplutonium elements. European Atomic Energy Community report EUR-2508.e. 254pp. (NSA 20: 1726).

Deutsches Hydrographisches Institut. 1968. Bibliographies in nuclear science and technology. Section 25: Maritime radioecology. Germany Zentralstelle für Atomkernenergie -- Dokumentation, Frankfurt/Main, report AED-C-25-02. viii, 35pp.

Dixon, M. N., et al. 1975. Indexed bibliography on tritium: its sources and projections, behavior, measurement and monitoring techniques, health physics aspects and waste management. Oak Ridge National Lab., U. S. AEC report ORNL-5057. vp. (NSA 32: 19707).

Edmundson, E., Jr., V. Schultz and A. W. Klement, Jr. 1969. Marine radioecology, a selected bibliography of non-Russian literature. U. S. AEC report TID-3917, -3917(Suppl.1). i, 127; i, 76pp.

Ensslen, B. (Comp.). 1970. Bibliographies in nuclear science and technology. Section 12: Nuclear science literature of eastern Countries. Germany Zentralstelle fur Atomkernergie-Dokumentation, Frankfurt/Main, report AEC-C-12-22. 71pp. (NSA 25: 2399).

Faust, R. A., et. al. (Comps. and eds.). 1975. Environmental aspects of the transuranics: a selected annotated bibliography. Oak Ridge National Lab. U. S. AEC report ORNL-EIS-75-21-5. 234pp. (NSA 32: 14289).

Fischer, E. (w/B. Strahlmann). 1973, 1974. Bibliographies in nuclear science and technology. Section 02. Contamination and decontamination of food. No. 11, 12, 13. F. R. Germany Zentralstelle fur Atomkernenergie-Dokumentation, Leopoldshafen and Karlsruhe, report ZAED-BIBL-02-11, -12, -13. 51, 41, 30pp. (In German) (NSA 28: 29902; 30: 29420, 32103).

Francis, C. W., S. S. Talmage and B. B. McMullin, 1975. Radionuclide movement in soils and uptake by plants. A selected annoted bibliography. Oak Ridge National Lab. U. S. AEC report ORNL-EIS-75-77. 371pp. (NSA 32: 20024).

Furchtgott, E. 1963. Behavioral effects of ionizing radiations: 1955-1961. Psychol. Bull. 60(2): 157-199. (NSA 17: 38896) (Literature review).

Gardner, H. L. 1974. USAEC translation list index, 1970-1973. U. S. AEC report TID-4028-S1-R1. 211pp. (NSA 30: 2208).

Garman, D. E. J., et al. 1970. Bibliography on isotope separation by ion exhange. Australia Atomic Energy Commission, Lucas Heights, report AAEC/LIB/Bib-273. 33pp. (NSA 25: 23640).

Garrett, K. H. 1975. Leaching of uranium ores: papers published from 1968. UK Atomic Energy Authority report AWRE/LIB/BIB 28. 9pp. (NSA 32: 16822). (Acid and bacterial leaching).

Gerber, C. R. and H. E. Voress. (Comps.). 1966. A selected, annotated bibliography of the civil, industrial and scientific uses for nuclear explosives. U. S. AEC report TID-3522(8th Rev.). v. 108pp.

German Federal Republic. 1969, 1973. Bibliographies in nuclear science and technology. Section 17, 25. Reactor safety enclosure. --- Maritime radiology. No. 03. --- No. 04. Zentralstelle-Dokumentation, Frankfurt/Main, report AED-C-1-17, -C-25-03, -C-25-04. 125, 68, 88pp. (In German) (NSA 24: 1610).

Goldmann, J. B. 1963. Radiation effects upon experimental animals, man and plants: an annotated bibliography. Vol. I. A-L., 441pp. Lockheed Missiles and Space Co., Sunnyvale, Calif., report SB-62-60 (Vol. 1) (AD-438178; NSA 18: 36949).

Hardin, N. E., C. R. Worley and C. B. Patterson. 1974. Indexed listing of information science documents cataloged by the Technical Information Center. U. S. AEC report TID-26625. 348pp. (NSA 31: 7961).

Hardin, N. E. and D. H. McGinnis. (Eds.). 1974. Index to conferences assigned CONF-numbers by the Techincal Information Center. U. S. AEC report TID-4045. 589pp. (NSA 29: 31543).

Harrison, E. A. 1975. Ecosystem models (a bibliography with abstracts). Report for 1964-October 1974. National Technical Information Service, Springfield, Va; report NTIS/PS-75/088. 149pp. (NSA 33: 643). (Incl. radioactivity).

Hoard, A. G., M. Eisenbud adn J. H. Harley. 1956. Annotated bibliography on fallout from nuclear explosions. Health and Safety Lab; U. S. AEC report NYO-4753, 3 supplements, 44, 15, 23, 28pp.

Holoway, C. F. W. A. Goldsmith and V. M. Eldridge. 1975. Uranium tailings bibliography. Oak Ridge National Lab., U. S. AEC report ORNL-5109. 97pp. (NSA 33: 11525).

Houser, B. L. 1972. Indexed bibliography on environmental monitoring for radioactivity. Oak Ridge National Lab., U. S. AEC report ORNL-NSIC-101. v, 319pp.

Hundemann, A. S. 1975. Power plant siting. A bibliography with abstracts. Search period covered: 1964-September 1975. National Technical Information Service, Sprinfield, Va, report NTIS/PS-75/748. 145pp. (NSA 33(10): 23306). (Incl. nuclear power).

Hutchinson, E. 1960. Bibliography on the biological effects of thorium. Univ. of Rochester, U. S. AEC report UR-563. 48pp. (NSA 14: 9271).

Ingram, M. 1966. Biological effects of ionizing radiation. An annotated bibliography covering the years 1898-1957. Univ. of Rochester, U. S. AEC report TID-3097. 1196pp. (NSA 21: 4843).

International Atomic Energy Agency. Vienna.

1959. Unprocessed and processed radioisotope preparations and special sources. '''Labelled compounds. Vol. 1, 2. International directory of radioisotopes. Publ. STI/PUB/5, /7. 290, 232pp. (NSA 13: 20782; 14: 6081).

1960-1971. List of bibliographies on nuclear energy. Publ. STI/PUB/11 (Vols. 1-9). vp. (NSA 14: 21210: 23: 11385, 45167; 24: 16397; 26: 17641).

1968, 1973. Isotope techniques in hydrology. Vol. 1, 2. Biblio. Ser. No. 32, 41; Publ. STI/PUB/21/32, /41, 228, 238pp. (NSA 24: 7503; 28: 18242).

1969. Radioisotopes and ionizing radiations in entomology. Vol. 4. (1966-1967). Biblio. Ser. No. 36, Publ. STI/PUB/36(Vol. 4). 822pp. (NSA 24: 4675)

(1970. Peaceful uses of nuclear explosives. Biblio. Ser. No. 38, Publ. STI/PUB/38. 475pp. (NSA 24: 36447).

India Bhabha Atomic Energy Research Centre. 1971. Bibliography on hydrological considerations in (i) water resources for agriculture (ii) ground disposal of radioactive wastes: Desalination and Effluent Engineering Division. Bombay, report BARC-573. 373pp. (NSA 26: 12025).

Jacobs, J. M., N. K. Smelcer and H. E. Voress. 1961. Bibliographies of interest to the atomic energy program. U. S. AEC report TID-3043(Rev. 1, Suppl. 2). (NSA 15: 24634).

Japan National Diet Library. 1959. A bibliography on radioisotopes. Tokyo. 25pp. (NSA 14: 12406).

Jarrett, J. H. 1971. Radioisitopes bibliography: reports and publications of General Electric Co., Hanford Atomic Products Operations (1960-1964) and Battelle-Northwest (1965-mid 1971). Battelle Northwest Lab., U. S. AEC report BNWL-1621. iv, 114pp. (Incl. ecology).

Johnson, M. E. 1958, 1959. A bibliography of biological applications of antoradiography. Univ. of California, Berkeley, U. S. AEC report UCRL-8400, -8901. 41, 23pp. (NSA 12: 16157; 14: 2261).

Johnson, M. E. 1961. A bibliography of biological applications of autoradiography, 1 July 1959 to 1 January. U. S. Naval Radiological Defense Lab., San Francisco, Calif., Reviews and Lectures No. 117. 37pp. (NP-10597).

Junod, E. 1972. Practical bibliography on activation analysis. Radiochem. Radioanal. Lett. 11(3/4):219-227. (In French) (NSA 26: 55670).

Kate, K. (Ed.). 1970. Bibliography on wastes of plutonium. Japan Atomic Research Institute, Tokyo, report JAERI-Memo 4153. 19pp. (In Japanese) (NSA 25: 13190).

Kay, F. D., Jr. and W. Hobson, Jr. 1959. A radiobiology guide. Part II RCA Service Co., Camden, New Jersey, U. S. Air Force, Wright Air Development Center, report WADC-TR-57-118(II). (NSA 14: 1717).

Kelly, R. C. (Comp.). 1972, 1974. Selected annotated bibliography of the civil, industrial, and scienfific uses for nuclear explosions. U. S. AEC report TID-3522(Rev. 9) (Suppl.1), -R9-52. 54, 125pp. (NSA 26: 58761; 30: 9463).

Kennedy, V. S. and J. A. Mihursky. 1967. Bibliography on the effects of temperature in the aquatic environment and addendum. Natural Resources Inst., Univ. of Maryland Contribution No. 326. 122pp. (NP-18965) (NSA 25: 49119).

Klement, A. W., Jr. 1965, 1970. Natural environmental radioactivity - an annotated bibliography. U. S. AEC report WASH-1061, -1061(Suppl.). iii, 115; i, 72pp.

Klement, A. W., Jr. and V. Schultz. Terrestrial and freshwater radioecology: a selected bibliography. (TID-3910, 1962 also: Bibliography A, pp. 649-928. In: Radioecology. (See Schultz and Klement, 1963).

Year Publ.	TID-3910(Suppl.) U. S. AEC Report	pp.
1962	-	ii, 79
1963	1	ii, 95
1964	2	iii, 123
1965	3	iii, 115
1966	4	i, 128
1968	5	i, 75
1970	6	ii, 201
1971	7	ii, 220
1972	8	ii, 146
1974	9	ii, 125
1975	10	ii, 120
1975	11	ii, 163
1978	12	iv, 137

Klement, A. W., Jr. and V. Schultz. 1972. Russian radioecology: A bibliography of Soviet publications with citations of English translations and abstracts. U. S. AEC report TID-3915(Suppl.1). iii, 66pp. (NSA 26: 53530) (See Klement, Lytle and Schultz, 1968; and Schultz, 1966; below).

Klement, A. W., Jr. and I. E. Wallen. 1960. A selected list of references on marine and aquatic radiobiology. U. S. AEC report TID-3903. 42pp. (Also: Appendix B, pp. 729-746. In: Radioecology. (See Schultz and Klement, 1963) (NSA 14: 23849).

Klement, A. W., Jr., C. F. Lytle and V. Schultz. 1968. Russian radioecology, a bibliography of Soviet publications with citations of English translations and abstracts. U. S. AEC report TID-3915. iv, 131pp.

Krusiewski, S. V. 1970. Selected bibliography on radioactive occurences in the United States. U. S. AEC report RME-4110. 136pp. (NSA 24: 36424).

Lattimer, J. M. 1970. Estuarine radioecology: a bibliography of report literature. Office of Library Services, U. S. Dept. of Interior, Washington. Bibliography Series No. 16. iii, 52pp.

Lazarus, H. D. and B. H. Levedahl. 1962. Effects of radiation on the mammalian eye (A literature survey). Univer. of California at Los Angeles, U. S. AEC report TID-3912. 520pp. (NSA 17: 35447).

Lehmann, E. J. 1975. Precipitation washout (a bibliography with abstracts). Period covered: 1964-June 1975. National Technical Inforamation Service, Springfield, Va., report NTIS/PS-75/517. 121pp. (NSA 32: 27934). (Incl. radionuclided).

Leikind, M. C. and M. H. Eller. 1948. Radioactive manganese, iron, cobalt, copper, zinc and strontium. A selected bibliography of recent literature with special reference to biological and medical applications. Library of Congress, Tech. Information Div., Washington. 19pp. (NF-232) (NSA 1: 388).

Lepin, T. K., L. F. Aleksandrov and A. M. Kuzin. 1962. Bibliography on the application of radioactive and stable iosotopes in biology from 1950-1958. Acad. Sci. USSR, Moscow. 408pp. (NSA 18: 3444) (References in original language).

Lowder, W. M. and L. R. Solon. 1956. Background radiation-a literature search. Health and Safety Lab., U. S. AEC report NYO-4712. 43pp. (NSA 11: 50)

Lutz, G. J. (Ed.). 1970. Oceanography: a bibliography of selected activation analysis literature. U. S. National Bureau of Standards, Washington, report NBS-TN-534. 36pp. (NSA 25: 49).

Lutz, G. J. (Ed.). 1970. Pollution analysis: a bibliography of the literature of activation analysis. U. S. National Bureau of Standards, Washington, report NBS-TN-532. 32pp. (NSA 25: 2129).

Lutz, G. J., et al. (Eds.). 1968, 1969, 1972. Activation analysis: Inst. for Materials Research, a bibliography. ---a bibliography through 1971. U. S. National Bureau of Standards, Boulder, Colo., report NBS-TN-467(Pt. 1 and 2); -467(Pt. 1)(Add 1), -467(Pt. 2) (App.); -467. 715, 420, 892pp. (NSA 23: 9298; 24: 29226; 26: 55650).

McCormick, J. A. 1958. Isotope techniques in biological sciences. A selected list of references. U. S. AEC report TID-3512. 44pp. (NSA 12: 10346).

McCormick, J. A. 1958. Isotopes in biochemistry and biosynthesis of labelled compounds. A selected list of references. U. S. AEC report TID-3513. 89pp. (NSA 12: 13701).

McCormick, J. A. 1958. Radioisotopes in animal physiology. A selected list of references. U. S. AEC report TID-3515. 118pp. (NSA 13: 11537).

McCormick, J. A. 1959. Radioiotopes in agriculture: animal husbandry, bacteriology, fertilizer uptake, plant physiology, photosynthesis, and entomology. A selected list of references. U. S. AEC report TID-3516. -3516(Suppl. 1. 67, 22pp. (NSA 13: 11538, 19780).

McCormick, J. A. 1959. Radiation detection. A literature search. U. S. AEC report TID-3518. 22pp. (NSA 13: 13455).

McCormick, J. A. 1959. Utilization of radioisotopes in physical and biological sciences-general topics. A literature search. U. S. AEC report TID-3519. 30pp. (NSA 13: 13150).

McCormick, J. A. 1960. Radioisotopes in agriculture: animal husbandry, fertilizer uptake, plant physiology, photosynthesis, and entomology. A selected bibliography. U. S. AEC report TID-3079. 35pp.

McDonald, C. J. 1974. Section I. Mediterranean fruit fly (Ceratitis capitata (Wied.) and Section II. Queensland fruit fly (Dacus tryoni (Frogg). (both Diptera: Tephritidae). Australia Atomic Energy Commission Research Establishment, Lucas Heights, report AAEC/LIB/Bib-418. 65pp. (NSA 30: 12999) (References pertinent to radiation control).

Mandl, A. M. 1964. The radiosensitivity of germ cells. Biol. Rev. Cambridge Phil. Soc. 39: 288-371. (NSA 19: 7057) (400 references).

Markham, O. D. 1973. National Reactor Testing Station environmentally related publications. Idaho Operations Office, U. S. AEC report IDO-12078. 11pp.

Martelli, M. and R. Chandebois. 1973. Functions of the Type I cell system and the problem of oncogenesis planarians. Oncology 28: 274-288. (NSA 30(5): 13003).

Martin, F. M. et al. (Comps.). 1974, 1975. Environmental aspects of the transuranics: a selected annotated bibliography. Oak Ridge National Lab., U. S. AEC report ORNL-EIS-74-21-3, -6. 236, 196pp. (NSA 31: 5969; 32: 24339).

Mayer, R., et al. 1972, 1974. Bioenvironmental effects assoicated with n nuclear power plants: a selected bibliography. Battelle Columbus Labs., U. S. AEC report TID-26164, BMI-X-654, -654(1st Add.) 292; iii, 399; 402pp. (NSA 26: 56092; 30: 25938).

McKown, D. A. (Comp.). 1964. Health Physics bibliography on tritium. Los Alamos Scient. Lab., U. S. AEC report LAMS-2946. 87pp. (NSA 18: 18066).

Nagarathna, R. A. 1971. Bibliography on radioactive fallout (1965-70). India Bhabha Atomic Research Center, Bombay. 233pp. (NIS-mf-142) (NSA 26: 17708).

Newland, E. A. 1972. Radium: its removal from uranium mill effluents. A bibliography. Leterature search covers the period 1948-1970. Australia Atomic Energy Commission, Lucas Heights, report AAEC/LIB/BIB/341. 10pp. (NSA 26: 33329).

Newland, E. A. 1973. Nuclear analysis in geology and geochemistry. A bibliography. Australia Atomic Energy Commission, Lucas Heights, report AAEC/LIB/Bib-378. 117pp. (NSA 28: 17842).

Oak Ridge National Laboratory. 1965, 1969, 1970. Indexed bibliography of current nuclear safety literature -1, -18, -19, -20, -21. U. S. AEC report ORNL-NSIC -8, -59, -62, -67, -70, -71. 147, 285, 353, 311, 311, 265pp. (NSA 19: 26578; 23: 23942, 33003, 45096, 52389; 24: 9134).

Oak Ridge National Laboratory. 1974. Environmental Sciences Division publications 1-500; an abstracted indexed bibliography. U. S. AEC report ORNL-TM-4545, x, 171pp.

Oak Ridge National Laboratory. 1972, 1973, 1974. Environmental aspects of plutonium: a selected bibliography. ---and other elements: a selected annotated bibliography. U. S. AEC report ORNL-EIS-72-21, -73,-21(Suppl. 1), -74-21(Supp. 2). vi, 387; ix, 476; x, 266pp.

Oen, C. 1972. A bibliography on radioecology. Oak Ridge National Lab., U. S. AEC report ORNL-EIS-71-10. vi, 225pp.

O'Farrell, T. P. and V. K. Christensen. (Eds.). 1971. A bibliography of environmental research: Ecosystems Department, 1952-1970. Battelle Northwest Lab., U. S. AEC report BNWL-1549. 104pp. (NSA 25: 24072).

Ohl, J. P. 1973. Bibliography of published reports by U. S. Geological Survey personnel on the geology and hydrology of the Amchitka supplementary test area, Aleutian Islands, Alaska, 1969-1972. U. S. Geological Survey U. S. AEC report USGS-474-164. 17pp. (NSA 28: 8197).

Ohl, J. P. 1973. Bibliography of published reports by U. S. Geological Survey perssonnel on the geology and hydrology of the Central Nevada Test Area, Nye County, Nevada, 1968-First quarter 1973. U. S. Geological Survey, U. S. AEC report USGS-474-165. 13pp. (NSA 28: 8198).

Perilli, M. L. 1969. Bibliography on underground nuclear explosions (1950-1968). Italy Comitato Nazionale per l'Energie Nucleare, Rome, report RT/GEN-(69)3. 39pp. (In English and Italina) (NSA 24: 29412).

Philippine Atomic Energy Commission. Manila.

1964. Applications of radioisotopes in hydrology. A literature search. Report PAEC(A) IN-646. 19pp. (NSA 19: 34378).

1964. Autoradiography. A literature search. Report PAEC(A) IN-645. 61pp. (NSA 19: 36753).

1967. A literature search. Iron-59(^{59}Fe). Report PAEC(A) IN-672. 76pp. (NSA 22: 6217).

1971. Chromium isotopes- ^{51}Cr. A literature search. Report PAEC(A)-719. 124pp. (NSA 26: 42967).

1971. Iodine isotopes 131(^{131}I): A literature search. Report PAEC(A) IN-716. 334pp. (NSA 26: 4467).

Philippine Atomic Energy Commission. (Continued)
1972. ^{60}Co. A literature search. Report PAEC(A)-727. 241pp. (NSA 27: 7708).

1973. Activation analysis: A literature search. Report PAEC(A) -7332 (Suppl. 1). 233pp. (NSA 30: 28919).

Phillips, M. D. and M. G. Nagelsen. 1951. A bibliography on the effects of x-rays on bacteria. Battellle Memorial Institute, Columbus, Ohio. 18pp. (NP-3610) (NSA 6: 1377).

Pierce, C. M. (Comp.). 1963. The effects of radiation and radioisotopes on the life processes. Annotated bibliography. U. S. AEC report TID-3098. Books 1, 2; viii, 1824pp. (Index by Bost, Ward, and Voress) (NSA 18: 17477).

Piper, H. B. 1972. Indexed bibliography on nuclear facility siting. Oak Ridge National Lab., U. S. AEC report ORNL-NSIC-105. 278pp. (NSA 26: 49940).

Polikarpov, G. G. (Ed.). 1974. Bibliographical index of publications of Institute of Biology of South Seas AS Ukr. SSR on the problem "radiation and chemical biology" (1957-1972). "Naukova Dumka", Kiev, 58pp. (In Russian).

Raleigh, H. D. (Comp.). 1969. Reactor safety. A leterature search. U. S. AEC report TID 3525(Rev. 5)(Suppl. 4), (Suppl. 6), (Suppl. 7). 155, 63, 75pp. (NSA 23(21): 45097; 25: 37047; 26: 49840).

Reichle, D. E., D. J. Nelson and P. B. Dunaway. 1971. Biological concentration and turnover of radionuclides in food chains: a bibliography. Oak Ridge National Lab., U. S. AEC report ORNL-NSIC-89; ORNL-TM-2492. 42pp. (NSA 25(18): 42315).

Robinson, R. F., M. D. Phillips and M. G. Nagelsen. 1951. A bibliogarphy on the effects of x-rays on bacteria. Oberlin College, O. 18pp. (NP-3610) (NSA 6: 1377).

Rossouw, S. F. (Comp.). 1966, 1967. Applications of radioisotopes and radiation in hydrology. A bibliography. South Africa Atomic Energy Board, Pelindaba, Pretaria, report P56-, -144. 6, 144pp. (NSA 20: 23322; 21: 28079).

Roule, L. 1963. Radioelements in human parisitology (1963). Centre Études Nucléaires de Saclay, Commissariat à l'Énergia Atomique, Bibliographie No. 31. 59pp. (Annotated, indexed) (In French).

Rudolph, A. W., T. E. Carroll and R. S. Davidson. 1971. Plutonium and its effects in the environment. A selected literature survey. Battelle Columbus Labs., U. S. AEC report TID-26130. 203pp. (NSA 26(20: 48058).

Rudolph, A. W., T. E. Carroll and R. S. Davidson. 1971. Tritium and its effects in the environment: a selected literature survey. Battelle Columbus Labs., U. S. AEC report BMI-171-203. 378pp. (NSA 26(21): 50612).

Schultz, V. 1963. Radionuclides and ionizing radiation in ornithology: a selected bibliography on wild and domestic birds. U. S. AEC report TID-17762, 29pp. (NSA 17: 15635).

Schultz, V. 1969. Ecological techniques utilizing radionuclides and ionizing radiation: a selected bibliography. Washington State Univ., U. S. AEC report RLO-2213-1. iv, 252pp.

Schultz, V. 1972. Ecological techniques utilizing radionuclides and ionizing radiation. A selected bibliography. Washington State Univ., U. S. AEC report RLO-2213-1(Suppl. 1). i, 129pp. (NSA 26: 56504).

Schultz, V. 1974. Ionizing radiation and wild birds. A selected bibliography. Washington State Univ., U. S. AEC report TID-3919. 24pp. (NSA 32: 1145).

Schultz, V. 1976. Russian radioecology: a bibliography of Soviet publica- with citations of English translations and abstracts. Washington State Univ., U. S. ERDA report TID-3912 (Suppl. 2). v, 80pp.

Schultz, V. and F. W. Whicker. 1971, 1975. Selected bibliography of terrestrial, freshwater and marine radiation ecology. Washington State Univ. and Colorado State Univ., U. S. AEC and U. S. ERDA report TID-25650, -25650, -25650-31. iv, 185, 244pp. (NSA 33: 14834).

Schwartz, F. J. 1961. A bibliography - effects of external forces in aquatic organisms. Chesapeake Biological Lab., Solomons, Md., Contribution 168. 85pp. (Contains sections on radiation).

Seagren, R. D. and L. B. Shappert. 1971. Indexed bibliography on transportation and handling of radioactive materials-2. Oak Ridge National Lab., U. S. AEC report ORNL-NSIC-84. 123pp. (NSA 25: 10798).

Shoup, C. S. 1974. Bibliography of the zoology of Tennessee and the Tennessee Valley Region. 251pp. (NP-19905). (NSA 29(8): 18383) (includes Oak Ridge).

Smith, J. E., C. C. Bertrand and K. Gschneidner, Jr. 1968. 25 years of rare earth research papers and reports published by the Ames laboratory from its founding to its 25th anniversary. Iowa State Univ., Ames, U. S. AEC report IS-RIC-2. 241pp. (NSA 23: 47423).

Smith, R. J., H. E. Voress and N. K. Smelcer. (Comps.). 1964, 1965. Bibliographies of interest to the atomic energy program. U. S. AEC report TID-3043 (Rev. 2)(Suppl. 2), (Supp. 3). 108pp, 84pp. (NSA 19: abstr. 10694; 20: 12547). (Revised periodically).

Sonnenblick, B. P. 1972. Low and very low dose influences of ionizing radiations of cells and organisms, including man: a bibliography. Bureau of Radiological Health, U. S. Food and Drug Administration, report BRH/DBE 72-1. vii, 325pp.

Sorrentino, R. 1963. Actual possibilities in the field of radiosensitization. Radiobiol. Latina 6: 399-413. (In Italian) (NSA 19: 5711) (Literature survey on animals; 154 references).

Sparrow, A. H., J. B. Binnington and V. Bond. 1958. Bibliography on the effects of ionizing radiations on plants, 1896-1955. Brookhaven National Lab., U. S. AEC report BNL-504 (L-103). v, 222pp. (NSA 13: 2634).

Spurny, Z. and J. Sulcova. 1973. Bibliography of thermoluminescent dosimetry (1968-1972). Health Physics 24(5): 573-587. (NSA 28: 2705).

Stahlhofen, W. (Comp.). Bibliography on radiation effects of living tissues and organisms. Serv. C, Sect. 4. Germany Zentralstelle fur Atomkernenergie-Kokumentation, Frankfurt/Main. (In German).

Year Publ.	Report No. AED-C-04	pp.	NSA
1962	03	154	-
1962	08	315	17:23074
1962	09	144	17:23075
1964	15	-	-
1964	16	178	19:21854
1965	17	265	19:45807
1965	18	148	20:1607
1965	19	314	20:1608
1965	21	92	20:5284
1965	22	98	20:12629

Stahlhofen, W. (Comp.). Bibliographies in nuclear science and technology. Ser. C, Sect. 4. Germany Zentralstelle fur Atemkernenergie-Dokumentation, Frankfurt/Main. (n several languages; or German and English) (See also Becker above).

Year Publ.	Report No. AED-C-04-	pp.	NSA
1966	23	116	20:16219
1966	24	92	20:16220
1966	25	110	20:16221
1967	27	125	21:28335
1967	29	174	21:41473
1967	30	125	22:4663
1968	31	182	22:51869
1970	32	151	25:16277

Stephens, S. V. and R. D. Boche. 1953. Annotated bibliography in radiobiology. Argonne Lab., U. S. AEC report ANL-5111. vii, 360pp. (NSA 8: 974).

Stoner, R. D. (w/M. W. Hess and V. P. Bond) (Comp.). 1965. Radiation and infection. An annotated bibliography. Brookhaven National Lab., U. S. AEC report-23990. ---Supplement 1. 272, 124pp. (NSA 22: 4565; 4566).

Stringham, W. S. 1972. Caron-13: a selected abstract bibliography. Mound Lab., U. S. AEC report MLM-1913. 205pp. (NSA 26: 45286) (A stable isotope of carbon).

Surosky, A. E. 1966. Nuclear bibliography and facilities survey (compiled by the Institute of Environmental Sciences Nuclear Radiation Committee). J. Environ. Sci. 9: 22-25. (NSA 20: 35188).

Thompson, R. C. 1973. Biology of the transuanium elements. A bibliography. Battelle Northwest Lab., U. S. AEC report BNWL-1782. 132pp. (NSA 28: 27669).

Union of Soviet Socialist Republics. 1960. Literature on meteorology, hydrology, and oceanography in 1959. Meteorol. gidrol. 10: 55-58. (JPRS-7471) (NSA 15: 9190).

United Kingdom.

1949. Some papers on the effects of radiations on bacteria and moulds. (A bibliography). Atomic Energy Authority, Harwell, report AERE-Inf/Bib-51. 7pp. (NSA 4: 1330).

1968. ^{90}Sr in Soil. Commonwealth Bureau of Soils, Harpendon, Biblio. No. 1034. 14pp. (NSA 22: 42863).

1968. Determination of soil moisture using neutron and gamma ray probes. Commonwealth Bureau of Soils, Harpendon, Biblio. No. 115. (NSA 22: 42862).

1968. Tritium in groundwater and as groundwater tracer. Commonwealth Bureau of Soils, Harpendon, Biblio. No. 759. 4pp. (NSA 22: 42864).

1969. U. K. AEA list of publications available to the public. Cumulation No. 13, January-December 1968. Atomic Energy Research Establishment, Harwell. 105pp. (NSA 23: 33137).

1974. Some references to radium beryllium, silver, gold, uranium, and thorium in soils and plants (1955-1973). Commonwealth Agricultural Bureau, Slough. 16pp. (NN-20217) (NSA 31: 16910).

1975. Tracer techniques (1974-1977). Commonwealth Bureau of Soils annotated bibliography. Commonwealth Agricultural Bureaux, Slough. 43pp. (NSA 32: 11869).

U. S. Atomic Energy Commission.

1958. Nuclear science and technology - a selected bibliography. Report TID-3085. xii, 236pp. (NSA 12: 13649).

1971. Savannah River Ecology Laboratory: list of publications. Savannah River Plant. 20pp.

1972-1975. Reports available in Plowshare open files. Nevada Operations Office, report NVO-86(Rev. 1, 3, 4, 5). 33, 44, 62, 49pp. (NSA 26: 45477; 28: 12828; 30: 18575; 32: 3435).

1973. Radioactive waste processing and disposal. Report TID-3311(Suppl. 4,5). 121, 92pp. (NSA 27: 15327; 28: 30707).

1973. Radioactive waste management. A bibliography of publicly available literature pertaining to the USAEC's Hanford, Washington, production site. ---Savannah River, South Carolina production site. --- National Reactor Testing Station Idaho. ---Oak Ridge, Tennessee, site. Report TID-3340, -3341, -3342, -3343. 157, 35, 56, 89pp. (NSA 28: 18922, 22130, 22131, 25275).

U. S. Atomic Energy Commission. (Continued).
1973, 1974. Reports available in the Alaskan information program. Nevada Operations Office, Las Vegas, Nev., U. S. AEC report NVO-103(Rev.3), -(Rev.4). 21, 24pp. (NSA 28: 89109; 29: 29263).

1973. Standards in nuclear science and technology. A bibliography. Report TID-3336. 239pp. (NSA 29: 253).

1973. Monitoring, control, and disposal of tritium. A selected bibliography. Report TID-3337. 103pp. (NSA 28: 5545).

1974. Title listing of power reactor docket information. Report PRDI-74-4, -8, -9, -11. 62, 66, 53, 54pp. (NSA 29: 31438, 30: 26080, 26081; 31: 2577).

1974. Radium: a bibliography. Report TID-3345. 635pp. (NSA 30: 6105).

1974. Serial titles cited in Nuclear Science Abstracts. Report TID-4579-R8. 506pp. (NSA 30: 12060).

U. S. Environmental Protection Agency. 1975. Research reports of the National Environmental Research Center-Las Vegas, January-December 1974. National Environmental Research Center, Las Vegas, Nev. 26pp. (NP-20601) (NSA 33: 635).

U. S. Energy Research and Development Administration. 1975. Environmental studies at the Savannah River Plant and immediate environs. A bibliography. U. S. ERDA report TID-3353. 99pp. (NSA 32: 14328).

U. S. Fish and Wildlife Service. 1971. Publications list. U. S. Dept. of Commerce. National Marine Fisheries Service Center for Menhaden Research, Beaufort, North Carolina. 22pp. (Contains publications concerning radiation ecology).

U. S. Geological Survey. 1970. Bibliography of published reports by U. S. Geological Survey personnel on the geology and hydrology of the Nevada Test Site, 1957-1969. U. S. AEC report NTS-220; USGS-474-88. 67pp. (NSA 25: 8355).

Ulrikson, G. U., C. D. Bopp and R. M. Carroll. 1971. Indexed bibliography on effects of radionuclides and ionizing radiation on ecological systems. Oak Ridge National Lab., U. S. AEC report ORNL-NSIC-95. 385pp. (NSA 25: 46763).

Van den Broeck, P. J. (Comp.). nd. Treatment of pollutants with ionizing radiations. A bibliography. Belgium Institut National des Radioelements, Brussels. 29pp. (NP-20478) (NSA 32: 11207).

Vaubert, B. and R. Bittel. 1970. Study of tritium hazards. France Commissariat à l'Énergia Atomique, Fontenay-aux-Roses, report CEA-Bib-182; EUR-4522. 47pp. (In French) (NSA 24: 50937).

Vaughan, B. E., et al. 1973. Bibliography of environmental research. Ecosystems Department, 1952-1973. Battelle Northwest Labs., U. S. AEC report BNWL-SA-4655. 108pp. (NSA 29: 352).

Voress, H. E., T. F. Davis and T. N. Hubbard, Jr. 1958. Radioactive waste processing and disposal. An annotated bibliography of selected literature. U. S. AEC report TOD-3311. x, 123pp.

Voress, H. E., R. J. Smith and J. M. Jacobs. 1959. Radioactive fallout, a bibliography of selected United States Atomic Energy Commission reports. U. S. AEC report TID-3087. 74pp. (NSA 14: 13953).

Wallace, R. 1958. Bibliography of technical reports on the effects of fallout. Univ. of California, Berkeley, U. S. AEC report UCRL-8412. 70pp. (NSA 13: 11750).

Ward, H. L. 1964. Insect control by radiation and radioisotopes: a literature search. U. S. AEC report TID-3579. 23pp. (NSA 19: 5628).

Ward, H. L. (Comp.). 1964. Radioisotopes in agriculture: analytical procedures, animal husbandry, entomology, fertilizer uptake, general studies, photosynthesis, plant genetics, and plant physiology. 24pp. U. S. AEC report TID-3078(Suppl. 1). (NSA 18: 31064).

Ward, H. L. (Comp.). 1965. Phsiological studies employing radioisotopes. U. S. AEC report TID-3515(Suppl. 1). 284pp. (NSA 20: 6727).

Ward, H. L. (Comp.). 1965. Radioisotopes in medicine. A selected literature search. U. S. AEC report TID-3077(Suppl. 1). 261pp. (NSA 20: 164).

Ward, H. L. 1965. Biological applications of gamma radiation. U. S. AEC report TID-3580. iv, 94pp. (An annotated bibliography).

Ward, H. L. 1965. Isotope techniques in the biological sciences. A literature search. U. S. AEC report TID-3512(Suppl. 1). iv, 42pp. (NSA 19: 25825).

Ward, H. L. 1965. Physiological studies employing radioisotopes. U. S. AEC report TID-3515(Suppl. 1). iv, 282pp.

Wasserman, R. H. and C. L. Comar. 1961. Annotated bibliography of strontium and calcium metabolism in man and animals. U. S. Dept. of Agr., Misc. Publ. 821. 135pp.

Wassink, E. C., J. A. J. Stolwijk and A. C. Heringa-Westerhof. 1956. A b bibliographical study on uptake and accumulation of nutrient elements by plants in view of the possible occurrence of radioactive isotopes of these elements among the waste of nuclear reactors. Wageningen, Netherlands. Landbhogesch. Meded. 56(16): 52pp.

West, R. G. and R. C. Kelly. (Comp.). 1969., 1971. A selected, annotated bibliography of the civil, industrial, and scientific uses for nuclear explosives. U. S. AEC report TID-3522(8th Rev.)(Suppl. 1), (Rev. 9). 43, 359pp. (NSA 23: 19876; 25: 57603).

Wilkinson, B. K., L. S. Corrill and E. D. Copenhaver. 1974. Environmental transport of chemicals bibliography. Oak Ridge National Lab., U. S. AEC report ORNL-EIS-74-68. 187pp. (NSA 30: 9527).

Wilson, B. J. 1968. Selected references to tracer techniques. Review No. 1. Revised edition. United Kingdom Atomic Energy Authority, Amersham. 17pp. (NSA 22: 43254).

Zeleny, V. nd. Estimating the consequences on environmental contamination with artifical radioactivity: experience of the U. N. Radiation Committee. Ceskoslovenska Akademie Ved, Prague. (INIS-mf-887, pp. 174-188). (In Czech) (NSA 29: 24477).

Zhilkina, M. I., N. V. Vasil'yeva and L. I. Gedeonov. 1973. Radiochemical determination of the radioisotopes ^{89}Sr and ^{90}Sr in samples from the environment. Radiokhimiya 15(4): 584-590. (Sov. Radiochemistry 15(4): 585-589) (NSA 30: 12250).